Cases of Selected Notifiable Diseases—*Continued*

Reporting Area	Meningococcal Infections	Mumps		Pertussis			Rubella			Syphilis (Primary & Secondary)		Toxic-shock Syndrome	Tuberculosis		Tularemia	Typhoid Fever	Typhus Fever (Tick-borne) (RMSF)	Rabies, Animal
	Cum. 1991	1991	Cum. 1991	1991	Cum. 1991	Cum. 1990	1991	Cum. 1991	Cum. 1990	Cum. 1991	Cum. 1990	Cum. 1991	Cum. 1991	Cum. 1990	Cum. 1991	Cum. 1991	Cum. 1991	Cum. 1991
UNITED STATES	1,998	44	4,031	53	2,575	4,450	11	1,372	1,093	41,006	48,867	274	23,543	23,973	188	456	635	6,486
NEW ENGLAND	153	-	28	2	280	508	-	4	8	1,085	1,629	15	664	708	5	33	9	201
Maine	13	-	-	-	52	24	-	-	1	3	7	4	33	18	-	1	-	-
N.H.	14	-	6	-	22	68	-	1	1	12	51	3	5	20	-	1	-	2
Vt.	16	-	4	-	5	8	-	-	-	2	2	-	12	13	-	-	-	-
Mass.	84	-	2	2	178	367	-	2	2	498	668	8	373	419	5	28	8	14
R.I.	3	-	4	-	-	10	-	-	1	57	26	-	83	75	-	-	-	-
Conn.	23	-	12	-	23	31	-	1	3	513	875	-	158	163	-	3	1	185
MID. ATLANTIC	216	4	296	18	274	557	10	585	11	7,041	9,473	44	5,627	5,659	2	105	25	2,298
Upstate N.Y.	109	2	107	18	172	332	-	539	10	219	924	21	333	392	1	19	14	957
N.Y. City	21	-	-	-	19	-	-	2	-	3,898	4,285	2	3,685	3,554	-	60	1	-
N.J.	42	-	65	-	12	37	-	1	-	1,213	1,483	-	917	952	1	18	6	976
Pa.	44	2	124	-	71	188	10	43	1	1,711	2,781	21	692	761	-	8	4	365
E.N. CENTRAL	341	2	419	2	382	1,074	-	319	164	4,940	3,701	49	2,301	2,233	9	41	43	182
Ohio	101	-	112	2	116	246	-	283	131	662	616	22	370	399	2	4	25	20
Ind.	50	-	9	-	71	161	-	2	-	179	114	-	230	232	1	-	10	29
Ill.	92	-	147	-	61	356	-	8	21	2,373	1,524	15	1,181	1,091	4	20	5	35
Mich.	74	2	121	-	37	87	-	25	9	1,136	996	12	418	423	2	12	3	33
Wis.	24	-	30	-	97	224	-	1	3	590	451	-	102	88	-	5	-	65
W.N. CENTRAL	117	1	128	-	214	231	-	19	44	939	528	41	529	636	54	6	39	842
Minn.	26	-	21	-	81	54	-	6	35	71	92	9	97	125	1	2	-	306
Iowa	15	1	24	-	26	20	-	6	4	68	75	7	69	71	-	-	1	155
Mo.	39	-	40	-	77	113	-	5	3	589	289	13	221	312	43	1	26	23
N. Dak.	1	-	2	-	4	5	-	1	1	-	1	-	8	18	-	-	-	107
S. Dak.	3	-	2	-	5	2	-	-	-	1	4	1	31	14	5	-	1	174
Nebr.	10	-	8	-	9	11	-	-	1	17	17	2	20	18	1	3	5	17
Kans.	23	-	31	-	12	26	-	1	-	193	50	9	83	78	4	-	6	60
S. ATLANTIC	361	34	1,559	10	263	364	1	11	22	11,882	15,321	27	4,393	4,478	4	72	289	1,493
Del.	5	-	7	-	-	9	-	-	-	185	190	1	34	37	-	-	-	183
Md.	35	1	251	2	63	96	-	1	2	972	1,200	2	417	393	-	10	26	564
D.C.	18	-	24	-	2	15	-	1	1	703	1,082	1	182	159	-	3	-	22
Va.	38	-	70	-	24	25	-	-	1	871	923	5	310	410	-	11	19	253
W. Va.	14	-	27	-	9	32	-	-	-	33	20	-	65	82	[illegible]	[illegible]	4	52
N.C.	58	19	269	2	41	79	-	2	1	1,967	1,755	[illegible]	[illegible]	[illegible]	[illegible]	[illegible]	159	23
S.C.	32	-	380	-	14	14	-	-	1	1,527	1,062	[illegible]	[illegible]	[illegible]	[illegible]	[illegible]	37	113
Ga.	79	-	86	-	50	41	-	-	1	2,868	3,893	[illegible]	[illegible]	[illegible]	[illegible]	[illegible]	40	253
Fla.	82	14	445	6	60	53	1	7	15	2,756	5,196	[illegible]	[illegible]	[illegible]	[illegible]	[illegible]	4	30
E.S. CENTRAL	138	-	232	2	101	180	-	100	4	4,422	4,557	[illegible]	[illegible]	[illegible]	[illegible]	[illegible]	107	153
Ky.	50	-	-	-	-	21	-	-	1	112	122	[illegible]	[illegible]	[illegible]	[illegible]	[illegible]	31	48
Tenn.	43	-	195	-	40	85	-	100	3	1,452	1,899	[illegible]	[illegible]	[illegible]	[illegible]	[illegible]	58	29
Ala.	42	-	15	2	57	66	-	-	-	1,573	1,391	[illegible]	[illegible]	[illegible]	[illegible]	[illegible]	16	76
Miss.	3	-	22	-	4	8	-	-	-	1,285	1,145	[illegible]	[illegible]	[illegible]	[illegible]	[illegible]	2	-
W.S. CENTRAL	132	2	341	-	168	221	-	9	91	7,531	8,553	15	2,792	2,781	56	29	113	604
Ark.	20	-	44	-	14	38	-	1	3	743	597	4	249	320	42	-	30	48
La.	36	-	41	-	17	34	-	1	-	2,692	2,650	-	301	276	-	5	-	7
Okla.	13	-	16	-	49	69	-	2	1	205	274	4	179	212	13	3	81	174
Tex.	63	2	240	-	88	80	-	5	87	3,891	5,032	7	2,063	1,973	1	21	2	375
MOUNTAIN	77	1	311	8	348	365	-	38	114	604	894	35	625	579	32	12	8	241
Mont.	10	-	-	-	6	36	-	11	15	6	-	1	10	30	9	-	6	41
Idaho	8	-	12	-	29	59	-	-	49	4	9	1	15	12	-	-	-	6
Wyo.	2	-	5	-	3	1	-	-	1	11	3	-	4	5	1	-	-	83
Colo.	18	1	135	8	145	128	-	3	4	87	56	6	69	50	10	2	2	25
N. Mex.	9	N	N	-	53	19	-	4	-	30	51	7	73	122	2	2	-	6
Ariz.	22	-	122	-	69	78	-	2	32	344	620	5	310	259	3	7	-	50
Utah	-	-	15	-	41	40	-	11	4	9	29	15	54	38	7	-	-	19
Nev.	8	-	22	-	2	4	-	7	9	113	126	-	90	63	-	1	-	11
PACIFIC	463	-	717	11	545	950	-	287	635	2,562	4,211	36	5,001	5,218	6	155	2	472
Wash.	68	-	171	3	136	230	-	8	-	178	385	5	302	302	2	10	1	1
Oreg.	59	N	N	-	67	125	-	5	77	84	137	-	132	138	2	7	1	5
Calif.	320	-	499	7	263	464	-	267	542	2,288	3,651	31	4,304	4,529	2	126	-	462
Alaska	10	-	17	-	13	18	-	1	-	4	18	-	61	66	-	-	-	3
Hawaii	6	-	30	1	66	113	-	6	16	8	20	-	202	183	-	12	-	1
Guam	-	U	-	U	-	1	U	-	-	1	2	-	8	40	-	-	-	-
P.R.	19	-	13	-	58	22	-	-	-	424	331	-	211	218	-	9	-	63
V.I.	-	U	10	U	-	-	U	-	-	93	44	-	3	4	-	-	-	-
Amer. Samoa	-	U	3	U	-	-	U	-	-	-	-	-	2	15	-	-	-	-
C.N.M.I.	-	U	-	U	-	4	U	-	-	5	5	-	18	57	-	-	-	-

Source: The Centers for Disease Control, Atlanta, GA.

Foundations in

MICROBIOLOGY

Foundations in
MICROBIOLOGY

Kathleen Talaro
Pasadena City College

Arthur Talaro
Pasadena City College

Contributions to Chapter 22 from:

Dr. Linda D. Caren
California State University–Northridge

Dr. Ann Wright
University of California–Los Angeles

WCB **Wm. C. Brown Publishers**

Book Team

Editor *Colin H. Wheatley*
Developmental Editor *Elizabeth M. Sievers*
Production Editor *Kennie Harris*
Designer *Kristyn Kalnes*
Art Editor *Miriam J. Hoffman*
Photo Editor *Carrie Burger*
Permissions Editor *Karen L. Storlie*
Art Processor *Joseph P. O'Connell*
Design Production Assistant *Kathleen Theis*
Visuals/Design Developmental Consultant *Marilyn Phelps*

Wm. C. Brown Publishers
A Division of Wm. C. Brown Communications, Inc.

Wm. C. Brown Communications, Inc.

Photo researcher *John D. Cunningham*

The credits section for this book begins on page 782 and is considered an extension of the copyright page.

Library of Congress Catalog Card Number: 91–70010

ISBN 0–697–00530–5

Printed in the United States of America by Wm. C. Brown Communications, Inc., 2460 Kerper Boulevard, Dubuque, IA 52001

10 9 8 7 6 5 4 3 2

To *Harold Benson,* for being a constant source of inspiration and for showing us the way,

To our students, past, present, and future, who have never failed to be interested and challenging, and

To our daughter, *Nicole,* for growing up normal and well-adjusted in a house literally immersed in microbiology.

Brief Contents

Expanded Contents

Chapter 4

Chapter 5

Chapter 6

Chapter 7

Chapter 8

Microbial Genetics 225

Chapter 9

Physical and Chemical Control of Microbes 265

Chapter 10

Drugs, Microbes, Host— The Elements of Chemotherapy 295

Chapter 11

Microbe–Human Interactions: Infection and Disease 325

Chapter 12

The Nature of Host Defenses 361

Chapter 13

The Acquisition of Specific Immunity and Its Applications 393

Chapter 14

Chapter 15

Chapter 16

Chapter 20

Chapter 21

Chapter 22

Preface

Why Study Microbiology?

To live a well-informed, safe existence in the modern world, it is becoming increasingly necessary to have a knowledge of tiny creatures called microorganisms and their effects on humans and the environment. There are several reasons that an exposure to microbiology can be useful and meaningful. Perhaps the most obvious is that it unlocks an invisible world that many humans never learn about or even suspect exists. Its unique perspective so increases our awareness of microorganisms and their roles in illness, industry, and ecology that it has the power to change people's lives.

A century ago, Louis Pasteur said, "Life would not long remain possible in the absence of microbes." The meaning behind this simple but pithy truism is one of many significant discoveries awaiting the budding microbiologist. Indeed, microorganisms have an impact upon the earth's ecosystems that rivals that of the sun, the oceans, and the activities of humans and other living things.

Although microbiology is primarily the study of microorganisms, it leads irrevocably to the study of humans and the human condition. We need only consider the arrival of AIDS on the world scene, the resurgent cholera epidemic in South America, new cancer therapies, and the explosion in biotechnology to understand the extent to which microbiology is woven into our everyday lives.

For most persons, a background in microbiology is also valuable because of its widely practical and sensible nature. Its applications become immediately pertinent to everyday situations involving food preparation and preservation, soil fertility, waste disposal, and the prevention and treatment of infectious diseases.

Microbiology students taking their first and sole microbiology course in preparation for a career in the medical and dental professions will immediately be struck by the additional relevance this science has for them. Allied health workers are actually *practical* microbiologists—it is they who are charged with preventing disease transmission among patients, giving drugs to treat infections and diseases, and caring for severely immunocompromised patients. Because these workers must be involved in daily hands-on care, assessment, and intervention, the ability to think from a microbiological perspective is a must. In fact, this is the current challenge of health care workers—to meet and master the demands of new technologies that promise to become a commonplace feature of patient care. Truly, science will be an integral part of their futures.

Content and Organization

When we first had the idea for this book, we wanted it to introduce microbiology in a way that left an exciting and indelible impression and made difficult concepts comprehensible. We also wanted it to be filled with innovative illustrations and aids to student learning. At the same time, we aimed for a style and organization that produced a readable, current, accessible, and attractive book. The fields encompassed by microbiology are so diverse and complex that no single book on the subject can offer everything to everyone. Our primary emphasis is on a survey of general topics needed by students entering careers in allied health. The book attempts to be thorough without being exhaustive and sufficiently broad in scope to appeal to any interested student desiring a background in the subject. A general science background is helpful but not required.

Organization of Subject Matter

This book is titled "foundations" because experience has shown that a background in science is developed by laying a foundation of elementary principles that gradually build upon one another to create a whole, working body of knowledge. This building process is carried out on at least two levels. Most chapters begin with a general description of the subject matter, followed by a more detailed description of each subtopic that flows naturally from one to the next. This plan is brought to the next level as well, so that concepts introduced in early chapters form a necessary foundation for understanding more complex topics in later chapters. This encourages students to develop an ever-expanding background of knowledge and improves their ability to understand increasingly sophisticated subjects.

The text is organized into 22 chapters grouped loosely into informal units of information. The first group of chapters is designed to develop the student's basic background and vocabulary. Chapter 1 discusses historical perspectives, the scientific method, classification of organisms, laboratory techniques, and the microscope. Chapter 2 surveys the basic concepts of chemistry and cells. It is written to serve students who have had no prior course and as a general review of chemistry for those needing a refresher. The next three chapters (3, 4, and 5) deal with the general structure, life cycle, classification, and miscellaneous characteristics of cellular microbes and viruses. The second group of chapters covers aspects of microbial function, including nutrition and growth (chapter 6), metabolism and energy procurement (chapter 7), and microbial genetics, synthesis, and genetic engineering (chapter 8). The chapters in the third unit rely on the unifying concepts from previous chapters to explain the control of microorganisms using physical and chemical agents (chapter 9) and antimicrobial chemotherapy (chapter 10).

Other topics that integrate multiple principles are included in the fourth segment of the book, which explores the interaction between infectious microorganisms and the human body. Chapter 11 discusses the harmful effects of microbes on the body, the transmission of diseases, and features of pathology and epidemiology. Chapter 12 describes the general defenses of the body against microbial infections, while chapter 13 covers the details of immune reactions, vaccinations, and immune testing. Chapter 14 is devoted to immune dysfunctions that occur in allergy, immunodeficiency diseases, and cancer.

The background of host-parasite relationships prepares the reader for a survey of typical infectious diseases, which begins with chapter 15. We have organized the pathogens according to their classification into microbial groups because we have found this to be the most successful system in our own courses. One advantage of this system is coherence: All characteristics of the microorganism and its diseases may be presented simultaneously and not in separate sections. Another is that students will already be familiar with this method of presentation from their laboratory work. This system is also effective because it emphasizes each organism and disease, rather than the site of infection. With the multiple infections that strike AIDS patients and the rise of new and old diseases, we feel it is necessary to be clear about how an infectious agent is isolated and identified, and about the epidemiology, symptoms, and treatment of diseases. Chapters 15, 16, and 17 cover bacterial diseases; chapter 18 surveys fungal diseases; chapter 19 deals with the parasites; and chapters 20 and 21 give information on viral diseases.

We finish up with chapter 22, which examines the beneficial roles of microorganisms. The first part of the chapter explores the roles of microorganisms in the aquatic and terrestrial environments of the earth, and the second part describes applications of microbial activities in food and industrial production.

Highlights of the Book Plan

Each chapter opens with a photograph or drawing that focusses interest on one of the thought-provoking subjects in the chapter.

The **chapter preview** briefly introduces the chapter and summarizes its primary emphasis and major concepts. Explanatory heads and subheads are liberally sprinkled through each chapter; these correspond with the expanded table of contents at the front of the book, which furnishes a functional overview of the content and flow of each chapter.

In planning the four-color illustration program, considerable attention was given to ensuring that color was used as meaningfully and memorably as possible. Because visual and mental pictures can be extremely helpful in understanding a process, structure, or abstraction, we have tried to develop unique, hand-tailored illustrations that will have a lasting impact. Characteristics of many diseases are presented by summary cycles to facilitate greater retention; illustrated tables, graphs, and flowcharts are used to condense information into a readily accessible format. Both color and black and white photographs have been selected for their pertinence to textual information.

In this course students will be stepping into a new world with its own language. Educators have shown that providing students with the origins of terminology greatly enhances their recall, usage, and proficiency with the subject matter. Key words appear in boldface and italics, and many of them are accompanied by pronunciation guidelines and basic roots of origin in the form of footnotes at the bottom of the page.

Numerous **feature boxes** present a wide spectrum of information. Many features cover topics that are integral to the subject of the chapter but are separated from the text to conveniently summarize or call attention to the topic. Other boxes discuss interesting historical vignettes, diseases, unusual microorganisms, current developments in the science, practical applications, laboratory correlates, and analogies.

Each chapter concludes with an encapsulated **review** into which **key terms** have been incorporated. These reviews are not mere summaries, but are actually the chapter in miniature form. They were intended as a tool for either quick review or more in-depth study of key concepts, and could even suffice as a preview of the subject matter if so desired.

As an additional study help, we have included a question section. The **true-false questions** are meant to be a quick but somewhat selective self-test. We suggest that students answer the questions and quiz themselves according to the directions—that is, by explaining what makes the false statements false. Appendix E contains the correct answers. The **concept questions** include essay-type questions that challenge comprehension of the principles in that chapter as well as matching or short-answer questions. The **practical/thought questions** are designed as brainteasers that challenge students to apply some of the principles in other contexts, to test their depth of understanding, to propose alternate theories, and to develop exercises and models.

Included at the end of the book are **appendix A,** Important Charts; **appendix B,** Exponents; **appendix C,** Methods for Testing Sterilization and Germicidal Processes; **appendix D,** Universal Blood and Body Fluid Precautions; and **appendix E,** Answers to True-False Questions. The **glossary** is a mini-dictionary that contains definitions for about 600 of the most common terms used in the text. We hope readers will also discover the endpapers, which include recent data and graphs (as of this printing) on infectious disease rates in the United States.

Note to the Student

The three factors that contribute most to an effective learning experience revolve around the dynamic relationship between student, instructor, and textbook.

Students learn most effectively when they have an inherent interest in the topic, approach it with an inquisitive and curious mind, and are well prepared and rested. Because you will not be in a position to judge the long-term usefulness or relevance of certain information, it is prudent to remain open-minded and not restrict your studies to only what may be required to pass an exam. The goals of study are threefold: to improve the retention and recall of factual information; to improve understanding; and to increase your ability to interpret and analyze information. Although there are probably as many different study strategies as there are students, a few general and specific activities can be recommended.

First, do not confuse reading and underlining with studying. Reading may acquaint you with the material in the chapter, but studying and learning are necessary to commit that key information to longer-term memory. Because you will usually be evaluated by means of exams, one helpful learning tactic is to become actively involved in self-testing. While reading, stop and test yourself on important concepts by closing the book and attempting to recreate a diagram, define a term, or summarize a concept. Write questions and terms on note or index cards that can be pulled out on a moment's notice to review ideas and terminology. Study in small groups that allow the free exchange of ideas and cross-quizzing.

Do not make the mistake of thinking that, just because concepts are described as basic, they are also easy. Becoming familiar with the basic ideas in microbiology and other sciences can be quite demanding and will require a considerable outlay of time, but the returns for study are proportionate to effort. The spacing of study time is also important, since devoting a couple of hours a day to study yields greater gains than cramming a month of notes into a two-day marathon.

Another player who fills a key part in an effective learning environment is your classroom instructor. The quest for knowledge can be thrilling and fulfilling, but for most beginners, the volume of available information is more than can be thoroughly covered in one semester or quarter. Your instructor will serve as the ultimate authority on which sections and topics will be covered and what will be omitted. Because he or she is the main source of guidance, interpretation, and course organization, you will want to follow his or her instructions concerning assignment of reading and study questions.

The third learning support is the textbook. As you can see, we have endeavored to prepare a tool that you can turn to for information and answers. We hope that much of what you encounter is new and challenging, and that at times, you will feel like a detective caught up in the thrill of discovery and exploration. We have purposely written the book to be comprehensive and up-to-date so that not only will it be a valuable source of information now, but will continue to be so long after the course is over.

Supplemental Material

Instructor's Manual with Test Item File The instructor's manual contains an extended lecture outline for every chapter. These outlines provide the instructor with the complete contents of each chapter, down to the D-head level. Instructors can decide at a glance what sections they wish to cover in a chapter.

The test item file, developed by Louis Giacinti, Milwaukee Area Technical College, tests students' comprehension of textual material, using true-false, multiple choice, matching, and short-answer objective questions. The answers to all questions are provided.

WCB TestPak The testpak is a computerized testing service available for generating examinations from the test item file. It provides a call-in/mail-in test preparation service. The complete test item file is also available on computer diskette for use with IBM (3.5- or 5.25-inch diskette), Apple IIe or IIc, or Macintosh computers.

Extended Lecture Outline The outline from the instructor's manual is also available on computer diskette for use with IBM (3.5- or 5.25-inch diskette), Apple IIe or IIc, or Macintosh computers. This allows instructors to organize their course syllabus on complete diskette.

Transparencies A set of 100 four-color transparencies of important illustrations from the text is available to all adopters.

Student Study Guide Written by Jackie Butler of Grayson County Community College, the study guide includes chapter objectives for each section of the chapter; questions (objective and essay) that test comprehension; activities to test students on terminology and help them organize textual information to better assimilate the textual presentation; and self-tests at the end of each chapter, with answers in an appendix.

Instructor's Resource Guide The resource guide contains additional information that instructors can use to enhance their lectures. In this guide the author provides additional boxed readings (similar to those appearing in the text), expanded information on concepts discussed in the text, and 50 transparency masters of unique line art not found in the text.

Laboratory Manual The new second edition of *Microbiology Laboratory Exercises* by Margaret Barnett has been correlated to this textbook to allow an integration of lecture and laboratory topics. This special edition provides a chart that ties the corresponding text information to the laboratory exercises. Instructors can easily assign readings from the text to support material that will be covered in each lab.

Instructor's Manual to the Laboratory Manual This instructor's manual contains all necessary information to set up and perform all the exercises, including names of supply houses, cultures, media used, stains and reagents, and notes on each lab exercise.

Acknowledgments

Writing this textbook has been somewhat like carrying a baby over a seven-year gestation period. It represents the diligent efforts of many capable and supportive individuals at Wm. C. Brown Publishers. Acknowledgments are due to our first editor, Ed Jaffe, whose patience and vision continued to sustain us and allowed us to see our dream through to fruition, and to our current editor, Colin Wheatley, who offered us direction and moral support during the difficult periods of final writing, rewriting, and editing. We are equally indebted to our developmental editor, Liz Sievers, whose enthusiasm for the book and special facility in biology were a constant source of encouragement and delight. Kennie Harris, our production editor, is truly a "woman for all seasons," who not only edits but keeps track of hundreds of miniscule production details. Our art editor, Miriam Hoffman, has handled the large and complicated art manuscript with perception and finesse. Photo editor Carrie Burger was proficient at the formidable task of labelling and organizing several hundred photographs. We thank editor Karen Storlie for her great care and thorough attention to permissions procurements. Credit also goes to Kristyn Kalnes, who added her special touch to the book design.

Others deserving special mention are Dr. John Cunningham of Visuals Unlimited who involved himself in long and laborious paper chases to obtain first-rate photographs; the artists who so beautifully translated our rough sketches into works of art; and our mother, Mrs. Grace Park, who spent many hours at the computer typing captions and master lists, and retyping parts of the manuscript.

No acknowledgments would be complete without a rousing salute to the following team of reviewers:

Shirley M. Bishel
Rio Hondo College

Dale DesLauriers
Life Science Division, Chaffey College

Warren R. Erhardt
Daytona Beach Community College

Louis Giacinti
Milwaukee Area Technical College

John Lennox
Penn State, Altoona Campus

Glendon R. Miller
Wichita State University

Joel Ostroff
Brevard Community College

Nancy D. Rapoport
Springfield Technical Community College

Mary Lee Richeson
Indiana University–Purdue University at Fort Wayne

Donald H. Roush
University of North Alabama

Dr. Pat Starr
Mt. Hood Community College

Pamela Tabery
Northampton Community College

The review process has been eye-opening, exhilarating, and occasionally painful, but through it all, we have gained tremendous respect and admiration for what reviewers do. It is only through reviews that the authors can examine the manuscript with a more critical eye, maintain balance, and decide what might be left out and what is necessary to include. All reviewers gave helpful suggestions and were supportive, but we are especially indebted to the careful and thorough suggestions and comments from Dale DesLauriers of Chaffey College, Alta Loma, California, and John E. Lennox, Pennsylvania State University. We would also like to recognize information in chapter 22 contributed by Dr. Linda Caren of California State University, Northridge, and Dr. Ann Wright of the University of California, Los Angeles. Although this book is a collaborative effort by dozens of persons in editing, production, design, and review, the authors take full responsibility for any errors of commission or omission, and would appreciate any comments or suggestions from our readers.

CHAPTER 1

The Main Themes of Microbiology

Louis Pasteur (1822–95), the great genius of microbiology. No other microbiologist in history can match the scope and impact of Pasteur's incredible contributions to the science.

Chapter Preview

This chapter introduces the central themes that are woven through microbiology like a fine web. First it discusses the nature and importance of microorganisms, the major scientific endeavors that concern microbiologists, and the historical highlights in microbiology. These topics are followed by a survey of the classification and evolutionary history of microorganisms. The chapter then concludes with a description of the basic laboratory tools used in visualizing, growing, handling, and identifying microorganisms. This chapter will prepare you for later chapters that examine the structure, physiology, and behavior of microbes and their roles in the natural scheme of things. It will also introduce you to the exciting, rapidly progressing, and practical sides of microbiology that we encounter in our everyday lives.

The World of Microorganisms

If all the microorganisms in the world were suddenly wiped out, humans would notice the effects immediately. Foods left unrefrigerated would not spoil, plaque would no longer develop on our teeth, and we would not have as many cleaning chores around the house. All in a single brilliant stroke, the world medical community's lofty goal of eradicating deadly infectious diseases—AIDS, tuberculosis, malaria, hepatitis B—would be a reality. Imagine it: Humans would never again have to fear contagious diseases. Of course, we would also have to do without certain favorite foods, like sauerkraut, yogurt, cheddar cheese, fine wine, and sourdough bread. At first, this seems a small price to pay, but are these slight inconveniences the only sacrifices to be expected? Within a few days, industrial processes that produce vitamins, hormones, and other drugs would lie silent and useless, and research labs would become barren and sterile. In a few months, compost containing dead animal and plant materials would build up around us, and in a few years and beyond, the very elements needed for life would be trapped in this *organic* muck, unavailable to the rest of the living world.

There is no denying that humans are greatly affected by microbes that act as *pathogens,* but we also depend upon them for many beneficial facets of life—one might even say for life itself. The paradoxical role that microbes play in the operation of the earth is constantly surprising (see feature 1.1, page 7). No one can emerge from a microbiology course without a changed view of the world and of themselves, and a greater understanding of the complex roles that microorganisms play in every aspect of existence.

The Scope of Microbiology

Microbiology is a specialized area of *biology* that studies living things ordinarily too small to be seen without magnification. Such **microscopic** organisms are collectively referred to as **microorganisms, microbes,** or numerous other terms, depending upon the purpose. Some persons may call them "germs" or "bugs" in reference to their role in infection and disease, but these terms have other biological meanings and perhaps place undue emphasis on the disagreeable reputation of microorganisms. Other terms that may be encountered in our study are **bacteria, viruses, fungi, protozoa,** and **algae;** these microorganisms are the major biological groups that microbiologists study (figure 1.1). The very nature of microorganisms makes them ideal subjects for study, often more attractive than **macroscopic** organisms. Among their attributes are relative simplicity, rapid reproduction, and adaptability.[1]

Microbiology is one of the largest and most complex of the biological sciences because, in addition to studying the natural history of microbes, it also deals with every aspect of microbe-human and microbe-environmental interaction, including infection, disease, drug therapy, immunology, genetic engineering, industry, agriculture, and ecology. The subordinate branches that come under the large and expanding umbrella of microbiology are presented in table 1.1.

Table 1.1 Branches of Microbiology

Science	Area of Study	Chapter Reference
Bacteriology	The bacteria—the smallest, simplest single-celled organisms	3
Mycology	The fungi, a group of organisms that includes both microscopic forms (molds and yeasts) and larger forms (mushrooms, puffballs)	4, 18
Protozoology	The protozoa—animal-like and mostly single-celled organisms	4
Virology	Viruses—minute, noncellular particles that parasitize living things	5
Parasitology	Parasitism and parasitic organisms—traditionally including pathogenic protozoa, helminth worms, and certain insects	4, 19
Phycology or algology	Simple aquatic organisms called algae, ranging from single-celled forms to large seaweeds	4
Microbial morphology	The detailed structure of microorganisms	3, 4, 5
Microbial physiology	Microbial function (metabolism) at the cellular and molecular levels	6, 7
Microbial taxonomy	Classification, naming, and identification of microorganisms	1, 3, 4
Microbial genetics, molecular biology	The function of genetic material and the biochemical reactions of cells involved in metabolism and growth	8
Microbial ecology	Interrelationships between microbes and the environment; the roles of microorganisms in the nutrient cycles of soil, water, and other natural communities	6, 22

The following areas of applied microbiology command considerable attention and interest:

Immunology studies the system of body defenses that protects against infection. This science includes *serology,* a discipline that tests the products of immune reactions in the blood serum and aids in diagnosis of

organic (or-gan'-ik) Gr. *organos.* In this sense, pertaining to substances derived from living things.

pathogens (path'-oh-jenz) Gr. *pathos,* disease, and *gennan,* to produce. Disease-causing agents.

biology Gr. *bios,* life, and *logos,* to study. The study of living organisms.

microscopic (my"-kroh-skaw'-pik) Gr. *mikros,* small, and *scopein,* to see.

microorganism (my"-kroh-or'-gun-izm) Short for microscopic organism. The term organism is synonymous with "living thing."

microbe (my'-krohb) Gr. *mikros,* small, and *bios,* life. This term is often used in the sense of disease-causing agents.

macroscopic (mak"-roh-skaw'-pik) Gr. *makros,* large, and *skopos,* to see. Visible with the naked eye.

1. Adaptability is the capacity of a living thing to change its structure or function in order to adjust to its environment.

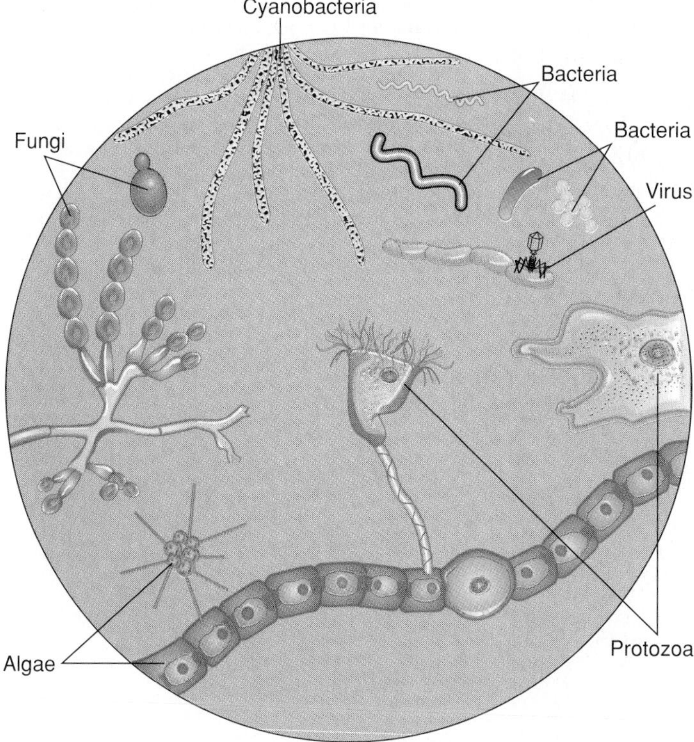

Figure 1.1 The diversity of the microbial world.

infectious diseases by this means, and *allergology,* the study of hypersensitive or allergic responses to ordinary, harmless materials (see chapters 12, 13, and 14).

Public health microbiology and **epidemiology** aim to monitor and control the spread of diseases in communities. The principal American and global institutions involved in this concern are the United States Public Health Service (USPHS) with its main agency, the Centers for Disease Control (CDC) located in Atlanta, Georgia, and the World Health Organization (WHO), the medical limb of the United Nations (see chapter 11).

Food microbiology, dairy microbiology, and **aquatic microbiology** examine the roles of microbes, both beneficial and detrimental, in consumable food and drink (see chapter 22).

Agricultural microbiology is concerned with the relationships between microbes and crops, with an emphasis on improving yield and combating plant diseases.

Biotechnology includes any process in which humans use organismic systems or processes to arrive at a desired product, ranging from bread making to gene therapy (see chapter 22).

Industrial microbiology is concerned with the uses of microbes to produce or harvest large quantities of useful and necessary materials such as vitamins, amino acids, solvents, drugs, and enzymes (see chapters 7 and 22).

Genetic engineering and **recombinant DNA technology** involve techniques that deliberately alter the genetic makeup of organisms to induce new compounds, different genetic combinations, and even unique organisms. This is the most powerful and rapidly growing area in modern microbiology (see chapters 8 and 22).

Each of the major disciplines in microbiology contains numerous subdivisions or specialties that in turn deal with a specific area or field. In fact, many areas of this science have become so specialized that it is not uncommon for a microbiologist to spend his or her whole life concentrating on a single group or type of microbe, biochemical process, or disease. On the other hand, rarely is one person a single type of microbiologist, and most can be classified in several ways. There are, for instance, bacterial physiologists who study industrial processes, molecular biologists who focus on the genetics of viruses, fungal taxonomists interested in agricultural pests, epidemiologists who are also nurses, and dentists who specialize in the microbiology of gum disease.

Studies in microbiology have led to greater understanding of many theoretical biological principles. For example, the study of microorganisms established universal concepts concerning the chemistry of life (see chapters 2 and 8), systems of inheritance (see chapter 8), and the global cycles of nutrients, minerals, and gases (see chapter 22). Microbiology is often serendipitous, meaning that basic research discoveries may later, quite by accident, lead to some new drug, therapy, product, food, or industrial process. For example, penicillin was discovered purely by a quirk of fate (see feature 10.1).

Infectious Diseases and the Human Condition Despite wonderful strides in understanding and treating infectious diseases, the world is still devastated by them. The World Health Organization (WHO), a medical agency of the United Nations, estimates that more than 20 million persons die each year worldwide from preventable infectious diseases (figure 1.2). Most of these diseases occur in developing countries where medical care may be unavailable or substandard. Not only are "new" diseases such as AIDS adding to this toll, but many older diseases, including tuberculosis, measles, and syphilis, are showing a dramatic resurgence. To complicate matters even further, increased numbers of patients with severe underlying health problems are being kept alive for extended periods. These individuals are subject to infections from common environmental microorganisms that do not affect healthy people. Even with miracle medical technology, microbes still appear to have "the last word," to quote Louis Pasteur.

The General Characteristics of Microorganisms

Cellular Organization

Two basic cell lines have appeared during evolutionary history. These lines, termed **procaryotic cells** and **eucaryotic cells,** differ primarily in the complexity of their cell structure (figure 1.3*a*). In general, procaryotic cells lack a special cell body called a nucleus, and eucaryotic cells have one. These two cell types and the organisms associated with them (called procaryotes and eucaryotes) will be covered in more detail in chapters 2, 3, and 4.

procaryotic (pro″-kar-ee-ot′-ik) Gr. *pro,* before, and *karyon,* nucleus.
eucaryotic (yoo″-kar-ee-ot′-ik) Gr. *eu,* normal, and *karyon,* nucleus.

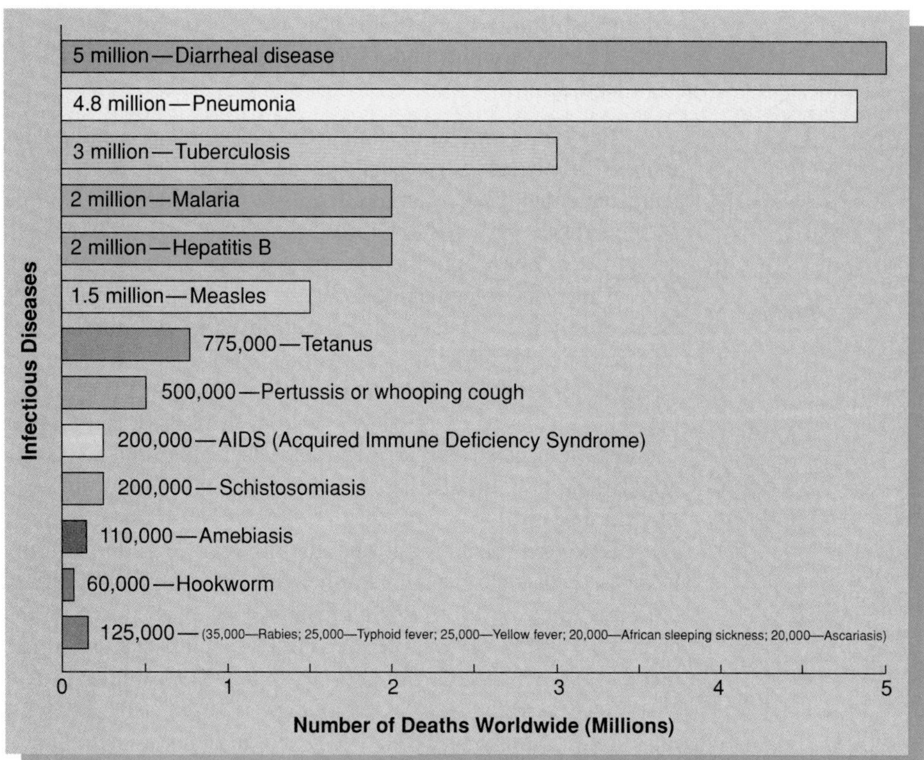

Figure 1.2 Worldwide mortality from infectious diseases for 1990, as estimated by the World Health Organization. (The bottom category combines statistics for several diseases.) The tragedy of this listing is that most of these diseases can be treated with drugs or prevented altogether through vaccination or improvements in health care and sanitation. Not included on this list are the millions of cases of disease that do not kill but severely debilitate the population. That figure, if available, would number in the billions.
Source: Data from World Health Organization 1990 reports.

All procaryotes are microorganisms, but only some eucaryotes are. The bodies of most microorganisms consist of either a single cell or just a few cells (figure 1.3*b, c*). Because of their role in disease, certain multicellular groups such as helminth worms and insects, many of which may be seen with the naked eye, are also considered in the study of microorganisms (figure 1.3*d*). Even in its seeming simplicity, the microscopic world is every bit as complex and diverse as the macroscopic one; in fact, microorganisms outnumber macroscopic organisms by a factor of several thousand.

A Note on Viruses Subject to intense study in microbiology are the viruses—small particles that exist at a level of complexity somewhere between large molecules and cells. Viruses are much simpler than cells, being composed essentially of a bit of hereditary material wrapped up in a protein covering. Some biologists refer to viruses as parasitic particles, while others consider them very primitive organisms.

Microbial Dimensions: How Small Is Small?

When we say that microbes are too small to be seen with the unaided eye, what sorts of dimensions are we talking about? This concept is best visualized by comparing microbial groups with the larger organisms of the macroscopic world and with the molecules and atoms of the molecular world (figure 1.4). Whereas the dimensions of macroscopic organisms are usually given in centimeters (cm) and meters (m), those of most microorganisms fall within the range of micrometers (μm) and to a lesser extent, nanometers (nm) and millimeters (mm). The size range of most microbes extends from the smallest viruses, measuring around 20 nm and actually not much bigger than a large molecule, up to protozoans measuring 3–4 mm and visible with the naked eye.

Life-styles of Microorganisms

The majority of microorganisms live freely in habitats such as soil and water where they are relatively harmless and often beneficial. A free-living organism can derive all required foods and other factors directly from the nonliving environment. A small number of microbes, termed **parasites,** are harbored and nourished by other living organisms, called **hosts.** Through its actions, a parasite causes damage to its host that is known as infectious disease. Most microbial parasites are some type of bacterium, fungus, protozoan, worm, or virus (but not alga). As we shall see later, a few microorganisms can exist on either free-living or parasitic levels.

The Historical Foundations of Microbiology

If not for the extensive interest, curiosity, and devotion of thousands of microbiologists over the last 300 years, we would know little about the microscopic realm that surrounds us. Each modern discovery in this science is based on the prior discoveries of men and women who toiled long hours in dimly lit laboratories with the crudest of tools. Each additional insight, whether large or small, has added to our current knowledge of living things and processes. This treatment of the early history of microbiology is not meant to be exhaustive, but will simply summarize the prominent discoveries made in the past 300 years:

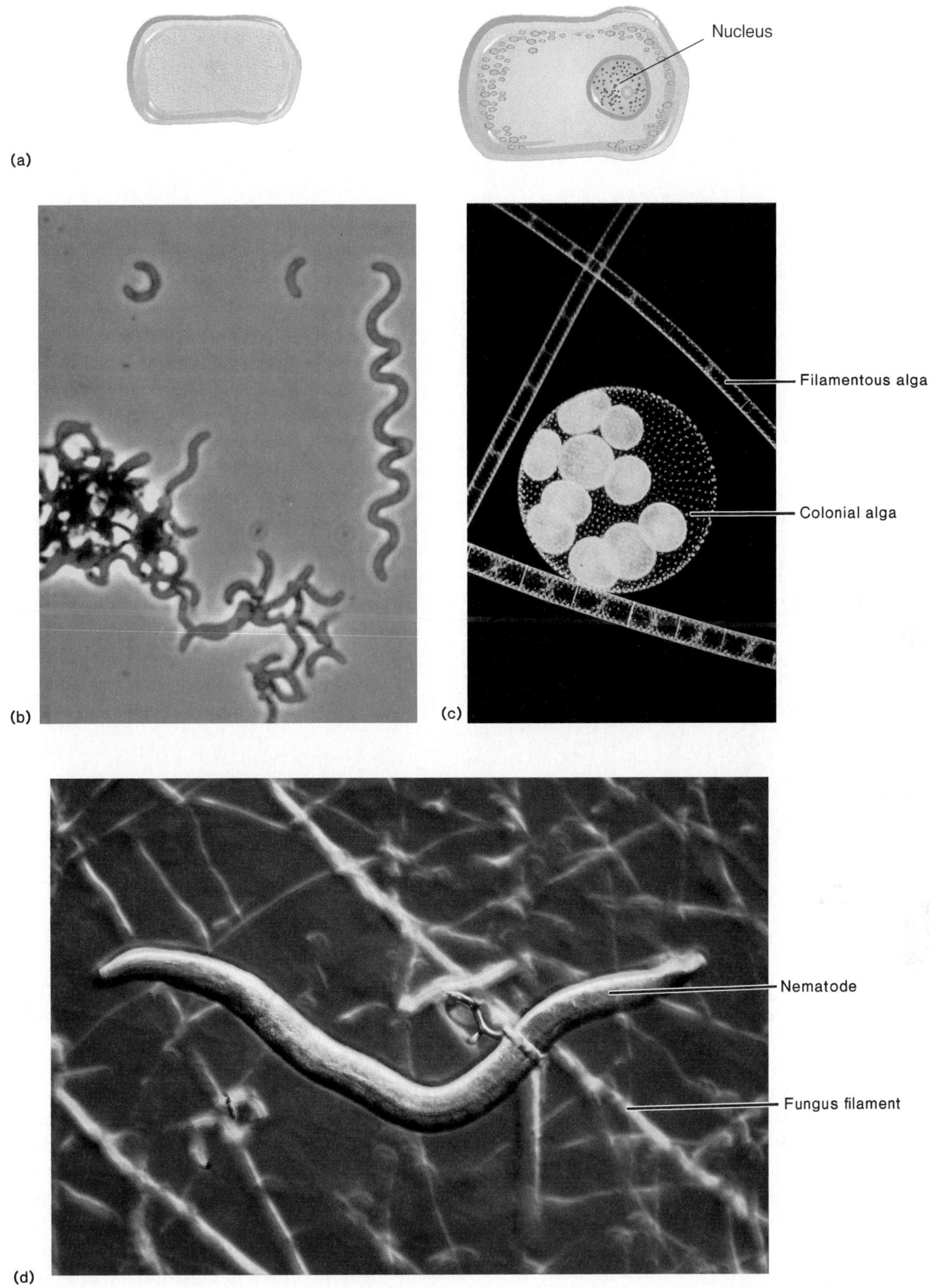

Figure 1.3 The organization of living things. (*a*) Microbial cells are of the small, relatively simple procaryotic variety (left) or the larger, more complex eucaryotic type (right). (*b*) A microscopic view of a photosynthetic bacterium, ***Rhodospirillum,*** indicates both the unicellular nature of these organisms and their relative simplicity. (*c*) Two eucaryotic organisms (algae) organized into more complex cell groupings. *Volvox* is a large, spherical colony composed of smaller colonies (white spheres) and cells (small green dots). *Spirogyra* is a filamentous alga composed of elongate cells joined end to end. (*d*) A nematode worm, a multicellular, microscopic animal, is shown being captured in the network of a killer fungus that preys upon it.

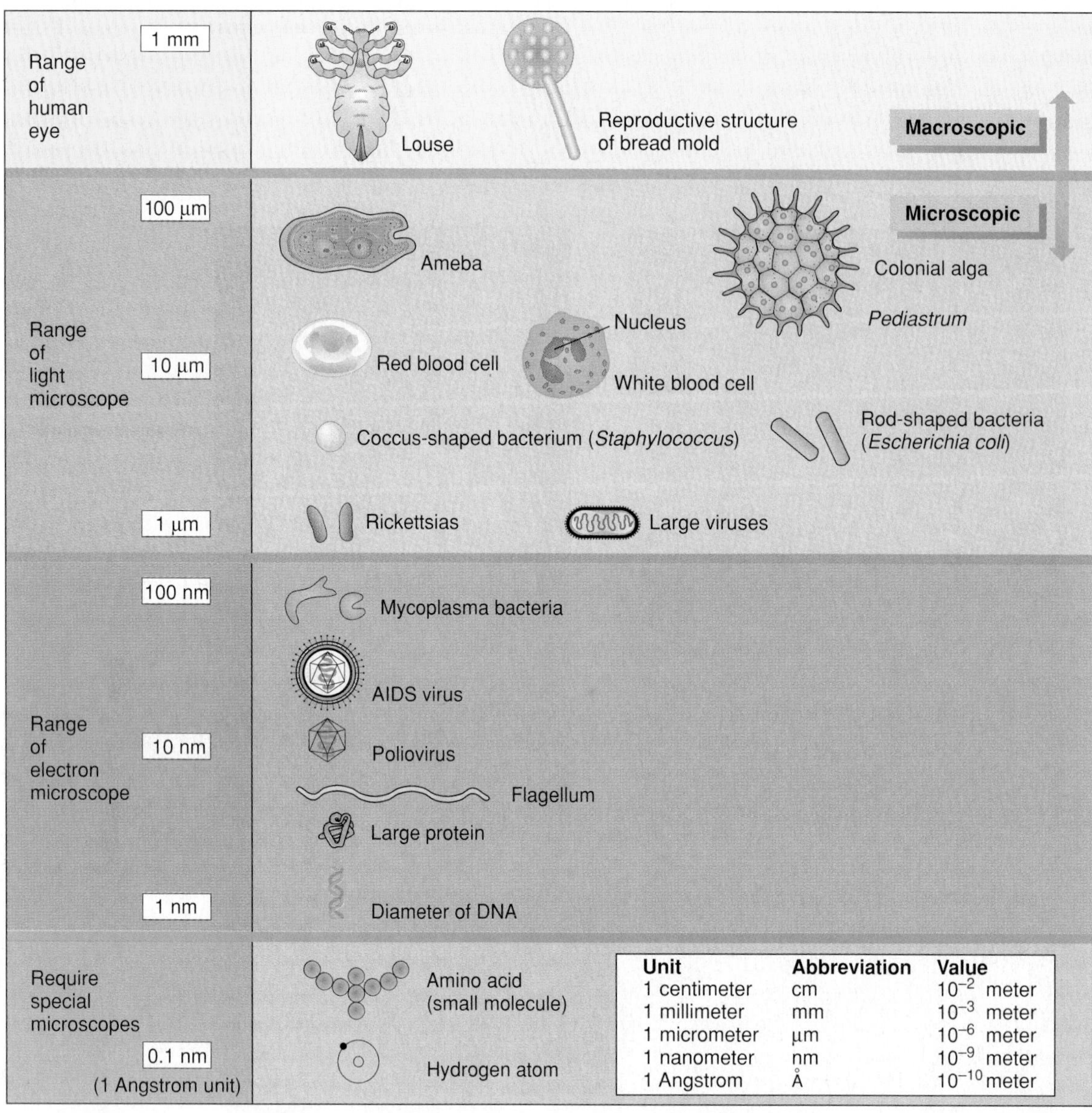

Unit	Abbreviation	Value
1 centimeter	cm	10^{-2} meter
1 millimeter	mm	10^{-3} meter
1 micrometer	µm	10^{-6} meter
1 nanometer	nm	10^{-9} meter
1 Angstrom	Å	10^{-10} meter

Figure 1.4 The size of things. Common measurements encountered in microbiology and a scale of comparison from the macroscopic to the microscopic, molecular, and atomic. Most microbes encountered in our studies will fall between 100 μm and 10 nm in overall dimensions.

microscopy, the rise of the scientific method, the germ theory, and the origins of modern microbiological techniques. Table 1.2 summarizes some of the pivotal events in microbiology, from its earliest beginnings to the present. Additional historical vignettes are integrated throughout this text to serve as background.

The Development of the Microscope: "Seeing Is Believing"

It was no doubt noted, even in early civilizations, that some spoiled foods became inedible or caused illness, whereas other foods became delicacies by the same process. And several centuries ago diseases such as the black plague and smallpox were already thought to be caused by some sort of transmissible matter. But these phenomena remained cloaked in mystery and were regarded with superstition, leading even well-informed scientists to believe in spontaneous generation (see feature 1.2, page 14). Humans would have remained ignorant of the large, diverse world of microorganisms were it not for the development of microscopes that revealed microbes as discrete entities sharing many of the characteristics of larger, visible plants and animals. Although individuals like Zaccharias Janssen, a Dutch spectacle maker, and Galileo, the Italian astronomer, fashioned magnifying lenses, their microscopes lacked the optical clarity needed for observing bacteria and other small, single-celled organisms. The first careful and exacting descriptions awaited the clever single-lens microscope hand-fashioned by Antony van Leeuwenhoek, a Dutch linen merchant and self-made microbiologist (figure 1.5).

Leeuwenhoek's wide-ranging investigations included observations of tiny organisms he called *animalcules* (little animals), blood and other human tissues (including his own tooth

Feature 1.1 Cherchez le microbe:[2] The Conflicting Character of Microorganisms

From earliest history, humans experienced a vague sense that "unseen forces" or "poisonous vapors" emanating from decomposing matter could cause disease. As the study of microbiology became more scientific and the invisible was made visible, the fear of such mysterious vapors was replaced by the knowledge and sometimes even the fear of "germs."

Microbial Threats About 110 years ago, the first studies by Robert Koch clearly linked a microscopic organism with a specific disease (anthrax). Since that time, microbiologists have conducted a continuous search for disease-causing agents. Over the past 20 years, new infectious agents, such as the human immunodeficiency virus (HIV) and other retroviruses, have been discovered, and newly emerging diseases such as Lyme disease and Legionnaire's disease have been linked definitively to microorganisms.

Medical microbiologists continue to turn up intriguing connections between microorganisms and diseases of unknown etiology to such an extent that Dr. Philip Paterson of Northwestern University Medical and Dental School believes that "all human disease has a microbial basis." Most recently, a correlation has been shown between type I diabetes and certain coxsackieviruses, between gastric ulcers and a novel bacterium that invades the stomach, and between rheumatoid arthritis and parvovirus infection. Although the precise involvement of infectious agents in many chronic diseases is yet to be determined, the possibility opens up entire new areas for potential study and treatment.

In addition to the health threat from some microorganisms, we are also waging a constant battle on another front. The adaptability and survival skills of microbes are so well developed that they readily occupy and even thrive in new habitats created by commercial materials including computer chips, jet fuel, paints, concrete, metal, plastic, paper, and cloth.

Microbial Allies At the same time that humans have been searching for the microbes in disease, they have also been seeking other microorganisms to solve various environmental, agricultural, and medical problems. Indeed, an explosion has occurred in nearly every area in which the beneficial nature of microbes is exploited. Consider, for instance, the new area of **bioremediation,** defined as the introduction of microbes to restore stability to disturbed or polluted environments. Increasingly, authorities are using microorganisms to clean up oil slicks or remove pollutants in lakes and water supplies. We rely on naturally occurring microbes to degrade the "biodegradable" materials poured into landfills, as well as to clean up sewage, extract minerals, and generally "recycle" nutrients in all ecosystems on earth.

The search for friendly microbes is a serious focus of **genetic engineering,** a field that tinkers with the very basis of life itself. Genetic engineers manipulate genetic material to produce new types of biological materials, microbes, and even plants and animals. Some of the greatest hopes of humankind—a vaccine for AIDS, self-fertilizing plants, miracle drugs, cures for genetic diseases, elimination of pollution, and solutions to world hunger—seem within the grasp of this "new microbiology."

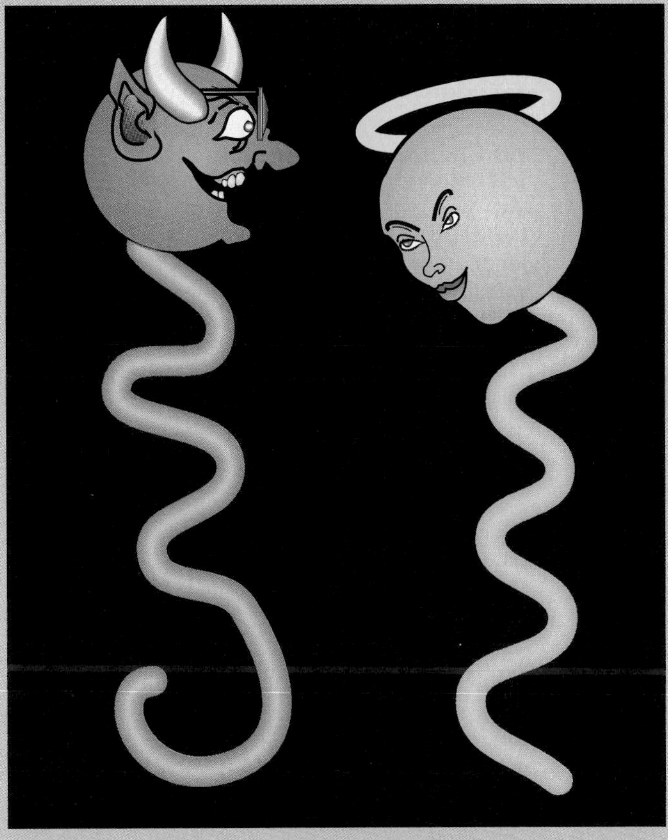

So, microorganisms pervade our lives in both an everyday, mundane sense and in a far wider view. We wash our clothes with detergents containing microbe-produced enzymes, eat food that derives flavor from microbial action, and in many cases, even eat microorganisms themselves. We are vaccinated with altered microbes to prevent diseases (that are caused by those very same microbes!); we treat various medical conditions with drugs produced by microbes; we dust our plants with insecticides of microbial origin; and we use microorganisms like tiny factories to churn out various industrial chemicals. Clearly, the relationship between humans and microbes is complex, and we can only benefit by our increased knowledge of it.

2. French for "Go find the microbe."

scrapings), insects, minerals, and plant materials. He constructed over 250 small, powerful microscopes that could magnify up to 300 times (figure 1.6). His descriptions of bacteria and protozoa were astute and precise, especially considering that he had no formal training in science and that he was the first person ever to observe this strange new world. Because of Leeuwenhoek's extraordinary contributions to microbiology, he is known as the father of bacteriology and protozoology.

From the time of Leeuwenhoek, microscopes evolved into more complex and improved instruments with the addition of refined lenses, a condenser, finer focussing devices, and built-in light sources. The prototype of the modern compound microscope, in use from about the mid-1800s, was capable of magnifications of 1,000 times or more. Even our modern laboratory microscopes are not greatly different in basic structure and function from those early microscopes. The technical characteristics of microscopes and microscopy will be discussed in a later section of this chapter.

The Rise of the Scientific Method

A serious impediment to the development of true scientific reasoning and testing was the tendency of early scientists to explain natural phenomena by a mixture of belief, superstition, and

Table 1.2 Significant Events in Microbiology

Date	Event and Its Importance
1361–1380	First widespread use of quarantine to control the spread of epidemic bubonic plague.
1481–1499	Elemental mercury given as a treatment for syphilis.
1590	Zaccharias Janssen, a Dutch spectacle maker, invents the first compound microscope.
1546	Italian physician Girolamo Fracastoro suggests that invisible organisms may be involved in disease.
1660	Englishman Robert Hooke explores various living and nonliving matter with a compound microscope that uses reflected light.
1676	Antony van Leeuwenhoek, a Dutch linen merchant, uses a simple microscope of his own design to observe bacteria and protozoa.
1668	Francesco Redi, an Italian naturalist, conducts experiments that demonstrate the fallacies in the spontaneous generation theory.
1776	An Italian anatomist, Lazzaro Spallanzani, conducts further convincing experiments that dispute spontaneous generation.
1796	English surgeon Edward Jenner introduces a vaccination for smallpox.
1838	Phillipe Ricord, a French physician, inoculates 2,500 human subjects to demonstrate that syphilis and gonorrhea are two separate diseases.
1839	Theodor Schwann, a German zoologist, and Matthias Schleiden, a botanist, formalize the theory that all living things are composed of cells.
1847–1850	The Hungarian physician Ignaz Semmelweis substantiates his theory that childbed fever is a contagious disease transmitted to women by their physicians during childbirth; he institutes the first use of antiseptics to reduce hand-borne disease.
1853–1854	John Snow, a London physician, demonstrates the epidemic spread of cholera through a water supply contaminated with human sewage.
1857	French bacteriologist Louis Pasteur shows that fermentations are due to microorganisms, and originates the process now known as pasteurization.
1858	Rudolf Virchow, a German pathologist, introduces the concept that all cells originate from preexisting cells.
1861	Louis Pasteur completes the definitive experiments that finally lay the theory of spontaneous generation to rest.
1867	The English surgeon Joseph Lister publishes the first work on antiseptic surgery, beginning the trend toward modern aseptic techniques in medicine.
1869	Johann Miescher, a Swiss pathologist, discovers the presence of complex acids in the cell nucleus, which he terms nuclein (DNA, RNA).
1876–1877	German bacteriologist Robert Koch* studies anthrax in cattle and implicates the bacterium *Bacillus anthracis* as its causative agent.
1881	Pasteur develops a vaccine for anthrax in animals.
	Koch introduces the use of pure culture techniques for handling bacteria in the laboratory.
	Walther and Fanny Hesse introduce agar-agar as a solidifying gel for culture media.
1882	Koch identifies the causative agent of tuberculosis.
1884	Koch outlines his postulates.
	Elie Metchnikoff,* a Russian zoologist, lays groundwork for the science of immunology by discovering phagocytic cells.
	The Danish physician Hans Christian Gram devises the Gram stain technique for differentiating bacteria.
1885	Pasteur develops a special vaccine for rabies.
1887	Julius Petri, a German bacteriologist, adapts two plates to form a container for holding media and culturing microbes.
1890	A German, Emil von Behring,* and a Japanese, Shibasaburo Kitasato, demonstrate the presence of antibodies in serum that neutralize the toxins of diphtheria and tetanus.

argument (see the discussion of abiogenesis in feature 1.2, page 14). The need for an experimental system that answered questions objectively and was not based on prejudice marked the beginning of true scientific thinking. These ideas gradually crept into the consciousness of the scientific community during the 1600s.

The general approach taken by scientists to explain a certain natural phenomenon is called the **scientific method.** The first step in this method is to formulate a **hypothesis,** a tentative explanation to account for what has been observed or measured. A good hypothesis must be capable of being either supported or falsified (discredited) by careful, systematic observation or experimentation. For example, the statement that "bees make honey from pollen" can be experimentally determined by the tools of science, but the statement that "bees buzz because they are happy" cannot.

The two types of reasoning commonly applied separately or in combination to develop and support hypotheses are **induction** and **deduction** (figure 1.7). In the **inductive** process, a scientist accumulates specific data or facts and then formulates a general hypothesis that accounts for these facts. The inductive approach asks, "Are various observed events best explained by this hypothesis or by another one?" In the **deductive** approach, a scientist constructs a hypothesis, tests its validity by outlining particular events that are predicted by the hypothesis, and then performs experiments to test for these events. The deductive process states: "If the hypothesis is valid, then certain specific events can be expected to occur."

Natural processes have numerous physical, chemical, and biological factors, or *variables,* that can hypothetically affect their outcome. To account for these variables, experiments are

Table 1.2—*Continued*

Date	Event and Its Importance
1892	A Russian, D. Ivanovski, is the first to isolate a virus (the tobacco mosaic virus) and show that it could be transmitted in a cell-free filtrate.
1895	Jules Bordet,* a Belgian bacteriologist, discovers the antimicrobial powers of complement.
1898	R. Ross* and G. Grassi demonstrate that malaria is transmitted by the bite of female mosquitos.
	Germans Friedrich Loeffler and P. Frosch discover that "filterable viruses" cause foot and mouth disease in animals.
1899	Dutch microbiologist Martinus Beijerinck further elucidates the viral agent of tobacco mosaic disease and postulates that viruses have many of the properties of living cells and that they reproduce within cells.
1900	The American physician Walter Reed and his colleagues clarify the role of mosquitos in transmitting yellow fever.
	An Austrian pathologist, Karl Landsteiner,* discovers the ABO blood groups.
1903	American pathologist James Wright and others demonstrate the presence of antibodies in the blood of immunized animals.
1905	Syphilis is shown to be caused by *Treponema pallidum,* through the work of German bacteriologists Fritz Schaudinn and E. Hoffman.
1906	August Wasserman, a German bacteriologist, develops the first serologic test for syphilis.
	Howard Ricketts, an American pathologist, links the transmission of Rocky Mountain spotted fever to ticks.
1908	The German Paul Ehrlich* becomes the pioneer of modern chemotherapy by developing salvarsan, an arsenic-based drug, to treat syphilis.
1910	An American pathologist, Francis Rous,* discovers viruses that can induce cancer.
1915–1917	British scientist F. Twort and French scientist F. D'Herelle independently discover bacterial viruses.
1928	Frederick Griffith lays the foundation for modern molecular genetics by his discovery of transformation in bacteria.
1929	A Scottish bacteriologist, Alexander Fleming,* discovers and describes the properties of the first antibiotic, penicillin.
1933–1938	Germans Ernst Ruska* and von Borries develop the first electron microscope.
1935	Gerhard Domagk,* a German physician, discovers the first sulfa drug and paves the way for the era of antimicrobic chemotherapy.
	Wendell Stanley is successful in inducing tobacco mosaic viruses to form crystals that still retain their infectiousness.
1941	Australian Howard Florey* and Englishman Ernst Chain* develop commercial methods for producing and purifying penicillin; this first antibiotic is tested and put into widespread use.
1944	Oswald Avery, Colin McLeod, and Maclyn McCarty show that DNA is the genetic material.
	Joshua Lederberg* and E. L. Tatum* discover conjugation in bacteria.
	The Russian Selman Waksman* and his colleagues discover the antibiotic streptomycin.
1953	James Watson,* Francis Crick,* Rosalind Franklin, and Maurice Wilkins* determine the structure of DNA.
1954	Jonas Salk develops the first polio vaccine.
1957	Alick Isaacs and Lindenmann discover the natural antiviral substance interferon.
1957	D. Carleton Gajdusek* discovers the underlying cause of slow virus diseases.
1959–1960	Gerald Edelman* and Rodney Porter* determine the structure of antibodies.
1972	Paul Berg* develops the first recombinant DNA in a test tube.
1973	Herb Boyer and Stanley Cohen clone the first DNA using plasmids.

*These persons were awarded Nobel prizes for their contributions to the field.

designed (1) to thoroughly test for or measure the consequences of each possible variable and (2) to accompany each variable with one or more *control groups.* A control group is designed exactly as the test group but omits only that variable being tested; thus, it may serve as a basis of comparison for the test group. The reasoning is that, if a certain experimental finding occurs only in the test group and not in the control group, the finding must be due to the variable being tested and not to some uncontrolled, untested factor that is not part of the hypothesis (figure 1.8).

A lengthy process of experimentation, analysis, and testing eventually leads to conclusions that either support or refute the hypothesis. If experiments do not uphold the hypothesis—that is, if it is falsified—the hypothesis or some part of it is rejected, and it is either discarded or modified to fit the results of the experiment. If the hypothesis is supported by the facts from the experiment, it is not (or should not be) immediately accepted as fact. It then must be tested and retested. Indeed, this is an important guideline in the acceptance of a hypothesis: The results must be published and repeated by other investigators. The history of science is littered with the carcasses of apparently ingenious hypotheses that simply could not be repeated and were thus ultimately refuted.

In time, as each hypothesis is supported by a growing body of data and survives rigorous scrutiny, it moves to the next level of acceptance—the **theory.** A theory is a collection of statements, propositions, or concepts that explains or accounts for a natural event. A theory is not the result of a single experiment repeated over and over again but is an entire body of ideas that expresses or explains many aspects of a phenomenon. It is not

Figure 1.5 An old painting of Antony van Leeuwenhoek sitting in his laboratory. J. R. Porter and C. Dobell have commented on the unique qualities Leeuwenhoek brought to his craft: "He was one of the most original and curious men who ever lived. It is difficult to compare him with anybody because he belonged to a genus of which he was the type and only species, and when he died his line became extinct."

a fuzzy or weak speculation, as is sometimes the popular notion, but a viable declaration that has stood the test of time and has yet to be disproven by serious scientific endeavors. Often, theories develop and progress through decades of research and are added to and modified by new findings. At some point, evidence of the accuracy and predictability of a theory is so compelling that the next level of confidence is reached and the theory becomes a **law** or principle. For example, although we may still refer to the "germ theory of disease," so little question remains that microbes can cause disease that it has clearly passed into the realm of law.

Science and its hypotheses and theories must progress along with technology. As advances in instrumentation allow new, more detailed views of living phenomena, old theories may be reexamined and altered and new ones proposed. But scientists do not take the stance that theories are ever absolutely proven. The characteristics that make scientists most effective in their work are curiosity, open-mindedness, skepticism, creativity, cooperation, and readiness to revise their views of natural processes.

The Development of Medical Microbiology

Early experiments on the sources of microorganisms led to the profound realization that, not only are air and dust full of microbes, but so is the entire surface of the earth, its waters, and

(a)

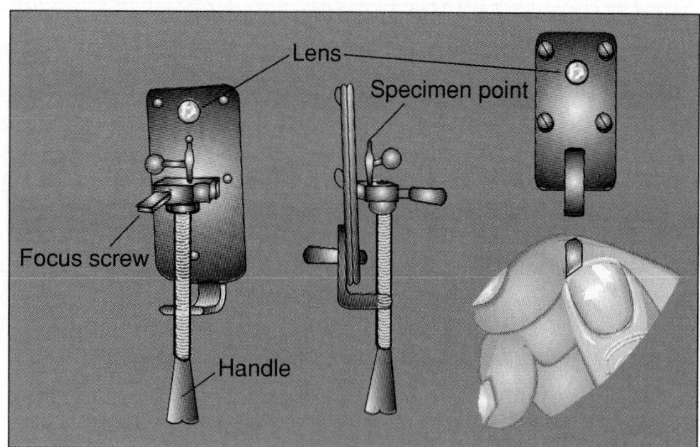

(b)

Figure 1.6 (*a*) A model of a Leeuwenhoek microscope. (*b*) Examples of bacteria drawn by Leeuwenhoek. He keenly observed, "I discovered living creatures in rain water which had stood but a few days in a new earthen pot, glazed blew within. This invited me to view this water with great attention, especially those little animals appearing to me ten thousand times less than those which may be perceived in the water with the naked eye." This is probably the first observation of bacteria.

all objects exposed to them. This discovery led to immediate applications in medicine, thus the seeds of medical microbiology were sown in the mid- to latter half of the nineteenth century with the introduction of the germ theory of disease and the resulting use of sterile, aseptic, and pure culture techniques.

The Study of Spores and Sterilization

Following Pasteur's inventive work with infusions (see feature 1.2), it was not long before English physicist John Tyndall demonstrated that heated broths would not spoil if stored in chambers completely free of dust. His studies provided the initial evidence that some of the microbes in dust and air have very high heat resistance, and that particularly vigorous treatment is required to destroy them. Later, the discovery and detailed description of heat-resistant bacterial endospores by Ferdinand Cohn, a German botanist, clarified why heat would sometimes

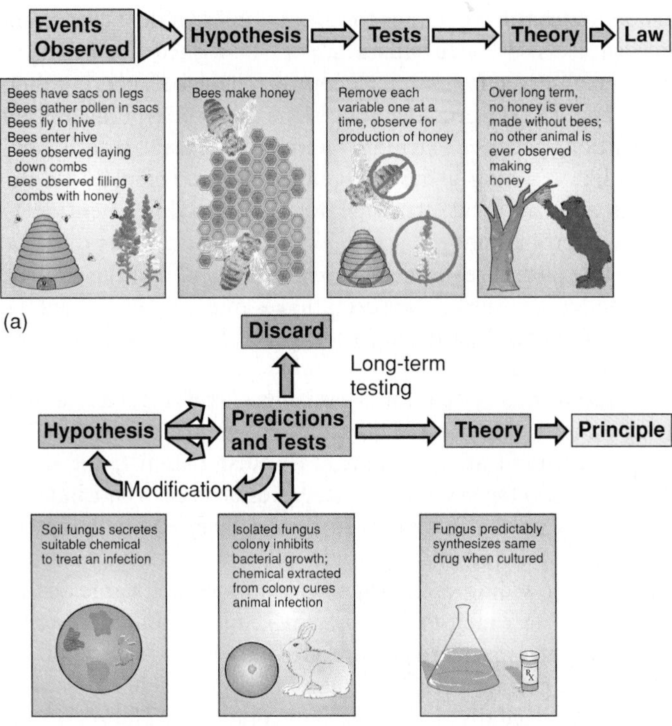

Figure 1.7 Comparing the inductive and deductive approaches to the scientific method. (*a*) The inductive process proceeds from specific observations to a general hypothesis. (*b*) The deductive process starts with a general hypothesis that predicts specific expectations.

Subject: Testing the factors responsible for dental caries

Hypothesis: Dental caries (cavities) involve dietary sugar or microbial action or both.

Variables:
Test animal: Requires a germ-free rat reared in total absence of microorganisms (in order to control this variable)

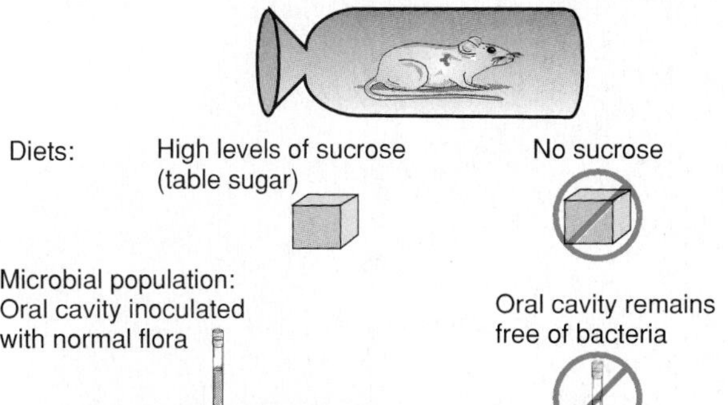

Experimental Protocol:

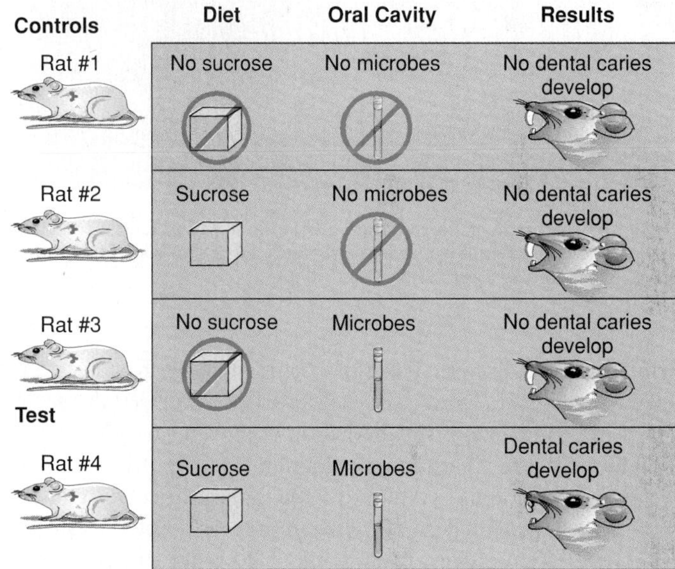

Conclusion: Dental caries will not develop unless both sucrose and microbial action are present. What other variables were not controlled?

Figure 1.8 An example of the use of controls in experiments.

fail to completely eliminate microorganisms. The modern sense of **sterile,** meaning completely free of all life forms, was established from that point (see chapter 9). The capacity to sterilize objects and materials is an absolutely essential part of microbiology, medicine, dentistry, and industry.

The Development of Aseptic Techniques

At the same time that abiogenesis was being hotly debated, a few budding microbiologists began to suspect that not only were spoilage and decay caused by microorganisms, but so were infectious diseases. It occurred to these rugged individualists that even the human body itself was a source of infection. An American physician, Oliver Wendell Holmes, observed that mothers experienced fewer infections when they gave birth at home, and the Hungarian Ignaz Semmelweis showed quite clearly that women became infected in the maternity ward following examinations by physicians coming directly from the autopsy room. Though some small notice was taken of these observations, it was the English surgeon Joseph Lister who first introduced **aseptic techniques** aimed at reducing microbes in a medical setting and preventing wound infections. The essence of his asepsis was to disinfect the hands and the air with strong antiseptic chemicals such as phenol prior to surgery. These techniques and the application of heat for sterilization became the bases for microbial control by physical and chemical methods still in use today (see chapter 9).

The Discovery of Pathogens and the Germ Theory of Disease

Two great founders of microbiology who introduced techniques that are still with us today were Louis Pasteur of France (see chapter opening illustration) and Robert Koch of Germany (figure 1.9). Pasteur made enormous contributions to our understanding of the microbial role in wine and beer formation. He also invented pasteurization and completed some of the first studies showing that diseases could arise when microbes entered the tissues and grew there, which became known as the

---

sterile (stair'-il) Gr. *steira,* barren.

aseptic (ay-sep'-tik) Gr. *a,* no, and *sepsis,* decay or infection. These techniques are aimed at reducing pathogens and do not necessarily sterilize.

Figure 1.9 Robert Koch, intent at his laboratory workbench and surrounded by the new implements of his trade: Petri plates, tubes and flasks filled with media, smears of bacteria, and bottles of stains.

germ theory of disease. Pasteur's contemporary, Koch, established *Koch's Postulates,* a series of proofs that verified the germ theory and could establish whether an organism was pathogenic and which disease it caused (see chapter 11). About 1875, Koch used this experimental system to show that anthrax was caused by a bacterium called *Bacillus anthracis.* So useful were his postulates that the causative agents of 20 other diseases were discovered between 1875 and 1900, and even today, they are the standard for identifying pathogens.

During this golden age of the 1880s, numerous other exciting creations emerged from Koch's prolific and probing laboratory work. He realized that study of the microbial world would require separating microbes from each other and growing them in culture. It is not an overstatement to say that he and his colleagues invented most of the techniques that are described in the last section of this chapter: inoculation, isolation, media, maintenance of pure cultures, and preparation of specimens for microscopic examination. Other highlights in this era of discovery are presented in later chapters on microbial control (see chapter 9) and vaccination (see chapter 13).

Taxonomy: Organizing, Classifying, and Naming Microorganisms

Students just beginning their microbiology studies are often struck by the seemingly endless array of new, unusual, and sometimes confusing names for groups and specific types of microorganisms. We once had a student who wished aloud that the cells she observed in her microscope carried labels bearing their name and correct pronunciation. One can well appreciate her frustration, because learning microbial **nomenclature** is very much like learning a new language, and occasionally its demands may be a bit overwhelming. But paying attention to proper microbial names is just like following a baseball game or a movie plot: It really helps to know the names of the players. Your understanding and appreciation of microorganisms will be greatly improved by learning a few general rules about why and how they are named.

The formal system for organizing, classifying, and naming living things is **taxonomy.** This science originated more than 250 years ago when Carl von Linné, a Swedish botanist, laid down the basic rules for taxonomic categories or **taxa.** Von Linné realized early on that a system for recognizing and defining the properties of living things would prevent chaos in scientific studies by providing each organism with a unique name and an exact "slot" in which to catalogue it. This classification would then serve as a means for future identification of that same organism and permit workers in many biological fields to know if they were indeed discussing the same organism. The von Linné system has served well in categorizing the two million or more different types of organisms that have been discovered since that time.

Taxonomy has several major aims, but its primary concerns are classification, nomenclature, and identification. These three areas are interrelated and play a vital role in keeping a dynamic inventory of the extensive array of living things. **Classification** is the orderly arrangement of organisms into groups, preferably in a format that shows evolutionary relationships. Nomenclature is the process of assigning names to the various taxonomic rankings of each microbial species. **Identification** is the process of discovering and recording the traits of organisms so that they may be placed in an overall taxonomic scheme. We will survey some general principles of identification in a later section of this chapter.

The Levels of Classification

The main taxa in a classification scheme are organized in seven descending ranks, beginning with a **kingdom,** the largest and most general, and ending with a **species,** the smallest and most specific. All the members of a kingdom share only one or a few general characteristics, while members of a species are all the same kind of organism—that is, they share the majority of their characteristics. The five taxa between the top and bottom levels

nomenclature (noh'-men-klay"-chur) L. *nomen,* name, and *calare,* to call. A system of naming.

taxonomy (tacks-on'-uh-mee) Gr. *taxis,* arrangement, and *nomos,* name.

taxa (tacks'-uh) sing. taxon.

species (spee'-sheez) L. *specere,* kind. In biology this term is always in the plural form.

Kingdom: Animalia
Phylum: Chordata
Class: Mammalia
Order: Primates
Family: Hominoidea
Genus: *Homo*
Species: *sapiens*

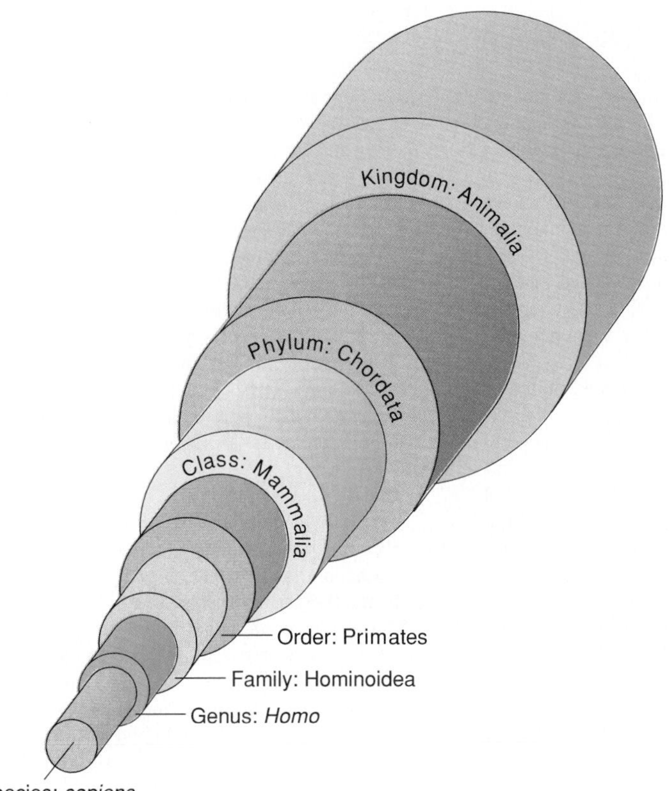

Figure 1.10 The levels in the classification of human beings from kingdom to species operate like a set of "Chinese boxes."

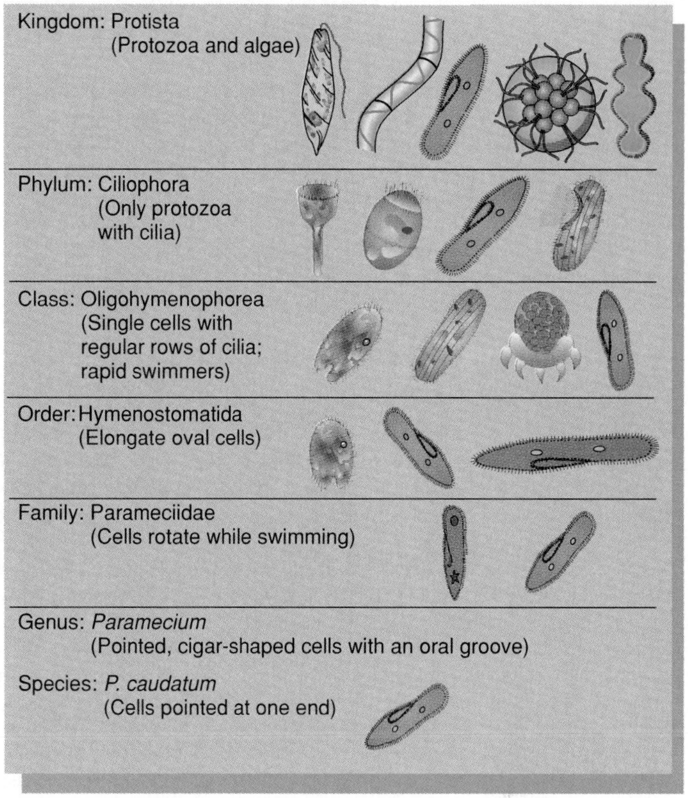

Kingdom: Protista
(Protozoa and algae)
Phylum: Ciliophora
(Only protozoa with cilia)
Class: Oligohymenophorea
(Single cells with regular rows of cilia; rapid swimmers)
Order: Hymenostomatida
(Elongate oval cells)
Family: Parameciidae
(Cells rotate while swimming)
Genus: *Paramecium*
(Pointed, cigar-shaped cells with an oral groove)
Species: *P. caudatum*
(Cells pointed at one end)

Figure 1.11 A common species of protozoan, *Paramecium caudatum*, traced through its taxonomic series. Note the gradual narrowing of the members, proceeding from general to specific levels.

are, in descending order: **phylum** or **division,**[3] **class, order, family,** and **genus.** Thus, each kingdom may be subdivided into a series of phyla, each phylum is made up of several classes, each class contains several orders, and so on. Since taxonomic schemes are to some extent artificial, certain groups of organisms do not exactly fit into the seven taxa. In this case, additional levels may be imposed immediately above (super) or below (sub) a taxon, giving us such categories as superphylum and subclass.

To illustrate the fine points of this system, we compare the taxonomic breakdowns of a human (figure 1.10) and a protozoan (figure 1.11). Humans belong to the Kingdom Animalia, and to emphasize just how broad this category is, ponder the fact that we belong to the same kingdom as sponges. Of the several phyla within this kingdom, humans belong to the Phylum Chordata (nerve cord-bearing animals), but even a phylum is rather all-inclusive, considering that humans share it with certain marine worms. The next level, Class Mammalia, narrows the field considerably by grouping only those vertebrates that have hair and suckle their young. The order that humans belong to is called the Primates, and it also includes apes, monkeys, and lemurs. Next comes the Family Hominoidea, containing just humans and apes. The final levels are our genus, *Homo* (all races of modern and ancient humans), and our species, *sapiens* (meaning wise). Notice that in both the human and the protozoan, the categories become less inclusive and the individual members more closely related. Other examples of classification schemes are provided in sections of chapters 3 and 4 and in several later chapters.

It would be well to remember that all taxonomic *hierarchies* are based on the judgment of scientists with certain expertise in a particular group of organisms, and that not all other experts may agree with the system being used. Consequently, no taxa are permanent to any degree; they are constantly being revised and refined as new information becomes available or new viewpoints become prevalent. The detailed taxonomy of many groups is subject to lively discussion, but since this text does not aim to emphasize details of taxonomy, we will usually be concerned with only the most general (kingdom, phylum) and specific (genus, species) levels.

phylum (fye'-lum) pl. phyla (fye'-luh) Gr. *phylon,* race.

3. The term phylum is used for animals and protozoa; the term division is used for bacteria, fungi, algae, and plants.

genus (jee'-nus) pl. genera (jen'-er-uh) L. birth, kind.

hierarchy (hy'-ur-ar-kee) L. *hierarchia,* levels of power. Things arranged in the order of rank.

Feature 1.2 Spontaneous Generation and the Battle of the Hypotheses

For thousands of years, it had been generally held that certain living things arose intact from vital forces inherent in nonliving or decomposing matter. This ancient belief in **spontaneous generation,** or abiogenesis, had evolved as people observed that meat left out in the open soon "produced" maggots, that mushrooms appeared on rotting wood, that rats and mice emerged from piles of litter, and other similar phenomena. On the surface, there is nothing incorrect about the observations themselves; it is the conclusions drawn from them that do not hold up to critical reasoning.

As early as 40 B.C., Vergil gave a recipe for growing bees artificially, and as late as 1550, von Helmont told how to produce mice by leaving piles of grain and cheese undisturbed in an attic. Even the discovery of single-celled organisms during the latter part of the mid-1600s did not kill the idea of abiogenesis, because some scientists assumed that microscopic beings were simply one stage in a process of forming larger organisms from small ones. Though some of these earlier ideas seem quaint and ridiculous in light of modern knowledge, we must remember that at the time, mysteries in life were accepted and the scientific method was not widely practiced. The eventual invalidation of the theory of abiogenesis was a very gradual process.

Two hypotheses attempted to explain the origin of the "simpler forms" of life: (1) They arose spontaneously by the massing of vital forces within nonliving matter (**abiogenesis**), or (2) they arose only from other living things of their same kind (**biogenesis**). If an investigator attempted to verify one, he was automatically capable of falsifying the other, because both cannot be true. Throughout history, serious proponents existed on both sides, but determining the "winning" hypothesis required 200 years of combined inquiries by microbiologists from several countries, each of whom dealt with a particular disputed part of the hypotheses.

Among the important variables to be considered in challenging the hypotheses were the effects of nutrients, air, and heat, and the presence of microbes in the environment. One of the first people to doubt the spontaneous generation theory was Francesco Redi of Italy. He conducted a simple test in which he placed meat in a jar and covered it with fine gauze. Flies gathering at the jar were blocked from entering and so laid eggs on the outside of the gauze. The maggots subsequently developed without access to the meat, indicating that maggots were the offspring of flies and did not arise from some "vital force" in the meat. This and related experiments laid to rest the idea that more complex animals such as insects and mice developed through abiogenesis, but it did not convince many scientists of the day that simpler organisms could not arise this way.

Frenchman Louis Jablot reasoned that even microscopic organisms must have parents, and his experiments with infusions (dried hay steeped in water) supported this. He divided an infusion that had been boiled to destroy any living things into a heat-treated container that was closed to the air and a heated container that was freely open to the air. Only the open vessel developed microorganisms, which he presumed had entered with dust-laden air. But the cause of biogenesis was temporarily set back by John Needham, an Englishman who did similar experiments using mutton gravy. His results were in conflict with Jablot's because both heated and unheated test containers teemed with microbes. Unfortunately, his experiments were done before the presence of heat-resistant endospores and the concept of true methods of sterility were widely known.

Not to let Needham's conclusion remain untested, an Italian naturalist, Lazzaro Spallanzani, took more extensive and sophisticated measures, which included long-term heating of the test vessels and infusions so as to destroy any lingering resistant forms. He found that a variety of liquid nutrients (broths, urine) thus treated and kept sealed in flasks from air would not develop growth. When he was accused of destroying the vegetative force of the nutrients by overheating, he showed that the heated nutrients could still grow microbes when exposed to air. Undaunted by this argument, other critics objected that Spallanzani's flasks were sealed from oxygen, a gas known to be required in the respiration of animals. Somehow, Leeuwenhoek's earlier discovery that free oxygen was not necessary for microbial growth was overlooked at the time.

Several types of experiments were applied in further defense of biogenesis. Franz Shultze and Theodor Schwann of Germany felt assured that air was the source of microbes and sought to prove this by passing air through strong chemicals or hot glass tubes into heat-treated infusions in flasks. When the infusions again remained devoid of living things, the supporters of abiogenesis claimed that the treatment of the air had made it harmful to the spontaneous development of life. These studies were followed up by Georg Schroeder and Theodor Van Dusch, who did not treat the air with heat or chemicals, but passed it through cotton wool to filter out microscopic organisms. Again, no microbes grew in the infusions. Though these experiments should have laid the arguments for spontaneous generation to rest, they did not.

Finally, in the mid-1800s, the great microbiologist Louis Pasteur entered the arena. He had recently been studying the roles of microorganisms in the fermentation of beer and wine, and it was clear to him that these processes were brought about by the activities of microbes introduced into the beverage from air, fruits, and grains. The methods he used to discount abiogenesis were simple yet brilliant. He repeated the experiments using cotton filters to trap dust from air and observed tiny objects (probably spores) in the filters. He also observed that these same filters would initiate growth in previously sterile broths. To further clarify that air was the source of microbes, he filled flasks with broth and fashioned their openings into elongate, swan neck-shaped tubes. The flasks' openings were freely open to the air but curved so that gravity would cause any airborne dust particles to deposit in the lower part of the necks. He heated the flasks to sterilize the broth and incubated them. As long as the flask remained intact, the broth remained sterile, but if the neck were broken off so that dust fell directly down into the container, microbial growth immediately commenced. Some of these ingenious little flasks are still on display at the Institute Pasteur in Paris in their original sterile form. Pasteur summed up his findings: "For I have kept from them, and am still keeping from them, that one thing which is above the power of man to make; I have kept from them the germs that float in the air, I have kept from them life."

---

abiogenesis (ah-bee″-oh-jen′-uh-sis) Gr. *a,* not, *bios,* living, and *gennan,* to produce.

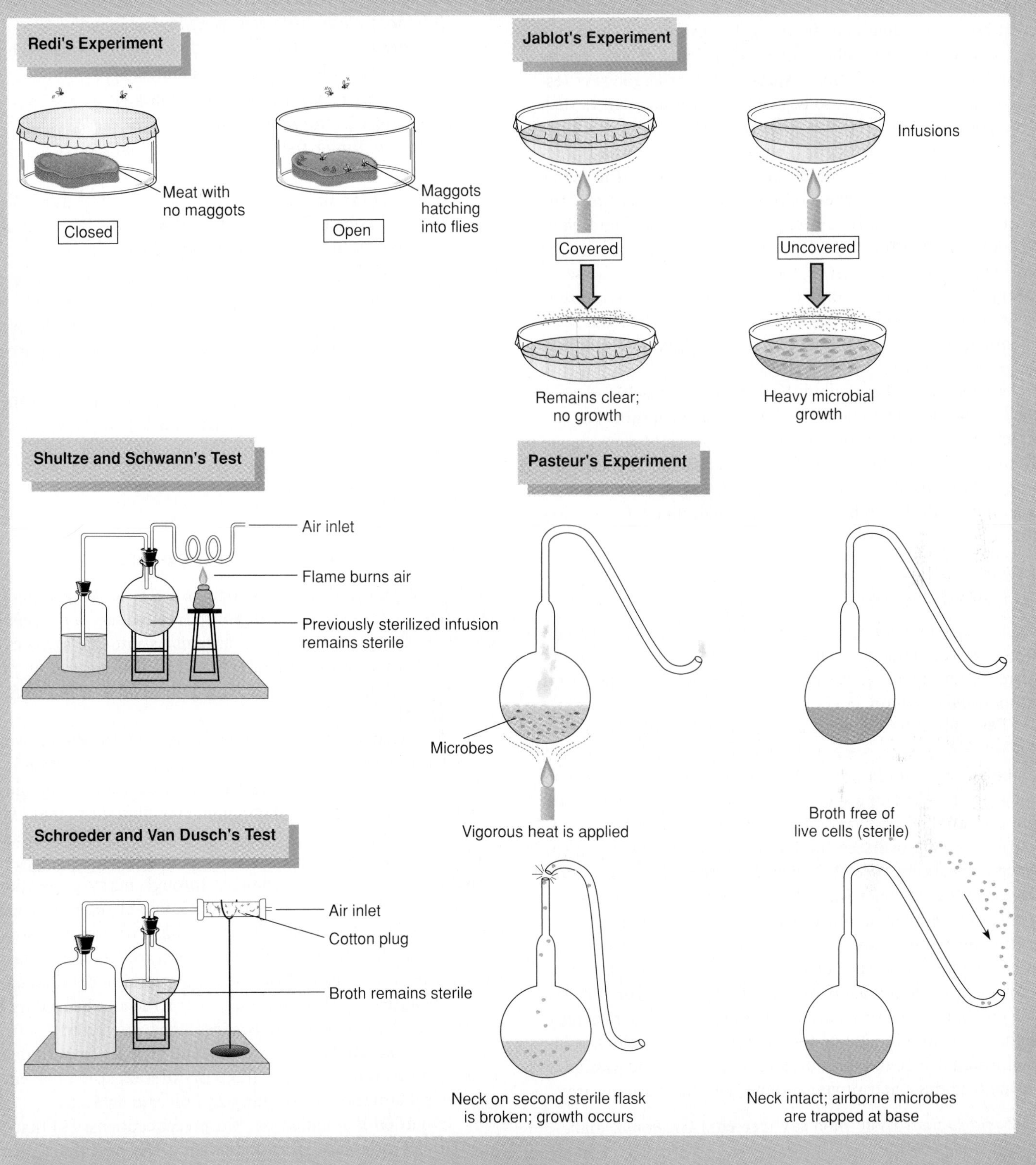

Redi's Experiment
Meat with
no maggots
Closed
Maggots
hatching
into flies
Open
Jablot's Experiment
Infusions
Covered
Uncovered
Remains clear;
no growth
Heavy microbial
growth
Shultze and Schwann's Test
Air inlet
Flame burns air
Previously sterilized infusion
remains sterile
Pasteur's Experiment
Microbes
Vigorous heat is applied
Broth free of
live cells (sterile)
Schroeder and Van Dusch's Test
Air inlet
Cotton plug
Broth remains sterile
Neck on second sterile flask
is broken; growth occurs
Neck intact; airborne microbes
are trapped at base

Assigning Species Names

Many larger organisms are known by a common name suggested by certain dominant features. For example, a bird species might be called a "blue-footed booby" or a flower species an "Indian paintbrush." Some species of microorganisms (especially pathogens) may also be called by informal names such as the gonococcus (*Neisseria gonorrhoeae*) or the leprosy bacillus (*Mycobacterium leprae*), but this is not the usual practice. If we were to adopt common names such as the "little yellow coccus" or the "club-shaped false diphtheria bacterium," the terminology would become even more cumbersome and challenging than scientific names. Even worse, common names are notorious for varying from region to region, even in the same country. A decided advantage of standardized nomenclature is that it provides a universal language, enabling scientists from Argentina, Japan, Turkey, or Outer Mongolia to freely exchange information.

The method of assigning the **scientific** or **specific name** is called a *binomial* (two-name) system of nomenclature. The scientific name is always a combination of the generic (genus) name followed by the species name. The generic part of the scientific name is capitalized, and the species part begins with a lowercase letter. Both should be italicized (or underlined if italics are not available), as follows:

Cryptococcus neoformans
genus species

Because other taxonomic levels are not expressed this way, one can always recognize a scientific name. To save space, an organism's scientific name is sometimes given in shorthand form, such as *C. neoformans.*

The source for nomenclature is usually Latin or Greek, but if other languages such as English or French are used, the endings of these words are revised to have Latin endings. The strategies for assembling species names vary, but in general, the name first applied to a species will be its stable name. An international group oversees the naming of every new organism discovered, making sure that standard procedures have been followed and that there isn't already an earlier name for the organism or another organism with that same name. The inspiration for names is extremely varied and often rather imaginative. A number of species have been named in honor of a microbiologist who originally discovered the microbe or who has made outstanding contributions to the field. Other names may designate a characteristic of the microbe (shape, color), a location where it is found, or a disease it causes. Some examples of specific names, their pronunciations, and their origins are:

1. *Homo sapiens* (hoh'-moh say'-pee-enz) Gr. *homo,* man, and *sapiens,* wise.
2. *Micrococcus luteus* (my''-kroh-kok'-us loo'-tee-us) Gr. *micros,* small, and *kokkus,* berry; L. *luteus,* yellow. "The little yellow coccus."
3. *Pseudomonas tomato* (soo''-doh-mon'-us toh-may'-toh) Gr. *pseudo,* false, *monas,* unit, and *tomato,* the fruit. A bacterium that infects the common garden tomato.
4. *Corynebacterium pseudodiphtheriticum* (kor-eye''-nee-bak-ter'-ee-yum soo-doh''-dif-ther-it'-i-kum) Gr. *coryne,* club, *bacterium,* little rod, *pseudo,* false, and *diphtheriticus,* relating to the disease diphtheria. "The club-shaped, false diphtheria bacterium."
5. *Cryptococcus neoformans* (krip''-toh-kok'-us nee'-oh-form''-anz) Gr. *cryptos,* hidden, *kokkos,* berry, and *neoforma,* tumor-forming. A fungus that causes a serious brain infection.
6. *Lactobacillus sanfrancisco* (lak''-toh-bass-ill'-us san''-fran-siss'-koh) L. *lacto,* milk, and *bacillus,* little rod. A bacterial species used to make sourdough bread.
7. *Vampirovibrio chlorellavorus* (vam-pye''-roh-vib'-ree-oh klor-ell-ah'-vor-us) F. *vampire;* L. *vibrio,* curved cell; *Chlorella,* a genus of green algae; and *vorus,* to devour. A small curved bacterium that sucks out the cell juices of *Chlorella.*
8. *Giardia lamblia* (jee-ar'-dee-uh lam'-blee-uh) for Alfred Giard, a French microbiologist, and Vilem Lambl, a Bohemian physician, both of whom worked on the organism, a protozoan that causes a severe intestinal infection.

The Origin and Evolution of Microorganisms

Earlier we indicated that taxonomists prefer to use a system of classification that shows the degree of relatedness of organisms, one that places closely related organisms into the same categories. This pattern of organization, called a *natural* or *phylogenetic* system, often uses selected observable traits to form the categories.

A phylogenetic system is based on the concept of evolutionary relationships among types of organisms. **Evolution** is an important theme that underlies all of biology, including microbiology. From its simplest standpoint, evolution states that living things change gradually through hundreds of millions of years, and that these evolvements are expressed in various types of structural and functional changes through many generations. The process of evolution is selective: Those changes that most favor the survival of a particular organism or group of organisms tend to be retained and those that are less beneficial to survival tend to be lost. Space does not permit a detailed analysis of evolutionary theories, but its occurrence is backed up by a tremendous amount of evidence from the fossil record and from the study of **morphology, physiology,** and **genetics** (inheritance). Evolution accounts for the millions of different species on earth and their adaptation to its many and diverse habitats.

Evolution is founded on two preconceptions: (1) that all new species originate from preexisting species, and (2) that closely related organisms have similar features due to having

---

evolution (ev-oh-loo'-shun) L. *evolutio,* to roll out.

morphology (mor-fol'-oh-jee) Gr. *morphos,* form, and *logos,* to study. The study of organismic structure.

physiology (fiz''-ee-ol'-oh-jee) Gr. *physis,* nature. The study of the function of organisms.

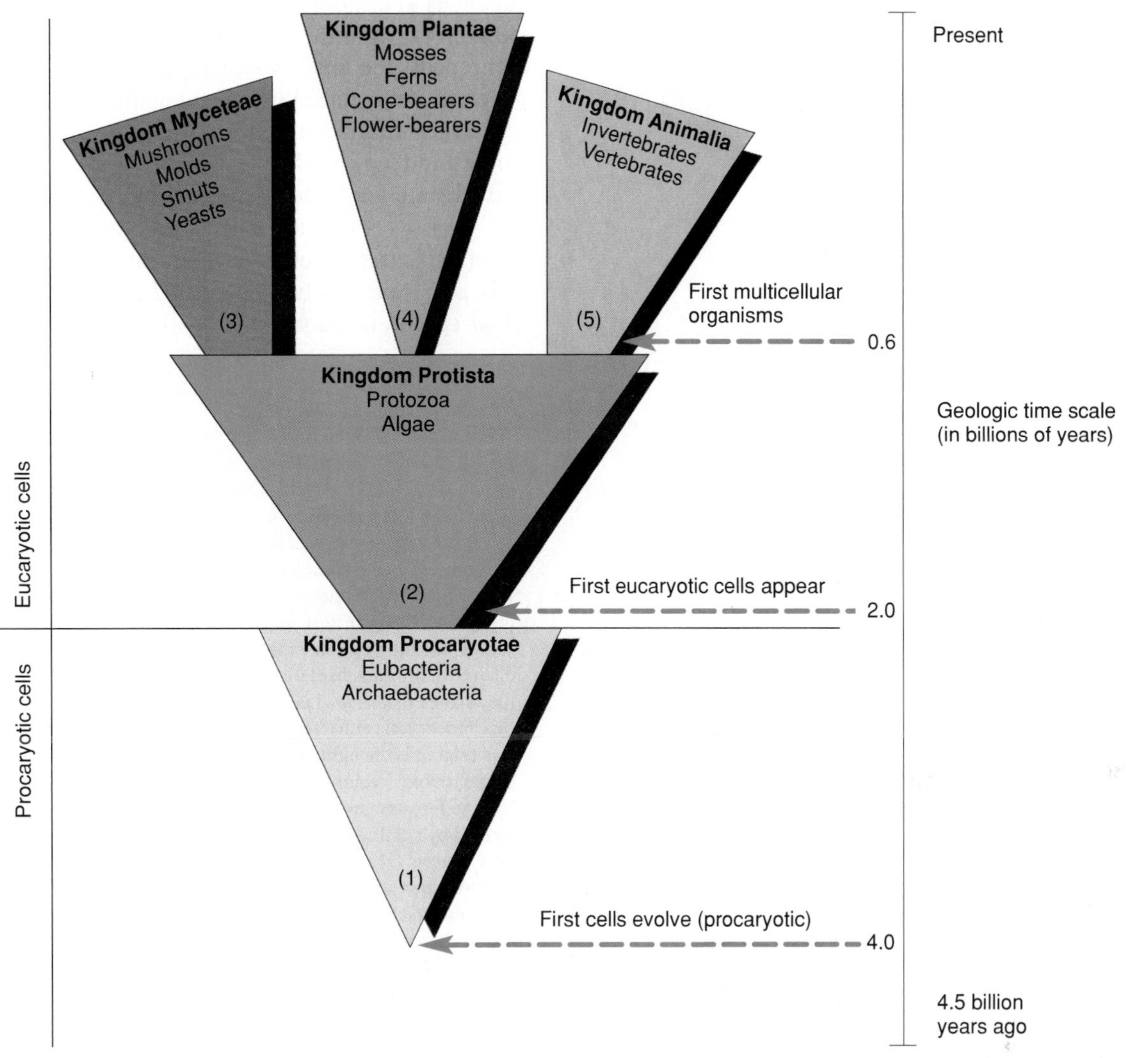

Figure 1.12 Kingdoms are based on cell structure and type, the nature of body organization, and nutritional type. (*1*) Procaryotae have procaryotic cells and are unicellular. (*2*) Protista have eucaryotic cells and are mostly unicellular. They may be photosynthetic (algae) or they may feed on other organisms (protozoa). (*3*) Myceteae have eucaryotic cells and are unicellular or multicellular; they have cell walls and are not photosynthetic. (*4*) Plantae have eucaryotic cells, are multicellular, have cell walls, and are photosynthetic. (*5*) Animalia have eucaryotic cells, are multicellular, do not have cell walls, and derive nutrients only from other organisms.

evolved from common ancestral forms. Usually, evolution progresses toward greater complexity, and evolutionary stages range from simple, more primitive forms that are closer to an ancestral organism to more complex, advanced forms. Although we use the terms primitive and advanced to denote the degree of change from the original set of ancestral traits, it is very important to realize that while all species presently residing on earth are modern, some have arisen more recently in evolutionary history than others.

The evolutionary patterns of organisms are often drawn as a family tree, with the trunk representing the main ancestral lines and the branches showing offshoots into specialized groups of organisms. This sort of arrangement places the more ancient groups at the bottom and the more recent ones at the top. The branches may also indicate closeness in relationships among the offshoot lines and, if a time scale is added, the beginning and length of an evolutionary period (figure 1.12).

The first phylogenetic trees of life were constructed on the basis of just two kingdoms (plants and animals). In time, it became clear that certain organisms did not truly fit either of these categories, so a third kingdom for simpler, single-celled organisms (protists) was recognized. Eventually, significant differences became evident even among the protists, thus Robert Whittaker proposed a fourth kingdom for the bacteria and a fifth one for the fungi. This currently accepted system places all living things in one of five basic kingdoms: (1) the Procaryotae or Monera, (2) the Protista, (3) the Myceteae or Fungi, (4) the Plantae, and (5) the Animalia (figure 1.12).

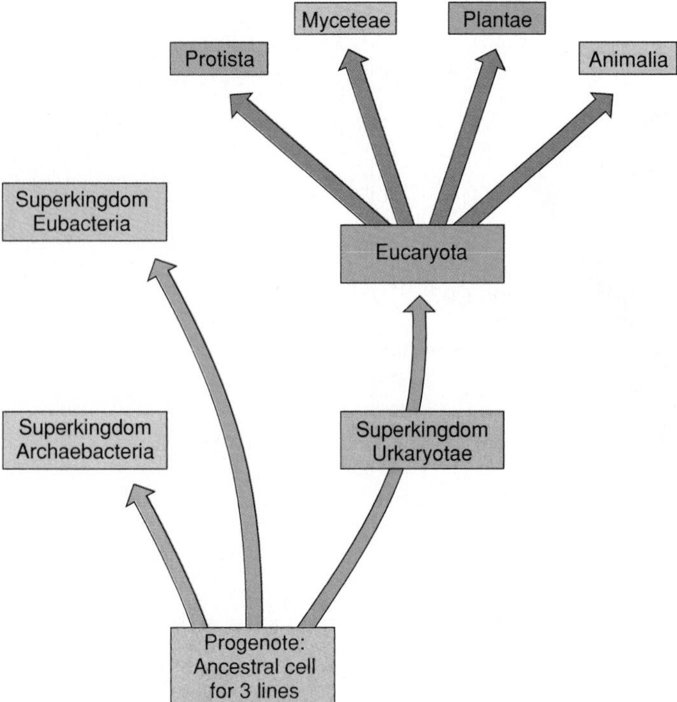

Figure 1.13 An alternate evolutionary system, suggesting that the major groups of living things evolved from three different cell lines (superkingdoms).

The simple, single-celled organisms at the base of the family tree are in the **Kingdom Procaryotae** (also called Monera). Since only those organisms with procaryotic cells are placed in it, cellular structure is the main defining characteristic for this kingdom. It includes all of the microorganisms commonly known as bacteria. Two main subgroups of the kingdom are the **eubacteria,** bacteria with characteristic procaryotic cell structure, and the **archaebacteria,** bacteria with atypical cell structure and function. Primitive bacteria were the earliest cells to appear on earth (see figure 3.27) and were the original ancestors of both more advanced bacteria and eucaryotic organisms. The Kingdom Procaryotae is a large and complex group that will be surveyed in more detail in chapter 3.

The other four kingdoms contain organisms composed of eucaryotic cells, whose probable origin from procaryotic cells is discussed in feature 4.1. The **Kingdom Protista** contains mostly single-celled microbes that lack more complex levels of organization, such as tissues. Its members include both the microscopic algae, defined as photosynthetic cells enclosed in walls, and the protozoans, which feed upon other live or dead organisms and lack cell walls. More information on this group's taxonomy is given in chapter 4. The **Kingdom Myceteae** contains the fungi, single- or multicelled eucaryotes that are encased in cell walls and absorb nutrients from other organisms (also see chapter 4). With the exception of certain infectious worms and arthropods, the final two kingdoms, **Animalia** and **Plantae,** are generally not included in the realm of microbiology because they are large, multicellular organisms. In general, animals move freely and feed on other organisms, while plants grow in an attached state and obtain nutrients by photosynthesis.

It is worth noting that viruses are not included in any of the classification or evolutionary schemes, because they are not cells and their position cannot be given with any confidence. Their special taxonomy will be discussed in chapter 5.

An Alternate View of Kingdoms and Cells Unquestionably, bacteria are the most ancient microbes on earth, and for the first two billion years of evolution, they were also the dominant life form (and still are in many ways!). Although biologists have found the system of five kingdoms and two basic cell types very useful, recent studies in molecular biology have forced them to reconsider its evolutionary accuracy. It has been determined that certain types of molecules in cells, called small ribosomal ribonucleic acid (rRNA), provide a "living record" of the evolutionary history of an organism. Analysis of this molecule in eubacteria, archaebacteria, and eucaryotic cells indicates that archaebacteria are so different from the other two groups that they should be included in a separate kingdom. These studies have also shown that procaryotic eubacterial cells and eucaryotic cells are not as different as was originally thought. These findings may lead to the possible redefinition of the current evolutionary scheme. A system proposed by Carl Woese and George Fox assigns all organisms to one of three superkingdoms, each described by a different type of cell: Archaebacteria, Eubacteria, and Urkaryotae (figure 1.13). It is believed that these three kingdoms arose from a common ancestor, called the progenote. Because this new system has not been universally embraced and is not yet in wide usage, we will continue to use five kingdoms and two basic cell types (procaryotic and eucaryotic) in this text. This example of a taxonomic disagreement emphasizes an important truism: Classification schemes can vary, but they don't change the characteristics or behavior of the microbes.

Tools of the Laboratory: The Methods of Studying Microorganisms

Microbiologists use five basic techniques to manipulate, examine, and characterize microorganisms: inoculation, incubation, isolation, inspection, and identification (the five "I's"; figure 1.14). Some or all of these activities are performed by microbiologists, whether the beginning student in the laboratory, the researcher attempting to isolate drug-producing bacteria from soil, or the clinical microbiologist working with a specimen from a patient's infection. These procedures make it possible to handle and maintain microorganisms as discrete and separate entities whose detailed biology can be studied and recorded.

Inoculation and Incubation

Persons studying large organisms such as animals and plants can, for the most part, readily see and differentiate their experimental subjects from the surrounding environment and from one another. In fact, they can use their senses of sight, smell, hearing, and touch to detect and evaluate major identifying

---

eubacteria (yoo″-bak-ter′-ee-uh) Gr. *eu,* true, and *bakterion,* little rod. All bacteria besides the archaebacteria.

archaebacteria (ark″-ee-bak-ter′-ee-uh) Gr. *archaios,* ancient.

Protista (pro-tiss′-tah) Gr. *protos,* the first.

Myceteae (my-cee′-tee-aye) Gr. *mycos,* the fungi.

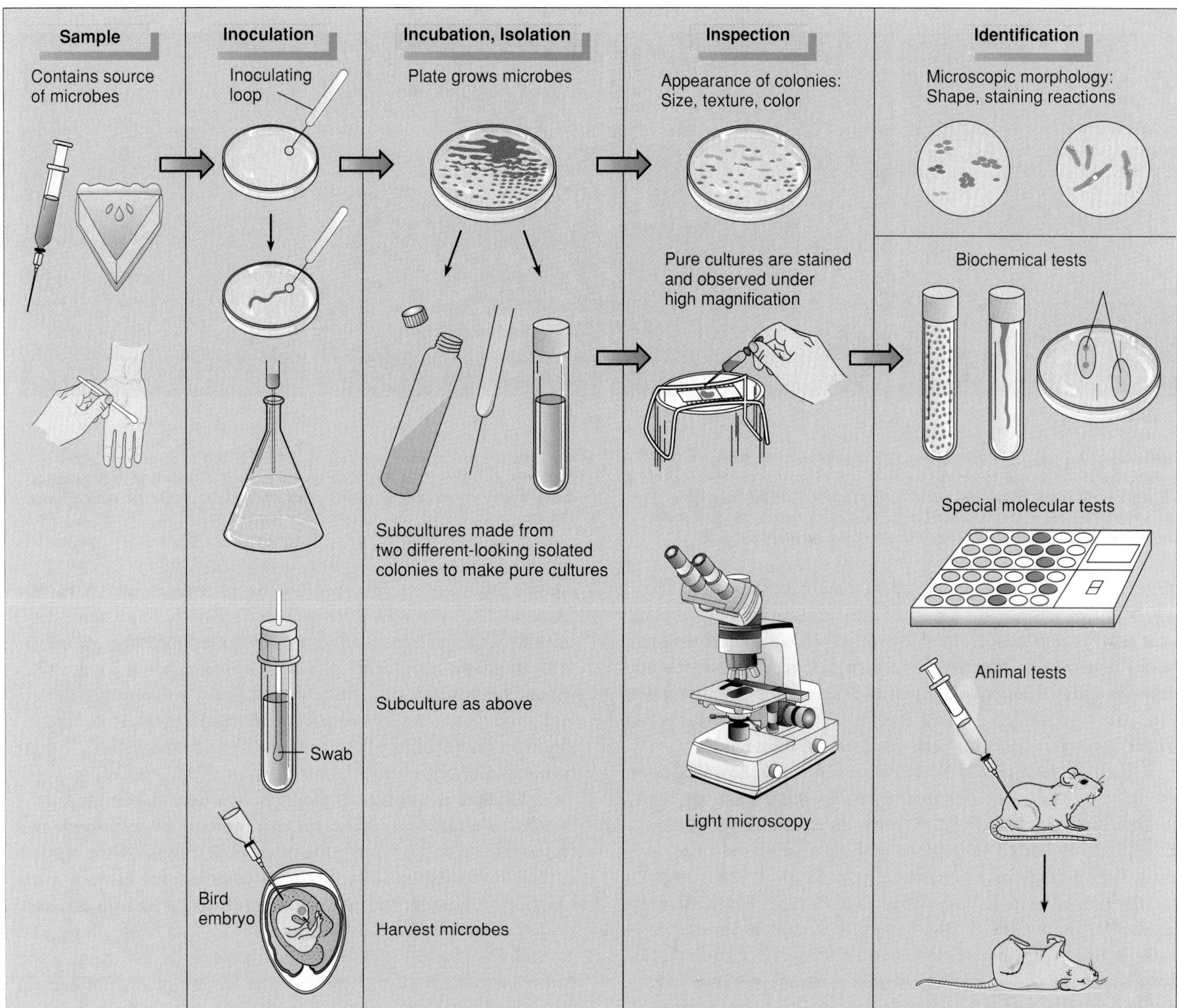

Figure 1.14 A summary of the general laboratory techniques carried out by microbiologists. It is not necessary to perform all the steps shown or to perform them exactly in this order, but all microbiologists participate in at least some of these activities. In some cases, one may proceed right from the sample to inspection, and in others, only inoculation and incubation on special media are required.

characteristics and to keep track of growth and developmental changes. But due to the minute nature of microorganisms and their dispersal throughout the environment, microbiologists are faced with some unique problems. First, to maintain and keep track of such elusive research subjects, it is often necessary to grow them in artificially prepared nutrient materials called culture **media.** Also, microbes in most habitats (such as the soil and the human mouth) exist in complex associations, thus to accurately assess the individual species, each must be pulled out of its mixed population and strictly separated from the other species. A third difficulty in working with microbes is that because they are invisible and widely distributed, undesirable ones can readily insinuate themselves into an experiment and destroy it. These impediments were largely responsible for the development of sterile, pure culture, and aseptic techniques.

The cultivation or **culturing** of microorganisms involves introducing a tiny sample of cells (the *inoculum*) into a container of nutrient medium and encouraging the cells to propagate. This process is called **inoculation.** The observable growth that appears in or on the medium is known as a **culture.** The

media (mee′-dee-uh) sing. medium; L. middle.

inoculation (in-ok″-yoo-lay′-shun) L. *in,* and *oculus,* bud.

culture (kull′-chur) Gr. *cultus,* to tend or cultivate.

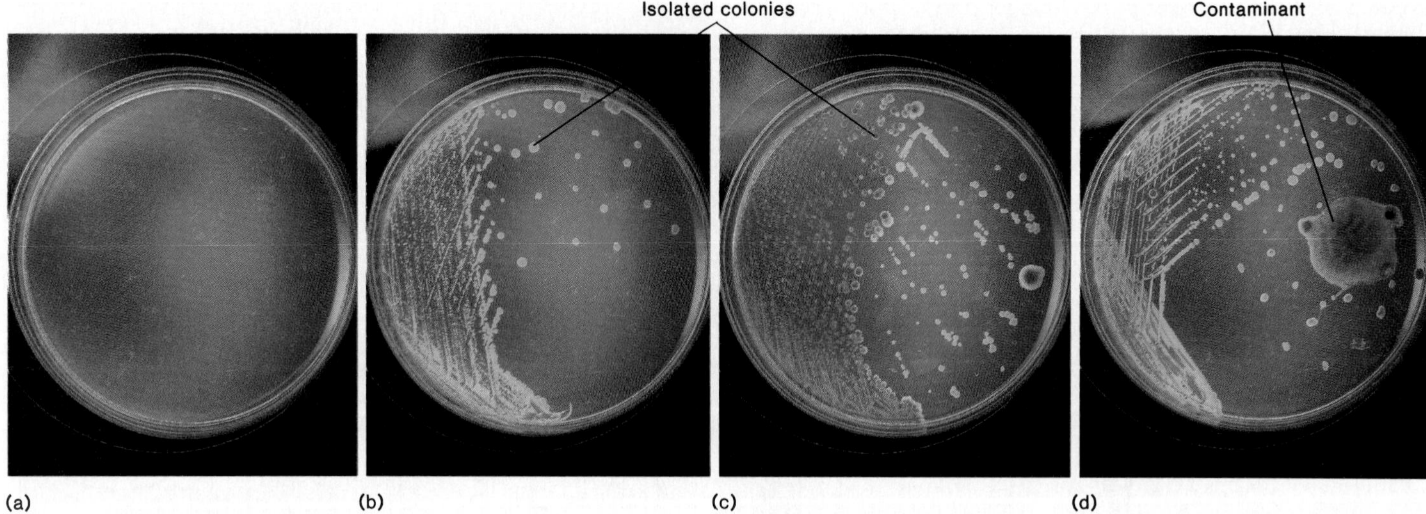

Figure 1.15 Various conditions of media and cultures. The plate in (*a*) contains sterile solid media, shows no visible growth, and is not yet a culture. All procedures in the culturing of microorganisms require sterile media. (*b*) A pure culture of *Micrococcus luteus*. It is evident that only this species is growing in the plate. (*c*) A mixed culture of *M. luteus* and *Serratia marcescens* species, which are readily differentiated by their contrasting pigments. (*d*) This plate was overexposed to room air, and it has developed distinctive fuzzy colonies of molds. Because these intruders are not desirable and are not identified, the culture is now contaminated.

nature of the sample depends upon the objectives of the analysis. Medical samples for determining the cause of an infectious disease are obtained from body fluids (blood, cerebrospinal fluid), discharges (sputum, urine, feces), or diseased tissue. Other samples frequently subject to microbiologic analysis are water, sewage, foods, air, and inanimate objects. Measures for proper specimen collection are discussed in chapter 15.

Culture media may be contained in test tubes, flasks, or petri dishes, and they are inoculated by such tools as loops, needles, pipettes, and swabs. As we will see in chapter 6, media are extremely varied in content and consistency and are specially formulated for a particular purpose. Culturing certain microorganisms (including viruses and certain bacteria) may require "living media" in the form of cell cultures or host animals. In order for an experiment to be properly controlled, the inoculation must start with a sterile medium and use inoculating tools with sterile tips. Measures must be taken to prevent introduction of nonsterile materials such as room air and fingers directly into the culture.

In some ways, culturing microbes is analogous to gardening on a microscopic scale. The cultures can be compared to carefully separated plots of tiny vegetables in which extreme care is taken to exclude weeds. A **pure culture** is a container of medium that grows only a single known species or type of microorganism (figure 1.15*b*). This is the usual state of microorganisms for laboratory study, because it allows the systematic examination and control of that microorganism by itself. Instead of the term pure culture some microbiologists prefer the term *axenic,* meaning that the culture is free of other living things except for the one being studied. A **mixed culture** (figure 1.15*c*) is a container that holds two or more *known,* easily differentiated species of microorganisms, not unlike a garden plot containing both carrots and onions. A **contaminated culture** (figure 1.15*d*) was once pure or mixed (and thus a known entity) but has since had *contaminants* (unwanted microbes of uncertain identity) introduced into it, like weeds into a garden. Because contaminants have the potential for causing disruption, vigilance is constantly required to exclude them from experiments in microbiology labs, as you will no doubt witness in your own experiences.

Once a container of medium has been inoculated, it is *incubated*—placed in a controlled chamber, an *incubator,* at a temperature that will facilitate microbial growth. Although microbes have adapted to growth at temperatures ranging from freezing to boiling, the usual temperatures used in laboratory propagation fall between 20° and 40°C. Incubators may also control the content of atmospheric gases such as oxygen and carbon dioxide that may be required for the growth of certain microbes. During the incubation period (ranging from a day to several weeks), the microbe multiplies and produces some visible manifestation of growth.

Inspection, Isolation, and Identification

It is common during various stages of incubation to evaluate a culture macroscopically[4] (with the naked eye). This is possible because, even though the individual microbes are too small to be seen individually, large masses of them are readily visible. Microbial growth in a liquid medium materializes as cloudiness, sediment, scum, or color. A common manifestation of growth on solid media, especially in bacteria and fungi, is the

axenic (ak-zee'-nik) Gr. *a,* no, and *xenos,* stranger.

contaminated (kon-tam'-ih-nay-tid) Gr. *con,* together, and L. *tangere,* to touch.

incubate (in'-kyoo-bayt) Gr. *incubatus,* to lie in or upon.

4. Macroscopically means to observe a microbe at the cultural, rather than the microscopic, level.

appearance of isolated colonies, clinging masses of cells in regular, discrete packets (figure 1.15*b, c*). Colony size, shape, color, and texture may help define certain characteristics of a microbe. Once isolation has occurred, it is standard practice to make a second-level culture, called a *subculture,* by picking out a tiny inoculum from a colony and inoculating it into a separate container of media, thereby preparing a pure culture.

Cultural inspection is inevitably tied to an even more direct visual method—microscopic inspection. Using a microscope lets one see firsthand the tiny cells that appear in cultures. It gives very tangible evidence of many characteristics of cell morphology, including size, shape, and details of internal and external structure. So essential is the microscope to accurate cell study that we include the general principles of microscopy in a subsequent section of this chapter.

How does one determine what sorts of microorganisms have been isolated in cultures? Certainly the combination of microscopic and macroscopic appearance can be very valuable in differentiating the smaller, simpler procaryotic cells from the larger, more complex eucaryotic cells. Appearance can be especially useful in identifying eucaryotic microorganisms to the level of genus or species because of their distinctive morphological features; however, bacteria are generally not identifiable by these methods because their morphologies are frequently similar. For them, we must resort to techniques that characterize their cell function or physiology. In general, these methods fall into the realm of biochemical tests because they can determine fundamental chemical characteristics such as nutrient requirements, products given off during growth, temperature and gas requirements, and mechanisms for deriving energy.

Several modern analytical and diagnostic tools that focus on the genetic and molecular characteristics can detect the exact nature of the genetic material of the cell, a molecule called deoxyribonucleic acid (DNA; see chapter 2). In the case of certain pathogens, further information on a microbe is obtained by inoculating a laboratory animal such as a rat, mouse, rabbit, or guinea pig. By compiling physiological testing results with macro- and microscopic traits, a profile is prepared, and this overview of the organism becomes the raw material used in final identification. In several subsequent chapters (3, 15, 16, and 17), we will present selected major methods of bacterial identification.

Maintenance and Disposal of Cultures

The fate of a laboratory culture depends upon the original purpose of isolation and identification. In some cases, the cultures and specimens constitute a potential hazard and will require immediate and proper disposal. Both steam sterilizing (see autoclave, chapter 9) and incineration (burning) are effective methods of destroying microorganisms. On the other hand, many teaching and research laboratories require a line of stock cultures, continuously maintained species that represent "living catalogues." The largest culture collection can be found at the American Type Culture Collection in Rockville, Maryland, which maintains a voluminous array of frozen and freeze-dried fungal, bacterial, viral, and algal cultures.

The Microscope: Window on an Invisible Realm

Imagine Leeuwenhoek's excitement and wonder when he first glimpsed in a drop of rainwater an amazing microscopic world that teemed with bizarre creatures. Beginning microbiology students still experience this sensation, and even experienced microbiologists never forget their first view. The microbial existence is indeed another world, but it would remain largely uncharted without an essential tool: the microscope. Your efforts in exploring will be more meaningful if you understand some essentials of **microscopy** and specimen preparation.

Magnification and Microscope Design

One of the first discoveries to spur the development of microbiology was that a clear, spherical piece of glass can serve as a lens to **magnify**—that is, appear to enlarge—small objects. Magnification in most microscopes results from a complex interaction between visible light waves and the curvature of the lens. When a beam or ray of light transmitted through air strikes and passes through the convex surface of glass, it experiences some degree of **refraction,** defined as the bending or change in the angle of the light ray as it passes into the lens (see figure 1.18). The greater the difference in the composition of the two substances the light passes between, the more pronounced is the refraction. When an object is placed a certain distance from the spherical lens and illuminated with light, an optical replica or **image** of it is formed by the refracted light. Depending upon the size and curvature of the lens, the image appears enlarged to a particular degree, which is called its power of magnification and is usually identified with a number combined with × (read "times"). This behavior of light is evident if one looks through an everyday object such as a glass ball, a water-filled flask, or a magnifying glass (figure 1.16). It is basic to the function of all **optical** or **light microscopes,** though many of them have additional features that define, refine, and increase the size of the image.

The first microscopes were **simple,** meaning they contained just a single magnifying lens and a few working parts. Examples of this type of microscope are a magnifying glass, a hand lens, and Leeuwenhoek's useful little tool shown in figure 1.6*a*. Among the refinements that led to the development of today's **compound** microscope were the addition of a second magnifying lens system, a lamp in the base that gives off visible light to *illuminate* the specimen, and a special lens called the *condenser* that converges or focusses the rays of light to a single point on the object. The fundamental parts of a modern compound light microscope are illustrated in figure 1.17.

The Principles of Light Microscopy

To be most effective, a microscope should provide adequate magnification, resolution, and clarity of image. Magnification

---

microscopy (mi-kraw'-skuh-pee) The science that studies microscope techniques.
magnify (mag'-nih-fye) L. *magnus,* great, and *ficere,* to make.
refract, refraction (ree-frakt', ree-frak'-shun) L. *refringere,* to break apart.
illuminate (ill-oo'-mih-nayt) L. *illuminatus,* to light up.

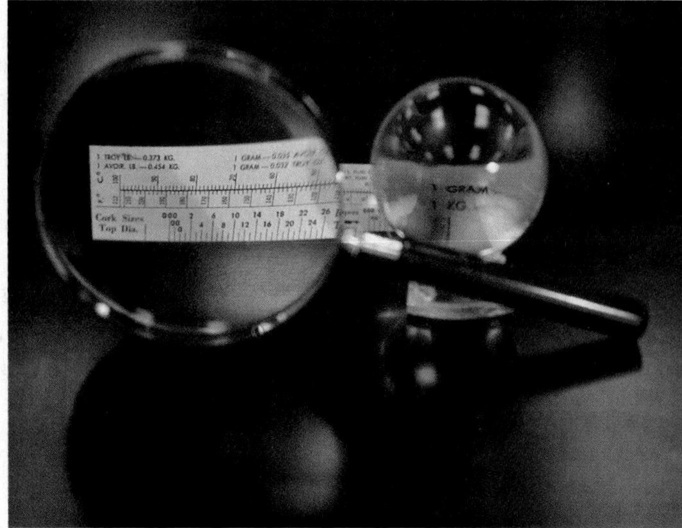
(a)

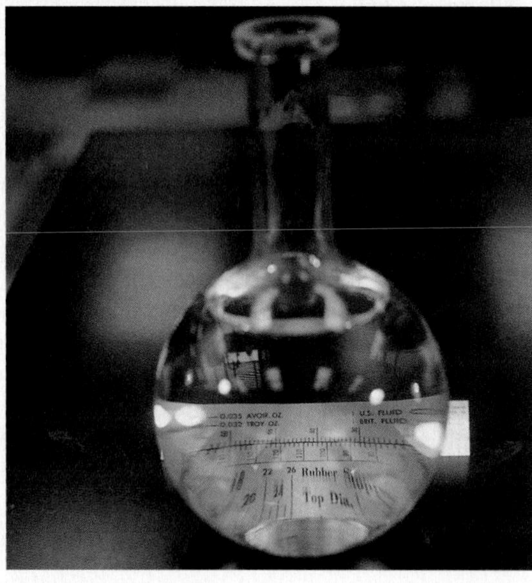
(b)

Figure 1.16 Demonstration of the magnification and image-forming capacity of clear glass and plastic "lenses," given a proper source of illumination. (*a*) A magnifying glass and a clear glass sphere; (*b*) a Florence flask containing water.

of the object or specimen by a compound microscope occurs in two phases. The first lens in this system (the one closest to the specimen) is the *objective lens,* and the second (the one closest to the eye) is the *ocular lens,* or eyepiece (figure 1.18). The objective forms the initial image of the specimen, called the *real image.* When this image is projected up through the microscope body to the plane of the ocular, a second or *virtual image* is formed. In essence, the virtual image is an image of an image, one that will be received by the eye and converted to a retinal and visual image. The magnifying power of the objectives alone usually ranges from 4× to 100×, and the power of the ocular

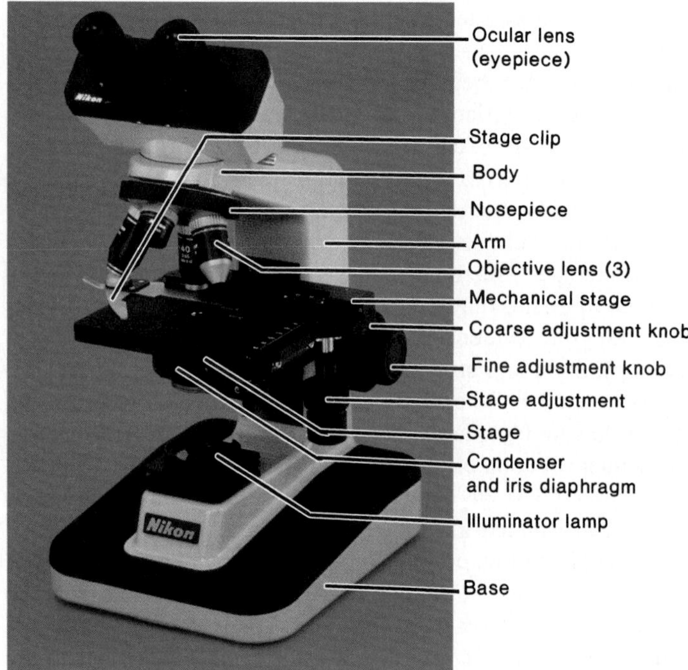

Figure 1.17 The parts of a student laboratory microscope. This microscope is a compound microscope with two oculars (called binocular). It has three objective lenses, a mechanical stage to move the specimen, a condenser, an iris diaphragm, and a built-in lamp.

alone ranges from 10× to 20×. The *total power of magnification* of the final image formed by the combined lenses is a product of the separate powers of each lens:

Power of objective ×	Power of ocular =	Total magnification
40× high dry objective	10× =	400×
100× oil immersion objective	10× =	1,000×
10× low power objective	20× =	200×

Microscopes are equipped with a nosepiece holding three or more objectives that can be rotated into position as needed. The power of the ocular usually remains constant for a given microscope. Depending on the power of the ocular, the total magnification of standard light microscopes can vary from 40× with the lowest power objective (called the scanning objective) to 2,000× with the highest power objective (the oil immersion objective).

Resolution: Distinguishing Magnified Objects Clearly

Despite the importance of magnification in visualizing tiny objects or cells, an additional property that is essential for seeing clearly is **resolution** or **resolving power.** Resolution is the capacity of an optical system to distinguish or separate two adjacent objects or points from one another. For example, at a certain fixed distance (about 10 inches), the lens in the human eye can resolve two small objects as separate points just as long as the two objects are no closer than 0.2 mm apart (see figure

resolve, resolution (re-zolv′, rez″-oh-loo′-shun) L. *resolvere,* to unbind.

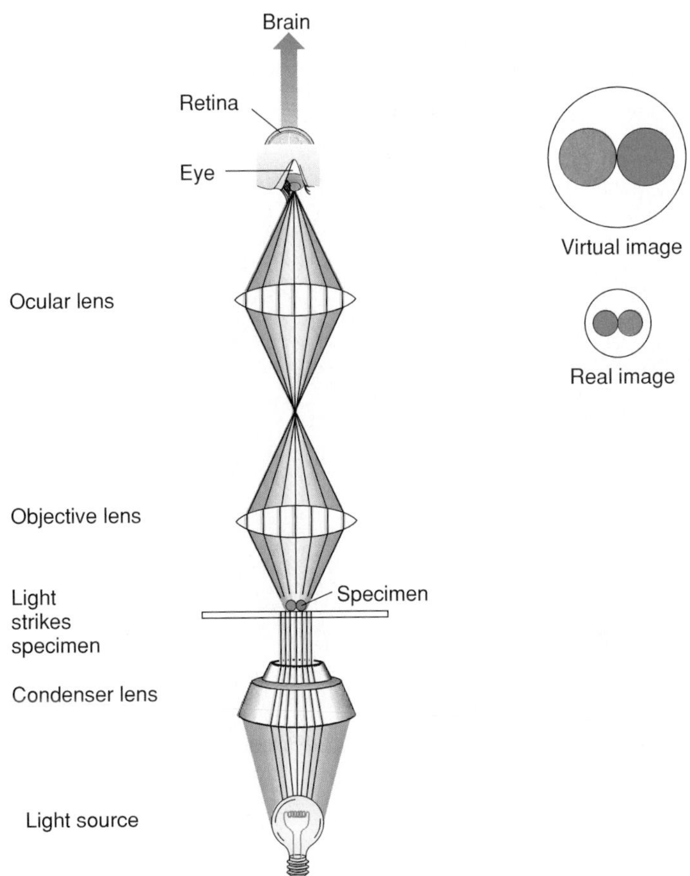

Figure 1.18 The pathway of light and the two stages in magnification of a compound microscope. As light passes through the condenser, it forms a solid beam that is focussed on the specimen. Light leaving the specimen that enters the objective lens is refracted so that an enlarged primary image (the real image) is formed. One does not see this image, but its degree of magnification is represented by the lower circle. The real image is projected through the ocular, and a second image (the virtual image) is formed by a similar process. This image is the final magnified image that will be received by the retina and perceived by the brain. Notice that the lens systems cause the image to be reversed.

1.19 to demonstrate this for yourself). The eye examination given by optometrists is in fact a test of the resolving power of the human eye for various-sized letters read at a distance of 20 feet. Because microorganisms are extremely small and usually very close together, they will not be seen with clarity or any degree of detail unless the microscope can resolve them.

A simple equation in the form of a fraction expresses the main determining factors in resolution:

$$\text{Resolving power (R.P.)} = \frac{\text{Wavelength of light in nm}}{2 \times \text{Numerical aperture of objective lens}}$$

From this equation it is evident that the resolving power is a function of the wavelength of light that forms the image along with certain characteristics of the objective. The light source for optical microscopes consists of a band of colored wavelengths in the visible spectrum (figure 9.7). The shortest visible wavelengths are in the violet-blue portion of the spectrum (400 nm), and the longest are in the red portion (750 nm). Because the wavelength must pass between the objects that are being resolved, shorter wavelengths (in the 400–500 nm range) will provide better resolution (figure 1.20). Some microscopes have a special blue filter placed over the lamp to limit the longer wavelengths of light from entering the specimen.

The other feature influencing resolution is the **numerical aperture,** a mathematical constant that describes the relative efficiency of a lens in bending light rays. Without going into the mathematical derivation of this constant, it is sufficient to say that each objective has a fixed numerical aperture reading that is determined by the microscope design and ranges from 0.1 in the lowest power lens to approximately 1.25 in the highest power (oil immersion) lens. The most important thing to remember about this number is that a higher numerical aperture will provide better resolution. In order for the oil immersion lens to arrive at its maximum resolving capacity, a drop of oil must be inserted between the tip of the lens and the specimen on the glass slide. Since oil has the same optical qualities as glass, it prevents refraction loss that would occur as peripheral light passes from the slide and into the air, and this effectively increases the numerical aperture (figure 1.21).

There is an absolute limitation to resolution in optical microscopes, and the nature of this limit can best be determined by calculating the resolution of the oil immersion lens using a blue-green wavelength of light:

$$\text{R. P.} = \frac{500 \text{ nm}}{2 \times 1.25}$$

$$= 200 \text{ nm (or } 0.2\ \mu\text{m)}$$

In practical terms, this means that the oil immersion lens can resolve any cell or cell part as long as it is at least 0.2 μm in diameter, and that it can resolve two adjacent objects as long as they are no closer than 0.2 μm (figure 1.22). In general, larger organisms such as fungi and protozoa and some of their internal structures are readily resolved, and most bacteria can be visualized because their diameter is greater than 1 μm. However, a few bacteria and most viruses are far too small to be resolved by the optical microscope and will require electron microscopy (discussed later). In summary, then, the factor that most limits the clarity of a microscope's image is its resolving power. Even if a light microscope were designed to magnify several thousand times, its resolving power could not be increased, and the image it produced would simply be enlarged and fuzzy.

Other constraints to the formation of a clear image are the structure of the lens, the quality of the light source, and lack of contrast in the specimen. No matter how carefully a lens is constructed, flaws remain. A typical problem is *spherical aberration,* a distortion in the image caused by irregularities in

---

aperture (ap′-uh-chur) The diameter of the lens opening through which light enters.

aberration (ab″-er-ay′-shun) L. *aberratio,* to go astray.

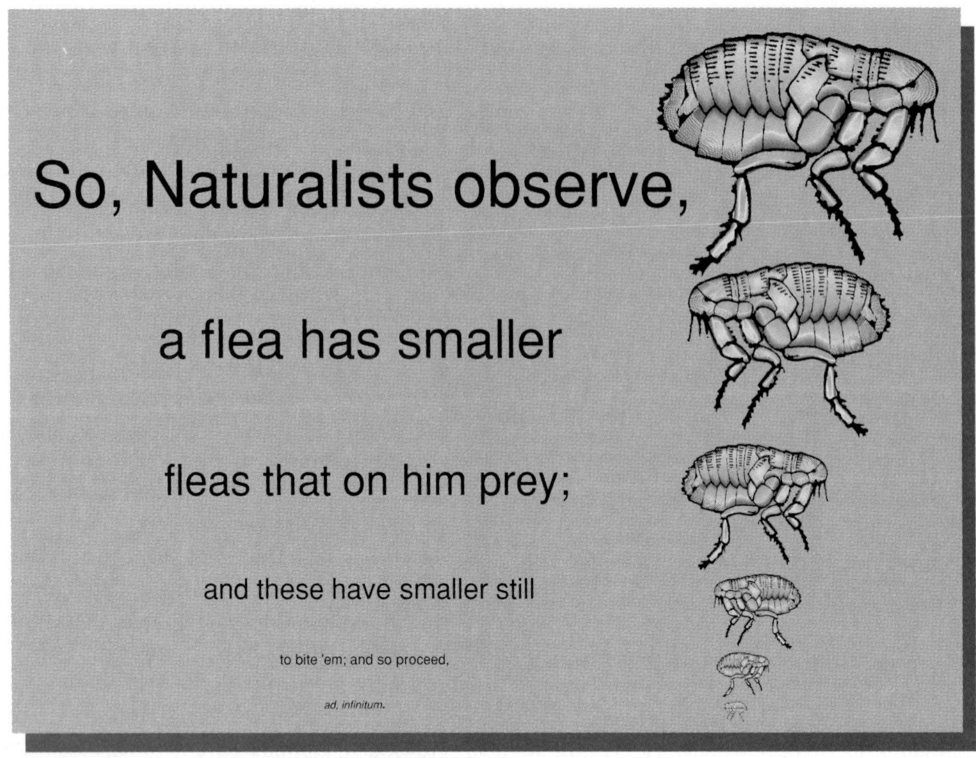

So, Naturalists observe, a flea
Has smaller fleas that on him prey;
And these have smaller still to bite 'em;
And so proceed, *ad infinitum.*
- *Jonathan Swift*

Figure 1.19 This is a test of your living optical system's resolving power. Prop your book against a wall about 20 inches away and determine the line that is no longer resolvable by your eye.
Source: Poem by Jonathan Swift.

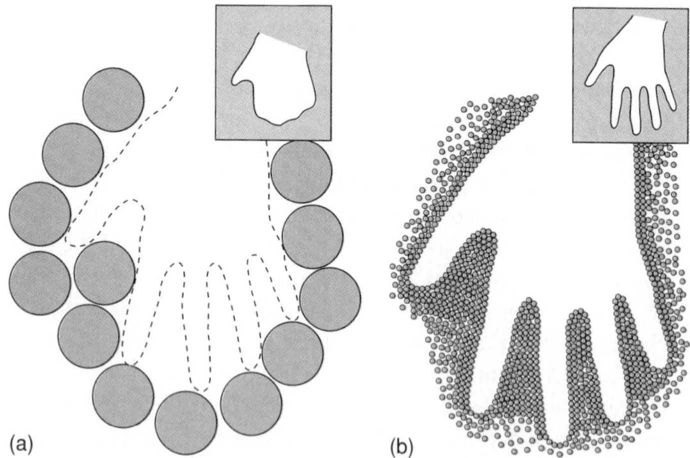

Figure 1.20 This simple model nicely demonstrates why the length of the light waves illuminating a specimen in part determines the capacity of a microscope to resolve. If the diameter of the circles represents the wavelengths of light and if the fingers of a hand represent the close objects to be resolved, then it is clear that a longer wavelength is too large to penetrate between small objects and produces a fuzzy, undetailed image (*inset a*). A shorter wave can insert itself between the individual components, and its image will be very defined (*inset b*).

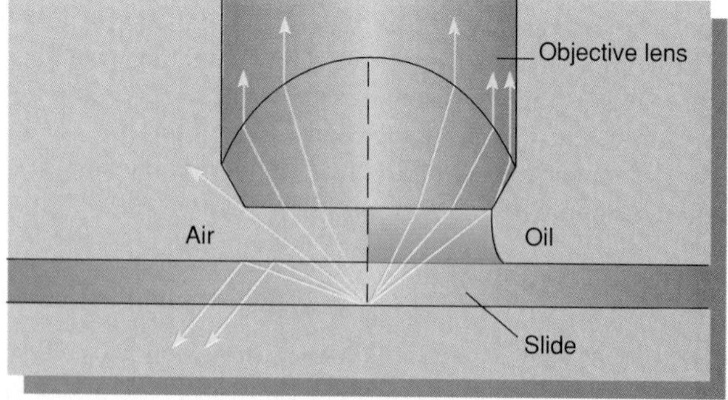

Figure 1.21 To maximize its resolving power, an oil immersion lens (the one with highest magnification) must have a drop of oil placed at its tip. This forms a continuous medium to transmit a beam of light from the condenser to the objective and effectively increase the numerical aperture. Without oil, some of the peripheral light that passes through the specimen is scattered into the air or onto the glass slide, and this decreases the numerical aperture.

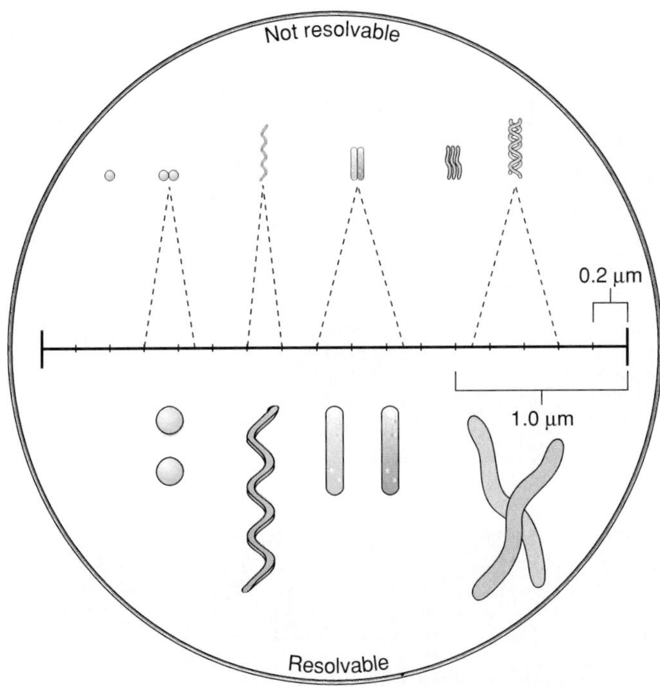

Figure 1.22 Comparison of cells that would not be resolvable, versus those that would be resolvable under oil immersion at 1,000× magnification. Note that, in addition to differentiating two adjacent things, good resolution also means being able to see an object clearly.

the lens, which creates a curved rather than flat image (see figure 1.16). Another is *chromatic aberration,* a blurry, rainbow-like image that results because a lens acts as a prism and separates visible light into its colored bands. To correct for these, the manufacturer incorporates several compensating lenses in the objective and ocular. Brightness and direction of illumination also affect image formation. Because too much light can reduce contrast and burn out the image, an iris diaphragm on most microscopes controls the amount of light entering the condenser. An additional factor is that cells and their internal contents (except pigmented ones) tend to be colorless and translucent, thus showing little contrast with the surrounding background. This lack of contrast may be compensated for by using special lenses (the phase-contrast microscope) and by adding dyes to cells.

Variations on the Optical Microscope

Optical microscopes that use visible light may be described by the nature of their *field,* meaning the circular area viewed through the ocular lens. With special adaptations in lenses, condenser, and light sources, four special types of microscopes can be described: bright-field, dark-field, phase-contrast, and interference. A fifth type of optical microscope, the fluorescence microscope, uses ultraviolet radiation as the illuminating source. Each of these microscopes is adapted for viewing specimens in a particular way, as described in the next sections and summarized in parts of table 1.3.

Table 1.3 Comparison of Types of Microscopy

Microscope	Maximum Practical Magnification	Resolution	Important Features
Visible Light As Source of Illumination			
Bright-field	2,000×	0.2 µm (200 nm)	Common multipurpose microscope for live and preserved stained specimens; specimen is dark, field is white; provides fair cellular detail
Dark-field	2,000×	0.2 µm	Best for observing live, unstained specimens; specimen is bright, field is black; provides outline of specimen with reduced internal cellular detail
Phase-contrast	2,000×	0.2 µm	Used for live specimens; specimen is contrasted against gray background; excellent for internal cellular detail
Differential interference	2,000×	0.2 µm	Provides brightly colored, highly contrasting, three-dimensional images of live specimens
Ultraviolet Rays As Source of Illumination			
Fluorescent	2,000×	0.2 µm	Specimens stained with fluorescent dyes or combined with fluorescent antibodies emit visible light; specificity makes this microscope an excellent diagnostic tool
Electron Beam Forms Image of Specimen			
Transmission electron microscope (TEM)	1,000,000×	0.5 nm	Sections of specimen are viewed under very high magnification; finest detailed structure of cells and viruses is shown; used only on preserved material
Scanning electron microscope (SEM)	100,000×	10 nm	Scans and magnifies external surface of specimen; produces striking three-dimensional image

Bright-field Microscopy

The **bright-field microscope** is the most widely used type of light microscope. Although we ordinarily view objects like the words on this page with light reflected off the surface, a bright-field microscope forms its image when light is transmitted through the specimen. The specimen, being denser and more opaque than its surroundings, absorbs some of this light, and the rest of the light is transmitted directly up through the ocular into the field. As a result, the specimen will produce an image that is darker than the surrounding brightly illuminated field. The bright-field microscope is a multipurpose instrument that may be used for both live, unstained and preserved, stained material. The bright-field image is compared with that of other microscopes in figure 1.23.

Dark-field Microscopy

A bright-field microscope may be adapted as a **dark-field microscope** by adding a special disk called a *stop* to the condenser. The stop blocks all light from entering the objective lens except peripheral light that is reflected off the sides of the specimen itself. The resulting image is a particularly striking one: brightly illuminated specimens surrounded by a dark (black) field (figure 1.23*b*). Some of Leeuwenhoek's more successful microscopes may have operated with dark-field illumination. The most effective use of dark-field microscopy is to visualize unstained living cells that would be distorted by drying or heat. It can outline the organism's shape and permit rapid recognition of swimming cells that might appear in dental and other infections (figure 1.24), but it does not reveal fine internal details.

Phase-Contrast and Interference Microscopy

If similar objects made of clear glass, ice, cellophane, or plastic were immersed in the same container of water, an observer would have difficulty telling them apart because they are all clear and colorless. Internal components of a live, unstained cell are likewise difficult to distinguish because they too lack contrast. But cell structures do differ slightly in density, enough that they may alter the light that passes through them in subtle ways. The **phase-contrast microscope** has been constructed to take advantage of this characteristic. This microscope contains devices that translate the subtle changes in light waves passing through the specimen into differences in light intensity. With this technique, for example, relatively denser cell parts will alter the pathway of light more than less dense regions, and this will create contrast between them. The amount of internal detail visible by this method is greater than by either bright-field or dark-field methods. The phase-contrast microscope is most useful for observing intracellular structures such as bacterial spores, granules, and organelles, as well as the locomotor structures of eucaryotic cells (figure 1.25*a*).

Like the phase-contrast microscope, the *differential interference contrast (DIC) microscope* provides a detailed view of unstained, live specimens by manipulating the light. But this microscope has additional refinements, including two prisms that add contrasting colors to the image and two beams of light rather

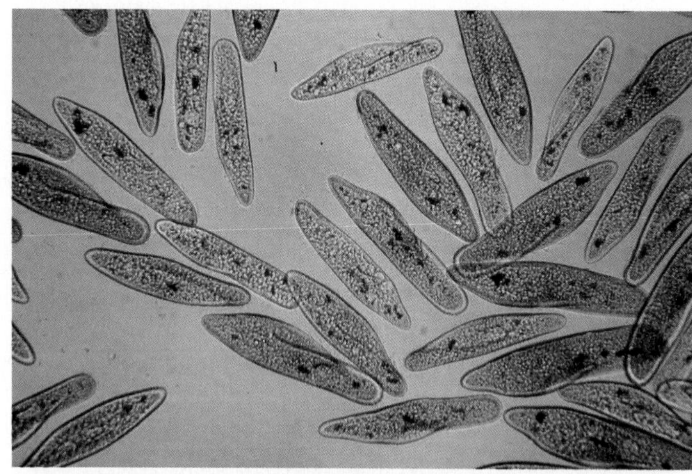

(a)

(b)

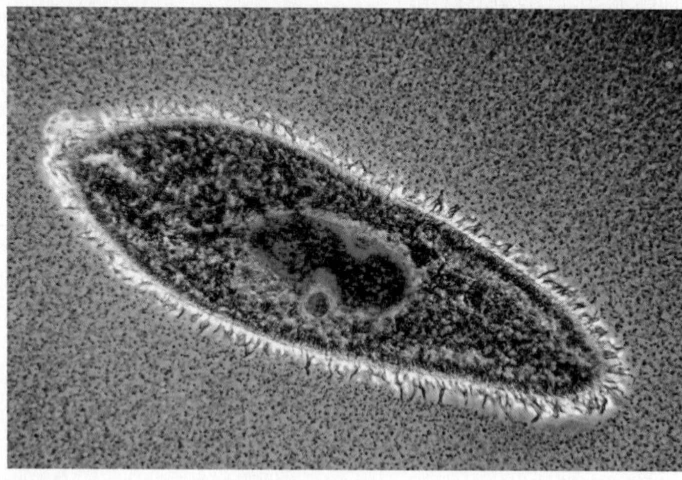

(c)

Figure 1.23 Live cells of *Paramecium* viewed with (*a*) bright-field (×100), (*b*) dark-field (×110), and (*c*) phase-contrast microscopy (×100). Note the difference in the appearance of the field and the degree of detail shown by each method of microscopy. Only in phase-contrast are the cilia (fine hairs) on the cells noticeable. Can you see the nucleus? The oral groove?

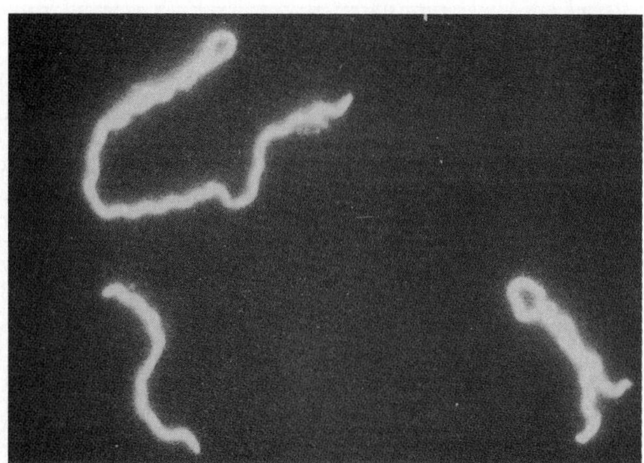

Figure 1.24 Dark-field photomicrograph of a spiral-shaped oral bacterium called *Treponema vincenti* (×1,100). This species lives in the space between the gums and the teeth and may be involved in a deteriorating infection called necrotizing gingivitis.

than a single one. DIC microscopes produce extremely well-defined images that are vividly colored and appear three-dimensional (figure 1.25*b*).

Fluorescence Microscopy

The **fluorescence microscope** is a specially modified compound microscope furnished with an ultraviolet (UV) radiation source and a filter that protects the viewer's eye from injury by these dangerous rays. The name for this type of microscopy originates from certain dyes (acridine, fluorescein) and minerals that possess the property of *fluorescence.* This means that they emit visible light when bombarded by shorter ultraviolet rays. For an image to be formed, the specimen must first be coated or placed in contact with a source of fluorescence. Subsequent illumination by ultraviolet radiation causes the specimen to give off light that will form its own image, usually an intense yellow, orange, or red against a black field.

This technique has its widest applications in diagnosing infections caused by certain bacteria, protozoans, and viruses. A staining technique with fluorescent dyes is commonly used to detect *Mycobacterium tuberculosis* (the agent of tuberculosis) in patients' specimens (see feature 16.4). In a number of diagnostic procedures, fluorescent dyes are affixed to specific antibodies (molecules produced by the immune system in response to an infection). These *fluorescent antibodies* can be used to detect the specific causative agents in such diseases as syphilis, chlamydiosis, trichomoniasis, herpes, and influenza (figure 1.26; see chapters 13, 17, 18, and 20). A fluorescence microscope can be handy for keeping track of microbes in complex mixtures because the fluorescence is so specific and the highlighted cells stand out.

fluorescence (floo″-oh-res′-ens) L. *flux,* a discharge.

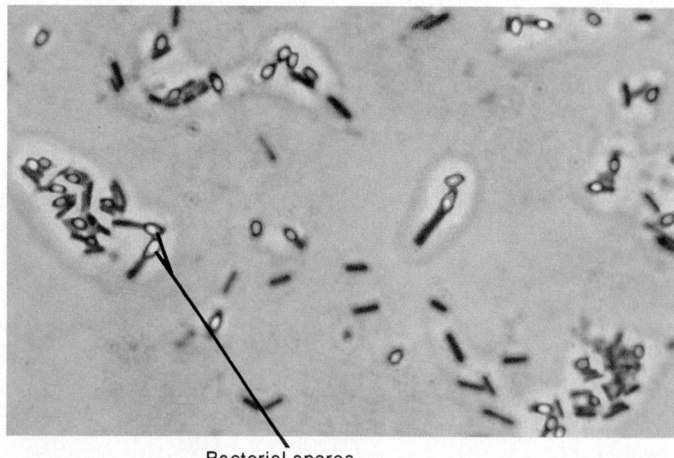

(a)

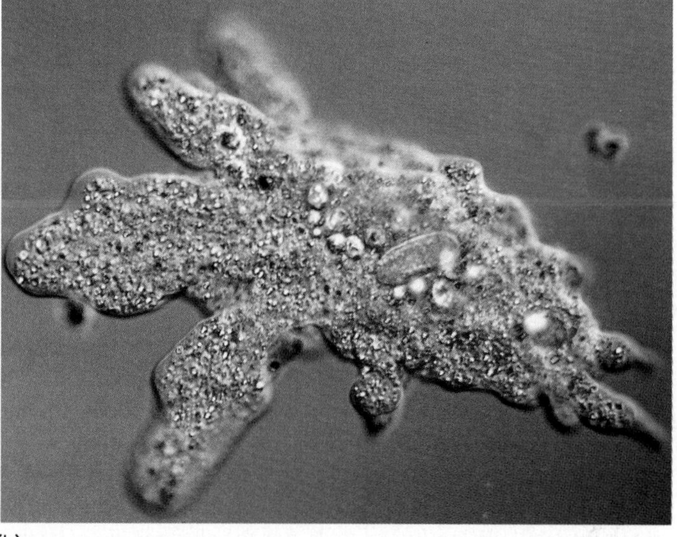

(b)

Figure 1.25 (*a*) Phase-contrast micrograph of a bacterium containing spores. The relative density of the spores causes them to appear as bright shiny objects against the darker cell parts (×600). (*b*) Differential interference micrograph of *Amoeba proteus,* a common protozoan. Note the outstanding internal detail, the depth of field, and the bright colors, which are not natural (×160).

Electron Microscopy

If conventional light microscopes are our windows on the microscopic world, the electron microscope (EM) is our window on the minutest details of that world. Although this microscope was originally conceived and developed for studying nonbiological materials such as metals and small electronics parts, biologists immediately recognized the importance of the tool and began to use it in the early 1930s. One of the most impressive features of the electron microscope is the resolution it provides.

Unlike the light microscope, which is limited by wavelength and numerical aperture, the electron microscope forms an image with a beam of electrons that behaves somewhat like a wave. Electrons are the smallest particles of an atom (see

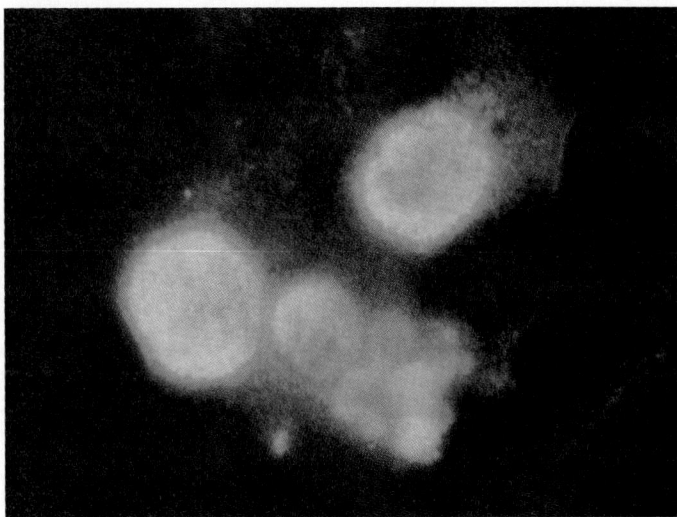

Figure 1.26 A fluorescent antibody stain of cells infected with influenza viruses. This technique does not allow one to actually see the viruses, but gives evidence of their presence. In this method, a special antibody specific to this virus is bound to a fluorescent dye, so that sites in the cell containing the virus will be turned this brilliant green color.

Table 1.4 Comparison of Light Microscopes and Electron Microscopes

Characteristic	Light or Optical	Electron (Transmission)
Useful magnification	2,000×	Over 1,000,000
Maximum resolution	200 nm	0.5 nm
Image produced by	Visible light rays	Electron beam
Image focussed by	Glass objective lens	Electromagnetic objective lenses
Image viewed through	Glass ocular lens	Fluorescent screen
Specimen placed on	Glass slide	Copper mesh
Specimen may be alive	Yes	No
Specimen requires special stains or treatment	Not always	Yes
Colored images produced	Yes	No

chapter 2) and travel in waves that are 100,000 times shorter than the waves of visible light. Because resolving power is a function of the wavelength, electrons have tremendous power to resolve minute structures. Indeed, it is possible to resolve atoms with an electron microscope, though the practical resolution for biological applications is approximately 0.5 nm. Because the resolution is so substantial, it follows that magnification can also be extreme—usually between 5,000× and 750,000× for biological specimens and up to 5,000,000× in some applications. Its capacity for magnification and resolution makes the EM an invaluable tool for seeing the finest structure (called the ultrastructure) of cells and viruses. If not for electron microscopes, our understanding of biological structure and function would still be in its early theoretical stages.

In fundamental ways, the electron microscope is a derivative of the compound microscope. It employs components analogous to, but not necessarily the same as, those in light microscopy (figure 1.27). For instance, it magnifies in stages by means of two lens systems, and it has a condensing lens, a specimen holder, and focussing apparatus. Otherwise, the two types have numerous differences (table 1.4). The electron microscope has an electron gun that aims the electron beam at the specimen through a vacuum and ring-shaped electromagnets rather than glass lenses to properly focus this beam. Specimens must be pretreated with dyes and other types of chemicals to increase contrast. The enlarged image is displayed on a viewing screen or photographed for further study rather than being directly observed. Since images produced by electrons lack color, electron micrographs (a micrograph is a microscopic photograph) are always shades of black, gray, and white. The color-enhanced ones that you see in this and other textbooks (figure 1.28*a*) have computer-added color (yes, they're colorized!).

Two general forms of EM are the transmission electron microscope (TEM) and the scanning electron microscope (SEM) (see table 1.3). **Transmission electron microscopes** are the method of choice for viewing the detailed structure of cells and viruses. This microscope produces its image by transmitting electrons through the specimen. Because electrons cannot readily penetrate thick preparations, the specimen must be sectioned into extremely thin slices (20–100 nm thick) and stained or soaked in metals that will increase image contrast. The darkest areas of TEM micrographs represent the thicker (denser) parts, and the lighter areas indicate the more transparent and less dense parts (figure 1.28*b*). The TEM may also be used to produce negative images and shadow casts of whole viruses (see figure 5.1).

The **scanning electron microscope** provides some of the most dramatic and realistic images in existence. The design of this instrument makes possible an extremely detailed three-dimensional picture of anything from a fly's eye to AIDS viruses escaping their host cells. To produce this image, the SEM does not transmit electrons through a thin section; it bombards the surface of a whole, metal-coated specimen with electrons, while scanning back and forth over it. A shower of electrons deflected from the surface is picked up with great fidelity by a sophisticated detector, and the electron pattern is displayed as an image on a television screen. The contours of the specimens resolved with scanning electron micrography are very revealing and often surprising. Areas that look smooth and flat with the light microscope display intriguing surface features with the SEM (figure 1.29). Improved technology has continued to refine electron microscopes and to develop variations on the basic plan. One of the most inventive relatives of the EM is the new scanning probe microscope (see feature 1.3).

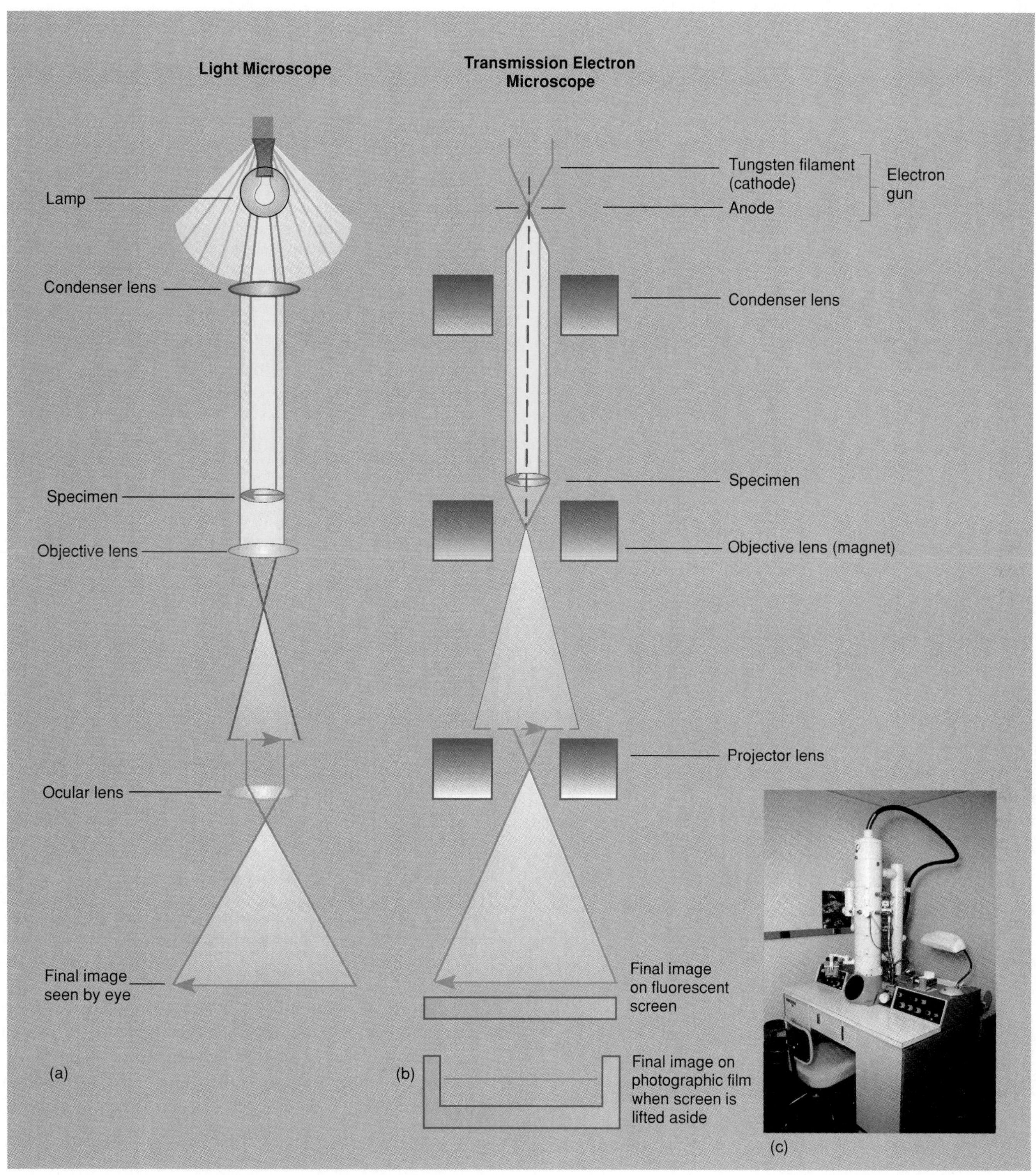

Figure 1.27 Comparison of (*a*) the light microscope and (*b*) one type of electron microscope (EM; transmission type). These diagrams are highly simplified, especially for the electron microscope, to indicate the common components. Note that the EM's image pathway is actually upside-down compared with that of a light microscope. (*c*) The EM is a larger machine with far more complicated working parts than most light microscopes.

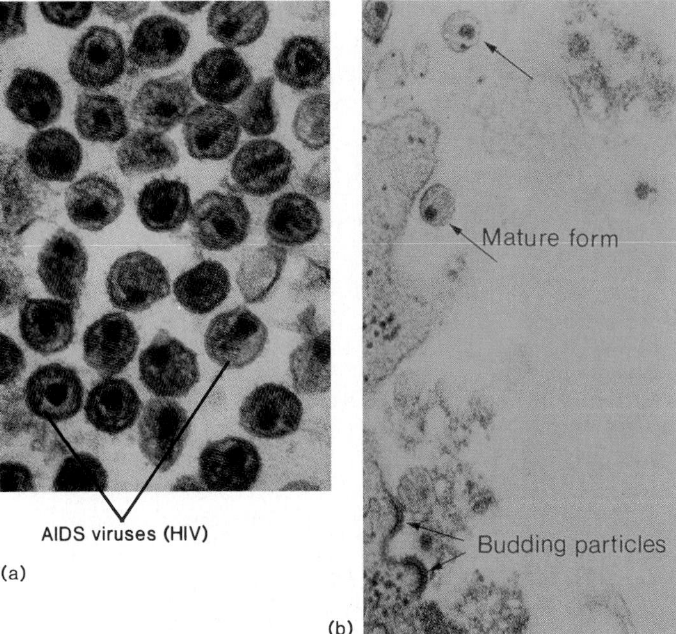

Figure 1.28 Transmission electron micrographs. (*a*) A false-color image of AIDS viruses entering a host cell (×200,000). (*b*) AIDS viruses leaving a host cell by budding off its surface (×50,000).

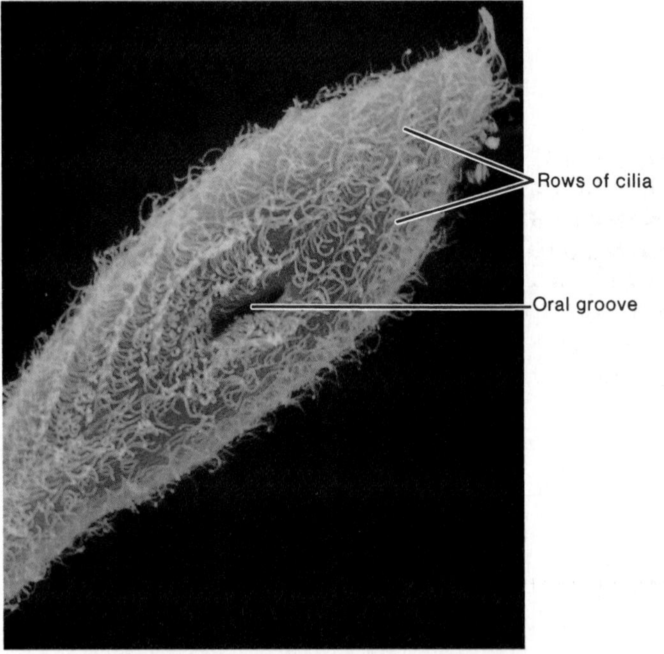

Figure 1.29 A scanning electron micrograph (SEM) of *Paramecium*, covered with what look like tiny, perfect rows of curls. Actually, these are its locomotor and feeding structures, the cilia. Compare this figure with figure 1.23 to appreciate the amazing degree of surface detail in an SEM.

Preparing Specimens for Optical Microscopes

A specimen for optical microscopy is generally prepared by mounting a sample on a suitable glass slide that sits on the stage between the condenser and the objective lens. How a slide specimen, or *mount*, is prepared depends upon (1) the condition of the specimen, either in a living or preserved state; (2) the aims of the examiner, whether to observe overall structure, identify the microorganisms, or see movement; and (3) the type of microscopy available, be it bright-field, dark-field, phase-contrast, or fluorescence.

Feature 1.3 A Revolution in Resolution: Probing Microscopes

Not very long ago, chemists, physicists, and biologists had to rely on indirect methods to provide information on the fine structures of the smallest molecules. But the mid-1980s witnessed the introduction of a new generation of microscopes that literally sees structure by "feeling" it. *Scanning probe microscopes* operate with a minute needle tapered to a tip that can be as narrow as a single atom! This probe scans over the exposed surface of a material and records an image of its outer texture. These revolutionary microscopes have such profound resolution that they have the potential to view single atoms (but not subatomic structure yet) and to magnify 100 million times. The scanning tunnelling microscope (STM) was the first of these microscopes. It uses a tungsten probe that hovers near the surface of an object and follows its topography while simultaneously giving off an electric signal of its pathway that is imaged on a screen. The STM is used primarily for detecting defects on the surfaces of electrical conductors and computer chips composed of silicon (see accompanying photo), but it recently provided the first incredible close-up views of DNA, the genetic material (see chapter 8 opening illustration). Another exciting new variant is the atomic force microscope (AFM), which gently forces a diamond and metal probe down onto the surface of a specimen like a needle on a record. As it moves along the surface, any deflection of the metal probe is detected by a sensitive device that relays the information to an imager. This microscope is very useful in viewing biological molecules (see chapter 2 opening illustration).

An image of perfectly arranged silicon atoms produced by a scanning tunnelling microscope (STM).

Fresh, Living Preparations

Live samples of microorganisms are placed in *wet* or *hanging drop mounts* so that they can be observed as much as possible in their natural state. The cells are suspended in a suitable fluid (water, broth, saline) that temporarily maintains viability (the living state) and provides space and a medium for locomotion. A wet mount consists of a drop or two of the culture placed on a slide and overlaid with a cover glass. Although this type of mount is quick and easy to prepare, the cover glass may damage larger cells, and the slide is very susceptible to drying and can contaminate the slide handler. A more satisfactory alternative is the hanging drop preparation made with a special concave (depression) slide, a vaseline adhesive/sealant film, and a coverslip from which a tiny drop of sample is suspended (see figure 3.4). These types of preparations, though workable for only a short term, provide the most accurate impression of the size, shape, arrangement, and (sometimes) color of cells. Used in conjunction with high definition microscopes like phase-contrast and interference microscopes, significant detail may be observed.

Fixed, Stained Smears

A more permanent mount for long-term study can be obtained by preparing fixed, stained specimens. One mainstay of the modern microbiology laboratory, the **smear,** was developed by Robert Koch more than 100 years ago. The technique consists of spreading a thin film made from a liquid suspension of cells on a slide and air drying it. During the step called *fixation,* the air-dried smear is usually heated gently, a process that simultaneously kills and secures the specimen to the slide. Another important action of fixation is to arrest various cellular components in a natural state without undue distortion. For some microbial cells, additional chemicals such as alcohol and formalin are necessary to preserve more delicate cell structures.

Like images on undeveloped photographic film, the unstained cells of a fixed smear are quite indistinct, no matter how great the magnification or how fine the resolving power of the microscope. "Developing" a smear is aimed at creating contrast and making indecipherable features stand out, and this is done by means of staining techniques. **Staining** is any procedure in which colored chemicals called *dyes* are added to smears. Dyes impart a color to cells or cell parts by becoming affixed to them through a chemical reaction. In general, dyes are classified as *basic dyes,* which have a positive charge, or *acidic dyes,* which have a negative charge. Because chemicals of opposite charge are attracted to each other, those cell parts that are negatively charged will attract basic dyes and those that are positively charged will attract acidic dyes (table 1.5). Many cells, especially those of bacteria, have numerous negatively charged acidic substances and thus stain more readily with basic dyes. Acidic dyes, on the other hand, tend to be repelled by cells, so they are good for negative staining (discussed in the next section). The chemistry of dyes and their staining reactions is covered in more detail in feature 2.2.

Negative Versus Positive Staining Two basic types of staining technique are used, depending upon how a dye reacts with the specimen (summarized in table 1.5). Most procedures involve a **positive stain,** in which the dye actually sticks to the specimen and gives it color. A **negative stain,** on the other hand, is just the reverse (like a photographic negative), because the dye does not stick to the specimen but settles around its outer boundary, forming a silhouette. In a sense, negative staining "stains" the glass slide to produce a dark field that highlights the cells in the specimen. Dyes used for this purpose are usually nigrosin (blue-black) and India ink (a black suspension of carbon particles). The cells themselves do not stain because these dyes are negatively charged and are repelled by the negatively charged surface of the cells. The value of negative staining is its relative simplicity and the reduced shrinkage or distortion of cells, as the smear is not heat-fixed. A quick assessment can thus be made regarding cellular size, shape, and arrangement. Negative staining is valuable in characterizing the capsule that surrounds certain bacteria and yeasts (see figure 1.31).

Table 1.5 Categories of Common Microbiological Stains

	Positive Staining	Negative Staining
Appearance of Cell	Colored by dye	Clear and colorless
Background	Not usually stained (generally white)	Stained (usually dark)
Dyes Employed	Basic dyes: Crystal violet Methylene blue Safranin Malachite green	Acidic dyes: Nigrosin India ink
Subtypes of Stains	Several types: Simple stain Differential stains Gram stain Acid-fast stain Spore stain Special stains Capsule Flagella Spore Granules Nucleic acid	Few types: Capsule

Simple Versus Differential Staining Positive staining methods are classified as simple, differential, or special (figure 1.30). Whereas **simple stains** require only a single dye and an uncomplicated procedure, **differential stains** use two different colored dyes, called the *primary dye* and the *counterstain,* to separate or distinguish cell types or parts. These stains tend to be more complex and may require additional chemical reagents to reveal a particular staining reaction.

Most simple staining techniques take advantage of the ready binding of bacterial cells to dyes like malachite green, crystal violet, basic fuchsin, and safranin. With simple stains, all cells appear the same color, regardless of type, but such bacterial characteristics as shape, size, and arrangement are revealed (figure 1.31*a*). One useful simple stain is Loeffler's methylene blue stain. The blue cells stand out against a relatively unstained background, so that size, shape, and grouping

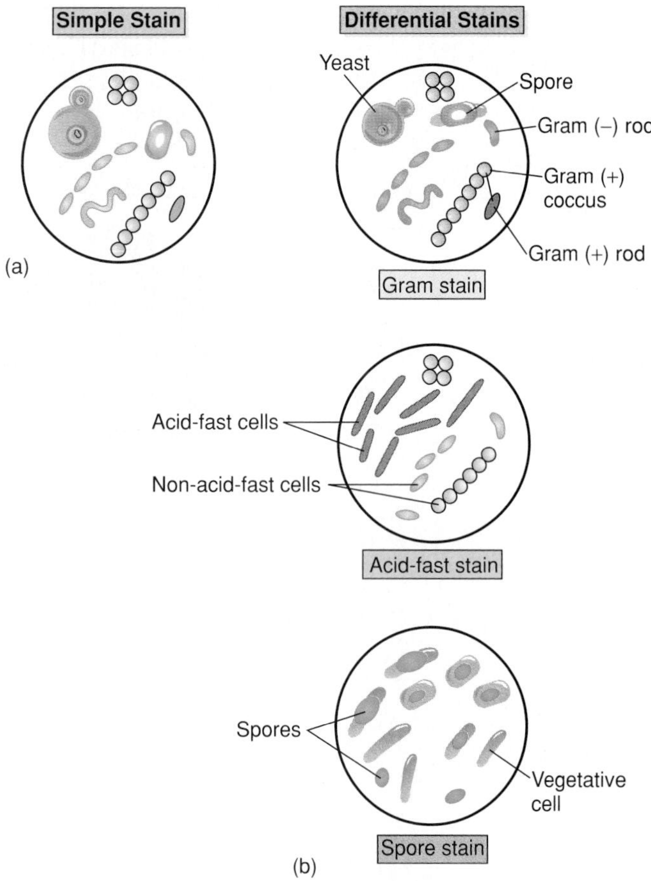

Figure 1.30 Comparison of simple and differential stains, using mixed samples. (*a*) In the simple stain, note that all cells are stained the same color, regardless of their shape. In differential stains, color is used to differentiate a minimum of two cell groups or types. (*b*) Gram stain differentiates by both color (purple, red) and shape. Acid-fast stains and spore stains indicate primarily two types of cells.

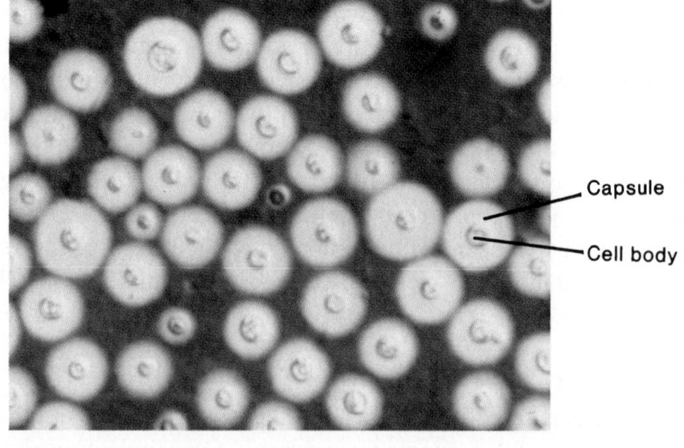

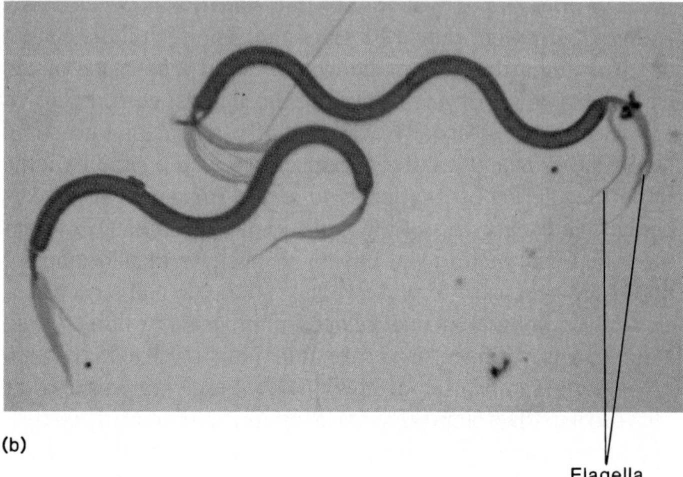

Figure 1.31 (*a*) India ink capsule stain of *Cryptococcus neoformans* (×150). (*b*) Flagellar stain of *Spirillum volutans* (×600). Note the paired flagella at each end.

show up easily. This method is significant because it may stain the internal granules of *Corynebacterium diphtheriae,* a bacterium that is responsible for diphtheria, thus it is also a type of special stain (see chapter 3).

Types of Differential Stains A satisfactory differential stain uses differently colored dyes to clearly contrast two different cell types or cell parts. Common combinations are red and purple, red and green, or pink and blue. Differential stains may also pinpoint other characteristics such as the size, shape, and arrangement of cells. Typical examples include Gram, acid-fast, and endospore stains.

Gram staining, a century-old method named for its developer, Hans Christian Gram, remains the most universal diagnostic staining technique for bacteria. It permits ready differentiation of major significant categories, based upon the color reaction of the cells: gram positive, which stain purple, and gram negative, which stain red (figure 1.30*b*). The Gram stain is the basis of several important bacteriological topics, including bacterial taxonomy, cell wall structure, and identification and diagnosis of infection; in some cases, it even guides the selection of the correct drug for an infection. Gram staining will be discussed in greater detail in chapter 3.

The *acid-fast stain,* like the Gram stain, is a multipurpose, differential stain. The basic principle of this procedure originated with Paul Ehrlich in 1882 as an improved method for the difficult-to-stain bacterium responsible for tuberculosis. He found that after the microbes had been stained with a certain heated dye (carbol fuchsin), they resisted the loss of that dye even when washed with a solution containing acid (figure 1.30*b*). This use of the term fast means to hold tightly or tenaciously. Later it was determined that these types of bacterial cells have a particularly impervious outer wall. Two other medically important bacilli that resist staining are the leprosy bacillus and *Nocardia,* an agent of lung or skin infections. Tuberculosis and leprosy may be identified by the presence of acid-fast bacteria in specimens (see chapter 16).

The *endospore stain* is similar to the acid-fast method in principle, in that a hot dye is forced into resistant bodies called spores or endospores (formation and significance to be discussed in chapter 3). This stain not only distinguishes between bacteria that produce spores and those that do not, but it also distinguishes between spores and the cells that form them (so-called *vegetative cells;* figure 1.30*b*). Of significance in medical microbiology are the gram-positive, spore-forming members of the genus *Bacillus* (the cause of anthrax) and *Clostridium* (the cause of botulism and tetanus)—both dramatic diseases with universal fascination that we shall consider later.

Special stains are used to bring out or emphasize certain cell parts such as flagella and spores that are not revealed by conventional staining methods. Some staining techniques (spore, capsule) fall into more than one category.

Capsule staining is the method for observing the microbial capsule, an unstructured protective layer surrounding the cells of some bacteria and fungi. The capsule does not readily accept stains, thus it is often easiest to form a negative image of it with India ink (figure 1.31*a*) or it may be demonstrated by positive stains. Because not all microbes exhibit capsules, this feature is useful for identifying pathogens in certain respiratory infections such as cryptococcosis, a common fungal infection in AIDS patients.

Flagellar staining is a method for revealing flagella, the tiny, slender filaments used by bacteria for locomotion (see chapter 3). Because the width of bacterial flagella lies beyond the resolving power of the light microscope, to be seen, they must be enlarged by depositing a coating on the outside of the filament and then staining it (figure 1.31*b*). This stain works best with fresh, young cultures, since flagella are delicate and may be lost or damaged on older cells. Their presence, number, and arrangement on a cell are taxonomically useful.

Chapter Review with Key Terms

Introduction to Microbiology

Whether microbes are good or evil depends upon one's point of view. But it is a fact that they are an integral part of nature and indispensable for normal, balanced life as we know it. Microbes are responsible for spoilage, decay, and disease, but they are also used in the manufacture of foods, beverages, and industrial chemicals, **bioremediation,** drug synthesis, and **genetic engineering.**

Microbiology is the biological study of **bacteria, viruses, fungi, protozoa,** and **algae,** which are collectively called **microorganisms** or **microbes.** In general, microorganisms are **microscopic** and, unlike **macroscopic** organisms that are readily visible, they require magnification to be seen. The unicellular simplicity, growth rate, and **adaptability** of microbes are some of the reasons that microbiology is so extensive and embraces many other areas of science and application. Early microbiology blossomed with the conceptual developments of **sterilization, aseptic techniques,** and the **germ theory of disease.**

Characteristics of Microorganisms

Organisms can be described according to their composition or way of life. For instance, the cells of **eucaryotic** organisms contain a nucleus, but those of **procaryotic** organisms do not. Some organisms are hosts, while others are **parasites;** some parasites are **infectious.**

Microbes may be measured in micrometers, dimensions too small to be seen without a microscope. With his simple microscope, Leeuwenhoek discovered organisms he called **animalcules.** As a consequence of his findings and the rise of the **scientific method,** the notion of **spontaneous generation** or **abiogenesis** was eventually abandoned for **biogenesis.** The scientific method applies **inductive** and **deductive** reasoning to develop rational **hypotheses** and **theories** that can be tested. Principles that withstand repeated scrutiny become **law** in time.

Classification of Microorganisms

Taxonomy is a hierarchy scheme for the **classification, identification,** and **nomenclature** of organisms, which are grouped in categories called **taxa,** based on features ranging from general to specific. Starting with the broadest category, the taxa are **kingdom, phylum (or division), class, order, family, genus,** and **species.** Organisms are assigned **binomial scientific names** consisting of their genus and species names.

The most widely adopted classification scheme is a five-kingdom system consisting of the following: **Kingdom Procaryotae (Monera),** containing the **eubacteria** and the **archaebacteria; Kingdom Protista,** comprised of primitive unicellar microbes like algae and protozoa; **Kingdom Mycetae,** containing the fungi; **Kingdom Animalia,** encompassing animals; and **Kingdom Plantae,** containing plants. Inspection of the **morphology, physiology,** and **genetics** of organisms reveals an ancestral **evolutionary** relationship among these kingdoms.

Laboratory Techniques

Laboratory steps routinely employed in microbiology are inoculation, incubation, isolation, inspection, and identification. To prepare a **culture** (visible growth specimen), a **medium** (nutrient substrate) is **inoculated** (implanted, seeded) and **incubated.** Cultures differ: A **pure culture** consists of a single type of microbe; a **mixed culture** deliberately contains more than one type; and a **contaminated culture** is tainted with some intruding microbe. Valuable cultures are maintained in culture collections.

Microscopy

Optical or **light microscopy** depends upon lenses that **refract** light rays, drawing them to a focus to produce a magnified **image.** A **simple microscope** consists of a single magnifying lens, and a **compound microscope** relies upon a series of two: the **ocular lens** and the **objective lens.** The objective is responsible for the **real image,** and the ocular forms the **virtual image.** The **total power of magnification** is calculated from the product of the ocular and objective magnifying powers.

Resolution, or the **resolving power,** relates not only to magnification but to optical clarity as well. Resolution is directly proportional to the wavelength of illumination but inversely related to the **numerical aperture** of the lens. Image distortion called **spherical aberration** derives from curvature flaws, especially along the periphery, whereas **chromatic aberration** is related to a mixture of wavelengths or colors.

Modifications in the lighting or the lens increase the versatility of the compound microscope system. Such variations give rise to the **bright-field, dark-field, phase-contrast, interference,** and **fluorescence microscopes.** Magnification and resolution in the electron microscope obey the same laws as in optical microscopes. Fundamental substitutions include electron emission for light, electromagnets for focussing the lens, and the proper apparatus to provide the necessary voltage and vacuum. Compared to optical microscopes and imagery, the **transmission electron microscope** (TEM) is analogous to the bright-field microscope, and the **scanning electron microscope** (SEM) corresponds to the dark-field microscope.

Specimen preparation of optical microscopy is governed by the condition of the specimen, the purpose of the inspection, and the type of microscope being used. **Wet** or **hanging drop mounts** permit examination of live organisms, but the resolution is poorer than in **dyed** preparations. A **stained smear** provides higher resolution by virtue of color contrasts, but the dyes used are often toxic. Moreover, a smear must be **heat-fixed** in order to adhere to the slide.

Staining can be controlled by choosing the dye. A **basic (alkaline) dye** carries a positive charge, and an **acidic dye** bears a negative charge, thus they selectively bind to oppositely charged surfaces. Bacterial surfaces are usually negatively charged and attract basic dyes. Although bacteria repel acid dyes, they are useful for background or **negative staining** because a glass slide tends to be positively charged.

Staining methods can be distinguished from one another. A **simple stain** uses an uncomplicated procedure and one dye, like methylene blue, malachite green, crystal violet, basic fuchsin, or safranin. A **differential stain** requires a primary dye and a contrasting counterstain, consequently the procedure is more involved. Classic examples of differential stains are the **Gram stain, acid-fast stain,** and **endospore stain. Special stains** are designed to bring out distinctive characteristics, as with the **spore stain, capsule stain,** and **flagella stain.**

Stains have practical diagnostic significance. The ability of the Gram stain to distinguish **gram-positive** bacteria and **gram-negative** bacteria is useful in deciding upon drug therapy. Finding internal granules by means of a simple stain is valuable in diphtheria diagnosis. The **acid-fast stain** is useful in diagnosing tuberculosis and leprosy. The **spore stain** is helpful in anthrax, botulism, and tetanus diagnosis. The **capsule stain** can assist in diagnosing respiratory infections.

True–False Questions

Determine whether the following statements are true (T) or false (F). If you feel a statement is false, explain why and reword the sentence so that it reads accurately.

____ 1. The world would function more effectively in the absence of microorganisms.

____ 2. Microorganisms are called microbes because they are all microscopic.

____ 3. Procaryotes differ from eucaryotes solely on the basis of a nucleus.

____ 4. Leeuwenhoek was instrumental in the development of the oil immersion lens.

____ 5. The notion of spontaneous generation was easy to verify but not to discredit.

____ 6. A hypothesis is temporary proof, but a theory is permanent.

____ 7. By microbiological definition, sterile denotes the inability to breed.

____ 8. Application of the germ theory of disease requires inductive reasoning.

____ 9. Phylum is to division as Monera is to Protista.

____ 10. A mixed culture is the same as a contaminated culture.

____ 11. A subculture is a colony growing beneath the media surface.

____ 12. Resolution is possible without magnification.

____ 13. A real image exists, but a virtual image does not.

____ 14. Motility is best seen with fixed, stained smears because of greater contrast.

____ 15. By convention, only positively charged dyes can yield a positive stain.

Concept Questions

1. How do you view disease, decay, and germs? What do you suppose the world would be like if there were cures for all infectious diseases and a means to destroy all microbes?
2. Identify the groups of microorganisms included in the scope of microbiology, and explain the rationale for encompassing such diversity.
3. What event, discovery, or invention would you consider the turning point that marks the birth of microbiology?
4. Why was the abandonment of the spontaneous generation theory so significant?
5. Using the scientific method, describe the steps you would take to address the notion that infectious disease is a necessary evil.
6. Is the germ theory of disease really a law?
7. What is the basis for a phylogenetic system of classification?
8. Name the five kingdoms, and give microbial examples of each.
9. Differentiate between microscopic and macroscopic methods in microbiology, citing specific examples.
10. What are some basic differences between inoculation and contamination?
11. What do pure, mixed, and contaminated cultures have in common, and how do they differ from each other?
12. On the basis of the formula for resolving power, explain why a smaller R. P. value is preferred to a larger one. What would a value of 1.0 signify? A value greater than 1.0?
13. Of what benefit is the dark-field microscope if the field is black?
14. Compare and contrast the optical compound microscope with the electron microscope.
15. Rationalize and rank the following preparations in terms of retaining microbial size, shape, and motility: spore stain, negative stain, wet mount, simple stain, hanging drop slide, and Gram stain.
16. Itemize the various staining methods, and briefly characterize each.

Practical/Thought Questions

1. List the major variables in abiogenesis outlined in feature 1.2, and explain how each was tested and controlled by the scientific method.
2. Construct the scientific name of a newly discovered species of bacterium, using your name, a pet's name, a place, or a unique characteristic. Be sure to use proper notation and endings.
3. When buying a microscope, what feature or features are most important to check? What is probably true of a $20 microscope that claims to magnify 1,000×?

CHAPTER 2

From Atoms to Cells: A Chemical Connection

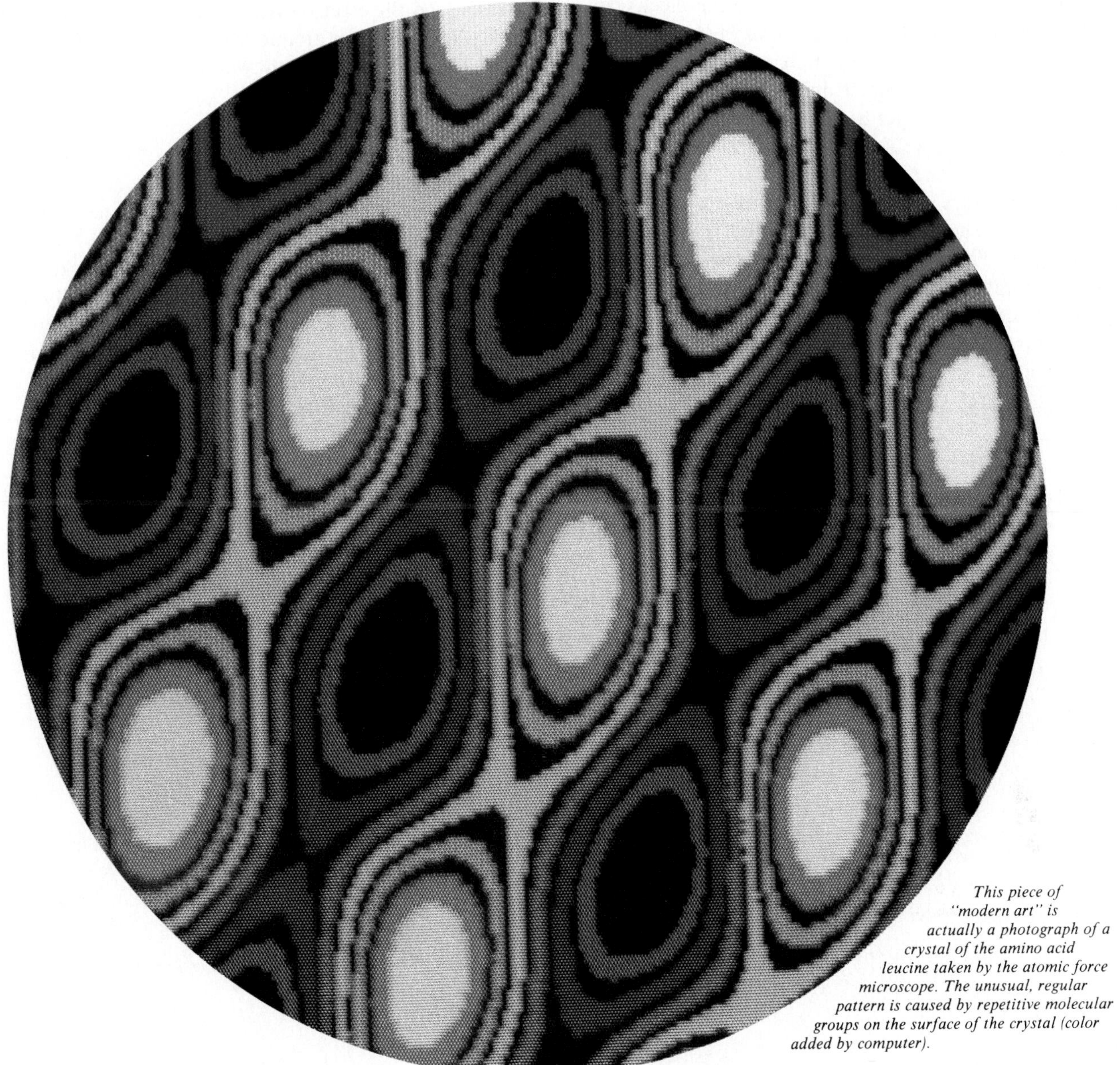

This piece of "modern art" is actually a photograph of a crystal of the amino acid leucine taken by the atomic force microscope. The unusual, regular pattern is caused by repetitive molecular groups on the surface of the crystal (color added by computer).

Chapter Preview

Because the characteristics of life ultimately stem from chemical events, the study of any biological science requires some familiarity with basic concepts of chemistry and biochemistry. Over the next 20 chapters you will encounter discussions of nutrition, metabolism, genetics, disinfectants, drugs, the immune system, the effects of diseases on the body, and the use of microbes to produce products as diverse as beer and hormones. The initial sections of this chapter are designed to give you a working knowledge of atoms, molecules, bonding, solutions, pH, and the structure of macromolecules that will serve you well in understanding these concepts and applying them in the future. The final section of this chapter contains an introduction to cell structure and a general comparison of procaryotic and eucaryotic cells in preparation for chapters 3 and 4.

Atoms, Bonds, and Molecules: Fundamental Building Blocks

All tangible materials that occupy space and have mass are called **matter.** The organization of matter—whether air or rocks or living things—begins with individual building blocks called atoms. An **atom** is a minute particle (an oxygen atom, for instance, is only 0.00000000066 mm in diameter) that cannot be further subdivided into a simpler unique particle by chemical means. Although we have yet to actually see it, the fine structure of atoms has been well established by extensive physical analysis. In general, atoms consist of subatomic particles called **protons** (p^+), which are positively charged, **neutrons** (n^o), which have no charge (are neutral), and **electrons** (e^-), which are negatively charged. The relatively larger protons and neutrons comprise a central core or **nucleus** that is surrounded by one or more smaller, constantly moving electrons (figure 2.1). The atomic structure is maintained and stabilized by the mutual attraction of the protons and the electrons. In addition, because the number of protons is exactly balanced by the number of electrons circling the nucleus, the opposing charges cancel each other out, and an isolated, intact atom theoretically carries no charge. The nucleus makes up the larger mass (weight) of the atom, and the electron region accounts for the greater volume. To illustrate, if an atomic nucleus were the size of a period on this page, the entire atom would be the size of a blimp.

atom (at′-um) Gr. *atomos,* not cut.

Elements and Their Properties

Although all atoms share the same fundamental structure, and all protons, neutrons, and electrons are identical, variations in the combination of these particles give rise to unique types of atoms called **elements.** Each type of element is known by a name and an abbreviated symbol. To date, 91 naturally occurring elements have been described, and 16 have been produced artificially by physicists. Table 2.1 lists some of the elements common to biological systems (*bioelements*), their atomic characteristics, and their main functions or uses.

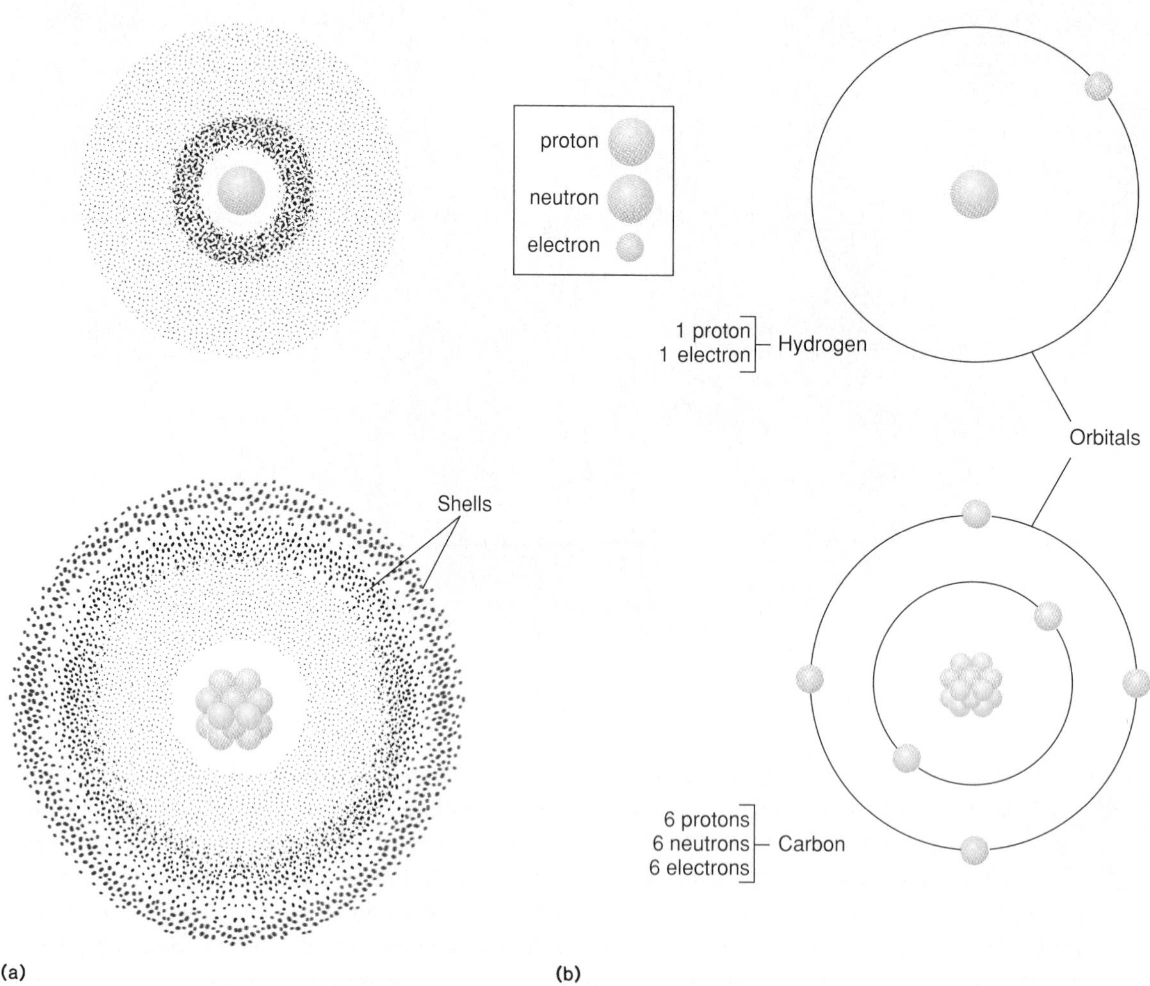

Figure 2.1 Diagrams of atoms. (*a*) Models of a hydrogen atom and a carbon atom depict the nucleus surrounded by a haze, indicating the probable position of electrons in orbitals and shells. (*b*) Working models of these same two atoms make it easier to visualize the numbers and arrangements of electrons, with orbitals shown as concentric circles.

The unique properties of each element result from the numbers of protons, neutrons, and electrons it has, and each element is identified by certain physical measurements. The **atomic weight** or **mass** of an element is equal to the sum of the numbers of its protons and neutrons. Atomic weight is not only a measure of weight but a relative scale for comparing atomic size, since larger, heavier elements have higher atomic weights. An element is also assigned an **atomic number,** based upon the number of protons alone. By knowing the atomic number, we also automatically know the number of electrons carried by each element. Because hydrogen has only a single proton, a single electron, and no neutron, it is the only element that has the same atomic weight and number.

Table 2.1 The Major Elements of Life and Their Primary Characteristics

Element	Atomic Symbol*	Atomic Number	Atomic Weight**	Ionized Form	Importance
Calcium	Ca	20	40.1	Ca^{++}	Part of outer covering of certain shelled amebae; stored within bacterial spores
Carbon	C	6	12.0	—	Principal structural component of biological molecules
		6	14.0	—	Isotope used in dating fossils
Chlorine	Cl	17	35.5	Cl^-	Component of disinfectants; used in water purification
Cobalt	Co	27	58.9	Co^{++}, Co^{+++}	Trace element needed by some bacteria to synthesize vitamins
		27	60	—	An emitter of gamma rays; used in food sterilization; used to treat cancer
Copper	Cu	29	63.5	Cu^+, Cu^{++}	Necessary to the function of some enzymes; Cu salts are used to treat fungal and worm infections
Hydrogen	H	1	1	H^+	Necessary component of water and many organic molecules; H_2 gas released by bacterial metabolism
		1	3	—	Tritium has 2 neutrons; radioactive; used in clinical laboratory procedures
Iodine	I	53	126.9	I^-	A component of antiseptics and disinfectants; contained in one reagent of the Gram stain
		53	131, 125		Radioactive isotopes for diagnosis and treatment of cancers
Iron	Fe	26	55.8	Fe^{++}, Fe^{+++}	Necessary component of respiratory enzymes; some microbes require it to produce toxin
Magnesium	Mg	12	24.3	Mg^{++}	A trace element needed for some enzymes; component of chlorophyll pigment
Manganese	Mn	25	54.9	Mn^{++}, Mn^{+++}	Trace elements for certain respiratory enzymes
Nitrogen	N	7	14.0	—	Component of all proteins and nucleic acids; the major atmospheric gas
Oxygen	O	8	16.0	—	An essential component of many organic molecules; molecule used in metabolism by many organisms
Phosphorus	P	15	31	—	A component of ATP, nucleic acids, cell membranes; stored in granules in cells
		15	32		Radioactive isotope used as a diagnostic and therapeutic agent
Potassium	K	19	39.1	K^+	Required for normal ribosome function and protein synthesis; essential for cell membrane permeability
Sodium	Na	11	23.0	Na^+	Necessary for transport; maintains osmotic pressure; used in food preservation
Sulfur	S	16	32.1	—	Important component of proteins; makes disulfide bonds; storage element in many bacteria
Zinc	Zn	30	65.4	Zn^{++}	An enzyme cofactor; required for protein synthesis and cell division

*Based on the Latin name of the element. The first letter is always capitalized; if there is a second letter, it is always lowercase.

**The atomic weight does not come out even because it is an average of the several isotopes.

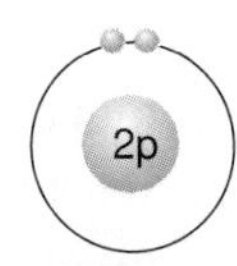

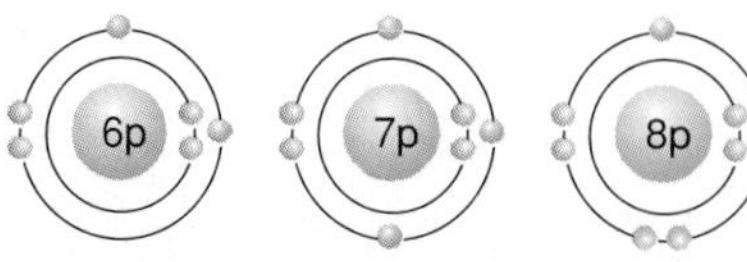

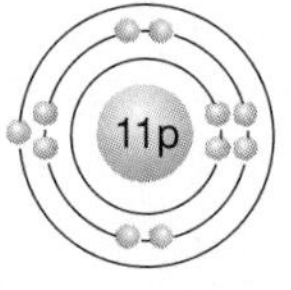

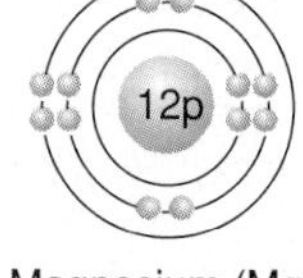

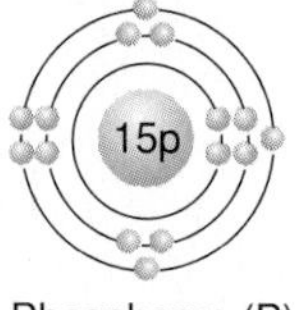

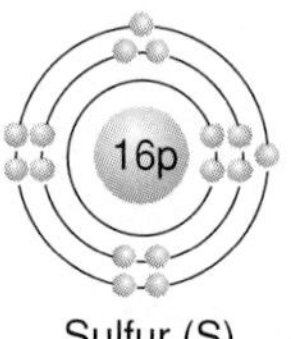

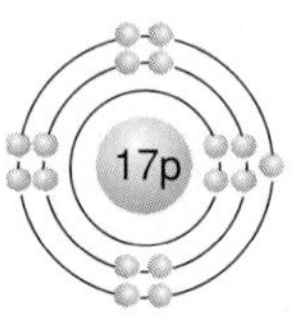

Figure 2.2 Models of several elements show how the shells are filled by electrons as the atomic numbers increase (numbers noted inside nuclei). Electrons tend to appear in pairs, except for certain elements with incompletely filled outer shells.

Table 2.1 also indicates that certain elements exist in more than one form, and close inspection will show that these have the same atomic number but a different atomic weight. These variant forms of an element, called **isotopes,** differ only in the number of neutrons in the nucleus. The nuclei of some isotopes are unstable and spontaneously break down, releasing energy in the form of radiation. Such *radioactive isotopes* are widely used in research, in treatment of disease, as tracers in certain types of immunologic testing, and in some forms of sterilization (see chapter 9).

Electron Orbitals and Shells

Models of atomic structure such as that in figure 2.1*b* make it appear as though electrons are small bodies orbiting the nucleus like planets around a sun. It is important to realize, however, that these diagrams are not meant to show exact structure but to present the characteristics of an atom in workable fashion. A model of an actual atom is a bit difficult to represent on paper because the electrons do not move about in perfect concentric circles. It is more accurate to use a three-dimensional cloud to show the volume in space through which an electron can be expected to move. The pathway of an electron's movement—the *orbital*—falls within an energy sphere called a *shell* (figure 2.1*a*). (The orbital is named for the superficial resemblance of early atomic models to a solar system, while the shell is named for the notion that shells fit together like the layers of an onion.) Each orbital has a finite number of electrons; each shell contains one or more orbitals; and the orbitals and shells vary in energy, from the lowest ones lying closest to the nucleus to the highest ones lying farthest away.

Electrons fill the orbitals in pairs according to a predictable pattern (figure 2.2). The first shell can contain one pair (2 electrons), the second shell can contain up to four pairs (8 electrons), the third can hold a maximum of nine pairs (18), and the fourth can contain up to sixteen pairs (32). The number of shells and how completely they are occupied depends upon the number of electrons a given element has. The shells are filled up from inside out, and those elements with higher atomic numbers will have more electrons and more shells. For example, helium with only 2 electrons has a single filled first shell, and oxygen with 8 electrons has a filled first shell (2 e^-) and a partially filled second shell (6 e^-). Selected elements and the distribution of their electrons are depicted in figure 2.2.

The Ties That Bind: Atomic Bonds and Molecules

Most elements do not exist in nature in pure, uncombined form, but rather in the form of molecules and compounds. A **molecule** is a distinct chemical substance that results from the combination of two or more atoms. A molecule may consist of atoms of the same element, as in oxygen gas (O_2) or nitrogen gas (N_2), or atoms of dissimilar elements, as in most of the molecules in living things. When a molecule is a combination of two or more different elements in a certain ratio, it is called a **compound.** Just as an atom has an atomic weight, a molecule has a molecular weight (MW), which is calculated from the sum of all of the atomic weights of the atoms it contains.

Chemical bonds are formed when two or more atoms share, donate (lose), or accept (gain) electrons (figure 2.3). The number of electrons in the outermost orbital, known as the *valence,* determines the different degrees of reactivity and the types of bonding among atoms. Those elements with a filled outer orbital are relatively stable because they have no "extra" electrons to share or donate to other atoms, while those with partially filled outer orbitals are less stable and more apt to form

molecule (moll'-ih-kyool) L. *molecula,* little mass.

valence (vay'-lents) L. *valentia,* strength. A numerical measure of binding capacity.

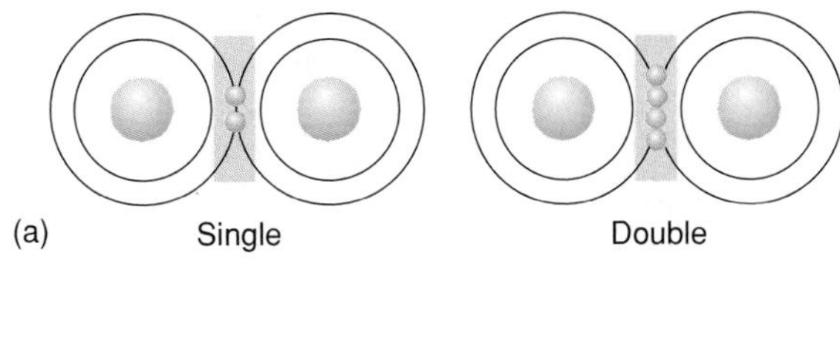

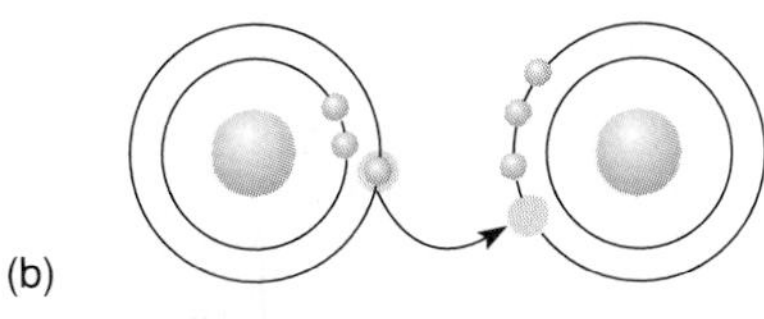

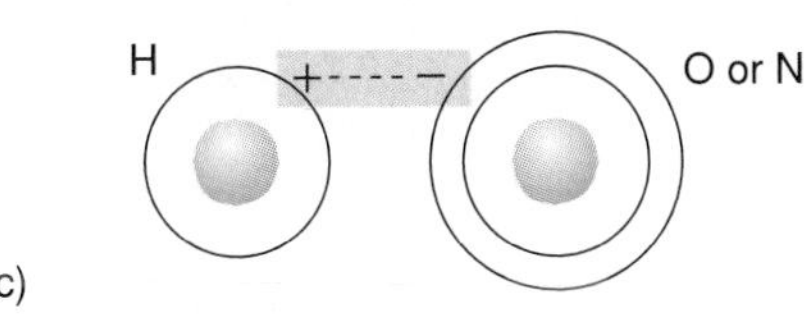

Figure 2.3 General representation of three types of bonding. (*a*) Covalent bonds, both single and double. (*b*) An ionic bond. (*c*) Hydrogen bonds. Note that hydrogen bonds are represented in models and formulas by dotted lines.

some sort of bond. For example, because helium has a single filled shell, it is an inert (nonreactive) gas. By comparison, an oxygen atom has 2 electrons less than a full outer shell and is highly reactive. In addition to reactivity, the number of electrons in the outer shell dictates the number of chemical bonds an atom can make. For instance, an oxygen atom can generally bind with up to two other atoms, and carbon can bind with four (see figure 2.13).

Many bonding interactions are based on the tendency of atoms with unfilled outer shells to gain greater stability by achieving, or at least approximating, a filled outer shell. For example, an atom such as oxygen that could gain 2 additional electrons bonds readily with atoms (such as hydrogen) that readily give up electrons. We will explore some additional examples of the basic types of bonding in the following section.

Covalent Bonds and Polarity

Covalent (cooperative valence) **bonds** form between atoms with valences that suit them to sharing electrons rather than to donating or receiving them. A simple example is hydrogen gas (H_2), which consists of two hydrogen atoms. Although a hydrogen atom has a single electron, the combination of two hydrogen atoms provides 2 electrons that can orbit around both nuclei, thereby approximating a filled orbital for both atoms (figure 2.4*a*). Covalent bonding also occurs in oxygen gas (O_2), but in this molecule, each atom has 2 electrons to share, and their combination forms a *double* rather than a *single* bond

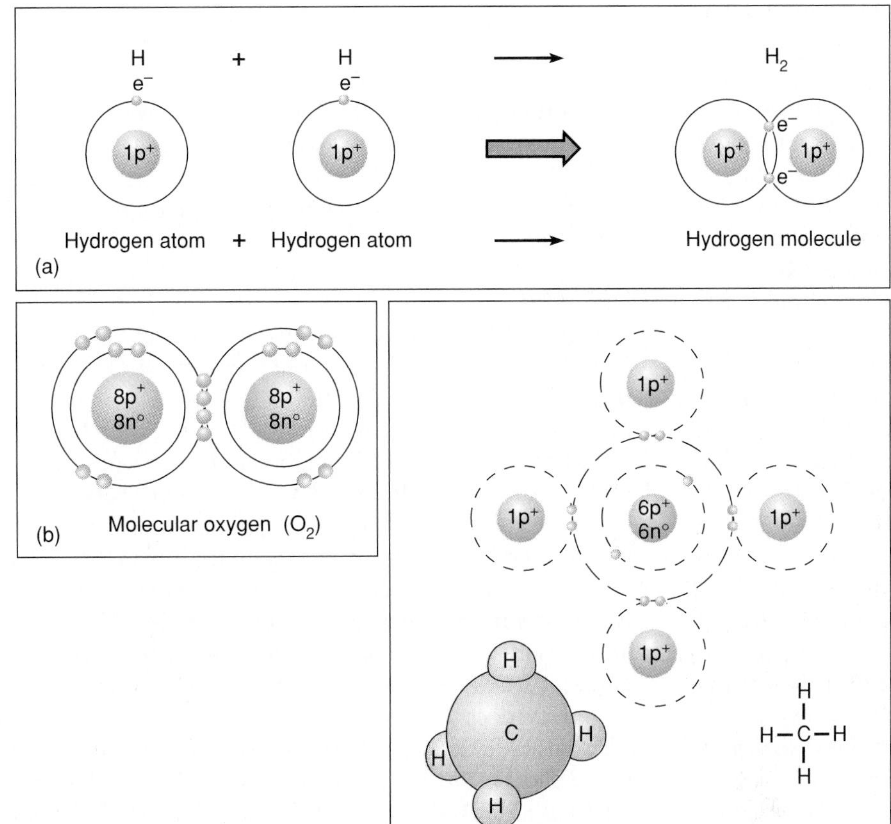

Figure 2.4 Examples of molecules with covalent bonding. (*a*) A hydrogen molecule is formed when two hydrogen atoms share their electrons and form a single bond. (*b*) In a double bond, the outer orbitals of two oxygen atoms overlap and permit the sharing of four electrons (one pair from each) and the saturation of the outer orbital for both. (*c*) Simple, working, and three-dimensional models of methane. Note that carbon has four electrons to share and hydrogens each have one, thereby completing the shells for all atoms in the compound.

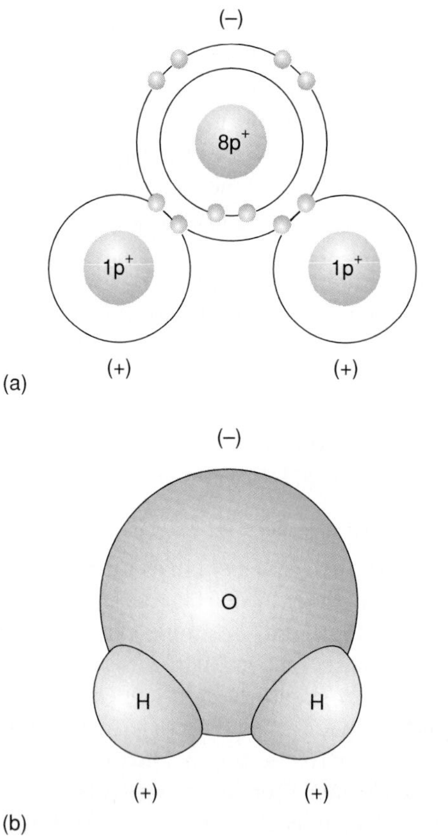

Figure 2.5 (*a*) Simple model and (*b*) three-dimensional model of a water molecule indicate the polarity, or unequal distribution, of electrical charge, which is caused by the pull of the shared electrons toward the oxygen side of the molecule.

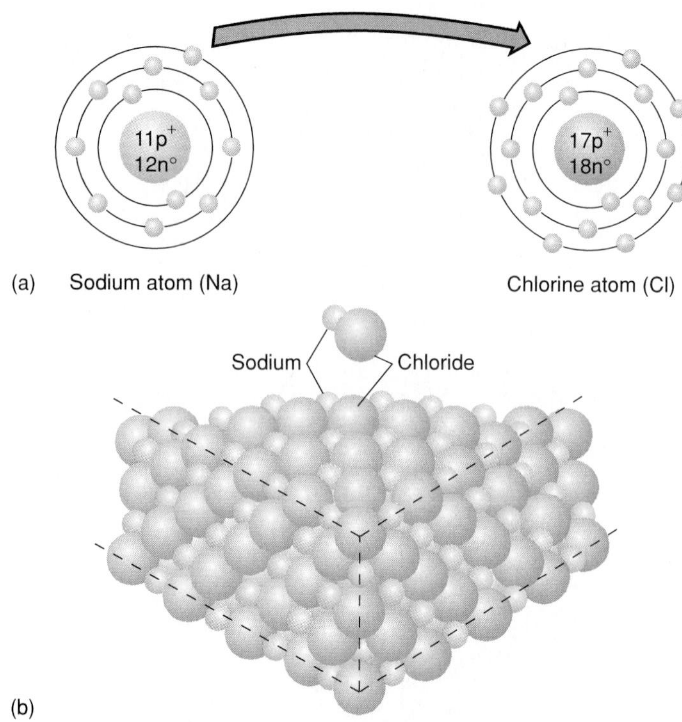

Figure 2.6 Ionic bonding between sodium and chlorine. (*a*) When the two elements are placed together, sodium loses its single outer orbital electron to chlorine, thereby saturating chlorine's outer orbital. (*b*) This reaction produces sodium and chloride ions that form large complexes or crystals in which the two atoms alternate in a definite, regular, geometric pattern.

(figure 2.4*b*). The majority of the molecules associated with living things are composed of covalent bonds between the most common biological elements (carbon, hydrogen, oxygen, nitrogen, sulfur, and phosphorus), which are discussed in more depth in chapter 6. A slightly more complex pattern of covalent bonding is shown for methane gas (CH_4) in figure 2.4*c*.

When covalent bonds are formed between two identical or similar atoms, the electrons are equally shared and distributed between the two, so that no part of the molecule has a greater attraction for the electrons. This sort of molecule is termed **nonpolar.** When two different atoms bind covalently, however, especially if one is of higher atomic number, the electrons are not shared equally and may be pulled more toward one atom than another. This causes one side of a molecule to be more negatively charged and the other side to be more positively charged. A molecule with such an asymmetrical distribution of charges is termed **polar** (having positive and negative poles), as exemplified by water (figure 2.5). The two hydrogens and the single oxygen are oriented so that the shared electrons are drawn with greater force towards the nucleus of the oxygen molecule. This causes the hydrogen side of the molecule to take on a positive charge (due to the protons in their nuclei) and the oxygen side to take on a negative charge (due to the electrons being attracted there). The polar nature of water plays an extensive role in a number of biological reactions to be discussed in a subsequent section. Polarity is a significant property of many of the large molecules in living systems and greatly influences both their reactivity and their structure.

Ionic Bonds

In many bonding reactions, electrons are transferred completely from one atom to another and are not shared. These reactions invariably occur between atoms with valances that complement each other, meaning that one atom is more stable by gaining electrons and one atom is more stable by losing electrons. A striking example is the reaction that occurs between sodium (Na) and chlorine (Cl) (figure 2.6). Elemental sodium is a soft, lustrous metal so potent that it can burn human flesh, and molecular chlorine is a very poisonous yellow gas. But when the two are combined, they form sodium chloride[1] (NaCl)—the familiar nontoxic table salt—a compound with properties quite different from either parent element.

How does this transformation occur? Since sodium has 11 electrons (2 in shell one, 8 in shell two, and only 1 in shell three), it is 7 short of having a complete outer shell. Chlorine has 17 electrons (2 in shell one, 8 in shell 2, and 7 in shell three), making

1. In general, when a salt is formed, the ending of the name of the negatively charged ion is changed to *-ide*.

Feature 2.1 Electron Transfer and Oxidation/Reduction

The property of atoms that causes them to gain or lose electrons is an important system for managing energy in living things, since electrons are a readily portable and constant source of energy (energy being the capacity to do work). The phenomenon in which electrons are transferred from one atom or molecule to another is termed an **oxidation and reduction** (shortened to **redox**) reaction. Though the term oxidation was originally adopted to refer to reactions with oxygen, the current usage refers to any reaction in which electrons are given up, regardless of whether oxygen is involved. By comparison, reduction is the tendency to receive electrons during a reaction, and all such redox reactions occur in pairs. There are a number of ways that such transfers can be made, but to analyze the phenomenon, let us again review the production of NaCl, this time from a different standpoint. Although it is true that these atoms form ionic bonds, the chemical combination of the two is also a type of redox reaction.

Redox pair   Oxidized   Reduced

$2Na + Cl_2 \rightarrow 2Na^+ + 2\ Cl^-$ (equation balanced for accuracy)

When these two atoms react to form sodium chloride, a sodium atom gives up an electron to a chlorine atom. During this reaction, sodium is oxidized (loses an electron), and chlorine is reduced (gains an electron). To take this definition further, an atom or molecule like sodium that can donate electrons and thereby reduce another molecule is a **reducing agent**; one that can receive extra electrons and thereby oxidize another molecule is an **oxidizing agent.** You may find this concept easier to keep straight if you think of them as agents that cause reduction (electron gain) in their partner molecule and agents that cause oxidation (electron loss). These reactions are involved in many of the metabolic processes discussed in chapter 7. In cellular metabolism, electrons alone may be transferred from one molecule to another as described here, but sometimes oxidation and reduction occur through the transfer of hydrogen atoms (which are a proton and an electron) from one compound to another. As we will see, many types of electron transfers in cells involve large carrier molecules that can pick up electrons, hydrogens, or both from one compound and donate them to another.

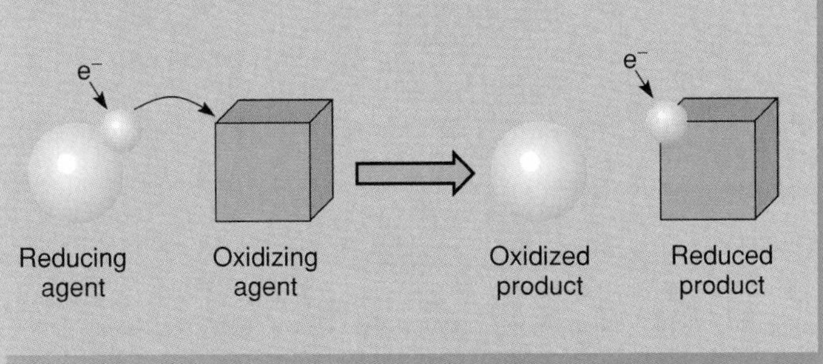

Simplified diagram of the exchange of electrons during an oxidation-reduction reaction.

it 1 short of a complete outer shell. These two atoms are very reactive with one another, since a sodium atom will readily donate its single electron and a chlorine atom will avidly receive it. (The reaction is slightly more involved than a single sodium atom combining with a single chloride atom (see feature 2.1), but this does not detract from the fundamental reaction as described here.) The result of this reaction is not a discrete in-

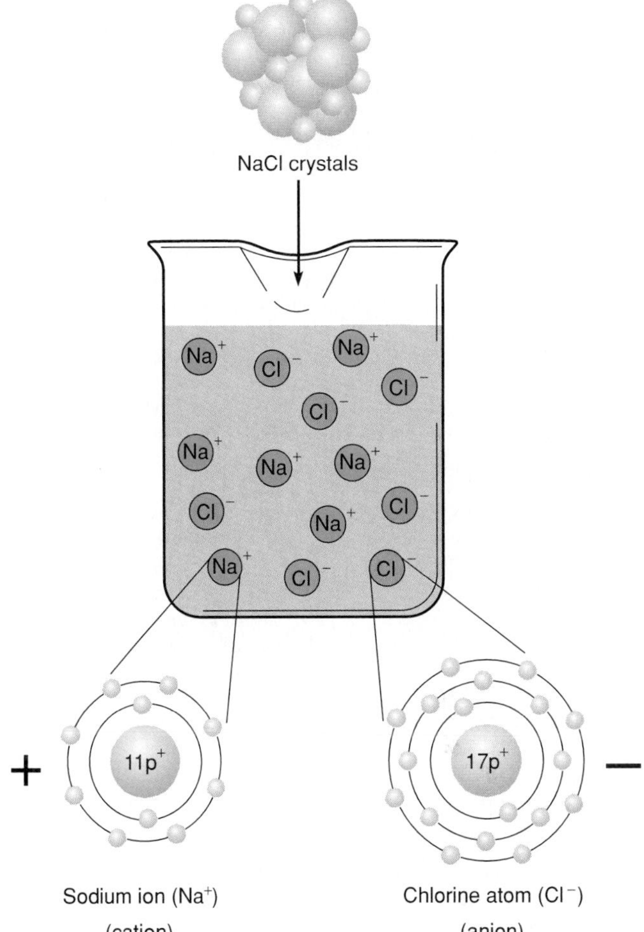

Figure 2.7 When NaCl in the crystalline form is added to water, the ions are released from the crystal as separate charged particles (cations and anions) into solution. (See also figure 2.11.)

dependent molecule of NaCl but a massive atomic linkage of these atoms organized into a perfect geometric solid or crystal.

Ionization When ionic compounds are dissolved in water, they **ionize.** This means that the ionic bond is broken, and the atoms dissociate (separate) into individual, nonattached, charged particles called **ions** (figure 2.7). To illustrate what imparts a charge to ions, let us look again at the reaction between sodium and chlorine. Having lost one electron through its reaction with chlorine, a separated sodium atom (ion) now has more protons than electrons, giving it a positive charge. Having gained one electron, chlorine now has more electrons than protons, and this gives it a negative charge.

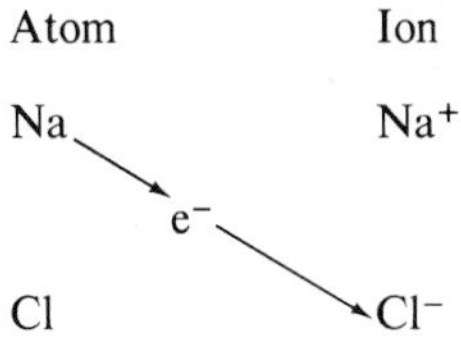

ion (eye'-on) Gr. *ion,* going.

Feature 2.2. The Chemistry of Dyes and Staining

Staining of microorganisms with dyes is an essential way to observe them and their structure in greater detail. Dyes are colored compounds synthesized from the common organic solvent benzene. When certain double-bonded groups (C=O, C=N, N=N) are attached to complex ringed molecules, the resultant compound gives off a specific color. Most dyes are in the form of a sodium or chloride salt of an acidic or basic compound that ionizes when dissolved in a compatible solvent. (This makes them an example of organic compounds that ionize.) The color-bearing ion, termed a **chromophore,** is thus charged and has an affinity for certain cell parts that are of the opposite charge.

Dyes that ionize into a negatively charged chromophore and a positively charged mineral ion are **acidic** or **anionic.** An example is sodium eosinate, a bright red dye that dissociates into eosin$^-$ and Na^+. Acidic dyes are attracted to the positively charged molecules of cells, but because bacterial cells have numerous acidic substances and carry a slight negative charge on their surface, they do not stain well with acidic dyes.

Basic or **cationic** dyes ionize into a positively charged chromophore and a negative mineral ion. For example, basic fuchsin gives off fuchsin$^+$ and Cl^-. Since these dyes are positively charged, they are attracted to negatively charged cell components such as nucleic acids and proteins. Because bacteria have a preponderance of negative ions, they stain readily with basic dyes, and most of the common dyes used in staining procedures (crystal violet, malachite green) are basic.

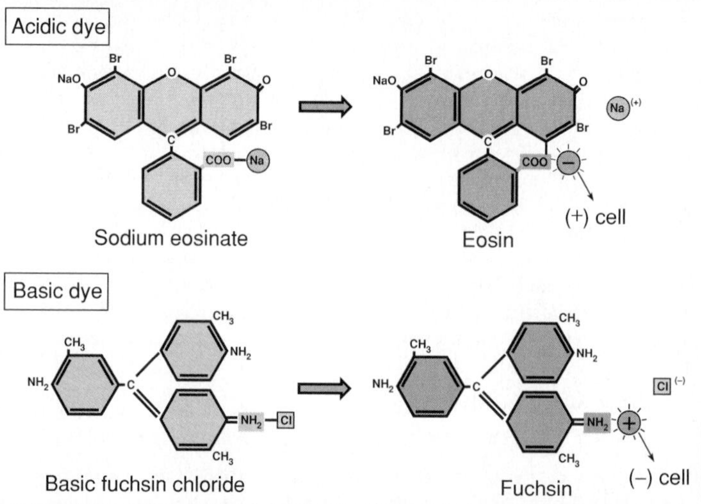

Positively charged ions such as Na^+ are termed **cations,** and negatively charged ions such as Cl^- are termed **anions.** Substances such as salts, acids, and bases (discussed in a later section), which release ions when dissolved in water, are termed **electrolytes** because they can conduct an electrical current. Since ions carry charges, they can interact with other molecules and atoms according to the principle that atoms of like charge repel each other and those of opposite charge attract each other (that is, they possess an *electrostatic* attraction). The interactions of charged and polar compounds are important in many cellular chemical reactions, in the formation of solutions, and in staining reactions (see feature 2.2) that will be covered in this chapter and others. Because electrons are a source of energy, their transferral from one molecule to another constitutes a significant mechanism by which biological systems handle energy (see feature 2.1).

Hydrogen Bonds

Some types of bonding involve neither sharing, losing, nor gaining electrons but are due to attractive forces between nearby molecules or atoms. A **hydrogen bond** is a weak type of bond that forms between a hydrogen covalently bonded to one molecule and an oxygen or nitrogen atom on the same or a different molecule. Because hydrogen in a covalent bond tends to be positively charged, it will attract a nearby negatively charged atom and form an easily disrupted bridge with it. This type of bonding is usually represented in molecular models with a dotted line. A simple example of hydrogen bonding is given for water (figure 2.8), and a more complex type that is partly responsible for the stabilization and structure of proteins and nucleic acids is given in figures 2.22*b* and 2.25*a*.

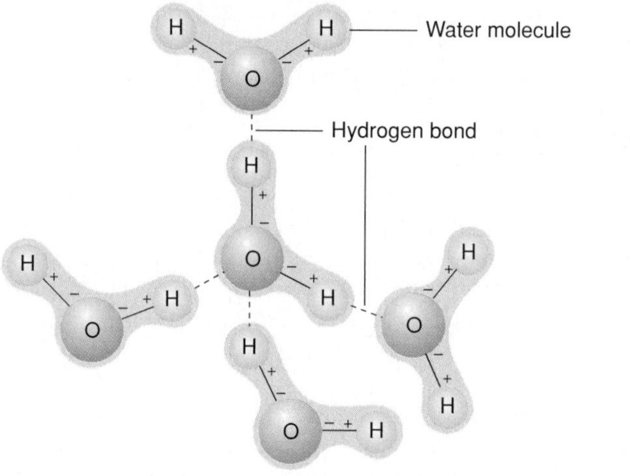

Figure 2.8 Hydrogen bonding in water. Due to the polarity of water molecules, the negatively charged oxygen end of one water molecule is weakly attracted to the positively charged hydrogen end of an adjacent water molecule.

Formulas, Models, and Chemical Equations

The atomic content of molecules can be represented by a few convenient *formulas.* We have already been exposed to one type, the molecular formula, which concisely gives the atomic

---

cation (cat'-eye-on) An ion that migrates toward the negative pole or cathode of an electrical field.

anion (an'-eye-on) An ion that migrates toward the positive pole or anode.

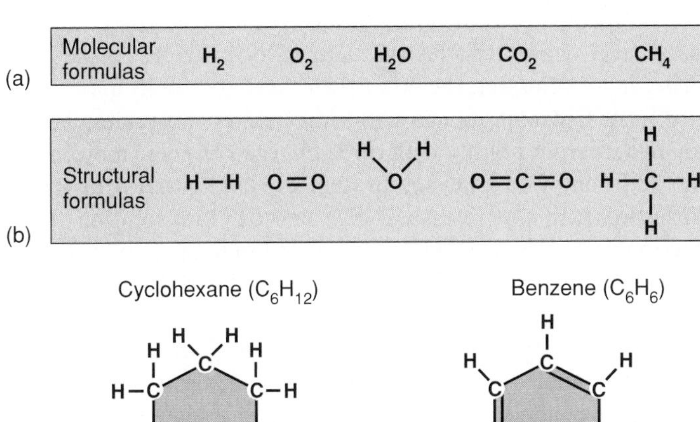

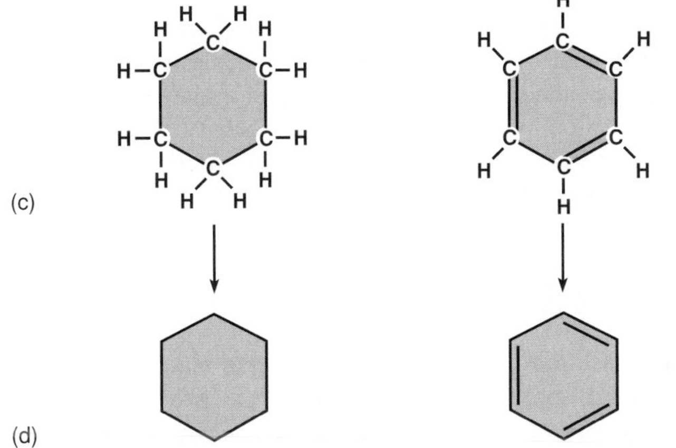

Figure 2.9 Comparison of molecular and structural formulas. (*a*) Molecular formulas provide a brief summary of the elements in a compound. (*b*) Structural formulas clarify the exact relationships of the atoms in the molecule, depicting single bonds by a single line and double bonds by two lines. (*c*) In structural formulas of organic compounds, cyclic or ringed compounds may be completely filled in, or (*d*) they may be presented in a shorthand form in which carbons are assumed to be at the angles and attached to hydrogens.

symbols, with the number of the elements involved in subscript (CO_2, H_2O). More complex molecules such as glucose ($C_6H_{12}O_6$) can also be symbolized this way, but this empirical formula is not unique since fructose and galactose also share it. Molecular formulas are useful, but they only summarize the atoms in a compound; they do not show the precise bonds between atoms. For this purpose, chemists use structural formulas illustrating the relationships of the atoms and the number and types of bonds (figure 2.9). Structural models represent the three-dimensional appearance of a molecule, illustrating the orientation of atoms (differentiated by color codes) and the molecule's overall shape (figure 2.10).

The printed page tends to make molecules appear static, but this picture is far from correct, because molecules are constantly changing through chemical reactions. To make it easier to trace chemical exchanges between atoms or molecules, and to derive some sense of the dynamic character of reactions, chemists use shorthand equations containing symbols, numbers, and arrows to simplify or summarize the major characteristics of a reaction. Molecules entering or starting a reaction are called **reactants,** and substances left by a reaction are called **products.** In most instances, summary chemical reactions do not give the details of the exchange, in order to keep the expression simple and save space.

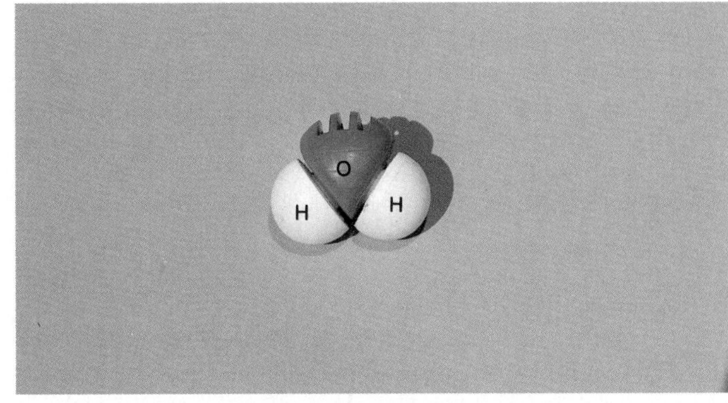

(a)

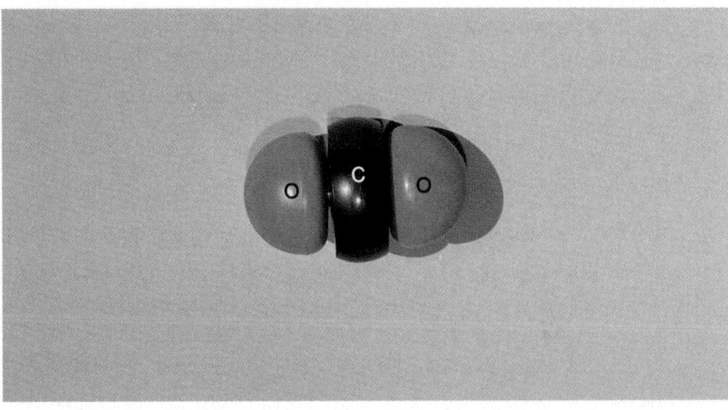

(b)

(c)

Figure 2.10 Three-dimensional or space-filling models of (*a*) water, (*b*) carbon dioxide, and (*c*) glucose. By convention, the red atoms are oxygen, the white ones hydrogen, and the black ones carbon.

In a **synthesis** reaction, the reactants bond together in a manner that produces an entirely new molecule (reactant A plus reactant B yields product AB). An example of this is the production of sulfur dioxide, a by-product of burning sulfur fuels and an important component of smog:

$$S + O_2 \rightarrow SO_2$$

synthesis (sin'-the-sis) Gr. *synthesis,* putting together.

Some synthesis reactions may not be such simple combinations. When water is synthesized, for example, the reaction does not really involve one oxygen atom combining with two hydrogen atoms, because elemental oxygen exists as O_2 and elemental hydrogen exists as H_2. A more accurate equation for this reaction is:

$$2H_2 + O_2 \rightarrow 2H_2O$$

The equation for this sort of reaction must be balanced—that is, the number of atoms on one side of the arrow must equal the number on the other side to reflect the actual participants in the reaction. Multiply the prefix number by the subscript number to arrive at the total number of atoms; if no number is given, it is assumed to be 1.

In **decomposition** reactions, the bonds on a more complex reactant are permanently broken to release simpler products. This occurs when large nutrient molecules are digested into smaller units:

$$ABC \rightarrow AB + C \text{ or } ABC \rightarrow A + B + C$$

During **exchange** reactions, the reactants trade portions between each other and release products that are combinations of the two. This type of reaction occurs between acids and bases when they form water and a salt:

$$AB + XY \leftrightarrow AX + BY$$

The reactions in biological systems are often **reversible,** meaning that reactants and products may be converted back and forth. These reversible reactions are symbolized with double arrows, each pointing in opposite directions, as in the exchange reaction just shown. Whether a reaction is reversible depends upon the proportions of these compounds, the amount of energy necessary for the reaction, and the presence of catalysts (substances that increase the rate of a reaction). Additional reactants coming from another reaction may also be indicated by arrows that enter or leave at the main arrow:

$$\overset{CD \quad C}{\smile}$$
$$X + Y \rightarrow XYD$$

Many of these principles will be encountered again in chapters 6 and 7.

Solutions: Homogeneous Mixtures of Molecules

A **solution** is a mixture of one or more substances called **solutes** uniformly dispersed in a dissolving medium called a **solvent.** An important characteristic of a solution is that the solute cannot be separated by filtration or ordinary settling. The solute may be gaseous, liquid, or solid, but the solvent is usually a liquid. Examples of solutions are salt or sugar dissolved in water and iodine dissolved in alcohol. In general, a solvent will dissolve a solute only if it has similar electrical characteristics as indicated by the rule of solubility, expressed simply as "like dissolves like." For example, water is a polar molecule and will readily dissolve a polar solute such as NaCl, yet a nonpolar solvent such as benzene will not dissolve NaCl.

Water is the most common solvent in natural systems and has several characteristics that suit it to this role. As we saw earlier in this chapter, the polarity of the water molecule causes it to form hydrogen bonds with other water molecules, but it can also interact readily with other charged or polar molecules. The addition of an ionic solute such as NaCl crystals to water causes ionization or release of Na^+ and Cl^- into solution. This reaction occurs because Na^+ is attracted to the negative pole of the water molecule and Cl^- is attracted to the positive pole; this way, they are drawn away from the crystal separately into solution. Eventually, each ion becomes **hydrated,** which means that it is surrounded by a sphere of water molecules (figure 2.11). Covalently bonded molecules of glucose or amino acids do not ionize like salts, but because they have polarity, they can also be hydrated in solution. Molecules having groups that attract water to their surface are termed *hydrophilic.* Nonpolar molecules such as benzene that repel water are considered *hydrophobic.* A third class of molecules (certain lipids) that have both hydrophilic and hydrophobic properties are called *amphipathic.*

Because most biological activities take place in aqueous (water-based) solutions, the concentration of these solutions can be very important (see chapter 6). The *concentration* of a solution is the amount of solute dissolved in a certain amount of solvent. It may be expressed by weight, volume, or percentage. Concentration by percentage is calculated as the weight of the solute, measured in grams (g), dissolved in a specified volume of solvent, measured in milliliters (ml). For example, dissolving 5 g of NaCl in 100 ml of water produces a 5% solution; dissolving 60 g in 100 ml produces a 60% solution; and dissolving 60 g in 1,000 ml (1 liter) produces a 6% solution. Solutions with a small amount of solute and a relatively greater amount of solvent are considered dilute or weak, and solutions with a large amount of solute dissolved in a small amount of solvent are considered concentrated or strong.

Acidity, Alkalinity, and the pH Scale

Another factor that has tremendous impact on living systems is the concentration of acid or base to which they are exposed. To understand how solutions develop acidity or basicity, we must again look at the behavior of water molecules. Hydrogens and oxygen tend to remain bonded by covalent bonds, but in certain instances, a single hydrogen can break away as the ionic form (H^+), leaving the remainder of the molecule in the form of an OH^- ion. Because an H^+ ion is essentially a proton molecule that has lost its electron, it is positively charged, while the OH^- remains in possession of that electron and is negatively charged. Ionization of water is constantly occurring, but in pure water containing no other ions, H^+ and OH^- are produced in equal amounts, and the solution remains neutral. A solution is **acidic**

---

hydrophilic (hy-dro-fil'-ik) Gr. *hydros,* water, and *philos,* to love.
hydrophobic (hy-dro-fo'-bik) Gr. *phobos,* to fear.

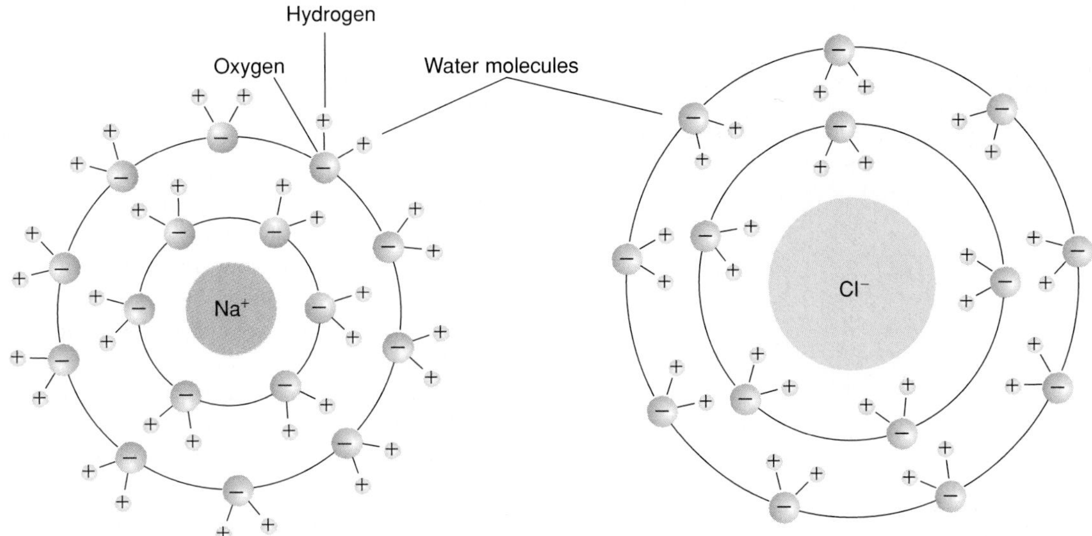

Figure 2.11 Hydration spheres formed around ions in solution. In this example, sodium cations attract the negatively charged end of water molecules, and chloride anions attract the positively charged region of water molecules. In both cases, the ions become covered with concentric layers of water molecules.

when a compound (acid) releases additional hydrogen ions (H^+) into a solution, and a solution is **basic** when a compound releases additional hydroxyl ions (OH^-), so that there is no longer a balance between the two ions.

Acid and base concentrations of solutions are measured on the **pH scale,** a graduated numerical scale that ranges from 0 (the most acidic) to 14 (the most basic). The pH readings of various substances are shown in figure 2.12. Although this scale is a useful and easily applied system for rating relative acidity and basicity, it is actually a mathematical derivation of the negative logarithm (reviewed in appendix B) of the concentration of H^+ ions in a solution, also represented as:

$$\text{pH} = \log \frac{1}{[H^+]\ (\text{Concentration of } H^+ \text{ in grams/liter})}$$

Acids have a greater concentration of H^+ than OH^-, and the number on the pH scale with greatest acidity is pH 0, which has a $[H^+]$ (hydrogen ion concentration) of 1.0 g/l. Each of the subsequent whole-number readings in the scale changes in $[H^+]$ by a tenfold reduction, so that pH 1 contains [0.1 g H^+/l], pH 2 contains [0.01 g H^+/l], and so on, continuing in the same manner up to pH 14, which contains [0.00000000000001 g H^+/l]. These same concentrations may be represented more manageably by exponents—that is, the $[H^+]$ for pH 2 is 10^{-2} and the $[H^+]$ for pH 14 is 10^{-14} (table 2.2). It is evident that the pH units are derived (by a method we will not go into) from the exponent itself. Even though the basis for the pH scale is $[H^+]$, it is important to note that, as the $[H^+]$ in a solution decreases, the $[OH^-]$ increases in direct proportion. At midpoint—pH 7 or neutrality—the concentrations are exactly equal and neither predominates, this being the pH of pure water previously mentioned.

In summary, the pH scale may be used to rate or determine the degree of acidity or basicity (also called alkalinity) of

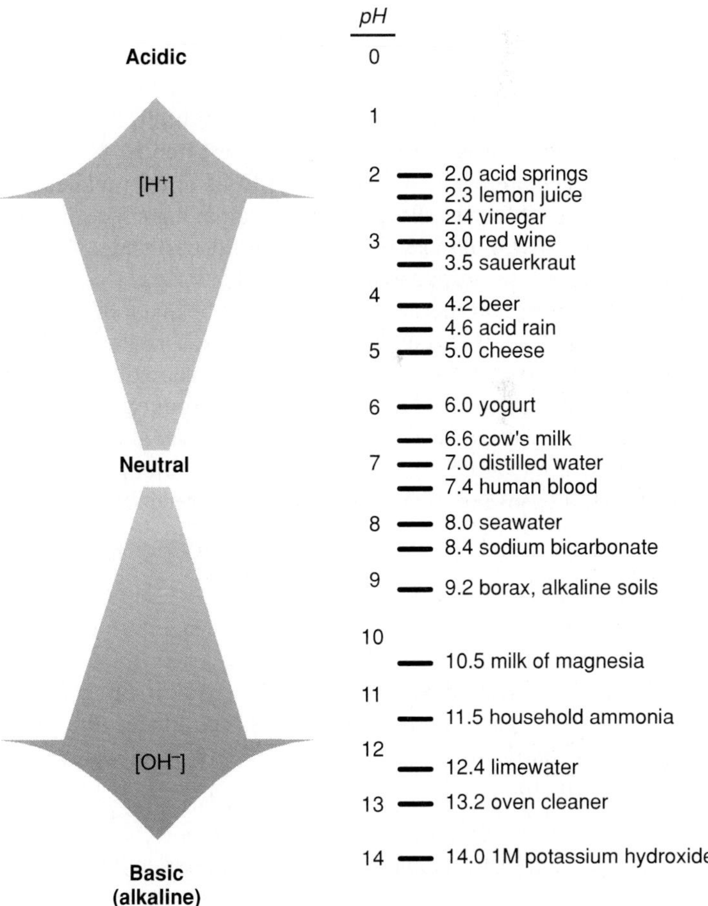

Figure 2.12 The pH scale, showing the relative degrees of acidity and basicity and the approximate pH readings for various substances and habitats.

Table 2.2 Hydrogen Ion and Hydroxyl Ion Concentrations at a Given pH

g/l of Hydrogen Ions	Logarithm	pH	g/l of OH^-
1.0	10^{-0}	0	10^{-14}
0.1	10^{-1}	1	10^{-13}
0.01	10^{-2}	2	10^{-12}
0.001	10^{-3}	3	10^{-11}
0.0001	10^{-4}	4	10^{-10}
0.00001	10^{-5}	5	10^{-9}
0.000001	10^{-6}	6	10^{-8}
0.0000001	10^{-7}	7	10^{-7}
0.00000001	10^{-8}	8	10^{-6}
0.000000001	10^{-9}	9	10^{-5}
0.000000000.1	10^{-10}	10	10^{-4}
0.0000000000.1	10^{-11}	11	10^{-3}
0.00000000000.1	10^{-12}	12	10^{-2}
0.000000000000.1	10^{-13}	13	10^{-1}
0.0000000000000.1	10^{-14}	14	10^{-0}

a solution. On this scale, a pH below 7 is acidic, and the lower the pH, the greater the acidity; a pH above 7 is basic, and the higher the pH, the greater the basicity. Incidentally, though pHs are given here in even whole numbers, more often, a pH reading exists in decimal form—for example, pH 4.5 or 6.8 (acidic) and pH 7.4 or 10.2 (basic). Because of the damaging effects of very concentrated acid or base, most cells operate best under neutral, weakly acidic, or basic conditions (see chapter 6).

An important factor that governs the behavior of aqueous solutions is that, when acids and bases coexist in the same solution, some of the H^+ and OH^- they release reacts to form water. This reaction, called *neutralization,* produces other neutral by-products as well. For instance, when equal amounts of solutions of hydrochloric acid (HCl) and sodium hydroxide (NaOH, a base) are mixed, the reaction proceeds as follows:

$$HCl + NaOH \rightarrow H_2O + NaCl$$

In this reaction, the acid and base ionize to H^+ and OH^- ions, which combine to form water, and other ions, Na^+ and Cl^-, which produce the sodium chloride salt. Any neutral product of this sort that arises when acids and bases react is called a salt. Many of the organic acids (lactic acid, succinic acid) that function in *metabolism* (see chapter 7) exist in the salt form (lactate, succinate).

The Chemistry of Carbon and Organic Compounds

Thus far we have discussed atoms and small, simple molecules that are known as **inorganic,** but the reactions in living things occur at the level of much larger molecules. Although inorganic atoms, ions, and molecules are important to living things, most of the actual structures and chemical reactions of cells involve **organic** chemicals, which are defined as compounds that contain carbon and hydrogen. Certain carbon compounds such as carbon dioxide (CO_2), carbonates (CO_3), and cyanide (CN) are inorganic, but the great majority are organic. Organic molecules vary in complexity from the simplest one, methane (CH_4; figure 2.4*c*), which has a molecular weight of 16, to certain antibody molecules (produced by an immune reaction) that have a molecular weight of nearly 1,000,000 and are among the most complex molecules on earth.

The role of carbon as the fundamental element of life can best be understood if we look at its chemistry and bonding patterns. The valence of carbon makes it an ideal atomic building block to form the backbone of organic compounds: It has 4 electrons in its outer orbital, which it can share with other atoms (including other carbons) through covalent bonding. As a consequence, it forms stable chains containing thousands of carbon atoms and still has bonding sites available for forming covalent bonds with numerous other atoms. The nature of the bonds it forms may be linear, branched, or ringed, and it may form four single bonds, two double bonds, or one triple bond (figure 2.13). The atoms with which carbon is most often associated in organic compounds are hydrogen, oxygen, nitrogen, sulfur, and phosphorus.

Functional Groups of Organic Compounds

One important advantage of carbon serving as the molecular skeleton for living things is that it is free to bind with an unending array of other molecules. These special molecular groups or accessory molecules that bind to organic compounds are called **functional groups.** Functional groups help define the chemical class of certain groups of organic compounds and confer unique reactive properties on the whole molecule (table 2.3). Since each type of functional group behaves in a distinctive manner, reactions of an organic compound can be predicted by knowing the kind of functional group or groups it carries. Many synthesis, decomposition, and transfer reactions rely upon functional groups such as $-OH$, $-C{=}O$, or $-NH_2$.

Macromolecules: Molecular Superstructures

The compounds of life are studied by the science of **biochemistry.** Biochemicals are organic compounds produced by (or components of) living things, and they include four main families: carbohydrates, lipids, proteins, and nucleic acids (table 2.4). The compounds in these groups are assembled from smaller molecular subunits or building blocks, and because they are often very large compounds, they are termed *macromolecules.* All macromolecules except lipids are formed by *polymerization,* a process in which repeating subunits termed *monomers* are bound

metabolism (me-tab′-o-lizm) A general term referring to all chemical and physical processes in the cell.

inorganic (in-or-gan′-ik) Any chemical substances that are not combinations of carbon and hydrogen.

organic (or-gan′-ik) Gr. *organikos,* instrumental.

monomer (mawn′-oh-mur) Gr. *mono,* one, and *meros* part.

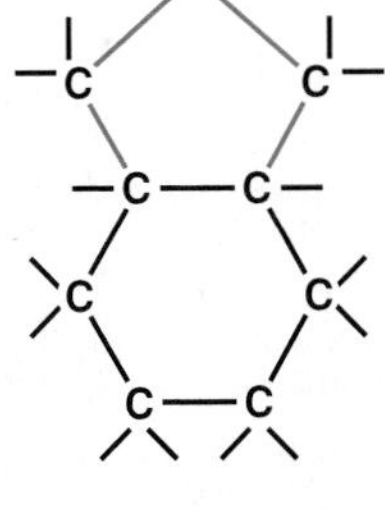

Figure 2.13 The versatility of bonding in carbon. In most compounds, carbon makes a total of four bonds. (*a*) Both single and double bonds may be made with other carbons, oxygen, and nitrogen; single bonds are made with hydrogen. Simple electron models show how the electrons are shared in these bonds. (*b*) Multiple bonding of carbons may give rise to long chains, branched compounds, and ringed compounds, many of which are extraordinarily large and complex.

Table 2.3 Representative Functional Groups and Classes of Organic Compounds

Formula of Functional Group	Name	Class of Compounds
R*— O—H	Hydroxyl	Alcohols, polysaccharides
R— C(=O)—OH	Carboxyl	Fatty acids, proteins, organic acids
R— C(H)(H)— NH_2	Amino	Proteins, nucleic acids
R— C(=O)—O—R	Ester	Lipids
R— C(H)(H)—SH	Sulfhydryl	Cysteine (amino acid), proteins
R— C(=O)—H	Carbonyl, terminal end	Aldehydes, polysaccharides
R— C(=O)—C—	Carbonyl, internal	Ketones, polysaccharides
R— O—P(=O)(OH)—OH	Phosphate	DNA, RNA, ATP

*The R designation on a molecule is shorthand for residue, and its placement in a formula indicates that what is attached at that site varies from one compound to another.

into chains of various lengths termed *polymers*. For example, proteins (polymers) are composed of a chain of amino acids (monomers) (see figure 2.22*a*). The large size and complex, three-dimensional shape of macromolecules enables them to function as structural components, molecular messengers, energy sources, enzymes (biochemical catalysts), nutrient stores, and sources of genetic information. In the next section and in subsequent chapters, we will consider numerous concepts relating to the roles of macromolecules in cells.

Carbohydrates: Sugars and Polysaccharides

The term **carbohydrate** originates from the resemblance of most members of this family to combinations of carbon and water. Although carbohydrates may be generally represented by the formula $(CH_2O)_n$, in which *n* indicates the number of units and

polymer (paul′-ee-mur) Gr. *poly,* many, and *meros,* part. Also the root for polysaccharide and polypeptide.

Table 2.4 Macromolecules and Their Functions

Macromolecule	Constituents	Function	Examples
Carbohydrates			
Monosaccharides	Glucose (dextrose)	Main sugar of metabolic reactions; building block of polysaccharides	
	Fructose (fruit sugar)	Alternative sugar in metabolism; building block of sucrose	
Disaccharides	Maltose (malt sugar)	Composed of two glucoses; an important breakdown product of starch	
	Lactose (milk sugar)	Composed of glucose and galactose	
	Sucrose (table sugar)	Composed of glucose and fructose	
Polysaccharides	Monosaccharides	Structures, food storage	Starch, cellulose, glycogen
Lipids			
Triglycerides	Fatty acids, glycerol	Major component of cell membranes; storage	Fats, oils
Phospholipids	Fatty acids, glycerol, phosphate	Membranes	
Waxes	Fatty acids, alcohols	Cell wall of mycobacteria	Mycolic acid
Steroids	Ringed structure (not a polymer)	Membranes of eucaryotes and some bacteria	Cholesterol; ergosterol
Proteins	Amino acids	Metabolic reactions; structural components	Enzymes; part of cell membrane, cell wall, ribosomes, antibodies
Nucleic Acids	Pentose sugar, phosphate, Purines: adenine, guanine Pyrimidines: cytosine, thymine, uracil		
Deoxyribonucleic Acid (DNA)	Deoxyribose sugar; all of the above, but thymine instead of uracil	Inheritance	Chromosomes; genetic material of viruses
Ribonucleic Acid (RNA)	Ribose sugar; all of the above, but uracil instead of thymine	Expression of genetic traits	Ribosomes; mRNA, tRNA

the atomic ratio of elements, some carbohydrates contain additional atoms of sulfur or nitrogen. In molecular configuration, the carbons form chains or rings with two or more hydroxyl groups and either an aldehyde or ketone group, thus they are technically referred to as *polyhydroxy aldehydes* or *polyhydroxy ketones* (figure 2.14).

Carbohydrates come in a variety of forms. The common term *sugar* (*saccharide*) refers to a simple carbohydrate such as a monosaccharide or disaccharide that has a sweet taste. A **monosaccharide** is a simple polyhydroxy aldehyde or ketone molecule containing from 3 to 7 carbons; a **disaccharide** is a combination of two monosaccharides; and a **polysaccharide** is a polymer of five or more monosaccharides bound in linear or branched chain patterns (figure 2.14). Monosaccharides and disaccharides are specified by combining a prefix that describes some characteristic of the sugar with the suffix *-ose*. For example, *hexoses* are composed of 6 carbons, and *pentoses* contain 5 carbons. **Glucose** (Gr. sweet) is the most common and universally important hexose; **fructose** is named for fruit (one of its sources); and xylose, a pentose, derives its name from the Greek word for wood. Disaccharides are named similarly: *Lactose* (L. milk) is an important component of milk; *maltose* means malt sugar; and *sucrose* (Fr. sugar) is common table or cane sugar.

The Nature of Carbohydrate Bonds

The subunits of disaccharides and polysaccharides are linked by means of *glycosidic* bonds, in which the carbons (each is assigned a number) on adjacent sugar units are bonded to the same oxygen atom like links in a chain (figure 2.15). For example, maltose is formed when the number one (#1) carbon on a glucose bonds to the #4 carbon on a second glucose; sucrose is formed when glucose and fructose bind between their #1 and #2 carbons; and lactose is formed when glucose and galactose connect by their #1 and #4 carbons. In order to form this bond, one carbon gives up its OH group and the other (the one contributing the oxygen to the bond) loses the H from its OH group. Because a water molecule is produced, this reaction is known as *dehydration synthesis,* a process common to most polymerization reactions (see proteins in a later section of this chapter). Many polysaccharides are polymers of glucose, although the precise nature of their bonding provides several variations in structure (figure 2.16). The synthesis and breakage of glycosidic bonds requires specific catalysts (enzymes) (see chapter 7).

saccharide (sak'-uh-ryde) Gr. *sakcharon,* sweet.

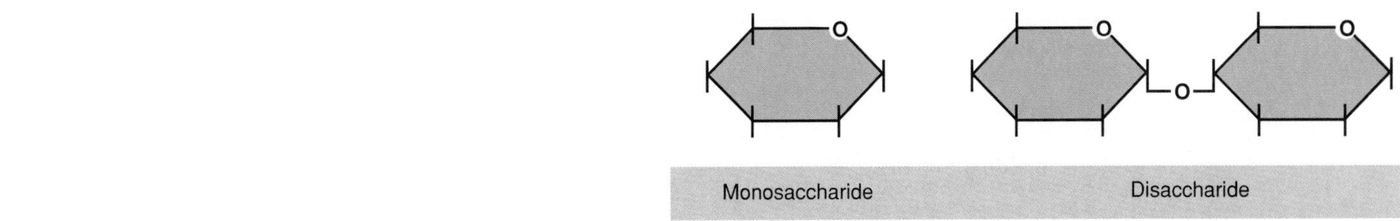

(a)

Polysaccharide

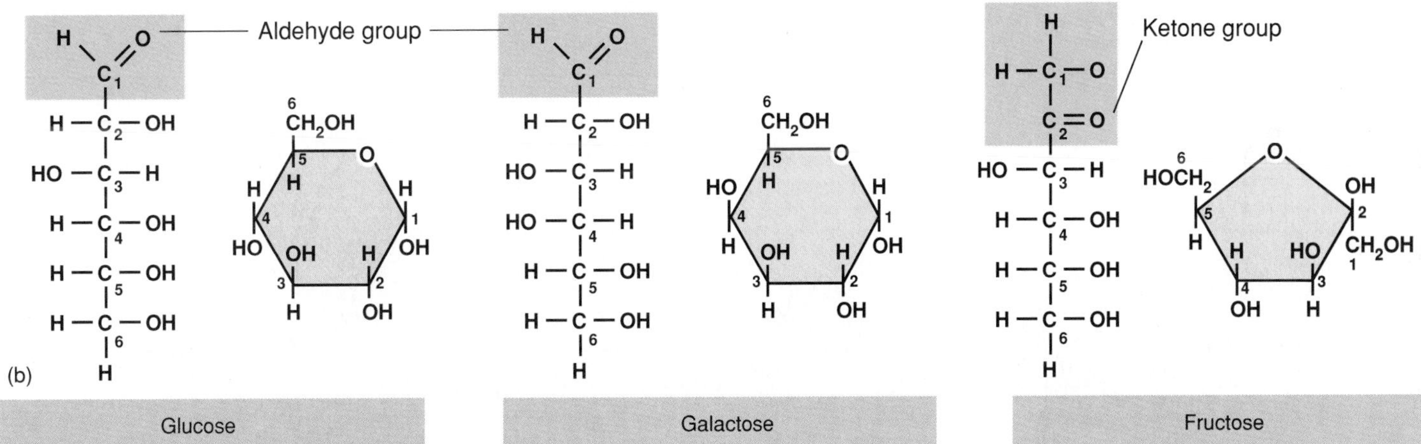

Figure 2.14 Common classes of carbohydrates. (*a*) Major saccharide groups, named for the number of sugar units each contains. (*b*) Structural formulas of three hexoses. Sugars may be represented in structural models in both linear and ring format. The linear form makes the aldehyde and ketone groups more evident, although the sugars are usually found in the ring form in solution. Note that the carbons are numbered so as to keep track of reactions within and between monosaccharides.

The Functions of Polysaccharides

Polysaccharides typically contribute to structural support and protection, and serve as nutrient and energy stores. The cell wall in plants and many microscopic algae derives its strength and rigidity from **cellulose,** a long, fibrous polymer (figure 2.16*a*). Because of this role, cellulose is probably one of most common organic substances on earth, yet it is digestible only by certain bacteria, fungi, and protozoa, which produce an enzyme that can attack the bonds (see figure 6.3). Other structural polysaccharides may be conjugated (chemically bonded) to amino acids, nitrogen bases, lipids, or proteins. **Agar-agar,** an indispensable polysaccharide used to solidify culture media, is a natural component of certain seaweeds. It is a complex polymer of galactose and sulfur-containing carbohydrates. The exoskeletons of insects and crustacea and the cell walls of certain fungi contain *chitin,* a polymer of glucosamine (a sugar with an amino functional group). One special class of compounds in which polysaccharides (glycans) are linked to peptide fragments (a short chain of amino acids) is **peptidoglycan** (pep″-tih-doh-gly′-kan), the component that gives rigidity to the bacterial cell wall. As we shall see in chapter 9, this is a target of the penicillin group of antibiotics. Another portion of the cell wall of certain bacteria is *lipopolysaccharide,* a complex of lipid and polysaccharide responsible for symptoms in certain infections (see chapters 11 and 16).

The outer surface of some cells has a fine "sugar coating" composed of polysaccharides bound in various ways to proteins (the combination is called mucoprotein or glycoprotein). This structure, called the *glycocalyx,* may function in attachment to other cells or as part of a *receptor*—which is a molecule that receives and responds to external stimuli. Small sugar molecules are the basis for the human blood types, and many molecules involved in immune reactions (antibodies) have a

---

glycocalyx (gly″-koh-kay′-lix) Gr. *glycos,* sweet, and *calyx,* covering.

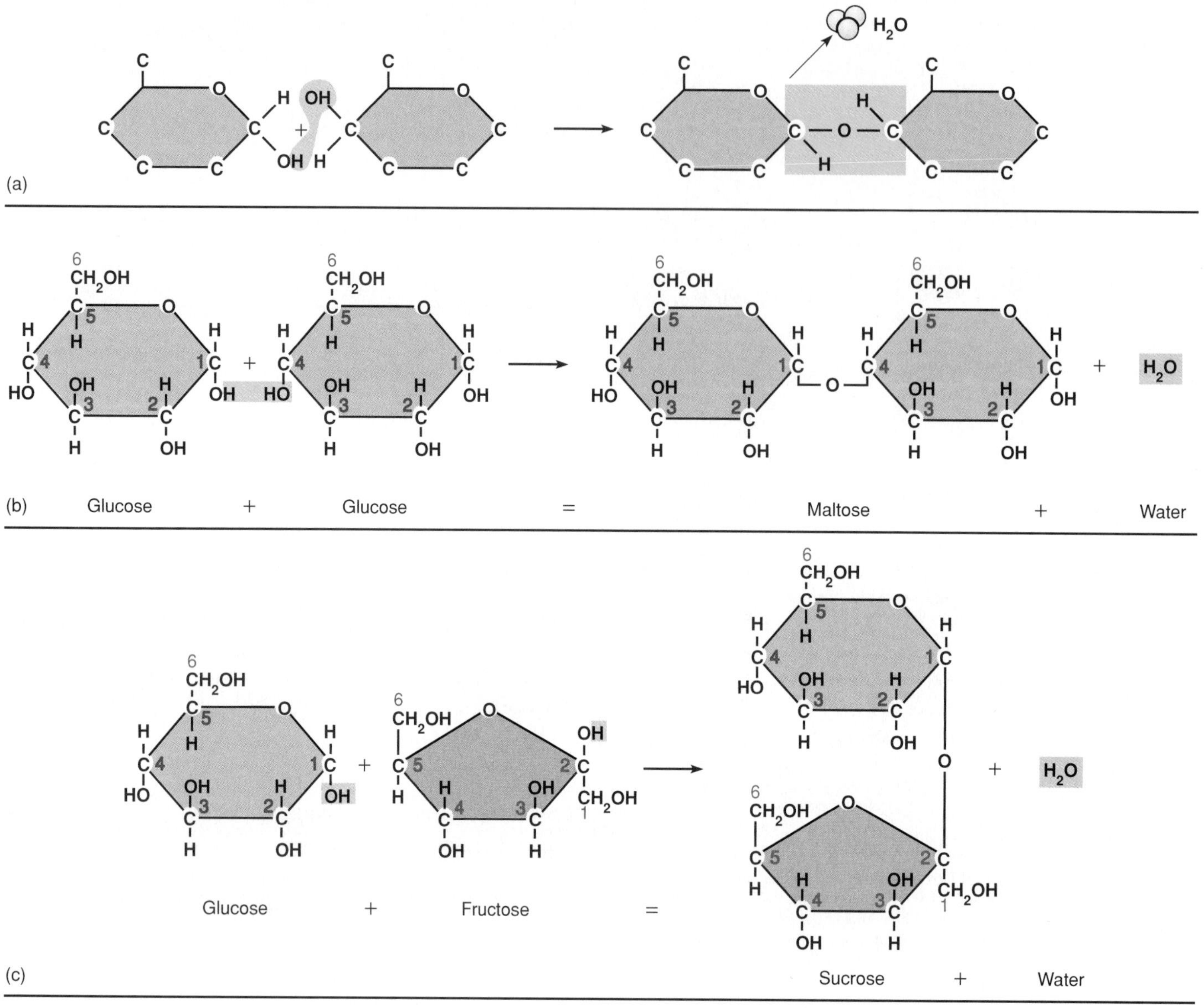

Figure 2.15 (*a*) General scheme in the formation of a glycosidic bond by dehydration synthesis. (*b*) Formation of the 1,4 bond between two glucoses to produce maltose. (*c*) Formation of the 1,2 bond between glucose and fructose to produce sucrose.

carbohydrate component. Many bacteria are coated by a protective polysaccharide capsule that contributes to their infectiousness (see figure 3.12), and some viruses have glycoproteins on their surface that function in invasion of cells.

Polysaccharides are usually stored in the form of compact polymers of glucose such as starch (figure 2.16*b*) or glycogen, but only organisms that have the appropriate digestive enzymes can break them down and use them as a nutrient source. Since breaking the bonds between the glucose molecules of these polysaccharides requires the addition of a water molecule, digestion is also termed *hydrolysis*. Whereas *starch* is stored by green plants, microscopic algae, and some fungi, *glycogen* (animal starch) is a prominent storage molecule in animals and some bacteria.

hydrolysis (hy-droll'-uh-sis) Gr. *hydros,* water, and *lyein,* to dissolve.

Lipids: Fats, Phospholipids, and Waxes

The term **lipid,** derived from the Greek word *lipos,* meaning fat, is not a chemical designation, but an operational term for a variety of substances that are not soluble in polar solvents such as water (recall that oil and water do not mix) but will dissolve in nonpolar solvents such as benzene, ether, or chloroform. This is largely because the substances we call lipids contain relatively long or complex C–H (hydrocarbon) chains that are nonpolar and thus hydrophobic. The main groups of compounds classified as lipids are triglycerides, phospholipids, steroids, and waxes.

(a)

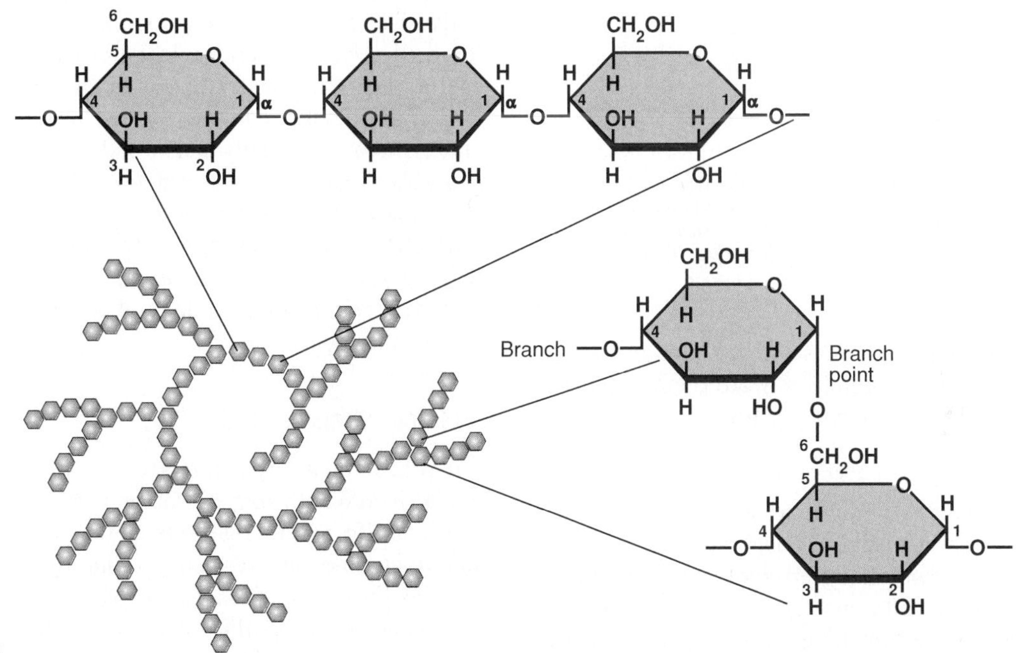

Figure 2.16 Polysaccharides. (*a*) Cellulose is composed of glucose bonded in β 1,4 bonds that produce linear, lengthy chains of polysaccharides—the typical structure of wood and cotton fibers. (*b*) Starch is composed of two glucose polymers: the main chain structure is amylose bonded in a 1,4 pattern, with side branches of amylopectin bonded by 1,6 bonds. The entire molecule is a compact, granular structure.

One important group of lipids are the **triglycerides,** a category that includes fats and oils. Triglycerides are composed of three fatty acids bound to a single molecule of glycerol by means of ester bonds (figure 2.17). **Glycerol** is a 3-carbon alcohol with three OH groups that serve as binding sites, and **fatty acids** are long chain hydrocarbon molecules with a carboxyl group (COOH) at one end that is free to bind to the glycerol. The bond that forms between the –OH group and the –COOH is defined as an **ester bond.** The hydrocarbon portion of a fatty acid can vary in length from 4 to 24 carbons, and depending on the fat, it may be saturated or unsaturated. If the carbons in the chain are single-bonded to 2 other carbons and 2 hydrogens, the fat is saturated, and if there is at least one C–C double bond in the chain, it is unsaturated.

The structure of fatty acids is what gives fats and oils (liquid fats) their greasy, insoluble nature. In general, solid fats (such as beef tallow) are more saturated, and oils (or liquid fats) are more unsaturated. In most cells, triglycerides are stored in long-term concentrated form as droplets or globules. When the ester linkage is acted on by digestive enzymes called lipases, the fatty acids and glycerol are freed to be used in metabolism. Fatty acids are a superior source of energy, yielding twice as much

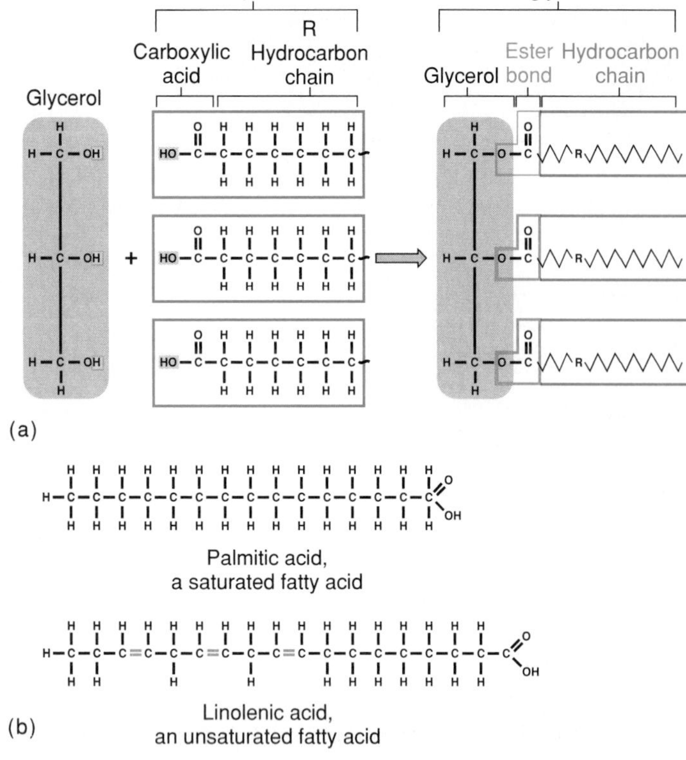

Figure 2.17 (*a*) Synthesis and structure of a triglyceride. Because a water molecule is released at each ester bond, this is another form of dehydration synthesis. The jagged lines and R symbol represent the hydrocarbon chains of the fatty acids, which are commonly very long. (*b*) Structural formulas for saturated and unsaturated fatty acids.

per gram as other storage molecules (starch). Soaps are K^+ or Na^+ salts of fatty acids whose qualities make them excellent grease removers and cleaners (see chapter 9).

Membrane Lipids

A class of lipids that serves as a major structural component of cell membranes are the **phospholipids.** Though phospholipids are similar to triglycerides in containing glycerol and fatty acids, they have some significant differences. Phospholipids contain only two fatty acids attached to the glycerol, while the third glycerol binding site holds phosphoric acid. The phosphoric acid is in turn bonded to an alcohol,[2] which varies from one phospholipid to another (figure 2.18*a*).

The arrangement of phospholipids accounts for some of their biologically significant properties. Due to a charge on the phosphoric acid/alcohol "head" of the molecule and the lack of a charge on the long "tail" of the molecule (formed by the fatty acids), these lipids have both a hydrophilic and a hydrophobic end. When exposed to an aqueous solution, they form single or double layers in which the hydrophilic heads are attracted to the water phase and the hydrophobic tails are repelled away from the water phase (figure 2.18*b*). The capacity of lipids to assume double layers has great biological importance, in that they create the primary framework of cell membranes. When two single layers of polar lipids come together to form a double layer, the natural tendency is for the outer hydrophilic faces of each single layer to be exposed on the outer surfaces and for the hydrophilic portions to be immersed in the core of the bilayer. The structure of lipid bilayers gives membranes several of their characteristics, such as their selective permeability and fluid nature (see feature 2.4).

Miscellaneous Lipids

Steroids are complex ringed compounds that are common components of cell membranes and animal hormones. The best known of these is the sterol (meaning a steroid with an OH group) called *cholesterol* (figure 2.19). Cholesterol apparently reinforces the structure of the cell membrane in animal cells and an unusual group of cell-wall-deficient bacteria called the mycoplasmas (see chapter 3). The cell membranes of fungi also contain a sterol called ergosterol. *Prostaglandins* are fatty acid derivatives found in trace amounts that function in inflammatory and allergic reactions, blood clotting, and smooth muscle contraction. Chemically, a *wax* is an ester formed between a long chain alcohol and a saturated fatty acid. This creates a material that is typically pliable and soft when warmed but hard and water-resistant when cold (paraffin, for example). Among living things, fur, feathers, fruits, leaves, human skin, and insect exoskeletons are naturally waterproofed with a coating of wax. Bacteria that cause tuberculosis and leprosy produce a wax (wax D) that repels ordinary stains and contributes to the pathogenicity of these bacteria.

Proteins: Shapers of Life

The predominant organic molecules in cells are **proteins,** a fitting term adopted from the Greek word *proteios,* meaning first or prime. To a very real extent, the structure, behavior, and unique qualities of each living thing are a consequence of the proteins they contain. To best explain the origin of the special properties and versatility of proteins, we must examine their general structure. The building blocks of proteins are amino acids, which exist in 20 different naturally occurring forms (table 2.5). Various combinations of these amino acids account for the nearly infinite variety of proteins.

All amino acids have a basic skeleton consisting of a carbon (called the α carbon) linked to an amino group (NH_2), a carboxyl group (COOH), and a hydrogen atom (H). The variations among the amino acids occur at the fourth group (R for residue), which is different in each amino acid and which imparts the unique characteristics to the molecule and to the proteins that contain it (figure 2.20). The amino group on one amino acid readily forms a covalent bond (called a **peptide bond**) with the carboxyl group on another amino acid (figure 2.21). As a result of peptide bond formation, it is possible to produce chains varying in length from two to thousands of amino acids.

2. Alcohols are hydrocarbons containing OH groups.

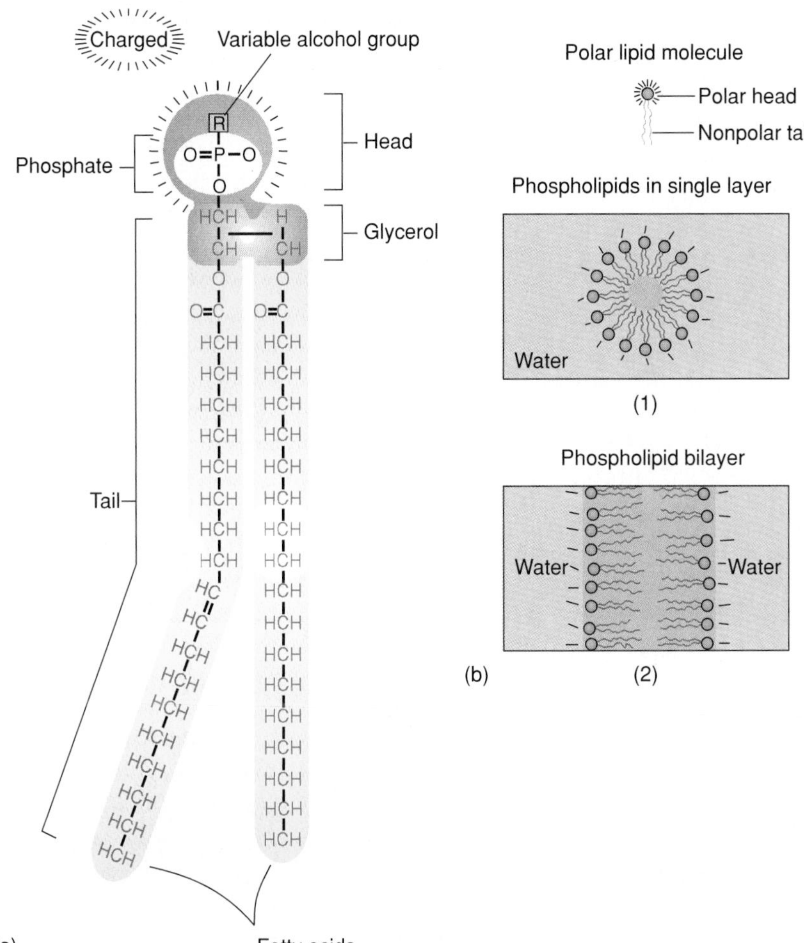

Figure 2.18 Phospholipids—amphipathic molecules. (*a*) A complex model of a single molecule of a phospholipid. The phosphate-alcohol head lends a charge to one end of the molecule; its long, trailing hydrocarbon chain is uncharged. (*b*) The behavior of phospholipids in water-based solutions causes them to become arranged (*1*) in single layers with the charged head oriented toward the water phase and the hydrophobic nonpolar tail buried away from the water phase, or (*2*) in double-layered phospholipid systems with the hydrophobic tails sandwiched between two hydrophilic layers.

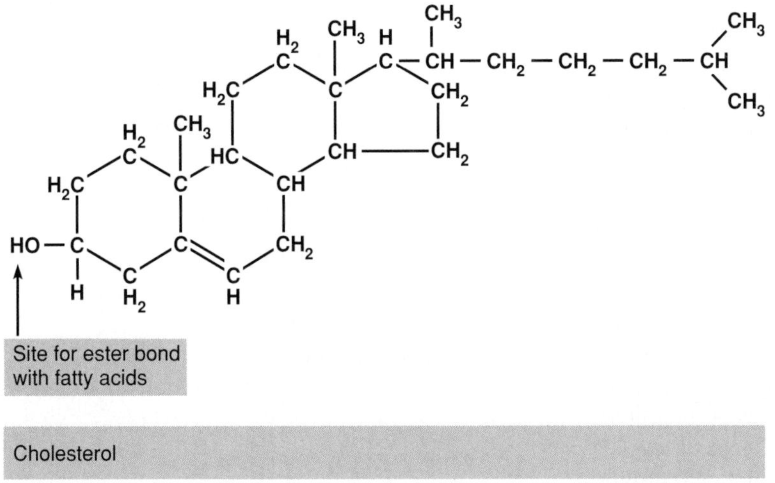

Site for ester bond with fatty acids

Cholesterol

Figure 2.19 Formula for cholesterol, an alcoholic steroid that is inserted in some membranes. Cholesterol may become esterified with fatty acids at its OH group, imparting a polar quality similar to that of phospholipids.

Various terms are used to denote the nature of compounds containing peptide bonds. *Peptide* usually refers to a molecule composed of short chains of amino acids, such as a dipeptide (two amino acids), a tripeptide (three) and a tetrapeptide (four). A *polypeptide* contains an unspecified number of amino acids, but usually has more than 20, and is often a smaller subunit of a protein. A *protein* is the largest of this class of compounds and usually contains a minimum of 50 amino acids. It is common

peptide (pep′-tyde) Gr. *pepsis,* digestion.

Table 2.5 The 20 Amino Acids and Their Abbreviations

Amino Acid	Abbreviation
Alanine	Ala
Arginine	Arg
Asparagine	Asp
Aspartic acid	Asp
Cysteine	Cys
Glutamic acid	Glu
Glutamine	Gln
Glycine	Gly
Histidine	His
Isoleucine	Ile
Leucine	Leu
Lysine	Lys
Methionine	Met
Phenylalanine	Phe
Proline	Pro
Serine	Ser
Threonine	Thr
Tryptophan	Trp
Tyrosine	Tyr
Valine	Val

for the terms polypeptide and protein to be used interchangeably, though not all polypeptides are large enough to be considered proteins. In chapter 8 we will see that protein synthesis is not just a random connection of amino acids—it is directed by information provided in DNA (described in the next section).

Protein Structure and Diversity

An important contributing factor in protein diversity is that a protein does not function in the form of a straight, extended chain (called its primary structure). It has a natural tendency to assume more complex levels of organization called the secondary, tertiary, and quaternary structures (figure 2.22). The **primary (1°) structure** is described as the type, number, and order of amino acids in the chain, which varies extensively from protein to protein. The **secondary (2°) structure** arises when various functional groups exposed on the outer surface of the molecule form weak interactions such as hydrogen bonds with one another. The result is that the amino acid chain twists into a coiled configuration called the α helix, or forms a flat, β-pleated sheet. Proteins in the secondary helix state undergo a third level of folding, or **tertiary (3°) structure,** which results in a compact, spherical mass. Proteins that assume such three-dimensional shapes are known as **globular** proteins. The tertiary level of structure is created by weak bonds between functional groups, but in proteins with the sulfur-containing amino acid cysteine (sis′-tuh-yeen), additional stability is achieved through strong covalent **disulfide bonds** between sulfur atoms on two different parts of the molecule (figure 2.22*c*). More complex proteins may have a **quaternary (4°) structure,** in which polypeptides in tertiary form combine into a large, multiunit protein. This is typical of antibodies (see chapter 13) and some enzymes that act in cell synthesis.

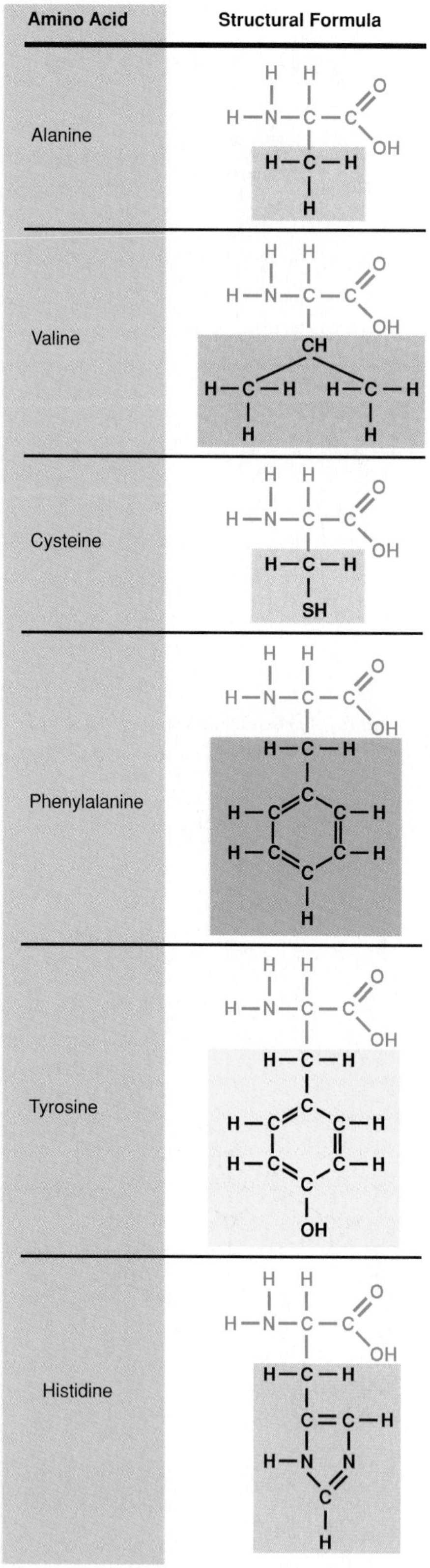

Amino Acid	Structural Formula
Alanine	
Valine	
Cysteine	
Phenylalanine	
Tyrosine	
Histidine	

Figure 2.20 Structural formulas of selected amino acids. The basic structure common to all amino acids is shown in blue, and the variable group, or R group, varies from a methyl (CH_3) group (as in alanine) to more complex ringed compounds (as in tyrosine).

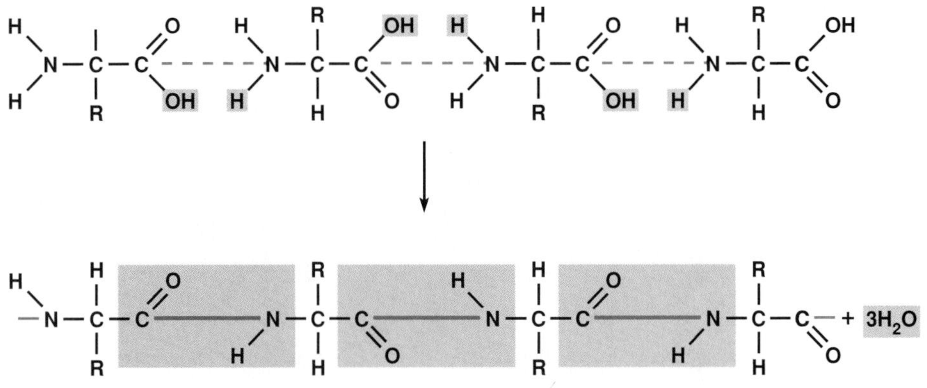

Figure 2.21 The formation of peptide bonds in a tetrapeptide.

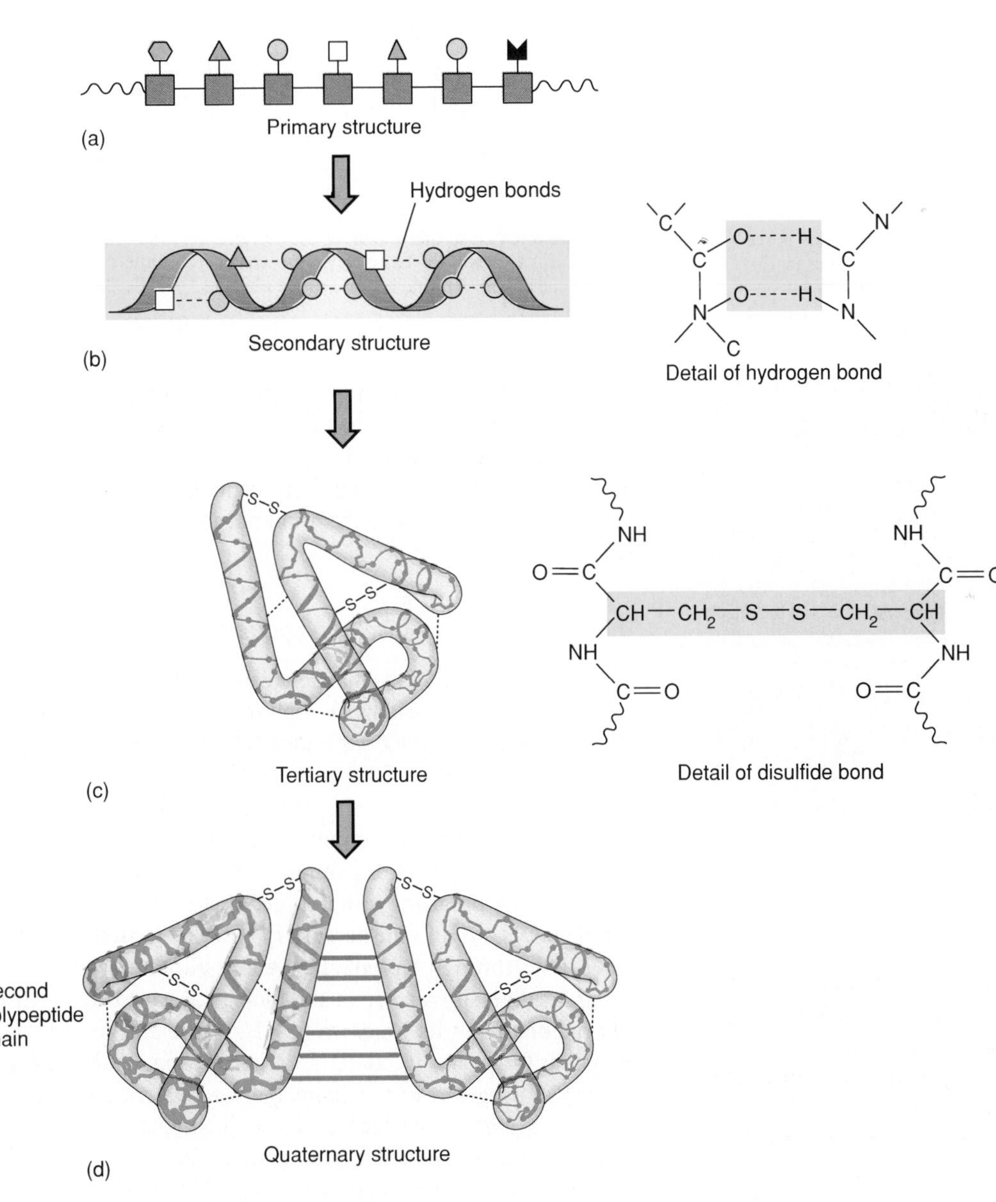

Figure 2.22 Stages in the formation of a functioning globular protein: (*a*) The primary structure, a series of amino acids in a long chain; (*b*) the secondary structure, twisting of the primary chain into an alpha helix; (*c*) the tertiary structure, folding of the molecule into a three-dimensional globule that is further stabilized by disulfide bonds; and (*d*) the quaternary structure. Some proteins are composed of more than a single polypeptide chain; these are attached by weak, reversible bonds.

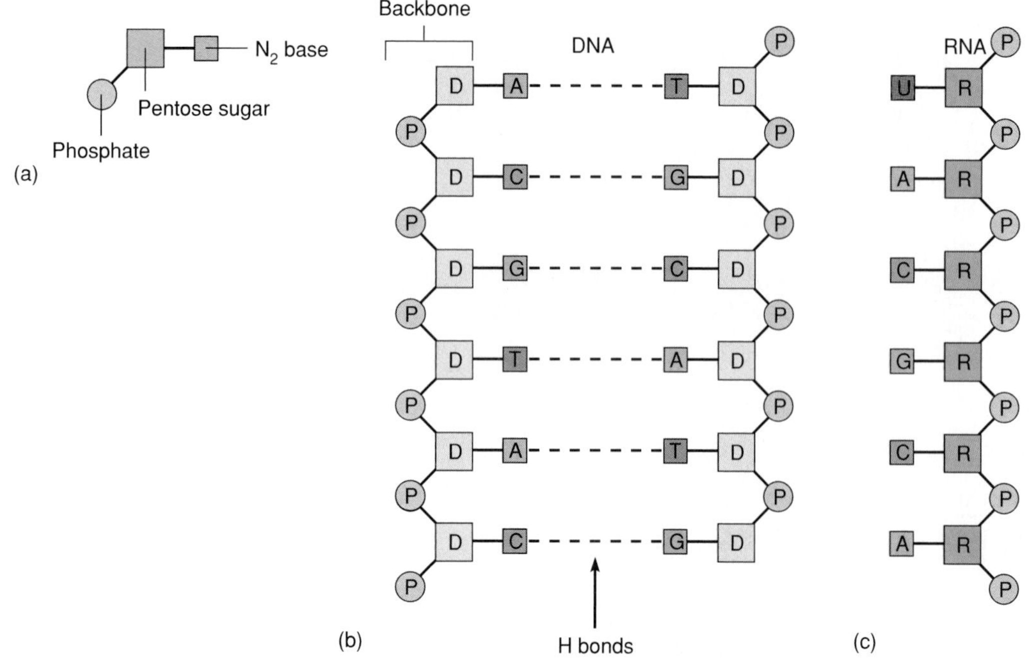

Figure 2.23 The basic structure of nucleic acids. (*a*) A nucleotide is the monomer of both DNA and RNA. (*b*) In DNA the basic polymer is composed of alternating deoxyribose (D) and phosphate (P) with nitrogen bases (A, T, C, G) attached to the deoxyribose. Two of these polynucleotide strands are oriented so that the bases are paired across the central axis of the molecule. (*c*) In RNA the basic polymer is composed of alternating ribose (R) and phosphate (P) attached to nitrogen bases (A, U, C, G); but it is only a single strand.

Feature 2.3 Proteins and Paper Airplanes

The levels of protein structure have steps in folding analogous to those of a handmade paper airplane. The flat sheet of paper one starts with is equivalent to the primary state. Initial folds made for wings, body, and tail produce a two-dimensional secondary configuration. When these folds are correctly pulled out or inserted, they yield a three-dimensional tertiary structure that is the airplane. If a more elaborate airplane were constructed, other parts such as wing and tail fins could be added, thus creating a quaternary level of structure. This analogy is more than just structural—it is also functional. The initial shape and size of the paper (1°) will dictate the nature of the folds (2°); these folds will determine the shape and size of the airplane (3°); and finally, the airplane will fly (function) only if the proper 3° state is formed.

The most important outcome of the intrachain bonding and folding process is that each different type of protein develops a unique shape, and its surface displays a pattern of pockets and bulges. This means that a protein can react only with molecules that complement or fit its particular surface features, and this results in a great deal of specificity. The differences in specificity that are possible provide the functional diversity required for many thousands of different cellular activities. **Enzymes,** for example, contain globular proteins and serve as the catalysts for all chemical reactions, and every such reaction will require a specific enzyme (see chapter 7). Antibodies also have sites that accommodate only certain molecules; toxins (products of certain infectious agents) composed of proteins act on one specific organ or tissue; and carrier proteins can transport only certain nutrients. The functional three-dimensional form of a protein is termed the *native state,* and if it is disrupted by some means, the protein is said to be *denatured.* Such agents as heat, acid, alcohol, and some disinfectants disrupt the stabilizing intrachain bonds and cause the molecule to unfold back to the primary (and nonfunctional) state (see figure 9.4).

The Nucleic Acids: A Cell Computer and Its Programs

The nucleic acids, **deoxyribonucleic acid (DNA)** and **ribonucleic acid (RNA),** were originally isolated from the cell nucleus. Shortly thereafter, they were also found in other parts of nucleated cells, in cells with no nuclei (bacteria), and in viruses. The universal occurrence of nucleic acids in all known cells and viruses emphasizes their important roles as informational molecules. DNA, the master computer of cells, contains a special coded genetic program with detailed instructions for biological function. DNA is also the hereditary material passed on to future generations through reproduction, and it ultimately accounts for the great diversity we see in the biological world. DNA transfers the details of its program to RNA, operator molecules responsible for carrying out DNA's instructions and translating the DNA program into a language for performing life functions. The most important of these functions is protein synthesis, a concept pursued in greater detail in chapter 8. For now, let us briefly consider the structure of DNA, RNA, and a close relative, adenosine triphosphate (ATP).

deoxyribonucleic, ribonucleic (dee-ox″-ee-rye″-boh- noo-klay′-ik) (rye″-boh-noo-klay′-ik) Named for their sugars.

Both nucleic acids are polymers of repeating units called **nucleotides,** each of which is composed of three smaller units: a *nitrogen base,* a *pentose* (5-carbon) sugar, and a *phosphate* (figure 2.23*a*). The nitrogen base is a cyclic compound that comes in two forms: *purines* (two rings) and *pyrimidines* (one ring). There are two types of purines, *adenine (A)* and *guanine (G),* and three types of pyrimidines, *thymine (T), cytosine (C),* and *uracil (U)* (figure 2.24). A characteristic that differentiates DNA from RNA is that DNA contains all of the nitrogen bases except uracil, and RNA contains all of the nitrogen bases except thymine. The nitrogen base is covalently bonded to the sugar—*ribose* in RNA and *deoxyribose* (meaning it has one oxygen less than ribose) in DNA. Phosphate (PO_4), a derivative of phosphoric acid (H_3PO_4), provides the final covalent bridge that connects sugars in series. Thus, the backbone of a nucleic acid strand is a chain of alternating phosphate-sugar-phosphate-sugar molecules, and the nitrogen bases branch off the side of this backbone (see figure 2.23*b,c*).

The Double Helix of DNA

DNA is a huge molecule formed by two very long polynucleotide strands bonded together by their nitrogen bases, which are directed towards the center of the molecule. The pairing of the nitrogen bases occurs according to a predictable pattern: Adenine always pairs with thymine, and cytosine always pairs with guanine, and no other combination ever occurs in DNA. These base pairs are held together by hydrogen bonds—two between A–T pairs and three between C–G pairs (figure 2.25*a*). For ease in understanding the structure of DNA, it is sometimes compared to a ladder, with the sugar-phosphate backbone representing the rails and the paired nitrogen bases representing the steps. Due to the manner of nucleotide pairing and stacking of the bases, the actual configuration of DNA is a *double helix* more reminiscent of a spiral staircase (figure 2.25). As is true of protein, the structure of DNA is intimately related to its function. DNA molecules are usually extremely long, which satisfies a requirement for storing genetic information. The hydrogen bonds between pairs can be disrupted when DNA is being copied, and the fixed complementary base pairing is essential in maintaining the genetic code.

RNA: Organizers of Protein Synthesis

RNA also consists of a long chain of nucleotides, but in addition to differing from DNA in the type of sugar and one nitrogen base, RNA is a single strand and generally not a double helix (see figure 2.23). Although all three main types of RNA are copies of the DNA program, each has a specific function in protein synthesis. Messenger RNA (mRNA) carries the genetic message of the order and type of amino acids in a protein; transfer RNA (tRNA) brings in the correct amino acids for protein assembly; and ribosomal RNA (rRNA) is a major component of ribosomes (see figure 3.19).

---

nucleotide (noo′-klee-oh-tyed) Gr. *nucis,* nut, and Fr. *acide,* acid.

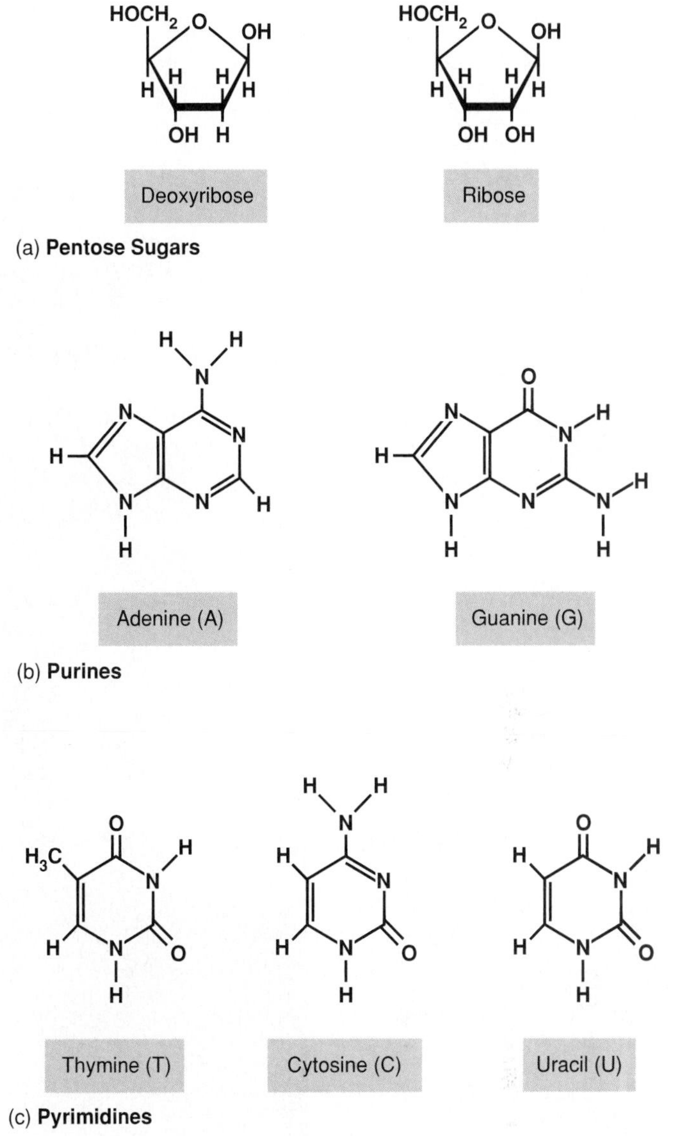

Deoxyribose

Ribose

(a) **Pentose Sugars**

Adenine (A)

Guanine (G)

(b) **Purines**

Thymine (T)

Cytosine (C)

Uracil (U)

(c) **Pyrimidines**

Figure 2.24 The sugars and nitrogen bases that make up DNA and RNA. (*a*) DNA contains deoxyribose, and RNA contains ribose. (*b*) A and G purines are found in both DNA and RNA. (*c*) C pyrimidine is found in both DNA and RNA, but T is found only in DNA and U is found only in RNA.

ATP: The Energy Molecule of Cells

A relative of RNA involved in an entirely different cell activity is **adenosine triphosphate (ATP)**. ATP is a nucleotide containing adenine, ribose, and three phosphates rather than one (figure 2.26). It belongs to a category of high-energy compounds (also including guanosine triphosphate—GTP) that give off energy when the bond is broken between the second and third (outermost) PO_4. The presence of these "high-energy bonds" makes it possible for ATP to give off and store energy for cellular chemical reactions. Breakage of the third phosphate not only releases energy to do cellular work but also generates adenosine diphosphate (ADP). ADP can be converted back to ATP when the third phosphate is restored, thereby serving as an energy depot. Carriers for oxidation-reduction activities (nicotinamide adenine dinucleotide [NAD], for instance) are also derivatives of nucleotides (see chapter 7).

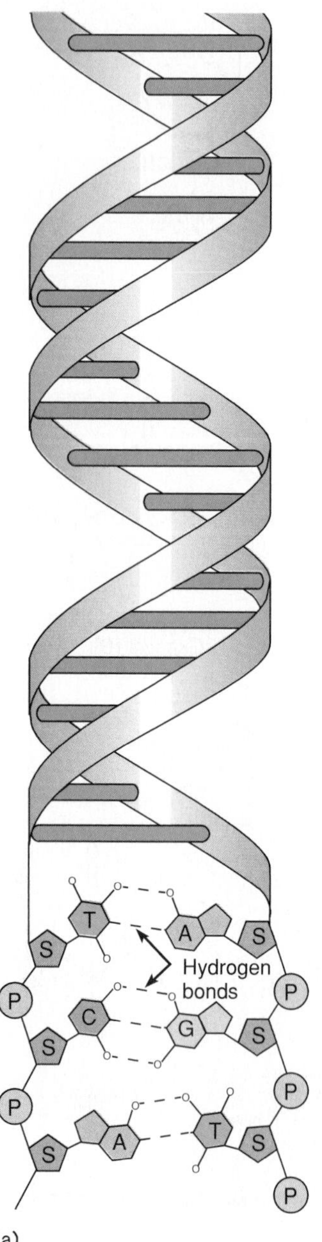

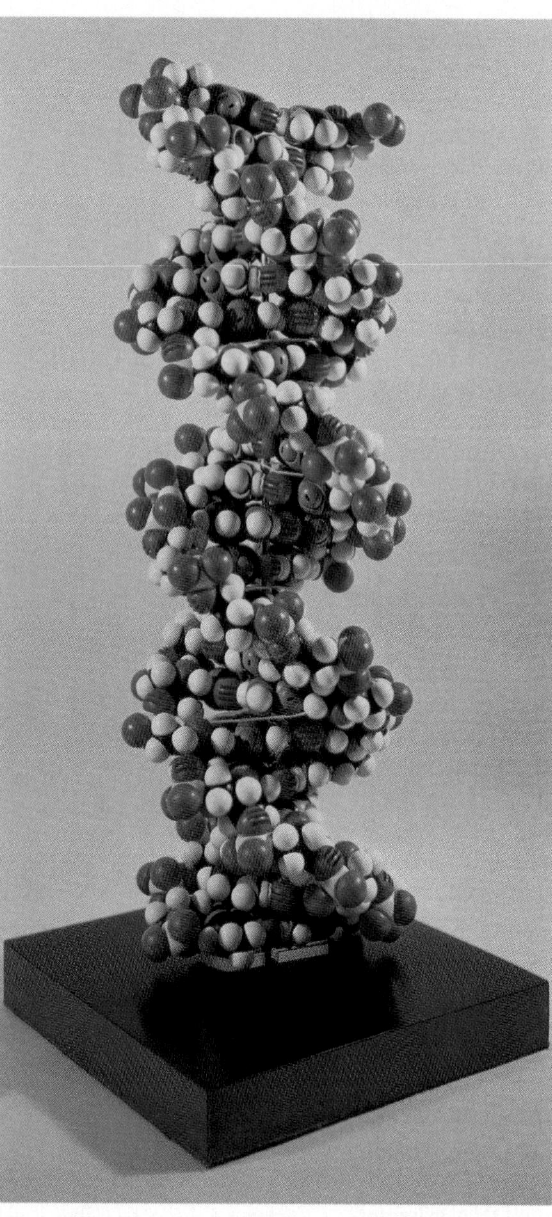

Figure 2.25 (*a*) A structural representation of the double helix of DNA shows the details of hydrogen bonds between the nitrogen bases of the two strands. (*b*) A space-filling model indicates the actual configuration of a small section of a DNA molecule.

Cells: Where Chemicals Come to Life

As we proceed in this chemical survey from the level of simple molecules to the increasingly complex levels of macromolecules, at some point we cross a line from the realm of lifeless molecules and arrive at the fundamental unit of life called a **cell.**[3] A cell is indeed a huge aggregate of carbon, hydrogen, oxygen, nitrogen, and many other atoms, and it follows the basic laws of chemistry and physics, but it is much more. The combination of these atoms produces characteristics, reactions, and products that can only be described as **living.**

3. The word was originally coined from an Old English term meaning "small room" because of the way plant cells looked to early microscopists.

Fundamental Characteristics of Cells

Some organisms such as bacteria and protozoa consist only of single cells, but others such as animals and plants have bodies containing trillions of cells. Regardless of the organism, all cells have a few common characteristics. They tend to be spherical, polygonal, cubical, or cylindrical in shape, and their protoplasm (internal cell contents) is enclosed by a cell or cytoplasmic membrane (see feature 2.4); they have chromosomes containing DNA and ribosomes for protein synthesis; and they are exceedingly complex in function. Beyond these similarities, cells belong to one of two fundamentally different lines (previously mentioned): the small, seemingly simple **procaryotic** cells, and the larger, structurally more complicated **eucaryotic** cells.

Adenine | Ribose | (1) (2) (3) Phosphates

Adenosine

Figure 2.26 The structural formula of an ATP molecule, the chemical form of energy transfer in cells. The wavy lines that connect the second and third phosphates represent bonds that release high energy when broken.

Feature 2.4 Membranes: Cellular Skins

The word **membrane** appears frequently in descriptions of cells in this chapter and in chapters 3 and 4. The word itself describes any lining or covering, including such multicellular structures as the mucous membranes of the body. From the perspective of a single cell, however, a membrane is a very thin sheet composed of specialized lipid molecules (averaging about 40% of membrane content) and protein molecules (averaging about 60%). Membranes appear in several areas of a cell, but their most prominent role is as a **cell** or **cytoplasmic membrane** enclosing the cell contents. Membranes are also components of eucaryotic organelles such as nuclei, mitochondria, and chloroplasts, and they appear in internal pockets of certain procaryotic cells. Even some viruses, which are not cells at all, have a membranous protective covering.

Cell membranes are so very thin—on the average, just 0.0070 μm thick—that they cannot actually be seen with an optical microscope. Even at magnifications made possible by electron microscopy (500,000×), very little of the precise architecture can be visualized, and a cross-sectional view looks something like railroad tracks. Following more detailed microscopic and chemical analysis, S. J. Singer and C. K. Nicholson proposed a simple and elegant theory for membrane structure called the **fluid mosaic model.** According to this theory, a membrane is a continuous bilayer formed by lipids that are oriented with the polar (hydrophilic) lipid heads toward the outside and the nonpolar (hydrophobic heads) toward the center of the membrane. Embedded at numerous sites in this bilayer are various-sized globular proteins. Some proteins are situated only at the surface, while others extend fully through the entire membrane. The configuration of the inner and outer sides of the membrane can be quite different due to the variations in proteins. Membranes have a dynamic, constantly changing character because the lipid phase flows and the proteins migrate freely about, somewhat like icebergs in a sea of lipids.

The structure of membranes accounts for their behavior and functions in cells. The lipid phase provides an impenetrable barrier to many substances, which accounts for the capacity of membranes to (1) control transport of molecules into and out of the cell, (2) be selectively *permeable,* and (3) segregate reactions within the cell itself. Membrane proteins serve many functions, including receiving molecular signals, binding (receptor) molecules, acting as enzymes, and serving as channels of transport (see chapter 6).

mosaic (moh-zay'-ik) An intricate design made up of many smaller pieces.
permeable (per'-mee-uh-bull) Capable of being passed through.

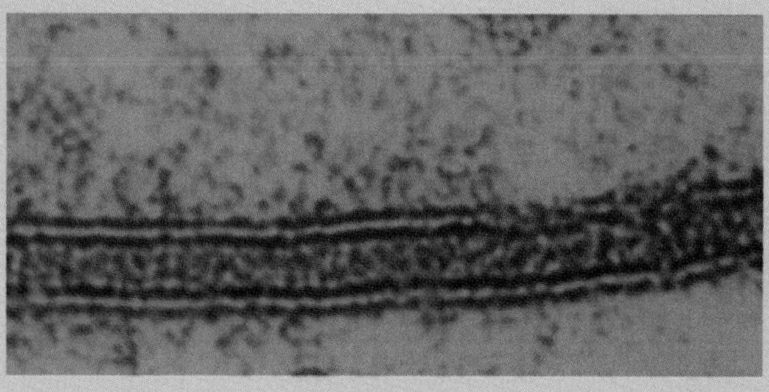

(a)

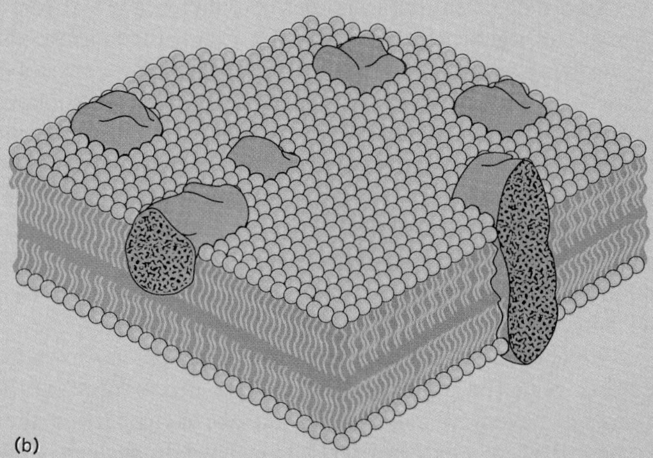

(b)

Extreme magnification of (*a*) a cross section of a cell membrane, which appears as double tracks. (*b*) A generalized version of the fluid mosaic model of a cell membrane indicates a bilayer of lipids with globular proteins embedded to some degree in the lipid matrix. This structure explains many characteristics of membranes, including flexibility, solubility, permeability, and transport.

Table 2.6 A General Comparison of Procaryotic and Eucaryotic Cells and Viruses *

Function or Structure	Characteristic	Procaryotic Cells	Eucaryotic Cells	Viruses**
Genetics	Nucleic acids	+	+	+
	Chromosomes	+	+	−
	True nucleus	−	+	−
	Nuclear envelope	−	+	−
Reproduction	Mitosis	−	+	−
	Production of sex cells	−/+	+	−
	Binary fission	+	+	−
Biosynthesis	Independent	+	+	−
	Golgi apparatus	−	+	−
	Endoplasmic reticulum	−	+	−
	Ribosomes	+***	+	−
Respiration	Enzymes	+	+	−
	Mitochondria	−	+	−
Photosynthesis	Pigments	+/−	+/−	−
	Chloroplasts	−	+/−	−
Motility/Locomotor Structures	Flagella	+/−***	+/−	−
	Cilia	−	+/−	−
Shape/Protection	Cell wall	+***	+/−	−
	Capsule	+/−	+/−	−
	Spores	+/−	+/−	−
Complexity of Function	—	+	+	+/−
Size (in general)	—	0.5–3μm	2–100μm	<0.2μm

*(+) means most members of the group exhibit this characteristic; (−) means most lack it; (+/−) means only some members have it.
**Viruses cannot participate in metabolic or genetic activity outside their host cells.
***The procaryotic type is functionally similar to the eucaryotic, but structurally unique.

Eucaryotes contain a number of complex internal structures called **organelles** that perform a useful function for the cell involving growth, nutrition, or metabolism. By convention, organelles are also defined as cell components that are enclosed by membranes (see feature 2.4). In addition to their specific functions, organelles also partition the eucaryotic cell into smaller compartments. The most visible organelle is the **nucleus,** a roughly ball-shaped mass of genetic material surrounded by a double membrane. Some of the other organelles include the Golgi apparatus, endoplasmic reticulum, vacuoles, and mitochondria (table 2.6).

Procaryotic cells are found exclusively in organisms called bacteria. Sometimes it may seem that the procaryotes are the microbial "have-nots" because, for the sake of comparison, they are described by what they lack. They have no nucleus or organelles. This apparent simplicity is misleading, because the ultramicroscopic structure of procaryotes is really quite elaborate. Overall, procaryotic cells can engage in every activity that eucaryotic cells can—and many can function in ways that eucaryotes cannot.

Processes That Define Life

To lay the groundwork for a detailed coverage of cells in chapters 3 and 4, this section provides an overview of cell structure and function and introduces the primary characteristics of life. The biological activities or properties that help define and characterize cells as living entities are: (1) the capacity to reproduce, (2) the capacity to metabolize, including cell synthesis and the release of energy, (3) movement, (4) mechanisms for transporting substances into and out of the cell, and (5) mechanisms for cell support and storage. Although eucaryotic cells have specific organelles to perform these functions, procaryotic cells must rely on a few simple, multipurpose cell components for these same functions. As we saw in chapter 1, viruses are not cells, are not generally considered living things, and show certain signs of life only when they invade a host cell. Table 2.6 indicates their relative simplicity compared to cells.

Reproduction

All cells have a *genome* composed of long strands of DNA. In eucaryotic cells this DNA is packed into chromosomes that are

organelle (or″gan-el′) L. *organella,* little organ.

genome (jee′-nome) A combination of the words gene and chromosome. Refers to an organism's entire set of hereditary factors.

entirely enclosed by a nuclear membrane. Procaryotic DNA is in a special type of circular *chromosome* that is not enclosed by a membrane of any sort.

Living things devote a portion of their life cycle to producing offspring that will carry on their particular genetic line for multiple generations. In sexual reproduction, offspring are produced through the union of sex cells from two different parents. In asexual[4] reproduction, offspring originate through the division of a parent cell into two daughter cells. Sexual reproduction, often of a complex nature, occurs in most eucaryotes, but eucaryotic cells also reproduce asexually by several processes. One is a type of cell division called binary fission, a simple process in which the cell splits equally in two. Another simple asexual method is budding, in which the division is unequal and one cell (the bud) is smaller. Many eucaryotic cells engage in *mitosis,* an orderly division of chromosomes that usually accompanies cell division (see figure 4.6). In contrast, procaryotic cells reproduce primarily by binary fission or budding. They have no mitotic apparatus, nor do they reproduce by typical sexual means.

Metabolism

The principal cell component involved in protein synthesis is a tiny particle called a *ribosome.* In eucaryotes, these may be scattered thoughout the cell or as part of membranous sacs known as the *endoplasmic reticulum* (see figure 4.8). Procaryotes also have ribosomes, but they are a smaller version than those of eucaryotes, and procaryotes have no endoplasmic reticulum. Energy in the form of ATP arises from chemical reactions in the *mitochondria* in eucaryotes, and from the cell membrane in procaryotes. Photosynthetic microorganisms (algae and some bacteria) trap solar energy by means of pigments and convert it to chemical energy in the cell. The algae (eucaryotes) have compact membranous bundles called *chloroplasts,* which contain the pigment and perform the photosynthetic reactions. Photosynthetic reactions and pigments in photosynthetic bacteria do not occur in chloroplasts, but in specialized internal membranes.

Motility

The capacity for true motility or self-propulsion is present in a great variety of cells (but not all of them). Eucaryotic cells can move by one of the following locomotor organelles: cilia—short, hairlike appendages, flagella—longer, whiplike appendages, or pseudopods—fingerlike extensions of the cell membrane. Procaryotes move by means of unusual, propeller-like flagella unique to bacteria or by special fibrils that produce a gliding form of motility. They have no cilia or pseudopods.

Protection and Storage

Many cells are supported and protected by rigid *cell walls,* which prevent them from rupturing and impart a constant shape. Among eucaryotes, cell walls occur in plants, microscopic algae, and fungi, but not in animals. The majority of procaryotes have cell walls, but theirs differ in composition from the eucaryotic varieties. As protection against depleted nutrient sources, many microbes store food intracellularly. Eucaryotes do this in membranous sacs called vacuoles, while procaryotes concentrate it in crystals called granules.

Transport

A cell needs to draw nutrients from its external environment and to remove waste and other metabolic products from its internal environment. This two-directional transport is accomplished in both eucaryotes and procaryotes by the cell membrane. This membrane, described in the following section, has a very similar structure in both eucaryotic and procaryotic cells. Eucaryotes have an additional organelle, the *Golgi apparatus,* that assists in packaging materials for transport and removal from the cell.

Table 2.6 summarizes the differences and similarities among procaryotes, eucaryotes, and viruses. You will probably want to refer to this table again after you have completed chapters 3 and 4.

chromosome (kro'-moh-some) Gr. *chroma,* colored, and *soma,* body. The name is derived from the fact that chromosomes stain readily with dyes.

4. Asexual refers to the absence of sexual union. (The prefix *a* or *an* in front of a word means not.)

mitosis (my-toh'-sis) Gr. *mitos,* thread, and *osis,* a condition. Often the term mitosis is used synonymously with eucaryotic cell division.

Chapter Review with Key Terms

Fundamentals of Chemistry

Basic Properties of Elements

All matter in the universe is composed of minute particles called **atoms**; an atom is the simplest form of matter not divisible into a simpler substance by chemical means. The behavior and identity of atoms conforms to known, predictable chemical principles.

Atoms are composed of smaller particles called **protons** (p^+), which are positive in charge; **neutrons** (n^0), which are uncharged; and **electrons** (e^-), which are negative in charge. Protons and neutrons comprise the center or **nucleus** of an atom, and electrons circle around the nucleus in spherical patterns called **orbitals** that occupy energy levels called **shells.**

Differences in the numbers of protons, neutrons, and electrons in an atom give rise to different **elements,** each known by a distinct name and symbol. Elements can be described by **atomic weight** (the total number of protons and neutrons in the nucleus) and **atomic number** (the number of protons alone). The electron number equals the proton number, thus a single unreacted atom is neutral because the opposite charges cancel one another out. Elements may exist in variant forms called **isotopes,** which differ in the number of neutrons they contain.

Electron Orbitals and Shells

Electrons fill the orbitals in a pattern of pairs, from the inner- to the outermost orbitals; the first orbital (shell) can accept 2 e^-, the second shell can hold 8 e^- (4 pairs), and the third shell can hold 18 e^- (9 pairs). Because each element has a distinct number of electrons, the result of orbital filling will be unique for it; most elements do not "come out even," and have an outermost orbital that becomes the focus of reactivity and bonding of an atom. In general, atoms with a filled outermost orbital (such as helium) are less reactive than atoms with unfilled ones (such as oxygen).

Atomic Bonds and Molecules

When outer orbitals of atoms interact in various ways, they form **chemical bonds.** The result of these bonds is a **molecule,** a combination of atoms forming a substance different from the atoms that made it. Member atoms of molecules may be the same element or different elements; if the member elements are different, the substance is a **compound** (CO_2, NH_3).

The nature of molecular bonds is dictated by the electron makeup, or **valence,** of the outer orbitals of the atoms in the molecule. **Covalent bonds** occur when the valences of the atoms are compatible so that electrons can remain with both atoms and orbit within the entire molecule. This is the common bond in molecules containing the **bioelements** (carbon, hydrogen, nitrogen, oxygen, phosphorus, and sulfur). When both members of the covalently bonded molecule are the same element, the electrons are equally shared and the molecule is **nonpolar;** if the elements are different in weight, one of the atoms will pull the electrons more towards it and create positively and negatively charged poles, called a **polar** molecule (H_2O).

Ionic bonds occur when an atom with a low number of valence electrons loses them to an atom that has a nearly filled outer orbital (for example, Na and Cl). Together they form a large crystalline aggregate of the two atoms (NaCl), rather than a discrete molecule. Ionic bonds are readily broken by dissolution, and when this occurs, the atoms separate or **ionize** into charged particles called **ions** that retain the electron makeup of the original bond. Ions that have lost electrons and have a positive charge are **cations** (Na^+), and those that have gained electrons and have a negative charge are **anions** (Cl^-). As a general rule, ions of the opposite charge are attracted to each other (positive attracts negative), and those with the same charges repel each other (positive versus positive or negative versus negative).

Redox Reactions: Chemicals may participate in a transfer of electrons called an **oxidation-reduction** or redox reaction. These reactions always occur between pairs of atoms or molecules. **Oxidation** is a reaction in which electrons are released, and **reduction** is a reaction in which these same electrons are received. Any atom or molecule that donates electrons to another atom or molecule is a **reducing agent,** and one that picks up electrons is an **oxidizing agent.**

Hydrogen bonds are caused by weak attractive forces between covalently bonded hydrogen and polarized atoms. Oxygen and nitrogen atoms on the same or nearby molecules bear a negative charge.

Formulas, Models, and Chemical Equations

Formulas of molecules and compounds may be presented as simple molecular formulas that summarize the atoms contained therein (H_2O, CO_2), or they can be expressed as structural formulas that give the details of bonding ($H-O-H$, $O=C=O$).

Chemical Equations: Equations summarize a chemical reaction by showing the reactants (starting chemicals) and the products (resulting chemicals), and indicate whether certain bonds are broken or created. **Synthesis** reactions: $A + B \rightarrow AB$; **decomposition** reactions: $XY \rightarrow X + Y$; **exchange** reactions: $AB + XY \rightarrow AX + BY$; reversible reactions: $MN \leftrightarrow M + N$.

Solutions: A solution is a mixture of a chemical termed a **solute** dissolved in a medium called a **solvent.** Solutes are atoms or molecules in solid, liquid, or gaseous form, and solvents are usually liquid molecules. For a solute to dissolve (be soluble) in a solvent, the two must have similar characteristics of polarity. Water-soluble solutes are either charged (ionic salts) or polar (certain organic molecules). The dissolved solute will become **hydrated**—surrounded by a shell of water molecules due to electrostatic attraction. Chemicals that attract water this way are **hydrophilic**; those that repel it are **hydrophobic.** The **concentration** of a solution is defined as the amount of solute dissolved in a given amount of solvent; the greater the amount of solute relative to solvent, the higher a solution's concentration.

Acids, Bases, and pH: An **acid** is a solution that contains a concentration of **hydrogen ions** $[H^+]$ greater than 0.0000001 grams /liter (1×10^{-7}), and a **base** is a solution that contains a concentration of $[H^+]$ below that amount. When a solution contains exactly that $[H^+]$, it is considered neutral—neither acidic nor basic. The relative acidity and basicity of solutions can be represented by a pH scale based on a standardized range of $[H^+]$; pH ranges from the most acidic reading of 0 to the most basic (alkaline) reading of 14; in the exact middle is pH 7, the reading of neutrality. Acidic and basic ions in aqueous solutions may interact to form water and a salt according to this reaction, called **neutralization**: $H^+X + OH^-Y \rightarrow HOH\ (H_2O) + XY$ (salt).

Inorganic compounds are composed of some combination of atoms other than carbon and hydrogen (H_2O, CO_2, NaCl); **organic compounds** contain carbon and hydrogen (and usually some others) in combination. Organic compounds constitute the most important molecules in the structure and function of cells. The capacity of carbon to form covalent bonds with other carbons and with atoms such as hydrogen, oxygen, nitrogen, and phosphorus gives it great versatility in creating the backbone of biological molecules. **Functional groups** such as carboxyl ($-COOH$), amino ($-NH_2$), and phosphate ($-PO_4$) are special accessory molecules that bind to carbon and provide the tremendous diversity and reactivity seen in organic compounds.

Biochemistry and Macromolecules

Biochemicals are organic compounds produced by cells; they are often very large, so are termed **macromolecules.** Many are assembled from individual smaller building blocks called **monomers** into multiunit chains called **polymers,** in a process known as **polymerization.**

Carbohydrates: Compounds composed of carbon and water in the combination (CH_2O). The presence of hydroxyl groups ($-OH$) and a carbonyl places them in the chemical class of aldehyde or ketone. Simple sugars, or **monosaccharides,** consist of three to seven carbons and have names that end in -ose. Monosaccharides may be joined through *glycosidic bonds* between the $-OH$ group on one sugar and a carbon on the second sugar ($C-O-C$); because water is removed during the formation of this bond, it is termed *dehydration synthesis.* **Disaccharides** are composed of two monosaccharides (**lactose, maltose, sucrose**); **polysaccharides** are chains of five or more monosaccharides.

Polysaccharides are the most important structural and storage carbohydrates; **cellulose** is a long chain glucose polymer that is a major component of the cell wall in plants and some microbes; **peptidoglycan** is a combination of glycans and peptides that reinforces the bacterial cell wall; **starch** and **glycogen** are compact storage forms of glucose in plant and animal cells, respectively. The bonds of polysaccharides are digested by specific enzymes and require a water molecule—thus their breakage is termed **hydrolysis.**

Lipids: Any organic compounds that are not soluble in water and other polar solvents due to their nonpolar, hydrophobic hydrocarbon chains. **Triglycerides,** including fats and oils, consist of a **glycerol** molecule bound by each of its three OH groups to a COOH group of a **fatty acid** molecule. The three bonds are called **ester bonds:**

```
  O   O
  |   ||
C—O—C—
```

If the carbons in the fatty acid are each bonded to four other atoms, the fat is saturated; if there is one or more $C=C$ bond, it is unsaturated. Triglycerides are storage lipids that yield considerable energy on hydrolysis.

Phospholipids are lipids composed of a glycerol bound to two fatty acids and a phosphoric acid, attached to some type of alcohol; the molecule has a polar quality because the PO_4/alcohol head is charged and hydrophilic and the long fatty acid chains are uncharged and hydrophobic; this property makes them form single or double lipid layers in the presence of water, with the hydrophilic end facing the water and the hydrophobic end repelling it. Phospholipids are important constituents of cell membranes.

Proteins: Highly complex biochemicals assembled from 20 different subunits called **amino acids** (aa). The amino acids are combined in a certain order (by dehydration synthesis) by **peptide bonds** that occur between the carboxyl group and amino group of two amino acids: —C — NH — O — C. A **peptide** is any short chain, such as a dipeptide with two aa's or a tripeptide with three; a **polypeptide** is usually 20–50 aa's in a chain; a **protein** contains more than 50 aa's. The predominant molecules in cells are these larger proteins.

The chain of amino acids is a protein's **primary structure,** but not its functional structure. The functional groups extending from the amino acids on this chain cause proteins to form additional levels of structure. Hydrogen and other weak bonds within the chain twist it into a helix configuration called the **secondary structure;** the secondary helix folds upon itself and forms strong **disulfide bonds** (between the sulfur atoms on nearby cysteines) and additional weak bonds, thus producing a three-dimensional globular shape, or **tertiary structure.** The surface configuration of the third level provides proteins with their specificity and makes possible the enormous diversity in enzymes, structural molecules, antibodies, and cell receptors. Huge proteins composed of two or more smaller subunits exist in the **quaternary level** of structure.

Nucleic Acids: Very complex molecules that carry, express, and pass on the genetic information of all cells and viruses. Their basic building block is a nucleotide, composed of a **nitrogen base,** a **pentose sugar,** and a **phosphate.** Nitrogen bases are the ringed compounds: *adenine* (A), *cytosine* (C), *thymine* (T), *guanine* (G), and *uracil* (U); pentose sugars may be deoxyribose or ribose. The sugar combines covalently with a nitrogen base and a phosphate; the polymerization of nucleotides produces a molecule having a repeating backbone of sugar-phosphate with the bases branching off the sides.

Both **deoxyribonucleic acid (DNA)** and **ribonucleic acid (RNA)** are polynucleotide strands. DNA contains **deoxyribose** sugar, has all of the bases except uracil, and occurs as a double-stranded helix with the bases hydrogen-bonded in pairs situated in the middle of the molecule; the pairs must mate according to the pattern A–T and C–G. DNA provides a master code for all life processes. RNA contains **ribose** sugar, has all of the bases except thymine, and is a single-stranded molecule that may come in three varieties. RNA is the interpreter of the DNA code and is directly involved in protein synthesis.

Adenosine triphosphate (ATP) is another nucleotide involved in the transfer and storage of energy in cells. It contains adenine, ribose, and three phosphates in a series. Splitting off the last phosphate in the triphosphate releases a packet of energy that may be used to do cell work.

Introduction to Cell Structure

Cells are huge aggregates of organic compounds organized to carry out complex processes described as **living.** All organisms consist of cells, which fall into one of two types: **procaryotic,** which are small, structurally simple bacterial cells that lack a **nucleus** and other **organelles,** and **eucaryotic,** which are larger, contain a nucleus and organelles, and are found in the cells of plants, animals, fungi, and protozoa. Viruses are not cells and are not generally considered living because they cannot function independently.

An organism must participate in certain processes in order to be considered alive. Eucaryotic cells carry out these processes using specific organelles; procaryotic cells use a few multipurpose structures. **Reproduction** involves producing offspring asexually (with one parent) or sexually (with two parents); **metabolism** refers to the chemical reactions in cells, including the synthesis of proteins on ribosomes and the release of energy (ATP); **motility** or movement originates from special locomotor structures such as flagella and cilia; **protective** external structures include capsules and cell walls; nutrient **storage** takes place in compact intracellular masses; and **transport** involves conducting nutrients into the cell and wastes out of the cell.

Important structures that surround and contain the protoplasm of all cells, create organelles, and divide the cytoplasm of eucaryotic cells into compartments are **membranes.** Membranes are continuous ultrathin sheets. They are comprised of phospholipids, which provide the flowing network of the membrane bilayer, and proteins, which are embedded in this bilayer. Lipids provide an impenetrable barrier that dictates cell **permeability,** and proteins serve as channels of transport and sites of recognition and chemical reactions.

True–False Questions

Determine whether the following statements are true (T) or false (F). If you feel a statement is false, explain why and reword the sentence so that it reads accurately.

___ 1. An atom is the smallest unit of matter whose behavior can be predicted.

___ 2. Protons are positively charged, neutrons are negatively charged, and electrons are neutral.

___ 3. Electrons circle around the nucleus of an atom in tracks called orbitals.

___ 4. The atomic number of an element is about twice as large as its atomic weight.

___ 5. A compound is a type of molecule in which the atoms are different.

___ 6. Water molecules (H_2O) are formed through ionic bonds between oxygen and hydrogen.

___ 7. A molecule has polarity when one side is different in charge from the other side.

___ 8. Hydrogen bonds form between two hydrogen atoms on separate molecules.

___ 9. Ions with the same charge will be repelled by each other, and ions with opposite charges will be attracted to each other.

___ 10. Oxidation is the tendency of an atom to give up electrons, and reduction is the tendency to receive electrons during reactions.

___ 11. In a solution of NaCl and water, NaCl is the solvent and water is the solute.

___ 12. An acid is a substance that releases H^+ into a solution.

___ 13. A solution with a pH of 2 has a higher $[H^+]$ than a solution with a pH of 8.

___ 14. Fructose is a type of monosaccharide.

___ 15. Bond formation in polysaccharides and polypeptides is accompanied by the removal of a water molecule.

___ 16. Cellulose is a structural polysaccharide, and starch is a storage polysaccharide.

___ 17. A phospholipid contains three fatty acids, a glycerol, and a phosphate.

___ 18. Twenty naturally occurring amino acids are used to make proteins.

___ 19. Disulfide bonds are special covalent bonds that form between two sulfur atoms in proteins.

___ 20. DNA is a double-stranded molecule that contains deoxyribose, phosphate, and four different nitrogen bases.

Concept Questions

1. How are the concepts of an atom and an element related? What causes elements to differ?
2. How are atomic weight and atomic number derived? Using data in table 2.1, give the electron number of nitrogen, sulfur, calcium, phosphorus, and iron. What is distinctive about isotopes of elements?
3. How is the concept of molecules and compounds related?
4. Why is an isolated atom neutral? Describe the concept of the atomic nucleus, electron orbitals, and shells. What causes atoms to form chemical bonds? Why do some elements not bond readily?
5. Distinguish between the general reactions in covalent, ionic, and hydrogen bonds.
6. Which kinds of elements tend to make covalent bonds? Distinguish between a single and a double bond. What is polarity? Why are some covalent molecules polar and others nonpolar? What is an important consequence of the polarity of water?
7. Which kinds of elements tend to make ionic bonds? Exactly what causes the charges to form on atoms in ionic bonds? Verify the proton and electron numbers for Na^+ and Cl^-. Differentiate between an anion and cation. What kind of an ion would you expect magnesium to make, based on its valence?
8. Differentiate between an oxidizing agent and a reducing agent.
9. Why are hydrogen bonds relatively weak?
10. Compare the three basic types of chemical formulas; review the types of chemical reactions and the general ways they may be expressed in equations.
11. Define solution, solvent, and solute. What properties of water make it an effective biological solvent, and how does a molecule like NaCl become dissolved in it? How is the concentration of a solution determined?
12. What determines whether a substance is an acid or a base? Briefly outline the pH scale. How can a neutral salt be formed from acids and bases?
13. What atoms must be present in a molecule for it to be considered organic? What characteristics of carbon make it ideal for the formation of organic compounds? What are functional groups? Differentiate between a monomer and a polymer. How are polymers formed?
14. What characterizes the carbohydrates? Differentiate between mono-, di-, and polysaccharides, and give examples of each. What is a glycosidic bond, and what is dehydration synthesis?
15. Draw simple structural molecules of triglycerides and phospholipids to compare their differences and similarities. What is an ester bond? How are saturated and unsaturated fatty acids different? What characteristic of phospholipids makes them essential components of cell membranes?
16. Describe the basic structure of an amino acid. What is a peptide bond? Differentiate between a peptide, a polypeptide, and a protein. Explain what causes the various levels of structure of a protein molecule. What functions do proteins perform in a cell?
17. Describe a nucleotide and a polynucleotide, and compare and contrast the general structure of DNA and RNA. Name the two purines and the three pyrimidines. Why is DNA called a double helix? What is the function of RNA? What is ATP, and what is its function in cells?
18. Outline the general structure of a cell, and describe those characteristics of cells that qualify them as living. Why are viruses not considered living? Compare the general characteristics of procaryotic and eucaryotic cells. What are cellular membranes, and what are their functions? Explain the fluid mosaic model of a membrane.

Practical/Thought Questions

1. The octet rule in chemistry helps predict the tendency of atoms to acquire or donate electrons from the outer shell. It says that those with fewer than 4 tend to donate electrons and those with more than 4 tend to accept additional electrons; those with exactly 4 can do both. Using this rule, determine what category each of the following elements falls into: N, S, C, P, O, H, Ca, Fe, and Mg. (You will need to work out the valence of the atoms.)
2. Predict the kinds of bonds that occur in ammonium (NH_3), phosphate (PO_4), disulfide (—S—S—), and $MgCl_2$. (Use simple models such as those in figure 2.3.)
3. Work out the following problems: What is the number of protons in helium? Will an H bond form between $H_3—C—CH=O$ and H_2O? Why, or why not? Predict whether the following compounds are polar or not: Cl_2, NH_3, CH_4, glucose, leucine. What is the pH of a solution with a concentration of 0.00001 g/ml of H^+? With a concentration of 0.00001 g/ml of OH^-?
4. Describe how hydration spheres are formed around cations and anions. Which substances will be expected to be hydrophilic and hydrophobic?
5. Name several inorganic compounds.
6. In what way are carbon-based compounds like children's Tinker Toys or Lego blocks?
7. Is galactose an aldehyde or a ketone sugar?
8. How many water molecules are released for each triglyceride that is formed?
9. How many peptide bonds are in a tetrapeptide?
10. Use pipe cleaners to help understand the formation of the 2° and 3° structure of proteins. Note the various ways that this object can be folded and the diversity of shapes that may be formed.
11. Looking at figure 2.25, can you see why adenine can form hydrogen bonds only with thymine and why cytosine can form them only with guanine? Why can no other combinations occur in DNA?

CHAPTER 3

Procaryotic Profiles: The Bacteria

Colorful floating bloom of purple sulfur bacteria and green algae in the Baltic Sea.

Chapter Preview

A bacterial cell appears deceptively simple when viewed with an ordinary microscope. Although procaryotic cells are the simplest of all cells, not until they are subjected to the scrutiny of the electron microscope and biochemical studies does their intricate and functionally complex nature become evident. They are among the most common and ubiquitous microorganisms on earth, and they play essential roles in the balance of nature and in the lives of humans. You will see the significance of these organisms verified daily in newspaper headlines, in this course, and in your career. A knowledge of the structure and behavior of procaryotic cells is fundamental to later studies in nutrition, genetics, drug therapy, infection, and disease. The primary topics to be covered in this chapter are elements of microscopic anatomy, physiology, classification, identification, methods of cultivation, and a survey of special bacterial groups.

Bacterial Form and Function

The evolutionary history of the bacteria extends back 3.5 billion years (see figure 3.27). For bacteria to have endured for so long in such a variety of habitats indicates that their cellular structure and function are amazingly adaptable. An analysis of the cellular organization of a procaryotic cell produces the following general pattern:

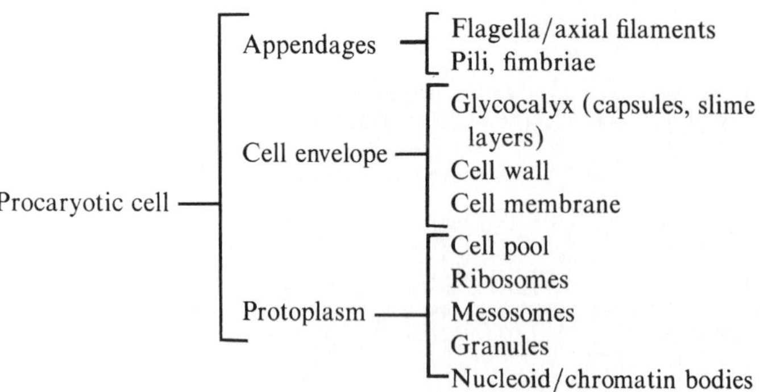

All bacterial cells invariably have a cell envelope, glycocalyx, cell membrane, cell pool, ribosomes, and a nucleoid; the majority have a cell wall. Although they are common to many species, flagella, pili, fimbriae, capsules, slime layers, and granules are *not* universal components of all bacteria.

A Highly Magnified View of a Generalized Procaryotic Cell

Before taking a closer look at the anatomy of a single procaryotic cell, it will be helpful to understand that much of our knowledge of the structures described here is derived from electron microscope studies. Most of you will not see these details from firsthand laboratory experiences. Though photomicrographs of cell structure have a flat, two-dimensional appearance, keep in mind that these structures exist in a three-dimensional configuration. The descriptions of bacterial structure, except where otherwise noted, refer to the **eubacteria,** a category of typical bacteria with peptidoglycan in their cell walls. Figure 3.1 contrasts the three-dimensional anatomy of a generalized (rod-shaped) bacterial cell alongside an electron micrograph of an actual cell. As we survey the principal anatomical features of this cell, we will perform a microscopic dissection of sorts, following a course that begins with the outer cell structures and proceeds to the internal contents.

Appendages: Structures for Swimming and Clinging

Several discrete types of accessory structures sprout from the surface of bacteria. These elongate molecules have been termed

Glycocalyx
Fimbriae
Cell membrane
Chromatin body
Cell wall
Pilus
Ribosomes
Mesosome
Granule
Flagellum

(a)

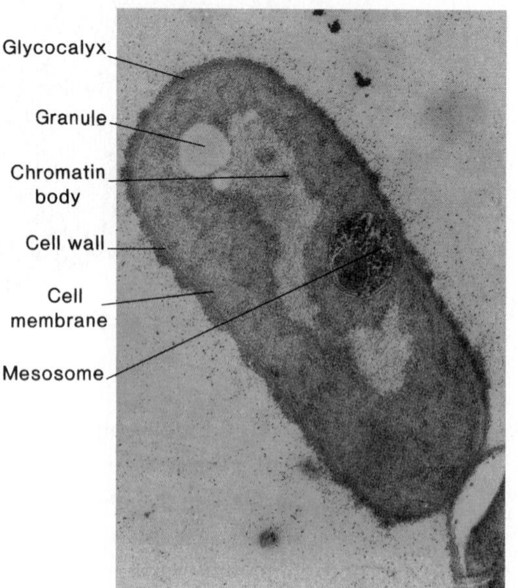

(b)

Figure 3.1 (*a*) Cutaway view of a typical rod-shaped bacterium, showing major structural features. Note that not all components are found in all cells. (*b*) Transmission electron micrograph of a cell (×50,000).

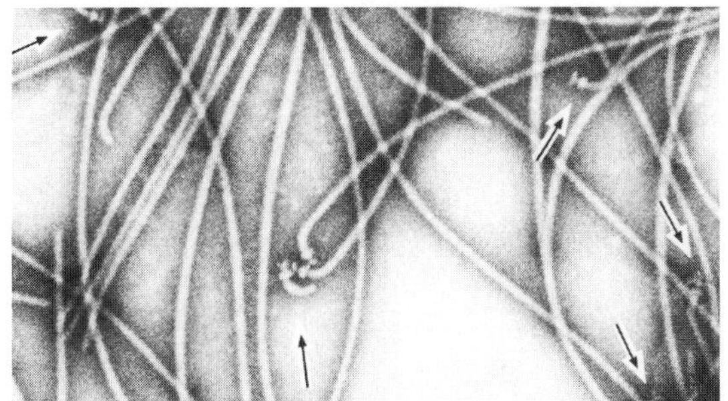

(a)

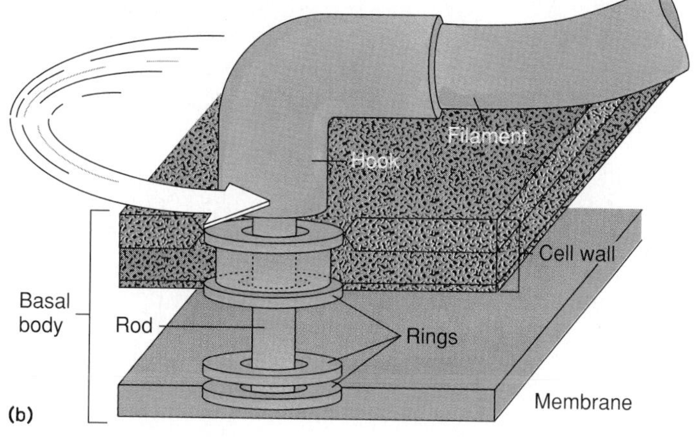

(b)

Figure 3.2 (*a*) Electron micrograph of basal bodies and filaments (at arrows) of bacterial flagella (×66,000). (*b*) Details of the basal body. The hook, rings, and rod function together as a tiny device that rotates the filament 360°.

appendages, and though common, they are not present on all species. Appendages can be divided into two major subgroups: those that provide motility (flagella and axial filaments) and those that provide attachments (fimbriae and pili).

Flagella

The **flagellum** in procaryotes is an appendage of truly amazing construction, certainly unique in the biological world. The primary function of flagella is to confer **motility** or self-propulsion—that is, the capacity of a cell to swim freely through an aqueous habitat. The extreme fineness of a bacterial flagellum necessitates high magnification to reveal its special architecture, which occurs in three distinct parts: the filament, the hook (sheath), and the basal body (figure 3.2). The *filament,* a helical structure composed of proteins that is approximately 20 nm in diameter and varies from 1 to 70 μm in length, is inserted into the *hook.* The curved, tubular hook is anchored to the cell by the basal body, a stack of rings fixed firmly through the cell wall, periplasmic space, and cell membrane. The mechanism of the hook/basal body articulation is very much like a ball in a socket. This arrangement permits the hook with its filament to rotate 360°, rather than undulating back and forth like a whip as was once thought. The rotation of the flagellum causes the cell body to spin in the opposite direction, and this gives it a forward motion (see figure 3.5).

One can generalize that all spirilla, about half of the bacilli, and a small number of cocci are flagellated. Flagella vary both in number and arrangement according to two general patterns: (1) In a *polar* arrangement, the flagella are attached at one or both ends of the cell. Three subtypes of this pattern include *monotrichous,* with a single flagellum; *lophotrichous,* with small bunches or tufts of flagella emerging from the same site; and *amphitrichous,* with flagella at both poles of the cell. (2) In a *peritrichous* arrangement, flagella are dispersed randomly over the surface of the cell (figure 3.3). The type of arrangement has some bearing on the swimming speed of the bacterium.

If identification of a bacterium requires detection of the actual number and placement of flagella, special stains or electron microscope preparations are required, as flagella are too minute to be seen in unstained live preparations with an ordinary light microscope. Often it is sufficient to know simply whether a bacterial species is motile. One way to detect motility is to place a tiny mass of cells into a soft (semisolid) medium. Growth spreading rapidly through the entire medium is indicative of motility (see figure 6.21). Alternately, cells can be observed microscopically with a hanging drop slide (figure 3.4). A truly motile cell will flit, dart, or wobble around the field, making some progress, whereas one that is nonmotile jiggles about in one place and does not make progress.

---

appendage (e-pen′-dij) Any external projection of a body.
flagellum (fle-jel′-em) pl. flagella; L. a whip.

monotrichous (maw-not′-rih-kus) Gr. *mono,* one, and *tricho,* hair.
lophotrichous (lo-fot′-rih-kus) Gr. *lopho,* tuft or ridge.
amphitrichous (am-fit′-rih-kus) Gr. *amphi,* on both sides.
peritrichous (pe-rit′-rih-kus) Gr. *peri,* around.

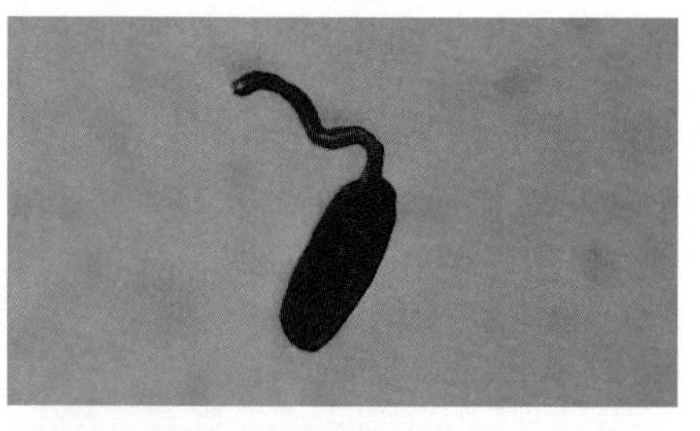

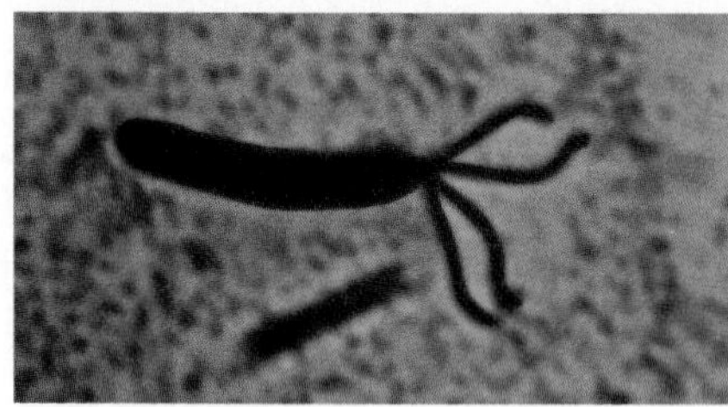

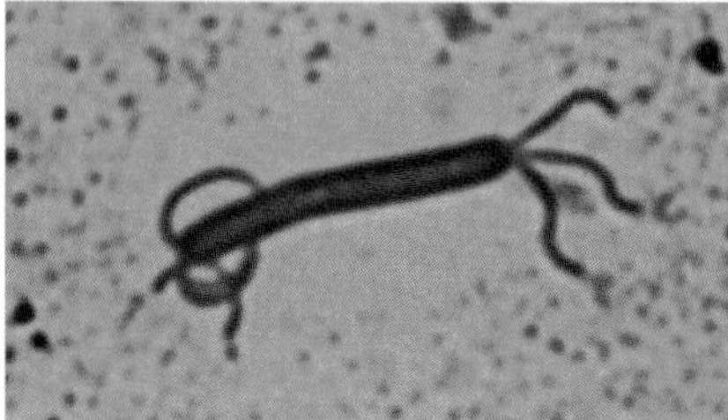

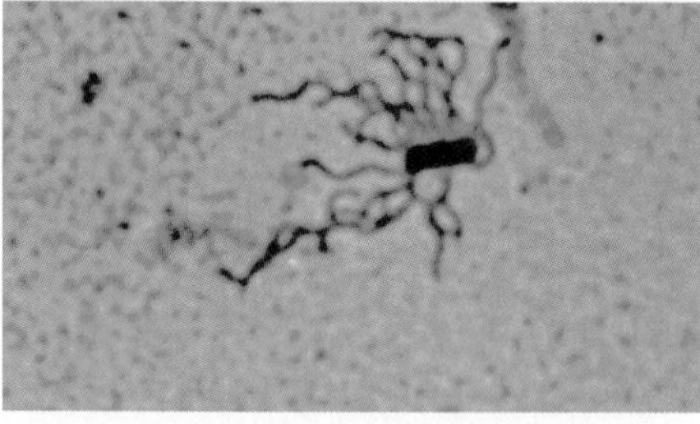

Figure 3.3 Flagellar arrangements. (*a*) Polar types: monotrichous, lophotrichous, and amphitrichous. (*b*) Peritrichous types: light micrograph and scanning electron micrograph.

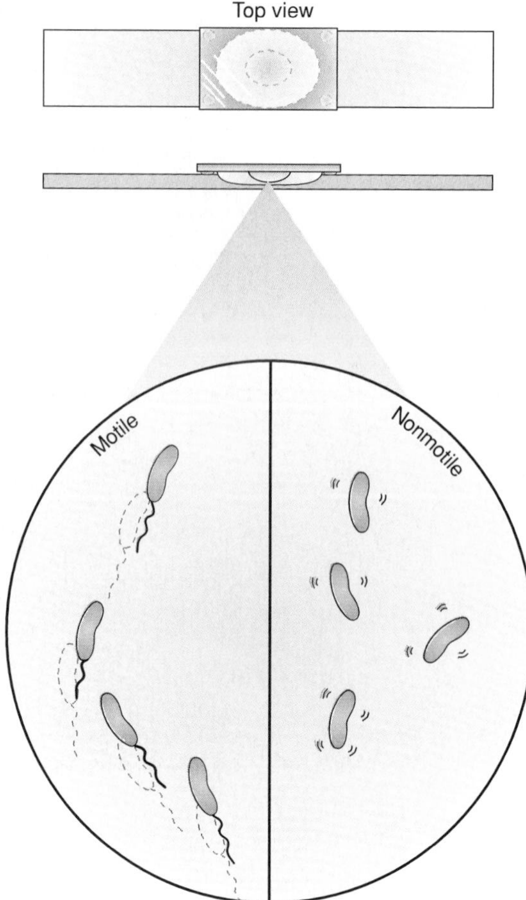

Figure 3.4 A hanging drop slide can be used to detect motility by observing the microscopic behavior of the cell. In true motility, the cell swims and progresses from one point to another. It is assumed that a motile cell has one or more flagella. Nonmotile cells oscillate in the same relative space due to bombardment by molecules, a physical process called Brownian movement.

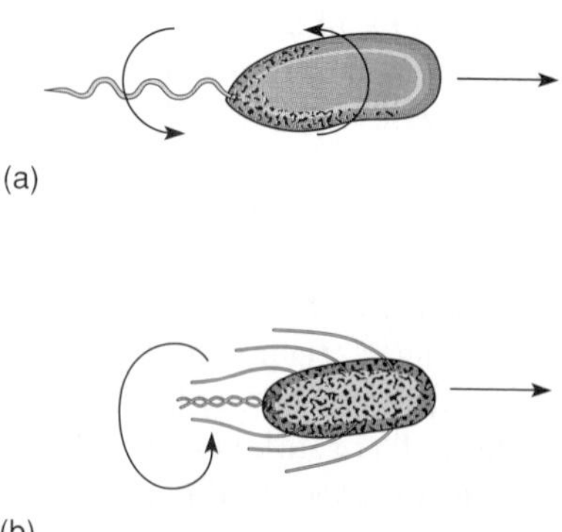

Figure 3.5 The operation of flagella and the mode of locomotion in bacteria with polar and peritrichous flagella. (*a*) In general, when the flagellum rotates in one direction, the cell rotates in the opposite direction. (*b*) In peritrichous forms, all flagella sweep toward one end of the cell and rotate as a single group.

Details of Flagellar Function Flagellated bacteria can perform some rather sophisticated feats. They can move toward potential food sources and away from adverse environmental conditions—a type of behavior called **chemotaxis.** Positive chemotaxis is movement in the direction of a favorable chemical stimulus (usually a nutrient); negative chemotaxis is movement of a cell away from a repellent (potentially harmful) chemical.

The flagellum is effective in guiding bacteria through the environment primarily because the system for detecting chemicals is linked to the mechanisms that drive the flagella. The cell membrane and periplasmic space at the site of flagellum insertion contain specific receptors[1] that bind sugars and amino acids. In the presence of a particular nutrient, a series of energy-requiring reactions set the flagellum in rotary motion. If several flagella are present, they bunch together and rotate simultaneously (figure 3.5). As a flagellum rotates counterclockwise, the cell itself is driven clockwise. This combination results in a *run,* defined as smooth, linear swimming in the direction of the stimulus. Runs are interrupted at various intervals by *tumbles,* during which the flagellum reverses direction and the cell turns away from a straight course. The type of chemotaxis shown at any one time is dependent upon the number of tumbles. It is believed that attractant molecules inhibit tumbles and permit progress toward the stimulus. Repellents cause numerous tumbles and avoidance of the stimulus (figure 3.6).

Not all bacteria travel at the same speed. The speediest forms are polar, flagellated rods such as *Thiospirillum,* which can zip along at 5.2 mm/minute, and *Pseudomonas aeruginosa,* which can swim 3.4 mm/minute. Peritrichous forms tend to be slower swimmers. For instance, the motility rate of *Escherichia coli* is 1 mm/minute, and of *Sporosarcina ureae* (a coccus), 1.7 mm/minute. Taking into account the small dimensions of these bacteria, such speeds are comparable to some protozoa and animals.

chemotaxis (ke″-moh-tak′-sis) pl. chemotaxes; Gr. *chemo,* chemicals, and *tax,* arrangement.

1. Cell-surface molecules that receive stimuli when other molecules bind to them.

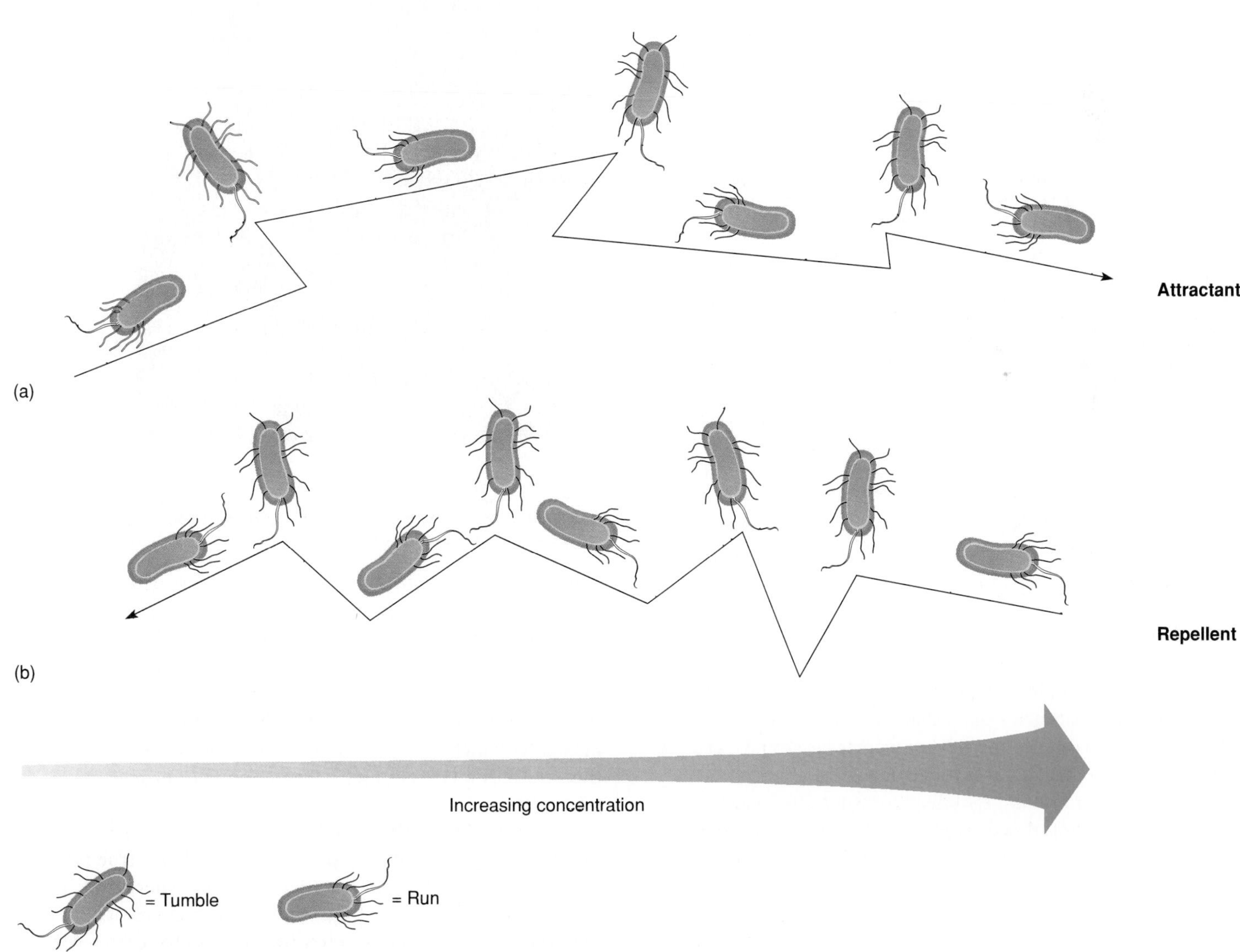

Figure 3.6 Chemotaxis in bacteria. The cell shows a primitive mechanism for progressing (*a*) toward positive stimuli and (*b*) away from irritants by swimming in straight runs or by tumbling.

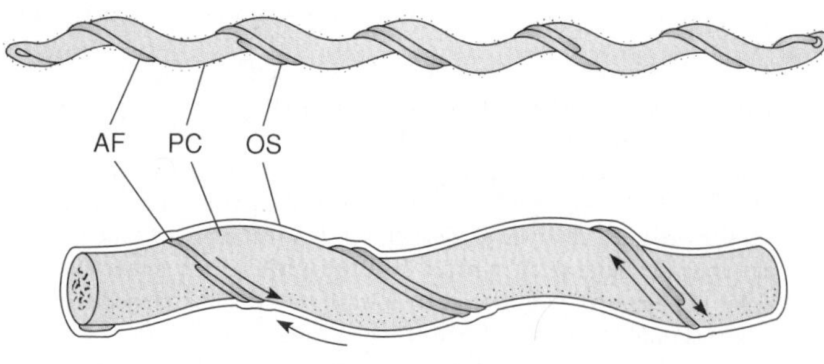

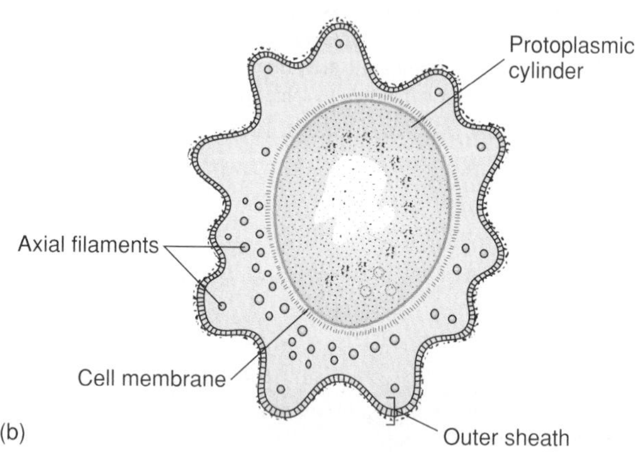

Figure 3.7 The orientation of axial filaments on the spirochete cell. (*a*) Longitudinal section; (*b*) cross section. Contraction of the filaments imparts a spinning and undulating pattern of locomotion.

Axial Filaments

Corkscrew-shaped bacteria called **spirochetes** show an unusual, wriggly mode of locomotion caused by two or more long, coiled threads, the axial filaments (fibers). Because it is indeed a type of modified flagellum, an axial filament consists of a long, thin microfibril inserted into a hook. Unlike flagella, however, the entire structure is enclosed in the space (periplasmic) between the cell wall and the cell membrane, and for this reason it is also called an *endoflagellum* (figure 3.7). The filaments curl closely around the spirochete coils and are free to rotate and impart a twisting or flexing motion to the cell. This form of locomotion must be seen in live cells to be truly appreciated.

Non-locomotor Appendages

Customarily, the terms **pilus** or **fimbria** have been used interchangeably to indicate any bacterial surface appendage not involved in motility. Since a distinction is helpful, we will use the term fimbriae to refer to the shorter, numerous strands, and pili to refer to the longer, sparser appendages.

Fimbriae are small bristlelike fibers sprouting off the surface of many bacterial cells (figure 3.8). Their exact composition varies, but most of them contain protein. Fimbriae have an inherent tendency to stick to each other and to surfaces. For example, they are often responsible for the mutual clinging of cells that leads to films and other thick aggregates of cells on the surface of liquids, and for the microbial colonization of inanimate solids such as rocks and glass. Some pathogens can colonize and infect host tissues because of a tight adhesion between their fimbriae and epithelial cells (figure 3.8*b*). This is also the means by which the gonococcus (agent of gonorrhea) invades the genitourinary tract and *Escherichia coli* invades the intestine.

A pilus (also called *sex pilus*) is an elongate, rigid tubular structure made of a special protein, *pilin.* So far, true pili have been found only on gram-negative bacteria, where they are involved primarily in a mating process between cells called *conjugation,* which involves partial transfer of DNA from one cell

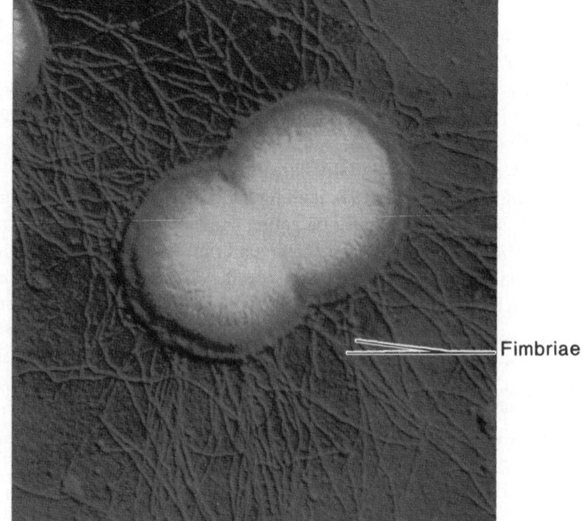

(b)

Figure 3.8 Bacterial fimbriae. (*a*) Fringe of fimbriae surrounding a *Streptococcus* cell. (*b*) Section of upper respiratory tract, showing *Streptococcus* attached to host cells by means of its fimbriae.

spirochete (spy′-roh-keet) Gr. *speira,* coil, and *chaite,* hair.
pilus (py′-lus) pl. pili; L. hair.
fimbria (fim′-bree-ah) pl. fimbriae; L. a fringe.

conjugation (kon-ju-gay′-shun) L. *conjugatus,* linked together.

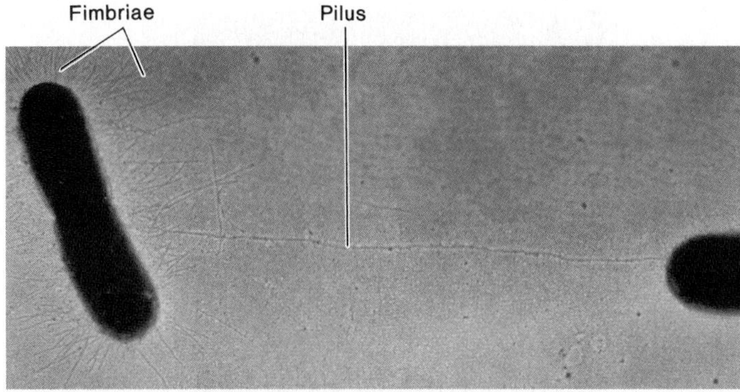

Figure 3.9 Bacteria in the process of conjugating. Clearly evident is a single sex pilus forming a conjugation bridge. Note also the numerous fine fimbriae on one cell.

to another (figure 3.9). A pilus from the donor cell unites with a recipient cell, thereby bringing the two into close proximity for making the transfer. Production of pili is controlled genetically, and conjugation takes place only between closely related species or genera. We will further explore the roles of pili and conjugation in chapter 8.

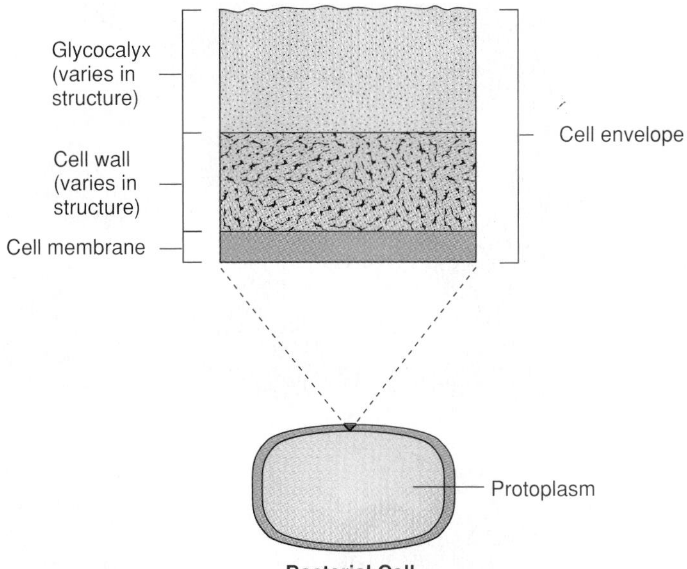

Figure 3.10 The relationship of the three layers of the cell envelope.

The Cell Envelope: The Outer Wrapping of Bacteria

The majority of bacteria have a chemically complex external covering that, for a long time, was simply termed the cell wall. More detailed study eventually revealed that the cell wall is not the outermost covering and that most cells have surface coatings of various types. The precise anatomy of the bacterial surface and wall is so diverse that we will collectively refer to the complex of layers external to the cell protoplasm as the **cell envelope.** The layers of the envelope are stacked one upon another and are often tightly bonded together like the casings of a peanut. The three basic layers that can be identified in electron micrographs are the glycocalyx, the cell wall, and the cell membrane (figure 3.10). Though the envelope layers each perform a distinct function, they also act together as a single protective unit. The envelope is extensive, and can account for one-tenth to one-half of a cell's bulk.

The Bacterial Surface

The bacterial cell surface is frequently exposed to severe environmental conditions. The **glycocalyx** develops as a coating of macromolecules to protect the cell and, in some cases, help it adhere to its environment. Glycocalyces differ among bacteria in thickness, organization, and chemical composition. Some bacteria are covered with a loose, soluble shield called a **slime layer** that evidently protects them from loss of water and nutrients (figure 3.11*a*). Certain bacteria produce **capsules** of repeating polysaccharide units, of protein, or of both (figure 3.11*b*). Unlike the slime layer, a capsule is bonded to the cell to some degree, and it has a thick, gummy consistency that gives a prominently sticky (mucoid) character to the colonies of most encapsulated bacteria (figure 3.12). Although capsules and other surface layers have various functions as described in the next section, they are not absolutely essential to bacterial survival.

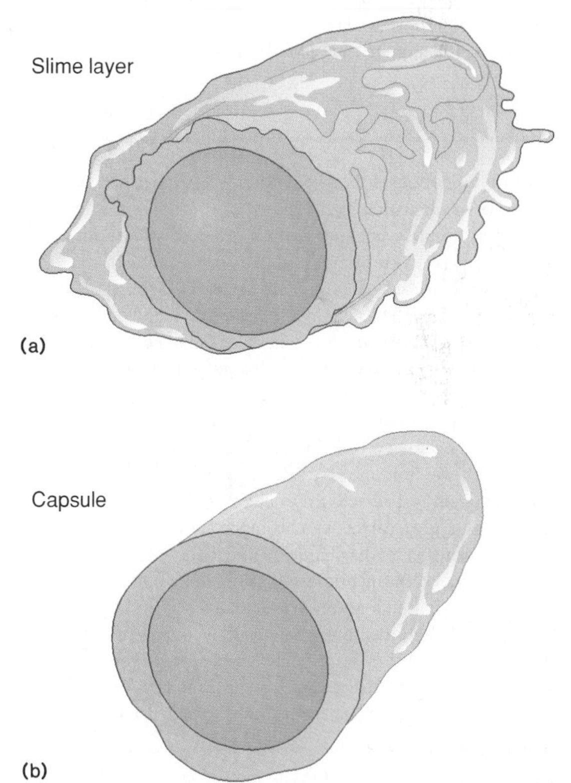

Figure 3.11 Bacterial cells sectioned to show the types of glycocalyces. (*a*) The slime layer is a loose structure that is easily washed off. (*b*) The capsule is a thick, structured layer that is not readily removed.

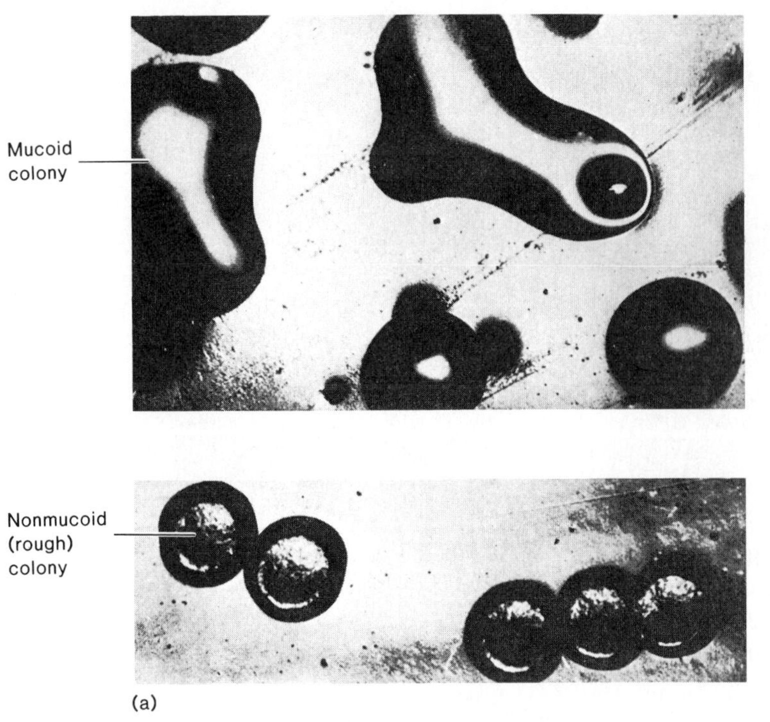

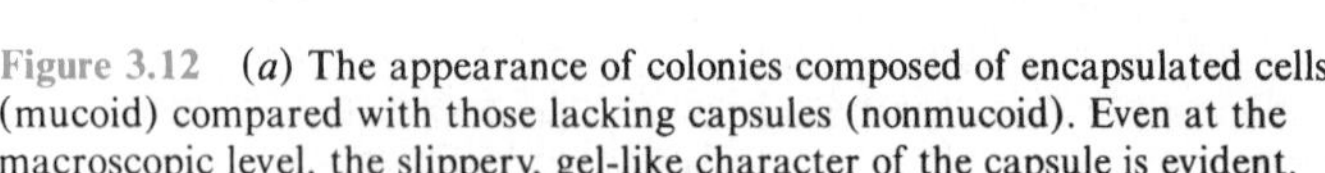

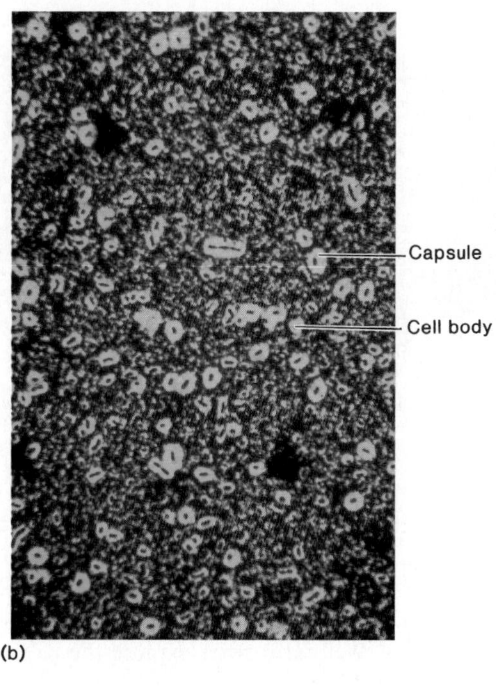

Figure 3.12 (*a*) The appearance of colonies composed of encapsulated cells (mucoid) compared with those lacking capsules (nonmucoid). Even at the macroscopic level, the slippery, gel-like character of the capsule is evident. (*b*) Staining reveals the microscopic appearance of a large, well-developed capsule.

Biomedical Significance of the Glycocalyx Capsules are formed by several pathogenic species, such as *Streptococcus pneumoniae* and *Klebsiella pneumoniae* (both cause pneumonia—an infection of the lung), *Haemophilus influenzae* (one cause of meningitis), and *Bacillus anthracis* (the cause of anthrax). Encapsulated bacterial cells generally have greater pathogenicity because capsules protect them against white blood cells called phagocytes. The texture of capsules effectively blocks phagocytosis, described as the engulfing and destroying action of these cells. By escaping phagocytosis, the bacteria are free to multiply and damage body tissues. Encapsulated bacteria may mutate to nonencapsulated forms, and when they do, they usually lose their pathogenicity. Because capsules can be antigenic (that is, they can stimulate specific immune responses), they form the basis of some vaccines and serological tests discussed in chapter 13.

Other types of glycocalyces can be important in colonization. The thick, white film that forms on teeth (plaque) is due to the surface slimes produced by certain streptococci in the oral cavity. This slime initially allows them to gain a foothold on the teeth and provides an avenue for other oral bacteria that in time may lead to dental caries. The glycocalyx of some bacteria is so highly adherent that it is responsible for persistent colonization of nonliving materials such as plastic catheters, intrauterine devices, and metal pacemakers that are in common medical use (figure 3.13).

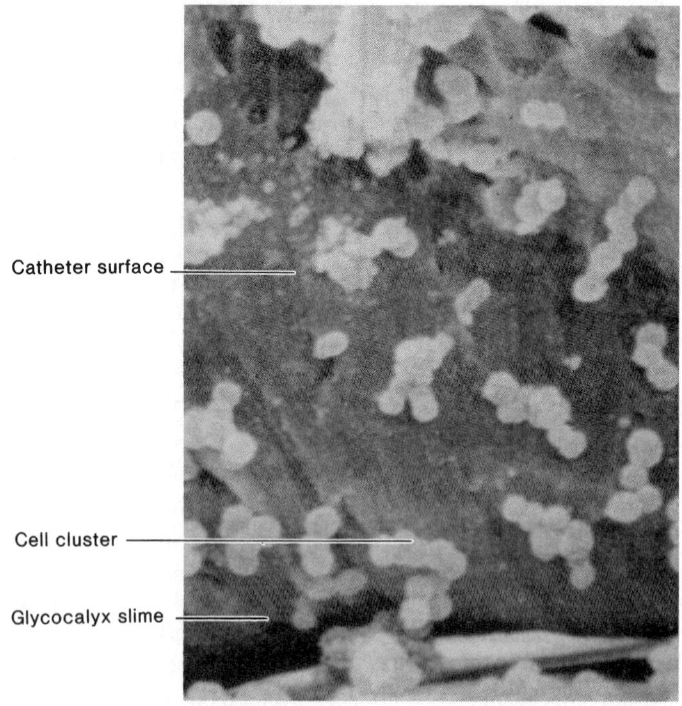

Figure 3.13 Scanning electron micrograph of *Staphylococcus aureus* attached to a catheter.

The Rigid Structure of the Cell Wall

Immediately below the glycocalyx lies a second layer, the **cell wall.** This structure accounts for a number of important bacterial characteristics. In general, it determines the shape of a bacterium, and it also provides the kind of strong structural

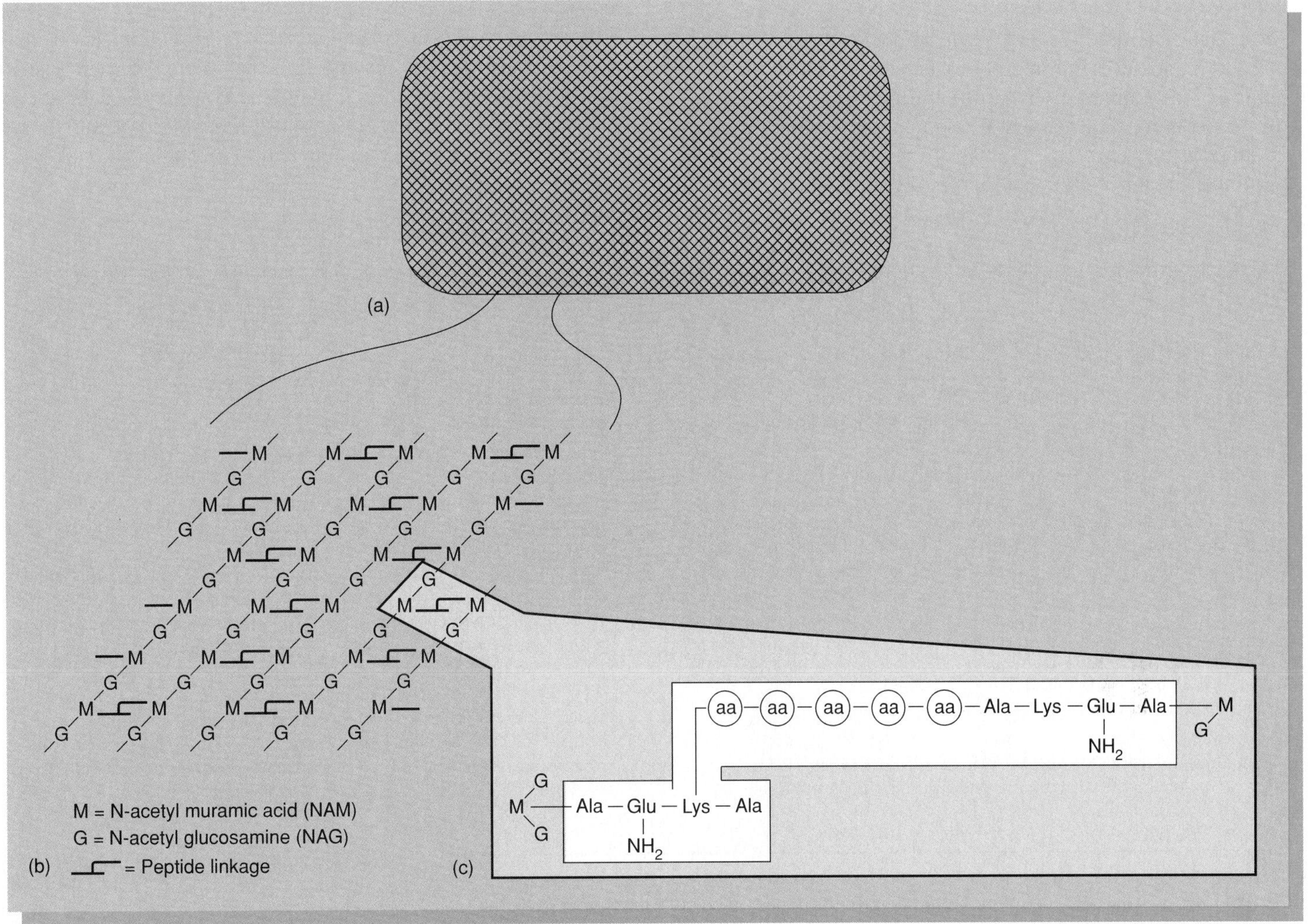

Figure 3.14 General structure of the peptidoglycan component of the cell wall. (*a*) An artist's interpretation of the peptidoglycan meshwork (not to scale). Such is the union of the molecules that the wall is actually one continuous and sometimes many-layered molecule that surrounds the cell like a finely woven basket. (*b*) The structure at the molecular level consists of a backbone of alternating glycan molecules (N-acetyl muramic acid [NAM] and N-acetyl glucosamine [NAG]). These chains are bound together at adjacent NAMs into a meshwork by means of a segment consisting of a short chain of amino acids (peptide). This sort of interlinking pattern confers rigidity yet flexibility. (*c*) The general molecular appearance of a peptide cross-bridge. Its precise structure varies among various groups.

support necessary to keep a bacterium from bursting or collapsing due to changes in pressure. In this way, the cell wall functions like a bicycle tire that maintains the necessary shape and prevents the more delicate innertube from bursting when it is filled with air. The relatively rigid, protective quality of the wall is due to a unique macromolecule called **peptidoglycan** (PG). This compound is a giant polymeric molecule composed of a repeating framework of long *glycan* chains cross-linked by short peptide fragments (figure 3.14). As we will see, peptidoglycan is not the only substance in cell walls, and its amount varies among the general bacterial groups.

Biomedical Significance of Peptidoglycan in the Cell Wall Because many bacteria live in habitats that contain a considerable amount of free water, they are constantly taking in excess water by osmosis. If it were not for the strength and relative rigidity of the peptidoglycan in the cell wall, this would soon exert a pressure capable of rupturing them. Understanding the function of this "weak link" has been a tremendous boon to the drug industry. Several types of drugs to treat infection (penicillin, cephalosporins) are effective because they prevent the peptidoglycan layer of bacteria from forming normally. With their cell walls incomplete or missing, such cells have very little protection from *lysis*. Some disinfectants (alcohol, detergents) also kill bacterial cells by damaging the cell wall. Lysozyme, a natural defense contained in tears and saliva, helps destroy certain bacteria by dissolving peptidoglycan.

lysis (ly'-sis) Gr. to loosen. A process of cell destruction, as occurs in bursting.

Differences in Cell Wall Structure

More than a hundred years ago, long before the detailed anatomy of bacteria could even remotely be known, a Danish physician named Hans Christian Gram developed a staining technique, the **Gram stain,** that delineates two generally different groups of bacteria (see feature 3.1). Though the reasons were not determined for many years, the contrasting staining reactions that occur with this stain are due entirely to some very fundamental differences in the morphology of bacterial cell walls. We call the two major groups shown by this technique the **gram-positive** and **gram-negative** bacteria. Since the Gram stain does not actually reveal the nature of these physical differences, we must turn to the electron microscope and to biochemical analysis.

Feature 3.1 The Gram Stain—A *Grand* Stain

In 1884, Hans Christian Gram discovered a staining technique that could be used to make bacteria in infectious specimens more visible. His technique consisted of a timed, sequential application of crystal violet (the primary dye), Gram's iodine (IKI, the mordant), an alcohol rinse (decolorizer), and safranin (the counterstain). In the finished product, bacteria that stained purple were called gram positive and those that stained red were called gram negative.

Although these staining reactions involve an attraction of the cell to a charged dye (see chapter 2), it is important to note that the terms gram "positive" and "negative" are used to indicate not the electrical charge of cells or dyes but whether or not a cell retains the primary dye-iodine complex after decolorization. There is nothing specific in the reaction of gram-positive cells to the primary dye or in the reaction of gram-negative cells to the counterstain. We now know that the chemical theory of the Gram stain involves differences in the structure of the cell wall and how it reacts to the sequence of reagents. In the first step, crystal violet stains all cells in a sample the same purple color. The addition of the mordant (intensifier) is a key differentiating step of the stain, because the mordant, Gram's iodine, forms large crystals with the dye, which are deposited in the peptidoglycan mesh-work of the cell wall. Because of the larger peptidoglycan layer in gram-positive cells, this reaction is far more extensive in gram-positive cells than in gram-negative ones. Furthermore, when the alcohol decolorizer is applied, the alcohol dissolves lipids in the outer membrane of gram-negative cells, permitting the dye to be removed from the thin cell wall. By contrast, the crystals of dye are trapped in the cell walls of gram-positive bacteria, where they are relatively inaccessible and resistant to removal. Because gram-negative cells will be colorless after the alcohol step, their presence is demonstrated by applying the counterstain safranin.

This century-old method remains the universal basis for bacterial taxonomy and identification. It permits differentiation of four major categories based upon color reaction and shape: gram-positive rods, gram-positive cocci, gram-negative rods, or gram-negative cocci (see table 3.5). Besides its application in classifying bacterial cells, the Gram stain is a practical aid in diagnosing infection and in guiding drug treatment. For example, gram staining a fresh urine or throat specimen may help pinpoint the possible cause of infection, and it may be possible to begin drug therapy on the basis of this stain. Even in this day of elaborate and expensive medical technology, the Gram stain remains an important and unbeatable first tool in diagnosis.

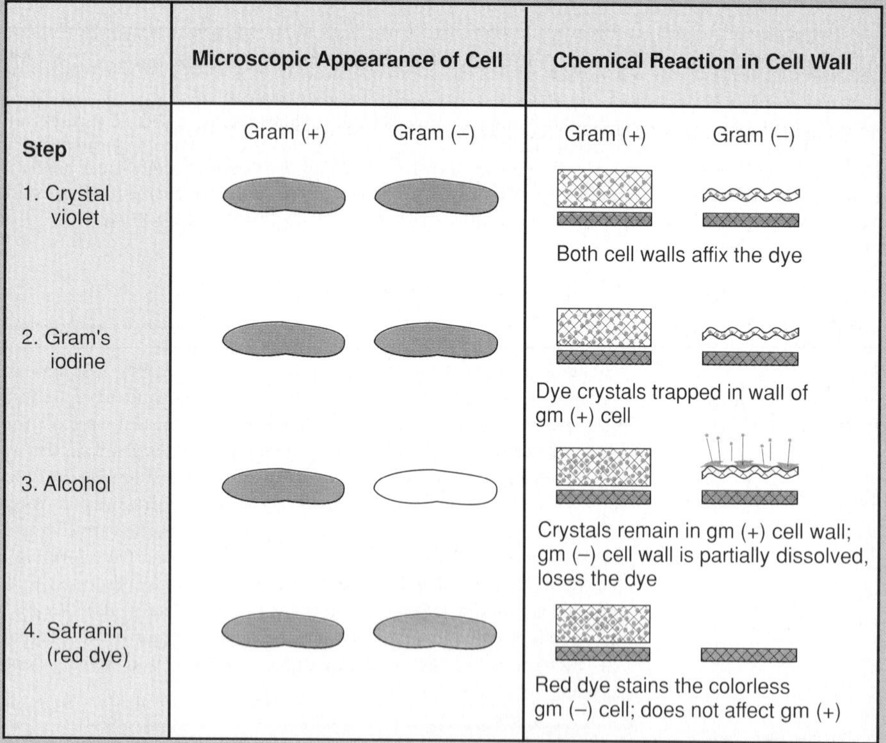

Step	Microscopic Appearance of Cell: Gram (+)	Microscopic Appearance of Cell: Gram (–)	Chemical Reaction in Cell Wall: Gram (+)	Chemical Reaction in Cell Wall: Gram (–)
1. Crystal violet			Both cell walls affix the dye	
2. Gram's iodine			Dye crystals trapped in wall of gm (+) cell	
3. Alcohol			Crystals remain in gm (+) cell wall; gm (–) cell wall is partially dissolved, loses the dye	
4. Safranin (red dye)			Red dye stains the colorless gm (–) cell; does not affect gm (+)	

Gram stain technique and theory.

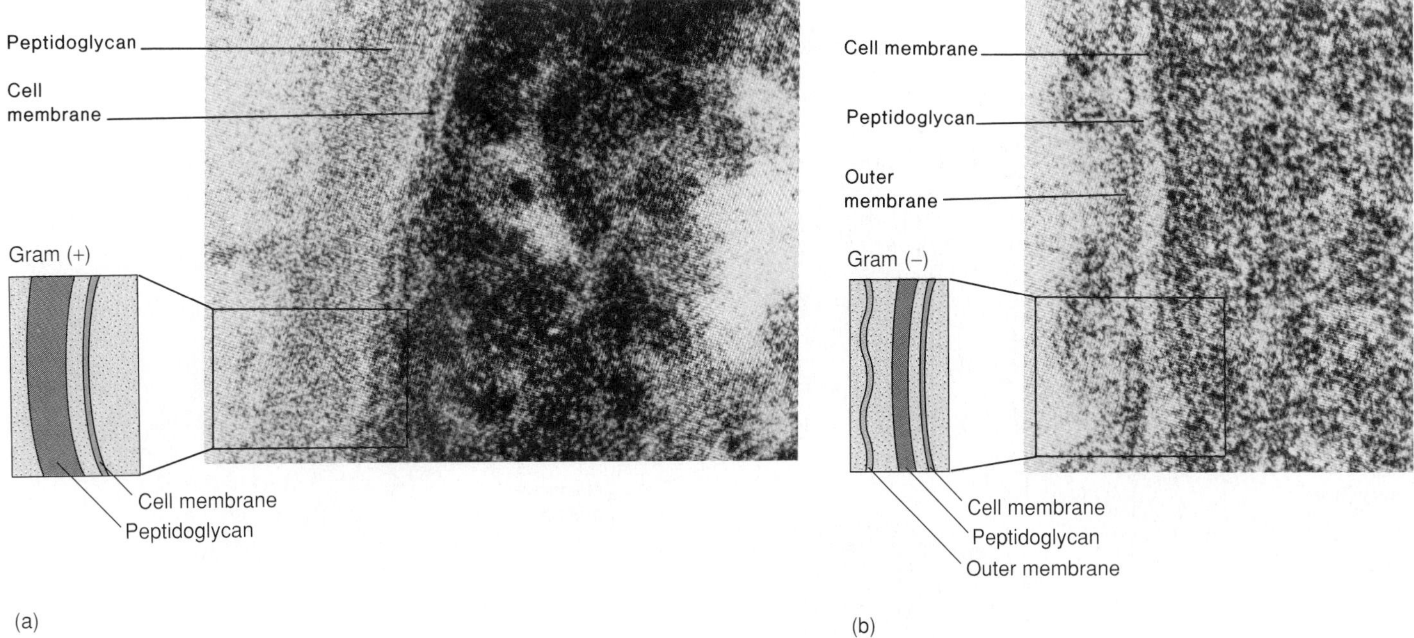

Figure 3.15 A comparison of the envelopes of gram-positive and gram-negative cells. (*a*) A photomicrograph of a gram-positive cell wall/membrane and an artist's interpretation of its sandwich-style layering. (*b*) A photomicrograph of a gram-negative cell wall/membrane and an artist's interpretation of its sandwich-style layering.

The extent of the differences between gram-positive and gram-negative bacteria is evident in the physical appearance of their cell envelopes (figure 3.15). In gram-positive walls, a microscopic section resembles an open-faced sandwich with two layers: the thick outer cell wall, composed primarily of peptidoglycan, and the cell membrane. A similar section of a gram-negative cell envelope shows a complete sandwich with three layers: the wall, composed of an outer membrane and a thin layer of peptidoglycan, and the cell membrane. With this structure in mind, let us examine the cell walls of the two cell types.

The Gram-Positive Cell Wall Though the bulk of the gram-positive cell wall is a thick sheath composed of numerous sheets of peptidoglycan, it also contains tightly bound acidic polysaccharides, including techoic acid and lipotechoic acid (figure 3.16). The function of techoic acid is in cell wall maintenance and enlargement during cell division, and it also contributes to the acidic charge on the cell surface. The cell wall of gram-positive bacteria is generally pressed tightly against the cell membrane with very little space between them.

The Gram-Negative Cell Wall The gram-negative cell wall is more complex in morphology because it contains an *outer membrane* (OM), a thin shell of peptidoglycan, and an extensive space between the PG and the cell membrane (figure 3.16). The outer membrane is somewhat similar in construction to the cell membrane, except that it contains additional types of lipids, polysaccharides, and proteins. The outer leaf of the OM consists of *lipopolysaccharide* (LPS), which is essentially polysaccharides integrated into a layer of lipid. The inner leaf of the OM consists of a lipid layer anchored by means of proteins to the peptidoglycan layer below. The outer membrane serves as a partial chemical sieve by allowing only relatively small molecules to penetrate. Access is provided by special membrane channels, the *porin proteins,* that span completely across the outer membrane. The size of these porins can be altered so as to block the entrance of chemicals, making them one defense of gram-negative bacteria against certain antibiotics (see chapter 10).

The bottom layer of the gram-negative wall is a single sheet of peptidoglycan. Though it acts as a somewhat rigid protective structure as previously described, its thinness gives gram-negative bacteria a relatively greater flexibility and sensitivity to lysis. Between the peptidoglycan and the cell membrane is a well-developed region called the *periplasmic space.* This space is an important receptacle and reaction site for a large and varied pool of substances that are entering and leaving the cell. It contains enzymes that are involved in digesting molecules entering the cell and special binding proteins that function in cell transport (see chapter 6).

---

periplasmic (per″-ih-plaz′-mik) Gr. *peri,* around, and *plasm,* molded.

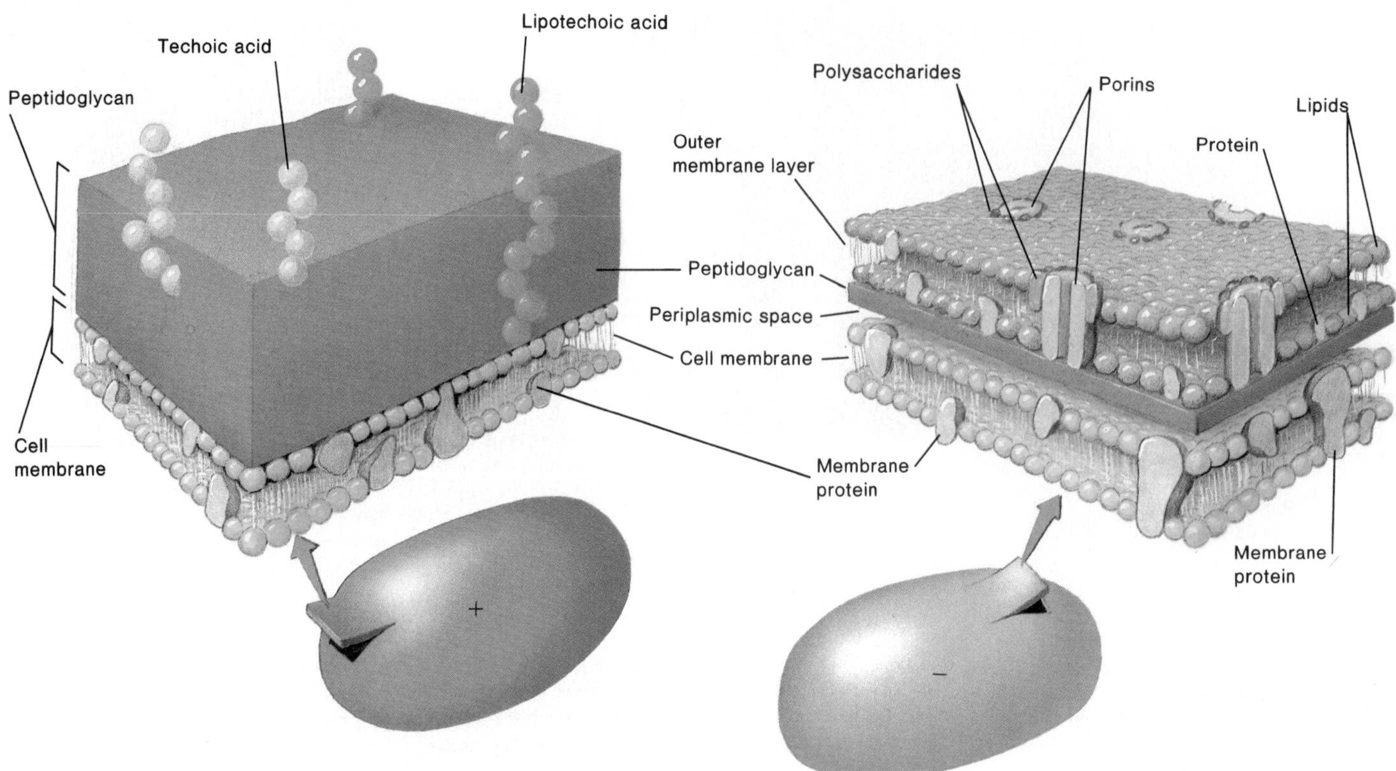

Figure 3.16 A comparison of the detailed structure of gram-positive and gram-negative cell walls.

Biomedical Significance of Differences in the Cell Walls The anatomy of the two gram reaction types contributes to other differences besides just the staining reaction. We have mentioned how the outer membrane contributes an extra barrier in gram-negative forms to slow or stop the entry of some antimicrobic chemicals. This imperviousness makes gram-negative bacteria generally more difficult to inhibit or kill than gram-positive ones. Treating infections by gram-negative bacteria often requires different drugs and more aggressive therapy than treating gram-positive ones. Gram-positive bacteria are also more sensitive to dyes and some types of disinfectants.

The cell wall or its parts can interact with human tissues and contribute to disease. Lipopolysaccharides in the outer membrane of gram-negative cells are a factor in some diseases due to their toxicity in humans and other mammals (see chapter 11). Proteins attached to the outer portion of the cell wall of several gram-positive species, including *Corynebacterium diphtheriae* (the agent of diphtheria) and *Staphylococcus aureus* (the cause of boils), also have toxic properties. The lipids in the cell walls of certain *Mycobacterium* spp. (notation for several different species) are harmful to human cells as well. Because most macromolecules in the cell walls are foreign to humans, they also stimulate the immune system to produce antibodies.

Exceptions in the Cell Wall Several bacterial groups lack the cell wall structure of gram-positive or gram-negative bacteria, and some bacteria have no cell wall at all. Although these exceptional forms may stain positive or negative in the Gram stain, examination of their fine structure and chemistry shows that they do not really fit the descriptions for typical gram-negative or -positive cells. For example, the cells of *Mycobacterium* and *Nocardia* contain peptidoglycan and stain gram positive, but the bulk of their cell wall is comprised of unique types of lipids. One of these is a very long chain fatty acid called *mycolic acid,* or cord factor, that contributes to the pathogenicity of this group (see chapter 16). The thick, waxy nature imparted to the cell wall by these lipids is also responsible for a high degree of resistance to certain chemicals and dyes. Such resistance is the basis for the *acid-fast stain* used to diagnose tuberculosis and leprosy. In this stain, hot carbol fuchsin dye becomes tenaciously attached (is held fast) to these cells so that an acid alcohol solution will not remove the dye (see chapter 16).

An unusual group of bacteria called the archaebacteria also have chemically distinct and complex cell walls. In some, the walls are composed almost entirely of polysaccharides, and in others, the walls are pure protein, but as a group, they all lack the true peptidoglycan structure described previously. Since a few archaebacteria and all mycoplasmas (discussed in a later section of this chapter) lack a cell wall entirely, their cell membrane must serve the function of protection as well as transport. These forms have a cell membrane stabilized and reinforced by special types of lipids.

Some bacteria that ordinarily have a cell wall may lose it during part of their life cycle. These wall-deficient forms are referred to as **L forms** or L-phase variants (for the Lister In-

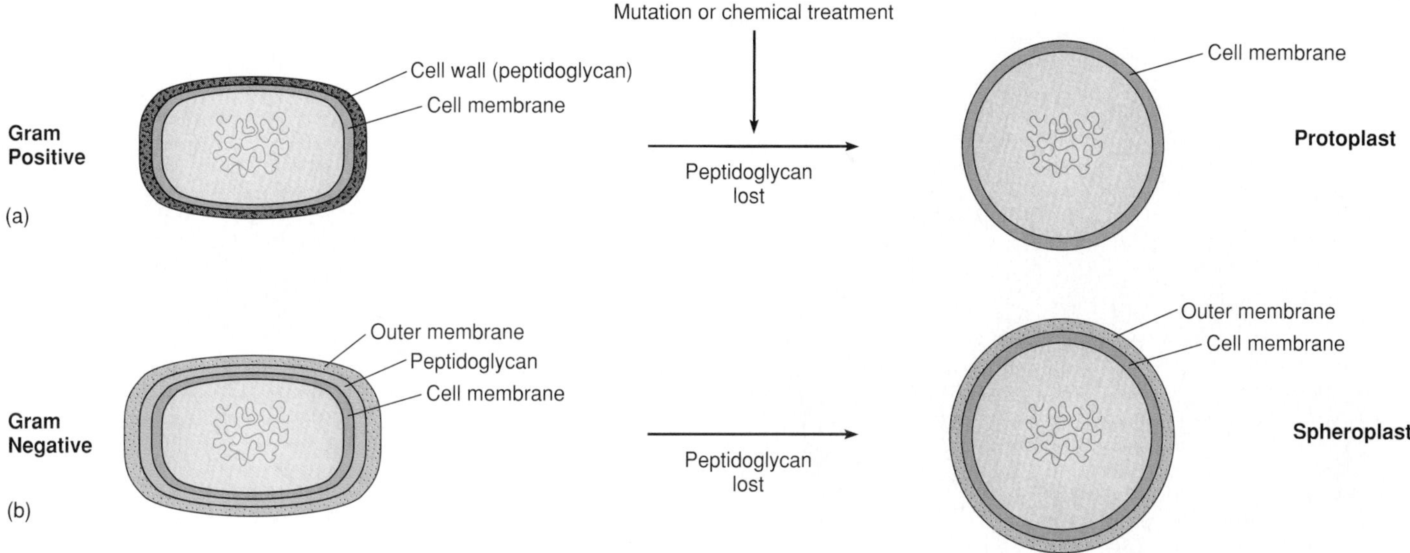

Figure 3.17 The conversion of walled bacterial cells to L forms.

stitute where they were discovered). L forms arise naturally from a mutation in the wall-forming genes, or they may be induced artificially by treatment with a chemical such as lysozyme or penicillin that disrupts the cell wall. When a gram-positive cell is exposed to either of these two chemicals, it will lose the cell wall completely and become a *protoplast,* a fragile cell bounded only by a membrane that is highly subject to lysis (figure 3.17*a*). A gram-negative cell exposed to these same substances loses its peptidoglycan but retains its outer membrane, leaving a less fragile, but nevertheless weakened, *spheroplast* (figure 3.17*b*). Evidence points to a role for L forms in certain infections (for example, cat-scratch disease; see chapter 17).

The Cell Membrane

The next layer to appear just beneath the cell wall is the cell membrane, a very thin (4–5 nm) flexible sheet completely surrounding the cell's internal contents. Its general composition was described in chapter 2 as a lipid bilayer with proteins embedded to varying degrees (see feature 2.4). Bacterial cell membranes have this typical structure, containing primarily phospholipids (making up about 30–40% of the membrane) and proteins (contributing 60–70%). Major exceptions to this description are mycoplasmal membranes that contain a high level of lipids called sterols, rigid molecules that function in membrane stabilization, and archaebacterial membranes, which contain unique fatty-acid free lipids.

In some locations the cell membrane folds up into fingerlike passages or pockets in the cytoplasm called **mesosomes** (see figure 3.1). These are prominent in gram-positive bacteria but harder to see in gram-negative bacteria because of their relatively small size. Mesosomes presumably increase the internal surface area available for membrane functions, although their actual functions continue to be a controversial topic.

Functions of the Cell Membrane and Mesosomes Since bacteria have no organelles, the cell membrane provides numerous functions that relate to transport, energy extraction, nutrient processing, and synthesis. Though the glycocalyx and cell wall may bar the passage of large molecules, regulating *transport* is not their primary function in procaryotes. It is the cell membrane that controls passage of nutrients into the cell and discharge of wastes. The cell membrane is also involved in *secretion,* or the discharge of a metabolic product into the extracellular environment. Although water and small solute molecules can diffuse across the membrane unaided, most substances will be handled by special carrier mechanisms (see chapter 6).

The membranes of procaryotes are an important site for a number of metabolic activities. It is here that most enzymes of respiration and other energy-processing activities are located (see chapter 7). Enzyme systems located in the cell membrane also synthesize structural macromolecules to be incorporated into the cell envelope and appendages. Other products (enzymes and toxins) are secreted by the membrane into the extracellular environment. A proposed function of mesosomes appears to be in guiding the duplicated chromatin bodies (see the following section) into the two daughter cells during cell division (see figure 6.15).

The Protoplasm: Internal Contents of the Cell

Within the bacterial envelope is a dense, gelatinous solution called **protoplasm,** which is a prominent site for the cell's biochemical and synthetic activities. Its major component, water (70–80%), serves as a solvent for a complex mixture of nutrients, including sugars, amino acids, and salts, called the

---

protoplast (proh'-toh-plast) Gr. *proto,* first, and *plastos,* formed.
spheroplast (sfer'-oh-plast) Gr. *sphaira,* sphere.
mesosome (mess'-oh-sohm) Gr. *mesos,* middle, and *soma,* body.

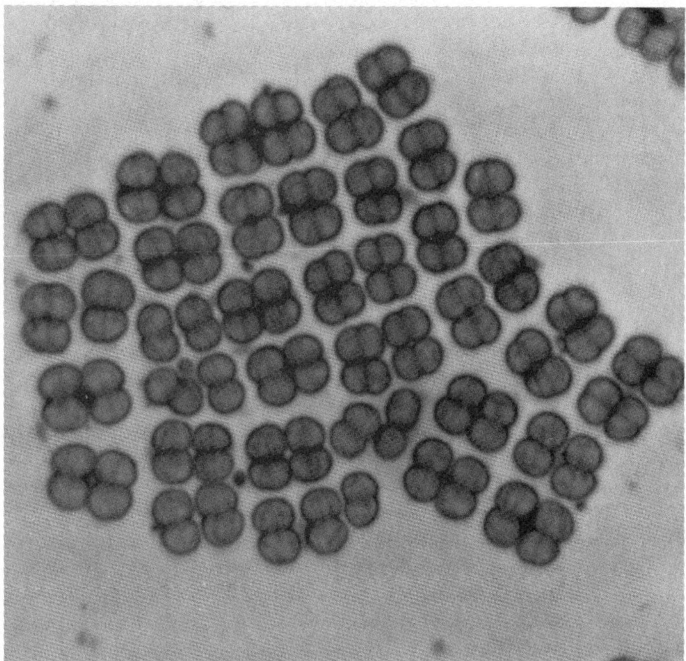

Figure 3.18 Chromatin bodies or nucleoids of *Deinococcus* are highlighted by a Giemsa stain. Note that this bacterium exists in a packet of four cells.

cell pool. The components of this pool serve as building blocks for cell synthesis or as sources of energy. The protoplasm also accommodates metabolic enzymes and larger discrete cell masses such as the chromatin body, ribosomes, mesosomes, and granules.

Chromatin Bodies / Plasmids

The hereditary material of bacteria exists in the form of a single circular strand of DNA designated as the **chromatin body** or bacterial chromosome (see figure 3.1). By definition, bacteria do not have a nucleus—that is, their DNA is not enclosed by a nuclear membrane, but instead is aggregated in a dense area of the cell called the **nucleoid.** The chromatin body is actually an extremely long molecule of DNA that is tightly coiled around special basic protein molecules so as to fit inside the cell compartment (see feature 8.3). Arranged along its length are genetic units (genes) that carry information required for bacterial maintenance and growth. When exposed to special stains or observed with an electron microscope, chromatin bodies have a granular or fibrous appearance (figure 3.18).

Although the chromatin body is the minimal genetic requirement for bacterial survival, many bacteria contain other, more disposable pieces of DNA called **plasmids.** These tiny, circular extrachromosomal strands often confer protective traits upon bacteria, such as resisting drugs and radiation and producing toxins and enzymes (see chapter 8). Because they can be readily manipulated in the laboratory and transferred from one bacterial cell to another, plasmids are an important agent in modern genetic engineering techniques.

chromatin (kroh'-mah-tin) Gr. *chromato,* color, and *in,* fiber.

nucleoid (noo'-klee-oid) L. *nucis,* nut, and *oid,* form or like.

plasmid (plaz'-mid) Gr. *plasma,* something molded, and *id,* connected with.

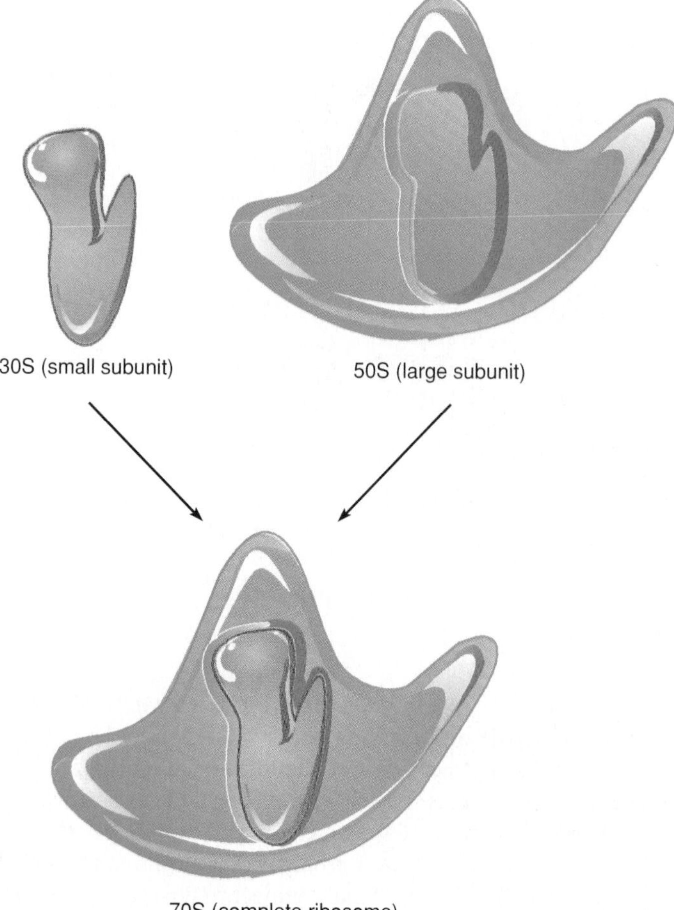

Figure 3.19 A model of a procaryotic ribosome, showing the small (30S) and large (50S) subunits, both separate and joined.

Ribosomes

A bacterial cell contains thousands of tiny discrete units, the **ribosomes.** When viewed even by very high magnification, ribosomes show up as fine, spherical specks dispersed through the cytoplasm that often occur in chains (polysomes). They are also attached to the cell membrane and mesosomes. Chemically, a ribosome is a combination of a special type of RNA called ribosomal or rRNA (about 60%) and protein (40%). One method of characterizing ribosomes is by S, or Svedberg[2] units, which rate the molecular sizes of various cell parts that have been spun down and separated by weight in a centrifuge. Heavier structures are assigned a higher S rating. Combining this method of analysis with very high resolution electron micrography has revealed that the procaryotic ribosome, which has an overall rating of 70S, is actually composed of two smaller subunits. The so-called 30S unit looks something like a heart; the 50S unit bears a resemblance to a crown (figure 3.19). They fit together to form a miniature platform upon which protein synthesis is performed. We will examine the more detailed functions of ribosomes in chapter 8.

ribosome (rye'-boh-sohm) Gr. *ribose,* a pentose sugar, and *some,* body.

2. Named in honor of T. Svedberg, the Swedish chemist who developed the ultracentrifuge.

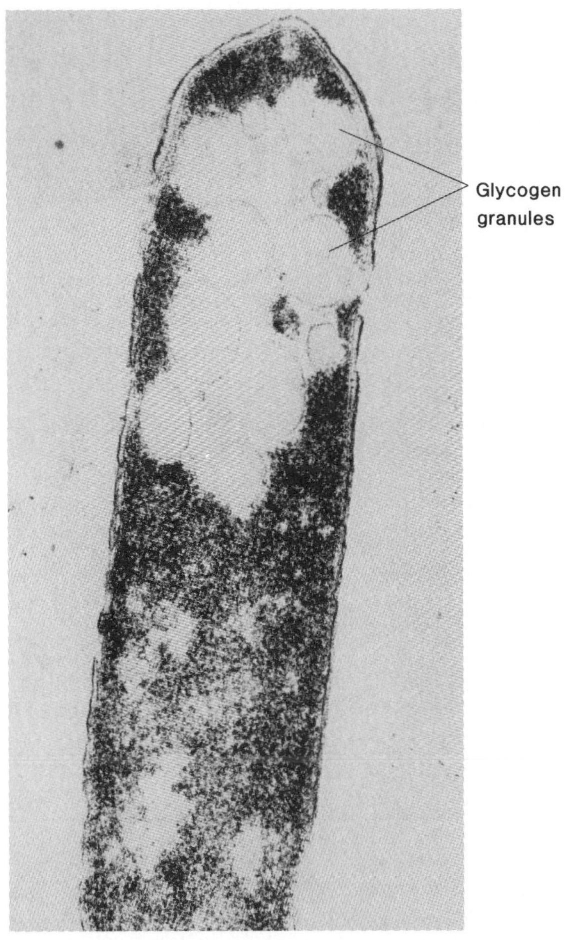

Figure 3.20 An example of a storage inclusion in a bacterial cell. Substances like glycogen may be stored in an insoluble, concentrated form that provides ample, long-term supply of that nutrient.

Granules or Inclusions

Most bacteria are exposed to severe shifts in the availability of food. To compensate for this, during periods of nutrient abundance, they may concentrate nutrients intracellularly in **inclusions** or **granules** of varying size, number, and content. As the environmental source of these nutrients is later depleted, the bacterial cell can gradually mobilize the stored source as required. Inclusions are composed of condensed, energy-rich organic substances, including glycogen and poly β-hydroxybutyrate (PHB), enclosed within special membranes (figure 3.20). Granules usually contain crystals of inorganic compounds and are not enclosed by membranes. Examples are the sulfur granules of photosynthetic bacteria (see figure 3.33) and the polyphosphate granules of *Corynebacterium* (see figure 3.22) and *Mycobacterium.* These are often termed **metachromatic** granules because they stain a contrasting color (red, purple) in the presence of certain blue dyes such as methylene blue. A unique type of inclusion found in some aquatic bacteria are gas vesicles that provide buoyancy and flotation.

inclusion (in-kloo'-zhun). Any substance held in the cell cytoplasm in an insoluble state.

metachromatic (met″-uh-kroh-mah'-tik) Gr. *meta,* other, and *chromo,* color.

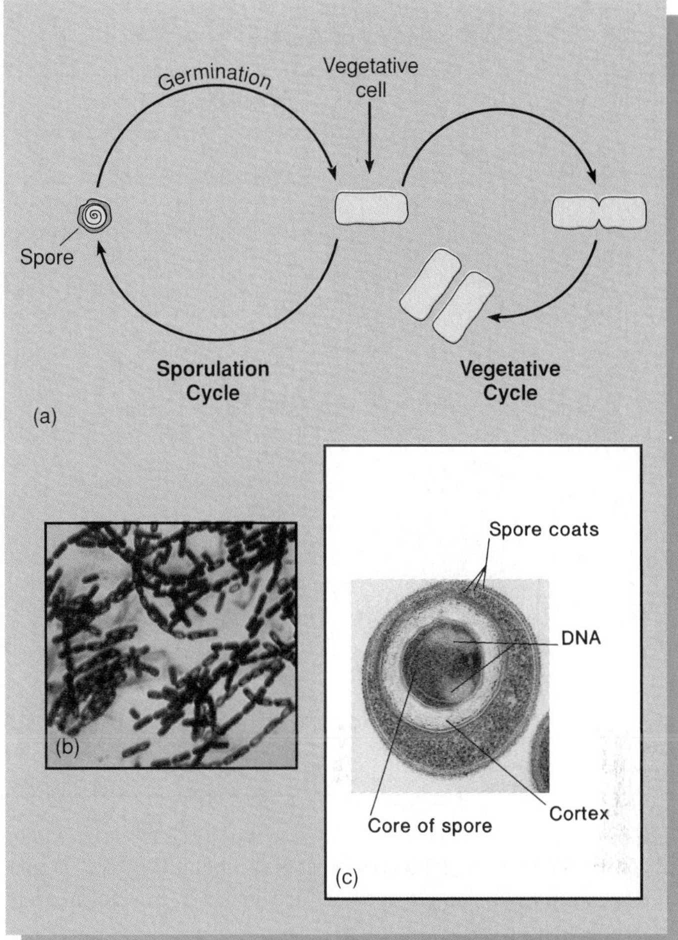

Figure 3.21 (*a*) The general life cycle of a spore-forming bacterium. Sporeformers can exist in an active, growing, vegetative state, or they can enter a dormant survival state through sporulation. Their actual fate depends upon the availability of nutrients. (*b*) A Gram stain of *Clostridium,* with central holes showing the position of spores. (*c*) A cross section of a single spore, revealing numerous layers.

Bacterial Endospores: Resistance in the Extreme

Ample evidence indicates that the anatomy of bacteria helps them adjust rather well to adverse habitats. But of all microbial structures, nothing can compare to the bacterial **endospore** (or simply spore) for withstanding hostile conditions and facilitating survival. Spores are dormant structures produced by members of the gram-positive genera *Bacillus* and *Clostridium* and several other genera. Spore formation, also called **sporulation** or sporogenesis, breaks the natural life cycle of these soil-inhabiting bacteria into two phases, termed the **vegetative** and **sporulative** stages. While the vegetative cell is a metabolically active and growing entity, the spore exists in an inert, resting condition (figure 3.21*a*). Spores can be readily shown with a spore stain or Gram stain (figure 3.21*b*). Because of their *refractility,* they show up well in a fresh hanging drop

endospore (en'-doh-spor) Gr. *endo,* inside, and *sporos,* seed.

refractility (ree-frak-til'-ih-tee). A property of causing light rays to bend, thus creating differences in contrast.

Table 3.1 General Stages in Endospore Formation

Stage	State of Cell	Process/Event
1	Vegetative cell	Cell in early stage of binary fission doubles chromatin body.
2	Vegetative cell becomes **sporangium** in preparation for sporulation	One chromatin body and a small bit of cytoplasm are walled off by a septum at one end of the cell. This **protoplast** or **core** of the spore contains the minimum structures and chemicals necessary for guiding life processes. During this time, the sporangium is actively synthesizing compounds required for spore formation.
3	Sporangium	The protoplast is engulfed by the sporangium to facilitate the formation of various protective layers around it.
4	Sporangium with **prospore**	Special peptidoglycan is laid down to form a cortex around the spore protoplast, now called the prospore; calcium and dipicolinic acid are deposited; core becomes dehydrated and metabolically inactive.
5	Sporangium with prospore	Three heavy and impervious protein spore coats are added.
6	Mature spore	Spore becomes thicker and heat resistance is complete; sporangium is no longer functional and begins to deteriorate.
7	Free spore	Complete loss of sporangium frees spore; it can remain dormant yet viable for thousands of years.

preparation as well. Features of spores, including size, shape, and position in the vegetative cell, are somewhat useful in identifying some species.

A Note on Terminology The term spore may have more than one usage in microbiology. It is a generic term that refers to any tiny cell-like bodies that are produced by vegetative or reproductive structures of microorganisms. Spores can be quite variable in origin, form, and function. The bacterial type is an endospore because it is produced inside a cell. It functions in survival, not reproduction, because neither cell division nor an increase in cell numbers is involved in its formation. In contrast, the fungi produce many different types of spores for both survival and reproduction (see chapter 4).

Endospore Formation and Resistance

The depletion of nutrients, especially an adequate carbon or nitrogen source, is the stimulus for a vegetative cell to begin spore formation. Once this stimulus has been received by the vegetative cell, it undergoes a conversion to a committed sporulating cell called a *sporangium.* Complete transformation of a vegetative cell into a sporangium and then into a spore requires 6 to 8 hours in most spore-forming species. Table 3.1 illustrates some major physical and chemical events in this process.

Bacterial endospores are the hardiest of all life forms, capable of withstanding extremes in heat, drying, freezing, radiation, and chemicals that would readily kill vegetative cells. Their survival under such harsh conditions is due to several factors. The heat resistance of spores has been linked to their high content of calcium and *dipicolinic acid,* though the exact role of these chemicals is not yet clear. We know, for instance, that heat destroys cells by inactivating proteins and DNA and that this process requires a certain amount of water in the protoplasm. Since the deposition of calcium dipicolinate in the spore removes water and leaves the spore very dehydrated, it will be less vulnerable to the effects of heat. Moreover, since the spore is already very water-free, it is also metabolically inactive and highly resistant to damage from further drying. Important contributions to radiation and chemical resistance also come from the thick, impervious cortex and spore coats (figure 3.21*b*). The longevity of spores is well documented. One record describes the isolation of viable spores from a 3,000-year-old archaeological specimen.

The Germination of Spores

After lying in a state of inactivity for a varying period of time, spores can be revitalized when favorable conditions arise. The breaking of dormancy, called germination, happens in the presence of water and a specific chemical or environmental stimulus (germination agent). Once initiated, it proceeds to completion quite rapidly (1½ hours). While the specific germination agent varies among species, it is generally a small organic molecule such as an amino acid or an inorganic salt. This agent stimulates the formation of hydrolytic (digestive) enzymes by the spore membranes. These enzymes digest the cortex and expose the core to water. As the core rehydrates and takes up nutrients, it begins to grow out of the spore coats. In time, it reverts to a fully active vegetative cell, resuming the vegetative cycle.

Biomedical Significance of Bacterial Spores Although most spore-forming bacteria are relatively harmless inhabitants of the soil, several bacterial pathogens are sporeformers. In fact, some aspects of the diseases they cause are related to the persistence and resistance of their spores. *Bacillus anthracis* is the agent of anthrax, a skin and lung infection of domestic animals transmissible to humans. The genus *Clostridium* includes even more pathogens, including *C. tetani,* the cause of tetanus (lockjaw), and *C. perfringens,* the cause of gas gangrene. When the spores of these species are embedded in a wound that contains dead tissue, they can germinate, grow, and release potent toxins. Another toxin-forming species, *C. botulinum,* is the agent of that deadly food poisoning, botulism.

Spores can be a constant problem in microbial control. Because they are in the soil and dust, they are easily carried into human habitats. Likewise, they can resist ordinary cleaning methods that use boiling water, soaps, and disinfectants. They present problems in the microbiology lab as frequent contaminants of cultures and media. Hospitals and clinics must take precautions to guard against the potential harmful effects of spores in wounds. Spore destruction is a particular concern of the food canning industry, since germinated spores can both spoil and poison food. As we will see in chapter 9, vigorous methods must be used to destroy them.

The Scope of Bacterial Types

For the most part, bacteria function as independent single-celled, or **unicellular,** organisms. While it is true that an individual bacterial cell may live attached to others in colonies or other such groupings, each one is fully capable of carrying out all necessary life activities, such as reproduction, metabolism, and nutrient processing (unlike the more specialized cells of a multicellular organism).

Shapes, Arrangements, and Sizes

Though a survey of the bacterial kingdom reveals a considerable variety in shape, size, and colonial arrangement, most bacteria can be described by one of three general shapes as dictated by the configuration of the cell wall (figure 3.22). These cells look rather two-dimensional and flat with traditional staining and microscope techniques, but they are seen to best advantage with a scanning electron microscope that emphasizes their striking three-dimensional nature (figure 3.23). If the cell wall is spherical or ball-shaped, the bacterium is described as a **coccus.** Cocci may be perfect spheres, but they may also be oval, bean-shaped, or even pointed. A cell that is cylindrical (longer than wide) is termed a **rod** or **bacillus.** There is also a genus named *Bacillus,* and in the early days of bacterial nomenclature, many rod-shaped bacteria were placed in this genus (see feature 3.2). As might be expected, rods too are quite varied in their actual form. Depending on the bacterial species, they may be blocky, spindle-shaped, round-ended, long and threadlike (filamentous), or even clubbed or drumstick-shaped. When a rod is short and plump, it is called a *coccobacillus,* and if gently curved, it is a **vibrio.**

An extreme variation on the curved or spiral-shaped cylinder is evident in the **spirillum,** a rigid helix, twisted twice or more along its axis (like a corkscrew). Another spiral cell mentioned earlier in conjunction with axial filaments is the **spirochete,** a more flexible form that resembles a spring or stretched Slinky. Refer to table 3.2 for a comparison of other features of the two helical bacterial forms.

It is rather common for cells of the same species to vary to some extent in shape and size. This phenomenon, called **pleomorphism** (figure 3.24*a*), is due to individual variations in cell wall structure caused by nutritional or slight hereditary differences. For example, though the cells of *Corynebacterium diphtheriae* are generally considered rod-shaped, in culture they display variations such as club-shaped, swollen, curved, filamentous, and coccoid. Pleomorphism reaches an extreme in the mycoplasmas, which entirely lack cell walls and so display extreme variations in shape (see figure 3.31).

Feature 3.2 A Note On Names

The scope of the bacterial world once seemed much simpler than it does now. During the mid to late 1800s, most known bacteria could be neatly covered by a very few generic names. Spherical bacteria were placed in the genera *Micrococcus, Streptococcus,* or *Staphylococcus;* and rod-shaped bacteria were assigned to either genus *Bacterium* or genus *Bacillus.* If bacteria had the shape of a curved rod, they were put in the genus *Vibrio,* and if they were spiral, they were placed in either genus *Spirillum* or genus *Spirochaeta.* Because very little was known then about the biochemical characteristics of these bacteria, and the Gram stain had not yet been applied as a general classification tool, their morphology was the primary method of identification. As a result, often bacteria that we now know are very different were placed in the same genus. Consider that *Escherichia coli* was once called *Bacterium coli, Pseudomonas aeruginosa* was *Bacterium aeruginosa,* and *Streptococcus lactis* was *Bacterium lactis,* even though the first two are rods and the third is a coccus.

Present classification schemes include over 1,000 genera, and bacterial identification is based on hundreds of characteristics. Though more than 5,000 species are presently known, new ones are discovered every year. The recent increase in new species is probably due in part to an improvement in identification techniques. Some bacteriologists feel that we have only scratched the surface, and that myriads of bacteria remain to be discovered.

You will notice that certain generic names appear capitalized and uncapitalized. For example, the word *Staphylococcus* as shown here refers to a particular genus of coccus-shaped bacteria with a particular set of characteristics. When the word appears uncapitalized, unitalicized, and in the plural (staphylococci), it is a general way of designating the members of that genus or of describing an arrangement or cell type. This same usage is true of several other genera, as in streptococci, micrococci, mycobacteria, pseudomonads, and clostridia. Likewise, *Bacillus* and *Spirillum* refer to genera, and bacillus and spirillum to shapes.

coccus (kok'-us) pl. cocci (kok'-sye) Gr. *kokkos,* berry.
bacillus (bah-sil'-lus) pl. bacilli (bah-sil'-eye) L. *bacill,* small staff or rod.

vibrio (vib'-ree-oh) L. *vibrare,* to shake.
spirillum (spye-ril'-em) pl. spirilla; L. *spira,* a coil.
pleomorphism (plee"-oh-mor'-fizm) Gr. *pleon,* more, and *morph,* form or shape.

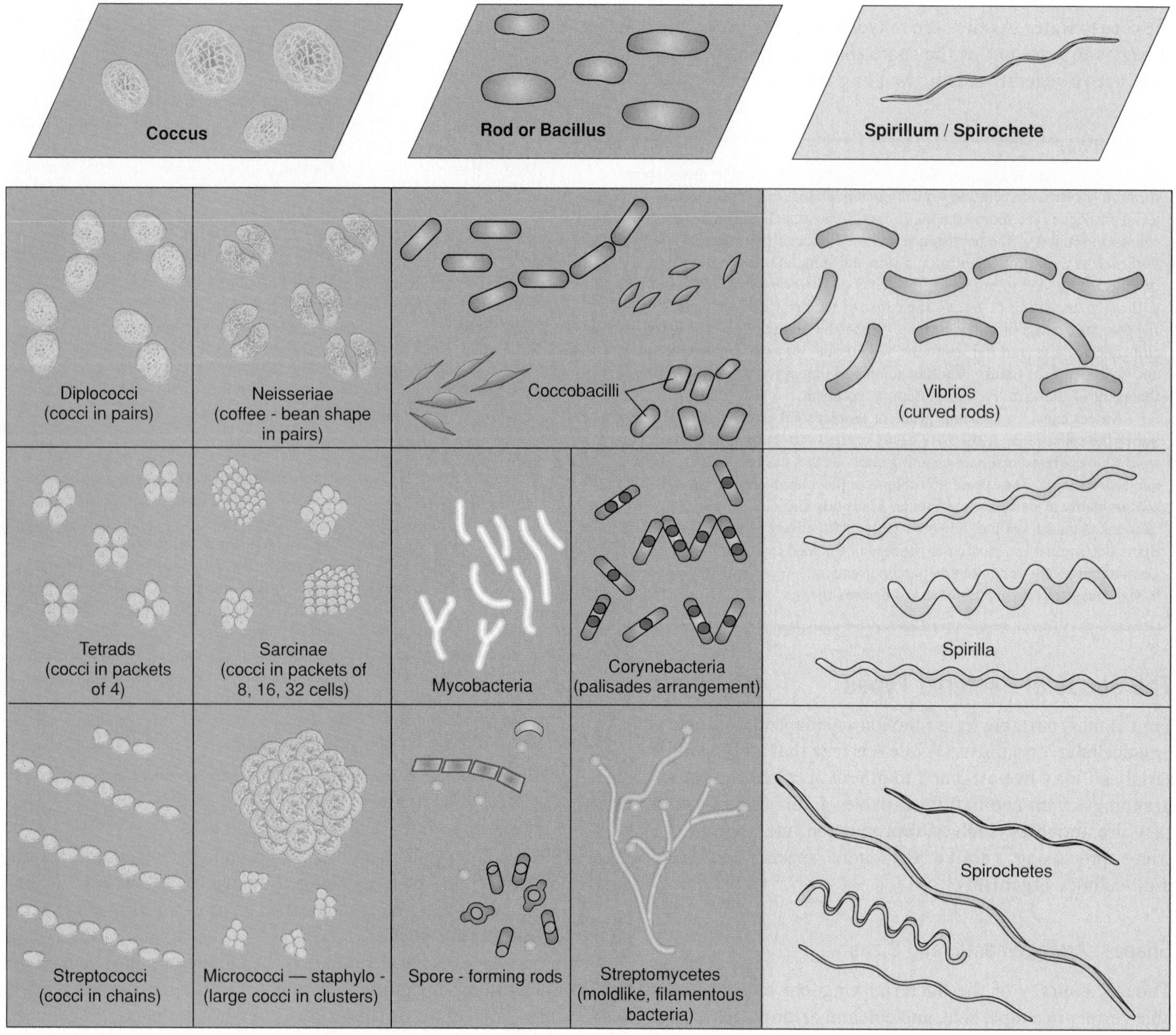

Figure 3.22 Bacteria arranged by basic shapes and arrangements, with specific examples of certain groups.

The cells of bacteria may also be categorized according to **arrangement,** or cell groupings (see figure 3.22). The main factors influencing the arrangement of a particular cell type are its pattern of division and whether the cells remain attached afterwards. The greatest variety in arrangement occurs in cocci, which may be single, paired (**diplococci**), in **tetrads** (groups of four), in irregular clusters (both **staphylococci** and **micrococci**), or in chains of four to hundreds of cells (**streptococci**). An even more complex grouping is a cubical packet of eight, sixteen, or more cells called a **sarcina** (sar′-sih-nah). These different coccal groupings are the result of a coccus dividing in either a single plane, two perpendicular planes, or several intersecting planes; after division, the resultant daughter cells remain attached (figure 3.25).

Bacilli are less varied in arrangement because they divide only in the transverse plane (perpendicular to the axis). They

diplococci; Gr. *diplo*, double.

staphylococci (staf″-ih-loh-kok′-sye) Gr. *staphyle*, a bunch of grapes.

micrococci; Gr. *mikros*, small.

streptococci (strep″-toh-kok′-sye) Gr. *streptos*, twisted.

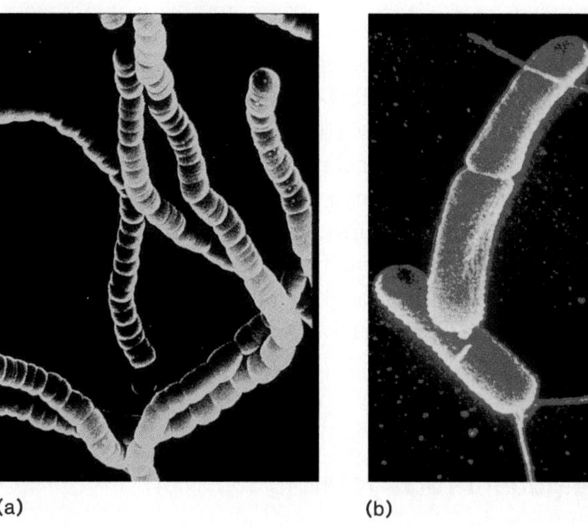

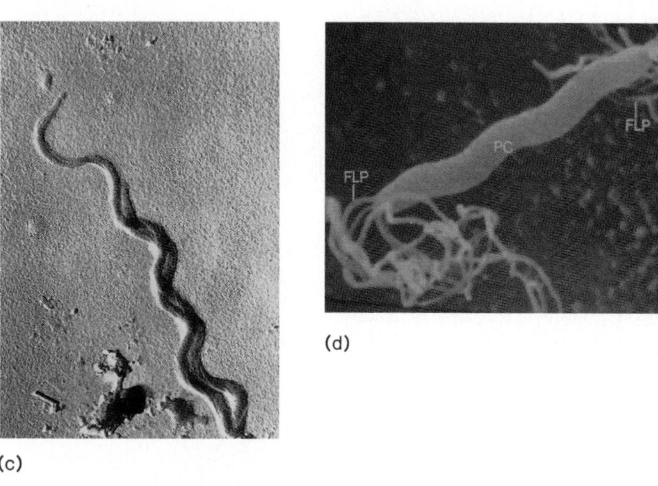

Figure 3.23 Basic bacterial shapes in three dimensions with surface features. (*a*) Cocci in chains. (*b*) A rod-shaped bacterium (*Escherichia coli*) with a convoluted surface. (*c*) A spirochete. Note the numerous coils. (*d*) A spirillum. Note the flagella (FLP) and spiral surface feature (PC).

Table 3.2 Comparison of the Two Spiral-Shaped Bacteria

	Spirilla	Spirochetes
Overall Appearance	Rigid helix	Flexible helix
	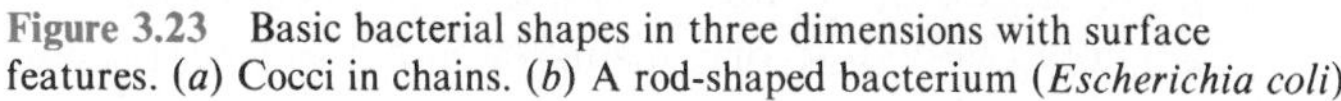	
Mode of Locomotion	Polar flagella—cells swim by rotating around like corkscrews; do not flex	Axial filaments within sheath; cells flex; can swim by rotation or by creeping on surfaces
	1 to several flagella; may be in tufts	2 to 100 axial filaments
Number of Helical Turns	Varies from ½ to 20	Varies from 3 to 70
Gram Reaction (Cell Wall Type)	Gram negative	Gram negative
Examples of Important Types	Most are harmless saprotrophs; one species, *Spirillum minor,* causes rat bite fever	*Treponema pallidum,* cause of syphilis; *Borrelia* and *Leptospira,* important pathogens

occur either as single cells or in some variation of an end-to-end grouping: diplobacilli (pairs) and streptobacilli (chains). A **palisades** arrangement, typical of the corynebacteria, is formed when the cells of a chain remain attached but only by a small hinge region at the ends. The cells tend to fold (snap) back upon each other, forming a row of cells oriented side by side (see figure 3.24*b*). The reaction can be compared to the behavior of boxcars on a jackknifed train, and the result looks something like

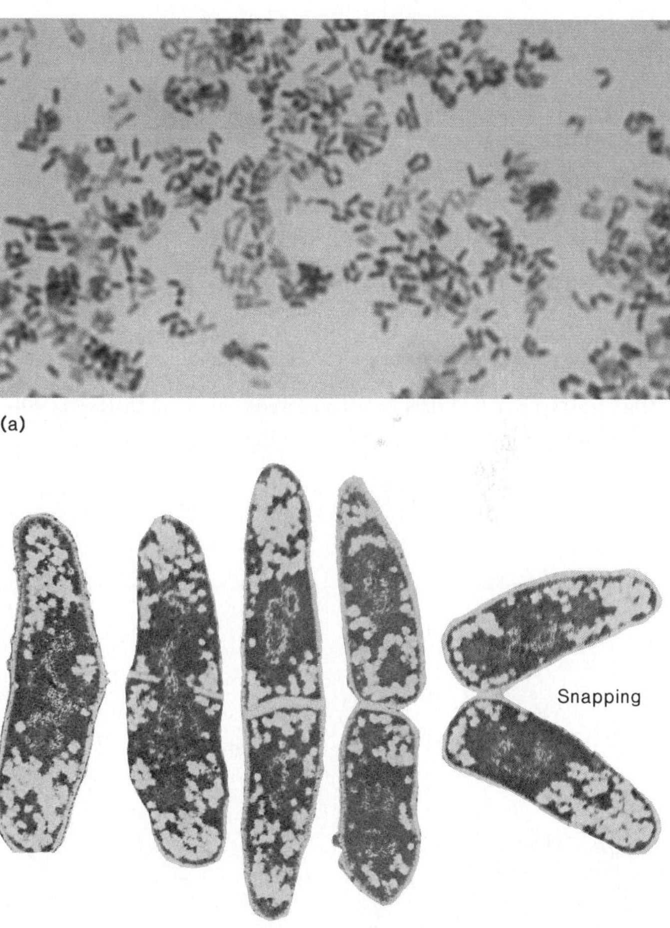

Figure 3.24 (*a*) Pleomorphism in *Corynebacterium.* Cells occur in a great variety of shapes and sizes. (*b*) This genus typically exhibits an unusual formation called a palisades arrangement that is caused by "snapping."

palisades (pal′-ih-saydz) L. *pale,* a stake. A fence made of a row of stakes.

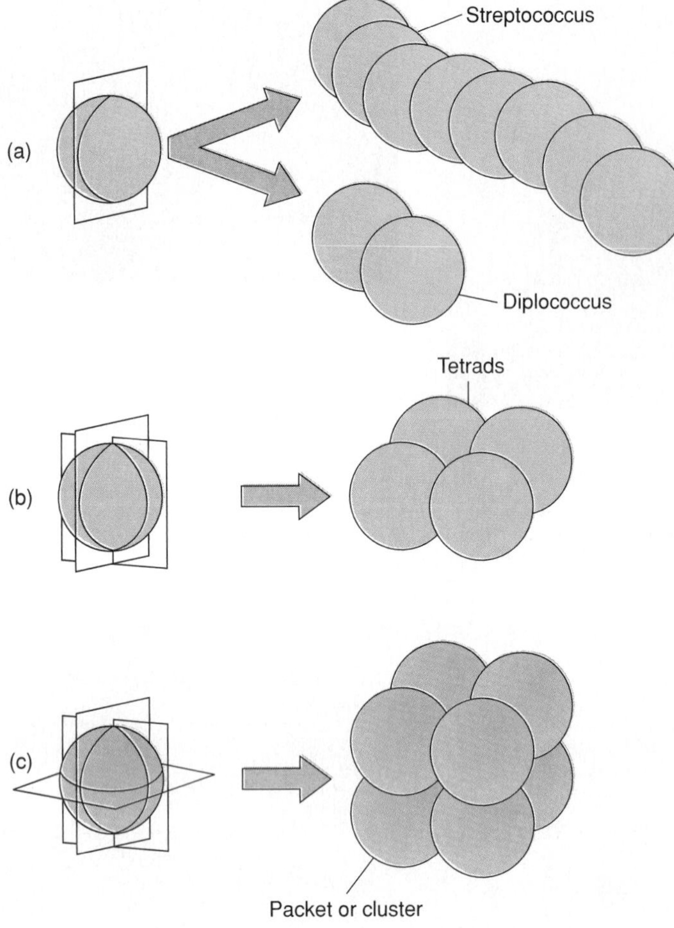

Figure 3.25 Arrangements of cocci resulting from different planes of cell division: (*a*) Division in one plane produces diplococci and streptococci. (*b*) Division in two planes at right angles produces a tetrad. (*c*) Division in three or more planes produces packets or clusters.

an irregular picket fence. Spirilla are occasionally found in short chains, but spirochetes rarely remain attached after division.

The dimensions of bacteria range from those just barely visible with light microscopy (0.2 μm) to those measuring a thousand times that size. Cocci measure anywhere from 0.5 to 3.0 μm in diameter; bacilli range from 0.2 to 2.0 μm in diameter and from 0.5 to 20 μm in length; vibrios and spirilla vary from 0.2 to 2.0 μm in diameter and from 0.5 to 100 μm in length. Spirochetes range from 0.1 to 3.0 μm in diameter and from 0.5 to 250 μm in length. Comparative sizes of typical cells are presented in figure 3.26.

Classification Systems in the Kingdom Procaryotae

Tracing the origins of bacteria has not been an easy task. As a rule, tiny, relatively soft organisms do not form fossils very readily. Several times within the past 30 years, however, scientists have discovered microscopic fossils of procaryotes that look very much like modern bacteria. Some of these rocks have been dated back billions of years (figure 3.27).

Bacterial Classification and Identification Systems

Classification systems serve both academic and practical purposes. While they aid in organizing microorganisms and studying their relationships and origins, they are also useful in differentiating and identifying unknown species in medical and applied microbiology. Since the classification of bacteria was started around 200 years ago, several thousand species have been identified, named, and catalogued (see feature 3.2). Early bacteriologists found it convenient to classify bacteria according to shape, variations in arrangement, growth characteristics, and habitat. As more species were discovered and as techniques for studying their biochemistry were developed, it soon became clear that similarities in cell morphology and arrangement do not automatically indicate relatedness. An example of this situation can be found in the gram-negative rods. Literally hundreds of

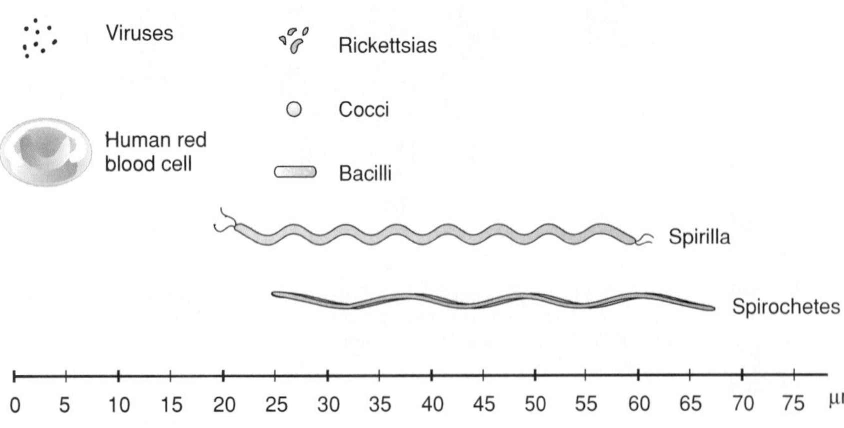

Figure 3.26 The range of bacterial sizes, to scale. The smallest cells of all three shapes are so tiny that they would be just barely discernible at 1,000× magnification. The sizes of human red blood cells and viruses are given for comparison.

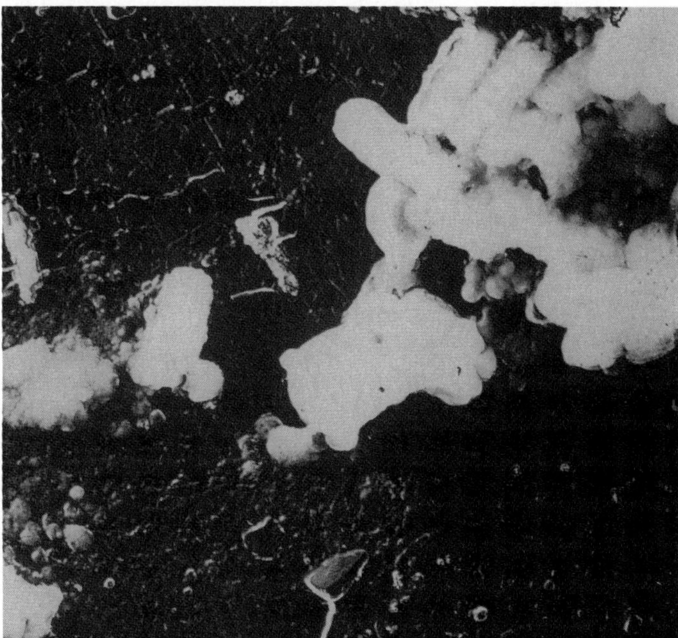

Figure 3.27 Electron micrograph of fossilized rod-shaped bacterial cells preserved in rock sediment. This specimen is more than two billion years old (×24,000).

Table 3.3 Major Taxonomic Groups of Bacteria per *Bergey's Manual*

Division I. Gracilicutes: Gram-Negative Bacteria

Class I. Scotobacteria: Gram-negative non-photosynthetic bacteria (examples in table 3.5)

Class II. Anoxyphotobacteria: Gram-negative photosynthetic bacteria that do not produce oxygen (purple and green bacteria)

Class III. Oxyphotobacteria: Gram-negative photosynthetic bacteria that evolve oxygen (cyanobacteria)

Division II. Firmicutes: Gram-Positive Bacteria

Class I. Firmibacteria: Gram-positive rods or cocci (examples in table 3.5)

Class II. Thallobacteria: Gram-positive branching cells (the actinomycetes)

Division III. Tenericutes

Class I. Mollicutes: Bacteria lacking a cell wall (the mycoplasmas)

Division IV. Mendosicutes

Class I. Archaebacteria: Bacteria with atypical compounds in the cell wall and membranes

From *Bergey's Manual of Systematic Bacteriology*, 9th ed. Copyright © the Williams & Wilkins Co., Baltimore. Reprinted by permission.

different species exist, with highly significant differences in biochemistry and genetics. If we had to rely on Gram stain and shape alone, we could not classify them below the level of class.

Although several schemes for organizing bacteria have been developed, new discoveries based on modern techniques of analysis (to be discussed later) keep them in a constant state of change. To be valuable, a scheme must (1) make use of current knowledge and show natural relationships, (2) be flexible enough to include new information on species and relationships, and (3) complement the several disciplines in microbiology. No one system of classification has been permanent or universally accepted. Of the current proposed systems, we will present two: (1) the general scheme of classification from the ninth edition of *Bergey's Manual of Systematic Bacteriology*, a manual of bacterial descriptions and classification published continuously since 1923 (table 3.3), and (2) a more detailed system that emphasizes the major, medically important bacterial families (see table 3.5).

The ninth edition of *Bergey's Manual* organizes the Kingdom Procaryotae into four major divisions. These are somewhat natural divisions, based upon the nature of the cell wall. The **Gracilicutes** have gram-negative cell walls and thus are thin-skinned; the **Firmicutes** have gram-positive cell walls that are thick and strong; the **Tenericutes** lack a cell wall and so are soft; and the **Mendosicutes** are primitive bacteria with unusual cell walls and nutritional habits. The first two divisions contain the greatest number of species, and the 200 or so species that cause human and animal diseases can be found in four classes: the Scotobacteria, Firmibacteria, Thallobacteria, and Mollicutes. The system used in *Bergey's Manual* further organizes bacteria into subcategories such as classes, orders, and families, but these are not available for all groups. An example of the entire classification of one bacterial species is shown in table 3.4.

Table 3.4 Taxonomic Rankings for *Borrelia burgdorferi*

Taxonomic Rank	Includes
Kingdom: Procaryotae	All bacteria
Division: Gracilicutes	All bacteria with gram-negative cell wall
Class: Scotobacteria	Non-photosynthetic gram-negative bacteria
Order: Spirochaetales	Helical, flexible, motile bacteria (spirochetes)
Family: Spirochaetaceae	Helical, motile bacteria lacking hooked ends
Genus: *Borrelia*	Tiny, loose, irregularly coiled spirochetes
Species: *Borrelia burgdorferi*	Causative agent of Lyme disease

Gracilicutes (gras″-ih-lik′-yoo-teez) L. *gracilus*, thin, and *cutis*, skin.

Firmicutes (fer-mik′-yoo-teez) L. *firmus*, strong.

Tenericutes (ten″-er-ik′-yoo-teez) L. *tener*, soft.

Mendosicutes (men-doh-sik′-yoo-teez) L. *mendosus*, to be false.

Many medical microbiologists prefer to use a working system that simply lists the major families and genera (table 3.5). This system is more applicable for their purposes because it is restricted to bacterial disease agents, it depends less on nomenclature, and it supports recognition and identification. It starts out with a similar split into gram-positive and gram-negative bacteria, then sets up subgroups according to cell shape, structure, or arrangement, and the following characteristics of oxygen usage: *Aerobic* bacteria use oxygen in metabolism; *anaerobic* bacteria do not use oxygen in metabolism; and facultative bacteria may or may not use oxygen. Though table 3.5 does not include species, they will be included later in the chapters on specific diseases.

Species and Subspecies in Bacteria

Among many organisms, the species level is a distinct, readily defined, and natural taxonomic category. In animals, for instance, a species is a distinct type of organism that can produce viable offspring only when it mates with others of its own kind. This definition does not work for bacteria, because they do not exhibit a typical mode of sexual reproduction, they can interbreed with unrelated forms, and they are notorious for their changeability. So it is necessary to hedge a bit when we define a bacterial species. Theoretically, it is a single kind of bacterium, all cells of which share an overall similar pattern of traits, in contrast to other groups whose pattern is different. The boundaries that separate two closely related species in a genus are in some cases very arbitrary, but this definition still serves as a way to separate the bacteria into various kinds that can be cultured and studied.

Because the individual members of given species can show variations, we must also define levels within species (subspecies) called **strains** and **types.** A strain or variety of bacteria is a culture that appears or behaves differently in some way from other cultures of that species. For example, there are pigmented and nonflagellated strains of *Pseudomonas fluorescens.* Types are subspecies that may also show differences in antigenic makeup (serotypes), in phage (virus) susceptibility (phage types), and in pathogenicity (pathotypes).

What's In a Name? It is well worth stating that it makes little difference to bacteria what we call them. The plague bacillus, whether named *Yersinia pestis* or *Pasteurella pestis* (its older name), is still quite capable of causing bubonic plague. In the medical microbiology lab, the public health lab, or the hospital, a scheme of classification becomes something much more than an orderly cataloguing of tongue-twisting names. It presents a foundation of characteristics to be used in identification. With this scheme, not only can technicians identify *Y. pestis* and scores of other pathogens, but they can distinguish these pathogens from their less injurious relatives.

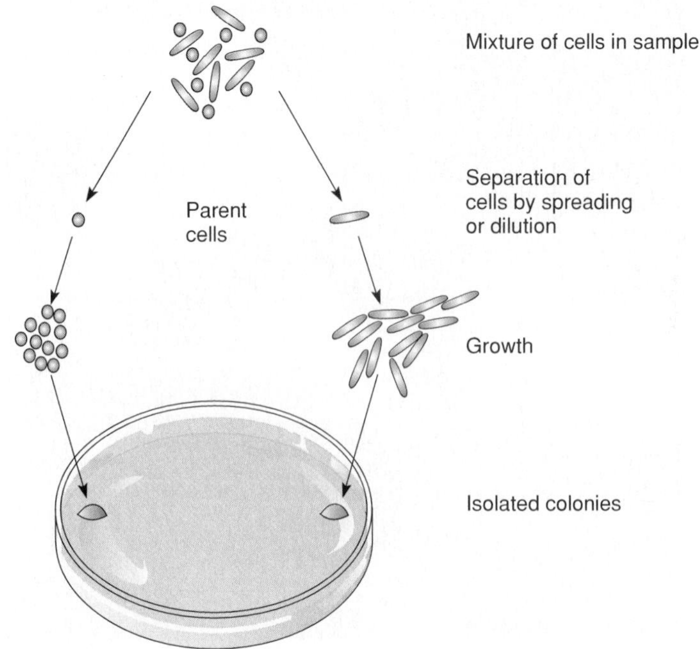

Figure 3.28 How isolated colonies form.

Laboratory Isolation, Cultivation, and Identification of Bacteria

Every pinch of soil, speck of saliva, and droplet of pond water harbors a great variety of bacteria. Though study of such mixed populations can be useful, in most laboratory studies, one must prepare pure cultures of the individual members of the population. This is especially critical in medical studies, in which an infectious agent must be identified and its individual characteristics determined so that proper treatment may be given. Isolation techniques are based on the concept that if an individual cell is separated from other cells and provided adequate space on a solid nutrient surface, it will grow into a macroscopic mass of cells called a colony (figure 3.28). Because it was formed from a single cell, a colony consists of just that single bacterial species and no other.

The main requirements for proper isolation are that a small number of cells be inoculated into a relatively large volume or area of growth medium. Media are extremely varied in formulation and usage, and may be chosen to grow a single type, a few types, or many types of bacteria. This topic and a detailed discussion of bacterial growth and media appear in chapter 6.

Both principal methods of isolation require the following materials: a medium that has been solidified by agar, *Petri plates* (clear, flat containers consisting of a smaller dish that holds the medium covered by a slightly larger dish), and an inoculating loop. In the **streak plate,** a small droplet of culture or sample is

Table 3.5. Medically Important Families and Genera of Bacteria, with Notes on Some Diseases*

I. Bacteria with Gram-Positive Cell Wall Structure

Cocci in clusters or packets that are aerobic or facultative

Family Micrococcaceae: *Staphylococcus* (members cause boils, skin infections)

Cocci in pairs and chains that are facultative

Family Streptococcaceae: *Streptococcus* (species cause strep throat, dental caries)

Anaerobic cocci in pairs, tetrads, irregular clusters

Family Peptococcaceae: *Peptococcus, Peptostreptococcus* (involved in wound infections)

Spore-forming rods

Family Bacillaceae: *Bacillus* (anthrax), *Clostridium* (tetanus, gas gangrene, botulism)

Non-spore-forming rods

Family Lactobacillaceae: *Lactobacillus, Listeria* (milk-borne disease), *Erysipelothrix* (erysipeloid)

Family Propionibacteriaceae: *Propionibacterium* (involved in acne)

Family Corynebacteriaceae: *Corynebacterium* (diphtheria)

Family Mycobacteriaceae: *Mycobacterium* (tuberculosis, leprosy)

Family Nocardiaceae: *Nocardia* (lung abscesses)

Family Actinomycetaceae: *Actinomyces* (lumpy jaw), *Bifidobacterium*

Family Streptomycetaceae: *Streptomyces* (important source of antibiotics)

II. Bacteria with Gram-Negative Cell Wall Structure

Family Neisseriaceae

Aerobic cocci

Neisseria (gonorrhea, meningitis), *Branhamella*

Aerobic coccobacilli

Moraxella, Acinetobacter

Anaerobic cocci

Family Veillonellaceae

Veillonella (dental disease)

Miscellaneous bacilli

Brucella (undulant fever), *Bordetella* (whooping cough), *Francisella* (tularemia)

Aerobic bacilli

Family Pseudomonadaceae: *Pseudomonas* (pneumonia, burn infections)

Miscellaneous: *Legionella* (Legionnaires' disease)

Facultative or anaerobic bacilli and vibrios

Family Enterobacteriaceae: *Escherichia, Edwardsiella, Citrobacter, Salmonella* (typhoid fever), *Shigella* (dysentery), *Klebsiella, Enterobacter, Serratia, Proteus, Yersinia* (one species causes plague)

Family Vibronaceae: *Vibrio* (cholera, food infection), *Campylobacter, Aeromonas*

Miscellaneous genera: *Chromobacterium, Flavobacterium, Haemophilus* (meningitis), *Pasteurella, Cardiobacterium, Streptobacillus*

Anaerobic bacilli

Family Bacteroidaceae: *Bacteroides, Fusobacterium* (anaerobic wound and dental infections)

Helical bacilli

Family Spirochetaceae: *Treponema* (syphilis), *Borrelia* (Lyme disease), *Leptospira* (kidney infection)

Obligate intracellular bacteria

Family Rickettsiaceae: *Rickettsia* (Rocky Mountain spotted fever), *Coxiella* (Q fever)

Family Bartonellaceae: *Bartonella*

Family Chlamydiaceae: *Chlamydia* (sexually transmitted infection)

III. Bacteria with No Cell Walls

Family Mycoplasmataceae: *Mycoplasma* (pneumonia), *Ureoplasma* (urinary infection)

*Details of pathogens and diseases in chapters 15, 16, and 17.

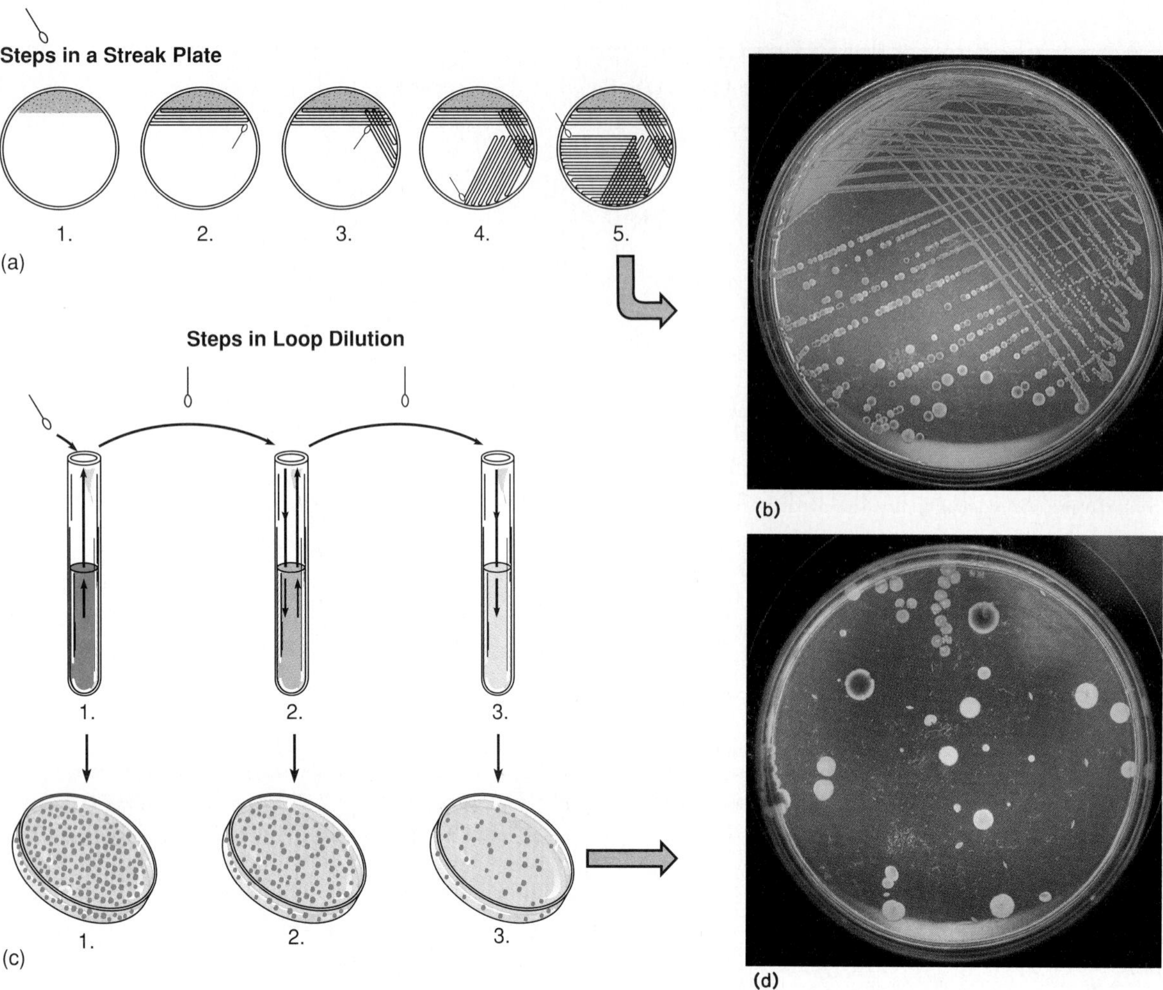

Figure 3.29 Methods for isolating bacteria. (*a*) Steps in a quadrant streak plate. (*b*) Resulting isolated colonies of bacteria. (*c*) Steps in the loop dilution method and (*d*) the end result.

spread over the surface of the medium according to a pattern that gradually thins the cells out and separates them spatially on several sections of the plate (figure 3.29*a*). Because of its effectiveness and economy of materials, the streak plate is the method of choice in most applications.

In the **loop dilution** method, the sample is inoculated serially into a series of cooled but still liquid agar tubes so as to dilute the number of cells in each successive tube in the series (figure 3.29*c*). Inoculated tubes are then poured into sterile petri dishes and allowed to harden. The result will be that the number of cells per volume is so decreased in the second or third tube that cells have ample space to grow into separate colonies. The difference is that in this technique some of the colonies will develop in the medium itself and not just on the surface.

Once colonies have developed, they are inspected for individual differences, and different colonies are subcultured. Since a colony consists of only one species of bacterium, this technique ensures that a pure culture will be made for further testing and identification. Some bacterial pathogens, such as rickettsias, chlamydias, the spirochete of syphilis, and the leprosy bacillus, cannot be grown in artificial media and must be provided with living animals or cultures of host cells, much like the viruses (see chapter 5).

Summary of Methods Used in Bacterial Identification

Methods the microbiologist uses to identify bacteria to the level of genus and species fall into the categories of morphology (microscopic and macroscopic), bacterial physiology or biochemistry, serological analysis, and genetic techniques. Data from the various methods of analysis are used to construct a profile of each bacterium. Final differentiation of the unknown species is accomplished by comparing its profile to profiles of known bacteria in tables, charts, and keys, and frequently by the assistance of a computer programmed to analyze many pieces of data simultaneously. Not all methods are used to identify all bacteria. Some bacteria may be identified by a few physiological tests and a Gram stain; others may require a whole spectrum of tests. The following list summarizes some of the general areas of bacteriologic testing.

Microscopic morphology: The shape and size of cells, Gram stain reaction, other differential stains, and special structures, including flagella, spores, granules, and capsules.

Macroscopic morphology: Appearance of colonies, including texture, size, shape, pigment; speed of growth, reactions with special media.

Physiological characteristics: Numerous diagnostic tests for determining the presence of specific enzymes and nutritional and oxygen requirements.

Serological analysis: Testing bacterial cultures with antibodies that are known to be specific for a given species or genus of bacterium.

Chemical analysis: Testing for the types of specific chemical substances that the bacterium has, such as proteins and fatty acids.

Genetic analysis: Testing the DNA for its total percentage of guanine and cytosine (the G/C content), which may be indicative of its relationship to other bacteria. This technique is used more often in classification than in identification. **DNA and RNA homologies**: Determining the degree of similarity of DNA molecules in cultures that would indicate genetic relationships. DNA probes (see chapters 8 and 22) have been developed for this purpose.

A more detailed analysis of these and other identification methods is presented in chapters 13, 16, 17, and 22.

Survey of Bacterial Groups with Unusual Characteristics

The bacterial world is so diverse that we cannot do justice to it in this introductory chapter. This variety extends into all areas of bacterial biology, including nutrition, mode of life, and behavior. Certain types of bacteria exhibit such unusual qualities that they deserve special mention. In this mini-survey, we will consider some medically important groups and some more remarkable representatives of bacteria living free in the environment that rarely do any harm to human beings. Many of the bacteria mentioned here do not have the morphology typical of those bacteria discussed in earlier sections.

Unusual Forms of Medically Significant Bacteria

Rickettsias

Rickettsias[3] are distinctive gram-negative bacteria that are among the tiniest of all cells (figure 3.30). Though they have a somewhat typical bacterial morphology, they are atypical in their life cycle and other adaptations. Most are pathogens that alternate between a mammalian host and blood-sucking arthropods,[4] such as fleas, lice, or ticks (see chapter 17). Since rickettsias cannot survive or multiply outside a host cell and cannot carry out metabolism completely on their own, they are considered obligate intracellular parasites (see feature 6.6). Several important human diseases are caused by rickettsias. Among these are Rocky Mountain spotted fever, caused by *Rickettsia rickettsii* (transmitted by ticks), and epidemic typhus, caused by *Rickettsia prowazekii* (transmitted by lice). An exceptional rickettsia, *Coxiella burnetii* (the cause of Q fever), is unusual in being resistant to the environment and transmitted through the air as well as by arthropods.

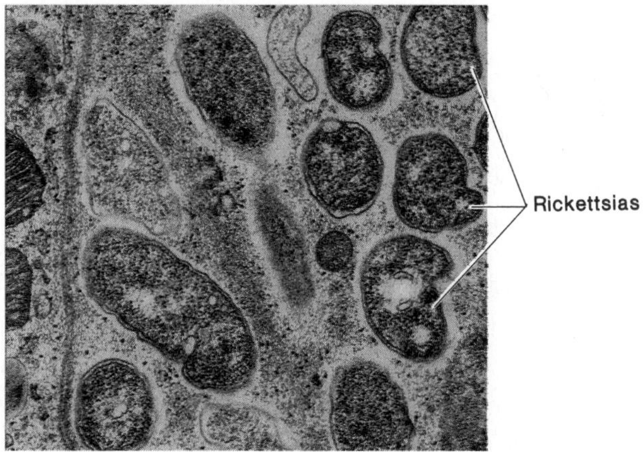

Figure 3.30 Rickettsias living inside a host cell (×59,000).

Chlamydias

Bacteria of the genus *Chlamydia* are very similar to the rickettsias in that they require host cells for growth and metabolism, but they are not carried by arthropods. Because of their tiny size and obligately parasitic life-style, they were at one time considered a type of virus. Later studies indicated that their structure was that of a typical gram-negative procaryotic cell and that their mode of cell division (binary fission) was clearly procaryotic. Two species that carry the greatest medical impact are *Chlamydia trachomatis,* the cause of both a severe eye infection (trachoma) that can lead to blindness and one of the most common sexually transmitted diseases (see chapter 17), and *Chlamydia psittaci,* the agent of ornithosis or parrot fever, a disease of birds that can be transmitted to humans.

Mycoplasmas and Other Cell-Wall-Deficient Bacteria

Mycoplasmas are bacteria that naturally lack a cell wall. Though other bacteria require an intact cell wall to protect the bursting of the cell, the mycoplasmal cell membrane is stabilized and resistant to lysis. These extremely tiny, pleomorphic cells have a minimum size of 0.1 μm and range in shape from filamentous to coccus to doughnut. They are *not* obligate parasites and can be grown on artificial media, though added sterols are required for the cell membranes of some species. Mycoplasmas are found in many habitats, including plants, soil, and animals. The most important medical species is *Mycoplasma pneumoniae* (figure 3.31), which adheres to the epithelial cells in the lung and causes an atypical form of pneumonia in humans (see chapter 17).

Unusual Bacteria Not Involved in Human Disease

Photosynthetic Bacteria

The nutrition of most bacteria is heterotrophic, meaning that they derive their nutrients from other organisms. Photosynthetic bacteria, however, are independent cells that can

3. Named for Howard Ricketts, who first worked with these organisms and later lost his life to typhus.

4. An invertebrate with jointed legs, such as an insect or spider.

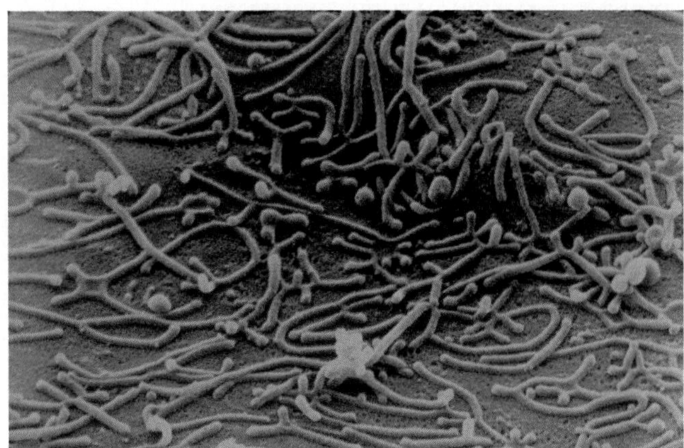

Figure 3.31 Scanning electron micrograph of *Mycoplasma pneumoniae* (×62,000). Cells like these that naturally lack a cell wall exhibit extreme pleomorphism.

synthesize all required nutrients from simple inorganic compounds in the presence of sunlight. The two general types of photosynthetic bacteria are those that evolve oxygen during photosynthesis and those that produce some other substance, such as sulfur granules or sulfates.

Cyanobacteria The cyanobacteria were called blue-green algae for many years and were grouped with the eucaryotic algae. Further study has verified that they are indeed bacteria with a gram-negative cell wall (Gracilicutes) and general procaryotic structure. These unusual bacteria range in size from 1 μm to 50 μm, and they can be unicellular, colonial, or filamentous (figure 3.32). Some species occur in packets surrounded by a gelatinous sheath (figure 3.32*c*). Among the special intracellular components of cyanobacteria are small sacs or inclusions containing gas that permit them to float on the water surface to increase their light exposure, cysts that fix nitrogen, and special photosynthetic membranes called *thylakoids* in which the granules of chlorophyll and other photosynthetic pigments are embedded. This group is sometimes called the blue-green bacteria in reference to their content of *phycocyanin* pigment that tints some members a shade of blue-green, though other members are colored yellow and orange. Cyanobacteria grow profusely in fresh water and seawater and may be responsible for periodic blooms that kill off fishes. The origin of an occasional pungent taste and odor in tap water may be from heavy growth of cyanobacteria in the local reservoir and their production of a chemical called geosmin. Other major contributions of this group are the addition of oxygen to the atmosphere and the conversion of gaseous nitrogen (N_2) into a form usable by plants (nitrogen fixation; see chapter 22).

Green and Purple Bacteria The green and purple bacteria (see chapter opening photo) are also photosynthetic and contain pigments on thylakoids, but they are different from the cyanobacteria in that they contain a special type of chlorophyll called bacteriochlorophyll, they do not give off oxygen, and they live in habitats that lack oxygen. These bacteria are named for their predominant colors, but they can also develop brown, pink, purple, blue, and orange coloration. They exist as single cells of many different shapes, and are frequently motile. Both groups utilize sulfur compounds in their metabolism, and some may deposit intracellular granules of sulfur or sulfates (figure 3.33). These organisms live in various aqueous habitats, including sulfur springs, freshwater lakes, and swamps.

phycocyanin (fye-koh-sye′-an-in) Gr. *phykos*, seaweed, and *cyan*, blue. A blue pigment unique to this group.

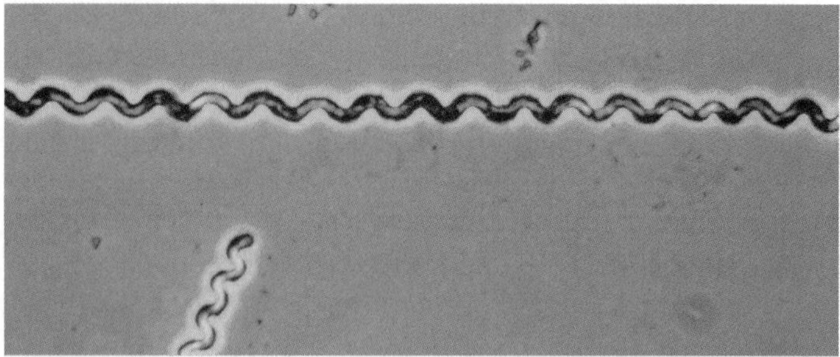

(a)

(b)

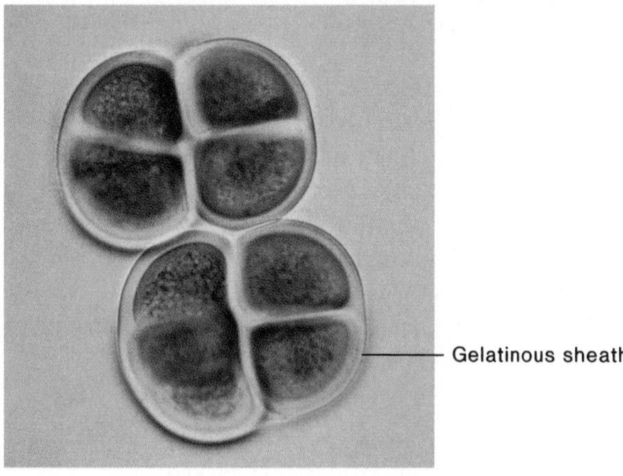

(c)

Figure 3.32 Morphological variations of cyanobacteria. (*a*) *Spirulina* (×300). (*b*) Two species of *Oscillatoria*, a gliding filamentous form (×100). (*c*) *Chroococcus*, a colonial form surrounded by a gelatinous sheath (×600).

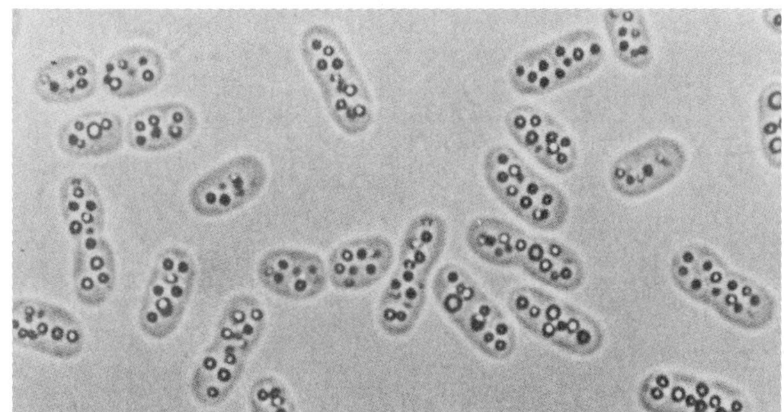

Figure 3.33 Photosynthetic purple sulfur bacteria, *Chromatium vinosum,* with internal sulfur granules.

Feature 3.3 Bacterial Predators

Despite their small size, bacteria are often subject to attack by smaller parasites (viruses) or predators (other bacteria). Among the most notable of the predatory bacteria are the **bdellovibrios.** These tiny, comma-shaped bacteria exhibit a peculiar behavior reminiscent of certain parasitic worms or fishes. Dashing through the water, they collide with their much larger prey cells (usually gram-negative rods). Upon impact, they attach to the cell and drill a hole in its outer wall by rapid rotation, then slip through this hole into the periplasmic space. Once inside their prey, bdellovibrios seal off the hole, kill the cell, and begin to feed on its protoplasm. During the next 4 hours, the miniscule predator elongates and subdivides into a number of new vibrios that are released when the dead shell of the prey bacterium breaks apart. Like the parent, these actively motile cells move out to seek and assault new prey.

bdellovibrio (del-oh-vib′-ree-oh) Gr. *bdella,* leech, and *vibrio,* a curved cell.

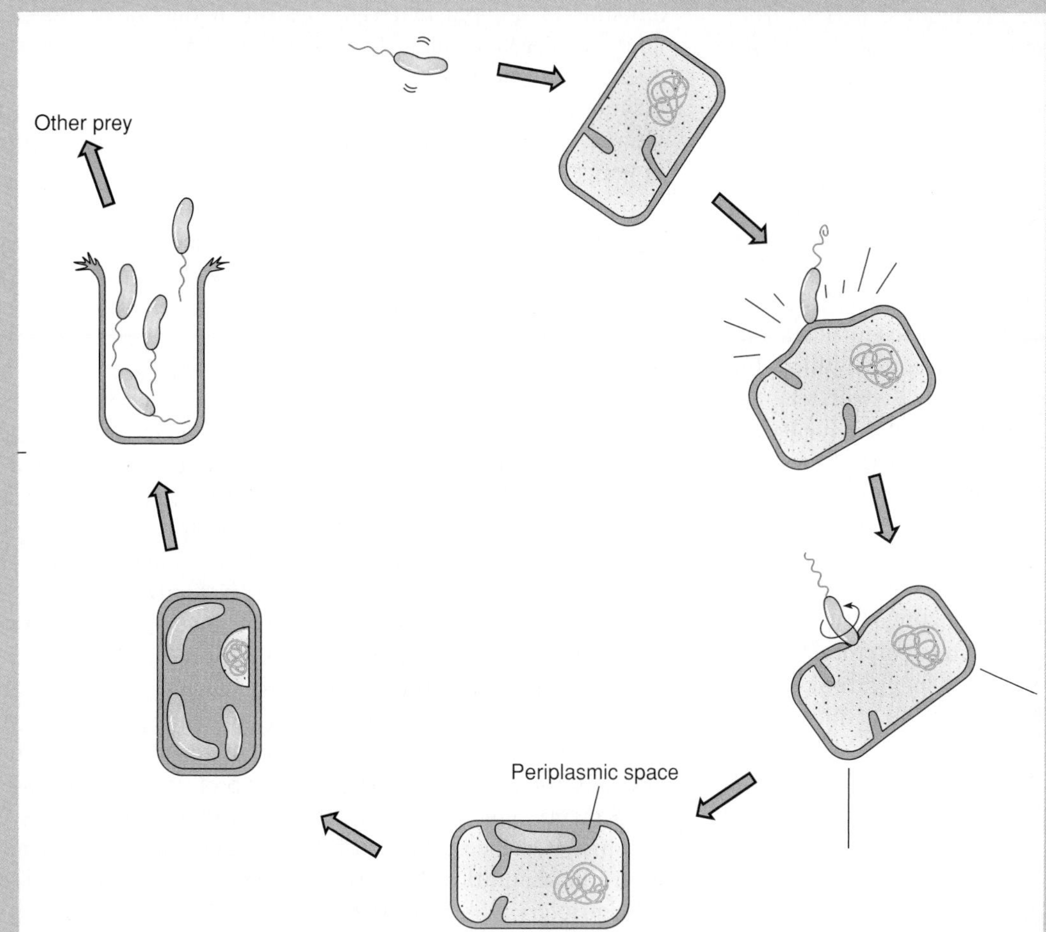

Bdellovibrio, a bacterial predator.

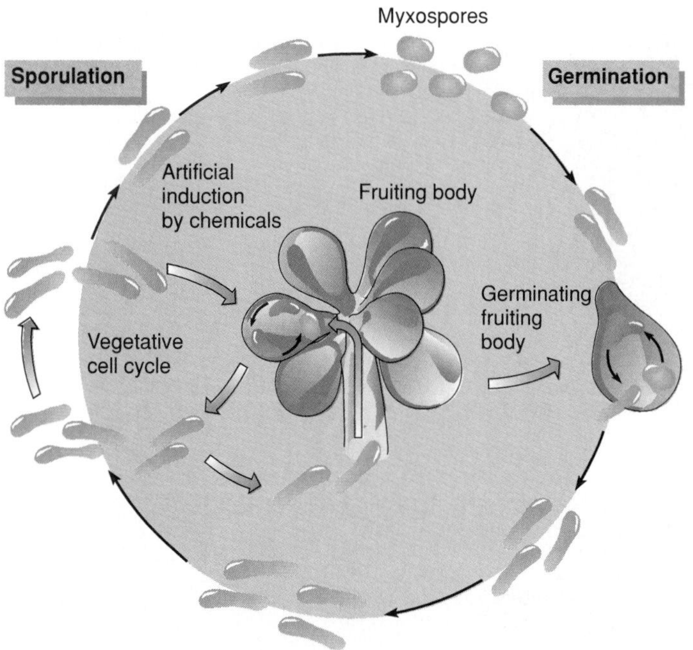

Figure 3.34 The life cycle of a myxobacterium (*Chondromyces*), showing its central mature fruiting body.

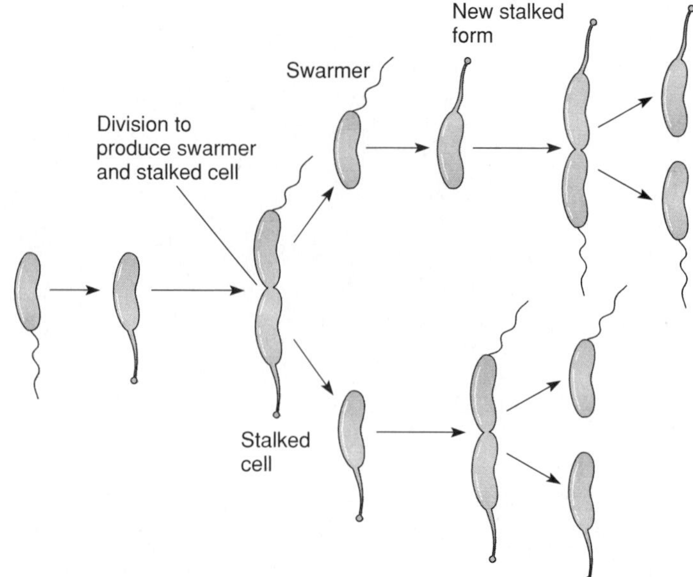

Figure 3.35 Budding bacteria. The life cycle of *Caulobacter,* showing the development of the flagellated swarm cell and stalked phases.

Gliding Bacteria

The **gliding bacteria** are a very diverse group of gram-negative bacteria that live in water and soil. The name is derived from the tendency of members to glide over moist surfaces. They have no flagella, and the gliding property evidently involves rotation of filaments or particles just under the outer membrane of the cell wall. There are several morphologic forms, including slender rods, long filaments, cocci, and some miniature, tree-shaped fruiting bodies. Probably the most intriguing and exceptional members of this group are the slime bacteria, or *myxobacteria* (figure 3.34). What sets the myxobacteria apart from other bacteria is the complexity and advancement of their life cycle. During this cycle, the vegetative cells swarm together and differentiate into a many-celled, colored structure called the fruiting body. The fruiting body is a survival structure that makes spores by a method very similar to that of the eucaryotic slime molds (see chapter 4). Though small, these fruiting structures can often be seen with the unaided eye on tree bark and plant debris.

Appendaged Bacteria

The appendaged bacteria are quite varied in their structure and life cycles, but all of them produce an extended process of the cell wall in the form of a bud, a stalk, or a long thread (figure 3.35). The stalked bacteria live attached to the surface of objects in aquatic environments. One type can even grow in distilled or tap water. The stalks evidently help them trap minute amounts of organic materials present in the water. Budding bacteria reproduce entirely by budding—that is, they develop a tiny bulb (bud) at the end of a thread. The bud breaks off, enlarges, develops a flagellum, and swarms to another area to start its own cycle (figure 3.35). These bacteria can also grow in very low-nutrient habitats.

The Mendosicutes or Archaebacteria Some of the most extreme habitats in nature are occupied by the **archaebacteria.** These bacteria are considered so exceptional in structure and physiology that some taxonomists would place them in a group entirely separate from procaryotic and eucaryotic cells. The group contains members that produce methane (methanogens) and those that live in environments that are very high in salt (halophilic or salt-loving bacteria; see chapter 6). One particularly distinct form are the *square bacteria,* found in extremely saline waters such as the Red Sea. These tiny cells are truly square, with straight sides and sharp corners, and they are arranged in block-shaped packets or sheets (figure 3.36).

Some archaebacteria are adapted to very high temperatures and thus are called thermophilic bacteria. In 1983, researchers sampling sulfur vents in the deep ocean discovered some of these bacteria flourishing at temperatures up to 250°C—150 degrees above the temperature of boiling water! Not only were these bacteria growing prolifically at this high temperature, but they were also living at 265 atmospheres of pressure. (On the earth's surface, pressure is about one atmosphere.) For additional discussion of the unusual adaptations of archaebacteria, see chapter 6.

---

myxobacteria (micks'-oh-bak-ter''-ee-uh) Gr. *myxa*, mucus or slime.

(a)

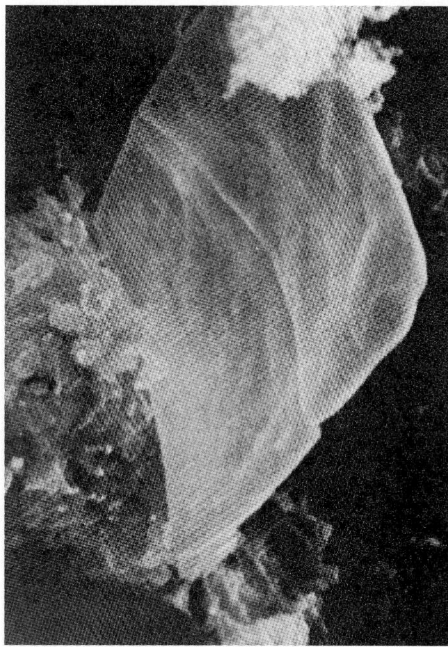

(b)

Figure 3.36 (*a*) A packet of square bacteria. (*b*) The actual configuration of an individual cell is something like a soda cracker.

Chapter Review with Key Terms

The Procaryotes

Alternate Names: General—**Archaebacteria** Monera, **Eubacteria,** bacteria, Schizomycetes (old).

Specific: See also classification.

Overall Morphology

Cellular: Procaryotic cells, lacking a nucleus, mitochondria, chloroplasts, or other membranous organelles. Outer covering is a **cell envelope** composed of **glycocalyx (slime layer, capsule)**, **cell walls** (in most), **cell membrane. Gram stain** differentiates two types of cells based on wall structure: **Gram-positive** cells have a thick layer of **peptidoglycan; gram-negative** cells have an outer membrane and a thin layer of peptidoglycan. Diameter of most bacteria is 1–5 μm. Special **appendages**: motility by bacterial **flagella** common; **fimbriae** important in cell adhesion; **pili** in genetic recombination. Other special structures: **endospores** that are highly resistant survival cells; **inclusions** and **granules** for storing nutrients.

Organismic: Most are **unicellular,** form colonies on solid nutrients. Multicellularity is rare. Have distinct shapes, including **cocci, bacilli, spirilla, spirochetes.** Certain groups may assume long filaments; some show variations in shape, or **pleomorphism.** Various **arrangements** based upon mode of cell division: **(diplo-, strepto-, staphylo-, palisades)**.

Nutritional/Habitat Requirements

A large number of bacteria are heterotrophs that require an organic carbon source derived from dead organisms by absorption or from a live host (parasites); several groups are capable of synthesizing nutrients using sunlight (photosynthetic).

Oxygen Requirements: Full range, including groups that require O_2 to grow (aerobic), can grow with or without O_2 (facultative), or must have O_2 excluded in order to grow (anaerobic).

Preferred Temperature for Growth: Most exist between 10° and 40°C. Some species live in very cold habitats, even at subfreezing temperatures, and some at high temperatures (up to 250°C).

Reproduction and Life Cycles

Bacteria generally have a single chromosome; reproduce primarily by asexual means, including budding and simple binary fission; no mitosis; genetic exchange may occur through conjugation; life cycle complex in sporeformers, gliding bacteria, appendaged bacteria, and archaebacteria.

Classification of Major Groups

The Kingdom Procaryotae is divided into four divisions, based on the type of cell wall.

Gracilicutes: The largest group; contains bacteria with gram-negative cell walls; includes several medically significant microbes, such as *Salmonella, Shigella,* and other intestinal pathogens, the Rickettsias, and the agents of gonorrhea and syphilis. The photosynthetic, gliding, sheathed, and appendaged bacteria are also in this group.

Firmicutes: Gram-positive cells; examples of pathogens are in the genera *Streptococcus, Staphylococcus, Mycobacterium* (tuberculosis and leprosy), and *Clostridium* (tetanus, gas gangrene).

Tenericutes: Cells without cell walls; primarily in the genus *Mycoplasma* (one type causes pneumonia).

Mendosicutes: Cells with very atypical cell walls; most species are free-living in extreme environments; includes the methanogens and the extreme heat- and salt-tolerant species. No known medically important species.

Identification

Through microscopic (Gram stain, shape, special structures) and macroscopic appearance, and through biochemical, enzyme, and genetic characteristics.

Where Found

Because they are highly adaptable as a group, nearly everywhere: water, soil, air, dust, food, deep sea, plants, animals, swamps, hot springs, north and south poles.

Ecological Importance

Important as decomposers of organic matter; play a role in the cycles of nitrogen, phosphorus, sulfur, and carbon. The cyanobacteria contribute oxygen to the atmosphere through photosynthesis.

Economic Importance

Beneficial: Bacteria are used in a number of industries, including food (needed for flavor and texture); drugs (production of antibiotics, vaccines, and other medicines); biotechnology (manipulation of bacteria for making large quantities of hormones, enzymes, and other proteins).

Adverse: Bacteria are responsible for spoilage of food and vegetables. Many plant pathogens adversely affect the agricultural industry.

Medical Importance

Bacteria are very common pathogens of humans. Approximately 200 species are known to cause disease in humans, and many normally inhabit the bodies of humans and other animals. Infections can be treated with antibiotics.

Unique Features

Smallest cells in existence; procaryotic structure; unique shapes, arrangements; lack of traditional sexual reproduction and mitosis; pleomorphism; cell wall of most contains peptidoglycan; mesosomes; single chromosome; 70S ribosomes.

True–False Questions

Determine whether the following statements are true (T) or false (F). If you feel a statement is false, explain why, and reword the sentence so that it reads accurately.

____ 1. All bacterial cells have a cell membrane, a nucleoid, and ribosomes.

____ 2. A bacterial flagellum is a long filament that moves the cell by a whiplike motion.

____ 3. In general, rods are motile and cocci are nonmotile.

____ 4. Pili are stiff, hollow rods that serve as a means for genetic exchange between bacteria.

____ 5. Examples of the glycocalyx in bacteria are pili and fimbriae.

____ 6. The bacterial cell wall functions in transport.

____ 7. Gram-positive cell walls have a thick layer of peptidoglycan, and gram-negative cell walls have an outer membrane and a thin layer of peptidoglycan.

____ 8. Mesosomes are internal extensions of the cell wall that function in cell metabolism and division.

____ 9. Granules are concentrated crystals of inorganic chemicals that function as stored nutrients.

____ 10. Bacterial endospores are reproductive bodies of bacteria.

____ 11. A micrococcus is arranged in packets of eight cells.

____ 12. A spirochete is rigid, and a spirillum is flexible.

____ 13. A bacterial colony is a discrete aggregate of cells that have arisen from an isolated parental cell.

Concept Questions

1. Name several general characteristics that could be used to define the procaryotes. What does it mean to say that bacteria are ubiquitous? In what habitats are they found? Give some general means by which bacteria derive nutrients. Do any other microbial groups besides bacteria have procaryotic cells?
2. Describe the structure of a flagellum and how it operates. What are the four main types of flagellar arrangement? Tell some direct and indirect ways that one may determine motility. How does the flagellum dictate the behavior of a motile bacterium? Differentiate between flagella and axial filaments.
3. What is the cell envelope? Explain the position of the glycocalyx. What are functions of slime layers and capsules? How is the presence of a slime observable even at the level of a colony?
4. Explain the roles of pili and fimbriae.
5. What is the function of peptidoglycan? To which part of the cell envelope does it belong? Give a simple description of its structure. What happens to a cell that has its peptidoglycan disrupted or removed?
6. What is the Gram stain? What is there in the structure of bacteria that causes some to stain purple and others to stain red? How does the precise structure of the cell walls differ in gram-positive and gram-negative bacteria? What other properties besides staining are different in gram-positive and gram-negative bacteria? What is the periplasmic space, and how does it function?
7. List five functions that the cell membrane performs in bacteria.
8. Of what are the chromatin body and plasmids composed? What are their functions?
9. What is unique about the structure of bacterial ribosomes? How do they function?
10. Explain the structure and function of inclusions and granules. What are metachromatic granules, and what do they contain?
11. What is the vegetative stage of a bacterial cell? What is an endospore, and how does it function? Why are endospores so difficult to destroy? Describe the endospore-forming cycle.
12. Draw the three bacterial shapes. How are spirochetes and spirilla different? What is a vibrio? A coccobacillus? What is pleomorphism?
13. What are the size ranges in bacteria according to shape? Where do they fit, size-wise, compared to viruses and eucaryotic cells in general?
14. What characteristics are used to classify bacteria? What are the most useful characteristics for categorizing bacteria into families? What is the species level in bacteria? What are subspecies?
15. What are L forms, and how are they important?
16. How are bacteria medically and ecologically important?

Practical/Thought Questions

1. What would happen if one stained a gram-positive cell with safranin? A gram-negative one with crystal violet? What would happen to the two types if the mordant were omitted? Speculate on what the gram reaction of your body cells would be.
2. What is required to kill endospores? How do you suppose archeologists were able to date some spores as being 3,000 years old?
3. Using clay, demonstrate how cocci may divide in several planes and show the outcome of this division. Show how the arrangements of bacilli occur, including palisades.
4. Use a corkscrew and a spring to compare the flexibility and locomotion in spirilla and spirochetes.
5. Under the microscope you see a rod-shaped cell that is swimming rapidly forward. What do you automatically know about that bacterium's structure? Can you think of another function of flagella besides locomotion?
6. Describe how you might go about isolating, cultivating, and identifying a pathogen from a urine sample.
7. Name a bacterium that has no cell walls.
8. Name a bacterium that is aerobic, gram-positive, and spore-forming.
9. Name a bacterium that is pleomorphic and has palisades arrangement and metachromatic granules.
10. Name an acid-fast bacterium.
11. Name two types of obligate intracellular parasitic bacteria.
12. Name a bacterium that lives in extremes of heat and salt.
13. Name a coccus that is gram positive and in chains.
14. Name a gram-negative diplococcus.
15. Name a bacterium that contains sulfur granules.
16. Name a bacterium that is photosynthetic.
17. Name a bacterium that preys on other bacteria.

CHAPTER 4

Eucaryotic Cells and Microorganisms

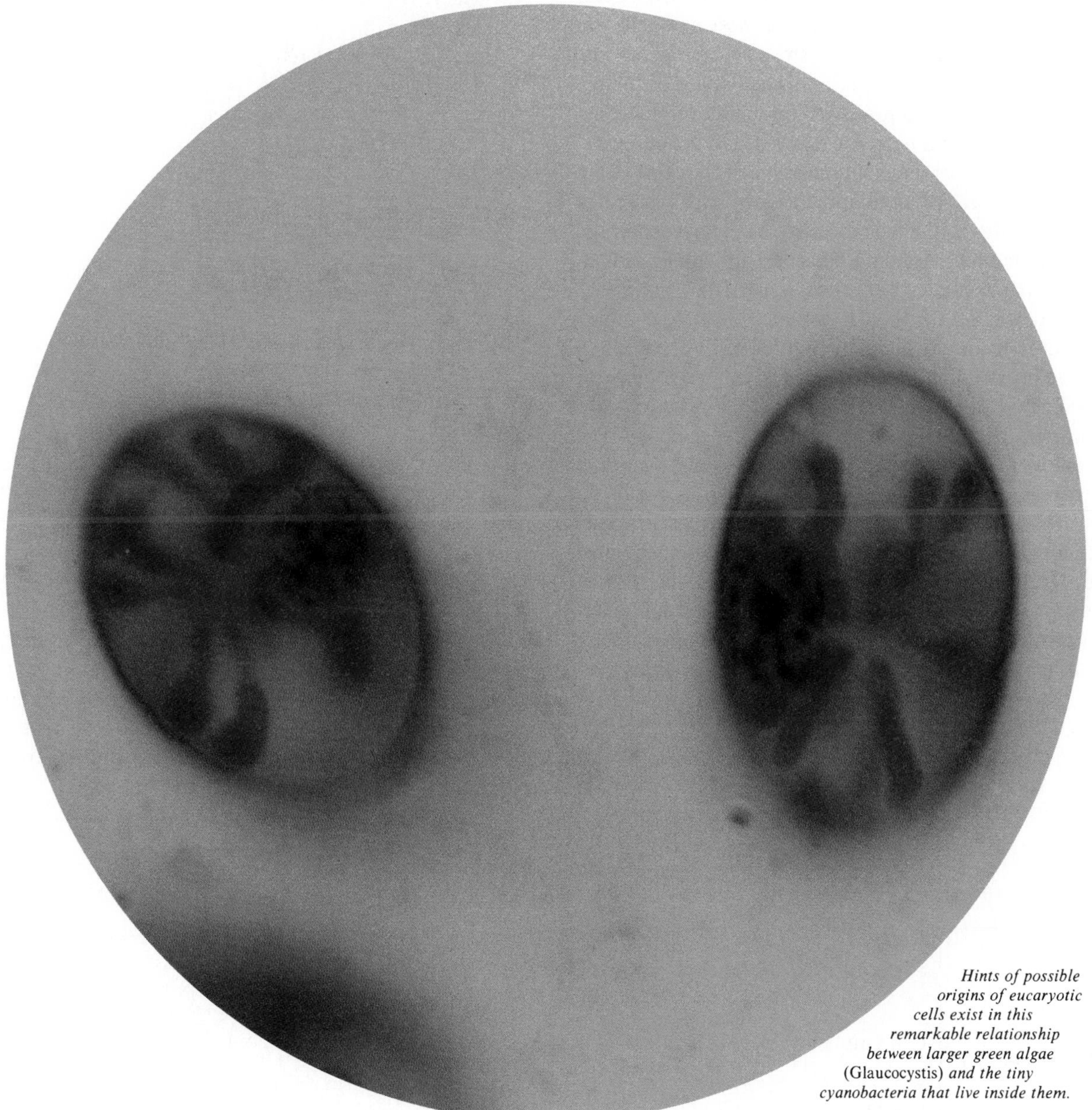

Hints of possible origins of eucaryotic cells exist in this remarkable relationship between larger green algae (Glaucocystis) *and the tiny cyanobacteria that live inside them.*

Chapter Preview

The eucaryotic cell is a complex, compartmentalized unit that differs from the procaryotic cell by containing a nucleus and several other specialized structures called organelles. Although exact cell structures differ somewhat among the several groups of eucaryotic organisms, the eucaryotic cell is the typical cell of certain microbial groups (fungi, algae, protozoans, and helminth worms) as well as all animals and plants. In this chapter we will examine the overall structure and function of eucaryotic cells in preparation for later chapters that deal with related microbiological concepts, including multiplication of viruses, metabolism, genetics, nutrition, drug therapy, immunology, and disease. Because of the tremendous variety of eucaryotic microorganisms and their practical importance in medicine, industry, and agriculture, this chapter will also cover the major characteristics of each group of microscopic eucaryotes.

The Nature of Eucaryotes

Evidence from paleontology indicates that the first eucaryotic cells appeared on earth approximately two billion years ago (figure 4.1). An appealing theory suggests that these cells evolved from procaryotic organisms by a process of intracellular *symbiosis* (see feature 4.1). The structure of these new cells was so versatile that eucaryotic microorganisms soon spread out into available habitats and adopted greatly diverse styles of living.

The first primitive eucaryotes were probably single-celled and independent, but over time, some forms began to aggregate, forming colonies. With further evolution, some of the cells within colonies became *specialized,* or adapted to perform a particular function advantageous to the whole colony, such as locomotion, feeding, or reproduction. At some point, as individual cells in the organism lost the ability to survive apart from the intact colony, the scene was set for the development of more complex multicellular organisms. Although a multicellular organism is composed of many cells, it is more than just an assemblage of cells like a colony. Rather, it is composed of distinct groups of cells that cannot exist independently of the rest of the body. The cell groupings of multicellular organisms that have a specific function are termed *tissues,* and groups of tissues make up *organs.*

Looking at modern eucaryotic organisms, we find examples of many levels of cellular organization. All protozoa, as well as numerous algae and fungi, live a unicellular or colonial existence. Truly multicellular organisms are found only among plants and animals and some of the fungi and algae. Only certain eucaryotes are small enough to fall into the realm of microbiology—namely, the protozoa and some of the algae, fungi, and animal parasites.

symbiosis (sim-bye-oh′-sis) Gr. *sym,* together, and *bios,* to live.

Figure 4.1 Fossilized eucaryotes, ancient eucaryotic cells dated around two billion years ago. Note the prominent nucleus and cell wall.

Form and Function of the Eucaryotic Cell

The cells of eucaryotic organisms are so varied that no one member can serve as a typical example. Figure 4.2 presents the generalized structure of typical algal, fungal, and protozoan cells. The following diagram shows the organization of a eucaryotic cell. (Only motile algae exhibit all possible eucaryotic organelles.)

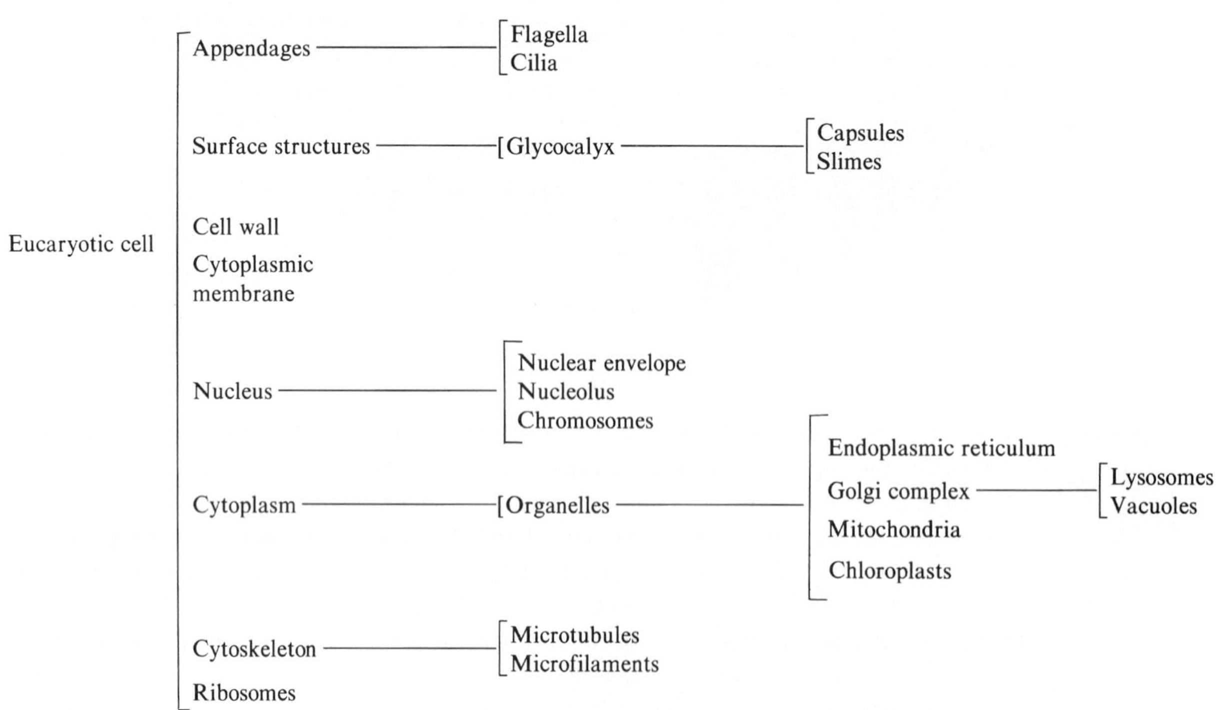

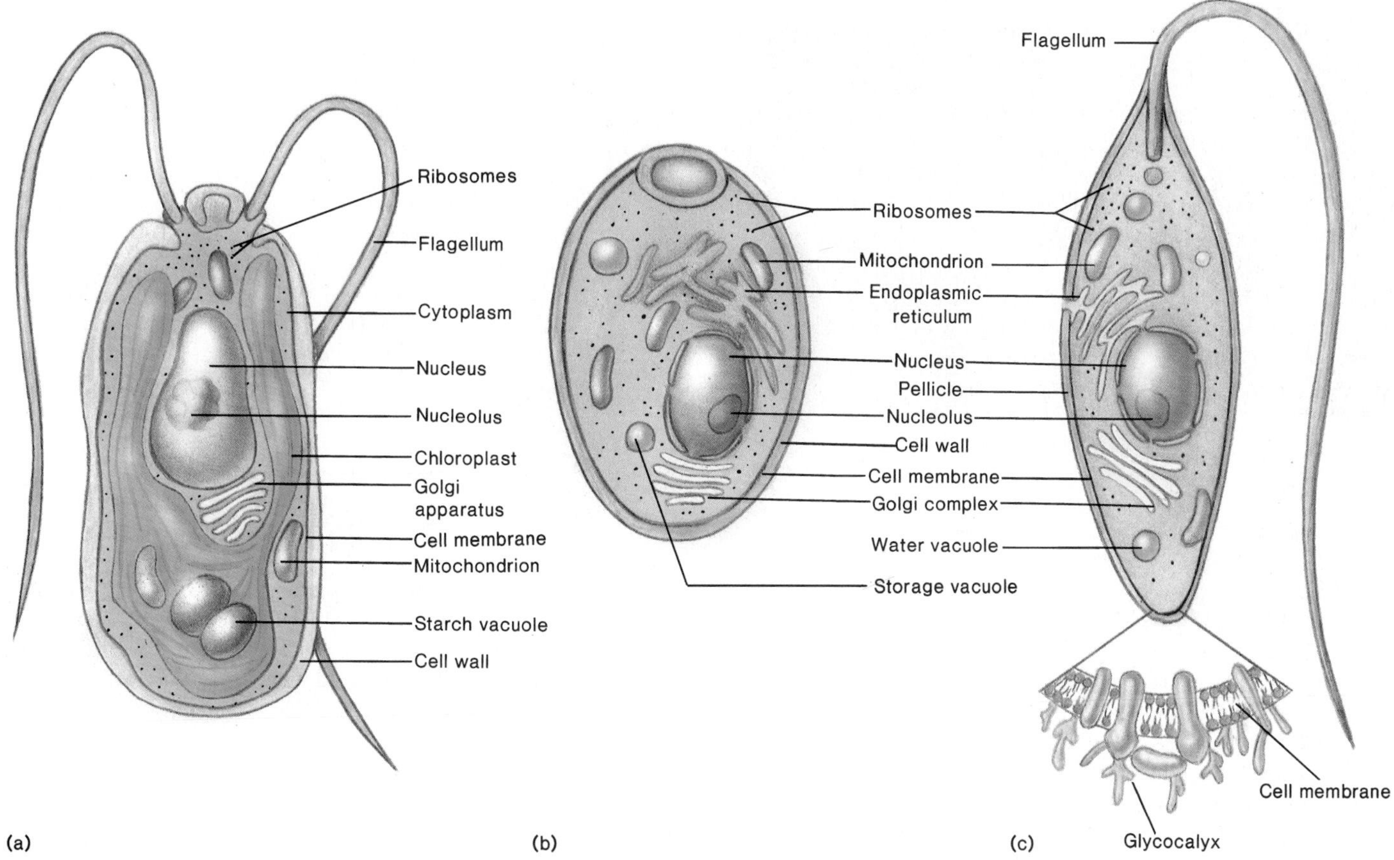

Figure 4.2 The structure of three representative eucaryotic cells. (*a*) *Chlamydomonas,* a unicellular motile alga. (*b*) A yeast cell (fungus). (*c*) A flagellated protozoan.

All eucaryotic microbial cells have a glycocalyx, cytoplasmic membrane, nucleus, mitochondria, endoplasmic reticulum, Golgi apparatus, vacuole, and cytoskeleton. Cell walls, locomotor appendages, and chloroplasts are found only in some groups. In the following sections we will cover the microscopic structure and functions of the eucaryotic cell. As with the procaryotes, we will begin on the outside and proceed inward through the cell.

Appendages: Locomotor Organelles

Motility allows a microorganism to locate life-sustaining nutrients and to migrate toward positive stimuli such as sunlight; it also permits avoidance of harmful substances and stimuli. Locomotion by means of flagella or cilia is a common property of protozoa, many algae, and a few fungal and animal cells.

Flagella

Although they share the same name, the structure and operation of eucaryotic flagella are much different from those of procaryotes. The eucaryotic flagellum is thicker (by a factor of 10), structurally more complex, and covered by an extension of the cell membrane. A single flagellum is a long, sheathed cylinder containing regularly spaced hollow tubules—the *microtubules*—that extend along its entire length (figure 4.3*a*). A cross section reveals nine pairs of closely attached microtubules surrounding a single central pair. This scheme, called the 9 + 2 arrangement, is a universal pattern of flagella and cilia. During locomotion, the paired microtubules slide past each other, which whips the flagellum back and forth. Although details of this process are too complex to discuss here, it involves expenditure of energy and a coordinating mechanism in the cell membrane. If the flagellum is located at the back end of the cell, it can push the cell forward like a boat's propeller; if oriented at the front end of the cell, it can pull it forward by a lashing or twirling motion (figure 4.3*b*). The placement and number of flagella are characteristics used to identify flagellated protozoa and certain algae.

Cilia

Cilia are very similar in overall architecture to flagella, but they are shorter and more numerous (some cells have several thousand). They are found only on a single group of protozoans and

---

cilia (sil′-ee-ah) sing. cilium; L. *siliam,* eyelid.

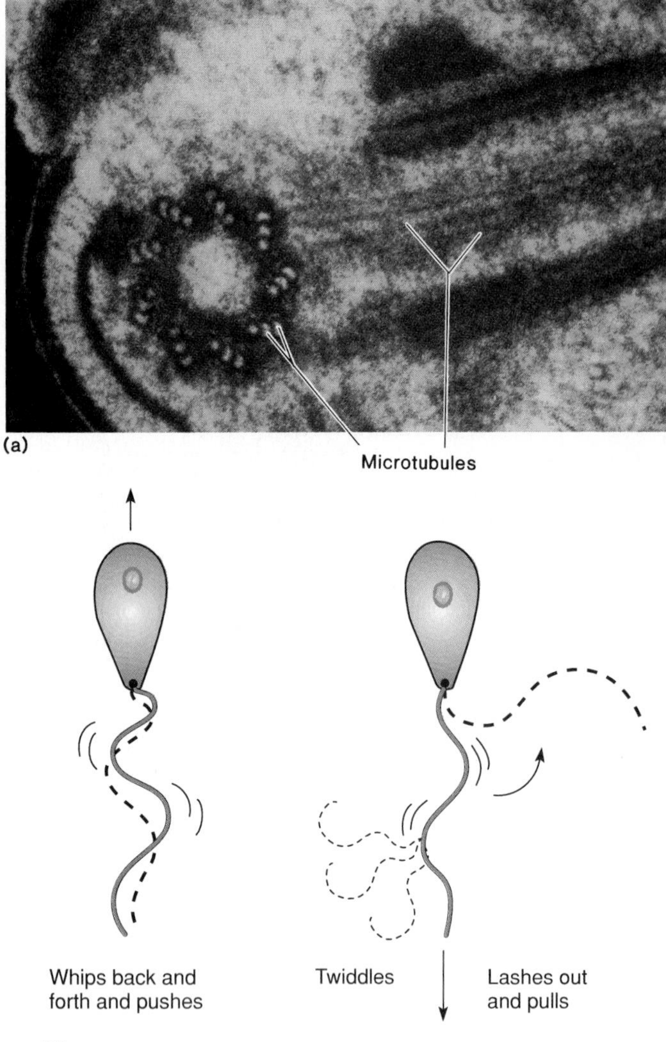

Figure 4.3 (*a*) Longitudinal section through a flagellum, showing microtubules. On the left is a cross section (circular area) that reveals the typical 9 + 2 arrangement found in both flagella and cilia. (*b*) Locomotor patterns seen in flagellates.

certain animal cells. In the ciliated protozoans, the cilia occur in rows over the cell surface, where they beat back and forth in regular oar-like strokes (see figure 4.31). Such protozoans are among the most rapid of all motile cells. The fastest ciliated protozoan may swim up to 2,500 μm/sec—that's a meter and a half per minute! On some cells, cilia also function as feeding and filtering structures. A notable example of the function of animal cilia occurs in the epithelial cells in the respiratory tract (see figure 12.2).

Surface Structures

Most eucaryotic cells have a *glycocalyx,* an outermost boundary that comes in direct contact with the environment (see figure 4.2). This structure is usually composed of polysaccharides and appears as a network of fibers, a slime layer, or a capsule much like the glycocalyx of procaryotes. Because of its positioning,

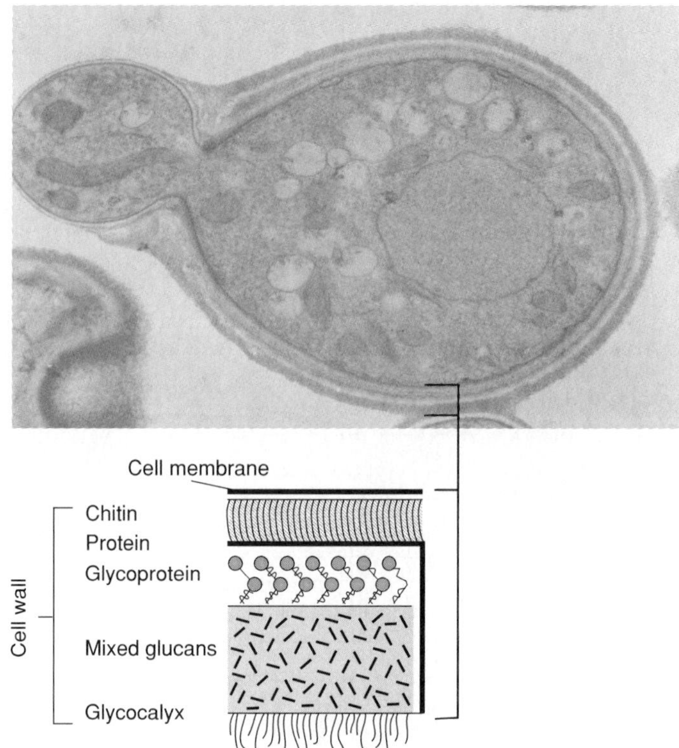

Figure 4.4 Structure of the fungal cell wall.

the glycocalyx contributes to protection, adherence of cells to surfaces, and reception of signals from other cells and from the environment. We will explore the function of receptors that are part of the glycocalyx in chapters 11 and 12 (host defenses and immune interactions). The nature of the layer supporting the glycocalyx varies among the several eucaryotic groups. Fungi and most algae have a thick, rigid cell wall, whereas protozoa, a few algae, and all animal cells lack a cell wall and have only a cytoplasmic membrane.

The Cell Wall

The cell walls of eucaryotic cells provide structural support and shape, but they are totally different in chemical composition from procaryotic ones. Fungal cell walls have a thick, inner layer of polysaccharide fibers composed of chitin or cellulose and a thin outer layer of protein (figure 4.4). The cell walls of algae are quite varied in chemical composition. Substances commonly found among various algal groups are cellulose, pectin,[1] mannans,[2] and minerals such as silicon dioxide and calcium carbonate.

1. A polysaccharide polymer of galacturonic acid.
2. A polymer of the sugar known as mannose.

Feature 4.1 The Marvelous Origin of Eucaryotic Cells

Biologists have grappled for years with the problem of how a cell as complex as the eucaryotic cell originated. One of the most fascinating theories is that of **endosymbiosis**, which proposes that eucaryotic cells arose when a smaller bacterial cell merged with a larger bacterial cell and, instead of either being harmed or destroyed, the two assumed a mutual relationship. As the smaller cells took up permanent residence, they came to perform specialized functions for the larger cell, such as food synthesis and oxygen utilization that enhanced the cell's versatility and survival. At some point, the relationship was obligatory, and the cell became a single entity. This phenomenon has been demonstrated in the laboratory with amebas, infected with bacteria, that gradually became dependent upon the bacteria for survival.

Although the theory of endosymbiosis has been greeted with some controversy, the biologist most responsible for its validation is Dr. Lynn Margulis. Using modern molecular techniques, she has accumulated convincing evidence of the relationships between the organelles of modern eucaryotic cells and the structure of bacteria. For example, the mitochondrion of eucaryotic cells is something like a tiny cell within a cell. It is capable of independent division, and contains its own DNA and ribosomes. Perhaps the most illuminating finding is that mitochondrial ribosomes are procaryotic and their DNA is very bacteria-like. The same holds true for the chloroplasts of algae and plants. These may have arisen from endosymbiotic cyanobacteria that provided their host cells with a built-in feeding mechanism (see chapter opening illustration). Dr. Margulis is even convinced that cilia and flagella are the consequence of endosymbiosis between spirochetes and the cell membrane of early eucaryotic cells.

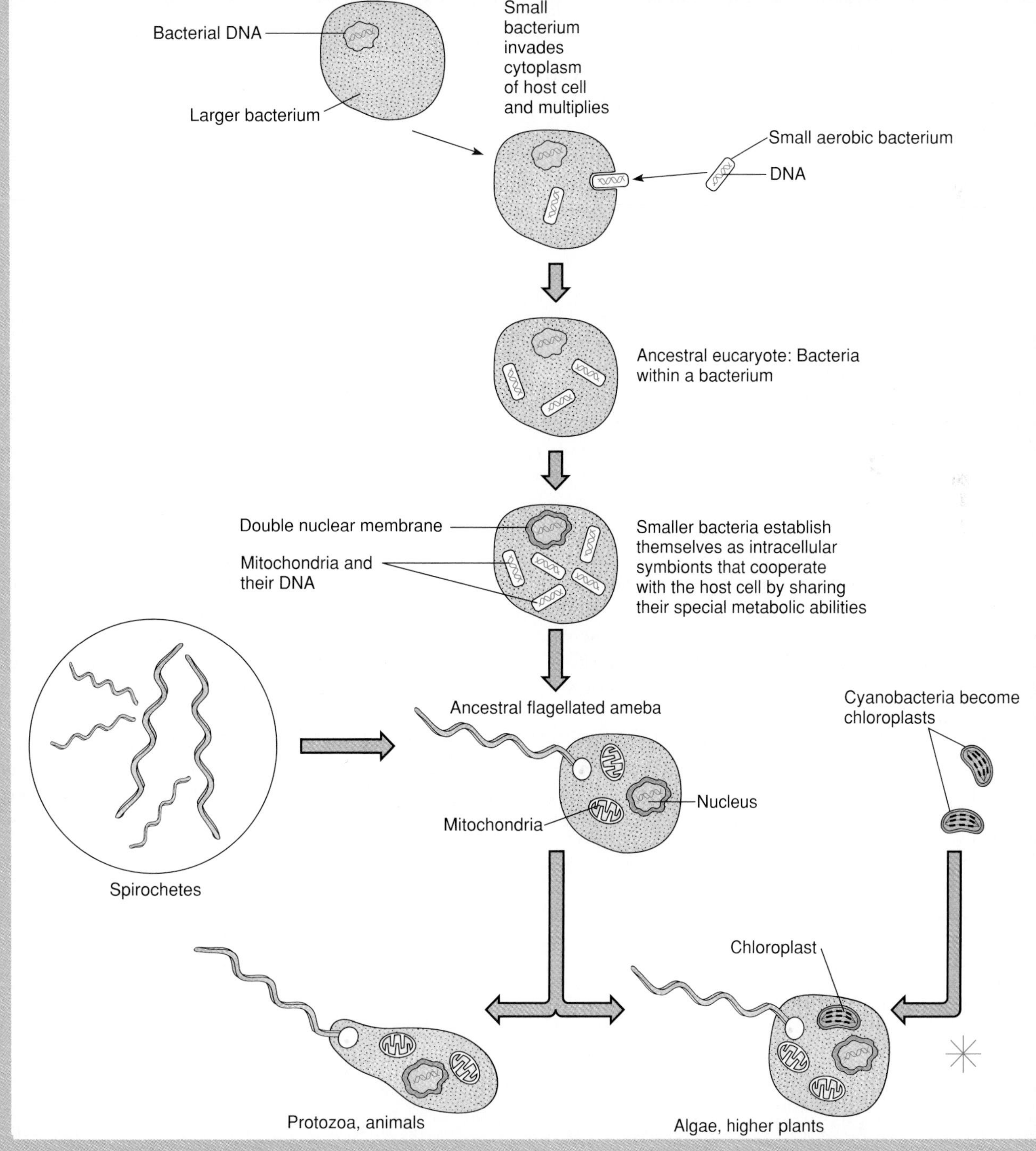

The Cytoplasmic Membrane

The cytoplasmic membrane of eucaryotic cells is a typical bilayer of lipids in which protein molecules are embedded. In addition to phospholipids, eucaryotic membranes also contain *sterols* of various kinds. Sterols are different from phospholipids in both structure and behavior (see figure 2.19). They are more rigid and less flexible, and in general, they help to stabilize eucaryotic membranes. This strengthening feature is extremely important in cells lacking a cell wall. Cytoplasmic membranes of eucaryotes are functionally similar to those of procaryotes: They are differentially permeable barriers that serve in transport (see chapter 6). Unlike procaryotes, eucaryotic cells also contain a number of individual membrane-bound organelles. In fact, the membranous component of a eucaryotic cell is so extensive that it comprises 60% to 80% of its volume.

Form and Function of Organelles

Eucaryotic cells do their cellular work with unique intracellular apparatuses called organelles. Each type of organelle carries out one or two functions, and it also partitions the cell into many small compartments, thereby serving to organize and separate various simultaneous events. Several functions of the eucaryotic cell proceed in an assembly-line fashion, with each organelle processing materials and then sending them on to another organelle in the series.

The Nucleus

The **nucleus** is a compact sphere that is the most prominent of a eucaryotic cell's organelles. It is separated from the cell cytoplasm by an external boundary called a **nuclear envelope.** Two unique architectural features of the envelope are: (1) It is composed of two parallel membranes separated by a narrow space, and (2) it is perforated with small, regularly spaced openings or pores formed at sites where the two membranes unite (figure 4.5). The nuclear pores are passageways through which macromolecules and large particles like ribosomes migrate from the nucleus to the cytoplasm and vice versa. Within the nucleus is a granular mass, the **nucleolus,** that may stain more intensely than the immediate surroundings because its high ribosomal RNA content attracts basic dyes. The role of the nucleolus in synthesizing the ribosomal subunits has been well established by numerous studies.

Another prominent feature of the nuclear contents in stained preparations is a network of fibers called **chromatin** because it has an affinity for certain dyes. At times in the cell's life cycle, the amount of chromatin changes in relation to the synthetic activity of the cell. In general, high degrees of cellular activity and growth are associated with reduced chromatin, and less active cells have more prominent chromatin.

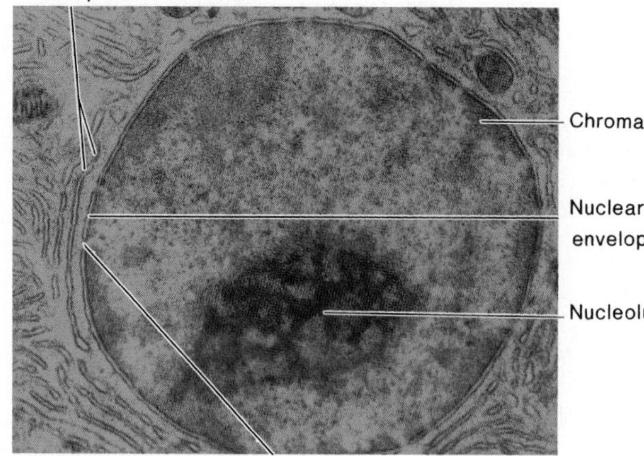

Figure 4.5 Electron micrograph section of an interphase nucleus, showing its most prominent features.

Analysis has shown that chromatin is composed of DNA and protein and that it actually comprises the eucaryotic **chromosomes,** central banks of genetic information in the cell. One does not actually see the chromosomes in the nucleus of most cells because they are long, linear chains of DNA bound in varying degrees to *histone*[3] proteins, and they are far too fine to be resolved as distinct structures without extremely high magnification. During the division cycle of a cell, however, it is essential that both the cell body and the chromosomes divide and separate equally into daughter cells. Division of the chromosomes occurs through **mitosis** (figure 4.6), and during this process, the structure of the nucleus and nucleolus is temporarily disrupted and the chromosomes themselves finally become visible. Additional levels of coiling and supercoiling of the DNA strand around the histones are responsible for the thick, X-shaped appearance of the chromosomes (figure 4.6). This orderly condensation of the DNA apparently prevents chromosomes from tangling as they are separated into new cells. This topic and other elements of genetics are covered in more detail in chapter 8.

nucleolus (noo-klee'-oh-lus) pl. nucleoli; L. diminutive of nucleus.

chromatin (kroh'-mah-tin) Gr. *chromos,* color.

3. Simple, basic proteins.

mitosis (my-toh'-sis) Gr. *mitos,* thread, and *osis,* a condition of.

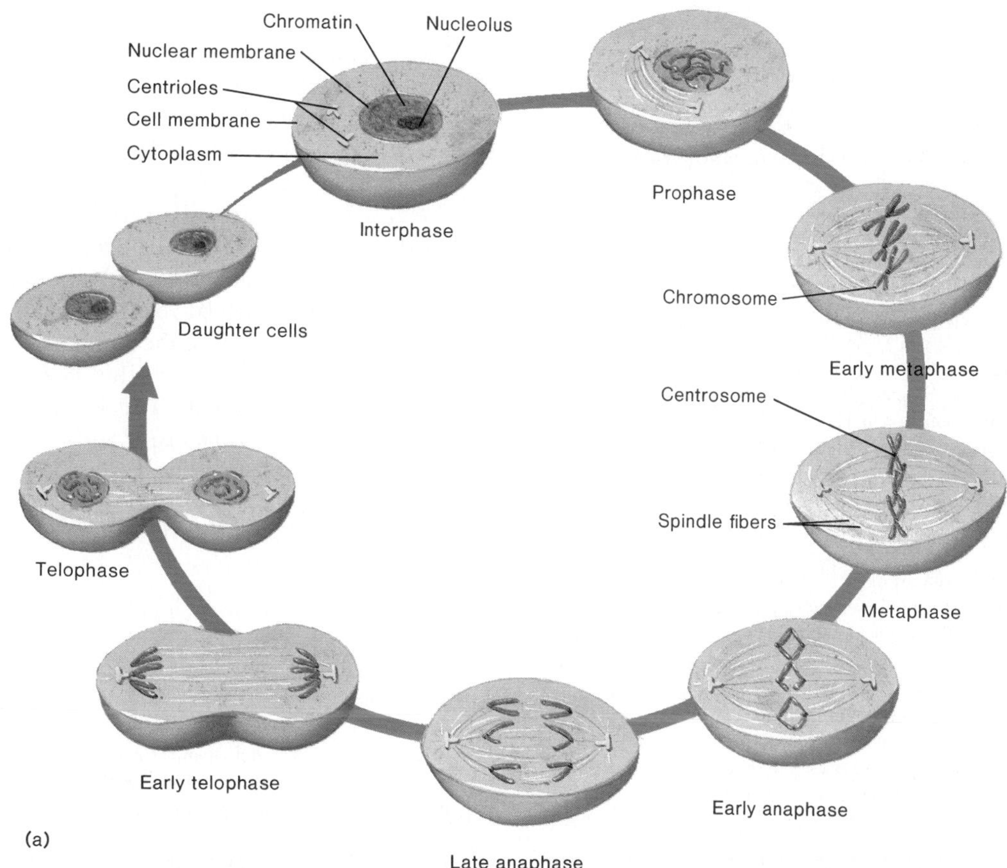

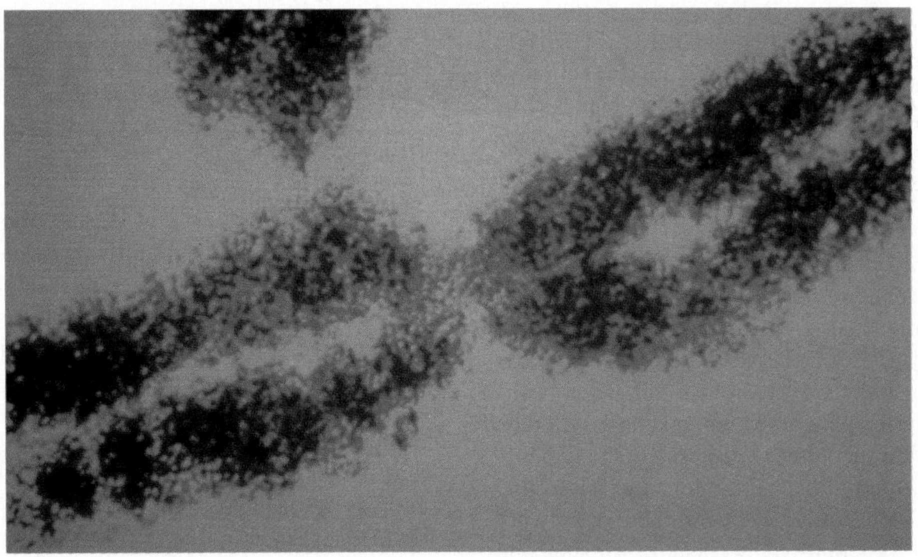

Figure 4.6 Changes in the cell and nucleus that accompany mitosis in a protozoan. (*a*) Prior to mitosis (at interphase), chromosomes are visible only as chromatin. As mitosis proceeds (early prophase), chromosomes take on a fine, threadlike appearance as they condense and the nuclear membrane and nucleolus are temporarily disrupted. (*b*) By metaphase, the chromosomes are fully visible as X-shaped structures. The shape is due to duplicated chromosomes attached at a central point, the centrosome. Spindle fibers attach to these and facilitate the separation of individual chromosomes during metaphase. Later phases serve in the completion of chromosomal separation and division of the cell proper into daughter cells.

Chromosome Number and Reproduction To ensure its continued existence, a eucaryotic cell must maintain in its nucleus a specified number of chromosomes that is typical for its species. These chromosomes may exist as single, unpaired (**haploid**) entities or in paired sets (**diploid**). Most fungi, many algae, and some protozoa are examples of organisms whose cells occur in the haploid state during most of the life cycle. For example, the chromosome number of one type of ameba is 25, and one species of the fungus *Penicillium* possesses 5 chromosomes. Regardless of the chromosome state of the adult organism, all gametes (sex cells) are haploid.

Most animal and plant cells (and some eucaryotic microbes) spend the greater part of their life cycle in the diploid state as a consequence of their origin from gamete fusion. When a female haploid gamete (egg) fuses with a male haploid gamete (sperm), this produces a double (diploid) set of chromosomes in the offspring. Examples are humans with 46 chromosomes (23 pairs of different chromosomes), a parasitic blood fluke with 16 chromosomes (8 pairs), and a redwood tree with 22 chromosomes (11 pairs).

During normal growth or cell division (asexual reproduction) of all eucaryotic organisms, the chromosome number, whether haploid or diploid, is maintained by mitosis (see figures 4.6 and 4.7). Sexual reproduction in diploid organisms, however, necessitates reduction of the diploid chromosome number to haploid, since gametes must be haploid for proper fertilization. This is accomplished through reduction division or **meiosis** (figure 4.7*b*). During this process, diploid cells in the sex organs undergo two sets of divisions that result in haploid sex cells. Formation of a diploid zygote then restores the chromosome number characteristic of that organism. Another example of the function of meiosis occurs in the sexual reproductive cycles of haploid fungi. Normal mitosis alone can produce their gametes, but after these gametes unite, they form a diploid zygote. In order to restore the fungus to the haploid state, the diploid nucleus undergoes meiosis and produces haploid offspring (see figure 4.7*a* and 4.20).

Endoplasmic Reticulum

The **endoplasmic reticulum** (**ER**) can be likened to a labyrinth (an intricate passageway) on a microscopic scale. In sections it appears as a parallel series of thin, flat pouches, bounded by membranes. Two kinds of endoplasmic reticulum are the **rough endoplasmic reticulum** (**RER**) (figure 4.8) and the **smooth endoplasmic reticulum** (**SER**). Electron micrographs show that the RER originates from the outer membrane of the nuclear envelope and extends in a continuous network through the cytoplasm, even out to the cell membrane. This architecture permits the RER to transport materials from the nucleus to the cytoplasm and ultimately to the cell's exterior. The RER appears rough because of large numbers of ribosomes embedded in the membranes, which underlines its other function as a synthetic organelle. Proteins synthesized on the ribosomes are shunted into the cavity of the reticulum and held there for later packaging and transport. The smooth ER is a closed tubular network without ribosomes that functions in nutrient processing and synthesis and in storage of nonprotein macromolecules such as lipids.

Golgi Complex

The **Golgi complex,** also called the Golgi body or apparatus, is a discrete organelle consisting of a stack of flattened, disc-shaped sacs, or *cisternae.* These sacs have outer limiting membranes and cavities like the endoplasmic reticulum, but they do not form a continuous network (figure 4.9). This organelle is always closely associated with the endoplasmic reticulum both in its location and function. At a site where it meets the Golgi complex, the endoplasmic reticulum buds off tiny membrane-bound packets of protein called *transitional vesicles* that are picked up by the forming face of the Golgi complex. Once in the complex itself, the proteins are often modified by the addition of polysaccharides and lipids. The final action of this apparatus is to pinch off finished *condensing vesicles* that will function in the cytoplasm (lysosomes) or be transported outside the cell (figure 4.10). The production of antibodies and receptors in human white blood cells involves this packaging and transport system (see chapter 13).

Nature's Assembly Line As the keeper of the eucaryotic genetic code, the nucleus ultimately governs and regulates all cell activities. But because the nucleus remains fixed in a specific cellular site, it must direct these activities through a structural and chemical network (figure 4.10). This network includes ribosomes, which originate in the nucleolus, and the rough endoplasmic reticulum, which is continuously connected with the nuclear envelope. Initially, a genetic code on DNA is copied and passed out through the nuclear pores directly to the ribosomes on the endoplasmic reticulum. Here, specific proteins are synthesized according to that particular code and deposited in the central pocket of the endoplasmic reticulum. After being picked up by a Golgi complex, the protein products are completed and packaged into vesicles that can be used by the cell in a variety of ways. Some of the vesicles contain enzymes to digest food inside the cell, and other vesicles are secreted to digest materials outside the cell.

Lysosomes and Vacuoles

A **lysosome** is one type of vesicle originating from the Golgi complex that contains digestive enzymes. So long as the membrane around a lysosome remains intact, the enzymes are inactive. In old or damaged cells, breakage of these "suicide bags" releases hydrolytic enzymes that can self-digest and destroy the cell. This is a natural mechanism for eliminating old or damaged cells in multicellular organisms. Other types of vesicles include *vacuoles,* which are membrane-bound sacs containing fluids or solid particles to be digested, excreted, or stored. They are formed in phagocytic cells (certain white blood cells and protozoans) in response to food and other substances engulfed through *endocytosis.* The contents of a food vacuole are digested through the merger of the vacuole with a lysosome (figure 4.11). Other types of vacuoles are used in storing reserve food

haploid (hap'-loyd) Gr. *haplos,* simple, and *eidos,* form.

diploid (dip'-loyd) Gr. *di,* two, and *eidos,* form.

meiosis (my-oh'-sis) Gr. *meiosis,* diminution. A process that permits paired chromosome sets to be reduced to single sets.

endoplasmic (en''-doh-plas'-mik) Gr. *endo,* within, and *plasm,* something formed; reticulum (rih-tik'-yoo-lum) L. *rete,* net or network.

Golgi (gol'-jee) Named for C. Golgi, an Italian histologist who first described the apparatus in 1898.

vesicle (ves'-ik-l) L. *vesios,* bladder. A small sac containing fluid.

lysosome (ly'-soh-sohm) Gr. *lysis,* dissolution, and *soma,* body.

vacuole (vak'-yoo-ohl) L. *vacuus,* empty. Any membranous space in the cytoplasm.

endocytosis (en''-doh-sy-toh'-sis) Gr. *endo,* in, and *kytos,* a hollow vessel.

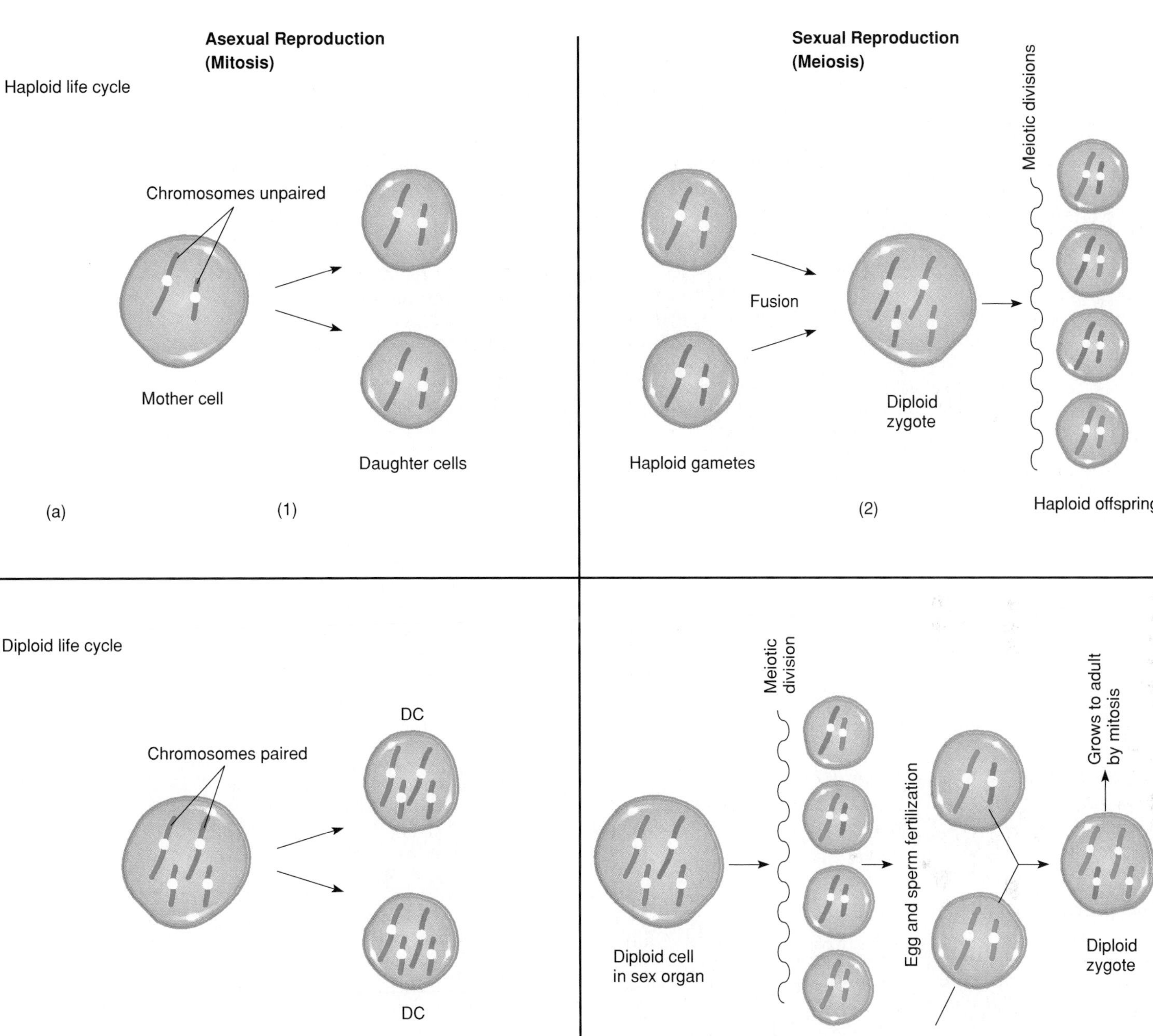

Figure 4.7 Schematic of haploid versus diploid life cycles, showing the roles of mitosis and meiosis in reproduction and maintaining an organism's chromosome number. (*a*) Cells with haploid life cycle (algae, fungi, many protozoa). Chromosomes are unpaired. (*1*) Asexual reproduction is accomplished through mitosis, which maintains the chromosome number in the daughter cells. (*2*) Sexual reproduction involves the fusion of specialized sex cells (already haploid), forming a diploid zygote. To restore the haploid state in the offspring, meiosis of the zygote must occur. Note that reassortment of chromosomes may produce haploid offspring unlike either original gamete. (*b*) Cells with a diploid life cycle (animals, plants, some algae and protozoa). Chromosomes occur in pairs. (*3*) Asexual reproduction through mitosis maintains the chromosome number. (*4*) In order to reproduce sexually, the normally diploid cells must produce haploid gametes through meiotic division. During fertilization, these haploid gametes (eggs, sperm) form a zygote that restores the diploid number. Depending upon which gametes fuse, combinations of chromosomes unlike the original parents are possible in offspring.

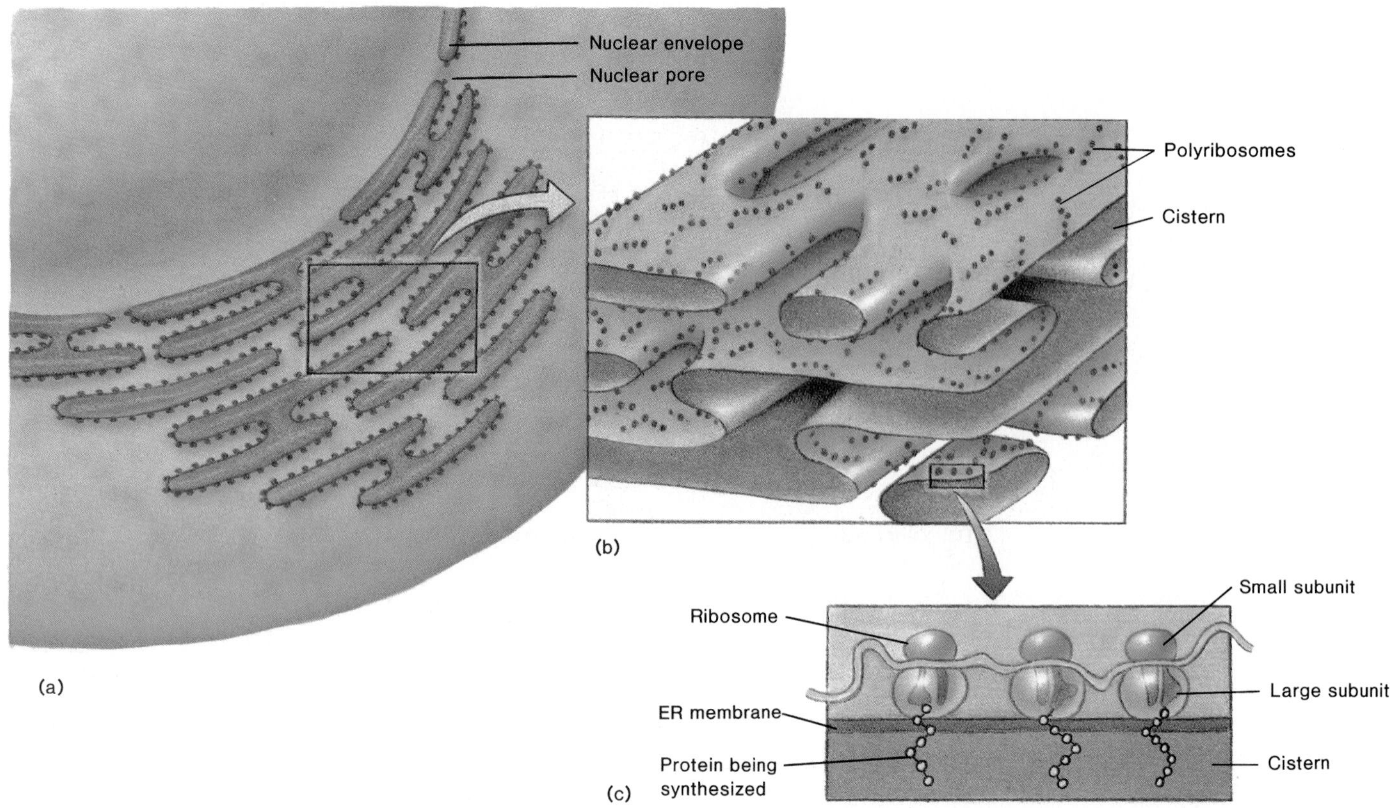

Figure 4.8 The origin and detailed structure of the rough endoplasmic reticulum (RER). (*a*) Schematic view of the origin of the RER from the outer membrane of the nuclear envelope. (*b*) Three-dimensional projection of the RER. (*c*) Detail of the orientation of a ribosome on the RER membrane.

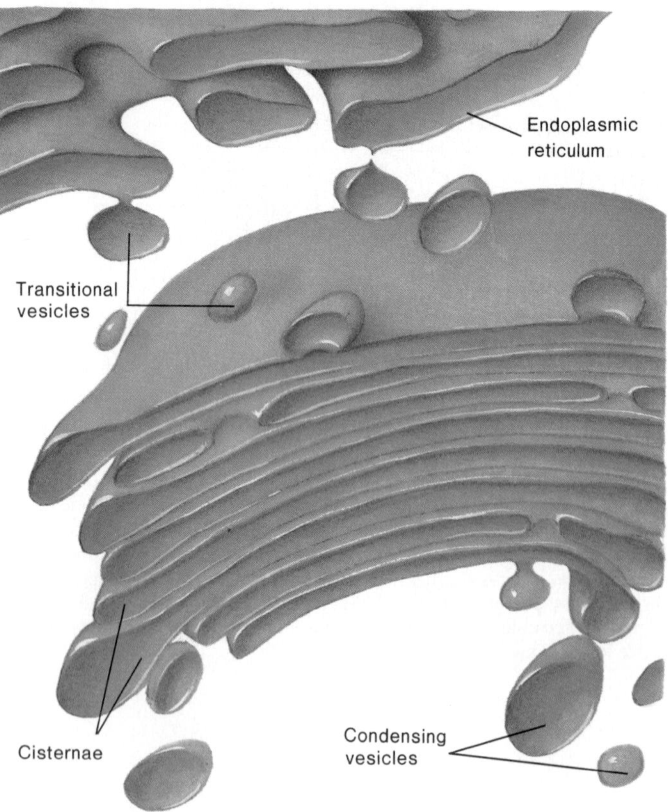

Figure 4.9 Detail of the Golgi complex, showing flattened layers of cisternae. Vesicles enter the upper surface and leave the lower surface of the complex.

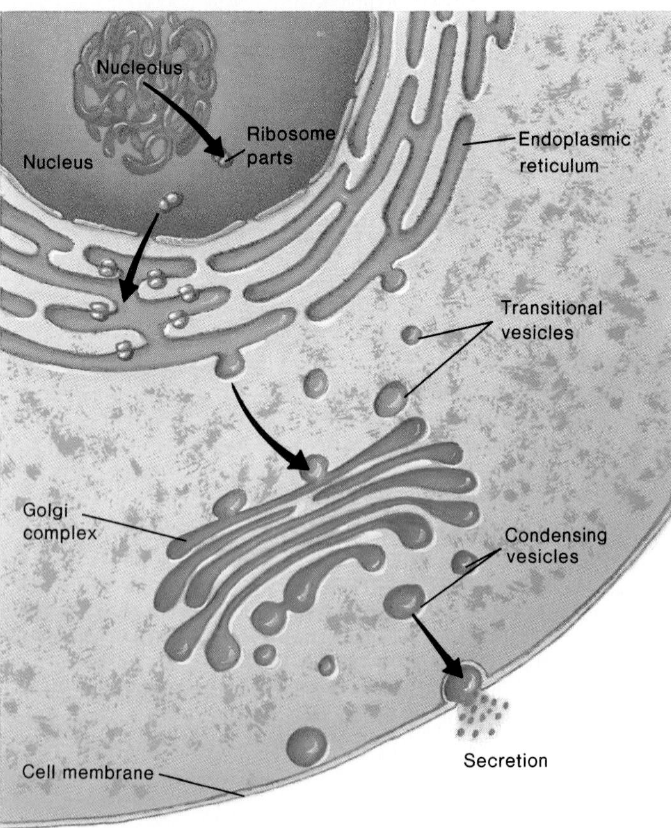

Figure 4.10 The cooperation of organelles in cell synthesis: nucleus → RER → Golgi complex → vesicles.

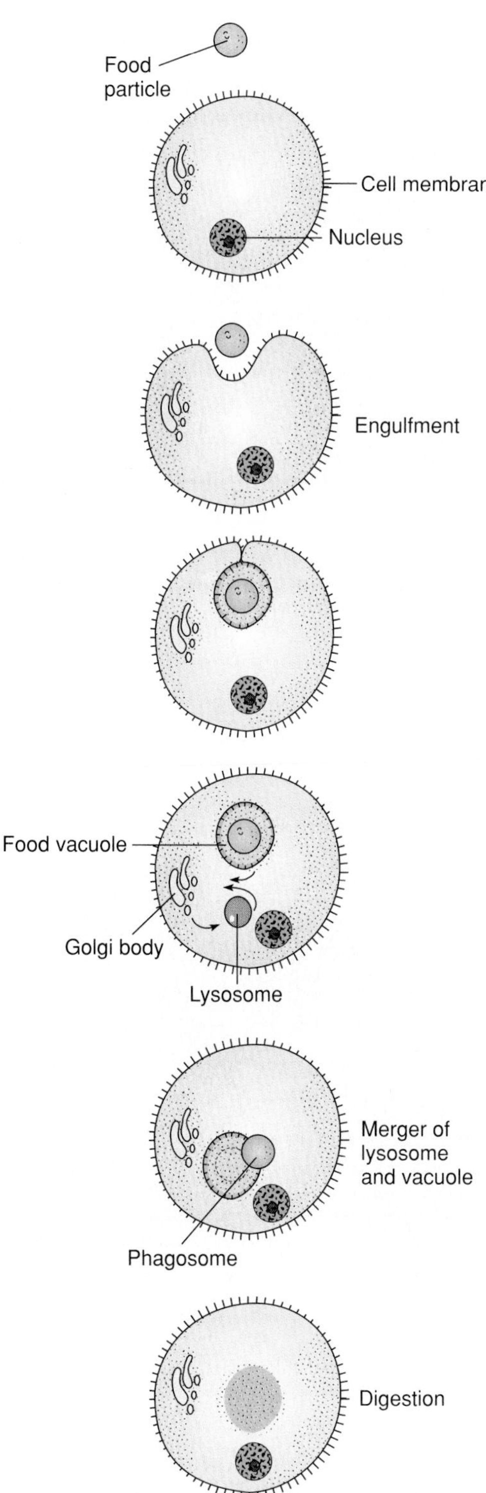

Figure 4.11 The origin and action of lysosomes in phagocytosis.

such as fats and glycogen. Protozoa living in freshwater habitats regulate osmotic pressure by means of contractile vacuoles, which regularly expel excess water that has diffused into the cell (see figure 4.30).

Mitochondria: Energy Generators of the Cell

Although the nucleus is the cell's control center, none of the cellular activities it oversees could proceed without a constant supply of energy, the bulk of which is generated in most eucaryotes by **mitochondria**. When viewed with light microscopy, mitochondria appear as round to oblong particles scattered through the cytoplasm. Closer scrutiny of the internal ultrastructure reveals that a single mitochondrion consists of two separate sets of membranes: (1) an outer membrane that is smooth and continuous, forming the external contour, and (2) an inner, folded membrane nestled neatly within the outer membrane (figure 4.12*a*). The folds on the inner membrane, called *cristae,* appear fingerlike in cross section, but are actually shelflike. Cristae provide an increased surface for metabolic activities while simultaneously partitioning the organelle into separate compartments. The spaces around the cristae are filled with a chemically complex fluid called the *matrix* (figure 4.12*b*).

The mitochondrial cristae are a major site of cellular respiration. This is an oxygen-using process that extracts chemical energy contained in nutrient molecules and stores it in the form of high-energy molecules, or ATP. The detailed functions of mitochondria will be covered in chapter 7. Mitochondria (along with chloroplasts) are unique among organelles in that they divide independently of the cell, they contain circular strands of DNA, and they have 70S (not 80S) ribosomes. As we saw earlier, these findings have prompted some intriguing speculations on their evolutionary origins (see feature 4.1).

Chloroplasts

Chloroplasts are remarkable packets found in algae and plant cells that are capable of converting the energy of sunlight into chemical energy through photosynthesis. The photosynthetic role of chloroplasts makes them the primary producers of nutrients upon which all other organisms (except certain bacteria) ultimately depend. Another important photosynthetic product of chloroplasts is oxygen gas. Chloroplasts are similar to mitochondria in overall structure, although chloroplasts are larger, contain special pigments, and are much more varied in overall shape.

Although there are differences among various algal chloroplasts, most are generally composed of two membranes, one enclosing the other. The even, outer membrane completely covers an inner membrane folded into small disclike sacs called *thylakoids* that are stacked upon one another into *grana.* These structures carry the green pigment chlorophyll and sometimes

---

mitochondria (my″-toh-kon′-dree-ah) sing. mitochondrion; Gr. *mitos,* thread, and *chondrion,* granule.

cristae (kris′-te) sing. crista; L. *crista,* a comb.

matrix (may′-triks) L. *mater,* mother or origin.

chloroplast (klo′-roh-plast) Gr. *chloros,* green, and *plastos,* to form.

thylakoid (thy′-lah-koid) Gr. *thylakon,* a small sac, and *eidos,* form.

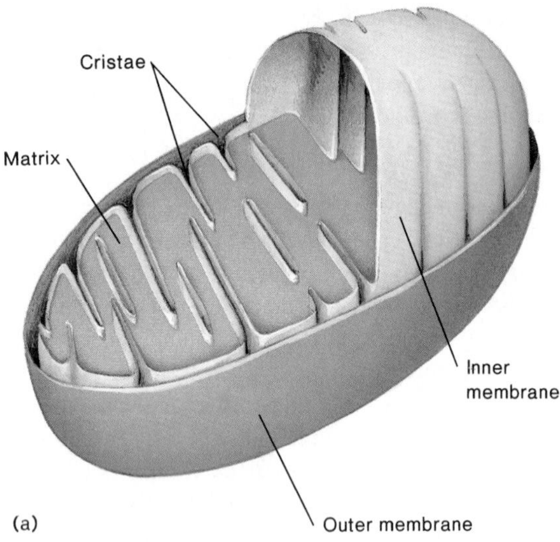

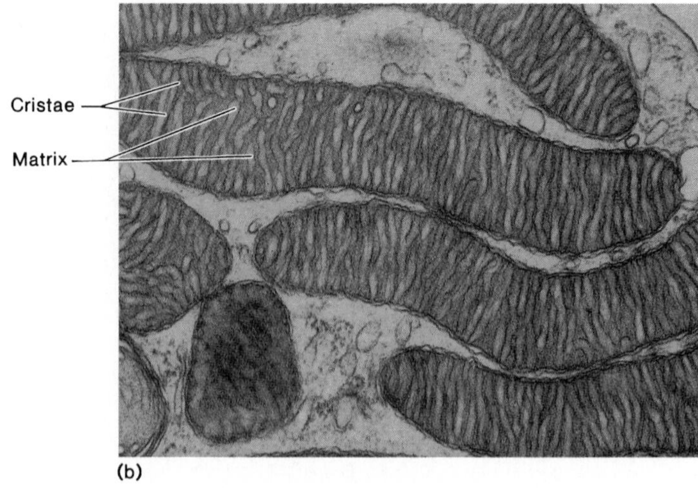

Figure 4.12 General structure of a mitochondrion as seen (*a*) in three-dimensional projection and (*b*) by electron micrograph. In most cells, mitochondria are elliptical or spherical, though in certain fungi, algae, and protozoa, they are long and filament-like.

additional pigments as well. Surrounding the thylakoids is a ground substance called the *stroma* (figure 4.13). The role of the photosynthetic pigments is to absorb and transform solar energy into chemical energy that is then converted by reactions in the stroma to carbohydrates. We will explore some important aspects of photosynthesis again in chapters 7 and 22.

The Cytoskeleton: The Network That Supports Cell Structure

The cytoplasm of a eucaryotic cell is much more than a flabby, unstructured gel. It is penetrated by a flexible framework of molecules called the cytoskeleton (figure 4.14). This framework appears to have several functions, such as anchoring organelles, providing support, and permitting shape changes and movement in some cells. The two main types of cytoskeletal elements are **microfilaments** and **microtubules.** Microfilaments are thin protein strands that attach to the cell membrane and form a network through the cytoplasm. Some microfilaments are responsible for movements of the cytoplasm, often made evident by the streaming of organelles around the cell in a cyclic pattern. Other microfilaments are active in *ameboid motion,* a type of movement typical of cells such as amebas and phagocytes that produce extensions of the cell membrane (pseudopods) into which the cytoplasm flows (see figure 4.30). Microtubules are long, hollow tubes made of spiralling rows of tubulin protein molecules. Their basic functions are to maintain the shape of eucaryotic cells without walls and to transport substances from one part of a cell to another. Microtubules are the spindle fibers that play an essential role in mitosis by attaching to chromosomes and separating them into daughter cells. As indicated earlier, they are also an essential component of cilia and flagella.

stroma (stroh′-mah) Gr. *stroma,* mattress or bed.

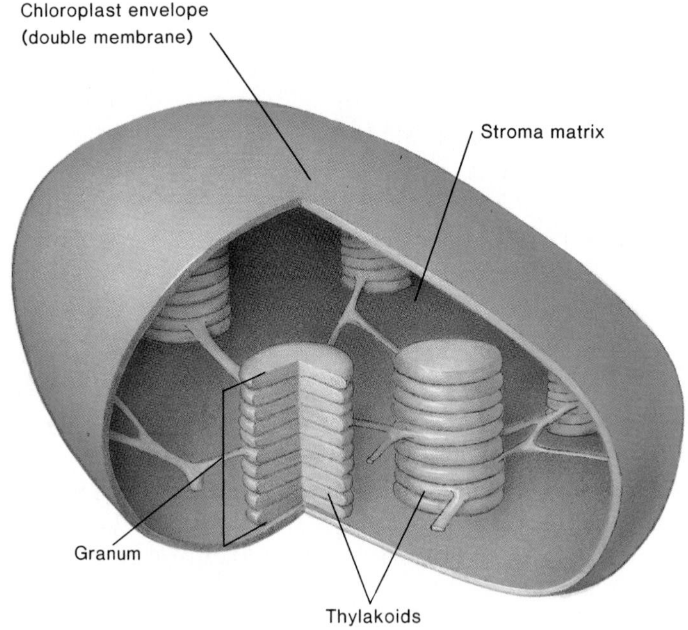

Figure 4.13 Detail of an algal chloroplast.

Ribosomes

In an electron micrograph of a eucaryotic cell, ribosomes are numerous, tiny particles that give a stippled appearance to the cytoplasm. Ribosomes are distributed in two ways: Some are scattered in the cytoplasm (free), while others are intimately associated with the rough endoplasmic reticulum as previously described. The basic structure of eucaryotic ribosomes is similar to that of procaryotes, both being composed of large and small subunits of ribonucleoprotein (see figure 4.8). By contrast, however, the eucaryotic ribosome is larger (80S) and originates from the nucleolus. As in the procaryotes, eucaryotic ribosomes are the staging areas for protein synthesis.

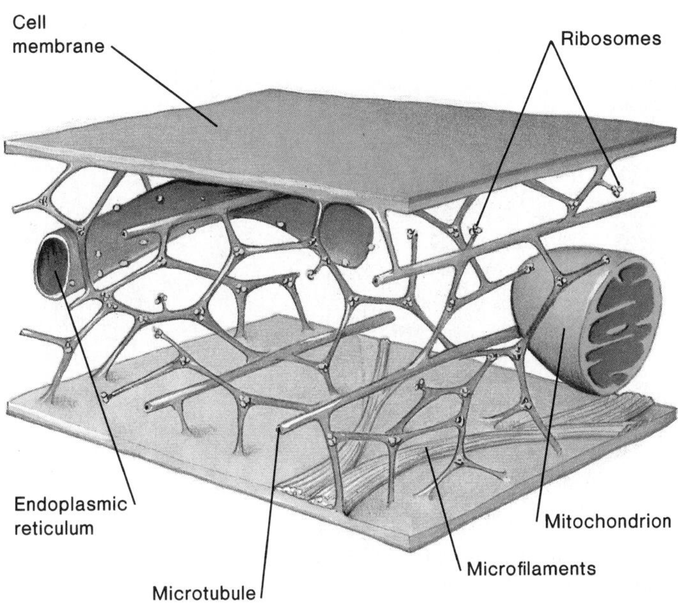

Figure 4.14 A model of the cytoskeleton, depicting the relationship between microtubules and microfilaments and organelles.

Survey of Eucaryotic Microorganisms

With the general structure of the eucaryotic cell in mind, let us next examine the amazingly wide range of adaptations that this cell type has undergone. The following sections contain a general survey of the principal eucaryotic microorganisms—fungi, algae, protozoa, and parasitic worms—while simultaneously introducing elements of their structure, life history, classification, identification, and importance to humans.

The Kingdom of the Fungi: Myceteae

The position of the *fungi* in the biological world has been debated for many years. They were originally classified with the green plants (along with algae and bacteria), and were later separated from plants and placed in a group with algae and protozoa (the Protista). Even at that time, many *mycologists* were struck by several unique qualities of fungi that warranted placement in a separate kingdom, and finally in 1969, this was achieved.

The **Kingdom Myceteae** is large and filled with forms of great variety and complexity. The approximately 100,000 species of fungi can be divided for practical purposes into the *macroscopic fungi* (mushrooms, puffballs, gill fungi) and the *microscopic fungi* (molds, yeasts). Although the majority of fungi are either unicellular or colonial, a few complex forms such as mushrooms and puffballs do have a limited form of cellular specialization.

Cells of the microscopic fungi exist in two basic morphological types: yeasts and hyphae. A **yeast** is any unicellular fungus with round to oval cells that reproduces asexually by giving off small cells called **buds.** Some species may form *pseudohyphae,* a string of yeast cells that have not separated after each new bud is formed (figure 4.15). A true **hypha** is a long, tubular or threadlike cell typical of the filamentous fungi or **molds** (figure 4.16). Some fungal cells exist only in a yeast form; others occur primarily as hyphae; and a few, called *dimorphic,* can exist as either, depending upon growth conditions. This changeability in growth form is particularly characteristic of some pathogenic molds (see chapter 18).

(a)

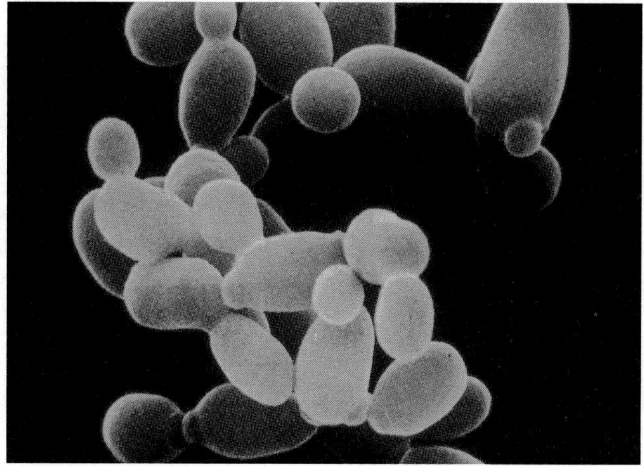

(b)

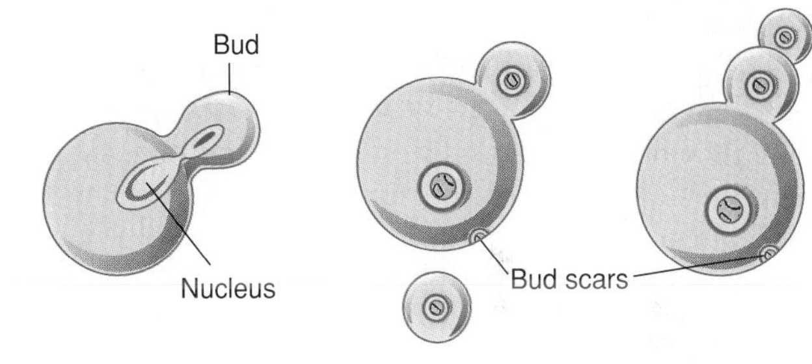

(c)

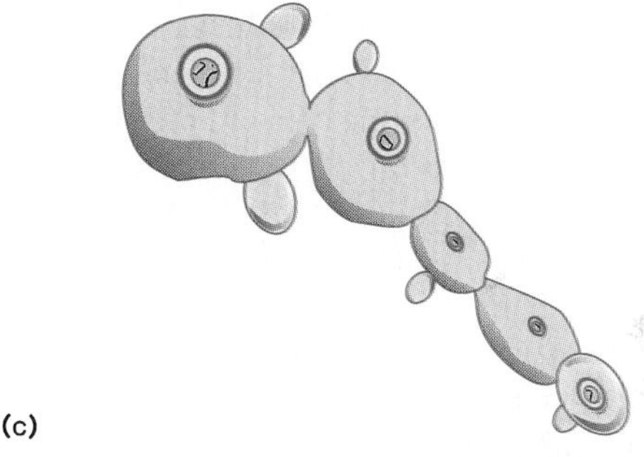

Figure 4.15 Microscopic morphology of yeasts. (*a*) Scanning electron micrograph of the brewer's or baker's yeast *Saccharomyces cerevisiae* (×21,000). (*b*) Formation and release of yeast buds. (*c*) Formation of pseudohypha (a chain of budding yeast cells).

fungi (fun'-jye) sing. fungus; Gr. *fungos,* mushroom.
mycologist (my-kol'-uh-jist) Gr. *mykes,* standard root for fungus. Scientists who study fungi.

pseudohyphae (soo''-doh-hy'-fee) sing. pseudohypha; Gr. *pseudo,* false, and *hyphe.*
hypha (hy'-fah) pl. hyphae (hy'-fee) Gr. *hyphe,* a web.
dimorphic (dy-mor'-fik) Gr. *di,* two, and *morphe,* form.

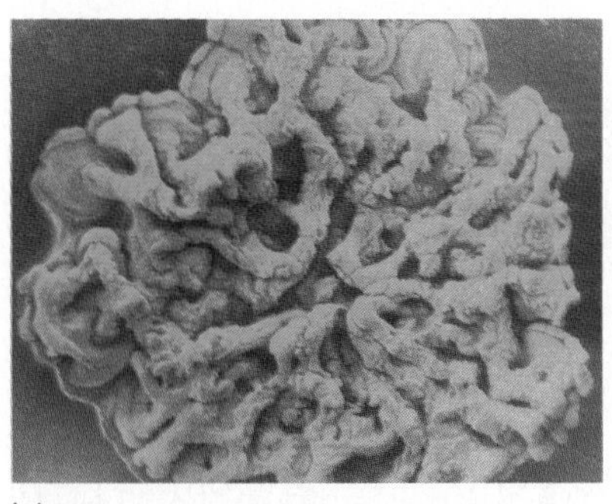

(a)

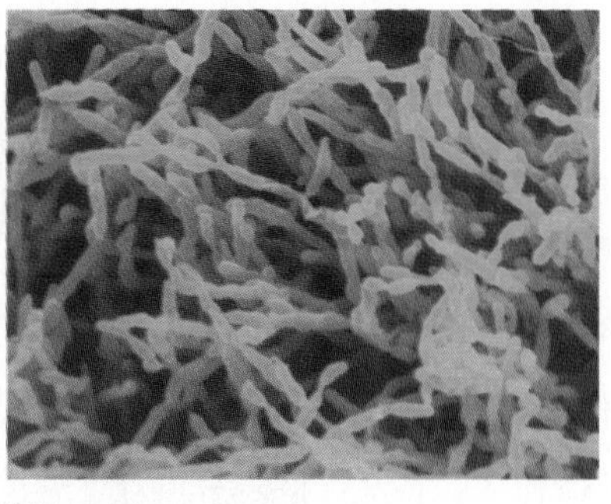

(b)

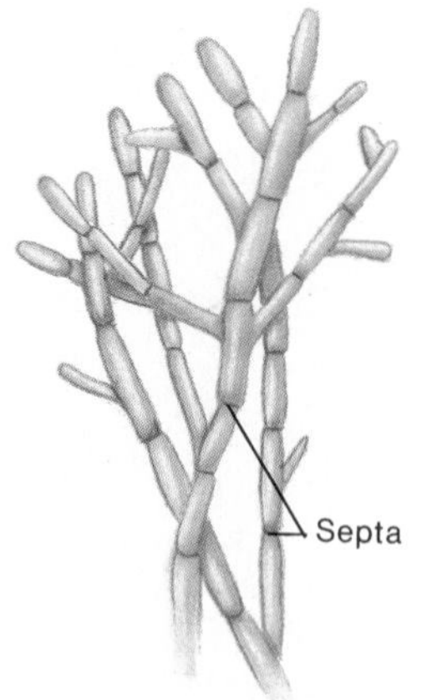

(c) Septate hyphae Nonseptate hyphae

Figure 4.16 Mold morphology from macroscopic to microscopic. (*a*) Mold colony arises from a fine network of intertwining hyphae (mycelium) (×70). (*b*) Scanning electron micrograph of hyphae making up the colony (×8,200). (*c*) Basic structural types of hyphae.

Fungal Nutrition

All fungi acquire nutrients from preformed organic materials called **substrates** (figure 4.17). Most fungi are *saprobes,* meaning that they obtain these substrates from the remnants of dead plants and animals in soil, fresh water, or salt water. For parasitic fungi, the substrate is actually the bodies of living animals or plants, though very few fungi absolutely require a living host. In general, the fungus penetrates the substrate and reduces it to small molecules that can be absorbed by the cells. Fungi have enzymes for digesting an incredible array of substrates, including feathers, hair, cellulose, petroleum products, wood, and rubber. It has been said that every naturally occurring organic material on earth can be attacked by some type of fungus. Fungi are often found in nutritionally poor or adverse environments. In our laboratory we occasionally discover fungi in mineral and weak acid solutions ("bottle imps") and on specimens preserved in formaldehyde. Various fungi thrive in substrates with high salt or sugar content, at relatively high temperatures, and even in snow and glaciers. Though only a few species (less than 300) can parasitize plants and animals, their medical and agricultural impact is extensive. Fungi often have bright colors, but these colors are not due to photosynthetic pigments.

Organization of Microscopic Fungi

The cells of most microscopic fungi grow in discrete and distinctive colonies. The colonies of yeasts are much like those of bacteria in that they have a soft, uniform texture and appearance. The colonies of filamentous fungi are noted for the striking cottony, hairy, or velvety textures that arise from their microscopic organization and morphology. The woven, intertwining mass of hyphae that makes up the body or colony of a mold is called a **mycelium** (see figure 4.16).

Although hyphae contain the usual eucaryotic organelles, they also have some unique organizational features. **Nonseptate** hyphae consist of one long, continuous cell not divided into individual compartments by **septa** or cross walls (see figure 4.16*c*). With this construction, the cytoplasm and organelles move freely from one region to another, and each hypha may have several nuclei. In more complex species, the hyphae are divided into segments by cross walls, a condition called **septate.** The nature of the septa varies from solid partitions with no communication between the compartments to partial walls with small pores that allow the flow of organelles and nutrients between adjacent compartments.

Hyphae can also be classified according to their particular function. **Vegetative hyphae** (mycelia) are responsible for the visible mass of growth that appears on the surface of a substrate and penetrates it to digest and absorb nutrients. During the development of a fungal colony, the vegetative hyphae ultimately give rise to structures called **fertile** (reproductive) or **aerial hyphae** that orient vertically from the vegetative mycelium. These hyphae are responsible for the production of spores. Other specializations of hyphae are pictured in figure 4.18.

Reproductive Strategies and Spore Formation

Fungi have many complex and successful reproductive strategies. Most can propagate by the simple outward growth of existing hyphae or by fragmentation, in which a separated piece of mycelium can generate a whole new colony. But the primary reproductive mode of fungi involves the production of various types of spores. Do not confuse fungal spores with the more resistant, nonreproductive bacterial spores. Fungal spores are responsible not only for multiplication but also for survival, variation, and dissemination. Due to their compactness and relatively light weight, spores are dispersed widely through the environment by air, water, and living things. Normal outdoor air may contain 10^5 fungal spores per cubic meter, and moist air containing nutrients may have as many as 10^9 spores per cubic meter. Upon encountering a favorable substrate, a spore will germinate and produce a new fungus colony in a very short time (see figure 4.18).

The fungi exhibit a marked diversity in spores, thus they are largely classified and identified by their spores and spore-forming structures. Although there are some elaborate systems for naming and classifying their spores, we will present a basic overview of the principal types. The most general subdivision is

saprobe (sap′-rohb) Gr. *sapros,* rotten, and *bios,* to live. An organism whose nutrient source is dead organisms or their parts.

mycelium (my-see′-lee-um) pl. mycelia; Gr. *mykes,* fungus, and *helos,* nail.

septa (sep′-tah) sing. septum; L. *seepio,* to fence or wall.

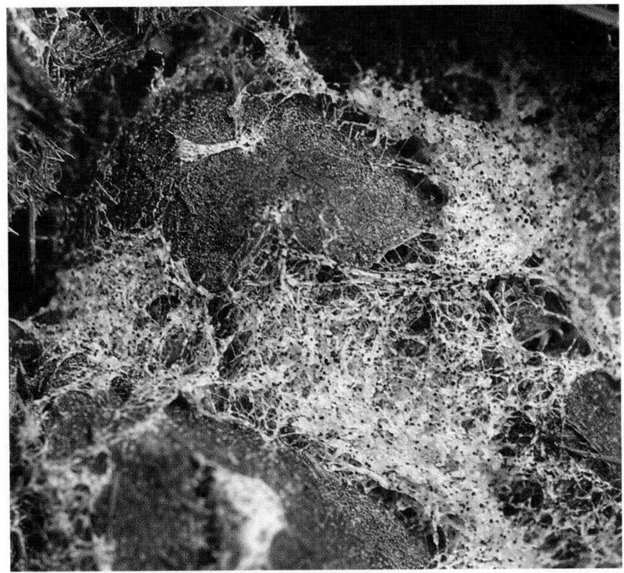

(a)

(b)

(c)

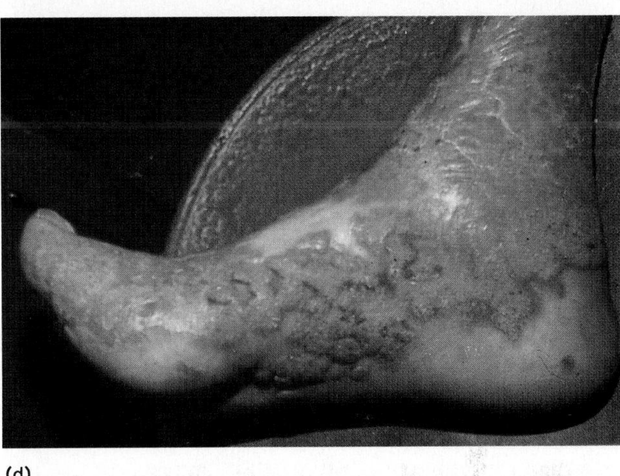

(d)

Figure 4.17 A small assortment of nutritional sources (substrates) of fungi. (*a*) A fungal mycelium growing in soil and forest litter (black spheres are spores). (*b*) A common mold, *Penicillium*, growing on oranges. (*c*) Cup fungus on tree bark. The cup is a fruiting body that holds the sexual spores. (*d*) The skin of the foot infected by a soil fungus, *Fonsecaea pedrosoi*.

based on the way the spores arise. **Asexual spores** are the products of mitotic division of a single parent cell, and **sexual spores** are formed through a process of two parental nuclei fusing, followed by meiosis.

Asexual Spore Formation Based upon the appearance of the fertile hypha and the origin of spores, there are two subtypes of asexual spores (figure 4.19):

1. **Sporangiospores** are formed by successive cleavages within a saclike head called a **sporangium**, which is attached to a stalk, the sporangiophore. These spores are initially enclosed but will be released when the sporangium ruptures.

2. **Conidia** (conidiospores) are free spores not enclosed by a special spore-bearing sac. They develop by being pinched off the tip of a special fertile hypha, or by the segmentation of a preexisting vegetative hypha. Conidia are the most common asexual spores, and they come in the following varieties:

   arthrospore (ar′-thro-spor) Gr. *arthron*, joint. A rectangular spore formed when a septate hypha fragments at the cross walls.

   chlamydospore (klam′ih-doh-spor) Gr. *chlamys*, cloak. A spherical conidium formed by the thickening of a hyphal

sporangium (spo-ran″-jee-um) sing. sporangia; Gr. *sporos*, seed, and *angeion*, vessel.

conidia (koh-nid′-ee-ah) sing. conidium; Gr. *konidion*, a particle of dust.

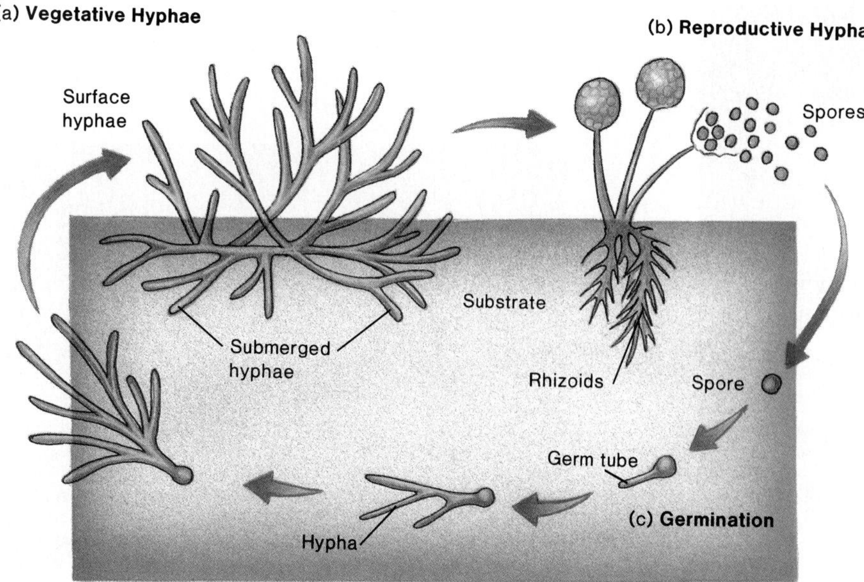

Figure 4.18 Functional types of hyphae in the mold *Rhizopus*. (*a*) Vegetative hyphae are those surface and submerged filaments that digest, absorb, and distribute nutrients from the substrate. Some species also have special anchoring structures called rhizoids. (*b*) Later, as the mold matures, it sprouts reproductive hyphae that produce asexual spores (as here) or sexual spores. (*c*) During the asexual life cycle, the dispersed mold spores settle on a usable substrate and send out germ tubes that elongate into hyphae. Through continued growth and branching, an extensive mycelium is produced. So prolific are the fungi that a single colony of mold can easily contain 5,000 spore-bearing structures. If each of these released 2,000 single spores and if every spore released from every sporangium on every mold were able to germinate, we would soon find ourselves in a sea of mycelia. Most spores do not germinate, but enough are successful to keep the numbers of fungi and their spores very high in most habitats.

cell. It is released when the surrounding hypha fractures and it serves as a survival or resting cell.

blastospore A spore produced by budding from a parent cell that may be a yeast or another conidium; also called a bud.

phialospore (fy'-ah-lo-spor) Gr. *phialos,* a vessel. A conidium that is budded from the mouth of a vase-shaped sporogenous cell called a phialide or sterigma, leaving a small collar.

microconidium and *macroconidium.* The smaller and larger conidia formed by the same fungus under varying conditions. Microconidia are one-celled, and macroconidia have two or more cells.

porospore. A conidium that grows out through small pores in the sporogenous cell; some are composed of several cells.

Figure 4.19*b* illustrates these types of conidia.

Strength in Variation Because fungi propagate very effectively with millions of asexual spores, one may ponder what function sexual reproduction serves for them. The answer lies in an important interchange that occurs when fungi of different genetic constitutions combine their genetic material. Just as in plants and animals, this linking of genes from two parents creates offspring with combinations of genes different from either parent. The offspring from such a union can have slight variations in form and function that are potentially advantageous in the adaptation and survival of their species. Since fungal cells spend most of their life cycle in the haploid state, their cells already exist in the correct state for compatible sexual union. In general, during sexual reproduction, the haploid nuclei from two compatible parental cells fuse, forming a diploid nucleus that subsequently participates in sexual spore development (see figure 4.7*a*). The actual spore-forming cycles vary a great deal, and many of them are very intricate.

Sexual Spore Formation Although a dominant portion of their life cycle is spent in the asexual phase, the majority of fungi produce sexual spores at some point. The nature of this process varies from the simple fusion of fertile hyphae of two different strains to a complex union of differentiated male and female structures and the development of special fruiting structures. Four different types of sexual spores have been identified, but we will consider the three most common: zygospores, ascospores, and basidiospores. These spore types provide an important basis for classifying the major fungal phyla.

Zygospores are sturdy diploid spores formed by the meeting of the hyphae of two opposite strains (called the plus and minus strains) (figure 4.20). Fusion of the hyphal tips creates a diploid cell or zygote that swells, becomes covered by strong, spiny walls, and is released into the environment. When its wall is disrupted and moisture and nutrient conditions are suitable, the zygospore germinates and forms a sporangium. Meiosis of diploid cells of the sporangium results in haploid nuclei that develop into sporangiospores. Both the sporangia and sporangiospores that arise from sexual processes are outwardly identical to the asexual type, but because the spores arose from the union of two separate fungal parents, they are not genetically identical.

In general, haploid spores called **ascospores** are created inside a special fungal sac or **ascus** (pl. asci) (figure 4.21). Although details may vary among types of fungi, the ascus and ascospores are formed when two different strains or sexes conjugate. In many species, the male sexual organ (the antheridium) fuses with the female sexual organ (the ascogonium). The end result is a number of terminal cells, each containing a diploid nucleus. Through differentiation, each of these cells enlarges to form an ascus, and its diploid nucleus undergoes meiosis (often followed by mitosis) to form four to eight haploid nuclei that will mature into ascospores. Late in the cycle, the asci burst and release the ascospores. Some species form an elaborate fruiting body to hold the asci, such as the cup fungus in figure 4.17*c*.

---

zygospores (zy'-goh-sporz) Gr. *zygon,* yoke, to join.
ascospores (as'-koh-sporz) Gr. *ascos,* a sac.

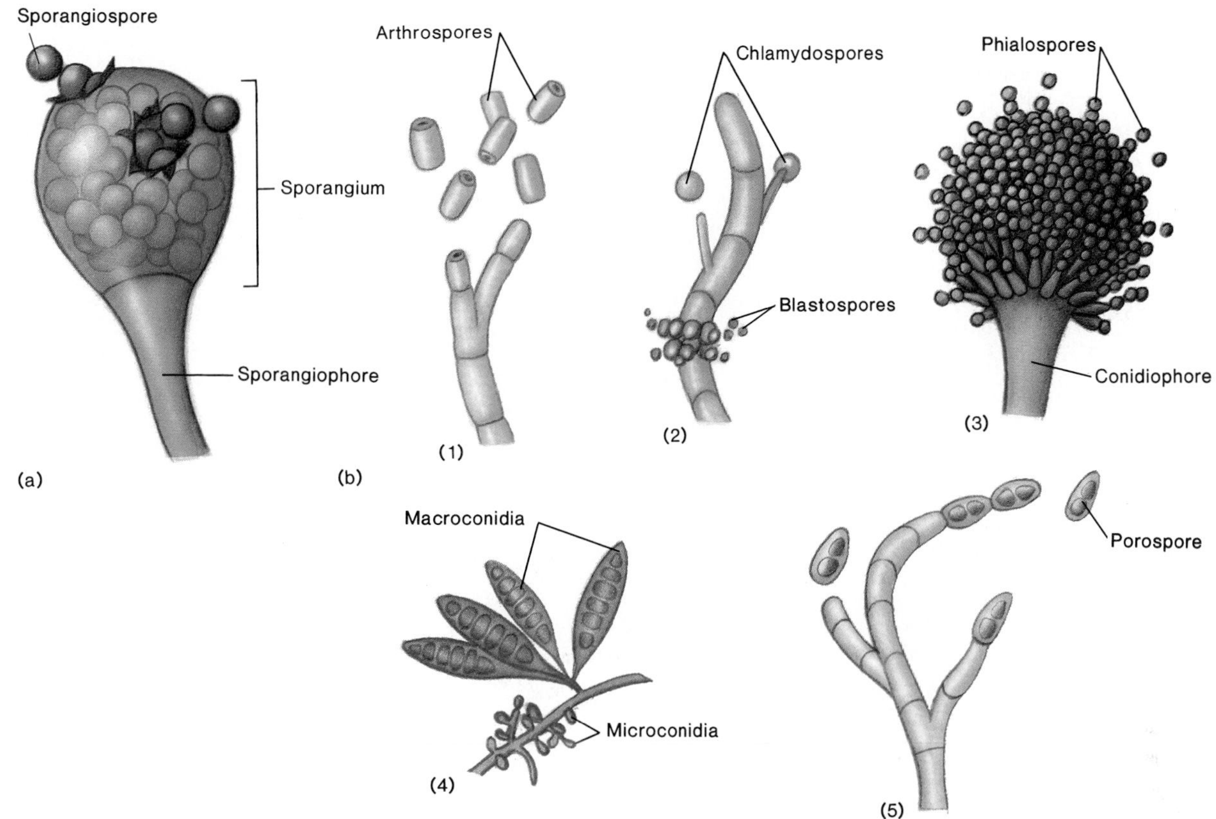

Figure 4.19 Types of asexual mold spores. (*a*) Sporangiospores (ex. *Absidia*). (*b*) Conidia: (*1*) arthrospores (ex. *Coccidioides*); (*2*) chlamydospores and blastospores (ex. *Candida albicans*); (*3*) phialospores (ex. *Aspergillus*); (*4*) macroconidia and microconidia (ex. *Microsporum*); and (*5*) porospores (ex. *Alternaria*).

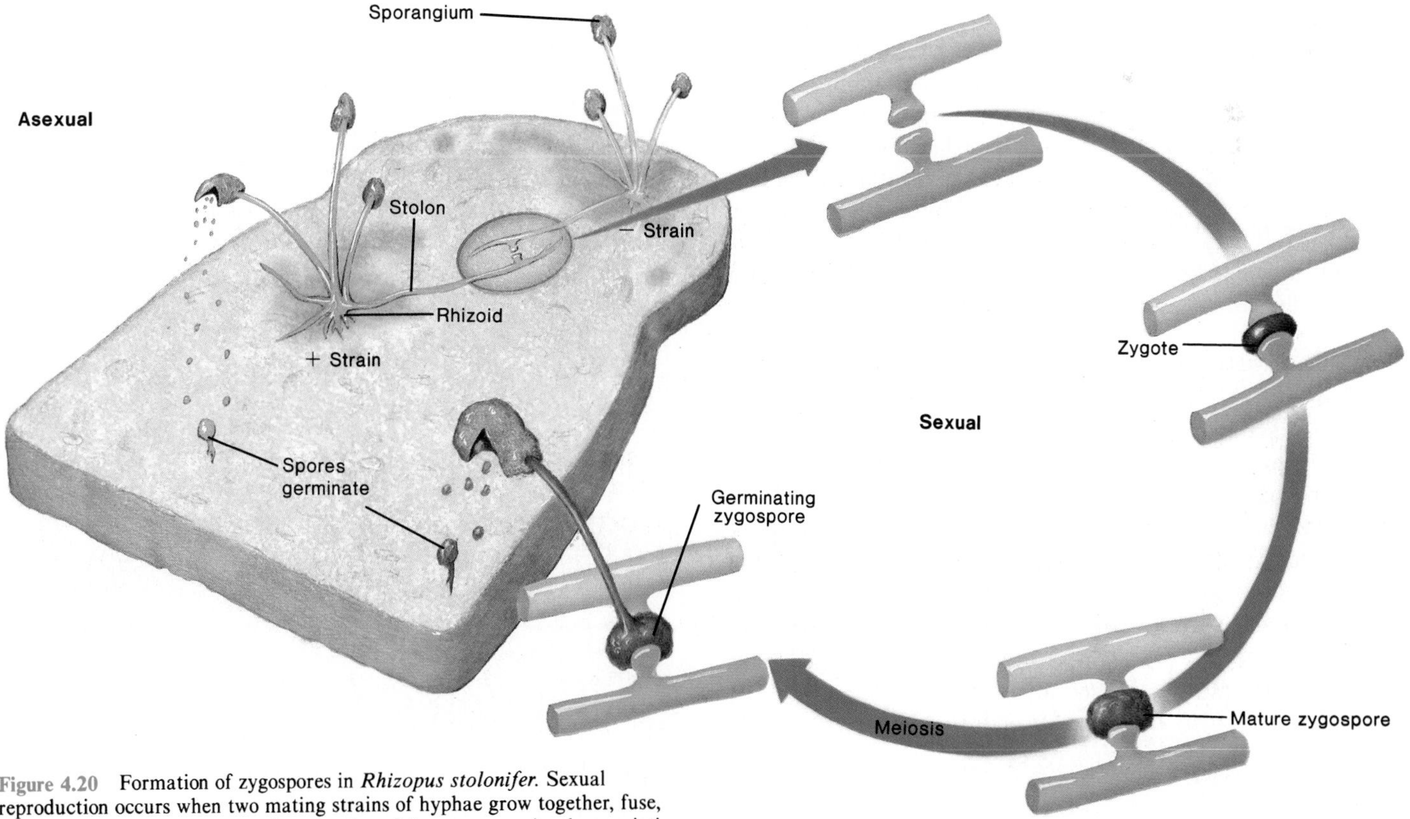

Sporangium
Asexual
Stolon
− Strain
Rhizoid
+ Strain
Spores germinate
Zygote
Sexual
Germinating zygospore
Mature zygospore
Meiosis

Figure 4.20 Formation of zygospores in *Rhizopus stolonifer.* Sexual reproduction occurs when two mating strains of hyphae grow together, fuse, and form a mature zygospore. Germination of the zygospore involves meiotic division and production of a haploid sporangium that looks just like the asexual one.

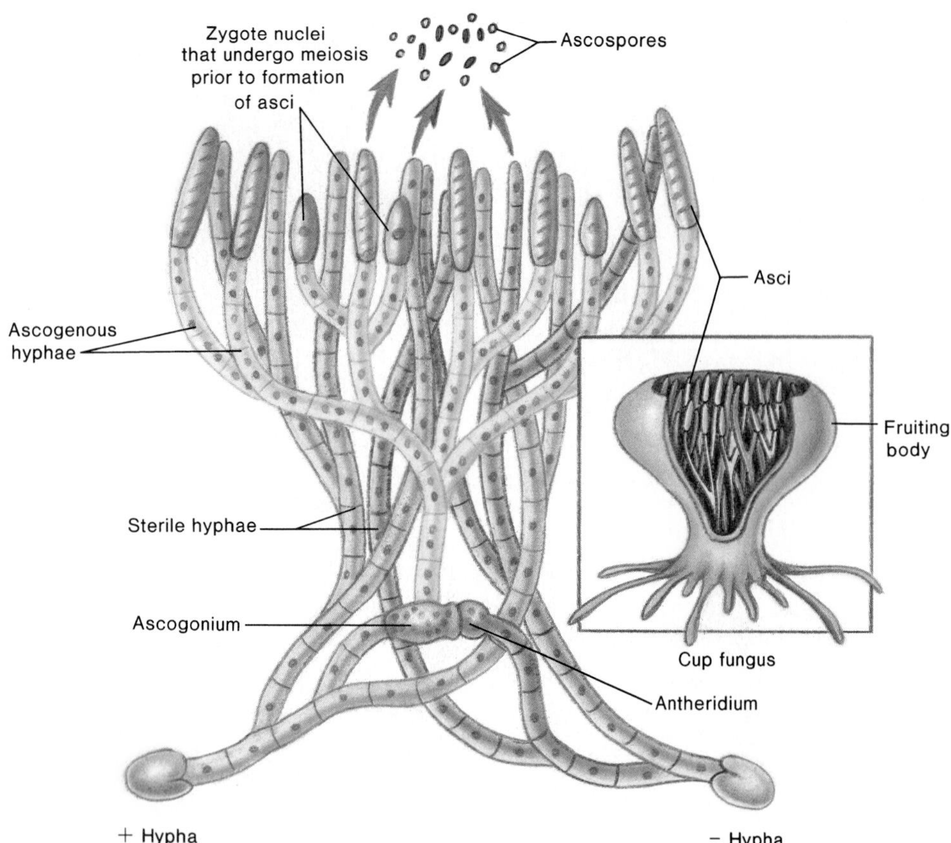

Figure 4.21 Production of ascospores in a cup fungus.

Basidiospores are haploid sexual spores formed on the outside of a club-shaped cell called a **basidium** (figure 4.22). In general, spore formation follows the same pattern of two mating types coming together, fusing and forming terminal cells with diploid nuclei. Each of these cells becomes a basidium, and its nucleus produces, through meiosis, four haploid nuclei. These nuclei are extruded through the top of the basidium, where they develop into basidiospores. Notice that the location of the basidia is along the gills in mushrooms, thus the spores are ejected downward. It may be a surprise to discover that the fleshy part of a mushroom is actually a fruiting body designed to protect and help disseminate its sexual spores.

Fungal Classification

It is frequently difficult for microbiologists to assign logical and/or useful classification schemes to microorganisms while at the same time preserving their relationships. This is because the organisms do not always perfectly fit the neat categories made for them, and even experts cannot always agree on the nature of the categories. Because the fungi are no exception, there are several ways to classify them. For our purposes, we will adopt a classification scheme with a medical mycology emphasis, in which the Kingdom Myceteae is subdivided into two subkingdoms, the Myxomycota (slime molds; see feature 4.2) and the

basidiospores (bah-sid′-ee-oh-sporz) Gr. *basidi,* a pedestal.

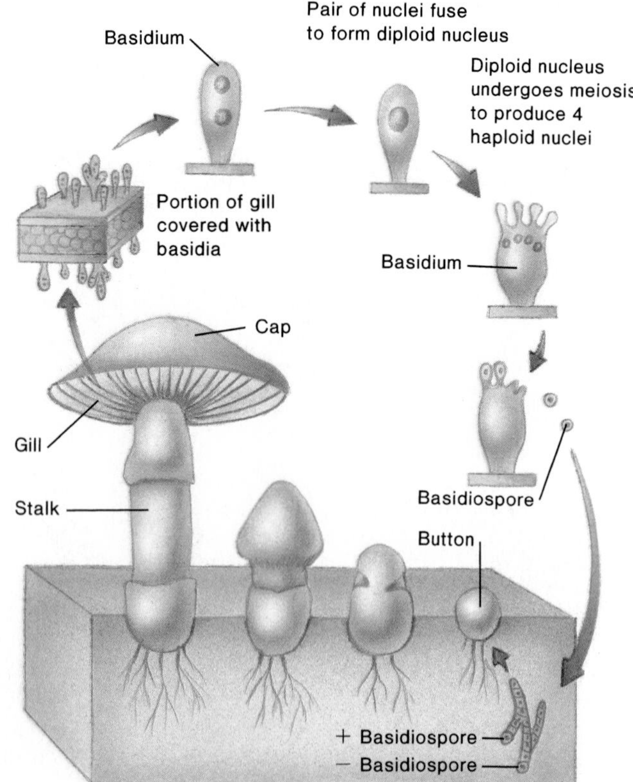

Figure 4.22 Formation of basidiospores in a mushroom.

Feature 4.2 Slime Molds: Fungi with a Difference

Among the strangest microorganisms are the **slime molds,** ephemeral creatures that inhabit the soil. The name is derived from their large, moist feeding stage and their superficial resemblance to the molds. Although we are including them taxonomically with the fungi, these organisms actually have complex life cycles that alternate between a protozoan-like, motile stage and a fungus-like, spore-forming stage. Without going into the details of their classification, the two basic types are acellular slime molds (featured here) and cellular slime molds.

Acellular slime molds have an astonishing feeding stage called the *plasmodium,* which is a huge multinucleate mass that creeps over the surface of leaves and logs. The plasmodium moves much like an ameba (there is no cell wall), and it feeds by slowly engulfing particles of organic matter. This stage is especially active during warm, humid weather, but it does not last very long. For many years, we have discovered dome-shaped foamy masses of plasmodia in moist areas of the garden. They make odd crackling noises and usually disappear as mysteriously as they came. Some of these pink plasmodia are a foot long. At some point, when the habitat dries up, the plasmodium undergoes a remarkable transformation into small mounds that are the beginnings of the sporangial stage. When the mature spores are released into a suitable environment, they germinate into single cells called myxamebas, which multiply and develop into flagellated swarm cells. As these swarm cells migrate together, their fusion creates the giant slime mass of the plasmodium—and the cycle has come full circle.

Slime molds are ecologically, but not medically, important. They are of interest mainly because of their unusual and beautiful forms and their use in experiments with chemical messages and chemotaxis. They release a chemical called cyclic AMP, which causes the individual cells to be attracted and migrate to the same location so that they can aggregate.

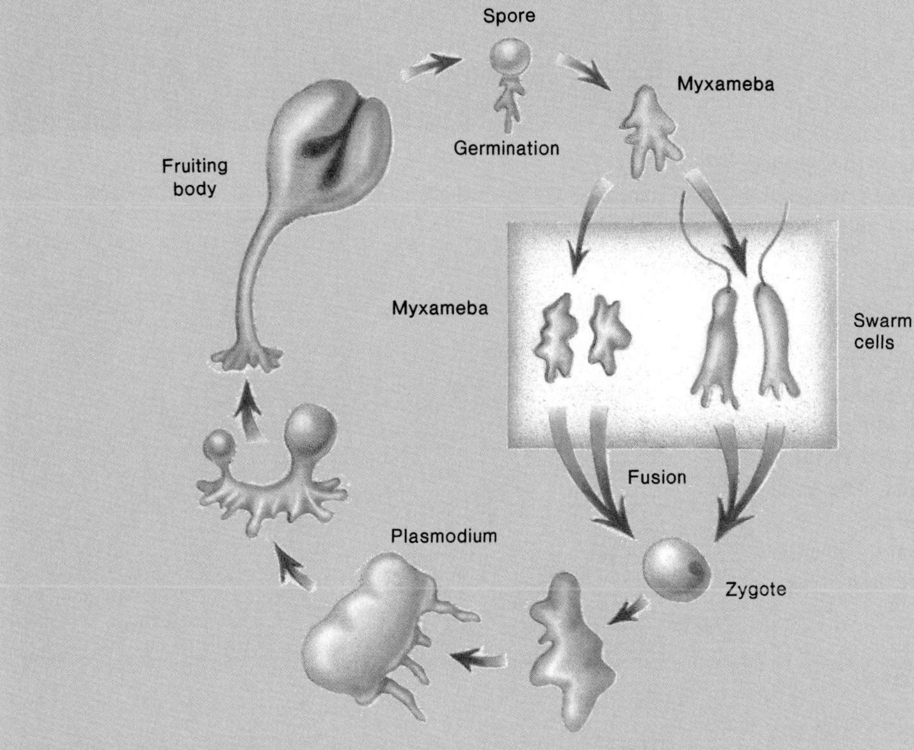

The life cycle of an acellular slime mold.

Eumycota (true fungi). The Eumycota contains the following four phyla, separated principally by their mode of sexual reproduction:

Fungi that produce sexual and asexual spores (perfect)

Phylum I—Zygomycotina (formerly Phycomycetes). Sexual spores: zygospores; asexual spores: mostly sporangiospores, some conidia. Hyphae are usually nonseptate. If septate, the septa are complete. Most species are free-living saprobes; some are animal parasites. May be obnoxious contaminants in the laboratory and on food and vegetables. Examples are mostly molds: *Rhizopus,* a black bread mold; *Mucor; Syncephalastrum; Circinella* (figure 4.23, asexual phase only).

Phylum II—Ascomycotina (formerly Ascomycetes). Sexual mode: produce asci and ascospores; asexual mode: many types of conidia, formed at the tips of conidiophores; hyphae with porous septa. Many important species. Examples: *Histoplasma,* the cause of Chicago disease; *Microsporum,* one cause of ringworm (a common name for certain fungal skin infections that often grow in a ringed pattern);

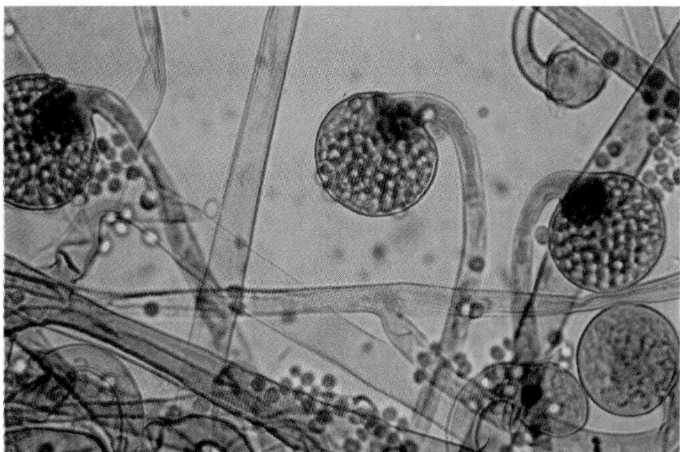
Figure 4.23 A representative Zygomycotine, *Circinella.* Note the sporangia and sporangiospores.

Penicillium, one source of antibiotics (figure 4.24, asexual phase only); and *Saccharomyces,* a yeast used in making bread and beer. Most of the species in this phylum are either molds or yeasts; it also includes many of the human and plant pathogens.

Phylum III—Basidiomycotina (formerly Basidiomycetes).

Sexual reproduction by means of basidia and basidospores; asexual spores: conidia, incompletely septate hyphae, some plant parasites, and one human pathogen; fleshy fruiting bodies are common. Examples: mushrooms, puffballs, bracket fungi, and plant pathogens called rusts and smuts. The one human pathogen, the yeast *Cryptococcus neoformans,* causes a serious infection in several organs, including the skin, brain, and lungs (see figure 18.25).

Fungi that produce asexual spores only (imperfect)

Phylum IV—Deuteromycotina.

Asexual spores: conidia of various types; hyphae septate; majority are yeasts or molds, some dimorphic; saprobes and a few animal and plant parasites. Examples: Several human pathogens were originally placed in this group, especially the imperfect states of *Blastomyces* and *Microsporum.* Other species are *Coccidioides immitis,* the cause of valley fever (figure 4.25); *Candida albicans,* the cause of various yeast infections; and *Sporothrix schenckii,* which causes sporotrichosis.

Representative types of the four phyla that comprise the Eumycota are shown in figures 4.23, 4.24, and 4.25.

Deuteromycotina (doo″-ter-oh-my-koh-tee′-nah) Gr. *deutero,* second, and *mykes,* fungus.

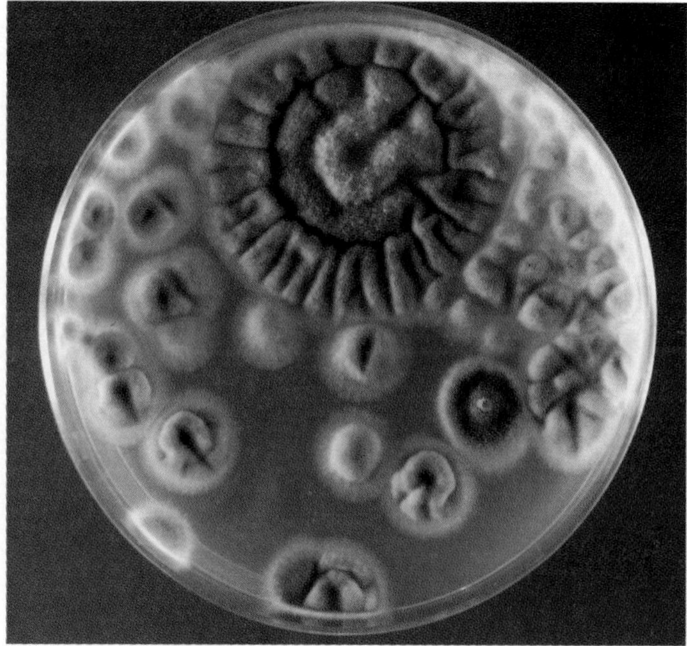
(a)

(b)

Figure 4.24 A common Ascomycotine, *Penicillium.* (*a*) Macroscopic view of a typical blue-green colony. (*b*) Microscopic view shows the brush arrangement of phialospores (×220).

A Taxonomic Dilemma From the beginnings of fungal classification, any fungus that lacked a sexual state was called "imperfect" and placed in a catchall category, the Fungi Imperfecti or Deuteromycotina. This category held that particular species until its sexual state was described (if ever). Gradually, many species of Fungi Imperfecti were found to make sexual spores, and they were assigned to the taxonomic grouping that best fit those spores. This created a quandary, because several very important species, especially pathogens, had to be regrouped and renamed. In many cases, the older names were so well entrenched in the literature that it was easier to retain them and assign a new generic name for their sexual stage. Thus, *Blastomyces* has a sexual state called *Ajellomyces; Histoplasma* is also called *Emmonsiella,* and *Penicillium* has a sexual state called *Talaromyces.*

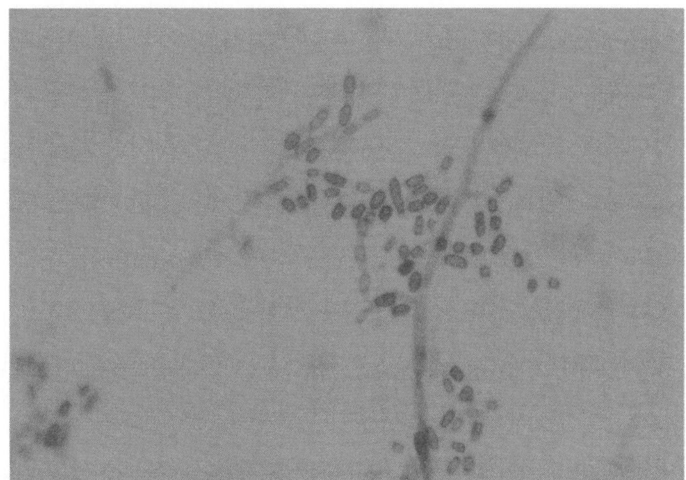

Figure 4.25 An important member of Fungi Imperfecti, *Coccidioides immitis,* is the mold that causes valley fever. Note arthrospores.

Fungal Identification and Cultivation

In order to identify fungi in medical specimens, they are first isolated on special types of media and then observed macroscopically and microscopically. Among the media for cultivating fungi are cornmeal, blood, and Sabouraud's agar. The latter medium is useful in isolating fungi from mixed samples because of its low pH, which inhibits the growth of bacteria but not most fungi. Because the fungi are classified into general groups by the presence and type of sexual spores, it would seem logical to identify them this way. In the medical setting, however, sexual spores are rarely if ever demonstrated, thus the asexual spore-forming structures and spores are used to identify to the level of genus and species. Other characteristics that may contribute to identification are hyphal type, colony texture and pigmentation, and, occasionally, physiological characteristics.

The Role of Fungi in Infection

Nearly all fungi are free-living and do not require a host to complete their life cycles. Even among those fungi that are pathogenic, most human infection occurs through accidental contact from an environmental source such as soil, water, or dust. Humans are generally quite resistant to fungal infection, except for two main types of fungal pathogens: the true pathogens, which can infect even healthy persons, and the opportunistic pathogens, which attack persons who are already weakened in some way. Chapter 18 is devoted to fungal infections, or **mycoses.**

Mycoses vary in the way the agent enters the body and the degree of tissue involvement (table 4.1). The list of opportunistic fungal pathogens has been increasing in the past few years because of newer medical techniques that keep compromised patients alive. Even so-called harmless species found constantly in the air and dust around us may be able to infect patients who already have AIDS, cancer, or diabetes. The problem of opportunistic fungal infections is and will continue to be a serious challenge to the health care worker.

Table 4.1 Major Fungal Infections of Humans

Degree of Tissue Involvement and Area Affected	Name of Infection	Name of Causative Fungus
Superficial (Not Deeply Invasive)		
Hair	Black piedra	*Piedraia hortae*
Outer epidermis	Pityriasis versicolor	*Malassezia furfur*
Epidermis, hair, and dermis may be attacked	Dermatophytosis, also called tinea or ringworm of the scalp, body, feet (athletes foot), toenails, and beard, depending on the area afflicted	*Microsporum, Trichophyton* and *Epidermophyton*
Mucous membranes, skin, nails	Candidiasis, or yeast infection	*Candida albicans*
Subcutaneous (Invades Just Beneath the Skin)		
Nodules beneath the skin; may invade lymphatics	Sporotrichosis	*Sporothrix schenckii*
Large fungal tumors of limbs and extremities	Mycetoma, madura foot	*Pseudoallescheria boydii, Madurella mycetomatis*
Systemic (Deep; Organism Enters Lungs; May Invade Other Organs)		
Lung	Coccidioidomycosis (San Joaquin Valley fever)	*Coccidioides immitis*
	North American blastomycosis (Chicago disease)	*Blastomyces dermatitidis*
	Histoplasmosis (Ohio Valley fever)	*Histoplasma capsulatum*
	Cryptococcosis (torulosis)	*Cryptococcus neoformans*
Lung, skin	Paracoccidioidomycosis (South American blastomycosis)	*Paracoccidioides brasiliensis*

Fungi are involved in other medical conditions besides infections. Fungal cell walls contain potent substances that can cause allergies. The toxins produced by poisonous mushrooms can induce neurological disturbances and even death. The mold *Aspergillus flavus* (see figure 18.27) makes a potentially lethal poison called aflatoxin,[4] which is the cause of a disease in domestic animals that have eaten grain infested with the mold and is also a suspected cause of liver cancer in humans.

Nonmedical Importance of Fungi

Fungi pose an everpresent economic hindrance to the agricultural industry. Not only are a number of species pathogenic to field plants such as corn and grain, but fungi also rot harvested fresh produce during shipping and storage. It has been estimated that as much as 40% of the yearly fruit crop is consumed not by humans but by fungi. On the positive side, fungi play an

4. From aspergillus, **flavus, toxin.**

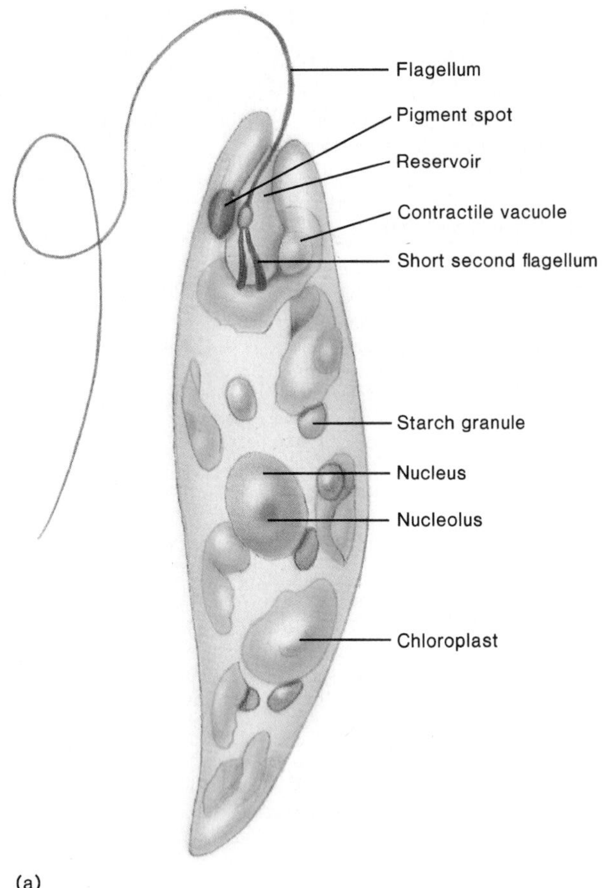

(a)

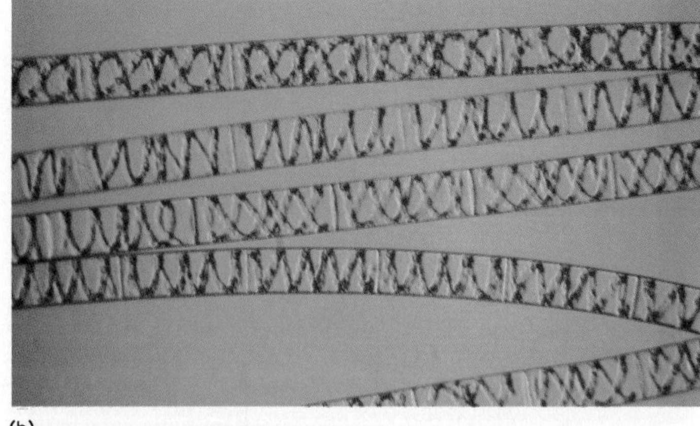

(b)

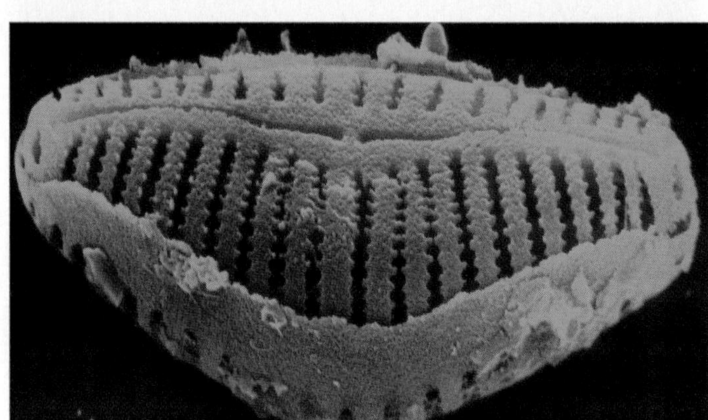

(c)

Figure 4.26 Representative microscopic algae. (*a*) *Euglena,* a unicellular motile example (×150). (*b*) *Spirogyra,* a colonial filamentous form with spiral chloroplasts. (*c*) The diatom *Trinacria regina,* a unicellular alga with finely detailed silica cell walls.

essential role in decomposing organic matter and returning essential minerals to the soil. Mankind has tapped the biochemical potential of fungi to produce industrial quantities of antibiotics, alcohol, organic acids, and vitamins, and some fungi are eaten or used to impart flavorings to food. The yeast *Saccharomyces* makes the alcohol in beer and wine, and it also produces the gas that causes bread to rise. Blue cheese, soy sauce, and cured meats derive their unique flavors from the actions of fungi (see chapter 22).

The Kingdom Protista

Although the fungi are placed in a separate kingdom, two groups of eucaryotic microbes are sufficiently similar in their characteristics to be combined into one. The Kingdom Protista contains the microscopic algae and the protozoa, which are both groups of mostly unicellular or colonial microorganisms that lack specialization into tissues.

The Algae

The **algae** are a group of photosynthetic organisms usually recognized by their larger members, such as seaweeds and kelps. In addition to being beautifully colored and diverse in appearance, they vary in length from a few μm to 200 meters. Microscopic forms are included with the protozoa in the Kingdom Protista (figure 4.26); macroscopic forms such as kelps are grouped with the plants. Algae occur in unicellular, colonial, and filamentous states, and the larger forms may have tissues. Table 4.2 lists the characteristics of the various subgroups of algae.

Algal cells as a group contain all of the eucaryotic organelles. The most noticeable of these are the chloroplasts, which contain, in addition to the green pigment chlorophyll, a number of other pigments that create the yellow, red, and brown coloration of some groups (table 4.2). Chloroplasts may have such extreme variations in shape and size that they are used in identification (for example, *Spirogyra,* figure 4.26*b*). Most algal cells are enclosed in an intricately patterned cell wall, which accounts for the distinctive appearance of unicellular members such as diatoms and dinoflagellates. The outermost structure in a group called the euglenoids (for example, *Euglena*) is a thick, flexible membrane called a *pellicle.* Motility by flagella or gliding is common among the algae, and many members contain tiny, light-sensitive areas (eyespots) that coordinate with the flagella so that the cell can swim toward the light it requires for photosynthesis.

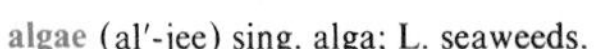

algae (al′-jee) sing. alga; L. seaweeds.

Table 4.2 Summary of Algal Characteristics

Division	Organization	Primary Habitat	Cell Wall	Pigmentation	Ecology/ Importance	Example
Microscopic Algae						
Euglenophyta (euglenoids)	Mainly unicellular; motile by flagella	Fresh water	None; pellicle instead	Chlorophyll, carotenoids, xanthophyll	Some are heterotrophic	*Euglena*
Chlorophyta (green algae)	Varies from unicellular, colonial, filamentous, to multicellular	Fresh water	Cellulose	Chlorophyll, carotenoids, xanthophyll	Precursor of higher plants	*Chlamydomonas, Spirogyra, Volvox*
Bacillariophyta (diatoms)	Mainly unicellular, flagellated; move by special gliding mechanism	Fresh water and marine	Silicon dioxide	Chlorophyll, fucoxanthin	Diatomaceous earth, major component of plankton	*Navicula*
Pyrrophyta (dinoflagellates, fire algae)	Unicellular, dual flagella	Marine plankton	Cellulose or atypical wall	Chlorophyll, carotenoids	Cause of "red tide"	*Gonyaulax*
Macroscopic Algae						
Phaeophyta (brown algae)	Multicellular, vascular system, holdfasts	Marine, subtidal forests	Cellulose, alginic acid	Chlorophyll, carotenoids, fucoxanthin	Source of an emulsifier, alginate	*Fucus, Sargassum*
Rhodophyta (red algae)	Multicellular	Marine, intertidal forests	Cellulose	Chlorophyll, carotenoids, xanthophyll, phycobilin	Source of agar and carrageenan, a food additive	*Gelidium*

Algae are common inhabitants of fresh and marine waters. They are one of the main components of the large floating community of microscopic organisms called *plankton.* Other algal habitats include the surface of soil, rocks, and plants. Several species are hardy enough to live in hot springs or snowbanks (see figure 6.11). Because algae are photosynthetic, they are not themselves infectious and usually pose no medical threat. Certain dinoflagellates overgrow during particular seasons of the year and impart a brilliant red color to the water referred to as a "red tide." These organisms produce toxins that cause death in fish and invertebrates and may cause food poisoning in humans who eat those animals (see figure 22.20).

Algae are often described by common names such as green algae, brown algae, golden brown algae, and red algae in reference to their predominant color. A more technical system divides the microscopic algae into four divisions, based on the types of chlorophyll and other pigments, the type of cell covering, and the nature of their stored foods. These groups are (1) Euglenophyta or euglenoids, (2) Chlorophyta or green algae, (3) Bacillariophyta or diatoms, and (4) Pyrrophyta or dinoflagellates (see table 4.2).

Algae reproduce asexually through fragmentation, binary fission, and mitosis, and some produce motile spores. Their sexual reproductive cycles may be highly complex, with cycles similar to fungi. The capacity of algae to photosynthesize is extremely important to the earth. They form the basis of aquatic food chains and play an essential role in the earth's oxygen and carbon dioxide balance. Some products of algae have industrial applications. For example, fossilized marine diatoms yield an abrasive powder called diatomaceous earth that is used in polishes, bricks, and filters, and certain seaweeds are a source of agar and algin used in microbiology, dentistry, and the food and cosmetics industries. We will consider the role of algae in microbial ecology in chapter 22.

Biology of the Protozoa

If a poll were taken to choose the most engrossing and vivid group of microorganisms, many people would choose the protozoans. Although their name comes from the Greek for "first animals," they are far from being simple, primitive organisms. The protozoans constitute a very large group (about 65,000 species) of creatures that, though single-celled, have startling properties when it comes to movement, feeding, and behavior. While most members of this group are harmless, free-living inhabitants of water and soil, a few species cause hundreds of millions of infections worldwide. Before we consider a few examples of important pathogens, let us examine some general aspects of protozoan biology.

Protozoan Form and Function Most protozoan cells are single units containing the major eucaryotic organelles except the cell wall and chloroplasts (see figure 4.2*c*). Their organelles may be highly specialized for feeding, reproduction, and locomotion. The cytoplasm is usually divided into a clear outer layer called the *ectoplasm* and a granular inner region called the *endoplasm.* Ectoplasm is involved in locomotion, feeding, and protection, and endoplasm houses the nucleus, mitochondria, and food and contractile vacuoles. Some ciliates and flagellates[5] even have organelles that work somewhat like a primitive nervous system to coordinate movement. The outer boundary of protozoan cells is a cell membrane that regulates the movement of food, wastes, and secretions. Cell shape may remain constant (as in most ciliates) or may change constantly (as in amebas). Certain amebas (foraminiferans) encase themselves in hard shells made of calcium carbonate. The size of most protozoan cells falls within the range of 3 to 300 μm. Some notable exceptions are giant amebas and ciliates that are large enough (3–4 mm in length) to be seen swimming in pond water.

Nutritional and Habitat Range Protozoa are **heterotrophic** and usually require their food in a complex organic form. Free-living species scavenge dead plant or animal debris and even graze on live cells of bacteria and algae. Some species have special feeding structures such as oral grooves, which sweep food particles into a passageway or gullet that packages the captured food into vacuoles for digestion. A remarkable feeding adaptation can be seen in the ciliate *Didinium,* which can easily devour another ciliate that is nearly its same size (see figure 4.31*c*). Some protozoa absorb food directly through the cell membrane. Parasitic species live on the fluids of their host, such as plasma and digestive juices, or they may actively feed on tissues.

Though protozoa have adapted to a wide range of habitats, their main limiting factor is the availability of moisture. Their predominant habitats are fresh and marine water, soil, plants, and animals. Even extremes in temperature and pH are not a barrier to their existence, since hardier species may be found in hot springs, ice, and habitats with low or high pH. Later we will discuss the propensity of many protozoans to convert to a resistant, dormant stage called a cyst.

Styles of Locomotion All but a few protozoa are motile by some means. Among the various protozoan groups are members that move strictly by **pseudopodia, flagella,** or **cilia.** A few species, however, have both pseudopodia (also called pseudopods) and flagella. Some unusual protozoans move by a gliding or twisting movement that does not appear to involve any of these locomotor structures. Pseudopodia may be blunt, branched, or long and pointed, depending on the particular ameba. As previously described, the flowing action of the pseudopods results in ameboid motion, and pseudopods also serve as feeding structures in many amebas (see figure 4.30). The structure and behavior of flagella and cilia were discussed in the first section of this chapter. Flagella vary in number from one to several, and in certain species they are attached along the length of the cell by an extension of the cytoplasmic membrane called the undulating membrane (see figure 4.29). In most ciliates, the cilia are distributed over the entire surface of the cell in characteristic patterns. Because of the tremendous variety in ciliary arrangements and functions, ciliates are among the most diverse and awesome cells in the biological world. In certain protozoans, cilia line the oral groove and function in feeding, and in others, they fuse together to form stiff props that serve as primitive rows of walking legs (see figure 4.31).

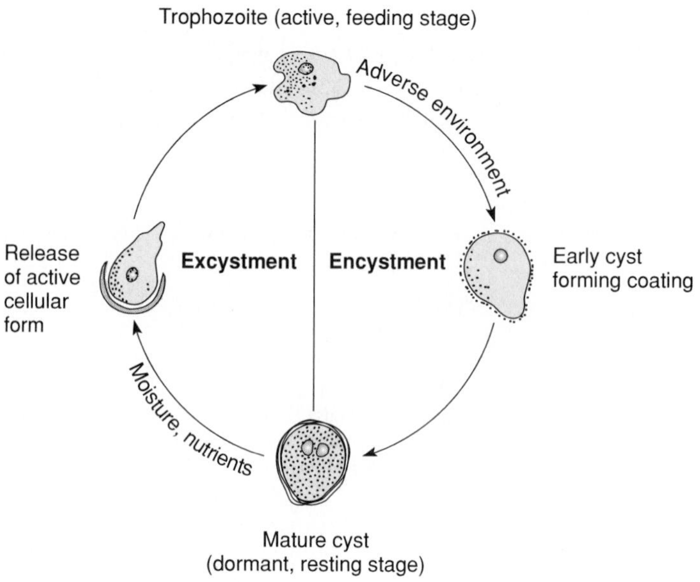

Figure 4.27 The general life cycle exhibited by many protozoa. (All protozoa have a trophozoite, but not all produce cysts.)

Life Cycles and Reproduction Most protozoans are recognized by a motile feeding stage called the **trophozoite** that requires ample food and moisture to remain active. A large number of species are also capable of entering into a dormant, resting stage called a **cyst** when conditions in the environment become unfavorable for growth and feeding. During *encystment,* the trophozoite cell rounds up into a sphere, and its ectoplasm secretes a tough, thick cuticle around the cell membrane (figure 4.27). Because cysts are more resistant than ordinary cells to heat, drying, and chemicals, they permit survival through adverse periods. They can be dispersed by air currents and may even be an important factor in the spread of diseases such as amebic dysentery. If provided with moisture and nutrients, a cyst breaks open and releases the active trophozoite.

The life cycles of protozoans vary from simple to complex. Several protozoan groups exist at all times in the trophozoite state. Many alternate between a trophozoite and a cyst stage, depending on the conditions of the habitat. The nature of the life cycle of a parasitic protozoan has direct impact upon its mode of infection. For example, *Trichomonas vaginalis,* a flagellate that causes a common sexually transmitted disease, does not form cysts and so must be transmitted through direct contact. In contrast, encysted intestinal pathogens such as *Entamoeba histolytica* and *Giardia lamblia* are readily transmitted in contaminated water and foods. Certain parasitic species go through a complex series of morphological changes as they alternate between different hosts.

5. The terms flagellate and cilate are common names of protozoan groups that move by means of flagella and cilia.

heterotrophic (het-ur-oh-troh'-fik) Gr. *hetero,* other, and *troph,* to feed. A style of nutrition that relies on an organic nutrient source.

pseudopodia (soo''-doh-poh'-dee-ah) sing. pseudopod; Gr. *pseudo,* false, and *pous,* feet.

trophozoite (trof''-oh-zoh'-yte) Gr. *trophonikos,* to nourish, and *zoon,* animal.

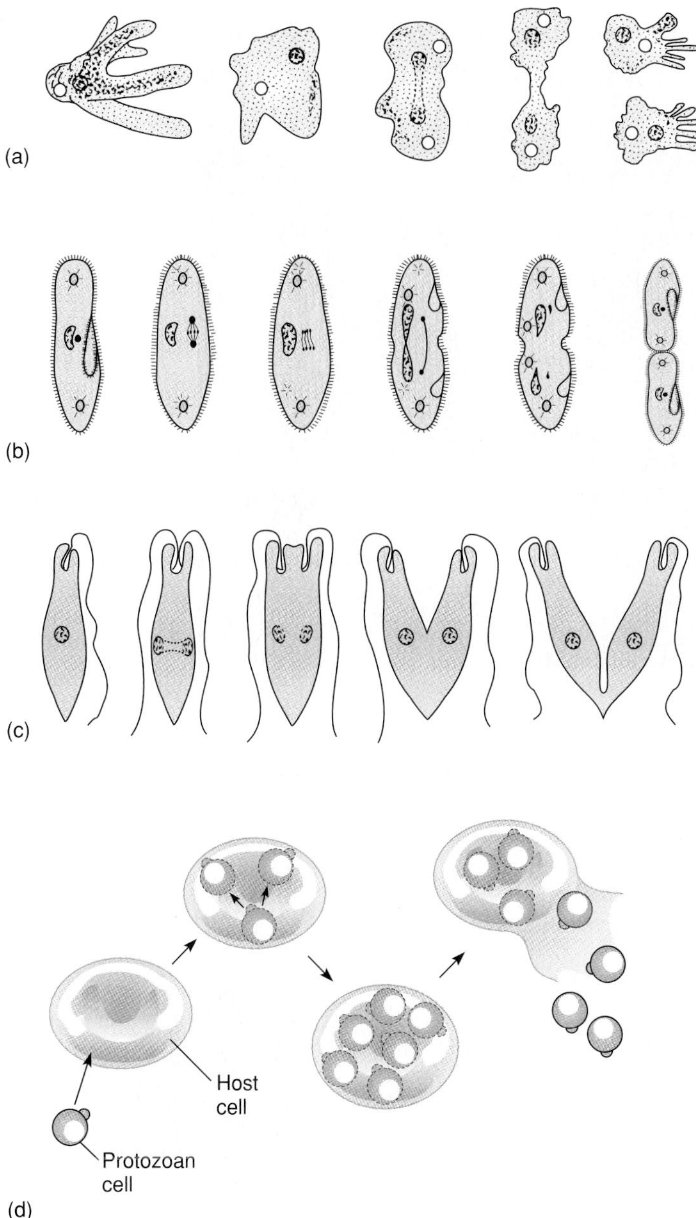

Figure 4.28 Asexual reproductive strategies in protozoa: (*a*) ameba; (*b*) ciliate; (*c*) flagellate. Note that in (*a–c*), the nucleus divides mitotically, followed by fission of the cell into two parts. (*d*) Multiple fission (schizogony) in sporozoa.

All protozoans reproduce by relatively simple, asexual methods, usually involving mitotic cell division (figure 4.28). Flagellates tend to divide longitudinally, and ciliates transversely. Several parasitic species, including the agents of malaria and toxoplasmosis, reproduce asexually inside a host cell by multiple fission (figure 4.28*d*). Sexual reproduction also occurs during the life cycle of most protozoans, though several groups of amebas and flagellates appear to reproduce only asexually. The simplest sexual mode is *syngamy,* in which two haploid motile gametes unite to form a diploid zygote. This zygote subsequently undergoes meiotic division to produce a number of haploid trophozoites. Ciliates participate in *conjugation,* a form of genetic exchange in which members of two different mating types fuse temporarily and exchange micronuclei. A micronucleus is the smaller of the two nuclei of ciliates that functions primarily in sexual recombination. Although the details of conjugation are too complicated to include here, suffice it to say that the exchange yields different genetic combinations that may be advantageous in evolution.

syngamy (sing'-ah-mee) Gr. *syn,* with, and *gamy,* marriage.

Classification of Selected Medically Important Protozoa The protozoa have not escaped problems in taxonomy. They, too, are very diverse, and frequently frustrate attempts to generalize or place them in neat groupings. The most recent system divides them into three phyla, with several subphyla and classes, but this method may be more complex than is necessary. We will simplify this system by dividing the Subkingdom Protozoa into four medically important groups, based on method of motility, mode of reproduction, and stages in the life cycle, as follows:

The Mastigophora (Flagellata)

Motility is primarily by flagella, with some ameboid species; single nucleus; sexual reproduction, when present, by syngamy; division by longitudinal fission. Several parasitic forms have no mitochondria and Golgi apparatus; most species form cysts and are free-living; the group also includes several parasites. Some species may be found in loose aggregations or colonies, but most are solitary. **Examples**: *Trypanosoma* and *Leishmania,* important blood pathogens with one flagellum; *Giardia,* an intestinal parasite with four flagella; *Trichomonas,* a parasite of the reproductive tract of humans with four to six flagella (figure 4.29).

The Sarcodina

Cell form is primarily an ameba; major locomotor organelles are pseudopodia, though some species may have flagella during reproductive states (figure 4.30). Asexual reproduction by fission; two groups have an external shell; mostly uninucleate; usually encyst. **Examples**: Most amebas are free-living and not infectious; *Entamoeba* is a pathogen or parasite of humans; shelled amebas called foraminifera and radiolarians are responsible for chalk deposits in the ocean.

The Ciliophora (Ciliata)

Trophozoites are motile by cilia; may have cilia in tufts for feeding and attachment (figure 4.31); most develop cysts; have both macronuclei and micronuclei; division by transverse fission; most have a definite mouth and feeding organelle; show relatively complicated and advanced behavior. **Example**: The majority of ciliates are free-living and harmless. One important pathogen, *Balantidium coli,* lives in vertebrate intestines and may infect humans.

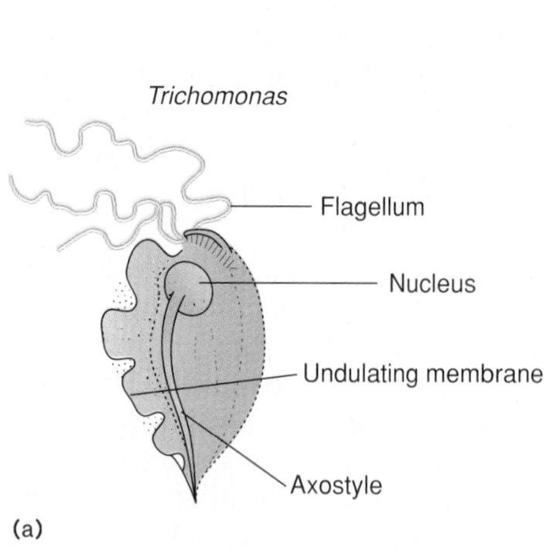

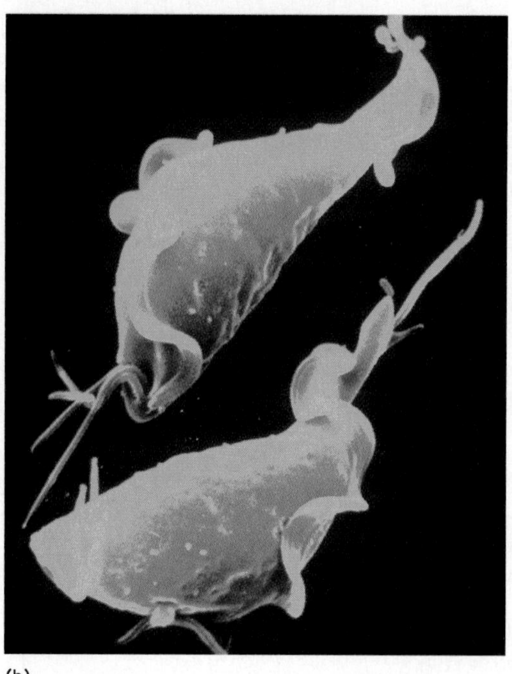

Figure 4.29 The structure of a typical mastigophoran, *Trichomonas vaginalis,* a genital tract pathogen, as shown in (*a*) a drawing, and (*b*) a scanning electron micrograph.

Figure 4.30 Examples of sarcodinians. (*a*) The structure of an ameba. (*b*) Radiolarian, a shelled ameba with long pointed pseudopods. Pseudopods are used in both movement and feeding.

The Sporozoa

Life cycles are complex, with well-developed asexual and sexual stages. Motility is not a prominent characteristic in most cells, but occasionally flagella or pseudopods are present. Sporozoans produce special sporelike cells called **sporozoites** following sexual reproduction, which are important in transmission of infections; most form cysts; entire group is parasitic (figure 4.32). **Examples**: *Plasmodium* is the cause of malaria and it is undoubtedly the most prevalent protozoan parasite, causing from 100 million to 300 million cases per year worldwide. It is an intracellular parasite with a complex cycle alternating between humans and mosquitos (see figure 19.10). *Toxoplasma gondii* causes an acute infection (toxoplasmosis) in humans, which is transmitted primarily by cats.

Protozoan Identification and Cultivation The unique appearance of most protozoans makes it possible for a knowledgeable

sporozoite (spor″-oh-zoh′-yte) Gr. *sporos,* seed, and *zoon,* animal.

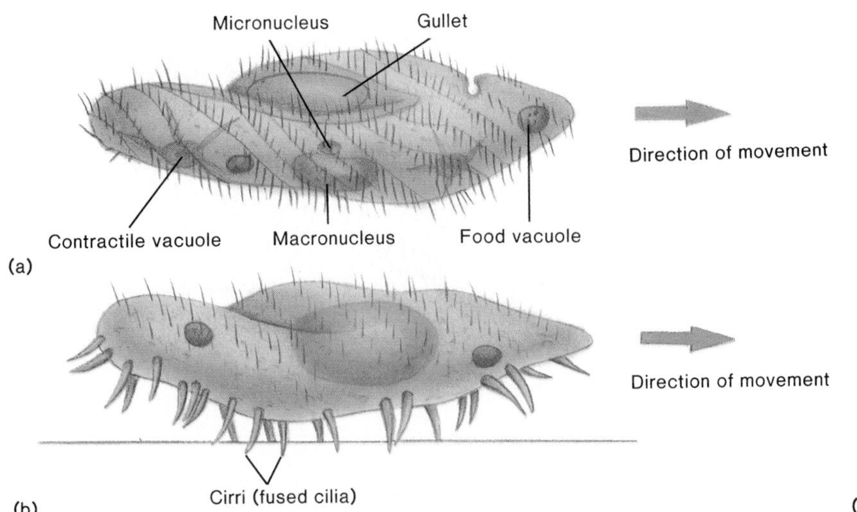

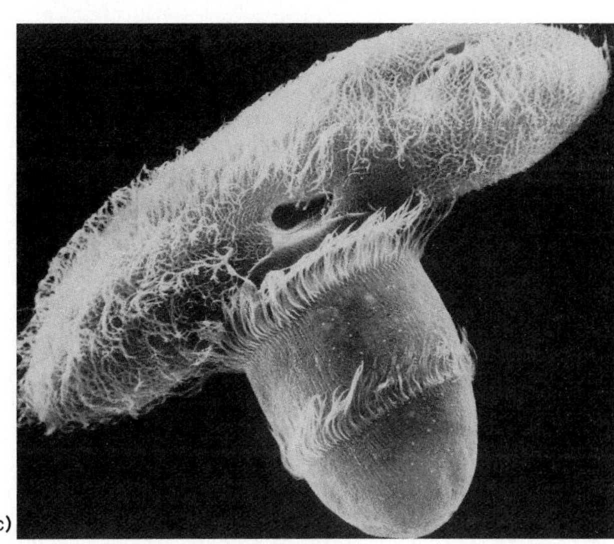

Figure 4.31 Selected ciliate representatives. (*a*) The structure of a typical representative, *Paramecium.* Cilia beat in coordinated waves, driving the cell forward and backward. Cilia are also used to sweep food particles into the gullet to form food vacuoles. (*b*) Fused cilia (cirri) of *Stylonychia* are used for walking over surfaces. (*c*) Scanning electron micrograph of a *Didinium* cell eating a *Paramecium* (×1,400). Note the number and distribution of cilia in each protozoan.

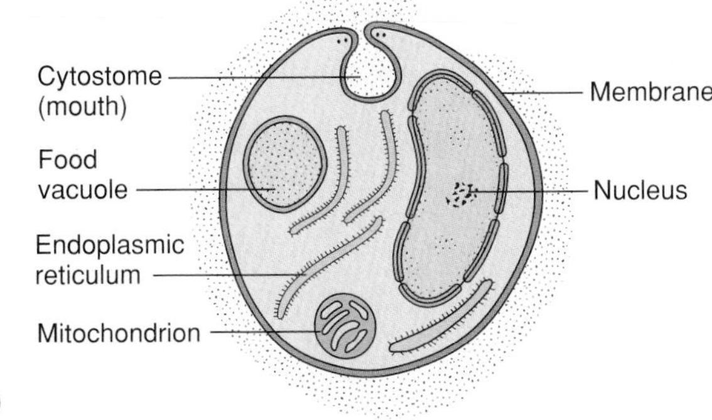

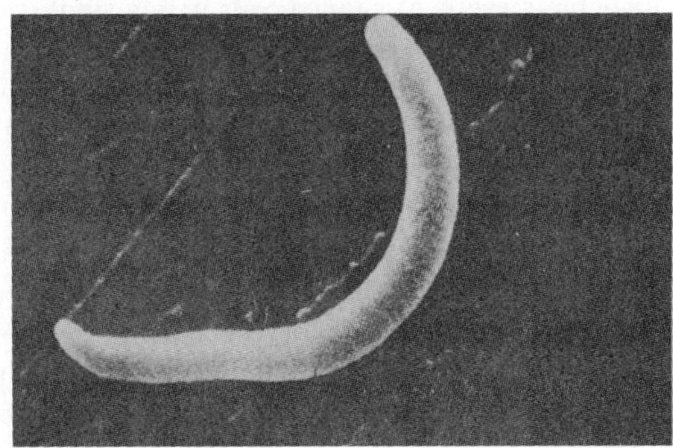

Figure 4.32 Sporozoan parasites. (*a*) General cell structure. Note the lack of specialized locomotor organelles. (*b*) Scanning electron micrograph of the infective stage of *Plasmodium,* the malaria parasite.

person to identify them to genus and often species by microscopic morphology alone. Characteristics to consider in identification include: the shape and size of the cell; the type, number, and distribution of locomotor structures; the presence of special organelles or cysts; and the number of nuclei. Medical specimens, taken from blood, sputum, cerebrospinal fluid, feces, or the vagina, are smeared directly onto a slide and observed with or without special stains. Occasionally, protozoans are cultivated on artificial media or in laboratory animals for further identification or study.

General Thoughts on the Parasitic Life-style

Parasitic protozoa are traditionally studied along with the helminths in the science of parasitology. While viruses and bacteria may also be parasitic, this science has historically been based on the study of larger animal or animal-like parasites. Although a parasite is generally defined as an organism that obtains food and other needs at the expense of a host, the range of host-parasite relationships can be very broad. At one extreme are the so-called "good" parasites, which occupy their host with little harm. An example would be certain amebas that live in the human intestine and feed off organic matter there. At the other extreme are parasites (*Plasmodium, Trypanosoma*) that invade and multiply in host tissues such as blood, brain, or muscles and cause severe damage and disease. Between these two extremes are parasites of varying pathogenicity, depending on their particular adaptations (see chapters 6 and 11).

The life cycle of most human parasites can be depicted in three general stages: (1) The microbe is transmitted to the human host from a source such as soil, water, food, other humans, or animals. (2) The microbe invades and multiplies in the host, producing more parasites that can infect other suitable hosts. The effects of this invasion on the human vary. (3) The microbe leaves the host in large numbers by a specific means and, to survive, must find and enter a new host. There are numerous variations on this simple theme. For instance, the microbe may invade more than one host species (an alternate host), and it may undergo several changes as it cycles through these hosts, such as sexual reproduction or encystment. The microbe may be spread from human to human by means of *vectors,* which

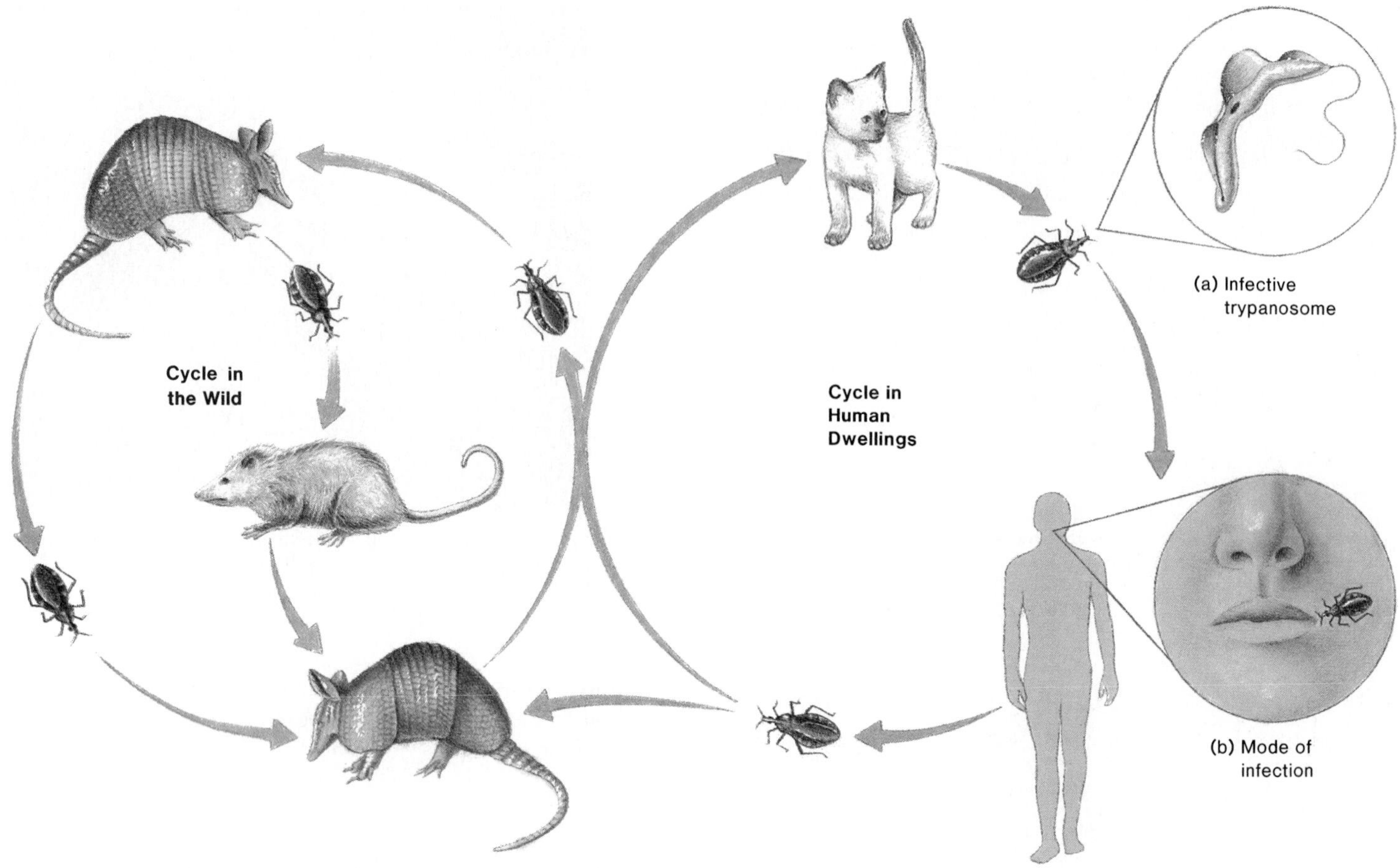

Figure 4.33 Cycle of transmission in Chagas' disease. Trypanosomes (inset *a*) may be transmitted among mammalian hosts and human hosts by means of a bite from the kissing bug (inset *b*).

are usually blood-sucking intermediate hosts, or it may be spread through bodily fluids and feces.

Selected Medically Important Protozoan Parasites

Although protozoan infections are very common, they are actually caused by only a small number of species. How do humans acquire various protozoan infections? One factor is the geographic area in which they live, because some pathogens are native only to a particular area of the world (such as the tropics). Another factor is the complexity of the parasite's life cycle. To introduce some concepts of the host-protozoan relationship, we will look at two specific examples. Details of several other medically important protozoa are given in chapter 19.

Trypanosomes Trypanosomes are parasites belonging to the genus *Trypanosoma.* The two most important representatives are *T. brucei* and *T. cruzi,* species that are closely related but geographically restricted. *Trypanosoma brucei* occurs in Africa, where it causes approximately 10,000 new cases of sleeping sickness per year. *Trypanosoma cruzi,* the cause of Chagas' disease,[6] is normally found only in South and Central America, where it infects approximately seven million people a year. Both species have long, crescent-shaped cells with a single flagellum that may be attached to the cell body by an undulating membrane (figure 4.33). Both occur in the blood during infection and are transmitted by blood-sucking vectors. We will use *T. cruzi* to illustrate the phases of a trypanosomal life cycle.

The trypanosome of Chagas' disease relies on the close relationship of two hosts: a warm-blooded mammal and an insect that feeds on the blood of that mammal. This multilevel parasitic existence is one of the profound yet common themes in parasitic relationships. The mammalian hosts are numerous, including man, dogs, cats, oppossums, armadillos, and foxes. The vector is the *reduviid* bug, an insect that lives in mammalian habitats, including the dwellings of humans. It is sometimes called the kissing bug because of its habit of biting its host at the corner of the mouth. Transmission occurs from bug to mammal and from mammal to bug, but usually not from mammal to mammal. The general and microscopic phases of this cycle are presented in figure 4.33.

vector (vek'-tur) L. *vectur,* one who carries.

Trypanosoma (try''-pan-oh-soh'-mah) Gr. *trypanon,* borer, and *soma,* body.

6. Named for Carlos Chagas, the discoverer of *T. cruzi.*

reduviid (ree-doo'-vee-id) A member of a large family of flying insects with sucking, beaklike mouths.

The trypanosome trophozoite multiplies in the intestinal tract of the reduviid bug and is harbored in the feces. During the night, the bug seeks a host and bites the mucous membranes, usually of the eye, nose, or lips. As it fills with blood, the bug soils the bite with feces containing the trypanosome. Ironically, the victims themselves inadvertently contribute to the entry of the microbe by scratching the wound. The trypanosomes ultimately become established and multiply in white blood and muscle cells. Periodically, these parasitized cells rupture, releasing large numbers of new trophozoites into the blood. Eventually, the trypanosome may spread to many systems, including the lymphoid organs, heart, liver, and brain. Manifestations of the resultant disease range from mild to very severe, and include fever, swelling, and heart and brain damage. In some instances, the disease persists for many years, and death from it is not uncommon. An effective treatment has yet to be developed. The insect phase of development is completed when a reduviid bug takes a blood meal from an infected mammal, thus establishing the microbe in the digestive tract of its insect host.

Infective Amebas Several species of amebas cause disease in humans, but probably the most common disease is amebiasis or **amebic dysentery,** caused by *Entamoeba histolytica.* This microbe is widely distributed in the world, from northern zones to the tropics, and is nearly always associated with humans. Amebic dysentery is the fourth most common protozoan infection in the world, with most cases confined to the tropics and subtropics. This microbe has a life cycle quite different from the trypanosomes in that it does not involve multiple hosts and a blood-sucking vector. It lives part of its cycle as a trophozoite and part as a cyst. In general, the cyst is more resistant and can survive in water and soil for several weeks, so it is the more important stage for transmission. The primary way that people become infected is by ingesting food or water contaminated with human feces.

Figure 4.34 shows the major features of the amebic dysentery cycle, starting with the ingestion of cysts. The viable, heavy-walled cyst passes through the stomach unharmed. Once inside the intestine, the cyst germinates to produce a large, multinucleate ameba that subsequently divides to form small amebas (the trophozoite stage). These trophozoites migrate to the large intestine, where they colonize the intestinal surface and begin to feed and grow. From this site, they may penetrate the lining of the intestine and invade the liver, lungs, and skin. Common symptoms include gastrointestinal disturbances such as nausea, vomiting, and diarrhea that often lead to weight loss and dehydration. Untreated cases with extensive damage to the organs experience a high death rate. The cycle is completed in the infected human when certain trophozoites in the feces begin to form cysts with several nuclei, and these cysts pass out of the body with fecal matter. Knowledge of the amebic cycle and the role of cysts has been a boon in controlling the disease. Important preventive measures include sewage treatment, curtailing the use of human feces as fertilizers, and adequate sanitation of food and water.

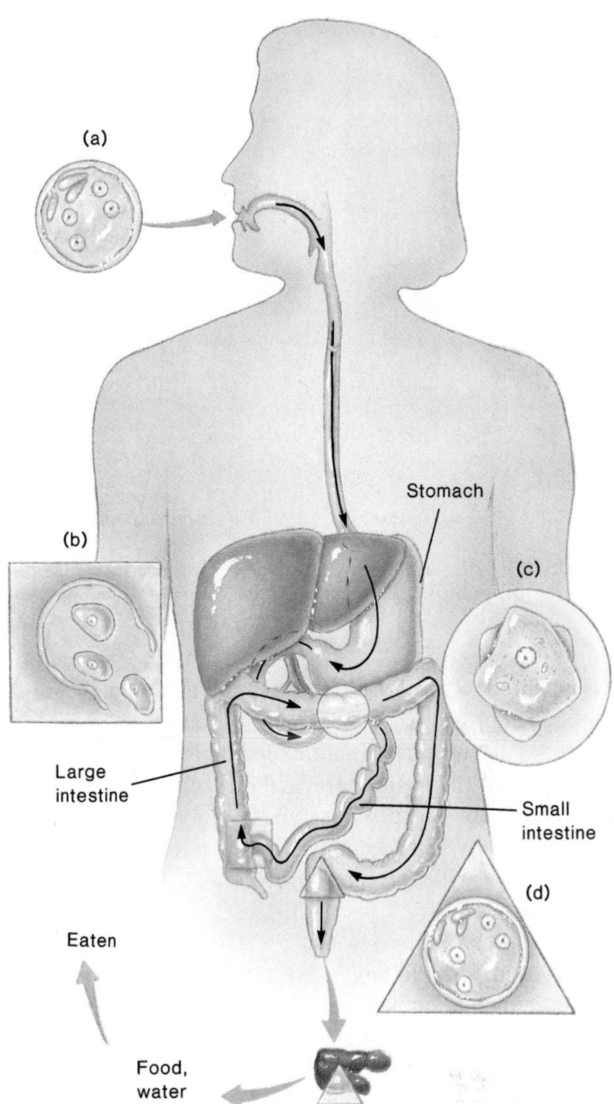

Figure 4.34 Stages in the infection and transmission of amebic dysentery, showing the route of infection and the appearance of *Entamoeba histolytica.* (*a*) Cysts are eaten. (*b*) Trophozoites (amebas) emerge from cysts. (*c*) Trophozoites invade the large intestinal wall. (*d*) Mature cysts are released in the feces.

The Parasitic Helminths

Tapeworms, flukes, and roundworms are collectively called **helminths,** from the Greek word meaning worm. Adult animals are usually large enough to be seen with the naked eye, and they range from the longest tapeworms, measuring up to 25 yards in length, to roundworms less than a millimeter in length. Nevertheless, they are included among microorganisms because the microscope is necessary to see the details of smaller helminths and to identify their eggs and larvae.

The two major groups of parasitic helminths, based upon morphological form, are the **flatworms,** with a very thin, often segmented body plan (figure 4.35), and the **roundworms** (also

dysentery (dis′-en-ter″-ee) Any inflammation of the intestine accompanied by bloody stools. It can be caused by a number of factors, both microbial and nonmicrobial.

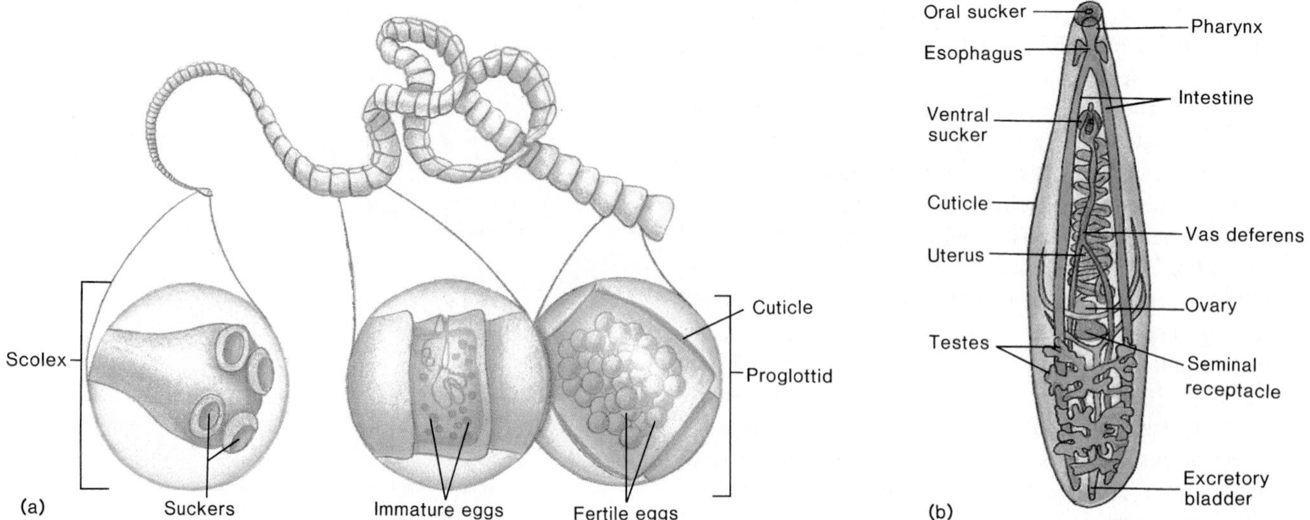

Figure 4.35 Parasitic flatworms. (*a*) A cestode (tapeworm), showing the scolex, long tapelike body, and magnified views of immature and mature proglottids. (*b*) The structure of a trematode (fluke). Note the suckers that attach to host tissue and the dominance of reproductive and digestive organs.

called **nematodes**), with an elongate, cylindrical, unsegmented body plan (figure 4.36). The flatworm group is subdivided into the **cestodes** or tapeworms, named for their long, ribbonlike arrangement, and the **trematodes** or flukes, characterized by flat, ovoid bodies. Not all flatworms and roundworms are parasites by nature—many live free in soil and water. Both here and in chapter 19, we will concern ourselves with the medically important worms.

General Worm Morphology All parasitic helminths are multicellular animals equipped to some degree with organs and organ systems, though the most developed organs are those of the reproductive tract, with some degree of reduction in the digestive, excretory, and nervous systems. In particular groups, such as the cestodes, reproduction is so dominant that the worms are reduced to little more than a series of flattened sacs filled with ovaries, testes, and eggs (see feature 4.3 and figure 4.35*a*). Not all worms have such extreme adaptations as cestodes, but most have a highly developed reproductive component, thick cuticles, and mouth glands for breaking down the host's tissue.

Life Cycles and Reproduction Many worms have very complex life cycles that alternate between hosts. Reproduction of individuals is primarily sexual, involving the production of eggs and sperm in the same worm or in separate male and female worms. Fertilized eggs are usually released to the environment and are provided with a protective shell and extra food to aid their development into larvae. Even so, most eggs and larvae are vulnerable to heat, cold, drying, and predators, and are destroyed or unable to reach a new host. To counteract this formidable mortality rate, certain worms have adapted a

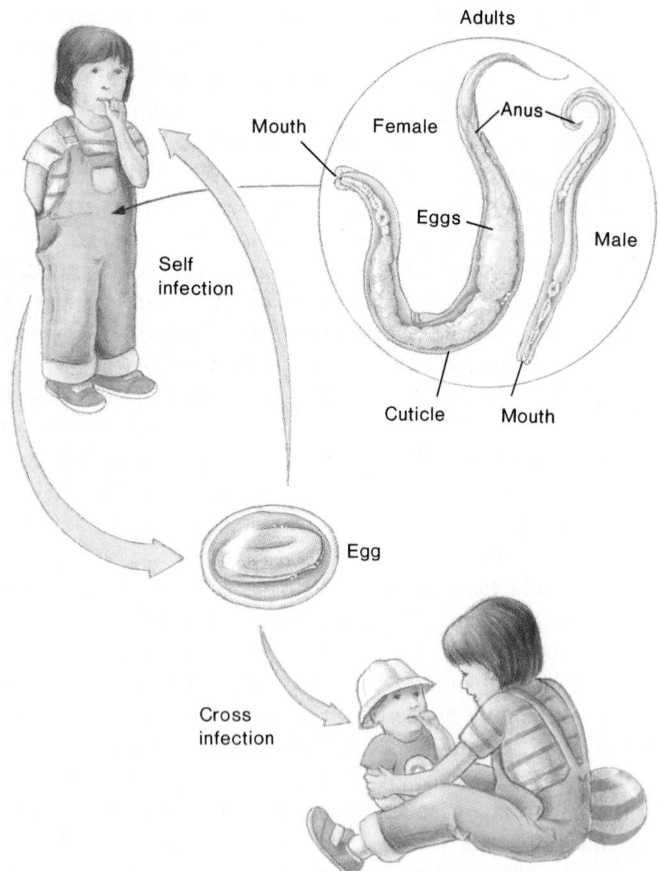

Figure 4.36 The life cycle of the pinworm. Eggs are the infective stage and are transmitted by unclean hands. Children frequently reinfect themselves and also pass the parasite on to others.

nematode (neem'-ah-tohd) Gr. *nemato,* thread, and *eidos,* form.

cestode (sess'-tohd) L. *cestus,* a belt, and *ode,* like.

trematode (treem'-a-tohd) Gr. *trema,* hole. Named for the appearance of having tiny holes.

Feature 4.3 Biological Adaptation in the Tapeworms

The adaptation of intestinal helminths serves as a useful model to illustrate the economy, opportunity, and even resourcefulness of parasitic worms. Deep within the dark, airless tunnel of the host's intestine, the worms experience no temperature extremes, no water shortage, no predators. They literally float in a sea of food that the host has already digested. Because these worms merely have to absorb nutrients and water through their skin, they do not need an extensive digestive tract of their own. Living this "life of Riley" is not without problems, however. First, the animals must complete their life cycles by having their eggs safely transmitted to another host. Second, they must avoid being digested or expelled by the host. The first problem is solved by ample reproductive organs, and the second is solved by developing a resistant covering or cuticle and by various hooks and suckers for attachment to host tissues.

reproductive capacity that borders on the incredible: A single female *Ascaris*[7] can lay 200,000 eggs per day, and a large female may contain over 25,000,000 eggs at varying stages of development! If only a tiny number of these eggs makes it to another host, the parasite will have been successful in completing its life cycle.

A Helminth Cycle To illustrate a helminth cycle in humans, we will use the example of a roundworm, *Enterobius vermicularis,* the pinworm or seatworm. This worm causes a very common infestation of the large intestine (figure 4.36). Worms range from 2 to 12 mm long, and have a tapered, curved cylinder shape. The condition they cause, enterobiasis, is usually a simple, uncomplicated infection that does not spread beyond the intestine.

A cycle starts when a person swallows microscopic eggs picked up from another infected person by hand-to-hand contact or by touching articles that person has touched. The eggs hatch in the intestine and release larvae that mature into adult worms in the appendix (within about one month). There, male and female worms mate, and the female migrates out to the anus to deposit eggs, which causes intense itchiness that is relieved by scratching. Herein lies a significant means of dispersal, because scratching contaminates the fingers, which in turn transfer eggs to bedclothes and other inanimate objects. This person becomes a host and a source of eggs, and can spread them to others in addition to reinfesting himself (figure 4.36). Enterobiasis occurs most often among families and in other close living situations. Its distribution is worldwide among all socioeconomic groups, but it seems to attack younger people more frequently than older ones and females three times more often than males.

7. *Ascaris* is a genus of parasitic intestinal roundworms.

Classification and Identification The helminths are classified according to their shape, size, the degree of development of various organs, the presence of hooks, suckers, or other special structures, the mode of reproduction, the kinds of hosts, and the appearance of eggs and larvae. They are identified in the laboratory by microscopic detection of the adult worm or its larvae and eggs, which often have distinctive shapes or external and internal structures (see chapter 19). Occasionally, they are cultured in order to verify all of the life stages.

Distribution and Importance About 50 species of helminths parasitize humans. They are distributed in all areas of the world that support human life. Some worms are restricted to a given geographical region, and many have a higher incidence in tropical areas. This knowledge must be tempered with the realization that jet age travel, along with human migration, are gradually changing the patterns of worm infections, especially of those species that do not require alternate hosts or special climatic conditions for development. The yearly estimate of worldwide cases numbers in the billions, and these are not confined to developing countries. A conservative estimate places 50,000,000 helminth infections in the United States alone. The primary targets are young, malnourished children, whose death rate can be increased in cases of gross infestation, multiple worm infections, and impaired host defenses.

Chapter Review with Key Terms

The Eucaryotes[8]

Structure: Appendages (cilia, flagella), glycocalyx, cell wall, cytoplasmic membrane, ribosomes, organelles (nucleus, nucleolus, endoplasmic reticulum, Golgi complex, mitochondrion, chloroplast, cytoskeleton, microfilaments, microtubules).

8. A review comparing the major differences between eucaryotic and procaryotic cells is provided in table 2.6, page 60.

The Kingdom Fungi (Myceteae)

Alternate Names: General—Mycophyta; common names of particular types—microscopic fungi (yeasts, molds); macroscopic fungi—mushrooms, bracket fungi, puffballs.

Overall Morphology: At the cellular (microscopic) level, typical eucaryotic cell, with thick **cell walls. Yeasts** are single cells that form **buds** and sometimes short chains called **pseudohyphae. Hyphae** are long, tubular filaments that may be septate or nonseptate and grow in a network called a **mycelium**; hyphae are characteristic of the filamentous fungi called **molds.**

Yeasts and molds appear macroscopically in colonial form; some fungi (mushrooms) produce multicellular structures such as fleshy fruiting bodies; largely **nonmotile,** except for some motile gametes.

Nutritional Mode/Distribution: All are **heterotrophic.** The majority are harmless **saprobes** living off organic **substrates** such as dead animal and plant tissues. A few are **parasites** living on the tissues of other organisms, but none are obligate. No chlorophyll, not photosynthetic. Most require oxygen in their habitat, though some yeasts can grow in the absence of it. Preferred temperature of growth is 10°–40°C, though a few prefer warmer or colder temperatures. Extremely widespread in many habitats, including soils, dust, fresh and salt water, on plants, food, detritus, bodies of animals.

Reproduction: Primarily through **spores** formed on special **fertile hyphae.** In **asexual** reproduction, spores are formed through budding, partitioning of a hypha, or in special sporogenous structures; examples are **conidia** and **sporangiospores.** In **sexual** reproduction, spores are formed through frequently elaborate means, involving fusion of male and female strains and the formation of a sexual hypha of some sort; sexual spores are basis for classification.

Major Groups: The four important subgroups among the true fungi, given with sexual spore type, are **Zygomycotina (zygospores), Ascomycotina (ascospores), Basidiomycotina (basidiospores)**, and **Deuteromycotina** (no sexual spores).

Importance: Ecologically essential as decomposers of cellulose and other plant materials in the environment. Fungi return valuable nutrients to the ecosystem. Economically beneficial as sources of antibiotics; fungi are used in making foods and in genetic studies. Harmful plant pathogens, attacking row and other crops; decomposition of fruits, vegetables, and other foods during storage; mildew of clothing, other objects. Medically, several fungi are important human pathogens. Cause superficial, and systemic **mycoses;** increasing number of opportunists attack severely compromised patients; some produce substances that are toxic if eaten.

Unique Features: Hyphae, spore-forming structures, fruiting bodies, types of spores, structure of cell walls.

The Kingdom Protista—The Algae

Alternate Names: General—plantlike protists, kelps, seaweeds. Specific—**euglenoids, green algae, diatoms, dinoflagellates.**

Overall Morphology: Eucaryotic, with **chloroplasts** containing **chlorophyll** and other pigments; cell wall; may or may not have flagella. Microscopic forms are unicellular, colonial, filamentous; macroscopic forms are colonial and multicellular.

Nutritional Mode/Distribution: **Photosynthetic;** most are free-living in the aquatic environment (common component of **plankton**); parasitism is extremely rare.

Reproduction: Similar to fungi.

Classification of Major Groups: Algae are divided into six groups according to types of pigments and cell walls. Microscopic algae: Euglenophyta are unicellular, green, flagellated algae with a **pellicle** and eyespot. Chlorophyta are predominantly green, walled algae that may be unicellular, colonial, multicellular, and flagellated. Bacillariophyta (also called diatoms) are golden-brown, unicellular, gliding algae with an intricate silica cell wall. Pyrrophyta (dinoflagellates) are red, unicellular, flagellated cells that cause red tides. Macroscopic algae: Phaeophyta and Rhodophyta are multicellular kelps and seaweeds.

Importance: Algae provide the basis of the food chain in most aquatic habitats, and they produce a large proportion of atmospheric O_2 through photosynthesis. Some are harvested as a source of cosmetics, food, and medical products.

The Protozoa

Alternate Names: General—Animal-like **protists;** unicellular animals. Specific— **ameba, flagellate, ciliate, sporozoan.**

Overall Morphology: Size is usually microscopic, with some exceptions. At the cellular level, eucaryotic; most have locomotor structures, such as **flagella, cilia, pseudopods**; special feeding structures may be present; lack a cell wall; cells vary in size from those large enough to be seen with the naked eye to those not much larger than a bacterium; extreme variations in shape. May exist in various forms: **trophozoite,** a motile, feeding stage or **cyst,** a dormant resistant stage. At the organismic level, most are unicellular, though some may be in colonies; no multicellular forms.

Nutritional Mode/Distribution: All are **heterotrophic.** A large number are free-living in fresh water, salt water, or soil; most feed on complex organic matter, bacteria, algae, and other microorganisms. A number are parasites on animals; several cause serious infectious diseases. May be spread from host to host by **insect vectors.** Most free-living forms require oxygen to grow (aerobic); some parasites can grow in the absence of oxygen (anaerobic).

Require a moist habitat (soil, water, plants, insects, vertebrates); preferred temperatures are 15°–40° C.

Reproduction: Asexual by binary fission and **mitosis,** budding; sexual by fusion of free-swimming gametes, conjugation.

Classification of Major Groups: Protozoans are subdivided into four groups, based upon mode of locomotion and type of reproduction: **Mastigophora,** the flagellates, motile by flagella; **Sarcodina,** the amebas, motile by pseudopods; **Ciliophora,** the ciliates, motile by cilia; **Sporozoa,** all parasites; motility not well developed; produce unique reproductive structures.

Importance: Ecologically important in regulating microbial populations in the environment, decomposing organic matter, serving as food for larger organisms. Medically significant in that every year hundreds of millions of people are afflicted with one of the many protozoan infections (malaria, trypanosomiasis, amebiasis, leishmaniasis), and a large number die from the severe effects. Protozoan infections may be geographically restricted.

Unique Features: Active, motile cells with complex, well-developed organelles; ability to encyst.

The Helminth Parasites

Other Names: Parasitic worms, **cestodes** (tapeworms), **trematodes** (flukes), **roundworms** (nematodes).

Overall Morphology: Animal cells; multicellular; individual organs specialized for reproduction, digestion, movement, protection, though some of these are reduced.

Nutritional Mode: Parasitize other animals; feed on host fluids, tissues; have mouth parts for attachment to or digestion of host tissues. Most are aerobic, but some may live in anaerobic conditions.

Reproduction and Life Cycle: Most have well-developed sex organs that produce eggs and sperm. Fertilized eggs go through larval period in or out of host body.

Major Groups: **Flatworms** have highly flattened body; no definite body cavity; digestive tract a blind pouch; male and female sex organs may or may not be present on the same worm. May have simple excretory and nervous systems. **Cestodes** or tapeworms consist of chains of rectangular segments; attach to host's intestine by hooked mouthpart. **Trematodes** or flukes are flattened, ovoid, nonsegmented worms with sucking mouth parts. **Roundworms (nematodes)** have round bodies in cross section; have a complete digestive tract; develop complex protective cuticle on their surface; motile by whiplike motion; may have spines and hooks on mouth; excretory and nervous systems poorly developed.

How Transmitted and Acquired: Through ingestion of larvae or eggs in food, soil, water; may be carried by insect vectors.

Importance: Afflict billions of humans. Because these worms may cause illness and death, they have medical and economic impact of staggering proportions.

True–False Questions

Determine whether the following statements are true (T) or false (F). If you feel a statement is false, explain why, and reword the sentence so that it reads accurately.

____ 1. Flagella and cilia are found primarily in protozoa.

____ 2. The nuclear envelope is a double membrane structure with pores that allow communication with the cytoplasm.

____ 3. If two haploid cells fuse, a diploid cell will result.

____ 4. The cell wall has the same structure in all eucaryotes.

____ 5. The roughness of the rough endoplasmic reticulum is due to ribosomes.

____ 6. The common names for types of microscopic fungi are yeasts and molds.

____ 7. Since the fungi photosynthesize, they are grouped with the higher plants.

____ 8. A hypha divided into compartments by cross walls is called septate.

____ 9. A conidium is a sexual spore, and a zygospore is an asexual spore.

____ 10. Most fungi are free-living and nonparasitic.

____ 11. All algae contain some type of chlorophyll.

____ 12. All protozoa have some means of locomotion.

____ 13. The trophozoite is the active feeding stage of protozoa, and the cyst is an inactive dormant stage.

____ 14. Sporozoans are all parasitic and nonmotile.

____ 15. Helminth parasites reproduce by means of eggs and sperm.

Concept Questions

1. Summarize the major similarities and differences between procaryotic and eucaryotic cells.
2. Which kingdoms contain eucaryotic microorganisms? Differentiate between the concepts of unicellular, colonial, and multicellular levels of organismic organization. Give examples of each level.
3. Describe the functional morphology of each of the major eucaryotic organelles. How are flagella and cilia similar? How are they different?
4. Trace the synthesis of cell products, their processing, and their packaging through the organelle network.
5. Describe the detailed structure of the nucleus. Why can one usually not see the chromosomes? When are the chromosomes visible? What causes them to be visible?
6. Define mitosis and explain its function. What happens to the chromosome number during this process, and how does a diploid organism remain diploid and a haploid organism remain haploid?
7. Define meiosis and explain its function. When does it occur in diploid organisms? When in haploid organisms?
8. Describe some of the ways that organisms use lysosomes.
9. Why would a cell need a skeleton?
10. Differentiate between the yeast and hypha types of fungal cell. What is a mold? What does it mean if a fungus is dimorphic?
11. How does a fungus feed? Where would one expect to find fungi?
12. Describe the functional types of hyphae. Describe the two main types of asexual fungal spores and how they are formed. What are some types of conidia? What is the reproductive potential of molds in terms of spore production?
13. Explain how sexual spores are important to fungi. Give the three main types, and explain with a simple diagram how each is formed.
14. How are fungi classified? Give an example of a member of each phylum of the fungi, and describe its structure and how it is important.
15. What is a mycosis? What kind of mycosis is athlete's foot? What kind is coccidioidomycosis?
16. What are protists?
17. Present the principal characteristics of algae. How are they important? What causes the many colors in the algae?
18. Explain the life cycle in protozoans. How do they feed? How do they reproduce?
19. Briefly outline the characteristics of the four protozoan groups. What is an important pathogen in each group?
20. Which protozoan group is the most complex in structure and behavior? In life cycle? What sets the sporozoans apart from the other protozoan groups?
21. Construct a chart that compares the four groups of eucaryotic microorganisms (fungi, algae, protozoa, helminths) in cellular structure. Address whether the group has a cell wall, chloroplasts, is motile, or some other unique feature. Compare the manner of nutrition and body plan (unicellular, colonial, multicellular).

Practical/Thought Questions

1. Suggest how one would go about determining if mitochondria and chloroplasts are a modified procaryotic cell.
2. Give the common name of a eucaryotic microbe that is unicellular, walled, non-photosynthetic, nonmotile, and bud-forming.
3. Give the common name of a microbe that is unicellular, nonwalled, motile with flagella, and has chloroplasts.
4. Give the common name of a microbe that has long, thin pseudopods and is encased in a hard shell.
5. Give the common name of a multicellular animal that is composed primarily of thin sacs of reproductive organs.
6. Give the name of a parasite that is transmitted in the cyst form.
7. You just found an old container of food in the back of your refrigerator. You open it and see a mass of multicolored fuzz. As a budding microbiologist, describe how you would determine what is growing on the food.
8. Here is a simple project using mushrooms as art. A unique print can be made by placing a mushroom cap, gill side down on a piece of paper and leaving it undisturbed for a few days. The spores will fall onto the paper, leaving a perfect pattern of the gills. If carefully lacquered, this can be preserved as a unique and beautiful design.
9. Explain why opportunistic mycoses are a growing medical problem.
10. How are bacterial endospores and cysts of protozoa alike?
11. You have gone camping in the mountains and plan to rely on water present in forest pools and creeks. Certain encysted pathogens often live in this type of water, but you don't discover this until you arrive at the campground. How might you treat the water to prevent becoming infected?
12. Explain the two levels of parasitism at work in Chagas' disease. Do you suppose the trypanosome has a parasite, too? If so, could its parasite have a parasite? What does this tell you about the prevalence of parasitism and its success as a mode of life? What is a potential weakness in parasitism?
13. Can you think of a way to determine if a child is suffering from pinworms?

CHAPTER 5

An Introduction to the Viruses

Beauty from disease. The variegations that impart the striking color patterns on these tulips are due to a viral infection (tulip mosaic disease), which is passed from generation to generation in the seeds and bulbs.

Chapter Preview

Virology is the study of an important group of unusual and intriguing particles whose major characteristics are so different from cellular microorganisms that they must be considered separately. The involvement of viruses in human diseases, ranging from AIDS and the common cold to influenza and cancer, challenges even the elementary microbiology student to have a basic grasp of their properties. A generous share of this chapter focusses on the relationship between the virus and its host cell because of the intimate connection between them. Along the way, we will also explore unique features of virus structure, physiology, multiplication, cultivation, and identification.

The Search for the Elusive Virus

The discovery of the light microscope made it possible to see firsthand the agents of many bacterial, fungal, and protozoan diseases. But the techniques developed for observing and cultivating these relatively large microorganisms were virtually useless for detecting viruses. For many years, the causes of viral infections such as smallpox and polio remained cloaked in mystery, even though it was clear that the diseases were transmitted from person to person. The French bacteriologist Louis Pasteur was certainly on the right track when he postulated that rabies was caused by a "living thing" smaller than bacteria, and in 1884 he was able to develop the first vaccine for rabies. It was also Pasteur who proposed that the term *virus* (L. poison) be used to denote this special group of infectious agents.

The first substantial revelations about the unique characteristics of viruses occurred in 1892, when D. Ivanovski isolated the tobacco mosaic virus in culture, and in 1898, when Friedrich Loeffler and Paul Frosch isolated the virus that causes foot-and-mouth disease in cattle. These early researchers found that when infectious fluids from host organisms were passed through porcelain filters designed to trap bacteria, the filtrate still remained infectious. This proved that an infection could be caused by a cell-free fluid containing agents smaller than bacteria, and thus first introduced the concept of a *filterable virus.*

Over the succeeding decades, a remarkable picture of the physical, chemical, and biological nature of viruses began to take form. Years of experimentation were required to show that viruses were not vague components of fluids, but actual particles, with a definite size, shape, and chemical composition, and that, like bacteria, they could be cultivated. By the 1950s, virology had grown into a multifaceted discipline that promised to provide much information on disease, genetics, and even life itself.

The Position of Viruses in the Biological World

Viruses are highly novel biological entities known to infect every type of cell, including those of microorganisms; there are bacterial, algal, fungal, protozoan, plant, and animal viruses. Although the emphasis in this chapter will be on viruses that cause human and animal disease, much credit for our knowledge must be given to work with bacterial and plant viruses. The exceptional and curious nature of viruses spurs numerous questions, including: (1) Are they organisms—that is, are they alive? (2) What are their distinctive biological characteristics? (3) How can particles so small, simple, and seemingly insignificant be capable of causing disease and death? and (4) What is the connection between viruses and cancer? In this chapter we will address these questions and many others.

The structure and behavior of viruses have invariably led to debates about their position in the microbial world. One viewpoint holds that, because viruses are unable to exist independently from the host cell, they are not living things, but are more akin to large, infectious molecules. Others propose that, even though viruses do not exhibit most of the life processes of cells (discussed in chapter 2), they can direct life processes of cells, and so are certainly more than inert and lifeless molecules. Depending upon the circumstances, both views are defensible. In medicine, this debate has greater philosophical than practical importance because viruses are agents of disease and must be dealt with through control, therapy, and prevention, whether we regard them as living or not. In keeping with their special position in the biological spectrum, it is best to describe viruses as infectious particles (rather than organisms) and as either active or inactive (rather than living or dead).

Viruses are different from their host cells in size, structure, behavior, and physiology. They are **obligate intracellular parasites,** meaning that they cannot multiply unless they invade a specific host cell and instruct its genetic and metabolic machinery to make and release quantities of new viruses. Because of this characteristic, viruses are capable of causing serious damage and disease.

The General Structure of Viruses

Size Range

As a group, viruses represent the smallest infectious agents (with some unusual exceptions to be discussed later). How small are they? Their size relegates them to the realm of the **ultramicroscopic.** This term means that most of them are so minute ($< 0.2\ \mu m$) that an electron microscope is necessary to detect them or examine their fine structure. Such extremely small size makes them **filterable,** able to pass through special filters that would trap most microbes. The way they are dwarfed by their host cells is also notable. Even as small as bacteria are, more than 2,000 bacterial viruses would fit into a single cell, and more than 50,000,000 polioviruses could be accommodated by an average human cell. Animal viruses range in size from the smallest parvoviruses[1] (around 20 nm [0.02 μm] in diameter) to the poxviruses[2] that are as large as small bacteria (up to 450 nm [0.4 μm] in length). Some cylindrical viruses are relatively long (800 nm [0.8 μm] in length), but so narrow in diameter (15 nm [0.015 μm]) that their visibility is still limited without the high magnification and resolution of an electron microscope. Table 5.1 compares the sizes of several viruses alongside procaryotic and eucaryotic cells and molecules.

Viral architecture is most readily observed through special stains in combination with electron microscopy (figure 5.1). Negative staining uses very thin layers of an opaque salt to outline the shape of the virus against a dark background and to enhance textural features on the viral surface. Internal details are revealed by positive staining of specific parts of the virus such as protein or nucleic acid. In *shadowcasting,* the virus preparation is attached to a surface and showered with a dense metallic vapor directed from a certain angle. The thin metal coating over the surface of the virus approximates its contours, and a shadow is cast on the unexposed side.

1. Naked DNA viruses that cause intestinal infections in humans.
2. A group of viruses, including smallpox, that create marked skin bumps called pox.

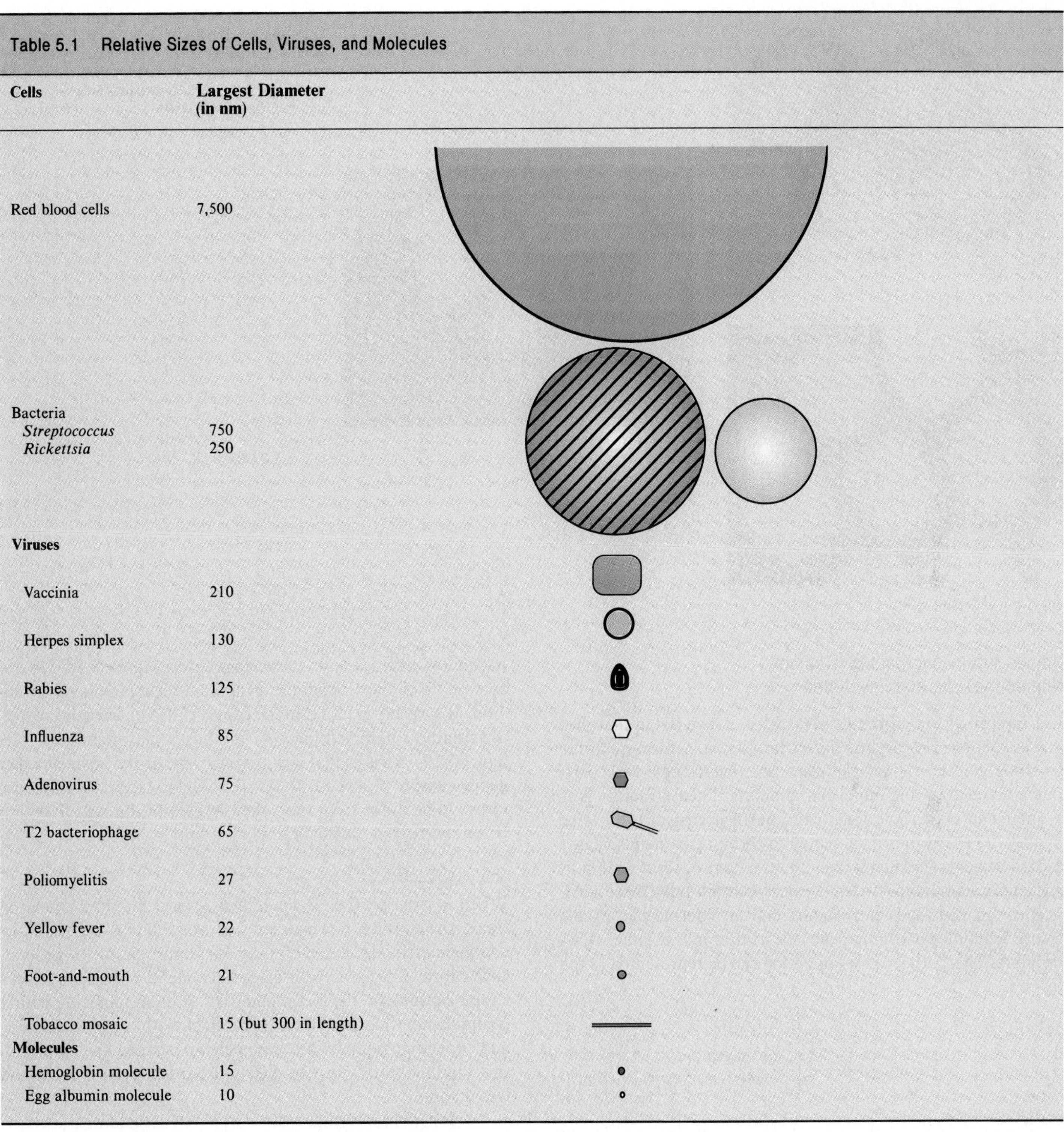

Table 5.1 Relative Sizes of Cells, Viruses, and Molecules

Cells	Largest Diameter (in nm)
Red blood cells	7,500
Bacteria	
Streptococcus	750
Rickettsia	250
Viruses	
Vaccinia	210
Herpes simplex	130
Rabies	125
Influenza	85
Adenovirus	75
T2 bacteriophage	65
Poliomyelitis	27
Yellow fever	22
Foot-and-mouth	21
Tobacco mosaic	15 (but 300 in length)
Molecules	
Hemoglobin molecule	15
Egg albumin molecule	10

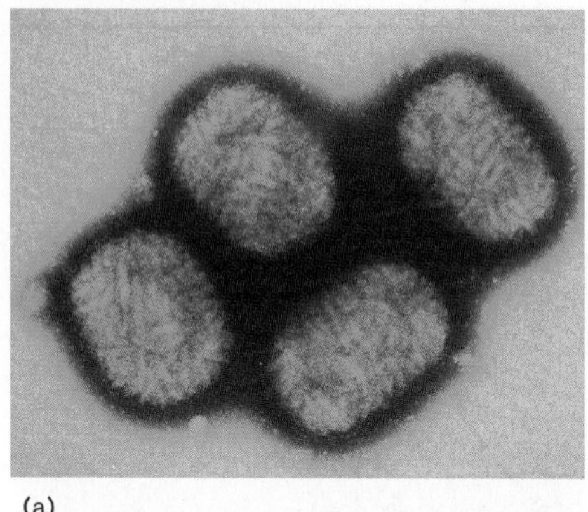
(a)

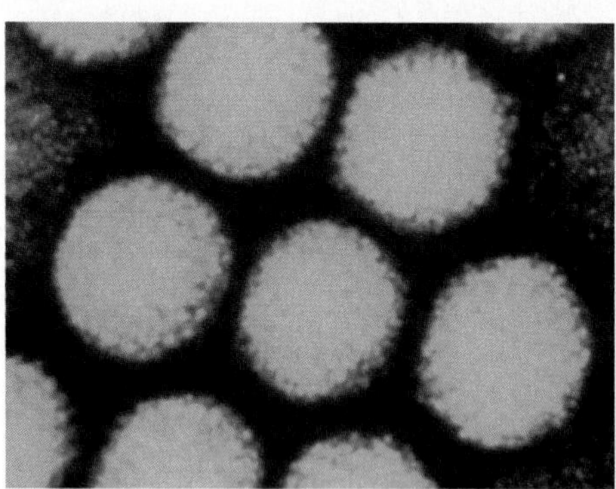
(b)

(c)

Figure 5.1 Methods of viewing viruses. (*a*) Negative staining of an orf virus, revealing details of its outer coat (×122,000). (*b*) Positive staining of a mass of adenovirus particles in a cell (×300,000). (*c*) Shadowcasting image of a vaccinia virus (×17,500).

Unique Viral Constituents: Capsids, Nucleocapsids, and Envelopes

It is important to realize that viruses are not cells and that they possess neither procaryotic nor eucaryotic structural qualities. Instead, they are large, complex macromolecules, with parts made up of repeating molecular subunits. Their structure is so regular and crystalline that many purified viruses form large aggregates or crystals if subjected to certain treatments (figure 5.2). The general plan of virus organization is exceptional in its simplicity and compactness. Viruses contain only those parts needed to invade and control a host cell: an external coating and a core containing one or more nucleic acid strands of either DNA or RNA. This pattern of organization can be represented as follows:

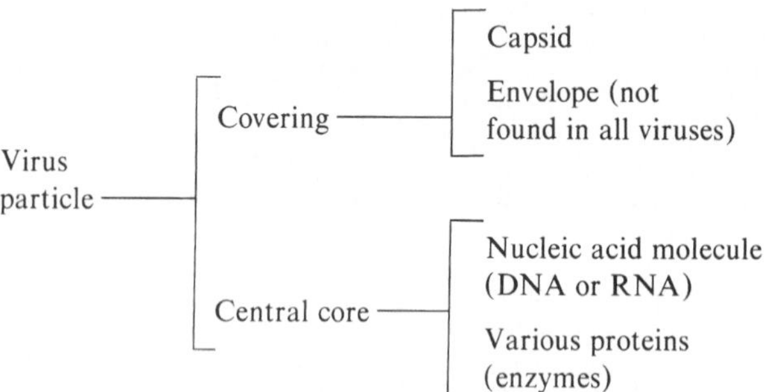

All viruses have a protein **capsid** or shell that surrounds the nucleic acid strand. Together, the capsid and the nucleic acid strand are referred to as the *nucleocapsid* (figure 5.3*a*). Members of 12 of the 17 families of animal viruses possess an additional covering external to the capsid called an **envelope,** which is actually a modified piece of the host's cell membrane (see figure 5.19). Viruses that lack this envelope are considered **naked nucleocapsids** (figure 5.3*b*). As we shall see later, the enveloped viruses also differ from the naked viruses in the way that they enter and leave a host cell.

The Viral Capsid

When a virus particle is magnified several hundred thousand times, the capsid is a prominent feature, often having the exact and geometric perfection of a crystal (figure 5.2*b*). In general, each capsid is constructed from smaller, identical building blocks called **capsomers**. Each capsomer is a protein molecule with a configuration that permits it to interlock with other capsomers, and depending on how the capsomers are shaped and arranged, this binding results in two different capsid types: helical and icosahedral.

Helical capsids have the simpler structure, with a single type of rod-shaped capsomer that bonds with other capsomers to form a hollow disc resembling a bracelet. During the formation of the nucleocapsid, these discs fuse with other discs to form a continuous helix, with the nucleic acid strand coiled tightly within (figure 5.4). In electron micrographs the appearance of a helical capsid varies with the type of virus. The nucleocapsids of naked helical viruses are very rigid and tight, thereby creating a gross cylindrical appearance (figure 5.5*a, b*).

capsid (kap'-sid) L. *capsa*, box.

capsomer (kap'-soh-meer) L. *capsa*, box, and *mer*, part.

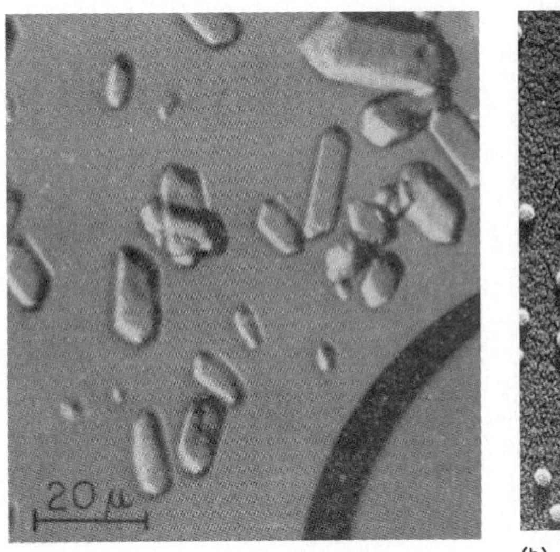

(a)

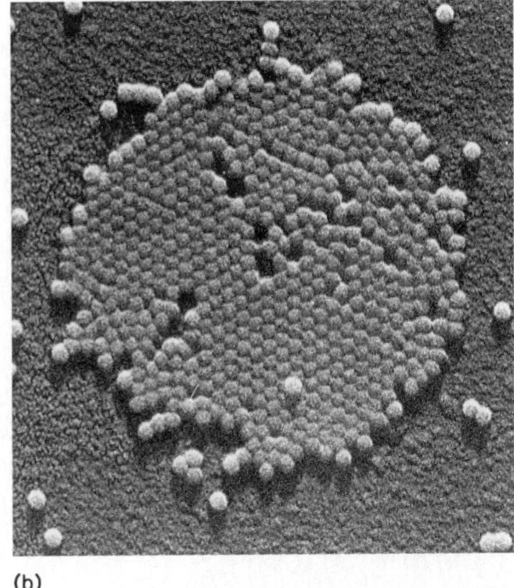

(b)

Figure 5.2 The crystalline nature of viruses. (*a*) Slightly magnified (×1,200) crystals of purified poliovirus particles. (*b*) Highly magnified (×150,000) electron micrograph of the capsids of this same virus, demonstrating their highly geometric nature.

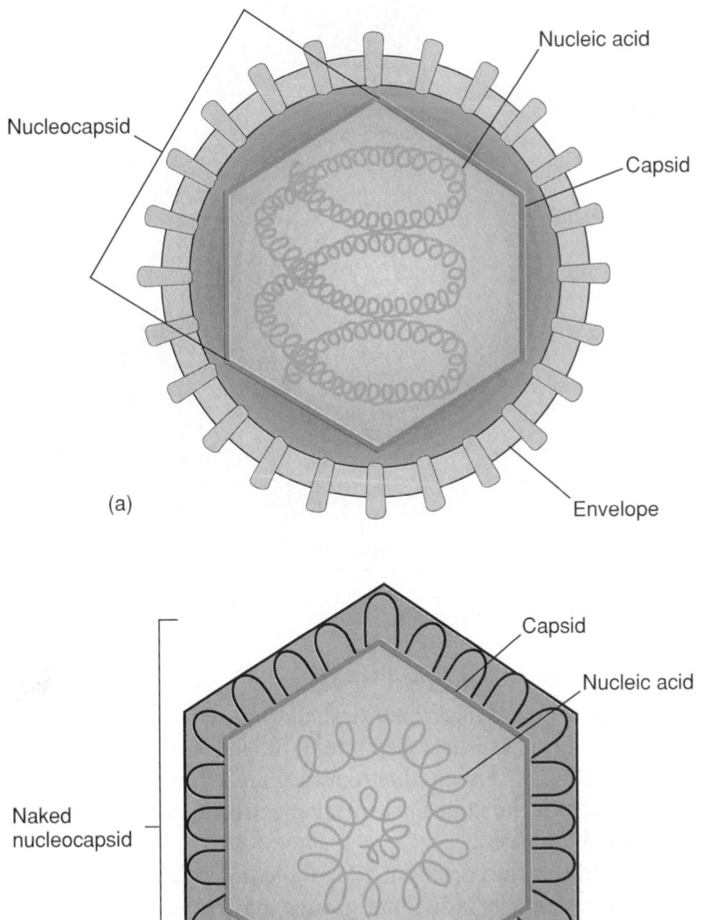

Figure 5.3 Generalized structure of viruses. (*a*) An enveloped virus composed of a nucleocapsid surrounded by a membrane or envelope. (*b*) A naked virus composed of a capsid surrounding nucleic acid (nucleocapsid).

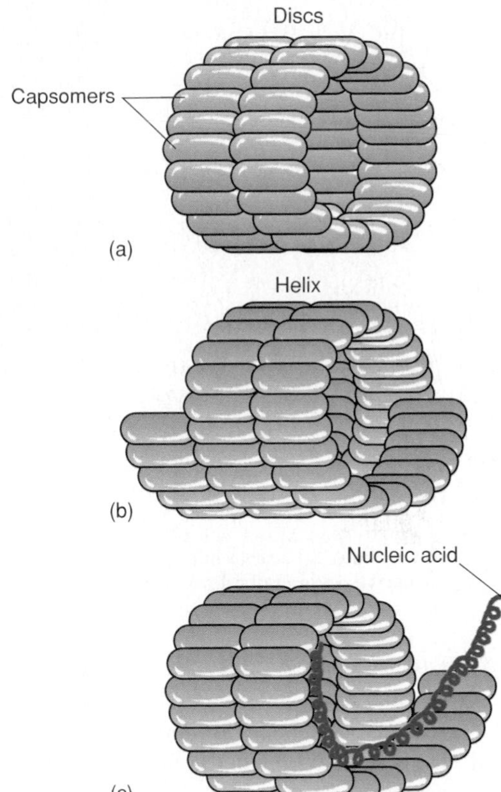

Figure 5.4 Stages in the assembly of helical nucleocapsids. (*a*) Individual capsomers are bound together into discs. (*b*) Several adjacent discs are assembled into a helix. (*c*) A nucleic acid strand is inserted into the core of the helix.

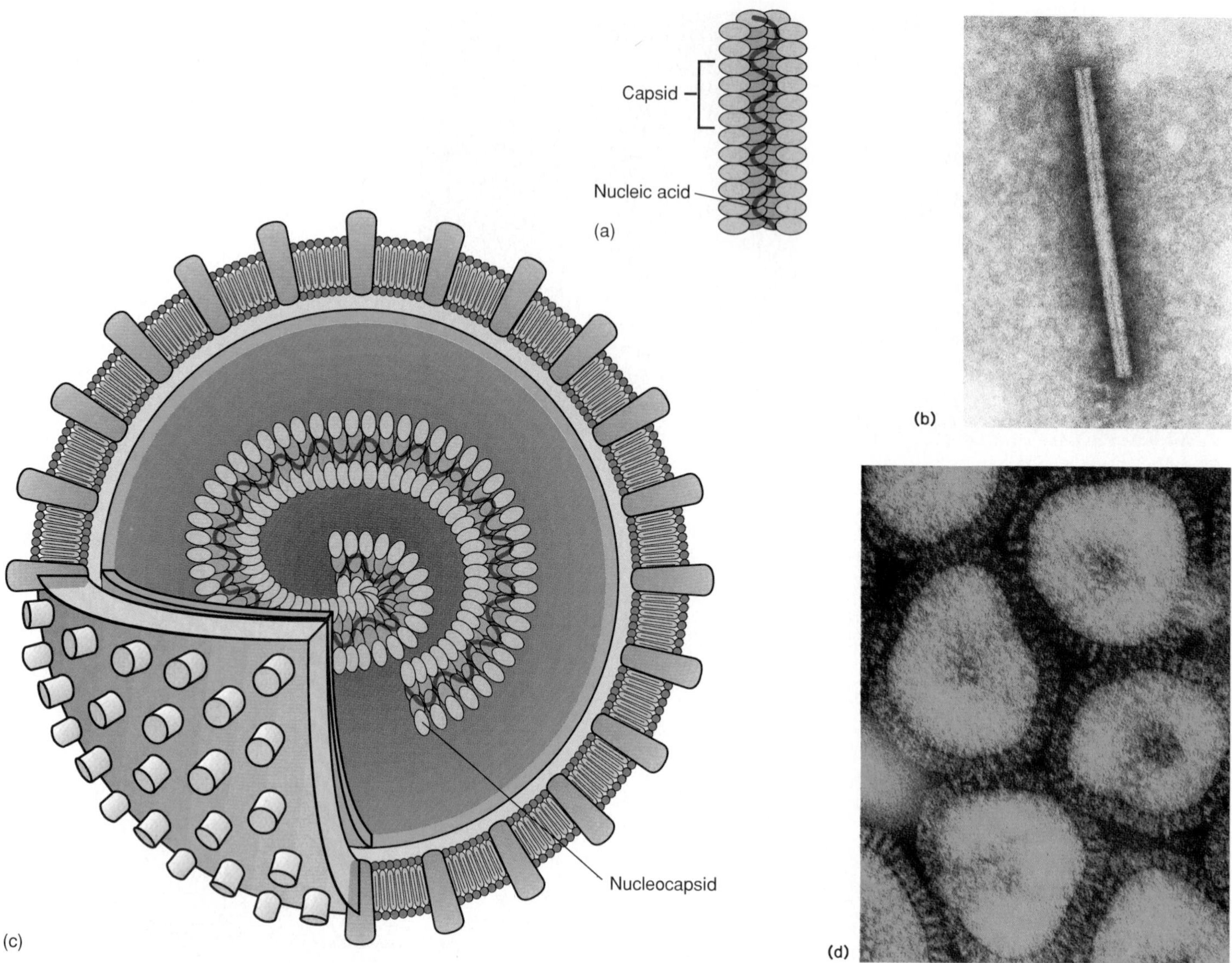

Figure 5.5 Typical variations of viruses with helical nucleocapsids. Naked helical (tobacco mosaic) virus: (*a*) a schematic view and (*b*) a greatly magnified micrograph. Note the overall cylindrical morphology. Enveloped helical (influenza) virus: (*c*) a schematic view and (*d*) an electron micrograph of the same virus (×350,000).

An example is the *tobacco mosaic virus* that attacks tobacco leaves and gives them a mottled appearance. Enveloped helical nucleocapsids, on the other hand, are more flexible and tend to be arranged as a loose helix within the envelope (figure 5.5*c, d*). This type of morphology is found in several enveloped human viruses, including those of influenza, measles, mumps, and rabies.

The capsids of a number of major virus families are arranged in a three-dimensional, many-sided figure generally known as a polyhedron. A specific example of this pattern is to be seen in the **icosahedral viruses.** An icosahedral virus is built of triangular capsomers that are very uniform and symmetrical and can become attached to one another along their three sides, which according to the rules of geometry and crystallography, results in a hollow icosahedral shell (figure 5.6). During assembly of the virus (discussed later), the nucleic acid is packed into the center of this icosahedron, forming a nucleocapsid. This arrangement is the same basic principle used in the construction of geodesic domes.

Although the capsids of all icosahedral viruses have a similar basic structure, individual groups may show variations. In photomicrographs the capsid may appear spherical or cubical, and individual capsomers may look either ring- or dome-shaped (figure 5.7). Another factor that alters the appearance of icosahedral viruses is whether they are naked or enveloped. The viruses of the common cold and polio (picornaviruses) have naked icosahedral nucleocapsids. Mumps viruses, influenza viruses, and herpesviruses (agents of chickenpox and cold sores) are enveloped.

---

icosahedral (eye''-koh-sah-hee'-drul) Gr. *eikosi*, twenty, and *hedra*, side. A icosahedron is a highly regular geometric figure composed of 20 triangular-shaped faces that fit together to form a 12-cornered unit.

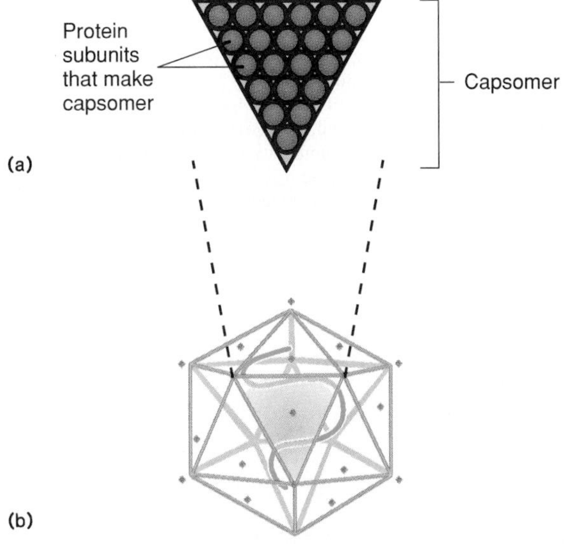

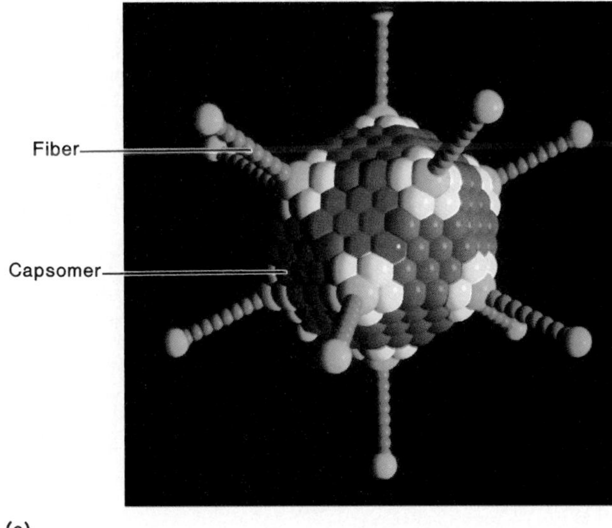

(c)

Figure 5.6 Formation and structure of icosahedral viruses. (*a*) The basic building block is a triangular-shaped capsomer composed of spherical proteins. In this example, each capsomer contains 21 protein units, although viruses may vary in this feature. (*b*) View of a completed capsid, shown transparent to reveal its 20 triangular panels and 12 corners. (*c*) Three-dimensional model of an adenovirus.

The Envelope

Following multiplication, enveloped viruses bud off the surface of the host cell, taking with them a bit of its membrane system (see figure 5.19). The viral envelope, like the cell membrane, contains lipids and proteins, but it is different in an important way. During development inside the host cell, enveloped viruses replace some or all of the regular membrane proteins with special viral proteins. Some proteins form a binding layer between the envelope and capsid of the virus, and others (glycoproteins) remain exposed on the outside of the envelope. These protruding molecules, called **spikes,** function in attachment and invasion of another host cell. Because the envelope is rather supple, enveloped viruses may take on shapes ranging from a sphere to a bullet to a filament. Figure 5.8 summarizes the morphological types of some common viruses.

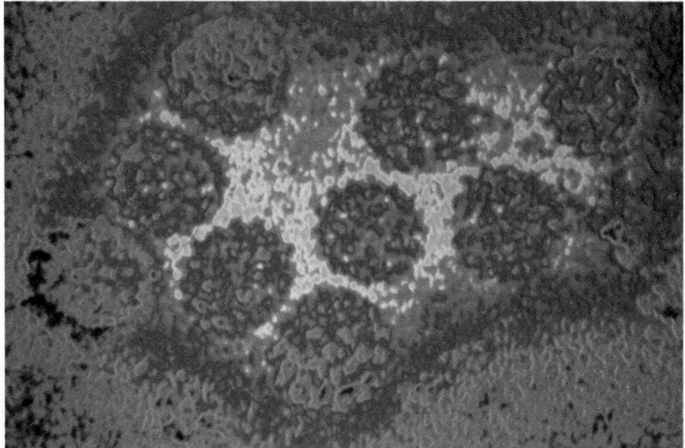

(a)

(b)

Figure 5.7 Two types of icosahedral viruses, highly magnified. (*a*) False-color micrograph of papilloma viruses with unusual, ring-shaped capsomers. (*b*) Herpes simplex virus, an enveloped icosahedron (×300,000).

Complex Viruses

Two special groups of viruses, termed **complex viruses** (figure 5.9), are more intricate in structure than those just described, and are not restricted to only helical, icosahedral, naked, or enveloped virus morphology. The **poxviruses** (including the agent of smallpox) are typical in that they contain a nucleic acid core, but they lack a regular capsid and have in its place several layers of lipoproteins and coarse fibrils. Large **bacteriophages,** which are viruses that parasitize bacteria, are likewise very complex, with a cubical head, a helical tail, and fibers for attachment to the host cell. Their mode of multiplication will be covered in a later section of this chapter.

---

bacteriophage (bak-teer′-ee-oh-fayj″) From *bacteria,* plus Gr. *phagein,* to eat. Also called a bacterial virus.

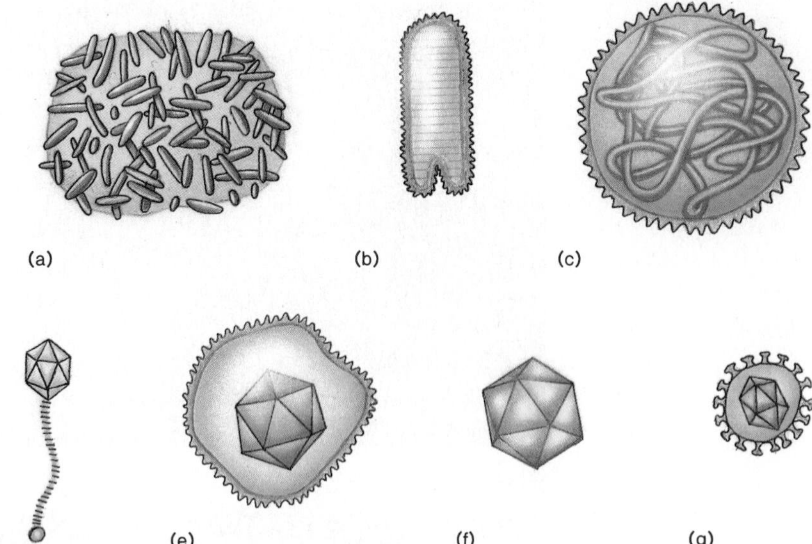

Figure 5.8 An array of virus types. (*a*) Poxvirus, a complex virus. (*b*) Rhabdovirus, a bullet-shaped enveloped virus. (*c*) Mumps virus, a spherical enveloped virus with helical nucleocapsid. (*d*) Flexible tailed bacteriophage. (*e*) Herpesvirus, a spherical, enveloped icosahedral nucleocapsid. (*f*) Papilloma (wart) virus, a naked icosahedral nucleocapsid. (*g*) AIDS virus, an enveloped icosahedral nucleocapsid.

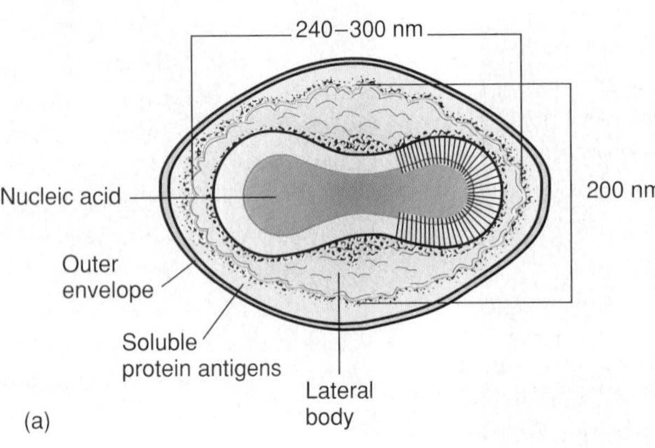

Figure 5.9 Detailed structure of complex viruses. (*a*) The vaccinia virus, a poxvirus. (*b*) Photomicrograph and (*c*) diagram of a T4 bacteriophage. (See text for functions.)

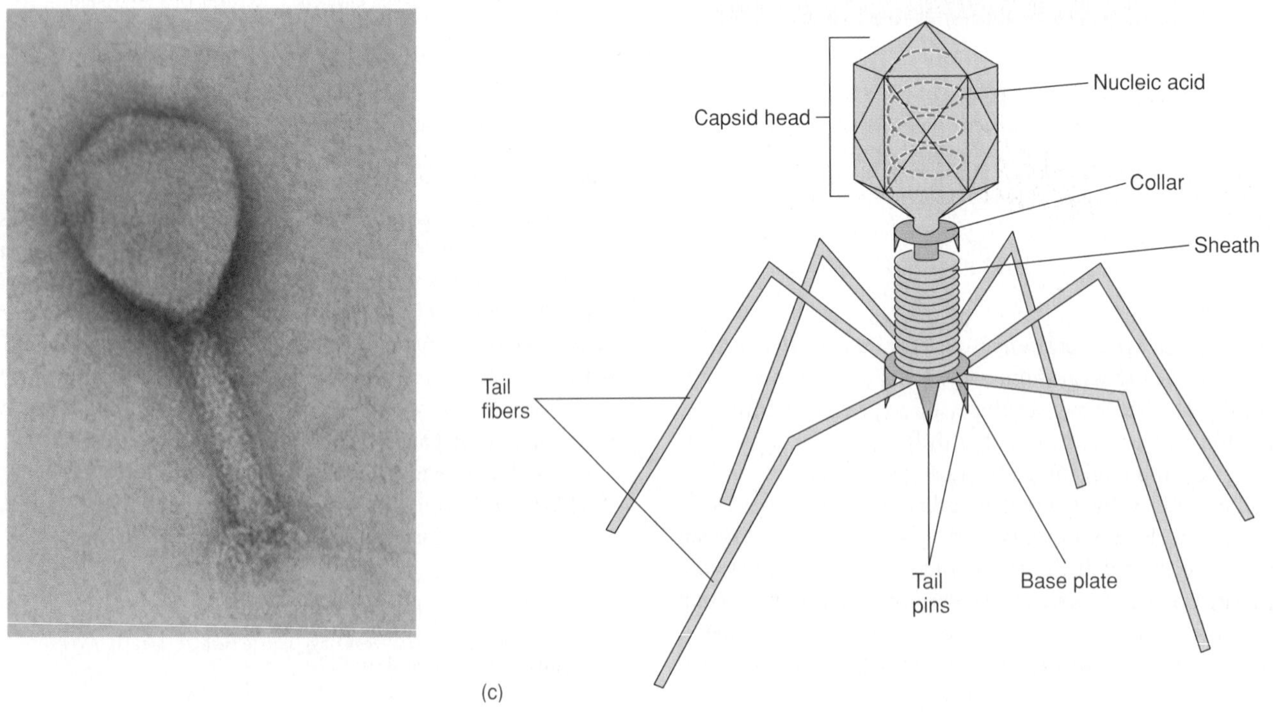

Biomedical Significance of the Viral Capsid/Envelope The outermost covering of a virus, whether a capsid (as in naked viruses) or an envelope, is indispensable to viral function, because it helps sustain the viral genetic material. Capsids, envelopes, and the outer shells of poxviruses protect the nucleic acid from the effects of various enzymes and chemicals when it is outside the host cell. For example, the capsids of enteric (intestinal) viruses such as polio and hepatitis A are resistant to the acid- and protein-digesting enzymes of the gastrointestinal tract. Capsids and envelopes are also responsible for helping to introduce the viral DNA or RNA into a suitable host cell, first by binding to the cell surface and then by assisting in genetic penetration (to be discussed in more detail later). In addition, parts of viral capsids and envelopes stimulate the immune system to produce antibodies. Antibodies can neutralize viruses and protect the host's cells against future infections. Antibodies in the blood are also the basis of various diagnostic tests for virus infections (see chapters 13, 20, and 21).

Nucleic Acids: The Core of a Virus

So far, one biological constant is that the genetic makeup of living things is encoded in macromolecules called nucleic acids. Viruses, though neither alive nor cells, are no exception to this rule—but with a significant difference. Unlike cells, which contain both DNA and RNA, viruses contain only one type: either DNA or RNA. Because viruses must pack into a tiny space all of the genes necessary to instruct the host cell to make new viruses, the size of a viral genome is rather small compared to that of a cell. It varies from five genes in the smallest viruses to dozens of genes in the largest ones. By comparison, the bacterium *Esherichia coli* has approximately 4,000 genes, while a human cell has approximately 100,000.

In chapter 2 we also learned that DNA usually exists in cells as a double-stranded molecule and that RNA is single-stranded. Although most viruses follow this same pattern, a few exhibit distinctive and exceptional forms. Notable examples are the parvoviruses (the cause of fifth disease), which contain single-stranded DNA, and reoviruses (a cause of respiratory and intestinal tract infections), which contain double-stranded RNA.

Biomedical Significance of the Viral Nucleic Acid Whether a virus has DNA or RNA in single or double strands, these tiny strands of genetic material hold the key to the behavior of that virus. In a very real sense, viruses are **genetic parasites,** because they cannot realize a full infectious cycle until their nucleic acid has reached the internal habitat of the host cell. Interestingly enough, viral nucleic acid alone has been shown to cause infection when experimentally introduced into a host cell.

Other Substances in the Virus Particle

In addition to the protein of the capsid, the proteins and lipids of envelopes, and the nucleic acid of the core, viruses may contain enzymes for specific operations within their host cell. Examples of these enzymes are *polymerases,* for synthesizing DNA and RNA, and enzymes for digesting host DNA and proteins. Viruses completely lack the genes for synthesis of metabolic enzymes. As we shall see, this lack has little consequence, because viruses have adapted to confiscate completely their hosts' metabolic resources.

How Viruses Are Classified and Named

Although viruses are not cells and thus are not included in the kingdoms discussed in chapter 1, they are diverse enough to require a classification scheme to aid in their study and identification. In a very informal and general way, we have already begun classifying viruses—as animal, plant, or bacterial viruses; enveloped or naked viruses; DNA or RNA viruses; and helical or icosahedral viruses. These introductory categories are certainly useful in organization and description, but the study of specific viruses requires a more standardized method of nomenclature. For many years, the animal viruses were classified mainly on the basis of their hosts and the kind of diseases they caused. This method gradually became less workable as more viruses were discovered and numerous areas of overlap emerged. For example, the same virus can have more than one host, and several different viruses may cause a similar type of disease. It was ultimately necessary to adopt a system that took into account the actual nature of the virus particles themselves, with only partial emphasis on host and disease. The main criteria presently used to group viruses are structure and chemical composition (including similarities in genetic makeup).

A widely used scheme for classifying animal viruses assigns them to *families* that represent major subgroups and to *genera* that correspond to each different viral type (tables 5.2 and 5.3). The two superfamilies of animal viruses are those containing DNA and those containing RNA. DNA viruses can be further divided into six families, and RNA viruses into eleven families. Virus families are given a name comprised of a Latin root followed by *-viridae.* Characteristics used for placement in a particular family include type of capsid, presence and type of envelope, overall viral size, and area of the host cell in which the virus multiplies.

Some virus families are named for their microscopic appearance (shape and size). Examples include *rhabdoviruses,* which have a bullet-shaped envelope, and *togaviruses,* which have a cloaklike envelope. Anatomic or geographic areas of isolation have also been used. For instance, *adenoviruses* were first discovered in adenoids (one type of tonsil), and bunyaviruses (bun-yah) were originally isolated in an area in Africa called Bunyamwera. Some names describe the effects on the host: *Herpesviruses* are named for the spreading character of herpes rashes. Acronyms made from blending several characteristics include *picornaviruses,* which are tiny RNA viruses, and reoviruses (or **r**espiratory **e**nteric **o**rphan viruses), which infect the respiratory tract and the intestine and are not closely related to any other virus.

polymerase (pol-im′-ur-ace) An enzyme that synthesizes a large molecule from smaller subunits.

rhabdovirus (rab″-doh-vy′-rus) Gr. *rhabdo,* little rod.
togavirus (toh″-gah-vy′-rus) L. *toga,* covering or robe.
adenovirus (ad″-uh-noh-vy′-rus) Gr. *aden,* gland.
herpesvirus (hur″-peez-vy′-rus) Gr. *herpes,* to creep.
picornavirus (py-kor″-nah-vy′-rus) Sp. *pico,* small, plus RNA.

Table 5.2 Animal Virus Families and Their Major Characteristics							
Family	**Nucleic Acid**	**Strand Type**	**Capsid Type**	**Envelope**	**Size (diameter in nm)**	**Additional Features**	**Common Name of Important Members***
DNA Viruses							
Poxviridae	DNA	Double	None	+	130–300	Complex virus; brick-shaped	Smallpox virus (Variola)
Iridoviridae	DNA	Double	Icosahedral	+/−	130	Spherical	None infect humans
Herpesviridae	DNA	Double	Icosahedral	+	150–200	Spherical; can become latent**	Herpes simplex virus, Varicella zoster virus, Epstein-Barr virus
Adenoviridae	DNA	Double	Icosahedral	−	70–90	Spherical	Human adenoviruses
Papovaviridae	DNA	Double	Icosahedral	−	45–55	Spherical; most produce tumors in host	Human papilloma virus
Hepadnaviridae	DNA	Single/double	Icosahedral	+	42	Spherical, with double shell	Hepatitis B virus
Parvoviridae	DNA	Single	Icosahedral	−	18–26	Spherical	Parvovirus B19
RNA Viruses							
Picornaviridae	RNA	Single	Icosahedral	−	20–30	Spherical	Hepatitis A virus; poliovirus, coxsackieviruses, rhinoviruses
Calciviridae	RNA	Single	Icosahedral	−	35–40	Spherical	Norwalk virus
Togaviridae	RNA	Single	Icosahedral	+	45–70	Spherical; most are carried by arthropods	Rubella virus; yellow fever virus; dengue fever virus
Bunyaviridae	RNA	Single	Helical	+	90–100	Spherical; transmitted by arthropods	Bunyamwera virus, Rift Valley fever virus
Reoviridae	RNA	Double	Icosahedral	−	60–80	Spherical	Human rotavirus, Colorado tick fever virus
Orthomyxoviridae	RNA	Single	Helical	+	80–120	Spherical (some filaments); envelope has glycoprotein spikes	Influenza viruses
Paramyxoviridae	RNA	Single	Helical	+	125–250	Spherical or filamentous with spiked envelope	Parainfluenza virus, mumps virus, measles virus
Rhabdoviridae	RNA	Single	Helical	+	60–75	Bullet-shaped	Rabies virus
Retroviridae	RNA	Single	Icosahedral	+	100	Spherical; responsible for cancers in several animal groups	Human immunodeficiency virus (AIDS), oncoviruses***
Arenaviridae	RNA	Single	?	+	50–300	Spherical; contain granules that appear to be ribosomes	Lassa virus; lymphocytic choriomeninigitis virus
Coronaviridae	RNA	Single	Helical	+	80–130	Spherical; envelope has large, clublike spikes	Human infectious bronchitis and corona viruses
CHINA	—	—	—	—	—	Chronic Infectious Neuropathic Agents (CHINA)****	

*This is only a partial list of the principal viruses that are important to humans. Table 5.3 enlarges on the characteristics of each viral genus.

**A term indicating that the virus is within the host but in a state of inactivity; the virus may become active at a later date.

***From the root *onco,* meaning mass. These are viruses that cause cancer.

****Infectious forms that appear to be viruslike and have only recently been observed microscopically. They all cause progressive, degenerative diseases of the nervous system.

Table 5.3 Important Human Viral Genera, Common Names, and Types of Diseases They Cause

Genus of Virus	Common Name of Genus Members	Name of Disease	Clinical Characteristics in Humans
DNA Viruses			
Orthopoxvirus	Variola major and minor	Smallpox	Pox—pustules on skin
Herpesvirus	Herpes simplex (HSV) I virus	Fever blisters, cold sores	Lesions on lips, eyes
	Herpes simplex (HSV) II virus	Genital herpes	Lesions on genitals; damage to newborn infants
	Varicella zoster virus (VZV)	Chickenpox, shingles	Generalized rash (pox)
	Human cytomegalovirus (CMV)	CMV infections	Congenital viral infections, some cases of mononucleosis
	Epstein-Barr virus (EBV)	Infectious mononucleosis	Sore throat, fever, enlargement of lymph glands, Burkitt's lymphoma tumors
Mastadenovirus	Human adenoviruses	Various acute respiratory, enteric, and eye infections	—
Papillomavirus	Human papilloma virus (HPV)	Several types of warts	Epidermal tumors on skin, mucous membranes
Polyomavirus	JC virus (JCV)	Progressive Multifocal Leukoencephalopathy (PML)	Generally fatal brain infection
None assigned	Hepatitis B virus (HBV or Dane particle)	Serum hepatitis	Progressive liver infections
RNA Viruses			
Enterovirus	Poliovirus	Poliomyelitis	Infection of nervous system, paralysis
	Coxsackievirus	Affects several systems: brain, upper respiratory tract, skin	—
	ECHO* viruses	Affects several systems: brain, upper respiratory tract, skin	—
	Hepatitis A virus (HAV)	Infectious hepatitis	Attacks liver cells; jaundice
Rhinovirus	Human rhinovirus	Common cold, bronchitis	Fever, cough, nasal congestion
Calicivirus	Norwalk virus	Viral diarrhea, Norwalk virus syndrome	Acute enteritis
Alphavirus	Eastern equine encephalitis virus	EEE	Fatal encephalitis**
	Western equine encephalitis virus	WEE	Encephalitis
	Ross River Virus	—	Rash, arthritis
Flavivirus	Dengue fever virus	Dengue fever	Fever, rash, hemorrhage
	Yellow fever virus	Yellow fever	Fever, hemorrhage, hepatitis
	St. Louis encephalitis virus	—	Encephalitis
Rubivirus	Rubella virus	Rubella (German measles)	Skin, lymph node, joint symptoms; damage to fetus
Bunyavirus	Bunyamwera viruses	Rift Valley fever	Fever and viremia***
	Bwamba C viruses	—	Fever and viremia
	California group	—	Encephalitis
Phlebovirus	Rift Valley fever	—	Fever, encephalitis, hemorrhages, blindness
Nairovirus	Crimean-Congo hemorrhagic fever virus (CCHF)	Crimean-Congo hemorrhagic fever	Fever, bleeding in intestine and skin
Orthoreovirus	Mammalian reovirus	—	Not yet associated with clinical disease
Orbivirus	Colorado tick fever virus	Colorado tick fever	Fever, encephalitis; attacks bone marrow
Rotavirus	Human rotavirus	—	Vomiting, diarrhea; dehydration in infants
Influenzavirus	Influenza virus, type A (Asian, Hong Kong, and Swine influenza viruses)	Influenza or "flu"	Acute infection of the nasopharynx, trachea and bronchi
	Influenza virus, type B	Influenza or "flu"	
Paramyxovirus	Parainfluenza virus, types 1–5	Parainfluenza	Respiratory infections, including croup, common cold
	Mumps virus	Mumps	Infects salivary glands, testes, ovaries, and brain
Morbillivirus	Measles virus	Measles (red)	Fever, rash, cough, nasal discharge
Pneumovirus	Respiratory syncytial virus (RSV)	Common cold syndrome	Pneumonia in infants
Lyssavirus	Rabies virus	Rabies (hydrophobia)	Fatal brain infection
Oncornavirus	Type-C oncovirus	—	—
	Human T-cell leukemia virus (HTLV)	T-cell leukemia	Cancer of T cells
Lentivirus	HIV (human immunodeficiency viruses I and II)	Acquired immunodeficiency syndrome (AIDS)	Loss of immune function
Arenavirus	Lassa virus	Lassa fever	Fever, severe hemmorrhage, shock
Coronavirus	Infectious bronchitis virus (IBV)	Bronchitis	Acute infection of upper respiratory tract
	Enteric corona virus	—	Intestinal infections

*An acronymn for **E**nteric **C**ytopathogenic **H**uman **O**rphan, denoting the origin and effect of these viruses on cells.

**An inflammation of the brain caused by an infectious agent or a toxic substance.

***Presence of viruses in the blood.

Viruses are assigned genus status according to their host, target tissue, and the type of disease they cause (table 5.3). Viral genera are denoted by a special Latinized root, followed by the suffix *-virus* (for example, *Enterovirus*). Because the use of standardized species names has not been widely accepted, the common or English vernacular names will predominate in discussions of specific viruses in this text (for example, poliovirus).

Modes of Viral Multiplication

Viruses are seldom far removed from their hosts. In addition to providing the viral habitat, the host cell is absolutely necessary for viral multiplication. The process of viral multiplication is an extraordinary biologic phenomenon. Viruses have often been aptly described as minute parasites that appropriate the synthetic and genetic machinery of cells. The nature of this cycle dictates viral pathogenicity, transmission, the responses of the immune defenses, and human measures to control viral infections. From these perspectives, we cannot overemphasize the importance of a working knowledge of the relationship between viruses and their host cells.

A Note on Terminology Although the terms virus or virus particle are interchangeable, virologists find it convenient to distinguish between the various states in which a virus may exist. A fully formed, extracellular particle that is virulent (able to establish infection in a host) is called a **virion** (vir′-ee-on). Once its genetic material has entered a host cell, a virus may undergo a vegetative stage, in which it multiplies at the expense of the host, or it may become quiescent or **latent** (lay′-tunt). If the multiplication cycle results in lysis of the host cell and release of more virions, it is a **lytic** (lih′-tik) **cycle.**

The multiplication cycles of viruses are similar enough that virologists use certain viruses (such as bacteriophages and enveloped animal viruses) as general models. Although a given cycle occurs continuously and not in discrete steps, it is helpful to demarcate the major events in the sequence. These events are **adsorption,** a recognition process between a virus and host cell that results in virus attachment to the external surface of the host cell; **penetration,** entrance of the virion (either a whole virus or just its nucleic acid) into the host cell; **replication,** expression of the viral genome at the expense of the host's synthetic equipment, resulting in the production of the various virus components; **maturation,** assembly of these individual viral parts into whole, intact virions; and **release,** escape from the host cell of the active, infectious viral particles (figure 5.10). We will present an overview of the genetic events of virus cycles here, but this topic is covered in greater detail in chapter 8. The following section compares and contrasts the multiplication stages for bacterial and human viruses.

adsorption (ad-sorp′-shun) L. *ad,* to, and *sorbere,* to suck. The attachment of one thing onto the surface of another.

replication (rep-lih-kay′-shun) L. *replicare,* to reply. To make an exact duplicate.

The Multiplication Cycle in Bacteriophages

When Frederick Twort and Felix d'Herelle discovered bacterial viruses in 1915, it first appeared that the bacterial host cells were being eaten by some unseen parasite, and this observation inspired the name bacteriophage. Most bacteriophages (often shortened to phage) contain double-stranded DNA, though single-stranded DNA and RNA types exist as well. So far as is known, every bacterial species is parasitized by various specific bacteriophages. Probably the most widely studied bacteriophages are those of *Escherichia coli,*[3] especially the T-even (for even-numbered type) phages. Their complex structure has been previously suggested: They have an icosahedral capsid head containing DNA, a central tube (surrounded by a sheath), collar, base plate, tail pins, and fibers, which in combination make an efficient package for infecting a bacterial cell (see figure 5.9*b*). Momentarily setting aside a strictly scientific and objective tone, it is tempting to think of these extraordinary viruses as minute spacecrafts landing on an alien planet, ready to unload their genetic cargo.

Adsorption to the Cell Surface

For a successful infection, a phage must meet and stick to a susceptible host cell. Adsorption takes place when certain molecules on the phage tail and fibers bind to specific molecules (receptors) on the cell envelope of the host bacterium (figure 5.10). Among the bacterial structures that serve as phage receptors are surface molecules of the cell wall and pili. Attachment is a function of correct fit and chemical attraction—that is, the phage's tail molecules must interact specifically with those of the bacterial receptor. Once the phage is firmly affixed to the cell, it is positioned for penetration.

Bacteriophage Penetration

Following adsorption, a phage is still outside its host and remains inactive unless it can gain entrance to the cytoplasmic compartment. The strong, rigid bacterial cell wall is quite an impenetrable barrier, and the entire virus particle is unable to cross it. But the T-even bacteriophages have an exquisite mechanism for injecting nucleic acid across this barrier and into the cell. Like a miniscule syringe, the sheath constricts, pushing the inner tube through the host's cell wall and membrane and dissolving a passageway for its nucleic acid into the interior of the cell (figure 5.11). Once inside, the viral nucleic acid is all that is necessary for completing the multiplication cycle. The spent virus shell (or ghost) has fulfilled its principal function—protection, adsorption, and injection of DNA—and it will remain attached to the outside of the cell even after the cell has disintegrated.

The Lytic Cycle

Entry of the nucleic acid into the cell results in sweeping changes in the bacterial cell's activities. Within a few minutes, synthesis of bacterial molecules stops and metabolism shifts to the

3. A common intestinal bacterium used in research and teaching laboratories.

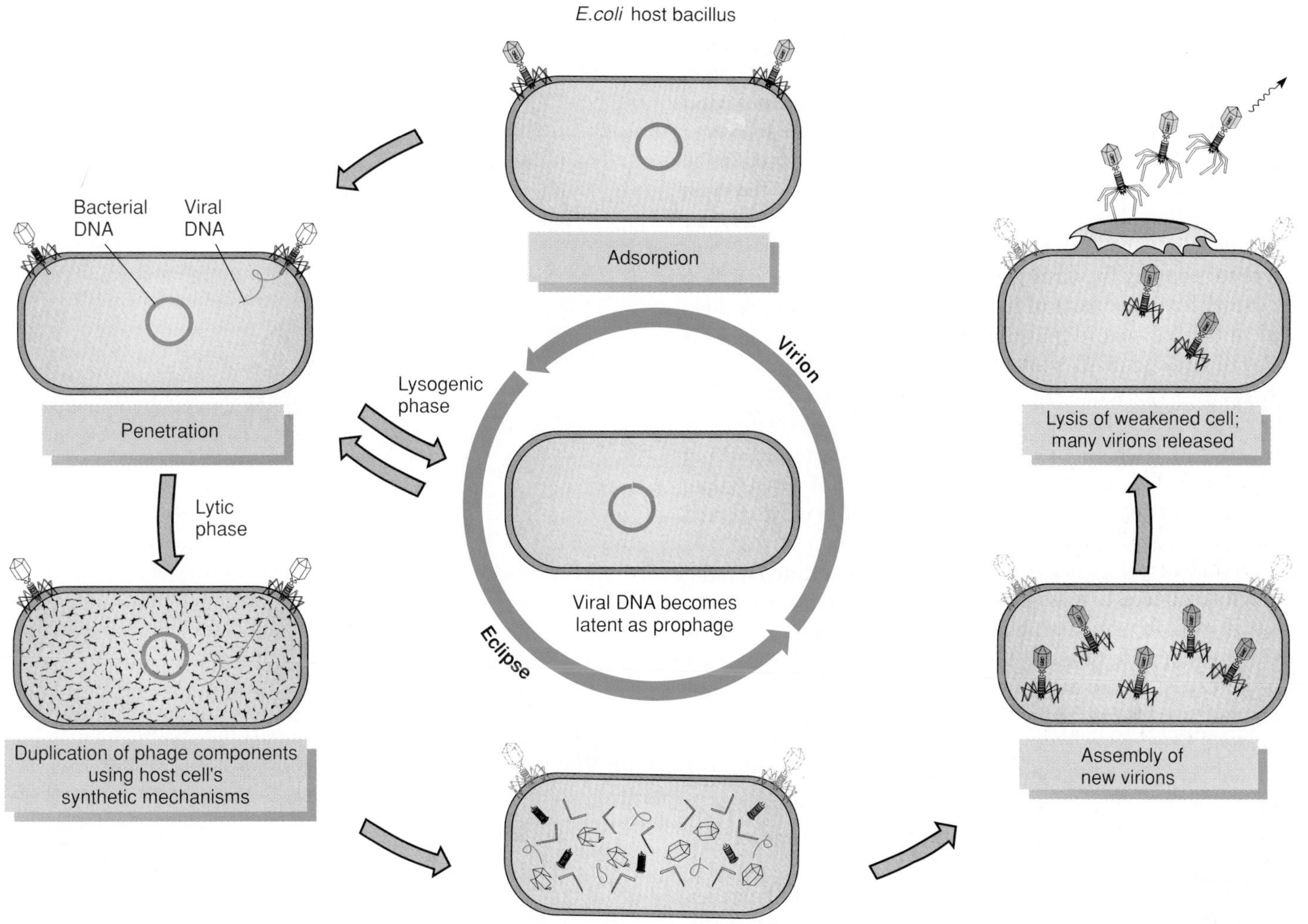

E.coli host bacillus

Adsorption

Bacterial DNA

Viral DNA

Penetration

Lysogenic phase

Virion

Viral DNA becomes latent as prophage

Eclipse

Lytic phase

Duplication of phage components using host cell's synthetic mechanisms

Assembly of new virions

Lysis of weakened cell; many virions released

Figure 5.10 Events in the multiplication cycle of T-even bacteriophages. The cycle is divided into the eclipse phase (during which the phage is developing but is not yet infectious) and the virion phase (when the virus is mature and capable of infecting a host). (Text provides further details of this cycle.)

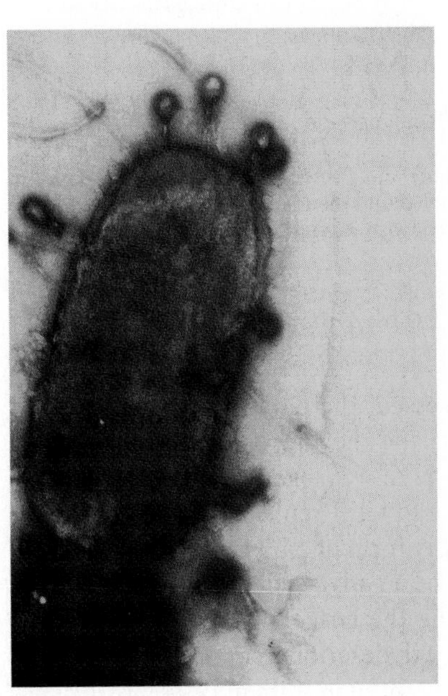

(a)

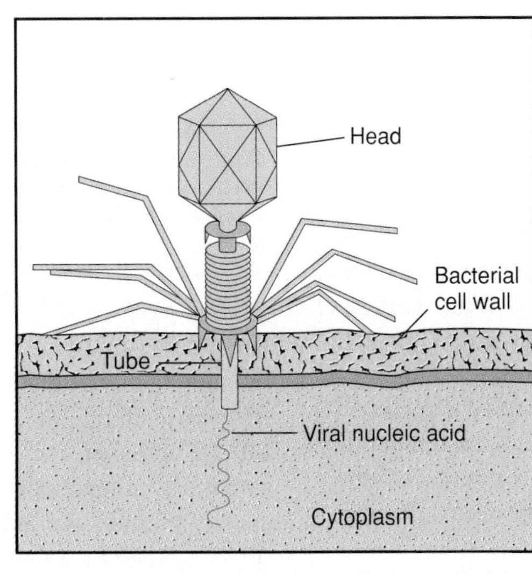

(b)

Figure 5.11 Penetration of a bacterial cell by a bacteriophage. (*a*) A single *E. coli* cell with several phages attached. Note that some heads are opaque and others are transparent due to having injected their nucleic acid. (*b*) Following adsorption, the phage plate becomes embedded in the cell wall, and the sheath contracts, pushing the tube through the cell wall and releasing the nucleic acid into the interior of the cell.

expression of genes on the viral nucleic acid strand. In T-even bacteriophages, the viral DNA redirects the genetic and metabolic activity of the cell, blocking the utilization of host DNA and ensuring that viral DNA is copied and synthesis of new viral components begins. As phage genes are expressed, the following viral molecules appear in the cell's cytoplasm: (1) proteins to "seal" the cell (because the host cell is punctured by the virus during injection, and it is advantageous for the virus to repair the hole before the host cell's integrity is destroyed); (2) enzymes for copying the virus genome; (3) proteins that make up the capsid head and parts of the tail; and (4) enzymes that help weaken the cell wall so that the phages can easily escape. This period of viral synthesis exploits the host's cytoplasmic nutrient resources, its ribosomes for synthesis, and its energy supplies. The early stages of viral replication are known as *eclipse,* a period during which no mature virions can be detected within the host cell. This corresponds with the period of penetration, replication, and early assembly, and if the cell were disrupted at this time, no infective virus would be released.

As the newly synthesized virus molecules accumulate, the host cell is induced to mass-produce large numbers of bacteriophage in assembly-line fashion (figure 5.12). First to be assembled are the capsid, tail, and fibers. The viral DNA is inserted into the capsid in the early stages of assembly, and later, the capsid takes on its finished angular form. The tail is formed by the union of the tube and sheath, and is completed by the addition of fibers. In final assembly these separate parts are bonded into mature phage particles. An average-sized *Escherichia coli* cell can contain up to 200 new phage units at the end of this period.

Eventually, the host cell becomes so packed with viruses that it **lyses**—splits open spontaneously, thereby liberating the mature virions (figure 5.13). One cause of lysis is believed to be the late release of virally induced enzymes, which have so weakened the cell envelope that it suddenly ruptures. This is due to enzymes that digest the cell membrane and the peptidoglycan of the cell wall. Upon release, the virulent phages are capable of spreading to other susceptible bacterial cells and beginning a new cycle.

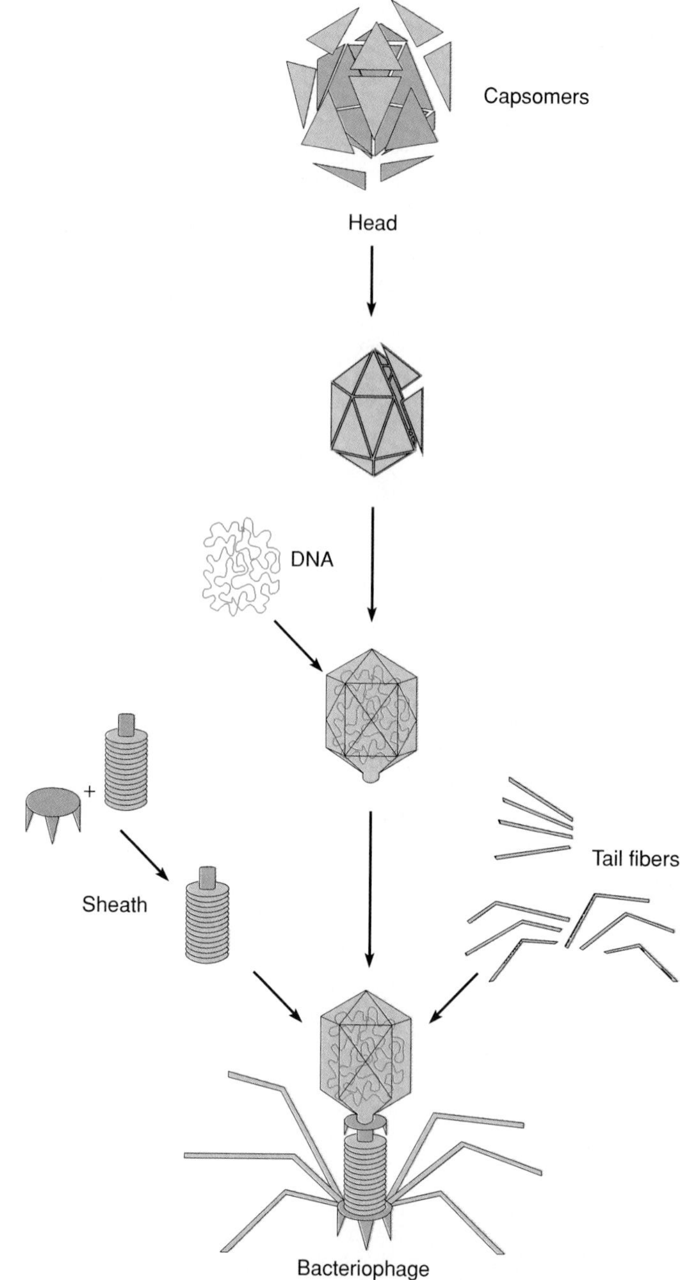

Figure 5.12 Bacteriophage assembly line. Protein subunits synthesized on the host cell's ribosomes are assembled into the major phage parts (head, sheath, tail fibers), which are subsequently brought together for the finished product.

Lysogeny: The Silent Virus Infection

The lethal effects of a virulent phage on the host cell present a dramatic view of virus-host interaction. Not all bacteriophages complete the lytic cycle, however. Special phages, called *temperate* phages, undergo adsorption and penetration in the bacterial host, but are not replicated or released. What happens instead is that the viral DNA enters an inactive *prophage* state where it is inserted into the bacterial chromosome. This viral DNA will be retained by the host cell and copied during cell division, so that the cell's progeny will also have the temperate phage DNA (figure 5.14). This condition of the host chromosome carrying bacteriophage DNA is termed **lysogeny.** Because the viral genome is not expressed, the bacterial cells carrying temperate phages do not lyse, and they appear entirely normal. But on occasion, the prophage in a lysogenic cell will be activated and progress directly into viral replication and the lytic cycle. Lysogeny is a less deadly form of parasitism than the full lytic cycle, and is thought to be an advancement that allows the virus to spread without killing the host. In a later section and in chapters 8, 14, and 20, we will describe a similar relationship that exists between certain viruses and human cells.

---

temperate (tem′-pur-ut) A reduction in intensity.
prophage (pro′-fayj) L. *pro,* before, plus phage.
lysogeny (ly-soj′-ehn-ee) The potential ability to produce phage.

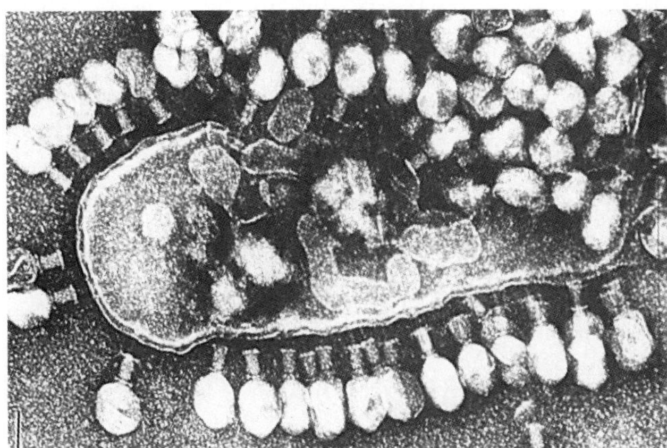

Figure 5.13 A weakened bacterial cell, crowded with viruses, has ruptured and released numerous virons that may attack nearby susceptible host cells.

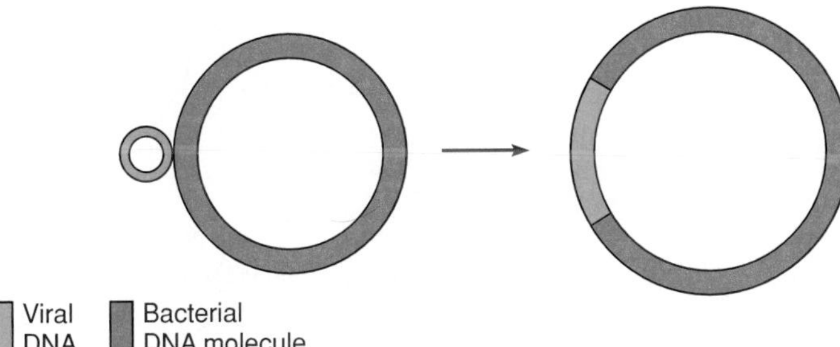

Figure 5.14 The lysogenic state in bacteria. A bacterial DNA molecule can accept and insert viral DNA molecules at specific sites on its genome. This additional viral DNA is duplicated along with the regular genome and may provide adaptive characteristics for the host bacterium.

Biomedical Significance/Considerations of Bacteriophages We have covered the cycle of bacterial viruses in some detail because they illustrate general features of viral multiplication in a very concrete and memorable way. It is most fascinating to realize that viruses are capable of lying "dormant" in their host cells, possibly becoming active at some later time. No mystery surrounds the role of bacteriophages in bacterial genetics. Because of the intimate association between the virus/host genetic material, phages occasionally serve as transporters of bacterial genes from one bacterium to another. This phenomenon, called transduction, is responsible for the toxic properties of several bacterial pathogens, including *Corynebacterium diphtheriae,* the cause of diphtheria, and *Streptococcus pyogenes,* the cause of scarlet fever (see chapter 8).

If you have been struck by the idea that bacteria (some of which are pathogens in their own right) are plagued by their own tiny parasites, you are certainly not alone. Not long after the destructive power of phages became known, some enterprising physicians attempted to use doses of bacteriophages to treat bacterial infections. They reasoned that the phages would selectively seek out the bacterial cells in the body, invade, and kill them, thereby ridding the patient of the infection. Unfortunately, it was determined early on that, for some reason, phage therapy was not effective. However, some physicians continued to use this unorthodox treatment for up to 20 years after it was found to be ineffective.

Table 5.4 Comparison of Bacteriophage and Animal Virus Multiplication

	Bacteriophage	Animal Virus
Adsorption	Precise attachment of special tail fibers to cell wall	Attachment of spikes, capsid, or envelope to cell surface receptors
Penetration	Injection of nucleic acid through cell wall; no uncoating of nucleic acid	Whole virus enters (is engulfed or fuses with cell membrane); uncoating of nucleic acid
Eclipse State	Occurs	Occurs
Replication and Maturation	Occurs in cytoplasm Cessation of host synthesis Viral DNA or RNA are replicated and begin to function Viral components synthesized	Occurs in cytoplasm and nucleus Cessation of host synthesis Viral DNA or RNA are replicated and begin to function Viral components synthesized
Viral Persistence	Lysogeny	Latency, chronic infection, cancer
Exit from Host Cell	Cell bursts when weakened by viral enzymes	Some cells lyse; enveloped viruses bud off host cell membrane
Cell Destruction	Immediate	Immediate; delayed in some

Multiplication Cycles in Animal Viruses

Now that we have a working concept of viral multiplication, let us turn to the animal viruses for yet another model system. Although the basic cycle is very similar to that of bacteriophages, several differences between both the host cells and their viruses contribute to dissimilarities in the details of the cycles (table 5.4). The phases in the cycle of animal viruses are eclipse, maturation, and release. The eclipse phase includes adsorption, penetration, and uncoating, while maturation and release include viral replication, assembly, and departure from the host cell. The length of the entire multiplication cycle varies from 6 hours in polioviruses to 36 hours in herpesviruses. See figure 5.15 for a comparison of the major phases of two types of animal viruses.

Adsorption and Host Range

Viral invasion begins when the virus encounters a susceptible host cell and adsorbs by means of specific surface structures to receptor sites on the cell membrane. The exact mode of attachment varies between the two general types of viruses. In enveloped forms such as influenza virus and human immunodeficiency virus (HIV or AIDS virus), glycoprotein spikes interlock with the cell surface molecules. But viruses with naked nucleocapsids (poliovirus, for example), possess a surface protein that adheres to membrane receptors (figure 5.16). The

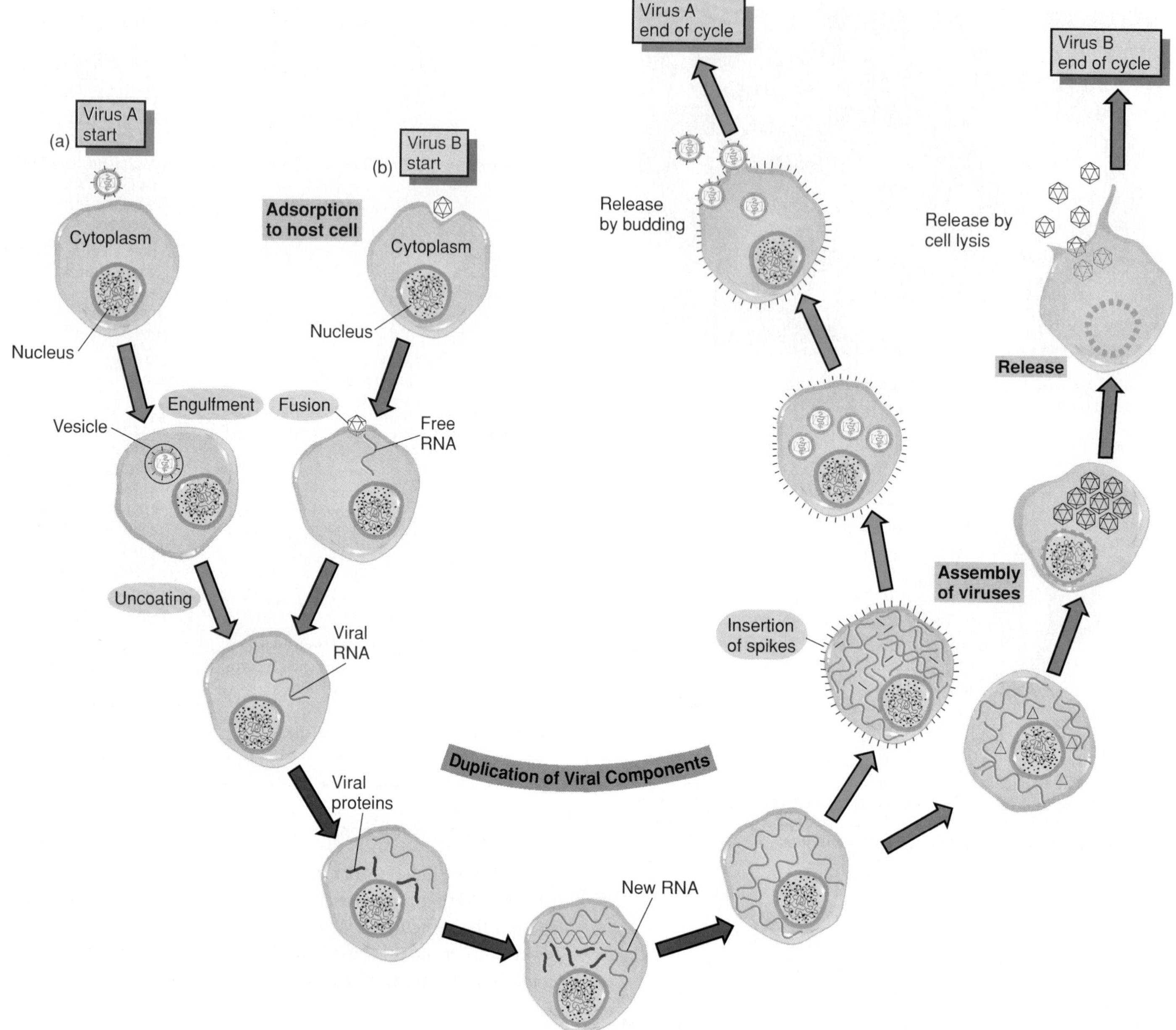

Figure 5.15 Features in the multiplication cycle of animal viruses, comparing two general types of RNA viruses. (*a*) An enveloped virus (Rubella virus). (*b*) A naked nucleocapsid (poliovirus). Both viruses adsorb to specific host receptors, but they penetrate the host cell by different mechanisms. Once in the cell, the free RNA molecule instructs the synthesis of new viral components (proteins for capsid and RNA). The two viruses also differ in the mode of release. Virus (*a*) inserts spikes into the host cell membrane and is budded off, taking with it an envelope. Virus (*b*) does not insert spikes and is released by cell lysis.

actual number of viruses adsorbed at any one time varies with the cell and the virus, but certain animal cells have as many as 100,000 viral receptors.

The specificity of virus-cell union limits the scope of hosts that a given virus can infect in a natural setting. This limitation, known as the *host range,* may be as restricted as that of the poliovirus, which can attach only to the intestinal and nerve cells of primates (humans, apes, and monkeys), or as broad as the range of the rabies virus, which can infect the cells of all mammals. Some cells are *resistant*—that is, they are incompatible with the virus and thus block adsorption. This explains why, for example, human cells resist infection with the canine hepatitis virus and dog cells are not naturally invaded by the human hepatitis A virus. It also explains why viruses tend to have a particular ***target tissue*** in the body that is the focus of infection and disease, as the hepatitis virus has for the liver or the mumps virus has for the salivary glands. However, some viruses can be induced to infect cells that they would not infect naturally, and this is what makes cultivation in the laboratory possible.

Penetration/Uncoating of Animal Viruses

Animal viruses exhibit some impressive mechanisms for entering a host cell. Unlike bacteriophages, they have no mechanism to inject their nucleic acids. Instead, the flexible cell

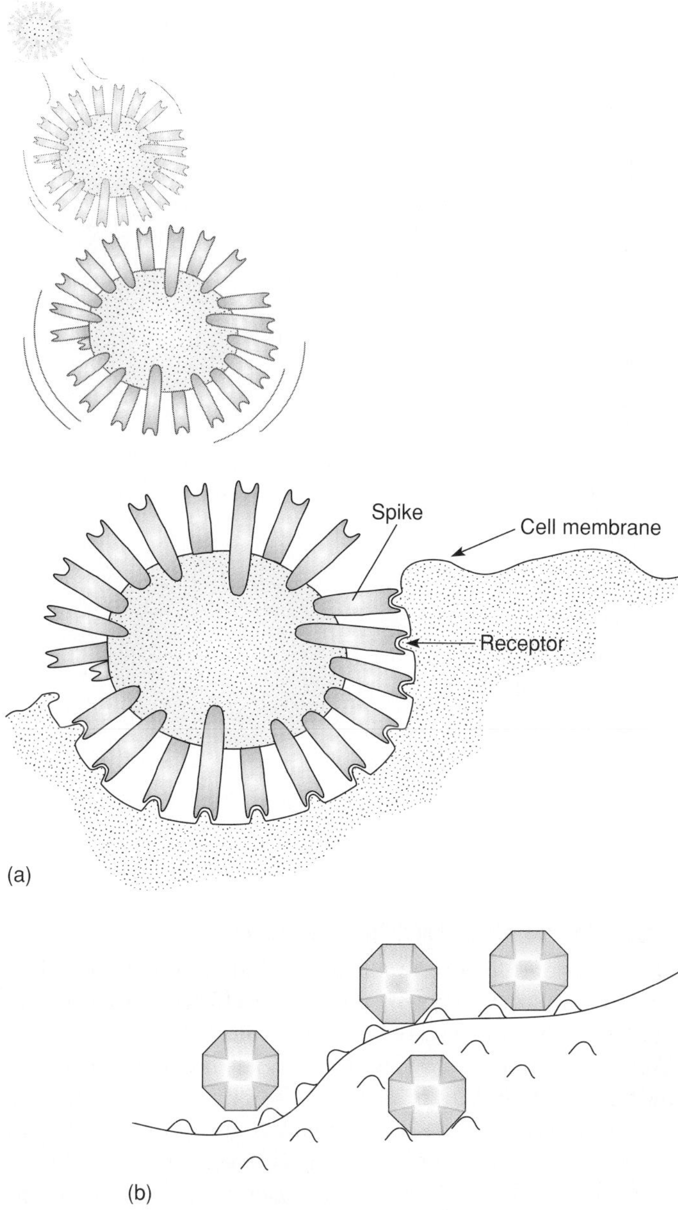

Figure 5.16 The mode by which animal viruses adsorb to the host cell membrane. (*a*) A virus with spikes. The configuration of the spike has a complementary fit for cell receptors. (*b*) Surface components of the naked capsid attach to membrane receptors.

membrane of the host is induced to admit the virus or its nucleic acid by one of three means (figure 5.17). In the case of endocytosis, the entire virus is engulfed by the cell and enclosed in a vacuole or vesicle. When enzymes in the vacuole dissolve its coatings (capsid, envelope), the virus is said to be **uncoated,** a process that releases the viral nucleic acid into the cytoplasm. The exact manner of uncoating varies, depending on whether the virus is naked, enveloped, or complex. Another means of entry involves direct fusion of certain enveloped viruses (influenza and mumps viruses) with the cell membrane of the host cell (figure 5.17*b*). In this form of penetration, uncoating and liberation of the nucleic acid occur during the process of fusion itself. A few nonenveloped viruses (such as poliovirus) enter by a means that is not well understood, in which the capsid adheres to the cell membrane and the nucleic acid is translocated into the cell.

Maturation of Animal Viruses: Host Cell As Factory

The synthetic and replicative phases of animal viruses are highly regulated and extremely complex at the molecular level. We will not attempt to cover these phases in any of their considerable detail. A more thorough coverage of viral multiplication and genetics is included in chapters 8, 20, and 21. As with bacteriophages, the free viral nucleic acid exerts control over the host's synthetic and metabolic machinery. How this proceeds will vary, depending on whether the virus is a DNA or RNA virus. In general, DNA viruses replicate in the host cell's nucleus, and RNA viruses remain in the cytoplasm.

The period of synthesis is divided into early and late phases. During the early phase, the viral nucleic acid alters the genetic expressions of the host and instructs it to begin synthesizing viral components. Some of the early viral proteins in the cell include enzymes for synthesizing viral nucleic acids. The synthetic period is a time of intense though largely invisible activity. Capsid proteins and viral nucleic acids are synthesized, and enzymes required for viral assembly are added to the growing pool of virus molecules. An event essential to the development of certain enveloped viruses is synthesis by the host cell of viral spikes and their insertion into its cell membrane (see figures 5.15 and 5.19).

Assembly and Release of Mature Viruses

Toward the end of the late phase, mature virus particles are constructed from the accumulated subunits. In most instances, the capsid is first laid down as an empty shell that will serve as a receptacle for the nucleic acid strand. Electron micrographs taken during this time show cells with masses of viruses, often in crystalline packets (figure 5.18). To complete the cycle, assembled viruses leave their host in one of two ways. Nonenveloped and complex viruses that reach maturation in the cell nucleus or cytoplasm are released by lysis of the cell. Enveloped viruses are liberated by **budding** or **exocytosis** from the membranes of the cytoplasm, nucleus, endoplasmic reticulum, or vesicles. In this process, which is somewhat like the reverse of adsorption, the nucleocapsid attaches to the inside of the membrane, combines with spikes (if they are present), and is pinched off with its envelope (figure 5.19). Enveloped viruses are not released simultaneously, but are shed gradually, without the immediate destruction of the cell. Most viral infections are ultimately lethal to the cell, regardless of how the viruses exit.

The number of viruses released by infected cells is variable, being controlled by factors such as the size of the virus and the health of the host cell. (An injured cell may be unable to synthesize viral components.) As "few" as three or four thousand virions are released from cells infected with poxviruses, while poliovirus-infected cells can release over 100,000 virions. If even a small number of these virions happens to meet another susceptible cell and infect it, the potential for viral proliferation is immense.

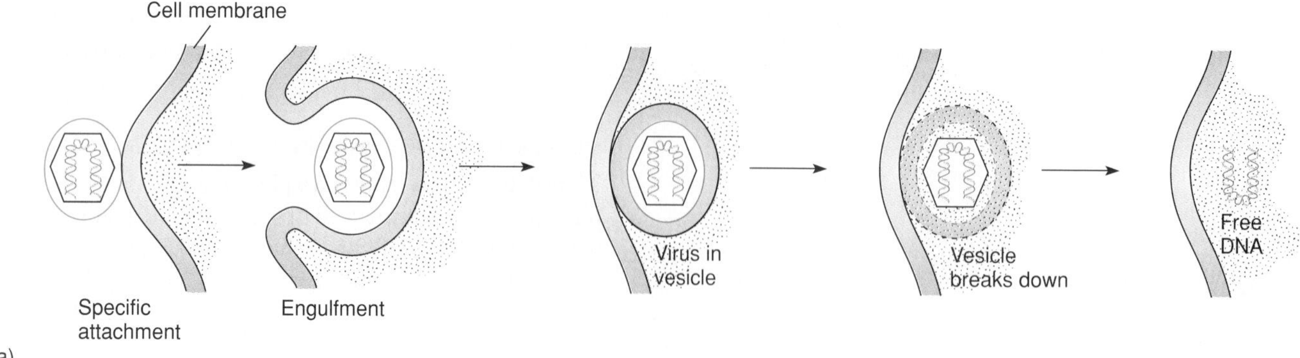

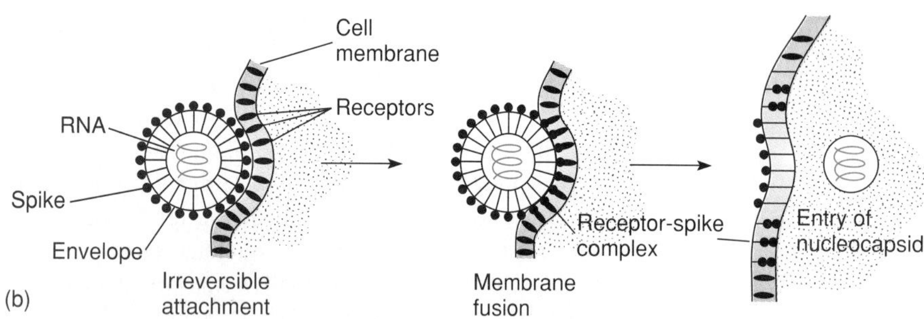

Figure 5.17 Two principal means by which animal viruses penetrate. (*a*) Endocytosis (engulfment) and uncoating of a herpesvirus. (*b*) Fusion of the mumps virus with the cell membrane.

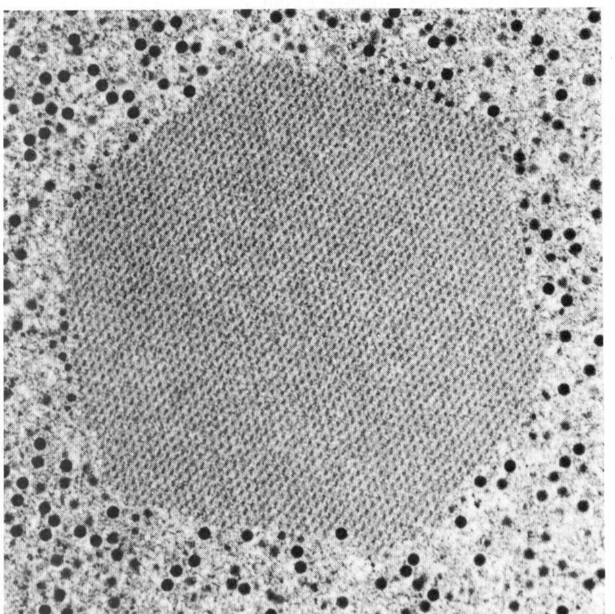

Figure 5.18 Nucleus of a cell, containing a crystalline mass of adenovirus (×35,000).

Damage to the Host Cell and Persistent Infections

The short- and long-term effects of viral infections on animal cells are well documented. **Cytopathic effects** (CPE) are defined as virus-induced damage to the cell that alters its microscopic appearance. Individual cells may become disoriented, undergo gross changes in shape or size, or develop intracellular changes (figure 5.20*a*). It is common to note *inclusion bodies,* or compacted masses of viruses or damaged cell organelles, in the nucleus and cytoplasm. Examination of cells and tissues for cytopathic effects is an important part of the diagnosis of viral infections. Table 5.5 summarizes some prominent cytopathic effects associated with specific viruses.

Although accumulated damage from a virus infection kills most host cells, some cells maintain a carrier relationship in which the cell harbors the virus and is not destroyed by it. These so-called *persistent infections* may last from a few weeks to lengthy periods (for life, in some cases). The most serious persistent viruses remain in a chronic latent state, periodically becoming reactivated. Examples of this are herpes simplex viruses (fever blisters and genital herpes) and herpes zoster virus (chickenpox and shingles), which can go into latency in nerve cells and later emerge under the influence of various stimuli to cause recurrent infections. Specific damage that occurs in viral diseases will be covered more completely in chapters 20 and 21.

cytopathic (sy″-toh-path′-ik) Gr. *cyto,* cell, and *pathos,* disease.

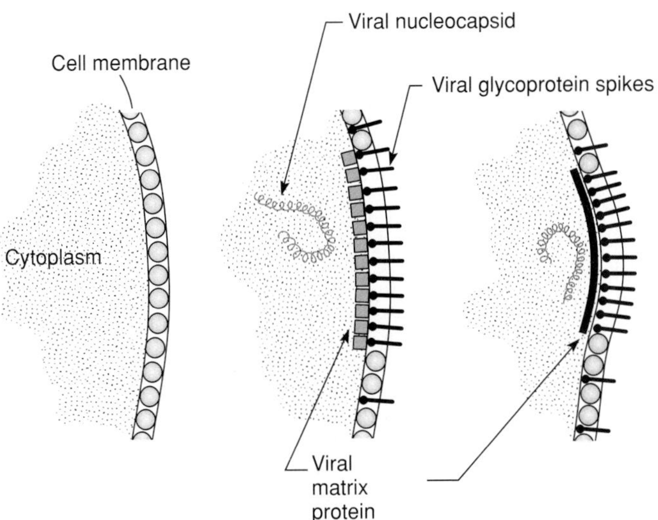

Figure 5.19 Stages in the maturation of an enveloped virus (influenza virus). As the virus is budded off the membrane, it simultaneously picks up an envelope and spikes.

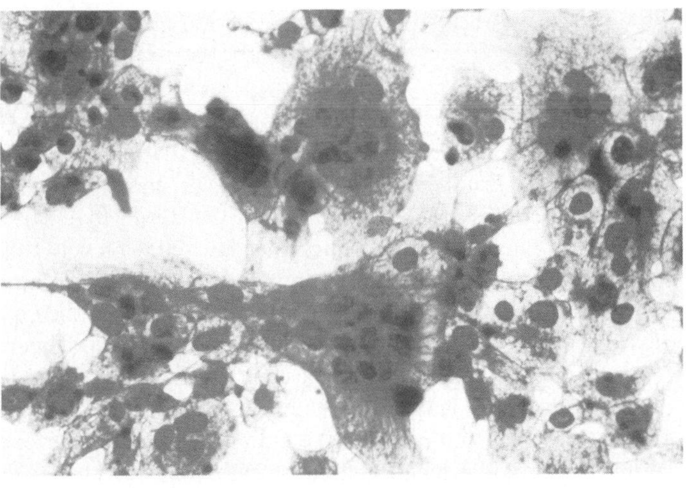

(a)

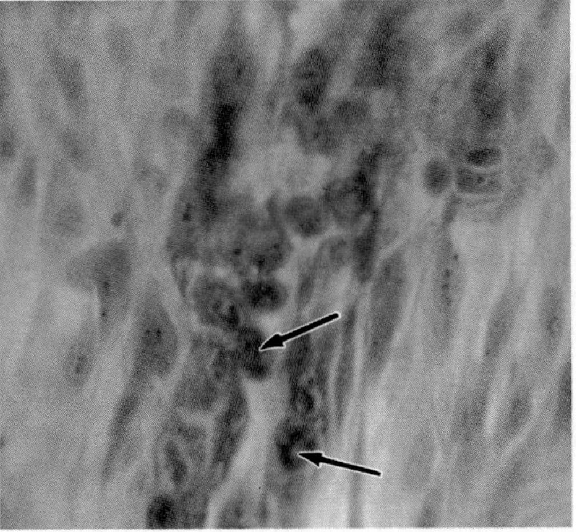

(b)

Figure 5.20 Cytopathic changes in cells and cell cultures infected by viruses. (*a*) Human epithelial cells infected by herpes simplex virus demonstrate the giant size of the cells and their multiple nuclei. (*b*) Human cells infected with cytomegalovirus. Note the inclusion bodies (arrows).

Table 5.5 Cytopathic Changes in Selected Viral-Infected Animal Cells

Virus	Response in Animal Cell
Smallpox virus	Cells round up; inclusions appear in cytoplasm
Herpes simplex	Cells become giant with multiple nuclei; nuclear inclusions
Adenovirus	Clumping of cells; nuclear inclusions
Poliovirus	Cell lysis; no inclusions
Togavirus	Cell lysis; no inclusions
Reovirus	Cell enlargement; vacuoles and inclusions in cytoplasm
Influenza virus	Cells round up; no inclusions
Rabies virus	No change in cell shape; cytoplasmic inclusions (Negri bodies)
HIV	Giant cells with numerous nuclei (multinucleate)

Some persistent viruses enter their host cell and alter its growth and metabolic patterns to such an extent that the cells become cancerous. These viruses are termed *oncogenic,* and their effect on the cell is called *transformation.* A startling feature of these viruses is that their nucleic acid is consolidated into the host DNA like a prophage. Changes acquired by transformed cells include: an increased rate of growth; chromosome alterations; changes in the cell's surface molecules; and the capacity to divide for an indefinite period, unlike normal animal cells. Mammalian viruses capable of initiating tumors are called *oncoviruses.* Some of these are DNA viruses such as papovaviruses (one type causes warts), herpesviruses, and adenoviruses. Retroviruses (RNA viruses that can program the synthesis of DNA using their single-stranded RNA as a template) have also been shown to cause cancer in humans. This finding has spurred a great deal of speculation on the possible involvement of viruses in cancers whose cause is still unknown. Additional information on the connection between viruses and cancer is found in chapters 8, 14, and 21.

Techniques in Cultivating and Identifying Animal Viruses

One problem hampering earlier animal virologists was their inability to propagate specific viruses routinely in pure culture and in sufficient quantities for their studies. Virtually all of the pioneering attempts at cultivation had to be performed in the animal or plant that was the usual host for the virus, even though using the intact host organism as a medium often left much to be desired. How could researchers have ever traced the stages of viral multiplication if they had been restricted to the natural host, especially in the case of human viruses? Fortunately, systems of cultivation with broader applications were developed, including *in vivo* inoculation of laboratory-bred animals and embryonic bird tissues and *in vitro* cell (or tissue) culture methods. Such use of substitute host systems permits greater control, uniformity, and wide-scale harvesting of viruses.

The primary purposes of viral cultivation in a medical setting are: (1) to isolate and identify viruses in clinical specimens; (2) to prepare viruses for vaccines; and (3) to do detailed research on viral structure, multiplication cycles, genetics, and effects on host cells. Table 5.6 summarizes some major human viruses and the systems used to culture them.

Table 5.6 Common Animal Viruses and Laboratory Methods for Cultivation

Virus	Cultivated In
Poliovirus	Human and primate tissue culture cells
Rhinovirus (cold virus)	Human embryonic kidney and lung tissue culture
Bunyavirus	Baby mice, mosquitos, human tissue culture
Rubella virus	Monkey cell culture
Influenza, mumps, measles virus	Human tissue culture, chicken embryos, monkey or calf kidney
Rhabdovirus (rabies)	Mice, human or hamster kidney tissue culture, chick embryo
HIV	Human lymphocyte cell culture, chimpanzees
Papilloma virus (warts)	Human fetal brain tissue culture
Herpesviruses	Human embryonic fibroblast culture
Poxviruses	Human cell culture, chicken embryo
Hepatitis A	Human cell culture
Hepatitis B	Primates (virus cannot yet be cultivated in cell culture)

Viral Cultivation Through Animal Inoculation

Specially bred strains of white mice, rats, hamsters, guinea pigs, and rabbits are the usual choices for animal cultivation of viruses. Invertebrates (mosquitos) or nonhuman primates are occasionally used as well. Because viruses may exhibit some host specificity, certain animals may propagate a given virus more readily than others. Depending on the particular experiment, tests can be performed on adult, juvenile, or newborn animals. The animal is exposed to the virus by injection of a viral preparation or specimen into the brain, blood, muscle, body cavity, skin, or footpads.

Use of Bird Embryos

An **embryo** is an early developmental stage of animals marked by rapid differentiation of cells. Birds undergo their embryonic period within the closed protective case of an egg, which makes an incubating bird egg a nearly perfect experimental system for viral propagation. It is an intact and self-supporting unit, complete with its own sterile environment and nourishment. Furthermore, it furnishes several embryonic tissues that readily support viral multiplication.

Chicken, duck, and turkey eggs are the most common choices for inoculation. Injecting an egg is a very delicate procedure, fraught with a series of difficulties. The opaque shell hinders viewing and presents a barrier to inoculation, so it must be penetrated, usually by drilling a hole or making a small window. Since this procedure can infect the sterile contents of the egg, rigorous sterile techniques must be used to prevent contamination by bacteria and fungi from the air and the outer surface of the shell. The site of inoculation depends upon the type of virus being cultivated and the needs of the experiment (figure 5.21). Targets include the allantoic cavity, a fluid-filled sac that functions in embryonic waste removal; the amniotic cavity, a sac that cushions and protects the embryo itself; the chorioallantoic membrane, which functions in embryonic gas exchange; the yolk sac, a membrane that mobilizes yolk for the nourishment of the embryo; and the embryo itself. The yolk sac develops earliest and is capable of supporting the growth of many different viruses. The amniotic cavity and chorioallantoic membrane are more developed in later embryos and are used to propagate viruses that require cells at a more advanced stage.

Viruses multiplying in embryos may or may not cause effects visible to the naked eye. The signs of viral growth include death of the embryo, defects in embryonic development, and localized areas of damage in the membranes, resulting in discrete, opaque spots called *pocks* (a variant of pox) (figure 5.22). If a virus does not produce overt changes in the developing embryonic tissue, virologists have other methods of detection. Embryonic fluids and tissues may be prepared for direct examination with an electron microscope. Certain viruses can also be detected by their ability to agglutinate[4] red blood cells or by their reaction with an antibody of known specificity that will affix to its corresponding virus, if it is present.

in vivo (in vee'-voh) L. *vivos*, life. Experiments performed in a living body.

in vitro (in vee'-troh) L. *vitros*, glass. Experiments performed in test tubes or other artificial environments.

4. To form large masses or clumps.

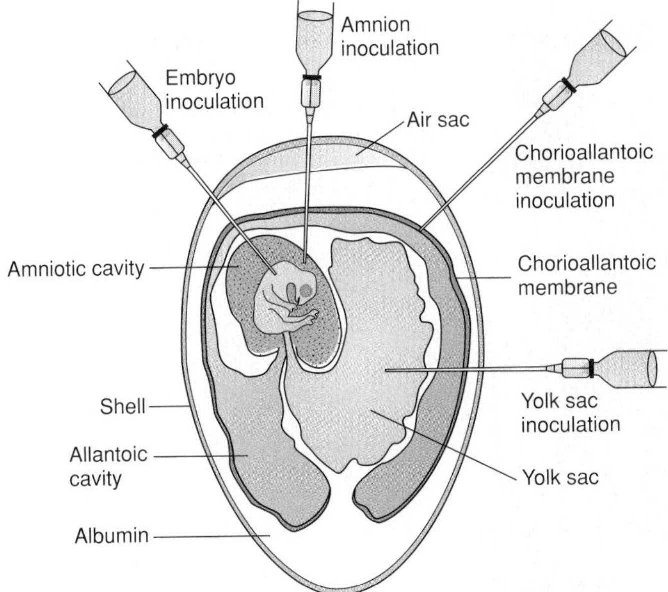

Figure 5.21 Cultivating animal viruses in a developing bird embryo. The shell is perforated using sterile techniques, and a virus preparation is injected into a site selected to grow the viruses.

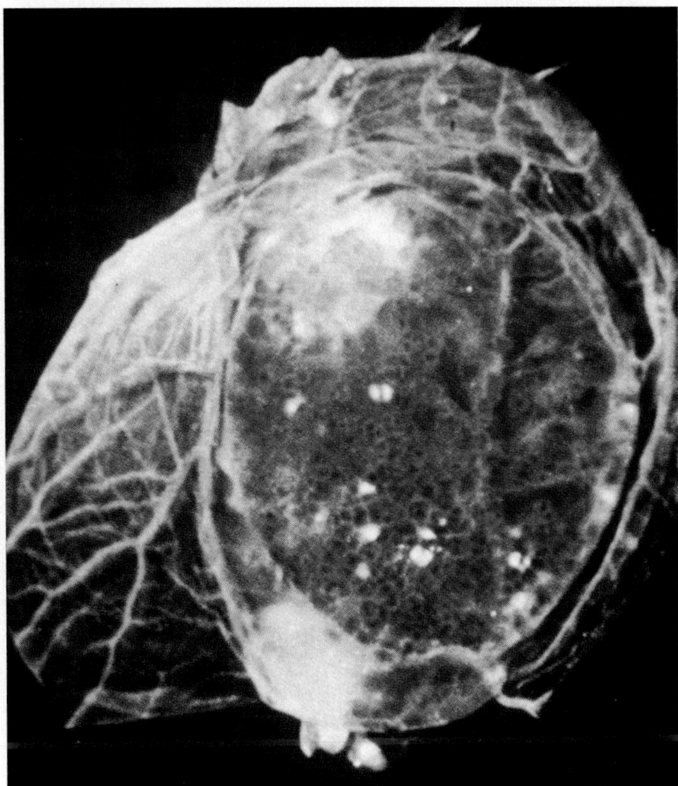

Figure 5.22 Pocks (white patches) developing on a chorioallantoic membrane removed from a chicken embryo. Several viral groups, including poxviruses and herpesviruses (shown here), cause pocks.

Cell (Tissue) Culture Techniques

The most important early discovery that led to cultivation of viruses was the development of a simple and effective way to grow populations of isolated animal cells in culture. These types of *in vitro* cultivation systems are termed **cell culture** or **tissue culture.** (Although these terms are used interchangeably, cell culture is probably a more accurate description.) So prominent is this method that most viruses are propagated in some sort of cell culture, and much of the virologist's work involves developing and maintaining these cultures. Animal cell cultures are grown in sterile chambers (petri dishes, bottles) with special media that contain the correct types and proportions of ions and nutrients that an animal cell requires to survive. The growth chamber is designed so that the cells can form a monolayer, or a single sheet of cells, to support viral growth (figure 5.23). The monolayer system permits close macroscopic and microscopic inspection for signs of viral multiplication.

Cultures of animal cells can be either primary or continuous in nature. *Primary cell cultures* are derived by removing a tissue from an animal, gently separating the cells, and then placing them in a growth chamber. Embryonic, fetal, adult, and even cancerous tissues from many organs have served as sources of primary cultures. The isolated cells undergo a series of mitotic divisions and form a thin, spreading layer. A primary culture retains several characteristics of the original tissue from which it was derived, but will be limited in the number of times it can be subcultured; eventually, it will die out or mutate into a different strain of cells. Such strains of cells derived from primary cell cultures are called *continuous cultures,* and are different in several ways. They may, for instance, have altered chromosome numbers, grow rapidly, or show changes in morphology, and they can be continuously subcultured for long periods of time. Such cultures or **cell lines** are immortal and will continue to divide, provided they are routinely transferred to fresh nutrient.

Because cell cultures contain nutrients that will readily support the growth of fungi and bacteria, strict aseptic techniques must be exercised to exclude those organisms during maintenance. Adding antibiotics (which do not inhibit viruses) also helps retard the growth of any contaminants that may occasionally be introduced. One very clear advantage of cell culture is that a specific cell line can be available for viruses with a very narrow host range. Strictly human viruses can be propagated in one of several primary or continuous human cell lines, such as embryonic kidney cells, fibroblasts, bone marrow, and heart cells.

One way to detect the growth of a virus in culture is to observe large-scale degeneration and lysis of infected cells. This can be shown with a special technique that overlays the monolayer of cells with agar so as to hold it in place. The areas where virus-infected cells have been destroyed show up as clear, well-defined patches in the cell sheet called *plaques* (figure 5.23*c*). This same technique is used to detect and count bacteriophages, because they also produce plaques when grown in soft agar cultures of their host cells (bacteria). A plaque develops when a virus infects a host cell and forms multiple progeny, which are released to nearby host cells in the medium. As new cells become

plaque (plak) Fr. *placke,* patch or spot.

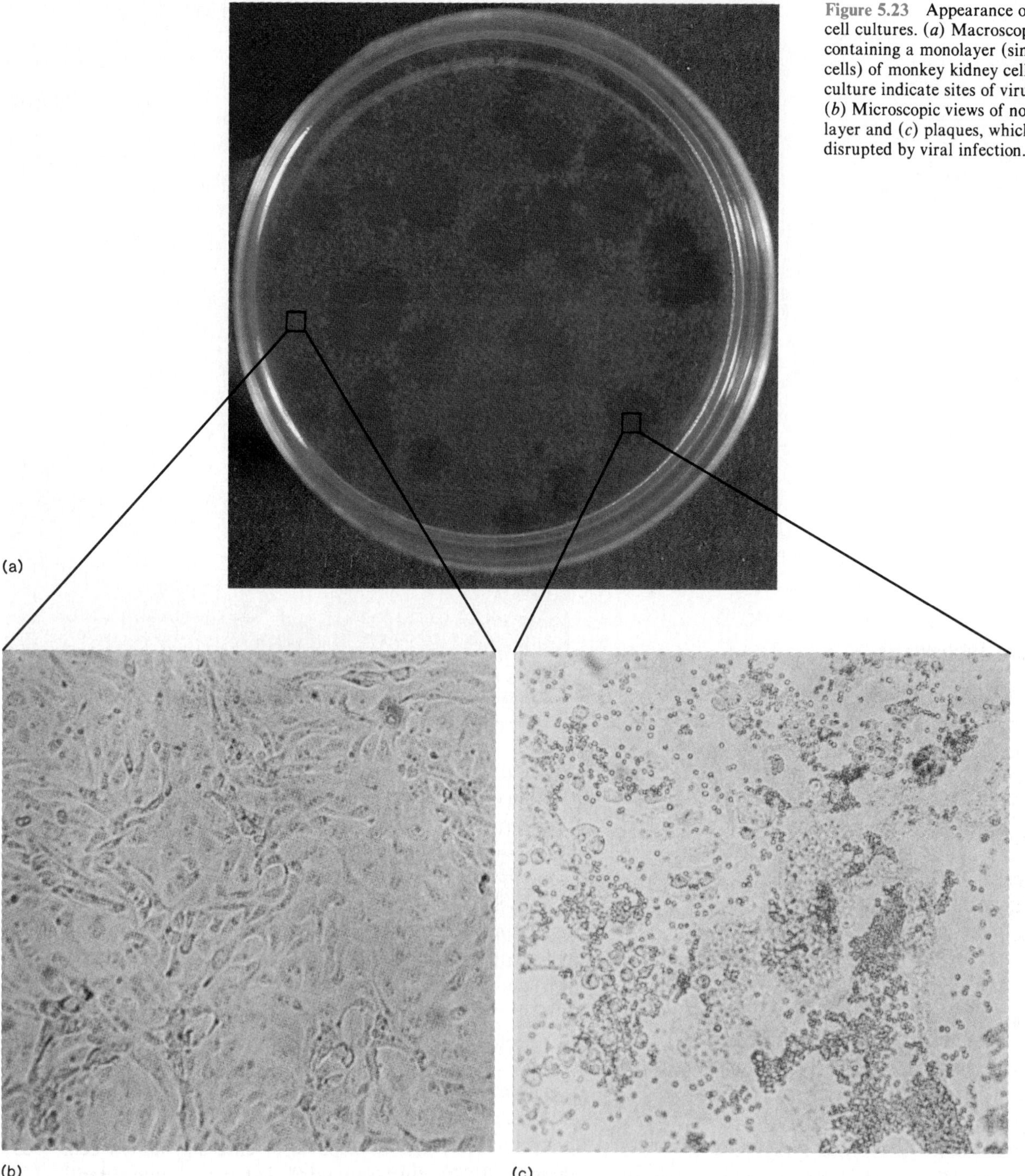

Figure 5.23 Appearance of normal and infected cell cultures. (*a*) Macroscopic view of a petri dish containing a monolayer (single layer of attached cells) of monkey kidney cells. Clear spaces in culture indicate sites of virus growth (plaques). (*b*) Microscopic views of normal, undisturbed cell layer and (*c*) plaques, which consist of cells disrupted by viral infection.

infected, they release more viruses, and so on. As this proceeds, the infection spreads gradually and symmetrically from the original point of infection, causing the macroscopic appearance of round, clear spaces that correspond to areas of dead cells.

Medical Importance of Viruses

The number of viral infections that occur on a worldwide basis is nearly impossible to predict accurately. Certainly, viruses are the most common cause of acute infections that do not result in hospitalization, especially when one considers widespread diseases such as colds, hepatitis, chickenpox, influenza, herpes, and warts. If one also takes into account prominent viral infections found only in certain regions of the world, such as Nile Valley fever, Rift Valley fever, and yellow fever, the total could easily exceed several billion cases per year. Although most viral infections do not result in death, some, such as rabies and AIDS, have very high mortality rates, and others may lead to long-term debility (polio, neonatal rubella). Current research is focussing on the possible connection of viruses to several chronic afflictions of unknown cause, such as diabetes and multiple sclerosis.

Special Viruslike Infectious Agents

Not all noncellular infectious agents have typical viral morphology. One group of most unusual forms, even smaller and simpler than viruses, appears to cause slow, persistent diseases in humans and animals. The term slow is used because of the long period of latency (usually several years) before the first clinical signs appear. Symptoms range from mental derangement to loss of muscle control, and the diseases are progressive and universally fatal.

Creutzfeldt-Jakob disease, a human condition that manifests as gradual degeneration of the central nervous system, afflicts several million persons per year worldwide. Its mode of transmission is unknown. Kuru is a related form of neural deterioration occurring among certain New Guinea tribes that observe the startling ritual of eating the brain tissue of dead tribesmen. The agent of kuru is thus spread remarkably—by cannibalism—but the disease is gradually becoming rare due to the eradication of such practices. Several animals (sheep, monkeys, mice) are victims of a similar disease called scrapie, which is readily transmissible from host to host. A common feature of these conditions is the deposition of distinct fibrils in the brain tissue (see figure 21.27). Some researchers have hypothesized that these fibrils are the agents of the disease and have named them *prions* (**pro**teinacious **in**fectious particles). Whether these fibrils actually represent an infectious agent is currently highly controversial. If continued study supports this hypothesis, the discovery that they are composed primarily of protein (no nucleic acid) will certainly revolutionize our ideas of what can constitute an infectious agent. (See chapter 21 for further discussion of this phenomenon.)

Plants are also parasitized by viruslike agents called **viroids.** Research has indicated that viroids differ from ordinary viruses by being very small (about one-tenth the size of an average virus) and very simple: They are nothing more than naked strands of RNA, lacking a capsid or any other type of coating. The chief significance of viroids is their ability to cause diseases in several economically important plants, including tomatoes, potatoes, cucumbers, citrus trees, and chrysanthemums.

Detection and Control of Viral Infections

Correct diagnosis can be crucial in detecting a potentially serious viral infection and in prescribing appropriate therapy. This is especially true for life-threatening viral infections such as AIDS, infections that can result in epidemics such as influenza, and infections that pose a serious risk to the fetus such as rubella (see feature 5.1). Viruses may be much more difficult to identify than cellular microbes. Physicians often use the overall clinical picture of the disease as one indicator, but in cases for which the virus must be identified, several other means are available. Direct, rapid methods include examining a specimen from the patient for the presence of the virus or signs of cytopathic changes (see herpesviruses, figure 20.9). In certain cases, the specimen must be cultivated in cell culture, embryos, or animals, but this method can be time-consuming and slower to give results. Sensitive tests to detect antibodies specific for a given virus in clinical specimens, such as blood, respiratory secretions, feces, throat, and vaginal secretions, are useful, and samples may be screened for the presence of indicator molecules from the virus itself. Details of these techniques are provided in chapters 13, 20, and 21.

Feature 5.1 Viruses, Fetuses, and Neonates

The common childhood viral infections such as measles, rubella, and chickenpox are usually mild and free of severe aftereffects—provided they occur in healthy, normal children whose immune defenses are active. But even otherwise innocuous viral infections can become life-threatening in highly vulnerable fetuses or newborns. One of the more dangerous of these is the rubella virus. If a mother develops rubella during pregancy, the virus can cross the placenta and cause abnormal development in the fetus. These intrauterine infections often result in congenital defects such as cardiac disease, cataracts, deafness, abnormal brain development, and even death (see figure 21.9). Fetal infections with herpesviruses such as chickenpox and cytomegalovirus may also permanently damage or kill a fetus.

Herpes simplex II, the cause of genital herpes in adults, does not usually cross the placenta and infect the fetus, but it may be picked up by the neonate through contact with the mother's birth canal. The disease it causes varies from a short-lived skin infection to extreme brain damage. Hepatitis B virus may also be acquired from the mother's birth canal. The potential risk is so great that pregnant mothers must be carefully monitored and screened for the presence of these viruses.

Treatment of Viral Infections

The novel nature of viruses has been a major impediment to effective therapy. Viruses are not cells, thus antibiotics aimed at disrupting various cell structures do not work on them. Most antiviral drugs, such as acyclovir for herpes infections and azidothymidine (AZT) for AIDS, prevent the virus from replicating and so must be capable of entering human cells (see chapter 10). Another compound that shows some potential for treating and preventing viral infections is a naturally occurring human cell product called *interferon* (see chapters 10 and 12). Vaccines that stimulate immunity are an extremely valuable tool for a small number of viral diseases (see chapter 13).

Chapter Review with Key Terms

The Viruses

Alternate or Common Names: General—**virions, viral particles.** Specific—**bacteriophage, poxvirus, herpesvirus,** etc.

Overall Morphology

Viruses are *not* **cellular**; have no organelles, no protoplasm, no locomotion of any kind; are large, complex molecules; may be **crystalline** in form. A virus particle is composed of a nucleic acid core (**DNA** or **RNA,** not both) surrounded by a protein shell or **capsid;** combination called a **nucleocapsid;** capsid is **helical** or **icosahedral** in configuration; many are covered by a membranous **envelope** containing protein **spikes;** complex viruses have additional external and internal structures.

Shapes/Sizes: Cuboidal, spherical, cylindrical, brick- and bullet-shaped. Smallest infectious forms range from the largest poxvirus (0.45 μm or 450 nm) to the smallest viruses (0.02μm or 20 nm).

Nutritional and Other Requirements

Viruses lack metabolism; have no enzymes for processing food or generating energy; are tied entirely to the host cell for all needs (**obligate intracellular parasites**).

Distribution/Host Range

Viruses are known to parasitize all types of cells, including bacteria, algae, fungi, protozoa, animals, and plants. Each viral type is limited in its **host range** to a single species or group, mostly due to specificity of adsorption of virus to specific host receptors.

Classification

The two types of viruses are DNA and RNA viruses. These are further subdivided into families, based on shape and size of capsid, presence or absence of an envelope, size of the nucleic acid, and antigenic similarities. **Animal viruses**: Common names of major DNA viruses include poxviruses (smallpox), herpesviruses, adenoviruses, papilloma (wart) virus, hepatitis B virus, parvoviruses. RNA viruses include poliovirus, hepatitis A virus, rhino (cold) virus, encephalitis viruses, yellow fever virus, rubella virus, influenza virus, mumps virus, measles virus, rabies viruses, and HIV (AIDS virus).

Multiplication Cycle

A unique method of multiplication occurs—no division involved. Viruses make multiple copies of themselves when they (1) attach or **adsorb** to a host cell, (2) **penetrate** a cell by means of whole virus or only nucleic acid, (3) assume control of the cell's genetic and metabolic components, thus (4) programming synthesis of new viral parts, and (5) directing **assembly** of these parts into new virus particles, and (6) **release** complete mature viruses. This may proceed through **lysis** of cell or **budding** of viruses. Bacteriophages and animal viruses differ in how they enter and leave the cell. Some viruses go into a **latent** or **lysogenic** phase in which they integrate into the DNA of the host cell and later may become active and produce a lytic infection.

Method of Cultivation

The need for an intracellular habitat makes it necessary to grow viruses in living cells, either in the intact host animal or plant or in isolated cultures of host cells (**cell culture**).

Importance

Medical: Viruses cause a variety of infectious diseases, ranging from mild respiratory illness (common cold) to destructive and potentially fatal conditions (rabies, AIDS). Some viruses may cause birth defects and cancer in humans and other animals.

Agricultural: Hundreds of cultivated plants undergo viral infections, often with adverse economic and ecologic repercussions.

Research: Because of their simplicity, viruses have become an invaluable tool for studying basic genetic principles.

Identification

By means of **cytopathic effects** in host cells.

Unique Features

Viruses exist at a level between living things and nonliving molecules. They consist of a capsid and nucleic acid, either DNA or RNA, not both; lack metabolism and respiratory enzymes; multiply inside host cells using the assembly line of the host's synthetic machinery; are genetic parasites; can pass through fine filters; are ultramicroscopic and crystallizable.

True-False Questions

Determine whether the following statements are true (T) or false (F). If you feel a statement is false, explain why, and reword the sentence so that it reads accurately.

___ 1. A virus is a special type of tiny infectious cell.

___ 2. Viruses infect all types of living things.

___ 3. The capsid is composed of protein subunits called capsomers.

___ 4. The envelope of an animal virus is derived from the cell wall of its host cell.

___ 5. The nucleic acid of viruses is either DNA or RNA, but not both.

___ 6. The general steps in the viral multiplication cycle are adsorption, penetration, duplication, assembly, and lysis.

___ 7. The virus replicates itself using the host's synthetic and metabolic machinery.

___ 8. A prophage is an early stage in the development of a bacteriophage.

___ 9. Animal viruses inject their nucleic acid into the host cell by a special mechanism.

___ 10. In general, an RNA virus multiplies in the nucleus of a cell and a DNA virus multiplies in the cytoplasm.

___ 11. Enveloped viruses pick up their envelope as they bud off the host cell.

___ 12. A latent virus persists in the cell for long periods and may cause recurrent disease.

___ 13. Viruses require their natural host cell in which to be cultivated.

Concept Questions

1. Describe 10 *unique* characteristics of viruses (may include structure, behavior, multiplication). After consulting table 2.6, what additional statements can you make about viruses, especially as compared with cells?
2. What does it mean to be an obligate intracellular parasite? What is another way to describe the sort of parasitism exhibited by viruses?
3. Characterize the viruses according to size range. What does it mean to say they are ultramicroscopic? Filterable?
4. Describe the general structure of viruses. What is the capsid, and what is its function? How are the two types of capsids constructed? What is a nucleocapsid? Give examples of viruses with the two capsid types. What is an enveloped virus, and how does the envelope arise? Give an example of a common enveloped human virus. What are spikes, how are they formed, and what is their function?
5. What are bacteriophages, and what is their structure? What is a tobacco mosaic virus? How are the poxviruses different from other animal viruses?
6. Since viruses lack metabolic enzymes, how can they synthesize necessary components? What are some enzymes that the virus does come equipped with?
7. How are viruses classified? What are virus families? How are generic and common names used? Looking at table 5.3, how many different viral diseases can you count?
8. Compare and contrast the main phases in the lytic multiplication cycle in bacteriophages and animal viruses. When is a virus a virion? What is necessary for adsorption? Why is penetration so different in the two groups? What is eclipse? In simple terms, what does the virus nucleic acid do once it gets into the cell? What is involved in assembly?
9. What is a prophage or temperate phage? What is lysogeny?
10. What dictates the host range of animal viruses? What are two ways that animal viruses penetrate the host cell? What is uncoating? Describe the two ways that animal viruses leave their host cell.
11. Describe several cytopathic effects of viruses. What causes the appearance of the host cell? How might it be used to diagnose viral infection?
12. What does it mean for a virus to be persistent or latent, and how are these events important? Briefly describe the action of an oncogenic virus.
13. Describe the three main techniques for cultivating viruses. What is the advantage of using cell culture? The disadvantage? What is a disadvantage of using live intact animals or embryos. What is a cell line? A monolayer? How are plaques formed?
14. What is the principal effect of the agents of Creutzfeldt-Jakob disease and kuru? How are the agents different from viruses?
15. Why are virus diseases more difficult to treat than bacterial diseases?

Practical/Thought Questions

1. What characteristics of viruses could be used to characterize them as life forms? What makes them more similar to lifeless molecules?
2. Comment on the possible origin of viruses. Is it not curious that the human cell welcomes a virus in and hospitably removes its coat? How do spikes play a part in the action of the host cell?
3. If viruses that normally form envelopes were prevented from budding, would they still be infectious?
4. The end result of most viral infections is death of the host cell. If this is the case, how can we account for such differences in the damage that viruses do (compare the effects of the cold virus with those of the rabies virus)? What is one adaptation of viruses that does not immediately kill the host cell?
5. Given that DNA viruses can actually be carried in the DNA of the host cell's chromosomes, comment on what this means in terms of inheritance in the offspring. Could some cancers be explained on the basis of inheritance from parents of a latent virus?
6. You have been given the problem of constructing a building on top of a mountain. It will be impossible to use huge pieces of building material. Using viruses as a model, can you think of a way to construct the building?
7. The AIDS virus attacks only specific types of human cells, such as certain white blood cells and nerve cells. Can you explain why this is true?
8. Consult table 5.2 to determine which viral infections you have had and which ones you have been vaccinated for. Which viruses could possibly be oncoviruses?
9. One early problem in cultivating the AIDS virus was the lack of a cell line that would sustain indefinitely *in vitro*, but eventually, one was developed. What do you expect were the stages in developing this cell line?
10. If you were involved in developing an antiviral drug, what would be some important considerations? (Can a drug "kill" a virus? How could multiplication be blocked?)
11. Generally speaking, the minimum requirement for a particle to be duplicated is that it have some type of nucleic acid. Can you think of a way that prions could multiply?
12. Why is an embryonic or fetal viral infection so harmful?

CHAPTER 6

Elements of Microbial Nutrition, Ecology, and Growth

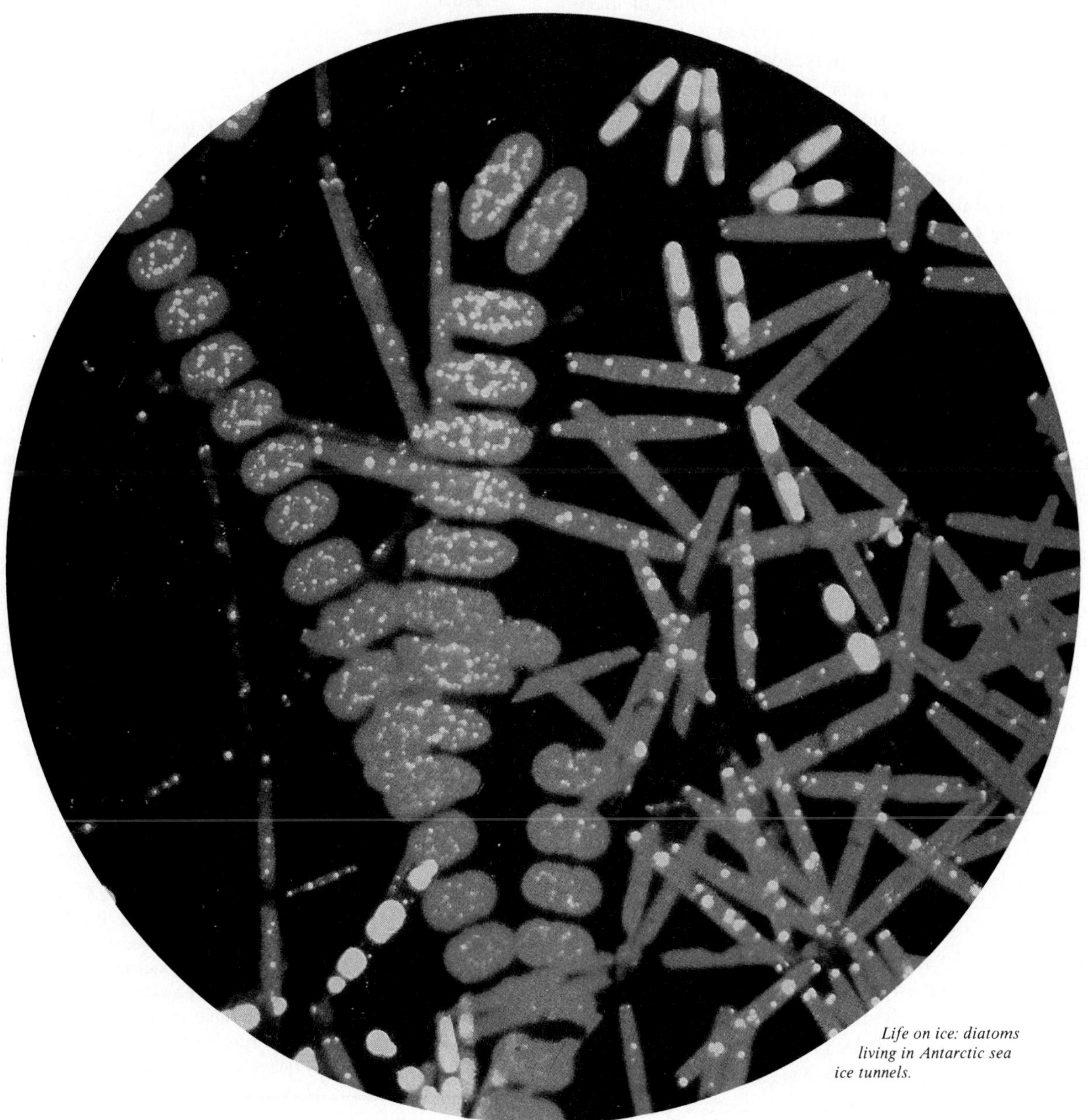

Life on ice: diatoms living in Antarctic sea ice tunnels.

Chapter Preview

In this chapter we explore the concepts that underlie microbial nutrition and nutrient transport, the effects of environmental factors on adaptation and growth, microbial and ecological interrelationships, and laboratory techniques for culturing microbes and studying their populations.

Conditions Affecting Microbial Life

With its widely fluctuating physical, chemical, and biological conditions, the earth provides an amazing array of habitats. The complex combination of environmental factors in these habitats dictates the survival strategies of its microbial residents. Those factors that exert the greatest influence are (1) nutrients and energy sources, (2) ambient temperature, (3) amount of moisture, (4) presence or absence of gases, (5) osmotic pressure, (6) pH (acidity or alkalinity), (7) presence or absence of light or radiation, and (8) other organisms. Some microbes are unable to survive the prevailing conditions in a habitat, some migrate to other habitats, and others remain and flourish. Microbes are so immensely adaptable and plastic that they have proliferated into every type of habitat and *niche,* from the most temperate to the most extreme (see feature 6.1).

Microbial Nutrition

Nutrition is a process by which chemical substances called **nutrients** are acquired from the surrounding environment and used in cellular activities such as metabolism and growth. With respect to nutrition, microbes are not really so different from humans. Bacteria living in mud on a diet of sulfur, or protozoa digesting wood in a termite's intestine, seem to be radical adaptations, but even these organisms require certain basic nutrients. These so-called **essential nutrients** are absorbed in some form from the environment, processed, and transformed into the chemicals of the cell (see feature 6.2, page 160).

The essential nutrients do not vary extensively from microbe to microbe at the level of the bioelements. You will recall from chapter 2 that cells need to receive some form of carbon, oxygen, nitrogen, hydrogen, phosphorus, sulfur, iron, potassium, magnesium, calcium, sodium, and other elements from their habitat. What varies is the ultimate source of a particular element, its chemical form, and how much of it the microbe needs (table 6.1).

Two categories of essential nutrients are **macronutrients** and **micronutrients** (table 6.2). Macronutrients are required in relatively large quantities and play principal roles in cell structure and metabolism. Examples of macronutrients are molecules that contain carbon, hydrogen, oxygen, and nitrogen. Micronutrients, sometimes called trace elements, are needed in much smaller amounts for enzyme and pigment structure and function. Examples of these are manganese and zinc. What constitutes a micronutrient can vary from one microbe to another and must be determined by experimentation in some cases.

niche (nitch) Fr. *nichier,* to nest. The role of an organism in its biological setting.
nutrition L. *nutrire,* nourishment.

Table 6.1 Nutrient Requirements and Sources for Two Ecologically Different Bacterial Species

	Thiobacillus thioxidans	*Mycobacterium tuberculosis*
Habitat	Sulfur springs	Human lung
Nutritional Type	Chemoautotroph (sulfur-oxidizing bacteria)	Chemoheterotroph (parasite; cause of tuberculosis in humans)
Required Element	**Provided Principally By**	
Carbon	Carbon dioxide (CO_2) in air	Minimum of L-glutamic acid (an amino acid), glucose (a simple sugar), albumin (blood protein), oleic acid (a fatty acid), and citrate
Hydrogen	H_2O, H^+ ions	Nutrients listed above, H_2O
Nitrogen	Ammonium (NH_4)	Ammonium, L-glutamic acid
Oxygen	Phosphate (PO_4), sulfate (SO_4)	Oxygen gas (O_2) in atmosphere
Phosphorus	PO_4	PO_4
Sulfur	Elemental sulfur (S), SO_4	SO_4
Miscellaneous Nutrients		
Vitamins	None required	Pyridoxine, biotin
Inorganic salts	Potassium, calcium, iron, chloride	Sodium, chloride, iron, zinc, calcium, copper, magnesium
Principal energy source	Oxidation of sulfur	Oxidation of glucose

A Dash of This, a Pinch of That The specific need for a given nutrient can be determined in the laboratory by empirical testing: Each nutrient is deliberately omitted from a growth medium to see if the microbe can grow in its absence. In the past, the need for a micronutrient was at times overlooked because the amount required could be provided by miniscule quantities on glassware and even in the air. Simply smoking tobacco in a room where media are being prepared can supply enough micronutrients for some microbes. We hardly advise this as a standard method, however.

Another way to categorize nutrients is according to their carbon content. An **inorganic nutrient** is a simple atom or molecule that has no carbon atoms in its structure (with one exception—CO_2). The natural reservoirs of inorganic compounds are mineral deposits in the crust of the earth, bodies of water, and the atmosphere. Examples include metals and their salts

Feature 6.1 Life in the Extremes

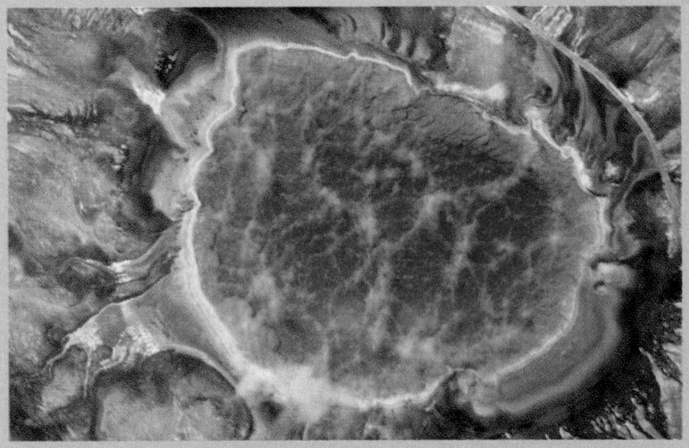

(a)

(b)

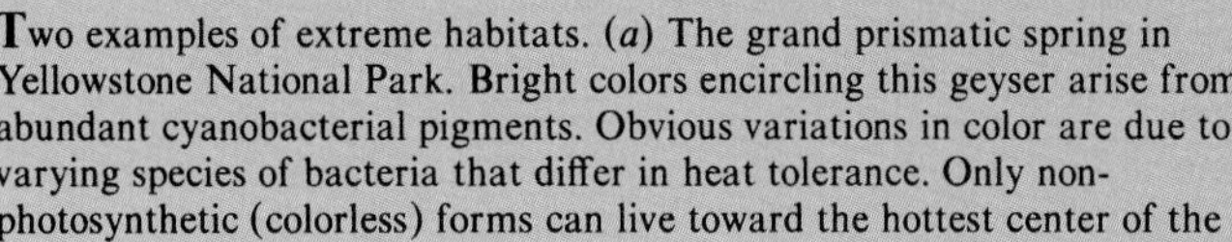

Two examples of extreme habitats. (*a*) The grand prismatic spring in Yellowstone National Park. Bright colors encircling this geyser arise from abundant cyanobacterial pigments. Obvious variations in color are due to varying species of bacteria that differ in heat tolerance. Only non-photosynthetic (colorless) forms can live toward the hottest center of the pool. (*b*) Evaporation ponds in the Great Salt Lake derive their red color from salt-loving bacteria. These microbes grow most abundantly where the salt concentration is highest. The pigment is also increased by direct exposure to sunlight.

Any extreme habitat—whether hot, cold, salty, acidic, alkaline, high in pressure, poor in nutrients, arid, oxygen-free, or toxic—is likely to harbor highly specialized microorganisms. Although in most instances the inhabitants are bacteria, certain fungi, protozoans, and algae are also capable of living in harsh habitats. Novel habitats produced by human activities are soon colonized by microbes as well.

Hot Habitats Perhaps the most extreme habitats are hot springs, geysers, and ocean vents, all of which support flourishing microbial populations. Temperatures in these regions range from 50°C to well above the boiling point of water, with some ocean vents even approaching 350°C. Most heat-adapted microbes are highly primitive organisms called archaebacteria, bacteria whose genetics and metabolism are extremely modified for this mode of existence. Some of the brilliant colors surrounding the sulfur springs and other hot regions of Yellowstone National Park are caused by growth of cyanobacteria. A completely unique ecosystem based on hydrogen sulfide-oxidizing bacteria exists in the hydrothermal vents lying along deep oceanic ridges (see feature 6.9). Heat-adapted bacteria even plague home water heaters and the heating towers and water-cooling equipment of power and industrial plants.

Chilly Climes A large part of the earth exists at cold temperatures. The Arctic, Antarctic, glacial mountains, snowbanks, and deepest parts of the ocean (except thermal vents) all hover near the freezing point of water (2°–10°C), nevertheless, microbes settle and grow in such regions. Several species of algae and fungi thrive on the surfaces of snow and glaciers (see figure 6.11). More surprising still is a strategy of bacteria and algae adapted to the sea ice of Antarctica. Although the ice appears to be completely solid, it is honeycombed by various-sized pores and tunnels filled with liquid water. These frigid microhabitats harbor a virtual microcosm of planktonic life, including predators (fish and shrimp) that live on photosynthetic algae and bacteria (see chapter opening photo).

Salty Situations Most microbial cells are inhibited by high amounts of salt; it is for this reason that salt is good for preserving some foods. In contrast, oceans, salt lakes, inland seas, and brines support whole communities of salt-dependent bacteria and algae, some of which flourish in a saturated salt (30%) solution. These microbes have demonstrable metabolic needs for high levels of minerals like sodium, potassium, magnesium, chlorides, or iodides. Because of their salt-loving nature, some species are pesky contaminants in salt processing plants, pickling vats, and salted fish.

pH Highly acidic or alkaline habitats are not common, but acidic bogs, lakes, and alkaline soils contain a specialized microbial flora. A few species of algae and bacteria can actually survive near the pH of concentrated hydrochloric acid. Not only do they require such a low pH for growth, but particular bacteria (for example, *Thiobacillus*) help maintain the low pH by releasing strong acid.

Deep in the Earth's Crust It was once thought that the region beneath the soil and upper crust of the earth's surface was sterile. New work with deep core samples (from 330 meters down) indicates a vast microbial population in these zones. A myriad of bacteria, protozoa, and fungi exist in this moist clay, which is rich in minerals and complex organic substrates. Bacteria ordinarily indigenous to the human intestine were found in some samples, leading to speculation that sewage gradually seeps down into these layers.

At the Microbial Fringes The predominant living things occupying the deepest part of the oceans (10,000 meters or below) are pressure- and cold-loving microorganisms. The lack of free oxygen or light is no hindrance to the numerous species that live in the depths of mud, swamps, and oceans. Nor does the lack of water inhibit microbes adapted to sand dunes and deserts. Even organic substances that accumulate in the earth (petroleum and coal) and mineral deposits containing copper, zinc, gold, and uranium have thriving bacterial populations. Although one may think that formaldehyde, alcohol, airplane fuel, methane, hydrogen sulfide gas, and carbolic acid are harmful to cells, these substances are food for some microbes. Moreover, some microbes are indigenous to asphalt, wet cement, and old rubber tires.

As a rule, once a microbe has adapted to an extreme habitat, it will die if placed in a moderate one. And, except for rare cases, none of the organisms living in these extremes are pathogens, because the human body is a hostile habitat for them.

Table 6.2 Analysis of the Chemical Composition of an *Escherichia coli* Cell

	% Total Weight	% Dry Weight
Organic Compounds		
Proteins	15	50
Nucleic acids		
RNA	6	20
DNA	1	3
Carbohydrates	3	10
Lipids	2	Not determined
Miscellaneous	2	Not determined
Inorganic Compounds		
Water	70	—
All others	1	3
Elements		
Carbon (C)	—	50
Oxygen (O)	—	20
Nitrogen (N)	—	14
Hydrogen (H)	—	8
Phosphorus (P)	—	3
Sulfur (S)	—	1
Potassium (K)	—	1
Sodium (Na)	—	1
Calcium (Ca)	—	0.5
Magnesium (Mg)	—	0.5
Chloride (Cl)	—	0.5
Iron (Fe)	—	0.2
Manganese (Mn), zinc (Zn), molybdenum (Mo), copper (Cu), cobalt (Co), zinc (Zn)	—	0.3

Note: Approximate number of different types of molecules making up the cell contents: 5,000. Basic group of nutrients allowing *E. coli* to synthesize these 5,000 chemicals: $(NH_4)_2SO_4$, $FeCl_2$, NaCl, trace elements, glucose, KH_2PO_4, $MgSO_4$, $CaHPO_4$, and water.

(magnesium sulfate, ferric nitrate), gases (oxygen, carbon dioxide) and water (table 6.3). The molecules of **organic nutrients** contain carbon atoms and are usually the products of living things. They range from the simplest organic molecule, methane (CH_4), to large polymers (carbohydrates, lipids, proteins, and nucleic acids). The utilization of nutrients is extremely varied: Some microbes (sulfur-reducing bacteria) obtain their nutrients strictly from inorganic sources, and others require a combination of organic and inorganic (*Escherichia coli;* see table 6.2). Microbes capable of invading and living in the human body derive all essential nutrients from host tissues, tissue fluids, secretions, and wastes.

Nutritional Analysis of a Microbe

To illustrate the nutrition of a representative microbe, let us analyze the contents of an average bacterial cell (the intestinal bacterium *Escherichia coli*) by means of table 6.2. Examination of this table should verify several important features of nutrition:

1. Water content is the highest of all the components (70%).
2. Proteins are the next most prevalent chemical.
3. A cell as "simple" as *E. coli* contains on the order of 5,000 different compounds, yet it needs to absorb only eight basic types of compounds to synthesize this great diversity.
4. About 97% of the dry cell weight is composed of organic macronutrients.

Feature 6.2 Dining with an Ameba

An ameba gorging itself on bacteria could be compared to a human eating a bowl of vegetable soup. Though the ameba may not eat from the four basic food groups, its nutrient needs and uses are fundamentally similar to a human's. Food is ingested, broken down by enzymes, and converted into the organism's own special brand of macromolecules through a series of stages. The nature of the ingested nutrients varies from small molecules that can be absorbed intact to food debris and large molecules that must be broken down into absorbable molecules. Once absorbed, these nutrients add to a dynamic pool of inorganic and organic compounds dissolved in the protoplasm. This pool supports cellular chemical reactions by providing raw materials for synthesis and a source of energy.

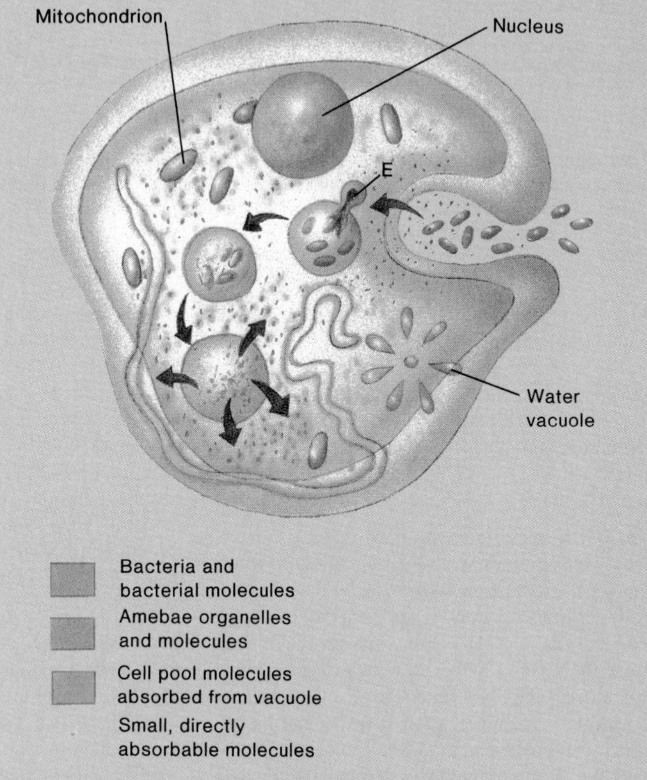

Steps in obtaining and incorporating nutrients. Food particles are surrounded and enclosed by pseudopods into a vacuole that receives digestive enzymes (E). Small nutritional "building blocks" that result from digestion move out of the vacuole into the cell pool and are used in the metabolic and synthetic activities of the cell.

Table 6.3 Principal Inorganic Reservoirs of Elements

Element	Inorganic Environmental Reservoir
Carbon	CO_2 in air; CO_3 in rocks and sediments
Oxygen	O_2 in air, certain oxides, water
Nitrogen	N_2 in air; NO_3, NO_2, NH_4 in soil and water
Hydrogen	Water, H_2 gas, mineral deposits
Phosphorus	Mineral deposits (PO_4, H_2PO_4)
Sulfur	Mineral deposits, volcanic sediments (SO_4, H_2S, S)
Potassium	Mineral deposits, the ocean (KCl, K_2PO_4)
Sodium	Mineral deposits, the ocean (NaCl) (NaSi)
Calcium	Mineral deposits, the ocean ($CaCO_3$, CaCl)
Magnesium	Mineral deposits, geologic sediments ($MgSO_4$)
Chloride	The ocean (NaCl, NH_4Cl)
Iron	Mineral deposits, geologic sediments ($FeSO_4$)
Manganese, molybdenum, cobalt, nickel, zinc, copper, other micronutrients	Various geologic sediments

5. About 96% of the cell is composed of the six bioelements (CHNOPS).
6. Bioelements are needed in the overall scheme of cell growth, but most of them are available to the cell as compounds and not pure elements (table 6.3).

Elements, Nutrients, and Cycles The elements that comprise nutrients ultimately exist in an environmental inorganic reservoir of some type. These reservoirs serve not only as a source of these elements, but can be replenished by the activities of organisms. Thus, elements cycle in a pattern from an inorganic form in an environmental reservoir to an organic form in organisms. Organisms may, in turn, serve as a continuing source of that element for other organisms. Eventually, the element is recycled to the inorganic form, usually by the actions of microorganisms. All in all, a tremendous variety of microorganisms are involved in processing the elements. From a larger perspective, the overall cycle of life on this planet depends upon the combined action of several interacting nutritional schemes, each performing a necessary step. In fact, as we shall see in chapter 22, the framework of microbial nutrition underlies the nutrient cycles of all life on earth.

Sources of Essential Nutrients

For convenience, the following section on nutrients is organized by element. You will no doubt notice that some categories overlap and that many of the compounds furnish more than one element.

Carbon Sources

Microorganisms need carbon for all cellular structures and processes. Those called **heterotrophs** obtain carbon principally in the form of organic matter from the bodies of other organisms and are consequently dependent on other life forms. Among the common organic molecules that can satisfy this requirement are proteins, carbohydrates, lipids, and nucleic acids. In most cases these nutrients provide several other elements as well. Some organic nutrients available to heterotrophs already exist in a form that is simple enough for absorption (for example, monosaccharides and amino acids), but many larger molecules must be digested by the cell prior to absorption. Moreover, heterotrophs vary in their capacities to use various organic carbon sources. Some are restricted to a few sources, whereas others (certain *Pseudomonas* species, for example) are so versatile that they can use hundreds of different ones.

The sole carbon source of microorganisms called **autotrophs** is carbon dioxide (CO_2), an inorganic gas that currently makes up about 0.35% of the earth's atmosphere. The percentage of this gas is gradually increasing due to the burning of hydrocarbons (gasoline and other fuel), which contributes to the gradual warming of the earth or the "greenhouse effect" (see chapter 22). Since autotrophs have the special capacity to convert CO_2 into organic compounds, they are not nutritionally dependent on other living things. In a later section we will further discuss nutritional types as based on carbon and energy sources.

A Clarification About Carbon It seems worthwhile to emphasize a point about the *extracellular source* of carbon as opposed to the *intracellular function* of carbon compounds. Although a distinction must be made between the forms of carbon brought into the cell as nutrients, the bulk of carbon compounds functioning *in* the cell are organic. In other words, autotrophs rely on CO_2 as their primary source of carbon, and heterotrophs must use organic carbon sources; but the intracellular operations of both types require organic carbon compounds. Furthermore, most heterotrophs use CO_2 in biosynthetic reactions such as fatty acid synthesis, even though it is not their primary source of carbon.

Nitrogen Sources

The main reservoir of nitrogen is nitrogen gas (N_2), which makes up about 79% of the earth's atmosphere. This element is indispensable to the structure of proteins, DNA, RNA, and ATP. Such nitrogenous compounds are the primary source for heterotrophs, but to be useful, they must first be degraded into their basic building blocks (protein into amino acids; nucleic acids into nucleotides). Some bacteria and algae utilize inorganic nitrogenous nutrients (NO_3, NO_2, or NH_3). A small number of bacteria can transform N_2 into compounds usable by living things through the process of nitrogen fixation (figure 6.1). It should be emphasized here that, regardless of the initial form in which the inorganic nitrogen enters the cell, it must first be converted to NH_3, the only form that can be directly combined with carbon to synthesize amino acids and other compounds.

heterotroph (het'-uhr-oh-trohf) Gr. *hetero,* other, and *troph,* feeder.

autotroph (aw'-toh-trohf) Gr. *auto,* self, and *troph,* to feed.

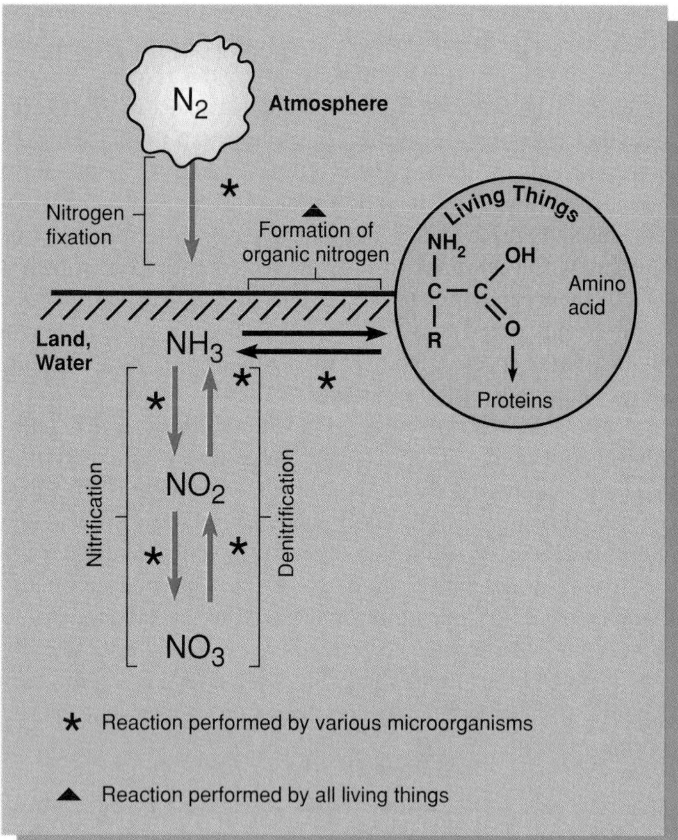

Figure 6.1 A simplified scheme of nitrogen compounds and their uses by the biologic community.

Oxygen Sources

Because oxygen is a major component of organic compounds such as carbohydrates, lipids, and proteins, it plays an important role in the structural and enzymatic functions of the cell. Oxygen is likewise a common component of inorganic salts such as sulfates, phosphates, and nitrates. The free gaseous state of oxygen in the atmosphere is absolutely essential to the metabolism of many organisms, as we shall see later in this chapter.

Hydrogen Sources

Hydrogen is a major element in all organic compounds and several inorganic ones, including water (H_2O), salts (H_2PO_4) and certain naturally occurring gases (H_2S, CH_4, and H_2). These gases are both used and produced by microbes. Hydrogen performs the following overlapping roles in the biochemistry of cells: (1) maintaining pH, (2) forming hydrogen bonds in macromolecules, and (3) acting as a prime force in the oxidation-reduction reactions of respiration (see chapter 8).

Phosphorus (Phosphate) Sources

The main inorganic source of phosphorus is phosphate (PO_4), derived from phosphoric acid (H_3PO_4) and found in rocks and oceanic mineral deposits. Phosphate is a key component of nucleic acids and is thereby essential to the genetics of cells and viruses. Because it is found in the nucleotide ATP, it also serves in cellular energy transfers. Other phosphate-containing compounds are phospholipids in cell membranes and coenzymes such as NAD and NADP (see chapter 8). Phosphate may be so scarce in the environment that it severely limits growth, which explains why certain bacteria concentrate it in metachromatic granules.

Feature 6.3 Metal Ions and Infectious Disease

A link exists between the course of an infection and the presence of certain metal ions. The ion best known for this effect is iron (Fe^{++}). Several fascinating studies have revealed that the balance of iron in a mammalian host's body can actually stimulate or inhibit infections. When the iron concentration is increased, faster growth has been observed in bacteria that cause gonorrhea, meningitis, enteric fevers, and gas gangrene; in the malaria protozoan; and in some yeasts. It has been suggested that some hosts have defense mechanisms for depriving an infectious agent of iron as a means of slowing the microbe's growth.

A low concentration of iron in the host has also been found to tip the balance in favor of the microbe, thus promoting infection. For example, the diphtheria bacillus produces its toxin in response to a low-iron environment, and the agent of bubonic plague (*Y. pestis*) is more virulent when iron-starved. The immune function of iron-starved animals can be so compromised that they are more susceptible to infection.

Other metals have also been implicated in infectious processes. Toxic shock syndrome is a disease in young women who use highly absorbent tampons. Evidently, when these tampons absorb large amounts of magnesium from the vaginal mucus, resident *Staphylococcus aureus* bacteria respond to the lower amount of this ion by producing a toxin with very destructive effects. Zinc plays a peripheral role in diseases caused by *Aspergillus* aflatoxins, in that these fungi require large amounts of zinc to produce their toxins.

Sulfur Sources

Sulfur is widely distributed throughout the environment in mineral form. Rocks and sediments (gypsum) may contain sulfate (SO_4), sulfides (FeS), hydrogen sulfide gas (H_2S), and elemental sulfur (S). The essential amino acids cysteine and methionine (meth-eye′-oh-neen) and certain vitamins are its principal organic forms. The contributions of sulfur to cell chemistry fall into the general areas of protein structure and enzyme function. The sulfur-containing amino acids contribute to the structural stability and shape of proteins by forming unique linkages called disulfide bonds (see figure 2.22).

Other Nutrients Important in Microbial Metabolism

The remainder of important elements are mineral ions. **Potassium** is essential to protein synthesis and membrane function. **Sodium** is important for some types of cell transport. **Calcium** is a stabilizer of the cell wall and endospores of bacteria. It is also combined with carbonate (CO_3) in the formation of shells by foraminiferans and radiolarians. **Magnesium** is a component of chlorophyll, a stabilizer of membranes and ribosomes, and a participant in cell energetics. **Iron** is an important component of the cytochrome pigments. Ions such as zinc, copper, cobalt, nickel, molybdenum, manganese, silicon, iodine, and boron are needed by some microbes but not others. A discovery with important medical implications is that metal ions can directly influence certain diseases by their effects on microorganisms (see feature 6.3).

Table 6.4 Nutritional Categories of Microbes by Carbon and Energy Source

Category	Carbon Source	Energy Source	Example
Autotroph	CO_2	Nonliving environment	Any "self-feeder"
Photoautotroph	CO_2	Sunlight	Photosynthetic organisms, such as algae, plants, cyanobacteria
Chemoautotroph	CO_2	Simple inorganic chemicals	Only certain bacteria, such as methanogens, vent bacteria
Heterotroph	Organic	Sunlight or other organisms	Any "other-feeder"
Photoheterotroph	Organic	Sunlight	Nonsulfur bacteria
Chemoheterotroph	Organic	Metabolic conversion of the protoplasm from other organisms	Protozoa, fungi, many bacteria, animals
Saprobe	Organic	Metabolizing the protoplasm of a dead organism	Fungi, bacteria (decomposers)
Parasite	Organic	Utilizing the tissues, fluids of a live host	Various parasites and pathogens; can be bacteria, fungi, protozoa, animals

Growth Factors: Essential Organic Nutrients

Few microbes are as synthetically versatile as *Escherichia coli,* the example used in table 6.2. Many (especially parasites) lack the genetic and metabolic mechanisms to synthesize all necessary organic compounds from raw materials. An organic compound such as an amino acid or vitamin that cannot be synthesized by an organism and must be provided as a nutrient is a **growth factor.** For example, though all cells require 20 different amino acids for proper assembly of proteins, many cells cannot synthesize them all. Those that must be obtained from food are called *essential amino acids.* A notable example of the need for growth factors occurs in *Haemophilus influenzae,* a bacterium that causes meningitis and respiratory infections in humans. It requires hemin (factor X), NAD (factor V), thiamine and pantothenic acid (vitamins), uracil, and cysteine. Microbes that require growth factors, complex nutrients, or other special conditions are termed **fastidious** (see section on media).

How Microbes Feed

The earth's limitless habitats and microbial adaptations are rivalled by an elaborate menu of nutritional schemes. Fortunately, most organisms show consistent trends and can be described by a few general categories (table 6.4) and a few selected terms (see feature 6.4). The main determiners of a microbe's nutritional type are its source of carbon and its source of energy. In a previous section we divided all microbes into autotrophs, which use CO_2, and heterotrophs, which use organic carbon compounds, but each of these can be further broken down into subtypes based on the energy source and, in the case of heterotrophs, the exact mode of obtaining organic nutrients.

fastidious (fass-tid'-ee-us) L. *fastidium,* loathing or disgust.

Feature 6.4 The Language of Microbial Adaptations

Although initially you may feel inundated with new words, there is really no mystery to the terminology of microbial adaptations. One can easily understand the ecological or nutritional description of a microbe by learning a few prefixes and suffixes. In terms of nutrition, the prefixes *auto-* (self), *hetero-* (other), *photo-* (light), *chemo-* (chemical), and the suffix *-troph* (to feed) can be put together in various ways to succinctly describe a microbe's carbon and energy sources (see table 6.4). Because an algae "feeds itself using the sun," it is a *photoautotroph,* while a parasitic worm that "feeds on the chemicals of others" is a *chemoheterotroph.* One can describe other nutritional patterns by using *sapro-* (rotten) with *-troph* and *halo-* (salt) with *-phile* (to love).

Other useful roots and words for characterizing adaptations are *thermo-* (heat), *psychro-* (cold), *meso-* (intermediate), *aero-* (air), *anaero-* (no air), or *baro-* (pressure) combined with suffixes *-phile* or *-bios* (living). Thus, a *thermophile* "loves heat" and an *aerobe* "lives in air."

Because microbes can vary in how they deal with growth conditions, it may be necessary to use modifying terms. Microbes that are *obligate* or *strict* can only live under certain conditions (nutrients, habitat). Thus, viruses are *obligate* intracellular parasites, and humans (and most animals) are *strict* aerobes—that is, they cannot survive without air.

Other microorganisms can adjust to a changing environment by adopting an alternate metabolic scheme or behavior. An all-around adjective for describing such a microbe is *facultative.* One must be careful in interpreting this term; it should be used in the sense of "can adapt to." For example, a *facultative halophile* can adapt to higher than normal salt concentrations, but it is not obligate nor does it normally live in a high-salt habitat. *Staphylococcus aureus,* a resident of the human nose, is an example of such a species. Likewise, a *facultative anaerobe* can metabolize and grow without air, but it is more efficient if air is present. The roots *-tolerant* and *-duric* denote survival but not growth under certain conditions. A *thermoduric* bacterium can survive heat for a short period, but it does not ordinarily grow at hot temperatures.

facultative (fak'uhl-tay-tiv) Gr. *facult,* capability or skill, and *tatos,* most.

Autotrophs

An autotroph derives energy from one of two possible nonliving sources. These are sunlight (photoautotrophs) and chemical reactions involving simple inorganic chemicals (chemoautotrophs). **Photoautotrophs** are photosynthetic—that is, they capture the energy of light rays and transform it into a chemical form that can be used in cell synthesis (see feature 6.5). Because photosynthetic organisms (algae, plants, some bacteria) trap the sunlight's energy and convert it into a form that can be used by themselves and heterotrophs, they form the basis for most food chains. Their role as primary producers will be discussed again in chapter 22.

Feature 6.5 Sunlight-Driven Organic Synthesis

Two equations sum up the reactions of photosynthesis in a simple way. The first equation shows a reaction that results in the production of oxygen:

$$CO_2 + H_2O \xrightarrow[\text{Chlorophyll}]{\text{Sunlight}} (CH_2O)_n{}^* + O_2$$

This oxygenic (oxygen-producing) type of photosynthesis occurs in plants, algae, and cyanobacteria (see figure 3.32). The function of chlorophyll is to capture light energy. Glucose produced by the reaction is a raw material that can be used by the cell to synthesize other cell components, and it is also an important nutrient for heterotrophs. The production of oxygen is vital to maintaining this gas in the atmosphere.

A second equation shows a photosynthetic reaction that does not result in the production of oxygen:

$$CO_2 + H_2S \xrightarrow[\text{Bacteriochlorophyll}]{\text{Sunlight}} (CH_2O)n + S + H_2O$$

This anoxygenic (no oxygen produced) type of photosynthesis is found in bacteria such as purple and green sulfur bacteria. Note that the type of chlorophyll (bacteriochlorophyll, a substance unique to these microbes), one of the reactants (hydrogen sulfide gas), and one product (elemental sulfur) are different from those in the first equation. These bacteria live in the anaerobic regions of aquatic habitats.

*$(CH_2O)_n$ is shorthand for a carbohydrate.

A significant group of bacteria called **chemoautotrophs** have such an extreme nutritional adaptation that they require neither sunlight nor organic nutrients. Some microbiologists prefer to call them *lithotrophs* (rock feeders) in reference to their total reliance on minerals. These bacteria derive energy in diverse and rather amazing ways. Putting it in very simple terms, they couple the oxidation of inorganic substrates such as hydrogen gas, hydrogen sulfide, sulfur, iron, or nitrite with the reduction of carbon dioxide. This reaction provides simple organic molecules and a modest amount of energy to drive the synthetic processes of the cell. Chemoautotrophic bacteria play an important part in recycling inorganic nutrients. For an example of chemoautotrophy and its importance to deep-sea communities, see feature 6.9.

One very unusual group of chemotrophs are the *methanogens* (figure 6.2), bacteria that produce methane (CH_4) from hydrogen gas and carbon dioxide:

$$4\ H_2 + CO_2 \rightarrow CH_4 + 2H_2O$$

Methane, sometimes called "swamp gas," is formed in anaerobic, hydrogen-containing microenvironments of soil, swamps, mud, and even the intestines of some animals. In their habitats methanogens associate with microbes that decompose organic detritus.[1] The methanogens use the metabolic breakdown products of these decomposers to produce methane. Because methane is a valuable and inexpensive energy source, this relationship has far-reaching practical applications. Farms around the world have adapted this technology for biogas generators—devices primed with a mixed population of microbes (including methanogens) and fueled with animal wastes. The constant supply of methane generated from these wastes is collected and burned to drive a steam generator. This may provide sufficient electricity and heat to make a farm energetically self-sufficient (see chapter 22). Some intriguing relationships of intestinal methanogens are discussed in feature 6.9.

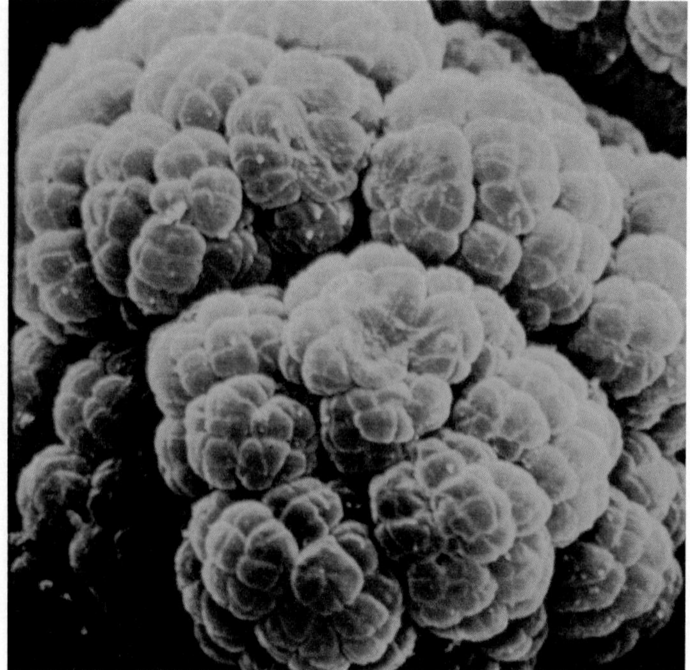

(a)

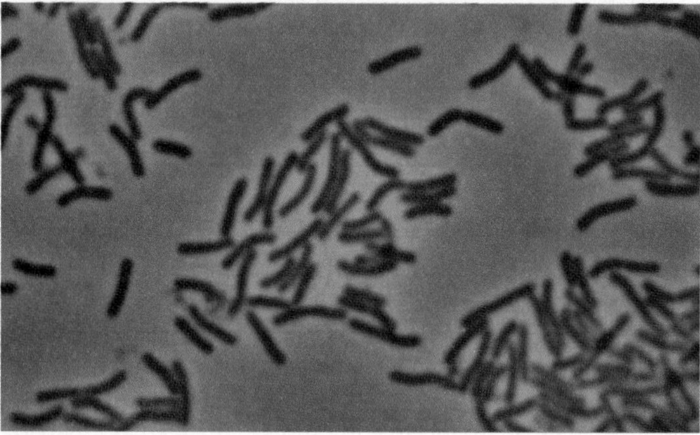

(b)

Figure 6.2 Methane-producing bacteria. Members of this group are primitive archaebacteria with unusual cell walls and membranes. (*a*) A small colony of *Methanosarcina*. (*b*) *Methanobacterium*, a fast-growing, heat-loving strain.

Heterotrophs

Chemoheterotrophic microorganisms derive both carbon and energy from organic compounds that originate from autotrophs

methanogen (meth′-ah-noh-gen) From *methane*, a colorless, odorless gas, and *gennan*, to produce.

1. Residue from dead plant parts or other organisms.

and other heterotrophs. Processing these organic molecules by respiration or fermentation releases energy in the form of ATP. An example of chemoheterotrophy is *aerobic respiration,* the principal energy-yielding reaction in all animals, most protozoa and fungi, and many bacteria. It can be simply represented by the equation:

$$\text{Glucose } [(CH_2O)n] + O_2 \rightarrow \text{ATP} + CO_2 + H_2O$$

You might notice that this reaction is complementary to photosynthesis. Here, sugar and oxygen are reactants, and carbon dioxide is given off. Indeed, the earth's balance of both energy and metabolic gases is greatly dependent on this relationship (see chapter 22).

Chemoheterotrophic microorganisms belong to one of two main categories that differ in their fundamental source of organic matter. **Saprobes**[2] are free-living microorganisms that feed on organic detritus from dead organisms, and **parasites** derive nutrients from the cells or tissues of a living thing (host). Most microbes of biomedical importance belong to these two categories.

Saprobes occupy a niche as decomposers of plant matter, animal carcasses, and even dead microbes. If not for the work of decomposers, the earth would gradually fill up with dead plant and animal material, and the nutrients they contain would not be recycled. Although some saprobes can phagocytose detritus particles and digest them intracellularly, saprobes with rigid cell walls (fungi and bacteria, which make up the majority) are incapable of engulfing large food particles. To compensate, they release enzymes to the extracellular environment and digest the food particles into smaller molecules that can pass freely into the cell (figure 6.3).

Examples of Saprobic Microbes Stalked, budding, and gliding bacteria feeding on detritus in natural aquatic environments would be considered *strict saprobes,* in being unable to adapt to the body of a live host. Apparently there are fewer of these strict species than we once thought, and many supposedly nonpathogenic saprobes can infect a susceptible host. When a saprobe infects a host, we call it a *facultative parasite,* and when it causes disease in a compromised individual, it is an *opportunistic pathogen.* For example, though its natural habitat is soil and water, *Pseudomonas aeruginosa* frequently causes infections in hospitals, and *Legionella,* which lives in moist air, can cause a deadly form of pneumonia (Legionnaire's disease). Most pathogenic fungi (*Histoplasma, Cryptococcus, Aspergillus*) are really facultative parasites because they normally do not live on a host.

The natural habitat of a parasitic microbe is the body of a host, to which it usually causes some degree of harm. Parasites range from viruses to helminths, and they may live on the body (ectoparasites), in the organs and tissues (endoparasites), or even within cells (intracellular parasites, the most extreme

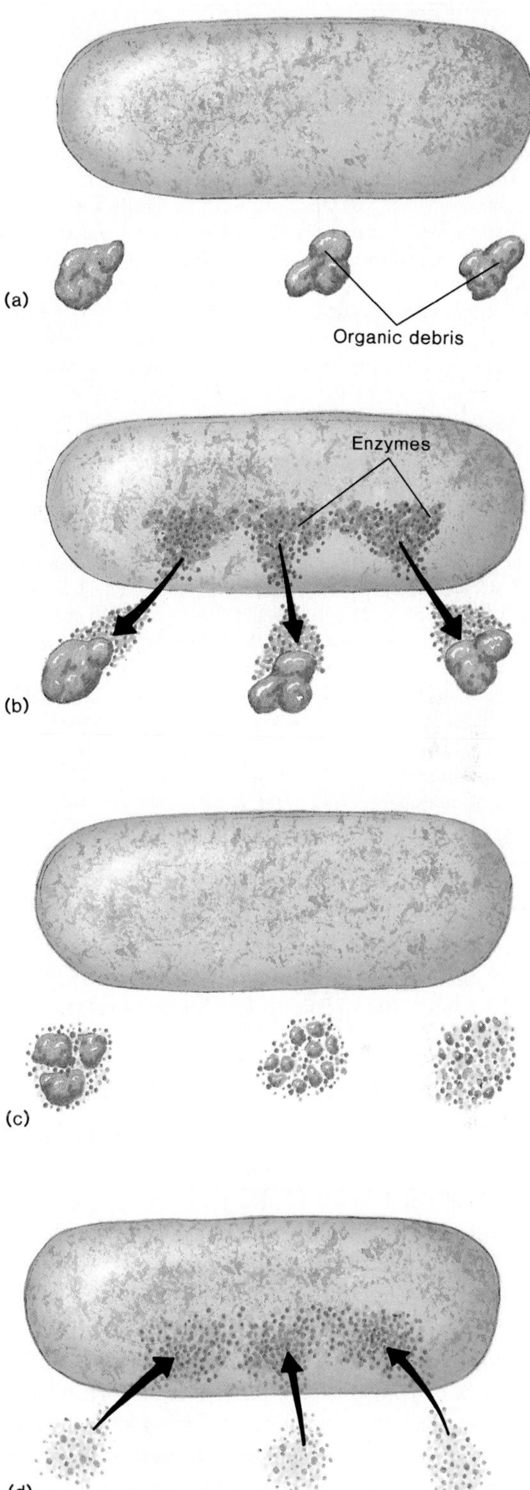

Figure 6.3 Extracellular digestion in a saprobe with a cell wall (bacterium, fungus). (*a*) A walled cell is inflexible and cannot engulf large pieces of organic debris. (*b*) In response to a usable substrate, the cell synthesizes enzymes that are transported across the wall into the extracellular environment. (*c*) The enzymes hydrolyze the bonds in the debris molecules. (*d*) Digestion produces molecules small enough to be transported into the cytoplasm.

2. Synonyms are saprotroph and saprophyte. We prefer saprobe or saprotroph because they are more consistent with other terminology and because saprophyte is a holdover from the time when bacteria and fungi were considered plants.

Feature 6.6 Degrees of Parasitism

Parasites inclined to cause damage to tissues (disease) or even death are called **pathogens.** In general, relatively new parasites (like HIV, the virus that causes AIDS) often have severe and deadly effects and are highly pathogenic, while milder pathogens give rise to diseases that mainly cause discomfort (the cold virus). Sometimes there is a fine line between pathogenicity and nonpathogenicity. Eventually, the host may even develop tolerance to a parasite, at which point we call the relationship symbiosis or commensalism (see "Microbial Interactions" and "Interrelationships Between Microbes and Humans" later in this chapter). Although *strict parasites* (for example, the leprosy bacillus, the spirochete of syphilis, and all viruses) cannot grow outside of a host, parasites termed *facultative saprobes* can grow in a nonliving habitat if provided with organic nutrients. Bacterial parasites such as *Streptococcus pyogenes* (the cause of strep throat), *Staphylococcus aureus,* and many other common pathogens behave as saprobes while growing in artificial media.

Life Inside the Cell Obligate intracellular parasitism is a unique but relatively common mode of life. Microorganisms that spend all or part of their life cycle inside a host cell include the viruses, a few bacteria (rickettsias, chlamydias), and certain protozoan parasites (sporozoans). Contrary to what one might think, the inside of a cell is not completely without hazards, and microbes must overcome some difficult challenges. They must find a way into the cell, keep from being destroyed, not destroy the host cell too soon, multiply, and find a way out of the cell to other cells. What an intracellular parasite obtains from the host cell varies from group to group. Viruses are at the extreme, parasitizing both genetic and metabolic machinery in order to multiply. Rickettsias are primarily energy parasites; chlamydia are metabolic and nutrient parasites; and the malaria sporozoan is a hemoglobin parasite.

type) (see feature 6.6). Although there are several degrees of parasitism, the more successful parasites generally have no fatal effects and eventually evolve to less harmful relationships.

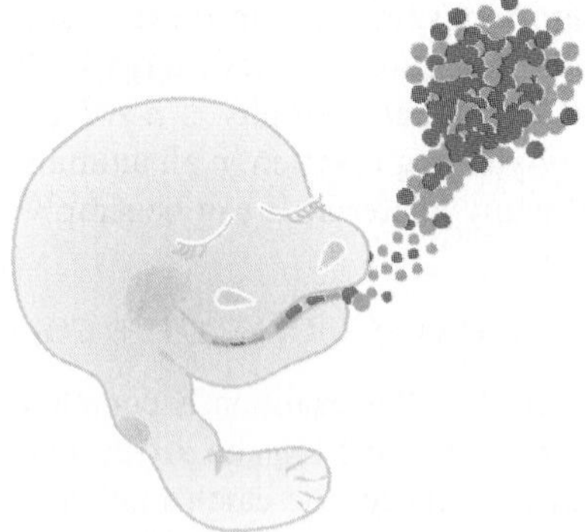

Passive Transport

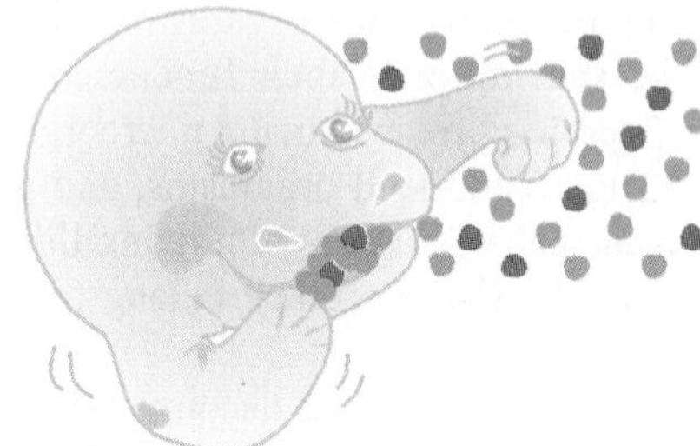

Active Transport

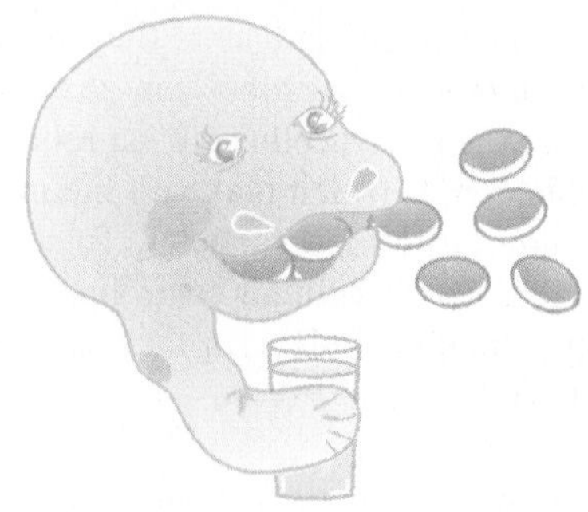

Bulk Transport

Figure 6.4 Simplified models of the three major types of transport. In passive transport, nutrients exist in a gradient from a high concentration outside the cell to a low concentration inside the cell. The molecules naturally diffuse into the cell without its having to expend energy. In active transport, the cell actively picks up nutrients from a solution in which the nutrients are not in a gradient (physical capture is required). In bulk transport, large solids or masses of liquids enter the cell intact by engulfment. The cell actively works during this process.

Absorption of Nutrients: Transport Mechanisms

A microorganism's habitat provides necessary nutrients—some in abundance, others scarce—but those nutrients must still be taken into the cell. Survival also requires that cells transport waste materials into the environment. Whatever the direction, transport occurs across the cell membrane, the structure specialized for this role. This is true even in organisms with cell walls (bacteria, algae, and fungi), because the cell wall is rather porous and does not block the entry of many molecules. Three general types of transport are **passive transport,** which follows physical laws that are not unique to living systems and do not generally require the work of the cell; **active transport,** which requires the activities of living membranes and the expenditure of energy; and **bulk transport,** the movement of large masses of material across membranes (figure 6.4).

Passive Transport: Diffusion, Osmosis

The driving force of passive transport is atomic and molecular movement—the natural tendency of atoms and molecules to be in constant random motion. The existence of this motion is evident in *Brownian movement* (see chapter 3) of small particles suspended in liquid. It can be sensed in a variety of other simple observations. A drop of perfume released into one part of a room is soon smelled in another part, or a lump of sugar in a cup of tea is soon spread through the whole cup without stirring. This phenomenon of molecular movement, in which atoms or molecules move in a gradient from an area of higher density or concentration to an area of lesser density or concentration, is **diffusion** (figure 6.5). Although it is passive and requires a gradient of the diffusing material, diffusion does help a cell obtain certain freely diffusable materials (oxygen, carbon dioxide, and water) and release its wastes to the environment.

diffusion (dih-few′-zhun) L. *dis,* apart, and *fundere,* to pour.

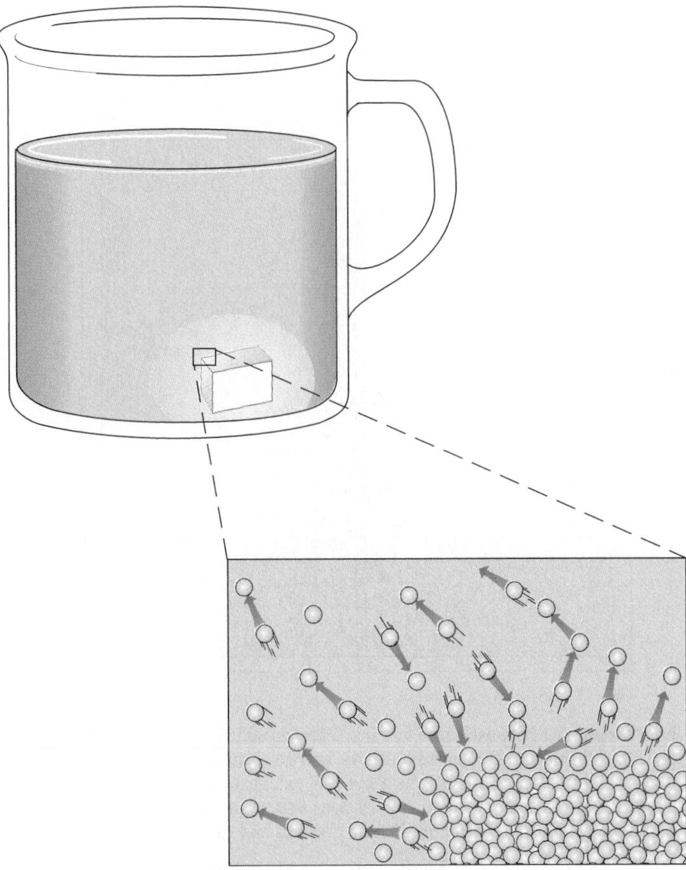

Figure 6.5 Diffusion of molecules in aqueous solutions. A heavy concentration of sugar exists in the cube at the bottom of the liquid. An imaginary molecular view of this area shows that sugar molecules are in a constant state of motion. Those at the edge of the cube diffuse from the concentrated area into more dilute regions. As diffusion continues, the sugar will spread evenly throughout the aqueous phase and eventually there will be no gradient. At that point, the system is said to be in equilibrium.

Diffusion of water through a membrane, a process called **osmosis**, is also a physical phenomenon that can be easily demonstrated in the laboratory with nonliving materials. It provides a model of how a cell deals with water (figure 6.6). In an osmotic system, the membrane is *selectively* or *differentially permeable,* having openings that allow the free movement of water but block the movement of larger molecules. When this membrane is placed between solutions of differing concentrations and the solute is not diffusable (protein, for example), then under the laws of diffusion, water will diffuse at a faster rate from the side that has more water (is a weaker solution) to the side that has less water (is a stronger solution). As long as the concentrations of the solutions differ, one side will experience a net loss of water and the other a net gain of water, until equilibrium is reached and the rate of diffusion is equalized.

osmosis (oz-moh'-sis) Gr. *osmos,* impulsion, and *osis,* a process.

Weak to Strong, the Work of Osmosis By definition, osmosis is the diffusion of solvent (water), thus the concentration and movement of the solvent are what should be followed when tracking osmosis. The solute concentration is also important, however, in that it tells us the strength and dictates how much water is present to diffuse. For example, a **strong** (70%) solution contains 70 parts of solute and 30 parts of water. By comparison, a **weak** (5%) solution contains 5 parts of solute and 95 parts of water. When these two solutions are placed on either side of a membrane, water will diffuse at a faster rate from the weaker solution into the stronger one.

Osmosis in living systems is similar to the model shown in figure 6.6, largely because living membranes are selectively permeable. They generally block the entrance and exit of larger molecules, and permit free diffusion of water. Because most cells are surrounded by some free water, the amount of water entering or leaving has a far-reaching impact on cellular activities and survival. This osmotic relationship is determined by the relative concentrations of the solutions on either side of the cell membrane—the protoplasm of the cell versus its external environment, for example (table 6.5). Such systems can be compared using the terms isotonic, hypertonic, or hypotonic.

Under **isotonic** conditions, the environment is equal in concentration to the cell's internal environment, and since diffusion of water proceeds at the same rate in both directions, there is no net change. Isotonic solutions are the most stable environments for cells, because they are already in osmotic equilibrium with the cell. Parasites living in host tissues are most likely to be living in isotonic habitats. Under **hypotonic** conditions, one solution is weaker than the solution to which it is being compared. Pure water provides the most hypotonic environment for cells. In such systems, the net direction of osmosis is into the cell, and cells without walls swell and may burst. **Hypertonic** conditions are also out of balance with the tonicity of the cell's protoplasm, but in this case, the environment is a stronger solution than the protoplasm. Because a hypertonic environment will cause water to diffuse out of a cell, it is said to have high *osmotic pressure* or potential. Salt water and concentrated sugar solutions are examples of such solutions, and food preservation with salt and sugar takes advantage of their hypertonic effect.

Let us now see how specific microbes have adapted osmotically to their environments. In general, isotonic conditions pose little stress on cells, so survival depends on counteracting the adverse effects of hypertonic and hypotonic environments. A bacterium and an ameba living in fresh pond water are examples of cells that live in constantly hypotonic conditions. The rate of water diffusing across the cell membrane into the protoplasm is rapid and constant, and they would die without compensatory mechanisms. The majority of bacterial cells have compensated by developing a cell wall that protects them from

isotonic (eye-soh-tawn'-ik) Gr. *iso,* same, and *tonos,* tension.
hypotonic (hy-poh-tawn'-ik) Gr. *hypo,* under, and *tonos,* tension.
hypertonic (hy-pur-tawn'-ik) Gr. *hyper,* above, and *tonos,* tension.

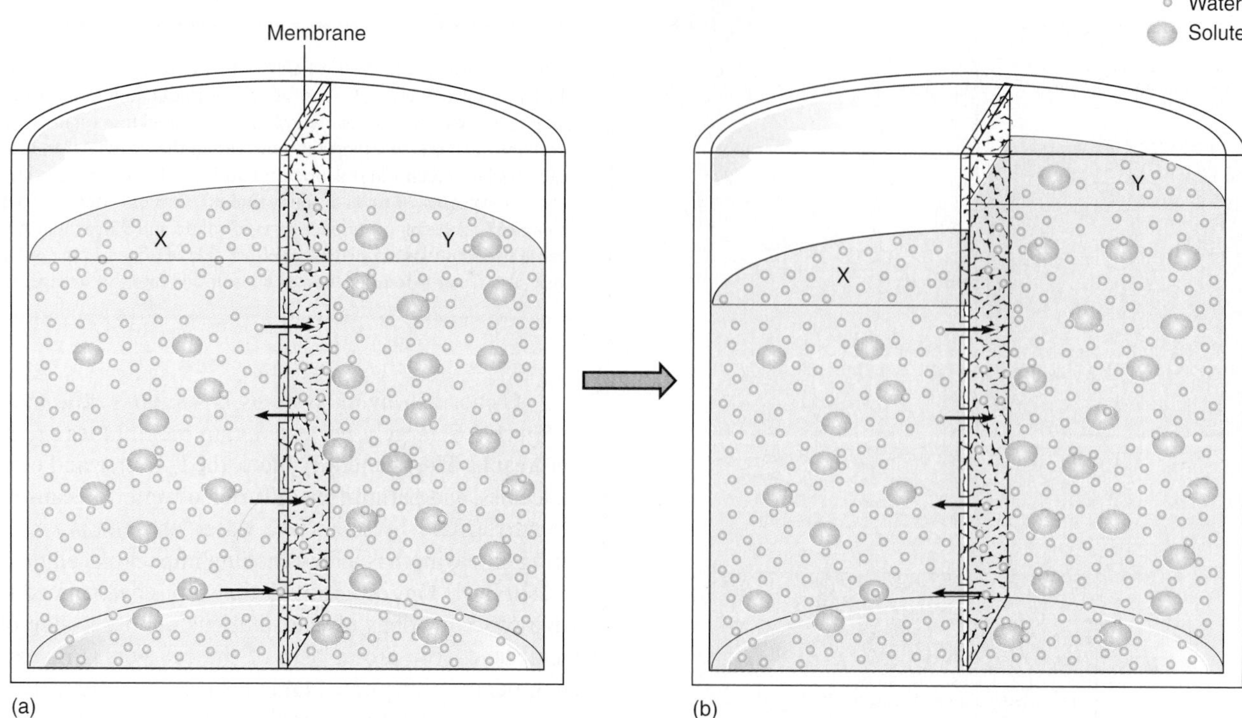

Figure 6.6 Osmosis, the diffusion of water through a selectively permeable membrane. (*a*) A membrane has pores that allow the ready passage of water but not large solute molecules from one side to another. Placement of this membrane between solutions of different strengths (X = weak and Y = strong) results in a diffusion gradient for water. Water behaves according to the law of diffusion and moves across the membrane pores in both directions. Because there is more water in solution X, the opportunity for a water molecule to successfully hit and go through a pore is greater than for solution Y. The result will be a net movement of water from X to Y. (*b*) The level of solution on the Y side rises as water continues to diffuse in. This process will continue until equilibration occurs and the rate of diffusion of water is equal on both sides.

bursting even as the protoplast becomes *turgid* with water. The ameba's adaptation is an anatomical and physiological one that requires the constant expenditure of energy. It has a water vacuole that siphons excess water back out into the habitat like a minute pump (see figure 4.30). A microbe living in a high-salt environment (hypertonic) has the opposite problem and must either restrict its loss of water to the environment or increase the salinity of its internal environment. Halobacteria living in the Great Salt Lake and the Dead Sea actually concentrate salt in their cells to become isotonic with the environment, thus they have a physiologic need for high salt.

A form of passive transport called **facilitated diffusion** (figure 6.7) also requires a concentration gradient of the transported molecule and does not expend energy, but it is more specific than other passive systems. Only molecules carried by specialized membrane proteins are transported. Through this mechanism, some yeasts and bacteria transport sugars, and spore-forming bacteria concentrate calcium in developing spores.

Active Transport: Bringing in Nutrients Against a Gradient

Free-living microbes exist under relatively nutrient-starved conditions and cannot rely on slow and rather inefficient passive transport mechanisms. To maintain their nutritional status, microbes must capture nutrients that are in extremely short supply and actively transport these nutrients into the cell. Features in-

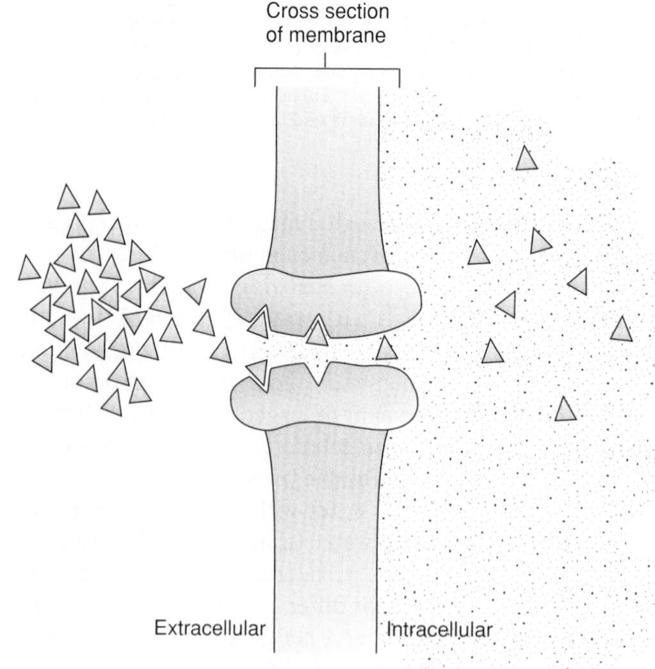

Figure 6.7 Facilitated diffusion involves the attachment of a molecule to a specific protein carrier, but because these molecules are carried to the cell by a diffusion gradient, it is still passive transport.

---

turgid (ter'-jid) A condition of being swollen or congested.

Table 6.5 Cell Responses to Solutions of Differing Osmotic Content

	Relative Tonicity of Solution External to the Cell		
	Isotonic	**Hypotonic**	**Hypertonic**
Cells with Cell Wall	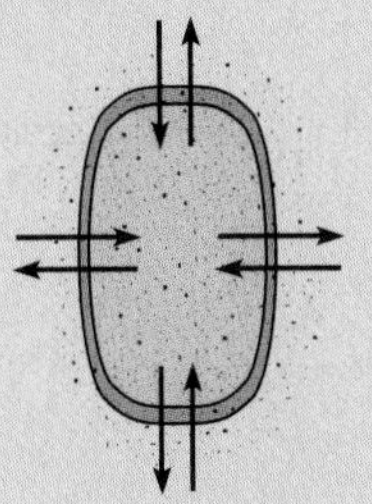	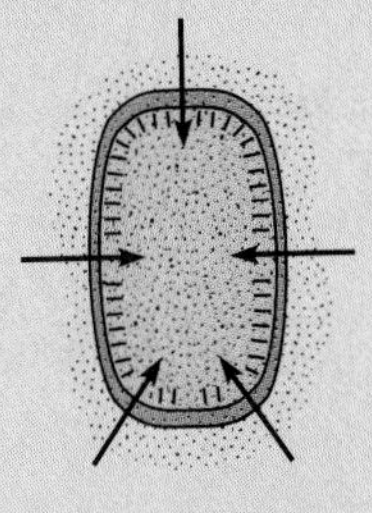	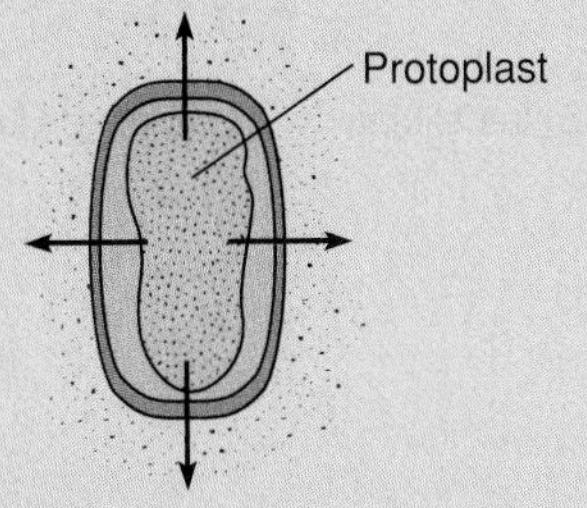 Protoplast
	Water concentration is equal inside and outside the cell, thus rates of diffusion are equal in both directions.	Net diffusion of water is into the cell; this swells the protoplast and pushes it tightly against the wall. Wall usually prevents cell from bursting.	Water diffuses out of the cell and shrinks the protoplast away from the cell wall; process is plasmolysis.
Cells Lacking Cell Wall	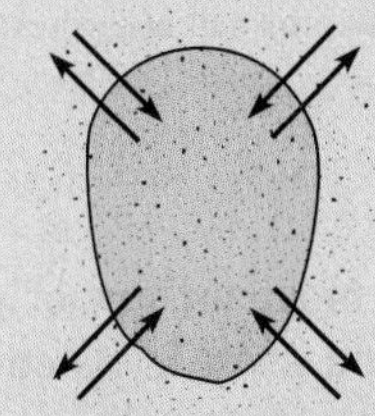	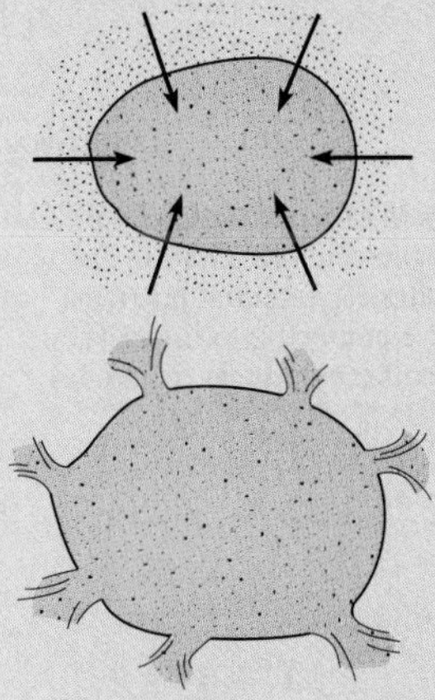 Early Late	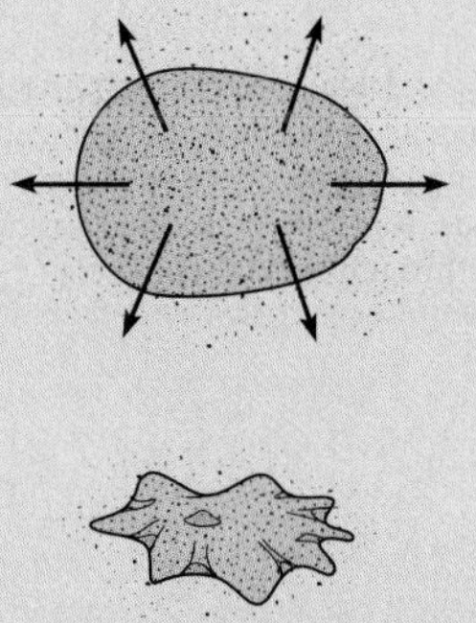 Early Late
	Rates of diffusion are equal in both directions.	Diffusion of water into the cell causes it to swell, and may burst it if no mechanism exists to remove the water.	Water diffusing out of the cell causes it to shrink and become distorted.

→ Direction of net water movement

herent in **active transport systems** are: (1) the transport of nutrients against the natural diffusion gradient or in the same direction as the natural gradient, but at a rate faster than by diffusion alone; (2) special membrane proteins (permeases and pumps, figure 6.8*a*); and (3) the expenditure of energy. Examples of substances transported actively are monosaccharides, amino acids, organic acids, phosphates, and metal ions. Some freshwater algae have such efficient active transport systems that an essential nutrient can be found in intracellular concentrations 200 times that of the habitat.

Active transport also occurs within intracellular membranes such as the mitochondrion, the chloroplast, and the endoplasmic reticulum. This behavior is particularly important in mitochondrial ATP formation (see the discussion of oxidative phosphorylation in chapter 7) and protein synthesis. One special type of active transport, **group translocation,** couples the transport of a nutrient with its conversion to a substance that is immediately useful to the cell (figure 6.8*b*).

Bulk Transport: Eating and Drinking by Cells

Some cells actively transport large molecules, particles, liquids, or even other cells across the cell membrane. These substances do not pass physically through the membrane but are carried into the cell by a process of membrane envelopment and vacuole or vesicle formation called **endocytosis** (figure 6.9). Amebae and certain white blood cells ingest large solid matter by *phagocytosis;* liquids, such as oils or large molecules in solution, enter the cell through *pinocytosis.*

endocytosis (en″-doh-cy-toh′-sis) Gr. *endo,* in; *cyte,* cell; and *osis,* a process.
phagocytosis (fag″-oh-sy-toh′-sis) Gr. *phagein,* to eat.
pinocytosis (pin″-oh-cy-toh′-sis) Gr. *pino,* to drink.

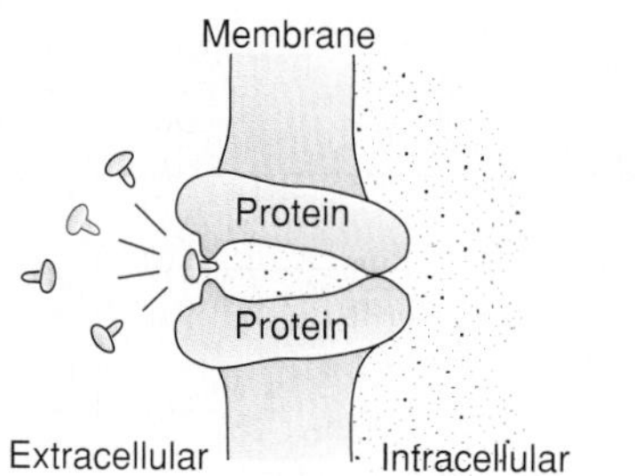

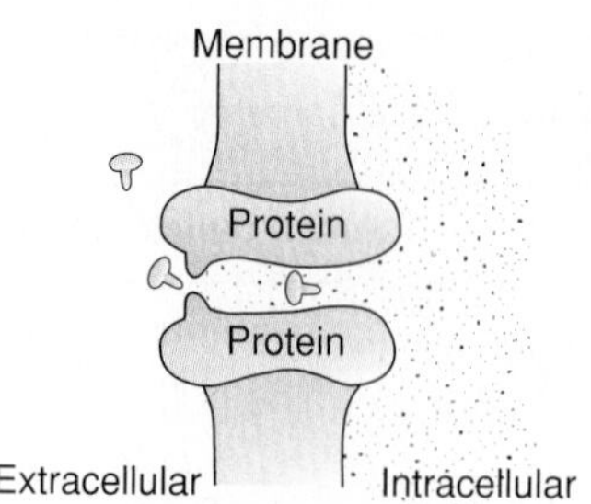

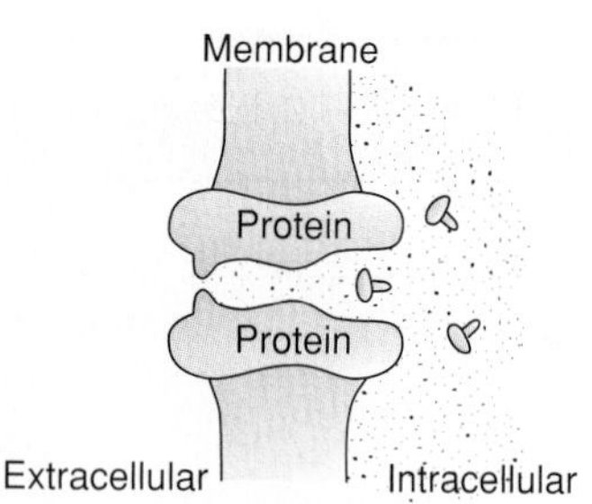

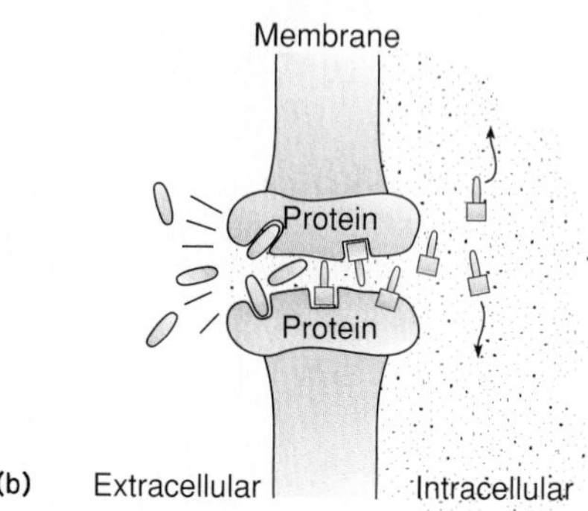

Figure 6.8 In active transport mechanisms, energy is expended to pick up the molecule and take it into the cell's interior. (*a*) Molecular channels. The membrane proteins (permeases) have attachment sites for essential nutrient molecules. As these molecules are captured, they are pumped into the cell's interior through special channels or pores in the membrane protein. Microbes have these systems for transporting various ions (sodium, iron) and small organic molecules. (*b*) In group translocation, the molecule is actively captured, but along the route of transport, it is chemically altered. By coupling transport with synthesis, the cell conserves energy.

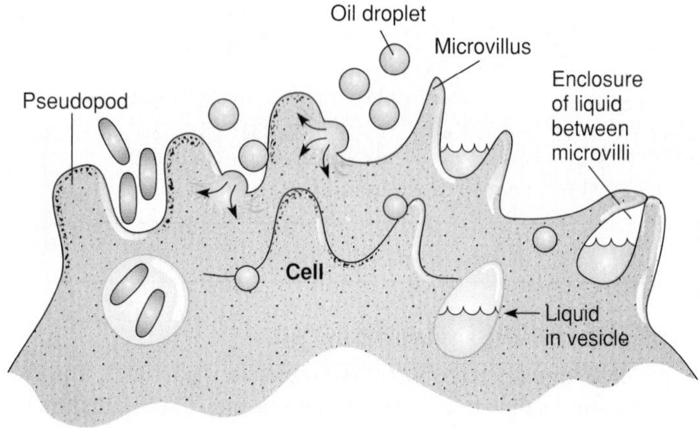

Figure 6.9 Endocytosis (phagocytosis and pinocytosis). Solid particles are phagocytosed by large cell extensions called pseudopods, and fluids are pinocytosed into vesicles by very fine cell protrusions called microvilli. Oil droplets fuse with the membrane and are released directly into the cell.

The Influence of Environmental Factors on Microbes

It should be clear that a microbe experiences a wide variety of environmental factors: heat, cold, gases, acid, radiation, moisture, osmotic and hydrostatic pressures, and even other microbes. A theme we will discuss repeatedly is how a microbe deals with or adapts to these factors. Adaptation is a complex long-term adjustment in biochemistry or genetics that enables an organism to survive and grow in its environment. Although the nature of these adjustments will become more evident in chapters 7 and 8, for now, it can be stated that environmental factors fundamentally affect the function of metabolic enzymes, and that survival in a changing environment is largely a matter of enzymatic flexibility. Incidentally, one must be careful to differentiate between growth in a given condition and tolerance, which implies survival without growth.

Temperature

Microbial cells assume the ambient temperature of their natural habitats and adapt to growth within a certain range of temperatures encountered in that habitat, called the *cardinal temperatures*. The **minimum temperature** is the lowest temperature that permits a microbe's continued growth and metabolism; below this temperature, its activities are generally inhibited. The **maximum temperature** is the highest temperature at which growth and metabolism can proceed. If the temperature rises slightly above maximum, growth will stop, but if it continues to rise beyond that point, the enzymes and nucleic acids will eventually become permanently inactivated and the cell will die. This is why heat works so well in microbial control. The **optimum temperature** encompasses a small range, intermediate between the minimum and maximum, which promotes the fastest rate of growth and metabolism (rarely is the optimum a single point).

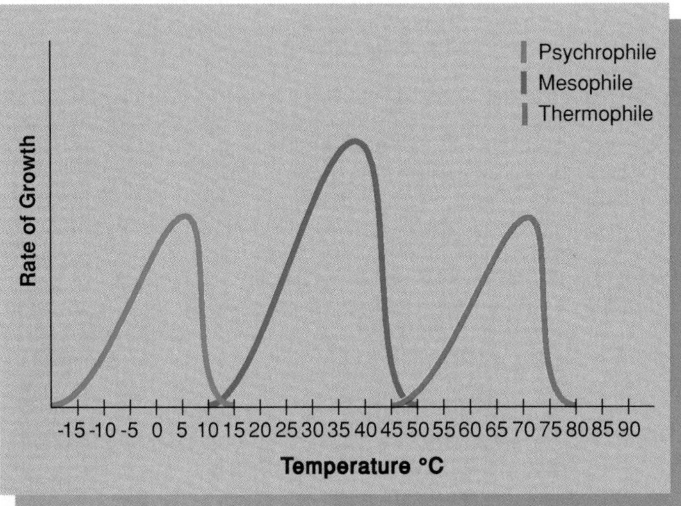

Figure 6.10 Ecological groups by temperature of adaptation. Psychrophiles can grow at or near 0°C and have an optimum below 15°C. As a group, mesophiles can grow between 10°C and 50°C, but their optima usually fall between 20°C and 40°C. Generally speaking, thermophiles require temperatures above 45°C and grow optimally between this temperature and 80°C. The cardinal temperatures are labelled for one group. Note that the extremes of the ranges can overlap to an extent.

(a)

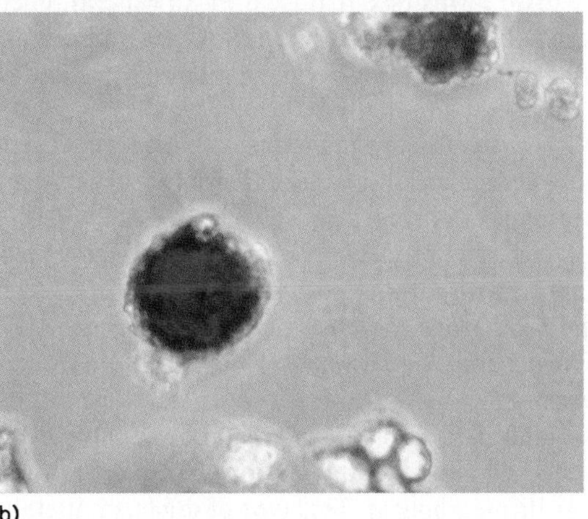
(b)

Figure 6.11 Red snow. (*a*) An early summer snowbank provides a perfect habitat for psychrophilic photosynthetic organisms like *Chlamydomonas nivalis*. (*b*) Microscopic view of this snow alga (actually classified as a "green" alga).

Examples of Temperature Ranges Depending on their natural habitats, some microbes have a broad cardinal range, others a narrow one. Consider, for instance, strict parasites that will not grow if the temperature varies more than a few degrees below or above the host's body temperature. The typhus rickettsia multiplies only in the range of 32°–38°C. Some pathogens are so strict in their temperature requirements that they will multiply only in certain body sites. The leprosy bacillus and the rhinovirus (the cause of the common cold) multiply successfully only in the cooler regions of the body. Many non-strict human parasites cannot grow beyond a range of ten degrees (for example, the gonococcus—30°–40°C, or the tubercle bacillus—30°–42°C). Other parasites are not so limited. Strains of *Staphylococcus aureus* grow within the range of 6°–46°C, and the intestinal bacterium *Enterococcus faecalis* grows within the range of 5°–50°C.

Depending upon its habitat, an organism may be adapted to temperature ranges described as cold, moderate, or hot. The terms that describe these ecological groups are psychrophile, mesophile, and thermophile (figure 6.10).

A **psychrophile** (sy'-kroh-fyl) is a microorganism that grows optimally below 15°C and is capable of growing at 0°C. It is obligate with respect to cold and generally cannot grow above 20°C. Lab work with true psychrophiles can be a real challenge. Inoculations have to be done in a cold room because room temperature may be lethal to the organism. Unlike most laboratory cultures, storage in the refrigerator incubates, rather than inhibits them. As one might predict, the habitats of psychrophilic bacteria, fungi, and algae are snowfields (figure 6.11), polar ice, and the deep ocean. Rarely, if ever, are they pathogenic. True psychrophiles must be distinguished from *facultative psychrophiles,* mesophilic organisms that grow slowly in cold but have an optimum temperature above 20°C.

The majority of microorganisms important to the medical microbiologist are **mesophiles** (mees'-oh-fylz), organisms that grow at moderate temperatures. Although an individual species may grow at the extremes of 10°C or 50°C, the optimum growth temperatures (optima) of most mesophiles fall into the range of 20°–40°C. Organisms in this group inhabit animals and plants as well as soil and water in temperate, subtropical, and tropical regions. Most human pathogens have growth optima somewhere between 30°C and 40°C (human body temperature is 37°C). *Thermoduric* microbes, which can survive short exposure to high temperatures but are normally mesophiles, are common contaminants of heated or pasteurized foods (see chapter 9). They are often cyst-forming, spore-forming, or thick-walled microbes.

A **thermophile** (thur'-moh-fyl) is a microbe that grows optimally at temperatures greater than 45°C. Such heat-loving microbes live in soil and water associated with volcanic activity and in habitats directly exposed to the sun. Thermophiles vary in heat requirements, with a general range of growth of 45°–80°C. Most eucaryotic forms cannot survive above 60°C, but a few thermophilic bacteria grow around 250°C (currently thought to be the temperature limit endured by protoplasm). Many thermophiles are spore-forming species of *Bacillus* and *Clostridium,* and a small number are pathogens.

Gas Requirements

The three atmospheric gases that most influence microbial growth are O_2, CO_2, and N_2. Of these, oxygen has the greatest impact on microbial adaptation. Not only is it an important respiratory gas, but it is also a powerful oxidizing agent that exists in many toxic forms (see feature 6.7). In general, microbes fall into these categories: those that use oxygen and can detoxify it; those that can neither use oxygen nor detoxify it; and those that do not use it but can detoxify it.

With respect to oxygen requirements, several general categories are recognized. An **aerobe** (aerobic organism) grows well in the presence of normal atmospheric oxygen and possesses the enzymes needed to process toxic oxygen products. An organism that cannot grow without oxygen is an *obligate aerobe.* Most fungi and protozoans as well as many bacteria (genera *Bacillus, Bdellovibrio,* and *Xanthomonas*) are strictly aerobic in their metabolism.

A **facultative anaerobe** is an aerobe capable of growth in the absence of oxygen—that is, oxygen is not absolutely required for its metabolism. This type of organism metabolizes by aerobic respiration when oxygen is present, but in the absence of oxygen, it adopts an anaerobic mode of metabolism such as fermentation. Facultative anaerobes usually possess catalase and dismutase. A large number of bacterial pathogens fall into this group (for example, enteric rods and staphylococci).

A **microaerophile** (myk″-roh-air′-oh-fyl) does not grow at normal atmospheric tensions, but requires a small amount of oxygen in metabolism. *Actinomyces israelii,* the cause of lumpy jaw in humans, and *Treponema pallidum,* the cause of syphilis, are examples of microaerophiles. Most organisms in this category live in a habitat (soil, water, or the human body) that provides small amounts of oxygen but is not directly exposed to the atmosphere.

An **anaerobe** (anaerobic microorganism) does not grow in normal atmospheric oxygen, and it lacks the metabolic enzyme systems for using oxygen in respiration. Because many anaerobes also lack the enzymes for processing toxic oxygen, they cannot tolerate any free oxygen in the immediate environment. Microbes killed or inhibited by oxygen are called *strict* or *obligate anaerobes.* Strict anaerobes live in highly reduced habitats, such as deep muds, lakes, oceans, soil, and even the bodies of animals (see feature 6.8). Among the more important anaerobic pathogens are some species of *Clostridium, Bacteroides, Fusobacterium,* and the protozoan *Trichomonas.* Growing anaerobic bacteria usually requires special media or incubation chambers that exclude oxygen (figure 6.13*a*).

Aerotolerant anaerobes do not utilize oxygen, but can survive in its presence. These organisms are not killed by oxygen, mainly because they possess alternate mechanisms for breaking down peroxides and superoxide. Certain lactobacilli and streptococci use manganese ions or peroxidases to perform this task. Determining the oxygen requirements of a microbe from a biochemical standpoint can be a very time-consuming process. Often it is illuminating to perform culture tests with reducing

Feature 6.7 Toxic Forms of Oxygen

As oxygen enters into cellular reactions, it is transformed into several toxic products. Singlet oxygen ($^1O_2$) is an extremely reactive molecule produced by both living and nonliving processes. Notably, it is one of the substances produced by phagocytes to kill invading bacteria (see chapter 12). The buildup of singlet oxygen and the oxidation of membrane lipids and other molecules can damage and destroy a cell. The highly reactive superoxide ion (O_2^-), peroxides, and hydroxides are other destructive metabolic products of oxygen. To protect themselves against damage from these chemicals, most cells have developed enzymes that go about the business of scavenging and neutralizing these products. The complete conversion of superoxide ion into harmless oxygen requires a two-step process and at least two enzymes:

$$\text{Step 1. } O_2^- + O_2^- + H_2 \xrightarrow{\text{Superoxide dismutase}} H_2O_2 \text{ (hydrogen peroxide)} + O_2$$

$$\text{Step 2. } H_2O_2 + H_2O_2 \xrightarrow{\text{Catalase or peroxidase}} 2H_2O + O_2$$

In this series of reactions (essential for aerobic organisms), the superoxide ion is first converted to hydrogen peroxide and normal oxygen by the action of an enzyme called superoxide dismutase. Since hydrogen peroxide is also toxic to cells (it is a disinfectant and antiseptic), it must be degraded by another enzyme (either catalase or peroxidase) into water and oxygen. If a microbe is not capable of dismantling toxic oxygen in this manner, it must live in habitats free of oxygen.

Feature 6.8 Anaerobic Infections

Even though human cells use oxygen, and oxygen is found in the blood and tissues, some body sites present anaerobic microhabitats, or pockets, where colonization or infection can occur. It is now believed that these areas are maintained in this anaerobic state by the coexistence of aerobic and facultative organisms that use up the oxygen. As long as oxygen users continue to grow, so can anaerobes. One region that is very supportive of this sort of microbial cooperation (synergism) is the oral cavity, where tooth crevices and the gingival sulcus are important sites of anaerobic infections. Dental caries are partly due to the complex actions of aerobic and anaerobic bacteria, and most gingival infections consist of similar mixtures of oral bacteria that have invaded damaged gum tissues. Another region that supports synergism is the large intestine. Because the majority of bacteria that inhabit the intestine are anaerobic, anaerobic infections often accompany abdominal surgery. Gas gangrene, tetanus, and infections arising from human or animal bites are also caused by anaerobes.

---

aerobe (air′-ohb) Though the prefix means air, it is used in the sense of oxygen.

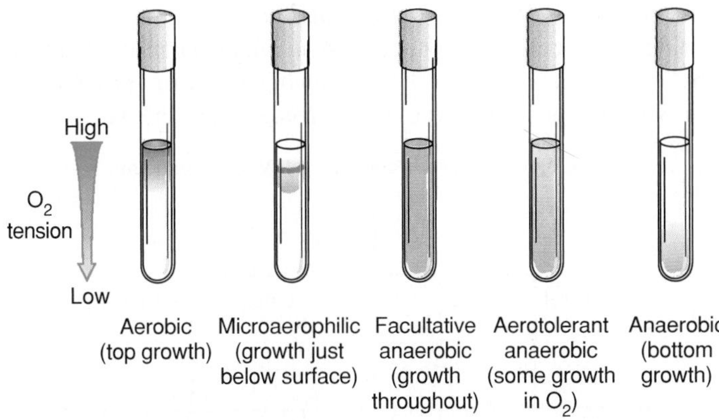

Figure 6.12 Use of thioglycollate broth to demonstrate oxygen requirements. Thioglycollate is a chemical that absorbs O_2 gas and renders it unavailable to bacteria. Oxygen from the air continues to be dissolved in the medium and absorbed; its progress can be shown by the red dye resazurin. When a series of tubes is inoculated with bacteria that differ in O_2 requirements, the relative position of growth provides some indication of their adaptations to oxygen use.

media (those that contain an oxygen-absorbing chemical). One such technique demonstrates oxygen usage by the location of growth in a tube of fluid thioglycollate (figure 6.12).

Although all microbes require some carbon dioxide in their metabolism, the *capnophiles* grow best at a higher CO_2 tension than is normally present in the atmosphere. This becomes important in the initial isolation of some pathogens from clinical specimens, notably *Neisseria* (gonorrhea, meningitis), *Brucella* (undulant fever), and *Streptococcus pneumoniae*. Incubation is carried out in a system that provides 3% to 10% CO_2—either a CO_2 incubator, special plastic bag, or candle jar (figure 6.13*b*).

Effects of pH

The degree of acidity or alkalinity (basicity) of a solution can be expressed by a chemist's rating called the pH scale, a series of numbers ranging from 0 to 14 (see chapter 2). The pH of pure water (7.0) is neutral, neither acidic nor basic. As the pH value decreases toward 0, the acidity increases, and as the pH increases toward 14, the alkalinity increases. The majority of organisms do not live or grow in high or low pH habitats, because acid and base can be highly damaging to enzymes and other cellular substances. The optimum pH range for most microorganisms is between 6 and 8, and most human pathogens are neutrophiles that grow optimally at a pH of 6.5 to 7.5.[3]

Some microorganisms live at pH extremes. Obligate *acidophiles* such as *Euglena mutabalis* grow in acid pools of between 0 and 1.0 pH. *Thermoplasma,* a bacterium that lacks a cell wall, lives in hot coal piles at a pH of 1 to 2 and will lyse if placed at pH 7. Because molds and yeasts may tolerate an acidic environment, they frequently contaminate pickled foods. Alkalinophiles live in hot pools and soils that contain high levels

Lockscrew
Outer lid
Inner lid
Catalyst chamber contains palladium pellets.
$2H_2 + O_2 \longrightarrow 2H_2O$
CO_2
H_2
Rubber gasket provides air-tight seal.
Gas Pak
BBL
Gas generator envelope (10 ml of water is added to chemicals in envelope to generate H_2 and CO_2. Carbon dioxide promotes more rapid growth of organisms.)
Anaerobic indicator strip (Methylene blue becomes colorless in absence of O_2.)
Reaction (Oxygen is removed from chamber by combining with hydrogen on surface of palladium pellets.)

(a)

(b)

Figure 6.13 Systems for growing anaerobes and capnophiles. (*a*) The anaerobic jar, or CO_2 incubator. A packet that generates hydrogen gas is added to the jar with cultures and tightly sealed. As H_2 is generated, it reacts with any O_2 present to form water, effectively producing anaerobiosis. A CO_2 packet may also be introduced if capnophilic growth is desired. (*b*) The candle jar. This simple yet useful method involves lighting a candle and immediately closing the jar. After a few moments, the flame goes out due to depleted O_2. Meanwhile, the combustion process has added CO_2 to the air as well.

of basic minerals (up to pH 10.0). Bacteria that decompose urine create alkaline conditions, since ammonium (NH_4^+, an alkaline ion) may be produced when urea (a component of urine) is digested.

Osmotic Pressure

Although most microbes exist under hypotonic or isotonic conditions, a few, called the **halophiles** (hay′-loh-fylz) or **osmophiles** (oz′-moh-fylz), live in solutions that have a high solute

capnophile (kap′-noh-fyl) Gr. *kapnos,* smoke.

3. The pH of human blood is normally 7.4.

concentration (are hypertonic). *Obligate halophiles* like *Halobacterium* and *Halococcus* inhabit strongly saline lakes and ponds. They grow optimally in solutions of 25% NaCl but require at least 15% NaCl (combined with other salts) for growth. These bacteria lyse in hypotonic habitats. Some bacteria, though not residents of high-salt environments, are remarkably resistant to salt. *Escherichia coli* maintains in its periplasmic space an osmotically active solution that responds to changes in the medium and prevents overt swelling or shrinkage of its protoplast. Although it is common to use high osmotic pressures in preserving food (jams, jellies, syrups, and brines) to prevent growth of contaminants, many osmophilic bacteria and fungi actually thrive under these conditions and can spoil food.

Miscellaneous Environmental Factors

Various forms of electromagnetic radiation (ultraviolet, infrared, visible light) stream constantly onto the earth from the sun. Some microbes (phototrophs) may use visible light rays as an energy source, but non-photosynthetic microbes tend to be damaged by the toxic oxygen products produced by contact with light. Some microbial species, such as *Staphylococcus aureus,* can produce a yellow carotenoid pigment that protects against the damaging effects of light by absorbing and dismantling toxic oxygen. Other types of radiation that can damage microbes are ultraviolet and ionizing rays (X rays and cosmic rays). In chapter 9 we will see just how these types of energy are applied in microbial control.

As a body descends into the ocean depths, it is subject to increasing hydrostatic pressure. Microbes (mostly bacteria) living there exist under pressures anywhere from a few times to over one thousand times the pressure of the atmosphere. **Barophiles** are bacteria so strictly adapted to high pressures that they will rupture when exposed to normal atmospheric pressure.

Due to the high water content of protoplasm, all cells require water from their environment to sustain growth and metabolism. Water is the solvent for cell chemicals, and it is needed for enzyme function and digestion. A certain amount of water on the external surface of the cell is required for the diffusion of nutrients and wastes. Even in apparently dry habitats, such as sand or dry soil, the particles retain a thin layer of water usable by microorganisms. Dormant, dehydrated cell stages (for example, spores and cysts) tolerate extreme drying because of the inactivity of their enzymes.

Microbial Interactions

Up to now we have considered the importance of nonliving environmental influences on the growth of microorganisms. Another profound influence comes from other organisms that share (or sometimes are) their habitats. In all but the rarest instances, microbes live in direct proximity to other life forms, and these mixed populations give rise to interrelationships of elaborate and fascinating scope. Some generalities that apply to such relationships are: (1) They may occur between microbes; (2) they may involve multicellular organisms such as animals or plants; and (3) they can have beneficial, neutral, or harmful effects on the organisms involved. The following diagram provides an overview of the major types of microbial interactions:

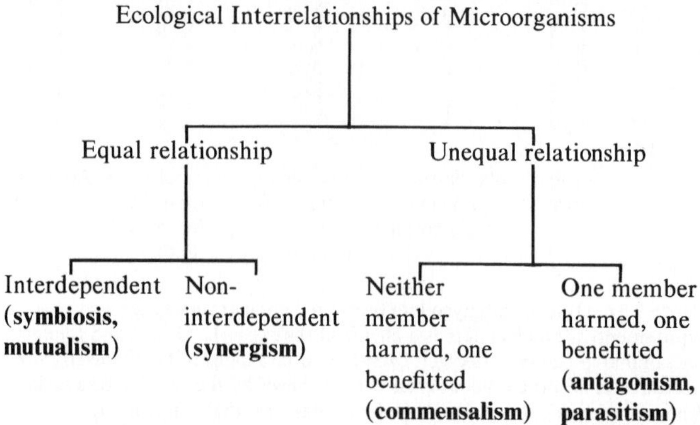

When two organisms live together in a close partnership, they are involved in **symbiosis** and are termed symbionts. Although symbiosis originally meant a simple cohabitation of organisms, it is sometimes used in the sense of a mutually beneficial obligate relationship, also known as *mutualism.* A symbiotic relationship often involves an interaction between two organisms that maintain a shared habitat or interchange nutrients. The examples in feature 6.9 illustrate this concept.

Synergism is a cooperative relationship between organisms that is beneficial to both members but not obligatory. It is a common sort of communal life that occurs when two organisms work together to accomplish a reaction that neither one could accomplish alone. Sometimes the relationship is nutritional (as in cross feeding), with the metabolic products enhancing the growth of both members. An example of synergism is the relationship between plants and soil bacteria, which mutually stimulates growth (see chapter 22). In synergistic infections, a combination of organisms (sometimes as many as 10 species) can produce tissue damage that a single organism could not cause alone. Gum disease, dental caries, tetanus, and gas gangrene all have a synergistic component.

Many interrelationships involve an imbalance in the contributions of the participants. In **commensalism** one member (A) is neither harmed nor benefitted, yet A provides benefits to the other member (B). In this situation, B is called the commensal of A. A classic example of commensalism between microorganisms is *satellitism,* a phenomenon based on nutritional or protective factors. In nutritional satellitism, microbe A provides a growth factor that microbe B needs (figure 6.14). In another form, microbe A breaks down a substance that would be toxic or inhibitory to microbe B. Relationships between humans and resident microorganisms that derive nutrients from the human body are discussed in the following section.

---

synergism (sin'-ur-jizm) Gr. *syn,* together; *erg,* work; and *ism,* process.
commensalism (kuh-men'-sul-izm) L. *com,* together, and *mensa,* table.

Hemophilus satellite colonies

Staphylococcus aureus growth

Figure 6.14 Satellitism, a type of commensalism between two microbes. In this example on blood agar, *Staphylococcus aureus* provides growth factors for *Hemophilus influenzae*, which grows as tiny satellite colonies near the streak of *Staphylococcus*. By itself, *Hemophilus* could not grow on blood agar.

The concept of parasitism as a harmful interrelationship has already been discussed. Another harmful association, **antagonism**, arises when members of a community compete. In this unequal interaction, one microbe secretes antimicrobial chemicals that inhibit or destroy another microbe in the same habitat. This gives the first microbe a competitive advantage. Interactions of this type are common in the soil, where mixed communities often compete for space and food. The most prominent examples are seen in the *antibiosis* (production of antibiotics) by fungi and bacteria and in the production of *bacteriocins* by bacteria. The natural inclination of microorganisms to produce antibiotics has been well exploited in the control of diseases (see chapter 10). Bacteriocins are proteins produced by gram-negative rods that are inhibitory to other closely related bacteria.

Interrelationships Between Microbes and Humans

The human body is a rich habitat for bacteria, fungi, and protozoa. The microbes that normally live on the skin, in the alimentary tract, and in other sites are called the *normal flora* (see chapter 11). These residents participate in symbiotic, synergistic, commensal, and parasitic relationships with their human hosts. For example, certain bacteria living symbiotically in the intestine produce some vitamins, and species of symbiotic *Lactobacillus* residing in the vagina help maintain an environment that protects against infection by other microorganisms. Hundreds of commensal species "make a living" on the body without either harming or benefitting it. Consider, for instance, *Staphylococcus epidermidis* and *Pityrosporum ovale* residing in the outer dead regions of the skin, oral microbes feeding on the constant flow of nutrients in the mouth, or the billions of bacteria that occupy the large intestine. Because the normal flora and the body are in a constant state of change, these relationships are not absolute, and a symbiont or commensal could convert to a pathogen by invading body tissues and causing disease.

---

antagonism (an-tag'-oh-nizm) Gr. *antagonistes*, an opponent.
antibiosis (an"-tee-by-oh'-sis) *anti*, against, and *bios*, life.
bacteriocin (bak-teer'-ee-oh-sin) Gr. *bakterion*, little rod, and *ios*, poison.

Feature 6.9 Life Together

Protozoa often contain symbiotic bacteria and algae in their cytoplasm. Not a great deal is known about the contribution of these symbionts, but it is possible that the protozoan derives growth factors from them, which in turn are provided a habitat by the protozoan. One peculiar ciliate propels itself by affixing symbiotic bacteria to its cell membrane to act as "oars." So obligatory is the relationship in some amebas and ciliates that their symbiotic bacteria are absolutely necessary for survival. This kind of relationship is especially striking in the case of the multiple symbiosis of termites, which harbor endosymbiotic protozoans with endosymbiotic bacteria, so that wood eaten by the termite gets processed by the microbes, and all three organisms succeed (sometimes too well).

Symbiosis Between Microbes and Animals Microorganisms carry on symbiotic relationships with animals as diverse as sponges, worms, and mammals. Bacteria and protozoa are essential in the operation of the rumen (a complex, four-chambered stomach) of cud-chewing mammals. These mammals produce no enzymes of their own to break down the cellulose in their diet, but the microbial population harbored in their rumens does. The complex food materials are digested through several stages, during which time the animal regurgitates and chews the partially digested plant matter (the cud) and occasionally burps methane produced by the microbial symbionts.

Thermal Vent Symbionts Another fascinating symbiotic relationship has been found in the deep thermal ridges (vents) in the seafloor where geologic forces spread the crustal plates and release heat and gas. These vents are a focus of tremendous biological and geological activity. Discoveries first made in the late 1970s demonstrated that the basis of the energy chain in this community is not the sun, because the vents are too deep for light to penetrate (2,600 meters). Instead, this ecosystem is based on a massive autotrophic bacterial population that oxidizes the abundant H_2S given off by the volcanic activity there.

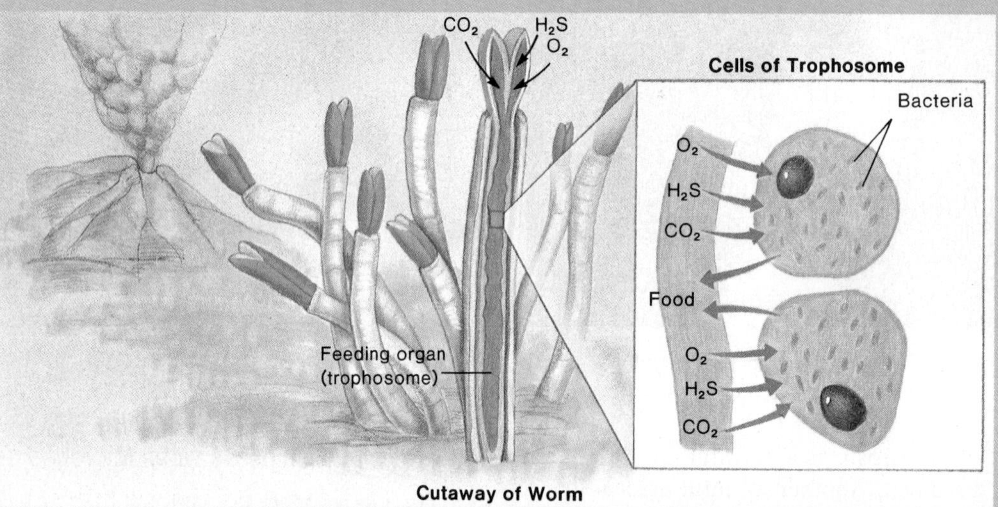

A view of a vent community based on mutualism and chemoautotrophy. The giant tube worm *Riftia* houses bacteria in its specialized feeding organ, the trophosome. Raw materials in the form of dissolved inorganic molecules are provided to the bacteria through the worm's circulation. With these, the bacteria produce usable organic food that is absorbed by the worm.

The Study of Microbial Growth

Microbes that have been provided with essential nutrients and the required environment are metabolically active and grow. Growth takes place on two levels. On one level, a cell builds up protoplasm and increases its size, and on the other level, the number of cells in the population increases. This capacity for multiplication—increasing the size of the population by cell division—has tremendous importance in antimicrobial control, infectious disease, and biotechnology. In the following section we will focus primarily on the characteristics of bacterial growth that are generally representative of microorganisms as a whole.

The Basis of Population Growth: Binary Fission

The division of a bacterial cell occurs mainly through **binary** or **transverse fission**; binary means that one cell becomes two, and transverse refers to the position of the division plane forming across the width of the cell. During binary fission, the parent cell enlarges, duplicates its chromosome, and forms a central transverse septum that divides the cell into two daughter cells. This is repeated at intervals by each new daughter cell in turn, and with each successive round of division, the population increases. The stages in this continuous process are shown in greater detail in figures 6.15 and 6.16.

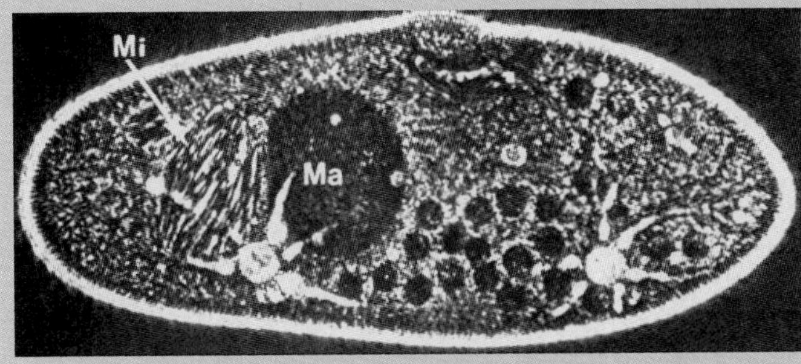

(a)

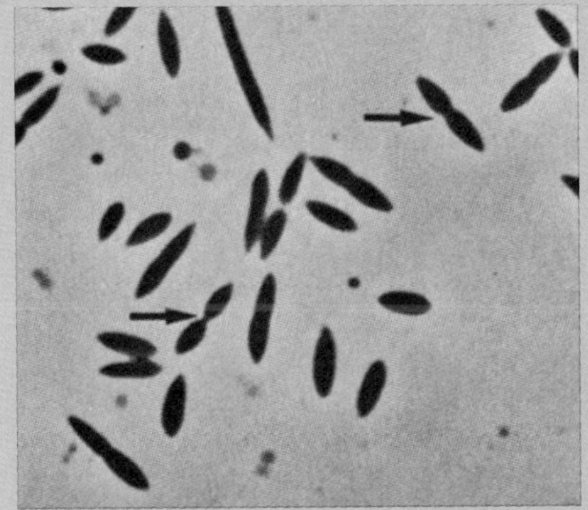

(b)

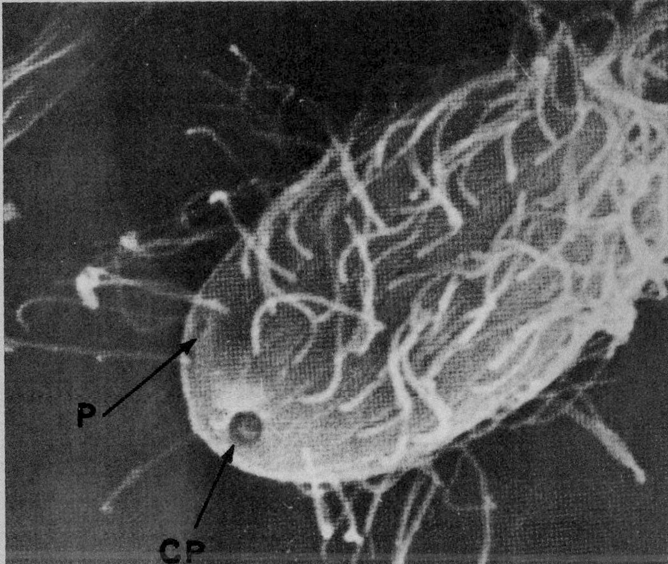

(c)

The endosymbiotic bacteria of *Paramecium*. (*a*) This familiar pond protozoan harbors a bacterium in its micronucleus (Mi) that swells it to 50 times its normal size and makes it as large as the macronucleus (Ma). (*b*) A magnified view of the fusiform-shaped bacteria. (*c*) A novel ciliated protozoan from the rumen of red cattle with unusual feeding structures—the cytopharynx (CP) and the pharyngeal rod (P).

The Rate of Population Growth

The time required for a complete fission cycle—from parent cell to two new daughter cells—is called the **generation** or **doubling time.** The term generation is used somewhat in the same sense as in humans—the period between an individual's birth and his reproductive maturity. In bacteria, it is a doubling process in which each new fission cycle or generation increases the population by a factor of 2. So, the parent stage consists of 1 cell, the first generation consist of 2 cells, the second of 4, the third of 8, then 16, 32, 64, and so on (figure 6.16). As long as the environment remains favorable, this doubling effect can continue at a constant rate, and with the passing of each generation, the population will double, over and over again. The length of the generation time is a measure of the growth rate of an organism (see feature 6.10).

Close analysis of figure 6.16 should make it evident that: (1) The cell population after the first division can be represented by the number 2 with an exponent (2^1, 2^2, 2^3, 2^4); (2) the exponent increases by one in each generation; and (3) the number of the exponent is also the number of the generation. This growth pattern is termed **exponential.** Population growth also conforms to a **geometric progression,** expressed on a graph as a constantly

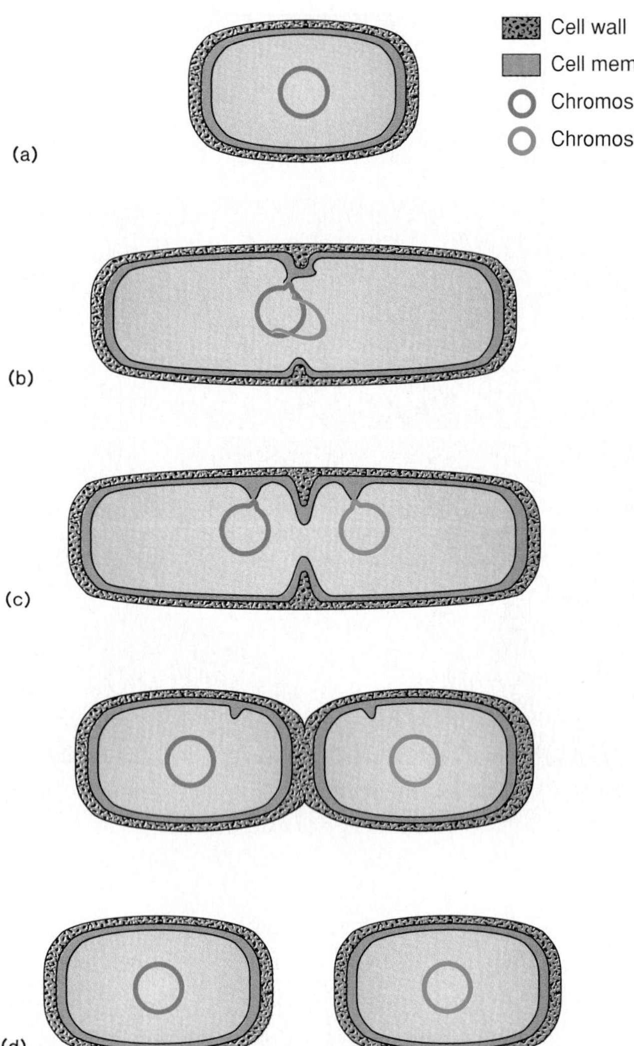

Figure 6.15 Steps in binary fission. (*a*) A parent cell prepares for division by enlarging its cell wall, cell membrane, and overall volume. (*b*) Midway in the cell, the wall develops notches that will eventually form the transverse septum, and the chromosome is duplicated, becoming affixed to mesosomes. (*c*) The wall septum grows inward, and the chromosomes are pulled toward opposite cell ends as the mesosome splits. Other cell components are also distributed (equally) to the two developing cells. (*d*) The septum is synthesized completely through the cell center, and the cell membrane patches itself so that there are two separate cell chambers. At this point the daughter cells are divided. Some species will separate completely as shown here, and others will remain attached.

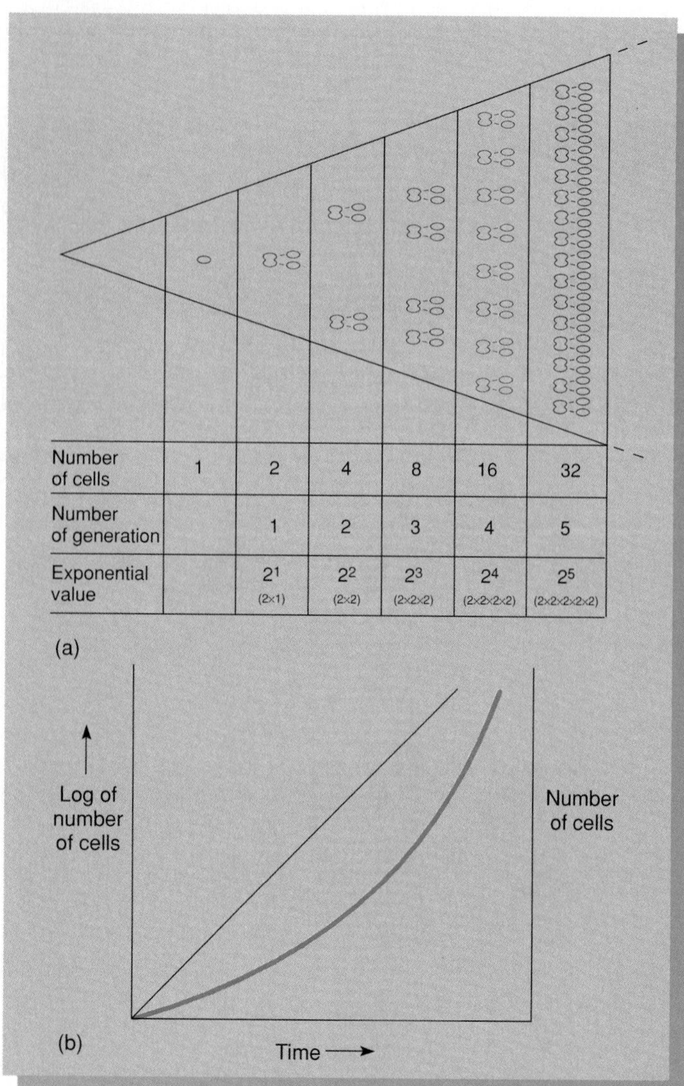

Figure 6.16 The mathematics of population growth. (*a*) Starting with a single cell, if each product of reproduction goes on to divide in a binary fashion, the population doubles with each new division cycle or generation. This process can be represented by simple numbers or by logarithms (2 raised to an exponent). (*b*) Plotting the cell numbers arithmetically gives a curved slope, whereas plotting the logarithm of the cells produces a straight line indicative of exponential growth.

increasing slope, and defined as a sequence of numbers in which each pair of adjacent numbers has the same ratio throughout the sequence, in this case 1 to 2. Because these populations often contain very large numbers of cells, it is useful to express them by means of exponents (see appendix B).

Predicting the number of cells that arise during a long growth period (yielding millions of cells) is based on a relatively simple concept. One could use the method of addition $2 + 2 = 4 + 4 = 8 + 8 = 16 + 16 = 32$, etc., or a method of multiplication (for example, $2^5 = 2 \times 2 \times 2 \times 2 \times 2$), but it is easy to see that for 20 or 30 generations, this calculation could take reams of paper and be very tedious. An easier way to calculate the size of a population over time is to use an equation such as:

$$N_f = (N_i)2^n$$

In this equation N_f is the total number of cells in the population at some point in the growth phase, N_i is the starting number, and 2^n is the number of generations. If we know any two of the values, the other values can be calculated. Let us use the example of *Staphylococcus aureus* in feature 6.10 to calculate how many cells (N_f) will be present in an egg salad sandwich

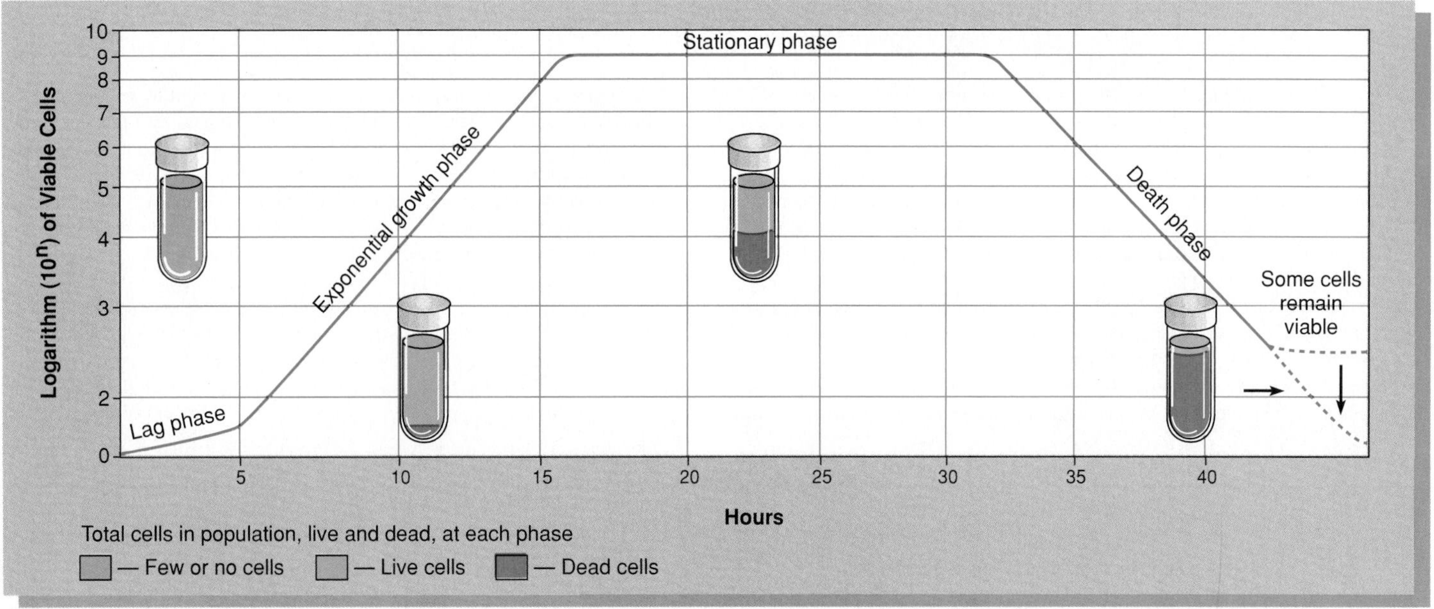

Figure 6.17 The growth curve. On this graph, the number of viable cells expressed as a logarithm (log) is plotted against time. See text for discussion of the various phases. Note that with a generation time of 30 minutes, the population has risen from 10 (10^1) cells to 1,000,000,000 (10^9) cells in only 16 hours. Cells are multiplying and dying all along the curve, but the relative rates of these events change as the curve proceeds.

Feature 6.10 How Fast Do Bacteria Grow?

Compared to the growth rates of most other living things, bacteria are notoriously speedy. The average generation time is 30–60 minutes under optimum conditions. At the extremes are species with very rapid growth rates and short generation times (5–10 minutes) and those with very slow growth rates and long generation times (measured in days). The longest generation times occur in species of *Mycobacterium,* particularly *M. leprae,* the cause of leprosy (10–30 days—as long as in some animals). Most pathogens have relatively shorter doubling times. *Salmonella enteriditis* and *Staphylococcus aureus,* both causes of food poisoning, double in 20 to 30 minutes, which is why leaving food at room temperature for "just a while" has caused many a person to be stricken by a nasty case of food poisoning. In a few hours, a population of these bacteria could easily grow from a few cells to a few million cells.

after it sits in a warm car for 4 hours. We will assume that N_i is 10 (number of cells deposited in the sandwich while being prepared). To derive *n,* we need to divide 4 hours (240 minutes) by the generation time (we will use 20 minutes). This comes out to 12, so 2^n is equal to 2^{12}. Referring to a table on the powers of numbers (available in various handbooks), 2^{12} is found to be 4,096.

$$\text{Final number} = 10 \times 4{,}096$$

$$= 40{,}960 \text{ cells in the sandwich}$$

This same equation, with modifications, is used to determine the generation time, a more complex calculation that requires knowledge of the number of cells at the beginning and end of a growth period. Such data are obtained through actual testing by a method discussed in the following section.

Aspects of Population Growth

In reality, a population of bacteria does not maintain its potential growth rate and does not double endlessly, because in most systems, numerous factors prevent the cells from dividing at their maximum rate. Quantitative laboratory studies indicate that a population typically displays a predictable pattern or *growth curve.*

The method commonly used to observe the population growth pattern is a viable count technique, in which the total number of live cells is counted over a period of time. In brief, this method entails the following sequence of events: (1) placing a tiny number of cells into a sterile liquid medium; (2) incubating this culture over a period of several hours; (3) sampling the broth at regular intervals during incubation; (4) plating each sample onto solid media; and (5) counting the number of colonies present after incubation. Feature 6.11 gives the details of this process.

Stages in the Normal Growth Curve

The system of batch culturing described in feature 6.11 is *closed,* meaning that nutrients and space are finite and there is no mechanism for the removal of waste products. The plotting of data from this type of system yields a curve with a series of phases termed the lag phase, the exponential (log) growth phase, the stationary phase, and the death phase (figure 6.17).

Feature 6.11 Steps in a Viable Plate Count—Batch Culture Method

To establish a growing population, a flask containing a known quantity of sterile liquid medium is inoculated with a few cells of a pure culture. The flask is placed in an incubator that favors the optimum rate of growth for this bacterium, and timing is begun. To assess the population size at any point in the growth cycle, a tiny measured sample of the culture is removed from the growth chamber and plated out on a solid medium to develop isolated colonies. This procedure is repeated at evenly spaced intervals (every hour for 24 hours).

Evaluating the samples involves a principle that is common and important in microbiology—one colony on the plate represents one cell or colony-forming unit (CFU) from the original sample. Since the CFU of some bacteria is actually composed of several cells (consider the clustered arrangement of *Staphylococcus,* for instance), using the colony count to estimate population size may underestimate the exact count by several orders. This is not a serious problem in that the CFU is the smallest functional unit of colony formation and dispersal in these bacteria. Multiplication of the number of colonies in a single sample by the container's volume gives a fair estimate of the total population size (number of cells) at any given point. To arrive at the growth curve, the number for each sample is graphed in sequence for the whole incubation period (see figure 6.17).

Because of the scarcity of cells in the early stages of growth, some samples may give a zero reading even if there are viable cells in the culture, and the sampling itself may remove enough viable cells to prejudice the tabulations. However, since the purpose here is to compare relative trends in growth, these factors do not greatly detract from the overall pattern.

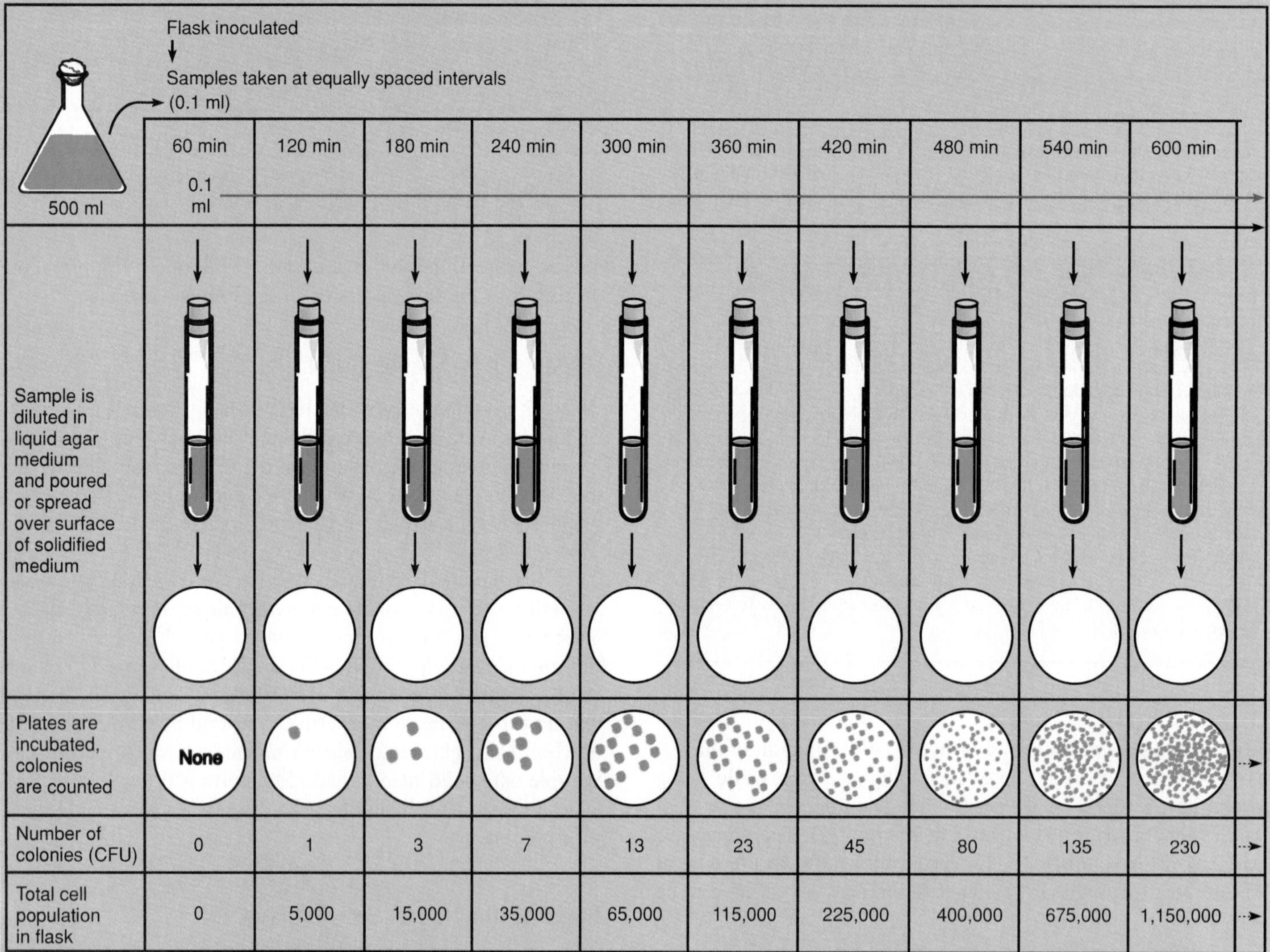

The viable plate count, a technique for determining population size and rate of growth.

The **lag phase** is a relatively "flat" period on the graph when the population appears not to be growing. It occurs because: (1) The newly inoculated cells require a period of adjustment, enlargement, and synthesis; (2) the cells are not yet multiplying at their maximum rate; and (3) the population of cells is so sparse or dilute that the sampling misses them. In time, the cells enter a period of maximum activity termed the **exponential growth phase,** a period during which the curve increases geometrically. This phase will continue as long as cells have adequate nutrients and the environment is favorable.

The **stationary growth phase** is marked by a decline in the growth rate and an increase in the death rate, caused by less optimal conditions such as depleted nutrients, excretion of organic acids and other biochemical pollutants into the growth medium, and an increased density of cells. Because cells are dying at the same overall rate as they are multiplying, the curve levels off. As the limiting factors intensify, cells begin to die in exponential numbers (literally perishing in their own wastes), or they are unable to multiply. The curve now dips downward as the **death phase** begins. How rapidly death occurs depends on the relative resistance of the species and how toxic the conditions are, but it is usually slower than the exponential growth phase. Viable cells often remain many weeks and months after this phase has begun. In the laboratory, refrigeration is used to slow the progression of the death phase.

Applications of the Growth Curve The tendency for populations to exhibit phases of rapid growth, slow growth, and death has important implications in microbial control, infection, food microbiology, and cultural technology. Although antimicrobial control agents rapidly accelerate the death phase in all populations, microbes in the exponential growth phase are far more susceptible to heat and disinfectants than those in the stationary or death phase because young, active cells are more sensitive.

Growth patterns in microorganisms can also account for the stages of infection (see chapter 11). Living microbes released during the exponential phase may be far more numerous and virulent than those released at a later stage of infection. So a person shedding cold virus in the early and middle stages of a cold by coughing and sneezing is more likely to spread infection. The course of an infection is also influenced by the relatively faster rate of multiplication of the microbe, which can overwhelm the slower growth rate of the host's own cellular defenses.

Understanding the stages of cell growth is crucial for work with cultures. For example, beginning students may reincubate a culture that has achieved maximum growth under the mistaken impression that nutrients are still present and that the microbe can continue to multiply. In most cases, it is unwise to continue incubating a culture beyond its exponential or stationary phase, as this will reduce the number of viable cells, and the culture could die out completely. It is preferable to do stains (an exception is the spore stain) and motility tests on young cultures, because the cells will show their natural size and correct reaction, and motile cells will have functioning flagella.

Other Methods of Measuring Population Growth

Turbidometry

Microbiologists have developed several alternate ways of analyzing bacterial growth qualitatively and quantitatively. One of the simplest methods for estimating the size of a population is through turbidometry. This technique relies on the simple observation that a tube of clear nutrient solution loses its clarity and becomes cloudy or **turbid** as microbes grow in it. In general, the greater the turbidity, the larger the population size, and this can be measured by means of sensitive instruments (figure 6.18).

Enumeration of Bacteria

Turbidity readings are useful for comparative growth studies, but if a more quantitative evaluation is required, the viable count described previously or some other enumeration (counting) procedure is necessary. The **direct** or **total cell count** involves counting the number of cells in a sample microscopically (figure 6.19*a*). This technique, very similar to that used in blood cell counts, employs a special microscope slide (cytometer) calibrated to accept a tiny sample that is spread over a premeasured grid. The cell count from a cytometer can be used to estimate the total number of cells in a larger sample (say, of milk or water). This method has one inherent inaccuracy: Because no distinction can be made between dead and live cells, both types are included in the count. Counting may be automated by sensitive devices (the Coulter counter or flow cytometer), which focus light into a tiny opening and electronically count each cell as it flows by (figure 6.19*b*). In some studies of population growth, a four-phase growth curve obtained from batch culturing is undesirable. The alternative is a continuous culture device, the *chemostat,* which stabilizes the growth rate and cell number, and can admit a steady stream of new nutrients and siphon off used media and old bacterial cells.

Media: Providing Nutrients in the Laboratory

A major stimulus to the rise of microbiology 100 years ago was the development of techniques for culturing microbes out of their natural habitats and in pure form in the laboratory. This milestone enabled the close examination of a microbe and its morphology, physiology, and genetics. It was evident from the very first that for successful cultivation, each microorganism had to be provided with all of its required nutrients. Such artificial nutrient preparations are called **media** (sing. medium).

We have seen how the nutritional requirements of microbes vary from a few very simple inorganic compounds to a complex list of specific inorganic and organic compounds. This tremendous diversity is also evident in the types of media that can be prepared. The *Difco Manual*[4] lists approximately 500 entries for media used in modern microbiology laboratories. Although most modern media are of the "instant" variety, prepared from dehydrated commercial powders, this was not true during the pioneering days of microbiology (see feature 6.12).

---

4. Difco, the Digestive Ferments Company, started in 1895 as a supplier of prepared artificial media.

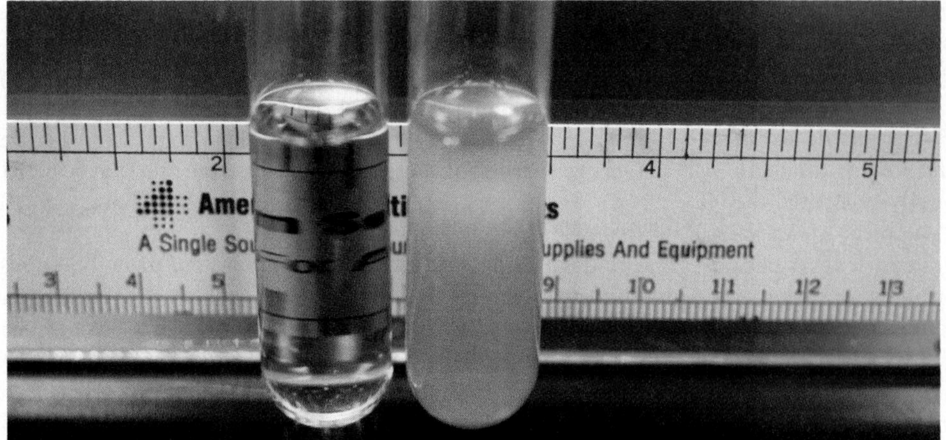

(a)

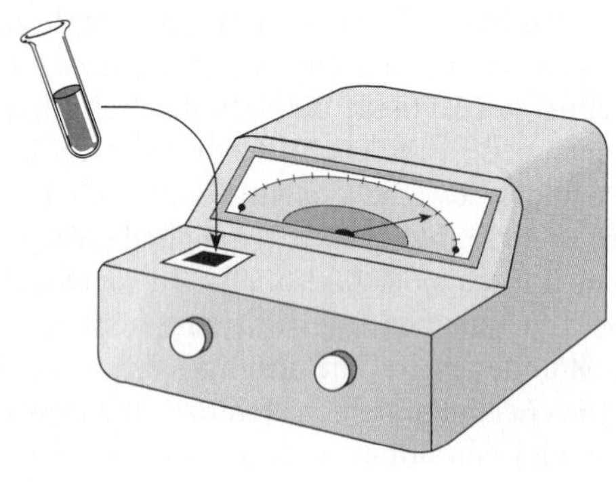
(b)

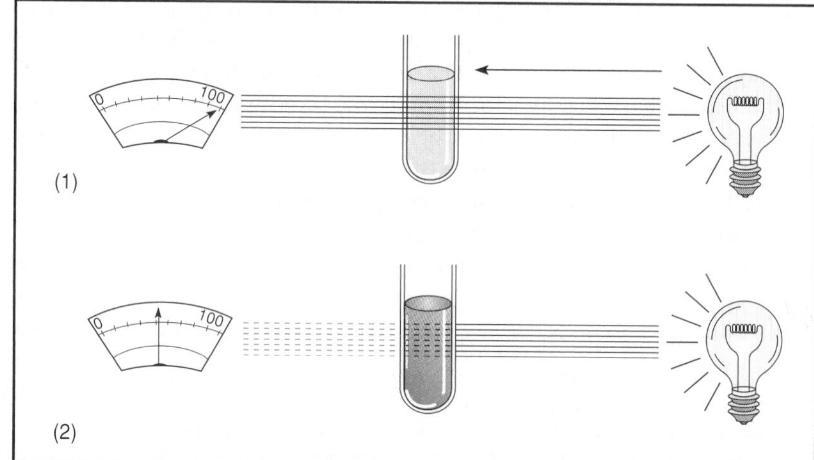

Figure 6.18 Turbidity measurements as indicators of growth. (*a*) Holding a broth to the light is one method of checking for gross differences in cloudiness (turbidity). The broth on the left is transparent, indicating no growth; the broth on the right is cloudy and opaque, indicating heavy growth. (*b*) Since the eye is not sensitive enough to pick up fine degrees in turbidity, more sensitive measurements may be made with a spectrophotometer. (*1*) A tube with less growth transmits more light and gives a higher reading, and (*2*) a tube with more growth transmits less light and gives a lower reading. This technique provides only a relative measure of growth; it cannot determine actual numbers or differentiate between dead and live cells.

Feature 6.12 The Way It Was: Media

With the ready availability of commercial products, media preparation in these times is quite straightforward and simple. But this was not the case many years ago. A media prep room might have looked like a cross between a kitchen and an alchemist's lab. A standard book reviewing culture media up to 1930* compiled 7,000 different recipes. Thumbing through this book, one gets a clear idea of the ingenuity and care that media formulators and preparers brought to their task. A fantastic assortment of raw materials were (and occasionally still are) added to preparations, and some formulae were extraordinarily complex. It was not unusual for a single medium to require 25 steps in preparation. The media compositions seem only to have been limited by the imagination of the "cook." Numerous unconventional animal products appear among the recipes: bone, cartilage, placenta, lung, lard, chopped beef, testicles, sperm, pus, feces, and urine. Plant materials, also a good source of nutrients, were ground, mashed, chopped, pureed, filtered, and cooked. A partial list reads like a garden: potatoes, carrots, spinach, onions, peas, fruits (bananas, prunes), grasses (straw, hay), and dozens of different seeds. Foods such as tomato juice, macaroni, coffee, spices, butter, and beer had their functions in certain concoctions. No conceivable source was overlooked, including such curious ingredients as soil, manure, brick, coal, sawdust, wood ashes, lint, rusty nails, asbestos, arsenic, and cyanide!

**A Compilation of Culture Media for the Cultivation of Microorganisms* (1930), Levine and Shoenlein.

Types of Media

Media may be classified on three primary levels: (1) physical form, (2) chemical characteristics, and (3) functional type (table 6.6).

Table 6.6 General Categories and Subcategories of Media

Physical State (Medium's Normal Consistency)	Chemical Composition (Type of Chemicals Medium Contains)	Functional Type (How Medium Is Used)
1. Liquid	1. Synthetic (chemically defined)	1. General purpose
2. Semisolid	2. Nonsynthetic (not chemically defined)	2. Enrichment
3. Convertibly solid		3. Selection
4. Permanently solid		4. Differentiation
		5. Anaerobic growth
		6. Specimen transport
		7. Assay
		8. Enumeration

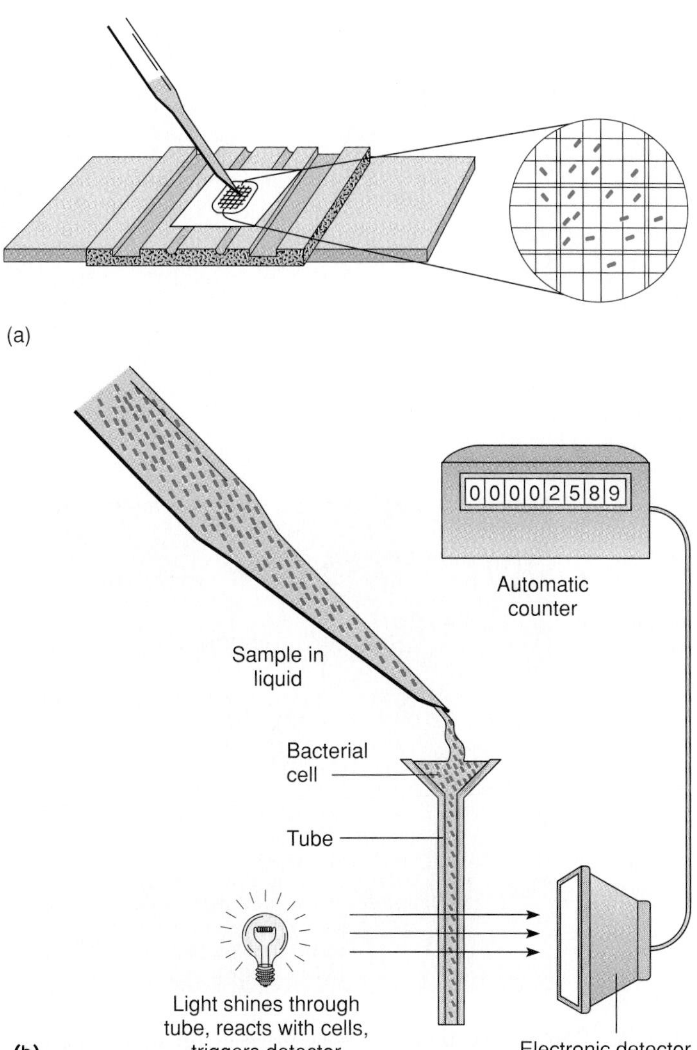

Figure 6.19 Direct counts of bacteria (*a*) Direct microscopic count. A small sample (0.02 mm²) is placed on the grid under a slide. Individual cells, both living and dead, are counted. (*b*) Coulter counter. As cells pass through this device, they trigger an electronic sensor that tallies their numbers. A variation on this machine, called a flow cytometer, may be used to differentiate between types of cells and particles as well.

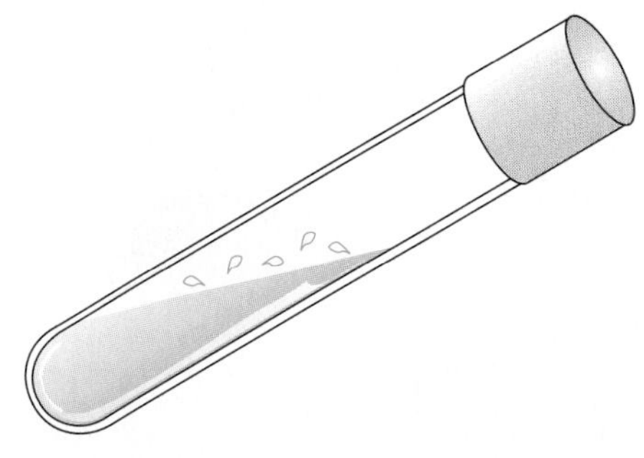

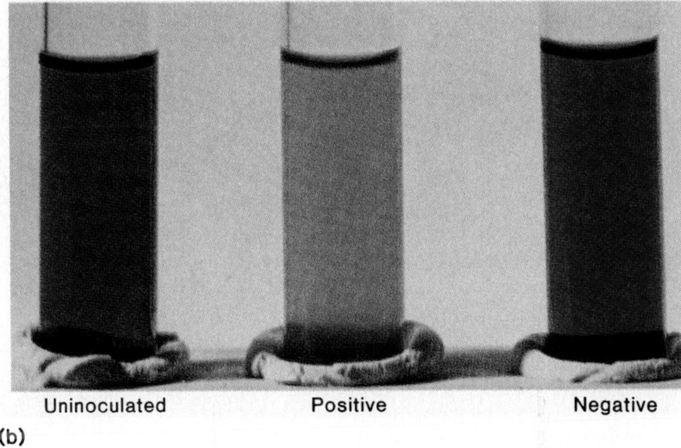

Figure 6.20 (*a*) Liquid media tend to flow freely when the container is tilted. (*b*) *Enterococcus faecalis* broth, used to identify this species.

Physical States of Media

Liquid media are defined as aqueous formulations that do not gel or solidify at temperatures above freezing and that tend to flow freely when the container is tilted (figure 6.20). These media, termed broths, milks, or nutrient solutions, are made by dissolving various solutes in distilled water. Growth occurs throughout the container and may present a dispersed cloudy or particulate appearance.

Examples of Liquid Media Sterile broths tend to be clear and transparent or translucent. A common laboratory medium, nutrient broth, contains beef extract and peptone dissolved in water. Methylene blue milk and litmus milk contain dyes and low-fat milk that render them relatively opaque. Fluid thioglycollate, a medium used for culturing anaerobes, contains a tiny amount of a thickening agent, but since it remains liquid, it is still a broth.

Semisolid media are not liquid at ordinary room temperature and exhibit a gelatinous consistency (figure 6.21). This is because they contain an amount of solidifying agent (agar or gelatin) that hardens them slightly but does not produce a firm substrate. Semisolid media are used to restrict slightly the movement of motile microbes, to grow microaerophiles or anaerobes, and to localize a reaction at a specific site. When heated, semisolid media liquefy.

Examples of Semisolid Media Motility test medium and SIM contain a small amount (0.3–0.5%) of **agar-agar** (see feature 6.13). The medium is stabbed carefully in the center and later observed for the pattern of growth around the stab line. In addition to motility, SIM is used to test for physiological characteristics (hydrogen sulfide production and indole reaction). Fermentation media such as cystine trypticase agar (CTA) and transport media are also semisolid.

Solid media that provide a firm surface on which cells can form discrete colonies are vital in isolating and subculturing bacteria and fungi (see chapters 1 and 3). They come in two forms: liquefiable and nonliquefiable. **Liquefiable solid media,** sometimes called reversible solid media, contain a solidifying agent that is **thermoplastic**—its physical properties change in

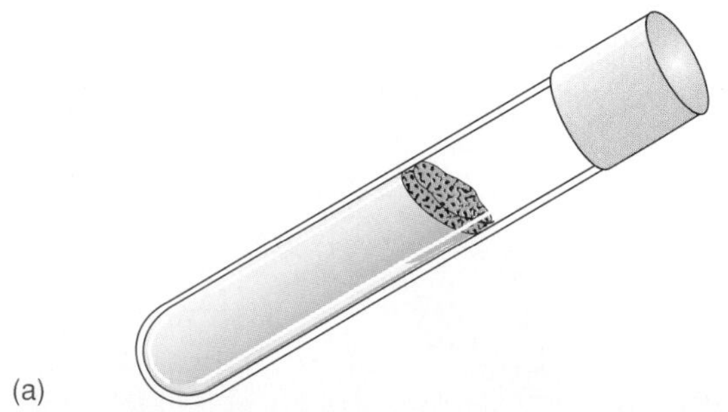

Figure 6.21 (*a*) Semisolid media have more body than liquid media but less body than solid media. They do not flow freely and have a soft, clotlike consistency. (*b*) Sulfur indole motility (SIM) medium. The growth patterns that develop in this medium may be used to determine various characteristics. The tube on the left shows no motility; the tube in the center shows motility; the tube on the right shows both motility and production of H_2S gas (black precipitate).

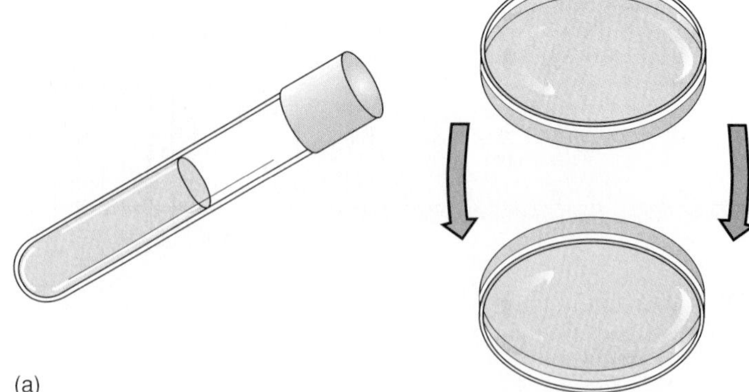

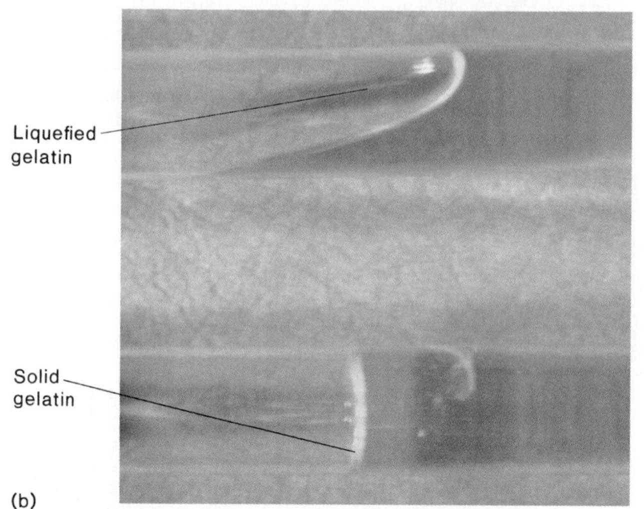

Figure 6.22 Solid media that are reversible to liquids. (*a*) Media containing 1–5% agar-agar are solid enough to remain in place when containers are tilted or inverted. They are liquefiable by heat but generally not by bacterial enzymes. (*b*) Nutrient gelatin contains enough gelatin (12%) to take on a solid consistency. Unfortunately, it may be liquefied by bacterial enzymes.

response to temperature. By far the most widely used and effective of these agents is **agar-agar** (usually shortened to agar), a complex polysaccharide isolated from the red alga *Gelidium* (see feature 6.13). The benefits of agar are numerous. It is solid at room and incubation temperature, and it melts (liquefies) at the boiling temperature of water (100°C). Once liquefied, agar does not resolidify until it cools to 42°C, allowing it to be inoculated and poured in liquid form at temperatures (45°–50°C) that will not harm the microbes or the handler. Agar is flexible and moldable, and it provides a basic framework to hold moisture and nutrients, though it is not itself a usable nutrient for the vast majority of microorganisms.

Examples of Liquefiable Solid Media The list of agars is very extensive. Any medium containing 1% to 5% agar usually has the word agar in its name. Nutrient agar is a common one. Like the broth, it contains beef extract and peptone, as well as 1.5% agar by weight. Many of the examples covered in the section on functional categories of media contain agar. Although gelatin is not nearly as satisfactory as agar, it will create a reasonably solid surface in concentrations of 10% to 15% (but it probably won't remain solid; see feature 6.13). Agar and gelatin media are illustrated in figure 6.22.

agar-agar (ah′-gur-ah′-gur) The Malay word for this seaweed product.

Nonliquefiable solid media that are not thermoplastic include materials such as rice grains (used to grow fungi), cooked meat media (good for anaerobes), and potato slices; all of these media start out solid and remain solid after heat sterilization. Egg media and some media containing serum start out liquid and are permanently coagulated or hardened by moist heat. Because these media are less versatile, they have fewer applications than agar media.

Chemical Content of Media

Media whose compositions are chemically defined are termed **synthetic.** Such media contain highly pure organic and inorganic compounds that do not vary from one source to another and have a molecular content specified by means of an exact formula. Synthetic media come in many forms. Some media for autotrophs contain nothing more than a few very simple inorganic compounds dissolved in water (see table 6.1). Others contain a variety of defined organic and inorganic chemicals (see table 6.7). Such standardized and reproducible media are most

Feature 6.13 Microbiology Discovers Agar-agar

Early bacteriologists recognized the advantages of growing bacteria in discrete colonies rather than solely in broths and infusions. For this, they needed a surface that was relatively solid and provided nutrients. Two old standbys were sterilized potato slices and gelatin, but these left much to be desired. Potatoes had a limited nutrient content, and gelatin lost its solidity in warmer temperatures and could be digested by many microorganisms. About 110 years ago, this problem was solved by a fortunate cultural exchange (of the human kind). The German physician and avid bacteriologist Walther Hesse had been studying airborne microbes and was frustrated by the messy outcome of gelatin cultures. Understanding his need for a gelatin substitute, his wife Fanny suggested that he try a substance her mother had discovered through Dutch friends from Java. This material, agar-agar, had traditionally been used by Malaysians to stabilize jellies and thicken soup, and it was a nearly perfect solidifying agent. Agar-agar could be liquefied and sterilized without adverse effect. Also, it was solid at room and incubation temperatures, very resistant to digestion, and relatively translucent. The success Hesse found with agar-agar quickly caught on among bacteriologists.

Dr. Walther and Frau Fanny Hesse.

useful in research and cell culture, where the exact nutritional needs of the test organisms are known. If even one component of a given medium is not chemically definable, the medium belongs in the next category. Many pathogens will not grow on synthetic media because they require more complex nutrients than can be provided by them.

Nonsynthetic or empirical media contain at least one ingredient that is *not* chemically definable—not a simple, pure compound and not representable by an exact chemical formula. Most of these substances are extracts of animals, plants, or yeasts, including such materials as ground-up cells, tissues, and secretions. Examples are blood, serum, and meat extracts or infusions. Infusions, made by soaking fresh meat in water and salvaging the liquid, are high in vitamins, minerals, proteins, and other organic nutrients. Other undefined ingredients are milk, yeast extract, soybean digests, complex plant carbohydrates, and peptone. Peptone is a partially digested protein, rich in amino acids, that is often used as a carbon and nitrogen source. Nutrient broth, blood agar, MacConkey agar, and mitis-salivarius

Table 6.7 Medium for the Growth and Maintenance of the Green Alga *Euglena*

Ingredient	Amount
Glutamic acid (aa)	6 g
Aspartic acid (aa)	4 g
Glycine (aa)	5 g
Sucrose (c)	30 g
Malic acid (oa)	2 g
Succinic acid (oa)	1.04 g
Boric acid	1.14 mg
Thiamine HCl (v)	12 mg
Monopotassium PO_4	0.6 g
Magnesium sulfate	0.8 g
Calcium carbonate	0.16 g
Ammonium carbonate	0.72 g
Ferric chloride	60 mg
Zinc sulfate	40 mg
Manganese sulfate	6 mg
Copper sulfate	0.62 mg
Cobalt sulfate	5 mg
Ammonium molybdate	1.34 mg

Note: These ingredients are dissolved in 1,000 ml of water. (aa) = amino acid; (c) = carbohydrate; (oa) = organic acid; (v) = vitamin; g = gram; mg = milligram.

agar, though very different in function and appearance, are all nonsynthetic. Nonsynthetic media present a rich mixture of nutrients for pathogens and other heterotrophs with complex nutritional needs.

Distinguishing Synthetic from Nonsynthetic Let us use a specific example to compare what differentiates a synthetic medium from a nonsynthetic one. The synthetic *Euglena* medium contains three known amino acids in known amounts, while nonsynthetic nutrient broth and MacConkey agar both contain peptone and, consequently, amino acids, but exactly which ones and in what quantities are unknown and variable. Inorganic salts and organic acids are added in pure salt form for *Euglena*, but these components are provided by undefined beef or yeast extract in many nonsynthetic media.

Functional Categories of Media

Microbiologists have many types of media at their disposal, with new ones being devised all the time. Depending upon what is added, a microbiologist can fine-tune a medium for nearly any purpose. As a result, only a few species of bacteria or fungi cannot yet be cultivated artificially. Media are used for primary isolation, to maintain cultures in the lab, to determine biochemical and growth characteristics, and for numerous other functions. Note that a single medium may serve multiple purposes; for example, it may be general purpose and enriched, selective and differential, or enriched and differential.

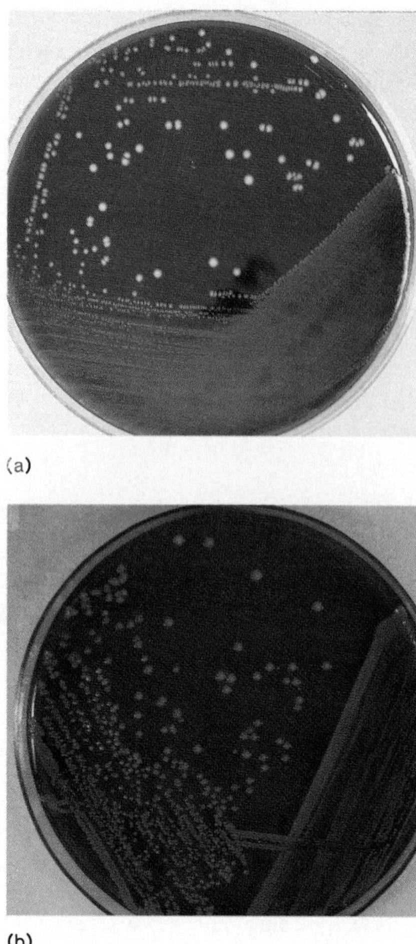

Figure 6.23 Examples of enriched media. (*a*) Blood agar growing *Enterococcus faecalis,* a common fecal bacterium. (*b*) Chocolate agar, a medium that gets its brown color from heated blood, not chocolate. It is commonly used to culture the fastidious gonococcus *Neisseria gonorrhoeae.*

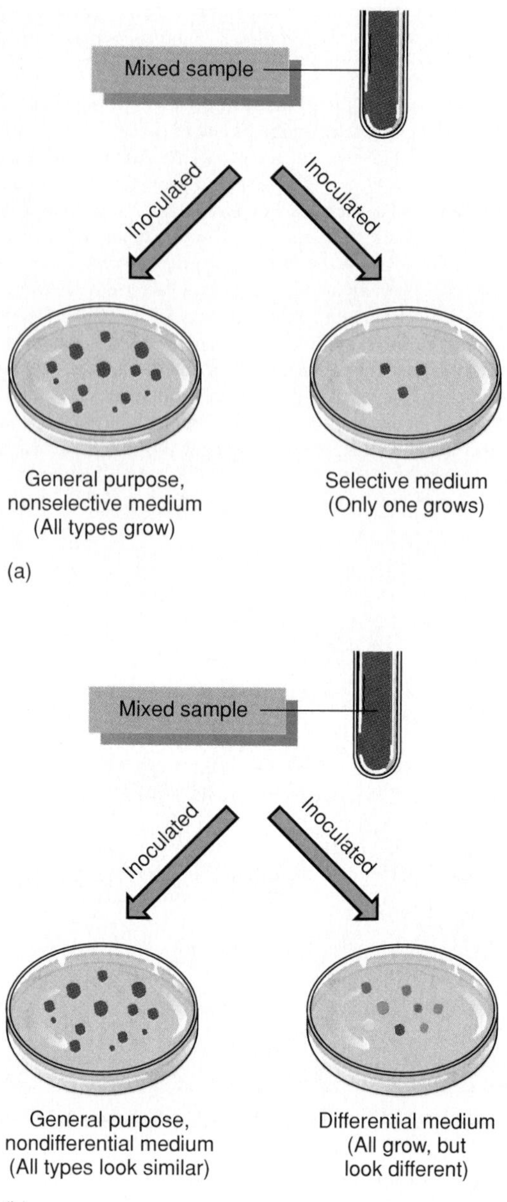

Figure 6.24 Comparison of selective and differential media with general purpose media. (*a*) The same mixed sample containing five different species is streaked onto plates of general purpose (nonselective medium) and selective medium. Note the results. (*b*) Another mixed sample containing three different species is streaked onto plates of general purpose (nondifferential medium) and differential medium. Note the results.

General Purpose Media

General purpose media are designed to grow as broad a spectrum of microbes as possible. As a rule, they are nonsynthetic and contain a mixture of nutrients that could support the growth of pathogens and nonpathogens alike. Examples include nutrient agar and broth, blood agar, brain-heart infusion, stock culture agar, and trypticase soy agar (TSA). TSA contains partially digested milk protein (casein), soybean digest, NaCl, and agar.

Enriched Media

An **enriched medium** contains basic nutrients similar to general purpose media, but it is enriched with blood, serum, hemoglobin, or individual growth factors needed to culture species that would ordinarily live on a host. Such fastidious species as *Streptococcus pneumoniae* are often grown on blood agar, which is made by adding sheep, horse, cow, or rabbit blood (taken aseptically) to a sterile isotonic agar base (figure 6.23*a*). Pathogenic *Neisseria* species are grown on Thayer-Martin medium or chocolate agar, which is essentially cooked blood agar (figure 6.23*b*). The requirements of *Hemophilus influenzae* for the growth factors discussed earlier in this chapter are met by Fildes medium, which contains digested sheep blood.

Selective and Differential Media

Some of the cleverest and most inventive media belong to the categories of selective and differential media (figure 6.24). Because these media are designed for special microbial groups, they have extensive applications in isolation and identification.

A **selective medium** (table 6.8) contains an agent (or more than one) that inhibits the growth of a certain microbe or mi-

Medium	Selective Agent	Used For
Mueller tellurite	Potassium tellurite	Isolation of *Corynebacterium diphtheriae*
Mitis-salivarius agar	Tellurite, crystal violet	Isolation of oral streptococci
Azide glucose broth	Sodium azide	Isolation and identification of streptococci
Phenylethanol agar	Phenylethanol	Isolation of staphylococci and streptococci
Tomato juice agar	Tomato juice, acid	Isolation of lactobacilli from saliva
Desoxycholate citrate agar	Desoxycholate, citrate	Isolation of gram-negative enterics
SS agar	Bile, citrate, brilliant green	Isolation of *Salmonella* and *Shigella*
Lowenstein-Jensen	Malachite green dye	Isolation and maintenance of *Mycobacteria*
Emmon's agar	Chloramphenicol (antibiotic)	Isolation of fungi—inhibits bacteria
Sabouraud's agar	pH of 5.6 (acid)	Isolation of fungi—inhibits bacteria

Table 6.8 Selective Media, Agents, and Functions

crobes (A, B, C) but not others (D), and thereby encourages or *selects* for growth of group D. Selective media are very important in primary isolation of a specific type of microorganism from samples containing a highly mixed population—for example, feces, saliva, skin, water, and soil. These media hasten isolation by suppressing the background organisms and favoring growth of the desired ones. They sometimes permit, in a single step, the preliminary identification of a genus or even a species.

Examples of Selective Media Mannitol salt agar (MSA) contains a concentration of NaCl (7.5%) that is quite inhibitory to most human pathogens. One exception is the genus *Staphylococcus*, which grows well in this medium and may consequently be isolated from infections or the environment in a single step. Although bile salts, a component of feces, are inhibitory to most gram-positive bacteria, they permit the growth of many gram-negative rods. Media for isolating enteric pathogens (MacConkey agar, eosin-methylene blue [EMB] agar) contain these salts as a selective agent. Dyes such as methylene blue and crystal violet also inhibit certain gram-positive bacteria. Other agents that have selective properties are antibiotics, acid, and oxygen (added or removed). (See also mitis-salivarius agar, figure 6.25.)

In the usual sense, an **enrichment medium** (not the same as enriched) is an extremely selective medium used primarily to favor the growth of a pathogen that would otherwise be overlooked because of its low numbers in a specimen. The enrichment agent inhibits nearly all competing organisms. Selenite and brilliant green dye are used in media to isolate *Salmonella* from feces, and potassium tellurite enriches the growth of *C. diphtheriae* from throat cultures.

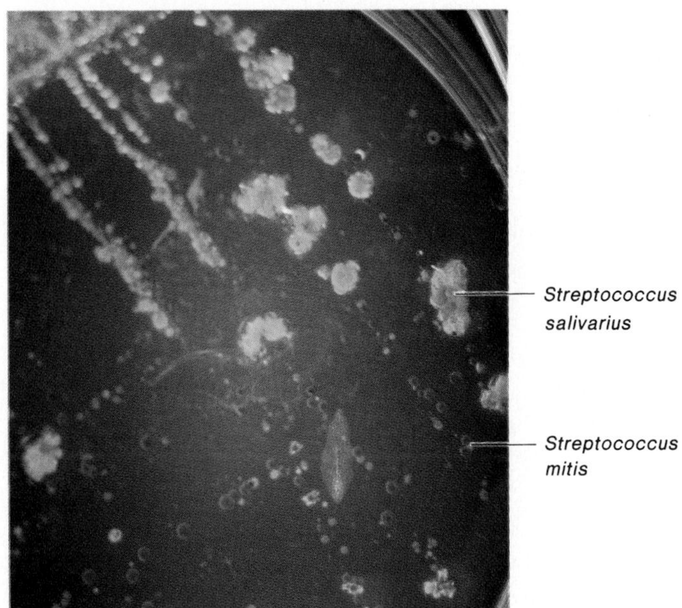

Figure 6.25 Example of a selective medium. Mitis-salivarius agar contains a selective agent called tellurite. If a saliva sample is streaked on this medium, only oral streptococci will grow; other oral microbes are inhibited. (Note that this medium is also differential.)

A **differential medium** can grow several types of microorganisms, but by virtue of their differing appearance on this medium, the eye can distinguish one microorganism from another. Differentiation manifests itself as variations in colony size and color, in media color changes, and in the formation of gas bubbles and precipitates. What causes these differences in appearance? The answer lies in the nature of the agents added to the medium. For example, when a substance is visibly altered by organism A but not by organism B, then A will appear different from B in the presence of that substance. The simplest differential media differentiate two types—a bacterium that acts on a sugar versus one that does not, or a dye that colors one type of colony and does not affect other colonies (table 6.9). Some media are sufficiently complex to show three or four different reactions (for example, TSIA agar).

Examples of Differential Media Dyes may serve as differential agents by changing color in response to alterations in pH (figure 6.26*a*). Such pH indicators are often used in differential media to demonstrate the production of acid or base. For example, MacConkey agar contains lactose and neutral red, a dye that is yellow when neutral and pink or red when acidic. A bacterium like *E. coli* that ferments lactose and produces acid develops red to pink colonies, and one like *Salmonella* that does not ferment lactose remains its natural color (off-white) (figure 6.26*b*). Mitis-salivarius agar differentiates oral *Streptococcus* species by means of sucrose. Whereas *S. salivarius* can use the sucrose to synthesize a slime layer and produce large "gumdrop" colonies, *S. mitis* does not do this and remains tiny and flat.

Medium	Contains	Differentiates Between
Blood agar	Intact red blood cells	Types of hemolysis
Mannitol salt agar	Mannitol and phenol red	Species of *Staphylococcus*
EMB agar	Eosin, methylene blue, lactose	*E. coli,* other lactose fermenters, and lactose nonfermenters
Simmons citrate	Citrate, bromthymol blue	Bacteria that can use citrate as a sole carbon source from those that cannot
SIM	Thiosulfate, iron	H_2S gas producers from nonproducers
TSI agar	Triple sugars, iron, and phenol red dye	Fermentation of sugars, H_2S production
XLD agar	Lysine, xylose, iron, thiosulfate, phenol red	*Enterobacter, Escherichia, Proteus, Providencia, Salmonella,* and *Shigella*
Birdseed agar	Special birdseed	*Cryptococcus neoformans* and other fungi

Table 6.9 Differential Media

Miscellaneous Media

A *reducing medium* contains a substance that absorbs oxygen or slows the penetration of oxygen in a medium, thus reducing the reduction-oxidation potential. Examples of reducing agents are thioglycollic acid, ascorbic acid, and cystine. Reducing media are important for growing anaerobic bacteria or determining oxygen requirements (see figure 6.12). *Carbohydrate fermentation media* contain sugars that can be converted to acids (fermented) and a pH indicator to show this reaction (figure 6.27). Media for determining other biochemical reactions number in the hundreds and collectively provide the basis for identifying many bacteria and fungi.

Transport media are used to maintain and preserve specimens that have to be shipped or held for a period of time prior to clinical analysis, or for delicate species that die rapidly if not held under stable conditions. Stuart's and Amies transport media contain salts, buffers, and absorbants to prevent cell destruction by enzymes, pH changes, and toxic substances, but they do not contain nutrients to support growth. *Assay media* are used by technologists to test the effectiveness of antimicrobial drugs (see chapter 10) and by drug manufacturers to assess the effect of disinfectants, antiseptics, cosmetics, and preservatives on the growth of microorganisms. *Enumeration media* are used by industrial and environmental microbiologists to count the numbers of organisms in milk, water, food, soil, and other samples.

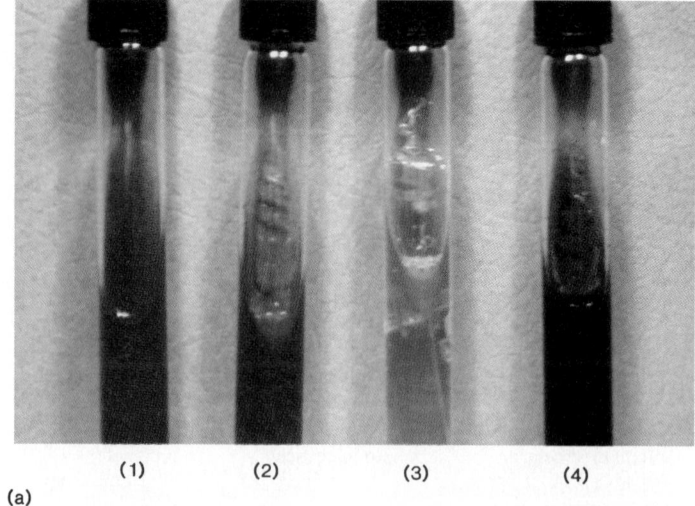

(a)

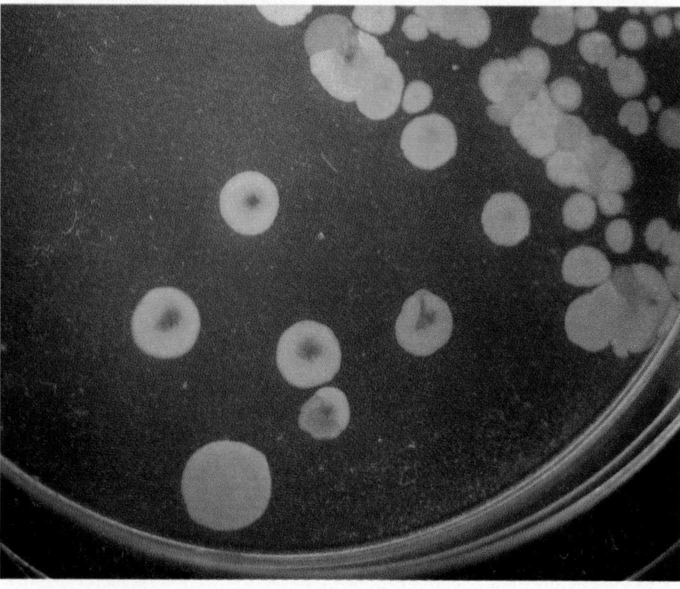

(b)

Figure 6.26 Media that differentiate characteristics. (*a*) Triple-sugar iron agar (TSIA) in a slant. This medium contains three fermentable carbohydrates, phenol red to indicate pH changes, and a chemical (iron) that indicates H_2S gas production. Reactions are: (*1*) An uninoculated (sterile) tube neutral (red); (*2*) a tube in which no acid production has occurred; (*3*) a tube in which acid has been produced (yellow) (cracks in agar are due to gas production); and (*4*) a tube in which H_2S has been produced, forming a black precipitate. (*b*) MacConkey agar differentiates between lactose-fermenting bacteria (indicated by a pink-red reaction in the center of the colony) and lactose-negative bacteria (indicated by an off-white colony with no dye reaction).

Figure 6.27 Carbohydrate fermentation in broths. This medium is designed to show fermentation (acid production) and gas formation by means of a small inverted Durham tube for collecting gas bubbles. The tube on the left is an uninoculated negative control; the center tube is positive for acid (yellow) and gas (open space); the tube on the right shows growth, but neither acid nor gas.

Chapter Review with Key Terms

Microbial Nutrition, Ecology, and Growth

Nutrition

Nutrition consists of taking in material, assimilating some of it into protoplasm and extracting energy from the rest. Substances that can only be obtained by feeding, as opposed to material that an organism can assemble from simpler components, are considered **essential nutrients.** Depending upon the quantity of intake, essential nutrients are considered as either **macronutrients** (C,H,N,O,P,S) or **micronutrients** (trace elements, vitamins, growth factors). Depending upon hydrocarbon content, they are classed as either **inorganic** or **organic.** In any case, a **growth factor** is an organic nutrient that cannot be synthesized and thus must be preformed. Principal growth factors are essential amino acids, hemin, NAD, vitamins, and uracil.

Nutrient Sources: Distinguishing among diverse carbon and energy source adaptations requires definitive terminology. An **autotroph** is a microorganism that depends on no more than carbon dioxide for its carbon needs. If its energy needs are met by light, it is a **photoautotroph,** but if it extracts energy from inorganic substances, it is a **chemoautotroph (lithotroph).** In contrast, a **heterotroph** acquires carbon from more complex organic molecules. A **saprobe** satisfies its nutritional needs by feeding upon the dead.

Environmental Influences on Microbes

Temperature: Each type of organism exhibits an **optimum temperature,** a narrow range at which it grows best. Growth diminishes as temperature rises or falls from the optimum, and ceases altogether with **minimum** or **maximum temperatures.** Organisms that cannot grow above 20°C but thrive below 15°C and continue to grow even at 0°C are known as **psychrophiles.** **Mesophiles** grow from 10°C to 50°C, having temperature optima from 20° to 40°C. The growth range of **thermophiles** is 45°–80°C. A distinction is made between growing at the optimum temperature and being able to briefly withstand temperature extremes. Hence, a **facultative psychrophile** grows slowly in the cold, although its optimum is above 20°C. Similarly, a **thermoduric** organism can tolerate short exposure to high temperature. (To indicate the nature of withstanding a particular condition, the roots **-tolerant** and **-duric** are used.)

Oxygen Requirements: Free oxygen (O_2) is potentially toxic as a consequence of spontaneous reactive **superoxide** ion and **peroxide** formation. All but **strict (obligate) anaerobes** can produce **superoxide dismutase** and either **catalase** or **peroxidase;** these enzymes degrade reactive oxides and permit safe exposure to O_2.

Oxygen dependence varies. **Aerobes** thrive in normal atmospheric oxygen, and organisms that will not grow without oxygen are called **obligate aerobes.** An aerobic organism capable of living without oxygen temporarily is a **facultative anaerobe.** An aerobe that prefers a small amount of oxygen but does not grow under anaerobic conditions is a **microaerophile.** Anaerobes are of two types: An **aerotolerant anaerobe** cannot use oxygen for respiration, yet is not injured by it. Organisms that are actually damaged or killed by oxygen are **strict** or **obligate anaerobes.**

Capnophiles thrive in the presence of carbon dioxide in concentrations up to about 10% greater than that found in the atmosphere.

Effects of pH: Acidity and alkalinity affect the activity and integrity of enzymes and the structural components of a cell. Optimum pH for most microbes ranges approximately from 6 to 8. **Acidophiles** prefer lower pH, and **alkalinophiles** prefer higher pH.

Other Environmental Factors: Electromagnetic radiation and barometric pressure affect microbial growth. While a phototroph relies upon light for photosynthesis, light and other electromagnetic rays can induce the formation of noxious oxygen radicals in that organism and in others as well. A **barophile** is adapted to life under high (usually hydrostatic) pressure (bottom dwellers in the ocean, for example).

Influences of Microbes on the Environment

Microbes are vital for recycling matter. **Methanogens** are special types of chemoautotrophs characterized by their ability to generate methane under anaerobic conditions. The remains of dead organisms furnish nutrients for decomposers or saprobes. **Parasites** feed upon live hosts, a relationship that may result in disease. Disease-causing parasites are **pathogens.**

Transport Mechanisms

A microbial cell is able to absorb material from its surroundings, relocate various materials within the cell, and secrete substances. Depending upon energy requirements, transport mechanisms are either **active** or **passive.** In active transport, substances are concentrated, a process that consumes energy. **Diffusion** is a form of passive transport directed down a concentration gradient. **Osmosis** is essentially diffusion of water through a selectively permeable membrane. A special form of passive transport that permits movement of selected substances is called **facilitated diffusion. Phagocytosis** and **pinocytosis** are forms of active transport in which bulk quantities of solid and fluid material, respectively, are taken into the cell.

Because microbes are enclosed in a cell membrane that permits passage of water but not larger molecules, they are subject to **osmotic pressure.** They face **hypotonic** stress when subjected to a medium of lower solute concentration and **hypertonic** stress when they come in contact with a medium of higher concentration. In an **isotonic** medium, solute concentrations are nearly the same inside and outside the cell, and there is no net movement of water in either direction. A **halophile (osmophile)** thrives in hypertonic surroundings, and an **obligate halophile** requires a salt concentration of at least 15%, but preferably 25%.

Microbial Interactions

Microbes coexist in varied relationships. **Symbiosis** pertains to cohabitation, but its usage often connotes **mutualism,** a reciprocal, obligatory, and beneficial relationship between two organisms. **Synergism** is a mutually beneficial but not obligatory coexistence. **Commensalism** exists when an organism is exploited but not harmed by the other organism in the relationship. **Satellitism** is a special form of commensalism. **Antagonism** entails competition, inhibition, and injury directed against the opposing organism. Special cases of antagonism are **antibiosis** and **bacteriocin** production.

Microbial Growth

The halving of a parent bacterial cell to form a pair of similar-sized daughter cells is known as **binary** or **transverse fission.** The duration of each division is called the **generation** or **doubling time.** Because a population theoretically doubles with each generation, the growth rate is **exponential,** and each cycle increases in **geometric progression.** A **growth curve** is a graphic representation of a closed population that changes in size and quality as environmental conditions deteriorate. Plotting a curve requires an estimate of live cells, called a **viable count.** The initial, seemingly quiet period of the curve is called the **lag phase.** This is followed by the **exponential growth phase** in which viable cells increase in geometric progression. As environmental conditions stagnate and the growth and death rates are the same, a plateau or **stationary growth phase** occurs. The **death phase** is typified by an adverse environment and death that exceeds growth. Should a census of all cells be necessary, a direct or total cell count is conducted. A microscope counting chamber or Coulter counter (flow cytometer) is used for this.

Growing Microbes in the Lab: Artificial nutrient media vary according to their physical form, chemical characteristics, and purpose. They may be **liquid** (broth, milk), **semisolid,** or **solid,** depending upon the absence or the quantity of solidifying agent (usually **agar-agar** or **gelatin**). A **synthetic medium** is any preparation that is chemically defined, but a medium containing a poorly identified component is **nonsynthetic.** A **general purpose medium** is used to raise a wide assortment of microbial types; when supplemented with blood or tissue infusion to culture **fastidious** species, it is called an **enriched medium.** A **selective medium** permits preferential growth of certain organisms in a mixture and inhibits others. An enrichment medium is a highly selective medium used expressly to favor the growth of an organism that is ordinarily sparse. A **differential medium** supports the growth of several types of microbes, but they can be distinguished by virtue of dissimilar features such as color, texture, and size. A **reducing medium** limits oxygen availability and is useful for cultivating anaerobes. The ability to utilize sugars can be determined with **carbohydrate fermentation media.** To convey fragile microbes, special **transport media** are needed to stabilize viability. **Assay media** are used to evaluate the effectiveness of antimicrobial agents. Environmental and industrial surveillance routinely call for **enumeration media** to determine the number of microbes in food, drinking water, soil, sewage, and other sources.

True–False Questions

Determine whether the following statements are true (T) or false (F). If you feel a statement is false, explain why, and reword the sentence so that it reads accurately.

___ 1. A growth factor is an essential nutrient that an organism cannot synthesize itself from raw materials.

___ 2. All bioelements exist in an inorganic environmental reservoir.

___ 3. A chemoautotroph can synthesize all its required organic components from simple raw materials using energy from the sun.

___ 4. An obligate halophile can live in fresh water.

___ 5. Photosynthetic autotrophs all contain pigments.

___ 6. A pathogen is a parasite that damages a host's tissues by its actions.

___ 7. Passive transport is not an important mechanism of transport in cells.

___ 8. A cell exposed to a hypertonic environment will lose water to its surroundings.

___ 9. Active transport requires the expenditure of ATP.

___ 10. Environmental factors such as temperature and pH exert their effect on the enzymatic activity of microbial cells.

___ 11. Mesophiles are killed by refrigeration.

___ 12. Superoxide ion is toxic to strict anaerobes because they have no catalase.

___ 13. The time required for binary fission to occur is called the generation time.

___ 14. In a viable plate count, each colony represents a cell or colony-forming unit from the sample population.

___ 15. During the stationary phase, no new cells are being added to the population.

___ 16. Agar-agar is superior to gelatin as a solidifying agent because it does not melt at room temperature.

Concept Questions

1. Differentiate between micronutrients and macronutrients.
2. Is there an organism that could grow on a medium that contains only the following compounds dissolved in water: $CaCO_3$, $MgNO_3$, $FeCl_2$, $ZnSO_4$, and glucose? Defend your answer.
3. Briefly describe the general function of the bioelements CHNOPS in the cell.
4. What is the function of metallic ions in cells?
5. Compare autotrophs and heterotrophs with respect to carbon and energy sources.
6. Describe the nutritional strategy of two types of lithotrophs described in the chapter.
7. Compare the effects of isotonic, hypotonic, and hypertonic solutions on an ameba and on a bacterial cell. If a cell is in a hypotonic environment, what is the condition of its protoplasm relative to the environment?
8. Why are most pathogens mesophilic?
9. What are the ecological roles of psychrophiles and thermophiles?
10. What is the habitat of a facultative parasite? A strict saprobe?
11. Name three groups of obligate intracellular parasites.
12. What might be the habitat of an aerotolerant anaerobe?
13. Where in the body are anaerobic habitats apt to be found?
14. Where do superoxide ions and hydrogen peroxide originate?
15. Differentiate between mutualism, commensalism, synergism, parasitism, and antagonism, using examples. How are parasitism and antagonism similar? How are they different? Are any of these relationships obligatory?
16. Why is growth called exponential? What makes it a geometric progression?
17. If an egg salad sandwich sitting in a warm car for 4 hours develops 40,960 bacterial cells, how many more cells would result with just one more hour of incubation? With 10 hours of incubation? What if the initial bacterial dose were 100? What does this tell you about using clean techniques in food preparation (other than aesthetic considerations)?
18. Tell what is happening at points A, B, C, D.

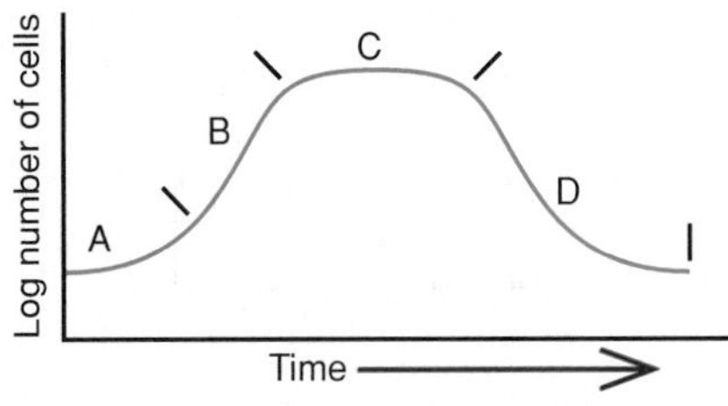

19. Why are we concerned with differentiating live versus dead cells in populations?
20. Predict the appearance of the growth curve in an *open* system.
21. Why is an older culture needed for spore staining?
22. Mitis-salivarius agar contains 10 g tryptose, 10 g proteose peptone, 1 g glucose, 50 g sucrose, 4 g dipotassium phosphate, two dyes, and 15 g agar. Though both this and *Euglena* agar contain exact amounts of specific sugar and inorganic salts, why is mitis-salivarius still considered a nonsynthetic medium? How could you make it synthetic?
23. Tell the two principal functions of dyes in media.
24. Differentiate among enriched, selective, and differential media.

Practical/Thought Questions

1. Describe how one might determine the nutrient requirements of a microbe from Mars. If, after exhausting the nutrient schemes, it still does not grow, what other factors might one take into account?
2. How can osmotic pressure and pH be used in preserving foods? How do they affect microbes?
3. How might one effectively treat anaerobic infections using gas? Explain how it would work.
4. How can you explain the observation that unopened milk will spoil even while refrigerated?
5. What would be the effect of a fever on a thermophilic pathogen?
6. Patients with acidotic diabetes are especially susceptible to fungal infections. Can you explain why?
7. Using the concept of synergism, can you describe a way to grow a fastidious microbe?
8. Describe a way to isolate an antibiotic-producing bacterium.
9. Why must blood agar be made with an *isotonic* base medium?
10. Check the list in table 6.2 (page 160). Using the "rule of essentials," are all needed elements present here? Where is the carbon?

CHAPTER 7

Microbial Metabolism: The Chemical Crossroads of Life

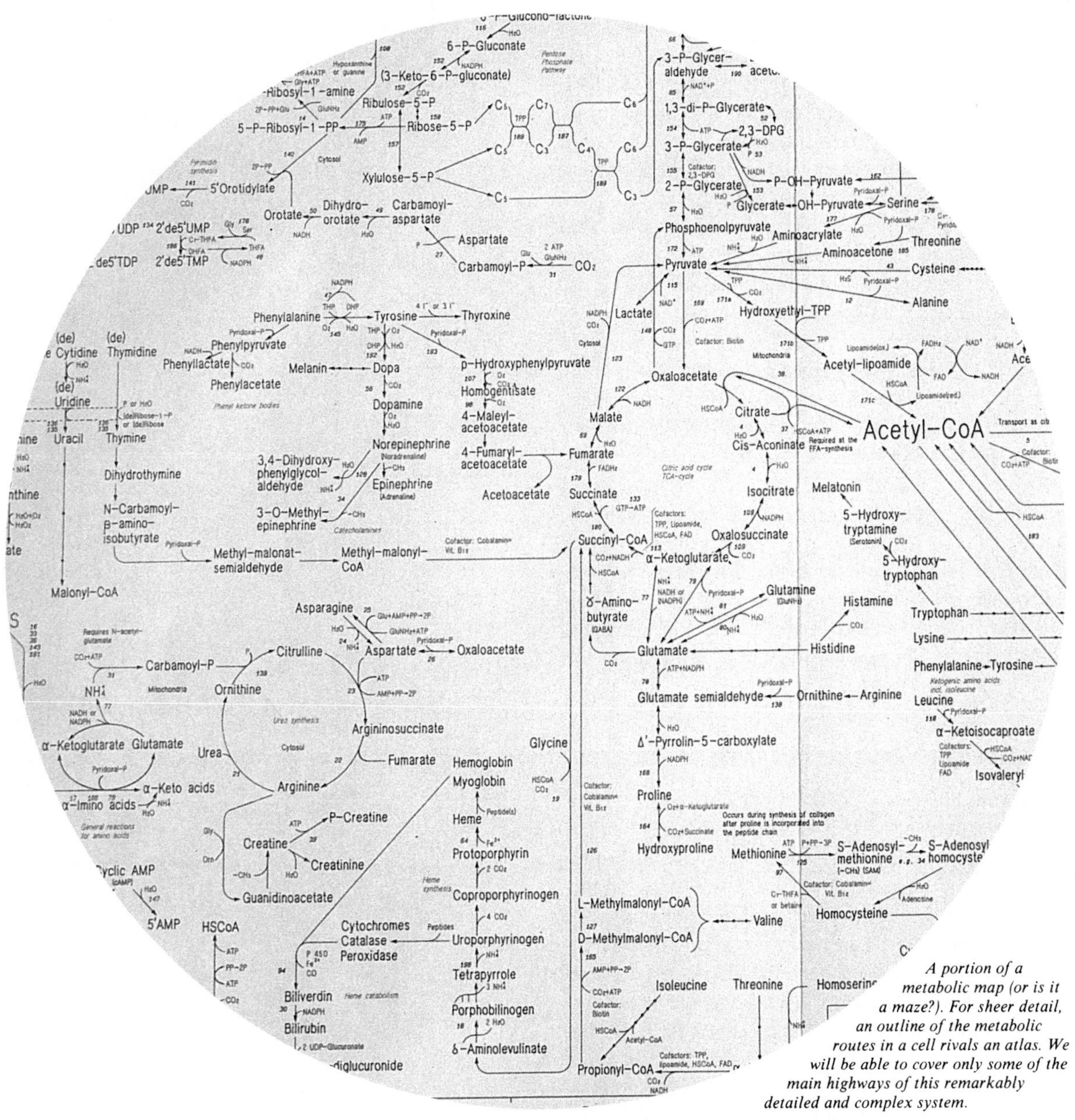

A portion of a metabolic map (or is it a maze?). For sheer detail, an outline of the metabolic routes in a cell rivals an atlas. We will be able to cover only some of the main highways of this remarkably detailed and complex system.

Chapter Preview

This chapter will discuss four interrelated concepts that are all part of metabolism: the nature of enzymes, the function of enzymes, the flow of energy in the cell, and the pathways that govern nutrient processing. Because a common metabolic thread passes among all organisms, a study of metabolism increases our understanding of life processes in general. Our present knowledge of human cell physiology and the success of modern drug therapy have resulted largely through studies of human and microbial metabolism. Such diverse areas as biotechnology and clinical diagnosis depend on microbial metabolism. The manufacture of antibiotics, organic acids, enzymes, alcohols, hormones, vitamins, amino acids, and detergents is based on the actions of microbes.

The Metabolism of Microbes

Metabolism, from the Greek term *metaballein,* meaning change, pertains to all chemical reactions and physical workings of the cell. Although metabolism entails thousands of different reactions, most of them fall into one of two general categories. **Anabolism,** sometimes also called biosynthesis, is any process that results in synthesis of cell molecules and structures. It is a building and bond-making process that forms larger molecules from smaller ones, and it usually requires the input of energy. **Catabolism** is the opposite, or complement, of anabolism. Catabolic reactions are degradative; they break bonds, convert larger molecules into smaller components, and often produce energy. The linking of anabolism to catabolism ensures the efficient completion of many thousands of cellular processes.

Metabolism is a cyclical self-regulatory process that maintains the stability of the cell (figure 7.1). Its actions provide a dynamic pool of chemical building blocks and give rise to enzymes and structural components of the cell. Metabolism of nutrients can extract energy in the form of adenosine triphosphate or ATP (previously described in chapter 2) that may be channeled into such processes as biosynthesis, transport, growth, and motility. As we will see, **metabolites**—compounds given off by the complex networks of metabolism—often serve multipurpose roles in the economy of the cell. Metabolism is highly organized and responsive to fine controls. For instance, there are mechanisms for diminishing or ceasing the production of a substance that is not in demand and for diverting overabundant nutrients into storage. The details of metabolic schemes and pathways are indeed complex, but only through such an intricate chemical organization can cells function and survive.

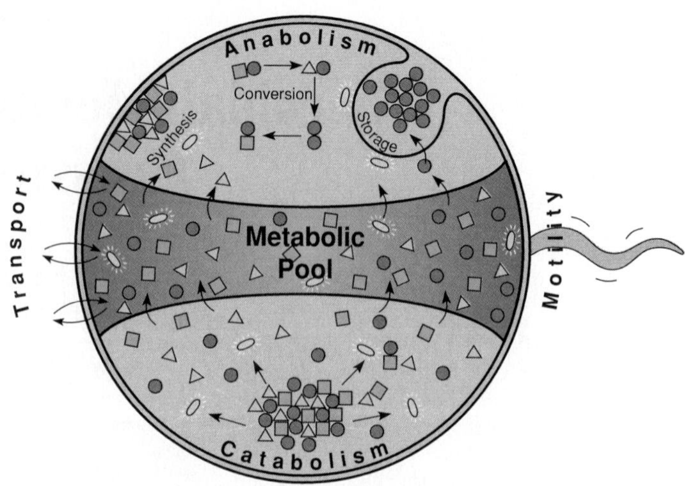

Figure 7.1 Summary of metabolic functions.

Enzymes: Catalyzing the Chemical Reactions of Life

A microbial cell could be viewed as a microscopic factory, complete with basic building materials, a source of energy, and a "blueprint" for running its extensive network of metabolic reactions. But the chemical reactions of life, even when highly organized and complex, cannot proceed without a special class of molecules called **enzymes.** Enzymes are a remarkable example of **catalysts,** chemicals that increase the rate of a chemical reaction without becoming part of the products. Do not make the mistake of thinking that an enzyme creates a reaction. Because of the greater energy of some molecules, a reaction would occur spontaneously at some point (figure 7.2). But such uncatalyzed metabolic reactions cannot occur fast enough to sustain life processes, and so enzymes are indispensable to life. Other characteristics of enzymes are summarized in table 7.1.

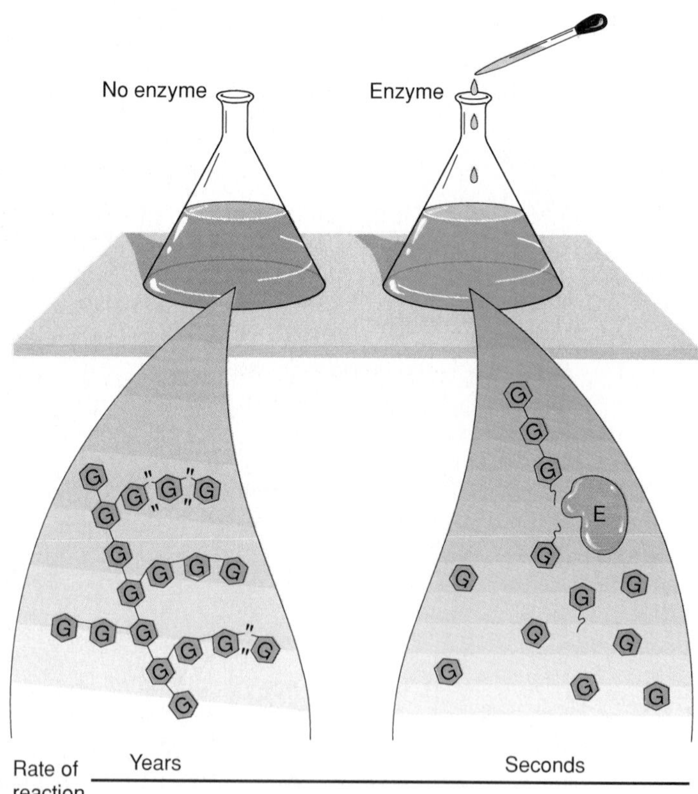

Figure 7.2 Comparison of a noncatalyzed reaction versus a catalyzed reaction. A starch molecule will spontaneously decompose into glucose, given enough time. An enzyme can cause the same reaction in a fraction of a minute.

---

anabolism (ah-nab'-oh-lizm) Gr. *anabole,* a throwing up.

catabolism (kah-tab'-oh-lizm) Gr. *katabole,* a throwing down.

enzyme (en'-zyme) Gr. *en,* in, and *syme,* leaven. Named for catalytic agents first found in yeasts.

catalyst (kat'-uh-list) Gr. *katalysis,* dissolution.

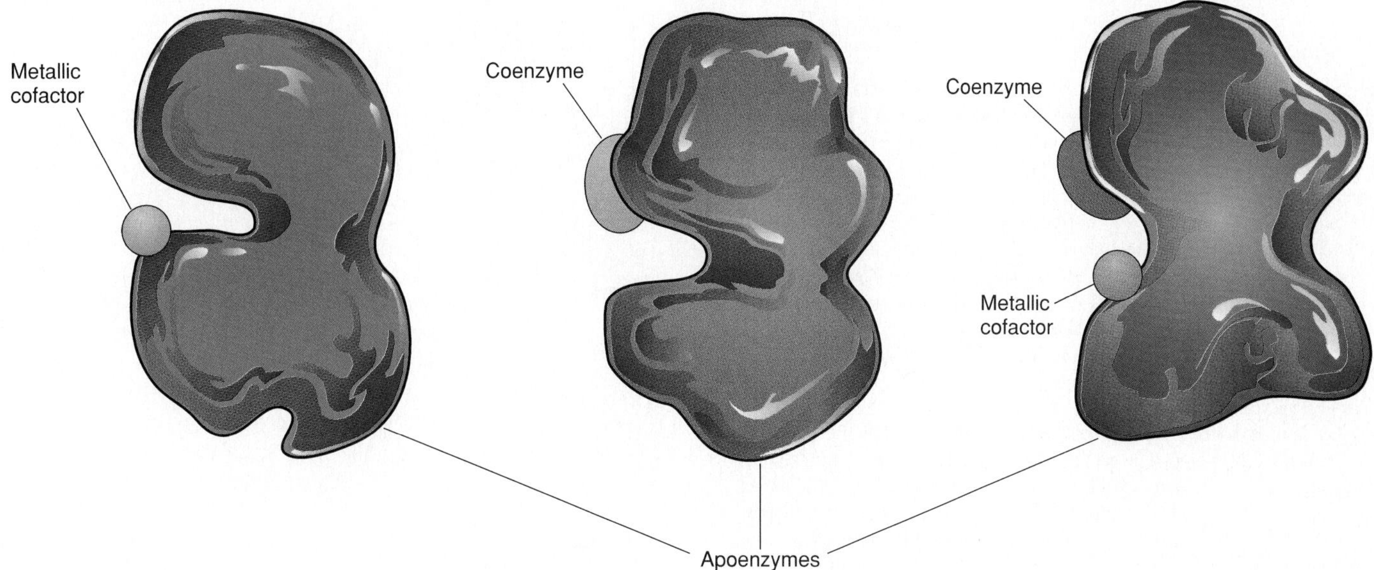

Figure 7.3 Conjugated enzymes are composed of various combinations. All have an apoenzyme (polypeptide or protein) component and may have one or more cofactors.

Table 7.1 Checklist of Enzyme Characteristics
Act as organic catalysts to speed up the rate of cellular reactions
Are composed of protein and may require cofactors
Enable metabolic reactions to proceed at a speed compatible with life, and also help regulate them
Provide a reactive site for target molecules called substrates
Associate closely with substrates, but do not become integrated into the reaction products
Are not used up or permanently changed by the reaction
Have unique characteristics such as shape, specificity, and function
Lower the activation energy required for a chemical reaction to proceed (see feature 7.1)
Work rapidly
Are reusable, thus function in extremely low concentrations
Are limited by particular conditions of temperature and pH
Can be regulated by feedback and genetic mechanisms

How Do Enzymes Work?

We have said that an enzyme speeds up the rate of a metabolic reaction, but just how does it do this? During a chemical reaction, reactants are converted to products by bond formation or breakage. A certain amount of energy is required for every such reaction, and this requirement limits its rate. This resistance to a reaction, which must be overcome for a reaction to proceed, is measurable and is called the **energy of activation** (see feature 7.1). In the laboratory, overcoming this initial resistance can be achieved (1) by increasing thermal energy (heating), which increases molecular velocity, (2) by increasing the concentration of reactants, or (3) by adding a catalyst. In living systems, the first two alternatives are not feasible, since elevating the temperature is potentially harmful and increasing the density of reactants is not practical. This leaves only the action of catalysts, and enzymes fill this need efficiently and potently.

At the molecular level, an enzyme physically promotes a reaction by serving as a physical site upon which one or more reactant molecules—the **substrate** (or substrates)—can be positioned for various interactions. Although an enzyme becomes physically attached to the substrate and participates directly in bonding, it does not become a part of the products, is not used up by the reaction, and can function over and over again. In fact, studies have shown that a single molecule of an enzyme can catalyze the reactions of several million substrate molecules. To further visualize the roles of enzymes in metabolism, we must next look at their structure.

Enzyme Structure

Enzymes can be classified as simple or conjugated. *Simple* enzymes consist of protein alone, while *conjugated* enzymes (figure 7.3) contain protein and other nonprotein molecules (with some exceptions; see feature 7.2). A conjugated enzyme, sometimes referred to as a **holoenzyme,** is a combination of a protein, now called the **apoenzyme,** and one or more **cofactors** (table 7.2). Cofactors are either organic molecules, called **coenzymes,** or inorganic elements (metal ions). In some enzymes the cofactor is loosely associated with the apoenzyme by noncovalent bonds, while in others it is tenaciously linked by covalent bonds.

holoenzyme (hol-oh-en′-zyme) Gr. *holos,* whole.

Feature 7.1 Enzymes As Biochemical Levers

An analogy will allow us to envision the relationship of enzymes to the energy of activation. A large boulder sitting precariously on a cliff's edge contains a great deal of potential energy, but it will not fall to the ground and release this energy unless something disturbs it. This might eventually happen spontaneously as the cliff erodes below it, but that could take a very long time. Nudging it with a crowbar (to overcome the resistance of the boulder's weight) will cause it to tumble down freely by its own momentum. If we relate this analogy to chemicals, the reactants are the boulder on the cliff, the activation energy is the energy needed to move the boulder, the enzyme is the crowbar, and the product is the boulder at the bottom of the cliff. Like the crowbar, an enzyme permits a reaction to occur rapidly by lowering the energy obstacle. (Though this analogy concretely illustrates the concept of the energy of activation, it is imperfect in that the enzyme, unlike the crowbar, does not actually add any energy to the system.)

This phenomenon can be represented by plotting the energy of the reaction against the direction of the reaction, which results in a hill-shaped curve. In the accompanying graph, the energy state of the reactants is higher than that of the product, and the energy of activation is the increasing slope of the curve. If a catalyzed reaction is compared with an uncatalyzed reaction, it is clear that the enzyme lowers the energy of activation. It does this by providing the reactant with an energy "shortcut" by which it can progress to its final state.

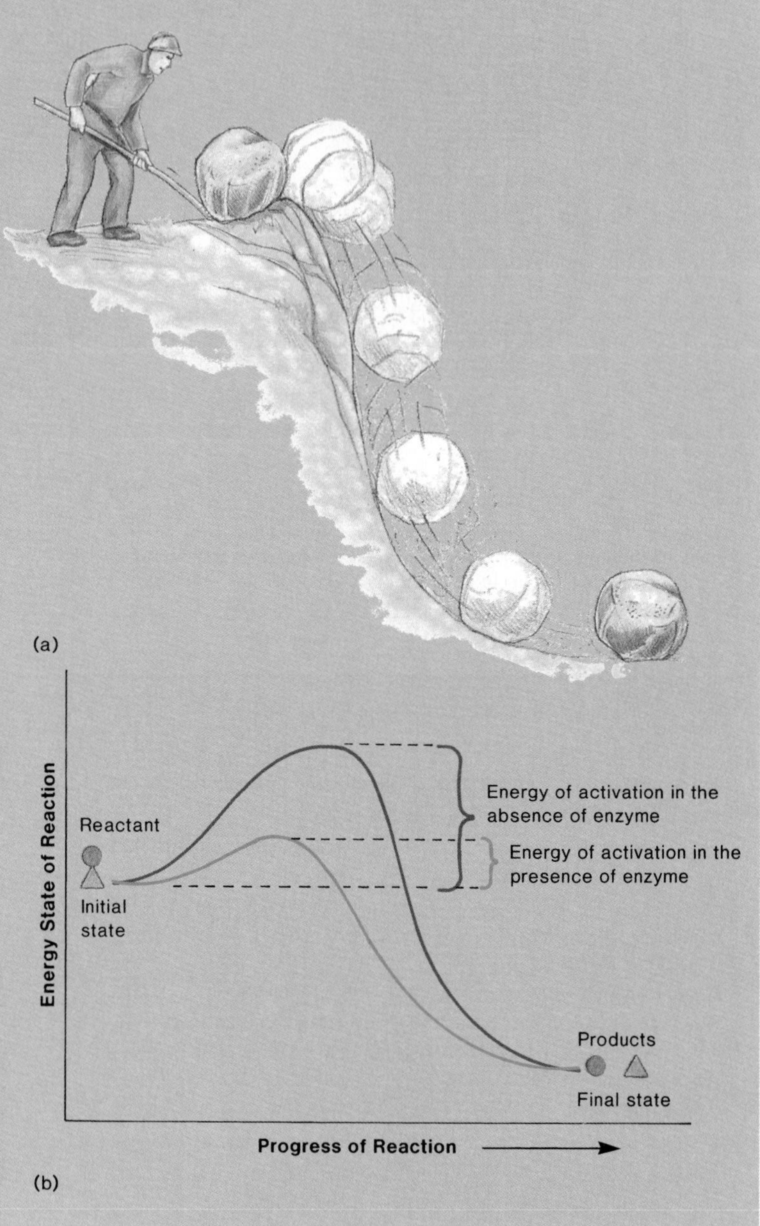

The influence of enzymes on chemical reactions. (*a*) Analogy in which the boulder represents potential energy. (*b*) Graph of chemical reaction, with and without an enzyme. Energy (called energy of activation) is required in both cases to convert a reactant molecule to products. But in an enzyme-catalyzed reaction, the enzyme significantly lowers this energy of activation and allows the reaction to proceed more readily and rapidly. Note that the products are at the same energy state with or without an enzyme.

Apoenzymes: Specificity and the Active Site

Apoenzymes range in size from small polypeptides with a hundred amino acids and a molecular weight of 12,000 to large polypeptide conglomerates with thousands of amino acids and a molecular weight of over a million. Like all proteins, an apoenzyme exhibits levels of molecular complexity called the primary, secondary, tertiary, and in larger enzymes, quaternary organization (figure 7.4). As we saw in chapter 2, the first three levels of structure arise from a single polypeptide chain that has undergone an automatic folding process and achieved stability by forming disulfide bonds. Folding causes the surface of the apoenzyme to acquire three-dimensional surface features. This relates to enzyme specificity in that: (1) The apoenzyme of each enzyme differs from others in its primary structure (the type and sequence of amino acids[1]); (2) nuances in polypeptide folding produce a particular three-dimensional configuration (tertiary structure); and (3) surface features of the tertiary structure provide a unique and specific site (usually a crevice or groove) for attachment of substrate molecules. The site that accepts a substrate is called the **active** or **catalytic site,** and there may be from one to several such sites (figure 7.4).

1. As specified by the genetic blueprint in DNA (see chapter 8).

Table 7.2 Selected Enzymes, Catalytic Actions, and Cofactors

Enzyme	Action	Cofactor
Catalase	Breaks down hydrogen peroxide	Iron (Fe)
Oxidase	Adds electrons to oxygen	Iron, copper (Cu)
Hexokinase	Transfers phosphate to glucose	Magnesium (Mg)
Urease	Splits urea into ammonium	Nickel (Ni)
Nitrate reductase	Reduces nitrate to nitrite	Molybdenum (Mo)
DNA polymerase complex	Synthesis of DNA	Zinc (Zn) and Mg
	Coenzyme and Its Associated Vitamin	
Glycine synthetase	Tetrahydrofolate requires folic acid	
Transaminase	Pyridoxal phosphate requires vitamin B_6	
Pyruvate dehydrogenase	Coenzyme A requires pantothenic acid	
Flavin dehydrogenase	Flavin adenine dinucleotide (FAD) requires riboflavin	
Various other dehydrogenases	Nicotinamide adenine dinucleotide (NAD) requires niacin	

Feature 7.2 Unconventional Enzymes

The secrets of the cell never cease to surprise biologists. Just when it seemed that the protein nature of biological catalysts and enzymes had been well established, researchers working on bacteria discovered a novel type of RNA that also acts as a catalyst. Even more startling is the fact that the target of these molecules, called *ribozymes,* is RNA itself. In their basic mechanisms, ribozymes are really very similar to protein-based enzymes: They have a very specialized active site that binds to a certain place on a substrate and changes that substrate in some way. It appears that the precise function of RNA-based enzymes is somewhat more restricted. They help process the genetic code by cleaving RNA molecules, but they do not handle the diversity of substrates that regular protein enzymes do.

Meanwhile, back at the lab, immunologists were studying enzymatic possibilities from yet another perspective. They were already well acquainted with the concept of a lock-and-key specific fit because of the way that antibodies lock onto their targets. This knowledge led to the inevitable question: Can antibodies, which are also proteins, act as enzymes, too? The answer is a qualified yes—if the antibody is carefully chosen. These catalytic antibodies, or *abzymes,* have extreme specificity for their substrate and can speed up a reaction, form an abzyme-substrate complex, participate in bond breaking, and release a product. What is very exciting about the use of abzymes is that, unlike enzymes, antibodies are very easy to fine-tune to fit almost any possible substrate configuration and can be manufactured in large amounts using monoclonal methods (see chapter 13). The applications for medicine and industry are considerable. For instance, an abzyme could be designed to do double duty—both attaching to a target molecule and enzymatically destroying it. This technology has the potential for treating cancers, infections, abnormal blood clots, and other medical problems very specifically. The possible use of abzymes to purify drugs and inactivate viruses is already being explored.

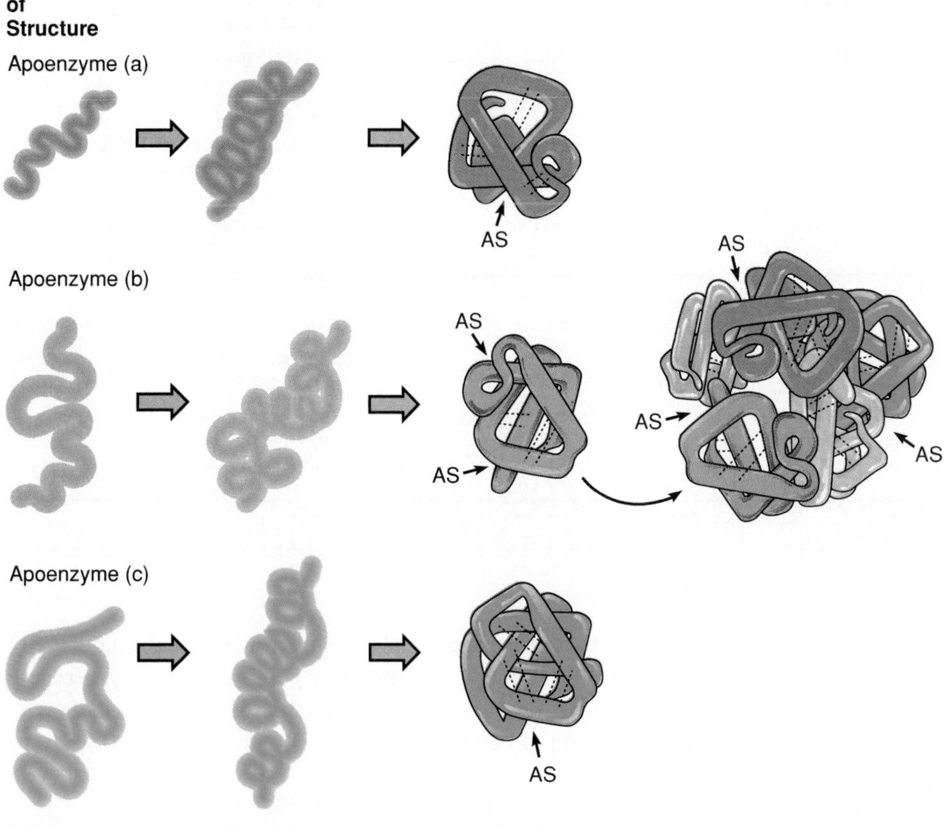

Figure 7.4 How the active site and specificity of the apoenzyme arise. As the polypeptide forms intrachain bonds and folds, it assumes a three-dimensional (tertiary) state with numerous surface features. Because each different polypeptide folds differently, each apoenzyme will have a differently shaped active site (AS). Some enzymes have more than one active site; others have sites to attach cofactors and regulatory compounds; and more complex enzymes have a quaternary structure consisting of several polypeptides bound by weak forces, as in (*b*).

Feature 7.3 The Enzyme Name Game

Most metabolic reactions require a separate and unique enzyme. Up to the present time, researchers have discovered and named over 2,000 of them. Given the complex chemistry of the cell and the extraordinary diversity of living things, several times that many enzymes probably remain to be discovered. Early workers did not have a consistent plan for naming enzymes, thus many enzymes had more than one name, different enzymes shared the same name, or the name was not particularly descriptive (ptyalin—salivary amylase, and pepsin—gastric protease being notable examples). As the list of enzymes began to grow, an international system of nomenclature was developed to standardize their naming and classification.

The naming of enzymes is really quite straightforward. A name is composed of two parts: a prefix or stem word derived from a certain characteristic—usually the substrate acted upon or the type of reaction catalyzed, or both—followed by the ending *-ase*.

The system assigns the enzyme to one of the following six classes, based on its general biochemical action: (1) *Oxidoreductases* transfer electrons from one substrate to another, and *dehydrogenases* transfer a hydrogen from one compound to another. (2) *Transferases* transfer C, N, P, or S groups from one substrate to another. (3) *Hydrolases* cleave bonds on molecules with the addition of water. (4) *Lyases* add groups to or remove groups from double-bonded substrates. (5) *Isomerases* change a substrate into its isomeric form. (6) *Ligases* catalyze the formation of bonds with the input of ATP and the removal of water.

Each enzyme is also assigned a systematic name that indicates the specific reaction it catalyzes, but some of these names are too cumbersome to be used routinely, and the common name, based on the substrate or some other notable feature, is preferred. Examples are *hydrogen peroxide oxidoreductase,* known more commonly as *catalase,* and *mucopeptide N-acetylmuramoylhydrolase,* called (thankfully) *lysozyme.* With this system, an enzyme that digests a carbohydrate substrate is a *carbohydrase;* more specifically, *amylase* acts on starch (amylose is a major component of starch). This enzyme could also be known generally as a *polysaccharidase.* A *disaccharidase* such as *maltase* digests 2-unit sugars such as maltose. An enzyme that hydrolyzes peptide bonds of a protein is a *proteinase, protease,* or *peptidase,* depending on the size of the protein substrate. Some fats and other lipids are digested by *lipases.* DNA is hydrolyzed by *deoxyribonuclease,* generally shortened to *DNase.* A *synthetase* or *polymerase* bonds together many small molecules into large molecules. Other examples of enzymes are presented in table 7.3. (See also table 7.2.)

Table 7.3 A Sampling of Enzymes, Their Substrates, and Their Reactions

Common Name	Systematic Name	Enzyme Class	Substrate	Action
Cellulase	1,4 β-glycosidase	Hydrolase	Cellulose	Cleaves 1, 4 β-glycosidic linkages
Lactase	β-D-galactosidase	Hydrolase	Lactose	Breaks lactose down into glucose and galactose
Penicillinase	Beta-lactamase	Hydrolase	Penicillin	Hydrolyzes beta-lactam ring
Lipase	Triacylglycerol acylhydrolase	Hydrolase	Triglycerides	Cleaves bonds between glycerol and fatty acids
DNA polymerase	DNA nucleotidyl-transferase	Transferase	DNA nucleosides	Synthesizes a strand of DNA using the complementary strand as a model
Hexokinase	ATP-glucose phosphotransferase	Transferase	Glucose	Catalyzes transfer of phosphate from ATP to glucose
Aldolase	Fructose diphosphate aldolase	Lyase	Fructose diphosphate	Catalyzes the conversion of the substrate to two 3-carbon fragments
Lactate dehydrogenase	Same as common name	Oxidoreductase	Pyruvic acid	Catalyzes the conversion of pyruvic acid to lactic acid
Oxidase	Cytochrome oxidase	Oxidoreductase	Molecular oxygen	Catalyzes the reduction (addition of electrons and hydrogen) to O_2

Enzyme-Substrate Interactions

For a reaction to take place, a temporary enzyme-substrate union must occur, fittingly described as a complementary lock-and-key fit, at the active site (figure 7.5). The bonds formed between the substrate and enzyme are weak and, of necessity, easily reversible. Once the temporary enzyme-substrate complex has formed, appropriate reactions occur on the substrate, often with the aid of a cofactor, and a product is formed and released. The enzyme can then attach to another substrate molecule and repeat this action. Although enzymes can potentially catalyze reactions in both directions, most examples in this chapter will depict them working in one direction only.

Cofactors: Supporting the Work of Enzymes

In chapter 6 we learned that microorganisms require specific metal ions called trace elements and certain organic growth

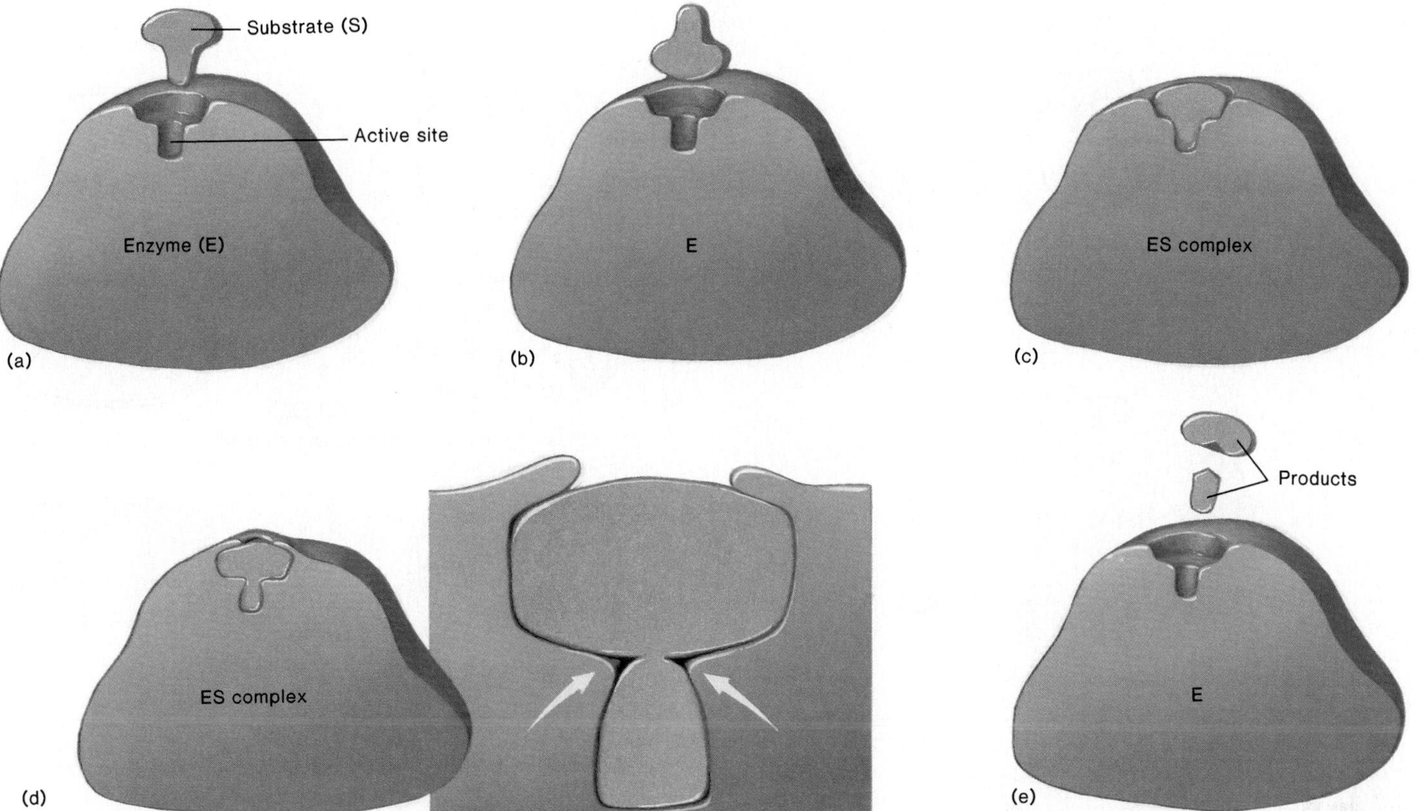

Figure 7.5 Enzyme-substrate reactions: fit, proximity, and orientation. When the enzyme and substrate come together in the transition state, the substrate (S) must be close and in correct position with respect to the enzyme (E). (*a*) and (*b*) show incorrect proximity and orientation, while (*c*) shows a correct fit, position, and proximity. (*d*) When the ES complex is formed, it enters a transition state. (*e*) During this temporary but tight interlocking union, the enzyme may actually flex itself around the substrate as it participates directly in breaking or making bonds. Once the reaction is complete, the enzyme releases the products.

factors. In many cases the need for these substances arises from their roles as cofactors. The **metallic cofactors,** including iron, copper, magnesium, manganese, zinc, cobalt, selenium, and many others, participate in precise functions between the enzyme and its substrate. In general, metals enhance catalysis, help bring the active site and substrate closely together, and participate directly in chemical reactions with the enzyme-substrate complex. Often, metallic cofactors are referred to as enzyme activators.

Coenzymes are organic compounds that work in conjunction with an apoenzyme to perform a necessary alteration of a substrate. The general function of a coenzyme is to remove a functional group from one substrate molecule and add it to another, thereby serving as a transient carrier of this group (figure 7.6). The specific activities of coenzymes are many and varied. In a later section of this chapter, we shall see that coenzymes carry and transfer hydrogen atoms, electrons, carbon dioxide, and amino groups. Because coenzymes are vitamins or contain vitamins, it becomes even clearer why vitamins are so important to nutrition and are required as growth factors in the majority of living things. Vitamin deficiencies prevent the complete holoenzyme from forming, which compromises both the chemical reaction and the structure or function dependent upon that reaction.

Classification of Enzyme Functions

Location and Regularity of Enzyme Action Enzymes may be classified according to their site of activity (figure 7.7). After initial synthesis in the cell, **exoenzymes** are transported extracellularly, where they break down (hydrolyze) large food molecules or harmful chemicals. Examples of exoenzymes are cellulase, amylase, and penicillinase. By contrast, **endoenzymes** are retained intracellularly and function there. Most enzymes of the metabolic pathways are of this variety.

In terms of their presence in the cell, enzymes are not all produced in equal amounts or at equal rates. Some, called **constitutive enzymes,** are always present and in relatively constant amounts, regardless of the amount of substrate. Because glucose is central to the metabolic pathways of organisms, the enzymes involved in glucose utilization are invariably constitutive. An **induced** (or inducible) enzyme is not constantly present and

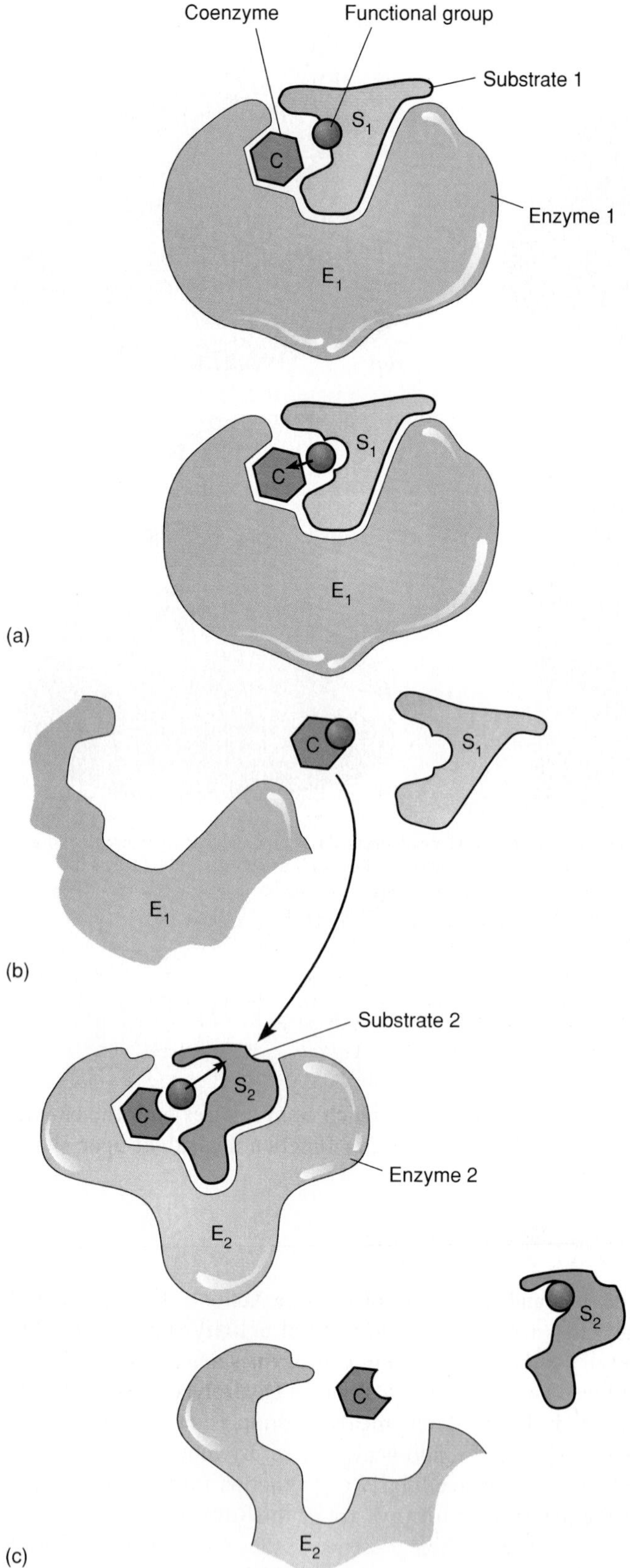

Figure 7.6 The carrier functions of coenzymes. (*a*) The coenzyme carrier in position to remove a functional group from substrate 1. (*b*) Coenzyme with the attached functional group and the substrate that donated the group are released. (*c*) The coenzyme carries the functional group to a second enzyme-substrate complex and passes the group to a second substrate. The now empty coenzyme and filled substrate are released.

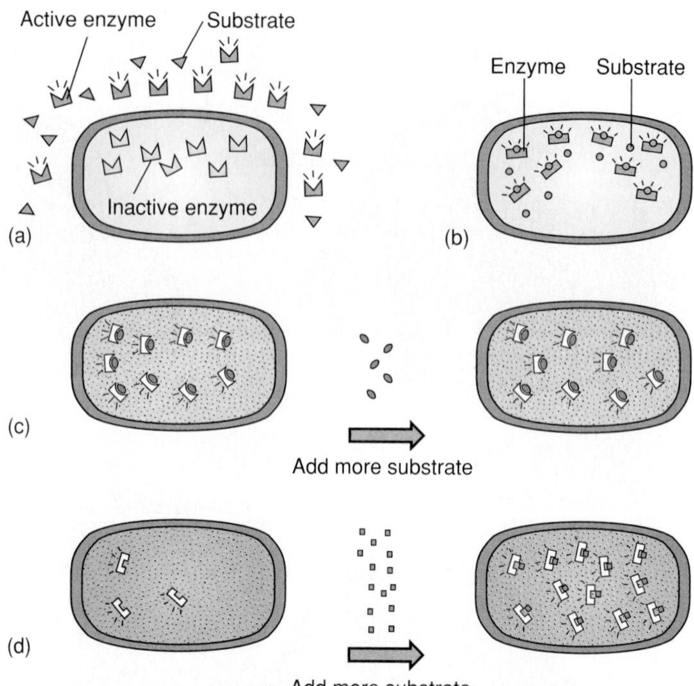

Figure 7.7 Types of enzymes, as described by their location of action and quantity. (*a*) Exoenzymes function extracellularly. (*b*) Endoenzymes function intracellularly. (*c*) Constitutive enzymes are present in constant amounts in a cell. The addition of more substrate does not increase the numbers of these enzymes. (*d*) Induced enzymes are normally present in trace amounts, but their quantity may be increased a thousandfold by the addition of substrate.

is produced only when its substrate is present. Induced enzymes are present in amounts ranging from a few molecules per cell to several thousand times that many, depending on the metabolic requirement. This property of selective synthesis of enzymes prevents a cell from wasting energy by making enzymes that will not be used immediately. The induction of enzymes constitutes an important metabolic control discussed later in this section of the chapter and again in chapter 8.

Synthesis and Hydrolysis Reactions A growing cell is in a frenzy of activity, constantly synthesizing proteins, DNA, and RNA, forming storage polymers such as starch and glycogen, and assembling new cell parts. Such anabolic reactions require enzymes (ligases) to form covalent bonds between smaller substrate molecules. Also known as *condensation* reactions, synthesis reactions require one ATP and release one water molecule for each bond made (figure 7.8*a*). Catabolic reactions involving energy transactions, remodeling of cell structure, and digestion of macromolecules are also very active during cell growth. These processes require enzymes that reduce substrates to a series of smaller molecules by breaking intramolecular bonds. Because the breaking of each bond requires the input of a water molecule, digestion is often termed *hydrolysis* (figure 7.8*b*).

---

hydrolysis (hy-drol′-uh-sis) Gr. *hydro*, water, and *lysis*, setting free.

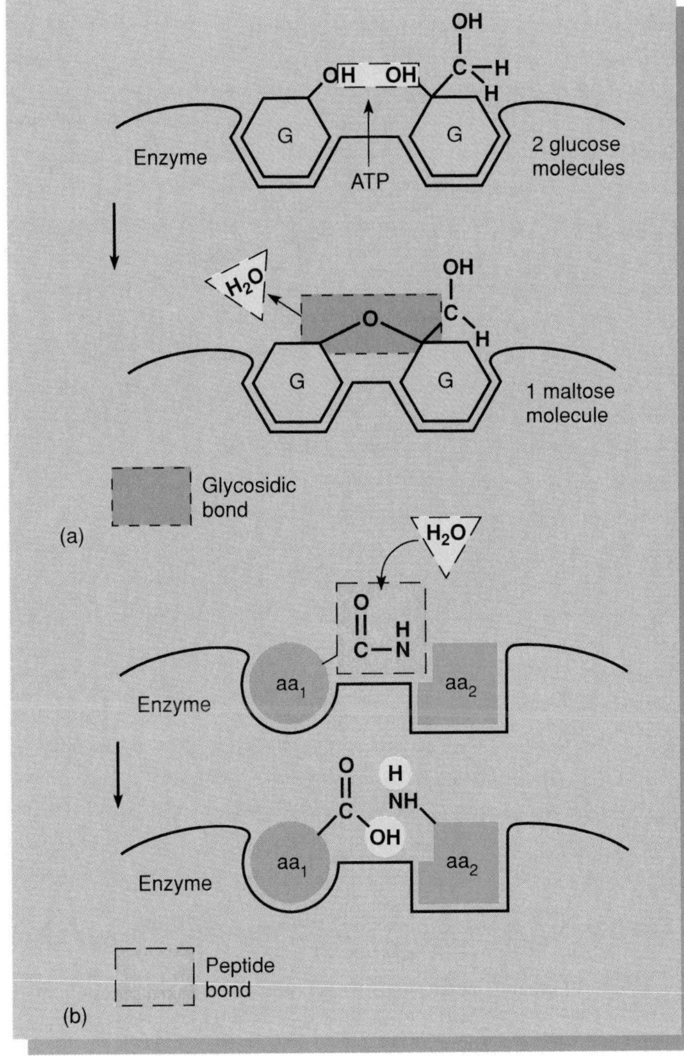

Figure 7.8 Examples of enzyme-catalyzed synthesis and hydrolysis reactions. (*a*) Condensation reaction: Forming a glycosidic bond between two glucose molecules to generate maltose requires the removal of a water molecule and energy from ATP. (*b*) Hydrolysis reaction: Breaking a peptide bond between two amino acids requires the insertion of a water molecule.

Transfer Reactions by Enzymes Other enzyme-driven processes that involve the simple addition or removal of a functional group are important to the overall economy of the cell. Oxidation-reduction and other transfer activities are examples of these types of reactions.

Some atoms and compounds readily give or receive electrons and participate in oxidation (the loss of electrons) or reduction (the gain of electrons). The compound that loses the electrons is **oxidized,** and the compound that receives the electrons is **reduced.** Such oxidation-reduction (redox) reactions are common in the cell and indispensable to the energy transformations discussed later in this chapter. Important components of cellular redox reactions are oxidoreductases, which remove electrons from one substrate and add them to another, and their coenzyme carriers, nicotinamide adenine dinucleotide (NAD; see figure 7.14) and flavin adenine dinucleotide (FAD).

Other enzymes play a role in the molecular conversions necessary for the economical use of nutrients by directing the transfer of functional groups from one molecule to another. For example, *aminotransferases* convert one type of amino acid to another by transferring an amino group (see figure 7.27); *phosphotransferases* participate in the transfer of phosphate groups, also a form of energy transfer; *methyltransferases* move a methyl (CH_3) group from substrate to substrate; and *decarboxylases* (also called carboxylases) catalyze the removal of carbon dioxide from organic acids in several metabolic pathways.

The Role of Microbial Enzymes in Disease Many bacterial pathogens secrete unique exoenzymes that help them avoid host defenses or promote their multiplication in tissues. Because these enzymes contribute to pathogenicity, they are referred to as virulence factors, or toxins in some cases. *Streptococcus pyogenes* (a cause of throat and skin infections) produces a streptokinase that digests blood clots and apparently assists in invasion of wounds. The lipases of *Staphylococcus aureus* (the cause of skin boils) increase the virulence of this species by promoting its invasion of oil-producing glands on the skin. *Pseudomonas aeruginosa,* a respiratory pathogen, produces elastase, which digests elastin, a protein common in many tissues. *Clostridium perfringens,* an agent of gas gangrene, synthesizes lecithinase C, a lipase that profoundly damages cell membranes and accounts for the tissue death associated with this disease. The collagenases of some parasitic worms digest the principal protein of connective tissue and promote invasion of tissues. Not all enzymes digest tissues; some, like penicillinase, inactivate penicillin and thereby protect a microbe from its effects.

Although viruses do not synthesize enzymes independently, they instruct the host cell to synthesize a small but impressive group of viral enzymes. Most of these are involved in viral replication, but some contribute to cell damage and viral escape. Bacteriophages instruct the synthesis of lysozyme, which digests and weakens the bacterial cell wall and facilitates lysis and release. Influenza virus contains spikes of neuraminidase that help liberate the virus from an infected cell. Perhaps the most intriguing enzyme of all is the reverse transcriptase present in retroviruses (AIDS virus is one type), which promotes conversion of viral RNA to viral DNA, thereby allowing the virus genes to be inserted into a host chromosome.

The Sensitivity of Enzymes to Their Environment

The activity of an enzyme is highly influenced by the cell's environment. In general, enzymes operate only under the natural temperature, pH, and osmotic pressure of an organism's habitat. When enzymes are subjected to changes in these normal conditions, they tend to be chemically unstable or **labile.** The enzymes of microbes living in moderate temperatures and pH are especially sensitive to environmental extremes. Low temperatures inhibit catalysis. High temperatures denature the apoenzyme—**denaturation** being a process by which the weak bonds that collectively maintain the native shape of the apoenzyme are broken. This disruption causes extreme distortion of the enzyme's shape and prevents the substrate from attaching to the active site (see figure 9.4). Such nonfunctional enzymes block metabolic reactions, which can lead to cell death. Low or high pH or certain chemicals (heavy metals, alcohol) are also denaturing agents.

Regulation of Enzymatic Activity and Metabolic Pathways

Metabolic reactions proceed in a systematic, highly regulated manner that maximizes the use of available nutrients and energy. The cell responds to environmental conditions by adopting those metabolic reactions that most favor growth and survival. Because enzymes are critical to these reactions, the regulation of metabolism is largely the regulation of enzymes by an elaborate system of checks and balances. Before we examine some of these control systems, let us take a look at some general features of metabolic pathways.

Metabolic Pathways

Metabolic reactions rarely consist of a single action or step. More often, they occur in a multistep series or pathway, with each step catalyzed by an enzyme. An individual reaction is shown in various ways, depending on the purpose at hand (figure 7.9). The product of one reaction is often the reactant (substrate) for the next, forming a linear chain reaction. Many pathways have branches that provide alternate methods for nutrient processing. Others take a cyclic form, in which the starting molecule is regenerated to initiate another turn of the cycle (for example, the TCA cycle; see figure 7.23). Pathways generally do not stand alone; they are interconnected and merge at many sites. Every pathway has one or more enzyme pacemakers (usually the slowest enzyme in the series) that sets the rate of a pathway's progression. These enzymes respond to various control signals and, in so doing, determine whether or not a pathway proceeds (see feature 7.4).

Regulation of pacemaker enzymes proceeds on two fundamental levels. Either the enzyme itself is directly inhibited or activated, or the amount of the enzyme in the system is altered (decreased or increased). Factors that affect the enzyme directly provide a means for the system to be finely controlled or tuned, whereas regulation at the genetic level (enzyme synthesis) is coarser.

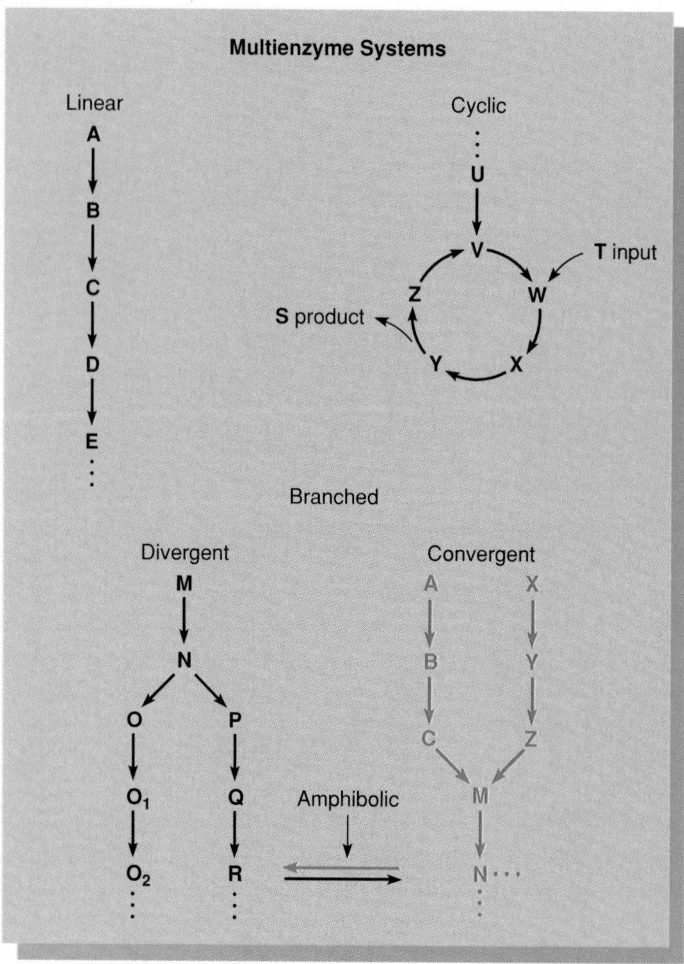

Figure 7.9 In general, metabolic pathways consist of a linked series of individual chemical reactions that produce intermediary metabolites and lead to a final product. These pathways occur in several patterns, including linear, cyclic, and branched. Anabolic pathways involved in biosynthesis are additive and result in a more complex molecule, each step adding on a functional group, whereas catabolic pathways generate energy and involve the dismantling of molecules. Virtually every reaction in a series involves a specific enzyme.

Direct Controls on the Behavior of Enzymes

Competitive Inhibition In **competitive inhibition,** other molecules with a structure similar to the normal substrate can occupy the enzyme's active site. Although these molecular mimics have the proper fit for the site, they cannot be further acted upon. This effectively prevents an enzyme from attaching to its usual substrate and blocks its activity, output, and possibly the rest of that pathway. This mechanism is common in natural metabolic pathways. It is also an important mode of action by some drugs for treating infections (see chapter 10).

Feedback Control The term *feedback* is borrowed from the field of engineering, where it describes a process in which the product of a system is *fed back* into the system to control it. Feedback is positive or negative, depending upon whether the process is stimulated (+) or inhibited (−) by the product. **Negative feedback** is common in electrical appliances such as hot plates or furnaces with automatically controlled temperatures. These devices possess a means of controlling the temperature (a heating coil or flame) and a temperature sensor (thermostat) that is set to a desired point and regulates the turning on or off of the heat source. In the case of a hot plate, when the set temperature has been reached, the thermostat switches the heating elements off. A subsequent fall below the set temperature causes the heating element to be turned on again.

Many enzymatic reactions are controlled by a similar negative mechanism, whereby the end product being fed back into the system *negates* (cancels) an enzyme's activity (figure 7.10*a*). At high levels of substrate and scant levels of end product, the enzyme works freely, but as the end product builds up, it stops the action of that enzyme. The mechanisms of negative feedback—that is, how the product actually stops the action of an enzyme—can best be understood if we look at the allosteric behavior of enzymes.

We have previously shown that globular proteins of enzymes are bulky and possess numerous surface features, and that

Feature 7.4 Enzymes—Headgates That Control Metabolism

One could liken the regulation of the biochemical pathways to that of an irrigation system consisting of interconnected canals and ditches whose water flow is regulated by headgates. As the demand for water downstream diminishes, ceases, or increases, the headgate may be adjusted to meet the demand. In this analogy, the canals correspond to the metabolic pathways and the headgates to the enzymes. This system lends itself to regulation, in that suppression of a key enzyme—say, one beginning a sequence—can essentially block a pathway and all reactions connected with that pathway down the line. Or, the system may be fine-tuned by repressing an intermediate enzyme farther down the pathway. So, too, may a pathway be activated when an enzyme is "turned on."

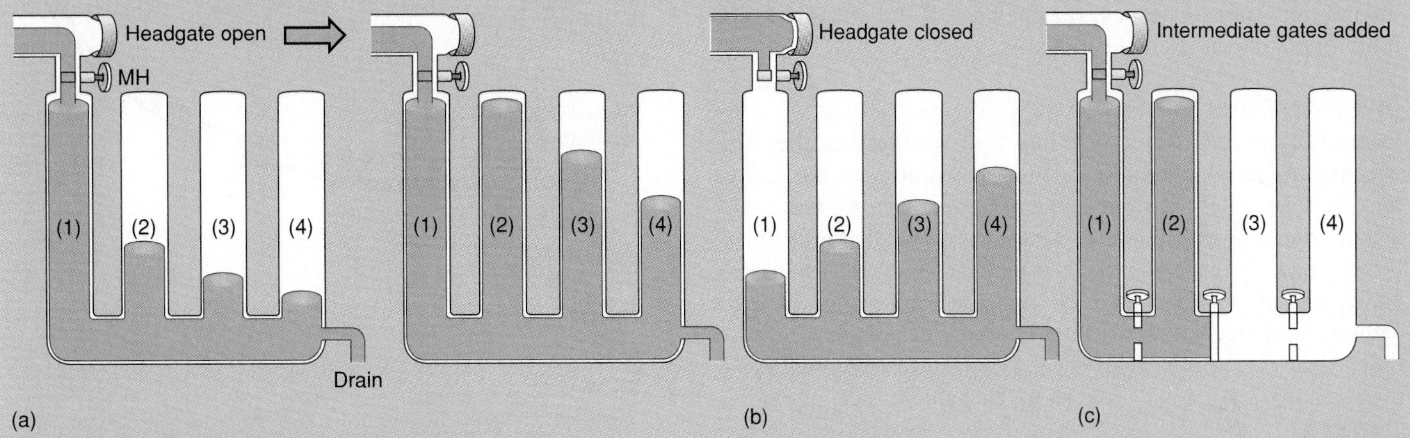

Enzymes as headgates in metabolic control are depicted in (*a*) and (*b*). In this type of system, the main headgate (MH) controls all water going into the system. This is comparable to a pathway in which an enzyme at the beginning of a sequence (*1*) controls the overall pathway. The turning on or off of (*1*) effects the events in (*2*), (*3*), and (*4*). (*c*) Gates placed between channels permit even finer control of a specific limb or portion.

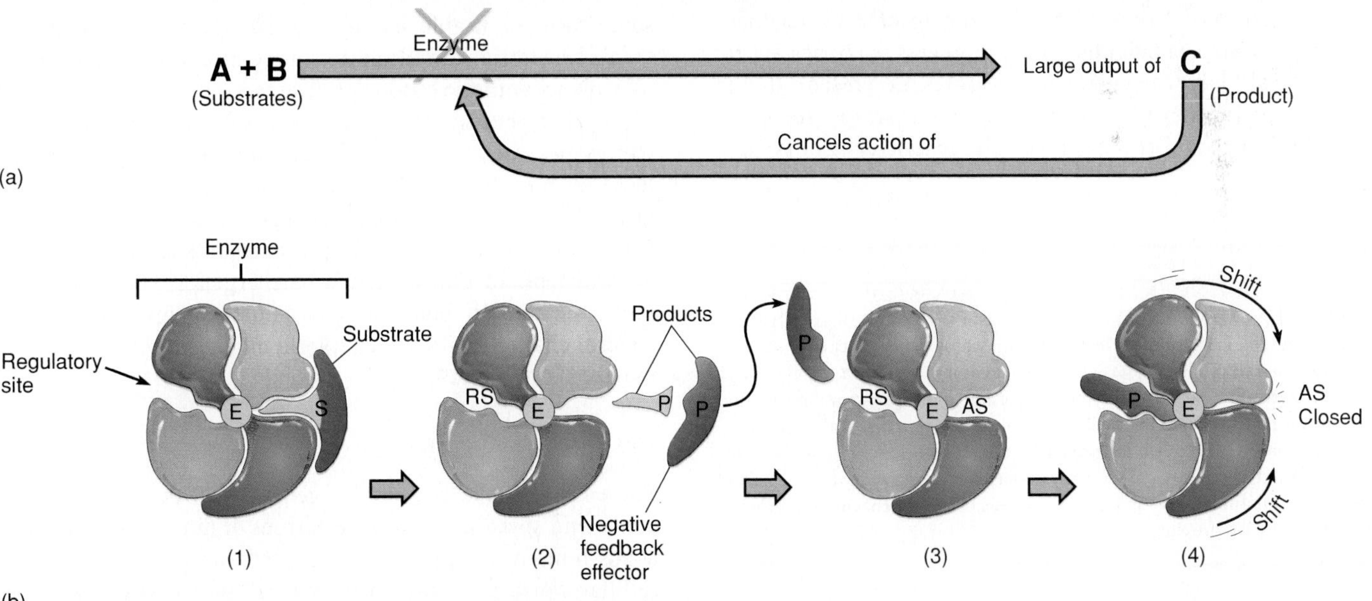

Figure 7.10 Enzyme control by negative feedback in system. (*a*) General equation: As end product builds up, it binds to the enzyme and stops its action. (*b*) The behavior of allosteric enzymes. (*1, 2*) The enzyme complex normally attaches to the substrate at the active site and releases products (P). (*3*) One product can function as a negative-feedback effector by fitting into a regulatory site (RS) on the enzyme. (*4*) The entrance of P into RS causes a confirmational shift of the enzyme that closes the active site (AS). The enzyme cannot for the time being catalyze further reactions with the AS.

one part of an enzyme's terrain (the active site) accommodates the substrate. **Allosteric** enzymes have an additional **regulatory site** for the attachment of molecules other than substrate. When an end product molecule fits into this regulatory site, the enzyme's active site is so distorted that it can no longer bind to its substrate. Although this does not denature the enzyme and is quite reversible, allosteric inhibitors can temporarily stop the action of that enzyme (figure 7.10*b*). This mechanism is termed **feedback inhibition.** When at some point the end product is used up and more of it is needed, the enzyme will be released from inhibition and can resume catalysis.

Controls on Enzyme Synthesis

Controlling enzymes by controlling their synthesis is effective because enzymes do not last indefinitely. Some wear out, some may be deliberately degraded, and others are diluted with each cell division. For catalysis to continue, enzymes eventually must be replaced. This works into the scheme of the cell, where replacement of enzymes can be regulated according to cell demand. The mechanisms of this system are genetic in nature—that is, they require regulation of DNA and the protein synthesis machinery, topics we shall encounter once again in chapter 8.

Enzyme repression is a means to stop further synthesis of an enzyme somewhere along its pathway. As the level of the end product from a given enzymatic reaction has built to excess, the genetic apparatus responsible for replacing these enzymes is automatically suppressed (figure 7.11). The response time takes longer than feedback inhibition, but its effects are more enduring. A response that resembles the inverse of feedback repression is **enzyme induction.** In this process, enzymes appear (are induced) only when suitable substrates are present; that is, the synthesis of enzyme is induced by its substrate (see figure 8.20). Both mechanisms are important genetic control systems in bacteria.

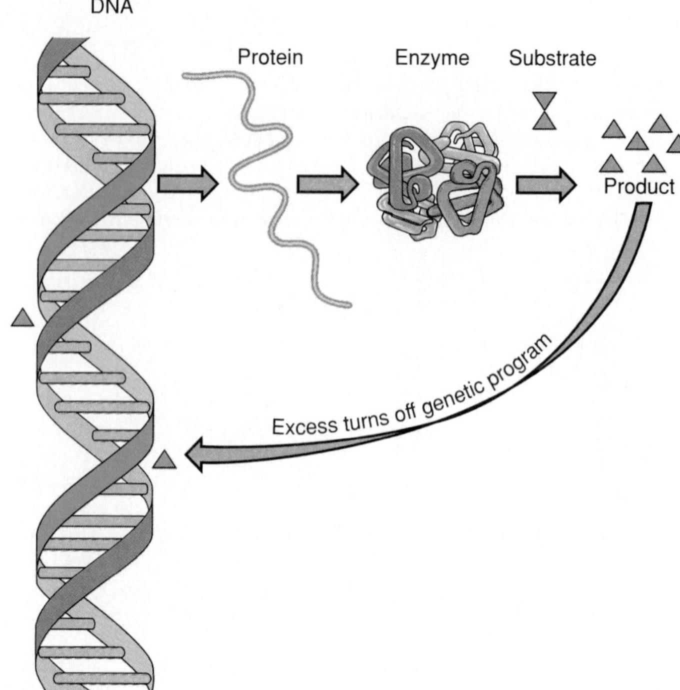

Figure 7.11 One type of genetic control of enzyme synthesis: feedback/enzyme repression. The enzyme is synthesized continuously until enough product has been made, at which time the excess product acts on the program for that enzyme's synthesis and blocks it.

Waste-Not, Want-Not A classic model of enzyme induction occurs in the response of *Escherichia coli* to certain sugars. For example, if a particular strain of *E. coli* is inoculated into a medium whose principal carbon source is lactose, it will produce lactase to hydrolyze it into glucose and galactose. If the bacterium is subsequently inoculated into a medium containing only sucrose as a carbon source, it will cease synthesizing lactase and begin synthesizing sucrase. Because this response enables the organism to adapt to a variety of nutrients, this type of enzyme has been termed *adaptive.* It also prevents a microbe from "spinning its wheels," making enzymes for which no substrates are present.

The Pursuit and Utilization of Energy

In chapter 6 we surveyed the constant supply of nutrients required for the cellular machine to function. But none of the major cell processes—biosynthesis, movement, transport, or growth—could proceed without the constant input and expenditure of some form of usable **energy.** Except for certain chemoautotrophic bacteria, the ultimate source of energy is the sun, but only photosynthetic autotrophs can tap this source directly. This capacity of photosynthesis to convert solar energy into chemical energy provides both a nutritional and an energy basis for all heterotrophic living things. Although energy exists in several general forms, only chemical energy can routinely operate cell transactions at the level of the protoplasm (see feature 7.5). The general topic of this section is bioenergetics—the oxidation of fuels, the role of redox carriers, and the origins of ATP. The general characteristics of energy in biological systems are summarized in table 7.4.

The Energy in Electrons

The energy of the cell resides generally in the bonds between atoms and specifically in the electrons of nutrients. When these bonds remain untapped or unchanged and are not spent doing cellular work, they have what we call **potential energy.** When bond energy is freed for cellular work, it is termed **kinetic energy.** Kinetic energy may be directed toward synthesis, movement, and growth.

Not all cellular reactions are equal with respect to energy. Some release energy, and others require it to proceed. If the cellular reaction:

$$X + Y \xrightarrow{\text{ATP}} Z$$

allosteric (al-oh-stair′-ik) Gr. *allos,* other, and *steros,* solid. Literally, "another space."

Table 7.4 Characteristics of Energy in Living Systems

General Function

Does work or affects changes in the cell

Specific Functions

Biosynthesis, movement, active transport, growth

Rules of Usage

1. Energy is not created, but released, expended, transferred, or stored
2. Energy can be transformed from one form to another

General Forms Usable by Cells

Primarily chemical and radiant;
electrical and mechanical to an extent

General Source

Derived from the sun in photoautotrophs and from chemicals in chemotrophs

Immediate Source in Heterotrophs

Bond breaking—redox reactions on organic compounds

Flow of Energy in Cell

Catabolic—Downhill; from high-energy compounds to low-energy; released electrons are shuttled from one chemical system to another, thus releasing energy used to form ATP

Anabolic—Uphill; simple molecules are bonded together to form more complex ones, which requires energy from ATP

Temporary Storage

High-energy bonds of ATP and GTP and other phosphorylated compounds

Long-term Storage

Stored carbohydrates, fats, proteins that can be broken down to release energy

proceeds with the release of energy, it is termed **exergonic.** If the reaction:

$$A + B \xrightarrow{\text{ATP}} C$$

requires the input of energy, it is **endergonic.**

Summaries of metabolism make it seem that cells "create" energy from nutrients, but they do not. What they actually do is extract chemical energy already present in nutrient fuels and apply that energy toward useful work in the cell. Indeed, a cell may be likened to a gasoline engine that releases energy as it burns the fuel and uses the energy to perform work. The engine does not actually produce energy, but it converts the potential energy of the fuel to kinetic energy of work.

The mechanisms by which cells handle energy are well understood, and most of them are explainable in basic biochemical terms. At the simplest level, cells possess specialized enzyme systems that entrap the energy present in the bonds of nutrients as they are progressively broken (figure 7.12). During exergonic reactions, electrons are transferred directly or through special

exergonic (ex-er-gon'-ik) Gr. *exo,* without, and *ergon,* work. In general, catabolic reactions are exergonic.

endergonic (en-der-gon'-ik) Gr. *endo,* within, and *ergon,* work. Anabolic reactions tend to be endergonic.

Feature 7.5 Energy in Biological Systems

Energy is the capacity to do work or cause change. Energy commonly exists in various forms: (1) thermal or heat energy from molecular motion; (2) radiant (wave) energy from visible light or other rays; (3) electrical energy from a flow of electrons; (4) mechanical energy from a physical change in position; (5) atomic energy from reactions in the nucleus of an atom; and (6) chemical energy present in the bonds of molecules. Cells are far too fragile to rely in any constant way on thermal or atomic energy for cell transactions. Since cells are largely chemical entities, it is only logical that chemicals are the basis of cellular energetics.

Fortunately for cells, energy is convertible from one form to another. Light energy is convertible to heat energy; heat is convertible to light; light to chemical; chemical to mechanical; and so on. This phenomenon allows the biological world to use energy effectively and efficiently. For example, photosynthetic microbes trap the energy in visible light and transform it into the chemical energy of nutrients. This chemical energy, in turn, can be transformed to the mechanical energy of flagellar movement, or it can even produce light. Conversions do not occur without some energy being lost as heat (in amounts compatible with life). But even heat serves a useful purpose by maintaining a temperature optimal for enzyme function.

(a)

(b)

Photobacteria contain a system of enzymes that expend energy to produce light. (*a*) When a compound called luciferin is acted on by luciferase, these bacteria emit a fair amount of visible light. The possible function of light emission in bacteria is very speculative. (*b*) Certain marine fishes carry a type of these bacteria symbiotically in a sac beneath the eye, which apparently serves as a light organ to attract other fish and to facilitate predation. This is one of the more extreme examples of biological energy conversion.

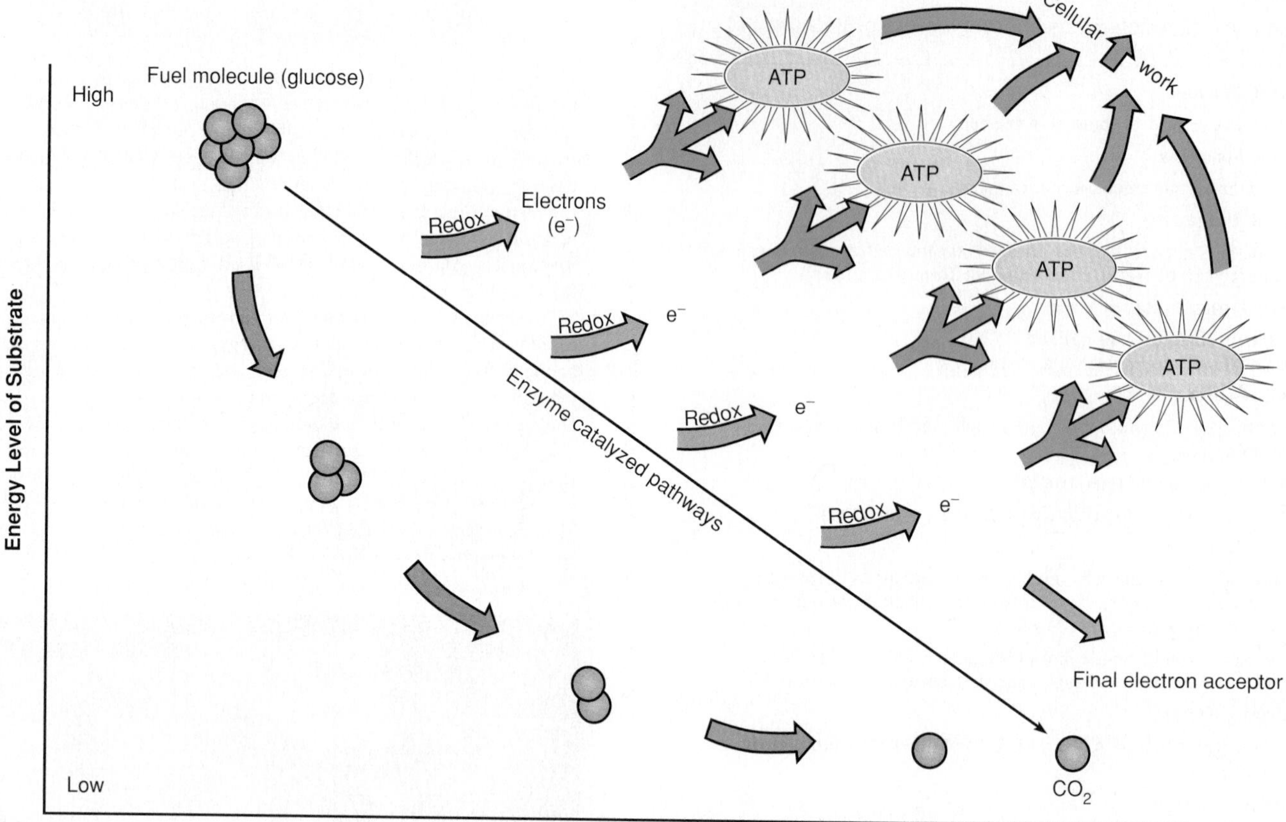

Figure 7.12 A simplified view of the cell's energy machine. The central events of cell energetics include the evolution of energy during the systemic dismantling of a carbon-based molecule. This is achieved by redox reactions that give off electrons and the shuttling of these electrons to sites in the cell where their energy may be transferred to ATP.

carriers to various high-energy phosphate molecules such as ATP. As we shall see, the ability of ATP to store and release the energy of chemical bonds fuels endergonic cell reactions. Before discussing ATP, let us examine the force behind energy extraction: redox reactions.

A Closer Look at Biological Oxidation and Reduction

We stated earlier that energy extraction in biological systems involves **redox reactions.** Such reactions always occur in pairs, with an electron donor and an electron acceptor. Together, the electron donor and acceptor constitute a *conjugate pair,* or *redox pair.* Written in equation form:

$$\text{Electron donor (oxidized)} \leftrightarrow e^- + \text{Electron acceptor (reduced)}$$
$$\downarrow$$
$$\text{ATP}$$

A temporary electron carrier may or may not be involved.

This process salvages electrons along with their inherent energy, but it also releases some kinetic energy, leaving the reduced compound with less energy than the oxidized one. The released energy can **phosphorylate,** or add an inorganic phosphate (P_i), to ADP or to some other compound, a process that captures this energy and stores it in a high-energy molecule (ATP, for example). In many cases, the cell does not handle electrons as discrete entities but rather as parts of an atom such as hydrogen. For simplicity's sake, we will continue to use the term electron transfer, but keep in mind that hydrogen ions are often part of the transfer process. The removal of hydrogens (one hydrogen atom consists of a single proton and a single electron) from a compound during a redox reaction is called **dehydrogenation.** The job of handling these hydrogens and electrons falls to one or more carriers, which function as short-term repositories for the electrons until they can be used in ATP synthesis (figure 7.13). As we shall see, dehydrogenations are an essential supplier of electrons for the respiratory electron transport system.

Electron Carriers: Molecular Shuttles

Electron carriers resemble shuttles that are alternately loaded and unloaded, repeatedly accepting and releasing electrons and hydrogens to facilitate the transfer of redox energy. Most carriers are coenzymes that transfer both electrons and hydrogens, but some transfer electrons only. The most common carrier is NAD, which carries hydrogens (and a pair of electrons) from dehydrogenation reactions (figure 7.14). Reduced NAD can be represented in various ways. Because 2 hydrogens are removed,

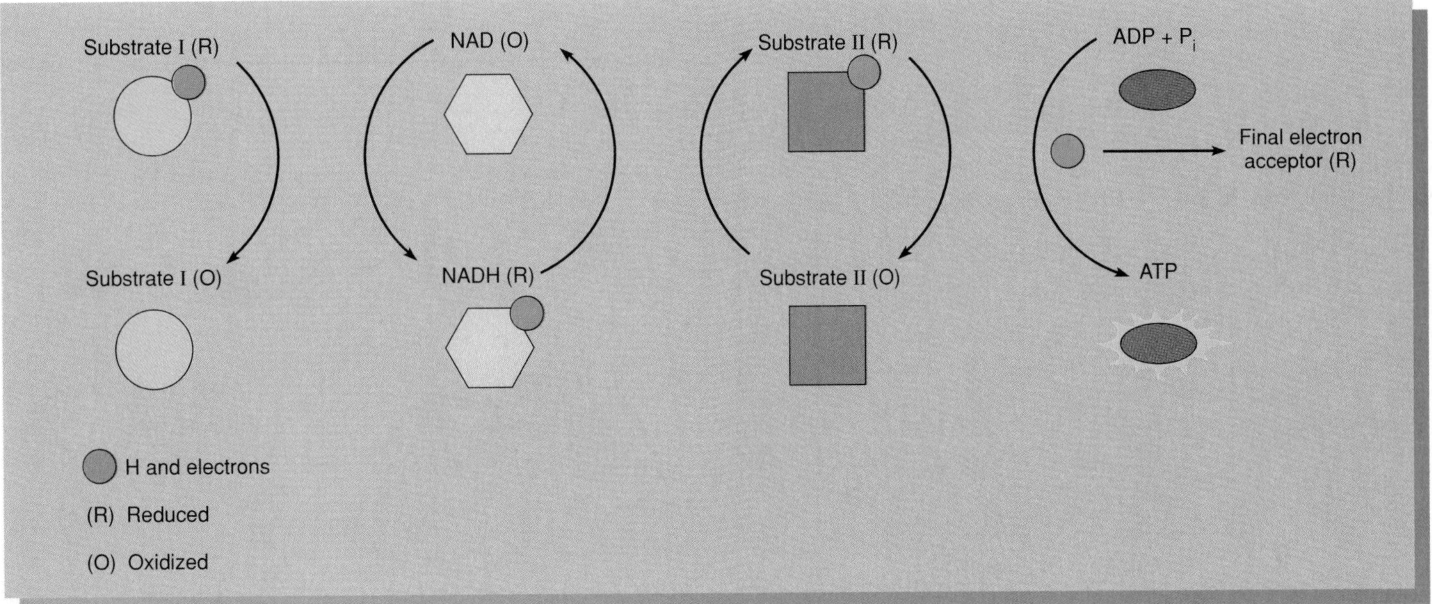

Figure 7.13 The role of electron carriers in redox reactions. Electrons and hydrogens are transferred by paired or coupled reactions in which the donor compound is oxidized (loses electrons) and the acceptor is reduced (gains electrons). A molecule such as NAD may serve as a transient intermediate carrier of the electrons between substrates. During this series of reactions, some energy given off by electron transfer is used to synthesize ATP. To complete the reaction, the electrons are passed to a final electron acceptor.

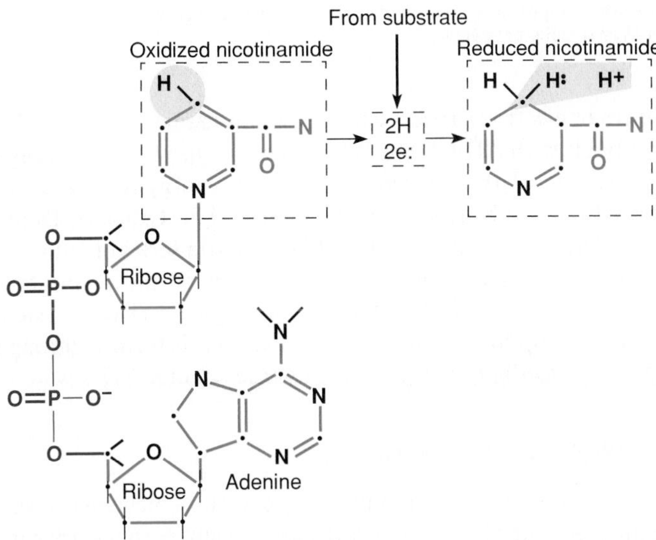

Figure 7.14 Details of NAD reduction. This coenzyme contains the vitamin nicotinamide and the purine adenine attached to double ribose phosphate molecules (a dinucleotide). The principal site of action is on the nicotinamide (boxed area). Hydrogens and electrons donated by a substrate interact with a carbon on the top of the ring (yellow circled areas). One hydrogen bonds there, carrying two electrons, and the other hydrogen is carried in solution as H^+ (a proton).

the actual carrier state is NADH + H^+, but this is somewhat cumbersome, so we will represent it with the shorter NADH. In catabolic pathways, electrons are extracted and carried through a series of redox reactions until the **final electron acceptor** at the end of a particular pathway is reached. In aerobes, this acceptor is molecular oxygen, and in anaerobes, it is some other inorganic or organic compound. Other common redox carriers are FAD, NADP (NAD phosphate), coenzyme A, and the compounds of the respiratory chain, which are fixed into membranes (see figure 7.24).

Adenosine Triphosphate: Metabolic Money

In what ways do cells extract chemical energy from electrons, store it, and then tap the storage sources? To answer these questions we must look more closely at the powerhouse molecule, adenosine triphosphate. ATP has also been described as metabolic money because it can be earned, banked, saved, spent, and exchanged. As a temporary energy repository, ATP provides a connection between energy-yielding catabolism and all other cellular activities that require energy. Some clues to its energy-storing properties lie in its unique molecular structure (figure 7.15).

The Molecular Structure of ATP ATP is a three-part molecule consisting of a nitrogen base (adenine) linked to a 5-carbon sugar (ribose), with a chain of three phosphate groups bonded to the ribose (figure 7.15; see figure 2.26). Breaking the bonds between two successive phosphates yields adenosine diphosphate (ADP) and adenosine monophosphate (AMP). It is worthwhile noting how economically a cell manages its pool of adenosine nucleotides. AMP derivatives help form the backbone of RNA and are also a major component of certain coenzymes (NAD, FAD, and coenzyme A).

The type, arrangement, and especially the proximity of atoms in ATP combine to form a compatible but unstable high-energy molecule. The high energy of ATP originates in the

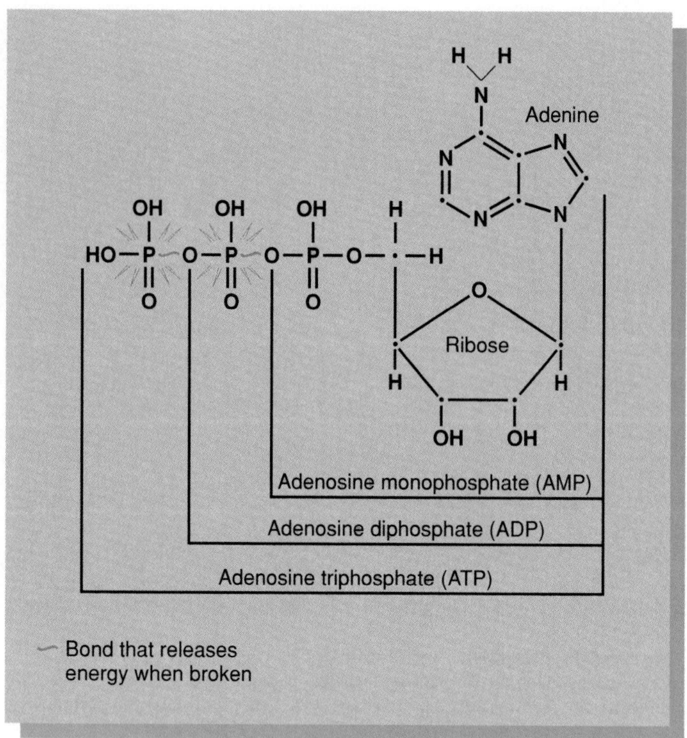

Figure 7.15 The structure of adenosine triphosphate (ATP) and its sister compounds, ADP and AMP.

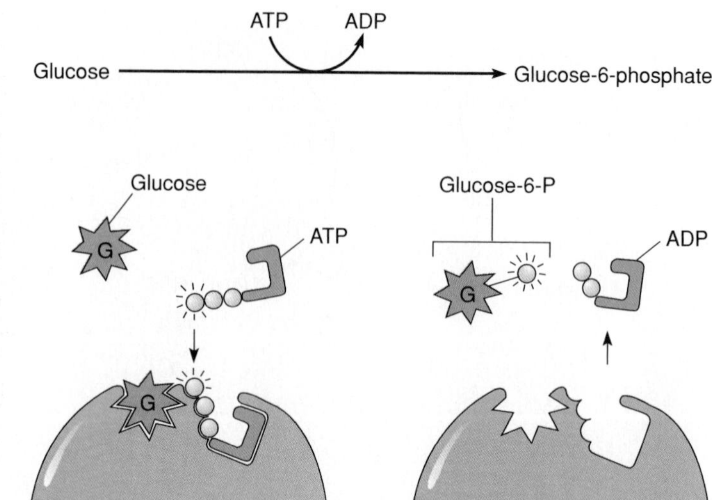

Figure 7.16 An example of ATP function: the activation of glucose. The first step in catabolizing glucose is the addition of a phosphate from ATP. This use of high-energy phosphate as an activator is a recurring feature of many metabolic pathways.

orientation of the phosphate groups, which are relatively bulky and carry negative charges. The proximity of these repelling electrostatic charges imposes a strain that is most acute on the bonds between the last two phosphate groups. When ATP is formed, this instability is accommodated by the bond stress being distributed through the rest of the molecule. The strain on the phosphate bonds accounts for the energetic quality of ATP, because removal of the terminal phosphates releases the bond energy.

The Metabolic Role of ATP Cellular reactions involving ATP are necessarily intertwined and balanced. ATP expenditure must inevitably be followed by ATP regeneration, and so on, in a never-ending cycle in an active cell. In many instances, the energy released during ATP hydrolysis powers biosynthesis by activating individual subunits before they are enzymatically linked together. This is a necessary part of protein synthesis, for example, where ATP donates a phosphate to each amino acid prior to the formation of a peptide bond. ATP is also used to prepare a molecule for catabolism such as the double phosphorylation of a 6-carbon sugar during the early stages of glycolysis (figure 7.16; see figure 7.19).

The formation of ATP occurs through a reversal of the process by which it was spent. In heterotrophs, the energy infusion that regenerates a high-energy phosphate comes from certain energy-yielding steps of catabolic pathways, in which nutrients such as carbohydrates provide a source of electrons. ATP is formed when redox reactions involving pairs of substrates or electron carriers energize an inorganic phosphate (P_i) that binds with ADP. Some ATP molecules are formed through *substrate level phosphorylation*—that is, energy is released directly from a substrate to ADP (figure 7.17). Other ATPs are formed through *oxidative phosphorylation,* a series of redox reactions occurring during the final phase of the respiratory pathway (see figure 7.25). Phototrophic organisms have a system of *photophosphorylation,* in which the ATP is formed through a series of sunlight-driven reactions (see chapter 22).

Pathways of Bioenergetics

One of the scientific community's greatest achievements was deciphering the biochemical pathways of cells. Initial work with bacteria and yeasts, followed by studies with animal and plant cells, clearly demonstrated metabolic similarities and strongly supported the concept of the universality of metabolism (see feature 7.6). The complexity of life's total metabolic scheme is staggering; even the chapter opening illustration, which presents a partial map of the major reactions, had to be simplified in order to fit on a single diagram. The major metabolic pathways include catabolic routes that gradually degrade nutrients and anabolic or biosynthetic routes that are involved in cell growth. Though these pathways are interconnected and interdependent and most individual reactions are reversible, anabolic pathways are not simply reversals of catabolic ones. At first glance it might seem more economical to use identical pathways, but having different enzymes and divided pathways

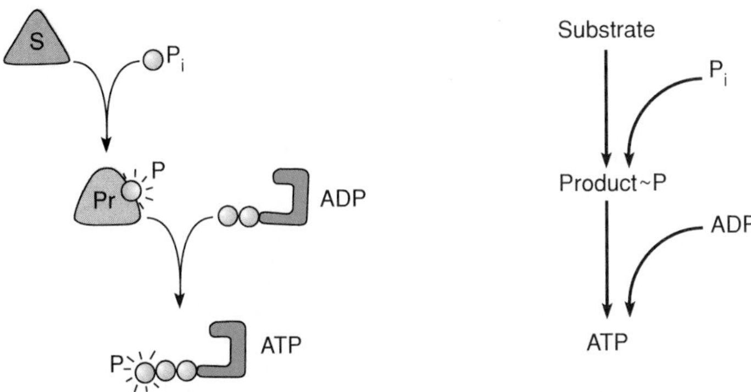

Figure 7.17 ATP formation at the substrate level. The inorganic phosphate (P_i) and the substrate form a bond that has high potential energy. In a reaction catalyzed by ATP synthetase, the phosphate is transferred to ADP, thereby producing ATP.

allows anabolism and catabolism to proceed simultaneously without interference. For simplicity, we shall focus our discussion on the most common catabolic pathways that will illustrate general principles of other pathways as well.

Catabolism: An Overview of Nutrient Breakdown and Energy Release

The primary catabolism of fuels that results in energy release in many organisms proceeds through a series of three coupled pathways: (1) **glycolysis,** also called the Embden-Meyerhof-Parnas (EMP) pathway, (2) the **tricarboxylic acid cycle** (TCA), also known as the citric acid or Krebs cycle,[2] and (3) the **respiratory chain** (electron transport and oxidative phosphorylation). Each segment of the pathway is responsible for a specific set of actions on various products of glucose. How these pathways interconnect in aerobic respiration is represented in figure 7.18, and their reactions are summarized in table 7.5. In the following sections, we will observe each of these pathways in greater detail.

Energy Strategies in Microorganisms

Nutrient processing is extremely varied, especially in bacteria, yet in most cases it is based on three basic catabolic pathways. In previous discussions, microorganisms were categorized according to their requirements for oxygen gas, and this relates directly to their mechanisms of energy release. As we shall see, **aerobic respiration** is a process in which glucose is completely degraded through glycolysis, the TCA cycle, and the respiratory chain. It relies on free oxygen as the final acceptor for electrons and hydrogens, and produces a relatively large amount of ATP. Aerobic respiration is characteristic of many bacteria, fungi, protozoa, and animals, and is the system we will emphasize here. Facultative and aerotolerant anaerobes may use only the glycolysis scheme to incompletely oxidize or **ferment** glucose. In this case, oxygen is not required, organic compounds are the final electron acceptors, and a relatively small amount of ATP is produced. Some strictly anaerobic microorganisms metabolize by means of **anaerobic respiration,** a system that involves the same three pathways as aerobic respiration, but does not use molecular oxygen as the final electron acceptor. Aspects of fermentation and anaerobic respiration will be covered in subsequent sections of this chapter.

glycolysis (gly-kol'-ih-sis) Gr. *glykys,* sweet, and *lysis,* a loosening.

2. The EMP pathway is named for the biochemists who first outlined its steps. TCA refers to the involvement of several organic acids containing three carboxylic acid groups, citric acid being the first tricarboxylic acid formed; Krebs is in honor of Sir Hans Krebs who, with F. A. Lipmann, delineated this pathway, an achievement for which they won the Nobel Prize in 1953.

Feature 7.6 Microbial Models of Metabolism

Looking back to the days of my first microbiology course, I recall the reverence with which the instructors spoke of the bacterium that was a constant part of our lab work and a common example in the lecture. *Escherichia coli* (everyone called it plain *E. coli*) was the laboratory workhorse and the research pet; at times, it began to seem like an old friend. It is no wonder that this species, named for the German bacteriologist Escherich and for the colon, where it is commonly found, has gained such a following.

A Tribute to *Escherichia coli* Due to ease in cultivation, metabolic versatility, and relative nonpathogenicity, *E. coli* was the species of choice from the earliest studies in metabolism and genetics and was a source of major discoveries in these areas. Many of the known biochemical pathways and metabolic enzymes were worked out with it. A picture that emerged from these and ensuing studies on other organisms was that the chemical activities of cells, ranging from *E. coli* to humans, were fundamentally the same. Because so much information was compiled on the physiology and genetics of *E. coli* over time, it automatically became the model for techniques in biotechnology and genetic engineering. Although bacterial species in the genera *Bacillus* and *Salmonella,* the yeast *Saccharomyces,* and cell cultures are more recent contenders for large roles in research and biotechnology, the colon bacillus continues to be a valuable subject.

Microbes—The Modern Guinea Pigs? So similar is the metabolism of certain fungi and bacteria to the metabolism of mammals that these microbes are now being used as models of mammalian metabolism. The organisms are used alone or in mixed cultures to decipher the fine details of pathways that are still obscure or to test the metabolic pathways of drugs and toxic substances. In the latter case, researchers are looking for metabolic breakdown products of new drugs and their possible adverse effects on humans. Fungi such as *Aspergillus* have pathways of drug processing remarkably similar to those of the human liver. Progress is also being made in determining how microbes process pollutants and cancer-causing chemicals. Among the greatest benefits of microbial modelling are that the number of experimental animals can be reduced, that the microbes are easy to grow in large numbers, and that metabolic by-products can be amassed in bulk amounts.

Aerobic Respiration

Aerobic respiration is a series of enzyme-catalyzed reactions in which electrons are transferred from fuel molecules to oxygen as a final electron acceptor. This pathway is the principal energy-yielding scheme for aerobic heterotrophs, and it provides both ATP and metabolic intermediates for many other pathways in the cell, including those of protein, lipid, and carbohydrate synthesis.

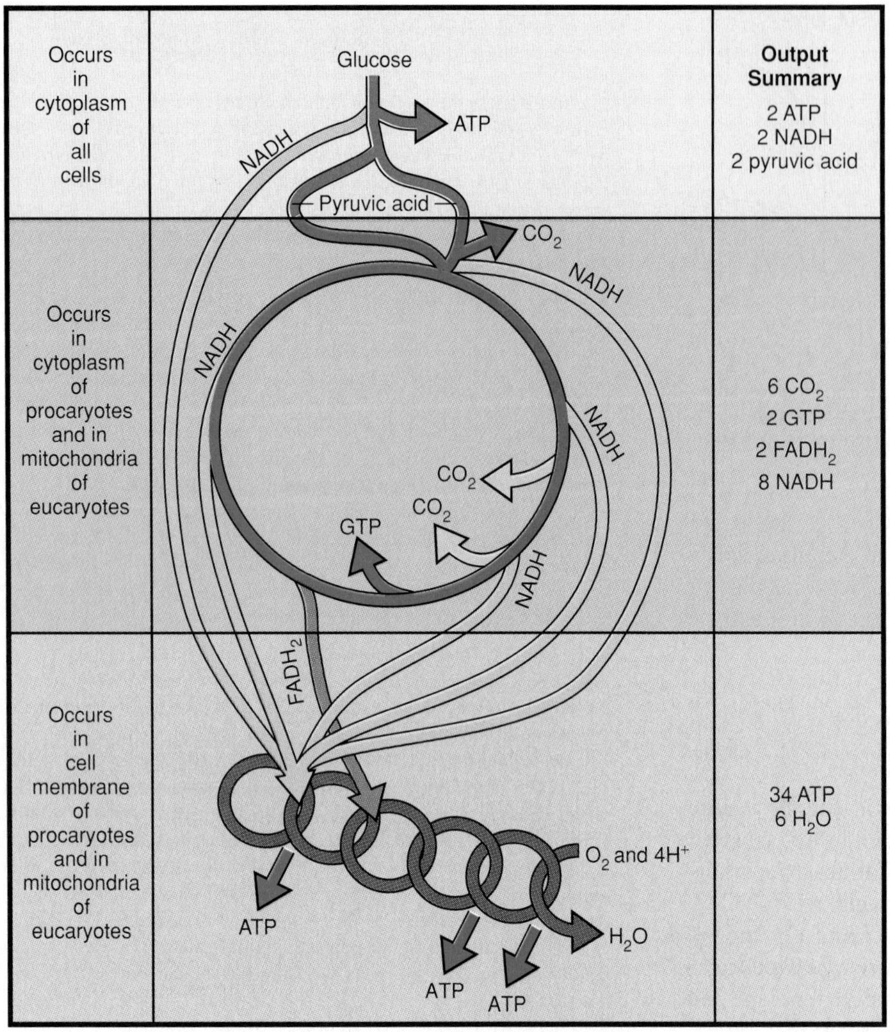

Figure 7.18 Overview of the flow, location, and products of pathways in aerobic respiration. Glucose is degraded through a gradual stepwise process to carbon dioxide and water. Glycolysis proceeds in the absence of oxygen, reduces the glucose to two 3-carbon fragments, and produces a small amount of ATP. The tricarboxylic acid cycle receives these 3-carbon fragments and processes them through redox reactions that extract the electrons and hydrogens. These are shuttled into the electron transport to be used in ATP synthesis. CO_2 is an important product of the TCA cycle. Transport of electrons in the third phase generates a large amount of ATP; in the final step, the electrons and hydrogens are received by oxygen, which forms water. Both TCA and electron transport require oxygen to proceed.

Table 7.5 Metabolic Strategies Among Heterotrophic Microorganisms

Scheme	Pathways Involved	Final Electron Acceptor	Net Products	Chief Microbe Type
Aerobic respiration	Glycolysis, TCA cycle, electron transport	O_2	38 ATP, CO_2, H_2O	Aerobes; facultative anaerobes
Anaerobic metabolism				
Fermentative	Glycolysis	Organic molecules	2 ATP, CO_2, ethanol, lactic acid*	Facultative, aerotolerant, strict anaerobes
Respiration	Glycolysis, TCA cycle, electron transport	Various inorganic salts (NO_3, SO_4, CO_3)	CO_2, ATP, organic acids, H_2S, CH_4	Anaerobes; some facultatives

*The products of microbial fermentations are extremely varied and include organic acids, alcohols, and gases.

Aerobic respiration can be summarized by an equation:

$$\text{Glucose } (C_6H_{12}O_6) + 6\ O_2 + 38\ \text{ADP} + 38\ P_i \rightarrow$$

$$6\ CO_2 + 6\ H_2O + 38\ \text{ATP}$$

The seeming simplicity of the equation for aerobic respiration conceals its complexity. Fortunately, we do not have to present all of the details to answer a few important questions concerning its reactants and products. The most important ideas to explore are: (1) the steps in the oxidation of glucose, (2) the involvement of coenzyme carriers and the final electron acceptor, (3) how molecules are shunted from pathway to pathway, (4) where and how ATP originates, (5) where carbon dioxide originates, (6) where oxygen is needed, (7) where water originates, and (8) connections with other pathways. The prominent elements in aerobic respiration are summarized in table 7.6.

Glucose: The Starting Compound Why are carbohydrates like glucose such good fuels? These compounds are highly reduced—that is, they are superior hydrogen donors. The enzymatic withdrawal of hydrogen from them also removes electrons that are unstable and rich in energy. The end products of the conversion of these carbon compounds are energy-rich ATP and energy-poor carbon dioxide and water. Polysaccharides (starch, glycogen) and disaccharides (maltose, lactose) are stored sources of glucose for the respiratory pathways. Although we use glucose as the main starting compound, other hexoses (fructose, galactose) and fatty acid subunits may enter the pathways of aerobic respiration as well (see figure 7.26). Each step of metabolism is catalyzed by a specific enzyme, but we will not mention it for most reactions.

Glycolysis: The Starting Lineup An anaerobic process called **glycolysis** enzymatically converts glucose into **pyruvic acid.** Depending on the organism and the conditions, it may be only the first phase of aerobic respiration, or it may serve as the primary metabolic pathway (fermentation). Glycolysis is significant for providing a means to synthesize a small amount of ATP anaerobically and also to generate pyruvic acid, an essential intermediary metabolite.

Steps in the Glycolytic Pathway Glycolysis proceeds along nine linear steps, starting with glucose and ending with pyruvic acid (figure 7.19). The first portion of the Embden-Meyerhof-Parnas pathway involves activation of the substrate, and the steps following involve oxidation reactions of the glucose fragments, the synthesis of ATP, and the formation of pyruvic acid. The following outline lists the principal steps of glycolysis:

1. **Glucose** is phosphorylated by means of an **ATP** to activate it for the subsequent oxidations. The product is **glucose-6-phosphate.** (Numbers in chemical names refer to the position of the phosphate on the carbon skeleton.)
2. Glucose-6-phosphate is converted to its isomer, **fructose-6-phosphate.**

Table 7.6 Reaction Checklist for Aerobic Respiration

Input
Glucose, O_2, ADP, P_i

Participants
Various intermediate metabolites, enzymes, electron donors, electron carriers, electron recipients

Flow of Metabolites and Electrons
Through glycolysis, TCA cycle, and electron transport

Compartments of the Cell Where It Occurs
Glycolysis—cytoplasm in all organisms; TCA cycle—mitochondria in eucaryotes and cytoplasm in procaryotes; electron transport—mitochondria in eucaryotes and cell membrane in bacteria

Ultimate Output
ATP, CO_2, H_2O

3. Another **ATP** is spent in phosphorylating the first carbon of fructose-6-phosphate, which yields **fructose-1,6-diphosphate.**

(To this point, no energy has been released, no oxidation-reduction has occurred, and in fact, 2 ATPs have been used. In addition, the molecules remain in the 6-carbon state.)

4. Now doubly activated, fructose-1,6-diphosphate is split into two 3-carbon fragments: **glycerol-3-phosphate (G-3-P)** and dihydroxyacetone phosphate (DHAP). These molecules are isomers, and DHAP is enzymatically converted to the metabolically more useful G-3-P.

(The effect of this step is to double every subsequent reaction, because where there was once a single molecule, there are now two to be fed into the remaining pathways.)

5. Each molecule of glyceraldehyde-3-phosphate becomes involved in the single oxidation-reduction reaction of glycolysis, a reaction that sets the scene for ATP synthesis. Two reactions occur simultaneously and are catalyzed by the same enzyme. First, the coenzyme **NAD** picks up or removes hydrogen from G-3-P, forming **NADH.** This is followed by the addition of an **inorganic phosphate** (P_i) to form a high-energy bond on the third carbon of the G-3-P substrate. The product of these reactions is **diphosphoglyceric acid (DPGA).**

(In aerobic organisms, the NADH formed during this step will undergo further reactions in the electron transport system, where the final H acceptor will be oxygen and additional ATPs will be generated [see figure 7.25]. In organisms that ferment glucose anaerobically, the NADH will be oxidized back to NAD, and the hydrogen acceptor will be an organic compound [see figure 7.21].)

6. One of the **high-energy phosphates** of DPGA is donated to **ADP,** resulting in a molecule of **ATP.** The product of this reaction is **3-phosphoglyceric acid.**

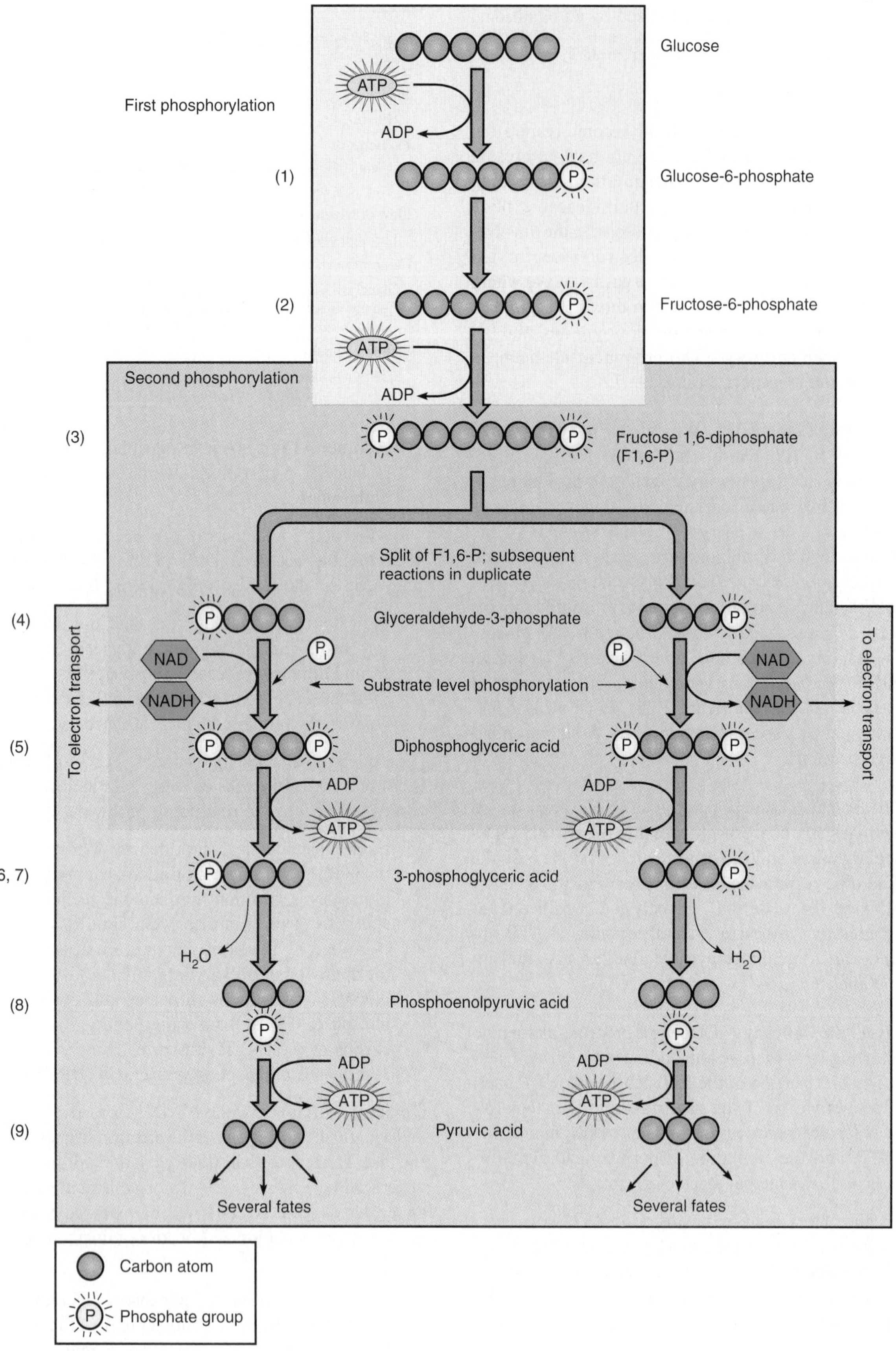

Figure 7.19 The reactions of the glycolysis system (Embden-Meyerhof-Parnas pathway).

7,8. A substrate for the synthesis of a second ATP is produced. First, 3-phosphoglyceric acid is converted to **2-phosphoglyceric acid** through the shift of a phosphate from the third to the second carbon. Subsequent removal of a water molecule from 2-phosphoglyceric acid produces **phosphoenolpyruvate** and generates another **high-energy phosphate** that will be donated to ADP in the next reaction.

9. In the final reaction of glycolysis, phosphoenolpyruvate gives up its high-energy phosphate to form a **second ATP,** and produces **pyruvic acid,** also called pyruvate, a compound with many roles in metabolism.

The 2 ATPs formed during steps 6 and 9 are examples of substrate-level phosphorylation, in that the high-energy phosphate is transferred directly from a substrate to ADP. These reactions are catalyzed by kinases, special types of transferase that can phosphorylate a substrate reversibly. Because both molecules that arose in step 4 undergo these reactions, this makes an overall total of 4 ATPs generated in the partial oxidation of a glucose to 2 pyruvic acids. However, 2 ATPs were expended for steps 1 and 3, so the net number of ATPs available to the cell from these reactions is 2.

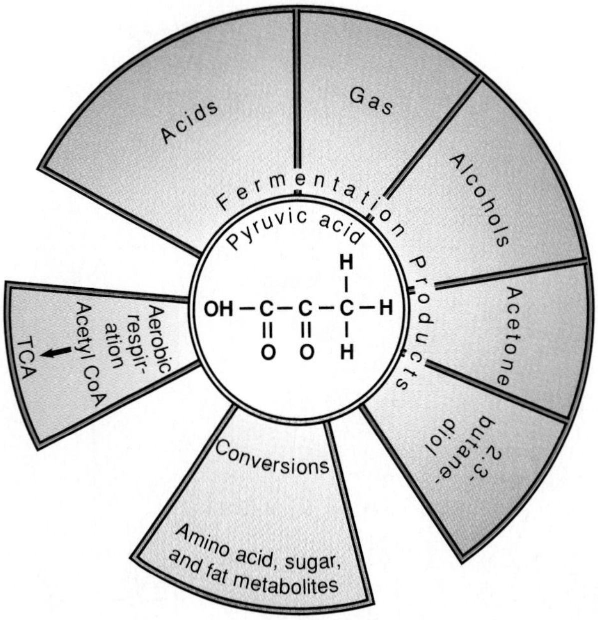

Figure 7.20 Pyruvic acid (pyruvate) is an important hub in the processing of nutrients by microbes. It may be fermented anaerobically to several end products or oxidized completely to CO_2 and H_2O through the TCA cycle and the electron transport system. It may also serve as a source of raw material for synthesizing amino acids and carbohydrates.

Pyruvic Acid—The Grand Central Station of Molecules

Pyruvic acid occupies an important position in several pathways, and different organisms handle it in different ways (figure 7.20). In strictly aerobic organisms and some anaerobes, pyruvic acid enters the TCA cycle for further processing and energy release. Facultative anaerobes may adopt a fermentative metabolism, in which glycolysis is followed by reduction of pyruvic acid and formation of fermentation products. Of all the results of pyruvate metabolism, probably the most varied is fermentation. Before we continue discussing aerobic respiration, let us take a short side trip to explore some important highlights of fermentation.

The Importance of Fermentation

Technically speaking, **fermentation** is the incomplete oxidation of glucose or other carbohydrates in the absence of oxygen, a process that uses organic compounds as the terminal electron acceptors and yields a small amount of ATP. Over time, the term fermentation has acquired several looser connotations. Originally, Pasteur called the microbial action of yeast during wine production *ferments* (see feature 7.7), and to this day, biochemists use the term in reference to the production of ethyl alcohol by yeasts acting on glucose and other carbohydrates. Fermentation is also what bacteriologists call the formation of acid, gas, and other products by the action of various bacteria on pyruvic acid. The process is a common metabolic strategy among bacteria. Industrial processes that produce chemicals on a massive scale through the actions of microbes are also called fermentations (see chapter 22). Each of these usages is acceptable for one application or another.

Fermentation extracts only meager amounts of energy from glucose and other carbohydrates. Despite this relative inefficiency, microbes that rely on fermentative metabolism gain sufficient energy to survive. From one standpoint, fermentation permits independence from molecular oxygen and allows colonization of anaerobic environments. It also enables microorganisms with a versatile metabolism to adapt to variations in the availability of oxygen. For them, fermentation provides a means to grow even when oxygen levels are too low for aerobic respiration.

Bacteria that digest cellulose in the rumens of cattle (as described in chapter 6) are largely fermentative. After initially hydrolyzing cellulose to glucose, they ferment the glucose to organic acids that are then absorbed as the bovine's principal energy source. Even human muscle cells may undergo a form of fermentation that permits short periods of activity after the oxygen supply in the muscle has been exhausted. Muscle cells convert pyruvic acid into lactic acid, which allows anaerobic production of ATP to proceed for a time. But this cannot go on indefinitely, and after a few minutes, the accumulated lactic acid causes muscle fatigue.

Products of Pyruvic Acid Fermentation

Alcoholic beverages (wine, beer, whiskey) are perhaps most prominent among fermentation products, as are solvents (acetone, butanol), organic acids (lactic, acetic), dairy products, and many other foods. Derivatives of proteins, nucleic acids, and

---

fermentation (fur-men-tay'-shun) L. *fervere,* to boil, or *fermenatum,* leaven or yeast.

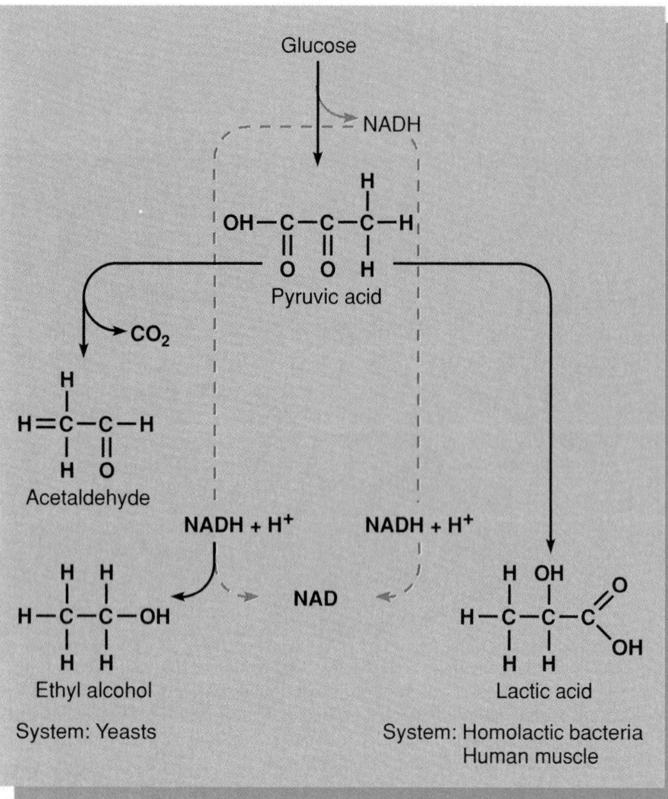

Figure 7.21 The chemistry of fermentation systems that produce acid and alcohol. In both cases the final electron acceptor is an organic compound. In yeasts, pyruvic acid is decarboxylated to acetaldehyde, and the NADH given off in the glycolytic pathway reduces it to ethyl alcohol. In homolactic fermentative bacteria, pyruvic acid is reduced by NADH to lactic acid.

other organic compounds are fermented to produce vitamins, antibiotics, and even hormones such as hydrocortisone.

Fermentation products may be grouped into two general categories—alcoholic fermentation products and acidic fermentation products (figure 7.21). **Alcoholic fermentation** occurs in yeast species that have metabolic pathways for converting pyruvic acid to ethanol. This process involves a decarboxylation of pyruvic acid to acetaldehyde, followed by a reduction of the acetaldehyde to ethanol. In oxidizing the NADH formed during glycolysis, this step returns a balance to the oxidation-reduction scheme. These processes are crucial in the production of beer and wine, though the actual techniques for arriving at the desired amount of ethanol and the prevention of unwanted side reactions are important tricks of the brewer's trade (see feature 7.7). Note that the products of alcoholic fermentation are not only ethanol but also CO_2, a gas that accounts for the bubbles in champagne and beer.

Alcohols other than ethanol may be produced during bacterial fermentation pathways. Certain clostridia produce butanol and isopropanol through a complex series of reactions. Although this process was once an important source of alcohols for industrial use, it has been largely replaced by a nonmicrobial petroleum process.

Feature 7.7 Pasteur and the Wine-to-Vinegar Connection

The microbiology of alcoholic fermentation was greatly clarified by Louis Pasteur after French wine makers hired him to uncover the causes of periodic spoilage in wines. Especially troublesome was the conversion of wine to vinegar and the resultant sour flavor. Up to that time, wine formation had been considered strictly a chemical process. After extensively studying beer making and wine grapes, Pasteur concluded that wine, both fine and not-so-fine, was the result of microbial action on the juices of the grape and that wine "disease" was caused by contaminating organisms that produced undesirable products such as acid. Although he did not know it at the time, the bacterial contaminants responsible for the acidity of the spoiled wines were likely to be *Acetobacter* or *Gluconobacter* introduced by the grapes, air, or wine-making apparatus. These common gram-negative genera further reduce ethanol to acetic acid and are presently used in commercial vinegar production. Pasteur's far-reaching solution to the problem is still with us today—mild heating, or *pasteurization,* of the grape juice to destroy the contaminants, followed by inoculation of the juice with a pure yeast culture. It also introduced the concept that a single microbe could be responsible for both a desired product and a disease. The topic of wine making is explored further in chapter 22.

The pathways of **acidic fermentation** are extremely varied (see figure 7.21). Lactic acid bacteria ferment glucose in the same way that humans do—by reducing it to lactic acid. If the product of this fermentation is *only* lactic acid, as in certain species of *Streptococcus* and *Lactobacillus,* it is termed *homolactic.* The souring of milk is due largely to the production of this acid by bacteria.

When glucose is fermented to a mixture of lactic acid, acetic acid, and carbon dioxide, as is the case with *Leuconostoc* and other species of *Lactobacillus,* the process is termed heterolactic fermentation. Many members of the family Enterobacteriaceae (*Escherichia, Shigella,* and *Salmonella*) possess enzyme systems for converting pyruvic acid to several acids simultaneously (figure 7.22). **Mixed acid fermentation** produces a combination of acetic, lactic, succinic, and formic acids and lowers the pH of a medium to about 4.0. *Propionibacterium* produces primarily propionic acid, which gives the characteristic flavor to Swiss cheese while fermentation gas produces the holes. Some members also further decompose formic acid completely to carbon dioxide and hydrogen gases. Because enteric bacteria commonly occupy the intestine, this fermentative activity accounts for the accumulation of some types of gas—primarily CO_2 and H_2—in the intestine. Some bacteria oxidize the organic acids and produce the neutral end product 2,3-butanediol (see feature 7.8).

We have provided only a brief survey of fermentation products, but it is worth noting that microbes may be harnessed to synthesize a variety of other substances by varying the raw materials provided them. In fact, so broad is the meaning of the word fermentation that the large-scale industrial syntheses by microorganisms often utilize entirely different mechanisms from those described here, and even occur aerobically. This is particularly true in antibiotic, hormone, vitamin, and amino acid production.

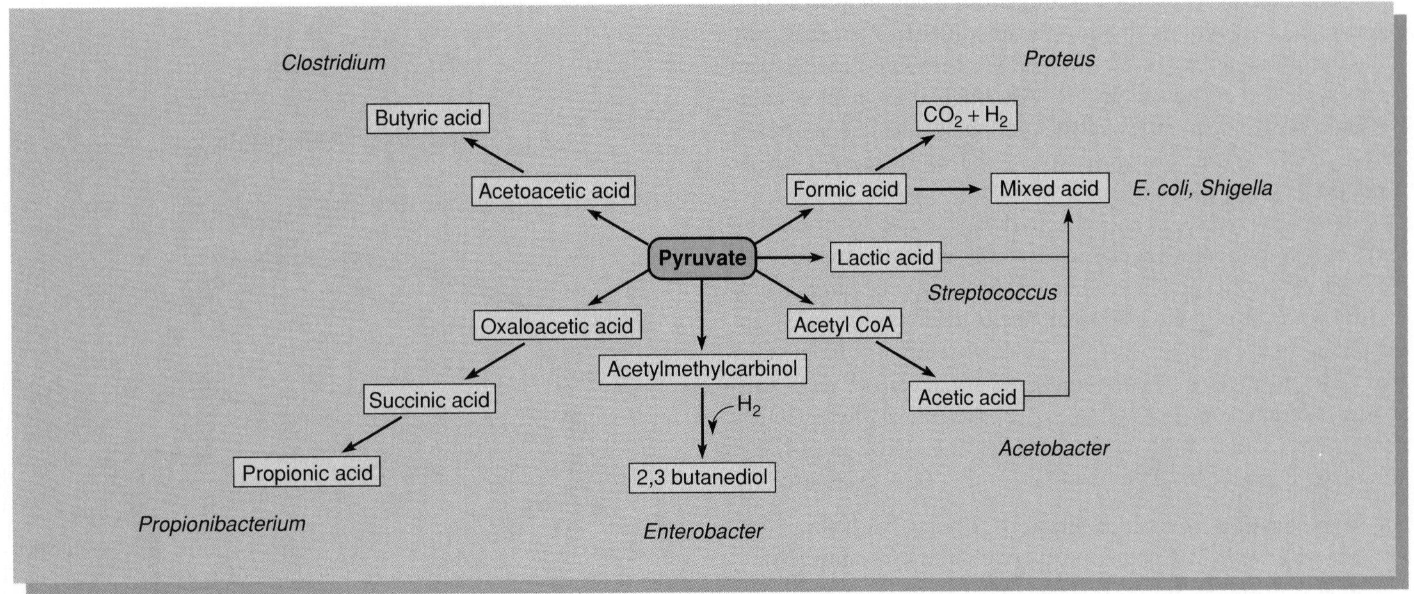

Figure 7.22 Miscellaneous products of pyruvate fermentation and the bacteria involved in their production.

Feature 7.8 Fermentation and Biochemical Testing

The knowledge and understanding of the fermentation products of given species are important not only in industrial production, but also in identifying bacteria by biochemical tests. Fermentation patterns in enteric bacteria, for example, are an important identification tool. Specimens are grown in media containing various carbohydrates, and the production of acid or acid and gas is noted (see figure 6.27). For instance, *Escherichia* ferments the milk sugar lactose, whereas *Shigella* and *Proteus* do not. On the basis of its gas production during glucose fermentation, *Escherichia* can be further differentiated from *Shigella,* which ferments glucose but does not generate gas. Other enteric bacteria are separated on the basis of whether glucose is fermented to mixed acids or to 2,3 butanediol. *Escherichia coli* produces mixed acids, while *Enterobacter* and *Serratia* form primarily 2,3-butanediol. These are part of the IMViC testing described in chapter 16. The M refers to the methyl red test for mixed acids, and the V refers to the Voges-Proskauer test for the butanediol pathway.

Continuation of Aerobic Respiration

As we have seen, the anaerobic oxidation of glucose yields a comparatively small amount of energy and leaves pyruvic acid. Pyruvic acid is still energy-rich, containing a number of extractable hydrogens and electrons to power ATP synthesis, but this can be achieved only through the work of the second and third phases of respiration (see figure 7.18). The tricarboxylic acid cycle begins by converting the 3-carbon pyruvic acid to a 2-carbon fragment. The fragment is then channeled through a circuit of organic acids (TCA) where it is degraded to carbon dioxide and where multiple redox reactions occur. This takes place in the mitochondrial matrix in eucaryotes and in the cytoplasm of bacteria. Coupled closely to the TCA cycle is the respiratory chain, a system that transports the hydrogens and electrons, siphons off their energy into ATP, and donates them to oxygen.

The Tricarboxylic Acid Cycle: A Carbon and Energy Wheel

In the first step of the **tricarboxylic acid** cycle, pyruvic acid is converted to a compound that can be fed into the TCA cycle proper (figure 7.23). This step involves the first oxidation-reduction reaction of this phase of respiration, and it also removes a carbon as carbon dioxide. During this very complex reaction, a cluster of enzymes and **coenzyme A** participates in the dehydrogenation (oxidation) of pyruvic acid, the reduction of NAD to NADH, and the decarboxylation of pyruvic acid to a 2- carbon **acetyl** group. The acetyl group remains attached to coenzyme A, forming a complex called **acetyl coenzyme A** (acetyl-CoA) that begins the TCA cycle. This is a huge enzyme complex. In *E. coli,* it contains four vitamins and has a molecular weight of six million—larger than a ribosome!

The NADH formed during this reaction will be shuttled into electron transport and used to generate ATP; its formation is one of five dehydrogenations associated with the TCA cycle and is what accounts for the relatively greater output of ATP in aerobic respiration. This is also the first of three occasions of CO_2 formation in this pathway. Be reminded that all reactions described actually happen twice for each glucose because of the two pyruvates that are released during glycolysis. An important incidental feature of this pathway is that acetyl groups from the breakdown of certain fats can enter the pathway at this same point.

The acetyl groups, the remaining fragments of glucose that started the glycolytic pathway, are subsequently dismantled by first combining with a 4-carbon oxaloacetate molecule to form citric acid, a 6-carbon molecule. Although this seems a cumbersome way to dismantle such a small molecule, it is necessary in biological systems, and it favors the extraction of larger amounts of energy.

As we have seen, a cyclic pathway is one in which the starting compound is regenerated at the end (see figure 7.9). The tricarboxylic acid cycle has eight steps, beginning and ending with the organic acid **oxaloacetic acid**[3] (figure 7.23). As we take a single spin around the TCA wheel, it will be instructive to keep track of (1) the numbers of carbons of each substrate reactant and product (#C), (2) reactions where CO_2 is generated, (3) the involvement of NAD and FAD, and (4) the site of ATP synthesis. The reactions in the TCA cycle are:

1. **Oxaloacetic acid** (oxaloacetate; 4C) reacts with the **acetyl group** (2C) on acetyl-CoA, thereby forming **citric acid** (citrate; 6C), and releasing coenzyme A to react with another acetyl.
2. Citric acid is converted to its isomer, **isocitric acid** (isocitrate; 6C), to prepare this substrate for the decarboxylation and dehydrogenation of the next step.
3. Isocitric acid is acted upon by an enzyme complex including NAD or NADP (depending on the organism), in a reaction that generates NADH or NADPH, splits off a carbon dioxide, and leaves **α-ketoglutaric acid** (α-ketoglutarate; 5C).
4. Alpha-ketoglutaric acid serves as a substrate for the last decarboxylation reaction and yet another redox reaction involving coenzyme A and yielding NADH. The product is the high-energy compound **succinyl-CoA** (4C).

(At this point, the cycle has converted the 3 original carbons of pyruvic acid to 3 CO_2's, yet it is only half over. As it turns out, the remaining steps are needed not only to regenerate the oxaloacetic acid to start the cycle again, but also to extract more energy from the intermediate compounds leading to oxaloacetic acid.)

5. Succinyl-CoA releases its high-energy bond, this time not to an ADP but to its relative and metabolic equivalent, GDP (guanosine diphosphate), forming GTP. GTP will later be converted to ATP. This is the single substrate level phosphorylation in the TCA cycle. This reaction leaves **succinic acid** (succinate; 4C).
6. Succinic acid is dehydrogenated, and the electrons are accepted by FAD, forming $FADH_2$. Like NADH, $FADH_2$ carries its electrons to the respiratory chain to drive ATP synthesis, though $FADH_2$ comes in at a different point. **Fumaric acid** (fumarate; 4C) is the product of this reaction.
7. The addition of water to fumaric acid (called hydration) results in **malic acid** (malate; 4C).

3. In biochemistry, the organic acids are represented interchangeably as the acid (oxaloacetic acid) or as its salt (oxaloacetate).

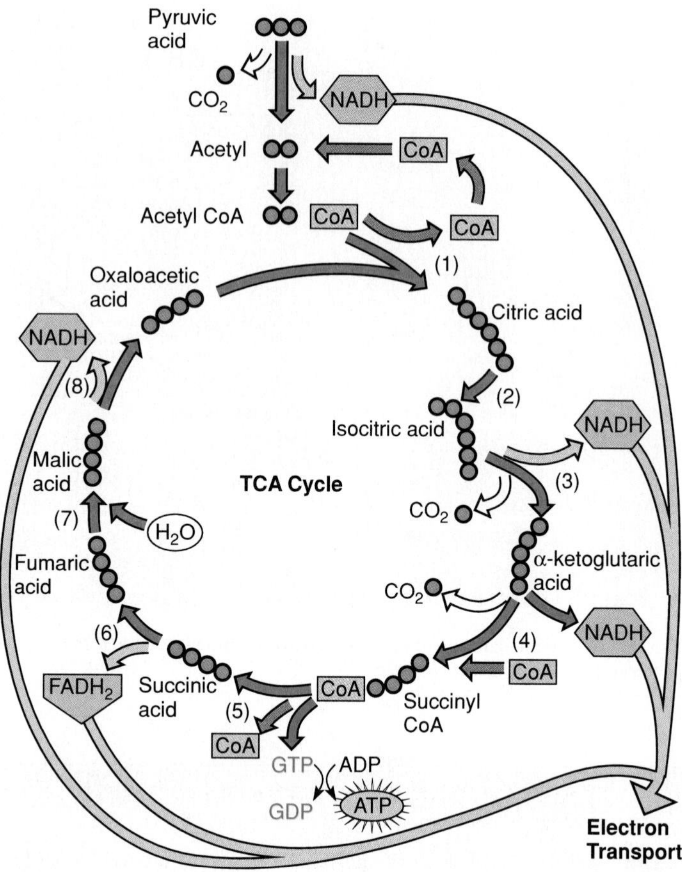

Figure 7.23 The reactions of a single turn of the TCA cycle. Each glucose will produce two spins.

8. Malic acid is dehydrogenated (with formation of a final NADH), and **oxaloacetic acid** is formed. This brings the cycle full circle and ready for another round.

The Respiratory Chain: Electron Transport and Oxidative Phosphorylation

We now come to the energy chain, which is the final "processing mill" for electrons and hydrogen and the major generator of ATP. Overall, this system consists of a chain of special redox carriers that receive electrons from reduced carriers (NADH, $FADH_2$) generated by the TCA cycle and glycolysis and shuttle them in a sequential and orderly fashion (see figure 7.19). The flow of electrons down this chain is highly energetic and gives off ATP at various points. At its end, an enzyme catalyzes the final acceptance of electrons and hydrogen by oxygen, producing water. Some variability exists from one organism to another, but the principal compounds that carry out these complex reactions are NADH dehydrogenases, flavoproteins, coenzyme Q (*ubiquinone*), and *cytochromes*. These complex compounds

ubiquinone (yoo-bik'-wih-nohn) L. *ubique*, everywhere. A type of chemical, similar to vitamin K, that is very common in cells.

cytochrome (sy'-toh-krohm) Gr. *cyto*, cell, and *kroma*, color. Cytochromes are pigmented, iron-containing molecules similar to hemoglobin.

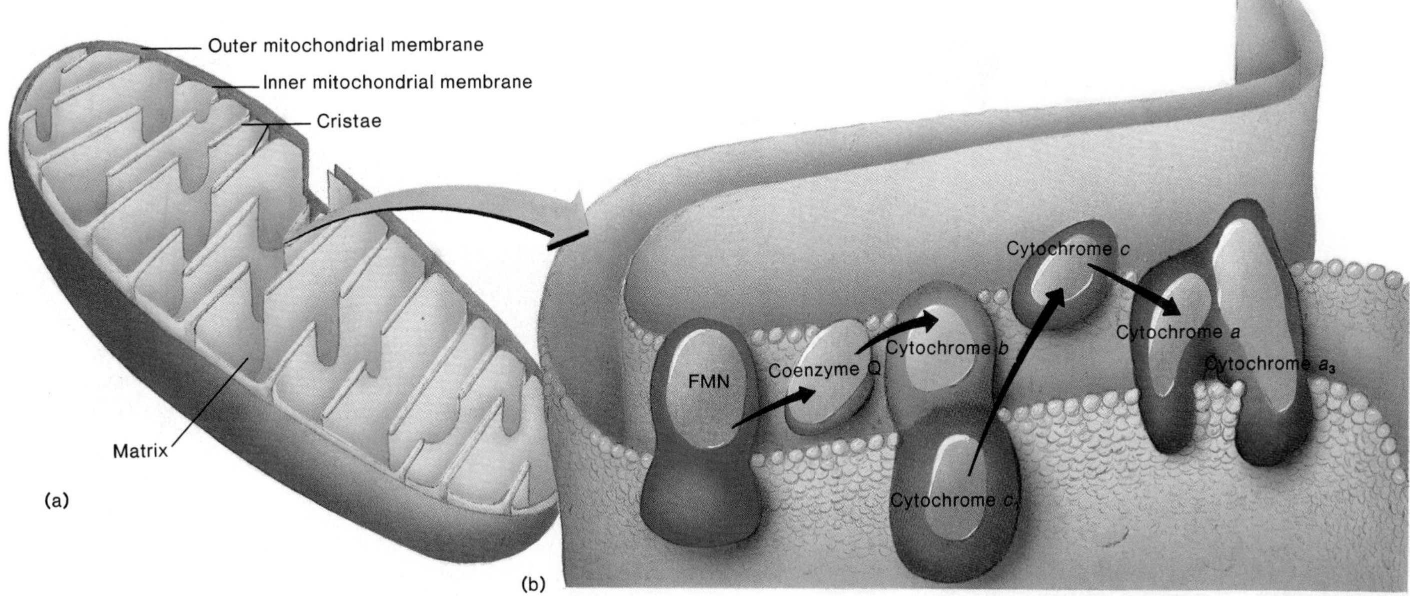

Figure 7.24 (*a*) Location and arrangement of respiratory carriers on the cristae of mitochondria. (*b*) The electron transport chain, showing actual orientation of electron carriers with respect to the inner mitochondrial membrane.

contain a metal that readily facilitates receiving and donating electrons (being reduced and oxidized). The highly compartmentalized structure of the respiratory chain is an important factor in its function. Note in figure 7.24 that the electron transport carriers and enzymes are embedded in the inner mitochondrial membranes in eucaryotes and in the cell membrane of bacteria.

Elements of Electron Transport: The Energy Cascade

The principal questions about the electron transport system are: How are the electrons passed from one carrier to another in the series; how is this coupled to ATP synthesis; and where and how is oxygen utilized? Although the biochemical details of this process are rather complicated, the basic reactions consist of a number of redox reactions now familiar to us. In general, the seven carrier compounds and their enzymes are arranged in linear sequence and are reduced and oxidized in turn as follows:

> Component A in the series receives a pair of electrons from NADH and is reduced to AH. AH then donates the electrons (is oxidized) to the adjacent compound B, forming BH (reduced). BH is then oxidized by C, forming CH. CH donates to D, and down the chain until the final reduction of oxygen.

The sequence of electron carriers in the respiratory chain of most aerobic organisms is (1) NADH dehydrogenase, (2) flavin mononucleotide (FMN), (3) coenzyme Q, (4) cytochrome *b*, (5) cytochrome c_1, (6) cytochrome *c*, and (7) cytochrome a/a_3. Conveyance of the NADHs from the TCA cycle and glycolysis to the first carrier sets in motion the remaining six steps. With each redox exchange, the energy level of the reactants is lessened. The released energy is captured and used by the ATP synthetase complex to produce 3 ATPs for each NADH that enters (figure 7.25). This coupling of ATP synthesis to electron transport is termed **oxidative phosphorylation.** Since FADH from the TCA cycle is metabolically equivalent to FMN, it side-steps ATP synthesis at the first site, resulting in one less ATP for its trip through the transport chain.

The Theory of ATP Formation by Oxidative Phosphorylation

What biochemical mechanisms are involved in coupling electron transport to the production of ATP? We have observed that the components of electron transport are embedded in a precise sequence along the respiratory membrane. This stations them essentially between two compartments: the inner mitochondrial matrix and the outer membrane and cytoplasm. According to a widely accepted concept called the *chemiosmotic hypothesis,* as the electron transport carriers shuttle electrons, they actively pump hydrogen ions (protons) into the outer compartment of the mitochondrion. This process sets up a concentration gradient of hydrogen ions, and it also generates a difference in charge between the outer membrane compartment (+) and the inner membrane compartment (−). The potential energy inherent in such a separation of charge can be harnessed to produce ATP. Although the exact mechanism is not completely understood, it appears special pores that penetrate the ATP synthetase complex transport hydrogen ions (protons) from the outside compartment back into the mitochondrial matrix. This process releases sufficient free energy for the synthesis of an ATP. This theory has been supported by tests showing that oxidative phosphorylation is blocked if the mitochondrial or bacterial cell membranes are disrupted.

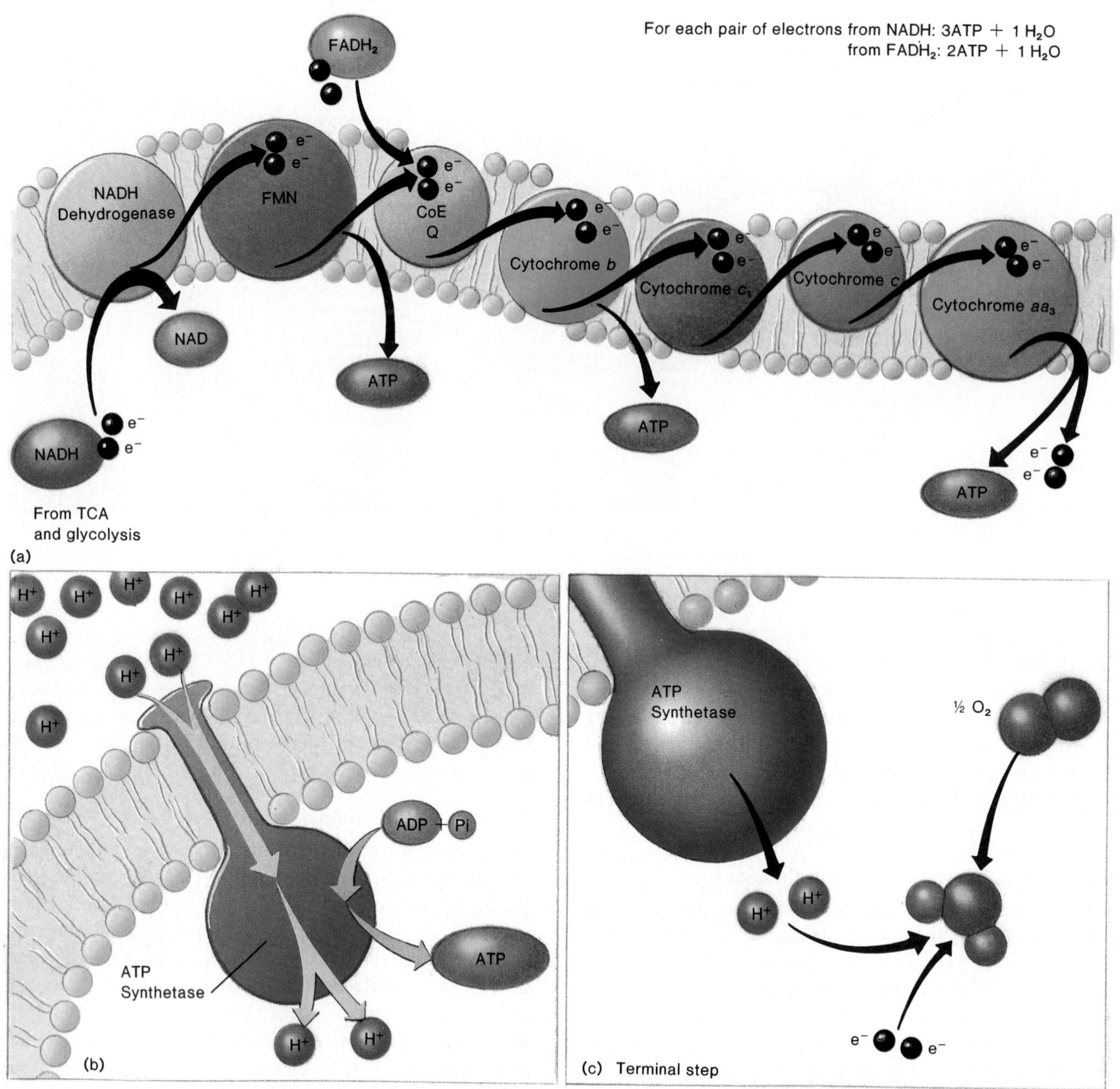

Figure 7.25 The electron transport system and oxidative phosphorylation. (*a*) Starting at NADH dehydrogenase, electrons brought in from the TCA cycle by NADH are passed along the chain of electron transport carriers. Each adjacent pair of transport molecules undergoes a redox reaction. At sites along this chain, this electron transport is coupled to ATP synthesis as shown in (*b*). The membrane actively transports H^+ ions (protons) to the cytoplasmic side of the mitochondrial membrane, a process that forms a charge gradient between the outer and inner compartments. This sort of electric potential distributed across the membrane is capable of generating energy for an ATP to be synthesized by the ATP synthetase. (*c*) At the terminal step, the acceptance of hydrogens and electrons by oxygen produces water.

Table 7.7 Summary of Aerobic Respiration for One Glucose Molecule

	Glycolysis	TCA Cycle	Respiratory Chain	Total Net Output
ATP Produced	2 × 2* = 4	1 × 2 = 2**	17 × 2* = 34	40 − 2 (used) = 38
ATP Used	2	0	0	—
NADH Produced	1 × 2 = 2	4 × 2 = 8	0	10
FADH Produced	0	1 × 2 = 2	0	2
CO_2 Produced	0	3 × 2 = 6	0	6
O_2 Used	0	0	3 × 2 = 6	—
H_2O Produced	1	0	3 × 2 = 6	7 − 1 (used) = 6
H_2O Used	0	1	0	—

*Products are multiplied by 2 because the first figure represents the amount for only one trip through the pathway, and two molecules make this trip per glucose.
**The product is actually GTP, but this will be converted to ATP.

Yield of ATPs from Oxidative Phosphorylation

The total of five NADHs (four from the TCA cycle and one from glycolysis) can be used to synthesize:

15 (5 × 3) ATPs per pair of electrons

and **15 × 2 = 30** ATPs per glucose

The single FADH produced during the TCA cycle results in:

2 ATPs per pair of electrons

and **2 × 2 = 4** ATPs per glucose

Table 7.7 summarizes the total of ATP and other products for the entire aerobic pathway.

Summary of Aerobic Respiration

Originally, we presented a summary equation for respiration. We are now in a position to tabulate the input and output of this equation at various points in the pathways and sum up the final ATPs (see table 7.7). Close examination of this table will review several important facets of aerobic respiration: (1) The total yield of ATP is 40: 4 from glycolysis, 2 from the TCA cycle, and 34 from electron transport. However, 2 ATPs were expended in early glycolysis, leaving a net of **38 ATPs.** (2) Six carbon dioxide molecules are generated during the TCA cycle. (3) Six oxygen molecules are consumed during electron transport. (4) Six water molecules are produced in electron transport and 1 in glycolysis, but because 1 is used in the TCA cycle, this leaves a net of 6.

The Terminal Step

The terminal step during which oxygen accepts the electrons is catalyzed by cytochrome aa_3, also called cytochrome oxidase. This large enzyme complex is specially adapted to receive electrons from cytochrome *c,* pick up hydrogens from solution, and react with oxygen to form a molecule of water. This reaction, though in actuality more complex, is summarized as follows:

$$2\,H + 2\,e^- + \tfrac{1}{2}O_2 \rightarrow H_2O$$

Most eucaryotic aerobes have a fully functioning cytochrome system, but bacteria exhibit wide-ranging variations in this part of the system. Some species lack one or more of the redox steps, while others have several alternative electron transport schemes. Because many bacteria lack cytochrome *c,* this variation can be used to differentiate among certain genera of bacteria. The oxidase test (see figure 15.29) is diagnostic of cytochrome *c.* It is used to help identify members of the genera *Neisseria* and *Pseudomonas* and some species of *Bacillus.* Another variation in the cytochrome system is evident in certain bacteria *(Klebsiella, Enterobacter)* that can grow even in the presence of cyanide because they lack cytochrome oxidase. Cyanide would ordinarily cause rapid death in humans and animals because it blocks cytochrome oxidase, thereby completely terminating aerobic respiration.

A potential side reaction of the respiratory chain in aerobic organisms is the incomplete reduction of oxygen to superoxide ion ($:O_2^-$) and hydrogen peroxide (H_2O_2). Because these toxic oxygen products (previously discussed in feature 6.7) can be very damaging to cells, aerobes have various neutralizing enzymes (*superoxide dismutase* and *catalase*). One exception is the genus *Streptococcus,* which can grow well in oxygen yet lacks both cytochromes and catalase. The tolerance of these organisms to oxygen can be explained by the neutralizing effects of the peroxidase it contains. The lack of cytochromes, catalase, and peroxidases in anaerobes as a rule limits their ability to process free oxygen and contributes to its toxic effects on them.

Alternate Catabolic Pathways

Certain bacteria follow a different pathway in carbohydrate catabolism than in aerobic respiration. The **phosphogluconate pathway** (also called the hexose monophosphate shunt) provides ways to anaerobically oxidize glucose and other hexoses, to release ATP, to produce large amounts of NADPH, and to process pentoses (5-carbon sugars). This pathway, common in heterolactic fermentative bacteria, yields various end products, including lactic acid, ethanol, and carbon dioxide. Furthermore, it is a significant intermediate source of pentoses for nucleic acid synthesis.

Anaerobic Respiration

Some bacteria have evolved an anaerobic respiratory system that functions like the aerobic cytochrome system except that it utilizes oxygen containing salts, rather than free oxygen, as the final electron acceptor. Of these, the nitrate (NO_3) and nitrite (NO_2) reduction systems are best known. The reaction in species such as *Escherichia coli* is represented as:

$$NO_3^- + H_2 \xrightarrow{\text{Nitrite reductase}} NO_2^- + H_2O$$

Reduction here is the removal of oxygen from nitrate. A test for this reaction is one of the battery of physiological tests used in identifying bacteria.

Some species of *Pseudomonas* and *Bacillus* possess enzymes that can further reduce nitrite to nitric oxide (NO), nitrous oxide (N_2O), and even nitrogen gas (N_2). This process, called *denitrification,* is a very important step in recycling nitrogen in the biosphere (see chapter 22). Other oxygen-containing nutrients reduced anaerobically by various bacteria are carbonates and sulfates.

Biosynthesis and the Crossing Pathways of Metabolism

Our discussion now turns from catabolism and energy extraction to anabolic functions and biosynthesis. In this section we will overview aspects of intermediary metabolism, including amphibolic pathways, the synthesis of simple molecules, and the synthesis of macromolecules.

The Frugality of the Cell—Everything Used, Nothing Wasted

It must be obvious by now that cells have mechanisms for careful management of carbon compounds. Rather than being dead ends, most catabolic pathways contain strategic molecular intermediates that can be diverted into anabolic pathways. The opposite is also true. In this way, a given molecule may serve multiple purposes, and the maximum benefit may be derived from all nutrients and metabolites of the cell pool. The property of a system to function in both catabolism and anabolism is **amphibolism.**

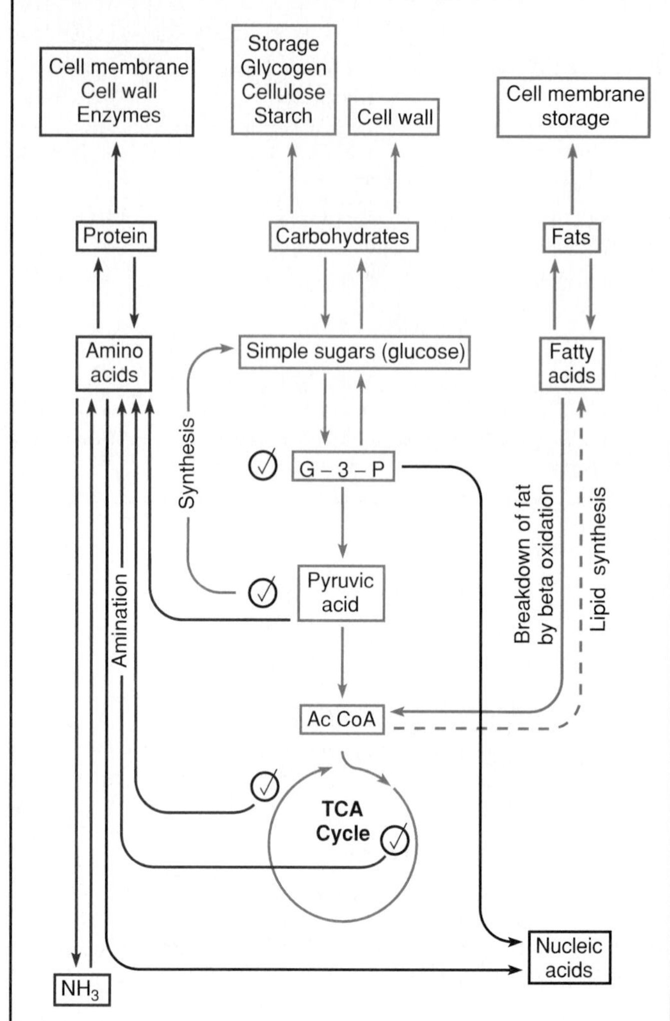

Figure 7.26 A summary of metabolic interactions. Many intermediate compounds serve as amphibolic branch points (✓). With comparatively small modifications, these compounds may be converted into other compounds and enter a different pathway. Note that catabolism of glucose (center) furnishes numerous intermediates for amino acid, fat, nucleic acid, and carbohydrate synthesis. Likewise, molecules arising from fat metabolism may be channeled into the TCA cycle.

The amphibolic nature of intermediary metabolism can be better appreciated if we look at some examples of the branch points and crossroads of several metabolic pathways (figure 7.26). The pathways of glucose catabolism are an especially rich "metabolic marketplace." The principal sites of amphibolic interaction occur during glycolysis (glyceraldehyde-3-phosphate and pyruvic acid) and the TCA cycle (acetyl coenzyme A and various organic acids).

Amphibolic Sources of Cellular Building Blocks

Glyceraldehyde-3-phosphate may be diverted away from glycolysis and converted into precursors for amino acid, carbohydrate, and triglyceride (fat) synthesis. (A precursor molecule is a compound that is the source of another compound.) Earlier we noted the numerous directions that pyruvic acid catabolism may take. In terms of synthesis, pyruvate also plays a pivotal

---

amphibolism (am-fee-bol′-izm) Gr. *amphi,* two-sided, and *bole,* a throw.

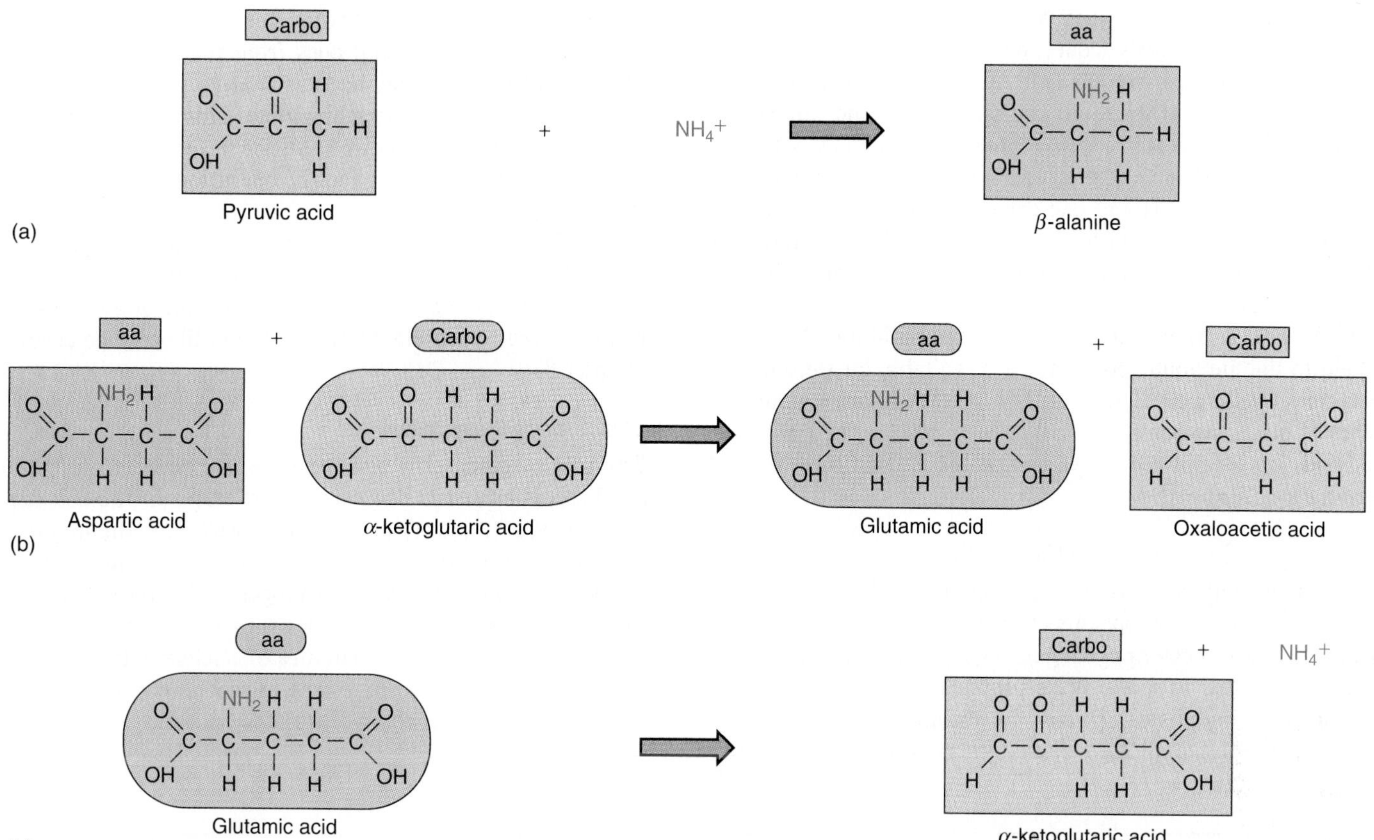

Figure 7.27 Reactions that produce and convert amino acids. (*a*) Through amination (the addition of an ammonium molecule [amino group]), a carbohydrate may be converted to an amino acid. (*b*) Through transamination (transfer of an amino group from an amino acid to a carbohydrate fragment), metabolic intermediates can be converted to amino acids that are in low supply. (*c*) Through deamination (removal of an amino group), an amino acid may be converted to a useful intermediate of carbohydrate catabolism. This is how proteins may be used to derive energy. Ammonium is one waste product.

role in providing intermediates for amino acids. In the event of an inadequate glucose supply, it serves as the starting point in glucose synthesis from various metabolic intermediates, a process called *gluconeogenesis*.

The acetyl group that starts the TCA cycle is another extremely versatile metabolite that may be fed into a number of synthetic cycles. This 2-carbon fragment can be converted as a single unit into one of several amino acids, or a number of these fragments can be condensed into hydrocarbon chains that are important bases for fatty acid and lipid synthesis. Note that the reverse is also true—fats may be degraded to acetyl and thereby enter the TCA cycle. This aerobic process, called *beta oxidation,* can provide a large amount of energy. Oxidation of a 6-carbon fatty acid yields 50 ATP, compared with 38 for a 6-carbon sugar.

Two metabolites of carbohydrate catabolism that the TCA cycle spins off, oxaloacetic acid and α-ketoglutaric acid, are essential intermediates in the synthesis of certain amino acids. This occurs through **amination,** the addition of an amino group to a carbon skeleton (figure 7.27). A certain core group of amino acids may then be used to synthesize others. Amino acids and carbohydrates may be interchanged through transamination (figure 7.27).

Pathways that synthesize the nitrogen bases (purines, pyrimidines), which are components of DNA and RNA, originate in amino acids, and so can be dependent on intermediates from the TCA cycle as well. Because the coenzymes NAD, NADP, FAD, and others contain purines and pyrimidines similar to the nucleic acids, their synthetic pathways are also dependent on amino acids. During times of carbohydrate deprivation, organisms may likewise convert amino acids to intermediates of the TCA cycle by deamination (removal of an amino group) and thereby derive energy from proteins. Deamination results in the formation of nitrogen waste products such as ammonium or urea (figure 7.27).

Formation of Macromolecules

Monosaccharides, amino acids, fatty acids, nitrogen bases, and vitamins—the building blocks that make up the various macromolecules and organelles of the cell—come from two possible sources. They can enter the cell from the outside, as nutrients, or they can be synthesized through amphibolic and other pathways. The degree to which an organism can synthesize its own

gluconeogenesis (gloo″-koh-nee″-oh-gen′-uh-sis) Literally, the formation of "new" glucose.

building blocks is determined by its genetic makeup, a factor that varies tremendously from group to group. In chapter 6 we found that autotrophs require only CO_2 as a carbon source, a few minerals to synthesize all cell substances, and no organic nutrients. Some heterotrophic organisms (*E. coli,* yeasts) are also very efficient in that they can synthesize all cellular substances from minerals and one organic carbon source such as glucose. Compare this with a strict parasite that has few synthetic abilities of its own and drains most precursor molecules from the host.

Whatever their source, once these building blocks are added to the metabolic pool, they are available for synthesis of polymers by the cell. The details of synthesis vary among the types of macromolecules, but all of them involve the formation of bonds by specialized enzymes and the expenditure of ATP (see figure 7.8*a*).

Amino Acids and Protein and Nucleic Acid Synthesis

Proteins account for a large proportion of a cell's constituents. They are essential components of enzymes, the cell membrane, the cell wall, and cell appendages. As a general rule, 20 amino acids are needed to make these proteins (see chapter 2). Although some organisms (*E. coli,* for example) have pathways that will synthesize all 20 of these, others (especially animals) lack some or all of the pathways for amino acid synthesis and must acquire the essential ones from their diets. Protein synthesis itself is a complex and fascinating process that involves far more than simply lining up amino acids and forming peptide bonds between them. As we shall see in the next chapter, it requires a genetic blueprint and the operation of intricate cellular machinery.

DNA and RNA are important for the hereditary continuity of cells and the overall direction of protein synthesis. Because nucleic acid synthesis is a major topic of genetics and is closely allied to protein synthesis, it will likewise be covered in chapter 8.

Carbohydrate Biosynthesis

The role of glucose in bioenergetics is so crucial that its biosynthesis is ensured. But certain structures in the cell depend on an adequate supply of glucose as well. It is the major component of the cellulose cell walls of some eucaryotes and of certain storage granules (starch, glycogen). Monosaccharides other than glucose are important in the synthesis of bacterial cell walls (figure 3.14) and in the structure of nucleic acids (ribose, deoxyribose). The formation of the microbial capsule and glycocalyx also involves carbohydrate biosynthesis.

Chapter Review with Key Terms

Microbial Metabolism

Metabolism is the sum of cellular chemical changes; it involves scores of reactions that interlink in linear or branched pathways. Metabolism is a twin process consisting of **anabolism,** the reactions that convert small molecules into macromolecules, and **catabolism,** in which large molecules are degraded. Intermediary molecules called **metabolites** are generated, and their levels are regulated.

Enzymes

Control is exercised by regulating the activity of organic **catalysts** or **enzymes.** Enzymes facilitate reactions by lowering the **energy of activation,** but rather than being consumed, they are retrieved and reused. The majority of cellular reactions are catalyzed, and each enzyme acts specifically upon its assigned metabolite called the **substrate.**

Enzyme Structure: Depending upon its composition, an enzyme is either **conjugated** or **simple.** A conjugated enzyme consists of a protein component called the **apoenzyme** and one or more activators called **cofactors.** Some cofactors are organic molecules called **coenzymes,** and others are inorganic elements, typically metal ions. To function, a conjugated enzyme must be a complete **holoenzyme** with all its parts. Simple enzymes are functional, though comprised solely of protein.

Enzyme-Substrate Interactions: Metallic cofactors impart greater reactivity to the enzyme-substrate complex. Coenzymes are transfer agents that pass components from one substrate to another. Coenzymes are usually **vitamins.** Substrate attachment occurs in a special pocket called the **active** or **catalytic site.** In order to fit, a substrate must conform to the unique shape, size, and binding features of the active site. This configuration is determined by the amino acid content, sequence, and folding of the apoenzyme. Thus, enzymes are **substrate-specific.**

Enzyme Classification: Enzyme names consist of a prefix and suffix; the former is derived from the type of reaction or the substrate, and the latter is **-ase.** Traditional names such as ptyalin, though less descriptive, are still used. By convention, enzymes may also be classified according to whether or not they are released. Thus, an **exoenzyme** is secreted, but an **endoenzyme** isn't. Moreover, a **constitutive enzyme** is regularly found in a cell, while an **induced enzyme** is synthesized only if its substrate is present.

Types of Enzyme Function: Metabolic reactions vary. Sometimes water is released in anabolism to forge new covalent bonds; these are called **condensation reactions.** Conversely, some **hydrolysis reactions** may occur in catabolism during bond breakage. **Functional groups** may be added, removed, or traded in many reactions. Reactions frequently involve relocating valence electrons, often accompanied by protons (H^+). Compounds yielding electrons are **oxidized,** whereas those gaining electrons are **reduced.**

Enzyme Sensitivity: Enzymes are **labile** (unstable) and function only within narrow operating ranges of temperature, pH, and osmotic pressure. By virtue of their large size and complexity, enzymes are especially vulnerable to **denaturation.** Enzymes are vital metabolic links, and thus constitute easy targets for many harmful physical and chemical agents.

Regulation of Enzymatic Activity

Regulation and interference are exerted upon key enzymes located at or near the beginning of a pathway. A substance that resembles the normal substrate and can occupy the same active site is said to exert **competitive inhibition.** In **feedback,** or **end-product control,** the concentration of the output at the end of a pathway exerts control over the key enzyme. A special mechanism called **feedback inhibition** operates through the plasticity of **allosteric enzymes.** Elsewhere on these large enzymes, secondary molecules govern enzyme activity by attaching to special **regulatory sites** away from the active site. The resultant distortion transmitted through the enzyme affects the specificity of the active site.

Rather than relying upon the suppression of enzyme activity, **feedback repression** operates through enzyme replacement. Depending upon the status of the end product under control, the replenishment of key enzymes is genetically restrained by limiting synthesis. The impact is slower but enduring. **Enzyme induction** is an economizing mechanism that conserves cell resources by producing enzymes only when the appropriate substrate is present.

How Energy Works

Potential and **kinetic energy** both have the capacity to perform work. Energy is consumed in **endergonic reactions** but released in **exergonic reactions.** The freed energy is associated with electrons that can be temporarily captured and transferred in **high-energy molecules** like ATP and GTP. Extracting energy requires a series of **electron carriers** arrayed in a downhill **reciprocal redox chain** between electron donors and electron acceptors. In **oxidative phosphorylation,** threshold levels of energy are applied to **inorganic phosphate,** causing it to form high-energy storage compounds like ATP.

Glycolysis and the TCA Cycle: Carbohydrates like glucose and other hydrocarbon sources like lipids are energy-rich because they are highly reduced and can yield a large number of electrons per molecule. Glucose is dismantled in two stages. **Anaerobic respiration** or **glycolysis** is the degradation of glucose to pyruvic acid, a process that does not require oxygen. Also, pyruvic acid is processed in **aerobic respiration** via the **tricarboxylic acid (TCA) cycle** and its associated respiratory chain. While oxygen is the **final electron acceptor** in aerobic respiration, other acceptors like sulfate, nitrate, or nitrite are employed in anaerobic respiration. Important intermediates in glycolysis are glucose-6-phosphate, fructose-1,6-diphosphate, glycerol-3-phosphate, diphosphoglyceric acid, phosphoglyceric acids, phosphoenolpyruvate, and pyruvic acid.

Fermentation: Anaerobic respiration in which both the electron donor and final electron acceptors are organic compounds. Although fermentation does not reap as much energy as aerobic respiration, this adaptation enables anaerobic and facultative microbes to exploit environments devoid of oxygen. Production of alcoholic brews, breads, pickles, cheese, vinegar, and certain industrial solvents relies upon fermentation. The flavors and aromas of pickled foods are owed to the production of one or several types of (mixed) organic acids. The **phosphogluconate pathway** is an alternate anaerobic pathway for hexose oxidation that also provides for the synthesis of NADPH and pentoses.

Products of Pyruvic Acid Oxidation: Acetyl-coenzyme A is a product of pyruvic acid processing. The net effect of pyruvic acid oxidation in the TCA cycle is the generation of ATP, some GTP, CO_2, and H_2O. Important intermediary metabolites are pyruvate, acetate, oxaloacetate, citrate, isocitrate, α-ketoglutarate, succinyl CoA, succinate, fumarate, and malate. The respiratory chain completes energy extraction. Important redox carriers of the electron transport system are NAD, FAD, coenzyme Q, and cytochromes. The **chemiosmotic hypothesis** is a conceptual model that explains the origin and maintenance of electropotential gradients across a semipermeable membrane, leading to ATP synthesis.

Versatility of Glycolysis and TCA Cycle: Glycolysis and the TCA cycle are bidirectional or **amphibolic pathways.** In addition to use as fuel, some metabolites of these pathways double as building blocks. Intermediates that are convertible into amino acids through **amination** contribute to peptide synthesis. Conversely, the same amino acids from peptide degradation can be **deaminated** and used for fuel. Components for purines and pyrimidines are derived from amino acid pathways. Two-carbon acetate molecules from pyruvate **decarboxylation** are units available for fatty acid synthesis. Glyceraldehyde-3-phosphate is a backbone for fatty acid attachment. The combination yields triglyceride, a typical storage fat, especially in times of carbohydrate abundance. Reversing the process, as in **beta oxidation** of fatty acid, fuel from storage sites is tapped.

True–False Questions

Determine whether the following statements are true (T) or false (F). If you feel a statement is false, explain why, and reword the sentence so that it reads accurately.

____ 1. Anabolism is another term for biosynthesis.

____ 2. Catabolism is a form of metabolism in which small molecules are converted into larger ones.

____ 3. An enzyme lowers the activation energy, thus causing a chemical reaction to occur.

____ 4. An enzyme donates part of itself to form the products of a reaction.

____ 5. An apoenzyme is the part of the enzyme where one finds the active site.

____ 6. Many coenzymes are vitamins.

____ 7. To digest cellulose in its environment, a fungus would produce an endoenzyme.

____ 8. Labile enzymes are easily denatured by an adverse environment.

____ 9. In negative feedback control of enzymes, a buildup of the amount of product increases the activity of the enzyme.

____ 10. The binding of an agent to the regulatory site of an allosteric enzyme distorts the shape of the enzyme and blocks its function.

____ 11. Energy in biological systems is primarily chemical.

____ 12. ATP carries energy from catabolic to anabolic reactions.

____ 13. Exergonic reactions are coupled to endergonic reactions.

____ 14. Cellular redox reactions often involve the removal of electrons and hydrogens.

____ 15. Aerobic organisms require oxygen to serve as the final hydrogen acceptor during respiration.

____ 16. The products of glycolysis are ATP, CO_2, and H_2O.

____ 17. Fermentation of a glucose molecule gives off only 2 ATPs, compared with 38 given off by aerobic respiration.

____ 18. The TCA cycle begins with the formation of acetyl coenzyme A and ends with the formation of oxaloacetic acid.

____ 19. The $FADH_2$ formed during the TCA cycle enters the electron transport system at the same site as NADH.

____ 20. Anabolic pathways are simply reversals of catabolic pathways.

Concept Questions

1. Show diagrammatically the interaction of a holoenzyme and its substrate, and general products that may be formed from a reaction.
2. Give the general name of the enzyme that: synthesizes ATP; digests RNA; carries out the transformation from DHAP to 3-G-P (step 4 in glycolysis); catalyzes the formation of acetyl from pyruvic acid (step 1 in the TCA cycle); reduces pyruvic acid to lactic acid; reduces nitrate to nitrite.
3. Two steps in glycolysis are catalyzed by allosteric enzymes. These are (1) phosphofructokinase, which catabolizes step 3, and (2) pyruvate kinase, which catabolizes step 9. Suggest what metabolic products might regulate these enzymes. How might one place these regulators in figure 7.12?
4. How can oxidation of a substrate proceed without oxygen?
5. In the following redox pairs, which compound is reduced and which is oxidized? NAD and NADH; $FADH_2$ and FAD; lactic acid and pyruvic acid; NO_3 and NO_2; ethanol and acetaldehyde.
6. Discuss the relationship of anabolism to catabolism; of ATP to ADP; of glycolysis to fermentation; of electron transport to oxidative phosphorylation.
7. How is substrate level phosphorylation different from oxidative phosphorylation?
8. Outline the basic steps in glycolysis, indicating where ATP is used and given off. Where does NADH originate, and what is its fate in an aerobe? What is its fate in a fermentative organism?
9. How many ATPs would be formed as a result of aerobic respiration if cytochrome oxidase were missing from the respiratory chain (as is the case with many bacteria)?
10. Compare the general equation for aerobic metabolism with the summary table (7.7), and verify that all figures balance.
11. Water is one of the end products of aerobic respiration. Where in the metabolic cycles is it used?
12. Briefly outline the use of certain metabolites of glycolysis and TCA in amphibolic cycles.
13. Summarize the chemiosmotic theory of ATP formation.

Practical/Thought Questions

1. Using the concept of fermentation, describe the microbial/biochemical mechanisms that cause milk to sour.
2. *Trichomonas vaginalis* is a protozoan agent of a sexually transmitted disease in humans. It lives on the vaginal or urinary mucous membranes, feeding off dead cells and glycogen. Astonishingly, this eucaryote completely lacks mitochondria. Predict what sort of metabolism it must have.
3. Beer production requires an early period of rapid aerobic metabolism of glucose by yeast. Given that anaerobic conditions are necessary to produce alcohol, can you explain why this step is necessary?
4. Microorganisms are being developed to control man-made pollutants and oil spills that are metabolic poisons to animal cells. A promising approach has been to genetically engineer bacteria to degrade these chemicals. What is actually being manipulated in these microbes?

CHAPTER 8

Microbial Genetics

The first direct glimpse at DNA's structure. This false-color scanning tunneling micrograph of calf thymus gland DNA (magnified ×2,000,000) brings out the well-defined ridges of the helix.

Chapter Preview

The study of modern genetics is really a study of the language of the cell, a special language found in deoxyribonucleic acid—DNA. In this chapter we shall investigate the structure and function of this molecule and explore how it is copied, how its language is interpreted into useful cell products, how it is controlled, how it changes, how it is transferred from one cell to another, and how it can be manipulated by biotechnology.

Introduction to Genetics and Genes: Unlocking the Secrets of Heredity

Genetics is the study of the inheritance or **heredity** of living things. It is a wide-ranging science that explores the transmission of biological properties (traits) from parent to offspring, the expression and variation of these traits, the structure and function of the genetic material, and how this material changes. The study of genetics exists on several levels (figure 8.1). Organismic genetics observes the heredity of the whole organism or cell; chromosomal genetics examines the behavior of chromosomes; and molecular genetics deals with the biochemistry of the genes. All of these levels are useful areas of exploration, but in order to understand the expressions of microbial structure, physiology, mutations, and pathogenicity, we need to examine the operation of genes at the cellular and molecular levels. The study of microbial genetics provides a greater understanding of human genetics and an increased appreciation for the astounding advances in genetic engineering we are currently witnessing.

The Nature of the Genetic Material

For a species to survive, it must have the capacity of self-replication. In single-celled microorganisms, reproduction involves the division of the cell by means of binary fission, budding, or mitosis, but these forms of reproduction involve a more significant activity than just simple cleavage of the cell mass. Because the genetic material is responsible for inheritance, it

genetics L. *genesis,* birth, generation.
heredity L. *hereditas,* heirship.

Feature 8.1 Deciphering the Structure of DNA

An important milestone in the history of genetics was the purification of DNA by Oswald Avery and his coworkers in 1945. The realization by the scientific world that DNA was the principal carrier of genetic information and the blueprint for life created an avalanche of questions, the most tantalizing of which concerned its molecular structure.

A few years later (1951), American biologist James Watson and English physicist Francis Crick collaborated on solving the DNA puzzle. Although they did little of the original research, they attacked the problem with brilliance, perception, and considerable energy. The existing data from experiments with DNA intrigued them: It had been determined by Erwin Chargaff that any model of DNA structure would have to contain deoxyribose, phosphate, purines, and pyrimidines arranged in a way that would provide variation and a simple way of copying itself. Watson and Crick spent long hours constructing models with cardboard cutouts, and kept alert for any and every bit of information that might give them an edge. Two English biophysicists, Maurice Wilkins and Rosalind Franklin, had been painstakingly collecting data on X-ray crystallographs of DNA for several years. With this technique, molecules of DNA bombarded by X rays produce a photographic image that can predict the three-dimensional structure of the molecule. After being allowed to view certain of Wilkins and Franklin's data privately, Watson and Crick noticed an unmistakable pattern: The molecule appeared to be a double helix. Gradually, the pieces of the puzzle fell into place, and a final model was assembled—a model that explained all of the qualities of DNA, including how it is copied. Although Watson and Crick were rightly hailed for the clarity of their solution, it must be emphasized that their success was due to the considerable efforts of a number of English and American scientists. This historical discovery showed that the tools of physics and chemistry have useful applications in biological systems, and it also spawned ingenious research in all areas of molecular genetics.

Since the discovery of the double helix, an extensive body of biochemical, microscopic, and crystallographic analysis has left little doubt that the model first proposed by Watson and Crick is correct. A new technique using scanning tunnelling microscopy produces three-dimensional images of DNA magnified two million times (see chapter opening illustration). These images verify the helical shape and twists of DNA represented by models.

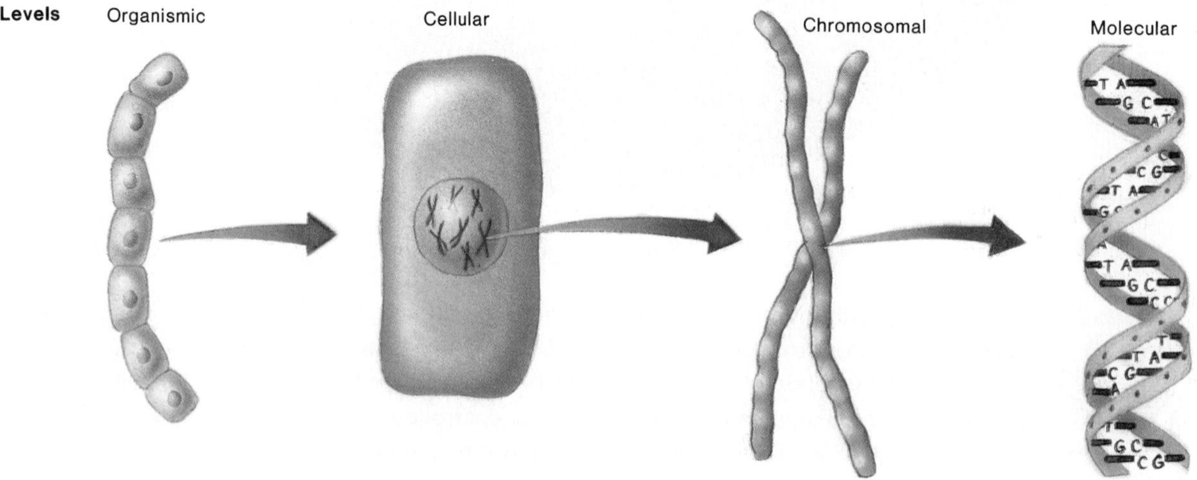

Figure 8.1 Levels of genetic study. The operations of genetics may be observed at the organismic, cellular, chromosomal (gene), and molecular (DNA) levels.

must be exactly duplicated and separated to each offspring during reproduction. In other words, when the parent cell divides, each chromosome must be exactly copied to ensure that each daughter cell has what it needs for normal function. This genetic material is a long, encoded molecule of DNA with several orders of structure. Before we look at how DNA is copied, let us explore the organization of this genetic material, proceeding from the general to the specific.

The Levels of Structure and Function of the Genome

The **genome** is the sum total of genetic material of a cell. Although most of the genome exists in the form of chromosomes, genetic material may appear in nonchromosomal sites as well (figure 8.2). For example, bacteria and fungi contain tiny extra pieces of DNA (plasmids), and certain organelles of eucaryotes (the mitochondria and chloroplasts) are equipped with their own genetic programs. Genomes of cells are composed exclusively of DNA, but viruses contain either DNA or RNA as the principal genetic material. Although the specific genome of an individual organism is unique, the general pattern of nucleic acid structure and function is similar among all organisms.

In general, a **chromosome** is a discrete cellular structure composed of a long, neatly packaged piece of DNA; however, the chromosomes of eucaryotes and bacterial cells differ in several respects. The structure of eucaryotic chromosomes consists of a DNA molecule tightly wound around histone proteins (see feature 8.3), whereas a bacterial chromosome (chromatin body) is condensed and secured into a packet by means of basic proteins. Other differences in eucaryotic chromosomes are: They are located in the nucleus; they vary in number from three to hundreds; they may occur in pairs (diploid) or singles (haploid); and they are linear. In contrast, the single chromosome of a bacterium exists freely in the cytoplasm in granules, is haploid, and maintains a closed circle configuration.

The chromosomes of all cells are subdivided into basic informational packets called genes. A **gene** can be defined from more than one perspective. In classical genetics, the term refers to the fundamental unit of heredity responsible for a given trait in an organism. In the molecular and biochemical sense, it is a site on the chromosome that provides information for a certain cell function. More specifically still, it is a certain segment of DNA that contains the necessary code to make a protein or RNA molecule. This last definition of a gene will be emphasized in this chapter. For an analogy that clarifies the relationship of the genome, chromosomes, genes, and DNA, see feature 8.2.

The Size of Genomes

Genomes vary greatly in size. The smallest viruses have 4 or 5 genes; the bacterium *E. coli* has a single chromosome containing about 3,000 genes, and a human cell packs about 100,000 genes into 46 chromosomes (though even the human genome is relatively modest in size compared to some plants and other animals). The total length of DNA relative to cell size is notorious: The chromosome of *E. coli* would measure about 1 mm if unwound and stretched out linearly, and yet this fits within a cell that measures only about 2 μm across, making the DNA 500

Cells

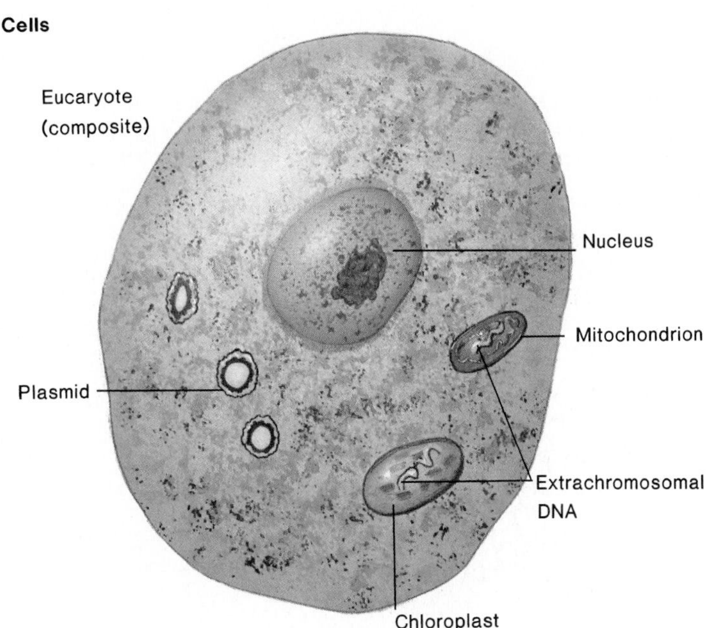

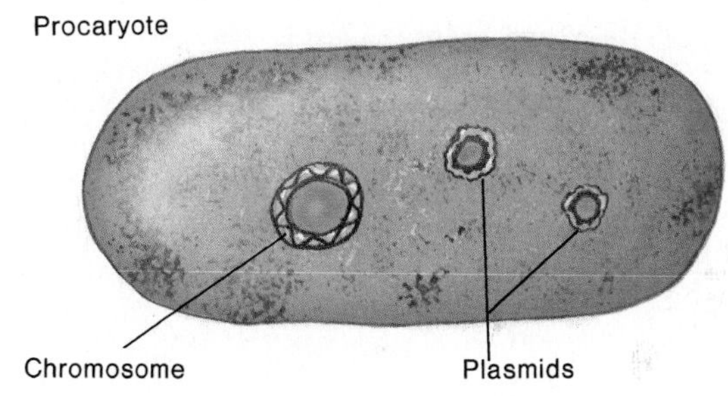

Viruses

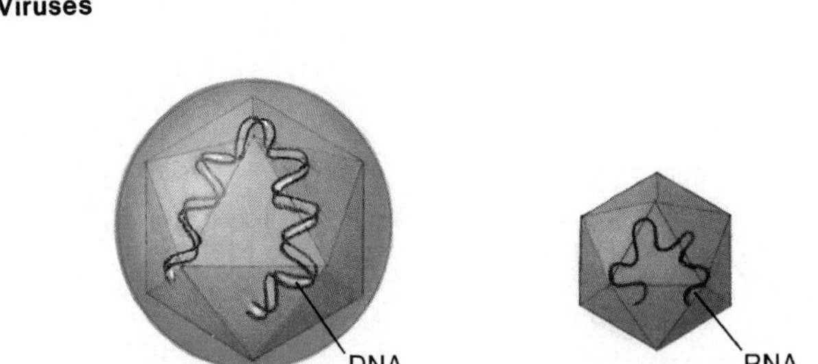

Figure 8.2 The general location and forms of the genome in microbes.

Feature 8.2 The Video Genome

The relationship of the genome, chromosomes, and genes to one another and to DNA is analogous to a collection of videotapes of family events. The whole library of tapes (genome) contains several individual cassettes (chromosomes); the tape on the cassette is divided sequentially into several separate events (genes), which may be selected and played to generate a picture on the television screen (product). This analogy works in several other ways:

1. At all levels (library, cassette, event), the basic informational unit is still the tape itself (the DNA molecule).
2. Like a tape, the DNA molecule can be copied, spliced, and edited.
3. Somewhat like the spools of a cassette, the long DNA of a chromosome is carefully wrapped up so that it is compact, easy to read, and will not get tangled (see feature 8.3).
4. For the tape (DNA code) to be translated into images (cell product), a tape player (special cell machinery) is necessary.
5. The tape, like the DNA molecule, makes sense only if played in a certain direction.
6. The entire collection of tapes, like the genome, is a store of information. Not all of it is being "played" at any one time.

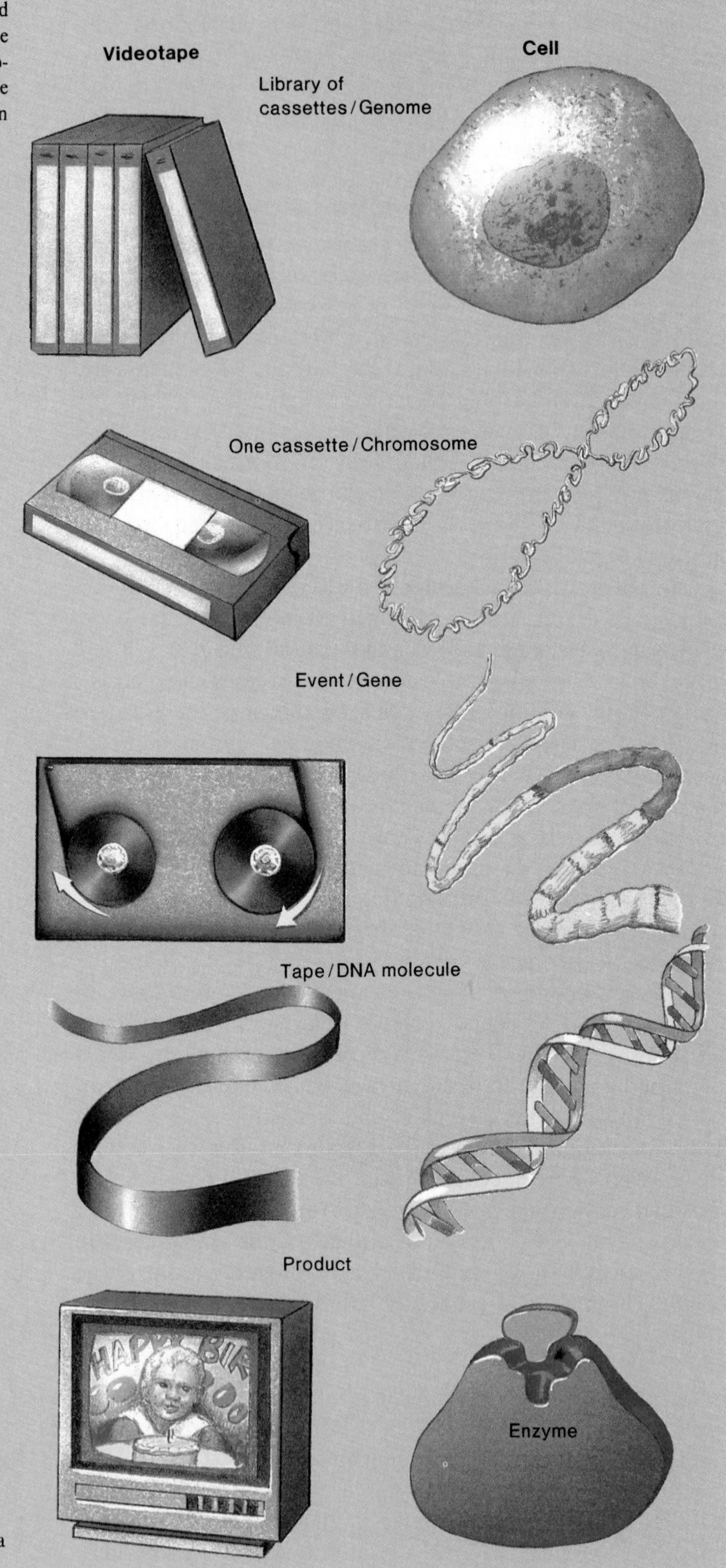

The relationships of the genome, chromosomes, and genes as compared to a videotape system.

Feature 8.3 The Packaging of DNA: Winding, Twisting, and Coiling

The analogy of DNA to a cassette tape is imperfect because the DNA molecule is not perfectly wound around a spool. Packing the mass of DNA into the cell involves further levels of DNA structure called supercoils or superhelices. In the simpler system of procaryotes, the chromosome, a continuous circle, is twisted around itself by the action of a special enzyme called a *topoisomerase* (specifically DNA *gyrase*). This enzyme folds, splices, and holds DNA, thereby introducing a reversible series of twists into the molecule. The system in eucaryotes is more complex, with three or more levels of coiling. First, the DNA molecule of a chromosome, which is linear, is wound twice around the histone proteins, creating a chain of *nucleosomes.* The nucleosomes fold in a spiral formation upon one another. Most experts believe that an even greater supercoiling occurs when this spiral arrangement further twists on its radius into a giant spiral with loops radiating from the outside. This extreme degree of compactness is what makes the eucaryotic chromosome visible during mitosis (see figure 4.6). In addition to reducing the volume occupied by DNA, supercoiling solves the problem of keeping the chromosomes from getting tangled during cell division, and it protects the code from massive disruptions due to breakage.

From a different perspective, highly coiled DNA is far too condensed to be available for cell activities. Another type of topoisomerase (also called a helicase or untwisting enzyme) uncoils the supercoils to permit copying and other functions of DNA during cell division and protein synthesis.

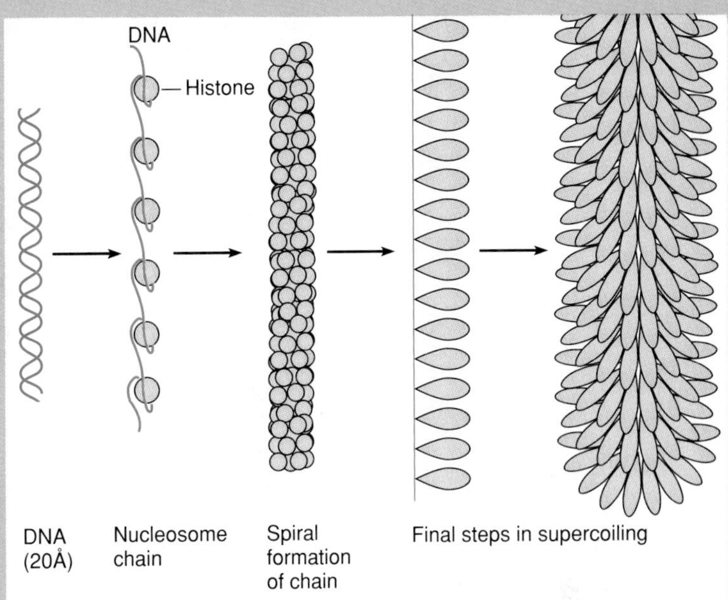

The packaging of DNA. Eucaryotic DNA undergoes several orders of coiling and supercoiling, which greatly condenses it.

topoisomerase (tah"-poh-eye-saw'-mur-aze) Any enzyme that changes the configuration of DNA.

gyrase (jy'-rayz) L. *gyros,* ring or circle. A bacterial enzyme that produces supercoils.

nucleosomes "Nucleus bodies" arranged like beads on a chain.

times longer than the cell (figure 8.3). Still the bacterial chromosome takes up only about one-third to one-half of the cell's volume. Likewise, if the sum of all DNA contained in the 46 human chromosomes were unravelled and laid end to end, it would measure about 6 feet. This means that the DNA is about 180,000 times longer than a cell 10 μm wide and a million times longer than the width of the nucleus. How can such elongated genomes fit into the miniscule volume of a cell, and in the case of eucaryotes, into an even smaller compartment, the nucleus? The answer lies in the regular coiling of the DNA chain (see feature 8.3).

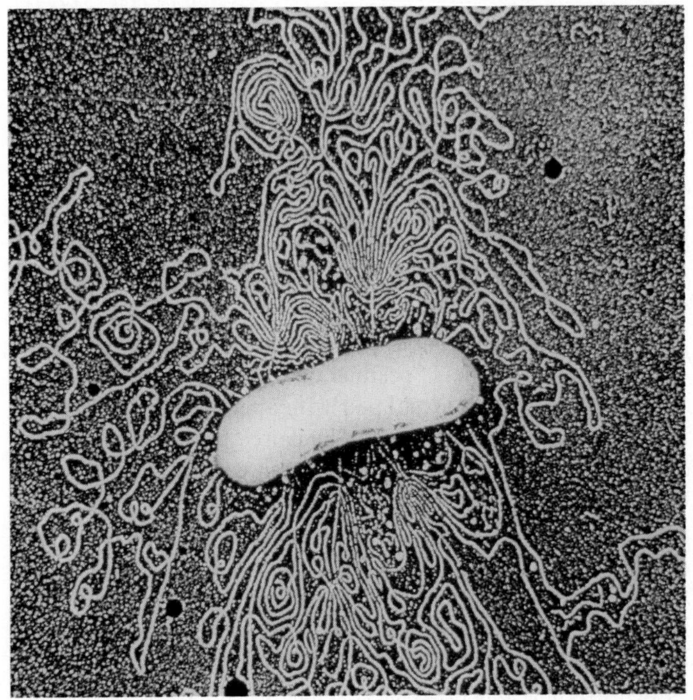

Figure 8.3 An *E. coli* bacillus disrupted to release its DNA molecule. The cell has spewed out its single, uncoiled DNA strand into the surrounding medium.

The DNA Code: A Simple Yet Profound Message

Examining the function of DNA at the molecular level—cracking its code—requires an even closer look at its structure. To do this we will imagine being able to magnify a small piece of a gene about five million times. (Although genes have not yet been viewed at quite this level, we are rapidly approaching this capability; see chapter opening illustration.) What such fine scrutiny will disclose is one of the great marvels, and for many years, mysteries, of biology. Our first view of DNA in chapter 2 revealed that it is a gigantic molecule, a type of nucleic acid, with two polynucleotide strands combined into a double helix (figure 2.25). This general structure is universal, regardless of

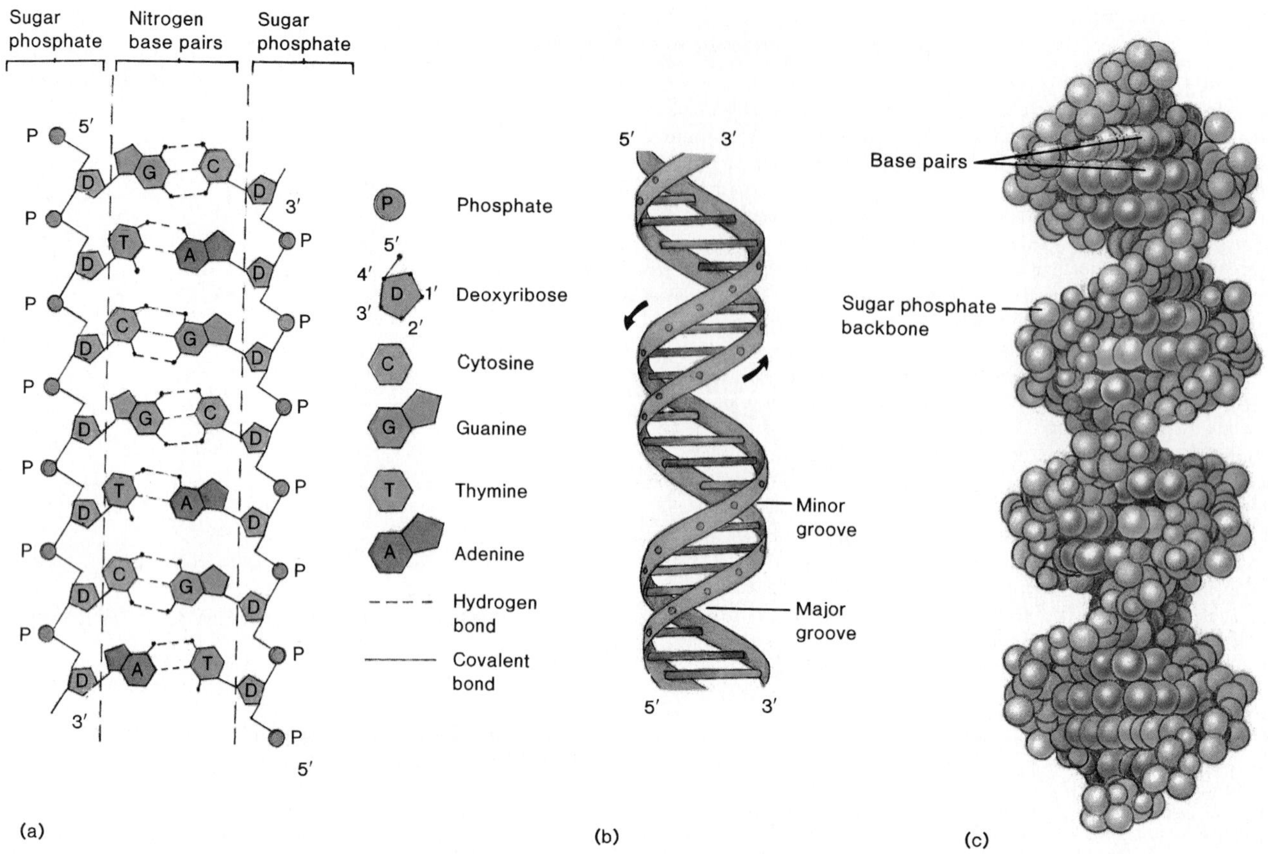

Figure 8.4 Three views of DNA structure. (*a*) A highly schematic nonhelical model, showing sugar-phosphate backbones, details of deoxyribose bonds with phosphate (5′ or 3′), and bonds with sugar (1′). Purine/pyrimidine base pairs are shown with hydrogen bonds; G/C have three of these bonds; A/T have two. (*b*) Simplified model that highlights the antiparallel arrangement and the major and minor grooves. (*c*) Space-filling model that more accurately depicts the structure of DNA.

the organism being viewed. It consists of a **deoxyribose sugar-phosphate** backbone attached to **nitrogenous base** crosspieces. The sugar and phosphate molecules alternate, creating a repetitive and regular molecular skeleton. Two phosphates are bonded covalently to one deoxyribose sugar by means of phosphodiester bonds. One of the bonds is to the number 5′ (read "five prime") carbon on deoxyribose, and the other is to the 3′ carbon, which confers a predictable order on the backbone of the molecule (figure 8.4).

The nitrogenous bases, **purines** and **pyrimidines,** attach by covalent bonds at the 1′ position of the sugar. They span the midline of the molecule and pair with appropriate complementary bases from the other side. The paired bases are so aligned as to be joined by hydrogen bonds. Such weak bonds are easily broken, allowing the molecule to be "unzipped" into its complementary strands. Later we will see that this feature is of great importance in gaining access to the information encoded in the nitrogenous base sequence. Pairing of purines and pyrimidines is not random, it follows unvarying rules dictated by the chemical and physical reality that a given base can form hydrogen bonds with only one other base. Thus, in DNA, the purine **adenine** (A) can pair only with the pyrimidine **thymine** (T), and the purine **guanine** (G) can pair only with the pyrimidine **cytosine** (C). This small and simple pairing rule is the guiding force of a cell's genetic language. Although the base pairing partners do not vary, the sequence of base pairs along the DNA molecule may assume any order, resulting in an infinite number of possible nucleotide sequences.

Other important considerations of DNA structure concern the nature of the double helix itself. The halves are not exactly parallel—that is, they are not oriented in the same direction. In order for the A–T and G–C pairs to meet properly, one side of the helix runs in the opposite direction of the other, in an *antiparallel* arrangement (figure 8.4*b*). To keep track of the direction of the helix, we assign numbers to the order of the phosphate attachment to the deoxyribose carbon positions on each strand. So one helix runs from the 5′ to 3′ direction and the other runs from the 3′ to 5′ direction. This characteristic has significant impact on the replication and decoding of the DNA program. As apparently perfect and regular as the DNA molecule may seem, it is not exactly symmetrical. The torsion in the helix and the stepwise stacking of the nitrogen bases produce two different-sized surface features, the major and minor grooves (figure 8.4*b*).

The Significance of DNA Structure The nitrogen bases influence DNA in two major ways:

1. **Maintenance of the code during reproduction.** The constancy of base pairing guarantees that the code will be retained during cell growth and division. When the two strands are separated, each one provides a **template** (pattern or mold) for the replication (exact copying) of a new molecule (figure 8.5). Because the sequence of one strand automatically gives the sequence of its partner, the code can be duplicated with fidelity.
2. **Providing variety.** The order of bases along the length of the DNA strand constitutes the genetic program—the language—of the DNA code. Adding to our earlier definitions, the message present in a gene is a precise arrangement of these bases, and the genome is the collection of all DNA bases that, in an ordered combination, are responsible for the unique qualities of each organism.

It is tempting to ask how such a seemingly simple code can account for the extreme differences among forms as diverse as a virus, *E. coli,* and a human. The English language, based on 26 letters, can create an infinite variety of words, but how can this complex genetic language be based on just four nitrogen base "letters"? Put in mathematical terms, for a segment of DNA only 1,000 bases long, there are $4^{1,000}$ different genetic combinations. Because the complexity of even the simplest bacterial genome is many times greater than this, one must conclude that the potential for variation is virtually infinite.

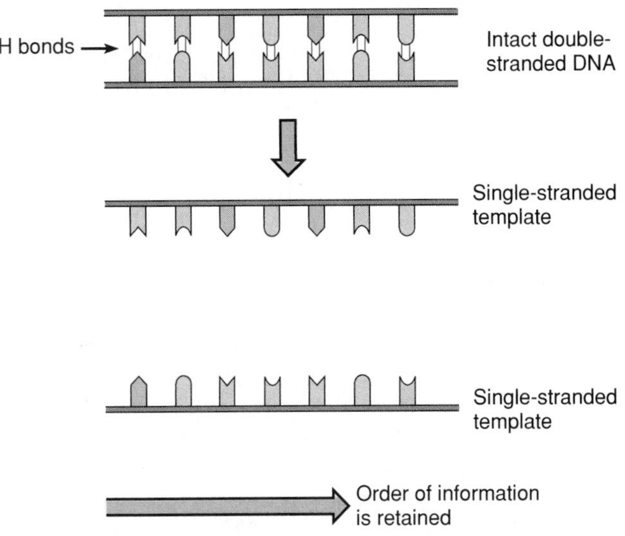

Figure 8.5 Implications of DNA's code and structure. The easy disruption of hydrogen bonds between base pairs is an important factor in the separation of the double strands and the formation of single-stranded templates. Each strand may now serve as the pattern for forming two double-stranded molecules. Yet the covalent bonds between molecules on one strand are not easily broken, and the order of bases, which is the language, remains intact.

DNA Replication: Preserving the Code and Passing It On

Because the sequence of bases along the length of a gene constitutes the language of DNA, a special mechanism enables this genetic program to be duplicated exactly in each new cell and preserved for hundreds of generations with minimum losses or additions to the genes. Our example will follow the process in bacteria, but with minor exceptions, it also applies to the process as it works in other procaryotes, eucaryotes, and some viruses. Early in the division cycle, the metabolic machinery of a bacterium responds to a message and initiates the duplication of the chromosome (figure 6.15). It is indicative of the speed of this process that the duplication of DNA must be completed during a single generation (around 20 minutes in *E. coli*).

Table 8.1 Some Enzymes Involved in DNA Replication and their Functions

Enzyme	Function
Helicase	Unzipping the DNA helix
Primase	Synthesizing an RNA primer
DNA polymerase III	Adding bases to the new DNA chain; proofreading the chain for mistakes
DNA polymerase I	Removing primer, closing gaps
Ligase	Final binding of nicks in DNA
Gyrase	Supercoiling

The Overall Replication Process

What features allow the DNA molecule to be exactly duplicated, and how is its integrity retained? DNA replication requires a careful orchestration of the actions of 30 different enzymes (partial list in table 8.1), which separate the strands of the existing DNA molecule, copy its template, and produce two complete daughter molecules. These actions, in turn, require that large amounts of raw materials and ATP be readily available from the cell pool. The **basic steps of replication** are shown in figure 8.6 and include the following: (1) uncoiling the parent DNA molecule, (2) unzipping the hydrogen bonds between the base pairs, thus separating the two strands and exposing the nucleotide sequence of the helix to serve as templates, and (3) synthesizing two double strands by attachment of the correct complementary nucleotides to each single-stranded template. It is worth noting that each daughter molecule will be identical to the parent in composition, but neither one is completely new: Half of each—the half that served as a template—is an original parental DNA strand. The preservation of the parent molecule in this way, termed *semiconservative replication,* helps explain the reliability and fidelity of replication.

Refinements and Details of Replication

The circular bacterial DNA molecule replicates by means of a special configuration called a **replicon.** Replication begins at a precise initiation site composed of a palindrome (see feature 8.4). Here, enzymes called helicases ("unzipping enzymes") come into play. They untwist the helix and break its hydrogen bonds, a process that requires ATP. This process is best seen in animation; the molecule does not unzip like a common clothing zipper—it rotates as it is untwisted.

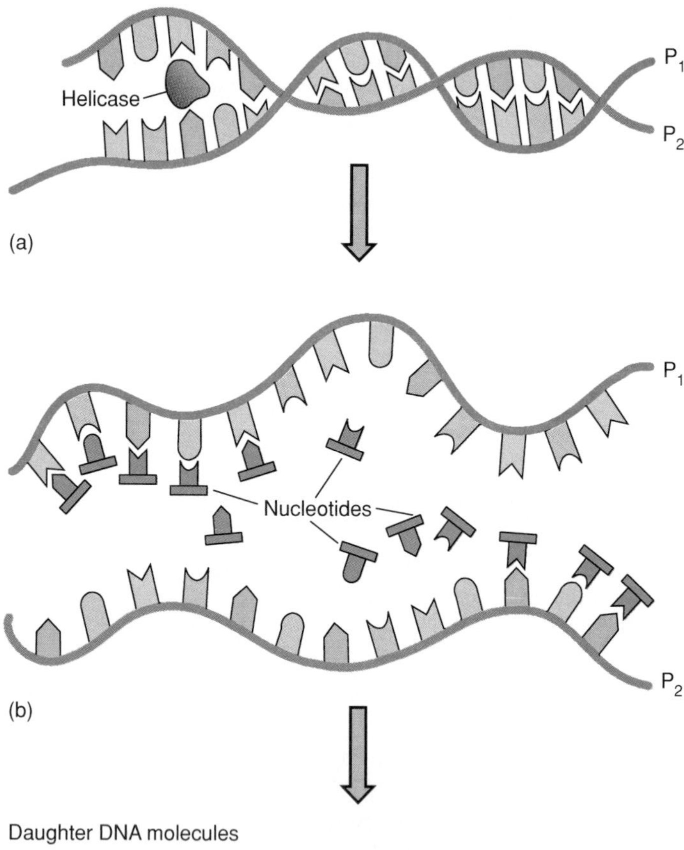

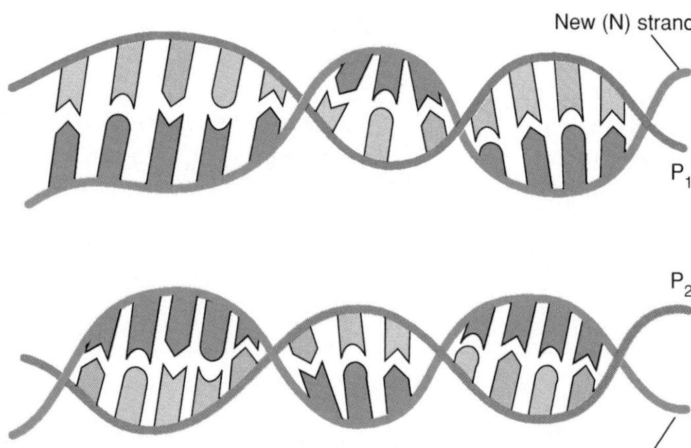

Figure 8.6 Simplified steps in semiconservative replication of DNA. (*a*) A helicase unwinds the double helix into two parent strands (P_1 and P_2). (*b*) A pool of preformed nucleotide building blocks enter and participate in the replication of a new complementary strand according to the template patterns through base pairing. (*c*) Completed daughter molecules each contain one strand that is newly synthesized and one of the original parent strands.

Replication begins when an RNA primer enters the initation site. The primer provides DNA polymerase with a "hook" to correctly position itself to begin synthesis. Because the DNA molecule is circular, there are two replication **forks,** each containing its own polymerase. As replication proceeds, these forks open up and gradually move apart (figure 8.7*a*). Although simple models show that further elongation of the molecules proceeds by the simple addition of complementary nucleotides to each parent strand continuously as the fork is opened up, this is not actually what happens on both strands. Because DNA polymerase is correctly oriented for synthesis *only* in the 5′ to 3′ direction of the parental template molecule, only this one strand, the *leading strand,* can be synthesized continuously. Given the antiparallel arrangement of DNA, the strand that runs in the 3′ to 5′ direction, termed the *lagging strand,* cannot be synthesized continuously. This dilemma is solved by the polymerase synthesizing the lagging strand in short segments in the direction away from the fork for a short distance. As the fork opens up a bit, the next segment is synthesized backwards to the point of the previous segment, a process repeated at both forks until synthesis is complete (figure 8.7*b*). This manner of synthesis leaves small pieces of DNA (100 to 1,000 bases long) called *Okazaki fragments* that will eventually have to be spliced together to form two complete molecules.

Elongation and Termination of the Daughter Molecules As replication proceeds along both forks, the replicon acquires the look of a half-open eye, or the Greek letter *theta* (θ), due to one duplicating strand that droops down (figure 8.8*a*). The addition of bases proceeds at an astonishing pace, estimated in some bacteria at 750 bases per second at each fork! When the replication forks come full circle and meet, enzymes move along the lagging strand to begin the initial linking of the fragments and complete synthesis and separation of the two circular daughter molecules (figure 8.8*b*).

Feature 8.4 Palindromes—Word Games with the Language of DNA

Intriguing components of DNA structure are the numerous palindromic sequences present throughout the molecule. In language, a palindrome is a word, phrase, or sentence that reads the same both forward or backward—for example, **radar; madam I'm Adam; too hot to hoot;** and **poor Dan is in a droop.** A DNA palindrome, also called an inverted repeat, might read as follows:

GCTAGC
CGATCG

Unlike words, a DNA palindrome occurs not in a single line but in the order of bases in the two complementary strands, with the top strand being read from left to right and its complement being read from right to left.

DNA palindromes vary in size from a few bases to several hundred, and appear to have a number of functions. They may, for instance, be regulatory, providing a starting site for DNA replication or serving as a binding site for enzymes and other molecules that govern genetic expression. Inverted sequences may also permit loops to form in supercoiling DNA, thereby relieving tension on the molecule. Of great significance is the discovery of special enzymes of bacterial origin called **restriction endonucleases,** which hydrolyze or *nick* DNA internally at sites of palindromic sequences. These enzymes recognize foreign DNA and are capable of snipping it apart at these sites. In the bacterial cell, this protects against the incompatible DNA of bacteriophages or plasmids. In the biotechnologist's lab, the enzymes can be used to cleave DNA at desired sites and are a must for the techniques of recombinant DNA technology.

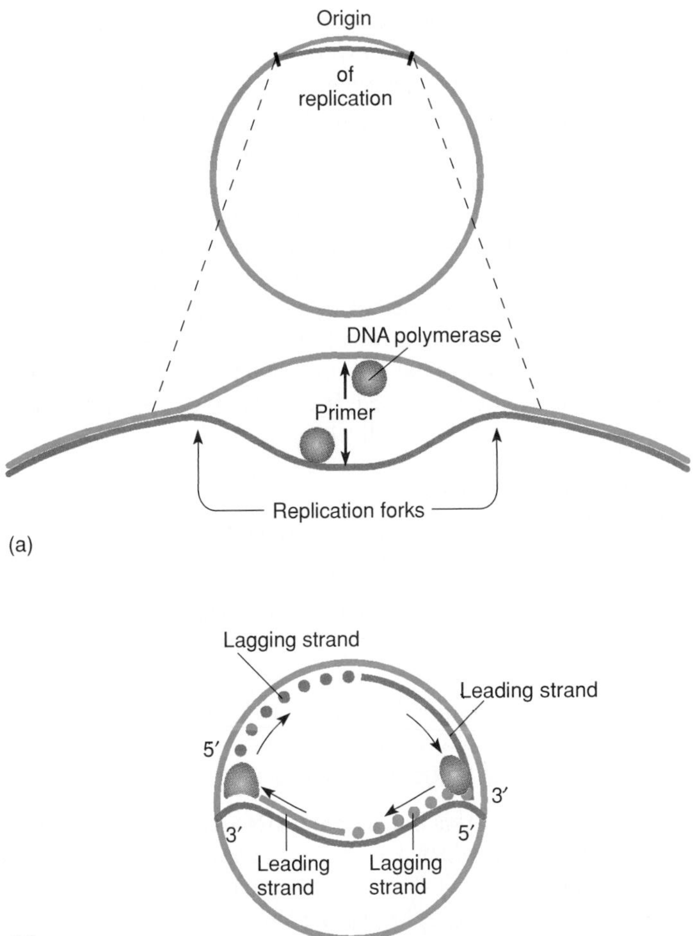

Figure 8.7 The bacterial replicon: a model for DNA synthesis. (*a*) Circular DNA has a special site where replication originates. When strands are separated, two replication forks form and a DNA polymerase enters at each fork. (*b*) Since DNA polymerase works only in the 5′ to 3′ direction, at each fork there will be two different patterns of replication. The leading strand, which orients in this direction, will be synthesized continuously, and the lagging strand, which orients in the opposite direction, will be synthesized in short pieces that are later linked together.

Other Variations on the Theme The bidirectional replicon as described for bacteria is also the mechanism of many viruses and some bacterial plasmids. Because of its linear structure, the eucaryotic chromosome is replicated simultaneously at numerous sites along the molecule, a process that speeds up duplication of the larger eucaryotic chromosome. Other aspects of the process, however, are very similar in all groups. A novel form of DNA synthesis called *rolling circle* occurs in some bacterial viruses and plasmids (figure 8.9). Although rather complicated in practice, it involves the synthesis of DNA in a single direction, with the new strand of DNA rolling out from the circle of DNA like paper from a roll of tissue.

Like any language, DNA is occasionally "misspelled" when an incorrect base is added to the growing chain. Studies have shown that such mistakes are made once in approximately 100,000 bases, but most of these are corrected. If not corrected, they become mutations and may lead to serious cell dysfunction and even death. Because continued cellular integrity is very dependent on perfect replication, DNA has developed its own proofreading function. The same enzyme complex that elongates the molecule can also detect incorrect, unmatching bases, excise them, and replace them with the correct base. This is yet another reason for the fidelity of DNA replication.

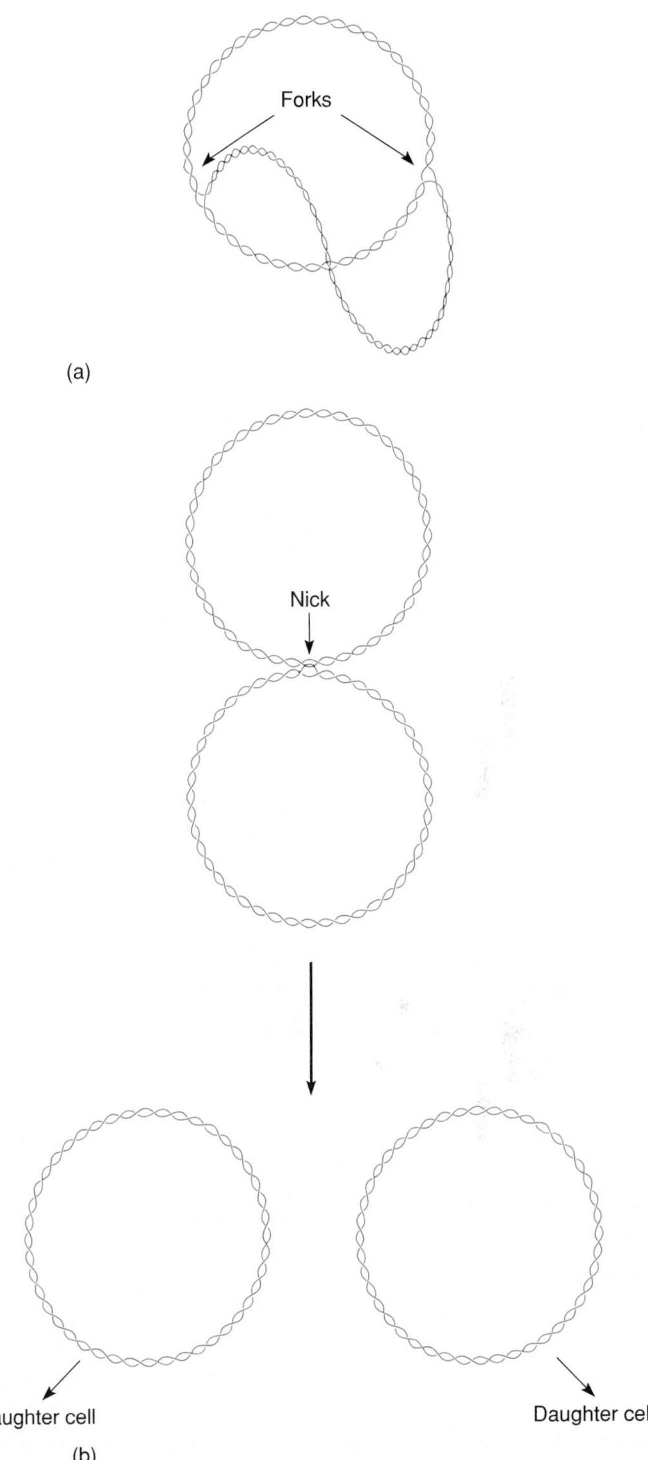

Figure 8.8 Completion of chromosome replication in bacteria. (*a*) The theta or "eye" stage of replication, in which one strand loops down as it grows in length. (*b*) Nicking, separation, repair, and release of two completed molecules that will be separated into daughter cells during binary fission.

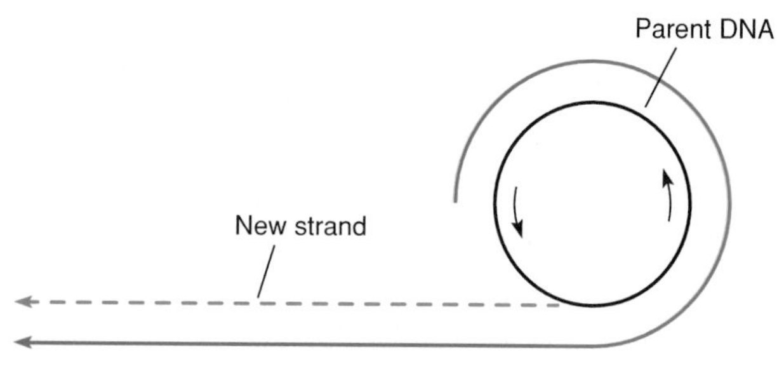

Figure 8.9 Simplified model of a rolling circle type of DNA replication as occurs in some viruses and plasmids. As the parent DNA rotates, a single strand (green) is synthesized on its template and this new strand comes off. A complementary strand (blue) is then formed in sections in accordance with this template. A large number of molecules may be generated continuously by this means.

Applications of the DNA Code: Transcription and Translation

Having explored how the genetic message in the DNA molecule is conserved through accurate replication, we must now consider the precise role of DNA in the cell. Given that the sequence of bases in DNA is a genetic code, just what is the nature of this code and how is it utilized by the cell? Although the DNA library is full of critical information, the molecule itself does not perform cell processes directly. Its stored information is conveyed to other molecules, which carry out instructions. The concept that genetic information flows from DNA to RNA to protein is a central theme of molecular biology (figure 8.10). More precisely, it states that the master code of DNA is first copied onto an RNA molecule called a messenger through **transcription,** and the RNA message is decoded by special cell components into proteins during **translation.** The principal exceptions to this pattern are found in RNA viruses, which convert RNA to other RNA, and in retroviruses, which convert RNA to DNA.

The Gene-Protein Connection

Genes fall into three basic categories: *structural genes* that code for proteins, genes that code for RNA, and *regulatory genes* that control gene expression. The sum of an organism's structural genes constitutes its distinctive genetic makeup or **genotype,** which when expressed in the individual, creates traits (certain structures or functions) referred to as the **phenotype.** Just as a person inherits a combination of genes (genotype) that gives a certain eye color or height (phenotype), a bacterium inherits genes that direct the formation of a flagellum, and a virus inherits genes for its capsid structure.

The Triplet Code

As we pursue the details of transcription and translation, several questions invariably arise concerning the relationship between genes and cell function. For instance, how does gene structure lead to the expression of traits in the individual, and what features of gene expression cause one organism to be so distinctly different from another? For answers we must turn to the correlation between gene and protein structure. Because each structural gene is a linear sequence of bases that codes for a protein and each protein is different, then each gene must differ somehow in its basic composition. In fact, the language of DNA exists in the order of groups of three consecutive bases called **triplets** (also codons) on one DNA strand (figure 8.11). Thus, one gene differs from another in its composition of triplets. An equally important part of this concept is that each triplet represents a code for a particular amino acid. When the triplet code is transcribed and translated, it dictates the type and order of amino acids in a polypeptide chain.

The Importance of Proteins

To further analyze the crucial relationship between the DNA code and protein, let us review two ideas concerning proteins:

1. A protein's primary structure—the order and type of amino acids in the chain—determines its characteristic shape and function.
2. Proteins ultimately determine phenotype—the expression of all aspects of cell function and structure. Put more simply, living things are what their proteins make them.

To complete this line of logic, DNA is mainly a blueprint that tells the cell which kinds of proteins to make and how to make them. Recalling that enzymes are proteinaceous presents a wonderful circular logic—somewhat akin to the chicken/egg riddle—that the enzymes needed in DNA replication, transcription, and translation are also made by these processes! As we shall see in the ensuing section, translating nucleic acid language into protein language is a very precise and elegant process.

The Major Participants in Transcription and Translation

The processes of transcription and translation have been so rigorously researched that the major elements and the orderly sequence of events are well delineated. Transcription and translation, like replication, are highly complex. A number of components participate—most prominently, messenger RNA, transfer RNA, ribosomes, several types of enzymes, and a storehouse of raw materials. After first examining each of these components, we shall see how they come together in the assembly line of the cell.

RNAs: Tools in the Cell's Assembly Line

Ribonucleic acid is an encoded molecule like DNA, but its general structure is different in several ways: (1) It is a **single-stranded molecule**—that is, it is a single helix. This is not to say that it does not form a secondary structure (hairpin loops) or a tertiary structure (see later discussion of transfer RNA), but these structures still arise from a single strand. (2) RNA contains **uracil,** instead of thymine, as the complementary base pairing mate for adenine. This does not change the inherent DNA code in any way because the uracil still follows the pairing

---

genotype (jee'-noh-type) Gr. *gennan,* to produce, and *typos,* type.

phenotype (fee'-noh-type) Gr. *phainein,* to show. The physical manifestation of gene expression.

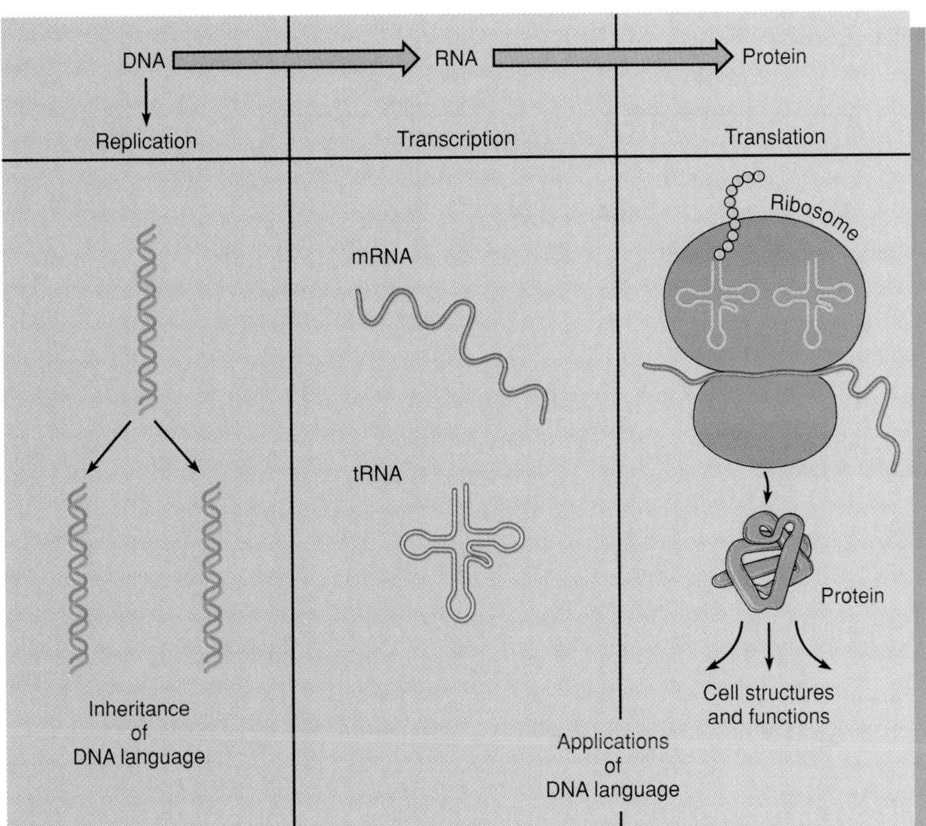

Figure 8.10 Summary of the flow of genetic information in cells. This biological theme, sometimes succinctly summed up as "DNA makes RNA makes protein," indicates the role of DNA as storehouse and distributor of genetic information and shows the involvement of other participants in carrying out its "orders."

Table 8.2 Types of Ribonucleic Acid

RNA Type	Contains Codes For	Function in Cell	Translated
Messenger (m)	Sequence of amino acids in protein	Brings codons to ribosome	Yes
Transfer (t)	A cloverleaf tRNA to carry amino acids	Brings amino acids to ribosome	No
Ribosomal (r)	Several large structural rRNA molecules	Forms the major part of a ribosome	No
Primer	An RNA that can begin DNA replication	Primes DNA	No
Small nuclear	Self-splicing RNA	Removes small pieces of RNA (acts as an enzyme)	No

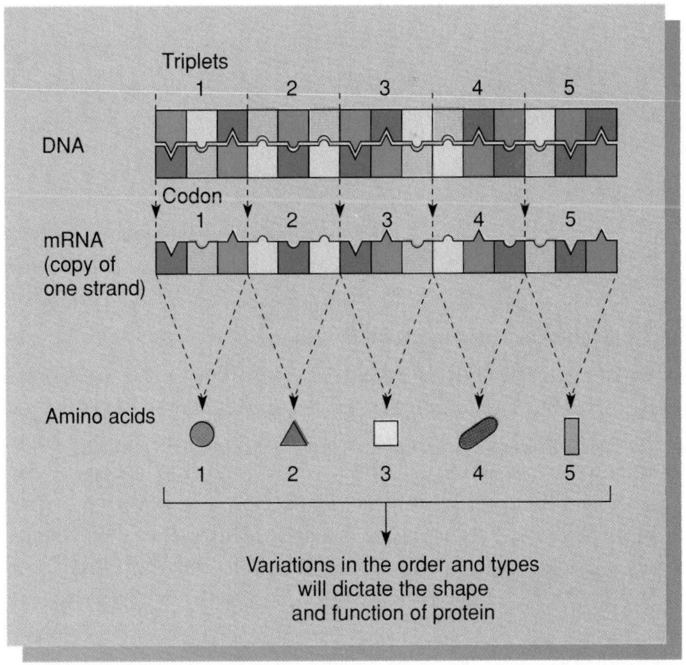

Figure 8.11 The DNA molecule is a continuous chain of base pairs, but the sequence must be interpreted in groups of three base pairs (a triplet). Each triplet as copied into mRNA codons will translate into one amino acid—that is, the ratio of base pairs to amino acids is 3:1.

rules. (3) Although RNA, like DNA, contains a backbone that consists of alternating sugar and phosphate molecules, the sugar in RNA is **ribose** rather than deoxyribose. The many functional types of RNA range from small regulatory pieces to large structural ones (table 8.2). All types of RNA are formed through transcription of a DNA gene, but only mRNA is further translated into another type of molecule (protein).

Messenger RNA (mRNA)

Codon 1 Codon 2 Codon 3

A U G C U G A C U

(a) (P) = Phosphate (R) = Ribose U = Uracil

Transfer RNA (tRNA)

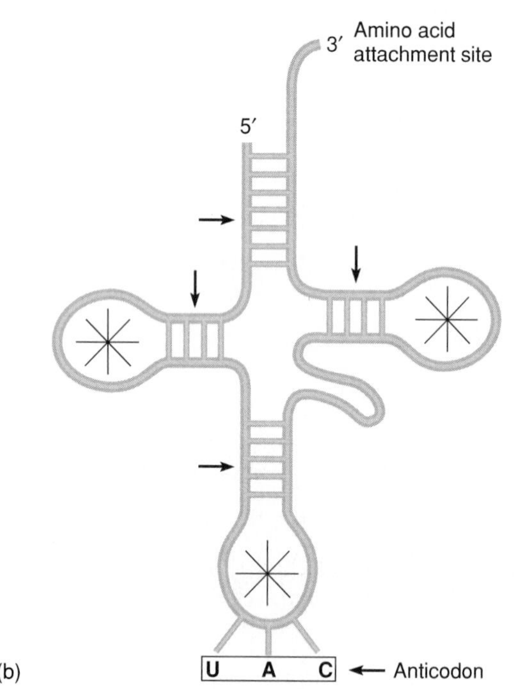

(b)

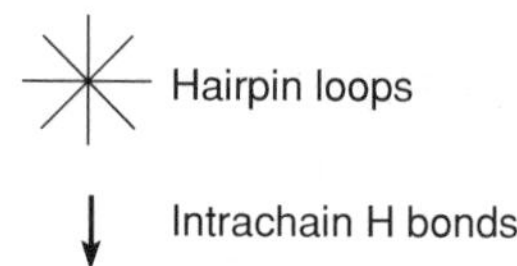

Figure 8.12 Characteristics of messenger and transfer RNA. (*a*) A short piece of mRNA illustrates the general structure of RNA: single-strandedness, repeating phosphate-ribose sugar backbone; single nitrogen bases with uracil instead of thymine. (*b*) Transfer RNA (tRNA) can loop back on itself to form intrachain hydrogen bonds. The result is a cloverleaf structure, shown here in simplified form. The tRNA is an important adaptor molecule. At its bottom is an anticodon that specifies the attachment of a particular amino acid at the 3′ trailing end.

Messenger RNA: Carrying DNA's Message

Messenger RNA (**mRNA**) is a transcript (copy) of a structural gene or genes complementary to DNA. Messenger RNA is synthesized by a process similar to DNA replication, and the complementary base pairing rules ensure that the code will be faithfully copied in the mRNA transcript. The message of this transcribed strand is displayed in a series of triplets called **codons** (figure 8.12*a*). The details of transcription and the function of mRNA will be covered shortly.

Transfer RNA: The Key to Translation

Transfer RNA (**tRNA**) is also a copy of a DNA code; however, it differs from mRNA. It contains sequences of bases that form hydrogen bonds with complementary sections of the same tRNA strand. At these areas, the molecule bends back upon itself into several *hairpin loops* (figure 8.12*b*), giving the molecule a secondary *cloverleaf* structure that folds even further into a complex, three-dimensional helix. This compact molecule is an adaptor that converts RNA language into protein language. The bottom loop of the cloverleaf exposes a triplet, the **anticodon,** that both designates the specificity of the tRNA and complements mRNA's codons. Trailing off the opposite end of the molecule is an acceptor position for the amino acid specified by that tRNA's anticodon. For each of the 20 amino acids (see table 2.5), there is at least one specialized type of tRNA to carry it. The charging of the tRNA takes place in two enzyme-driven steps: First an ATP activates the amino acid (aa-AMP), and then this group binds to the acceptor end of the tRNA. Because tRNA is the molecule that will convert the master code on mRNA into a protein, the accuracy of this step is crucial.

The Ribosome: A Mobile Molecular Factory for Translation

The procaryotic (70S) ribosome is a particle composed of tightly packaged ribosomal RNA and protein. Electron microscope and biochemical studies have revealed that the ribosomal particle is made up of two subunits that, when joined together, form a special niche to hold the components of protein synthesis (see figure 8.14). Ribosomes contribute both enzymatic and attachment functions for mRNA and tRNA. A metabolically active bacterial cell can accommodate 20,000 of these miniscule factories—all actively engaged in reading the genetic program, taking in raw materials, and emitting proteins at an impressive rate.

Transcription: The First Stage of Gene Expression

How does DNA present its code? At a segment corresponding to a gene, the DNA helices are separated. This exposes the nitrogen base triplets so that they may serve as a template for synthesis of an mRNA strand. Only one strand of the DNA—the *sense strand*—contains meaningful instructions for synthesis of a functioning polypeptide. To produce a strand of mRNA that is a replica of the sense strand, its complementary strand—the *antisense strand*—will be copied.

Transcription proceeds through several stages, directed by a huge and very complex enzyme system, **RNA polymerase.** During initiation, RNA polymerase recognizes a segment of the DNA called the *promoter region* that lies near the beginning of the gene segment to be transcribed (figure 8.13). At this site, the polymerase begins building the mRNA chain. During elongation, which proceeds in the 5′ to 3′ direction, nucleotide building blocks are assembled in accordance with the DNA template, except that uracil (U) is placed as adenine's complement. As elongation continues (at the rate of 40 nucleotides added per second), the part of DNA already transcribed is rewound into its original helical form. At termination, the polymerases recognize another code that signals the separation and release of the mRNA strand. How long is the mRNA? Just as long as the gene (or genes) that encodes a given protein. The very smallest mRNA might consist of 100 bases, an average-sized mRNA might consist of 1,200 bases, and a large one, of several thousand.

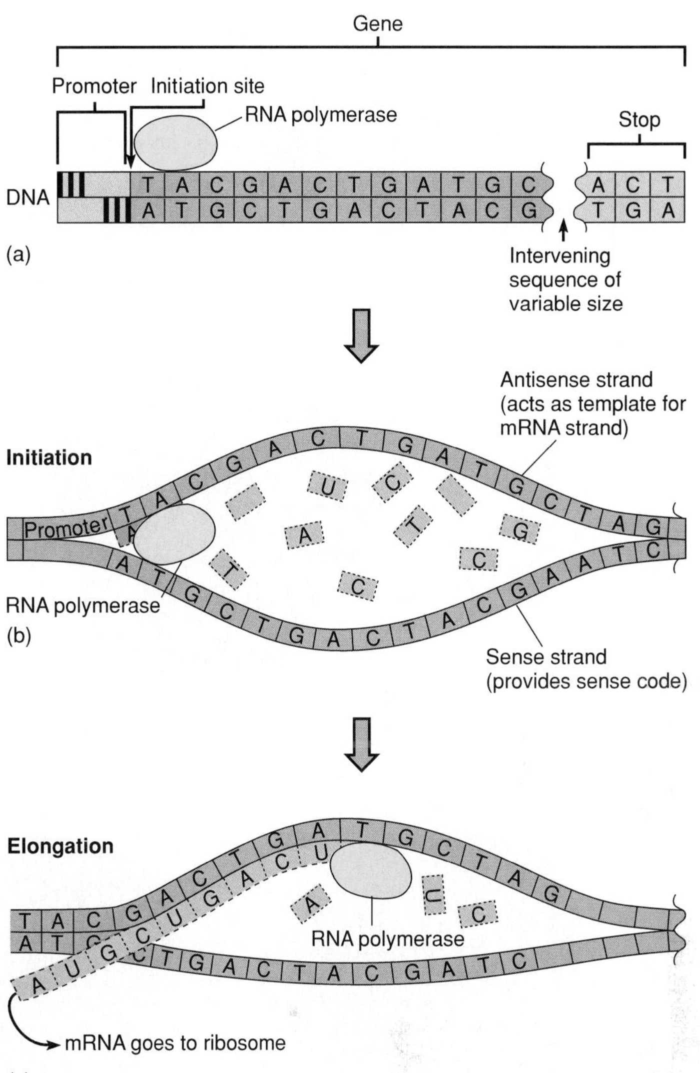

Figure 8.13 The steps in mRNA synthesis or transcription. When polymerase reaches the stop message, mRNA transcript terminates and is released to a ribosome to be translated.

Translation: The Second Stage of Gene Expression

In translation, all of the elements needed to synthesize a protein, from the mRNA to the amino acids, are brought together on the ribosome (figure 8.14*a*). The process is divided into the stages of initiation, elongation, termination, and protein folding and processing.

Initiation of Translation

The mRNA molecule leaving the DNA transcription site is short-lived and must immediately be translated. In fact, the mRNA is usually translated as it is being transcribed, which greatly speeds up the entire process. Ribosomal subunits nearby are ready to assemble, accept the mRNA, and bind to its 5′ end (figure 8.14*b*). One way to envision the attachment of the RNA is to return to our videotape analogy and imagine the end of the mRNA being strung through the ribosome somewhat like a tape through the head of a tape player, so that it will be read on the proper frame. The mRNA usually has a short leader that does not encode part of the protein (like the first part of a tape before the movie begins). As the ribosome begins to scan the mRNA, it encounters the *start codon,* which is almost always **AUG** (and rarely GUG).

With the mRNA message in place on the assembled ribosome, the next step in translation involves entrance of tRNAs with their amino acids. The pool of cytoplasm around the region contains a complete array of tRNAs, previously charged by having the correct amino acid attached. The step in which the complementary tRNA meets with the mRNA code is guided by the two sites on the large subunit of the ribosome called the **P site** (left) and the **A site** (right).[1] Think of these sites as shallow, scooped-out regions in the larger subunit of the ribosome, each of which accommodates a tRNA.

1. P stands for peptide site; A stands for aminoacyl site (an aminoacyl group is the charged amino acid).

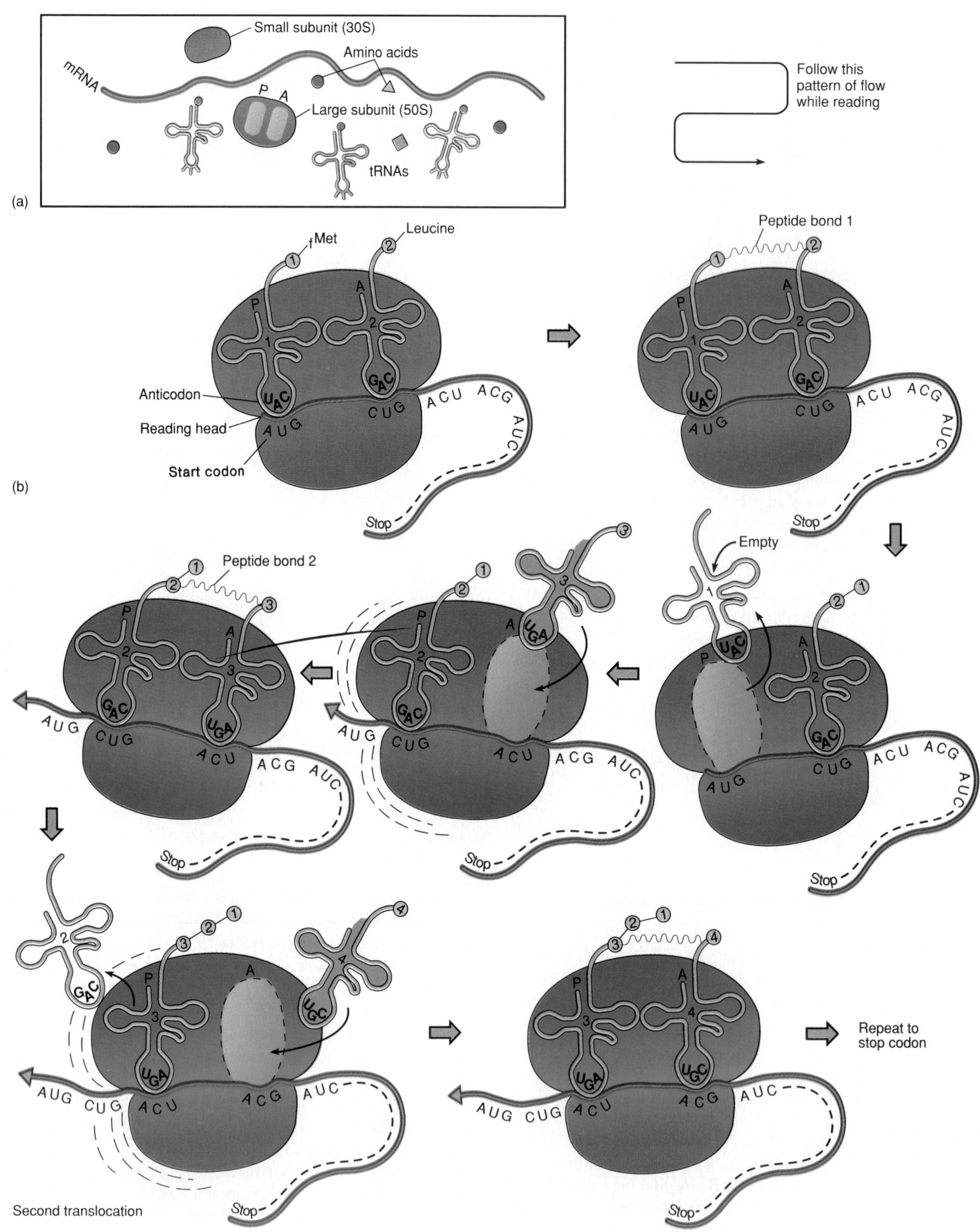

Figure 8.14 (*a*) Participants in translation. (*b*) The events in protein synthesis.

Table 8.3 The Genetic Code

First Position (5′ End)	Second Position: U		C		A		G		Third Position (3′ End)
U	UUU	Phe	UCU	Ser	UAU	Tyr	UGU	Cys	U
	UUC	Phe	UCC	Ser	UAC	Tyr	UGC	Cys	C
	UUA	Leu	UCA	Ser	UAA	STOP**	UGA	STOP**	A
	UUG	Leu	UCG	Ser	UAG	STOP**	UGG	Trp	G
C	CUU	Leu	CCU	Pro	CAU	His	CGU	Arg	U
	CUC	Leu	CCC	Pro	CAC	His	CGC	Arg	C
	CUA	Leu	CCA	Pro	CAA	Gln	CGA	Arg	A
	CUG	Leu	CCG	Pro	CAG	Gln	CGG	Arg	G
A	AUU	Ile	ACU	Thr	AAU	Asn	AGU	Ser	U
	AUC	Ile	ACC	Thr	AAC	Asn	AGC	Ser	C
	AUA	Ile	ACA	Thr	AAA	Lys	AGA	Arg	A
	AUG START	fMet*	ACG	Thr	AAG	Lys	AGG	Arg	G
G	GUU	Val	GCU	Ala	GAU	Asp	GGU	Gly	U
	GUC	Val	GCC	Ala	GAC	Asp	GGC	Gly	C
	GUA	Val	GCA	Ala	GAA	Glu	GGA	Gly	A
	GUG	Val	GCG	Ala	GAG	Glu	GGG	Gly	G

*This codon initiates translation.

**For these codons, which give the order to stop translation, there are no corresponding tRNAs and no amino acids.

With mRNA serving as a template, the stage is finally set for actual protein assembly. The correct **tRNA** (#1) enters the P site and fits itself against the start codon presented by the mRNA. Rules of pairing dictate that the anticodon of this tRNA must be complementary to the mRNA codon AUG, thus the tRNA with anticodon UAC is the one to first occupy site P. It happens that the amino acid carried by this tRNA is formyl *methionine* ($_f$Met; table 8.3), though in many cases, it may not remain a permanent part of the finished protein.

Elongation and Termination of the Protein Chain

Elongation of the chain now begins in earnest. This process is so dynamic that it is best presented in animation, but close study of a model (figure 8.14) will clarify its basic pattern. To keep track of protein assembly, you will want to remain aware that the ribosome and its "reading head" shift to the right along the mRNA from one codon to the next. As each new codon is brought into the head, a complementary tRNA is brought to the A position, a peptide bond is formed between the amino acids on the P and A tRNAs, and the polypeptide grows in length.

The first step in elongation occurs with the filling of the A site by a second **tRNA** (#2 on figure 8.14). The identity of this tRNA and its amino acid is dictated by the second mRNA codon. By convention, the master genetic code is represented by the mRNA codons and the amino acids they specify (table 8.3). Except in a very few cases, this code is universal, whether for procaryotes, eucaryotes, or viruses. It is worth noting that once the triplet code on mRNA is known, the original DNA sequence, the complementary tRNA code, and the types of amino acids in the protein are automatically known (figure 8.15). One cannot predict with any certainty in the opposite direction, however.

Table 8.3 shows that more than one mRNA codon (and thus more than one tRNA anticodon) specifies a given amino acid. Because there are 64 different triplet codes,[2] and only 20 different amino acids, it is not surprising that some amino acids are represented by several codons. For example, leucine and serine can each be represented by any of six different triplets, and only tryptophan and methionine are represented by a single codon. When a coded system of this type has more than one code for each translated element, it is termed *degenerate,* and it prevents us from using protein structure alone to determine genetic code. Fortunately, highly technical machines called sequencers can determine the order of codons on DNA and mRNA and the order of amino acids in proteins.

Once tRNA #2 has been brought to the A site in accordance with the second codon on mRNA, the two adjacent tRNAs and their amino acid (aa) cargo are now in favorable proximity for a peptide bond to form. This joining of the $_f$Met and **aa #2** results in a *dipeptide.* Following this step, the dipeptide immediately separates from tRNA #1 while remaining attached to tRNA #2.

2. $64 = 4^3$ (the 4 different codons in all possible combinations of 3).

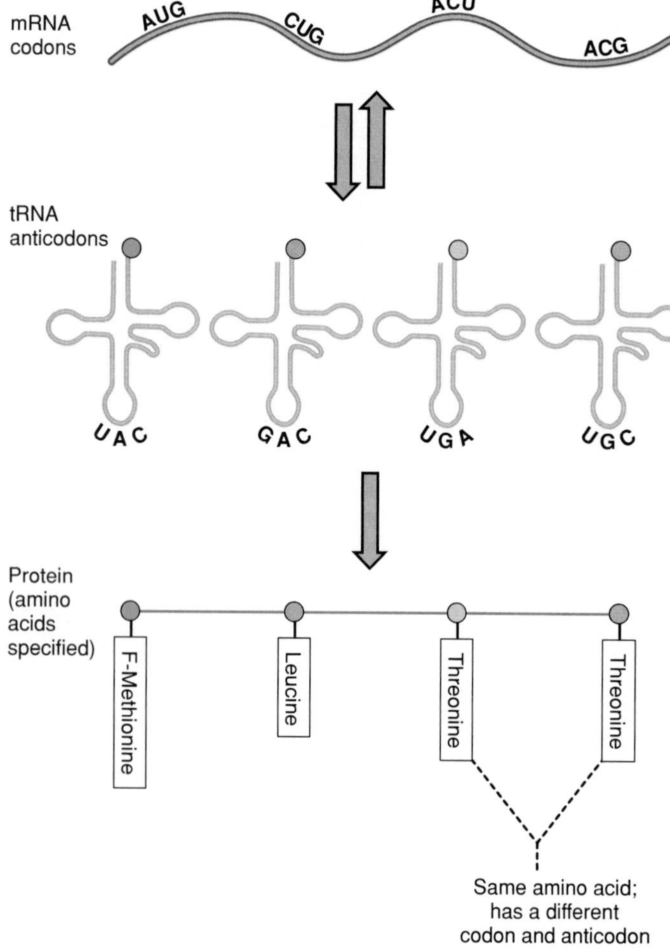

Figure 8.15 If a codon is known, the anticodon is automatically known, and vice versa. But because of the degeneracy of the code, it is not possible to determine the exact codon or anticodon from protein structure.

For the next step in elongation to proceed, some room must be made on the ribosome, and the next codon in sequence must be brought into position for reading. This is accomplished by *translocation,* the enzyme-directed shifting of the ribosome to the right along the mRNA strand, which causes: (1) the blank tRNA (#1) to be discharged from the ribosome, (2) tRNA #2 with the dipeptide attached to move into P position, leaving site A temporarily empty, and (3) the next codon (#3) to be brought into place on the ribosome (figure 8.14). The mechanism of this process remains unsettled. The tRNA that has been released is now free to drift off into the cytoplasm and become recharged with an amino acid for later additions to this or another protein. The stage is now set for the insertion of **tRNA#3** at site A as directed by the the third mRNA codon. This is followed once again by peptide bond formation between the dipeptide and **aa3** (making a tripeptide), splitting of the peptide from tRNA#2, and translocation (release of tRNA #2, shifting of mRNA, movement of tRNA #3 to position P, and entry of codon #4 into place). From this point on, peptide elongation proceeds repetitively by this basic series of actions out to the end of the mRNA.

The termination of protein synthesis is not simply a matter of reaching the last codon on mRNA. It is brought about by the presence of at least one novel codon near the end of the mRNA, occurring just after the codon for the last amino acid. These blank codons—**UAA, UAG,** or **UGA**—do not translate into either a tRNA or an amino acid. Although they are often called **nonsense codons,** they carry a necessary and useful message: *Stop* here. These codons were originally called nonsense triplets because, when first seen in a mutant strain of *E. coli,* they occurred out of place and caused premature arrest of peptide synthesis. Later it was determined that they normally occur near the end of all mRNAs. When this codon is reached, a special enzyme (release factor) breaks the bond between the final tRNA and the finished polypeptide chain, releasing it from the ribosome.

Before newly made proteins can carry out their structural or enzymatic roles, they may require finishing touches. These so-called posttranslational modifications vary from one protein to another. Even before the peptide chain is released from the ribosome, it begins folding upon itself to achieve its biologically active tertiary conformation. Some proteins must have the starting amino acid (formyl methionine) clipped off; proteins destined to become complex enzymes have cofactors added; and some join with other completed proteins to form quaternary levels of structure.

Further Thoughts on Protein Synthesis The machinelike operation of transcription and translation is overwhelming in its precison. Protein synthesis in bacteria is both efficient and rapid. At 37°C, 12–17 amino acids per second are added to a growing peptide chain. An average protein consisting of about 400 amino acids requires less than half a minute for complete synthesis. Further efficiency is gained when the translation of mRNA starts in the middle of transcription (figure 8.16). A single mRNA is long enough to be fed through more than one ribosome, thus several polypeptides may be synthesized from the same mRNA transcript arrayed in tandem procession along a chain of ribosomes. This **polyribosomal complex** is indeed an assembly line for mass production of proteins.

Protein synthesis consumes an enormous amount of energy. The equivalent of 3 ATPs is used in forming each peptide bond. An ATP is needed to charge a tRNA; one GTP is needed for the binding of a tRNA; and another GTP is spent on shifting the mRNA. Nearly 1,200 ATPs are required just for an average-sized protein.

Protein synthesis in eucaryotes is very similar in many respects, but it appears to be slower. In developing red blood cells, only two amino acids are added per second at 37°C. Polyribosomes operate in eucaryotes, but with fewer ribosomes in the complex. Other differences in the genetic expression of eucaryotes are discussed next.

Eucaryotic Transcription and Translation: Similar Yet Different

Although the fundamentals of procaryotic and eucaryotic gene expression are similar, there are differences, primarily in the cellular location of transcription and in how the mRNA is handled. Because eucaryotic DNA lies in the nucleus, that is where mRNA originates. To be translated, it has to pass through the pores in the nuclear envelope to the ribosomes. Simultaneous transcription and translation as exhibited in procaryotes is not possible.

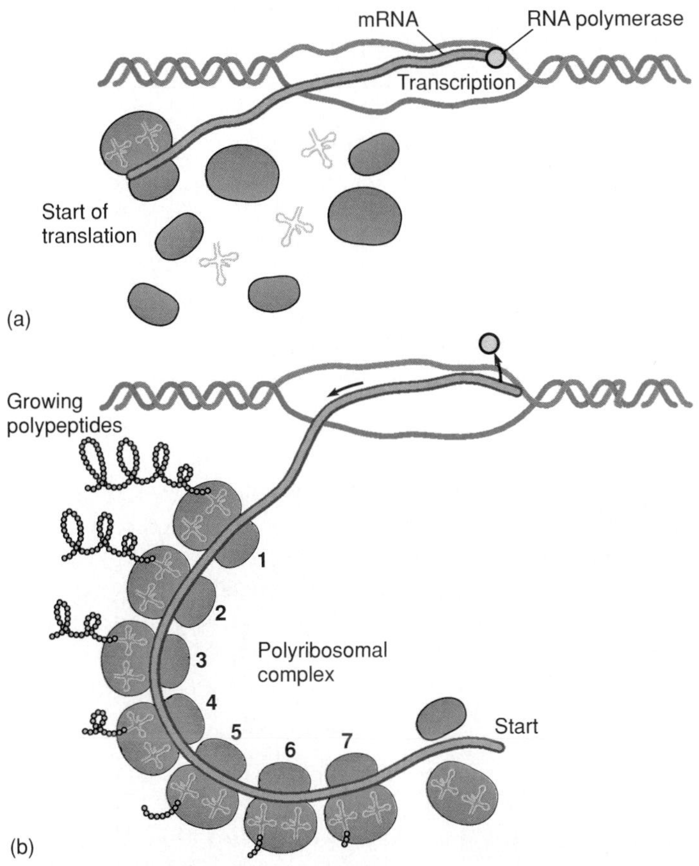

Figure 8.16 Speeding up the protein assembly line in bacteria. (*a*) The mRNA transcript encounters ribosomal parts immediately as it leaves the DNA. (*b*) The ribosomal factories assemble along the mRNA in a chain, each ribosome reading the message and translating it into protein. Many products will thus be well along the synthetic pathway before transcription has even terminated.

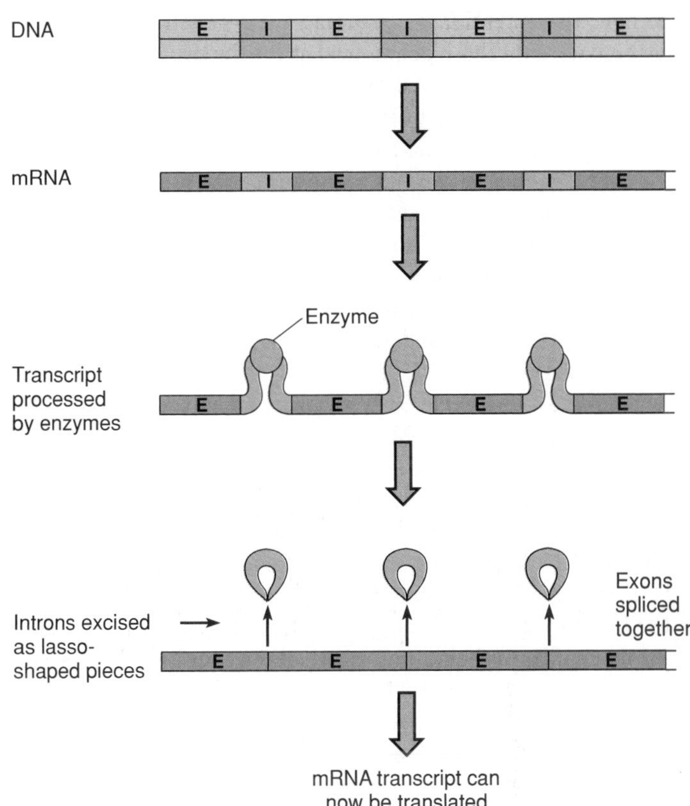

Figure 8.17 The split gene of eucaryotes. Eucaryotic genes have an additional complicating factor in their translation. Their coding sequences, or exons (E), are interrupted at intervals by segments called introns (I) that are not part of that protein's code. Introns are transcribed but not translated, which necessitates their removal by enzymes prior to translation.

We have given the simplified definition of a gene that works well for procaryotes, but a eucaryotic gene in most cases does not actually follow this colinear[3] pattern—that is, it is *not* an uninterrupted series of triplets coding for a protein. A eucaryotic gene contains the code for a protein, but located along this code are one to several intervening sequences of bases, called **introns,** that do not code for product. Introns are interspersed between coding regions, called **exons,** that will be translated (figure 8.17). Using words as examples, a short section of colinear procaryotic gene might read TOM SAW OUR DOG DIG OUT, while a eucaryotic gene that codes for the same portion would read TOM SAW XZKP FPL OUR DOG QZWVP DIG OUT. The recognizable words are the exons, and the gibberish represents the introns.

This unusual genetic architecture, sometimes called a *split gene,* requires further processing prior to translation (figure 8.17). Transcription of the entire gene, with both exons and introns, occurs first. Then a group of self-splicing RNAs (ribozymes, see feature 7.2) capable of cleaving the points between exons and introns goes to work. The splicer loops the introns into lasso-shaped pieces, excises them, and joins the exons end-to-end. This produces a strand of mRNA with no intron material that can proceed to the cytoplasm to be translated. At first glance, this system seems to be a cumbersome way to make a transcript, and the value of this extra genetic baggage is still the subject of much debate. Several different types of introns have been discovered, some of which do code for proteins. One particular intron discovered in yeast gives the code for a reverse transcriptase, leading to speculation that cells have their own mechanisms for producing DNA from RNA. Some experts hypothesize that introns serve as a "sink" for extra bits of genetic material that could be available for splicing into existing genes, thus promoting genetic change and evolution.

The Genetics of Animal Viruses

Chapter 5 mentioned the genetic nature of viruses. It was noted that they essentially consist of one or more pieces of DNA or RNA enclosed in a protective coating. Above all, they are genetic parasites that require access to their host cell's genetic and metabolic machinery to be replicated, transcribed, and translated. Although viruses may damage or kill their host cells, they

3. Colinearity means that the base sequence can be read directly into a series of amino acids.

also have the potential for changing them genetically. Because they contain only those genes needed for the production of new viruses, the genomes of viruses tend to be very compact and economical. In fact, this very simplicity makes them excellent subjects for the study of gene function.

The genetics of viruses is quite diverse (see chapters 20 and 21). In many viruses, the nucleic acid is linear in form, while in others it is circular. The genome of most viruses exists in a single molecule, though in a few, it is segmented into several smaller molecules. Most viruses contain normal double-stranded (ds) DNA or single-stranded (ss) RNA, but entirely novel genetic patterns occur in ssDNA viruses, dsRNA viruses, and retroviruses that work backwards by making dsDNA from ssRNA. In some instances, viral genes overlap one another, and at times, both strands of DNA contain a sense message.

A few generalities may be stated about viral genetics. In all cases, the viral nucleic acid penetrates the cell and is introduced into the host's gene-processing machinery at some point. In successful infection, an invading virus instructs the host's machinery to synthesize large numbers of new virus particles by a mechanism specific to a particular group. Replication of the DNA molecule of DNA animal viruses occurs in the nucleus, where the cell's DNA replication machinery lies (except in the pox viruses); the genome of most RNA viruses is replicated in the cytoplasm (except in the orthomyxoviruses [influenza]). In all viruses, viral mRNA is translated into viral proteins on host cell ribosomes using host tRNA. In the next section we will briefly observe the patterns of genetic replication in the major viral groups as summarized in table 8.4.

Table 8.4 Patterns of Genetic Flow in Viruses

DNA Viruses
dsDNA → dsDNA (semiconservative)
ssDNA → dsDNA → ssDNA
RNA Viruses
(+) ssRNA → (−) ssRNA → (+) ssRNA
(−) ssRNA → (+) ssRNA → (−) ssRNA
ssRNA → ssDNA → dsDNA → ssRNA (retroviruses)
dsRNA → ssRNA → dsRNA (conservative)

Replication, Transcription, and Translation of dsDNA Viruses

Replication of dsDNA viruses is divided into phases (figure 8.18). During the early phase, viral DNA enters the nucleus, and a large section of it, several genes long, is transcribed into a messenger RNA. This transcript moves into the cytoplasm to be translated into viral proteins (enzymes) needed to replicate the viral DNA, which occurs in the nucleus. The host cell's own DNA polymerase is often involved, though some viruses (herpes, for example) may have their own. During the late phase, other parts of the viral genome are transcribed and translated into proteins required to form the capsid and other structures. The new viral genomes and capsids are assembled, and the mature viruses are released by budding or cell disintegration.

DNA viruses interact directly with the DNA of their host cell. In some viruses the viral DNA becomes silently *integrated* into the host's genome by insertion at a particular site on the host genome (figure 8.18). This integration may later lead to the transformation[4] of the host cell into a cancer cell and the production of a tumor. Several DNA viruses, including hepatitis B (HBV), the herpesviruses, papilloma (warts) viruses, and adenoviruses are known or suspected to be initiators of cancers and are thus termed *oncogenic*. The mechanisms of transfor-

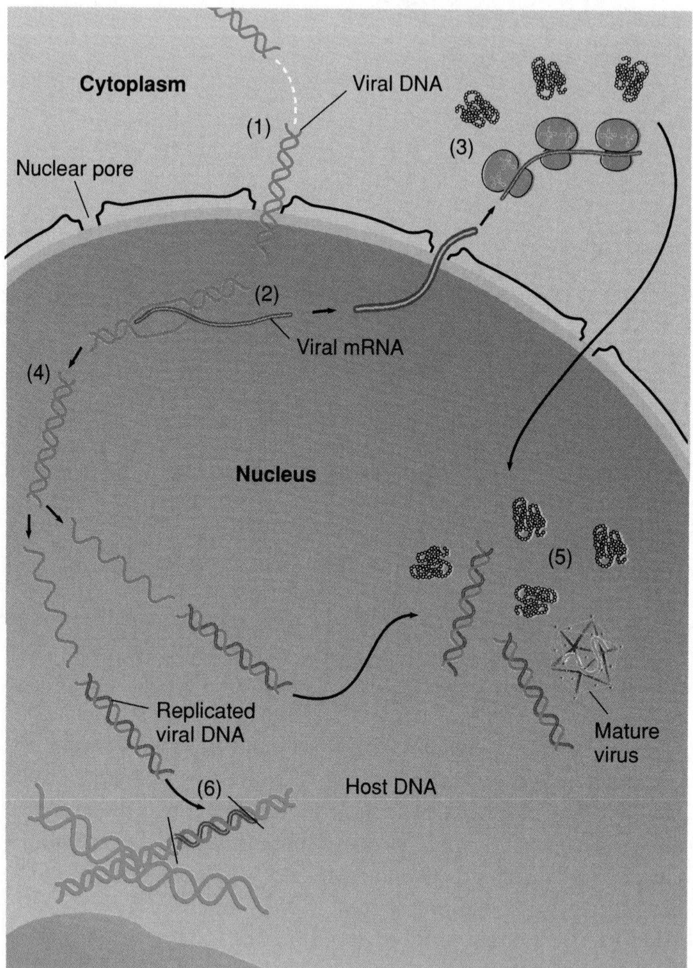

Figure 8.18 Genetic stages in the multiplication of DNA viruses. The virus penetrates the host cell and releases DNA, which (*1*) enters the nucleus and (*2*) is transcribed. Other events are: (*3*) Viral mRNA is translated into structural proteins; proteins enter the nucleus. (*4*) Viral DNA is replicated repeatedly in the nucleus. (*5*) Viral DNA and proteins are assembled into mature virus in the nucleus. (*6*) Being double-stranded, viral DNA may insert into host DNA (latency).

4. The process of genetic change in a cell, leading to malignancy.

oncogenic (awn″-koh-jen′-ik) Gr. *onkos*, mass, and *gennan*, to produce. Refers to any cancer-causing process. Viruses that do this are termed oncoviruses.

mation and oncogenesis appear to involve special genes called oncogenes that can regulate cellular genomes (see chapter 14).

Replication, Transcription, and Translation of RNA Viruses

RNA viruses exhibit several differences from DNA viruses. Their genomes are smaller and less stable, they enter the host cell already in an RNA form, and the virus cycle occurs entirely in the cytoplasm for most viruses. An RNA virus brings with it one of the following genetic messages: a positive-sense genome (+) that comes ready to be translated into proteins, a positive-sense genome (+) that can be converted to DNA, a negative-sense genome (−) that must be converted to positive prior to translation, or a dsRNA genome.

Positive-Sense Single-Stranded RNA Viruses Examples of positive-sense single-stranded viruses are poliovirus and hepatitis A virus (HAV). Shortly after the virus uncoats in the cell, its positive strand is translated into a large protein that is soon cleaved into individual functional units, one of which is a polymerase that initiates the replication of the viral strand (figure 8.19). Replication of a single-stranded molecule poses a special problem. Given that the mature viruses will each require a new positive strand, how is this going to be made starting from a positive template? It is done in two steps. First, a negative strand is synthesized using the parental positive strand as a template by the usual base pairing mechanism. The resultant negative strand becomes a master template against which numerous positive daughter strands are made. Further translation of the viral genome produces large numbers of structural proteins for final assembly and maturation of the virus.

RNA Viruses with Reverse Transcriptase: Retroviruses A most unusual class of viruses has a unique capability to reverse the order of the flow of genetic information. Thus far in our discussion, all genetic entities have shown the patterns DNA → DNA, DNA → RNA, or RNA → RNA. Retroviruses, including HIV, the cause of AIDS, and HTLV I, a cause of one type of leukemia, synthesize DNA using their RNA genome as a template (see figure 21.14). They accomplish this by means of an enzyme, **reverse transcriptase,** that comes packaged with each virus particle. This enzyme assembles a single-stranded DNA against the viral RNA template and then directs the formation of a complementary strand of this ssDNA, resulting in a double strand of viral DNA (see figure 21.14). The dsDNA strand enters the nucleus, where it may be integrated into the host genome. There, it is perfectly situated to be transcribed by the usual mechanisms into new viral ssRNA. Translation of the viral RNA yields viral proteins that combine with the RNA genome during final virus assembly. The capacity for a retrovirus to become inserted into the host's DNA as a provirus has several possible consequences. In some cases, these viruses are oncogenic and are known to transform cells and produce tumors. On the other hand, the AIDS virus appears to remain latent in an infected cell until a stimulus activates it to continue a lytic cycle. These processes and their outcome are described more fully in chapters 14 and 21.

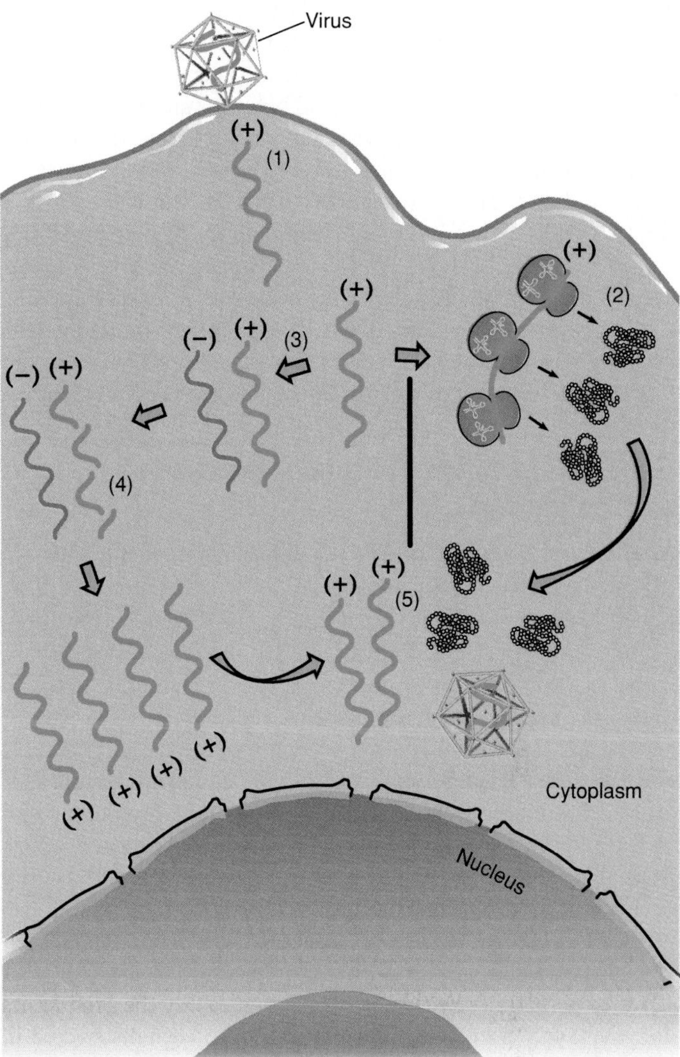

Figure 8.19 Replication of positive-sense RNA viruses. In general, these viruses do not enter the nucleus. (*1*) Penetration and uncoating of viral RNA. (*2*) Since it is positive in sense and single-stranded, the RNA can be directly translated on host cell ribosomes into various necessary viral proteins. (*3*) To produce large numbers of positive genomes for final assembly, first a negative genome is synthesized against the positive template. (*4*) The negative template may be used to produce a series of positive replicates. (*5*) RNA strands and proteins assemble into mature viruses.

Negative-Strand ssRNA Viruses Most RNA viruses including those of rabies, mumps, measles, and influenza, possess negative-sense strands. Having a negative strand imposes some complexities on viral synthesis. In the primary stage the negative strand becomes a template to synthesize positive strands of RNA. This occurs immediately, because the virus brings with it a preformed viral polymerase for this job. During a second stage of synthesis, the positive strands serve as the master templates for forming large numbers of negative strands. These negative strands serve as the genomes for new viruses or may be used to synthesize a number of secondary positive strands. Positive strands will also serve as mRNAs for translation into proteins needed in the assembly of mature viruses.

The Genetics of dsRNA Viruses Reoviruses are distinct among the viruses in having a dsRNA genome arrayed on several linear molecules. Human pathogens include the agent of Colorado tick fever and rotavirus diarrhea. A dsRNA genome cannot be replicated in the semiconservative manner of DNA because the two RNA strands do not completely separate. In an alternate strategy, one strand of the parent RNA is transcribed into a positive mRNA that serves as a template for a complementary negative mRNA strand. These two strands then remain attached and become the genome for new viruses. This replication is conservative in that the parent strand remains intact and the progeny RNA molecule is completely new. The positive strand also directs the synthesis of viral products needed for assembly.

Genetic Regulation of Protein Synthesis and Metabolism

In chapter 7 we surveyed the metabolic reactions in cells and the enzymes involved in these reactions. At that time we mentioned types of metabolic regulation that are genetic in origin. These fine control mechanisms ensure that not all genes are active at all times. In this way enzymes will be produced to appropriately reflect nutritional status and prevent the waste of energy and materials in dead-end synthesis. Genetic function is regulated by a specific segment of DNA called an **operon.** Operons contain various control genes (regulators, promoters, and operators) that govern the operation of related structural genes. Such gene regulation responds to external stimuli and usually occurs at the level of transcription. Operons act in one of two ways: (1) The operon may be turned on (*induced*) by the substrate of the enzyme for which the structural genes code, or (2) the operon may be turned off (*repressed*) by the product its enzymes synthesize. Although operons have been discovered in bacteria, comparable devices probably govern the much more complicated eucaryotic metabolism as well.

The Lactose Operon: A Model for Inducible Gene Regulation in Bacteria

The best understood cell system for explaining control through genetic induction is the **lactose (*lac*) operon.** This concept, first postulated in 1961 by François Jacob and Jacques Monod, accounts for the regulation of lactose metabolism in *Escherichia coli.* Several other operons with similar modes of action have since been identified, and together they furnish convincing evidence that the environment of a cell may have great impact on gene expression.

The lactose operon is made up of three segments or *loci* of DNA: (1) the **regulator,** composed of the gene that codes for a protein capable of repressing the operon (a **repressor**); (2) the **control** locus, composed of two genes, the **promoter** (identified as a palindrome) and the **operator,** where transcription of the structural genes is initiated; and (3) the **structural** locus, made up of three genes, each coding for a different enzyme (β-galactosidase, permease, and transacetylase) needed to catabolize lactose (figure 8.20). In bacteria, structural genes needed for the metabolism of a nutrient tend to be arranged in procession, an efficient strategy that permits genes for a particular metabolic pathway to be induced or repressed in unison. The enzymes of the operon are of the inducible sort mentioned in chapter 7. The promoter, operator, and structural components lie in contiguous order, but the regulator may be at a distant site.

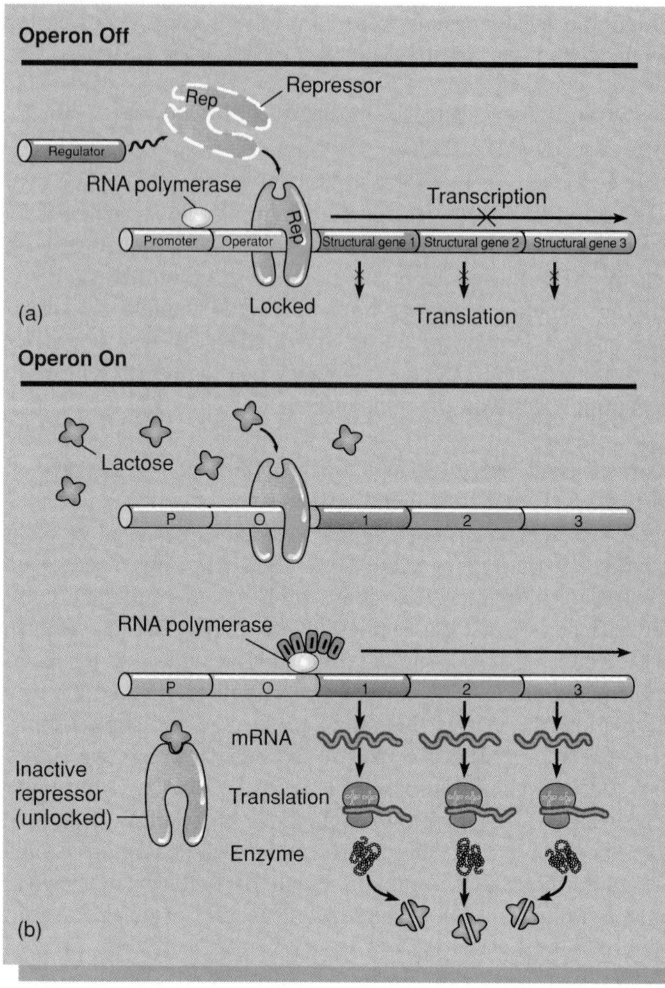

In inductive systems like the *lac* operon, the operon is normally in an *off mode* and does not initiate enzyme synthesis when the appropriate substrate is absent (figure 8.20*a*). How is

Figure 8.20 The lactose operon in bacteria: how inducible genes are controlled by substrate. (*a*) Operon off. In the absence of lactose, a repressor protein (the product of a regulatory gene) attaches to the operator gene of the operon. This effectively locks the operator and prevents any transcription of structural genes downstream (to its right). Suppression of transcription (and, consequently, translation) prevents the unnecessary synthesis of enzymes for processing lactose. (*b*) Operon on. Upon entering the cell, the substrate (lactose) becomes a genetic inducer by attaching to the repressor, which loses its grip and falls away. The operator is now free to initiate transcription, and enzymatic products of translated genes perform the necessary reactions on their lactose substrate.

operon (op'-ur-on) L. *opera,* exertion.

loci (loh'-sy) sing. locus (loh'-kus) L. *locus,* a place. The site on a chromosome occupied by a gene.

the operon maintained in this mode? The key is in the repressor protein that is coded by the regulatory locus. This relatively large and pliant molecule has allosteric binding conformations, one for the operator and another for lactose. In the absence of lactose, this repressor binds with the operator locus, thereby blocking the transcription of the structural genes lying downstream. Think of the repressor as a lock on the operator, and if the operator is locked, the structural genes cannot be transcribed.

If lactose is added to the cell's environment, it triggers several events that turn the operon *on.* Because lactose is ultimately responsible for stimulating protein synthesis, it is called an *inducer.* The preferential binding of lactose to the repressor protein institutes a conformational change in the repressor that dislodges it from the operator segment (figure 8.20*b*). The control segment that was previously static is now unlocked. Located at the promoter site is an RNA polymerase ready to begin transcription at the operator site. The structural genes are transcribed in a single unbroken transcript coding for all three enzymes, and this mRNA is also translated as a unit. After synthesis, the enzymes are cleaved into individual units that participate in the catabolism of lactose.

As lactose is depleted, further enzyme synthesis is not necessary, so the order of events reverses. Because at this point there is no longer sufficient lactose to inhibit the repressor, it is again free to attach to the operator. The operator is locked, and transcription of the structural genes and protein synthesis related to lactose both stop.

A Repressible Operon

Bacterial systems for amino acid, purine, and pyrimidine synthesis work on a slightly different principle—that of repression. Similar factors such as repressor proteins, operators, and a series of structural genes exist for this operon, but with some important differences. A repressible operon governs anabolism, the synthesis of an important cell nutrient. Unlike the *lac* operon, this operon is normally in the on mode and will be turned off only when this nutrient is no longer required. The nutrient plays an additional role as a **corepressor** needed to block the action of the operon.

A growing cell that needs the amino acid arginine (arg) effectively illustrates the operation of a repressible operon. Under these conditions, the *arg* operon is set to *on,* and arginine is being actively synthesized through the action of its enzymatic products (figure 8.21*a*). In an active cell, the arginine will be used up and will not accumulate. In this instance, the repressor will remain inactive because there is too little free arginine to activate it. As the cell's metabolism begins to slow down, however, the synthesized arginine will no longer be used up and will accumulate. Now it is freer to act as a corepressor by attaching to the repressor. This reaction produces a functioning repressor that locks the operator and stops further transcription and arginine synthesis (figure 8.21*b*).

Analogous gene control mechanisms in eucaryotic cells are not as well understood, but it is known that genes can be turned on and off by intrinsic regulatory segments similar to operons.

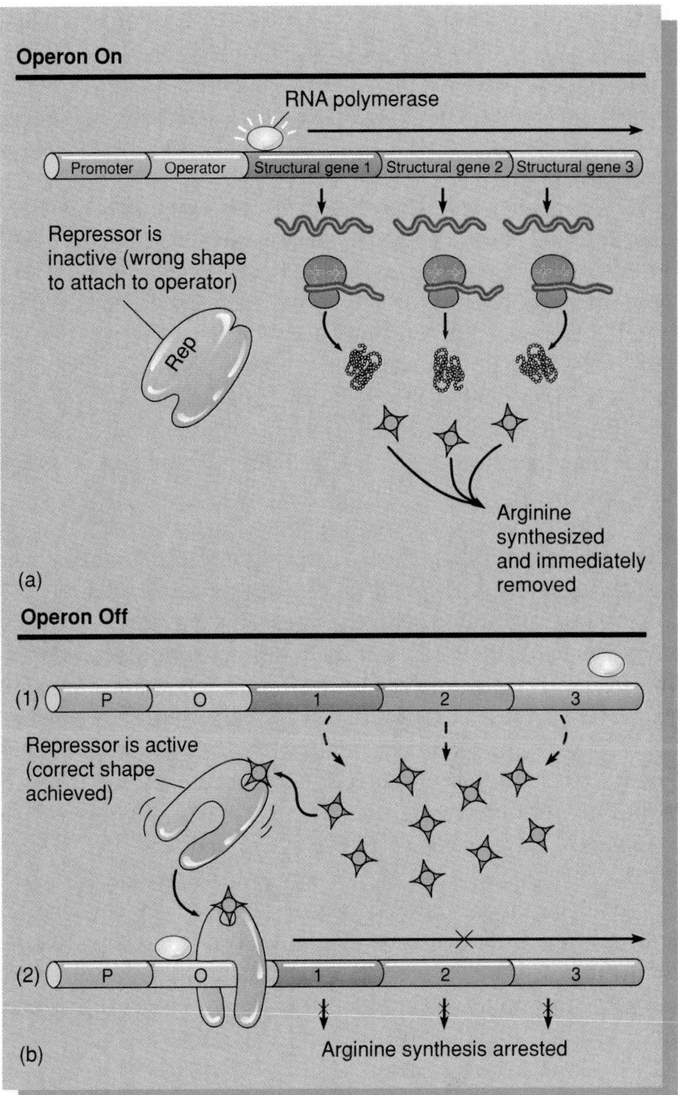

Figure 8.21 Repressible operon: control of a gene through excess nutrient. (*a*) Operon on. A repressible operon remains on when its nutrient products (here, arginine) are in great demand by the cell. This is because the repressor remains inactive at low nutrient levels. (*b*) Operon off. The operon is repressed when (*1*) arginine builds up and, serving as a corepressor, reacts with the repressor, thereby activating it, and (*2*) the repressor complex affixes to the operator and blocks further transcription of genes for arginine synthesis.

Evidence of certain kinds of regulation (or its loss) can be detected in cancer cells, where genes called oncogenes abnormally override built-in genetic controls. An oncogene may exert its control by causing the formation of a chemical that permanently activates a part of the genome controlling cell growth. A cell with no constraints on the number of cell divisions may grow out of all normal bounds into tumors and leukemias. The fascinating genetics of cancer are discussed further in chapter 14.

Antibiotics That Affect Transcription and Translation

Naturally occurring cell nutrients are not the only agents capable of modifying gene expression. Some infection therapy is based on the concept that certain antibiotics react with DNA,

RNA, and ribosomes and thereby alter genetic expression (see chapter 10). Treatment with such drugs is based on an important premise: that growth of the infectious agent will be inhibited by blocking its protein synthesizing machinery selectively, without disrupting the cell synthesis of the patient receiving the therapy.

Drugs that inhibit protein synthesis exert their influence on transcription or translation. For example, the rifamycins used in therapy for tuberculosis bind to RNA polymerase, blocking the initiation step of transcription, and are selectively more active against bacterial RNA polymerase than the corresponding eucaryotic enzyme. Actinomycin D binds to DNA and halts mRNA chain elongation, but its mode of action is not selective for bacteria. For this reason, it is very toxic and never used to treat bacterial infections, though it can be applied in tumor treatment.

The ribosome is a frequent target of antibiotics that inhibit ribosomal function and ultimately protein synthesis. The value and safety of these antibiotics again depend upon the differential susceptibility of procaryotic and eucaryotic ribosomes. One problem with drugs that selectively disrupt procaryotic ribosomes is that the mitochondria of humans contain a procaryotic type of ribosome, and these drugs may inhibit the function of the host's mitochondria (chapter 10). One group of antibiotics (including erythromycin and spectinomycin) prevents translation by interfering with the attachment of mRNA to ribosomes. Chloramphenicol, lincomycin, and tetracycline bind to the ribosome in a way that prevents the elongation of the polypeptide, and aminoglycosides (such as streptomycin) inhibit peptide initiation and elongation. It is interesting to note that these drugs have served as important tools to explore genetic events because they can arrest specific stages in these processes.

Changes in the Genetic Code: Mutations and Intermicrobial Exchange and Recombination

As precise and predictable as the rules of genetic expression seem, permanent changes do occur in the genetic code. Indeed, genetic change is the driving force of evolution. In microorganisms, such changes are often phenotypic, being first observed as a trait that abruptly appears or disappears. For example, a normally colored bacterium loses its ability to form pigment, or a strain of the malarial parasite develops resistance to a drug. Phenotypic changes of this type are ultimately due to changes in the genotype. Any permanent, inheritable change, large or small, in the genetic information of the cell is a **mutation.** On a strictly molecular level, a mutation is an alteration (loss, addition, rearrangement) in the nitrogen base sequence of DNA.

A microorganism that exhibits a natural, nonmutated characteristic is known as a *wild type* or wild strain. If it bears a mutation, it is called a *mutant strain.* Mutant strains show variance in morphology, nutritional characteristics, genetic control mechanisms, resistance to chemicals, temperature preference, and nearly any type of enzymatic function. Mutant strains

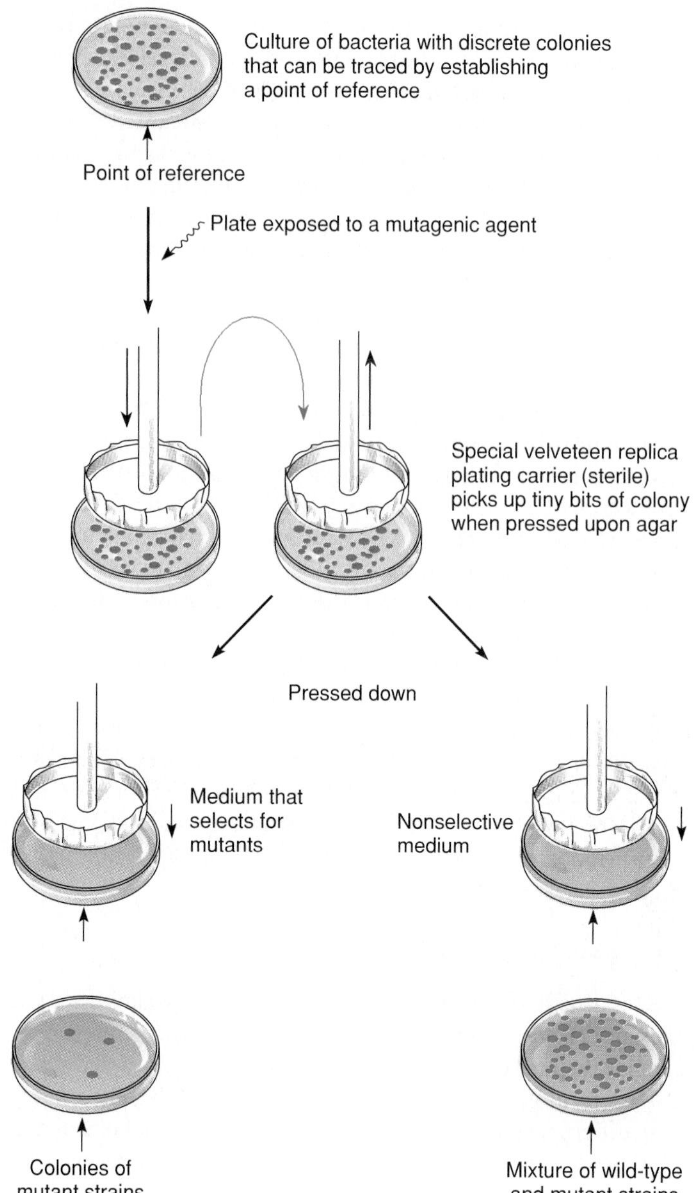

Figure 8.22 The general basis of replica plating, a method for detecting and isolating mutant strains of microorganisms.

are very useful for tracking genetic events, unravelling genetic organization, and pinpointing genetic markers. A classic method of detecting and isolating mutant strains is through nutritional manipulation of a culture. For example, in a culture of a wild-type bacterium that is lactose-positive (meaning it has the necessary enzymes for fermenting this sugar), a small number of mutant cells are probably lactose-negative, having lost the capacity to ferment this sugar. If the culture is plated on a medium containing indicators for fermentation, each colony could be observed for its fermentation reaction, and the negative strain isolated. Another standard method of detecting and isolating microbial mutants is by replica plating (figure 8.22).

Agent	Effect
Chemical	
Nitrous acid, bisulfite	Remove an amino group from some bases
Mustard gas	Causes cross-linkage of DNA strands
Acridine dyes	Cause frameshifts due to insertion between base pairs
Nitrogen base analogs	Compete with natural bases for sites on replicating DNA
Radiation	
Ionizing (gamma rays, X rays)	Form free radicals that cause single or double breaks in DNA
Ultraviolet	Causes cross-links between adjacent pyrimidines

Table 8.5 Selected Mutagenic Agents and Their Effects

Causes of Mutations

A mutation is described as spontaneous or induced, depending upon its origin. A *spontaneous mutation* is a random change in the DNA arising from mistakes in replication or the detrimental effects of natural background radiation (cosmic rays) on DNA. The frequency of spontaneous mutations has been measured for a number of different organisms. Mutation rates vary from one mutation in 10^5 replications (a high rate) to one mutation in 10^{10} replications (a low rate). The rapid rate of bacterial reproduction allows these mutations to be observed more readily in bacteria than in animals.

Induced mutations result from exposure to known **mutagens,** which are primarily physical or chemical agents that interact with DNA in a disruptive manner (table 8.5). The carefully controlled use of mutagens has proved a useful way to induce mutant strains of microorganisms for study.

Chemical mutagenic agents such as acridine dyes insert completely across the DNA helices between adjacent bases to produce a frameshift mutation (see feature 8.5) and distort the helix (figure 8.23). Analogs[5] of the nitrogen bases (5-bromodeoxyuridine and 2-aminopurine, for example) are chemical mimics of natural bases that are incorporated into DNA during replication. Insertion of these abnormal bases leads to mistakes in base pairing. Many chemical mutagens are also carcinogens or cancer-causing agents (see the discussion of the Ames test in a later section of this chapter).

5. An analog is a chemical structured very similarly to another chemical except for minor differences in functional groups.

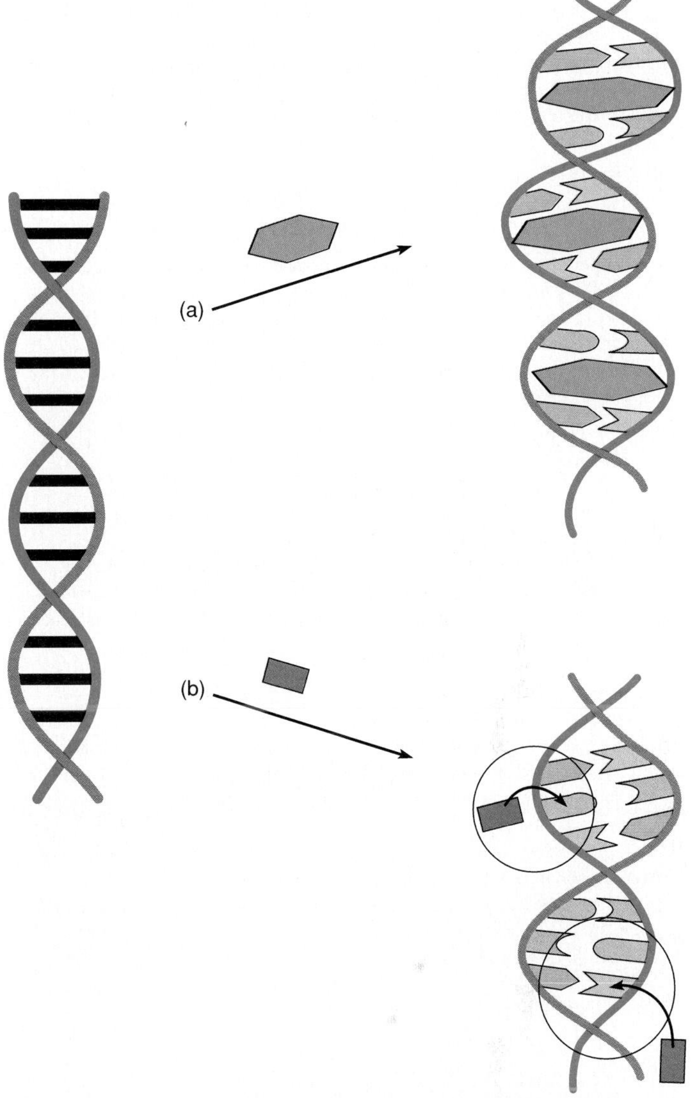

Figure 8.23 Modes of action of chemical mutagenic agents. (*a*) Some mutagens insert into the helix and interfere with its function. (*b*) Other chemicals act as "imposters." Due to their structural similarity to a regular nitrogenous base, they will insert in its place, but they cannot function like the regular base.

Physical agents that alter DNA are primarily types of radiation. High-energy gamma rays and X rays create reactive free radicals in the cell. Because of its role as a genetic repository, DNA is a vulnerable target of these radicals, and it accumulates breaks that may not be repairable. Ultraviolet (UV) radiation induces abnormal bonds between adjacent pyrimidines that prevent normal replication (see figure 9.11). The overall effects of radiation depend upon the length of exposure and its intensity. Exposure to large doses of radiation can be fatal, which is why it is so effective in microbial control; it can also be carcinogenic in animals. Ionizing radiation such as X-ray therapy has been linked to thyroid cancer, and exposure to nuclear radiation is a known cause of leukemia.

Feature 8.5 Mutations: Mistakes in DNA Language

The nature of mutations and their effects on the product can best be demonstrated by a sentence analogy, with letters representing the bases. Let us say that the message of DNA reads:

DIDPAMSEETADSUP?

Or, if set in word triplets (like DNA) to make sense:

DID PAM SEE TAD SUP?

Two kinds of mutation can move the reading frame of the code. Such **frameshift mutations** disrupt the natural order of the message by shifting the frame forward or back, thus setting the whole sequence off. In **deletion,** one or more bases is removed and the bases all shift left of the deletion point to form new triplets. In our example, removing an E in SEE, creates the following frameshift:

E
|
DID PAM SET ADS UP-?

In this example, the message still means something. This sort of missense mutation on DNA may have some use in the cell.

If a new base or bases are added to the DNA message, as during **insertion,** the base combinations must shift right of the insertion point. Thus, if X is inserted before SEE in the original sentence, the following sentence results:

DID PAM XSE ETA DSU P?

This change removes the sense from both the words and the sentence, and is an example of a frameshift that could lead to a useless protein and possibly alter cell function. The location of a frameshift can be very significant in that it determines how many amino acids will be changed. In general, one occurring near the beginning of a gene will have greater effect than one near the end because it shifts more bases, changes more amino acids, and increases the chance for a nonsense mutation to occur.

Other types of mutations do not necessarily shift the frame in the manner of our previous examples because they are limited to one base or codon and do not affect those in surrounding positions. In an **inversion,** for example, the order of bases is switched, which alters only one or possibly two codons. Consider inversion of the letters in SUP:

DID PAM SEE TAD USP?

With base **substitution,** one base is removed and replaced with another. This, again, can cause a different codon and amino acid to be added to the protein during translation. If an N were substituted for the final P in the original sentence, the code would read:

|
DID PAM SEE TAD SUN?

Inversions and substitutions generally have subtler effects than frameshifts. For more work with mutations, see Practical/Thought Questions at the end of this chapter.

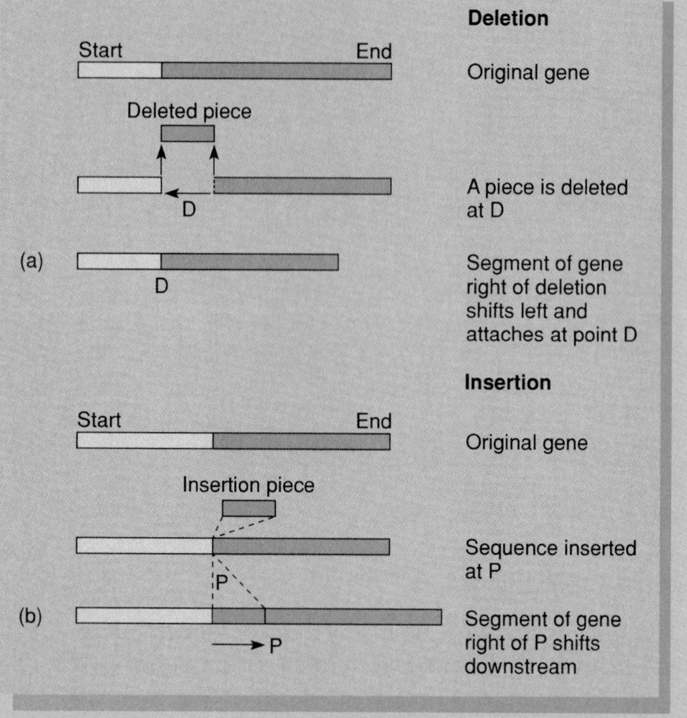

A diagrammatic scheme of two types of point mutations that shift the frame of reference. (*a*) Deletion; (*b*) insertion. (Double-stranded DNA is represented here as a bar, for easier visualization.)

Categories of Mutations

Mutations range from large mutations, in which whole chromosomes are lost or large genetic sequences are inserted (see later section on transposons), to small ones that affect only a single base on a gene. These latter mutations, which involve addition, removal, or substitution of a few bases, are called **point mutations.** The effects of some point mutations are described in feature 8.5.

To understand how a change in DNA influences the cell, remember that the DNA code appears in a particular order of triplets (three bases) that is transcribed into mRNA codons, each of which specifies an amino acid. A permanent alteration in the DNA that is copied faithfully into mRNA and translated can change the structure of the protein. A change in a protein may likewise change the morphology or physiology of a cell.

Most mutations have a harmful effect on the cell, leading to cell dysfunction or death; these are called lethal mutations. Neutral mutations produce neither adverse nor helpful changes. A small number of mutations are beneficial in that they provide the cell with a useful change in structure or physiology. A change in the code that leads to placement of a different amino acid is called a **missense mutation** (see feature 8.5). A missense mutation can do one of the following: (1) create a faulty, nonfunctional protein, (2) produce a different but functional protein, or (3) cause no significant alteration in protein function.

A **nonsense mutation,** on the other hand, changes a normal codon into a stop codon that does not code for an amino acid and stops the production of the protein wherever it occurs. A nonsense mutation almost always results in a nonfunctional protein. A *silent mutation* alters a base, but does not change the amino acid and thus has no effect. Because of the degeneracy of the code, ACU, ACC, ACG, and ACA all code for threonine, so a mutation that changes only the last base will not alter the sense of the message in any way. A *back-mutation* occurs when a gene that has undergone mutation reverses (mutates back) to its original base composition.

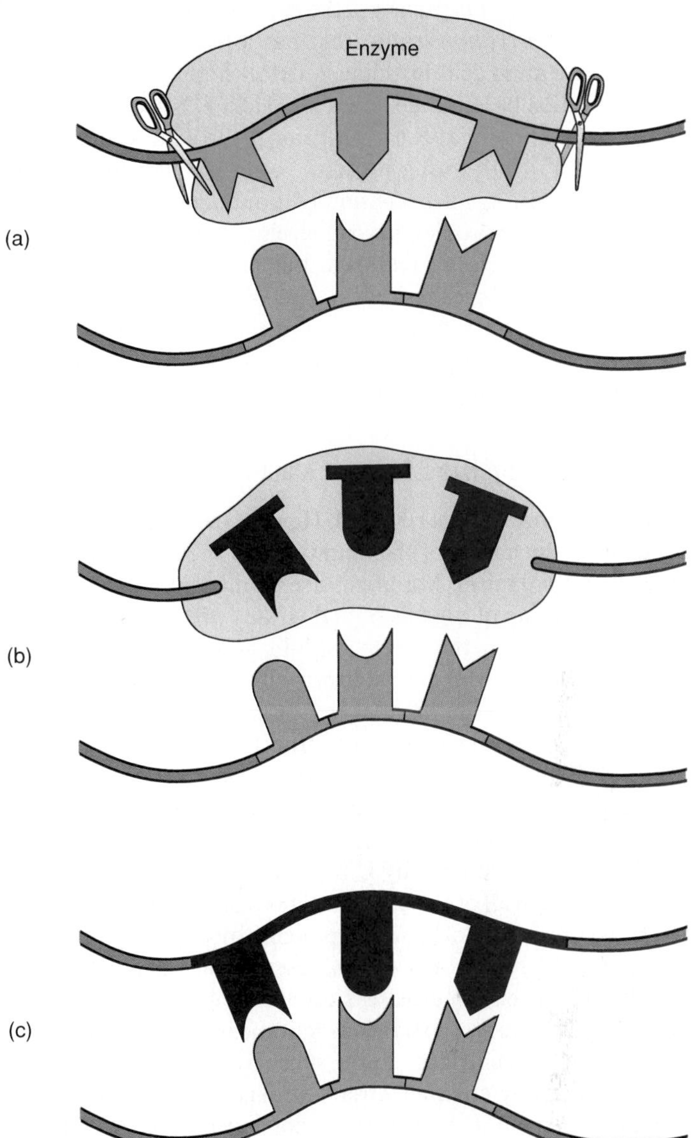

Figure 8.24 Excision repair of mutation by an enzyme. (*a*) The enzyme recognizes a section of incorrect bases and removes it. (*b*) The repair enzyme places correct bases and seals the gap. (*c*) Repaired DNA.

Repair of Mutations

Earlier we indicated that DNA has a proofreading mechanism to repair mistakes in replication that might otherwise become permanent. Because mutations are potentially life threatening, the cell has additional systems for finding and repairing DNA that has been damaged by various mutagenic agents and processes. Most ordinary DNA damage is resolved by enzymatic systems specialized for ferreting out and rectifying such defects.

DNA that has been damaged by ultraviolet radiation can be healed by *photoactivation* or *light repair.* This repair mechanism requires visible light and a light-sensitive enzyme, DNA photolyase, which can detect and attach to the damaged areas (sites of abnormal pyrimidine binding). When this enzyme absorbs visible light, it derives sufficient energy to break the abnormal bonds. Ultraviolet repair mechanisms are successful only for a relatively small number of UV mutations. Cells cannot repair severe, widespread damage and will die. Another scheme involves the *suicide enzyme* that repairs mutated guanine by plucking off the mutagenic group. What is most unusual about this enzyme is that, unlike others, it destroys itself after function and cannot be used again.

Light Repair of DNA Photolyase is found in all normal cells. In humans, the genetic disease *xeroderma pigmentosa* is due to nonfunctioning genes for this enzyme. Persons suffering from this rare disorder develop severe skin cancers; this provides strong evidence for a link between cancer and mutations.

Mutations may be excised by a series of enzymes that remove the incorrect bases and add the correct ones. This process is known as excision repair. First, enzymes break the bonds between the bases and the sugar-phosphate strand at the site of the error. A different enzyme subsequently removes the defective bases one at a time, leaving a gap that will be filled in by regular DNA replication (figure 8.24). A repair system can also locate **mismatched** bases that were missed during proofreading—for example, C mistakenly paired with A, or G with T. The base must be replaced soon after the mismatch is made, or it will soon become lost in the complex maze of codons.

The Ames Test

New agricultural, industrial, and medicinal chemicals are constantly being added to the environment, and exposure to them is widespread. The discovery that many such compounds are mutagenic and that up to 83% of these mutagens are linked to cancer is significant. Although animal testing has been a standard method of detecting chemicals with carcinogenic potential, a more rapid screening system, called the **Ames test**[6] is also commonly used. In this ingenious test, the experimental subjects are bacteria whose gene expression and mutation rate can be readily observed and monitored. The premise is that any chemical capable of mutating bacterial DNA can similarly mutate mammalian (and thus human) DNA and is therefore potentially hazardous.

6. Named for its creator, Bruce Ames.

One indicator organism in the Ames test is a mutant strain of *Salmonella typhimurium*[7] that has lost the ability to synthesize the amino acid histidine, a defect highly susceptible to back-mutation because the strain also lacks DNA repair mechanisms. Mutations that cause reversion to the wild strain, which is capable of synthesizing histidine, occur spontaneously at a low rate. A test agent is considered a mutagen if it enhances the rate of back-mutation beyond levels that would occur spontaneously. One variation on this testing procedure is outlined in figure 8.25. The Ames test has proved invaluable for screening an assortment of environmental and dietary chemicals for mutagenesis and carcinogenicity without resorting to more expensive and time-consuming animal studies.

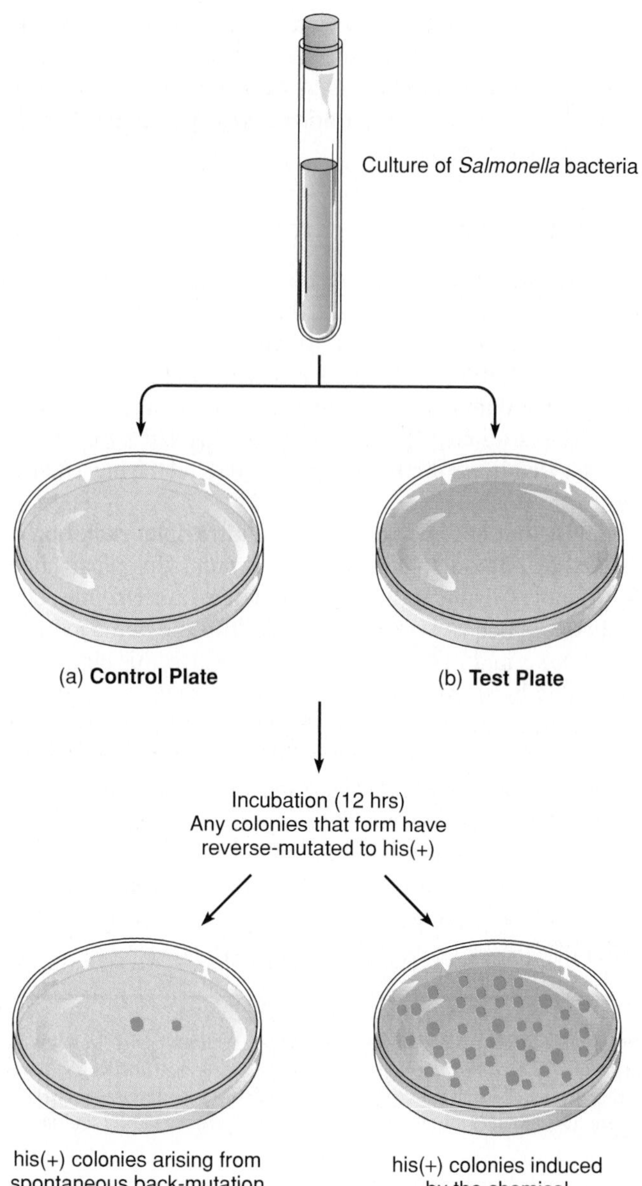

Figure 8.25 The Ames test uses a strain of *Salmonella typhimurium* that cannot synthesize histidine (his −), lacks the enzymes to repair DNA so that mutations show up readily, and has leaky cell walls that permit the ready entrance of chemicals. Because many potential carcinogens (benzanthracene and aflatoxin, for example) are mutagenic agents only after being acted on by mammalian liver enzymes, an extract of these enzymes may be added to the test medium. (*a*) In the control setup, bacteria are plated on a histidine-free medium containing liver enzymes but lacking the test agent. (*b*) The experimental plate is prepared the same way except that it contains the test agent. After incubation, plates are observed for colonies. Any colonies developing on the plates are due to a back-mutation in a cell, which has reverted it to a his(+) strain. By comparing the number of colonies growing on the control plate with the test plate, the degree of mutagenicity of the chemical agent can be calculated. Chemicals that produce an increased incidence of back-mutation are considered carcinogens.

Positive and Negative Effects of Mutations

Many mutations are not repaired. How the cell copes with them depends on the nature of the mutation and the strategies available to that organism. Mutations are permanent and heritable, and the offspring of organisms and viruses will inherit any mutations acquired by the parent, with long-lasting effects. Some mutations are harmful to organisms, while others provide adaptive advantages.

If a mutation leading to a nonfunctional protein occurs in a gene for which there is only a single copy, as in haploid or simple organisms, the cell will probably die. This happens when certain mutant strains of *E. coli* acquire mutations in the genes needed to repair damage by UV radiation. Adverse mutations are damaging to all organisms. Mutations of the human genome affecting the action of a single protein (mostly enzymes) are responsible for more than 400 diseases by last count. Feature 8.6 discusses one such disease, sickle-cell anemia.

Although most spontaneous mutations are not beneficial, a small number contribute to the success of the individual and the population by creating variant strains with alternate ways of expressing a trait. Microbes are not "aware" of this advantage and do not direct these changes; they simply respond to the environment they encounter. Those organisms with beneficial mutations can more readily adapt, survive, and reproduce. In the long-range view, mutations and the variations they produce are the raw materials for change in the population and thus, for evolution.

Mutations that create variants occur frequently enough that any population contains mutant strains for a number of characteristics, but as long as the environment is stable, the population will remain unchanged. When the environment changes, however, it can become hostile for the survival of certain individuals, and only those microbes bearing protective mutations can **adapt** to the new environment and survive. In this way, the environment **naturally selects** certain mutant strains that will reproduce, give rise to subsequent generations, and in time, be the dominant strain in the population. Through these means, any change that confers an advantage during selection pressure will be retained by the population. One of the clearest models for this sort of selection and adaptation is acquired drug resistance in bacteria (see chapter 10). One of the fascinating mechanisms bacteria have developed for increasing their adaptive capacity is genetic exchange (called genetic recombination).

7. *S. typhimurium* inhabits the intestine of poultry and causes food poisoning in humans. It is used extensively in genetic studies of bacteria.

Feature 8.6 The Magnitude of a Mutation: Sickle-Cell Anemia

Sickle-cell anemia provides insight into the dramatic repercussions of a single base change in the DNA molecule. This change is magnified across all levels of the biological spectrum: A codon is changed, an amino acid is substituted, the protein is abnormal, the cells containing it become abnormal, the tissues are harmed, the organs malfunction, the individual is drastically impaired, and even the population itself is affected.

Persons with this severe and life-threatening illness have inherited two recessive genes (homozygous) for an abnormal type (S) of hemoglobin (Hb), the principal oxygen-carrying protein of the blood. What is wrong with this abnormal hemoglobin, how did it get that way, and what is its effect on the person possessing the defective genes?

In certain populations at some time in the past, a point mutation has occurred on the gene that encodes one polypeptide of the hemoglobin molecule. When this mutation is germinal (that is, when it affects the gametes), it is passed on to offspring, where it will be found in every cell. The red blood forming cells in which the hemoglobin genes are active are most affected initially. The exact site of this mutation is the sixth codon of the gene, specifically GAA, which would ordinarily cause the insertion of glutamic acid. The mutation causes substitution of a single base (T) in the center, yielding GTA (or a GUA transcript), which is the code for valine. Insertion of valine in place of glutamic acid during translation causes the polypeptide to behave differently.

Such a change seems insignificant—only one base—yet the consequences are profound. Hemoglobin S bears "sticky" spots that cause mutual attraction among the individual hemoglobin molecules. They crystallize into long chains that build up in the red blood cells. Although red blood cells are ordinarily disc-shaped, these chains distort them into elongate, crescent- or sickle-shaped cells. The fine capillary network becomes plugged by these oddly shaped cells, blocking circulation in the tissues and precipitating a crisis that severely damages vital organs. Fragile sickle cells are subject to bursting, which instigates anemia and other side effects. These symptoms first appear in infancy and recur throughout life, which is generally shortened.

The highest incidence of the HbS mutation exists in the populations of Central Africa, though it is also found among Mediterranean, Arabic, and Indian peoples. Sickle-cell anemia occurs in about 1 in 400 births among Afro-Americans. Why such a debilitating mutation was spread so effectively in the population is probably due to its relationship to malaria, a prominent infectious disease of those regions. Carriers of a single HbS gene (heterozygous) do not develop a full case of sickle-cell anemia, but in fact, actually have resistance to malaria, giving them a survival advantage over normal noncarriers. Thus, a mechanism that maintains the carriers simultaneously conserves the sickle-cell anemia gene, and as long as there are carriers, there is the chance that some offspring will inherit two abnormal genes.

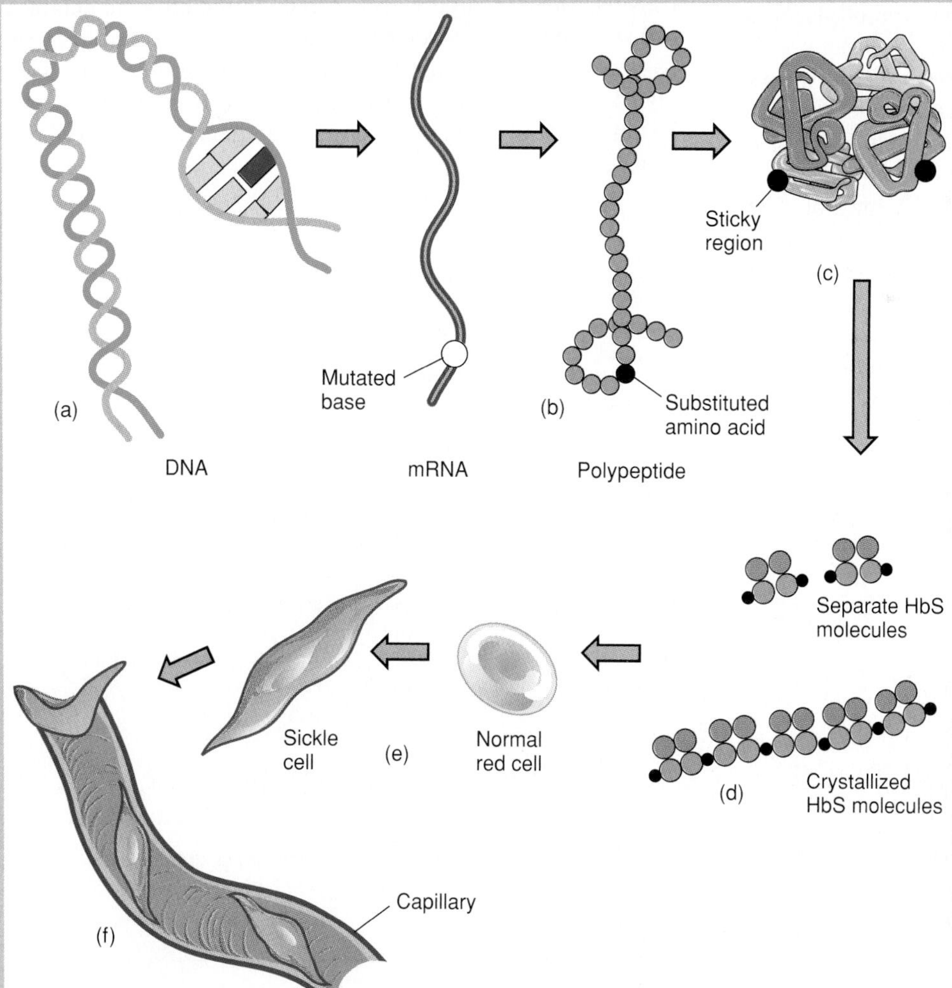

The scope of mutational effects in organisms: sickle-cell anemia in humans. (*a*) Substitution of a single base creates a mutation on the hemoglobin gene that causes a different amino acid to be added during hemoglobin synthesis. (*b*) Abnormal HbS polypeptide folds differently, and (*c*) in its quaternary structure, exposes sticky regions. (*d*) HbS molecules form long crystalline strands instead of remaining separate. (*e*) Distortion of RBCs into sickle cells. (*f*) Sickle cells get hung up and cannot circulate in capillaries, leading to organ stress.

Intermicrobial DNA Transfer and Recombination

Genetic recombination, or hybridization through sexual reproduction, is an important foundation of genetic variation in eucaryotes. Although bacteria have no exact equivalent to sexual reproduction, they exhibit a primitive counterpart whereby parts of their genome are exchanged. An event in which one bacterium donates DNA to another bacterium is a type of biologically induced mutation termed **intermicrobial transfer.** The properties of bacteria that account for such exchanges are: (1) Their genetic material is not found solely in chromosomes; (2) bacteria are naturally suited to sharing odds and ends of genetic material among themselves; and (3) bacteria harbor a repertory of potentially interchangeable genetic information. The end result of recombination is a new strain different from both the donor and the original recipient strain. Genetic exchanges have tremendous effects on the genetic characteristics of bacteria, and unlike spontaneous mutations, are generally beneficial to them. They provide additional genes for resistance to drugs and metabolic poisons, new nutritional and metabolic capabilities, and increased virulence and adaptation to the environment in general.

Genetic Recombination Involving Extrachromosomal DNA

DNA transmitted between bacteria is in the form of **plasmids,** closed circular molecules of DNA separate from chromosomes, or other small DNA fragments that have escaped from a lysed cell. Genes in extrachromosomal DNA are not necessary to basic bacterial survival, but they may be genetically useful and confer greater versatility or adaptability on the recipient cell. In addition to being transferrable, exchanged DNA may be integrated into the chromosome of the recipient and replicated and transmitted to progeny during cell division. Plasmids and integrated genes are transcribed and translated along with the regular chromosome. Intermicrobial exchanges occur at a low rate in natural populations, and they may also be induced in the laboratory. In a later section on genetic engineering, we will see that bacteria, viruses, and some yeasts are capable of receiving and using the DNA of other organisms that is grafted into a plasmid.

Depending upon the mode of transmission, these means of genetic exchange are called conjugation, transformation, and transduction. **Conjugation** requires the attachment of two related species through a pilus and the presence of a special plasmid. **Transformation** entails the transfer of naked DNA and requires no special vehicle. **Transduction** is DNA transfer mediated through the action of a bacterial virus (table 8.6).

Conjugation: Bacterial Sex Conjugation is an unconventional mode of sexual mating in which a plasmid or other genetic material is transferred by a donor to a recipient cell via a specialized appendage. It occurs primarily in gram-negative bacteria. The donor cell, also called the male, possesses a plasmid (**fertility** or **F factor**) that allows it to synthesize a **sex pilus** or **conjugative pilus.** The recipient cell, or female, is a closely related strain or species that has a recognition site on its surface. A cell's role in conjugation is denoted by F^+ for the male, which has the F factor, and F^- for the female, which has none. Contact is made when a sex pilus grows out from the male cell, attaches to the surface of the female cell, contracts, and draws the two cells together (see figure 3.9). Although the physical details of transmission are still somewhat obscure, the genetic material passes across or through the bridge formed by the pilus. Conjugation is a very conservative process, in that the donor bacterium generally retains a copy of the genetic material being transferred.

conjugation (kawn″-jew-gay′-shun) L. *conjugatus,* yoked together.
transformation (trans-for-may′-shun) L. *trans,* across, and *formatio,* to form. This term is also used to mean the cancerous (malignant) conversion of cells.
transduction (trans-duk′-shun) L. *transducere,* to lead across.

Table 8.6 Types of Intermicrobial Exchange

Mode	Requirements	Direct or Indirect*	Genes Transferred
Conjugation	Sex pilus on donor Fertility plasmid in donor Both donor and recipient live Closely related species	Direct	Drug resistance; resistance to metals, toxins, enzymes
Transformation	Free donor DNA (fragment) Live, competent recipient cell	Indirect	Polysaccharide capsule; unlimited with cloning techniques
Transduction	Donor is lysed bacterial cell Defective bacteriophage is carrier of donor DNA Live, competent recipient cell	Indirect	Toxins; sugar fermentation; drug resistance

*Direct means the donor and recipient are in contact during exchange; indirect means they are not.

According to findings with *E. coli,* conjugative transfer exhibits two patterns:

1. The donor (F^+) cell makes a copy of its F factor and transmits this to a recipient (F^-) cell. This changes the F^- cell into an F^+ cell capable of producing a sex pilus and conjugating with other cells (figure 8.26*a*). No additional donor genes are transferred at this time.
2. In high-frequency recombination (Hfr) donors, the fertility factor has been integrated into the F^+ bacterial chromosome. The term high-frequency recombination was adopted to denote that an integrated F factor is more rapidly spread through the population because it is replicated right along with the rest of the chromosome and passed on to progeny. This is an important event for future conjugative episodes in that it enables the F factor to direct a more comprehensive transfer of part of the donor chromosome to a recipient cell. This occurs through a rolling circle mechanism, with one strand of DNA retained by the donor and the other strand transported across to the recipient cell (figure 8.26*b*). The transfer of an entire chromosome takes about 90 minutes, but the pilus bridge between cells is ordinarily broken before this time, and rarely is the entire genome of the donor cell transferred.

Conjugation has great biomedical importance. Special **resistance (R) plasmids** or **factors** that bear genes for resisting antibiotics and other drugs are commonly shared among bacteria through conjugation. Transfer of R factors can confer multiple resistance to tetracycline, chloramphenicol, streptomycin, sulfonamides, and penicillin. The details of this phenomenon are presented in chapter 10. Other types of R factors carry genetic codes for resistance to heavy metals (nickel and mercury) or for synthesizing virulence factors (toxins and enzymes) that increase the pathogenicity of the bacterial strain. Conjugation studies have also provided an excellent way to map the bacterial chromosome.

Transformation: Capturing DNA from Solution One of the cornerstone discoveries in microbial genetics was made in the late 1920s by the English biochemist Frederick Griffith working with *Streptococcus pneumoniae* and laboratory mice. The pneumococcus exists in two major strains based on the presence of the capsule, colonial morphology, and pathogenicity. Encapsulated strains bear a smooth (S) colonial appearance and are virulent, while strains lacking a capsule have a rough (R) appearance and are nonvirulent (figure 3.12). (Recall that the capsule protects a bacterium from the phagocytic host defenses.) To set the groundwork, Griffith showed that when a mouse was injected with a live, virulent (S) strain, it soon died

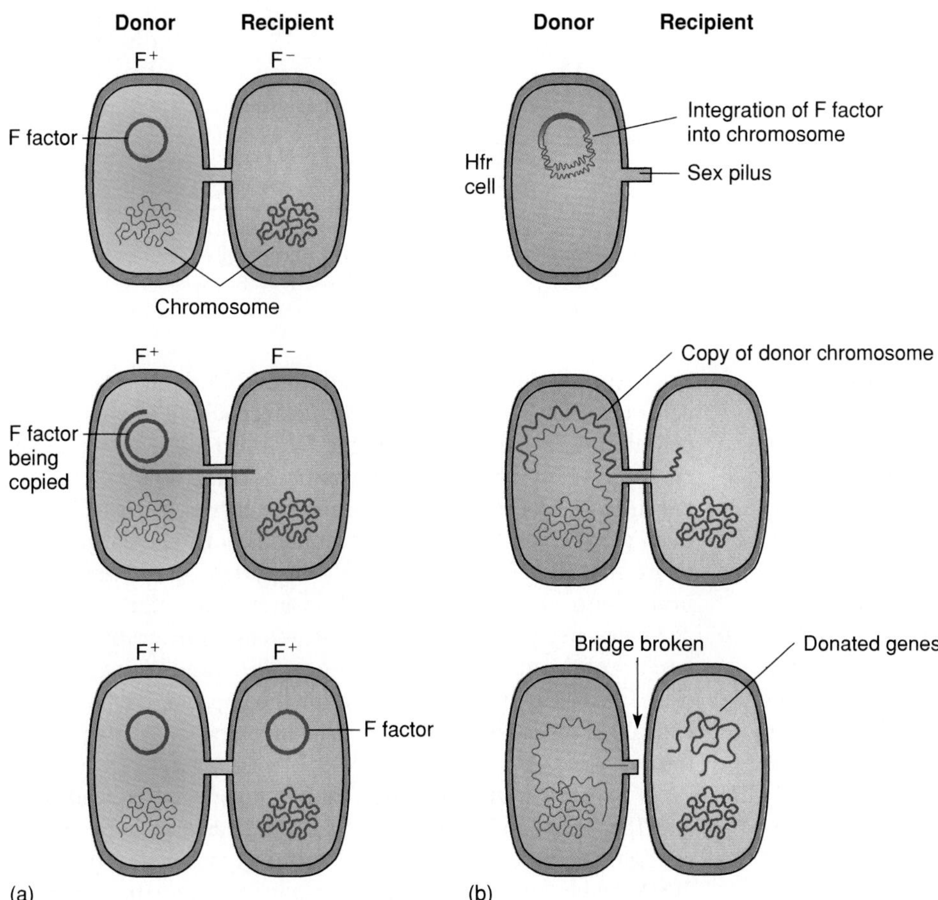

Figure 8.26 Conjugation: genetic transmission through direct contact. The sex pilus forms a connection between the cells (donor on the left, recipient on the right). Whether the donated genes actually pass through the bridge remains controversial. (*a*) Transfer of the F^- factor or conjugative plasmid. A cell must have this to transfer chromosomal genes. (*b*) High-frequency (Hfr) transfer involves transmission of chromosomal genes from a donor to a recipient cell.

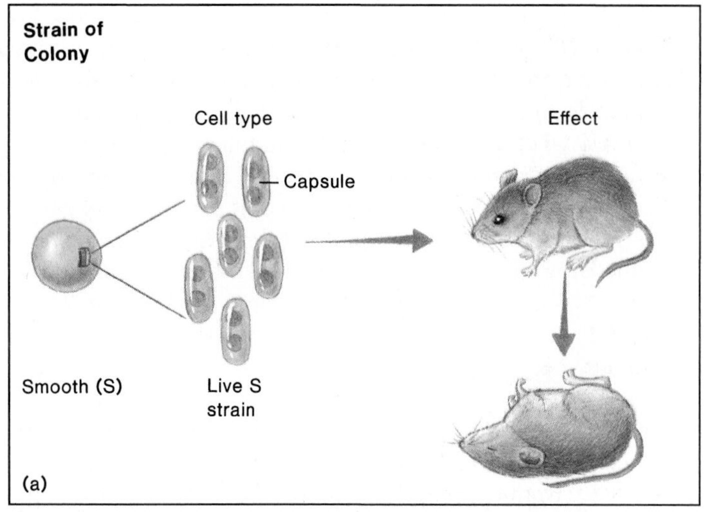

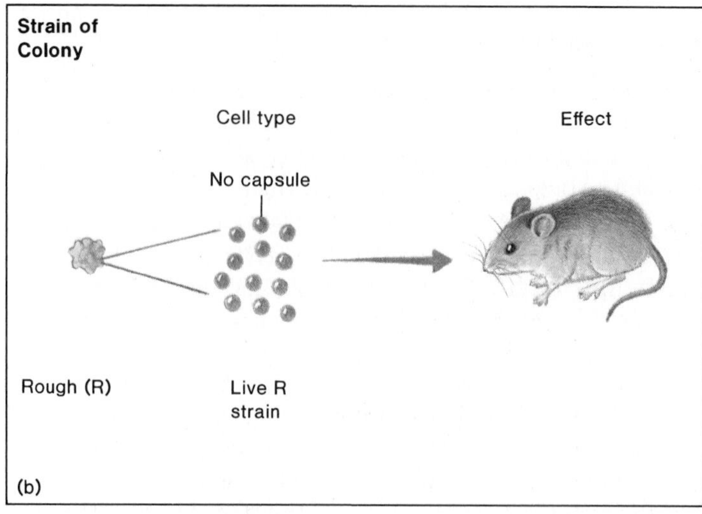

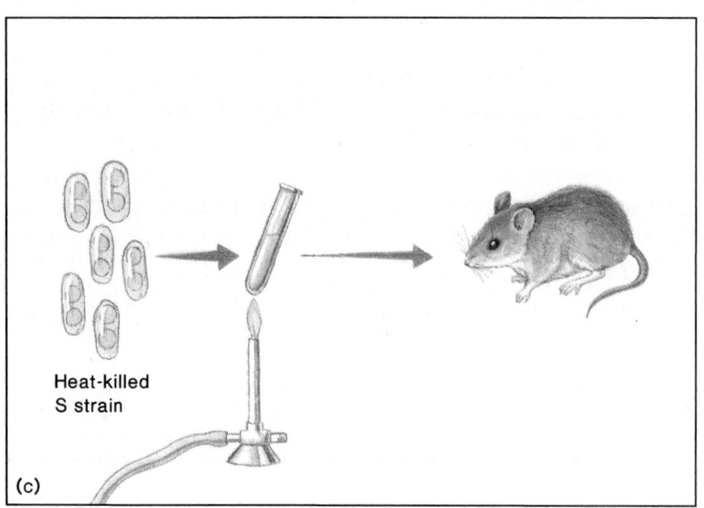

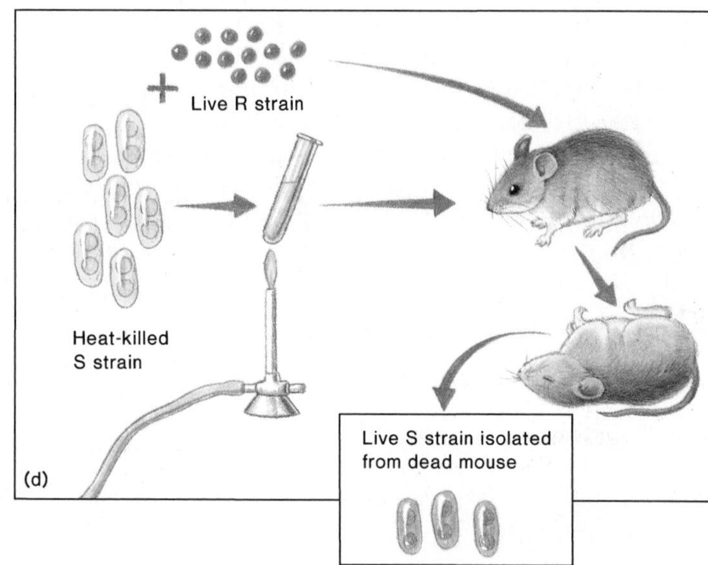

Figure 8.27 Griffith's classic experiment in transformation. In essence, this experiment proved that DNA released from a killed cell can be acquired by a live cell. The cell receiving this new DNA is genetically transformed, in this case, from a nonvirulent strain to a virulent one.

(figure 8.27*a*). When another mouse was injected with a live, nonvirulent (R) strain, the mouse remained alive and healthy (figure 8.27*b*). Next he tried a variation on this theme. First, he heat-killed an S strain and injected it into a mouse, which remained healthy (figure 8.27*c*). Then came the ultimate test: Griffith injected both dead S cells and live R cells into a mouse, with the result that the mouse died from a pneumococcal blood infection (figure 8.27*d*). Because killed bacterial cells do not "come back to life" and the nonvirulent strain was harmless, how could this have happened? Although he did not know it at the time, Griffith had shown that during their sojourn in the body of the mouse, the dead S cells broke open and released some of their DNA (by chance, that part containing the genes for making a capsule), and a few of the live R cells subsequently picked up this loose DNA and were **transformed** by it into virulent, capsule-forming strains.

Later studies supported the concept that a chromosome released by a lysed cell breaks into fragments small enough to be accepted by a recipient cell, and that DNA, even from a dead cell, retains its code. This nonspecific acceptance by a bacterial cell of small fragments of soluble DNA from the surrounding environment is termed **transformation.** Transformation is apparently facilitated by small openings occurring in the cell wall during cell enlargement that accommodate a fairly large fragment of DNA. This fragment is transported by the cell membrane into the cytoplasm and inserted into the bacterial chromosome. It is a natural event among both gram-positive streptococci and gram-negative *Hemophilus* and *Neisseria* species. In addition to genes coding for the capsule, bacteria also exchange genes for antibiotic resistance and bacteriocin synthesis in this way.

Because transformation requires no special appendages, and the donor and recipient cells do not have to be in direct contact, the process is useful for certain types of recombinant DNA technology. With this technique, foreign genes from a completely unrelated organism are inserted into a plasmid, which is then introduced into a bacterial cell of *E. coli* or *Bacillus subtilis* through transformation. These recombinations can be carried out easily in a test tube, and human genes can be experimented upon and even expressed outside the human body by placing them in a microbial cell. This same phenomenon in eucaryotic cells, termed *transfection,* is an essential aspect of genetically engineered yeasts, bird embryos, and mice, and it has been proposed as a future technique for curing genetic diseases in humans. These topics are covered in more detail later in this chapter and in the biotechnology section of chapter 22.

Transduction: The Case of the Piggyback DNA When we last encountered bacteriophages (bacterial viruses), it was to discuss their role as destructive bacterial parasites. Infection by a virus does not always kill the host cell, however, and viruses may indeed serve as genetic vectors (an entity that can bring foreign DNA into a cell). The process by which a bacteriophage serves as the carrier of DNA from a donor cell to a recipient cell is **transduction.** Although it occurs naturally in a broad spectrum of bacteria, the participating bacteria in a single transduction event must be the same species because of the specifity of viruses for host cells. The events in transduction are shown in figure 8.28.

There are two versions of transduction. In *generalized transduction,* random fragments of disintegrating host DNA are taken up by the phage during assembly. Virtually any gene may be transmitted through this means. In *specialized transduction,* a highly specific part of the host genome is regularly incorporated into the virus. This specificity is explained by the prior existence of a temperate phage that acts as a prophage inserted in a fixed site on the bacterial chromosome. When activated, the prophage separates from the bacterial chromosome, carrying a small segment of host genes with it. During the lytic cycle, these specific viral-host gene combinations are incorporated by the viral particles and carried to another bacterial cell.

Several cases of specialized transduction have biomedical importance. The virulent strains of bacteria such as *Corynebacterium diphtheriae, Clostridium* spp., and *Streptococcus pyogenes* all produce toxins with profound physiological effects, whereas nonvirulent strains do not produce toxins. It turns out that virulence is due entirely to transduction of genes that code for these toxins by a bacteriophage, and only those bacteria infected with a temperate phage are toxin formers. Other instances of transduction are seen in staphylococcal transfer of drug resistance and in the transmission of gene regulators in gram-negative rods (*Escherichia, Salmonella*).

Transposons: "This Gene is Jumpin"

One sort of genetic transferral of great interest involves **transposons.** These elements have the distinction of shifting from one part of the genome to another, and so are termed "jumping genes." When their existence was first postulated in corn plants, it was greeted with some skepticism, since it had long been believed that the locus of a given gene was set, and that a gene did not or could not move to another one. Now it is evident that jumping genes are widespread among procaryotic and eucaryotic cells and viruses.

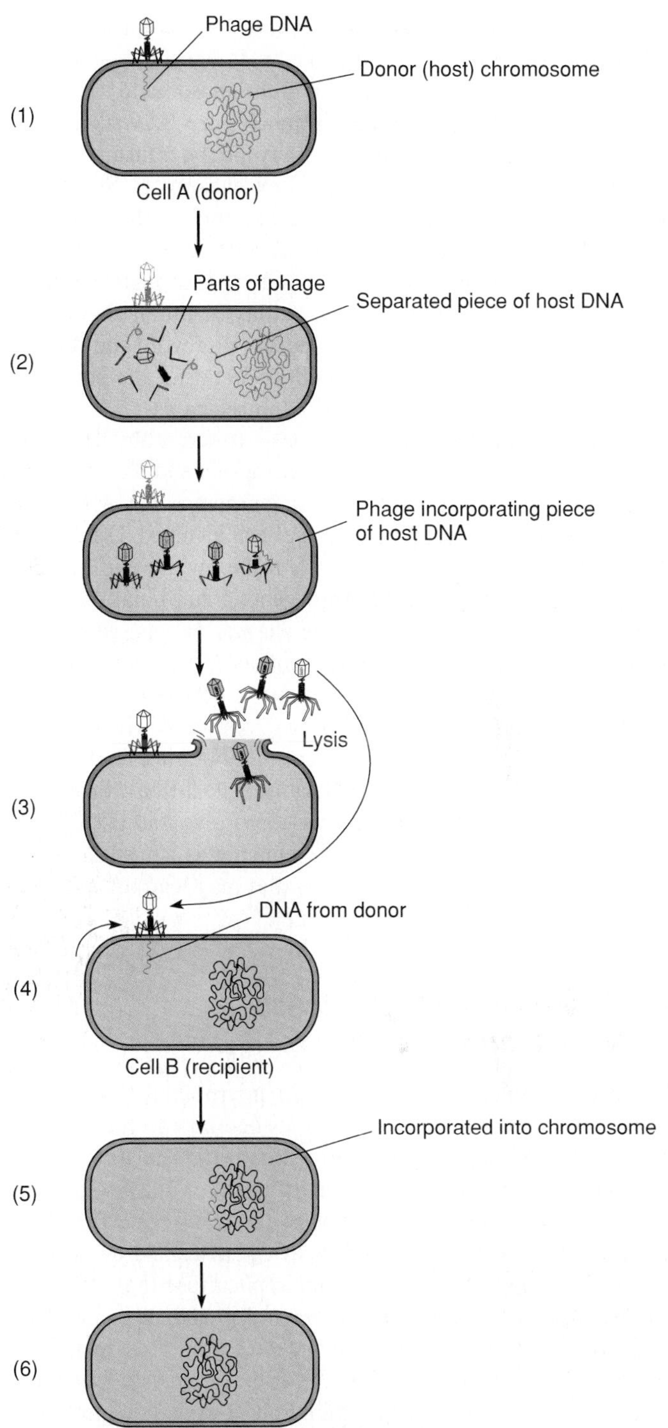

Figure 8.28 Transduction: genetic transfer by means of a virus carrier. (*1*) A phage infects cell A (the donor cell) by normal means. (*2*) During replication and assembly, a phage particle incorporates a segment of bacterial DNA by mistake. (*3*) Cell A then lyses and releases the mature phages, including the genetically altered one. (*4*) The altered phage adsorbs to and penetrates another host cell (cell B), injecting the DNA from cell A rather than viral nucleic acid. (*5*) Cell B receives this donated DNA, which recombines with its own DNA. (*6*) Since the virus is defective (biologically inactive as a virus), it is unable to complete a lytic cycle. The transduced cell (B) survives and may use this new genetic material.

All transposons share the general characteristic of travelling from one location to another on the genome—from one chromosomal site to another, from a chromosome to a plasmid, or from a plasmid to a chromosome (figure 8.29*a*). Because transposons occur in plasmids, they may also be transmitted from one cell to another in bacteria and a few eucaryotes. Some transposons replicate themselves before jumping to the next location, and others simply move without replicating first.

Transposons may be recognized by the palindromic sequences at each end, some of which may be hundreds of bases long (figure 8.29*b*). These regions permit discovery and removal of the transposon sequence as well as dictating the site of insertion into DNA when the transposon relocates.

The overall effect of transposons—to scramble the genetic language—can be beneficial or adverse, depending upon such variables as where insertion occurs in a chromosome, what kinds of genes are relocated, and the type of cell involved. On the beneficial side, transposons are known to be involved in (1) gene rearrangement to create different genetic combinations, as in mammalian immunoglobulins (see chapter 13); (2) changes in traits such as colony morphology, pigmentation, pili, and antigenic characteristics; (3) replacement of damaged DNA; and (4) the intermicrobial transfer of drug resistance (in bacteria). On the negative side, rearrangement of DNA that leads to mutations such as deletions, insertions, translocations, inversions, and chromosome breakage may be disruptive and even lethal. Retroviruses like the AIDS virus behaving as transposons randomly insert in the genome in ways that may lead to severe cell dysfunction (see chapters 14 and 21).

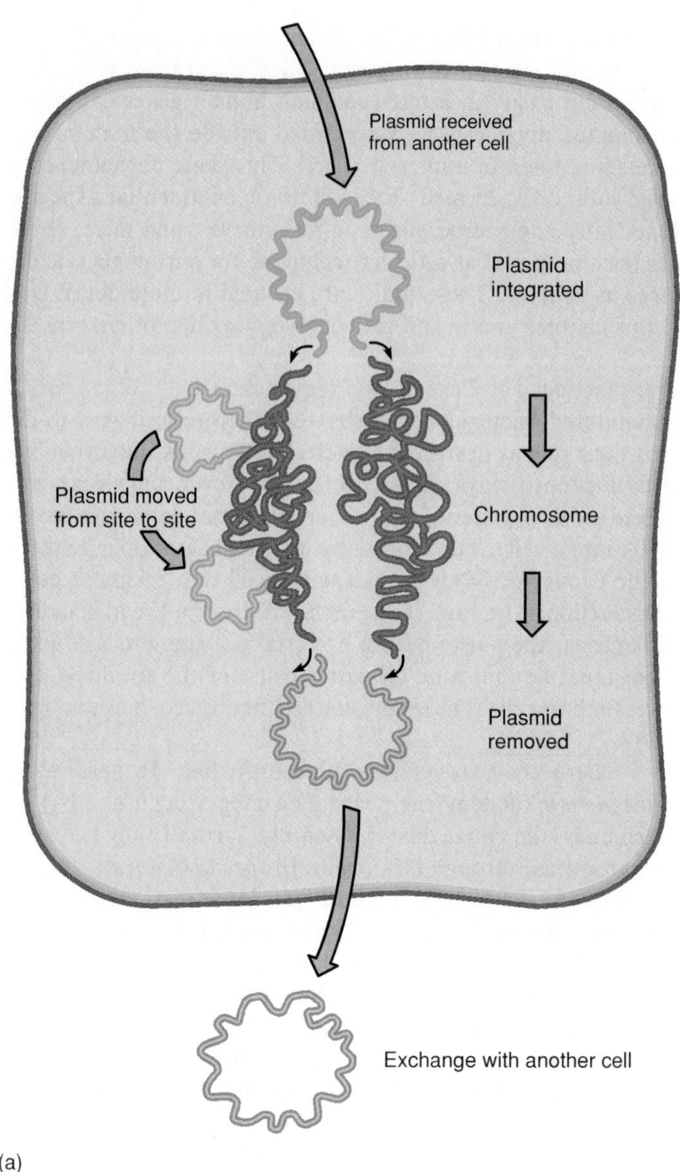

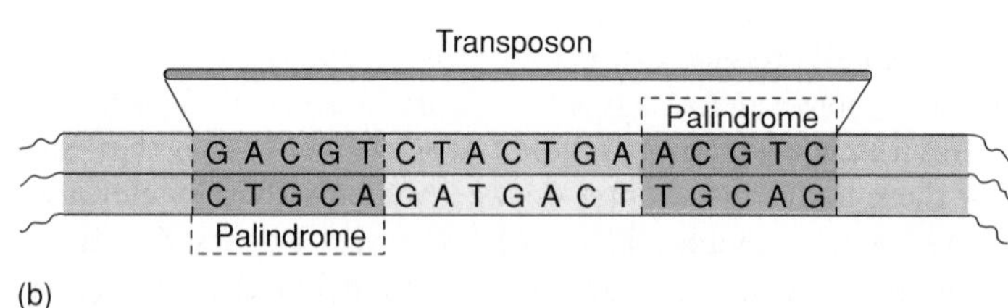

Figure 8.29 Transposons: shifting segments of the genome. (*a*) Potential mechanisms in the movement of transposons in bacterial cells. (*b*) A transposon circumscribed by palindromic sequences.

The Applications of Discoveries in DNA: Gene Technology

The intense and creative research into microbial DNA has been responsible for a deluge of discoveries and new techniques that rivals any other period of scientific discovery. This new body of knowledge has presented the scientific community with some powerful and exciting tools to explore and manipulate DNA (see feature 8.7), and it has placed us at the threshold of fulfilling a wishlist of medical and industrial applications that were once only dreamed of. In this section we will introduce some of the basic concepts of **genetic engineering,** techniques in which DNA is manipulated artifically to identify and derive useful genes and genetic products. The most prominent of the technologies in this field involve *recombinant DNA* and gene cloning.

Imitating Nature: Recombinant DNA Techniques

Recombinant DNA technology, which unites DNA sequences from different organisms, originated from the clever tricks that bacteria were found to do with bits of extra DNA such as plasmids, transposons, and proviruses. Scientists reasoned that because bacteria are naturally good at accepting, replicating, and expressing DNA from other organisms, they could be genetically engineered to mass-produce "foreign" genes and gene products that were otherwise difficult to come by—for example, hormones, enzymes, vaccines, and agricultural products.

The cornerstone of recombinant methods is the **cloning** of a gene. A clone has been defined as a genetically identical strain or organism that originated from the same parent (see chapter 3). In molecular biology, cloning refers to the duplication of a gene isolated from another organism by the cells of a microbial host. The microbe, or **cloning host,** must have particular characteristics (table 8.7). *Escherichia coli* has been the traditional subject, but this species has a potential for virulence (endotoxin) and could become established as part of the normal flora

Table 8.7 Desirable Features in a Cloning Host
Rapid overturn, fast growth rate
Grows readily *in vitro*
Nonpathogenic
Relatively simple genome
Genetic makeup well delineated (mapped)
Capable of accepting plasmid vectors
Maintains foreign genes through multiple generations
Will secrete a high yield of proteins from expressed foreign genes

of humans. For this reason, the common brewer's yeast *Saccharomyces cerevisiae,* which fulfills most of the criteria, is widely used in industry.

At first, a number of problems arose in recombining the DNA of other organisms with bacteria. One of these was locating the exact site of a particular gene on the donor organism's chromosome, and another was isolating it. The complexity of the human genome made locating specific genes impossible for many years. Eventually, this limitation was overcome, and several genes were pinpointed. Details of the techniques for isolating a gene are far beyond the level of this text, but there are essentially three strategies: (1) The genome can be separated into fragments by endonucleases, each of which is screened for a certain genetic expression. This is a laborious process, since each fragment of DNA must be examined for the cloned gene. Finding the proverbial needle in a haystack would probably be easier. (2) DNA is synthesized from RNA transcripts, using a reverse transcriptase. (3) DNA is synthesized artificially on a machine. Although listing these methods seems very straightforward, they require long and tedious work. The reward has been that once a gene is isolated and cloned, it can be maintained as such just like a microbial pure culture. *Genomic libraries,* which are like culture collections, now exist for hundreds of genes. Another extremely useful offshoot of gene isolation has been the development of genomic maps for humans (discussed in chapter 22).

Cloning techniques require some sort of **vector** to carry the foreign DNA into the cloning host. Most vectors are plasmids or bacteriophages. One long-standing vector is an *E. coli* plasmid that carries genetic markers for resistance to antibiotics, but it can carry only a small amount of foreign DNA. A type of phage vector, the *Charon*[8] phage, is a modified virus having large sections of missing genome that may be replaced with foreign genes. It can carry and clone a fairly large segment of foreign DNA. A hybrid vector that combines both a plasmid and a phage, called a *cosmid,* is also capable of carrying large genomic sequences. Plasmids are inserted into cloning hosts by transformation and phage-based vectors are inserted by transduction.

8. Named for the mythical boatman in Hades who carried souls across the River Styx.

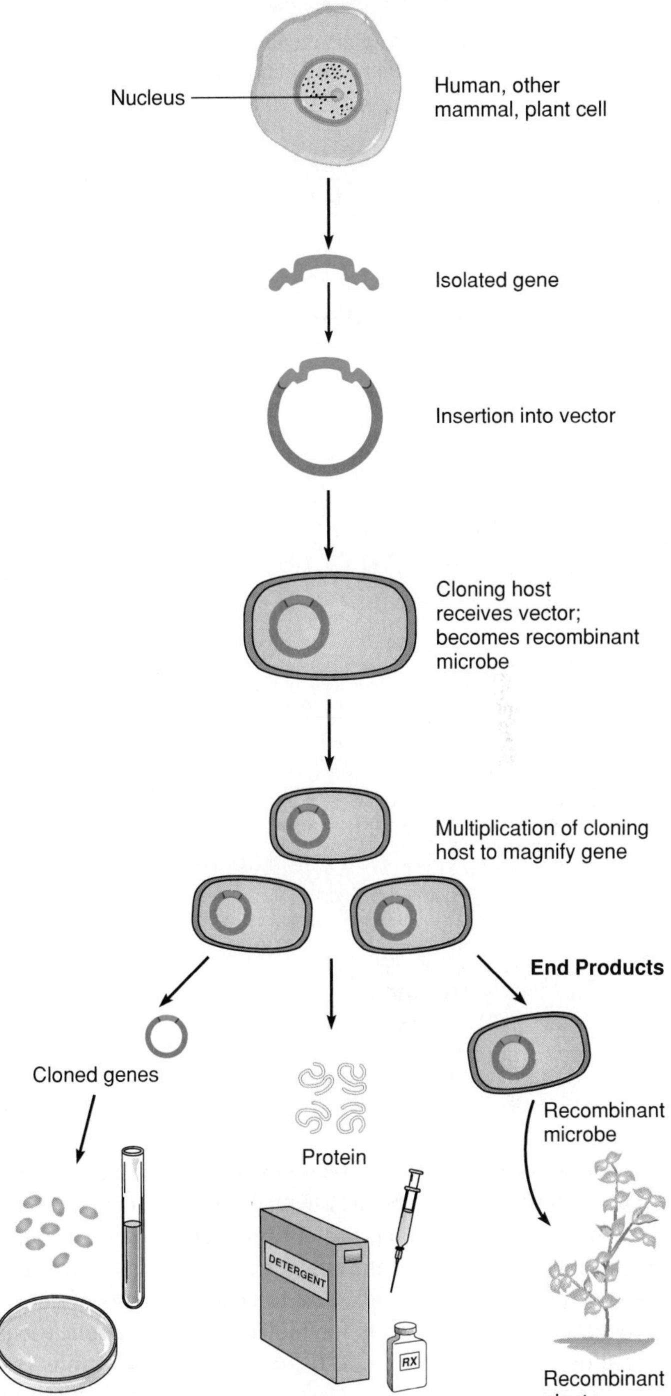

Figure 8.30 Strategy for the applications of gene cloning in genetic engineering.

Technical Aspects of Recombinant DNA and Gene Cloning

Recombinant DNA techniques lend themselves to an unlimited number of applications in human genetics, industry, and agriculture. Although the techniques have slight variations, the general events include isolating a desired gene from an animal, plant, virus, or bacterium, inserting the gene into a vector, cloning the vector with its foreign gene in a cloning host, and isolating an end product (figure 8.30).

Table 8.8 Partial List of Applications for Recombinant DNA Technology

Protein Products

Hormones and Factors

- Insulin
- Growth hormone
- Erythropoietin (EPO)
- Interleukin-2
- Interferon
- Factor VIII (for hemophilia treatment)
- CD4 to treat AIDS patients

Vaccines

- AIDS capsid protein (current tests)
- Hepatitis B surface antigen
- Malaria surface antigen

Antibiotics

Enzymes

- Clot-digesting drugs
- Detergent additives

Cloned Genes

- DNA probes
- RNA probes
- Gene libraries for mapping, research, and biotechnology
- Genes for gene therapy

Recombinant Organisms

- Frost-free bacteria
- Recombinant viruses for vaccines
- Recombinant viruses to act as cloning vectors
- Pesticide-resistant plants
- Virus-resistant plants
- Tobacco plants that synthesize drugs
- Plants that synthesize monoclonal antibodies
- Oil-eating bacteria
- Fungicide-producing bacteria
- Pollutant-degrading bacteria
- Tobacco plants that glow in the dark

A summary of the applications of recombinant DNA is presented in table 8.8. The desired ends of these techniques are threefold: to mass-produce protein products such as enzymes and hormones, to increase the number of cloned genes to be used for gene probes and research, and to create cultures of genetically recombined organisms for biotechnological applications. These designer organisms constitute entirely new strains that would never exist otherwise. Because they are products of research and development, they can even be patented.

So fruitful have these studies been, that commercial exploitation has already begun in earnest, and several large companies, earning billions of dollars, are based entirely on products obtained through genetic engineering. The list of new applications grows longer every year. One innovative but very controversial application involves the recombinant soil bacterium *Pseudomonas syringae,* which has been engineered to block frost formation on plants. Plants sprayed with this product (*Frostban*) are protected from frost, but environmentalists are concerned about the outcome of releasing the recombinant bacteria into the soil.

Feature 8.7 DNA—The Marvelous Molecular Toy

The nature of DNA is mind boggling. We have already seen some of its surprises—a structure like a child's building blocks, a language like the most challenging word game, the springiness of an elastic band—but these are only rudiments of the ways it may be manipulated in the laboratory with a number of inventive techniques.

Molecular biologists have found that melting (denaturing) DNA with mild heat causes the double helix to separate longitudinally into two strands, each of which can be examined, compared, and tested. When left alone to cool slowly, the two strands come back together at the original complementary sites. DNA may also be cut crosswise at chosen positions to pave the way for splicing and grafting. The endonucleases have been a real boon for clipping DNA into smaller pieces. One group (restriction endonucleases) selectively degrades some DNA but not others, so that sites of cutting may be finely controlled. A restriction endonuclease has the unique property of recognizing palindromes, which it nicks across in a blunt or staggered manner. Staggered cuts are sometimes referred to as "sticky ends" due to their tendency to insert at open sites on plasmids or chromosomes through complementary base pairing. Linear DNA may become circularized when its sticky ends are attracted to each other. This is how plasmids and some genomes maintain their circular form. Hundreds of restriction nucleases are presently known, each type having a particular genomic palindrome as its target. Endonucleases are named by combining the first letter of the genus, the first two letters of the species, and the endonuclease number. Thus, Eco R I is the first endonuclease found in *E. coli,* and Hind II is the second endonuclease discovered in *Hemophilus influenzae* type d.

Fragments of DNA produced by endonucleases may be placed in compartments on a special agar gel and subjected to an electric current (gel electrophoresis). The electricity causes the DNA pieces to migrate toward the positive pole and sorts them according to size. The largest fragments move more slowly and remain nearer the top of the gel, whereas the smallest fragments migrate faster and are found at the bottom. The positions of DNA on the gels may be determined by developing them with special stains or photographic film (if radioactive isotopes have been used) and comparing them against a known standard.

Through these techniques, we are moving closer to the era of grafting genes into chromosomes to repair mutations or correct genetic defects. The combination of cutting DNA into shorter fragments with endonucleases and melting it to separate the two strands has spawned the techniques of nucleic acid hybridization and probes. In hybridization, an unknown DNA is denatured and exposed to small pieces of radioactive, labelled, single-stranded DNA or RNA that will react with the sites on the test DNA that are complementary. Endonucleases and DNA melting have likewise made it possible to make DNA "fingerprints"—a pattern of fragmented DNA that is peculiar to each individual organism. Mapping DNA and locating specific genetic loci have been greatly facilitated by having well-defined chromosome fragments. These specialized areas are discussed in greater detail in chapter 22.

Recombinant technology has also been applied in treating human diseases caused by the lack of a certain hormone, such as diabetes and dwarfism. Traditional therapy involves replacing the missing hormone. Porcine or bovine insulin is used for diabetes, even though such animal products always carry with them the chance for reactions to foreign animal protein.

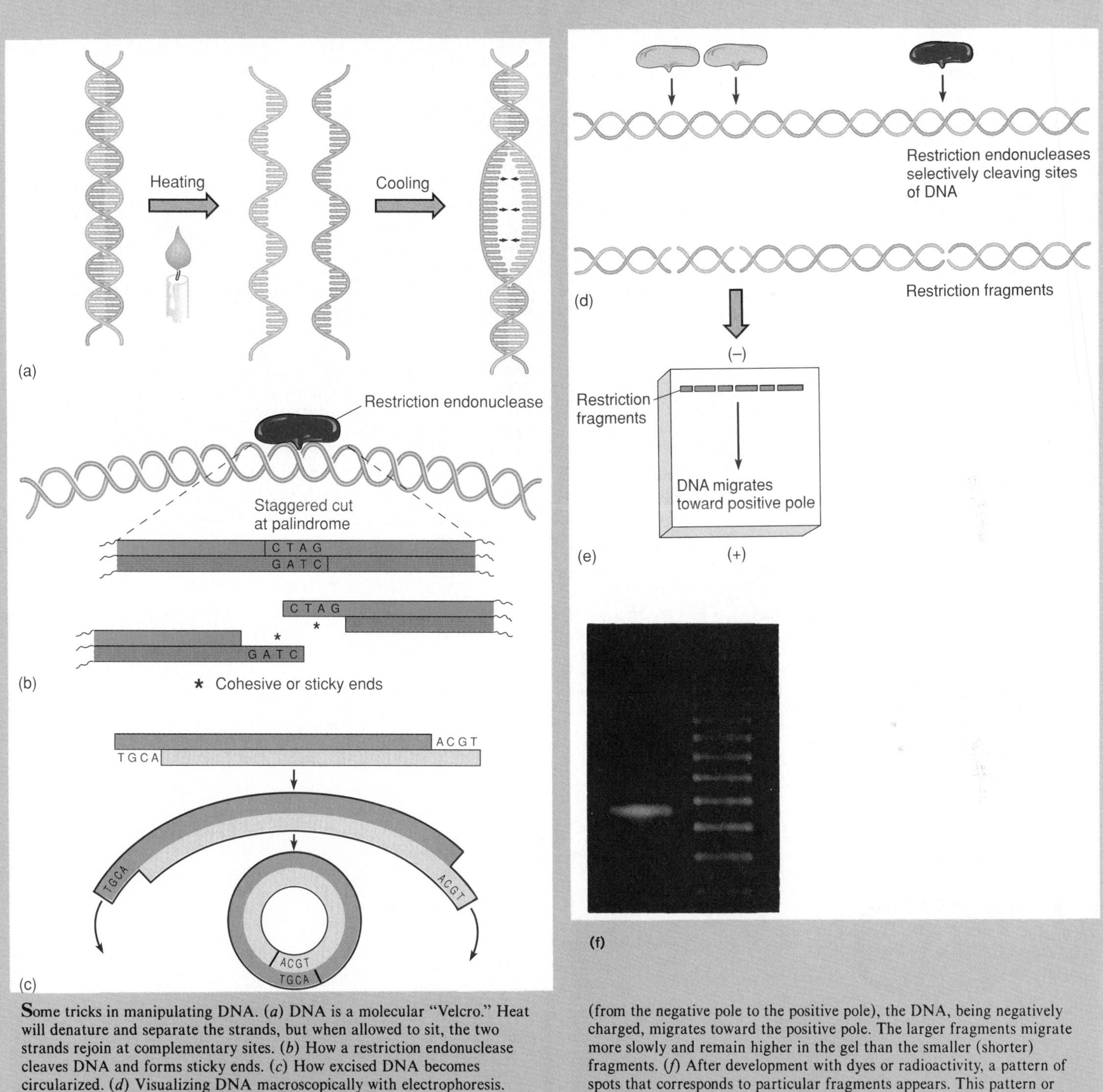

Some tricks in manipulating DNA. (*a*) DNA is a molecular "Velcro." Heat will denature and separate the strands, but when allowed to sit, the two strands rejoin at complementary sites. (*b*) How a restriction endonuclease cleaves DNA and forms sticky ends. (*c*) How excised DNA becomes circularized. (*d*) Visualizing DNA macroscopically with electrophoresis. After cleavage into fragments, DNA is placed into sites on one end of a block of special agarose gel. (*e*) When an electric current is passed through the gel (from the negative pole to the positive pole), the DNA, being negatively charged, migrates toward the positive pole. The larger fragments migrate more slowly and remain higher in the gel than the smaller (shorter) fragments. (*f*) After development with dyes or radioactivity, a pattern of spots that corresponds to particular fragments appears. This pattern or fingerprint is highly unique for a given DNA.

Dwarfism cannot be treated with growth hormones from other mammals, so human growth hormone (HGH) had to be obtained from the pituitaries of cadavers. At one time, not enough HGH was available to treat the thousands of children in need.

Recombinant technology changed the face of these two conditions. Through exhaustive research, the genes that code for insulin and growth hormone were isolated from human chromosomes. The principal steps involved in the recombinant DNA technique are insertion of the hormone gene into an appropriate vector, introduction of the vector into a compatible cloning host, and growth of this cloned host cell in a medium. The vector DNA replicates right along with other cell DNA, so all progeny contain the cloned gene too. This is a simple yet incredibly powerful way to magnify a gene, going from a few recombined cells to billions in a short time. The next step is to stimulate the recombined cell to transcribe and translate the foreign gene, synthesize the hormone, and secrete it into the growth medium. After synthesis, the cloning cells and other chemical and microbial impurities are removed from the medium, yielding the human hormonal product (figure 8.31). This technology has the potential for producing large quantities of an otherwise rare human product in a short time. As for the reality, diabetics can now take Humulin, human insulin, and children with dwarfism or failure to grow are being prescribed Protropin, a recombinant growth hormone.

Other offshoots of biotechnology that have direct medical applications in humans are gene mapping, DNA fingerprinting, and gene therapy, all discussed in chapter 22.

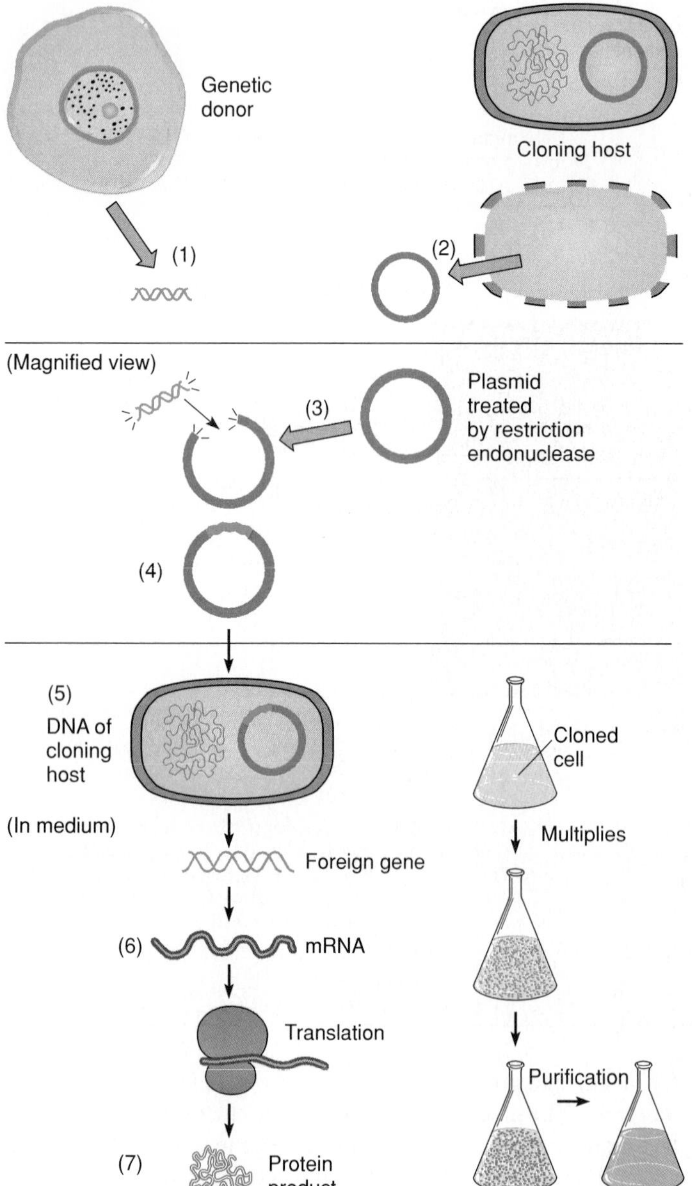

Figure 8.31 The general steps in recombinant DNA, gene cloning, and product retrieval are as follows: (*1*) The required gene is isolated from a genetic donor. (*2*) Cloning plasmids are retrieved from the cloning host by digesting it with lysozyme. (*3*) A restriction endonuclease is applied to the plasmid and foreign gene to produce sticky ends. (*4*) Incubation of the two types of DNA hastens the insertion of the foreign gene into the plasmid. (*5*) The recombinant plasmid is introduced into an intact host cell by transformation. (*6*) The cell transcribes and translates the foreign gene. (*7*) The protein product from translation of the foreign gene is recovered and purified.

Feature 8.8 The Implications of Fooling Mother Nature

Aside from the immediate benefits of genetic engineering and its apparently bright future, what might be some of its longer term and potentially harmful effects? Some people feel that manipulating the genes of humans and other organisms is tantamount to playing God, and they predict dire consequences that range from sociological and religious to ecological and medical. Critics argue that creating entirely new organisms and releasing them into the world may upset the balance of nature. And, they ask, if it is possible to engineer plants that resist herbicides and bacteria that make insulin, is it not also possible to create a deadly virus that could insert into the human genome and cause cancer or some other disease? Just as recombinant DNA methods lend themselves to beneficial uses, they can also be applied to biological warfare.

Is the world heading toward a time when normal-sized children will be given genetically engineered growth hormone to create super athletes? Or, will tests to screen for genetic disease be demanded of all persons seeking a job or applying for insurance? Eventually, as more is known about the human genome, the results of such testing may be available to all businesses. Such possibilities will give rise to numerous ethical and legal questions. It will eventually be necessary for representatives of many disciplines—law, medicine, business, government, education, even sports—to come together and set some basic moral and medical rules for using this new technology. These stunning tools—with such potential for benefit—will eventually force some difficult choices on all of us.

Chapter Review with Key Terms

Microbial Genetics

Genes and the Genetic Material

Genetics is the study of **heredity,** and the **genome** is the sum total of genetic material of a cell. A **chromosome** is comprised of DNA in all organisms; **genes** are specific segments of this linear molecule. Genomic sizes range from 4–5 genes in viruses, to several million genes in eucaryotic cells. Compactness of DNA comes from compound winding, twisting, and coiling. Genes code for specific peptides such as enzymes, antibodies, or structural proteins.

Gene Structure and Replication

A gene consists of DNA, **antiparallel** strands of repeating deoxyribose-sugar-phosphate units attached to nitrogenous bases of purine or pyrimidine. Complementary base pairing (adenine–thymine and cytosine–guanine) ensures genetic fidelity in DNA synthesis called **replication.** Replication is **semiconservative** and requires enzymes like **helicase, primase, polymerase, ligase,** and **gyrase.** These components, in conjunction with the chromosome being duplicated, constitute a **replicon.** Helicase severs hydrogen bonds, exposing the nitrogen bases of the unzipped strands, which function as **templates.** Synthesis proceeds along two **replication forks,** each with a **leading strand** and a **lagging strand.** In bacteria, the replicon resembles **theta,** and in certain viruses, the replicon is called a **rolling circle.**

Gene Function

Genetic information processing is **unidirectional**—from DNA to RNA and to protein. RNA synthesis is called **transcription,** and protein synthesis is called **translation.** An organism's genetic makeup, or **genotype,** dictates all its traits, called its **phenotype.** The genetic code is organized into **triplets** of nitrogen bases, and each triplet (codon) corresponds to a particular amino acid. Thus, DNA encodes the type, number, and sequence of amino acids in proteins, molecules that ultimately express traits.

Types of RNA: Unlike DNA, RNA is **single-stranded** and contains **uracil** instead of thymine and **ribose** instead of deoxyribose. Chief forms are messenger, transfer, and ribosomal RNA (**mRNA, tRNA,** and **rRNA**). On mRNA, triplets of bases called **codons** convey genetic information from DNA. The corresponding **anticodon** on tRNA ensures delivery of the appropriate amino acid. Ribosomes are the staging sites where all translation components come together.

Transcription and Translation: Transcription begins with the recognition of a **promoter** region of DNA by **RNA polymerase,** and elongation proceeds in the 5′ to 3′ directon. Only one DNA strand (**antisense**) is copied. The end result is a single long mRNA strand that will proceed to a ribosome for translation. Translation begins at the **AUG** or **start codon** on mRNA. The assembly of a protein begins by the occupation of P and A sites on the ribosome by tRNAs with anticodons complementary to the codons of the mRNA. Peptide bonds are formed between amino acids on adjacent tRNAs. This proceeds in repetitive fashion until the **nonsense codons** on mRNA are reached. These triplets have no corresponding amino acids. On the other hand, the genetic code is considered **degenerate** because some codons correspond to more than one amino acid.

Eucaryotic Gene Expression: Eucaryotes have **split genes.** Chromosomes are interrupted by noncoding regions called **introns** that separate coding segments called **exons.** The synthesis of mRNA requires splicing to delete stretches that correspond to introns.

The Genetics of Viruses

The genomes of viruses are diverse. Some are linear, others circular. Some consist of a single segment while others have several. Some are made of DNA, yet others of RNA. There are double-stranded (ds) viruses, single-stranded (ss) DNA viruses, ssRNA viruses, and dsRNA viruses. Retroviruses synthesize dsDNA from ssRNA. The replication of dsDNA viruses is a complex cycle of early and late events. Some viral DNA may be silently integrated into the host's genome. Integration by **oncogenic** viruses may lead to **transformation** of the host cell into an immortal cancerous cell. RNA viruses do not exhibit early and late phases, but their replication is complicated by strand polarity (positive- or negative-sense genome) and double-strandedness. Additionally, a retrovirus must undertake preliminary DNA synthesis before it begins replication.

Genetic Regulation of Proteins

Protein synthesis and metabolism are regulated by **gene induction** or **repression,** as controlled by an **operon.** An operon is a DNA unit of **regulatory genes** (made up of **regulators, promoters,** and **operators**) that controls the expression of **structural genes** (which code for enzymes and structural peptides). **Inducible operons** like the **lactose operon** are normally *off* but are turned *on* by lactose, an inducer. **Repressible operons** govern anabolism, and unlike the lactose operon, are usually *on,* but can be shut *off* when the end product is no longer needed.

Certain antibiotics affect transcription and translation. Adverse modes of action include interference with RNA polymerase, RNA elongation, and ribosomal activity.

Gene Mutation

Genome changes in microbes come from **mutations** and intermicrobial genetic exchanges. The term **wild type** or strain denotes the original form, while **mutant** strain refers to the altered version. Mutations are **spontaneous** if they occur randomly and **induced** if they are due to directed chemical or physical agents called **mutagens. Point mutations** entail addition, removal, or substitution of a few bases. A **missense mutation** leads to amino acid substitution, and a **nonsense mutation** arrests peptide synthesis without amino acid insertion. A **silent mutation** is the consequence of genetic code "degeneracy" that allows for base substitution without amino acid substitution. A **back-mutation** is a reversion to the original base composition.

Repair of Mutations: Enzymes that can identify and repair some mutations exist. Some enzymes locate **mismatched** bases and engage in repairs. A **suicide enzyme** mends damaged guanine and deactivates itself. DNA damaged by ultraviolet light may be corrected by **photoactivation** or **light repair.** The **Ames test** is a mutation-screening method based on the susceptibility of mutant *Salmonella typhimurium* to back-mutation. A mutagen will cause the strain to resume histidine synthesis, a genetic marker.

Effects of Mutations: Mutations may furnish adaptive advantage and be beneficial, but some are lethal. Intermicrobial transfer and genetic recombination permit gene sharing between bacteria. **Conjugation** entails plasmid (**fertility** or **F factor**) transfer apparently via a **sex** or **conjugative pilus.** Conjugation resulting in transfer of **resistance** or **R plasmids** is biomedically important. Naked DNA is acquired in bacterial **transformation** without agents. This type of transfer was discovered in smooth and rough pneumococci. **Transduction** is mediated by bacteriophages. Random hitchhiking of disintegrating host DNA is called **generalized** transduction, but transmission of specific fragments is known as **specialized** transduction. Specialized transduction is involved in *Corynebacterium diphtheriae, Clostridium,* and *Streptococcus pyogenes* toxigenicity.

Transposons are large base sequences that regularly depart and reinsert into other chromosomal sites.

Gene Technology

Genetic engineering is feasible through **recombinant DNA** techniques that manipulate the genes through plasmid transfer, **vectors,** transformation, and transduction. Vectors enable the cloning of desired foreign genes in a host microorganism and the induction of transcription and translation of that gene into products such as hormones, drugs, nucleic acid probes, and genetically altered microbes, plants, or animals.

True–False Questions

Determine whether the following statements are true (T) or false (F). If you feel a statement is false, explain why, and reword the sentence so that it reads accurately.

____ 1. A dividing bacterial cell receives half of its DNA from the parent cell and synthesizes the rest itself.

____ 2. A chromosome is composed of units of DNA called genes.

____ 3. The nitrogen bases in DNA are bonded to the phosphate.

____ 4. DNA replication is semiconservative because the template parent strand is retained by the daughter molecules.

____ 5. In DNA, adenine is the complementary base for thymine, and cytosine is the complement for guanine.

____ 6. The bonds between base pairs are covalent.

____ 7. Messenger RNA is formed by translation of a gene on one unzipped DNA strand.

____ 8. Transfer RNA carries the anticodon and amino acid to the ribosome for protein synthesis.

____ 9. The *lac* operon is activated by a repressor molecule.

____ 10. A mutation is an inheritable change in the DNA code.

____ 11. A plasmid is a circular piece of DNA that is needed for normal function of a bacterial cell.

____ 12. Transduction is a form of intermicrobial gene transfer that involves a virus carrier.

____ 13. In recombinant DNA technology, plasmids containing foreign genes are cloned in microorganisms.

Concept Questions

1. Briefly describe how DNA is packaged to fit inside the cell.
2. Give the base sequence of the complementary strand of the following strand of DNA: 5′-ATCGGCTACGTTCAC-3′
3. Describe what is meant by the antiparallel arrangement of DNA.
4. Name several characteristics of DNA structure that enable it to be replicated with such great fidelity generation after generation.
5. Explain the following relationship: DNA makes RNA makes protein.
6. What message does a gene provide? How is the language of the gene expressed?
7. If a protein is 3,300 amino acids long, how many nucleotide pairs long is the gene sequence that codes for it?
8. What about a palindrome makes it recognizable by a restriction endonuclease? Draw a short sequence of a palindrome.
9. Compare the structure and functions of DNA and RNA.
10. What are the sense and antisense strands of DNA?
11. Compare and contrast the actions of DNA and RNA polymerase.
12. What are the functions of start and nonsense codons?
13. The following sequence represents triplets on DNA:
    TAC CAG ATA CAC TCC CCT GCG ACT
    Give the mRNA codons and tRNA anticodons that correspond with this sequence, and then give the sequence of amino acids in the polypeptide. What alternate code can be used to synthesize this protein?
14. Explain how bacterial and eucaryotic cells differ in gene structure, transcription, and translation.
15. Compare DNA viruses with RNA viruses in their general methods of nucleic acid synthesis and viral replication.
16. Compare and contrast the *lac* operon with a repressible operon system. How does the *lac* system relate to the feedback control of enzymes mentioned in chapter 7?
17. What is the premise of the Ames test?
18. Describe the principal types of mutations.
19. Compare conjugation, transformation, and transcription on the basis of general method, nature of donor, and nature of recipient. List some examples of genes that may be transferred by intermicrobial transfer.
20. By means of a flowchart, show the possible jumps that a transposon can make.
21. What characteristics of bacteria make them good cloning hosts?
22. Outline the steps in cloning a gene. Once cloned, what may this gene be used for?

Practical/Thought Questions

1. A simple test you can do to demonstrate the coiling of DNA in bacteria is to open a large elastic band, stretch it taut, and twist it. First it will form a loose helix, then a tighter helix, and finally, to relieve stress, it will twist back upon itself. Further twisting will result in a series of knotlike bodies; this is how bacterial DNA is condensed.
2. To envision the replicon of bacteria, take a very thick rubber band and nick it longitudinally at some point. This will be the site of transcription origin. Slice the rubber band for a few millimeters and label the orientation of the two beginning strands (5′ to 3′ and 3′ to 5′). Label one strand leading and one strand lagging. This will also produce the two forks of replication and allow you to play with the production of the two new strands. Snip toward one fork to imagine how the polymerase will enter to add the correct bases. To represent the bases, add one paperclip to the leading strand in sequence toward the fork on the leading strand as each snip is made. Add clips to the lagging strand in the appropriate direction for that strand, but show that it is discontinuous. Do the same on the other fork, and continue around until you have two separate strands. About halfway through, look for the theta configuration.

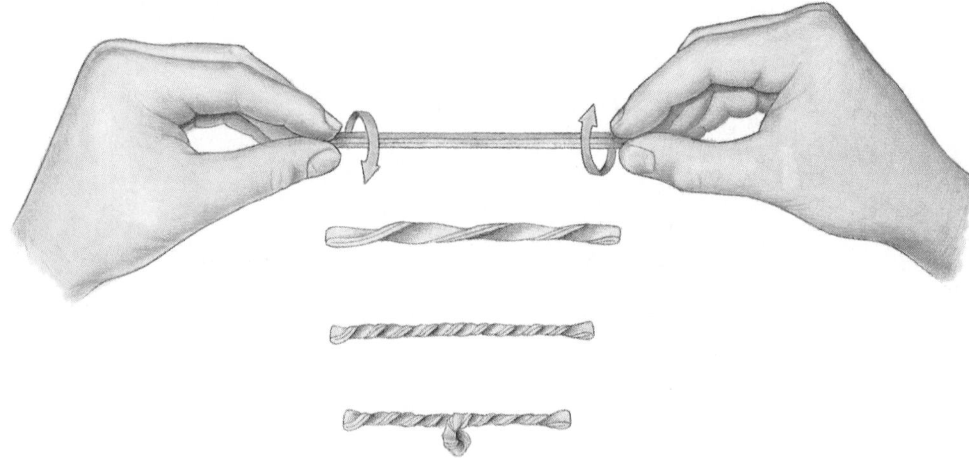

3. Knowing that retroviruses operate on the principle of reversing the direction of transcription from RNA to DNA, propose a drug that might possibly interfere with their replication.
4. Using the piece of DNA in concept question 13, show a deletion, an insertion, a substitution, and an inversion. Which ones are frameshift mutations? Are any of your mutations nonsense? Missense? (Use the universal code to determine this.)
5. Briefly explain how one might genetically engineer a plant to be frost-free.
6. Describe how a virus might be genetically engineered to make it highly virulent.
7. Give two possible methods of treating or curing sickle-cell anemia using biotechnology.

CHAPTER 9

Physical and Chemical Control of Microbes

Illustration of protective clothing used by doctors in the 1700s to avoid exposure to plague victims. The beaklike portion of the hood contained volatile perfumes to protect against foul odors and possibly inhaling "bad air."

Chapter Preview

The natural condition of humanity is to share surroundings with a large, diverse population of microorganisms. The complete exclusion of microbes from the environment is not only impossible but of questionable value. However, certain circumstances require constant and concerted efforts to exclude them. Such daily activities as cleaning, refrigeration, and cooking are, by nature, antimicrobial processes. The medical, dental, and commercial methods that prevent the spread of infectious agents, hinder spoilage, and make products safe are routine and very broad in scope. These methods for destroying, removing, and inhibiting microbes are the subjects of this and the following chapter.

Controlling Microorganisms

Much of the time in our daily existence, we take for granted tap water that is drinkable, food that is not spoiled, shelves full of products to eradicate "germs," and drugs to treat infections. Controlling our degree of exposure to potentially harmful microbes is a monumental concern in our lives, and it has a long and eventful history (see feature 9.1).

General Considerations in Microbial Control

The methods of microbial control belong to the general category of *decontamination* procedures, in that they destroy or remove contaminants. In microbiology, contaminants are microbes present at a given place and time that are undesirable or unwanted. Most decontamination methods employ either **physical** agents such as heat or radiation or **chemical** agents such as disinfectants, antiseptics, or drugs. This separation is convenient, but the categories may overlap in some cases—for instance, radiation may cause damaging chemicals to form, or chemicals may work by generating heat. The flowchart in figure 9.1 summarizes the major applications and aims in microbial control.

Relative Resistance of Microbial Forms

The primary targets of antimicrobial control are microorganisms capable of causing infection or spoilage that are constantly present in the external environment and on the human body. This targeted population is rarely simple or uniform; in fact, it often contains mixtures of microbes with extreme differences in resistance and harmfulness. Contaminants that can have far-reaching effects if not adequately controlled include bacterial vegetative cells and endospores, fungal hyphae and spores, yeasts, protozoan trophozoites and cysts, worms, insects and their eggs, and viruses. The following scheme compares the relative resistance of these forms to physical and chemical methods of control:

Highest resistance
Bacterial endospores.

Moderate resistance
Protozoan cysts; some fungal sexual spores (zygospores); some viruses. In general, naked viruses are more resistant than enveloped forms. Among the most resistant viruses are the hepatitis B virus and the poliovirus. Particular vegetative bacteria that have higher resistance are the *Mycobacterium tuberculosis, Staphylococcus aureus,* and *Pseudomonas* species.

Least resistance
Most bacterial vegetative cells; ordinary fungal spores and hyphae; enveloped viruses, yeasts, trophozoites.

Actual comparative figures on the requirements for destroying various groups of microorganisms are shown in table 9.1. Since bacterial endospores are the most resistant microbial entities, their destruction constitutes the goal of processes that sterilize (see definition in following section), because any process that kills them will invariably kill all less resistant forms. Other methods of control (disinfection, antisepsis) act primarily upon microbes that are less hardy than spores.

Feature 9.1 Microbial Control in Ancient Times

No one knows for sure when humans first applied methods that could control microorganisms, but perhaps the discovery and use of fire in prehistoric times was the starting point. We do know that records describing simple measures to control decay and disease appear from civilizations that existed several thousand years ago. We know, too, that these ancient people had no concept that germs caused disease, but they did have a mixture of religious beliefs, skills in observing natural phenomena, and, possibly, a bit of luck. This combination led them to carry out simple and sometimes rather hazardous measures that contributed to the control of microorganisms.

Salting, smoking, pickling, and drying foods, and exposure of food, clothing, and bedding to sunlight were prevalent practices among early civilizations. The Egyptians showed surprising sophistication and understanding of decomposition by embalming the bodies of their dead with strong salts and pungent oils. They introduced filtration of wine and water as well. The Greeks and Romans burned clothing and corpses during epidemics, and they stored water in copper and silver containers. The armies of Alexander the Great reportedly boiled their drinking water and buried their wastes. Burning sulfur to fumigate houses and applying sulfur as a skin ointment also date from this approximate era.

During the great plague pandemic of the Middle Ages, it was commonplace to bury corpses in mass graves, burn the clothing of plague victims, and ignite aromatic woods in the houses of the sick in the belief that fumes would combat the disease. In a desperate search for some sort of protection, survivors wore peculiar garments (see chapter opening illustration) and anointed their bodies with herbs, strong perfume, and vinegar. These attempts may sound foolish and antiquated, but it now appears that they may actually have had some benefits. Burning wood releases formaldehyde, which could have acted as a disinfectant; herbs, perfume, and vinegar contain mild antimicrobial substances. Each of these early methods, although somewhat crude, laid the foundations for microbial control methods that are still with us today.

Table 9.1 Relative Microbial Resistance to Physical and Chemical Agents

Method	Spores*	Vegetative Forms*	Relative Resistance**
Heat (moist)	120°C	80°C	1.5×
Radiation (X ray) dosage	0.4 Mrad	0.1 Mrad	4×
Ultraviolet rays (exposure time)	1.5 hrs	10 min	9×
Sterilizing gas (ethylene oxide)	1,200 mg/l	700 mg/l	1.7×
Sporicidal liquid (2% glutaraldehyde)	3 hrs	10 min	18×

*Values are based on methods (concentration, exposure time, intensity) that are required to destroy the most resistant pathogens in each group.
**The greater resistance of spores versus vegetative cells given as an average figure.

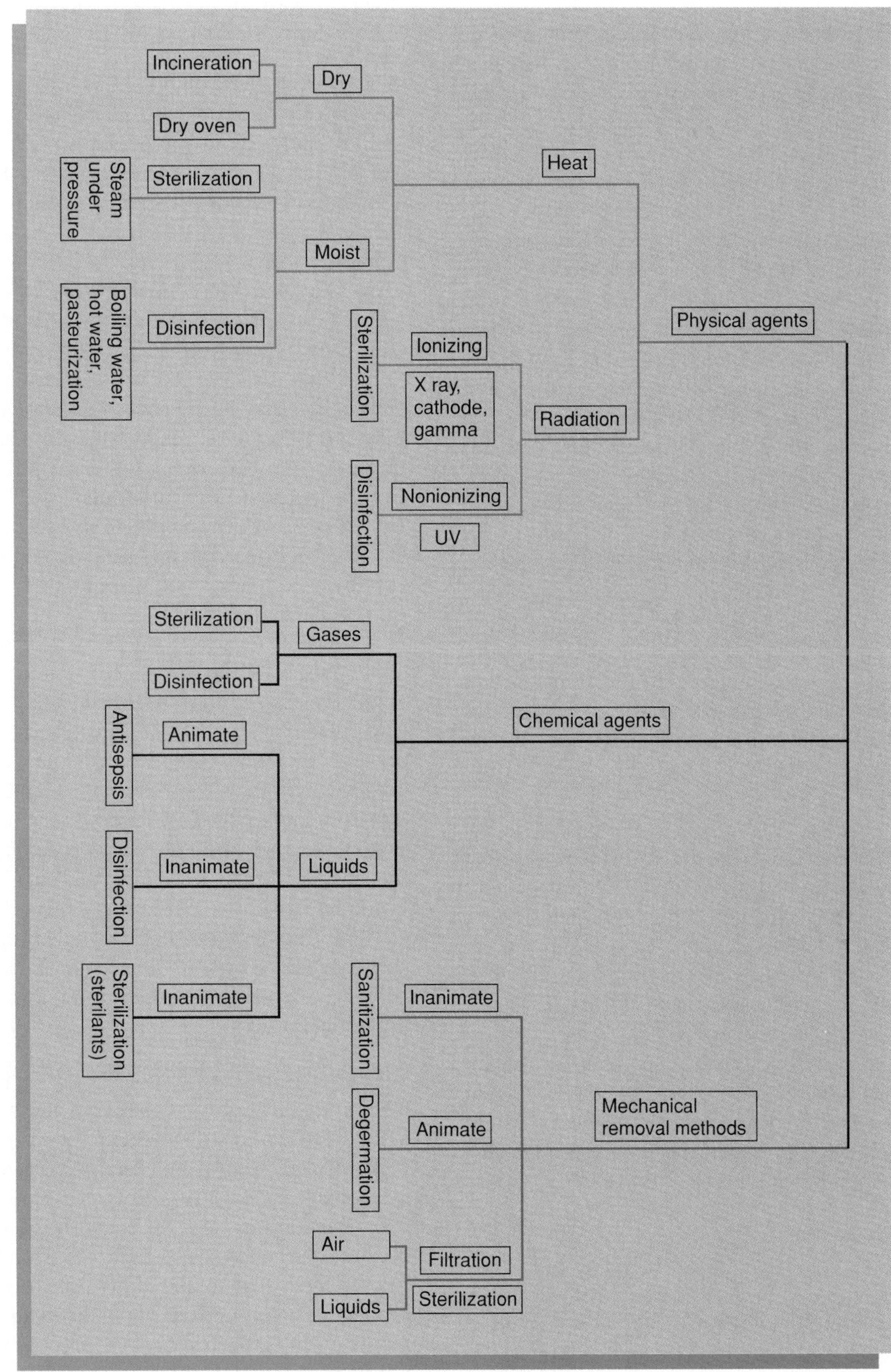

Figure 9.1 Flow summary for microbial control measures.

Terminology and Methods of Microbial Control

Through the years, a growing terminology has emerged for describing and defining measures that control microbes. To complicate matters, the everyday use of some of these terms may at times be vague and inexact. For example, occasionally one may be directed to "sterilize" or "disinfect" a patient's skin, even though this usage does not fit the technical definition of either term. To lay the groundwork for the concepts in microbial control to follow, we present here a series of concepts, definitions, and usages in antimicrobial control.

Sterilization

Sterilization is a process that eliminates, through destruction or removal, all viable microorganisms, including viruses. Any material that has been subjected to this process is said to be **sterile.** These terms should be used only in the strictest sense for methods that have been proved to sterilize. An object cannot be slightly sterile or almost sterile—it is either sterile or not sterile. Control methods that sterilize are generally used on inanimate objects, because sterilizing parts of the human body or live animals would call for such harsh treatment that it would be highly dangerous and impractical. As we shall see in a subsequent chapter, many internal parts of the body—the brain, muscles, and liver, for example—are naturally free of microbes.

Sterilized products—surgical instruments, syringes, and commercially packaged foods, just to name a few—are frequently essential to human well-being. Although most sterilization is performed with a physical agent such as heat, a few chemicals called *sterilants* can be classified as sterilizing agents because of their ability to destroy spores.

At times, sterilization is neither practicable nor necessary, and only certain groups of microbes need to be controlled. Some antimicrobial agents eliminate only the susceptible vegetative states of microorganisms but do not destroy the more resistant endospore and cyst stages. Keep in mind that the destruction of spores is not always a necessity, because most of the infectious diseases of humans and animals are caused by non-spore-forming microbes.

Microbicidal Agents

The root *-cide,* meaning to kill, may be combined with other terms to define an antimicrobial agent aimed at destroying a certain group of microorganisms. For example, a **bactericide** is a chemical that destroys bacteria except for those in the endospore stage. It may or may not be effective on other microbial groups. A **fungicide** is a chemical that can kill fungal spores, hyphae, and yeasts. A **virucide** is any chemical known to inactivate viruses, especially on living tissue. A **sporicide** is an agent capable of destroying bacterial endospores. A sporicidal agent may also be a sterilant because it can destroy the most resistant of all microbes. A **biocide** is a substance that kills all living things, but particularly microorganisms.

Agents That Cause Microbistasis

The Greek words *stasis* and *static* mean to stand still. They can be used in combination with various prefixes to denote a condition in which microbes are temporarily prevented from multiplying but are not killed outright. Although killing or permanently inactivating microorganisms is the usual goal of antimicrobial control, microbistasis does have meaningful applications. **Bacteriostatic** agents prevent the growth of bacteria on tissues or on objects in the environment, and *fungistatic* chemicals inhibit fungal growth. Materials used to control microorganisms in the body (antiseptics and drugs) have microbistatic effects because microbicidal compounds would generally be too toxic to human cells.

Germicides, Disinfection, Antisepsis

A **germicide,** also called a microbicide, is any chemical agent that kills pathogenic microorganisms. A germicide may be used on inanimate (nonliving) materials or on living tissue, but it ordinarily cannot kill resistant microbial cells. Any physical or chemical agent that kills "germs" is said to have **germicidal** properties.

The related term, **disinfection,** refers to the use of a physical process or a chemical agent (a **disinfectant**) to destroy vegetative pathogens but not bacterial endospores. It is important to note that disinfectants are normally used only on inanimate objects because, in the concentrations required to be effective, they can be toxic to human and other animal tissue. Disinfection processes also remove the harmful products of microorganisms (toxins) from materials.

Examples of disinfection include: (1) Applying a solution of 5% bleach to an examining table; (2) boiling food utensils used by a sick person; and (3) immersing thermometers in tincture of iodine between uses.

In modern usage, **sepsis** is defined as the growth of microorganisms or the presence of microbial toxins in blood and other tissues. The term **asepsis** refers to any practice that prevents the entry of infectious agents into sterile tissues and thus prevents infection. Aseptic techniques commonly practiced in health care range from sterile methods that exclude all microbes to **antisepsis.** In antisepsis, chemical agents called **antiseptics** are applied directly to exposed body surfaces (skin and mucous membranes), wounds, and surgical incisions to destroy or inhibit vegetative pathogens.

Examples of antisepsis include preparing the skin prior to surgical incisions with iodine compounds, swabbing an open root canal with hydrogen peroxide, and ordinary handwashing with a germicidal soap.

Methods That Reduce the Numbers of Microorganisms

Several applications in commerce and medicine do not require actual sterilization, disinfection, or antisepsis but are based on reducing the levels of microorganisms so that the possibility of infection or spoilage is greatly decreased.

Restaurants, dairies, breweries, and other food industries consistently handle large numbers of soiled utensils and quantities of foods that could readily become sources of infection and spoilage. These industries must keep microbial levels to a minimum during preparation and processing. **Sanitization** is any cleansing technique that mechanically removes microorganisms (along with food debris) to reduce the level of contaminants. A **sanitizer** is a compound such as soap or detergent used to perform this task.

---

sterile (ster′-ill) Gr. *steira,* barren. (This has another, older meaning that connotes the inability to produce offspring.)

germicide (jer′-mih-syd) L. *germen,* germ, and *caedere,* to kill. Germ is a common term for a pathogenic microbe.

disinfection (dis″-in-fek′-shun) L. *dis,* apart, and *inficere,* to corrupt.

asepsis (ay-sep′-sis) Gr. *a,* no or none, and *sepsis,* decay.

antisepsis *anti,* against.

sanitization (san″-ih-tih-zay′-shun) L. *sanitas,* health.

Cooking utensils, dishes, glass bottles, cans, and used clothing that have been washed and dried may not be completely free of microbes, but they are considered safe for normal use (sanitary). Air sanitization with ultraviolet lamps reduces airborne microbes in hospital rooms, veterinary clinics, and laboratory installations. It is important to note that some sanitizing processes (such as dishwashing machines) are rigorous enough to sterilize objects, but this is not true of all sanitization methods.

It is often necessary to reduce the numbers of microbes on the human skin for surgical preparations, drawing blood, and in treating wounds. Reduction is achieved by scrubbing the skin and/or immersing it in chemicals. This procedure, called **degermation,** emulsifies oils that lie on the outer cutaneous layer and mechanically removes potential pathogens on the outer layers of the skin.

Examples of degerming procedures are the surgical handscrub, the application of alcohol wipes to the skin, and the cleansing of a wound with germicidal soap and water. The concepts of antisepsis and degermation clearly overlap, since a degerming procedure can simultaneously create antisepsis, and vice versa.

What Is Microbial Death?

Death is a phenomenon that involves the permanent termination of an organism's vital processes. Signs of life in complex organisms such as animals are self-evident, and death is made clear by loss of nervous function, respiration, or heartbeat. In contrast, death in microscopic organisms that are composed of just one or a few cells is often hard to detect, because they reveal no conspicuous vital signs to begin with. Lethal agents (such as radiation and chemicals) do not necessarily alter the overt appearance of microbial cells. Even the loss of movement in a motile microbe cannot be used to indicate death. This fact has made it necessary for workers in the field of microbial control to develop special qualifications that define and delineate microbial death.

The destructive effects of chemical or physical agents occur at the level of a single cell. As the cell is continuously exposed to an agent such as intense heat or toxic chemicals, various cell structures become dysfunctional, and the entire cell may sustain irreversible damage. At present, the most practical way to detect this damage is to determine if a microbial cell can still reproduce when exposed to a suitable environment. If the microbe is so disturbed metabolically and structurally that it is unable to multiply, then it is no longer viable. The permanent loss of reproductive capability, even under optimum growth conditions, has become the accepted microbiological definition of death.

Factors in Microbial Control

The ability to define microbial death has tremendous theoretical and practical importance. With this concept, workers in the field of microbial control have defined the conditions required to destroy microorganisms and have pinpointed the manner by which antimicrobial agents kill cells. Standards of sterilization and disinfection in medicine, dentistry, and industry have been established largely by means of rigorous experimentation. Literally hundreds of testing procedures have been developed for evaluating physical and chemical agents. Appendix C presents a general breakdown of certain types of antimicrobial tests, along with the kinds of variables that must be controlled.

The cells of a pure culture show marked variations in susceptibility to a given microbicidal agent. Death of the whole population is not instantaneous, but requires a certain time of exposure (figure 9.2*a*). The most susceptible cells (younger, actively growing cells) die immediately, whereas less susceptible cells (older, inactive ones) have greater resistance and require longer exposure times. At some point, the chance that any live cells remain is severely diminished; this point is called sterilization. In most actual circumstances of disinfection and sterilization, the target population is not a single species of microbe but a mixture of bacteria, fungi, spores, and viruses, presenting an even greater spectrum of microbial resistance.

The effectiveness of a particular agent is a function of several factors besides time. The following additional factors influence the action of antimicrobial agents:

1. The number of microorganisms (figure 9.2*b*). A higher load of contaminants requires more time to destroy.
2. The nature of the microorganisms in the population (figure 9.2*c*). Are highly resistant forms present?
3. The temperature and pH of the environment.
4. The concentration (dosage, intensity) of the agent.
5. The mode of action of the agent (figure 9.2*d*).
6. The presence of solvents, interfering organic matter, and inhibitors.

The influence of these factors will be discussed in greater detail in subsequent sections.

Detecting Microbistasis

Permanent cessation of reproduction is the accepted criterion for microbial death; however, differentiating a microbistatic effect from a microbicidal one can present a problem. This is particularly true of some disinfectants and antiseptics that only retard cell growth. Cells kept from multiplying long enough may eventually become unable to do so, and may therefore be dead by definition. One feature of microbicidal testing is to reduce the potential for mistaking "static" effects for "cidal" effects (figure 9.2*d*).

How Antimicrobial Agents Work: Their Modes of Action

An antimicrobial agent's adverse effect on cells is known as its *mode* (or *mechanism*) *of action.* Determining an agent's precise mode of action may be very difficult, and we do not yet completely understand how agents such as radiation and heat achieve their effects. Current evidence indicates that various parts of the cell accumulate injuries until it ceases to function. Many antimicrobial agents affect more than one cellular target and may inflict both primary and secondary damages that eventually lead to cell death. Agents can be classified according to the following degrees of selectiveness: (1) agents that are less selective in their scope of destructiveness inflict severe damage on

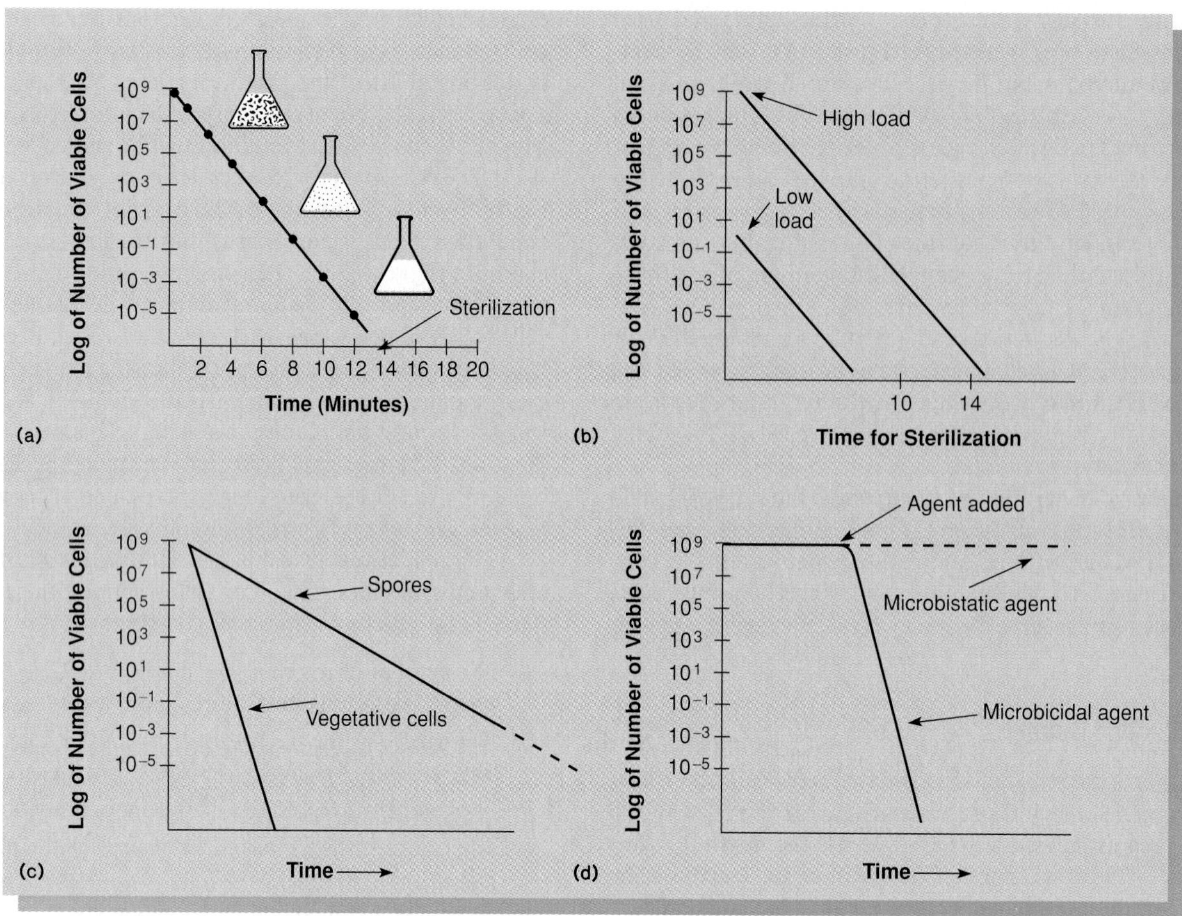

Figure 9.2 Factors that influence the rate at which microbes are killed by antimicrobial agents. (*a*) Length of exposure to the agent. During exposure to a chemical or physical agent, all cells of a microbial population, even a pure culture, do not die simultaneously. Over time, the number of viable organisms remaining in the population decreases logarithmically, giving a straight line relationship on a graph. The point at which the number of survivors is infinitesimally small is considered sterilization. (*b*) Effect of the microbial load. (*c*) Relative resistance of spores versus vegetative forms. (*d*) Action of the agent, whether destructive or inhibitory.

many cell parts, and are generally very biocidal (heat, radiation, some disinfectants); (2) moderately selective agents with intermediate specificity (certain disinfectants and antiseptics); and (3) more selective agents (drugs) whose target is usually limited to a specific cell structure or function and whose effectiveness is restricted only to certain microbes.

The cellular targets of physical and chemical agents fall into four general categories: (1) the cell wall, (2) the cell membrane, (3) cellular synthetic processes (DNA, RNA), and (4) proteins.

The Effects of Agents on the Cell Wall

The cell wall maintains the structural integrity of bacterial and fungal cells. Several types of chemical agents damage the cell wall by blocking its synthesis, digesting it, or breaking down its surface. A cell deprived of a functioning cell wall becomes fragile and is lysed very easily. Examples of this mode of action include some antimicrobial drugs (penicillins) that interfere with the synthesis of the cell wall in bacteria (see chapter 10). The enzymes lysozyme and lysostaphin digest specific bonds in the cell wall of gram-positive bacteria, eroding and weakening it, and one effect of detergents and alcohol is to disrupt cell walls (figure 9.3*a*).

How Agents Affect the Cell Membrane

All microorganisms have a cell membrane composed of lipids and proteins, and even some viruses have an outer membranous envelope. As we learned in previous chapters, a cell's membrane provides a two-way system of transport that also regulates osmotic pressure. If this membrane is disrupted, a cell loses its selective permeability and can neither prevent the loss of vital molecules nor bar the entry of damaging chemicals, which leads to cell death. Detergents called **surfactants** work as microbicidal agents because they lower the surface tension of cell membranes. To envision how this works, think of the lipid part of the membrane as a fluid interface, with a cohesive tension at its surface caused by mutual attraction of the hydrophilic lipid heads. Surfactants are polar molecules with hydrophilic and hydrophobic regions that can physically bind to the lipid layer and penetrate the internal hydrophobic region of membranes. In effect, this process "opens up" the once tight interface, leaving leaky spots that allow injurious chemicals to seep into the cell and important ions to seep out (figure 9.3*b*).

---

surfactant (sir-fak'-tunt) A word derived from **surface-active agent.**

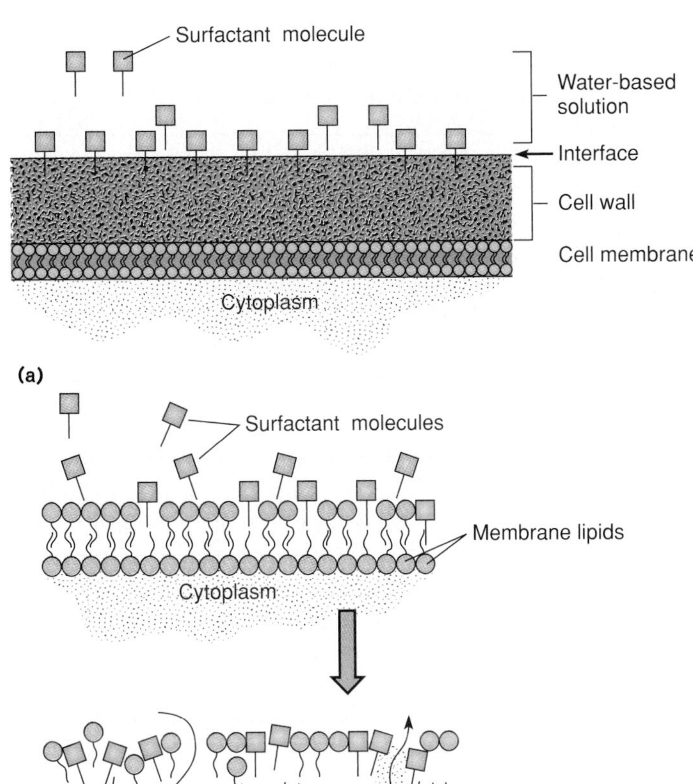

Figure 9.3 Mode of action of surfactants. (*a*) Effect on cell wall. Surfactants are attracted to the surface of the cell wall, where they attach and weaken its interfacial tension. This results in the wetting and easier penetration of the cell wall. (*b*) Effect on the cell membrane. Surfactants inserting in the lipoidal layers disrupt it and create abnormal channels that alter permeability and cause leakage both into and out of the cell.

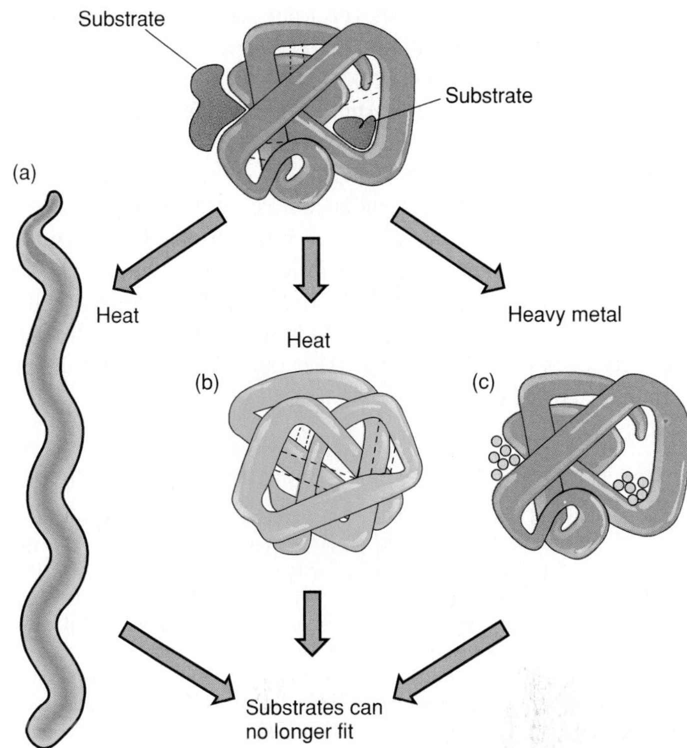

Figure 9.4 Modes of action affecting protein function. The functional three-dimensional or native state maintained by specific bonds creates active sites to react with substrate. Some agents denature the protein by breaking all or some secondary and tertiary bonds. Results are (*a*) complete unfolding or (*b*) random bonding and incorrect folding. (*c*) Some agents react with protein functional groups and can block the active site or interfere with bonding.

Agents Affecting Protein and Nucleic Acid Synthesis

Microbial life depends upon an orderly and continuous supply of proteins to function as enzymes and structural molecules. As we saw in chapter 8, these proteins are synthesized on the ribosomes through a complex process. Any agent that interferes with accurate translation also prevents synthesis of a complete, functioning protein. For instance, the antibiotic chloramphenicol binds to the ribosomes of bacteria in a way that stops peptide bonds from forming. In its presence, many bacterial cells are inhibited from forming proteins required in growth and metabolism, and are thus inhibited from multiplying. Most of the agents that block protein synthesis are drugs used in antimicrobial therapy. These drugs will be discussed in greater detail in chapter 10.

The nucleic acids are likewise necessary for the continued functioning of microbes. DNA must be regularly replicated and transcribed in growing cells, and any agent that either impedes these processes or changes the genetic code is potentially antimicrobial. Some agents bind irreversibly to DNA, preventing both transcription (formation of RNA) and translation, while others are mutagenic agents. Gamma, ultraviolet, or X radiation causes mutations that result in permanent inactivation of DNA. Chemicals such as formaldehyde and ethylene oxide also interfere with DNA and RNA function.

Agents That Alter Protein Function

A microbial cell contains large quantities of proteins that function properly only if they remain in a normal three-dimensional configuration called the *native state*. The antimicrobial properties of some agents arise from their capacity to disrupt or **denature** proteins. In general, denaturation occurs when the bonds that maintain the secondary and tertiary structure of the protein are broken. This will cause the protein to unfold or create random, irregular loops and coils (figure 9.4). One way that proteins may be denatured is through coagulation by moist heat (the same reaction seen in the irreversible solidification of the white of an egg when boiled). Chemicals such as strong organic solvents (alcohols, acids) and phenolics also coagulate proteins. Other antimicrobial agents, such as metallic ions, attach to the active site of the protein and prevent it from interacting with its correct substrate. Regardless of the exact mechanism, such losses in normal protein function can promptly arrest metabolism.

Practical Concerns in Microbial Control

Numerous considerations govern the selection of a workable method of microbial control. The following are among the most pressing concerns: (1) Does the application require sterilization, or is disinfection adequate? In other words, must spores be destroyed, or is it necessary to destroy only vegetative pathogens? (2) Is the item to be reused or permanently discarded? If it will be discarded, then the quickest and least expensive method should be chosen. (3) If it will be reused, can the item withstand heat, pressure, radiation, or chemicals? (4) Is the control method suitable for a given application? (For example, ultraviolet radiation is a good sporicidal agent, but it will not penetrate solid materials.) Or, in the case of a chemical, will it leave an undesirable residue? (5) Will the agent penetrate to the necessary extent? (6) Is the method cost- and labor-efficient, and is it safe?

Management of Microbial Control A remarkable variety of substances may require sterilization. They run the gamut from durable solids like rubber to sensitive liquids like serum, and from air to tissue grafts. Hundreds of situations requiring sterilization confront the network of persons involved in health care, be it technician, nurse, doctor, or manufacturer, and no universal method works well in every case. Considerations such as cost, effectiveness, and ecological impact are all important. For example, the disposable plastic items such as catheters and syringes that are used in invasive medical procedures have the potential for infecting the tissues. These must be sterilized during manufacture by a nonheating method (gas or radiation), because heat can damage delicate plastics. After these items have been used, it is often necessary to destroy or decontaminate them before they are discarded because of the potential risk to the handler (needle sticks). Steam sterilization, which is quick and sure, is a sensible choice at this point, because it doesn't matter if the plastic is destroyed. But what happens to this nonbiodegradable medical waste when it is finally disposed of? We have already seen the unfortunate results of burying it in landfills or dumping it in oceans, lakes, and rivers. One solution has been to burn some of these materials in an incinerator, but this just adds another type of pollution to our already polluted world.

Methods of Physical Control

Microorganisms have adapted to the tremendous diversity of habitats the earth provides, even severe conditions of temperature, moisture, pressure, and light. For microbes that normally withstand extreme physical conditions (heat, cold, drying, radiation), our attempts at control would probably have little effect. Fortunately, the vast majority of microbes that must be controlled are not adapted to such extremes and are readily controlled by abrupt changes in environment. Most prominent among antimicrobial physical agents is heat. Other less widely used agents include radiation, filtration, ultrasonic waves, and even cold. The following sections will examine some of these methods and explore their practical applications in medicine, commerce, and the home.

Heat As an Agent of Microbial Control

A sudden departure from a microbe's temperature of adaptation, either an increase or a decrease, is likely to have a detrimental effect on it. Increasingly elevated temperatures (exceeding the maximum) are microbicidal, whereas lower temperatures (below the minimum) tend to have inhibitory or static effects. The two physical states of heat used in microbial control are moist and dry. **Moist heat** occurs in the form of hot water, boiling water, or steam (vaporized water). In practice, the temperature of moist heat usually ranges from 60°C to 135°C. As we shall see, the temperature of steam can be regulated by adjusting its pressure in a closed container. The expression **dry heat** denotes air with a low moisture content that has been heated by a flame or electric heating coil. In practice, the temperature of dry heat ranges from 160°C to several thousand degrees C.

Table 9.2 Comparison of Times and Temperatures to Achieve Sterilization with Moist and Dry Heat

	Temperature	Time to Sterilize
Moist Heat	121°C	15 min
	125°C	10 min
	134°C	3 min
Dry Heat	150°C	150 min
	160°C	120 min
	170°C	60 min

Mode of Action and Relative Effectiveness of Heat

In addition to their physical state, moist and dry heat differ in their efficiency: At a given temperature, moist heat works several times faster than dry heat (table 9.2), and at the same length of exposure, moist heat kills cells at a lower temperature than dry heat. The mechanisms by which microorganisms are destroyed by heat are rather complex. It is clear that both forms of heat disrupt important cell components, but it also appears that their specific modes of action are different. Exposure to moist heat generally coagulates and therefore denatures cytoplasmic proteins. The greater efficacy of moist heat has been attributed to this particular mode of action, in that protein (enzyme) denaturation occurs more rapidly and at a lower temperature if moisture is present. Cell structures such as the membrane, ribosomes, DNA, and RNA are also damaged by moist heat.

Dry heat also has several effects on microbes. For instance, it can oxidize cells, and at extremely high temperatures, reduce them to ashes. It may dehydrate cell components, thereby concentrating the protoplasm. In time, it also denatures proteins and DNA, but because proteins are more stable in dry heat, higher temperatures are required to inactivate them.

Heat Resistance and Thermal Death: Spores

Bacterial endospores exhibit the greatest resistance, and vegetative states of bacteria and fungi are the least resistant to both moist and dry heat. Destruction of spores usually requires temperatures above boiling (table 9.3), although resistance varies widely even among sporeformers. In boiling water (100°C), the spores of *Bacillus anthracis* (the agent of anthrax) can be destroyed in a few minutes, whereas the spores of some thermo-

Table 9.3 Thermal Death Times of Spores

Organism	Temperature	Time of Exposure to Kill Spores
Moist Heat		
Bacillus stearothermophilis	121°C	12 min
	100°C	7.5 hrs
Bacillus anthracis	100°C	10 min
Clostridium botulinum	120°C	10 min
	100°C	5 hrs
Clostridium tetani	105°C	10 min
Dry Heat		
Bacillus stearothermophilis	140°C	5 min
Bacillus xerothermodurans	200°C	139 hrs
Bacillus anthracis	180°C	3 min
Clostridium botulinum	120°C	2 hrs
	180°C	15 min
Clostridium tetani	100°C	60 min
	140°C	15 min

Table 9.4 Average Thermal Death Times of Microorganisms*

Microbial Type	Temperature	Time (Min)
Non-spore-forming pathogenic bacteria	58°C	28
Non-spore-forming nonpathogenic bacteria	61°C	18
Vegetative stage of spore-forming bacteria	58°C	19
Fungal spores	76°C	22
Yeasts	59°C	19
Heat inactivation of viruses		
Nonenveloped	57°C	29
Enveloped	54°C	22
Protozoan trophozoites	46°C	16
Protozoan cysts	60°C	6
Worm eggs	54°C	3
Worm larvae	60°C	10

*Not including bacterial endospores.

philic and anaerobic species may require 10 hours or more. Differences in the resistance of spores to moist and dry heat are also evident: A suspension of spores of *Bacillus subtilis,* a common soil bacterium, is destroyed in 1 minute with moist heat at 121°C, but requires up to 120 minutes with dry heat at 121°C. Other comparative data on spore destruction with moist and dry heat are shown in table 9.3.

Heat Resistance and Thermal Death: Vegetative Stages

The thermal death requirements of vegetative cells also vary, though not to the same extent as for spores (table 9.4). Among bacteria, the destruction range with moist heat ranges from 50°C for 3 minutes (*Neisseria gonorrhoeae*) to 60°C for 60 minutes (*Staphylococcus aureus*). It is worth noting that vegetative cells of sporeformers are just as susceptible as vegetative cells of non-sporeformers, and that pathogens are neither more nor less susceptible than nonpathogens. Other microbes, including fungi (yeasts, molds, and some of their spores), protozoa, and worms, are rather similar in their sensitivity to heat. Viruses are surprisingly resistant to heat, with a tolerance range extending from 55°C for 2–5 minutes (adenoviruses) to 60°C for 600 minutes (hepatitis virus). For practical purposes, all nonresistant forms of bacteria, yeasts, molds, protozoa, worms, and viruses are destroyed when exposed to 80°C for 10 minutes.

Practical Concerns in the Use of Heat

Adequate sterilization requires that both temperature and length of exposure be considered. As a general rule, higher temperatures allow shorter exposure times, and lower temperatures require longer exposure times. A combination of these two variables constitutes the **thermal death time,** or TDT, defined as the shortest length of time required to kill all microbes at a specified temperature. The TDT has been experimentally determined for the microbial species that are common or important contaminants in various heat-treated materials. Another way to compare the susceptibility of microbes to heat is the thermal death point (TDP), defined as the lowest temperature required to kill all microbes in a sample in 10 minutes.

Applications of the TDT Many perishable substances are processed with moist heat. Some of these products are intended to remain on the shelf at room temperature for several months or even years. The chosen heat treatment must render the product free of agents of spoilage or disease. At the same time, the quality of the product and the speed and cost of processing must be considered. For example, in the commercial preparation of canned green beans, one of the cannery's greatest concerns is to prevent growth of the agent of botulism. From several possible TDTs for *Clostridium botulinum* spores, the cannery must choose one that kills all spores but does not turn the beans to mush. Out of these many considerations emerges an optimal TDT for a given processing method. Commercial canneries heat low-acid foods at 121°C for 30 minutes, a treatment that sterilizes these foods. Because of such strict controls in canneries, most cases of botulism are contracted from home-canned vegetables and meats.

Common Methods of Moist Heat Control

The four ways that moist heat is employed to sterilize or disinfect are (1) steam under pressure, (2) live, nonpressurized steam, (3) boiling water, and (4) pasteurization.

Steam Under Pressure The highest temperature that steam can reach is 100°C under normal atmospheric pressure at sea level, a pressure equal to 15 pounds per square inch (psi), or 1 atmosphere. In order to raise its temperature above this point, steam must be pressurized in a closed chamber. This phenomenon is explained by the physical principle that governs the behavior of gases under pressure. When a gas is compressed, its temperature rises in direct relation to the amount of pressure.

So, when the pressure is increased to 5 psi above normal atmospheric pressure, the temperature of steam rises to 109°C. When the pressure is increased to 10 psi, its temperature will be 115°C, and at 15 psi (2 atmospheres), it will be 121°C. It is not the pressure by itself that is killing microbes, but the increased temperature it produces.

Such pressure-temperature combinations can only be achieved with a special device that subjects pure steam to pressures greater than 1 atmosphere. Health and commercial industries use an *autoclave* for this purpose, and a comparable home appliance is the pressure cooker. The autoclave comes in various forms, all of which have a fundamentally similar plan—a cylindrical metal chamber with an airtight door on one end, racks to hold materials, and a steam-filled jacket around the sterilizing chamber (figure 9.5). Its construction includes a maze of valves, pressure and temperature gauges, and ducts for regulating and measuring pressure and conducting the steam into the chamber. Sterilization is achieved when the steam condenses against the objects in the chamber and gradually raises them to its temperature.

Several details require special attention in autoclaving. Experience has shown that the most efficient pressure/temperature combination to sterilize is 15 psi, which yields 121°C. It is possible to use higher pressure to reach higher temperatures (for instance, increasing the pressure to 3 atmospheres raises the temperature 11°C), but this will not significantly reduce the exposure time and may harm the items being sterilized. Because this temperature will develop only if saturated steam is fed into the inner chamber, the modern autoclave is engineered to automatically evacuate the air in the chamber and fill it with steam. It is also important to avoid overpacking or haphazardly loading the chamber, which prevents steam from circulating freely around the contents and impedes the full contact that is necessary. The duration of the process is adjusted according to the bulkiness of the items in the load (thick bundles of material or large flasks of liquid) and how full the chamber is. The range of holding times varies from 10 minutes for light loads to 40 minutes for heavy or bulky ones; the average time is 20 minutes.

Applications of the Autoclave The autoclave is an excellent device to sterilize heat-resistant materials such as glassware, cloth (surgical dressings), rubber (gloves), metallic items (surgical and dental instruments), liquids, paper, some media, and some heat-resistant plastics. If the items are heat-sensitive (plastic petri dishes) but will be discarded, the autoclave is still a good choice. However, the autoclave is ineffective for sterilizing substances that repel moisture (oils, waxes, powders).

Intermittent Sterilization Autoclaving sterilizes by a one-shot method and is fast and efficient for most heat-resistant items. However, some materials cannot withstand the high temperature. If no other method is available, these materials may be subjected to **intermittent sterilization,** also called *tyndallization.*[1] This technique requires a chamber to hold the materials and a reservoir for boiling water. Items in the chamber are exposed to free-flowing steam for 30–60 minutes. This temperature is not sufficient to reliably kill spores, so a single exposure will not suffice. Assuming that surviving spores will germinate into less resistant vegetative cells, the items are incubated at appropriate temperatures for 23–24 hours, and then again subjected to steam treatment. This cycle is repeated for three days in a row. Because the temperature never gets above 100°C, highly resistant spores that do not germinate may survive even after three days of this treatment.

Applications of Tyndallization Intermittent sterilization is used most often to process heat-sensitive culture media, such as those containing sera, egg, or carbohydrates (which may break down at higher temperatures), and some canned foods. It is probably not effective in sterilizing items such as instruments and dressings that provide no environment for spore germination, but it certainly would disinfect them (see next section).

Boiling Water: Disinfection A simple boiling water bath or chamber can quickly decontaminate items in the clinic and home. Because a single processing at 100°C will not kill all resistant cells, this method is, strictly speaking, disinfection rather than sterilization. When materials are exposed to boiling water for 30 minutes, all non-spore-forming pathogens will be killed, including resistant species such as the tubercle bacillus and staphylococci. Probably the greatest disadvantage with this method is that the items may be easily recontaminated when removed from the water. Boiling is also a recommended method of disinfecting unsafe drinking water.

Applications in Disinfection Medical and dental offices sometimes employ a special bath for disinfecting metal and rubber instruments. Chemicals may be added to the water to prevent the edges of scalpels and needles from corroding. In the home, boiling water is a fairly reliable way to sanitize and disinfect materials for babies, food preparation, and utensils, bedding, and clothing from the sickroom.

Pasteurization: Disinfection of Beverages Fresh beverages such as milk, fruit juices, beer, and wine are easily contaminated during collection and processing. Because microbes have the potential for spoiling these foods and causing illness, it may be necessary to use heat to reduce the microbial load and destroy pathogens. **Pasteurization** (see feature 7.7) is a technique in which heat is applied to liquids to kill potential agents of infection and spoilage, while at the same time retaining flavor and food value. This process is considered a form of moist heat because of the high moisture content of the beverage. Ordinary pasteurization techniques require special vats and heat exchangers that expose the liquid to 63°–66°C for 30 minutes (batch method) or 71.6°C for 15 seconds (flash method). The second method is preferable because it is less likely to change flavor and nutrient content, and it is more effective against certain resistant pathogens such as *Coxiella* and *Mycobacterium.* Although these treatments inactivate most viruses and destroy the

1. Named for the British physicist John Tyndall who did early experiments with sterilizing procedures.

(a)

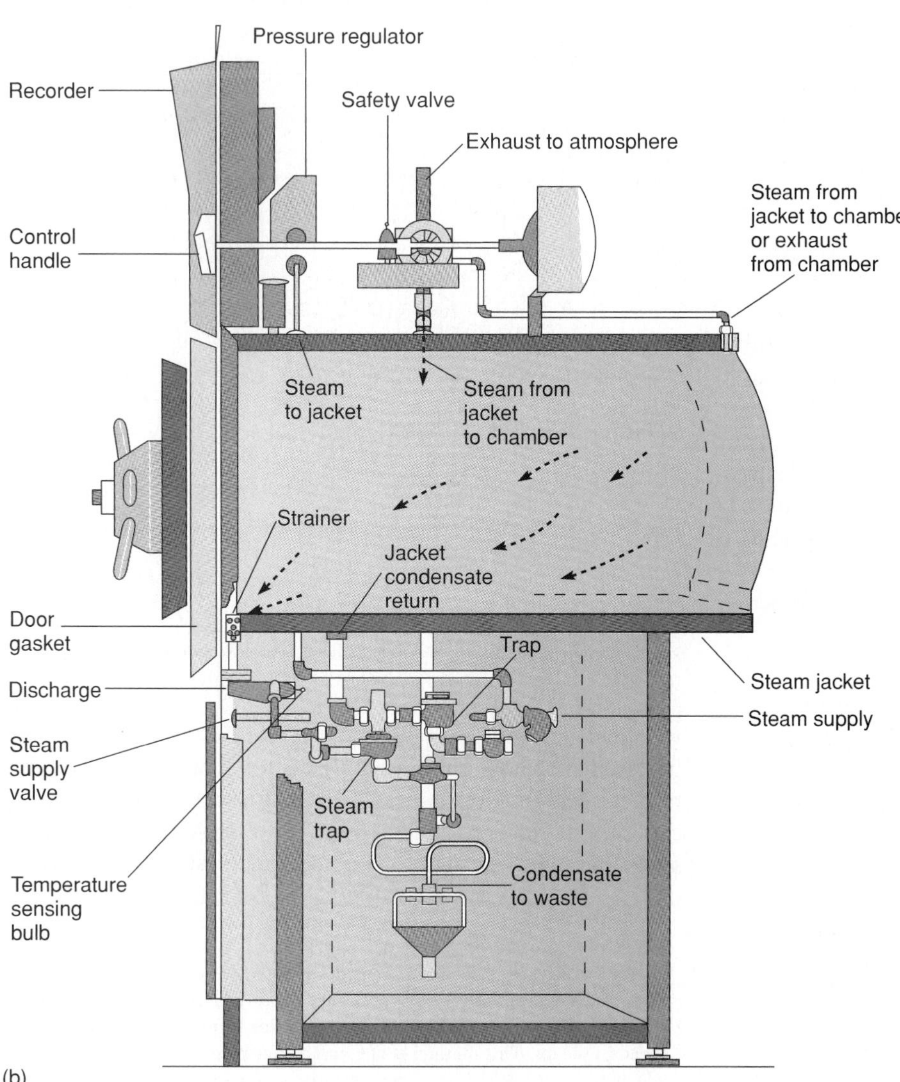

(b)

Figure 9.5 Steam sterilization with the autoclave. (*a*) A large automatic autoclave used in sterilization by drug companies. (*b*) Cutaway section, showing autoclave components.

(*b*) From John J. Perkins, *Principles and Methods of Sterilization in Health Science,* 2d ed., 1969. Courtesy of Charles C Thomas, Publisher, Springfield, Illinois.

vegetative stages of 97–99% of bacteria and fungi, they do not kill endospores or *thermoduric* species (mostly nonpathogenic lactobacilli, micrococci, and yeasts). In recent times, containers labelled *sterile milk* have appeared on supermarket shelves. This milk, which has a storage life of 3 months (unopened and unrefrigerated), has been processed with ultrahigh temperature (UHT)—134°C for 1–2 seconds (see chapter 22).

Applications of Pasteurization Pasteurization is most often employed to prevent the transmission of milk-borne diseases from infected cows or milk handlers. The primary targets of pasteurization are non-spore-forming pathogens: *Salmonella* species (a common cause of food infection), *Campylobacter jejeuni* (acute intestinal infection), *Listeria monocytogenes* (listeriosis), *Brucella* species (undulant fever), *Coxiella burnetii* (Q fever), *Mycobacterium bovis* and *M. tuberculosis* (tuberculosis), *Streptococcus* species, and several enteric viruses. Milk is not sterile after regular pasteurization. In fact, it may contain 20,000 microbes per milliliter or more, which explains why even an unopened carton of milk will eventually spoil. In 1985, a California epidemic of listeriosis afflicted hundreds of people and resulted in 40 deaths, mostly among infants and elderly persons. Epidemiologists searching for the infection source were first led to a particular type of fresh, soft cheese that had been eaten by the patients and was contaminated with *Listeria*. Later evidence linked the outbreak to milk taken from infected cows and inadequately pasteurized prior to cheesemaking. Although modern dairy maintenance and milk processing technologies have reduced the probability of milk-borne infection, pasteurization is still a valuable preventative measure that also extends milk storage time. Some wineries and breweries still pasteurize wine and beer to kill unwanted contaminants and to increase shelf life, but other methods such as filtration are gradually replacing heat.

Dry Heat: Hot Air and Incineration

Dry heat is not as versatile or widely used as moist heat, but it has several important sterilization applications. The temperatures and times employed in dry heat vary according to the particular method, but in general, they are greater than with moist heat. **Incineration** in a flame or electric heating coil is perhaps the most rigorous of all heat treatments. The flame of a Bunsen burner reaches 1,870°C at its hottest point, and furnace/incinerators operate at temperatures of 800°–6,500°C. Direct exposure to such intense heat ignites and reduces the microbes, and sometimes their vehicles, to ashes and gas.

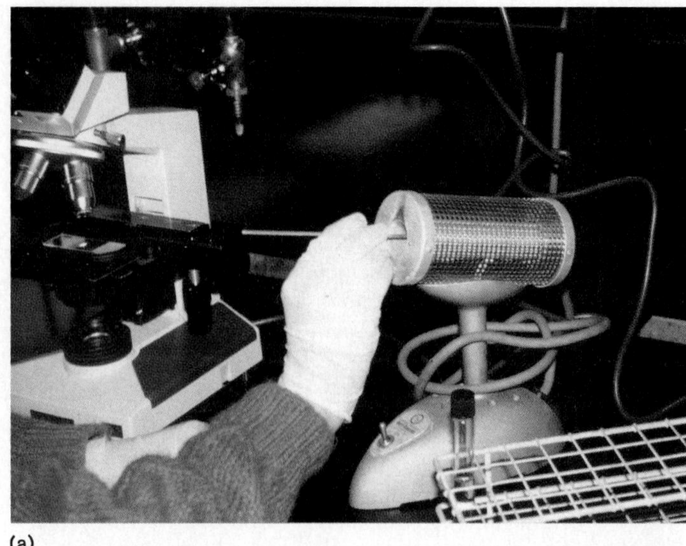
(a)

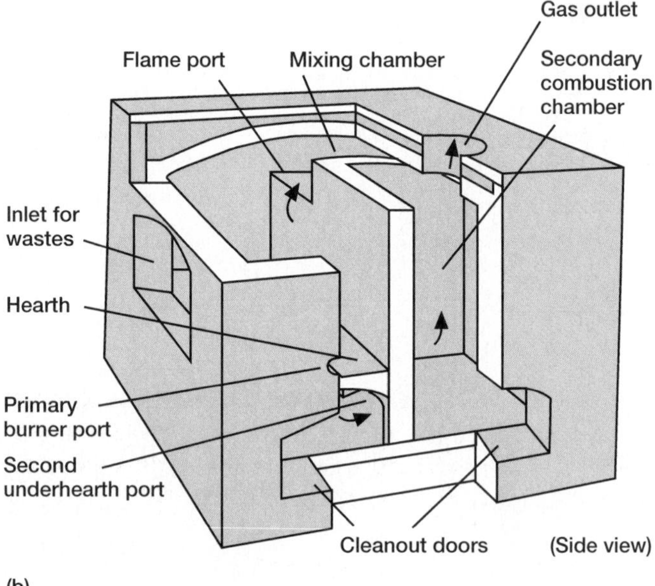

(b)

Figure 9.6 Methods of incineration. (*a*) Infrared incinerator with shield to prevent spattering of microbial samples during flaming. (*b*) Simplified diagram of an incinerator used to burn infectious wastes.
Source: (*b*) Danielsen, *Air Pollution Engineering Manual*, Environmental Protection Agency, 1973.

Applications of Incineration Incineration of microbial samples on inoculating loops and needles is a very common practice in the microbiology laboratory. No microbe on earth, sporeformer or otherwise, could possibly survive direct contact with the hottest part of a Bunsen burner flame. This method is fast and effective, but it is also limited to metals and heat-resistant glass materials. Incinerators (figure 9.6) are regularly employed in hospitals and research labs for complete destruction and disposal of infectious materials such as syringes, needles, cultural materials, dressings, bandages, bedding, animal carcasses, wastes, and pathology samples. However, furnace incineration poses some disadvantages: It contributes to air pollution, requires the use of disposable supplies, and sometimes does not completely combust large packets of materials.

The hot air oven provides another means of dry heat sterilization. The so-called **dry oven** is usually electric (occasionally gas) and has coils that radiate heat within an enclosed compartment. Heated, circulated air contacts the items, transferring its heat in the process. Sterilization is accomplished by exposure of items to 150°–180°C for 2 to 4 hours, which ensures thorough heating of the objects and destruction of spores.

Applications of the Dry Oven The dry oven is used in laboratories and clinics for heat-resistant items that do not sterilize well with moist heat. Substances appropriate for dry ovens are glassware, powders and oils that steam does not penetrate well, and metallic instruments that may be corroded by steam. This method is not suitable for plastics, cotton, and paper, which may burn at the high temperatures, or for solutions, which will dry out. Household ovens can be used to sterilize various metal utensils and glassware. The major limitation to this method is the time and temperature involved.

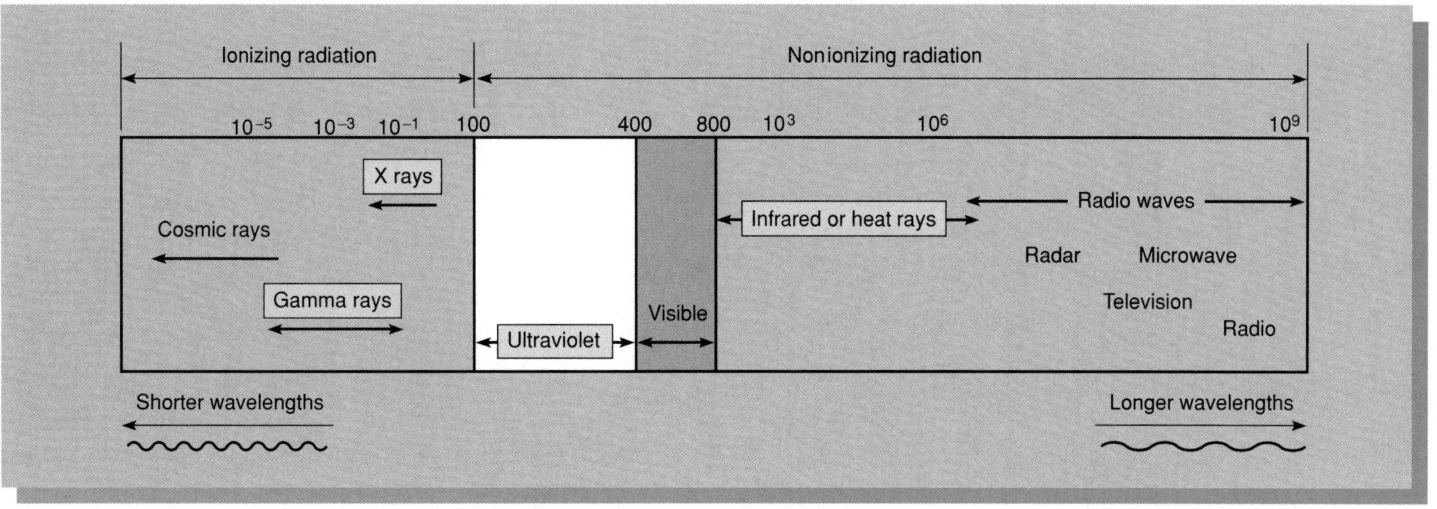

Figure 9.7 Electromagnetic radiations used in chemical control. Short wavelengths of gamma and X rays are the most manageable and practical forms of ionizing radiation. Nonionizing radiation, such as ultraviolet rays in the range of 240 to 280 nm and infrared (heat) rays, are fair microbicidal agents. Other waves are not routinely used in microbial control.

Miscellaneous Methods: Cold and Desiccation

The principal benefit of cold treatment is to slow microbial growth in the microbiology laboratory and in food during processing and storage. It must be emphasized that cold merely retards the activities of most microbes. While it is true that some microbes may be killed by cold temperatures, still others are not adversely affected by gradual cooling, long-term refrigeration, or deep-freezing. In fact, freezing temperatures, ranging from −70°C to −135°C, provide an environment that can preserve cultures of bacteria, viruses, and fungi for long periods. Some microbes (psychrophiles) grow very slowly even at freezing temperatures and may continue to secrete toxic products. Unawareness of these facts is probably responsible for numerous cases of food poisoning from frozen foods that have been defrosted at room temperature and then inadequately cooked. Pathogens able to survive several months in the refrigerator are *Staphylococcus aureus, Clostridium* species (sporeformers), *Salmonella* species, *Streptococcus* species, and several types of yeasts, molds, and viruses.

Vegetative cells directly exposed to normal room air gradually become dehydrated or *desiccated.* More delicate pathogens such as the pneumococcus, the spirochete of syphilis, and the gonococcus die after a few hours of air drying, but many others are not killed by drying and some are even preserved. Endospores of *Bacillus* and *Clostridium* are viable for thousands of years under extremely arid conditions. Vegetative forms show varying degrees of resistance to drying, depending on the substrate carrying them. Staphylococci and streptococci, if protected by dried secretions (pus, saliva), and the tubercle bacillus, when surrounded by droplets of sputum, can remain viable in air and dust for lengthy periods. Many viruses (especially nonenveloped) and fungal spores can also withstand long periods of desiccation. Desiccation can be a valuable way to preserve foods such as meats, vegetables, and fruits because it greatly reduces the amount of water available to support microbial growth.

It is interesting to note that a combination of freezing and drying—*lyophilization*—is a common method of preserving microorganisms and other cells in a viable state for many years. Pure cultures are frozen instantaneously on dry ice and exposed to a high vacuum that rapidly removes the water (it goes right from the frozen state into the vapor state). This avoids the formation of ice crystals that would damage the cells. Although not all cells will survive this process, enough of them do to permit future reconstitution of that culture.

As a general rule, chilling, freezing, and desiccation should not be construed as methods of disinfection or sterilization because their antimicrobial effects are erratic and uncertain, and one cannot be sure that they kill pathogens.

Radiation As a Microbial Control Agent

Another type of energy that exerts antimicrobial effects is radiation. For our purposes, **radiation** is defined as any form of energy emitted from atomic activities that travels at high velocity through matter or space. The two major forms of radiation used in microbial control are electromagnetic and particulate. **Electromagnetic radiation** is wavelike in its behavior, with a spectrum ranging from high-energy, short wavelength gamma rays at one extreme to low-energy, very long radio waves at the other (figure 9.7). For the most part, only electromagnetic radiations at the gamma ray, X ray (also called roentgen ray), and ultraviolet ray (UV) levels are suitable for microbial control. **Particle radiation** consists of subatomic particles such as electrons, protons, and neutrons that have been freed from the atom. The only particle radiation feasible for antimicrobial applications is the high-speed electron (also called

desiccate (des'-ih-kayt) To dry at normal environmental temperatures.

lyophilization (ly-off''-il-ih-za'-shun) Gr. *lyein,* to dissolve, and *philein,* to love.

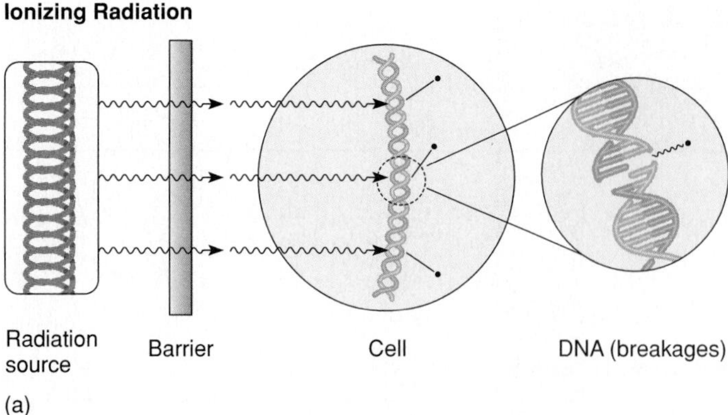

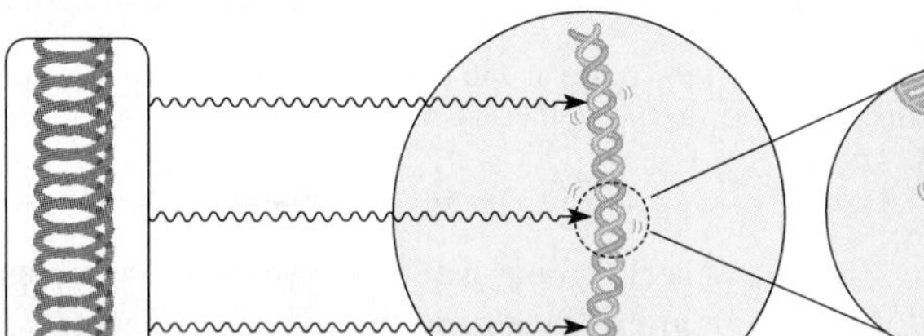

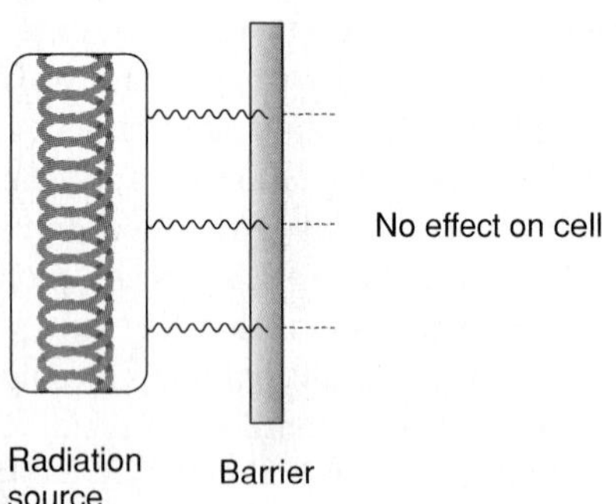

Figure 9.8 Cellular effects of irradiation. (*a*) Ionizing radiation can penetrate a solid barrier, bombard a cell, enter it, and dislodge electrons from molecules. Breakage of DNA creates massive mutations. (*b*) Nonionizing radiation enters a cell, strikes molecules, and excites them. The effect on DNA is mutation by formation of abnormal bonds. (*c*) A solid barrier cannot be penetrated by nonionizing radiation.

a β particle or cathode ray). Both electromagnetic and particle radiation exist naturally in the universe and can also be produced through artificial means.

Visible light does exert some antimicrobial effects, but it is not predictable enough in its effects to control all microbes. Infrared rays may be destructive by virtue of the heat they produce, but these rays have fewer applications than other heating methods. Microwaves also apparently kill microorganisms through heat production, but the antimicrobial effects of long radio waves are minimal.

Modes of Action of Ionizing Versus Nonionizing Radiation

To understand the actual physical effects of radiation on microbes, it will be helpful to visualize the process of *irradiation,* or bombardment with radiation, at the cellular level (figure 9.8). When a cell is bombarded by certain waves or particles, its molecules absorb some of the available energy, leading to one of two consequences: (1) If the radiation disrupts the atomic structure of molecules by dislodging and ejecting orbital electrons, the result is an electrical imbalance and the formation of ions. This is *ionizing radiation.* One of the most sensitive targets for ionizing radiation is the DNA molecule, which can sustain mutations on a broad scale. Secondary lethal effects appear to be chemical changes in organelles and the production of toxic substances. Gamma rays, X rays, and high-speed electrons are all ionizing in their effects. (2) *Nonionizing radiation,* best exemplified by UV, excites atoms by raising them to a higher energy state, but it does not ionize them. This atomic excitation, in turn, leads to abnormal linkages within molecules such as DNA, and is thus a source of mutations (see chapter 8).

Ionizing Radiation: Gamma Rays, X Rays, and Cathode Rays in Microbial Control

Over the past several years, ionizing radiation has become safer and more economical to use, and its applications have mushroomed. It is a highly effective alternative for sterilizing materials that are sensitive to heat or chemicals. Because it sterilizes in the absence of heat, radiation is sometimes termed *cold ster-*

ilization. Devices that emit ionizing rays include gamma ray machines containing radioactive cobalt, X-ray machines similar to those used in medical diagnosis, and cathode ray machines that operate like the vacuum tube in a television set. Items are placed in these machines and irradiated for a short time with a carefully chosen dosage. The dosage is measured in rads (**r**adiation **a**bsorbed **d**ose). Depending on the application, exposure ranges from 0.5 to 5 megarads (Mrads; a megarad is equal to 1,000,000 rads). Although all ionizing radiations can penetrate solids and liquids, gamma rays are most penetrating, X rays intermediate, and cathode rays least penetrating. The microbial forms with the greatest resistance to radiation are certain gram-positive cocci (*Deinococcus*), because of extremely efficient DNA repair mechanisms, and bacterial spores, due to their density. Viruses are second in resistance, followed by yeasts and molds. Vegetative bacterial cells, protozoa, worms, and insects are the most sensitive to radiation (table 9.5).

Table 9.5 Radiation Dosages Required to Kill Selected Microorganisms

Types of Microbe	Dosage (in Mrads)
Sporeformers	
Clostridium botulinum	0.33
Bacillus sphaericus	1.00
Vegetative Bacteria	
Salmonella typhimurium	0.02
Staphylococcus aureus	0.02
Dienococcus radiodurans	1.5
Miscellaneous Microbes	
Saccharomyces (yeast)	0.05
Penicillium (mold)	0.02
Vaccinia virus	0.17
Foot-and-mouth virus	1.3
Trichinella (worm)	0.05

Applications of Ionizing Radiation For 40 years, various agencies have been experimenting with ionizing radiation to disinfect and sterilize food. Two problems that inevitably occur with irradiated food are changes in flavor and nutrition and the possibility of introducing harmful by-products of chemical reactions. These disadvantages have been reduced by applying minimal dosages and by carrying out the process in very cold, oxygen-free chambers (figure 9.9). Foods that are sterilized using this method include cured meats and some spices and seasonings. More often, irradiation is used to cut down the microbial load, disinfect foods, and kill insects. Fresh pork may be irradiated to destroy *Trichinella* worms (the cause of trichinosis); chicken carcasses, to control *Salmonella*; wheat, to kill insect eggs; and fresh fruits and vegetables, to cut down on surface microbes that can increase the rate of spoilage.

Sterilizing medical products with ionizing radiation is a rapidly expanding field. Drugs, vaccines, ointments, powders, medical instruments (especially plastics), syringes, sutures, sponges, surgical gloves, and tissues such as bone, cartilage, skin, and heart valves for grafting all lend themselves to this mode of sterilization. Its main advantages include speed, high penetrating power (it can sterilize materials through outer packages and wrappings), and the absence of heat. Its main disadvantages are potential dangers to machine operators from exposure to radiation and possible damage to some materials. Contrary to popular belief, radiation used this way does not render food and other materials radioactive.

Figure 9.9 Irradiation machine for controlling fruit flies and bacteria on grapefruit. The effectiveness of this technique in preventing spoilage will make it a satisfactory substitute for pesticides.

Nonionizing Radiation: Ultraviolet Rays

Sunlight is the natural source of ultraviolet rays, and this accounts for its microbicidal effects, as demonstrated by the simple experiment in figure 9.10. Absorption of most of the sun's UV waves by the atmosphere prevents its full intensity from being felt and thus limits its practical use. Ultraviolet radiation ranges in wavelength from approximately 100 nm to 400 nm. It is most lethal from 240 nm to 280 nm (with a peak at 260 nm). In everyday practice, the source of UV radiation is the germicidal lamp, which generates radiation at 254 nm. Due to its lower energy state, UV radiation is not as penetrating as ionizing radiation. Because UV radiation passes readily through air, slightly through liquids, and only poorly through solids, the object to be disinfected must be directly exposed to it for full effect.

As UV radiation passes through a cell, it is initially absorbed by DNA, and this brings reproduction to a halt. Specific molecular damage occurs on the pyrimidine bases (thymine and

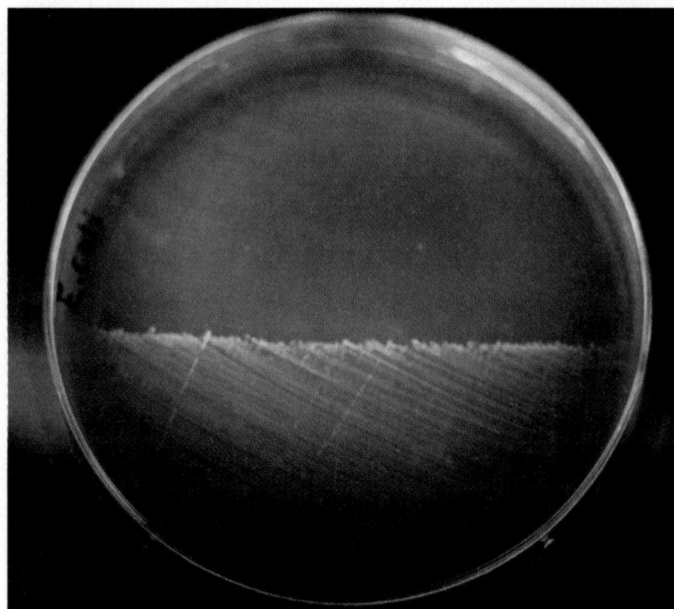

Figure 9.10 Effect of sunlight on bacterial growth. The plate was streaked on all parts with *E. coli.* The cover was removed, and half of the plate was exposed to direct sunlight for 30 minutes while the other half remained covered. Very few cells survived on the exposed side.

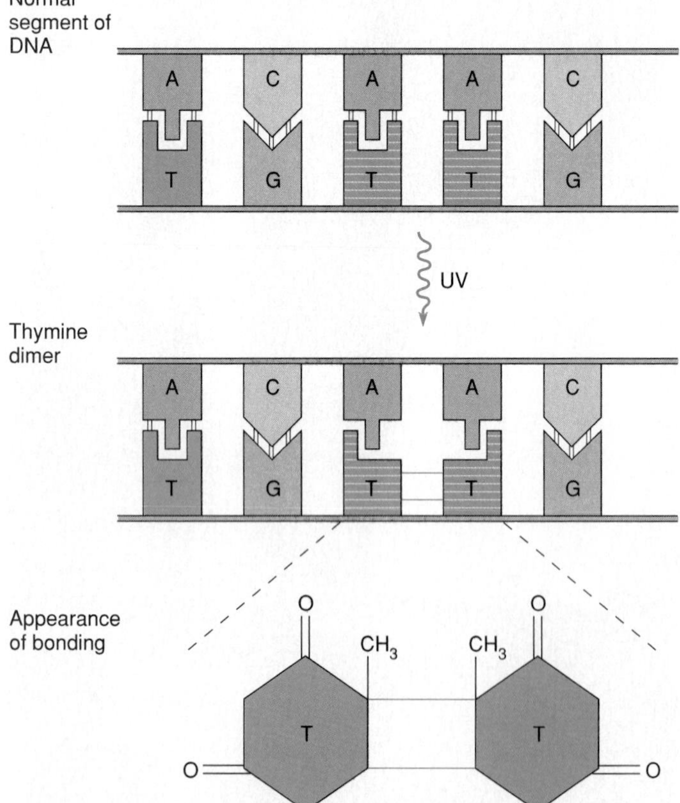

Figure 9.11 Formation of pyrimidine dimers by the action of ultraviolet (UV) radiation. This shows what occurs when two adjacent thymine bases on one strand of DNA are induced by UV rays to bond laterally with each other. The result is a thymine dimer. These dimers can also occur between adjacent cytosines and the thymine and cytosine bases. If they are not repaired, dimers can prevent that segment of DNA from being correctly replicated or transcribed. Massive dimerization is lethal to cells.

cytosine), which are so altered by the UV rays that they form abnormal linkages with each other called *pyrimidine dimers* (figure 9.11). The fact that they occur between adjacent bases on the same DNA strand and interfere with normal DNA replication and transcription accounts for the corresponding inhibition of growth and cellular death. In addition to altering DNA, UV radiation also disrupts cells by generating toxic photochemical products (free radicals). The universal damage inflicted by UV rays makes it a powerful tool for destroying fungal cells and spores, bacterial vegetative cells, protozoa, and viruses. Bacterial spores are about 10 times more resistant to radiation than vegetative cells, but they can be killed by increasing the time of exposure.

Applications of UV Radiation Control methods using UV radiation are directed primarily at disinfection rather than sterilization. They are used to cut down on airborne contaminants in hospital rooms, operating rooms, schools, food preparation areas, nursing homes, and military housing. UV disinfection of air has proved effective in reducing postoperative infections, in preventing the transmission of infections by respiratory droplets, and in curtailing the growth of microbes in food processing plants and slaughterhouses. In this procedure, germicidal lamps are positioned on the walls or ceiling to prevent undue exposure of room occupants; ultraviolet lamps can also be located in heating and air conditioning ducts to treat incoming air. Using this method, the concentration of airborne microbes is decreased as much as 99% of the original number. UV irradiation of liquids requires special equipment to spread the liquid into a thin, flowing film that is exposed directly to a lamp. This method may be used to treat drinking water (in place of chlorination) and to purify other liquids (milk, fruit juices, wine, and beer) as an alternative to heat. UV treatment has proved effective in freeing vaccine antigens and plasma from contaminants. The surfaces of solid, nonporous materials such as walls, floors, counters, chemical crystals, meat, nuts, tissues for grafting, and drugs have also been successfully disinfected with UV.

One major disadvantage of UV relates to its poor powers of penetration—it does not adequately penetrate opaque materials such as glass, metal, cloth, plastic, and even paper. Another drawback to UV is the damaging effect of overexposure on human tissues, including sunburn, retinal damage, cancer, and skin wrinkles.

Sound Waves in Microbial Control

High-frequency sound (sonic) waves beyond the sensitivity of the human ear are known to disrupt cells. These frequencies range from 15,000 to more than 200,000 cycles per second (supersonic to ultrasonic). Such vibrations transmitted through a water-filled chamber (sonicator) induce pressure changes and minute, bubblelike cavities in the liquid, a phenomenon called **cavitation.** The short-lived bubbles swell and collapse, creating intense points of turbulence that can stress and burst cells in the vicinity (figure 9.12). Gram-negative rods are most sensitive to ultrasonic vibrations, and gram-positive cocci, fungal spores, and bacterial spores are most resistant to them. Sonication not only helps destroy some microbes, but it also forcefully dislodges foreign matter from objects. Heat generated by sonic waves (up to 80°C) also appears to contribute to the antimicrobial action. Ultrasonic devices are used in dental and some medical offices to clear debris and saliva from instruments before sterilization and to clean dental restorations, but most

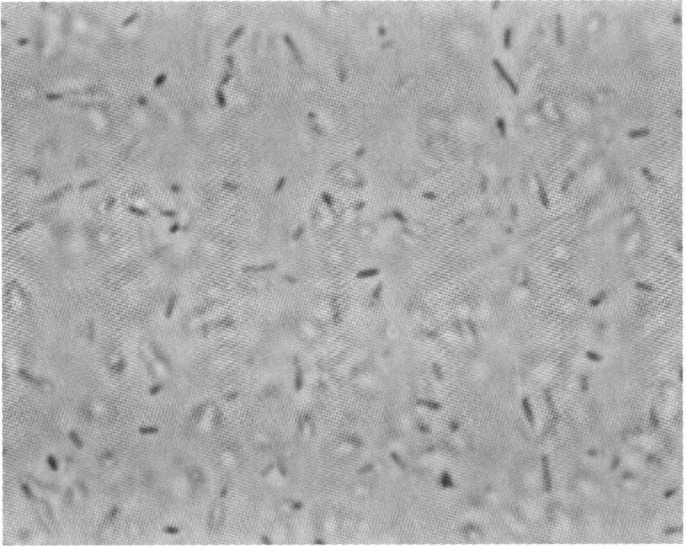

(a)

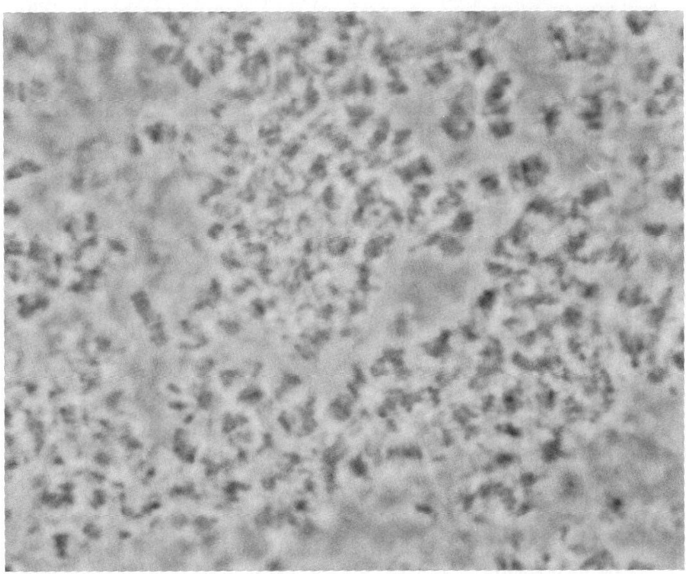

(b)

Figure 9.12 (*a*) Phase-contrast microscopic views of *E. coli* cells before being subjected to ultrasonic vibrations (sonication). (*b*) Masses of fragmented bacteria after sonication.

sonic machines are not predictable enough to be used in disinfection or sterilization. Other types of ultrasonic devices are available for medical diagnosis and for removing plaque and calculus from teeth.

Sterilization by Filtration: A Technique for Removing Microbes

In the search for means to control microbes in air and liquids, technology has not overlooked mechanical methods that remove instead of destroy them. The basis for this technique is simple: A fluid (air or liquid) is strained through a layer of material with openings large enough for the fluid to pass through but too small for microorganisms to pass through (figure 9.13).

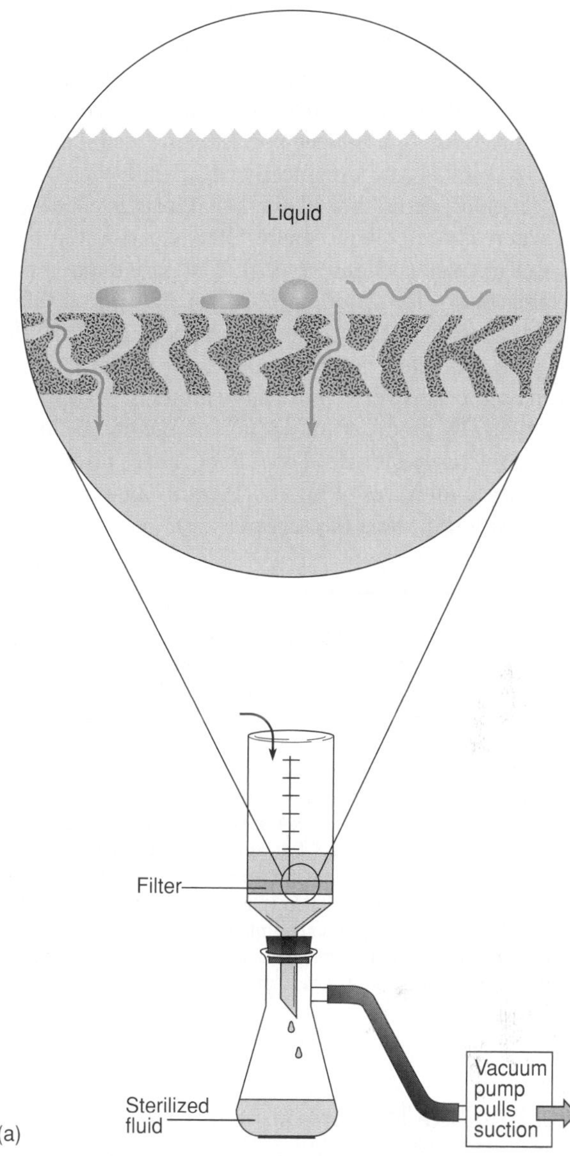

(a)

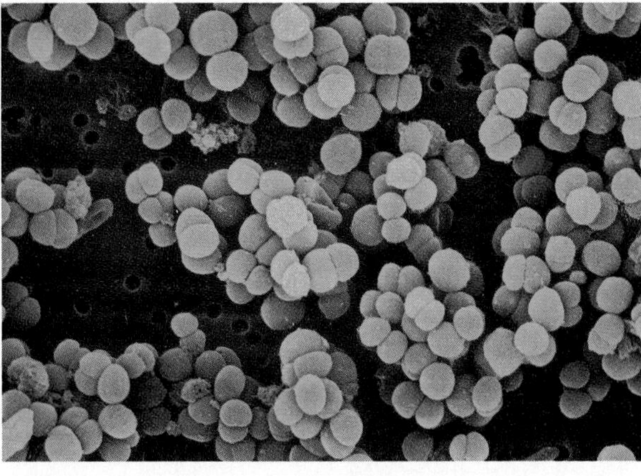

(b)

Figure 9.13 Membrane filtration. (*a*) Vacuum assembly for achieving filtration of liquids through suction. Inset shows filter as seen in cross section, with tiny passageways (pores) too small for the microbial cells to enter, but large enough for liquid to pass through. (*b*) Scanning electron micrograph of filter, showing relative size of pores and cocci trapped on its surface (×5,000).

Most modern microbiological filters are thin membranes of cellulose acetate, polycarbonate, and a variety of plastic materials (Teflon, nylon) whose pore size can be carefully controlled and standardized. Ordinary substances like charcoal, diatomaceous earth, or unglazed porcelain are also used in some applications. Viewed microscopically, the cellulose acetate filter consists of regularly cross-linked fibers that form irregular spaces (pores), whereas the polycarbonate filter is perforated by very precise, uniform holes (figure 9.13*b*). The pore diameters vary from coarse (8 μm) to ultrafine (0.02 μm), permitting selection of the minimum particle size to be trapped. Those with the smallest pore diameters permit true sterilization by removing viruses, and some will even remove large proteins. A sterile liquid filtrate is typically produced by suctioning the liquid through a sterile filter into a presterilized container. These filters are also used to separate mixtures of microorganisms and to enumerate bacteria in water analysis (see chapter 22).

Applications of Sterilization by Filtration The field of filtration sterilization is very broad. This method is used to prepare liquids that cannot withstand heat, including serum and other blood products, vaccines, drugs, IV fluids, enzymes, and media. Filtration has been employed to sterilize milk and other heat-labile beverages and to reduce the microbial load of water. It is also a valuable way to filter out particulate impurities (crystals, fibers, etc.) that can cause severe reactions in the body, though it has the disadvantage of not removing soluble molecules (toxins) that can cause disease and it may permit viruses to escape. Filtration is also an efficient means of removing airborne contaminants that are a common source of infection and spoilage. Fiberglass or cellulose filters are placed in air ducts to mechanically sieve out microbes and particulate matter as the air is continuously circulated by means of a fan. Particulate filters are also used to sterilize respiratory and anaesthetic gases and to provide a steady flow of sterile air in sterile rooms.

Chemical Agents in Microbial Control

Chemical control of microbes probably emerged as a serious science in the early 1800s, when physicians used chloride of lime and iodine solutions to treat wounds and wash hands prior to surgery. At the present time, approximately 10,000 different antimicrobial chemical agents are manufactured; probably 1,000 of these are utilized routinely in the allied health sciences and the home. Despite the genuine need to avoid infection and spoilage, the abundance of products available on the drugstore shelf to "kill germs, disinfect, antisepticize, clean and sanitize, deodorize, fight plaque, or purify the air" indicates a preoccupation with eliminating microbes from the environment that, at times, seems excessive.

Antimicrobial chemicals occur in the liquid, gaseous, or even solid state, and vary from disinfectants and antiseptics to sterilants and preservatives (chemicals that inhibit the deterioration of substances). Liquid agents generally have a base of water, alcohol, or a mixture of the two, into which various solutes are dissolved. Solutions containing pure water as the solvent are termed *aqueous*; whereas those with pure alcohol or water-alcohol mixtures are termed *tinctures*.

Choosing a Microbicidal Chemical

The choice and appropriate use of antimicrobial chemical agents is of constant concern in medicine and dentistry. Although actual clinical practices of chemical decontamination vary widely, some desirable qualities in a germicide have been identified, including: (1) rapid action even in low concentrations, (2) solubility in water or alcohol and long-term stability, (3) broad-spectrum microbicidal action without being toxic to human and animal tissues, (4) penetration of inanimate surfaces to sustain a cumulative or persistent action, (5) resistance to becoming inactivated by organic matter, (6) noncorrosive or nonstaining properties, (7) sanitizing and deodorizing properties, and (8) inexpensiveness and ready availability. As yet, no chemical can completely fulfill all of these requirements, but several do approach this ideal. At the same time, we should question the rather overinflated claims of certain commercial agents such as mouthwashes and disinfectant air sprays.

Germicides are evaluated in terms of their effectiveness in destroying microbes on medical and dental materials. The three levels of chemical decontamination procedures are ***high, intermediate,*** and *low* (table 9.6). High-level germicides kill endospores, and if properly used, are sterilants. Materials that necessitate high-level control are medical devices—for example, catheters, heart-lung equipment, and implants—that are not heat-sterilizable and are intended to enter body tissues during medical procedures. Intermediate-level germicides kill fungal (but not bacterial) spores, resistant pathogens such as the tubercle bacillus, and viruses. They are used to disinfect items (respiratory equipment, endoscopes) that come in intimate contact with the mucous membranes but are noninvasive. Low levels of disinfection eliminate only vegetative bacteria, vegetative fungal cells, and some viruses. They are required for materials such as electrodes, straps, and furniture that may touch the skin surfaces but not the mucous membranes.

Factors That Affect the Germicidal Activity of Chemicals

Factors that control the effect of a germicide include the nature of the microorganisms being treated, the nature of the material being treated, the degree of contamination, the time of exposure, and the strength and chemical action of the germicide (table 9.7). Standardized procedures for testing the effectiveness of germicides are summarized in appendix C. The modes of action of most germicides involve the cellular targets discussed in an earlier section of this chapter: proteins, nucleic acids, the cell wall, and the cell membrane.

A chemical's strength or concentration is expressed in various ways, depending upon convention and the method of preparation, and many chemical agents can be expressed by more than one notation. In dilutions, a small volume of the liquid chemical (solute) is diluted in a larger volume of water to achieve a certain ratio. For example, a common laboratory phenolic disinfectant such as Amphyl is usually diluted 1:200—that is, one part of chemical has been added to 200 parts of water by volume. Solutions such as chlorine that are effective in very high dilutions are expressed in parts per million (ppm). In percent so-

Table 9.6 Qualities of Chemical Agents Used in Allied Health

Agent	Target Microbes	Level of Activity	Toxicity	Comments
Chlorine	Sporicidal (slowly)	Intermediate	Gas is highly toxic; solution irritates skin	Inactived by organics; unstable in sunlight
Iodine	Sporicidal (slowly)	Intermediate	Can irritate tissue; toxic if ingested	Iodophors* are milder forms
Phenolics	Some bacteria, viruses, fungi	Intermediate to low	Can be absorbed by skin; can cause CNS damage	Poor solubility; expensive
Alcohols	Most bacteria, viruses, fungi	Intermediate	Toxic if ingested; a mild irritant; dries skin	Inflammable, fast-acting
Hydrogen peroxide,* stabilized	Sporicidal	High	Toxic to eyes; toxic if ingested	Greater stability; works well in organic matter
Quaternary ammonium compounds	Some bactericidal, virucidal, fungicidal activity	Low	Irritating to mucous membranes; poisonous if taken internally	Weak solutions can support microbial growth; easily inactivated
Soaps	Certain very sensitive species	Very low	Very bland; few if any toxic effects	Used for removing soil
Mercurials	Only weakly microbistatic	Low	Highly toxic if ingested, inhaled, absorbed	Easily inactivated
Silver nitrate	Bactericidal	Low	Toxic, irritating	Discolors skin
Glutaraldehyde*	Sporicidal	High	May irritate skin; toxic if absorbed	Not inactivated by organic matter; unstable
Formaldehyde	Sporicidal	High to intermediate	Very irritating; fumes damaging, carcinogenic	Slow rate of action
Ethylene oxide gas*	Sporicidal	High	Very dangerous to eyes, lungs; carcinogenic	Explosive in pure state; good penetration; materials must be aerated
Dyes	Weakly bactericidal, fungicidal	Low	Low toxicity	Stain materials, skin
Chlorhexidine*	Most bacteria, some viruses, fungi	Low to intermediate	Low toxicity	Fast-acting, mild, has residual effects

*These forms approach the ideal by having some of the following characteristics: broad spectrum, low toxicity, fast action, penetrating abilities, residual effects, stability, potency in organic matter, and solubility.

lutions, the solute is added to water by weight or volume to achieve a certain percentage in the solution. Alcohol, for instance, is used in percentages ranging from 50% to 95%. In general, solutions of low dilution or high percentage have more of the active chemical (are more concentrated) and tend to be more germicidal, but expense and potential toxicity may necessitate using the minimum strength that is effective.

Another factor that contributes to germicidal effectiveness is the length of exposure. Most compounds require adequate contact time to allow the chemical to penetrate and to act on the microbes present. The composition of the material being treated must also be considered. Smooth, solid objects are more reliably disinfected than those with pores, pockets, or spaces that can trap soil. An item contaminated with common biological matter such as serum, blood, saliva, pus, tissue fluid, fecal material, or urine presents a problem in disinfection. Large amounts of organic material can hinder the penetration of a disinfectant and, in some cases, form bonds that reduce its activity. Germicides containing chlorine, mercurials, dyes, and some detergents are inactivated to some extent in the presence of a high load of organic substances.

Nothing Substitutes for Good, Clean Techniques Each decontamination practice in a hospital or clinic must be critically evaluated on the basis of the potential risk for patient infection. For example, sterilants (ethylene gas) used to sterilize syringes, needles, and surgical instruments must be applied for an adequate time. Quick exposure will not ensure the degree of destruction needed to sterilize. Even intermediate-level disinfection with iodine and low-level disinfection with detergents require methodical and scientific practices. Of considerable concern at all levels is that items be cleaned prior to their chemical treatment. Removal of microbial contaminants on instruments and other reusable materials ensures that the germicide or sterilant will better accomplish the job for which it was chosen. Simply dousing heavily contaminated materials with stronger chemicals will not ensure any level of disinfection and, consequently, can never replace competent techniques.

Germicidal Categories According to Chemical Group

Several general groups of chemical compounds are widely used for antimicrobial purposes in medicine and commerce (table 9.6). Prominent agents include halogens, heavy metals, alcohols, phenolic compounds, oxidizers, aldehydes, detergents, and

Table 9.7 Required Concentrations and Times for Chemical Destruction of Selected Microbes

Organism	Concentration	Time
Agent: Aqueous Iodine		
*Staphylococcus aureus**	2%	2 min
*Escherichia coli***	2%	1.5 min
Enteric viruses	2%	10 min
Agent: Chlorine		
*Mycobacterium tuberculosis**	50 ppm	50 sec
Entamoeba cysts (protozoa)	0.1 ppm	150 min
Hepatitis A virus	3 ppm	30 min
Agent: Phenol		
Staphylococcus aureus	1:85 dil	10 min
Escherichia coli	1:75 dil	10 min
Agent: Ethyl Alcohol		
Staphylococcus aureus	70%	10 min
Escherichia coli	70%	2 min
Poliovirus	70%	10 min
Agent: Hydrogen Peroxide		
Staphylococcus aureus	3%	12.5 sec
Neisseria gonorrhoeae	3%	0.3 sec
Herpes simplex virus	3%	12.8 sec
Agent: Quaternary Ammonium Compound		
Staphylococcus aureus	450 ppm	10 min
*Salmonella typhi***	300 ppm	10 min
Agent: Silver Ions		
Staphylococcus aureus	8 μg/ml	48 hrs
Escherichia coli	2 mg/ml	48 hrs
Candida albicans (yeast)	14 mg/ml	48 hrs
Agent: Glutaraldehyde		
Staphylococcus aureus	2%	$<$ 1 min
Mycobacterium tuberculosis	2%	$<$ 10 min
Herpes simplex virus	2%	$<$ 10 min
Agent: Ethylene Oxide Gas		
Streptococcus faecalis	500 mg/l	2–4 min
Influenza virus	10,000 mg/l	25 hrs
Agent: Chlorhexidine		
Staphylococcus aureus	1:10 dil	15 sec
Escherichia coli	1:10 dil	30 sec

*Gram-positive vegetative bacterium.
**Gram-negative vegetative bacterium.

gases. These groups will be surveyed in the following section from the standpoint of each agent's specific forms, modes of action, indications for use, and limitations.

The Halogen Antimicrobial Chemicals

The **halogens** are fluorine, bromine, chlorine, and iodine, a group of nonmetallic elements that commonly occur in minerals, seawater, and salts. Although they may exist in either the ionic (halide) or nonionic state, most halogens exert their antimicrobial effect primarily in the nonionic state, not the halide (chloride, iodide, for example). Because fluorine and bromine are difficult and dangerous to handle and no more effective than chlorine and iodine, only the latter two are used routinely in germicidal preparations. These elements are highly effective components of disinfectants and antiseptics because they are microbicidal and not just static, and they are sporicidal with longer exposure. It is no wonder that these halogens are the active ingredients in nearly one-third of all antimicrobial chemicals currently marketed.

halogens (hay'-loh-jenz) Gr. *halos,* salt, and *gennan,* to produce.

Chlorine and Its Compounds Chlorine has been used for disinfection and antisepsis for approximately 200 years. The major forms used in microbial control are liquid and gaseous chlorine (Cl_2), hypochlorites (OCl), and chloramines (NH_2Cl). In solution, these compounds combine with water and release a very active substance, hypochlorous acid (HOCl). This substance oxidizes the sulfhydryl (S–H) group on the amino acid cysteine and interferes with disulfide (S–S) bridges on numerous enzymes. The resulting denaturation of the enzymes is permanent and suspends metabolic reactions. Chlorine kills not only bacterial cells and endospores, but also fungi and viruses. The major limitations of chlorine compounds are: (1) They are ineffective if used at an alkaline pH; (2) excess organic matter can greatly reduce their activity; and (3) they are relatively unstable, especially if exposed to light.

Applications of Chlorine Pure **elemental chlorine** is an extremely toxic gas that must be stored in steel cylinders during transport and handling. Gaseous and liquid chlorine are used almost exclusively for large-scale disinfection of drinking water, sewage, and waste water from such sources as agriculture and industry. To ensure that water is safe to drink, it is chlorinated to a concentration of 0.6 to 1.0 parts of chlorine per million parts of water. This rids the water of pathogenic vegetative microorganisms without unduly affecting its taste (some persons may debate this).

Hypochlorites are perhaps the most extensively used of all chlorine compounds. These are available as sodium and calcium salts dissolved in water to a concentration of 0.5–70%. The scope of applications is broad, including sanitization and disinfection of food equipment in dairies, restaurants, and canneries, and treatment of swimming pools, spas, drinking water, and even fresh foods. Hypochlorites are used in the allied health areas to treat wounds, irrigate root canals, and disinfect equipment, bedding, and instruments. Common household bleach is a weak solution (5%) of sodium hypochlorite that serves as an all-around disinfectant, deodorizer, and stain remover.

Chloramines (dichloramine, halazone) are being employed more frequently as an alternative to pure chlorine in treating water supplies. Because standard chlorination of water is now believed to produce unsafe levels of cancer-causing substances such as trihalomethanes, some water districts have been directed by federal agencies to adopt chloramine treatment of water supplies. This changeover could initially create some difficulties for kidney dialysis patients and tropical fish enthusiasts because, though chloramines do not produce trihalomethanes, they have toxic effects if absorbed by the blood during hemodialysis and they damage fish gills. Chloramines also serve as sanitizers and disinfectants and for treating wounds and skin surfaces.

Iodine and Its Compounds Iodine is a pungent black chemical that forms brown-colored solutions when dissolved in water or alcohol. Soon after its discovery in 1812, it was rapidly adopted by medical fields as an antiseptic and later as a disinfectant. The two primary iodine preparations are *free iodine* in solution (I_2) and *iodophors*. Iodine rapidly penetrates the cells of microorganisms, where it apparently disturbs a variety of metabolic functions by interfering with the hydrogen and disulfide bonding of proteins (similar to chlorine). All classes of microorganisms are killed by iodine if proper concentrations and exposure times are used. Iodine activity is not as adversely affected by organic matter and pH as chlorine is.

Applications of Iodine Solutions Free iodine solutions occur in three forms. **Aqueous iodine** contains 2% iodine and 2.4% sodium iodide; it is used as a topical antiseptic prior to surgery and occasionally as a treatment for burned and infected skin. Strong iodine solution (5% iodine and 10% potassium iodide) is used primarily as a disinfectant because of its potency. It is an effective disinfectant for plastic items, rubber instruments, cutting blades, catheters, and thermometers. **Iodine tincture** is a 2% solution of iodine in sodium iodide and dilute alcohol. It is a powerful skin germicide, especially when a high degree of asepsis is needed. Because iodine can be extremely irritating to the skin and toxic when absorbed, strong aqueous solutions and tinctures (5–7%) are no longer considered safe for routine antisepsis. Further disadvantages include iodine's discoloring, corrosive, and malodorous qualities. Drinking water and wastewater are occasionally treated with iodine compounds, though this practice is too costly for widespread use. Iodine tablets are available for disinfecting water during emergencies or destroying pathogens in impure water supplies.

Iodophors are complexes of iodine and a neutral polymer such as a polyvinylalcohol. This formulation allows the slow release of free iodine and increases its degree of penetration. These compounds have largely replaced free iodine solutions in medical antisepsis because they are less prone to staining or irritating tissues. Common iodophor products marketed as Betadine, Povidone (PVP), and Isodine contain 2–10% of available iodine and are used as medical and dental antiseptics, degerming agents, and disinfectants. Applications include preparing skin for surgery, treating burns and vaginal infections, mouthwashing, preparing skin and mucous membranes for injections, surgical handscrubbing, and disinfecting equipment and surfaces in general.

Phenol and Its Derivatives

Phenol (carbolic acid) is an acrid, poisonous compound derived from the distillation of coal tar. First adopted by Joseph Lister in 1867 as a surgical germicide, phenol was the major antimicrobial chemical until other phenolics with fewer toxic and irritating effects were developed about 50 years later. Solutions of phenol are now used only in certain limited cases, but it remains one standard against which other phenolic disinfectants are rated. Substances chemically related to phenol are often referred to as phenolics. Hundreds of these chemicals are now available.

Phenolics consist of one or more aromatic carbon rings with added functional groups (figure 9.14). Among the most important are alkylated phenols (cresols), chlorinated phenols, and bisphenols. The mode of action of these compounds has been thoroughly studied. In high concentrations, they are cellular poisons, rapidly disrupting cell walls and membranes and precipitating proteins, whereas in lower concentrations, they inactivate certain critical enzyme systems. The phenolics are strongly microbicidal, and will destroy vegetative bacteria (in-

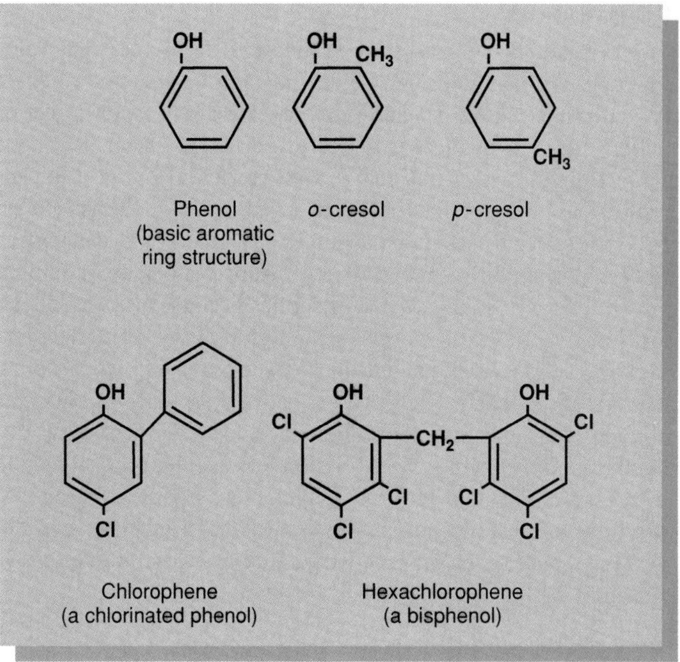

Figure 9.14 Some phenolics. All have a basic aromatic ring, with added groups such as Cl and CH_3.

cluding the tubercle bacillus), fungi, and most viruses (not the hepatitis B virus), but they are not reliably sporicidal. Their continued activity in the presence of organic matter and their detergent actions contribute to their usefulness. Unfortunately, the toxicity of many of the phenolics makes them too dangerous to use as antiseptics. And because they are only weakly soluble in water, many preparations must be mixed with soap for maximum effect.

Applications of Phenolics Phenol itself is still used for general disinfection of drains, cesspools, and animal quarters, but it is seldom applied as a medical disinfectant or antiseptic. The cresols are simple phenolic derivatives that are combined with soap for intermediate or low levels of disinfection in the hospital. Lysol and creolin, in a 1–3% emulsion, are common household versions of this type.

The bisphenols are also widely employed in commerce, clinics, and the home. One type, orthophenyl phenol, is the major ingredient in disinfectant aerosol sprays. This same phenolic is also found in some proprietary compounds (Amphyl, O-syl) often used in hospital and laboratory disinfection. One particular bisphenol, hexachlorophene, is known to possess high germicidal activity, especially against gram-positive pathogens in the genera *Staphylococcus* and *Streptococcus*. Up until 1972, this compound was added to numerous cleansing soaps. The germicidal soaps pHisoHex and Hexagerm (3% hexachlorophene) had been used for about 10 years in surgical handscrubbing, to prevent infections in burn units, and in bathing newborn infants to prevent staphylococcal infections. When it was discovered that hexachlorophene is absorbed through the skin and can cause neurological damage, it was no longer sold over the counter in cosmetics or soaps, used indiscriminately on burn patients, or applied in the newborn nursery. Although it has been more or less replaced by chlorhexidine, hexachlorophene is periodically employed in the hospital for hand cleaning or controlling outbreaks of skin infections.

Chlorhexidine

The compound chlorhexidine (Hibiclens, Hibitane) is a complex organic base containing chlorine and two phenolic rings. It is somewhat related to the cationic detergents in its mode of action, being a surfactant and a protein denaturant. At moderate to high concentrations, it is bactericidal for both gram-positive and gram-negative bacteria, but inactive against spores. Its effects on viruses and fungi vary. At decreased concentrations, chlorhexidine is merely microbistatic. It possesses distinct advantages over many other antiseptics because of its mildness, low toxicity, and rapid action. Although it binds to the skin surface and shows a residual antimicrobial effect over several hours, it is not absorbed into deeper tissues to any extent. Alcoholic or aqueous solutions of chlorhexidine are now commonly used for handscrubbing, preparing skin sites for surgical incisions and injections, and whole body washing. Chlorhexidine solution also serves as an obstetric antiseptic, a neonatal wash, a wound degermer, a mucous membrane irrigant, and a preservative for eye solutions.

Alcohols As Antimicrobial Agents

Alcohols are colorless hydrocarbons with one or more —OH functional groups. Of several alcohols available, only ethyl (2 carbons) and isopropyl (3 carbons) are suitable for microbial control. Methyl alcohol (1 carbon) is not very microbicidal, and longer chain alcohols are either poorly soluble in water or too expensive for routine use. Alcohols are employed alone in aqueous solutions or as solvents for other antimicrobial chemicals (mercurials and iodine, for example). Alcohol's mechanism of action depends in part upon its concentration. Concentrations of 50% and above dissolve membrane lipids, disrupt cell surface tension, and compromise membrane integrity. Alcohol that has entered the protoplasm denatures proteins through coagulation, but only in alcohol-water solutions of 50–95%. Absolute alcohol (100%) dehydrates cells and inhibits their growth, but is generally not a protein coagulant.

Though useful in intermediate- to low-level germicidal applications, alcohol does not destroy bacterial spores at room temperature. There are numerous examples in the literature of spores surviving for months in alcoholic solutions. Alcohol can, however, destroy resistant vegetative forms, including tubercle bacilli and fungal spores, provided the time of exposure is adequate. Alcohol is generally more effective in inactivating enveloped viruses than the more resistant nonenveloped viruses such as poliovirus and hepatitis B virus.

Applications of Alcohols Ethyl alcohol, also called ethanol or grain alcohol, is known for its relatively germicidal, nonirritating, nontoxic, and inexpensive characteristics. Solutions of 70% to 95% are routinely used as skin degerming agents because the surfactant action removes skin oil, soil, and even some microbes sheltered in deeper skin layers. Ethyl alcohol is occasionally used to disinfect electrodes, face masks, and thermometers, which are first cleaned and then soaked in alcohol for 15–20 minutes. Isopropyl alcohol, sold as rubbing alcohol, is even more microbicidal and less expensive than ethanol, but these benefits must be weighed against its toxicity. It must be used with caution in disinfection or skin cleansing, because inhalation of its vapors may adversely affect the nervous system.

Hydrogen Peroxide and Related Germicides

Hydrogen peroxide (H_2O_2) is a colorless, caustic liquid that decomposes in the presence of light, metals, or catalase into water and oxygen gas. Peroxide solutions have been used for about 50 years, but early formulations were unstable and inhibited by organic matter. These flaws earned it a reputation as an unsatisfactory germicide. However, manufacturing methods now permit synthesis of H_2O_2 so stable that even dilute solutions retain activity through several months of storage.

The germicidal effects of hydrogen peroxide were originally attributed to the oxidation of sulfhydryl bonds. However, current thought suggests that hydrogen peroxide is more likely to participate in the formation of hydroxyl free radicals (.OH), which, like the superoxide radical (see chapter 6), are highly reactive and toxic to cells. Although most microbial cells produce catalase to inactivate the small amounts of hydrogen peroxide produced normally during their own metabolism, this small quantity of catalase cannot neutralize the amount of hydrogen peroxide entering the cell during disinfection and antisepsis. Hydrogen peroxide is bactericidal, virucidal, and fungicidal, and in higher concentrations, also sporicidal.

Applications of Hydrogen Peroxide As an antiseptic, 3% hydrogen peroxide serves a variety of needs, including skin and wound cleansing, bedsore care, and mouthwashing. It is especially useful in treating infections by anaerobic bacteria because of the lethal effects of oxygen on these forms. Hydrogen peroxide also has versatile applications for disinfecting inanimate materials: soft contact lenses, surgical implants, plastic equipment, utensils, bedding, and room interiors. Solutions of 6% to 25% hydrogen peroxide are potent enough to sterilize and are being considered for sterilization of food packaging equipment and even the interiors of spaceships.

Other compounds that have effects similar to those of hydrogen peroxide are: (1) ozone (O_3), occasionally used to disinfect air and water; (2) peracetic acid, an extremely potent oxidizer that is employed to control fungal growth on fruits and to sterilize pacemakers and isolation chambers for animals; and (3) potassium permanganate ($KMnO_4$), a bright pink solution, strongly microbicidal even when very dilute. Its manganese content makes it rather toxic, so it is now used mainly as an algicide.

Chemicals with Surface Action: Detergents

Detergents are complex organic substances that, as a group, are often termed surfactants. Chemical types are nonionic, anionic, and cationic detergents. The nonionic (uncharged) detergents are not really effective germicides. Most anionic (negatively charged) detergents have limited microbicidal power. Soaps belong to this group. Cationic detergents are the most effective of the three, especially the quaternary ammonium compounds (usually shortened to *quats*).

All detergents have a general molecular structure that includes a long chain hydrocarbon residue and a highly charged or polar group (figure 9.15). Given this configuration, detergents operate by lowering cellular surface tension. This can have several effects, but chief among them is the disruption of the cell membrane and the loss of its selective permeability. Qua-

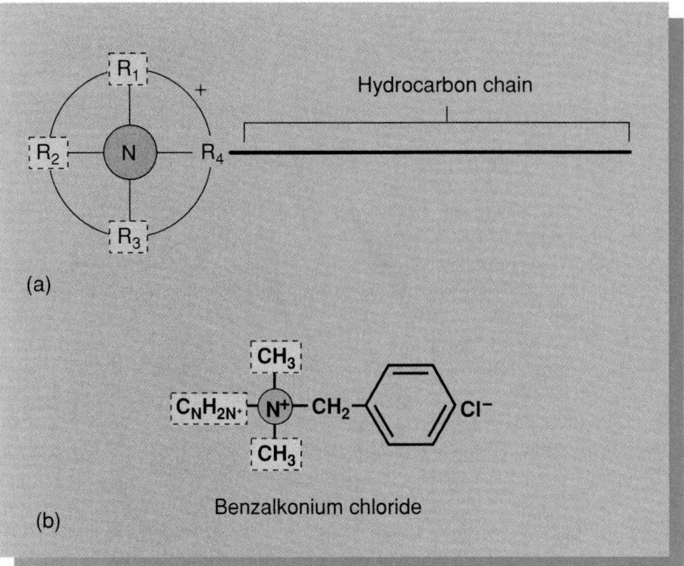

Figure 9.15 (*a*) In general, detergents are polar molecules with a positively charged head and at least one long, uncharged hydrocarbon chain. The head contains a central nitrogen nucleus with various alkyl (R) groups attached. (*b*) The structure of a common quaternary ammonium detergent, benzalkonium chloride.

ternary ammonium compounds kill microbes by causing leakage of microbial protoplasm, precipitating proteins, and inhibiting metabolism. Because of their ability to interact with surfaces, detergents make good wetting agents, cleansing agents, and emulsifiers.

The range of activity of detergents is varied. Quaternary ammonium compounds, if used at medium concentrations, are effective against some gram-positive bacteria, viruses, fungi, and algae. In low concentrations, they have only microbistatic effects. The quats are ineffective against the tubercle bacillus, hepatitis virus, *Pseudomonas,* and spores at any concentration. Their activity can be greatly reduced in the presence of organic matter, and they function best in alkaline solutions. As a result of their limitations, quats are rated only for low-level disinfection in the clinical setting.

Applications of Detergents Quaternary ammonium compounds include benzalkonium chloride, Roccal, Zephiran, and cetylpyridinium chloride (Ceepryn). In dilutions ranging from 1:100 to 1:1,000, these are commonly mixed with cleaning agents to simultaneously disinfect and clean hospital floors, walls, furniture, equipment surfaces, and public restrooms. Although these compounds were once used to disinfect surgical and dental instruments, their level of disinfection is far too low for this application. Their detergent properties and low toxicity make quats among the best sanitizers; they are used to clean restaurant eating utensils, food processing equipment, dairy equipment, and clothing. Benzalkonium chloride is frequently incorporated as a preservative for ophthalmic solutions and cosmetics.

Soaps are alkaline compounds made by combining the fatty acids in oils (coconut, castor, cottonseed, linseed) with sodium or potassium salts. In usual practice, soaps are only weak microbicides, and they destroy only highly sensitive forms such as the gonococcus, the meningococcus, and the spirochete of syphilis. The common hospital pathogen *Pseudomonas* is so unaffected by soap that various species grow abundantly in soap dishes. Soaps function primarily as cleansing agents and sanitizers in industry and the home. The superior sudsing and wetting properties of soaps help to mechanically remove large amounts of surface soil, greases, and other debris that contains microorganisms. Soaps gain greater germicidal value when mixed with agents such as alcohol (called green soap), chlorhexidine, or iodine, and they may be used for cleaning instruments prior to heat sterilization, degerming patients' skin, routine handwashing by medical and dental personnel, and preoperative handscrubbing. Vigorously brushing the hands with germicidal soap over a 15-minute period is an effective way to remove dirt, oil, and surface contaminants as well as some resident microbes, but it will never sterilize the skin (see feature 9.2 and figure 9.16).

Heavy Metal Compounds

Various forms of the metallic elements mercury, silver, gold, copper, arsenic, and zinc have been applied in microbial control over several centuries. These are often referred to as heavy metals because of their relatively high atomic weight. However, from this list, only preparations containing mercury and silver still have any significance as germicides. Although some metals (zinc, iron) may actually be needed in small concentrations for the metabolic processes of microorganisms, metals with a higher molecular weight (mercury, silver, gold) have no beneficial cellular function and are in fact very toxic, even in minute quantities (parts per million). This property of having antimicrobial effects in exceedingly small amounts is called an *oligodynamic action* (figure 9.17). Heavy metal germicides contain either an inorganic or organic metallic salt and come in the form of aqueous solutions, tinctures, ointments, or soaps.

Mercury, silver, and most other metals exert microbicidal effects by forming ions that complex with several functional groups, including sulfhydryls, hydroxyls, amines, and phosphates. The resulting inactivation of proteins rapidly brings metabolism and growth to a standstill (see figure 9.4). This mode of action can destroy many types of microbes, including vegetative bacteria, fungal cells and spores, algae, protozoa, and viruses (but not endospores).

Unfortunately, there are several drawbacks to using metals in microbial control: (1) Metals are very toxic to humans if ingested, inhaled, or absorbed through the skin, even in small quantities, for the same reasons that they are toxic to microbial cells; (2) they commonly cause allergic reactions; (3) large quantities of biological fluids and wastes neutralize or depress their actions; and (4) microbes may develop resistance to metals. Health and environmental considerations have dramatically reduced the use of metallic antimicrobial compounds in medicine, dentistry, commerce, and agriculture.

oligodynamic (ol″-ih-goh-dy-nam′-ik) Gr. *oligos,* little, and *dynamis,* power.

Feature 9.2 The Quest for Sterile Skin

More than a hundred years ago, before sterile gloves were a routine part of medical procedures, the hands remained bare during surgery. Realizing the danger from microbes, medical practitioners deemed that sterilizing the hands of surgeons and their assistants would prevent surgical infections. Several stringent (and probably very painful) techniques involving strong chemical germicides and vigorous scrubbing were practiced. Here are a few examples:

In Schatz's method, the hands and forearms were first cleansed by brisk scrubbing with liquid soap for 3–5 minutes, then soaked in a saturated solution of permanganate at a temperature of 110°F until they turned a deep mahogany brown. Next, the limbs were immersed in saturated oxalic acid until the skin became decolorized. Then, as if this were not enough, the hands and arms were rinsed with sterile limewater and washed in warm bichloride of mercury for 1 minute.

Or, there was Park's method (more like a torture). First, the surfaces of the hands and arms were rubbed completely with a mixture of cornmeal and green soap to remove loose dirt and superficial skin. Next, a paste of water and mustard flour was applied to the skin until it began to sting. This potion was rinsed off in sterile water, and the hands and arms were then soaked in hot bichloride of mercury for a few minutes, during which the solution was rubbed into the skin.

Another method once earnestly suggested for getting rid of microorganisms was to expose the hands to a hot-air cabinet to "sweat the germs" out of skin glands. Pasteur himself advocated a quick flaming of the hands to maintain asepsis.

The old dream of sterilizing the skin was finally reduced to some basic realities: The microbes entrenched in the epidermis and skin glands cannot be completely eradicated even with the most intense efforts, and the skin cannot be sterilized without also seriously damaging it. Because this is true for both medical personnel and their patients, the chance always exists that infectious agents can be introduced during invasive medical procedures. Fortunately, the Goodyear Tire and Rubber Company developed rubber gloves in 1890, making it possible to place a sterile barrier around the hands. Of course, this did not mean that skin cleansing and antiseptic procedures were abandoned or downplayed. A thorough scrubbing of the skin, followed by application of an antiseptic, is still needed to remove the most dangerous source of infections, the superficial contaminants picked up by touching the environment.

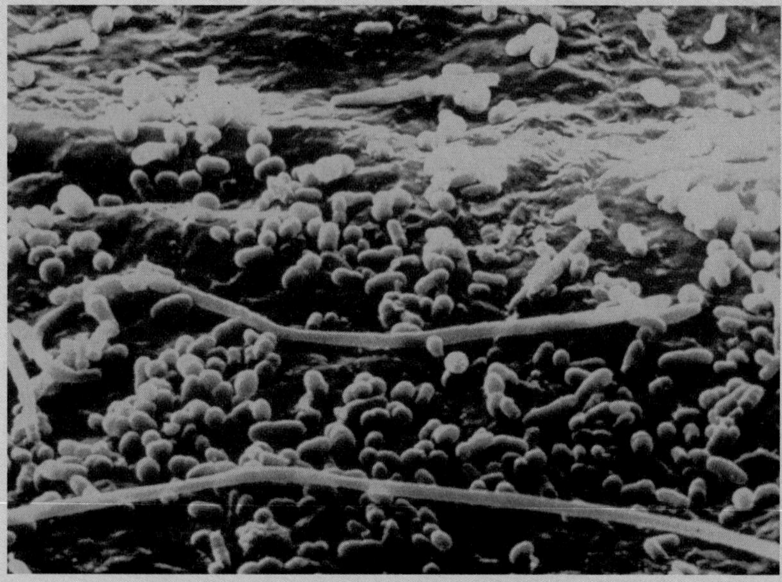

(a)

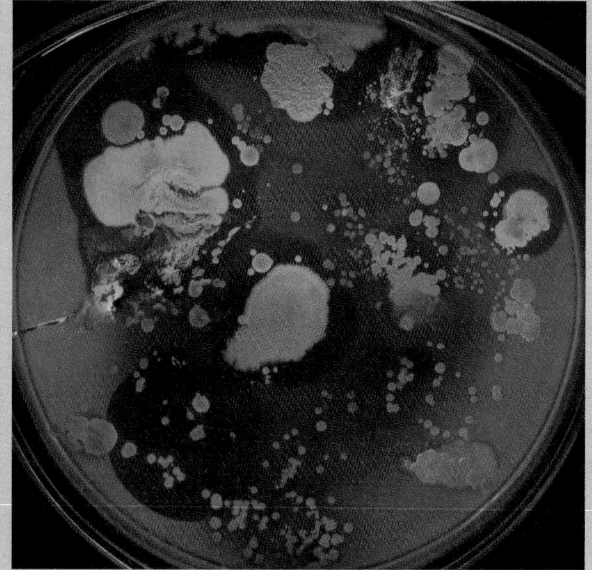

(b)

Microbes on normal unwashed hands. (*a*) A scanning electron micrograph of a fingertip shows clusters of bacteria perched atop a fingerprint ridge (×47,000). (*b*) Heavy growth of microbial colonies on a plate of blood agar. This culture was prepared by passing an open sterile plate around a classroom of 30 students, and having each one touch its surface. After incubation, a mixed population of bacteria and fungi appeared. Clear zones in agar may be indicative of pathogens.

Applications of Heavy Metals Weak (0.001% to 0.2%) organic mercury tinctures such as thimerosal (Merthiolate) and nitromersol (Metaphen) are fairly effective antiseptics and infection preventatives, but they should never be used on broken skin because they are harmful and can delay healing. Ointments are still employed on a limited basis for treating eye infections, ringworm, and other parasitic skin infections. The organic mercurials also serve as preservatives in cosmetics and ophthalmic solutions. Mercurochrome, that old staple of the medicine cabinet, is now considered among the poorest of antiseptics.

The silver compound that remains in common usage is silver nitrate ($AgNO_3$). Crede introduced it in the late nineteenth century for preventing gonococcal infections in the eyes of newborn infants who had been exposed to an infected birth canal. With Crede's technique, drops of a 1% solution are instilled onto the conjunctival surface for a short period, followed by a rinse with saline to reduce irritation. This preparation is not used as often as it once was because other infectious agents (*Chlamydia,* for example) that pose a danger for neonatal eye infections are resistant to $AgNO_3$.

Solutions of silver nitrate are used as topical germicides on mouth ulcers and to disinfect cavities on teeth surfaces, though this can discolor teeth. Silver sulfadiazine ointment, when added to dressings, effectively prevents infection in second- and third-degree burn patients. Colloidal silver preparations (Argyn and Neosilvol) contain silver salts complexed to protein. Because they gradually release silver ions, they are much milder and less toxic than inorganic salts, but are restricted to use as mild germidical ointments or rinses for the mouth, nose, eyes, and vagina.

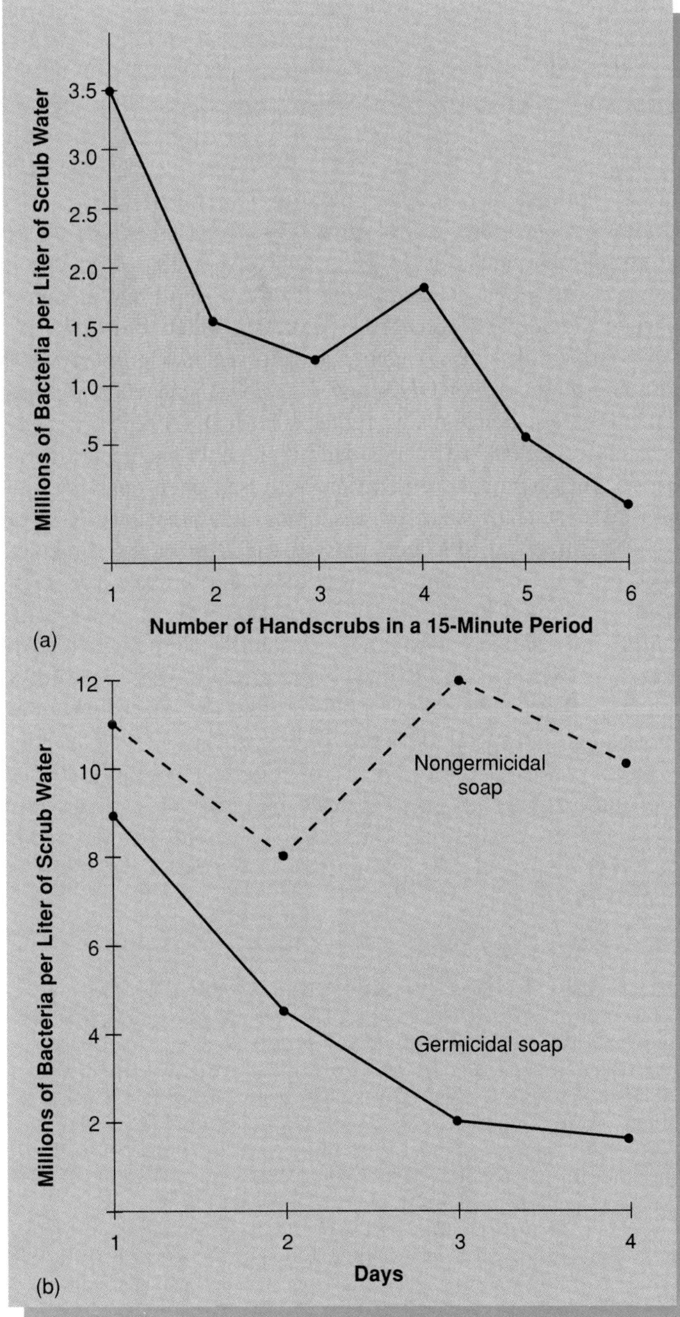

Figure 9.16 Graphs showing effects of handscrubbing. (*a*) Reduction of microbes on the hands during a single (15 min) episode of handwashing with nongermicidal soap. The increase in the fourth scrub is due to a high level of resident microbes uncovered in that skin layer. (*b*) Comparison of scrubbing over several days with a nongermicidal soap versus a germicidal soap. Germicidal soap has persistent effects on skin over time, keeping the microbial count low. Without germicide, soap does not show this sustained effect.

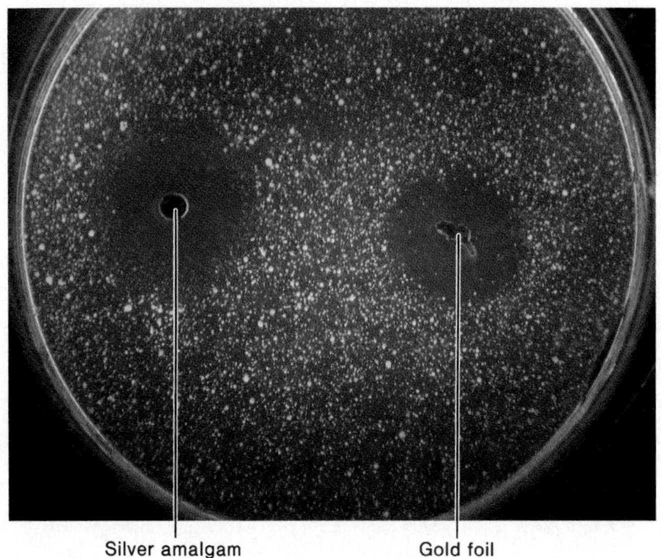

Figure 9.17 Demonstration of the oligodynamic action of heavy metals. A pour plate inoculated with saliva has small fragments of heavy metals pressed lightly into it. During incubation, clear zones indicating growth inhibition developed around both fragments. The slightly larger zone surrounding the amalgam probably reflects the synergistic effect of the two metals (silver and mercury) it contains.

Aldehydes As Germicides

Organic substances bearing a —CHO functional group (a strong reducing group) on the terminal carbon are called aldehydes. Several common substances such as sugars and some fats are technically aldehydes. The two aldehydes used most often in microbial control are **glutaraldehyde** and **formaldehyde.**

Glutaraldehyde is a yellow acidic liquid with a mild odor. The molecule's two aldehyde groups favor the formation of polymers. The mechanism of activity against microbes is not completely understood, but glutaraldehyde appears to cross-link protein molecules on the cell surface through alkylation of amino acids, a process in which a hydrogen atom on an amino acid is replaced by the glutaraldehyde molecule itself (figure 9.18). It can also irreversibly disrupt the activity of enzymes within the cell. Glutaraldehyde is a rapid, broad-spectrum antimicrobial chemical and one of the few chemicals officially accepted as a sterilant and high-level disinfectant. It kills spores in 3 hours and fungi and vegetative bacteria (even *Mycobacterium* and *Pseudomonas*) in a few minutes. Viruses, including the most resistant forms, appear to be inactivated after relatively short exposure times. Glutaraldehyde has other advantages besides these. It retains its potency even in the presence of organic matter, is noncorrosive, does not damage plastics, and is less toxic or irritating than formaldehyde. Its principal disadvantage is that it is somewhat unstable, especially with increased pH and temperature.

Formaldehyde is a sharp, irritating gas that readily dissolves in water to form an aqueous solution called *formalin.* Full saturation of formaldehyde (37%) produces a solution of 100% formalin. The chemical is microbicidal through its attachment to nucleic acids and functional groups of amino acids. Formalin is an intermediate- to high-level disinfectant, although it acts

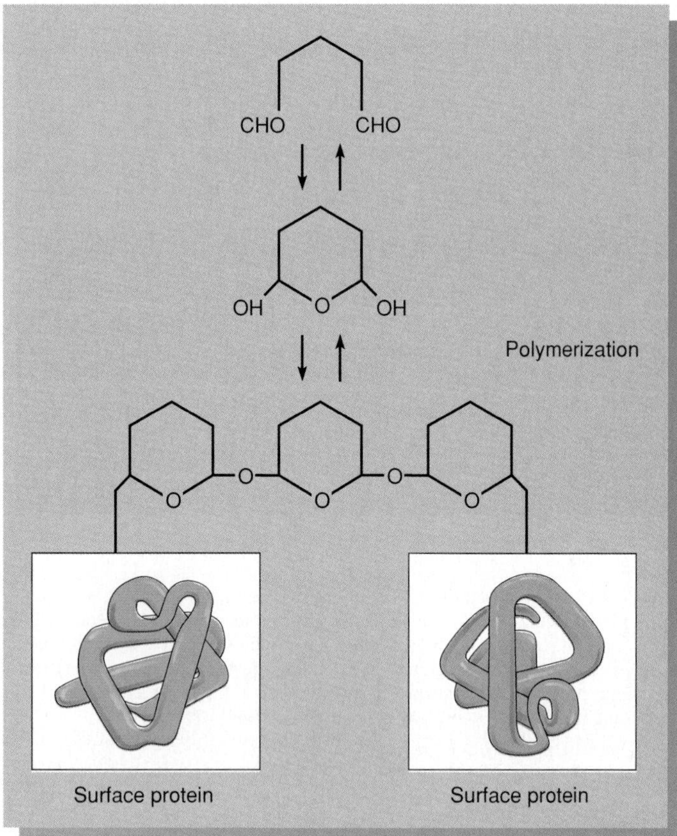

Figure 9.18 Actions of glutaraldehyde. The molecule forms a closed ring that can polymerize. When these strong alkylating agents react with amino acids, they cross-link and inactivate proteins.

more slowly than glutaraldehyde. Formaldehyde's extreme toxicity (it is classified as a carcinogen) and irritating effects on the skin and mucous membranes greatly limit its clinical usefulness.

Applications of the Aldehydes Glutaraldehyde was introduced in 1963 as a substitute for formalin in chemically sterilizing materials that are damaged by heat. Commercial products (Cidex, Glutarol) are diluted to 2% and activated to sterilize respiratory therapy equipment, hemostats, fiberoptic endoscopes (laparoscopes, arthroscopes), and anesthetic devices. Glutaraldehyde is employed in dental offices for practical disinfection of instruments (when heat sterilization is not possible or necessary) because of its ability to destroy the hepatitis B virus. It is an effective, though expensive, alternative environmental disinfectant in medicine and industry. Glutaraldehye can also be used to preserve vaccines, sanitize poultry carcasses, and degerm cow's teats.

Formalin can be diluted in alcohol (8%) or water (1–8%) for various applications. Formalin tincture has limited use as a disinfectant for surgical instruments, and 1% formalin solution is still used to disinfect some reusable kidney dialysis instruments. Any object that will come in intimate contact with the body must be thoroughly rinsed with sterile water to remove the formalin residue. It is, after all, one of the active ingredients in embalming fluid.

Gaseous Sterilants and Disinfectants Processing inanimate substances with chemical vapors and aerosols provides a unique and versatile alternative to heat or liquid chemicals. A hundred years ago, highly dangerous and reactive compounds such as sulfur dioxide and chlorine gases were employed to fumigate large spaces. Discovery of their adverse effects on people and materials led to the introduction of other products. Currently, those gases having the broadest applications are ethylene oxide (ETO), propylene oxide, and betapropiolactone (BPL).

Ethylene oxide is a colorless substance that exists as a gas at normal temperatures. It is very explosive in air, a feature that can be eliminated by combining it with a high percentage of carbon dioxide or fluorocarbon. Like the aldehydes, ETO is a very strong alkylating agent, and it reacts vigorously with guanine molecules of DNA and functional groups of proteins. Through these mechanisms, it blocks both DNA replication and enzymatic actions. Ethylene oxide is the only gas generally accepted for chemical sterilization because, when employed according to strict procedures, it is a biocide. A specially designed ETO sterilizer called a chemiclave, a variation on the autoclave, is equipped with a chamber, gas ports, and temperature, pressure, and humidity controls (figure 9.19). ETO is rather penetrating but relatively slow acting. Depending on the temperature and gas mixture used, sterilization requires from 90 minutes to 3 hours. Some items absorb ETO residues and must be aerated for several hours after exposure to ensure dissipation of as much residual gas as possible. For all of its effectiveness, ETO has some unfortunate features. Its explosiveness makes it dangerous to handle; it can damage the lungs, eyes, and mucous membranes if contacted directly; and it is rated as a carcinogen by the government.

Applications of Gases and Aerosols Ethylene oxide was first used to disinfect sugar, spices, and tobacco. It was later adopted for treating drugs and foodstuffs and was even considered as a possible means to sterilize blood, serum, and culture media. ETO is currently used primarily to sterilize and disinfect plastic materials and delicate instruments in hospitals and industries. ETO in concentrations of 450–800 mg/l (Carboxide, Cryoxide) can safely sterilize prepackaged heart pacemakers, artificial heart valves, surgical supplies, syringes, and disposable petri dishes. It is still used as a sterilant for spices and dried foods.

Propylene oxide is a close relative of ETO, with similar physical properties and mode of action, although it is less toxic. Because it breaks down into a relatively harmless substance, it is safer than ETO for sterilization of foods (nuts, powders, starches, spices) and for disinfectants.

Betapropiolactone (BPL) is a nonexplosive liquid that also alkylates DNA. Although it is even more microbicidal than the oxides, it is not nearly as penetrating. This feature and its high toxicity to animals greatly restrict its applications. Betapropiolactone is used for disinfecting buildings and sterile rooms, disinfecting instruments, sterilizing bone and arterial grafts, and inactivating viruses in vaccines.

Dyes As Antimicrobial Agents

Dyes are important in staining techniques and as selective and differential agents in media; they are also a primary source of certain drugs used in chemotherapy. Because aniline dyes such as crystal violet and malachite green are very active against gram-positive species of bacteria and various fungi, they are incorporated into solutions and ointments to treat skin infections (ringworm, for example). The yellow acridine dyes, acriflavine

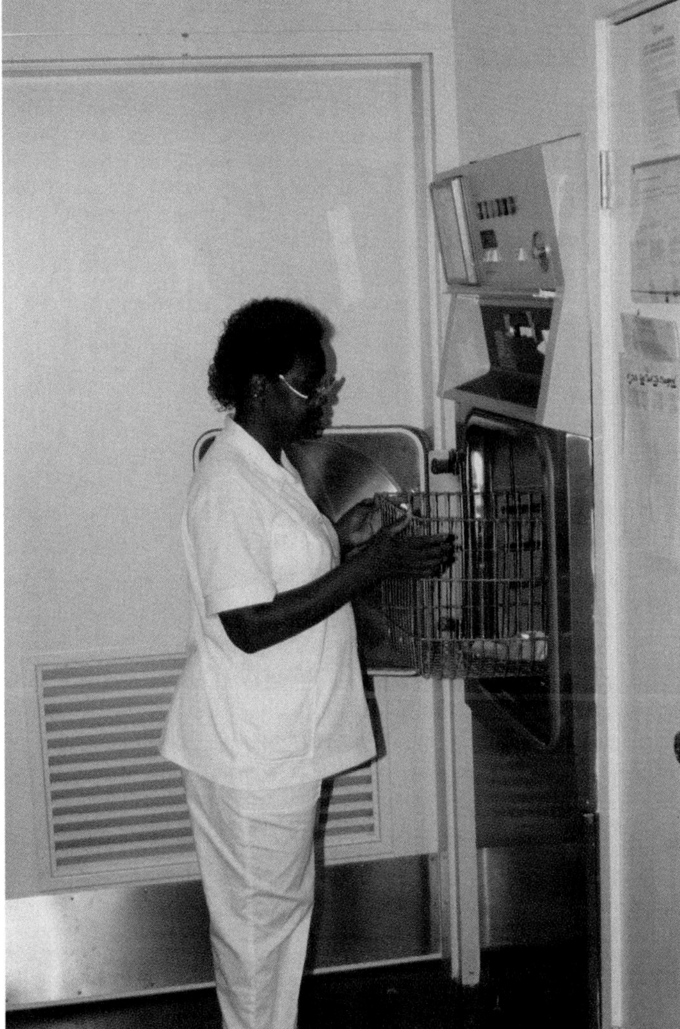

(a)

Air intake
Air filter
Vacuum pump
Chamber
Door
Gas mixer
ETO cylinder
CO_2 cylinder

(b)

Figure 9.19 (*a*) An automatic ethylene oxide sterilizer. (*b*) The machine is equipped with gas canisters containing ethylene oxide (ETO) and carbon dioxide, a chamber to hold items, and mechanisms that evacuate gas and introduce air.

Feature 9.3 How Can You Sterilize the World?

Most human beings manage to remain healthy despite the fact that they live in continual intimate contact with microorganisms. We really don't have to be preoccupied with microbes every minute or feel overly concerned that the things we touch, drink, or eat are sterile, as long as they are somewhat clean and free of pathogens. For most of us, resistance to infection is well maintained by our numerous host defenses; but such is not the case for everyone.

In the last 20 years, largely due to increasingly sophisticated technological advances, more babies with severe combined immune deficiency disease (SCID) have survived. SCID children are unable to develop natural immune cells and antibodies, thus they are in constant mortal danger from infections by common, even "innocuous" microbes that exist all around us (see chapter 14). Just so they can be kept alive, they are delivered by cesarian section (to maintain sterility) and then isolated in a sterile environment.

Many SCID children have received successful bone marrow transplants, a procedure that offers them freedom and a normal life—if they can overcome one more problem. For many months after the transplant, as the children are gradually developing immunities, they cannot be exposed to the usual levels of microbial contaminants. At first they are kept in strict isolation, and all personnel and visitors are antiseptically scrubbed and gowned before entering the room. Later, the children are allowed to go home, but with many restrictions and preventative routines.

The parents of SCID children are under tremendous pressure to keep their surroundings as germ-free as possible; at times, they must feel compelled to sterilize the world. Everything that passes the child's lips, touches the hands or skin, or could be inhaled in the air requires evaluation. It is frequently necessary for both parents and child to be gowned and masked. Direct physical contact, outside visitors, and playmates must be discouraged. Toys and hands are fastidiously and frequently treated with germicides, and eating utensils, as well as all food and water, are sterilized. The house requires constant and thorough disinfection and vacuuming, and even then, the child cannot crawl on the floors. Every little sniffle can be a cause for panic. In the next few years, we will continue to see dramatic changes in life support systems and means of keeping critically ill persons alive, not the least of which will be the discovery of new modes of antimicrobial control that may help severely compromised patients survive.

and proflavine, are sometimes utilized for antisepsis and wound treatment in medical and veterinary clinics. Overall, though, dyes will probably continue to have limited applications because they stain and have a narrow spectrum of activity.

Acids and Alkalies

Conditions of very low or high pH can destroy or inhibit microbial cells; however, most inorganic acids and alkalies are far too corrosive, caustic, and hazardous to use in microbial control. Aqueous solutions of ammonium hydroxide remain a common component of detergents, cleansers, and deodorizers. Organic acids are widely used in food preservation because they prevent spore germination and bacterial and fungal growth, and because they are generally regarded as safe to eat. Acetic acid (in the form of vinegar) is a pickling agent that inhibits bacterial growth; propionic acid is commonly incorporated into breads and cakes to retard molds; lactic acid is added to sauerkraut and olives to prevent spoilage by anaerobic bacteria (especially the clostridia), and benzoic and sorbic acids are added to beverages, syrups, and margarine to inhibit yeasts.

Chapter Review with Key Terms

Physical Control of Microorganisms: Antimicrobial and Decontamination Methods

Moist Heat

Hot water or steam; may be used in **disinfection** or **sterilization.** Mode of action: Denaturation of proteins, destruction of membranes and DNA.

For Sterilization: (1) **Autoclave** uses steam under pressure (15 psi/121°C/10–40 min). Applied on heat-resistant materials that steam can penetrate, including cloth, glassware, metallic items, media, rubber, certain foods. Not recommended for oils, powders, heat-sensitive fluids, plastics (that melt).

(2) **Intermittent sterilization** uses free-flowing, unpressurized steam, 100°C applied for 30–60 minutes on three successive days. Used for substances that cannot be autoclaved, especially media with egg, serum, sugars, and certain canned foods.

For Disinfection: (1) **Boiling** water baths at 100°C used to destroy pathogens (not spores) on dishes, clothing, utensils, instruments, baby's articles, glassware, bedding; does not kill spores; materials may become recontaminated.

(2) **Pasteurization** is application of heat less than 100°C to liquids; batch method in bulk tanks heated to 62°C for 30 minutes; flash method in thin layers heated to 71°C for 15 seconds. Used to destroy non-spore-forming milk-borne pathogens, such as *Salmonella* and *Listeria* and/or to reduce spoilage by lowering the overall microbial count; also applied to beer, wine, fruit juices. Only reduces microbial content; many thermodurics survive. Can change flavor, nutritional content of food.

Dry Heat

Hot air for sterilization or **decontamination.**

Mode of Action: Depending on the temperature, combustion to ashes, oxidation, dehydration, coagulation of proteins.

(1) **Incineration**: Open flame—materials are placed for a few seconds in a very hot (800°–1,800°C) flame. Used to flame tips of inoculating implements, sterilize test tubes in a Bunsen burner.

(2) Incineration in furnace: Incinerator chamber equipped with very high temperature flame (600°–1,200°C) that burns items and microbes to ashes. Used for disposing hospital and industrial wastes containing heavy amounts of contaminants. Must be used only on disposable items; may add pollutants to atmosphere.

(3) **Dry Oven**: Materials exposed to 150°–180°C for 2–4 hours. Used for sterilizing empty glassware, metal instruments, needles, oils, waxes, powders. Too vigorous for liquids containing water, plastics, paper, cloth; takes too long compared to autoclaving.

Cold Temperatures

Refrigeration 0°–15°C) or freezing (below 0°C), for microbial inhibition or preservation.

Mode of Action: **Microbistasis** slows the growth rate of most microorganisms. Used for maintenance of foods, increasing shelf life of drugs, chemicals; for preserving the viability of microbial cultures. Not a mode of disinfection or sterilization; many pathogens survive chilling and freezing.

Drying/Desiccation

Gradual withdrawal of water from cells by exposure to room air.

Mode of Action: Concentration of cellular solutes, leading to metabolic inhibition. Used in preparing dehydrated foods. Does not kill pathogens, especially if embedded in protective vehicles such as saliva, pus, feces; not really effective for infection control.

Radiation/Irradiation

Energy in the form of waves (**electromagnetic**) and particles that can be transmitted through space. Used in **cold sterilization.**

Ionizing: High-energy, short waves and particles that can dislodge electrons from atoms. Types are gamma rays, X rays, cathode rays (high-speed electrons) used in **irradiation.** Mode of action: Primary—direct damage to DNA chain, causing breaks and mutations. Secondary—formation of substances that poison the cell. Used to sterilize heat-sensitive medical materials such as plastics, drugs, instruments, tissue grafts, vaccines; also to disinfect and increase storage time of fresh meats, poultry, fish, fruits, vegetables; very penetrating. May alter flavor and nutrition; still somewhat more expensive than other methods.

Nonionizing: Moderate energy, medium-length waves that excite, do not ionize, atoms. Type is ultraviolet; irradiation source is germicidal lamp. Mode of action: Acts on DNA molecule, forming **dimers** between **adjacent pyrimidines**; production of toxic products. Used primarily for disinfection; reduction of microbial load of air in clinics, hospitals, schools, military housing, industry; destruction of microbes in water, sera, vaccines, drugs; disinfection of solid surfaces in food preparation. Does not penetrate glass, paper, plastic, wood, metal; can cause damage to human tissues (skin, eyes).

Sound Waves

Very high-frequency waves—super and ultrasonic. Mode of action: Creates strong vibration and massive turbulence; disrupts cells. Uses are limited; mainly for cleaning and disinfection of instruments.

Filtration

Physical removal of microbes from liquids and air by means of fine filters; filter contains pores that permit passage of the fluid but trap microbes; used to sterilize or disinfect. Filters include cellulose acetate, glass, plastics; contain pores that allow medium to pass. Used for sterilizing heat-sensitive liquids such as serum, vaccines, blood products, IV drugs, some media, milk, water; also for removing microbes from air in hospital rooms, isolation units, drug and food preparation areas. Does not filter out toxic secretions of microbes.

Chemical Control of Microorganisms

General Uses: Disinfectants, **antiseptics, sterilants,** preservatives, **sanitizers, degermers.**

Physical States: Solutions come in **aqueous** form (chemicals dissolved in water) and **tincture** form (chemicals dissolved in alcohol); solutions expressed as dilutions, percents, and parts/million. In gases, chemical is in vapor or aerosol phase.

Halogens

Chlorine: Forms are Cl_2, hypochlorites, chloramines. Mode of action: Denaturation of proteins by disrupting disulfide bonds. Spectrum of action: All microbial types, including endospores, with adequate time. Applications: Elemental chlorine (1ppm) is added to water and wastewater to control pathogenic vegetative pathogens; hypochlorites (chlorine bleach) are used extensively for disinfection and sanitization in medicine, dentistry, the food industry, and the home; chloramines are clinical disinfectants and antiseptics and alternative water disinfection agents. Limitations: Chemical action may be retarded by high levels of organic matter; may be unstable at basic pHs and in light.

Iodine: Forms: Free iodine (I_2) and iodophors (iodine complexed to organic polymers). Mode of action: Interference with protein interchain bonds, causing denaturation. Spectrum of action: Broad—sporicidal, though usually not used for sterilization. Applications: Weak solutions—topical antiseptic for prepping skin; iodine tincture, 2%, for high degree of surgical asepsis; strong solutions (5–7%)—disinfection of plastic, rubber, glass, some metal instruments. **Iodophors**—aqueous solutions with 2–10% iodine (Betadine, Povidone) are milder, yet effective components of medical and dental degerming (handscrubbing, skin prepping) agents, disinfectants, and ointments. Limitations: Iodine solutions stain, corrode, and have a strong odor. Strong solutions are too irritating and toxic to use on tissue.

Phenolics:

Forms: Cresols and bisphenols. Mode of action: Disruption of cell membrane, protein precipitators. Spectrum of action: **Bactericidal, fungicidal, virucidal,** but not **sporicidal.** Applications: Mainly as disinfectants. All forms remain active even in biological fluids, wastes; phenol is an old disinfectant that is too toxic for most uses; 1–3% cresol in soap (Lysol) is a common housekeeping disinfectant and cleaner. Orthophenyl phenol is a milder phenolic used in air sprays and some disinfectants (Amphyl). Limitations: Most are too toxic for use as antiseptics and are relatively insoluble.

Chlorhexidine

Forms: Hibiclens, Hibitane. Mode of action: Surfactant, protein denaturant. Spectrum of activity: Bactericidal, some antiviral and antifungal effects, not sporicidal. Applications: In an aqueous or alcoholic solution, is a skin degerming agent for handscrubbing and preoperative preps; also a wash for newborns, wounds, and burns. Product is relatively mild, nontoxic, and fast-acting.

Alcohols

Forms: Ethyl and Isopropyl, usually in solutions of 50–95%. Modes of action: Dissolution of membrane lipids (surfactant) and protein coagulation; pure (100%) alcohol slowly dehydrates cells. Spectrum of action: Narrow; mainly on bacterial vegetative cells and fungi; not sporicidal. Applications: Ethyl alcohol (70–95%) is a germicide in the clinic, laboratory, and home for skin degerming and sometimes disinfection and sanitization of equipment; isopropyl alcohol has similar uses, though it is more toxic and less safe. Limitations: Does not inactivate polio and hepatitis B virus.

Hydrogen Peroxide

Forms: Weak to strong hydrogen peroxide solutions. Mode of action: Produces highly active hydroxyl-free radical that damages proteins and DNA molecules; decomposes to water and O_2 gas, which can be toxic. Spectrum of action: Broad; strong solutions are sporicidal. Applications: 3% hydrogen peroxide is used for skin and wound antisepsis, care of mucous membrane infections; also for disinfection of equipment, utensils, implants; 6–25% can be used if sterilization of equipment is required. Limitations: Decomposes in presence of light and catalase.

Detergents

Forms: Cationic (**quaternary ammonium compounds—quats**) and soaps. Mode of action: **Surfactant**—lower surface tension of cells, disrupt membranes, and alter permeability; quats also precipitate proteins. Spectrum of activity: Somewhat erratic; have limited bactericidal and fungicidal action; are not sporicidal. Applications: Quats (benzalkonium and cetylpyridinium chlorides), diluted 1:100–1:1000, are environmental disinfectants and cleansers in clinics and laboratories, sanitizers in the food industry, and preservatives for pharmaceuticals; soaps are not very microbicidal but function in the mechanical removal of soil on skin, utensils, and environmental surfaces. Limitations: do not destroy the tubercle bacillus, hepatitis B virus; are inactivated by large quantities of organic matter.

Heavy Metal Compounds

Mercury and silver, inorganic and organic solutions and tinctures. Mode of action: **Oligodynamic action,** precipitate proteins. Spectrum of activity: Bactericidal, fungicidal, virucidal, not sporicidal. Applications: Organic mercurial tinctures (thimerosal, nitromersol) are used as skin antiseptics and to treat infections. Solutions of silver nitrate of 1–2% are used on newborn eyes to prevent infections and as a topical treatment for mucous membranes; silver sulfadiazine ointment is a burn infection preventative; colloidal silver preparations are used for mouth and eye rinses. Limitations: Metal solutions are highly toxic, cause allergies, and are neutralized by organic substances; organic preparations are not widely microbicidal.

Aldehydes

Forms: **Glutaraldehyde** and **formaldehyde** solutions (formalin). Mode of action: Alkylation of amino and nucleic acids. Spectrum of activity: Very broad; glutaraldehyde may be used as a sterilant. Applications: Glutaraldehyde in 2% solutions (Cidex) is used for sterilizing heat-sensitive instruments in the medical and dental office, also for environmental disinfection, but usually not antisepsis; formalin in 1–8% has limited use in disinfection of some instruments, rooms, and as a preservative. Limitations: Glutaraldehyde is somewhat unstable; formaldehyde is toxic and irritating.

Gases

Forms: **Ethylene oxide** (ETO), propylene oxide, and betapropiolactone. Mode of action: Alkylation of nucleic acids and proteins. Spectrum of activity: All are sporicidal, but only ETO is considered a sterilant. Applications: Ethylene oxide mixed with carbon dioxide (Carboxide) is the gas of choice to sterilize plastic materials for medical and industrial usage and to treat spices and dried foods; propylene oxide can disinfect food (nuts, starch); betapropiolactone is used to disinfect rooms and some biological materials. Limitations: ETO is somewhat dangerous to use because of its explosiveness and toxicity; it is slow-acting. Betapropiolactone is very toxic, with poor penetration.

Dyes

Forms: Aniline (crystal violet) and acridine (acriflavine). Mode of action: Probably binding to nucleic acids and proteins. Spectrum of activity: Mostly inhibitory for gram-positive bacteria and fungi. Applications: Primarily as treatments for skin infections and components of media. Limitations: Discolor; are narrow spectrum.

Acids and Alkalies

Mode of action: Very low or high pHs disrupt biological molecules. Applications: Organic acids (benzoic, propionic, acetic) are food preservatives; sodium and ammonium hydroxide are detergents and cleaning agents. Limitations: corrosive and caustic.

True–False Questions

Determine whether the following statements are true (T) or false (F). If you feel a statement is false, explain why, and reword the sentence so that it reads accurately.

____ 1. Autoclaving a flask of media for half the usual time will partially sterilize it.

____ 2. Microbial control methods that kill bacterial endospores will sterilize.

____ 3. Disinfection is any process that destroys the non-spore-forming contaminants on inanimate objects.

____ 4. Sanitization is a process by which objects are made sterile with chemicals.

____ 5. Agents that lower the surface tension of cells alter the selective permeability of the cell membrane and cause the leakage of cell contents.

____ 6. High temperatures are microbicidal, and low temperatures inhibit microbial growth.

____ 7. The general operating scheme for an autoclave is 121°F for 10–40 minutes and 15 psi.

____ 8. Following the flash method of pasteurization, milk is sterile.

____ 9. Ionizing radiation ionizes an atom, whereas nonionizing radiation changes an atom's energy level.

____ 10. Filtration is the method of choice for sterilizing heat-sensitive liquids.

____ 11. Iodine tincture is the antiseptic of choice for wound treatment.

____ 12. Phenolics are all good sporicides.

____ 13. Silver nitrate solution is instilled into the eyes of some newborn infants to prevent gonococcal infection.

____ 14. Detergents are ineffective germicides because of their narrow spectrum of action.

Concept Questions

1. Explain what a tuberculocide does. What about a pseudomonicide? A virustatic agent?
2. Compare sterilization with disinfection and sanitization. Describe the relationship of the concepts of sepsis, asepsis, and antisepsis.
3. Briefly explain how the type of microorganisms will influence the exposure to antimicrobial agents. Explain how the numbers of contaminants influence the nature of exposure.
4. Why does a population of microbes not die instantaneously when exposed to an antimicrobial agent?
5. Why are antimicrobial processes inhibited in the presence of organic matter?
6. Describe four modes of action of antimicrobial agents, and give a specific example of how each works.
7. Summarize the nature, mode of action, and effectiveness of moist and dry heat. Compare their effects on vegetative cells and spores.
8. How can the temperature of steam be raised above 100°C? Explain the relationship involved.
9. What do you see as a basic flaw in tyndallization? In boiling water devices? In incineration? In ultrasonic devices?
10. What are several microbial targets of pasteurization?
11. Explain why desiccation and cold are not reliable methods of disinfection.
12. What are some advantages of ionizing radiation as a method of control? Some disadvantages?
13. What is the precise mode of action of ultraviolet radiation? What are some disadvantages to its use?
14. What are the superior characteristics of iodophors over free iodine solutions?
15. Name one chemical for which the general rule that a higher concentration is more effective is *not* true.
16. Name the principal sporicidal chemical agents.
17. Why is hydrogen peroxide solution so effective against anaerobes?
18. Give the uses and disadvantages of the heavy metal chemical agents, glutaraldehyde, and the sterilizing gases.

Practical/Thought Questions

1. What is wrong with the statement, "The patient's skin was sterilized or disinfected with alcohol"? What would be the more correct wording?
2. For each item on the following list, give a reasonable method of sterilization. You cannot use the same method more than three times; the method must sterilize, not just disinfect; and the method must not destroy the item or render it useless. After considering a workable method, think of a method that would not work.

room air	cloth dressings
inside of a refrigerator	disposable syringes
blood	leather thrift shop shoes
wine	talcum powder
serum	a cheese sandwich
a jar of vaseline	milk
a pot of soil	human hair (for wigs)
a child's toy	orchid seeds
plastic petri dishes	a flask of nutrient agar
fruit	metal instruments
heat-sensitive drugs	the world
rubber gloves	

3. Graph the data on table 9.3, plotting the time on the Y axis and the temperature on the X axis for three or four species. What conclusions can you draw regarding the effects of time and temperature on thermal death? Is there any difference between the graph for a sporeformer versus a non-sporeformer?
4. Can you think of situations in which the same microbe would be considered a serious contaminant in one case and completely harmless in another?
5. What microbial control methods around the house require high levels of decontamination? Moderate levels? Low levels?
6. Devise an experiment that will differentiate between bacteriocidal and bacteriostatic effects.
7. There is quite a bit of concern that chlorine used as a water purification chemical presents serious dangers. Can you think of some alternative methods to purify huge water supplies and yet keep them safe from contamination?
8. The shelf life and keeping qualities of fruit and other perishable foods are greatly enhanced through irradiation, saving industry and consumers billions of dollars. How would you personally feel about eating a piece of irradiated fruit? How about spices that had been sterilized with ETO?
9. The microbial levels in saliva are astronomical ($> 10^6$ cells per ml, on average). What effect do you think a mouthwash can have against this high number? Comment on the claims made for such products.
10. Can you think of some innovations the health care community can use to deal with medical waste and its disposal that prevent infection but are ecologically sound?

CHAPTER 10

Drugs, Microbes, Host—The Elements of Chemotherapy

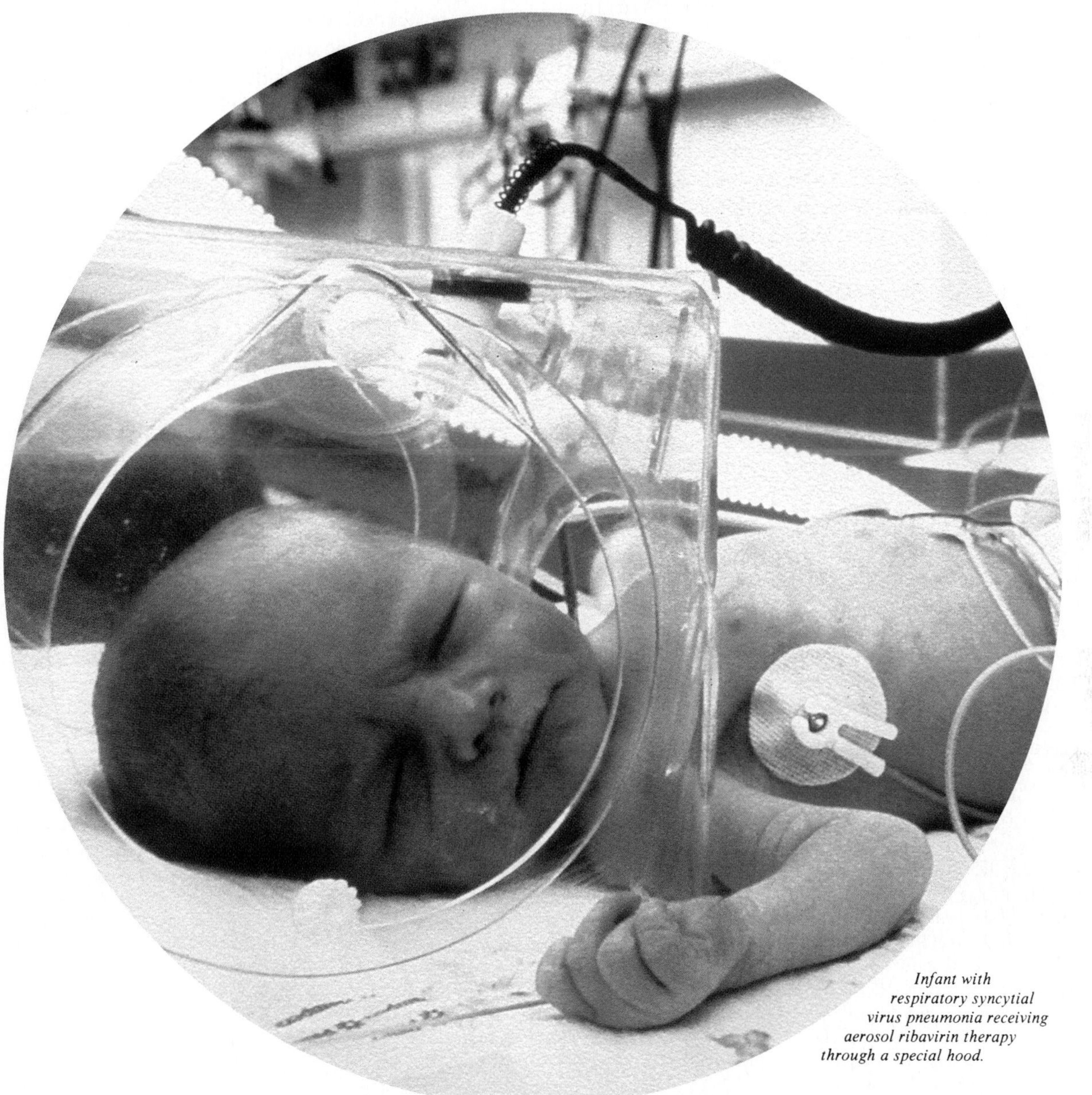

Infant with respiratory syncytial virus pneumonia receiving aerosol ribavirin therapy through a special hood.

Chapter Preview

The use of chemicals to control or prevent infection—antimicrobial chemotherapy—is the subject of this chapter. As you will see, the administration of a drug has impact beyond simply the action of the drug on the microbe. It also includes any effects microbes exert against the drug and, because drugs work inside the body, interactions between the drug and the patient's tissues. The principal areas to be covered here are general concepts of chemotherapy; mechanisms of drug action; drug resistance; antibacterial, antifungal, antihelminth, and antiviral drugs; adverse human reactions to drugs; methods of drug testing; and factors in drug selection.

Principles of Antimicrobial Therapy

A hundred years ago in the United States, one out of three children was expected to die of an infectious disease before the age of five. Early death or severe lifelong debilitation from scarlet fever, diphtheria, tuberculosis, meningitis, and many other bacterial diseases was a fearsome yet undeniable fact of life to most of the world's population. The introduction of modern drugs to control infections in the 1930s began a medical revolution that has added significantly to the lifespan and health of humans. It is no wonder that for many years, antibiotics in particular were regarded as the miracle cure-all for infectious diseases. In later discussions we will evaluate this misconception in light of the shortcomings of chemotherapy. Although antimicrobic drugs have greatly reduced the incidence of certain infections, they have definitely not eradicated infectious disease and probably never will. In fact, in many parts of the world, mortality rates from infectious diseases are as high as before the arrival of antimicrobic drugs. Nevertheless, humans have been taking medicines to try to control diseases for thousands of years (see feature 10.1).

The Terminology of Chemotherapy

Any chemical used in treatment, relief, or **prophylaxis** of disease is defined as a **chemotherapeutic drug** or agent. When chemotherapeutic drugs are given as a means to control infection, the practice is termed **antimicrobial chemotherapy.** Antimicrobial drugs (also termed anti-infective drugs) are a special class of compounds capable even in high dilutions of destroying

prophylaxis (proh''-fih-lak'-sis) Gr. *prophylassein,* to keep guard before. A process that prevents infection or disease in a person at risk.

chemotherapy (kee''-moh-ther'-uh-pee) Gr. *chemieia,* chemistry, and *therapeia,* service to the sick. Use of drugs to treat disease.

Feature 10.1 From Witchcraft to Wonder Drugs

Early human cultures relied on various types of primitive medications such as potions, salves, poultices, and mudplasters. Many were concocted by medicine men from plant, animal, and mineral products that had been found, usually through trial and error or accident, to have some curative effect upon ailments and complaints. In one ancient Chinese folk remedy, a fermented soybean curd was applied to skin infections. The Greeks used wine and plant resins (myrrh and frankincense), rotting wood, and various mineral salts to treat diseases. Some folk medicines were occasionally effective, but most of them were probably witches' brews that either had no effect or were even harmful. It is interesting that the Greek word *pharmakeutikos* originally meant the practice of witchcraft. These ancient remedies were handed down from generation to generation, but it was not until the Middle Ages that a specific disease was first treated with a specific chemical. Dosing syphilitic patients with inorganic arsenic and mercury compounds may have proved the ancient axiom: Graviora quaedum sunt remedia periculus ("Some remedies are worse than the disease").

In the 1600s Jesuit priests discovered that an extract of the bark of the South American cinchona tree could curb the symptoms of malaria. They brought this knowledge back to Europe, but it took another 200 years before a regular supply of this bark was available to the European population. Of course, we now know that the magic ingredient in this potion was quinine, which remained the major therapy for malaria until the pressures of World War II made it necessary to develop other sources of antimalarial drugs.

An enormous breakthrough in the science of drug therapy came with the proof of the germ theory of infection by Robert Koch (see chapter 1). This allowed disease treatment to focus on a particular microbe, which in turn opened the way for Paul Ehrlich to formulate the first theoretical concepts in chemotherapy in the late 1800s. Ehrlich had observed that specific dyes often affixed themselves to specific microorganisms and not to animal tissues. This led to the profound idea that if a drug were properly selective in its actions, it would zero in and destroy a microbial target and leave human cells unaffected. He decided to test this theory in treating syphilis. First, he worked out the structure of an arsenic-based drug that was very toxic to the spirochete of syphilis but, unfortunately, to humans as well. Ehrlich gradually and systematically altered this parent molecule, creating numerous derivatives. Finally, on the 606th try, he arrived at a compound he called *salvarsan.* This drug had some therapeutic merit and was used for a few years, but it eventually had to be discontinued because it was still not selective enough in its toxicity. Ehrlich's work had laid important foundations for many of the developments to come.

Another pathfinder in early drug research was Gerhard Domagk, whose discoveries in the 1930s launched a breakthrough in therapy that marked the true beginning of broad-scale usage of antimicrobial drugs. Experimenting with numerous synthetic dyes, Domagk showed that the red dye prontosil was active against certain bacterial infections in animals, even though it was inactive against that same infectious agent in a test tube. In time, this peculiar observation was explained: Prontosil was chemically changed by the body into an entirely different compound, with specific activity against bacteria. This new substance was sulfonamide—the first *sulfa* drug. In a short time, the structure of this drug was determined, and it became possible to synthesize it on a wide scale and to develop scores of other sulfonamide drugs. Although these drugs had immediate applications in therapy (and still do), still another fortunate discovery was needed before the golden age of antibiotics could really blossom.

The story of penicillin dramatically demonstrates how developments in science and medicine often occur through a combination of accident, persistence, collaboration, and vision. In the London laboratory of Alexander Fleming in 1928, a plate of *Staphylococcus aureus* became contaminated with the mold *Penicillium notatum.* Observing these plates, Fleming noted that the colonies of *Staphylococcus* were evidently being destroyed by some activity of the nearby *Penicillium* colonies. Struck by this curious phenomenon, he extracted from the fungus a compound he called penicillin, and showed that it was responsible for the inhibitory effects. Although he recognized the therapeutic potential of penicillin, he was unable to devise a method for its purification, stabilization, and large-scale industrial production.

Fleming's monumental discovery languished for a decade until the pressures of impending war spurred Howard Florey and Ernst Chain to develop methods for industrial production of penicillin in England. (At that time, war fatalities due to infectious diseases were higher than for other causes.) Clinical trials of penicillin, conducted in 1941, ultimately proved its effectiveness, and cultures of the mold were brought to the United States for an even larger-scale effort. When penicillin was made available to the world's population, it seemed a godsend at first, but in time, due to extreme overuse and misunderstanding of its capabilities, it also became the model for one of the most serious drug problems—namely, drug resistance. Fortunately, by the 1950s the pharmaceutical industry had entered an era of drug research and development that soon made penicillin only one of a large assortment of antimicrobial drugs.

or inhibiting microorganisms. The origin of modern antimicrobial drugs is varied. Some, called **antibiotics,** are substances produced by the natural metabolic processes of microorganisms that can inhibit or destroy other microorganisms. Other antimicrobic substances, termed **synthetic** drugs, are derived in the laboratory from dyes or other organic compounds. Although separation into these two categories has been traditional, they tend to overlap, because many antibiotics are now chemically altered in the laboratory (*semisynthetic*). The current trend is to use the term **antimicrobic** for all antimicrobial drugs, regardless of origin.

Antimicrobial drugs vary in their scope of activity. The so-called **narrow-spectrum** agents are effective against a limited array of different microbial types. Examples are bacitracin, an antibiotic whose inhibitory effects extend mainly to certain gram-positive bacteria, or griseofulvin, which is used chiefly in fungal skin infections. **Broad-spectrum** agents are active against a wider range of different microbes. The targets of antibiotics in the tetracycline group, for example, are a variety of gram-positive and gram-negative bacteria, rickettsias, and mycoplasmas. The concept of spectrum of activity has become blurred to some extent, because an agent can be induced by artificial means to change its spectrum.

Drug, Microbe, Host—Some Basic Interactions

The broad concept of anti-infective therapy revolves around three interacting factors: the drug, the microorganism, and the infected host. To put it simply, the drug should destroy the infectious agent without harming the host's cells. This process can be visualized more tangibly as a series of stages (figure 10.1): (1) The drug is administered to the host via a designated route—

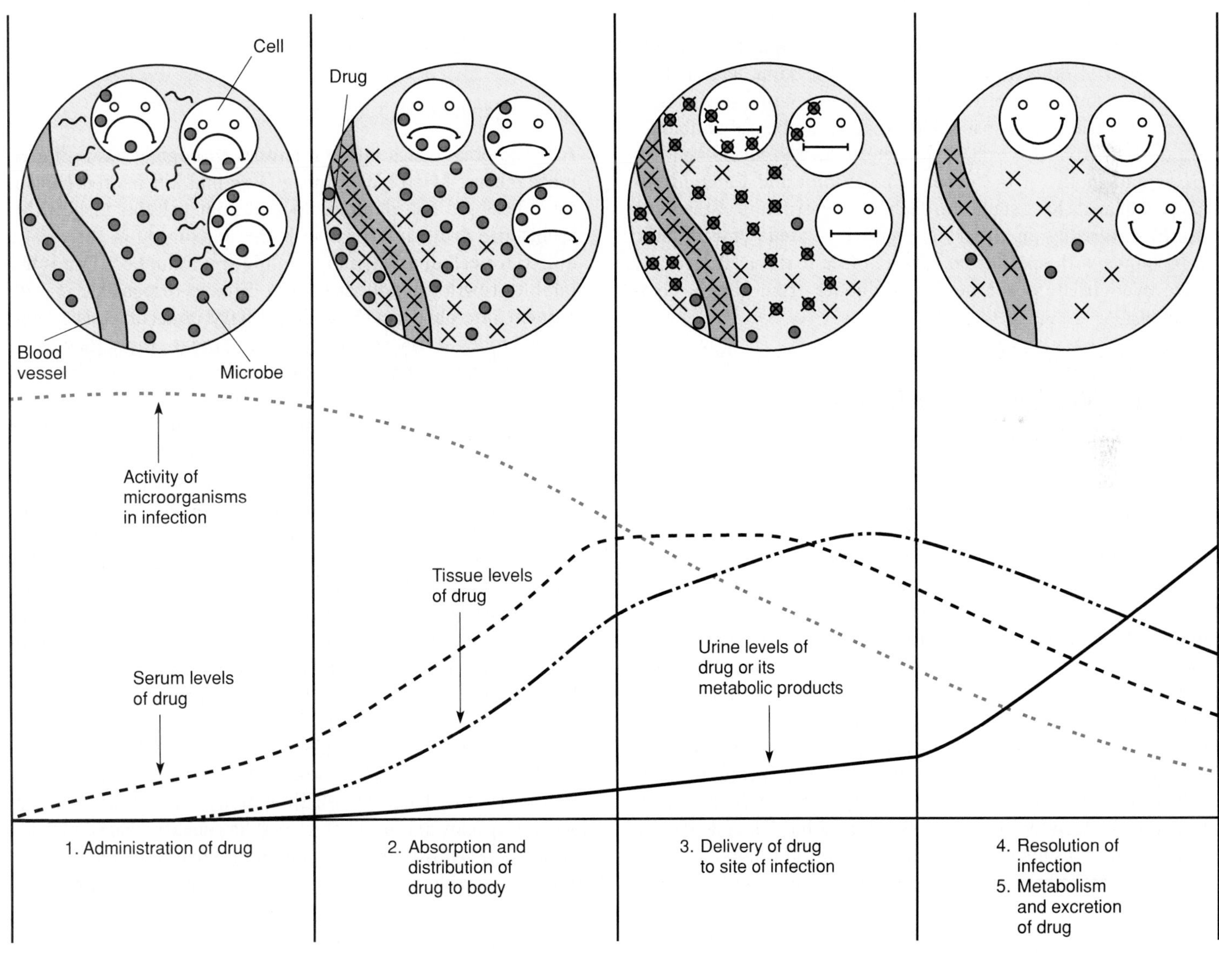

Figure 10.1 The course of events in chemotherapy: interaction between patient, drug, and infectious agent.

primarily by mouth (oral), by injection into a vein (intravenous), by injection into a muscle (intramuscular), by applying to the skin surface (topical), and occasionally, by injection into the skin or body cavity. (2) The drug is absorbed by and dissolved in body fluids or cells. (3) The drug is delivered to the infected area (extracellular or intracellular). (4) The drug destroys the infectious agent or inhibits its growth. (5) The drug is eventually excreted or broken down by the host's organs, ideally without harming them. Among the factors that affect the outcome of drug therapy are those that aid or block drug action and those that influence the behavior of microbe or host tissues. Favorable features of antimicrobial drugs in relation to the infectious agent and host are presented in table 10.1.

Table 10.1 Characteristics of the Ideal Antimicrobic Drug
Selectively toxic to the microbe but nontoxic to vertebrate cells
Microbicidal rather than microbistatic
Relatively soluble and functions even when highly diluted in body fluids
Remains potent long enough to act and is not broken down or excreted prematurely
Not subject to the development of antimicrobial resistance
Complements or assists the activities of the host's defenses
Remains active despite the presence of large volumes of organic materials
Does not disrupt the host's health by causing allergies or predisposing the host to other infections

The Origins of Antimicrobial Drugs

Nature is undoubtedly the most prolific producer of antimicrobial drugs. Antibiotics, after all, are common metabolic products of aerobic spore-forming bacteria and fungi. By inhibiting the growth of other microorganisms in the same habitat (antagonism), antibiotic producers presumably enjoy less competition for nutrients and space. The greatest numbers of antibiotics are derived from bacteria in the genera *Streptomyces* and *Bacillus* and molds in the genera *Penicillium* and *Cephalosporium.* Chemists have sought to decipher the structure of the more useful antibiotics in order to produce related semisynthetic compounds with specialized features (see feature 10.2). They have even tried to synthesize antibiotics from simple compounds.

Characteristic Interactions Between Drug and Microbe

The primary effect of antimicrobial drugs is to disrupt the cell processes or structures of bacteria, fungi, and protozoa, or to inactivate viruses. Most of the drugs used in chemotherapy interfere with the function of enzymes required to synthesize or assemble macromolecules, or they destroy structures already formed in the cell. Preferably, such drugs should be **selectively toxic,** which means they produce adverse effects on microbial cells without simultaneously damaging host tissues. This concept of selective toxicity is central to chemotherapy, and the best drugs are those that block the actions or synthesis of molecules in microorganisms but not in vertebrate cells. Examples of drugs with selective toxicity are those that block the synthesis of the cell wall in bacteria (penicillins). They have low toxicity and few direct effects on human cells because human cells lack a wall and are thus neutral to this action of the antibiotic. Among the most toxic to human cells are drugs that act upon a common structure such as the cell membrane (amphotericin B, for example). In a later section we will address the circumstances in which the ideal of selective toxicity is not met.

Mechanisms of Drug Action

Antimicrobial drugs injure microbes through several different mechanisms. Microbicidal drugs lyse and kill microorganisms by inflicting direct damage upon specific cellular targets. Microbistatic drugs interfere with the machinery in the cell required for cell division, and so inhibit reproduction. Drugs that inhibit growth are not considered directly responsible for subsequent microbial death; their primary importance is to keep the microbes at bay and prevent their spread, thus allowing the host defenses an opportunity to destroy and remove the infectious agent.

Antimicrobial drugs function specifically in one of the following ways: (1) They inhibit cell wall synthesis; (2) they inhibit nucleic acid synthesis; (3) they inhibit protein synthesis; or (4) they interfere with the function of the cell membrane. These categories are not completely discrete, and some effects may overlap. In chapter 9 we presented an overall picture of antimicrobial mechanisms; here, we include selected examples of antimicrobics that will illustrate the general ways that drugs act upon microbial cells. It is worth noting that much knowledge of cellular structure and function has emerged as an added bonus from research on the effects of drugs.

Antimicrobial Drugs That Affect the Bacterial Cell Wall

The cell walls of most bacteria contain a rigid girdle of peptidoglycan. This structure, which is many layers thick in gram-positive species and quite thin in gram-negative ones, protects the cell against rupture from hypotonic environments. Cells actively engaged in enlargement or binary fission must constantly

Feature 10.2 A Modern Quest for Designer Drugs

Extraordinary time and effort were expended in the discovery, testing, and marketing of the first truly significant antibiotic (penicillin). Once this monumental event had transpired, the world immediately witnessed a scientific scramble to find more antibiotics. This search was advanced on several fronts. Hundreds of investigators began the laborious task of screening samples from soil, dust, muddy lake sediments, rivers, estuaries, oceans, plant surfaces, compost heaps, sewage, skin, and even the hair and skin of animals for antibiotic-producing bacteria and fungi. This intense effort has paid off over the past 50 years, because more than 10,000 antibiotics were eventually discovered (though surprisingly, only a relatively small number have been applicable to chemotherapy). Finding a new antimicrobic substance is only a first step. The complete pathway of drug development from discovery to therapy takes at least nine years.

Antibiotics are products of a fermentation pathway that occurs in nearly all spore-forming bacteria and fungi. The true role of antibiotics in the lives of these microbes continues to be somewhat mysterious, although the evolutionary preservation of genes for antibiotic production does point to an important function for them. Some experts theorize that antibiotic-releasing microorganisms can inhibit or destroy nearby competitors or predators, while others feel that these compounds play a part in spore formation. Whatever benefit the microbes derive, the existence of these compounds has been extremely profitable for humans. Every year, the pharmaceutical industry farms vast quantities of microorganisms and harvests their products to treat diseases caused by other microorganisms. Researchers have facilitated the work of nature by selecting mutant species that yield more abundant or useful products, by varying the growth medium, or by altering the procedures for large-scale industrial production (see chapter 22).

Another approach in the drug quest is that of organic synthesis—the chemical manipulation of molecules by adding or removing functional groups. Drugs produced in this way are designed to have advantages over other, related drugs. With the **semisynthetic** method, a natural product of the microorganism is joined with various preselected functional groups. The antibiotic is reduced to its basic molecular framework (called the nucleus), and to this nucleus specially selected side chains (R groups) are added. A case in point is the metamorphosis of the semisynthetic penicillins. The nucleus is an inactive penicillin derivative called aminopenicillanic acid, which has an opening on the number 6 carbon for addition of R groups. A particular carboxylic acid (R group) added to this nucleus can "fine-tune" the penicillin, giving it special characteristics. For instance, some R groups will make the product resistant to penicillinase (methicillin), some confer broader spectrum (ampicillin), and others make the product acid-resistant (penicillin V). Cephalosporins and tetracyclines have likewise been subjected to semisynthetic conversions.

In the search for new antimicrobic drugs, scientists have also taken notice of unusual animal and plant sources. Recently, very powerful antimicrobial peptides were isolated from the skin of the African clawed frog and from the white blood cells of rabbits. Equally surprising was the discovery of an effective antibiotic in bee glue, the substance that bees use to build and repair their hives. A particularly fascinating methodology is being followed by pharmaceutical anthropologists. Having observed that apes treat their illnesses in the wild by using certain plants, these researchers are sampling several of these plants in the hope of isolating compounds with antimicrobial properties. The potential for using bioengineering techniques to design drugs seems almost limitless, and indeed, several drugs have already been produced by manipulating the genes of antibiotic-producers.

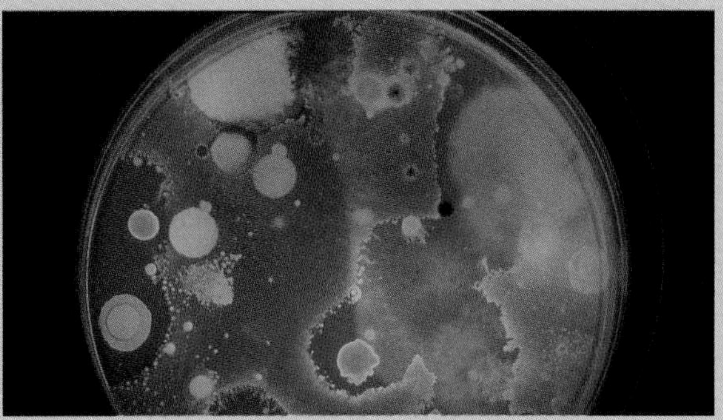

A plate with several discrete colonies of soil bacteria was sprayed with a culture of *Escherichia coli* and incubated. Zones of inhibition (clear areas with no growth) surrounding several colonies indicate species that produce antibiotics.

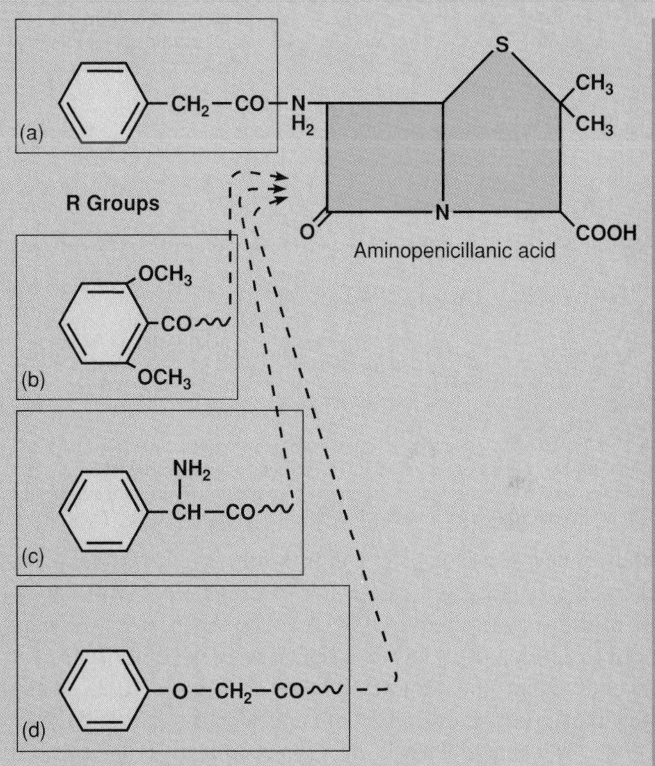

Synthesizing new types of penicillins. (*a*) The original penicillin G molecule is a fermentation product of *Penicillium chrysogenum* that appears somewhat like a split-level house with a removable carport on one end. This house without the carport is the basic nucleus called aminopenicillanic acid. (*b–d*) Various new fixtures (R groups) may be added in place of the carport according to need. These R groups will produce different penicillins: (*b*) methicillin; (*c*) ampicillin; and (*d*) penicillin V.

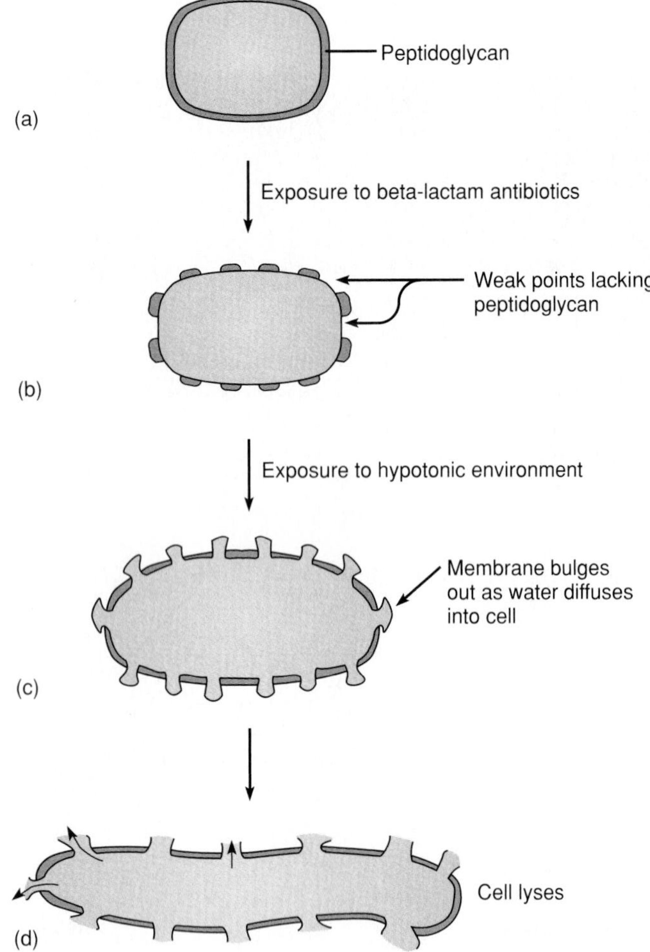

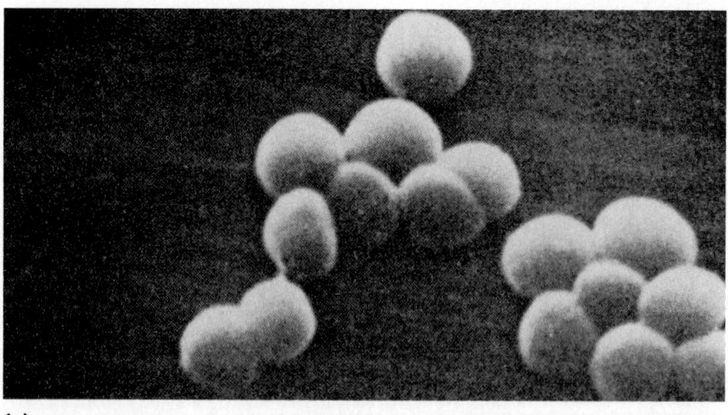
(e)

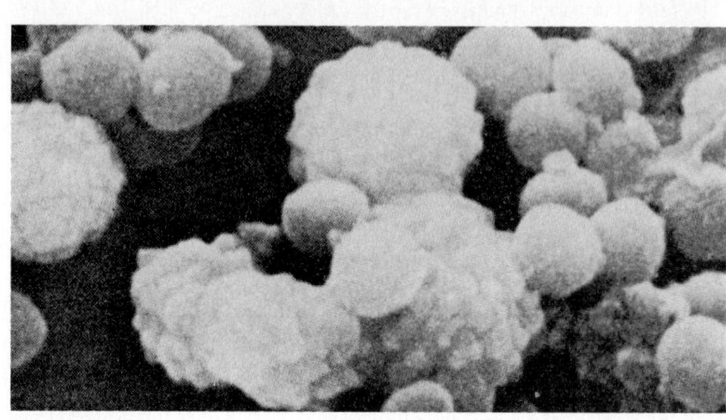
(f)

Figure 10.2 The consequences of exposing a growing cell to antibiotics that prevent cell wall synthesis. (*a*) Coccus is exposed to beta-lactam agent (cephalosporin). (*b*) Weak points develop where peptidoglycan is incomplete. (*c*) The weakened cell is exposed to a hypotonic environment. (*d*) The cell lyses. (*e*) Scanning electron micrograph of the cell in its normal state (×10,000). (*f*) SEM of the same cell in the drug-affected state, showing surface bulges (×10,000).

synthesize new peptidoglycan and transport it to its proper place in the cell envelope. Several groups of drugs react with one or more of the enzymes required to complete this process, causing the cell to develop peptidoglycan-deficient weak points at growth sites and to become osmotically fragile (figure 10.2). Antibiotics that produce this effect are considered bactericidal, because the weakened cell will be subject to lysis. It is essential to note that most of these antibiotics are active only in young, actively growing cells, since old, inactive, or dormant cells are not synthesizing peptidoglycan. (One exception is a new class of antibiotics called the penems.)

Drugs That Block Cell Wall Synthesis Cycloserine inhibits the formation of the basic peptidoglycan subunits; vancomycin hinders the elongation of the peptidoglycan; and bacitracin blocks the transport of the peptidoglycan subunits to their position in the wall. The beta-lactams (penicillins and cephalosporins) bind to peptidases that are essential to cross-link the glycan molecules, thereby interrupting the completion of the cell wall (figure 10.3). Some penicillins are less effective against gram-negative bacteria because the outer membrane of their cell envelope blocks passage of the drug into the cells, but broad-spectrum synthetic penicillins and cephalosporins do have activity against gram-negative species.

Antimicrobial Drugs That Affect Nucleic Acid Synthesis

As you will recall from chapter 8, the metabolic pathway that generates DNA and RNA molecules is a long, enzyme-catalyzed series of reactions. Like any complicated process, it is subject to breakdown at many different points along the way, and inhibition at any point in the sequence can block subsequent events. Antimicrobial drugs interfere with nucleic acid synthesis essentially by blocking synthesis of the raw materials, inhibiting replication, or stopping transcription. Because functioning DNA and RNA are required for proper translation as well, the effects on microbial metabolism can be far-reaching.

One of the more extensively studied modes of action is that of the **sulfonamides** (sulfa drugs). These synthetic drugs are termed **antimetabolites** because they interfere with an essential metabolic process in bacteria. They also represent a model for **competitive inhibition,** because they are structural or **metabolic analogs** that mimic the natural substrate of an enzyme and vie for its active site. More specifically, a sulfa drug is

analog (an′-uh-log) A compound whose configuration closely resembles another compound required for cellular reactions.

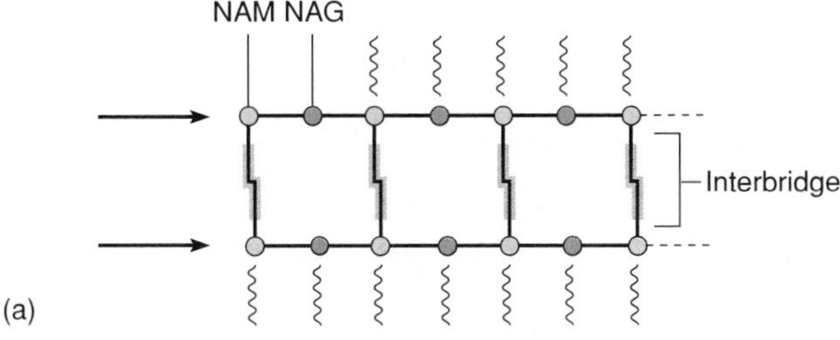

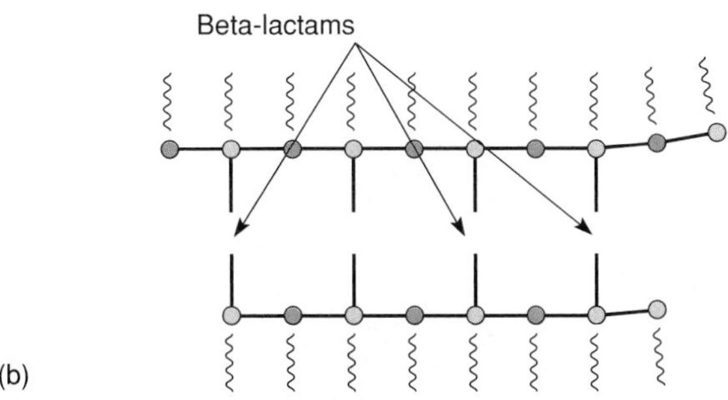

Figure 10.3 The mode of action of beta-lactam antibiotics on the bacterial cell wall. (*a*) Intact peptidoglycan has alternating NAM and NAG glycans cross-linked by peptide interbridges. (*b*) The beta-lactam antibiotics block the peptidases required to attach the interbridges, thereby greatly weakening the cell wall meshwork.

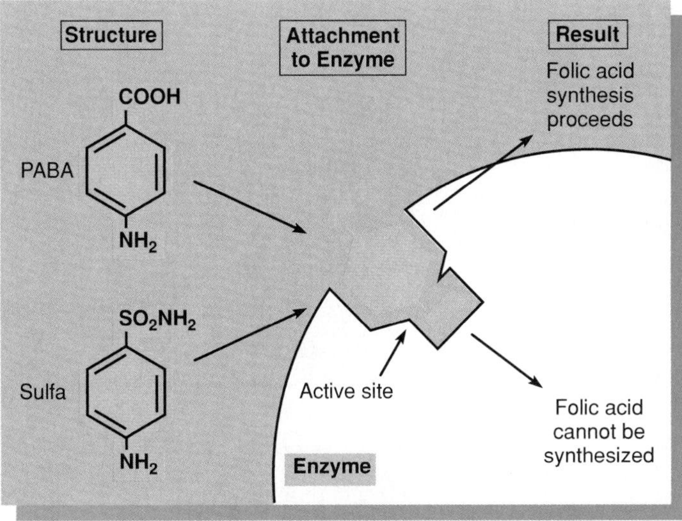

Figure 10.4 Competitive inhibition of a pathway in nucleic acid synthesis. Sulfonamides are structural analogs of PABA, a chemical required to synthesize folic acid. The similar configuration of the two means that sulfonamides can bind to the active site on an enzyme involved in folic acid synthesis. Although it binds, sulfa still cannot complete the required synthesis.

very similar to the natural metabolic compound PABA (para-aminobenzoic acid) that is needed by bacteria to synthesize folic acid. Folic acid, in turn, is required as a component of the coenzyme tetrahydrofolic acid that participates in the synthesis of purines and certain amino acids. A sulfonamide molecule has extreme affinity for the PABA site on the enzyme that synthesizes folic acid, thus it can successfully compete in a "chemical race" with PABA for the same sites (figure 10.4).

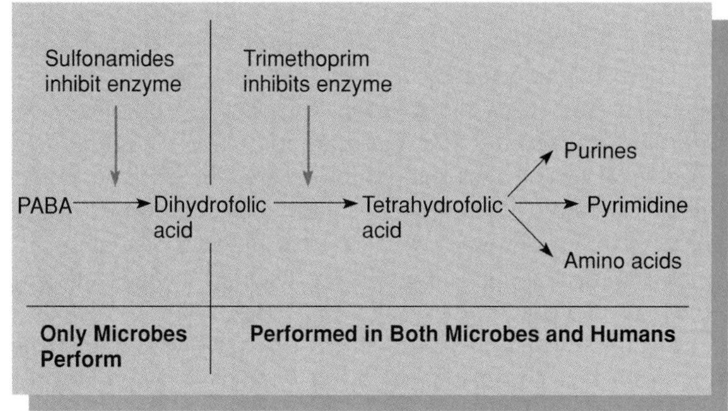

Figure 10.5 A scheme for the synthesis and function of folic acid and points of inhibition by antimetabolites. Sulfa drugs (sulfonamides) inhibit the first step, and trimethoprim blocks the second; together, they have a synergistic action that inhibits the entire pathway leading to nitrogen base and amino acid synthesis.

Sulfonamides ultimately cause an inadequate supply of folic acid for purine production, which invariably halts nucleic acid synthesis. The metabolic changes brought on by sulfonamides and other metabolic analogs prevent bacterial cells from multiplying, so such drugs are primarily bacteriostatic in their effects.

Sulfonamides are valuable in therapy because they act on bacteria and certain protozoa, but not on mammalian cells. This one-sided inhibition stems from a basic nutritional difference between humans and microorganisms. Although humans require folic acid for nucleic acid synthesis as much as bacteria do, humans cannot synthesize folic acid; it is an essential nutrient (vitamin) that must come from the diet. Because human cells lack this special enzymatic system for incorporating PABA into folic acid, their metabolism cannot be inhibited by sulfa drugs.

Examples of Nucleic Acid Inhibitors Para-aminosalicylic acid (PAS) and the sulfones are other structural analogs of PABA that have a mode of action similar to sulfonamides. Trimethoprim is a metabolic analog that competes for a site on the enzyme needed during a subsequent step in folic acid synthesis (figure 10.5), thus blocking later steps leading to DNA and RNA. Trimethoprim is often given simultaneously with sulfonamides to achieve a **synergistic** effect. In pharmacology, this refers to an additive or cooperative effect achieved by two drugs working together, thus requiring a lower dose of each.

Other antimicrobics inhibit DNA synthesis. Chloroquine (an antimalarial drug) binds and cross-links the double helix. Novobiocin and newer broad-spectrum quinolones inhibit DNA unwinding enzymes or gyrases. Antiviral drugs that are analogs of purines and pyrimidines (including idoxuridine and acyclovir) insert in the viral nucleic acid and block further replication. Rifampin prevents the correct function of RNA polymerase and stops synthesis of mRNA. This, of course, also terminates protein synthesis (see the next section).

Drugs That Block Translation

Most inhibitors of translation, or protein synthesis, react with the ribosome-mRNA complex. Although human cells also have ribosomes, the ribosomes of eucaryotes are different in size and structure from those of procaryotes, so for the most part, these antimicrobics have a selective action against bacteria. One potential therapeutic consequence of drugs that bind to the procaryotic ribosome is the damage they may do to eucaryotic mitochondria, which contain a procaryotic type of ribosome. Two possible targets of ribosomal inhibition are the 30S subunit and the 50S subunit (figure 10.6). Aminoglycosides (streptomycin, gentamicin, for example) insert on sites on the 30S subunit and cause the misreading of the mRNA, leading to abnormal proteins. Tetracyclines block the attachment of tRNA on the A acceptor site and effectively stop further synthesis. Other antibiotics attach to sites on the 50S subunit in a way that prevents the formation of peptide bonds (chloramphenicol) or inhibits translocation of the subunit during translation (erythromycin).

Drugs That Disrupt Cell Membrane Function

A cell with a damaged membrane invariably dies from metabolic insufficiency or lysis, and does not even have to be actively dividing to be destroyed. The three main antibiotic classes that damage cell membranes are the polymyxins, the polyenes, and the imidazoles. Each of these has specificity for a particular microbial group, based on differences in the types of lipids in their cell membranes. Unfortunately, this selectivity is not exacting, and the universal presence of membranes in microbial and animal cells alike means that most of these antibiotics are quite toxic to humans.

Examples of Membrane-Active Drugs Polymyxins interact with membrane phospholipids, distort the cell surface, and cause leakage of proteins and nitrogen bases, particularly in gram-negative bacteria (figure 10.7). The polyene antifungal antibiotics (amphotericin B and nystatin) form complexes with the sterols on fungal membranes, which causes abnormal openings and seepage of small ions. It was recently shown that the lethal effects of drugs known as aminoglycosides (streptomycin, for example) are due to a combination of mechanisms. First, the misreading of the mRNA code causes abnormally folded proteins to be inserted in the cell membrane; then, the incorrect fit of these proteins leaves holes in the membrane.

The Acquisition of Drug Resistance

The wide-scale use of antimicrobics soon led to microbial **drug resistance,** an adaptive response in which microorganisms become able to tolerate an amount of drug that would ordinarily be inhibitory. The development of mechanisms for circumventing or inactivating antimicrobic drugs is a logical outgrowth of drug therapy, given the genetic versatility and adaptability of microbial populations. The property of drug resistance may be intrinsic as well as acquired. Intrinsic drug resistance exists when the natural structure or physiology of a species or group is protective, not acquired through specific genetic changes. In our context, the term drug resistance will mean resistance acquired by microbes that had normally been sensitive to the drug.

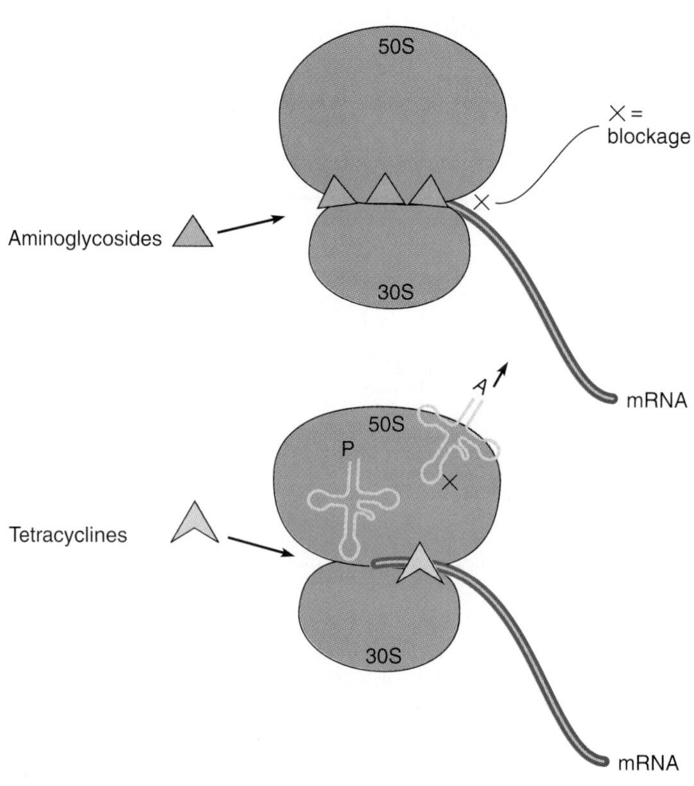

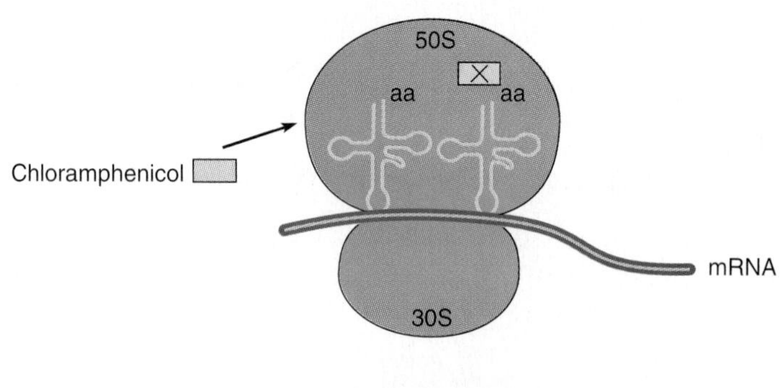

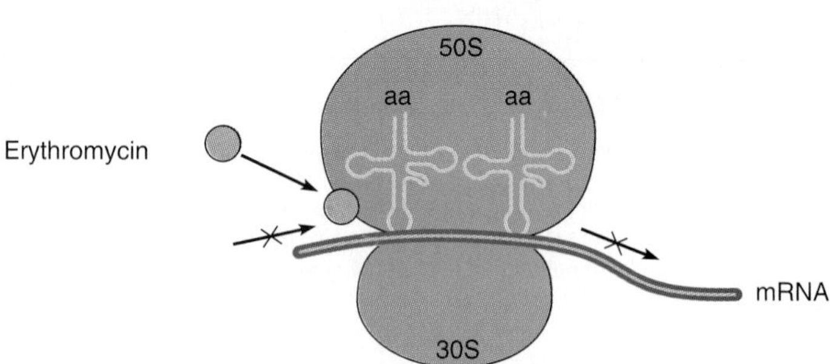

Figure 10.6 Sites of inhibition on the procaryotic ribosome and major antibiotics that act on these sites. All have the general effect of blocking protein synthesis. Blockage sites are indicated by X.

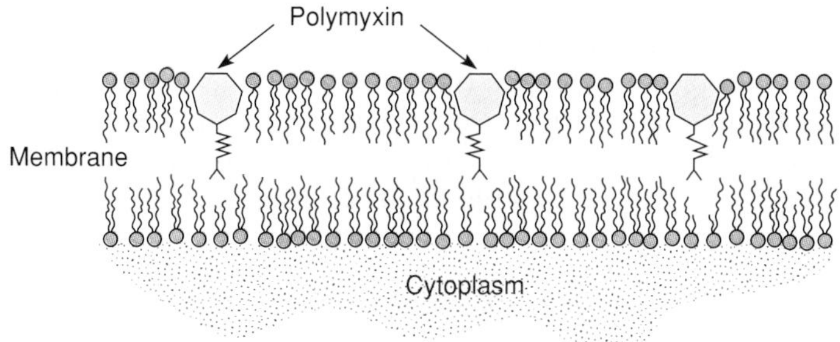

Figure 10.7 The detergent action of polymyxin. After passing through the cell wall of gram-negative bacteria, polymyxin binds to the cell membrane and disrupts its structure.

How Does Drug Resistance Develop?

The genetic events most often responsible for drug resistance are either chromosomal mutations or transfer of extrachromosomal DNA from a resistant species to a sensitive one. Chromosomal drug resistance results from spontaneous random mutations in bacterial populations. The chance that such a mutation will be advantageous is minimal, and the chance that it will confer resistance to a specific drug is lower still. Nevertheless, given the huge numbers of microorganisms in any population and the constant rate of mutation, such mutations do occur. The end result varies from slight changes in microbial sensitivity, which can be overcome by larger doses of the drug, to complete loss of sensitivity.

Resistance associated with intermicrobial transfer originates from plasmids called *resistance factors* or *R factors* that are transferred through conjugation, transformation, or transduction. Studies have shown that plasmids encoded with drug resistance, like mutations occurring on chromosomes, are naturally present in microorganisms before they have been exposed to the drug. Such traits are "lying in wait" for an opportunity to be expressed and to confer adaptability on the species. Many bacteria also maintain transposable drug resistance sequences (transposons) that are duplicated and inserted from one plasmid to another or from a plasmid to the chromosome. Transposons apparently enable drug resistance genes to spread within and between cells. Chromosomal genes and plasmids containing codes for drug resistance are faithfully replicated and inherited by all subsequent progeny. This sharing of resistance genes accounts for the rapid proliferation of drug-resistant species (figure 10.8).

Specific Mechanisms of Drug Resistance

In general, a microorganism loses its sensitivity to a drug by expressing genes that prevent the drug from completing its mechanism of action. Gene expression takes the form of (1) synthesis of enzymes that inactivate the drug, (2) decrease in cell permeability and uptake of the drug, (3) change in the number or affinity of the drug receptor sites, or (4) modification of an essential metabolic pathway. Some bacteria can become resistant indirectly by lapsing into dormancy or, in the case of penicillin, by converting to a cell-wall-deficient form (L form) that penicillin cannot affect.

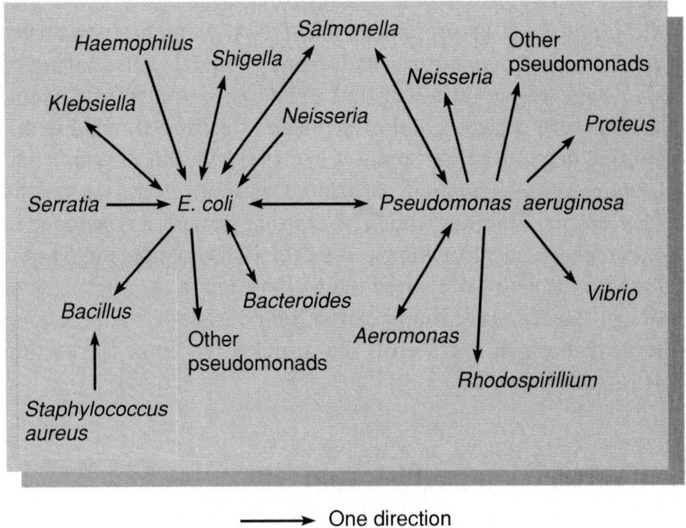

Figure 10.8 The promiscuous exchange of drug resistance is shown by this network of bacterial genera that have received or donated plasmids for drug resistance. (All are gram negative, except *Bacillus* and *Staphylococcus*.) This phenomenon is responsible for the growing numbers of drug-resistant microbes.

Source: Data from Young and Mayer, *Review of Infectious Diseases* 1:55, 1979.

Drug Inactivation Mechanisms Microbes inactivate drugs by producing enzymes that permanently alter drug structure. One example, bacterial exoenzymes called **beta-lactamases,** hydrolyze the *beta-lactam* ring structure of some penicillins and cephalosporins. Two beta-lactamases—*penicillinase* and *cephalosporinase*—disrupt the structure of certain penicillin or cephalosporin molecules (figure 10.9*a*). So many strains of *Staphylococcus aureus* produce penicillinase that regular penicillin is rarely a possible therapeutic choice. Now that some strains of *Neisseria gonorrhoeae,* called PPNG,[1] have also acquired penicillinase, alternate drugs are often required to treat gonorrhea. A large number of other gram-negative species are inherently resistant to some of the penicillins and cephalosporins because of naturally occurring beta-lactamases. Resistance to aminoglycoside antibiotics and chloramphenicol is due to enzymes that add functional groups to the antibiotic molecules, thereby preventing drug attachment to the bacterial ribosomes.

Decreased Permeability to a Drug The resistance of some bacteria may be due to a mechanism that prevents the drug from entering the cell and acting on its target. For example, the outer membrane of the cell wall of certain gram-negative bacteria is a natural blockade for some of the penicillin drugs. Resistance to the tetracyclines may arise from plasmid-encoded proteins that pump the drug out of the cell. Resistance to the aminoglycoside antibiotics is a special case in which microbial cells have lost the capacity to transport the drug intracellularly.

---

beta-lactam (bay′-tuh-lak′-tam) Molecular structure shown in figure 10.11.

1. Penicillinase-producing *Neisseria gonorrhoeae.*

Change of Drug Receptors Because most drugs act on a specific target such as protein, RNA, DNA, or membrane structure, microbes may circumvent drugs by altering the nature of this target. In bacteria resistant to rifampin and streptomycin, the structure of key proteins has been altered so that these antibiotics can no longer bind. Erythromycin, lincomycin, and clindamycin resistance is associated with an alteration on the 50S ribosomal binding site. The types of penicillin resistance in *Streptococcus pneumoniae* and methicillin resistance in *Staphylococcus aureus* are related to an alteration in the binding proteins in the cell wall. Fungi can become resistant by decreasing their synthesis of ergosterol, the principal receptor for certain antifungal drugs.

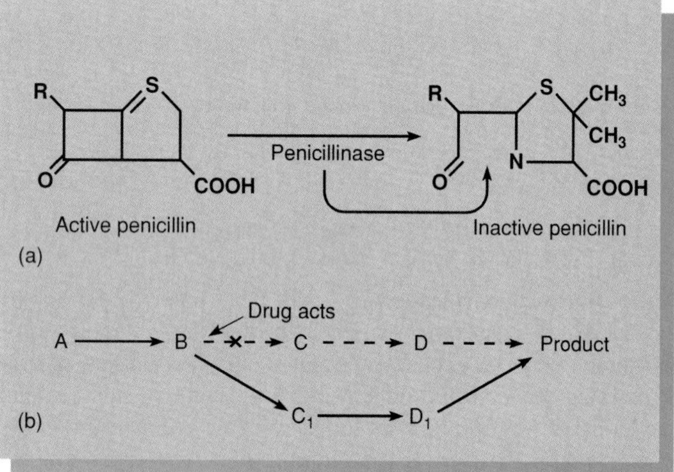

Figure 10.9 Examples of mechanisms of acquired drug resistance. (*a*) Inactivation of a drug like penicillin by penicillinase, an enzyme that cleaves a portion of the molecule and renders it inactive. (*b*) The drug has blocked the usual metabolic pathway, so the microbe circumvents it by using an alternate, unblocked pathway that achieves the required outcome.

Changes in Metabolic Patterns The action of antimetabolites can be circumvented if a microbe develops an alternative metabolic pathway or enzyme (figure 10.9*b*). Sulfonamide and trimethoprim resistance develops when microbes deviate from the usual patterns of folic acid synthesis. Fungi may acquire resistance to flucytosine by competely shutting off certain metabolic activities.

Natural Selection and Drug Resistance

So far, we have been considering drug resistance at the cellular and molecular levels, but its full impact will be felt only if this resistance occurs throughout the cell population. Let us examine how this might happen and its longer-term therapeutic consequences. Recall that any large population of microbes is likely to contain a few individual cells that are already drug resistant because of prior mutations or transfer of plasmids (figure 10.10*a*). As long as the drug is not present in the habitat, the numbers of these resistant forms will remain low because they have no particular growth advantage. But if the population is subsequently exposed to this drug (figure 10.10*b*), sensitive individuals are inhibited or destroyed, and resistant forms survive and proliferate. During subsequent population growth, all offspring of these genetically altered microbes will inherit this drug resistance. In time, the replacement population will have a preponderance of the drug-resistant forms and may even become completely resistant (figure 10.10*c*). In ecological terms, the environmental factor (in this case, the drug) has put selection pressure on the population, thus the "fitter" microbe (the drug-resistant one) survived, and the population has evolved to a condition of drug resistance. Natural selection for drug-resistant forms is apparently a common phenomenon. It takes place most frequently in various natural habitats, laboratories, and medical environments, but it occasionally occurs within the bodies of humans and animals during drug therapy.

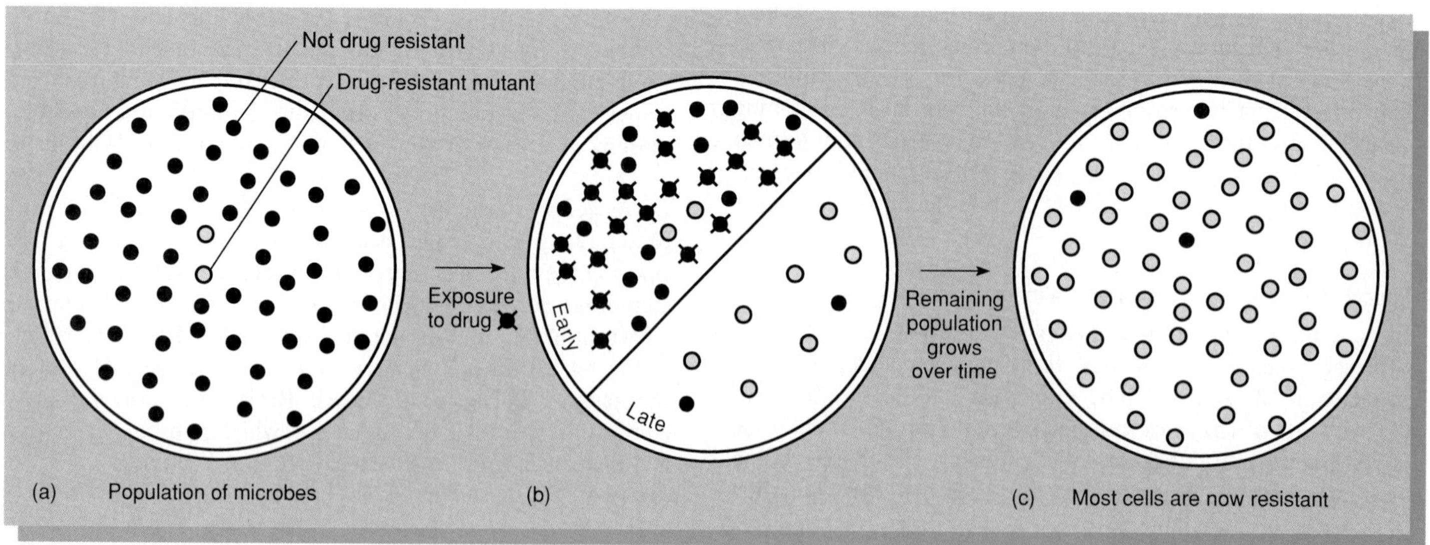

Figure 10.10 Natural selection and drug resistance. (*a*) Populations of microbes may harbor some members with a prior mutation that confers drug resistance. (*b*) Environmental pressure (here, the presence of the drug) selects for survival of these mutants so that (*c*) they eventually become the dominant members of the population.

Biomedical Significance of Drug Resistance The far-reaching effects of drug resistance are best appreciated by considering some practical examples. The medical setting is one situation destined to bring microbial populations into frequent and intimate contact with antimicrobic drugs. Drugs are inevitably released into the hospital environment during therapy, and a hospital is also something of a magnet for pathogens, thus conditions are ripe for the emergence of drug-resistant strains of pathogens. During the early years of penicillin therapy, most strains of *Staphylococcus aureus* were sensitive to penicillin. As this antibiotic was used increasingly and indiscriminately, the first penicillin-resistant strains appeared, and within a few years, the majority of hospital strains were resistant. Now *S. aureus* shows a wide range of resistance to many antibiotics, often making selection of an appropriate antibiotic difficult.

The widespread presence of plasmid-mediated drug resistance in most members of the Family Enterobacteriaceae and *Pseudomonas* seems to be associated with extensive oral antibiotic therapy. The once common practice of spiking farm animal feed with penicillin or tetracycline to help prevent intestinal infections and increase animal growth rate had the unfortunate side effect of selecting for resistant strains, especially of intestinal pathogens. These strains ultimately made their way into humans (see feature 11.8). It now appears that normal flora such as *Escherichia coli* can maintain a gene pool of resistance plasmids that it readily shares with numerous other intestinal bacteria, creating strains of *Shigella* and *Salmonella* with resistance to several drugs.

The progressive development of drug-resistant species means that microbial sensitivity or susceptibility to drugs cannot be assumed. Infectious agents must be tested for possible resistance to prevent failure in therapy. It almost seems that medical science is in a race with the microbes—and that we are barely a few jumps ahead of them. Who will triumph in this race, only time will tell. In the meantime, several procedures are advisable for minimizing and curbing the spread of drug resistance (table 10. 2).

Table 10.2 Strategies to Limit Drug Resistance by Microorganisms

Drug Usage

Dosage: Maintenance of sufficient amounts of circulating drug diminishes the opportunity for selection of mutants with resistance to low drug levels. Patients must comply with and carefully follow the physician's guidelines for therapy. It is important for the patient to take the whole course of therapy to ensure that the infectious agent has been completely eliminated and not merely inhibited.

Combined Therapy: Administering two or more drugs simultaneously can ensure that at least one of the drugs will be effective, and that a microbial strain that is resistant to any of the drugs being administered will not persist. The basis for this method lies in the unlikelihood of simultaneous resistance to both drugs.

Drug Research

Pharmaceutical companies continue to seek new antimicrobial drugs with structures that are not as readily inactivated by microbial enzymes, or drugs with modes of action that are not readily circumvented.

Drug Controls

Antimicrobics (especially those for bacterial infections) are overproduced, overprescribed, and used inappropriately on a very wide scale. Proposals to reduce the abuses in the United States range from educational programs for physicians to requiring written justification from the physician on all antibiotics prescribed. Especially valuable antimicrobics may be restricted in their use to only one or two types of infections. The addition of antimicrobics to animal feeds has been curtailed in most parts of the United States.

Survey of Major Antimicrobial Drug Groups

Scores of antimicrobial drugs are marketed in the United States, and new ones are being discovered or developed every year (see feature 10.2). Although the medical and pharmaceutical literature contains a dizzying array of names for antimicrobics, most of them are variants of a small number of drug families. About 250 different antimicrobial drugs currently classified are grouped in 18 drug families. Drug reference books may give the impression that there are ten times that many because various drug companies assign different trade names to the very same generic drug. Ampicillin, for instance, is sold as Ampen, Amcill, Omnipen, PenA, Principen, and nearly 50 other names. Most antimicrobics are organic cyclic (ring or aromatic) compounds comprised of some combination of peptides, amino acid derivatives, sugars, or nucleotides. Most are useful in controlling bacterial infections, though we shall also consider a number of antifungal, antiviral, and antiprotozoan drugs. Table 10.3 summarizes some major infectious agents, the diseases they cause, and the drugs indicated to treat them (drugs of choice).

Antibacterial Drugs

Penicillin and Its Relatives

The **penicillin** group of antibiotics, named for the parent compound, is a large, diverse group of compounds, most of which end in the suffix *-cillin*. Although penicillins could be completely synthesized in the laboratory from simple raw materials, it is more practical and economical to obtain natural penicillin through microbial fermentation. The natural product may then be used either in unmodified form or to make semisynthetic derivatives. *Penicillium notatum* was the species from which penicillin was first isolated, but it was later replaced by the species *P. chrysogenum,* an even more productive source of the drug. All penicillins consist of three parts: a thiazolidine ring, a beta-lactam ring, and a variable side chain that dictates its microbicidal activity (figure 10.11).

Subgroups and Uses of Penicillins The characteristics of certain penicillin drugs are shown in table 10.4. Penicillins G and V are the most important natural forms. Penicillin G is considered the drug of choice for infections by known sensitive, gram-positive cocci (most streptococci, pneumococci) and some gram-negative bacteria (most strains of gonococci, meningococci, and the spirochete of syphilis). It is usually administered by injection with a repository compound—a chemical that holds the drug at the injection site and releases it slowly. Penicillin V is given orally for many of the same types of infectious agents. It is one of a very few antibiotics approved for antimicrobic prophylaxis and is currently suggested as a daily therapy for children with sickle-cell anemia who are threatened by systemic pneumococcal infections.

Certain semisynthetic penicillins such as ampicillin, carbenicillin, and amoxicillin have broader spectra and so can be used to treat infections by gram-negative enteric rods. Penicillinase-resistant penicillins such as methicillin, nafcillin, and cloxacillin are useful in treating infections caused by some penicillinase-producing bacteria. Mezlocillin and azlocillin, the newest penicillins, have such an extended spectrum that they may be substituted for combinations of antibiotics.

Table 10.3 Selected Survey of Chemotherapeutic Agents in Infectious Diseases

Infectious Agent	Typical Infection	Drug of Choice	Alternate Drug*
Bacteria			
Gram-Positive Cocci			
Staphylococcus	Abscess, skin	Penicillinase-resistant penicillin	Vancomycin, cefoxitin, erythromycin
Streptococcus	Sepsis, erysipelas, pneumonia	Penicillin	Cephalosporin, erythromycin
Gram-Positive Rods			
Bacillus	Anthrax	Penicillin	Erythromycin
Clostridium	Gas gangrene, tetanus	Cephalosporin	Erythromycin, penicillin
Corynebacterium	Diphtheria	Erythromycin	Penicillin
Acid-Fast Rods			
Mycobacterium	Tuberculosis	(INH, rifampin, ethambutol)**	—
M. leprae	Leprosy	(Dapsone, rifampin, clofazimine)**	—
Nocardia	Nocardiosis	Sulfonamide	—
Gram-Negative Cocci			
Neisseria gonorrhoeae	Gonorrhea	Penicillin	Cefoxitin, quinolone, spectinomycin
N. meningitidis	Meningitis	Penicillin	Chloramphenicol
Gram-Negative Rods			
Bordetella	Pertussis	Erythromycin	Ampicillin
Brucella	Brucellosis	Tetracycline	Chloramphenicol
Escherichia coli	Sepsis, diarrhea, urinary tract infection	Aminoglycoside	Ampicillin
Francisella	Tularemia	Streptomycin	Tetracyclines
Hemophilus influenzae	Meningitis	Chloramphenicol	Ampicillin
Legionella	Legionnaire's disease	Erythromycin	Rifampin
Pseudomonas	Opportunistic lung, burn infections	Aminoglycoside	Tetracyclines, carbenicillin, polymyxin
Salmonella	Typhoid fever, salmonellosis	Chloramphenicol, ampicillin	Ampicillin, amoxicillin, quinolone
Shigella	Dysentery	Ampicillin	SxT***, quinolone
Vibrio cholerae	Cholera	Tetracyclines	SxT
Yersinia pestis	Plague	Streptomycin	Tetracyclines
Spirochetes			
Treponema pallidum	Syphilis	Penicillin	Tetracyclines
Borrelia	Lyme disease	Tetracyclines	Penicillin
Mycoplasmas	Pneumonia, urinary infections	Tetracyclines	Erythromycin
Rickettsiae	Spotted fevers	Tetracyclines	Chloramphenicol
Chlamydiae	Urethritis, vaginitis	Tetracyclines, penicillin	Sulfonamides, quinolones
Fungi			
Superficial Mycoses (Dermatophytoses)			
Epidermophyton	Athlete's foot	Topical miconazole	—
Microsporon	Ringworm	Topical clotrimazole	—
Trichophyton	Athlete's foot	Oral griseofulvin	—
Candida albicans	Superficial candidiasis, intestinal candidiasis	Miconazole (topical), nystatin (oral)	—
Systemic Mycoses			
Aspergillus	Aspergillosis	Amphotericin B and/or Flucytosine	—
Blastomyces	Blastomycosis	Amphotericin B	Flucytosine, hydroxystilbamidine
Candida albicans	Candidiasis	Amphotericin B and/or Flucytosine	—
Coccidioides immitis	Valley fever	Amphotericin B	—
Cryptococcus neoformans	Cryptococcosis	Amphotericin B and/or Flucytosine	—
Sporothrix schenckii	Sporotrichosis	Iodides	Amphotericin B

Table 10.3—*Continued*

Infectious Agent	Typical Infection	Drug of Choice	Alternate Drug*
Protozoa			
Balantidium coli	Dysentery	Oxytetracycline	Iodoquinol
Entamoeba histolytica	Amebiasis	Metronidazole, chloroquine	Emetine plus iodoquinol
Giardia lamblia	Giardiasis	Quinacrine	Metronidazole
Plasmodium	Malaria	Chloroquine, primaquine	Quinine
Pneumocystis carinii	Pneumonia	SxT, pentamidine	—
Trichomonas vaginalis	Trichomoniasis	Metronidazole	—
Trypanosoma cruzi	Chagas' disease	Nitrifurimox****	—
T. brucei	Sleeping sickness	Suramin****	Pentamidine
Helminths			
Ascaris	Ascariosis	Mebendazole, pyrantel	Piperazine
Cestodes	Tapeworm	Niclosamide	Praziquantel
Schistosoma	Schistosomiasis	Praziquantel	Metrifonate
Various fluke infections	—	Praziquantel	Tetrachlorethylene
Various roundworm infections	—	Mebendazole, thiabendazol	Piperazine
Viruses			
Orthomyxovirus	Type A influenza	Amantadine	
Herpesvirus	Genital herpes, oral herpes, shingles	Acyclovir, vidarabine	Idoxuridine
HIV virus	AIDS	AZT	DDI, DDC

*Given in the case of patient allergy, microbial resistance, or some other medical concern.
**( ) Usually given in combination.
***A combination of sulfamethoxazole and trimethoprim.
****Available in the United States only from the Drug Service of the Centers for Disease Control.

Table 10.4 Characteristics of Selected Penicillin Drugs

Name	Spectrum of Action	Uses, Advantages	Disadvantages
Penicillin G	Narrow	Best drug of choice when bacteria are sensitive; low cost; low toxicity	Can be hydrolyzed by penicillinase; allergies occur
Penicillin V	Narrow	Good absorption from intestine; otherwise, similar to penicillin G	Hydrolysis by penicillinase; allergies
Oxacillin, dicloxacillin	Narrow	Not susceptible to penicillinase; good absorption	Allergies; expensive
Methicillin, nafcillin	Narrow	Not usually susceptible to penicillinase	Poor absorption; allergies; growing resistance
Ampicillin	Broad	Works on gram-negative bacilli	Can be hydrolyzed by penicillinase; allergies; only fair absorption
Amoxicillin	Broad	Gram-negative infections; good absorption	Hydrolysis by penicillinase; allergies
Carbenicillin	Broad	Same as ampicillin	Poor absorption; used only parenterally
Azlocillin, mezlocillin	Very broad	Effective against *Pseudomonas* species; low toxicity compared to aminoglycosides	Allergies; susceptible to many beta-lactamases

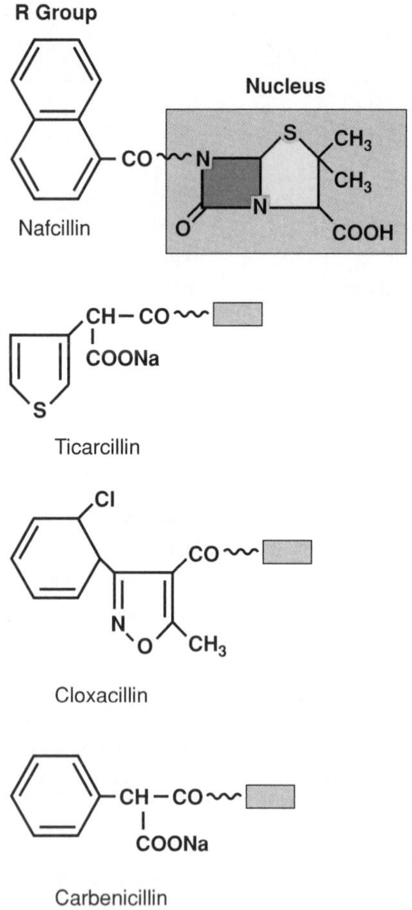

R Group

Nucleus

CO N S CH3 CH3 N O COOH

Nafcillin

CH–CO COONa S

Ticarcillin

Cl CO N O CH3

Cloxacillin

CH–CO COONa

Carbenicillin

Figure 10.11 Chemical structure of penicillins. All penicillins contain a thiazolidine ring (yellow) and a beta-lactam ring (red), but each differs in the nature of the side chain (R group), which is also responsible for differences in biologic activity.

The Cephalosporin Group of Drugs

The **cephalosporins** are a comparatively new group of antibiotics that currently account for a majority of all antibiotics administered. The first compounds in this group were isolated in the late 1940s from the mold *Cephalosporium acremonium.* It wasn't until 20 years later that drug researchers discovered a mutant strain of this species that could be used in drug manufacture. Like penicillin, cephalosporins possess a beta-lactam ring structure and lend themselves to synthetic alterations (figure 10.12). The generic names of these compounds are often recognized by the presence of the root "cef" in their names.

Subgroups and Uses of Cephalosporins The major assets of the cephalosporins stem from their versatility: They are relatively broad spectrum, resistant to penicillinases, and cause fewer allergic reactions. There are presently three generations of cephalosporins, based upon their antibacterial activity. First-generation cephalosporins such as cephalothin and cefazolin are most effective against gram-positive cocci and are mildly active against a few gram-negative bacteria. Second-generation forms include cefaclor and cefonacid, which are more effective than the first-generation forms in treating infections by certain gram-negative bacteria such as *Enterobacter, Proteus,* and *Hemophilus.* Third-generation cephalosporins, represented by cephalexin and cefadizime, are relatively broad spectrum, but have especially well-developed activity against enteric bacteria that produce beta-lactamases. Some third-generation compounds such as cefoperazone are especially good anti-*Pseudomonas* agents. Cefoxitin is a new semisynthetic drug that shows great promise in treating difficult gram-positive and gram-negative infections and anaerobes. Although some cephalosporins are given orally, many are poorly absorbed from the intestine and must be administered **parenterally,** by injection into a muscle or vein.

parenterally (par-ehn'-tur-ah-lee) Gr. *para,* beyond, and *enteron,* intestine. A route of drug administration other than the gastrointestinal tract.

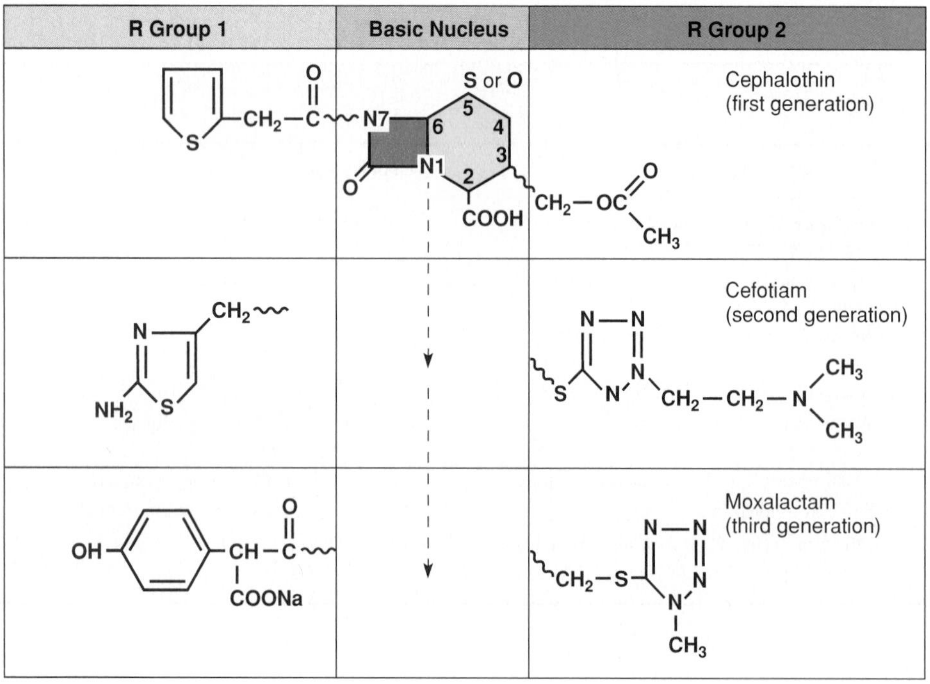

R Group 1	Basic Nucleus	R Group 2
CH_2–C(=O) (thiophene)	S or O; N7, 6, 5, 4, 3, 2, N1; COOH	CH_2–OC(=O)CH_3 — Cephalothin (first generation)
CH_2 (aminothiazole, N, S, NH_2)		S–(tetrazole N—N, N, N)–CH_2–CH_2–N(CH_3)$_2$ — Cefotiam (second generation)
OH–(phenyl)–CH(COONa)–C(=O)		CH_2–S–(tetrazole N—N, N, N–CH_3) — Moxalactam (third generation)

Figure 10.12 The structure of cephalosporins. Like penicillin, these have a beta-lactam ring (red), but a different main ring (yellow). Also, unlike penicillins, they have two sites for placement of R groups (at positions 3 and 7). This makes possible several generations of molecules with greater versatility in function and complexity in structure.

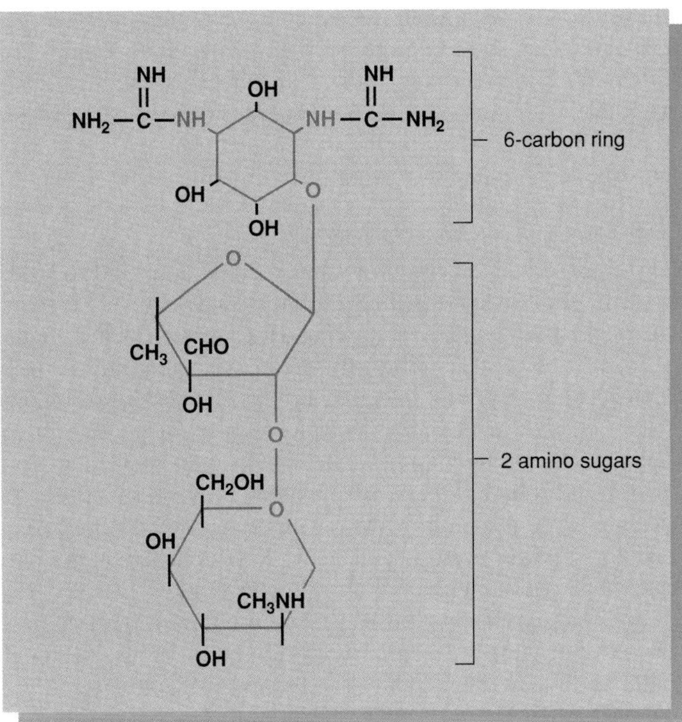

Figure 10.13 The structure of streptomycin, showing the general arrangement of an aminoglycoside (colored portions of molecule).

The New Beta-Lactam Antibiotics

Within the past 15 years, some very powerful and effective beta-lactam-based antibiotics have been discovered. Imipenem has proved to be an excellent broad-spectrum drug for infections with gram-positive and gram-negative aerobic and anaerobic pathogens. Not only is it active in very small concentrations, but it may be taken by mouth and appears to have low human toxicity. Aztreonam, isolated from the purple bacterium *Chromobacterium violaceum,* is a newer narrow-spectrum drug with special effectiveness in pneumonia, septicemia, and urinary tract infections by *Pseudomonas* and other gram-negative aerobic bacilli.

The Aminoglycoside Drugs

Antibiotics composed of two or more amino sugars and an aminocyclitol (6-carbon) ring are referred to as **aminoglycosides** (figure 10.13). These complex compounds are exclusively the products of various species of soil *actinomycetes* in the genera *Streptomyces* (figure 10.14) and *Micromonospora.*

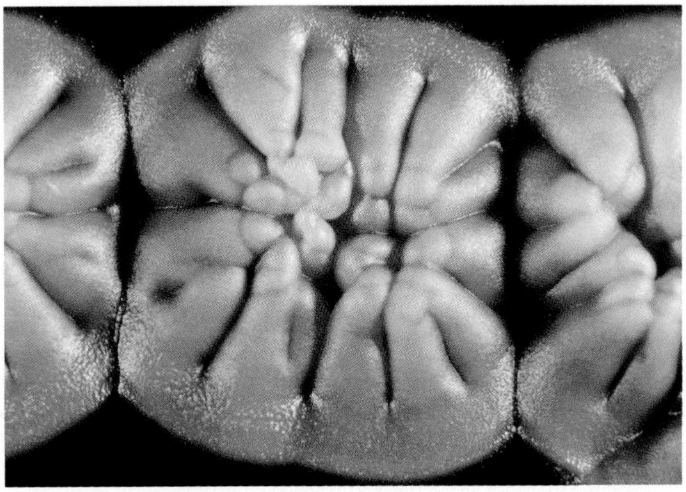

actinomycetes (ak″-tin-oh-my-see′-teez) Gr. *actinos,* ray, and *myces,* fungus. A group of filamentous, funguslike bacteria.

Figure 10.14 A colony of *Streptomyces,* one of nature's most prolific antibiotic producers.

Subgroups and Uses of Aminoglycosides As a group, aminoglycosides have a relatively broad antimicrobial spectrum, but they are especially useful in treating infections caused by aerobic gram-negative rods and certain gram-positive bacteria. Streptomycin is among the oldest of the drugs and has gradually been replaced by newer forms with less mammalian toxicity. It is still the antibiotic of choice for treating bubonic plague, tularemia, and brucellosis, and is a considered a good antituberculosis agent. Gentamicin is less toxic and is widely administered for infections caused by gram-negative rods (*Escherichia coli, Proteus, Klebsiella, Pseudomonas, Salmonella,* and *Shigella*). Two relatively new aminoglycosides, tobramycin and amikacin, are also used for gram-negative bacillary infections and have largely replaced kanamycin. A relative of the aminoglycosides, spectinomycin, is used to treat infections with penicillin-resistant *Neisseria gonorrhoeae.*

Tetracycline Antibiotics

In 1948, a yellow colony of *Streptomyces* isolated from a soil sample gave off a substance, aureomycin, with strong antimicrobic properties. Later isolates (terramycin) were chemically altered to produce the antibiotic called tetracycline. These natural parent compounds and semisynthetic derivatives are called collectively, the **tetracyclines** (figure 10.15*a*). All of the tetracycline drugs are very broad spectrum and appear to enter mammalian cells readily during therapy.

Subgroups and Uses of Tetracyclines The major tetracyclines differ primarily in their stability. Tetracycline and oxytetracycline are short-acting compounds; methacycline and demeclocycline are intermediate in activity; the newer compounds doxycycline and minocycline are long-acting. The scope of microorganisms inhibited by tetracyclines includes gram-positive and gram-negative rods and cocci, aerobic and anaerobic bacteria, mycoplasmas, rickettsias, and spirochetes. Tetracycline compounds are administered orally to treat several sexually transmitted diseases, Rocky Mountain spotted fever, typhus, *Mycoplasma* pneumonia, cholera, leptospirosis, acne, and even some protozoan infections. Although generic tetracycline is low in cost and easy to administer, its side effects—namely, gastrointestinal disruption and deposition in hard tissues—can limit its use (table 10.5).

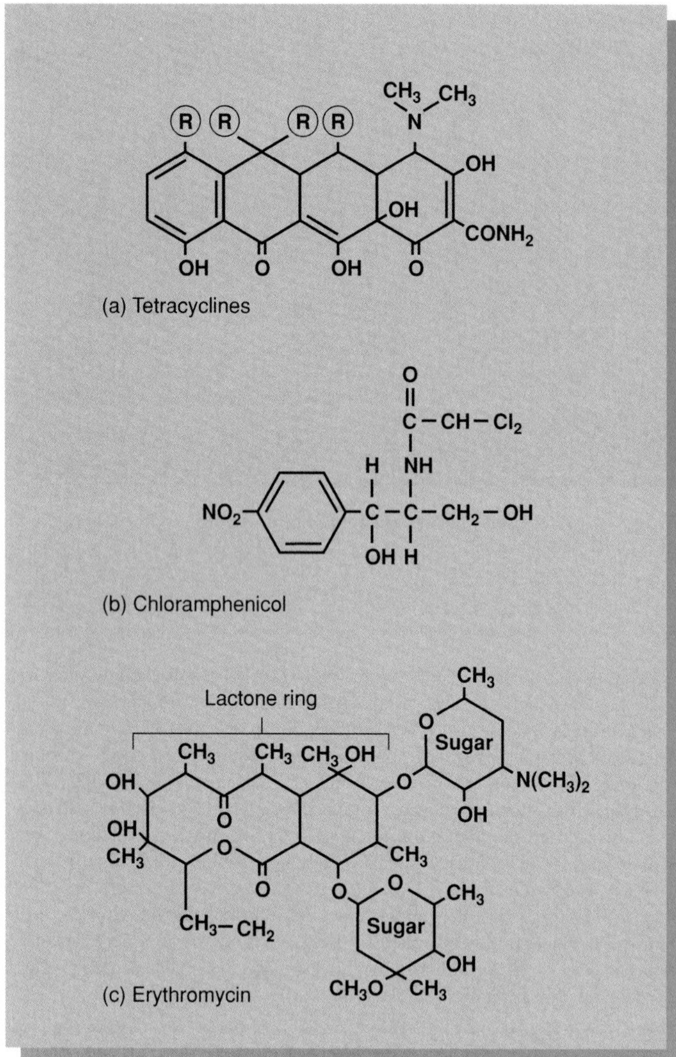

Figure 10.15 Structures of miscellaneous broad-spectrum antibiotics. (*a*) Tetracyclines. These are named for their regular group of four rings. The several types vary in structure and activity by substitution at the four R groups. (*b*) Chloramphenicol. (*c*) Erythromycin. Its central feature is a large lactone ring to which two hexose sugars are attached.

Chloramphenicol

Originally isolated in the late 1940s from *Streptomyces venezuelae,* **chloramphenicol** is a potent broad-spectrum antibiotic with a unique nitrobenzene structure (figure 10.15*b*). It is one type of antibiotic that is no longer derived from the natural source but is entirely synthesized through chemical processes. In the United States chloramphenicol is marketed under the trade name Chloromycetin. Although this drug is fully as broad spectrum as the tetracyclines, evidence of its extreme toxicity began to mount within a short time after its introduction, and by the mid-1950s, restrictions were placed on its use. One out of 24,000–40,000 people undergoing long-term therapy with this drug incurs irreversible damage to the bone marrow that usually results in a fatal form of aplastic anemia.[2] Its administration is now limited to typhoid fever, brain abscesses, eye infections, some forms of meningitis, and certain life-threatening infections for which an alternate therapy is not available. Chloramphenicol should never be given in large doses repeatedly over a long period of time, and the patient's blood must be monitored during therapy.

Erythromycin, Clindamycin, Vancomycin

Erythromycin is a macrolide antibiotic first isolated in 1952 from a strain of *Streptomyces* in a Philippine soil sample. Its structure consists of a large lactone ring with sugars attached (figure 10.15*c*). This drug is relatively broad spectrum and of fairly low toxicity. It is administered orally as the drug of choice for *Mycoplasma* pneumonia, legionellosis, *Chlamydia* infections, pertussis, and diphtheria, and as a prophylactic drug prior to intestinal surgery. It also offers a useful substitute for dealing with penicillin-resistant streptococci and gonococci, and for treating syphilis and acne. Erythromycin is the only macrolide antibiotic with clinical applications.

Clindamycin is a relatively new antibiotic synthesized from lincomycin, another *Streptomyces*-derived antibiotic. Although both antibiotics have relatively broad antibacterial spectra, clindamycin is the more active of the two. The tendency of clindamycin to cause adverse reactions in the gastrointestinal tract limits its applications to: (1) serious infections in the large intestine and abdomen due to anaerobic bacteria that are unresponsive to other antibiotics (*Bacteroides* and *Clostridium*), (2) infections with penicillin-resistant staphylococci, and (3) acne medications applied to the skin.

Vancomycin is a narrow-spectrum antibiotic most effective in treating staphylococcal infections in cases of penicillin and methicillin resistance or in patients with an allergy to penicillins. It has also been chosen to treat *Clostridium* infections in children and endocarditis (infection of the lining of the heart) caused by *Enterococcus faecalis.* Because it is very toxic and hard to administer, vancomycin should be used only in the most serious, life-threatening conditions.

Rifampin

Another product of the prolific *Streptomyces* is rifamycin, which is altered chemically into **rifampin.** This dark red antibiotic has a somewhat limited spectrum because the molecule is too large to pass through the cell envelope of many gram-negative bacilli. It is, however, very effective in treating infections by several gram-positive rods and cocci and a few gram-negative bacteria. Rifampin figures most prominently in treating mycobacterial infections, especially tuberculosis and leprosy, but it is usually given in combination with other drugs to prevent development of resistance. Rifampin is recommended for prophylaxis in *Neisseria meningitidis* carriers and their contacts, and it is occasionally used to treat *Legionella, Brucella,* and *Staphylococcus* infections.

The *Bacillus* Antibiotics: Bacitracin and Polymyxin

Bacitracin is a narrow-spectrum peptide antibiotic produced by a strain of *Bacillus subtilis* isolated from the infected compound fracture of a girl named Tracy. Since then, its greatest

2. A failure of the blood-producing elements that results in very low levels of red and white blood cells.

claim to fame has been as a major ingredient in a common drugstore antibiotic ointment (Neosporin) for combating superficial skin infections by streptococci and staphylococci. For this purpose, it is usually combined with neomycin (an aminoglycoside) and polymyxin.

Bacillus polymyxa is the source of the **polymyxins,** narrow-spectrum peptide antibiotics with a unique fatty acid component that contributes to their detergent activity (figure 10.7). Only two polymyxins—B and E (also known as colistin)—have any routine applications, and even these are limited by their toxicity to the kidney. Either drug may be indicated to treat drug-resistant *Pseudomonas aeruginosa* and severe urinary tract infections caused by other gram-negative rods. Polymyxin B alone is commonly added to topical skin ointments.

Synthetic Antibacterial Drugs

The synthetic antimicrobics as a group do not originate from bacterial or fungal fermentations. Some were developed from aniline dyes and others were originally isolated from plants. Although they have been largely supplanted by antibiotics, several types are still useful.

The Sulfonamides, the Sulfones, and Trimethoprim

The very first modern antimicrobic drugs were the **sulfonamides** or sulfa drugs, named for para-aminobenzenesulfonamide (sulfanilamide; figure 10.16). Sulfanilamide was the first sulfa drug characterized and the first drug to be used in a clinical trial in the United States. Although thousands of sulfonamides have been formulated, only a few have gained any importance in chemotherapy. Because of its solubility, sulfisoxazole is the best agent for treating nocardiosis, acute urinary tract infections, and certain protozoan infections. Silver sulfadiazine ointment and solution are prescribed for treatment of burns and eye infections. In many cases, sulfamethoxazole is given in combination with trimethoprim (Septra, Bactrim) to take advantage of the synergistic effect of the two drugs.

Sulfones are compounds chemically related to the sulfonamides but lacking their broad-spectrum effects. This does not diminish their importance as key drugs in treating leprosy. The most active form is dapsone, usually given in combination with rifampin and clofazamine (an antibacterial dye) over long periods. **Trimethoprim** is a synthetic pyrimidine with well-developed activity against many gram-positive and gram-negative bacteria. It is usually combined with sulfonamides to treat infections of the urinary, respiratory, and gastrointestinal tracts, and has proved an effective medication for some sexually transmitted diseases and for *Pneumocystis* pneumonia in AIDS patients.

Miscellaneous Antibacterial Agents **Isoniazid** (INH) has been in use since 1952 to treat tuberculosis. It is bactericidal to *Mycobacterium tuberculosis,* but only against growing cells. Oral doses are indicated for both active tuberculosis and prophylaxis in cases of a positive TB test. Ethambutol, a closely related

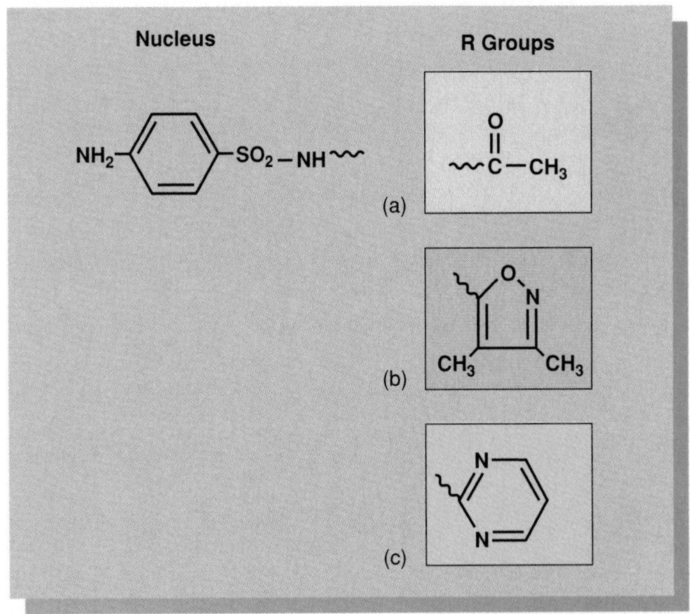

Figure 10.16 The structures of some sulfonamides: (*a*) sulfacetamide; (*b*) sulfadiazine; (*c*) sulfisoxazole.

compound, is effective in treating the early stages of tuberculosis.

Nitrofurantoin, a derivative of furan sugars, is a fairly broad-spectrum synthetic drug most effective in gram-negative urinary tract infections because it is excreted by the kidney and is concentrated in the urine.

Much excitement has been generated by the new class of synthetic drugs chemically related to quinine called **fluoroquinolones.** These drugs exhibit several ideal traits, including potency and broad spectrum. Even in minimal concentrations, quinolones inhibit nearly every gram-positive and gram-negative bacterial species tested. They are readily absorbed from the intestine and less subject to microbial resistance than other drugs. The principal quinolones, norfloxacin and ciprofloxacin, have been successful in therapy for urinary tract infections, sexually transmitted diseases, gastrointestinal infections, prophylaxis in immunosuppressed patients, osteomyelitis, respiratory infections, and soft tissue infections. However, early evidence points to harmful effects on the central nervous system and intestinal tract that could limit applications of these drugs.

Agents to Treat Fungal Infections

The cells of fungi, being eucaryotic, present special problems in chemotherapy. For one, the great majority of chemotherapeutic drugs useful in treating bacterial infections are generally ineffective in combating fungal infections, and for another, the similarities between fungal and human cells often mean that drugs toxic to fungal cells are likely to harm human tissues. A small battery of agents with special antifungal properties has been developed for treating systemic and superficial fungal infections. Four main drug groups currently in use are the macrolide polyene antibiotics, griseofulvin, the synthetic imidazoles, and flucytosine (figure 10.17).

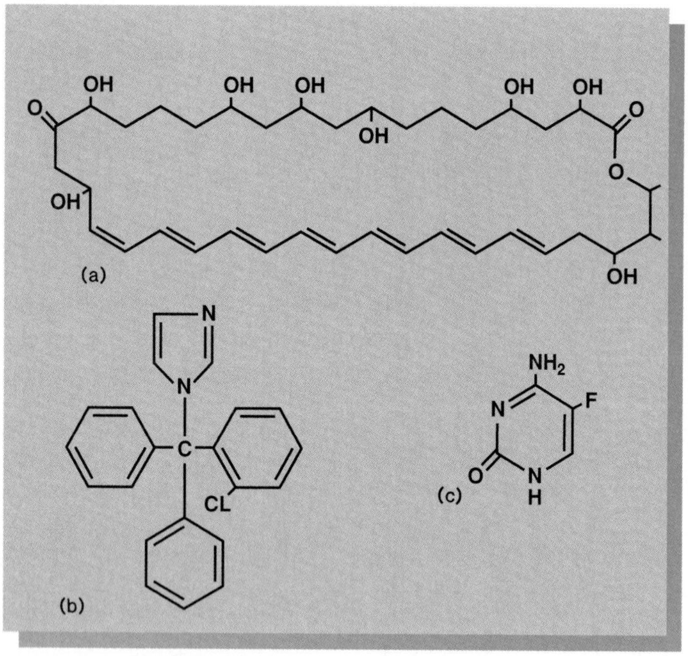

OH OH OH OH OH
O
OH
O
O
OH
(a)
OH
N
N
NH$_2$
N
F
C
O
N
H
CL
(c)
(b)

Figure 10.17 Some antifungal drug structures. (*a*) Polyenes. The example shown is amphotericin B (proposed structure), a complex steroidal antibiotic that inserts into fungal cell membranes. (*b*) Clotrimazole, one of the imidazoles. (*c*) Flucytosine, a structural analog of cytosine that contains fluoride.

Macrolide polyenes, represented by **amphotericin B** (named for its acidic and basic—amphoteric—properties) and **nystatin** (for New York State, where it was discovered), have a structure that mimics the lipids in some cell membranes. Both compounds were originally isolated from species of *Streptomyces.* Amphotericin B (Fungizone) is by far the most versatile and effective of all antifungals. Not only does it work on most fungal infections, including skin and mucous membrane lesions caused by *Candida albicans,* but it is presently the only drug that can be injected to treat systemic fungal infections. The current increase in the number of patients with these life-threatening infections makes amphotericin B an invaluable drug. Nystatin is used only topically or orally to treat candidiasis of the skin and mucous membranes, but it is not useful for subcutaneous or systemic fungal infections or for ringworm.

Griseofulvin is an antifungal product of the mold *Penicillium.* An interesting property of this drug makes it especially active in certain dermatophyte infections such as athlete's foot. Oral doses are absorbed into the blood and deposited in the epidermis, nails, and hair as they grow. Fungi infecting these areas will be inhibited and eventually destroyed and sloughed off with these tissues. Because complete eradication requires several months and griseofulvin is relatively nephrotoxic, this therapy is given in only the most extreme cases.

The **imidazoles** are broad-spectrum antifungal agents with a complex ringed structure. The three most effective drugs are ketoconazole, which is used orally and topically for cutaneous mycoses, vaginal and oral candidiasis, and some systemic mycoses, and clotrimazole and miconazole, which are applied in creams and solutions to infections in the skin, mouth, and vagina.

Flucytosine is an analog of cytosine that was first developed for tumor therapy in mammals. Although not an effective anticancer drug, it turned out to be useful in combating fungi. Its best features are its rapid absorption after oral therapy and its property of entering the blood and cerebrospinal fluid. Alone, it may be used to treat *Cryptococcus* meningitis and candidiasis. Because many fungi are resistant to flucytosine, it must be combined with amphotericin B to effectively treat other systemic mycoses.

Antiparasitic Chemotherapy

The enormous diversity among protozoan and helminth parasites and their corresponding therapies reaches far beyond the scope of this textbook; however, a few of the more common drugs will be surveyed here and again in chapter 19. Presently, a small number of approved and experimental drugs are used to treat malaria, leishmaniasis, trypanosomiasis, amebic dysentery, and helminth infections, but the need for new and better drugs has spurred considerable research in this area.

Antimalarial Drugs: Quinine and Its Relatives

For hundreds of years, quinine, extracted from the bark of the cinchona tree, was the principal treatment for malaria. After World War II, it was replaced by the synthesized quinolines, mainly chloroquine and primaquine, which had less toxicity to humans. But because there are several species of *Plasmodium* (the malaria parasite) and many stages in its life cycle, no one drug is universally effective for every species and stage, and each drug is restricted in application. Primaquine eliminates the liver phase of infection, and chloroquine suppresses acute attacks associated with infection of red blood cells. The drugs are taken together for malarial prophylaxis and cure. Although quinine chemotherapy was abandoned for a time because of its toxicity, the recent development of chloroquine-resistant *Plasmodium* in South America and Southeast Asia restored quinine to a role in treating drug-resistant infections (see chapter 19).

Chemotherapy for Other Protozoan Infections

A widely used amebicide, metronidazole (Flagyl), is effective in treating mild and severe intestinal infections and hepatic disease caused by *Entamoeba histolytica.* Given orally, it also has applications for infections by *Giardia lamblia* and *Trichomonas vaginalis.* Another antidote for amebic infections is dihydroemetine, a drug whose action is to induce emesis (vomiting), which helps eliminate the pathogen.

Antihelminthic Drug Therapy

Treating helminth infections has been one of the most difficult and challenging of all chemotherapeutic tasks. Flukes, tapeworms, and roundworms are much larger parasites than other microorganisms and have greater similarities to human physiology, so drugs that merely block their reproduction are usually not successful in eradicating the adult worms. The most effective drugs immobilize, disintegrate, or inhibit the metabolism of the worms.

Mebendazole and thiabendazole are broad-spectrum antiparasitic drugs used in several roundworm and tapeworm intestinal infestations. Working locally within the intestine, these drugs inhibit the function of the cellular microtubules of adult worms, thereby interfering with their glucose utilization and disabling them. Egg and larval stages are also killed by these drugs. The compounds pyrantel and piperazine paralyze the muscles of intestinal roundworms. Niclosamide destroys the scolex and the adjoining proglottids of tapeworms, thereby loosening the worm's holdfast. In these forms of therapy, the worms are unable to maintain their grip on the intestinal wall and are expelled along with the feces by the normal peristaltic action of the bowel. One of the newer antihelminth drugs, praziquantel, appears to successfully treat various tapeworm and fluke infections.

Antiviral Chemotherapeutic Agents

The treatment of viral infections with chemotherapeutic agents is still in its infancy. Traditionally, viral infections have been prevented by vaccination, but a useful vaccine has yet to be developed for many viruses. Recent epidemics of AIDS, genital warts, and measles, not to mention the ever present common cold, continue to emphasize the need for effective antiviral drugs. The major limitations of the small number of existing antiviral agents are that most of them have an extremely narrow spectrum and act only on viruses inside the cell. Most compounds halt the viral multiplication cycle by (1) barring complete penetration of the virus into the host cell, (2) blocking the transcription and translation of viral molecules, or (3) preventing the maturation of viral particles. Although antiviral drugs protect uninfected cells by keeping viruses from being synthesized and released, most are unable to destroy extracellular viruses or those in a latent state.

The mode of action of several antiviral agents is to mimic the structure of nucleotides and compete for sites on replicating DNA. The incorporation of these synthetic nucleotides inhibits further DNA synthesis. **Acyclovir** (Zovirax) is a synthetic purine compound that blocks DNA synthesis in a small group of viruses, particularly the herpes simplex viruses (HSV). In the topical form, it is most effective in controlling the primary attack of facial or genital herpes. Intravenous or oral acyclovir therapy has proved successful in reducing the severity of primary and recurrent genital HSV episodes. This form is also used to treat shingles and chickenpox caused by the herpes zoster virus. *Idoxuridine* is an analog of thymine that inhibits the synthesis of DNA in certain viruses; at present, it is prescribed only for eye infections by herpes simplex. An analog of adenine, *vidarabine,* is also effective against the herpesviruses. **Ribavirin** is a guanine analog used in aerosol form to treat life-threatening infections by the respiratory syncytial virus (RSV) in infants (see chapter opening illustration). It has been one of the large battery of experimental drugs tested by AIDS patients, though it does not appear to have any therapeutic benefit in that disease.

Azidothymidine (**AZT** or Retrovir) is a thymine analog used exclusively to treat AIDS patients. This drug is specific for HIV by preventing the natural action of the viral reverse transcriptase and blocking further DNA synthesis and viral replication (see chapter 21). It is indicated for patients with overt AIDS and for those with AIDS-related complex (ARC). It also appears to benefit patients in the early asymptomatic stage of HIV disease. Although it is not a cure, AZT does slow the course of the disease in most instances.

Amantadine and its relative, rimantidine, are amines restricted almost exclusively to treating infections by influenza A virus. Because their action appears to inhibit the uncoating of the viral RNA, these drugs must be given rather early in an infection. Amantadine is also used prophylactically to reduce the incidence of influenza A infections in the elderly. In addition to these antiviral drugs, dozens of other agents are under investigation. For a novel approach to antiviral therapy, see feature 10.3.

At one time, great hope was held for the antiviral promise of interferon. **Interferon** is a carbohydrate-containing protein produced naturally by infected animal cells. In humans, interferon is produced primarily by fibroblasts and leukocytes in response to various infectious stimuli. The discovery of interferon's potential application in viral infections and cancer therapy

Feature 10.3 Household Remedies—From Apples to Zinc

Who would have thought that drinking a glass of apple juice, eating a clove of garlic, or a sneezing into a facial tissue might nip a viral infection in the bud? A series of research findings from the past few years seems to point to a possible role for these and other humble medicinal aids. Apple juice, fruit juices, and even tea contain natural antiviral substances, thought to be tannic or other organic acids, that kill the poliovirus and coxsackievirus outside the body. Thus, drinking beverages that contain these substances may help prevent the passage of those viruses into the intestine (their usual site of entry). Could this be a reason that "an apple a day keeps the doctor away"?

Specialists in human rhinoviruses suggest that the most important route of transmission of cold viruses is through hand-to-hand contact. Researchers with Kimberly-Clark used this information to develop a special tissue impregnated with iodine and citric acid. If used by patients to catch sneezes, coughs, and secretions, these tissues proved quite effective in retarding the spread of the cold virus and even in reducing the severity of the infection in some cases. The micronutrient zinc shows some promise in inhibiting rhinoviruses and herpesviruses, especially if the patient sucks on a zinc lozenge periodically during the early stages of infection. Although large doses of ascorbic acid or vitamin C have been cited for years as a superior cold remedy, this claim is still considered controversial by the medical community, and most scientific tests have been inconclusive.

The therapeutic benefits of certain foods are often surprising. Yogurt made with live cultures has recently been shown to contain a natural antibiotic. This may explain the benefits of eating yogurt to control yeast infections of the gastrointestinal tract and vagina. Research indicates that garlic extract also contains active ingredients that clearly inactivate several types of animal viruses. If all else fails, one should not overlook the recuperative powers of chicken soup, sometimes known as "Jewish penicillin." This, too, has been found in controlled scientific tests to shorten the length and relieve the symptoms of colds, though the active ingredients have not been isolated. It appears that a timely trip to the kitchen cabinet could be as beneficial as one to the medicine cabinet. Might this be what is meant by "feeding a cold and starving a fever"?

prompted a tremendous push for research on its biological effects. We now know that it is a versatile part of animal host defenses, having a broad spectrum of activities and great import in natural immunities. (Its mechanism will be discussed in chapter 12.)

The first investigations of interferon's antiviral activity were limited by the extremely minute quantities that could be extracted from human blood. As techniques in recombinant DNA technology made it possible to produce larger quantities of interferon, extensive clinical trials to test its effectiveness in viral infections and cancer were undertaken, with inconclusive and somewhat disappointing results. Some major conclusions regarding interferon therapy may be summarized as follows: (1) It reduces the time of healing and some of the complications in certain infections (mainly of herpesviruses) but does not cure them; (2) it may prevent some symptoms of cold and papilloma (warts) viruses; (3) it is somewhat toxic if given in large doses; and (4) it apparently slows the progress of a few cancers, including bone and breast cancer and certain leukemias and lymphomas. A form of interferon is approved for treating a rare cancer called hairy-cell leukemia. It is presently being tested as a nasal spray to control cold virus, as a treatment for hepatitis C (a liver infection spread by blood transfusions), and as one experimental therapy for AIDS. Whether interferon has the potential to be an important all-purpose antiviral or anticancer resource is still not clear.

Characteristics of Host/Drug Reactions

Although selective antimicrobial toxicity is the ideal constantly being sought, chemotherapy by its very nature involves contact with foreign chemicals that can harm human tissues. In fact, estimates indicate that at least 5% of all persons taking an antimicrobic drug experience some type of serious adverse reaction to it. The major **side effects** of drugs fall into one of three categories: direct damage to tissues through toxicity, allergic reactions, and disruption in the balance of normal microbial flora. The damage incurred by antimicrobial drugs may be short-term and reversible or permanent, and it ranges in severity from cosmetic to lethal. Table 10.5 summarizes drug groups and their major side effects.

Toxicity to Organs

Certain drugs adversely affect the following organs: the liver (hepatotoxic drugs), kidneys (nephrotoxic drugs), gastrointestinal tract, cardiovascular system and blood-forming tissue, nervous system (neurotoxic drugs), respiratory tract, skin, bones, and teeth.

Because the liver is responsible for metabolizing and detoxifying foreign chemicals in the blood, it may be damaged by a drug or its breakdown products. Injury to liver cells can result in enzymatic abnormalities, fatty liver deposits, hepatitis, and jaundice, and it may even lead to fatalities. The kidney, too, excretes drugs and their metabolites in its work as a blood filter. Some drugs irritate the nephron tubules, creating changes that interfere with their filtration abilities. Drugs such as sulfonamides crystallize in the kidney pelvis and form stones that can obstruct the flow of urine.

The most common complaint associated with oral antimicrobial therapy is diarrhea, which can progress to severe intestinal irritation or colitis. Although some drugs directly irritate the intestinal lining, the usual gastrointestinal complaints are caused by disruption of the intestinal microflora (discussed in a subsequent section).

Many drugs given for parasitic infections are toxic to the heart, causing changes in blood pressure, dysrhythmias, and even cardiac arrest in extreme cases. Chloramphenicol can cause severe depression of blood-forming cells in the bone marrow, resulting in either a reversible or permanent (fatal) anemia. Some drugs hemolyze the red blood cells, some reduce white blood cell counts, and still others damage platelets or interfere with their formation, thereby inhibiting blood clotting.

Certain antimicrobics act directly on the brain and cause seizures. Others, such as aminoglycosides, damage nerves (very commonly, the 8th cranial nerve), leading to dizziness, vertigo, deafness, or motor and sensory disturbances. When drugs block the transmission of impulses to the diaphragm, respiratory failure can result.

The skin is a frequent target of drug-induced side effects. The skin response can be a symptom of drug allergy or a direct toxic effect. Some drugs interact with sunlight to cause photodermatitis, a skin inflammation. Tetracyclines are contraindicated (not advisable) for children from birth to eight years of age because they bind to the enamel of the teeth, creating a permanent gray to brown discoloration (figure 10.18). Pregnant women should avoid tetracyclines because they cross the placenta and can be deposited in the developing fetal bones and teeth.

Allergic Responses to Drugs

Probably the most frequent drug reaction is heightened sensitivity, or *allergy*. This reaction occurs because the drug acts as an antigen (a foreign material capable of stimulating the immune system) and stimulates an immune response. This response may be provoked by the intact drug molecule or by substances that develop from the body's metabolic alteration of the drug. In the case of penicillin, for instance, it is not the penicillin molecule that causes the allergic response, but a major product, *benzylpenicilloyl*. Allergic reactions have been reported for every major type of antimicrobic drug, but the penicillins account for the greatest number of antimicrobic allergies, followed by the sulfonamides.

Table 10.5 Major Adverse Toxic Reactions to Common Drug Groups		
Antimicrobic Drug	**Tissue Affected**	**Damage or Abnormality Produced**
Antibacterials		
Penicillin G	Brain	Seizures*
Carbenicillin	Platelets	Abnormal bleeding
Ampicillin	GI tract	Diarrhea and enterocolitis**
Cephalosporins	Platelet function	Inhibition of prothrombin synthesis
	Brain	Seizures*
	Kidney	Nephritis
Tetracyclines	GI tract	Diarrhea and enterocolitis
	Liver	Damage to hepatocytes*
	Teeth, bones	Gray to brown discoloration of tooth enamel in fetuses and children from birth through age eight.
	Skin	Reactions to sunlight
Chloramphenicol	Bone marrow	Injury to red and white blood cell precursors; blood cell deficiencies
Aminoglycosides (streptomycin, kanamycin, gentamicin, amikacin)	GI tract, hair cells in cochlea, vestibular cells, neuromuscular, kidney tubules	Diarrhea and enterocolitis; malabsorption; loss of hearing, dizziness, vertigo; respiratory failure; loss of filtration ability
Isoniazid	Liver	Hepatitis
	Brain	Seizures
	Skin	Dermatitis
Sulfonamides	Kidney, pelvis	Formation of crystals; blockage of urine flow
	Red blood cells	Hemolysis
	Platelets	Reduction in number
Polymyxin	Kidney	Damage to membranes of tubule cells
	Neuromuscular	Weakened muscular responses
Rifampin	Liver	Damage to hepatic cells
	Skin	Dermatitis
Antifungals		
Amphotericin B	Kidney	Disruption of tubular filtration
Flucytosine	White blood cells	Decreased number
Antiprotozoan Drugs		
Metronidazole	GI tract	Nausea, vomiting
Chloroquine	GI tract	Vomiting
	Brain	Headache
	Skin	Itching
Antihelminthics		
Niclosamide	GI tract	Nausea, abdominal pain
Pyrantel	GI tract	Irritation
	Brain	Headache, dizziness
Antivirals		
Acyclovir	Brain	Seizures, confusion
	Skin	Rash
Amantadine	Brain	Nervousness, lightheadedness
	GI tract	Nausea
AZT	Marrow	Immunosuppression, anemia

*A very rare reaction.
**An inflammation of the intestinal tract.

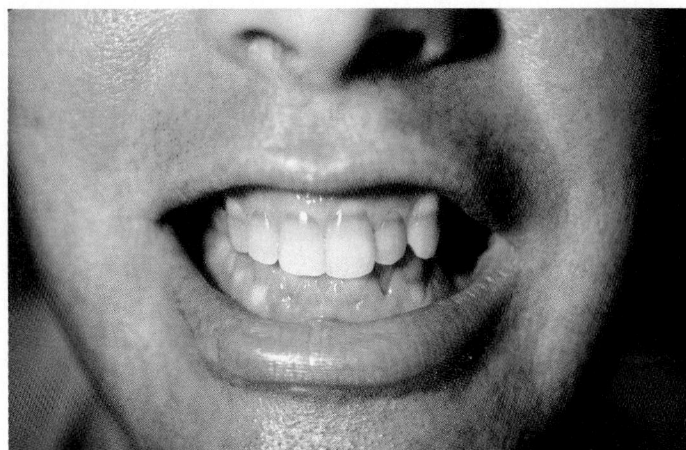

Figure 10.18 An adverse effect of tetracycline is the permanent discoloration of tooth enamel.

Manifestations of Drug Allergies Allergic persons become sensitized to a drug during the first contact, usually without symptoms. Once the immune system is sensitized, a second exposure to the drug can lead to an immediate reaction such as a skin rash (hives), respiratory inflammation, and rarely, anaphylaxis, an acute, overwhelming allergic response that comes on rapidly and may be fatal. More commonly, however, the response is delayed, with symptoms developing several days after therapy. Persons with this form of hypersensitivity develop cutaneous reactions and fever. (This topic is discussed in greater detail in chapter 14.)

Suppression and Alteration of the Microflora by Antimicrobics

Most normal, healthy body surfaces, such as the skin, large intestine, outer openings of the urogenital tract, and oral cavity, provide numerous habitats for a virtual "garden" of microorganisms. These normal colonists or residents, called the **flora** or microflora, consist mostly of harmless or beneficial bacteria, but some may be potential pathogens. In this coexistence with humans, the flora remains relatively stable and in a balance necessary to the well-being of both parties. Although we shall defer a more detailed discussion of this topic to chapter 11, let us now focus on the general effects of drugs on this population.

Significant Effects of Antimicrobics on Microbial Flora

The mixed population of microbial colonists normally exists in a state of ecological balance that tends to hold pathogenic members in check. If a broad-spectrum antimicrobic is introduced into the host to treat infection, it will destroy microbes regardless of their roles in the ecologic balance, affecting not only the targeted infectious agent, but many others in sites far removed from the original infection (figure 10.19). The result of this therapy may be the destruction of beneficial resident species and the subsequent survival and overgrowth of potentially harmful residents or contaminants, a complication called a **superinfection.**

Some common examples will show how a disturbance in microbial flora leads to replacement flora and superinfection. A broad-spectrum cephalosporin used to treat a urinary tract infection by *Escherichia coli* will cure the infection, but it will also kill off the lactobacilli in the vagina that normally maintain a protective acidic environment there. The drug has no effect, however, on *Candida albicans,* a yeast that also resides in normal vaginas. Released from the inhibitory pH normally provided by lactobacilli, the yeasts proliferate and cause an infection. *Candida* may cause similar superinfections of the oropharynx (thrush) and the large intestine.

Oral therapy with tetracyclines, clindamycin, and broad-spectrum penicillins and cephalosporins is associated with a serious and potentially fatal condition known as *antibiotic-associated colitis* (pseudomembranous colitis). This condition is due to the overgrowth in the bowel of *Clostridium difficile,* a spore-forming anaerobic species that is resistant to the antibiotic. It invades the intestinal lining and releases toxins that induce diarrhea, fever, and abdominal pain. Excessive use of broad-spectrum antimicrobial drugs leads to the selection for drug-resistant strains of the microflora, and these can be an ongoing source of superinfections in hospital patients.

Considerations in Selecting an Antimicrobial Drug

Before actual antimicrobial therapy can begin, it is important that at least three factors be known: (1) the nature of the microorganism causing the infection; (2) the degree of the microorganism's susceptibility (also called sensitivity) to various drugs; and (3) the overall medical condition of the patient.

Identifying the Agent

Identification of infectious agents from body specimens should be attempted as soon as possible. It is especially important that such specimens be taken before the antimicrobic drug is given, because the drug may bar the isolation of the infectious agent. Direct examination of gram-stained specimens of body fluids, sputum, or stool is still a valuable method for rapidly detecting and perhaps even identifying bacteria or fungi. A doctor often begins the initial therapy on the basis of such immediate findings. The choice of drug will be based on drugs that are known from past experience to be effective against the microbe; this is called the "informed best guess." For instance, if a sore throat appears to be caused by *Streptococcus pyogenes,* the physician might prescribe penicillin, because this species seems to be universally sensitive to it so far. If the infectious agent is not or cannot be isolated, epidemiological statistics may be required to predict the most likely agent in a given infection. For example, *Hemophilus influenzae* accounts for the majority of cases of meningitis in children, followed by *Streptococcus pneumoniae* and *Neisseria meningitidis.* Microbiologic statistics of this sort are available for many diseases and can be a valuable aid.

---

flora (flor'-ah) Gr. *flora,* the goddess of flowers. The microscopic life present in a particular locale.

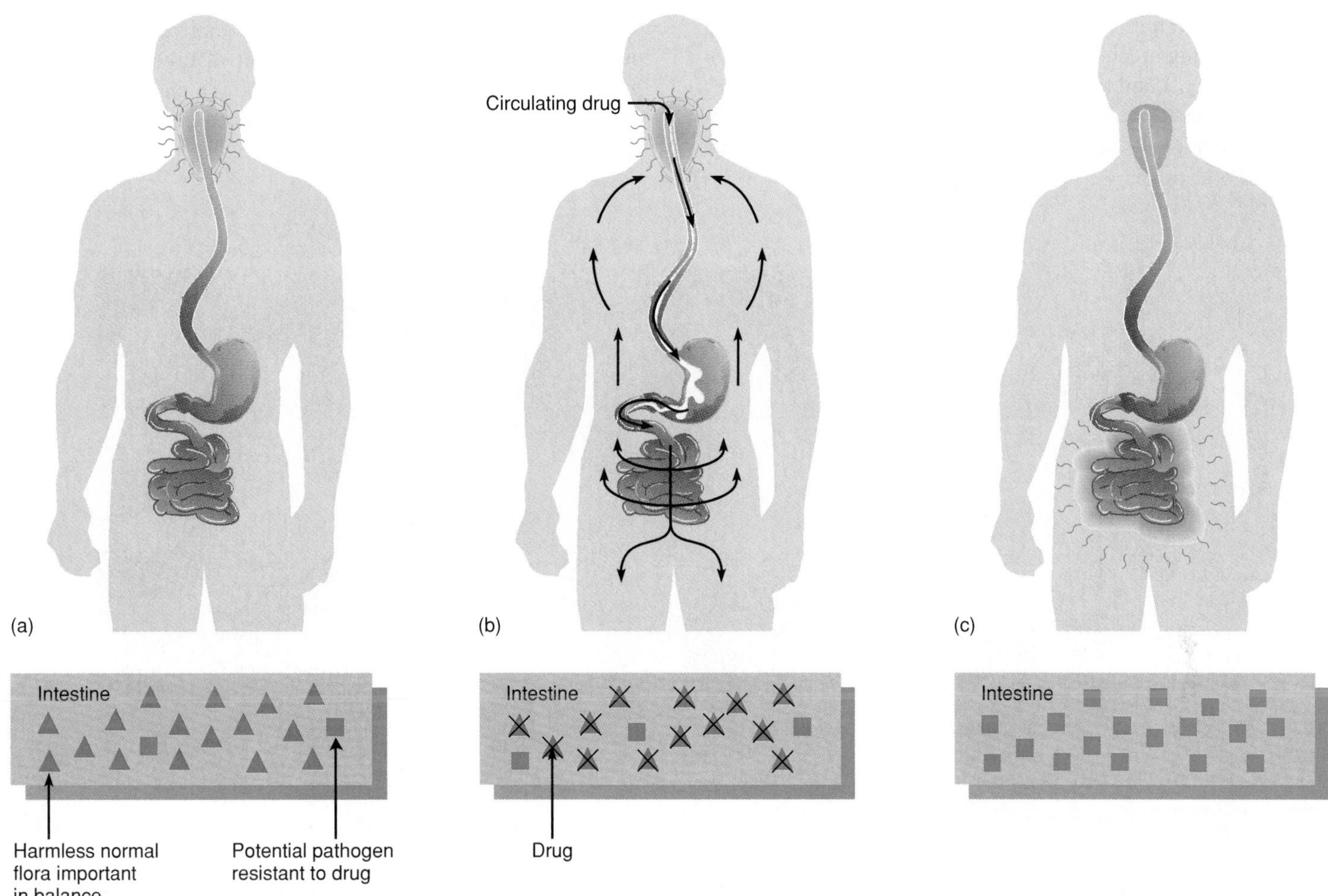

Figure 10.19 The role of antimicrobics in disrupting microbial flora and causing superinfections. (*a, b*) A primary infection in the throat is treated with an oral antibiotic that enters the intestine and is absorbed into the circulation. (*c*) The primary infection is cured, and the intestine is superinfected by drug-resistant pathogens that survived and proliferated in the intestine.

Testing for the Drug Susceptibility of Microorganisms

Not all infectious agents require antimicrobial sensitivity testing. Drug testing in fungal or protozoal infections is difficult and may be unnecessary. And because certain groups, such as group A streptococci and all anaerobes (except *Bacteroides*), have so far been uniformly susceptible to penicillin G, testing may not be necessary unless the patient is allergic to penicillin. But testing is essential in those groups commonly showing resistance, primarily *Staphylococcus* species, *Neisseria gonorrhoeae,* certain streptococci (*S. pneumoniae* and *S. faecalis*), and the aerobic gram-negative enteric bacilli.

Selection of a proper antimicrobial agent begins by demonstrating the *in vitro* activity of several drugs against the infectious agent by means of standardized methods. In general, these tests involve exposing a pure culture of the isolated bacterium to several different drugs and observing it macroscopically for the effects of the drugs on growth. The simplest method, the *Kirby-Bauer* technique, is an agar diffusion test that provides useful semiquantitative or qualitative data on antimicrobic susceptibility. In this test, the surface of a plate of special medium is seeded carefully with the isolated test bacterium, and small discs containing a premeasured amount of antimicrobic are dispensed onto the bacterial lawn. After 18 to 24 hours of incubation at 37 °C, the zone of inhibition surrounding the discs is measured and compared with a standard for each drug (figure 10.20 and table 10.6). The profile of antimicrobic sensitivity, or antibiogram, provides data for drug selection. An advantage of the Kirby-Bauer procedure is that many drugs can be tested simultaneously in a single plate; however, it is less effective for anaerobic, highly fastidious, or slow-growing bacteria. Another disadvantage is that it does not provide precise quantitative information on the proper dosage of drug required to treat the infection.

More sensitive and quantitative results can be obtained with tube dilution tests. First the antimicrobic is diluted serially in containers of broth, and then each tube is inoculated with a small uniform sample of pure culture. After incubation for 18–24 hours, the tubes are examined for growth (turbidity). The smallest concentration of drug that visibly inhibits growth is called the **minimum inhibitory concentration,** or MIC. The MIC is useful in determining the smallest effective dosage of a drug and in providing a comparative index against other antimicrobics (figure 10.21 and table 10.7). It can be informative to carry this test further by selecting tubes with no visible growth, subculturing them, and retesting this population for susceptibility. The lowest drug concentration that completely and permanently suppresses growth in this subset of tubes is called the *minimum lethal concentration,* or MLC. In many clinical laboratories, these antimicrobic testing procedures are automated (figure 10.21*b*).

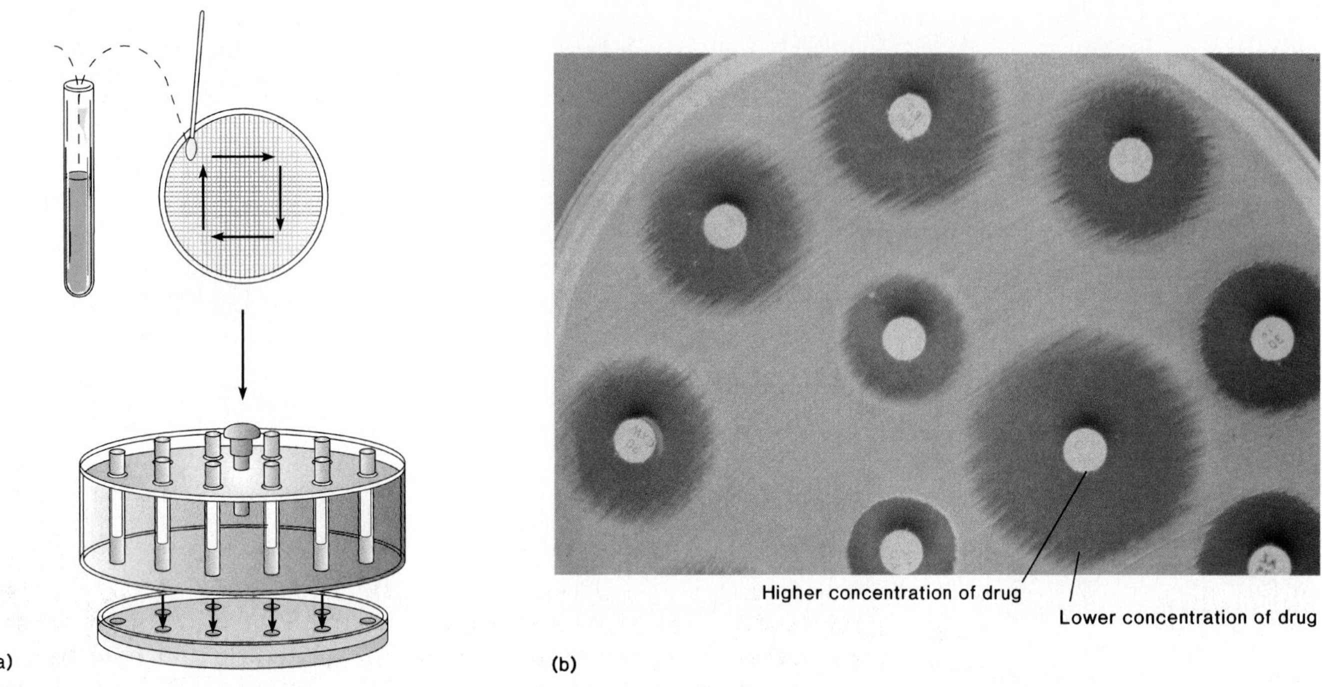

Figure 10.20 Method and interpretation of disc diffusion. (*a*) Steps in plate preparation and measurement. During incubation, antimicrobics become increasingly diluted as they diffuse out of the disc into the medium. (*b*) Interpretation of results. If the test species is sensitive to a drug, it will be inhibited and a zone of inhibition (no growth) develops. The larger this zone is, the greater is the sensitivity to the drug (as the bacterium is inhibited by lower concentrations). The zones' diameters are measured and compared to a standardized chart (see table 10.6). If a species is drug resistant, there will be little or no zone of inhibition around the disc.

Table 10.6 Results of Kirby-Bauer Test

Drug	Zone Sites (in mm) Required For: Susceptibility (S)	Resistance (R)	Actual Result for Lab Strain of *Staphylococcus aureus*	Evaluation
Bacitracin	> 13	< 8	15	S
Chloramphenicol	> 18	< 12	20	S
Erythromycin	> 18	< 13	2	S
Gentamicin	> 13	< 12	16	S
Kanamycin	> 18	< 13	20	S
Neomycin	> 17	< 12	12	R
Penicillin G	> 29	< 20	10	R
Polymyxin B	> 12	< 8	10	R
Streptomycin	> 15	< 11	11	R
Vancomycin	> 12	< 9	15	S
Tetracycline	> 19	< 14	25	S

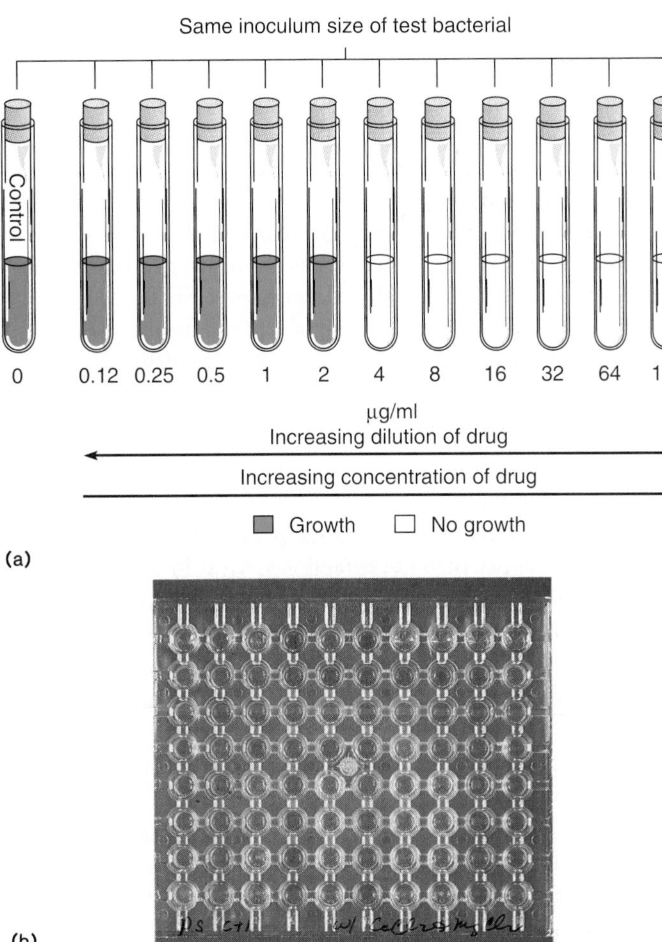

Figure 10.21 Tube dilution test for determining the minimum inhibitory concentration (MIC). (*a*) The antibiotic is diluted serially through tubes of liquid nutrient from right to left. All tubes are inoculated with an identical sample of a test bacterium and then incubated. The first tube on the left is a control containing only the microbe and nutrient. The turbidity of each successive tube is compared with the control. The dilution of the first tube in the series that shows no growth (no turbidity) is the MIC. (*b*) An automated method for determining the MIC of an isolated pathogen. A machine automatically dilutes and dispenses the drug and inoculates the test microbe into a multiple-chambered plate. After incubation, the clear wells are those showing inhibition; turbid wells indicate growth and lack of inhibition by the drug at that dilution.

Applying the Results of Drug Susceptibility Tests The results of antimicrobial sensitivity tests guide the physician's choice of a suitable drug. If therapy has already commenced, it is imperative to determine if the tests bear out the use of that particular drug. If this drug shows little or no activity *in vitro,* a modification of the therapy will be necessary. Once therapy has begun, it is important to observe the patient's clinical response, because the *in vitro* activity of the drug may not always correlate with its *in vivo* effect. The failure of antimicrobic treatment results from such factors as: (1) the infectious agent being inaccessible because the drug cannot diffuse into that body compartment (the brain, joints, skin); (2) a few resistant cells in the culture that were not seen in the sensitivity test; and (3) the infection being caused by more than one pathogen (mixed), some of which are resistant to the drug. If therapy does fail, a different drug, combined therapy, or different method of administration will have to be considered.

Medical Considerations in Anti-infective Chemotherapy

Many factors influence the choice of an antimicrobial drug besides microbial sensitivity to it. The nature and spectrum of the drug, its potential adverse effects, and the condition of the patient can be critically important. When several antimicrobic drugs are available for treating an infection, final drug selection advances to a new series of considerations. In general, it is better to choose the narrowest spectrum drug of those that are effective. This decreases the potential for superinfections and other adverse reactions. Because drug toxicity is of concern, it is best to choose the one with high selective toxicity for the infectious agent and low human toxicity. This concept describes the **therapeutic index,** the ratio of a drug's toxic dose to its minimum effective (therapeutic) dose. This index really compares the circulating drug level that is expected to cause toxic reactions in the patient with the minimum circulating level that will inhibit the microbe *in vivo.* The closer these two figures are (the smaller the ratio), the greater is the potential for toxic drug reactions. For example, a drug that has a therapeutic index of:

$$\frac{10\ \mu g/ml\text{: toxic dose}}{9\ \mu g/ml\ (MIC)} = 1.1$$

Table 10.7 Comparative MICs in μg/ml for Five Common Drugs and Seven Pathogens

Bacterium	Penicillin G	Ampicillin	Sulfamethoxazole	Tetracycline	Cefaclor
Staphylococcus aureus	4.0	0.05	3.0	0.3	4.0
Enterococcus faecalis	3.6	1.6	100.0	0.3	60.0
Neisseria gonorrhoeae	0.5	0.5	5.0	0.8	2.0
Escherichia coli	100.0	12.0	3.0	6–50.0	3.0
Pseudomonas aeruginosa	>500.0	>200.0	—	>100.0	—
Salmonella spp.	12.0	6.0	10.0	1.0	0.8
Clostridium	0.16	—	—	3.0	12.0

is a riskier choice than one with a therapeutic index of:

$$\frac{10\ \mu g/ml}{1\ \mu g/ml} = 10$$

Drug companies recommend dosages that will inhibit the microbe but not adversely affect patient cells. When a series of drugs being considered for therapy have similar MICs, the drug with the highest therapeutic index has the widest margin of safety.

The physician must also take a careful history of the patient to discover any preexisting medical conditions that will influence the activity of the drug or the response of the patient. A history of allergy to a certain class of drugs should preclude the administration of that drug. This is true even in cases of two closely related drug families such as the penicillins and the cephalosporins. Underlying liver or kidney disease will ordinarily necessitate the modification of drug therapy, because these organs play such an important part in metabolizing or excreting the drug. Very young, very old, or pregnant patients demand special precautions. Age can diminish gastrointestinal absorption and organ function, and most antimicrobial drugs cross the placenta and could affect fetal development.

Because drugs can interact with one another, the patient's current intake of other drugs must be carefully scrutinized. Drug incompatibilities can result in increased toxicity or failure of one or more of the drugs. For example, the combination of aminoglycosides and cephalosporins increases nephrotoxic effects; antacids reduce the absorption of isoniazid; and the interaction of tetracycline or rifampin with oral contraceptives may abolish the contraceptive's effect. Some drugs (penicillin, some aminoglycosides, amphotericin B, and flucytosine) act synergistically so that reduced doses of each may be used in combined therapy. Other concerns in choosing drugs include any genetic or metabolic abnormalities in the patient, the site of infection, the route of administration, and the cost of the drug.

An Antimicrobial Drug Dilemma

We began this chapter with a view of the exciting strides made in chemotherapy during the past few years, but we must end it on a note of qualification and caution. There is now a worldwide problem in the management of antimicrobial drugs, which rank second only to some nervous system drugs in overall usage. The remarkable progress in treating many infectious diseases has spawned a view of antimicrobics as the answer to infections as diverse as the common cold and acne. And, while it is true that nothing is as dramatic as curing an infectious disease with the correct antimicrobic drug, in many instances, drugs have no effect or may even be harmful. The depth of this problem can perhaps be appreciated better with a few statistics:

1. Roughly 200 million prescriptions for antibiotics are written in the United States every year. It has been estimated that about half of these prescriptions are inappropriate because the infection is viral in origin. Many drugs are also misprescribed as to type, dosage, or length of therapy. The estimated cost of these wasted drugs is in the billions of dollars.
2. The overuse of antimicrobials in hospitals for surgical prophylaxis is thought to have brought on increased antimicrobial resistance without really benefitting the patient in many cases.
3. There is a tendency to use a "shotgun" antimicrobial therapy for non-life-threatening infections, which involves administering a broad-spectrum drug instead of a more specific narrow-spectrum one. This may lead to superinfections as well as toxic reactions. Tetracyclines and chloramphenicol are still prescribed routinely for infections that would be treated more effectively with narrower spectrum, less toxic drugs.

The Art and Science of Choosing an Antimicrobic Drug Even when all the information is in, the final choice of a drug is not always easy or straightforward. Consider the case of an elderly alcoholic patient with pneumonia caused by *Serratia marcescens* and complicated by diminished liver and kidney function. All drugs must be given parenterally because of prior damage to the gastrointestinal lining and poor absorption. Drug tests show that the infectious agent is sensitive to third-generation cephalosporins, gentamicin, amikacin, and azlocillin. The patient's history shows previous allergy to the penicillins, so these would be ruled out. Drug interactions occur between alcohol and the cephalosporins, and these drugs are also associated with serious bleeding in elderly patients, so this is perhaps not a good choice. The aminoglycosides are nephrotoxic and poorly cleared by damaged kidneys. On the surface, it would appear that none of the drugs is a good choice, but on closer examination, it turns out that amikacin is less nephrotoxic than gentamicin.

In the case of a cancer patient with severe systemic *Candida tropicalis* infection, there will be fewer criteria to weigh. Intravenous amphotericin B alone or in combination with flucytosine is about the only possibility—despite the drug's nephrotoxicity and other possible adverse side effects.

A pregnant woman with Rocky Mountain spotted fever also presents a real medical emergency for which there is little choice. Regardless of the potential harm to the fetus, she must be promptly treated with tetracycline or chloramphenicol, though perhaps at lower doses. In a life-threatening situation, where a dangerous chemotherapy is perhaps the only chance for survival, the choices are reduced, the priorities different. At the other extreme, a mild gram-negative urinary tract infection or a superficial streptococcal or staphylococcal skin infection in an otherwise healthy adult may present several possible therapeutic choices.

---

4. Drugs are often prescribed without benefit of culture or susceptibility testing, even when such testing is clearly warranted.
5. More expensive drugs are chosen when a less costly one would be just as effective. Among the most expensive drugs are the cephalosporins and the longer-acting tetracyclines, yet these are among the most commonly prescribed antibiotics.
6. Tons of excess antimicrobial drugs produced in this country are exported to other countries, where controls are not as strict. Nearly 200 different antibiotics are sold over-the-counter in Latin American and Asian countries. Many residents of these countries have gotten into the habit of medicating themselves with periodic doses of antimicrobics. Drugs used in this way are largely ineffectual, but worse yet, they may inadvertently account for the emergence of some drug-resistant bacteria that subsequently cause epidemics.

The medical community recognizes that most physicians are motivated by important and prudent concerns, such as the need for immediate therapy to protect a sick patient and for defensive medicine to provide the very best care possible, but many experts feel that more education is needed for both physicians and patients concerning the proper occasions for prescribing antibiotics. In the final analysis, every allied health professional should be critically aware not only of the admirable and utilitarian nature of antimicrobics, but also of their limitations.

Chapter Review with Key Terms

Antimicrobial Chemotherapy

Purposes of Chemotherapeutic Drugs: Treatment of infections, control of microbes in the body.

Chemical Nature: Aromatic peptides, sugars, amino acids, nucleotides.

Categories of Antimicrobics: **Antibiotics** are chemicals derived from bacteria and molds, natural or **semisynthetic** (part natural, part synthetic); **synthetic drugs** are derived completely from industrial processes. Drugs may be **narrow spectrum** or **broad spectrum,** microbicidal or microbistatic.

Scope: Antibacterial (largest number), antifungal, antiprotozoan, antihelminthic, antiviral.

Ideal Qualities of Antimicrobics: **Selective toxicity,** meaning high toxicity to microorganisms, low toxicity to vertebrates; potency unaltered by dilution; stability and solubility in tissue fluids; lack of disruption to host's immune system or microflora; exempt from drug resistance.

Route of Administration: By mouth; **parenteral** (injection into vein, muscle); topical on skin, mucous membranes.

Major Adverse Results of Chemotherapy: Toxicity ranging from slight, short-term damage to permanent damage, debilitation, death; allergic reactions; disruption of normal microbial flora of body; **superinfections** by resistant species; **drug resistance** (selection of strains of microorganisms genetically resistant through mutation or intermicrobial transfer of resistance factors).

Special Clinical Approaches: **Prophylaxis,** administering antimicrobic drugs to prevent infections in highly susceptible persons; combined therapy, administering two or more antimicrobics simultaneously to circumvent drug resistance or to achieve **synergism,** the additive or magnified effectiveness of certain drugs working together.

Stages in Selection of Proper Drug: Identification of microbe ***in vitro;*** testing antimicrobial sensitivity or susceptibility (determining the **MIC**); assessing the **therapeutic index;** assessing patient condition. Final drug selection weighs potential effectiveness, toxicity, ***in vivo*** effects, spectrum, and the medical condition of the patient.

Perspectives in Antimicrobic Abuse: Drugs are overprescribed, overproduced, and used inappropriately on a worldwide basis, with unfortunate medical and economic consequences.

Drug Categories: 18 drug families and more than 250 drugs.

Antibacterial Antibiotics

Penicillins

Types/Source: **Beta-lactam**-based drugs; from ***Penicillium chrysogenum*** mold; natural form is penicillin G; semisynthetic forms (ampicillin, carbenicillin, methicillin, nafcillin) vary in specific therapeutic applications.

Mode of Action: Bactericidal; blocks the completion of the cell wall, causes weak points and cell rupture.

Specific Uses/Spectra: Penicillins G and V are narrow spectrum—mostly gram-positive and a few gram-negative (for example, gonococcus) bacteria. Semisynthetics are moderate to broad in spectrum, effective in infections by specified gram-positive and gram-negative species.

Problems in Therapy: Not highly toxic, but common cause of allergic reactions; some semisynthetics can cause superinfections; bacterial resistance to some forms of penicillin occurs through **beta-lactamase** (for example, **penicillinase**).

Cephalosporins

Types/Source: Natural and semisynthetic forms from *Cephalosporium acremonium* mold.

Mode of Action: Similar to penicillins; inhibit peptidoglycan synthesis.

Specific Uses/Spectra: First generation is narrow spectrum, against gram-positive cocci; second generation is narrow spectrum, for some gram-negative rods; third generation is broad spectrum, especially for gram-negative enteric rods and gram-positive cocci.

Problems in Therapy: Adverse blood and kidney reactions; superinfections; allergic reactions; bacterial resistance through **cephalosporinases.**

Aminoglycosides

Types/Source/Spectrum: Mostly narrow spectrum; from ***Streptomyces.***

Mode of Action: Interference with the bacterial ribosome, inhibition of protein synthesis.

Specific Uses/Spectra: Streptomycin, narrow spectrum, not in common usage, except for tuberculosis therapy; gentamicin, standard therapy for enteric bacillary infections; tobramycin and amikacin, replacement drugs for kanamycin in gram-negative infections.

Problems in Therapy: Toxic reactions to acoustic and vestibular apparatus of ear, kidney damage, intestinal disturbances; drug resistance.

Tetracyclines and Chloramphenicol

Source/Spectrum: Very broad-spectrum drugs originally from species of *Streptomyces* (now semi- or fully synthetic).

Mode of Action: Interfere with translation (protein synthesis).

Specific Uses: Tetracyclines (ex. tetracycline and minocycline) differ in length of potency; used for rickettsial infections, *Mycoplasma* pneumonia, cholera, acne, and some sexually transmitted diseases. Chloramphenicol (Chloromycetin) limited by toxicity; indicated for serious infections where there is no alternative.

Problems in Therapy: Tetracycline may lead to hepatotoxicity, gastric disturbance, discoloration of tooth enamel in children, superinfections; chloramphenicol may damage bone marrow.

Bacitracin and Polymyxin

Spectrum/Source: Narrow spectrum, from *Bacillus.*

Mode of Action: Bacitracin prevents synthesis of the cell wall of gram-positive bacteria; polymyxin has detergent action that disrupts cell membrane of gram-negative bacteria.

Specific Uses: Bacitracin used in ointments with neomycin for skin infections; polymyxins used to treat *Pseudomonas* infection or in skin ointments.

Problems in Therapy: Polymyxin can cause nephrotoxic and neuromuscular reactions; bacitracin is useful only for topical applications.

Miscellaneous

Erythromycin, clindamycin, vancomycin, rifampin—streptomycete products.

Modes of Action: Erythromycin and clindamycin disrupt protein synthesis; vancomycin interferes with early cell wall synthesis; rifampin inhibits RNA synthesis.

Specific Uses/Spectra: Erythromycin is broad spectrum, for *Legionella, Chlamydia, Mycoplasma,* and some penicillin-resistant cocci; clindamycin used for intestinal infections by anaerobes; vancomycin applied in life-threatening staphylococcal infections; rifampin used for tuberculosis and leprosy.

Problems in Therapy: Vancomycin is neurotoxic; clindamycin and erythromycin can harm the GI tract; rifampin is hepatotoxic; resistant bacteria occur for all.

Antibacterial Synthetic Drugs

Sulfonamides (Sulfa Drugs)

Types/Spectrum: Originally derived from prontosil; all types have similar basic structure; commonest in use is sulfisoxazole; relatively broad spectrum.

Mode of Action: Acts as an **antimetabolite,** a **metabolic analog** that causes **competitive inhibition,** resulting in blockage in nucleic and amino acid synthesis.

Specific Uses: Urinary tract infections, nocardiosis, burn and eye infections; often combined with trimethoprim.

Problems in Therapy: Formation of crystals in kidney and allergy.

Miscellaneous

Trimethoprim, medication for urinary, respiratory, and gastrointestinal infections; a competitive inhibitor in nucleic acid synthesis; can cause bone marrow damage. Dapsone, a major antileprosy drug, combined with rifampin to block drug resistance. **Isoniazid** (INH), an antituberculosis drug; blocks synthesis of cell wall of mycobacteria; may damage liver. *Fluoroquinolones,* promising new broad-spectrum drugs.

Drugs for Fungal Infections

Polyenes: Amphotericin B, nystatin, antibiotics that disrupt cell membrane by detergent action. Amphotericin is a key drug in systemic fungal infections; nystatin is used for skin and mucous membrane candidiasis. Both are commonly nephrotoxic.

Imidazoles: Synthetic drugs that interfere with membrane synthesis; ketoconazole, miconazole, and clotrimazole for cutaneous and membrane infections; can cause liver damage.

Flucytosine: Synthetic inhibitor of DNA synthesis; used alone or in combination with amphotericin for systemic mycoses; may lower WBC count; fungal resistance.

Drugs for Protozoan Infections

Quinines: Dominant types are chloroquine, primaquine, and quinine for managing malaria; choice depends upon sensitivity and stage in cycle of *Plasmodium*; can cause intestinal symptoms and eye disturbances; resistance and complexity of life cycle are main hurdles.

Others: Metronidazole (Flagyl) for amebiasis, giardiasis, trichomonas infections; suramin, melarsoprol, indicated in treatment of African trypanosomiasis; nitrifurimox, for acute South American trypanosomiasis.

Drugs for Helminth Infections

Mebendazole, thiabendazole, and praziquantel, all-purpose agents in treating intestinal roundworm, tapeworm, and some fluke infestations; pyrantel and piperazine, primarily for intestinal roundworms; niclosamide, for tapeworms. Taken orally, cure occurs only upon incapacitation or death of worms and eggs followed by their expulsion in feces.

Drugs for Viral Infections

Acyclovir, idoxuridine, and vidarabine, synthetic nitrogen bases that block synthesis of viral components in herpesviruses; **amantadine,** restricted to treating influenza A infections; **AZT,** an anti-AIDS drug; **interferon,** a naturally occurring protein that seems useful in reducing symptoms of some viral infections and treating a few cancers. Most antivirals function intracellularly to block virus multiplication; main drawbacks are lack of diversity and toxicity.

True–False Questions

Determine whether the following statements are true (T) or false (F). If you feel a statement is false, explain why, and reword the sentence so that it reads accurately.

____ 1. An antibiotic is a compound produced by the metabolism of bacteria or fungi that inhibits the growth of or destroys other microbes.

____ 2. It is desirable for an antimicrobial drug to be broken down or excreted rapidly.

____ 3. Drugs that prevent the formation of the bacterial cell wall are effective on spores.

____ 4. Sulfonamide drugs inhibit bacterial growth by competing with PABA for a site on the enzyme that synthesizes folic acid.

____ 5. Microbial resistance to drugs is acquired through conjugation, transformation, and transduction of R factors.

____ 6. An antibiotic that disrupts the normal flora can cause a superinfection.

____ 7. Most antihelminth drugs function by weakening the worms so they can be flushed out by the intestine.

____ 8. An antimicrobic drug with a high therapeutic index is safer to use than one with a low index.

____ 9. The minimum inhibitory concentration is the highest dilution of a drug that is required to inhibit growth of a microbe.

____ 10. The best drug to choose for treating an infection is the one with the largest zone of inhibition in the disc diffusion method.

Concept Questions

1. Differentiate between antibiotics and synthetic drugs.
2. Differentiate between narrow-spectrum and broad-spectrum antibiotics. Can you determine why some drugs have narrower spectra than others? (Hint: Look at their mode of action.) How might one determine whether a particular antimicrobic is broad or narrow spectrum?
3. What is the major source of antibiotics? What appears to be the natural function of antibiotics?
4. Using the following diagram as a guide, briefly explain how the three factors in drug therapy interact. What drug characteristics will make the cure most effective? What are the major aims of new antimicrobic drugs? Which one do you think is perhaps the most important?
5. Explain the major modes of action of antimicrobial drugs, and give an example of each. What is an antimetabolite? What is the basic reason that an analog molecule can inhibit metabolism? What are the long-term effects of drugs that block transcription? Why would a drug that blocks translation on the ribosomes of bacteria also affect human cells? Why do drugs that act on membranes generally have greater toxicity?
6. Explain the phenomenon of drug resistance from the standpoint of microbial genetics. How can one test for drug resistance?
7. Multiple drug resistance is becoming increasingly common in microorganisms. Explain how one bacterium can acquire resistance to several drugs.
8. Explain four general ways that microbes evade the effects of drugs. What is the effect of beta-lactamase?
9. What causes mutated or plasmid-altered strains of drug-resistant microbes to persist in a population?
10. Review the major groups of antibacterial drugs (antibiotics and synthetics), antifungal drugs, antiparasitic drugs, and antiviral drugs.
11. Explain why there are so few antifungal, antiparasitic, and antiviral drugs. What effect do nitrogen base analogs have upon viruses?
12. Summarize the biological actions of interferon.
13. Generally overview the adverse effects of antimicrobic drugs on the host. On what basis can one explain allergy to drugs?
14. Describe the stages in a superinfection.
15. Outline the steps in antimicrobic susceptibility testing. How is the MIC used? What is the therapeutic index, and how is it used?

Practical/Thought Questions

1. Describe the events that occur when an oral drug is taken to treat (1) a skin infection or (2) meningitis (an infection of the meninges of the brain).
2. Occasionally, one will read that a microbe has become "immune" to a drug. What is a better way to explain what is happening?
3. Can you think of additional ways that drug resistance can be prevented? What can health care workers do? What can one do on a personal level?
4. Drugs are often given to surgical patients, to dental patients with heart disease, or to healthy family members exposed to contagious infections. What word would you use to describe this use of drugs? What is the purpose of this form of treatment?
5. (*a*) Your pregnant neighbor has been prescribed a daily dose of oral tetracycline for acne. Do you think this therapy is advisable for her? Why, or why not? (*b*) A woman has been prescribed a broad-spectrum oral cephalosporin for a strep throat. What are some possible consequences in addition to cure of the infected throat? (*c*) A man has a severe case of gastroenteritis that is negative for bacterial pathogens. A physician prescribes an oral antibacterial drug in treatment. What are your opinions of this therapy?
6. You have been directed to take a sample from a growthless portion of the zone of inhibition in the Kirby-Bauer test and inoculate it onto a plate of nonselective medium. What does it mean if growth occurs on the new plate? What if there is no growth?
7. In cases where it is not possible to culture or drug test an infectious agent (such as middle ear infection), how would the appropriate drug be chosen?
8. Using the results in table 10.6 and reviewing drug characteristics, choose an antimicrobic for each of the following situations (explain your choice): (*a*) for an adult patient suffering from mycoplasma pneumonia; (*b*) for a child with meningitis (drug must enter into cerebrospinal fluid); (*c*) for a patient with allergy to erythromycin.
9. What might be the mechanism of drug synergism?
10. How would you personally feel about being told by a physician that your infection cannot be cured by an antibiotic, and that the best thing to do is go home, drink a lot of fluids, and take aspirin or other symptom-relieving drugs?
11. Refer to the tube dilution test shown in figure 10.21*a*, and give the MIC of the drug being tested.

CHAPTER 11

Microbe-Human Interactions: Infection and Disease

Painting of Dr. Jesse Lazear exposing the arm of James Carroll to a mosquito infected with the yellow fever virus. Human volunteers contributed significantly to our understanding of the transmission of yellow fever and many other diseases.

Chapter Preview

The human body exists in a state of dynamic equilibrium with microorganisms. In the healthy individual, this balance is maintained as peaceful coexistence and lack of disease. But on occasion, the balance tips in favor of the microorganism, and an infection or disease results. In this chapter we will systematically explore each component of the host-parasite relationship, beginning with the nature and function of normal flora, moving to the stages of infection and disease, and closing with a view of population medicine and the spread of disease. These topics will set the scene for the next two chapters, which deal with the ways the host defends itself against assault by microorganisms.

The Human Host

In previous chapters we considered several of the basic interrelationships between humans and microorganisms. Most of the microbes inhabiting the human body benefit from the nutrients and protective habitat it provides. But from the human point of view, these relationships run the gamut from mutualism to commensalism to parasitism, and may be beneficial, neutral, or harmful (see page 174). A common characteristic of all microbe-human relationships, regardless of where they lead, is that they begin with contact.

Contact, Infection, Disease—A Continuum

The body surfaces are constantly barraged by microbes—some of them become implanted there as colonists (normal flora), some are rapidly lost (transients), and others invade the tissues. Such intimate contact with microbes inevitably leads to **infection,** a condition in which pathogenic microorganisms penetrate the host defenses, enter the tissues, and multiply. When the cumulative effects of the infection damage or disrupt the tissues, an **infectious disease** results. A disease is literally any deviation from health. Hundreds of different diseases are caused by derangements in organ function, nutrition, genetics, hormones, aging, or other factors. But in this chapter, our discussion will be confined to infectious disease—the malfunction of a tissue or organ caused by microbes or their products.

The pattern of the host-parasite relationship is a continuum—beginning with contact, progressing to infection, and ending in disease—but the continuation of this process is not inevitable (see feature 11.1). When a potentially infectious microbe is present on the body without yet invading, it is called a **contaminant**; thus, to be contaminated is not the same as to be infected. This host-microbe interaction has numerous points of departure, so that not all contaminations lead to infection and not all infections lead to disease. In fact, contamination without infection and infection without disease are the rule. Before we consider further details of infection and disease, let us examine the fascinating relationship between humans and their resident flora.

Resident Flora: The Human As a Habitat

With its constant source of nourishment and moisture, relatively stable pH and temperature, and extensive surfaces upon which to settle, the human body provides a favorable habitat for an abundance of microorganisms. In fact, it is so favorable that, cell-for-cell, microbes outnumber human cells ten to one! The large and mixed collection of microbes adapted to the body have been variously called the normal **resident flora, indigenous flora,** or **microflora,** though some microbiologists prefer to use the terms microbiota, commensals, amphibionts, and associates. The normal microflora is comprised of an array of bacteria, fungi, protozoa, and, to a certain extent, viruses and arthropods.

Acquiring Resident Flora

Like a virgin planet being settled by alien beings, humans begin to acquire microflora from their earliest contact with the animate and inanimate world. Development of the flora proceeds

indigenous (in-dih'-juh-nus) Belonging or native to.

Feature 11.1 Hovering at the Crossroads—The Potentials of Microbe-Human Associations

1. **Contact** with microbe (contamination) → Colonization by flora; Loss; Allergy; → 2. **Infection** → Cure, immunity; Carrier state; → 3. **Disease** → Mortality; Cure, immunity; Morbidity; Carrier state

1. Microbes are first acquired on the exposed areas of the body through contact with other living things or the environment. The instances and consequences of microbe-human contact are largely a matter of happenstance. The fate of acquired microbes varies with the type of microbe, the condition of the body, and the actions of the host:

a. After initial contact, some microbes take up residence on the body as part of the normal flora, a diverse yet relatively stable collection of microbes that are important in maintaining the host equilibrium. (Note that even normal flora may under certain circumstances cause infection.)
b. Some microbes with greater infectious potential may evade the host defenses, infiltrate the body, and cause an infection—that is, they act as pathogens. Many factors determine whether an infection takes place, but in general, the greater the virulence of the microbe, the more likely it is that infection will take place.
c. After initial contact, some microbes (the transients) stay on the body for only a short time and are destroyed by the host defenses or removed by host behavior (such as cleaning or antisepsis).
d. Some microbes or microbial products result in hypersensitivity reactions or allergy.

2. In the event of infection, two outcomes are possible:

a. The immune system may arrest the infection before injury to tissues and organs occurs.
b. The infectious agent is not arrested and becomes entrenched in tissues where its **pathological** effects cause some degree of damage.

3. If infection proceeds to disease, several outcomes are possible:

a. The immune system may eventually arrest the microbe and stop the disease process.
b. Damaged tissues and organs may lead to dysfunction or **morbidity.**
c. Damage may be severe enough to cause death or mortality.
d. The microbe may be harbored (carried) inconspicuously for varying lengths of time.

pathological (path''-uh-loj'-ih-kul) A disease state caused by structural and functional damage to tissues.

Table 11.1 Sites That Harbor a Normal Flora
Skin and its contiguous mucous membranes
Upper respiratory tract
Alimentary canal (various parts)
Outer opening of urethra
External genitalia
Vagina
External ear canal
External eye (lids, conjunctiva)

Table 11.2 Sterile (Microbe-Free) Anatomical Sites and Fluids
All Internal Tissues and Organs
Heart and circulatory system
Liver
Kidneys and bladder
Lungs
Brain
Spinal cord
Muscles
Ovaries
Testes
Glands
Bone marrow
Sinuses
Middle and inner ear
Internal eye
Fluids Within an Organ or Tissue
Blood
Urine in kidneys, ureters, bladder
Cerebrospinal fluid
Saliva prior to entering the oral cavity
Semen prior to entering the urethra
Amniotic fluid surrounding the embryo and fetus

in a relatively orderly and characteristic way. Of the seemingly limitless influx of microbes, only some are capable of persisting on the body, and the vast majority are removed or destroyed by host defenses. Those that remain are capable of evading the defenses and adapting to a particular microhabitat—one that fills special requirements for food, moisture, presence or lack of oxygen, layers of dead cells, and even other microorganisms. It is safe to say that most anatomical surfaces directly exposed to the environment can and do harbor microbes. As indicated in table 11.1, these surfaces are mainly the skin and mucous membranes, parts of the inner surface of the alimentary tract, and openings to the cutaneous surface from the urinary, respiratory, and reproductive tracts. Moist areas through which food passes (the alimentary tract) tend to have the largest and most diverse populations. By contrast, organs and fluids inside the body cavity and the central nervous system remain free of normal flora during life and are maintained in a sterile state by various defense mechanisms (table 11.2). Although microorganisms may transiently enter these sites, they do not normally become established there.

Interactions with indigenous flora have developed over millions of years into a complex and dynamic relationship that directly and indirectly influences many aspects of human life. Although its functions are far from completely understood, the useful effects of normal flora are mainly protective or nutritional. Once established, the flora modifies its microhabitats by (1) altering pH and oxygen tension, (2) excreting chemicals such as fatty acids, gases, alcohol, and antibiotics, or (3) creating physical obstacles. These factors favor colonization by a select group of microbes while simultaneously repulsing newcomers and restricting their population size and activity through antagonism and competition for nutrients. Certain enteric bacteria release residual vitamins, creating a supplementary dietary source, though the significance of this remains in question. Other theorized functions of the normal flora are discussed in a later section on germ-free animals.

Although relatively stable, the flora fluctuates to a limited extent with the seasons, with age, and with variations in diet, hygiene, hormones, drug therapy, and general health. Most species are not serious pathogens and remain harmless as long as they do not penetrate superficial skin and mucosal surfaces. A few species have the potential to become medically important if the balance of the flora shifts. We observe this in the overgrowth of *Clostridium difficile* or *Candida albicans* in patients undergoing antibiotic therapy. After being temporarily disturbed, the flora is restored to its former levels and composition by surviving residents harbored in protected areas and by contact with the environment. Agents in the flora may also become infectious if the health of the host is compromised (as by cancer or AIDS), and certain agents (for example, the pneumococcus and *Pneumocystis*) are sufficiently infectious to overcome the weakened host defenses (see table 11.5). It is notable that certain organisms harbored by some persons can be pathogenic to other persons. The flora constitutes an important reservoir for bacterial pathogens such as *Staphylococcus aureus, Corynebacterium diphtheriae,* and *Neisseria meningitidis.*

Initial Colonization of the Newborn

The uterus and its contents are normally sterile during embryonic and fetal development and remain essentially germ-free until just prior to birth (except in in utero infections, which often lead to damage or death of the fetus). The event that first exposes the infant to microbes is the breaking of the fetal membranes, at which time microbes from the mother's vagina may enter the womb. A wholesale exposure occurs during the birth process itself, when the baby unavoidably comes in intimate contact with the birth canal (figure 11.1). Within 8 to 12 hours after delivery, the newborn typically has been settled by bacteria such as streptococci, staphylococci, and lactobacilli, acquired primarily from its mother. The nature of the flora initially colonizing the large intestine depends upon whether the baby is bottle- or breast-fed. Bottle-fed infants (receiving milk or a milk-based formula) tend to acquire a mixed population of coliforms,

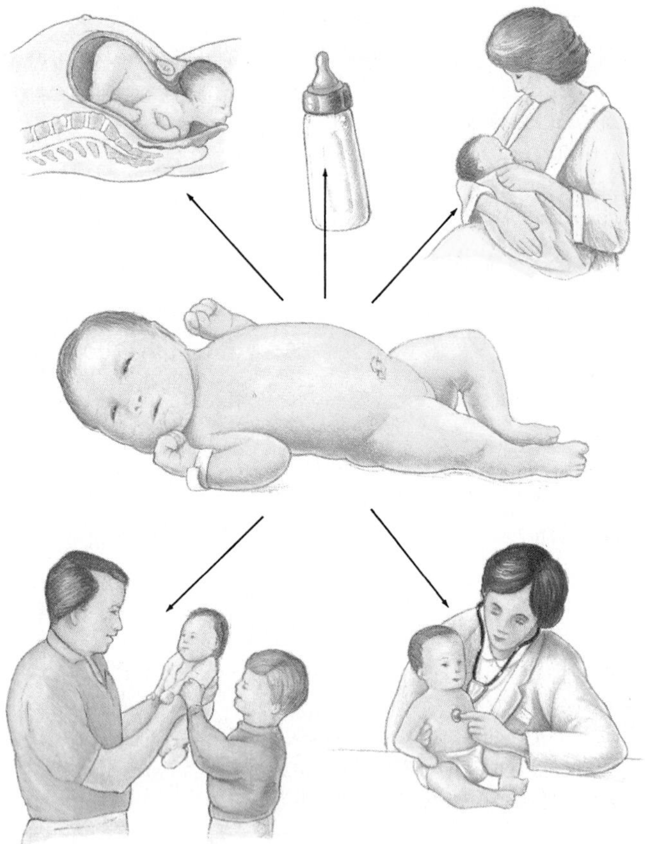

Figure 11.1 The origins of flora in newborns.

lactobacilli, enteric streptococci, and staphylococci, while the intestinal flora of breast-fed infants is very simple, consisting primarily of *Bifidobacterium* species. The skin, gastrointestinal tract, and portions of the respiratory and genitourinary tract all exhibit a similar pattern of colonization as contact continues with family members, pediatric personnel, the environment, and food.

Milestones that contribute to development of the adult pattern of flora are eruption of teeth, weaning, and introduction of the first solid food. Although exposure to microbes is unavoidable and even necessary for the maturation of the infant's flora, contact with pathogens is dangerous, because the neonate is not yet protected by a full complement of flora, and due to its immature immune defenses, is extremely susceptible to infection. Neonatal infections by *Neisseria gonorrhoeae, Chlamydia trachomatis, Candida, Staphylococcus aureus, Clostridium botulinum,* and herpesvirus, to name a few, may lead to complications and even death.

Indigenous Flora of Specific Regions

Although we tend to speak of the flora as a single unit, it is a complex mixture of hundreds of species, differing somewhat in quality and quantity from one individual to another. Studies of the flora have shown that most persons harbor certain specially adapted bacteria, fungi, and protozoa (table 11.3).

Flora of Human Skin The skin is the largest and most accessible of all organs. Its major layers are the epidermis, an outer layer of dead cells continually being sloughed off and replaced, and the dermis, which lies atop the subcutaneous layer of tissue (figure 11.2*a*). Depending on its location, skin also contains hair follicles and several types of glands, and the outermost surface is covered with a protective, waxy cuticle that may help microbes adhere. The normal flora resides only in or on the dead cell layers, and except for the base of the follicles and glands, the flora does not extend into the dermis or subcutaneous levels. The nature of the population varies according to site. Oily, moist skin supports a more prolific flora than dry skin. Humidity, occupational exposure, and clothing habits also influence its character. Transition zones where the skin joins with the mucous membranes of the nose, mouth, and perineum harbor a particularly rich flora.

Ordinarily there are two cutaneous populations. The **transient population,** or exposed flora, clings to the skin surface, but does not ordinarily grow there. It is acquired by routine contact, and it varies markedly from person to person, from site to site, and even from moment to moment. The transient flora could include any microbe a person has picked up, even species that do not ordinarily live on the body, and it is greatly influenced by the hygiene of the individual. For example, the hands of a child who has been playing in dirt will show vast differences in quality and quantity of transients from a dentist's hands, which may be washed as often as 150 times a day.

The **resident population,** or sheltered flora, lives and multiplies in deeper layers of the epidermis and in glands and follicles (figure 11.2*b*). Because the resident flora is regular and relatively stable, it is more predictable and less influenced by hygiene. The normal skin residents consist primarily of species of *Staphylococcus, Corynebacterium, Propionibacterium,* and yeasts. Chronically moist skin folds, especially between the toes, tend to harbor fungi, while lipophilic mycobacteria and staphylococci are prominent in sebaceous[1] secretions of the axilla, external genitalia, and external ear canal. One species, *M. smegmatis,* lives in the cheesy secretion or *smegma* on the external genitalia of men and women.

Flora of the Alimentary Tract The alimentary tract or canal receives, moves, digests, and absorbs food and removes waste. It encompasses the oral cavity, esophagus, stomach, small intestine, large intestine, rectum, and anus. Saying that the alimentary canal harbors flora may seem to contradict our earlier statement that internal organs are sterile, but it is not an exception to the rule. How can this be true? In reality, the alimentary tract is a long, hollow tube (with numerous pockets and curves), bounded by the mucous membranes of the oral cavity on one extreme and those of the anus on the other. Because the innermost surface of this tube is exposed to the environment, it

1. From sebum, a lipid material secreted by glands in the hair follicles.

Table 11.3 Life on Humans: Sites Containing Well-Established Flora and Representative Examples

Anatomic Sites	Common Genera	Remarks
Skin	**Bacteria:** *Staphylococcus, Micrococcus, Corynebacterium, Propionibacterium*	Microbes live only in upper dead layers of epidermis, glands, and follicles; dermis and layers below are sterile.
	Fungi: *Pityrosporum* yeast	—
	Arthropods: *Demodix* mite	Present in sebaceous glands and hair follicles
Alimentary Tract		
Oral Cavity	**Bacteria:** *Streptococcus, Neisseria, Veillonella, Staphylococcus, Fusobacterium, Lactobacillus, Bacteroides, Corynebacterium, Actinomyces, Eikenella, Treponema, Hemophilus*	Colonize the epidermal layer of cheeks, gingiva, pharynx; surface of teeth; found in saliva in huge numbers
	Fungi: *Candida* species	—
	Protozoa: *Trichomonas tenax, Entamoeba gingivalis*	Frequent the gingiva of persons with poor oral hygiene
Large Intestine and Rectum	**Bacteria:** *Bacteroides, Fusobacterium, Eubacterium, Bifidobacterium, Clostridium,* fecal Streptococci, *Peptococcus, Lactobacillus,* coliforms (*Escherichia, Enterobacter*)	Sites of lower alimentary canal other than large intestine and rectum have sparse or nonexistent flora. Flora are predominantly strict anaerobes; others are aerotolerant or facultative.
	Fungi: *Candida*	—
	Protozoa: *Entameba coli, Trichomonas hominis*	—
Upper Respiratory Tract	Microbial population exists in the nasal passages, throat, and pharynx; due to close proximity, flora is similar to that of oral cavity	Trachea and bronchi have a sparse population; smaller breathing tubes and alveoli have no normal flora and are essentially sterile
Genital Tract	**Bacteria:** *Lactobacillus, Streptococcus, Corynebacterium, Escherichia, Mycobacterium*	In females, flora occupy the external genitalia and vaginal and cervical surfaces; internal reproductive structures normally remain sterile. Flora responds to hormonal changes during life.
	Fungi: *Candida*	—
Urinary Tract	*Staphylococcus, Streptococcus,* coliforms	In females, flora exist only in the first portion of the urethral mucosa; the remainder of the tract is sterile. In males, the entire reproductive and urinary tract is sterile except for a short portion of the anterior urethra.

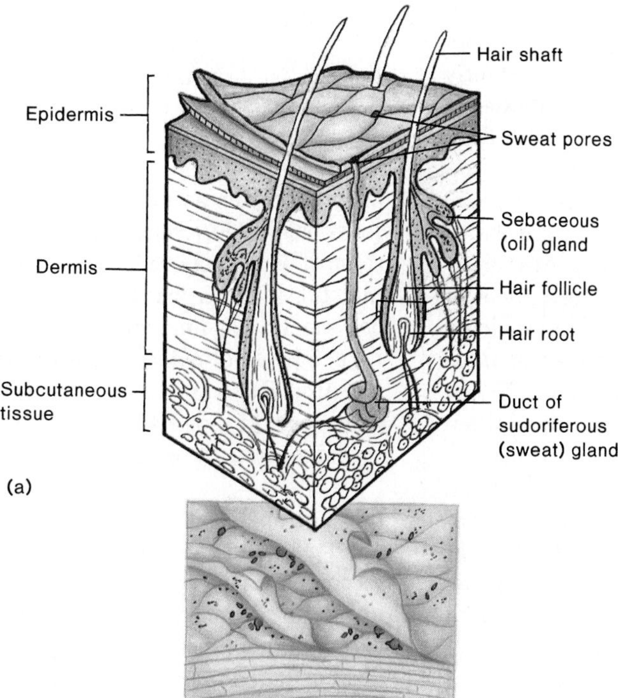

Figure 11.2 The landscape of the skin. (*a*) The epidermis and associated glands and follicles (colored) provide rich and diverse habitats. Noncolored regions (dermis and subcutaneous layers) are free of microbial flora. (*b*) A highly magnified view of the skin surface reveals beds of rod- and coccus-shaped bacteria and yeasts beneath peeled-back skin flakes.

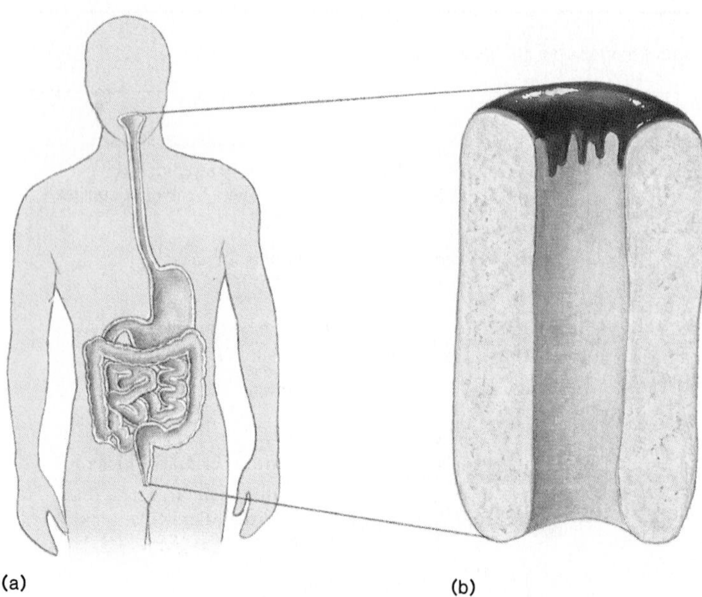

Figure 11.3 (*a*) The body is like a donut, perforated from one end to the other by a long tube, the alimentary tract. (*b*) Things inside the alimentary lumen are not inside the body, just as something inside the donut hole is not inside the donut itself.

is topographically outside the body, so to speak (figure 11.3). This architecture permits ingested materials to be screened before they cross the mucosal epithelium, and at the same time, it creates numerous niches for microbes.

The shifting conditions of pH, peristalsis, oxygen tension, and microscopic anatomy in the alimentary tract are reflected by the variations in or distribution of the flora (figure 11.4). Some microbes remain attached to the mucous epithelium or its associated structures, and others dwell in the *lumen.* Although the abundance of nutrients invites microbial growth, the only areas that harbor appreciable permanent flora are the oral cavity, large intestine, and rectum. The esophagus undergoes wavelike contractions (peristalsis), a process that constantly flushes microorganisms; the stomach acid inhibits most microbes; and peristalsis and digestive enzymes help exclude flora from all but the terminal segment of the small intestine.

Flora of the Mouth The oral cavity has a unique flora that is among the most diverse and abundant of the body. Microhabitats, including the cheek epithelium, gingiva, tongue, floor of the mouth, and tooth enamel, provide numerous adaptive niches for hundreds of different species to colonize. The most common residents are aerobic *Streptococcus* species—*S. sanguis, S. salivarius, S. mitis*—that colonize the smooth superficial epithelial surfaces. Two species, *S. mutans* and *S. sanguis,* make a major contribution to dental caries by forming sticky dextran slime layers in the presence of simple sugars. The adherence of dextrans to the tooth surface establishes a medium (plaque) that attracts other bacteria (see figure 17.30). Over time, acidic products given off by the growing bacteria etch the enamel.

Once the teeth have erupted, an anaerobic habitat is established in the gingival crevice that favors the colonization of

lumen (loo′-men) The space within a tubular structure.

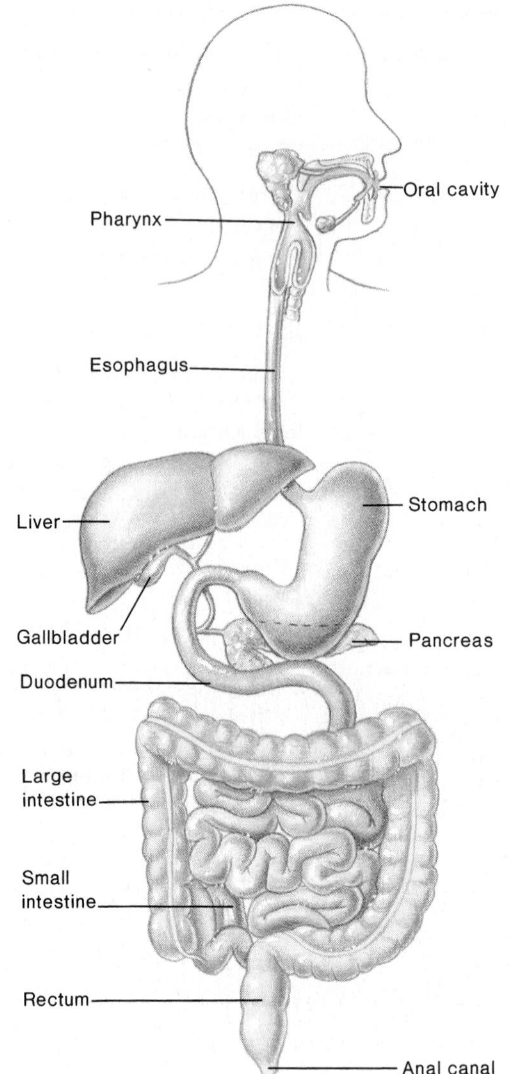

Figure 11.4 Areas of the alimentary tract that shelter resident flora are highlighted in color. Noncolored areas do not regularly harbor residents.

anaerobes such as *Lactobacillus,* spirochetes (*Borrelia*), fusiform bacilli (*Fusobacterium*), *Bacteroides,* and *Actinomyces,* some of which may be involved in dental caries and periodontal[2] infections. The instant that saliva is secreted from ducts into the oral cavity, it is immediately laden with resident and transient flora. Saliva normally has one of the highest bacterial counts of any secretion (up to 5×10^9 cells per milliliter), a fact that tends to make mouthwashes rather ineffective and a human bite very dangerous (see feature 11.2).

Flora of the Large Intestine Much is yet to be learned about the flora of the intestinal tract, but observations to date verify its complex and profound interactions with the host. The large intestine (cecum and colon) and the rectum harbor a huge population of microbes (10^8–10^{11} per gram of feces) (figure 11.5). So abundant and prolific are these microbes that they comprise

2. Situated or occurring around the tooth.

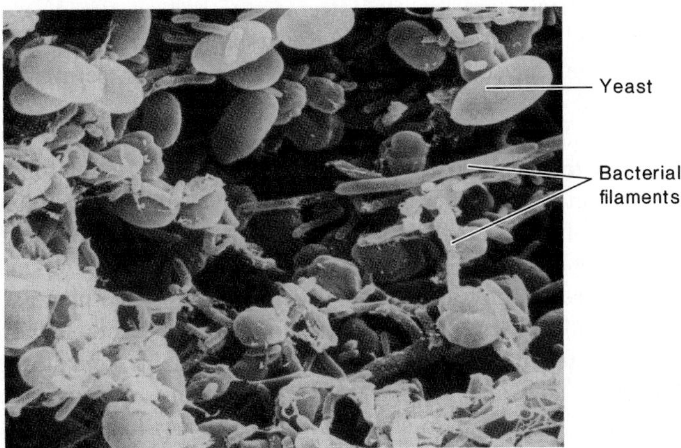

Figure 11.5 Electron micrograph of the mucous membrane of the large intestine. Present are microcolonies of yeasts and long, filamentous bacteria located in pockets.

10–30% of the fecal volume. One measure of the contribution of bacteria to fecal mass is that even an individual on a long-term fast passes feces consisting primarily of bacteria.

Owing to the state of the intestinal environment, the predominant fecal flora are strictly anaerobic bacteria (*Bacteroides, Bifidobacterium, Fusobacterium,* and *Clostridium*). *Coliforms* such as *Escherichia coli, Klebsiella, Enterobacter,* and *Citrobacter* are present in smaller numbers. Many species ferment waste materials in the feces, generating vitamins (B_{12}, vitamin K, pyridoxine, riboflavin, and thiamine) and acids (acetic, butyric, and propionic acids) of potential value to the host. Occasionally significant are bacterial digestive enzymes that convert disaccharides to monosaccharides or promote steroid metabolism.

The Lack of Lactase Late in childhood, members of certain ethnic groups lose the ability to secrete the enzyme lactase. When they ingest milk or other lactose-containing dairy products, lactose is acted upon instead by intestinal bacteria, and severe intestinal distress may result. The recommended treatment for this deficiency is avoiding these foods or eating various lactose-free substitutes.

Feature 11.2 Bites and Needles—A Traumatic Portal of Entry

That venerable tale of "man bites dog" turns the tables in more ways than one, because a human bite is probably far more dangerous than a dog bite. The main reason for its potency is the presence in human saliva of huge numbers of relatively virulent bacteria. A bite also creates a ragged, deep wound favorable to anaerobes. The most common lacerating injuries occur in fistfights, when the knuckles of the hitter are bashed against the teeth of his opponent. The inoculated wound rapidly advances to a deep ulcer that can spread and cause damage to the joints if left untreated. Infections also occur in bite wounds inflicted by children or psychiatric patients, in persons who self-inflict a wound by nervously chewing the insides of their cheeks, and in dental personnel accidentally inoculated with saliva (see figure 11.23).

Though rare, a shark's bite carries hidden dangers. Only one-fourth of shark bite victims die from the trauma of the attack itself, while others escape with varying degrees of raking lacerations or deep bites. These survivors are still at extreme risk from massive wound infection by unidentified bacteria from the normal flora of the shark's teeth and mouth. Without prompt and rigorous treatment (drugs and surgery), the bite infection may prove fatal.

A tragic and growing societal problem is the drug abuser who frequently and willingly forms a portal of entry into his own skin and veins with a hypodermic syringe. There is no denying that those who inject heroin, cocaine, or amphetamines are in dire jeopardy from the drug itself, but many drug-abuse practices increase chances for infection and are just as life-threatening. A fiend looking for the perfect system to transmit infectious agents might well use the behavioral patterns of the drug addict as a model. In some large cities (particularly those with inclement weather), drug users congregate in abandoned buildings to buy drugs and drug paraphernalia (the "works") and to "shoot up." Drugs may be injected into the skin (skin popping) or into the veins (IV). Some IV users "boot" the drug by drawing out a small amount of blood and mixing it with drugs before reinjecting it—a sure way to contaminate the syringe. Most serious of all, the works are often shared and even rented with meager or no attempts at disinfection. Considering that a single "shooting gallery" may play host to hundreds of users every day, and that many participants hop from one gallery to another, the potential for spread of infection is tremendous. Recklessness is a disastrous part of the addiction, in that the need for the drug is greater than the fear of infection. Programs have been adopted by many cities to distribute kits with sterile syringes and bleach for disinfecting them.

Users who inject drugs are predisposed to a disturbing list of well-known diseases: hepatitis, AIDS, tetanus, tuberculosis, osteomyelitis, and malaria. A resurgence of some of these infections is directly traceable to drug use. These individuals are also very vulnerable to infections that reflect a history of chronic drug use. Contaminated needles often contain bacteria from the skin or environment that induce heart disease (endocarditis), lung abscesses, and chronic infections of the injection site.

A contribution from intestinal bacteria that is at once engrossing, curious, and obnoxious are odoriferous chemicals (indole, *skatole,* and amines) and gases (CO_2, H_2, CH_4, and H_2S). Intestinal gas is known in polite circles as *flatus,* and the expulsion of it as *flatulence.* Gas arises in essentially three ways: (1) air is swallowed (mainly N_2 and O_2, relieved by belching); (2) gases diffuse into the GI tract from the blood (much of the soluble gas is reabsorbed and eliminated by the lungs or used in metabolism); and (3) bacteria produce an average of 7 to 10 liters daily, though only about one-tenth is actually ejected as flatus. Gas production is profoundly affected by diet. Vegetables such as cabbage, cauliflower, corn, onion, and the notorious beans contain carbohydrate residues that are not attacked by human digestive enzymes, yet are readily fermented by gas-producing bacteria. Combustible gases occasionally form an explosive mixture in the presence of oxygen that has reportedly ignited during intestinal surgery and ruptured the colon!

Flora of the Respiratory Tract The upper respiratory tract (nasal passages and pharynx) acquires flora within a few hours after birth. The first microorganisms are predominantly oral streptococci. Inhaled air regularly deposits microbes that are

coliform (koh'-lih-form) L. *colum,* a sieve. Gram-negative, facultatively anaerobic, and lactose-fermenting flora.

skatole (skat'-ohl) Gr. *skatos,* dung. One chemical that gives feces its characteristic stench.

filtered out, destroyed, or expelled, although some can adapt to specific regions of this habitat. *Staphylococcus aureus* preferentially resides in the nasal entrance, nasal vestibule, and anterior nasopharynx, and *Neisseria* species take up residence in the mucous membranes of the nasopharynx behind the soft palate (figure 11.6). Lower still, in the vicinity of the tonsils and lower pharynx, are assorted streptococci and species of *Hemophilus*. Conditions lower in the respiratory tree (bronchi, bronchioles, alveoli) are unfavorable habitats for permanent residents.

Figure 11.6 Colonized regions of the respiratory tract. The moist mucous blanket of the nasopharynx has a well-entrenched flora (pink). Some colonization occurs in the upper trachea, but lower regions of bronchi, bronchioles, and lungs lack residents.

Flora of the Genitourinary Tract The parts of the genitourinary tract that harbor microflora are the vagina and outer opening of the urethra in females and the anterior urethra in males (figure 11.7). The internal reproductive organs are kept sterile through physical barriers such as the cervical plug and other host defenses. The remainder of the urinary tract—kidney, ureter, bladder, and upper urethra—is presumably kept sterile by urine flow and regular bladder emptying. Because the urethra in women is so short (about 1.5 inches long), it may form a stairway for bacteria to the bladder, leading to frequent urinary tract infections. The principal residents of the urethra are nonhemolytic streptococci, staphylococci, corynebacteria, and occasionally, coliforms.

The vagina presents a notable example of how changes in physiology may greatly influence the composition of the normal flora. An important factor influencing these changes in women is the hormone estrogen. Estrogen normally stimulates the vaginal mucosa to secrete the carbohydrate glycogen, which certain bacteria (primarily *Lactobacillus acidophilus*) ferment, thus lowering the pH to about 5. Before puberty, a girl produces little estrogen, little glycogen, and has a vaginal pH of about 7. These conditions favor the establishment of diphtheroids,[3] staphylococci, streptococci, and some coliforms. As hormone levels rise at puberty, the vagina begins to deposit glycogen and the flora shifts to the acid-producing lactobacilli. It is thought that the acidic pH of the vagina during this time is protective, preventing the establishment and invasion of microbes with potential for harming a developing fetus. The estrogen-glycogen effect continues, with minor disruptions, throughout the child-

3. Any nonpathogenic species of *Corynebacterium*.

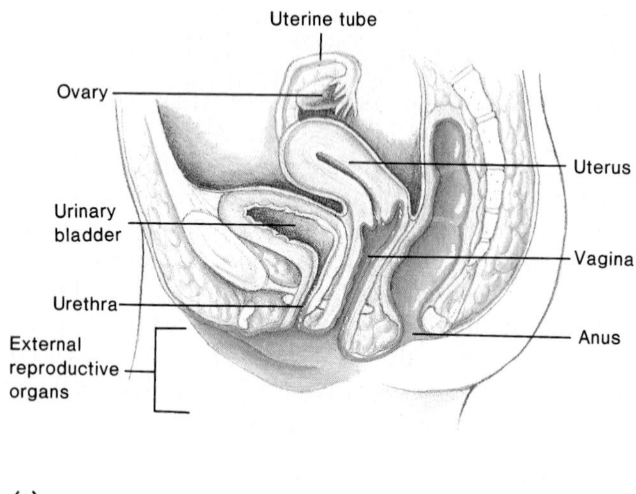

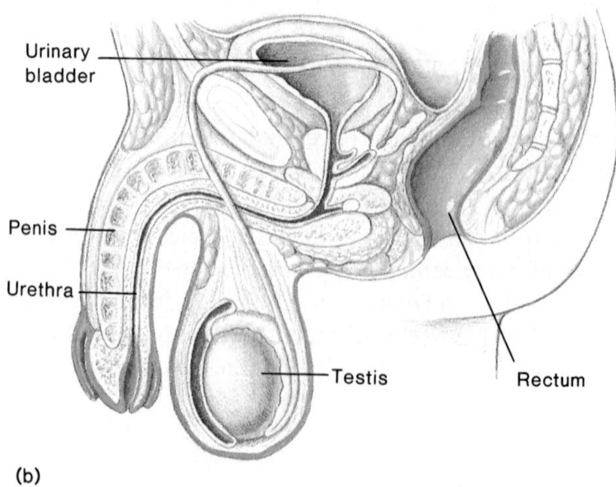

Figure 11.7 Location of (*a*) female and (*b*) male genitourinary flora (indicated by color).

bearing years until menopause, when the flora returns to a mixed population similar to that of prepuberty. These transitions are not abrupt, but occur over several months to years.

Studies with Germ-Free Animals

For years, questions lingered about how essential the microbial flora is to normal life and what functions various members of the flora might serve. The need for animal models to further investigate these questions led eventually to development of laboratory strains of **germ-free** or **axenic** mammals and birds. The techniques and facilities required for producing and maintaining a germ-free colony are exceptionally rigorous. After the young mammals are taken from the mother aseptically by cesarian section, they are immediately transferred to a sterile isolator or incubator (figure 11.8). The newborns must be fed by hand through gloved ports in the isolator until they can eat on their own, and all materials entering their chamber must be sterile. One great help in maintaining the colonies is that, once germ-free parents exist, they will continue to produce germ-free young without human intervention. Rats, mice, rabbits, guinea pigs, monkeys, dogs, hamsters, and cats are some of the mammals raised in the germ-free state.

Experiments with germ-free animals are of two basic varieties: (1) general studies on how the lack of normal microbial flora influences the nutrition, metabolism, and anatomy of the animal, and (2) *gnotobiotic* studies, in which the germ-free subject is inoculated with a single type of microbe to determine its individual effect, or with several known microbes to determine interrelationships. Results are validated by comparing the germ-free group with a conventional, normal control group. Table 11.4 summarizes some major conclusions arising from studies with axenic animals.

A dramatic characteristic of germ-free animals is that they live longer and have fewer diseases than normal controls, as long as they remain in a sterile environment. From this standpoint, it is clear that the flora is not needed for survival and may even be the source of infectious agents. But more significant is the other side of this coin: The flora contributes significantly to the development of the immune system. When germ-free animals are placed in contact with normal control animals, they gradually develop a flora similar to the controls. However, germ-free subjects are less able to tolerate some of the colonists and may die from infections by relatively harmless species. This is due to the immature character of the immune system of germ-free animals. These animals have smaller lymph nodes, a reduced number of certain types of white blood cells, and slower antibody response.

Gnotobiotic experiments have clarified the dynamics of several infectious diseases. Perhaps the most striking discoveries were made in the case of oral diseases. For a number of years, the precise involvement of microbes in dental caries had been ambiguous. Studies with germ-free rats, hamsters, and beagles confirmed that caries development is influenced by heredity, a diet high in sugars, and poor oral hygiene, but even when all these predisposing factors are present, germ-free animals still remain free of caries unless they have been inoculated with specific bacteria. Further discussion on dental diseases can be found in chapter 17.

The ability of known pathogens to cause infection can also be influenced by normal flora, sometimes in opposing ways. Studies have indicated that germ-free animals are highly susceptible to experimental infection by the enteric pathogens *Shigella* and *Vibrio,* whereas normal animals are less susceptible, presumably because of their protective flora. In marked contrast, *Entamoeba histolytica* (the agent of amebic dysentery)

gnotobiotic (noh″-toh-by-aw′-tik) Gr. *gnotos,* known, and *biota,* the organisms of a region.

Figure 11.8 Sterile enclosure for rearing and handling axenic laboratory animals.

Table 11.4 Effects and Significance of Experiments with Germ-Free Subjects

Effect in Germ-Free Animal	Significance
Enlargement of the cecum; other degenerative diseases of the intestinal tract of rats, rabbits, chickens	Microbes are needed for normal intestinal development
Vitamin deficiency in rats	Microbes are a significant nutritional source of vitamins
Underdevelopment of immune system in most animals	Microbes are needed to stimulate development of certain host defenses
Absence of dental caries and periodontal disease in dogs, rats, hamsters	Normal flora are essential in caries formation and gum disease
Heightened sensitivity to enteric pathogens (*Shigella, Salmonella, Vibrio cholerae*) and to fungal infections	Normal flora are antagonistic against pathogens
Lessened susceptibility to amebic dysentery	Normal flora facilitate the completion of the life cycle of the ameba in the gut

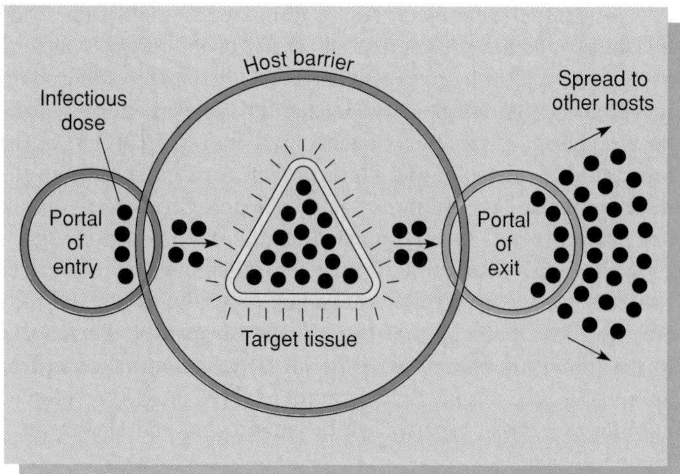

Figure 11.9 An overview of the events in infection. An adequate dose of an infectious agent overcomes a defense barrier and enters the sterile tissues of a host through a portal of entry. From here, it moves or is carried to a specific organ or tissue, where it multiplies and causes some degree of damage. The exit of the pathogen through the same or another portal may facilitate its transmission to another host.

is more pathogenic in the normal animal than in the germ-free animal. One explanation for this phenomenon is that *E. histolytica* must feed on intestinal bacteria to complete its life cycle. These gnotobiotic studies have provided invaluable information on the host-parasite relationship, but they have also emphasized that maintaining a germ-free condition would be extremely difficult, expensive, and highly impractical for humans.

The Anatomy of an Infection

When the association between host and parasite tilts in the direction of infection and disease, a particular series of events is set into motion. The microbe enters the body, attaches itself, invades (crosses host barriers), multiplies in a target tissue, and finally is released to the exterior by various pathways (figure 11.9).

The type and severity of an infection depend on numerous factors, most of which relate to the pathogenicity of the microbe and the condition of the host. *Pathogenicity,* we learned earlier, is a broad concept that characterizes a microorganism's actual and comparative potential to cause an infection or disease. Pathogenic microbes have been traditionally divided into two categories, depending upon the nature of the microbe-host relationship. **True pathogens** (primary pathogens) are capable of causing infection and disease in healthy persons with normal immune defenses. Examples include the influenza virus, plague bacillus, rabies virus, malarial protozoan, and other microorganisms with well-developed qualities of **virulence** (discussed next). Primary pathogens are generally associated with a distinct, recognizable disease. The degree of pathogenicity of true pathogens ranges from weak to potent—that is, infection with the cold virus is mild and nonfatal, while infection with the rabies virus is nearly 100% fatal.

virulence (veer'-yoo-lents) L. *virulentia,* virus, poison.

Table 11.5 Factors That Weaken Host Defenses and Increase Susceptibility to Infection *

Old age
Extreme youth (infancy, prematurity)
Malnutrition
Genetic defects in immunity
Acquired defects in immunity (AIDS)
Dysfunctional host defenses
Physical and mental stress
Organ transplant
Cancer
Chemotherapy
Anatomical defects
Diabetes
Liver disease
Surgery

*These conditions compromise defense barriers or immune responses.

Opportunistic pathogens (secondary pathogens) mainly infect persons whose host defenses (immunities) are compromised. They are not generally considered pathogenic to the normal, healthy person and do not have well-developed virulence properties. It could be argued that most microbes have some capacity for pathogenicity—for example, if a severely compromised host is challenged by a large inoculum in a vulnerable site. Another side of the argument is that anyone who becomes infected is compromised to an extent at that moment. Factors that greatly **predispose** a person to opportunistic infections are shown in table 11.5.

Recognizing that classifying pathogens as either true or opportunistic may be an oversimplification, the Centers for Disease Control has adopted a system of biosafety categories for pathogens based on their degree of pathogenicity and the relative danger in handling them. With this system, presented in appendix D, pathogens are assigned a level or class. Microbes not known to cause disease in humans (such as *Micrococcus luteus*) are given class 1 status; moderate risk agents such as *Staphylococcus aureus* are assigned to class 2; readily transmitted and virulent agents such as *Mycobacterium tuberculosis* are in class 3; and the highest risk microbes (dangerous, deadly pathogens such as rabies and Lassa fever viruses) are in class 4.

Pathogenicity and Virulence Factors

Differences in pathogenicity are more explicitly explained on the basis of virulence. Although this term is sometimes used interchangeably with pathogenicity, it is actually not a synonym. Whereas pathogenicity is a general term for describing microbial infectiousness, virulence defines microbial invasiveness and toxigenicity. These properties, which contribute to a pathogen's capacity to infect and damage host tissues, are called **virulence factors.** Virulence may be due to single or multiple factors. In some microbes, the causes of virulence are clearly established, but in others, they are not. In the following section we will examine the effects of virulence factors while simultaneously outlining the stages in the progress of an infection.

The Portal of Entry: Gateway to Infection

To initiate an infection, a microbe enters the tissues of the body by a characteristic route, the **portal of entry,** usually a cutaneous or membranous boundary. The source of the infectious agent can be **exogenous,** originating from a source outside the body (the environment or another person or animal), or **endogenous,** already existing on or in the body (normal flora or latent infection).

For the most part, the portals of entry are the same anatomical regions that also support normal flora: the skin, alimentary tract, respiratory tract, and urogenital tract. In fact, this easy access by the normal flora often helps it penetrate a portal at a time of lowered host resistance. The majority of pathogens have adapted to a specific portal of entry, one that provides a habitat for further growth and spread. This adaptation can be so restrictive that if certain pathogens enter the "wrong" portal, they will not be infectious. For instance, inoculation of the nasal mucosa with the influenza virus invariably leads to the flu, but if this virus contacts only the skin, no infection will result. Likewise, contact with athlete's foot fungi in small cracks in the toe webs can induce an infection, but inhaling the fungus spores will not. Occasionally, an infective agent can enter by more than one portal. For instance, the gonococcus can enter and infect the throat and anus in addition to the urogenital tract; *Mycobacterium tuberculosis* enters through both the respiratory and gastrointestinal tracts; *Corynebacterium diphtheriae* may infect by way of the throat and the skin; and *Francisella tularensis,* the cause of tularemia, is known to enter by every portal except the urogenital tract.

Infectious Agents That Enter the Skin

The skin is such a large and accessible target that it is a very common portal of entry. The actual sites of entry are usually nicks, abrasions, and punctures (many of which are tiny and inapparent) rather than smooth, unbroken skin. *Staphylococcus aureus* (the cause of boils), *Streptococcus pyogenes* (an agent of impetigo), the fungal dermatophytes, and agents of gangrene and tetanus gain access through damaged skin. The viral agent of cold sores (herpes simplex, type 1) enters through the skin of the lips.

Some infectious agents create their own passageways into the skin using digestive enzymes. For example, certain helminth worms burrow through the skin directly to gain access to the tissues. Other infectious agents enter through bites. The bites of insects, ticks, and mites offer an avenue to a variety of viruses, rickettsias, and protozoa. Bites by larger animals are a portal of entry for rabies and various nonspecific infections. An artificial means for breaching the skin barrier is the use of contaminated hypodermic needles by intravenous drug abusers (see feature 11.2).

Although the conjunctiva, the outer protective covering of the eye, is ordinarily a relatively good barrier to infection, bacteria such as *Hemophilus aegyptius* (pinkeye), *Chlamydia trachomatis* (trachoma), and *Neisseria gonorrhoeae* have special affinity for this membrane. Foreign bodies in the eye or minor injuries may serve as a portal for herpes simplex and *Acanthamoeba* (a protozoan that may contaminate homemade contact lens solutions).

The Gastrointestinal Tract As Portal

Pathogens with a portal of entry in the gastrointestinal tract begin their journey in the mouth with food, drink, and other materials that are swallowed. Along this journey, they must survive the acidity of the stomach and the bile of the intestine. Most enteric pathogens possess specialized mechanisms for entering and localizing in the mucosa of the small or large intestine. The best-known enteric agents of disease are gram-negative rods in the genera *Salmonella, Shigella, Vibrio,* and certain strains of *Escherichia coli* (see chapter 16). Viruses that enter through the gut are poliovirus, hepatitis A virus, echovirus, and rotavirus. Important enteric protozoans are *Entamoeba histolytica* (amebiasis) and *Giardia lamblia* (giardiasis). Although the anus is not a typical portal of entry, it becomes one in people who practice anal sex. About 20 different bacterial, protozoan, and fungal pathogens are involved in anorectal infections (those involving the anus and rectum).

The Respiratory Portal of Entry

The oral cavity is also the gateway to the respiratory tract, the route by which the largest number of pathogens enter. The extent to which an agent will be carried into the respiratory tree is based primarily on its size. In general, small cells and particles are inhaled more deeply than larger ones. Infectious agents deposited on the membrane of the nasopharynx include the bacteria of streptococcal sore throat, meningitis, diphtheria, and whooping cough, and the viruses of influenza, measles, mumps, rubella, chickenpox, and the common cold. Pathogens inhaled into the lower regions of the respiratory tract invade the epithelium of the bronchioles and lungs and elicit **pneumonia,** an inflammatory condition of the lung with multiple causes. Bacteria (the pneumococcus, *Klebsiella, Legionella, Mycoplasma*), fungi (*Cryptococcus*), and protozoa (*Pneumocystis, Cryptosporidium*) are a few of the agents involved in pneumonias. All types of pneumonia are on the increase due to the greater susceptibility of AIDS patients to them. Other agents causing unique recognizable lung diseases are *Mycobacterium tuberculosis* and fungal pathogens such as *Histoplasma.*

Urogenital Portals of Entry

The majority of reproductive and urinary tract infections are contracted by sexual means (intercourse or intimate direct contact). The older term, venereal disease, has been replaced by the more descriptive **sexually transmitted disease** (STD; see feature 11.3). The microbes of STDs enter the skin or mucosa of the penis, external genitalia, vagina, cervix, and urethra. Some can penetrate an unbroken surface, while others require a cut or abrasion. The once predominant sexual diseases syphilis and gonorrhea have been supplanted by a large and growing list of STDs headed by genital warts, chlamydia, and herpes. Evolving sex practices have increased the incidence of STDs that were once uncommon, and diseases that were not originally

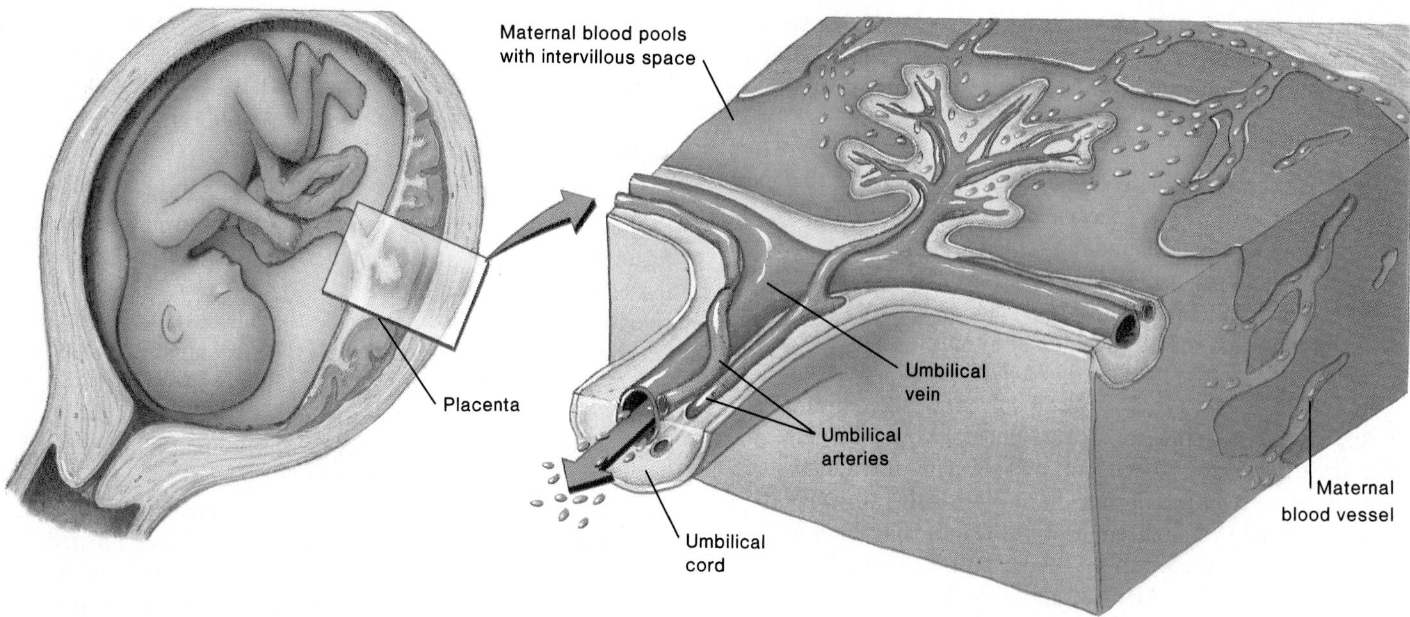

Figure 11.10 Transplacental infection of the fetus. Microbes move from the maternal blood vessels into the blood pool. From here, they seed the fetal circulation by way of the umbilical vein.

Feature 11.3 What's Love Got to Do with It?

The word venereal is derived from Venus, the Latin name of the goddess of love. For centuries, sexually oriented disease was considered a curse from God, with the greater share of blame attributed to women. During biblical times, so great was the condemnation of women afflicted with sex-associated diseases that many of them were executed. Attempts to treat or control infection during the early twentieth century were squelched by clergymen protesting that, "If God had wanted to eradicate venereal disease, he would not have given it to women in the first place." Posters from the first and second world wars and dating even into the 1950s flaunt the role of women in these diseases and depict them in the most debasing ways. It is clear that STDs were and are stigmatizing, just as it is now obvious that men are also profoundly involved in their transmission. The term sexually transmitted disease recognizes this fact.

considered STDs are now so classified.[4] Other common sexually transmitted agents are *Trichomonas* (a protozoan), *Candida albicans* (a yeast), *Hemophilus ducreyi* (the cause of chancroid), hepatitis B virus, pubic mites, and HIV (AIDS virus).

The Placental Portal of Entry

The placenta is an exchange organ formed by maternal and fetal tissues that separates the blood of the developing fetus from that of the mother, yet permits diffusion of dissolved nutrients and gases to the fetus. Because the fetal and maternal circulation systems are not directly connected, the placenta is ordinarily an effective barrier against microorganisms in the maternal circulation. However, a few microbes can cross the placenta, enter the umbilical vein, and spread by the fetal circulation into the fetal tissues (figure 11.10). The outcome of placental penetration depends upon the stage of pregnancy and the type of microbe involved. The most serious complications are spontaneous abortion, congenital malformations, fatal brain damage, prematurity, and stillbirths. Agents most associated with injurious transplacental infections are rubella virus, cytomegalovirus, HIV, the spirochete of syphilis, *Listeria monocytogenes, Campylobacter fetus,* and the protozoan *Toxoplasma gondii.*[5]

The Size of the Inoculum

Another factor crucial to the course of an infection is the quantity of microbes in the inoculating dose. For most agents, infection will proceed only if a minimum number, called the *infectious dose* (ID), is present. This number has been determined experimentally for many microbes. In general, microorganisms with smaller infectious doses have greater virulence. The ID varies from the astonishing number of 1 cell in Q fever to about10 virus particles in rabies, tuberculosis, giardiasis, tularemia, and coccidioidomycosis, and from 1,000 cells for gonorrhea and 10,000 cells in typhoid fever to 1,000,000,000 cells in cholera. If the size of the inoculum is smaller than the infectious dose, infection will generally not progress. If the quantity is far in excess of the ID, the onset of disease may be more rapid. Even weakly pathogenic species may be rendered more virulent with a large inoculum.

4. Amebic dysentery, scabies, salmonellosis, and *Strongyloides* worms are examples.

5. Several diseases that are transmitted fetally or at birth are represented by the acronym STORCH, which includes syphilis, toxoplasmosis, other diseases (such as hepatitis B, listeriosis, HIV, and CMV), rubella, chlamydia, and herpes simplex virus.

Mechanisms of Invasion and Establishment of the Pathogen

Following entry of the pathogen, the next stage in infection requires that the pathogen (1) bind to the host, (2) penetrate the epithelial boundary, and (3) become established in the tissues. How the pathogen achieves these ends greatly depends upon its specific biochemical characteristics.

How Pathogens Attach

Adhesion is a process by which microbes gain a more stable foothold at the portal of entry. Because this often involves a specific interaction between surface molecules on the microbial surface and receptors on the host cell, adhesion may determine the specificity of a pathogen for its host organism and in some cases the specificity of a pathogen for a particular cell type. Once attached, the pathogen cannot be readily dislodged and is poised advantageously to invade the sterile body compartments. Bacterial pathogens attach most often by mechanisms such as fimbriae (pili), flagella, and adhesive slimes or capsules, while viruses attach by means of specialized receptors (figure 11.11). Protozoa may infiltrate by means of their organelle of locomotion; for example, *Balantidium coli* is said to penetrate by the boring action of its cilia. Parasitic worms are mechanically fastened to the portal of entry by suckers, hooks, and barbs (see chapter 19). Adhesion methods of various microbes and the diseases they lead to are shown in table 11.6.

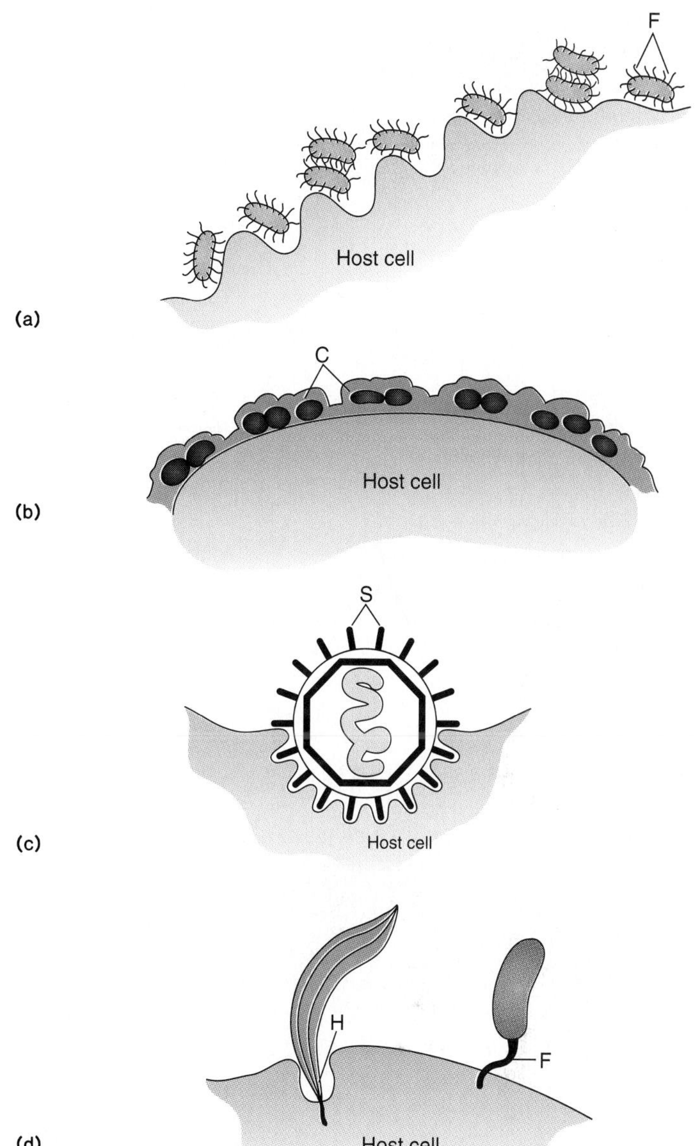

Figure 11.11 Mechanisms of adhesion by pathogens (see table 11.6 for specific examples). (*a*) Fimbriae (F), minute bristlelike appendages. (*b*) Adherent extracellular capsules (C), slime, or other sticky substance. (*c*) Viral envelope spikes (S). (*d*) Specialized cell hooks (H) or filaments (F).

How Virulence Factors Contribute to Tissue Damage

The behavior of a microbe during entry and invasion is a balancing act of sorts. While seeking a habitat that will nourish and sustain it, the parasite must not destroy itself or the host. Well-adapted parasites such as the cold virus or dermatophytic fungi invade the tissues and multiply with relatively minor effects. But unfortunately, many parasites (the tetanus bacillus and AIDS virus, for instance) cause diseases that can severely damage and even kill the host. For convenience, we will divide the virulence factors into exoenzymes, toxins, and antiphagocytic factors (figure 11.12). Although this distinction is useful, there is often a very fine line between enzymes and toxins because many substances called toxins actually function as enzymes.

Extracellular Enzymes Many pathogenic bacteria, fungi, protozoa, and worms secrete **exoenzymes** that break down and inflict damage on epithelial structures. The following enzymes dissolve the host's defense barriers and promote the spread of microbes to deeper tissues:

1. Mucinase, which digests the protective coating on mucous membranes, is produced by *Vibrio cholerae* and *Entamoeba histolytica.*
2. Keratinase, which digests the principal component of skin and hair, is produced by dermatophytic fungi.
3. Collagenase, which digests the principal fiber of connective tissue, is produced by *Clostridium* species and certain worms.
4. Hyaluronidase digests the ground substance that cements animal cells together (hyaluronic acid). This enzyme is an important virulence factor in staphylococci, clostridia, streptococci, and pneumococci. Because such destruction facilitates dissemination of microbes, this enzyme has been called the spreading factor.

Some enzymes react with components of the blood. Coagulase, an enzyme produced by pathogenic staphylococci and the plague bacillus, causes clotting of blood or plasma. By contrast, the bacterial kinases (streptokinase, staphylokinase) do just the opposite, dissolving fibrin clots and expediting the invasion of damaged tissues.

Table 11.6 Adhesive Factors in Microbes		
Microbe	**Disease**	**Adherence Mechanism**
Neisseria gonorrhoeae	Gonorrhea	Fimbriae attach to genital epithelium
N. menigitidis	Meningitis	Fimbriae cling to membrane of pharynx
Escherichia coli	Diarrhea	Well-developed K antigen capsule
Shigella	Dysentery	Fimbriae may attach to intestinal epithelium (not known for sure)
Campylobacter	Diarrhea	Flagellum and specialized cell tip embed into cells
Vibrio	Cholera	Glycocalyx anchors microbe to intestinal epithelium
Treponema	Syphilis	Tapered hook embeds in host cell
Mycoplasma	Pneumonia	Specialized filament at ends of bacteria fuses tightly to lung epithelium
Pseudomonas aeruginosa	Burn, lung infections	Fimbriae and slime layer
Corynebacterium diphtheriae	Diphtheria	Surface protein adheres to mucosa of pharynx
Streptococcus pyogenes	Pharyngitis, impetigo	Lipotechoic acid and M-protein anchor cocci to epithelium
Streptococcus mutans, S. sobrinus	Dental caries	Dextran slime layer glues cocci to tooth surface
Influenza virus	Influenza	Viral spikes react with receptor on cell surface
Poliovirus	Polio	Capsid proteins attach to receptors on susceptible cells

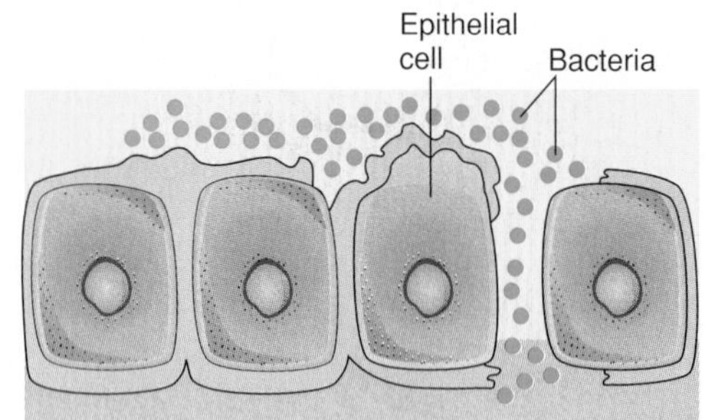

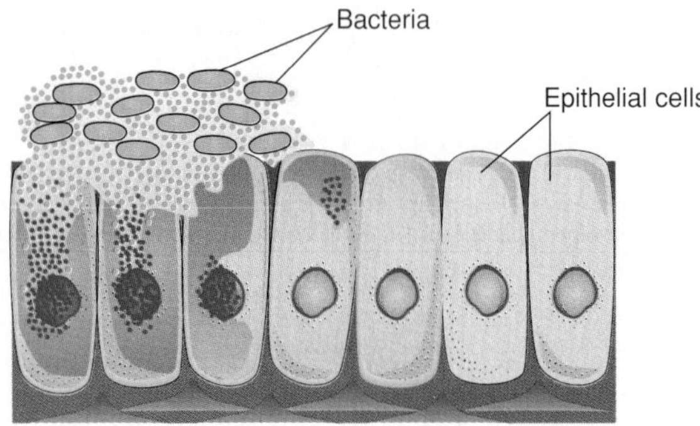

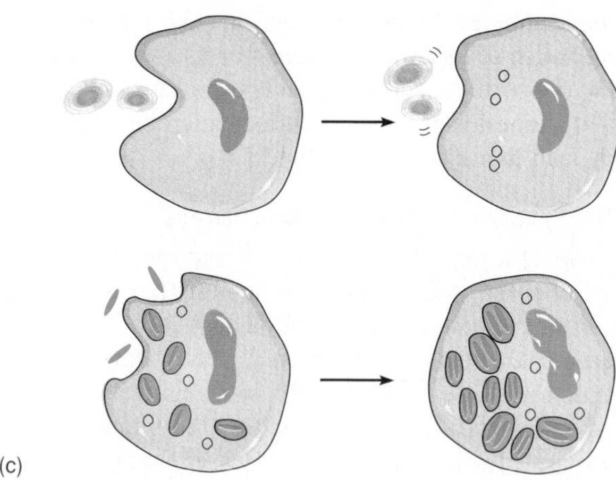

Figure 11.12 The function of exoenzymes, toxins, and phagocyte blockers in invasiveness. (*a*) Exoenzymes. Bacteria produce extracellular enzymes that dissolve extracellular barriers and penetrate through or between cells to underlying tissues. (*b*) Toxins (primarily exotoxins) secreted by bacteria diffuse to target cells, which they poison and disrupt by lysis or antimetabolic processes. (*c*) Blocked (top) or incomplete (bottom) phagocytosis. Phagocytosis may be ineffective for several reasons. For example, the pathogens may have a protective coating that makes them hard to engulf, or the phagocyte may ingest but not completely destroy them.

Bacterial Toxins: A Potent Source of Cellular Damage A **toxin** is a specific chemical product of microbes, plants, and some animals that is poisonous to other living things. **Toxigenicity,** the power to produce toxins, is a genetically controlled characteristic of many species and is responsible for the adverse effects of a variety of diseases generally called **toxinoses.** A type of toxinosis in which the toxin is spread by the blood from the site of infection is a **toxemia** (tetanus and diphtheria, for example), while one caused by ingestion of toxins is an **intoxication** (botulism). A toxin is named according to its specific target of action: Neurotoxins act on the nervous system; enterotoxins act on the intestine; hemotoxins lyse red blood cells; and nephrotoxins damage the kidneys.

A more traditional scheme classifies toxins according to their origins (figure 11.13). An unbound toxin molecule secreted by a living bacterial cell into the infected tissues is an **exotoxin.** A toxin that is not secreted but released only after the cell is damaged or lysed is an **endotoxin.** Other important differences between the two groups, summarized in table 11.7, are generally chemical and medical in nature.

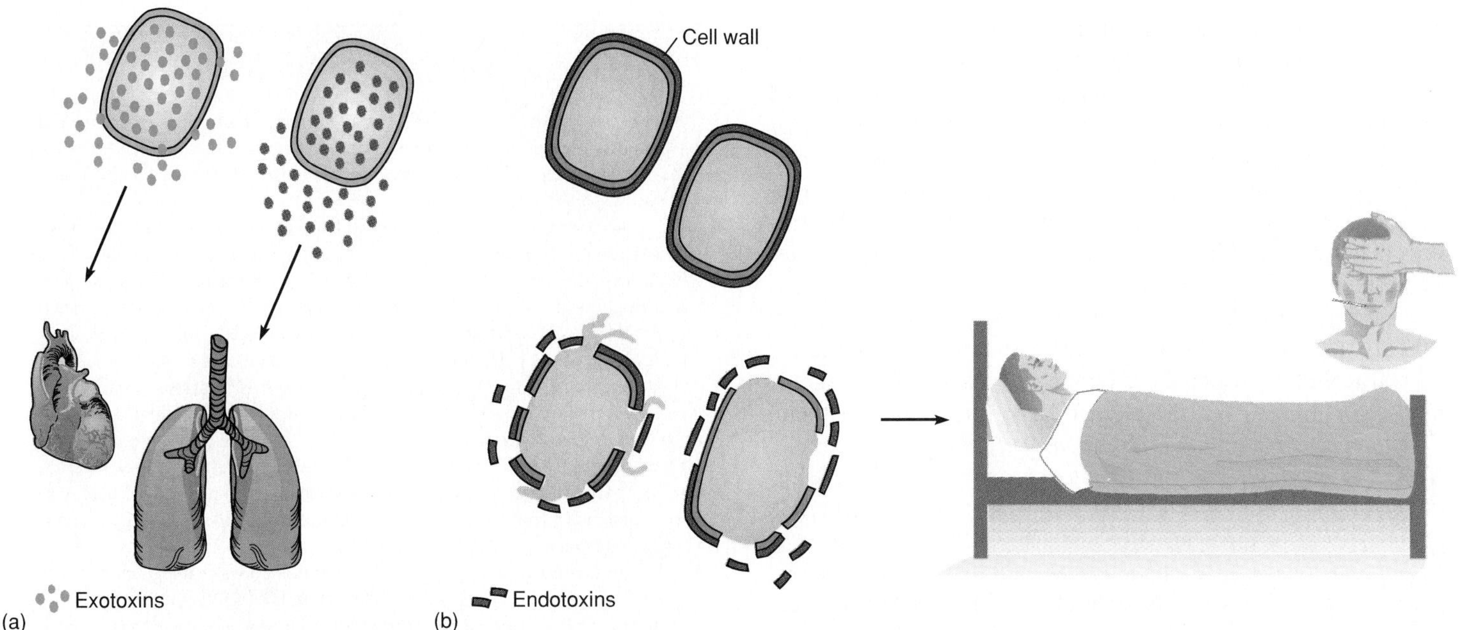

Figure 11.13 The origins and effects of circulating exotoxins and endotoxins. (*a*) Exotoxins, given off by live cells, have highly specific targets and physiological effects. (*b*) Endotoxins, given off when the cell wall of gram-negative bacteria disintegrates, have more generalized physiological effects.

Table 11.7 Differential Characteristics of Exotoxins and Endotoxins

Characteristic	Exotoxins	Endotoxins
Toxicity	Strong (damaging in minute amounts)	Weaker
Effects on the Body	Specific to a tissue	Generalized, nonspecific
Chemical Composition	Polypeptides	Lipopolysaccharide of cell wall
Heat Denaturation at 60°C	Unstable	Stable
Toxoid Formation	Convert to toxoid*	Do not convert to toxoid
Immune Response	Stimulate antitoxins**	Do not stimulate antitoxins
Fever Stimulation	No	Yes
Manner of Release	Secreted, no lysis	Released by lysis
Typical Sources	A few gram-positive and gram-negative bacteria	All gram-negative bacteria

*A toxoid is a neutralized toxin.
**An antitoxin is an antibody that reacts specifically with a toxin.

Exotoxins are proteins with a strong specificity for a target cell and extremely powerful, sometimes deadly, effects. Exotoxins generally affect cells by damaging the cell membrane and initiating lysis or by disrupting intracellular function. **Hemolysins** are a class of bacterial exotoxin that disrupts the cell membrane of red blood cells (and some other cells too). This damage causes the red blood cells to **hemolyze**—that is, to burst and release hemoglobin pigment. Hemolysins that increase pathogenicity include the streptolysins of *Streptococcus pyogenes,* the alpha (α) and beta (β) toxins of *Staphylococcus aureus,* and the soluble hemolysin of *Listeria monocytogenes.* When colonies of bacteria growing on blood agar produce hemolysin, distinct zones appear around the colony. The pattern of hemolysis is often used to identify bacteria and determine their degree of pathogenicity (see chapter 15).

The toxins of diphtheria, tetanus, and botulism, among others, attach to a particular target cell, become internalized, and interrupt an essential cell pathway. The consequences of cell disruption depend upon the target. One toxin of *Clostridium tetani* blocks the action of certain spinal neurons; the toxin of *Clostridium botulinum* prevents the transmission of nerve-muscle stimuli; pertussis toxin inactivates the respiratory cilia; and cholera toxin provokes profuse water and salt loss from intestinal cells. More details of the pathology of exotoxins can be found in the chapters that treat the specific diseases.

Endotoxins belong to a class of chemicals called lipopolysaccharides (LPS), which are an integral part of the cell walls of gram-negative bacteria. When gram-negative bacteria cause infections, some of them eventually lyse and release small bits of the cell wall into the infection site or into the circulation. Depending upon the amounts present, endotoxins have a variety of nonspecific effects, including fever, cardiovascular shock (in large amounts), hemorrhage, and tissue death. A blood infection by gram-negative bacteria such as *Salmonella, Shigella, Neisseria meningitidis,* and *Escherichia coli* is particularly dangerous, in that it can lead to fatal endotoxic shock (see feature 16.5).

hemolysin (hee-mawl'-uh-sin) Gr. *haima,* blood, and *lysis,* dissolution. A substance that causes hemolysis.

How Microbes Escape Phagocytosis Phagocytes are important cellular predators that block the advancement of microbes into the tissues. Through phagocytosis, pathogens are engulfed and destroyed by powerful enzymes and poisons (see chapter 12). Challenged by this efficient and widespread host defense, many microorganisms have adapted mechanisms (antiphagocytic factors) that circumvent some part of the phagocytic process (see figure 11.12*c*). The most aggressive strategy involves bacteria that kill phagocytes outright. Species of both *Streptococcus* and *Staphylococcus* produce **leukocidins,** substances that are toxic to white blood cells. Some microorganisms secrete an extracellular surface layer (slime or capsule) that makes it physically difficult for the phagocyte to get a hold on them. *Streptococcus pneumoniae, Salmonella typhi, Yersinia pestis, Neisseria meningitidis,* and *Cryptococcus neoformans* are notable examples. Some bacteria are well adapted to survival inside phagocytes after ingestion. For instance, pathogenic species of *Legionella, Listeria, Mycobacterium,* and many rickettsias are readily engulfed but are capable of avoiding further destruction. Surviving intracellularly in phagocytes has special significance because it provides a place for the microbes to hide, grow, and be spread throughout the body.

Establishment, Spread, and Pathological Effects

Aided by virulence factors, microbes eventually come to rest within a particular target organ and there continue to cause damage. The type and scope of injuries inflicted during this process account for the typical stages of an infection (see feature 11.4), the patterns of the infectious disease, and its signs and symptoms.

In addition to the adverse effects of enzymes, toxins, and other factors, multiplication by the parasite frequently undermines the structural integrity of the host. Accumulated cellular debris, change in pH, competition for oxygen, and excretion of waste products all may weaken the host tissues. Physical obstruction of tubular structures such as blood vessels, lymphatic channels, the intestinal lumen, or a bile duct adds further insult. Eventually, the tissues may be so destroyed as to cause **necrosis** (lysis of the host cells and death of tissue). Although viruses do not produce toxins or destructive enzymes, they are fully capable of destroying cells by multiplying within them. Many of the cytopathic effects of viral infection arise from the impaired metabolism and death of cells (see chapter 5).

Varied Patterns of Infection Just as host-parasite relationships come in many varieties, so do patterns of infection (figure 11.14). In the simplest situation, the target tissue is the same as the point of entry. An infection that remains confined to a specific body site is called a **localized infection.** Examples of localized infections are boils, fungal skin infections, and warts. Many infectious agents do not remain local but spread away from the initial site of entry to other tissues. In fact, this is necessary for pathogens whose target tissue is some distance from the site of entry, as in rabies and hepatitis A. The rabies virus travels from a bite wound to its target in the brain, and the hepatitis A virus goes from the intestine to the liver. The spread of a microbe from the site of entry to its target is usually *hematogenous*—that is, within the host's circulatory system.

Feature 11.4 The Classic Stages of Clinical Infections

As the body of the host responds to the invasive and toxigenic activities of a parasite, it passes through four distinct phases of infection and disease—the incubation period, the prodromium, the period of invasion, and the convalescent period.

The **incubation period** is the time from initial contact with the infectious agent (at the portal of entry) to the appearance of the first symptoms. (In a way, the term is misleading because it gives the impression that this is the only period of microbial growth, when in fact, the microbe is multiplying during all of the periods.) During the incubation period, the agent is multiplying at the portal of entry but has not yet caused enough damage to elicit symptoms. Although the incubation period is relatively well defined and predictable for each microorganism, it does vary according to host resistance, degree of virulence, and distance between the target organ and the portal of entry (the farther apart, the longer the incubation period). Overall, an incubation period may range from several hours in pneumonic plague to several years in leprosy. The majority of infections have incubation periods ranging between 2 and 30 days.

The earliest notable symptoms of infection appear as head and muscle aches, fatigue, upset stomach, and general malaise (a vague feeling of discomfort). This short period (1–2 days) is the **prodromium.** The infectious agent next enters a **period of invasion,** during which it multiplies at high levels, exhibits its greatest toxicity, and becomes well established in its target tissue. This period is often marked by fever and other prominent and more specific symptoms, which may include cough, rashes, diarrhea, loss of muscle control, swelling, jaundice, discharge of exudates, or severe pain, depending on the particular infection. The length of this period is extremely variable.

In most cases, the patient begins to respond to the infection, and the symptoms decline—sometimes dramatically, other times slowly. Hollywood has been particularly fond of the image of the pioneer woman tending a sick child all night until finally the fever "breaks" and the child is miraculously better. During the recovery that follows, called the **convalescent period,** the patient's strength and health gradually return due to the healing nature of the immune response. But sometimes the patient does not recover. An infection that results in death is **terminal.** This term is particularly applicable to an infection that is the immediate cause of death in a patient already suffering from a degenerative disease or cancer. Thus, an alcoholic might succumb to terminal pneumococcal pneumonia, or an AIDS patient might actually die of pneumocystis pneumonia.

The transmissibility of the microbe during these four stages must be considered on an individual basis. A few agents are released mostly during incubation (measles, for example); many are released during the invasive period (*Shigella*); and others can be transmitted during all of these periods (hepatitis B).

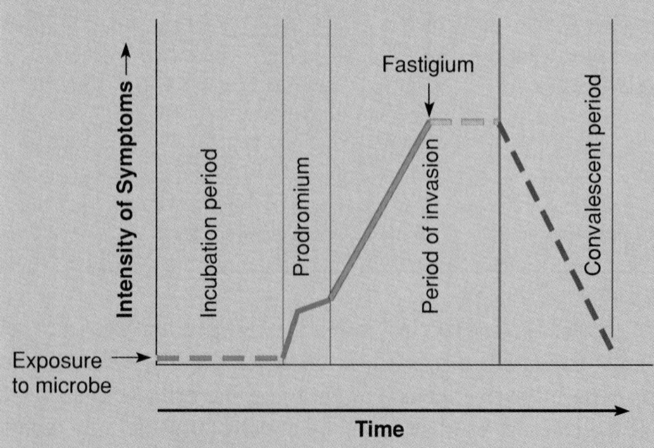

Stages in the course of infection and disease. Dashed lines represent periods with a variable length; the fastigium is the height of the disease.

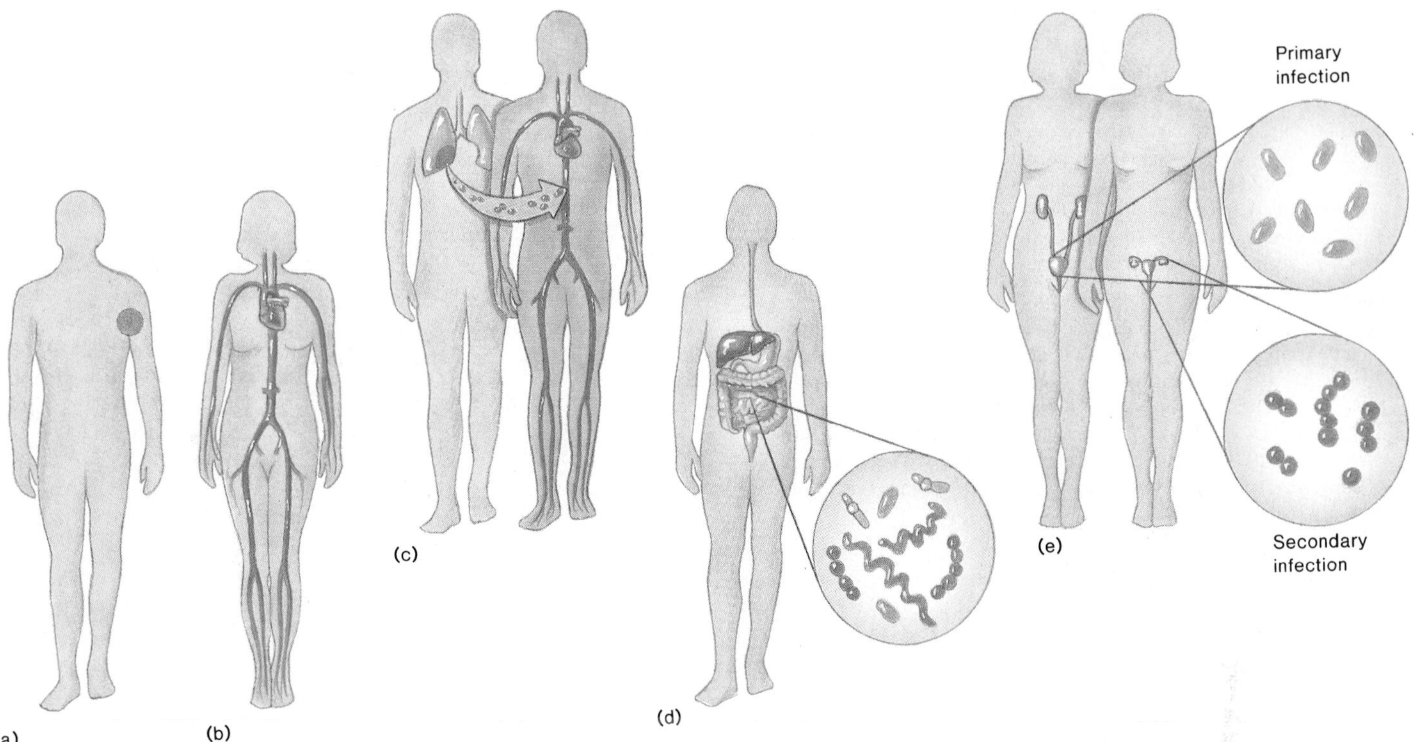

Figure 11.14 The occurrence of infections with regard to location, type of microbe, and space of time. (*a*) A localized infection, in which the pathogen is restricted to one specific site. (*b*) Systemic infection, in which the pathogen spreads through circulation to many sites. (*c*) A focal infection occurs initially as a local infection, but circumstances cause the microbe to be carried to other sites systemically. (*d*) A mixed infection, in which the same site is infected with several microbes at the same time. (*e*) In a primary/secondary infection, the same person has an initial infection that is complicated by a second one in a different location (usually) and caused by a different microbe.

When an infection spreads to several sites and tissue fluids, it is generalized or **systemic.** Examples of systemic infections are viral diseases (measles, rubella, chickenpox, and AIDS); bacterial diseases (brucellosis, anthrax, typhoid fever, and syphilis); and fungal diseases (histoplasmosis and cryptococcosis). Infectious agents may also travel to their targets by means of nerves (as in rabies) or cerebrospinal fluid (as in meningitis).

In **focal** infections, the infectious agent breaks loose from a local infection and is carried into other tissues. This pattern is exhibited by tuberculosis or by streptococcal pharyngitis that gives rise to scarlet fever. In a toxemia,[6] the infection itself remains localized at the portal of entry and does not spread into other tissues. While multiplying locally, the microbe gives off toxins that are carried by the blood to the actual target tissue. In this way, the target of the bacterial cells may be different from the target of their toxin (see discussions of tetanus and diphtheria in chapter 16).

An infection is not always caused by a single microbe. In a **mixed infection,** several agents establish themselves simultaneously at the infection site. In some mixed or synergistic infections, the microbes cooperate in breaking down a tissue. In other mixed infections, one microbe creates an environment that enables another microbe to invade. Gas gangrene, wound infections, dental caries, and human bite infections tend to be mixed.

6. Not to be confused with toxemia of pregnancy, which is a metabolic disturbance and not an infection.

Some diseases are described according to a time frame. When an initial or **primary infection** is followed by another infection caused by a different microbe, the second infection is termed a **secondary infection.** This pattern often occurs in a child with chickenpox (primary infection) who may scratch his pox and infect them with *Staphylococcus aureus* (secondary infection). The secondary infection need not be in the same site as the primary infection, and it usually indicates altered host defenses. Infections that come on rapidly, with severe but short-lived effects, are called **acute** infections. Infections that progress and persist over a long period of time are **chronic. Subacute** infections do not come on as rapidly as acute infections or persist as long as chronic ones. Feature 11.5 illustrates other common terminology used to describe infectious diseases.

Signs and Symptoms: Warning Signals of Disease

When the cumulative effects of an infection lead to disease, a variety of structural and functional changes, or **pathological** effects, characteristically manifest themselves as signs and symptoms. A **sign** is any objective, measurable evidence of disease as noted by an observer; a **symptom** is the subjective evidence of disease as felt or seen by the patient. In general, signs are more precise than symptoms, though both may have the same underlying cause. For example, an infection of the brain might present with the symptom of headache and the sign of bacteria in the spinal fluid. Or a respiratory infection might produce a cough (symptom) and abnormal chest sounds (sign). Disease

Feature 11.5 A Quick Guide to the Terminology of Infection and Disease

Words in medicine have great power and economy. A single technical term often serves for a whole phrase or sentence. A well-chosen word in a care plan or on a patient's chart can save time and space. But at first, the student may feel overwhelmed by what seems like a mountain of new words. Knowing a few root words and a fair amount of anatomy can help you learn many of these words and even deduce the meaning of unfamiliar ones.

The "itises" The suffix *-itis* means an inflammation of some body part, and when affixed to the end of an anatomical term, it indicates an inflammatory condition in that location. Thus, meningitis is an inflammation of the tissues surrounding the brain; encephalitis is an inflammation of the brain itself; hepatitis involves the liver; vaginitis, the vagina; gastroenteritis, the intestine; and otitis media, the middle ear. Pneumonitis is a condition of the lungs usually given the older and more common name pneumonia. Although not all inflammatory conditions are caused by infections, many infectious diseases inflame their target organs.

The "emias" The suffix *-emia* is derived from the Greek word *haeima*, meaning blood. When added to a word, it means "associated with the blood." Thus, septicemia means sepsis (infection) of the blood; bacteremia, bacteria in the blood; viremia, viruses in the blood; and fungemia, fungi in the blood. There are any number of other general or specific prefixes (toxemia, gonococcemia, spirochetemia).

The "oses" The suffix *-osis* means "a disease or morbid process." It is frequently added to the names of pathogens to indicate the disease they cause—for example, listeriosis, histoplasmosis, toxoplasmosis, shigellosis, salmonellosis, and borreliosis. A variation of this suffix is *-iasis*, as in trichomoniasis and candidiasis.

The "omas" The suffix *-oma* comes from the Greek word *onkomas* (swelling) and means tumor. Although the root is often used to describe cancers (sarcoma, melanoma), it is also applied in some infectious diseases that cause masses or swellings (tuberculoma, leproma).

Table 11.8 Common Signs and Symptoms of Infectious Diseases

Signs
Fever
Septicemia
Microbes in tissue fluids
Various "itises"
Chest sounds
Skin eruptions
Leukocytosis
Leukopenia
Swollen lymph nodes
Granuloma
Abscesses
Tachycardia (increased heart rate)
Antibodies in serum
Symptoms
Fever, chills
Pain, ache, soreness, irritation, redness
Drainage from body opening or sore
Malaise, fatigue
Cough
Rashes
External swelling
Diarrhea
Vomiting
Abdominal cramps
Anorexia (lack of appetite)
Pus

indicators such as fever and rashes can be sensed by the patient and measured by the physician, thus are both signs and symptoms. When a disease can be identified or defined by a certain complex of signs and symptoms, it is termed a *syndrome*. Signs and symptoms with considerable importance in diagnosing infectious diseases are shown in table 11.8.

Signs and Symptoms of Inflammation

The earliest symptoms of disease result from the activation of a body defense called the **inflammatory response,** a system of cells and chemicals that responds nonspecifically to disruptions in the tissue. This subject will be discussed in greater detail in chapter 12, but it is worth noting here that symptoms of infection and disease such as fever, chills, pain, soreness, redness, and irritation are caused by the mobilization of this system. Some signs of inflammation include: **edema,** the accumulation of fluid in an afflicted tissue; inflammatory conditions in specific tissues (see "itises," feature 11.5); *granulomas* and *abscesses,* walled-off collections of inflammatory cells and microbes in the tissues; and *lymphadenitis,* swollen lymph nodes. The skin is also a common focus of signs and symptoms (see feature 11.6).

Blood Signs

Changes in the number of circulating white blood cells, as determined by special counts, may indicate infection. *Leukocytosis* is an increase in the level of white blood cells, whereas *leukopenia* is a decrease. Other signs of disease revolve around the occurrence of a microbe or its products in the blood. The clinical term for blood infection, **septicemia,** refers to a general state in which microorganisms are multiplying in the blood and are present in large numbers. When small numbers of bacteria or viruses are found in the blood, the correct terminology is **bacteremia** or **viremia,** which means that these microbes are present in the blood but are not necessarily multiplying.

Signs of a Specific Immune Reaction

The normal host will invariably show signs of an immune response in the form of antibodies in the serum or some type of sensitivity to the microbe. This tendency is the basis for several serological tests used in diagnosing infectious diseases such as AIDS or syphilis. Such specific immune reactions account for

inflammatory (in-flam′-uh-tor″-ee) L. *inflammatio,* to set on fire.

edema (uh-dee′-muh) Gr. *oidema,* swelling.

leukocytosis (loo″-koh-sy-toh′-sis) From *leukocyte,* a white blood cell, and the suffix *osis.*

leukopenia (loo″-koh-pee′-nee-uh) From *leukocyte* and *penia,* a loss or lack of.

Feature 11.6 Out Damned Spot: When Infection Erupts in the Skin

The skin is often the canvas upon which the signs and symptoms of infection are sketched. Rashes and other skin eruptions are common in many diseases, and because they tend to mimic each other, it can be difficult to differentiate among diseases on this basis alone. The general term for any damaged or dysfunctional body area is **lesion.** Skin lesions form when a microbe directly invades the skin, when it exits via the skin, or when an inflammatory reaction occurs. Skin lesions may be restricted to the epidermis and its glands and follicles, or they may extend into the dermis and subcutaneous regions. The lesions of some infections undergo characteristic changes in appearance during the course of disease and thus fit more than one category (see figure 20.1).

A small, flat-colored skin lesion is a *macule*. The color may be red, tan, or white. Crops of these spots are early signs of measles and rubella. A raised lesion composed of solid tissue is a *papule.* A papule varies in texture from smooth to rough and may present any number of colors. Often, papules are a later stage in the development of a macule, producing the so-called maculopapular rash. A thicker papule that penetrates into the dermis is a *nodule.* Leprosy and several fungus infections produce nodular lesions.

Many lesions contain some sort of liquid or semiliquid material called an exudate. A *vesicle* is a small blister that contains clear, yellowish fluid, and a *bulla* is a large, fluid-filled blister. Vesicles are characteristic of herpes simplex, impetigo, fungal infections, and chickenpox (a pock is one type of vesicle); bullae are seen in some types of staphylococcal skin infections. If the space in the lesion is filled with pus (a mixture of fluid, white blood cells, tissue debris, and bacteria), it is a *pustule.* Acne lesions are types of pustules. Vesicles, bullae, and pustules have a fragile surface that may burst and release the exudate, which dries and forms a *crust.* Depending upon how deep they go, these lesions may or may not leave permanent scars.

When a lesion results from skin being sloughed, the process is known as *erosion* or *ulceration.* Erosion is a superficial loss of the epidermis, while an ulcer is a deeply penetrating open sore. The chancre of syphilis and the eschar of anthrax are erosive types of lesions; large, draining ulcers appear in leishmaniasis and some worm infections.

Some skin manifestations are due to blood or bleeding. *Erythema* is a confluent reddening of the skin due to increased blood flow. The rash of scarlet fever and the peculiar migrating skin eruption of Lyme disease are types of erythema. Hemorrhaging into the skin produces a brown to purple discoloration called *purpura,* a serious complication in bubonic plague and meningococcal meningitis.

---

lesion (lee'-zhun) L. *laesio,* to hurt.
macule (mak'-yool) L. *maculatus,* spotted.
papule (pap'-yool) L. *papula,* pimple.
nodule (nawd'-yool) L. *nodulus,* little knot.
vesicle (ves'-ik-ul) L. *vesicula,* small bladder.
bulla (byoo'-lah) L. *bulla,* bubble.
pustule (pust'-yool) L. *pustula,* blister or pimple.
erythema (air-ih-thee'-mah) Gr. *erythema,* flush upon the skin.
purpura (pur'-pur-ah) L. *purpura,* purple.

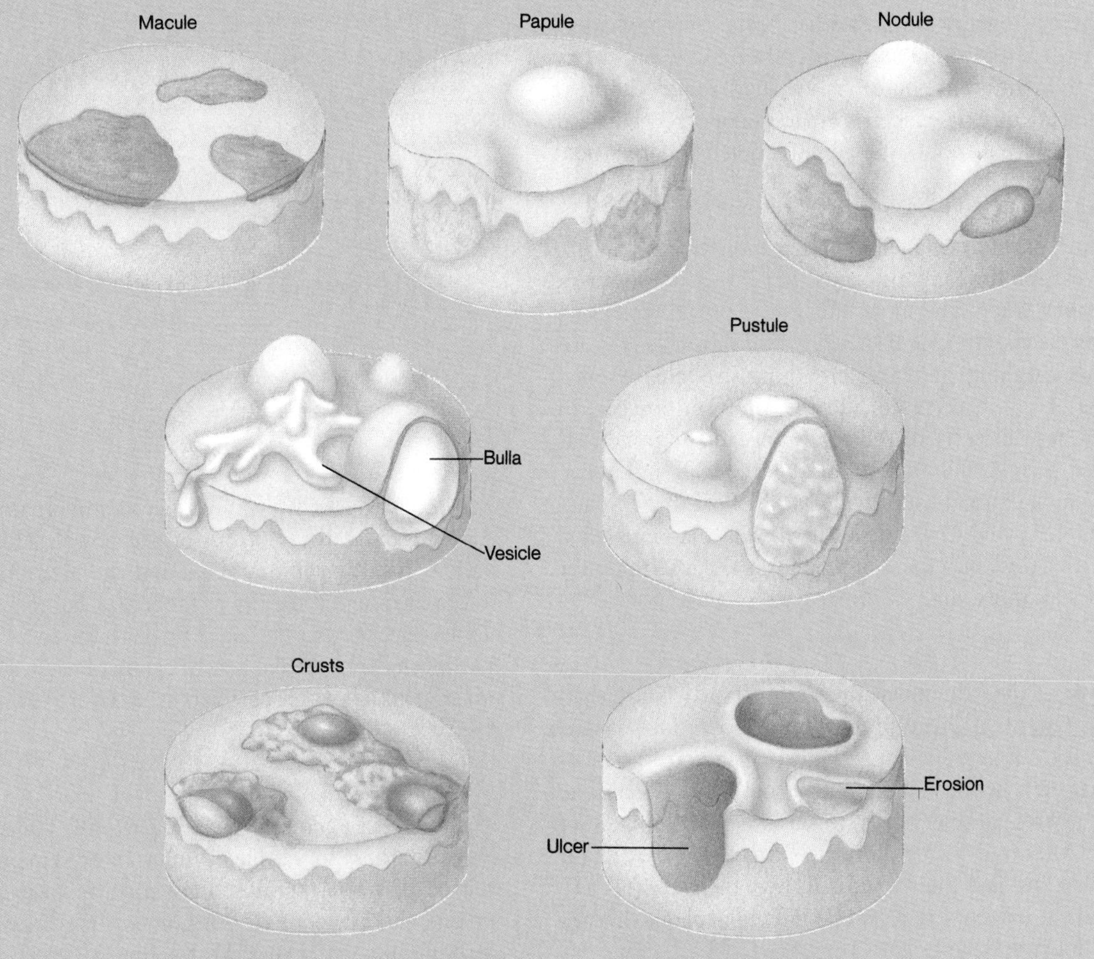

Cutaway views of the skin comparing various lesions.

any long-term protection against reinfection by that microbe. We will concentrate on this role of the host defenses in the next two chapters.

Infections That Go Unnoticed

It is rather common for an infection to produce no noticeable symptoms, even though the microbe is active in the host tissue. In other words, though infected, the host does not manifest the disease. Infections of this nature are known as **subclinical, inapparent,** or **asymptomatic,** because the patient experiences no symptoms or disease and does not seek medical attention. However, it is important to note that most infections are attended by some sort of sign. In the section on epidemiology, we will further address the significance of subclinical infections in the transmission of infectious agents.

The Portal of Exit: Vacating the Host

Earlier, we introduced the idea that a parasite is considered *unsuccessful* if it kills its host. A parasite is equally unsuccessful if it does not have a provision for leaving its host and moving to other susceptible hosts. With few exceptions, pathogens depart by a specific avenue called the **portal of exit** (figure 11.15). In most cases, the pathogen is shed or released from the surface of the body through secretion, excretion, discharge, or sloughed tissue. The number of infectious agents in these materials is usually very high, which increases both virulence and the likelihood that the pathogen will reach other hosts. In many cases, the portal of exit is the same as the portal of entry, but a few pathogens seek a different route. As we shall see in the next section, the portal of exit concerns epidemiologists because it greatly influences the dissemination of infection in a population.

Respiratory and Salivary Portals

Mucus, sputum, nasal drainage, and other moist secretions are the media of escape for the pathogens that infect the lower or upper respiratory tract. The most effective means of releasing these secretions is coughing and sneezing (see figure 11.22), although talking, laughing, and mere breathing also release some. Tiny particles of liquid released into the air form aerosols or droplets, and these directly or indirectly spread the infectious agent to other people. The agents of tuberculosis, influenza, measles, and chickenpox most often leave the host through airborne droplets. Microbes that infect the salivary gland are shed right along with saliva into the mouth. This is the exit route for mumps, herpes simplex, and rabies viruses.

Skin Scales

The outer layer of the skin and scalp are constantly being shed into the environment. A large proportion of household dust is actually composed of skin scales. A single person may shed several billion skin cells per day, and some persons, called shedders, disseminate massive numbers of bacteria into their immediate surroundings. Although normal flora are shed, so too are certain pathogens. Skin lesions, their exudates, and their crusts may serve as portals of exit in warts, fungal infections, boils, herpes simplex, smallpox, and syphilis.

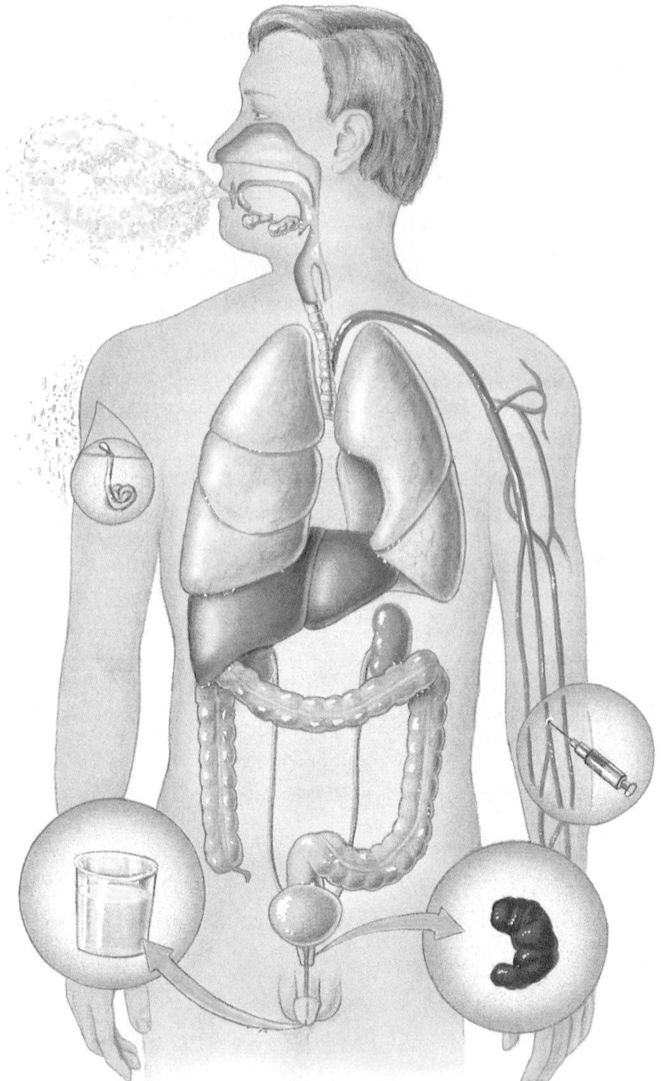

Figure 11.15 Major portals of exit of infectious diseases.

Fecal Exit

Feces are a very common portal of exit. Some intestinal pathogens grow in the intestinal mucosa and create an inflammation that increases the motility of the bowel. This speeds up peristalsis, resulting in diarrhea, and the more fluid stool provides a more rapid exit for the pathogen. A number of metazoan parasites release cysts and eggs through the feces. Feces containing pathogens are a public health problem when allowed to contaminate drinking water or when used to fertilize crops (see chapter 19).

Urogenital Tract

A number of agents involved in sexually transmitted infections leave the host in vaginal discharge or semen. This is also the source of neonatal infections such as herpes simplex, *Chlamydia,* and *Candida albicans,* which infect the infant as it passes through the birth canal. Less commonly, certain pathogens that infect the kidney are discharged in the urine—for instance, the agents of leptospirosis, typhoid fever, tuberculosis, and schistosomiasis.

Removal of Blood or Bleeding

Although the blood does not have a direct route to the outside, it can serve as a portal of exit when it is removed or released through the site of an injury. This may occur by natural or artificial means. Blood-feeding insects such as mosquitos and fleas are common transmitters of pathogens (see feature 17.5). The AIDS and hepatitis viruses are transmitted by shared needles or through small gashes in a mucous membrane caused by sexual intercourse. Blood donation is also a means for certain microbes to leave the host, though this is rather unusual because of close monitoring of the donor population and blood used for transfusions.

The Persistence of Microbes and Pathological Conditions

The apparent recovery of the host does not always mean that the microbe has been completely removed or destroyed by the host defenses. Following the initial symptoms in certain chronic infectious diseases, the infectious agent retreats into a dormant state, a condition called **latency.** Throughout this latent state, the microbe may periodically become active and produce a recurrence of disease. The viral agents of herpes simplex, herpes zoster, hepatitis B, AIDS, and Epstein-Barr persist in the host for long periods of time. The agents of syphilis, typhoid fever, tuberculosis, and malaria also enter into latent stages. The person harboring a persistent infectious agent may or may not shed it during the latent stage. If it is shed, such persons are chronic carriers who serve as sources of infection for the rest of the population.

The Aftermath of Infection

The final outcome of infection and disease is best represented as a continuum, with effects ranging from inapparent to mild, moderate, severe, or fatal. Most infections are inapparent, mild, or moderate in severity, and only a small number are severe or fatal. This last group includes diseases that leave **sequelae** or *residua,* long-term or permanent pathologic states such as deafness (meningitis), blindness (trachoma), heart valve damage (streptococcal pharyngitis), arthritis (Lyme disease), endocarditis (Q fever), and paralysis (polio), to name a few.

Epidemiology: The Study of Disease in Populations

So far, our discussion has revolved primarily around the impact of an infectious disease at the level of a single individual. Let us now turn our attention to the effects of diseases on the community (see feature 11.7); this is the realm of **epidemiology.** Although this term literally means the study of the occurrence of disease, epidemiology has evolved into a science that looks at every aspect of the determinants and distribution of disease. It involves many disciplines—not only microbiology, but anatomy, physiology, immunology, medicine, psychology, sociology, ecology, and statistics—and it considers many diseases other than infectious ones, including heart disease, cancer, drug addiction, and mental illness. The epidemiologist is a medical sleuth of sorts, collecting clues as to the causative agent, pathology, sources of infection, and modes of transmission, and tracking the agent's numbers and distribution in the community. In fulfilling these demands, the epidemiologist asks, who, when, where, how, what, and why? about diseases. The outcome of these studies helps public health departments develop prevention and treatment programs and establish a basis for predictions.

sequelae (suh-kwee′-lee) L. *sequi,* to follow.
residua (ruh-zid′-yoo-uh) L. *residuum,* remainder.
epidemiology (ep″-ih-dee-mee-awl′-uh-gee) Gr. *epidemios,* prevalent.

Feature 11.7 The Impact of Epidemics

The AIDS pandemic has been frightening to witness. For many young people born in the era of vaccines and antibiotics, it is their very first confrontation with a communicable infection that so far lacks a cure. Through the ages, many diseases of this nature have erupted in a similar manner and thrown the population into a state of panic. The bubonic plague of the Middle Ages ultimately killed about one out of every three persons living in Europe, and tuberculosis was responsible for one-fourth of the deaths of all persons living there in the 1800s. A pandemic of influenza in the early part of this century killed somewhere on the order of 50 million persons. Even today, diseases such as malaria, measles, and tuberculosis have a worldwide mortality rate far exceeding that of AIDS. Although noninfectious diseases such as heart dysfunction and cancer present a more imminent danger to the populus, the greatest fear and concern are reserved for diseases that are communicable.

The dream of modern medicine has been to arrest the spread of disease, but in reality, worldwide outbreaks of infectious disease are still very common. Much of the reason for this is that microorganisms are formidable at responding and adapting to alterations in the individual and community. Just when one infectious disease has been eradicated or greatly reduced through vaccination or chemotherapy, a new one moves in to fill the vacancy. In some cases, changing epidemiologic factors increase the prevalence of a disease that was once sporadic. An important contributing factor in pandemics nowadays is our increased personal freedom and rate of travel. A person may become infected in one country and be home for several days before the disease appears. Along the way, the pathogen may have been transmitted to large numbers of other persons.

Who, When, and Where? Occurrence of Disease in the Population

Epidemiologists are concerned with all of the factors covered earlier in this chapter: virulence, portals of entry and exit, and the course of disease. But they are also interested in **surveillance**—that is, collecting, analyzing, and reporting data on the rates of occurrence, mortality, morbidity, and transmission of infections. This involves keeping data for a large number of diseases seen by the medical community and reported to public health authorities (table 11.9). By law, many diseases (termed **reportable** or notifiable) must be reported to authorities, while others are reported on a voluntary basis.

Table 11.9 Reportable Diseases in the United States*

AIDS	Lymphogranuloma venereum
Amebiasis	Malaria
Anthrax	Measles (rubeola)
Arboviral infections	Meningococcal infections
Aseptic meningitis	Mumps
Botulism	Pertussis
Brucellosis	Plague
Chancroid	Poliomyelitis
Chickenpox	Psittacosis
Cholera	Rabies
Diphtheria	Rheumatic fever
Encephalitis	Rubella
Enterovirus	Salmonellosis
Gonorrhea	Shigellosis
Hepatitis A	Syphilis
Hepatitis B	Tetanus
Hepatitis, other	Trichinosis
Influenza	Tuberculosis
Legionellosis	Tularemia
Leprosy	Typhoid fever
Leptospirosis	Typhus
Lyme disease	Yellow fever

*Depending on the state, some of these diseases are reported only if they occur at epidemic levels; others must be reported on a case-by-case basis.

A well-developed network of individuals and agencies at the local, district, state, national, and international levels keeps track of infectious diseases. Physicians and hospitals report all notifiable diseases that are brought to their attention. This is done as a case report (on one individual) or a collective report (on several individuals). It is very important to maintain the right of privacy of the persons in these reports. Local public health agencies first receive the case data and determine how they will be handled. In most cases, health officers investigate the history and movements of patients to trace their prior contacts and to control the further spread of the infection as soon as possible through drug therapy, immunization, and education. In sexually transmitted diseases, patients are asked to name their partners so that these persons may be notified, checked, and treated. The principal government agency responsible for keeping track of infectious diseases nationwide is the Centers for Disease Control (CDC) in Atlanta, Georgia, which is a part of the United States Public Health Service. The CDC publishes a weekly notice of diseases (the *Morbidity and Mortality Report*) that provides weekly and cumulative summaries of the case rates and deaths for about 36 notifiable diseases, highlights important and unusual diseases, and presents data concerning disease occurrence in the major regions of the United States (see the endsheets of this book). Ultimately, the CDC shares its statistics on disease with the World Health Organization (WHO) for worldwide tabulation and control.

morbidity (mor-bid′-ih-tee) L. *morbidus*, sick. A state of disease.

Epidemiological Statistics: Case Rates Over Time

The **prevalence** of a disease is the total number of existing cases with respect to the entire population. It is a cumulative statistic, usually represented as the percentage of the population having a particular disease at any given time. Disease **incidence** measures the number of new cases over a certain time period, as compared to the general healthy population. This statistic, also called the case or morbidity rate, indicates both the rate and the risk of infection. The equations used to figure these rates are:

$$\text{Prevalence} = \frac{\text{Total number of cases in population}}{\text{Total number of persons in population}} \times 100 = \%$$

$$\text{Incidence} = \frac{\text{Number of new cases}}{\text{Number of healthy persons}} = \text{Ratio}$$

As an example, let us use a classroom of 50 students exposed to a new strain of influenza. Prior to exposure, the prevalence and incidence in this population are both zero (0/50). If in one week, 5 out of the 50 people contract the disease, the prevalence is $5/50 = 10\%$, and the incidence is 1 in 9 (5 cases compared to 45 healthy persons). If after two weeks, 5 more students contract the flu, the prevalence becomes $5 + 5 = 10/50 = 20\%$, and the incidence becomes 1 in 8 (5/40). When dealing with large populations, the incidence is usually given in numbers of cases per 1,000 or 100,000 population.

The changes in incidence and prevalence are usually followed over a seasonal, yearly, and long-term basis and are helpful in predicting trends (figure 11.16). Statistics of concern to the epidemiologist are the rates of disease with regard to sex, race, or geographical region. Also of importance is the mortality rate, which measures the total number of deaths in a population due to a certain disease. Over the past century, the overall death rate from infectious diseases has dropped, although the number of persons afflicted with infectious diseases (the morbidity rate) has remained relatively high.

Monitoring statistics also makes it is possible to define the frequency of a disease in the population (figure 11.17). An infectious disease that exhibits a relatively constant number of cases over a long period of time in a particular geographic locale is **endemic.** For example, Lyme disease is endemic to certain areas of the United States where the tick vector is found. A certain number of new cases are expected in these areas every year. With a **sporadic** disease, a few isolated cases are reported in widespread locales in an unpredictable manner. Rabies and polio are reported sporadically in the United States (fewer than 10 cases per year). When statistics indicate that the prevalence of an endemic or sporadic disease is increasing beyond what is expected for that population, the pattern is described as an **epidemic.** (See figure 11.16 for an idea of how epidemics appear graphically.) The existence of an epidemic is determined by an increasing trend, and it is not based on a particular population size (it can be anything from a hospital to a nation). There is

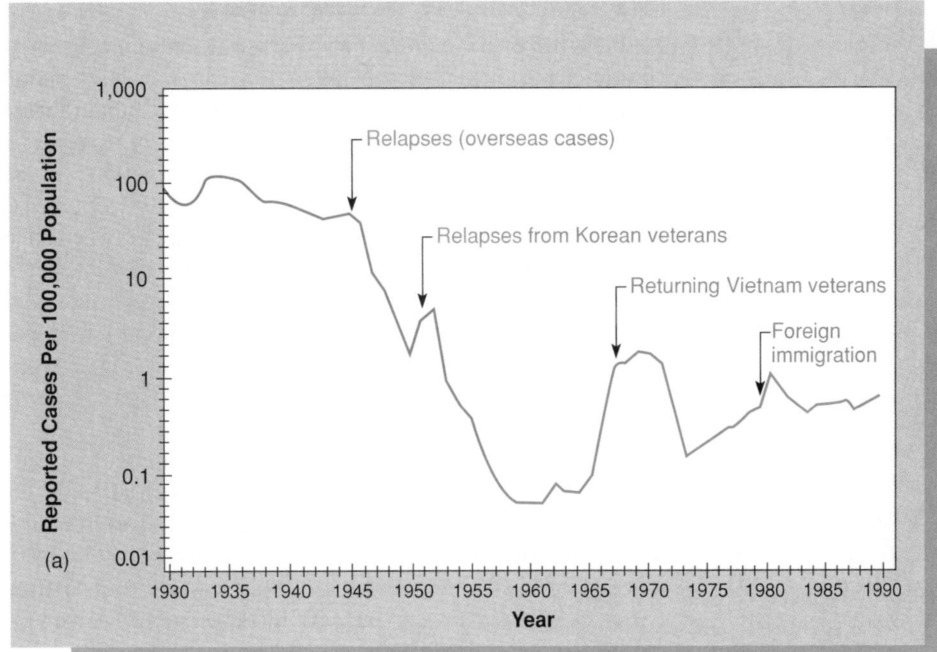

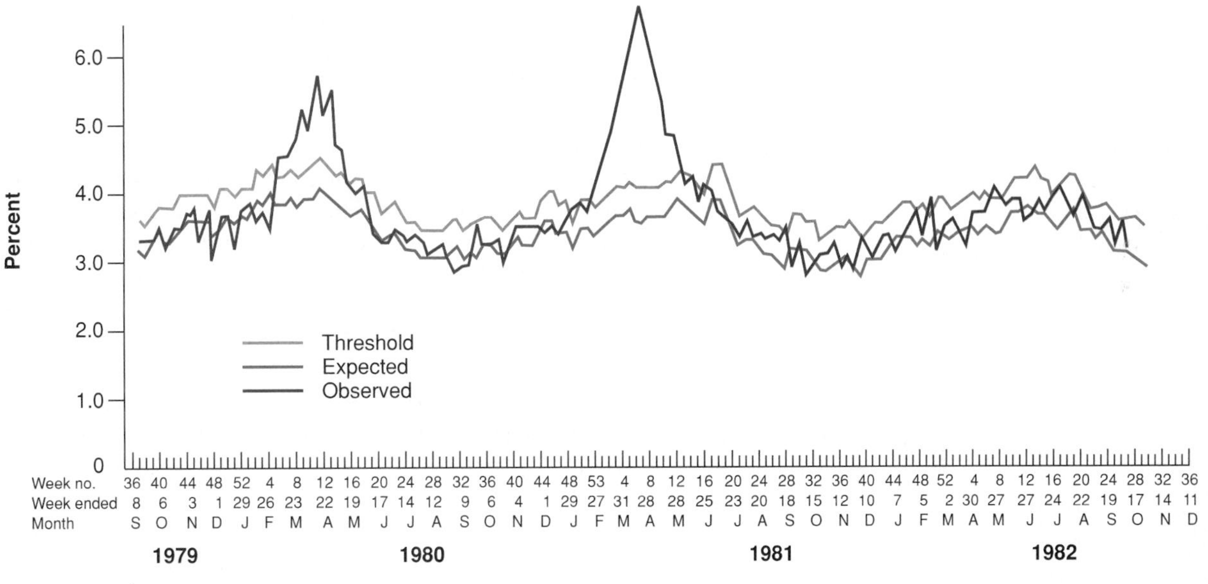

Figure 11.16 Examples of the use of epidemiologic statistics. (*a*) A graph showing the incidence of malaria in the United States over a 60-year period. The trend during the first 30 years was a gradual decrease; the case rate has risen since then. Factors contributing to increases and epidemics are returning servicemen in times of war and immigrants from endemic regions.
(*b*) Mortality rates of influenza pneumonia on a weekly basis over three years in 121 cities. One graph line shows the actual death rate, while the other shows what was estimated by the Centers for Disease Control (CDC).

Source: Data from the Centers for Disease Control.

also no defined time period (it can range from hours in food poisoning to years in syphilis), nor is an exact percentage of increase needed before an outbreak can qualify as an epidemic. Each situation is evaluated separately. For example, from 1981 to 1985, the yearly incidence of measles in the United States remained a stable 1,000 to 3,000 cases, but beginning in 1986, a new epidemic erupted that had reached nearly 25,000 cases by 1990. This epidemic was blamed on the fact that children had not been properly vaccinated. The spread of an epidemic across continents is a **pandemic,** as exemplified by AIDS and influenza (see chapter 21).

One important epidemiological truism might be called the "iceberg effect." Regardless of case reporting and public health screening, a large number of cases of infection in the community go undiagnosed and unreported. Take salmonellosis, for instance. Although approximately 40,000 cases are reported per year, this is believed to be only 1% to 10% of actual cases, which more likely range between 400,000 and 4,000,000 (see figure 16.35). The iceberg effect may be even more lopsided for sexually transmitted diseases or for commonplace infections such as influenza and colds.

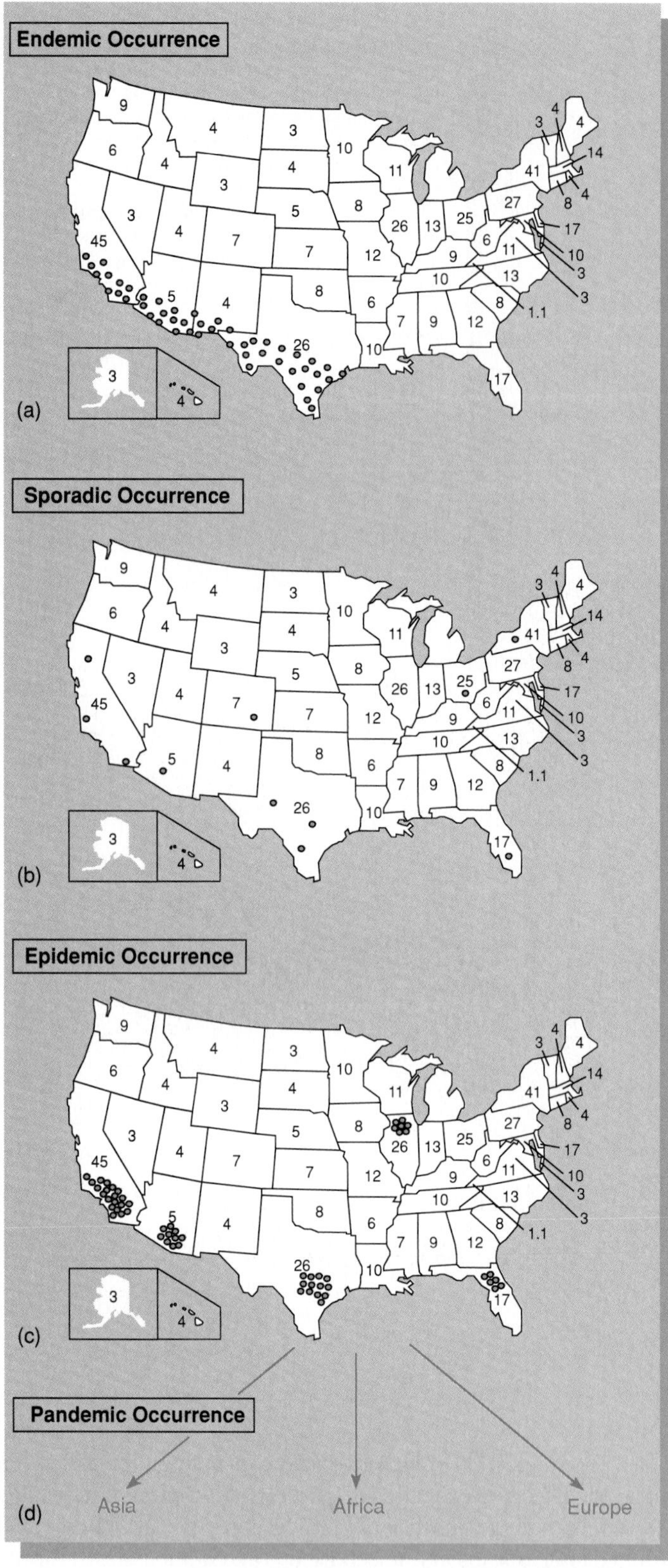

Figure 11.17 Patterns of infectious disease occurrence. (*a*) In endemic occurrence, cases are concentrated in one area at a relatively stable rate. (*b*) In sporadic occurrence, a few cases are dispersed over a wide area. (*c*) An epidemic is an increased number of cases that often appear in geographic clusters. (*d*) Pandemic occurrence means that an epidemic ranges over more than one continent.

Investigative Strategies of the Epidemiologist

At first, the evidence of a new disease or an epidemic in the community comes in piecemeal. A few cases are seen sporadically by physicians and are eventually reported to authorities. It may take several reports before any alarm is registered. In a way, epidemiologists and public health departments are initially working blind. With only odds and ends of data from a series of apparently unrelated cases, they must work backwards to reconstruct the epidemic pattern. A completely new disease requires even greater preliminary investigation, because the infectious agent must be isolated and linked directly to the disease (see the discussion of Koch's postulates in a later section of this chapter). All factors possibly impinging on the disease are scrutinized. Investigators search for *clusters* of cases indicating spread between persons or a public (common) source of infection; they also look at possible contact with animals, contaminated food, water, and public facilities, and at human interrelationships or changes in community structure. Out of this maze of case information, the investigators hope to recognize a pattern that indicates the source of infection so they can quickly move to control it (see feature 11.8).

Reservoirs: Where Pathogens Persist

For an infectious agent to continue to exist and be spread, it must have a place to reside. A **reservoir** is any long-term animate or inanimate receptacle that serves as a habitat and a focus of dissemination for an infectious agent. Often the reservoir is a human or animal carrier, although soil, water, and plants are also reservoirs. The reservoir may be distinguished from the **source** of infection, the individual or object from which an infection is actually acquired. In diseases such as syphilis, the reservoir and the source are the same (the human body). In the case of hepatitis A, the reservoir (a human carrier) may be different from the source (contaminated food).

Living and Nonliving Reservoirs

Many pathogens continue to exist and spread because they are harbored by members of a host population. Unlike persons or animals with frank or clinical symptomatic infection, a **carrier** is, by definition, an individual that *inconspicuously* shelters a pathogen and unknowingly spreads it to others. Although human carriers are occasionally detected through routine screening (blood tests, cultures) and other epidemiologic devices, they are unfortunately very difficult to discover and control. As long as a pathogenic reservoir is maintained by the carrier state, the disease will continue to exist in that population, and the potential for epidemics will be a constant threat. The duration of the carrier state may be short- or long-term, and may or may not involve actual infection of the carrier.

Several situations can produce the carrier state. **Asymptomatic** or apparently **healthy carriers** are indeed infected, but as previously indicated, they show no symptoms (figure 11.18*a*). A few asymptomatic infections (gonorrhea and genital warts, for instance) may carry out their entire course without overt manifestations. Other asymptomatic carriers, called **incubation carriers,** spread the infectious agent during the incubation period. For example, AIDS patients may harbor and spread the

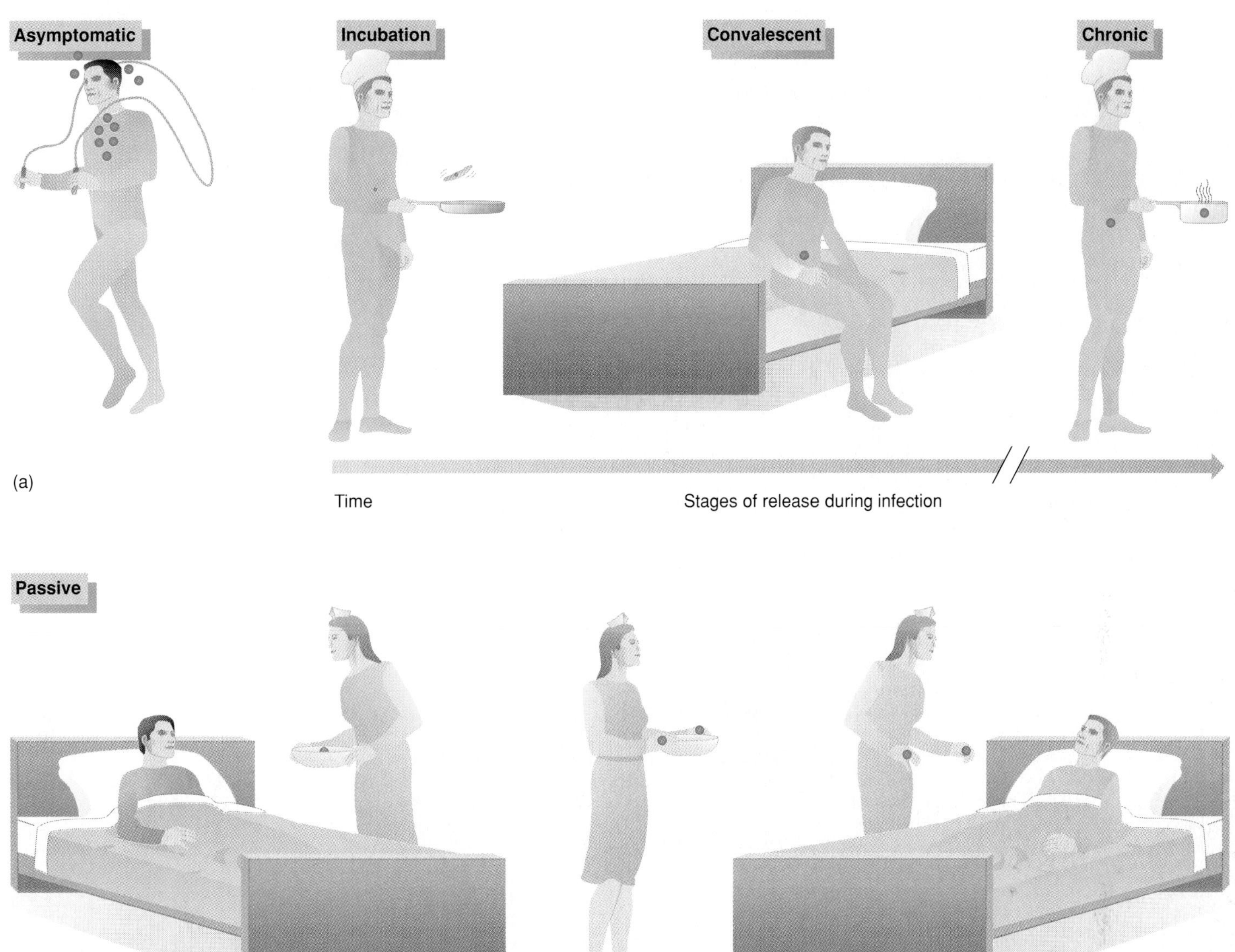

Figure 11.18 Types of carriers. (*a*) An asymptomatic carrier is infected without symptoms. Incubation carriers are in the early stages of infection; convalescent carriers are in late stages of recovery; chronic carriers sequester the microbe for long periods after the infection is over. (*b*) A passive carrier is contaminated but not infected.

virus for months and years before their first symptoms appear. Recuperating patients without symptoms are considered **convalescent carriers** when they continue to shed viable microbes and convey the infection to others. In diphtheria, patients spread the microbe for up to 30 days after the disease has subsided.

An individual who shelters the infectious agent for a long period after recovery due to the latency of the infectious agent is a **chronic carrier.** Patients who have recovered from tuberculosis, hepatitis, and herpes infections frequently carry the agent chronically. About one in 20 victims of typhoid fever continues to harbor *Salmonella typhi* in the gallbladder for several years, and sometimes for life. The most infamous of these was "Typhoid Mary," a cook who created an epidemic in the early 1900s (see feature 16.7).

A carrier state of great concern may occur during patient care (see a later section on nosocomial infections). Medical and dental personnel who must constantly handle materials that are heavily contaminated with patient secretions and blood are at risk for picking up pathogens mechanically and accidentally transferring them to other patients. When an individual transmits infectious agents from an infected patient to other patients, he is acting as a **passive** or **contact carrier** (figure 11.18*b*). Proper handwashing, handling of contaminated materials, and sterile techniques greatly reduce this likelihood.

Feature 11.8 Chronology of a Mystery Epidemic*

In the mid-1980s, CDC epidemiologist Scott Holmberg was caught up in an engrossing and circuitous search for the source of an epidemic of salmonellosis in the Midwest. Although he encountered numerous false clues, dead ends, and the slimmest of leads, several key discoveries eventually helped him solve the mystery. Here is a diary of the investigative pathway Holmberg took:

Incident: Outbreak of food infection, a cluster of 11 cases—**Cluster #1**
Where: Southern Minnesota.
Etiologic agent: *Salmonella newport.*
Unusual factors: All persons were infected with a species not usually found in that part of the Midwest. The bacteria were resistant to amoxicillin and penicillin.
Patient condition: Very ill, requiring hospitalization.
Patient history: All but two patients were unrelated; all had been taking oral amoxicillin or penicillin just prior to the attack.
Investigative questions and course:

1. Was antibiotic contaminated? Patients' leftover drugs and the pharmacy supply were tested thoroughly for contamination; a single colony of *Salmonella* was found in one sample. Should medical authorities and the public be warned? Conclusion: Capsule was probably contaminated by fingers of pill taker—**Dead end #1.**
2. What other factors may be involved? Recent histories of patients were taken, but no common travel experience was discovered, and no remarkable food habits were noted—**Dead end #2.**
3. In the field of epidemiology, dead ends are a way of life, so Dr. Holmberg dug deeper. He made a personal visit to the homes of stricken persons, interviewed them in depth, and inspected their homes closely. Again, no common source was discovered—**Dead end #3.**
4. Next Holmberg reexamined the interacting facts of which he was certain:
   Salmonella newport infection . ? → antibiotic therapy.
   The association between the infection and the antibiotic therapy appeared to be an important clue. Did the microbe become resistant in the patient? What was its source? Clearly, there was a third unknown factor.
5. New technology was applied. A DNA analysis of *S. newport* isolates indicated that all strains were exactly the same; the source, whatever it was, had to be a common one.
6. The Big Breakthrough. A series of inquiries to state health departments in neighboring states uncovered a cluster of cases of drug-resistant *S. newport* reported from South Dakota, with one death—**Cluster #2.** All patients were relatives, but they evidently had not shared a common source. The inquiry revealed one important bit of evidence: One patient was a dairy farmer.
7. The farm was closely scrutinized. A history of epidemic calf diarrhea was disclosed. State officials had isolated the same strain of *Salmonella newport* from one calf. But it was concluded that milk could not have been a source, because other victims in the cluster had not gotten milk from this dairy—**Dead end #4?**
8. Further discussion with the farmer revealed that his uncle on an adjacent farm had a beef herd which supplied meat (ground beef) to the families striken by infection. Further investigation revealed that dairy and beef herds had mingled through a broken fence and that the beef cattle had been fed feed mixed with antibiotics that would have selected for antibiotic-resistant strains of *Salmonella.*
9. The Final Connection. Dr. Holmberg next followed the course of the cattle connection: The South Dakota farmer had sold his beef cattle to a broker who subsequently sold them to a slaughterhouse. The slaughterhouse made ground beef from some of the meat, and this was shipped to markets in Minnesota from which the patients in Cluster #1 had bought ground beef. Eventually, 18 cases were linked to this one source.
10. The Loose Ends. The ultimate source of *S. newport* in the cattle was never discovered. But several loose ends were tied up. It was concluded that:

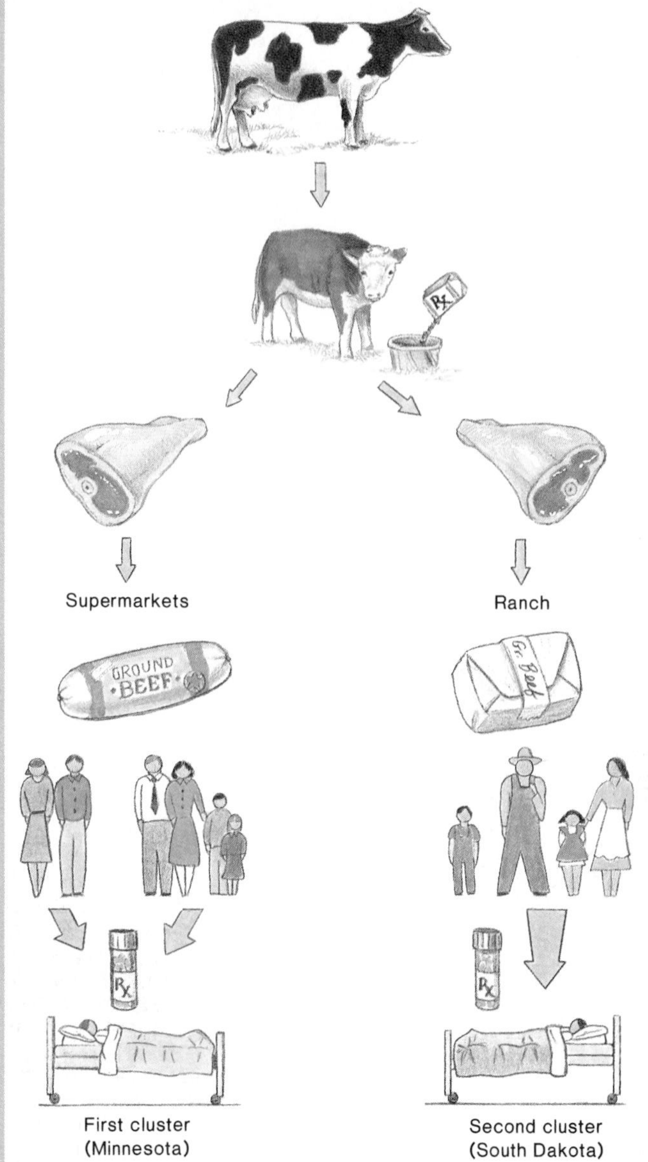

Reconstruction of an epidemic of salmonellosis.

Low doses of antibiotics helped maintain drug-resistant *Salmonella* in the herd.

During slaughter, meat contaminated with the bacterium was made into ground beef, which served as a very effective dispersal agent because meat from several cows was mixed.

Ground beef was not adequately cooked before consumption. The microbe came to be harbored in the intestines of the patients asymptomatically.

Antibiotic therapy for an unrelated infection created a selective environment that led to the overgrowth of *Salmonella* and development of disease.

Besides being a brilliant bit of epidemiologic detective work, this discovery created a furor among agriculture and health officials, because it clearly revealed that feeding antibiotics to animals could actually favor strains of resistant species that could be transmitted to humans.

*For a thorough discussion of this subject, see *Science,* 5 October, 1984, p. 80.

Animals As Reservoirs and Sources

Up to now, we have lumped animals with humans in discussing living reservoirs or carriers, but animals deserve special consideration as vectors of infections. The word **vector** is used by epidemiologists to indicate a live animal that transmits an infectious agent from one host to another. (The term is sometimes misused to include any object that spreads disease.) The majority of vectors are arthropods such as fleas, mosquitos, flies, and ticks, although larger animals may also spread infection—for example, mammals (rabies), birds (psittacosis), or lower vertebrates (salmonellosis).

By tradition, vectors are placed into one of two categories, depending upon the animal's relationship with the microbe (figure 11.19). A **biological vector** actively participates in a pathogen's life cycle, serving as a site where it can multiply or complete its life cycle. A biological vector communicates the infectious agent to the human host by biting, aerosol formation, or touch. In the case of biting vectors, the animal may (1) inject infected saliva into the blood (as does the mosquito), (2) defecate around the bite wound (the flea), or (3) regurgitate blood into the wound (the tsetse fly). More detailed discussions of the roles of biological vectors are in chapters 16, 17, 19, and 21.

Mechanical vectors are not necessary to the life cycle of an infectious agent, and indeed, are passive participants in the transmission process. The external body parts of these animals become contaminated through mechanical contact with a pathogenic source. The agent is subsequently transferred to humans indirectly by an intermediate such as food or, occasionally, by direct contact (as in certain eye infections). Houseflies have habits that suit them to the role of mechanical vector. Their mouthparts are adapted for feeding on decaying garbage and feces, which facilitates contamination of their feet and mouth parts. They also regurgitate juices onto food to soften and digest it. Despite these obnoxious habits, the fly is a pretty amazing creature—it can fly a distance of 20 miles and generate several hundred progeny every 14 days. Unfortunately, flies also spread more than 20 bacterial, viral, protozoan, and worm infections. Other nonbiting flies transmit tropical ulcers, yaws, and trachoma (see chapter 17). Cockroaches, which have similar unsavory habits, are known to carry about 40 different species of infectious agents and may play a role in the mechanical transmission of fecal pathogens.

Many vectors and animal reservoirs spread their infections to humans. An infection indigenous to animals but naturally transmissible to humans is a **zoonosis.** In these types of infections, the human is essentially a dead-end host and does not contribute to the natural persistence of the microbe. Some zoonotic infections (rabies, for instance) may have multihost involvement, and others may have very complex cycles in the wild (see plague in chapter 16). Zoonotic spread of disease is promoted by close associations of humans with animals, and people in animal-oriented or outdoor professions are at greatest risk. At least 150 zoonoses exist worldwide; the most common ones are listed in table 11.10. It is worth noting that zoonotic infections are impossible to completely eradicate without also eradicating the animal reservoirs. This has been attempted on mosquitos and certain rodents, but it is inconceivable that such extreme measures would ever be tried on game animals in Africa (to eliminate trypanosomiasis) or on migratory birds (to control certain viruses).

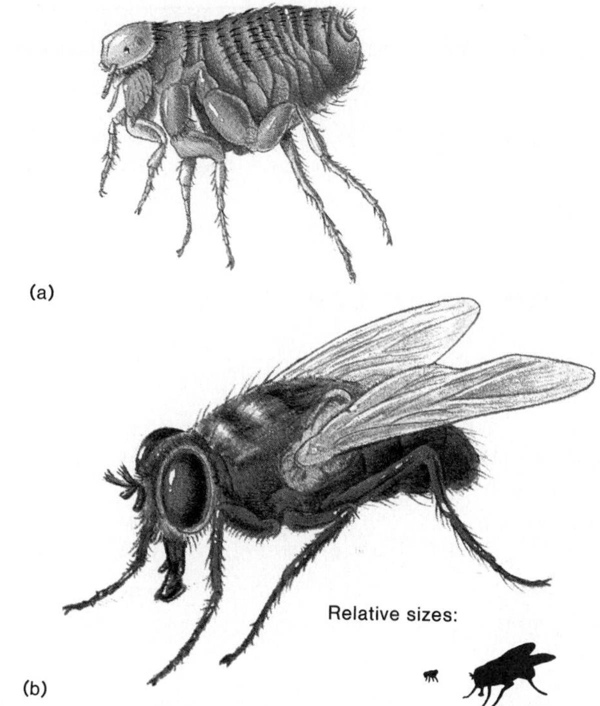

Figure 11.19 Two types of vectors. (*a*) Biological vectors serve as hosts during pathogen development and include the flea, a carrier of bubonic plague and murine typhus. (*b*) Mechanical vectors include nonbiting filth eaters such as the housefly, a transmitter of enteric infection.

Nonliving Reservoirs

It is clear that microorganisms have adapted to nearly every habitat in the biosphere. They thrive in soil and water and often find their way into the air. Although most of these microbes are saprobic and cause little harm and considerable benefit to humans, some are opportunists and a few are regular pathogens. Because the human host is in regular contact with these environmental sources, acquisition of pathogens from natural habitats is of diagnostic and epidemiologic importance.

Soil harbors the vegetative forms of bacteria, protozoa, helminths, and fungi, as well as their resistant or developmental stages such as spores, cysts, ova, and larvae. Soil also serves as a repository for animal and plant wastes and resident flora from dead hosts. Regular bacterial pathogens include the anthrax bacillus and species of *Clostridium* that are responsible for gas gangrene, botulism, and tetanus. Pathogenic fungi in the genera *Coccidioides* and *Blastomyces* are spread by spores in the soil and dust. The invasive stages of the hookworms *Necator* and *Ancyclostoma* occur in the soil. Natural bodies of water carry fewer nutrients than soil, but still support a number of pathogenic species such as *Legionella, Naegleria* (a pathogenic ameba), and *Giardia.*

zoonosis (zoh″-uh-noh′-sis) Gr. *zoion,* animal, and *nosos,* disease.

Table 11.10 Animal Hosts and Their Common Zoonotic Infections

Disease	Domestic Animals					Wild Animals				
	Cats	*Dogs*	*Cattle*	*Horses*	*Poultry*	*Arthropods*	*Birds*	*Rodents*	*Primates*	*Carnivores*
Viruses										
Rabies	+	+	+	—	—	+	—	+	+	+
Yellow fever	—	—	—	—	—	+	—	+	+	—
Viral fevers	—	—	—	—	—	+	—	+	—	—
Newcastle disease	—	—	—	—	+	—	+	—	—	—
Influenza	—	—	—	—	+	—	+	—	—	—
Bacteria										
Q fever	—	—	+	—	—	—	+	+	—	—
Rocky Mountain spotted fever	—	+	—	—	—	+	—	—	—	+
Psittacosis	—	—	—	—	+	—	+	—	—	—
Leptospirosis	+	+	+	+	—	—	—	+	+	+
Anthrax	+	+	+	+	—	—	—	+	—	+
Brucellosis	—	+	+	—	—	—	—	—	—	—
Listeriosis	—	+	+	+	—	—	+	+	—	+
Plague	+	+	—	—	—	+	—	+	—	+
Salmonellosis	+	+	+	+	+	+	+	+	+	+
Tularemia	+	+	—	+	—	+	+	+	—	+
Miscellaneous										
Ringworm	+	+	+	+	—	—	—	+	+	+
Toxoplasmosis	+	—	+	—	—	—	+	+	—	+
Trypanosomiasis	+	+	+	—	—	+	—	+	+	—
Larval migrans	+	+	—	—	—	—	—	—	—	—
Trichinosis	—	—	—	+	—	—	—	—	—	+
Tapeworm	—	—	+	—	—	—	—	—	—	—
Scabies (mange)	+	+	+	+	+	—	—	—	—	—

How and Why? The Acquisition and Transmission of Infectious Agents

Infectious diseases may be categorized on the basis of how they are acquired. A disease is **communicable** when an infected host can transmit the infectious agent to another host and establish infection in that host. (Although this is standard terminology, one must realize that it is not the disease that is communicated, but the microbe. Also be aware that the word infectious is sometimes used interchangeably with communicable, but that this is not precise usage.) The transmission of the agent may be direct or indirect, and how readily the disease is transmitted varies considerably from one agent to another. If the agent is highly transmissible, especially through direct contact, the disease is **contagious.** Influenza and measles move readily from host to host and thus are contagious, whereas leprosy is only weakly communicable. Because they can be spread through the population, communicable diseases will be our main focus in the following sections.

In contrast, a **non-communicable** infectious disease does **not** arise through transmission of the infectious agent from host to host. The infection and disease are acquired through some other, special circumstance. Non-communicable infections occur primarily when a compromised person is invaded by his own microflora (as with certain pneumonias, for example) or when an individual has accidental contact with a facultative parasite that exists in a nonliving reservoir (figure 11.20). Cases in point are mycoses, acquired through inhalation of fungal spores (from dust), and tetanus, in which *Clostridium tetani* spores from a soiled object enter a cut or wound. Persons thus infected do not become a source of disease to others.

Patterns of Transmission in Communicable Diseases

The routes or patterns of disease transmission are many and varied. Spread is by direct or indirect contact with animate or inanimate objects, and may be horizontal or vertical. The term *horizontal* means the disease is spread through a population from one infected individual to another; *vertical* signifies transmission from parent to offspring via the ovum, sperm, placenta, or milk. The extreme complexity of transmission patterns among microorganisms makes it very difficult to generalize, but for easier organization, we will divide them into two major groups, as shown in figure 11.21: transmission by direct contact or transmission by indirect routes (with the latter category divided into vehicle and airborne transmission).

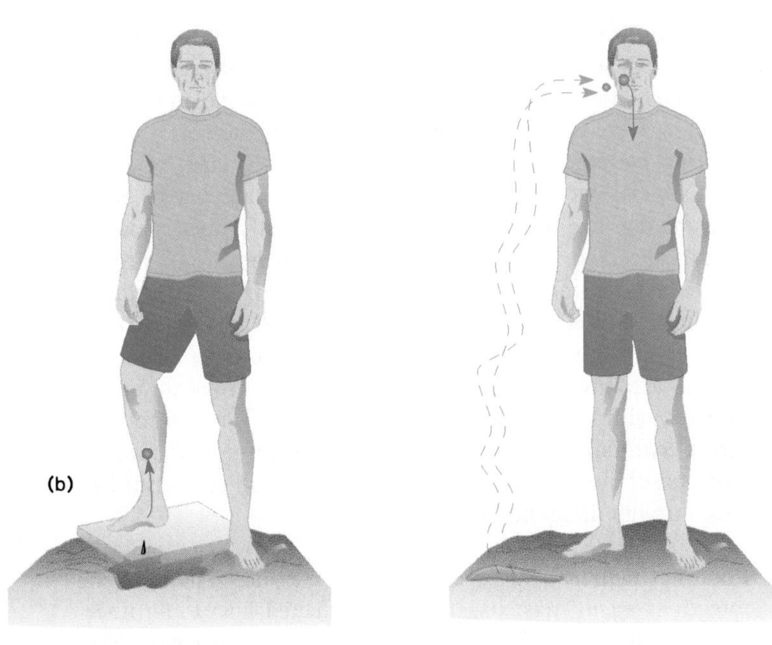

Figure 11.20 Non-communicable infections may be acquired (*a*) from normal flora and (*b*) from contact with microbes that live in the soil and water. Note that these infections are not acquired from another infected person, either directly or indirectly.

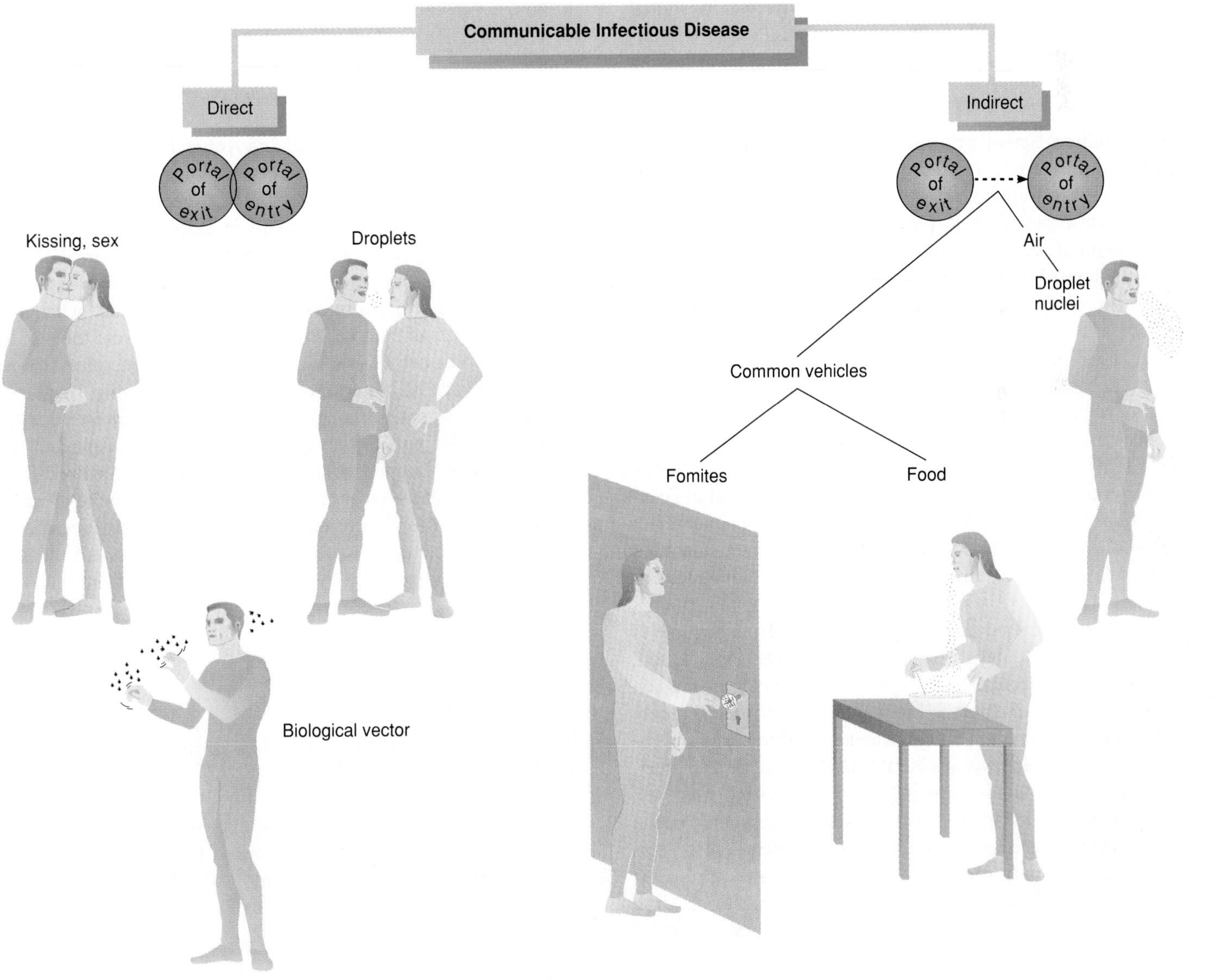

Communicable Infectious Disease

Direct

Portal of exit

Portal of entry

Indirect

Portal of exit

Portal of entry

Kissing, sex

Droplets

Air

Droplet nuclei

Common vehicles

Fomites

Food

Biological vector

Figure 11.21 Summary of how communicable infectious diseases are acquired.

Modes of Direct Transmission For microbes to be directly transferred, some type of contact must occur between the skin or mucous membranes of the infected person and that of the infectee. It may help to think of this route as the portal of exit meeting the portal of entry without the involvement of an intermediate object or entity. Included in the category of direct transmission are fine droplets sprayed directly upon a person during sneezing or coughing (as distinguished from droplet nuclei that are transmitted some distance by air). Most sexually transmitted diseases are spread directly. Infections that result from kissing, nursing, placental transfer, or bites by biological vectors are direct. When a person eats meat from an animal infected with salmonellosis or tularemia, this is also a form of direct communication. Most obligate parasites are far too sensitive to survive for long outside the host and can be transmitted only through direct contact.

Figure 11.22 The explosiveness of a sneeze. Special photography dramatically captures droplet formation in an unstifled sneeze. Even the merest attempt to cover a sneeze with one's hand will reduce this effect considerably. When such droplets dry and remain suspended in air, they are droplet nuclei.

Routes of Indirect Transmission To be considered indirect, the infectious agent must pass from an infected host to an intermediate object or substance and then on to another host. This form of communication occurs when the infected individual contaminates inanimate objects, food, or air through his activities. The transmitter of the infectious agent may be either overtly infected or a carrier.

Vehicles: Indirect Spread by Contaminated Materials The term **vehicle** specifies any inanimate material commonly used by humans that can transmit infectious agents. A *common vehicle* is a single material that serves as the source of infection for many individuals. Some specific types of vehicles are food, water, various biological products (such as blood, serum, and tissue), and fomites. A **fomite** is an inanimate (nonliving) object that harbors pathogens because it is subject to constant contamination. The list of possible fomites is as long as your imagination allows. Probably highest on the list would be public facilities such as doorknobs, telephones, push buttons, and faucet handles, because the hands are very frequently the source of infectious agents. Shared bed linens, handkerchiefs, toilet seats, nursery school toys, eating utensils, clothing, personal articles, and syringes are other examples. Although paper money is impregnated with a disinfectant to inhibit microbes, pathogens may still be isolated from a fairly large proportion of bills as well as coins.

Outbreaks of food poisoning often result from the role of food as a common vehicle. The source of the agent may be soil, the handler, or a mechanical vector. In the type of transmission termed the *oral-fecal route,* a fecal carrier with inadequate personal hygiene contaminates food during handling, and an unsuspecting person ingests it. Hepatitis A, amebic dysentery, shigellosis, and typhoid fever are often transmitted this way. Other food-associated illnesses may be due to inoculation of the food by soil (botulism) or human flora (staphylococcal food poisoning). Because milk provides a rich growth medium for microbes, it is a significant means of transmitting pathogens from diseased animals, infected milk handlers, and environmental sources of contamination. The agents of brucellosis, tuberculosis, Q fever, salmonellosis, listeriosis, and streptococcal infections are transmitted by contaminated milk. Water that has been contaminated by feces or urine may introduce pathogens such as *Salmonella, Shigella, Vibrio,* and viruses (hepatitis A, polio).

Airborne Route: Droplet Nuclei and Aerosols Unlike soil and water, outdoor air cannot provide nutritional support for microbial growth and seldom transmits airborne pathogens. On the other hand, indoor air (especially in a closed space) can serve as an important medium for the suspension and dispersal of certain respiratory pathogens via droplet nuclei and aerosols. **Droplet nuclei** are dried microscopic residues created when microscopic pellets of mucus and saliva are ejected from the mouth and nose. They are generated forcefully in an unstifled sneeze or cough (figure 11.22) or mildly during nose blowing, speaking, laughing, or singing. Although the larger beads of moisture settle rapidly, smaller particles evaporate and remain suspended for longer periods. Droplet nuclei are implicated in the spread of hardier pathogens such as the tubercle bacillus and the measles virus. **Aerosols** are suspensions of fine dust or moisture particles in the air that contain live pathogens. Q fever is spread by dust from animal quarters, and psittacosis by aerosols from infected birds.

Nosocomial Infections: The Hospital As a Source of Disease

Infectious diseases acquired during a hospital stay are known as **nosocomial** infections. This category of disease seems incongruous at first thought, because we think of a hospital as a place to get a cure, not a place to get a disease. However, it is not unusual for a surgical patient's incision to become infected or a burn patient's treatment to be complicated by pneumonia. The rate of nosocomial infections may be as low as 0.1% or as high

fomite (foh'-myt) L. *fomes,* tinder.

nosocomial (nohz''-oh-koh'-mee-al) Gr. *nosos,* disease, and *komeion,* to take care of. Originating from a hospital or infirmary.

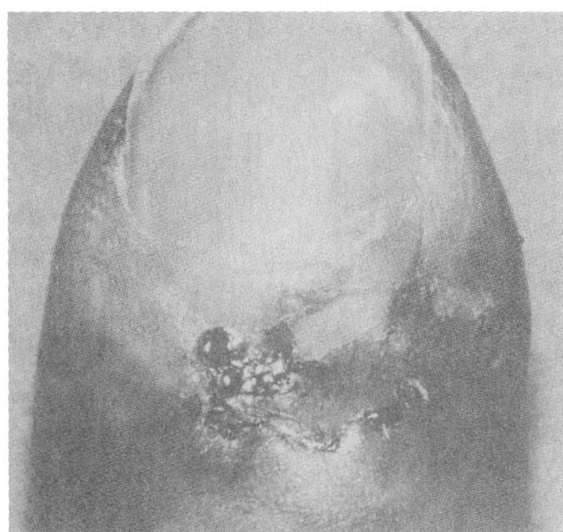

Figure 11.23 Nosocomial infection in dentistry. The finger of a dentist was accidently penetrated by a saliva-contaminated periodontal probe. This shows the infectious potential of human saliva.

as 20% of all admitted patients at any one time, but the average is about 5%. This results in increased mortality, morbidity, and cost (up to several billion dollars per year).

The infectious agent may originate from resident flora, or it may be acquired directly or indirectly from the air, fomites, other patients, or medical personnel. So many factors unique to the hospital environment are tied to nosocomial infections that a certain amount of infection is virtually unavoidable. After all, the hospital both attracts and creates compromised patients, and it concentrates virulent pathogens in a single place. Furthermore, the health care process itself makes it possible for microbes to be carried from one patient to another. Treatments using instruments, respirators, and nebulizers constitute a possible source of infectious agents. Indwelling devices such as catheters, prosthetic heart valves, grafts, shunts, joints, and tracheostomy tubes form a ready portal of entry or focus of growth for microbes. An additional factor is the tendency for drug-resistant strains of microorganisms to develop in hospitals, thereby further complicating treatment.

The most common nosocomial infections involve the urinary tract, surgical wounds, the lungs, and the bloodstream. In general, gram-negative rods (*Escherichia coli, Klebsiella, Pseudomonas*) are most often cultured from patient specimens, followed by staphylococci and streptococci. Not to be minimized are nosocomial infections in patient caretakers. Needle sticks (a type of inoculation infection), handling infectious materials, and direct contact with the patient are among the common routes of infection in clinical workers (figure 11.23).

The potential seriousness and medical impact of nosocomial infections have forced most hospitals to develop a special committee to monitor epidemiology. A critical part of this team is the infection control nurse or nurse-epidemiologist. This person has the fascinating but challenging job of overseeing the infection situation in the entire hospital. This includes keeping track of infection in the wards, determining possible epidemic situations, relaying this information to the rest of the committee, seeking out breaches in asepsis in nursing care or surgery, and training others in aseptic techniques. Among the control procedures brought to bear on the problem of nosocomial infections are monitoring housekeeping and contamination, isolating patients, using antibiotic prophylaxis, enforcing strict sterile and disinfection procedures, restricting infected personnel, immunizing personnel, and emphasizing proper handwashing techniques.

Using Koch's Postulates to Determine Etiology

An essential aim in the study of infection and disease is determining the precise **etiologic,** or causative, agent. In our modern technological age, we take for granted that a certain infection is caused by a certain microbe, but such has not always been the case. More than a century ago, Robert Koch realized that, in order to prove the germ theory of disease, he would have to develop a standard for determining causation that would stand the test of scientific scrutiny. Out of his experimental observations on the transmission of anthrax in cows came a series of proofs, called **Koch's postulates,** that established the principal criteria for etiologic studies (figure 11.24). These postulates direct an investigator to (1) find evidence of a particular microbe in every case of a disease, (2) isolate that microbe from an infected subject and cultivate it artificially in the laboratory, (3) inoculate a second susceptible healthy subject with the laboratory isolate and observe the resultant disease, and (4) reisolate the agent from the second subject.

When applying Koch's postulates, several details are of great importance: Each isolated culture must be pure, observed microscopically, and identified by means of characteristic tests; the first and second isolate must be identical; and the pathological effects, signs, and symptoms of the disease in the first and second subject must be the same. Once established, these postulates were rapidly put to the test, and within a short time, they had helped determine the causative agents of tuberculosis, diphtheria, and plague. Today, few infectious diseases have not been directly linked to a known infectious agent.

Koch's postulates continue to play an essential role in modern epidemiology. Every decade, new diseases challenge the scientific community and require application of the postulates. Prominent examples are toxic shock syndrome, AIDS, Lyme disease, and Legionnaire's disease (named for the American Legion members who first contracted a mysterious lung infection in Philadelphia).

Despite the reliability of Koch's postulates, once in a while they cannot be completely fulfilled. One reason is that some infectious agents are not readily isolated or grown in the laboratory. Another is that a good animal model is lacking for some human diseases. If one cannot elicit a similar infection by inoculating it into an animal, it is much more difficult to prove the etiology. In the past, scientists have attempted to circumvent this problem by using human subjects (see feature 11.9). A small but vocal group of critics has claimed that the postulates have not been adequately carried out for AIDS, and thus, that HIV cannot be claimed as the cause of it. Perhaps the necessary studies can now be completed with some recently developed models using embryonic mice infected with HIV genes and mice implanted with a human immune system (see chapters 13 and 14).

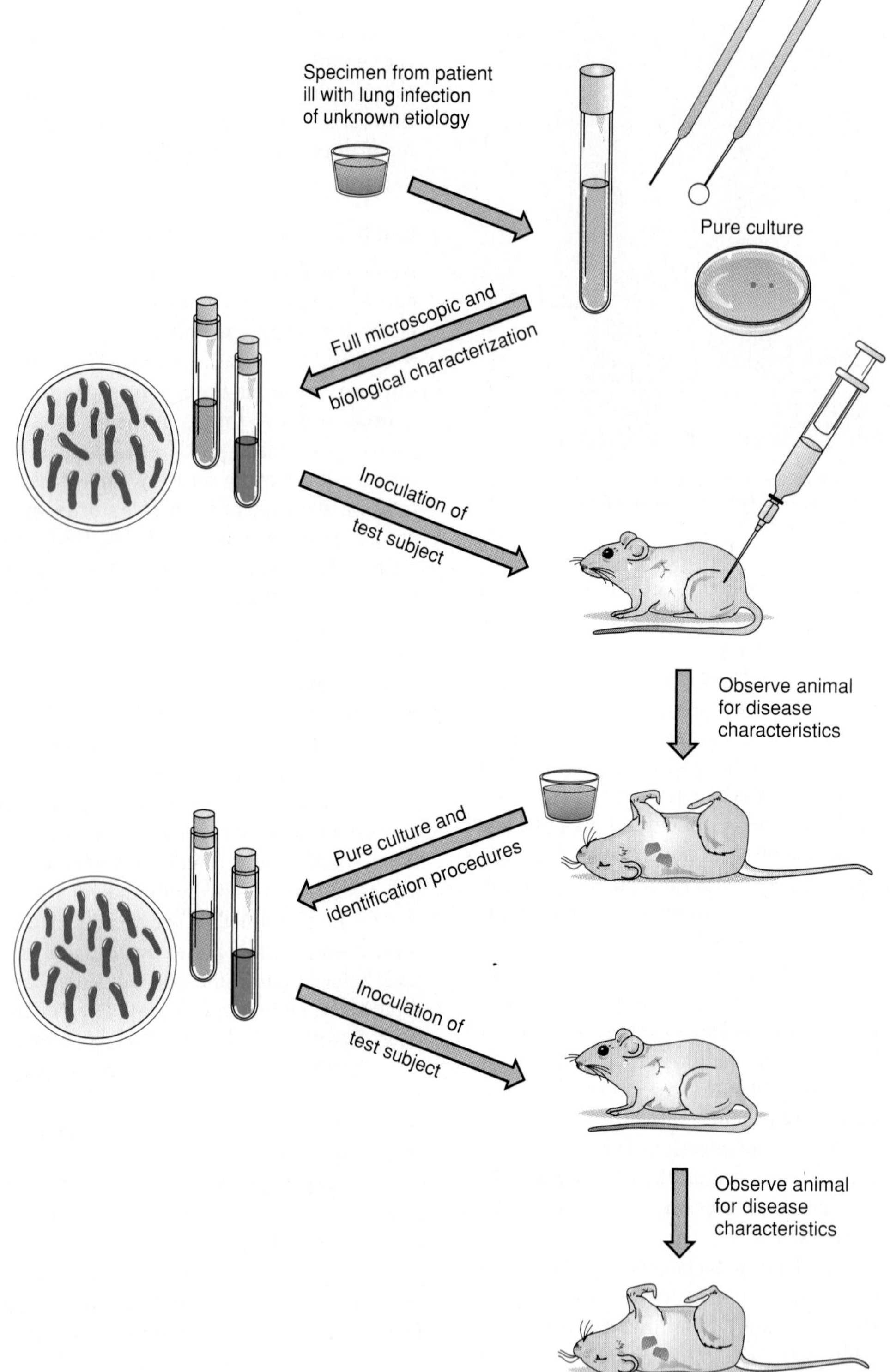

Figure 11.24 Koch's postulates: Is this the etiologic agent? Both the microbe in the initial and second isolation and the disease in the two experimental animals must be identical for the postulates to be satisfied.

Feature 11.9 Humans As Guinea Pigs: "Those Daring Young Men"

In this day and age, human beings are not used as subjects for determining the cause of infectious disease, but this has not always been true. A long tradition of human experimentation dates well back into the eighteenth century, with the subject oftentimes being the experimenter himself. In some studies, mycologists inoculated their own skin and even that of family members with scrapings from fungal lesions to demonstrate that the disease was transmissible. And in the early days of parasitology, it was not uncommon for a brave researcher to swallow worm eggs in order to study the course of his infection and the life cycle of the worm.

By contrast, a German colleague of Koch's named Max von Petenkofer believed so strongly that cholera was *not* caused by a bacterium that he and his assistant swallowed cultures of the vibrio. Fortunately for them, they did not acquire serious infection. Many self-experimenters have not been so lucky.

One of the most famous cases is that of Jesse Lazear, a Cuban physician who worked with Walter Reed on the etiology of yellow fever in 1900 (see chapter opening illustration). Dr. Lazear was convinced that mosquitos were directly involved in the spread of the agent of yellow fever, and by way of proof, he allowed himself and two volunteers to be bitten by mosquitos infected with the blood of yellow fever patients. Although all three became ill as a result of this exposure, Dr. Lazear's sacrifice was the ultimate one—he died of yellow fever. Years later, paid volunteers were used to completely fulfill the postulates.

Dr. Lazear was not the first martyr in this type of cause. Fifteen years previously, a young Peruvian medical student, Daniel Carrion, attempted to prove that a severe blood infection, Oroya fever, had the same etiology as *verucca peruana,* an ancient disfiguring skin disease. He inoculated himself with fluid from a skin lesion, and in three weeks developed the acute fever and died, becoming a national hero. In his honor, the disease now identified as bartonellosis is also sometimes called Carrion's disease. Eventually, the microbe (*Bartonella bacilliformis*) was isolated, a monkey model was developed, and the sandfly was shown to be the vector for this disease.

Not all such self-experimentation has advanced science. Consider the story of John Hunter, whose inoculations led to the erroneous conclusion that gonorrhea and syphilis were the same disease (see feature 15.3). The impetus for self-experimentation has continued—but present-day researchers are more likely to test experimental vaccines than dangerous microbes. Jonas Salk injected himself with his polio vaccine before allowing it to be used on others, and the French researcher Daniel Zagury tested his version of an AIDS vaccine on himself. We must not forget that the final trials of vaccine and drug testing are totally dependent on human volunteers.

Chapter Review with Key Terms

Microbe-Human Interactions

Contact-Infection-Disease: The Host-Parasite Relationship

The human body is constantly in **contact** (**contaminated**) with microbes. Some microbes (transients) are rapidly lost; some commensals colonize the body as residents; other microbes are pathogens that may cause an **infection** by circumventing the host defense system, entering normally sterile tissues, and multiplying there. When infections lead to a disruption in tissues, **infectious disease** results. The outcome is highly variable, but most contacts do not result in infection, and most infections do not lead to disease.

The Body As a Habitat

The **resident flora** or **microflora** is a huge and rich mixed population of microorganisms residing on body surfaces exposed to the environment, including the skin, mucous membranes, parts of the alimentary tract, urinary tract, reproductive tract, and upper respiratory tract. Anatomical sites lying within the body cavity (organs) and fluids (blood, urine) in those sites do not harbor flora.

Colonization: Begins just prior to birth and continues over an individual's life; variations occur in response to individual differences in age, diet, hygiene, and health.

Role of Flora: Bacteria may maintain a balance in the normal conditions; some produce vitamins; studies with **axenic** animals (free of any normal flora) show that flora contribute to the development of the immune and gastrointestinal systems, and also to some diseases (dental caries). Normal flora are sometimes agents of infection.

Factors Affecting the Course of Infection and Disease

Pathogenicity and Virulence: **Pathogenicity** is the property of microorganisms to cause infection and disease. **Virulence** is the precise means by which the microbe invades and damages host tissues; it helps define the degree of pathogenicity. Pathogenicity varies with a microbe's ability to invade or harm host tissues and with the condition of host defenses. A **true pathogen** produces **virulence factors** that allow it to readily evade host defenses and to harm host tissues. True pathogens can infect normal, healthy hosts with intact defenses. An **opportunistic pathogen** is not highly virulent but can invade and cause disease in persons whose host defenses are compromised by **predisposing conditions** such as age, malnutrition, genetic defects, medical procedures, and underlying organic disease.

Mechanisms of Infection and Disease: Entry, adherence, invasion, multiplication, and disruption of target tissues. The **portal of entry** is the route by which microbes enter the tissues, primarily via skin, alimentary tract, respiratory tract (**pneumonia**), urogenital tract (**sexually transmitted diseases**), or placenta. Pathogens that come from outside the body are **exogenous**; those that originate from normal flora are **endogenous.** The size of the infectious dose is of great importance. In the process of **adhesion,** a microbe attaches to the host cell by means of fimbriae, flagella, capsule, or other receptors; this puts it in advantageous position for invasion.

Virulence Factors: Enzymes, **toxins, antiphagocytic factors. Exoenzymes** digest host epithelial tissues; disrupt tissues; aid invasion. **Toxigenicity** is a microbe's capacity to produce toxins at site of multiplication; may affect local or distant targets. **Toxinoses** are diseases caused by toxins acting on particular target organs; toxins damage structure or function of host cells; **toxemia** refers to toxins absorbed into the blood; **intoxication** means ingestion of toxins. An **exotoxin** is a protein secreted by living bacteria with powerful effects on a specific organ. Examples are **hemolysins,** toxins of tetanus and diphtheria. An **endotoxin** is the lipopolysaccharide portion of a gram-negative cell wall released when a bacterial cell dies; more generalized and weaker in its toxicity; one cause of fever. Antiphagocytic chemicals include leukocidins (white blood cell poisons) and capsules.

Effects on Target Organ/Spread of Infection

Patterns of Infection: Stages in infection/disease are **incubation period,** period from contact with infectious agent until appearance of first symptoms; **prodromium,** short period of initial, vague symptoms; **period of invasion,** a variable period during which microbe multiplies in high numbers and affects the function of target organ (accompanied by severest symptoms); **convalescent period,** period of recovery, with decline of symptoms.

Types of Infections/Diseases: **Localized infection,** microbe remains in isolated site; **systemic infection,** microbe is spread through the tissues by circulation; **focal infection,** microbe spreads from local site to entire body (systemic); **mixed infection,** several microbes cause one type of infection simultaneously; **primary infection,** the initial infection in a series; **secondary infection,** a second infection by a different microbe that complicates a primary infection; ***-itis,*** root indicating an infection that causes inflammation of an organ (gastroenteritis, meningitis); ***-emia,*** root that refers to microbes or their products in the blood (**septicemia,** bacteremia); **acute infection,** appears suddenly, has a short course, is relatively severe; **chronic infection,** persists over a long period of time; **subacute infection,** length is in between acute and chronic.

Signs and Symptoms: Manifestations of disease, indicators of **pathologic** effects on target organs. A **sign** is objective, measurable evidence noted by an observer. Examples include septicemia, change in number of white blood cells; skin **lesions;** inflammation; **necrosis,** lysis or death of tissue. A **symptom** is a subjective effect of disease as sensed by patient. Examples are pain, cough, fatigue, redness, swelling. A **syndrome** is a disease that manifests as a predictable complex of symptoms; infections that do not show symptoms are called **subclinical, inapparent,** or **asymptomatic.** Through the **portal of exit,** microbe is released with bodily secretions and discharges so that it may have access to new host; portals include respiratory types (droplets from sneezing, coughing); saliva; skin; feces; urogenital tract (urine, mucus, semen); blood. A microbe may become dormant (**latent**) and cause recurrent infections. Damage to organs and tissues that remains after infection is a **sequela.**

Epidemiology

Science that determines the factors influencing causation, frequency, and distribution of disease in a community; epidemiologists are involved in **surveillance** of reportable diseases, consider measures to protect the public health; are concerned with disease statistics such as **prevalence** (the total number of cases in a community), **incidence** (the number of new cases in a population), **morbidity** (general health of the population), and **mortality** (death).

Frequency of Disease: **Endemic,** a disease constantly present in a certain geographical area; **sporadic,** disease occurs occasionally with no predictable pattern; **epidemic,** sudden outbreak of disease in which numbers increase beyond expected trends; **pandemic,** worldwide epidemic.

Origin of Pathogens

The **reservoir** is a place where the pathogen ultimately originates, its habitat; **source** of infection refers to the immediate origin of an infectious agent; **carrier** is an individual that inconspicuously shelters a pathogen and spreads it to others; **asymptomatic carrier** is infected but without symptoms. Examples are **incubation carriers,** carriage early in disease; **convalescent carriers,** carriage in last phases of recovery; **chronic carriers,** carriage for long periods after recovery; **passive or contact carriers** are uninfected but convey infectious agents from infected persons to uninfected ones by hand and instrument contact.

Vectors/Zoonoses

A **vector** is an animal that transmits pathogens; **biological vector,** an alternate animal host (mosquito, flea) that assists in completion of life cycle of microbe; often transmits by biting; **mechanical vector,** animal not host in microbial life cycle, transmits by contaminated body parts (housefly); **zoonosis,** an infection for which animals are natural reservoir and host; can be transmitted to humans.

Acquisition of Infection

Communicable infectious disease occurs when pathogen is transmitted from host to host directly or indirectly; **contagious diseases** are readily transmissible through direct contact; **non-communicable diseases** are not spread from host to host; acquired from one's own flora (pneumonia) or from a nonliving environmental reservoir (tetanus).

Direct Transmission: Infectious agent is spread through direct contact of portal of exit with portal of entry (STDs, herpes simplex).

Indirect Transmission: A material (**vehicle**) contaminated with pathogens serves as intermediate source of infection; **fomite,** inanimate object contaminated with pathogens (public facilities, personal items); food may be a vehicle; **droplet nuclei,** airborne dried particles containing infectious agents, formed by sneezing, coughing.

Nosocomial Infections: Infectious diseases that originate in the hospital or clinical setting. Common among surgical and chronically ill patients; may occur in health care workers.

Koch's Postulates: A series of criteria that must be followed to determine the **etiologic** (causative) agent of disease.

True–False Questions

Determine whether the following statements are true (T) or false (F). If you feel a statement is false, explain why, and reword the sentence so that it reads accurately.

____ 1. The resident flora are all harmless commensals.

____ 2. Resident flora are commonly found in the stomach.

____ 3. The degree of pathogenicity is increased by virulence factors.

____ 4. Exotoxins are heat-sensitive secreted proteins, and endotoxins are heat-tolerant parts of the cell membrane of bacteria.

____ 5. Hemolysins are toxins that lyse white blood cells.

____ 6. A patient may be infectious during the period of incubation, the prodromium, the period of invasion, and the period of convalescence.

____ 7. Bacteremia is the presence of bacteria in the blood.

____ 8. A subclinical infection is acquired in a hospital.

____ 9. A mechanical vector is a passive animal transmitter of pathogens.

____ 10. One can acquire tetanus from a patient with tetanus.

____ 11. A disease may be simultaneously infectious, communicable, and contagious.

____ 12. Examples of portals of entry include the mouth, skin, wounds, and intestine.

____ 13. Examples of portals of exit include spinal fluid, urine, saliva, and blood.

Concept Questions

1. Differentiate between contamination, infection, and disease. What are the possible outcomes in each of the three situations?
2. How are infectious diseases different from other diseases?
3. Name the general body areas that are sterile. Why is the inside of the intestine not sterile like many other organs?
4. What causes variations in the flora of the newborn intestine?
5. What factors influence how the flora of the vagina develops?
6. Why must axenic young be delivered by cesarian section?
7. Explain several ways that true pathogens differ from opportunistic pathogens.
8. Distinguish between pathogenicity and virulence; what are virulence factors?
9. Describe the course of infection from contact with pathogen to its exit from host.
10. Why are most microbes so limited to a single portal of entry? For each portal of entry, give a vehicle that carries the pathogen and the course it must travel to invade the tissues.
11. Differentiate between exogenous and endogenous infections.
12. What factors possibly affect the size of the infectious dose? Name five factors involved in microbial adhesion.
13. Which body cells or tissues are affected by hemolysins, leukocidins, hyaluronidase, kinases, tetanus toxin, pertussis toxin, and enterotoxin?
14. Compare and contrast: endotoxins versus exotoxins; systemic versus local infections; primary versus secondary infections; infection versus intoxication.
15. What is the difference between signs and symptoms? (First put yourself in the place of a patient with an infection and then in the place of a physician examining you. Describe what you would feel and what the physician would detect upon examining the affected area.)
16. What is important about the portal of exit?
17. Outline the science of epidemiology and the work of an epidemiologist. Which of the following statistics, based on number of reported cases, show sporadic frequency? Endemic? Epidemic? Pandemic?

United States Regions	**Coccidioidomycosis**	
	1989	**1991**
Western	350	295
Midwest	10	6
Eastern seaboard	1	0
Northeast	0	2
Southeast	2	3

	St. Louis Encephalitis	
	1989	**1991**
Western	2	3
Midwest	7	5
Eastern seaboard	0	1
Northeast	2	4
Southeast	13	95

	Diphtheria	
	1989	**1991**
Western	4	5
Midwest	0	0
Eastern seaboard	2	1
Northeast	0	0
Southeast	1	0

18. Distinguish between mechanical and biological vectors, using examples.
19. Describe what it means to be a carrier of infectious disease. Describe four ways that people may be carriers. What is epidemiologically and medically important about carriers in the population?
20. What is the precise difference between communicable and non-communicable infectious diseases? Between direct and indirect modes of transmission?
21. List the main features of Koch's postulates. Why is it so difficult to prove them for some diseases?

Practical/Thought Questions

1. Consider the relationship between the vaginal residents and the colonization of the newborn. Can you think of some serious medical consequences of this relationship? Why would normal flora cause some infections to be more severe and other infections to be less severe?
2. The following patient specimens produce positive cultures when inoculated and grown on appropriate media. For each, tell whether this result is to be expected or not:

Urine	Blood
Throat	Cerebrospinal fluid
Lung biopsy	Urine from bladder
Feces	Liver biopsy
Saliva	Semen

What are the important clinical implications of positive blood or cerebrospinal fluid?

3. Pretend that you have been given the job of developing a colony of germ-free cockroaches. What will be the main steps in this process? What possible experiments can you do with these animals?
4. If healthy persons are resistant to infection with opportunists or weak pathogens, what is the expected result if a compromised person is exposed to a true pathogen?
5. Why can exotoxins be used in vaccination, but not endotoxins?
6. Explain how the endotoxin gets into the bloodstream of a patient with endotoxic shock.
7. You are a physician following the course of an infection in a small child who has been exposed to scarlet fever. Describe the main events that occur, and tell what is happening during each stage and what causes its signs and symptoms.
8. Use technical words to describe each of the following infections (may fit more than one category):
   Caused by needle stick in dental office
   Pneumocystis pneumonia in AIDS patient
   Bubonic plague from rat flea bite
   Diphtheria
   Acute necrotizing gingivitis
   Syphilis of long duration
   Large numbers of gram-negative rods in the blood
   A boil on the back of the neck
   An inflammation of the meninges
   Scarlet fever
9. Using figure 11.16, determine the approximate prevalence and incidence of malaria for the years 1975 to 1985.
10. Name 10 fomites that you touched today.
11. Can you tell what parts of Koch's postulates were unfulfilled by Dr. Lazear's experiment described in feature 11.9?

CHAPTER 12

The Nature of Host Defenses

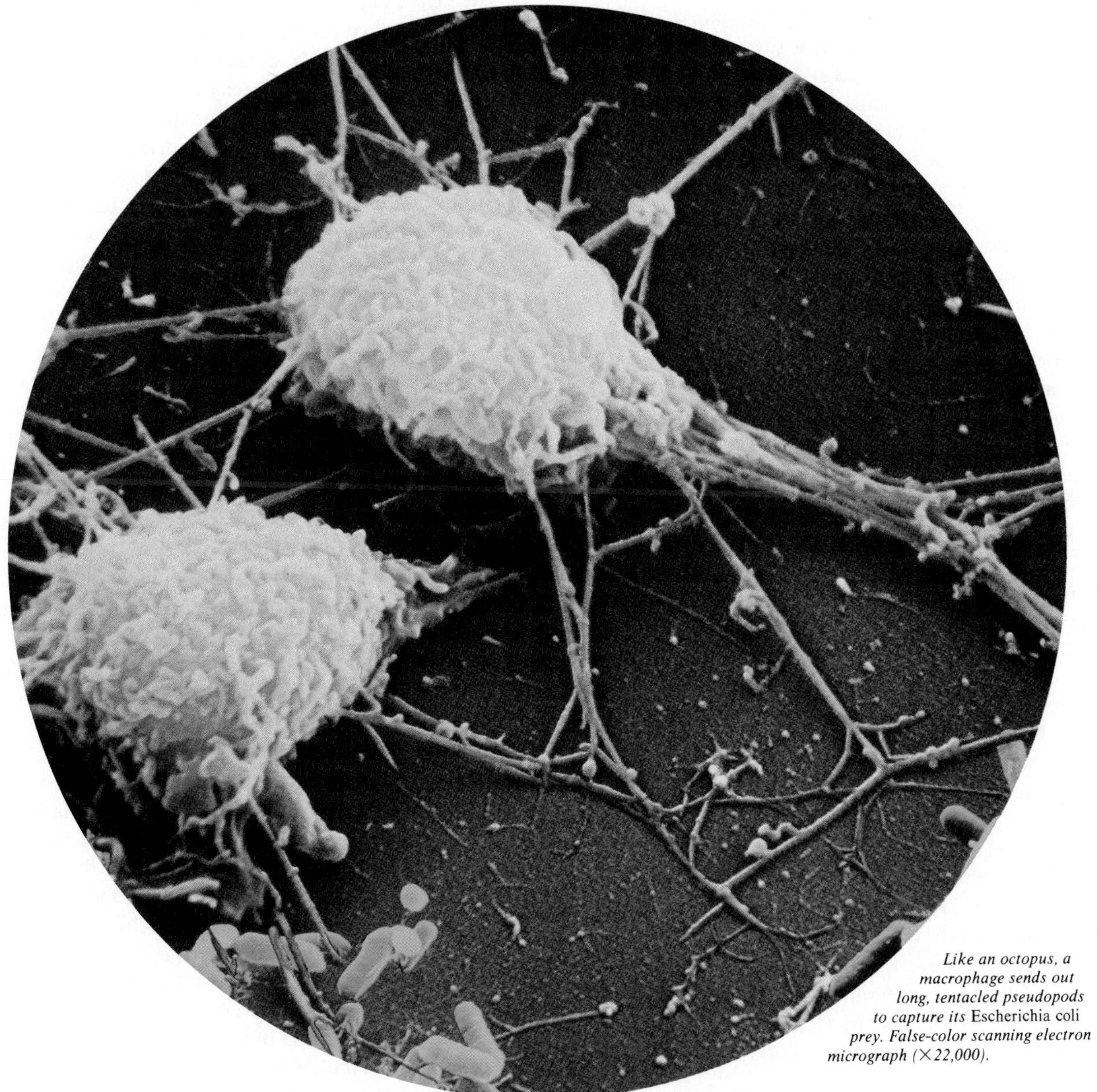

Like an octopus, a macrophage sends out long, tentacled pseudopods to capture its Escherichia coli *prey. False-color scanning electron micrograph (×22,000).*

Chapter Preview

The survival of the host depends upon an elaborate network of defenses that bar the penetration of harmful microbes and other foreign materials and, should they penetrate, prevent them from becoming established in tissues. Defenses range from nonspecific to specific and from active to passive, and they involve barriers, cells, and chemicals. This chapter will introduce the nonspecific host defenses intrinsic to all humans; the anatomical and physiological components of surveillance, recognition, and destruction of foreign substances; and the general concepts of the adaptive responses that account for an individual's long-term resistance to infection and disease.

Defense Mechanisms of the Host in Perspective

In chapter 11 we explored the host-parasite relationship, with emphasis on the role of microorganisms in disease. In this chapter we will examine the other side of the relationship—that of the host defending itself against microorganisms. As previously stated, in light of the unrelenting contamination and colonization of humans by microbes, it is something of a miracle that we are not constantly infected and diseased. But we are **not** because of a remarkable, fascinating, and amazingly complex system of defenses. In the war against all sorts of invaders, microbial and otherwise, the body puts up a series of barriers, sends in an army of cells, and emits a flood of chemicals to protect tissues from harm.

The host defenses embrace a multilevel network of inborn, nonspecific components and specific **immunities** referred to as the first, second, and third lines of defense (figure 12.1). The interaction and cooperation of these three levels of defense normally provide complete protection against infection. The **first line of defense** includes any barrier that blocks invasion at the portal of entry. This mostly nonspecific line of defense limits access to the internal tissues of the body. It is not truly part of the immune system because it does not involve recognition of or response to a specific foreign substance. The **second line of defense** is a slightly more sensitive system of protective cells and fluids that includes inflammation and phagocytosis. It acts rapidly at both the local and systemic levels once the first line of defense has been circumvented. The highly specific **third line of defense** is acquired on an individual basis as each foreign substance is encountered by specially adapted cells called lymphocytes. The reaction with each different microbe produces unique protective substances and cells that come into play if that microbe is encountered again. The third line of defense provides long-term immunity.

The human body is armed with various levels of defense, but these defenses do not operate in a completely separate fashion; most defenses overlap and are even redundant in some of their effects. This "immunological overkill" literally bombards microbial invaders with an entire assault force, making their survival unlikely. Because of the interwoven nature of host defenses, it is difficult to introduce one part without also discussing another, so we will set the scene by presenting general concepts and then tackle the specifics in the latter part of this chapter and in chapter 13.

Barriers at the Portal of Entry: A First Line of Defense

A number of normal defenses that are an integral part of the anatomy and physiology of the body combine to impede the entry of microbes at the site of first contact. These natural, inborn, nonspecific defenses may be divided into physical, chemical, and genetic barriers (figure 12.2).

Physical or Anatomical Barriers at the Body's Surface

The skin and mucous membranes of the respiratory and digestive tracts have several built-in defensive features. The outermost layer (stratum corneum) of the skin is composed of epithelial cells that have been cornified and keratinized, meaning that they have become compacted, cemented together, and impregnated with an insoluble protein, keratin. The result is a thick, tough layer that is highly impervious and waterproof. Few pathogens can penetrate this unbroken barrier, especially in regions such as the soles of the feet or palms of hands, where the stratum corneum is much thicker than on other parts of the body. The constant flaking of skin also provides a means to rid the host of potential pathogens. Although hair follicles, sebaceous glands, and sweat glands could provide access into the body, several adaptations reduce this possibility. The hair shaft is periodically extruded, and the follicle cells *desquamated.* The flushing effect of sweat glands also helps remove microbes.

The mucocutaneous membranes of the digestive, urinary, and respiratory tracts, and the eye are thin, moist, permeable surfaces. Despite the normal wear and tear upon these epithelia, damaged cells are rapidly replaced. The mucous coat on the free surface of some membranes impedes the entry of bacteria. Blinking and *lacrimation* flush the eye's surface with tears and rid it of irritants. The constant flow of saliva helps carry microbes into the harsh conditions of the stomach. Vomiting and defecation also evacuate noxious substances or microorganisms from the body.

The respiratory tract is constantly guarded from infection by elaborate and highly effective adaptations. Nasal hair traps larger particles, and the turbulence of inhalation causes suspended particles to be flung against the sticky, mucus-lined walls. Rhinitis, the copious flow of mucus and fluids that occurs in allergy and colds, exerts a flushing action. In the respiratory tree (primarily the trachea and bronchi), a ciliated epithelium conveys foreign particles entrapped in mucus toward the pharynx either to be expelled or swallowed (figure 12.3). This so-called ciliary "escalator" propels entrapped particles at a rate of 10 to 30 mm/hr. Irritation of the nasal passage reflexly initiates a sneeze, which expels a large volume of air at high velocity. Similarly, the acute sensitivity of the bronchi, trachea, and larynx to foreign matter ensures that the cough reflex is readily triggered and that irritants are explosively ejected.

---

immunity (im-yoo′-nih-tee) Gr. *immunis,* free, exempt. A state of resistance to infection.

desquamate (des′-kwuh-mayt) L. *desquamo,* to scale off. The casting off of epidermal scales.

lacrimation (lak″-rih-may′-shun) L. *lacrimatio,* tears.

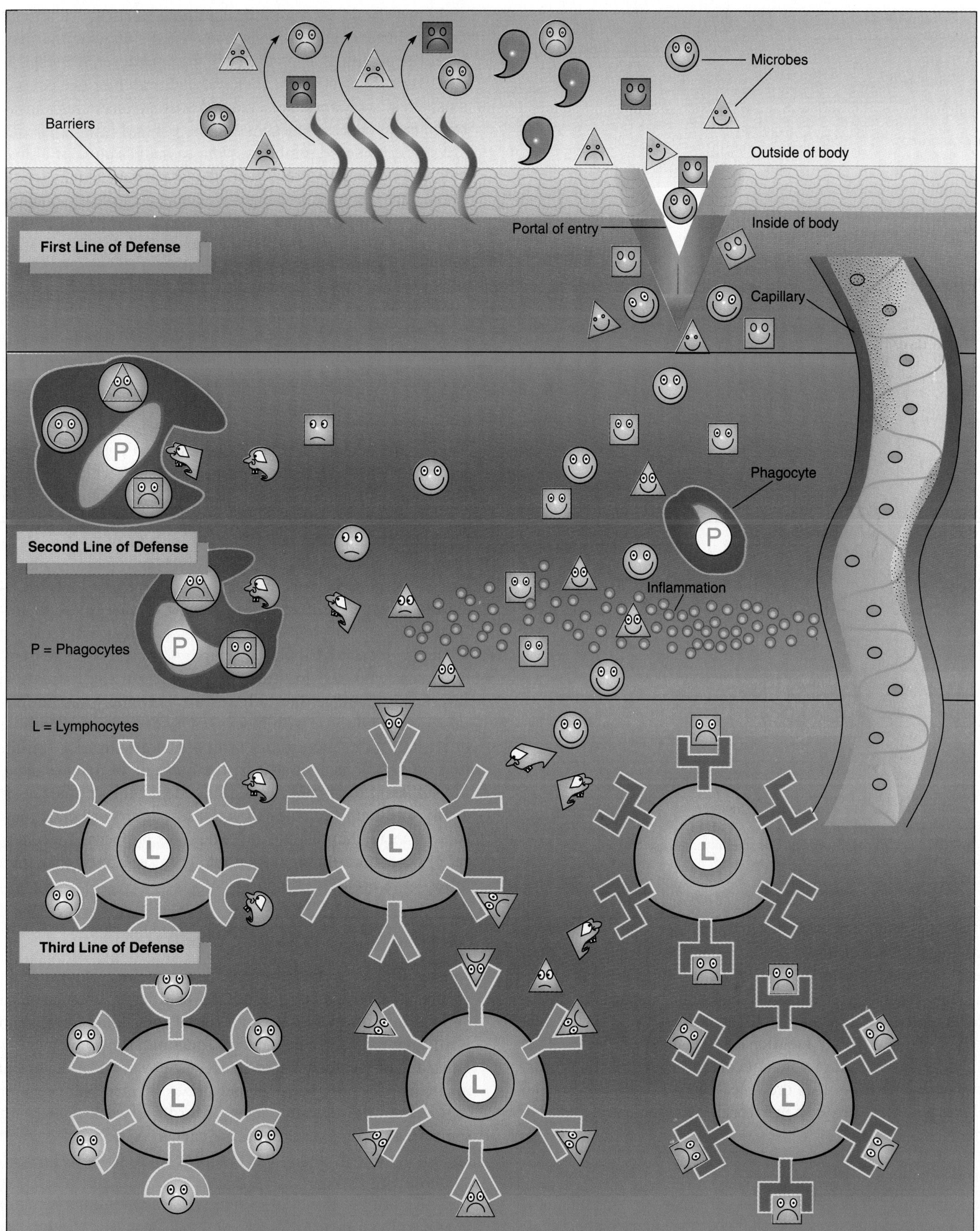

Figure 12.1 The levels of host defense. The first and second lines of nonspecific defense can limit or destroy microbes without regard to their differences. The third line is discriminating and reacts specifically to each different microbe.

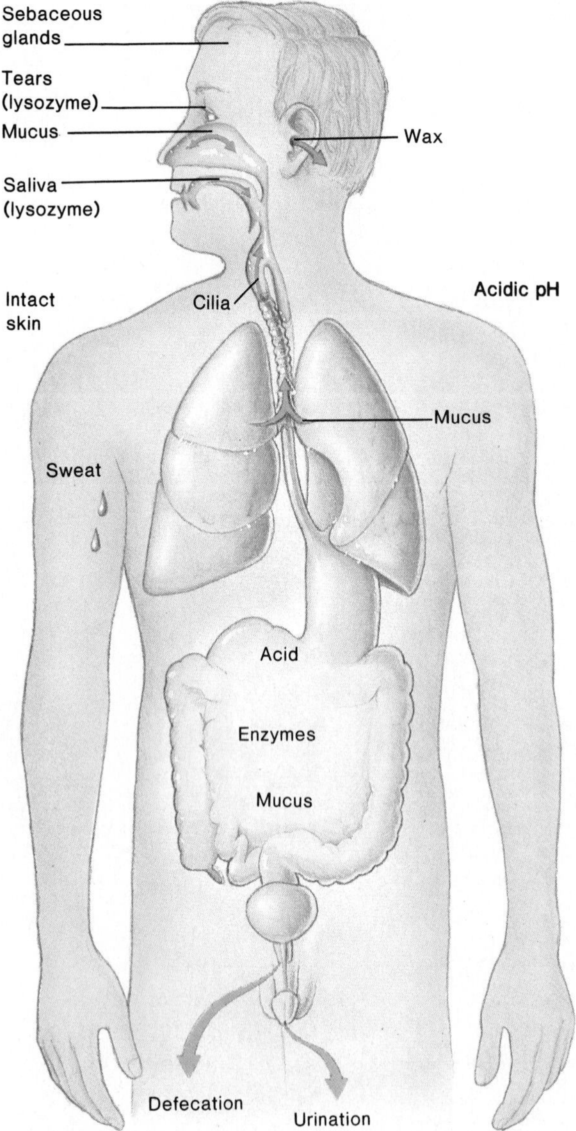

Figure 12.2 The primary physical and chemical defense barriers.

The genitourinary tract derives partial protection from the continuous trickle of urine through the ureters and from periodic bladder emptying that flushes the urethra.

The composition and protective effect exerted by microflora were discussed in chapter 11, thus no further detail is necessary, except to briefly mention its beneficial influence. Even though the resident flora does not constitute an anatomical barrier, its presence may block the access by pathogens to epithelial surfaces and may create an unfavorable environment for pathogens.

Nonspecific Chemical Defenses

The skin and mucous membranes offer a variety of chemical defenses. Sebaceous secretions that coat the emerging hair shaft exert an antimicrobial effect, just as specialized sebaceous glands (the meibomian glands of the eyelids) lubricate the conjunctiva and inner eyelid with an antimicrobial secretion. An important defense in tears and saliva is **lysozyme,** an enzyme that hydrolyzes the peptidoglycan in the cell wall of bacteria. Perspiration can attain high concentrations of sodium chloride, potassium ions, urea, and lactic acid that effectively discourage invasion by many microbes. The skin's acidic pH and fatty acid content are also considered inhibitory. The stomach's content of hydrochloric acid renders protection against pathogens that are swallowed, and the intestine's digestive juices and bile are potentially destructive to microbes. Semen contains an antimicrobial chemical (spermine) that inhibits bacteria, while in the vagina, the acidic pH maintained by normal flora is protective.

Genetic Defenses

Some defenses exist in the negative sense—that is, a host is protected by the lack of something rather than by its presence. For example, in a species defense, some pathogens have such great specificity for one host species that they are incapable of infecting other species. Reduced to everyday terms, "Humans can't acquire distemper from cats, and cats can't get mumps from humans." This is particularly true of viruses, which can invade only by attaching to a specific host receptor. On the other hand, zoonotic infectious agents may attack a broad spectrum of animals. Genetic differences in susceptibility may also exist within members of one species. Humans carrying a gene or genes for sickle-cell anemia are resistant to malaria (see feature 8.6). Racial or ethnic differences also exist in susceptibility to tuberculosis, leprosy, and certain systemic fungal infections.

The vital contribution of barriers is clearly demonstrated in persons who have lost them or never had them. Patients with severe skin damage due to burns are extremely susceptible to infections, and those with blockages in the salivary glands, tear ducts, intestine, and urinary tract are at greater risk for infection. But as important as it is, the first line of defense alone is not sufficient to protect against infection. Many pathogens find a way to circumvent the barriers using their virulence factors (discussed in chapter 11), and when this happens, a whole new set of defenses—inflammation, phagocytosis, and other immunities—are brought into play.

Introducing the Immune System

Immunology is the study of immune phenomena. Although once a purely descriptive discipline, the field has mushroomed into an exciting, precedent-setting area of molecular biology, constantly enlarging its boundaries, manipulating biological systems, and dominating the progress of many areas of biology and medicine. Several recent Nobel prizes in medicine were presented for work in immunology. Hardly a week goes by that some breakthrough in the areas of cancer, AIDS, or therapeutic applications does not emerge from the immunological research community.

A study of immunities necessitates examining a number of interrelated concepts. For example, because immune reactions actually happen at the molecular level, we must understand the concepts of foreignness, surveillance, recognition, and

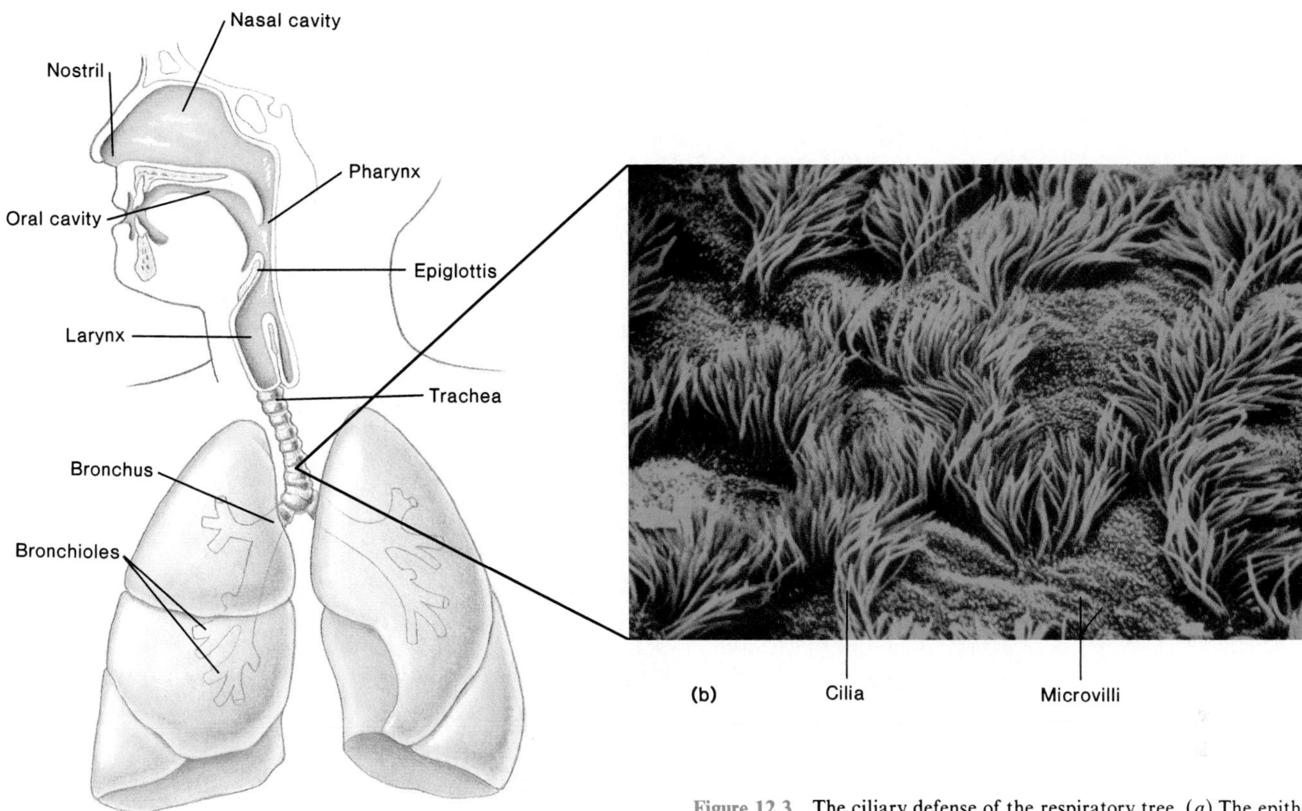

Figure 12.3 The ciliary defense of the respiratory tree. (*a*) The epithelial lining of the airways contains a brush border of cilia to entrap and propel particles upward toward the pharynx. (*b*) Tracheal mucosa (×5,000).

specificity. In order to discuss phagocytosis, we must know something about blood cells; in order to discuss inflammation, we must understand the roles of body fluids and chemical mediators; and in order to understand specific immunities, we must know something about all of these and much more.

An Expedition into the Molecular Domains of the Immune System

Whenever foreign material such as an infectious agent enters the tissues, the cells of the immune system are rapidly enlisted in a formidable molecular interchange. The main players in these reactions are molecules called **markers**[1] that stick out of the cell surface like miniscule signposts announcing that cell or molecule's identity. The identity of a given marker is based on its **specificity,** a unique configuration that dictates the kinds of immune responses it can elicit. Markers can also be **receptors,** molecules that bind specifically with complementary molecules in ways that signal, communicate, and trigger reactions inside the cell.

1. The term marker is also employed in genetics in a different sense—that is, to denote a detectable characteristic of a particular genetic mutant. A genetic marker may or may not be a surface marker.

How Are Markers Detected?

The body has a fleet of white blood cells strategically located to grope continually through its tissues, compartments, and fluids like billions of tiny fingers, feeling and evaluating what is there and searching for any markers that are new or different. This process of scouting the tissues for foreign receptors and other possibly threatening particles is **surveillance.** One could say that surveillance is a sort of immunologic investigation concerned with discovering and evaluating any miniscule pieces of matter that find their way into the body (figure 12.4). To be completely effective, however, it must be accompanied by recognition.

Recognition of Self and Nonself

Surveillance cells evaluate and differentiate the molecules they discover by a process of **recognition.** Inherent in recognition is the capacity of the probing cells to sort out the **natural markers of the body** (that is, **self**) from the **foreign markers** (or **nonself**) (see feature 12.1). This is an essential step, since the immune system is programmed to react as if anything foreign is potentially harmful and should be earmarked for destruction, whereas self will not. In this hunt for things that are not part of self, the immune system has a "sense of touch" so profound and exquisite that it often boggles the mind. When nonself is discovered and recognized, a whole battalion of responses is brought into play to **entrap and destroy** (kill, neutralize) the invading offender (figure 12.4).

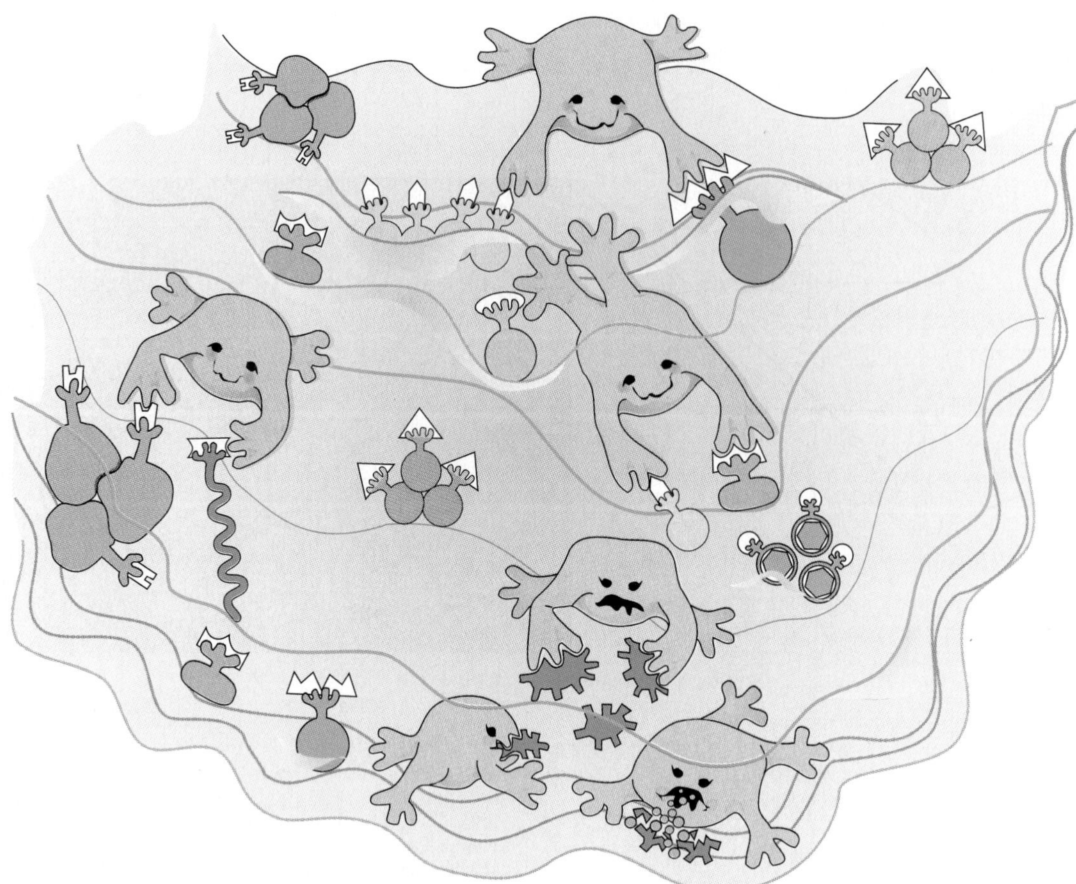

Figure 12.4 Search, recognize, and destroy is the mandate of the immune system. White blood cells are equipped with a very sensitive sense of "touch." As they sort through the tissues, they feel surface markers that help them determine what is self and what is not. When self markers are recognized, no response occurs. When nonself is detected, a reaction to destroy it is mounted.

Eliminating "Old" Self White blood cells have the amazing capacity to detect fine differences between self and nonself and to react against nonself. Yet, consider this: The immune system is also intimately involved in eradicating old or damaged self (red blood cells, tumors). It would appear then, that these abnormal cells become altered in such a way that the immune system treats them as nonself. The change is probably in the surface markers.

Systems Involved in Immune Defenses

Unlike many systems, the **immune system** is not a single, well-defined site; rather, it encompasses a large, complex, and diffuse network of cells and fluids that permeate every organ and tissue. The nature of this anatomic and physiologic design facilitates the processes of surveillance and recognition previously mentioned.

The Compartmentalized Systems of the Body

The body is partitioned into intracellular, extracellular, lymphatic, cerebrospinal, and blood compartments, all fluid-filled regions that help keep functions somewhat separate for greater efficiency. Although these compartments are separated, they are far from closed, because part of their architecture involves extensive interchange and communication (figure 12.5). Four main body compartments that dominate in immune function are: (1) the *reticuloendothelial system (RES),* (2) the spaces surrounding tissue cells that contain *extracellular fluid (ECF),* (3) the *bloodstream,* and (4) the *lymphatic system.* In the following section, we shall consider how these main compartments interact, their anatomy and functions, and how they operate in the second and third lines of defense.

The Merging Body Compartments

For effective immune responsiveness, the activities in one fluid compartment must have impact on other compartments. Let us see how this occurs by viewing a piece of tissue at the microscopic level (figure 12.5*a*). At this level clusters of tissue cells are in direct contact with the reticuloendothelial system (RES) and the extracellular fluid (ECF). Other compartments (vessels) that penetrate at this level are the capillaries and the lymphatic system. Cells and chemicals that originate in the RES and ECF can diffuse or migrate into the blood and lymphatics and vice versa. Likewise, any products of a lymphatic reaction will eventually be transmitted directly into the blood through the connection between these two systems. Furthermore,

Feature 12.1 What Is Foreign? What Is Not?

When beginning to discuss the immune system, it is essential to introduce the concept of **foreignness.** In the most general sense, a foreign material is something that may be recognized or distinguished as not being a natural part of an organism's body. The properties of recognition and identification of nonself are traits of primitive origin, present to a limited extent in most organisms, and often essential to survival. When an ameba seeks out and ingests bacteria, it is showing a primitive capacity to distinguish that which is foreign (and edible). A similar recognition occurs in the life cycle of slime molds, whose separate ameboid cells migrate toward one another and aggregate into a large, many-celled stage. Even sponges, the most primitive animals, have a recognition system sufficiently developed to distinguish the cells of other species. If two species of sponges are finely divided and mixed, the cells reaggregate only with cells from their original species. This property is even more refined among the higher invertebrates and vertebrates, which have specialized blood cells to conduct surveillance and defense. Regardless of evolutionary status, one requirement is universal: the ability to discriminate between the harmless and the harmful, between self and nonself.

An even more specific term for a foreign, nonself molecule is **antigen.** An antigen is a substance, often a surface marker, that evokes a specific immune response. As we shall see later in this chapter and in the next, antigens are responsible for eliciting specific reactions that occur during infection, allergies, and vaccination.

antigen (an'-tih-jen) From *antibody* + *gen,* to produce. Originally, any substance that elicits the production of antibodies, but the term is now used to include other kinds of reactions.

certain cells and chemicals originating in the blood can move through the vessel walls into the extracellular spaces, and from there migrate into the lymphatic system. The flow of events among these systems depends on where an infectious agent or foreign substance first intrudes. A typical progression might begin in the extracellular spaces and RES, move to the lymphatic circulation, and ultimately end up in the bloodstream. Regardless of which compartment is first exposed, an immune reaction in any one of them will eventually be felt in the others through exchanges at the microscopic level. An obvious benefit of such an integrated system is that no cell of the body is far removed from competent protection, no matter how isolated. Let us next view each of these compartments in detail.

Immune Functions of the Reticuloendothelial System

The tissues of the body are permeated by a support network of connective tissue fibers, or a *reticulum,* that originates in the cellular basal lamina, interconnects nearby cells, and meshes with the massive connective tissue network surrounding all organs. The reticular network is intrinsic to the immune function because it (1) participates in surveillance, (2) provides a route of passage for phagocytes, and (3) forms a continuous network through which materials may enter the extracellular fluid surrounding tissues (figure 12.6). The term **reticuloendothelial system** has traditionally been used, though some

reticuloendothelial (reh-tik″-yoo-loh-en″-doh-thee′-lee-al) L. *reticulum,* a small net, and *endothelium,* lining of the blood vessel.

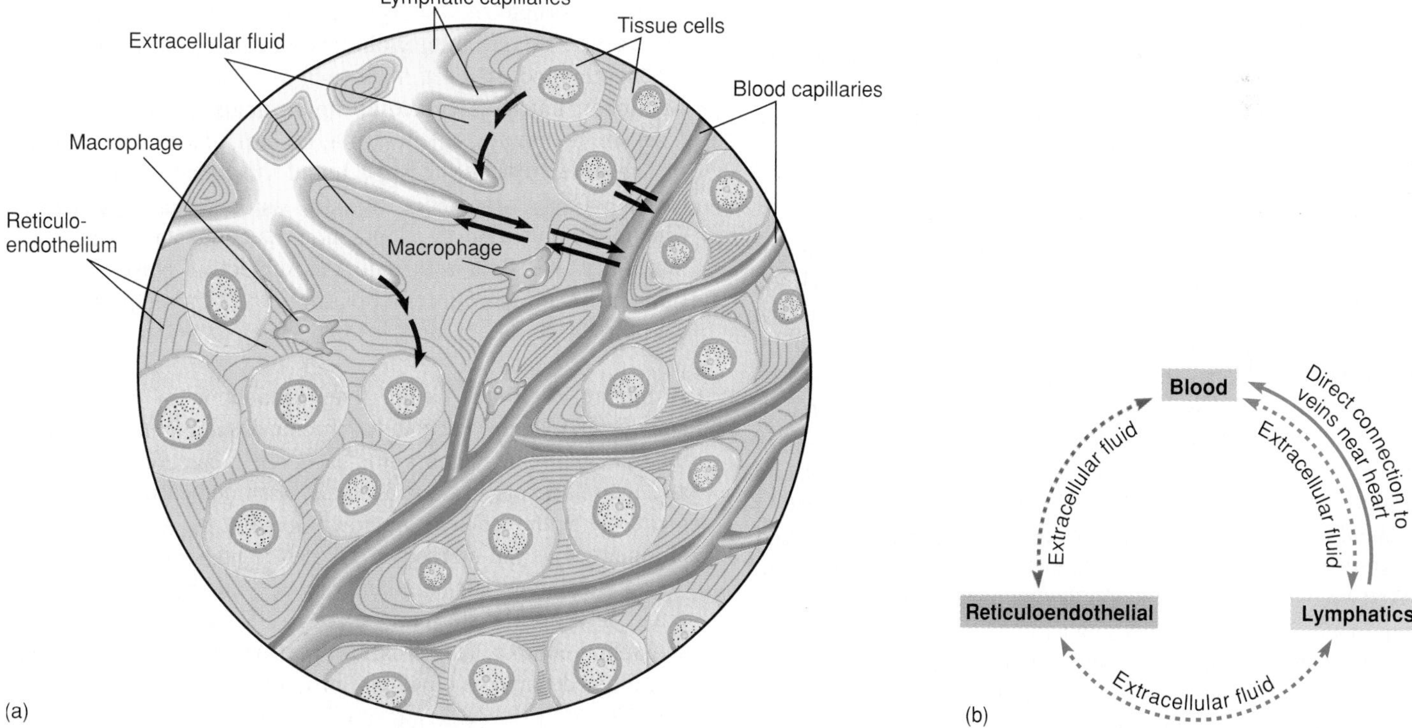

Figure 12.5 The body compartments are separate but connected. (*a*) The meeting of the major fluid compartments at the microscopic level. (*b*) Schematic view of the main fluid compartments and how they form a continuous cyclic system of exchange. Reactions in one section are rapidly broadcast to the others.

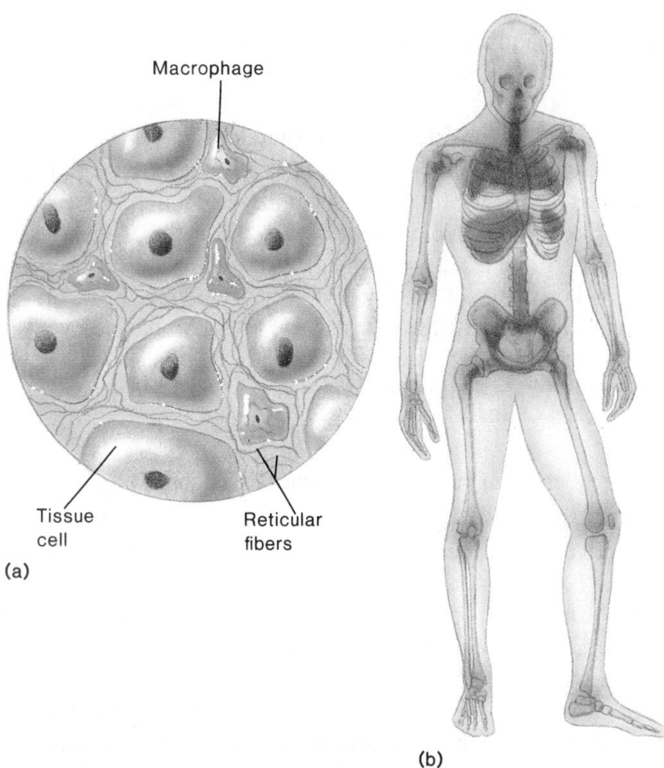

Figure 12.6 The reticuloendothelial system occurs as a pervasive, continuous connective tissue framework throughout the body. (*a*) This system begins at the microscopic level with a fibrous support network (reticular fibers) enmeshing each cell. This web connects one cell to another within a tissue or organ, and provides a niche for macrophages, which can crawl within and between tissues using the fibers as guide lines. (*b*) The degrees of shading in the body indicate variations in phagocyte concentration (darker = greater).

persons prefer to emphasize the system's phagocytic properties and have termed it the **mononuclear phagocyte system.** The system is heavily endowed with macrophages waiting to attack passing foreign intruders as they arrive in the skin, lungs, liver, lymph nodes, spleen, and bone marrow.

Origin, Composition, and Functions of the Blood

The circulatory system consists of two separate but linked systems. The circulatory system proper consists of the heart, arteries, veins, and capillaries that circulate the blood, and the lymphatic system consists of lymphatic vessels and lymphatic organs (lymph nodes) that circulate lymph (see figure 12.15). As we shall see, these two circulations parallel, interconnect, and complement one another.

The substance that courses through the arteries, veins, and capillaries is **whole blood,** a liquid connective tissue consisting of **blood cells** (formed elements) suspended in **plasma** (ground substance). One can visualize these two components with the naked eye when a tube of *unclotted* blood is allowed to sit or is spun in a centrifuge. The cells' density causes them to sediment into an opaque layer at the bottom of the tube, leaving the plasma, a clear, yellowish fluid, on top (figure 12.7). The terms plasma and serum are sometimes mistakenly interchanged, and though these substances have the same basic origin, they differ in one important way. Because **serum** is the fluid extruded from

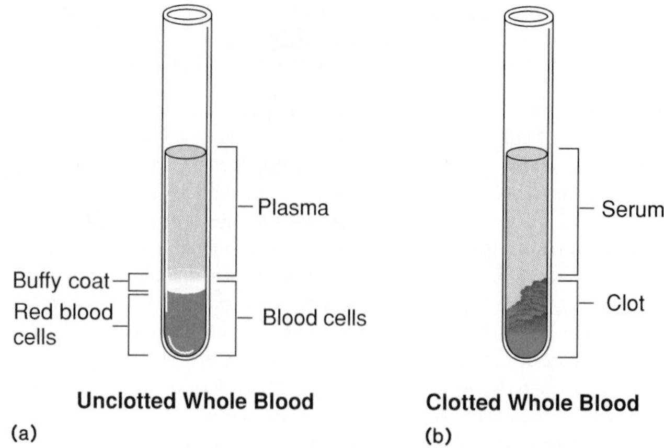

Figure 12.7 The macroscopic composition of whole blood. (*a*) When blood containing anticoagulants is allowed to sit for a period, it stratifies into a clear layer of plasma, a thin layer of off-white material called the buffy coat (which contains the white blood cells), and a layer of red blood cells in the bottom, thicker layer. (*b*) When blood is drawn and allowed to clot, a clear layer of serum is formed above the clot.

clotted blood and the clotting proteins enter into the formation of the clot, plasma contains the clotting proteins and serum does not. An easy way to remember this is: "Plasma can clot and serum cannot." Serum does contain all other substances regularly found in plasma, however.

Fundamental Characteristics of Plasma Plasma contains hundreds of different chemicals produced by the liver, white blood cells, endocrine glands, and nervous system, and absorbed from the digestive tract. The main component of this fluid is water (92%), and the remainder consists of proteins and miscellaneous organic and inorganic substances. The main solutes in plasma are: albumin; globulins (including antibodies) and other immunochemicals; fibrinogen and other clotting factors; hormones; nutrients (glucose, amino acids, fatty acids); ions (sodium, potassium, calcium, magnesium, chloride, phosphate, bicarbonate); dissolved gases (O_2 and CO_2); and waste products (urea). These substances support the normal physiological functions of nutrition, development, protection, homeostasis, and immunity. We shall return to the subject of plasma and its function in immune interactions later in this chapter and in chapter 13.

A Survey of Blood Cells The production of blood cells, **hemopoiesis,** begins early in embryonic development in the yolk sac (an embryonic membrane). Later it is taken over by the liver and lymphatic organs, and is finally assumed entirely and permanently by the red bone marrow (figure 12.8). Although much of a newborn's red marrow is devoted to hemopoietic function, the active marrow sites gradually recede, and by the age of four years, only the ribs, sternum, pelvic girdle, flat bones of the skull and spinal column, and proximal portions of the humerus and femur are devoted to blood cell production.

hemopoiesis (hee″-moh-poy-ee′-sis) Gr. *haima,* blood, and *poiesis,* a making. Also called hematopoiesis.

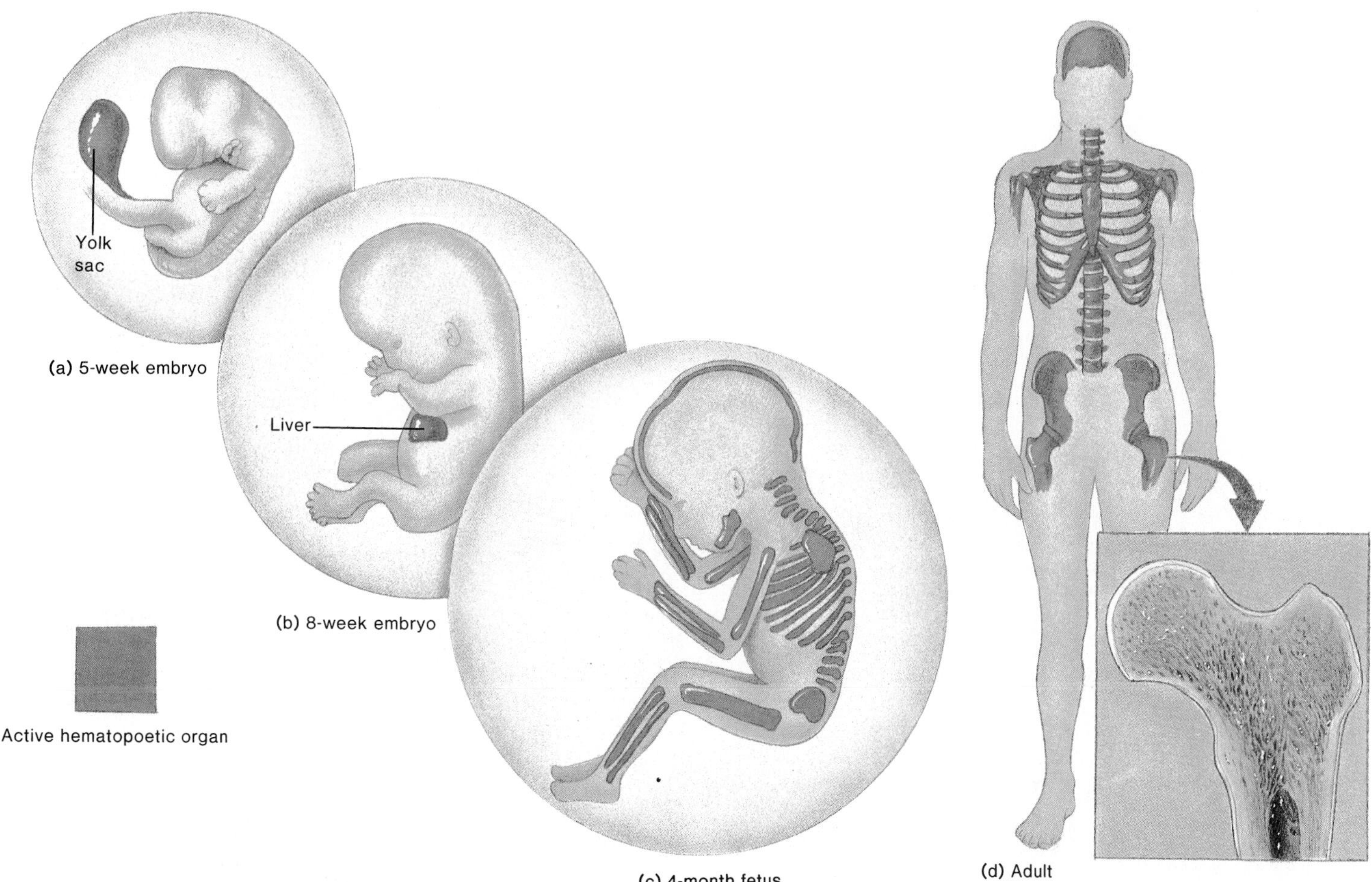

Figure 12.8 Stages in hemopoiesis. The sites of blood cell production change as development progresses from (*a, b*) yolk sac and liver in the embryo, to (*c*) extensive bone marrow sites in the fetus and (*d*) selected bone marrow sites in the child and adult. (*Inset*) Red marrow occupies the spongy bone.

The relatively short life of blood cells demands a rapid turnover that is continuous throughout a human lifespan. The ultimate precursor of new blood cells is a pool of undifferentiated cells called pluripotential[2] **stem cells,** which is maintained in the marrow. During development, these stem cells proliferate and *differentiate*—meaning that immature or unspecialized cells develop the specialized form and function of mature cells. The primary lines of cells that arise from this process produce red blood cells (RBCs or erythrocytes), white blood cells (WBCs or leukocytes), and platelets (thrombocytes). As we will see, the white blood cell lines are programmed to develop into several secondary lines of cells during the final process of differentiation (figure 12.9). These mature white blood cells are largely responsible for immune function.

The **white blood cells (leukocytes)** are traditionally evaluated by their reactions with Wright stain, a mixture of dyes that differentiates cells by color and morphology (figure 12.10). When this stain was first used on blood smears evaluated by the light microscope, the leukocytes appeared either with or without noticeable colored granules in the cytoplasm, and were subsequently divided into groups called **granulocytes** or **agranulocytes.** Later, greater magnification revealed that even the agranulocytes have tiny granules in their cytoplasm, leading some hematologists to recategorize leukocytes on the basis of the appearance of the nucleus. **Polymorphonuclear leukocytes (PMN,** or granulocytes) have a lobed nucleus, and **mononuclear leukocytes** (agranulocytes) have an unlobed, rounded nucleus. Note that the cells fall into the same groups regardless of the system used (figure 12.9).

Granulocytes Types of polymorphonuclear leukocytes present in the bloodstream are neutrophils, eosinophils, and basophils. All three are known for prominent cytoplasmic granules that stain with some combination of acidic dye (eosin) or basic dye

2. Having many potentials; unipotential means having one potential.

leukocyte (loo′-koh-syte) Gr. *leukos,* white, and *kytos,* cell. The whiteness of unstained WBCs is best seen in the white layer or buffy coat of sedimented blood.

polymorphonuclear Gr. *poly,* many; *morph,* shape; and *nuclear,* nucleus.

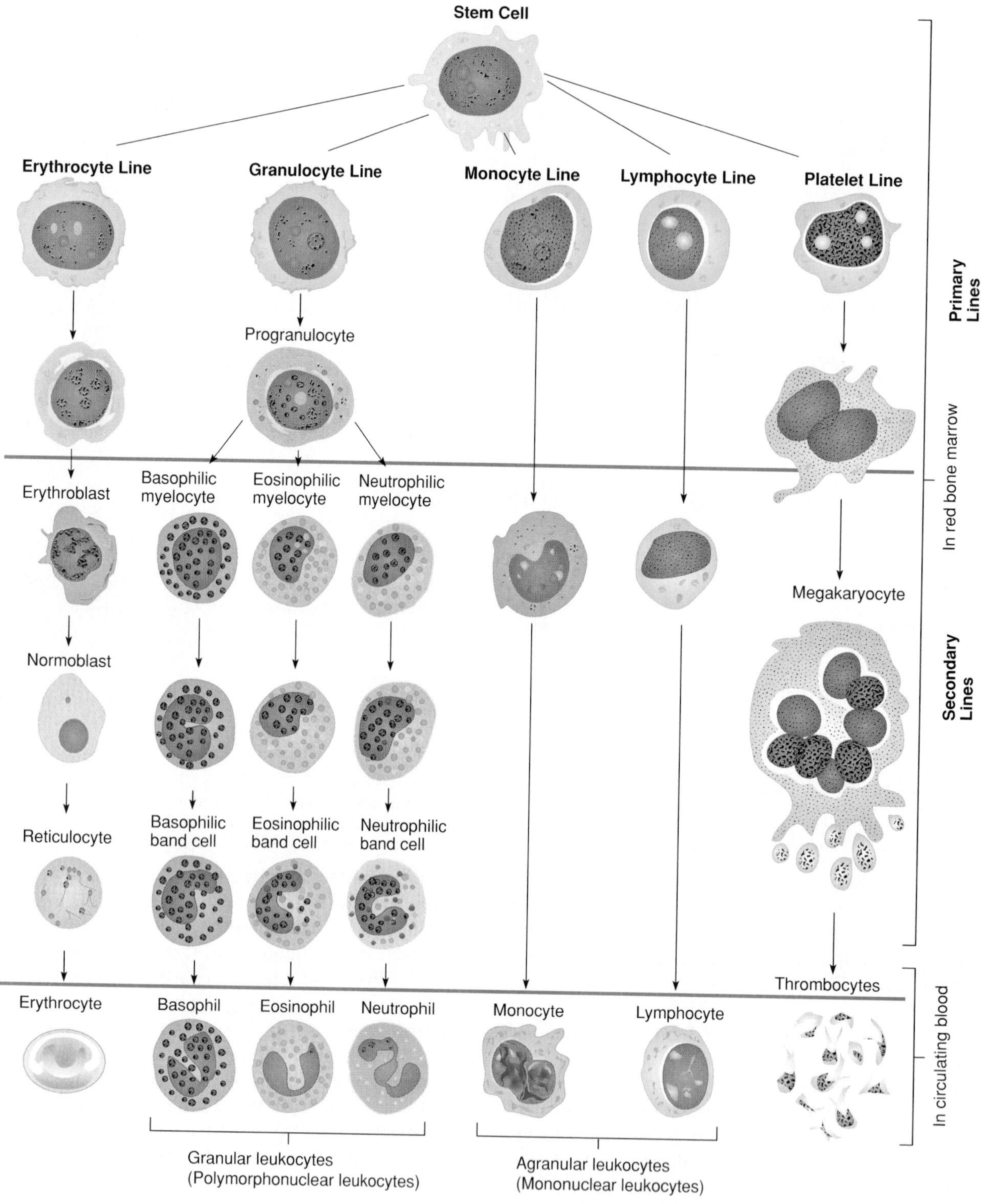

Figure 12.9 The development of blood cells and platelets. Each cell type in circulating blood (bottom row) is ultimately derived from an undifferentiated stem cell in the red marrow. During differentiation, the stem cell gives rise to several cell lines that become more and more specialized. Mature cells are released into the circulatory system.

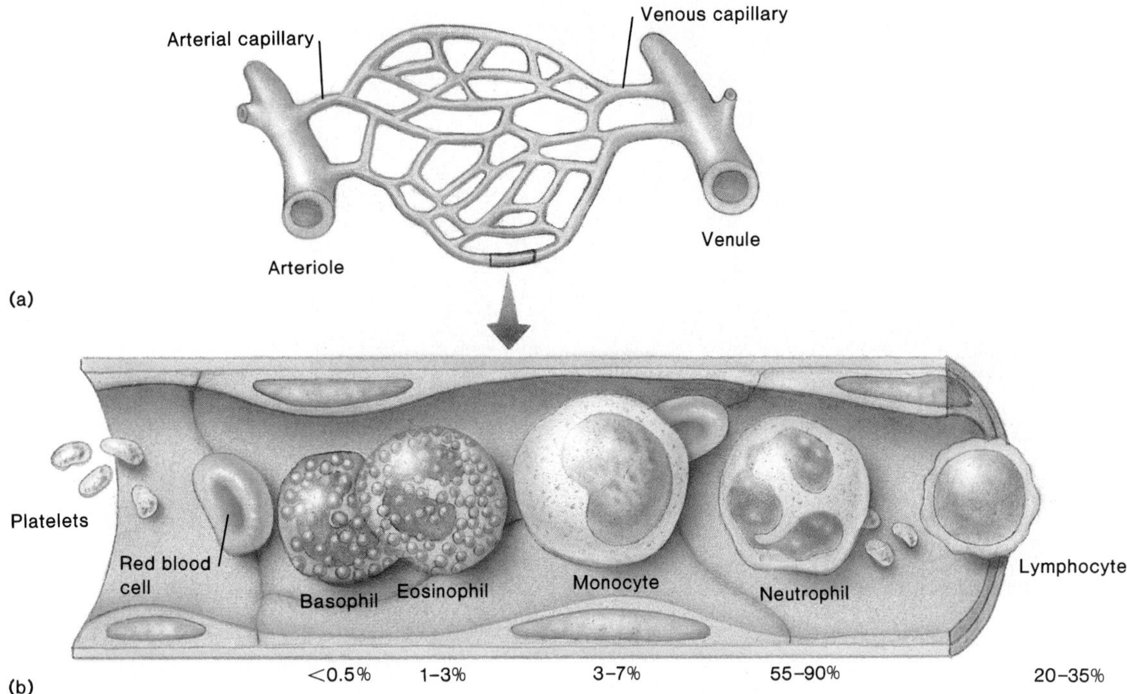

Figure 12.10 The microanatomy and circulating cells of the bloodstream. (*a*) The circulatory system in tissues consists of arterioles that deliver blood to the capillary bed and venules that carry the blood away from it. (*b*) A cutaway view of a capillary reveals a histological picture of thin, tile-like endothelial cells that form a thin, one-cell-thick conduit. The cell types and relative proportions of circulating blood cells are also shown.

(methylene blue). Although these granules are useful diagnostically, they also function in numerous physiologic events.

Neutrophils are distinguished from other leukocytes by their conspicuous, lobed nuclei connected by thin strands and by their fine, pale lavender granules. In cells newly released from the bone marrow, the nuclei are horseshoe-shaped, but as they age, they form multiple lobes (up to five). Polymorphonuclear neutrophils (PMN) make up 55% to 90% of the circulating leukocytes, comprising about 25 billion cells in the circulation at any given moment. The main work of the neutrophils is in phagocytosis.[3] Their high numbers in both the blood and tissues suggest a constant challenge from resident microflora and environmental sources. Most of the cytoplasmic granules carry digestive enzymes and other chemicals that degrade the phagocytosed materials (see the discussion of phagocytosis later in this chapter). The average neutrophil lives only about eight days, spending most of this time in the tissues and only about 6–12 hours in circulation.

Eosinophils have about the same diameter as neutrophils but are readily distinguished in a Wright stain preparation by their larger, orange to red (eosinophilic) granules and bilobed nucleus. They are much more numerous in the bone marrow and the spleen than in the circulation, contributing only 1% to 3% of the total WBC count in peripheral blood. A variety of inflammatory substances stimulates the release of eosinophils from bone marrow and attracts them to a particular site. The role of the eosinophil in the immune system is not fully defined, though several functions have been suggested. Their granules contain peroxidase, lysozyme, and other digestive enzymes. Eosinophils appear to be weakly phagocytic for bacteria, foreign particles, and antigen-antibody complexes. High levels of eosinophils are observed in infections by helminth parasites and fungi, suggesting that they are directed against larger eucaryotic parasites (figure 12.11). Much evidence is accumulating on the role of eosinophils as mediators in immediate allergies such as asthma and anaphylaxis (see chapter 14).

Basophils are also about the size of neutrophils, but they are characterized by pale-stained, constricted nuclei and very prominent dark blue to black granules. They are the scarcest type of leukocyte, comprising only 0.5% of the total circulating WBCs in a normal individual. Basophils share some morphological and functional similarities with widely distributed tissue cells, called **mast cells.** Although these two cell types were once regarded as identical, they differ in motility, origin, and location. Mast cells are nonmotile elements bound to connective tissue around blood vessels, nerves, glands, and epithelia. Basophils are motile elements derived from bone marrow. With

neutrophil (noo′-troh-fil) L. *neuter,* neither, and *philos,* to love. The granules are neutral and do not react markedly with either acidic or basic dyes. In clinical reports they are often called "polys" or PMNs for short.

3. The neutrofil is sometimes called a microphage or "small eater."

eosinophil (ee″-oh-sin′-oh-fil) Gr. *eos,* dawn, rosy, and *philos,* to love. Eosin is a red, acidic dye attracted to the granules.

basophil (bay′-soh-fil) Gr. *basis,* foundation. The granules are attractive to basic dyes.

mast From Ger. *mast,* food. Early cytologists thought these cells were filled with food vacuoles.

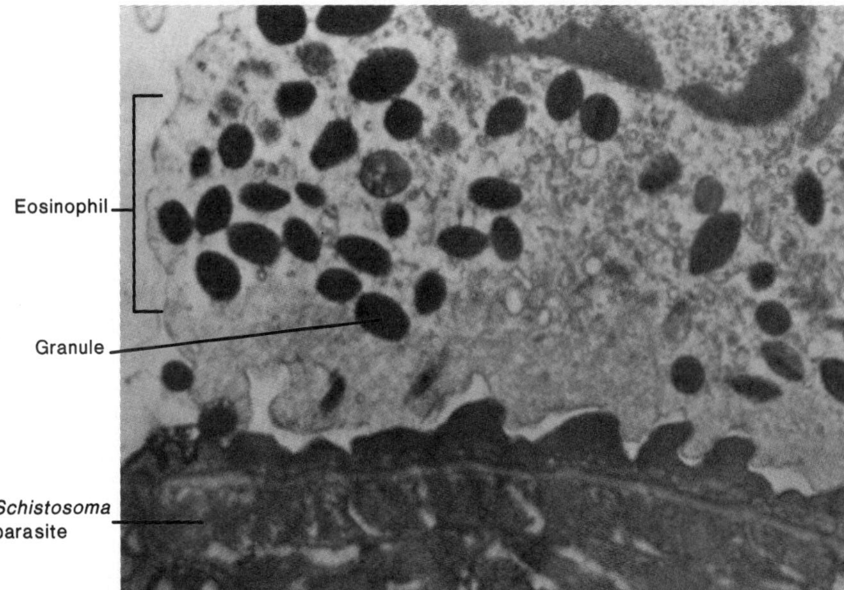

Figure 12.11 An electron micrograph section of an eosinophil attacking a parasitic worm (*Schistosoma*). The eosinophil releases the contents from its prominent granules; this initially disrupts the worm's outer surface and later destroys it completely.

regard to immune function, both types act primarily in immediate allergy and inflammation. Their granules release chemicals such as histamine, serotonin, heparin, and several enzymes with pronounced physiologic effects. As we shall see in chapter 14, many allergy symptoms are directly attributable to the effects of these chemicals on tissues and organs.

Agranulocytes Agranular leukocytes have globular, nonlobed nuclei and lack prominent cytoplasmic granules when viewed with the light microscope. The two general types are monocytes and lymphocytes.

Monocytes are generally the largest of all white blood cells and the third most common in the circulation (3–7%). As they mature, the nucleus becomes oval or kidney-shaped—indented on one side, off-center, and often contorted with fine, prunelike wrinkles. The pale blue cytoplasm holds many fine granules containing digestive enzymes. Monocytes are discharged by the bone marrow into the bloodstream and spend about a day and a half in the peripheral blood before leaving the circulation to undergo final differentiation into **macrophages** (see figure 12.22). Because monocytes are the precursors of macrophages, the two are rather similar in anatomy and function. Unlike many other WBCs, the monocyte-macrophage series is relatively long-lived and retains an ability to multiply. Macrophages are among the most versatile and important of cells. Most immune reactions involve them either directly or indirectly. In general, they are responsible for: (1) many types of specific and nonspecific phagocytic and killing functions (they assume the job of cellular "scullery maids," mopping up the messes created by infection and inflammation); (2) processing foreign molecules and presenting them to lymphocytes; and (3) secreting biologically active compounds that assist, mediate, attract, focus, and inhibit immune cells and reactions. We shall touch upon these functions in several ensuing sections.

Lymphocytes are the second most common WBC in the blood, comprising 20% to 35% of the total circulating leukocytes. The fact that their overall number throughout the body is among the highest of all cells indicates how important they are to immunity. One estimate suggests that about one-tenth of all adult body cells are lymphocytes, exceeded only by erythrocytes and fibroblasts. The entire collection of lymphocytes for a 150-pound human would tip the scales at about 2.2 pounds. This is equal in weight to another vital organ, the brain. In a stained blood smear, most lymphocytes are small, spherical cells with a uniformly dark blue, rounded nucleus surrounded by a thin fringe of clear, pale blue cytoplasm, although in tissues, they can become much larger and may even mimic monocytes in appearance (see figure 12.9). Lymphocytes exist as two functional types—the bursal-equivalent or **B lymphocytes (B cells,** for short) and the thymus-derived or **T lymphocytes (T cells,** for short). B cells were first demonstrated in and named for a special lymphatic gland of chickens called the *bursa of Fabricius,* the site for their maturation in birds. Humans lack a bursa, but some unidentified equivalent organ, possibly the bone marrow, liver, or GALT,[4] is responsible for their maturation. T cells mature in the thymus gland in all birds and mammals. Both populations of cells are transported by the bloodstream and lymph, and move about freely between lymphoid organs and connective tissue.

monocyte (mon'-oh-syte) From *mono,* one, and *cytos,* cell.

macrophage (mak'-roh-fayj) Gr. *macro,* large, and *phagein,* to eat. They are the "large eaters" of the tissues.

4. GALT is an acronym for gastrointestinal-associated lymphoid tissue, part of the lymphatic system.

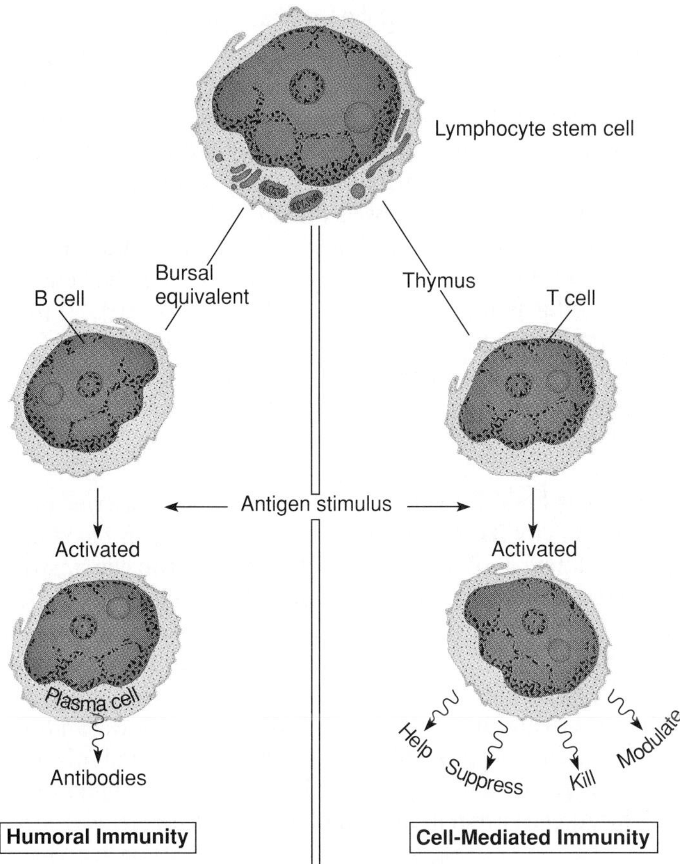

Figure 12.12 Summary of the general development and functions of lymphocytes, which are the cornerstone of specific immune reactions. Both B cells and T cells arise from the same stem cell but later diverge into two cell lines. Their appearances are similar, and one cannot differentiate them on the basis of a simple stain. Note the relatively large nucleus/cytoplasm ratio and the lack of granules.

Lymphocytes are the key cells of the third line of defense and the specific immune response (figure 12.12). When stimulated by foreign substances (antigens), lymphocytes are transformed into activated cells that neutralize and destroy that foreign substance. The contribution of B cells is mainly in **humoral immunity,**[5] defined as protective molecules carried in the fluids of the body. When activated B cells divide, they form specialized *plasma cells,* which produce **antibodies,** large protein molecules that interlock with an antigen and participate in its destruction. Activated T cells engage in a spectrum of immune functions characterized as **cell-mediated immunity** (CMI) in which T cells help, suppress, and modulate immune functions and kill foreign cells. The action of both classes of lymphocytes accounts for the recognition and memory typical of immunity. So important are lymphocytes to the defense of the body that we will devote a large portion of the next chapter to their reactions.

5. In reference to the humors, the liquids of the body. Humoral immunity includes antibodies, complement, and interferon.

antibody (an'-tih-bahd"-ee) Gr. *anti,* against, and O.E. *bodig,* body.

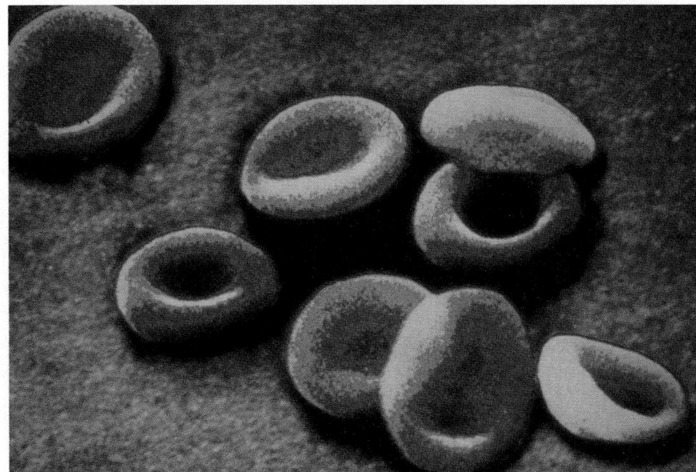

Figure 12.13 Red blood cells. A colorized scanning electron micrograph highlights the biconcave, indented centers.

Erythrocyte and Platelet Lines The **erythrocyte** line of cells goes through a process in which the nucleus is gradually extruded. The resultant red blood cells or corpuscles are simple, biconcave discs—little more than sacs of hemoglobin that transport oxygen and CO_2 to and from the tissues (figure 12.13). These are the most numerous of circulating blood cells, appearing in Wright stains as small pink circles. Red blood cells do not ordinarily have immune function, though they may be the target of immune reactions (see chapter 14).

Platelets, or *thrombocytes,* are formed elements in circulating blood that are *not* whole cells. They originate in the bone marrow when a giant multinucleate cell called a *megakaryocyte* disintegrates into numerous tiny, irregular-shaped pieces, each containing bits of the cytoplasm and nucleus (see figure 12.9). In Wright stains, platelets are blue-gray with fine red granules and are readily distinguished from cells by their smaller size (1–3 μm in diameter) and greater number (150,000 to 450,000 per mm^3 of blood). Platelets function primarily in hemostasis (plugging broken blood vessels to stop bleeding) and in releasing chemicals that act in blood clotting and inflammation.

Unique Dynamic Characteristics of White Blood Cells Many lymphocytes and phagocytes make regular journeys from the blood and lymphatics to the tissues and back again to the circulation as part of the constant surveillance of the compartments. In order for these WBCs to complete this circuit, they are not regularly propelled about in the blood, but instead, adhere to the inner walls of the smaller blood vessels. From this position, they are poised to emigrate out of the blood into the tissue spaces by a process called **diapedesis.** Upon contact with the small spaces between the patchwork of endothelial cells that line blood vessels, these WBCs extrude themselves between the

erythrocyte (eh-rith'-roh-syte) Gr. *erythros,* red. The red color comes from hemoglobin.

platelet (playt'-let) Gr. *platos,* flat, and *let,* small.

thrombocyte (throm'-boh-syte) Gr. *thrombos,* clot.

diapedesis (dye"-ah-puh-dee'-sis) Gr. *dia,* through, and *pedan,* to leap.

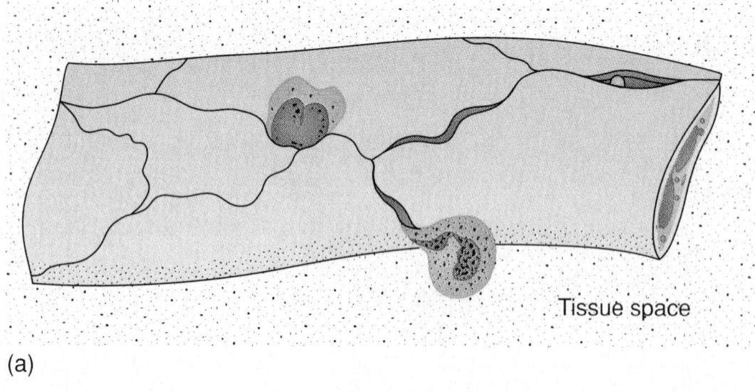

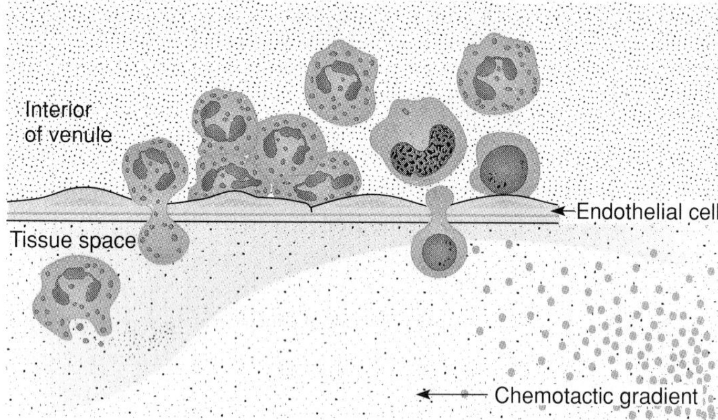

Figure 12.14 Diapedesis, the creepy, crawly character of leukocytes. (*a*) Three-dimensional view of a venule depicts white blood cells engaged in this phenomenon. (*b*) Cross section through a venule shows the marginated pool of leukocytes adhering to the endothelial wall. Two white blood cells are partially inserted between the endothelial junctions so as to slip into the tissue space. Shading indicates a chemical gradient to which the cells are responding.

spaces and slip into the nearby extracellular region (figure 12.14). (Their actively motile nature and plastic shape enable them to do this.) This process occurs primarily in postcapillary venules (little veins) whose endothelial cells have loose junctions between them, and it happens without injury to the cells or vessel wall.

An important factor in the migratory habits of these WBCs is **chemotaxis,** the tendency of cells to migrate in the direction of a specific chemical stimulus given off at a site of injury or infection (see inflammation and phagocytosis later in this chapter). Through this means, cells swarm from many compartments to the site of infection and remain there to perform general and specific immune functions. These basic properties are absolutely essential for the sort of intercommunication and deployment of cells required for most immune reactions.

chemotaxis (kee-moh-tak'-sis) NL *chemo*, chemical, and *taxis*, arrangement.

Components and Functions of the Lymphatic System

The lymphatic system is a compartmentalized network of vessels, cells, and specialized accessory organs (figure 12.15). Structurally, it begins in the farthest reaches of the tissues as tiny capillaries that transport a special fluid (lymph) through an increasingly larger tributary system of vessels and filters (lymph nodes) to major vessels that drain back into the regular circulatory system. The three major functions of the lymphatic system are: (1) to provide an auxiliary route for the return of extracellular fluid to the circulatory system proper;[6] (2) to act as a "drain-off" system for the inflammatory response; and (3) to render surveillance, recognition, and protection against foreign materials through a system of lymphocytes, phagocytes, and antibodies.

Lymphatic Fluid Lymph is a thick yellow liquid carried by the lymphatic circulation that is formed when certain blood components filter out of the blood vessels into the extracellular spaces and diffuse or migrate into the lymphatic capillaries. Thus, the composition of lymph parallels that of serum in many ways. It is made up of water, dissolved salts, and 2% to 5% protein (especially antibodies and albumin). Like blood, it also transports numerous white blood cells (especially lymphocytes) and miscellaneous materials such as fats, cellular debris, and infectious agents that have gained access to the tissue spaces. Unlike blood, red blood cells are not normally found in lymph.

Lymphatic Vessels The system of vessels that transports lymph is constructed along the lines of regular blood vessels (figure 12.16). The tiniest ones, lymphatic capillaries, accompany the blood capillaries and permeate all parts of the body except the central nervous system and certain organs such as bone, placenta, or thymus. Lymphatic capillaries are thin-walled and permeable, so as to readily admit extracellular fluid. The density of lymphatic vessels is particularly high in the hands, feet, and around the areolae of the breasts. Unlike the bloodstream, the lymphatic system does not circulate cyclically, but in one direction—toward the heart (figure 12.16). The lymphatic capillary network feeds into a series of fewer but larger vessels that, in turn, drain finally into two main ducts (the thoracic duct and the right lymphatic duct), which empty the lymph into the large veins at the base of the neck. In the larger lymphatic vessels, the flow of fluid is facilitated by muscle contraction, gravity, and one-way valves similar to the check valves found in veins.

Lymphoid Organs and Tissues Other organs and tissues that perform lymphoid functions are the lymph nodes (glands), thymus, spleen, and clusters of tissues in the gastrointestinal tract (GALT) and the pharynx (the tonsils, for example). A trait common to these organs is a loose connective tissue framework that houses aggregations of lymphocytes, the important class of white blood cells mentioned previously.

6. The importance of this function is most evident in cases of impaired lymphatic drainage such as patients with filariasis, a roundworm infection (see figure 19.20).

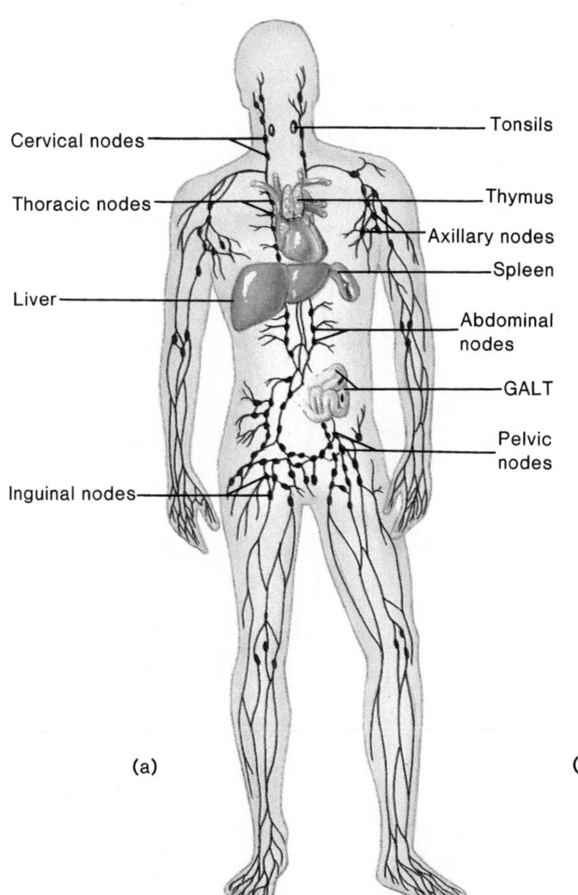

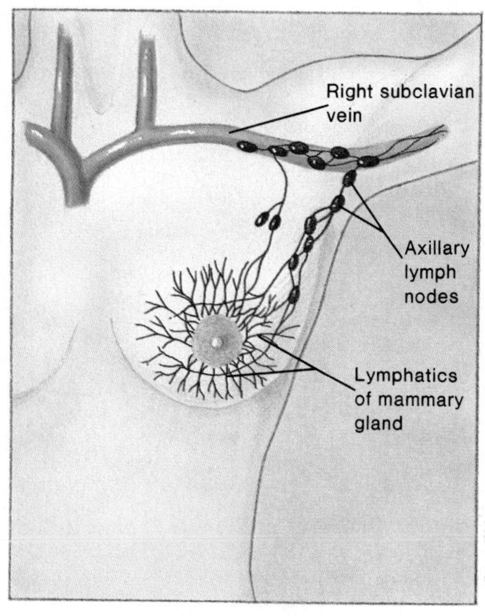

Figure 12.15 General components of the lymphatic system. (*a*) This system consists of an anastomosing, branching network of vessels that permeate most tissues of the body. Note the especially high concentration of this network in the hands, feet, and breasts. (*b*) Organs that form part of this system are lymph nodes, clustered at major drainage points (armpit, groin, intestine). Other lymphatic structures are the spleen, part of the small intestine (GALT), the thymus, and the tonsils.

Lymph Nodes Lymph nodes are small, encapsulated, bean-shaped organs stationed, usually in clusters, along lymphatic channels and large blood vessels of the thoracic and abdominal cavities (figure 12.15). Major aggregations of nodes occur in the loose connective tissue of the armpit (axillary nodes), groin (inguinal nodes), and neck (cervical nodes). Both the location and architecture of these nodes clearly specialize them for filtering out materials that have entered the lymph and providing appropriate cells and niches for immune reactions to take place.

A view of a single sectioned lymph node reveals its filtering and cellular response systems (figure 12.17). Incoming lymphatic vessels transport the lymph into a channel called a central sinus, which flows into smaller sinuses. Within the core of the node, sinuses converge and become tributaries of a single main lymphatic vessel that leaves the node via an indentation. Along its course, lymph percolates through the sections of the node that contain segregated populations of lymphocytes. The central zone, or *medulla* is the location of T cells, and the surrounding germinal centers in the *cortex* are packed with B cells. This system of sinuses and discrete lymphocyte zones filters out particulate materials (microbes, for instance) and contributes WBCs to the lymph as it passes through. Many of the initial encounters between lymphocytes and microbes that result in specific immune responses occur in the lymph nodes.

Spleen The spleen is a lymphoid organ in the upper left portion of the abdominal cavity. It is somewhat similar to the lymph nodes in its basic structure and function, except that the spleen circulates blood instead of lymph (figure 12.18). It consists of a connective tissue network of vascular sinuses, the *red pulp,* which is rich in macrophages, polymorphonuclear lymphocytes, and erythrocytes. Nestled within the red pulp are compact regions of lymphocytes called the *white pulp.* B cells and T cells occupy separate predetermined areas of the white pulp. The spleen filters blood and tissue fluids, serves as an important station for phagocytosis of foreign matter and immune reactions against bacteria, and removes and breaks down worn erythrocytes. Although adults whose spleens have been surgically removed may live a relatively normal life, asplenic children are extremely immunocompromised.

medulla (meh-dul'-ah) L. *medius,* middle or marrow.
cortex (kor'-teks) L. *cortex,* bark, rind, shell.

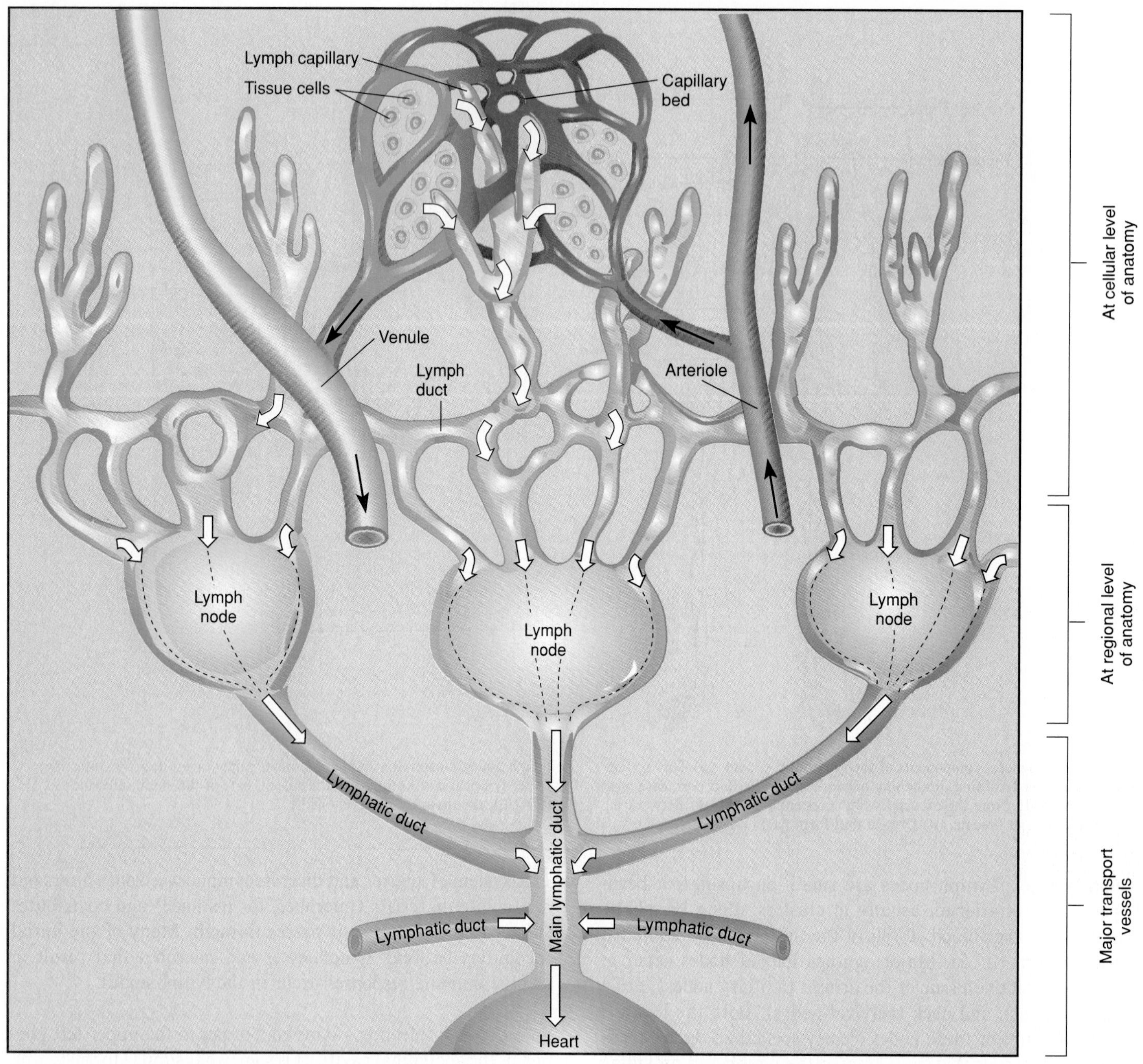

Figure 12.16 The circulatory scheme of the lymphatic vessels. The lymphatic capillaries originate at the finest level of the tissues as tiny, fingerlike projections that absorb excess tissue fluid and transport it into a network of fewer and larger vessels (ducts). These ducts drain their fluid (lymph) into the lymph nodes, where it is filtered and transported into a larger series of drainage vessels. These vessels empty their load of lymph into large main ducts near the heart. By this means, lymphatic chemicals and cells may be returned to the blood.

The Thymus: Site of T-Cell Maturation The **thymus** is the first lymphoid organ to appear in the human embryo, originating as two lobes in the pharyngeal region that fuse into a triangular-shaped structure. The size of the thymus is greatest proportionally at birth (figure 12.19), and it continues to exhibit high rates of activity and growth until puberty, after which it begins to shrink gradually through adulthood. During the last few decades of life, its function is greatly diminished, because its primary work has been completed. The thymus is sectioned into a medulla, composed of special epithelial cells, and a cortex, containing undifferentiated lymphocytes called thymocytes (figure 12.19). Under the influence of thymic hormones, thymocytes develop specificity and are released into the circulation as mature T cells. The T cells subsequently migrate to and settle in other lymphoid organs (for example, the lymph nodes and spleen), where they occupy the specific sites described previously. By adulthood, these lymphatic organs have received enough mature T cells that thymic function can be reduced without adverse effect. However, babies born without a functional thymus are vulnerable to a disease called DiGeorge syndrome (see chapter 14).

---

thymus (thigh'-mus) Gr. *thymos*, soul, mind.

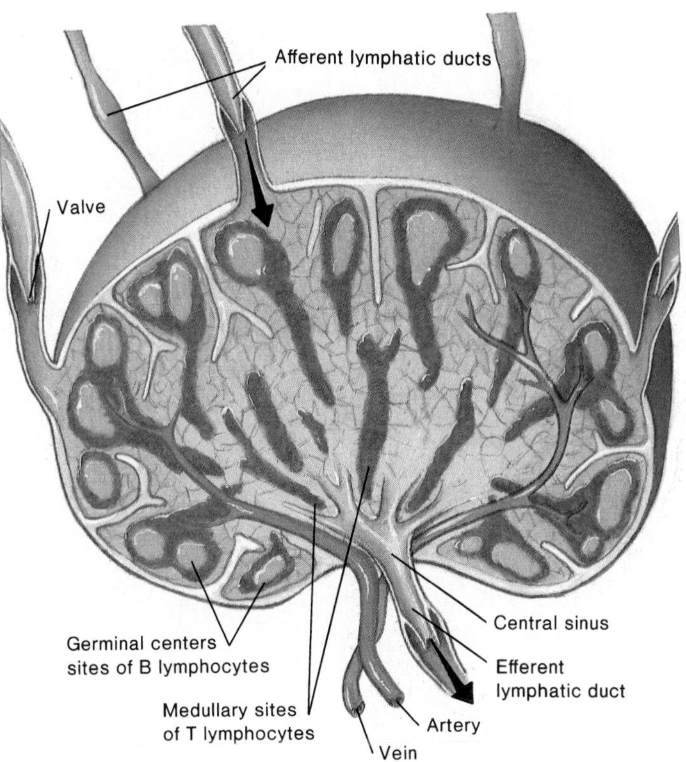

Figure 12.17 The anatomy of a lymph node. Section of a node reveals afferent lymphatic ducts that penetrate into the sinuses. As the lymph percolates through these sinuses, it is filtered and exposed to populations of B and T lymphocytes that are segregated into specific sites. Filtered lymph is delivered to a central sinus and leaves via an efferent lymphatic duct. Note that the node is penetrated by a blood circulation that provides an opportunity for further blood-lymph exchange.

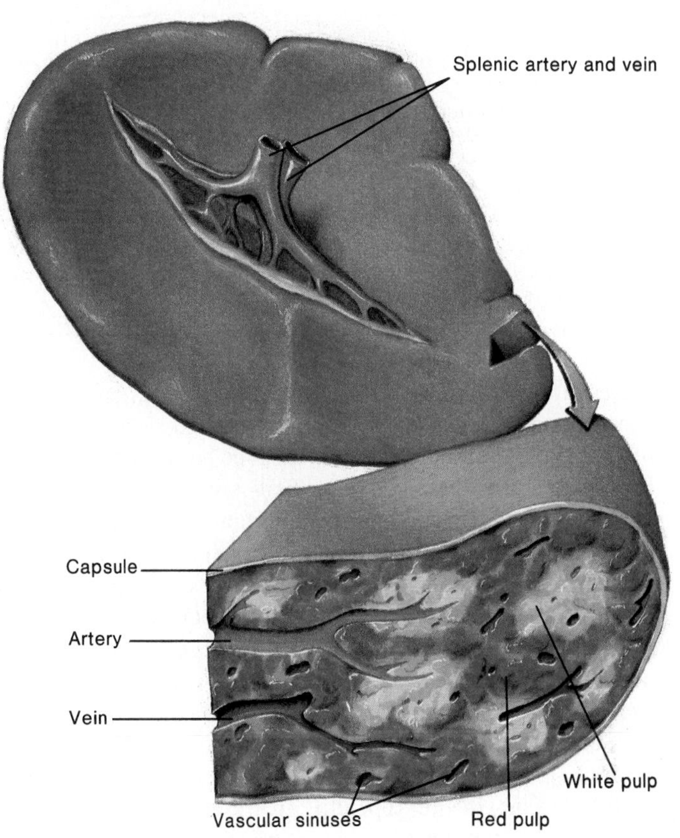

Figure 12.18 The anatomy of the spleen.

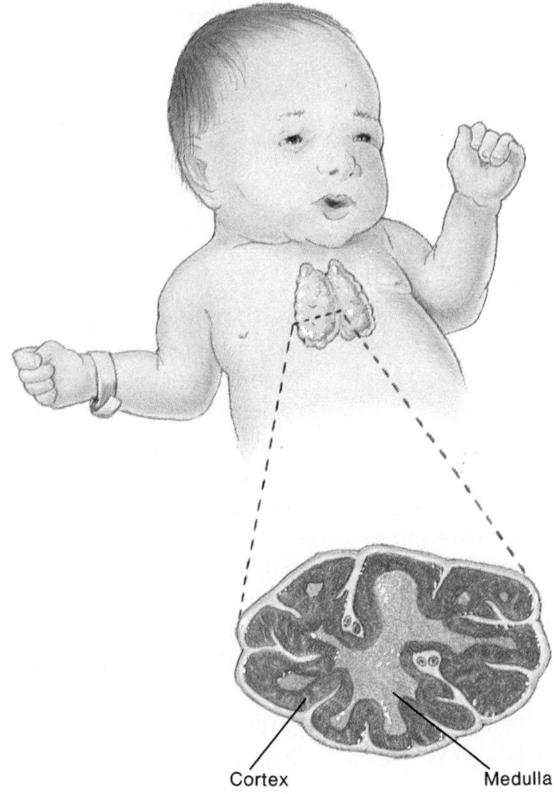

Figure 12.19 The thymus gland. Immediately after birth, the thymus is a large organ that nearly fills the region over the midline of the upper thoracic region. In the adult, it is proportionately smaller (to compare, see figure 12.15*a*). Section shows the main anatomical regions of the thymus.

The Misunderstood Thymus The thymus gland has an interesting history. Up until the early 1960s, its function was mysterious, and it was thought to be something of a vestigial organ. Medical science was so mistaken about its importance that children's necks were sometimes irradiated to "cure" a condition called "enlarged thymus." Experiments on rodents finally clarified its link to lymphocyte development. If the thymus is surgically removed in neonates, they become severely immunodeficient and fail to thrive. Thymectomized adults, however, experience less life-threatening and more subtle effects. Do not confuse the thymus with the thyroid gland, which is located nearby, but has an entirely different function.

Miscellaneous Lymphoid Tissue At many sites on or just beneath the mucosa of the gastrointestinal and respiratory tracts lie discrete bundles of lymphocytes. The positioning of this diffuse system provides an effective first-strike potential in regions that are constantly contaminated with microbes and other foreign materials in food and air. In the pharynx, a ring of tissues called the **tonsils** provides an active source of lymphocytes. Removal of the tonsils to control chronic infections does not appear to compromise immunity, though this surgery is less common nowadays due to the availability of broad-spectrum antibiotics. Collections of lymphocytes also underlie the mucosa of the trachea and bronchi. Perhaps the greatest concentration of this sort of tissue is in the gastrointestinal tract (the GALT). Immune functions exist in the appendix, in the lacteals (special lymphatic vessels stationed in each intestinal villus), and in *Peyer's patches,* compact aggregations of lymphocytes in the ileum of the small intestine. GALT is essential in immunities against intestinal pathogens and may be important in the development of B cells. The breasts of pregnant and lactating women also become temporary sites of antibody-producing lymphoid tissue (see colostrum, feature 12.7).

Nonspecific Immune Reactions of the Body's Compartments

Now that we have introduced the principal anatomical and physiological framework of the immune system, let us address some mechanisms that play an important part in host defenses: inflammation, phagocytosis, interferon, and complement. Because of the generalized nature of these defenses, they contain elements that are nonspecific, but they also support and interact with the specific immune responses described at the end of this chapter and in chapter 13.

The Inflammatory Response: A Complex Concert of Reactions to Injury

At its most general level, the inflammatory response is a reaction to any traumatic event in the tissues. It is so very commonplace that most of us manifest inflammation in some way every day. It appears in the nasty flare of a cat scratch, the blistering of a burn, the painful lesion of an infection, and the symptoms of allergy. It is readily identifiable by a classic series of signs and symptoms (see feature 12.2). The eliciting factors of inflammation include trauma from infection (the primary emphasis here), tissue injury or necrosis due to physical or chemical agents, and specific immune reactions. Although the details of inflammation are very complex, its chief functions can be summarized as follows: (1) to mobilize and attract immune components to the site of the injury, (2) to set in motion mechanisms to repair tissue damage and localize and clear away harmful substances, and (3) to destroy microbes and block their further invasion (figure 12.20). The inflammatory response is a powerful defensive reaction, a means for the body to maintain stability and restore itself after an injury. But it also has the potential in chronic situations to actually *cause* tissue injury, destruction, and disease (see feature 12.4).

Feature 12.2 Rubor, Calor, Tumor, Dolor—Gross Signs of Inflammation

The classic signs and symptoms of inflammation have been known for centuries and are characterized succinctly by four Latin terms: *rubor, calor, tumor,* and *dolor.* Rubor (redness) is caused by increased circulation and vasodilation in the injured tissues; calor (warmth) is the heat given off by the increased flow of blood; tumor (swelling) is caused by increased fluid escaping into the tissues; and dolor (pain) is caused by the stimulation of nerve endings. Combined, these events often cause the temporary loss of function of the afflicted tissue (*functio laesa*). Many of these reactions are unpleasant, and at times may seem more harmful than protective. It is clear, however, that they serve both to warn that injury has taken place and to set in motion responses that save the body from further injury.

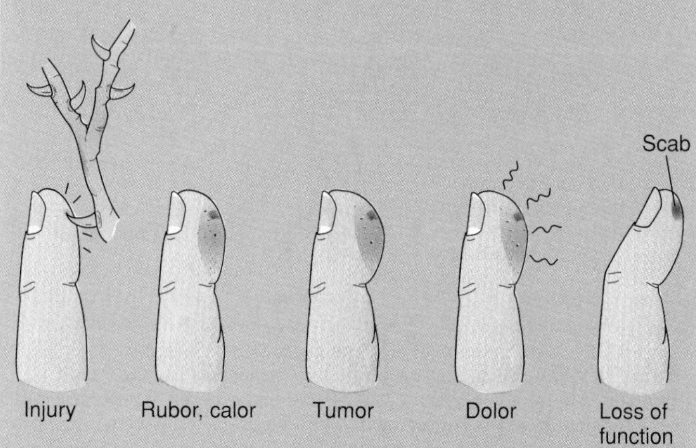

The response to injury. This classic checklist encapsulates the reactions of the tissues to an assault and is still very applicable because each of the events is an indicator of one of the mechanisms of inflammation described in succeeding sections of this chapter.

The Stages of Inflammation

The process leading to inflammation is a dynamic, predictable sequence of events that may be acute, lasting from a few minutes or hours, to chronic, lasting for days, weeks, or years. Once the initial injury has occurred, a chain reaction takes place at the site of damaged tissue, summoning beneficial cells and fluids into the injured area. In order to visualize this process, it is helpful to view an injury at the microscopic level, looking over

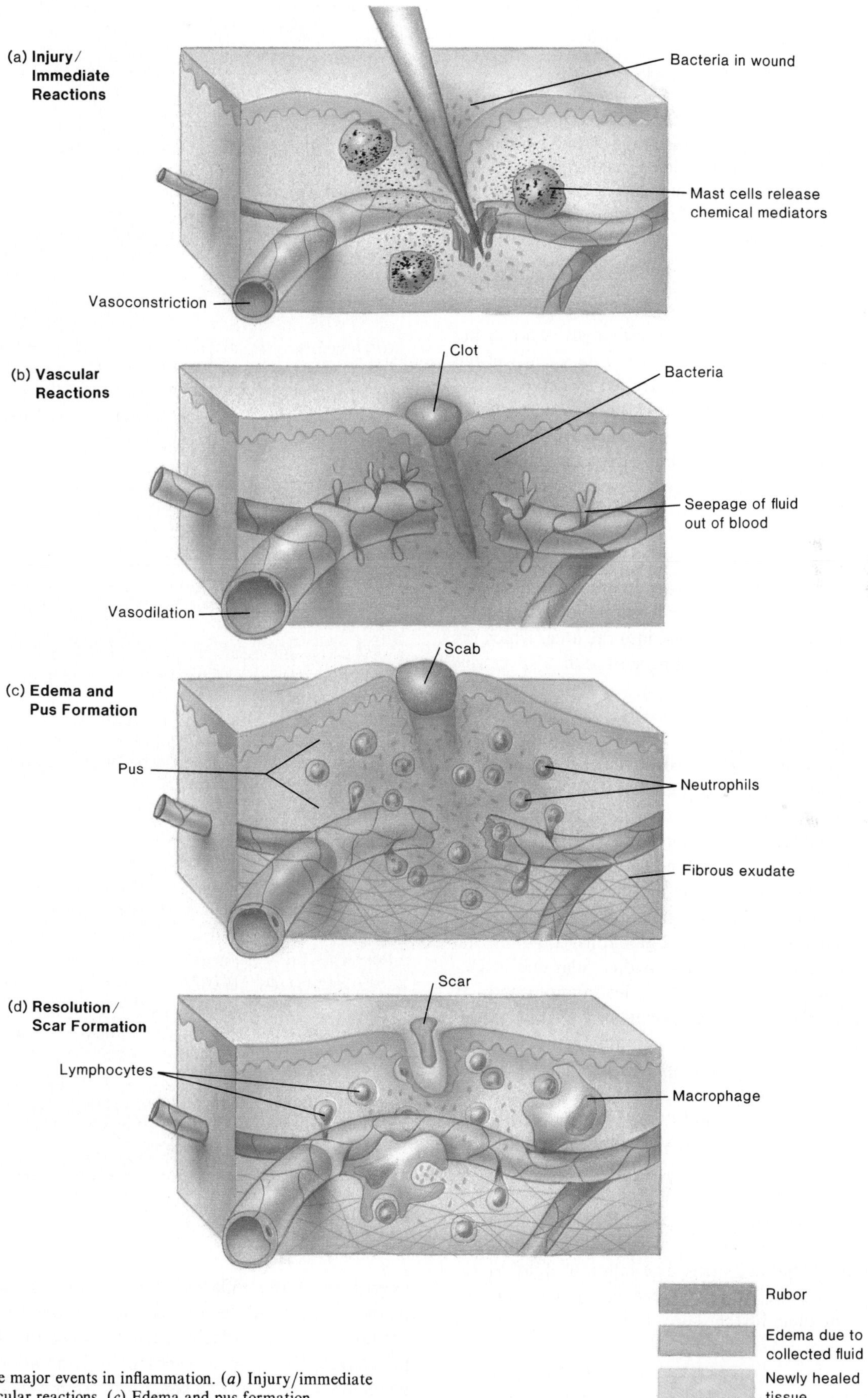

Figure 12.20 The major events in inflammation. (*a*) Injury/immediate reactions. (*b*) Vascular reactions. (*c*) Edema and pus formation. (*d*) Resolution/scar formation.

each major participant and reaction (figure 12.20). The flow of major events in inflammation can be represented as follows:

> Injury → Reflex narrowing of the blood vessels (vasoconstriction) lasting for a short time → Increased diameter of blood vessels (vasodilation) → Increased blood flow → Increased vascular permeability → Leakage of fluid (plasma) from blood vessels into tissues (exudate formation) → Edema → Infiltration of site by neutrophils → Infiltration by macrophages and lymphocytes → Repair, either by complete resolution and return of tissue to normal state or by formation of scar tissue.

Vascular Changes: Early Inflammatory Events

Following an injury, some of the earliest changes occur in the vasculature (arterioles, capillaries, venules) in the vicinity of the damaged tissue. These changes are controlled by nervous stimulation and **chemical mediators** or **cytokines** released by blood cells, tissue cells, and platelets in the injured area. Some of these mediators are *vasoactive*—that is, they affect the endothelial cells and smooth muscle cells of blood vessels; some are chemotactic factors for white blood cells; and some influence other immune reactions (see feature 12.3 and figure 12.21). Although the constriction of arterioles is stimulated first, it lasts for only a few seconds or minutes, and is followed in quick succession by the opposite reaction, vasodilation. The overall effect of vasodilation is to increase the flow of blood into the area, which facilitates the influx of immune components and also causes redness and warmth.

Edema: Leakage of Fluid into Tissues

One result of the actions of some vasoactive substances is that, in a matter of minutes, the endothelial cells surrounding postcapillary venules contract and form gaps through which blood-borne components may exude into the extracellular spaces. The fluid part that escapes, called the *exudate,* accumulates in the tissues and causes local swelling and hardness, a reaction known as **edema.** The edematous exudate contains varying amounts of plasma proteins, such as globulins, albumin, the clotting protein fibrinogen, blood cells, and cellular debris. Depending upon its content, the exudate varies in appearance. It may be clear and cell-free (serous); or it may contain a preponderance of either red blood cells (serosanguinous) or white blood cells (purulent or pus-containing). In some types of edema, the fibrinogen is converted to fibrin threads that enmesh the injury site. Within an hour, multitudes of neutrophils responding chemotactically to special signalling molecules converge on the injured site (see figure 12.20*c*).

The Benefits of Edema and Chemotaxis Both the formation of edematous exudate and the infiltration of neutrophils are physiologically beneficial activities. The influx of fluid dilutes toxic substances, and the fibrin clot can effectively trap microbes and prevent their further spread. The neutrophils that aggregate in the inflamed site are immediately involved in phagocytosing and destroying bacteria, dead tissues, and particulate matter (by mechanisms discussed in a later section on phagocytosis). In some types of inflammation, accumulated phagocytes contribute to **pus,** a whitish mass of cells, liquefied cellular debris, and bacteria. Certain bacteria (streptococci,

---

Feature 12.3 The Dynamics of the Cellular Communications Network

Many of the cells of the immune system store powerful cytokines that regulate, stimulate, and limit immune reactions. Cytokines serve as a chemical communications network that helps the body mount cooperative campaigns against foreign materials. There are several sources of cytokines: (1) monokines, produced by monocytes and macrophages; (2) lymphokines, produced by lymphocytes; (3) inflammatory peptides, produced by neutrophils; (4) vasoactive amines, which come from platelets and mast cells; and (5) miscellaneous substances. Some cytokines act during inflammation and allergy, while others function as part of the specific immune response. Many cytokines and their reactions are too complex for the scope of this book, but we will present here an overview of the most important ones:

Histamine is a vasoactive mediator produced by mast cells and basophils that causes vasodilation, increased vascular permeability, and mucus production. It functions primarily in inflammation and allergy.

Serotonin is a mediator produced by platelets and intestinal cells that participates in vasoconstriction, blood coagulation, and smooth muscle contraction.

Bradykinin is a vasoactive amine from the blood or tissues that stimulates smooth muscle contraction and increases vascular permeability, mucus production, and pain. It is particularly active in allergic reactions.

Chemotactic factors are common agents that induce white blood cells to migrate; they include complement C5A, lymphokines, and factors produced by mast cells.

Arachidonic acid metabolites come from any injured cell. Arachidonic acid is a natural tissue fatty acid that may be metabolically converted to several active compounds:

Prostaglandins, produced by most body cells, are among the most potent of all chemical mediators. There are several types, many of which have opposing effects (for example, dilation or constriction of blood vessels) and are powerful stimulants of inflammation and pain.

Leukotrienes stimulate the contraction of smooth muscle and enhance vascular permeability. They are implicated in the more severe manifestations of immediate allergies (constriction of airways).

Interferon is an antiviral immune stimulant produced by lymphocytes and certain tissue cells (its mechanism of action will be discussed later in this chapter).

Interleukin-1 is a monokine that stimulates T cells and B cells and acts as a pyrogen. *Interleukin-2* is a lymphokine that stimulates the proliferation and killing activities of T cells (see chapter 13).

Complement is a complex system of chemicals with numerous biological effects, some of which are inflammatory, killing, and chemotactic (figure 12.21).

Platelet-activating factor, a substance released from basophils, causes the aggregation of platelets and the release of other chemical mediators during immediate allergic reactions.

---

cytokine (sy′-toh-kyne) Gr. *cytos,* cell, and *kinein,* to move. A chemical stimulant released from one tissue that induces activity in another tissue.

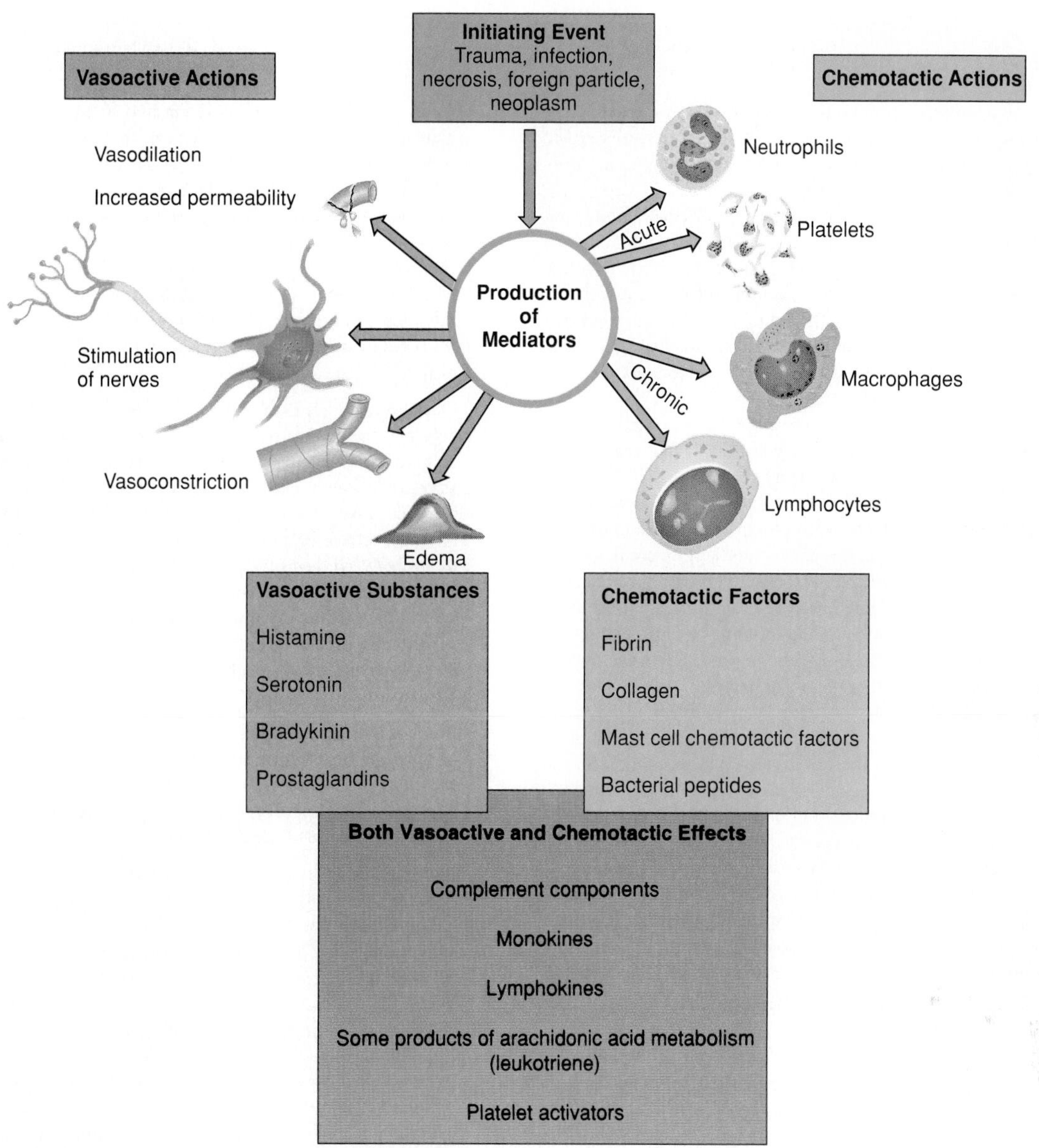

Figure 12.21 Chemical mediators of the inflammatory response and their effects.

staphylococci, gonococci, and meningococci) are especially powerful attractants for neutrophils and are thus termed **pyogenic** or pus-forming bacteria.

Late Reactions of Inflammation A mild inflammation may be resolved by the nonspecific reactions described so far. After a period of time (about 12 hours), cells that react more slowly (monocytes, lymphocytes, and macrophages) aggregate at the reaction site. The concluding stage in the inflammatory response now turns to cleanup and healing. Clearance of pus, cellular debris, dead neutrophils, and damaged tissue falls to the macrophages, the only cells that can engulf and dispose of such large masses. At the same time, B lymphocytes react with foreign molecules and cells by producing specific antimicrobial proteins (antibodies), and T lymphocytes kill intruders directly. As this process is going on, the regenerative powers of the tissues are brought into play, and the tissue is completely repaired, if possible, or replaced by connective tissue (a scar; figure 12.20*d*). If the inflammation cannot be relieved or resolved this way, chemical mediators of chronic inflammation enter and often create a long-term pathological condition (see feature 12.4).

Fever: An Adjunct to Inflammation

The inflammatory mechanisms discussed thus far are somewhat localized, but inflammation also has an important systemic component—**fever,** defined as an abnormally elevated body temperature. Although fever is a nearly universal symptom of infection, it is also associated with certain allergies, specific

Feature 12.4 When Inflammation Gets Out of Hand

Not every aspect of inflammation is protective or results in the proficient resolution of tissue damage. As one looks over a list of diseases, it is rather striking how many of them are due in part or even completely to an overreactive or dysfunctional inflammatory response.

Some "itis" reactions mentioned in chapter 11 are a case in point (see feature 11.5). Inflammatory exudates that build up in the brain in African trypanosomiasis, cryptococcosis, and other brain infections may be so injurious to the nervous system that impairment is permanent. Frequently, an inflammatory reaction that causes the pathogen to be walled off gives rise to an abscess, a swollen mass of neutrophils and dead, liquefied tissue that often harbors live pathogens in the center. Abscesses are a prominent feature of staphylococcal, amebic, and enteric infections.

Other pathological manifestations of chronic diseases—for example, the tubercles of tuberculosis, the gummas of syphilis, the disfiguring nodules of leprosy, and the cutaneous ulcers of leishmaniasis are due to an aberrant tissue response called *granuloma formation* (see figure 16.23). Granulomas develop not only in response to microbes, but also in response to inanimate foreign bodies (sutures and mineral grains that are difficult to break down). This condition is initiated when neutrophils ineffectively and incompletely phagocytose the pathogens or materials involved in an inflammatory reaction. The macrophages enter to clean up and attempt to phagocytose the now dead neutrophils and foreign substances, but they also fail to completely manage them. They respond by storing these ingested materials in vacuoles and becoming inactive. Over a period of time, large numbers of adjacent macrophages fuse into giant, inactive multinucleate cells called foreign body giant cells. These sites are further infiltrated with lymphocytes. The resultant collections make the tissue appear granular—hence, the name. A granuloma may exist in the tissue for months, years, or even a lifetime.

The inflammatory response also plays a major role in adverse immune reactions such as allergy and autoimmunity (see chapter 14). The damage to the heart valves and kidneys characteristic of rheumatic fever and glomerulonephritis has a strong inflammatory component. Although acute inflammatory reactions may increase the white blood cell count, those of long standing may ultimately suppress hematopoiesis and cause leukopenia.

immune responses, cancers, and other organic illnesses. Fevers whose causes are unknown are called fevers of unknown origin, or FUO.

The body temperature is normally maintained by a special control center in the hypothalamus that regulates the body's heat production and heat loss and sets the core temperature at around 37°C (98.6°F), with slight fluctuations (1°F) during a daily cycle. Fever is initiated when a circulating substance called **pyrogen** resets the hypothalamic thermostat to a higher setting, which in turn signals the musculature to increase heat production and the peripheral arterioles to decrease heat loss through vasoconstriction (see feature 12.5). Fevers range in severity from low grade (37.7°–38.3°C or 100°–101°F) to moderate (38.8°–39.4°C or 102°–103°F) to high (40.0°–41.1°C or 104°–106°F). Fevers of 106°F and above are very dangerous, but they are rare and are usually due to organic disease, not infectious agents. Pyrogenic substances are described as *exogenous* (coming from outside the body) or *endogenous* (originating internally). Exogenous pyrogens are products of infectious agents such as viruses, bacteria, protozoa, and fungi. One well-characterized exogenous pyrogen is endotoxin, the lipopolysaccharide found in the cell walls of gram-negative bacteria. Blood, blood products, vaccines, or injectable solutions may contain exogenous pyrogens. Products labelled "nonpyrogenic" have been found through testing to contain no fever-causing contaminants. Endogenous pyrogens are liberated by monocytes, neutrophils, and macrophages during the process of phagocytosis, and appear to be a natural part of the immune response. One of the most potent of these is interleukin-1, a product of macrophages that have been stimulated by viruses, endotoxin, or chemical mediators.

pyrogen (py'-roh-jen) Gr. *pyr,* fire, and *gennan,* produce. As in funeral pyre and pyromaniac.

Feature 12.5 The Enigma of Fever

Fever is such a prevalent reaction that it takes several pages of a medical dictionary just to tabulate the diseases that involve it. For thousands of years, people believed fever was part of an innate protective response. Hippocrates offered the idea that it was the body's attempt to burn off a noxious agent. Sir Thomas Sydenham wrote in the seventeenth century: "Why, fever itself is Nature's instrument!" So widely held was the view that fever could be therapeutic that pyretotherapy (treating disease by inducing an intermittent fever) was once used to treat syphilis, gonorrhea, leishmaniasis (a protozoan infection), and cancer. This attitude fell out of favor when drugs for relieving fever (aspirin) first came into use in the early 1900s, and an adverse view of fever began to dominate.

Our Changing Views of Fever In recent times, the medical community has returned to the original concept of fever as more healthful than harmful. Experiments with vertebrates indicate that fever is a universal reaction, even in cold-blooded animals such as lizards and fish. After all, why would such a response be retained in evolution unless it was physiologically beneficial? A study with *febrile* mice and frogs indicated that fever increases the rate of antibody synthesis. Work with tissue cultures showed that increased temperatures stimulate the activities of T cells and increase the effectiveness of interferon. Artificially infected rabbits and pigs allowed to remain febrile survive at a higher rate than those given suppressant drugs. Fever appears to enhance phagocytosis of staphylococci by neutrophils in guinea pigs and humans.

Modern Use of Fever Therapy Fever may be induced by injecting pyrogens or applying heat (artificial fever). The latter has been effective in treating the cutaneous lesions of sporotrichosis, paracoccidioidomycosis, and leishmaniasis. The discovery that some types of cancer cells are sensitive to heat has led to the use of heat on tumors.

Hot and Cold. Why Do Chills Accompany Fever? Fever almost never occurs as a single response; it is usually accompanied by chills. What causes this oddity—that a person flushed with fever periodically feels cold and trembles uncontrollably? The explanation lies in the natural physiological interaction between the thermostat in the hypothalamus and the temperature of the blood. As long as the blood's temperature is detected by the hypothalamus as cooler than the set point of the thermostat, mechanisms for increasing the body temperature are brought into play. For example, if the thermostat has been set (by pyrogen) at 102°F, but the blood temperature is 99°F, the muscles are stimulated to contract involuntarily (shivering) as a means of producing heat. In addition, the vessels in the skin constrict, creating a sensation of cold, and the piloerector muscles in the skin cause "goose bumps" to form. As long as the two temperatures are different, the chills will continue. They can be equalized either by allowing the fever to continue or by suppressing it, and when the two temperatures are the same, the chills will abate.

febrile (fee'-bril) L. *febris,* fever. Feverish.

Tape Here

Tape Here

NO POSTAGE NECESSARY IF MAILED IN THE UNITED STATES

BUSINESS REPLY MAIL

FIRST- CLASS MAIL PERMIT NO. 47 EDWARDSVILLE, KS

POSTAGE WILL BE PAID BY ADDRESSEE

MEDI-SIM, INC.
P.O. BOX 13267
EDWARDSVILLE KS 66113-9989

Benefits of Fever The association of fever with infection strongly suggests that it serves a beneficial role, a view still being debated but gaining acceptance. Aside from its practical and medical importance as a sign of a medical problem, increased body temperature also appears to have some physiologic benefits: (1) Fever inhibits multiplication of temperature-sensitive microorganisms such as the poliovirus, cold viruses, herpes zoster virus, systemic and subcutaneous fungal pathogens, *Mycobacterium* species, and the syphilis spirochete. (2) Fever impedes the nutrition of bacteria by reducing the availability of iron. It has been demonstrated that during fever, the macrophages stop releasing their iron stores, and this could retard several enzymatic reactions needed for bacterial growth. (3) Fever increases metabolism and stimulates immune reactions and naturally protective physiological processes. It speeds up hematopoiesis, phagocytosis, and specific immune reactions.

Treatment of Fever With this revised perspective on fever, whether to suppress it or not can be a difficult decision. Some advocates feel that a slight to moderate fever in an otherwise healthy person should be allowed to run its course, in light of its potential benefits and minimal side effects. Others believe that the possible benefits of fever do not outweigh its discomfort and potential harm. All medical experts do agree that high and prolonged fevers, or fevers in patients with cardiovascular disease, seizures, and respiratory ailments, are risky and must be treated immediately with suppressant drugs. The classic therapy for fever is an *antipyretic* drug such as aspirin or acetaminophen (Tylenol) that lowers the setting of the hypothalamic center and restores normal temperature. Any physical technique that increases heat loss (cold water or alcohol baths, for example) can also help reduce the core temperature.

Phagocytes—The Ever Present Busybodies of Inflammation and Specific Immunity

By any standard, a phagocyte represents an impressive piece of living machinery, meandering through the tissues to seek, capture, and destroy a target. The general activities of phagocytes are: (1) to survey the tissue compartments and discover microbes, particulate matter (dust, carbon particles, antigen-antibody complexes), and injured or dead cells; (2) to ingest and eliminate these materials; and (3) to extract immunogenic information (antigens) from foreign matter (see chapter 13). It is generally accepted that all cells have some capacity to engulf materials, but the *professional phagocytes* do it for a living. The three main types of phagocytes are neutrophils, monocytes, and macrophages.

Granulocytic Phagocytes: Neutrophils and Eosinophils

As previously stated, neutrophils are general-purpose phagocytes that react early in the inflammatory response to bacteria and other foreign materials and to damaged tissue (figure 12.20).

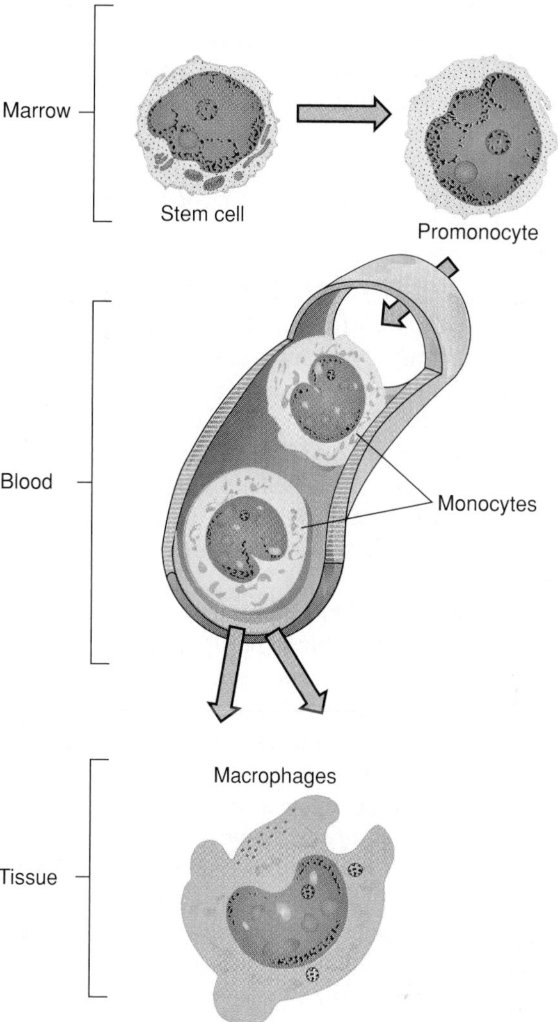

Figure 12.22 The developmental stages of monocytes/macrophages. The cells progress through maturational stages in the bone marrow and peripheral blood. Once in the tissues, macrophages may remain nomadic or take up residence in a specific organ.

A common sign of bacterial infection is a high neutrophil count in the blood (neutrophilia), and neutrophils are also a primary component of pus. Eosinophils are attracted to sites of parasitic infections and antigen-antibody reactions, though they play only a minor phagocytic role.

Macrophage: King of the Phagocytes

After emigrating out of the bloodstream into the tissues, monocytes are transformed by various inflammatory mediators into macrophages. This process is marked by an increase in size and by enhanced development of lysosomes and other organelles (figure 12.22). At one time, macrophages were classified as either fixed (adherent to tissue) or wandering, but this terminology can be misleading. All macrophages retain the capacity to move about. Whether they reside in a specific organ or wander depends upon their stage of development and the immune stimuli they receive. Specialized macrophages called *histiocytes* migrate to a certain tissue and remain there during their lifespan.

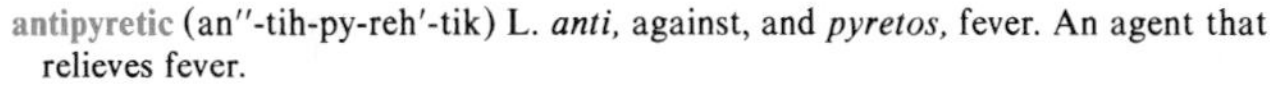

antipyretic (an″-tih-py-reh′-tik) L. *anti*, against, and *pyretos*, fever. An agent that relieves fever.

Examples are alveolar (lung) macrophages, the Kupffer cells in the liver, Langerhans cells in the skin (figure 12.23), and macrophages in the spleen, lymph nodes, bone marrow, kidney, bone, and brain. Other macrophages do not reside permanently in a particular tissue and drift nomadically throughout the RES. Not only are macrophages dynamic scavengers, but they also process foreign substances and prepare them for reactions with B and T lymphocytes (see chapter 13).

Mechanisms of Phagocytic Discovery, Engulfment, and Killing

Although the term phagocytosis literally means the engulfment of particles by cells, phagocytes actually endocytose both particulate and liquid substances. Because phagocytes unleash a veritable medicine chest of antimicrobial chemicals to attack and destroy the engulfed material, phagocytosis is also more than just the physical process of engulfment. The events in phagocytosis include chemotaxis, ingestion, phagolysosome formation, destruction, and excretion (figure 12.24).

Chemotaxis and Ingestion Phagocytes move into a region of inflammation with a deliberate sense of direction, attracted by a gradient of stimulant products from the parasite and host tissue at the site of injury. Phagocytes on the scene of an inflammatory reaction are primed to make immediate physical contact, often by cornering the target against the fibrous network of connective tissue or upon the wall of blood and lymphatic vessels. Somewhat more elegant is *opsonization* (discussed again in chapter 13), a process in which compounds such as antibodies coat the surface of microorganisms, thereby facilitating recognition and engulfment. Once the phagocyte has made contact with its prey, it extends pseudopods that enclose the cells or particles in a pocket and internalize them in a vacuole called a *phagosome.*

Phagolysosome Formation and Killing In a short time, *lysosomes* migrate to the scene of the phagosome and fuse with it to form a *phagolysosome.* Other granules containing antimicrobial chemicals are released into the phagolysosome, forming a potent brew designed to poison and then dismantle the ingested material. The destructiveness of phagocytosis is evident by the death of bacteria within 30 minutes after contacting this battery of antimicrobial substances.

Destruction and Elimination Systems Two separate systems of destructive chemicals await the microbes in the phagolysosome. One system requires active, aerobic metabolism, and the other does not. The oxygen-dependent system elaborates several substances that were described in chapters 6 and 9. Myeloperoxidase, an enzyme found in granulocytes, forms halogen ions (OCl^-) that are strong oxidizing agents. Other products of oxygen metabolism such as hydrogen peroxide, the superoxide anion (O_2^-), activated or so-called singlet oxygen ($^1O_2$), and the hydroxyl free radical (.OH) separately and together have formidable killing power. Mechanisms that come into play in the absence of oxygen are the liberation of lactic acid and lysozyme. Cationic proteins that injure bacterial cell membranes and a

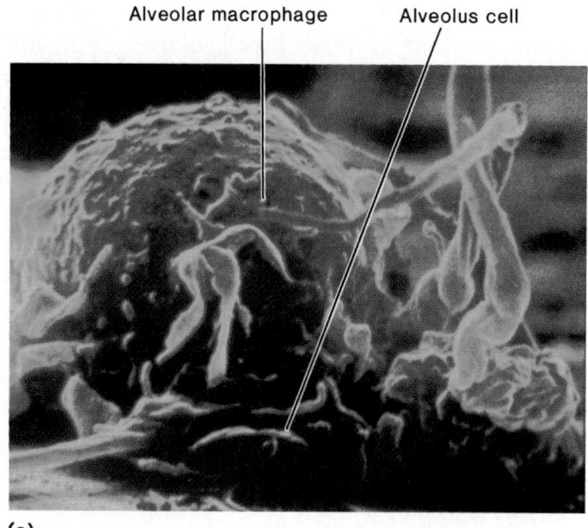

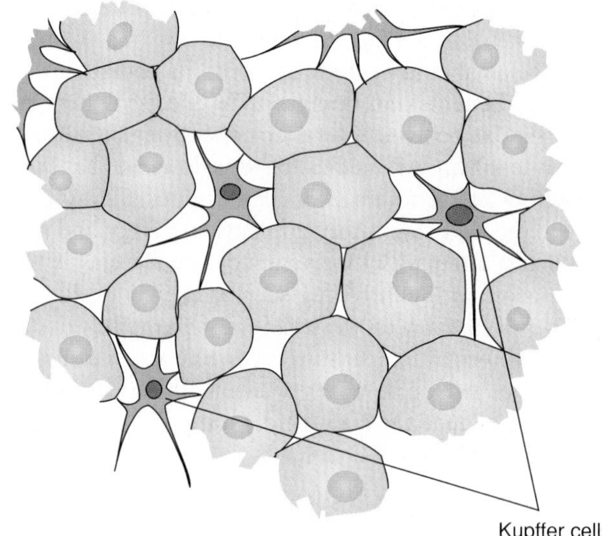

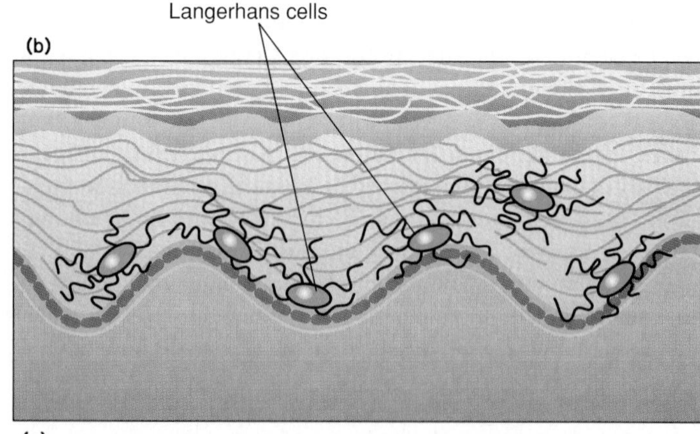

Figure 12.23 Sites containing macrophages. (*a*) Scanning electron micrograph view of a lung with an alveolar macrophage. (*b*) Liver tissue with Kupffer cells. (*c*) Langerhans cells deep in the epidermis.

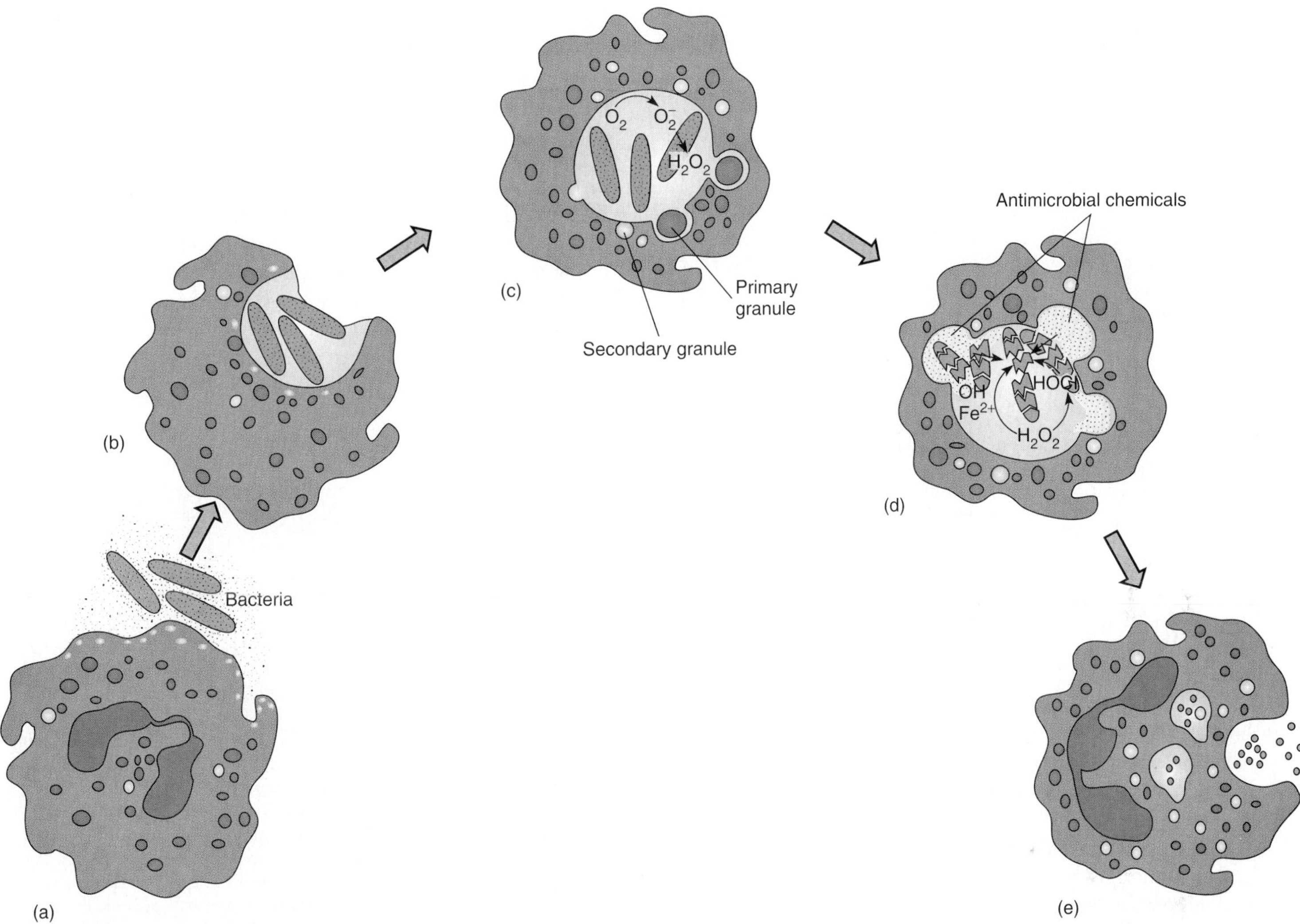

Figure 12.24 The phases in phagocytosis. (*a*) Chemotaxis. (*b*) Contact and ingestion (forming a phagosome). (*c*) Formation of phagolysosome (granules fuse with phagosome). (*d*) Killing, digestion of the microbe. (*e*) Release of debris.

number of proteolytic, lipolytic, and other hydrolytic enzymes complete the job of destruction and dismantling. The small bits of undigestible debris are exocytosed and released.

Contributors to the Body's Chemical Immunity

Interferon: Antiviral Compound and Immune Stimulant

Interferon (IF) was described in chapter 10 as a small protein produced naturally by certain white blood and tissue cells that has sometimes been used in therapy against viral infections and cancer. Although the interferon system was originally thought to be directed exclusively against viruses, it is now known to be involved also in defenses against other microbes and in immune regulation and intercommunication. Three major types are: *alpha interferon,* a product of lymphocytes that is induced by infection with viruses, bacteria, and other agents; *beta interferon,* a product of fibroblasts, epithelial cells, and macrophages in response to viruses; and *gamma interferon,* a product of T cells that functions in immune regulation. The first two types are important in nonspecific suppression of viral infection, while gamma interferon is part of a specific response to antigens.

Characteristics of Antiviral Interferon The binding of a virus or other inducer to the receptors of an infected cell sends a signal into the cell nucleus that activates the genes coding for interferon (figure 12.25). As interferon is synthesized, it is rapidly secreted by the cell into the extracellular spaces. The action of antiviral interferon is indirect—it does not kill or inhibit the virus directly. After diffusing to nearby, uninfected cells and entering them, IF activates a gene complex that codes for another protein. This second protein, not interferon itself, interferes with the multiplication of viruses. Interferon is not virus-specific, so IF raised to one type of virus will also protect against other types. Because this second protein is the direct inhibitor of virus, it could be a valuable treatment for AIDS and other virus infections (see chapter 21).

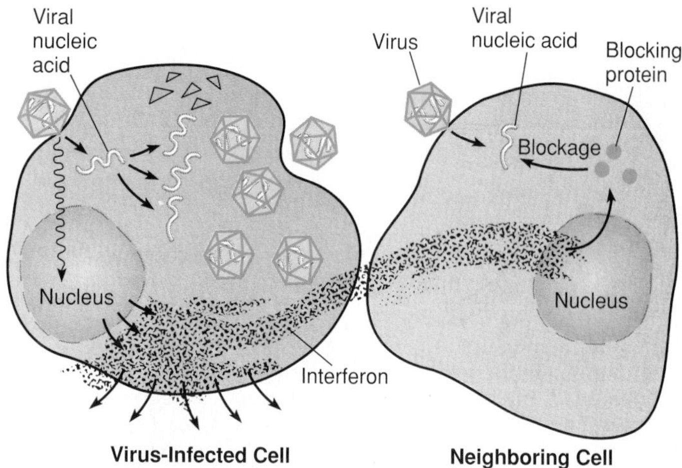

Figure 12.25 The antiviral activity of interferon. When a cell is infected, its nucleus is triggered to transcribe and translate the interferon (IF) gene. The interferon diffuses out of the infected cell into nearby (uninfected) cells where it enters the nucleus. Here IF activates a gene for synthesizing a peptide that blocks viral replication. Note that the original cell is not protected by IF and that IF does not prevent viruses from invading the protected cells.

Other Roles of Interferon Interferons are also important cytokines. Alpha interferon produced by T lymphocytes activates a subset of cells called natural killer (NK) cells. One type of beta interferon plays a role in the maturation of B and T lymphocytes and in inflammation. Gamma interferon inhibits cancer cells, stimulates B lymphocytes, and aids in a number of other specific immune reactions.

Complement: A Versatile Backup System

Among its many overlapping functions, the immune system has another complex and multiple-duty system called **complement (C factor)** that, like inflammation and phagocytosis, is brought into play at several levels. The complement system, named for its property of "completing" immune reactions, consists of 20 blood proteins that work in concert to destroy bacteria and certain viruses. The sources of complement factors are liver hepatocytes, lymphocytes, and monocytes. Some knowledge of this important system will be crucial to several later topics.

To understand how complement functions, it is helpful to be acquainted with the concept of a cascade reaction. A cascade reaction is a sequential physiologic response like that of blood clotting, in which the first substance in a chemical series activates the next substance, which activates the next, and so on, until a desired end product is reached. Complement exhibits two schemes, the *classical pathway* and the *alternative pathway,* which differ in the stimuli that initiate them and in several reactions. The two pathways merge at the final stages into a common pathway with a similar end result (figure 12.26). These factors were numbered C1–C9 in order of their discovery, but unfortunately, the numbering of the earlier proteins (C1–C4) does not necessarily coincide with their order of activation (table 12.1).

Overall Stages in the Complement Cascade In general, the complement cascade includes the three stages of *initiation, amplification,* and *membrane attack.* At the outset, an initiator (microbes or parts of microbes, chemical mediators, and antibodies; table 12.2) reacts with the first complement chemical, and this propels the reaction on its cascading course. All C factors except for the last five exist in an inactive (proenzyme) state and must be activated by an enzyme that cleaves them into two fragments. According to this pattern, activating the first factor in the series turns it into an enzyme that cleaves the second factor. This converts the second factor into an active enzyme that cleaves the third factor, and so on (figure 12.26*a*). Details of the pathways differ, but whether classical or alternative, the functioning end product is a large, multienzyme *membrane attack complex* that can digest holes in the cell membranes of bacteria, cells, and enveloped viruses, thereby destroying them (figure 12.26*a*). The two pathways are described in feature 12.6.

Feature 12.6 How Complement Works

The **classical pathway** is a part of the specific immune response covered in chapter 13. It is initiated when antibody, called complement-fixing antibody, attaches to antigen on the surface of a membrane. The first chemical, C1, is a large complex of three molecules, C1q, C1r, and C1s (see table 12.1). When the C1q subunit has recognized and bound two or more antibodies, the C1r subunit cleaves the C1s proenzyme, and an activated enzyme emerges. During amplification, the C1s enzyme has as its primary targets proenzymes C4 and C2. Through the enzyme's action, C4 is converted into C4a and C4b, and C2 is converted into C2a and C2b. C4b and C2a fragments remain attached as an enzyme, C3 convertase, whose substrate is factor C3. The cleaving of C3 yields subunits C3a and C3b. C3b has the property of binding strongly with the cell membrane in close association with the component C5, and it also forms an enzyme complex with C4b–C2a that converts C5 into two fragments, C5a and C5b. C5b will form the nucleus for the membrane attack complex. This is the point at which the two pathways merge. From this point on, C5b reacts with C6 and C7 to form a stable complex inserted in the membrane. Addition of C8 to the complex causes the polymerization of several C9 molecules into a giant cylindrical membrane attack complex that bores ring-shaped holes in the membrane, which lyse the target cell or destroy the virus (figure 12.26*b*).

The **alternative pathway,** sometimes called the **properdin pathway,** is not specific to a particular microbe, because it can be initiated by a wide variety of microbes, tumors, and cell walls (but not antibodies). It requires a different group of serum proteins—factors B, D, and P (properdin), C3b, and magnesium in the initiation and amplification phases rather than C1, C2, or C4 components (see table 12.1). The remainder of the steps occur as in the classic pathway. The principal function of the alternative system is to provide a rapid method for lysing foreign cells (especially gram-negative bacteria) and viruses in the absence of specific immunity.

You will notice that at many of the steps of the classical pathway, two molecules are given off. One of these continues in the formation of the membrane attack complex, and the other (C2b, C24a, C3a, or C5a) goes on to become a chemical mediator or stimulant of inflammation and other immune reactions. Complement components also augment phagocytosis (the opsonizing type), cause immune adherence or "stickiness" of erythrocytes, leukocytes, and platelets, and are useful in chemotaxis. Complement may participate in inflammation and allergy by causing liberation of vasoactive substances from mast cells and basophils. C3a and C5a are so potent in this response that injecting only one quadrillionth of a gram elicits an immediate flare-up at the site.

properdin (proh'-pur-din) L. *pro,* before, and *perdere,* to destroy.

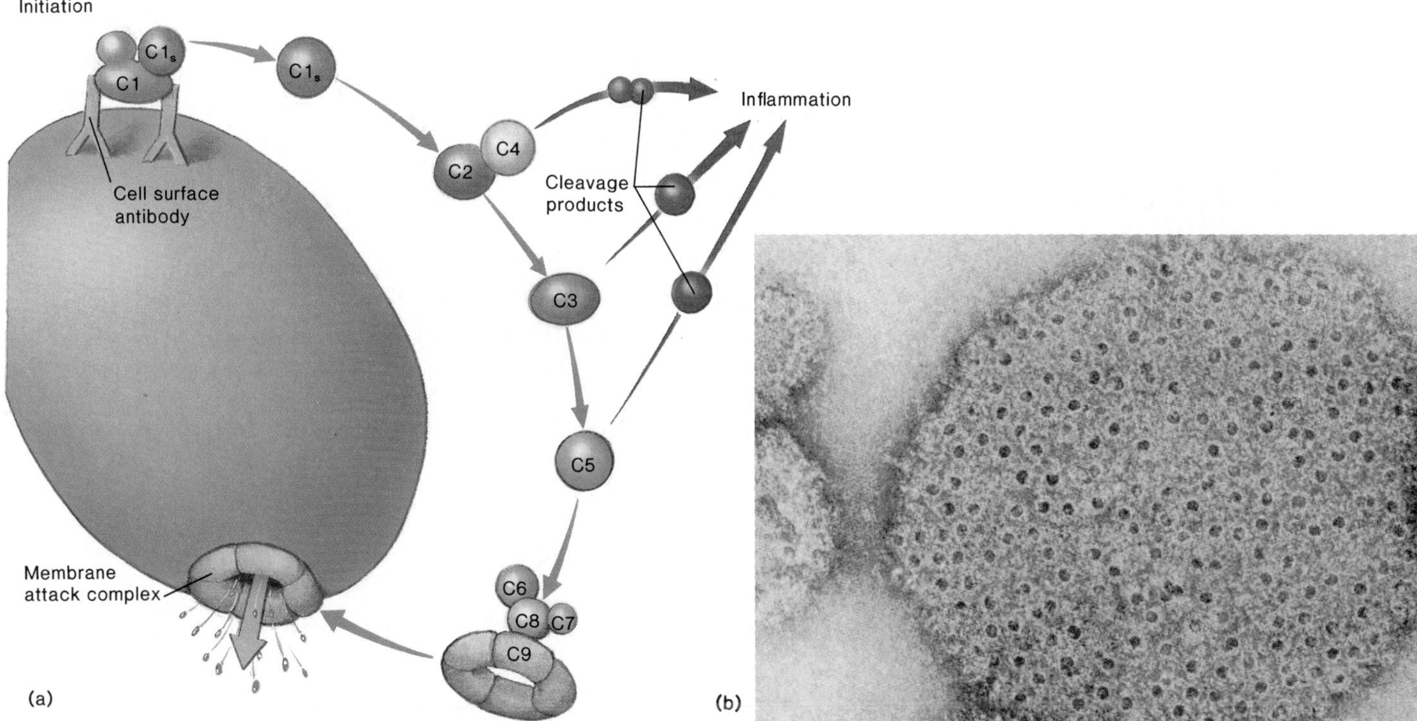

Figure 12.26 The cascade reaction of the classic complement pathway. (*a*) This fine-tuned molecular chain reaction involves about 20 factors that exist in the inactive (proenzyme) state. When a factor is cleaved by an enzyme (another C component), it converts to an active counterpart that itself cleaves the next component. The final product of the enzyme cascade is a large, donut-shaped enzyme complex that punctures through the cell membrane of the microbe and causes its lysis. (*b*) An electron micrograph (×187,000) of a cell reveals multiple puncture sites over its surface. The lighter, ringlike structures are the actual enzyme complex.

Table 12.1 Complement Proteins

Pathway	Component
Classic (Initial Portion)	C1q C1r C1s C4 C2 C3
	Membrane Attack Components (Common to Both Pathways) C5 C6 C7 C8 C9
Alternative or Properdin (Initial Portion)	Properdin Factor B Factor D Factor C3b and Mg

Table 12.2 Substances That Activate the Complement Pathways

Activators in the Classical Pathway	Activators in the Alternative Pathway
Complement-fixing antibodies: IgG, IgM	Cell wall components; e.g., yeast and bacteria
	Viruses; e.g., influenza virus
Bacterial lipopolysaccharide	Parasites; e.g., *Schistosoma*
Pneumococcal C-reactive protein	Fungi; e.g., *Cryptococcus*
Retroviruses	Some tumor cells
Polynucleotides	X-ray opaque media, dialysis membranes
Mitochondrial membranes	

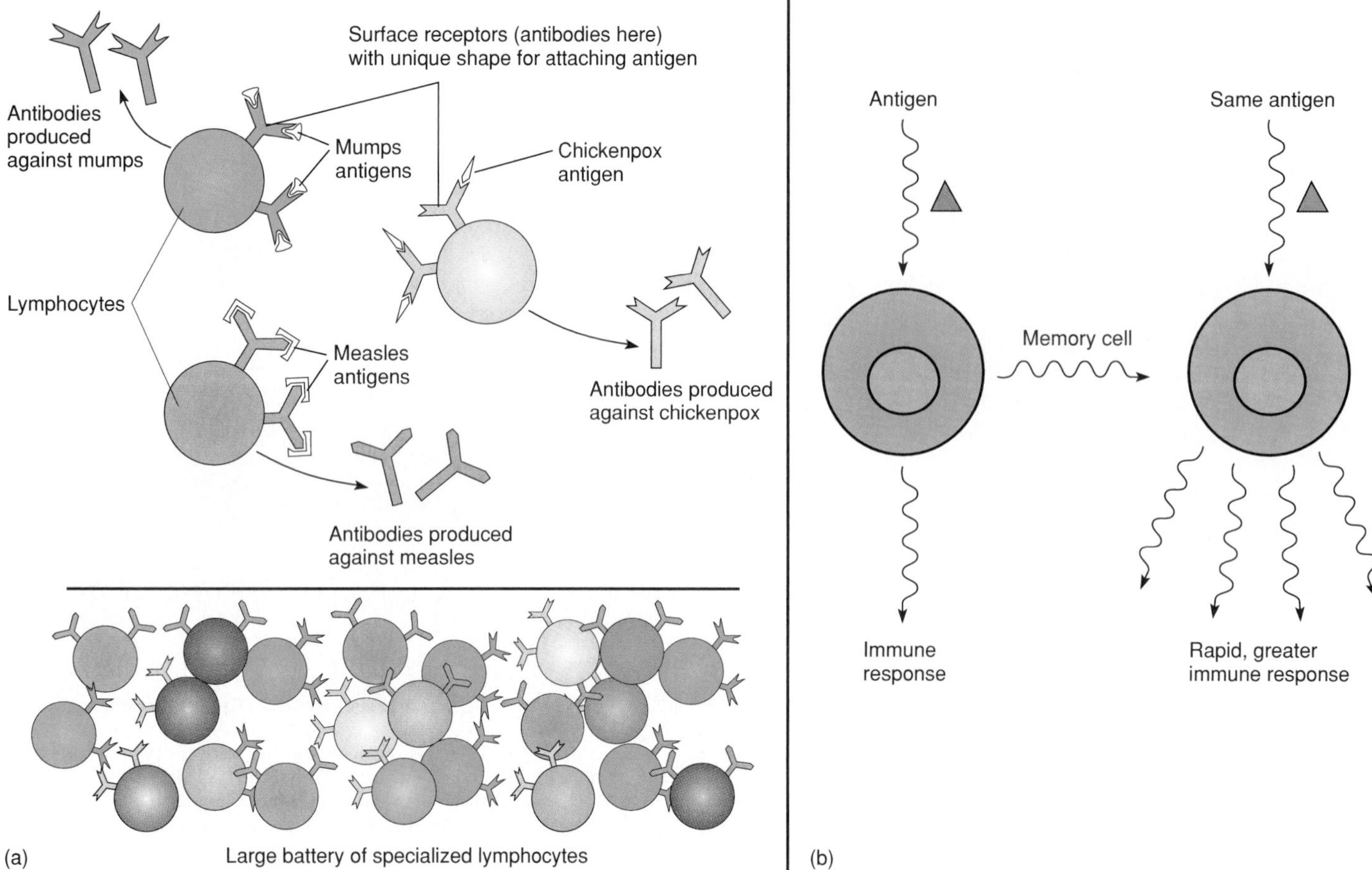

Figure 12.27 The basic characteristics of acquired immunity. (*a*) Specificity of lymphocyte receptors to antigen. (*b*) Memory: First contact with antigen creates a memory and quick recall upon second and other future contacts with that antigen.

Specific Immunities—The Third and Final Line of Defense

In instances where host barriers and nonspecific defenses fail to control an infectious agent, a person with a normal functioning immune system has an extremely substantial mechanism to resist the pathogen—the third, specific line of immunity. This aspect of immunity is the resistance developed after contracting childhood ailments such as chickenpox or measles that provides long-term freedom from future attacks. This sort of immunity is not innate, but adaptive; it is acquired only after an immunizing event such as an infection. The absolute need for acquired or adaptive immunity is impressively documented in children who genetically lack this system or in AIDS patients who have lost it. Even with heroic measures to isolate the patient, combat infection, or restore lymphoid tissue, the victim is constantly vulnerable to life-threatening infections.

Acquired specific immunity arises from a dual system that we have previously mentioned—the B and T lymphocytes. During fetal development, these lymphocytes undergo a selective process that specializes them for reacting only to one specific antigen. During this time, **immunocompetence,** the ability to react with myriads of different foreign substances, develops. An infant is born with the theoretical potential to acquire millions of different immunities.

Two features that most characterize this third line of defense are **specificity** and **memory.** Unlike mechanisms such as anatomic barriers or phagocytosis, acquired immunity is highly selective. For example, the antibodies produced during an infection against the chickenpox virus will function only against that virus and not against the measles virus (figure 12.27*a*). The property of memory pertains to the brisk mobilization of lymphocytes that have been programmed to "recall" their first engagement with the invader, and rush to the attack once again (figure 12.27*b*). The next section will summarize the highlights of this complex and fascinating response, which will be covered more extensively in chapter 13.

The General Scheme of Adaptive Immunity

The means by which humans acquire immunities can be nicely encapsulated within four interrelated categories: active, passive, natural, and artificial.

Active immunity occurs when an individual receives an immune stimulus (antigen) that activates the B and T cells and causes that person to produce immune substances such as antibodies. Active immunity is marked by several characteristics: (1) It is an essential attribute of an immunocompetent individual; (2) it

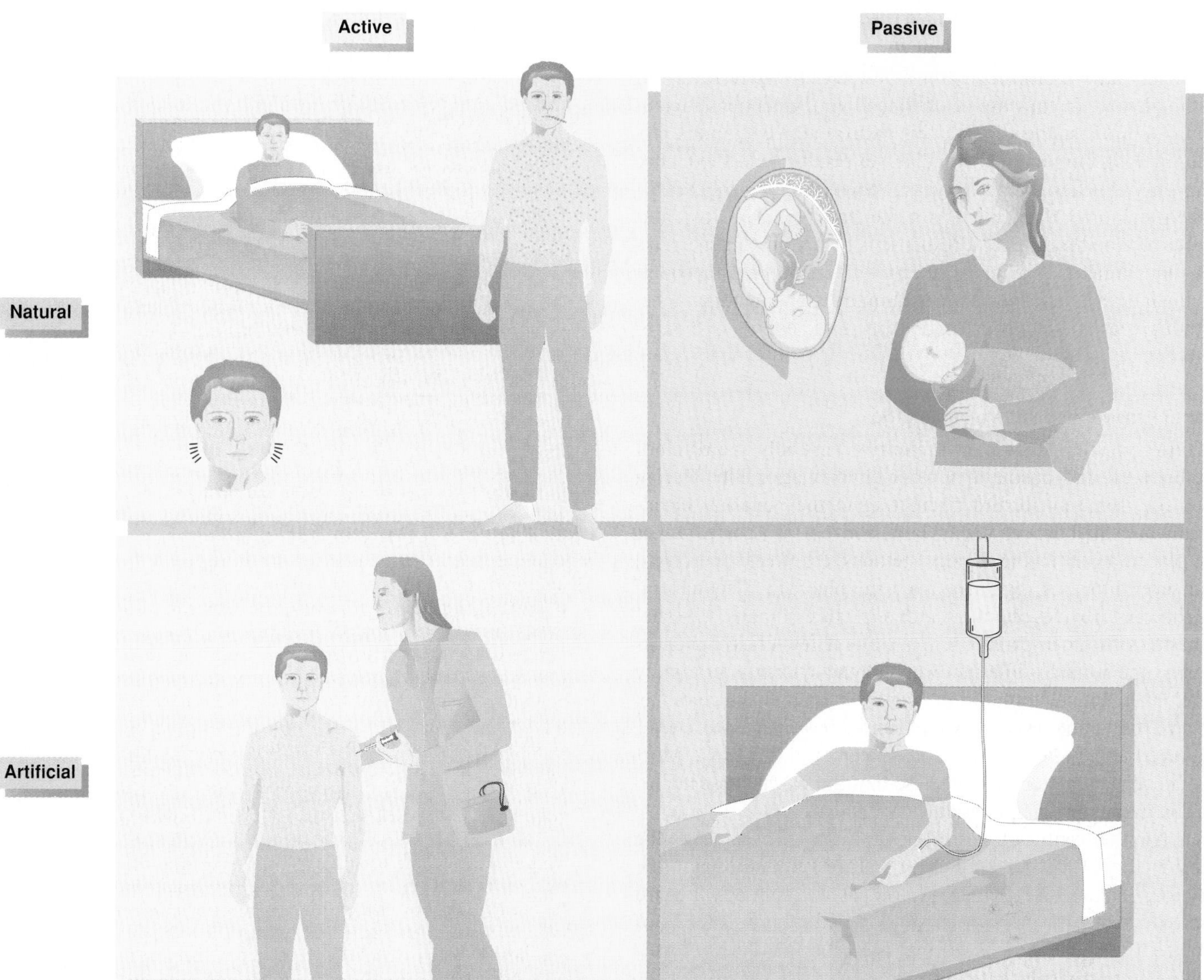

Figure 12.28 Categories of acquired immunities. Natural immunities, which occur during the normal course of life, may be either active (through acquiring an infection and recovering) or passive (antibodies donated from mother to child). Artificial immunities are acquired through medical practices and are either active (vaccinations that stimulate one's own immunities by exposure to antigens) or passive immune therapy, in which a serum containing antibodies from another source is given to the patient.

creates a memory that renders the person ready for quick action upon reexposure to that same antigen; (3) it requires several days to develop; and (4) it lasts for a relatively long time, sometimes for life. Active immunity may be stimulated by natural or artificial means.

Passive immunity occurs when an individual receives immune substances (antibodies) that were produced actively in the body of another human or animal (donor). The recipient is protected for a time even though he or she has not had prior exposure to the antigen. It is characterized by: (1) lack of memory to the original antigen, (2) lack of production of new antibodies against that disease, (3) immediate onset of protection, and (4) short-term effectiveness, because antibodies have a limited period of function, and ultimately, the recipient's body disposes of them. Passive immunity may also be natural or artificial in origin.

Natural immunity encompasses any immunity acquired during the normal biological experiences of an individual, not involving medical intervention.

Artificial immunity is protection from infection obtained through medical procedures. This type of immunity is induced by immunization with vaccines and immune serum.

Figure 12.28 illustrates the various possible combinations of acquired immunities.

Natural Active Immunities: Getting the Infection

After recovering from infectious disease, a person may be actively resistant to reinfection for a period that varies according to the disease. In the case of childhood viral infections such as measles, mumps, and rubella, this natural active stimulus provides lifelong immunity. Other diseases result in a less extended immunity of a few months to years (pneumococcal pneumonia, shigellosis), and reinfection is quite possible. One fortunate aspect of the natural active response is that a subclinical infection can stimulate protective immunity. This probably accounts for the fact that some persons are immune to an infectious agent without ever having been noticeably infected or vaccinated with it.

Natural Passive Immunity: Mother to Child

Natural, passively acquired immunity occurs only as a result of the prenatal and postnatal mother-child relationship. During fetal life, some antibodies circulating in the maternal bloodstream are small enough to pass or be actively transported across the placenta. Antibodies against tetanus, diphtheria, pertussis, and several viruses regularly cross the placenta. This natural mechanism provides an infant with a mixture of many maternal antibodies that can protect it for the first few critical months outside the womb, while its own immune system is gradually developing active immunities. Depending upon the microbe, passive protection lasts anywhere from a few months to a year, but eventually, the infant's body clears the antibody. Most childhood vaccinations are timed so that there is no lapse in protection against common childhood infections.

Another source of natural passive immunity comes to the baby by way of mother's milk (see feature 12.7). Although the human infant acquires 99% of natural passive immunity in utero and only about 1% through nursing, the milk-borne antibodies provide a special type of intestinal protection that is not forthcoming from transplacental antibodies.

Artificial Immunity: Immunization

Immunization is any clinical process that produces immunity in a subject. Because it is often used to give advance protection against infection, it is also called immunoprophylaxis. The use of these terms is sometimes imprecise, thus it should be stressed that active immunization, in which a person is administered antigen, is synonymous with vaccination, and that passive immunization, in which a person is given antibodies, is a type of immune therapy.

Vaccination: Artificial Active Immunization

The term **vaccination** originated from the Latin word *vacca* (cow), because the cowpox virus was used in the first preparation for active immunization against smallpox (see feature 13.6). Vaccination exposes a person to a specially prepared microbial (antigenic) stimulus, which then triggers the immune system to produce antibodies and lymphocytes to protect the person upon future exposure to that microbe. As with natural active immunity, the degree and length of protection varies. Commercial vaccines are currently available for about 24 diseases. Methods of vaccine antigen selection and modes of vaccination are discussed more fully in the applications section of chapter 13.

Feature 12.7 Breast-Feeding: The Gift of Antibodies

An advertising slogan from the past claims that cow's milk is "nature's most nearly perfect food." One could go a step further and assert that human milk is nature's *perfect* food for young humans. Clearly, it is loaded with the required nutrients, not to mention being available on demand from a readily portable, hygienic container that does not require refrigeration. But there is another, perhaps even greater, benefit. During lactation, the breast becomes a site for the proliferation of lymphocytes which produce IgA, a special class of antibody that protects the mucosal surfaces from local invasion by microbes. The very earliest secretion of the breast, a thin, yellow milk called *colostrum,* is very high in IgA. These antibodies form a protective coating in the gastrointestinal tract of a nursing infant that guards against infection by a number of enteric pathogens (*E. coli, Salmonella,* poliovirus, rotavirus). Protection at this level is especially critical because an infant's own IgA and natural intestinal barriers are not yet developed. As with immunity in utero, the necessary antibodies will be donated only if the mother herself has active immunity to the microbe through a prior infection or vaccination.

The benefits of nursing have been known since time immemorial. In the Middle Ages, women of privilege who did not want to nurse often bore huge families of 20 or more children and dispatched each to the care of a wet-nurse for its first two years of life. In an era when infant mortality was extremely high, the children often had little contact with the birth mother until their survival was assured. In more recent times, the ready availability of artificial formulas and the changing life-styles of women have reduced the incidence of breast feeding. Where adequate hygiene and medical care prevail, bottle-fed infants get through the critical period with few problems, because the foods given them are relatively sterile and they have received protection against some childhood infections in utero. A compromise solution for some women is to nurse for the first month or two when it is most beneficial to the infant. Mothers in developing countries with untreated water supplies or poor medical services are strongly discouraged from using prepared formulas, because mother's milk has the double benefit of providing protective antibodies and being less subject to contamination.

Pediatricians have sought a way to provide intestinal antibodies to infants whose mothers cannot nurse them. Preparations of sterile cow colostrum have been tried, but the sterilization process destroyed the antibodies. Certain organizations advocate that lactating mothers extract excess breast milk and freeze it for mothers who cannot nurse. It has been shown that preemies (premature infants) are extremely susceptible to necrotizing enterocolitis and that formula with added antibodies reduced the incidence of this disease.

Immunotherapy: Artificial Passive Immunization

In **immunotherapy,** a patient at risk for acquiring a particular infection is administered a preparation that contains specific antibodies against that infectious agent. In the past, these therapeutic substances were obtained by vaccinating animals (horses in particular), then taking blood, and extracting the serum. Horse serum is now used only in limited situations because of the potential for hypersensitivity to it. More often used are pooled human serum from donor blood (gamma globulin) and **immune serum globulins** containing high quantities of antibodies. Immune serum globulins are used to protect persons who have been exposed to hepatitis, measles, and rubella. More specific immune serum, obtained from patients recovering from a recent infection, is useful in preventing and treating hepatitis B, rabies, pertussis, and tetanus (see chapter 13).

Chapter Review with Key Terms

Host Defenses

The Three Levels of Host Defenses

The **first line of defense** is comprised of barriers that block the pathogen at portal of entry; the **second line of defense** includes protective cells and fluids in tissues; and the **third line of defense** includes specific immune reactions with microbes. Each comes into play as necessary; the third line of defense is required for survival.

First Line Barriers: Physical barriers are skin, mucous membranes, tears, cilia, coughing, diarrhea; chemical barriers are fatty acids, lysozyme in tears, saliva, gastric acidity, sebaceous secretions; genetic barriers involve lack of susceptibility to an infectious agent due to genetic specialization of animal and microbe.

Immunity/Immunology

Immunity: The specific resistance one acquires to an infectious agent. Immune system involved in very specific functions: in **surveillance,** specialized **white blood cells (WBCs)** sift through tissues and fluids of the body to evaluate the state of the **markers** or receptors (molecules on surface of cells) there; in **recognition,** WBCs are responsible for detecting and differentiating those markers that are normal (**self**) and those that are **foreign** or **antigens (nonself)**; WBCs mount a campaign to destroy and remove the foreign material, while self is usually unaffected.

Immune System: A diffuse network of cells, fibers, chemicals, fluids, tissues, and organs that permeates the entire body. At the cellular level, various separate but interdigitating compartments come into play: **reticuloendothelial system (mononuclear phagocyte),** a system of fibers and phagocytes that forms a continuous network around all tissues and organs; **extracellular fluid (ECF),** a liquid environment in which all cells are bathed; **lymphatic system,** a series of vessels and organs that carry lymph away from tissues; and the **bloodstream,** which circulates blood in all organs. The constant communication among the compartments ensures that a reaction in one will soon be felt in another.

Circulatory System: Blood and Lymphatics

Composition of Whole Blood: **Plasma** is a clear liquid separated from blood cells (formed elements); plasma is a mixture of numerous nutrients, ions, gases, hormones, antibodies, albumin, and waste products dissolved in water. **Serum** is plasma minus the clotting factors. **Blood cells** are formed by **hemopoiesis** in particular bone marrow sites. From **stem cells,** three main lines of cells are differentiated—white blood cells (**leukocytes**), red blood cells (**erythrocytes**), and megakaryocytes that give rise to **platelets.**

Functions of Leukocytes: Leukocytes are most important in immune function; the two lines are: **granulocytes (polymorphonuclear leukocytes)** and **agranulocytes (mononuclear leukocytes)**.

Granulocytes were named according to their appearance in Wright stain, showing distinct granules in cytoplasm; **neutrophils,** the most numerous WBCs in circulation, are phagocytes; **eosinophils,** the fourth most common in circulation, function in worm and fungal infections; **basophils,** the least common in circulation, are involved in mediation of allergy, along with tissue **mast cells.**

Agranulocytes show no noticeable granules in Wright stain; **monocytes** are large, third most common in circulation, important as blood phagocytes, give rise to **macrophages** in tissues; **lymphocytes,** second most common in circulation, important in specific immune reactions. Two main types of lymphocytes are **B cells,** which produce **antibodies** as part of **humoral immunities,** and **T cells,** which participate in **cell-mediated immunities** (CMI).

Leukocyte Behavior: Motile by ameboid motion; can insert between the endothelial cells of small blood vessels (**diapedesis**) and enter surrounding tissues. Respond to tissue injury or infection by migrating toward chemical signals (**chemotaxis**).

Lymphatic System

One-way system of vessels; begins as fine capillaries in tissues, gradually joining together in larger vessels that eventually drain into the blood system; transport a special fluid, **lymph,** that contains serum components and white blood cells; organs include **lymph nodes,** special lymph filters that contain large numbers of lymphocytes and are common sites of immune challenge; **spleen,** lymphatic organ that filters blood and serves as source of immunologic cells; **thymus,** site of T-cell maturation; miscellaneous tissue, including tonsils, gastrointestinal-associated lymphoid tissue (**GALT**), and Peyer's patches, an important local immune response system.

Generalized Immune Reactions

Inflammatory Response: Complex response system activated by tissue injury (infection, burn, allergy); general functions are to mobilize immune system against pathogens, repair damage, and clear infection; symptoms are redness, heat, swelling, and pain.

Stages of Inflammatory Response: Changes in vascular bed—vessels narrow and dilate in response to **chemical mediators (cytokines),** released by injured tissues and white blood cells; **edema** is swelling due to increased tissue fluid, which may block spread of microbe; by chemotaxis, mediators attract neutrophils to site of injury to engulf debris and microbes; **pus** may form; late in inflammation, macrophages and lymphocytes arrive to clean up residue of inflammation and undergo specific antimicrobial reactions; sometimes long-term inflammation results in injury and disease (granuloma); **fever** is increase in body temperature above normal (37°C or 98.6°F) due to body's reaction to **pyrogens,** substances that alter the hypothalamus setting. Fever may be beneficial by slowing microbial multiplication and stimulating the immune response.

Phagocytosis: General process of engulfment and destruction of foreign materials by certain white blood cells; neutrophils, early phagocytes that engulf small particles, microbes, molecules; macrophages, largest phagocytes that do most generalized scavenging and extraction of antigenic information; some linked to a specific tissue or organ (liver, lung, skin); others free and wandering. Phagocytes engulf materials into vacuole or **phagosome,** and **lysosomes** containing powerful chemicals fuse with these and digest and dismantle foreign material.

Generalized Chemical Immunities

Interferon: Small protein produced by lymphocytes and fibroblasts in response to infection or cancer; three types; immunological importance includes blockage of viral replication and immune stimulation and regulation.

Complement: Complex chemical defense system that destroys certain pathogens and acts as an immune mediating system; involves 20 chemicals, the **complement** or **C factors;** operates on principle of cascade reaction—one component activates the next in line, which activates the next. Two major pathways occur: Classical involves activation of complement by specific antibody; alternative involves nonspecific reaction to infections. The result of complement activation is a huge enzyme, the **membrane attack complex,** that can kill cells and viruses by digesting holes in their outer membranes.

Introduction to Specific Immunities

Immunocompetent individuals possess a third line of immunity that confers an ability to acquire specific immunities; involves individualized response to the **specificity** of each microbe and **immunologic memory,** which provides protection in event of reexposure to same pathogen.

Categories of Acquired Immunity: **Natural,** acquired as part of normal life experiences; **artificial,** acquired through medical procedures such as **immunization; active,** individual is challenged with antigen that stimulates production of protective substances (antibodies); creates memory, takes time, is lasting; **passive,** individual is given protective substances (antibodies) produced in the body of another individual; no memory, is immediate, and short-term.

Four combinations of acquired immunity are **natural active,** acquired upon infection and recovery; **natural passive,** transfer of antibodies from mother to child through placenta and breast milk; **artificial active (vaccination),** person inoculated with carefully selected antigen that will stimulate immunity and prevent future infections; **artificial passive,** person is injected with specific **immune serum** or globulins as prophylactic or therapeutic measure.

True–False Questions

Determine whether the following statements are true (T) or false (F). If you feel a statement is false, explain why, and reword the sentence so that it reads accurately.

___ 1. Active and passive immunities are part of the first line of defense.

___ 2. Recognition is a process through which the immune system distinguishes self from nonself.

___ 3. Neutrophils function principally as phagocytes.

___ 4. The spleen, thyroid gland, lymph nodes, and GALT are lymphoid tissues.

___ 5. Complement is a complex series of biochemical reactions that forms a barrier to infection.

___ 6. Monocytes are agranular leukocytes that develop into macrophages.

___ 7. Immunity arising from a carefully administered antigen is artificial passive.

___ 8. Basophils occur infrequently in the normal circulation because they function in allergic reactions.

___ 9. Inflammatory mediators stimulate vasodilation and chemotaxis.

___ 10. Endogenous pyrogens are fever-causing agents from microorganisms.

Concept Questions

1. Explain the function of the three lines of defense. Which is the most important to life?
2. How is surveillance of the tissues carried out? What is responsible for it? What is the importance of recognition of foreign materials or antigens?
3. Trace the complete cycle of a bacterium through the immune compartments, starting with the blood, the RES, the lymphatics, and the ECF. (You will start and end at the same point.) What is the ultimate connection between the bloodstream and the lymphatic circulation?
4. What are the main components of the reticuloendothelial system? Why is it also called the mononuclear phagocyte system?
5. Prepare a simplified outline of the cell lines of hemopoiesis.
6. Differentiate between granulocytes and agranulocytes.
7. What is the principal function of lymphocytes? Differentiate between the two lymphocyte types and between humoral and cell-mediated immunity.
8. Why are platelets called formed elements?
9. Explain diapedesis and chemotaxis. Can you tie in the impact of these processes with question 3?
10. How is lymph formed? Why are white cells but not red cells found in it? What are the functions of the lymphatic system? Explain the filtering action of a lymph node. What is GALT, and what does it do?
11. Differentiate between pyogenic and pyrogenic processes, using examples.
12. Briefly account for the origins and actions of inflammatory mediators (cytokines).
13. Explain how macrophages arise. What types are there, and what are the principal functions?
14. In what ways is a phagocyte a tiny container of disinfectants?
15. Briefly describe the biological effects of interferon.
16. Describe the general complement reaction in terms of a cascade. What is the end result of complement activation? What are some other functions of complement components?
17. What are immunocompetence, immunological specificity, and immunological memory?
18. Evaluate this statement for accuracy: In active immunity, antigens are given; in passive immunity, antibodies are given.

Practical/Thought Questions

1. Children born without a functioning lymphocyte system are lacking what?
2. What is the most important component extracted for bone marrow transplants?
3. Which would be used to perform a clotting test, plasma or serum? If one wishes to give antibody immunotherapy, which is better to use, or does it matter?
4. A patient's chart shows an extensive eosinophilia. What does this cause you to suspect?
5. How can adults continue to function relatively normally after surgery to remove the thymus, tonsils, spleen, or lymph nodes?
6. What part of the inflammatory and immune defense accounts for swollen lymph nodes and leukocytosis? What is pus? How can edema be beneficial?
7. Patients with a history of tuberculosis sometimes show residua in the lungs and even recurrent infection. Account for this on the basis of the inflammatory response.
8. Account for the several symptoms that occur in the injection site when one has been vaccinated against influenza.
9. Knowing that fever is potentially both harmful and beneficial, what are some possible guidelines for deciding whether to suppress it or not? What is the specific target organ of a fever-suppressing drug?
10. For passive immunity to exist, what must also occur? With this knowledge, can you think of an effective way to immunize a fetus? Is this safe? What would be an even safer way to ensure that fetuses get necessary antibodies?
11. What type of host defense or immune response does each of the following represent? (In answering this question, you should use terms such as acquired, inborn, active, passive, natural, artificial, nonspecific barriers, inflammation, specific immunity, etc.)

vaccination for tetanus	injection of gamma globulin
lysozyme in tears	action of neutrophils
immunization with horse serum	edema
in utero transfer of antibodies	inability to get canine distemper
booster injection for diphtheria	stomach acid
recovery from mumps	cilia in trachea
colostrum	asymptomatic chickenpox
interferon	complement

CHAPTER 13

The Acquisition of Specific Immunity and Its Applications

A large, crablike cancer cell is attacked by killer T cells, which pour it full of lymphotoxins and gradually incapacitate the monster.

Chapter Preview

The specific reactions of the human immune system represent some of the most exciting, useful, and challenging concepts in microbiology. A detailed exploration of the inner workings of this remarkable, finely tuned system will reveal not only how it defends against microorganisms and destroys cancer cells, but also how it is involved in diseases as diverse as hay fever, diabetes, and multiple sclerosis. Understanding the machinery of the immune response has an added practical benefit, because the concepts are directly applicable to medical procedures such as vaccination, organ transplants, blood transfusions, blood tests, cancer therapy, and allergy management.

Further Explorations into the Immune System

In chapter 12 we explored general characteristics and components of the immune system. We described the capacity of the immune system to survey, recognize, and react to foreign cells and molecules, and we overviewed the characteristics of nonspecific host defenses, blood cells, phagocytosis, inflammation, and complement. We introduced the concepts of specificity and acquired immunity, and examined the differences between active and passive immunity. In this chapter we will take a closer look at acquired immunity and its practical applications.

The elegance and complexity of immune function are largely due to lymphocytes working "hand in hand" with macrophages. To simplify and clarify the network of immunologic development and interaction, we present it here as a series of five stages, with each stage covered in a separate section (figure 13.1). The principal stages are: I. lymphocyte development and differentiation; II. the processing of antigens; III. the challenge of B and T lymphocytes by antigens; IV. the production and activities of antibodies; and V. T-lymphocyte responses. Notice that as each stage is covered, we will simultaneously be following the sequence of an immune response.

The Dual Nature of the Immune Response

In the following overview, Roman numerals correspond with the subsequent text sections that cover these topics in more detail.

I. Development of the Dual Lymphocyte System

Lymphocytes are at the center of the immune system's universe. They undergo a sequential development that begins in the embryonic yolk sac and shifts to the liver and bone marrow. Although all lymphocytes arise from the same basic stem cell type, at some point in development, they diverge into two distinct types. Final maturation of B cells occurs in a site believed to be the bone marrow or gastrointestinal-associated lymphoid tissue (GALT), and that of T cells occurs in the thymus. This process commits each individual B cell or T cell to one specificity. Both cell types subsequently migrate to precise, separate areas in the lymphoid organs (nodes, spleen, as described in chapter 12) to serve as a lifelong continuous source of immune responsiveness.

II. Entrance and Processing of Antigens and Clonal Selection

When foreign cells or particles enter a fluid compartment of the body, they immediately encounter a network consisting of the lymphatics, blood, and the reticuloendothelial system (RES). In any of these sites, the antigenic materials encounter a battery of cells that work together to screen, entrap, and eliminate them. Certain specialized macrophages are usually the first cells to recognize and react. They ingest and process antigens and present them to the lymphocytes that are specific for that antigen. This selection process is the trigger that activates the lymphocytes. In most cases, the response of B lymphocytes also requires the additional assistance of special classes of T cells.

III B and III T. Activation of Lymphocytes and Clonal Expansion

When challenged by antigen, both B cells and T cells further differentiate and proliferate. The multiplication of a particular lymphocyte creates a clone, or group of genetically identical cells, some of which are memory cells that will ensure future reactiveness against that antigen. Because the B-cell and T-cell responses depart notably from this point in the sequence, they will be summarized separately.

IV. Activated B Lymphocytes and Their Products—Humoral Immunity

The active progeny of a dividing B-cell clone are called plasma cells. These cells are programmed to synthesize and secrete large protein molecules (antibodies) into the tissue fluids. When these antibodies attach to the antigen for which they are specific, the antigen is marked for destruction or neutralization. Because secreted antibodies circulate freely in the tissue fluids, lymph, and blood, the immunity they provide is humoral.

V. Activated T Lymphocytes and Cell-Mediated Immunity (CMI)

T-cell types and responses are extremely varied. When activated (sensitized) by antigen, a T cell gives rise to one of several types of progeny, each involved in a cell-mediated immune function. The four main classes of T cells are (1) helper cells that assist in immune reactions, (2) suppressor cells that suppress immune reactions, (3) cytotoxic or killer cells that destroy specific target cells, and (4) delayed hypersensitivity cells that function in certain allergic reactions. Although T cells secrete chemicals called lymphokines that help destroy antigen or regulate immune responses, they do not produce antibodies.

Essential Preliminary Concepts for Understanding Immune Reactions of Sections I–V

Before discussing lymphocyte development and function, antigens, and antibodies, we need to integrate a number of very important concepts that have been covered previously. Past chapters dealt with the unique structure of molecules (especially proteins), the characteristics of cell surfaces (membranes and envelopes), the ways that genes are expressed, and immune recognition and identification of self and nonself. Let us now examine how these basic ideas come together in one working system. Ultimately, the shape and function of protein receptors and markers protruding from the surfaces of cells are the result of genetic expression, and these molecules are responsible for specific immune recognition and thus, immune reactions.

Markers on Cell Surfaces Involved in Recognition of Self and Nonself

Chapter 12 touched on the fundamental idea that cell markers or receptors confer specificity and identity. A given cell may

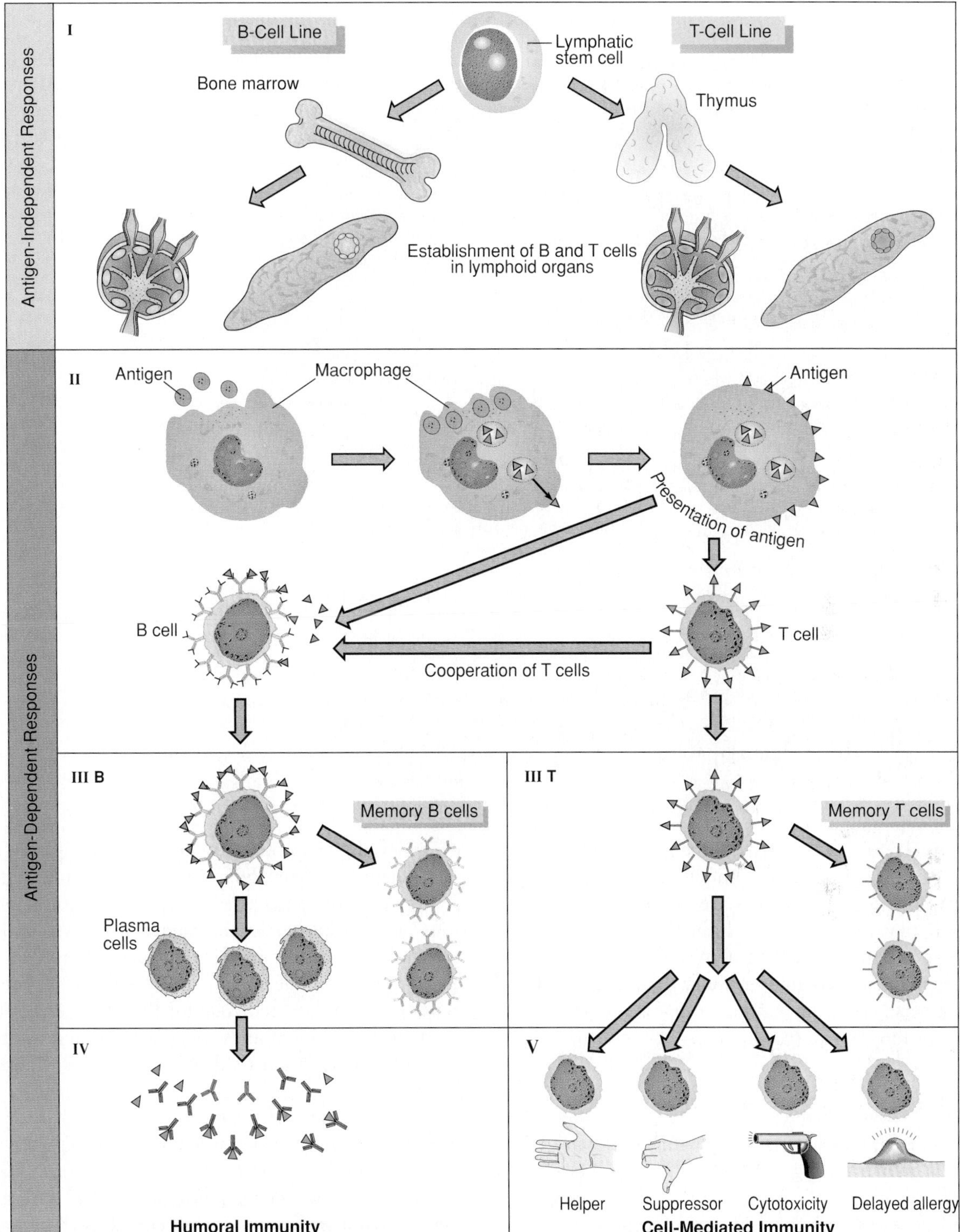

Figure 13.1 Overview of the stages of lymphocyte development and function. I. Development of B- and T- lymphocyte specificity and migration to lymphoid organs. II. Antigen processing by macrophage and presentation to lymphocytes; assistance to B cells by T cells. III B and III T. Lymphocyte activation, clonal expansion, and formation of memory B and T cells. IV. Humoral immunity. B-cell line produces antibodies to react with the original antigen. V. Cell-mediated immunity. Activated T cells perform various functions on the original antigen.

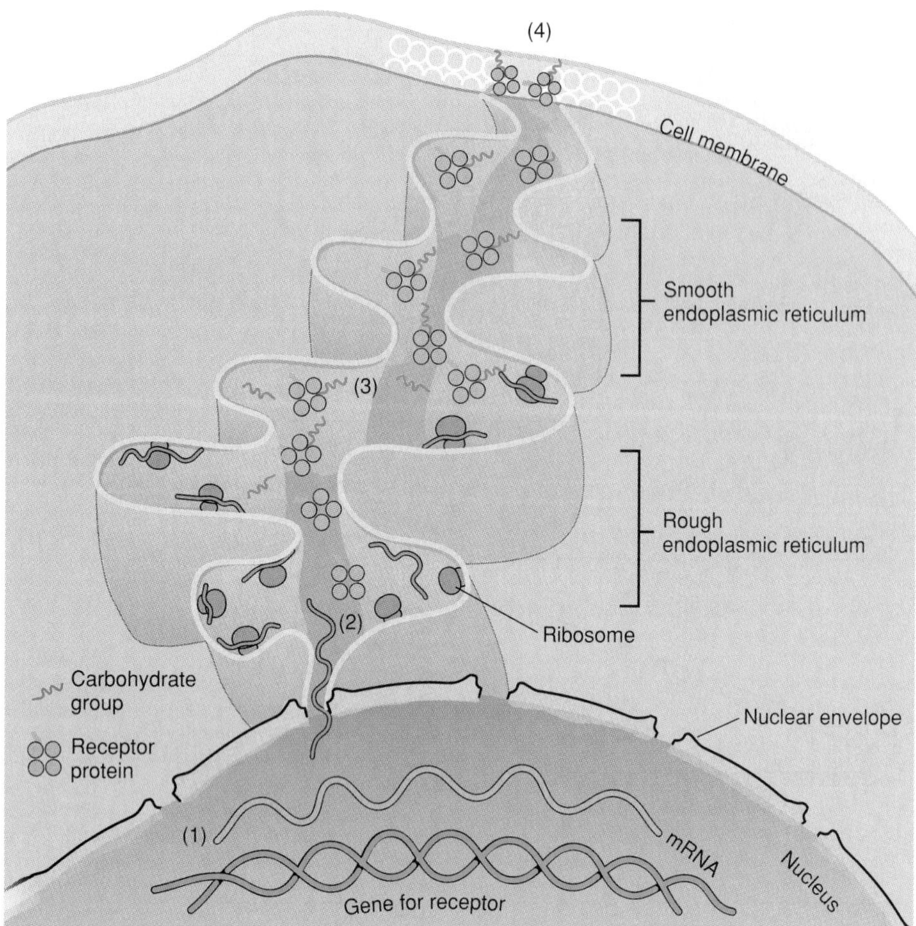

Figure 13.2 Receptor formation in a developing cell. (*1*) Gene coding for the receptor is transcribed. (*2*) In the endoplasmic reticulum mRNA is translated into the protein portion of the receptor. (*3*) A carbohydrate side chain is added, forming glycoprotein. (*4*) The finished receptor is transported to and inserted in the cell membrane.

express several different receptors, each type playing a distinct and significant role in detection, recognition, and cell communication. Major functions of receptors are (1) to perceive and attach to nonself (antigens), (2) to help in the recognition of self, (3) to receive and transmit chemical messages among other cells of the system, and (4) to aid in cellular development. Because of their importance in the immune response, we will concentrate here on the major receptors of lymphocytes and macrophages.

How Are Receptors Formed?

The nature of cell receptors is dictated by which particular genetic program is active during development. As a cell matures—be it liver or brain cell, lymphocyte or macrophage—certain selected genes that code for the cell receptors will be transcribed and translated into protein products with a distinctive shape, specificity, and function. This receptor is transported to the surface of the cell and inserted there to be accessible for interactions with antigens and other cells (figure 13.2). The receptors of many cells, including lymphocytes and macrophages, are glycoproteins with additional carbohydrate fragments added.

Major Histocompatibility Complex

One set of genes that codes for human cell receptors is the **major histocompatibility complex (MHC).** This gene complex gives rise to a series of glycoproteins (called MHC antigens) found on all cells except red blood cells. Because these markers were first identified in humans on the surface of white blood cells, the MHC is also known as the **human leukocyte antigen (HLA)** system. This receptor complex plays a vital role in recognition of self by the immune system and in rejection of foreign tissues. Genes that regulate and code for the MHC of humans are located on the sixth chromosome, clustered in a multigene complex of three subgroups called class I, class II, and class III (figure 13.3).

The functions of the three MHC groups have been identified. Class I genes code for markers that regulate acceptance or rejection of tissue grafts, which is how the term *histo-* (tissue) *compatibility* (acceptance) originated. The system is rather complicated in its details, but in general, each human being inherits a particular combination of class I MHC (HLA) genes in a relatively predictable fashion. Although millions of different combinations and variations of these genes are possible among humans, the closer the relationship, the greater the probability for similarity in MHC profile (see figure 14.19). Individual differences in the exact inheritance of MHC genes, however, make it very likely that even closely related persons will express MHC receptors that their relatives do not have. This important observation will be addressed again later: Although humans are all the same species, most cells express mol-

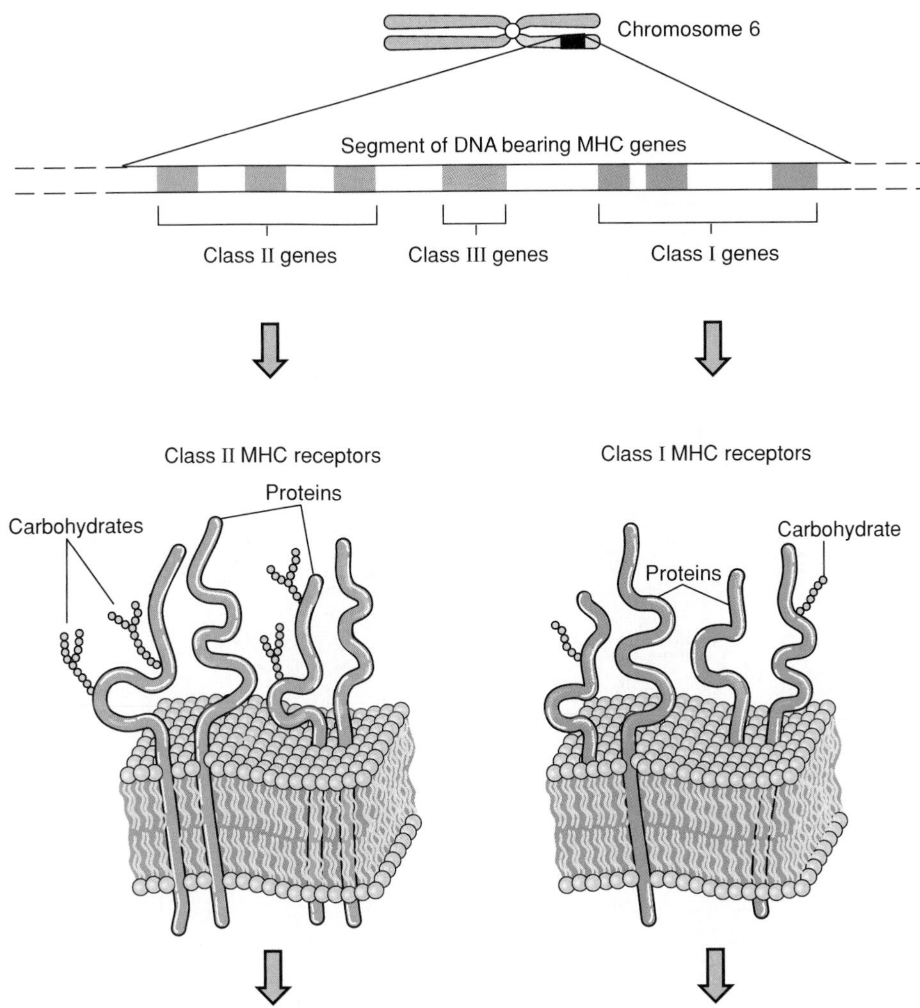

Figure 13.3 Glycoprotein receptors of the human major histocompatibility (human leukocyte antigen) gene complex (MHC).

ecules that are foreign (antigenic) to other humans. This becomes so important in organ transplant procedures and transfusions that the HLA marker antigens between donor and recipient must be tested (see graft rejection, chapter 14).

Class II MHC genes regulate immune responses. This system of genes and receptors is located primarily on macrophages and B cells, and it functions in cooperative immune responses to antigens mounted by macrophages, T cells, and B cells. Unlike the other two classes, which code for molecules that are inserted into cell surfaces, class III MHC genes code for certain secreted complement components—C2, C4, and factor B (described in chapter 12).

Lymphocyte Receptors and Development of Specificity to Antigen

The part lymphocytes play in immune surveillance and recognition emphasizes the essential role of their receptors. Although they possess MHC antigens for recognizing self, they also carry receptors on their membranes that recognize nonself or antigens. One rule of thumb underlying all lymphocyte-antigen interactions is that one uniquely specific receptor must exist in the immune repertoire for each different antigen. The diversity of foreign materials a body can potentially come in contact with is practically boundless. It includes not just microorganisms, but an awesome array of organic and inorganic compounds in the air, food, and household products. Looking at proteins alone, if the 20 different amino acids were arranged in all possible assorted numbers and combinations, billions of different antigenic shapes could potentially be produced. Some of the most fascinating questions in immunology have revolved around how the lymphocyte receptors can be varied to fit this large number of different antigens. A related question is: How can a cell accommodate enough genetic information to respond to millions or even billions of antigens? Furthermore, when, where, and how does the capacity to distinguish native from foreign tissue arise? Answers to all of these questions lie in a theory of genetic diversity called the clonal selection theory.

The Origin of Variety and Specificity in the Immune Response

The Clonal Selection Theory

In order to explain how lymphocytes use only about 500 genes to produce a tremendous variety of specific receptors, we must look at the early stages of lymphocyte development. Although

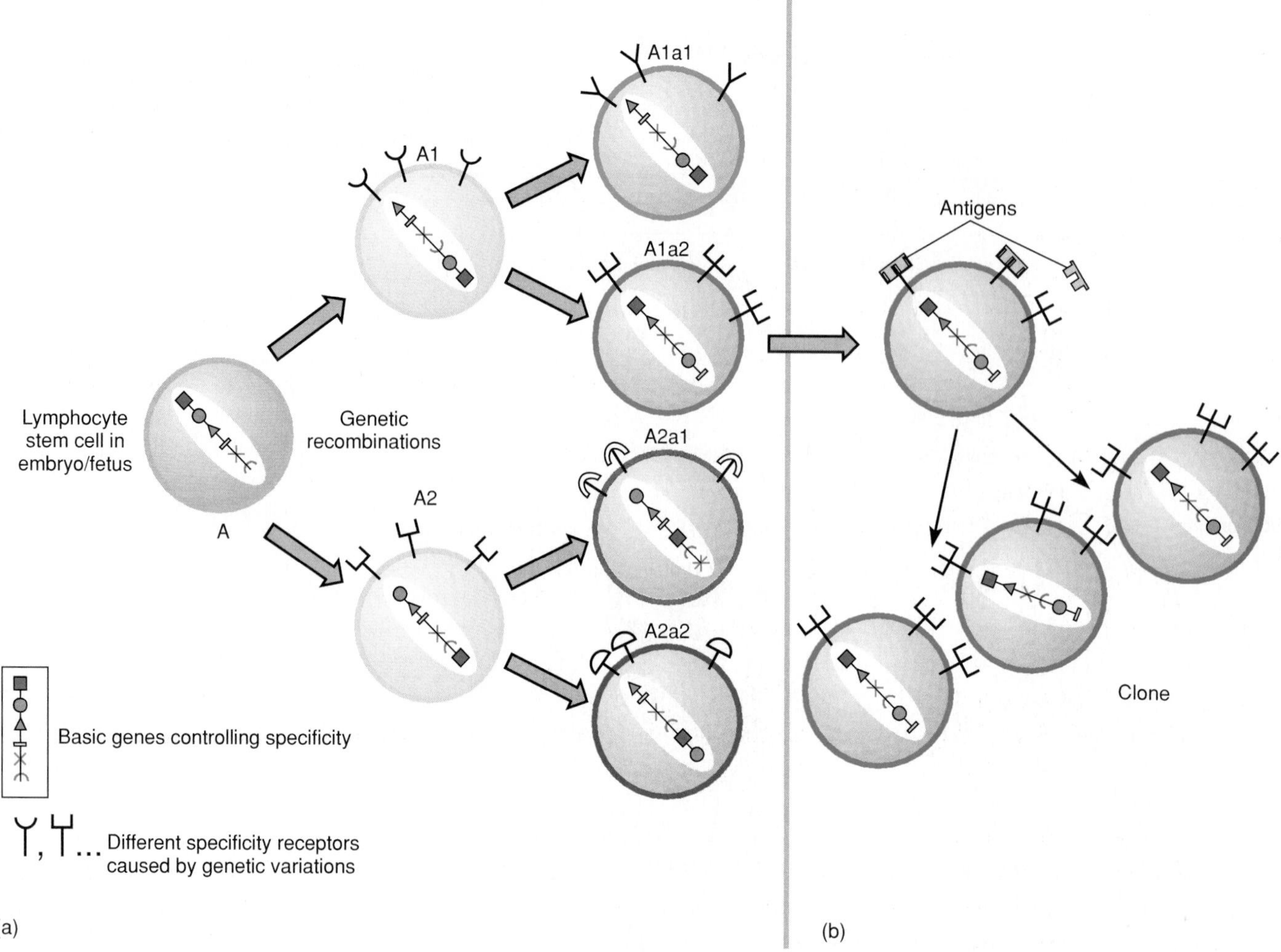

Figure 13.4 Clonal selection theory, the origins of lymphocyte specificity. (*a*) During the development of stem cells, the genes that program for specificity to antigen are rearranged so that lymphocyte daughter cells inherit differing genetic programs. These alterations are expressed in the nature of the protein and the shape of the receptor. The potential for genetic and receptor variation is so great that millions of genetically different clones can be formed. The receptor announces the lymphocyte's specificity while simultaneously providing a recognition and reaction site for an antigen. This part of development does not require the antigen to be present. (*b*) Clonal selection and expansion (antigen is present). The complete repertoire of lymphocytes has developed by the time a child is born, so that any antigen entering the system will automatically encounter and select its prespecified lymphocyte and stimulate it to expand into a genetically identical group or clone, each of which can respond to that same antigen. Note that all antigen receptors on a single lymphocyte are identical.

B cells and T cells have different kinds of receptors on their surfaces, the general principles of genetics and acquisition of specificity are the same for both. According to the most widely accepted **clonal selection theory,** as undifferentiated lymphocytes proliferate during the embryonic stage of development, their genetic programs are susceptible to random recombinations and mutations (figure 13.4). Stem lymphocytes start out with the same genetic program, but during development, they become altered genetically so that their expression is extremely varied. The mechanism might be pictured as follows: As an immature, undifferentiated lymphocyte (A) divides, its receptor genes recombine randomly to produce differing genetic codes in the two daughter cells. Because of gene shuffling, daughter cell A1 receives a different combination of genes and a different specificity from daughter cell A2. By the same random shuffling, the progeny of A1 and A2 will likewise receive a different genetic program, and so on, for millions of different lymphocytes. The ultimate impact of these genetic variations is that the amino acid composition of the receptor, and thus its shape and specificity, will be different for each lymphocyte type. Each genetically distinct group of lymphocytes that possesses the same specificity is called a **clone.**

As a result of gene reassortment on such a massive scale, the lymphoid system eventually develops (at least in theory) perhaps a billion different clones of lymphocytes, so that at birth, the ability to react with a tremendous variety of antigens is already in place in the lymphoid tissue. Two important generalities of immune responsiveness one can derive from the clonal selection theory are that (1) lymphocyte specificity is preprogrammed, existing in the genetic makeup before an antigen has ever entered the system, and (2) each genetically different type of lymphocyte expresses only a single specificity.

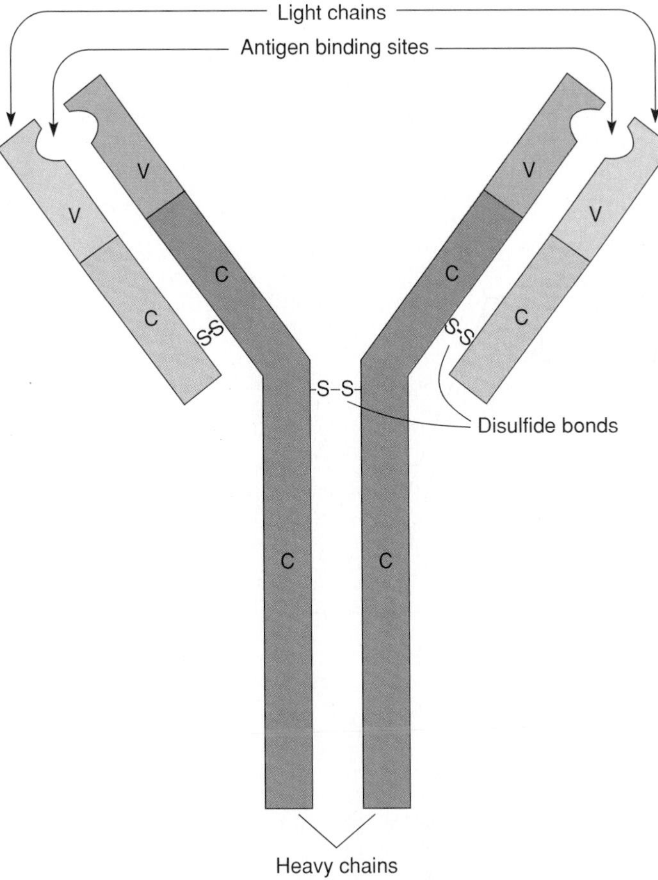

Figure 13.5 Simplified structure of an immunoglobulin molecule. The main components are four polypeptide chains—two identical light chains and two identical heavy chains bound by disulfide bonds as shown. Each chain consists of a variable region (V) and a constant region (C). The variable regions of light and heavy chains form a binding site for antigen.

According to another part of the clonal selection theory, any lymphocyte that could possibly mount a harmful response against *self* molecules is eliminated or suppressed. Immunologists call this **tolerance to self.** Some immune diseases (called autoimmune) are believed to arise when the immune system loses this tolerance and attacks self (see chapter 14). The final part of the clonal selection theory proposes that the first introduction of each distinct type of antigen into the immune system selects a genetically distinct lymphocyte and causes it to expand into a clone of cells that can react to that antigen.

The Specific B-Cell Receptor: An Immunoglobulin Molecule

In the case of B lymphocytes, the receptor genes that undergo the recombination described are those governing **immunoglobulin (Ig)** synthesis. Immunoglobulins are large glycoprotein molecules that serve as the specific receptors of B cells and as antibodies. The basic immunoglobulin molecule is a composite of four polypeptide chains: a pair of identical heavy (H) chains and a pair of identical light (L) chains (figure 13.5). One light chain is bonded to one heavy chain, and the two heavy chains are bonded to one another with disulfide bonds, creating a symmetrical, Y-shaped arrangement. The ends of the forks formed by the light and heavy chains contain pockets, called the **antigen binding sites.** It is essential that these sites be varied in shape to fit a myriad of antigens. This need is accommodated by the presence of **variable regions (V),** sites where amino acid composition is highly varied from one clone of B lymphocytes to another. The remainder of the light chains and heavy chains consist of **constant regions (C),** so named because their amino acid content does not vary greatly from one antibody to another. Although we will subsequently discuss immunoglobulins and their function as antibodies, for now, we must concentrate on their genetics in order to understand the origins of lymphocyte specificity.

The genes that code for immunoglobulins lie on three different chromosomes. An undifferentiated lymphocyte has about 150 different genes that code for the variable region of light chains and a total of about 250 genes for the variable and **diversity regions (D)** of the heavy chains. It has only a few genes coding for the constant regions and a few for the joining regions (J) that join segments of the molecule together. Only a few of these genes will ever be expressed in a given lymphocyte because, during the genetic recombination of development, certain genes are randomly selected and active in the mature cell, and the rest are inactive (figure 13.6). Factors involved in this recombination will be discussed next.

Development of the Receptors During Lymphocyte Maturation

It is helpful to think of the immunoglobulin genes as blocks of information that code for a polypeptide lying in sequence along a chromosome. During development of each lymphocyte, the genetic blocks are independently segregated, randomly rearranged, and assembled as follows:

1. For a heavy chain, a variable region gene and a diversity region gene are randomly selected from among the hundreds available and spliced to one joining region gene and one constant region gene.
2. For a light chain, one variable, one joining, and one constant gene are spliced together.
3. After transcription and translation of each gene complex into a polypeptide, a heavy chain combines with a light chain to form half an immunoglobulin; two of these combine to form a completed monomer (figure 13.6).

Once synthesized, the immunoglobulin product is transported to the cell membrane and inserted there to act as a receptor that expresses the specificity of that cell and to react with an antigen. The first receptor on most B cells is a small form of IgM, one of the classes of immunoglobulins (see table 13.2). It is notable that for each lymphocyte, the genes that were selected for the variable region, and thus for its specificity, will be locked in for the rest of the life of that lymphocyte and its progeny. We will discuss the different classes of Ig molecules further in section IV.

---

immunoglobulin (im″-yoo-noh-glawb′-yoo-lin) The technical name for an antibody.

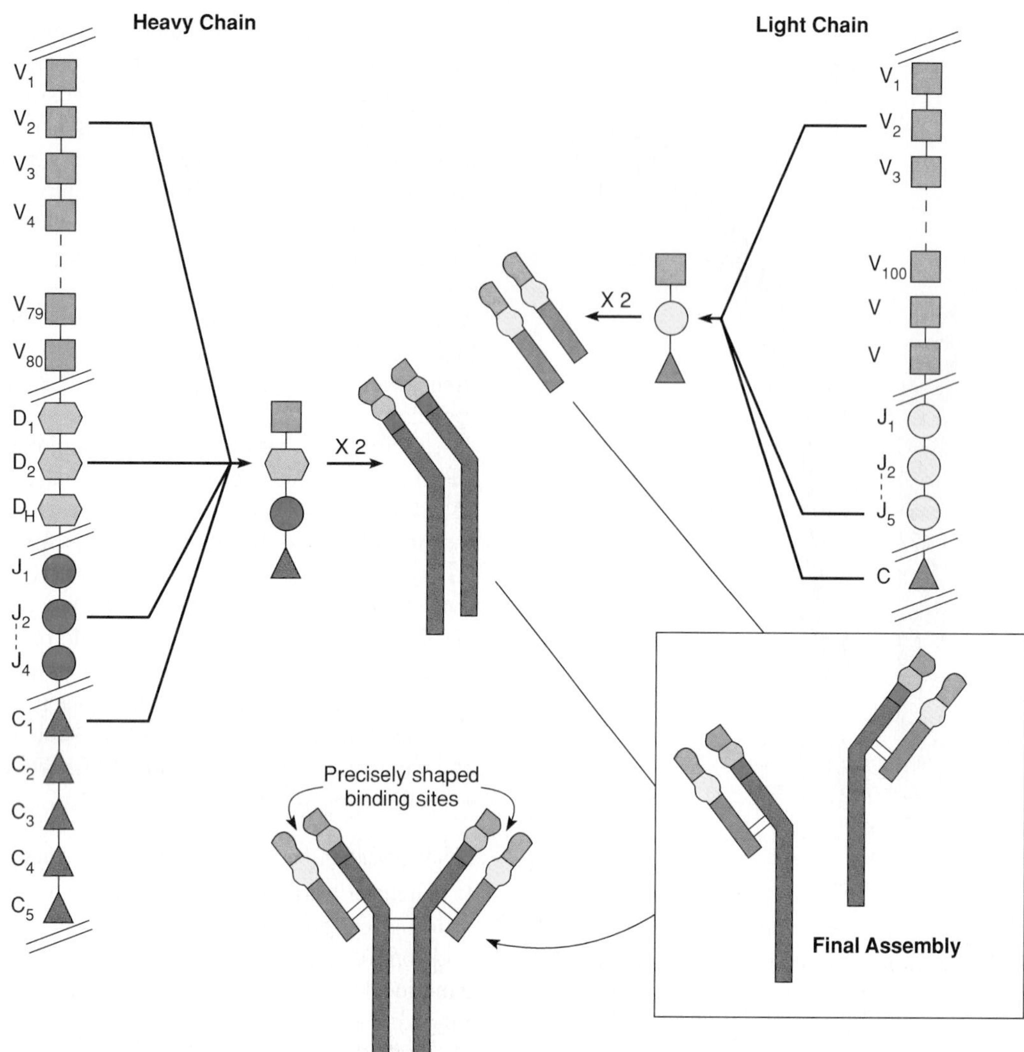

Figure 13.6 A simplified look at immunoglobulin genetics. The phenomenon of B-cell differentiation is something like an immunological "cut and paste," in which the final gene that codes for a heavy or light chain is assembled by splicing blocks of genetic material from several regions. Left: The heavy chain gene is composed of genes from four separate segments (V, D, J, and C) that are transcribed and translated to form a single polypeptide chain. Right: The light chain genes are put together like heavy ones, except that the final gene is spliced from three gene groups (V, J, and C). During final assembly, first the heavy and light chains are bound, and then the heavy/light combinations are connected to form the immunoglobulin molecule.

T-Cell Receptors

The T-cell receptor for antigen belongs to the same immunoglobulin superfamily as the B-cell receptor. It is similar to B cells in being formed by genetic modification, having variable and constant regions, being inserted in the membrane, and having an antigen-combining site formed from two parallel polypeptide chains (figure 13.7). Unlike the immunoglobulins, the T-cell receptor is relatively small and does not appear to have a humoral function. The several other receptors of T cells are described in a later section.

The Lymphocyte Response System in Depth

Now that you have a working knowledge of some factors in specific immune function, let us look at each stage of an immune response.

I. The Stages in Lymphocyte Origin, Differentiation, and Maturation

Development generally follows a similar general pattern in both types of lymphocytes (see figure 13.1). Starting in embryonic and fetal stages, stem cells in the yolk sac, liver, and bone marrow give rise to immature lymphocytes that are released into the circulation. What happens to these lymphocytes next is something akin to their getting an education. At some specific anatomical location, they undergo developmental changes (are educated) so as to become specific and responsive to antigen. This maturation occurs along two separate lines that will characterize all future responses (summarized in table 13.1). Lymphocyte differentiation and immunocompetence are basically complete by the late fetal or early neonatal period.

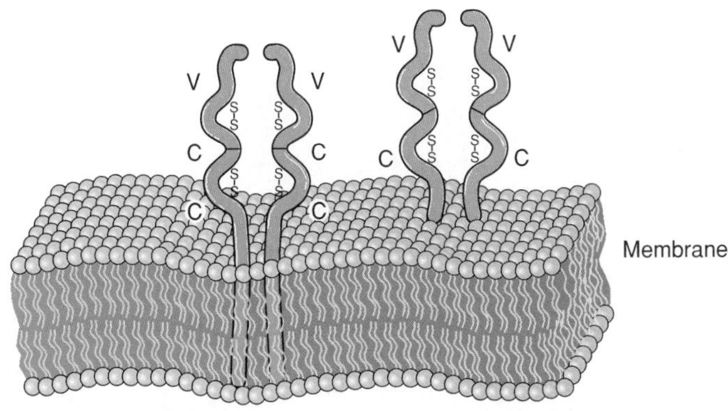

Figure 13.7 Proposed structure of the T-cell receptor for antigen.

Table 13.1 Contrasting Properties of B-Cell and T-Cell Lines

	B Cells	T Cells
Site of Maturation	Fetal bone marrow, liver, GALT	Thymus
Nature of Surface Markers	Immunoglobulin	Several types
Texture of Surface	Rough	Smoother
Circulation in Blood	Low numbers	High numbers
Rosette Formation with Plain Sheep RBCs	No	Yes
Distribution in Lymphatic Organs	Cortex (in follicles)	Paracortical (interior to the follicles)
Product of Antigenic Stimulation	Plasma cells	Sensitized lymphocytes of several types
General Functions	Production of antibodies	Cells function in helping, suppressing, killing, delayed hypersensitivity

Specific Happenings in B-Cell Maturation

Although the site of B-cell maturation is well established in birds, its site in humans and other mammals is not as clear. The most likely candidates for this function are the liver, bone marrow, and GALT, but the factors that initiate development in these organs are still to be discovered. As a result of gene modification and selection, hundreds of millions of distinct B cells are released to sites in the lymph nodes, spleen, and other lymphoid organs, where they will come in contact with antigens throughout life. In addition to having immunoglobulins as surface receptors, a fully differentiated B cell has a distinctively rough appearance under high SEM magnification, with numerous microvillus projections (figure 13.8).

Specific Happenings in T-Cell Maturation

The maturation of T cells and the development of their specific receptors are directed by the thymus gland and its hormones.

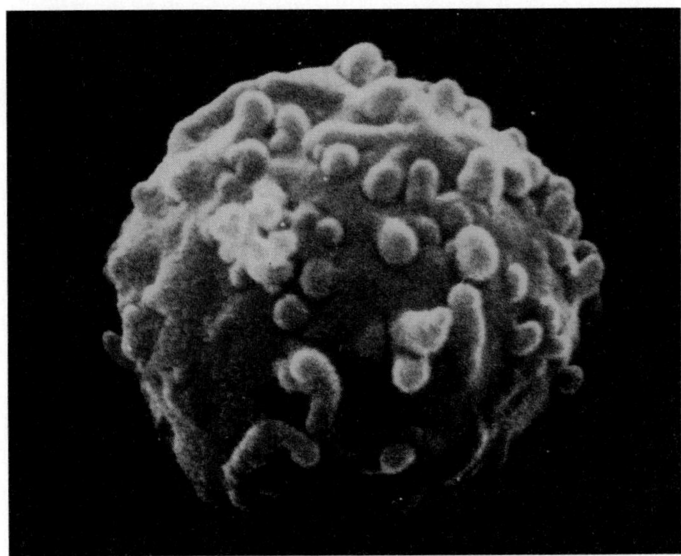
Figure 13.8 Appearance of a B cell by scanning electron microscopy (×16,000). Note the well-developed projections that represent receptors.

Owing to the complexity of T-cell function, there are at least seven classes of T-cell receptors or markers, termed the CD cluster (common determinant—CD1, CD2, etc.). Some receptors (figure 13.7) recognize antigen molecules, while others recognize self molecules on B cells, other T cells, and macrophages. Additional characteristics typical of mature T cells are the lack of microvilli under high SEM magnification and the property of rosette formation when mixed with normal sheep red blood cells (see figure 13.34).

II. The Nature of Antigens and the Host's Response to Them

Having reviewed the characteristics of lymphocytes, let us now examine the properties of antigens, the substances that cause them to react. As we reported in chapter 12, an **antigen (Ag)**[1] is a substance that provokes an immune response in specific lymphocytes. The property of behaving as an antigen is called **antigenicity.** The term *immunogen* is another term of reference for a substance that can elicit an immune response. To be perceived as an antigen or immunogen, a substance must meet certain requirements in foreignness, shape, size, and accessibility.

Characteristics of Antigens

In chapter 12, we discovered that an antigen is perceived by lymphocytes and certain phagocytes as being nonself or **foreign,** meaning that it is not a normal constituent of the body. Whole microbes or their parts, cells, or substances that arise from other humans, animals, plants, and various molecules all possess this quality of foreignness, and thus are potentially antigenic to the immune system of an individual (figure 13.9). Molecules of

1. Originally, **anti**body **gen**erator.

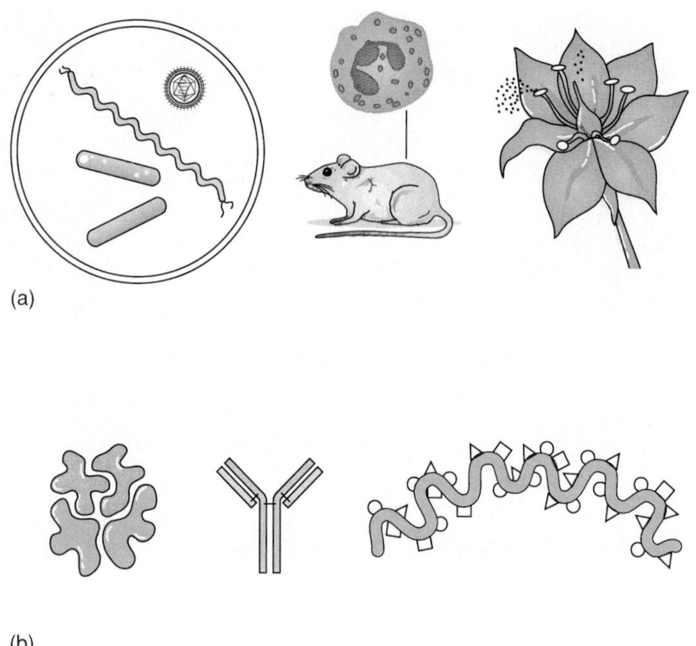

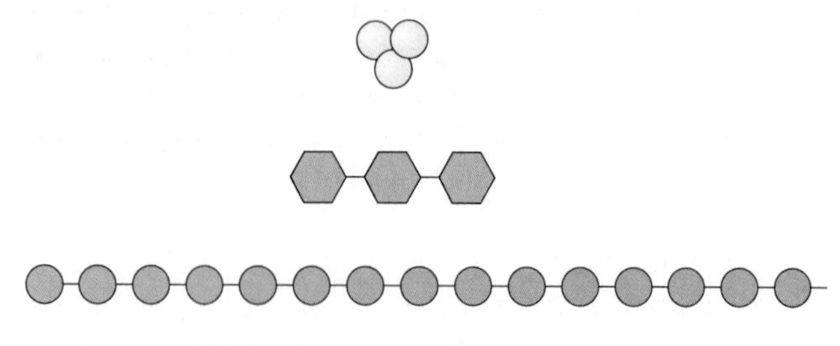

Figure 13.9 Characteristics of antigens. (*a*) Whole cells and viruses make good immunogens. (*b*) Complex molecules with several antigenic determinants make good immunogens. (*c*) Poor immunogens include small molecules by themselves, simple molecules, or large but repetitive molecules.

complex composition such as proteins and protein-containing compounds prove to be more immunogenic than repetitious polymers comprised of a single type of unit. In summary, most materials that serve as antigens fall into the following chemical categories:

1. Proteins and polypeptides (enzymes, albumin, antibodies, hormones, exotoxins)
2. Lipoproteins (cell membranes)
3. Glycoproteins (blood cell markers)
4. Nucleoproteins (DNA complexed to proteins, but pure DNA is not)
5. Polysaccharides (certain bacterial capsules) and lipopolysaccharides

Effects of Molecular Shape and Size To initiate an immune response, a substance must also be large enough to "catch the attention" of the lymphocytes. Molecules with a molecular weight (MW) of less than 1,000 are seldom complete antigens, and those between 1,000 MW and 10,000 MW are weakly so. Complex macromolecules approaching 100,000 MW are the most immunogenic, a category also dominated by large proteins. Note that large size alone is not sufficient for antigenicity; glycogen, a polymer of glucose with a highly repetitious structure, has an MW over 100,000 and is not antigenic, whereas insulin, a protein with an MW of 6,000, is antigenic.

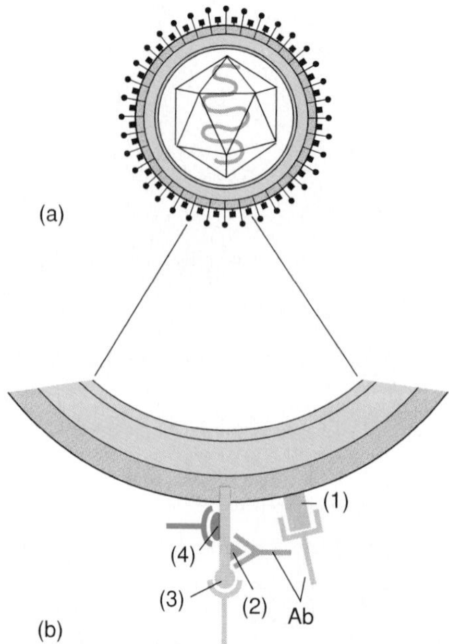

Figure 13.10 Mosaic antigens. (*a*) Microbes such as viruses present various sites that serve as separate antigenic determinants. (*b*) Inset indicates that each determinant (*1,2,3,4*) will stimulate a different lymphocyte and antibody response.

A lymphocyte's capacity to discriminate differences in molecular shape is so fine that it recognizes and responds to only a portion of the antigen molecule. A minute piece of the antigen molecule called the **antigenic determinant** signals that the molecule is indeed foreign (figure 13.10). This determinant has a particular tertiary structure and shape that must conform like a key to the receptor "lock" of the lymphocyte, which then responds to it. Certain amino acids accessible at the surface of proteins or protruding carbohydrate side chains are typical examples. Many foreign cells and molecules are very complex antigenically, with numerous determinants, each of which will elicit a separate and different lymphocyte response. Examples of these multiple or *mosaic antigens* include bacterial cells containing cell wall, membrane, flagellar, capsular, and toxin antigens, and viruses (figure 13.10).

Small foreign molecules that consist only of a determinant group and are too small by themselves to elicit an immune response are termed **haptens.** However, if such an incomplete antigen is linked to a larger carrier molecule, the entire composite becomes immunogenic (figure 13.11). The carrier group contributes to the size of the complex and enhances the proper spatial orientation of the determinative group, while the hapten serves as the antigenic determinant. Haptens include such molecules as drugs, metals, and ordinarily innocuous household, industrial, and environmental chemicals. Many haptens develop antigenicity in the body by combining with large carrier molecules such as serum proteins (see allergy in chapter 14).

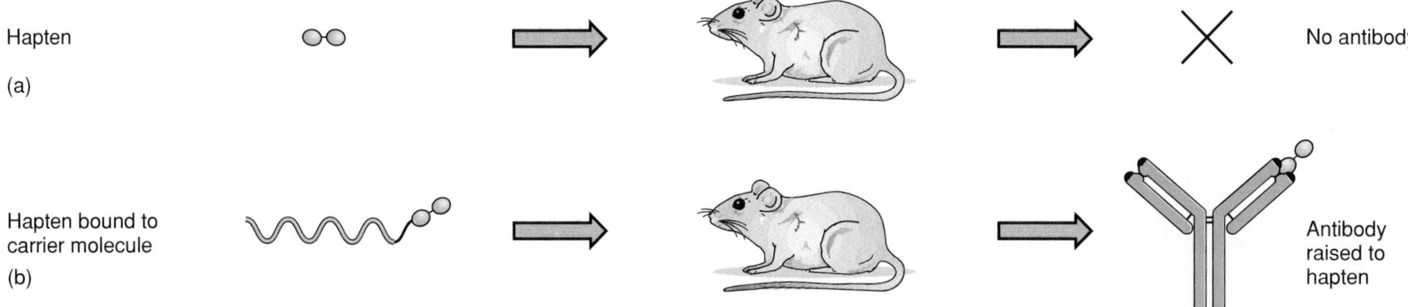

Figure 13.11 The hapten-carrier phenomenon. (*a*) Haptens are too small to be discovered by an animal's immune system; no response. (*b*) A hapten bound to a large molecule will serve as an antigenic determinant and stimulate a response and an antibody that is specific for it.

Special Types of Antigens Up to now, we have emphasized the role of microbial antigens, but tissues and proteins (including enzymes and antibodies) from other humans and animals are also antigenic. On occasion, even a self substance can take on the character of an antigen. How can this be so? During lymphocyte differentiation, immune tolerance to self tissues occurs, but a few anatomic sites may contain sequestered (hidden) molecules that escape this assessment. Such molecules, called **autoantigens,** occur in tissues (eye, thyroid gland, for example) that are walled off early in embryonic development before the surveillance system is in complete working order. Tolerance to these substances has not been established, so they may subsequently be mistaken as foreign, and this mechanism appears to account for some types of immune diseases (see autoimmune diseases in chapter 14).

Because each human being is genetically and biochemically unique (except for identical twins), the proteins and other molecules of one person may be antigenic to another. **Alloantigens (isoantigens)** are cell surface markers and molecules that occur in some members of the same species but not in others. Alloantigens are the basis for an individual's blood group (see chapter 14) and major histocompatibility profile, and they are responsible for incompatibilities that can occur in blood transfusion or organ grafting.

On the other hand, different organisms may possess some molecules with the same or a similar determinant group. These **heterophile** or heterogenetic antigens stimulate a response from the same lymphocyte clone. Antibodies raised against a heterophile antigen from one organism will **cross-react** with a similar or identical antigen from another source. Examples of antigens of different origins with very similar antigenic determinants are carbohydrate residues on the surfaces of bacteria and red blood cells, the antigens of group A streptococci and human heart tissue, and cardiolipin, a phospholipid present in a wide assortment of living things. Heterophile antigens may be partly responsible for diseases such as rheumatic fever and for false positive diagnostic tests (as in syphilis).

alloantigen, isoantigen (al'-oh, eye'-soh) Gr. *allos,* other, and *iso,* same.
heterophile (het'-ur-oh-fyl) Gr. *hetero,* other, and *phile,* to love.

Antigens that evoke allergic reactions, called **allergens,** will be characterized in detail in chapter 14.

Host Response to Antigens: A Cooperative Affair

The early events in the encounter between antigens and cells of the immune system are currently being clarified. Microbes and other foreign substances enter most often through the respiratory or gastrointestinal mucosa and less frequently through other mucous membranes, the skin, or across the placenta. When introduced intravenously, antigens become localized in the liver, spleen, bone marrow, kidney, and lung. If introduced by some other route, antigens are carried in lymphatic fluid and concentrated by the lymph nodes. The lymph nodes and spleen are important in concentrating the antigens and circulating them thoroughly through all areas populated by lymphocytes so that they come in contact with the proper clone. To help imagine this, see feature 13.1.

The Role of Macrophages—Antigen Processing and Presentation

In most immune reactions, the antigen must be further acted upon and formally presented to lymphocytes by special macrophages or other cells called **antigen-processing cells** (APCs). These cells engulf the antigen and alter it by partial degradation or addition of molecules, presumably to increase immunogenicity and recognition by lymphocytes. After processing is complete, the antigen is moved to the surface of the APC and bound to the MHC receptor so that it will be accessible to the lymphocytes during presentation (figure 13.12).

Presentation of Antigen to the Lymphocytes and Its Early Consequences

Various requirements must be met before an appropriate response to the macrophage-bound antigen will occur. **T-cell-dependent antigens** require that certain recognition steps occur between the macrophage, antigen, and lymphocytes. The first cells on the scene to assist the macrophage in activating B cells and other T cells are a special class of **helper T cells (T_H)**. This class of T cell bears a receptor that binds simultaneously with the class II MHC receptor on the macrophage and with one site on the antigen (figure 13.12). Once identification has occurred,

Feature 13.1 An Immunological Bargain Basement

Realizing the great variety and quantity of antigens that find their way into the body, one might wonder how this microscopic flotsam and jetsam can ever meet up with the right lymphocyte. After all, the body's fluid compartments are relatively large, and antigens are constantly entering through some portal or another. The events of this awesome sorting process could be likened to a rummage sale in a microscopic department store. The store (lymph node) is crowded with large numbers of all types of customers (lymphocytes), and it contains all sorts of merchandise (antigens). Picture the mounds of antigenic merchandise arrayed on tables and thousands of lymphocyte customers milling around, pawing through the piles of materials with their receptor hands to find just the right specificity (size, fit, color, shape, and brand). The sales personnel (macrophages and T-helper cells) display the merchandise, help in selection, and expedite the sale (contact of the antigen with its matching lymphocyte). Like customers, some lymphocytes become "experienced shoppers." On subsequent shopping sprees, they retain a memory of previous sales and are even more efficient.

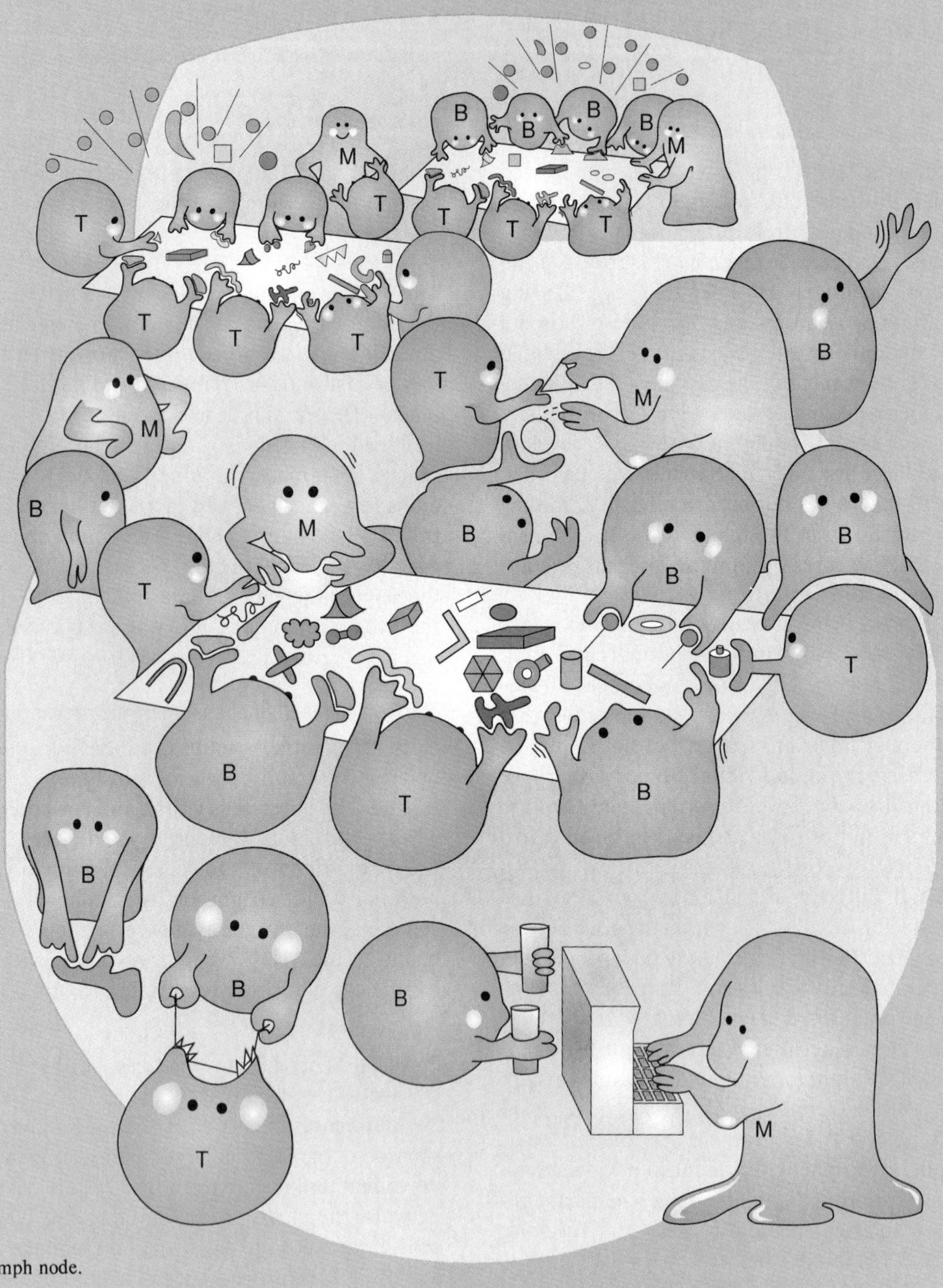

Rummage sale in a lymph node.

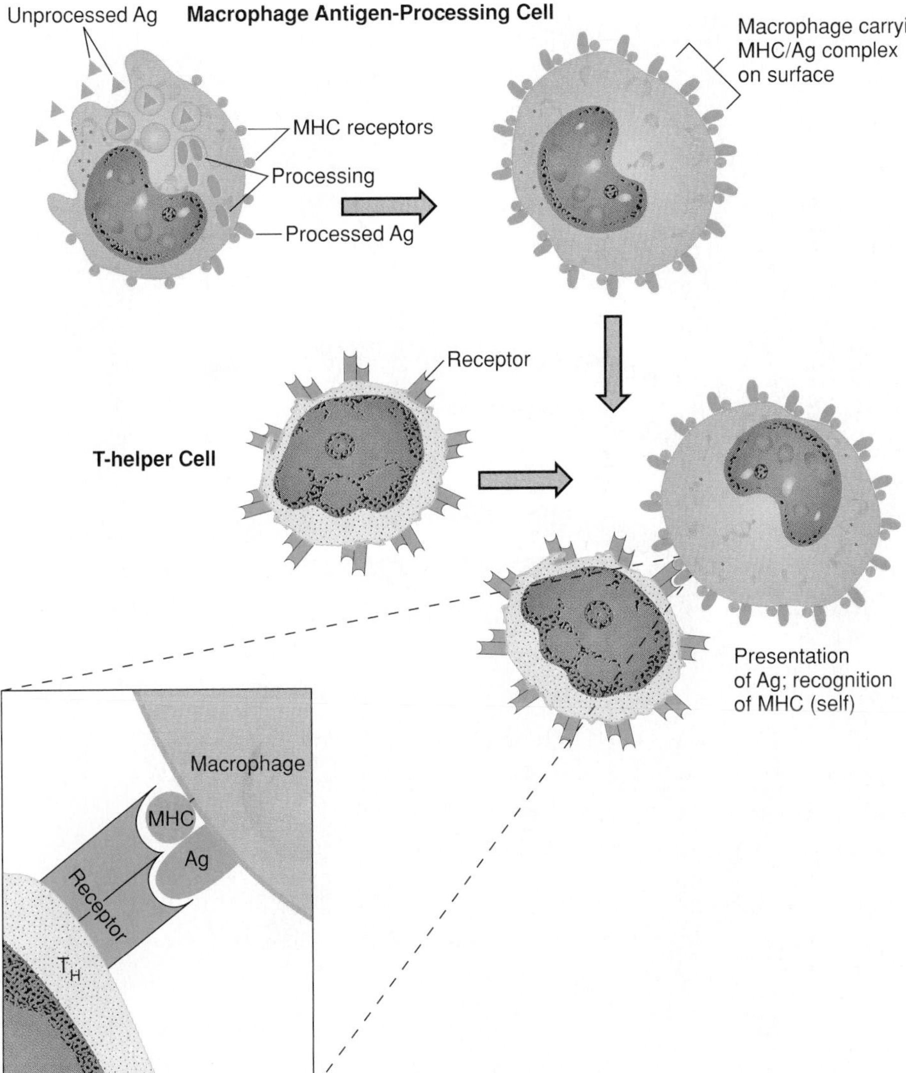

Figure 13.12 Cell cooperation between a macrophage and a T cell during the initial response to antigen. Inset of macrophage-lymphocyte identification shows that the macrophage-receptor complex fits precisely with the receptor of the T_H cell so that there is recognition of both self and nonself.

a monokine, **interleukin-1 (IL-1),** produced by the macrophage, activates this T-helper cell. The T_H cell, in turn, produces a lymphokine, **interleukin-2 (IL-2),** that stimulates a general increase in activity of committed B and T cells (see feature 13.5). The manner in which B and T cells subsequently become activated by the macrophage/T-helper cell complex and their individual responses to antigen will be addressed separately in sections III B and IV, and III T and V.

A few antigens can trigger a general B-lymphocyte response without the cooperation of macrophages or T-helper cells. These **T-cell-independent** antigens are usually simple molecules such as carbohydrates with many repeating and invariable determinant groups. Examples include lipopolysaccharide from the cell wall of *Escherichia coli* and polysaccharide from the capsule of *Streptococcus pneumoniae.* Because so few antigens are of this type, most B-cell reactions require T cells.

interleukin (in″-tur-loo′-kin) A chemical that carries signals between white blood cells.

III B. The Response of B Lymphocytes to Antigenic Challenge: Clonal Selection, Expansion, and Antibody Production

The immunologic activation of most B cells requires a series of events (figure 13.13):

1. **Presentation of antigen and clonal selection.** In this case, a precommitted B cell of a particular clonal specificity is presented with the antigen by the macrophage complex so that the antigen binds to the B-cell receptors. Further recognition of self, involving the binding of a B-cell MHC receptor to the T_H cell, also occurs at this point.
2. **Instruction by chemical mediators.** The B cell receives developmental signals secreted by the macrophage (interleukin-1) and T cells (interleukin-2 and various B-cell growth and differentiation factors).
3. The combination of these stimuli on the membrane receptors causes a signal to be transmitted internally to the B-cell nucleus.

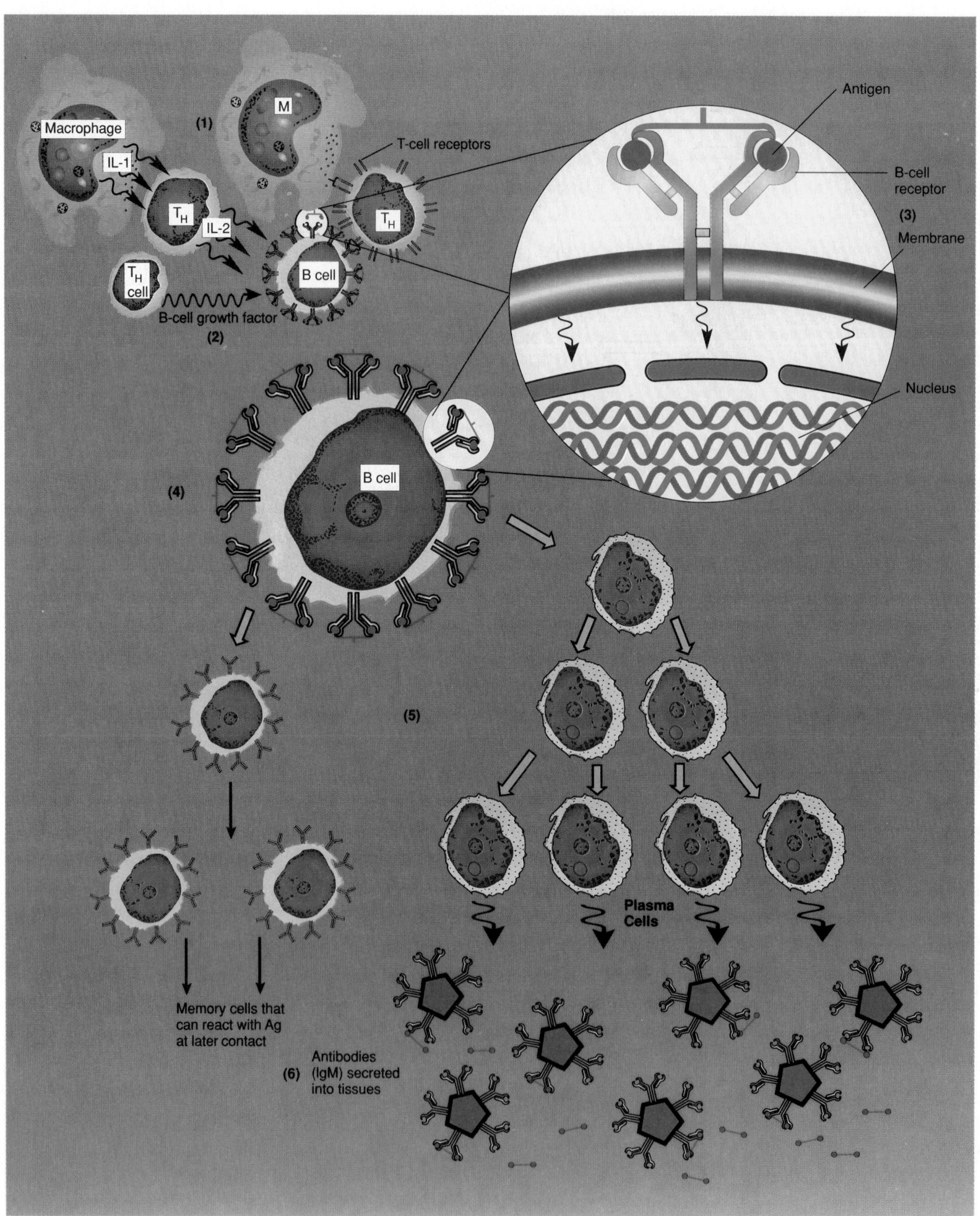

Figure 13.13 Events in B-cell activation. (*1*) Clonal selection. Antigen is presented to the B cell by macrophage; activated T cells facilitate recognition by binding to both the macrophage and the B cell and (*2*) by producing growth factors that stimulate the B cell. (*3*, inset) The binding of antigen on the immunoglobulin receptors of the B cell transmits a signal to the cell nucleus to begin activation of the B cell. (*4*) The B cell is transformed into a blast cell. (*5*) The blast cell undergoes mitotic divisions and clonal expansion into memory and plasma cells. (*6*) Plasma cells secrete antibodies.

4. This event triggers the cell to begin **blast transformation.** A blast is an enlarged, highly active cell whose DNA synthesis and organelle bulk are greatly increased in preparation for mitotic divisions.
5. A B cell transformed into the blast stage multiplies through successive mitotic divisions and produces a large population of genetically identical daughter cells. Some cells that stop short of becoming fully differentiated are **memory cells,** which remain in the immune repertoire should that same antigen enter again at a later time. This reaction also increases the clone size, so the next time that same antigen enters, even more cells will have that specificity. By far the most numerous progeny are large, specialized, terminally differentiated B cells called **plasma cells.**
6. Plasma cells have one purpose: to secrete into the surrounding tissues copious amounts of antibodies with the same specificity as the original receptor (figure 13.13). Although an individual plasma cell can produce around 2,000 antibodies per second, this does not continue indefinitely because of regulation from the T-suppressor (T_S) class of cells. The plasma cells do not survive for long, and the making of antibodies is something of a last suicidal act.

IV. Characteristics of Antibodies: Structure, Function, Classes

Antibody Design: The Structure of "Magic Bullets"[2]

Earlier we saw that a basic immunoglobulin molecule contains four polypeptide chains connected by disulfide bonds. Let us view this structure once again, using an **IgG molecule** as a model. Two functionally distinct segments called fragments can be differentiated. The two "arms" that bind antigen are termed **antigen binding fragments (Fab),** and the rest of the molecule is the **crystallizable fragment (Fc),** so called because it has been crystallized in pure form. The distal end of each Fab fragment (consisting of the variable regions of the heavy and light chains) folds into a groove that will accommodate one antigenic determinant. The Fc fragment is involved in binding to various cells and molecules of the immune system itself. Although the overall molecular model is Y-shaped, it can also conform to a T shape because of a special *hinge* region at the site of attachment between the Fab and Fc fragments. Swivelling at the hinge permits the Fab fragments to change their angle to accommodate nearby antigen sites that may vary slightly in distance and position. Figure 13.14 shows three views of antibody structure, and feature 13.2 suggests a working model.

2. The words of Paul Erhlich, a German bacteriologist, who first described the remarkable behavior of antibodies.

Feature 13.2 The Headless Antibody

A vivid (and we hope memorable) analogy compares basic immunoglobin structure to a human body standing with the legs together and the arms outspread. It is a fair model of the structure if you pretend that the head is gone and realize that no body part corresponds to a section of the light chains. The arms (Fab fragments) are connected to the body by hinges (the hinge regions) that may move the arms up and down. The body and legs (Fc fragments) may be planted solidly in one spot. The structure is symmetrical—that is, one side is a mirror image of the other. The model is also functionally dynamic. It reminds us that the basic immunoglobin molecule has two antigen binding sites (hands, which could be likened to the hypervariable region that binds antigen). Using the fingers and thumb, the interior three-dimensional shape of a hand can be varied (like Ag binding sites) to conform to and hold a variety of materials (just think of the infinite shapes of the items one's hand can precisely grip: paper, pen, steering wheel, utensils, book, handlebars, basketball, and other hands).

Antibody-Antigen Interactions and the Function of the Fab

The site on the antibody where the antigenic determinant inserts is composed of a *hypervariable region* whose amino acid content may be extremely varied. Antibodies differ somewhat in the exactness of this groove for antigen, but a certain complementary fit is necessary for the antigen to be held effectively (figure 13.15). The specificity of antigen binding sites for antigens is very similar to enzymes and substrates (in fact, some antibodies are used as enzymes; see feature 7.2). So specific are some immunoglobulins for antigen that they can distinguish between a single functional group of a few atoms. Because of its pair of identical Fab sites, an IgG molecule can bind two sites on one cell, or it may bind sites on two separate cells and thereby link them.

The principal activity of an antibody is to unite with, immobilize, call attention to, or neutralize the antigen for which it was formed. Antibodies called **opsonins** stimulate **opsonization,** a process in which microorganisms or other particles are coated with specific antibodies so that they will be more readily recognized by phagocytes, which dispose of them (figure 13.16). Opsonization has been cleverly likened to putting handles on a slippery object to provide phagocytes a better grip. The interaction of an antibody with complement can result in the specific rupturing of cells and some viruses. The capacity for antibodies to aggregate or agglutinate antigens is also one principle behind highly sensitive and specific laboratory tests to be discussed later. In **neutralization** reactions, antibodies fill the surface receptors on a virus or the active site on a molecule to prevent it from functioning normally. **Antitoxins** are a special type of antibody that neutralize bacterial exotoxins. It should be noted that not all antibodies are protective—some neither benefit nor harm, and a few actually cause diseases (see chapter 14).

opsonization (awp″-son-uh-zay′-shun) Gr. *opsonein,* to prepare food.

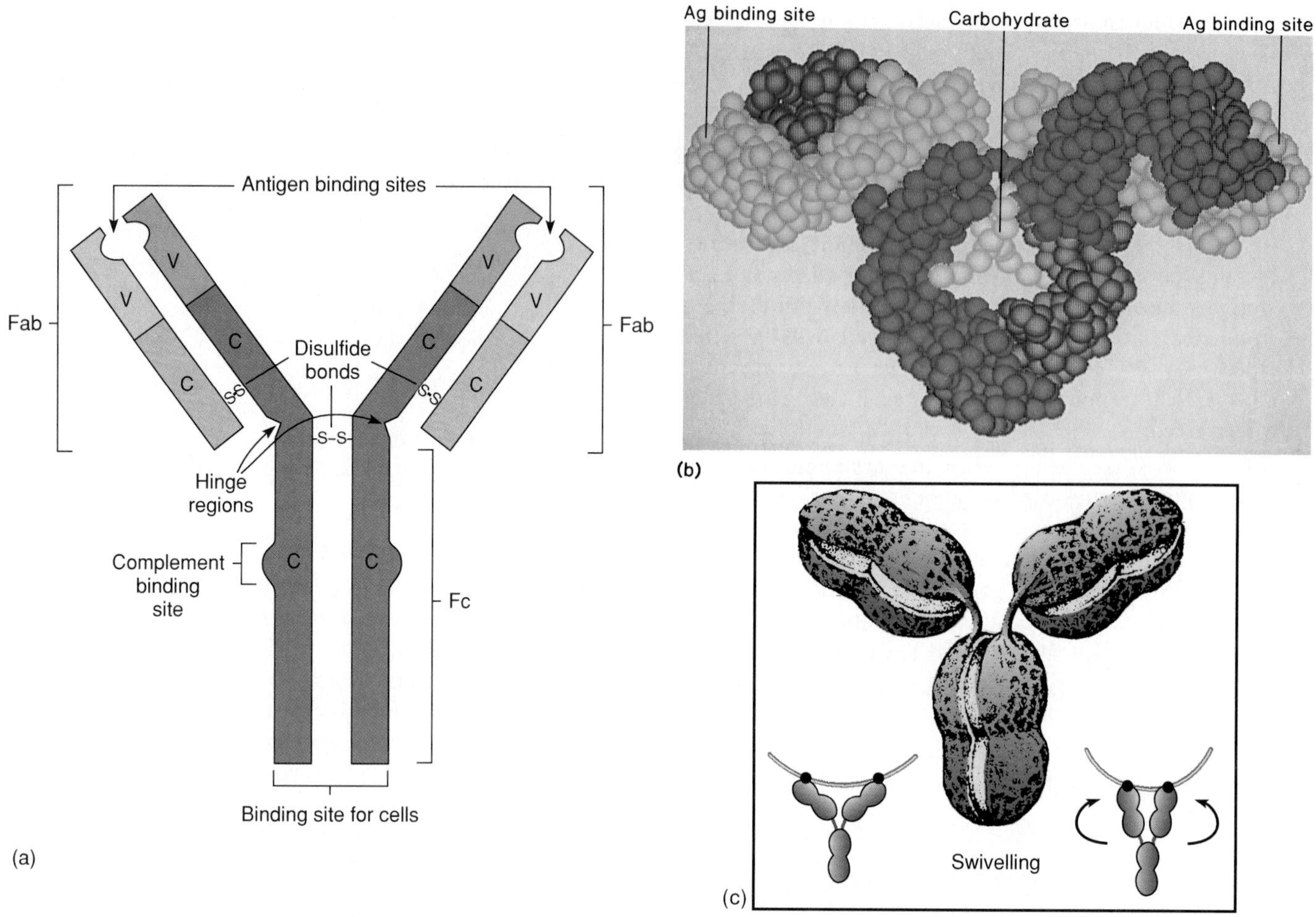

Ag binding site
Carbohydrate
Ag binding site

(b)

Antigen binding sites
Fab
V
V
C
C
Disulfide bonds
S-S
Hinge regions
Complement binding site
C
C
Fc
Binding site for cells
(a)

Swivelling
(c)

Figure 13.14 Working models of antibody structure. (*a*) Diagrammatic view of IgG depicts the principal functional areas (Fab and Fc) of the molecule. (*b*) Realistic model of immunoglobulin shows the tertiary and quaternary structure achieved by additional intrachain and interchain bonds and the position of the carbohydrate component. (*c*) The "peanut" model of IgG helps illustrate swivelling of Fabs relative to one another and to Fc.

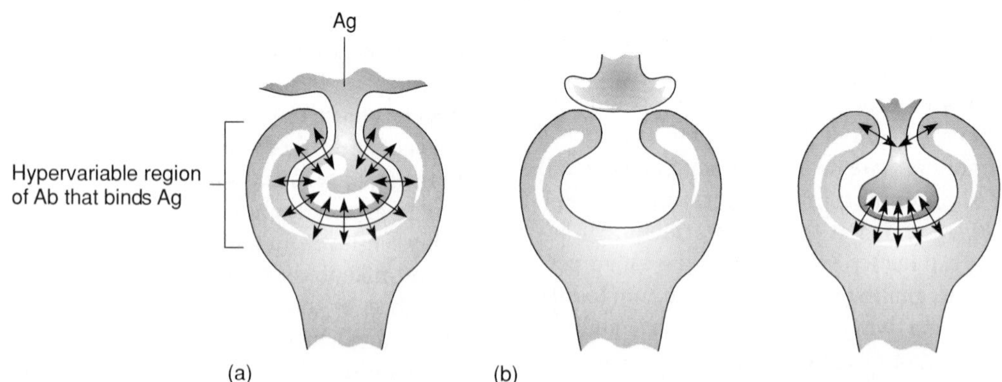

Figure 13.15 Antigen-antibody binding. (*a*) The union of Ab and Ag is characterized by a certain degree of fit and supported by weak linkages such as hydrogen bonds and electrostatic attraction. In a snug fit such as that shown here, there is greater opportunity for attraction and strong attachment. The strength of this union confers high affinity. (*b*) Examples of potential interactions of other antigens with this same antibody. The first Ag clearly cannot be accommodated, whereas the second antigen has a near fit that could effectively bind the Ab. This illustrates how cross reactivity can occur among Ags and Abs.

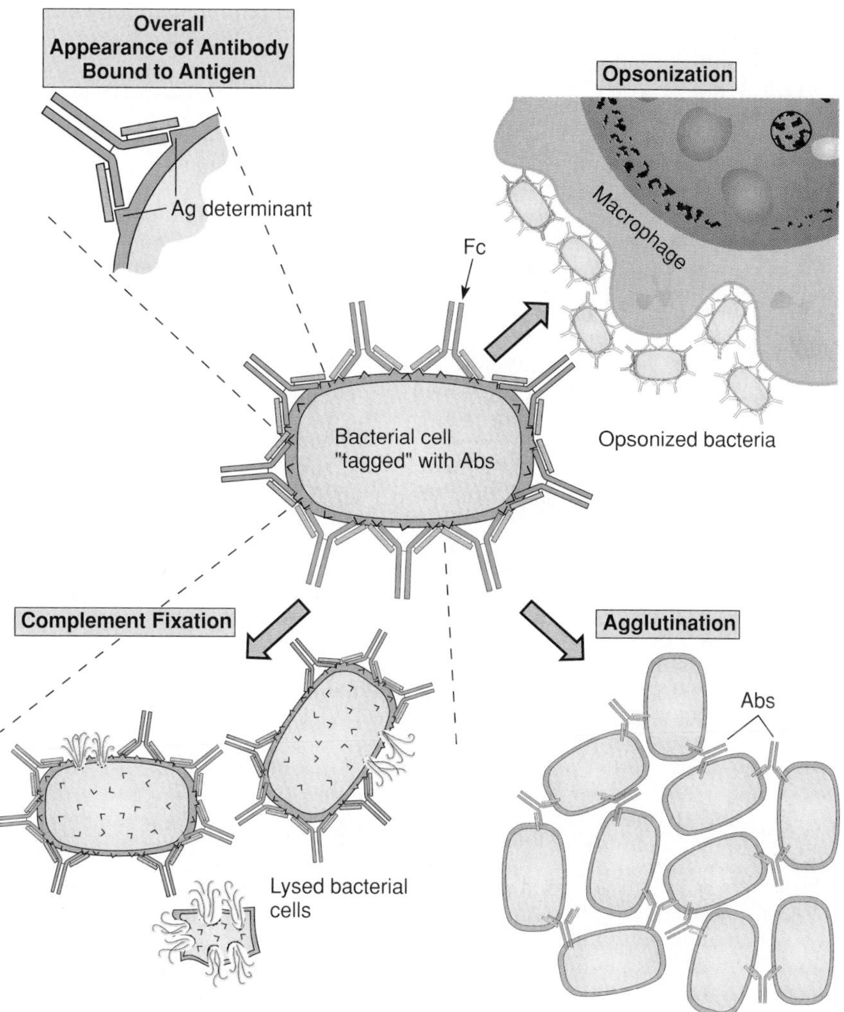

Figure 13.16 Summary of antibody functions.

Functions of the Crystallizable Fragment: Interactions with Self

Although the Fab fragments bind antigen, the Fc fragment has a different binding function. In most classes of immunoglobulin, the proximal end of Fc contains an effector molecule that can bind to the Fc receptors on the membrane of one's own cells, including macrophages, neutrophils, eosinophils, mast cells, basophils, and lymphocytes. The effect of an antibody's Fc fragment binding to a cell receptor depends upon that cell's role. In the case of opsonization, the antibody coats foreign cells and viruses in such a way that the Fc fragments are exposed to neutrophils and macrophages. Certain antibodies have receptors on the Fc portion for fixing complement, and in some immune reactions, the binding of Fc causes the release of chemical mediators. For example, the antibody of allergy (IgE) binds to basophils and mast cells, which causes the release of allergic mediators such as histamine (see chapter 14). The size and amino acid composition of Fc also determine an antibody's permeability, its distribution in the body, and its class.

Accessory Molecules on Immunoglobulins

All antibodies contain molecules in addition to the basic polypeptides. Varying amounts of carbohydrates are affixed to the constant regions in most instances (table 13.2). Two additional accessory molecules are the *J chain* named both for its shape and for its function in joining polymeric Igs, and the *secretory component,* which helps move Ig across mucous membranes. These proteins occur only in certain immunoglobulin classes

The Classes of Immunoglobulins

Immunoglobulins exist as structural and functional classes called *isotypes* (compared and contrasted in table 13.2). The differences in these classes are due primarily to variations in the Fc fragment and its accessory molecules. The classes are differentiated with shorthand names (Ig, followed by a letter: IgM, IgG, IgA, IgD, IgE).

IgM (M for *macro*) is a huge molecule composed of five monomers (making it a pentamer) attached by the Fc receptors

Table 13.2 Characteristics of the Immunoglobulin (Ig) Classes

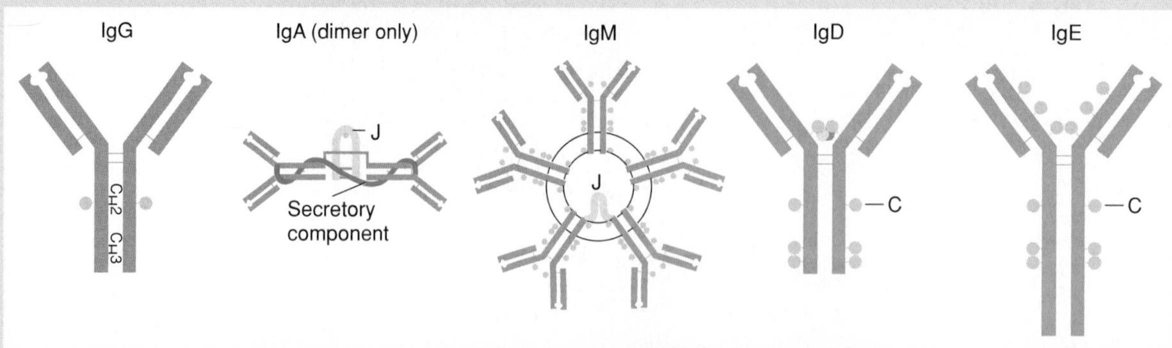

	IgG	IgA (dimer only)	IgM	IgD	IgE
	Monomer	**Dimer, Monomer**	**Pentamer**	**Monomer**	**Monomer**
Number of Antigen Binding Sites	2	4 2	10	2	2
Molecular Weight	150,000	170,000–385,000	900,000	180,000	200,000
Percent of Total Antibody in Serum	80%	13%	6%	1%	0.002%
Average Life in Serum (Days)	23	6	5	3	2.5
Crosses Placenta?	Yes	No	No	No	No
Fixes Complement?	Yes	No	Yes	No	No
Fc Binds To	Phagocytes	Phagocytes	B lymphocytes	B lymphocytes	Mast cells and basophils
Biological Function	Long-term immunity; memory Abs	Secretory Ab; on mucous membranes	Produced at first response to Ag; can serve as B-cell receptor	Receptor on B cells	Antibody of some types of allergy; worm infections

C = Carbohydrate
J = J chain

to a central J chain. With its 10 binding sites, this molecule has tremendous avidity for antigen (avidity means the capacity to bind antigens). It is the first class expressed, both as a receptor in the educated B cell and as an antibody released by a plasma cell following initial encounter with antigen. Its complement-fixing and opsonizing qualities make it an important antibody in many immune reactions. It circulates mainly in the blood and is far too large to cross the placental barrier.

The structure of **IgG** has already been presented. It is a monomer produced by memory cells responding the second time to a given antigenic stimulus. It is by far the most prevalent antibody circulating throughout the tissue fluids and blood. It has numerous functions: It neutralizes antitoxins, opsonizes, fixes complement, and is the only antibody capable of crossing the placenta.

The two forms of **IgA** are: (1) a monomer that circulates in small amounts in the blood and (2) a dimer that is a significant component of the mucous and serous secretions of the salivary glands, intestine, nasal membrane, breast, lung, and genitourinary tract. The dimer, called **secretory IgA,** is formed in a plasma cell by two monomers attached by a J piece. To facilitate the transport of IgA across membranes, a secretory piece is later added by the gland cells themselves. IgA coats the surface of these membranes and appears free in saliva, tears, colostrum, and mucus. It confers the most important specific local immunity to enteric, respiratory, and genitourinary pathogens. Its contribution in protecting newborns who derive it passively from nursing has been mentioned (see feature 12.7).

The function of **IgD** has been something of a mystery. It is a monomer found in miniscule amounts in the serum, and it does not fix complement, opsonize, or cross the placenta. What is known is that it is found on immature B cells as a receptor for antigen, usually along with IgM. It seems to be the triggering molecule for B-cell activation, and it may also play a role in immune suppression.

IgE is also an uncommon blood component unless one is allergic or has a parasitic worm infection. Its Fc region interacts with receptors on mast cells and basophils. Its biological significance appears to be in stimulating an inflammatory response through the release of potent physiological substances by the basophils and mast cells. Because inflammation would enlist blood cells such as eosinophils and lymphocytes to the site of

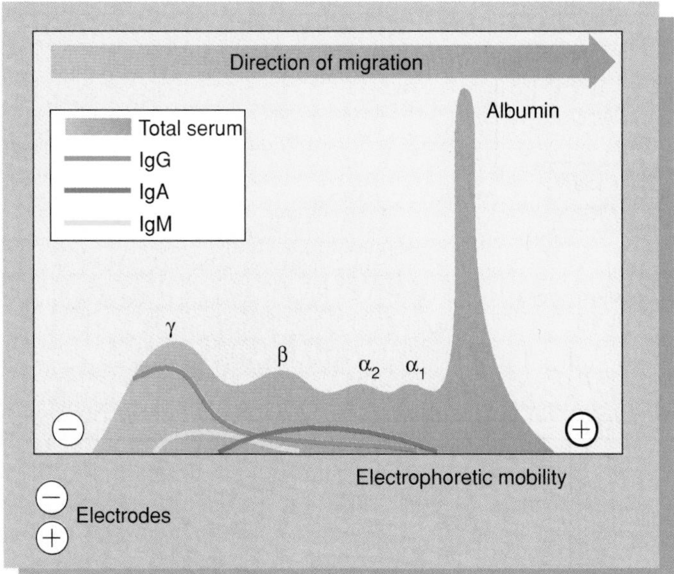

Figure 13.17 Pattern of human serum following electrophoresis. When antiserum is subjected to electric current, the various proteinaceous components are separated into bands. The relative sizes of the molecules are somewhat indicated here if one knows that heavier molecules migrate more slowly than light ones.

infection, it would certainly be one defense against parasites. Unfortunately, IgE has another, more insidious effect—that of mediating anaphylaxis, asthma, and certain other allergies.

Evidence of Antibodies in Serum

Regardless of the site where antibodies are first secreted, a large quantity eventually ends up in the blood by way of the body's communicating networks. If one submits a sample of **antiserum** (serum containing specific antibodies) to electrophoresis, the major groups of proteins migrate in a pattern consistent with their mobility and size (figure 13.17). The albumins show up in one band, and the globulins in four bands called alpha-1 (α_1), alpha-2 (α_2), beta (β), and gamma (γ) globulins. Most of the globulins represent antibodies, which explains how the term immunoglobulin was derived. **Gamma globulin** is composed primarily of IgG, whereas β and α_2 globulins are a mixture of IgG, IgA, and IgM. As we will see, the gamma globulin fraction of serum is important in passive immune therapies.

Monitoring Antibody Production over Time: Primary and Secondary Responses to Antigens

We can learn a great deal about how the immune system reacts to an antigen by studying the levels of antibodies in serum over time (figure 13.18). A quantitative way to express this level is the **titer,** or concentration of antibodies. Invaded for the first time by a given antigen, the system undergoes a **primary response.** Even though the earliest part of this response, the *latent period,* is marked by a lack of antibodies for that antigen, much activity is occurring. During this time, the antigen is being concentrated in lymphoid tissue, taken up by macrophages, and presented to the correct clones of B lymphocytes. As plasma cell progeny synthesize antibodies, the serum titer increases to a certain plateau and then tapers off to a low level over a few weeks or months. When the class of antibodies produced during this response is tested, an important characteristic of the response is uncovered. It turns out that, early in the primary response, most of the antibodies are of the IgM type, which is the first class to be assembled by B cells. Later, the class of the antibodies (but not their specificity) is switched to IgG.

When the immune system is exposed again to the same immunogen within weeks, months, or even years, the response is amplified. The rate of antibody synthesis, the peak titer, and the length of antibody persistence are greatly increased over the primary response. The rapidity and strength of this **secondary response** are attributable to the memory B cells that were formed during the primary response (an elite corps of "veterans" previously trained to attack that antigen). Because of its association with recall, the secondary response is also called the **anamnestic response.** The advantage of this response is evident: It provides a quick and potent strike against subsequent exposures to infectious agents. This memory effect has been well used in providing artificial active immunity by giving **boosters**—additional doses of vaccine—to increase the serum titer.

Monoclonal Antibodies: A New Technology from Cancer Cells

The value of antibodies as tools for locating or identifying antigens is well established. For many years, antiserum extracted from human or animal blood was the main source of antibodies for tests and therapy, but most antiserum has a basic problem: It is **polyclonal,** containing a mixture of antibodies with multiple specificities derived from several clones. This is to be expected, because several immune reactions may be occurring simultaneously, and even a single species of microbe can stimulate several different types of antibodies. For therapy, testing, and research, a pure preparation of **monoclonal antibodies** (MABs), originating from a single clone and having a single specificity, has greater desirability.

Until recently, producing monoclonal antibodies in reasonably large quantities had not been possible. Owing to the manner in which antibodies are formed in the living body, it is very difficult to stimulate only one clone of lymphocytes *in vivo* and then to collect the resulting antibodies. On the surface, it might seem that one could use cell sorting techniques to isolate a single B cell, expose it to antigen in a culture dish, and collect the antibodies. The problem is that B cells transform to antibody-secreting plasma cells, and plasma cells cannot be grown indefinitely in cell culture. A technology involving cell hybridization that was introduced in the 1970s solved this problem (figure 13.19). A central part of this technique arose from the discovery that tumors isolated from multiple *myelomas* in mice consist of identical plasma cells. These monoclonal

titer (ty'-tur) Fr. *titre,* standard. One method for determining titer is shown in figure 13.26.

anamnestic (an-am-ness'-tik) Gr. *anamnesis,* a recalling.

myeloma (my-uh-loh'-muh) Gr. *myelos,* marrow, and *oma,* tumor. A malignancy of the bone marrow.

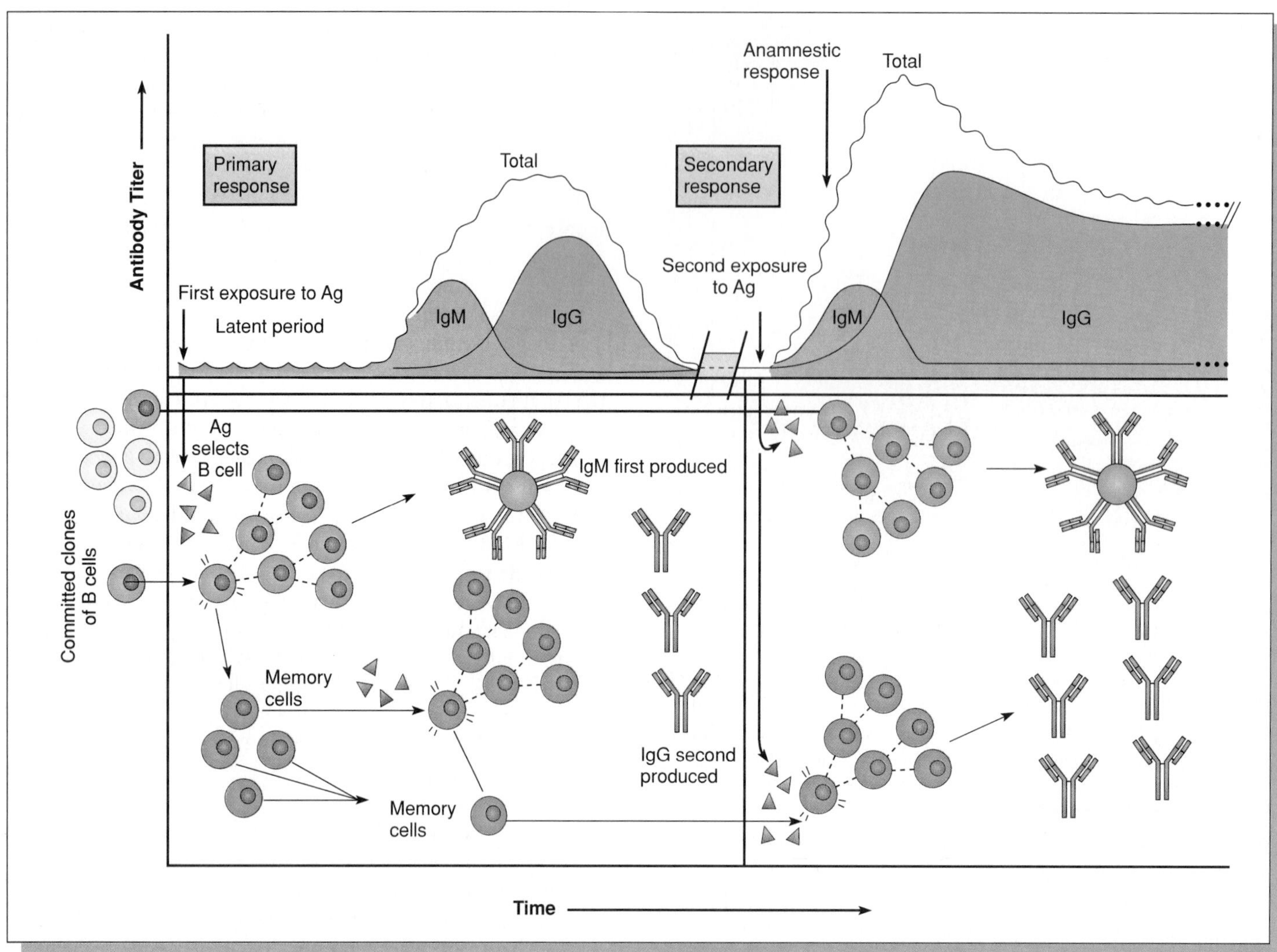

Figure 13.18 Primary and secondary responses to antigens. Top: The pattern of antibody titer and subclasses as monitored during initial and subsequent exposure to the same antigen. Bottom: A view of the B-cell responses that account for the pattern. Depicted are clonal selection, clonal expansion, production of memory cells, and the predominant Ab class occurring at first and second contact with Ag.

plasma cells were found to secrete a strikingly pure form of antibodies with a single specificity and to continue to divide indefinitely. Unfortunately, however, the antibodies were defective. Immunologists recognized the potential in these plasma cells and devised a **hybridoma** approach to creating MABs. The basic idea behind this approach is to hybridize or fuse a myeloma cell with a normal plasma cell from a mouse spleen to create an immortal cell that secretes a supply of functional antibodies with a single specificity.

The introduction of this technology opened up numerous biomedical applications. Monoclonal antibodies have provided immunologists with excellent standardized tools for studying the immune system and for expanding disease diagnosis and treatment. Most of the successful applications thus far use MABs in *in vitro* diagnostic testing and research. Although injecting monoclonal antibodies to treat human disease is an exciting prospect, so far this therapy has been stymied because most MABs are of mouse origin, and many humans are hypersensitive to them. The development of human MABs and other novel approaches is currently underway (see feature 13.3).

III T and V. The Nature of Cell-Mediated Immunity (CMI): How T Cells Respond to Antigen

All the time that B cells have been actively responding to antigens, the T-cell limb of the system has been similarly engaged. The responses of T cells, however, are **cell-mediated** immunities, which require the direct involvement of T lymphocytes throughout the course of the reaction. These reactions are among the most complex and diverse in the immune system, and involve several subsets of T cells whose particular actions are dictated by CD receptors. All mature T cells have CD2 (causes rosetting) and CD3, but **CD4** and **CD8** are found only on certain classes (table 13.3). T cells are restricted—that is, they require some type of MHC (self) recognition before they can be

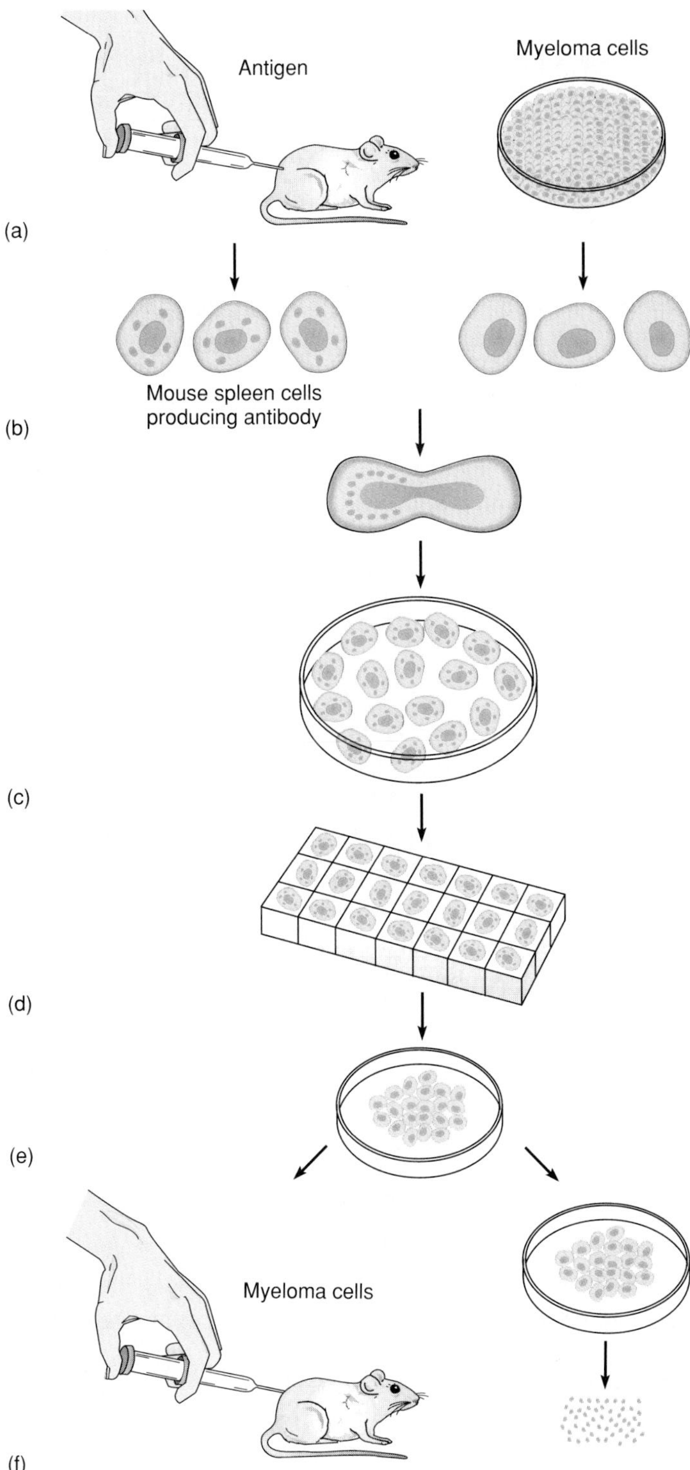

Figure 13.19 Summary of the technique for producing monoclonal antibodies from hybridized myeloma and normal plasma cells. (*a*) A normal mouse is inoculated with selected antigen; plasma cells are isolated from its spleen. (*b*) A strain of mouse with myeloma provides tumor cells. (*c*) The two cell populations are mixed with polyethylene glycol, which causes some cells in the mixture to fuse and form hybridomas. (*d*) Surviving cells are cultured and separated into individual wells. (*e*) Tests are performed on each hybridoma to determine its antibody specificity. (*f*) A hybridoma with the desired specificity is grown in tissue culture; antibodies are isolated and purified. The isolated hybridoma is maintained in a myeloma-susceptible mouse or frozen for future use.

Barrett, et al., *Biology*, © 1986, p. 338. Reprinted by permission of Prentice-Hall, Inc., Englewood Cliffs, NJ.

Table 13.3 Characteristics of Subsets of T Cells

Subset	Shorthand Designations	Functions/ Important Features
T-helper or inducer cells	T_H, T_4	Assist B cells in recognition of antigen; assist other subsets of T cells in recognition and reaction to Ag; identified by CD4 receptors
T-suppressor	T_s, T_8	Regulate immune reactions; cells limit the extent of antibody production; block some T-cell activity; identified by CD8 receptors
Cytotoxic killer cells	T_K, T_{CTL}	Destroy a target foreign cell by lysis; important in destruction of complex microbes, cancer cells, virus-infected cells; graft rejection; allergy
Delayed hypersensitivity	T_D, T_{DTH}	Responsible for allergies occurring several hours or days after contact; skin reactions as in tuberculin test

activated, and all produce chemical mediators or lymphokines with a spectrum of biological effects (see feature 13.4).

The Activation of T Cells and Their Differentiation into Subsets

The educated T cells in lymphoid organs are primed to react with antigens that are processed and presented to them by macrophages. A T cell is initially **sensitized** when antigen is bound to its receptor. By mechanisms not yet fully characterized, sensitization leads to the final differentiation of the cell into one of the four functionally specialized subsets: T-helper, T-suppressor, cytotoxic, or delayed hypersensitivity (table 13.3 and figure 13.20). As with B cells, activated T cells transform into lymphoblasts in preparation for mitotic divisions, and they divide into effector cells and memory cells that can interact with the antigen upon subsequent contact. Memory T cells are some of the longest lived blood cells known (70 years in one well-documented case).

T-Helper Cells Helper or inducer cells play a central role in immune reactions to antigens, including those of B cells and other T cells. They do this directly by receptor contact, and indirectly through various lymphokines (interleukin-2, B-cell growth factor) that recruit or help differentiate lymphocytes. T-helper cells are the most prevalent type of T cell in the blood and lymphoid organs, making up about 65% of this population. The severe depression of this class of T cells (with CD4 receptors) by HIV is what largely accounts for the immunopathology of AIDS.

Feature 13.3 Monoclonal Antibodies—Variety Without Limit

Imagine releasing millions of tiny homing pigeons into a molecular jungle and having them home directly to their proper roost, and you have some sense of what monoclonal antibodies can do. Laboratories use them to identify antigens, receptors, and antibodies; to differentiate cell types (T cells versus B cells) and cell subtypes (different sets of T cells); to diagnose diseases such as cancer and AIDS; and to identify bacteria and viruses.

A number of promising techniques have been directed toward using monoclonals as drugs. A newer method combining genetic engineering with hybridoma technology has opened up the potential for producing antibodies of almost any desired specificity and makeup. For instance, *chimeric* MABs that are part human and part mouse have been produced through splicing antibody genes. It is even possible to design an antibody that has antigen binding sites with different specificities. With this technology, extra molecular groups may be added to increase antibody sensitivity or to give antibodies destructive powers. Monoclonals can be hybridized with plant or bacterial toxins to form **immunotoxins,** complexes that seek out and attach to a cell and poison it. The most exciting prospect of this therapy is that it can destroy a specified cancer cell and not harm normal cells. Such antibodies could also be used to suppress allergies, autoimmunities, and graft rejection. One such drug, Orthoclone, is currently used to prevent the rejection of kidney transplants by incapacitating cytotoxic T cells. Now that researchers have genetically engineered plants to produce human MABs, there literally will be no limitations on future therapeutic uses.

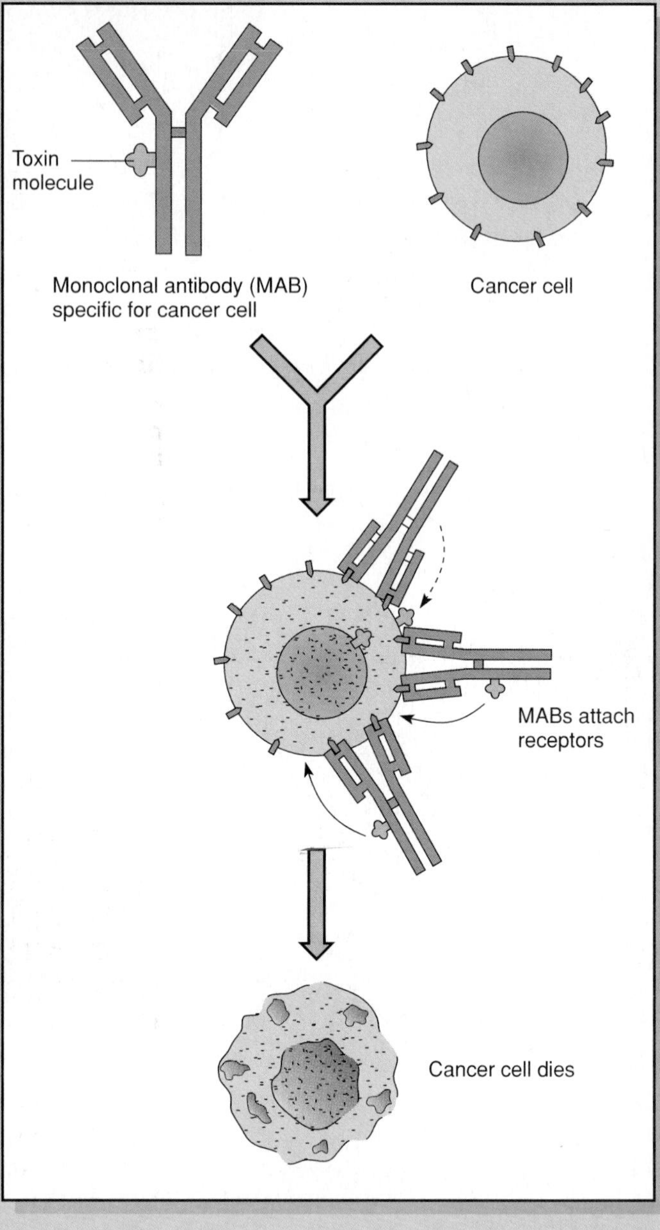

The mechanism of an immunotoxin. A potential therapy for cancer and immune dysfunctions is the use of a monoclonal antibody specific for a certain tumor that carries a potent toxin molecule. This antibody would circulate to cancer cells and deliver the toxin to them. Normal body cells would be unharmed.

T-Suppressor Cells An essential part of the immune mechanism is to restrict rampant, uncontrolled immune responses that could be inappropriate or destructive. Although immunosuppression reactions are not completely understood, they are known to involve T-suppressor cells (those that bear the CD8 receptor).

Cytotoxic T Cells: Cells That Kill Other Cells **Cytotoxicity** is the capacity of certain T cells to kill a specific target cell. It is a fascinating and powerful property that accounts for much of our immunity to foreign cells and cancer, and yet, under some circumstances, it can lead to disease. In essence, an activated **killer T cell** recognizes its target cell through receptors and

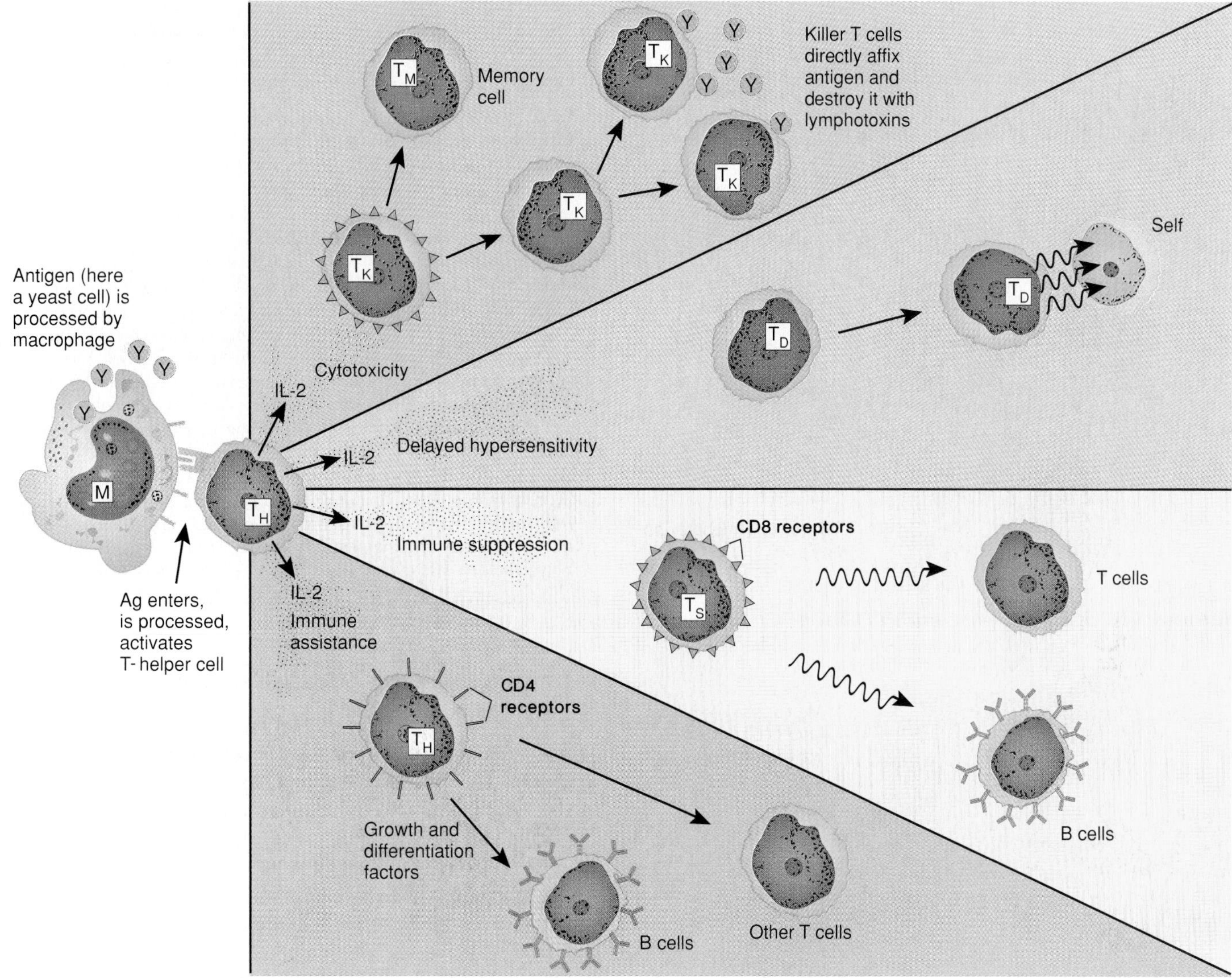

Figure 13.20 Overall scheme of T-cell activation and differentiation into various subsets.

mounts a direct toxic attack upon it. Examples of foreign cells that effectively provoke cytotoxic T cells are:

1. Fungi, protozoa, and complex bacteria (mycobacteria).
2. Virally infected cells (figure 13.21). Killer cells recognize these because of telltale virus receptors expressed on their surface. Cytotoxic defenses are an essential protection against viruses.
3. Cancer cells. T cells constantly survey the tissues and immediately attack any abnormal cells they encounter. (See the chapter opening illustration for a dramatic image of this behavior.) The importance of this function is clearly demonstrated in the susceptibility of T-cell-deficient persons to cancer (chapter 14).
4. Cells from other animals and humans. Cytotoxic CMI is the most important factor in **graft rejection.**
5. Reactions that destroy one's own cells (some types of hypersensitivity; see chapter 14).

Other Types of Killer Cells Some killer cells are not restricted to a single type of antigen. **Natural killer** (**NK**) cells appear to be relatives of T cells that occur in the spleen, blood, and lungs. These cells are endowed with natural, nonspecific cytotoxic powers against various types of cancer cells (leukemia and carcinoma for example). When NK cells are stimulated by interferon, they show greater activity against virus-infected cells. A novel immunotherapy for cancer involves lymphokine-activated killer cells (LAK), which are NK cells incubated with the immune stimulant interleukin-2.

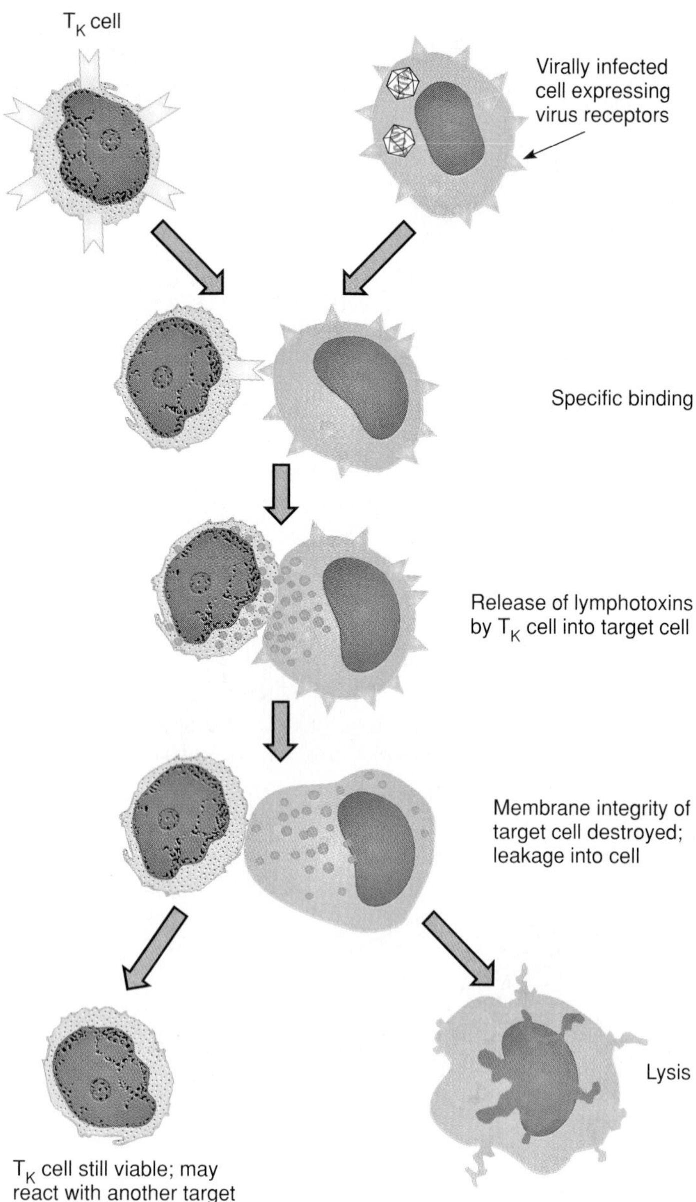

Figure 13.21 Stages of cell-mediated cytotoxicity and the action of lymphotoxins on target cells.

Delayed Hypersensitivity T Cells Although immediate allergies such as hay fever and anaphylaxis are mediated by antibodies, certain delayed responses to allergens (the tuberculin reactions for example) are brought on by special T cells. These reactions will be discussed in chapter 14.

Practical Applications of Immunological Function

Knowing how the immune system reacts to antigens has led to extremely valuable biomedical applications in two major areas: (1) use of antiserum and vaccination to provide artificial protection against disease, and (2) diagnosis of disease through immunologic testing.

Feature 13.4 Lymphokines—Chemical Products of T Cells

Although the immunities of T cells are usually thought of as cell-mediated, one must not overlook the fact that T cells are also prolific chemical factories. The products of sensitized T cells that communicate with and act upon other cells are **lymphokines.** Lymphokines have diverse effects. Some (interferon and interleukin) regulate immune reactions; some mediate inflammation (chemotactic factors); and others (lymphotoxins) kill whole cells. In the previous chapter, we mentioned the role of **gamma interferon** in regulating B cells and T cells, activating natural killer cells (NK), and stimulating macrophages. **Interleukin-2** was discussed in conjunction with T-helper cells and its role as a growth promoter and general stimulus for lymphocyte activity. **Interleukin-3** is a powerful stimulus for stem cell development in the bone marrow. Other lymphokines (B-cell growth factor and B-cell differentiation factor) are important activators of B cells.

For years it has not been clear just how cytotoxic cells destroy their whole cell targets. Now it appears that they are very predatory, directly contacting the foreign cell's membrane, secreting disruptive compounds into it, and causing it to burst. The compounds that kill the target cell are termed **lymphotoxins.** In one mechanism, the lymphotoxin appears to disrupt the cell membrane by forming pores. In another, the cytotoxic T cell secretes the lymphotoxin into the target cell and poisons it (figure 13.21).

Immunization: Methods of Manipulating Immunity for Therapeutic Purposes

We previously encountered the concept of artificial immunity in chapter 12. Both the active and passive forms are widely used in the health care field to treat and prevent disease.

Immunotherapy: Artificial Passive Immunity

The first attempts at passive immunization by transferring blood products from immune individuals to nonimmune individuals occurred in the early 1900s. At that time, horse serum containing antitoxins was used to prevent tetanus and to treat patients who had been exposed to diphtheria. Now antisera produced in mammals are used less often because of the chance of hypersensitivity to animal proteins contained in them. More often, products of human origin that function with various degrees of specificity are used. **Immune serum globulin (ISG),** sometimes called gamma globulin, contains immunoglobulin extracted from the pooled blood of at least 1,000 human donors. Each lot of serum presents a broad cross section of IgG (and some IgM) antibodies, and the content varies from lot to lot. The method of processing ISG concentrates the antibodies to increase potency and eliminates potential pathogens (such as the hepatitis B and HIV viruses). It is a treatment of choice in preventing measles and hepatitis A and in replacing antibodies in immunodeficient patients. Most forms of ISG are injected intramuscularly to minimize adverse reactions, and the protection it provides lasts two to three months.

A preparation called **specific immune globulin (SIG)** is derived from a more defined group of donors. Companies that prepare SIG obtain serum from patients who are convalescing and in a hyperimmune state after such infections as pertussis, rabies,

Feature 13.5 The Wisdom of the Body

The involvement of receptors and chemical stimuli in the actions of the immune system affirm what medical scientists have long suspected: that the cells of the nervous system and immune system communicate. This discovery enables biologists to explain how emotions can influence the state of health through their effect on the immune system, as well as how the immune system "converses" with the nervous system. This linkup is partially chemical. For example, various cells of the immune system respond to transmitter substances such as norepinephrine and acetylcholine released by nerve fibers. Moreover, the brain and certain parts of the peripheral nervous system produce interleukin (IL-1), an important immunoregulatory chemical. Lymphocytes and macrophages may also release chemicals that stimulate or inhibit nervous function. For example, the lymphokine IL-2 acts on the hypothalamus and causes fever, headaches, muscle pains, and other symptoms associated with infection.

The nervous and immune systems are also anatomically connected. Researchers have demonstrated extensive innervation of the bone marrow, thymus, spleen, lymph nodes, and blood vessels. This neurological wiring even extends to the level of single T cells! One theory suggests that this circuitry provides information to the brain about the status of immune reactions. This would allow the brain to regulate and modulate these reactions by means of neurotransmitters.

Evidence accumulating in the new area of *psychoneuroimmunology* indicates that the mental state of a person influences the condition of his or her immune system. Although medical science has acknowledged for a long time that certain diseases appear to be psychosomatic (a disease caused by or influenced by mental state), recent research has furnished greater credence to the basis for these syndromes. For instance, it has been shown that certain diseases of immune dysfunction (asthma, arthritis, and colitis) are aggravated by brain chemicals. In addition, researchers have been able to quantify the degree to which stress can depress immune function. Clinically depressed patients show moderate to severe inhibition of immune function. Otherwise-healthy medical students were shown to have diminished degrees of T-helper activity while under the stress of final exams. Some scientists feel that the psychological outlook of AIDS and cancer patients could be a significant factor in their prognoses. Preliminary observations indicate that more aggressive, hopeful, and less depressed patients appear to survive longer. Although the details of the brain-immune system connection are currently sketchy, future advances could offer numerous therapeutic approaches for a broad spectrum of diseases.

tetanus, chickenpox, and hepatitis B. These globulins are preferable to ISG because they contain higher titers of specific antibodies obtained from a smaller pool of patients. Although useful for prophylaxis in persons who have been exposed or may be exposed to infectious agents, these sera are often limited in availability.

Alternatives to human immune globulin (used only in cases where a human product is not available) are antisera and antitoxins produced in a suitable mammal, usually the horse. Equine sera available for passive immunization are diphtheria and botulism antitoxins and antivenoms for spider and snake bites. Unfortunately, the presence of numerous horse antigens may lead to acute allergic reactions, serum sickness, or anaphylaxis (see chapter 14). Although donated immunities only last a relatively short time, they act immediately and protect patients for whom no other useful medication or vaccine exists.

Most passive immunization involves the administration of antibodies, but occasionally T cells are given passively. For example, after closely matching the HLA antigens between the donor and recipient, sensitized T cells called *transfer factor* can be provided for immunodeficient patients chronically infected with *Candida albicans.*

Artificial Active Immunity: Vaccination

Active immunity may be conferred artificially by **vaccination**—exposing a person to material that is antigenic but not pathogenic. The discovery of vaccination was one of the farthest reaching and most important developments in medical science (see feature 13.6). It profoundly reduced the prevalence and impact of many infectious diseases that were once common and often deadly. In this section, we will survey the principles of vaccine preparation and important considerations surrounding vaccination in a community. (Vaccines are also given specific consideration in later chapters on bacterial and viral diseases.)

Principles of Vaccine Preparation A vaccine must be considered from the standpoints of antigen selection, effectiveness, ease in administration, safety, and cost. In natural immunity, an infectious agent stimulates appropriate B and T lymphocytes and creates memory clones. In artificial active immunity, the objective is to obtain this same response with a modified version of the microbe or its components. A safe and effective vaccine should mimic the natural protective response, not cause a serious infection or other disease, have long-lasting effects in a few doses, and be easy to administer. Most vaccine preparations contain one of the following antigenic stimulants (figure 13.22): (1) killed whole cells or inactivated viruses, (2) live, *attenuated* cells or viruses, (3) parts of cells or viruses (subunit vaccines), or (4) genetically engineered microbes or microbial parts.

Large, complex antigens such as intact cells or viruses are very effective immunogens. Depending on the vaccine, these are either killed or attenuated. **Killed, whole vaccines** are prepared by cultivating the correct strain or strains of a bacterium or virus and treating them with formalin, radiation, or some other agent that does not destroy antigenicity. Current vaccines for the bacterial diseases pertussis and typhoid fever are of this type (see chapter 16). Salk polio vaccine and rabies vaccine contain killed whole viruses. Because the microbe does not multiply, killed vaccines often require a larger dose and more boosters to be effective. Although one might think that vaccines containing dead microorganisms would be quite safe, they are not without adverse side effects.

A number of vaccines are prepared from **live, attenuated** microbes. **Attenuation** is any process that substantially lessens or negates the virulence of viruses or bacteria. It is usually achieved by modifying the growth conditions or manipulating microbial genes in a way that eliminates virulence factors. Long-term cultivation *in vitro,* selection of mutant strains that grow

attenuated (ah-ten′-yoo-ayt-ed) L. *attenuare,* to thin. Able to multiply, but nonvirulent.

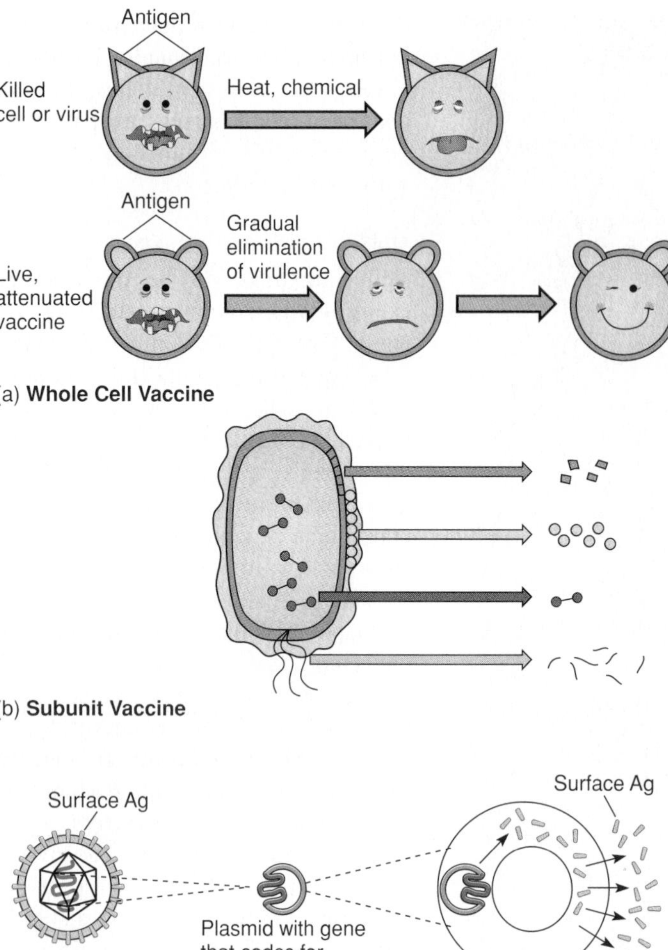

Figure 13.22 General strategies in vaccine preparation. (*a*) Whole cells, killed or attenuated. (*b*) Subunit vaccines, in which disruption of the cell (or virus) releases various molecules or cell parts that can be isolated and purified. (*c*) A recombinant gene coding for a surface or other molecule is isolated from the pathogen (here, a hepatitis virus). Insertion of the gene into the cloning vector (yeast) results in the production of large amounts of viral surface antigen to be used in vaccine.

at colder temperatures (cold mutants), or passage of the microbe through unnatural hosts or tissue culture have all proved successful methods. The vaccine for tuberculosis (BCG)[3] was obtained after 13 years of subculturing the agent of bovine tuberculosis (see chapter 16). Vaccines for measles, mumps, polio (Sabin), and rubella contain live, nonvirulent viruses. The advantages that favor live preparations are: (1) Viable microorganisms can multiply and produce infection (but not disease) like the natural organism; (2) they confer long-lasting protection; and (3) they usually require fewer doses and boosters than other types of vaccines. Disadvantages of using live microbes in vaccines are that they require special storage facilities, can be transmitted to other persons, and can mutate back to a virulent strain (see polio, chapter 21).

If the exact antigenic determinants that stimulate immunity are known, it is possible to produce a vaccine based on a selected part of a microorganism. These are called **subunit** or **cell-free vaccines.** Examples of subunit antigens are the capsules of the pneumococcus and meningococcus, the protein surface antigen of anthrax, and the surface receptors of hepatitis B virus. A special type of vaccine is the **toxoid,** which consists of a purified exotoxin that has been chemically denatured. By eliciting the production of antitoxins that can neutralize the natural toxin, toxoid vaccines provide protection against toxinoses such as diphtheria and tetanus. A survey of the major licensed vaccines and their uses is presented in table 13.4.

New Vaccine Strategies

One reason many infectious diseases lack practical vaccines is that a suitable antigen preparation has yet to be developed. Currently, much attention is being focussed on newer strategies for antigen preparation that employ synthesis, recombinant DNA, gene cloning technology, and other laboratory manipulations.

When the exact composition of an antigenic determinant is known, it is possible to synthesize it. This permits preserva-

3. BCG, **B**acille-**C**almette-**G**uerin, for the word bacillus and the names of its French developers.

toxoid (tawks′-oyd) Toxinlike.

Feature 13.6 The Lively History of Active Immunization

The basic notion of immunization has existed for thousands of years. It probably stemmed from the observation that persons who had recovered from certain communicable diseases rarely if ever got a second case. No doubt the earliest crude attempts involved bringing a susceptible person into contact with a diseased person or animal. The first recorded attempt at immunization occurred in sixth century China. It consisted of drying and grinding up smallpox scabs and blowing them with a straw into the nostrils of vulnerable family members. By the tenth century, this practice had changed to the deliberate inoculation of dried pus from the smallpox pustules of one patient into the arm of a healthy person, a technique later called **variolation** (variola is the smallpox virus). This method had been used in parts of the Far East for centuries before it was brought to England in 1721 by Lady Mary Montagu. Although the principles of the technique had some merit, unfortunately, many recipients and their contacts died of smallpox. This vividly demonstrates a cardinal rule for a workable vaccine: It must contain an antigen that will provide protection but not cause the disease. Variolation was so controversial that any English practitioner caught doing it was charged with a felony.

Eventually, this human experimentation paved the way for the first really effective vaccine, developed by the English physician Edward Jenner in 1796 (see chapter 20). Jenner conducted the first scientifically controlled study, one that had a tremendous impact on the advance of medicine. His work gave rise to the words **vaccine** and **vaccination** (from L. *vacca,* cow), which now apply to any immunity obtained by inoculation with selected antigens. Jenner was inspired by the case of a farmer, Benjamin Jesty, who had been infected by cowpox (a related virus that afflicts cows but causes only a mild infection in humans) and was thereafter free of smallpox. Jesty even went so far as to inoculate his family with cowpox, which protected them as well. In his initial test of this cross-resistance, Jenner prepared material from human cowpox lesions and inoculated a young boy. When challenged two months later with an injection of crusts from a smallpox patient, the boy proved immune. Jenner's discovery—that a less pathogenic agent could confer protection against a more pathogenic one—is especially remarkable in view of the fact that microscopy was still in its infancy and the nature of viruses was unknown. When his method proved successful, word of its significance spread widely, and it was adopted in many other countries. At various times, the method of producing the vaccine was changed, and somewhere along the line, the original virus mutated into a unique strain (*vaccinia* virus). But the essence of Jenner's method—scratching the vaccine into the skin with a sharp tool—remained. Now that smallpox is no longer a threat, smallpox vaccination has been essentially discontinued.

Other historical developments in vaccination included using heat-killed bacteria in vaccines for typhoid fever, cholera, and plague, and techniques for using neutralized toxins for diphtheria and tetanus. Throughout the history of vaccination, there have been vocal opponents and minimizers, but numbers do not lie: Whenever a vaccine has been introduced, the prevalence of that disease has declined.

The quest for vaccines has never ceased; greater understanding and improved technologies have made it possible to develop new vaccines and to increase the safety and effectiveness of older preparations. The world health community's current goal is to exterminate polio, measles, rubella, pertussis, and diphtheria worldwide by the early 1990s. In addition to vaccinating children, this program also targets young women of childbearing age to ensure that passive immunities will protect children at the most vulnerable time of life. The success of this formidable campaign remains to be seen.

---

tion of antigenicity while greatly increasing antigen purity and concentration. Because such molecules may be rather small, their immunogenicity may have to be increased by adding a carrier or adjuvant.

Some of the genetic engineering concepts introduced in chapter 8 offer novel approaches to vaccine development. These methods are particularly effective in designing vaccines for obligate parasites that are difficult or expensive to culture, such as the syphilis spirochete or the malaria parasite. This technology provides a means of isolating the genes that encode various microbial antigens, inserting them into plasmid vectors, and cloning them in appropriate hosts. The outcome of recombination can be varied as desired. For instance, the cloning host can be stimulated to synthesize and secrete a protein product (antigen), which is harvested and purified (figure 13.22*c*). This is how one type of hepatitis B vaccine is made, and it is the basis for several AIDS vaccines undergoing clinical trials. Antigens from the agents of syphilis, malaria, and influenza have been similarly isolated and cloned, and are currently being considered as potential vaccine material.

Another ingenious technique using genetic recombination has been nicknamed the *Trojan horse* or piggyback vaccine. The first term derives from an ancient legend in which the Greeks sneaked soldiers into the fortress of their Trojan enemies by hiding them inside a large, mobile wooden horse. In the microbial equivalent, genetic material from a selected microorganism is inserted into a live carrier microbe that is nonpathogenic. In theory, the recombinant microbe will multiply and express the piggyback genes, and the vaccine recipient will be immunized against the piggyback antigens. *Vaccinia,* the virus originally used to vaccinate for smallpox, has proved a practical agent for this technique. It is used as the carrier in one of the many new experimental vaccines for AIDS, herpes simplex 2, leprosy, and tuberculosis. Another group of researchers has had preliminary success with an antimalarial vaccine made from an attenuated strain of *Salmonella typhimurium* genetically transformed by *Plasmodium* (malaria) genes. Other diseases for which vaccines are being developed and tested are gonorrhea, syphilis, hepatitis A, rotavirus, salmonellosis, shigellosis, and schistosomiasis.

Vaccine effectiveness relies, in part, on the production of antibodies that closely fit the natural antigen. Realizing that such reactions are very much like a molecular jigsaw puzzle, researchers have proposed an entirely new concept in vaccines. The *anti-idiotypic vaccine* is based on the principle that the antigen binding (variable) region, or *idiotype,* of a given antibody (A) can be antigenic to a genetically different recipient and can cause that recipient's immune system to produce antibodies (B; also called anti-idiotypic antibodies) specific for antibody A (figure 13.23). What follows is that the variable region on these

idiotype (id'-ee-oh-type) Gr. *idios,* own, peculiar. Another term for the antigen binding site.

Table 13.4 Currently Approved Vaccines

Disease	Route of Administration	Recommended Usage/Comments
Contain Killed Whole Bacteria		
Cholera	Subcutaneous (SQ) injection	For travelers; effect not long-term
Pertussis*	Intramuscular (IM) injection	For newborns and children; possible adverse side effects
Typhoid	SQ and IM	For travelers only; efficacy variable
Plague	SQ	For exposed individuals and animal workers; variable protection
Contain Live, Attenuated Bacteria		
Tuberculosis (BCG)	Intradermal (ID) injection	For high-risk occupations only; protection variable
Subunit Vaccines (Capsular Polysaccharides)		
Meningitis (meningococcal)	SQ	For protection in high-risk infants, military recruits; short duration
Meningitis (*H. influenzae*)	IM	For infants and children; some controversy as to its effectiveness
Pneumococcal pneumonia	IM or SQ	Important for persons at high risk: the young, elderly, and immunocompromised; moderate protection
Toxoids (Formaldehyde-Inactivated Bacterial Exotoxins)		
Diphtheria	IM	A routine childhood vaccination; highly effective
Tetanus	IM	A routine childhood vaccination; highly effective
Botulism	IM	Only for exposed individuals such as laboratory personnel
Contain Killed Whole Viruses		
Hepatitis B	IM	For medical, dental, laboratory personnel and other persons at risk
Poliomyelitis (Salk)	IM	Routine childhood vaccine; effective
Rabies	IM	For victims of animal bites or persons otherwise exposed; effective
Contain Live, Attenuated Viruses (Mostly RNA Viruses)		
Adenovirus infection	Oral	For immunizing military recruits
Measles (rubeola)	SQ	Routine childhood vaccine; very effective
Mumps (parotitis)	SQ	Routine childhood vaccine; very effective
Poliomyelitis	Oral	Routine childhood vaccine; very effective
Rubella	SQ	Routine childhood vaccine; very effective
Chickenpox (varicella)	SQ	For children with cancer; protection not long-term; some risk in giving live DNA viruses
Yellow fever	SQ	Travelers, military personnel in endemic areas
Subunit Viral Vaccines		
Influenza*	IM	For high-risk populations; requires constant updating for new strains; immunity not durable
Recombinant Vaccines		
Hepatitis B	IM	Medical, dental, and laboratory personnel

*Alternate cell- or virus-free vaccine being tested.

anti-idiotypic antibodies will mimic (have the same configuration as) the antigen for which antibody A was specific. Antibody B could be used in vaccines, because the antibodies it stimulates in the vaccine recipient will react with the natural antigen. The vaccine recipient is not actually given microbial antigen, thus such vaccines have less potential for dangerous side effects. In addition, the exact nature of the microbial antigen need not be known in order for a workable vaccine to be developed. Using monoclonal antibodies, this approach has been used to mimic the surface antigen of hepatitis B virus and *Trypanosoma* with some success.

Route of Administration and Side Effects of Vaccines

Most vaccines are injected by subcutaneous, intramuscular, or intradermal routes. Oral vaccines are available for only two diseases (table 13.4), but have some distinct advantages. An oral dose of a vaccine can stimulate protection (IgA) on the mucous membrane of the portal of entry. Oral vaccines are also easier to give, more readily accepted, and well tolerated. An influenza vaccine given intranasally is another method that shows promise. Some vaccines (especially those that are weak immunogens) require the addition of a special binding substance or *adjuvant*. An adjuvant is any compound that enhances immunogenicity and prolongs antigen retention at the injection site. The adjuvant precipitates the antigen and holds it in the tissues so that it will be released gradually. This presumably facilitates contact with macrophages and lymphocytes. Common adjuvants are

adjuvant (ad'-joo-vunt) L. *adjuvare*, to help.

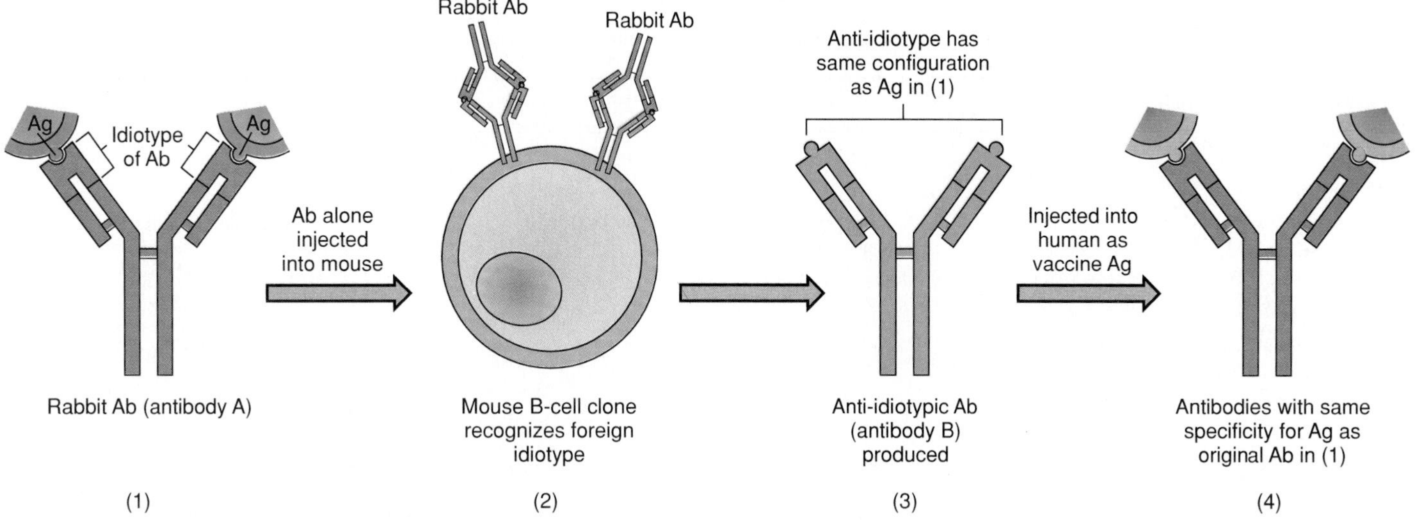

Figure 13.23 Anti-idiotype vaccines use antibodies as vaccine antigens.

alum (aluminum hydroxide salts), Freund's adjuvant (emulsion of mineral oil, water, and extracts of mycobacteria), and beeswax.

Vaccines must go through many years of trials in experimental animals and humans before they are licensed for general use. Even after they have been approved, like all therapeutic products, they are not without complications. The most common of these are local reactions at the injection site, fever, allergies, and other adverse reactions. Relatively rare reactions (about 1 case out of 220,000 vaccinations) are panencephalitis (from measles vaccine), back-mutation to a virulent strain (from polio vaccine), disease due to contamination with dangerous viruses or chemicals, and neurological effects of unknown cause (from pertussis and swine flu vaccines). Some patients experience allergic reactions to the vaccine medium (eggs or tissue culture), rather than to vaccine antigens. Professionals involved in giving vaccinations must understand their inherent risks but also realize that the risks from the infectious disease almost always outweigh the chance of an adverse vaccine reaction. The greatest caution must be exercised in giving live vaccines to immunocompromised and pregnant patients, the latter because of possible risk to the fetus.

To Vaccinate: Why, Whom, and When?

Several advantages accompany vaccination. Certainly it confers long-lasting, sometimes lifetime, protection in the individual, but an equally important reason is to protect the public health. Vaccination is an effective method of establishing **herd immunity** in the population. According to this concept, individuals immune to a communicable infectious disease will not be carriers, which reduces the occurrence of that microbe; therefore, the larger the number of immune individuals in a population (herd), the less likely that an unimmunized member of the population will encounter the agent. In effect, collective immunity through mass immunization confers indirect protection on the nonimmune (such as children). Herd immunity maintained through immunization is an important force in averting epidemics.

Vaccination is recommended for all typical childhood diseases for which a vaccine is available and for people in certain special circumstances (health workers, travelers, military personnel). Table 13.5 outlines a general schedule for childhood immunization and the indication for special vaccines. Not only are some vaccines mixtures of antigens, but on occasion several vaccines are administered simultaneously. Sometimes the question arises, Can a person be challenged with too many immunogens at once? Consider the case of military recruits who line up for as many as 15 injections within a few minutes, or children who receive boosters for DPT and polio at the same time they receive the MMR. Experts doubt that immune interference (inhibition of one immune response by another) is a significant problem in these instances, and the mixed vaccines are carefully balanced to prevent this eventuality. The main problem with simultaneous administration is that side effects can be amplified.

Vaccination in Decline Diligence in vaccinating the public is not just a concern in developing countries. It is true that the occurrence of childhood diseases in the U.S. has dropped significantly in the past several years, but so have the number of vaccinations. As a direct result, the level of herd immunity has also decreased, making the population vulnerable to epidemics. This accounts for the recent sudden and unexpected outbreaks of measles, rubella, mumps, and pertussis. Furthermore, vaccines for such diseases as influenza, pneumococcal pneumonia, and hepatitis B, which should be administered to persons at special risk, often go ignored or unused. An estimated 20 million citizens become ill from preventable infections per year, and 70,000 of them die. Several factors may account for this unfortunate situation. Parents may not comply with vaccination because (1) they fear adverse side effects such as those reported with pertussis, (2) they mistakenly believe that a disease has been eradicated, or (3) they are immigrants who have not had access to medical care. In some cases, the physician fails to inform the patient of a vaccine, or the patient refuses his recommendations. The reduction in federal funds to underwrite immunization for preschool children seems to be an important factor. What remains an important reality is that the prevention costs for a disease are only a fraction of the treatment costs.

Table 13.5 Recommended Regimen for Individual and Community Immunizations

Used in Normal Children, Adults

Vaccine	Ages at Time of Vaccination*	Comments
Mixed Vaccines		
Polio (OPV) Diphtheria (D) Pertussis (P) Tetanus (T)	6–10 weeks, 4–5 months, 6–7 months, 15–18 months, 4–6 years	OPV and DPT administered simultaneously with 1 initial dose and 4 boosters; final OPV and DPT given preferably before child first enters school
Tetanus-diphtheria (Td) or tetanus toxoid (TT)	Over age 7	Repeated every 10 years; for prophylaxis in traumatic injuries
Measles, mumps, rubella (MMR)	15 months	MMR administered simultaneously; booster recommended
Measles alone	Various ages	May be given at 12 months of age in areas of high incidence; booster advisable for children prior to entering school and for anyone vaccinated before 1970
Single Vaccines		
Hemophilus influenzae B (HiB)	2–18 months	Especially for young children; not recommended for children over 5 years

Used in Cases of Specific Risk Due to Occupational or Other Exposure

Group Targeted	Vaccine
Hospital, laboratory, health care workers	Hepatitis B, influenza, polio, tuberculosis
People whose jobs involve contact with animals (veterinarians, ranchers, forest rangers)	Rabies, plague
People with specific life-styles (homosexual males, drug abusers)	Hepatitis B
Travelers to endemic regions, including military recruits (varies with geographic destination)	Cholera, hepatitis B, measles, yellow fever, meningococcal meningitis, polio, rabies, typhoid, plague

* May vary slightly from region to region, depending on the incidence of disease.

Serological and Immune Tests: Measuring the Immune Response *in Vitro*

The antibodies formed during an immune reaction are important in combating infection, but they hold additional practical value. Characteristics of antibodies (such as their quantity or specificity) can reveal the history of a patient's contact with microorganisms or other antigens. This is the underlying basis of **serological testing. Serology** is the branch of immunology that traditionally deals with *in vitro* diagnostic testing of serum. Serological testing is based on a now-familiar concept. Antibodies have extreme specificity for antigens, so that when a particular antigen is exposed to its specific antibody, it will fit like a hand in a glove. The ability to visualize this interaction in some way provides a powerful tool for detecting, identifying, and quantifying antibodies—or for that matter, antigens. The scheme works both ways, depending on the situation. One may detect or identify an unknown antibody using a known antigen, or if the antigen is unknown, one may use an antibody of known specificity to help detect or identify it (figure 13.24). Modern serological testing has grown into a field that tests more than just serum. Urine, cerebrospinal fluid, whole tissues, and saliva may also be used to determine the immunologic background of patients. These and other immune tests are helpful in confirming a suspected diagnosis or in screening a certain population for disease.

General Features of Immune Testing

The strategies of immunologic tests are diverse, and they underline some of the brilliant and imaginative ways that antibodies and antigens may be used as tools. We will summarize them under the headings of agglutination, precipitation, immunodiffusion, complement fixation, fluorescent antibody tests, and immunoassay tests. Let us first overview the general characteristics of immune testing, and then look at each type separately.

Characteristics aimed for in a serological test are specificity and sensitivity (figure 13.25) **Specificity** is the property of a test to focus upon only a certain antibody or antigen and not to react with unrelated or distantly related ones. **Sensitivity** refers to the capacity of the test to detect even very small amounts of antibodies or antigens that are the targets of the

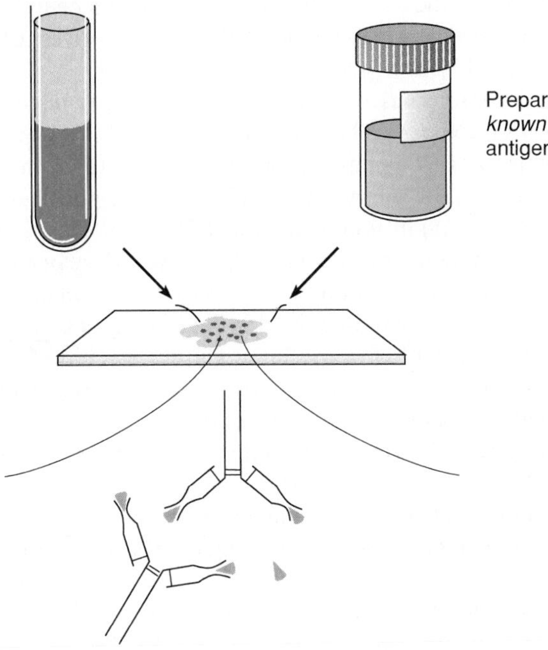

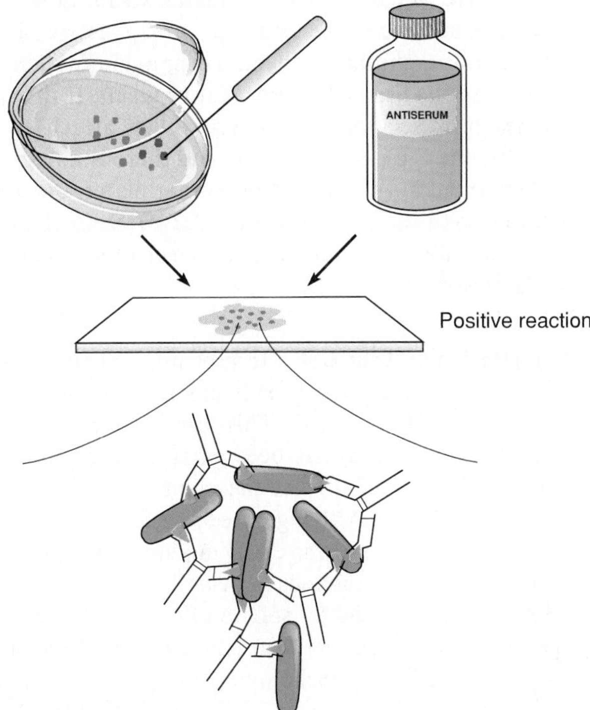

Figure 13.24 Basic principles of testing using antibodies and antigens. (*a*) In serological diagnosis of disease, a blood sample is scanned for the presence of antibody using an antigen of known specificity. A positive reaction is usually evident in some readily visible reaction, such as color change or clumping, that indicates a specific interaction between antibody and antigen. (The reaction at the molecular level is rarely observed.) (*b*) Identification of an unknown microbe using serum containing antibodies of known specificity, a procedure known as serotyping. Microscopically or macroscopically observable reactions indicate correct match between Ab and Ag.

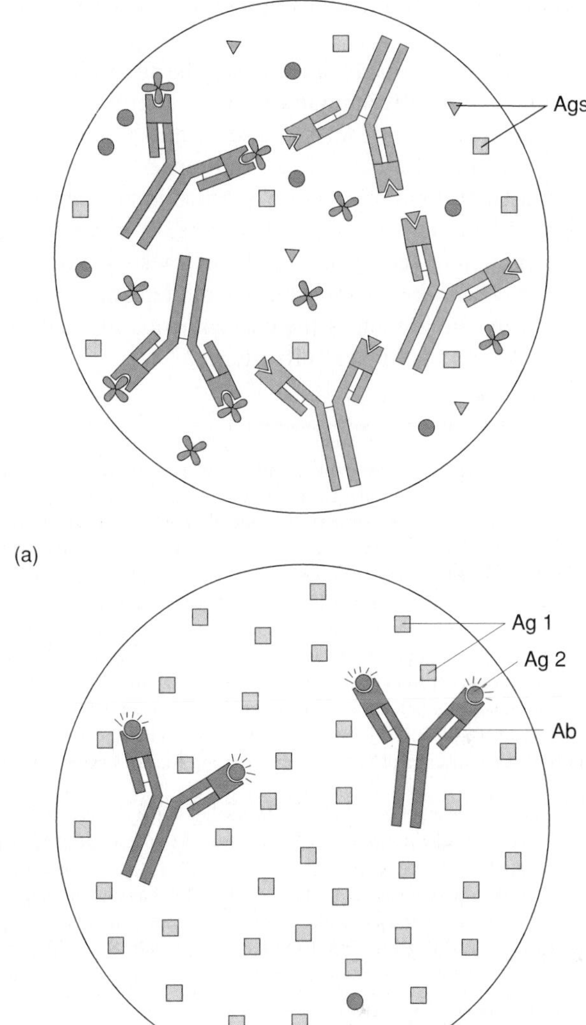

Figure 13.25 Specificity and sensitivity in immune testing. (*a*) This test shows specificity, in that Ab attaches with great exactness with only one type of Ag. (*b*) Sensitivity is demonstrated by the fact that Ab can locate Ag even when Ag is greatly diluted.

test. New systems using monoclonal antibodies have greatly improved specificity, and those using radioactivity, enzymes, and electronics have improved sensitivity.

Visualizing Antigen-Antibody Interactions We have seen that, at the molecular level, an antibody attaches to a specific molecular site on an antigen. Because this reaction cannot be readily seen without an electron microscope, all tests involve some type of endpoint reaction visible to the naked eye or with regular magnification that tells whether the test is positive or negative. In the case of large antigens such as cells, the binding of Ag to Ab forms large clumps or aggregates that are visible macroscopically or microscopically. Smaller Ag-Ab complexes that do not result in readily observable changes will require special indicators in order to be visualized. Endpoints are often revealed by dyes or fluorescent reagents that can tag molecules of interest. Similarly, radioactive isotopes incorporated into antigens or antibodies constitute sensitive tracers that are detectable with photographic film.

An antigen-antibody reaction may be quantified by means of a **titer** reading, a technique that gives the relative quantity or concentration of antibodies and so permits different samples to be standardized and compared. In titer tests, a sample is serially diluted in tubes or a multiple-welled microtiter plate and mixed with antigen (figure 13.26*b*). Titer is expressed as the highest dilution of serum that produces a visible reaction with an antigen. The more a sample may be diluted and yet still react with antigen, the greater is the concentration of antibodies in that sample and the higher is its titer.

When Positive Is Negative: How to Interpret Serological Test Results

What if a patient's serum gives a positive reaction—is seropositive—in a serological test? In most situations, it means that antibodies raised against a particular microbe have been detected in the sample. But one must be cautious in proceeding to the next level of interpretation. The mere presence of antibodies does not necessarily indicate that the patient has a disease, but only that he or she has had contact with a microbe or its antigens through infection (even subclinical) or vaccination. In screening tests for determining a patient's history (rubella, for instance), knowing that a certain titer of antibodies is present can be significant, because it shows that the person has some protection. However, when the test is being used to diagnose current disease, a series of tests to show a rising titer of antibodies is necessary.

Another important consideration in testing is the occasional appearance of biological *false positives*. These are results in which a patient's serum shows a positive reaction, even though, in reality, he is not or has not been infected by the microbe. False positives such as those in syphilis and AIDS testing arise when antibodies or other substances present in the serum **cross-react** with the test reagents, producing a positive result. Such false results may require retesting with a test that greatly minimizes cross reactions.

Agglutination and Precipitation Reactions

The essential differences between agglutination and precipitation are in size, solubility, and location of the antigen. In agglutination, the antigens are whole cells such as red blood cells or bacteria with determinant groups on the surface. In precipitation, the antigen is a cell-free molecule in solution. In both instances, when Ag and Ab are optimally combined so that neither is in excess, one antigen is interlinked by several antibodies to form an insoluble, three-dimensional aggregate so large that it cannot remain suspended and it settles out (figure 13.26*a*).

Agglutination Testing Agglutination is discernible because the antibodies called *agglutinins* cross-link the antigens or *agglutinogens* to form visible clumps. Agglutination tests are performed routinely by blood banks to determine ABO and Rh (Rhesus) blood types in preparation for transfusions. In this test, antisera containing antibodies against the blood group antigens on red blood cells are mixed with a small sample of blood and read for the presence or absence of clumping (see figure 14.9). The *Widal test* is an example of a tube agglutination test for diagnosing typhoid fever, other salmonelloses, and undulant fever. In this test, a serum sample is serially diluted with saline in small tubes, incubated with antigen, and centrifuged. In addition to detecting specific antibody, it also gives the serum titer (figure 13.26*b*).

Numerous variations of agglutination testing exist. The VDRL (Venereal Disease Research Lab) is a common test for antibodies to syphilis. Because the reagents for the test include a heterophile antigen (cardiolipin), false positives may occur. The cold agglutinin test, named for antibodies that react only at lower temperatures (4°–20°C), was developed to diagnose *Mycoplasma* pneumonia. The *Weil-Felix reaction* is an agglutination test sometimes used in diagnosing rickettsial infections.

In some tests, special agglutinogens have been prepared by affixing antigen to the surface of an inert particle. In *latex agglutination* tests, the inert particles are tiny latex beads. Kits using latex beads are available for assaying pregnancy hormone in the urine, identifying the yeast *Candida* and bacteria (staphylococci, streptococci, and gonococci), and diagnosing rheumatoid arthritis. In *viral hemagglutination* testing, the particle is a red blood cell that reacts naturally with certain viral antigens. The RBCs are mixed with a known virus and a patient's serum of unknown content (figure 13.27). The test is interpreted differently than other agglutination tests because it is based on a competition between the RBCs and the antibodies for the virus antigens. If the patient's serum does **not** contain antibodies specific to the virus, the virus reacts with the RBCs instead and agglutinates them. Thus, agglutination here indicates no antibodies and a seronegative result. If the specific antibodies for the virus are present, they attach to the virus particles, the RBCs remain free, and there is no agglutination. As performed in the wells of microtiter plates, the difference is apparent to the naked eye (figure 13.27*c*). Several viral diseases (measles, rubella, mumps, mononucleosis, and influenza) may be diagnosed with this test.

Precipitation in Agar Gel In precipitation reactions, the soluble antigen (*precipitinogen*) is precipitated (made insoluble) by an antibody (*precipitin*). This reaction is observable in a test tube where antiserum has been carefully laid over an antigen solution (figure 13.28*a*). At the point of contact, a cloudy or opaque zone forms. Although precipitation is a useful detection tool, the precipitates are so easily disrupted in liquid media that most precipitation reactions are carried out in agar gels. These substrates are sufficiently soft to allow the reactants (Ab and Ag) to freely diffuse, yet firm enough to hold the Ag-Ab precipitate in place. One technique with applications in microbial identification and diagnosis of disease is the **double diffusion** (Ouchterlony) method. It is called double diffusion because it involves diffusion of both antigens and antibodies. The test is performed by punching a pattern of small wells into an agar medium and filling them with test antigens and antibodies. A band forming between two wells indicates that antibodies from

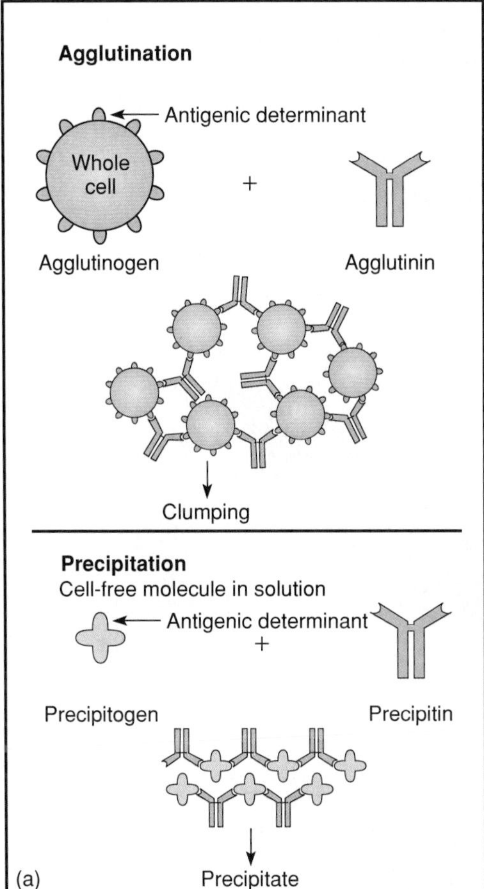

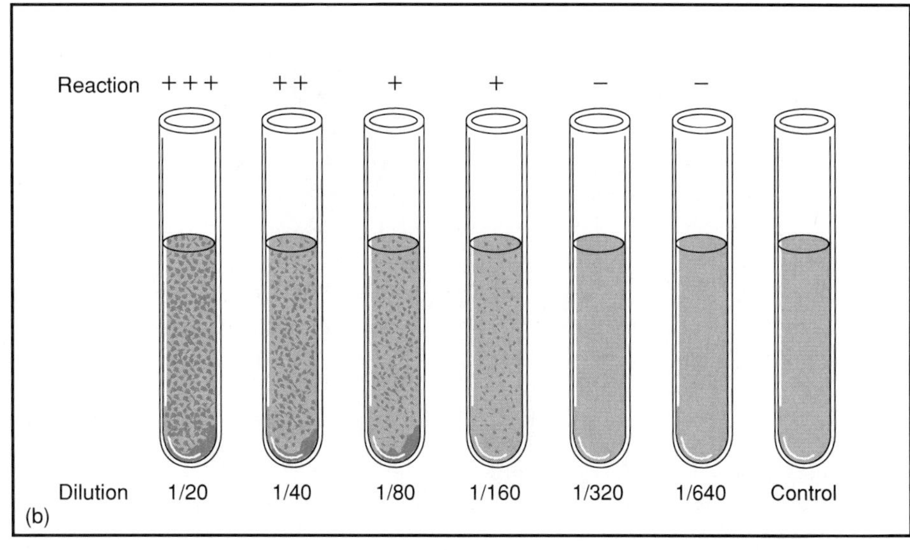

Figure 13.26 (*a*) Cellular/molecular view of agglutination and precipitation reactions. (Though IgG is shown as the Ab, IgM is also involved in these reactions.) (*b*) The tube agglutination test. A sample of patient's serum is diluted with saline through a series of tubes. The dilution is binary, so that the number of antibodies in the serum is halved in each subsequent tube. An equal amount of the antigen (here, dead bacterial cells) is added to each tube. The control tube has antigen but no serum. After incubation and centrifugation, each tube is examined for flakes (agglutination) as compared to the control, which will be cloudy and flake-free. The titer is defined as the dilution of the last tube in the series that shows agglutination.

one well have met and reacted with antigens from the other well. Variations on this technique provide several results (figure 13.28*b*): (1) One may place antigen in the center well and use it to assess the content of sera. (2) One may place antiserum in the center well to help identify antigens placed in the outer wells. (3) One may read identification reactions between the contents of the outer wells.

Immunoelectrophoresis constitutes yet another refinement of diffusion and precipitation in agar. With this method, a serum sample is first electrophoresed to separate the serum proteins as previously shown (see figure 13.17). After antibodies that can react with these serum samples are placed in wells parallel to the direction of migration, arc patterns specific for blood proteins develop (figure 13.29). This test is widely used to detect disorders in the production of antibodies. *Counterimmunoelectrophoresis,* which uses an electric current to speed up the migration of antibody and antigen, is a newer technique for identifying bacterial and viral antigens in blood.

The Western Blot

A variation of immunoelectrophoresis is the basis for the **Western blot** test. A counterpart of the Southern blot test for identifying DNA, this test is a very specific and sensitive way to identify or verify a particular antibody or antigen in a sample (figure 13.30). First, the test material is electrophoresed in a gel, which separates out particular bands. The gel is then transferred to a special blotter substrate that binds the reactants in place. To this, a known antigen or antibody is added, and the preparation is incubated. To demarcate antigen that has bound to any antibodies on the blotter, a radioactive or enzymatic indicator is added, and the blot is developed. The pattern of bands that appears is compared to known positive and negative samples. This is currently the verification test for persons who are antibody-positive for HIV in the ELISA test (described in a later section), because it is more specific for this antibody and less subject to misinterpretation. The technique has enormous possibilities, because it is applicable for detecting microbes and their antigens in specimens.

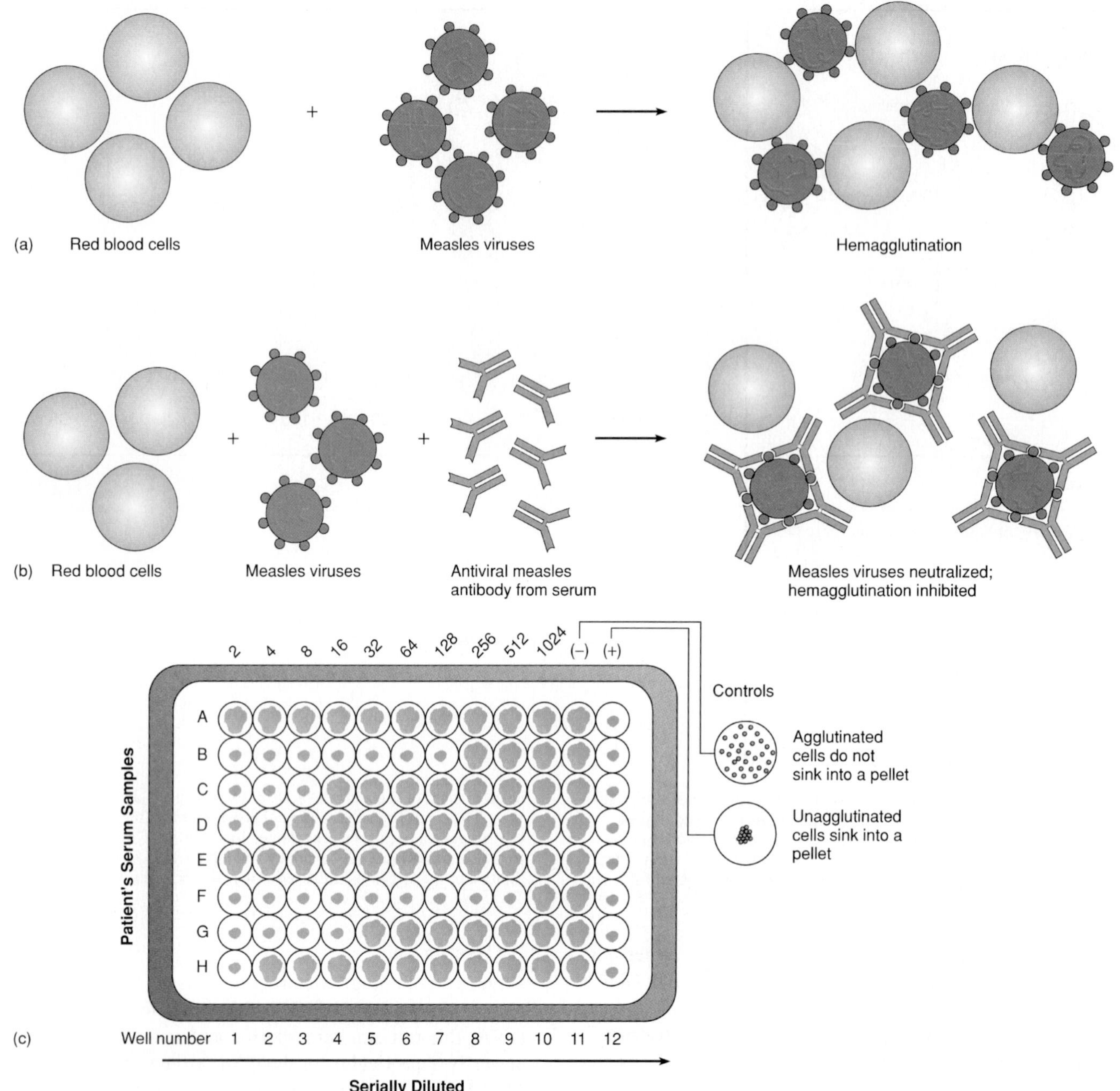

Figure 13.27 Theory and interpretation of viral hemagglutination. (*a*) In an antibody-free system, the natural tendency of measles virus to combine and interlink red blood cells causes an agglutination reaction. (*b*) In a test system in which antibodies specific to the measles virus are present, these Abs will react with the viruses and prevent them from binding to red blood cells; this blocks agglutination. Thus, no agglutination means a positive reaction for antibody. (*c*) In an actual test, a multiwell microtiter plate is set up to run dilutions of several patients' sera. In a negative test, there will be agglutination, which appears as complexes covering the bottom of the well; in a positive test, the unagglutinated red blood cells fall into a small pile (pellet) in the well's bottom. (This test allows one to read titer as well.)

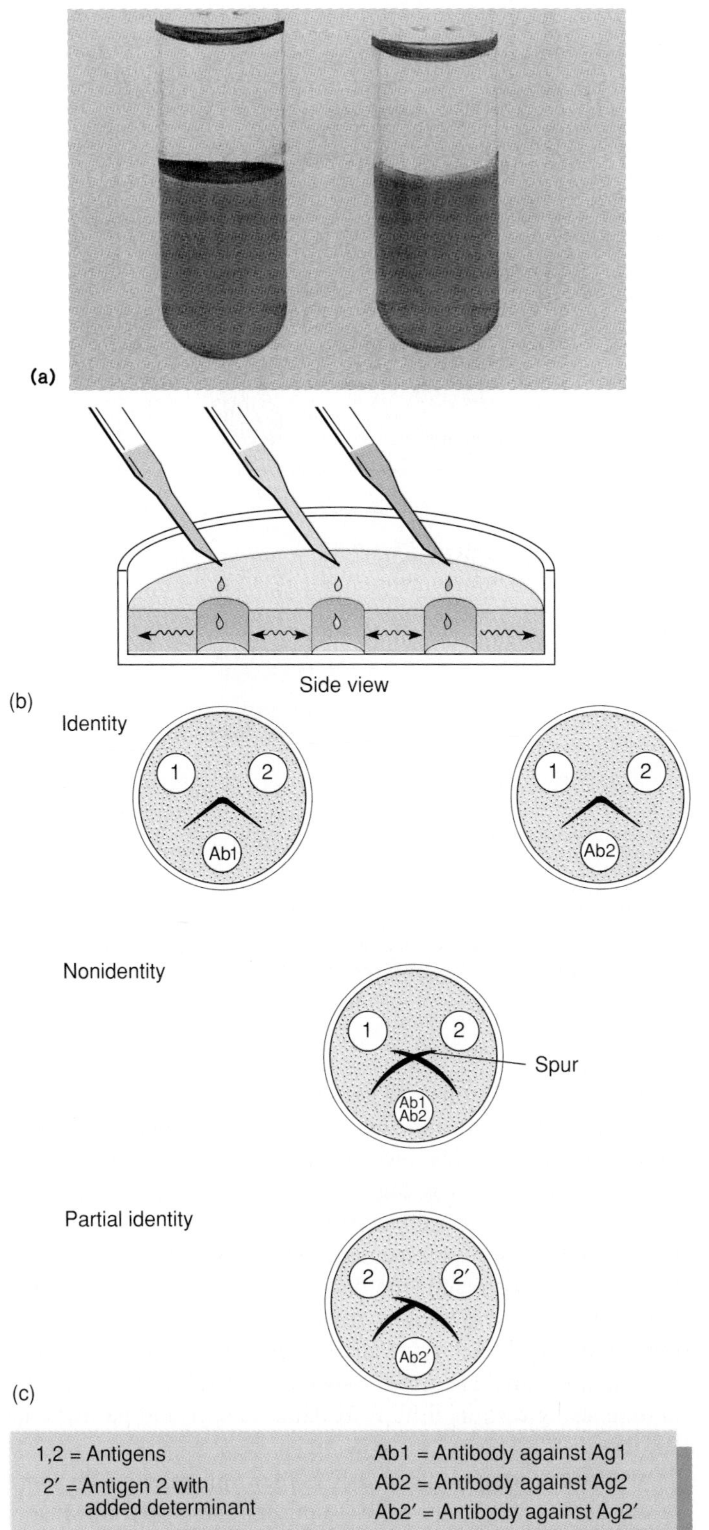

Figure 13.28 Precipitation reactions. (*a*) A tube precipitation test for streptococcal group antigens. Specific antiserum has been placed in the bottom of the tubes and antigen solution carefully overlaid to form a zone of contact. The right-hand tube has developed a heavy band of precipitate indicative of a positive reaction between Ab and Ag. The left-hand tube is negative. (*b*) In one method of setting up a double diffusion test, wells are punctured in soft agar, and Abs and Ags are added in a pattern. As the contents of the wells diffuse toward each other, a number of reactions can result. When a well containing Ab is placed so as to react with two Ags, lines of precipitate that form between the Ab/Ag wells are indicative of various characteristics of the test antigens. A line indicates a distinct Ag/Ab reaction between the two wells; a spur indicates lack of identity. (*c*) In the top example, the smooth V between wells 1 and 2 and the Ab show that they are the same antigen. When two spurs cross each other as in the middle figure, this indicates complete nonidentity of Ags 1 and 2. If a spur occurs on one side only (bottom figure), there is partial serological similarity between the two antigens.

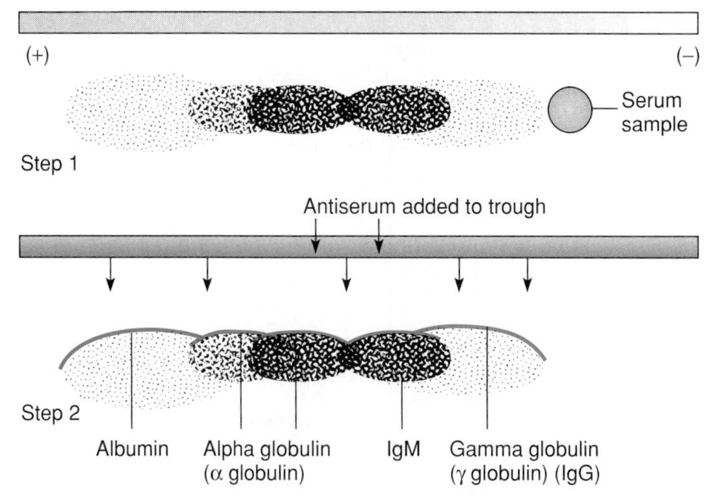

Figure 13.29 Immunoelectrophoresis of normal human serum. Step 1. Proteins are separated by electrophoresis on a gel. Step 2. To identify the bands and increase visibility, antiserum containing antibodies specific for serum proteins is placed in a trough and allowed to diffuse toward the bands. This produces a pattern of numerous arcs representing major serum components.

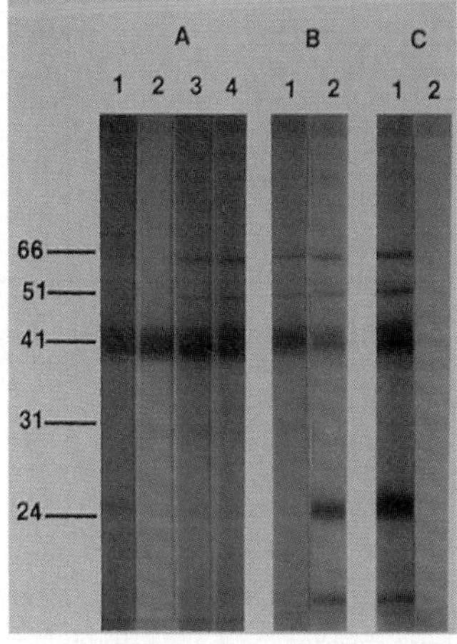

Figure 13.30 The Western blot procedure. This method tests for antibodies to specific HIV antigens. First HIV antigens are separated by electrophoresis and transferred or blotted onto special strips (primary HIV Ags are numbered on the left). Test strips are incubated with test sera, exposed to radioactive markers, and developed with X-ray film. Areas where HIV Ags have focussed antibodies show up as bands. Patients' sera are compared with a positive control strip (Cl) containing antibodies for all HIV Ags. In this series, A strips are from AIDS patients, B strips are from ARC patients, and strip C2 is HIV negative (no antibodies).

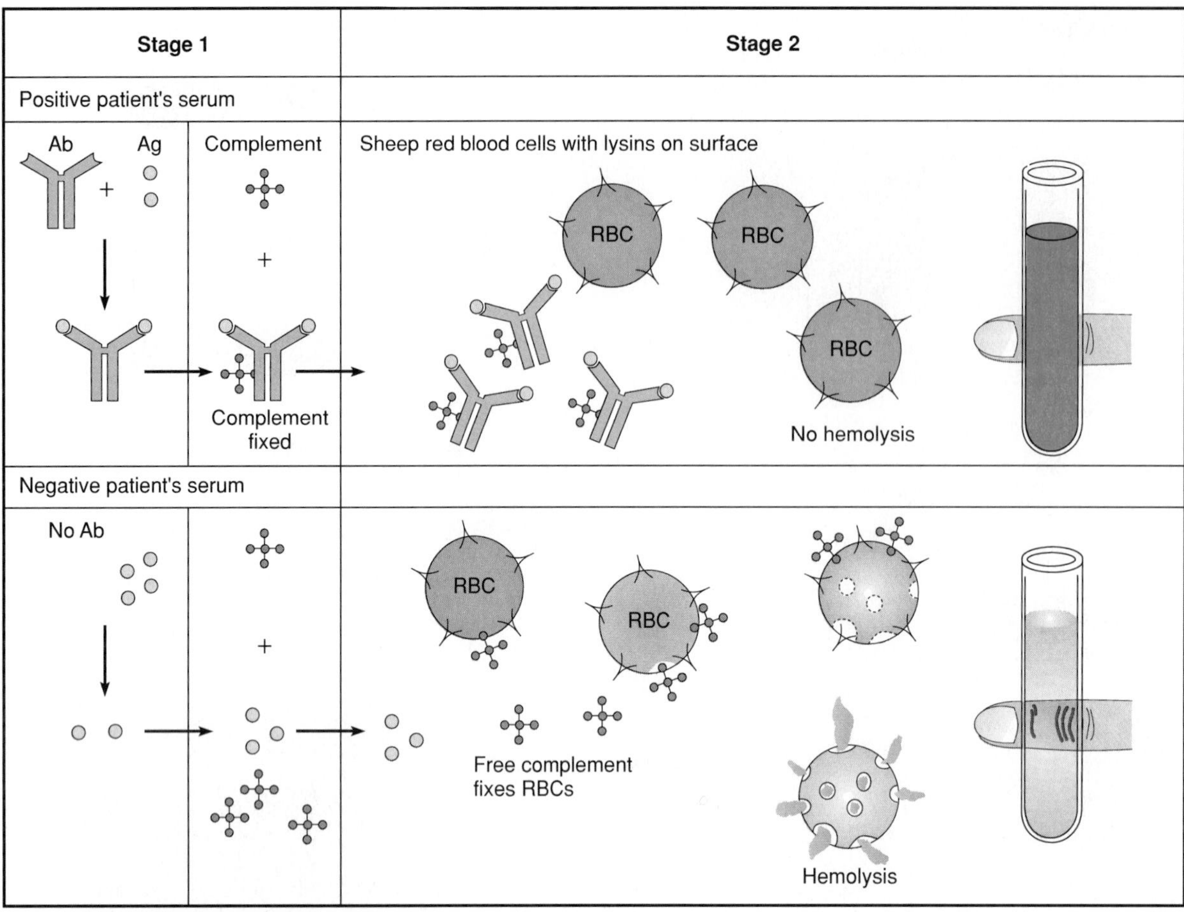

Figure 13.31 Complement fixation test. In this example, a patient's serum is being tested for antibodies to a certain infectious agent. Note that macroscopically, unhemolyzed red blood cells present a turbid solution that one cannot see through. Hemolysis creates a transparent pink solution.

Complement Fixation

An antibody that requires (fixes) complement to complete the lysis of its antigenic target cell is a **lysin** or **cytolysin.** When lysins act in conjunction with the intrinsic complement system on red blood cells, the cells hemolyze (lyse and release their hemoglobin). This lysin-mediated hemolysis is the basis of a group of tests called **complement fixation,** or CF (figure 13.31). In general, CF testing is carried out with four components: antibody, antigen, complement, and sensitized sheep red blood cells, and it is conducted in two phases. In the first phase, the test antigen is allowed to react with the test antibody (one must know the identity of at least one of them) in the absence of complement. If the Ab-Ag are specific for each other, they form complexes. To this mixture, purified complement proteins from guinea pig blood are added. If antibody and antigen have complexed during the previous step, they attach or **fix** the complement to them, thus preventing it from participating in further reactions. The extent of this complement fixation is determined in the second phase. In this phase, sheep RBCs are sensitized by mixing them with sheep-specific lysins. This will produce an indicator complex that can also fix complement. Contents of the stage 1 tube are mixed with the stage 2 tube and observed for hemolysis (which can be observed with the naked eye as a clearing of the solution). If hemolysis *does not* occur, it means that the complement was used up by the first stage Ab-Ag complex and that the unknown antigen or antibody was indeed present, thus the test is considered positive. If hemolysis does occur, it means that unfixed complement from tube 1 reacted with the RBC complex instead, thereby causing lysis of the sheep RBCs. With this result, the test is negative for the antigen or antibody that is the basis of the test. The experimental procedure may also be set up to determine antibody titer. Complement fixation tests are somewhat complicated, yet are invaluable in diagnosing certain viral, rickettsial, fungal, and parasitic infections.

The antistreptolysin O (ASO) titer test measures the levels of antibody against the streptolysin toxin, an important hemolysin of group A streptococci. It employs a technique related to complement fixation. A serum sample is exposed to known suspensions of streptolysin and then allowed to incubate with RBCs. Lack of hemolysis indicates antistreptolysin antibodies in the patient's serum that have neutralized the streptolysin and prevented hemolysis. This is an important verification procedure for scarlet fever, rheumatic fever, and other related streptococcal syndromes (see chapter 15).

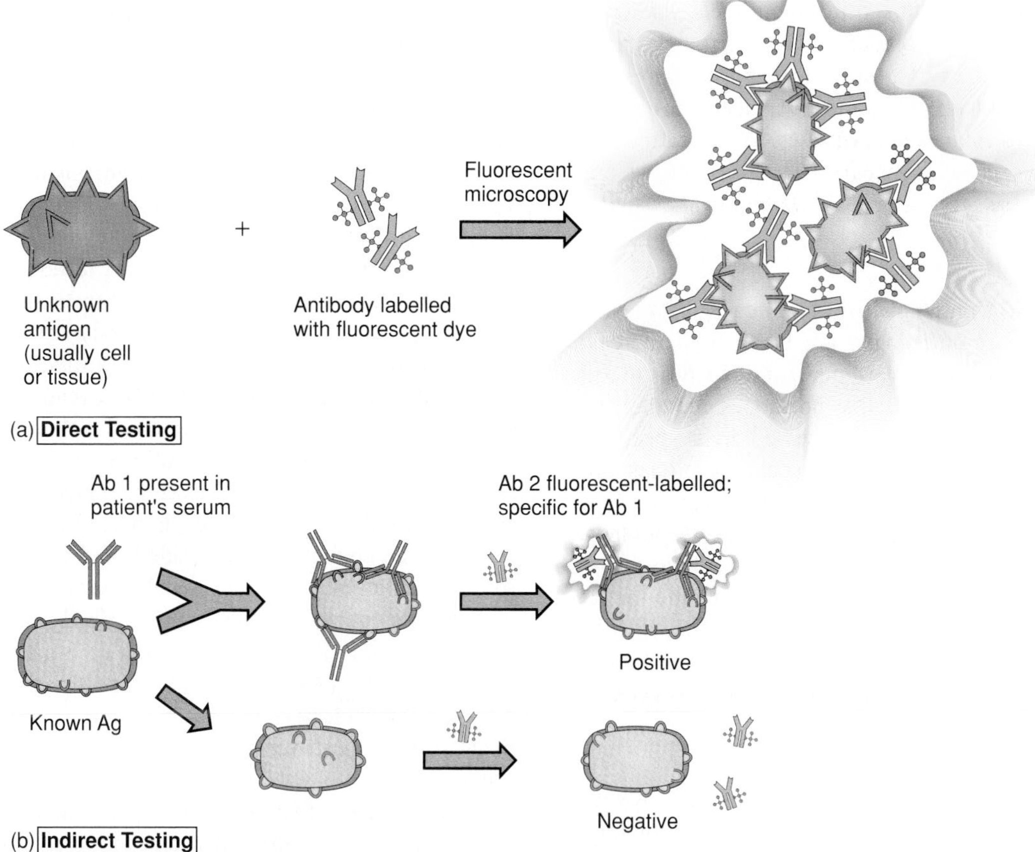

Figure 13.32 Immunofluorescence testing. The success of these techniques is contingent upon the accumulation of sufficient labelled Ab in one site that it will show up as fluorescing cells or masses. (*a*) Direct: Unidentified Ag is directly tagged with fluorescent Ab. (*b*) Indirect: Ag of known identity is used to assay unknown Ab; a positive reaction occurs when the second Ab (with fluorescent dye) affixes to the first Ab.

Miscellaneous Serological Tests

A test that relies on changes in cellular activity as seen microscopically is the *Treponema pallidum immobilization* (TPI) test for syphilis. The impairment or loss of motility of the *Treponema* spirochete in the presence of test serum and complement indicates that the serum contains anti-*Treponema pallidum* antibodies (see chapter 17). In *toxin neutralization* tests, a test serum is incubated with the microbe that produces the toxin. If the serum inhibits the growth of the microbe, the serum does contain antitoxins.

Serotyping is an antigen-antibody technique that allows a microbiologist to identify, classify, and subgroup certain bacteria into categories called serotypes, using antisera raised to various cell antigens such as the capsule, flagellum, and cell wall. It is widely used in typing *Salmonella* species and strains, and is the basis for the numerous serotypes of streptococci. The Quellung test that identifies serotypes of the pneumococcus involves a precipitation reaction in which antibodies react with the capsular polysaccharide. Although the reaction makes the capsule seem to swell, it is actually creating a zone of Ab-Ag complex on the cell's surface (see chapter 15).

Fluorescent Antibodies and Immunofluorescence Testing

The property of dyes such as fluorescein and rhodamine to emit visible light in response to ultraviolet radiation was discussed in chapter 1. This valuable property has found numerous applications in diagnostic immunology. The fundamental tool in immunofluorescence testing is a fluorescent antibody, a monoclonal antibody labelled by a fluorescent dye (fluorochrome).

The two ways that fluorescent antibodies may be used are shown in figure 13.32. In *direct testing,* an unknown test specimen or antigen is fixed to a slide and exposed to a fluorescent antibody solution of known composition. If the antibodies are complementary to antigens in the material, they will bind to it. After the slide is rinsed to remove unattached antibodies, it is observed with the fluorescent microscope. Fluorescing cells or specks indicate the presence of Ab-Ag complexes and a positive result. These tests are valuable for identifying and locating antigens on the surfaces of cells or in tissues and in identifying the disease agents of gonorrhea, chlamydiosis, whooping cough, Legionnaires' disease, plague, trichomoniasis, meningitis, and listeriosis.

Indirect testing with fluorescent antibodies is a bit more complex, because the fluorescent antibody does not react directly with the test antigen. In this scheme, the fluorescent antibodies are *anti-isotypic* antibodies—that is, they are antibodies raised to the Fc region of another antibody (remember that antibodies may be antigenic). An antigen of known character (a bacterial cell, for example) is combined with a test serum of unknown antibody content. To visualize whether the serum contains antibodies that have affixed to the antigen, a fluorescent antibody solution made to react with the unknown antibody is applied and rinsed off. A positive test shows fluorescing aggregates or cells, indicating that the fluorescent antibodies have combined with the unlabelled antibodies. In a negative test, no fluorescent complexes will appear. This technique is frequently used to diagnose syphilis (FTA-ABS) and various viral infections.

Immunoassays: Tests of Great Sensitivity The elegant tools of the microbiologist and immunologist are being used increasingly in athletics, criminology, government, and business to test for trace amounts of substances such as hormones, metabolites, and drugs. But finding a chemical "needle" in a specimen "haystack" would be impossible with serological techniques that rely only on agglutination or precipitation. This changing face of immunological testing has led to **immunoassay technology,** extremely sensitive methods that permit rapid and accurate measurement of trace antigen or antibody. Some of the means of detecting an antigen or antibody in minute quantities include radioactive isotope labels, enzyme labels, or sensitive electronic sensors. Exquisitely sensitive monoclonal antibody procedures have been enlisted to carry out many of these tests.

Radioimmunoassay (RIA) Antibodies or antigens labelled with a radioactive isotope can be used to pinpoint minute amounts of a corresponding antigen or antibody. Although very complex in practice, the tests are essentially based on comparing a test system for the amount of radioactivity before and after incubation with a known, labelled antigen or antibody. The labelled substance competes with its natural, nonlabelled one for a reaction site. Large amounts of a bound radioactive component indicate that the unknown test substance was not present. The amount of radioactivity is measured with an isotope counter or a photographic emulsion (autoradiograph). RIA has been employed to measure the levels of insulin and other hormones and to diagnose allergies, chiefly by the radioimmunosorbent test (RIST) for measurement of IgE in allergic patients and the radioallergosorbent test (RAST) to standardize allergenic extracts (see chapter 14).

Enzyme-Linked Immunosorbent Assay (ELISA) The **ELISA test** is named for its main elements: an antibody that is simultaneously linked to an antigen and to an enzyme-antibody complex that produces a color change. The enzymes used most often are horseradish peroxidase and alkaline phosphatase, both of which release a dye (chromogen) when exposed to their substrate. This technique also relies on a solid support such as a plastic microtiter plate that can *adsorb* (attract on its surface) the reactants (figure 13.33).

As with immunofluorescence, this technique is applicable for both direct and indirect situations. In *direct ELISA* or sandwich tests, a known antibody adsorbed to the bottom of a well is incubated with a solution containing unknown antigen. After rinsing off excess unbound components, an enzyme-antibody complex that can react with the antigen is added. Any bound antigen will attract enzyme-antibody and hold it in place. Next, the substrate to the enzyme is placed in the wells and incubated. If enzyme has been affixed to Ag, it hydrolyzes the substrate and releases a colored dye. Thus, any color developing in the wells is a positive result. Lack of color means that the antigen was not present and that the subsequent rinsing removed the enzyme-antibody complex.

The *indirect ELISA* test is used to test a patient's serum for antibodies. As with other indirect tests, this one indicates a positive reaction by means of an antibody-antibody reaction, and like the direct ELISA, the indicator antibody is complexed to an enzyme, and a positive reaction is read as a color change (figure 13.33). The starting reactant is a known antigen that is adsorbed to a surface in a well of a plate. To this, an unknown serum is added. After rinsing, an enzyme-Ab reagent that can react with unlabelled antibody is placed in the well. The substrate to the enzyme is then added, and the wells are scanned for color changes. Any color development indicates that all the components reacted and that the antibody was present in the patient's serum. This is the common screening test for the antibodies to HIV (AIDS virus), various rickettsial species, *Salmonella,* the cholera vibrio, certain bacterial toxins, and viruses. Because false positives may occur, a verification test may be necessary.

Some newer technology uses electronic monitors that directly read out antibody-antigen reactions. Without belaboring the technical aspects, these systems contain computer chips that sense the minute changes in electric current given off when an antibody binds to antigen. The potential for sensitivity is extreme—it is thought that amounts as small as 12 molecules of a substance can be detected in a sample. New immunoassays using bioluminescent molecules and enzymes (firefly luciferin and luciferase) that spontaneously release light can also be adapted to measure antigen and antibody reactions.

Tests That Differentiate T Cells and B Cells

So far we have concentrated on tests that identify antigens and antibodies in samples, but numerous techniques can differentiate between B cells and T cells and quantify subsets of each. Knowing about the types and numbers of lymphocytes in blood and other samples has been very important in evaluating immune dysfunctions such as those in AIDS, immunodeficiencies, and cancer. The simplest method for identifying T cells is to mix them with untreated sheep red blood cells. Receptors on the T cell bind the RBCs into a flowerlike cluster called a *rosette formation* (figure 13.34*a*). Rosetting can also occur in B cells if one uses Ig-coated bovine RBCs or mouse erythrocytes.

Indirect fluorescent antibodies have been developed that can differentiate not only between T cells and B cells but can subgroup them (figure 13.34*b*). These subgroup tests utilize monoclonal antibodies produced in response to specific cell markers. B-cell tests categorize different stages in B-cell development and are very useful in characterizing B-cell cancers. T-cell tests help differentiate the CD4, CD8, and other subsets important in AIDS and other immunodeficiency diseases.

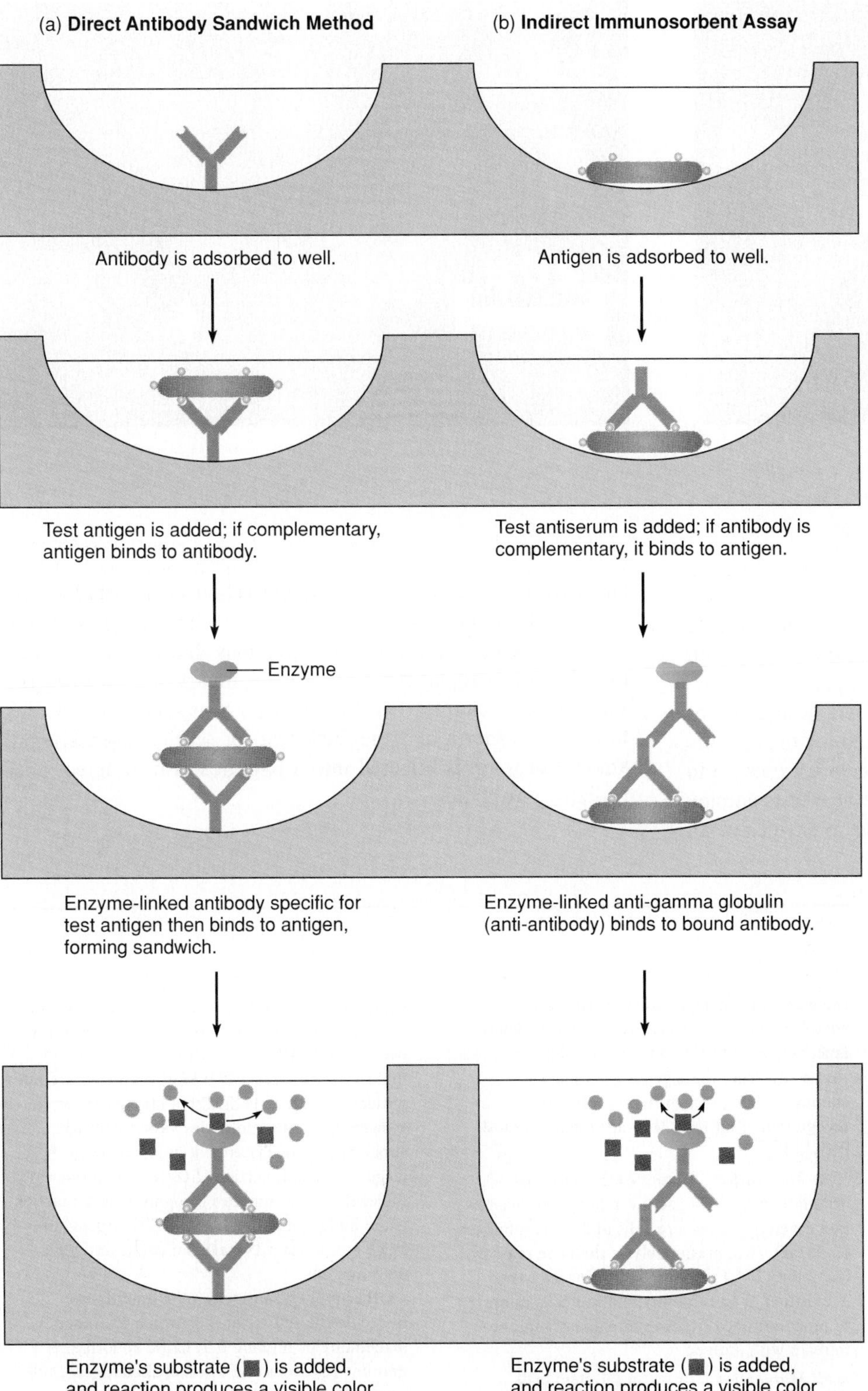

Figure 13.33 Methods of ELISA testing. The success of these tests depends upon adequate rinsing between steps to remove unreacted or nonspecific components. (*a*) Direct: antibody sandwich method. (*b*) Indirect: immunosorbent assay. Current screening for HIV is performed as in (*b*).

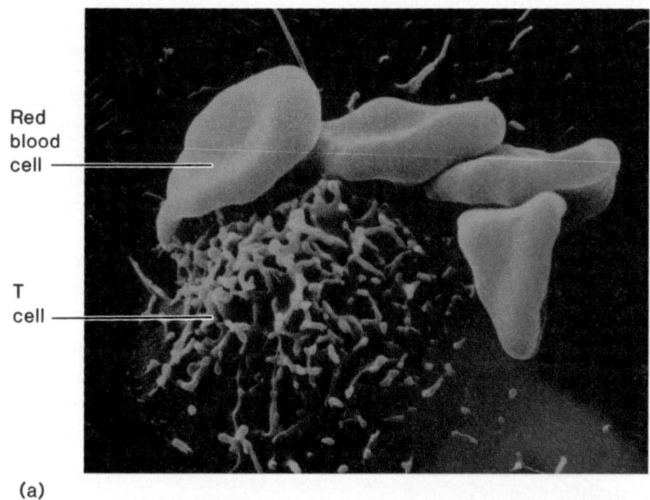

(a)

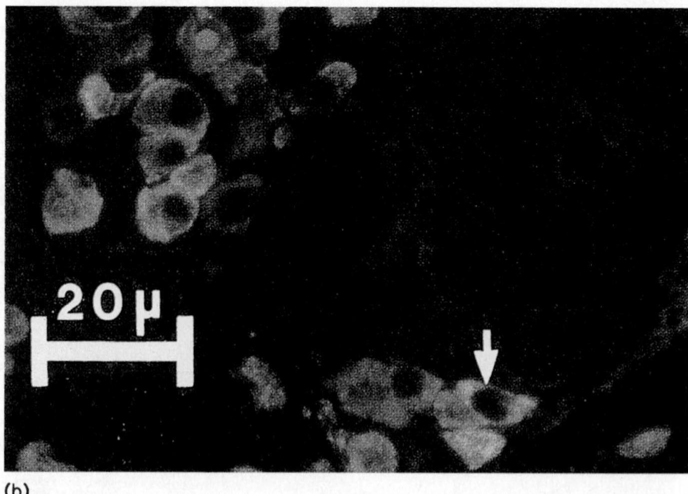

(b)

Figure 13.34 Tests for characterizing T cells and B cells. (*a*) Photomicrograph of rosette formation that identifies T cells. (*b*) Plasma cells highlighted by fluorescent antibodies.

In Vivo Testing

Probably the first immunologic tests were performed not in a test tube, but on the body itself. A classic example of one such technique is the *tuberculin test,* in which a small amount of purified protein from *Mycobacterium tuberculosis* is injected or pricked into the skin and periodically inspected for the appearance of a red, raised lesion that indicates previous exposure to tuberculosis (see feature 16.4). In practice, *in vivo* tests employ principles similar to serological tests, because an antigen or an antibody is introduced into a patient to elicit some sort of visible reaction. Like the tuberculin test, some of these diagnostic skin tests are useful for evaluating infections due to fungi (coccidioidin and histoplasmin tests, for example) or allergens (see skin testing in chapter 14). In sensitivity tests such as the Schick test for diphtheria and the Dick test for scarlet fever, a small amount of toxin is injected into a patient's arm to detect neutralizing antibodies.

Chapter Review with Key Terms

Specific Immunity and Its Applications[4]

Development of Lymphocyte Specificity/Receptors

Acquired immunity involves the reactions of B and T lymphocytes to antigens. Each lymphocyte must undergo differentiation into its final functional type. This involves the development of unique protein receptors for antigen, the specificity of which is genetically controlled and unique for each type of lymphocyte.

The **clonal selection theory** explains that genetic recombination during embryonic and fetal proliferation of lymphocytes produces millions of different lymphocyte clones, each bearing a different kind of antigen receptor. This provides a huge repertoire for reactions with the millions of antigens that can be contacted during life. **Tolerance to self,** the elimination of any lymphocyte clones that can attack self, occurs during this time.

Receptors on B cells are **immunoglobulin** (antibody) molecules, and receptors on T cells are smaller glycoprotein molecules. Other receptors needed in recognition are governed by the **major histocompatibility** (**MHC**) gene complex, which is also referred to as the **human leukocyte antigen** (**HLA**) complex. The receptors that result from these genes govern cell communication, recognition of self, and recognition of antigen; they are found on most body cells.

B-Cell Maturation: Immature B stem cells originate in the yolk sac, liver, and bone marrow and develop into mature cells under the influence of certain sites, presumably in the bone marrow, liver, and GALT; differentiation involves the acquisition of Ig receptors; mature cells migrate to predetermined sites in lymphoid organs, ready to react with antigen.

T-Cell Maturation: Immature T stem cells originate in the same areas of the embryo and fetus but mature in the thymus gland; specificity is acquired through addition of CD receptors; mature cells migrate to different sites in lymphoid organs.

Introduction of Antigens/Immunogens

Ability to respond to antigens (Ag) is present at birth. **Requirements for antigenicity,** the ability to stimulate an immune reaction, include foreignness (recognition as nonself), size, complexity of cell or molecule. Foreign cells and large complex molecules (proteins, lipoproteins over 10,000 MW) are most antigenic; foreign molecules less than 1,000 MW (**haptens**) are not antigenic unless attached to a larger carrier molecule; simpler molecules with repeating subunits (glycogen) are not antigenic even if large. The small portion of the entire antigen molecule that stimulates recognition and reaction is the **antigenic determinant.** Cells, viruses, and large molecules may have numerous antigenic determinants.

Other types of antigens are **autoantigens,** molecules on self tissues for which tolerance is inadequate or missing that cause abnormal immune reactions and disease; **alloantigens,** cell surface markers (HLA, red blood cell) of one member of a species that are antigens to another of that same species; **heterophile antigens,** molecules in two completely unrelated species with an identical or similar antigenic determinant; **allergen,** the antigen that provokes allergy.

Cooperation in Immune Reactions to Antigen

Antigens enter portal of entry and are concentrated in the lymphoid compartments.

4. Also review figure 13.1.

T-cell-dependent antigens are first phagocytosed and processed by special macrophages, the **antigen-processing cells**; these cells alter the antigen and attach it to their MHC receptor for presentation to lymphocytes. Antigen presentation involves a precise, three-way collaboration among the macrophage, a T-helper (T_H) cell, and a selected B cell. Interaction requires direct matching of MHC receptors, specific attachment of antigen, and chemical activation of the T cell by **interleukin-1** from the macrophage and activation of the B cell with **interleukin-2** from the T cell.

B-Cell Activation/Blast Formation/Antibody Production

B-cell receptors receive the antigen from the cooperative complex, and this event, plus B-cell growth and differentiation factors, provide the final stimuli. B cell undergoes **blast formation** in preparation for mitosis and **clonal expansion**; numerous divisions form **plasma cells** that secrete antibodies and **memory cells** that can react to that same antigen later.

Nature of Antibodies: A monomer is composed of large proteins or immunoglobulins (Ig) consisting of four polypeptide chains that form a Y-shaped molecule; contains two identical limbs (**Fab**), the ends of which have a pocket or active site that binds antigen; the shape of the pocket varies among antibodies so as to accommodate different antigens; the solitary leg of the antibody (**Fc**) binds to self.

Antigen-Antibody (Ag-Ab) Reactions: In **opsonization,** antibodies tag the antigen to make it more readily discovered and phagocytosed; in **neutralization,** antibodies attach to an active site on a toxin (**antitoxin**) or a receptor on a virus; antibodies may aggregate antigens; some antibodies fix complement and cause destruction of cells; the Fc portion can bind to various body cells and mediate inflammation and allergy.

Antibody Classes: Five classes exist, differing in size and function: **IgM,** the first Ab to be formed in response to Ag, a huge, five-part molecule with 10 binding sites; **IgG,** major Ab in circulation, produced by memory cells, crosses the placenta; **IgA,** secretory Ab, dimer present on mucous membranes and secretions of them, provides first-strike potential in local entry of Ag; **IgD,** primarily a receptor on B cells; **IgE,** antibody of atopic allergy and anaphylaxis.

Antibodies in Serum (Antiserum): Quality may be determined by electrophoresis, and quantity may be determined by examining **titer** (levels of antibodies) over time; first introduction of an Ag to the immune system produces a **primary response,** with a gradual increase in Ab titer that gradually recedes; second contact with same Ag produces rapid response, larger titer called the **secondary** or **anamnestic response,** which is due to the memory cells produced during initial response.

Monoclonal Antibodies: Artificially induced antibodies formed by fusing a B cell with a cancer cell; are not mixtures of Abs, but have a single specificity to Ag; used in diagnosis of disease, identification of microbes, therapy.

Cell-Mediated Immunity (CMI)/T Cells

T cells react in a direct frontal attack upon a variety of foreign cells and cell markers. Following presentation of Ag by a macrophage, the sensitized T cell goes into blast formation, and depending upon its **CD** receptor type, will become one of four functional types: (1) **T-helper cells** (T_H) assist other T cells and B cells, directly and by means of mediators; (2) **T-suppressor cells** (T_S) limit the actions of other T cells and B cells; (3) **cytotoxic** or **killer T cells** (T_K) seek out and destroy large, complex foreign or abnormal cells (microbes, cancer, grafted tissues, virus-infected cells); (4) **delayed hypersensitivity cells** (T_D) whose reactions with foreign antigens cause a form of allergy. Long-lasting memory T cells remain after response; T cells secrete a series of chemicals called **lymphokines (interleukin, interferon, lymphotoxins)** that destroy antigen or stimulate reactions.

Biomedical Applications of Specific Immunity

Immunization: Producing immunity by medical intervention. Passive immunotherapy includes administering immune serum globulin and specific immune globulins pooled from donated serum to prevent infection and disease in those at risk; antisera and antitoxins from animals are occasionally used.

Active immunization is synonymous with **vaccination**; provides an antigenic stimulus that does not cause disease but produces long-lasting, protective immunity. **Vaccines** contain (1) **killed** whole cells or inactivated viruses that do not reproduce but are antigenic; (2) **live, attenuated** cells or viruses that are able to reproduce but have lost virulence; (3) **subunit** or parts of microbes such as surface antigen or neutralized toxins (**toxoids**); and (4) genetic engineering techniques, including cloning of antigens, and recombinant attenuated microbes. Vaccines are given by injection or orally. Boosters (additional doses) are often required. Vaccination increases **herd immunity,** protection provided by mass immunity in a population.

Serological/Immune Testing: The basis of **serology** is that persons with infectious disease will have antibodies specific to the microbe in the serum. The basis of tests to diagnose disease is that Abs specifically bind to Ag; if an Ag of known identity is added to an unknown serum sample, it will locate and affix to the antibody; the reverse is also true—known antibodies can be used to discover and type antigens; these Ag-Ab reactions are made visible by cross-linking Ag with Ab to form obvious clumps and precipitates, with dyes, or with radioactivity. Tests are performed *in vitro;* result is read as positive or negative; desirable properties of tests are **specificity** and **sensitivity.**

Types of Tests: In **agglutination** tests, antibody cross-links whole cell antigens, forming three-dimensional complexes that settle out and form visible clumps in the test chamber; examples are tests for blood type, some bacterial diseases, viral diseases. **Double diffusion precipitation** tests involve the diffusion of Ags and Abs in a soft agar gel, forming zones of precipitation where they meet.

In **immunoelectrophoresis,** migration of serum proteins in gel is combined with precipitation by antibodies. **Western blot** test involves electrophoresis of Ags or Abs in a gel to separate discrete biochemical bands. Gel is affixed to a blotter, reacted with a test specimen, and developed by radioactivity or with dyes. **Complement fixation** test detects **lysins,** antibodies that fix complement and may lyse target cells; involves first mixing test Ag and Ab with complement and then with sensitized sheep RBCs; if the complement is fixed by the Ag-Ab, the RBCs remain intact and the test is positive; hemolysis of RBCs by the complement indicates lack of specific antibodies.

In **direct assay,** known marked Ab is used to detect unknown Ag (microbe) without an intermediate step. In **indirect testing,** known Ag reacts with unknown Ab; reaction is made visible by attachment of a second marked Ab to first Ab. In **immunofluorescence testing, fluorescent antibodies** (antibodies tagged with fluorescent dye) are used directly and indirectly to form microscopically visible fluorescent cells or aggregates.

Immunoassay is highly sensitive testing for Ag and Ab. In **radioimmunoassay,** Ags or Abs are labelled with radioactive isotopes and traced. The **enzyme-linked immunosorbent assay (ELISA)** is used directly or indirectly to detect unknown Ag or Ab; positive result is visualized when enzyme releases colored product. Tests are also available to differentiate B cells from T cells and subsets of T cells. With *in vivo* testing, Ags are introduced into the body directly to determine the patient's immunologic history.

True–False Questions

Determine whether the following statements are true (T) or false (F). If you feel a statement is false, explain why, and reword the sentence so that it reads accurately.

____ 1. The B-cell receptor is an immunoglobulin molecule.

____ 2. B cells mature in the bone marrow and T cells in the thymus.

____ 3. The best antigens are small, foreign, simple molecules.

____ 4. A virus is a mosaic antigen because it bears several antigenic determinants.

____ 5. Plasma cells are the actual secretors of antibodies.

____ 6. Opsonization is a process in which microbes are cross-linked by antibody.

____ 7. The greatest concentration of antibodies is found in the gamma fraction of the serum.

____ 8. The antibody titer is a measurement of the relative amount of Ab in the serum.

____ 9. The anamnestic response is the short lag period that occurs before the primary response.

____ 10. T-helper cells assist in the specific functions of certain B cells and other T cells.

____ 11. Natural killer cells are general, nonspecific white blood cells that attack cancers.

____ 12. An attenuated microbe is one that has been lysed so as to release its several antigens.

____ 13. Immune serum globulin provides short-term protection to patients at risk.

____ 14. In agglutination reactions, the antigen is a large molecule; in precipitation reactions, it is a small molecule.

____ 15. A person with a high titer of antibodies to an infectious agent has greater protection than a person with a low titer.

____ 16. Direct immunofluorescence tests can identify a microbe in a specimen, whereas indirect immunofluorescence can identify unknown antibodies.

Concept Questions

1. What function do receptors play in specific immune responses? How can receptors be made to vary so widely?
2. Describe the major histocompatibility complex, and explain how it participates in immune reactions.
3. Evaluate the following statement: Each unique lymphocyte type must have a different receptor to react with antigen. How many different Ags might one be expected to meet up with during life?
4. What constitutes a clone of lymphocytes? Explain the clonal selection theory of antibody specificity and diversity. When during development is Ag not needed, and why is it not needed? When is Ag needed? Why must the body develop tolerance to self?
5. Trace the development of the B-cell receptor from gene to cell surface. What is the structure of the receptor? What is the function of the variable regions?
6. Trace the origin and development of B lymphocytes; of T lymphocytes. What is happening during lymphocyte education?
7. Compare and contrast B cells and T cells.
8. What is an antigen or immunogen? The antigenic determinant? How do foreignness, size, and complexity contribute to antigenicity? Why are haptens by themselves not antigenic, even though foreign? How can they be made immunogenic?
9. Differentiate among autoantigens, alloantigens, and heterophile antigens.
10. What is the advantage of having lymphatic organs screen the body fluids, directly and indirectly?
11. Trace the immune response system, beginning with the entry of a T-cell-dependent antigen, antigen processing, presentation, and the cooperative response among the macrophage and lymphocytes.
12. On what basis is a particular B-cell clone selected? How are B cells activated? What happens when they are activated? What are the functions of plasma cells, clonal expansion, and memory cells?
13. Describe the structure of immunoglobulin. What is the function of the Fab and Fc portions?
14. What are some possible outcomes of the attachment of Abs to Ags? (What eventually happens to the Ags?)
15. Multiple matching. Place all possible matches in the space at the left.

    ______ IgM
    ______ IgG
    ______ IgA
    ______ IgD
    ______ IgE

    a. Found in mucous secretions
    b. A monomer
    c. A dimer
    d. Has greatest number of Fabs
    e. Is primarily a surface receptor for B cells
    f. Major Ig of primary response to Ag
    g. Major Ig of secondary response
    h. Crosses the placenta
    i. Fixes complement
16. Contrast primary and secondary response to Ag. What causes the latent period? The anamnestic response?
17. Contrast monoclonal and polyclonal antibodies.
18. Why are the immunities involving T cells called cell-mediated? How do T cells become sensitized? Summarize the function of each category of T cell. How do cytotoxic cells kill their target? Why would the immune system naturally require suppression?
19. Describe the strategies for developing vaccines. What is the advantage of a killed vaccine, a live, attenuated vaccine, a subunit vaccine, a recombinant vaccine? Explain by means of an outline how an inoculation with tetanus toxoid will protect a person the next time they step on a dirty piece of glass.
20. What is the basis of serology and serological testing? Differentiate between specificity and sensitivity; how are Ag-Ab reactions detected? Why is titer important?
21. How are agglutination and precipitation alike? How are they different? Give examples of the two types of testing procedures.
22. Briefly describe and give an example of a specific test using immunoelectrophoresis, Western blot, complement fixation, fluorescence testing (direct and indirect), immunoassays (ELISA—direct and indirect), and *in vivo* testing.

Practical/Thought Questions

1. Double-stranded DNA is a large, complex molecule, but it is not immunogenic by itself. Can you think of a reason why?
2. Tell the relationship between an antitoxin, a toxoid, and a toxin.
3. At least three boosters are given for DPT vaccines. Explain what each subsequent booster does, and why more than one is needed.
4. How would you go about producing monoclonal antibodies that would participate in the destruction of cancer cells but would not kill normal human cells?
5. At the cellular/microscopic level, describe the events in the immune system of AIDS patients that result in opportunistic infections and cancers.
6. Design a vaccine that: would induce protective IgA in the intestine; would give immunity to dental caries; would protect against the liver phase of the malaria parasite; is derived from a single microbe and would immunize against two different infectious diseases.
7. Can you think of a reason that it could be risky to administer a live DNA virus vaccine? A vaccine made from the blood of carriers?
8. Determine the vaccines you have been given; ones you may need in the future; ones for which you may require periodic boosters.
9. Using three colors of clay, fashion Abs to show how the anti-idiotype vaccine is made and how it works to immunize.
10. When traders and missionaries first went to the Hawaiian Islands, the natives there experienced severe disease and high mortality rates from smallpox, measles, and certain STDs. Can you explain why?
11. Why does rising titer of Abs indicate infection?
12. Why do some tests for Ab require backup verification?
13. Looking at figure 13.26*b*, what is the titer as shown? What would it mean as far as the condition of the patient if it had occurred in 1/40? What if no agglutination had occurred in any tube? From figure 13.27*c* can you tell which patients have measles antibody and which do not? What are the titers of the Ab-positive patients?
14. Why does positive hemolysis in the complement fixation test mean negative for the test substance?

CHAPTER 14

Disorders in Immunity

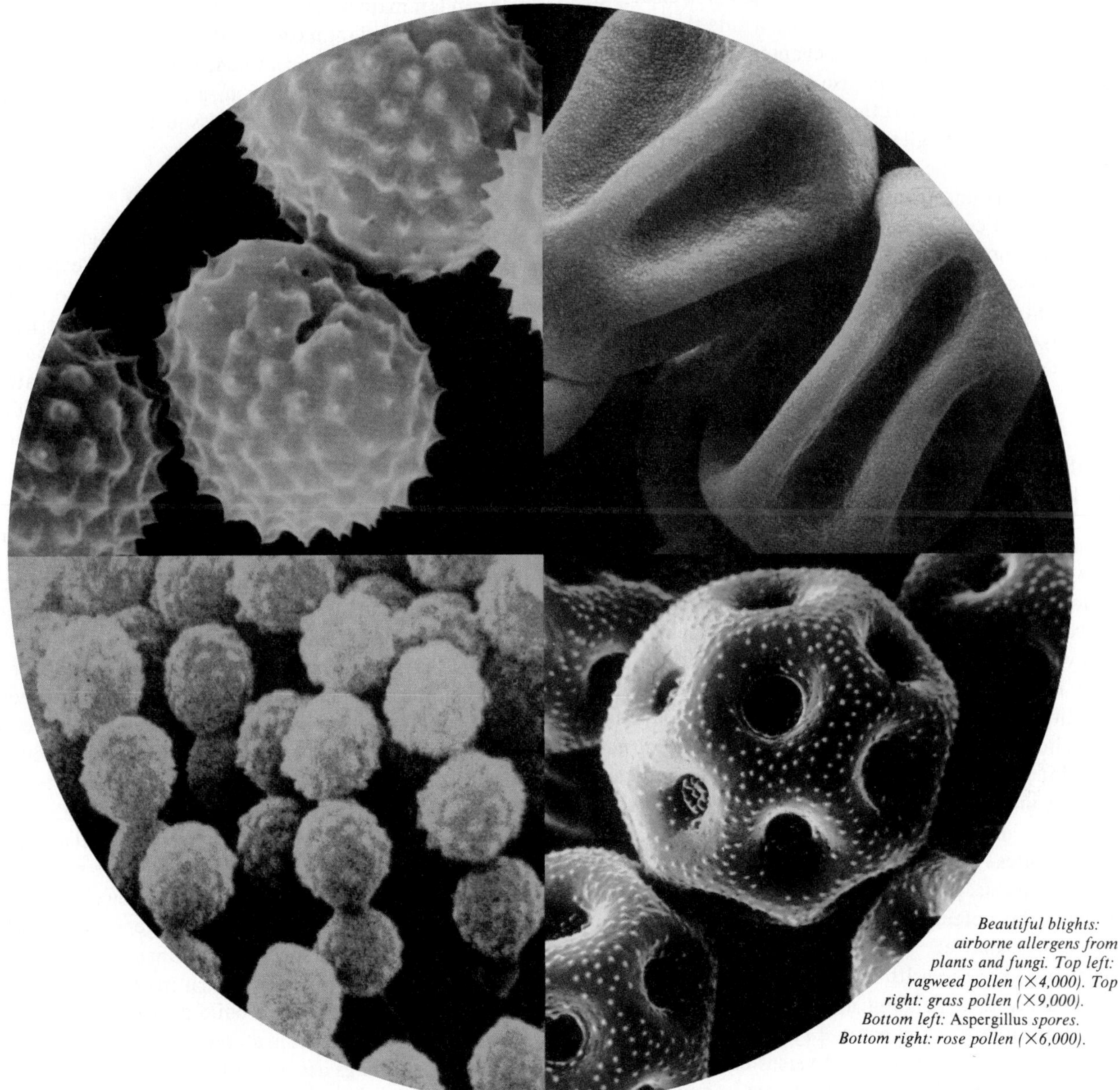

Beautiful blights: airborne allergens from plants and fungi. Top left: ragweed pollen (×4,000). Top right: grass pollen (×9,000). Bottom left: Aspergillus *spores. Bottom right: rose pollen (×6,000).*

Chapter Preview

Humans possess a powerful and intricate system of defense, which by its very nature, also carries the potential for causing injury and disease. In most instances, a defect in immune function is expressed in commonplace but miserable symptoms such as those of hay fever and dermatitis. But abnormal or undesirable immune functions are also actively involved in debilitating or life-threatening diseases such as asthma, anaphylaxis, rheumatoid arthritis, graft rejection, and cancer.

The Immune Response—A Two-Sided Coin

With few exceptions, our previous discussions of the immune response have centered around its numerous beneficial effects. The precisely coordinated system that seeks out, recognizes, and destroys an unending array of foreign materials is clearly protective, but it also presents another side—a side that promotes rather than prevents disease. In this chapter, we will survey **immunopathology,** the study of disease states associated with too much or too little immunity at the wrong place and time (figure 14.1). In the cases of **allergies** and **autoimmunity,** the tissues are innocent bystanders attacked by out-of-control immunological functions. In **grafts** and **transfusions,** a recipient reacts to the foreign tissues and cells of another individual. In **immunodeficiency disease,** immune function is incompletely developed or suppressed. **Cancer** falls into a special category, being both a cause and an effect of immune dysfunction. As we shall see, one fascinating by-product of studies of immune disorders has been our increased understanding of the basic workings of the immune system.

Overreactions to Antigens: Allergy/Hypersensitivity

The term **allergy** was first adopted to denote any state in which the tissues exhibit altered reactivity to a substance. According to its original meaning, allergy included both protective and hypersensitive responses. Usage has narrowed its meaning until now, allergy is used interchangeably with **hypersensitivity,** a term describing an exaggerated immune reaction that is injurious. Allergic individuals are acutely sensitive to repeated contact with antigens, called **allergens,** that do not noticeably affect nonallergic individuals. Although the general effects of hypersensitivity are detrimental, we must be aware that it involves the very same types of immune reactions as those at work in protective immunities—humoral and cell-mediated immunity, inflammatory response, phagocytosis, and complement. Thus, all humans have the potential to develop hypersensitivity under particular circumstances.

Originally, hypersensitivities were defined as either immediate or delayed, depending upon the time lapse between contact with the allergen and onset of symptoms. Subsequently, they were differentiated as humoral versus cell-mediated. But as information on the nature of the allergic immune response accumulated, it became evident that, although useful, these schemes oversimplified what is really a very complex spectrum of reactions. The most widely accepted classification, first introduced by immunologists P. Gell and R. Coombs, includes four major categories: type I (atopic allergies), type II (cytotoxic autoimmunities), type III (immune complex), and type IV (delayed hypersensitivity) (table 14.1). It should be noted that some allergies are combinations of more than one type. In general, types I, II, and III involve a B-cell–immunoglobulin response, and type IV involves a T-cell response. The antigens that elicit hypersensitivity can be exogenous, originating from outside the body (microbes, pollen grains, and foreign cells and proteins), or endogenous, arising from self tissue.

One of the reasons allergies are easily mistaken for infections is that both involve damage to the tissues and thus trigger the inflammatory response (see figure 12.20). Many symptoms and signs of inflammation (redness, heat, skin eruptions, edema, and granuloma) are prominent features of allergies. But unlike inflammation from infection, allergic inflammation is uncontrolled and usually escalates tissue damage.

Type I Allergic Reactions: Atopy and Anaphylaxis

All type I allergies share a similar physiological mechanism, are immediate in onset, and are associated with exposure to specific antigens, but it is convenient to recognize two subtypes: atopy and anaphylaxis. **Atopy** is any chronic local allergy such as hay fever or asthma. **Anaphylaxis** is a systemic, often explosive reaction that involves airway obstruction and circulatory collapse. In the following sections, we will consider the epidemiology of type I allergies, allergens and routes of inoculation, mechanisms of disease, and specific syndromes.

Epidemiology and Modes of Contact with Allergens

Allergies exert profound medical and economic impact. Allergists (physicians who specialize in treating allergies) estimate that about 10% to 30% of the population is prone to atopic allergy. It is generally acknowledged that self treatment with over-the-counter medicines accounts for significant underreporting of cases. The 40 million people afflicted by hay fever (15–20% of the population) spend about half a billion dollars annually for medical treatment. The monetary loss due to employee debilitation and absenteeism is immeasurable. Although many type I allergies are relatively mild, hospitalization and deaths, especially from asthma and anaphylaxis, are not rare.

The predisposition for type I allergies is inherited. Be aware that what is hereditary is a generalized *susceptibility,* not the allergy to a specific substance. For example, a parent who is allergic to ragweed pollen may have a child allergic to cat hair. The prospect of a child developing atopic allergy is at least 25% if one parent is atopic, increasing up to 50% if grandparents or siblings are also afflicted. The actual basis for atopy appears to be the inheritance of a genetic program that leads to IgE synthesis, the production of IgE-specific receptors on mast cells, and the susceptibility of target tissue to allergic mediators (see discussion of mechanisms later in this chapter). Allergic persons often exhibit a combination of syndromes, such as hay fever, eczema, and asthma.

Other factors that affect the presence of allergy are age, infection, imbalance in the nervous system, and geographic locale. New allergies tend to crop up throughout an allergic person's life, especially as new exposures occur after moving or changing life-style. In some persons, atopic allergies last for a

allergy (al′-er-jee) Gr. *allos,* other, and *ergon,* work. Some allergists refer to immediate reactions (hay fever) as allergies and to delayed responses (tuberculin reaction) as hypersensitivities.

atopy (at′-oh-pee) Gr. *atop,* out of place.

anaphylaxis (an″-uh-fih-lax′-us) Gr. *ana,* excessive, and *phylaxis,* protection.

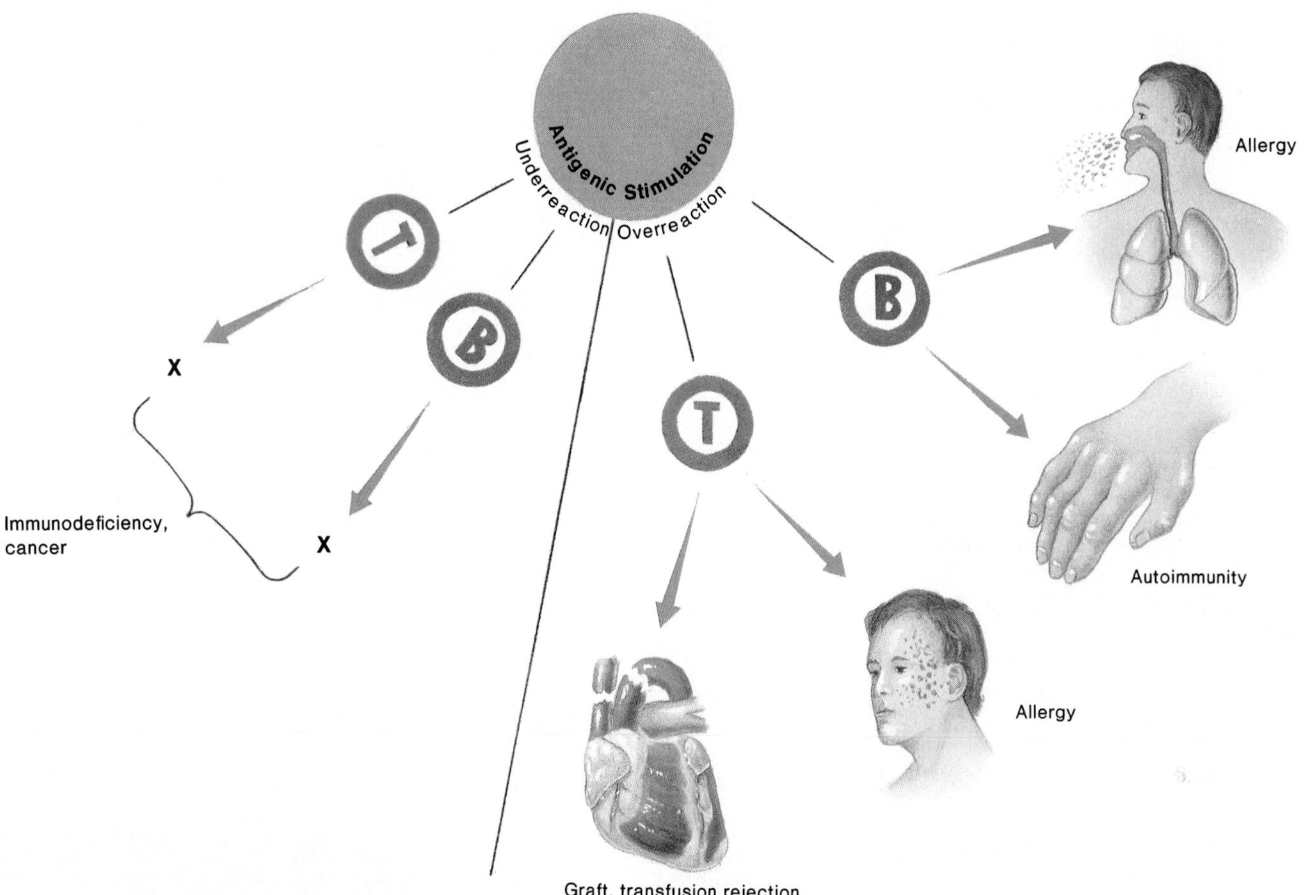

Figure 14.1 The immune system in health and disease. Just as the system of T cells and B cells provides necessary protection against infection and disease, the same system can cause serious and debilitating conditions by overreacting and underreacting to immune stimuli.

Table 14.1 Hypersensitivity States

Type	Antigen Source	Mechanism	Examples
I	Exogenous	Immediate allergy; IgE-mediated; involves mast cells, basophils	Anaphylaxis, atopic allergies such as hay fever, asthma
II	Exogenous, endogenous	IgG, IgM antibodies act upon cells and cause cell lysis; includes some autoimmune diseases	Blood group incompatibility; pernicious anemia
III	Exogenous, endogenous	Immune complex; antibody-mediated inflammation; circulating IgG complexes deposited in basement membranes of target organs; includes some autoimmune diseases	Systemic lupus erythematosus; rheumatoid arthritis; serum sickness
IV	Exogenous	Delayed hypersensitivity, mediated by T cells, NK cells, and macrophages; granulomas and skin reactions	Infection reactions; contact dermatitis; graft rejection

lifetime; others "outgrow" them, and still others suddenly develop them later in life. Some features of allergy have never been completely explained.

The Nature of Allergens and Their Portals of Entry

As with other antigens, allergens have certain immunogenic characteristics. Not unexpectedly, proteins are more allergenic than carbohydrates, fats, or nucleic acids. Many allergens are haptens, non-proteinaceous substances with a molecular weight of less than 1,000 that can form complexes with carrier molecules in the body (see figure 13.11). Organic and inorganic chemicals found in industrial and household products, cosmetics, food, and drugs are commonly of this type. Thousands of commercially synthesized chemicals are potential allergens, and dozens of new agents are developed each year. Table 14.2 lists a number of common allergenic substances.

Table 14.2. Common Allergens, Organized by Portals of Entry

Inhalants	Ingestants	Injectants	Contactants
Pollen	Food (chocolate, wheat, eggs, milk, nuts, strawberries, fish)	Hymenopteran venom	Drugs
Dust	Food additives	Drugs	Cosmetics
Mold spores	Drugs (aspirin, penicillin)	Vaccines	Heavy metals
Dander		Serum	Detergents
Animal hair		Enzymes	Formalin
Insect parts		Hormones	Rubber
Formalin			Glue
Drugs			Solvents
Enzymes			Dyes

Feature 14.1 Odd Allergies

When we think of allergies, most of us conjure up a picture of someone who reacts to bee stings, sneezes around cats, or can't eat a certain food without breaking out in a rash. But not all allergies can be traced to a single substance, and physicians report some surprising and somewhat bizarre causes of allergy. Consider the small number of persons who are actually allergic to exercise. This condition, called exercise-induced asthma, is apparently triggered by loss of moisture from the respiratory tract during strenuous exercise. Other atypical allergic disorders involve sensitivity to environmental factors such as temperature and pressure changes. Cold temperatures may cause swelling of the hands or lips, or a rash in certain sensitive individuals. Heat and sunlight occasionally stimulate local and systemic allergy symptoms, and even a rare allergy to water has been reported. The mechanisms of these peculiar allergies are not well understood.

An encounter with an allergen typically occurs through epithelial portals of entry in the respiratory tract, gastrointestinal tract, and skin. The mucosal surfaces of the gut and respiratory system present a thin, moist surface that is normally quite penetrable. The dry, tough keratin coating of skin is less permeable, but access still occurs through intact skin, glands, and hair follicles. It is worth noting that the organ of allergic expression may or may not be the same as the portal of entry.

Airborne environmental allergens such as pollen, house dust, dander (shed skin scales), or fungal spores are termed *inhalants.* Each geographic region harbors a particular combination of airborne substances that varies with the season and humidity (figure 14.2*a*). Pollen (see chapter opening illustration), the most common offender, is given off seasonally by the reproductive structures of pines and flowering plants (weeds, trees, and grasses). Unlike pollen, mold spores are released throughout the year and are especially profuse in moist areas of the home and garden. Airborne mammalian hair and dander, feathers, and the saliva of dogs and cats are common sources of allergens. The component of house dust that appears to account for most dust allergies is not soil, lint, or other debris, but the decomposed bodies of tiny mites that commonly live in this dust (figure 14.2*b*). Some people are allergic to their work, in the sense that they are exposed to allergens on the job. Florists,

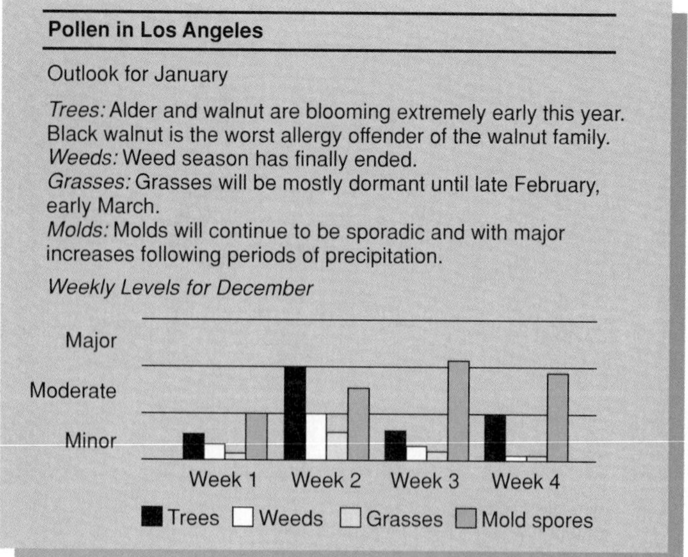

(a)

(b)

Figure 14.2 Monitoring airborne allergens. (*a*) The air in heavily vegetated places with a mild climate is especially laden with allergens. Though the seasons change, pollen and mold spores are present year-round. For example, in just one month in southern California, weed pollen subsided to near zero and mold spore levels doubled. (*b*) Because the dust mite *Dermatophagoides* feeds primarily on human skin cells in house dust, they are found in abundance in bedding and carpets. Airborne mite feces and particles from their bodies are an important source of allergies.

bakers, beauty operators, woodworkers, food processors, veterinarians, farmers, bookbinders, drug processors, leather tanners, welders, and plastics manufacturers work in environments that exacerbate inhalant allergies.

Allergens that enter by mouth, called *ingestants,* often cause food allergies (see a subsequent section of this chapter). *Injectant* allergies are an important adverse side effect of drugs or other substances used in diagnosing, treating, or preventing disease. A natural source of injectants is venom from stings by hymenopterans, a family of insects that includes honeybees and wasps. *Contactants* are allergens that enter through the skin. Many contact allergies are of the type IV, delayed variety discussed later in this chapter.

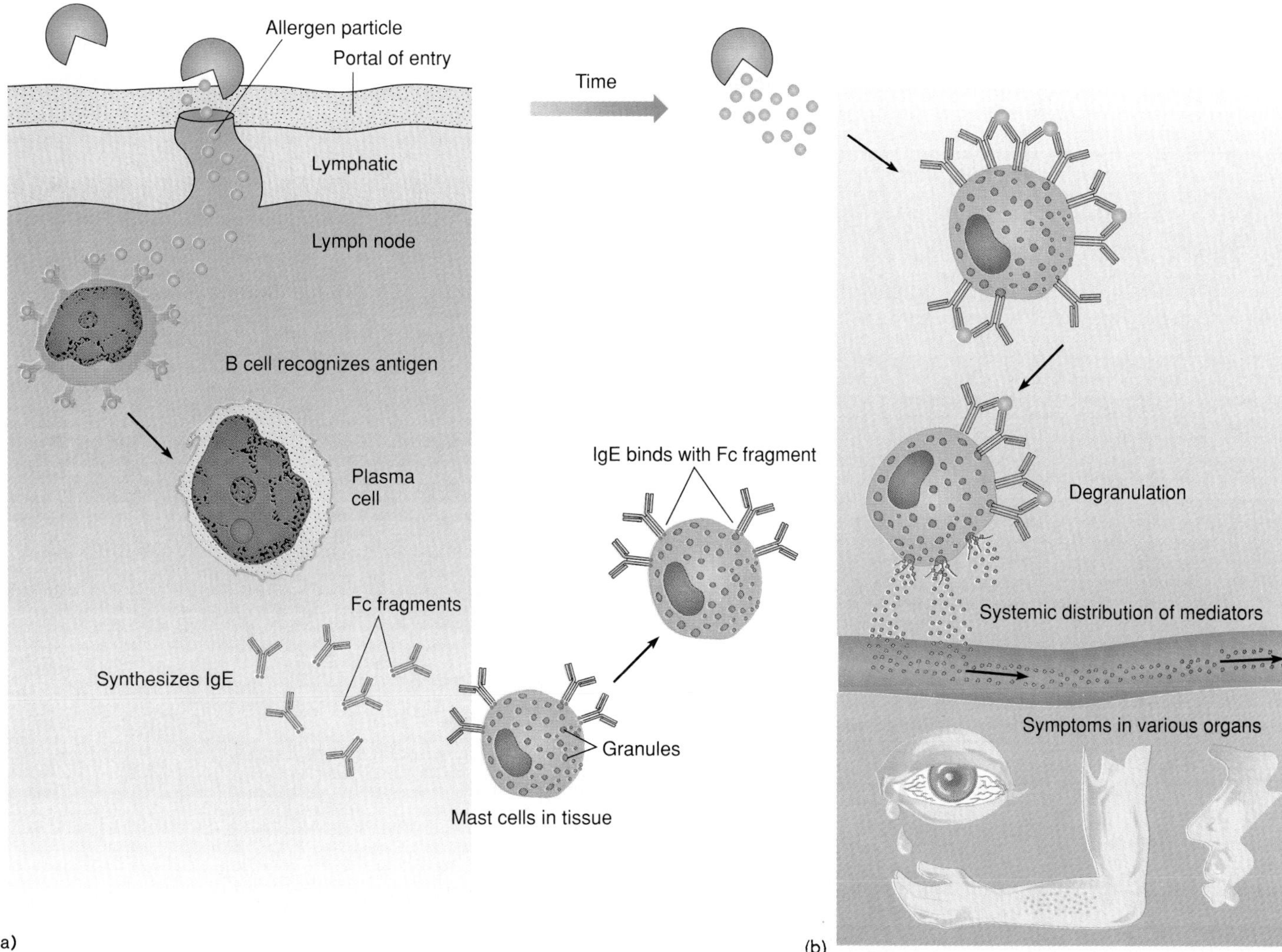

Figure 14.3 A schematic view of cellular reactions during the allergic response. (*a*) Sensitization (initial contact with sensitizing dose). (*b*) Provocation (later contacts with provocative dose).

Mechanisms of Type I Allergy: Sensitization and Provocation

What causes some people to sneeze and wheeze every time they step out into the spring air, while others suffer no ill effects? This hyperactivity helps define allergy, but it does not address what occurs in the tissues of the allergic individual that does not occur in the normal person. In general, type I allergies develop in stages (figure 14.3). The initial encounter with an allergen provides a **sensitizing dose** that primes the immune system for a subsequent encounter with that allergen, but elicits no symptoms. The memory cells and immunoglobulin are then ready to react with a subsequent **provocative dose** of the same allergen. This second contact precipitates the initial symptoms of allergy. Despite numerous anecdotal reports of persons becoming allergic upon first contact with an allergen, it is generally believed that these individuals unknowingly had contact at some previous time. Fetal exposure to allergens from the mother's bloodstream is one possibility, and foods may be a prime source of "hidden" allergens such as penicillin.

The Physiology of Sensitization and Provocation

During primary contact, the allergen penetrates the portal of entry (figure 14.3*a*). When large particles like pollen grains, hair, and spores encounter a moist membrane, they release molecules of allergen that pass into the tissue fluids and lymphatics. The lymphatics carry the allergen to the lymph nodes, where specific clones of B cells discover it (with the help of T-helper cells), are activated, and proliferate into plasma cells. So far, the response duplicates an infection, but what occurs next does not. These plasma cells produce **immunoglobulin E** (IgE), the antibody of allergy also known as *reagin.* IgE is different from other immunoglobulins in having an Fc receptor region with great affinity for mast cells and basophils. The binding of IgE to these cells in the tissues sets the scene for the reactions that occur upon repeat exposure to the same allergen (see figures 14.3*b* and 14.6).

---

reagin (ree′-ah-jin) Derived from the terms *react* or *reaction.*

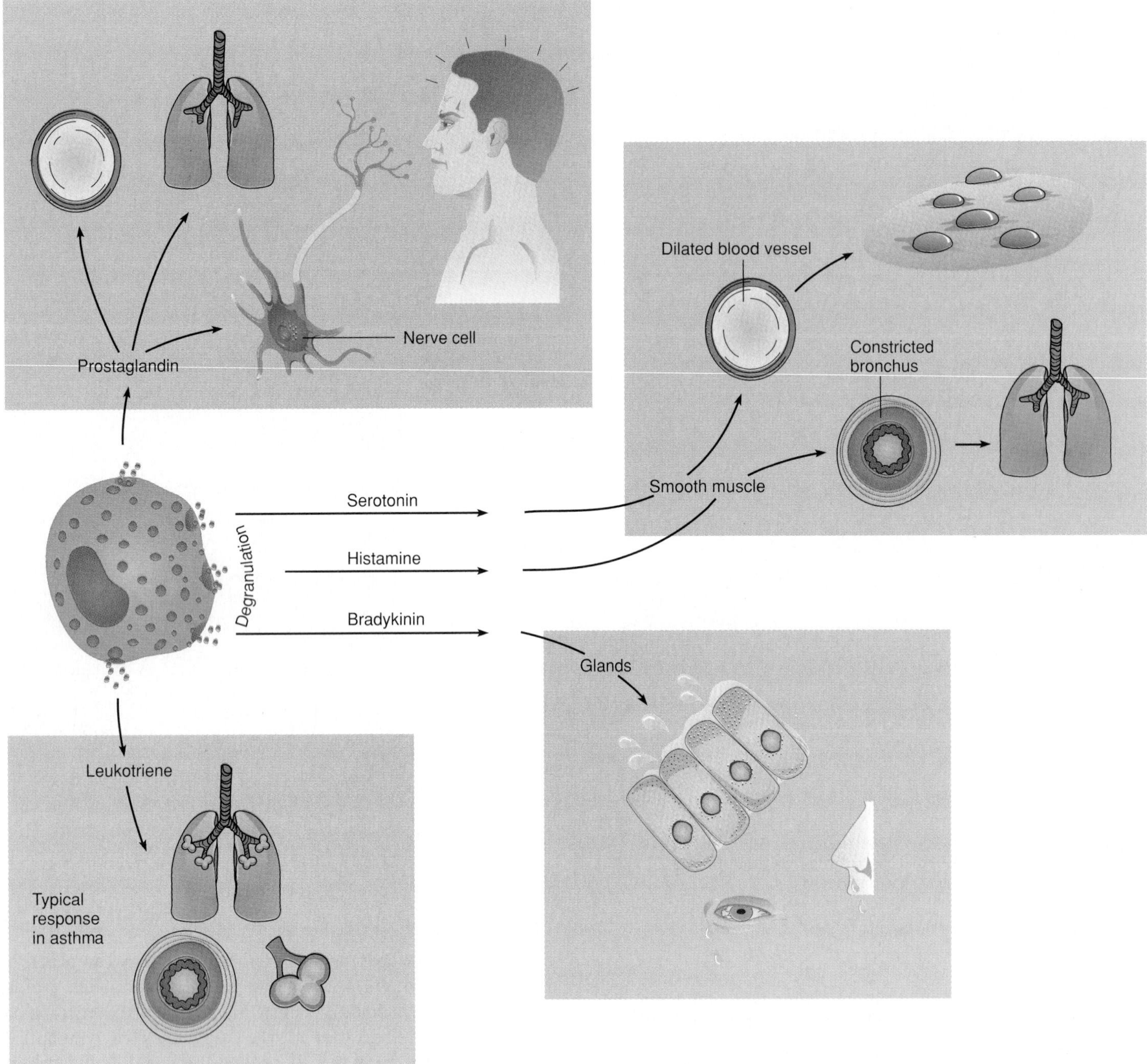

Figure 14.4 The spectrum of reactions to inflammatory mediators and the common symptoms they elicit in target tissues and organs. Note the extensive overlapping effects.

The Role of Mast Cells and Basophils

The characteristics of mast cells and basophils were covered in chapter 12. The most important features of these cells include:

1. Their ubiquitous location in tissues. Mast cells are located in the connective tissue of virtually all organs, but particularly high concentrations exist in the lungs, skin, gastrointestinal tract, and genitourinary tract. Basophils circulate in the blood, but migrate readily into tissues.
2. Their capacity to bind IgE during sensitization (figure 14.3). Each cell carries 30,000 to 100,000 cell receptors that attract 10,000 to 40,000 IgE antibodies.
3. Their cytoplasmic granules (secretory vesicles), which contain physiologically active chemical mediators (histamine, serotonin—introduced in chapter 12).
4. Their tendency to **degranulate** (figures 14.3 and 14.4) or release the contents of the granules into the tissues when properly stimulated by allergen.

Let us now see what occurs when the immune system is challenged with allergen a second time.

The Second Contact with Allergen

After sensitization, the IgE-primed mast cells may remain in the tissues for years. Even if a person goes long periods without contact, he retains the capacity to react immediately upon reexposure. The next time allergen molecules enter the sensitized system, they bind across adjacent receptors and stimulate the mast cell (or basophil) to degranulate. The diffusion of chemical mediators into the tissues and bloodstream accounts for numerous local and systemic reactions, many of which come on with surprising speed. The symptoms of allergy are not caused by the direct action of allergen on tissues but by the physiological effects of mast cell mediators on target organs.

> **Feature 14.2 Of What Value Is Allergy?**
>
> Why would humans and other mammals evolve a response capable of doing so much harm and even causing death? It is unlikely that this limb of immunity exists merely to make people miserable—it must have a role in protection and survival. What are the underlying biological functions of IgE, mast cells, and the array of potent mediators? Analysis has revealed that, although allergic persons have higher levels of IgE, trace quantities are present even in the sera of nonallergic individuals, just as mast cells and inflammatory mediators are also part of normal human physiology. It is generally believed that one important function of this system is to defend against helminth worms that are ubiquitous human parasites. In chapter 12 we learned that mediators serve valuable inflammatory functions, such as increasing blood flow and vascular permeability to summon essential immune components to an injured site. They are also responsible for increased mucous secretion, gastric motility, sneezing, and coughing, which help expel noxious agents. The difference is that in allergic persons, the quantity and quality of these reactions are excessive and uncontrolled.

Mediators, Target Organs, and Allergic Symptoms

About two dozen physiologically active substances that mediate allergy (and inflammation) have been identified. The principal mast cell and basophil mediators are histamine, serotonin, leukotriene, platelet-activating factor, prostaglandins, and bradykinin (figure 14.4). These chemicals, acting alone or in combination, account for the tremendous diversity in allergic manifestations. Tissues lining the body's surface—the skin, upper respiratory tract, gastrointestinal tract, and conjunctiva—are common local targets of allergic symptoms. The general responses of these organs include rashes, itching, redness, rhinitis, sneezing, diarrhea, and profuse lacrimation. Systemic targets include smooth muscle, glands, and nervous tissue. Because smooth muscle is responsible for regulating the size of blood vessels and respiratory passageways, changes in its activity can profoundly alter blood flow, blood pressure, and respiration. Pain, anxiety, agitation, lethargy, and other changes in perception or behavior are also attributable to the effects of mediators on the nervous system.

Histamine is the most profuse and the fastest-acting of the mediators. It is a potent stimulator of smooth muscle, glands, and eosinophils. Histamine's actions on smooth muscle vary with location. It constricts the smooth muscle layers of the small bronchi and intestine, thereby causing labored breathing and increased intestinal motility. In contrast, histamine relaxes vascular smooth muscle and dilates arterioles and venules. This gives rise to a *wheal and flare* reaction (see figure 14.6*b*), pruritis (itching), flushed skin, and headache. More severe reactions (such as anaphylaxis) may be attended by edema and vascular dilation, which lead to hypotension, tachycardia, circulatory failure, and frequently, shock. Salivary, lacrimal, mucous, and gastric glands are also histamine targets.

Although the role of **serotonin** in human allergy is uncertain, its effects appear to complement those of histamine. In experimental animals, serotonin increases vascular permeability, capillary dilation, smooth muscle contraction, intestinal peristalsis, and respiratory rate, but diminishes central nervous system activity.

Before the specific types were identified, **leukotriene** was known as the "slow-reacting substance of anaphylaxis," or SRS-A, for its property of inducing gradual contraction of smooth muscle. This type of leukotriene is responsible for the prolonged bronchospasm, vascular permeability, and mucous secretion of the asthmatic, which do not respond to antihistamine therapy. Other leukotrienes stimulate the activities of polymorphonuclear leukocytes.

Platelet-activating factor is a lipid released by basophils, neutrophils, monocytes, and macrophages that causes platelet aggregation and lysis. The physiological response to stimulation by this factor is similar to that of histamine, including increased vascular permeability, pulmonary smooth muscle contraction, pulmonary edema, hypotension, and a wheal and flare response in the skin.

Prostaglandins are a group of powerful inflammatory agents. Normally, these substances regulate smooth muscle contraction (for example, they stimulate uterine contractions during birth). In allergic reactions, they are responsible for vasodilation, increased vascular permeability, increased sensitivity to pain, and bronchoconstriction. Certain anti-inflammatory drugs work by preventing the synthesis of prostaglandins.

Bradykinin is related to a group of plasma and tissue peptides known as kinins that participate in blood clotting and chemotaxis. In allergy, it causes prolonged smooth muscle contraction of the bronchioles, dilatation of peripheral arterioles, increased capillary permeability, and increased mucous secretion.

histamine (his'-tah-meen) Gr. *histio,* tissue, and amine.

wheal (weel) A smooth, slightly elevated, temporary skin welt.

serotonin (ser''-oh-toh'-nin) L. *serum,* whey, and *tonin,* tone.

leukotriene (loo''-koh-try'-een) Gr. *leukos,* white blood cell, and *triene,* a chemical suffix.

prostaglandin (pross''-tah-glan'-din) From prostate gland. The substance was originally isolated from semen.

bradykinin (brad''-ee-kye'-nin) Gr. *bradys,* slow, and *kinein,* to move.

Specific Diseases Associated with IgE and Mast Cell-Mediated Allergy

The mechanisms just described are basic to hay fever, allergic asthma, food allergy, drug allergy, eczema, and anaphylaxis. In this section we will cover the main characteristics of these conditions, followed by methods of detection and treatment.

Atopic Diseases

Hay fever is a generic term for **allergic rhinitis,** a seasonal reaction to inhaled plant pollen or molds or a chronic, year-round reaction to a wide spectrum of airborne allergens or inhalants, (table 14.2). The targets are typically respiratory membranes, and the symptoms include nasal congestion, sneezing, coughing, profuse mucous secretion, itchy, red and teary eyes, and mild bronchoconstriction. Psychological factors can significantly influence an individual's perception of allergic discomfort.

Asthma is a respiratory disease characterized by episodes of impaired breathing due to bronchoconstriction. The airways of asthmatic persons are exquisitely responsive to minute amounts of inhalant allergens, food, or other stimuli, such as infectious agents. The symptoms of asthma range from occasional, annoying bouts of difficult breathing to fatal suffocation. Labored breathing, shortness of breath, wheezing, cough, chest tightness, and ventilatory *rales* are present to one degree or another. The respiratory tract of an asthmatic person is chronically inflamed and severely overreactive to allergic mediators, especially leukotrienes and serotonin from pulmonary mast cells. Other pathological components are thick, mucous plugs in the air sacs and lung damage that can result in long-term respiratory compromise (figure 14.4). Although an imbalance in the nervous control of the respiratory smooth muscles is apparently involved in asthma, the mechanisms are obviously complex and yet to be fully explained. Asthmatic episodes are highly responsive to the psychological state of the person, which strongly supports a neurological connection.

There are about 9.7 million asthmatics in the United States, with nearly one-third of them children. For reasons that are not completely understood, asthma is on the increase, and deaths from it have doubled in the past 10 years, even though effective agents to control it are more available now than they have ever been before (see discusssion of therapy later in this chapter).

Atopic dermatitis is an intensely itchy inflammatory condition of the skin, sometimes also called **eczema.** Sensitization occurs through ingestion, inhalation, and occasionally, skin contact with allergens. It usually begins in infancy with reddened, vesicular, weeping, encrusted skin lesions, and it progresses in childhood and adulthood to a dry, scaly, thickened (lichenified) skin condition (figure 14.5). Lesions may occur on the face, scalp, neck, and inner surfaces of the limbs and trunk. The itchy, painful lesions cause considerable discomfort, and they are often predisposed to secondary bacterial infections. An anonymous writer once aptly described eczema as "the itch that rashes" or "one scratch is too many but one thousand is not enough."

Figure 14.5 Atopic dermatitis, or eczema. Vesicular, encrusted lesions are typical in afflicted infants. This condition is prevalent enough to account for 1% of pediatric care.

Food Allergy

The ordinary diet contains a vast variety of compounds that are potentially allergenic. It is generally held that food allergies are due to a digestive product of the food or to an additive (preservative or flavoring). Although the mode of entry is intestinal, food allergies can also affect the skin and respiratory tract. Gastrointestinal symptoms include vomiting, diarrhea, and abdominal pain. In severe cases, nutrients are poorly absorbed, leading to growth retardation and failure to thrive in young children. Other manifestations of food allergies include eczema, hives, rhinitis, asthma, and occasionally, anaphylaxis. Classic food hypersensitivity involves IgE and degranulation of mast cells, but not all reactions involve this mechanism (see feature 14.3). The most common food allergens come from peanuts, fish, cow's milk, eggs, shellfish, and soybeans.

Drug Allergy

Modern chemotherapy has been responsible for many medical advances. Unfortunately, it has also been hampered by the fact that drugs are foreign compounds capable of stimulating allergic reactions. In fact, allergy to drugs is one of the most common side effects of treatment (present in 5–10% of hospitalized patients). Depending upon the allergen, route of entry, and individual sensitivities, virtually any tissue of the body may be affected, and reactions range from mild atopy to fatal anaphylaxis. Compounds implicated most often are antibiotics (penicillin is number one in prevalence), synthetic antimicro-

rhinitis (rye-nye'-tis) Gr. *rhis,* nose, and *itis,* inflammation.
asthma (az'-muh) The Greek word for gasping.
rales (rails) Abnormal breathing sounds.
eczema (eks'-uh-mah; also ek-zeem'-uh) Gr. *ekzeo,* to boil over.

Feature 14.3 "Please Hold the Additives"

A number of adverse reactions to food additives can provoke the symptoms of allergy. In some cases, a true allergy to the substance exists; in others, the reaction is an unusual intolerance of unknown origin.

Sodium metabisulfite (sulfite) is sometimes added to wine to prevent spoilage or to vegetables in salad bars to prevent browning. Sensitive persons exposed to this agent undergo severe, asthma-like attacks and occasionally even anaphylaxis (salad bar sickness). "Hot dog headache" may be experienced by some persons after ingesting nitrate and nitrite preservatives commonly used in processed meats (bacon, sausages). Tartrazine yellow dye 5 used to color foods, pills, tablets, and capsules may cause symptoms of asthma, rhinitis, or hives. Monosodium glutamate flavor enhancer is responsible for the so-called "Chinese restaurant syndrome," a sensation of burning, tightness, or numbness in the chest, neck, and face that begins shortly after the first few bites and lasts two to three hours.

Food sensitivity may also arise from microbial contaminants. Poisoning from negligently handled and stored seafood has been traced to bacterial contaminants that produce histamine-like compounds. Certain plants naturally contain small amounts of active compounds that exert a pharmacologic effect. Coffee has caffeine, tea has theophylline, and cocoa has theobromine. In larger quantities, these may produce insomnia, headache, nervousness, tachycardia, nausea, abdominal pain, and diarrhea. Some fresh fruits and vegetables and processed foods (chocolate, cheese, and dried meats) contain vasoactive amines (histamine, epinephrine, norepinephrine) that may cause headache and other symptoms in sensitive individuals.

bics (sulfa drugs), aspirin, opiates, and anaesthetics. In the United States, about 80 million infections are treated annually with penicillin or its semisynthetic derivatives. The actual allergen is not the intact drug itself but a hapten given off when the liver processes the drug. Penicillin allergy has also been traced to contamination of drugs, meat, milk, and other foods. Exposure to *Penicillium* mold in the environment is another suspected means of sensitization.

Anaphylaxis: An Overpowering Systemic Reaction

The term **anaphylaxis** or **anaphylactic shock** was first used to denote a reaction of animals injected with a foreign protein. Although the animals showed no response during the first contact, upon reinoculation with the same protein a few days later, they exhibited acute symptoms—itching, sneezing, difficult breathing, prostration, and convulsions—and many died in a few minutes. Two clinical types of anaphylaxis are distinguished in humans. *Cutaneous anaphylaxis* is the wheal and flare inflammatory reaction to the local injection of allergen. *Systemic anaphylaxis* is characterized by sudden respiratory and circulatory disruption that may be fatal. In humans, the allergen and route of entry are individually variable, though bee stings and injections of antibiotics or serum are implicated most often. Bee venom is a complex material containing several allergens and enzymes that contribute to the severity of the allergic response. Some persons have such extreme sensitivity that they retain it for decades.

The underlying physiological events in systemic anaphylaxis parallel those of atopy, but the concentration of mediators and the strength of the response are greatly amplified. The immune system of a sensitized person exposed to a provocative dose of allergen responds with a sudden, massive outpouring of allergic mediators into the tissues and blood, which act rapidly on the target organs. Anaphylactic persons have been known to die in 15 minutes from complete airway blockage.

Diagnosis of Allergy

Because allergy mimics infection and other conditions, it is important to determine if a person is actually allergic. If possible or necessary, it is also helpful to identify the specific allergen or allergens. Allergy diagnosis involves several levels of tests, including nonspecific, specific, *in vitro,* and *in vivo* methods.

A new test that can distinguish whether a patient has experienced an allergic attack measures elevated tryptase levels in the blood. The amount of tryptase, an enzyme released by mast cells, increases during an allergic response and remains high for several hours. Several types of specific *in vitro* tests can determine the allergic potential of a patient's blood sample. The leukocyte histamine-release test measures the amount of histamine released from the patient's basophils when exposed to a specific allergen. Serological tests that use radioimmune assays (see chapter 13) to reveal the quantity and quality of IgE are also clinically helpful. A general test for measuring the concentration of IgE in serum is the paper radioimmunoabsorbent test (PRIST). The radioallergosorbent test (RAST) is an extremely sensitive way to measure allergen-specific IgE, but it can be very expensive if a person has multiple allergies.

Skin Testing

A useful *in vivo* method to detect precise atopic or anaphylactic sensitivities is skin testing. With this technique, a patient's skin is injected, scratched, or pricked with a small amount of a pure allergen extract. Hundreds of these allergen extracts are obtainable from pharmaceutical companies, and they include common airborne allergens (plant and mold pollen) and more unusual allergens (mule dander, theatre dust, parakeet feathers). In patients with numerous allergies, the allergist maps the skin on the inner aspect of the forearms or back and injects the allergens intradermally according to this predetermined pattern (figure 14.6*a, b*). Approximately 20 minutes after antigenic challenge, each site is appraised for a wheal response indicative of histamine release. The diameter of the wheal is measured and rated on a scale of 0 (no reaction) to 4 (greater than 15 mm). Figure 14.6*c* shows skin test results for a person with extreme inhalant allergies.

Food Allergies: A Special Challenge

Pinpointing the offending ingredient in the patient's diet is a difficult but vital first step in treating and preventing food allergy. The best indicators of food allergy are allergic symptoms each time a particular item is ingested. In cases lacking direct correlation, identification of the allergen is by trial and error, a task greatly complicated by the complexity of foods and the presence of contaminants and additives. Patients may be instructed to compile a daily diary of their intake, eliminate suspected foods, watch for abatement of symptoms, and then

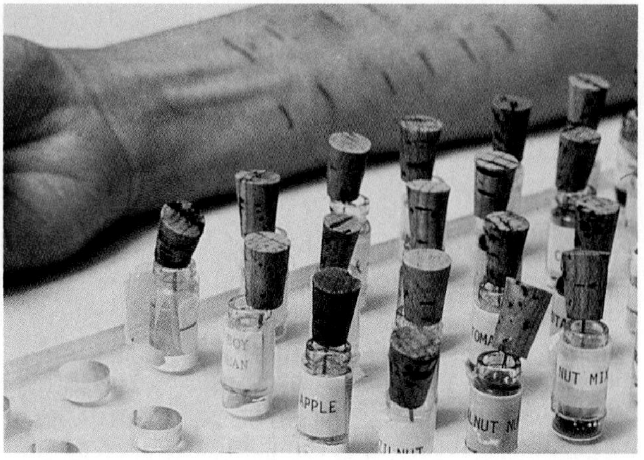

(a)

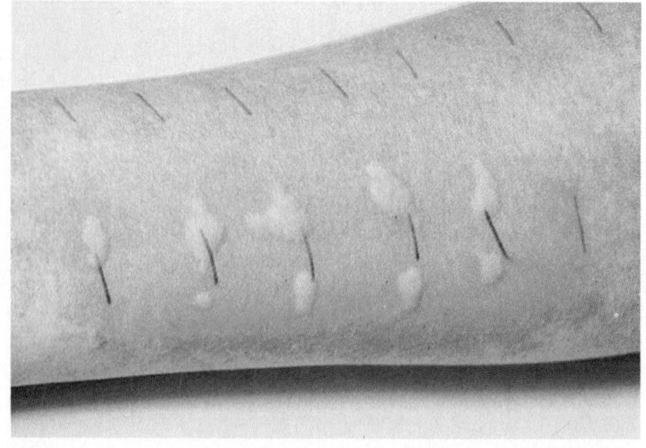

(b)

Environmental Allergens

No. 1 Standard Series

No.	Allergen	ID 8/85
1.	Acacia gum	+++
2.	Cat dander	+++
3.	Chicken feathers	++++
4.	Cotton lint	++++
5.	Dog dander	++
6.	Duck feathers	+
7.	Glue, animal	+
8.	Horse dander	++
9.	Horse serum	X
10.	House dust #1	+++
11.	Kapok	+
12.	Mohair (goat)	+
13.	Paper	+
14.	Pyrethrum	++++
15.	Rug pad, ozite	+++
16.	Silk dust	+
17.	Tobacco dust	+
18.	Tragacanth gum	+
19.	Upholstery dust	+++++
20.	Wool	+++
21.	....	

No. 2 Airborne Insect Particles

No.	Allergen	ID 8/85
1.	Ant	+++
2.	Aphis	+++++
3.	Bee	++++
4.	Housefly	++++
5.	House mite	X
6.	Mosquito	+++
7.	Moth	++++
8.	Roach	+++
9.	Wasp	++
10.	Yellow Jacket	0
	Airborne Mold Spores	
11.	*Alternaria*	++
12.	*Aspergillus*	+++
13.	*Cladosporium*	++
14.	*Hormodendrum*	+++
15.	*Penicillium*	0
16.	*Phoma*	+
17.	*Rhizopus*	+++
18.	....	

(c)

Figure 14.6 A method for conducting skin test allergies. (*a*) The forearm (or back) is mapped prior to injection with a selection of allergen extracts. The allergist must be very aware of potential anaphylaxis attacks triggered by these injections. (*b*) Closeup of skin wheals showing a moderately positive reaction (dark lines are measurer's marks). (*c*) Results of standardized skin tests for allergies to some common environmental allergens.

reintroduce the food to see if symptoms recur. Although the skin is a target in food allergy, skin testing with suspensions of food extracts is plagued with false-positive or false-negative reactions, so skin tests do not reliably verify food allergies.

Treatment and Prevention of Allergy

In general, the methods of treating and preventing type I allergy involve (1) avoiding the allergen, though this may be very difficult in many instances, (2) taking drugs that block the action of lymphocytes, mast cells, or mediators, and (3) undergoing desensitization therapy.

It is not possible to completely prevent initial sensitization, because there is no way to tell in advance if a person will develop an allergy to a particular substance. The practice of delaying the introduction of solid foods apparently has some merit in preventing food allergies in children, though even breast milk may contain allergens ingested by the mother. Rigorous cleaning and air conditioning can reduce contact with airborne allergens, but it is not feasible to isolate a person from all allergens, and that is why drugs are so important in control.

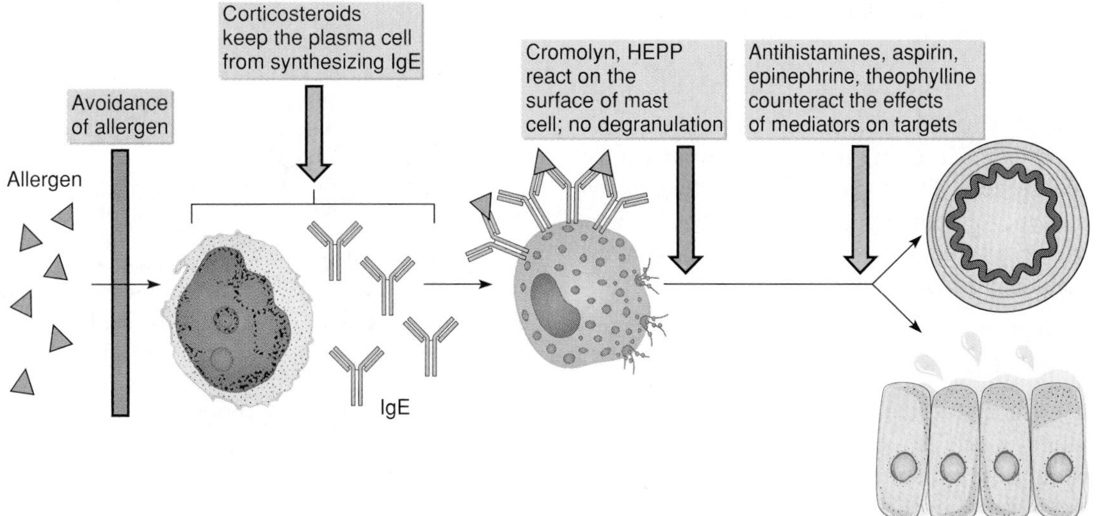

Figure 14.7 Strategies for circumventing allergic attacks.

Therapy to Counteract Allergies

The aim of antiallergy medication is to block the progress of the allergic response somewhere along the route between IgE production and the appearance of symptoms (figure 14.7). Oral anti-inflammatory drugs such as corticosteroids inhibit the activity of lymphocytes and thereby reduce the production of IgE, but these drugs have dangerous side effects and should not be taken for prolonged periods. Asthmatic patients can benefit from a corticosteroid inhaler. Some drugs block the degranulation of mast cells and reduce the levels of inflammatory mediators. The most effective of these are diethylcarbamazine, cromolyn, and a new synthetic drug called HEPP.

The most prevalent medications for preventing symptoms of atopic allergy are *antihistamines*, the active ingredients in most over-the-counter allergy control drugs. These drugs interfere with histamine activity by binding to histamine receptors on target organs. Drawbacks to antihistamines are that (1) the body develops a tolerance to them, requiring higher doses; (2) they do not prevent the many symptoms of allergy caused by other mediators; and (3) they induce drowsiness because of their effect on the brain. A newer antihistamine, terfenadine (Seldane), lacks this last side effect because it does not cross the blood-brain barrier. Other drugs that relieve inflammatory symptoms are aspirin and acetaminophen, which reduce pain by preventing prostaglandin synthesis, and theophylline, a bronchodilator that reverses spasms in the respiratory smooth muscles. Persons who suffer from anaphylactic attacks are urged to carry at all times a solution of injectable epinephrine (adrenalin) and an identification tag indicating their sensitivity. An aerosol inhaler containing epinephrine can also provide rapid relief. The effects of epinephrine are to reverse constriction of the airways and slow the release of allergic mediators. In an emergency, a tracheostomy may be necessary to bypass the constricted breathing passages.

Approximately 70% of allergic patients benefit from controlled injections of specific allergens, which are determined by skin tests. This technique, called **desensitization** or **hyposensitization,** is a therapeutic way to prevent reactions between allergen, IgE, and mast cells. The allergen preparations contain pure, preserved suspensions of plant antigens, venoms, dust mites, dander, and molds (but so far, hyposensitization for foods has not proved very effective). The immunologic mechanism of this treatment is open to differences in interpretation. One theory suggests that the quality and quantity of injected antigen stimulate the formation of high levels of IgG (figure 14.8). It has been proposed that these IgG *blocking antibodies* attach to the allergen and remove it from the system before it can bind to IgE, thus preventing the degranulation of mast cells. Other evidence points to the possibility that antigen delivered in this fashion combines with IgE *before* the IgE has a chance to react with the mast cells. It has been further postulated that the therapy induces specific clones of suppressor T cells that prevent B cells from producing IgE.

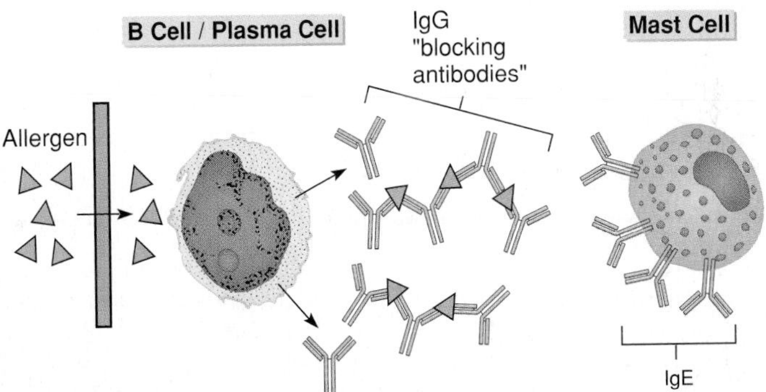

Figure 14.8 The blocking antibody theory for allergic desensitization. An injection of allergen causes IgG antibodies to be formed instead of IgE; these blocking antibodies cross-link and effectively remove the allergen before it can react with the mast cell.

Type II Hypersensitivities: Reactions That Lyse Foreign Cells

The diseases termed type II hypersensitivities are a complex group of syndromes that involve complement-assisted destruction (lysis) of cells by antibodies (IgG and IgM) directed against those cells' surface antigens. This category includes transfusion reactions and some types of autoimmunities (discussed in a later section). Often the cells targeted for destruction are red blood cells, but other cells may be involved.

Human Blood Types

Chapters 12 and 13 described the functions of unique surface receptors or markers on cell membranes. Ordinarily, these receptors play essential roles in cell recognition and development, but they become medically important when the tissues of one person are placed into the body of another person. Blood transfusions and organ donations (described later in this chapter) present an opportunity for the isoantigens (molecules that differ in the same species) on donor cells to be recognized by the lymphocytes of the recipient. These reactions are not really immune dysfunctions like allergy and autoimmunity. The immune system is in fact working normally, but it is not equipped to distinguish between the desirable foreign cells of a transplanted tissue and the undesirable ones of a microbe.

The Discovery of Blood Types Blood transfusions have been practiced since antiquity. At first, transfusions were performed between animals and humans, and then later, from one human to another. Occasionally these transfusions "took," but more often, they ended fatally, causing several countries to ban the procedure. Nonetheless, replacing lost blood was too alluring an idea to remain permanently suppressed, and in 1904 Karl Landsteiner finally discovered the source of this baffling incompatibility. His careful work first demonstrated that the serum of one person could clump or agglutinate the red blood cells of another. From several agglutination patterns, he identified the four blood types that came to be called the ABO blood groups (table 14.3).

The Basis of Human ABO Isoantigens and Blood Types

Like the MHC antigens on white blood cells, the ABO isoantigen markers on red blood cells are genetically determined and comprised of glycoproteins. These ABO antigens are inherited as two (one from each parent) of three alternative *alleles:* A, B, or O. A and B alleles are dominant over O and codominant with one another. As table 14.3 indicates, this mode of inheritance gives rise to four blood types (phenotypes), depending on the particular combination of genes. Thus, a person with an *AA* or *AO* genotype has **type A** blood; genotype *BB* or *BO* gives **type B;** genotype *AB* produces **type AB;** and genotype *OO* produces **type O.** Some important points about the blood types are: (1) They are named for the dominant antigen(s); (2) the RBCs

Table 14.3 Characteristics of ABO Blood Groups

Genotype	Phenotype A or B* RBC Antigen	Prevalence in Population**	Serum Content of Antibodies
OO	Neither	Most common	Both anti-a and anti-b
AA, AO	A	Second most common	Anti-b
BB, BO	B	Third most common	Anti-a
AB	AB	Least common	Neither antibody

*Capital letters generally denote Ag; lowercase denotes Ab.

**True of most large populations of mixed racial and ethnic groups.

of type O persons have receptors—they just don't have A and B receptors; and (3) tissues other than RBCs carry A and B antigens. Aspects of the molecular biology of the RBC markers are discussed in feature 14.4.

Antibodies Against A and B Antigens

Although an individual does not normally produce antibodies in response to his or her own RBC antigens, the serum may contain antibodies that can react with blood of another antigenic type, even though contact with this other blood type has *never* occurred. These preformed antibodies account for the immediate and intense quality of transfusion reactions. As a rule, type A blood contains antibodies (anti-b) that react against the antigens on type B and AB red blood cells. Type B blood contains antibodies (anti-a) that react with type A and AB red blood cells. Type O blood contains antibodies against both A and B antigens. Type AB blood does not contain antibodies against either A or B antigens[1] (table 14.3). The question that immediately arises is: Where do the anti-a and anti-b antibodies originate? The answer apparently lies in heterophile antigens, certain surface molecules distributed ubiquitously throughout the cellular world (see chapter 13). The surface polysaccharides of intestinal bacteria and some plant cells mimic the structure of A and B isoantigens.[2] Exposure to these natural sources during early childhood stimulates antibodies capable of reacting with the A and B isoantigens.

Clinical Concerns in Transfusions

With this information in mind, what are the clinical concerns in giving blood transfusions? First, the individual blood types of donor and recipient must be determined. Using a standard technique, a drop of blood is mixed with antisera that contain antibodies against the A and B antigens and then observed for the evidence of agglutination (figure 14.9).

allele (ah-leel') Gr. *allelon,* of one another. An alternate form of a gene for a given trait.

1. Why would this be true? The answer lies in the first sentence of the paragraph.
2. Evidence for this comes from germ-free chickens that do not have antibodies against the isoantigens of blood types, whereas normal chickens do.

Feature 14.4 The Origin of ABO Antigens

The A and B genes each code for an enzyme that adds a terminal carbohydrate to RBC receptors during maturation. RBCs of type A contain an enzyme that adds N-acetylgalactosamine to the receptor; RBCs of type B have an enzyme that adds D-galactose; RBCs of type AB contain both enzymes that add both carbohydrates; and RBCs of type O lack the genes and enzymes to add a terminal molecule.

The genetics of ABO antigens were once used to rule out paternity. For example, if a man is type A, the mother type O, and the child, type B, we know this man could not have fathered this child. However, this same logic cannot prove paternity. If the child is type A instead, it is possible for the man to be the father, but so could some other man with blood type A. Highly sensitive methods based on specific and variable MHC antigens and DNA fingerprinting have been developed to more precisely gather evidence of paternity or maternity (in cases of kidnapping or adoption, for instance).

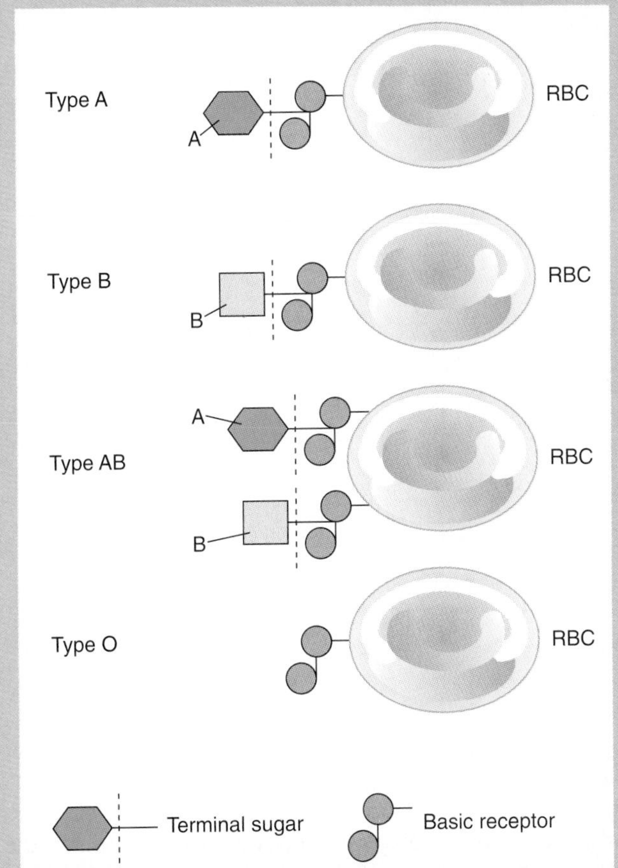

The genetic/molecular basis for the A and B antigens (receptors) on red blood cells. In general, persons with blood types A, B, and AB inherit a gene for the enzyme that adds a certain terminal sugar to the basic RBC receptor. Type O persons do not have such an enzyme and lack the terminal sugar.

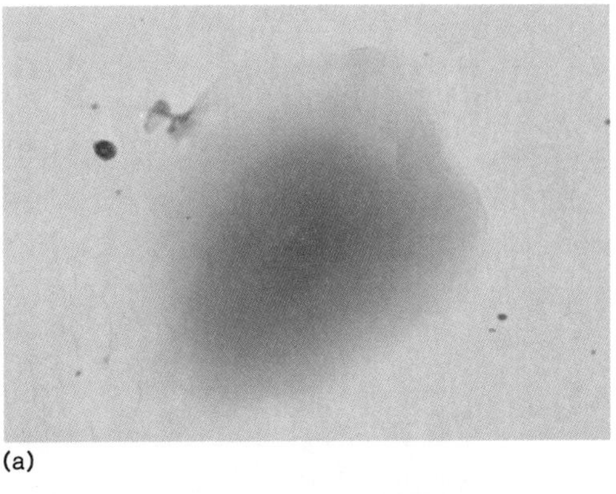

(a)

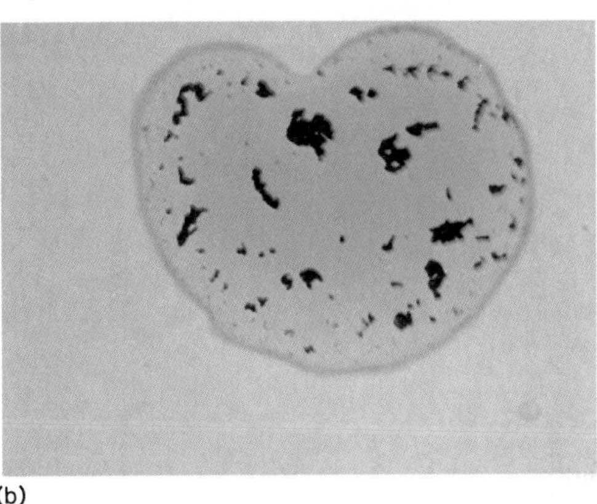

(b)

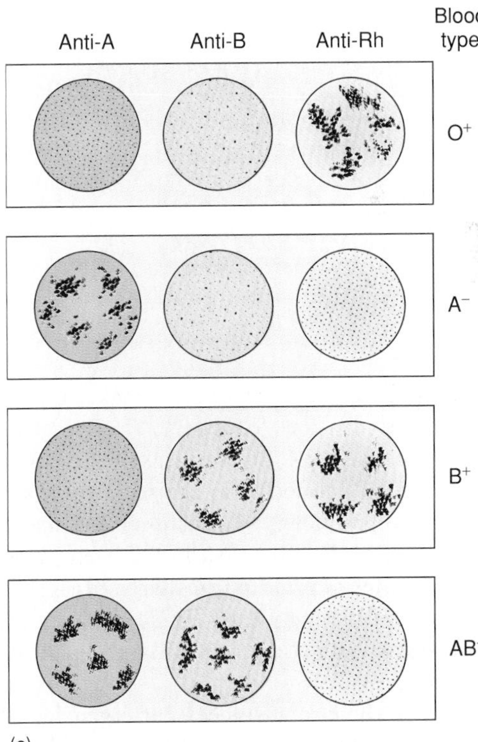

(c)

Figure 14.9 Interpretation of blood typing. In this test, a drop of blood is mixed with a specially prepared antiserum known to contain antibodies against the A, B, or Rh antigens. (*a*) If that particular Ag is not present, the RBCs in that droplet do not agglutinate. (*b*) If that Ag is present, agglutination occurs and the RBCs form visible clumps. (*c*) Several patterns and their interpretations. Anti-A, anti-B, and Rh are shorthand for the antiserum applied to the drops. (In general, O^+ is the most common blood type, and AB^- is the rarest.)

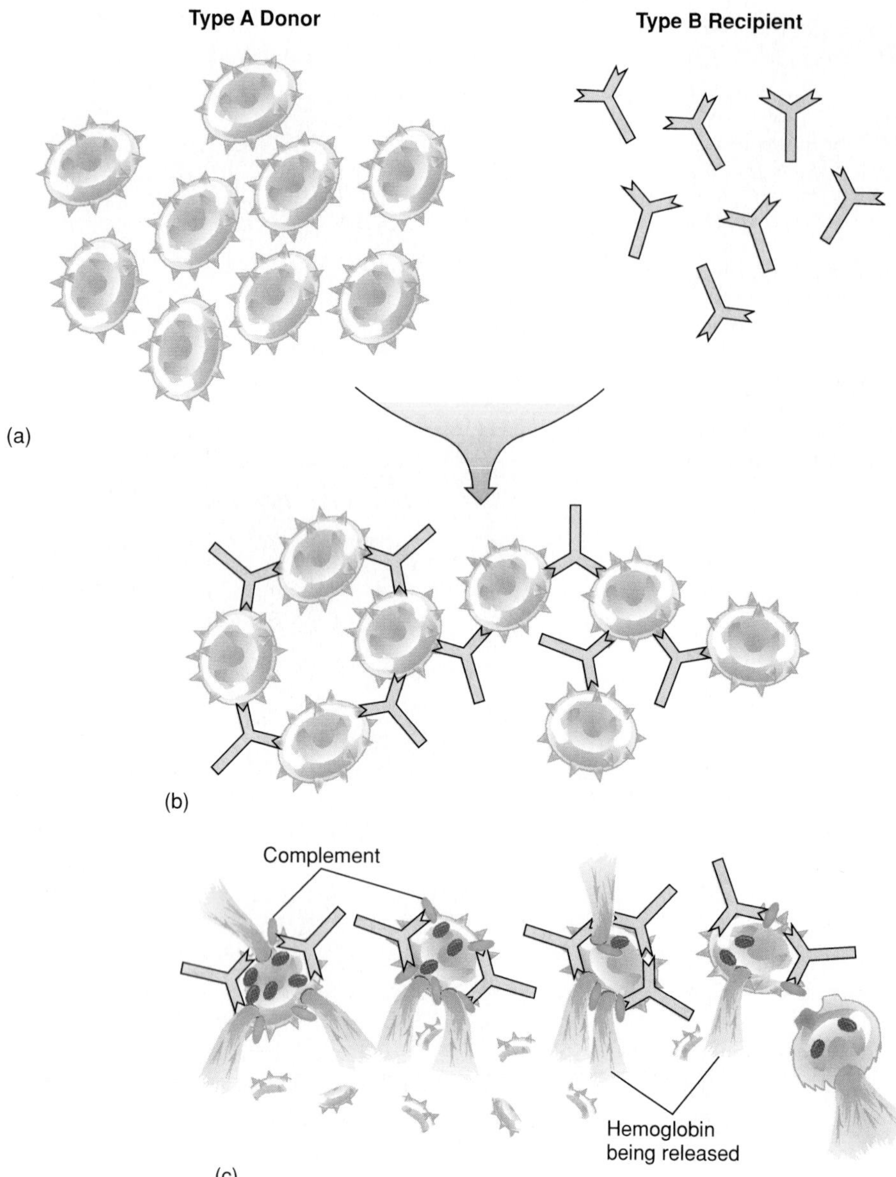

Figure 14.10 Microscopic view of a transfusion reaction. (*a*) Incompatible blood. The red blood cells of the type A donor contain antigen A, while the serum of the type B recipient contains anti-a antibodies that can agglutinate donor cells. (*b*) Agglutination particles can block the circulation in vital organs. (*c*) Activation of the complement by Ab on the RBCs can cause hemolysis and anemia. This sort of incorrect transfusion is very rare because of the great care taken by blood banks to ensure a correct match.

Knowing the blood types involved makes it possible to determine which transfusions are safe to do. The general rule of compatibility is that the RBC antigens of the donor must not be agglutinated by antibodies in the recipient's blood (figure 14.10). The ideal practice is to transfuse blood that is a perfect match (A to A, B to B). But even in this event, blood samples must be cross-matched prior to transfusion because other blood group incompatibilities may exist. This test involves mixing the blood of the donor with the serum of the recipient to check for agglutination.

Under certain circumstances (emergencies, the battlefield), the concept of universal transfusions may be used. To appreciate how this works, we must apply the rule stated in the previous paragraph. Type O blood lacks A and B antigens and will not be agglutinated by other blood types, so it could theoretically be used in any transfusion. A person with this blood type is called a *universal donor.* Because type AB blood lacks agglutinating antibodies, an individual with this blood could conceivably receive any type of blood. Type AB persons are consequently called *universal recipients.* Although both types of transfusions involve antigen-antibody incompatibilities, these are of less concern because of the dilution of the donor's blood in the body of the recipient. Additional RBC markers that may be significant in transfusions are the Rh, MN, and Kell antigens (see next section).

Transfusion of the wrong blood type causes various degrees of adverse reaction. The most severe reaction is massive hemolysis when the donated red blood cells react with recipient antibody and trigger the complement cascade (figure 14.10). The resultant destruction of red cells leads to systemic shock and kidney failure brought on by the blockage of glomeruli (blood filtering apparatus) by cell debris. Death is a common outcome. Other reactions caused by RBC destruction are fever, anemia, and jaundice. A transfusion reaction is managed by im-

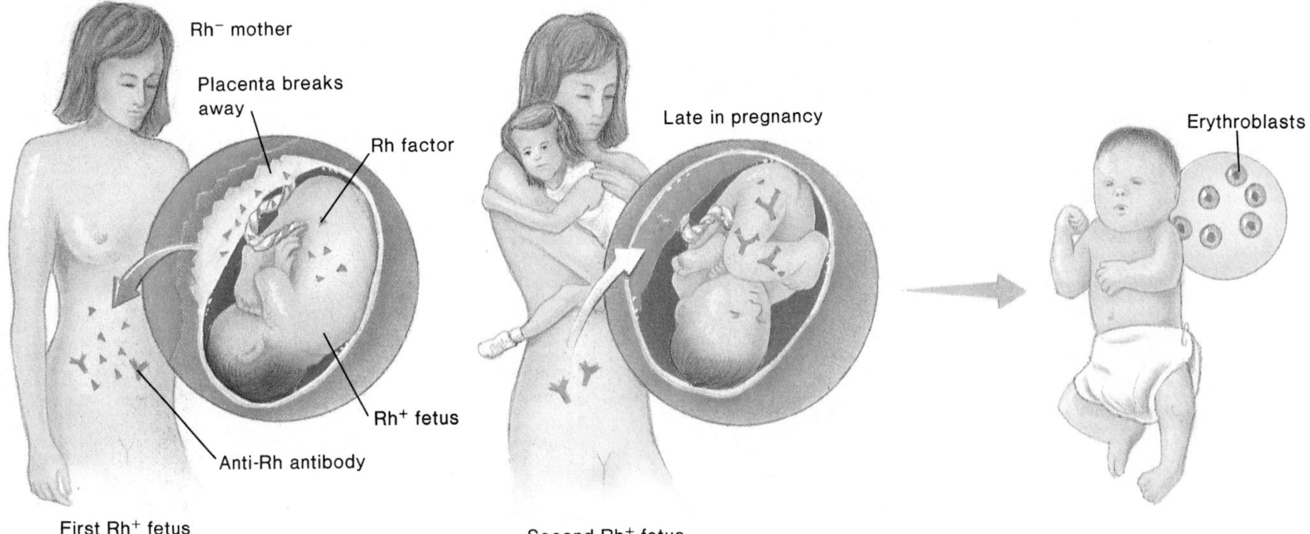

Figure 14.11 The development and aftermath of Rh sensitization. Initial sensitization of the maternal immune system to fetal Rh factor occurs during birth, when the placenta tears away. The child will escape disease in most instances, but the mother, now sensitized, will be capable of an immediate reaction to a second Rh^+ fetus and its antigen. At that time, her anti-Rh antibodies pass into the fetal circulation and elicit severe hemolysis in the fetus and neonate.

mediately halting the transfusion, administering drugs to remove hemoglobin from the blood, and beginning another transfusion with red cells of the correct type.

The Rh Factor and Its Clinical Importance

Another RBC isoantigen of major clinical concern is the **Rh factor** (or D antigen). This factor was first discovered in experiments exploring the genetic relationships among animals. When rabbits were inoculated with the RBCs of rhesus monkeys (the origin of *Rh*), the rabbits produced an antibody that also reacted with human RBCs. Further tests showed that this rhesus monkey antigen was present in about 85% of humans and absent in about 15%. The details of Rh inheritance are more complicated than those of ABO, but in simplest terms, a person's Rh type results from a combination of two possible alleles—a dominant one that produces the factor and a recessive one that does not. A person inheriting at least one Rh gene will be Rh^+; only those persons inheriting two recessive genes are Rh^-. (This factor is typed in a fashion similar to the ABO blood groups and is denoted by a symbol above the blood type, as in O^+ or B^-) (see figure 14.9*c*). Unlike the ABO antigens, exposure to normal flora does not sensitize Rh^- persons to the Rh factor. The only ways one can develop antibodies against this factor are through placental sensitization or transfusion.

Hemolytic Disease of the Newborn and Rh Incompatibility

The potential for placental sensitization occurs when a mother is Rh^- and her unborn child is Rh^+. The obvious intimacy between mother and fetus makes it possible for fetal RBCs to leak into the mother's circulation during childbirth, when the detachment of the placenta creates extensive avenues for fetal blood to enter the maternal circulation. The mother's immune system detects the foreign Rh factors on the fetal RBCs and is sensitized to them by producing antibodies and memory B cells. The first Rh^+ child is usually not affected because the process begins so late in pregnancy that the child is born before maternal sensitization is completed. However, the mother's immune system has been strongly primed for a second contact with this factor in a subsequent pregnancy (figure 14.11).

In the next pregnancy with an Rh^+ fetus, fetal blood cells escape into the maternal circulation late in pregnancy and elicit a rapid response. The fetus is at risk when the maternal anti-Rh antibodies cross the placenta into the fetal circulation, where they affix to fetal RBCs and cause complement-mediated lysis. The outcome is a potentially fatal **hemolytic disease of the newborn** (HDN) called *erythroblastosis fetalis* (eh-rith″-roh-blas-toh′-sis fee-tal′-is). Ordinarily, the bone marrow does not release RBCs into the circulation until they have matured (lost their nuclei). In this condition, immature nucleated RBCs called erythroblasts are released into the infant's circulation to compensate for the massive destruction of RBCs by maternal antibodies. Additional symptoms are severe anemia, jaundice, and enlarged spleen and liver.

Maternal-fetal incompatibilities are also possible in the ABO blood group, but adverse reactions occur less frequently than with Rh sensitization because the IgM antibodies to these blood group antigens are too bulky to cross the placenta in large numbers. In fact, the maternal-fetal relationship is a fascinating instance of foreign tissue not being rejected, despite the extensive potential for contact (see feature 14.5).

Preventing Hemolytic Disease of the Newborn

Prevention of initial sensitization of Rh^- women is by far the best control, because once sensitization has occurred, any other Rh^+ fetuses will be at risk for this disease. A careful family

Feature 14.5 Why Doesn't a Mother Reject Her Fetus?

Think of it: Even though mother and child are genetically related, the father's genetic contribution guarantees that the fetus will contain molecules that are antigenic to the mother. In fact, with the recent practice of implanting one woman with the fertilized egg of another woman, the surrogate mother is carrying a fetus that has no genetic relationship to her. Yet, even with this essentially foreign body inside the mother, dangerous immunological reactions like Rh incompatibility are rather rare. In what ways do fetuses avoid the surveillance of the mother's immune system? The answer appears to lie in the placenta and embryonic tissues. The fetal components that contribute to these tissues are not strongly antigenic, and they form a barrier that keeps the fetus isolated in its own antigen-free environment. The placenta is coated with a thick layer that prevents the passage of maternal cells, and it can also function as a sponge to sop up, remove, and inactivate circulating antigens.

history of an Rh^- pregnant woman can predict the likelihood that she is already sensitized or is carrying an Rh^+ fetus. It must take into account other children she has had, their Rh types, and the Rh status of the father. If the father is also Rh^-, the child will be Rh^- and free of risk, but if the father is Rh^+, the probability that the child will be Rh^+ is 50% or 100%. In some instances, the blood type of the child can be determined through fetal testing. In any event, if there is any possibility that the fetus is Rh^+, the mother must be passively immunized with antiserum containing antibodies against the Rh factor (*Rh_o (D) immune globulin* or *RhoGAM*[3]). This antiserum, injected at 28–32 weeks and again immediately following delivery, reacts with any fetal RBCs that have escaped into the maternal circulation, thereby preventing the sensitization of the mother's immune system to Rh factor (figure 14.12). RhoGAM must be given with each pregnancy that involves an Rh^+ fetus. It is ineffective if the mother has already been sensitized by a prior Rh^+ fetus or an incorrect blood transfusion, which can be determined by a serological test. As in ABO blood types, the Rh factor should be matched for a transfusion, although it is acceptable to transfuse Rh^- blood if the Rh type is not known.

Other RBC Antigens

Although the ABO and Rh systems are of greatest medical significance, other minor red cell isoantigen groups have been discovered. Examples of these are the *MN, Ss, Kell,* and *P* blood groups. Transfusion reactions due to incompatibilities in these blood types are rare, and typing of these blood groups is not routinely performed. The study of these blood antigens (as well as ABO and Rh) has led to some very interesting applications. For example, they may be useful in forensic medicine (crime detection), studying ethnic ancestry, and tracing prehistoric migrations in anthropology. Many blood cell antigens are remarkably hardy and may be detected in dried blood stains, semen, and saliva. Even the 2,000-year-old mummy of King Tutankhamen has been typed A_2MN!

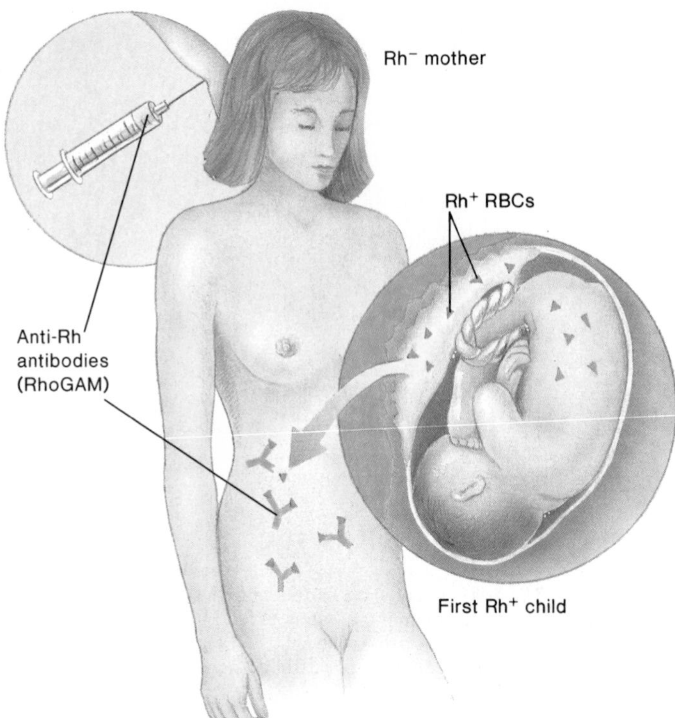

Figure 14.12 Prevention of erythroblastosis fetalis with anti-Rh immune globulin (RhoGAM). Injecting a mother who is at risk with RhoGAM during her first Rh^+ pregnancy helps to inactivate and remove the fetal Rh factor before her immune system can react with it and develop sensitivity.

Type III Hypersensitivities: Immune Complex Reactions

Type III hypersensitivity is similar to type II, because it involves the production of IgG and IgM antibodies after repeated exposure to antigens and the activation of complement. In type III, however, the antigens are small, soluble molecules not attached to the surface of a cell. The interaction of these antigens with antibodies produces large, free-floating complexes that may be deposited in the tissues, resulting in an **immune complex reaction** or disease. This category includes therapy-related disorders (serum sickness and the Arthus reaction) and a number of autoimmune diseases (such as glomerulonephritis and lupus erythematosus).

Mechanisms of Immune Complex Disease

Following initial exposure to a profuse amount of antigen, the immune system produces large quantities of antibodies that circulate in the fluid compartments. When this antigen enters the system a second time, it reacts with the antibodies to form antigen-antibody complexes (figure 14.13). These complexes summon various inflammatory mediators such as complement

3. Immunoglobulin fraction of human anti-Rh serum, prepared from pooled human sera.

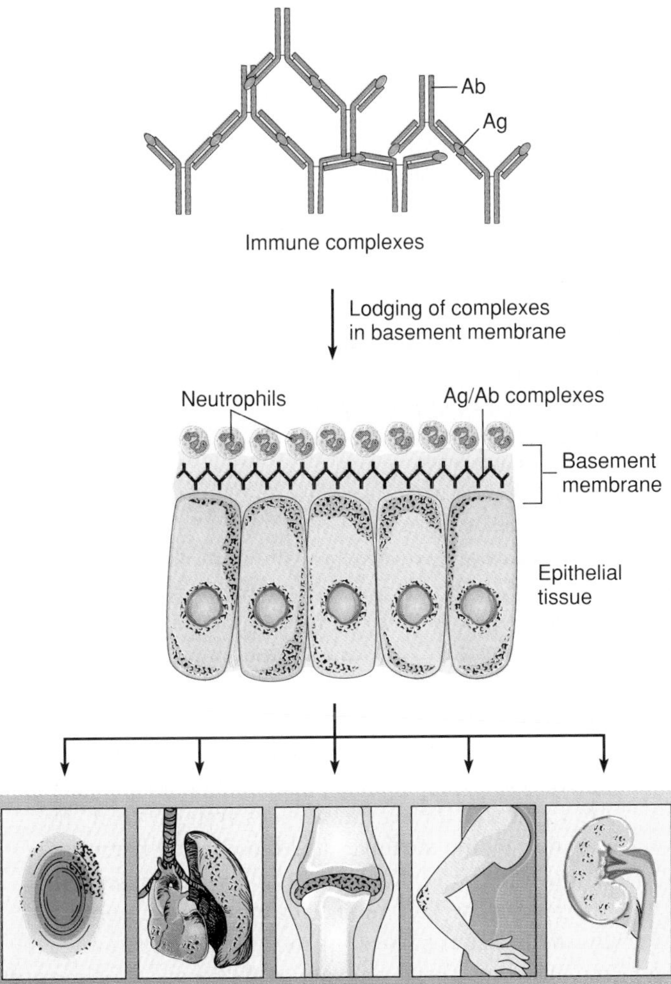

Figure 14.13 The background of immune complex disease. In general, circulating immune complexes become lodged in the basement membrane of the epithelia and cause vascular damage and organ malfunction.

and neutrophils, which ordinarily eliminate Ag-Ab complexes as part of the normal immune response. In an immune complex disease, however, these complexes are so abundant that they precipitate in the basement membranes[4] of epithelial tissues and are inaccessible. When the neutrophils are "frustrated" in their function, they release lysosomal granules into the tissues, causing chronic destructive inflammation. The symptoms of class III hypersensitivities are due in great measure to this pathological state.

Types of Immune Complex Disease

During the early tests of immunotherapy using animals, hypersensitivity reactions to serum and vaccines were common. In addition to anaphylaxis, two syndromes, the **Arthus reaction**[5] and **serum sickness,** were identified. But despite its serious side effects, antiserum therapy was so promising that inevitably it was applied to humans. Not surprisingly, it was immediately apparent that humans administered certain types of passive immunization (especially with animal serum) are prone to the very same complications as other animals.

Like anaphylaxis, both serum sickness and the Arthus reaction occur after sensitization and require preformed antibodies. Several characteristics set them apart, however. Arthus and serum sickness: (1) depend upon IgG, IgM, or IgA (precipitating antibodies) rather than IgE; (2) require large doses of antigen (not a miniscule dose as in anaphylaxis); and (3) have delayed symptoms (a few hours to days). The Arthus reaction and serum sickness differ from each other in some important ways. The Arthus reaction is a *localized* dermal injury due to inflamed blood vessels in the vicinity of injected antigen. Serum sickness is a *systemic* injury initiated by antigen-antibody complexes that circulate in the blood and settle into membranes at various sites.

The Arthus Reaction

The Arthus reaction is usually a response to a second injection of vaccines (boosters) or drugs at the same site as the first injection. In a few hours, the area becomes red, hot to the touch, swollen, and very painful. These symptoms are mainly due to the destruction of tissues in and around the blood vessels and the release of histamine from mast cells and basophils. Although the reaction is usually self-limiting and rapidly cleared, intravascular blood clotting can occasionally cause necrosis and loss of tissue.

Serum Sickness

Serum sickness was named for a condition that appeared in soldiers following repeated injections of horse serum to treat tetanus. It may also be caused by injections of animal hormones and drugs. The immune complexes that form enter the circulation, are carried throughout the body, and are eventually deposited in blood vessels of the kidney, heart, skin, and joints (figure 14.13). Depending on the organ involved, symptoms may include enlarged lymph nodes, rashes, painful joints, swelling, fever, and renal dysfunction.

An Inappropriate Response Against Self: Autoimmunity

The immune diseases we have covered so far are all caused by foreign antigens. However, in the incongruous case of **autoimmunity,** an individual actually develops hypersensitivity to himself. This pathological process accounts for **autoimmune diseases,** in which **autoantibodies** and, in certain cases, T cells mount an abnormal attack against self antigens. The scope of autoimmune diseases is extremely varied. In general, they may be differentiated as *systemic,* involving several major organs, or *organ-specific,* involving only one organ or tissue. They usually fall

4. Basement membranes are basal partitions of epithelia that normally filter out circulating antigen-antibody complexes.

5. Named after Maurice Arthus, the physiologist who first identified this localized inflammatory response.

Table 14.4 Selected Autoimmune Diseases

Disease	Target	Characteristics
Systemic lupus erythematosus (SLE)	Systemic	Vasculitis of many organs; antibodies against RBC, WBC, platelets, clotting factors, nucleus
Rheumatoid arthritis and ankylosing spondylitis	Systemic	Vasculitis; frequent target is joint lining; antibodies against other antibodies (rheumatoid factor)
Scleroderma	Systemic	Excess collagen deposition in organs; antibodies formed against many intracellular organelles
Hashimoto's thyroiditis	Thyroid	Destruction of the thyroid follicles
Graves' disease	Thyroid	Antibodies against TSH receptors
Pernicious anemia	Stomach lining	Antibodies against receptors prevent transport of vitamin B_{12}
Myasthenia gravis	Muscle	Antibodies against the acetyl choline receptors on the nerve-muscle junction alter function
Type I diabetes	Pancreas	Antibodies stimulate destruction of insulin-secreting cells
Type II diabetes	Insulin receptor	Antibodies block attachment of insulin
Multiple sclerosis	Myelin	T cells and antibodies sensitized to myelin sheath destroy neurons
Goodpasture's syndrome	Kidney	Antibodies to basement membrane of the glomerulus damage kidneys
Rheumatic fever	Heart	Antibodies to group A *Streptococcus* cross-react with heart tissue

into the categories of type II or type III hypersensitivity, depending upon how the autoantibodies bring about injury. The major autoimmune diseases, their targets, and basic pathology are presented in table 14.4.

Genetic and Sexual Correlation in Autoimmune Disease

In most cases, the precipitating cause of autoimmune disease remains obscure, but we do know that susceptibility is influenced by genetics and sex. Cases cluster in families, and even unaffected members tend to show the autoantibodies for that disease. If one identical twin has autoimmune disease, it will invariably occur in the other. More direct evidence comes from studies of the major histocompatibility gene complex. Particular genes in the class I and II major histocompatibility complex (see figure 13.3) coincide with certain autoimmune diseases. For example, autoimmune joint diseases such as arthritis and ankylosing spondylitis are more common in persons with the B-27 HLA type; SLE, Graves' disease, and myasthenia gravis are associated with the B-8 HLA antigen. Why autoimmune diseases (except ankylosing spondylitis) afflict more females than males also remains a mystery. Females are more susceptible during childbearing years than before puberty or after menopause, suggesting a possible hormonal relationship.

The Origins of Autoimmune Disease

The presence of autoantibodies in healthy individuals, although in much lower titers than in autoimmune patients, suggests some normal function for them. Although at first it seems contradictory, a moderate, regulated amount of autoimmunity is probably required to dispose of old cells and cellular debris. Disease apparently arises when this regulatory or recognition apparatus goes awry. Attempts to explain the origin of autoimmunity include the following theories:

The sequestered antigen theory explains that during embryonic growth, some tissues are *immunologically privileged*—that is, they are sequestered behind anatomical barriers and cannot be scanned by the immune system (figure 14.14*a*). Examples of these sites are regions of the central nervous system, which are shielded by the meninges and blood-brain barrier; the lens of the eye, which is enclosed by a capsule; and antigens in the thyroid and testes, which are protected by an epithelial barrier. As long as these self antigens remain sequestered, no reaction will occur. As time goes by and the antigen is exposed by means of infection, trauma, or deterioration, the immune system mistakes these self tissues for nonself and attacks them.

According to another concept, the immune system of a fetus develops tolerance by eradicating all lymphocyte clones (forbidden clones) that can react with self, while those that react to foreign antigens are retained. In autoimmunity, some of these forbidden clones may survive this tolerance process, and when stimulated at a later time, they are liable to attack self.

The theory of immune deficiency proposes that mutations in normal lymphocytes render them reactive to self, or that a general breakdown in the normal T-suppressor function sets the scene for inappropriate immune responses. Some autoimmune diseases may be caused by anti-idiotypic antibodies (antibodies produced in response to other antibodies). It is likely that certain autoimmune disorders (type I diabetes, multiple sclerosis, for example) are triggered by infection. Viruses, in particular, can change the nature of human cell receptors to the extent that the immune system will attack the cells bearing these viral receptors (figure 14.14*b*).

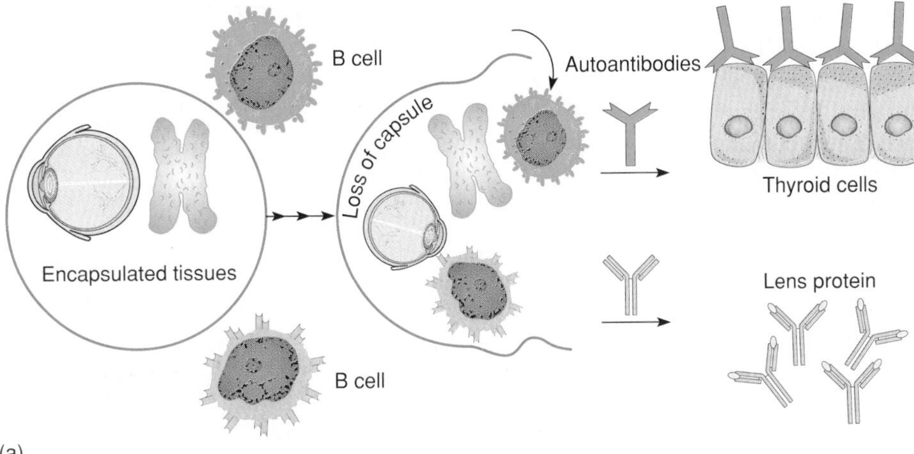

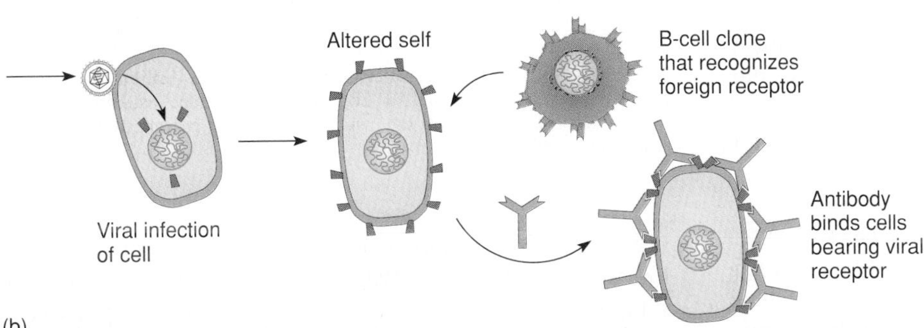

Figure 14.14 Possible explanations for autoimmunity. (*a*) Self is sequestered and is later incorrectly identified as an antigen. (*b*) Self is altered by viral infection.

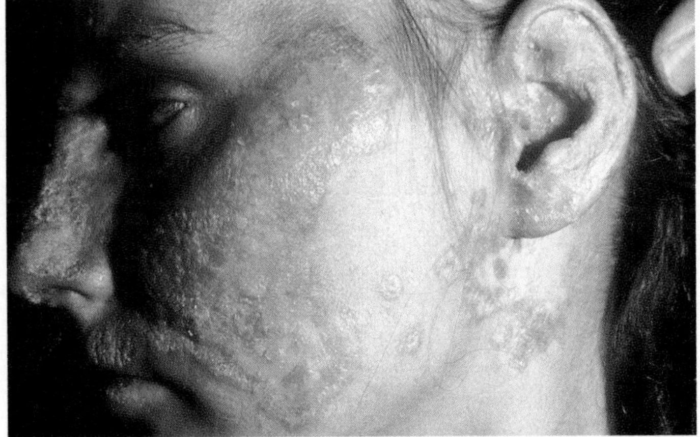
(a)

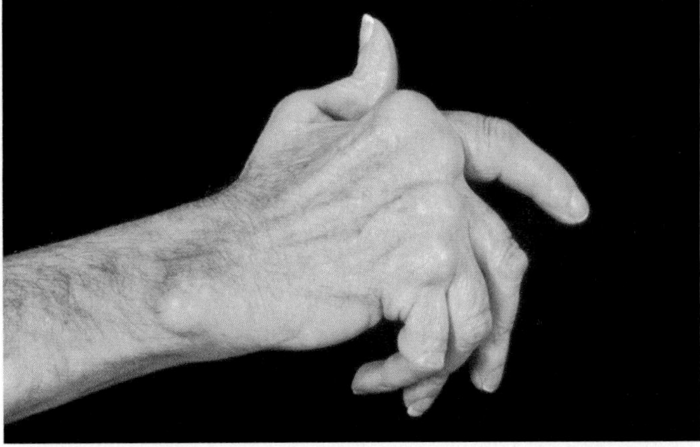
(b)

Figure 14.15 Common autoimmune diseases. (*a*) Systemic lupus erythematosus. One symptom is a prominent rash across the bridge of the nose and on the cheeks. These papules and blotches can also occur on the chest and limbs. (*b*) Rheumatoid arthritis commonly targets the synovial membrane of joints. Over time, chronic inflammation causes thickening of this membrane, erosion of the articular cartilage, and fusion of the joint, which severely limits motion and may swell and distort the joints.

Examples of Autoimmune Disease

Systemic Autoimmunities One of the most severe chronic autoimmune diseases is **systemic lupus erythematosus** (SLE or lupus). This name originated from the characteristic rash that spreads across the nose and cheeks in a pattern suggesting the appearance of a wolf (figure 14.15*a*). Although the manifestations of the disease vary considerably, all patients produce autoantibodies against a great variety of organs and tissues. The organs most involved are the kidneys, bone marrow, skin, nervous system, joints, muscles, heart, and GI tract. Antibodies to intracellular materials such as the nucleoprotein of the nucleus and mitochondria are also common.

In SLE, autoantibody/autoantigen complexes appear to be deposited in the basement membranes of various organs. Kidney failure, blood abnormalities, lung inflammation, myocarditis, and skin lesions are the predominant symptoms. One

systemic lupus erythematosus (sis-tem′-ik loo′-pis air″-uh-theem-uh-toh′-sis) L. *lupus,* wolf.

form of chronic lupus (called discoid) is influenced by exposure to the sun and afflicts primarily the skin, not other organs. The etiology of lupus is still a puzzle. It is not known how such a generalized loss of self-tolerance arises, though viral infection or loss of T-cell suppressor function are suspected. The fact that women of childbearing years account for 90% of cases indicates hormones may be involved. The diagnosis of SLE can usually be made with blood tests. Antibodies against the nucleus (ANA) and various tissues (detected by indirect fluorescent antibody or radioimmune assay techniques) are common, and a positive test for the lupus factor (an anticoagulant factor) is also very indicative of the disease. Therapy involves anti-inflammatory drugs such as aspirin and cortisone.

Rheumatoid arthritis, another systemic autoimmune disease, incurs progressive, debilitating damage to the joints. In some patients, the lung, eye, skin, and nervous system are also involved. In the joint form of the disease, autoantibodies form immune complexes that bind to the synovial membrane of the joints and activate complement. Chronic inflammation leads to scar tissue and joint destruction. The joints in the hands and feet are affected first, followed by the large joints (figure 14.15*b*). The precipitating cause in rheumatoid arthritis is not known, though infectious agents have been suspected. The most common feature of the disease is the presence of an IgM antibody, called rheumatoid factor (RF), directed against other antibodies. Treatment for this very difficult disease involves anti-inflammatory agents, immunosuppressive drugs, and sometimes gold salt injections.

Autoimmunities of the Endocrine Glands On occasion, the thyroid gland is the target of autoimmunity. The underlying cause of **Graves' disease** is the attachment of autoantibodies to receptors on the follicle cells that secrete the hormone thyroxin. The abnormal stimulation of these cells causes the overproduction of this hormone and the symptoms of hyperthyroidism. In **Hashimoto's thyroiditis,** the autoantibodies have the reverse effect. One type inactivates thyroxin, and the other initiates the destruction of follicle cells. As a result of these reactions, the levels of hormone are greatly reduced, and the patient suffers from hypothyroidism.

The pancreas and its hormone, insulin, are other autoimmune targets. Insulin, secreted by islet cells of the pancreas, regulates and is essential to the utilization of glucose by cells. **Diabetes mellitus** is caused by the dysfunction of this system (figure 14.16). Type I diabetes (also termed juvenile-onset or insulin-dependent) is associated with autoantibodies formed against the islet cells. A complex inflammatory reaction leading to lysis of these cells greatly reduces the amount of insulin secreted. In a form of type II diabetes (maturity-onset or insulin-independent), sufficient insulin is usually produced, but cellular receptors for insulin are reduced in number or availability. This may be due to autoantibodies that compete with insulin for receptor binding sites.

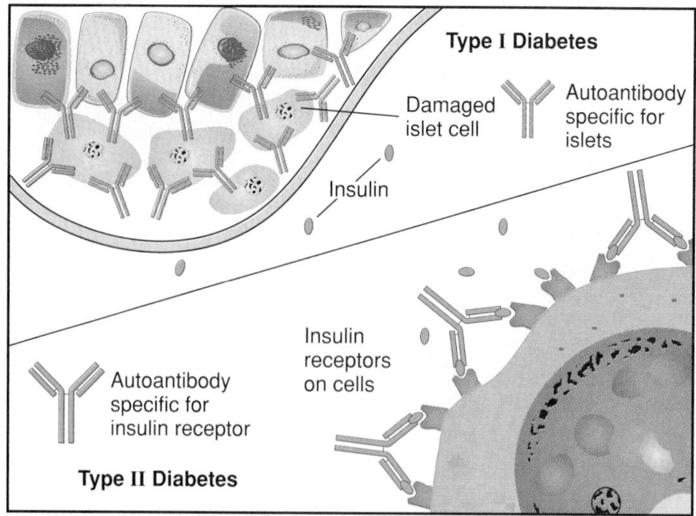

Figure 14.16 The autoimmune component in diabetes mellitus. In type I, autoantibodies injure the islets of Langerhans and reduce insulin synthesis. In type II, autoantibodies cover the insulin receptor and prevent insulin from attaching to cells.

Neuromuscular Autoimmunities **Myasthenia gravis** is named for the pronounced muscle weakness that is its principal symptom. Although the disease afflicts all skeletal muscle, the first effects are usually felt in the muscles of the eyes and throat. It may progress to complete loss of muscle function and death. The classic syndrome is caused by autoantibodies binding to the receptors for acetylcholine, a chemical required to transmit a nerve impulse across the synaptic junction to a muscle (figure 14.17). The immune attack so severely damages the postsynaptic membrane that transmission is blocked and paralysis ensues. Present treatment may include immunosuppressive drugs and therapy to remove the autoantibodies from the circulation. Experimental therapy using immunotoxins to destroy lymphocytes that produce autoantibodies shows some promise.

Multiple sclerosis (MS) is a paralyzing neuromuscular disease associated with lesions in the insulating myelin sheath that surrounds neurons. The neurons in the white matter of the central nervous system (brain, spinal cord) are the primary sites of attack. The underlying pathology involves damage to the sheath by both T cells and autoantibodies that severely compromises the capacity of neurons to send impulses. The principal motor and sensory symptoms are muscular weakness and tremors, difficulties in speech and vision, and some degree of paralysis. Most MS patients first experience symptoms as young adults, and they tend to experience remissions (periods of relief) alternating with recurrences throughout their lives. Studies with animals reveal that treating the disease passively with monoclonal antibodies or actively by vaccination may be possible in the near future.

---

rheumatoid arthritis (roo'-muh-toyd ar-thry'-tis) Gr. *rheuma*, a moist discharge, and *arthron*, joint.

myasthenia gravis (my''-us-thee'-nee-uh grah'-vis) Gr. *myo*, muscle, *astheneia*, weakness, and *gravida*, heavy.

sclerosis (skleh-roh'-sis) Gr. *sklerosis*, hardness.

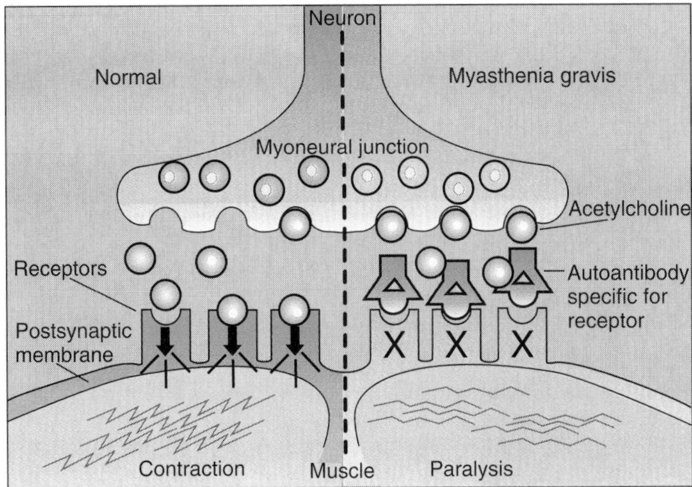

Figure 14.17 Proposed mechanisms for involvement of autoantibodies in myasthenia gravis.

Type IV Hypersensitivities: Cell-Mediated (Delayed) Reactions

The adverse immune responses we have covered to this point are explained primarily by B-cell involvement and antibodies. But type IV hypersensitivity involves primarily the T-cell limb of the immune system. Type IV immune dysfunction has traditionally been known as delayed hypersensitivity or allergy because the symptoms arise one to several days following the second contact with an antigen. In general, type IV diseases result when T cells are directed at self tissues or transplanted foreign cells. Examples of type IV hypersensitivity include delayed allergic reactions to infectious agents, contact dermatitis, and graft rejection.

Delayed-Type Allergies to Microorganisms

A classic example of a delayed-type allergy occurs when a person sensitized by tuberculosis infection is injected with an extract (tuberculin) of the bacterium *Mycobacterium tuberculosis*. The so-called **tuberculin reaction** is an acute skin inflammation at the injection site appearing within 24 to 48 hours. So useful and diagnostic is this technique for detecting present or prior tuberculosis that it is the chosen screening device (see feature 16.4). Other infections for which similar skin testing may be used are leprosy, syphilis, histoplasmosis, and toxoplasmosis. This form of hypersensitivity arises from time-consuming cellular events involving the T_D class of cells. After these cells receive processed microbial antigens from macrophages, they release broad-spectrum lymphokines that attract inflammatory cells to the site—particularly mononuclear cells, fibroblasts, and other lymphocytes. In a chronic infection (tertiary syphilis, for example), extensive damage to organs may occur through granuloma formation.

Contact Dermatitis

The most common delayed allergic reaction, **contact dermatitis,** is caused by exposure to poison ivy or poison oak (see feature 14.6), to simple haptenic allergens in household and personal articles (jewelry, cosmetics, elasticized undergarments), and to certain drugs. Like immediate atopic dermatitis, the reaction to these allergens requires a sensitizing and a provocative dose. The allergen first penetrates the outer skin layers, is processed by Langerhans cells (skin macrophages), and is presented to T cells. When subsequent exposures attract lymphocytes and macrophages to this area, their intense response toward the antigen destroys epidermal cells in the immediate vicinity. This accounts for the intensely itchy papules and blisters that are the early symptoms (figure 14.18). As healing progresses, the epidermis is replaced by a thick, horny layer. Depending upon the dose and the sensitivity of the individual, these events, from contact to healing, may last a week to 10 days.

Feature 14.6 Pretty, Prickly, Potentially Poisonous Plants

"Measles make you bumpy,
And mumps'll make you lumpy.
And chickenpox'll make you jump and twitch.
A common cold'll fool ya,
And whoopin' cough can cool ya.
But poison ivy, Lord'll make you itch.
You're gonna need an ocean
Of calamine lotion.
You'll be scratchin' like a hound—
The minute you start to mess around
Poison ivy, poison ivy."

From *Poison Ivy*, words and music by Jerry Leiber and Mike Stoller; recorded by the Coasters, 1959.

As a cause of allergic contact dermatitis (affecting about 10 million persons a year), nothing can compare with a single family of beautiful but pesky plants. At least one of these plants—either poison ivy, poison oak, or poison sumac—flourishes in the forests, woodlands, or along the trails of most regions of America. The allergen in these plants, an oil called urushiol, has such extreme potency that a pinhead-sized amount could spur symptoms in 500 people, and it is so long-lasting that botanists must be careful when handling 100-year-old plant specimens. Although degrees of sensitivity vary among individuals, it is estimated that half of all Americans are potentially hypersensitive to this compound. Some persons are so acutely sensitive that even the most miniscule contact, such as handling pets or clothes that have touched the plant or standing next to it, can trigger an attack. Probably the greatest danger arises from forest fires, when the burning plants release smoke laden with vaporized urushiol. Breathing this potent material can cause an overwhelming, and occasionally fatal, allergic response in sensitive fire fighters.

Humans first become sensitized by contact during childhood. Individuals at great risk (fire fighters, backpackers) are advised to determine their degree of sensitivity using a skin test, so that they may be adequately cautious and prepared. Over the years, various preventatives and treatments have been sought. Remedies include unusual skin potions such as horse urine, bleach, gasoline, buttermilk, ammonia, mustard, hair spray, and meat tenderizer. Drinking milk from goats that have grazed on poison ivy is believed by some to be an effective desensitizing method. Allergy researchers are capitalizing on this idea by testing oral vaccines containing a form of urushiol, which seem to work in experimental animals. Others have devised a compound to be rubbed on the skin that blocks the entry of the urushiol for up to 48 hours.

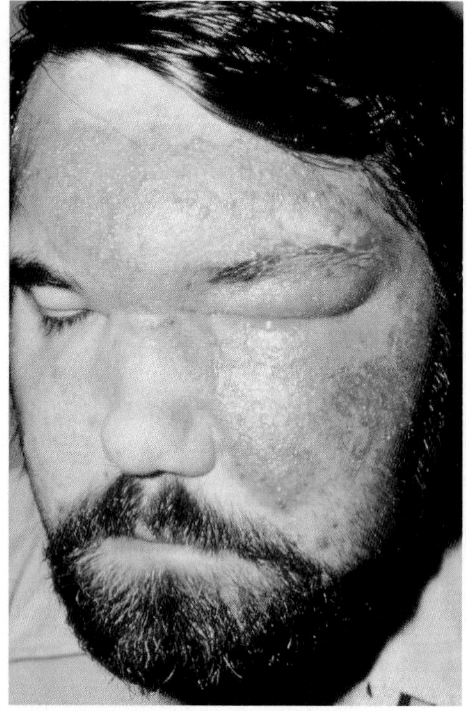
(a)

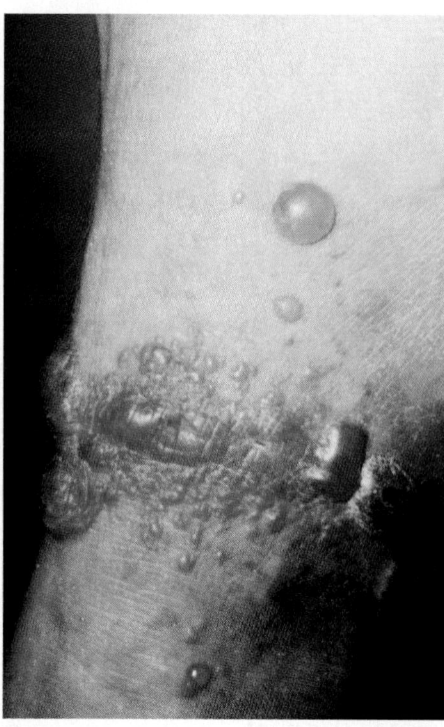
(b)

Figure 14.18 Contact dermatitis from poison oak, showing various stages of involvement: blisters, scales, thickened patches, and swollen eyelids. (*a*) Facial rash; (*b*) closeup of blisters on the wrist.

T Cells and Their Role in Organ Transplantation

Transplantation or grafting of organs and tissues is a common medical procedure. Although it is life-giving, this technique is plagued by the natural tendency of lymphocytes to seek out foreign tissues and mount a campaign to reject them. The bulk of the damage that occurs in graft rejections can be attributed to expression of cytotoxic T cells and other killer cells. In this section we will consider the mechanisms involved in graft rejection, tests for transplant compatibility, reactions against grafts, prevention of graft rejection, and types of grafts.

The Genetic and Biochemical Basis for Graft Rejection

In chapter 13 we discussed the role of major histocompatibility (MHC or HLA) genes and receptors in immune function. In general, the genes and receptors in MHC classes I and II are extremely important in recognizing self and in regulating the immune response. These receptors also set the events of graft rejection in motion. The MHC genes of humans are inherited from among a large pool of genes, so the cells of each person exhibit a relatively individual pattern of cell surface molecules (figure 14.19). The pattern is identical in different cells of the same person and may be very similar in related siblings and parents, but the further apart the relationship, the less likely that the MHC genes and receptors will be similar. When donor tissue (a graft) displays surface receptors of a different MHC class, the T cells of the recipient (called the host) will recognize its foreignness and react against it.

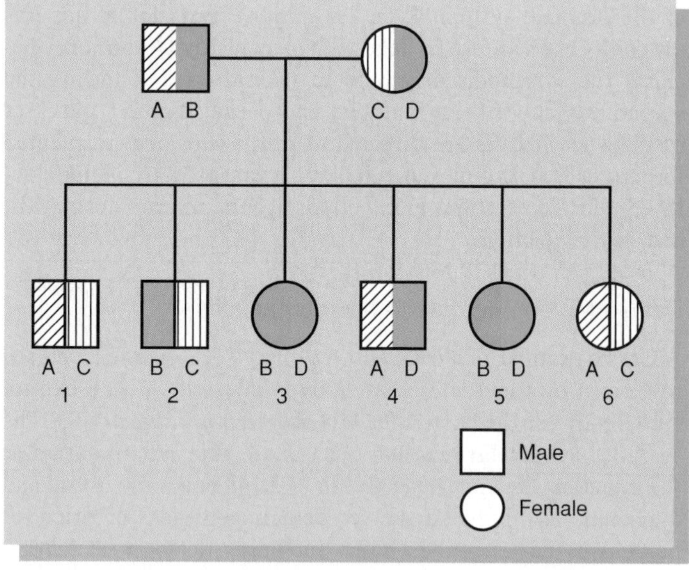

Figure 14.19 The pattern of inheritance of MHC (HLA) genes. A simplified version of the human leukocyte antigen (HLA) complex in a family. In this example, there are two genes in the complex, and each parent has a different set of genes (A/B and C/D). A child can inherit one of four different combinations. Out of six children, two sets (1 and 6, 3 and 5) have identical HLA genes and are good candidates for exchange grafts. Children sharing one gene (1 and 2, 3 and 4, 2 and 6) are close matches, but two pairs of children (2 and 4, 3 and 6) do not match at all.

T-Cell-Mediated Recognition of Foreign MHC Receptors

Host Rejection of Graft When the T cells of a host recognize foreign class II MHC receptors on the surface of grafted cells, they release interleukin-2 as part of a general immune mobilization. Receipt of this stimulus triggers killer (T_K or cytotoxic) T cells specific to the foreign antigens on the donated cells. T_K cells bind to the grafted tissue and secrete cytotoxic lymphokines that begin to reject it within two weeks of transplantation (figure 14.20*a*). Late in this process, the antibodies also

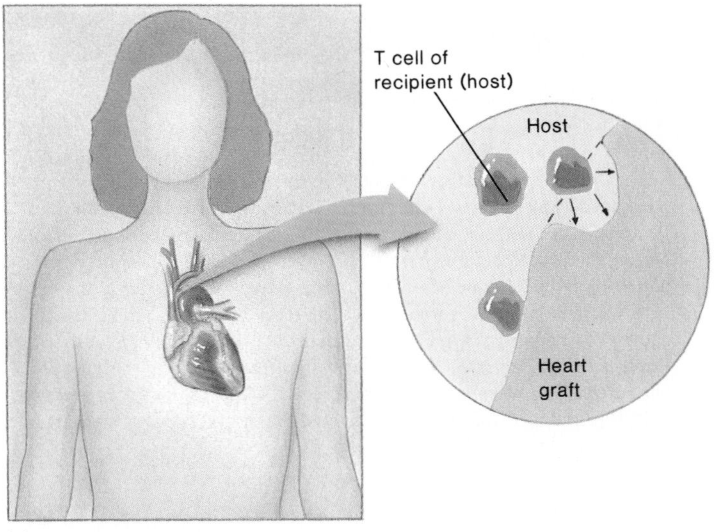

Figure 14.20 Potential reactions in transplantation. (*a*) The host's immune system (primarily cytotoxic T cells) encounters the cells of the donated organ (heart) and rejects the organ. (*b*) Grafted tissue contains endogenous T cells that recognize the host's tissues as foreign and mount an attack. The recipient will develop symptoms of graft versus host disease.

formed against the graft tissue speed its loss. A final blow is the destruction of the vascular supply, promoting death of the grafted tissue.

Graft Rejection of Host Under certain circumstances, the host cannot or does not reject a graft; but this failure may not protect the host from serious damage, because graft incompatibility is a two-way phenomenon. Some grafted tissues (especially bone marrow) contain an indigenous population of lymphocytes. This makes it quite possible for the graft to reject the host, causing **graft versus host disease (GVHD)** (figure 14.20*b*). Because any host tissue bearing MHC receptors foreign to the graft can be attacked, the effects of GVHD are widely systemic and toxic. A papular, peeling skin rash is the most common symptom. Other organs to be affected are the liver, intestine, muscles, and mucous membranes. GVHD occurs in approximately 30% of bone marrow transplants within 100–300 days of the graft. A relatively high percentage of recipients die from its effects.

Classes of Grafts

Grafts are generally classified according to how closely the donor and recipient tissue are related. Tissue transplanted from one site of an individual to another is known as an **autograft.** Typical examples are skin replacement in burn repair and the use of a vein to fashion a coronary artery bypass. In an **isograft (syngraft),** tissue from an identical twin is used. Because isografts do not contain foreign antigens, they are not rejected, but this type of grafting has obvious limitations. **Allografts (homografts),** the most common type of grafts, are exchanges between genetically different individuals belonging to the same species (two humans). A close genetic correlation is sought for most allograph transplants (see next section). A **xenograft** is a tissue exchange between individuals of different species. Until rejection can be better controlled, most xenografts are experimental.

Avoiding and Controlling Graft Incompatibility

Graft rejection can be averted or lessened by directly comparing the tissue of the recipient with that of potential donors. Several **tissue matching** procedures are used. In one method, the white blood cells of a donor and recipient are incubated together in a tissue culture medium. If, after a specified time, a number of cells have died, a mismatch is indicated. In the *mixed lymphocyte reaction (MLR),* lymphocytes of the two individuals are mixed and incubated. If an incompatibility exists, some of the cells will undergo transformation into activated cells. *Tissue typing* is similar to blood typing, except that specific antisera are used to disclose the HLA antigens on the surface of lymphocytes. In most grafts (one exception is bone marrow transplants), the ABO blood type must also be matched. Although a small amount of incompatibility is tolerable in certain grafts (liver, heart, kidney), a closer match is more likely to be successful, so the closest match possible is sought.

Drugs That Suppress Allograph Rejection Despite a nationwide computerized hotline for matching recipients with donors and a greater availability of viable organs than in the past, an ideal match between donor and recipient is still the exception rather than the rule, and some sort of immunosuppressive therapy to overcome rejection is usually required. Rejection can be controlled with agents such as methotrexate, prednisone, and antihuman thymocyte immunoglobulin. Not unexpectedly, intervention with these drugs is imprecise, and it is complicated by general suppression of the immune system (especially T cells) and frequent opportunistic infections.

A miracle drug used since 1983 to suppress organ rejection is the polypeptide *cyclosporine A,*[6] which is isolated from

6. Marketed as Sandimmune.

a fungus. It has revolutionized the science of transplantation by dramatically improving the survival rate of allograph patients (kidney, heart, liver, and bone marrow) and reducing the incidence of fatal infections. Although its action is not entirely understood, cyclosporine appears to block the activation of T-helper cells and interfere with the release of lymphokines. What makes this drug so valuable is that its effects are selective—that is, other important lymphoid cells and phagocytes are not inhibited, and the body is better able to ward off infections. It does cause the adverse effects of kidney toxicity and increased blood pressure, but these can be reduced by adjusting the dose and monitoring blood levels of the drug. In the future it is hoped that cyclosporine can be used to treat autoimmune diseases such as type I diabetes and rheumatoid arthritis.

Types of Transplants

We tend to think of tissue and organ grafting as a modern phenomenon, but skin grafts have been performed for more than a thousand years. Today, transplantation is a recognized medical procedure whose benefit is reflected in several thousand transplants each year. It has been performed on every major organ, including parts of the brain. The most frequent transplant operations involve skin, heart, kidney, coronary artery, cornea, and bone marrow. The sources of organs and tissues are live donors (kidney, skin, bone marrow, liver), recently dead cadavers (heart, kidney, cornea), and fetuses (a controversial technique banned under some circumstances). In the past decade, we have witnessed some unusual types of grafts. Fetal pancreas has been implanted as a potential treatment for diabetes, and fetal brain tissues for Parkinson's disease. Part of a liver has been transplanted from a live parent to a child, and a baboon's heart has been grafted into a newborn baby.

Bone marrow transplantation is a rapidly growing medical procedure. More than 2,000 transplants have been performed for patients with immune deficiencies, aplastic anemia, leukemia and other cancers, and radiation damage. This procedure is extremely expensive, costing up to $200,000 per patient. Before bone marrow from a closely matched donor can be infused (see feature 14.7), the patient is pretreated with chemotherapy and whole-body irradiation, a procedure designed to destroy his own blood stem cells and thus prevent rejection of the new marrow cells. Within two weeks to a month after infusion, the grafted cells are established in the host. Because donor lymphoid cells can still cause GVHD, antirejection drugs may be necessary. An amazing consequence of bone marrow transplantation is that a recipient's blood type may change to the blood type of the donor. Since blood cells are the primary target of HIV, bone marrow replacement is also being tested as a possible treatment for AIDS patients.

Feature 14.7 The Mechanics of Bone Marrow Transplantation

In some ways, bone marrow is the most exceptional form of transplantation. It does not involve invasive surgery in either the donor or recipient, and it permits the removal of tissue from a living donor that is fully replaceable. While the donor is sedated, the bone marrow sample (semiliquid) is aspirated by inserting a special needle into an accessible marrow cavity. The most favorable sites are the crest and spine of the ilium (major bone of the pelvis). During this procedure, which lasts one to two hours, 3% to 5% of the donor's marrow is withdrawn in 20 to 30 separate extractions. The donor may experience some pain and soreness, but there are rarely any serious complications. In a few weeks, the depleted marrow will naturally replace itself. Implanting the harvested bone marrow is rather convenient, because it is not necessary to place it directly into the marrow cavities of the recipient. Instead, it is dripped intravenously into the circulation, and the new marrow cells automatically settle in the correct sites.

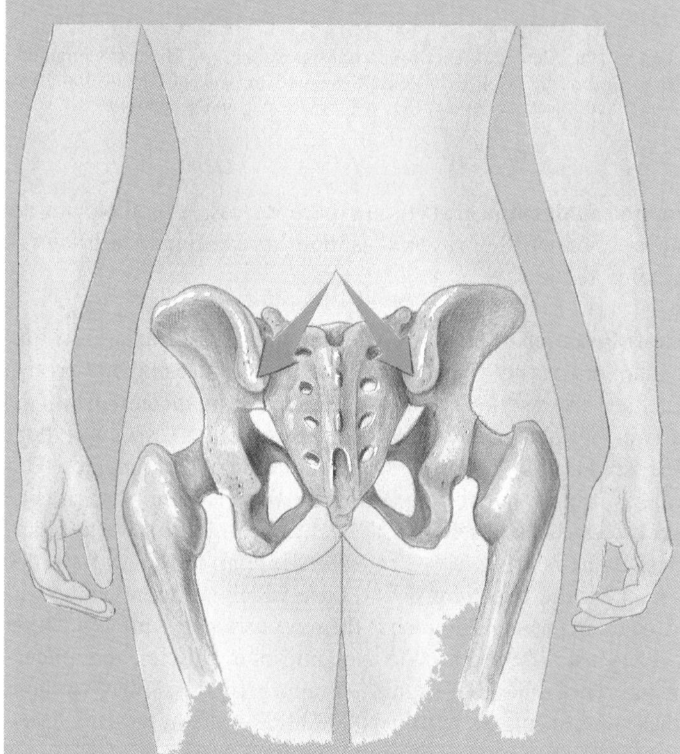

Removal of a bone marrow sample for transplantation. One site into which a needle is inserted is the iliac spine (arrows). (The ilium is a prolific producer of bone marrow.)

Immunodeficiency Diseases: Hyposensitivity of the Immune System

It is a marvel that the development of the immune system proceeds as normally as it does in human beings. On occasion, however, an error occurs and a person is born with a deficient immune repertoire. In many cases, these very "experiments" of nature have provided penetrating insights into the exact functions of certain cells, tissues, and organs because of the specific symptoms shown by the immunodeficient individuals. The predominant consequences of immunodeficiencies are recurrent, overwhelming infections, often with microbes that are not pathogenic to the immunocompetent person. Immunodeficiencies fall into two general categories: *primary diseases,* present at birth (congenital) and usually stemming from genetic errors, and *secondary diseases,* acquired after birth and caused by natural or artificial factors (table 14.5).

Table 14.5 General Categories of Immunodeficiency Diseases

Primary Immune Deficiencies (Genetic)
B-Cell Defects (Low levels of antibodies)
Agammaglobulinemia (X-linked, non-sex-linked)
Hypogammaglobulinemia
Selective immunoglobulin deficiencies
T-Cell Defects (Lack of cytotoxic T cells and graft rejection)
Thymic aplasia (DiGeorge syndrome)
Chronic mucocutaneous candidiasis
Combined B-Cell and T-Cell Defects (Usually caused by lack or abnormality of lymphoid stem cell)
Severe combined immunodeficiency disease (SCID)
Adenosine deaminase deficiency (ADA)
Wiskott-Aldrich syndrome
Ataxia-telangiectasia
Phagocyte Defects
Chédiak-Higashi syndrome
Chronic granulomatous disease of children
Lack of surface glycoprotein
Complement Defects
Lacking one of C components
Hereditary angioedema
Associated with rheumatoid diseases
Associated with other immune deficiencies
Secondary Immune Deficiencies (Acquired)
From Natural Causes
Infection: AIDS, leprosy, tuberculosis, measles
Other disease: cancer, diabetes
Protein calorie malnutrition
Stress
Pregnancy
Aging
From Immunosuppressive Agents
Irradiation
Severe burns
Steroids (cortisones)
Drugs to treat graft rejection and cancer

Primary Immunodeficiency Diseases

Deficiencies affect both specific immunities such as antibody production and nonspecific ones such as phagocytosis. Consult figure 14.21 to overview the places in the normal sequential development of lymphocytes where defects can occur and the possible consequences. In many cases, the deficiency is due to an inherited abnormality, though the exact nature of the abnormality is not known for a number of diseases. Because the development of B cells and T cells departs at some point, an individual may lack one or both cell lines. It must be emphasized, however, that some deficiencies affect others. For example a T-cell deficiency may affect B-cell function because of the role of T-helper cells. In some deficiencies, the lymphocyte in question is completely absent or present at very low levels, whereas in others, lymphocytes are present but do not function normally.

Clinical Deficiencies in B-Cell Development or Expression

Genetic deficiencies in B cells usually appear as an abnormality in immunoglobulin expression. In some instances, only certain immunoglobulin classes are absent; in others, the level of all types of immunoglobulins is reduced. A significant number of B-cell deficiencies are X-linked recessive traits, meaning that the gene occurs on the X chromosome and the disease appears primarily in male children.

The term **agammaglobulinemia** literally means the absence of gamma globulin, the primary fraction of serum that contains immunoglobulins. It was originally thought that no immunoglobulin at all is present, but it is very rare for Ig to be completely absent, and some physicians prefer the term **hypogammaglobulinemia.**[7] Both sex-linked and autosomal[8] recessive types of this rare syndrome occur. In both cases, mature B cells are absent, the level of antibodies is greatly reduced, and lymphoid organs are incompletely developed. T-cell function in these patients is normal. The symptoms of recurrent, serious bacterial infections usually appear about 6 months after birth. The bacteria most often implicated are pyogenic cocci, *Pseudomonas,* and *Hemophilus influenzae,* and the most common infection sites are the lungs, sinuses, meninges, and blood. Although antibodies are not as important in warding off viruses and protozoa, many Ig-deficient patients may have recurrent infections with these agents too. Patients often manifest a wasting syndrome and have a reduced lifespan, but modern therapy has improved their prognosis. The current treatment for this condition is passive immunotherapy with IgG preparations and continuous antibiotic therapy.

The lack of a particular class of immunoglobulin is a relatively common condition. Although genetically controlled, its underlying mechanisms are not yet clear. **IgA deficiency** is the most prevalent, occurring in about one person in 600. Such persons have normal quantities of B cells and other immunoglobulins, but they are unable to synthesize IgA. Consequently, they lack protection against local microbial invasion of the mucous membranes, and suffer recurrent respiratory and gastrointestinal infections. The usual treatment using Ig replacement will not work, because conventional preparations are high in IgG, not IgA.

IgM deficiency associated with cancer of the lymphoid organs often results in life-threatening septicemia by pyogenic cocci. IgG deficiencies can also be associated with increased pyogenic infections.

Clinical Deficiencies in T-Cell Development or Expression

Owing to their critical role in immune defenses, a genetic defect in T cells results in a broad spectrum of disease, including severe opportunistic infections, wasting, and cancer. In fact, a dysfunctional T-cell line is usually more devastating than a defective B-cell line because T-helper cells are required for most specific immune reactions. The deficiency can occur anywhere along the developmental spectrum, from thymus to mature, circulating T cells.

7. Denoting a lowered level of Ig.

8. Meaning that the gene is located on a chromosome other than a sex chromosome.

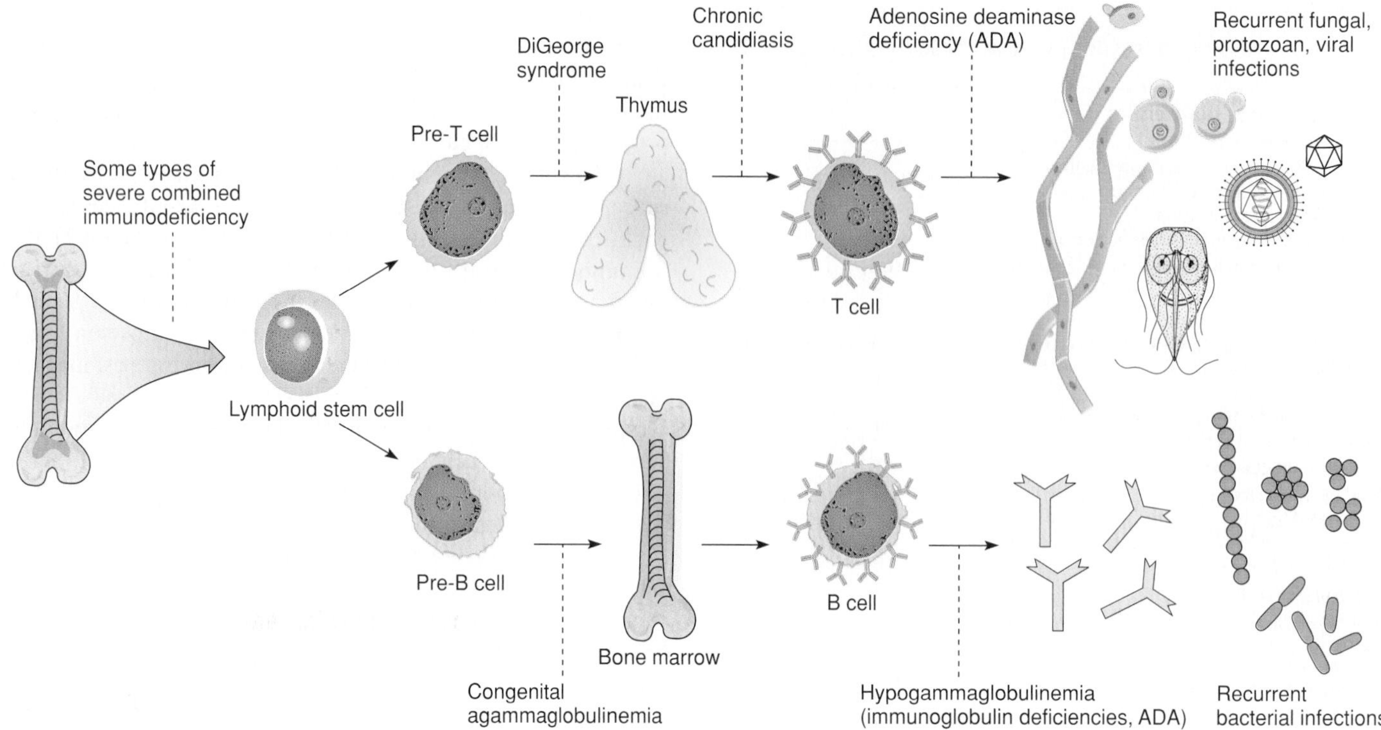

Figure 14.21 The stages of development and the functions of B cells and T cells whose failure causes immunodeficiencies.

Abnormal Development of the Thymus The most severe of the T-cell deficiencies involve the congenital absence or immaturity of the thymus gland. Thymic aplasia, or **DiGeorge syndrome,** results when the embryonic 3rd and 4th pharyngeal pouches fail to develop. Some cases are associated with a deletion in chromosome 22. Children with this syndrome not only lack thymus activity, but also manifest congenital heart defects and hypoparathyroidism (figure 14.22). The accompanying lack of cell-mediated immunity makes them highly susceptible to persistent infections by fungi, protozoa, and viruses. Common, usually benign childhood infections such as chickenpox, measles, or mumps can be overwhelming and fatal in these children. Even vaccinations using attenuated microbes pose a danger. Other symptoms of thymic failure are growth retardation, wasting of the body (from intestinal infections), and an increased incidence of lymphatic cancer. These children may have reduced antibody levels due to lack of T-helper cells, and they are unable to reject transplants. The major therapy for these children is a transplant of thymus tissue. Some individuals who experience a milder form of DiGeorge syndrome, possessing a small rudimentary (hypoplastic) thymus, can eventually recover T-cell function.

Chronic mucocutaneous candidiasis is a very selective T-cell deficiency in which the T cells that would ordinarily react to pathogenic *Candida* (a yeast) are defective. The cause is not known, though possibly, the T-cell clones that would ordinarily react to this microbe are absent. The symptoms, which begin in young children, are chronic infections of the skin, nails, and mucous membranes.

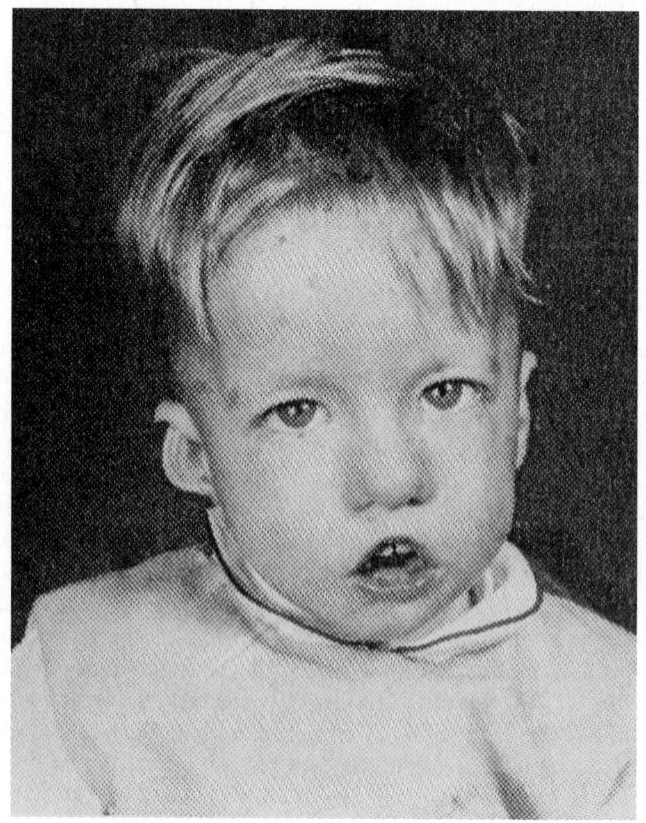

Figure 14.22 Facial characteristics of a child with DiGeorge syndrome. Typical defects include: low-set, deformed earlobes; wide-set, slanted eyes; a small, bow-like mouth; and the absence of a philtrum (the vertical furrow between the nose and upper lip).

Severe Combined Immunodeficiencies—Dysfunction in B and T Cells

Severe combined immunodeficiencies (SCID) are the most dire and potentially lethal of the immunodeficiency diseases because they involve dysfunction in both lymphocyte systems. Some SCIDs are due to the complete absence of the lymphocyte stem cell in the marrow, while others are attributable to the dysfunction of B cells and T cells later in development. This debilitating condition appears to have several forms. In the two most common ones, Swiss-type agammaglobulinemia and thymic alymphoplasia, the numbers of all types of lymphocytes are extremely low, the blood antibody content is greatly diminished, and the thymus and cell-mediated immunity are poorly developed. Although their genetic basis differs (Swiss condition being an autosomal recessive trait and alymphoplasia being sex-linked), both diseases are due to a defect in the lymphoid cell line. Infants with SCID usually manifest the T-cell deficiencies first (due to passive maternal antibody protection). Within days after birth, they may develop candidiasis, sepsis, pneumonia, or systemic viral infections.

An unusual form of SCID, known as **adenosine deaminase (ADA) deficiency,** is caused by an autosomal recessive defect in the metabolism of adenosine. In this case, lymphocytes develop, but a metabolic product builds up abnormally and selectively destroys them. Infants with ADA deficiency are subject to the same clinical symptoms as occur in other SCID syndromes—recurrent infections and severe wasting.

Because of their profound lack of specific adaptive immunities, SCID children require the most rigorous kinds of aseptic techniques to protect them from opportunistic infections (see feature 9.3). They must be isolated in a microbe-free environment until a bone marrow graft is available.

Correcting SCID David Vetter, the most famous SCID child, lived all but the last two weeks of his life in a sterile environment to isolate him from the microorganisms that would have quickly ended his life (figure 14.23). Medical tests performed before birth indicated that David might inherit this disease, so as a precaution, he was delivered by cesarian section and immediately placed in a sterile isolette. From that time, he lived in various plastic chambers—ranging from room-size to a special suit that allowed him to walk outside. Remarkably, he developed into a well-adjusted child, even though his only physical contact with others was through special rubber gloves. When he was 12, his doctors decided to attempt a bone marrow transplant that might allow him to leave his bubble prison. David was given a transplant of his sister's marrow, but the marrow harbored Epstein-Barr virus. Because he lacked any form of protective immunities against this oncogenic virus, a cancer spread rapidly through his body. Despite the finest medical care available, he died a short time later from the cancer.

Aside from life in a sterile plastic bubble, the only serious option for survival of SCID children is total replacement or correction of dysfunctional lymphoid cells. Some infants can benefit from fetal liver or bone marrow grafts. Although transplanting compatible bone marrow has been about 50% successful in curing the disease, it is complicated in many cases by graft versus host disease. The condition of some ADA patients has been partly corrected by periodic transfusions of red blood cells containing large amounts of adenosine deaminase. The future hope for ADA lies in gene therapy—insertion of a gene to replace the defective gene (see figure 22.46).

Figure 14.23 David Vetter—the boy in the plastic bubble.

Secondary Immunodeficiency Diseases

Secondary, or acquired deficiencies in B cells and T cells are caused by one of four general agents: (1) infection, (2) organic disease, (3) chemotherapy, or (4) radiation.

The most notorious infection-induced immunodeficiency is **AIDS.** This syndrome is caused when several types of immune cells, including T-helper cells, monocytes, and macrophages, are infected by the human immunodeficiency virus (HIV). It is generally thought that the depletion of T-helper cells ultimately accounts for the symptoms of cancers and opportunistic protozoan, fungal, and viral infections associated with this disease. (Chapter 21 contains an extensive section on AIDS.) Other infectious agents that may deplete immunities are Epstein-Barr virus, measles virus, and the leprosy bacillus.

Cancers that target the bone marrow or lymphoid organs can be responsible for extreme malfunction of both humoral and cellular immunity (see next section). In leukemia, the massive number of cancer cells compete for space and literally displace the normal cells of the bone marrow and blood. Plasma cell tumors produce large amounts of nonfunctional antibodies, and thymus gland tumors cause severe T-cell deficiencies. Organic diseases such as Hodgkin's disease, lymphoma, rheumatoid arthritis, and multiple sclerosis are often associated with extreme lymphopenia.

An ironic outcome of life-saving medical procedures is the possible suppression of a patient's immune system. For instance, some immunosuppressive drugs that prevent graft rejection by T cells may likewise suppress beneficial immunities. Although radiation and anticancer drugs are the first line of therapy for many types of cancer, both agents are extremely damaging to the bone marrow and other body cells.

Cancer: The Crab That Grows Within

The term **cancer** comes from the Latin word for crab and is presumably derived from the appendage-like projections that a spreading tumor develops. The disease is also known by the synonyms *neoplasm* and tumor, or by more specific terms that usually end in the suffix *-oma*. **Oncology,** the field of medicine that specializes in cancer, is growing with such great rapidity that we can do justice here only to its principal concepts as they relate to immunology, including the basic characteristics, classification, and origins of cancer.

Characteristics and Classification of Cancers

The feature that most distinguishes cancer from other diseases is an uncontrolled growth of abnormal cells that have arisen from normal cells. Most cancer cells show other characteristics, including (1) disorganized behavior and independence from surrounding normal tissues, (2) permanent loss of cell differentiation, and (3) expression of special markers on their surface. As a general rule, cancer originates in cells such as skin and bone marrow that have retained the capacity to divide, while mature cells that have lost this power (neurons, for instance) do not commonly become cancerous.

An abnormal growth or tumor is generally characterized as benign or malignant. A **benign tumor** is a self-contained mass within an organ that does not spread into adjacent tissues. The mass is usually slow-growing, rounded, and not greatly different from its tissue of origin. Benign tumors do not ordinarily cause death unless they grow into a critical space such as the heart valves or brain ventricles. On the other hand, the cells of a **malignant tumor (cancer)** do not remain encapsulated, but tend to spread both locally and distantly to other organs. As the initial (primary) tumor grows, it invades nearby tissues, enters lymphatic and blood vessels, and establishes secondary tumors in remote sites. This property of spreading is called **metastasis,** and the cancer is said to *metastasize*. Malignant tumors range in effects from those, like pancreatic cancer, that spread rapidly and are almost always fatal, to others, like basal cell carcinoma of the skin, that are much less aggressive and more treatable.

Another tumor-classifying scheme relies on the tissue or cell of primary origin. Although cancers are commonly referred to simply as bladder cancer, breast cancer, or liver cancer, their technical names more clearly define the tumor origin. Generally, cancers originating from epithelial tissues are called **carcinomas,** and those originating from mesenchyme (embryonic connective tissue) are **sarcomas.** Combining these terms with the tissue of origin supplies the full descriptive name. For example, adenocarcinoma develops in glandular tissue such as the pancreas or thyroid; squamous cell carcinoma arises in the skin epidermis; retinoblastoma occurs in the retina; lymphosarcoma in the lymph nodes; hepatoma (hepatic sarcoma) in the liver; and melanoma in the skin melanocytes. A special name, leukemia, denotes cancer of the blood-forming tissues. Considering the large number of cell types in a given organ, the great diversity of tumors (more than 100 clinical types) is to be expected.

---

neoplasm (nee′-oh-plazm) Gr. *neo*, new, and *plasm*, formation.

carcinoma (kar″-sih-noh′-mah) Gr. *karkinos*, crab.

sarcoma (sar-koh′-mah) Gr. *sarcos*, flesh, and *oma*, tumor. Sometimes referred to as soft-tissue tumors.

Epidemiology of Cancer

Cancer is the second leading cause of death in humans (following heart and circulatory disease). Annually, approximately 850,000 new cases are diagnosed, and 450,000 persons die from various cancers. Cancer rates also vary according to sex, race, ethnic group, age, geographic region, occupational exposure to carcinogens, diet, and life-style. For example, the most prevalent cancer among white males is lung cancer, followed by prostate, colon and rectal, and urinary cancers. The most prevalent cancer in white females occurs in the breast, followed by colon and rectal, lung, and uterine cancers. Lung cancer accounts for the largest number of deaths in both sexes. Cancers currently on the increase are melanoma (presumably due to increased levels of UV radiation entering the atmosphere), Kaposi's sarcoma (in AIDS patients), thyroid cancer, and esophageal cancer.

It is increasingly evident that susceptibility to certain cancers is inherited. At least 30 neoplastic syndromes that run in families have been described. A particularly stunning discovery is that cancer in some animals is actually due to endogenous viral genomes passed from parent to offspring in the gametes. Many cancers reflect an interplay of both hereditary and environmental factors. An example is lung cancer, which is associated mainly with tobacco smoking. But not all smokers develop lung cancer, and not all lung cancer victims smoke tobacco.

Proposed Mechanisms of Cancer

Cancer is clearly a complex disease associated with numerous physical, chemical, and biological agents (table 14.6). For years, there was no clear unifying explanation for how so many different factors could cause normal cells to become cancerous. But recent research findings link all cancers to genetic damage or inherited genetic predispositions that alter the function of certain genes normally found in all cells. The evidence for the interrelationship between genes and cancer is derived from the following observations: (1) Cancer cells often have damaged chromosomes; (2) a specific alteration in a gene can lead to cancer; (3) the predisposition for some cancers is inherited; (4) rates of cancer are higher in individuals who cannot repair damaged DNA; (5) mutagenic agents cause cancer; (6) cells contain genes that can be transformed to cancer-causing oncogenes; and (7) tumor suppressor genes (anti-oncogenes) exist in the normal genome. These discoveries and other advances in cancer research have provided the seeds for a general unifying theory for cancer.

Oncogene: A Normal Gene Gone Haywire? Cancer is another experiment of nature that continues to inform us about normal cell function. In the section on operons in chapter 8, we learned that normal cell operations are strictly controlled by regulatory genes. Extracellular chemical signals are received by receptors, which transmit them to a regulatory complex in the nucleus that switches the appropriate genes on or off. During the rapid growth of a young individual, cells are extremely active, and the switch that initiates cell division is operating much of the time. By contrast, most adult cells exist in a nondividing state, except for divisions involved in repair and maintenance. This state of bal-

Table 14.6 Selected Cancer-Triggering Agents and Their Diseases

Agent	Type of Exposure and Cancer Type
Physical Agents	
Ionizing radiation	Occupational exposure, therapy; many types
UV radiation (sunlight)	Sunlight, suntanning lamps; skin cancer
Asbestos	Manufacturing, building insulation; lung cancer
Chemical Agents	
Tobacco	Smoking, chewing; oral, lung cancers
Alcohol	Drinking; oral, pharyngeal, esophageal, liver
Arsenic	Manufacturing, mining; lung, skin, liver
Benzene	Various industrial processes; leukemia
Polycyclic hydrocarbons (coal tar derivatives)	Medicines, industrial use; lung, skin
Vinyl chloride	Manufacture of plastic; liver
Aflatoxin	Fungus-contaminated foods; liver
Aromatic amines	Manufacturing processes; bladder
Heavy-metal dust	Industry; lung and nasal
Anabolic steroids	Medication; liver
Estrogens	Medication; vagina, endometrium, liver
Biological Agents (Acquired Through Infection)	
Retrovirus	T-cell leukemia
Papilloma viruses	Cervical cancer
Polyoma viruses	Various tumors in mice and hamsters
Epstein-Barr virus	Burkitt's lymphoma; nasopharyngeal carcinoma
Hepatitis B virus	Liver cancer

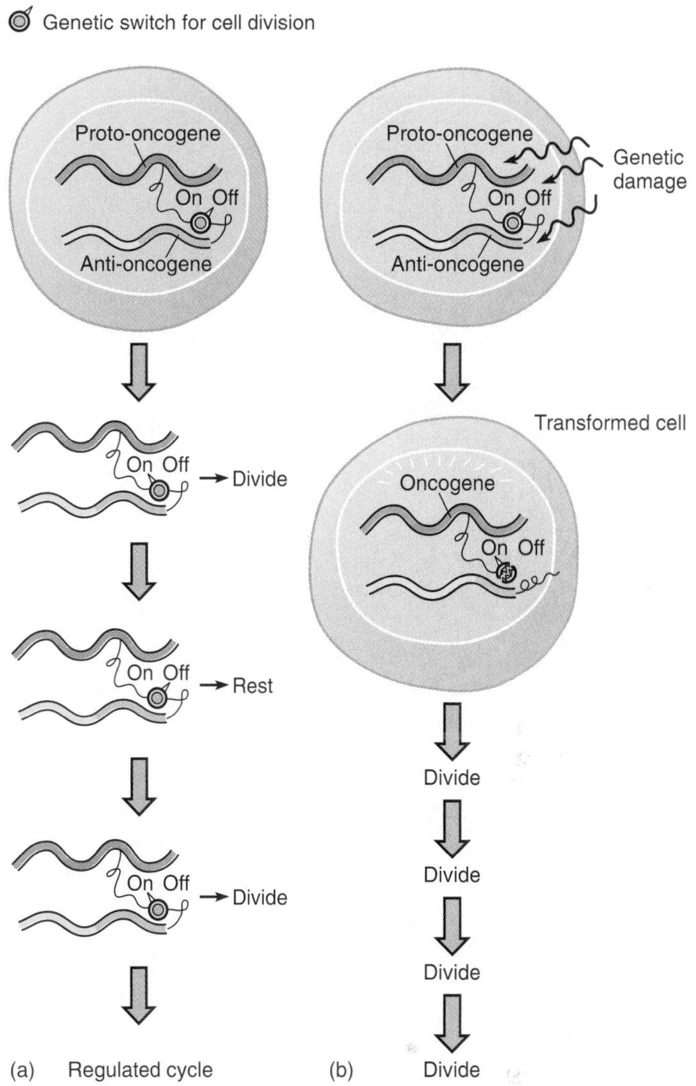

Figure 14.24 Common pathway for neoplasias. (*a*) Normal cell cycle and genetic controls. (*b*) Event that stimulates transformation into a cancer cell. Any agent that can alter the genetic control mechanisms can cause a cell to divide at incorrect times and in incorrect places.

ance ensures that cells divide at a rate compatible with normal development and function.

Cancer results when the system that keeps cell division quiescent is somehow short-circuited. Current theories explain that the gene complex controlling cell division contains a gene termed a **proto-oncogene** that regulates the onset of mitosis. The proto-oncogene is in turn regulated by another gene, the **anti-oncogene** (also called a tumor-suppressor gene), that prevents the proto-oncogene from acting continuously (figure 14.24*a*). It appears that any disruption of the proto-oncogene or anti-oncogene causes the proto-oncogene to remain permanently switched on. In this state of continuous activity, the gene, now called an **oncogene,** overrides the normal cell division cycle and causes the cell to divide without control (figure 14.24*b*). This process of cell immortalization is termed **transformation.** A number of happenings might explain how an oncogene goes permanently out of control and cannot be turned off. It may be that the anti-oncogene system is missing or has been inactivated by mutations, or that new genetic material has been added. Another possibility is that the oncogene products have been altered by mutation and no longer respond to regulation (remember that genes code for protein products). Some genes express their oncogenic potential after jumping from one site on a chromosome to another. Up to now, about 50 oncogenes have been discovered, and their modes of action are gradually being deciphered.

The Role of Viruses in Cancer One documented oncogenic change occurs when a cell is infected by a retrovirus (figure 14.25; see chapter 21). Once inside a host cell, these RNA viruses utilize an enzyme called reverse transcriptase to synthesize viral DNA that is subsequently inserted at a particular site on a host chromosome. Some retroviruses, such as the Rous

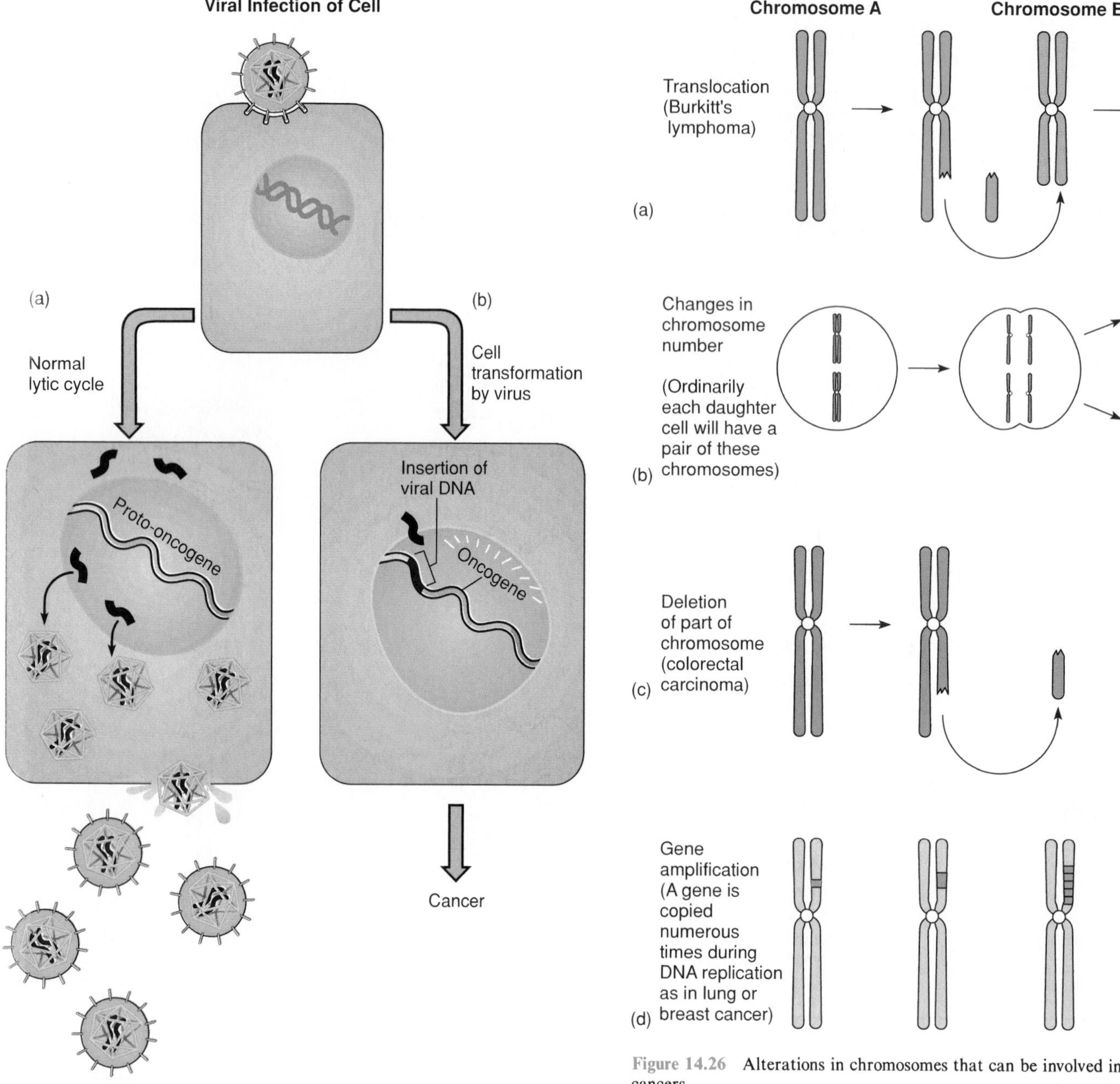

Figure 14.25 Outcomes of viral infection. (*a*) A virus may go through the normal lytic cycle and destroy its host cell. (*b*) A possible mechanism for viral induction of cancer. DNA viruses and retroviruses may insert DNA into the host's genome. If it inserts into a sensitive site (near a proto-oncogene), the cell can be converted into a tumor cell.

Figure 14.26 Alterations in chromosomes that can be involved in some cancers.

sarcoma virus of chickens, carry viral oncogenes whose products cause transformation of host cells into cancer cells. Other retroviral genomes insert on host chromosomes at regulatory sites, which may lead to activation of the cell's own proto-oncogenes. Only one type of cancer in humans, T-cell leukemia, is attributable to insertional mutagenesis by a retrovirus. DNA viruses are also involved in cancer. Human papilloma virus (HPV), which causes genital warts and is strongly implicated in cervical carcinoma, inserts its genome directly into a chromosome in a way that overrides the usual cellular growth controls. Epstein-Barr virus (EBV), a cause of infectious mononucleosis and Burkitt's lymphoma, appears to induce a chromosomal alteration called a translocation (see next section).

Chromosomal Damage and Cancer A number of cancers have been associated with gross chromosomal changes or other aberrations that occur during mitotic divisions (figure 14.26). One form of chromosomal disruption is called *translocation.* In translocation, a segment of one chromosome is transferred to another chromosome that is not its mate. The abnormal placement of the translocated genes is believed to act as a strong oncogenic stimulus. Burkitt's lymphoma (see figure 20.13), a malignant tumor of lymphocytes, is associated with the translocation of part of chromosome 8 to chromosome 14. Most cases of myelogenous leukemia are due to the translocation of a proto-oncogene on chromosome 9 to a different site on chromosome 22.

Certain cancers appear to result from a mistake in mitosis that produces cells bearing three rather than the usual pair of a particular chromosome. Such *trisomies* occurring with chromosome 8 and 12 may lead to leukemia, and an extra copy of chromosome 7 is associated with melanoma. Several types of leukemia may be traced to the *loss* of a whole chromosome. The loss of part of a chromosome during cell division (chromosomal deletion) is also an important factor in some cancers. A large number of colorectal carcinomas manifest a deletion of a small part of chromosome 17, and retinoblastoma occurs through deletions on chromosome 13 (see feature 14.8).

Sufficient evidence indicates that some cells become cancerous because of *gene amplification,* a process whereby numerous extra copies of a gene are produced during DNA replication. When the copied gene is a proto-oncogene, the likelihood of oncogenic transformation is greatly increased. Certain tumors arising in squamous cell tissue or tissue of the lung, brain, and breast are linked to gene amplification.

The Role of Carcinogens What is the true etiologic role of chemical and physical carcinogenic agents (table 14.6)? It appears that carcinogens genetically alter or damage DNA and trigger the activation or deregulation of a proto-oncogene, transforming it into an oncogene. Although physical agents such as X rays and UV radiation have well-developed powers to mutate or damage DNA (see figure 9.8), most chemical carcinogens operate in a stepwise fashion that requires at least two different stimuli. The first chemical, the *initiating stimulus,* is metabolized by the liver into a more reactive chemical that permanently alters the DNA of a particular target cell. This is the carcinogenic chemical targeted by the Ames test (see chapter 8). Before the altered cell becomes neoplastic, it must be acted upon by a second chemical stimulus, called a *promoter* or *co-carcinogen,* that completes the transformation to cancer.

Progression of Cancer

Transformation is marked by the appearance of aberrant histology and physiology in the malignant cell and its progeny. Cancer cells are frequently pleomorphic. They may appear as giant cells with bizarre shapes, an abnormal, vacuolated cytoplasm, and enlarged, multiple nuclei. They express surface markers that may or may not be normal. Some markers are surface receptors for receiving stimuli for *growth factors,* small polypeptides involved in cellular development that stimulate cancerous growth.[9] Some tumors secrete chemical factors that stimulate the development of additional circulatory vessels, an effect not unlike "self-feeding." An important feature of malignant cells is the loss of cohesiveness that keeps normal cells aggregated, so that they readily become dislodged from a tumor, metastasize into the circulation, and are carried to distant sites. Such invasive tumor cells also have the property of binding to and disrupting connective tissue barriers surrounding tissues and organs.

9. It is felt that certain oncogenes are the source of these growth factors.

Feature 14.8 The Retinoblastoma Gene—A Model Tumor Suppressor?

The recently discovered retinoblastoma (RB) gene has substantially increased our understanding of cancer genetics. Studies have shown that in normal humans the RB genes effectively prevent retinal cancer. Individuals who have inherited defective genes or whose normal genes have been damaged by carcinogens may develop multiple retinal tumors. The importance of the RB gene does not stop there. Soon after the discovery of this gene's function, it was determined that persons who develop retinoblastoma are also susceptible to osteosarcoma (a form of bone cancer), breast cancer, and a form of lung cancer. While attempting to trace the common link between these tumor types, researchers made some fascinating and momentous discoveries. Various studies found that the non-retinoblastoma cancers also lacked functioning RB genes, and that transplanting chromosome 13 or the isolated RB gene into cultured osteosarcoma cells converted them to normal cells. Molecular geneticists, intrigued by these amazing findings, began to study the protein product of the RB gene and determined that this protein indeed binds to a specific site on DNA. Subsequent studies showed that this gene plays an important role as an anti-oncogene whose usual function is to block cellular growth and prevent tumor formation. It is now evident that several groups of viruses are tumorigenic, because they override the action of this important regulatory gene.

Most metastatic cancer cells require a special environment in which to become established. Although the explanation for this phenomenon has been elusive, it is clear that not every type of metastatic cancer can take up residence and thrive in every tissue; a process of selection is involved. Experience thus makes it possible to predict likely organs of metastasis. For example, prostate cancer disseminates primarily to bone; melanoma to the lung, brain, and liver; and breast cancer to the lung, liver, and bone.

The Function of the Immune System in Cancer

What role does the immune system play in controlling cancer? Reaching back to earlier concepts, we find that **immune surveillance** is an attractive explanation for how cancer cells are eliminated. Many experts postulate that cells with cancer-causing potential arise constantly in the body, but that the immune system ordinarily discovers and destroys these cells, thus keeping cancer in check. Experiments with mammals have amply demonstrated that components of cell-mediated immunity interact with tumors and their antigens. The primary types of cells that operate in surveillance and destruction of tumor cells are cytotoxic T cells, natural killer (NK) cells, lymphokine-activated killer cells (LAK), and macrophages. It appears that these cells recognize abnormal or foreign surface markers on the tumor cells and destroy them by mechanisms shown in figure 13.21. Antibodies help destroy tumors by interacting with macrophages and natural killer cells. This involvement of the immune system in destroying cancerous cells has inspired several new therapies (see feature 14.9).

How do we account for the commonness of cancer in light of this powerful range of defenses against tumors? To an extent, the answer is simply that, as with infection, the immune system can and does fail. In some cases, the cancer may not be immunogenic enough; it may retain self markers and not attract the attention of the surveillance system. In other cases, the tumor antigens may have mutated to escape detection. As we saw earlier, patients with immunodeficiencies such as AIDS and SCID are more susceptible to various cancers because they lack essential T-cell or cytotoxic functions. It may turn out that most cancers are associated with some sort of immunodeficiency, even a slight or transient one.

Feature 14.9 Experimental Cancer Treatments Using Immunotherapy

The traditional methods of treating cancer have been surgical removal, radiation therapy, and chemotherapy. Disadvantages of these methods are that they are nonspecific or can severely damage normal tissues and organs; also, cancer cells may escape or become resistant to them. An exciting development in cancer therapy is the new area of immunotherapy—stimulating the patient's own immune defenses against a tumor. One technique involves removing a patient's lymphocytes, incubating them with interleukin-2, and reinfusing these cells along with additional interleukin-2. This therapy stimulates the formation of lymphokine-activated killer cells (LAKs) that will attack the cancer. A later variation on this basic technique involves white blood cells called tumor-infiltrating-lymphocytes (TIL), which are 100 times more aggressive than LAKs. A small piece of the patient's tumor is ground up and incubated with lymphocytes and interleukin-2, then replaced in the patient. This stimulates the lymphocytes to find, infiltrate, and destroy that specific type of tumor. This therapy has been especially valuable for melanoma and for colorectal, kidney, and lung cancers. Other promising immunotherapies involve tumor necrosis factor, interferon, and monoclonal antibodies (immunotoxins; see feature 13.4).

Chapter Review with Key Terms

Immune Disorders

Immunopathology

The study of disease states involving the malfunction of the immune system is called **immunopathology. Allergy,** or **hypersensitivity,** is an exaggerated and adverse expression of certain protective immune responses that occurs in some individuals. Abnormal responses to foreign antigens are characteristic of **immune complex disease**; undesirable reactions to foreign tissues are graft reactions. **Autoimmunity** involves abnormal responses to self antigens. A deficiency or loss in immune function is called **immunodeficiency.**

Allergy/Hypersensitivity

An allergic reaction is a heightened immune response to antigens involving misdirected but natural mechanisms of humoral immunity, cellular immunity, phagocytosis, and inflammation. The four types of immune reactions have been classified according to the type of lymphocyte involved, type of antigen, and nature of damage.

Type I Hypersensitivities

Immediate-onset allergies involve contact with **allergens,** antigens that affect only certain persons and not others; susceptibility is inherited; allergens may enter through four portals: Inhalants are breathed in (pollen, dust); ingestants are swallowed (food, drugs); injectants are inoculated (drugs, bee stings); contactants react on the skin surface (cosmetics, glue). Symptoms vary with the site of entry.

Mechanism: Sensitization occurs on first contact with allergen: Specific B cells react with allergen, form a special antibody class called **IgE,** which affixes by its Fc receptor to **mast cells, basophils.** This **sensitizing dose** primes the allergic response system. Upon subsequent exposure, (the **provocative dose**), the same allergen binds to the IgE-mast cell complex, which causes **degranulation,** release of intracellular granules containing chemical mediators (**histamine,** serotonin, leukotriene, prostaglandin), with powerful physiological effects on targets such as smooth muscle (vasodilation, bronchoconstriction) and glands; attendant symptoms are rash, itching, redness, increased mucous discharge, pain, swelling, and difficulty in breathing.

Specific Diseases: An **atopic allergy** is a local reaction to an allergen. **Allergic rhinitis (hay fever)** is a seasonal respiratory allergy; **asthma** is a chronic respiratory condition in which the airways are extremely sensitive to allergic mediators; **atopic dermatitis (eczema)** is characterized by an intensely itchy skin rash; **food allergy** involves respiratory, cutaneous, and skin reactions to common foodstuffs; **drug allergy** is hypersensitivity to drugs that occurs through contact, injection, or ingestion.

Systemic anaphylaxis is an acute, extreme reaction to allergens that results in severe respiratory and circulatory symptoms; death may occur through compromised respiration and circulatory collapse.

Diagnosis of Allergy: Histamine release test on basophils; serological assays for IgE; in skin testing, small amounts of allergen are injected directly into the skin, and the degree of skin reaction is then documented.

Control of Allergy: Drugs are used to control the action of histamine, inflammation, and release of mediators from mast cells; desensitization therapy involves the administration of purified allergens to prevent provocation of the allergic response.

Type II Hypersensitivities

Type II reactions involve the interaction of antibodies, foreign cells, and complement, leading to lysis of the foreign cells. They include transfusion reactions, in which humans may become sensitized to special isoantigens on the surface of the red blood cells of other humans. The **ABO blood groups** are genetically controlled: Type A blood has A antigens on the RBCs; type B has B antigens; type AB has both A and B antigens; and type O has neither antigen. Contact with heterophile antigens on bacteria and plants apparently causes a person to produce antibodies against any A or B antigens they are lacking (that are foreign to them); these antibodies can react with these antigens if the wrong blood type is transfused without regard to compatibility. **Rh factor** is another RBC antigen that becomes a problem if a mother lacking the factor is sensitized by a fetus having the factor; a second fetus can receive antibodies she has made against the factor and develop **hemolytic disease of the newborn.** Prevention involves therapy with **Rh immune globulin.**

Type III Immune Complex Reactions

Exposure to a large quantity of soluble foreign molecules (serum, drugs) stimulates antibodies (IgG, IgM) that react with the antigens to produce large, insoluble Ag-Ab complexes; the trapping of these **immune complexes** in various organs and tissues incites a damaging inflammatory response. **Arthus reaction** is a local reaction to a series of injected antigens in the same body site; may lead to tissue destruction, necrosis. **Serum sickness** is a systemic disease resulting from repeated injections of foreign proteins; high levels of circulating immune complexes are deposited in various organs (kidney, skin, joints) and cause tissue damage.

Autoimmunity: Sometimes, in types II and III hypersensitivity, the immune system has lost tolerance to self molecules (autoantigens) and forms antibodies (**autoantibodies**) and sensitized T cells to them; antibodies to self disrupt normal organ function; diseases are systemic or organ-specific and genetically determined and most common in females.

Systemic lupus erythematosus (**SLE**) is a chronic, severe systemic disease in which antibodies are deposited in the kidney, skin, lungs, and heart; difficult to manage. **Rheumatoid arthritis** is a chronic systemic autoimmunity that centers on the joints; appears to be associated with immune complexes that cause chronic inflammation and scar tissue. **Endocrine autoimmunities** include Graves' disease, Hashimoto's thyroiditis, and types I and II **diabetes mellitus. Myasthenia gravis** is an immune attack upon the myoneural junction, with muscle paralysis. In **multiple sclerosis,** T cells and antibodies damage the myelin sheath of nerve cells; accompanied by severe motor and sensory loss.

Type IV Cell-Mediated Hypersensitivity

A delayed response to antigen involving the activation of and damage by T cells.

Delayed Allergic Response: Skin response to allergens, including infectious agents. Example is **tuberculin reaction;** contact dermatitis is caused by exposure to poison plants (ivy, oak) and simple environmental molecules (metals, cosmetics); cytotoxic T cells acting on allergen elicit a severe skin reaction.

Graft Rejection: Reaction of cytotoxic T cells directed against the foreign cells of a grafted tissue; involves recognition of foreign HLA by T cells and rejection of tissue. Host may reject graft; graft may reject host. Types of grafts include: **autograft,** from one part of body to another; **isograft,** grafting between identical twins; **allograft,** between two members of same species; **xenograft,** between two different species. All major organs (heart, kidney, bone marrow, skin) may be successfully transplanted. Allografts require close **tissue match** (HLA antigens must correspond); rejection is controlled with drugs.

Immunodeficiency Diseases: Components of the immune response system are absent; involve B and T cells, phagocytes, complement. **Primary immunodeficiency** is genetically based, congenital; defect in inheritance leads to lack of B-cell activity, T-cell activity, or both. **B-cell defect** is called **agammaglobulinemia**; patient lacks antibodies; serious recurrent bacterial infections result. In Ig deficiency, one of the classes of antibodies is missing or deficient. In **T-cell defects,** the thymus is missing or abnormal. In **DiGeorge syndrome,** the thymus fails to develop; afflicted children experience recurrent infections with eucaryotic pathogens and viruses; immune response is generally underdeveloped. In **severe combined immunodeficiency** (**SCID**), both limbs of the lymphocyte system are missing or defective; no adaptive immune response exists; fatal without replacement of bone marrow or special facilities. **Secondary (acquired) immunodeficiency** is due to damage after birth (infections, drugs, radiation). AIDS is the most common of these; T-helper cells are main target; deficiency manifest in numerous opportunistic infections and cancers.

Cancer and the Immune System

Cancer is characterized by overgrowth of abnormal tissue, also known as a neoplasm. It appears to arise from malfunction of immune surveillance. **Tumors** can be **benign** (a nonspreading local mass of tissue) or **malignant** (a cancer) that spreads (metastasizes) from the tissue of origin to other particular sites by means of the circulation; malignant tumors may be **carcinomas,** originating from epithelial tissue, or **sarcomas,** originating from embryonic connective tissue. Cancers occur in nearly every cell type (except mature, nondividing cells).

Cancer cells appear to share a common basic mechanism involving some type of gene alteration that turns a normal gene (**proto-oncogene**) into an **oncogene**; this **transforms** the cell and triggers uncontrolled growth and abnormal structure and function. Gene alterations may be due to chromosome alterations during cell division, viral infection, or the actions of chemical and physical **carcinogens.** Malignant cancer cells display special markers, respond to growth factors, and lose their cohesiveness. Treatment is with surgery, chemicals, radiation, and immune therapy.

True–False Questions

Determine whether the following statements are true (T) or false (F). If you feel a statement is false, explain why, and reword the sentence so that it reads accurately.

___ 1. Allergy is the state in which a heightened response is shown to common environmental substances.

___ 2. B cells are responsible for type I allergies, while T cells are responsible for types II, III, and IV hypersensitivities.

___ 3. All allergies have the potential to be manifest in the skin, regardless of the route of entry of the allergen.

___ 4. Atopic allergies and anaphylaxis involve production of IgE and degranulation of mast cells.

___ 5. Type AB blood may be transfused into a type O person, because one lacks antigens and the other lacks antibodies.

___ 6. Sensitization of an Rh^- mother by an Rh^+ fetus can cause hemolytic disease in later children.

___ 7. In type II hypersensitivities, the antigens form large complexes with antibodies that clog epithelial tissues.

___ 8. In autoimmunity, the body attacks self tissues.

___ 9. Systemic lupus erythematosus involves many organs, and myasthenia gravis primarily involves muscle function.

___ 10. The tuberculin skin test is an example of a delayed-type allergy.

___ 11. The rejection of a transplant is due to cytotoxic T cells that attack host tissue.

___ 12. In SCID, children lack primarily the B-cell limb of lymphocyte immunity.

___ 13. Cancer is caused by a genetic defect or change that leads to uncontrolled growth of cells.

___ 14. Metastasis is the tendency of cancer cells to display abnormal anatomy and physiology.

Concept Questions

1. What is allergy or hypersensitivity? What accounts for the reactions that occur in these conditions? What does it mean when a reaction is immediate or delayed? Which allergies are immediate; which are delayed?
2. Describe several factors that influence allergies.
3. How are atopic allergies similar to anaphylaxis? How are they different?
4. How do allergens gain access to the body? What are some examples of allergens that enter by these portals?
5. Trace the course of a pollen grain through sensitization and provocation in type I allergies. What is the role of mast cells and basophils? Of IgE? Of allergic mediators? Briefly describe the effects of histamine, prostaglandin, and leukotriene in allergy. Briefly outline the target organs and symptoms of the principal atopic diseases. How are they diagnosed and treated?
6. What is anaphylaxis? What are its usual causes? How is it diagnosed and treated? Explain how hyposensitization works.
7. What is the mechanism of type II hypersensitivity? Why are the tissues of some persons antigenic to others? Would we be concerned about this problem if it were not for transfusions? What is the actual basis of the four ABO blood groups? Where do we derive our natural hypersensitivities to the A or B antigens that we do not possess?
8. Explain the rules of transfusion. Illustrate what will happen if type A blood is accidently transfused into a type B person.
9. Contrast type II and type III hypersensitivities with respect to type of antigen, antibody, and manifestations of disease. What is immune complex disease? Differentiate between the Arthus reaction and serum sickness.
10. In a few words, what is autoimmunity? What factors are involved in development of autoimmunity? Describe four major types of autoimmunity, comparing target organs and symptoms.
11. Compare and contrast type I (atopic) and type IV (delayed) hypersensitivity as to mechanism, symptoms, eliciting factors, and allergens.
12. What is the basis for a host rejecting the graft tissue? What is the basis for a graft rejecting the host? Compare the four types of grafts. What does it mean to say that two tissues constitute a close match? Describe the procedure involved in a bone marrow transplant.
13. In general, what causes primary immunodeficiencies? Secondary immunodeficiencies? Why can T-cell deficiencies have greater impact than B-cell deficiencies? What kinds of symptoms accompany a B-cell defect? A T-cell defect? Combined defects? Give examples of specific diseases that involve each type of defect.
14. Why would a patient be given immunosuppressive drugs? What is the outcome?
15. Define cancer, and give some synonyms. Differentiate between a benign tumor and a malignant tumor, and give examples. How are cancers technically differentiated?
16. Describe a possible mechanism to account for transformation of a normal cell to a cancer cell. Describe the possible genetic and environmental factors that increase oncogenesis.
17. Relate how the immune system is involved in the development (or prevention) of cancer.

Practical/Thought Questions

1. Discuss why the immune system is sometimes called a double-edged sword. What possible function does allergy have?
2. Can you think of a reason that some humans are so hypersensitive to hymenopteran stings?
3. A three-week-old neonate develops severe eczema after being given penicillin therapy for the first time. Can you explain what has happened?
4. Can you explain why a person is allergic to strawberries when he eats them, but shows a negative skin test to them?
5. Where in the course of type I allergies do antihistamine drugs work? Cortisone? Desensitization?
6. Although we call persons with type O blood universal donors and those with type AB blood universal recipients, what problems might accompany this procedure? What is the truly universal donor and recipient blood type? (Hint: Include Rh type.) Can you explain how a person could have a genotype for type A or B blood but a phenotype for type O?
7. Why would it be necessary for an Rh^- woman who has had an abortion or an ectopic pregnancy to be immunized against the Rh factor?
8. Can you explain how persons with autoimmunity could develop antibodies against intracellular organelles such as the nucleus?
9. Would a person show allergy to poison oak the first time he contacted it? Why?
10. Looking at figure 14.19, predict which family members would be good allograft pairs and which would be incompatible. Why might a graft be rejected even between closely matched siblings? What kind of graft is done when a baboon heart is transplanted into a baby?
11. Why are primary immunodeficiencies experiments of nature?
12. Why are defective genes found on the X chromosome expressed most often in male children?
13. Why don't babies with agammaglobulinemia show symptoms of pyogenic infections until about 6 months after birth?
14. SCID children cannot reject grafts. Why? What would be the major problem in most bone marrow transplantations for these children?
15. In what ways is cancer both a cause and a symptom of immunodeficiency?
16. What features of cancer cells account for metastasis?
17. In what ways could cancer be inherited?
18. If a chemical is found to be a mutagen by the Ames test, what other factors will be required to induce cancer?
19. Most cancer therapy removes or destroys the metastatic cells. Can you think of a genetic prevention for cancer?

Introduction to Medical Bacteriology

Preface to the Survey of Microbial Diseases

The next seven chapters of the textbook will cover the most clinically significant bacterial, fungal, parasitic, and viral diseases in the world, with an emphasis on those most prevalent in North America. This survey will encompass the morphology, physiology, and virulence of the microbes and the epidemiology, pathology, treatment, and prevention of the diseases they cause. The bacteria are divided into the gram-positive and gram-negative cocci (chapter 15); the gram-positive and gram-negative bacilli (chapter 16); miscellaneous bacteria, and dental infections (chapter 17); fungal, protozoan, and helminth parasites (chapters 18 and 19); DNA viruses (chapter 20); and RNA viruses (chapter 21). Prior chapters with useful references include chemotherapy (chapter 10) and host-parasite relations and epidemiology (chapter 11). Particular attention is called to feature 11.5 on the terminology of disease, feature 11.6 on skin lesions, and the discussion of vaccines and serological tests (last section of chapter 13).

The investigation and diagnosis of these diseases depends to a considerable degree on a series of basic laboratory techniques. This section summarizes the methods used by clinical microbiologists and other medical personnel to collect, isolate, and identify infectious agents.

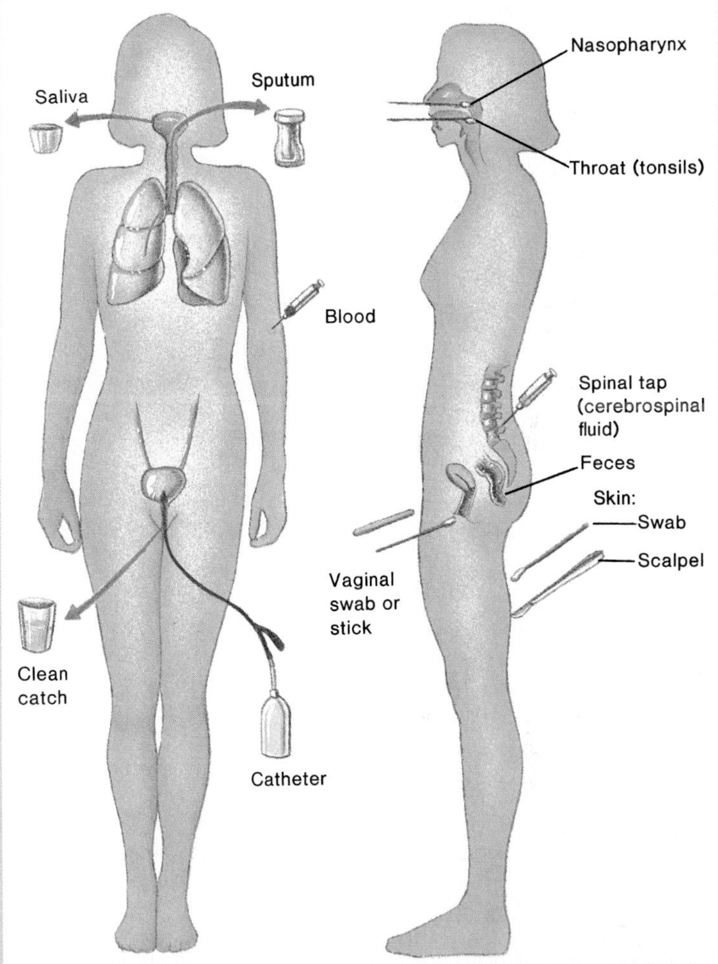

Figure A Sampling sites and methods of collection for clinical laboratories.

On the Track of the Infectious Agent: Specimen Collection

Regardless of the method of diagnosis, specimen collection is the common point that guides the health care decisions of every member of a clinical team. Indeed, the success of identification and treatment depends on how specimens are collected, handled, and stored. Specimens may be taken by a medical technologist, nurse, physician, or even by the patient himself. However, it is imperative that general aseptic procedures be used, including sterile sample containers and other tools to prevent contamination from the environment or the patient. Figure A delineates the most common sampling sites and procedures.

In sites that normally contain resident microflora, care should be taken to sample only the infected site and not surrounding areas. For example, throat and nasopharyngeal swabs should not touch the tongue, cheeks, or saliva. Saliva is an especially undesirable contaminant because it contains millions of bacteria per milliliter, most of which are normal flora. Saliva samples are occasionally taken for dental diagnosis by having the patient expectorate into a container. Depending on the nature of the lesion, skin may be swabbed or scraped with a scalpel to expose deeper layers. The mucous lining of the vagina, cervix, or urethra may be sampled with a swab or applicator stick.

Urine may be taken aseptically from the bladder with a thin tube called a catheter or by catching the flow midstream after washing the urethra (known as a "clean catch"). The latter method inevitably incorporates a few normal flora into the sample, but these can usually be differentiated from pathogens in an actual infection. Sputum, the mucous secretion that coats the lower respiratory surfaces, especially the lungs, is discharged by coughing or taken by catheterization to avoid contamination with saliva. Sterile materials such as blood,

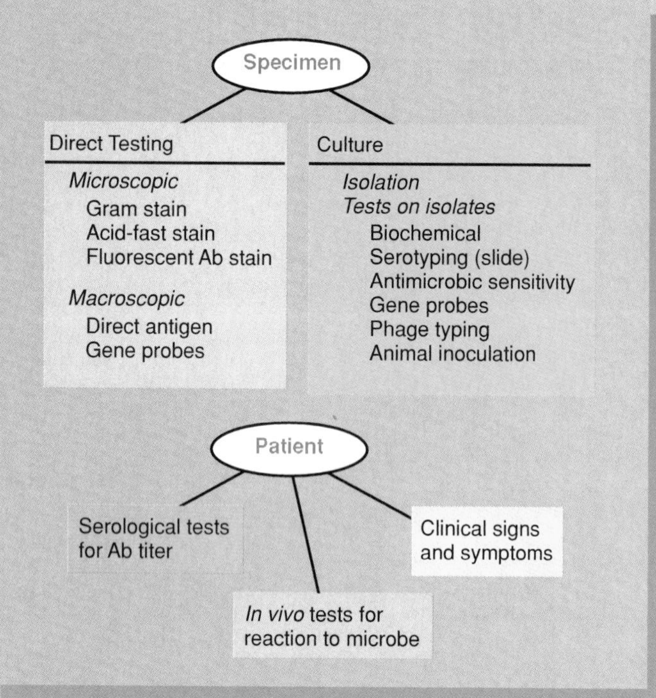

Figure B A scheme of specimen isolation and identification.

cerebrospinal fluid, and tissue fluids must be taken by sterile needle aspiration. Antisepsis and degermation of the puncture site are extremely important in these cases. Additional sources of specimens are the eye, ear canal, nasal cavity (all by swab), and diseased tissue that has been surgically removed by instrumentation, as in a biopsy.

After proper collection, the specimen is promptly transported to a lab and stored appropriately (usually refrigerated) if it must be held for a time. Nonsterile samples in particular, such as urine, feces, and sputum, are especially prone to deterioration at room temperature. Special swab and transport systems are designed to take the specimen and maintain it exactly as it was at the time of collection. These devices contain nonnutritive maintenance media (so the microbes do not grow), a buffering system, and an anaerobic environment to prevent possible destruction of oxygen-sensitive bacteria.

Overview of Laboratory Techniques

Two routes are taken in specimen analysis: (1) direct tests using microscopic, immunologic, or other specific methods that provide immediate clues as to the identity of the microbe or microbes in the sample, and (2) cultivation, isolation, and identification of pathogens using a wide variety of general and specific tests (figure B). Most test results fall into two categories: **presumptive data,** which place the isolated microbe (isolate) in a preliminary category such as a genus, and more specific, **confirmatory data,** which provide more definitive evidence of a species. Some tests are more important for some groups of bacteria than others. The total time required for analysis ranges from a few minutes in a streptococcal sore throat to weeks in tuberculosis.

Some diseases may be diagnosed without processing or identifying specimens. Serological tests on a patient's serum can detect signs of an antibody response, and skin testing can pinpoint a delayed allergic reaction to a microorganism (see chapters 13 and 14). These tests are also important in screening the general population for exposure to an infectious agent such as rubella or tuberculosis. Because diagnosis is both a science and an art, the ability of the practitioner to interpret signs and symptoms of disease can be very important. AIDS, for example, is often diagnosed by serological tests and a complex of signs and symptoms without ever isolating the virus.

A Pathogen or Not a Pathogen? Questions that can be difficult but necessary to answer in this era of debilitated patients and opportunists are: Is an isolate clinically important, and how do you decide whether an isolate is a contaminant or just part of the normal flora? The number of microbes in a sample is one useful criterion. For example, a few colonies of *Escherichia coli* in a urine sample may simply indicate normal flora, whereas several hundred may mean active infection. In contrast, the presence of a single colony of a true pathogen such a *M. tuberculosis* in sputum or an opportunist in sterile sites such as cerebrospinal fluid or blood is highly suggestive of its role in disease. Furthermore, the repeated isolation of a relatively pure culture of any microorganism may mean it is an agent of disease, though care must be taken in this diagnosis. Another problem facing the laboratory technician is that of differentiating a pathogen from species that are similar in morphology from their more virulent relatives. For this reason, identification usually requires a series of laboratory tests.

Immediate Direct Examination of Specimen

Direct microscopic observation of a fresh or stained specimen is one of the most rapid methods of determining presumptive and sometimes confirmatory characteristics (see figures 15.26 and 18.30). Stains most often employed for bacteria are the Gram stain (see feature 3.1) and the acid-fast stain (see figure 16.14). But although these ordinary stains are somewhat useful, they do not work with certain organisms. For bacteria such as the syphilis spirochete, direct fluorescence antibody (DFA) tests highlight the presence of the microbe in patient specimens by means of labelled antibodies (figure C). These tests are particularly useful for bacteria that are not readily cultivated in the laboratory.

Another way that specimens can be analyzed is through *direct antigen testing,* a technique similar to direct fluorescence in that known antibodies are used to identify antigens on the surface of bacterial isolates. But in direct antigen testing, the reactions can be seen with the naked eye. Quick test kits that greatly speed clinical diagnosis are available for *Staphylococcus aureus, Streptococcus pyogenes* (see figure 15.14), *Neisseria gonorrhoeae, Hemophilus influenzae,* and *Neisseria meningitidis.* However, when the microbe is very sparse in the specimen, direct testing is like looking for a needle in a haystack, and more sensitive methods are necessary.

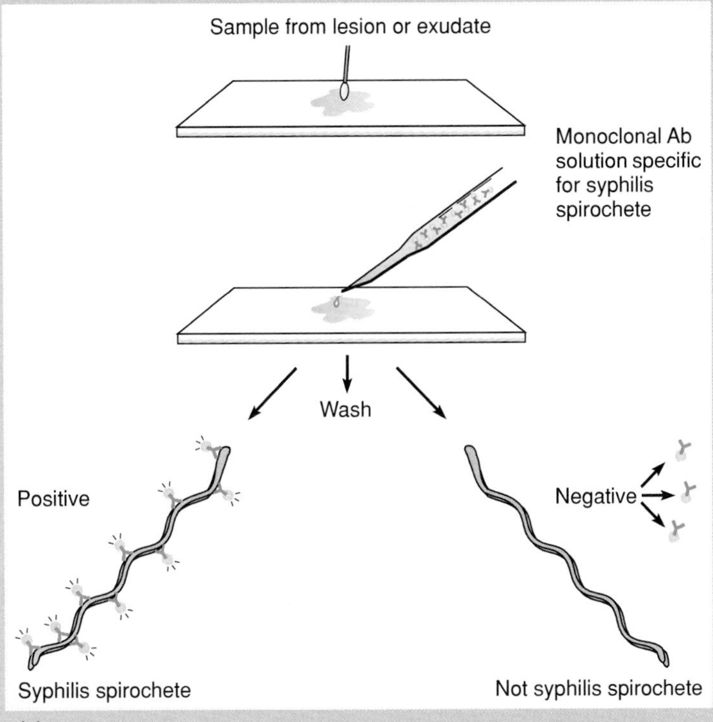

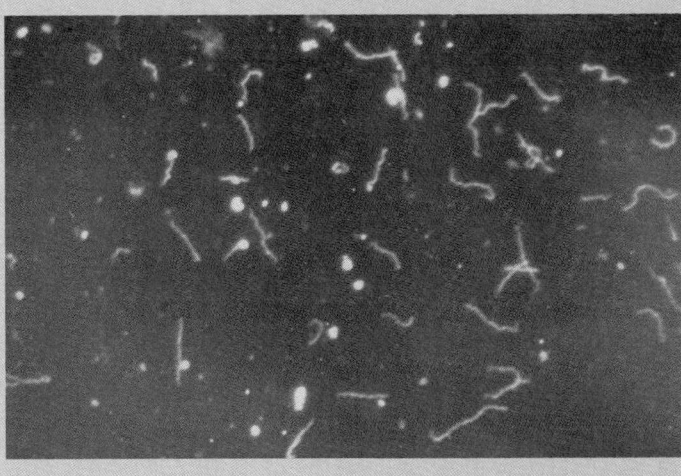

Figure C (*a*) Direct fluorescent antigen test results for *Treponema pallidum,* the syphilis spirochete, and an unrelated spirochete. (*b*) Photomicrograph of a blood sample from a syphilitic patient.

Cultivation of Specimen

Isolation Media Such a wide variety of media exist for microbial isolation that a certain amount of preselection must occur, based on the nature of the specimen. In cases where the suspected pathogen is present in small numbers or is easily overgrown, the specimen may be initially enriched with specialized media. However, specimens such as urine and feces with high bacterial counts and a diversity of species are initially plated onto selective media (see figure 16.30). In most cases, specimens are streaked onto solid differential media that define such characteristics as reactions in blood (blood agar) and fermentation patterns (mannitol salt and MacConkey agar). A patient's blood is usually cultured in a special bottle of broth that can be periodically sampled for growth. Numerous other examples of isolation, differential, and biochemical media were presented in chapter 6 and appear in various sections of chapters 15, 16, 17, and 18. In order for the subsequent steps in identification to be as accurate as possible, all work must be done from isolated colonies or pure cultures, because working with a mixed or contaminated culture gives misleading and inaccurate results. From such isolates, clinical microbiologists obtain information about a pathogen's microscopic morphology and staining reactions, cultural appearance, motility, oxygen requirements, and biochemical characteristics.

Biochemical Testing The physiological reactions of bacteria to nutrients and other substrates provide excellent indirect evidence of the types of enzyme systems present in a particular species. Many of these tests are based on the following scheme:

Test performed
to detect this
↓
Microbe + Substrate → End product
↓
If (+), microbe
has enzyme
If (−), microbe
lacks enzyme

The microbe is cultured in a medium with a special substrate and then tested for a particular end product. The presence of the end product indicates that the enzyme is expressed in that species, while its absence means it lacks the enzyme for utilizing the substrate in that particular way. These types of reactions are particularly meaningful in bacteria, which are haploid and generally express their genes for utilizing a given nutrient.

Among the prominent biochemical tests are carbohydrate fermentation (acid and/or gas); hydrolysis of gelatin, starch, and other polymers; and enzyme actions such as catalase, oxidase, and coagulase. Examples of these tests will be presented in chapters 15, 16, and 17. Many are presently performed with rapid, miniaturized systems that can simultaneously determine up to 20 characteristics in small individual cups or spaces (see figures 15.7 and 16.31). An important plus, given the complexity of biochemical profiles, is that such systems are readily adapted to computerized analysis.

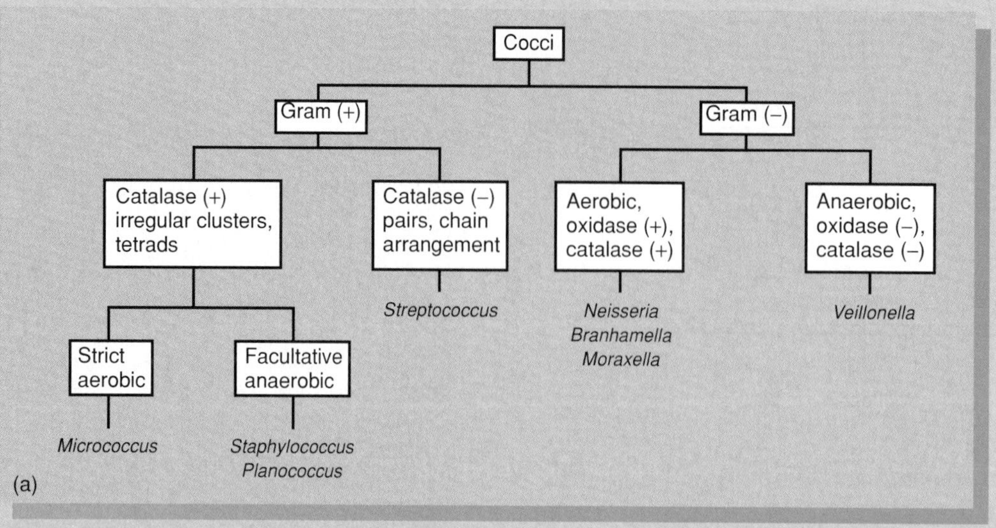

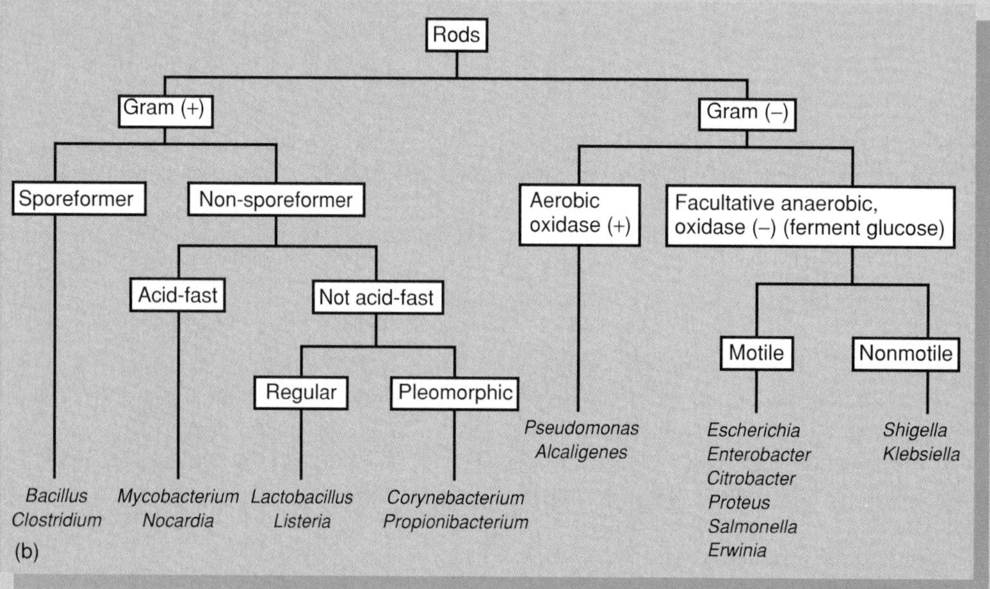

Figure D Flowchart to separate the genera of gram-positive and gram-negative (*a*) cocci and (*b*) rods involved in human diseases.

Common schemes for identifying bacteria are somewhat artificial but convenient. They are based on easily recognizable characteristics such as motility, oxygen requirements, Gram stain reactions, shape, spore formation, and various biochemical reactions. Schemes may be set up as flowcharts (figure D) or keys that trace a route of identification by offering pairs of opposing characteristics (positive versus negative, for example) from which to select. Eventually, an endpoint is reached, and the name of a genus or species that fits that particular combination of characteristics appears. Diagnostic tables that provide more complete information are preferred by many laboratories because variations from the general characteristics used on the flowchart may be misleading. Both systems are used in this text.

Miscellaneous Tests When morphological and biochemical tests are insufficient to complete identification, other tests come into play. Commercial **serotyping** kits are important for identifying isolates in the genera *Salmonella, Shigella,* and *Streptococcus.* In general, a bit of culture is mixed with specific, animal-derived antisera or monoclonal antibodies that agglutinate the cells if they are complementary (technique shown in figure 13.24*b*).

Antimicrobic sensitivity tests are not only important in determining the drugs to be used in treatment (see figure 10.20), but their patterns of sensitivity may be used in presumptive identification of some species of *Streptococcus* (see figure 15.14*a*), *Pseudomonas,* and *Clostridium.* Antimicrobics are also used as selective agents in many media.

Bacteria host viruses called bacteriophages that are very species- and strain-specific. Such selection by a virus for its host is useful in typing some bacteria, primarily *Staphylococcus* and *Salmonella.* The technique of **phage typing** involves inoculating a lawn of cells onto a petri dish, mapping it off into blocks, and applying a different phage to each block. Cleared areas corresponding to lysed cells indicate sensitivity to that phage. Phage typing is chiefly used for tracing strains of bacteria in epidemics.

Animals must be inoculated to cultivate bacteria such as *Mycobacterium leprae* and *Treponema pallidum,* while avian embryos are used to grow rickettsias, chlamydias, and viruses. Animal inoculation is also occasionally used to test the virulence of a strain of bacterium or fungus.

Modern molecular biology has provided some extremely specific and sensitive systems for identifying microbes. The value of such molecular testing is that it measures similarities or differences in the actual molecular and genetic structure of microbes, not just in morphological appearance or physiological reactions. Although this area is expanding beyond the scope of this text, one particular test that is currently gaining importance will be mentioned here. You will recall that DNA and RNA are composed of specific arrangements of nitrogen bases that are unique to each species or type of organism. This concept has been influential in developing **gene probe** and **hybridization** technology. Gene probes consisting of small segments of single-stranded DNA or RNA from known species may be mixed with single-stranded DNA or RNA from an unknown microbe. If the nucleic acid of the probe is complementary to the nucleic acid of the unknown microbe, the two strands will align, bind, and hybridize (see figure 22.47). Gene probes are currently used to identify *Escherichia coli* and a number of other gram-negative enteric bacteria, *Mycobacterium, Mycoplasma, Legionella, Chlamydia,* and a wide variety of viruses.

Review of Clinical Laboratory Technology with Key Terms

Concerns of the Clinical Laboratory

The lab provides technical support in determining the causative agents of disease and the antimicrobics to be used for treatment. The **method of specimen collection** is crucial; must be done aseptically; swabs used for nasopharynx, vagina, skin; saliva and urine collected as free flow; sterile urine may be taken by catheter; sputum is coughed up or aspirated from lung; blood and spinal fluid taken by puncture of sterile site with needle.

Handling

Sample transported to lab; those containing fragile pathogens must be placed in special environment, processed rapidly, or refrigerated for short time.

Laboratory Protocols

Identification of microbe and diagnosis of disease require a wide array of procedures, including (1) direct specimen testing, (2) cultivation, isolation, and examination of biochemistry, antigenicity, and genetics of microbe, and (3) examination of patient. Collected data falls into categories of presumptive evidence or confirmatory evidence of a species.

Testing

Direct tests include microscopic examination of stained specimen, direct fluorescent antibody tests, and macroscopic antigen tests, all of which provide rapid clinical data. **Cultivation** is usually required for confirmation.

Initial isolation of pathogen is performed on some types of selective, enrichment, or differential media, chosen according to the expected nature of the specimen and pathogen; cultures are examined for number and types of colonies; decision is made as to whether the isolates are normal flora, contaminants of sampling, or actual cause of infection.

Isolated colonies are examined for microscopic and macroscopic morphology, motility, staining reactions, and the presence of enzyme systems for processing nutrients; hundreds of biochemical tests are available to determine these characteristics; the most important ones are selected in accordance with evidence of a given group or genus; miniaturized methods are in common use; computerized analysis of results is helpful; collected data is processed through flowcharts or tables to arrive at species that most closely fits the unknown isolate.

Other tests used in determining species are **serotyping, antimicrobic sensitivity, gene probes**; use of animals may be necessary to isolate and confirm some microbes. **Immunological tests** are performed on patient serum or body; serological tests give evidence of antibodies in serum; skin testing (such as tuberculin) shows reactions of body to proteins derived from a pathogen; both are helpful in screening for past or current infection; diagnosis also requires observation of clinical signs and symptoms.

Review Questions

1. Why do specimens need to be taken aseptically when nonsterile sites are being sampled and selective media are to be used?
2. Explain the general principles in specimen collection.
3. What is involved in direct specimen testing? In presumptive and confirmatory tests? In cultivating and isolating the pathogen? In biochemical testing? In gene probes?
4. Differentiate between the serological tests used to identify isolated cultures of pathogens and those used to diagnose disease from patients' serum.
5. Why is it important to prevent microbes from growing in specimens?
6. Why is speed so important in the clinical laboratory?

CHAPTER 15

The Cocci of Medical Importance

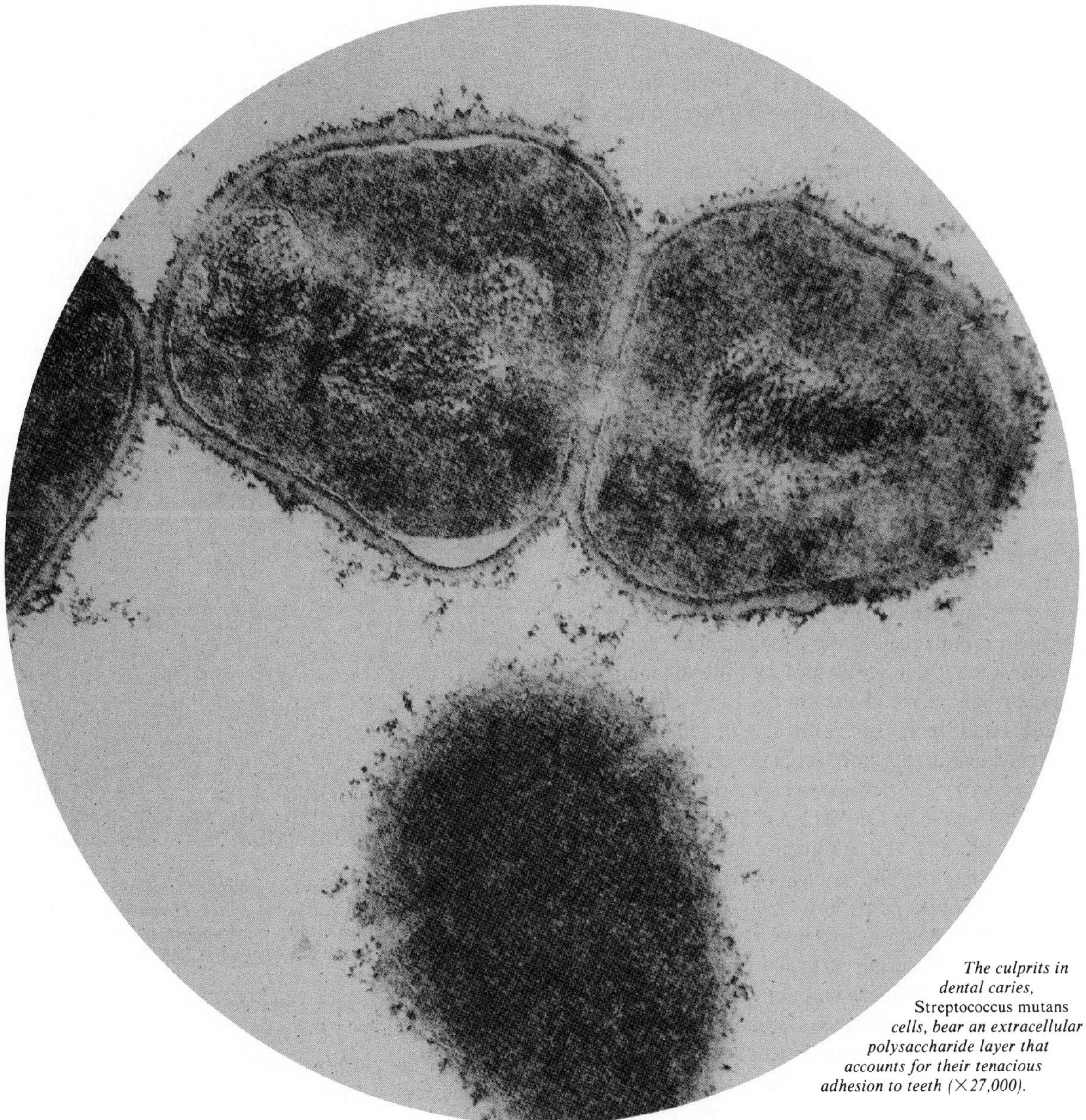

The culprits in dental caries, Streptococcus mutans *cells, bear an extracellular polysaccharide layer that accounts for their tenacious adhesion to teeth (×27,000).*

Chapter Preview

Gram-positive and gram-negative cocci are among the most significant infectious agents of humans. Because these bacteria tend to stimulate pus formation, they are often referred to collectively as the **pyogenic cocci.** The most common infectious species in this group belong to three genera: *Staphylococcus* (Family Micrococcaceae), *Streptococcus* (Family Streptococcaceae), and *Neisseria* (Family Neisseraceae).

General Characteristics of the Staphylococci

The genus ***Staphylococcus*** is a common inhabitant of the skin and mucous membranes that accounts for a considerable proportion of human infections (often called "staph" infections). Its spherical cells are arranged primarily in irregular clusters and occasionally in short chains and pairs (figure 15.1). Though typically gram positive, stains from older cultures and clinical specimens sometimes do not stain true. As a group, the staphylococci lack spores and flagella, and are only occasionally encapsulated.

Bergey's Manual lists 19 species in the genus *Staphylococcus,* but the most important human pathogens are (1) *S. aureus,*[1] (2) *S. epidermidis,* once called *S. albus* for its white colonies, and (3) *S. saprophyticus.* Of the three, *S. aureus* is considered the most serious pathogen, although the other species have become increasingly associated with opportunistic infections and can no longer be regarded as harmless commensals. From this point we will focus on the significant characteristics of each species separately.

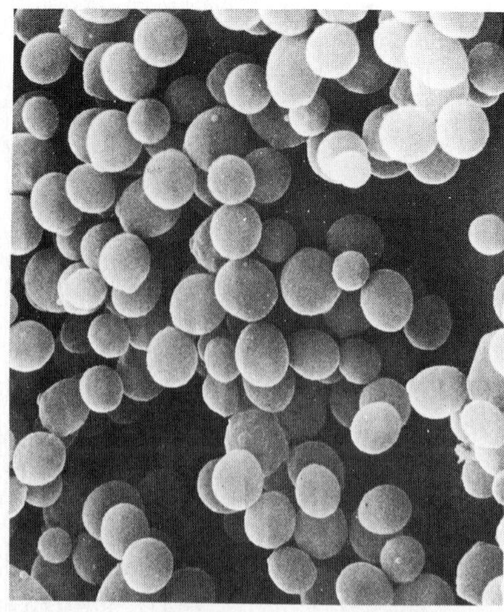

Figure 15.1 Scanning electron micrograph of *Staphylococcus aureus* (×7,500). This genus owes its name to the grapelike (Gr. *staphyle*) appearance of the clusters.

Growth and Physiological Characteristics of *Staphylococcus aureus*

Staphylococcus aureus grows in large, round, opaque colonies (figure 15.2) at an optimum of 37°C, though it can grow anywhere between 10°C and 46°C. The species is a facultative anaerobe, and growth is enhanced in the presence of O_2 and CO_2. Its nutrient requirements can be satisfied by routine laboratory media, and most strains are metabolically versatile—that is, they can digest proteins and lipids, and ferment a variety of sugars. This species is considered the most resistant of all non-spore-forming pathogens, with well-developed capacities to withstand high salt (7.5–10%), extremes in pH, and high temperatures (up to 60°C for 60 minutes). It also remains viable after months of air drying, and resists the effects of many disinfectants and antibiotics. These properties contribute to the reputation of *S. aureus* as a troublesome hospital pathogen.

Perhaps no other bacterial pathogen produces as many virulence factors to facilitate invasion of host tissues or evasion of various defenses as does *S. aureus.* Because known virulent strains lack one or more of the known virulence factors, no single factor accounts totally for virulence, leading some experts to theorize that overall staphylococcal pathogenicity is due to a combination of factors. The following section will deal with the major determinants of pathogenicity that fall into the categories of enzymes and toxins.

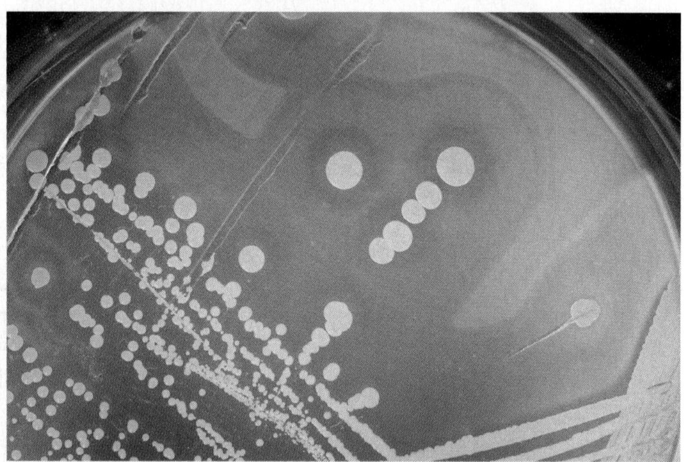

Figure 15.2 Blood agar plate growing *S. aureus.* Some strains show two zones of hemolysis. The relatively clearer inner zone is caused by α-toxin, whereas the outer zone is fuzzy and appears only if the plate has been refrigerated. This is the β or "hot-cold" hemolysin.

The Enzymes of *S. aureus*

Pathogenic *S. aureus* typically produces **coagulase,** an enzyme that coagulates plasma and blood (see figure 15.6*b*). The precise importance of coagulase to the disease process remains obscure. One theory proposes that coagulase causes a fibrin coating to develop around the cells that helps retard host defenses or that it may aid in adherence to tissues. Because 97% of all human isolates of *S. aureus* produce this enzyme, its presence is considered the most diagnostic species characteristic.

An enzyme that appears to promote invasion is hyaluronidase, which has been described as the "spreading factor" because it digests the ground substance (hyaluronic acid) around host cells. Other enzymes associated with *S. aureus* include staphylokinase, which digests blood clots; a nuclease capable of attacking DNA (DNase); and lipases that help bacteria colonize oily skin surfaces. A majority of strains are now resistant to certain forms of penicillin through their production of penicillinases.

The Toxins of *S. aureus*

Most of the toxic products of this species are extracellular proteins that injure a specific target tissue or cell. In general, there

1. From L. *aurum,* gold. Named for the tendency of some strains to produce a golden-yellow pigment, though this characteristic is not common enough to be used as a criterion for identification.

are blood cell toxins (hemolysins and leukocidins), intestinal toxins, and epithelial toxins. The activity of staphylococcal blood cell toxins is directed toward the cell membrane. The **hemolysins,** for example, lyse red blood cells by disrupting their membranes. All of the staphylococcal hemolysins produce a zone of hemolysis in blood agar (figure 15.2). The most far-reaching in its biological effects is a form of hemolysin called α (alpha)-toxin. This powerful substance lyses the red blood cells of various mammals and damages leukocytes, skeletal muscle, heart, and renal tissue as well. It is regarded as a significant contributor to the pathological process. Other powerful hemolysins isolated from human strains include β (beta)-toxin; δ (delta)-toxin, and γ (gamma)-toxin. (Note that the Greek letters used for the staphylococcal hemolysins do not correlate with the Greek letters describing general patterns of hemolysis in chapter 11. For example, α-toxin gives β-hemolysis, while β-toxin produces α-hemolysis.)

Staphylococcal **leukocidin** alters the permeability of neutrophils and macrophages, leading to their lysis. Because only virulent staphylococci produce this toxin, it probably helps incapacitate the phagocytic line of defense. Exotoxins that act upon the gastrointestinal tract of humans are **enterotoxins,** of which five distinct types exist. A few strains produce an **exfoliative** or epidermolytic toxin that separates the epidermal layer from the dermis and causes the skin to peel away. This toxin is responsible for staphylococcal scalded skin syndrome, in which the skin looks burned (see figure 15.5). The most recent toxin brought to light by research on *S. aureus* is **toxic shock syndrome toxin I** (TSSI). The presence of this toxin in victims of toxic shock syndrome indicates its probable role in the systemic pathologies of this dangerous condition.

Epidemiology and Pathogenesis of *S. aureus*

It is surprising that a bacterium with such great potential for virulence as *Staphylococcus aureus* is a common, intimate human associate. The microbe is present in most environments frequented by humans and is readily isolated from fomites. Colonization of the infant begins within hours after birth and continues throughout life. The carriage rate for normal healthy adults varies anywhere from 30% to 50%, and the pathogen tends to be harbored intermittently rather than chronically. Carriage occurs mostly in the anterior nares and, to a lesser extent, in the skin, nasopharynx, and intestine. Usually this colonization is not associated with symptoms, nor does it ordinarily lead to disease in carriers or their contacts. Circumstances that predispose an individual to infection include poor hygiene and nutrition, tissue injury, preexisting primary infections, diabetes mellitus, and immunodeficiency states. Nosocomial infections by *S. aureus* are especially problematic in the newborn nursery and in surgical wards. The so-called "hospital strains" can readily spread in an epidemic pattern within and outside the hospital.

The Scope of Clinical Staphylococcal Disease

Depending on the degree of invasion or toxin production by *S. aureus,* disease ranges from localized to systemic. A local staphylococcal infection often presents as an inflamed, fibrous lesion enclosing a core of pus called an **abscess** (figure 15.3). Toxigenic disease may present itself as a toxemia due to production of toxins in the body or as food intoxication, the ingestion of preformed toxin in food.

Localized Cutaneous Infections

Staphylococcus usually invades the skin through wounds, follicles, or skin glands. The most common infection is a mild, superficial inflammation of hair follicles (**folliculitis**) or glands (hiradenitis). Although these lesions are usually resolved with no complications, they may lead to infections of subcutaneous tissues. A **furuncle** (boil) results when the inflammation of a single hair follicle or sebaceous gland progresses into a large, red, and extremely tender abscess or pustule (figure 15.3). Furuncles often occur in clusters (furunculosis) on parts of the body such as the buttocks, breasts, axillae, and back of the neck, where skin rubs against other skin or clothing. A **carbuncle** is a larger (sometimes as big as a baseball) and deeper lesion created by aggregation and interconnection of a cluster of furuncles. It is usually found in areas of thick, tough skin such as on the back of the neck (figure 15.3*c*). Carbuncles are extremely painful and can even be fatal in elderly patients when they give rise to systemic disease. One staphylococcal skin infection not confined to follicles and skin glands is the bullous type of **impetigo.** It is characterized by bubblelike epidermal swellings that may break and peel away, and as such, is considered a localized form of scalded skin syndrome (see figure 15.5).

Miscellaneous Systemic Infections

Most systemic staphylococcal infections are focal in nature, beginning as a local cutaneous infection in a wound or a furuncle that seeds the blood and is carried to other sites, a common one being bone (figure 15.4). In **osteomyelitis,** the pathogen is established in the highly vascular metaphysis of a variety of bones (often the femur, tibia, ankle, or wrist). Abscess formation in the affected area results in an elevated, tender lump and necrosis of the bone tissue (figure 15.4*b*). Symptoms of osteomyelitis include fever, chills, pain, and muscle spasm. This form of osteomyelitis is seen most frequently in growing children, adolescents, and intravenous drug abusers. Another type, secondary osteomyelitis, develops after traumatic injury (compound fracture) or surgery in cancer and diabetes patients.

No organ can remain untouched by a systemic staphylococcal infection. Because these bacteria inhabit the nasopharynx, they may be aspirated into the lungs and cause a form of pneumonia involving multiple lung abscesses and symptoms

---

exfoliative (eks-foh′-lee-ay″-tiv) L. *exfoliatio,* a falling off in layers.

furuncle (fur′-unkl) L. *furunculus,* little thief.

carbuncle (car′-bunkl) L. *carbunculus,* little coal.

impetigo (im-puh-ty′-goh) L. *impetus,* to attack. Another type is caused by group A streptococci.

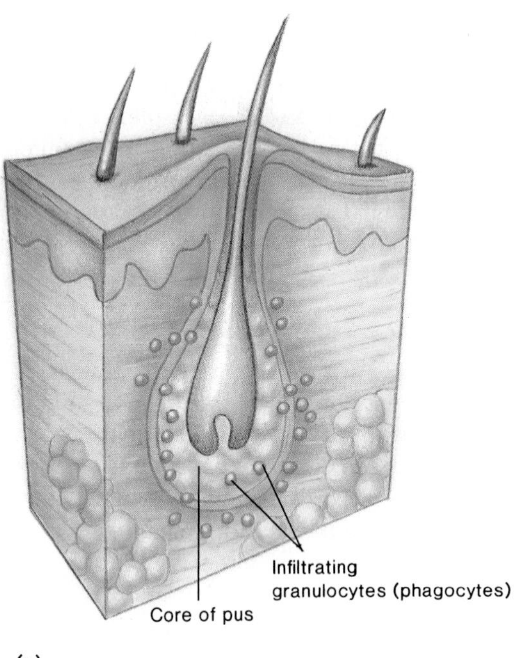

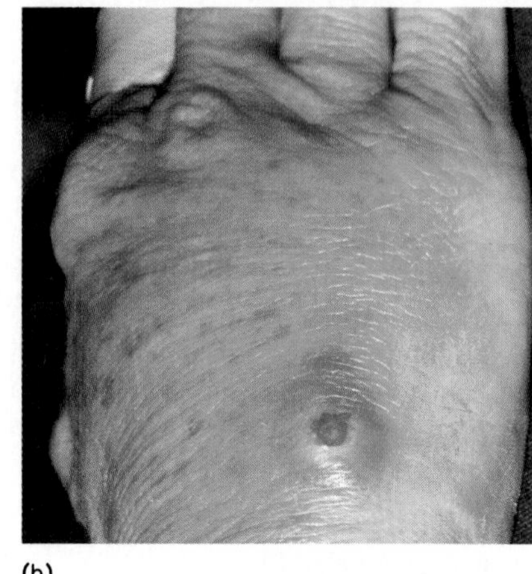

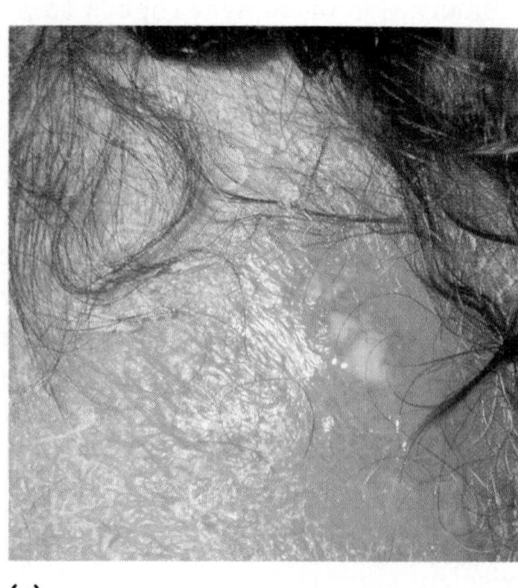

Figure 15.3 Cutaneous lesions of *S. aureus*. Fundamentally, all are skin abscesses that vary in size, depth, and degree of tissue involvement. (*a*) Sectional view of a boil or furuncle, a single pustule that develops in a hair follicle or gland and is the classic lesion of the species. The inflamed infection site becomes abscessed when masses of phagocytes, bacteria, and fluid are walled off by fibrin. (*b*) A furuncle on the back of the hand. (*c*) A carbuncle on the back of the neck. Carbuncles are massive deep lesions that result from multiple, interconnecting furuncles. Swelling and rupture into the surrounding tissues can be marked.

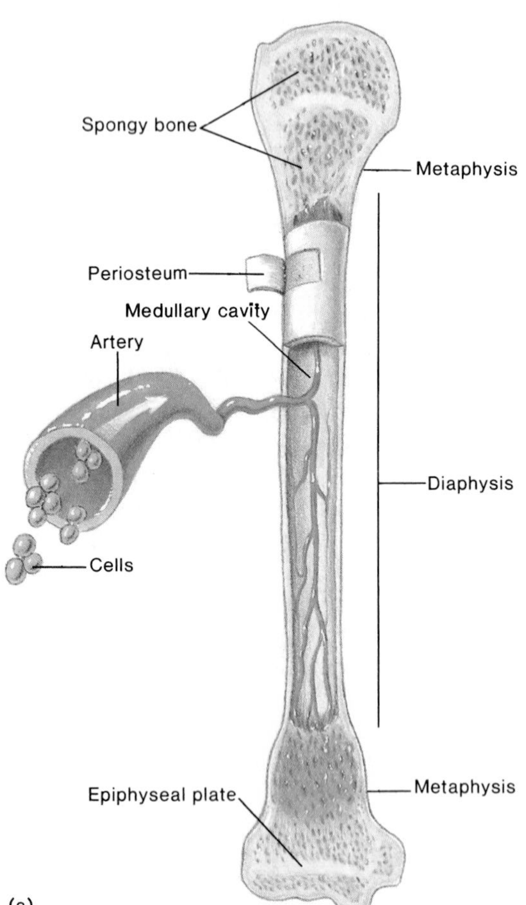

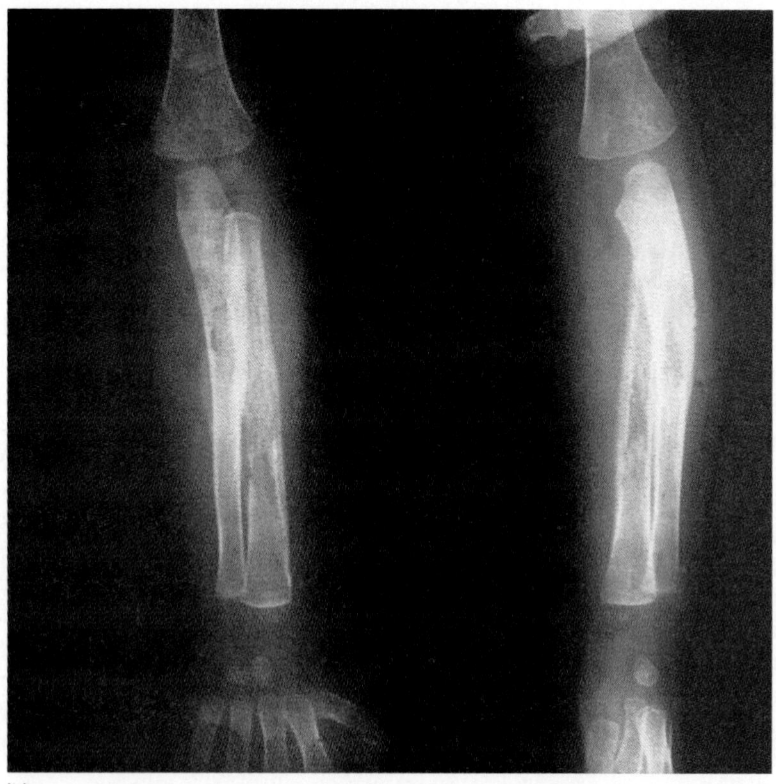

Figure 15.4 Staphylococcal osteomyelitis in a long bone. (*a*) In the most common form, the bacteria spread hematogenously from some other infection site, enter the artery, and lodge in the small vessels in bony pockets of the metaphysis or diaphysis. As they multiply, intense soft tissue damage begins, with swelling and necrosis. (*b*) X-ray photographs of a ruptured forearm bone caused by osteomyelitis.

of fever, chest pain, and bloody sputum. Although staphylococci account for only a small proportion of such pneumonia cases, their fatality rate is 50%. Most cases occur in infants and children suffering from cystic fibrosis and measles, and this form of pneumonia is also one of the most serious complications of influenza and certain lung diseases.

Staphylococcal bacteremia causes a high mortality rate among hospitalized patients with chronic disease. Its primary origin is bacteria that have broken loose from cutaneous and lung infections or from colonized medical devices (catheters, shunts). Circulating bacteria transported to the kidneys, liver, spleen, and muscles often form abscesses and contribute to a serious toxemia. One consequence of staphylococcal bacteremia is a fatal form of endocarditis associated with the colonization of the heart's lining, cardiac abnormalities, and rapid destruction of the valves. Infection of the joints may produce a deforming arthritis (pyoarthritis). A severe form of meningitis (accounting for about 15% of cases) occurs when *S. aureus* invades the cranial vault.

Toxigenic Staphylococcal Disease

Disorders due strictly to the toxin production of *S. aureus* are food intoxication, scalded skin syndrome (SSSS), and toxic shock syndrome (see feature 15.1). **Enterotoxins** produced by certain strains are responsible for the most common type of food poisoning in the United States (see chapter 22). This illness is associated with eating foods such as custards, sauces, cream pastries, processed meats, chicken salad, or ham that have been contaminated by handling and then left unrefrigerated for a few hours. Because of the high salt tolerance of *S. aureus,* even foods containing salt as a preservative are not exempt. The toxins produced by the multiplying bacteria do not noticeably alter the food's taste or smell. Enterotoxins are heat-stable (inactivation requires 100°C for at least 30 minutes), so heating the food after toxin production may not prevent disease. The ingested toxin acts upon the gastrointestinal lining and its nerves, with acute symptoms of cramping, nausea, vomiting, and diarrhea that appear in 2 to 6 hours. Recovery is rapid, usually within 24 hours.

Children with infection of the umbilical stump or eyes are susceptible to a toxemia called **staphylococcal scalded skin syndrome** (SSSS). Upon reaching the skin, this toxin induces a painful, bright red flush over the entire body that first blisters and then causes desquamation of the epidermis (figure 15.5). The vast majority of SSSS cases have been described in infants and children under the age of four. This same toxin causes the local reaction in bullous impetigo, which can afflict persons of all ages.

Host Defenses Against *S. aureus*

Despite regular close contact throughout life, humans have a well-developed resistance to staphylococcal infections. Studies performed on human volunteers established that an injection of several hundred thousand staphylococcal cells into unbroken skin was not sufficient to cause abscess formation. Yet, when sutures containing only a few hundred cells were sewn into the skin, a

Feature 15.1 Tampons and TSS

Toxic shock syndrome (TSS) was first identified as a discrete clinical entity in 1978 in young women using vaginal tampons. At first, the precise link between TSS and the use of tampons was a mystery. Then Harvard Medical School researchers discovered that ultra-absorbent brands of tampons strongly bind magnesium ions, and that the resultant low magnesium concentrations in the vaginal fluid can trigger TSS toxin production by the *S. aureus* population that is sometimes part of the normal vaginal flora. TSS toxemia causes a series of reactions, including fever, vomiting, rash, and renal, liver, blood, and muscle involvement, that are sometimes fatal. Cases of TSS have also been reported in children, men, and nonmenstruating women, but these are very rare and arise from skin or lung infections. The evidence implicating ultra-absorbent tampons as a major factor in the disease was so compelling that they were taken off the market in 1981. From the highest yearly count of 880 TSS cases in 1981, the number has gradually decreased to about 250 in 1991.

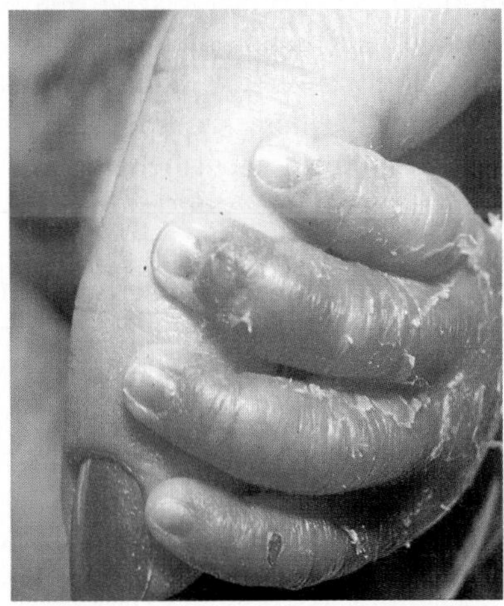

(a)

Epidermis

Dermis

(b)

Figure 15.5 Staphylococcal scalded skin syndrome (SSSS) in a newborn child. (*a*) Exfoliative toxin produced in local infections causes blistering and peeling of the skin that exposes an angry red underlayer.
(*b*) Photomicrograph of a segment of skin affected with SSSS. The line of desquamation is at the dermis. The lesions will heal well because the level of separation is so superficial.

classic lesion rapidly formed, indicating the aggravating effect of a foreign body in infection. Specific antibodies are produced against most of the staphylococcal antigens and intracellular substances, though none of these antibodies appears to effectively immunize a person against reinfection for long periods (except for the antitoxin to SSSS toxin). The most powerful defense lies in the phagocytic response by neutrophils and macrophages. When aided by the opsonic action of complement, phagocytosis effectively disposes of staphylococci that have gained access to lymphatics and other tissues. Inflammation and cell-mediated immunity that stimulate abscess formation also help contain staphylococci and prevent their further spread.

The Other Staphylococci

Because they lack the enzyme coagulase, other species in the genus *Staphylococcus* are called the **coagulase-negative staphylococci.** Several species are included in this group, some of human and others of nonhuman, mammalian origin. Although this group was once considered clinically insignificant, its importance has greatly increased over the past 20 years. Coagulase-negative staphylococci currently account for a large proportion of nosocomial and opportunistic infections in immunocompromised patients.

The normal habitat of ***Staphylococcus epidermidis*** is the skin, hair follicles, and mucous membranes. From this site, the bacterium is positioned to enter breaks in the protective skin barrier. Infections usually occur after surgical procedures, such as insertion of shunts, catheters, and various prosthetic devices that require incisions through the skin. It appears that these inert objects provide a substrate for the colonization of bacteria and support their survival and proliferation. It is particularly perturbing that the rate of these infections has increased due to technologic advances in maintaining patients. Although *Staphylococcus epidermidis* is not as invasive or toxic as *S. aureus,* it can also cause endocarditis, bacteremia, and urinary tract infections.

Less is known about the distribution and epidemiology of ***S. saprophyticus.*** It is an infrequent resident of the skin, lower intestinal tract, and vagina, but its primary importance is in urinary tract infections. For unknown reasons, urinary *S. saprophyticus* infection is found almost exclusively in sexually active adolescent women; in fact, it is the second most common cause of urinary infection in this group.

Identification of *Staphylococcus* in Clinical Samples

Staphylococci are frequently isolated from pus, tissue exudates, sputum, urine, and blood. To prevent inadvertently culturing staphylococci from the environment or from normal residents, great care in collection must be exercised. Primary isolation is achieved by inoculation on sheep or rabbit blood agar or, in heavily contaminated specimens, on selective media such as mannitol salt agar. Because isolation can take several hours, the specimen may be gram stained and observed for irregular clusters of gram-positive cocci. But differentiating among gram-positive cocci is not possible using colonial and morphological characteristics alone, and other presumptive tests are required (see figure D in "Introduction to Medical Bacteriology"). The production of catalase, an enzyme that breaks down the toxic hydrogen peroxide accumulated during oxidative metabolism, can be used to differentiate the staphylococci that produce it from the streptococci that do not (figure 15.6*a*). The property of *Staphylococcus* to grow anaerobically and to ferment sugars separates it from *Micrococcus,* a nonpathogenic genus that is a common specimen contaminant.

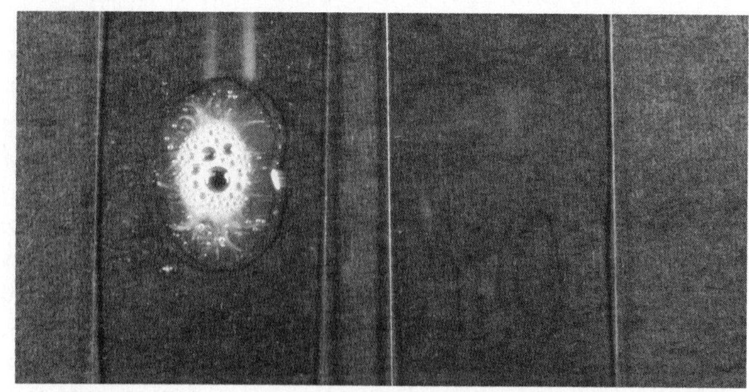
(a)

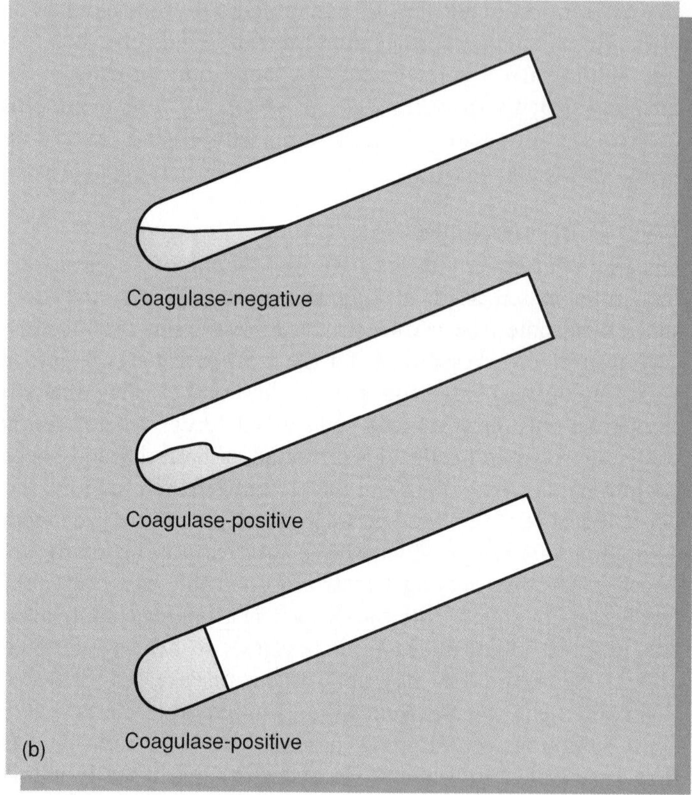

(b)

Figure 15.6 Tests for differentiating the genus *Staphylococcus* and the species *aureus.* (*a*) In the slide catalase test, adding 3% solution of hydrogen peroxide to a small sample of isolate will cause the sample to bubble vigorously if catalase is present and remain undisturbed if catalase is lacking. (*b*) Staphylococcal coagulase is an enzyme that reacts with factors in plasma to initiate fibrin or clot formation. If a tube of plasma inoculated with an isolate remains liquid and flows, the test is negative. If the plasma develops a lump or becomes completely clotted, the test is positive.

One key technique for separating *S. aureus* from other species of *Staphylococcus* is the coagulase test (figure 15.6*b*).

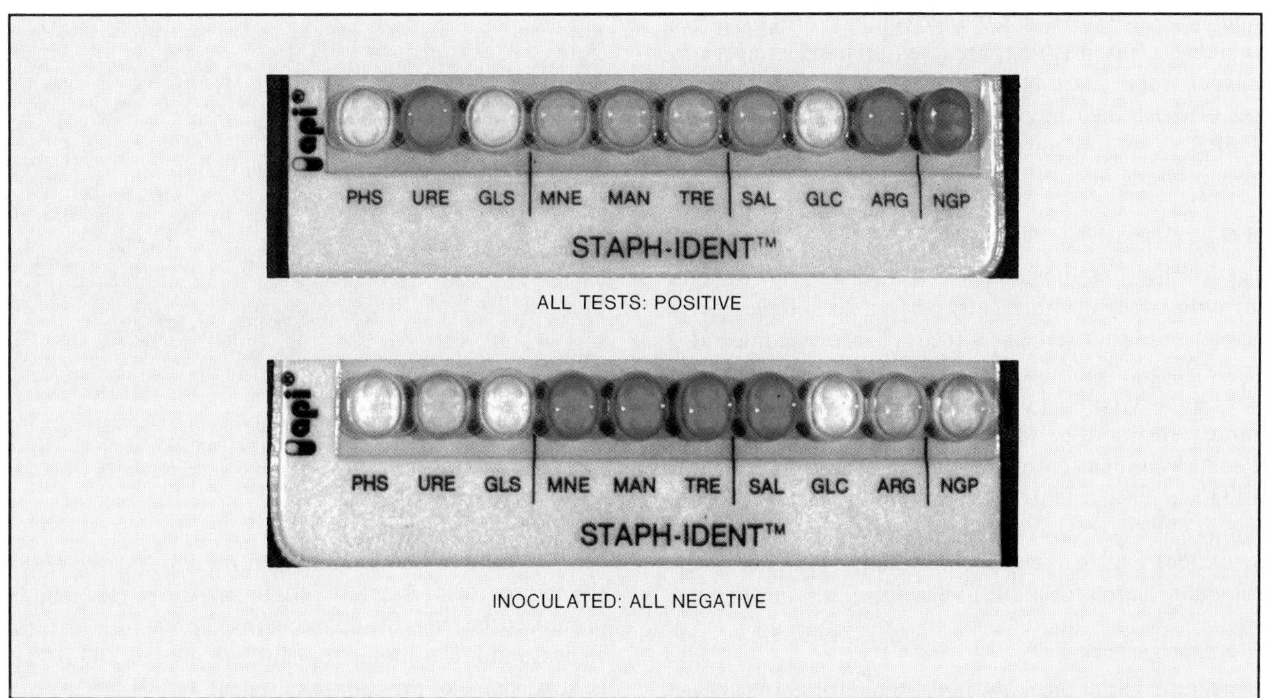

Figure 15.7 Miniaturized test system that fine-tunes the identification of *Staphylococcus* isolates. Some authors identify 19 species with this system. The cupules contain substrates that detect phosphatase production (PHS), urea hydrolysis (URE), glucosidase production (GLS), mannose fermentation (MNE), mannitol fermentation (MAN), trehalose fermentation (TRE), salicin fermentation (SAL), glucuronidase production (GLC), arginine hydrolysis (ARG), and galactosidase production (NGP).

Table 15.1 Separation of Clinically Important Species of *Staphylococcus*

Test	*S. aureus*	*S. epidermidis*	*S. saprophyticus*
Coagulase*	+**	−	−
Growth in Anaerobic Media	+	+	W
Mannitol Fermentation			
Aerobic	+	W	W
Anaerobic*	+	−	−
β-hemolysis by α Toxin	+	−	−
Produces DNase and RNase*	+	−	−
Sensitive to Lysostaphin*	+	−	−
Susceptible to Novobiocin*	+	+	−
Pathogenicity	Primary	Opportunistic	Opportunistic

*Tests most useful in differentiation.
**A few strains test negative for this.
W = May occur, but weakly.

By definition, any isolate that coagulates plasma is *S. aureus,* and all others are coagulase-negative. Rapid multitest systems are used routinely to collect other physiological information (figure 15.7). Table 15.1 lists some of the characteristics that further separate the three species most often involved in human infections. Differentiating the two common coagulase-negative staphylococci from one another relies primarily upon the novobiocin resistance of *S. saprophyticus* and the anaerobic fermentation of glucose by *S. epidermidis.*

Clinical Concerns in Staphylococcal Infections

The extensive distribution and resistance of staphylococci continue to defy attempts at medical control. Most strains of *S. aureus* have acquired genes for penicillinase, which makes them resistant to penicillin and ampicillin. Various strains demonstrate tolerance to erythromycin, tetracycline, chloramphenicol, aminoglycosides (gentamicin, kanamycin), cephalosporins, and even methicillin and oxacillin that are usually unaffected by penicillinase. Since resistance to such common drugs is likely, antimicrobial susceptibility tests are essential in selecting a correct therapeutic agent (see chapter 10). Strains with multiple resistance, often termed methicillin-resistant *Staphylococcus aureus* (MRSA), may require management with vancomycin, quinolones, or clindamycin.

Treatment of Staph Infections

Clinical experience has shown that abscesses will not respond to therapy unless they are surgically perforated and cleared of pus. Foreign bodies that serve as a focus of infection must also be removed. Severe systemic conditions such as endocarditis, septicemia, osteomyelitis, pneumonia, and toxemia respond slowly and require intensive, lengthy therapy by oral or injected drugs or some combination of the two. Attempts at immunization using bacteria or toxoids have proved largely unsuccessful, and in some cases, have even caused damage. No current recommendations exist for any form of staphylococcal vaccination, though research for a useful vaccine is still in progress.

Prevention of Staph Infections

The principal reservoir of the pathogenic staphylococci can never be eliminated. As long as there are humans, there will be carriers and infections. It is difficult to block the colonization of the human body, but one can minimize the opportunity for infection by careful hygiene and adequate cleansing of wounds and burns. Because the incidence of pathogenic staphylococci increases in the hospital, a major concern is reduction of nosocomial infections and epidemics, especially among persons at greatest risk—surgical patients and neonates. Patients or hospital personnel who are asymptomatic carriers or have frank infections constitute the most common source. The burden of effective deterrence again falls to meticulous handwashing, proper disposal of infectious dressings and discharges, isolation of persons with open lesions, and attention to indwelling catheters and needles.

When possible, other preventative measures may be instituted. For example, hospital workers who are known nasal carriers of *S. aureus* may be barred from nurseries, operating rooms, and delivery rooms. Occasionally, carriers are treated for several months with a combination of antimicrobic drugs (rifampin and dicloxacillin, for example). One curious and startling experimental technique is *interference therapy,* the deliberate insertion of a small inoculum of a less virulent strain of *S. aureus* into the nasal mucosa or umbilicus of a newborn. If successful, this procedure promotes colonization by this strain and inhibits establishment of more virulent staphylococci, but it must be used with caution.

General Characteristics of the Streptococci and Relatives

The genus ***Streptococcus*** includes a large and varied group of bacteria, some of which are normal residents or agents of disease in humans and animals and others that are free-living in the environment. A notable characteristic of this genus are cocci organized in long, beadlike chains. The length of these chains varies, and it is common to find them in pairs (figure 15.8). The general shape of the cells is spherical, but they may also appear ovoid or rodlike, especially in actively dividing young cultures.

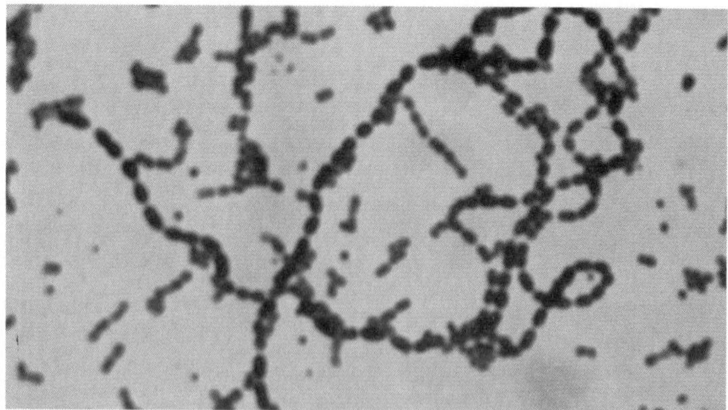

Figure 15.8 One of the truly great sights in microbiology, a freshly isolated *Streptococcus* with its long, intertwining chains. These occur only in liquid media and are especially well-developed if nutrients are in limited supply (×1,000).

Streptococci are non-spore-forming and nonmotile (except for an occasional flagellated strain), and they may be encapsulated. They are facultative anaerobes that ferment a variety of sugars, usually with the production of lactic acid (homofermentative). Streptococci are catalase-negative, but they do have a peroxidase system for inactivating hydrogen peroxide, which is why they survive even in the presence of oxygen. Most parasitic forms have elaborate nutritional requirements, including enriched media for cultivation. Colonies are usually small, nonpigmented, and glistening. Most members of the genus are quite sensitive to drying, heat, and disinfectants, and seldom develop drug resistance, though pneumococci and enterococci are exceptions.

Species of *Streptococcus* have traditionally been classified according to a system developed by Rebecca Lancefield in the 1930s. She discovered that the cell wall carbohydrates (antigens) of various cultures stimulated formation of antibodies with differing specificities. She characterized these different groups using an alphabetic system (A, B, C, . . .). Another convenient division for species is based on their reaction in blood agar (figure 15.9). Those producing a zone of β-hemolysis on sheep blood agar are members of Lancefield groups A, B, C, and certain strains of D, and those producing α-hemolysis are *Streptococcus pneumoniae* and the *viridans* streptococci. Several species of nonhemolytic cocci exist, but they are usually of little clinical significance. Table 15.2 summarizes the streptococcal groups, their habitats, and their pathogenicity.

Despite the large number of streptococcal species, human disease is most often associated with ***S. pyogenes, S. agalactiae,*** the **viridans streptococci,** ***S. pneumoniae,*** and ***Enterococcus***

Streptococcus Gr. *streptos,* winding, twisted. The chain arrangement is the result of division in only one plane.

viridans (vih'-rih-denz) L. *viridis,* green. This term comes from the green color that develops around colonies that produce α-hemolysis.

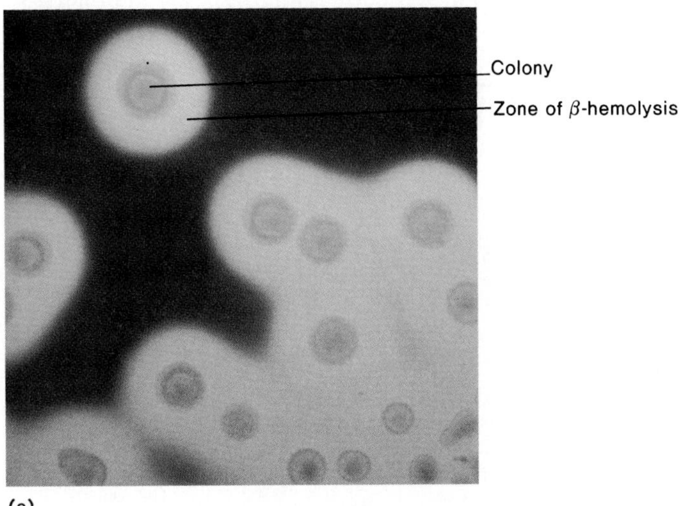

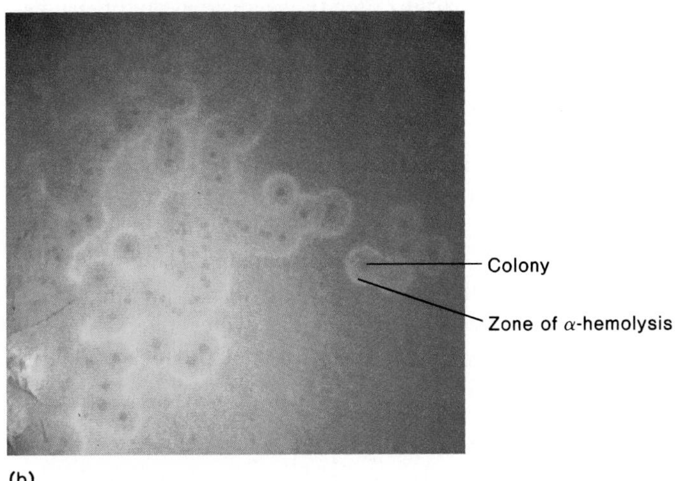

Figure 15.9 Hemolysis patterns on blood agar may be used in preliminary identification of streptococcal groups. (*a*) Colonies of *S. pyogenes* with β-hemolysis, in which RBCs are completely lysed. (*b*) Colonies of *S. pneumoniae* demonstrate the greening of blood agar typical of α-hemolysis.

Table 15.2 Major Species of *Streptococcus* and Related Genera

Species	Lancefield Group	Habitat	Pathogenicity to Humans
S. pyogenes	**A**	**Human throat**	**Skin, throat infections**
S. agalactiae	**B**	**Human vagina, cow udder**	**Neonatal, wound infections**
S. equisimilis	C	Swine, cows, horses	Pharyngitis, endocarditis
S. equi	C	Various mammals	Rare, in abscesses
S. zooepidemicus	C	Various mammals	Rare, in abscesses
S. dysgalactiae	C	Cattle	Rare
Enterococcus faecalis	**D**	**Human, animal intestine**	**Endocarditis, UTI***
E. faecium	D	Human, animal intestine	Similar to *E. faecalis*
E. durans	D	Human, animal intestine	Similar to *E. faecalis*
S. bovis	D	Cattle	Subacute endocarditis, bacteremia
S. equinus	D	Horses	Rare
S. anginosus	F, G, L	Humans, dogs	Endocarditis, URT** infections
S. sanguis	**H**	**Human oral cavity**	**Endocarditis, dental caries**
S. salivarius	K	Human saliva	Endocarditis
Lactococcus lactis	N	Dairy products	Very rare
L. cremoris	N	Dairy products	Not reported
S. suis	R	Swine	Rare, meningitis
S. mutans	**NI*****	**Human oral cavity**	**Dental caries**
S. uberis	NI	Domestic mammals	Rare
S. mitis	O, M	Human oral cavity	Tooth abscess, endocarditis
S. milleri	F	URT	Endocarditis, organ abscess
S. acidominimus	NI	Domestic mammals	Rare
S. pneumoniae	**NI**	**Human URT**	**Bacterial pneumonia**

*Urinary tract infection.
**Upper respiratory tract.
***No group C carbohydrate identified.
Note: Species in bold type are the most significant sources of human infection and disease.

faecalis. The latter species was formerly called *Streptococcus faecalis* until DNA hybridization tests indicated that it is a distinct genus.

β-Hemolytic Streptococci: *Streptococcus pyogenes*

By far the most serious streptococcal pathogen of humans is *Streptococcus pyogenes,* the only representative of group A. It is a relatively strict parasite, inhabiting the throat, nasopharynx, and occasionally the skin of humans. The involvement of this species in severe disease is partly due to the substantial array of surface antigens, toxins, and enzymes it generates, although even highly virulent strains do not produce all of the toxins and enzymes.

Cell Surface Antigens and Virulence Factors

A capsule is formed by most *S. pyogenes* strains, but it remains attached to the cell surface only during the rapid growth phase of a culture and is probably not as important to virulence as other factors (figure 15.10). C-carbohydrates, specialized polysaccharides or techoic acids found on the surface of the cell wall, are the basis for Lancefield groups. Apparently, their role in pathogenesis is to protect the bacterium from being dissolved by the lysozyme defense of the host. One cell surface molecule that accounts for the adherence of *S. pyogenes* to epithelial cells in the skin or pharynx is lipotechoic acid (LTA). Another type-specific molecule is the *M-protein,* of which about 80 different subtypes exist. This substance is the main component of fimbriae, the spiky surface projections seen in electron micrographs. The M-protein contributes to virulence by being both an antiphagocytic and adherence factor.

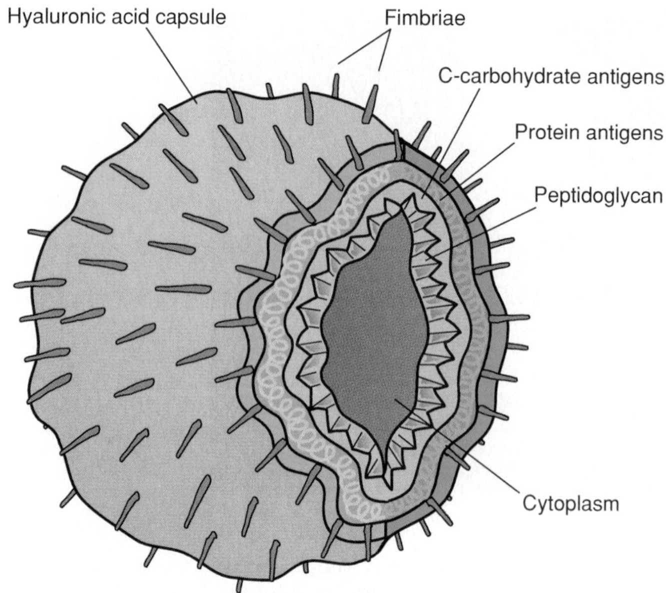

Figure 15.10 Cutaway view of group A *Streptococcus.* The outermost fringe consists of fimbriae composed in part of M-protein, a group-specific substance. Other layers making up the cell envelope are the capsule, the protein antigens, and the C-carbohydrate antigens.

Major Extracellular Toxins Group A streptococci owe several pathologic properties to the effects of hemolysins called **streptolysins.** The two types are streptolysin O (SLO) and streptolysin S (SLS).[2] Although both types cause β-hemolysis of sheep blood agar, SLO produces hemolysis only in deep (anaerobic) colonies, while SLS produces surface (and consequently, major) hemolysis. In the body, SLO is rapidly injurious to the cell membranes of many cell types, and it probably accounts for some of the damage in rheumatic fever. SLS has multiple cytotoxic effects on leukocytes, liver, and heart muscle.

A key toxin in the development of scarlet fever (discussed in a later section of this chapter) is the **erythrogenic** or **pyrogenic toxin,** names that underline its biologic activities. This toxin is responsible for the bright red rash typical of this disease, and it also induces fever by acting upon the temperature regulatory center. Only lysogenic strains of *S. pyogenes* that contain genes from a temperate bacteriophage can synthesize this toxin.

Major Extracellular Enzymes Several enzymes for digesting macromolecules are given off by the group A streptococci, though their role in pathogenesis is somewhat obscure. Streptokinase, similar to staphylokinase, activates a pathway leading to the digestion of fibrin clots and may play a role in invasion. Hyaluronidase breaks down the intercellular glue and promotes spreading of the pathogen into the tissues, while streptodornase (DNase) liquefies purulent discharges by hydrolyzing DNA.

Epidemiology and Pathogenesis of *S. pyogenes*

Streptococcus pyogenes has traditionally been linked with a diverse spectrum of infection and disease. Before the era of antibiotics, it accounted for a major portion of serious human infections and deaths from such diseases as rheumatic fever and puerperal sepsis. Although its importance has diminished in the past 50 years, recent epidemics have reminded us of its potential for sudden and serious illness (see feature 15.2).

Humans are the only significant reservoir for *S. pyogenes.* Healthy or subclinical carriers of virulent strains comprise about 5–15% of the population. The infection is generally transmitted through direct contact, droplets, and occasionally through food or fomites. The bacteria invade at periods of lowered host resistance, primarily through the skin and pharynx. The incidence and types of infections are altered by climate, season, and living conditions. Skin infections occur more frequently during the warm temperatures of summer and fall, and pharyngeal infections increase in the winter months. Children 5–15 years of age are the predominant group affected by both types of infections. In addition to local cutaneous and throat infections, *S. pyogenes* can give rise to a variety of systemic infections and progressive sequelae if not properly treated.

2. In SLO, "O" stands for oxygen because the substance is inactivated in oxygen. SLO is produced by most strains of *S. pyogenes.* In SLS, "S" stands for serum because the substance has an affinity for serum proteins. SLS is oxygen-stable.
erythrogenic (eh-rith″-roh-jen′-ik) Gr. *erythros,* red, and *gennan,* to produce.

Feature 15.2 *Streptococcus pyogenes*— The Return of a Killer

The winter of my eighth year, I came down with a case of scarlet fever that was soon complicated by rheumatic fever. I remember lying in bed in a totally darkened room, burning with fever, my mind hallucinating with strange voices, and my ears ringing. I recall even today the angry red flush of my skin and later, the aching joints and odd heart palpitations that felt like a butterfly was loose in my chest. Most young people now have never heard of scarlet fever or rheumatic fever, let alone experienced it, yet those diseases are still very much with us. Although their incidence has greatly declined during the era of antibiotics, in 1985 clusters of cases began appearing again for reasons not completely understood. The bacteria isolated from outbreaks in Utah, Montana, Nevada, Idaho, Pennsylvania, and Ohio appeared to be a new, more virulent strain that causes a disease similar to toxic shock syndrome. This disease was responsible for the death of muppet creator Jim Henson. An increased number of other *S. pyogenes* ailments have been reported, including strep throat, skin infections, and pneumonia. The resurgence of these diseases has provided another humbling lesson on the unpredictable nature of the host-parasite relationship.

Skin Infections When virulent streptococci invade a nick in the skin or the mucous membranes of the throat, an inflammatory primary lesion is produced. Pyogenic infections appearing after local invasion of the skin are pyoderma or erysipelas, while those developing in the throat are pharyngitis or tonsillitis.

Pyoderma, or streptococcal impetigo, is marked by burning, itchy papules that break and form a highly contagious yellow crust (figure 15.11*a*). Impetigo often occurs in epidemics among school children, and it is also associated with insect bites, poor hygiene, and crowded living conditions. A slightly more invasive form of skin infection is **erysipelas.** The pathogen usually enters through a small wound or incision on the face or extremities and eventually spreads to the dermis and subcutaneous tissues. Early symptoms are edema and redness of the skin near the portal of entry, fever, and chills. The lesion begins to spread outward, producing a slightly elevated edge that is noticeably red, hot, and often vesicular (figure 15.11*b*). Depending on the depth of the lesion and how the infection progresses, cutaneous lesions may remain superficial or produce long-term systemic complications. Severe cases involving large areas of skin are occasionally fatal.

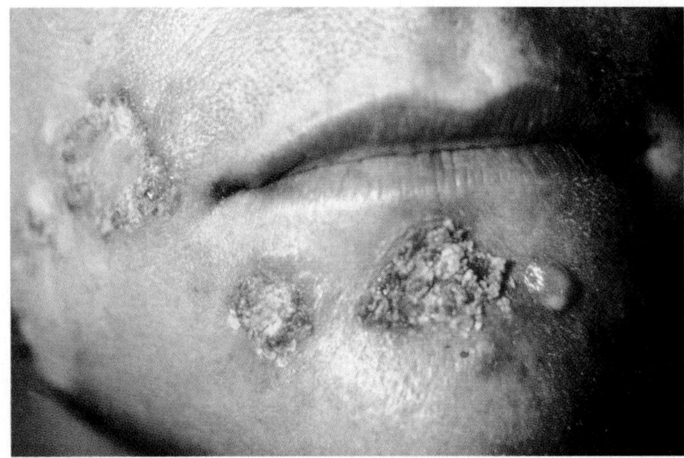

(a)

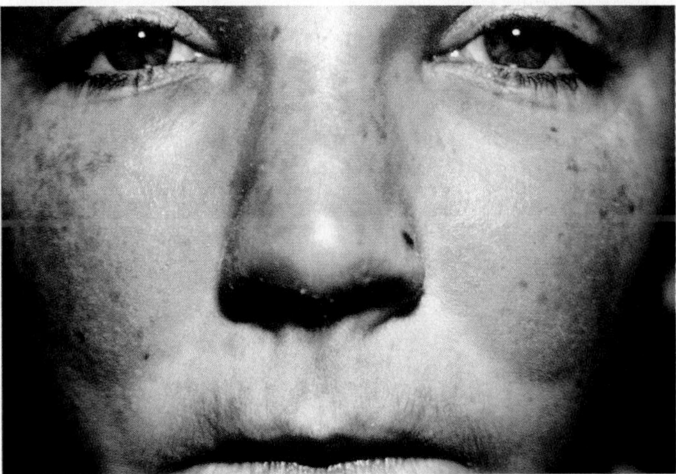

(b)

Figure 15.11 Streptococcal skin infections. (*a*) Impetigo lesions on the face. (*b*) Erysipelas of the face. Although it is a superficial infection, the inflammatory reaction spreads horizontally from the initial point of entry. Tender, red, puffy lesions have a sharp border that may burst and release fluid.

Most people associate streptococci with the condition called strep throat or, more technically, **streptococcal pharyngitis** (tonsillitis). Estimates indicate that most humans will acquire this particular infection at some time. The organism multiplies in the tonsils or pharyngeal mucous membranes, causing redness, edema, enlargement, and extreme tenderness that makes swallowing difficult and painful (figure 15.12). These symptoms may be accompanied by fever, headache, nausea, and abdominal pain. Other signs include a *purulent* exudate over the tonsils, swollen cervical lymph nodes, and occasionally white, pus-filled nodules on the tonsils.

Throat infection may lead to **scarlet fever** (scarlatina) when it involves a strain of *S. pyogenes* carrying a prophage that codes for erythrogenic toxin. Systemic spread of this toxin results in high fever and a bright red, diffuse rash over the face, trunk, inner arms and legs, and even the tongue. Within 10 days, the rash and fever usually disappear, often accompanied by desquamation (sloughing) of the epidermis. Many cases of pharyngitis and even scarlet fever are mild and uncomplicated, but on occasion, they elicit an inflammatory reaction that leads to severe sequelae.

erysipelas (er″-ih-sip′-eh-las) Gr. *erythros,* red, and *pella,* skin.

purulent (puh′-roo-lent) L. *purulentus,* inflammation. In reference to pus.

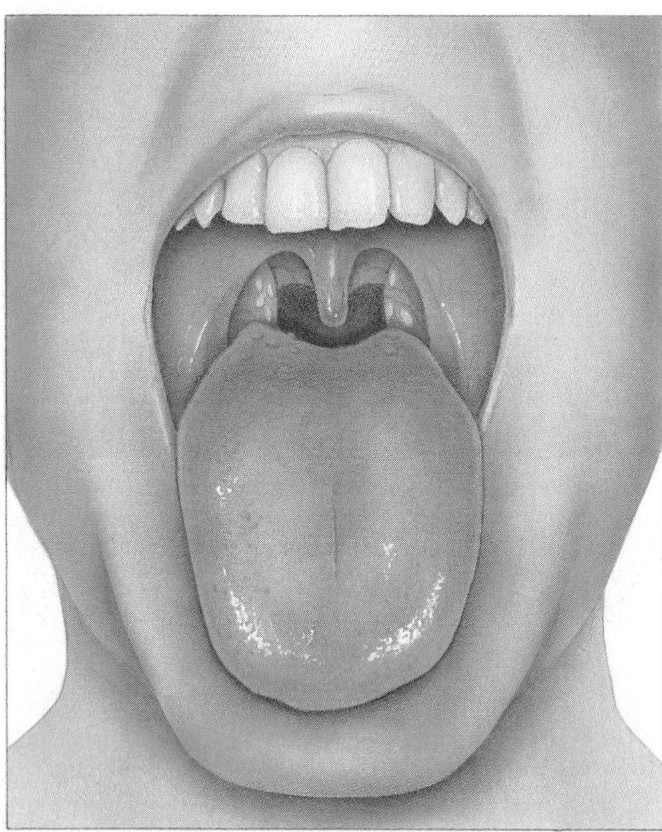

Figure 15.12 The appearance of the throat in pharyngitis and tonsillitis. The tongue, pharynx, and tonsils (if present) become bright red and suppurative. Whitish pus nodules may also appear on the tonsils.

Systemic Infections The dissemination of streptococci into the lymphatics and the blood may give rise to septicemia, an infrequent complication limited to persons who are already weakened by underlying disease. Another relatively rare infection is *S. pyogenes* pneumonia, which accounts for less than 5% of bacterial pneumonias and is usually secondary to influenza or some other pulmonary weakness. At one time, a major cause of maternal death was childbed or *puerperal* fever caused by *S. pyogenes* introduced into the vagina by contaminated hands. From this site, it spread rapidly into uterine tissue disrupted by childbirth and caused massive abdominal sepsis. Modern obstetric techniques and antimicrobic therapy have greatly reduced this cause of postpartum disease. Presently, most uterine infections occurring after childbirth are due to normal vaginal flora and not exogenous pathogens.

Long-Term Complications of Group A Infections

The principal sequelae that appear within a few weeks after group A streptococcal infections are (1) **rheumatic fever (RF)**, a delayed inflammatory condition of the joints, heart, and subcutaneous tissues, and (2) **acute gomerulonephritis (AGN),** a disease of the kidney glomerulus and tubular epithelia. The actual mechanisms for these streptococcal sequelae have yet to be completely explained, though several theories have been proposed. It is possible that tissue injury in both diseases is tied directly to streptococcal invasion of tissues or to the action of toxins such as SLO or SLS. The theory currently favored by many experts, however, indicates an autoimmune response to various streptococcal antigens. According to this theory, antibodies formed against the bacterial cell wall and membranes cross-react with molecules in the host that are similar to these bacterial antigens. In the heart, these antibody-antigen reactions trigger inflammation and injury of cardiac tissues, and in the kidney, immune complexes deposited in the epithelia inflame and damage the filtering apparatus (see chapter 14).

Rheumatic fever usually follows an overt or subclinical case of streptococcal pharyngitis or tonsillitis in children. Its major clinical features are carditis, an abnormal electrocardiogram, painful arthritis, chorea, nodules under the skin, and fever.[3] The course of the syndrome extends from three to six months, usually without lasting damage. In patients with severe carditis, however, extensive damage to the heart valves may occur (figure 15.13). Although the degree of permanent damage does not usually reveal itself until middle age, it is often extensive enough to require the replacement of damaged valves. Heart disease due to RF is on the wane because of medical advances in diagnosing and treating streptococcal infections, but it is still responsible for about 15,000 deaths per year in the United States and many times that number in the rest of the world.

In AGN, the kidney cells are so damaged that they cannot adequately filter blood. The first symptoms are nephritis (appearing as swelling in the hands and feet and low urine output), increased blood pressure, and occasionally heart failure. Urine samples are extremely abnormal, with high levels of red and white blood cells and protein. AGN may clear up spontaneously, may become chronic and lead to kidney failure later, or may be immediately fatal.

Although many types of antibodies are produced in response to streptococcal antigens, only two give long-term protection. One is the type-specific antibody produced in response to the M-protein, though it will not prevent infections with a different strain (which explains why one may have recurring cases of strep throat). The other is a neutralizing antitoxin against the erythrogenic toxin that prevents the fever and rash of scarlet fever. Serological tests can reveal anti-streptococcal antibodies in the serum within one to two weeks after infection, but this occurs too late for prompt diagnosis and treatment. Such tests are used primarily to determine whether kidney, heart, and joint disease is of streptococcal origin, to detect continuing or recent infection, and to judge a patient's susceptibility to rheumatic fever and glomerulonephritis.

---

puerperal (poo-er′-per-al) L. *puer*, child, and *parere*, to bear. From puerperium, the period of confinement after labor and birth.

rheumatic (roo-mat′-ik) Gr. *rheuma*, flux. Involving inflammation of joints, muscles, and connective tissues.

3. Carditis is inflammation of heart tissues; chorea is a nervous disorder characterized by involuntary, jerky movements.

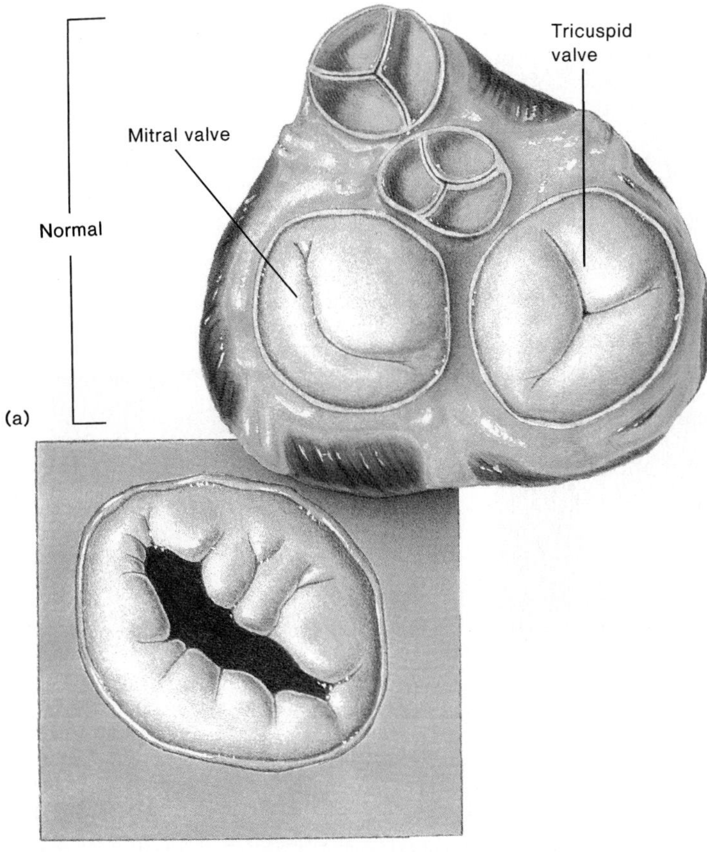

Figure 15.13 The cardiac complications of rheumatic fever. Pathologic processes of group A streptococcal infection can extend to the heart. In this example, it is believed that cross reactions between streptococcal-induced antibodies and heart proteins have a gradual destructive effect on the atrioventricular (especially the mitral) or semilunar valves. Scarring and deformation change the capacity of the valves to close and shunt the blood properly. (*a*) Normal valves, viewed from above. (*b*) Inset reveals scar tissue on a damaged mitral valve.

***In Vivo* Tests for Scarlet Fever** Immunity to the erythrogenic toxin is the basis of two tests that were formerly used in clinical assessment of scarlet fever. The Dick test determines the patient's susceptibility and degree of immunity to the erythrogenic toxin by injecting a small amount of the toxin into the skin. If a rash develops at that site, there is no immunity to the toxin, but if the skin shows no reaction, there are circulating antitoxins and thus immunity to the toxin. The counterpart to this is the Shultz-Carlton test used to diagnose a preexisting rash. The technique involves injecting tiny samples of erythrogenic antitoxin into an area of rash. If the redness fades at that site within a few hours, the rash is scarlet fever; if the rash does not fade, some other disease is indicated. Although newer methods of clinical diagnosis have largely replaced these *in vivo* tests, the principles behind them are still very useful.

Group B: *Streptococcus agalactiae*

Several other species of β-hemolytic *Streptococcus* in groups B, C, and D live among the normal flora of humans and other mammals, and may be isolated in clinical specimens from diseased human tissue. The group B streptococci (GBS), represented by the species *S. agalactiae,* demonstrate clearly how the distribution of a parasite can change in a relatively short time. Within the past several years, this species has shifted from its primary habitat on cows' udders, where it causes mastitis,[4] to regular residence in the human vagina, pharynx, and large intestine. The major results of this association with humans have been a sudden increase in serious infections in newborns and compromised persons.

Streptococcus agalactiae has been chiefly implicated in neonatal, puerperal, wound, and skin infections, and in endocarditis. Persons suffering from diabetes and vascular disease are particularly susceptible to wound infections. Because of its location in the vagina, GBS can be transferred to the infant during delivery, sometimes with dire consequences. In some newborns, these bacteria become part of the normal flora without infection, but infants with inadequate phagocytic defenses or deficient passive immunity are vulnerable to two forms of clinical disease. The early-onset type that develops a few days after birth is accompanied by sepsis, pneumonia, and high mortality (50%). The late-onset disease comes on in two to six weeks, with symptoms of meningitis—fever, vomiting, and seizures. Because most cases occur in the hospital, personnel must be aware of the risk of passively transmitting this pathogen, especially in the neonatal and surgical units.

Group D: Enterococci and Streptococci

Enterococcus faecalis, E. faecium, and *E. durans* are called the enterococci because they are normal colonists of the human large intestine. Two other members of group D, *S. bovis* and *S. equinus,* are nonenterococci that colonize other animals and occasionally humans. Infections caused by *Enterococcus faecalis* arise most often in elderly patients undergoing surgery and affect the urinary tract, wounds, blood, the endocardium, the appendix, and other intestinal structures. Infections by nonenterococci include bacteremia and endocarditis, especially in patients with gastrointestinal tumors.

Laboratory Identification Techniques

The failure to recognize group A streptococcal infections (even very mild ones) can have devastating effects. Thus, there is pressure for rapid cultivation and diagnostic techniques that will ensure proper treatment and prevention measures. Several companies have developed rapid diagnostic test kits to be used in clinics or offices to detect group A streptococci from pharyngeal swab samples. These tests, based on monoclonal antibodies that react with the C-carbohydrate of group A, are highly specific and sensitive (figure 15.14).

Complete identification usually necessitates cultivating a specimen on sheep blood agar plates and occasionally enrichment media. Group A streptococci are by far the most common β-hemolytic isolates in human lesions, but lately an increased number of infections by group B streptococci and the existence

4. Inflammation of the mammary gland.

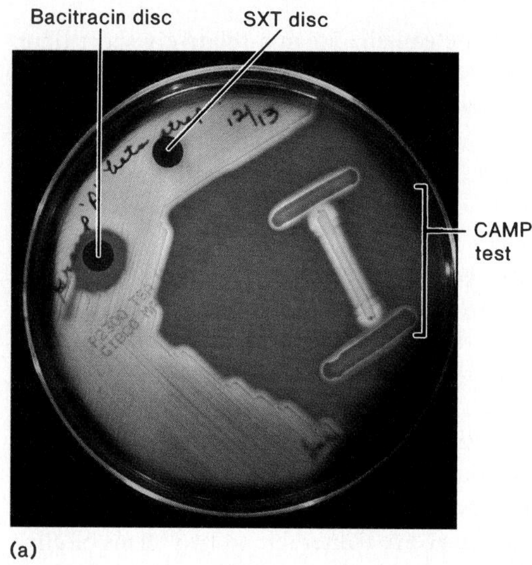

(a)

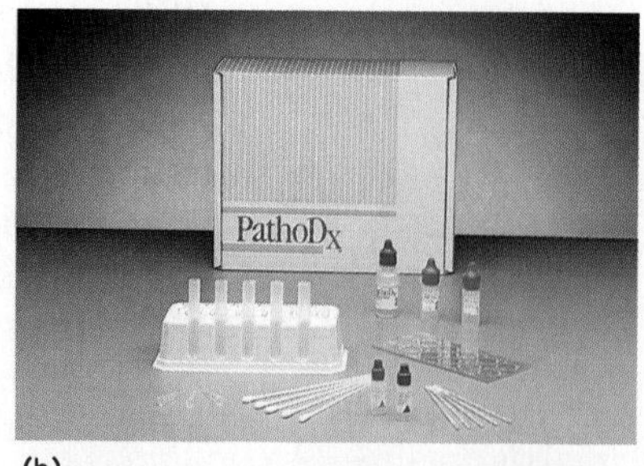

(b)

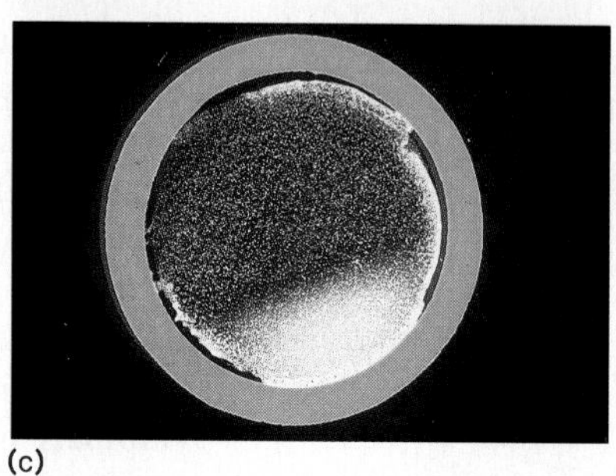

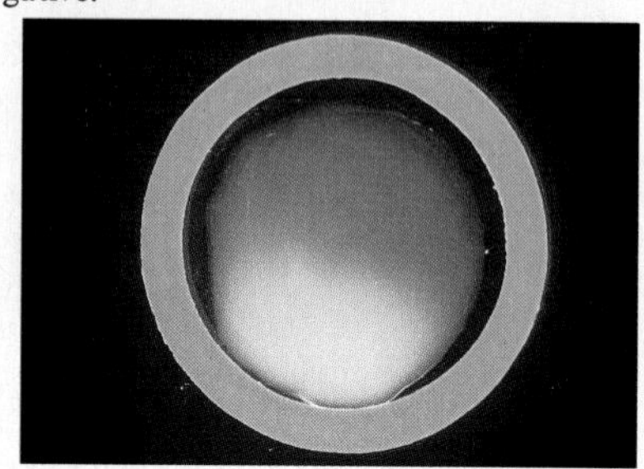

(c)

Figure 15.14 Streptococcal tests. (*a*) Bacitracin disc test. With very few exceptions, only *S. pyogenes* is sensitive to a minute concentration (0.02 μg) of bacitracin. Any zone of inhibition around the B disc is interpreted as a presumptive indication of this species. (Note: Group A streptococci are negative for SXT sensitivity and the CAMP test.) (*b*) A rapid, direct test kit for diagnosis of group A infections. (*c*) With this method, a patient's throat swab is introduced into a system composed of latex beads and monoclonal antibodies. Left: In a positive reaction, the C-carbohydrate on group A streptococci produces visible clumps. Right: A smooth, milky reaction is negative.

Table 15.3 Scheme for Differentiating β-Hemolytic Streptococci

Characteristic	Group A (*S. pyogenes*)	Group B (*S. agalactiae*)	Group C (*S. equisimilis*)	Group D *Enterococcus*
Bacitracin Sensitivity	+	−	−	−
CAMP Factor*	−	+	−	−
Esculin Hydrolysis in 40% Bile**	−	−	−	+
SXT Sensitivity***	−	−	+	−
Growth at 45° C	−	−	−	+
Growth in 6.5 % Salt	−	+	−	+

*Name is derived from the first letters of its discoverers. CAMP is a diffusable substance of group B, which lyses sheep red blood cells in the presence of staphylococcal hemolysin.
**A sugar that can be split into glucose and esculetin by the group D streptococci.
***Sulfa and trimethoprim. The test is performed (like bacitracin) with discs containing this combination drug.

of β-hemolytic enterococci have made it important to use differentiation tests. (Groups C and G are also β-hemolytic, but they are most often associated with infections in other mammals and only infrequently found in humans.) A positive bacitracin disc test (figure 15.14*a*) provides important evidence of group A. Other important characteristics that separate the various β-hemolytic streptococci are presented in table 15.3.

Group B streptococci are differentiated from groups A and D by the CAMP reaction and hippurate hydrolysis, both of which are positive for group B and negative for groups A and D (figure 15.15). *Enterococcus faecalis* (the predominant enterococcus) differs from the streptococci in its resistance to heat, 6.5% salt, and low concentrations of penicillin G. For these reasons, certain strains can be confused with *S. aureus,* but a Gram stain and catalase test would help differentiate these two. Because *E. faecalis* is sometimes found in the pharynx, strains may also be mistaken for *S. pyogenes,* but the ability of *E. faecalis* to hydrolyze esculin in 40% bile salt and its resistance to bacitracin help distinguish it. Not all of the streptococci can be typed with the Lancefield serological system, and it is not used for routine identification methods.

Treatment and Prevention of Streptococcal Infections

Antimicrobic therapy is aimed at curing infection and preventing complications. All strains of *S. pyogenes* continue to be very sensitive to penicillin or one of its derivatives. In cases of pharyngitis, small children receive an intramuscular injection of 600,000 units of benzathine penicillin to achieve the necessary circulating levels for a period of 10 days; larger children receive 900,000 units, and adults 1.2 million units. An alternate therapy often used for impetigo is oral penicillin V taken for 10

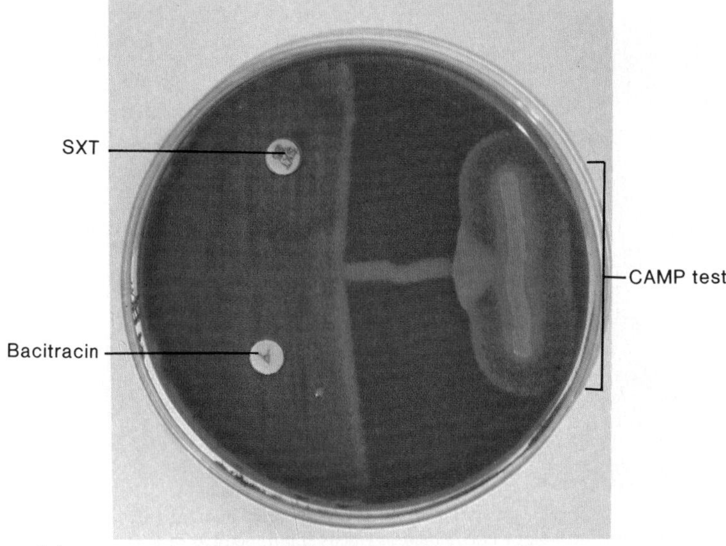

(a)

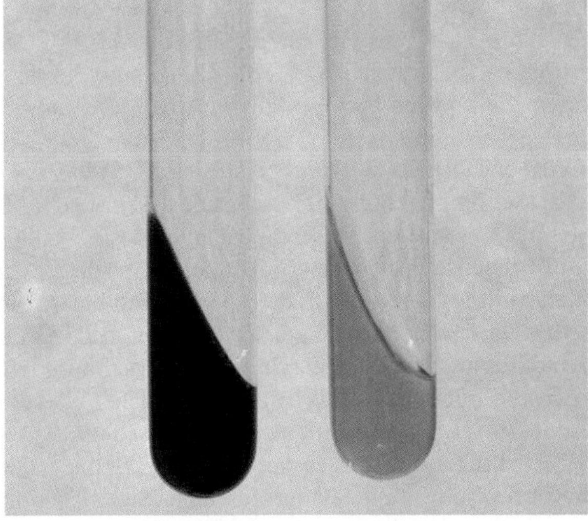
(b)

Figure 15.15 Tests for characterizing other β-hemolytic streptococci. (*a*) In the CAMP test, a special strain of β-hemolytic *S. aureus* that reacts synergistically with group B *Streptococcus* (GBS) is streaked between two smears of the sample. If the isolate is GBS, increased areas of hemolysis appear where the smears meet. (Note that group B streptococci are negative for bacitracin and SXT tests.) (*b*) Group D streptococci are identified in part by growing in bile and hydrolyzing esculin. One of the by-products of this reaction causes the precipitation of an iron salt that is brown to black in color. Here, the tube on the left is positive, and the one on the right is negative.

days. If a patient is allergic to the penicillins, erythromycin (if the strain is sensitive) or a cephalosporin may be used.

The deeper involvement of pneumonia, erysipelas, and wound infections requires higher dosages and longer treatment. Occasionally, surgery is needed to remove the affected tissue. The only certain way to arrest rheumatic fever or acute glomerulonephritis is to treat the preceding infection, because once these two pathological states have developed, there are no specific treatments. Work aimed at developing a safe vaccine for group A streptococci is still proceeding. The antibodies produced against a particular M-type can protect against further infection and disease by that strain, but the presence of 70–80 different M-types complicates vaccine development. Furthermore, the M-proteins used in vaccines must be separated from other antigens that cause undesirable toxic side effects.

Streptococcal Diseases: When in Doubt—Culture Because streptococcal disease may come on rapidly, it is beneficial to routinely culture throat and skin infections of children. Some physicians recommend that persons with a history of rheumatic fever or recurring strep throat receive continuous, long-term penicillin prophylaxis. Mass prophylaxis has also been indicated for other high-risk groups, primarily young people exposed to epidemics in boarding schools, military camps, and other institutions. Some controversial preventative measures include removing the tonsils of persons with recurrent tonsillitis and treating carriers with antimicrobics. Tonsillectomy has been used extensively in the past, but with questionable effectiveness, because these bacteria can infect other parts of the throat, and it appears that the carrier state is not easy to eradicate simply by drug therapy. In hospitals, known carriers of *S. pyogenes* should not be allowed to work with surgical, obstetric, and immunocompromised patients. Patients with group A infections must be isolated, and high-level precautions must be practiced in handling infectious secretions.

The antibiotic treatment of choice for group B streptococcal infection is penicillin G; alternatives are erythromycin and cephalosporins. Some physicians advocate routine penicillin prophylaxis in colonized mothers and infants, but others fear that this may increase the rate of allergies and infections by resistant strains. Artificial passive immunization with human immunoglobulins is currently being considered for treatment and prevention in high-risk groups. Once enterococcal infection has been verified through testing, it may be effectively treated by combined therapy with high doses of ampicillin and an aminoglycoside such as gentimicin, which work synergistically.

α-Hemolytic Streptococci: The Viridans Group

The viridans category encompasses a large and complex group of human streptococci not entirely groupable by Lancefield serology. They are the most numerous and widespread residents of the oral cavity (gingiva, cheeks, tongue, saliva) and are also found in the nasopharynx, genital tract, and skin. These species can cause serious systemic infections, although most of them are opportunists and lack the full complement of toxins and enzymes that occur in group A. The one characteristic shared by all species (*S. mitis, S. mutans, S. milleri, S. salivarius, S. sanguis*) is α-hemolysis. Other characteristics used to distinguish them are too numerous to include here.

Because viridans streptococci are not highly invasive, their entrance into tissues usually occurs through dental or surgical instrumentation and manipulation. These organisms are constant inhabitants of the gums and teeth, and even chewing hard candy or brushing the teeth can provide a portal of entry for them. Dental procedures can lead to bacteremia, meningitis, abdominal infection, sinusitis, and wound infections. But the most important complication of all viridans streptococcal infections is **subacute endocarditis.** In this condition, blood-borne bacteria settle on areas of the heart lining or valves that have

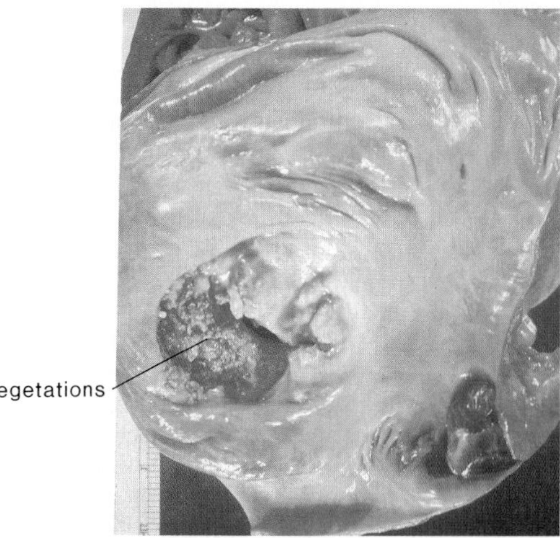

Figure 15.16 A view of the heart in subacute bacterial endocarditis. The tricuspid valve has nodular vegetations on the leaflets that constantly release bacteria into the circulation.

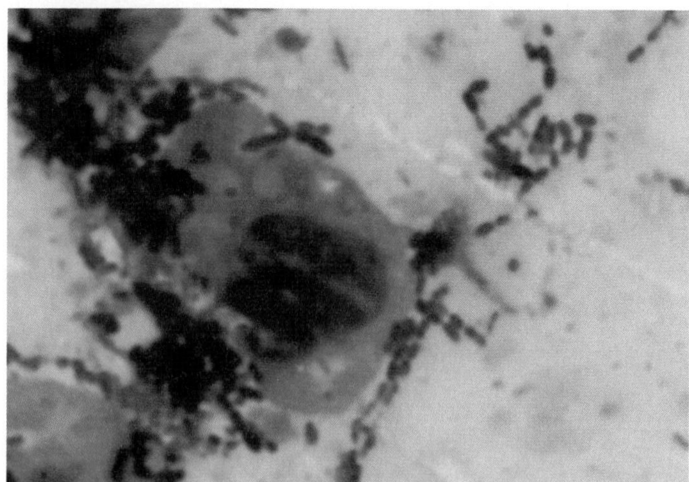

Figure 15.17 This Gram stain of sputum from a pneumonia patient indicates the morphology of *S. pneumoniae* as small, pointed diplococci. Large cells interspersed among the pneumococci are phagocytes (×600).

previously been injured by rheumatic fever, valve surgery, or the like. Colonization of these surfaces leads to tissue deformations called vegetations (figure 15.16). As the disease progresses, vegetations increase in size, constantly releasing masses of bacteria into the circulation. Because of its insidious and somewhat concealed course, endocarditis is considered subacute rather than acute or chronic. Symptoms and signs range from fever, heart murmur, and emboli to weight loss and anemia.

Endocarditis is diagnosed almost exclusively by blood culture, and repeated blood samples positive for bacteremia are highly suggestive of it. The goal in treatment is to completely destroy the microbes in the vegetations, and this can be accomplished by long-term therapy with penicillin G. Because persons with preexisting heart conditions are at high risk for this disease, they usually receive prophylactic antibiotics prior to dental procedures.

Another very common dental disease involving viridans streptococci is dental caries. In the presence of sugar, *S. mutans* and *S. sanguis* produce slime layers made of glucose polymers that adhere tightly to tooth surfaces (see chapter opening illustration). These sticky polysaccharides are the basis for plaque, the sticky white material on teeth that fosters dental diseases (see chapter 17).

Streptococcus pneumoniae: The Pneumococcus

High on the list of significant human pathogens is ***Streptococcus pneumoniae,*** a unique species that was formerly classified as *Diplococcus pneumoniae* until its genetic similarity to the streptococci was demonstrated. Because it causes 80% of all bacterial pneumonias, *S. pneumoniae* is also referred to as the **pneumococcus.** Gram stains of sputum specimens from pneumonia patients reveal small lancet-shaped cells arranged in pairs and short chains (figure 15.17). All pathogenic strains form rather large capsules, this being their major virulence factor. Cultures require complex media such as blood or chocolate agar, and they produce dome-shaped colonies with smooth or mucoid textures and α-hemolysis (see figures 3.12 and 15.9*b*). Growth is improved by the presence of 5–10% CO_2, and cultures die in the presence of oxygen because they lack catalase and peroxidases.

Only encapsulated (smooth) strains of *S. pneumoniae* cause disease; rough strains lack a capsule and are nonvirulent. Capsules help the streptococci escape phagocytosis, which is the major host defense in pyogenic infections. Capsules contain a polysaccharide antigen called the specific soluble substance (SSS) that varies chemically among the pneumococcal types and stimulates antibodies of varying specificity. So far, 84 different capsular types (specified by numerals 1, 2, 3, . . .) have been identified, using a technique called the *Quellung* test or capsular swelling reaction (see figure 15.20). Some strains of pneumococci also contain cell surface molecules that promote adhesion to the nasopharynx, but none of the pneumococcal toxins or enzymes have been linked to the disease process.

Epidemiology and Pathology of the Pneumococcus

From 5% to 50% of all persons carry *S. pneumoniae* as part of the normal flora in the nasopharynx. Although infection is often acquired endogenously from one's own flora, it may occur after direct contact with respiratory secretions or droplets from carriers. *S. pneumoniae* is very delicate and does not survive long out of its habitat. Factors that favor development of pneumonia are (1) old age, (2) the season (rate of infection is higher in the winter), (3) other diseases (persons with underlying lung disease or viral infections have weakened defenses), and (4) living in institutions, which increases the chance of contact with infected persons.

The Pathology of Pneumonia People commonly inhale microorganisms into the respiratory tract without serious consequences due to the host defenses present there. But pneumonia occurs when mucus containing a heavy load of bacterial cells is aspirated from the pharynx into lungs whose defense mechanisms have been temporarily or permanently compromised. If

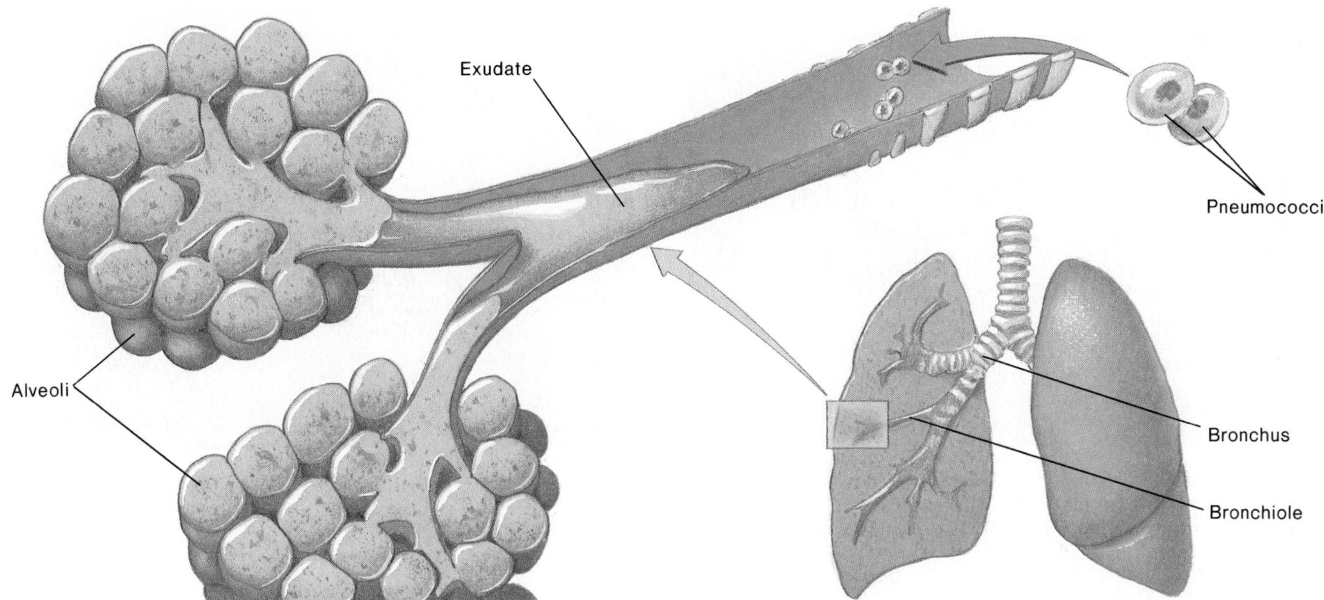

Figure 15.18 The course of bacterial pneumonia. As the pneumococcus traces a pathway down the respiratory tree, it provokes intense inflammation and exudate formation. The blocking of the bronchioles and alveoli by consolidation of inflammatory cells and products is evident.

pneumococci enter the bronchioles and lungs and multiply, they induce an overwhelming inflammatory response, marked by the release of a torrent of edematous fluid. In a form of pneumococcal pneumonia termed **lobar pneumonia,** this fluid accumulates in the alveoli along with red and white blood cells and spreads rapidly through the lung. In this instance, the patient can actually "drown" in his own fluid. If this mixture of exudate, cells, and bacteria solidifies in the air spaces, a condition known as *consolidation* occurs (figure 15.18). In infants and the elderly, the areas of infection are usually spottier and centered more in the bronchi than in the alveoli (bronchial pneumonia).

Symptoms of pneumococcal pneumonia are chills, shaking, rapid breathing, and fever. The patient may experience severe pain in the chest wall, cyanosis (due to compromised oxygen exchange), a cough that produces rusty-colored (bloody) sputum, and abnormal breathing sounds. Among the systemic complications of pneumonia are pleuritis and endocarditis, but pneumococcal bacteremia and meningitis are the greatest danger to the patient.

In young children, *S. pneumoniae* is a common agent of upper respiratory tract infections that can spread to the meninges and cause meningitis. It is even more common for this agent to gain access to the chamber of the middle ear by way of the eustachian tube and cause a middle ear infection called **otitis media.** Otitis media is the third most common childhood disease in the United States, and the pneumococcus accounts for the majority of cases. The severe inflammation in the small space of the middle ear induces acutely painful earaches and sometimes even temporary deafness (figure 15.19).

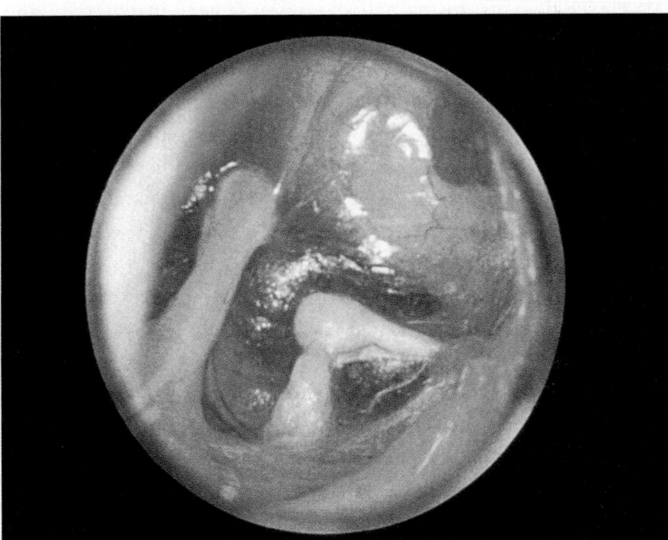

Figure 15.19 Pneumococcal otitis media. The acute inflammation and internal pressure of a middle ear infection are evident in the bulging of the tympanum as seen during otoscope examination. Occasionally, the eardrum breaks.

Healthy persons have high natural resistance to the pneumococcus. The natural mucous and ciliary responses of the respiratory tract help flush out transient organisms. Respiratory phagocytes are also essential in the eradication process, but this system works only if the capsule is coated by opsonins in the presence of complement. After recovery, the individual will be immune to future infections by that particular pneumococcal type. This immunity is the basis for a successful vaccine (see subsequent section on prevention).

Laboratory Cultivation and Diagnosis A specimen is usually necessary before diagnosing pneumococcal pneumonia, because

---

consolidation L. *consolidatio,* to make firm.

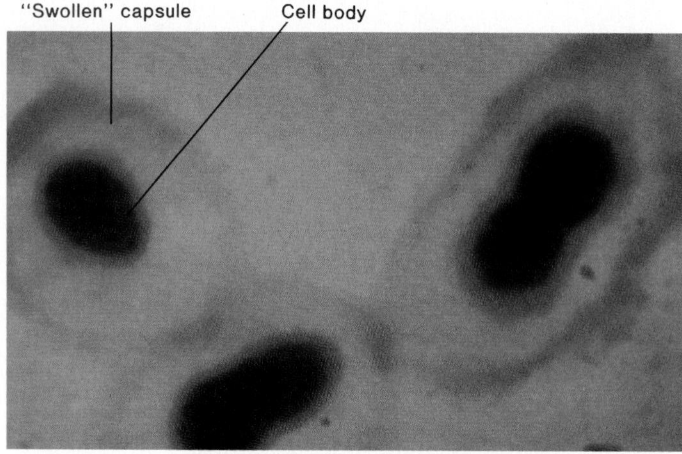

Figure 15.20 A positive Quellung, or capsular swelling test with anticapsular precipitins, is confirmatory for *S. pneumoniae.* It may also be used to identify the precise capsular serotype. The reaction of antibodies with the capsular polysaccharide intensifies the capsule.

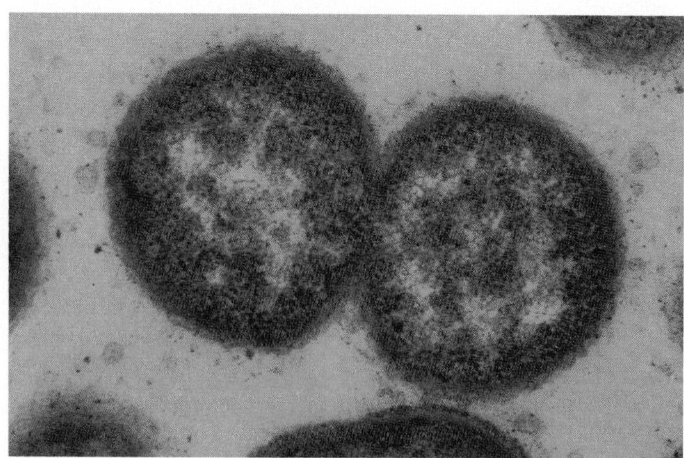

Figure 15.21 This transmission electron micrograph of *Neisseria* (×52,000) clearly indicates the manner of cell division.

a number of other pneumonia agents exist. Blood cultures, sputum, pleural fluid, and spinal fluid are common specimens. If the patient is unable to deliver sputum, the trachea may be catheterized. A gram-stained specimen is very instructive, and presumptive identification can often be made by this method alone. Also highly definitive is the Quellung reaction, a serological test in which sputum is mixed with various antisera and observed microscopically. Precipitation of antibodies on the surface of the capsule gives the appearance of swelling (figure 15.20). Alpha-hemolysis helps differentiate the pneumococcus from all streptococcal groups except the viridans. Diagnosis may be confirmed by checking the agent's sensitivity to the drug optochin and by positive bile solubility and inulin fermentation tests.

Treatment and Prevention of Pneumococcal Pneumonia

Treating pneumococcal infections and preventing their more serious complications require early administration of penicillin. In cases of microbial resistance or patient allergy to penicillin, erythromycin, clindamycin, or cephalosporin can be substituted. Daily penicillin prophylaxis has recently been suggested to protect children with sickle-cell anemia against recurrent pneumococcal infections. Untreated cases in these children may result in up to 30% mortality.

Although antibiotics arrest the course of pneumonia, active immunity is important in preventing recurrences. This is the one streptococcal disease for which effective vaccination is available. Polyvalent vaccines (Pneumovax and Pnu-immune) that contain capsular antigens of 12 to 23 of the most frequently encountered serotypes are indicated for patients at particularly high risk, including those with sickle-cell anemia, lack of a spleen, congestive heart failure, lung disease, diabetes, kidney disease, and advanced age. These vaccines are effective for about five years in 80–95% of those vaccinated. Immunization is not indicated in normal, healthy infants or in seriously immunosuppressed individuals.

The Family Neisseriaceae: Gram-Negative Cocci

Members of the Family Neisseriaceae are residents of the mucous membranes of warm-blooded animals. Most species are relatively innocuous commensals, but two are primary human pathogens with far-reaching medical impact. The genera contained in this group are *Neisseria, Moraxella,* and *Acinetobacter.* Of these, *Neisseria* has the greatest clinical significance.

A distinguishing feature of the ***Neisseria*** is their cellular morphology. Rather than being perfectly spherical, the cells are bean-shaped and paired, with their flat sides touching (figure 15.21). None develop flagella or spores, but capsules can be found on the pathogens. The cells are typically gram negative, possessing an outer membrane in the cell wall and, in many cases, pili.

Most *Neisseria* are strict parasites that do not survive long outside the host, particularly where hostile conditions of drying, cold, acidity, or light prevail. *Neisseria* species are aerobic or microaerophilic and have an oxidative form of metabolism. They have catalase, enzymes for producing acid (mostly acetic) from various carbohydrates, and a special cytochrome enzyme (oxidase) that can be used at a general level in identification. The pathogenic species, *N. gonorrhoeae* and *N. meningitidis,* require complex enriched media and grow best in an atmosphere containing additional CO_2. We shall concentrate on other features of the pathogenic *Neisseria* in the following sections.

Neisseria gonorrhoeae—The Gonococcus

Gonorrhea has been known as a sexually transmitted disease since ancient times. Its name originated with the Greek physician Claudius Galen, who thought that it was caused by an excess flow of semen. For a fairly long period in history, gonorrhea was confused with syphilis (see feature 15.3). In 1879

gonorrhea Gr. *gonos,* seed, and *rhein,* to flow.

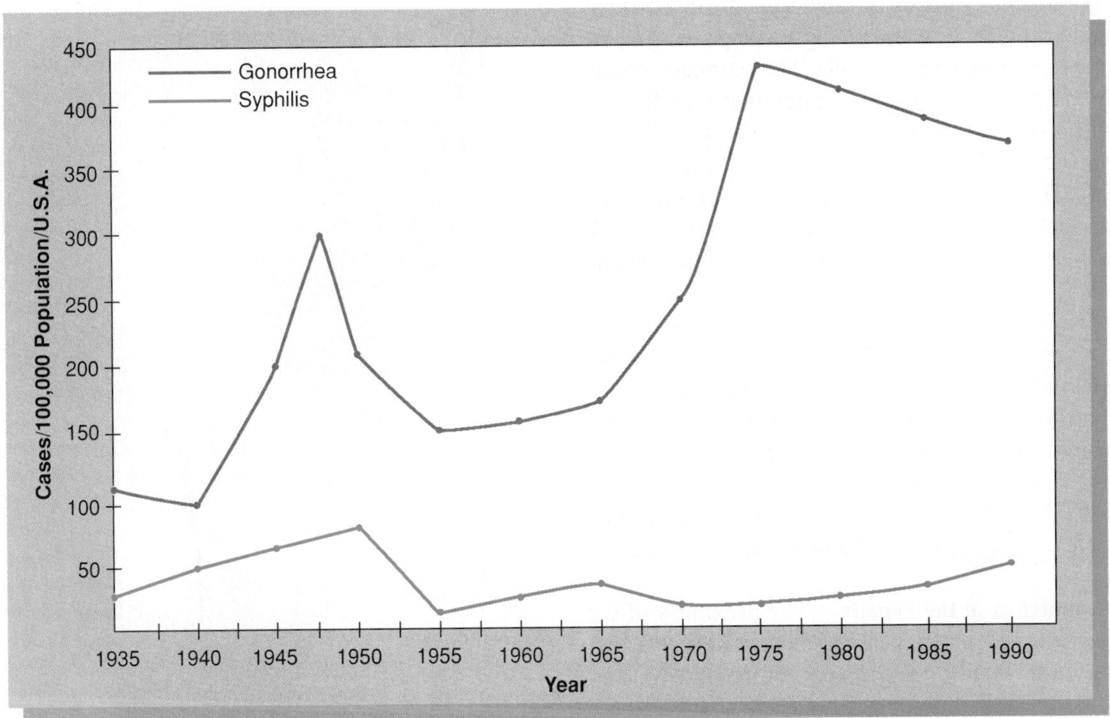

Figure 15.22 Comparative graph of two reportable infectious STDs. Gonorrhea was once the most common reportable STD in the United States. Notice the rise in cases (epidemics) corresponding with 1940–46 (World War II) and 1965–75 (the Vietnam War). This pattern correlates with times of social upheaval and changing sexual mores. Epidemiologists also theorize that increased use of oral contraceptives contributed to the second epidemic period. The current trend for gonorrhea is a slight decline in the number of cases. Although syphilis declined over the same period, a recent epidemic has increased the case level to nearly that of World War II.

Feature 15.3 An Unfortunate Self-Experiment

Throughout the Middle Ages, gonorrhea and syphilis were thought to be different manifestations of the same disease. In 1767, a dedicated and daring English physician named John Hunter attempted to determine if this were so with a shocking and unfortunate experiment. He played the parts of both guinea pig and scientist by inoculating himself with pus from a gonorrhea patient. However, Dr. Hunter did not know that the patient was simultaneously infected with gonorrhea and syphilis. Not only did the poor doctor acquire both infections, but his findings continued to foster the old and totally incorrect belief that the two diseases were one and the same.

Albert Neisser, a German physician, first viewed the bacterium that was later named for both him and the disease. Other workers went on to cultivate *N. gonorrhoeae,* also known as the **gonococcus,** and to prove conclusively that it alone was the etiologic agent of gonorrhea.

Factors Contributing to Gonococcal Pathogenicity

The virulence of the gonococcus is due chiefly to the presence of pili that promote mutual attachment of cocci to each other and invasion and infection of epithelial tissue. In addition to their role in adherence, pili also seem to slow phagocytosis by macrophages and neutrophils. Another contributing factor in pathogenicity is a protease that cleaves the secretory antibody (IgA) on mucosal surfaces and prevents its action. Whether the gonococcal capsule or the cell wall lipopolysaccharide are virulence factors remains unclear.

Epidemiology and Pathology of Gonorrhea

Gonorrhea is a strictly human infection that occurs worldwide and ranks among the top five sexually transmitted diseases. Although 600,000 to 700,000 cases are reported in the United States each year, it is estimated that the actual number may be closer to several million, counting subclinical infections. All races and ages are susceptible, but most cases occur in urban dwellers 18–24 years old who are poor and promiscuous. Figures on the prevalence of gonorrhea over the past 60 years show a fluctuating pattern, apparently corresponding to periods of social and political upheaval when promiscuity tends to increase (figure 15.22). Although gonorrhea has long occurred in epidemic proportions, present trends indicate a slight decline in the number of cases.

For various reasons, not every person coming in contact with a virulent strain of the gonococcus becomes infected. Studies done with male volunteers revealed that an infectious dose can range from 100 to 1,000 colony-forming units. Because *Neisseria gonorrhoeae* does not survive for more than one or two hours out of the body and must be transferred directly to a suitable mucous membrane, it is extremely unlikely that it would be picked up in a viable state from a fomite. Except for neonatal infections, the gonococcus spreads through some form of sexual contact. Men and women who carry the pathogen asymptomatically serve as the principal reservoir. The pathogen comes in contact with an appropriate portal of entry that may be genital or extragenital (rectum, eye, or throat). After attaching to the epithelial surface by pili, the bacteria invade the underlying connective tissue. In two to six days, this process results in an inflammatory reaction that may or may not be noticed. The following sections will survey the several categories of gonorrhea.

Genital Gonorrhea in the Male Infection in males usually occurs in the urethra, where it elicits urethritis, painful urination, and a yellowish discharge, though a relatively large number of cases may be asymptomatic. In most cases, infection is limited to the distal urogenital tract, but it may occasionally spread from the urethra to the prostate gland and epididymis (figure 15.23). Scar tissue formed in the spermatic ducts during healing of an invasive infection may render the individual infertile.

Genitourinary Gonorrhea in the Female The proximity of the genital and urinary tract openings increases the likelihood that both organs can be infected during sexual intercourse. A mucopurulent or bloody vaginal discharge occurs in about half the cases, along with painful urination if the urethra is affected. Major complications occur when the infection ascends from the vagina and cervix to higher reproductive structures such as the uterus and fallopian tubes (figure 15.24). One disease resulting from this progression is **salpingitis,** also known as **PID** (pelvic inflammatory disease), a condition characterized by fever, abdominal pain, and tenderness. It is not unusual for the microbe to spread into the peritoneum and become involved in mixed infections with anaerobic bacteria. Scar tissue from this infection may disfigure and block the fallopian tubes, causing sterility and ectopic pregnancies.

Extragenital Gonococcal Infections in Adults Extragenital sexual transmission and carriage of the gonococcus are not uncommon. Anal intercourse may lead to proctitis, and oral copulation may result in pharyngitis and gingivitis. These complications appear most often in homosexual males, but cases can occur in women. Careless personal hygiene may account for self inoculation of the eyes and a serious form of conjunctivitis. In 1% of gonorrhea cases, the gonococcus enters the bloodstream and is disseminated to the joints and skin. Involvement of the wrist and ankle may lead to chronic arthritis and a painful, sporadic, papular rash on the limbs. Rare complications of gonococcal bacteremia are meningitis and endocarditis.

Gonococcal Infections in Children Infants born to mothers harboring the gonococcus vaginally are in danger of being infected as they pass through the birth canal. Gonococcal eye infections are very serious and often manifest sequelae such as keratitis, ophthalmia, and even blindness (figure 15.25).

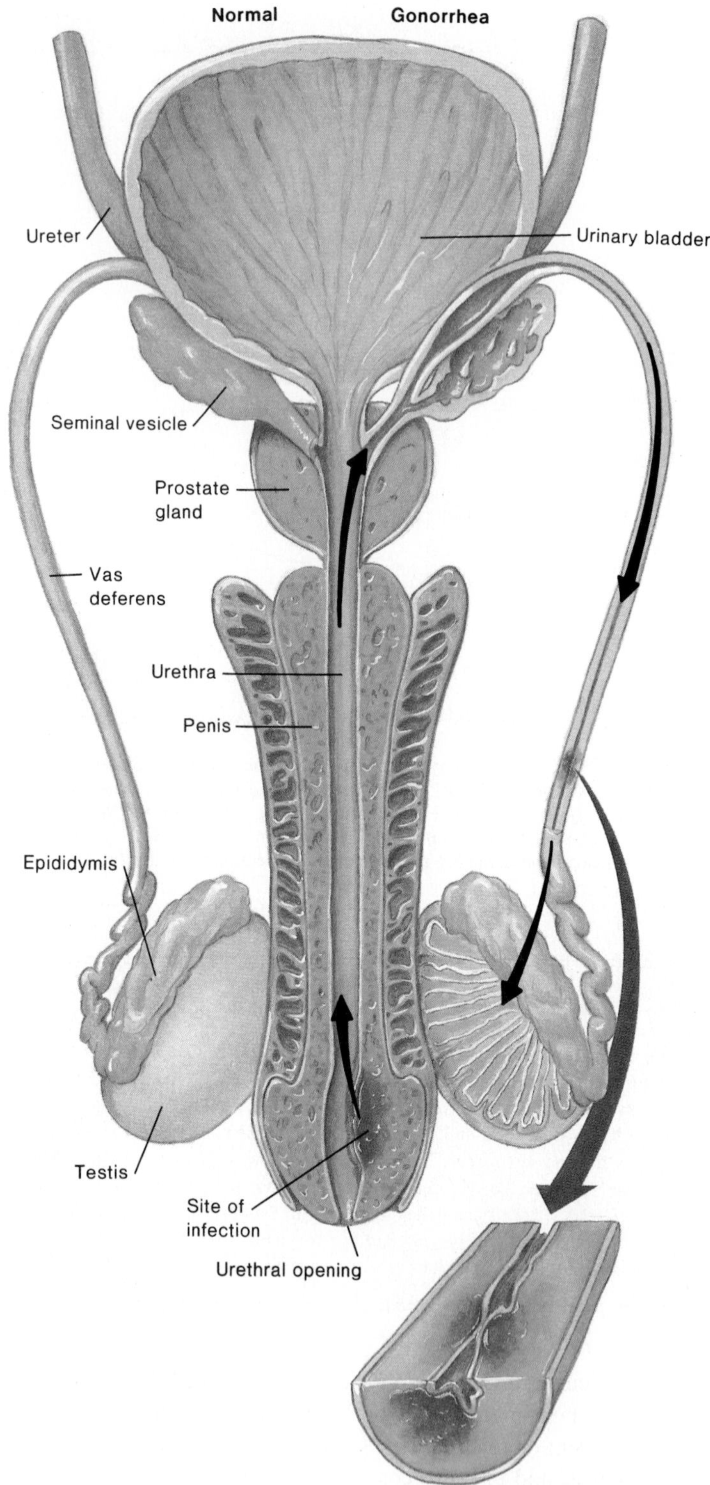

Figure 15.23 Left: Frontal view of the male reproductive tract in its normal, uninfected state. The sperm-carrying ducts are continuous from the testis to the urethral opening. Right: The route of ascending gonorrhea complications. Infection begins at the tip of the urethra, ascends the urethra through the penis, and passes into the vas deferens. Occasionally, it may even enter the epididymides and testes. Inset: Damage to the ducts carrying sperm may create scar tissue and blockage, which reduces sperm passage and may even lead to sterility.

---

salpingitis (sal″-pin-jy′-tis) Gr. *salpinx,* tube, and *itis,* inflammation. An inflammation of the fallopian tubes.

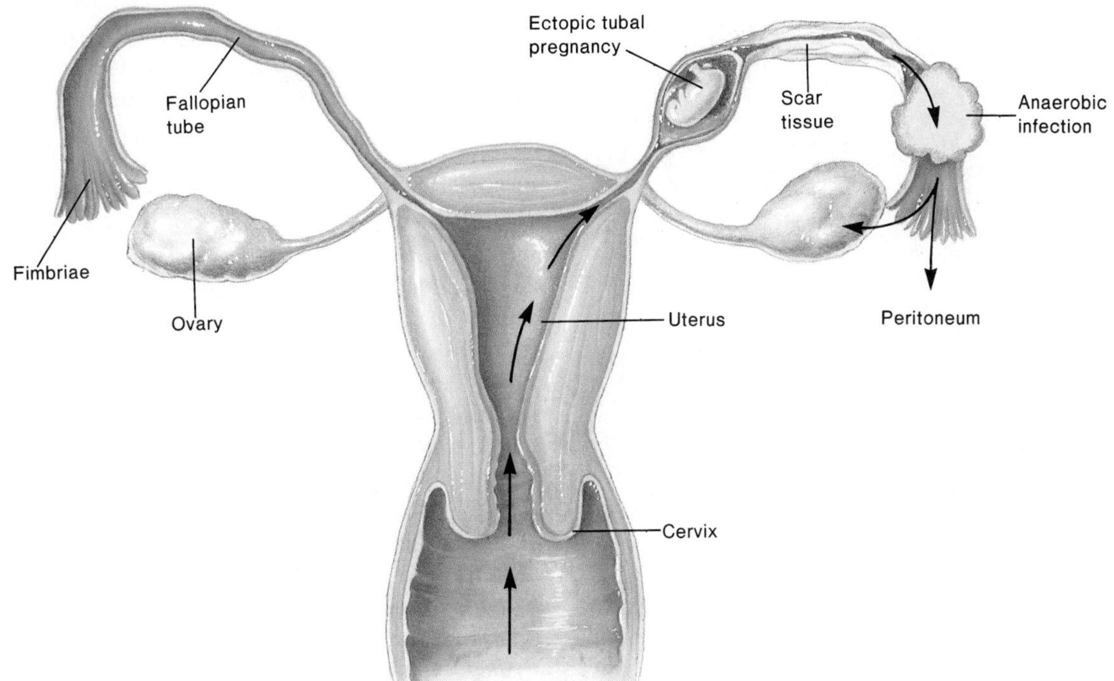

Figure 15.24 Invasive gonorrhea in women. Left: Normal state. Right: In ascending gonorrhea, the gonococcus is carried from the cervical opening up through the uterus and into the fallopian tubes. On rare occasions, it can escape into the peritoneum and invade the ovaries, causing peritonitis. Pelvic inflammatory disease (PID) is a serious complication that may lead to scarring in the fallopian tubes, ectopic pregnancies, and mixed anaerobic infections.

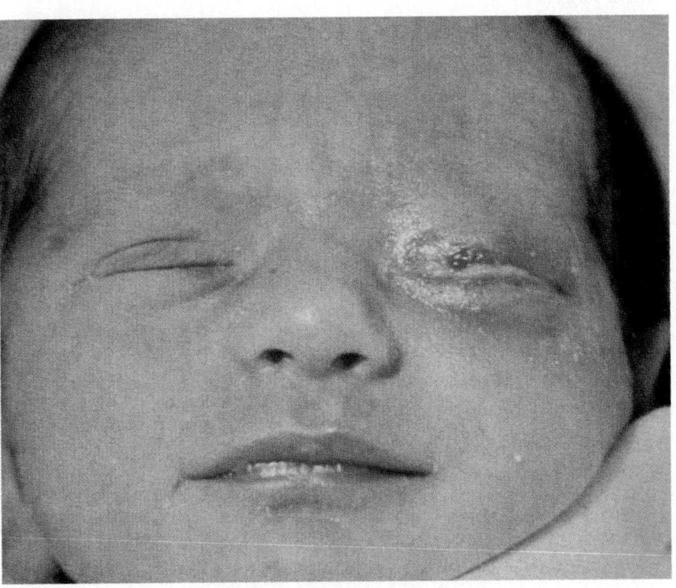

Figure 15.25 Gonococcal ophthalmia neonatorum in a week-old infant. The infection is marked by intense inflammation and edema, and if allowed to progress, it causes damage that can lead to blindness. Fortunately, this infection is completely preventable and treatable.

A universal precaution to prevent these complications is the instillation of antibiotics or silver nitrate into the conjunctival sac of newborn babies. Because of the potential harm to the fetus, physicians usually screen pregnant mothers for the gonococcus. Finding gonorrhea in children other than neonates is strong evidence of sexual abuse by infected adults, a condition that is being reported with increased frequency unfortunately.

Clinical Diagnosis of Gonorrhea

Diagnosing gonorrhea requires several lab tests. The presence of gram-negative diplococci in neutrophils from urethral, vaginal, cervical, or eye exudates is especially diagnostic because gonococci tend to be engulfed and remain viable within phagocytes (figure 15.26). This simple procedure can provide at least presumptive evidence of gonorrhea. It is most successful in males and least successful in asymptomatic infection. Other tests used to identify *Neisseria gonorrhoeae* and differentiate it from related species are discussed in a later section.

Host Defenses, Treatment, and Prevention of Gonococcal Infections

Although gonococcal infections stimulate local production of antibodies and activate the complement system, these responses do not produce lasting immunity, and some persons experience recurrent infections. Two accepted regimens for treating active cases of genital and extragenital infections are (1) intramuscular injection of 4.8 million units of procaine penicillin G, given with

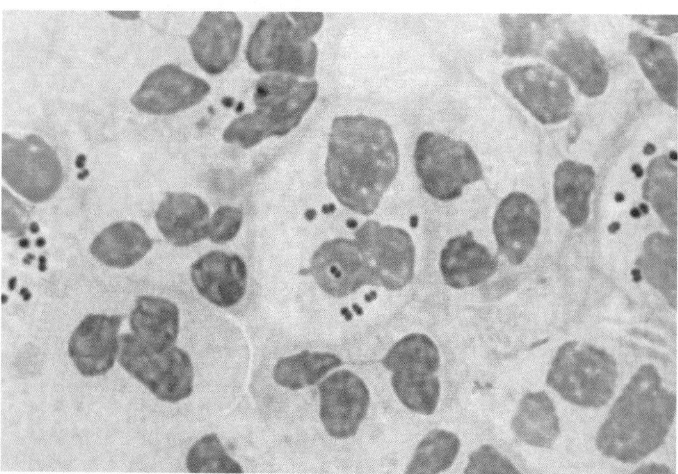

Figure 15.26 Gram stain of urethral pus from a patient with gonorrhea (×1,000). Note the intracellular (phagocytosed) gram-negative diplococci in polymorphonuclear lymphocytes (PMNs).

oral probenecid, a drug that slows the kidney's excretion of penicillin, or (2) a single large combined dose (3.5 grams) of oral ampicillin and probenecid. Complications such as PID and prostatitis require combined therapy with penicillin and tetracycline. Neonatal eye infections are treated with smaller doses of intravenous penicillin along with irrigation of the eye. Since more penicillinase-producing strains of *N. gonorrhoeae* (PPNG) have developed and penicillin allergy is so common, drugs such as broad-spectrum cephalosporins are sometimes substituted. First reported in the Far East in the 1970s, PPNG strains have steadily increased in incidence and now may account for as many as 10% of all cases reported to the CDC.

Gonorrhea is a reportable infectious disease, which means that any physician diagnosing it must forward the information to a public health department. The follow-up to this finding involves tracing sexual partners to offer prophylactic antibiotic therapy. There is a pressing need to seek out and treat asymptomatic carriers and their sexual contacts, but complete control of this group is nearly impossible. Other control measures include nonjudgmental education programs that emphasize the immediate and long-term effects of all STDs and promote safer sexual practices. Two developments, both dependent upon immunological technology, would revolutionize the management of this disease. The first would be a serological screening test to detect antibodies to the gonococcus in a patient's serum (much like that used for syphilis), thus making it possible to detect asymptomatic carriers. The second would be the perfection of a pilus vaccine to immunize the high-risk population. The latter possibility is the subject of intensive research, but so far the gonococcus has proved a slippery target.

Neisseria meningitidis: The Meningococcus

Another serious human pathogen is ***Neisseria meningitidis,*** a bacterium known commonly as the **meningococcus** and usually associated with epidemic cerebrospinal meningitis. Important factors in meningococcal invasiveness are a polysaccharide capsule, pili, and IgA protease. Although nine different strains of capsular antigens exist, serotypes A, B, and C are responsible for most cases of infection. Another virulence factor with potent pathological effects is the lipopolysaccharide (endotoxin) released from the cell wall.

Epidemiology and Pathogenesis of Meningococcal Disease

The diseases of *N. meningitidis* have a sporadic or epidemic incidence in late winter or early spring. The continuing reservoir of infection is humans who harbor the pathogen in the nasopharynx. The carriage state, which may last from a few days to several months, exists in 3% to 10% of the civilian adult population and may exceed 90% in military personnel. The scene is set for transmission when carriers closely associate with nonimmune individuals, as might be expected in families, day-care facilities, and military barracks. The highest-risk groups are young children (6–24 months old) and older children and young adults (10–20 years old). *Neisseria meningitidis* is the second most frequent cause of **meningitis** in young children, after *Hemophilus influenzae* (see chapter 16).

Because this bacterium does not survive long in the environment, meningococci must be acquired through close contact with secretions or droplets. Upon reaching its portal of entry in the nasopharynx, the meningococcus attaches there with pili. In many persons, this may result in simple asymptomatic colonization. In the more vulnerable individual, however, the cocci enter an incubation period of a few days, which then culminates in pharyngitis.

The most serious complications of meningococcal pharyngitis are due to meningococcemia (figure 15.27). Bacteria entering the blood vessels rapidly permeate the meninges and produce symptoms of meningitis, the most common complication in children. It is marked by fever, weakness, headache, stiff neck, convulsions, and vomiting. The presence of the pathogen and its endotoxin in the generalized circulation also gives rise to crops of *petechiae* on the trunk and appendages. In a small number of cases, meningococcemia becomes a fulminant disease with a high mortality rate. It has a violent onset, with fever (higher than 40°C), chills, delirium, severe widespread *ecchymoses,* shock, and coma (figure 15.28). Generalized intravascular clotting, cardiac failure, damage to the adrenal glands, and death may occur within a few hours.

Clinical Diagnosis of Meningococcal Disease

Suspicion of bacterial meningitis constitutes a medical emergency, and differential diagnosis must be done with great haste and accuracy, since complications can come on so rapidly and with such lethal consequences. Cerebrospinal fluid, blood, or nasopharyngeal samples are stained and observed directly for the typical gram-negative diplococci. Occasionally, skin lesions are punctured to obtain infectious material. Cultivation may be necessary to differentiate the bacterium from other species (see

---

petechiae (pee-tee′-kee-ee) Small, nonraised, round purple spots caused by hemorrhage into the skin. Larger spots are called **ecchymoses** (ek″-ih-moh′-seez).

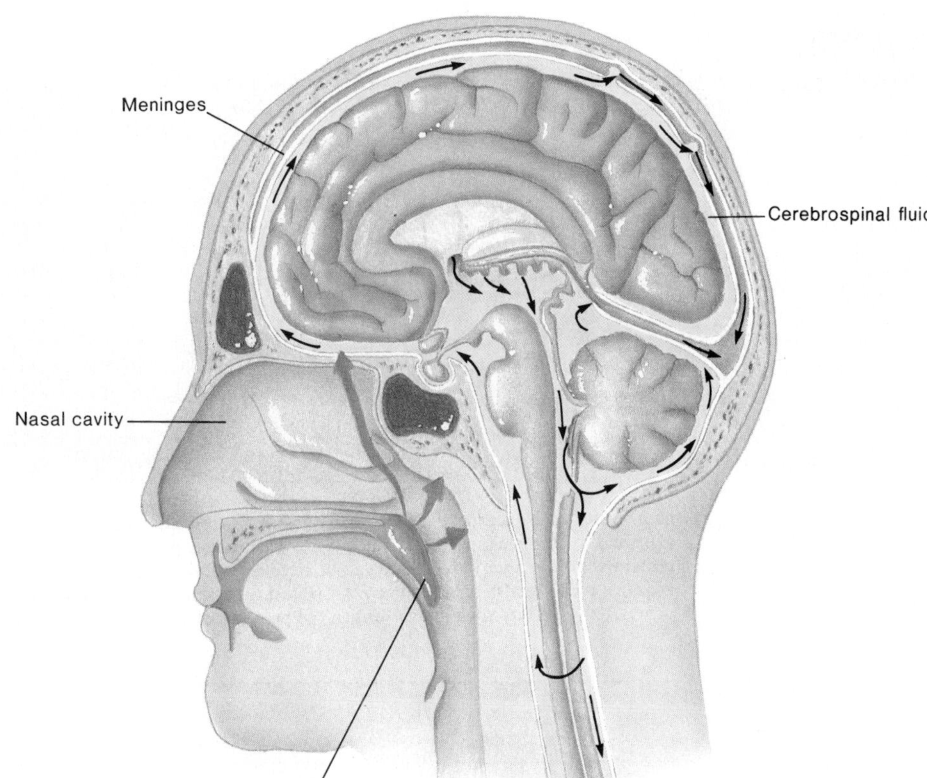

Figure 15.27 Dissemination of the meningococcus from a nasopharyngeal infection. Bacteria spread to the roof of the nasal cavity, which borders a highly vascular area at the base of the brain. From this location, they may enter the blood and escape into the cerebrospinal fluid. Infection of the meninges leads to meningitis and an inflammatory purulent exudate over the brain surface.

subsequent section on laboratory diagnosis). Specific rapid tests are also available for detecting the capsular polysaccharide or the cells directly from specimens without culturing.

Immunity, Treatment, and Prevention of Meningococcal Infection

Well-developed natural immunity to the meningococcus appears to be the rule among most individuals. The infection rate of exposed persons is only about 1%, so susceptibility is likely due to poorly mounted immune reactions. Immunity to the organism is due to opsonic antibodies that develop against the capsular polysaccharides in groups A and C and against membrane antigens in group B. Because untreated meningococcal disease has a mortality rate of up to 85%, it is vital that chemotherapy begin as soon as possible with one or more drugs. It may even be given while tests for the causative agent are underway. Penicillin G is the most potent of the drugs available for meningococcal infections; it is generally given in high doses intravenously. If the patient cannot tolerate penicillin, intravenous chloramphenicol is the second choice. Patients may also require treatment for shock and intravascular clotting.

When family members, medical or military personnel, or children in day care have come in close contact with infected persons, preventative therapy with rifampin or tetracycline may be warranted. Meningococcal vaccines that contain specific purified capsular antigens are available to protect high-risk groups, especially during epidemics. Group A vaccine protects all ages, but group C vaccine is useful only for individuals over two years of age, and a group B vaccine is not yet available.

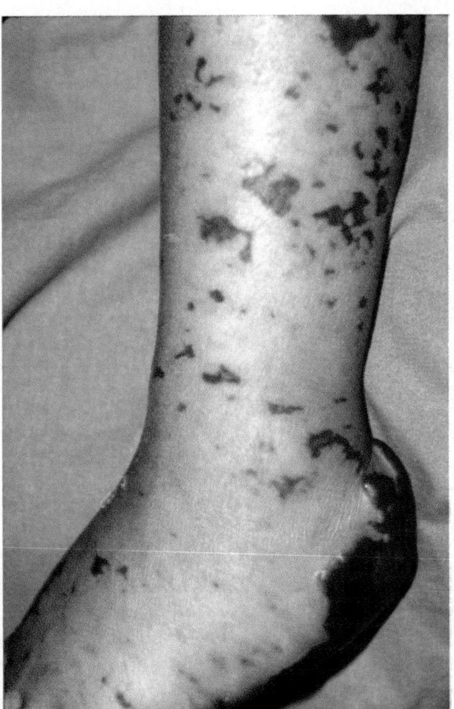

Figure 15.28 The appearance of meningococcemia. The generalized ecchymotic, purpuric blotches on the leg and foot of an adult male are due to subcutaneous hemorrhage. This condition can occur anywhere on the body, including the mucous membranes and conjunctiva. Endotoxins released during blood infection are thought to be largely responsible for this pathologic state.

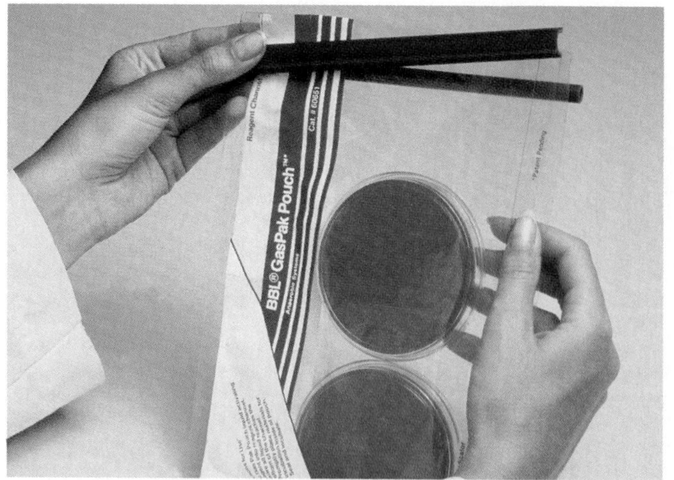

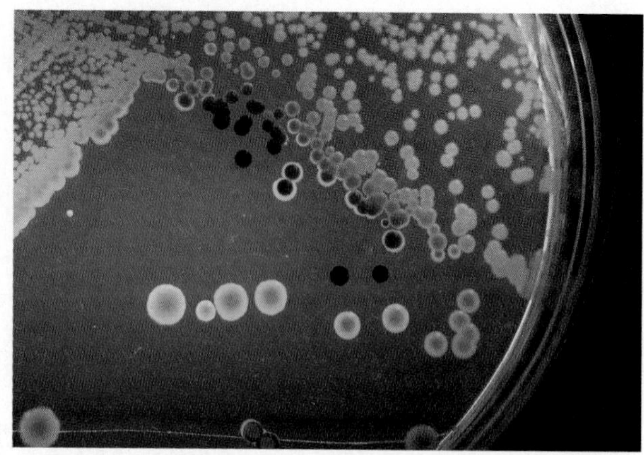

Figure 15.29 Selected laboratory techniques for *Neisseria* spp. (*a*) Special kits are available for collecting, culturing, and incubating a specimen. The kit shown contains plates of media and small, sealable bags for maintaining a high CO_2 atmosphere. These kits prevent destruction of the very sensitive pathogenic species. (*b*) The oxidase test. A drop of oxidase reagent is placed on a suspected *Neisseria* or *Branhamella* colony. If the colony reacts with the chemical to produce a purple to black color, it is oxidase positive; those that remain white to tan are oxidase negative. Because several species of gram-negative rods are also oxidase positive, this test is presumptive for these two genera only if a Gram stain has verified the presence of gram-negative cocci.

Table 15.4 Scheme for Differentiating Gram-Negative Cocci and Coccobacilli

Characteristic or Test	*N. gonorrhoeae*	*N. meningitidis*	*N. lactamica**	*N. sicca*	*Branhamella catarrhalis*	*Moraxella* spp.	*Acinetobacter* spp.
Oxidase	+	+	+	+	+	+	−
Growth on MTM	+	+	+	−	+/−	−	−
Growth on Nutrient Agar	−	−	−	+	+	+	+
Growth on Blood Agar at 22°C	−	−	+/−	+/−	+	+	+
Yellow Pigment	−	−	+	+/−	−	−	−
Require Increased CO_2 During Isolation	+	−	−	−	−	−	−
Sugar Fermentation							
Maltose	−	+	+	+	−	−	+/−
Sucrose	−	−	−	+	−	−	+/−
Lactose	−	−	+	−	−	−	+/−
Nitrate Reduction	−	−	−	−	+	+/−	−
Nitrite Reduction	−	+/−	++	+	+	−	−
Capsule	+	+	−	+/−	+	+/−	+/−
Human Pathogenicity	+	+	+	UDC**	Opportunistic	UDC	Opportunistic

*A weak pathogen, found in the nasopharynx of children and easily mistaken for *N. meningitidis*.
**Stands for Usually Dismissed as Contaminants, because pathogenicity is very rare.

Differentiating Pathogenic from Nonpathogenic *Neisseria*

It may become important to differentiate true pathogens from normal *Neisseria* that also live in the human body and may be present in infectious fluids. Immediately after collection, specimens are streaked on Modified Thayer-Martin (MTM)[5] or Martin-Lewis medium and incubated in a high CO_2 atmosphere (figure 15.29*a*). Presumptive identification of the genus is obtained by a Gram stain and oxidase testing on isolated colonies (figure 15.29*b*). Further testing may be necessary to differentiate the two pathogenic species from one another, from other oxidase-positive infectious species, and from normal flora of the oropharynx and genitourinary tract that can be confused with the pathogens. Sugar fermentations, growth patterns, nitrate reduction, and pigment production are useful differentiation tests at this point (table 15.4). Several rapid method identification kits have been developed for this purpose.

5. MTM is a selective medium that contains vancomycin to inhibit gram-positive cocci, colistin to inhibit gram-negative rods, and nystatin to inhibit yeasts.

Other Gram-Negative Cocci and Coccobacilli

The genera *Branhamella, Moraxella,* and *Acinetobacter* are included in the same family as *Neisseria* because of morphological and biochemical similarities. Most species are either relatively harmless commensals of humans and other mammals or saprobes living in soil and water. In the past few years, however, one species in particular has emerged as a significant opportunist in hosts with disturbed immune functions. This species, *Branhamella catarrhalis,* is found in the normal human nasophaynx and can cause purulent disease. It is associated with several clinical syndromes such as meningitis, endocarditis, otitis media, bronchopulmonary infections, and neonatal conjunctivitis. Adult patients with leukemia, alcoholism, malignancy, diabetes, or rheumatoid disease are the most susceptible.

The appearance of *B. catarrhalis* in specimens is so reminiscent of both the gonococcus and the meningococcus that the laboratory technician must use some other means to discriminate it from the pathogenic *Neisseria.* Key characteristics that allow this are the complete lack of carbohydrate fermentation and positive nitrate reduction in *B. catarrhalis.* Treatment of infection by this organism should be with erythromycin or cephalosporins, because so many strains of this species produce penicillinase.

Moraxella species are short plump rods rather than cocci, and some exhibit a twitching motility. These bacteria are widely distributed on the mucous membranes of domestic mammals and humans and are generally regarded as weakly pathogenic or nonpathogenic. Although some species are implicated in infections of cattle, goats, and horses, infections are rare, even in patients with impaired defenses.

The genus *Acinetobacter* is similar morphologically to *Moraxella.* It is a small, paired, gram-negative cell that varies in shape from a true rod to a coccus. However, it is quite different from the members of this family in several ways. For example, its habitat is soil, water, and sewage; it is nonfastidious; and it is oxidase-negative. *Acinetobacter* is so widely distributed that it is a common contaminant in clinical specimens and is ordinarily not considered clinically significant. In rare cases, however, it is clearly an agent of disease. Most infections are nosocomial in origin, affecting traumatized or debilitated persons with indwelling catheters or other instrumentation. Septicemia, meningitis, endocarditis, pneumonia, organ abscess, and urinary tract infections have been reported.

Branhamella catarrhalis (bran″-hah-mel′-ah cah-tahr-al′-is) After Sarah Branham, a bacteriologist working on the genus *Neisseria,* and L. *catarrhus,* to flow.

Moraxella (moh-rak-sel′-uh) From Victor Morax, a Swiss ophthalmologist who worked with the genus.

Acinetobacter (ass″-ih-nee-toh-bak′-tur) Gr. *a,* none, *kinetos,* movement, and *bacterion,* rod.

Chapter Review with Key Terms

The Medically Important Cocci

Genus *Staphylococcus*

Nonmotile, non-spore-forming cocci arranged in irregular clusters; facultative anaerobes; fermentative; salt-tolerant, catalase-positive. ***Staphylococcus aureus*** produces a number of virulence factors. Enzymes include **coagulase** (the confirmatory characteristic), hyaluronidase, staphylokinase, nuclease, and penicillinase. Toxins are β-hemolysins, leukocidin, enterotoxin, scalded skin syndrome toxin, toxic shock syndrome toxin. Microbe carried in nasal vestibule, nasopharynx, skin.

Infections: Target is skin; local **abscess** occurs at site of invasion of hair follicle, gland; manifestations are **folliculitis, furuncle, carbuncle,** and **bullous impetigo.** Other common infections are **osteomyelitis,** a focal infection of bone, bacteremia, leading to endocarditis, and pneumonia.

Toxic Disease: **Food intoxication** due to enterotoxin; **scalded skin syndrome** (**SSSS**), a skin condition that causes desquamation; **toxic shock syndrome,** toxemia in women due to infection of vagina, associated with wearing tampons.

Principal Coagulase-Negative Staphylococci: ***S. epidermidis,*** a normal resident of skin and follicles; an opportunist and one of the most common causes of nosocomial infections, chiefly in surgical patients with indwelling medical devices or implants; ***S. saprophyticus*** is a urinary pathogen.

Treatment: *S. aureus* has multiple resistance to antibiotics, especially penicillin, ampicillin, and methicillin; drug selection requires sensitivity testing; cephalosporins often used; abscesses require debridement and removal of pus; extreme resistance of staphylococci to harsh environmental conditions makes control difficult, requiring high level of disinfection and antisepsis; no vaccines available.

Streptococci

A large, varied group of bacteria (about 25 species), containing the genera ***Streptococcus*** and ***Enterococcus.*** Cocci are in chains, ranging from a few to thousands; nonmotile, non-spore-forming; often encapsulated; fermentative; catalase-negative; most pathogens fastidious and sensitive to environmental exposure. Classified into **Lancefield groups** (A–R) according to the type of serological reactions of the cell wall carbohydrate; also characterized by type of hemolysis. The most important sources of human disease are β-hemolytic ***S. pyogenes*** (group A), ***S. agalactiae*** (group B), ***Enterococcus faecalis*** (group D) and α-hemolytic ***S. pneumoniae*** and the **viridans streptococci.**

β-Hemolytic Streptococci: ***S. pyogenes*** is the most serious pathogen of family; produces several virulence factors, including C-carbohydrate, M-protein (fimbriae), streptokinase, hyaluronidase, DNase, **hemolysins** (SLO, SLS), **erythrogenic toxin.**

Microbe resides in nasopharynx of carriers; transmitted through close contact; invades skin and mucous membranes. **Skin infections** include **pyoderma** (strep **impetigo**), **erysipelas** (deeper, spreading skin infection). Systemic conditions include **strep throat** or **pharyngitis,** severe inflammation of throat membranes; may lead to toxemia, called scarlet fever—generalized flushing of skin and high fever due to erythrogenic toxin; pneumonia; puerperal fever.

Sequelae caused by immune response to streptococcal toxins include **rheumatic fever,** a delayed allergy that damages heart valves, joints. In **glomerulonephritis,** severe tissue injury leads to malfunction or destruction of kidney tubules.

Streptococcus agalactiae (group B), once a cow pathogen, is increasingly found in the human vagina; causes neonatal, puerperal, wound, skin infections, particularly in debilitated persons. ***Enterococcus faecalis*** and other enteric group D species are normal flora of intestine; cause opportunistic urinary, wound, and surgical infections. Group A and group B are treated primarily with some type of penicillin (G is very effective); sensitivity testing may be necessary for enterococci; no vaccines available.

α-Hemolytic Streptococci: The **viridans streps** *S. mitis, S. salivarius, S. mutans,* and *S. sanguis* constitute oral flora in saliva; principal infections are **subacute endocarditis,** mass colonization of heart valves following dental procedures; *mutans* and *sanguis* species attach to teeth leading to plaque and dental disease.

S. pneumoniae, the **pneumococcus,** has heavily encapsulated, lancet-shaped diplococci; capsule important virulence factor—84 types; reservoir is nasopharynx of normal healthy carriers.

The pneumococcus is the most common cause of bacterial pneumonia; attacks patients with weakened respiratory defenses; entrance of bacteria into lungs initiates acute, massive inflammatory response that fills lungs (and bronchioles) with fluid; **consolidation** of fluid leads to **lobar pneumonia**; respiration severely compromised, oxygen exchange inefficient; other symptoms include fever/chills, cyanosis, cough; **otitis media,** inflammation of middle ear, is common in children; vaccination available for patients at risk.

Gram-Negative Cocci

Primary pathogens are in the genus ***Neisseria,*** common residents of mucous membranes. *Neisseria* species are bean-shaped diplococci that may be encapsulated and are piliated and oxidase-positive, non-spore-forming, nonmotile; pathogens are fastidious and do not survive long in the environment.

N. gonorrhoeae: The **gonococcus,** cause of **gonorrhea**; microbe invades mucous membranes by attaching with **pili**; tends to be located intracellularly in pus cells; third or fourth most common STD; may be transmitted from mother to newborn; asymptomatic carriage is common in both sexes.

Symptoms of **gonorrhea in males** are urethritis, discharge; may infect deeper reproductive structures; causes scarring and infertility. Symptoms of **gonorrhea in females** include vaginitis, urethritis. Ascending infection may lead to **salpingitis (PID),** mixed anaerobic infection of abdomen; common cause of sterility and ectopic tubal pregnancies due to scarred fallopian tubes. **Extragenital infections** may be anal, pharyngeal, conjunctivitis, septicemia, arthritis.

Infections in newborns cause eye inflammation, occasionally infection of deeper tissues and blindness; can be prevented by prophylaxis immediately after birth.

Preferred treatment is penicillin G, but **PPNG** strains are increasing; no vaccine exists; safe sex practices, lack of promiscuity are important controls.

Neisseria meningitidis: The **meningococcus,** most common cause of **meningitis** in U.S. Agent is a common resident of the nasopharynx; seems to invade when resistance is lowered; is spread by close contact; bacterium invades by capsule and pili. Disease begins when bacteria enter bloodstream, pass into the cranial circulation, multiply in meninges; very rapid onset; initial symptoms are neurologic; **endotoxin** released by pathogen causes rash and shock; can be fatal; treated with penicillin and/or chloramphenicol; vaccines exist for groups A and C.

Other Gram-Negative Cocci and Coccobacilli: ***Branhamella catarrhalis,*** common member of throat flora; is an opportunist in cancer, diabetes, alcoholism. ***Moraxella,*** short rods that colonize mammalian mucous membranes. ***Acinetobacter,*** gram-negative coccobacilli that occasionally cause nosocomial infections.

True–False Questions

Determine whether the following statements are true (T) or false (F). If you feel a statement is false, explain why, and reword the sentence so that it reads accurately.

____ 1. All pyogenic cocci are gram positive.

____ 2. The coagulase test is used to differentiate *Staphylococcus aureus* from other staphylococci.

____ 3. Scalded skin syndrome is the same disease as scarlet fever.

____ 4. A penicillin-resistant *Staphylococcus* may be treated with methicillin without sensitivity testing.

____ 5. β-hemolysis is the partial lysis of red blood cells due to bacterial hemolysins.

____ 6. The source of pneumococci in infections is the environment.

____ 7. Viridans streptococci commonly cause subacute endocarditis.

____ 8. The main difference between the species of *Neisseria* is the portal of entry for disease.

____ 9. Gonorrhea infections in adults remain localized in the reproductive tract.

____ 10. The rash typical of meningitis is caused by endotoxemia.

Concept Questions

1. Differentiate between the pathologies in staphylococcal infections and toxinoses.
2. What is the action of hemolysins? of leukocidin? of kinases? of hyaluronidase?
3. What is an abscess? Distinguish between the four main skin infections of *Staphylococcus*.
4. What does it mean to say osteomyelitis is a focal infection?
5. What conditions favor staph food poisoning?
6. What conditions favor toxic shock syndrome?
7. What is the principal role of the coagulase-negative staphylococci in disease?
8. Compare streptococcal and staphylococcal impetigo.
9. Describe the major group A streptococcal infections.
10. Discuss the apparent pathology in rheumatic fever and acute glomerulonephritis.
11. What is the claim to fame of group B streptococci? How is the genus *Enterococcus* different from the genus *Streptococcus?*
12. How does lobar pneumonia arise? How is it different from the bronchial type? What causes the patient to get a blue tinge to his mucous membranes?
13. What features of male and female gonorrhea are similar? What features are different?
14. For which pyogenic cocci are there vaccines? For which pyogenic cocci is penicillin not a good choice for treatment?
15. Which genus of bacteria has pathogens that can cause blindness and deafness?
16. How does *N. meningitidis* get from the nasopharynx to the brain?
17. Give the three species of pyogenic cocci most implicated in neonatal disease. How do infants acquire such diseases?
18. Which infection of those covered could most likely be acquired from a contaminated doorknob? Why is this?
19. Single matching. Only one description in the right-hand column fits a word in the left-hand column.

___ furuncle	a. Complete red blood cell lysis
___ osteomyelitis	b. Substance involved in heart valve damage
___ coagulase	c. Dissolves blood clots
___ erythrogenic toxin	d. Enzyme of pathogenic *S.aureus*
___ rheumatic fever	e. Cutaneous infection of group A streps
___ β-hemolysis	f. Solidification of pockets in lung
___ consolidation	g. Unique pathologic feature of *N. gonorrhoeae*
___ viridans streptococci	h. A boil
___ erysipelas	i. Cause of tooth abscesses
___ endocarditis	j. Focal infection of long bones
___ streptolysin	k. Heart colonization by viridans streps
___ streptokinase	l. Cause of scarlet fever
	m. Long-term sequelae of strep throat

20. Multiple matching. Match the bacterium in the left-hand column with its characteristic. More than one characteristic may fit.

___ *Staphylococcus aureus*	a. Bacitracin sensitivity
___ *Streptococcus pyogenes*	b. Diplococcus
___ *Streptococcus pneumoniae*	c. Causes pneumonia
___ *Neisseria gonorrhoeae*	d. Complication is gomerulonephritis
___ *Neisseria meningitidis*	e. Infects female reproductive tract
___ *Enterococcus faecalis*	f. Resident of nasopharynx
	g. Is in clusters
	h. Is in chains
	i. Has a capsule
	j. Catalase production

Practical/Thought Questions

1. What is the probable significance of isolating large numbers of *Streptococcus mitis* from a throat swab sample? What is the significance of isolating five colonies of *Streptococcus pyogenes* in a throat culture? What is the significance of isolating three *Staphylococcus epidermidis* colonies from a swab culture of an open wound? How about 100 colonies from a swab from an indwelling catheter? How important is the isolation of 10 colonies of *N. meningitidis* from the nasopharynx? From the spinal fluid? What would be the possible explanation for finding *Enterococcus faecalis* in the blood?
2. Why is it so important to differentiate *S. aureus* from coagulase-negative staphylococci?
3. You have been handed the problem of diagnosing gonorrhea from a single test. Which one will you choose and why?
4. How do the gram-positive and gram-negative diplococci differ in exact morphology?
5. Why do you suppose there are no useful vaccines for most of the pyogenic cocci?
6. You have been called upon to prevent outbreaks of SSSS in the nursery of a hospital. What will be your main concerns and tactics?
7. Explain why pneumonia occurs most often in the elderly and the immunosuppressed.
8. How would an adult get gonococcal conjunctivitis?
9. You have been given the assignment of obtaining some type of culture that could help identify the cause of otitis media in a small child. What area might you sample? Why is this disease so common in children?
10. Explain why it is so important to take a detailed cardiovascular and infection history of dental patients. What is the policy for patients with heart disease?
11. An elderly man with influenza acquires a case of pneumonia. Gram-positive cocci isolated from his sputum give β-hemolysis on blood agar; the infection is very difficult to treat; it is unreactive to the Quellung test. Later it is shown that the man shared a room with a patient with osteomyelitis. Isolates from both infections were the same. What is the probable species? Why do you say this? What treatment will be necessary?
12. A person with an inflamed cut on his head went to a physician, where it was discovered that microbes had invaded the clot and spread into underlying tissues. Patchy areas developed around the lesion, leading to redness and edema. After a brief hospitalization, the pateint died. What was the probable condition, and which species and virulence factors were involved? What should the treatment have been?

CHAPTER 16

The Bacilli of Medical Importance

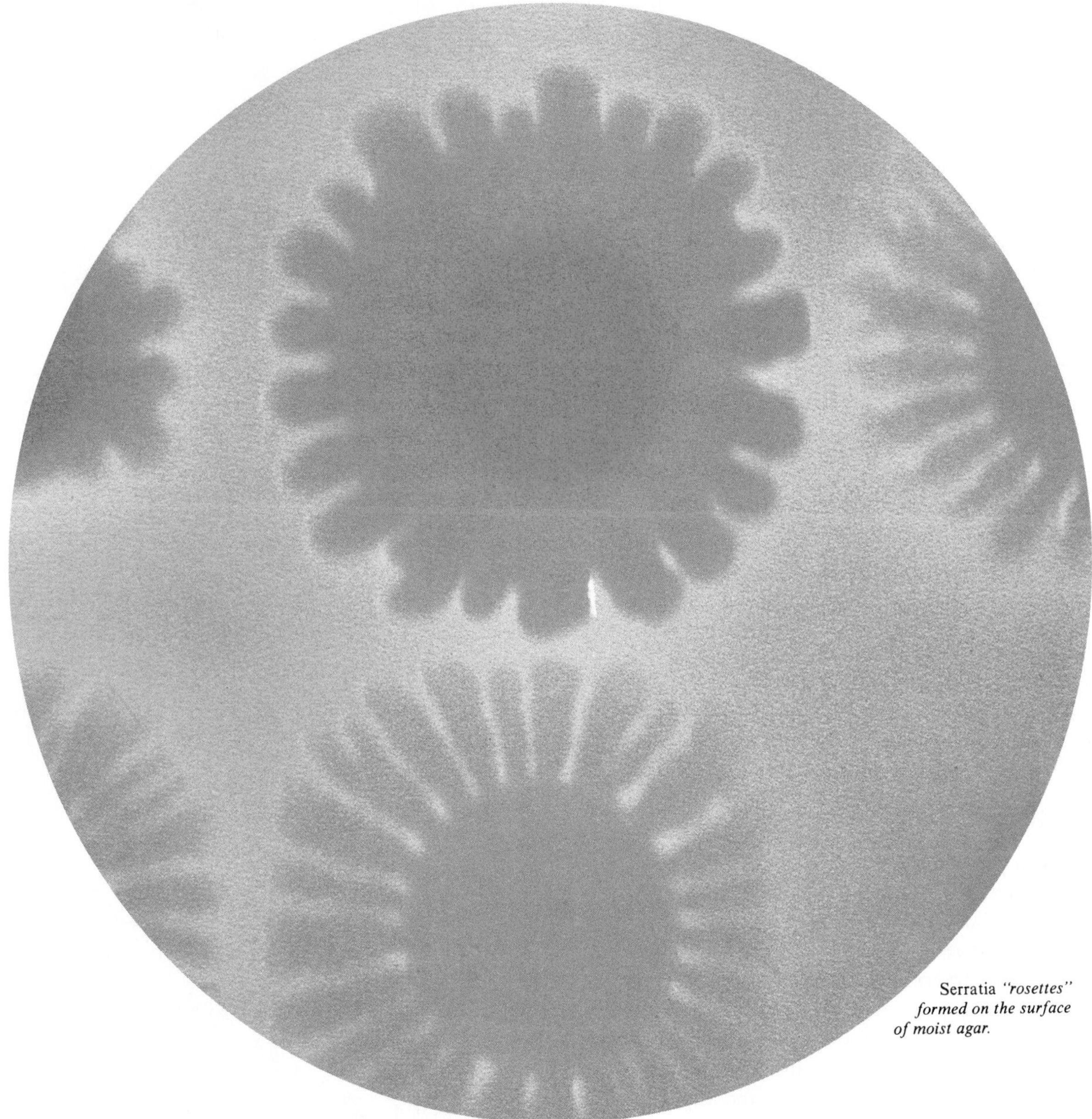

Serratia *"rosettes" formed on the surface of moist agar.*

Chapter Preview

In terms of sheer numbers, rod-shaped bacteria dominate the field of pathogenic bacteriology. Included among the diseases caused by bacilli are ancient, deadly, and fascinating ones such as bubonic plague, anthrax, tetanus, and leprosy, as well as newly emerging ones such as listeriosis and Legionnaires' disease. To conveniently characterize and differentiate the numerous genera and species, we are using a simple modification of the ninth edition of *Bergey's Manual of Systematic Bacteriology,* which divides the bacilli by gram reaction and subgroups them by morphological characteristics and oxygen utilization (see figure D, page 474). As in chapter 15, we will emphasize unique properties of the clinically significant species and the epidemiology, pathology, and control measures of the principal diseases.

Medically Important Gram-Positive Bacilli

The gram-positive bacilli may be subdivided into three general groups based on the presence or absence of endospores and the characteristic of acid-fastness. Further levels of separation correspond to oxygen requirements and cell morphology. This scheme may be organized as follows:

Endospore-forming
- Aerobic: *Bacillus*
- Anaerobic: *Clostridium*

Non-endospore-forming
- Regular in morphology
  - *Listeria*
  - *Erysipelothrix*
- Irregular in morphology
  - Aerobic: *Corynebacterium*
  - Anaerobic: *Propionibacterium*

Acid-fast bacilli
- *Mycobacterium*
- *Nocardia*

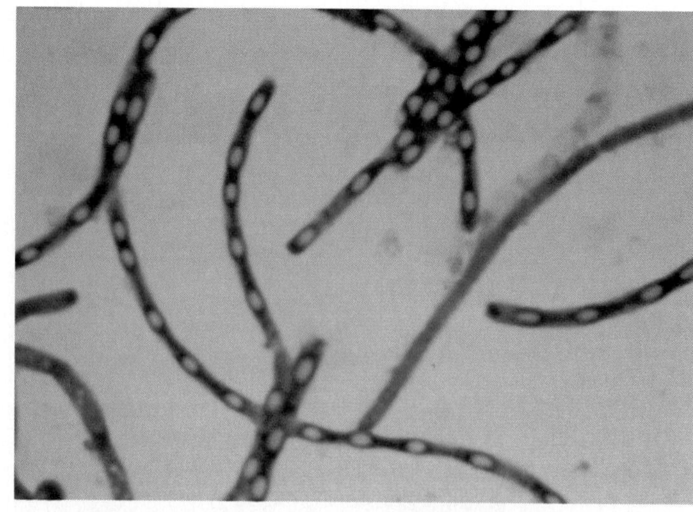

(a)

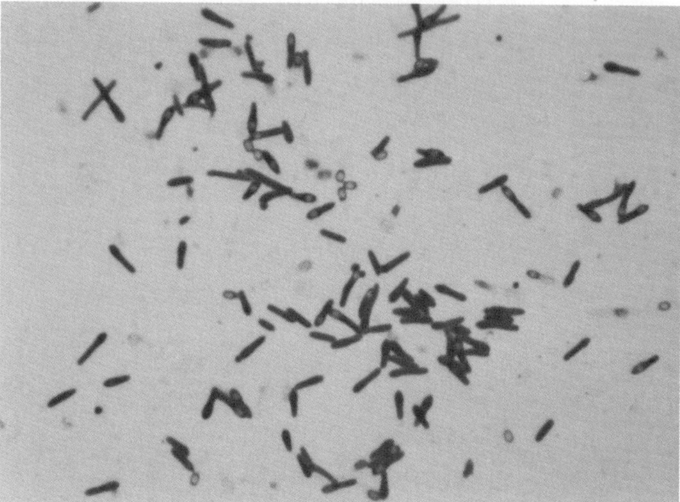

(b)

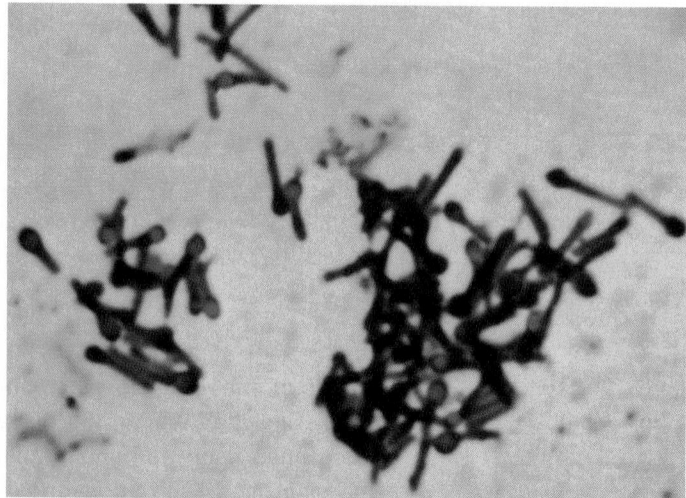

(c)

Figure 16.1 Examples of endospore-forming pathogens. (*a*) The morphological appearance of *Bacillus anthracis,* showing centrally placed endospores (×600). (*b*) A smear of *Clostridium perfringens* with central to terminal spores (×410). (*c*) *Cl. tetani* (×710). Its typical tennis racquet morphology is created by terminal spores that swell the sporangium.

Gram-Positive Spore-Forming Bacilli

Most endospore-forming bacteria are gram-positive, motile, rod-shaped forms in the genera *Bacillus, Clostridium,* and *Sporolactobacillus.* An endospore is a dense survival unit that develops in a vegetative cell in response to nutrient deprivation (figure 16.1; see figure 3.21). The extreme resistance to heat, drying, radiation, and chemicals accounts for the survival, longevity, and ecological niche of sporeformers, and it is also relevant to their pathogenicity.

General Characteristics of the Genus *Bacillus*

The genus *Bacillus* includes a large assembly of mostly saprobic bacteria widely distributed in the earth's habitats. *Bacillus* species are aerobic and catalase positive, and though they have varied nutritional requirements, none is fastidious. The group is noted for its versatility in degrading complex macromolecules, and it is also a common source of antibiotics. Because the primary habitat of many species is the soil, spores are continuously dispersed by means of dust into water and onto the bodies of plants and animals. Despite their ubiquity, the only two species with medical importance are *B. anthracis,* the cause of anthrax, and *B. cereus,* the cause of one type of food poisoning.

Bacillus anthracis and Anthrax

Bacillus anthracis is among the largest of all bacterial pathogens, being composed of block-shaped, angular rods 3–5 μm long and 1–1.2 μm wide, and central spores that develop under all growth conditions except in the living body of the host (figure 16.1*a*). Its virulence factors include a capsule and exotoxins that in varying combinations produce edema and cell death. For centuries, *anthrax* has been known as a zoonotic disease of

anthrax (an′-thraks) Gr. *anthrax,* carbuncle.

Feature 16.1 A Bacillus That Could Kill Us

Mechanisms of biological warfare are frequently based on the pathogenic potential of microorganisms. Because spore-forming bacteria like *Bacillus* are so hardy, they have held a particular attraction to scientists as research subjects and, in one case, as the basis of a deadly bomb.

Late in World War II, the mounting outrage over German bombings of London and the pressure for retaliation forced the British and American governments to collaborate on a powerful means of striking back. Given the knowledge that pneumonic anthrax was 100% fatal in a short period of time, a highly secret council designed a bomb containing the spores of *B. anthracis*. In the original plan, the detonation of these deadly projectiles over six German cities would have caused a shower of spores to be inhaled by humans and animals. Some estimates predicted that this could have wiped out the population over hundreds of square miles.

According to documents released in 1986, a plant in Indiana actually began to manufacture the bombs, but by this time, the war had started to wind down. How close the Allies came to dropping these bombs was not disclosed, but one undeniable legacy of the plan remains. In the early 1940s, a British bacteriological unit conducting preliminary tests released anthrax spores onto Gruinart, a small Scottish island. The pathogen became so entrenched in the soil that the island is uninhabitable to this day.

Twenty years later, the U.S. Army's Special Operations Division conducted a different yet chilling experiment using *Bacillus subtilis*. The underlying purpose of this study was to trace the spread of microbes in aerosol form among passengers in a crowded Washington D.C. airport and bus terminal. Operatives carried special suitcases equipped to spray huge numbers of these bacteria into the air. Other suitcases were rigged to sample the air and detect its spore concentration. By design, these tests simulated an actual plan for germ warfare using smallpox virus and were to provide evidence of how rapidly and successfully this mode of transmission could carry the microbe across the country. Hundreds of unsuspecting persons inhaled "large and acceptably uniform" doses and carried them to their destinations, presumably, with no ill effects.

This study and several others like it were based on the unfortunate and disturbing assumption that *B. subtilis* is a harmless air contaminant. It should be clear to any student of microbiology that many species of bacteria are opportunists and are capable of causing disease, especially if present in large numbers. In fact, *B. subtilis* can cause lung and blood infections in persons with weakened immunity. Equally troubling is the lack of a follow-up study by the investigators to determine any repercussions. Although flawed, the test did fulfill its original purpose. It was determined that if smallpox virus had been used instead of *B. subtilis*, 300 persons in 93 cities would have become infected, and that similar exposures at terminals in several other large cities would have spread the disease rapidly to thousands of persons. It was also shown that this contamination procedure could be accomplished easily by a secret agent without challenge or detection.

domesticated livestock (sheep, cattle, goats). It has an important place in the history of medical microbiology because it was Robert Koch's model for developing his postulates in 1877, and later, Louis Pasteur used the disease to prove the usefulness of vaccination.

The anthrax bacillus is a facultative parasite that undergoes its cycle of vegetative growth and sporulation in the soil. Infection occurs when animals grazing on contaminated grasslands ingest spores. The pathogen is returned to the soil in animal excrement or when the animals die. There it sporulates and serves as a long-term reservoir of infection for the animal population (see feature 16.1). The majority of anthrax cases are reported in livestock from Africa, Asia, and the Middle East. Most recent cases in the United States have occurred in textile

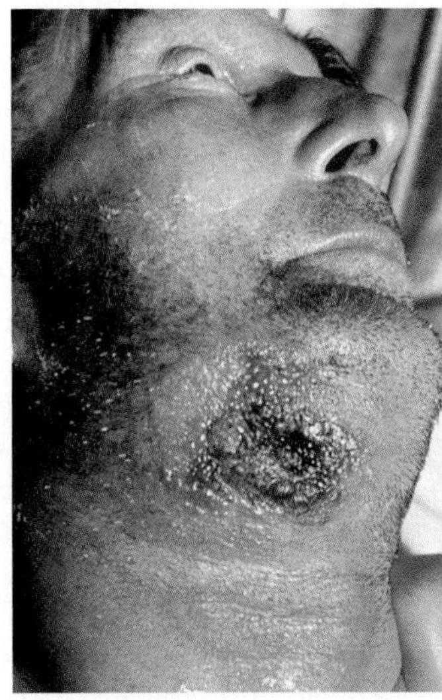

(a)

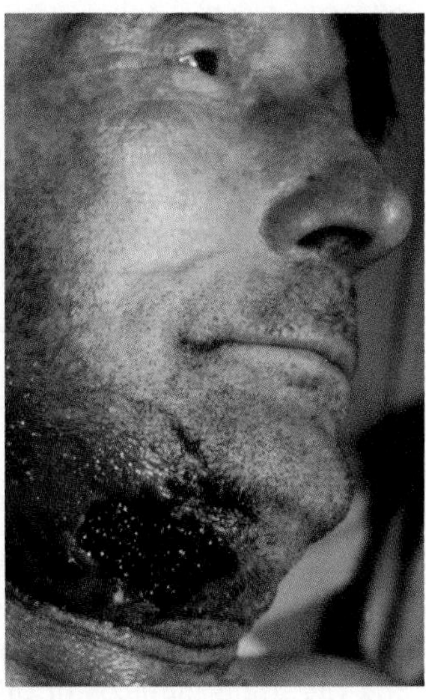

(b)

Figure 16.2 Cutaneous anthrax. (*a*) In early stages, the tissue around the site of invasion is inflamed and edematous. (*b*) A later stage reveals a thick necrotic lesion, the eschar. This will usually separate and heal spontaneously.

workers handling imported animal hair or hide or products made from them. Persons working with infected animals (veterinarians, livestock handlers) are also at slight risk. Because of effective control procedures, the number of cases in the United States is extremely low (fewer than 10 per year).

The circumstances of human infection depend upon the portal of entry. The most common and least dangerous of all forms is **cutaneous anthrax,** caused by spores entering the skin through small cuts and abrasions of the face, neck, or arms. Germination and growth of the pathogen in the skin is marked by the production of a papule that becomes increasingly inflamed and later ruptures to form a painless, black *eschar* (figure 16.2). Handlers of raw wool or hides are at risk of acquiring **pulmonary anthrax** (woolsorter's disease) by inhaling airborne spores. Bacilli growing in the lungs and releasing exotoxins produce a toxemia with wide-ranging pathological effects, including capillary thrombosis and cardiovascular shock. The disease is so deadly that it can be fatal in a few hours. Gastrointestinal anthrax acquired from contaminated meat is another rare but fatal form of the disease.

Methods of Anthrax Control Active cases of anthrax are treated with penicillin or tetracycline, but therapy does not lessen the effects of toxemia, so persons may still die. A vaccine containing live spores and a toxoid prepared from a special strain of *B. anthracis* are used to protect livestock in areas of the world where anthrax is endemic. Vaccination of humans with the purified toxoid alone is indicated only for persons who have occupational contact with livestock or products such as hides and bone. Animals that have died from anthrax must be burned or chemically decontaminated before burial to prevent establishing the microbe in the soil, and imported items containing animal hides, hair, and bone should be gas sterilized.

eschar (ess'-kar) Gr. *eschara*, scab.

Other *Bacillus* Species Involved in Human Disease

Bacillus cereus is a common airborne and dust-borne contaminant that multiplies very readily in cooked foods such as rice, potato, and meat dishes. The spores survive short periods of cooking and reheating, and when the food is stored at room temperature, the spores germinate and release enterotoxins. Ingestion of toxin-containing food causes nausea, vomiting, abdominal cramps, and diarrhea. There is no specific treatment, and the symptoms usually disappear within 24 hours.

For many years, most common airborne *Bacillus* species were dismissed as harmless contaminants with weak to nonexistent pathogenicity. However, infections by these species are increasingly reported in immunosuppressed patients, in those equipped with indwelling prosthetic devices, and in drug addicts who do not use sterile needles and syringes. An important contributing factor is that spores are abundant in the environment and the usual methods of disinfection and antisepsis are powerless to control them.

The Genus *Clostridium*

Another genus of gram-positive, spore-forming, heat-resistant rods that is widely distributed in nature is ***Clostridium.*** It is differentiated from *Bacillus* on the basis of oxygen requirements and catalase production. The large genus (nearly 100 species) is extremely varied in its habitats. Saprobic members reside in soil, sewage, vegetation, and organic debris, and commensals inhabit the bodies of humans and other animals. Infections caused by pathogenic species are not normally communicable, but occur when spores contaminate injured skin.

Clostridia cells produce oval or spherical spores that often swell the vegetative cell, and they occur in single, paired, or chain arrangements (see figure 16.1*b* and *c*). Their nutrient requirements are complex, and their metabolism is well adapted to decomposing a variety of substrates and synthesizing organic acids, alcohols, and other solvents through fermentation. This capacity makes some clostridial species essential tools of the chemical industry (see chapter 22). Other extracellular products, primarily exotoxins, play an important role in various clostridial diseases. Only six species of *Clostridium* are implicated in serious human disease. For a summary of some important species and their medical and economic importance, see table 16.1.

The Role of Clostridia in Infection and Disease

Clostridial disease can be divided into (1) wound and tissue infections, including myonecrosis, antibiotic-associated colitis, and tetanus, and (2) food intoxication of the perfringens and botulism varieties. Most of these diseases are caused when soluble exotoxins, some of which are highly potent, act on specific cellular targets (see feature 16.2).

Clostridium (klaw-strid′-ee-um) Gr. *closter,* spindle.

Table 16.1 Important Species of *Clostridium*

Species	Chief Importance to Humans	Description of Role
Cl. perfringens	Pathogen	Principal cause of gas gangrene and myonecrosis; common agent in enterotoxigenic food poisoning
Cl. novyi	Pathogen	Second most frequent cause of gas gangrene
Cl. septicum	Pathogen	Third most frequent cause of gas gangrene
Cl. tetani	Pathogen	Cause of tetanus
Cl. botulinum	Pathogen	Cause of botulism
Cl. difficile	Opportunist	Involved in antibiotic-associated colitis
Cl. iodophilum	Industrial uses	Produces organic acids and alcohols for commercial use
Cl. acetobutylicum	Industrial uses	Produces acids, alcohols, and benzene
Cl. butyricum	Industrial uses	Produces butyric acid in butter and cheese
Cl. cellobiofavum	Industrial uses	Digests cellulose

Feature 16.2 Toxicity in the Extreme

Imagine a substance so poisonous that a single microgram of it is a mass lethal dose for 200,000 mice, and a cup of it in concentrated form is capable of killing all the humans on the planet. This toxin is 100,000 times more powerful than rattlesnake venom and a million times more potent than strychnine. Although this formidable chemical might sound like the science fiction creation of a mad scientist, it actually exists as a product of the common soil bacterium *Clostridium botulinum.* Curiously, the genus *Clostridium* is the source of two of the most deadly toxins known, botulin and tetanospasmin. These two toxins share several other characteristics. For instance, both are polypeptides coded by plasmids and acquired through transduction by a virus. Both are neurotoxins that act on nerve cell processes involved in muscular activity, and both cause loss of voluntary muscle control (paralysis).

But here the similarities end. The diseases these two toxins cause are acquired through generally different modes. In tetanus, the toxin is formed in the tissues, and in the prominent form of botulism, the toxin is formed in food and then ingested. The mechanism by which the two toxins cause paralysis is also different. Tetanospasmin blocks the activity of inhibitory neurons in the central nervous system, and botulin acts on motoneurons in the peripheral nervous system. Tetanospasmin causes spastic (rigid) paralysis due to hyperactive muscles and sustained contraction (see figure 16.6). Botulin causes flaccid paralysis by interfering with the transmission of impulses to muscles and blocking contraction (see figure 16.8).

Despite the extreme toxicity of these compounds, they have some surprising medicinal applications. Botulin has been approved as a treatment to correct eye disorders such as cross-eye or walleye that are due to overactivity in one group of external eye muscles. Injecting a very diluted dose of the drug Oculinum into the eye muscles temporarily paralyzes them and allows the opposing muscles to correct the eye position. Tetanus toxin is such an excellent brain cell binding agent that medical scientists are adapting a nontoxic fragment of it to deliver antibodies or drugs across the blood-brain barrier and into the brain and spinal cord.

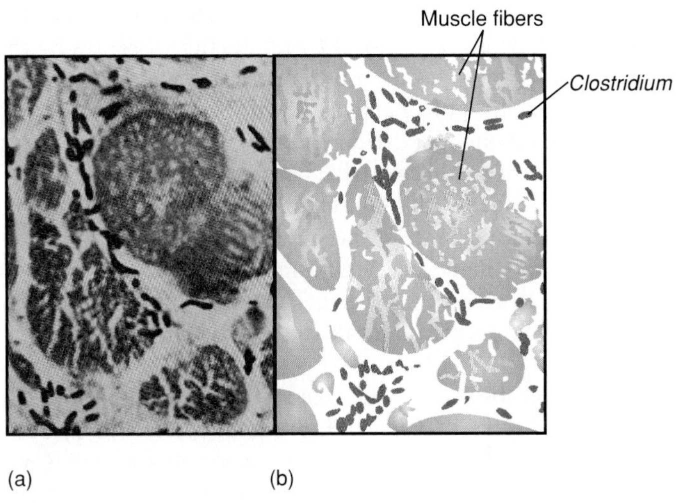

Figure 16.3 (*a*) A microscopic picture of clostridial myonecrosis, showing a histologic section of gangrenous skeletal muscle in cross section. (*b*) A schematic drawing of the same section. The growth of *Cl. perfringens* (plump rods) has caused gas formation and separation of the fibers.

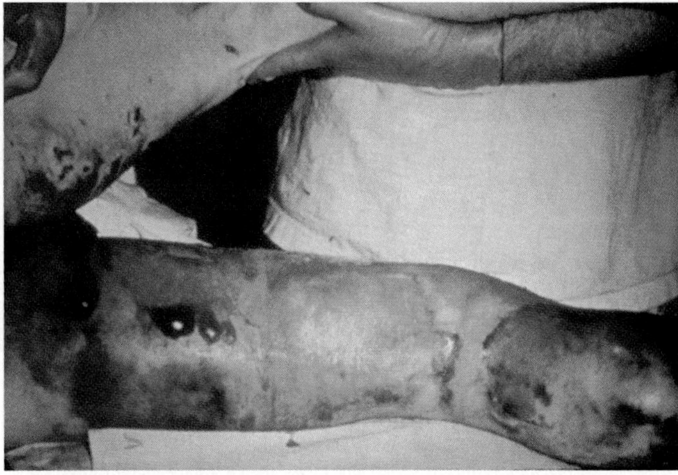
Figure 16.4 The clinical appearance of myonecrosis in a compound fracture of the leg. Necrosis has travelled from the main site of the break to other areas of the leg. Note the explosive nature of the infection, with blackening, general tissue destruction, and bubbles on the skin caused by gas formation in underlying tissue.

Gas Gangrene

The majority of clostridial soft tissue and wound infections are caused by ***Clostridium perfringens,*** *Cl. novyi,* and *Cl. septicum.* The spores of these species can be found in soil, on human skin, and in the human intestine and vagina. The disease they cause has the common name **gas gangrene** in reference to the gas produced by the bacteria growing in the tissue. It is technically termed anaerobic cellulitis or *myonecrosis.* The conditions that predispose a person to gangrene are surgical incisions, malignancies, compound fractures, diabetic sores, septic abortions, puncture and gunshot wounds, and crushing injuries contaminated by spores from the body or the environment.

Because clostridia are not highly invasive, infection requires damaged or dead tissue that supplies not only nutrients for growth but an anaerobic environment. The low oxygen tension results from an interrupted blood supply and the presence of aerobic bacteria that use up oxygen. Such conditions stimulate spore germination, rapid vegetative growth in the dead tissue, and synthesis and release of exotoxins. *Cl. perfringens* produces several physiologically active toxins; the most potent one, *alpha toxin,* causes RBC rupture, edema, and tissue destruction (figure 16.3).

Extent and Symptoms of Infection Two forms of gas gangrene have been identified. In **anaerobic cellulitis,** previously injured necrotic tissue is invaded, producing toxin and gas, but the infection remains localized and does not spread into healthy tissue. The pathology of true **myonecrosis** is more destructive. Toxins produced in injured tissue, often large muscles such as the thigh, shoulder, and buttocks, diffuse into nearby healthy tissue and cause local necrosis there. This damaged tissue then serves as a focus for continued clostridial growth, toxin formation, and gas production. The disease can progress through an entire limb or body area, destroying tissues as it goes (figure 16.4). Initial symptoms of pain, edema, and a bloody exudate in the lesion are followed by fever, tachycardia, and necrosis, which appears as blackened tissue filled with bubbles of gas. Gangrenous infections of the uterus due to septic abortions and clostridial septicemia are particularly serious complications. If treatment is not initiated early, the disease is invariably fatal.

Treatment and Prevention of Gangrene *Debridement* of diseased tissue eliminates the conditions that promote the spread of infection. This is most difficult in the intestine or body cavity, where only limited amounts of tissue can be removed. Surgery is supplemented by large doses of a broad-spectrum cephalosporin such as cefoxitin to control infection. Passive immunization with polyvalent antitoxin may also be attempted, even though its effectiveness is questionable. Hyperbaric oxygen therapy, in which the affected part is exposed to pure oxygen gas in a pressurized chamber, can also lessen the severity of infection. The increased oxygen content of the tissues presumably blocks further bacterial multiplication and toxin production. Extensive myonecrosis of a limb may call for surgical removal or amputation.

One of the most effective ways to prevent clostridial wound infections is immediate, thorough, and rigorous cleansing and surgical repair of deep wounds, decubitus ulcers (bedsores), compound fractures, and infected incisions, along with prophylactic antibiotic therapy. Because of so many different antigen subtypes in this group, active immunization is not possible.

---

gangrene (gang′-green) Gr. *gangraina,* an eating sore. A necrotic condition associated with the release of gases.

myonecrosis (my″-oh-neh-kroh′-sis) Gr. *myo,* muscle, and *necros,* dead.

debridement (dih-breed′-ment) Surgical removal of dead or damaged tissue.

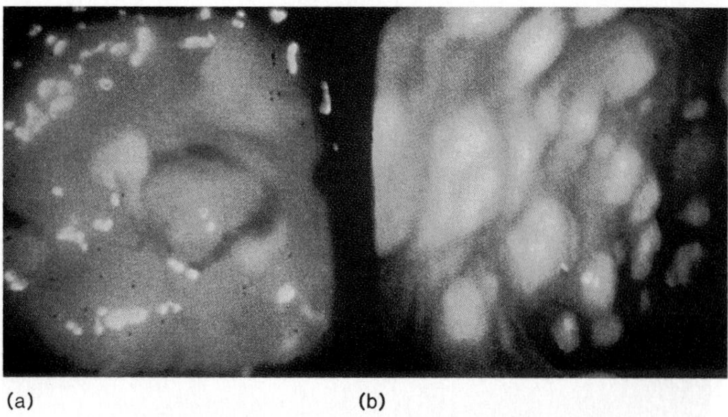

Figure 16.5 Antibiotic-associated colitis, as revealed by a sigmoidoscope, an instrument capable of photographing the interior of the colon. (*a*) A milder form with diffuse, inflammatory patches. (*b*) Heavy yellow plaques or pseudomembranes typical of more severe cases.

Antibiotic-Associated Colitis

A nosocomial infection called **antibiotic-associated colitis** is the second most common intestinal infection after salmonellosis in industrialized countries. The disease is caused by ***Clostridium difficile,*** a minor but normal resident of the intestine that was once considered relatively harmless. In most instances, this infection is traced to therapy with broad-spectrum antibiotics such as ampicillin, clindamycin, and cephalosporins.

Although the mechanisms of its pathogenesis are not completely understood, drug-resistant *Cl. difficile* seems to superinfect the intestine when the normal flora have been disrupted by antibiotics. Although this bacterium is relatively noninvasive, it produces toxins that injure the intestinal lining. The predominant symptom is diarrhea commencing late in therapy or even after therapy has stopped. More severe cases exhibit abdominal cramps, fever, and leukocytosis. The colon is inflamed and gradually sloughs off loose, membranelike patches called pseudomembranes consisting of fibrin and cells (figure 16.5). If the condition is not arrested, cecal perforation and death may result.

Mild, uncomplicated cases respond to withdrawal of antibiotics and replacement therapy for lost fluids and electrolytes. More severe infections are treated with oral vancomycin for several weeks until the intestinal flora returns to normal. Because infected persons often shed large numbers of spores in their stools, increased precautions are necessary to prevent spread of the agent to other patients who may be on antimicrobic therapy.

Tetanus or Lockjaw

Tetanus is a neuromuscular disease whose alternate name **lockjaw** refers to an early effect of the disease on the jaw muscle. The etiologic agent, ***Clostridium tetani,*** is a common resident of cultivated soil and the gastrointestinal tracts of humans and animals. Spores usually enter the body through accidental puncture wounds, burns, umbilical stumps, and (occasionally) medical procedures. Although tetanus is sometimes connected with stepping on a rusty nail, the introduction of spores by the soiled nail is of greater concern than the injury itself.

The incidence of tetanus is low in North America. Most cases occur among geriatric patients and intravenous drug abusers. In Third World countries, neonatal tetanus is common—predominantly the result of an infected umbilical stump or circumcision. The incidence of neonatal tetanus is higher in cultures that apply dung, ashes, and mud to these sites to arrest bleeding or as a customary ritual. The disease accounts for several hundred thousand infant deaths per year.

The Course of Infection and Disease The risk for tetanus occurs when spores of *Clostridium tetani* are forced into injured tissue. But the mere presence of spores in a wound is not sufficient to initiate infection because the bacterium is unable to invade damaged tissues readily. It is also a strict anaerobe, and the spores cannot become established unless tissues at the site of the wound are necrotic and poorly supplied with blood, conditions that favor germination.

As the vegetative cells grow, various metabolic products are released into the infection site. Of these, the most serious is **tetanospasmin,** a potent neurotoxin that accounts for the major symptoms of tetanus. The toxin spreads to nearby motor nerve endings in the injured tissue, binding there and travelling via axons to the ventral horns of the spinal cord (figure 16.6). In the spinal column, the toxin binds to specific target sites on the spinal neurons that are responsible for inhibiting skeletal muscle contraction. Only a small amount of toxin is required to initiate the disease. The incubation period varies from four to ten days, and shorter incubation periods signify a more serious condition.

When tetanospasmin blocks the release of neuroinhibitors necessary for regulating muscular contraction, the muscles are released from inhibition and begin to contract uncontrollably. Powerful muscle groups are most affected, and the first symptoms are clenching of the jaw, followed in succession by extreme arching of the back, flexion of the arms, and extension of the legs (figure 16.7). Lockjaw confers the bizarre appearance of *risus sardonicus* (sarcastic grin) that looks eerily as though the person is smiling (see figure 16.6*c*). Not only are these contractions spasmodic and extremely painful, but they are so powerful that they may even break bones, especially the vertebrae. Death is most often due to paralysis of the respiratory muscles and respiratory collapse. The fatality rate, ranging from 10% to 70%, is highest in cases involving delayed medical attention, a short incubation time, or head wounds. Full recovery requires a few weeks, and other than transient stiffness, no permanent damage to the muscles usually remains.

Treatment and Prevention of Tetanus Tetanus treatment is aimed at deterring the degree of toxemia and infection and maintaining patient homeostasis. A patient with a clinical appearance suggestive of tetanus should immediately receive either human tetanus immune globulin (TIGH) from pooled sera or heterologous tetanus antitoxin from horses (less acceptable because of possible allergic reactions). Although the antitoxin inactivates circulating toxin, it will not counteract the effects of

---

tetanus (tet′-ah-nus) Gr. *tetanos,* to stretch.

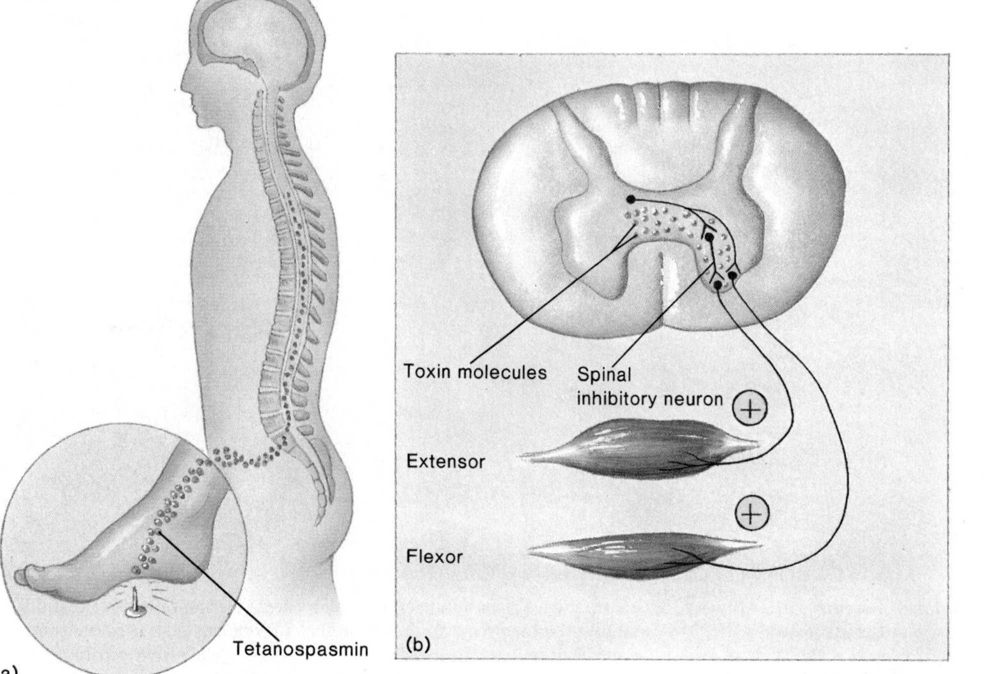

Figure 16.6 The events in tetanus. (*a*) Following traumatic injury, bacilli infecting the local tissues secrete tetanospasmin, which is absorbed by the peripheral axons and carried to the target neurons in the spinal column. (*b*) In the spinal cord, the toxin attaches to the junctions of regulatory neurons that inhibit inappropriate contraction. Released from inhibition, the muscles receive constant stimuli and contract uncontrollably, even opposing members of a muscle group. (*c*) Muscles contract spasmodically, without regard to regulatory mechanisms or conscious control.

Figure 16.7 Baby with neonatal tetanus, showing spastic paralysis of the paravertebral muscles, which locks the back into a rigid, arched position. Also note the abnormal flexion of the arms and legs.

toxin already bound to neurons. Other treatment methods include thoroughly cleansing and removing the afflicted tissue, controlling infection with penicillin or clindamycin, and administering muscle relaxants. The patient may require the assistance of a respirator, and a tracheostomy[1] is sometimes performed to prevent respiratory complications such as aspiration pneumonia or lung collapse.

Tetanus is one of the world's most preventable diseases, chiefly because of an effective vaccine containing tetanus toxoid. During World War II, only 12 cases of tetanus occurred among 2,750,000 wounded soldiers who had been previously vaccinated. The recommended vaccination series for one- to three-month-old babies consists of three injections given one month apart, followed by booster doses one and four years later. Children thus immunized probably have protection for 10 years. Additional protection against neonatal tetanus may be achieved by vaccinating pregnant women whose antibodies will be passed to the fetus. Toxoid should also be given to injured persons who have never been immunized, have not completed the series, or whose last booster was received more than 10 years previously. Passive immunization with human globulin is indicated primarily for individuals whose injuries are dirty, unattended, or necrotic. It is usually given in combination with the vaccine to achieve immediate and long-term protection.

Clostridial Food Poisoning

Two *Clostridium* species are involved in food intoxication, a type of food poisoning caused by ingesting a toxin produced by microorganisms multiplying in food. This is different from food infection, which occurs when ingested microbes infect and multiply in the intestine (see feature 16.6 and figure 22.27). ***Clostridium perfringens,*** type A, accounts for a mild illness that is the second most common form of food poisoning worldwide, whereas ***Clostridium botulinum*** produces a rarer but more toxic disease.

1. The surgical formation of an air passage by perforation of the trachea.

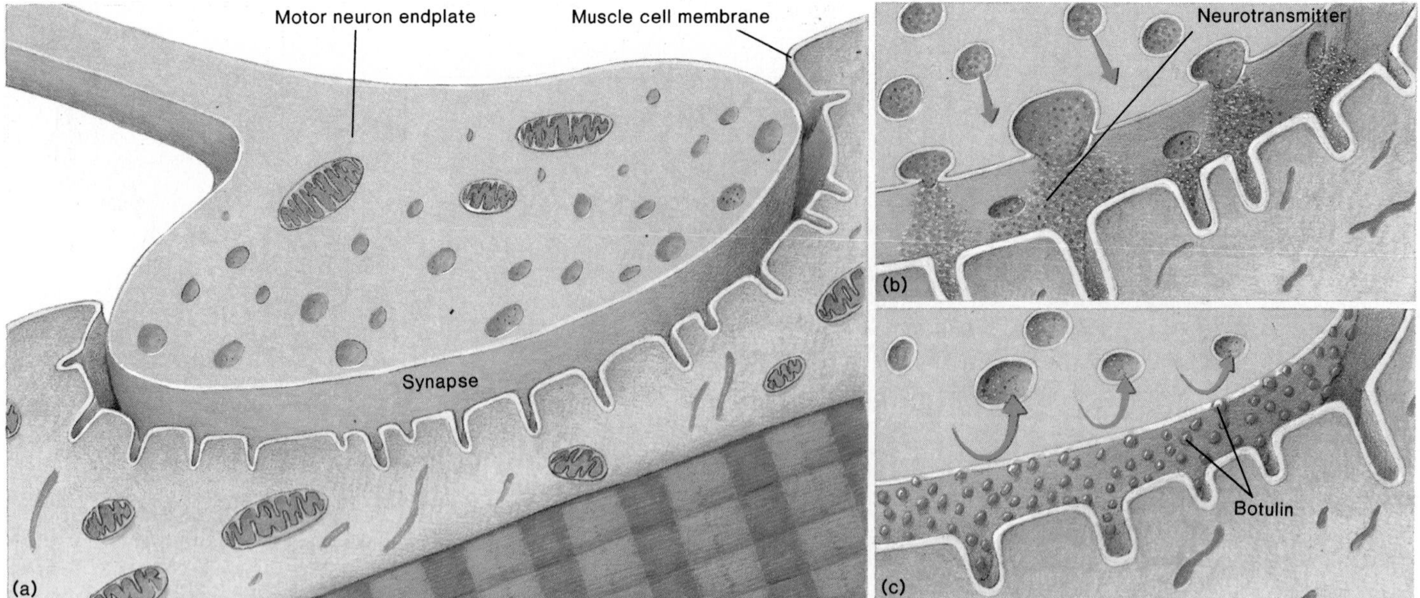

Figure 16.8 The physiological effects of botulism toxin (botulin). (*a*) The relationship between the motor neuron and the muscle at the myoneural junction. (*b*) In the normal state, a neurotransmitter released at the synapse crosses to the muscle and creates an impulse that stimulates muscle contraction. (*c*) In botulism, the toxin enters the motor end plate and attaches to the presynaptic membrane. This blocks release of the transmitter, prevents impulse transmission, and keeps the muscle from contracting.

Perfringens Food Poisoning *Clostridium perfringens* spores contaminate many kinds of food, but those most frequently involved in disease are meat, poultry, and fish that have not been cooked thoroughly enough to destroy the spores. When these foods are cooled, spores germinate, and the germinated cells multiply. If the food is eaten without adequate reheating, *Cl. perfringens* enters the small intestine of its victim and synthesizes enterotoxin. The toxin, acting upon epithelial cells, initiates acute abdominal pain and diarrhea, and occasionally nausea, in 8 to 24 hours. Recovery is rapid, and deaths are extremely rare.

Botulinum Food Poisoning **Botulism** is an intoxication associated with eating poorly preserved foods, though it may occasionally occur as an infection. Up until recent times, it was relatively prevalent and commonly fatal, but modern techniques of food preservation and medical treatment have reduced both its incidence and its fatality rate. However, botulism is a common cause of death in livestock that have grazed on contaminated food and in aquatic birds that have eaten decayed vegetation.

Clostridium botulinum is a spore-forming anaerobe that commonly inhabits soil and water, and occasionally the intestinal tract of animals. It is distributed worldwide, but occurs more often in the northern hemisphere. The species has eight distinctly different types, (designated A, B, C1, C2, D, E, F, and G), each of which varies in distribution among animals, regions of the world, and type of exotoxin. Human disease is usually associated with types A, B, E, and F, and animal disease with types A, B, C, D, and E.

botulism (boch'-oo-lizm) L. *botulis,* sausage. The disease was originally linked to spoiled sausage.

Botulism Food Intoxication There is a high correlation between cultural dietary preferences and food-borne botulism. In the United States, the disease is more often associated with low-acid vegetables (green beans, corn), fruits, and occasionally home-cured meats and fish. Because commercially canned foods are specifically heated to 121 °C for 20 or more minutes to destroy the botulinum spores, they are rarely implicated in botulism.

The factors in food processing that lead to botulism are very much a matter of circumstances. Spores are present on the vegetables or meat at the time of gathering and are difficult to remove completely by washing. When contaminated food is bottled and steamed in a pressure cooker that does not reach reliable pressure and temperature, some spores may survive (botulinum spores are highly heat resistant). At the same time, the pressure is sufficient to evacuate the air and create anaerobic conditions. Thus, when the bottles are stored at room temperature, the spores germinate and begin vegetative growth. One of the products of metabolism is **botulin,** the most potent microbial toxin known (see feature 16.2).

Bacterial growth may not be evident in the appearance of the bottle or can or in the food's taste or texture, and may give off only minute amounts of toxin. Swallowed toxin enters the small intestine and is absorbed into the lymphatics and circulation. From there, it travels to its principal site of action, the myoneural junctions of skeletal muscles (figure 16.8). The usual time before onset of symptoms is 12–36 hours, depending on the size of the dose. Neuromuscular symptoms first affect the muscles of the head, and include double vision and difficulty in swallowing and speaking, but there is no sensory or mental lapse. Although nausea and vomiting may occur at an early stage, they are not common. Later symptoms are descending muscular paralysis and respiratory compromise. In the past, death resulted from stoppage of respiration, but mechanical respirators have now greatly reduced the fatality rate.

Infant and Wound Botulism In rare instances, *C. botulinum* causes infection and toxemia. In these special circumstances, called infant and wound botulism, the spores infect a susceptible tissue and produce toxin there rather than in food.

Infant botulism was first described in the late 1970s in children between the ages of two weeks and six months who had ingested spores. The exact food source is not always known, although raw honey has been implicated in some cases. Apparently, the immature state of the neonatal intestine and microbial flora allows the spores to gain a foothold. As in adults, babies exhibit flaccid paralysis, usually with a weak sucking response, generalized loss of tone (the "floppy baby syndrome"), and respiratory complications. Although adults may also ingest botulinum spores in contaminated vegetables and other foods, the adult intestinal tract normally inhibits this sort of infection. In **wound botulism,** the spores enter a wound or puncture much as in tetanus, but the symptoms are similar to those of food-borne botulism. Increased cases of this form of botulism are being reported in intravenous drug abusers.

Treatment and Prevention of Botulism To differentiate botulism from other neuromuscular conditions, food samples should be retained, the patient's blood sampled, and the gastrointestinal tract pumped. The Centers for Disease Control provides a source of type A, B, and E trivalent horse antitoxins that must be administered early for greatest effectiveness. Patients are also managed with respiratory and cardiac support systems. Infectious botulism is treated with penicillin to control the microbe's growth and toxin production. Improved clinical management has increased the survival rate for food botulism to 85%.

Botulism will always be a potential threat for persons consuming home-preserved foods. Preventing it thus depends on educating the public about the proper methods of preserving and handling canned foods. Pressure cookers should be tested for accuracy in sterilizing, and home canners should be aware of the types of foods and conditions likely to cause botulism. Although acidic foods (tomatoes, fruits) are traditionally thought to inhibit the microbe, recent findings indicate that acid content alone may not be sufficient to prevent bacterial growth and toxin production. Other effective preventatives include disposal of bulging cans or bottles that look or smell spoiled and boiling all home-bottled foods for 10 minutes prior to eating, since the toxin is heat sensitive and rapidly inactivated at 100°C.

Differential Diagnosis of Clostridial Species

Although clostridia are common isolates, their clinical significance is not always immediately evident. Diagnosis frequently depends on the microbial load, the persistence of the isolate on resampling, and the condition of the patient. Laboratory differentiation relies on testing morphological and cultural characteristics, exoenzymes, carbohydrate fermentation, reaction in milk, and toxin production and pathogenicity. Some laboratories use a sophisticated method of gas chromatography that analyzes the chemical differences among species. Other valuable procedures are toxicity testing in mice or guinea pigs and serotyping with antitoxin neutralization tests.

Gram-Positive Regular Non-spore-forming Rods

The non-spore-forming gram-positive rods are a mixed group of genera subdivided on the basis of morphology and staining characteristics. One loose aggregate of seven genera is characterized as **regular** because they stain uniformly and do not assume pleomorphic shapes. Regular genera include *Lactobacillus, Listeria, Erysipelothrix, Kurthia, Caryophanon, Bronchothrix,* and *Renibacterium.* Medically significant representatives are *Listeria monocytogenes* and *Erysipelothrix rhusiopathiae.*

An Emerging Food-Borne Pathogen: *Listeria monocytogenes*

Listeria monocytogenes ranges in morphology from coccobacilli to long filaments in palisades formation (see table 16.2). Cells are actively motile with one to four flagella and do not produce capsules or spores. *Listeria* is not fastidious and is resistant to cold, heat, salt, pH extremes, and bile.

Epidemiology and Pathology of Listeriosis

The distribution of *L. monocytogenes* is so broad that its reservoir has been difficult to determine. It has been isolated all over the world from water, soil, plant materials, and the intestines of healthy mammals (including humans), birds, fish, and invertebrates. Apparently, the primary reservoir is soil and water, while animals, plants, and food are secondary sources of infection. Most cases of **listeriosis** are associated with ingesting contaminated milk, cheeses, ice cream, and meat. Except in cases of pregnancy, human-to-human transmission is probably not a significant factor, though humans may acquire infection from cattle, poultry, and shellfish. A predisposing factor in listeriosis seems to be the weakened condition of host defenses in the intestinal mucosa, since studies have shown that immunocompetent individuals are rather resistant to infection.

Listeriosis in normal adults is often a mild or subclinical infection with nonspecific symptoms of fever, diarrhea, and sore throat. Listeriosis in immunocompromised patients, fetuses, and neonates is associated with high morbidity and mortality. For reasons that are not well understood, pregnant women are highly susceptible to a mild form of the disease, which is transmitted to the infant prenatally when the microbe crosses the placenta or postnatally through the birth canal (figure 16.9). Intrauterine infections are widely systemic and usually result in premature abortion and fetal death. Infections in neonates localize in the meninges and cause extensive damage to the nervous system if not treated at an early stage.

Listeria monocytogenes (lis-ter′-ee-ah) For Joseph Lister, the English surgeon who pioneered antiseptic surgery; (mah″-noh-sy-toj′-uh-neez) For its effect on monocytes.

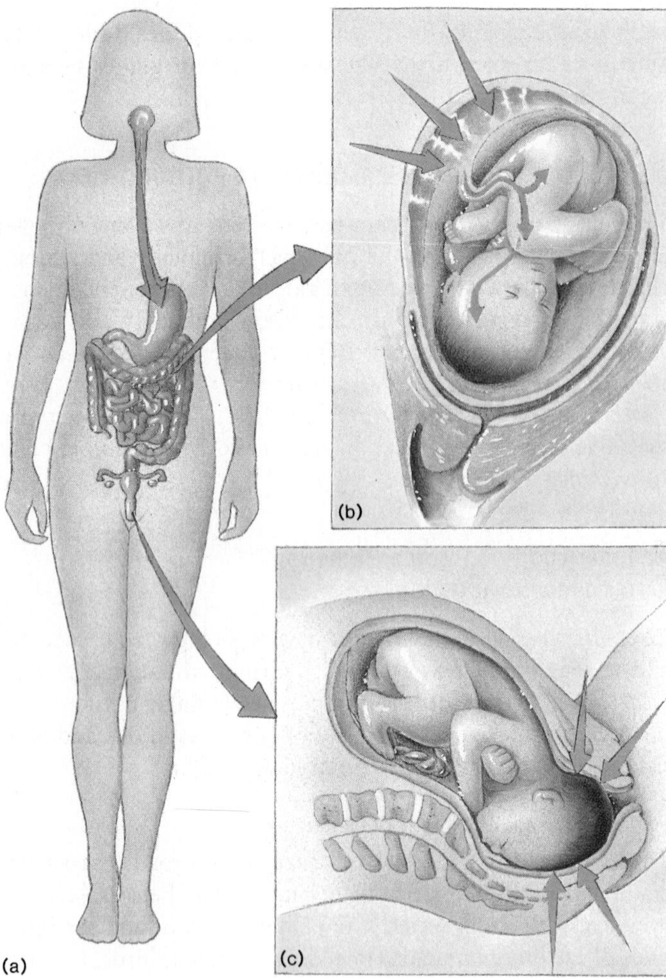

Figure 16.9 Principal routes of infection in listeriosis. (*a*) *Listeria monocytogenes* enters by the gastrointestinal route of adults and infects the intestine. From this site it may gain access to the circulation. (*b*) In pregnant women one complication occurs when the pathogen crosses the placenta and initiates a serious fetal infection or abortion. (*c*) An asymptomatic vaginal infection transmitted to the neonate at the time of birth creates symptoms of meningitis within a few hours to days.

Listeria in Dairy Products When an outbreak of listeriosis cropped up in California in 1985, almost every case was traced back to a fresh, soft cheese made with contaminated milk that had not been properly pasteurized. Of the 48 deaths, 30 occurred among fetuses and newborns and 18 among adults.

This epidemic, along with sporadic cases in several other states, spurred an in-depth investigation into the prevalence of *L. monocytogenes* in dairy products. Both domestic and imported cheeses were subjected to detailed bacteriologic analysis. The pathogen was isolated in 1% to 3.5% of the milk, cheeses, and ice creams tested. Cheeses made from raw milk and aged for several months are of special concern because *Listeria* bacilli can readily survive such processing and storage. Another area of current study is milk pasteurization, since *Listeria* is relatively resistant to ordinary pasteurization processes. Food and agriculture officials are intensifying their inspections of cheese producers and commercial cattle herds to avert a repeat epidemic.

Diagnosis and Control of Listeriosis

Diagnosing listeriosis is hampered by the difficulty in isolating it. The chances of isolation may be improved by using a procedure called cold enrichment, in which the specimen is held at 4°C and periodically plated onto media, but this may take four weeks. *L. monocytogenes* may be differentiated from the nonpathogenic *Listeria* and from other bacteria to which it bears a superficial resemblance by characteristics listed in table 16.2. Antibiotic therapy should be started as soon as listeriosis is suspected. Penicillin and ampicillin are the first choices, followed by erythromycin. Prevention focusses on adequately pasteurizing milk and cooking foods that may be contaminated with animal manure or sewage. Cold storage is not an effective control measure because the microbe can grow at refrigeration temperature.

Erysipelothrix rhusiopathiae: A Zoonotic Pathogen

Epidemiology, Pathogenesis, and Control

Erysipelothrix rhusiopathiae is a gram-positive rod widely distributed in animals and the environment. Its primary reservoir appears to be the tonsils of healthy pigs. It is also a normal flora of other vertebrates, and is commonly isolated from the excrement or dead carcasses of mammals, birds, and fish. It can persist for long periods in sewage, seawater, farm soils, smoked ham, and even pickled foods. The pathogen causes sporadic infections in wild and domestic animals and epidemics of swine erysipelas in pigs. Humans at greatest risk for infection are those who handle animals, carcasses, and meats, such as slaughterhouse workers, butchers, veterinarians, farmers, and fishermen.

The common portal of entry in human infections is a scratch or abrasion on the hand or arm. Upon entering the superficial tissues, the microbe multiplies locally to produce a disease known as **erysipeloid,** characterized by swollen, inflamed, dark red lesions that burn and itch (figure 16.10). Although the lesions usually heal without complications, rare cases of septicemia and endocarditis do arise. Inflamed red sores on the hands of persons in high-risk occupations may suggest erysipeloid, but the lesions must be cultured for confirmatory diagnosis. The condition is treated with penicillin or erythromycin. Swine erysipelas can be prevented by administering a vaccine, but the same vaccine does not protect humans. Animal handlers can lower their risk by wearing protective gloves.

Gram-Positive Irregular Non-spore-forming Rods

The **irregular,** non-spore-forming rods tend to be pleomorphic and to stain unevenly. Of the 20 genera in this category, *Corynebacterium, Mycobacterium,* and *Nocardia* have the greatest clinical significance. These three genera are grouped together because of similar morphological, genetic, and

Erysipelothrix rhusiopathiae (er″-ih-sip′-eh-loh-thriks) Gr. *erythros,* red, *pella,* skin, and *thrix,* filament. (ruz″-ee-oh-path′-ee-eye) Gr. *rhusios,* red, and *pathos,* disease.

Table 16.2 Characteristics for Differentiating *L. monocytogenes* from Similar Gram-Positive Bacteria

	Shape	Arrangement	Motility (20°–25°C)	Catalase	Morphology
L. monocytogenes	Rods, coccobacilli	Single, in chains	+	+	
Listeria spp.	Rods, coccobacilli	Single, in chains	+	+	
Streptococcus (esp. group D)	Cocci (oval)	Pairs, in chains	–	–	
Corynebacterium	Pleomorphic rods	Palisades, single	–	+	
Lactobacillus	Straight rods	Single, in chains	–	–	
Erysipelothrix	Long, slender rods	Filaments	–	–	

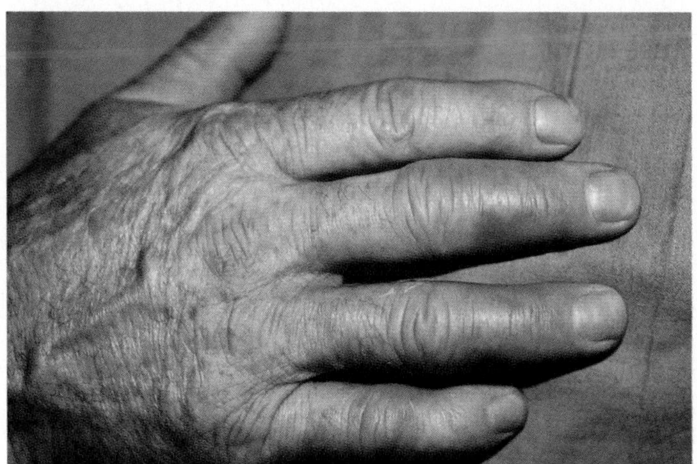

Figure 16.10 Erysipeloid on the finger.

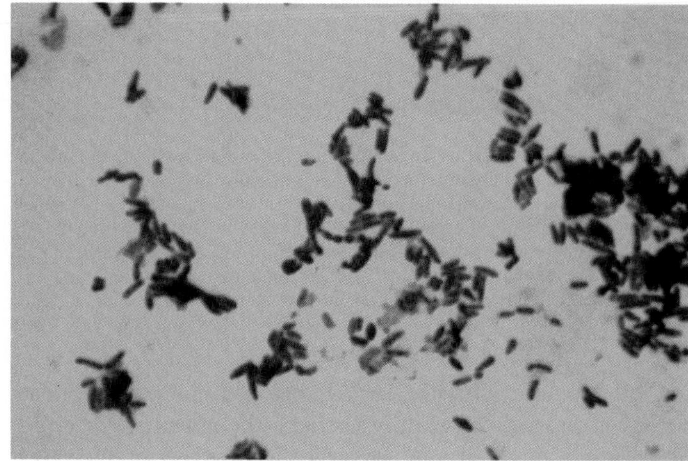

Figure 16.11 A photomicrograph of *C. diptheriae,* showing pleomorphism (especially club forms) and metachromatic granules (×600).

biochemical traits. They produce catalase and possess mycolic acids and a unique type of peptidoglycan in the cell wall. The following sections discuss the primary diseases associated with *Corynebacterium,* a similar species called *Propionibacterium,* and *Mycobacterium. Nocardia* is discussed with mycotic diseases in chapter 18.

Corynebacterium diphtheriae

Although several species of *Corynebacterium* are important, most human disease is associated with ***C. diphtheriae.*** In general morphology, this bacterium is a straight or somewhat curved rod that tapers at the ends, but it may have several pleomorphic variants because of thin spots that develop in the cell wall and distort cells into club, filamentous, and swollen shapes. Older cells are filled with polyphosphate granules and may occur in a palisades arrangement (figure 16.11).

Epidemiology of Diphtheria

For hundreds of years, **diphtheria** was one of the most devastating diseases, but in the last 50 years, both the number of cases and the fatality rate have steadily declined throughout the world. The current rate for the entire United States is 0.01 cases per

Corynebacterium (kor-eye″-nee-bak-ter′-ee-um) Gr. *koryne,* club, and *bakterion,* little rod.

diphtheria (dif-thee′-ree-ah) Gr. *diphthera,* membrane.

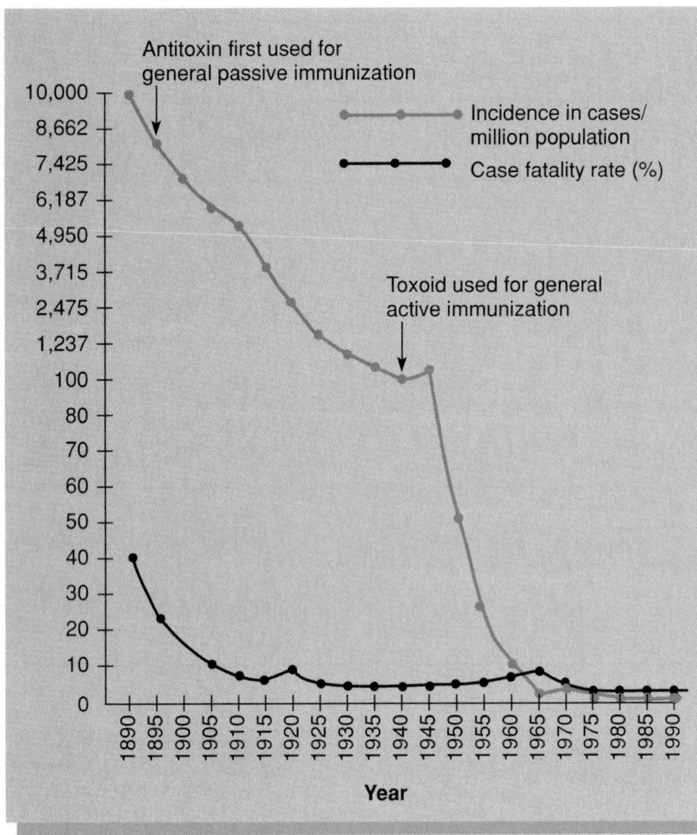

Figure 16.12 The incidence and case fatality rates for diphtheria in the United States during the last 100 years. This graph clearly documents the steady decline in cases, starting with the routine use of antitoxin in 1895 and the widespread use of the toxoid vaccine that began in 1945. The percentage of fatalities also dropped dramatically, long before antimicrobic drugs were available. The current residual level of cases is due to a small number of unimmunized carriers in the population at the present time.

million population (figure 16.12). Because many populations harbor a reservoir of healthy carriers, the potential for diphtheria is constantly present, as evidenced by the sporadic cases that crop up in every large U.S. city. Those at greatest risk are nonimmunized children from one to ten years of age living in crowded, unsanitary situations. Cutaneous diphtheria is a mild ulcerative condition of the skin that is most common among vagrant and homeless persons.

Pathology of Diphtheria

Exposure to the diphtheria bacillus usually results from close contact with skin crusts shed by human carriers or diseased persons, with droplet nuclei, or occasionally, with fomites or contaminated milk. The clinical disease proceeds in two stages: (1) local infection by *Corynebacterium diphtheriae,* and (2) toxin production and toxemia. The most common location of primary infection is in the upper respiratory tract (tonsils, pharynx, larynx, and trachea). The bacterium becomes established by means of virulence factors that assist in its attachment and growth. The cells are not ordinarily invasive and usually remain localized at the portal of entry.

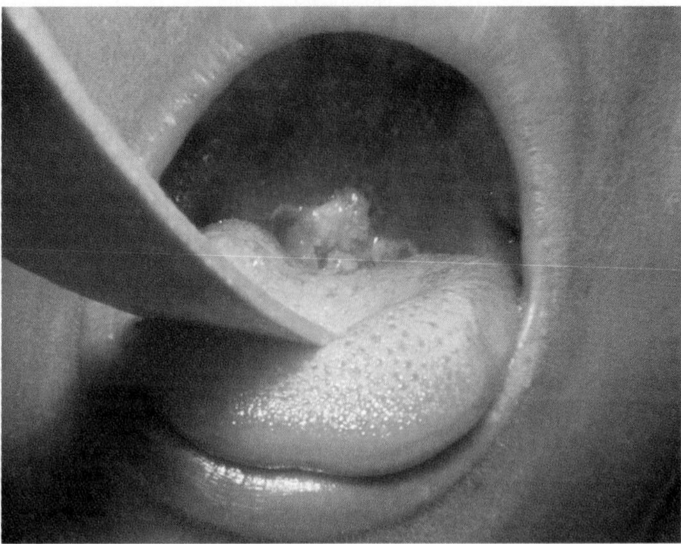

Figure 16.13 The clinical appearance of diphtheritic patients includes gross inflammation of the pharynx and tonsils marked by grayish patches (a pseudomembrane) and swelling over the entire area.

Diphtherotoxin and Toxemia Although infection is necessary for disease, the cardinal determinant of pathogenicity is the production of **diphtherotoxin.** According to studies, this exotoxin is produced only by toxigenic strains of *C. diphtheriae* that carry the structural gene for toxin production acquired from bacteriophages during transduction (see chapter 8). The toxin is a two-part protein consisting of one polypeptide fragment that enters mammalian target cells in the heart and nervous system and another fragment that arrests protein synthesis.

The toxin affects the body on two levels. Locally, it produces an inflammatory reaction, low-grade fever, sore throat, nausea, vomiting, enlarged cervical lymph nodes, and severe swelling in the neck. One life-threatening complication is the **pseudomembrane,** a greenish-gray sheet consisting of solidified fibrous exudate, cells, and fluid that develops in the pharynx (figure 16.13). The pseudomembrane is so leathery and tenacious that attempts to pull it away result in bleeding. If it develops in the respiratory tract, it can cause asphyxiation.

Systemically, the most dangerous complication is **toxemia,** which occurs when the toxin is absorbed from the throat and carried by the blood to certain target organs, primarily the heart and nerves. The action of the toxin on the heart causes myocarditis, fatty degeneration, and failure. Cranial and peripheral nerve involvement may progress to paralysis. Although toxic effects are usually reversible, patients with inadequate treatment often die from asphyxiation, respiratory complications, or heart damage.

Diagnostic Methods for the Corynebacteria

Diphtheria has such great potential for harm that often the physician must make a presumptive diagnosis and begin treatment before the bacteriologic analysis is complete. A gray membrane and swelling in the throat are somewhat indicative of diphtheria, although several diseases present a similar appearance. A Gram stain of a membrane or throat specimen could help rule out diphtheria. Epidemiologic factors such as living conditions, travel history, and immunologic history (a positive Schick test) can also aid in initial diagnosis.

Laboratory isolation can be completed on Loeffler's medium, Pai egg slants, or blood agar. A simple stain of *C. diphtheriae* colonies with alkaline methylene blue reveals cells with marked pleomorphism and granulation. Biochemical tests help differentiate *C. diphtheriae* from "diphtheroids"—similar species often present in clinical materials that are not primary pathogens. *Corynebacterium xerosis* normally lives in the eye, skin, and mucous membranes, and is an occasional opportunist in eye and postoperative infections. *Corynebacterium pseudodiphtheriticum,* a normal inhabitant of the human nasopharynx, may colonize natural and artificial heart valves.

Treatment and Prevention of Diphtheria

The adverse effects of toxemia are treated with diphtheria antitoxin (DAT) derived from horses. Prior to injection, the patient must be tested for allergy to horse serum and be desensitized if possible. The infection is treated with antibiotics such as penicillin or erythromycin. Bed rest, heart medication, and tracheostomy or bronchoscopy to remove the pseudomembrane may be indicated. Diphtheria can be easily prevented by a series of vaccinations with toxoid, usually given as part of a mixed vaccine against tetanus and pertussis called the DPT. Currently recommended are three vaccinations, starting at ten weeks of age, followed by a booster at one year and at school age. Older children and adults who are not immune to the toxin may be immunized with two doses of diphtheria-tetanus (DT) vaccine.

The Genus *Propionibacterium*

Propionibacterium resembles *Corynebacterium* in morphology and arrangement, but it differs by being aerotolerant or anaerobic and lacking mycolic acids. The most prominent species is ***P. acnes,*** a common resident of the pilosebaceous glands of human skin and occasionally the upper respiratory tract. The primary importance of this bacterium is its relationship with the familiar **acne vulgaris** lesions of adolescence. Acne is a complex syndrome influenced by genetic and hormonal factors as well as by the structure of the epidermis, but it is also an infection (see feature 16.3).

xerosis (zee-roh'-sis) Gr. *xerosis,* parched skin.

pseudodiphtheriticum (soo''-doh-dif-ther-it'-ih-kum) Gr. *pseudes,* false, and *diphtheriticus,* of diphtheria.

Propionibacterium (pro''-pee-on''-ee-bak-tee'-ree-um) Gr. *pro,* before, *prion,* fat, and *bakterion,* little rod. Named for its ability to produce propionic acid.

acnes (ak'-neez) Gr. *akme,* a point. (The original translators incorrectly changed an m to an n.)

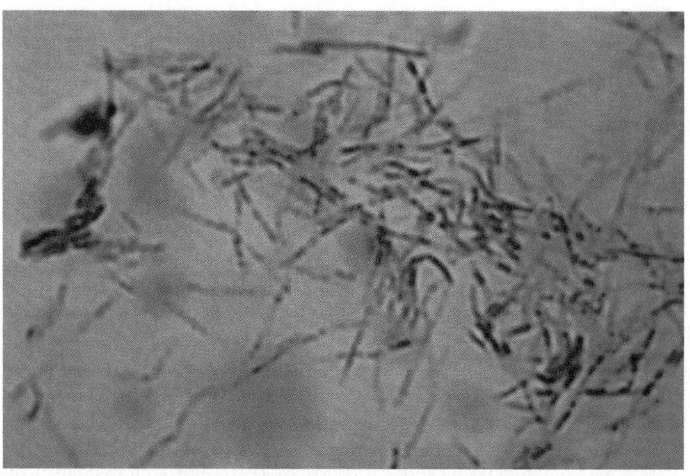

Figure 16.14 The microscopic morphology of mycobacteria (*M. tuberculosis* shown here), as illustrated by an acid-fast stain of sputum from a tubercular patient. Note the irregular morphology and the tendency for filaments to be formed (×700).

Mycobacteria: Acid-Fast Bacilli

The genus ***Mycobacterium*** is distinguished by its complex layered structure composed of high molecular weight mycolic acids and waxes. This high lipid content imparts the characteristic of **acid-fastness** and is responsible for the resistance of the group to drying, acids, and various germicides. The cells of mycobacteria are long, slender, straight, or curved rods with a slight tendency to be filamentous or branching (figure 16.14). Although they usually contain granules and vacuoles, they do not form capsules, flagella, or spores.

Most mycobacteria are strict aerobes that grow well on simple nutrients and media. Compared with other bacteria, the growth rate is generally slow, with generation times ranging from two hours to several days. Some members of the genus exhibit colonies containing yellow, orange, or pink carotenoid pigments that may require light for development, while others are nonpigmented. Many of the 50 mycobacterial species are saprobes living free in soil and water, and several are highly significant human pathogens (table 16.3). Worldwide, millions of persons are afflicted with the mycobacterial diseases tuberculosis and leprosy. Mycobacterial infections by species other than *M. tuberculosis* are called MOTT and are an increasing complication in immunosuppressed patients.

Mycobacterium tuberculosis: The Tubercle Bacillus

The *tubercle* bacillus is a long, thin rod that grows in sinuous masses or strands called cords. Unlike many bacteria, it produces no exotoxins or enzymes that could contribute to infectiousness. Most strains contain complex waxes and a cord factor

Mycobacterium (my''-koh-bak-tee'-ree-um) Gr. *myces,* fungus, and *bakterion,* a small rod.

tubercle (too'-ber-kul) L. *tuberculum,* little swelling.

Feature 16.3 Acne: The Microbial Connection

Normally, each pilosebaceous unit is a self-contained system for protecting, softening, and lubricating the skin (*1*). It contains one or more sebaceous glands that continuously release an oily secretion called sebum into the hair follicle. As hair and skin grow, the dead epidermal cells and sebum work their way upward and are discharged from the pore onto the skin surface.

Persons disposed to pimples and acne have inherited a skin structure that traps the mass of sebum and dead cells and clogs the pores. If the skin at the surface swells over the pore's entrance, a closed comedo or whitehead results (*2*); if the pore remains open to the surface but is blocked with a plug of sebum, it is called an open comedo or blackhead (*3*). The dark tinge of the blackhead is caused by the accumulation of the pigment melanin, not to uncleanliness. Further disruption to the follicle occurs when the sebaceous gland is stimulated by hormones (especially male) to secrete even more sebum into the blocked system. *Propionibacterium* bacteria in the follicle attempt to digest this surplus of oil by releasing lipases. The free fatty acids and bacterial antigens create a local inflammatory edema that bursts the follicle below the skin surface (*4*). In time, the lesion erupts as a papule or pustule (*5*). In a severe form of acne, secondary bacterial infections cause angry, deeply scarring lesions. Because acne depends in part on an infection, it can be suppressed with topical and oral antibiotics such as clindamycin, erythromycin, or tetracycline. Other types of therapy involve chemicals that enhance skin removal (benzoyl peroxide) and slow the production of sebum (Retin A and Accutane).

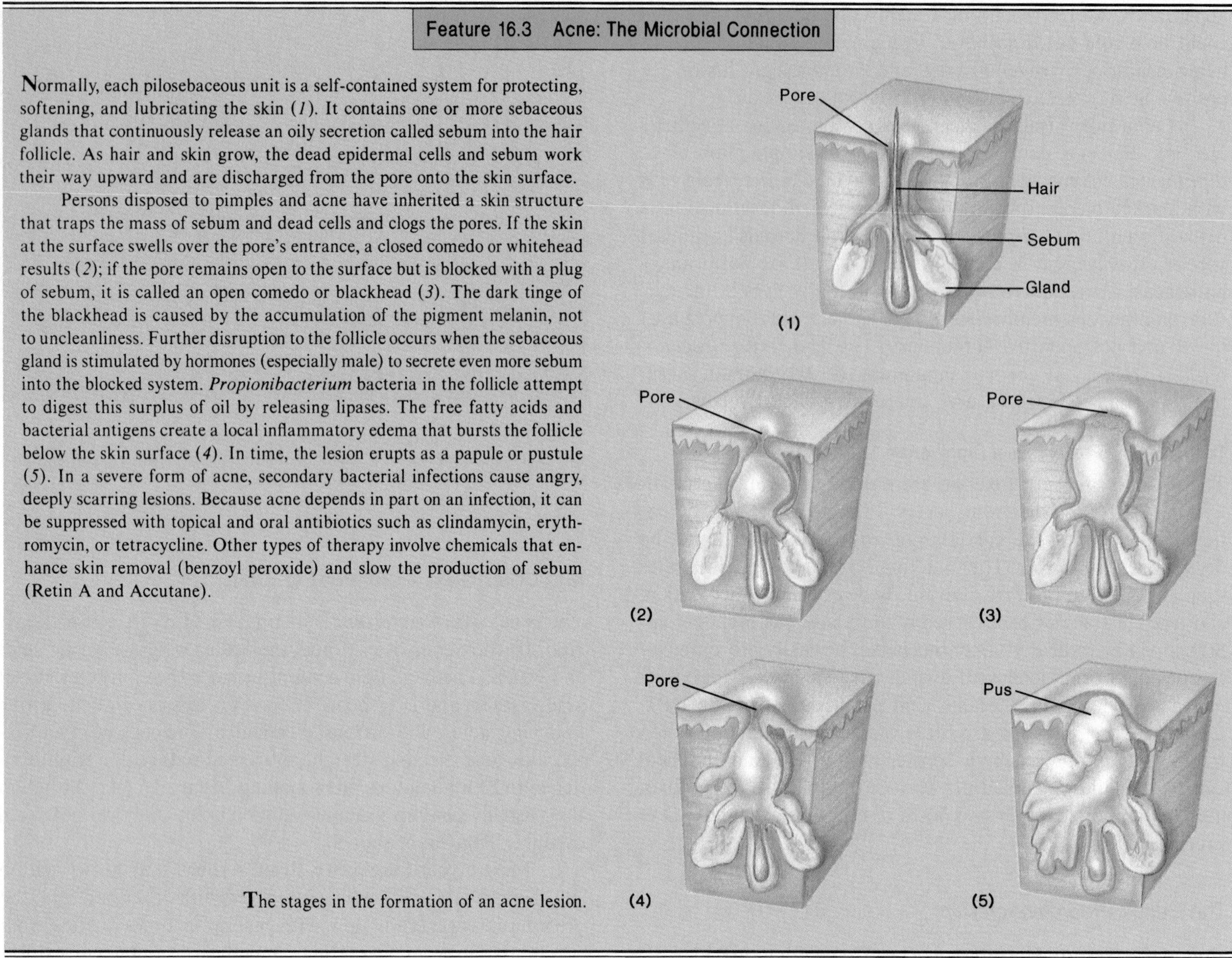

The stages in the formation of an acne lesion.

(figure 16.15), though the roles of these substances in pathogenesis are yet to be completely explained. It is currently thought that the antigens in the cell wall protect the mycobacteria from being destroyed by macrophages and that this contributes to their invasion and persistence.

Epidemiology and Transmission of Tuberculosis

Mummies from the Stone Age, ancient Egypt, and Peru provide unmistakable evidence that tuberculosis (TB) is one of the oldest human diseases. Just 100 years ago, tuberculosis was such a prevalent cause of death that it was called "Captain of the Men of Death" and "White Plague." Its epidemiologic patterns vary with the living conditions in a community or an area of the world. Factors that significantly affect a person's susceptibility to tuberculosis are poverty, inadequate nutrition, unsanitary living conditions, underlying debilitation of the immune system, lung damage, and genetics. It has been estimated that at least 20–25 million humans have active TB worldwide (with approximately three million deaths per year). Persons in developing countries are often infected as infants and harbor the microbe for many years until the disease is reactivated in young adulthood.

Cases in the United States show a strong correlation with the age, sex, and recent immigration history of the patient. The highest case rates occur in nonwhite males over 30 years of age and nonwhite females over 60. New immigrants from certain areas of Indochina, Central and South America, and Africa have a high rate of tuberculosis carriage. For several decades, the rates of tuberculosis had been steadily declining, but over the past five years, this trend has reversed, and TB is on the rise again due to higher rates of infection among homeless people, AIDS patients, and drug addicts.

The agent of tuberculosis is transmitted almost exclusively by fine dried mucous droplets that remain suspended in the air for several hours. The tubercle bacillus is very resistant and can survive for eight months in these aerosol particles. Some

Table 16.3 Differentiation of Important Mycobacteria

Species	Primary Habitat	Disease in Humans	Treatment	Rate of Growth*	Pigmentation*
M. tuberculosis	Humans	Tuberculosis (TB)	See text	S	NP
M. bovis	Cattle	Tuberculosis	Same as TB	S	NP
M. ulcerans	Humans	Skin ulcers	Surgery, grafts	S	NP
M. kansasii	Not clear	Opportunistic lung infection	Difficult, similar to TB	S	PP
M. marinum	Water, fish	Swimming pool granuloma	Tetracycline, rifampin	S	PP
M. scrofulaceum	Soil, water	Scrofula	Removal of lymph nodes	S	PS
M. avium-M. intracellulare complex	Birds	Opportunistic AIDS infection; lung infection like TB	Various drugs	S	NP
M. fortuitum-M. chelonae complex	Soil, water, animals	Wound abscess; postsurgical infection	4–6-drug regimen; surgery	R	NP
M. phlei	Sputum, soil	Not pathogenic	None	R	PS
M. smegmatis	Smegma, soil	Not pathogenic	None	R	Usually NP
M. leprae	Strict parasite of humans	Leprosy	See text	S	Cannot be grown in artificial media

*The mycobacteria are grouped into categories by their pigment production on special media or by their rate of growth. Photochromogens (PP) develop yellow to dark orange pigment in the presence of light; scotochromogens (PS) synthesize pigment in darkness; and nonpigmented forms (NP) have no color. Growth rate is rapid (R), occurring in less than seven days, or slow (S), occurring in more than seven days.

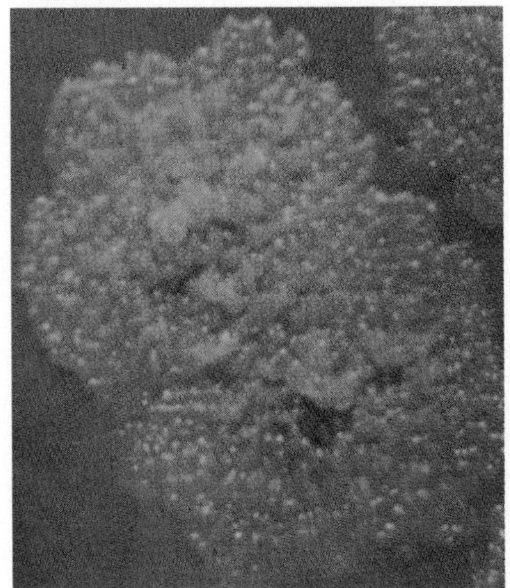

(a)

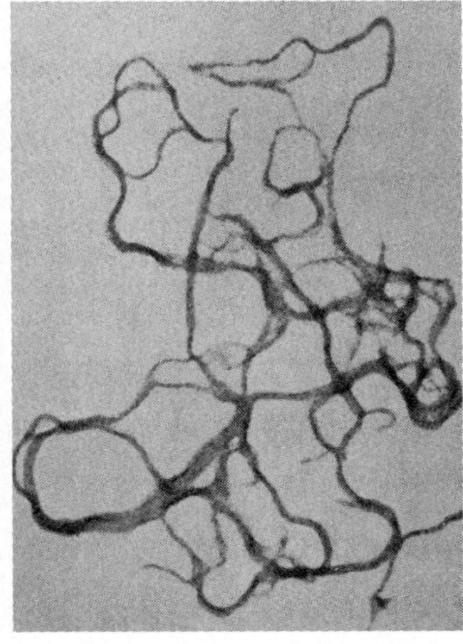

(b)

Figure 16.15 The cultural appearance of *M. tuberculosis*. (*a*) Colonies with a typical granular, waxy pattern of growth. (*b*) Magnified, the edge of a colony reveals thin, serpentine strands or cords typical of all virulent strains.

particles may be trapped in the mucus of the nasal passages and removed, but the tinier ones are inhaled into the bronchioles and alveoli. Persons sharing closed, small rooms with limited access to sunlight and fresh air are especially at risk.

The Course of Infection and Disease

A clear-cut distinction can be made between infection with the tubercle bacillus and the disease it causes. In general, humans are rather easily infected with the bacillus but resistant to the disease. Estimates project that only about 5% of infected persons actually develop a clinical case of tuberculosis. Untreated tuberculosis progresses slowly and is capable of lasting a lifetime, with periods of health alternating with episodes of morbidity. The majority (85%) of TB cases are centered in the lungs, even though disseminated tubercle bacilli may give rise to tuberculosis in any organ of the body. Clinical tuberculosis is divided into primary tuberculosis, secondary (reactivation or reinfection) tuberculosis, and disseminated tuberculosis.

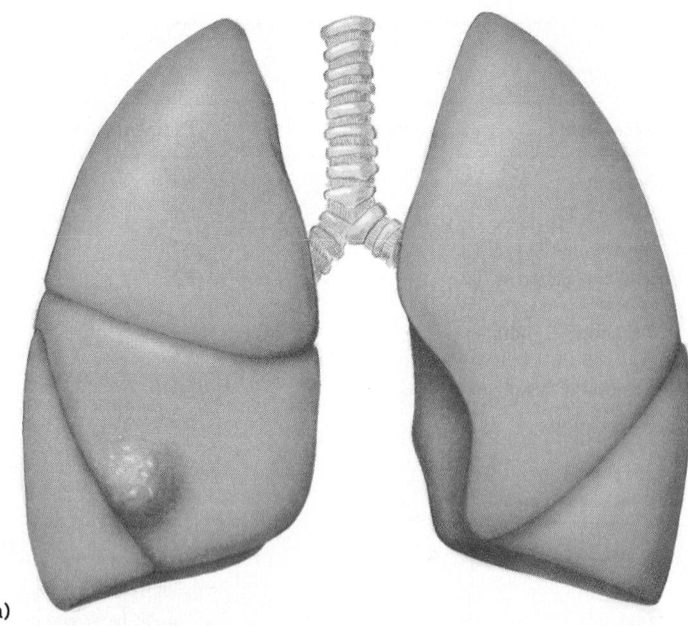

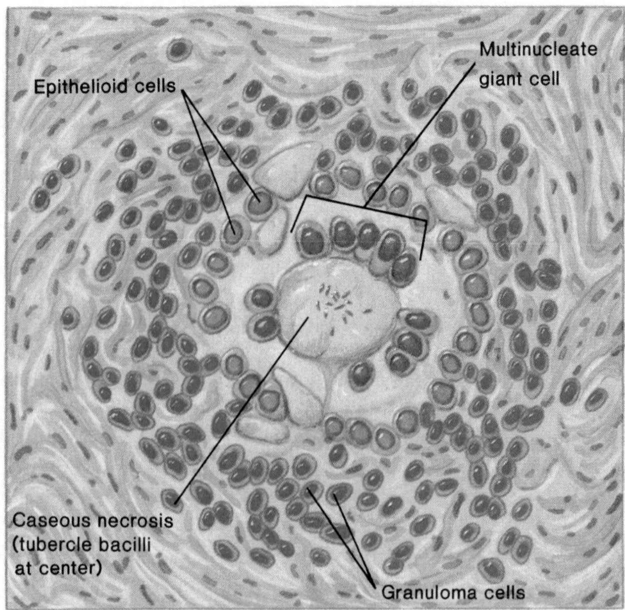

Figure 16.16 Tubercle formation. (*a*) A tubercle located in the lower right lung. (*b*) In this schematic drawing, the massive granuloma infiltrate has obliterated the alveoli and set up a dense collar of fibroblasts and lymphocytes (granuloma cells), and epithelioid cells. The core of this tubercle is caseous, with a cheesy material containing the bacilli.

Primary Tuberculosis The minimum infectious dose for lung infection is around 10 cells. The bacilli are phagocytosed by alveolar macrophages and multiply in this highly oxygenated environment. This period of hidden infection is asymptomatic or accompanied by mild fever, but some cells escape from the lungs into the blood and lymphatics. After three to four weeks, the immune system mounts a complex, cell-mediated assault against the bacilli. The large influx of mononuclear cells into the lungs plays a part in the formation of specific infection sites called **tubercles.** Tubercles are a type of granuloma that consist of a central core containing TB bacilli and enlarged macrophages and an outer wall made of fibroblasts, lymphocytes, and neutrophils (figure 16.16). Although this response checks further spread of infection and helps prevent the disease, it also carries a potential for damage. Frequently, the centers of tubercles break down into necrotic, *caseous* lesions that gradually heal by calcification (normal lung tissue is replaced by calcium deposits). The response of T cells to *M. tuberculosis* proteins also causes a delayed hypersensitivity response evident in the **tuberculin reaction,** a valuable diagnostic and epidemiologic tool (see feature 16.4).

Secondary Tuberculosis Although the majority of TB patients recover more or less completely from the primary episode of infection, live bacilli can remain dormant and become reactivated at a later time, especially in persons with weakened immunity. In chronic tuberculosis, tubercles filled with masses of bacilli expand and drain into the bronchial tubes and upper respiratory tract. Gradually, the patient experiences more severe symptoms, including violent coughing, greenish or bloody sputum, low-grade fever, anorexia, weight loss, extreme fatigue, night sweats, and chest pain.

A Very Fashionable Disease The older term for tuberculosis, **consumption,** conveys a sense that the TB patient (tubercular) wastes away, but during the eighteenth and early nineteenth centuries, many people actually believed that consumption generated great genius and creative powers. After all, many writers, poets, and composers of the time, including Frederic Chopin, John Keats, Elizabeth Barrett Browning, and Robert Louis Stevenson, died of the disease. The physical appearance of the tubercular—pale, emaciated, feverish—even became fashionable. Such was the sense of romance about this condition that Lord Byron was heard to say, "I should like to die of consumption . . . because the ladies would say, 'Look at that poor Byron. How interesting he looks in dying.' " And Henry Thoreau said, "Decay and disease are often beautiful, like the pearly tear of the shellfish and the hectic glow of consumption."

Extrapulmonary Tuberculosis During the course of secondary TB, the bacilli disseminate rapidly to sites other than the lungs. Organs most commonly involved in **extrapulmonary TB** are the regional lymph nodes, kidneys, long bones, genital tract, brain, and meninges. Because of the debilitation of the patient and the high load of tubercle bacilli, these complications are usually grave.

Renal tuberculosis results in necrosis and scarring of the renal medulla and the pelvis, ureters, and bladder. This damage is accompanied by painful urination, fever, and the presence of blood and the TB bacillus in urine. Genital tuberculosis in males damages the prostate gland, epididymis, seminal vesicles, and testes, and in females, the fallopian tubes, ovaries, and uterus. It often affects reproductive function in both sexes.

caseous (kay'-see-us) L. *caseus,* cheese. The material formed resembles cheese or curd.

Feature 16.4 Clinical Tests for Tuberculosis

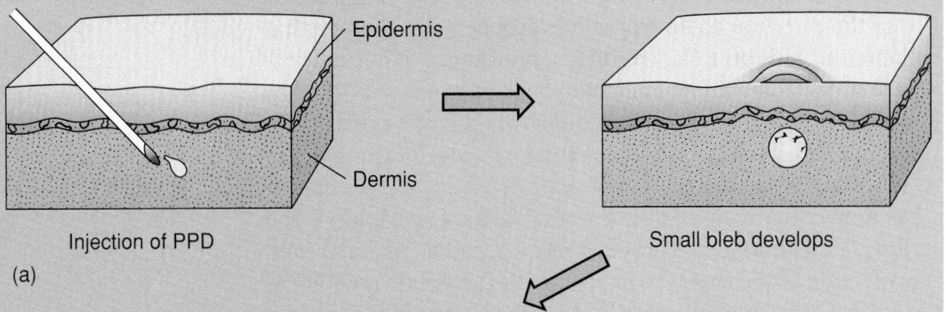

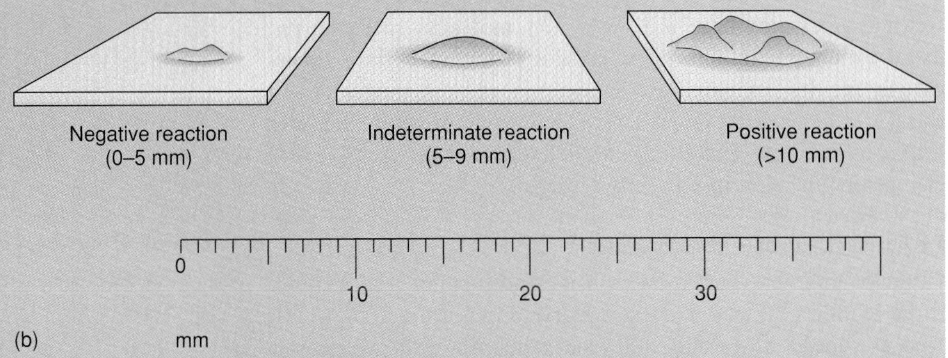

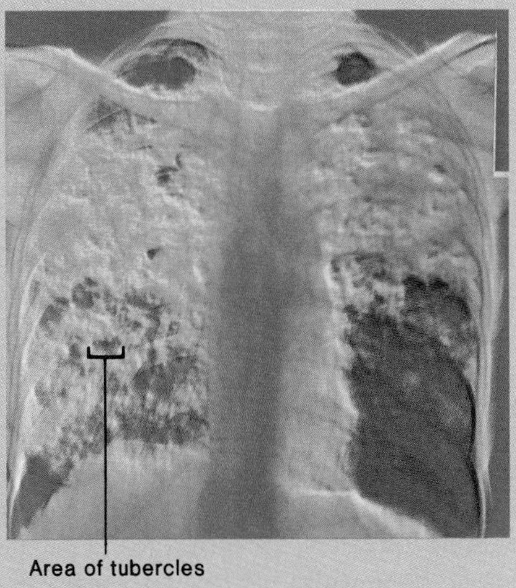

Because hypersensitivity to tuberculoproteins may persist throughout life, testing for them is an effective way to screen school children, public employees, and health care workers. A positive reaction to these proteins in an unimmunized individual is fairly definitive evidence of recent or past infection or disease. Because vaccination for TB also stimulates delayed hypersensitivity, tuberculin screening is mainly used in countries such as the United States where widespread vaccination is not carried out. Physicians should determine a patient's past history before giving the tuberculin test.

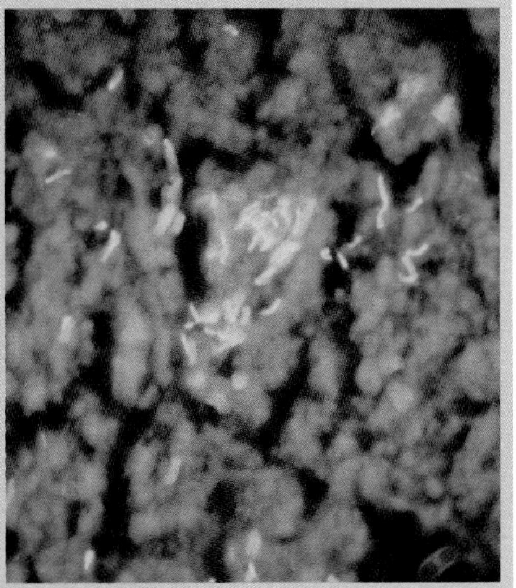

Tuberculin Testing The test itself involves exposure to a small amount of purified protein derivative (PPD), a standardized solution obtained from culture filtrates of *M. tuberculosis* (figure *a*). In the **Mantoux test,** 0.1 ml of PPD (equivalent to 5 tuberculin units) is injected intradermally into the forearm to produce an immediate small bleb. The skin is observed and measured for induration after 48 hours and 72 hours. The three reaction levels include (1) negative reactions, with 5 mm or less induration; (2) intermediate reactions, with 5 mm to 9 mm induration, which indicates doubtful sensitivity and requires retesting; and (3) positive reactions, with 10 mm or more induration (figure *b*). An alternate but less standardized method is the **tine test,** in which a small, sharp-pronged device is used to puncture PPD into the skin. Because tuberculin can cause a systemic reaction in sensitive individuals, it should not be injected into known tuberculin reactors, as these persons often experience severe ulceration and necrosis at the test site.

A positive reaction indicates prior contact with mycobacterial proteins, but not necessarily the presence of active disease. A false-positive reaction may be due to cross sensitivity to other mycobacteria or to prior vaccination. A false-negative reaction is defined as nonreactivity in a person who has the infection or disease. Nonreactivity of this type may be due to improper methods of injection, outdated PPD, or immunosuppressive disease.

Roentgenography and Tuberculosis Chest X rays or roentgenographs can help verify TB when other tests have given indeterminate results. X-ray films reveal abnormal radiopaque patches whose appearance and location can be very indicative. Primary tubercular infection presents the appearance of fine areas of infiltration and enlarged lymph nodes in the lower and central areas of the lungs. X rays of secondary tuberculosis show more extensive involvement, with masses of infiltration in the upper lungs and bronchi and marked tubercles (figure *c*). Scars from older infections often show up on X rays and may furnish a basis for comparison with which to identify newly active disease.

Testing for tuberculosis. (*a,b*) The Mantoux test. Tuberculin is injected into the dermis. A small bleb develops immediately but will be absorbed in a short time. After 48 to 72 hours, the skin reaction is rated by the degree of (size of) the reaction. (Views are magnified ×3.) (*c*) X ray showing a secondary tubercular infection. (*d*) A fluorescent acid-fast stain of *M. tuberculosis* from a patient sample.

The Acid-Fast Stain The diagnosis of tuberculosis in persons with positive skin tests or X rays may be backed up by acid-fast staining of sputum or other specimens. Several variations on the acid-fast stain are currently in use. The Ziehl-Neelsen stain produces bright red acid-fast bacilli (AFB) against a blue background, while fluorescence staining shows luminescent yellow-green bacilli against a dark brown background (figure *d*). The fluorescent acid-fast stain is becoming the method of choice because it is easier to read and provides a more striking contrast. Smears are evaluated in terms of the number of AFB seen per field. This quantity is then applied to a scale ranging from 0 to 4+, 0 being no AFB observed and 4+ being more than 9 AFB per field.

Tuberculosis of the bone and joints is a common complication. The spine is a frequent site of infection, though the hip, knee, wrist, and elbow may also be involved. Advanced infiltration of the vertebral column produces degenerative changes that collapse the vertebrae (figure 16.17), resulting in abnormal curvature of the thoracic region (humpback or kyphosis) or of the lumbar region (swayback or lordosis). Neurologic damage stemming from compression on nerves can cause extensive paralysis and sensory loss.

Tubercular meningitis is the result of an active brain lesion seeding bacilli into the meninges. Over a period of several weeks, the infection of the cranial compartments can create mental deterioration, permanent retardation, blindness, and deafness. Untreated tubercular meningitis is invariably fatal, and even treated cases may have a 30% to 50% mortality rate. An unusual type of reactivation tuberculosis termed **miliary tuberculosis** results from massive septicemia and dissemination. In this form of the disease, discrete, uniform tubercular nodules that look macroscopically like millet seeds (about the size of a rice grain) develop in a variety of organs.

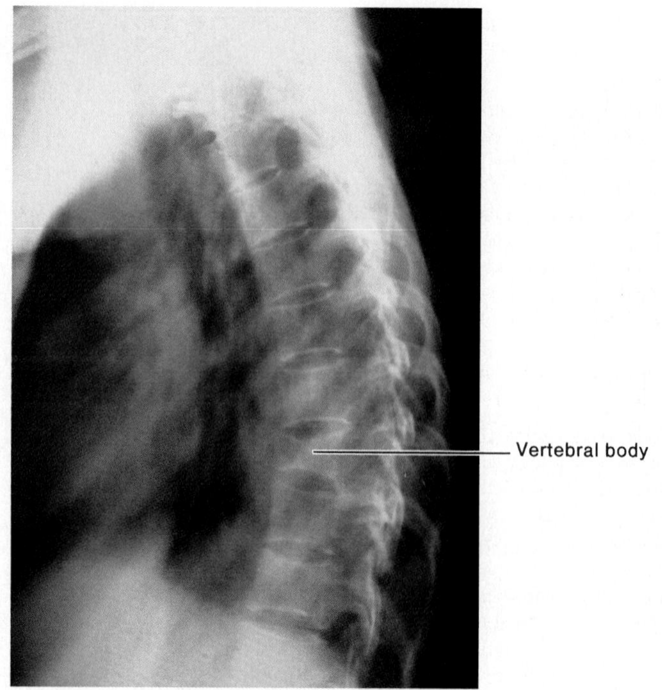

Figure 16.17 Tuberculosis of the spinal column (Pott's disease) as seen by an X ray.

Preliminary Diagnosis of Tuberculosis

Clinical diagnosis of tuberculosis traditionally relies upon three techniques, which are summarized in feature 16.4: (1) *in vivo* or tuberculin testing, (2) roentgenography (X rays), and (3) direct identification of acid-fast bacilli (AFB) in sputum or some other specimen.

Laboratory Cultivation and Diagnosis

M. tuberculosis infection is most accurately diagnosed by isolating and identifying the causative agent in pure culture. Because of the specialized expertise and technology required, this is not done by most clinical laboratories as a general rule. In the United States the handling of specimens and cultures suspected of containing the pathogen is strictly regulated by federal laws.

Diagnosis that differentiates between *M. tuberculosis* and other mycobacteria must be accomplished as rapidly as possible so that appropriate treatment and isolation precautions can be instituted. Because specimens are often contaminated with rapid-growing bacteria that will interfere with the isolation of *M. tuberculosis,* they are pretreated with chemicals to remove contaminants and plated onto selective egg-potato base media (such as Middlebrook 7H11 or Lowenstein-Jensen media). Cultures are incubated under varying temperature and lighting conditions to clarify thermal and pigmentation characteristics, and are then observed for signs of growth over eight weeks. The latest technologies differentiate the mycobacteria on the basis of chromatographic analysis of fatty acids and DNA probes. These tests are so rapid that a physician can receive a culture report in a few days instead of weeks.

Management of Tuberculosis

Treatment of TB involves administering drugs for a sufficient period of time to kill the bacilli in the lungs, organs, and macrophages, usually 6–24 months. To avoid drug resistance, the therapeutic regimen involves combined therapy with at least two drugs selected from a list of eleven, including isoniazid (INH), rifampin, ethambutol, streptomycin, pyrazinamide, thioacetazone, or para-aminosalycylic acid (PAS). The choice depends upon such considerations as effectiveness, adverse side effects, cost, and special medical problems of the patient. With so many choices, if one combination is not working well because of toxicity, drug resistance, or hypersensitivity, a reasonably effective replacement is usually available. The presence of a negative culture or a gradual decrease in the number of AFB on a smear indicates success. Cure will not occur if the patient stops taking the medication prematurely, and this is responsible for many relapses.

In the past, patients with active tuberculosis were quarantined or isolated in specialized hospitals (sanatoriums) to prevent transmission and provide special care. However, modern drugs for tuberculosis have changed this picture, and even infected persons can continue to live and work in the community without spreading the infection to others. Drugs have been so successful in eradicating the bacilli that surgical removal of all or part of a lung is rarely necessary now.

Prevention and Control of Tuberculosis

Although it is essential to identify and treat persons with active TB, it is equally important to seek out and treat persons who are in the early stages of infection or at high risk of becoming infected. Treatment groups are divided into tuberculin-positive "converters," who appear to have a reactivated infection, and tuberculin-negative persons in high-risk groups such as laboratory workers and the contacts of tubercular patients. The standard prophylactic treatment is a daily dose of isoniazid for a year.

A vaccine based on the attenuated "bacille Calmet-Guerin" (BCG) strain of *M. bovis* is given routinely to young children in countries that have high rates of tuberculosis. Studies have shown that anywhere from 20% to 80% of vaccinations give protection for several years. Because the United States does not experience as high a case rate, BCG vaccination is not generally recommended except among certain health professionals, military personnel, government workers, and missionaries.

Mycobacterium leprae: The Leprosy Bacillus

Mycobacterium leprae, the cause of leprosy, was first detected in 1873 by a Norwegian physician named Gerhard Hansen, and it is sometimes called Hansen's bacillus in his honor. The general morphology and staining characteristics of the leprosy bacillus are similar to other mycobacteria, but it is exceptional in two ways: (1) It is a strict parasite that has not been grown in artificial media or human tissue cultures, and (2) it is the slowest growing of all the species. *M. leprae* multiplies within host cells in large packets called globi at an optimum temperature of 30°C.

Leprosy is a chronic, progressive disease of the skin and nerves known for its extensive medical and cultural ramifications. In ancient times, it was accompanied by stigmatization, ignorance, and leprophobia. This was due to the severe disfigurement of the disease and to the belief that it was a divine curse. As if the torture of the disease were not enough, leprosy patients once suffered terrible brutalities, including imprisonment in dungeons under the most gruesome conditions. The modern view of leprosy is more enlightened. We know that it is not readily communicated and that it should not be accompanied by social banishment. Because of the unfortunate connotations associated with the term *leper* (a person who is shunned or ostracized), more sensitive terms such as leprosy patient, Hansen's patient, or *leprotic* are preferred.

Epidemiology and Transmission of Leprosy

Reliable statistics on the worldwide incidence of leprosy are difficult to obtain, but current estimates by WHO are in excess of 15 million cases, most of them are concentrated in tropical and subtropical areas of Asia, Africa, Central and South America, and the Pacific Islands. Leprosy is not restricted to warm climates, since it is also reported in Siberia, Korea, and northern China. The disease is endemic to a few limited locales in the United States, including parts of Hawaii, Texas, Louisiana, Florida, and California. The total number of reported cases nationwide is between 2,000 and 3,000, many of which are associated with recent immigrants.

Although the human body was long considered the sole host and reservoir of the leprosy bacillus, it is now clear that armadillos harbor a mycobacterial species indistinguishable from *M. leprae* and may develop a granulomatous disease similar to leprosy. Whether humans can acquire leprosy from armadillos is not yet known, but it is tempting to speculate that the infection is actually a zoonosis. The mechanism of transmission among humans is yet to be fully verified. Theories propose that the bacillus is directly inoculated into the skin through contact with a leprotic, that mechanical vectors are involved, or that inhalation of droplet nuclei is a factor.

Because the leprosy bacillus has very low virulence, most persons coming in contact with it do not develop clinical disease. Indeed, numerous persons have lived their entire lives among leprotics without acquiring leprosy. As with tuberculosis, it appears that health and living conditions influence susceptibility and the course of the disease. One known predisposing factor is an inherited or acquired defect in cell-mediated immunity. Mounting evidence also indicates that some forms of leprosy are associated with a specific genetic marker. Long-term household contact with leprotics, poor nutrition, crowded conditions, and inadequate hygiene are all contributing factors. Many persons become infected as children and harbor the microbe through adulthood.

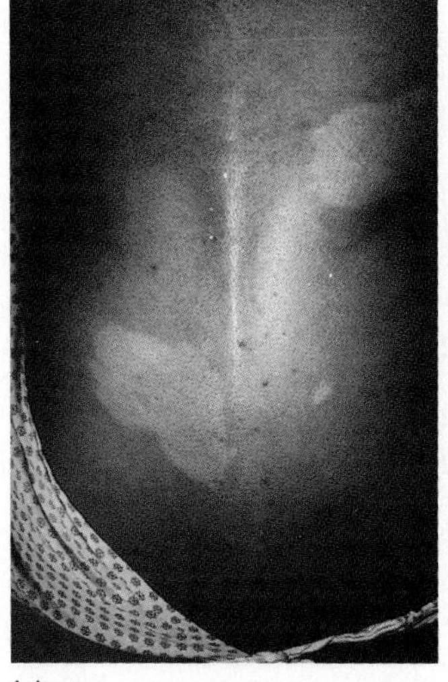
(a)

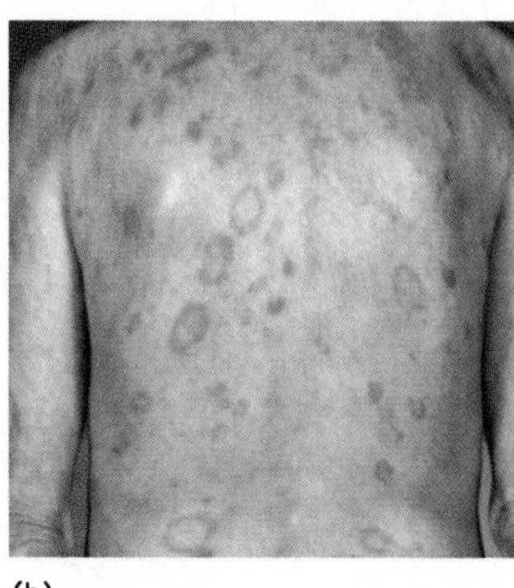
(b)

Figure 16.18 Early lesions in leprosy. (*a*) The initial infection in dark-skinned persons manifests as hypopigmented patches or macules. (*b*) In light-skinned individuals, it appears as reddish patches or papules.

The Course of Infection and Disease

Once *M. leprae* has entered a portal of entry, it is immediately phagocytosed by macrophages. In most infected persons, the macrophages successfully destroy the bacilli and there are no manifestations of disease. But in 4% to 12% of cases, the macrophage response is slow or weak, leading to intracellular survival of the pathogen. The usual incubation period varies from two to five years, with extremes of three months to 40 years. The earliest signs of leprosy appear on the skin of the trunk and extremities as small, spotty lesions colored differently from the surrounding skin (figure 16.18). In untreated cases, the bacilli grow slowly in the skin macrophages and Schwann cells of peripheral nerves, and the disease progresses to one of several outcomes.

A useful rating system for leprosy was developed by D. S. Ridley and W. H. Jopling in 1966. At the two extremes are

leprosy (lep′-roh-see) Gr. *lepros,* scaly or rough. Translated from a Hebrew word that refers to uncleanliness.

Table 16.4 The Two Major Clinical Forms of Leprosy	
Tuberculoid Leprosy	**Lepromatous Leprosy**
Few bacilli in lesions	Many bacilli in lesions
Few, shallow skin lesions in many areas	Numerous, deeper lesions concentrated in cooler areas of body
Local anesthesia in lesions	Sensory loss more generalized; occurs late in disease
No skin nodules	Gross skin nodules
Occasional mutilation of extremities	Mutilation of extremities common
Reactive to lepromin	Not reactive to lepromin
Lymph nodes not infiltrated by bacilli	Lymph nodes massively infiltrated by bacilli

tuberculoid leprosy (TT) and lepromatous leprosy (LL) (table 16.4), and in between are borderline tuberculoid (BT), borderline (BB), and borderline lepromatous (BL). Patients may have more than one form of leprosy simultaneously, and one type may progress to another.

Tuberculoid leprosy, the most superficial form, is characterized by one to three asymmetric, shallow skin lesions containing very few bacilli (figure 16.19). Microscopically, the lesions appear as thin granulomas and enlarged dermal nerves. Damage to these nerves usually results in local loss of pain reception and feeling. This form has fewer complications and is more easily treated than other types of leprosy.

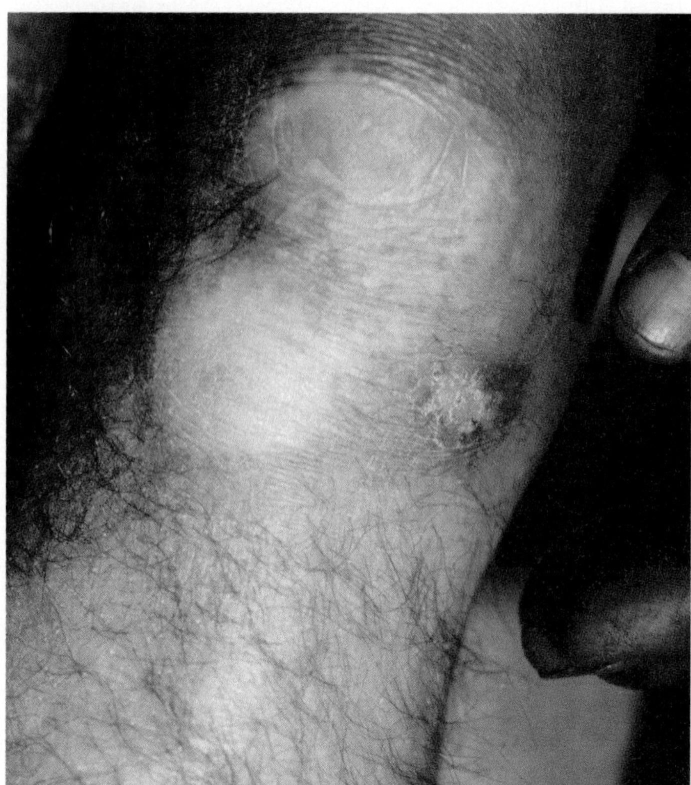

Figure 16.19 The clinical presentation of tuberculoid leprosy (TT). Skin lesions are shallow, painless, and solitary.

Lepromatous leprosy is responsible for the disfigurations commonly associated with the disease. It is marked by chronicity and severe complications due to widespread dissemination of the bacteria. Leprosy bacilli grow primarily in the cooler regions of the body, including the nose, ears, eyebrows, chin, and testes. As growth proceeds, the face of the afflicted person develops folds and granulomatous thickenings, called **lepromas,** which are caused by massive intracellular overgrowth of *M. leprae* (figure 16.20). Advanced LL causes a loss of sensitivity that predisposes the patient to trauma and mutilation, secondary infections, blindness, and kidney or respiratory failure.

The condition of **borderline leprosy** patients can progress either direction along the scale, depending upon their treatment and immunological competence. The most severe effect of intermediate forms of leprosy is early damage to nerves that control the muscles of the hands and feet. The subsequent wasting of the muscles and loss of control produces drop foot and claw hands (figure 16.21). Sensory nerve damage may lead to trauma and loss of fingers and toes.

Diagnosing Leprosy

Leprosy is diagnosed by a combination of symptomology, microscopic examination of lesions, and patient history. The feather test is one of the simplest and most effective tools for field diagnosis (figure 16.22). A skin area that has lost sensation and does not itch may be an early symptom. Numbness in the hands and feet, loss of heat and cold sensitivity, muscle weakness, thickened earlobes, and chronic stuffy nose are additional evidence. Laboratory diagnosis relies upon the detection of acid-fast bacilli in smears of skin lesions, nasal discharges, and tissue samples. Knowledge of a patient's prior contacts with leprotics also supports diagnosis. Laboratory isolation of the leprosy bacillus is difficult and not ordinarily attempted.

Treatment and Prevention of Leprosy

Leprosy infection may be controlled by drugs, but therapy is most effective when started before permanent damage to nerves and other tissues has occurred. Dapsone (DDS) is a safe and relatively inexpensive drug that was originally used alone until resistant stains developed; now, combined therapy is the rule. Tuberculoid leprosy can be managed with rifampin and dapsone for six months. Lepromatous leprosy requires a combination of rifampin, dapsone, and clofazimine until the number of AFBs in skin lesions has been substantially reduced (requiring up to two years). Then dapsone can be taken alone for an indeterminate period (up to 10 years or more). In the United States, many newly diagnosed leprosy patients are treated at the National Hansen's Disease Center operated by the U.S. Public Health Service in Carville, Louisiana.

Preventing leprosy requires constant surveillance of high-risk populations to discover early cases, chemoprophylaxis of healthy persons in close contact with leprotics, and isolation of leprosy patients. The WHO is currently sponsoring a trial of a vaccine containing killed leprosy bacilli, but its success will not

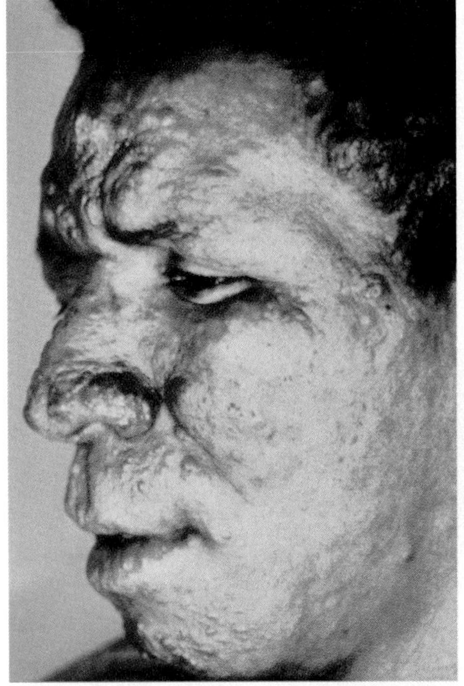

(a)

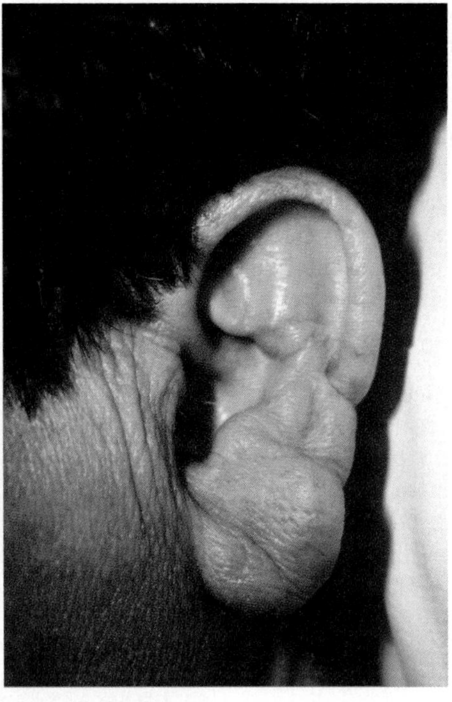

(b)

Figure 16.20 A clinical picture of lepromatous leprosy (LL). (*a*) Infection of the nose, lips, chin, and brows produces moderate facial deformation (leonine facies). (*b*) Detailed view of effects on the ear. Later, there will be hair loss and severe damage to the eyes.

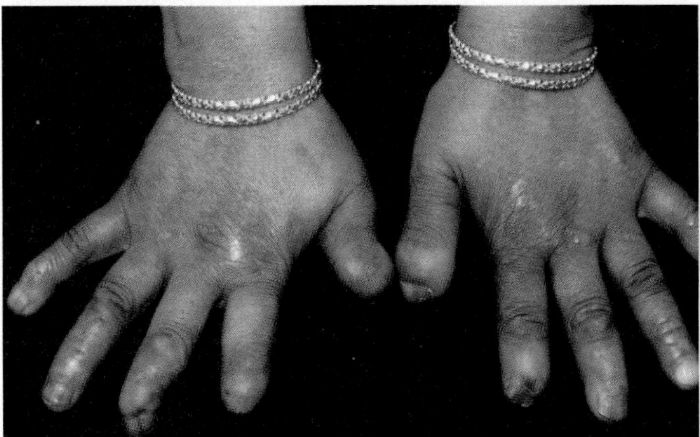

Figure 16.21 Deformation of the hands caused by borderline leprosy. The clawing and wasting are chiefly due to nerve damage that interferes with musculoskeletal activity. Individuals in a later phase of the disease may lose their fingers.

be known until the mid-1990s. Ultimately, the main barriers to the long-term eradication of this ancient and debilitating disease are poverty, overpopulation, ignorance, and fear.

Infections by Other Mycobacteria

For many years, most mycobacteria were thought to have low pathogenicity for humans (see table 16.3). Saprobic and commensal species are isolated so frequently in soil, drinking water, swimming pools, dust, air, raw milk, and even the human body, that both contact and asymptomatic infection appear to be widespread. However, the recent rise in opportunistic and nosocomial mycobacterial infections has demonstrated that many species are far from harmless.

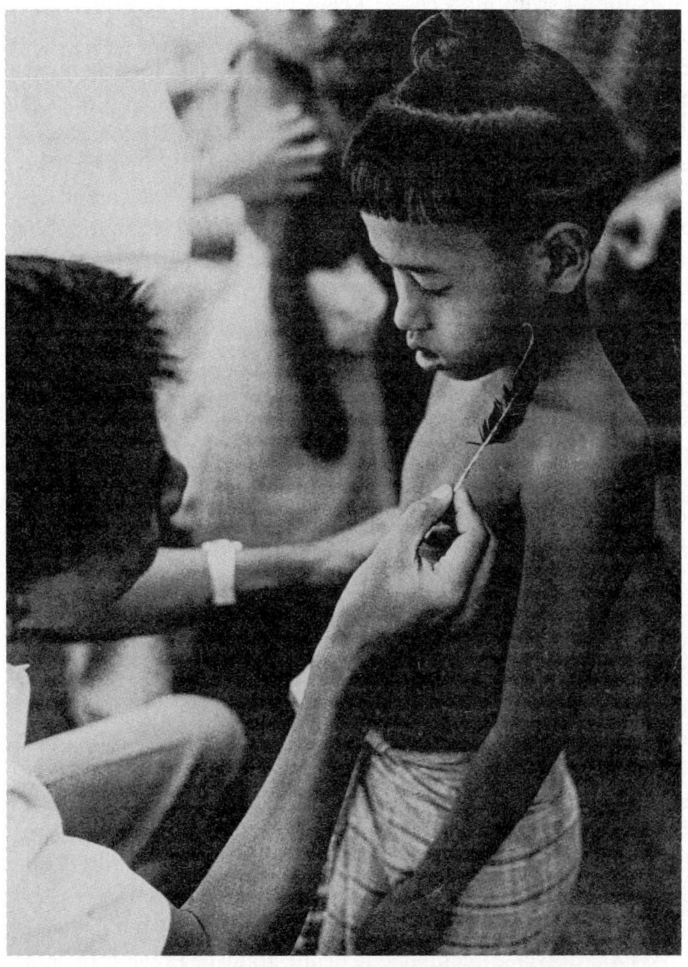

Figure 16.22 The feather test for leprosy. This Burmese child is being tickled over her face and trunk to determine loss of fine sensitivity to touch. This simple technique can provide evidence early in infection.

Disseminated Mycobacterial Infection in AIDS Bacilli of the *Mycobacterium avium-intracellulare* complex (MAI) frequently cause secondary infections in AIDS patients. In fact, MAI is the third most common cause of death in these patients, after pneumocystis pneumonia and cytomegalovirus infection. These common soil bacteria usually enter through the respiratory tract, multiply, and rapidly disseminate. Lacking an effective immune counterattack, the bacilli flood the body systems, especially the blood, bone marrow, bronchi, intestine, kidney, and liver. Even with a six-drug treatment regimen, the prognosis is poor.

Nontuberculous Lung Disease Pulmonary infections caused by commensal mycobacteria are similar to tuberculosis, though milder and not communicable. *Mycobacterium kansasii* infection is endemic to urban areas in the midwestern and southwestern United States and in parts of England. It occurs most often in adult white males who already have emphysema or bronchitis. *Mycobacterium fortuitum* complex causes pulmonary complications in immunosuppressed patients.

Skin, Lymph Node, and Wound Infections An infection by *M. marinum* has been labelled "swimming pool granuloma," because it is a hazard of scraping against the rough concrete surfaces lining swimming pools. The disease starts as a localized nodule, usually on the elbows, knees, toes, or fingers, which then enlarges, ulcerates, and drains (figure 16.23). The granuloma may clear up spontaneously, but it can also persist and require long-term treatment. *Mycobacterium scrofulaceum* causes an infection of the cervical lymph nodes in children living in the Great Lakes region, Canada, and Japan. The bacterium apparently infects the oral cavity and invades the lymph nodes when ingested with food or milk. In most cases, the infection is without complications, but certain children develop *scrofula,* in which the affected lymph nodes ulcerate and drain.

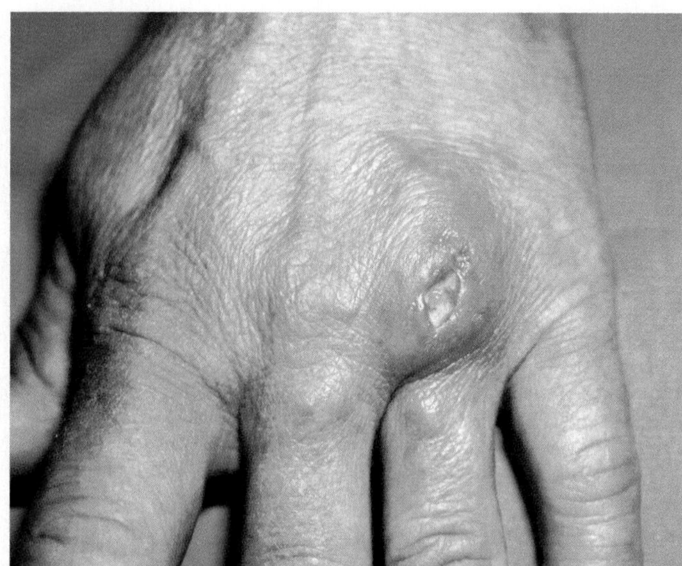

Figure 16.23 A chronic swimming pool granuloma of the hand.

> **Feature 16.5 Gram-Negative Sepsis and Endotoxic Shock**
>
> One serious consequence of infection with gram-negative bacteria, even if the species involved is not a frank or virulent pathogen, is septicemia. In a typical infection of this type, large numbers of bacteria are lysed by host defenses and other factors, and this liberates a large, heat-stable endotoxin into the blood. Being pyrogenic, endotoxin stimulates white blood cells and alters the temperature regulatory center of the brain, inducing fever. The activation of complement and the abnormal release of inflammatory mediators give rise to hypotension, intravascular coagulation, and shock. **Shock** is a pathological state of low blood pressure accompanied by a reduced amount of blood circulating to vital organs, particularly the brain, heart, lungs, and kidneys. Its major symptoms and signs are nausea, tachycardia, cold clammy skin, and weak pulse. Damage to the organs can bring on respiratory failure, coma, heart failure, and death in a few hours. Although the endotoxins of all gram-negative bacteria can cause shock, the most common clinical cases are due to gram-negative enteric rods.

Medically Important Gram-Negative Bacilli

The **gram-negative bacilli** are a large group of non-spore-forming bacteria adapted to a wide range of habitats and modes of life. The genera in this category are highly diverse in metabolism and pathogenicity. Because the group is so large and complex, and many of its members are not medically important, we have included only those that are human pathogens. To simplify coverage, they are organized into three subgroups according to oxygen requirements (table 16.5). A large number of representative genera are inhabitants of the gut (enteric); some are zoonotic; some are adapted to the human respiratory tract; and still others live in soil and water. It is useful to distinguish between the frank (true) pathogens (*Salmonella, Yersinia pestis, Bordetella,* and *Brucella,* for example), which are infectious to the general population, and the opportunists (*Pseudomonas* and coliforms), which are resident flora and cause infection in persons with weakened host defenses.

One universal component of all gram-negative rods is a complex cell wall with an outer membrane containing a lipopolysaccharide termed **endotoxin** (see chapter 11). Because endotoxin in the blood can have severe and far-reaching pathophysiologic effects, gram-negative septicemia, a common *iatrogenic* or nosocomial infection, is a cause for great concern (see feature 16.5).

Aerobic Gram-Negative Nonenteric Bacilli

The aerobic gram-negative bacteria that are medically important include a loose assortment of genera including *Pseudomonas,* an opportunistic pathogen; the zoonotic pathogens *Brucella* and *Francisella;* and *Bordetella* and *Legionella,* which are mainly human pathogens.

---

iatrogenic (eye-at''-troh-jen'-ik) Gr. *iatros,* doctor. An infection occurring as a result of medical treatment.

Pseudomonas (soo''-doh-moh'-nas) Gr. *pseudes,* false, and *monas,* a unit.

Table 16.5 Survey of Gram-Negative Pathogens

Oxygen Requirements	Genus	Species	Disease/Commentary
Aerobes			
	Pseudomonas	*aeruginosa, cepacia*	Opportunistic pathogens; invade surgical wounds, burns, lungs
	Brucella	*abortus, melitensis*	Brucellosis (undulant fever)
	Francisella	*tularensis*	Tularemia (rabbit fever)
	Bordetella	*pertussis*	Pertussis (whooping cough)
	Legionella	*pneumophilia*	Legionellosis; spread by environmental aerosols
Facultative Anaerobes			
The Family Enterobacteriaceae (Oxidase Negative)			
	Escherichia	*coli*	Most strains benign; pathogenic strains cause diarrhea or urinary tract infections
	Edwardsiella	*tarda*	Gastroenteritis; opportunistic
	Citrobacter, *Klebsiella*, *Enterobacter*, *Hafnia*, *Serratia*, *Proteus*, *Providencia*, *Morganella*		Pathogenic to the immunocompromised, causing opportunistic or secondary infection
	Salmonella		Salmonellosis
		typhi	Typhoid fever
		cholera-suis	Septicemia
		enteritidis	Gastroenteritis; dysentery
	Shigella	*dysenteriae*	Bacillary dysentery (shigellosis)
		flexneri	Bacillary dysentery
		boydii	Bacillary dysentery
		sonnei	Bacillary dysentery
	Yersinia	*pestis*	Bubonic plague
		pseudotuberculosis	Gastroenteritis; opportunistic
		enterocolitica	Gastroenteritis; opportunistic
The Family Pasteurellaceae (Oxidase Positive)			
	Pasteurella	*multocida*	Skin abscess; septicemia
	Haemophilus	*influenzae*	Meningitis, epiglottitis
Genera Not Affiliated with a Family			
	Gardnerella	*vaginalis*	Nonspecific vaginitis
	Eikenella	*corrodens*	Gingival infections
	Streptobacillus	*moniliformis*	Rat bite fever
	Calymmatobacterium	*granulomatis*	Donovanosis or granuloma inguinale
Obligate Anaerobes			
	Bacteroides	*fragilis*	Dominant species in intestine; cause of abdominal infections
		melaninogenicus	Oral soft tissue infections

Pseudomonas: The Pseudomonads

The pseudomonads are a large group of free-living bacteria whose primary niches are soil, seawater, and fresh water. They also colonize plants and animals, and are frequent contaminants in homes and clinical settings. These small, gram-negative rods have a single polar flagellum (figure 16.24), produce oxidase and catalase, and do not ferment carbohydrates. Although they ordinarily obtain energy aerobically through oxidative metabolism, some species may grow anaerobically if provided with a salt such as nitrate. Many species produce green, brown, red, or yellow pigments that diffuse into the medium and change its color.

Pseudomonas species can adapt to diverse, even hostile habitats, and manage to extract needed energy from miniscule

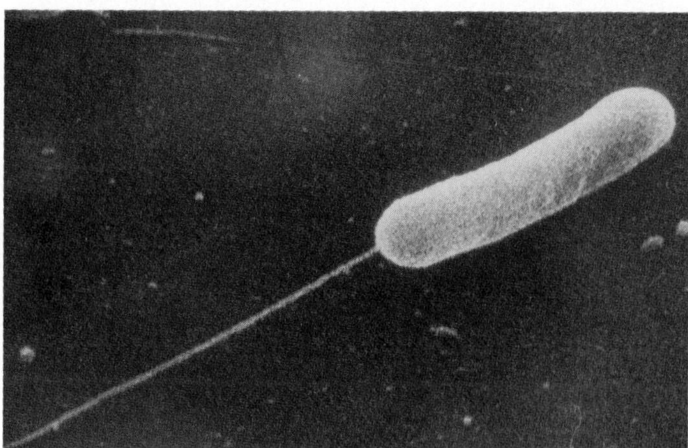

Figure 16.24 This electron micrograph of *Pseudomonas aeruginosa* emphasizes its single, polar flagellum.

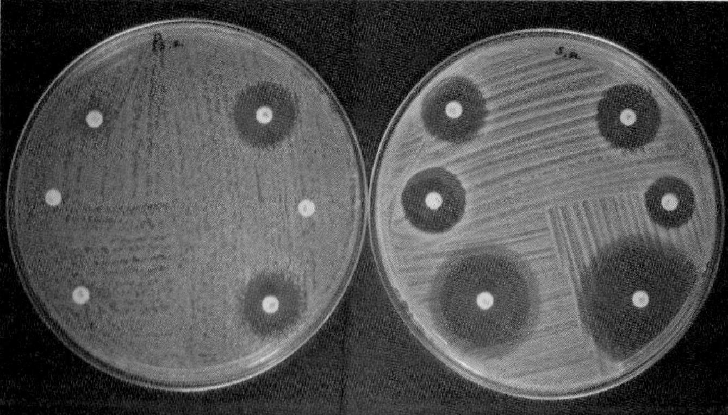

Figure 16.25 An antimicrobic sensitivity test of *Pseudomonas aeruginosa* reveals traits typical of the species. The plate on the left shows its multiple resistance to drugs and the blue-green pigment that diffuses into the medium. A plate of *Staphylococcus aureus* is shown on the right for comparison.

amounts of dissolved nutrients. Most members can grow in a medium containing minerals and one simple organic compound. This adaptability accounts for their constant presence in the environment and their capacity to thrive upon a host. These bacteria also have versatile properties for degrading extracellular substances by means of protease, amylase, pectinase, cellulase, and numerous other enzymes.

The Versatile Pseudomonads Pseudomonads might be thought of as the Dr. Jekyll and Mr. Hydes of the bacterial world because their impact on ecology, agriculture, and commerce is both adverse and beneficial. Consider, for instance, species that can metabolize fossil fuels, thereby constituting a contamination problem in manufacturing petroleum products but a boon in cleaning up oil spills (see chapter 22). In agriculture, pseudomonads rank among the most important plant pathogens, but recombinant forms are being used to protect plants from frost. Various species have roles in such diverse activities as food spoilage, recycling organic materials, and making artificial snow.

Pseudomonas aeruginosa is a common inhabitant of soil and water and an intestinal resident in about 10% of normal people. On occasion, it can be isolated from saliva or even a moist armpit or groin. Because the species is resistant to soaps, dyes, quaternary ammonium disinfectants, drugs, drying, and temperature extremes, it is a chronic nosocomial pathogen that is difficult to control. It is a frequent contaminant of humidifiers, ventilators, intravenous solutions, and anesthesia and resuscitation equipment. Even disinfected instruments, utensils, bathroom fixtures, and mops have been incriminated in hospital outbreaks.

In a pattern similar to that of the enteric bacteria discussed in a later section, *Ps. aeruginosa* is a typical opportunist. It is unlikely to cross healthy, intact anatomical barriers, thus its infectiousness results from invasive medical procedures or weak host defenses. The conditions that most predispose a person to infection are debilitating illness, injury, immunosuppressant medication, or intravenous injections. Once in the tissues, *Ps. aeruginosa* expresses virulence factors including exotoxins, a phagocytosis-resistant slime layer, and various enzymes and hemolysins that degrade host tissues. It also causes endotoxic shock.

The most common nosocomial *Pseudomonas* infections occur in patients with severe burns, neoplastic disease, and cystic fibrosis. Complications include pneumonia, urinary tract infections, abscesses, otitis, and corneal disease. *Pseudomonas* septicemia may give rise to diverse and grave conditions such as endocarditis, meningitis, and bronchopneumonia that have a high fatality rate (80%), even with treatment. Healthy persons are subject to outbreaks of skin rashes and urinary tract infections from community whirlpool baths, hot tubs, and swimming pools. This demonstrates that neither the temperatures nor the chlorine levels in these enclosures can inhibit the growth of this hardy pathogen. Wearers of contact lenses are also vulnerable to eye infections from contaminated storage and disinfection solutions.

Unusual characteristics of *Ps. aeruginosa* infections are a sweetish odor and the noticeable color that appears in tissue, pus, or other exudate ("blue pus"). The color is due to the bacterium's production of a blue-green or greenish-yellow pigment (pyocyanin) that fluoresces in ultraviolet radiation (figure 16.25). Lab identification relies on a battery of tests similar to those given for the Enterobacteriaceae (see figure 16.30). The notorious multiple drug resistance of *Pseudomonas aeruginosa* makes testing for the drug sensitivity of most isolates imperative. Drugs found effective in controlling infections are the third-generation cephalosporins, aminoglycosides, carbenicillin, polymyxin, quinolones, and sulfa-trimethoprim.

Other common hospital-associated species involved in opportunistic infections are *Ps. cepacia, Ps. fluorescens,* and *Ps. (Xanthomonas) maltophilia.*

Brucella and Brucellosis

Malta fever, undulant fever, and **Bang's disease** are synonyms for **brucellosis,** a zoonosis transmitted to humans from infected

aeruginosa (uh-roo″-jih-noh′-suh) L. *aeruginosa,* full of blue-green copper rust.

animals or contaminated animal products harboring ***Brucella.*** The three main species of these tiny, gram-negative coccobacilli are *B. melitensis* (from sheep and goats), *B. abortus* (from cattle), and *B. suis* (from pigs). Animal brucellosis manifests itself as an infection of the placenta and fetus that causes abortion, as typified by the illness in cattle called *Bang's*[2] disease. Humans infected with any of these three agents experience a severe febrile illness, but not abortion.

Brucellosis occurs worldwide, with concentrations in Europe, Africa, India, Mexico, and Central and South America. It is associated predominantly with occupational contact that occurs in slaughterhouses, livestock handling, and the veterinary trade. Infection takes place through contact with blood, urine, vaginal discharges, aborted fetuses, placentas, and aerosols. Human-to-human transmission has not been reported. About 10% of the recent cases arose from the consumption of raw milk or imported cheeses. Brucellosis is also a common disease of wild herds of bison and elk, in which it serves as a population controller. Cattle sharing grazing land with these wild herds often suffer severe outbreaks of Bang's disease, and contaminated herds may have to be destroyed.

Brucella enters through damaged skin or mucous membranes of the digestive tract, the conjunctiva, and the respiratory tract. Virulence appears to center upon some undefined factor that prevents the destruction of the bacillus by phagocytes. Carried in the lymphatic fluid to regional lymph nodes, the infected neutrophils release bacteria into the bloodstream, creating focal lesions in the liver, spleen, bone marrow, and kidney. The cardinal manifestation of human brucellosis is an undulating pattern of fever, which is the origin of one of its common names, undulant fever (figure 16.26). It is also accompanied by chills, profuse sweating, headache, muscle pain and weakness, malaise, and weight loss. Case fatalities are not common, though the syndrome may last for a few weeks to a year, even with treatment.

The patient's history can be very helpful in diagnosis, as are serological tests of the patient's blood. Tetracycline and streptomycin given for several weeks are usually effective in controlling infection. Prevention is effectively achieved by testing and elimination of infected cattle, quarantine of imported animals, vaccination, and pasteurization of milk.

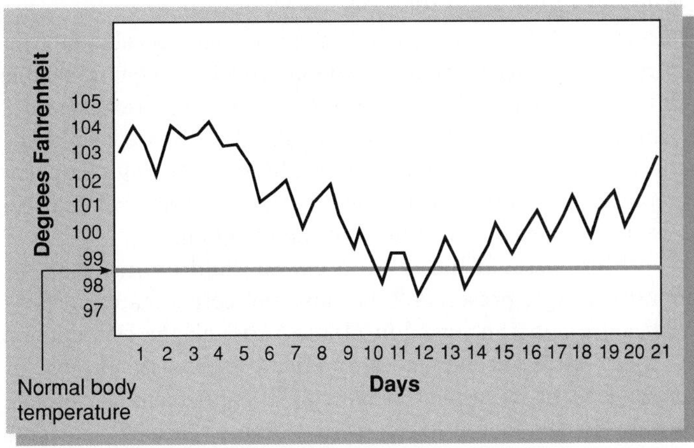

Figure 16.26 The temperature cycle in classic brucellosis. Body temperature undulates between day and night, and fluctuates between fever, normal, and subnormal.
Source: Data from A. Smith, *Principles of Microbiology,* 10th ed., 1985.

Francisella tularensis and Tularemia

Francisella tularensis is the causative agent of **tularemia,** a zoonotic disease of assorted mammals endemic to the northern hemisphere. In several characteristics, it is similar to *Yersinia pestis,* the cause of plague (discussed later in this chapter), and the two species were once included in the same genus, *Pasteurella.* Because the disease has been associated with outbreaks of disease in wild rabbits, it is sometimes called "rabbit fever."

Tularemia is abundantly distributed through numerous animal reservoirs and vectors in the USSR, Europe, Asia, and North America, but it does not appear in the tropics. This disease is noteworthy for its complex epidemiology and spectrum of symptoms. Although rabbits and rodents (muskrats especially) are the chief reservoirs, other wild animals (skunks, beavers, foxes, opossums) and some domestic animals are implicated as well. Of the more than 50 bloodsucking arthropod species known to carry *F. tularensis,* about half are vectors for humans. Ticks are most often involved, followed by biting flies, mites, and mosquitos.

Tularemia is strikingly varied in its portals of entry and disease manifestations. In a majority of cases, infection results when the skin or eye is inoculated during contact with infected animals, animal skins, and meats. Bites by vectors are also a frequent source of infection. Cases have been attributed to such diverse circumstances as ingesting undercooked or raw meat, drinking contaminated water, inhaling contaminated dust, or even being splattered in the eye by droplets, but the disease is not communicated from human to human. With an estimated infective dose of between 10 and 50 organisms, *Francisella tularensis* is often considered one of the most infectious of all bacteria.

After an incubation period ranging from a few hours to three weeks, acute symptoms of headache, backache, fever, chills, malaise, and weakness appear. Depending upon the original portal of entry, the clinical complications are an ulcerative skin lesion, swollen lymph glands, conjunctival inflammation, sore throat, intestinal disruption, systemic symptoms, and pulmonary involvement. The death rate in systemic and pulmonic forms is 10%, but proper treatment reduces it to almost zero. Streptomycin is the treatment of choice, and tetracycline and chloramphenicol are effective alternatives. Because the intracellular persistence of *F. tularensis* can lead to relapses, antimicrobial therapy must not be discontinued prematurely. Protection for occupational risk groups is afforded by live, attenuated vaccines and protective gloves, masks, and eye wear.

Brucella (broo-sel'-uh) After David Bruce, who isolated the bacterium.

2. After B. L. Bang, a Danish physician.

Francisella tularensis (fran-sih-sel'-uh too-luh-ren'-sis) After Edward Francis, one of its discoverers, and Tulare County, California, where the agent was first identified.

Bordetella pertussis and Whooping Cough

Bordetella pertussis is a minute, encapsulated coccobacillus responsible for **pertussis** or **whooping cough,** a communicable childhood affliction that was once practically a household word. Contrary to the common impression that the disease is mild and self-limiting, it often causes severe, life-threatening complications in babies. *Bordetella parapertussis* is a closely related species that causes a milder form of the infection.

Most cases of pertussis occur in children less than six months of age, presumably because protective maternal antibodies are not transferred in utero. Although the introduction of an effective vaccine has dramatically decreased the prevalence of pertussis in many countries, it is far from an obsolete disease. In developing countries that cannot routinely vaccinate and industrialized countries with lax vaccination, pertussis is still a major cause of childhood sickness and death. Around 2,000 to 4,000 cases are reported annually in the United States, with fewer than 10 deaths. The primary source of infection is other children afflicted with clinical disease; carriers or animals are not involved. Transmission is by direct contact with droplets or inhalation of infectious aerosols.

The primary virulence factors of *B. pertussis* are (1) receptors that specifically recognize and bind to ciliated respiratory epithelial cells and (2) toxins that destroy and dislodge ciliated cells (a primary host defense). The loss of the ciliary mechanism leads to buildup of mucus and blockage of the airways. The initial phase of pertussis is the *catarrhal stage,* marked by nasal drainage and congestion, sneezing, and occasional coughing. As symptoms deteriorate into the *paroxysmal stage,* the child experiences recurrent, persistent coughing. After fits of 10 to 20 abrupt, hacking coughs, the need for oxygen stimulates a deep inspiration that draws air swiftly through the narrowed larynx and gives off a "whoop." Death, if it occurs, is usually due to compromised respiration or complications of secondary bacterial or viral pneumonia.

The preferred control measure is early vaccination with a suspension of killed bacteria. Pertussis vaccine is combined with the diphtheria and tetanus vaccines (DPT), and the inoculation protocol is the same, except that boosters are not necessary. The vaccine occasionally causes severe reactions in older children and adults, and is not recommended for anyone over age seven. Other measures to protect susceptible persons during epidemics include erythromycin therapy and isolation of cases.

Legionella and Legionellosis

Legionella is a novel bacterium unrelated to other strictly aerobic gram-negative genera. Although the organisms were originally described in the late 1940s, they were not clearly associated with human disease until 1976. The incident that brought them to the attention of medical microbiologists was an explosive and mysterious epidemic of pneumonia that afflicted 200 American Legion members attending a convention in Philadelphia and killed 29 of them. After six months of painstaking analysis, epidemiologists isolated the pathogen and traced its source to contaminated air conditioning vents in the legionnaires' hotel. The news media latched onto the name **Legionnaires' disease,** and the experts named their new isolate *Legionella* (lee″-jun-ell′-uh).

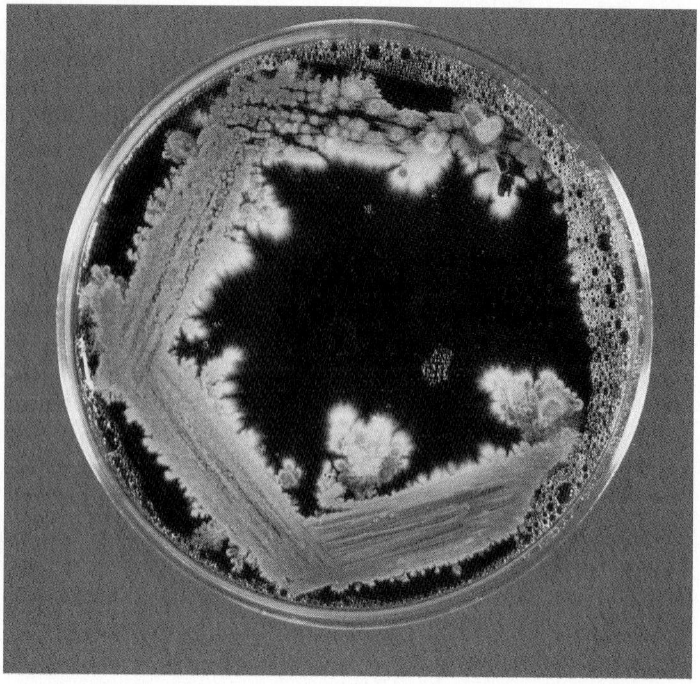

Bordetella pertussis (bor-duh-tel′-uh pur-tus′-is) After Jules Bordet, a discoverer of the agent, and L. *per,* severe, and *tussis,* cough.

Figure 16.27 *Legionella pneumophila* colonies on selective charcoal extract medium.

Legionellas are weakly gram-negative pleomorphic rods that range in morphology from cocci to filaments. They have fastidious nutrient requirements and can be cultivated only in special media (figure 16.27) and cell cultures. Several species or subtypes have been characterized, but *L. pneumophila* (lung-loving) is the one most frequently isolated in infections.

The epidemiology of legionellosis is best understood by considering the ecology of the bacterium. It is widely distributed in natural and man-made aquatic habitats and moist surroundings. It has been isolated from hot and cold tap water, drinking water supplies, heating towers, air conditioning systems, evaporative coolers, and showers. It can survive for a year in standing water and is heat tolerant. Some of the more unusual sources of infections include sprayers used to moisten produce in supermarkets and fallout from the eruption of Mt. St. Helen's.

Studies since 1976 have determined that Legionnaires' disease is not a new disease and that earlier epidemics occurred but went undiagnosed at the time. The disease is now known to be worldwide and is prevalent in males over 50 years of age. Most community outbreaks occur sporadically in conjunction with accidental exposure to a common environmental source of the agent. Nosocomial infections usually target elderly patients hospitalized with diabetes, malignant disease, transplants, alcoholism, and lung disease. The disease is not communicable and is acquired solely through inhaling moist, contaminated air.

Two major clinical forms of the disease are *Legionnaires' pneumonia* and *Pontiac fever.* The symptoms of both are a rising fever (up to 41°C), cough, diarrhea, and abdominal pain. Legionnaires' pneumonia is the more severe disease, progressing to lung consolidation and impaired respiration and organ function. It has a fatality rate of 10–20%. Pontiac fever does not lead to pneumonia and rarely causes death. Legionellosis is diagnosed by symptomology, patient's history, and fluorescent antibody staining of specimens. It is treated with erythromycin alone or in combination with rifampin. Because of *Legionella's* wide dispersal in aquatic habitats, control is difficult, though chlorination and regular cleaning of man-made sources are of some benefit.

Identification and Differential Characteristics of the Enterobacteriaceae

One large family of gram-negative bacteria that exhibits a considerable degree of relatedness is the Enterobacteriaceae. Although many members of this group inhabit soil, water, and decaying matter, they are also common occupants of the large bowel of humans and animals. The rods look very much alike with ordinary light microscopy, being small (the average is 1 μm by 2–3 μm) and non-spore-forming. The bacteria in this group grow best in the presence of air, but they are facultative and can ferment carbohydrates by an alternate anaerobic pathway. This group is probably the most common one isolated in clinical specimens—sometimes as normal flora, sometimes as agents of disease.

Enteric pathogens are the most frequent cause of **diarrheal illnesses** (see feature 16.6) that account for an annual mortality rate of five million persons worldwide—the second most common cause of death after cardiovascular illness. Counting the total number of cases that receive medical attention, along with estimates of unreported, subclinical cases, enteric illness is probably responsible for more morbidity than any other disease. Prominent pathogenic enterics include *Salmonella, Shigella,* and strains of *Escherichia,* many of which are harbored by human carriers.

Enterics (along with *Pseudomonas* species) also account for more than 50% of all isolates in nosocomial infections (figure 16.28). Their widespread involvement in hospital-acquired infections can be attributed to their constant presence in the hospital environment and their survival capabilities. Enterics are commonly isolated from soap dishes, sinks, and invasive devices such as fiberoptic probes, tracheal cannulas, and indwelling catheters. The most important enteric opportunists are *E. coli, Klebsiella, Proteus, Enterobacter, Serratia,* and *Citrobacter.* Interestingly, the diseases caused by these agents usually involve systems other than the gastrointestinal tract, such as the lungs and the urinary tract.

The genera in this family are traditionally divided into two subcategories. The **coliforms** include *Escherichia coli* and other gram-negative normal enteric flora that ferment lactose

Feature 16.6 Diarrheal Disease

Diarrhea is an acute syndrome of the intestinal tract in which the volume, fluid content, and frequency of bowel movements increase. It is usually a symptom of **gastroenteritis** (inflammation of the lining of the stomach and intestine) and may be accompanied by severe abdominal pain. Diarrhea is generated by several pathological states—most commonly, infection, intestinal disorders, and food poisoning. Although the human large intestine ordinarily harbors a huge microbial population, most bacterial, protozoan, and viral agents of diarrhea are not members of this normal gut flora but are acquired through contaminated food or water.

Infectious diarrhea has two basic mechanisms. In the toxigenic type of disease, bacteria release **enterotoxins** onto the surface of the small intestine. These toxins disrupt the physiology of epithelial cells and cause increased secretion of electrolytes and water loss, a condition called **secretory diarrhea.** The organism itself does not invade the tissues. Secretory diarrhea is characterized by its large volume, and there is little blood in the stool. This is the mechanism of cholera (see chapter 17) and some types of *E. coli* and *Shigella* poisoning.

In another form of diarrheal disease, the microbe invades the wall of the small or large intestine and disrupts its architecture, leading to gross injury. This form is attended by smaller fecal volume, pain in the rectum, blood in the stool, and ulceration of the inner mucosal lining. *Salmonella,* other strains of *Shigella* and *E. coli, Campylobacter,* and *Entamoeba histolytica* are responsible for this type of intestinal disease. Regardless of the cause, the loss of fluid that accompanies diarrhea results in severe dehydration and sometimes death. Infants are especially vulnerable to diarrheal illness because of their smaller fluid reserves and inadequate immunities. For a discussion of a simple but highly effective treatment for diarrheal symptoms, see feature 17.3.

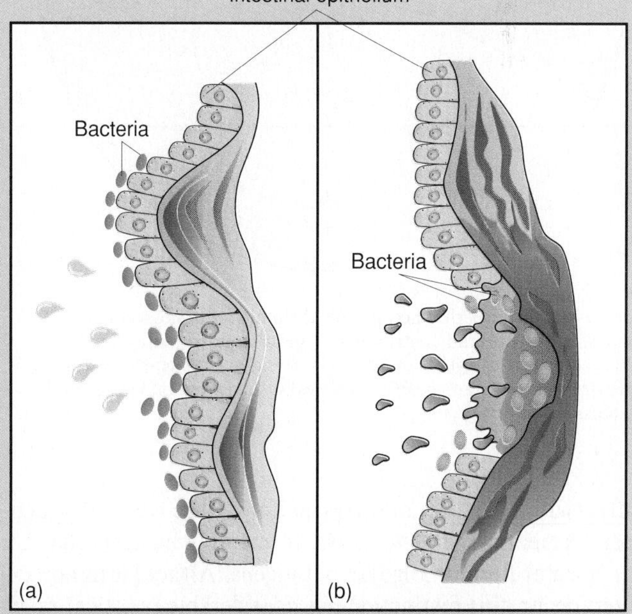

Mechanisms of infectious diarrhea. (*a*) In a toxigenic infection, the microbe remains on the surface of epithelial cells and secretes toxin into the cells. (*b*) In an invasive infection, the microbe breaks down epithelial cells and forms ulcerations on the intestinal lining.

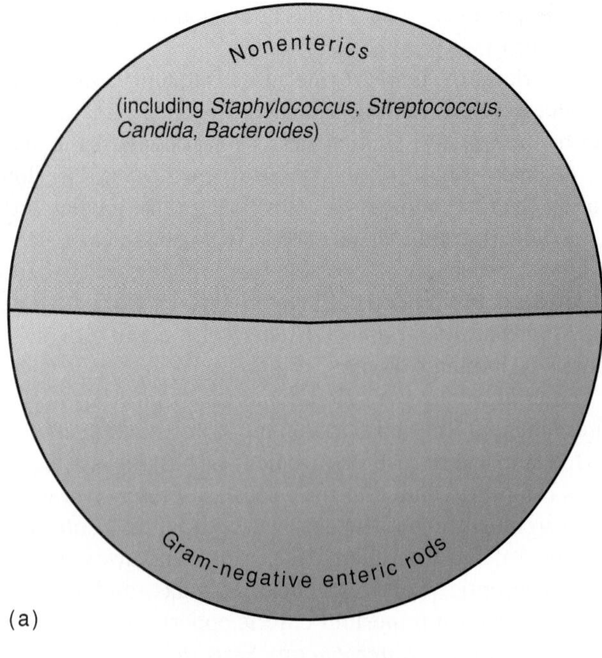

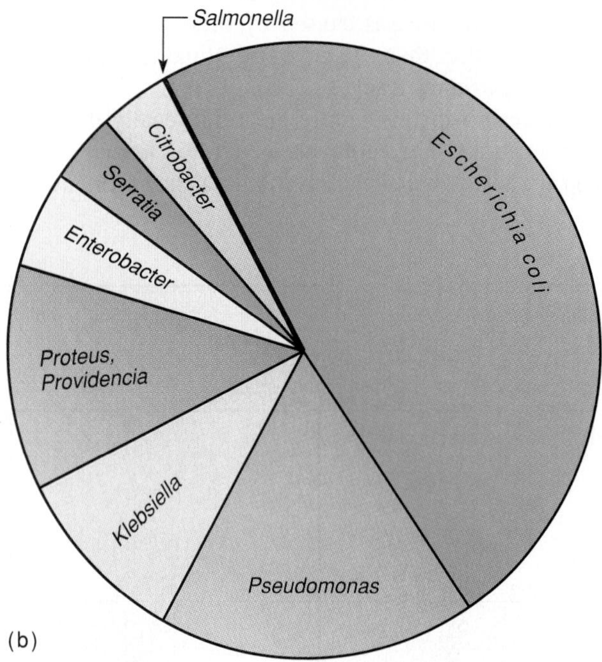

Figure 16.28 Bacteria that account for the majority of nosocomial infections. (*a*) The overall contribution of gram-negative rods versus nonenteric bacteria as agents in hospital infections. (*b*) Comparison of the relative proportion of each genus or species isolated from infections by the gram-negative enteric group.

rapidly (within 48 hours). **Noncoliforms** are generally nonlactose-fermenting or slow lactose-fermenting bacteria that are either normal flora or regular pathogens. Although minor exceptions occur, this distinction has considerable practical merit in laboratory, clinical, and epidemiologic terms. The following outline provides a framework for organizing the discussion of the Enterobacteriaceae:

Coliforms: Rapid lactose-fermenting enteric bacteria that are normal flora and opportunistic (some strains of *E. coli* are true pathogens)
- *Escherichia coli*
- *Klebsiella*
- *Enterobacter*
- *Hafnia*
- *Serratia*
- *Citrobacter*

Noncoliforms: Lactose negative; may or may not be normal flora
- Opportunistic, normal gut flora
  - *Proteus*
  - *Morganella*
  - *Providencia*
  - *Edwardsiella*
- Pathogenic enterics
  - *Salmonella typhi, S. cholerae-suis, S. enteritidis, Arizona hinshawii*
  - *Shigella dysenteriae, Sh. flexneri, Sh. boydii, Sh. sonnei*
  - *Yersinia enterocolitica*
  - *Yersinia pseudotuberculosis*
- Pathogenic nonenterics
  - *Yersinia pestis*

Morphology and staining characteristics are insufficient in themselves to identify the enteric bacilli. They share similar characteristics, being non-spore-forming, straight rods that are often peritrichously flagellated (exceptions are *Shigella* and *Klebsiella*), catalase positive, and oxidase negative. A large battery of biochemical tests is commonly used to identify the principal genera and species. Although the details of this process are beyond the scope of this text, a summary will overview the major steps (see figure 16.30). Because fecal specimens contain such a high number of normal flora, they may be inoculated first into enrichment media such as selenite or GN (gram-negative) broth that inhibit the normal flora and favor the growth of pathogens. Enrichment and nonfecal specimens are streaked onto several plates of selective, differential media such as MacConkey, EMB, or Hektoen enteric agar and incubated for 24–48 hours. Colonies developing on these media will provide the initial separation into lactose-fermenters and non-lactose-fermenters (figure 16.29; see figure 6.26).

After noting the reactions of isolated colonies, single colonies are subcultured on triple sugar iron (TSI) agar slants for further analysis. As is evident in figure 16.30, several other tests are required to identify to the level of genus (table 16.6). One series of reactions, the **IMViC** (indole, methyl red, Voges-Proskauer, *i* for pronounciation, and citrate), is a traditional panel that can be used to differentiate among several genera (table 16.7). Commercial identification systems such as the enterotube (figure 16.31) provide a rapid and compact means of acquiring these and additional data needed in species identification. In general, whether an isolate is clinically significant depends upon the specimen from which it came. For example,

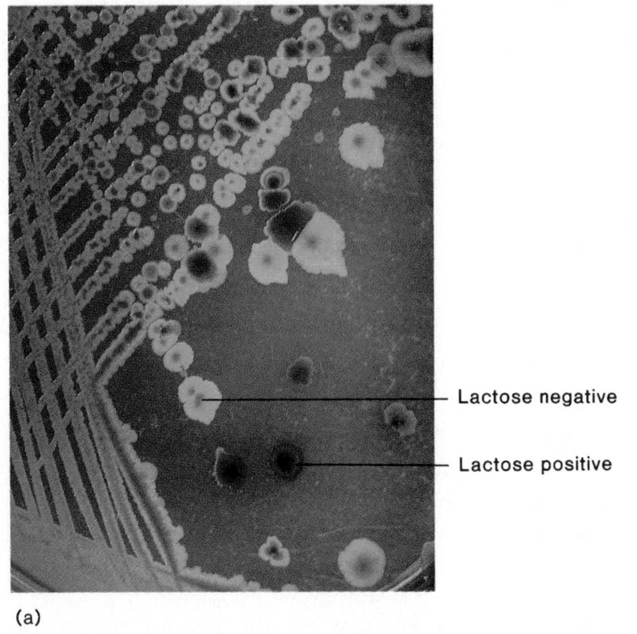

(a)

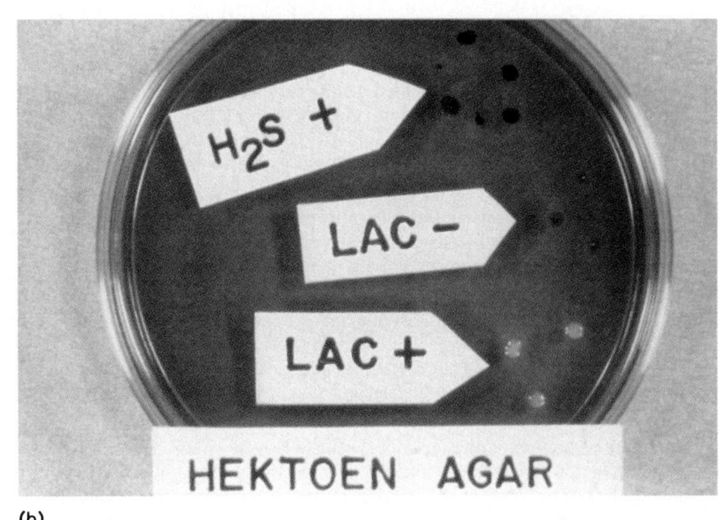

(b)

Figure 16.29 Isolation media for enterics, showing differentiating reactions. (*a*) Levine's Eosin methylene blue (EMB) agar. (*b*) Hektoen enteric agar. (See table 16.6.)

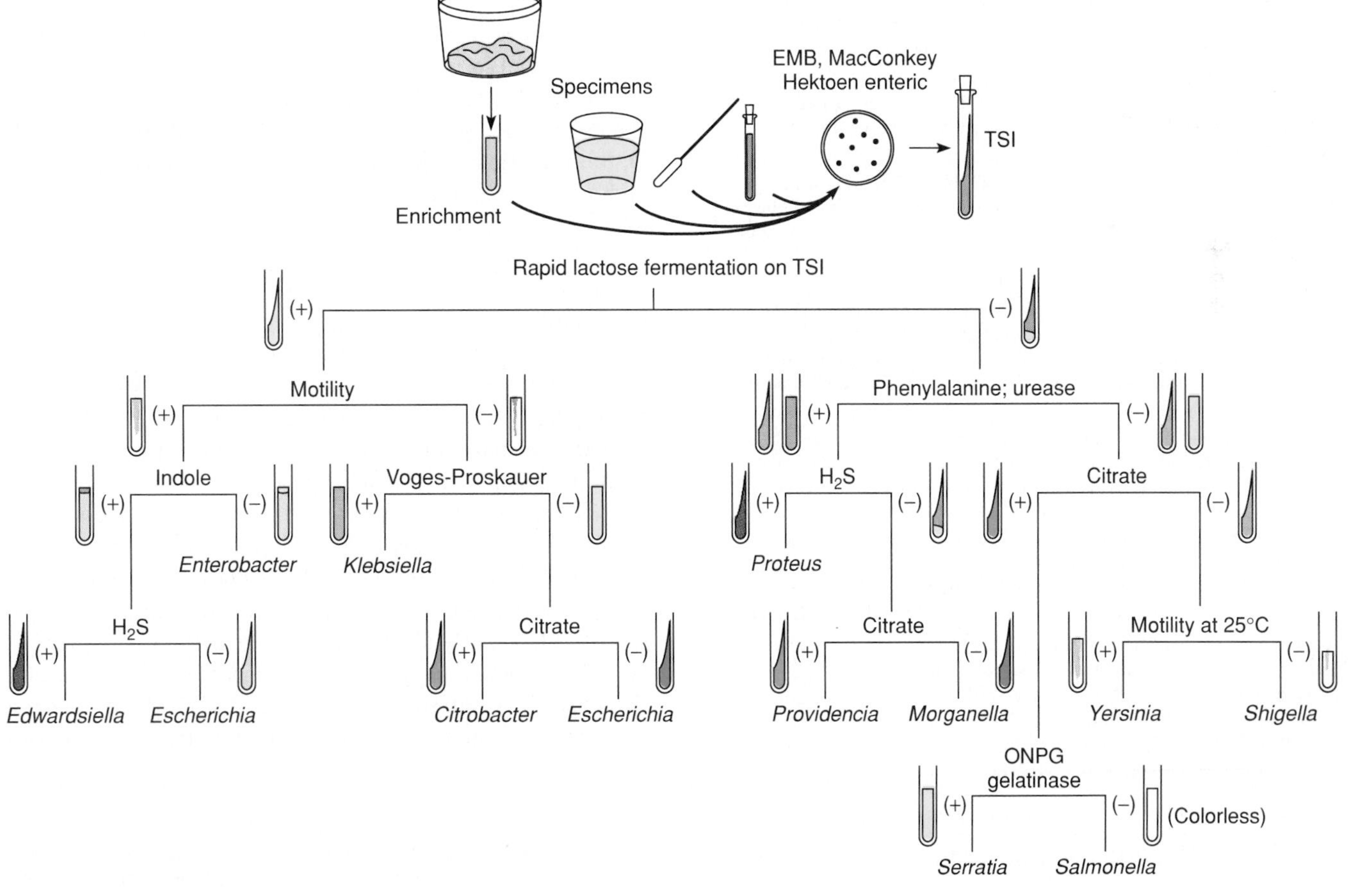

Specimens
EMB, MacConkey
Hektoen enteric
TSI
Enrichment
Rapid lactose fermentation on TSI
(+) (−)
Motility
(+) (−)
Phenylalanine; urease
(+) (−)
Indole
(+) (−)
Enterobacter
Voges-Proskauer
(+) (−)
Klebsiella
H_2S
(+) (−)
Proteus
Citrate
(+) (−)
H_2S
(+) (−)
Edwardsiella *Escherichia*
Citrate
(+) (−)
Citrobacter *Escherichia*
Citrate
(+) (−)
Providencia *Morganella*
Motility at 25°C
(+) (−)
Yersinia *Shigella*
ONPG
gelatinase
(+) (−) (Colorless)
Serratia *Salmonella*

Figure 16.30 Procedures for isolating and identifying selected enteric genera.

Table 16.6 Protocol in Isolating and Identifying Selected Enteric Genera *

MacConkey agar contains bile salts and crystal violet to inhibit gram-positive bacteria. It contains lactose and neutral red dye to indicate pH changes and differentiate the rapid lactose-fermenting bacteria from lactose non-fermenters or slow fermenters. Lactose (+) colonies are red to pink from the accumulation of acid acting on the neutral red indicator. Lactose (−) colonies are not colored in this way (see figure 6.26).

Levine's EMB agar also contains bile salts, plus eosin and methylene blue dyes that cause lactose fermenters to develop a dark nucleus and sometimes a metallic sheen over the surface. Non-fermenters are pale lavender and non-nucleated.

Hektoen enteric agar also contains bile and is especially good for isolating pathogens. The bromthymol blue and acid fuchsin indicators differentiate the lactose fermenters (salmon pink to orange) from the non-fermenters (green, blue-green), and also detect the production of H_2S gas, which turns the center of a colony black.

Utilization of lactose, a disaccharide, requires two enzymes—a permease to bring lactose into the cell and β-galactosidase to split it into the monosaccharides glucose and galactase. Rapid fermenters on the initial isolation plates and TSI have both enzymes. Slow fermenters have only galactosidase. **ONPG** is chemical shorthand for a test that detects these slow lactose fermenters, which turn a special ONPG medium yellow. Non-fermenters do not react in either test.

TSI is a nonselective medium that indicates some combination of fermentation reactions (primarily lactose and glucose, and with results from isolation medium, sucrose) by means of a phenol red indicator dye. It also reveals gas and hydrogen sulfide (H_2S) production (see figure 6.21). Hydrogen sulfide is a metabolic product of the reduction of an inorganic or organic sulfur source. H_2S is indicated by a reaction with iron salts to form a black precipitate of ferric sulfide.

The indole test indicates the capacity of an isolate to cleave a compound called indole off the amino acid tryptophan. If this occurs, Kovac's reagent forms a bright red ring at the surface of the tube; if it is negative, the color remains yellow.

The methyl red (MR) test indicates a form of glucose fermentation in which large amounts of mixed acids accumulate in the medium. The pH is lowered to around 5, so that methyl red dye remains red when added to the tube. MR-negative bacteria do not lower the pH to this degree, and the tube is yellow to orange in the presence of the dye.

The Voges-Proskauer (VP) test determines whether the product of glucose fermentation is a neutral metabolite called acetylmethycarbinol (acetoin). This substance reacts with Barritt's reagent to form a pink to rosy-red tinge in the medium. With negative results, the tube remains brown to yellow.

Citrate media contain citrate as the only usable carbon source and a pH indicator, bromothymol blue. If an isolate can utilize this carbon source, it grows and produces alkaline by-products, resulting in a conversion of the medium's color from green (neutral) to blue (alkaline).

Urea media contain 2% urea, a nitrogenous waste product of mammals, and phenol red pH indicator. Some bacteria produce the enzyme urease, which hydrolyzes urea into two molecules of ammonium. Ammonium raises the pH of the medium, and the indicator turns bright pink.

Phenylalanine (PA) deaminase, an enzyme produced primarily by the Proteeae group, removes an amino group from the amino acid phenylalanine to produce phenylpyruvic acid. This reaction is visualized by adding ferric chloride to the tube, which turns olive green when it reacts with the phenylpyruvic acid.

Motility is tested by a hanging drop slide or motility test medium (see figure 6.21).

*The metabolic basis for some tests is discussed in chapter 7.

Table 16.7 The IMViC Tests for Differentiating Common Opportunistic Enterics

Genus	Indole	Methyl Red	Voges-Proskauer	Citrate
Escherichia	+	+	−	−
Citrobacter	+	+	−	+
Klebsiella/ Enterobacter	−	−	V*	+
Serratia	−	V*	+	+
Proteus	+	−	−	+
Providencia	+	+	−	+
Pseudomonas	−	−	−	V*

*V = Species variable.

normal flora isolated from stool specimens do not usually signify infection in that site, but the same organisms isolated from extraintestinal sites in sputum, blood, urine, and spinal fluid are considered likely agents of opportunistic infections.

Antigenic Structures and Virulence Factors

The gram-negative enterics have complex surface antigens that are important in pathogenicity and are also the basis of immune responses (figure 16.32). By convention, they are designated **H,** the **flagellar antigen; K,** the **capsule** and/or **fimbrial antigen;** and **O,** the **somatic** or **cell wall antigen.**[3] Not all species carry the H and K antigens, but all have O, a lipopolysaccharide that produces endotoxic shock. Most species of gram-negative enterics exhibit a variety of subspecies or serotypes caused by slight variations in the chemical structure of the HKO antigens. These are usually identified by serotyping, a technique in which specific antibodies are mixed with a culture to measure the degree of agglutination or precipitation (see chapter 13). Not only is serotyping helpful in classifying some species, but it also helps pinpoint the source of outbreaks by a particular serotype, making it an invaluable epidemiologic tool.

The pathogenesis of enterics may be due to endotoxins (see feature 16.5), exotoxins, and to the pathogen's capacity to overcome host defenses and multiply in the tissues and blood. Although most enterics are commensals, most also have a latent potential to rapidly alter their pathogenicity. In chapters 8 and 10, we learned that gram-negative bacteria freely exchange chromosomal or plasmid-borne genes that can mediate drug resistance, toxigenicity, and other adaptive traits. If this recom-

3. H comes from the German word *hauch* (breath) in reference to the spreading pattern of motile bacteria grown on moist solid media. K is derived from the German term *kapsel.* O is for *ohne-hauch,* or nonmotile.

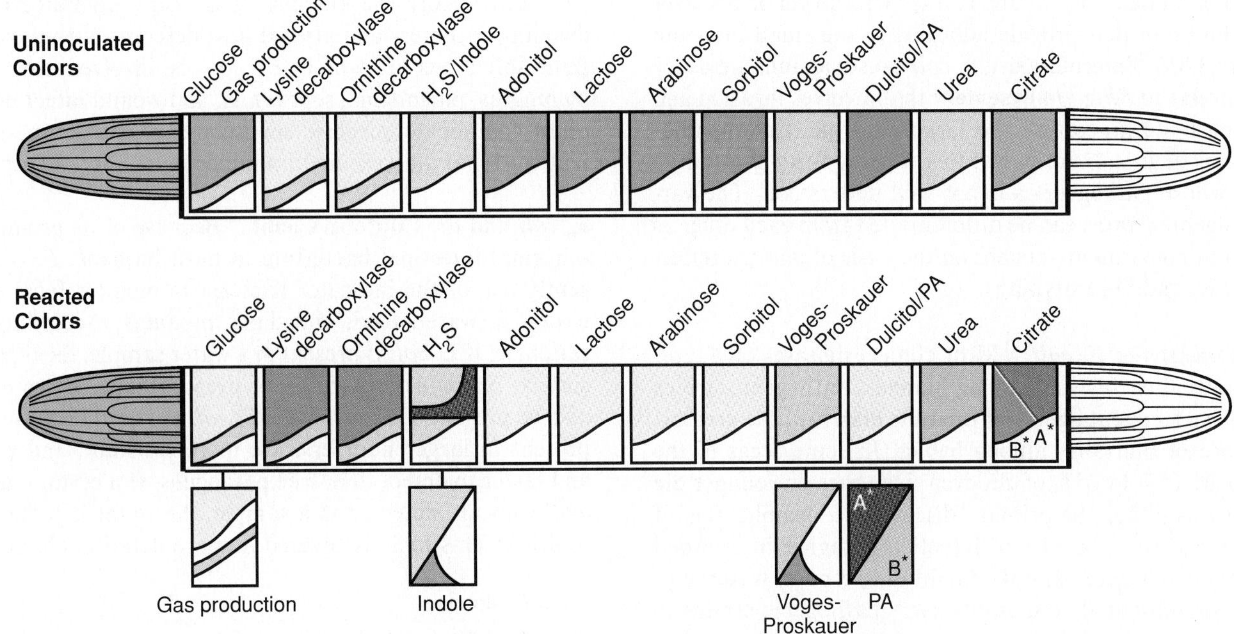

Figure 16.31 Enterotube II (Roche diagnostics), a miniaturized, multichambered tube used for rapid biochemical testing of enterics. An inoculating rod pulled through the length of the tube carries an inoculum to all chambers. Here, an uninoculated tube is compared with one that gives all positive results.

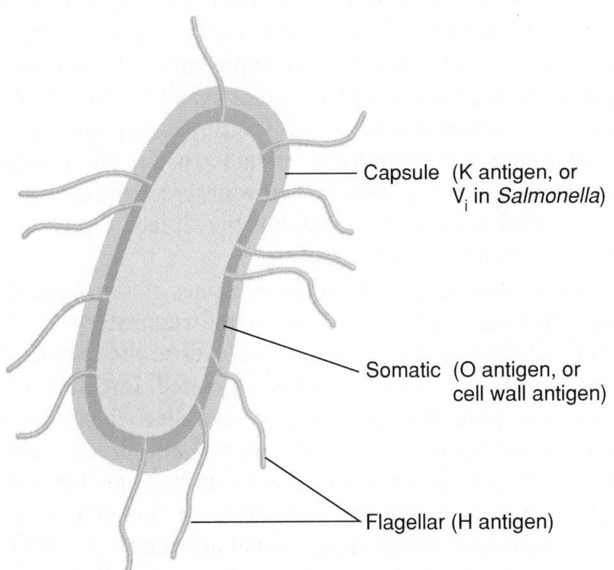

Figure 16.32 Antigenic structures in gram-negative enteric rods. Variations in the composition of these antigens provide the basis for the serological types found in most genera.

bination occurs in mixed populations of fecal residents (for example, in the intestines of carriers), genes that code for virulence factors may "make the rounds" among these bacteria, and it is possible for nonpathogenic or less pathogenic residents to acquire greater virulence (see *E. coli* in the next section). The most common virulence genes transferred code for enterotoxins, capsules, hemolysins, and fimbriae that promote colonization of the intestinal or urinary epithelia.

Coliform Organisms and Diseases

Escherichia coli: The Unpredictable Enteric Bacillus

Escherichia coli is the best known coliform, largely because of its use as a subject for laboratory studies. Although called the colon bacillus and sometimes regarded as the predominant species in the intestine of humans, *E. coli* is actually outnumbered 9 to 1 by the strictly anaerobic bacteria of the gut (*Bacteroides* and *Bifidobacterium*). Its prevalence in clinical specimens and infections is due to its being the most common aerobic and nonfastidious bacterium in the gut. Another mistaken view is that *E. coli* is a harmless commensal. Although most of the more than 100 strains are not infectious, some develop greater virulence through plasmid transfer, and others attack patients whose defenses are weakened.

Pathogenic Strains of *E. coli* **Enterotoxigenic** *E. coli* causes a severe diarrheal illness brought on by two exotoxins, termed heat-labile toxin (LT) and heat-stable toxin (ST), that stimulate heightened secretion and fluid loss. In this way it mimics

Escherichia (ess-shur-eek′-ee-uh) After the German physician Theodor Escherich.

the vibrio of cholera (see figure 17.14). This strain of bacteria also has fimbriae that provide adhesion to the small intestine (see figure 3.8*b*). **Enteroinvasive** *E. coli* causes an inflammatory disease similar to *Shigella* **dysentery** that involves invasion and ulceration of the mucosa of the large intestine. **Enteropathogenic** strains of *E. coli* are linked to a wasting form of infantile diarrhea whose pathogenesis is not well understood. The various pathogenic strains can be differentiated from each other as well as from nonvirulent versions on the basis of antigen differences (H, K, and O serotyping).

Clinical Diseases of *E. coli* Most clinical diseases of *E. coli* are transmitted exclusively among humans. Pathogenic strains of *E. coli* are frequent agents of **infantile diarrhea,** the greatest single cause of mortality among babies. In some areas of the world, about 15% to 25% of children five years or younger die of diarrhea as either the primary disease or a complication of some other illness. The rate of infection is higher in crowded tropical regions where sanitary facilities are poor, water supplies are contaminated, and adults carry pathogenic strains to which they have developed immunity. The immature, nonimmune neonatal intestine has no protection against these pathogens. A factor that increases the likelihood of infantile diarrhea is feeding the baby unsanitary food or water. The practice of preparing formula from dried powder mixed with contaminated water is tantamount to inoculating the infant with a pathogen. Many times, the babies are already malnourished, and further loss of body fluids and electrolytes is rapidly fatal. Mothers in third world countries especially are being urged to nurse their infants so that the children get better and safer nutrition as well as gastrointestinal immunity (see feature 12.7).

Traveler's diarrhea usually strikes persons visiting tropical countries and sampling the local food or drink. Despite the popular belief that "Montezuma's revenge," "Delhi belly," and other travel-associated gastrointestinal diseases are caused by exotic pathogens, a large proportion of cases are due to an enterotoxigenic strain of *E. coli*. Travelers encounter new strains to which the local population has developed immunity. The symptoms, occurring within 5 to 15 days, are profuse, watery diarrhea, low-grade fever, nausea, and vomiting. Some travelers develop more chronic and lasting enteroinvasive dysentery that resembles *Shigella* or *Salmonella* infection. The ability of antimicrobials to prevent or control infection is still unclear. Over-the-counter preparations that slow gut motility (kaolin, pectin, or *Lomotil*) afford symptomatic relief, but they may also serve to retain the pathogen longer and are probably not as effective as bismuth-salicylate mixture (Pepto-Bismol), which apparently counteracts enterotoxin.

Escherichia coli often invades sites other than the intestine. It causes 50% to 80% of **urinary tract infections** (UTI) in healthy persons. UTIs usually result when the urethra is invaded by its own endogenous bacterial colonists. The infection is more common in women because their relatively short urethras promote ascending infection to the bladder (cystitis) and occasionally the kidneys. It is also a complication of indwelling catheters and altered host defenses. Other extraintestinal infections in which *E. coli* is involved are neonatal meningitis, pneumonia, septicemia, and wound infections. These often complicate surgery, endoscopy, tracheostomy, catheterization, renal dialysis, and immunosuppressant therapy.

***E. coli* and the Coliform Count** Because of its prominence as a normal intestinal bacterium in most humans, *E. coli* is currently one of the indicator bacteria to monitor fecal contamination in water, food, and dairy products. According to this rationale, if *E. coli* is present in a water sample, fecal pathogens such as *Salmonella,* viruses, or even pathogenic protozoa may also be present. Coliforms like *E. coli* are used because they are present in larger numbers, are more resistant, and are easier and faster to detect than true pathogens. If a certain number of coliforms are detected in a sample, the water is judged unsafe to drink. This topic is covered in more detail in chapter 22.

Other Coliforms

Other coliforms of clinical importance mostly as opportunists are *Klebsiella, Enterobacter, Serratia,* and *Citrobacter.* Although these bacteria are enterics, most are infectious when introduced into sites other than the intestine. They cause infection primarily when the natural host defenses fail, often in conjunction with medical procedures that facilitate invasion. Various members of this group may infect nearly any organ, but the most serious complications involve septicemia and endotoxemia. Opportunists may be controlled to some extent by rigorous sterile and aseptic procedures along with antimicrobial therapy, but the rapid development of drug resistance in the group necessitates sensitivity testing. Due to their widespread occurrence and the poor medical state of the patients they infect, complete control will probably never be possible.

In addition to inhabiting the intestines of humans and animals, ***Klebsiella*** are also found in the respiratory tracts of normal individuals. This colonization leads to the chronic lung infections with which the species is associated. Infection by this organism is promoted by the large capsule, which prevents phagocytosis (figure 16.33). Some strains also produce toxins. The most important species in this group is ***K. pneumoniae,*** a frequent secondary invader and cause of nosocomial pneumonia, meningitis, bacteremia, wound infections, and UTIs.

Enterobacter and a closely related genus, *Hafnia,* commonly inhabit soil, sewage, and dairy products. Urinary tract infections account for the majority of clinical diseases associated with *Enterobacter.* The bacterium is also isolated from surgical wounds, spinal fluid, sputum, and blood. Although sometimes regarded as a minor pathogen, this organism is lethal when introduced into the bloodstream, as was shown in the 1970s when an epidemic caused by contaminated intravenous fluids infected 150 persons and killed 9. ***Citrobacter,*** a regular inhabitant of soil, water, and human feces, is an occasional opportunist in urinary tract infections and bacteremia in debilitated persons.

---

dysentery (dis'-en-ter"-ee) Gr. *dys,* bad, and *entera,* bowels.

Klebsiella (kleb-see-el'-uh) After Frederich Klebs, a German bacteriologist.

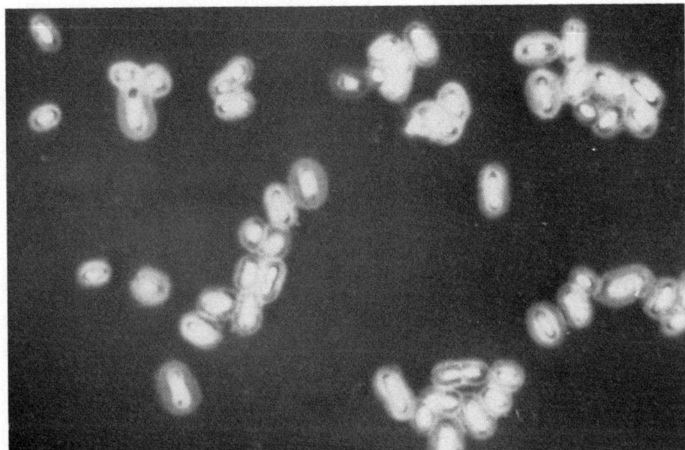

Figure 16.33 A capsule stain of *Klebsiella pneumoniae* (×1,500).

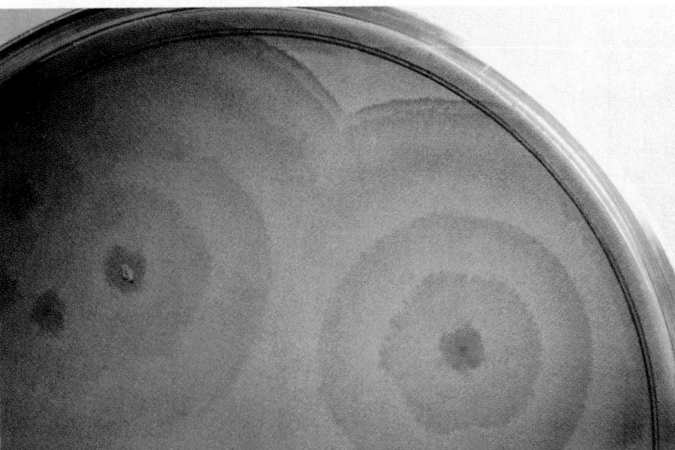

Figure 16.34 The wavelike swarming pattern of *Proteus mirabilis* is typical of this genus. Although this pattern is one of the most striking sights in the microbiology lab, it makes isolation of colonies in mixtures containing *Proteus* very difficult.

Serratia occurs naturally in soil and water as well as in the intestine. One species, *Serratia marcescens,* produces an intense red pigment (prodigiosin) when grown at room temperature. These bacteria are relatively innocuous in the healthy host, but since 1960 they have been found in nosocomial infections with increasing frequency. *Serratia* pneumonia, particularly prevalent in alcoholics, is transmitted by contaminated respiratory care equipment. The organism is often implicated in burn and wound infections, as well as in fatal septicemia and meningitis in immunosuppressed patients. To make matters worse, most strains have resistance to several classes of antibiotics.

Serratia—The Miracle Bug The brilliant red pigment formed by *Serratia marcescens* is striking and beautiful (see chapter opening illustration). Imagine how startled eighteenth-century Italians were when bright red dots appeared on the surfaces of cornmeal and communion wafers. In their haste to explain the phenomenon, they proclaimed that the spots were blood and the event was a miracle. When local scientist B. Bizio was called in to solve the mystery, he observed tiny red "fungi" under the microscope that he called *Serratia* (sur-at'-ee-uh) in memory of the great physicist S. Serrati and *marcescens* (mar-sess'-uns) for "fading away." This species was once considered so benign that microbiologists used it to trace the movements of air currents in hospitals and over cities, and even to demonstrate transient bacteremia following dental extraction. Of course, detecting these bacteria is greatly simplified by the obvious red colonies isolated from dust, air, or blood samples. Although 75% of the strains in clinical isolates are nonpigmented, infection with pigmented *Serratia* may cause streaks of red in the sputum, giving the false impression of blood.

Noncoliform Lactose-Negative Enterics

Opportunists: *Proteus* and Its Relatives

The three closely related genera *Proteus, Morganella,* and *Providencia* are saprobes in soil, manure, sewage, and polluted water, and commensals of humans and other animals. Despite a ubiquitous distribution, they are ordinarily harmless to a healthy individual. *Proteus* species swarm on the surface of moist agar in an unmistakable concentric pattern (figure 16.34). They are commonly involved in urinary tract infections, wound infections, pneumonia, septicemia, and occasionally infant diarrhea. *Proteus* urinary infections appear to stimulate renal stones and damage due to the urease they produce and the rise in urine pH this causes. *Morganella morganii* and *Providencia* species are involved in similar infections. One opportunist, *P. stuartii,* is a frequent invader of burns. Antibiotic therapy may be hindered by resistance to several antimicrobial agents.

Noncoliform Pathogens

The salmonellae and shigellae are distinguished from the coliforms and the *Proteus* group by having well-developed virulence factors, being primary pathogens, and not being normal flora of humans. The illnesses they cause—called **salmonelloses** and **shigelloses**—show some gastrointestinal involvement and diarrhea but often affect other systems as well.

***Salmonella:* Typhoid Fever and Other Salmonelloses** The most important pathogen in the genus *Salmonella* is *S. typhi,* the cause of the severe systemic disease typhoid fever. The other members of the genus are divided into two subgroups: *S. cholerae-suis,* a zoonosis of swine, and *S. enteritidis,* a large superspecies that includes around 1,700 different serotypes, based on variations on the major O, H, and V_i antigens (V_i being the virulence or capsule antigen in *Salmonella*). Each serotype has a species designation that reflects its source—for example, *S. typhimurium* (found in rats and mice) and *S. newport* (the city of isolation). The genus is so complex that most species determinations are performed by specialists at the Centers for Disease Control.

Salmonellae are motile; they ferment glucose with acid and sometimes gas; and most of them produce H_2S, but not

Proteus (pro'-tee-us) After *Proteus,* a Greek sea god who could assume many forms; ***Morganella*** (mor-gan-ell'-uh) after T. H. Morgan; and ***Providencia*** (prah-vih-den'-see-uh) after Providence, R.I.

Salmonella (sal-moh-nel'-uh) After Daniel Salmon, an American pathologist.

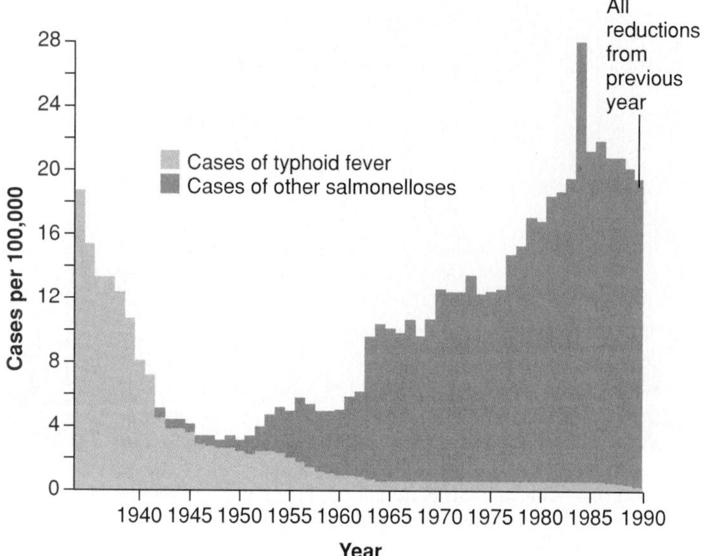

Figure 16.35 Data on the prevalence of typhoid fever and other salmonelloses from 1940 to 1990. (Nontyphoidal salmonelloses did occur prior to 1940, but the statistics are not available.)
Source: Data from *Science,* 21:965, November 1986.

Feature 16.7 The Infamous Typhoid Mary

Early in this century, a middle-aged Irishwoman named Mary Mallon immigrated to New York State to work as a cook in the houses of the rich. What followed has gone down in history as the classic example of a chronic carrier state. Seven of the eight families for which Mallon worked over a seven-year period were stricken by typhoid fever, and several persons died. A physician who specialized in diagnosing the disease spent months attempting to trace the source of the epidemic. Water, milk, food, and the environment were all tested for the presence of the typhoid bacillus, but to no avail.

Eventually, the doctor began to consider the possibility that the cook had been the culprit, and he set out to find her. But when finally tracked down, the wary and uncooperative Mary ran away. Due to the serious threat she posed to public health, the police were enlisted to capture her and take her to a hospital, where she was imprisoned. When tests for the pathogen proved positive, she was offered two ways to gain her freedom—have her gallbladder removed or stop working as a cook. Mary rejected both choices. She was released after three years on the condition that she not ply her trade, but she went right back to kitchen work, shedding her bacilli and causing more disease and death. After she was at last apprehended, she spent the remainder of her life as a ward of the hospital, and went to her grave stubbornly denying that she had ever been "Typhoid Mary."

urease. They are not particularly fastidious and grow readily on most laboratory media. They survive outside the host for long periods in inhospitable environments such as fresh water and freezing temperatures. These pathogens are resistant to chemicals such as bile and dyes (the basis for isolation on selective media) and do not lose virulence after long-term artificial cultivation.

Typhoid fever derives its name from its supposed symptomatic resemblance to typhus, a rickettsial disease, even though the two diseases are otherwise very different. In the United States, the incidence of typhoid fever has been steadily declining since the 1930s, and it now occurs primarily as sporadic cases in large cities (figure 16.35). Nationwide, about 500 cases are reported annually, roughly half of them imported from endemic regions. In other parts of the world, typhoid fever is still a serious health problem, responsible for 25,000 deaths each year and probably millions of cases.

The typhoid bacillus usually enters the alimentary canal along with water or food contaminated by the feces of overtly infected persons or carriers, although it is occasionally spread by close personal contact. Because humans are the exclusive hosts for *S. typhi,* asymptomatic carriers are important in perpetuating and spreading typhoid. Even six weeks after convalescence, the bacillus is still shed by about half of recovered patients. A small number of persons chronically carry the bacilli for longer periods in the gallbladder, from which they are constantly released into the intestine and feces. This carrier state, combined with unclean personal habits, provides a recipe for disaster (see feature 16.7).

The size of the *S. typhi* dose that must be swallowed to initiate infection is between 200,000 and a million bacilli. At the mucosa of the small intestine, the bacilli adhere and initiate a progressive, invasive infection that leads eventually to septicemia. The typhoid bacillus infiltrates the mesenteric lymph nodes and the reticuloendothelial network of the liver and spleen. Although macrophage uptake at these sites is prompt and extensive, the bacteria resist destruction and multiply intracellularly. The periodic escape of bacilli from infected lymphoid tissue gives rise to bacteremia and establishment in organs. Symptoms are fever, diarrhea, and abdominal pain. As the enteric phase progresses, the lymphoid tissue of the small intestine develops areas of ulceration that are vulnerable to hemorrhage, and in a few patients, deep erosion, perforation, and peritonitis occur (figure 16.36). The urinary tract and the liver may develop typhoidal abscesses. Untreated cases last a month, with about a 10% to 15% mortality rate, but with prompt treatment, very few patients die.

A preliminary diagnosis of typhoid fever can be based upon the patient's history and presenting symptoms, supported by a rising Widal agglutination titer (see chapter 13). Definitive diagnosis requires isolation of the typhoid bacillus. Chloramphenicol and ampicillin are the drugs of choice, though some resistant strains occur. Antimicrobials are usually effective in treating the chronic carrier, but surgical removal of the gallbladder may be necessary in individuals with chronic gallbladder inflammation. A killed whole-cell vaccine that provides temporary protection is available for travelers and military personnel.

Nontyphoidal Zoonotic Salmonelloses Salmonelloses other than typhoid fever are termed enteric fevers, *Salmonella* food poisoning, and gastroenteritis. These diseases are usually less severe than typhoid fever and are ascribed to one of the many serotypes of *Salmonella enteritidis*—most frequently, *S. paratyphi A, S. schottmulleri* (*paratyphi B*), *S. hirschfeldii* (*paratyphi C*), and *S. typhimurium.* A close relative, *Arizona*

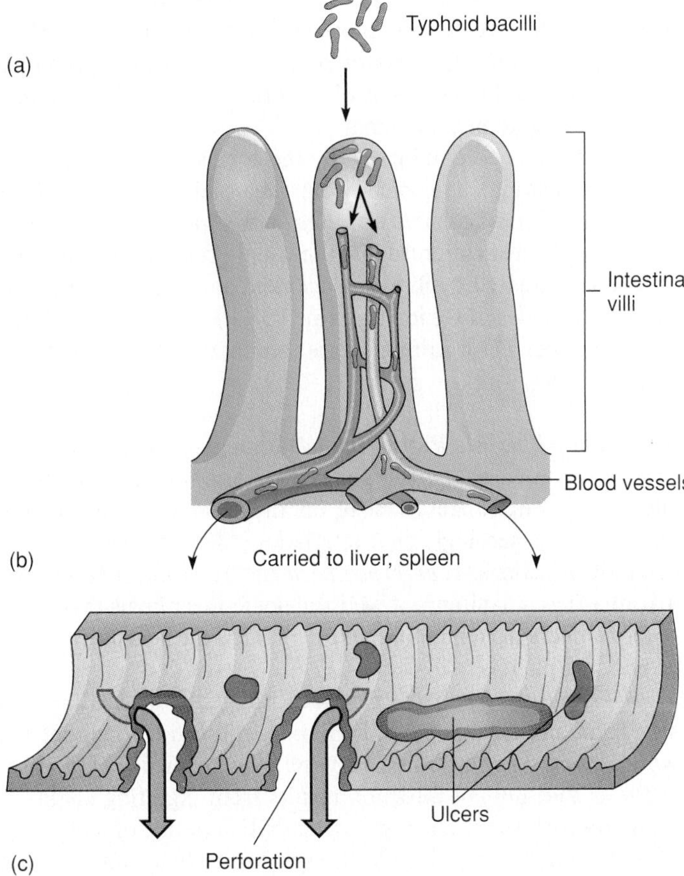

Figure 16.36 The phases of typhoid fever. (*a*) Ingested cells invade the small intestinal lining. (*b*) From this site, they enter the bloodstream, causing septicemia and endotoxemia. (*c*) Infection of the lymphatic tissue of the small intestine can produce varying degrees of perforation of the intestinal wall.

hinshawii,[4] is a pathogen found in the intestines of reptiles that parallels the salmonellae in its characteristics and diseases.

Nontyphoidal salmonelloses are more prevalent than typhoid fever and are currently on the increase (see figure 16.35). Unlike typhoid, all strains are zoonotic in origin, though humans may become carriers under certain circumstances. The primary reservoirs of these pathogens are cattle, rodents, poultry, and reptiles, and most cases of disease occur through a common food source. A notable example was the 1985 epidemic in a six-state area of the Midwest, which was due to contaminated milk from the same dairy and caused 14,000 cases of food infection. A smaller epidemic in the same area was caused by an antibiotic-resistant strain in ground beef (see feature 11.8). Another memorable outbreak occurred at the Circus-Circus casino in Las Vegas, Nevada, when a single food handler transmitted the pathogen to several hundred persons.

Animal Origins of *Salmonella* *Salmonella* are normal intestinal flora in many vertebrates. Because meat and milk can be readily contaminated during slaughter, collection, and processing, there are inherent risks in eating poorly cooked beef or unpasteurized fresh or dried milk, ice cream, and cheese. A particular concern is the contamination of foods by rodent feces. Several outbreaks of infection have been traced to unclean food storage or to food processing plants infested with rats and mice.

It is estimated that one out of every three chickens is contaminated with *Salmonella,* and other poultry such as ducks and turkeys are also affected. Eggs are a particular problem because the bacteria may actually enter the egg while the shell is being formed in the chicken. In general, one should always assume that poultry and poultry products harbor *Salmonella* and handle them accordingly, with clean techniques and adequate cooking.

Turtles and snakes are also carriers of the bacterium, thus it is essential to handle these animals carefully. Sales of small painted turtles are now outlawed in many areas due to the risk of salmonellosis.

The symptoms, virulence, and prevalence of *Salmonella* infections may be described by means of a pyramid, with typhoid fever at the peak, enteric fevers and gastroenteritis at intermediate levels, and asymptomatic infection at the base. In typhoid and enteric fevers, elevated body temperature and septicemia are much more prominent than gastrointestinal disturbances. Gastroenteritis is manifest by symptoms of vomiting, diarrhea, fluid loss, and mucosal lesions. Depending upon the organ or tissue involved, the lungs, nervous system, bones and joints may be sites for local infections. In otherwise healthy adults, symptoms spontaneously subside after two to five days; death is infrequent except in debilitated persons.

Serotyping is expensive, complicated, and not necessary before administering treatment. Therefore, it is not usually performed in the clinical setting and is attempted primarily when the source of infection must be traced, which is often a difficult bit of detective work. Treatment for complicated cases of gastroenteritis is similar to that for typhoid fever; uncomplicated cases are handled by fluid and electrolyte replacement. Because the pathogen has an animal reservoir and lacks a practical vaccine, total eradication of gastroenteritis caused by nontyphoidal salmonellae seems unlikely, but certain measures can greatly reduce its incidence (see feature 16.8).

***Shigella* and Bacillary Dysentery** *Shigella* causes a common but often incapacitating dysentery called shigellosis, which is marked by crippling abdominal cramps and frequent defecation of watery stool filled with mucus and blood. The etiologic agents (*Shigella dysenteriae, Sh. flexneri, Sh. boydii,* and *Sh. sonnei*) are primarily human parasites, though they may infect apes. All produce a similar disease that can vary in intensity. They are nonmotile, nonencapsulated, and not fastidious, and do not produce H_2S or urease. The shigellae resemble the enteropathogenic *E. coli* so closely that they are placed in the same subgroup.

4. Named for the state.

Shigella (shih-gel'-uh) After K. Shiga, a Japanese physician.

Feature 16.8 Avoiding Gastrointestinal Infections

Most food infections or other enteric illnesses caused by *Salmonella, Shigella, Vibrio, E. coli,* and other enteric pathogens are transmitted by the **4 F's—food, fingers, feces,** and **flies**—as well as by water. Individuals can protect themselves from infection by taking precautions when preparing food and traveling, and communities can protect their members by following certain public health measures.

In the Kitchen Confucius say, "A dirty cook gives diarrhea quicker than rhubarb." To which we might add, a little knowledge and good common sense go a long way. Food-related illnesses may be prevented by:

1. Use of good sanitation methods while preparing food. This includes preventing contamination by mechanical vectors like flies and cockroaches, disinfecting surfaces, sanitizing utensils, and liberal handwashing.
2. Use of temperature. Sufficient cooking of meats, eggs, and seafood is essential. These microbes are not heat-resistant and will be killed in 10 minutes at 100°C. Refrigeration prevents multiplication of bacteria in fresh food that is not to be cooked (but remember, it does not kill them).

While Traveling

1. It is well worth remembering that the 4 F's can be combated by the **4 B's—bread, beer, bananas,** and **bottled water.** In other words, choose food that is less likely to be contaminated with enteric pathogens.
2. Do not brush teeth or wash fruit or vegetables in tap water.
3. Go armed with antidiarrheal medicines and water disinfection tablets.

Community Measures

1. Water purification. Most enteric pathogens are destroyed by ordinary chlorination.
2. Pasteurization of milk.
3. Detection and treatment of carriers.
4. Restriction of carriers from food handling.
5. Constant vigilance during floods and other disasters that result in sewage spillage and contamination of food and drink.

Although *Sh. dysenteriae* causes the most severe form of dysentery, it is uncommon in the United States and occurs primarily in the eastern hemisphere. In the past decade, the prevalent agents in the United States have been *Sh. sonnei* and *Sh. flexneri.* Children between one and 10 account for more than half of the cases. In addition to the usual oral route, shigellosis is also acquired through direct person-to-person contact, largely because of the small infectious dose required (200 cells). The disease is mostly associated with lax sanitation, malnutrition, and crowding, and is spread epidemically in prisons, mental institutions, nursing homes, and boot camps. The so-called "gay bowel syndrome" is acquired through an atypical mode, namely unprotected anal sexual intercourse. As in other enteric infections, a chronic carrier period of weeks to months occurs in some persons.

Shigellosis is different from salmonellosis in that *Shigella* invades the villus cells of the large intestine, rather than the small intestine. In addition, it is not as aggressive as *Salmonella,* and does not perforate the intestine or invade the blood. The symptoms are caused by the release of toxins at the site of multiplication. Endotoxin causes fever, while enterotoxin brings on inflammation of the underlying gut wall layer, degeneration of the villi, and local erosion that causes bleeding and heavy mucous secretion (figure 16.37). Abdominal cramps and pain are caused by the disruption of the muscular function of the intestine. *Shigella dysenteriae* produces a heat-labile exotoxin that acts on the blood vessels of the brain and may lead to neurological symptoms and coma.

Diagnosis is complicated by the coexistence of several alternative candidates for bloody diarrhea such as *E. coli* and the protozoans *Entamoeba histolytica* and *Giardia lamblia.* Isolation and identification follow the usual protocols for enterics. Infection is treated by fluid replacement and oral drugs such as ampicillin and sulfa-trimethoprim (SXT). Prevention follows the same lines as for salmonellosis (see feature 16.8), and there is no vaccine.

The Enteric *Yersinia* Pathogens Although formerly classified in a separate category, the genus *Yersinia* has been placed in the family Enterobacteriaceae on the basis of cultural, biochemical, and serologic characteristics. All three species cause zoonotic infections. *Y. enterocolitica* and *Y. pseudotuberculosis* are intestinal inhabitants of wild and domestic animals that cause enteric infections in humans, and *Y. pestis* is the nonenteric agent of bubonic plague.

Yersinia enterolitica has been isolated from healthy and sick farm animals, pets, wild animals, and fish throughout the world. Its presence on fruits and vegetables and in drinking water suggests that human infection results from ingesting contaminated food or drink. During the incubation period of about 4 to 10 days, the bacteria invade the small intestinal mucosa, and some cells enter the lymphatics. Inflammation of the ileum and mesenteric lymph nodes gives rise to severe abdominal pain that mimics appendicitis. In fact, during an outbreak of *Yersinia* gastroenteritis and enterocolitis involving about 200 children a few years ago, a dozen children were mistakenly operated on for appendicitis. *Yersinia pseudotuberculosis* shares many characteristics with *Y. entercolitica,* though infections by the former are more benign and center upon lymph node inflammation rather than mucosal involvement.

Nonenteric *Yersinia pestis* and Plague

The word **plague**[5] conjures up visions of death and morbidity unlike any other infectious disease. Although pandemics of plague have probably occurred since antiquity, the first one that was reliably chronicled killed an estimated 100 million persons in the sixth century. Another one arose in the medieval 1340s, annihilating 40% of the population in some large cities. The last great pandemic occurred in the late 1800s and was transmitted around the world, primarily by rat-infested ships. Eventually, the disease was brought to the United States through the port of San Francisco. The chance circumstance of the 1906 earthquake and fire had far-reaching consequences, since infected rats abandoned the city and mingled with native populations of rodents. By this means, the pathogen gradually spread eastward to the Great Plains, and it exists today in many populations of

Yersinia (yur-sin'-ee-uh) After Alexandre Yersin, a French bacteriologist.

5. From the Latin *plaga,* meaning to strike, infest, or afflict with disease, calamity, or some other evil.

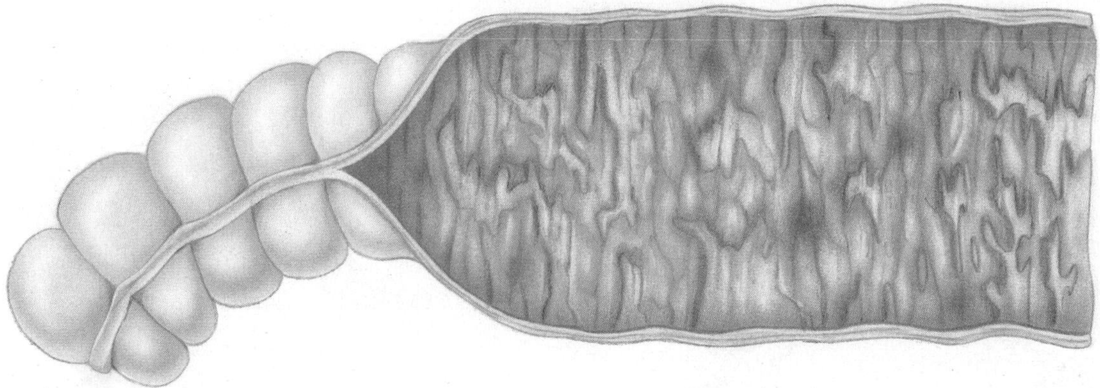

Figure 16.37 The appearance of large intestinal mucosa in *Shigella* (bacillary) dysentery. Note the patches of blood and mucus, the erosion of the lining, and the absence of perforation.

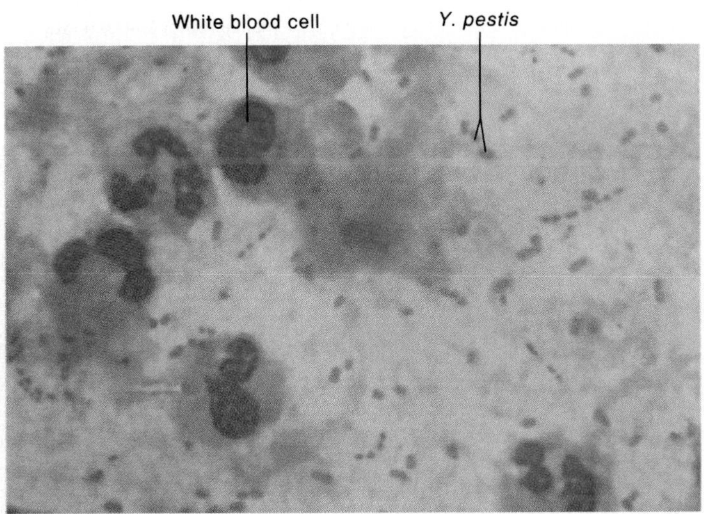

Figure 16.38 A Gram-stained preparation of *Yersinia pestis* in the blood of an infected mouse. The cells exhibit a distinctive bipolar morphology reminiscent of a safety pin.

rodents. The cause of this dread disease is a rather harmless-looking gram-negative rod with unusual bipolar staining and capsules called ***Yersinia pestis,*** formerly *Pasteurella pestis* (figure 16.38).

Virulence Factors Strains of the plague bacillus are just as virulent today as in the Middle Ages. Some of the virulence factors that coincide with its high morbidity and mortality are capsular and envelope proteins, which protect against phagocytosis and foster intracellular growth. The bacillus also produces coagulase, which clots blood and is involved in clogging the esophagus in fleas and obstructing blood vessels in humans. Other factors that contribute to pathogenicity are endotoxin and a highly potent murine toxin.

The Complex Epidemiology and Life Cycle of Plague

The plague bacillus exists naturally in many animal hosts, and its distribution is extensive, though the incidence of disease has been reduced in most areas. Plague still exists endemically in large areas of Africa, South America, the Mideast, Asia, and the USSR, where it sometimes erupts into epidemics. Cases occurring among soldiers during the Vietnam War reminded public health officials that plague is still a serious problem in many areas. In the United States, sporadic cases (single or small clusters) occur in 15 western states, mostly the result of contact with wild and domestic animals. No cases of human-to-human transmission have been recorded since 1924. Persons most at risk for developing plague are veterinarians, campers, backpackers, and those living near woodlands and forests.

The epidemiology of plague is among the most complex of all diseases. It involves several different types of vertebrate hosts and flea vectors, and its exact cycle varies from one region to another. A general scheme of the cycle is presented in figure 16.39. Humans may develop plague through contact with wild animals (**sylvatic plague**), domestic or semidomestic animals (**urban plague**), or infected humans.

The Animal Reservoirs The plague bacillus occurs in 200 different species of mammals. The primary long-term *endemic reservoirs* (foci) are various rodents such as mice and voles that harbor the organism but do not develop the disease. These hosts spread the disease to other mammals called *amplifying hosts* that become infected with the bacillus and experience massive die-offs during epidemics. These hosts, including the brown rat, black rat, ground squirrel, wood rat, chipmunk, prairie dog, and rabbit, are the sources of human plague. Which mammal is most important in this process depends on the area of the world. Other mammals (camels, sheep, coyotes, deer, dogs, and cats) may also be involved in the transmission cycle.

Flea Vectors The principal agents in the transmission of the plague bacillus from reservoir hosts to amplifying hosts to humans are fleas. These tiny, blood-sucking insects (see figure 11.19 and feature 17.5) have a special relationship with the bacillus. After an uninfected flea ingests a blood meal from a plague-ridden animal, the bacilli multiply in its gut. In a flea

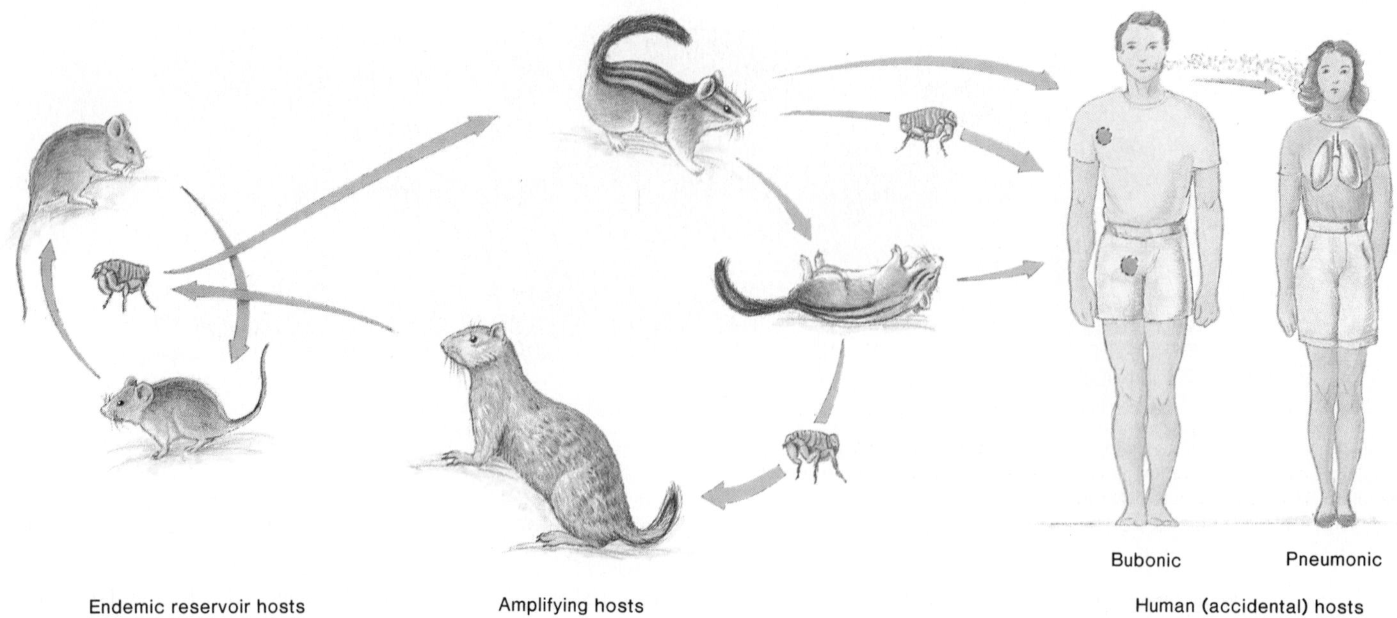

Figure 16.39 The infection cycle of *Y. pestis,* simplified for clarity.

that is a good transmitter of the bacillus, the combined bacterial growth and clotted blood eventually block the esophagus. Being unable to feed properly, the flea becomes ravenous and jumps from animal to animal in a futile attempt to get nourishment. During this process, regurgitated infectious material is inoculated into the bite wound.

When the flea's natural host is available, the flea will remain with that species and transmit the bacillus within that population. But many fleas are indiscriminate, and if a natural host species is not available or if it dies, they will seek other species, even humans. Depending on the weather conditions, fleas containing viable bacilli can survive up to three years in animal habitats. Fleas of rodents such as rats and squirrels are most often vectors in human plague, though occasionally, the human flea is involved. Humans are also infected by handling infected animals, animal skins, or meat, and by inhaling droplets.

Pathology of Plague

The number of bacilli required to initiate a plague infection is small—perhaps 3 to 50 cells. The manifestations of infection lead to bubonic, septicemic, or pneumonic plague. In **bubonic plague,** the plague bacillus multiplies in the flea bite, enters the lymph, and is filtered by the local lymph nodes. This causes an inflammatory swelling of the node called a **bubo,** typically in the groin and less often in the axilla or neck (figure 16.40). The incubation period lasts two to eight days, ending abruptly with the onset of fever, chills, headache, nausea, malaise, weakness, and extreme tenderness of the bubo.

Cases of bubonic plague often progress to massive bacterial growth in the blood termed **septicemic plague.** The release of virulence factors causes disseminated intravascular coagulation, circulatory stagnation, subcutaneous hemorrhage, and purpura that may degenerate into necrosis and gangrene. Due to visible darkening of the skin, the plague has often been called the "black death." In **pneumonic plague,** a dreaded complication, infection is lodged in the lungs and is highly contagious through sputum and pharyngeal aerosols. The fatality rate in untreated cases ranges from 50% to 75% in bubonic plague and approaches 100% in septicemic or pneumonic plague. Treated plague has a 90–95% survival rate.

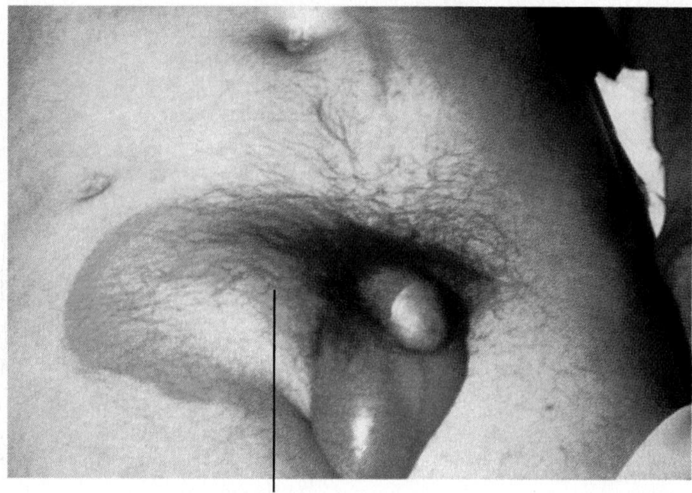

Figure 16.40 A classic inguinal bubo of bubonic plague. This hard nodule is very painful and may rupture onto the surface. (The brown discoloration is from iodine.)

Diagnosis, Treatment, and Prevention Because death may ensue as quickly as two to four days after the appearance of symptoms, prompt diagnosis and treatment of plague are

bubo (byoo′-boh) G. *boubon,* the groin.

imperative. The patient's history, including recent travel to endemic regions, symptoms, and laboratory findings from bubo aspirates, helps establish a diagnosis. Streptomycin, tetracycline, and sulfa drugs are satisfactory treatments.

The menace of plague is proclaimed by its status as one of the four internationally quarantinable diseases (the others are smallpox, cholera, and yellow fever). In addition to quarantine during epidemics, plague is controlled by trapping and poisoning rodents near urban and suburban communities and by dusting rodent burrows with insecticide to kill fleas. These methods do nothing to control the entire animal reservoir, so the potential for plague will always be present, especially as humans encroach more into the habitats of reservoir animals. A killed or attenuated vaccine that protects against the disease for a few months is given to military personnel and veterinarians.

Oxidase-Positive Nonenteric Pathogens

Pasteurella multocida *Pasteurella* is a zoonotic genus that occurs as normal flora in animals and is of concern to the veterinarian. Of the six recognized species, *P. multocida* is responsible for the broadest spectrum of opportunistic infections. For example, poultry and wild fowl are susceptible to cholera-like outbreaks, and cattle are especially prone to epidemic outbreaks of hemorrhagic septicemia or pneumonia known as "shipping fever." The species has also adapted to the nasopharynx of the household cat and is normal flora in the tonsils of dogs. Because many hosts are domesticated animals with relatively close human contact, zoonotic infections are an inevitable and serious complication.

Animal bites or scratches, usually from cats and dogs, cause a local abscess that can spread to the joints, bones, and lymph nodes. Patients with weakened immune function due to liver cirrhosis or rheumatoid arthritis are at greater risk for septicemic complications involving the central nervous system and heart. Patients with chronic bronchitis, emphysema, pneumonia, or other respiratory debility are vulnerable to pulmonary failure. Contrary to the antibiotic resistance of many gram-negative rods, *P. multocida* and other related species are susceptible to penicillin, and tetracycline is an effective alternative.

***Hemophilus:* The Blood-Loving Bacilli** *Hemophilus*[6] cells are tiny (0.5 × 0.8 mm) gram-negative pleomorphic rods sometimes confused with the genus *Neisseria* in clinical samples. The members of this group tend to be fastidious and sensitive to drying, temperature extremes, and disinfectants. Even though their name means blood-loving, none of these organisms can grow on blood agar alone without special techniques. Some *Hemophilus* species are normal colonists of the upper respiratory tract or vagina, and others (primarily *H. aegyptius, H. parainfluenzae,* and *H. ducreyi*) are virulent species responsible for conjunctivitis, childhood meningitis, and chancroid.

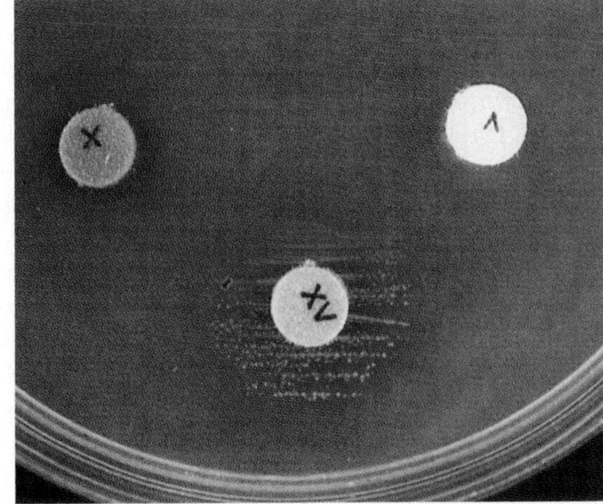

Figure 16.41 Discs containing separate or combined X- and V-factors are used to identify or type *Hemophilus* species. *H. influenzae* will grow only around the disc that has both factors. Other species (not shown here) require X- or V-factor but not both.

The characteristic that makes the hemophili fastidious is their requirement for X and/or V growth factors from blood to complete their metabolic syntheses (figure 16.41). The X-factor, hemin, is a part of the hemoglobin molecule and a necessary component of cytochromes, catalase, and peroxidase. The V-factor, nicotinamide adenine dinucleotide (NAD or NADP), is a coenzyme of certain dehydrogenases. These factors are made available to the hemophili in media such as chocolate agar (a form of cooked blood agar; see chapter 6) and Fildes medium. *Hemophilus* may also be grown on standard blood agar in conjunction with a catalase-positive organism like *Staphylococcus* that provides the necessary factors. The phenomenon of satellite colonies was illustrated in figure 6.14.

Hemophilus influenzae was originally named after it was isolated from patients with "flu" about 100 years ago. For over 40 years, it was erroneously proclaimed the causative agent until the real agent, the influenza virus, was discovered. The potential of this species to act as a pathogen remained in question until it was clearly shown to be the agent of **acute bacterial meningitis** in humans. This severe form of meningitis occurs chiefly in children between three months and five years of age, and the majority of cases are caused by the B serotype. In the past decade, about 20,000 cases have been reported in the United States annually, and the incidence appears to be increasing in some areas. In contrast to *Neisseria* meningitis, *Hemophilus* meningitis is not associated with epidemics in the general population, but tends to occur as sporadic cases or clusters in day-care and family settings. It is transmitted by close contact and nose and throat discharges. Most adults are immune to the agent, possibly due to a previous subclinical infection, and adult carriers are the usual reservoirs of the bacillus.

Hemophilus meningitis is very similar to meningococcal meningitis (see chapter 15), with symptoms of fever, vomiting, stiff neck, and neurological impairment. Untreated cases have a fatality rate of nearly 90%, but even with prompt diagnosis

Pasteurella multocida (pas"-teh-rel'-uh mul-toh-see'-duh) From Louis Pasteur, plus L. *multi,* many, and *cidere,* to kill.

6. Also spelled *Haemophilus.*

and aggressive treatment, 33% of children sustain residual disability, and about 5% must be placed in institutions for the mentally ill. Other important diseases caused by *H. influenzae* are an intense form of epiglottitis common in older children and young adults that may require immediate intubation or tracheostomy to relieve airway obstruction. This species is also an agent of otitis media, sinusitis, pneumonia, and bronchitis.

Hemophilus infections are usually treated with a combination of chloramphenicol and ampicillin. Outbreaks of disease in families and day-care centers may necessitate rifampin prophylaxis for all contacts. Vaccination with Hib vaccine, which is based on the capsular polysaccharide, is recommended for all children between the ages of two months and five years who enter day care.

Hemophilus aegyptius (**Koch-Weeks bacillus**) is an agent of acute communicable **conjunctivitis,** sometimes called **pinkeye.** The subconjunctival hemorrhage that accompanies infection imparts a bright pink tinge to the sclera (figure 16.42). The disease occurs primarily in children, is distributed worldwide, and is spread through contaminated fingers and shared personal items, as well as mechanically by gnats and flies. It is treated with antibiotic eyedrops.

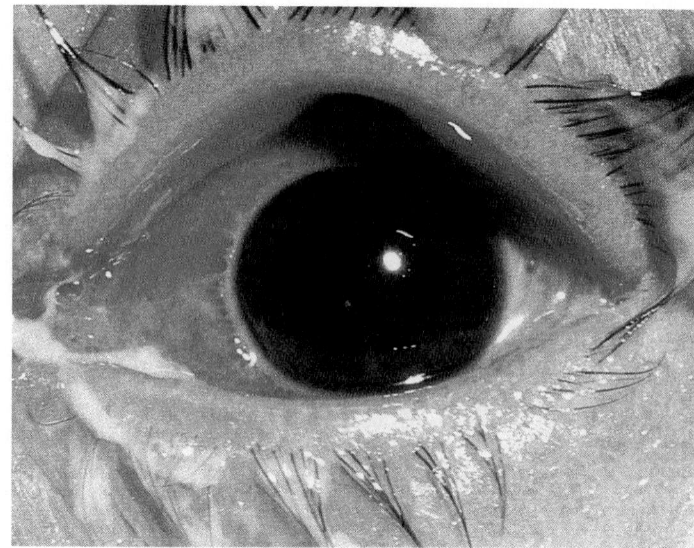

Figure 16.42 Acute conjunctivitis, or pinkeye.

Hemophilus ducreyi is the agent of **chancroid** (soft chancre), a sexually transmitted disease prevalent in the tropics and subtropics that afflicts mostly males. It is transmitted by direct contact with infected lesions and is favored by sexual promiscuity and unclean personal habits. After an incubation period lasting 2 to 14 days, lesions develop in the genital or perianal areas. First to appear is an inflammatory macule that evolves into a painful, necrotic ulcer similar to those found in lymphogranuloma venereum and syphilis (see chapter 17). Often the regional lymph nodes develop into bubo-like swellings that burst open. Although cotrimoxazole and other antimicrobials are effective, infection often recurs.

Hemophilus parainfluenzae and ***H. aphrophilus,*** members of normal oral and nasopharyngeal flora, are involved in infective endocarditis in adults who have underlying congenital or rheumatic heart disease. Such infections typically stem from routine dental procedures, periodontal disease, or some other oral injury.

Chapter Review with Key Terms

The Bacilli of Medical Importance

Gram-Positive Rods

Endosporeformers: Highly resistant bodies are produced. The aerobic genus ***Bacillus anthracis*** causes **anthrax,** a zoonosis of herbivorous animals; the bacterium inhabits the soil, where it is picked up by grazing animals that become infected and return it to the soil; humans are infected by contact with animals, their hides, or other products; the cutaneous form is manifest by **eschar,** a black ulcer at the site of entrance; the pulmonic form, acquired by inhaling spores, is highly fatal. Anthrax is controlled by antibiotics and animal vaccination. A type of food poisoning is caused by *B. cereus.*

The anaerobic genus ***Clostridium*** is common in soil and spread throughout most habitats, even the human body; wound and tissue diseases are due to introduction of the bacterium into damaged tissues; food intoxications arise when pathogenic clostridia release toxin into food; the basis for disease is powerful **exotoxins** that act upon specific target tissues. ***Cl. perfringens*** causes **gas gangrene,** a soft-tissue and muscle infection (**myonecrosis**) in areas of damage caused by punctures, broken limbs, surgery, diabetes, or crushing injuries; spores germinate in anaerobic, stagnant tissues and produce gas bubbles; may invade nearby healthy tissue; toxin destroys muscle, necrotizes tissue; may progress through whole limb; treatment by careful tissue **debridement** and cleansing, amputation, hyperbaric oxygen, antibiotics. ***Cl. difficile*** causes **antibiotic-associated colitis,** an acute diarrheal disease associated with disruptions in the GI tract flora due to antimicrobic therapy. ***Cl. tetani*** is the cause of **tetanus** or **lockjaw,** a disease associated with dirty puncture wounds, burns, umbilical stumps, or needles; spores in anaerobic areas of the wound grow and produce toxin called **tetanospasmin,** which travels to target cells in the CNS; affects muscular coordination by causing severe uncontrollable spasms of large muscles; loss of control may interfere with breathing, cause death; treatment is by antitoxin; control is through vaccination with tetanus toxoid.

Clostridial food poisoning occurs when *Cl. perfringens* is eaten with inadequately cooked meats and fish, producing toxin in the intestine; marked by severe diarrhea, vomiting, not usually fatal. **Botulism,** caused by ***Cl. botulinum,*** is associated with improperly home-canned foods; spores withstand food processing and grow in stored food; **botulin** is released. Botulism is a true food intoxication; ingested toxin enters circulation and acts on myoneural junctions; blocks muscle contraction, leads to flaccid paralysis; treated by antitoxin, maintenance on respirator; control by proper canning, heating of foods prior to eating. **Infant** and **wound botulism** are unusual types of infectious botulism contracted like other clostridial infections.

Non-spore-forming Rods: Straight, regular rods and irregular rods.

Straight, nonpleomorphic rods stain evenly. Genera that are regular in morphology include *Listeria monocytogenes* and *Erysipelothrix rhusiopathiae.* ***L. monocytogenes*** are widely distributed resistant bacteria that cause **listeriosis;** primary habitat appears to be water, soil, and the intestines; most cases are food infections associated with contaminated dairy products and meats; disease is mild, self-limited in young adults but may be severe and complicated in **fetuses,** neonates, the elderly, or the immunocompromised; treated with penicillin; for prevention, proper pasteurization of milk and dairy products is essential. ***E. rhusiopathiae*** is carried by animals (swine) and widely dispersed into the environment; cause of **erysipeloid,** a disease common among occupations that handle animals or animal products; symptom is inflamed red sores on fingers at portal of entry.

Irregular, non-spore-forming rods include *Corynebacterium* and *Propionibacterium*. ***C. diphtheriae*** is a highly pleomorphic aerobic rod with metachromatic granules and palisades arrangement; agent of *diphtheria,* a disease whose reservoir is healthy human carriers; spread by droplets; occurs first as an infection of the throat that remains localized and causes pharyngeal symptoms; symptoms include a **pseudomembrane** in throat that may suffocate; later symptoms are due to a classic **toxemia**; toxin spreads in blood to targets; may damage heart and CNS; treated by antitoxin, penicillin (erythromycin); controlled by vaccination (DPT). ***Propionibacterium acnes*** is an anaerobic rod found in the skin of humans; is regularly associated with **acne vulgaris** lesions; acts on skin oils to create inflammatory condition that erupts into pimples.

Acid-Fast Bacilli (AFB): Contain large amounts of mycolic acids and waxes; are strict aerobes; slender filamentous rods; widely distributed; tend to be resistant to environmental conditions. ***Mycobacterium tuberculosis*** causes **tuberculosis** (TB), a common lung infection; tubercle bacillus is spread by droplets in close quarters; most susceptible are persons with weakened immunities, poor nutrition, unhealthful living circumstances. The disease is divided into (1) **primary pulmonary** infection, often marked by formation of granulomas in the lungs called **tubercles;** (2) **secondary TB,** a severe lung complication that occurs in persons whose primary disease has been reactivated; and (3) **disseminated extrapulmonary** infections due to blood-borne TB bacilli carried to bone, kidneys, lymph nodes, and brain; may be fatal; TB diagnosed by tuberculin testing, chest X rays, acid-fast stain for bacilli in sputum; managed by combined drug therapy, BCG vaccine in some countries.

M. leprae causes **leprosy,** a chronic disease that begins in skin and mucous membranes and progresses into nerves; very prevalent in endemic regions of the world; spread through direct inoculation from **leprotics**; two forms are (1) **tuberculoid,** a superficial infection that does not produce skin nodules but damages nerves and causes loss of pain perception, and (2) **lepromatous,** a deeply nodular, disfiguring form that occurs primarily on the cooler regions of the body; in both forms, mutilation of parts may occur due to loss of pain receptors; treatment is by long-term combined therapy.

MOTT (typical mycobacteria other than the TB bacillus) are common environmental species that may be involved in disease. For example, MAI causes a disseminated disease of AIDS patients.

Gram-Negative Rods

A large group of loosely affiliated families and genera, all of which are non-spore-forming; cell walls of most species contain lipopolysaccharide with **endotoxic** properties, which causes fever, cardiovascular disruptions, and shock, and is one of the gravest complications of gram-negative septicemia.

Aerobic Rods: ***Pseudomonas*** species are among the most widely distributed bacteria; exist in most natural habitats and may be normal flora of humans; thrive even in hostile conditions or where nutrients are scarce; very versatile and may be beneficial; medically, ***Ps. aeruginosa*** is a common opportunist of medical treatments, especially where normal defenses are compromised; may attack lungs, skin, burns, urinary tract, eyes, ears; may also infect healthy persons; drug resistance limits treatment choices.

Brucella is a zoonotic genus that causes abortions in cattle, pigs, and goats, and **brucellosis** or **undulant fever** in humans; bacterium is transmitted through direct contact with infected animals or contaminated animal products and ingestion of raw milk; infection occurs in several systems; marked by organ abscess, fluctuating fever, and severe, long-term symptoms; treated with tetracycline therapy; controlled through animal vaccination, pasteurization.

Francisella tularensis causes **tularemia** or "rabbit fever," a zoonosis of rabbits, rodents, and other wild mammals that spreads to humans through direct contact with animals, bites by vectors (ticks), ingestion of contaminated food or water, or inhalation; very contagious infection; symptoms depend upon portal of entry and organ involved, but include skin, lymph nodes, lungs, intestine; treated with tetracycline; vaccine available for risk groups.

Bordetella pertussis causes a strictly human disease called **pertussis** or **whooping cough**; very contagious infection that afflicts mostly children under six months and may be fatal; infectious droplets are carried to respiratory tract; attachment of pathogen and toxin production destroy cilial defense and produce complications such as cough and bronchial inflammation. As disease escalates, so does coughing, which comes in several bursts followed by a sudden inspiration that rushes through the larynx; death may be caused by secondary infections; vaccine (DPT) very effective preventative in children.

Legionella pneumophila causes **legionellosis,** commonly called **Legionnaires' disease**; unusual agent is a wide-ranging inhabitant of natural water that also survives for months in human aquatic environments (cooling towers, air conditioners, sewers, taps) and can cause serious lung disease when accidentally inhaled; community infections occur from common environmental sources; hospital epidemics occur due to contaminated air and water supplies.

Anaerobic Rods: The **Family Enterobacteriaceae** is the largest group of gram-negative bacteria; small, often motile, fermentative rods occurring in many habitats but often found in animal intestines, thus are called **enterics;** informal division of family is made between **coliforms,** normal flora that are rapid lactose fermenters, and **noncoliforms,** genera that do not or only weakly ferment lactose; enterics are predominant bacteria in clinical specimens; some species cause **diarrheal disease** due to **enterotoxins** acting on the intestinal mucosa or to invasion and disruption of the mucosa; other bacteria are agents of nonintestinal opportunistic infections; some are true pathogens; some are opportunists only; identification of group includes series of biochemical and serological tests; primary **antigens** are flagellar (H), cell wall (O), and capsular (K or Vi); drug resistance is well entrenched through plasmids; treatment with drugs requires sensitivity testing.

The best known coliform diseases are caused by ***Escherichia coli,*** the most common enteric; exists in several forms; **pathogenic strains** have acquired virulence factors and invasive, toxigenic capacities: cause **infantile diarrhea,** a complication of malnourished babies fed unsanitary food or water; **traveler's diarrhea** occurs in people who pick up an odd toxigenic strain from water and food in other countries. *E. coli* is the usual cause of **urinary tract infections**; may be primary or opportunistic infection from normal flora; also causes nosocomial pneumonia and septicemia.

Other coliforms are ubiquitous in the hospital environment. *Klebsiella, Enterobacter, Serratia,* and *Citrobacter* account for a broad spectrum of nosocomial infections associated with instrumentation, including tracheostomies, respiratory care equipment, endoscopes, catheters; pneumonia, and burn and incision infections are common.

Noncoliform infections are caused by opportunists, pathogenic enterics, and pathogenic nonenterics. **Opportunists** include *Proteus, Morganella,* and *Providencia,* which cause nosocomial infections such as urinary tract infections, wound infections, pneumonia, sepsis, and diarrhea.

True enteric pathogens include ***Salmonella*** and ***Shigella.*** *Salmonella* causes **salmonelloses;** most severe disease is **typhoid fever,** caused by ***S. typhi;*** agent is spread only by humans via unclean food or water; often chronically carried in gallbladder; bacillus crosses wall of small intestine into circulation, is carried to liver, other organs where it may form abscesses; symptoms include fever, diarrhea, septicemia, and ulceration of small intestine that may lead to perforation into body cavity; treatment with chloramphenicol; vaccine available. Other species belong to a serotype of *S. enteriditis,* such as *S. typhimurium* or *S. paratyphi A;* bacilli are common flora of cattle, poultry, rats, mice; contaminate meat, milk, and eggs; diseases somewhat milder but more prevalent than typhoid fever; all cause zoonotic food infections called enteric fever or gastroenteritis, depending on severity of symptoms; treatment with antibiotics, oral rehydration.

Shigella causes **shigellosis,** a **bacillary dysentery** characterized by acute painful diarrhea with bloody, mucus-filled stools;

Sh. dysenteriae causes most severe form, but *Sh. sonnei* and *flexneri* cause similar milder disease; all are primarily human parasites; spread by **fingers, feces, food,** and **flies**; bacteria invade large intestine, but remain local and do not produce septicemia; damage to intestinal villi causes erosion, bleeding; treated with oral antimicrobics and rehydration therapy.

Enteric disease can be prevented through cleanliness in procuring and processing food, adequate cooking and refrigeration, keeping flies away, awareness of animal carriers, proper toilet habits, control of water and sewage, monitoring carriers, not ingesting questionable food or water.

Yersinia causes **yersinioses,** zoonoses that are spread from mammals to humans; *Y. enterocolitica* and *Y. pseudotuberculosis* cause food infection with appendicitis-like symptoms. ***Yersinia pestis*** is a nonenteric agent of the **plague,** an ancient virulent disease; epidemiology is very complex; agent is maintained by relationship between **endemic hosts** (mice) and **amplifying hosts** (rats, squirrels) and **flea vectors** that carry the bacillus between them as a result of blood-sucking habits; humans enter this cycle by accident, usually through flea bite or contact with infected animal; infected humans may pass the agent to other humans. Forms are (1) **bubonic plague,** in which multiplication at site of bite creates regional lymphatic swelling or **bubo**; (2) **septicemic plague,** a deadly complication marked by disseminated coagulation and hemorrhage; and (3) **pneumonic plague,** a lung infection spread by aerosols. Treatment includes streptomycin, tetracycline; control measures are vector and reservoir control and a vaccine.

Nonenteric pathogens include: (1) ***Pasteurella multocida,*** a zoonosis of cattle, poultry, cats, and dogs; spread by bites, scratches, other contact; usual disease is abscess and lymph node swelling. (2) ***Hemophilus influenzae,*** is the most frequent cause of **acute bacterial meningitis** in children between three months and five years; sporadic infection occurring primarily in day-care and similar settings; passed from adult carriers to children in discharges; disease is typical meningitis, with acute neurological complications, high morbidity and mortality, and mental sequelae; treated with chloramphenicol, ampicillin; vaccination with Hib recommended for children at risk. (3) ***H. aegyptius*** causes a form of conjunctivitis called pinkeye. (4) ***H. ducreyi*** causes the STD known as **chancroid.**

True–False Questions

Determine whether the following statements are true (T) or false (F). If you feel a statement is false, explain why, and reword the sentence so that it reads accurately.

___ 1. Endospore-forming bacteria that are agents of disease are widely distributed in soil and dust.

___ 2. Many clostridial diseases require anaerobic conditions for their development.

___ 3. *Clostridium perfringens* cannot invade healthy tissues.

___ 4. Tetanus causes spastic paralysis, and botulism causes flaccid paralysis.

___ 5. Listeriosis is a zoonosis spread from cattle to humans.

___ 6. Erysipeloid is a lung infection of swine that is transmitted to humans.

___ 7. The most severe symptoms of diphtheria are caused by toxemia.

___ 8. TB is spread by contaminated fomites.

___ 9. Leprosy is not highly communicable.

___ 10. Brucellosis causes abortion in host animals but not in humans.

___ 11. Tularemia is the most infectious bacterial disease known.

___ 12. *Shigella* infections lead to septicemia, but *Salmonella* infections do not.

___ 13. Typhoid fever is a human infection, whereas other salmonelloses are zoonoses.

___ 14. The bubo of bubonic plague is an enlarged lymph node.

___ 15. *Hemophilus influenzae* is the agent of influenza.

Concept Questions

1. What is the role of spores in infections?
2. Briefly outline the epidemiology of anthrax; what is the main manifestation of most infections? Why is pneumonic anthrax so deadly?
3. What characteristics of *Clostridium* contribute to its pathogenicity? Compare the toxigenicity of tetanospasmin and botulin. What predisposes the patient to clostridial infections?
4. Outline the epidemiology of the major wound infections and food intoxications of *Clostridium*. What is the origin of the gas in gas gangrene? How does hyperbaric oxygen treatment work? Why is amputation necessary in some cases? What is debridement, and how does it prevent some clostridial infections?
5. What is the mechanism of antibiotic-associated colitis?
6. What causes the jaw to lock in lockjaw? Why do patients with no noticeable infection sometimes present with tetanus?
7. Compare the symptomology of botulism and tetanus. How are they alike and different? What is the difference between food botulism and infant and wound botulism?
8. Describe the epidemiology and infection in listeriosis. Why is listeriosis a serious problem even with refrigerated foods?
9. Why is erysipeloid an occupation-associated infection?
10. What are the distinctive morphological traits of *Corynebacterium?* Differentiate between diphtheria infection and toxemia. How can the pseudomembrane be life-threatening? What is the ultimate origin of diphtherotoxin?
11. Give the unique characteristics of *Mycobacterium*. What is the epidemiology of TB? Differentiate between TB infection and TB disease. What are tubercles? What is the course of disseminated disease? What are the principles of tuberculin testing, chest X rays, and acid-fast staining? Why does tuberculosis require combined therapy?
12. What makes *M. leprae* different from other mycobacteria? Differentiate between tuberculoid and lepromatous leprosy. What causes the deformations? What causes the mutilation of extremities? What is the importance of the MOTT?
13. Why are bacteria like *Pseudomonas* and coliforms so often involved in nosocomial infections?
14. Briefly describe the human infections caused by *Pseudomonas, Brucella,* and *Francisella.*
15. What is the pathological effect of whooping cough? Why is it a baby killer?
16. What is unusual about *Legionella?* What is the epidemiological pattern of the disease?
17. What causes endotoxic shock, and what are the symptoms? Differentiate between toxigenic diarrhea and infectious diarrhea.
18. What is an enteric? A coliform? A noncoliform? Which bacteria are true enteric pathogens? Which are opportunists?

19. Briefly describe how to identify enterics. What is the basis of serological tests, and what is their main use for enterics?
20. How did *E. coli* develop pathogenicity? For what kinds of infections is it primarily responsible? Describe the roles of other coliforms in infections.
21. What is salmonellosis? What is the pattern of typhoid fever? How does the carrier state occur? What is the main source of the other salmonelloses?
22. What causes the blood and mucus in dysentery?
23. How does an individual avoid enteric infection and disease?
24. Briefly trace the epidemiologic cycle of plague. Compare the portal of entry of bubonic plague with that of pneumonic plague. Why is plague called the black death?
25. Briefly describe the epidemiology and pathology of *Hemophilus* meningitis.
26. Compare the types of food-related illness discussed in this chapter according to (1) the gram reaction of the agent, (2) whether it is a food infection or intoxication, and (3) the kinds of foods involved.
27. List the bacteria for which general, routine vaccines are given. For which special groups of bacteria are there vaccines? For which bacteria are there none? Why are there no vaccines for these?
28. Briefly outline the zoonotic infections in this chapter and describe how they are spread to humans.
29. Give the portal of entry and target tissues for tuberculosis, leprosy, gas gangrene, diphtheria, plague, pertussis, legionellosis, and shigellosis. Which are primary pulmonary pathogens?
30. Multiple matching. Match the zoonosis in the left-hand column with its primary biological vector in the right-hand column. (More than one answer may be possible.)

___ pasteurellosis	a. poultry
___ salmonellosis	b. fleas
___ gastroenteritis	c. herbivores
___ tularemia	d. rodents
___ anthrax	e. swine
___ brucellosis	f. dogs
___ erysipeloid	g. cats
___ plague	h. carnivores
___ yersiniosis	i. rabbits
	j. ticks

Practical/Thought Questions

1. What is the main clinical strategy in preventing gas gangrene? Why does it work?
2. Why is it unlikely that diseases like tetanus and botulism will ever be completely eradicated? Name some bacterial diseases that could be completely eradicated and explain how.
3. Why is the cause of death similar in tetanus and botulism?
4. Why does botulin not affect the senses? Why does botulism not commonly cause intestinal symptoms?
5. Account for the fact that boiling does not inactivate botulism spores but does inactivate botulin.
6. Adequate cooking is the usual way to prevent food poisoning. Why doesn't it work for *perfringens* and *bacillus* food poisoning?
7. Why do patients who survive tetanus and botulism often have no sequelae?
8. What would be the likely consequence of diphtheria infection alone without toxemia?
9. How can one tell that acne involves an infection?
10. Do you think the spittoons of the last century were effective in controlling tuberculosis? Why?
11. What could it mean to say that TB and leprosy are "family diseases"?
12. Why is *E. coli* the focus of tests to detect fecal contamination of water?
13. Given that so many infections are caused by gram-negative opportunists, what is the future medical prediction as the number of compromised patients increases?
14. An infectious dose of a million cells seems like a lot. In terms of the size and abundance of microbes, is it really that large? Could you see a cluster containing that many cells with the naked eye? Referring back to chapter 6 on microbial growth cycles, how long would it take an average bacterial species to reach that number if allowed to multiply?
15. Students in our classes sometimes ask how it is possible for a single enteric carrier to infect 1,000 people at a buffet or for a box turtle to expose someone to food infection. We always suggest that they use their imagination. Can you explain the course of events that causes these types of infections?
16. A farm worker sustained a crushing injury to his hand and was taken to a hospital where he received treatment and tetanus toxoid. During successive weeks, he had several operations to repair the damaged hand and was given antibiotics. After a time, he lost muscle tone and had difficulty talking. Finally, his hand required amputation, and he was given antitoxin. What do you think the disease was? Why did the earlier treatments not work? What other disease is this like?
17. A woman living near a wooded area in a western state discovered her cat carrying a sick mouse. She discarded the mouse, but a neighbor's dog found it and carried it home. In a few days, one child in the neighbor's family got a case of febrile illness that responded to antibiotics, while the cat died from an open sore on its neck. Later, the veterinarian who treated the cat developed a fatal pneumonia. What disease is possible here? Where did it originate?
18. Name five bacteria from this chapter that could be used in biological warfare. What are some possible ways they could be used to wage war?
19. Several persons working in an exercise gym acquired an acute disease characterized by fever, cough, pneumonia, and headache. Treatment with erythromycin cleared it up. The source was never found, but an environmental focus was suspected. What do you think might have caused the disease? People in a different gym got skin infections from sitting in a redwood hot tub. What could have caused that?

CHAPTER 17

Miscellaneous Bacterial Agents of Disease

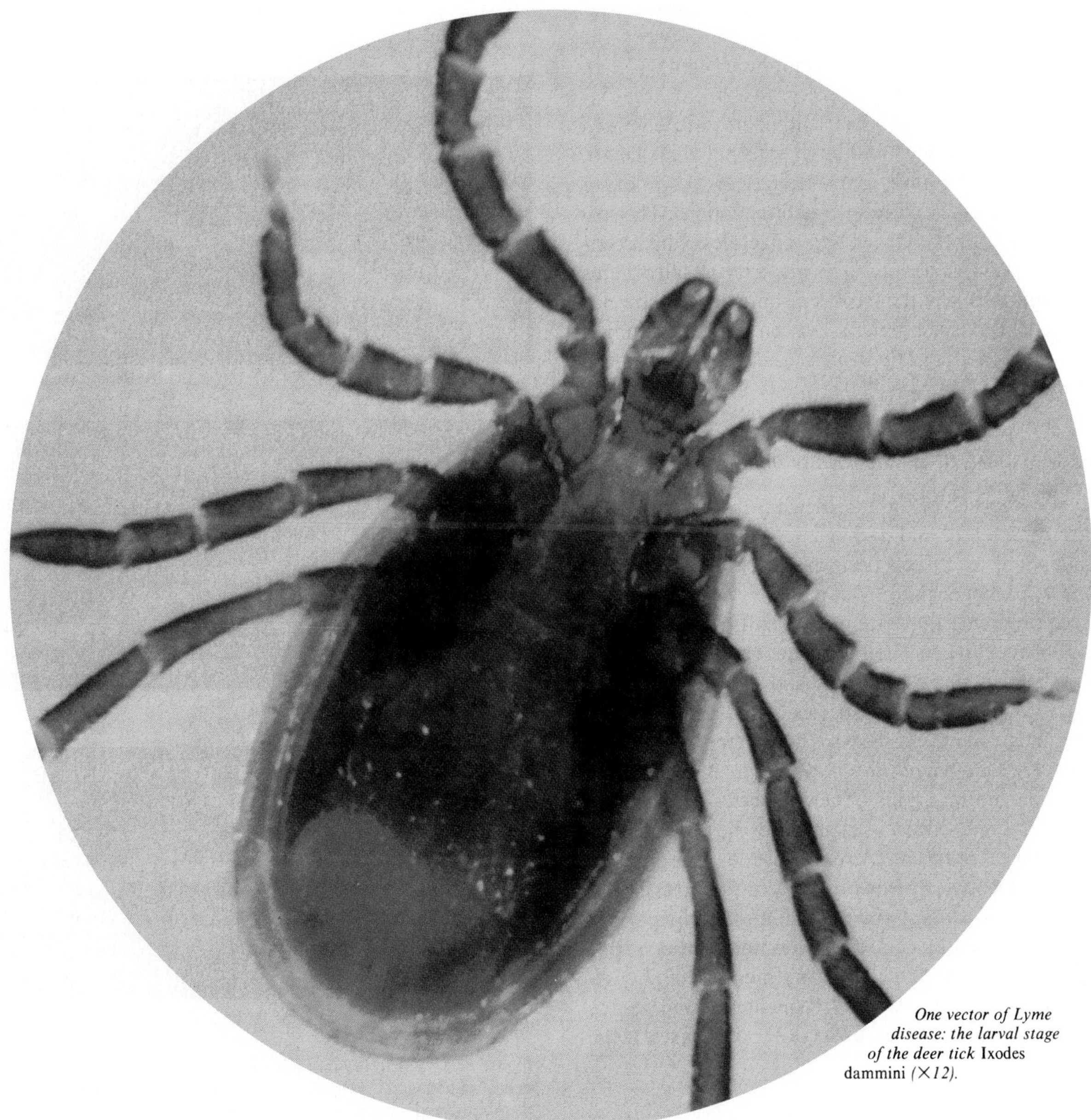

One vector of Lyme disease: the larval stage of the deer tick Ixodes dammini *(×12).*

Chapter Preview

A number of agents of bacterial infections do not fit the usual categories of gram-positive or gram-negative rods or cocci. This group includes spirochetes, obligate intracellular parasites such as rickettsiae and chlamydiae, and mycoplasmas. This chapter covers not only those agents, but also the vectors that many of them are borne by—ticks, lice, and other arthropods. The last section of the chapter surveys oral ecology, dental diseases, and the mixed bacterial infections that are responsible for them.

The Spirochetes

Bacteria called spirochetes have a helical form and a mode of locomotion that appear especially striking in live, unstained preparations using the dark-field or phase-contrast microscope (see figure 17.7). Other traits include a typical gram-negative cell wall and a well-developed periplasmic space between the underlying peptidoglycan and the cell membrane (figure 17.1*a*). Contrary to regular flagellated gram-negative bacilli, the flagella (also called axial filaments) of spirochetes are enclosed within the periplasmic space. Although internal flagella are constrained somewhat like limbs in a sleeping bag, their flexing propels the cell by rotation and even crawling motions. The spirochetes are classified in the Order Spirochaetales, which contains two families and five genera. The majority of spirochetes are free-living saprobes or commensals of animals and are not primary pathogens. But three genera that contain major human pathogens are *Treponema, Leptospira,* and *Borrelia* (figure 17.1*b, c,* and *d*).

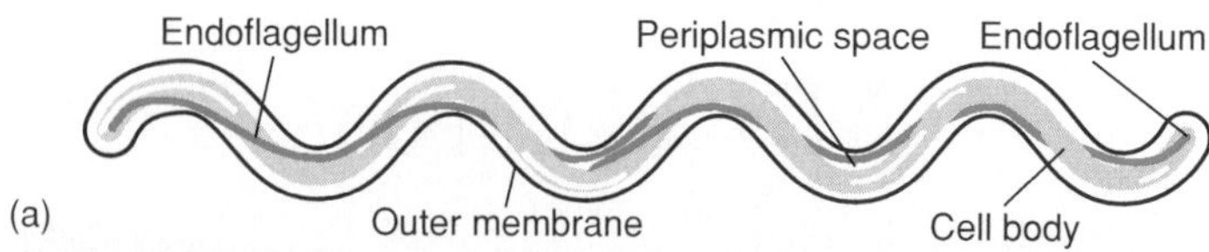

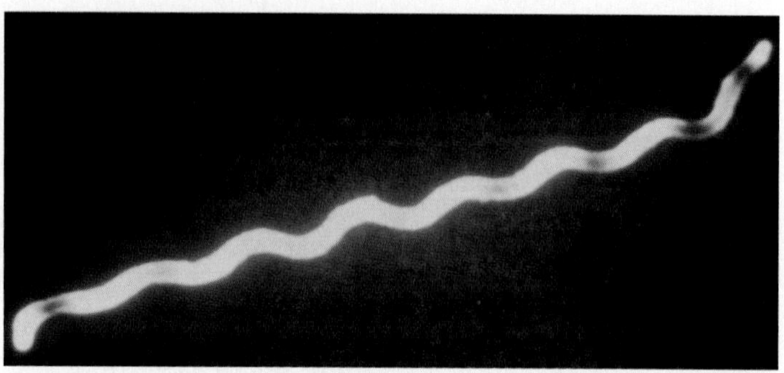

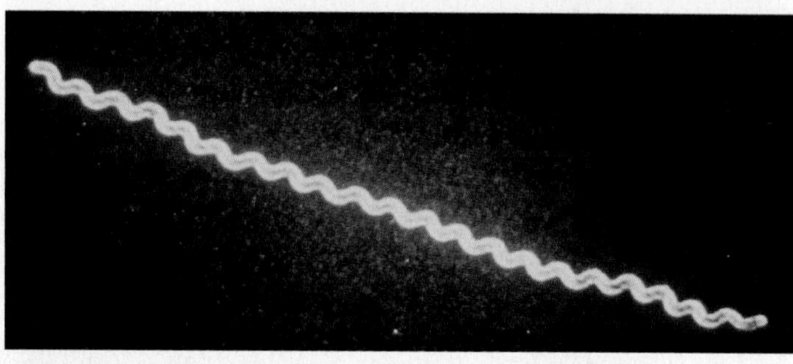

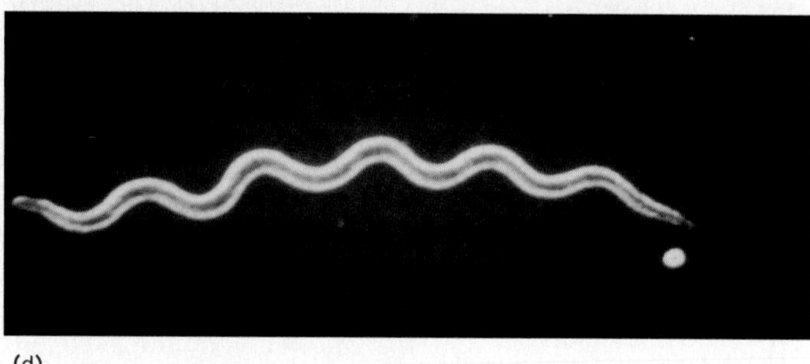

Figure 17.1 (*a*) Representation of general spirochete morphology with a pair of endoflagella inserted in the opposite poles, lying beneath the outer membrane and within the periplasmic space. In some spirochetes, the free ends overlap, as shown here. (*b-d*) Variations of the basic helical form. (*b*) *Treponema* has 8–20 evenly spaced coils. (*c*) *Leptospira* has numerous fine, regular coils and one or both curved ends. (*d*) *Borrelia* has 3–10 loose, irregular coils.

Treponemes: Members of Genus *Treponema*

Treponemes are composed of tight, somewhat regular, spiralled cells. They are commensal and pathogenic species living in the oral cavity, intestinal tract, and perigenital regions of humans and animals. The pathogens are strict parasites with such complex growth requirements that attempts to culture them on artificial media have failed. Their oxygen requirements are minimal, some being strictly anaerobic and others microaerophilic. Pathogenic species and subspecies of the genus *Treponema* are the etiologic agents of diseases called **treponematoses.** The subspecies *Treponema pallidum pallidum* is responsible for venereal and congenital syphilis; the subspecies *T. p. endemicum* causes nonvenereal endemic syphilis, or bejel; and *T. p. pertenue* causes yaws. *Treponema carateum* is the cause of pinta. Infection by the treponemes begins in the skin, progressing to other tissues in gradual stages, and is often marked by periods of healing interspersed with relapses. The major portion of this discussion will center on syphilis, and any mention of *T. pallidum* refers to this subspecies. Other treponemes of importance are involved in infections of the gingiva (see oral diseases at the end of the chapter).

Treponema pallidum: The Spirochete of Syphilis

The origin of syphilis is an obscure yet intriguing topic of speculation. The disease was first recognized at the close of the fifteenth century in Europe, a period coinciding with the return of Columbus from the West Indies, which led some medical scholars to conclude that syphilis was introduced to Europe from the New World. However, a more probable explanation contends that the spirochete evolved from a related subspecies, perhaps the endemic treponeme already present in the Mediterranean basin. In any case, the combination of the immunologically naive population of Europe, the European wars, and sexual promiscuity set the stage for worldwide transmission of syphilis that continues to this day.

Treponema pallidum (trep''-oh-nee'-mah pal'-ih-dum) Gr. *trepo,* turn, and *nema,* thread; L. *pallidum,* pale. The spirochete does not stain with the usual bacteriologic methods.

Synonyms for Syphilis Syphilis has been called many things over the centuries. Mention of the disease first appeared in a poem entitled "Syphilis sive Morbus Gallicus" by Fracastorius (1530), which was about a mythical shepherd whose name eventually became synonymous with the disease he suffered from. Another early Latin name for it was *lues venerea*—literally, the plague of love. Attempting to distance themselves from the disease, various peoples at various times have called syphilis the "Italian disease" or the "French disease." The name the Great Pox was adopted in the Middle Ages to distinguish syphilis from smallpox. Another nickname, the Great Imitator, points out that the complex stages of syphilis can easily be mistaken for infectious and noninfectious diseases.

Epidemiology and Virulence Factors of Syphilis

Although infection can be provoked in laboratory animals such as chimpanzees and rabbits, the human is evidently the sole natural host and source of *T. pallidum.* It is an extremely fastidious and sensitive bacterium that cannot survive for long outside the host, being rapidly destroyed by heat, cold, drying, disinfectants, soap, high oxygen tension, and pH changes. It survives a few minutes to hours when protected by body secretions and about 36 hours in stored blood. Research with human subjects has demonstrated that the risk of infection from sexual intercourse is 12% to 30%. Less common modes of transmission are passage to the fetus in utero and laboratory or medical accidents. Syphilitic infection through blood transfusion or exposure to fomites is rare.

Syphilis, like other STDs, has experienced periodic increases during times of social disruption (see figure 15.22). It is currently experiencing a sudden and dramatic outbreak nearly as great as that of the Second World War. This epidemic is concentrated in larger metropolitan areas and seems to occur most often among prostitutes, their contacts, and intravenous drug abusers. In some areas, the case rate has doubled. Because many cases go unreported, the actual incidence is likely to be several times higher than these reports show. Syphilis in Third World countries, especially Africa and Asia, is also on the rise. Persons with syphilis often suffer concurrent infection with other STDs such as gonorrhea or HIV; the latter is an especially deadly combination with a rapidly fatal course.

Pathogenesis and Host Response

Brought into direct contact with mucous membranes or abraded skin, *T. pallidum* binds avidly to the epithelium (figure 17.2). The number of cells required to establish infection is not known, though it may be as few as five. At the binding site, the spirochete multiplies and penetrates the capillaries by dissolving the hyaluronic acid between endothelial cells. Within a short time, it moves into the circulation, and the body is literally transformed into a large receptacle for incubating the pathogen—virtually any tissue is a potential target. The slow generation time of this bacterium (about 30 hours) accounts for the slow but progressive nature of the disease.

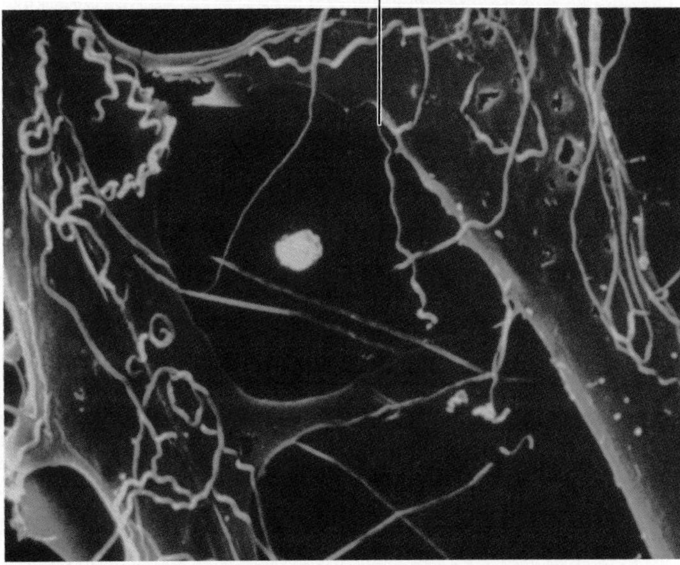

Figure 17.2 Electron micrograph of the syphilis spirochete attached to cells. Notice the hooklike nature of the specialized tip of the spirochete.

Specific factors that account for the virulence of the syphilis spirochete are yet to be discovered. It produces no toxins and does not appear to kill cells directly. Studies have shown that although phagocytes seem to act against it and several types of antitreponemal antibodies are formed, cell-mediated immune responses seem inadequate to contain it. The primary lesion occurs when the spirochetes invade the spaces around arteries and stimulate an inflammatory response. Organs are damaged when granuloma cells accumulate at these sites and obliterate circulation.

Clinical Manifestations Untreated syphilis is marked by distinct clinical stages designated as primary, secondary, and tertiary syphilis (table 17.1). It also has latent periods of varying duration during which the disease is quiescent. The spirochete appears in the lesions and blood during the primary and secondary stages, and thus is communicable at these times. It is largely non-communicable during the tertiary stage, though syphilis may be transmitted during early latency.

Primary Syphilis The earliest indication of syphilis infection is the appearance of a **chancre** at the site of inoculation, after an incubation period that varies from nine days to three months (figure 17.3). The chancre begins as a small, red, hard bump that enlarges and breaks down, leaving a shallow crater with firm margins. The base of the ulcer beneath the encrusted surface swarms with spirochetes. Chancres are single or multiple,

---

chancre (shang'-ker) Fr. for canker; from L. *cancer,* crab. An injurious sore.

Table 17.1 Syphilis: Stages, Symptoms, Diagnosis, and Control

Stage	Average Duration	Clinical Setting	Diagnosis	Treatment
Incubation	3 weeks	No lesion; treponemes adhere and penetrate the epithelium; after multiplying, they disseminate	Asymptomatic phase	
Primary	4–6 weeks	Initial appearance of chancre at inoculation site; intense treponemal activity in body; chancre later disappears	Dark-field microscopy; VDRL, FTA-ABS, MHA-TP testing	Benzathine penicillin G, 2×10^6 units; aqueous benzyl or procaine penicillin G, 4.8×10^6 units
Primary latency	4–8 weeks	Healed chancre; little scarring; treponemes in blood; few if any symptoms	Serological tests (+)	As above
Secondary	6 weeks after chancre leaves	Skin, mucous membrane lesions; hair loss; patient highly infectious; fever, lymphadenopathy; symptoms may persist for months	Dark-field testing of lesions; serological tests	Double doses of penicillins listed above
Latency	6 months–8 years	Treponemes quiescent unless relapse occurs; lesions may reappear; seropositivity	Microscopy useless	As above
Tertiary	Variable, up to 20 years	Neural, cardiovascular symptoms; gummas develop in organs; seropositivity	Treponeme not demonstrated	As above

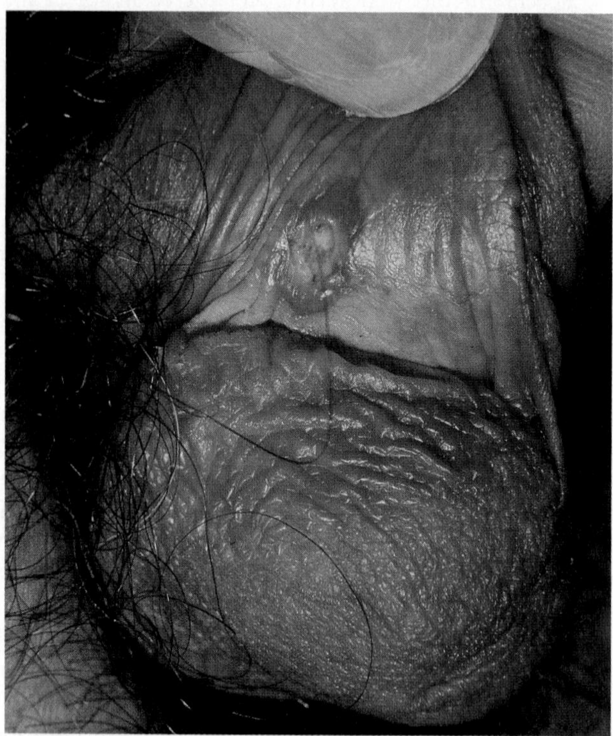

Figure 17.3 A chancre, the lesion of primary syphilis.

with most appearing on the internal and external genitalia. About 20% occur on the lips, oral cavity, nipples, fingers, or around the rectum. Genital lesions tend to be painless, and in the female, internal ones may even escape notice. Lymph nodes draining the affected region become enlarged and firm, but systemic symptoms like fever or headache are virtually absent. The chancre heals spontaneously without scarring in three to six weeks, but this healing is deceptive, because the spirochete has escaped into the circulation and is entering a period of tremendous activity.

Secondary Syphilis About three weeks to six months (average is six weeks) after the chancre heals, the secondary stage appears. By now, many systems of the body have been invaded, and the symptoms are more profuse and intense. Initially there is fever, headache, and sore throat, followed by lymphadenopathy and a peculiar red or brown rash that breaks out on all skin surfaces, including the palms and the soles (figure 17.4). Like the chancre, the lesions contain viable spirochetes and disappear spontaneously. While no system is spared, the major complications develop in the bones, hair follicles, joints, liver, eyes, brain, and kidneys. In most individuals, the symptoms of secondary syphilis disappear in a few weeks, though certain symptoms may linger for months and years.

Latency and Tertiary Syphilis After resolution of the symptoms of secondary syphilis, about 30% of persons infected enter a highly varied latent period that may last for 20 years or longer. Latency is divisible into early and late phases, and though antitreponeme antibodies are readily detected, the parasite itself is not. The final stage of disease, late or **tertiary syphilis,** is quite rare today because of widespread use of antibiotics to treat other infections. By the time a patient reaches this phase, the combined action of the latent infection and the body's response to it produce severe pathological complications. Cardiovascular syphilis results from damage to the small arteries in the aortic wall. As the fibers in the wall weaken, the aorta is subject to distension and fatal rupture. The same pathologic process can

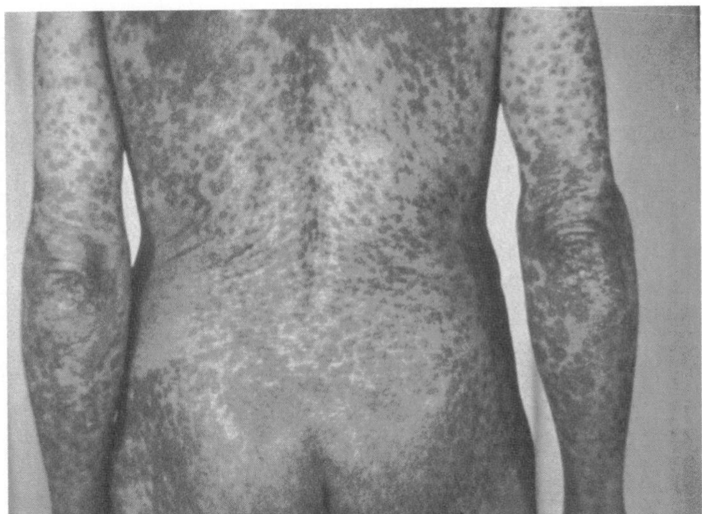

Figure 17.4 The skin rash in secondary syphilis may form on the trunk, arms, and even palms and soles (this latter feature is particularly diagnostic). The rash does not hurt or itch and may persist for months.

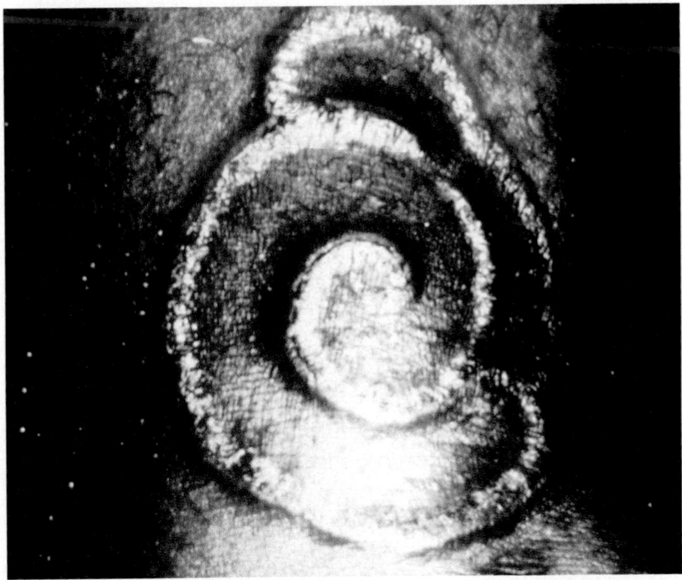

Figure 17.5 The pathology of late or tertiary syphilis. A ring-shaped, erosive gumma appears on the arm of this patient. Other gummas may be internal.

damage the aortic valves, resulting in insufficiency and heart failure.

In one form of tertiary syphilis, painful swollen syphilitic tumors called **gummas** develop in tissues such as the liver, skin, bone, and cartilage (figure 17.5). Gummas are usually benign and only occasionally lead to death, but they can impair function. **Neurosyphilis** can involve any part of the nervous system, but it shows particular affinity for the blood vessels in the brain, cranial nerves, and dorsal roots of the spinal cord. The diverse reactions include severe headaches, convulsions, mental

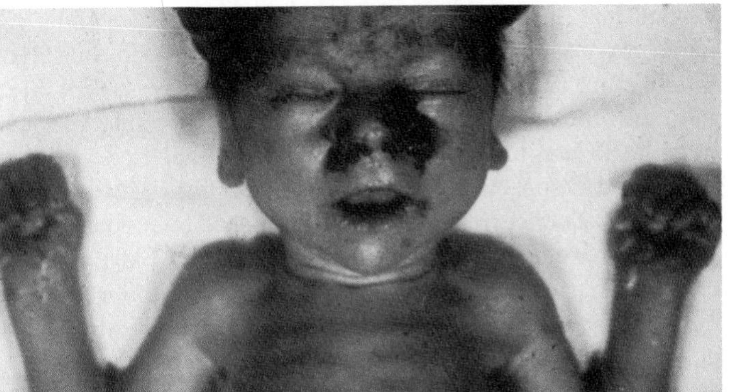

(a)

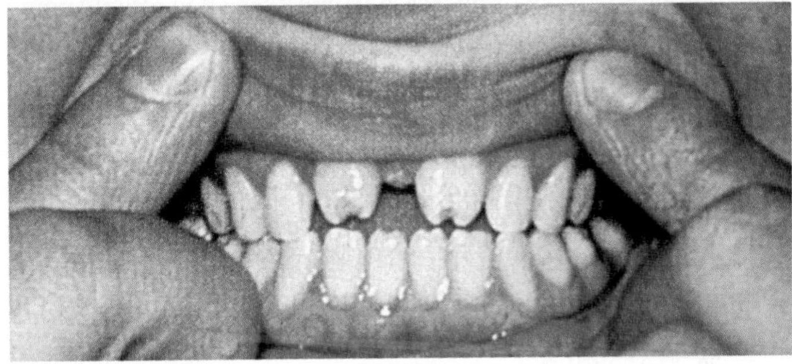

(b)

Figure 17.6 Congenital syphilis. (*a*) An early symptom is snuffles, a profuse nasal discharge that obstructs breathing. (*b*) A common characteristic of late congenital syphilis is notched, barrel-shaped incisors (Hutchinson's teeth).

derangement, the Argyll Robertson pupil,[1] atrophy of the optic nerve and blindness, and destruction of parts of the spinal cord that may lead to muscle wasting and loss of activity and coordination.

Congenital Syphilis When *Treponema pallidum* passes from the circulation of a syphilitic mother into the placenta, enters the umbilical vein, and is carried throughout the fetal tissues, it may give rise to **congenital syphilis.** Infection may occur in any of the three trimesters, though it is most common in the second and third. As it becomes established in the fetal tissues, the pathogen inhibits cell growth and disrupts critical periods of development with varied consequences, ranging from mild to the extremes of spontaneous abortion or stillbirth. Two clinical forms are recognized. Early congenital syphilis encompasses the period from birth to two years of age and is usually first detected three to eight weeks after birth. Infants present with nasal discharge (figure 17.6*a*), skin eruptions and loss, bone deformation, and nervous system abnormalities. The late form gives rise to an unusual assortment of stigmata in the bones, eyes, inner ear, joints, and teeth (figure 17.6*b*). The number of cases of congenital syphilis has been increasing in conjunction with the epidemic in adults, and because it is sometimes not diagnosed, some children will die or sustain lifelong disfiguring disease.

gumma (goo'-mah) L. *gummi*, gum. A soft tumorous mass containing granuloma tissue.

1. Perhaps the most common sign still seen today, this condition is caused by adhesions along the inner edge of the iris that fix the pupil's position into a small, irregular circle.

Clinical and Laboratory Diagnosis The pattern of syphilis imposes many complications on diagnosis. Not only do the stages mimic other diseases, but their appearance can be so separated in time as to seem unrelated. The chancre and secondary lesions must be differentiated from bacterial, fungal, and parasitic infections, tumors, and even allergic reactions. Overlapping symptoms of concurrent, sexually transmitted infections such as gonorrhea or chlamydiosis can further complicate diagnosis. The clinician must weigh presenting symptoms, patient history, and microscopic and serologic tests in rendering a definitive diagnosis.

For specific diagnosis of primary, early congenital, and to a lesser extent, secondary syphilis, no laboratory test is more valuable than dark-field microscopy of a suspected lesion (figure 17.7). The lesions must be carefully wiped with saline solution (not antiseptic) and squeezed or gently scraped to extract clear serous fluid. A wet mount prepared from the exudate is observed for the characteristic size, shape, and motility of *T. pallidum.* A single negative test is insufficient to exclude syphilis, since the patient may have removed the organism by washing, so two follow-up tests without prior washing are recommended. Another useful test for discerning the spirochete directly in samples is immunofluorescence staining with monoclonal antibodies (see figure C, page 473).

If dark-field or direct antigen tests are negative, serologic tests provide valuable though indirect diagnostic support. These tests are based upon detection of antibody formed in response to *T. pallidum* infection (table 17.1). Several of the tests (VDRL, Rapid Plasma Reagin (RPR), Kolmer) are variations on the original test developed by Wasserman using cardiolipin, a natural constituent of many cells, as the antigen. Although anticardiolipin antibodies are not specific for syphilis, the test is an effective way to screen the population for persons who may be infected.

Testing Blood for Syphilis Premarital blood tests required by most states are partly for detecting syphilis antibodies. Blood tests are also commonly applied to high-risk groups such as homosexuals, male and female prostitutes, persons with other STDs, and pregnant women. In the case of a positive result, it is important that a series of serologic tests be carried out to detect an elevated antibody titer indicative of active infection, since a single positive test may be due to a prior cured infection. Because the VDRL and its related tests rely on reactions to a substance found normally in human tissue, biological false positives can occur, especially in patients with autoimmune diseases or impaired immunity. A more specific test is needed for persons who are suspected of having a false-positive result.

Tests based on antigens specific to *Treponema* are more accurate than the VDRL test. Typical of these is the *T. pallidum* microhemagglutination assay (MHA-TP), which employs red blood cells that have been coated with treponemal antigen. Agglutination of the cells by serum indicates antitreponemal antibodies and infection. Another standard test is an indirect immunofluorescent method called the FTA-ABS (Fluorescent Treponemal Antibody Absorbance) test. The test serum is first absorbed with treponemal cells and reacted with antihuman globulin antibody labelled with fluorescent dyes. If antibodies to the treponeme are present, the fluorescence on the outside of these cells is highly visible with a fluorescent microscope (see figure 13.32*b*). The most sensitive and specific of all is the *T. pallidum* immobilization (TPI) test, in which live syphilis spirochetes are mixed with the test serum and observed microscopically for loss of motility. This has been a particularly expensive and difficult test because of the need to maintain cultures of live treponemes. In the past, rabbit testes have served this purpose, but now that live virulent strains have been cultivated in rabbit epithelial cell culture, this test will be more manageable.

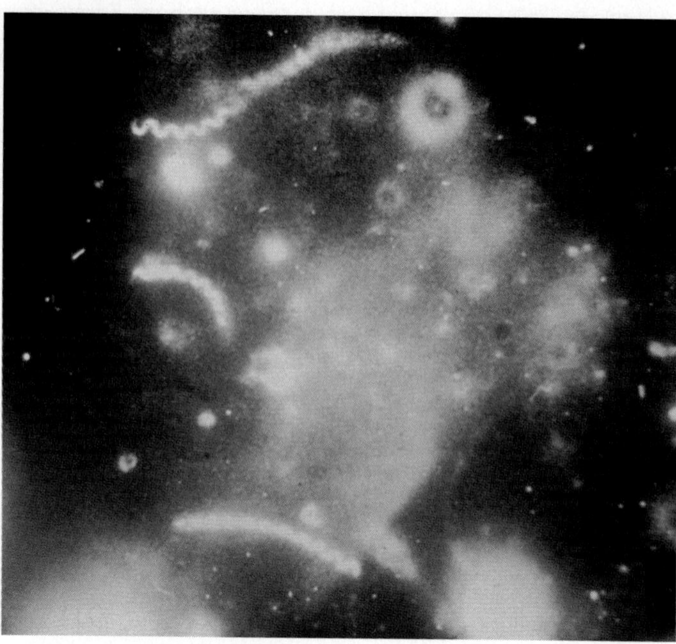

Figure 17.7 *Treponema pallidum* from a syphilitic chancre, viewed with dark-field illumination. Spirochetes contrast sharply with the red blood cells and tissue cells.

Treatment and Prevention Penicillin G retains its status as a wonder drug in the treatment of all stages and forms of syphilis. It is given parenterally in very large doses with benzathine or procaine to maintain a blood level lethal to the treponeme for at least seven days (table 17.1). Alternate drugs (tetracycline and erythromycin) are less effective and are indicated only if penicillin allergy has been documented. It is important that all patients be monitored for compliance or possible treatment failure.

The core of an effective prevention program depends upon detection and treatment of the sexual contacts of syphilitic patients. Public health departments and physicians are charged with the task of questioning the patients and tracing their contacts. All individuals identified as being at risk, even if they show no signs of infection, are given immediate prophylactic penicillin in a single, long-acting dose. The barrier effect of a condom provides superior protection, but washing after sexual intercourse alone is not an adequate protective measure. Protective immunity apparently does arise in humans and in experimentally infected rabbits, and this raises the prospect of an effective immunization program in the future. The recent cloning of treponemal surface antigens using recombinant DNA technology promises to provide a ready supply of purified antigen for developing vaccines and new diagnostic testing methods.

Nonsyphilitic Treponematoses

The other treponematoses are ancient diseases that closely resemble syphilis in their effects, though they are rarely transmitted sexually or congenitally. These infections, known as bejel, yaws, and pinta, are endemic to certain tropical and subtropical regions of the world, especially rural areas with unsanitary living conditions. The treponemes that cause these infections are nearly indistinguishable from those of syphilis in morphology and behavior. The diseases are slow and progressive, and involve primary, secondary, and tertiary stages. They begin with local invasion by the treponeme into the skin or mucous membranes and its subsequent spread to subcutaneous tissues, bones, and joints. Drug therapy with penicillin, erythromycin, or tetracycline remains the treatment of choice for these treponematoses.

Bejel Bejel is also known as endemic syphilis and nonvenereal childhood syphilis. The pathogen, a subspecies of *T. pallidum,* is harbored by a small reservoir of nomadic and seminomadic humans in arid areas of the Middle East and North Africa. It is a chronic, inflammatory childhood disease transmitted by direct contact or shared household utensils and other fomites, and is facilitated by minor abrasions or cracks in the skin or a mucous membrane. Often the infection begins as small moist patches in the oral cavity (figure 17.8*a*) and spreads to the skin folds of the body and to the palms.

Yaws Yaws is a West Indian name for a chronic disease known by the regional names bouba, frambesia tropica, and patek. It is endemic to warm, humid, tropical regions of Africa, Asia, and South America. The microbe is readily spread by direct contact with skin lesions or fomites, thus crowded dwellings and poor community or personal hygiene are contributing factors. Infection usually begins early in life and probably occurs only at damaged skin sites. The earliest sign is a large, abscessed papule called the "mother yaw," usually on the legs or lower trunk. After the initial lesion has healed, a secondary crop of moist nodular tumors appears, which erode the skin, periosteum, and bones, but do not penetrate to the viscera (figure 17.8*b*). Late stage yaws may result in mutilating ulcerations of the face and extremities. Yaws can be prevented by correcting predisposing conditions, shielding minor skin injuries from mechanical insect vectors, mass treatment, and surveillance for new cases.

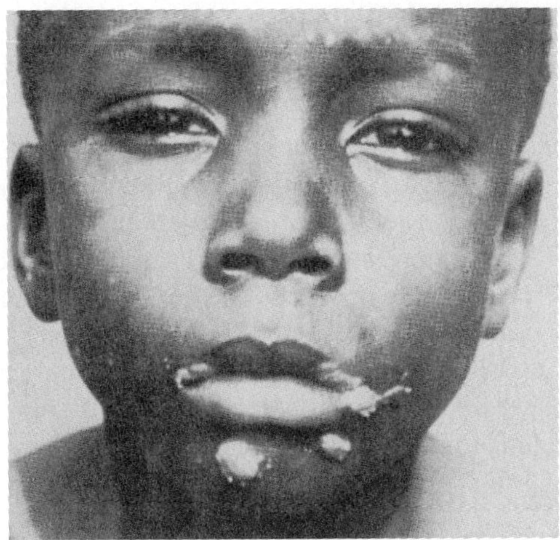

(a)

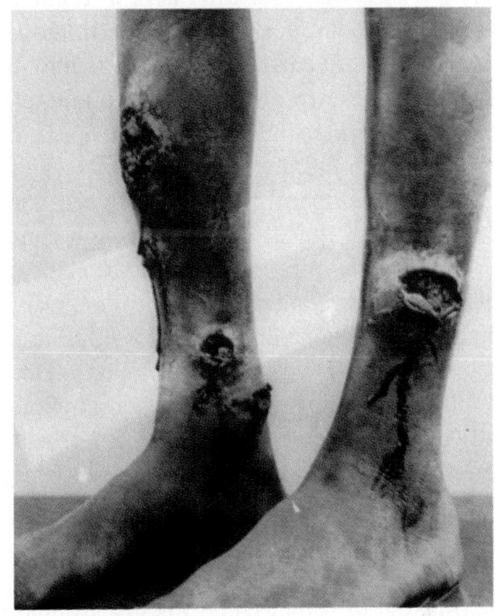

(b)

Figure 17.8 Endemic treponematoses. (*a*) Skin and membrane nodules in a young boy with endemic syphilis (bejel). (*b*) The clinical appearance of yaws. The large draining sores are "mother yaws" that give rise to new lesions.

Pinta The names mal del pinto and carate are regional synonyms for pinta, a chronic skin infection caused by *T. carateum.* Transmission evidently requires several years of close personal contact accompanied by poor hygiene and inadequate health facilities. Even though pinta is not currently widespread, the disease is still found in isolated populations inhabiting the tropical forest and valley regions of Mexico and Central and South America. Infection begins in the skin with a dry, scaly papule reminiscent of psoriasis or leprosy. In time, pigmented secondary macules and blanched tertiary lesions appear. Pinta is not life-threatening, but it often creates scars on the afflicted area.

---

carateum (kar-uh'-tee-um) From *carate,* the South American name for pinta.

Leptospira and Leptospirosis

Leptospires are typical spirochetes marked by tight, regular, individual coils with a bend or hook at one or both ends (figure 17.9). There are only two species in the genus: *Leptospira interrogans,* which causes **leptospirosis** in humans and animals, and *L. biflexa,* a harmless, free-living saprobe. The single hook of *L. interrogans* suggests the appearance of a question mark and helps differentiate it from *L. biflexa,* which has flexible hooks at both ends. The two species are serologically, genetically, and physiologically distinct. *L. interrogans* demonstrates nearly 200 serotypes distributed among various animal groups, which accounts for the extreme variations in leptospirosis among humans.

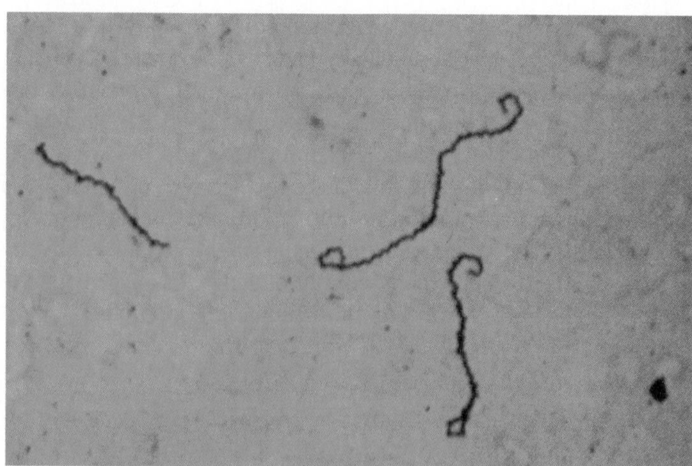

Figure 17.9 *Leptospira interrogans,* the agent of leptospirosis. Note the curved hook at one end of the spirochete.

Epidemiology and Transmission of Leptospirosis

Leptospirosis is a zoonosis associated with wild animals such as rodents, skunks, raccoons, foxes, and some domesticated animals, particularly dogs, cattle, pigs, sheep, and horses. Although these reservoirs are distributed throughout the world, the disease is concentrated mainly in the tropics. Leptospires shed in the urine of an infected animal can survive for several months at moderate temperature in neutral or alkaline soil or water. Infection occurs almost entirely through contact of skin abrasions or mucous membranes with animal urine or some environmental source containing urine. Transmission does not seem to occur through ingestion, animal bites, inhalation, or from human to human. In the United States, most of the relatively few cases (100–200) each year are reported among older male children and young adults who have swum or bathed in polluted water. Soldiers involved in jungle training are also at high risk for exposure.

Pathology of Leptospirosis and Host Response

Despite the amassed information on the disease and successful techniques for growing *L. interrogans* in artificial media, knowledge of its virulence mechanisms is fragmentary. Disease proceeds in two phases, and its principal targets are the kidneys, liver, brain, and eyes. During the early or leptospiremic phase, the pathogen appears in the blood and cerebrospinal fluid. Symptoms are sudden high fever, chills, headache, muscle aches, conjunctivitis, and vomiting. During the second or immune phase, the blood infection disappears, and the fever lessens in response to circulating IgM antibodies. This period is marked by milder fever, headache due to leptospiral meningitis, and *Weil's syndrome,* a state characterized by kidney invasion, hepatic disease, jaundice, anemia, and neurologic disturbances. Long-term disability and even death may ensue from injury to the kidneys and liver, but this occurs primarily with virulent strains and in elderly patients.

Diagnosis, Treatment, and Prevention

A history of environmental exposure, along with presenting symptoms, may support initial diagnosis of leptospirosis, but definitive diagnosis relies on the results of *Leptospira* culture and serologic tests. Isolation is accomplished by inoculating a specimen into special media or laboratory animals. Because leptospiral infection stimulates a strong humoral response, it is possible to test the patient's serum for its antibody titer. A fast, specific, and effective test called the macroscopic slide agglutination test is most often employed for routine screening. It is based on reacting live or formalinized *L. interrogans* with the patient's serum and observing agglutination or lysis with a darkfield microscope under low power.

Leptospires are sensitive to many classes of antibiotics; however, treatment must begin early in the infection. Penicillin or tetracycline given by the fourth day of illness rapidly reduces symptoms and shortens the course of disease, but after this time, therapy is less effective. Strain-specific human and animal vaccines made from killed cells are available, but with so many serotypes, these can confer protection only in areas where exposure to a specific endemic strain is anticipated. Vaccination is aimed at persons at greatest risk such as combat troops training in jungle regions and animal care and livestock workers. Due to the large animal reservoir and the readiness with which soil, water, and animal products become contaminated with urine, the best controls are to wear protective footwear and clothing and to avoid swimming or wading in livestock watering ponds.

Borrelia: Arthropod-Borne Spirochetes

Members of the genus *Borrelia* are morphologically distinct from other pathogenic spirochetes. They are comparatively larger, ranging from 0.2 to 0.5 μm in width and from 10 to 20 μm in length, and they contain 4 to 30 irregularly spaced and loose coils (see figure 17.1*d*) with an abundance (30 to 40) of periplasmic flagella. The nutritional requirements of *Borrelia* are so complex that the bacterium can be grown in artificial media only with difficulty.

Leptospira (lep″-toh-spy′-rah) Gr. *leptos,* slender or delicate, and *speira,* a coil.

Borrelia (boh-ree′-lee-ah) Named after Amédée Borrel, a French bacteriologist.

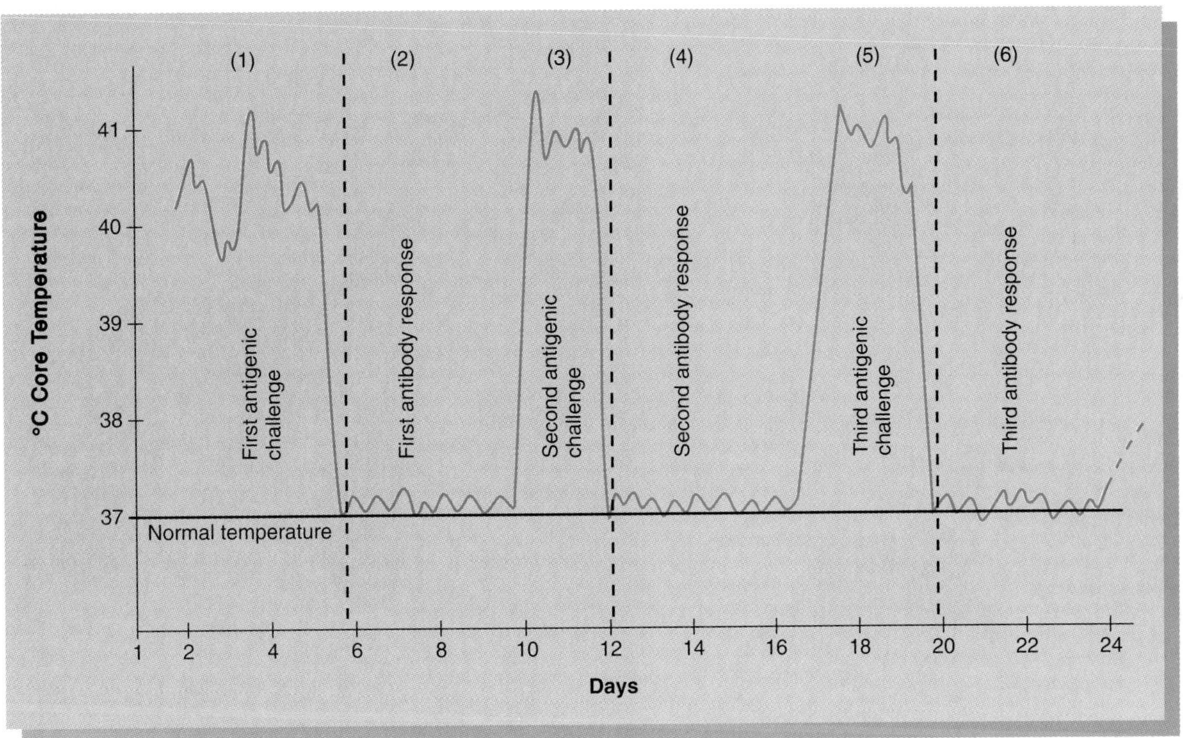

Figure 17.10 The pattern in relapsing fever: (*1*) Primary infection and fever; (*2*) initial antibody response with concurrent reduction in symptoms; (*3*) reinfection with a new antigenic type, causing renewed symptoms; and (*4*) a second antibody response, producing a second remission. (*5, 6*) This pattern may continue for up to four relapses.

Epidemiology of Relapsing Fever

Human infections with *Borrelia,* termed **borrelioses,** are all transmitted by some type of arthropod vector, usually ticks or lice. One significant disease is known as **relapsing fever.**

Borrelia hermsii, the cause of tick-borne relapsing fever, is carried by soft ticks (see feature 17.5) of the genus *Ornithodoros.* The mammalian reservoirs of this zoonosis are squirrels, chipmunks, and other wild rodents, and the human is generally an accidental host. The spirochetes mature and persist in the salivary glands and intestines of the tick, and both the bite itself and the subsequent scratching initiate infection. Because ticks can survive in the wild for many years, a single tick is capable of infecting a variety of host species and more than one individual. Tick-borne relapsing fever occurs sporadically in the United States, usually in campers, backpackers, and forestry personnel who frequent the higher elevations of western states. The incidence of infection is higher in endemic areas of the tropics, especially where rodents have easy access to dwellings.

Whenever famine, war, mass migrations, or natural disasters are coupled with poor hygiene, crowding, and inadequate medical attention, epidemics of louse-borne relapsing fever occur. Such conditions favor the survival and spread of the louse vector *Pediculus humanus* (see feature 17.5), which harbors the spirochete *B. recurrentis* in its body cavity. A host will not become infected unless the louse is smashed and accidentally scratched into a wound or the skin. Although only killed lice can transmit the disease and a single louse can infect only once, the abundance of lice usually compensates for this peculiar mode of epidemic spread. Louse-borne fever is most common in parts of China, Afghanistan, and Africa.

Pathogenesis and the Nature of Relapses

The pathological manifestations are similar in both tick- and louse-borne borrelioses. Even though the blood level may reach an imposing 500,000 borrelias per milliliter, there is little sign of disease during the 2- to 15-day incubation period, and even the site of entrance shows virtually no local inflammatory reaction. The incubation period ends abruptly with the onset of high fever, shaking chills, headache, and fatigue. Later features of the disease include anorexia, nausea, vomiting, muscle aches, and abdominal pain. Extensive damage to the liver, spleen, heart, kidneys, and cranial nerves occurs in many cases. Half of the patients hemorrhage profusely into organs, and some develop a rash on the shoulders, trunk, and legs. Untreated cases are often lengthy and debilitating, and are attended by 5% to 40% mortality.

As the name relapsing fever indicates, the fever follows a fluctuating course that is explained by changes in the parasite and the attempts of the immune system to control it (figure 17.10). *Borrelia* have adopted a remarkable strategy for tricking the immune system and avoiding destruction. They are genetically programmed to change surface antigens during growth, so that in time, the initial antibodies become useless against the antigenically altered cells. These masked cells survive, multiply, and cause a second wave of symptoms. Eventually, the immune system forms new antibodies, but it is soon faced with yet

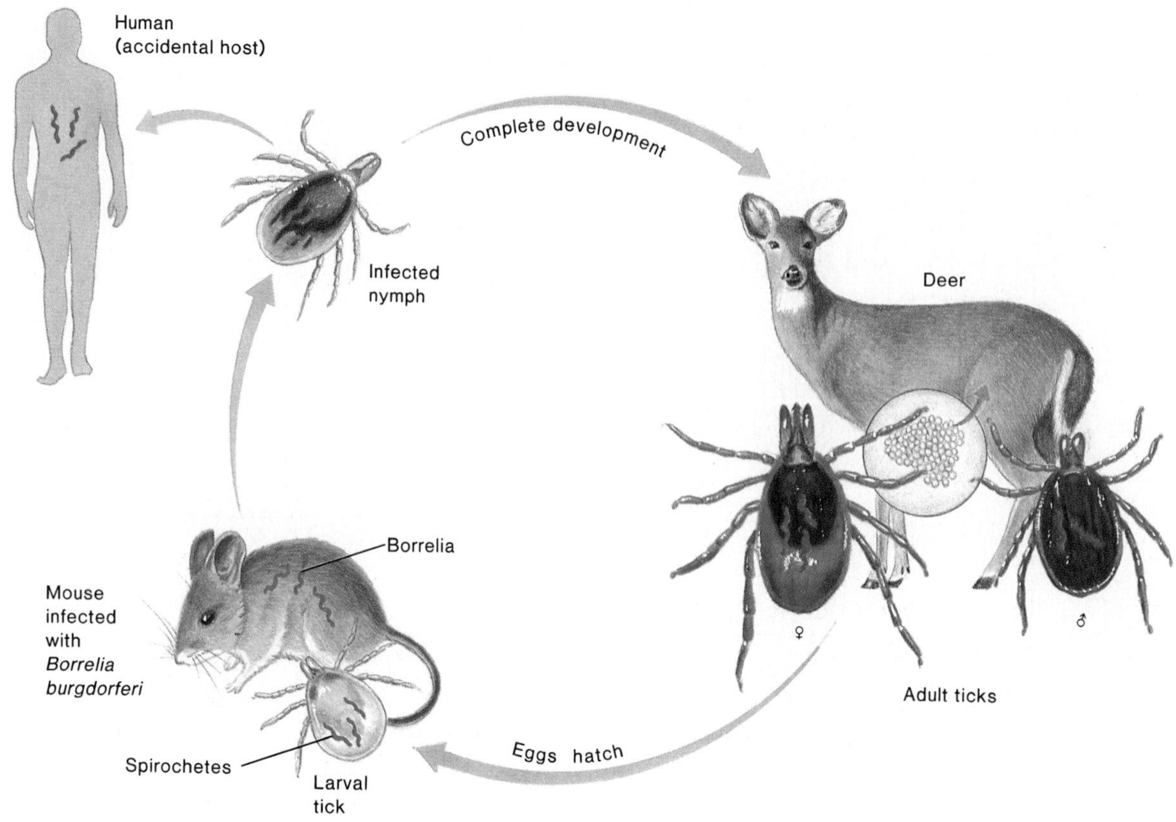

Figure 17.11 The cycle of Lyme disease.

another antigenic form. A single strain has been known to generate 24 distinct serological types. Eventually, cumulative immunity against the menu of antigens develops, and complete recovery may occur.

Diagnosis, Treatment, and Prevention A patient's history of exposure, clinical symptoms, and the presence of *Borrelia* in blood smears are very definitive evidence of borreliosis. Except for pregnant women and children under seven years old, tetracycline is the treatment of choice. Chloramphenicol, erythromycin, and streptomycin are also effective antimicrobial agents. Because vaccines are not available, prevention of relapsing fever has depended upon controlling rodents and avoiding tick bites. Louse-borne relapsing fever is effectively arrested by improving the standards of individual and community hygiene.

Borrelia burgdorferi and Lyme Disease

Lyme disease is a newly described syndrome that has become the most important borreliosis in the United States (see feature 17.1). Its spirochetal agent, ***Borrelia burgdorferi,*** is transmitted primarily by hard ticks of the genus *Ixodes* (see chapter opening illustration). This tick passes through a complex two-year cycle that involves two principal hosts (figure 17.11). As a larva or nymph, it feeds on the white-footed mouse, where it picks up the infectious agent. The nymph is relatively nonspecific and will try to feed on nearly any type of vertebrate, thus this is the form most likely to bite humans. The adult reproductive phase of the cycle is completed on deer.

Lyme disease is nonfatal, but often evolves into a slowly progressive syndrome that mimics neuromuscular and rheumatoid conditions. The most prominent early symptom is a rash at the site of a tick bite. The lesion, called *erythema chronicum migrans,* looks something like a bull's-eye, with a raised erythematous ring that gradually spreads outward and a pale central region (figure 17.12). Other early symptoms are fever, headache, stiff neck, and dizziness. If not treated or if treated too late, the disease may advance to the second stage, during which cardiac dysrhythmias and neurological symptoms such as facial palsy develop. After several weeks or months, a crippling polyarthritis may attack the joints, especially those of the knees. Some persons acquire chronic neurological complications that are severely disabling.

The greatest concentrations of Lyme disease are in areas having high mouse and deer populations. Most of the several thousand cases that have been recorded since 1978 have occurred in New Jersey, Massachusetts, Rhode Island, New York, Wisconsin, and Minnesota, though the number in northern California is growing. Highest risk groups include hikers, backpackers, and persons living in newly developed communities near wilderness areas. Peak seasons are the summer and early fall.

Diagnosis of Lyme disease is aided by the presence of ring-shaped lesions, isolation of spirochetes from the patient, and serological testing, though false-positive reactions are common.

burgdorferi (berg-dor'-fer-eye) Named for its discoverer, Dr. Willy Burgdorfer.

Feature 17.1 The Disease Named for a Town

In the 1970s, an enigmatic cluster of arthritis cases appeared in the town and surrounding suburbs of Old Lyme, Connecticut. This phenomenon caught the attention of nonprofessionals and professionals alike, whose persistence and detective work ultimately disclosed the unusual nature and epidemiology of Lyme disease. The story of its unraveling began in the home of Polly Murray, who along with her family, was beset for years by recurrent bouts of stiff neck, swollen joints, malaise, and fatigue that seemed vaguely to follow a rash from tick bites. When Mrs. Murray's son was diagnosed as having juvenile rheumatoid arthritis, she became skeptical. Conducting her own literature research, she began to discover inconsistencies. Rheumatoid arthritis was described as a rare, noninfectious disease, yet over an eight-year period, she found that 30 of her neighbors had experienced similar illnesses. Mrs. Murray carefully recorded these cases and reported to the state health authorities, but to no avail.

After similar observations were reported by Judith Mensch, another resident of the same region, the findings came to the attention of Dr. Allen Steere, a rheumatologist with a CDC background. He was able to forge the vital link between Mrs. Murray's case histories, the disease symptoms, and the presence of unique spirochetes in ticks preserved by some of the patients. These same spirochetes had been previously characterized in 1981 by Dr. Willy Burgdorfer, though he did not realize their importance at the time.

Early treatment with penicillin and tetracycline has been effective, and other antibiotics are being tested. Because dogs can also acquire the disease, a vaccine has been marketed to protect dogs living in certain areas of the United States. Avoiding endemic areas in season, wearing protective clothing, using insect sprays, and fastidiously removing ticks can prevent Lyme disease and other tick-borne infections as well (see preventative measures for Rocky Mountain spotted fever later in this chapter).

Other Curviform Gram-Negative Bacteria of Medical Importance

Three genera of curved or short spiral rods prominent in medical bacteriology are *Vibrio, Campylobacter,* and *Spirillum.* Vibrios are comma-shaped, curved rods with a single polar flagellum. Campylobacters are short spirals that have one or more flagella and a corkscrew type of motility. Spirilla are inflexible coiled rods that move by multiple polar flagella. *Vibrio* is a member of the Family Vibrionaceae, and *Campylobacter* and *Spirillum* are in the Family Spirillaceae.

The Biology of *Vibrio cholerae*

A freshly isolated specimen of ***Vibrio cholerae*** reveals quick, darting cells slightly resembling a weiner or a comma (figure 17.13*a*). *Vibrio* shares many cultural and physiological characteristics with members of the Enterobacteriaceae, a closely

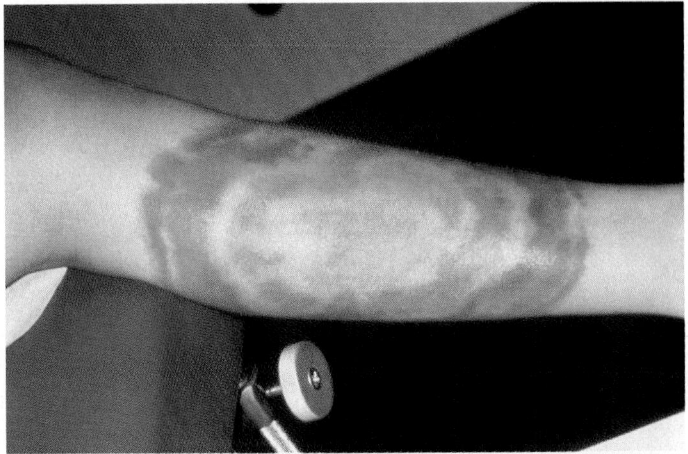

Figure 17.12 Lesions of Lyme disease on the leg. Note the flat, reddened, ringlike form. Primary lesions often give rise to large numbers of secondary lesions in other locations.

Feature 17.2 The Mysterious Epidemiology of Cholera

Cholera is a strictly human disease, and there are no demonstrable animal reservoirs. The microbe's natural habitat appears to be the soil and water of two large Indian Rivers, the Ganges and the Brahmaputra. In these endemic areas, transmission of cholera is associated with pilgrimages to the river and religious practices involving mass bathing. Periodically during the past two centuries, this vibrio has migrated by means of waterways and human carriers to areas of Asia, Africa, Europe, the Middle East, and South America, and in the process, it has caused at least seven pandemics. The most recent epidemics that started in 1991 in Peru rapidly spread to several other South American countries. In a few months, several hundred thousand persons were affected, and thousands died. Epidemiologists traced many cases to ingestion of contaminated raw seafood.

The pattern of cholera transmission and the onset of epidemics are greatly influenced by the season of the year and the climate. Cold, acidic, dry environments inhibit the migration and survival of *Vibrio,* whereas warm, monsoon, alkaline, and saline conditions favor them. The most persistent pandemic, which began in 1961 and is still continuing, is due to the sudden predominance of the hardier and more virulent *El Tor* biotype. This strain infects a higher number of persons, causes a lengthier disease, and is more likely to be chronically carried.

related family. They are fermentative and grow on ordinary or selective media containing bile at 37°C. They possess unique O (somatic) antigens, H (flagella) antigens, and membrane receptor antigens that provide some basis for classifying members of the family.

Epidemiology of Cholera

Epidemic cholera, or Asiatic cholera, has been a devastating disease for centuries. Although the human intestinal tract was once thought to be the primary reservoir, it is now known that the parasite is free-living in certain endemic regions (see feature 17.2). In nonendemic areas such as North America, the microbe is spread by water and food contaminated by asymptomatic carriers. Cholera is relatively uncommon in the United States, though sporadic outbreaks have occurred along the Gulf

Vibrio (vib'-ree-oh) L. *vibrare,* to shake.

Campylobacter (kam''-pih-loh-bak'-ter) Gr. *campylo,* curved, and *bacter,* rod.

cholerae (kol'-ur-ee) Gr. *chole,* bile. The bacterium was once named *V. comma* for its comma-shaped morphology.

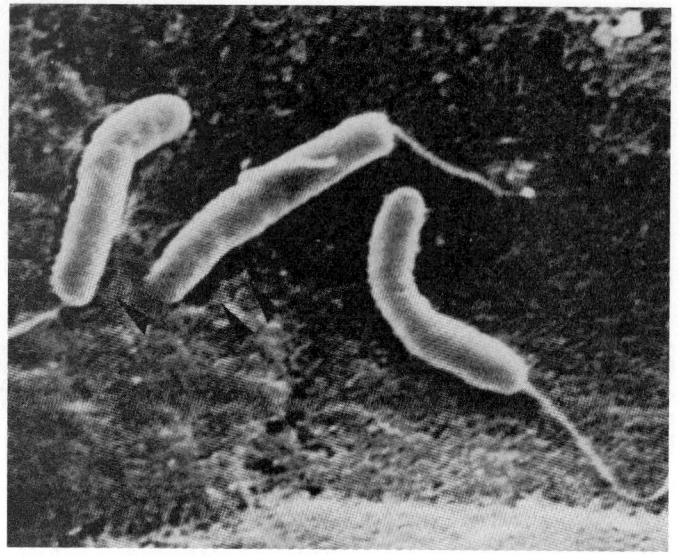

(a)

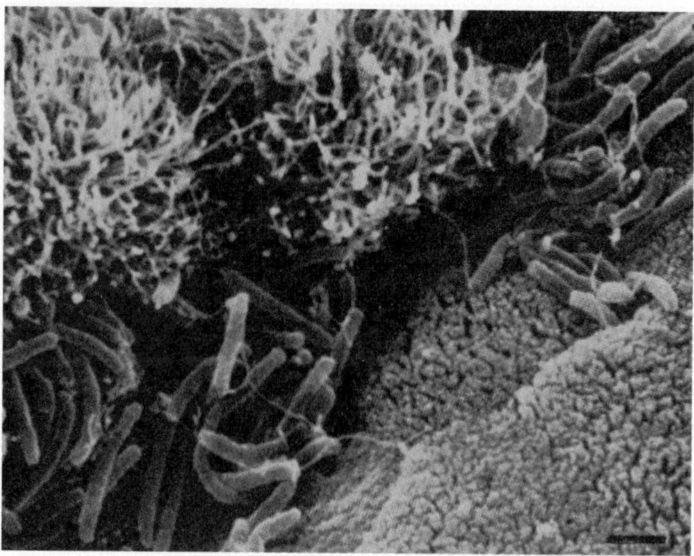

(b)

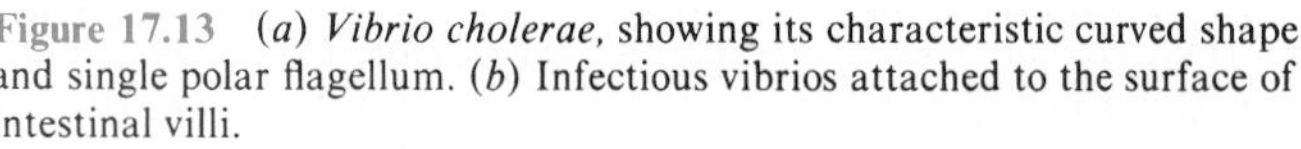

Figure 17.13 (*a*) *Vibrio cholerae,* showing its characteristic curved shape and single polar flagellum. (*b*) Infectious vibrios attached to the surface of intestinal villi.

of Mexico. Most cases of cholera are mild or self-limited, but in children and weakened individuals, the disease can explode with tremendous force.

Pathogenesis of Cholera

Following ingestion with food or water, *V. cholerae* encounters the potentially destructive acidity of the stomach. This hostile environment influences the size of the infectious dose (10^8 cells), though certain types of food also shelter the pathogen more readily than others. At the mucosa of the duodenum and jejunum, the surviving vibrios penetrate the mucous barrier using their flagella, adhere to the outside of the epithelium, and multiply there (figure 17.13*b*). The cells are strictly epipathogens in that they do not enter the cells or invade the mucosa. The virulence of *V. cholerae* is due entirely to an exotoxin called *choleragen* (CT) that disrupts the normal physiology of intestinal cells. This toxin stimulates the cells to shed large amounts of electrolytes into the intestine, an event that is accompanied by severe water loss (figure 17.14).

Following an incubation period of a few hours to a few days, symptoms begin abruptly with vomiting, followed by copious watery feces called **secretory diarrhea.** This voided fluid is odorless and contains flecks of mucus, hence the description "rice-water stool." Fluid losses of nearly one liter per hour have been reported in severe cases, and an untreated patient can lose up to 50% of body weight during the course of the disease. The diarrhea causes loss of blood volume, acidosis from bicarbonate loss, and potassium depletion that predispose the patient to muscle cramps, severe thirst, flaccid skin, sunken eyes, apathy, and in young children, coma, convulsions, and fever. Secondary circulatory consequences may include hypotension, weak pulse, tachycardia, cyanosis, and collapse from shock within 18 to 24 hours. If cholera is left untreated, death may occur in less than 48 hours and the mortality rate approaches 55%.

Diagnosis and Remedial Measures

During epidemics, clinical evidence is usually sufficient to diagnose cholera. But confirmation of the disease is often required for epidemiologic studies and detection of sporadic cases. Because patients in the acute phase can shed 10^6–10^8 organisms/ml in the stool, *V. cholerae* can be readily isolated and identified in the laboratory from stool samples. Direct dark-field microscopic observation of fresh feces can reveal characteristic curved cells with brisk, darting motility as confirmatory evidence. Immobilization or fluorescent staining of feces with group-specific antisera is supportive as well. Difficult or elusive cases can be traced by detecting a rising antitoxin titer in the serum.

The key to cholera therapy is prompt replacement of water and electrolytes, since their loss accounts for the severe morbidity and mortality. This can be accomplished by various rehydration techniques that replace the lost fluid and electrolytes (see oral rehydration [ORT] in feature 17.3).

Cases in which the patient is unconscious or has complications from severe dehydration require intravenous replenishment as well. The solution for this therapy is very similar to ORT, but administering it does require clinical facilities. Only after the patient has been treated for physiologic disruption can antibiotics to control the growth of the microbe come into play. Oral antibiotics such as tetracycline and drugs such as trimethoprim-sulfa can terminate the diarrhea in 48 hours, and they also promote recovery and diminish the period of vibrio excretion.

Prompt and aggressive detection and treatment alone are insufficient to control cholera. Effective prevention is contingent

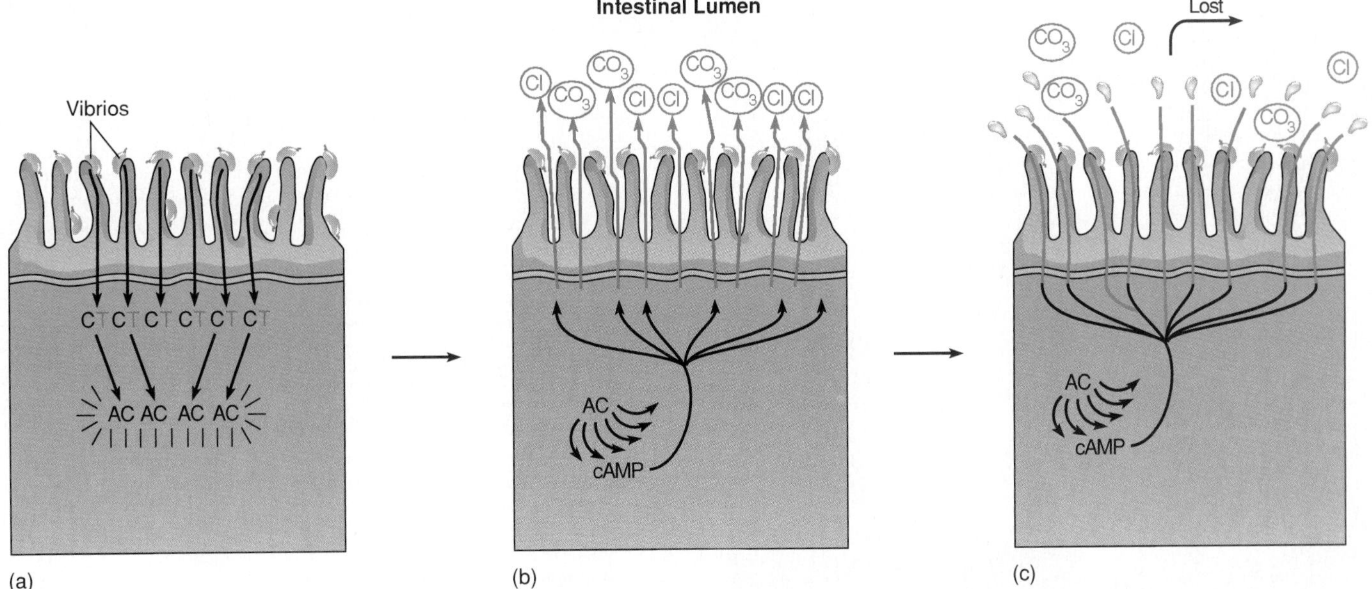

Figure 17.14 (*a*) The specific action of choleragen (CT) upon the intestinal epithelial cells heightens the activity of an enzyme called adenyl cyclase (AC). (*b*) This enzyme stimulates abnormally high levels of cAMP (cyclic adenosine monophosphate), a chemical messenger that normally mediates the action of hormones on cells but in higher concentrations promotes removal of anions by the cell membrane. (*c*) Under the constant action of cAMP, the cells begin to secrete large quantities of chloride (Cl) and bicarbonate (CO_3) ions into the intestinal lumen. Electrolyte loss is followed by water loss from epithelial cells, and the body experiences massive fluid depletion.

upon proper sewage disposal and water purification. Detecting and treating carriers with mild or asymptomatic cholera is a serious goal, but one that is frequently difficult to attain because of inadequate medical provisions in those countries where cholera is endemic. To date, attempts at vaccination have been lackluster. A worthwhile vaccine should protect for at least two years and be easily administered, but the principal vaccine developed so far containing killed cholera vibrios protects for only six months or less and does not stop the continued shedding of vibrios by infected individuals. Encouraging results have been reported for an oral vaccine using live, attenuated vibrios.

Vibrio parahaemolyticus: A Pathogen Carried by Seafood

Vibrio parahaemolyticus is a halophilic inhabitant of estuaries and coastal fishing waters. In this habitat, it associates with zooplankton and colonizes the chitinous exoskeleton of crustacea. Apparently, the ecologic role of this bacterium is to degrade chitin and recycle the resultant breakdown products. In temperate zones, vibrios survive over the winter by settling into the ocean sediment, and when resuspended by upwelling during the warmer seasons, they become incorporated into the food chain, eventually growing on fish, shellfish, and other edible seafood. *V. parahaemolyticus* shares with *V. cholerae* the general characteristics of morphology, culturing, motility, staining, and antigenicity.

Features of *V. parahaemolyticus* Gastroenteritis

Vibrio parahaemolyticus food infection, an acute form of gastroenteritis, was first described in Japan more than 30 years ago. The great majority of cases appear in individuals who have eaten raw, partially cooked, or unrefrigerated seafood. The vehicles most often implicated are squid, mackerel, sardines, crabs, tuna, shrimp, oysters, and clams. Outbreaks tend to be concentrated along coastal regions during the summer and early fall. Occasionally, the bacterium infects wounds and burns of swimmers, dock workers, seafood cooks, and other individuals exposed to seawater. The incubation period of nearly 24 hours is followed by explosive, watery diarrhea accompanied by nausea, vomiting, headache, abdominal cramps, and sometimes fever. Symptoms normally last about 72 hours but can persist for 10 days. Treatment of severe cases may require fluid and electrolyte replacement and occasionally antibiotics, but because the disease is self-limiting, hospitalization is not usually indicated. Control measures are focussed on keeping the bacterial count in all seafood below the infective dose by continuous refrigeration during transport and storage, sufficient cooking temperatures, and prompt serving.

Diseases of the *Campylobacter* Vibrios

Campylobacters are slender, curved or spiral bacilli propelled by polar flagella at one or both poles, often appearing in S-shaped or gull-winged pairs (figure 17.15). These bacteria tend to be microaerophilic inhabitants of the intestinal tract, genitourinary tract, and oral cavity of humans and animals. Species of *Campylobacter* most significant in medical and veterinary practice are *C. jejuni* and *C. fetus.* A close relative, *Helicobacter* (formerly *Campylobacter*) *pylori,* has recently been implicated in certain types of stomach ulcers (see feature 17.4).

Feature 17.3 Oral Rehydration Therapy

A special oral rehydration therapy (ORT) using glucose-electrolyte solution appears to be a miracle cure for cholera. This relatively simple formulation, developed by the World Health Organization, consists of a mixture of the electrolytes sodium chloride, sodium bicarbonate, potassium chloride, and glucose or sucrose dissolved in water (table 17.2). When administered early in amounts ranging from 750 to 1,500 ml/hour, the solution can restore a patient in four hours, often literally bringing him back from the brink of death. It works even if the individual has diarrhea, because the particular combination of ingredients is well tolerated and rapidly absorbed. Infants and small children who once would have died now survive so often that the mortality rate for treated cases of cholera is near zero. This therapy has several advantages, especially for countries with few resources. It requires no medical facilities, no high technology equipment, and no complex medication protocols. It is also inexpensive, noninvasive, fast-acting, and useful for a number of diarrheal diseases besides cholera.

Note the sunken eyes in this child suffering from severe dehydration.

Table 17.2 Formulation for Oral Rehydration Therapy*

	Glucose	NaCl	$NaHCO_3$	KCl
Oralyte (suitable for infants if used in 2:1 regimen and for adult noncholera or pediatric cholera patients if administered without extra water)	20 g/l	3.5 g/l	2.5 g/l	1.5 g/l
Adult formula (suitable for adult diarrheas and for cholera patients of all ages; extra water allowed *ad libitum*)	20 g/l	4.2 g/l	4.0 g/l	1.8 g/l
Infant formula (suitable for noncholera infantile diarrheas; could be used without extra water)	20 g/l	2.3 g/l	3.3 g/l	2.6 g/l

*Similar to a commercial product called Pedialyte. Packets containing these substances are to be dissolved in thoroughly boiled, cooled water.

From Strickland, *Hunter's Tropical Medicine,* 6th ed. Copyright © W. B. Saunders, Philadelphia, PA. Reprinted by permission.

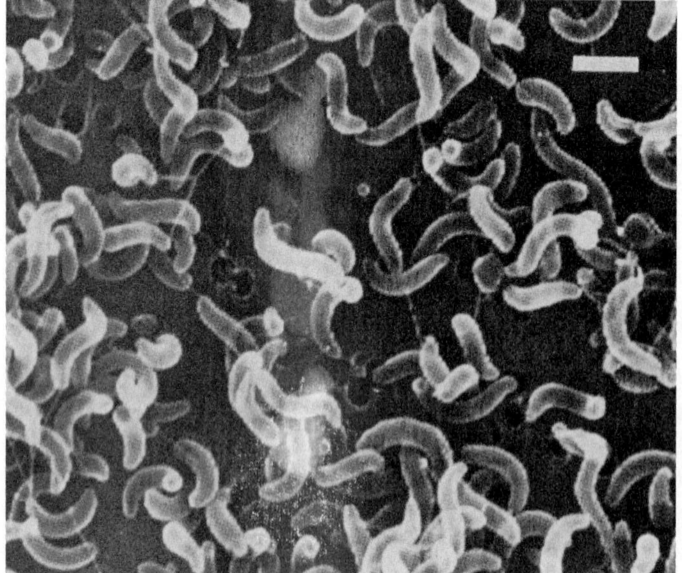

Figure 17.15 Scanning micrograph of *Campylobacter jejuni,* showing comma, S, and spiral forms.

Campylobacter jejuni Enteritis

Campylobacter jejuni has recently emerged as a pathogen of such imposing proportions that it is considered one of the most important causes of bacterial gastroenteritis worldwide. Early results of epidemiologic and pathologic studies have shown that this species is a primary pathogen transmitted through contaminated beverages and food, especially water, milk, meat, and other animal products. The source of a Vermont outbreak of enteritis was traced to a water storage reservoir that supplied most of the community and eventually caused disease in 20% of the population. Another epidemic was associated with a water tank exposed to bird feces. Cases are extremely common in developing countries due to frequent contact with animal reservoirs and unreliable sanitation.

When ingested *C. jejuni* cells reach the mucosa at the last segment of the small intestine (ileum) near its junction with the colon, they adhere by means of receptors on the cell envelope

jejuni (jee-joo′-nye) L. *jejunum.* The small section of intestine between the duodenum and the ileum.

Feature 17.4 *Helicobacter pylori* and Stomach Disease

Although the human stomach is usually regarded as a hostile habitat, researchers have recently uncovered a vibrio, *Helicobacter pylori*, that appears to reside there. Because this microbe can also be isolated from certain types of ulcers, some microbiologists feel that it is an important contributor to gastric disease. Also supporting this link is the fact that most patients with peptic and duodenal ulcers have readily detectable levels of antibodies directed against this species in their serum and gastric mucosa. Human volunteers who ingested a culture of *H. pylori* developed abnormal stomach histology that lingered for a time, and one epidemic of gastritis has been traced to persons accidentally exposed to it. Whether it is a causative agent, a contributing factor, or just normal flora is currently the object of much research.

and penetrate by burrowing. After an incubation period of one to seven days, symptoms of headache, fever, abdominal pain, and bloody or watery diarrhea commence. The mechanisms of pathology appear to involve a heat-labile enterotoxin called CJT that stimulates a secretory diarrhea like that of cholera.

Diagnosis of *C. jejuni* enteritis requires isolating it from fecal samples and occasionally from blood samples. Rapid presumptive diagnosis relies on direct examination of feces with a dark-field microscope, which accentuates the characteristic curved rods and darting motility. Resolution of infection occurs in most instances with simple, nonspecific rehydration and electrolyte balance therapy. In more severely affected patients, it may be necessary to administer erythromycin, tetracycline, aminoglycosides, or chloramphenicol. Because vaccines are yet to be developed, prevention depends upon rigid sanitary control of water and milk supplies and careful hygiene.

Traditionally of interest to the veterinarian, *C. fetus* (subspecies *venerealis*) causes a sexually transmitted disease of sheep, cattle, and goats. Its role as an agent of abortion in these animals has considerable economic impact on the livestock industry. About 40 years ago, the significance of *C. fetus* as a human pathogen was first uncovered, though its exact mode of transmission in humans is yet to be clarified. This bacterium appears to be an opportunistic pathogen that attacks debilitated persons or women late in pregnancy. Diseases to which *C. fetus* has been linked are meningitis, pneumonia, arthritis, fatal septicemic infection in the newborn, and occasionally, sexually transmitted proctitis in adults.

Spirillum minus: A Cause of Rat Bite Fever

Rat bite fever or *sodoku* is caused by a bacterium called *Spirillum minus* (*minor*) that has two to three rigid coils and lophotrichous flagella. This relatively rare zoonosis occurs mostly in Japan and the Far East and is more common in urban areas with poor sanitation and a high rat population. The primary host for the pathogen is the rat, in which it causes septicemia, eye infections, and lung infections. It is transmitted to human hosts almost exclusively by bites from infected wild rodents or their predators. Fleas or other arthropods do not serve as vectors, and direct human-to-human or fomite transmission has not been demonstrated.

In humans, the initial wound heals readily, but after an incubation period of two weeks, the site becomes swollen, inflamed, and painful, turns purple, and ulcerates. The local lymph nodes enlarge and become tender. Fever, malaise, headache, and a rash radiating from the wound site develop but then progressively diminish. Diagnosis by isolation and laboratory growth is not feasible because *S. minus* cannot be cultivated on artificial media. Direct dark-field visualization of fresh blood smears, lymphatic exudate, and lesion samples is a suitable alternative. The infection responds to penicillin, streptomycin, and tetracycline. Another form of rat bite fever caused by *Streptobacillus moniliformis* is more common in the United States, where several cases have occurred in biomedical personnel who handle laboratory rats.

sodoku (soh-doh'-koo) Jap. *so,* rat, and *doku,* poison.

Medically Important Bacteria of Unique Morphology and Biology

Bacterial groups that exhibit atypical morphology, physiology, and behavior are (1) rickettsias and chlamydias, obligately parasitic gram-negative coccobacilli, and (2) mycoplasmas, highly pleomorphic cell-wall-deficient bacteria.

Family Rickettsiaceae: The Rickettsias

The family Rickettsiaceae contains three genera pathogenic to humans: *Rickettsia, Coxiella,* and *Rochalimaea.* These organisms are known commonly as **rickettsiae** or rickettsias, and the diseases they cause are called **rickettsioses.** In most instances, the rickettsial life cycle is passed within arthropod or other invertebrate vectors, and all require live cells to be cultivated except for *Rochalimaea.* Table 17.3 summarizes the major diseases, vectors, and distribution of rickettsial species.

Morphologic and Physiologic Distinctions of Rickettsias

For many years, the rickettsiae were thought to be related to viruses because they multiplied only within host cells. That they are indeed bacteria, possessing a gram-negative cell wall, binary fission, metabolic pathways for synthesis and growth, and both DNA and RNA, eventually became evident. They are among the smallest cells, ranging from 0.3 to 0.6 μm wide and from 0.8 to 2.0 μm long, and are nonmotile pleomorphic rods or coccobacilli (figure 17.16).

The precise nutritional requirements of the rickettsiae have been difficult to demonstrate because of a close association with host cell metabolism. Their obligate parasitism originates from

Coxiella (kox"-ee-el'-ah) For H. R. Cox, an American bacteriologist who first isolated this rickettsia in the U.S.

Rochalimaea (roh-chah"-lih-may'-ah) From H. da Rocha-Lima, an early investigator of rickettsial diseases.

Table 17.3 Characteristics of Rickettsias Important in Human Disease

Disease Group	Species	Disease	Vector	Primary Reservoir	Mode of Transmission to Humans	Where Found
Typhus						
	R. prowazekii	Epidemic typhus	Body louse	Humans	Louse feces rubbed into bite; inhalation	Worldwide
	R. typhi (mooseri)	Murine typhus	Flea	Rodents	Flea feces rubbed into skin; inhalation	Worldwide
Spotted Fever						
	R. rickettsii	Rocky Mountain spotted fever	Tick	Small mammals	Tick bite; aerosols	North and South America
	R. akari	Rickettsialpox	Mite	Mice	Mite bite	Worldwide
Scrub Typhus	*R. tsutsugamushi*	—	Chigger	Rodents	Chigger bite	Asia, Australia, Pacific Islands
Q Fever	*Coxiella burnetii*	—	Tick	Cattle, sheep, goats	Airborne; contact with ticks	Worldwide
Trench Fever	*Rochalimaea quintana*	—	Body louse	Humans	Louse feces scratched into bite	Africa, Mexico, Europe

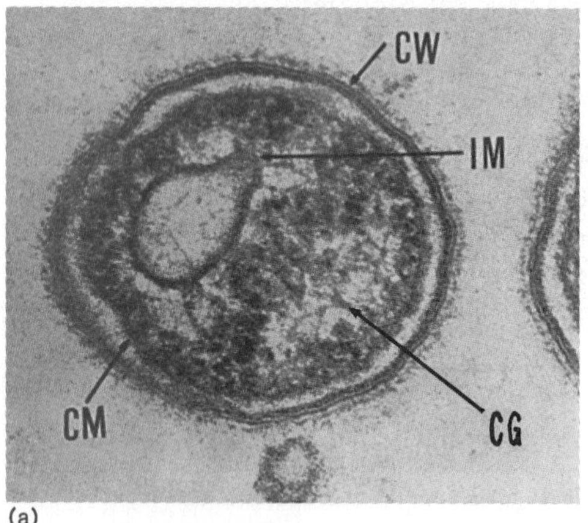

(a)

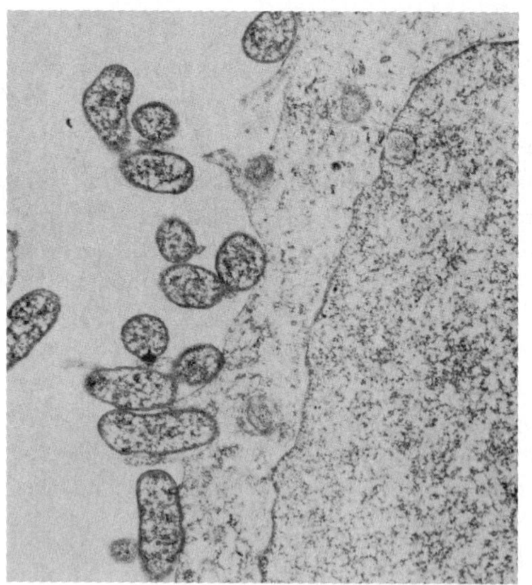
(b)

Figure 17.16 (*a*) The morphology of *Rickettsia*. Several features, including the cell wall (CW), cell membrane (CM), chromatin granules (CG), and even the ribosomes, identify these as tiny, pleomorphic, gram-negative bacteria (×185,000). (*b*) *Rochalimaea quintana* adhering to the surface of a mouse tissue culture cell.

an inability to metabolize AMP, an important precursor to ADP and ATP, which they must obtain from the host. Rickettsiae are generally sensitive to environmental exposure, although *R. typhi* can survive several years in dried flea droppings, and *Coxiella burnetii* shows extreme resistance to physical and chemical agents because of an endospore-like survival cell.

Distribution and Ecology of Rickettsial Diseases

The Role of Arthropod Vectors The life cycle of many rickettsiae depends upon a complex exchange between blood-sucking arthropod[2] hosts and vertebrate hosts, including humans (see feature 17.5). Humans accidentally enter the zoonotic life cycles through occupational contact with the animals except in the case of louse-borne typhus and trench fever. Because humans are often incidental or "dead end" hosts, they are not a regular source of infection. Most vectors apparently harbor rickettsiae

2. Ticks and mites are technically classified as arachnids, and lice and fleas are insects.

Feature 17.5 Of Mice and Mites and Lice and Bites: The Arthropod Vectors of Infectious Disease

Distributed the world over are 810 species of ticks, all of them ectoparasites and about 100 of them vectors of infectious disease. Two families of ticks are differentiated—soft (argasid) and hard (ixodid)—based on the absence or presence of a dorsal shield called the scutum. Argasids seek specific, sheltered habitats in caves, barns, burrows, or even bird cages, and they parasitize wildlife during nesting seasons and domesticated animals throughout the year. Endemic relapsing fever is transmitted by this type of tick.

In contrast, the ixodids prefer confined habitats, and many have adapted to a wide-ranging life-style, hitchhiking along as their hosts wander through forest, savannah, or desert regions. Depending upon the species, ticks feed during larval, nymph, and adult metamorphic stages. The longevity of ticks is formidable; metamorphosis can extend for two years and adults can survive for four years away from a host without feeding. Many tick-borne agents cause mild infection in their primary wildlife hosts, but serious infection in livestock and humans. Ixodid ticks are implicated in several rickettsial infections, most particularly, Rocky Mountain spotted and Q fevers, and they also play host to borreliae and certain viruses.

Mites So abundant are mites that the 30,000 classified species are estimated to represent less than one-tenth of those in existence. Most species are free-living, and only a handful of the parasitic ones are of direct medical importance. The house mouse mite, *Liponyssoides sanguineus,* is a vector for the etiologic agent of rickettsialpox. Although domestic mice and rats are the natural hosts for this mite, it is not specific in its food getting and will readily nibble on humans if the opportunity arises. Larval mites called chiggers develop in the undergrowth of jungles, grasslands, coniferous forests, cultivated fields, gardens, and riverbanks that are frequented by humans. The larvae may escape notice at first, since they are minute and the bite is initially painless. But the bite can become intensely itchy in sensitive individuals. Once the larval stage is passed, the nymph and adult feed on nonvertebrate hosts. Chiggers are the primary reservoir and carriers of *Rickettsia tsutsugamushi* (scrub typhus).

Fleas Fleas are laterally flattened, wingless insects with well-developed jumping legs and prominent probosci for piercing the skin of warm-blooded animals. They go through a complex metamorphosis on the host or in the environment, and are known for their extreme longevity and resistance. Many fleas are notorious in their nonspecificity, passing with ease from wild or domesticated mammals to humans. In response to mechanical stimulation and warmth, fleas jump onto their targets and crawl about, feeding as they go. A well-known vector is the oriental rat flea *Xenopsylla cheopis* that transmits *Rickettsia typhi,* the cause of murine typhus (see figures 11.19 and 16.39). The flea harbors the pathogen in its gut, and periodically contaminates the environment with virulent rickettsias by defecating. This same flea is involved in the transmission of plague (see chapter 16).

Lice Lice are small, flat insects equipped with biting or sucking mouthparts. The lice of humans usually occupy head and body hair (*Pediculus humanus*) or pubic, chest, and axillary hair (*Phthirus pubis*). They feed by gently piercing the skin and sucking blood and tissue fluid. Infection develops when the louse or its feces are inadvertently crushed and rubbed into wounds, skin, eyes, or mucous membranes. Rickettsiae transmitted by lice are *R. prowazekii* (epidemic typhus) and *Rochalimaea quintana* (trench fever).

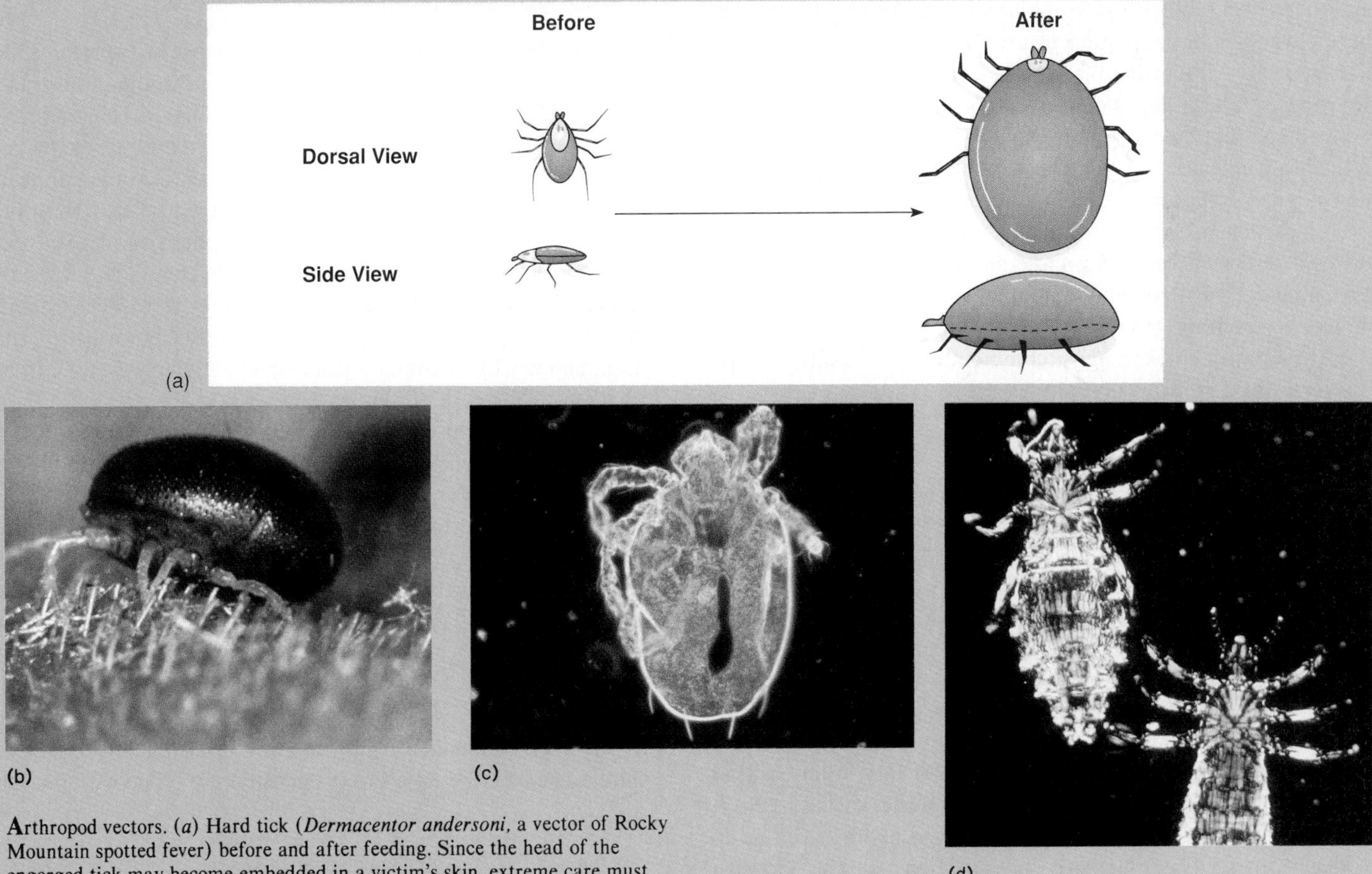

Arthropod vectors. (*a*) Hard tick (*Dermacentor andersoni,* a vector of Rocky Mountain spotted fever) before and after feeding. Since the head of the engorged tick may become embedded in a victim's skin, extreme care must be taken in its removal. (*b*) An engorged soft tick, represented here by *Ornithodoros* (the vector of relapsing fever). (*c*) A chigger mite, the vector of scrub typhus. (*d*) The body louse, a human parasite that transmits epidemic typhus and one form of relapsing fever.

with no ill effect, but others, like the human body louse, die from typhus infection and so cannot continuously harbor the pathogen. In certain vectors such as the tick of Rocky Mountain spotted fever, rickettsiae are transferred transovarially, from the infected female to her eggs. This event has profound impact, since the microbe is inherited continuously through multiple generations of ticks, creating a long-standing reservoir.

These arthropods feed on the blood or tissue fluids of their mammalian hosts, but not all of them transmit the rickettsial pathogen through direct inoculation with saliva. Ticks and mites directly inoculate the skin lesion as they feed, but fleas and lice harbor the infectious agents in their intestinal tract. During their stay on the host's body, these latter insects defecate or are smashed, thereby releasing the rickettsias onto the skin or into a wound. Ironically, scratching the bite helps the pathogen invade deeper tissues. One rickettsial species, *Coxiella burnetii,* is spread to humans primarily by nonvector means such as food, air, fomites, and animal products, and less commonly, by ticks.

General Factors in Rickettsial Pathology and Isolation

How rickettsiae cause injury to humans is less clearly understood than the clinical expression of their diseases. A common target in most infections is the endothelial lining of the small blood vessels. The bacteria somehow recognize, enter, and multiply within endothelial cells, causing necrosis. Part of the damage to cells is believed to be mediated by an endotoxin. In an apparent effort to replace the injured lining, the proliferating endothelial cells block the vascular lumen. Among the immediate pathological consequences are vasculitis, perivascular infiltration by inflammatory cells, vascular leakage, and thrombosis. These pathologic effects may be attended by skin rash, edema, hypotension, and gangrene. Intravascular clotting in the brain accounts for the stuporous mental changes and other neurological symptoms that sometimes occur.

Isolation of most rickettsiae from clinical specimens requires a suitable live medium and specialized laboratory facilities, including controlled access and safety cabinets. The usual choices for routine growth and maintenance are the yolk sacs of embryonated chicken eggs, chick embryo cell cultures, and to a lesser extent, mice and guinea pigs.

Specific Rickettsioses

Rickettsioses may be differentiated on the basis of their clinical features and epidemiology as: (1) classical, epidemic, or louse-borne typhus; (2) murine, endemic, or flea-borne typhus; (3) tropical, scrub, or mite-borne typhus; (4) spotted fevers; (5) trench fever; and (6) Q fever.

Epidemic Typhus and *Rickettsia prowazekii*

Epidemic or louse-borne typhus has been a constant accompaniment to war, poverty, and famine. Historically, it has caused more human suffering than any other bacterial disease, and it has altered the course of political and social history even up to modern times. The extensive investigations of Dr. Howard Ricketts and Stanislas von Prowazek in the early 1900s led to the discovery of the vector and the rickettsial agent, but not without mortal peril, since both men died of the very disease they investigated. *Rickettsia prowazekii* was named in honor of their pioneering efforts.

Epidemiology of Epidemic Typhus Humans are the sole hosts of human body lice and the only reservoirs of *R. prowazekii.* A louse acquires infection by drawing a blood meal from a rickettsemic human into its intestine. While feeding on another person, the louse defecates, and the feces become implanted into the bite or other breaks in the skin. Infection of the eye or respiratory tract may take place by direct contact or inhalation of dust containing dried louse feces, but this is a rarer mode of transmission.

Infected lice die from their infection in one to three weeks, without transmitting the rickettsiae to their eggs. The entire metamorphic cycle (egg to adult) can take place upon an infested person's garments, which provide an optimum developmental temperature of about 20°C. Crowding, infrequently changing clothing, or exchanging clothing greatly favor transmission. The overall incidence of epidemic typhus in the United States is very low, with no epidemics since 1922. Recent sporadic cases involving transmission from the flying squirrel indicate a possible animal reservoir. Although no longer common in regions of the world with improved standards of living, epidemic typhus presently persists in regions of Africa, Central America, and South America.

Disease Manifestations and Immune Response in Typhus Once rickettsias enter the circulation, they pass through an intracellular incubation period of 10 to 14 days. The first clinical manifestations are sustained high fever, chills, frontal headache, and muscular pain. Within seven days, a generalized rash appears, initially on the trunk and then spreading to the extremities, except the palms and soles. Personality changes, oliguria (low urine output), hypotension, and gangrene complicate the more severe cases. Untreated disease continues for three weeks, ending in either rapid recovery or death. Mortality is lowest in children, about 10% in young adults, and as high as 40–60% in patients over 50 years of age.

Recovery confers resistance to typhus for a long time. In some cases, the rickettsiae are not completely eradicated by the immune response and enter into latency. After several years, a milder recurring form of the disease known as *Brill-Zinsser disease* may appear. This disease is seen most often in persons who have immigrated from endemic areas and is of concern mainly because it provides a continuous reservoir of the etiologic agent.

Treatment and Prevention of Typhus The standard chemotherapy for typhus is tetracycline or chloramphenicol. But despite antibiotic therapy, the prognosis may be poor in patients

---

typhus (ty'-fus) Gr. *typhos,* smoky or hazy, underlining the mental deterioration seen in this disease. Typhus is commonly confused with typhoid fever, an unrelated enteric illness caused by *Salmonella typhi.*

with advanced circulatory or renal complications. Eradication of epidemic typhus is theoretically possible by exterminating the vector. Widespread dusting of human living quarters with insecticides has provided some environmental control, and individual treatment with an anti-louse shampoo or ointment is also effective. Vaccination against the rickettsial agent provides still another effective method of control. The vaccine, prepared in infected yolk sacs, is administered in paired doses a month apart, followed by booster doses every 6 to 12 months during conditions of exposure.

Epidemiology and Clinical Features of Endemic Typhus

The agent for **endemic typhus** is *Rickettsia typhi* (*R. mooseri*), which shares many characteristics with *R. prowazekii* except for pronounced virulence. Synonyms for this rickettsiosis are endemic typhus, murine (mouse) typhus, and flea-borne typhus. The disease occurs in certain parts of Central and South America and in the Southeast, Gulf Coast, and Southwest regions of the United States, where *R. typhi* is regularly harbored by mice and rats. Transmission to the human population is chiefly through infected rat fleas that inoculate the skin, though disease is occasionally acquired by inhalation. In the United States, most reported cases arise sporadically among persons working in rat-infested facilities such as shipyards, grain elevators, warehouses, and markets. Many more cases probably occur, but they escape notice because they are scattered or not severe enough to warrant medical attention.

Felines and Typhus In the mid-1980s residents of certain areas of southern California were warned that they might have been exposed to murine typhus by someone living in their house—their cat. How felines became involved in the cycle is unknown, but health officials believe that typhus-infected fleas from backyard opossums and rats have abandoned their natural hosts in favor of the cat. The cat perpetuates the fleas in and around the house, where they may then attack humans. Years of high flea infestation seem to have increased the number of cases as well.

In contrast to epidemic typhus, the clinical manifestations of endemic typhus are briefer, milder, and cause fewer complications. Following an incubation period of one to two weeks, fever, headache, muscle aches, and malaise begin abruptly. After five days, a skin rash, transient in milder cases, begins on the trunk and radiates toward the extremities. Symptoms dissipate in about two weeks. Tetracycline and chloramphenicol are effective therapeutic agents, while various pesticides are available for vector and rodent control.

Scrub Typhus or Tsutsugamushi Disease

Scrub typhus, also called *tsutsugamushi disease,* was first described nearly two centuries ago in Japan. The illness, caused by *R. tsutsugamushi,* is endemic to Southeast Asia, Australia, and India. Current incidence of this disease is difficult to trace, but it appears to occur sporadically in endemic areas.

Scrub typhus rickettsias live naturally in the bodies of chigger mites. Small mammalian and avian hosts that become infected help perpetuate the reservoir. Although humans are not their natural hosts, chiggers will jump upon and bite them as they pass through fields, forests, and riverbanks. At the site of the chigger bite, a distinctive black scab called an *eschar* develops in one to three weeks. Early symptoms of fever, headache, and muscle aches resemble those of endemic typhus, and in about half the cases, a generalized rash originates on the trunk and radiates peripherally. Severe cases are accompanied by mental confusion, delirium, pneumonia, and circulatory collapse. Mortality in untreated disease reaches 50%, but with prompt diagnosis and adequate therapy, fatalities can be virtually reduced to zero. Chemotherapy and prevention are nearly identical to that of endemic typhus.

Rocky Mountain Spotted Fever: Epidemiology and Pathology

The rickettsial disease with greatest impact on persons living in North America is **Rocky Mountain spotted fever** (RMSF), named for the place it was first seen—the Rocky Mountains of Montana and Idaho. Ricketts identified the etiologic agent *Rickettsia rickettsii* in smears from infected animals and patients and later discovered that it was transmitted by ticks. Despite its geographic name, this disease occurs infrequently in the western United States. The majority of cases are concentrated in the southeast and eastern seaboard regions (figure 17.17). Infections occur most frequently among children living in rural and mountain areas, generally in the spring and summer months when the tick vector is most active. The prevalence of RMSF is more than 1,000 cases per year, and it has been steadily rising in recent years because of increased recreation in wilderness areas and the gradual encroachment of suburban housing developments into tick habitats.

The principal reservoir and vector of *R. rickettsii* is a hard tick such as the wood tick (*Dermacentor andersoni*), the American dog tick (*D. variabilis,* among others), or the Lone Star tick (*Amblyomma americanum*). The dog tick is probably most responsible for transmission to humans, since it is the major vector in the southeastern United States (figure 17.18). The tick passes the pathogen to its offspring, to various domestic and wild mammals, and to humans. Unengorged ticks, picked up as a person brushes against low shrubs or other habitats, can go unnoticed. After wandering on the body, the tick eventually embeds its mouthparts into the skin, feeds, and sheds rickettsias into the bite.

Pathogenesis and Clinical Manifestations Although the severity of spotted fever ranges from mild to rapidly fatal, we will discuss the fully developed disease. After two to four days incubation, the first symptoms are a sustained fever, chills, headache, and muscular pain. The distinctive spotted rash usually comes on within two to four days after the prodromium (figure 17.19). It starts on the wrist and ankles, moves to the arms and

---

tsutsugamushi (soot″-soo-gah-moo′-shee) Jap. *tsutsuga,* small and menacing, and *mushi,* a creature.

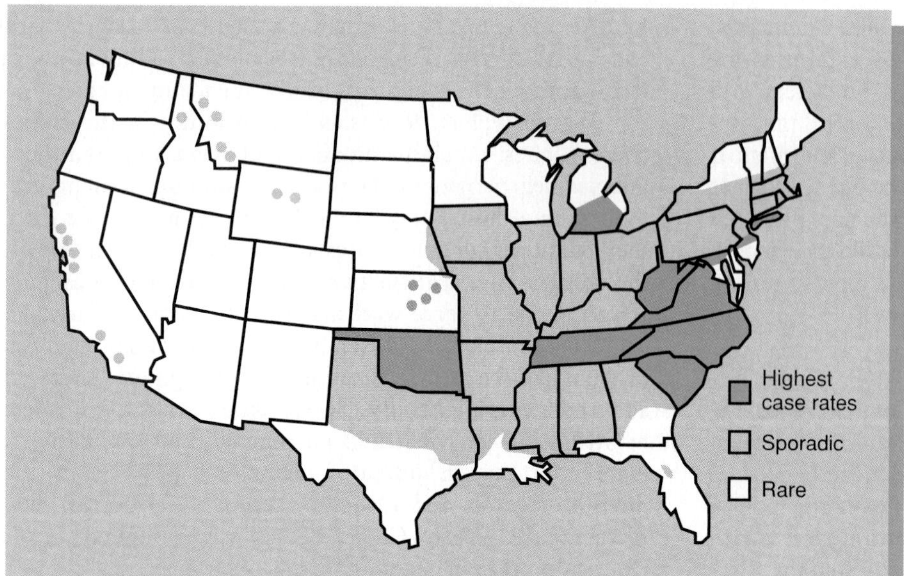

Figure 17.17 The prevalence of Rocky Mountain spotted fever during the last decade. Most cases are reported in states on the eastern seaboard and southeastern United States.

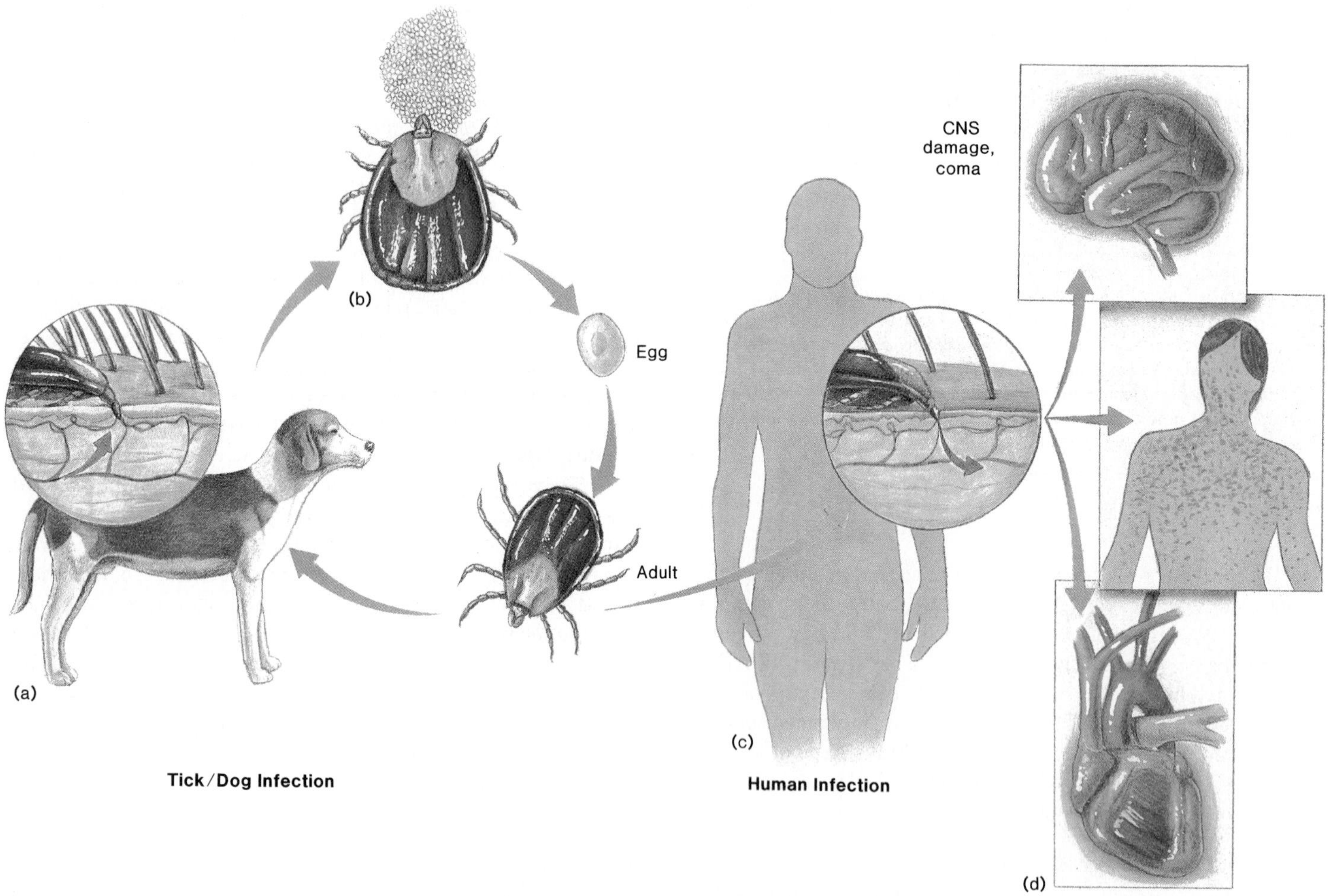

Figure 17.18 The transmission cycle in Rocky Mountain spotted fever. (*a*) Dog or wood ticks are the principal vectors. Ticks are infected from a mammalian reservoir during a blood meal. (*b*) Transovarial passage of *R. rickettsia* to tick eggs serves as a continual source of infection within the tick population. Infected eggs produce infected adults. (*c*) A tick attaches to a human, embeds its head in the skin, feeds, and sheds rickettsias into the bite. (*d*) Systemic involvement includes severe headache, fever, rash, coma, and circulatory disruption.

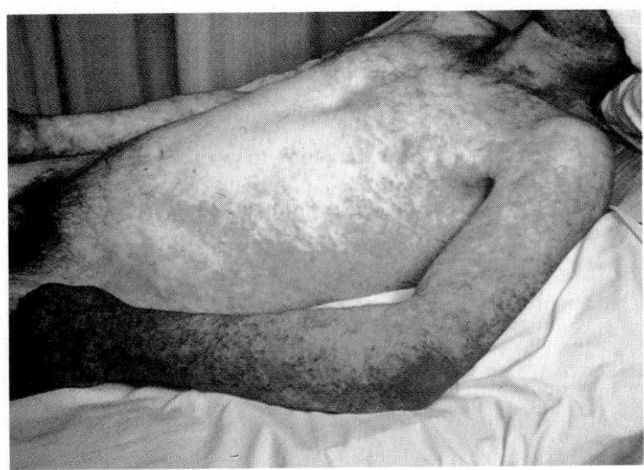

Figure 17.19 Late generalized rash of Rocky Mountain spotted fever. In some cases lesions become hemorrhagic and predispose to gangrene of the extremities.

legs, converges toward the chest, and eventually covers the entire body. Early lesions are slightly mottled like measles, but later ones are macular, maculopapular, and even petechial. In the most severe, untreated cases, the enlarged lesions merge and may become necrotic, predisposing to gangrene of the toes or fingertips.

Other grave manifestations of disease are cardiovascular disruption, including hypotension, thrombosis, and hemorrhage. Conditions of restlessness, delirium, convulsions, tremor, deafness, and coma are signs of the often overwhelming effects on the central nervous system. Fatalities occur in an average of 20% of untreated cases and 5–10% of treated cases. Although death usually occurs late in the disease, there have been instances of mortality within three days of onset of symptoms.

Diagnosis, Treatment, and Prevention of Spotted Fever Any case of Rocky Mountain spotted fever is a cause for great concern and requires immediate treatment even before laboratory confirmation. Careful clinical observation and patient history are thus essential. Indications sufficiently suggestive to start antimicrobial therapy are (1) a cluster of symptoms, including sudden fever, headache, and rash, (2) recent contact with ticks or dogs, and (3) possible occupational or recreational exposure in the spring or summer. A recent boon to early diagnosis is a method for staining rickettsias directly in a tissue biopsy using fluorescent antibodies. Isolating rickettsiae from the patient's blood or tissues is still desirable, but it is expensive and requires specially qualified personnel and laboratory facilities.

Because antibodies appear only after two or three weeks, a change in serum titer is used retrospectively to confirm a presumptive diagnosis. The drug of choice for suspected and known cases is tetracycline (doxycycline) administered every day for one week. Chloramphenicol is used if diagnosis has failed to differentiate between spotted fever and meningococcal meningitis, since this drug is effective in both cases. So widespread are endemic areas and so deeply entrenched is *R. rickettsii* in reservoir animals that anyone involved in outdoor activities should wear protective clothing, boots, leggings, and insect repellent containing DEET. Individuals exposed to heavy infestation should routinely inspect their bodies (and their dogs!) for ticks and remove them gently without crushing, preferably with forceps or fingers protected with gloves, because it is possible to become infected by tick feces or body fluids. A new vaccine, made with rickettsiae grown in chick embryo cell cultures, shows some promise in controlling the disease in humans who are at high risk.

Other Rickettsioses

Rickettsialpox is a benign, self-limited disease caused by *Rickettsia akari.* It is transmitted in the United States by a mite that is ectoparasitic on the common house mouse. The disease appears most frequently among persons living in crowded urban habitats that may harbor many mites. Rickettsialpox causes little morbidity and is not reportable, so the actual incidence is not known, though sporadic outbreaks have been reported in various eastern cities.

Trench fever first appeared during World War I, when it afflicted at least a million persons. Since 1945, cases have occurred infrequently in endemic regions of Europe, Africa, and Asia. The causative agent is *Rochalimaea quintana,* a species that, like epidemic typhus, cycles between humans and lice, but unlike the typhus rickettsia, does not multiply intracellularly and does not kill the louse vector. Symptoms are highly variable, but they generally include a five-to-six-day fever (hence five-day or quintana fever); leg pains, especially in the tibial region (shinbone fever); headache; chills; and muscle aches. A macular rash may also occur. The microbe may persist in the blood long after convalescence and is responsible for later relapses.

Q fever[3] was first described in Queensland, Australia. Its origin was mysterious for a time, until Cox working in Montana and Burnet in Australia discovered the agent later named ***Coxiella burnetii.*** Since this discovery, *Coxiella* has proved to be a unique and complex microbe in several ways. It is highly resistant, due to its production of an unusual type of spores (figure 17.20), and it is apparently harbored by a wide assortment of vertebrates and arthropods, especially ticks. Arthropod vectors play an essential role in transmitting the rickettsiae between wild and domestic animals, but transmission to humans is largely by means of environmental contamination and airborne spread. Sources of infectious material include urine, feces, milk, and the afterbirth of infected animals, which may produce airborne and dust-borne particles. The primary portals of entry are the lungs, skin, conjunctiva, and gastrointestinal tract.

Coxiella burnetti has been isolated from most regions of the world. California and Texas have the highest case rates in the United States, though most cases probably go undetected. Persons at highest risk are farm workers, meat cutters, wool and

3. For "query," meaning to question, or of unknown origin.

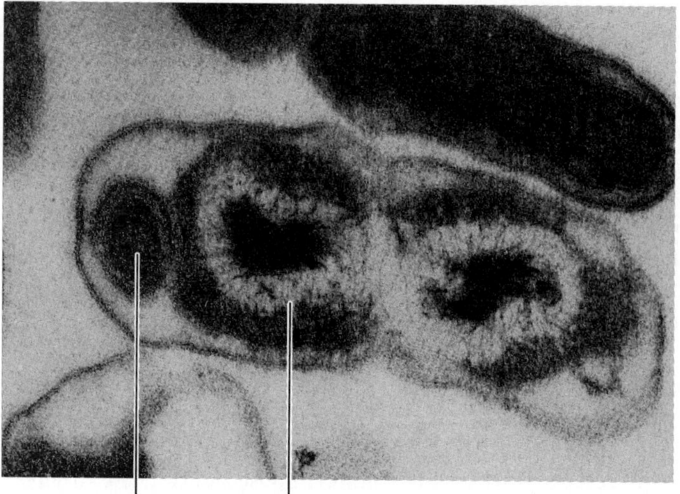

Figure 17.20 The vegetative cells of *Coxiella burnetii* produce unique endospores that are released when the cell disintegrates. Free spores survive outside the host and are important in transmission.

leather workers, laboratory technicians, and consumers of raw milk. The clinical manifestations typical of rickettsial infection are abrupt onset of fever, chills, head and muscle ache, and malaise. In contrast to most rickettsioses, skin rashes are seldom seen. Disease may also be accompanied by pneumonitis, hepatitis, or delayed endocarditis, which are occasionally fatal. Mild or subclinical cases recover spontaneously, and more severe cases respond to tetracycline therapy. Prevention and control are hampered by the lack of a satisfactory vaccine and the difficulty of eradicating the agent from livestock and their environment. In areas of endemic Q fever, cow and sheep milk should be pasteurized or boiled before consumption.

Other Obligate Parasitic Bacteria: The Chlamydiaceae

Like rickettsias, the chlamydias are obligate parasites that depend on certain metabolic constituents of host cells for growth and maintenance. They show further resemblance to the rickettsias with their small size, gram-negative cell wall, and pleomorphic morphology, but they are markedly different in several aspects of their life cycle. The two species of greatest medical significance are *Chlamydia trachomatis,* a very common pathogen involved in sexually transmitted, neonatal, and ocular disease (trachoma), and *Chlamydia psittaci,* a zoonosis of birds and mammals that causes ornithosis in humans.

The Biology of *Chlamydia*

Chlamydias alternate between two distinct stages: (1) a small, metabolically inactive, infectious form called the **elementary body** that is released by the infected host cell, and (2) a larger, noninfectious, actively dividing form called the **reticulate body** that grows within the host cell vacuoles (figure 17.21). Reticulate bodies ultimately differentiate into elementary bodies. Elementary bodies are dense spheres with an average diameter of 0.2–0.4 μm. They are not spores, although they are shielded by a rigid, impervious envelope that ensures survival outside the eucaryotic host cell. Reticulate bodies range in diameter from 0.5 μm to 1.5 μm, are finely granulated, and have a thin cell wall. Studies of the reticulate bodies indicate that they are energy parasites, entirely lacking enzyme systems for catabolizing glucose and other substrates and for synthesizing ATP, though they do possess ribosomes and mechanisms for synthesizing proteins, DNA, and RNA.

Chlamydia trachomatis (klah-mid'-ee-ah trah-koh'-mah-tis) Gr. *chlamys*, a cloak, and *trachoma*, roughness.

psittaci (sih-tah'-see) Gr. *psittacus*, a parrot.

Diseases of *Chlamydia trachomatis*

The reservoir of pathogenic strains of *Chlamydia trachomatis* is the human body. The microbe shows astoundingly broad distribution within the population, often being carried with no symptoms. Elementary bodies are transmitted in infectious secretions, and although infection can occur in all age groups, disease is most severe in infants and children. The two human strains are the **trachoma** strain, which attacks the squamous or columnar cells of mucous membranes in the eyes, genitourinary tract, and lungs, and the **lymphogranuloma venereum** (LGV) strain, which invades the lymphatic tissues of the genitalia.

Chlamydial Diseases of the Eye The two forms of chlamydial eye disease, trachoma and inclusion conjunctivitis, differ in their patterns of transmission and ecology. **Ocular trachoma,** an infection of the epithelial cells of the eye, is an ancient disease and a major cause of blindness in certain parts of the world. Although a few cases occur yearly in the United States, it is endemic to parts of Africa and Asia, where several million cases occur every year. Transmission is favored by contaminated fingers, fomites, flies, and a hot, dry climate.

The first signs of infection are a mild conjunctival exudate and slight inflammation of the conjunctiva. This is followed by marked infiltration of lymphocytes, monocytes, and macrophages into the infected area. As these cells build up in the follicles and lymphatics, they impart a pebbled (rough) appearance to the inner aspect of the upper eyelid (figure 17.22*a*). In time, a vascular pseudomembrane of exudate and inflammatory leukocytes forms over the cornea, a condition called *pannus* that lasts a few weeks and usually heals. Complications contributing most to corneal damage and impaired vision are repeated infections, secondary bacterial infection, and deformation of tear ducts and eyelashes, which predispose to dry eyes and corneal abrasion. Early treatment of this disease with tetracycline or sulfa drugs is highly effective and will prevent all of the complications. It is truly unfortunate in this day of preventative medicine that millions of children will develop blindness for want of a few dollars worth of antibiotics.

Inclusion conjunctivitis is usually acquired through contact with secretions of an infected genitourinary tract. Infantile conjunctivitis comes on 5 to 12 days after a baby has passed through the birth canal of its infected mother, and is the most

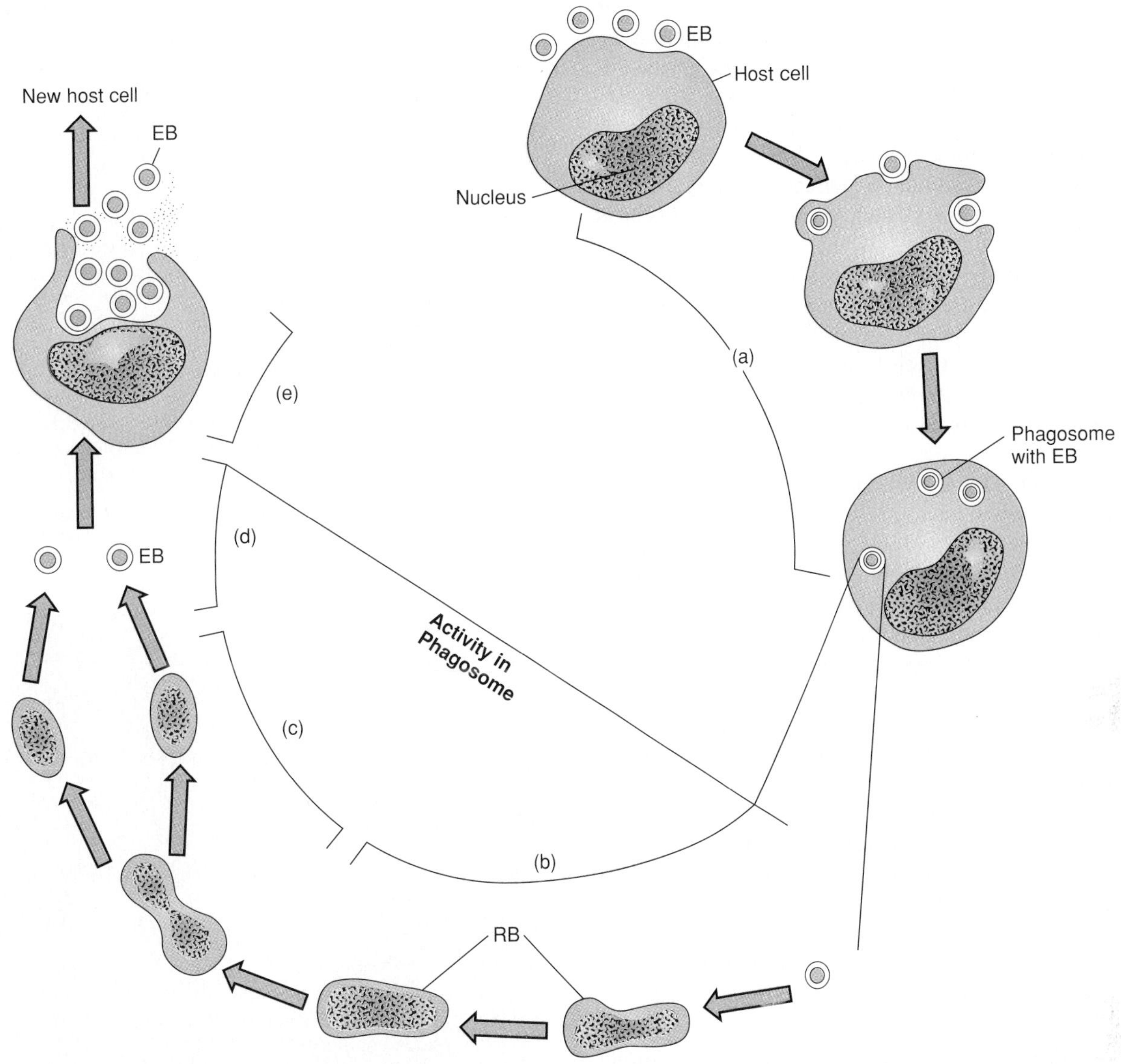

Figure 17.21 The life cycle of *Chlamydia.* (*a*) The infectious stage, or elementary body (EB), is taken into phagocytic vesicles by the host cell. (*b*) In the phagosome, each elementary body develops into a reticulate body (RB). (*c*) Reticulate bodies multiply by regular binary fission. (*d*) Mature RBs become reorganized into EBs. (*e*) Completed EBs are released from the host cell.

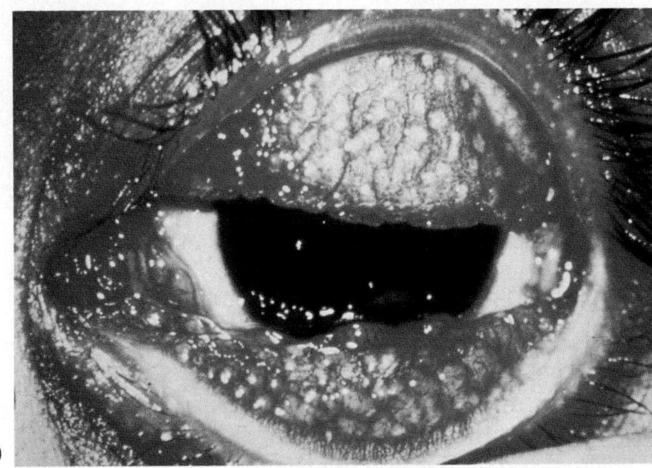

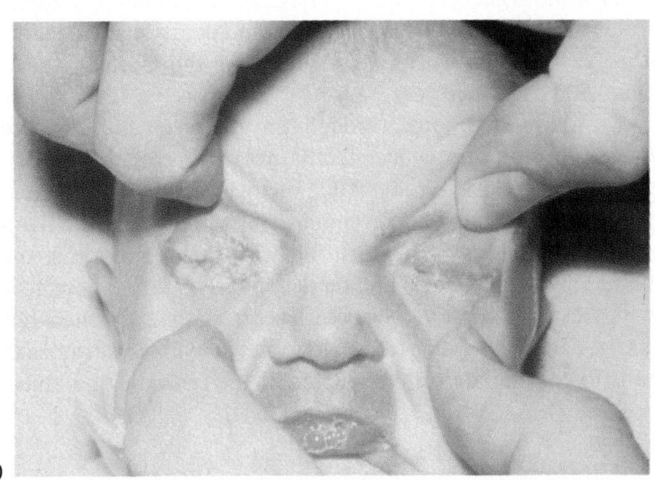

Figure 17.22 The pathology of primary ocular chlamydial infection. (*a*) Trachoma, an early, pebblelike inflammation of the conjunctiva and inner lid in a child. (Note that the eyelid is being pulled away by the examiner to reveal the lesion.) (*b*) Inclusion conjunctivitis in a newborn. Within five to six days, an abundant, watery exudite collects around the conjunctival sac. This is currently the most common cause of ophthalmia neonatorum.

prevalent form of conjunctivitis (100,000 cases per year). The initial signs are conjunctival irritation, a profuse adherent exudate, redness, and swelling (figure 17.22*b*). Although the disease is generally self-limited and heals spontaneously, trachoma-like scarring occurs often enough to warrant routine prophylaxis of all newborns (as for gonococcal infection). Because the traditional silver nitrate solution is ineffective against *C. trachomatis,* antibiotics such as erythromycin and tetracycline must be instilled into the eyes.

Inclusion conjunctivitis in adults has been traced to swimming pools, shared towels, and finger-to-eye contact. Most afflicted persons have concurrent genital infections, thus the majority of cases appear to result from self-inoculation or contamination during sexual intercourse. The infection is similar to trachoma and may cause corneal scarring if untreated.

Sexually Transmitted Chlamydial Diseases It has been estimated that *C. trachomatis* is carried in the reproductive tract of up to 10% of all persons, with even higher rates among the promiscuous. About 70% of infected women harbor it asymptomatically on the cervix, while 10% of infected males show no signs or symptoms. Depending on the condition of the host, the bacterium is capable of causing a whole spectrum of diseases, many of which resemble gonorrheal infections (see feature 17.6).

A syndrome appearing among males with chlamydial infections is an inflammation of the urethra called **nongonococcal urethritis** (NGU). This diagnosis is derived from the symptoms that mimic gonorrhea, yet do not involve gonococci. *Chlamydia* is the most frequent cause of this syndrome. Women with symptomatic chlamydial infection have cervicitis accompanied by a white drainage, endometritis, and salpingitis (pelvic inflammatory disease). As is often the rule with sexually transmitted diseases, chlamydia frequently appears in mixed infections with the gonococcus and other genitourinary pathogens, thereby greatly complicating treatment.

When a particularly virulent strain of *Chlamydia* chronically infects the genitourinary tract, the result is a severe, often disfiguring disease called **lymphogranuloma venereum.**[4] The external genitalia, anus, rectum, and the inguinal lymph nodes are the typical targets of invasion. The disease is endemic to regions of South America, Africa, and Asia, but occasionally occurs in other parts of the world. Its incidence in the United States is about 500 cases per year.

Chlamydiae enter through tiny nicks or breaks in the perigenital skin or mucous membranes and form a small, painless vesicular lesion that often escapes notice. Other acute symptoms are headache, fever, and muscle aches. As the lymph nodes near the lesion begin to fill with granuloma cells, they enlarge and become firm and tender (figure 17.23). These nodes or buboes may burst and heal with scarring that obstructs lymphatic channels. Long-term blockage of lymphatic drainage leads to chronic, deforming edema of the genitalia and anus.

4. Also called tropical bubo or lymphogranuloma inguinale.

Feature 17.6 New Threats from *Chlamydia*

Chlamydiosis is one of the most prevalent of all sexually transmitted diseases, probably exceeding 10 million cases in the United States annually. Medically and socioeconomically, its clinical significance now eclipses gonorrhea, herpes simplex II, and syphilis. Far from being simple and non-life-threatening, chlamydial infections, even inapparent ones, have the potential for great harm. When infection ascends to the uterus, fallopian tubes, and peritoneum, it may damage the interior of the reproductive organs. The result is that more than 10 thousand young women and men become sterilized every year, while thousands of women suffer interrupted tubal pregnancies. An unusual form of arthritis caused when the pathogen disseminates to the joints has been reported in both men and women. Cervical infections of mothers may lead to prematurity and morbidity in the fetus, chlamydial pneumonitis in neonates (30 thousand cases per year), and eye disease.

A different *Chlamydia* pathogen, *C. pneumoniae,* has been linked to severe respiratory illness in asthma patients. It now appears that infection with this agent accounts for the rise in fatal asthmatic attacks (see chapter 14).

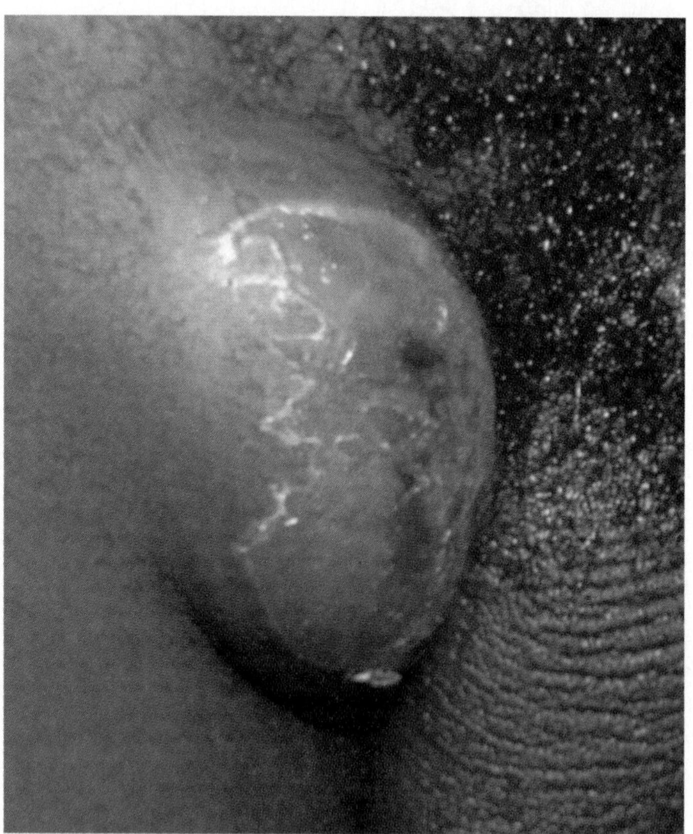

Figure 17.23 The clinical appearance of advanced lymphogranuloma venereum in a man. A chronic local inflammation blocks the lymph channels, causing swelling and distortion of the external genitalia.

Identification, Treatment, and Prevention of Chlamydiosis Because chlamydias reside intracellularly, specimen sampling must be sufficiently forceful to dislodge some of the cells from the mucosal surface. Genital samples are taken with a swab inserted a few centimeters into the urethra or cervix, rotated, and removed. Although the surest diagnosis comes from culture in

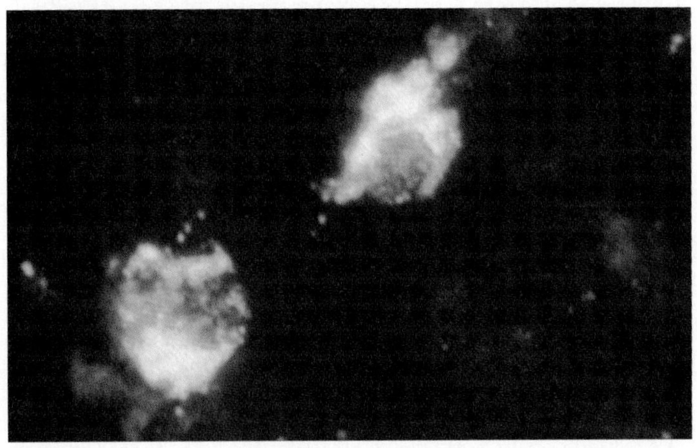
(a)

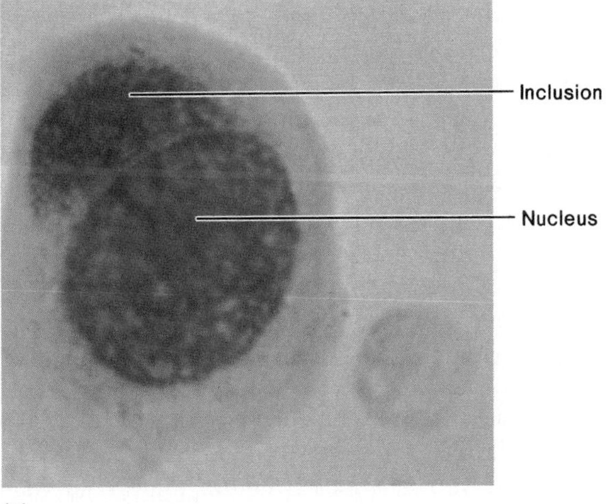

(b)

Figure 17.24 Direct diagnosis of chlamydial infection. (*a*) A patient's urethral, cervical, rectal, or conjunctival specimen is stained with monoclonal antibodies bearing a fluorescent dye. Infected cells glow a bright apple green. (*b*) Direct Giemsa stain of the eyelid scraping of a patient suffering from inclusive conjunctivitis. Note that the enlargement of the developing inclusion has displaced the nucleus to one side.

the yolk sacs of chicken embryos or mice, this procedure is too costly and time-consuming to be used routinely in STD clinics; however, it is an essential part of diagnosing neonatal infections. The fastest and most sensitive and specific test currently available is a direct assay of specimens using immunofluorescence (figure 17.24*a*). Methods useful in diagnosing inclusion conjunctivitis are Giemsa or iodine stains (figure 17.24*b*), but they are not recommended for urogenital specimens because of low sensitivity and the possibility of obtaining false-negative results in asymptomatic patients.

Treatment of urogenital chlamydial infections requires a drug that works intracellularly to block infection and multiplication. Drugs that meet this criterion most effectively are the tetracyclines, erythromycin, and rifampin. Penicillin and the aminoglycosides are not effective and must not be used. Because of the high carriage rate and the difficulty in detection, prevention of chlamydial infections has been and will continue to be very difficult. As a general rule, sexual partners of infected persons should be treated with drug therapy to prevent reinfection, and sexually active persons can achieve some protection with a condom. Screening pregnant women and treating those who test chlamydia-positive helps prevent fetal or neonatal disease. But relying on this technique alone can fail if carriers do not seek prenatal care, thus routine ocular prophylaxis of newborns is still necessary.

Chlamydia psittaci and Ornithosis

The term psittacosis was adopted to describe a pneumonia-like illness contracted by persons working with imported parrots and other psittacine birds in the last century. As outbreaks of this disease appeared in areas of the world having no parrots, it became evident that other birds could carry and transmit the microbe to humans and other animals. In light of this, the generic term **ornithosis** has been suggested as a replacement.

Ornithosis is a worldwide zoonosis that is carried in a latent state in wild and domesticated birds but becomes active under stressful conditions such as overcrowding. In the United States, poultry have been subject to extensive epidemics that killed as many as 30% of flocks. Infection is communicated to other birds, mammals, and humans by contaminated feces and other discharges that become airborne and are inhaled. Most human cases in the United States occur among poultry and pigeon handlers, and infection is probably more common than the 150 cases per year that are currently reported. Transmission is not exclusively zoonotic, because occasional outbreaks have been reported among humans having no contact with birds.

The symptoms of ornithosis mimic those of influenza and pneumococcal pneumonia. Early manifestations are fever, chills, frontal headache, and muscle aches, and later ones are coughing and lung consolidation. Unchecked, infection may lead to systemic complications involving the meninges, brain, heart, or liver. Although most patients respond well to tetracycline or erythromycin therapy, recovery is often slow and fraught with relapses. Control of the disease is usually attempted by quarantining and giving antimicrobial therapy to imported birds and by taking precautions in handling birds, feathers, and droppings.

Mollicutes and Other Cell-Wall-Deficient Bacteria

Bacteria in the Class Mollicutes, also called the **mycoplasmas,** are the smallest self-replicating microorganisms. All of them naturally lack a cell wall (figure 17.25*a*), and except for one genus, all species are parasites of animals and plants. The two most clinically important genera are *Mycoplasma* and *Ureaplasma.* Disease of the respiratory tract has been primarily associated with *Mycoplasma pneumoniae,* while *M. hominis* and *Ureaplasma ureolyticum* are implicated in urogenital tract infections.

---

ornithosis (or″-nih-thoh′-sis) Gr. *ornis,* bird. The number of birds that harbor *C. psittaci* exceeds 90 species.

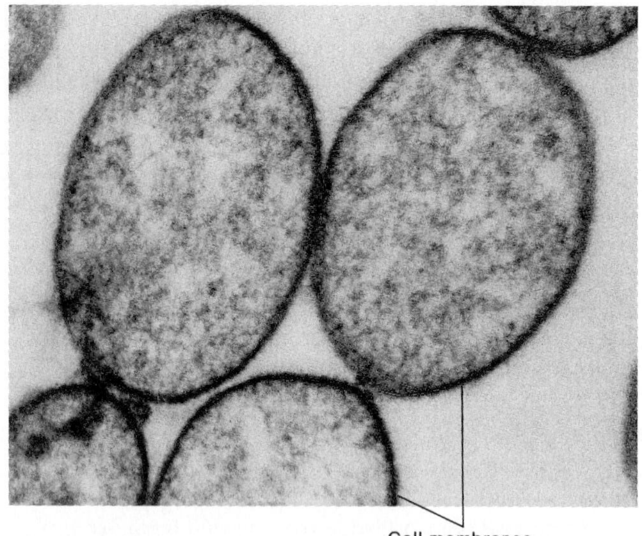

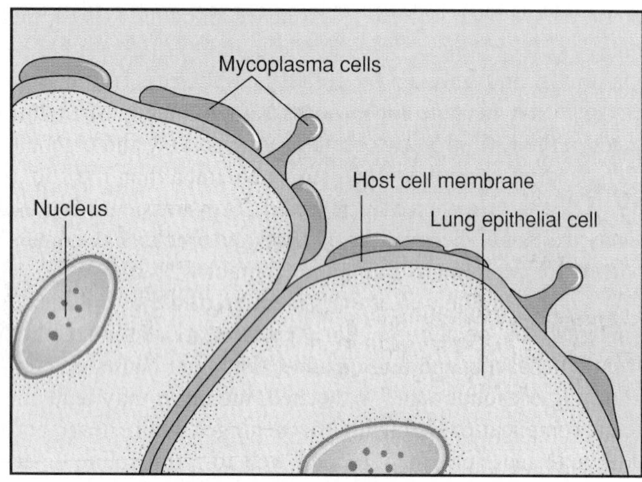

Figure 17.25 The morphology of the mycoplasmas. (*a*) A transmission electron micrograph of coccoid mycoplasma cells (×95,000) with single trilaminar cell membranes and no cell walls. (*b*) The pathology of mycoplasmal pneumonia. *M. pneumoniae* is a membrane parasite that adheres tightly and fuses with the host cell membrane. This makes its destruction and removal very difficult.

Biological Characteristics of the Mycoplasmas

Without a rigid cell wall to delimit their shape, mycoplasmas are exceedingly pleomorphic. The small (0.3–0.8 μm) flexible cells assume a spectrum of shapes, ranging from cocci and filaments to doughnuts, clubs, and helices (see figure 3.31). Mycoplasmas are not strict parasites, and they can grow in cell-free media, generate metabolic energy, and synthesize proteins with their own enzymes. However, most are fastidious and require complex media containing sterols, fatty acids, and preformed purines and pyrimidines. Animal species of mycoplasmas are sometimes referred to as membrane parasites because they form intimate associations with membranes of the respiratory and urogenital tracts (figure 17.25*b*). Because mycoplasmas bind to specific receptor sites on cells and adhere so tenaciously that they are not easily removed by usual defense mechanisms, infections are chronic and difficult to eliminate. Mycoplasmas are inhibited by antibiotics that repress protein synthesis but are resistant to penicillin and cephalosporins that block cell wall synthesis.

Mycoplasma pneumoniae and Atypical Pneumonia

Mycoplasma pneumoniae is a human parasite that is the most common agent of **primary atypical pneumonia (PAP)**. This syndrome is atypical in that its symptoms do not resemble those of pneumococcal pneumonia. PAP may also be caused by rickettsias, chlamydias, respiratory syncytial viruses, and adenoviruses. Mycoplasmal pneumonia is transmitted by aerosol droplets among persons confined in close living quarters, especially families, students, and the military. Community resistance to this pneumonia is high; only 3–10% of those exposed develop it, and fatalities are rare.

M. pneumoniae selectively binds to specific receptors of the respiratory epithelium and inhibits ciliary action. As the bacteria grow and spread over the next two to three weeks, both the cilia and the epithelium are gradually disrupted, even though most mycoplasmas are neither invasive nor particularly toxigenic. The first symptoms—fever, malaise, sore throat, and headache—are not suggestive of pneumonia. A cough is not a prominent early symptom, and when it does appear, it is mostly unproductive. As the disease progresses, nasal symptoms, chest pain, and earache may develop. The lack of acute illness in most patients has given rise to the nickname "walking pneumonia."

Diagnosis Because a culture can take two or three weeks, early diagnosis of mycoplasma pneumonia is difficult and relies chiefly on close clinical observation. Stains of sputum appear devoid of bacterial cells, leukocyte counts are within normal limits, and X-ray findings are nonspecific. Some physicians find the appearance of a small blister on the tympanic membrane a useful sign. Others use the patient's history to rule out other bacterial and viral agents. Serological tests based on complement fixation, immunofluorescence, and indirect hemagglutination are useful later in the disease.

Tetracycline and erythromycin inhibit mycoplasmal growth and help to rapidly diminish symptoms, but they do not stop the shedding of viable mycoplasmas. Patients frequently experience relapses if treatment is not continued for 14 to 21 days. Preventive measures include controlling contamination of fomites, avoiding contact with droplet nuclei, and reducing aerosol dispersion.

Other Mycoplasmas

Mycoplasma hominis and *Ureoplasma urealyticum* are regarded as weak, sexually transmitted pathogens. They are frequently encountered in samples from the urethra, vagina, and cervix of newborns and adults. These species initially colonize an infant at birth and subsequently diminish through early and late childhood. A second period of colonization and persistence is initiated by the onset of sexual intercourse. Evidence linking genital mycoplasmas to human disease is substantial and growing every year. *Ureaplasma urealyticum* is implicated in some types of nonspecific or nongonococcal urethritis and prostatitis. *Mycoplasma hominis* is more often associated with vaginitis, pelvic inflammatory disease, and kidney inflammation. Evidence suggests that mycoplasmas are involved in postpartum or postabortion fever, in male and female infertility, and in disruptions to the fetus and newborn.

The Most Deadly *Mycoplasma* of All? A newly discovered *Mycoplasma* species has stunned the scientific community and promises to change our view of their pathogenicity. This intracellular pathogen, tentatively named *M. incognitus*, was isolated from the spleen, liver, brain, and blood of AIDS patients. At first it appeared to be a potent cofactor accounting for rapid deterioration and cell death in some AIDS patients; however, researchers have now found that it is capable of causing a fatal infection on its own. This mycoplasma has been associated with aggressive flu-like illness in persons who are negative for HIV infection, and it appears to suppress the immune system on its own. Microbiologists from around the world are focussing their attention on this novel and deadly agent.

Bacteria That Have Lost Their Cell Walls

Exposure of typical walled bacteria to certain drugs (penicillin) or enzymes (lysozyme) may result in wall-deficient bacteria called L forms or L-phase variants (see figure 3.17). L forms are induced or occur spontaneously in numerous species and may even become stable and reproduce themselves, but they are not naturally related to mycoplasmas.

L Forms and Disease

The role of certain L forms in human and animal disease is a distinct possibility, but proving etiology has been complicated because infection cannot be followed in accordance with Koch's postulates. One theory proposes that antimicrobic therapy with cell-wall active agents induces certain infectious agents to become L forms. In this wall-free state, they resist further treatment with these drugs and remain latent until the therapy ends, at which time, they reacquire walls and resume their pathogenic behavior.

Infections with L-phase variants of group A streptococci, *Proteus*, and *Corynebacterium* have been reported, though they are uncommon. In a number of chronic pyelonephritis and endocarditis cases, cell-wall-deficient bacteria have been the only isolates. Research on persons with a chronic intestinal syndrome called Crohn's disease has uncovered a strong association with wall-deficient mycobacteria. The etiologic agent of **cat-scratch disease** (CSD), an infection connected with a cat scratch or some other contact with cats, has long been a mystery. Infection begins as a raised red nodule at the site of inoculation and goes on to produce long-term regional lymphadenopathy (figure 17.26*a*). If the portal of entry is the eye, a special type of conjunctivitis may be initiated. Microbiologists at the Armed Forces Institute of Pathology have isolated and tentatively identified the causative bacterium, a tiny pleomorphic gram-negative rod (the CSD bacillus; figure 17.26*b*). Closer examination of this bacillus indicates that it goes through an L-phase variation during infection, but connecting the final link, including verification by Koch's postulates, is the object of continuing research.

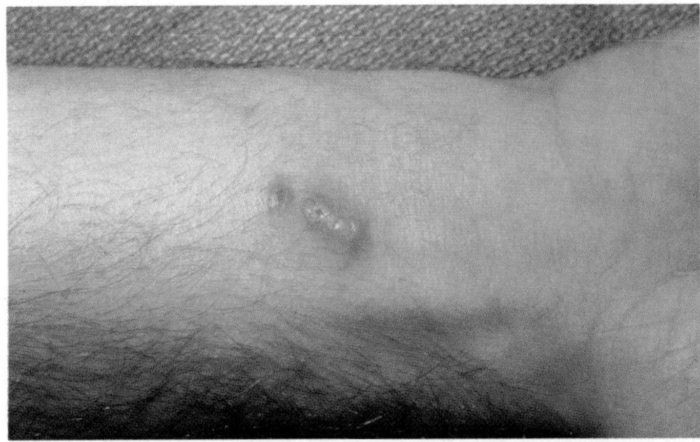

(a)

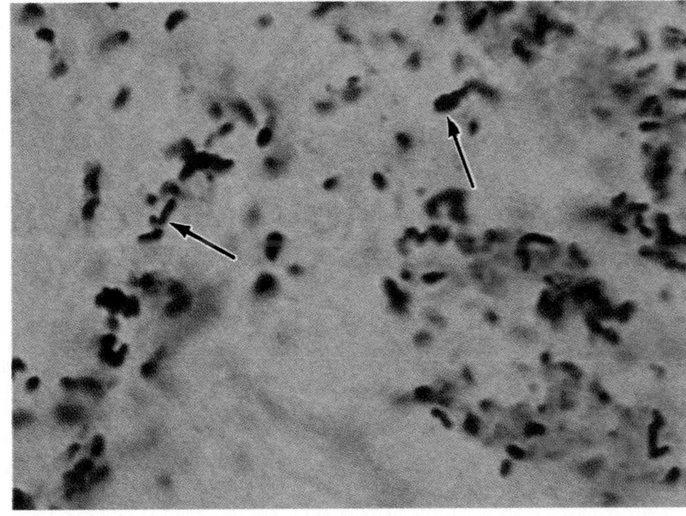

(b)

Figure 17.26 Cat-scratch disease. (*a*) A primary nodule appears at the site of the scratch in about 21 days. In time, large quantities of pus collect, and the regional lymph nodes swell. (*b*) Biopsy samples stained by the Warthin/Starry method show pleomorphic bacilli (arrows) having the characteristics of L forms.

Bacteria in Dental Disease

The relationship between humans and their oral microflora is so complex that, on a small scale, it is a type of ecosystem. The mouth contains a diversity of surfaces for colonization, including the tongue, teeth, gingiva, palate, and cheeks, and it provides numerous aerobic, anaerobic, and microaerophilic microhabitats for the estimated 1,000 different oral species with which humans coexist. This oral ecosystem is greatly enriched by the periodic infusion of food. In most humans, the oral environment remains in balance with little adverse effect, but in persons with poor or nonexistent oral hygiene, it teeters constantly on the brink of disease.

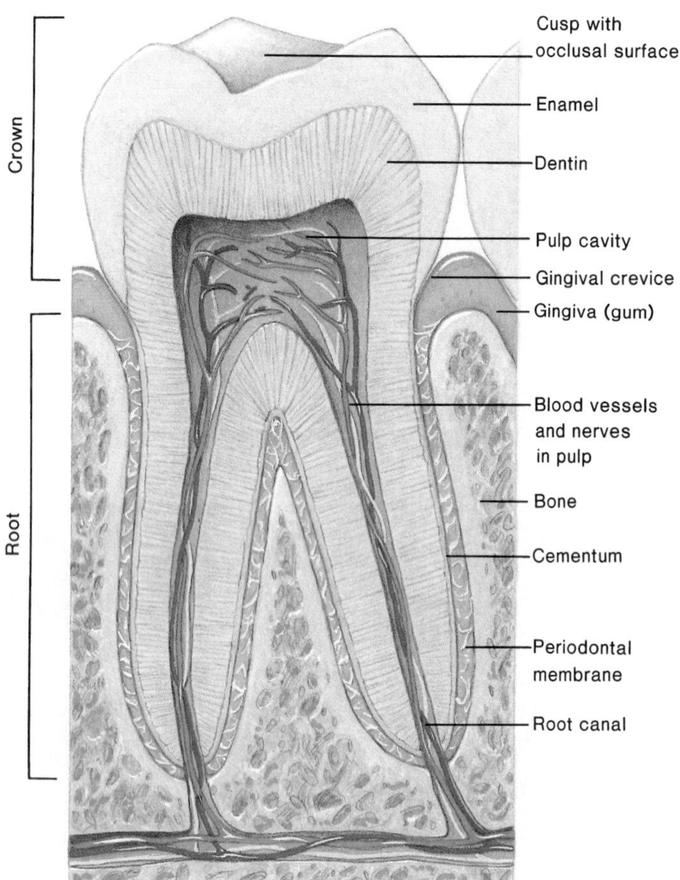

Figure 17.27 The anatomy of a tooth.

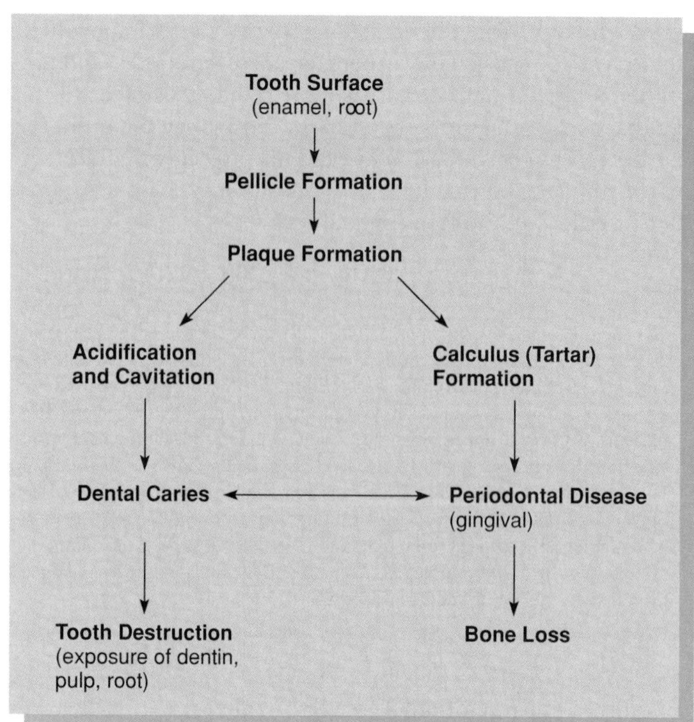

Figure 17.28 Summary of the events leading to dental caries, periodontal disease, and bone loss.

The Structure of Teeth and Associated Tissues

Most dental diseases are focussed on the **tooth** and its surrounding supportive structures or **periodontium** (per″-ee-oh-don′-shee-um) as shown in figure 17.27. A tooth is composed of a **crown** that protrudes above the gum and a **root** that is inserted into a bony socket. The outer surface of the crown is protected by a dense coating of **enamel,** an extremely hard, noncellular material composed of tightly packed rods of $CaPO_4$ that cannot be replaced once the tooth has matured. The root is surrounded by a layer of cementum, which anchors the tooth to the periodontal membrane that lines the socket. The major portion of the tooth inside the crown and root is composed of a highly regular calcified material called dentin, and the very core contains a pulp cavity that supplies the living tissues with blood vessels and nerves. The root canal is the portion of the pulp that extends into the roots. Places where the bone protrudes toward the crown are covered by connective tissue and a mucous membrane called the **gingiva** or gum. Important sites for initial dental infections are the enamel, especially the cusps, and the crevice or sulcus formed where the gingiva meets the tooth.

Dental pathology generally affects both hard and soft tissues (figure 17.28). Although both categories of disease are initiated when microbes adhere to the tooth surface and produce dental plaque, their outcomes vary. In the case of dental caries, the gradual breakdown of the enamel leads to invasive disease of the tooth itself, whereas in soft-tissue disease, calcified plaque damages the soft gingival tissues and predisposes them to bacterial invasion. Both diseases are responsible for the loss of teeth, though dental caries are usually implicated in children, and periodontal infections in adults.

Dental Caries: Hard-Tissue Disease

Dental caries (kar′-eez) is the most common human disease. It is a complex mixed infection of the dentition that gradually destroys the enamel and often lays the groundwork for the destruction of deeper tissues. It occurs most often on tooth surfaces that are less accessible and harder to clean and on those that provide pockets or crevices where bacteria can cling. Caries commonly develop on enamel pits and fissures, especially those of the occlusal surfaces, though they may also occur on the smoother crown surfaces and subgingivally on the roots.

Over the years, several views have been put forth to explain how dental caries originate. At various times it has been believed that sugar, microbes, or acid cause teeth to rot; however, studies with germ-free animals have now shown that no single factor can account for caries (see figure 1.8 and chapter 11). Caries development occurs in many phases and requires multiple interactions involving the anatomy, physiology, diet, and bacterial flora of the host. The principal stages in the formation of dental caries are pellicle formation, plaque formation, acid production and localization, and enamel etching (figure 17.29*a*).

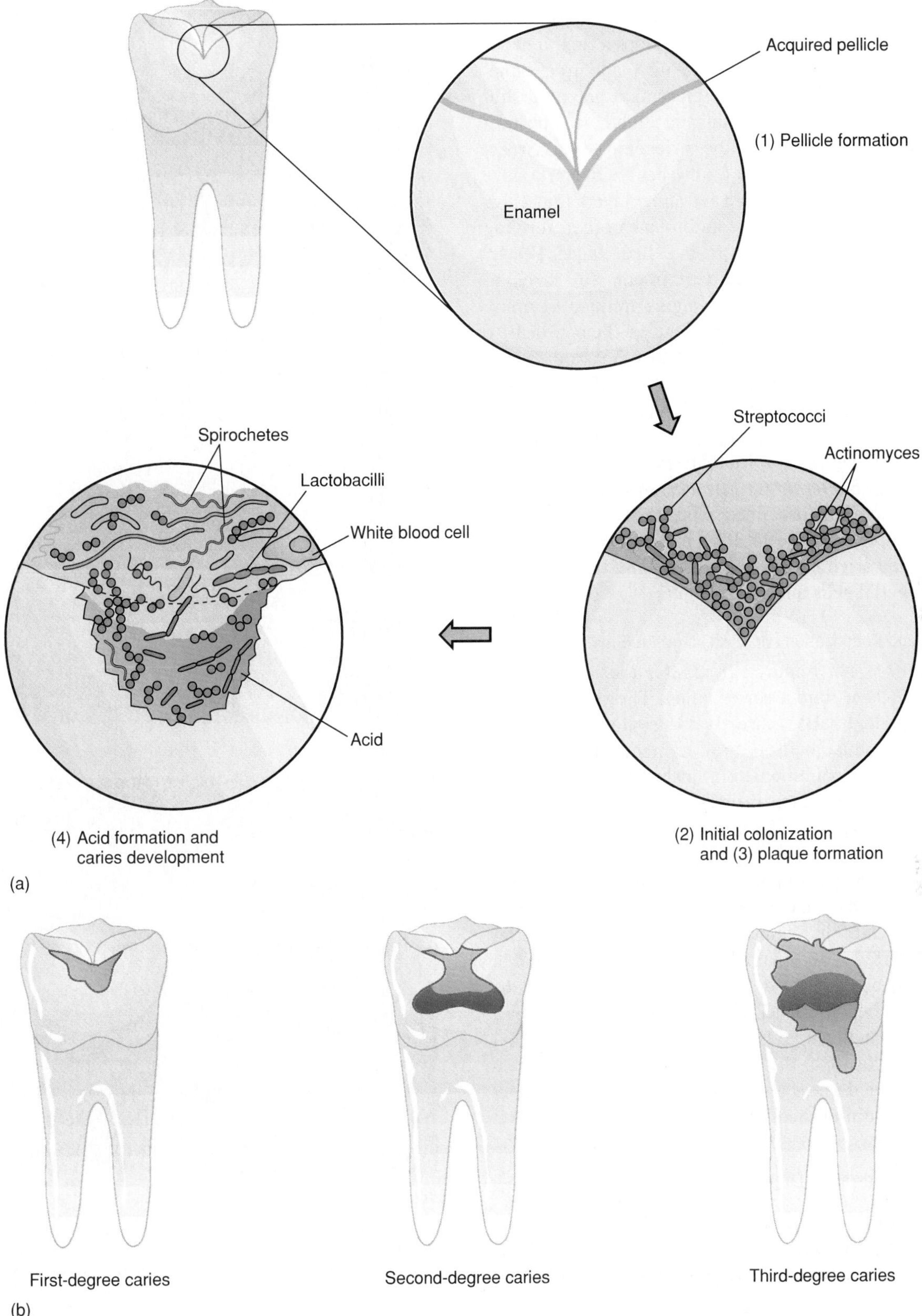

Figure 17.29 Stages in plaque development and cariogenesis. (*a*) A microscopic view of pellicle and plaque formation, acidification, and destruction of tooth enamel. (*b*) Progress and degrees of cariogenesis.

Plaque Formation

A freshly cleaned tooth is a perfect landscape for colonization by microbes. Within a few moments, it develops a thin, mucous coating called the **acquired pellicle,** which is made up of adhesive salivary proteins. This structure presents a potential substrate upon which certain bacteria first gain a foothold. The most prominent pioneering colonists are the cariogenic genera *Streptococcus* and *Actinomyces.* These gram-positive bacteria have adhesive receptors such as fimbriae and slime layers that allow them to cling to the tooth surfaces and to each other, forming a foundation for the dense whitish mass called **plaque.** Fed by a diet high in sucrose, glucose, and certain complex carbohydrates, *Streptococcus mutans* and *S. sanguis* produce a gummy polymer of glucose called dextran that helps them attach to smooth enamel surfaces, holds the plaque together, and adds to its bulk (see chapter 15 opening illustration). As these primary invaders continue to grow and plaque begins to build up, the scene is set for the invasion and aggregation of additional species of bacteria. Among the species that are secondary invaders are *Lactobacillus, Bacteroides, Fusobacterium,* and *Treponema.* Plaque is especially prominent when the teeth are stained with disclosing tablets (figure 17.30*a*), and a microscopic view reveals a rich and varied network of bacteria and their products along with epithelial cells and fluids (figure 17.30*b*).

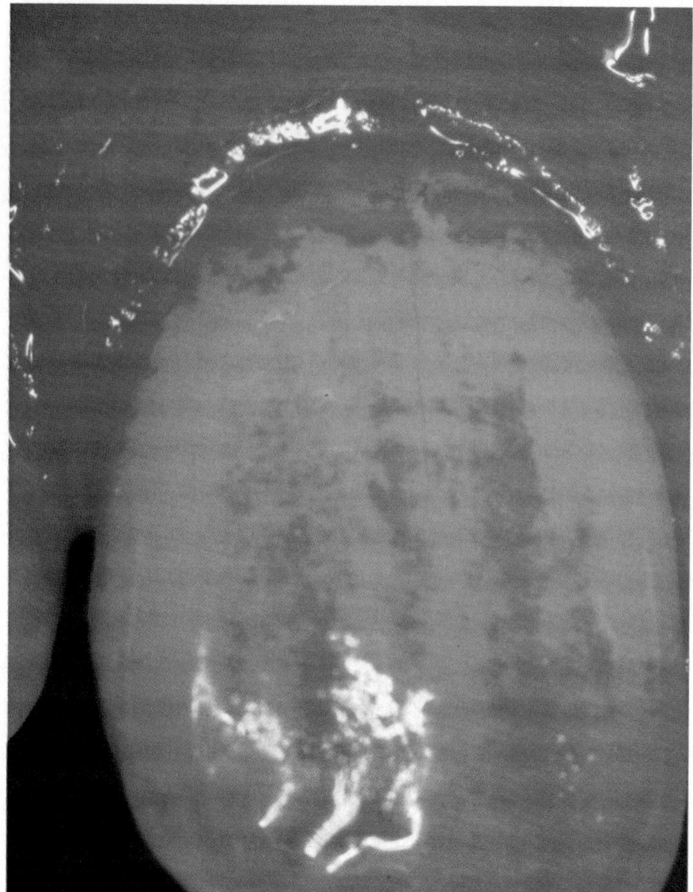

(a)

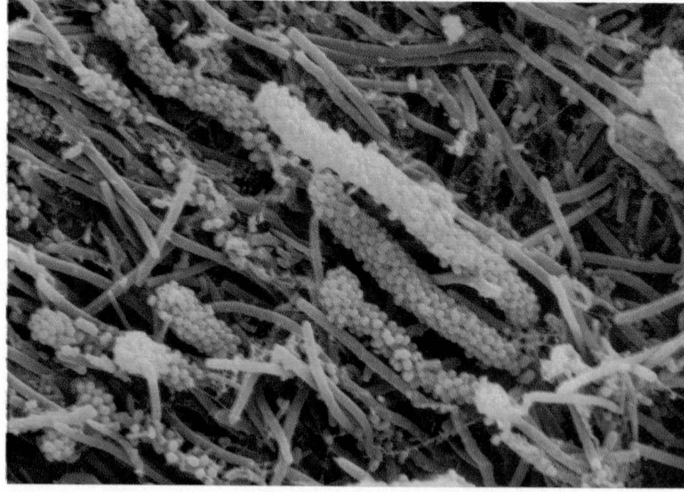

(b)

Figure 17.30 The macroscopic and microscopic appearance of plaque. (*a*) Tablets containing vegetable dye stain heavy plaque accumulations at the junction of the tooth and gingiva. (*b*) Scanning electron micrograph of plaque with long filamentous forms and "corn cobs" that are mixed bacterial aggregates.

Acid Formation and Localization and Etching of the Enamel

If mature plaque is not removed from sites that readily trap food, it usually evolves into a caries lesion. The role of plaque in caries development relates directly to streptococci and lactobacilli, which produce acid as they ferment dietary carbohydrates. If this acid is immediately flushed from the plaque and diluted in the mouth, it has little effect. However, the dense nature of plaque holds the acid against the enamel surface and prevents it from diffusing out. Eventually, acid accumulation lowers the pH of the environment adjacent to the enamel to below 5, which is acidic enough to begin to dissolve (decalcify) the inorganic $CaPO_4$ of the enamel in that spot. This initial lesion may remain localized in the enamel (first-degree caries) and be repaired with various inert materials (fillings). Once the deterioration has reached the level of the dentin (second-degree caries), tooth destruction speeds up, and the tooth may be rapidly destroyed. Exposure of the pulp (third-degree caries) is attended by severe tenderness and toothache, and the chance of saving the tooth is diminished (see figure 17.29*b*).

Soft-Tissue (Periodontal) Disease

Periodontal disease is so common that 97–100% of the population has some degree of it by age 45. Most kinds are due to bacterial colonization and varying degrees of inflammation that

---

plaque (plak) Fr. a patch.

occur in response to gingival damage. The most common predisposing condition occurs when the plaque becomes mineralized (calcified) with calcium and phosphate crystals. This process produces a hard, porous substance called **calculus** on both the sub- and supra-gingival tooth surfaces (figure 17.31). Supra-gingival calculus is usually not harmful and may even protect the tooth from caries, but its presence in the relatively sheltered gingival sulcus initiates a pathological lesion that leads to periodontal disease.

Calculus accumulating in the gingival sulcus irritates the delicate gingival membrane, and the chronic trauma to this area causes a pronounced inflammatory reaction. This event in turn creates an avenue of invasion for a variety of normal flora. These bacteria are primarily anaerobic and gram negative, and include *Veillonella* (cocci), *Bacteroides melaninogenicus* (rods), *Fusobacterium* (spindle-shaped rods), and numerous spirochetes. In response to the mixed infection, the damaged area becomes infiltrated by neutrophils and macrophages and later by lymphocytes, which cause further inflammation and tissue damage (figure 17.32). The initial signs of **gingivitis** are swelling, loss of normal contour, patches of redness, and increased bleeding of the gingiva. Spaces or pockets of varying depth also develop between the tooth and the gingiva. If this condition persists, a more serious disease called **periodontitis** results. This is the natural extension of the disease into the periodontal membrane and cementum. The deeper involvement increases the size of the pockets and can cause bone resorption severe enough to loosen the tooth in its socket. If allowed to progress, the tooth may be lost.

Another serious periodontal disease is **acute necrotizing ulcerative gingivitis** (ANUG), formerly called trench mouth or Vincent's disease. This disease is a synergistic infection involving *Treponema vincentii* and other spirochetes and fusobacteria that are sufficiently aggressive to invade the gingival tissues and severely damage them (figure 17.32*d*). It comes on rapidly and is associated with severe pain, bleeding, pseudomembrane formation, and necrosis. This condition usually results from poor oral hygiene, altered host defenses, or prior gum disease, and is not communicable. It responds well to broad-spectrum antibiotics.

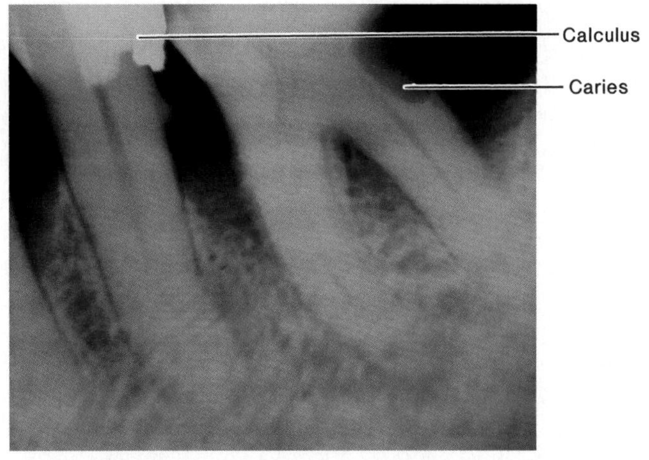

(a)

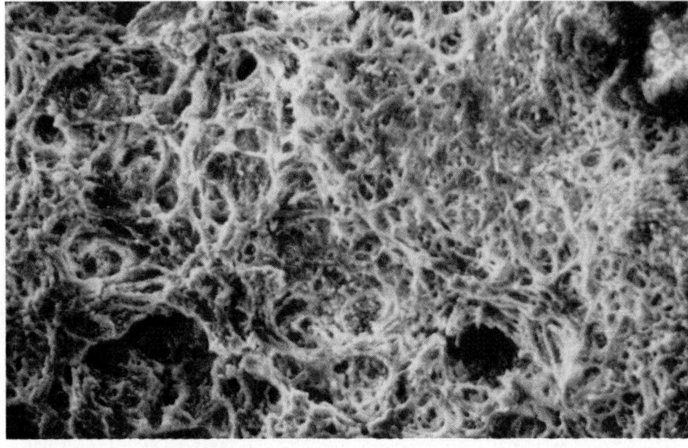

(b)

Figure 17.31 The nature of calculus. (*a*) Radiograph of mandibular premolar and molar, showing calculus on the top and a caries lesion on the right. Bony defects caused by periodontitis affect all three teeth. (*b*) Scanning electron micrograph revealing the rough, dense, and porous nature of calculus (×500).

Factors in Dental Disease

Nutrition and eating patterns have a profound effect on oral diseases. Persons whose diet is high in refined sugars (sucrose, glucose, and fructose) tend to have more caries, especially if these foods are eaten constantly throughout the day without brushing the teeth. The practice of putting a baby down to nap with a bottle of fruit juice or formula may lead to rampant dental caries ("nursing bottle caries"). In addition to diet, numerous anatomic, physiologic, and hereditary factors influence oral diseases. The structure of the tooth enamel may be influenced by genetics and by environmental factors such as fluoride, which strengthens the enamel bonds. The amount of calcium available to developing fetuses and neonates can also affect tooth development. Saliva that is more fluid in consistency inhibits oral colonization, and its content of inhibitory factors such as antibodies and lysozyme can also help prevent dental disease.

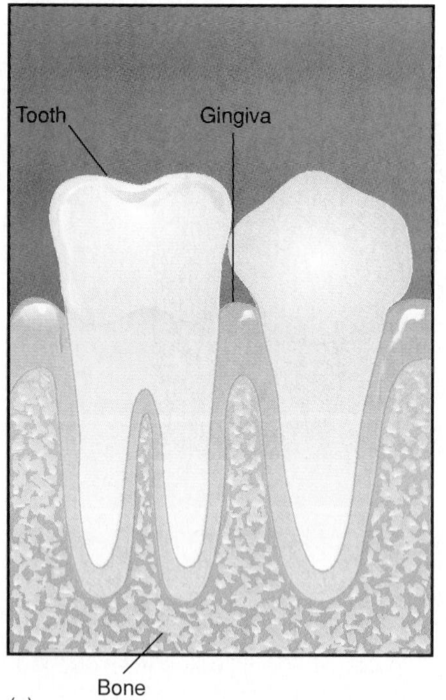

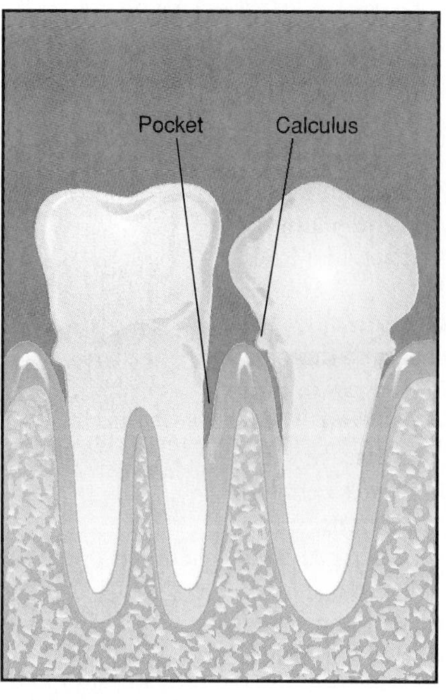

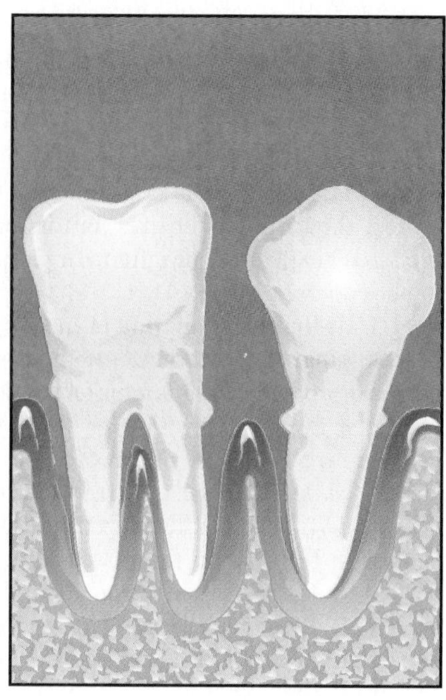

(d)

Figure 17.32 Stages in soft-tissue infection and gingivitis. (*a*) The relationship of tooth, gingiva, and bone in the nondiseased state. (*b*) Calculus buildup and early gingivitis. (*c*) Late stage disease, tissue destruction, deep pocket formation, and bone loss. (*d*) A sample of exudate from a gingival pocket stained with crystal violet (×560). Note the numerous spirochetes and fusiform bacilli.

The greatest control of dental diseases is preventive dentistry, including regular brushing and flossing to remove plaque, because stopping plaque buildup automatically reduces caries and calculus production. Once calculus has formed on teeth, it cannot be removed by brushing but can only be dislodged by special mechanical procedures (scaling) in the dental office. One of the most exciting prospects is the possibility of a vaccine to protect against the primary colonization of the teeth. Some success in inhibiting plaque formation has been achieved in experimental animals with vaccines raised against the whole cells of *S. mutans* and the fimbriae of *Actinomyces viscosus.*

Effects of Mouthwashes Mouthwashes are relatively ineffective in controlling plaque formation because of the high bacterial content of saliva and the relatively short acting time of the mouthwash. The federal government has recently blocked advertisements for certain mouthwashes that claim to prevent plaque formation. Dentists are also concerned about the high alcohol content of some mouthwashes, which could predispose patients to oral cancer.

Chapter Review with Key Terms

Miscellaneous Bacterial Infections

Spirochetes

Spirochetes are helical, flexible bacteria that move by axial filaments. Several ***Treponema*** species are obligate parasitic spirochetes with 8–12 regular spirals; best observed under dark-field microscope; cause **treponematoses;** direct observation of treponeme in tissues and blood tests important in diagnosis; cannot be cultivated in artificial media; treatment by large doses of penicillin or tetracycline.

Major Treponematoses

Syphilis: ***Treponema pallidum*** is agent; very sensitive pathogen that causes complex progressive disease in adults and children. **Sexually transmitted syphilis** is acquired through close contact with an infected cutaneous surface; untreated sexual disease occurs in stages over long periods; many symptoms mimic other diseases. At site of entrance, multiplying treponemes produce a **primary lesion,** a hard ulcer or **chancre,** which disappears as the microbe becomes systemic. **Secondary syphilis** occurs when spirochete infects many organs; marked by skin rash, fever, damage to mucous membranes. Both primary and secondary syphilis are communicable; after latent period of months to years, the pathogen is established in tissues. Final, non-communicable **tertiary stage** is marked by tumors called **gummas** and life-threatening cardiovascular and neurological effects.

Congenital syphilis is acquired transplacentally; disrupts embryonic and fetal development; survivors may have respiratory, skin, bone, teeth, eye, and joint abnormalities if not treated.

Nonsyphilitic Treponematosis: Slow progressive cutaneous and bone diseases endemic to specific regions of tropics and subtropics; usually transmitted under unhygienic conditions in rural areas. **Bejel** is a deforming childhood infection of the mouth, nasal cavity, body, and hands; **yaws** occurs from invasion of skin cut, causing a primary ulcer that seeds a second crop of lesions; **pinta** is superficial skin lesions that depigment and scar the skin.

Leptospira interrogans

L. interrogans has very regular coils and a prominent hook; causes **leptospirosis,** a worldwide zoonosis acquired through contact with urine of wild and domestic animal reservoirs. Spirochete enters cut, multiplies in blood and spinal fluid, causes muscle aches, headache; second phase is marked by Weil's syndrome, which involves kidneys and liver; requires early treatment with penicillin or tetracycline.

Borrelia

Borrelia is a loose, irregular spirochete that causes **borreliosis**; infections are vector-borne (mostly by ticks). *B. hermsii* is zoonotic (in wild rodents) and carried by soft ticks; *B. recurrentis* has a strictly human reservoir and is carried by lice; both species cause relapsing fever. Borrelias are introduced into blood and multiply; fever and other symptoms recur because the spirochete repeatedly changes antigenically and forces the immune system to keep adapting; this may occur several times.

Borrelia burgdorferi: A zoonosis carried by mice and spread by a hard tick (*Ixodes*) that lives on deer and mice; causes **Lyme disease,** a syndrome that occurs endemically in several regions of the U.S. Tick bite leads to fever and a prominent ring-shaped rash that migrates; if allowed to progress, may cause cardiac, neurological, and arthritic symptoms; can be controlled by antibiotics and by preventing tick contact.

Curved Bacteria (Vibrios)

Vibrios are short spirals or sausage-shaped cells with polar flagella.

Vibrio cholerae: Causes **epidemic cholera,** a human disease that originated in Asia but is now distributed worldwide. Organism is free-living in soil and water and is ingested in contaminated food and water; microbe infects the surface of epithelial cells in small intestine, is not invasive; severity of cholera due to potent **choleragen** toxin that causes the cells to actively release electrolytes and passively lose water, resulting in **secretory diarrhea**; electrolyte loss and dehydration lead to muscle, circulatory, and neurologic symptoms and death; treatment with oral rehydration (electrolyte and fluid replacement) completely restores stability; vaccine available.

Vibrio parahaemolyticus: Causes food infection associated with seafood; organism lives in seawater and is prevalent during warm months; symptoms similar to mild cholera; for prevention, food must be well cooked and refrigerated during storage.

Campylobacter Species: Vibrios with two polar flagella. ***C. jejuni*** is a common cause of severe **gastroenteritis** worldwide; acquired through food or water contaminated by animal feces; enteritis is due to enterotoxin with symptoms like cholera; ***C. fetus*** causes diseases in pregnant women and fatal septicemia in neonates; ***Helicobacter pylori*** may be etiologically involved in diseases of the stomach lining such as gastritis and ulcers.

Obligate Intracellular Parasitic Bacteria

Rickettsiae and **chlamydiae** are tiny, gram-negative rods or cocci that require certain metabolic elements of the host cell to multiply; diseases treatable with tetracycline and chloramphenicol. ***Rickettsia, Coxiella, Rochalimaea*** are genera that cause **rickettsioses;** most are zoonoses spread by arthropod vectors.

Characteristics of Ectoparasitic, Blood-sucking Arthropod Vectors: Transmitters of parasites between various vertebrate reservoirs; life cycles may be complex; human usually not a major participant in life cycle; vectors may pass pathogen to offspring transovarially or to hosts through bite, feces, or mechanical injury and scratching.

Ticks are arachnids; hard ticks have a hard shield and soft ticks lack it; ticks devoid of blood are small, can live in environment for many years; cling to host, feed, and inoculate the host with saliva containing the pathogen; carry Rocky Mountain spotted fever, Q fever, borrelioses, viral fevers.

Mites and chiggers are tiny arachnids that live on wild vegetation and nibble on various hosts, including humans if given the chance; bite is painless, causes itchiness; carry rickettsialpox and scrub typhus.

Fleas are flattened insects with jumping legs and a blood-probing mouth; extremely resistant and nonspecific to host; infect by defecating on injured skin; carry murine typhus, bubonic plague.

Lice are insects that cling to body hair and gently pierce skin; infection occurs when louse is crushed by scratching and rubbed into skin; carry epidemic typhus, trench fever, and relapsing fever.

Epidemic Typhus: Caused by ***Rickettsia prowazekii;*** carried by lice; associated with overcrowding; probably not a zoonosis; disease starts with high fever, chills, headache; rash occurs; Brill-Zinsser disease is a chronic, recurrent form.

Endemic (Murine) Typhus: Zoonosis caused by ***R. typhi,*** harbored by mice and rats; occurs in areas of high flea infestation; not common; like epidemic form, but symptoms milder.

Scrub Typhus: Caused by ***R. tsutsugamushi,*** a zoonosis of Japan and Asia; transmitted by chiggers; causes fever, rash; often fatal.

Rocky Mountain Spotted Fever: Etiologic agent is ***R. rickettsii;*** zoonosis carried by dog and wood ticks (*Dermacentor*), which live in blood of various mammals; most cases on eastern seaboard; infection causes distinct spotted, migratory rash; acute reactions include heart damage, CNS damage, gangrene; can be fatal; disease prevented by blocking ticks' access to body and inspecting for and carefully removing them.

Q Fever: Caused by ***Coxiella burnetii***; agent has unusual life cycle, with a resistant spore form that can survive out of host; a zoonosis of domestic animals; transmitted by air, dust, milk, ticks; risks include working with animals or drinking unpasteurized milk; usually inhaled, causing pneumonitis, fever, hepatitis, but no rash.

Chlamydia

Organisms in the genus *Chlamydia* pass through a transmission phase involving a hardy **elementary body** and an intracellular **reticulate body** that has pathological effects.

Chlamydia trachomatis: A strict human pathogen that causes eye diseases and STDs. **Ocular trachoma** is a severe infection that deforms the eyelid and cornea and may cause blindness. **Conjunctivitis** occurs in babies following birth; is associated with maternal infection; prevented by ocular prophylaxis after birth. *C. trachomatis* is a very common bacterial STD pathogen; causes nongonococcal urethritis in males, and cervicitis, salpingitis (PID), infertility, and scarring in females; causes **lymphogranuloma venerum,** a disfiguring disease of the external genitalia and pelvic lymphatics.

Chlamydia psittaci: Causes **ornithosis,** a zoonosis transmitted to humans through respiratory discharges and feces of bird vectors; highly communicable among all birds; pneumonia or flu-like infection with fever, lung congestion.

Mollicutes/Mycoplasmas

Mycoplasmas are naturally cell-wall-deficient bacteria; highly pleomorphic cells; not obligate parasites but require special lipids; fuse tightly to host membranes during infection, are difficult to dislodge; diseases treated with tetracycline, erythromycin.

Mycoplasma pneumonia: Causes **primary atypical pneumonia**; pathogen slowly spreads over interior respiratory surfaces; symptoms not pronounced, may be fever and chest pain, sore throat, no cough; *M. hominis* and *Ureaplasma urealyticum* are weak STDs; normal colonists of most persons; may cause urethritis, PID, and other reproductive tract diseases.

L Forms: Bacteria that normally have cell walls but have transiently lost them through drug therapy; may be involved in **cat-scratch disease,** a zoonosis of cats.

The Role of Mixed Infections in Dental Disease

The oral cavity contains hundreds of microbial species that participate in an ecosystem involving interactions between themselves, the human host, and the nutritional role of the mouth; it is continuously vulnerable to infection and disease.

Hard-Tissue Disease

Dental caries is a slow, progressive infection of irregular areas of enamel surface; begins with colonization of tooth by dextran-forming species of *Streptococcus* and cross adherence with *Actinomyces*; process forms layer of thick, whitish material called **plaque**; this complex bed of material harbors dense masses of bacteria and extracellular substances; acid formed by agents in plaque is held in close proximity against enamel and dissolves the inorganic salts; if unabated, this produces a deep caries lesion, which may invade dentin and root canal and destroy tooth.

Soft-Tissue Disease

Periodontal disease involves **periodontium** (**gingiva** and surrounding tissues); also begins when plaque forms on tooth (root surface below gingiva) and is mineralized to a hard concretion called **calculus**; this irritates tender gingiva; inflammatory reaction and swelling create **gingivitis** and pockets between tooth and gingiva invaded by bacteria (spirochetes and gram-negative bacilli); immune reaction further damages site; tooth socket may be involved (**periodontitis**), and the tooth may be lost.

True–False Questions

Determine whether the following statements are true (T) or false (F). If you feel a statement is false, explain why, and reword the sentence so that it reads accurately.

____ 1. *Treponema* can be cultured on blood agar.
____ 2. The treponematoses are zoonoses.
____ 3. A gumma is a syphilitic tumor.
____ 4. Penicillin is no longer used to treat syphilis because of drug resistance.
____ 5. Lyme disease is a borreliosis spread by hard ticks.
____ 6. Relapsing fever is a borreliosis spread mostly by lice.
____ 7. *Vibrio cholerae* is a highly invasive pathogen.
____ 8. The best therapy for cholera is oral antibiotics.
____ 9. Rickettsias and chlamydias are obligate intracellular parasites.
____ 10. Ornithosis is caused by a chlamydia carried in the saliva of cats.
____ 11. Mycoplasmas develop close associations with the cell membranes of host cells.
____ 12. Dental diseases are mixed infections.

Concept Questions

1. Describe the characteristics of *T. pallidum* that relate to its transmission; what is responsible for the current epidemic of syphilis?
2. Describe the stages of untreated syphilis infection. Where does the chancre occur, and what is in it? What is happening as the chancre disappears? Which stages are symptomatic, and which are infectious? What is latency? For which tissues does the spirochete have an affinity?
3. Describe the conditions leading to congenital syphilis and the long-term effects of the disease.
4. Describe the nonspecific and specific tests for syphilis. What do they test for? Why are they so important?
5. Give the general characteristics of bejel, yaws, and pinta.
6. Describe the epidemiology and pathology of leptospirosis.
7. Describe the two major borrelioses. How are arthropods involved? Why are there relapses in relapsing fever? What does it mean for a patient to be borrelemic or rickettsemic? How do these conditions relate to the role of vectors in the spread of disease?
8. Trace the route of the infectious agent from a tick bite to infection. Do the same for lice.
9. Overview the natural history of Lyme disease and its symptoms.
10. Briefly, what is the natural history of cholera? What is its principal pathological feature?
11. Why does a cholera patient lose so much water? What is secretory diarrhea? How does oral rehydration therapy work?
12. Briefly describe the food infection by *V. parahaemolyticus* and the diseases of *Campylobacter*.
13. What do rickettsias and chlamydias derive from the host? How do antibiotics work to control them?
14. Describe the life cycles of *Rickettsia prowazekii, Rickettsia rickettsii,* and *Coxiella burnetii.* What are the predisposing factors for each type of disease? What makes *Coxiella* unique among rickettsias? Why are dogs so important in transmission of RMSF? What are the general symptoms of rickettsial infections?
15. Compare the elementary body and the reticulate body of chlamydias. How are chlamydias transmitted? Describe the major complications of eye infections and STDs.
16. Why are the symptoms of mycoplasma pneumonia so mild? Why doesn't penicillin work on mycoplasma infection?
17. In what ways are dental diseases mixed infections? Discuss the major factors in the development of dental caries and periodontal infections.
18. Matching. Match each disease in the first column with its vector (or vectors) in the second column.

Disease	Vector
___ Leptospirosis	a. wild animals
___ Lyme disease	b. flea
___ Murine typhus	c. tick
___ Ornithosis	d. birds
___ Relapsing fever	e. louse
___ Lymphogranuloma venereum	f. mite
___ Cat-scratch disease	g. domestic animals
___ Epidemic typhus	h. none of these
___ Rocky Mountain spotted fever	
___ Q fever	
___ Scrub typhus	
___ Cholera	

19. Matching. Match each disease in the first column with its portal of entry in the second column.

Disease	Portal of Entry
___ Q fever	a. skin
___ Ornithosis	b. mucous membrane
___ Dental caries	c. respiratory
___ ANUG	d. urogenital
___ Mycoplasma	e. eye
___ Syphilis	f. oral
___ Leptospirosis	g. gastrointestinal
___ Lymphogranuloma venereum	
___ Cholera	
___ Lyme disease	
___ Trachoma	
___ *Campylobacter* infection	

Practical/Thought Questions

1. Why is it so difficult to trace the historical origin of disease, as in syphilis?
2. Why does syphilis have such profound effects on the human body? Why is long-term immunity so difficult to achieve?
3. How can congenital syphilis be prevented?
4. How are the nonsyphilitic treponematoses similar to syphilis?
5. In view of the fact that cholera causes the secretion of electrolytes into the intestine, explain what causes the loss of water. What are the principles of osmosis behind this phenomenon?
6. What would be the best type of vaccine for cholera?
7. Explain the general relationships of the vector, the reservoir, and the agent of infection.
8. Humans are accidental hosts in many vector-borne diseases. What does this say about the vector and the microbial agent?
9. Why can a louse cause infection only once? Why must it be crushed in order to cause infection? What kind of vector can infect many individuals? How do these vectors infect their hosts?
10. Why is arthropod vector control so difficult? Summarize the methods of preventing arthropod-borne disease.
11. Which bacteria presented in this chapter can be cultivated on artificial media? Which require embryos or cell culture?
12. Name four bacterial diseases for which the dark-field microscope is an effective diagnostic tool.
13. Explain how L forms could be involved in disease.
14. In what way is the oral cavity an ecological system? What makes it go out of balance? What are some logical ways to prevent dental disease besides removing plaque?
15. A woman journalist returning from a trip experienced severe fever, vomiting, chills, and muscle aches, followed by symptoms of meningitis and kidney failure. Early tests were negative for septicemia; throat cultures were negative; and penicillin was an effective treatment. Doctors believed the patient's work in the jungles of South America was a possible clue to her disease. What do you think might have been the cause?
16. A woman went for a hike in the mountains of New York State and later developed fever and a rash. What two totally different diseases might she have contracted, and what could have been the circumstances of infection?

CHAPTER 18

Fungal Diseases

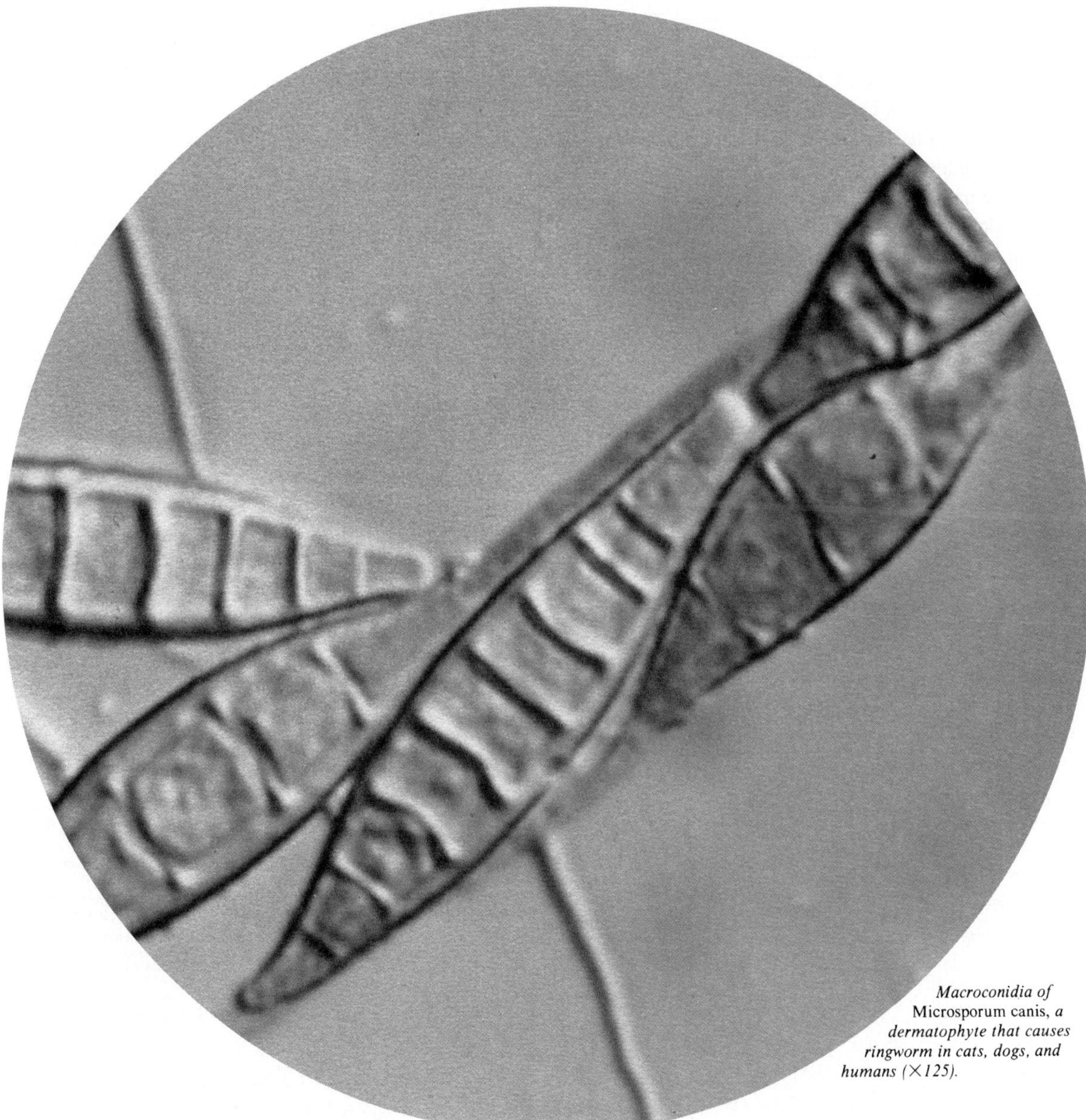

Macroconidia of Microsporum canis, *a dermatophyte that causes ringworm in cats, dogs, and humans (×125).*

Chapter Preview

The eucaryotic microbes collectively called fungi were introduced in chapter 4. The profound importance of fungi stems primarily from their role in the earth's ecological balance and their impact on agriculture. To a lesser—but not minor—extent, the fungi are also medically significant, as agents in human disease, allergies, and mycotoxicoses (intoxications due to ingesting fungal toxins). Diseases resulting from fungal infections, primarily by yeasts and molds, are termed mycoses. Because of their prevalence, mycotic infections will be the main focus of this chapter, with the less common respiratory allergies to fungi and mycotoxicoses briefly discussed at the end.

Fungi As Infectious Agents

Molds and yeasts are so widely distributed in air, dust, fomites, and even among the normal flora that humans are incessantly exposed to them. The fact that the planet's surface is totally dusted with spores led W. B. Cooke to christen it "our moldy earth." Fortunately, because of the relative resistance of humans and the comparatively nonpathogenic nature of fungi, most exposures do not lead to overt infection. Of the estimated 100,000 fungal species, only about 300 have been linked to disease in animals, though among plants, fungi are the most common and destructive of all pathogens. Human mycotic disease (*mycosis*) is associated with true or primary fungal pathogens that exhibit some degree of virulence, or with opportunistic or secondary pathogens that take advantage of defective resistance (table 18.1 and table 18.2).

True Versus Opportunistic Pathogens

A **true fungal pathogen** is a species that can invade and grow in a healthy, noncompromised animal host. This behavior is contrary to the metabolism and adaptation of fungi, most of which are inhibited by the relatively high temperature and low oxygen

mycosis (my-koh'-sis) pl. mycoses; Gr. *mykos*, fungi, and *osis*, a disease process.

Table 18.2 Comparison of True and Opportunistic Fungal Infections

	True Pathogenic Infections	Opportunistic Infections
Degree of Virulence	Well developed	Limited
Condition of Host	Resistance high or low	Resistance low
Primary Portal of Entry	Respiratory	Respiratory, mucocutaneous
Nature of Infection	Usually primary, pulmonary, and systemic; usually asymptomatic	Varies from superficial skin to pulmonary and systemic; usually symptomatic
Nature of Immunity	Well-developed, specific	Weak, short-lived
Infecting Form	Primarily conidial	Conidial or mycelial
Shows Thermal Dimorphism	Strongly	Not usually
Habitat of Fungus	Soil	Varies from soil to flora of humans and animals
Geographical Location	Restricted	Distributed worldwide

Table 18.1 Representative Fungal Pathogens, Degree of Pathogenicity, and Habitat

Microbe	Disease/Infection*	Primary Habitat and Distribution
I. Primary Pathogens		
Histoplasma capsulatum	Histoplasmosis	Soils high in bird guano; Ohio/Mississippi valleys of U.S.; Central, South America; Africa
Blastomyces dermatitidis	Blastomycosis	Presumably soils, but isolation has been difficult; southern Canada; Midwest, Southeast, Appalachia in U.S.; along drainage of major rivers
Coccidioides immitis	Coccidioidomycosis	Highly restricted to alkaline desert soils in southwestern U.S. (California, Arizona, Texas, and New Mexico)
Paracoccidioides brasiliensis	Paracoccidioidomycosis	Soils of rain forests in South America (Brazil, Columbia, Venezuela)
II. Pathogens with Intermediate Virulence		
Sporothrix schenckii	Sporotrichosis	In soil and decaying plant matter; widely distributed
Genera of dermatophytes *Microsporum, Trichophyton, Epidermophyton*	Dermatophytosis (various ringworms or tineas)	Human skin, animal hair, soil throughout the world
III. Secondary Pathogens		
Cryptococcus neoformans	Cryptococcosis	Pigeon roosts and other living sites (buildings, barns, trees); worldwide distribution
Candida albicans	Candidiasis	Normal flora of human mouth, throat, intestine, vagina; also normal in other mammals, birds; ubiquitous
Aspergillus spp.	Aspergillosis	Soil, decaying vegetation, grains; common airborne contaminants; extremely pervasive in environment
Genera in Mucorales *Rhizopus, Absidia, Mucor*	Mucormycosis	Soil, dust; very widespread in human habitation

*Specific mycotic infections are usually named by adding *-mycosis*, *-iasis*, or *-osis* to the generic name of the pathogen.

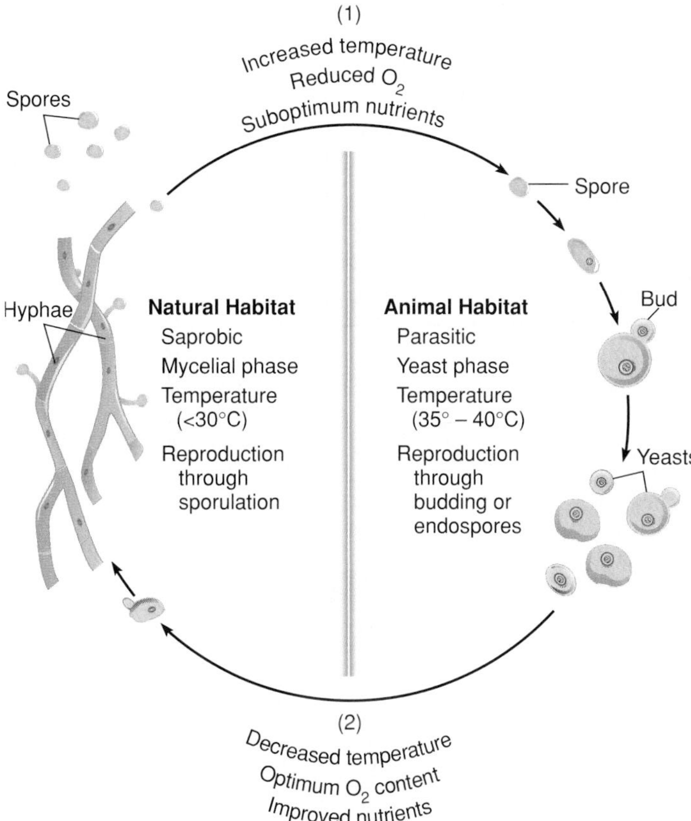

Figure 18.1 The general changes associated with thermal dimorphism, using generic hyphae, spores, and yeasts as examples. (*1*) When fungal spores from the environment gain entrance to a warm-blooded animal, they germinate into yeasts and remain in this phase in the host. (*2*) Yeast cells leaving the animal and returning to the environment revert to the sporulating hyphal state. These conversions can be demonstrated on artificial media in the laboratory.

Feature 18.1 Opportunistic Mycoses As Diseases of Medical Progress

At one time, opportunistic fungal infections in hospitalized patients were rather unusual. Common airborne fungi such as *Rhizopus, Geotrichum,* and *Rhodotorula* were rarely isolated as etiologic agents 25 years ago. Textbooks from the past describe these agents as common contaminants with weak if any pathogenic potential, and infections were considered extreme deviations from the norm.

Older ideas concerning these so-called harmless contaminants were rarely challenged, because immunodeficient and debilitated patients often died from their afflictions long before fungal infections could take hold. However, the advent of innovative surgeries, drugs, and other therapies that maintain such patients for extended periods has dramatically altered our perspective. Such advances in medical technology have significantly increased the survival rates of debilitated patients, but they have also increased the numbers of compromised patients. One clinical dilemma that cannot be completely eliminated, even with rigorous disinfection, is the exposure of such patients to potential fungal pathogens from the air, fomites, and other humans. Up to 4% of all nosocomial infections are caused by opportunistic fungi.

Pathologic states most likely to predispose a person to opportunistic fungal infections are:

1. Immune system disorders caused by cancer. Lymphoma and leukemia are commonly complicated by fungal infections.
2. Immune system disorders due to concurrent infections. AIDS and tuberculosis are notable examples.
3. Immune or endocrinological disruptions arising from therapy for cancer, transplants, and other infections. Immunosuppressive drugs, antibiotics, antineoplastic agents, and radiation are significantly implicated.
4. Organ damage (especially chronic), seen in diabetes, heart disease, kidney failure, and third-degree burns, all of which increase the patient's vulnerability.
5. The presence of indwelling and prosthetic devices such as catheters and heart valves.
6. Malnutrition, alcoholism, and drug abuse.

tensions of a warm-blooded animal's body. But a small number of fungi have acquired the morphological and physiological adaptations required to survive and grow in this habitat. By far their most striking adaptation is a switch from hyphal cells typical of the saprobic phase to yeast cells typical of the parasitic phase (figure 18.1). This biphasic characteristic of the life cycle is termed *thermal dimorphism* because it is initiated by increased temperature.

An **opportunistic fungal pathogen** is different from a true pathogen in several ways (table 18.2). An opportunist has weak to nonexistent invasiveness or virulence, and the host's defenses must be impaired to some degree for the microbe to gain a foothold. Although some species show both mycelial and yeast stages in their life cycles, thermal dimorphism is not a uniform occurrence. Opportunistic pathogens vary in their manifestations from superficial and benign colonizations to deep, chronic systemic disease that is rapidly fatal. Mycoses due to opportunists are an increasingly serious problem (see feature 18.1).

Some fungal pathogens exist in a category between true pathogens and opportunists. These species are not inherently invasive but can grow when inoculated into the skin wounds or abrasions of healthy persons. Examples are *Sporothrix,* the agent of a subcutaneous infection, and the *dermatophytes,* which cause ringworm and athlete's foot. Some mycologists believe these fungi are gradually becoming true pathogens.

Epidemiology of the Mycoses

The majority of fungal pathogens do not require a host to complete their life cycles, and the infections they cause are not communicable. Notable exceptions are some dermatophyte and *Candida* species that naturally inhabit the human body and are transmissible. For the remainder of the pathogens, exposure and infection occur when a human happens upon fungal spores in the environment (usually air, dust, or soil). Unlike the opportunists, true fungal pathogens are distributed in a predictable pattern that coincides with the pathogen's adaptation to the specific climate, soil, or other factors of a relatively restricted

dermatophyte (der-mah'-toh-fyte") Gr. *dermos,* skin, and *phyte,* plant. Fungi were once included in the plant kingdom.

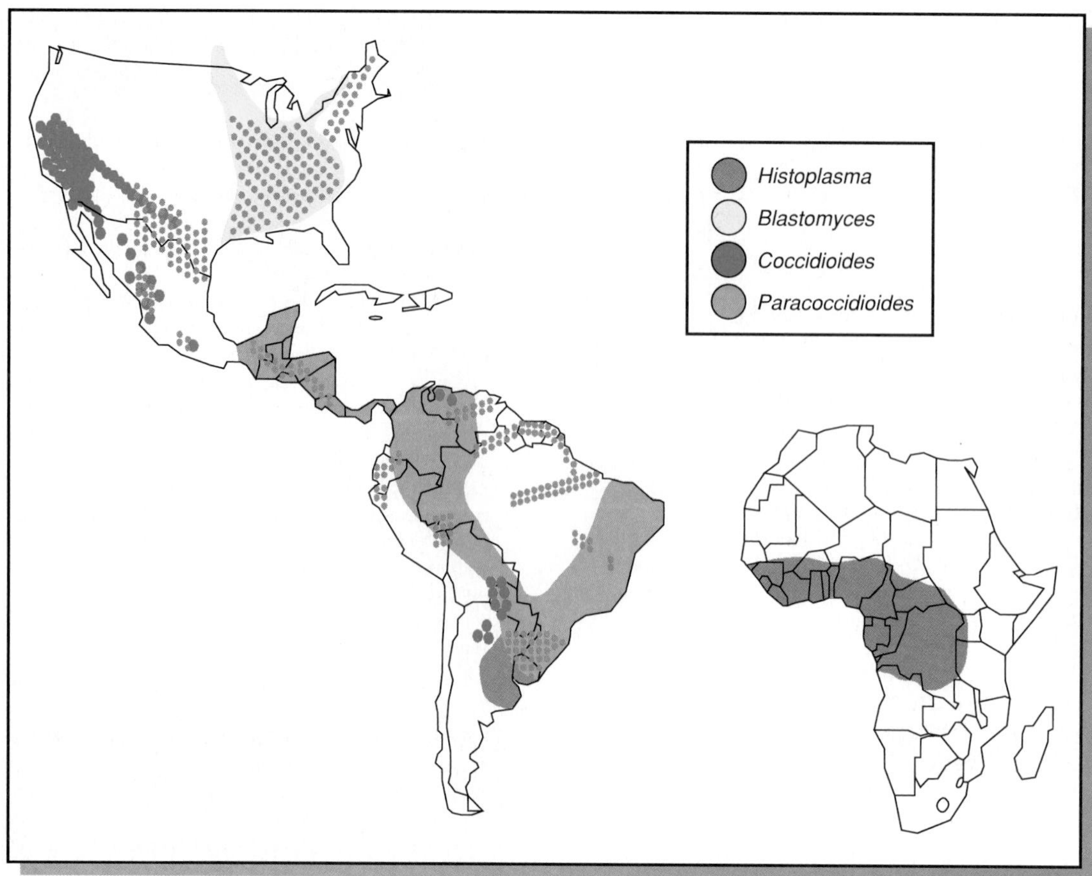

Figure 18.2 Distribution of the four fungal pathogens. The exposure of humans to a given pathogen is greatly influenced by where they live or travel. *Histoplasma* and *Blastomyces* occupy somewhat the same areas of North America, and their incidence correlates with the course of the Mississippi River and its tributaries. A unique form of *Histoplasma* occurs in Africa and is the only true fungal pathogen on a continent other than the Americas.

geographic region (figure 18.2). Infection with these agents requires human activity in that region at a time coinciding with sporulation.

Because case reporting is not mandatory, the incidence of all fungal infections has traditionally been difficult to measure. Dermatophytoses are probably the most prevalent, and it is thought that at least 90% of all humans will acquire ringworm or athlete's foot at least once. Estimates provided by routine skin testing indicate that millions of persons have experienced true mycoses, though most cases probably go undiagnosed or misdiagnosed.

Epidemics of mycotic infections can occur following mass exposure to a common source. Memorable incidents include spelunkers who were stricken by histoplasmosis after exploring bat caves, and South African mine workers who rubbed against contaminated wood planks and came down with sporotrichosis. Coccidioidomycosis is a particular hazard of construction workers and persons living in the paths of aberrant windstorms. Some epidemics arise from host-to-host transmission. Dermatophytoses are transmitted readily through shared personal articles, public facilities such as schools, swimming pools, and gymnasiums, and contact with infected animals. Candidiasis can be communicated during sexual contact and from mother to child at birth.

Pathogenesis of the Fungi

Mycoses involve complex interactions among the portal of entry, the nature of the infectious dose, the virulence of the fungus, and host resistance. Fungi enter the body mainly via respiratory, mucous, and cutaneous routes. In general, the agents of primary mycoses have a respiratory portal (spores inhaled from the air); subcutaneous agents enter through inoculated skin (trauma); and cutaneous and superficial agents enter through contamination of the skin surface. Spores, hyphal elements, and yeasts may all be infectious, but spores are most often involved due to their durability and abundance. Spore size is a factor in respiratory infections, because smaller spores are likely to be inhaled more deeply into the respiratory tract, and a large infecting dose increases the severity of disease. Thermal dimorphism greatly increases virulence by enabling fungi to tolerate the relatively high temperatures and low O_2 tensions of the body. Fungi in the yeast form seem more invasive because they grow more rapidly and spread through tissues and blood, while those producing hyphae tend to localize along the course of blood vessels and lymphatics. Specific factors contributing to fungal virulence are the subject of much research. Toxinlike substances have been isolated from several species, but how these substances actually damage the tissues remains obscure. Fungi

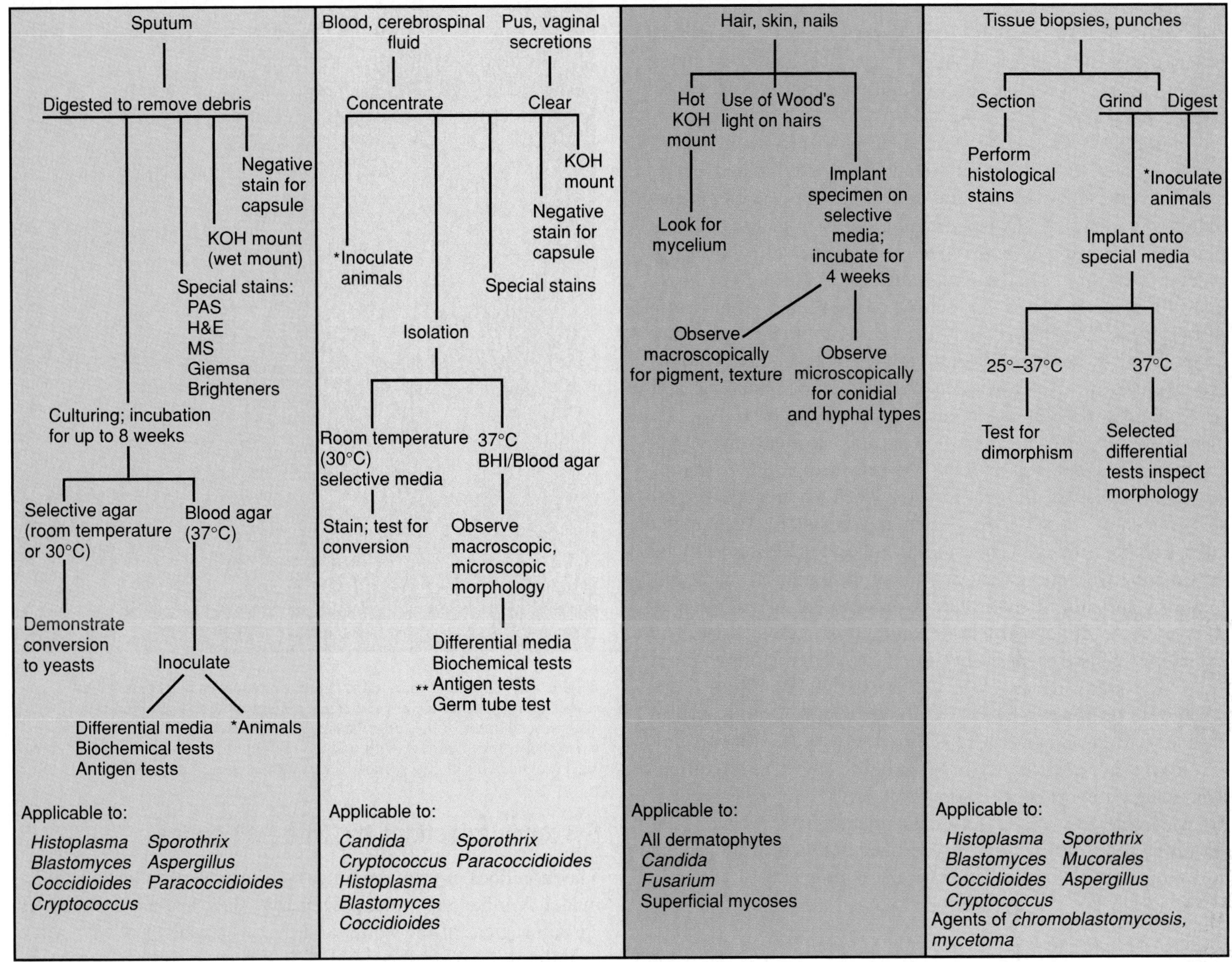

Figure 18.3 Methods of processing specimens in fungal disease. *Animal inoculation is performed only to help diagnose systemic mycoses when other methods are unavailable or indeterminant. **Some yeasts when incubated in serum for two to four hours sprout tiny hyphal tubes called germ tubes. *Candida albicans* is identified by this characteristic.

also produce various adhesion factors, hydrolytic enzymes, inflammatory stimulants, and allergens, all of which generate strong host responses.

The human body is extremely resistant to establishment of fungi. One of its numerous antifungal defenses is the normal integrity of the skin, mucous membranes, and respiratory cilia, but the most important defenses are cell-mediated immunity, phagocytosis, and the inflammatory reaction. Long-term protective immunity can develop for some of the true pathogens, but for the rest, reinfection is a distinct possibility.

Diagnosis of Mycotic Infections

Satisfactory diagnosis of fungal infections depends mainly upon isolating and identifying the pathogen in the laboratory. Accurate and speedy diagnosis is especially critical to the immunocompromised patient who must have prompt antifungal chemotherapy. For example, a patient with systemic *Candida* infection can die if the infection is not detected and treated within five to seven days. Because therapy may be species-specific, identification of the pathogen to that level is often necessary.

A suitable specimen may be obtained from sputum, skin scrapings, skin punches, cerebrospinal fluid, blood, tissue exudates, biopsies, urine, or vaginal samples, as determined by the patient's symptoms. Routine laboratory procedures include isolation, microscopic and macroscopic examination, histologic stains, serology, and animal inoculation (figure 18.3).

Because culture may require several days, the immediate direct examination of fresh samples is recommended. Wet mounts can be prepared by mixing a small portion of the sample on a slide with saline, water, or potassium hydroxide to clear the specimen of background debris. The relatively large size and unique appearance of many fungi help them stand out. Large round or oval budding cells are evidence of yeasts, while thick branching strands suggest hyphae. Nonspecific fluorescent stains or whiteners are valuable for highlighting fungi in tissue samples (figure 18.4).

Samples can also be stained histologically with methenamine silver, periodic acid Schiff, hematoxylin and eosin, Wright, Giemsa, or Gram stains to clearly delineate fungal elements (see figure 18.23). Isolating the fungal pathogen requires planting the specimen onto three or four types of solid media recommended for fungal isolation such as Sabouraud's dextrose agar, mycosel agar, inhibitory mold medium, and brain-heart-infusion agar. Depending on the desired result, these media may be modified by adding one of the following agents: blood to increase the growth of fastidious species, chloramphenicol and gentamicin to inhibit bacteria, or cyclohexamide to slow the growth of fungal contaminants from the specimen or the air. Cultures are incubated at room temperature, at 30°C, and at 37°C, and observed daily for growth, which may require several weeks in some species. Gross colonial morphology, such as the color and texture of the colony's surface and underside, may be very distinctive. Initial identification of the pathogen can be followed by specialized confirmatory procedures, including physiological and serological tests. Examples of these tests will be given during discussions of specific diseases later in this chapter.

Although various *in vitro* tests can detect antifungal antibodies in serum (see figure 18.10), serological testing may be inconsistent as a diagnostic tool. Skin testing, the *in vivo* testing of delayed hypersensitivity to fungal antigens, is mostly of epidemiologic import, since it does not verify ongoing infection. Considerable cross reactivity exists among *Histoplasma, Blastomyces,* and *Coccidioides.* A negative test in a healthy person, however, can rule out infection by these fungi.

Control of Mycotic Infections

Antimicrobic therapy for fungal infections is covered in chapter 10 and on a disease-by-disease basis in this chapter. Although immunization for most fungal diseases is not considered feasible at this time, T-cell infusions may be used successfully to treat children with candidiasis, and research on a vaccine for *Coccidioides* is proceeding. Few specific prevention measures exist for fungal infections.

Organization of Fungal Diseases

Fungal infections may be presented in several schemes, none of which is completely satisfactory. Traditional methods are based on taxonomic group, location of infection, and type of pathogen. Chapter 4 presents a taxonomic breakdown of the fungi, and this chapter treats them in the following categories according to the type and level of infection they cause and their degree of pathogenicity: (1) **systemic** true, **subcutaneous, cutaneous,** and **superficial** cutaneous mycoses (the four levels of infection depicted in figure 18.5), and (2) opportunistic mycoses (see table 18.1).

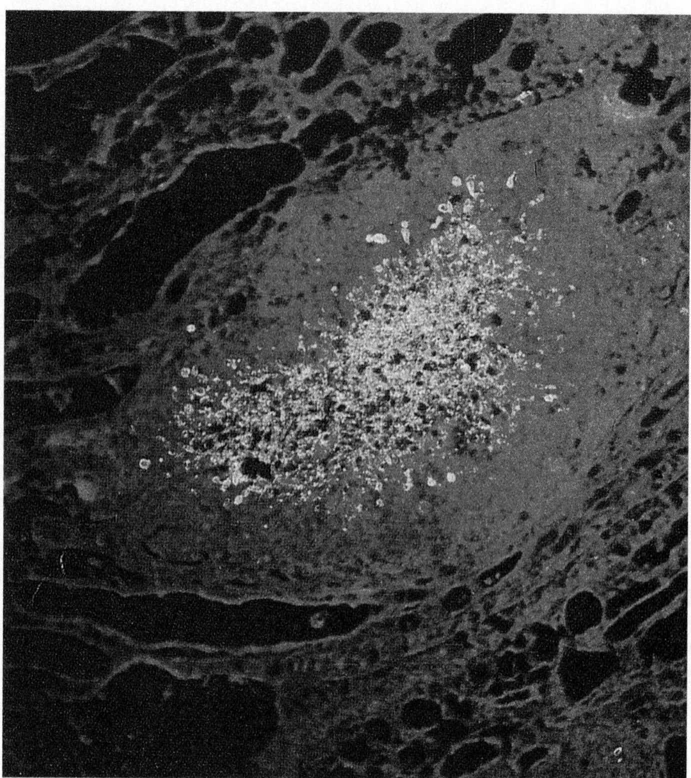

Figure 18.4 Special tissue stains called whiteners or brighteners may amplify the presence of fungal elements in specimens. They bind tightly to the carbohydrates of the fungal surface, and when viewed by a fluorescent microscope, they fluoresce with a blue-green light in contrast with the yellow background. This kidney section of *Candida* is stained with tinopal.

Systemic Infections by True Pathogens

The infections of primary fungal pathogens all follow a similar model. We have previously seen that they live endemically in certain regions of the world, and that when soil or other matter containing the fungal conidia is disturbed, airborne spores can be inhaled into the lower respiratory tract. In the lungs, the spores germinate into yeasts or yeastlike cells and produce an asymptomatic or mild **primary pulmonary infection (PPI)** that parallels tuberculosis. In some hosts, this infection becomes systemic and creates severe, chronic lesions. Less commonly, spores are inoculated into the skin where they form localized granulomatous lesions. All diseases result in immunity that may be long-term and that manifests itself clinically as an allergic reaction to fungal antigens.

Histoplasmosis

The most common true pathogen is *Histoplasma capsulatum,* the cause of histoplasmosis. This disease has probably afflicted humans since antiquity, though it was not described until 1905 by Dr. Samuel Darling. Through the years, it has been known

---

Histoplasma capsulatum (his″-toh-plaz′-mah kap″-soo-lay′-tum) Gr. *hist,* tissue, and *plasm,* shape; L. *capsula,* small box.

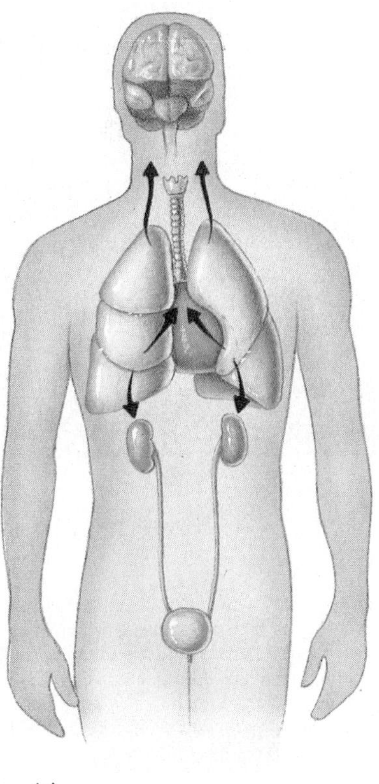
(a)

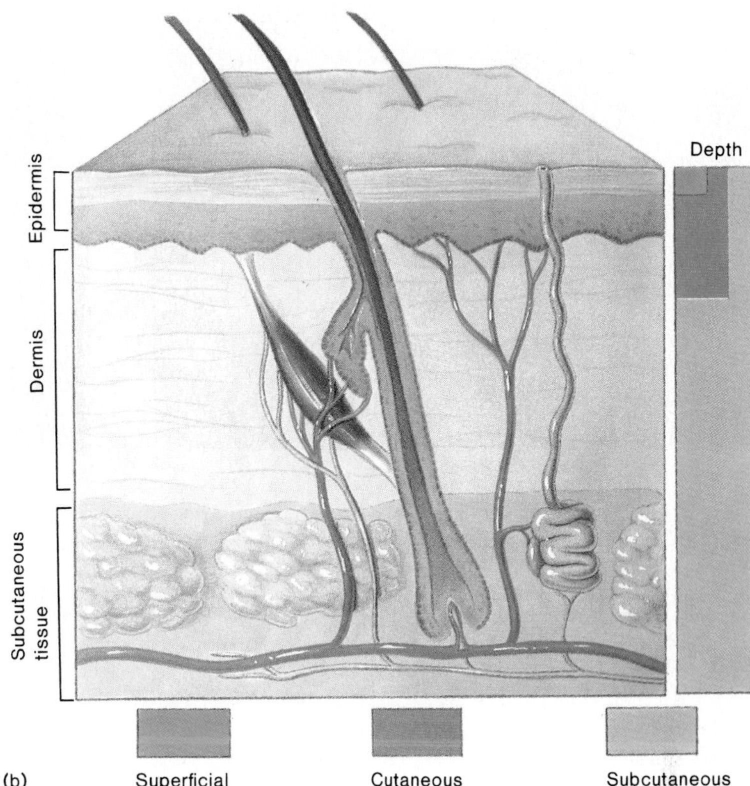

(b)

Figure 18.5 Levels of invasion by fungal pathogens. (Some species may invade more than one level.) (*a*) In systemic (deep) mycoses, fungemia causes the fungus to disseminate to the internal organs. Some pathogens show a particular affinity for one organ over another. (*b*) The skin and its attendant structures provide many potential sites for invasion, including the scalp, smooth skin, hair, and mucous membranes. Differing depths of involvement are: superficial, consisting of extremely shallow epidermal colonizations; cutaneous, involving the stratum corneum and occasionally the upper dermis; and subcutaneous, occurring after a puncture wound has introduced the fungus deep into the subcutaneous tissues.

by various synonyms—Darling's disease, Ohio Valley fever, and reticuloendotheliosis. The impact of histoplasmosis throughout history is not really known. Certain aspects of its current distribution and epidemiology suggest that it has been an important disease as long as humans have practiced agriculture. Some have gone so far as to speculate that histoplasmosis was the "mummy's curse" that caused sudden deaths in an archeological expedition unearthing King Tutankhamen's tomb in the early 1920s.

Biology and Epidemiology of *Histoplasma capsulatum*

Histoplasma capsulatum is typically dimorphic. Growth on media below 35°C is characterized by a white or brown, hairlike mycelium, and growth at 37°C on blood agar produces a white, smooth colony (figure 18.6).

H. capsulatum is discontinuously distributed on all continents except Australia, though its areas of greatest endemicity are the eastern and central regions of the United States (the Ohio Valley). This fungus appears to grow most abundantly in soils high in nitrogen content, especially those supplemented by bird and bat *guano,* and in regions with adequate moisture and moderate temperatures.

A useful tool for determining the distribution of *H. capsulatum* is to inject a fungal extract called **histoplasmin** into the skin and monitor an allergic reaction. Application of this test has verified the extremely widespread distribution of the fungus. In high prevalence areas such as southern Ohio, Illinois, Missouri, Kentucky, Tennessee, Michigan, Georgia, and Arkansas, 80% to 90% of the population show signs of prior infection. Histoplasmosis prevalence in the United States is estimated at about 500,000 cases per year, with several thousand of them requiring hospitalization and a small number resulting in death.

The spores of the fungus are probably dispersed by the wind and, to a lesser extent, animals. The most striking outbreaks of histoplasmosis occur when concentrations of spores have been dislodged by humans working in parks, bird roosting areas, buildings, and the like. People of both sexes and all ages incur infection, but adult males experience the majority of cases. The oldest and youngest members of a population are most likely to develop serious disease.

Infection and Pathogenesis of *Histoplasma*

Histoplasmosis presents the most formidable array of manifestations of any mycosis. It can be benign or severe, acute or chronic, and it can show pulmonary, systemic, or cutaneous lesions. Inhaling a small dose of microconidia into the deep recesses of the lung establishes a primary pulmonary infection that

guano (gwan′-oh) Sp. *huanu,* dung. An accumulation of animal manure.

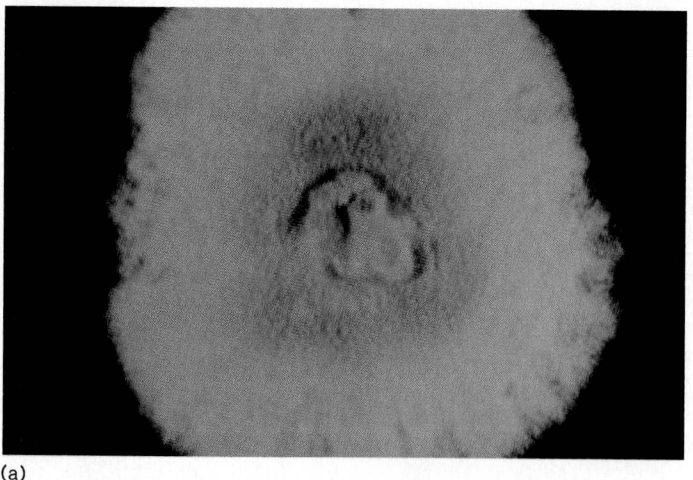

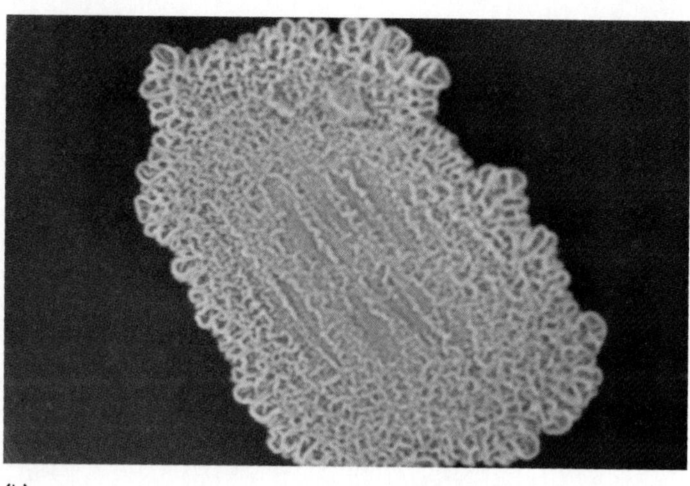

Figure 18.6 Cellular and cultural characteristics of *Histoplasma capsulatum*. (*a*) A colony at 25° C produces an abundant mycelium. (The spores are shown in figure 18.7.) (*b*) A yeast colony (37°C) is dense and may have a lacy texture.

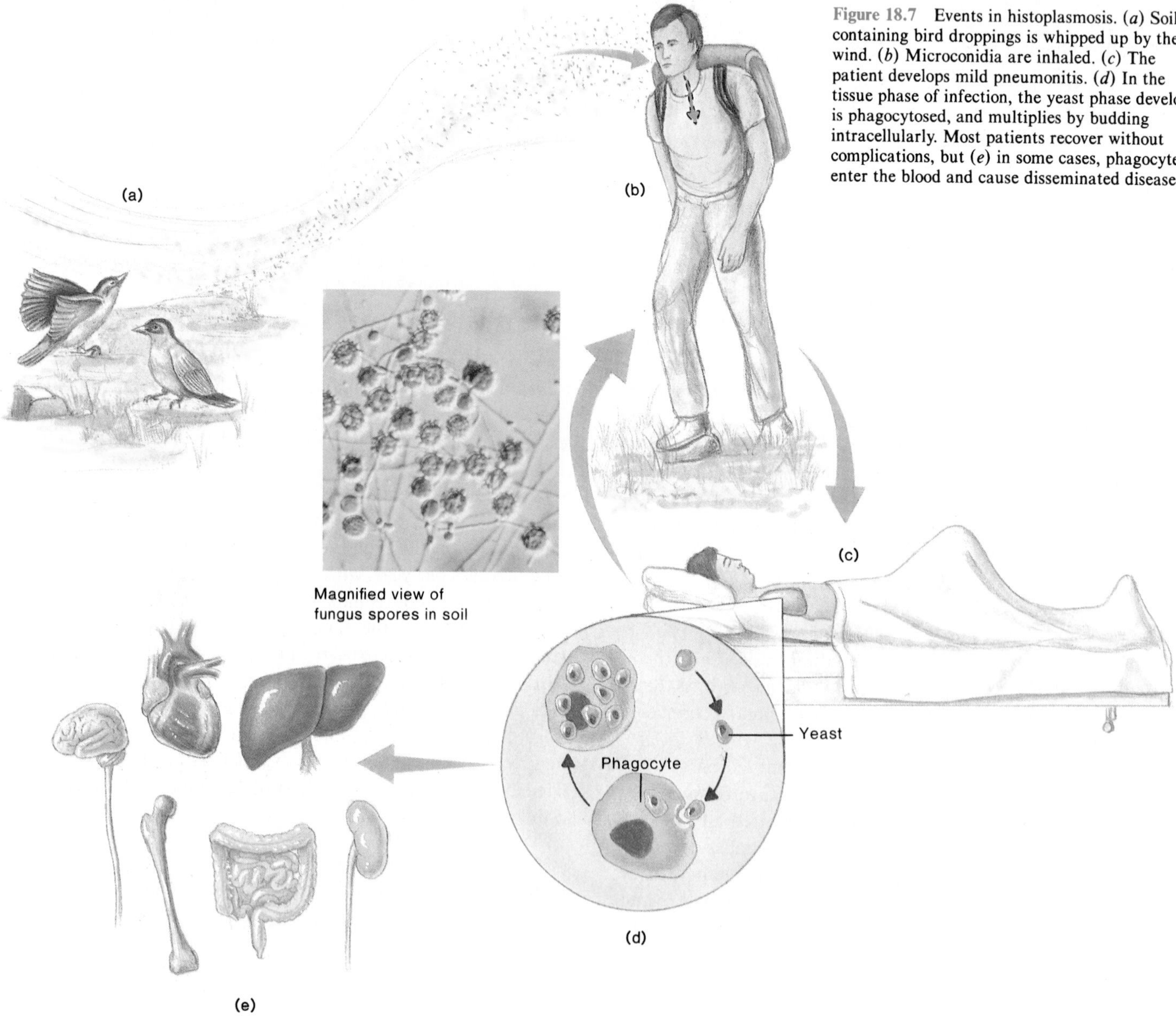

Figure 18.7 Events in histoplasmosis. (*a*) Soil containing bird droppings is whipped up by the wind. (*b*) Microconidia are inhaled. (*c*) The patient develops mild pneumonitis. (*d*) In the tissue phase of infection, the yeast phase develops, is phagocytosed, and multiplies by budding intracellularly. Most patients recover without complications, but (*e*) in some cases, phagocytes enter the blood and cause disseminated disease.

in most cases is benign or asymptomatic. Some persons experience mild symptoms such as aches, pains, and coughing, but a few develop more severe symptoms, including fever, night sweats, and weight loss (the fungus flu). Except for lung calcifications and a positive skin test, the benign, acute forms leave no residual effect. Primary cutaneous histoplasmosis, in which the agent enters via the skin, is very rare.

A more serious chronic form of histoplasmosis arises when patients lack an important component of host defenses and the fungus disseminates from the primary site of infection, often hidden within macrophages (figure 18.7). In children, this can lead to hepatosplenomegaly, anemia, circulatory collapse, and death. Adults with systemic disease may acquire lesions in the brain, intestine, adrenal gland, heart, liver, spleen, lymph nodes, and bone marrow. Occasionally, dissemination to the skin and mucous membranes results in swelling and lesions in the mouth and on the external genitalia. Persistent colonization of patients with emphysema and bronchitis causes *chronic pulmonary histoplasmosis,* a complication that has signs and symptoms similar to those of tuberculosis.

Diagnosis and Control of Histoplasmosis

Discovering *Histoplasma* in clinical specimens is a substantial diagnostic indicator. Usually it appears as spherical, "fish-eye" yeasts intracellularly in macrophages, and occasionally as free yeasts in samples of sputum and cerebrospinal fluid. An essential confirmatory step is to isolate the agent and demonstrate dimorphism, but this may require up to 12 weeks. The histoplasmin test does not indicate concurrent infection, thus it is not useful in diagnosis. However, complement fixation and immunodiffusion serological tests may support a diagnosis by showing a rising antibody titer.

Undetected or mild cases of histoplasmosis resolve without medical management, but chronic or disseminated disease calls for systemic chemotherapy. The principal drug, amphotericin B, is administered in daily intravenous doses for a few days to a few weeks. Under some circumstances, ketoconazole or miconazole are the drugs of choice. Surgery to remove affected masses in the lungs or other organs may also be effective.

Coccidioidomycosis: Valley Fever, San Joaquin Fever, California Disease

Coccidioides immitis is the etiologic agent of **coccidioidomycosis.** Although the fungus has probably lived in soil for millions of years, human encounters with it are relatively recent and coincide with the encroachment of humans into its habitat. This unique and fascinating fungus may be the most virulent of all mycotic pathogens.

Biology and Epidemiology of *Coccidioides*

The morphology of *C. immitis* is very distinctive. At 25°C, it forms a moist, white to brown colony with abundant, branching, septate hyphae. These hyphae fragment into thick-walled, blocky **arthroconidia** at maturity (inset photo, figure 18.8). On special media incubated at 37°–40°C, an arthrospore germinates into the parasitic phase, a small, spherical cell (**spherule**). This structure swells into a giant sporangium that cleaves internally to form numerous endospores (figure 18.8*c*).

Coccidioides immitis occurs endemically in various natural reservoirs and casually in areas where it has been carried by wind and animals. Conditions favoring its settlement in a given habitat include high carbon and salt content and a semiarid, relatively hot climate. The fungus has been isolated in such regions from soils, plants, and a large number of vertebrates (mammals, birds, and even reptiles and amphibians). The natural history of *C. immitis* follows a cyclic pattern—a period of dormancy in winter and spring, followed by growth in summer and fall. Natural dispersal of the arthrospores is aided by windstorms, dust storms, drainoff water, and even burrowing animals.

Skin testing has disclosed that the highest incidence of coccidioidomycosis (approximately 100,000 cases per year) is in the southwestern United States, though it also occurs in Mexico and parts of Central and South America. Especially concentrated reservoirs exist in the San Joaquin Valley of California and in southern Arizona. Occasional epidemics are reported in conjunction with soil agitation by humans (digging, excavating, and farming). Apparently, *Coccidioides* has adapted well to the ancient dwellings of Native Americans. Archeologists digging in these sites are constantly exposed to it and have been common victims of the disease. All persons inhaling the arthrospores probably develop some degree of infection, but adult males, dark-skinned persons, and pregnant women are more likely to acquire a severe form.

Infection and Pathogenesis of Coccidioidomycosis

The arthrospores of *C. immitis* are lightweight and readily inhaled. In the lung they convert to spherules, which swell, sporulate, burst, and release spores that continue the cycle. In 60 patients out of 100, this primary pulmonary infection is inapparent, while in the other 40, it is accompanied by coldlike symptoms such as fever, chest pain, cough, headaches, and malaise. In uncomplicated cases, complete recovery and lifelong immunity are the rule.

In about five out of a thousand cases, the primary infection does not resolve, and progresses with varied consequences. Chronic progressive pulmonary disease is manifest by nodular growths called *fungomas* and cavity formation in the lungs that compromise respiration. Dissemination of the endospores into major organs occasionally takes place in persons with impaired

Coccidioides immitis (kok-sid″-ee-oy′-deez ih′-mih-tis) From *coccidia,* a sporozoan, and L. *immitis,* fierce. The original discoverers thought the microbe looked like a protozoan.

fungoma (fun-joh′-mah) A fungus tumor or growth.

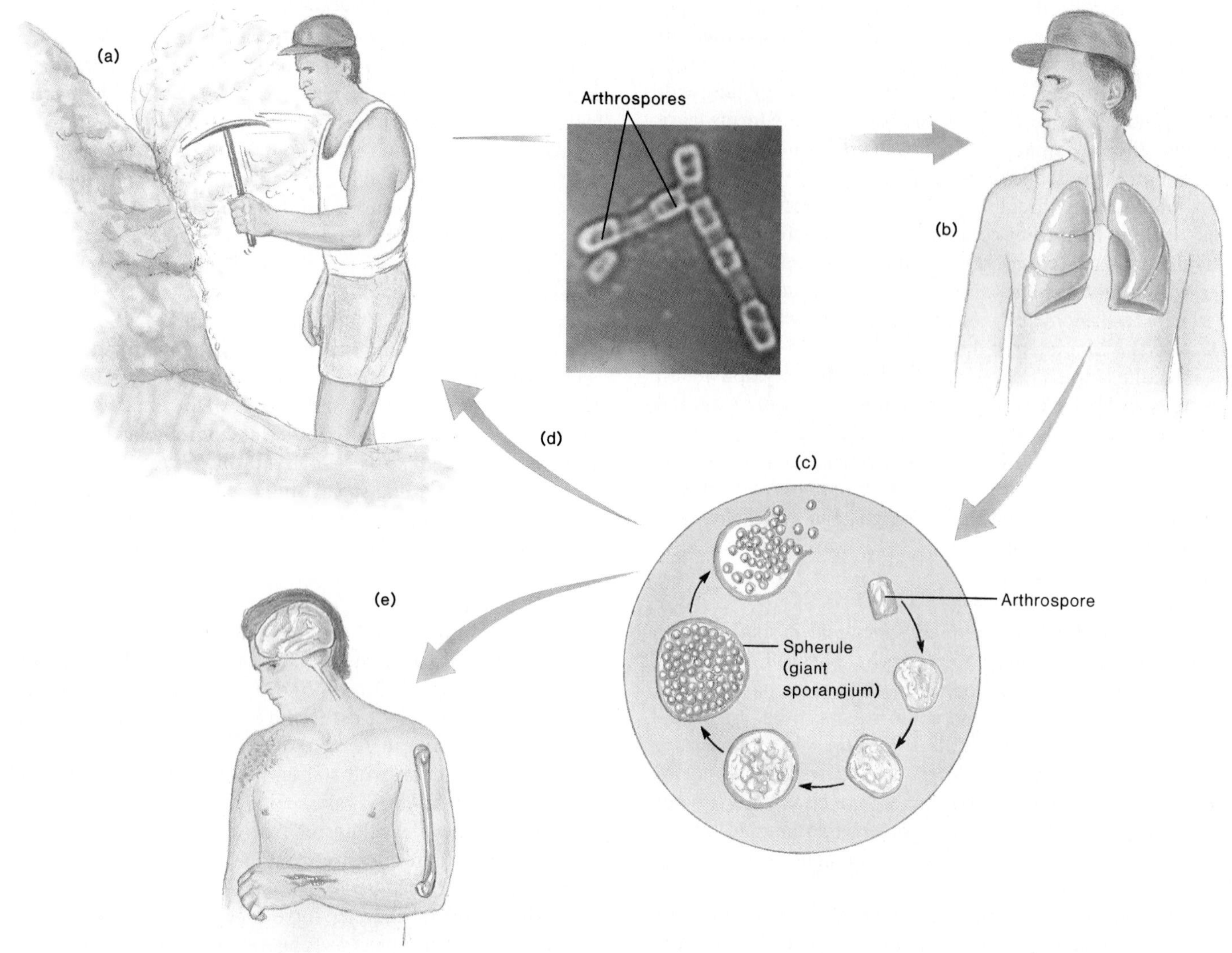

Figure 18.8 Events in coccidioidomycosis. (*a*) A person digging in soil produces aerosol. Inhaled arthrospores (inset photo) establish (*b*) a lung infection. (*c*) Arthrospores develop into spherules containing endospores; endospores are released in the lungs. (*d*) Immunocompetent persons effectively fight infection and return to health. (*e*) Compromised persons acquire severe complications in the brain, bone, and skin.

cell-mediated immunity. Such disseminated disease impressively demonstrates that when degree of virulence and immunodeficiency coexist in the same infection, there may be multiorgan involvement with rapid, explosive, and fatal results (figure 18.9).

Diagnosis and Control of Coccidioidomycosis

Diagnosis of coccidioidomycosis is straightforward when the highly distinctive spherules are found in sputum, tissue exudates, and biopsies. This finding is further supported if the typical mycelium and spores are isolated on Sabouraud's agar and if spherules are induced to form. Newer specific antigen tests have been a great boon in identifying and differentiating *Coccidioides* from other fungi. So great is the potential for contamination and infection that experts recommend growing cultures in tubes or bottles rather than on plates, and either opening them in a biological containment hood or killing them prior to inspection. Immunodiffusion (figure 18.10) and latex agglutination tests on serum samples are excellent screens for detecting

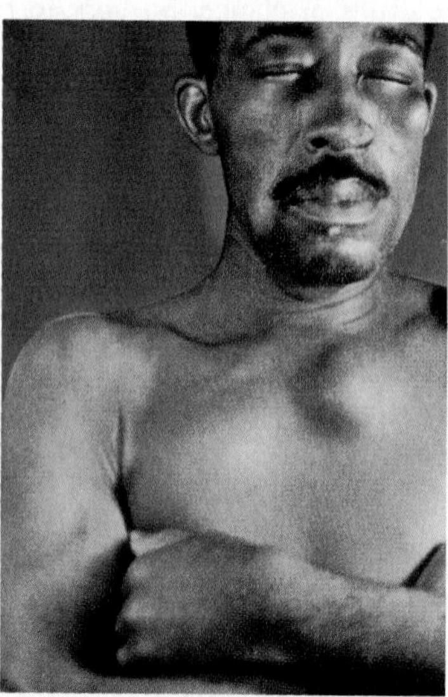

Figure 18.9 Disseminated coccidioidomycosis manifested by subcutaneous abscesses in the chest.

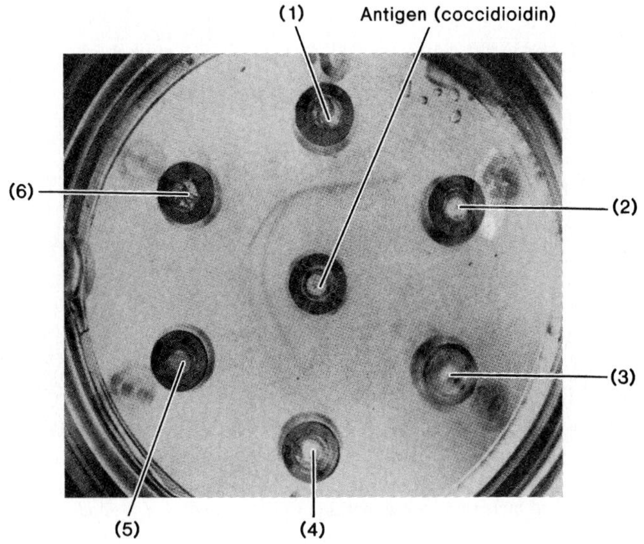

Figure 18.10 Immunodiffusion testing for active coccidioidomycosis. Small wells are punched into agar, and patients' sera are placed in numbered wells; the center well contains a known fungal antigen such as coccidioidin. Lines of precipitation forming between the outer well and the inner well indicate identity and antibodies for the disease. In this test, reactions occurring for sera 1, 5, and 6 will be interpreted as positive evidence of infection. Subtle differences in these lines can show qualitative differences in the severity of the disease. Wells 2, 3, and 4 indicate the absence of antibodies and infection.

early infection. Skin tests using an extract of the fungus (coccidioidin or spherulin) are of primary importance in epidemiologic studies.

The majority of coccidioidomycosis patients do not require treatment. However, in persons with disseminated disease, amphotericin B is administered intravenously. Imidazoles can have some benefit, but they may require a longer term of therapy. Minimizing contact with the fungus in its natural habitat has been of some value. For example, oiling roads and planting vegetation help reduce spore aerosols, and using dust masks while excavating soil prevents workers from inhaling spores.

Biology of *Blastomyces dermatitidis*

Blastomyces dermatitidis, the cause of **blastomycosis,** is another fungal pathogen endemic to the United States. Alternate names for this disease are Gilchrist's disease, Chicago disease, and North American blastomycosis, though discoveries of blastomycosis in areas other than North America make the latter name less appropriate. The dimorphic morphology of *Blastomyces* follows the pattern of other pathogens. Colonies of the saprobic phase are uniformly white to tan, with a thin, septate mycelium and simple ovoid conidia (figure 18.11*a*). Temperature-induced conversion results in a wrinkled, creamy-white colony that yields large, heavy-walled yeasts with buds nearly as large as the mother cell.

Blastomyces dermatitidis (blas″-toh-my′-seez der″-mah-tit′-ih-dis) Gr. *blastos,* germ, *myces,* fungus, *dermato,* skin, and *itis,* inflammation.

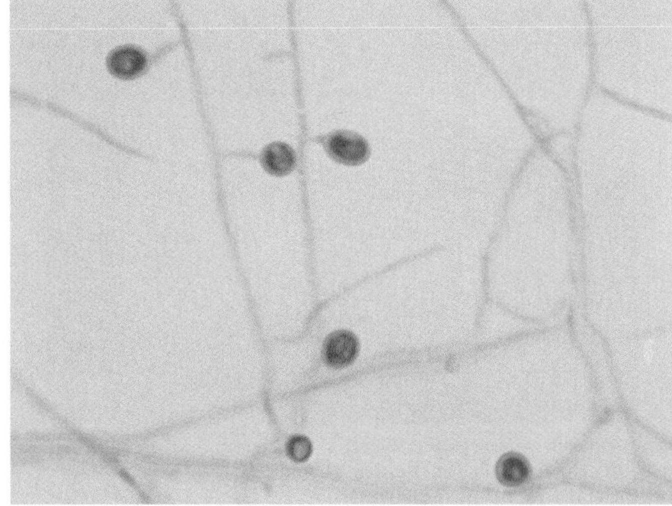

(a)

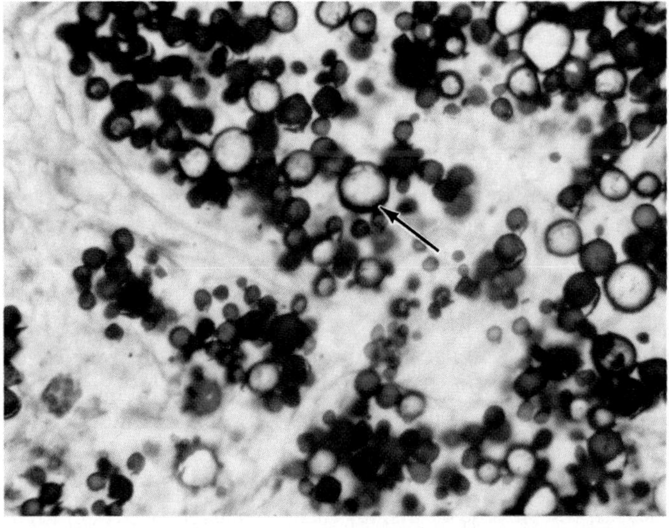

(b)

Figure 18.11 The dimorphic nature of *Blastomyces dermatitidis.* (*a*) Hyphal filaments bear conidia that look like tiny lollipops. (*b*) The tissue phase as seen in a sputum sample. The arrow points out the very thick cell wall typical of this species.

The distribution of *B. dermatitidis* is less delineated than that of other true pathogens. Studies indicate that it inhabits areas high in organic matter such as forest soil, decaying wood, animal manure, and abandoned buildings, but it has been difficult to isolate with regularity. Its life cycle features dormancy during the warmer, dryer times of the year and growth and sporulation during the colder, wetter seasons. In general, blastomycosis occurs from southern Canada to southern Louisiana and from Minnesota to the Carolinas and Georgia. Cases have also been reported in Central America, South America, Africa, and the Middle East. Humans, dogs, cats, and horses are the chief targets of infection, which is usually acquired by inhaling conidia-laden dust from living quarters, farm buildings, or forest litter. According to the few statistics available, the frequency of blastomycosis is highest among the middle-aged, males, blacks, and pregnant women.

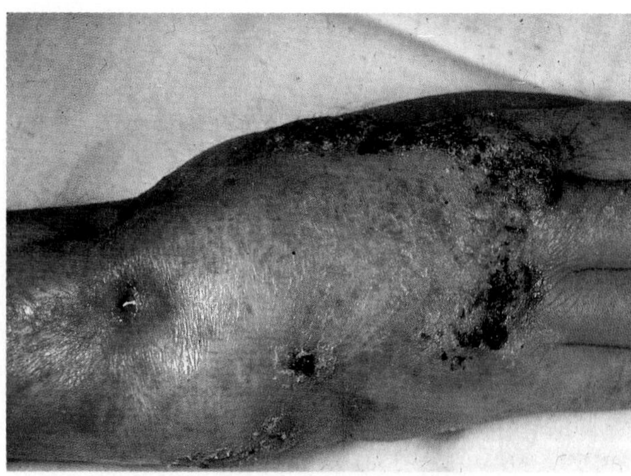

Figure 18.12 Cutaneous blastomycosis in the hand and wrist as a complication of disseminated infection. Note the darkly colored, tumorlike vegetations and scar tissue on the hand.

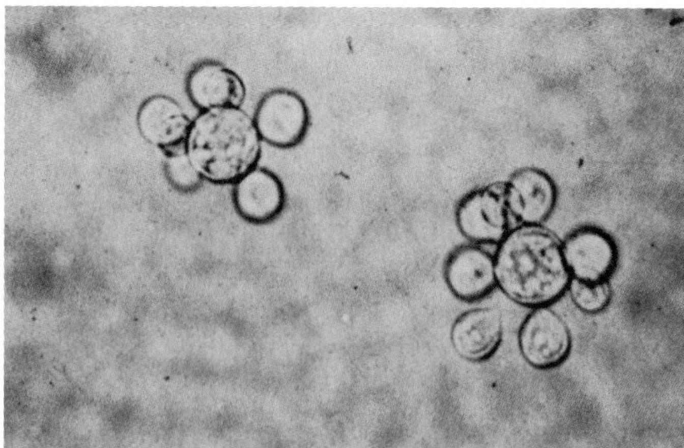

Figure 18.13 The morphology of *Paracoccidioides*. A Gridley stain from a skin lesion (×1,200) reveals the central round mother cell with a series of narrow-necked buds that look something like the handles of a pilot's wheel.

Infection and Pathology in Blastomycosis

The primary portal of entry of *B. dermatitidis* is the respiratory tract, though it also may enter through accidental inoculation. Inhaling only 10 to 100 conidia is enough to initiate infection. As conidia convert to yeasts and multiply, they encounter macrophages in the lung. What ensues is an inflammatory response and the formation of alveolar granulomas. Whether a large proportion of primary pulmonary infections are inapparent and quickly resolved is not known, but medical mycologists feel most cases are symptomatic to some degree.

Mild disease is accompanied by cough, chest pain, hoarseness, and fever. More severe progressive and chronic blastomycosis can involve the lungs, skin, and numerous other organs. Nodules, abscesses, and tumorlike growths developing in the lungs are often mistaken for cancer. Compared to other fungal diseases, the chronic cutaneous form of blastomycosis is rather common. It frequently begins on the face, hand, wrist, or leg as a subcutaneous nodule that erodes to the skin surface (figure 18.12). Dissemination of the yeasts into the skeleton induces symptoms of arthritis and osteomyelitis, and involvement of the central nervous system leads to headache, convulsions, coma, and mental confusion. The yeast may also spread to the reproductive tract, spleen, liver, and kidneys. Chronic systemic blastomycosis lasting for weeks to years frequently overwhelms the host defenses and kills the patient.

Laboratory Diagnosis and Therapy

Microscopic smears of specimens showing large ovoid yeasts with broad-based buds can provide the most reliable diagnostic evidence of blastomycosis (see figure 18.11*b*). Cultures prepared on selective and enriched media and incubated at room temperature may require several weeks to develop. Further confirmation of the overall microscopic and macroscopic findings requires verification of dimorphism or a positive immunodiffusion test. Most skin tests for *Blastomyces* are not useful for general testing because of the frequency of false-positive and false-negative results. Although disseminated infections were once nearly 100% fatal, modern drugs have greatly improved the prognosis. The drug of choice for severe disseminated and cutaneous disease is amphotericin B, and milder cases may respond to several months of dihydroxystilbamidine therapy.

Paracoccidioidomycosis

The remaining dimorphic fungal pathogen would be a front-runner in a competition for the most tongue-twisting scientific term. ***Paracoccidioides brasiliensis*** causes **paracoccidioidomycosis,** also known as paracoccidioidal granuloma and South American blastomycosis. Due to its relatively restricted distribution, this disease is the least common of the primary mycoses. *Paracoccidioides* forms a small, nondescript colony with scanty, undistinctive spores at room temperature, but it develops an unusual yeast form at 37°C. The large mother cells sprout small, narrow-necked buds that may radiate around the periphery like spokes on a pilot wheel (figure 18.13).

The natural history of *Paracoccidioides* has not yet been completely clarified. It has been isolated from the cool, humid soils of tropical and semitropical regions of South and Central America, particularly Brazil, Colombia, Venezuela, Argentina, and Paraguay. People most often afflicted with paracoccidioidomycosis are rural agricultural workers and plant harvesters. A possible animal role in transmission has been postulated but not demonstrated. Other factors believed to influence disease are an altered hormonal state, poor nutrition, and impaired host resistance. Infection is initiated when fungal spores enter the lungs or, occasionally are inoculated into the skin. Most infections are probably benign, self-limited events that go completely unnoticed. In the minority of patients who do incur progressive systemic disease, the lungs, skin and mucous membranes (especially of the head), and lymphatic organs are frequently involved.

Paracoccidioides brasiliensis (pair″-ah-kok-sid″-ee-oy′-deez brah-sil″-ee-en′-sis) Named for its superficial resemblance to *Coccidioides* and its prevalence in Brazil.

Paracoccidioidomycosis is diagnosed by standard procedures previously described for the other mycoses in this chapter. Closely scrutinizing fresh or stained clinical specimens for the distinctive yeast phase is important, and a cultural follow-up with conversion media is essential. Serological testing (mainly immunodiffusion) can be instrumental in diagnosing and monitoring the course of infection. Principal treatment drugs for disseminated disease, in order of choice, are ketoconazole, amphotericin B, and sulfa drugs.

Subcutaneous Mycoses

When certain fungi are transferred from soil or plants directly into traumatized skin, they can invade the damaged site. Such infections are termed subcutaneous because they involve tissues within and just below the skin. Most species in this group are greatly inhibited by the higher temperatures of the blood and viscera, and only rarely do they disseminate. Nevertheless, these diseases are progressive and can destroy the skin and associated structures. Mycoses in this category are sporotrichosis, chromoblastomycosis, phaeohyphomycosis, and mycetoma.

The Natural History of Sporotrichosis: Rose-gardener's Disease

The cause of **sporotrichosis, *Sporothrix schenckii,*** is a very common saprobic fungus that decomposes vegetative matter in soil and humus and exhibits both mycelial and yeast phases (figure 18.14; see figure 18.16). *Sporothrix* resides in warm, temperate, and moist areas of the tropics and semitropics, though the incidence of sporotrichosis is highest in Africa, Australia, Mexico, and Latin America. A prick by a rose thorn is the classic origin of the cutaneous form of infection, and has given rise to its common name. Most episodes of human infection follow contact with thorns, wood, sphagnum moss, bare roots, grass, bark, and other vegetation. Horticulturists, foresters, gardeners, farmers, and basket weavers incur the majority of cases. Overt disease is uncommon unless a person has inadequate immune defenses or is exposed to a large inoculum or a virulent strain. Other mammals, notably horses, dogs, cats, rats, and mules, are also susceptible to sporotrichosis.

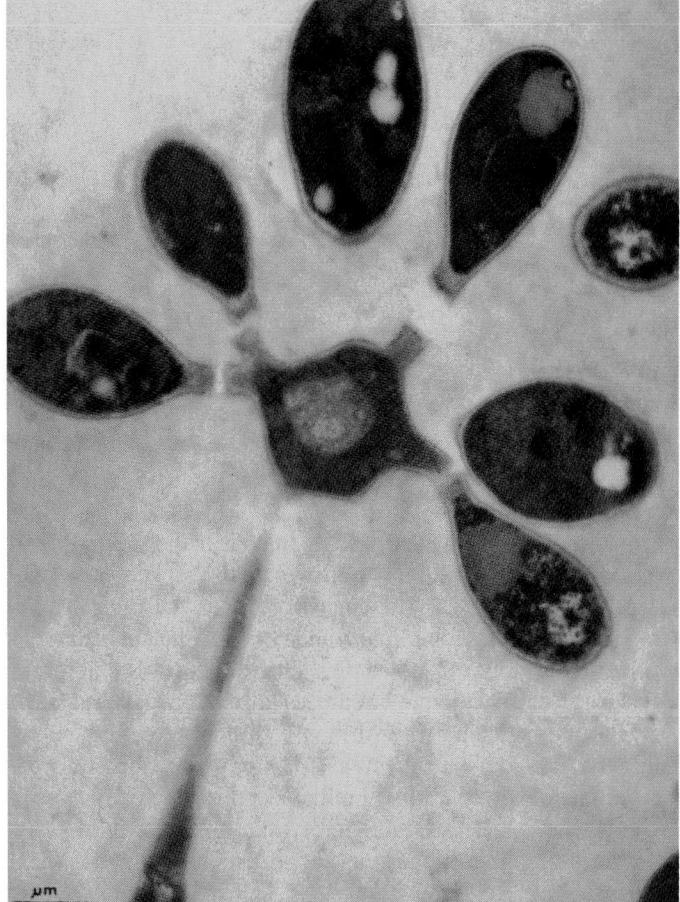

Figure 18.14 The microscopic morphology of *Sporothrix schenckii*. Spores develop as floral clusters borne on conidiophores or as single spores direct from the hyphae.

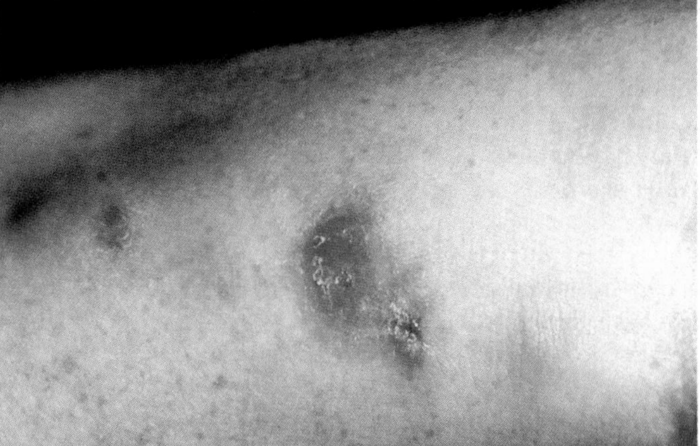
Figure 18.15 The clinical appearance of sporotrichosis. A primary sore is accompanied by a series of nodules running along the lymphatic channels of the arm.

Pathology, Diagnosis, and Control of Sporotrichosis

In **lymphocutaneous sporotrichosis,** the fungus grows at the site of penetration and develops a small, hard, nontender nodule within a few days to months. This subsequently enlarges into a bubo, becomes necrotic, breaks through to the skin surface, and drains (figure 18.15). Infection often progresses along the regional lymphatic channels, leaving a chain of lesions at various stages. If left untreated, this condition persists for several years, though it is contained by the regional lymph nodes and does not spread. When *Sporothrix* conidia are drawn into the lungs, primary pulmonary sporotrichosis can result. Although this infection was once thought to be rare, it is increasingly reported among hospitalized chronic alcoholics.

Discovery of the agent in tissue exudates, pus, and sputum is difficult because there are usually so few cells in the infection.

Sporothrix schenckii (spoh'-roh-thriks shenk'-ee-ee) Gr. *sporos,* seed, and *thrix,* hair. Named for B. R. Schenck, who first isolated it.

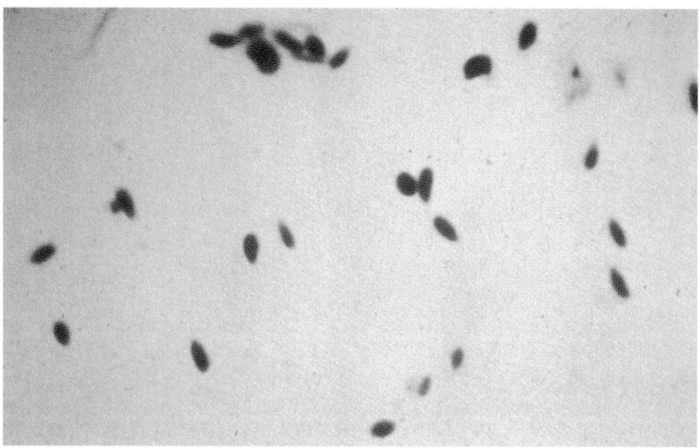

Figure 18.16 A scattering of tiny, plump *S. schenckii* yeasts with tapered ends.

To enhance its appearance, enzymes that clear the specimen and special stains that highlight the tiny, cigar-shaped yeasts may be applied (figure 18.16). Demonstration of the typical macroscopic and microscopic morphology on artificial media and verification of dimorphism confirm the diagnosis. Reliable serological tests are available but are usually not necessary. However, the sporotrichin skin test may be used to determine prior infection. An older but effective drug for sporotrichosis is potassium iodide, given orally in milk or applied topically to open lesions. Amphotericin B and flucytosine may assist in unresponsive cases. The fungus cannot withstand heat, thus local applications of heat packs to lesions have helped resolve some infections. As a preventive measure, persons with occupational exposure should shield their bare limbs.

Chromoblastomycosis and Phaeohyphomycosis: Diseases of Pigmented Fungi

Chromoblastomycosis is a progressive subcutaneous mycosis characterized by highly visible *verrucous* lesions. The principal etiologic agents are a collection of widespread soil saprobes containing large amounts of dark pigments. *Fonsecaea pedrosoi, Phialophora verrucosa,* and *Cladosporium carrionii* are among the most common causative species. **Phaeohyphomycosis,** a closely related infection described in feature 18.2, differs primarily in the causative species and the appearance of the infectious agent in tissue. Chromoblastomycotic agents produce very large, thick, yeastlike bodies called sclerotic cells, whereas the agents of phaeohyphomycosis remain typically hyphal.

The fungi associated with these infections are of low inherent virulence, and none exhibit thermal dimorphism. Infection occurs by happenstance when body surfaces, especially the legs and feet, are penetrated by soiled vegetation or inanimate objects. Chromoblastomycosis occurs throughout the world, with peak incidence in the American subtropics and tropics. Most vulnerable to infection are men with rural occupations who go barefoot. After a very long (2–3-year) incubation period, a small, colored, warty plaque, ulcer, or papule appears. Because this lesion is generally not painful, many patients do not seek treatment and may go for years watching it progress to a more advanced state (see figure 4.17*d*). Unfortunately, the patient may scratch the nodules, which aggravates the spread of infection on the body and provokes secondary bacterial infections.

Chromoblastomycosis is frequently confused with cancer, syphilis, yaws, and blastomycosis. Diagnosis is based on the clinical appearance of the lesions, microscopic examination of biopsied lesions, and the results of culture. Therapy for the condition includes heat, drugs, surgical removal of early nodules, and amputation in advanced cases. Combined topical amphotericin B and thiabendazole or systemic flucytosine have shown some success in arresting the disease.

chromoblastomycosis (kroh″-moh-blas″-toh-my-koh′-sis) Gr. *chroma,* color, and *blasto,* germ. The fungal body is highly colored *in vitro.*

verrucous (ver-oo′-kus) Tough, warty.

phaeohyphomycosis (fy″-oh-hy″-foh-my-coh′-sis) Gr. *phaeo,* brown, and *hypho,* thread.

Feature 18.2 Freakish Fungi

The infectious diseases termed phaeohyphomycoses are among the strangest oddities of medical science. The etiologic agents are soil fungi or plant pathogens with brown-pigmented mycelia. Victims of the disease are usually highly compromised patients who have become inoculated with these fungi, which are so widespread in household and medical environments that contact is inevitable. The disease process can be extremely deforming, and the microbe is sometimes hard to identify. Some genera of fungi commonly involved in these mycoses are *Alternaria, Aureobasidium, Curvularia, Dreschlera, Exophiala, Phialophora,* and *Wangiella.*

The fungi grow slowly from the portal of entry through the dermis and create enlarged, subcutaneous cysts. In patients with underlying conditions such as endocarditis, diabetes, and leukemia, the fungi may spread into the body proper (bones, brain, lungs). Perhaps the most bizarre case ever recorded was "the celery boy," a child who developed a small nodule on his cheek that eventually formed extensive thick, warty masses over his entire face and nasal cavity. After numerous attempts, physicians isolated a fungus that normally causes brown rot in celery plants. Over a period of years, the boy was literally eaten up by the infection.

Mycetoma: A Complex Disfiguring Syndrome

Another disease elicited when soil microbes are accidently implanted into the skin is **mycetoma,** a mycosis usually of the foot or hand that looks superficially like a tumor. It is also called *madura foot* for the Indian region where it was first described. About half of all mycetomas are caused by fungi in the genera *Pseudallescheria* or *Madurella,* though filamentous bacteria called actinomycetes (discussed later in this chapter) are often etiologically implicated. Mycetomas are endemic to equatorial Africa, Mexico, Latin America, and the Mediterranean, and cases occur sporadically in the United States. Infection begins when a bare foot, head, hand, or torso is pierced by a thorn, sliver, leaf, or other type of sharp plant debris. First to appear is a localized abscess in the subcutaneous tissues, which

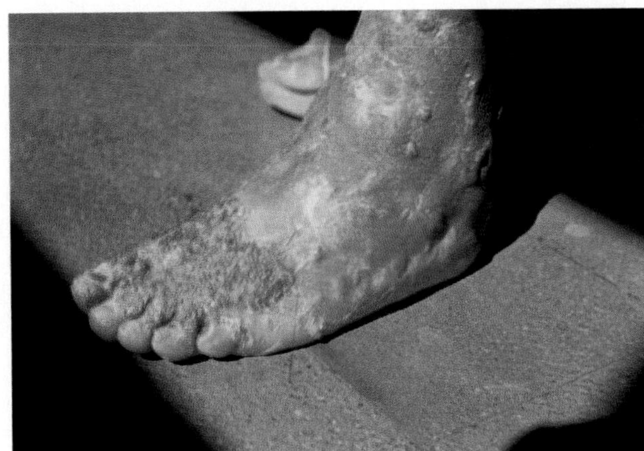

Figure 18.17 Early stages of mycetoma of the foot, caused by *Madurella*. Ulceration, swelling, and scarring are visible.

gradually swells and drains. Untreated cases that spread to the muscles and bones may cause pain and loss of function in the affected body part (figure 18.17). Mycetoma has an insidious, lengthy course and is exceedingly difficult to treat.

Cutaneous Mycoses

Fungal infections strictly confined to the nonliving epidermal tissues (stratum corneum) and its derivatives (hair and nails) are termed **dermatophytoses.** These important diseases are commonly called **ringworm,** because they tend to develop in circular, scaly patches (see figure 18.19), and **tineas,** because early observers thought worms caused them. Dermatophytoses are collectively attributed to 39 species in the genera *Trichophyton, Microsporum,* and *Epidermiphyton.* The causative agent of a given type of ringworm differs from person to person and from place to place, and is not restricted to a particular genus or species (table 18.3).

Table 18.3 The Dermatophyte Genera and Diseases

Genus	Name of Disease	Principal Targets	How Transmitted
Trichophyton	Ringworm of scalp Ringworm of body Ringworm of beard Ringworm of nail Athlete's foot	Hair, skin, nails	Human to human, animal to human
Microsporum	Ringworm of scalp Ringworm of skin	Scalp hair Skin; not nails	Animal to human, soil to human, human to human
Epidermophyton	Ringworm of groin Ringworm of nail	Skin, nails; not hair	Strictly human to human

Characteristics of Dermatophytes

The dermatophytes are so closely related and morphologically similar that they can be difficult to differentiate. Various species exhibit unique macroconidia, microconidia, and unusual types of hyphae. In general, *Trichophyton* produces thin-walled, smooth macroconidia and numerous microconidia; *Microsporum* produces thick-walled, rough macroconidia (see chapter opening photo) and sparser microconidia; and *Epidermiphyton* has ovoid, smooth, clustered macroconidia and no microconidia (figure 18.18).

Epidemiology and Pathology of Dermatophytoses

The natural reservoirs of dermatophytes are other humans, animals, and the soil (see feature 18.3). Infection is promoted by the following: shedding hairs, skin cells, and other epidermal tissues; the hardiness of the dermatophyte spores (they can last

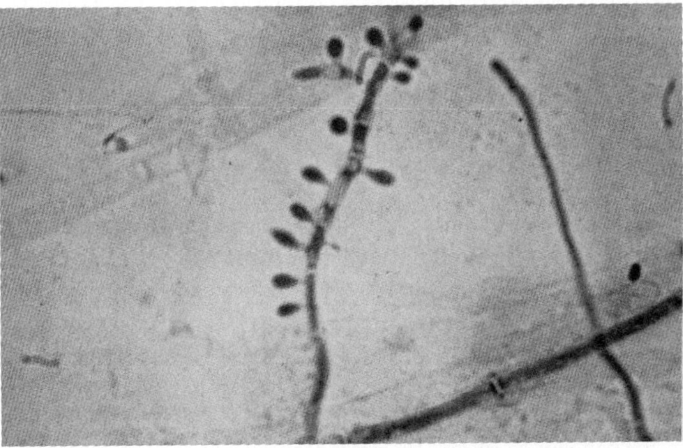

(a)

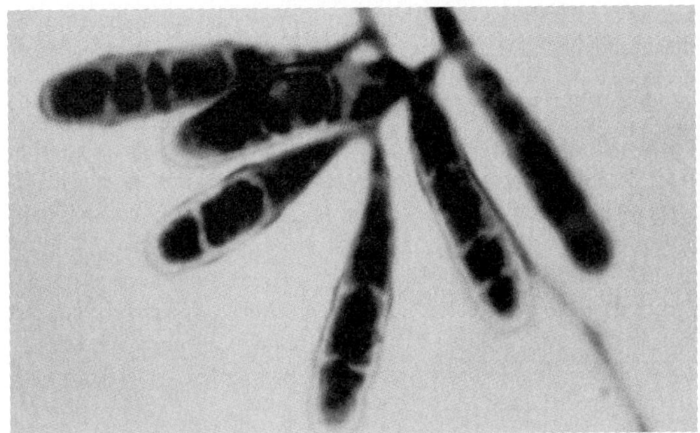

(b)

Figure 18.18 Examples of dermatophyte spores. (*a*) Regular, numerous microconidia of *Trichophyton.* (*b*) Smooth-surfaced macroconidia in clusters characteristic of *Epidermophyton.*

tinea (tin′-ee-ah) L. a larva or worm.

Feature 18.3 The Keratin Lovers

The dermatophytic fungi are especially well adapted to breaking down keratin, the primary protein of the epidermal tissues of vertebrates (skin, nails, hair, feathers, and horns). Their affinity for this compound gives them the name keratophiles. A study of dermatophyte ecology reveals a gradual evolutionary trend from saprobic soil forms that digest keratin but do not parasitize animals, to soil forms that occasionally parasitize animals, to species that are dependent on live animals. Some species can infect a broad spectrum of animals, and others are specific to one particular animal or region of the body. One adaptive challenge faced by relatively new fungal parasites is that they are likely to cause more severe reactions in the host's skin and to be attacked by the host defenses and eliminated. Thus, the more successful fungi equilibrate with the host by reducing their activity (growth rate, sporulation) to reduce the inflammatory response. Eventually, these dermatophytes become such "good parasites" that they colonize the host for life. A striking example is *Trichophyton rubrum,* a fungus that causes a form of athlete's foot. It has such a tenacious hold and is so hard to cure that its carriers might even be called the "T. rubrum people."

for years on fomites); hot, sweaty, chafed body parts; and intimate contact. Most infections exhibit a long incubation period (months), followed by localized inflammation and allergic reactions to the fungal proteins. As a general rule, infections acquired from animals and soil cause more severe reactions, and infections eliciting stronger immune reactions are resolved faster.

Dermatophytic fungi exist throughout the world. Although some species are endemic, they tend to spread rapidly to nonendemic regions through jet age travel. Dermatophytoses endure today as a serious health concern, not because they are life-threatening, but because of the extreme discomfort, stress, pain, and unsightliness they cause. In many cultures, ringworm and other skin mycoses are regarded as a social disgrace or a sign of uncleanliness.

An Atlas of Dermatophytoses

In the following section, the dermatophytoses are organized according to the area of body they affect, their mode of acquisition, and their pathologic appearance. Both the common English name and body site and its equivalent Latin name (tinea) are given.

Ringworm of the Scalp (Tinea Capitis) This mycosis results from the fungal invasion of the scalp and the hair of the head, eyebrows, and eyelashes (figure 18.19*a, b*). Very common in children, tinea capitis is acquired from other children and adults or from domestic animals. Manifestations range from small, scaly patches (gray patch), to a severe inflammatory reaction (kerion), to destruction of the hair follicle and permanent hair loss. Unless infected hairs are controlled, reinfection is common.

Ringworm of the Beard (Tinea Barbae) This tinea, also called "barber's itch," afflicts the chin and beard of adult males. Although once a common aftereffect of unhygienic barbering, it is now contracted mainly from animals.

Ringworm of the Body (Tinea Corporis) This extremely prevalent infection of humans can appear nearly anywhere on the body's glabrous (smooth and bare) skin. The principal sources are other humans, animals, and soil, and it is transmitted primarily by direct contact and fomites (clothing, bedding). The infection usually appears as one or more scaly reddish rings on the trunk, hip, arm, neck, or face (figure 18.19*c*). The ringed pattern is formed when the infection radiates from the original site of invasion into the surrounding skin. Depending on the causal species and the health and hygiene of the patient, lesions vary from mild and diffuse to florid and pustular.

Ringworm of the Groin (Tinea Cruris) Sometimes known as "jock itch," crural ringworm occurs mainly in males on the groin, perianal skin, scrotum, and occasionally, the penis. The fungus thrives under conditions of moisture and humidity created by profuse sweating or tropical climates. It is transmitted primarily from human to human and is pervasive among athletes and persons living in close situations (ships, military quarters).

Ringworm of the Foot (Tinea Pedis) Tinea pedis is known by a variety of synonyms, including athlete's foot and jungle rot. The disease is clearly connected to wearing shoes, since it is uncommon in cultures where the people customarily go barefoot (but as you have seen, bare feet have other fungal risks!). Any situation that encases the foot in a closed, warm, sweaty environment facilitates infection. Because tinea pedis is a known hazard in shared facilities such as shower stalls, public floors, and locker rooms, many public facilities have rules against bare feet. Infections begin with small blisters between the toes that burst, crust over, and may spread to the rest of the foot and nails (figure 18.20*a*).

Ringworm of the Hand (Tinea Manuum) Infection of the hand by dermatophytes is nearly always associated with concurrent infection of the foot. The fingers and palms of one hand only are the usual sites of lesions, which vary from white and patchy to deep and fissured.

Ringworm of the Nail (Tinea Unguium) Fingernails and toenails, being masses of keratin, are often sites for persistent fungus colonization. The first symptoms are usually superficial white patches in the nail bed. A more invasive form causes thickening, distortion, and darkening of the nail (figure 18.20*b*). Nail problems caused by dermatophytes are on the rise as more women wear artificial fingernails, which may provide a portal of entry into the nail bed.

Diagnosis of Ringworm

Dermatologists are most often called upon to diagnose dermatophytoses. Occasionally, the presenting symptoms are so dramatic and suggestive that no further testing is necessary, but in most cases, direct microscopic examination and culturing are needed. Diagnosis of tinea of the scalp is sometimes aided by a Wood's light, a special long-wave ultraviolet lamp that causes infected hairs to fluoresce with a greenish glow (see figure

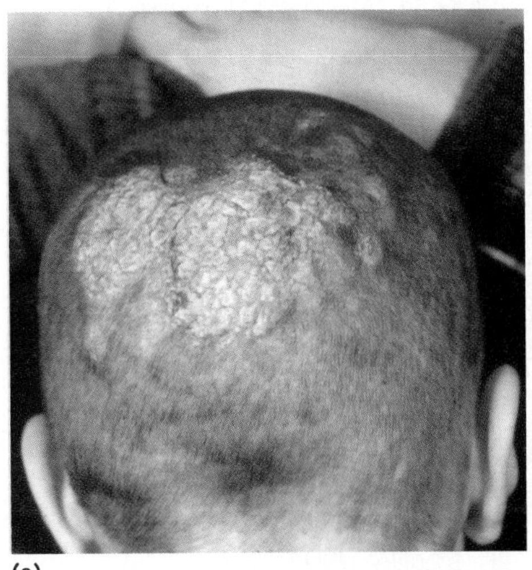
(a)

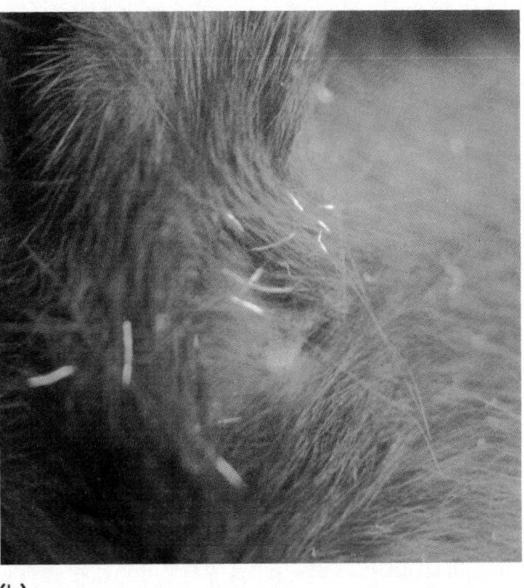
(b)

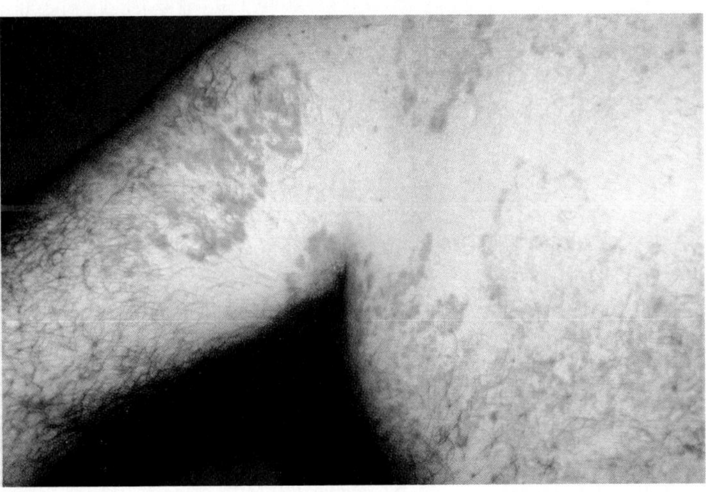
(c)

Figure 18.19 Ringworm lesions on the scalp and body vary in appearance. (*a*) Kerion, with deep exudative involvement and complete hair loss in the affected region. (*b*) A close-up of an infected hair fluorescing under a Wood's light. (*c*) Wide-scale lesions over the arm and shoulder have a dramatic ringed appearance that results from the gradual spread of inflammation from the center to the newest area of invasion in a circumferential pattern.

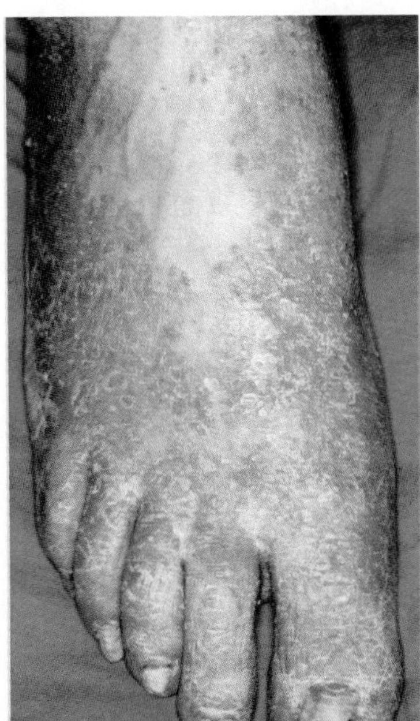
(a)

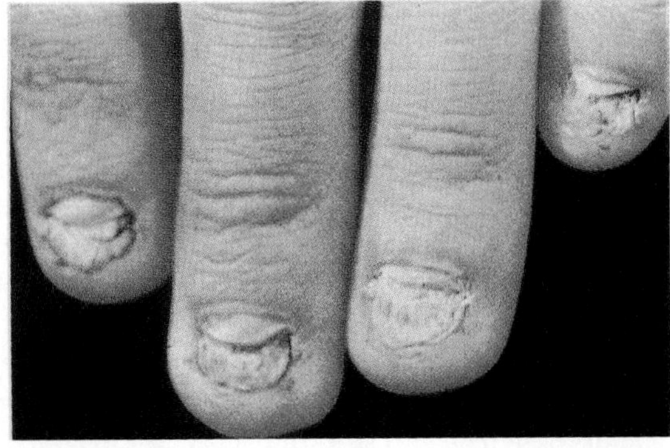
(b)

Figure 18.20 Ringworm of the extremities. (*a*) *Trichophyton* infection spreading over the foot in a "moccasin" pattern. The chronicity of these tineas is attributed to the lack of fatty-acid-forming glands in the feet and hands. (*b*) Ringworm of the nails. Invasion of the nail bed causes some degree of thickening, accumulation of cheesy debris, cracking, and discoloration; nails may be separated from underlying structures as shown here. Severe unchecked cases can result in permanent loss of the nails.

18.19*b*). Samples of hair, skin scrapings, and nail debris treated with heated KOH show a thin, branching fungal mycelium if infection is present. Culturing specimens on selective media and identifying the species also aid diagnosis in many cases.

Treatment of the Dermatophytoses

Ringworm therapy is based on the concept that the dermatophyte is feeding on dead epidermal tissues. These regions undergo constant replacement from living cells deep in the epidermis, so if multiplication of the fungus can be blocked, the fungus will eventually be sloughed off along with the skin or nail. Unfortunately, this takes time. Gentle debridement of skin and ultraviolet light treatments may be beneficial, but by far the most satisfactory choice for therapy is a topical antifungal agent. Over-the-counter ointments containing tolnaftate, imidazoles, haloprogin, logamel, or thiabendazine are applied regularly for several weeks. Some drugs work by speeding up loss of the outer skin layer. In intractable infections, a systemic drug such as griseofulvin can be given, but placing a patient on this hepatotoxic and nephrotoxic drug for up to two years is probably too risky in most cases.

Superficial Mycoses

Agents of **superficial mycoses** occupy the outer epidermal surface and are ordinarily neither inflammatory nor invasive. These diseases are innocuous and have mainly cosmetic repercussions. **Tinea versicolor**[1] is caused by the yeast *Malassezia furfur,* a normal inhabitant of human skin. For reasons that are not clear, its growth in some persons elicits mild, chronic scaling and interferes with pigmentation. The trunk, face, and limbs take on a mottled appearance. The disease is most pronounced in young persons living in the tropics who are frequently exposed to the sun. *Piedras* are marked by a tenacious, colored concretion forming on the outside surface of hair shafts (figure 18.21). In **white piedra,** caused by *Trichosporon beigelii,* a white to yellow adherent mass develops on the shaft of scalp, pubic, or axillary hair. At times, the mass is invaded secondarily by brightly colored bacterial contaminants, with startling results. **Black piedra,** caused by *Piedraia hortae,* is characterized by dark-brown to black gritty nodules, mainly on scalp hairs. Neither piedra is common in the United States.

Opportunistic Mycoses

Earlier in this chapter, we introduced the concept of opportunistic fungal infections and their predisposing factors (see feature 18.1 and table 18.2). The prevailing opportunistic pathogens of humans are the yeasts *Candida* and *Cryptococcus* and a small number of filamentous fungi, primarily *Aspergillus* and certain zygomycetes.

1. Versicolor refers to the color variations this mycosis produces in the skin.
piedra (pee-ay'-drah) Sp. stone. Hard nodules of fungus formed on the hair.

Hair shaft
White piedra
(a)

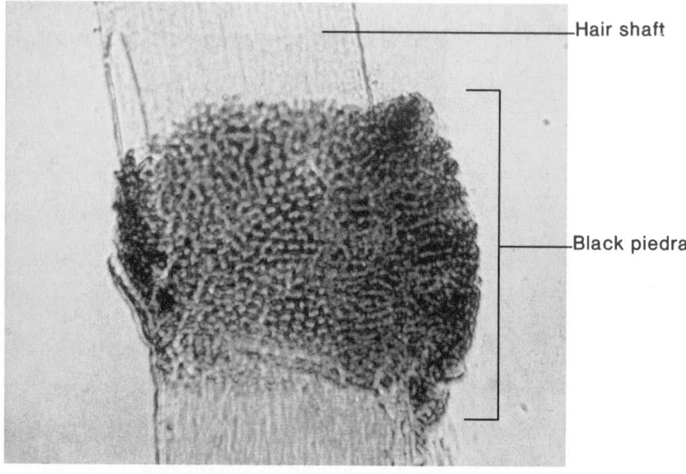

(b)

Figure 18.21 Examples of superficial mycoses. (*a*) Light-colored mass on a hair shaft with white piedra. (*b*) Dark, hard concretion enveloping the hair shaft in black piedra (×200).

Infections by *Candida:* Candidiasis

Candida albicans, an extremely widespread yeast, is the major cause of candidiasis (also called candidosis or moniliasis). Other *Candida* species are occasionally encountered in infected transplant patients and other persons with altered host defenses. Manifestations of infection run the gamut from short-lived, superficial skin irritations to overwhelming, fatal systemic diseases. Microscopically, *C. albicans* has budding cells of varying size that may be elongated and attached in a pseudomycelium (see figure 18.23*a*), and macroscopically, it is a white- to cream-colored, pasty colony with a yeasty odor.

Epidemiology of Candidiasis

Candida albicans occurs as normal flora in the oral cavity, genitalia, large intestine, or skin of 20% of humans. A healthy person

Candida albicans (kan'-dih-dah al'-bih-kanz) L. *candidus,* glowing white, and *albus,* white.

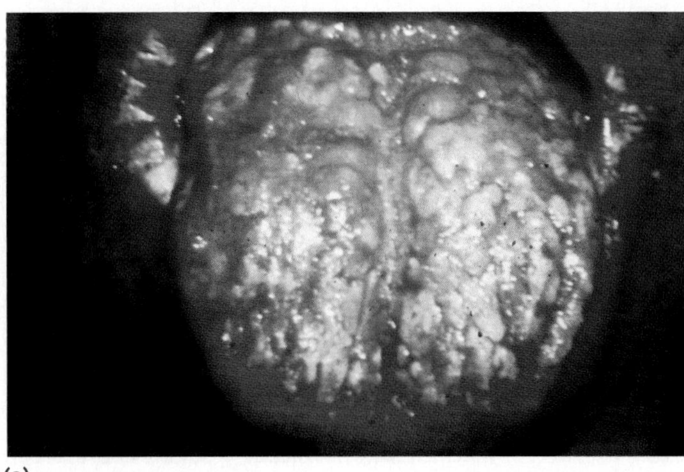

(a)

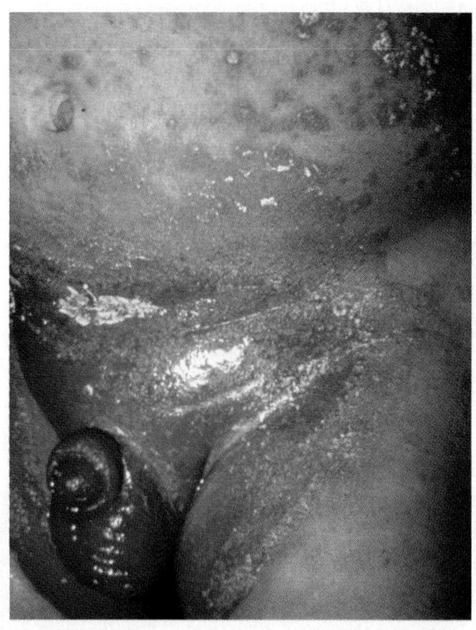

(b)

Figure 18.22 Common infections by *Candida albicans*. (*a*) Chronic thrush of an adult male with diabetes. (*b*) Candidiasis involving the genitalia and the skin of the lower abdomen.

with no breach in defenses can hold it in check, but the risk of invasion increases with extreme youth, pregnancy, use of birth control or antimicrobial drugs, nutritional and organic disease, immunodeficiency, the presence of invasive devices, and trauma. Any situation that maintains the yeast in a warm, moist environment against the skin provides an avenue for infection. Although candidiasis is usually endogenous and not contagious, it can be spread in nurseries or through surgery, childbirth, and sexual contact.

Pathology of *C. albicans*

Candida albicans produces local infections of the mouth, pharynx, vagina, skin, alimentary canal, and lungs, and it may also disseminate to internal organs. Mucous membranes most frequently involved are the oral cavity and vagina. **Thrush** is a white, adherent, patchy infection affecting the membranes of the mouth, gums, cheeks, or throat, usually in newborn infants and elderly, debilitated patients (figure 18.22*a*). **Vulvovaginal candidiasis** (**VC**), known more commonly as **yeast infection,** has widespread occurrence in adult women, especially those who are on antibiotics or oral contraceptives, have diabetes, or are pregnant (see feature 18.4). These conditions are thought to increase the glucose content of vaginal secretions, which favors candidal overgrowth and disrupts the balance of normal flora. Binding, impervious undergarments (panty hose, panties) are probably contributing factors. The chief symptoms of VC are a yellow to white milky discharge, inflammation, painful ulcerations, and itching. Infection of the penis occasionally occurs in males. The most severe cases spread from the vagina and vulva to the perineum and thighs (figure 18.22*b*).

Of all the areas of the gastrointestinal tract, *Candida* infects the esophagus and the anus most commonly. Esophageal candidiasis, a frequent complication in AIDS, causes painful, bleeding ulcerations, perforations, nausea, and vomiting. Infections of the stomach or intestine are relatively uncommon.

Feature 18.4 Pregnancy, Candidiasis, and the Neonate

Hormonal changes contribute to a high rate of candidiasis in pregnant women (up to 90% of women in the third trimester). Aside from the extreme discomfort of the symptoms, candidal vaginitis poses a threat to the newborn. Most cases of neonatal thrush are traced to contact with the mother's vagina during birth, and cutaneous infection is another common complication. Given the prevalence of vaginal infection during pregnancy, screening for *Candida* is yet another essential prenatal test. Fortunately, treating the mother with an intravaginal drug cream containing an imizadole or polyene is effective and does not jeopardize the fetus.

Candidal attack of keratinized structures such as skin and nails, called **onychomycosis,** is often brought on by predisposing occupational and anatomical factors. Persons whose occupations require their hands or feet to be constantly immersed in water are at risk for finger and nail invasion. *Intertriginous* infection occurs in moist areas of the body where skin rubs against skin, as beneath the breasts, in the armpit, and between folds of the groin. **Cutaneous candidiasis** may also complicate burns and produce a scaldlike rash on the skin of neonates.

Candidal blood infection in patients chronically weakened by surgery, organ transplants, cancer, or intravenous drug addiction is a grave sign, and it is usually followed by systemic invasion. Although the mechanisms of virulence are not well understood, the presence of *C. albicans* in the blood is such a serious assault that it causes more human mortalities than any other fungal pathogen. Principal targets of systemic infections

onychomycosis (awn″-ih-koh-my-koh′-sis) Gr. *onyx,* nail. Infection of the nails by *Candida* or *Aspergillus.*

intertriginous (in″-ter-trij′-ih-nus) L. *inter,* between, and *trigo,* to rub.

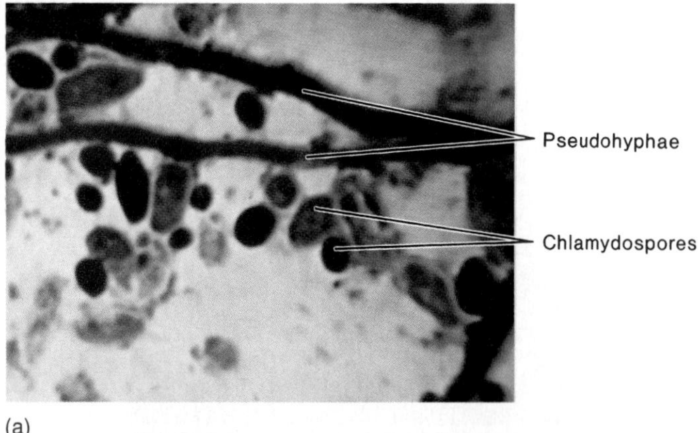

(a)

(b)

Figure 18.23 (*a*) A Gram stain of *Candida albicans* in a vaginal smear reveals gram-positive chlamydospores and pseudohyphae. In many cases of vaginal candidiasis, infection is detected during a routine Pap smear. (*b*) Rapid yeast identification system using biochemical reactions to 12 test substances.

are the urinary tract, endocardium, and brain. Patients with valvular disease of the heart or indwelling prosthetic devices are vulnerable to candidal endocarditis, usually caused by other species (*C. tropicalis* and *C. parapsilosis*). *Candida krusei* has reportedly caused terminal infections in bone marrow transplant patients and in recipients of anticancer therapy.

Laboratory Techniques in the Diagnosis of Candidiasis

A presumptive diagnosis of *Candida* infection is made if budding yeast cells and pseudohyphae are found in specimens from localized infections (figure 18.23*a*). Specimens are cultured on standard media incubated at 30°C. Identification is complicated by the numerous species of *Candida* and other look-alike yeasts, though confirmatory evidence of *C. albicans* can be obtained by the germ tube test, the presence of chlamydospores, and multiple panel systems that test for biochemical characteristics (figure 18.23*b*).

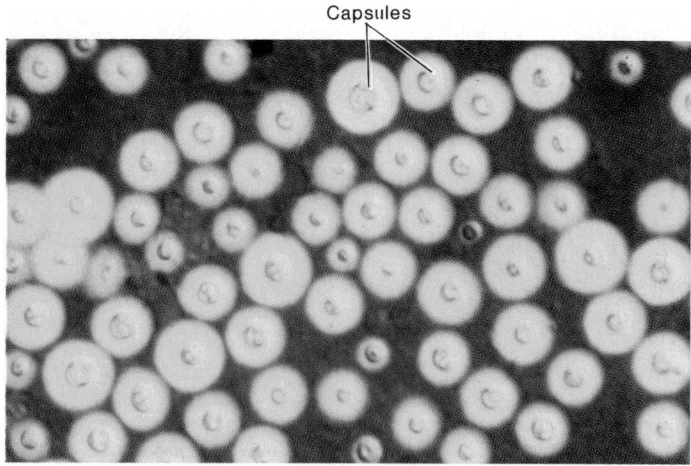

Figure 18.24 *Cryptococcus neoformans*. A negative stain using India ink clearly delineates the large capsules. This method must be performed with care because the capsule is fragile and easily destroyed (×150).

Treatment of Candidiasis

Because candidiasis is almost always opportunistic, it will return in many cases if the underlying disease is not also treated. Therapy for superficial mucocutaneous infection consists of topical antifungal agents (imidazoles and polyenes). Amphotericin B with or without flucytosine is the only choice available for systemic infections. Recurrent bouts of vulvovaginitis are managed by topical ketoconazole ointment, now available as an over-the-counter drug. Vaginal candidiasis is sexually transmissible, thus it is very important to treat both sex partners to avoid reinfection.

Cryptococcosis and *Cryptococcus neoformans*

Another widespread resident of human habitats is the fungus *Cryptococcus neoformans*. This yeast has a spherical to ovoid shape, with small, constricted buds and a large capsule that is important in its pathogenesis (figure 18.24). Its role as an opportunist is supported by evidence that healthy humans have strong resistance to it and that frank infection occurs primarily in debilitated patients. Most cryptococcal infections (*cryptococcoses*) center around the respiratory, central nervous, and mucocutaneous systems.

Epidemiology of *C. neoformans*

The primary ecologic niche of *Cryptococcus neoformans* revolves around birds. It is prevalent in urban areas where pigeons congregate, and it proliferates in the high nitrogen content of

---

cryptococcosis (krip''-toh-kok-oh'-sis) Older names are torulosis and European blastomycosis.

Cryptococcus neoformans (krip''-toh-kok'-us nee''-oh-for'-manz) Gr. *kryptos*, hidden, *kokkos*, berry, *neo*, new, and *forma*, shape.

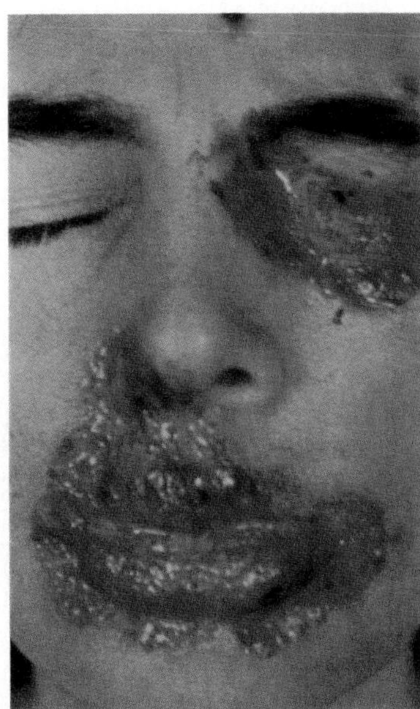

Figure 18.25 A late disseminated case of cutaneous cryptococcosis in which fungal growth produces a gelatinous exudate. The texture is due to the capsules.

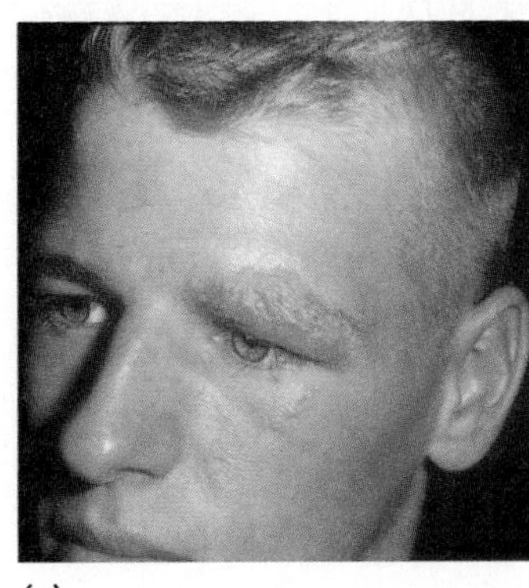

(a)

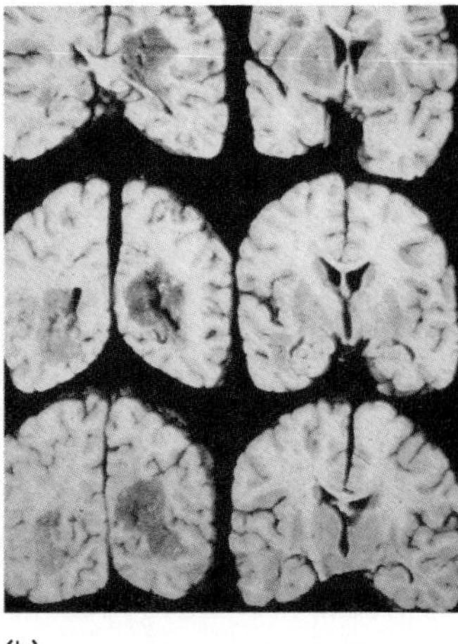

(b)

Figure 18.26 Clinical aspects of aspergillosis. (*a*) Generalized conjunctival infection accompanied by intense and painful inflammation. (Long-term contact lens wearers must guard against this infection.) (*b*) Brain abscesses (darkened areas) consistent with disseminated disease.

droppings that accumulate on pigeon roosts. As the masses of yeast cells dry, they are readily scattered into the air and dust. The species is sporadically isolated from dairy products, fruits, and the healthy human body (tonsils and skin), where it is considered a contaminant. The highest case rate of cryptococcosis occurs among AIDS patients and persons undergoing steroid therapy, though diabetic, leukemic, cancer, and SLE patients are also inclined to infection. The disease exists throughout the world, with highest prevalence in the United States. It is not considered communicable from human to human, though outbreaks have occurred in workers demolishing buildings where pigeons have roosted.

Pathogenesis of Cryptococcosis

The primary portal of entry for *C. neoformans* is respiratory, and most infections are subclinical and rapidly resolved. A few patients with pulmonary cryptococcosis develop fever, cough, and nodules in their lungs. The escape of the yeasts into the blood is intensified by weakened host defenses and attended by severe complications. Sites for which *Cryptococcus* shows an extreme affinity are the brain and meninges. The tumorlike masses formed in these locations may cause headache, mental changes, coma, paralysis, eye disturbances, and seizures. Because of the underlying illness of the host, many cases result in progressive deterioration and death. Some disseminated infections involve the skin, bones, and viscera (figure 18.25).

Diagnosis and Treatment of Cryptococcosis

The first step in diagnosis is negative staining of specimens to detect encapsulated budding yeast cells that do not occur as pseudohyphae. Colony development on blood-enriched or selective media requires two to three days. Two quick screening tests that presumptively differentiate *C. neoformans* from the seven other cryptococcal species are the urease and phenoloxidase tests. Confirmatory results include a negative nitrate assimilation, pigmentation on birdseed agar, and fluorescent antibody tests. Cryptococcal antigen may be detected in a specimen by means of serological tests. Systemic cryptococcosis requires immediate treatment with amphotericin B, often combined with flucytosine or miconazole, over a period of weeks or months.

Aspergillosis: Diseases of the Genus *Aspergillus*

Molds of the genus *Aspergillus* are possibly the most pervasive of all fungi. The estimated 600 species are isolated from soil, vegetation, leaf detritus, food, and compost heaps, and because of their omnipresence in air and dust, they are a serious pest in the microbiology laboratory. Of the eight species involved in human diseases, the thermophilic fungus *A. fumigatus* accounts for the most cases. Aspergillosis is almost always an opportunistic infection, lately posing a serious threat to AIDS, leukemia, and organ transplant patients. It is also involved in allergies and toxicoses (see feature 18.5).

Aspergillus infections usually occur in the lungs. Persons breathing clouds of conidia from the air of graineries, barns, and silos are at greatest risk. In heavily exposed healthy persons, the spores simply germinate in the lungs, forming *fungus balls*. Other benign, noninvasive infections include colonization of the sinuses, ear canals, eyelids, and conjunctiva (figure 18.26*a*). In the invasive form of aspergillosis, the fungus produces a necrotic pneumonia and disseminates to the brain (figure 18.26*b*), heart, skin, and a wide range of other organs. Systemic aspergillosis usually occurs in very ill hospitalized patients and has a very poor prognosis.

Aspergillus (as″-per-jil′-us) L. *aspergere*, to scatter.

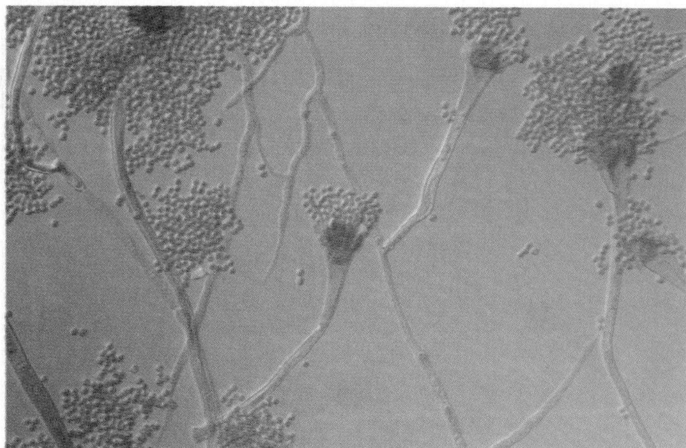

Figure 18.27 The microscopic appearance of *Aspergillus* in specimens (×400). Large, branched septate mycelia are the most common finding in tissues, but the characteristic fruiting heads are also present in some specimens. These molds are not dimorphic and exist only in the hyphal stage.

An obvious branching septate mycelium with characteristic conidial heads in a specimen is presumptive evidence of aspergillosis (figure 18.27). A culture of the specimen that produces abundant colonies and growth at 37°C will rule out its being only an air contaminant. A radioimmunoassay test for rapidly detecting antigen has been effective in diagnosing some cases. Noninvasive disease can be treated by surgical removal of the fungus tumor or by local drug therapy of the lung using amphotericin or nystatin. Combined therapy of amphotericin B and other antifungal agents is the only effective treatment for systemic disease.

Mucormycosis: Zygomycete Infection

The zygomycetes are extremely abundant saprobic fungi found in soil, water, organic debris, and food. The large, prolific sporangia of these fungi release multitudes of lightweight spores that literally powder the living quarters of humans and usually do little harm besides rotting fruits and vegetables. But an increasing number of critically ill patients are contracting a disease called *mucormycosis*. The genera most often involved in mucormycoses are *Rhizopus, Absidia,* and *Mucor.*

Underlying debilities that seem to increase the risk for mucormycosis are acidotic diabetes, leukemia, burns, and malnutrition. Uncontrolled diabetes can lead to reduced pH in the blood and tissue fluids. It is believed that this lowered pH, high temperature, and the glucose content of tissues enhance establishment of the fungus in the body. The specific pathology depends on the portal of entry of the fungus, which can be nasal, pulmonary, cutaneous, or gastrointestinal. In rhinocerebral mucormycosis, infection begins in the nasal passages or pharynx, and then grows in the blood vessels and up through the palate into the eyes and brain (figure 18.28). It progresses so swiftly that death may occur in a week or less. Invasion of the lungs occurs most often in leukemics, and gastrointestinal mucormycosis is a complication in children suffering from protein malnutrition (kwashiorkor). Contaminated dressings applied to burn and surgical patients may lead to skin infection, septicemia, and thrombosis. Unfortunately, the outlook for most of these patients is not hopeful.

mucormycosis (mew″-kor-my-koh′-sis) Any infection caused by members of the Order Mucorales. Also called phycomycosis and zygomycosis.

Feature 18.5 Fungi, Food, and Toxins

Ingesting fungi (mycophagy) is a common and often pleasant experience. In some cases, one eats them by choice, as with gourmet mushrooms, and quite frequently, they are unintentionally consumed in dairy products such as yogurt and cottage cheese, fruits, vegetables, and many other foods.

Although most of the fungi we eat accidently are harmless, poisonous ones cause toxic illness. A number of mushrooms and molds contain mycotoxins that act on the central nervous system, liver, gastrointestinal tract, bone marrow, or kidney. The number one cause of **mycotoxicosis** is the consumption of wild mushrooms, a centuries-old practice that has taken the life of many an avid mushroom hunter. The deadliest of all mushrooms is *Amanita phalloides,* the so-called death- or destroying angel whose cap contains enough toxin to kill an adult, though this happens rarely. Recently, some California and Oregon mushroom fanciers nearly lost their lives after eating a meal of *Amanita.* Clearly, collecting and eating wild mushrooms is recommended only for the most expert mycophagist, and the rest of us should be content to gather mushrooms in the grocery store.

Certain poisonous fungi have been significant in history, and others are tied to ancient religious practices. Rye plants infested by *Claviceps purpurea* develop black masses on the grains known as ergot. When such rye grains are inadvertently incorporated into food and ingested, absorption of the ergot alkaloids causes *ergotism.* People thus intoxicated exhibit convulsions, delusions, a burning sensation in the hands (St. Anthony's fire), and symptoms similar to LSD poisoning. This latter symptom occurs because one of the alkaloids in *Claviceps* and other mind-altering fungi contains lysergic acid, also a component of LSD. In the Middle Ages, millions of people died from ergotism, but now the disease is very rare. In fact, the active agent in ergot is extracted and marketed to treat (in minute amounts) migraine headaches and to induce uterine contractions.

Some mushrooms have played a part in religious rituals through the ages, and others have been taken for their hallucinogenic effects. European priests used the "fly agaric" mushroom to induce trances, visions, and other derangements in the senses. New World tribes ate *Psilosybe*—"magic mushrooms"—to bring on hallucinations and mild trances. Even today, a mushroom-induced higher consciousness is being sought by some people.

Mycotoxicosis is also caused by food contaminated by fungal toxins. One of the most potent of these toxins is *aflatoxin,* a product of *Aspergillus flavus* growing on grains, corn, and peanuts. Although the mold itself is relatively innocuous, the toxin is carcinogenic and hepatotoxic. It is especially lethal to turkeys, ducklings, pigs, calves, and trout that have eaten contaminated feed. In humans, it appears to be a cofactor in liver cancer. The potential hazard aflatoxin poses for domesticated animals and humans necessitates constant monitoring of peanuts, beer, grains, nuts, vegetable oils, animal feed, and even milk. A quick exposure of seeds to ultraviolet radiation is a useful screening test because they glow when contaminated by the toxin. During times of drought, the aflatoxin content of cereal grains and corn tends to increase. Evidently as the plant heads dry, the kernels split open and the ubiquitous soil fungus grows in them. The recent American drought precipitated a crisis when large amounts of inedible feed grains, corn, and even milk had to be destroyed or buried in landfills.

aflatoxin (af″-lah-toks′-in) Acronym for *Aspergillus flavus* toxin.

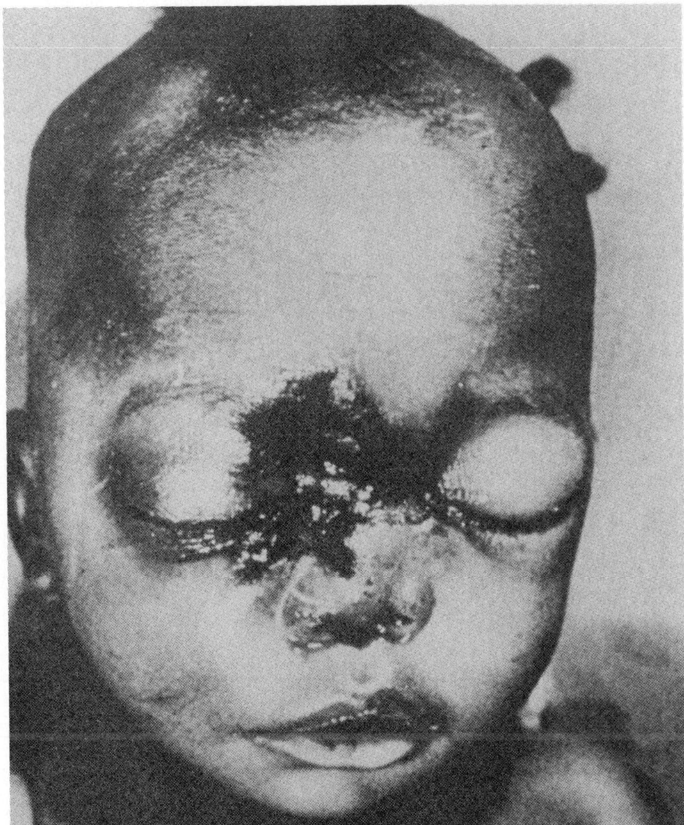

Figure 18.28 Mucormycosis in a malnourished child. This infection occurs in individuals so weakened that they usually have little chance for survival.

Diagnosis of mucormycosis rests on clinical grounds and biopsies or smears. Very large, thick-walled, nonseptate hyphae scattered through the tissues of a gravely ill diabetic or leukemic patient are very suggestive of mucormycosis. Attempts to isolate the molds from specimens are frequently met with failure, and confusion of the isolates with air contaminants is also a problem. Treatment, which is effective only if commenced early in the infection, consists of surgical removal of infected areas, accompanied by high doses of amphotericin.

Miscellaneous Opportunists

Geotrichosis is a rare mycosis caused by *Geotrichum candidum* (figure 18.29*a*), a mold commonly found in soil, in dairy products, and on the human body. The mold is not highly aggressive and is primarily involved in secondary infection in the lungs of tubercular or highly immunosuppressed patients. Species of *Fusarium* (figure 18.29*b*), a common soil inhabitant and plant pathogen, occasionally infect the eyes, toenails, and burned skin. Mechanical introduction of the fungus into the eye (for example, by contaminated contact lenses) may induce a severe mycotic ulcer. It can also cause a patchy infection of the nail bed somewhat like tinea unguium. *Fusarium* colonization of patients with whole-body burns has occasionally led to mortalities. Unusual opportunistic infections have been reported for other commonplace fungi such as *Penicillium, Paecilomyces,* and the red yeast *Rhodotorula.*

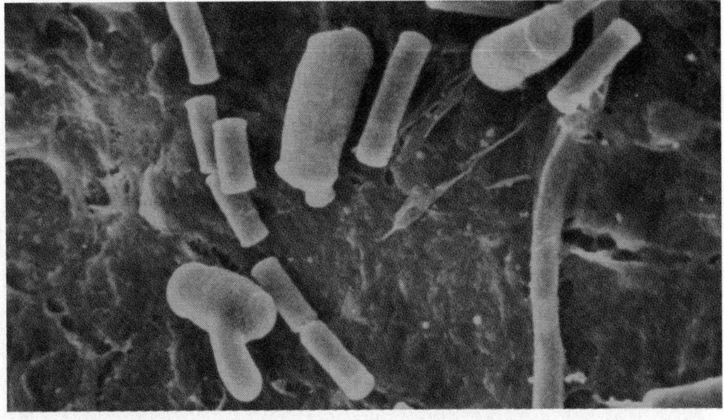

(a)

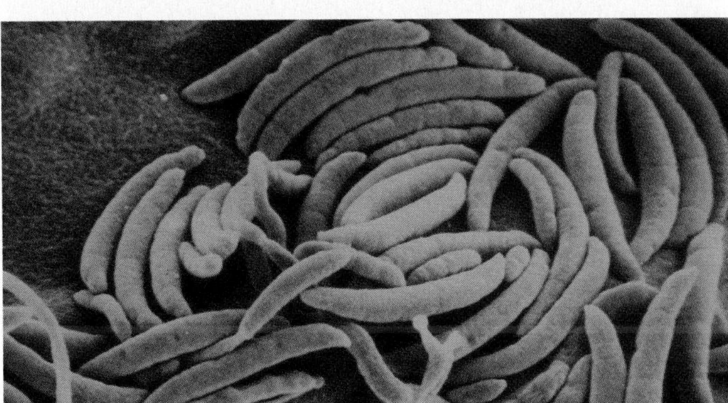

(b)

Figure 18.29 Common fungi that can cause uncommon infections. (*a*) *Geotrichum candidum* with arthrospores (×2,000). (*b*) *Fusarium* with crescent-shaped conidia (×750).

Fungal Allergies and Intoxications

It should be amply evident by now that fungi are constantly present in the air we breathe. In some persons, inhaled fungal spores are totally without effect; in others, they instigate infection; and in still others, they stimulate hypersensitivity. So significant is the impact of fungal allergens that airborne spore counts are performed by many public health departments as a measure of potential risk (see figure 14.2). Among the significant fungal allergies are (1) asthma, often occurring in seasonal episodes; (2) farmer's lung, a chronic and sometimes fatal allergy of agricultural workers exposed to moldy grasses; (3) teapicker's lung; (4) bagassosis, a condition caused by inhaling moldy dust from processed sugar cane debris; and (5) bark stripper's disease, caused by inhaling spores from logs. The role of fungi as toxic agents is discussed in feature 18.5.

Actinomycetes: Funguslike Bacteria

One group of bacteria traditionally included in a study of medical mycology are the pathogenic **actinomycetes.** The precedent

actinomycetes (ak″-tih-noh-my′-seets) Gr. *actinos,* a ray, and *myces,* fungi. Members of the Order Actinomycetales.

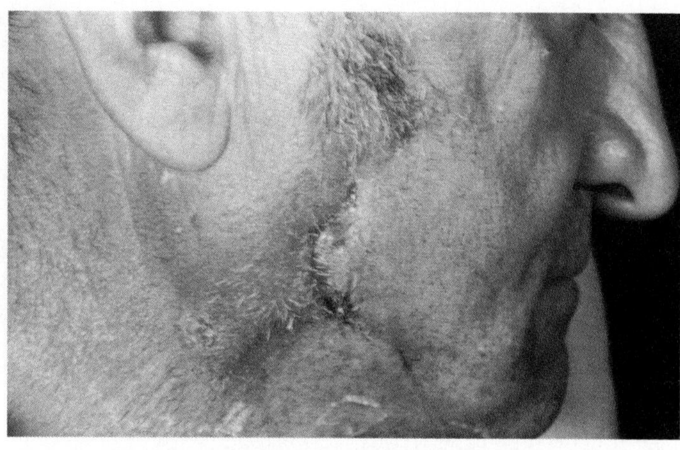

(a)

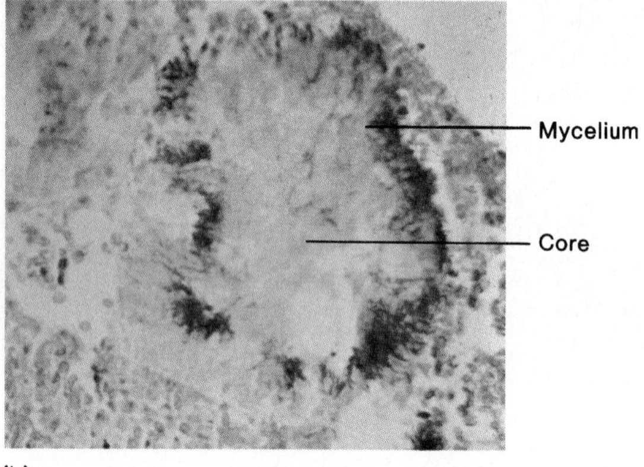

(b)

Figure 18.30 Clinical symptoms and microscopic signs of actinomycosis. (*a*) Early periodontal lesions work their way to the surface in cervicofacial disease. The lesions eventually break out and drain. (*b*) A section of crushed sulfur granule shows the mycelium surrounding a sulfur-yellow core.

for including these decidedly procaryotic organisms with the eucaryotic fungi was established by a century-old investigation that showed a superficial resemblance between them. This resemblance includes the formation of spores by some actinomycetes, the production of a branching, fragmented mycelium (figure 18.30*b*), and the chronic, granulomatous, disfiguring diseases they cause. In time, it became evident that actinomycetes were not fungi or even ancestors of fungi as had once been thought. In structure and physiology, the actinomycetes are filamentous, gram-positive bacteria related to the mycobacteria. The main genera of actinomycetes involved in human disease are *Actinomyces* and *Nocardia* (table 18.4).

Actinomycosis

Actinomycosis is an endogenous infection of the cervicofacial, thoracic, or abdominal regions by species of *Actinomyces* living normally in the human oral cavity, tonsils, and intestine. One form of disease, **lumpy jaw,** was once a common complication of tooth extraction, poor oral hygiene, and carious teeth. The lungs, abdomen, and uterus are also sites of infection.

Table 18.4 Selected Actinomycetes, Diseases, and Identifying Characteristics

	Actinomyces	*Nocardia*
Name	*Actinomyces israelii*	*N. asteroides* *N. brasiliensis*
Diseases	Actinomycosis (lumpy jaw)	Nocardiosis Mycetoma
Principal Characteristics	Branching filaments that fragment to rodlike cell	Filaments fragment to coccoid or elongate cells; may be acid-fast, with aerial spores
Respiration	Anaerobic, microaerophilic	Aerobic
Epidemiology	Worldwide; microbe is part of oral flora, invades through trauma	Worldwide; microbe lives in soil, water, grasses; spores inhaled
Diagnosis	Sulfur granules; short, branching gram-positive filaments	No sulfur granules; acid-fast, coccoid, and bacillary arthrospores

In cervicofacial involvement, the bacteria enter a damaged area of the oral mucous membrane and begin to multiply there. Diagnostic signs are swollen, tender nodules in the neck or jaw that give off a discharge containing macroscopic (1–2 mm) sulfur granules (figure 18.30). In most cases, infection remains localized, but bone invasion and systemic spread may occur in persons with poor health. Thoracic actinomycosis is a necrotizing lung disorder that can project through the chest wall and ribs. Abdominal actinomycosis is a complication of burst appendices, gunshot wounds, ulcers, or intestinal damage. Uterine actinomycosis has been increasingly reported in women using intrauterine contraceptive devices. Infections are treated with surgical drainage and antibacterial antibiotics (penicillin, erythromycin, tetracycline, and sulfonamides).

Compelling evidence indicates that oral actinomyces play a strategic role in the development of plaque and dental caries. Studies of the oral environment show that *A. viscosus* and certain oral streptococci are the very first microbial colonists of the tooth surface. Both groups have specific tooth-binding powers and can adhere to each other and to other species of bacteria (see chapter 17).

Nocardiosis

Nocardia is a genus of bacteria widely distributed in the soil. Most species are not infectious, but one, *N. brasiliensis,* is a primary pulmonary pathogen, and others, *N. asteroides* and *N. caviae,* are opportunists. Nocardioses fall into the categories of

Nocardia (noh-kar'-dee-ah) After Edmund Nocard, the French veterinarian who first described the genus.

pulmonary, cutaneous, or subcutaneous. Most cases in the United States are reported in patients with deficient immunity, but a few occur in normal individuals.

Pulmonary nocardiosis is a form of bacterial pneumonia with pathology and symptoms similar to tuberculosis. The lung develops abscesses and nodules, and may consolidate. Often the lesions extend to the pleura and chest wall, and disseminate to the brain, skin, and kidneys (figure 18.31). *Nocardia brasiliensis* is also one of the chief etiologic agents in mycetoma, discussed earlier in this chapter. Microbiologists at the University of California have recently reported a possible connection between a severe neurologic disorder, Parkinson's disease, and brain infection by *Nocardia asteroides*. Further studies dealing with the immunologic responses to the microbe promise to shed more light on this interesting discovery.

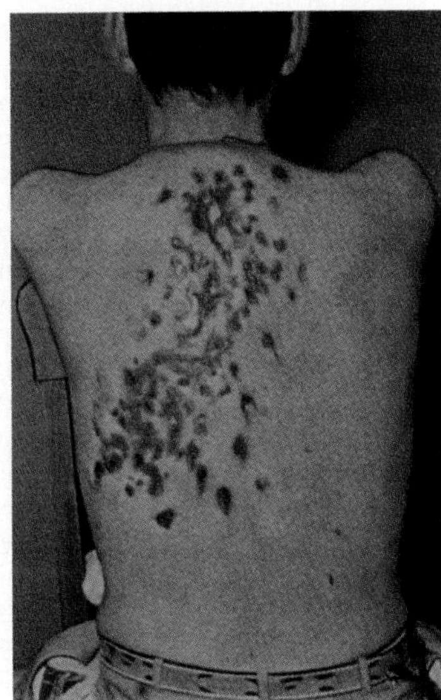

Figure 18.31 Nocardiosis, a case of pulmonary disease that has extended through the chest wall and ribs to the cutaneous surface.

Chapter Review with Key Terms

Fungal Infections

General Principles

Fungal diseases are called **mycoses,** and their names end in **-osis,** or **-iasis.** Infectious fungi occur in groups based upon the virulence of the pathogen and the level of involvement, whether **systemic, subcutaneous, cutaneous,** or **superficial.**

True or **primary pathogens** have virulence factors that allow them to invade and grow in a healthy host. They are also **thermally dimorphic,** occurring as hyphae in their natural habitat and converting to yeasts while growing as parasites at body temperature (37°C). When spores are inhaled, they initiate **primary pulmonary infection (PPI).** Systemic and cutaneous complications are common, especially in weakened patients. Disease is accidental, not transmissible, and causes a long-term allergic reaction to fungal proteins.

Opportunistic fungal pathogens may be normal flora or common inhabitants of the environment. They lack well-developed virulence and thermal dimorphism, but can attack patients whose immunity has been compromised by surgery, cancer, AIDS, or organic disease. Infections may be local and cutaneous or systemic. Other fungal pathogens such as the **dermatophytes** (skin fungi) are intermediate in virulence. They will grow if they come in contact with the body of a healthy person, but are not highly invasive.

Epidemiology: True pathogens are endemic to certain geographic areas; other fungal pathogens are widely distributed; dermatophytoses are the most prevalent; infections by the true pathogens are common in endemic regions; opportunistic infections are increasing due to improved medical technology that maintains immunocompromised patients.

Immunity: Primarily nonspecific barriers, inflammation, and cell-mediated; antibodies may be used to detect disease in some cases.

Diagnosis/Identification: Requires microscopic examination of fresh, stained specimens, culturing of pathogen in selective and enriched media, and specific biochemical and *in vitro* serological tests. Skin testing for true and opportunistic pathogens can determine prior disease, but is not useful in diagnosis.

Control Measures: Drugs include intravenous amphotericin B and/or flucytosine for local and systemic disease and imidazoles and nystatin for skin and membrane infections. No useful vaccines exist.

Systemic Mycoses Caused by True Pathogens

Histoplasmosis (Ohio Valley Fever): The agent is ***Histoplasma capsulatum,*** a dimorphic fungus with macro- and microconidia and small budding "fish-eye" yeasts; distributed worldwide, but most prevalent in eastern and central regions of the U.S.; infection common where land or animal excreta have been disturbed and blown by wind. Inhaled spores produce PPI, with flu-like symptoms; may be asymptomatic; may progress in patients with weak defenses to systemic involvement of a variety of organs and chronic lung disease.

Coccidioidomycosis (Valley Fever): ***Coccidioides immitis*** is the cause. The free-living stage has blocky **arthroconidia**; parasitic lung stage forms swollen **spherules** containing endospores; agent lives in alkaline soils and semiarid, hot climates, and is endemic to the southwestern U.S.; infection occurs when soils are disturbed by digging or wind and arthrospores are inhaled. Arthrospores convert to spherules that burst, reinfect lung, and form nodules; dissemination into major organs may cause fatal disease.

Blastomycosis (Chicago Disease): The agent is ***Blastomyces dermatitidis.*** Free-living form has simple conidia; parasitic yeast form is ovoid cell with large bud; distributed in soil, organic matter, and manure of a large section of the midwestern and southeastern U.S.; inhaled conidia convert to yeasts and multiply in lungs; inflammatory response produces cough, fever; may progress to chronic cutaneous nodules and abscesses and bone and nervous system infection.

Paracoccidioidomycosis (South American Blastomycosis): ***Paracoccidioides brasiliensis*** has an unusual, flowerlike yeast form; it is distributed in various regions of Central and South America; infection occurs through inhalation or inoculation of spores; systemic disease not common.

Subcutaneous Mycoses

Diseases of tissues in or below the skin.

Sporotrichosis (Rose-gardener's Disease): Caused by ***Sporothrix schenckii,*** a common saprobic fungus found worldwide that infects the appendages and lungs. The **lymphocutaneous** variety occurs when contaminated plant matter penetrates the skin and the pathogen forms a local nodule, then spreads to nearby lymph nodes.

Chromoblastomycosis: A deforming, deep infection of the tissues of the legs and feet with pigmented fungi following trauma.

Mycetoma: A progressive, tumorlike disease of the hand or foot due to chronic fungal infection; may lead to loss of the body part.

Cutaneous Mycoses

Infections confined to the skin.

Dermatophytoses: Commonly called tineas, ringworm, athlete's foot. Caused by dermatophytes, primarily species in genera *Trichophyton, Microsporum,* and *Epidermophyton.* Nearly any keratinized site may be attacked (skin, hair, nails). Some forms are communicable from other humans, animals; some are soil forms, found worldwide; infection facilitated by broken, moist, chafed skin and by hardiness of spores; reactions to fungi are inflammatory, causing itching and pain.

Ringworm of scalp (tinea capitis) affects scalp and hair-bearing regions of head; hair may be lost. Ringworm of beard (tinea barbae) afflicts the chin and beard of adult males. **Ringworm of body** (tinea corporis) occurs as inflamed red ring lesions anywhere on smooth skin. **Ringworm of groin** (tinea cruris) affects groin and scrotal regions exposed to constant moisture; readily transmissible in close living conditions. **Ringworm of foot** and **hand** (tinea pedis and tinea manuum) are caused by enclosing feet in hot sweaty shoes (athlete's foot), with hands becoming involved due to self inoculation; also spread by exposure of feet to public floors; occurs between toes, soft tissues and on nails. **Ringworm of nails** (tinea unguium) is a persistent colonization of the nails of the hands and feet that distorts the nail bed.

Superficial Mycoses

Infections of outermost superficial epidermal structures. **Tinea versicolor** causes mild scaling, mottling of skin; **white piedra** is whitish or colored masses on the long hairs of the body; **black piedra** is dark, hard concretions on scalp hairs.

Opportunistic Mycoses

Candidiasis: Diseases of ***Candida albicans*** and other *Candida* species. Yeast forms are chlamydospores and pseudohyphae; is normal resident of mouth, vagina, intestine, skin; infection occurs in cases of lowered resistance (babies, pregnancy, drug therapy, AIDS, trauma); disease acquired endogenously, though may be spread from mother to child or sexually. Diseases include **thrush,** invasion of mucous membranes of mouth and throat that produces thick, white adherent growth, often in neonates; **vulvovaginal yeast infection,** a painful inflammatory condition of the internal and external reproductive structures that causes ulceration and whitish discharge; **onychomycosis,** infection of keratinized structures due to long-term occupational exposure; **cutaneous candidiasis,** occurring in moist areas of skin that rub and in neonates and burn patients.

Cryptococcosis: Caused by ***Cryptococcus neoformans,*** a yeast with a well-developed capsule; widespread resident of urban soils, especially in association with pigeon roosts; dust-borne and inhaled; common infection of AIDS, cancer, or diabetes patients; pulmonary disease causes cough, fever, lung nodules; dissemination to meninges and brain infection can cause severe neurological disturbance and death; skin may be involved.

Aspergillosis: Aspergillus is a very common soil fungus that occurs on plants, in air; inhalation of clouds of spores causes fungus balls in lungs; invasive form occurs in highly debilitated patients and is often fatal.

Mucormycosis: Caused by large molds. *Rhizopus* and *Mucor* invade persons with acidosis (diabetics, malnutrition), which allows these ordinarily harmless air contaminants to grow in the mucous membranes of the nose, eyes, and brain, with deadly consequences.

Mycoses are increasingly caused by common unaggressive fungi such as *Geotrichum, Fusarium, Rhodotorula,* and *Penicillium.*

Fungal Allergies and Mycotoxicoses

Airborne fungal spores are common sources of atopic allergies. Ingestion of fungal toxins leads to **mycotoxicoses**; most commonly caused by eating poisonous or hallucinogenic mushrooms or food contaminated with fungal toxins such as ergot or **aflatoxin.**

Diseases Caused by Actinomycetes

The genera ***Actinomyces*** and ***Nocardia*** are filamentous bacteria, but the diseases they cause mimic those of fungi, and they are traditionally included with the mycoses.

Actinomycosis: A. israelii occurs in the human oral cavity, tonsils, intestine; oral disease or surgery may result in infection called **lumpy jaw**; also abdominal, thoracic, uterine complications.

Nocardiosis: N. brasiliensis causes pulmonary disease similar to TB, with abscesses and nodules that may spread to chest wall; one agent of mycetoma.

True–False Questions

Determine whether the following statements are true (T) or false (F). If you feel a statement is false, explain why, and reword the sentence so that it reads accurately.

___ 1. In thermal dimorphism, a fungus alternates between hyphal and yeast phases in response to temperature.

___ 2. In pathogens, the yeast phase is more virulent than the hyphal phase.

___ 3. Opportunistic fungi are more virulent than primary pathogenic fungi.

___ 4. True pathogens are endemic to specific geographic areas.

___ 5. Most systemic mycoses begin as pulmonary infections following the inhalation of spores from an environmental reservoir.

___ 6. Histoplasmosis is endemic to the Ohio Valley, and coccidioidomycosis is endemic to the southwestern United States.

___ 7. The common name for coccidioidomycosis is Valley fever and for blastomycosis, Chicago disease.

___ 8. Skin testing is a useful diagnostic procedure for blastomycosis.

___ 9. A mycetoma is a fungal tumor of the lung.

___ 10. Ringworm is a communicable fungal infection of the skin, hair, and nails.

___ 11. Dermatophytes invade and grow in the lungs and other organs.

___ 12. Candidiasis is a local and systemic infection readily transmitted to others through casual contact.

___ 13. Cryptococcosis is associated with contact with pigeon droppings.

___ 14. Actinomycetes are bacteria that produce funguslike diseases.

Concept Questions

1. Explain how true pathogens differ from opportunists in physiology, virulence, types of infection, and distribution. Give examples of each. Explain the adaptation and importance of thermal dimorphism.
2. Why are most fungi considered facultative parasites and fungal infections considered non-communicable? Give examples of two communicable mycoses.
3. What factors are involved in the pathogenesis of fungi? What parts of fungi are infectious? What are the primary defenses of humans against fungi?
4. What is the role of skin testing in tracing fungal infections? For which groups is it most useful?
5. Overview the major steps in fungal identification. What is one advantage over bacterial identification? How are *in vitro* and *in vivo* tests different? Describe an immunodiffusion test; the histoplasmin test.
6. What is true about treating fungal diseases with drugs?
7. Differentiate between systemic, subcutaneous, cutaneous, and superficial infections. What is the initial infection site in true pathogens? What does it mean if an infection disseminates?
8. Briefly outline the life cycles of *Histoplasma capsulatum, Coccidioides immitis,* and *Blastomyces dermatitidis.* In general, what is the source of each pathogen? Is disease usually overt or subclinical? To which organs do these pathogens disseminate? What causes cutaneous disease? How are cutaneous diseases treated?
9. What are the most common subcutaneous infections? Why are these organisms not more invasive? Describe the life cycle and the disease caused by *Sporothrix.* How could one avoid it? Differentiate between chromoblastomycosis and phaeohyphomycosis.
10. What are the dermatophytoses? What is meant by the term keratophile? What are the fungi actually feeding upon? What do the terms ringworm and tinea refer to? What are the three main sources of dermatophytes?
11. What are the five major types of ringworm, their common names, symptoms, and epidemiology? What conditions predispose an individual to these infections?
12. What is the difficulty in ridding the epidermal tissues of certain dermatophytes? What is necessary in treatment?
13. Briefly describe three superficial mycoses.
14. What is the relationship of *Candida albicans* to humans? What medical or other conditions may predispose a person to candidiasis? Describe the principal infections of women, adults of both sexes, and neonates. Why should women be screened for candidiasis? What are the treatments for localized and systemic candidiasis? If a woman tests positive for infection, besides treating her, what else should be done?
15. Describe the background of cryptococcosis. What specimens could be used to diagnose it? Which group of people is currently most vulnerable to this disease? What makes the fungus somewhat unique?
16. Briefly list the major diseases caused by *Aspergillus* and zygomycetes. Why are fungus balls not an infection?
17. Describe the bacteria in the actinomycete group, and explain why they are similar to fungi. Briefly describe two of the common diseases caused by this group.

Practical/Thought Questions

1. Why are humans not constantly battling fungal infections, considering how prevalent fungi are in the environment?
2. Name several medical conditions that compromise the immunites of patients. Can you think of some solutions to the problems of opportunistic infections in extremely compromised patients? Is there any such thing as a harmless contaminant with these patients? Can medical personnel do anything to lessen their patients' exposure to fungi?
3. Describe the consequences if a virulent fungal pathogen infects a severely compromised patient.
4. Using the general guidelines presented in the "Introduction to Medical Bacteriology," page 471, how can you tell for sure if a fungal isolate is the actual causative agent or merely a contaminant?
5. Why are some dermatophytes considered good parasites? Why would infections from soil and animal species be more virulent than ones from fungi adapted to humans?
6. Why are diabetics so susceptible to fungal infections?
7. What is the function of mushroom and other fungal toxins? How are these toxins harmful, and how are they used to our advantage?
8. A patient is admitted to a hospital with severe chest pain, cough, and fever. Because lab tests show tiny cocci-like cells in sputum, he is treated for bacterial pneumonia and released. However, the patient does not respond to antibiotics, and the lung condition worsens. After a few days, skin bumps begin to appear on the face and chest. A lung biopsy indicates huge sporangia filled with tiny spores, and the patient's history reveals a recent trip to Arizona. On this basis, how would you diagnose the infection? What other fungal disease might it suggest? What drug should have been given?
9. A male researcher wanting to determine the pathogenesis of a certain fungus that often attacks women did the following experiment on himself. He taped a heavily inoculated piece of gauze to the inside of his thigh and left it there for several days, periodically moistening it with sterile saline. After a time, he experienced a severe skin reaction that resembled a burn and peeled away. What do you suppose was the fungus, and what does this experiment tell you about the factors in pathogenesis related to environment and sex?
10. A man involved in demolishing an older downtown building is the victim of a severe lung infection. Without any lab data, what two fungal infections do you suspect? What is your reasoning?

CHAPTER 19

The Parasitic Diseases of Humans

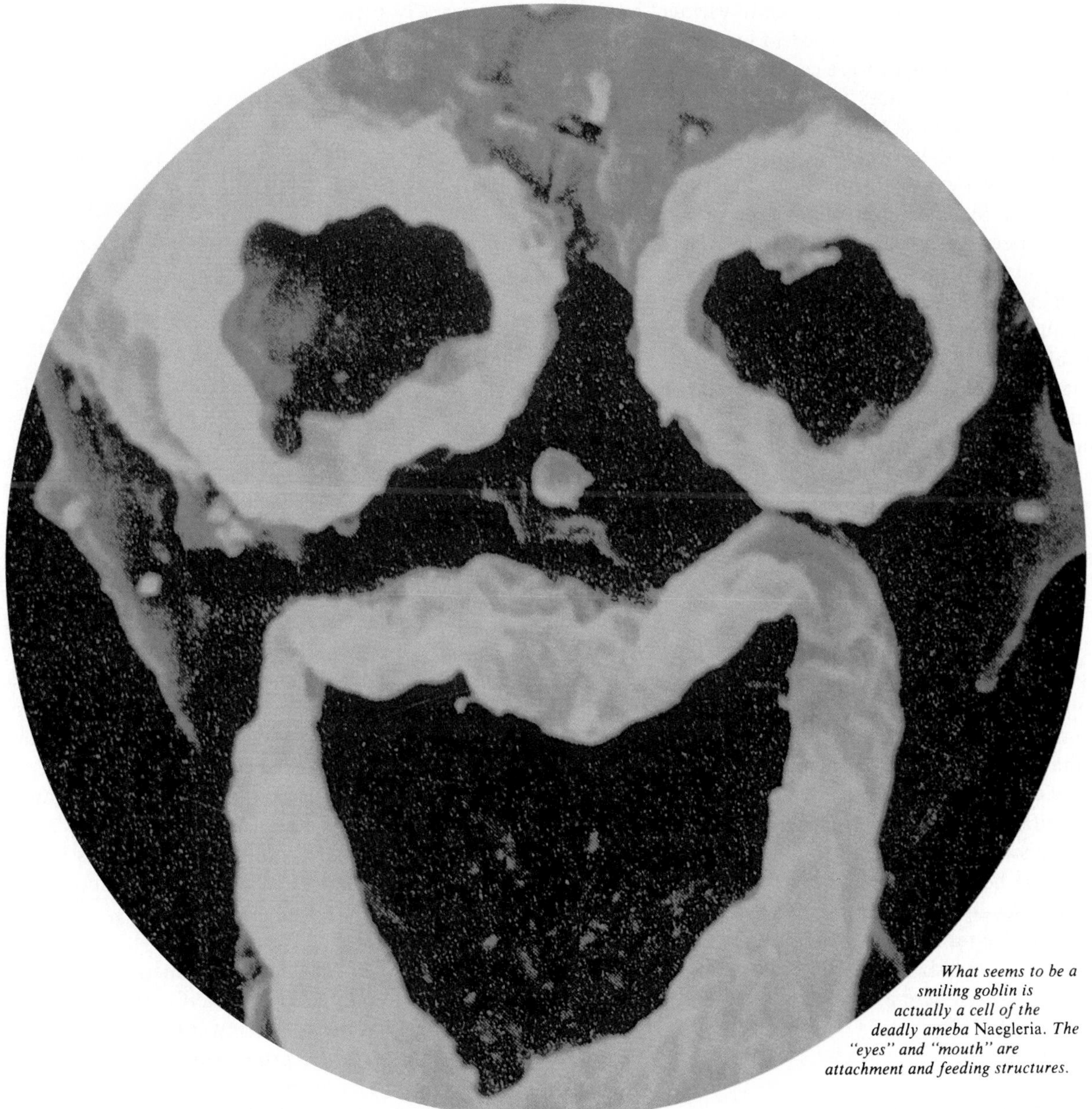

What seems to be a smiling goblin is actually a cell of the deadly ameba Naegleria. *The "eyes" and "mouth" are attachment and feeding structures.*

Chapter Preview

Parasitology is traditionally the study of eucaryotic parasites. It encompasses the protozoa (unicellular animals) and the helminths (worms), which are sometimes collectively called the macroparasites, as opposed to the microparasites (bacteria and viruses). Although this study is inherently fascinating and practical, we can afford only a compact survey of clinically important protozoa and helminths. This chapter covers elements of their morphology, life cycle, epidemiology, pathogenic mechanisms, and methods of control. Arthropods, which are often included in the field of parasitology, have been previously discussed as vectors for certain viral and bacterial infections in chapters 11, 16, and 17. Other chapters containing information on parasites are chapter 6 (nutrition) and chapter 10 (chemotherapy).

The Parasites of Humans

Due to their larger size and greater visibility, parasites and their diseases have been known since ancient times. Even so, the science of parasitology is relatively young and still in the early stages of definition at the molecular level. One cannot underestimate the impact of parasites on the everyday lives of humans. A 1990 WHO listing of the most prevalent tropical diseases indicated that seven out of eight were caused by parasites. In industrialized countries, we are less aware of parasites, but nevertheless, they infest most humans, in both hidden and overt forms at some time in their lives. The distribution of parasitic diseases is influenced by many factors, including rapid jet travel, immigration, and the increased number of immunocompromised patients. Because parasites can literally travel around the world, it is not uncommon for exotic parasitic diseases to be discovered even in the United States. Another occurrence that has given certain parasites new notoriety is their involvement in the secondary infections of AIDS patients.

Humans play host to dozens of different parasites. Some are indigenous to humans alone and are spread to other humans through direct contact or contaminated food and water. Others are transmitted by arthropod vectors from human to human. And still others are zoonoses that humans acquire from contact with pets or other animals.

Typical Protozoan Pathogens

When we first met the protozoa in chapter 4, they were described as single-celled, animal-like microbes usually having some form of motility. The four commonly recognized groups are sarcodinians (amebae), flagellates, ciliates, and sporozoans. Although the life cycles of pathogenic protozoa can vary a great deal, many of them propagate by simple asexual cell division of the active feeding cell, called a **trophozoite,** or by alternating between a trophozoite and **cyst,** a form that can survive for periods outside the host and in many cases also functions as an infective body (see figure 4.27). Others have a more complex life cycle that alternates between asexual and sexual phases (see later discusssion of sporozoans). The protozoan parasites and diseases to be surveyed here are listed in table 19.1.

Infective Amebas

Entamoeba histolytica and Amebiasis

Amebas are widely distributed in aqueous habitats and are frequent parasites of animals, but only a small number of them have the necessary virulence to invade tissues and cause serious pathology. The most significant pathogenic ameba is ***Entamoeba histolytica.*** The relatively simple life cycle of this parasite alternates between a large (20–40 μm) trophozoite that is actively motile by means of pseudopods and a smaller (10–15 μm), compact, nonmotile cyst (figure 19.1). The trophozoite lacks most of the organelles of other eucaryotes such as mitochondria, Golgi apparatus, endoplasmic reticulum, and lysosomes, and it has a large single nucleus that contains a prominent nucleolus called a *karyosome.* Amebas from fresh specimens are often packed with food vacuoles containing digested or whole erythrocytes and bacteria. The mature cyst is encased in a thin, yet tough wall and contains four nuclei as well as distinctive bodies called *chromatoidals,* which are actually dense clusters of ribosomes.

Table 19.1 Pathogenic Protozoa, Infections, and Primary Sources

Protozoan	Reservoir/Source
Ameboid Protozoa	
Amebiasis: *Entamoeba histolytica*	Human/water and food
Brain infection: *Naegleria, Acanthamoeba*	Free-living in water
Ciliated Protozoa	
Balantidiosis: *Balantidium coli*	Zoonotic in pigs
Flagellated Protozoa	
Giardiasis: *Giardia lamblia*	Zoonotic/water and food
Trichomoniasis: *Trichomonas tenax, T. hominis, T. vaginalis*	Human
Hemoflagellates	
Trypanosomiasis: *Trypanosoma brucei, T. cruzi*	Zoonotic/vector-borne
Leishmaniasis: *Leishmania donovani, L. tropica, L. brasiliensis*	Zoonotic/vector-borne
Sporozoan Protozoa	
Malaria: *Plasmodium vivax, P. falciparum, P malariae*	Human/vector-borne
Toxoplasmosis: *Toxoplasma gondii*	Zoonotic/vector-borne
Cryptosporidiosis: *Cryptosporidium*	Free-living
Pneumocystis pneumonia: *Pneumocystis carinii*	Free-living(?)

Epidemiology of Amebiasis Humans are the primary hosts of *E. histolytica.* Infection is usually acquired by ingesting food or drink contaminated with cysts from an asymptomatic carrier. Among the parasitic protozoan diseases, amebic dysentery is surpassed only by malaria in its capacity to cause morbidity and mortality. The ameba is thought to be carried in the intestine of one-tenth of the world's population, and it kills up to 110,000 persons per year. Its geographic distribution is partly due to varying amounts of soil contamination and local sewage disposal practices. Occurrence is highest in tropical regions (Africa, Asia, and Latin America), where night soil[1] or sewage is used on crops, drinking water supplies are contaminated, and sanitation may be substandard. Although the prevalence of the disease is lower in the United States, as many as 10 million persons may harbor the agent.

Entamoeba histolytica (en''-tah-mee'-bah his''-toh-lit'-ih-kuh) Gr. *ento,* within, *histos,* tissue, and *lysis,* a dissolution.

1. Human excrement used as fertilizer.

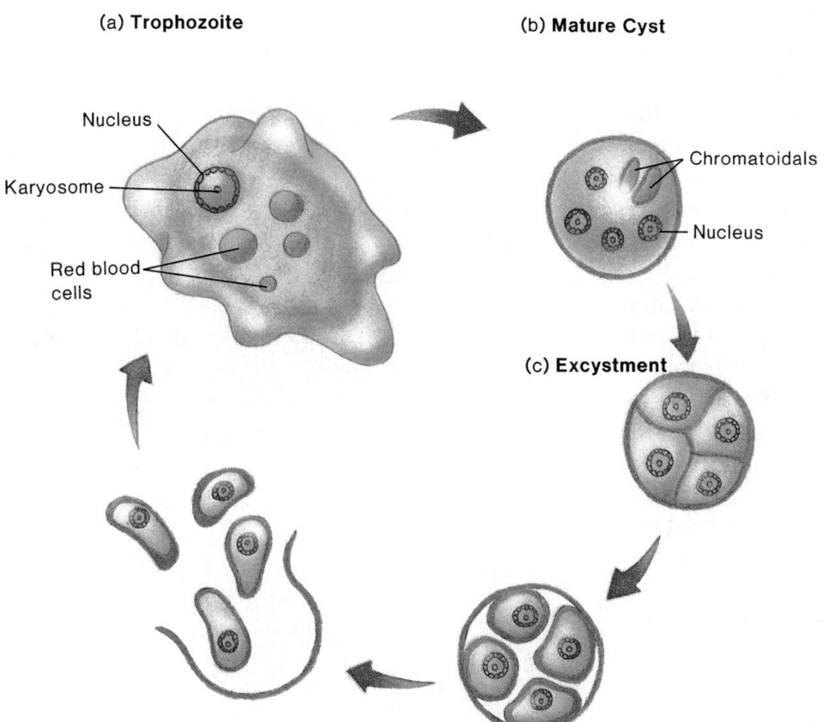

Figure 19.1 Cellular forms of *Entamoeba histolytica.* (*a*) A trophozoite containing a single nucleus, a karyosome, and red blood cells. (*b*) A mature cyst with four nuclei and two blocky chromatoidals. (*c*) Stages in excystment. Divisions in the cyst create four separate cells or metacysts that differentiate into trophozoites and are released.

Epidemics of amebiasis are infrequent, but have been documented in prisons, mental hospitals, juvenile care institutions, and communities where water supplies are polluted. In one of the more bizarre cases, patients in a chiropractic clinic were exposed to contaminated equipment while undergoing a type of colon flushing or irrigation treatment, and several died from the disease. Amebic dysentery can also be transmitted by oral-anal sexual contact as might occur among homosexuals, and up to 30% of U.S. homosexuals may be infected.

Life Cycle, Pathogenesis, and Control of *E. histolytica* Amebiasis typically begins when viable cysts survive the oral and gastric secretions and arrive in the small intestine (see figure 4.34). The alkaline pH and digestive juices of this environment stimulate excystment, and from the cyst emerge four trophozoites—one for each nucleus (figure 19.1*c*). Although freed in the small intestine, the trophozoites do not attach there but are swept along with the intestinal contents into the *cecum* and large intestine, attaching firmly to the mucosa with fine pseudopods. There, the trophozoites mature, multiply by binary fission, and actively move about and feed. In about 90% of patients, infection is asymptomatic or very mild, and the trophozoites do not invade beyond the most superficial layer. What actually determines whether an infection will lead to more severe disease is not completely understood. It is generally accepted that there are several strains of the parasite, some with greater virulence and aggressiveness and others more benign. Inoculation size, normal microbial flora, and host resistance also play some part in the degree of involvement.

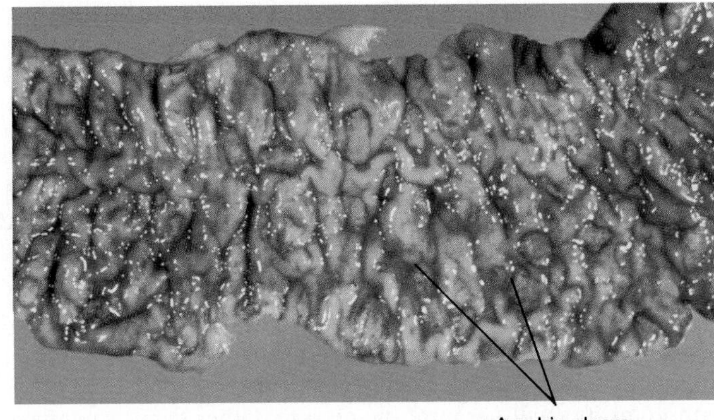

Figure 19.2 Intestinal amebiasis and dysentery of the cecum. Amebic ulcers dot the mucosa in this autopsy specimen.

As hinted by its species name, tissue damage is one of the formidable characteristics of untreated *E. histolytica* infection. Clinical amebiasis exists in intestinal and extraintestinal forms. The initial targets of intestinal amebiasis are the cecum, appendix, colon, and rectum. The ameba secretes enzymes that dissolve tissues, and actively eats its way into deeper layers of the mucosa, leaving small ulcerations (figure 19.2). Symptoms associated with this phase of disease are dysentery (bloody, mucus-filled stools), abdominal pain, fever, diarrhea, fatigue,

cecum (see′-kum) L. *caecus,* blind. The pouchlike anterior portion of the large intestine near the appendix.

and weight loss. The most life-threatening manifestations of intestinal infection are hemorrhage, perforation, appendicitis, and tumorlike growths called amebomas.

Extraintestinal infection occurs when amebas invade the viscera of the peritoneal cavity. The most common site of secondary infection is the liver (amebic hepatitis), where various-sized abscesses composed of necrotized liver cells and fluid develop at the sites of invasive trophozoites. Another common complication is the direct penetration of the diaphragm, leading to pulmonary amebiasis. Less frequently, the parasite invades other ectopic sites such as the spleen, adrenals, kidney, skin, and brain. Severe forms of the disease result in about a 10% fatality rate.

Amebiasis is often spread by chronic healthy carriers in whose intestines the encystment stage of the life cycle is completed. Cyst formation cannot occur in active dysentery because the feces are so rapidly flushed from the body. In the normal stools of carriers, however, the amebas encyst and are continuously shed in feces.

Amebic dysentery has symptoms and signs common to other forms of dysentery, especially bacillary dysentery. Initial diagnosis involves examining a saline suspension of a fecal smear for trophozoites or cysts under high-power magnification and differentiating them from nonpathogenic intestinal amebas that are part of the normal flora (figure 19.3). More definitive identification requires an oil immersion examination of a stained smear. It may be necessary to supplement laboratory findings with clinical data, including symptoms, serological tests, X rays, and the patient's history, which might indicate possible geographic, occupational, or sexual exposure.

The primary treatment for amebic dysentery is some combination of amebicidal drugs to destroy the parasite both in the feces and the tissues. Drugs such as iodoquinol act in the feces, whereas metronidazole, dehydroemetine, and chloroquine are used as systemic therapies. In combined intestinal and extraintestinal infections, both iodoquinol and metronidazole are given, while liver abscess alone is treated with chloroquine. Dehydroemetine (a less toxic form of emetine) is used to control symptoms but will not cure the disease. Other drugs are given to relieve diarrhea and cramps, while lost fluid and electrolytes are replaced by oral or intravenous therapy. Extreme forms of intestinal or liver amebiasis may require surgical removal of the diseased tissue to eliminate the parasite completely and cure the disease. Infection with *E. histolytica* provokes antibody formation against several antigens, but none appear to relieve amebiasis or prevent reinfection. Because neither natural nor artificial active immunity is presently practical, prevention depends upon concerted efforts similar to those for other enteric diseases (see feature 16.8). The regular chlorination of water supplies is insufficient to kill the cysts, and more rigorous methods such as boiling or iodine are required if contamination has occurred.

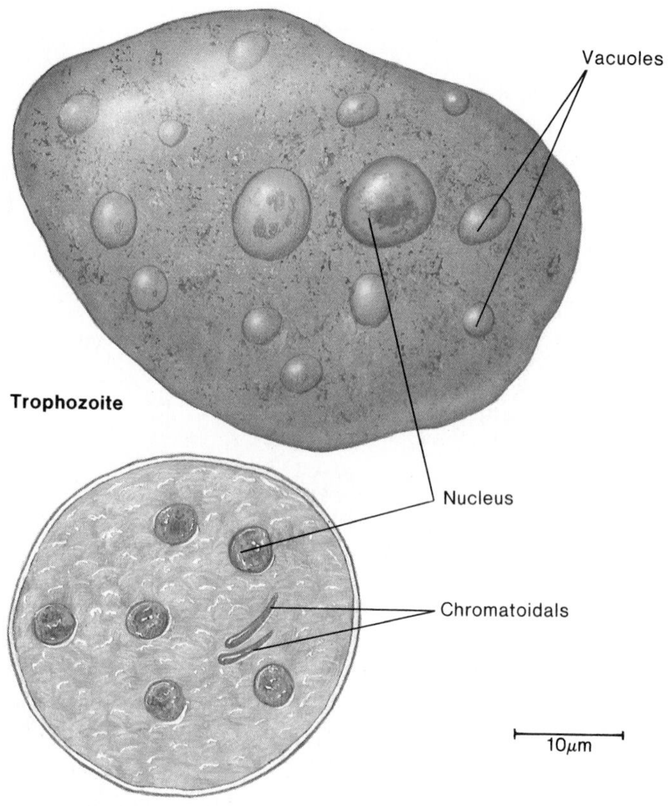

Figure 19.3 *Entamoeba coli,* a nonpathogenic intestinal ameba, may be mistaken for *E. histolytica,* but it is larger and less motile, with short, blunt pseudopodia. Its cyst has spiny chromatoidals and from one to eight nuclei.

Amebic Infections of the Brain

Two common free-living protozoans that cause a rare and usually fatal infection of the brain are ***Naegleria fowleri*** and ***Acanthamoeba.*** Both genera are accidental parasites that invade the body only under unusual circumstances. The trophozoite of *Naegleria* is a small, flask-shaped ameba that moves by means of a single broad pseudopod and has prominent feeding structures, or amebostomes, that lend it a facelike appearance in scanning electron micrographs (see chapter opening illustration). It forms a rounded, thick-walled, uninucleate cyst that is resistant to temperature extremes and mild chlorination. *Acanthamoeba* has a large ameboid trophozoite with spiny multiple pseudopods and a double-walled cyst.

Both *N. fowleri* and *Acanthamoeba* ordinarily inhabit standing fresh or brackish water, lakes, puddles, ponds, hot springs, moist soil, and even swimming pools and hot tubs. They

Naegleria fowleri (nay-glee'-ree-uh fow'-ler-eye) After F. Nagler and M. Fowler.
Acanthamoeba (ah-kan''-thah-mee'-bah) Gr. *acanthos,* thorn.

are especially abundant in warm water with a high bacterial count. Most cases of *Naegleria* meningoencephalitis reported worldwide occur in healthy young persons who have been swimming in warm, natural bodies of fresh water. One epidemic in Australia killed six persons after a public water supply was inadequately disinfected, and in Belgium, an outbreak was reported among persons who had bathed in polluted canal water.

Infection may begin when amebas are forced into human nasal passages as a result of swimming, diving, or other aquatic activities. Being suddenly inoculated into the abnormal but favorable habitat of the nasal mucosa, the ameba burrows in, multiplies, migrates up the olfactory nerves, crosses the cribiform plate, and enters the brain and surrounding structures. The result is **primary acute meningoencephalitis,** a rapid, massive destruction of brain and spinal tissue that causes hemorrhage and coma and invariably ends in death within a week or so.

Unfortunately, *Naegleria* meningoencephalitis advances so rapidly that treatment usually proves futile. Studies have indicated that early therapy with amphotericin B, sulfadiazine, tetracycline, or ampicillin, alone or in some combination, may be of some benefit. Because of the wide distribution of the ameba and its hardiness, no general means of control exists. Public swimming pools and baths should contain high levels of chlorine and be checked periodically for the ameba.

Acanthamoeba differs from *Naegleria* in its portal of entry, invading broken skin, the conjunctiva, and occasionally the lungs and urogenital epithelia. It also varies in the kinds of persons it attacks. Most cases occur in persons with debilitating diseases or trauma. Although it causes a meningoencephalitis somewhat similar to that of *Naegleria,* the course of infection is lengthier. Persons at special risk for infection are those with traumatic eye injuries or contact lens wearers who have small abrasions on the eye surface. A series of ocular infections that destroyed the eye in some cases and led to brain infection in others was traced to this organism. Eye infections can be avoided by carefully tending to injured eyes and using sterile solutions to store and clean contact lenses.

An Intestinal Ciliate: *Balantidium coli*

Most ciliates are free-living in various aquatic habitats, where they assume a role in the food chain as scavengers or predators upon other microbes. A few, such as the protozoan that digests wood (cellulose) in the termite's gut, have established relationships with animals, but only one, *Balantidium coli,* is pathogenic to humans. This giant ciliate (about the size of a period on this page) exists in both trophozoite and cyst states (figure 19.4). Its surface is covered by orderly oblique rows of cilia, and a shallow depression, the cytostome (cell mouth), is at one end.

Balantidium (bal″-an-tid′-ee-um) Gr. A small sac.

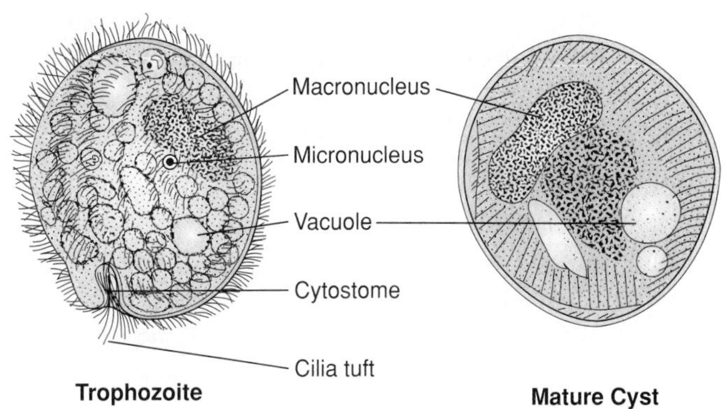

Figure 19.4 The morphologic anatomy of a trophozoite and a mature cyst of *Balantidium coli.*

These parasites require the anaerobic conditions and bacterial flora of the host to grow.

The large intestine of pigs and, to a lesser extent, sheep, cattle, horses, and primates constitutes the natural habitat of *B. coli.* Large cysts that remain infective for up to a week at room temperature are shed in the feces of these animals. Pigs are the most common source of human infection, though balantidiosis has also been spread by contaminated water and from person to person in institutional settings. These ciliates ordinarily scavenge epithelial and food debris from the large intestine, and healthy humans and animals are resistant to infection. In persons or animals with weakened defenses, however, the cilia and enzymes of the ciliate provide a means for invading the superficial layers of the colon.

Like cysts of *E. histolytica,* those of *B. coli* resist digestion in the stomach and small intestine, and liberate trophozoites that immediately burrow into the epithelium. The resultant erosion of the intestinal mucosa produces varying degrees of irritation and injury, leading to nausea, vomiting, diarrhea, dysentery, and abdominal colic. The protozoan rarely penetrates the intestine or enters the blood. The treatment of choice is oral tetracycline, but if this therapy fails, iodoquinol, nitrimidazine, or metronidazole may be given. Preventative measures are similar to those for controlling amebiasis, with additional precautions to prevent food and drinking water from being contaminated with pig manure.

The Flagellates (Mastigophorans)

The natural history of flagellated protozoans spans the ecological spectrum from free-living to commensalistic to parasitic. Pathogenic flagellates play an extensive role in human diseases, ranging from troublesome but usually self-limited infections (trichomoniasis and giardiasis) to serious and debilitating vector-borne diseases (trypanosomiasis and leishmaniasis). The major flagellate genera and species and their diseases are given in table 19.2.

Table 19.2 Flagellate Diseases and Target Organs

Disease	Major Lesion Site	Flagellate
Trichomoniasis		*Trichomonas* species
Vaginitis, urethritis	Vagina, urethra	*T. vaginalis*
Gingivitis	Opportunist in gums	*T. tenax*
Giardiasis		*Giardia*
Intestinal disease	Duodenum, jejunum	*G. lamblia*
Trypanosomiasis		*Trypanosoma* species
African sleeping sickness	Skin, viscera, brain	*T. brucei*
Chagas' disease	Skin, lymphatics, heart	*T. cruzi*
Leishmaniasis		*Leishmania* species
Oriental sore	Skin	*L. tropica*
Kala-azar	Viscera	*L. donovani*
Espundia	Skin, mucous membranes	*L. brasiliensis, L. mexicana*

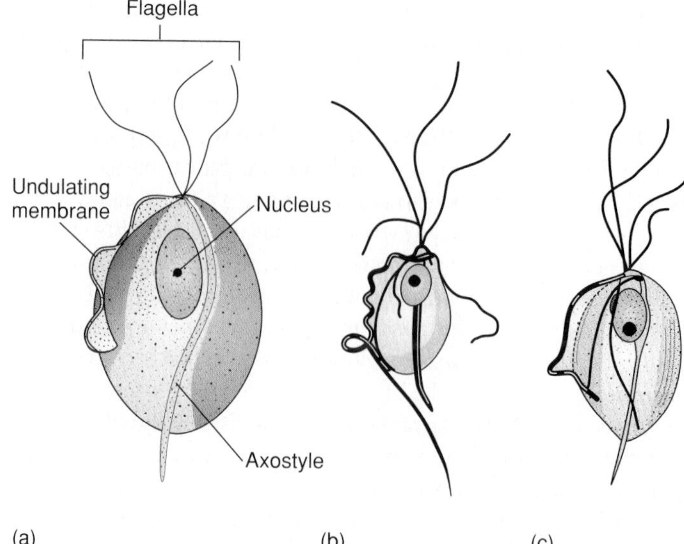

Figure 19.5 The trichomonads of humans. (*a*) *T. vaginalis*, a urogenital pathogen. (*b*) *T. tenax*, a gingival form (infection is rare). (*c*) *T. hominis*, an intestinal species (infection is rare).

Trichomonads: *Trichomonas* Species

Trichomonads are small, pear-shaped protozoa with four anterior flagella and an undulating membrane that together produce a characteristic twitching motility in live preparations. They exist only in the trophozoite form and do not produce cysts. Three trichomonads intimately adapted to the human body are *Trichomonas vaginalis, T. tenax,* and *T. hominis* (figure 19.5).

The most important species, ***Trichomonas vaginalis,*** is a pathogen of the reproductive tract that causes a sexually transmitted disease called **trichomoniasis.** The reservoir for this protozoan is the human urogenital tract, and about 50% of infected persons are asymptomatic carriers. Because *T. vaginalis* has no protective cysts, it is a relatively strict parasite that does not survive for long out of the host. Transmission is primarily through contact between genital membranes, though it may on rare occasions be transmitted through communal bathing, public facilities, and from mother to child. The incidence of infection is highest among promiscuous young women already infected with other sexually transmitted diseases such as gonorrhea or chlamydia. Approximately 3% to 15% of the U.S. adult population has either clinical or subclinical infection.

Increased vaginal acidity favors trichomonad growth and makes infection more likely. Symptoms and signs of trichomoniasis in the female include a frothy, foul-smelling, green-to-yellow vaginal discharge, vulvitis, cervicitis, and urinary frequency and pain. Severe inflammation of the infection site causes tenderness, edema, chafing, and itching. Males with overt clinical infections often experience irritating persistent or recurring urethritis, with a thin milky discharge and occasionally prostate infection.

Trichomonas (trik"-oh-moh'-nus) Gr. *thrix*, hair, and *monas*, unit.

Infection is diagnosed by demonstrating the active swimming trichomonads in a wet film of exudate. Detecting asymptomatic infection may require methods such as a Pap smear or culture of the parasite. The infection has been successfully treated with oral and vaginal metronidazole (Flagyl) for one week, and it is essential to treat both sex partners to prevent "ping-pong" reinfection.

Trichomonas tenax is a small trichomonad that ordinarily resides in the oral cavity of 5–10% of humans. Harborage is more common in persons with poor oral hygiene or dental disease. *T. tenax* is the only flagellate in the oral cavity, where it frequents the gingival crevices and areas of calculus, feeding on food debris and sloughed cells. It is transmitted by droplets, kissing, and occasionally by fomites such as eating utensils. Most dental experts agree that it is not a true pathogen but more likely an opportunist in lesions of gingivitis and periodontal pockets. *Trichomonas hominis* is a resident of the cecum of a small percentage of humans and great apes, but is believed to be a harmless commensal and is not associated with disease.

A Note on Names Many flagellates are named in honor of one of their discoverers—for example, *cruzi*, O. Cruz; *leishmania*, W. Leishman; *donovani*, C. Donovan; *Giardia*, A. Giard and *lamblia*, V. Lambl; and *brucei*, D. Bruce. Geographic location (*gambiense, rhodesiense, tropica, brasiliensis, mexicana*) is another basis for naming. Names may also be derived from a certain characteristic of the organism—for example, *tenax*, tenacious; *hominis* (L. *homo*, man); and *vaginalis*, vaginal habitat.

Giardia lamblia and Giardiasis

Giardia lamblia is a pathogenic flagellate first observed by Antonie van Leeuwenhoek in his own feces. For 200 years, it was considered a harmless or weak intestinal pathogen, and only in the last 40 years has its prominence as a cause of diarrhea been recognized. In fact, it is the most common flagellate isolated in clinical specimens. Observed straight on, the trophozoite has a unique, symmetrical heart shape with organelles positioned to resemble a face (figure 19.6*a*). Four pairs of flagella emerge from the ventral surface, which is concave and acts like a suction cup for attachment to the substrate. *Giardia* cysts are small, compact, and multinucleate.

Giardiasis has a general epidemiologic pattern similar to other protozoan intestinal infections. The protozoan has been isolated from beavers, cattle, coyotes, cats, and human carriers, but the precise reservoir is unclear at this time. Cysts are taken in with water and food or swallowed after close contact with infected persons or unhygienic fomites. In a study with prison volunteers, the cysts were shown to be highly contagious—only 10–100 were required for infection. Giardiasis is most common in children living in warm climates; however, in developed countries with improved sanitation, adults are infected as frequently as children. Instances of outbreaks have been so many and varied that a partial list can only hint at the possible modes of transmission. Not only are contaminated fresh waters a serious concern, but infection is also a problem for swimmers, homosexuals, and children in day-care centers. *Giardia* is also a well-known cause of travelers' diarrhea, which sometimes does not appear until the person has returned home. Cases of food-borne illness have been traced to carriers who contaminate food through unclean personal habits.

Water, water everywhere,
Nor any drop to drink.

Samuel Coleridge
The Rime of the Ancient Mariner

Unlike other pathogenic flagellates, *Giardia* is very hardy, and its cysts may survive for two months in cool water. Numerous infections have been reported in hikers and campers who used what they thought was clean water from ponds, lakes, and streams in remote mountain areas. Of course, checking water for purity by its appearance is unreliable, because the cysts are too small to be detected.

Wild mammals such as muskrats and beavers are probably intestinal carriers that contaminate the water with feces. Community water supplies in areas throughout the United States have also been implicated as common vehicles of infection, though it is not always possible to trace the exact source. *Giardia* epidemics have broken out in resorts with pristine mountain streams like Aspen, Colorado, and they have even been traced to chlorinated municipal water supplies in Massachusetts, California, and Washington.

Ingested *Giardia* cysts excyst in the duodenum and travel on to the jejunum where they feed and multiply. Some trophozoites remain on the surface, while others invade the glandular crypts to varying degrees. The outcome of infection ranges from asymptomatic to severe chronic giardiasis. Superficial invasion by the trophozoites causes damage to the epithelial cells, edema, and infiltration by white blood cells, but these effects are reversible. Typical symptoms include diarrhea, abdominal pain, flatulence, and muscular weakness. Stools containing large amounts of unabsorbed fat (a condition called steatorrhea) are probably the result of impaired absorption. Although both trophozoites and cysts escape in the stool, the cysts play a greater role in transmission. The infection is eradicated with quinacrine or metronidazole. Now that this parasite is apparently on the increase, many water agencies have had to rethink their policies on water maintenance and testing. The agent is killed by boiling, ozone, and iodine, but unfortunately, the usual chlorine content of treated water is inadequate to control it. Persons needing to use water from remote sources should assume that it is contaminated and boil or filter it.

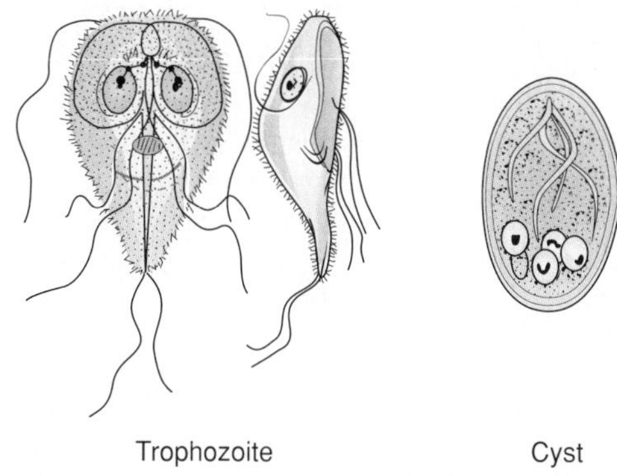

(a)

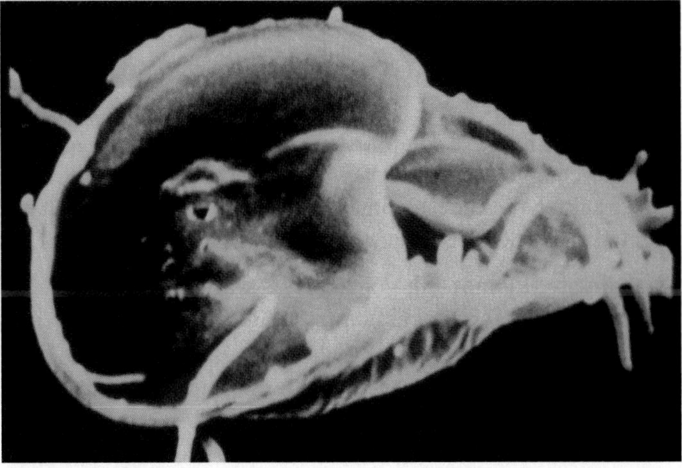

(b)

Figure 19.6 Trophozoites and cysts of *Giardia lamblia*. (*a*) The face of a trophozoite with eyes (paired nuclei). Side view shows the ventral depression and origin of the flagella. Cysts have two or four nuclei. (*b*) A scanning electron micrograph shows the fishlike appearance and flattened base (×9,800).

Hemoflagellates

Other parasitic flagellates are called hemoflagellates because of their propensity to live in the blood and tissues of the human host. Members of this group, which includes species of ***Trypanosoma*** and ***Leishmania,*** have several distinctive characteristics in common. All are obligate parasites that cause life-threatening and debilitating zoonoses; all are spread by blood-sucking vectors that serve as intermediate hosts; and all tend to be exotic tropical species that are rare in the United States.

Table 19.3 Cellular and Infective Stages of the Hemoflagellates

Genus/Species	Amastigote	Promastigote	Epimastigote	Trypomastigote
Leishmania	Intracellular in human macrophages	Found in sand fly gut; infective to humans	Does not occur	Does not occur
Trypanosoma brucei	Does not occur	Does not occur	Present in salivary gland of tsetse fly	In biting mouthparts of tsetse fly; transferred to humans
Trypanosoma cruzi	Intracellular in human macrophages, liver, heart, spleen	Occurs	Present in gut of reduviid (kissing) bug	In feces of reduviid bug; transferred to humans

Hemoflagellates have complicated life cycles and undergo morphologic changes between the vector and the human host. To keep track of these developmental stages, parasitologists have adopted the following set of terms:

amastigote (ay-mas′-tih-goht) Gr. *a-*, without, and *mastix,* whip. The form lacking a free flagellum.

promastigote (proh-mas′-tih-goht) Gr. *pro-,* before. The stage bearing a single, free anterior flagellum.

epimastigote (ep″-ih-mas′-tih-goht) Gr. *epi-,* upon. The flagellate stage, which has a free anterior flagellum and an undulating membrane.

trypomastigote (trih″-poh-mas′-tih-goht) Gr. *trypanon,* auger. The large, fully formed stage characteristic of *Trypanosoma.*

These stages suggest a metamorphic or evolutionary transition from the simple, cystlike amastigote form to the highly complex trypomastigote form. All four stages are present in *T. cruzi,* but *T. brucei* lacks the amastigote and promastigote forms, while *Leishmania* has only the amastigote and promastigote stages (table 19.3). In general, the vector host serves as the site of differentiation for one or more of the stages, and the parasite is infectious to humans only in the fully developed stage.

Trypanosoma Species and Trypanosomiasis

Trypanosoma species are distinguished by their infective stage, the trypomastigote, an elongate, spindle-shaped cell with tapered ends that is capable of eel-like, sinuous motility. Two types of **trypanosomiasis** are distinguishable on a geographic basis. *Trypanosoma brucei* is the agent of **African sleeping sickness,** and *T. cruzi* is the cause of **Chagas' disease,** which is endemic to Central and South America. Both exhibit a biphasic life-style alternating between a vertebrate and an insect host. Because the life cycle of *T. cruzi* was discussed in chapter 4, only that of sleeping sickness will be illustrated here.

***Trypanosoma brucei* and Sleeping Sickness** Trypanosomiasis has greatly affected the living conditions of Africans since ancient times. Even today, at least 250,000 persons are afflicted with it, and 20,000 die every year. It imposes an additional hardship when it attacks domestic and wild mammals. The two variants of sleeping sickness are the Gambian (West African) strain, caused by the subspecies *T. b. gambiense,* and the Rhodesian (East African) strain, caused by *T. b. rhodesiense* (figure 19.7*a*). These geographically isolated types are associated with different ecological niches of the principal tsetse fly vectors. In the West African form, the fly inhabits a niche in the dense vegetation along rivers and forests typical of that region, whereas the East African form is adapted to savanna woodlands and lakefront thickets.

The cycle begins when a tsetse fly becomes infected after feeding on an infected reservoir host, such as a wild animal (antelope, pig, lion, hyena), domestic animal (cow, goat), or human (figure 19.7*b*). The trypanosome is taken into the gut with a blood meal, where it multiplies, migrates to the salivary glands, and develops into the infectious mastigote. A single fly lives up to three months and is capable of delivering a mass of 50,000 cells when it bites a fresh host, even though only about 500 cells are required for infection. At the site of release, the trypanosome undergoes a series of divisions and produces a sore called the primary chancre. From there, the pathogen moves into the lymphatics and the blood. Although an immune response occurs, it is counteracted by an unusual adaptation of the trypanosome (see feature 19.1). The onset, severity, and length of the disease vary in the two geographical forms. The Rhodesian form is acute and advances to brain involvement in three to four weeks and to death in a few months. The Gambian form has a longer incubation period, is usually chronic, and may not affect the brain for several years.

The principal pathology of trypanosomiasis is directed at lymphatics and areas surrounding blood vessels. Symptoms are variable, but usually include intermittent fever, enlarged spleen,

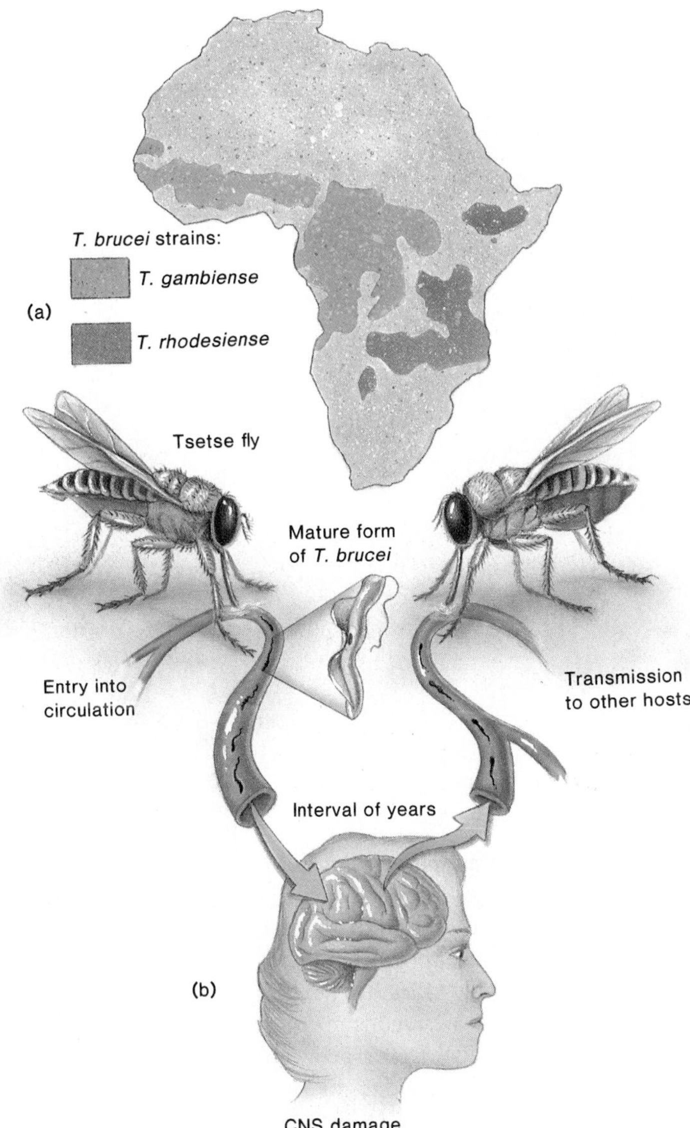

Figure 19.7 (*a*) The distribution of African trypanosomiasis. (*b*) The generalized cycle between humans and the tsetse fly vector. The saliva of a fly infected with *T. brucei* inoculates the human bloodstream. The parasite matures and invades various organs. In time, its cumulative effects cause central nervous system (CNS) damage. The trypanosome is spread to other hosts through another fly in whose alimentary tract the parasite completes a series of developmental stages.

swollen lymph nodes, and joint pain. In both forms, the central nervous system is affected, the initial signs being personality and behavioral changes that are at first subtle but progress to indifference, lassitude, and sleep disturbances. The disease is commonly called sleeping sickness, but in fact, uncontrollable sleepiness occurs primarily in the day and is followed by sleeplessness at night. Signs of advancing neurological deterioration are muscular tremors in the face, tongue, and limbs, shuffling gait, slurred speech, epileptic-like seizures, and local paralysis. Death results from coma, malnutrition, or secondary infections, and cardiac arrest may occur in the Rhodesian form.

Feature 19.1 The Trypanosome's Bag of Tricks

As in most infections, the immune system first responds to trypanosome surface antigens by producing immunoglobulins of the IgM type. These antibodies carry out their usual function of tagging the parasite to make it easier to destroy. If this were the end of the matter, sleeping sickness would be a mild infection with few fatalities, but such is not the case. Instead, the trypanosome strikes back with a fascinating genetic strategy in which surviving cells switch the structure of their surface glycoprotein antigens. This change in specificity renders the existing IgM ineffective so that the parasite eludes control and multiplies in the blood. When the host fights back by producing IgM against this new antigen, the parasite alters its surface again, the body produces corresponding different IgM, the surface changes again, and so on, in a repetitive cycle that may go on for months or even years. Eventually, the host becomes exhausted and overwhelmed by repeated efforts to catch up with this trypanosome masquerade. This cycle has tremendous impact on the pathology and control of the disease. The presence of the trypanosome in the blood and the severity of symptoms follow a wavelike pattern reminiscent of borreliosis. Development of a vaccine is hindered by the requirement to immunize against at least 100 different antigenic variations.

Sleeping sickness may be suspected if the patient has been bitten by a large fly (usually quite painful) while living or travelling in an endemic area. Trypanosomes are readily demonstrated in the blood, spinal fluid, or lymph nodes. Chemotherapy for the disease has been attempted with the relatively toxic drugs suramin and pentamidine, but it is most successful if administered prior to nervous system involvement. Brain infection is treated with melarsoprol.

Control of trypanosomiasis is possible in western Africa where humans are the main reservoir hosts. It involves eliminating tsetse flies by applying insecticides, using fly traps, or destroying vegetation that provides shelter and breeding sites. In eastern regions where cattle herds and large wildlife populations are reservoir hosts, control is far more difficult because it is impractical and unacceptable to eradicate mammalian hosts, and flies are less concentrated in specific sites. Giving prophylactic drugs to tourists or workers entering infested areas has met with little success.

***Trypanosoma cruzi* and Chagas' Disease** Chagas' disease accounts for millions of cases, thousands of deaths, and untold economic loss and human suffering in Latin America. The life cycle of its etiologic agent, *Trypanosoma cruzi* (see chapter 4), parallels that of *T. brucei,* except that the insect hosts are "kissing" or reduviid bugs that harbor the trypanosome in the hind gut and discharge it in feces. These bugs live throughout Central and South America and often share the living quarters of human and mammalian hosts. Infection occurs when the bug defecates near the site of its own bite, and the human inoculates this wound by rubbing bug feces into it. Infection is also possible when bug feces fall onto the mucous membranes or conjunctiva. The disease may even be spread by sexual intercourse, transplacental exchange, and blood transfusion in some instances. The course of the illness is somewhat similar to African

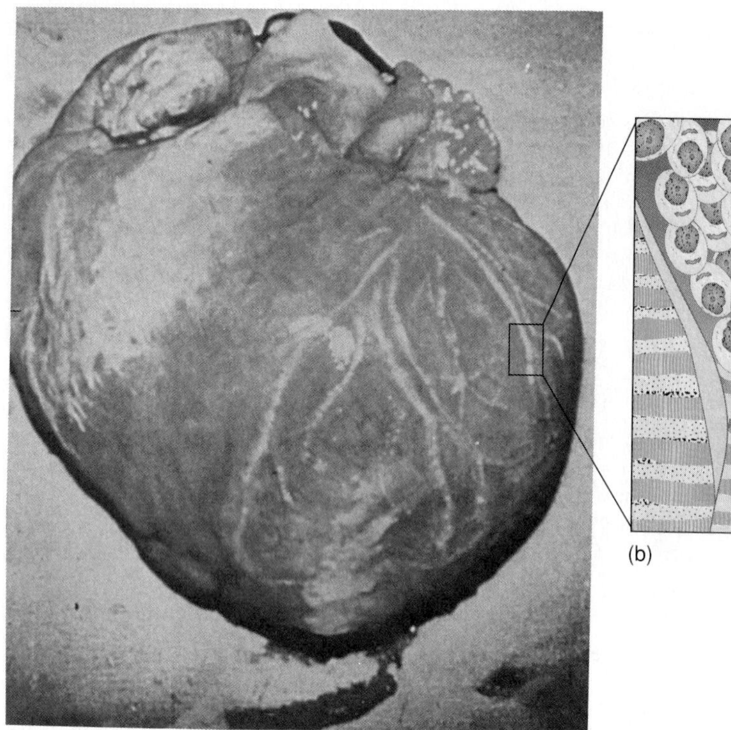

Figure 19.8 Heart pathology in Chagas' disease. (*a*) A greatly enlarged ventricular muscle. (*b*) Section of a heart muscle with a cluster of amastigotes nestled within the fibers.

trypanosomiasis, being marked by a local lesion, fever, and swelling of the lymph nodes, spleen, and liver. Particular targets are the heart muscle and large intestine, both of which become greatly enlarged and severely disrupted in function (figure 19.8). Chronically infected persons with cardiac complications usually die within two years.

In addition to microscopic examination of blood and serological testing, diagnosis of Chagas' disease is sometimes made using a clever technique called xenodiagnosis, in which a germ-free reduviid bug is allowed to feed on the patient and is then observed for signs of infection in a few weeks. Although treatment is sometimes attempted with two drugs, nifurtimox and benzonidazole, they are not effective if the disease has progressed too far, and they have damaging side effects. It has been shown that the prevalence of Chagas' disease can be reduced by vector control. Insecticides are of some benefit, but these bugs are very tough and capable of resisting doses that readily kill flies or mosquitoes. Mass collections of reduviid bugs and housing improvements to keep bugs out of human dwellings have also been effective. Vaccination has been contemplated, but the development of a successful vaccine lies in the indefinite future.

Leishmania Species and Leishmaniasis

Leishmaniasis is a zoonosis transmitted among various mammalian hosts by small female **phlebotomine** flies called sand flies that require a blood meal to produce mature eggs. ***Leishmania*** species have adapted to the insect host by timing their own development with periods of blood-taking (figure 19.9*a*). The protozoa enter the gut with blood and multiply and differentiate into promastigotes that migrate to the esophagus and proboscis of the fly. From there they are injected into the capillary of a host when the female takes her next blood meal.

Leishmaniasis is endemic to regions that lie on or near the equator, essentially wherever geographic and biologic conditions are favorable for sand flies to complete their life cycle. The disease can be transmitted by over 50 species of sand fly, and numerous wild and domesticated animals compose the rest of the reservoir (dogs, rodents, or wild carnivores depending on the region). Although humans are usually accidental hosts in the epidemiologic cycle, the fly is said to find human blood particularly delicious. Persons at particular risk are travelers or immigrants who have never had contact with the parasite and lack specific immunities.

The species of *Leishmania* pathogenic to humans are indistinguishable in appearance, even though they are geographically isolated and produce distinct types of diseases. In all cases, infection begins when an infected fly injects promastigotes while feeding. Although nearby macrophages capture the parasite, its virulence factors help it resist destruction, and this allows it to convert to an amastigote and multiply intracellularly in the macrophage. The differences in the nature of the infection depend on the pathway of the macrophage. If infected macrophages remain fixed, the infection is localized in the skin or mucous membranes, but if the infected macrophages migrate, systemic disease occurs.

Cutaneous leishmaniasis (oriental sore, Baghdad boil) is a localized infection of the capillaries of the skin caused by

---

phlebotomine (flee-bot'-oh-meen) Gr. *phlebo*, vein, and *tomos*, cutting.

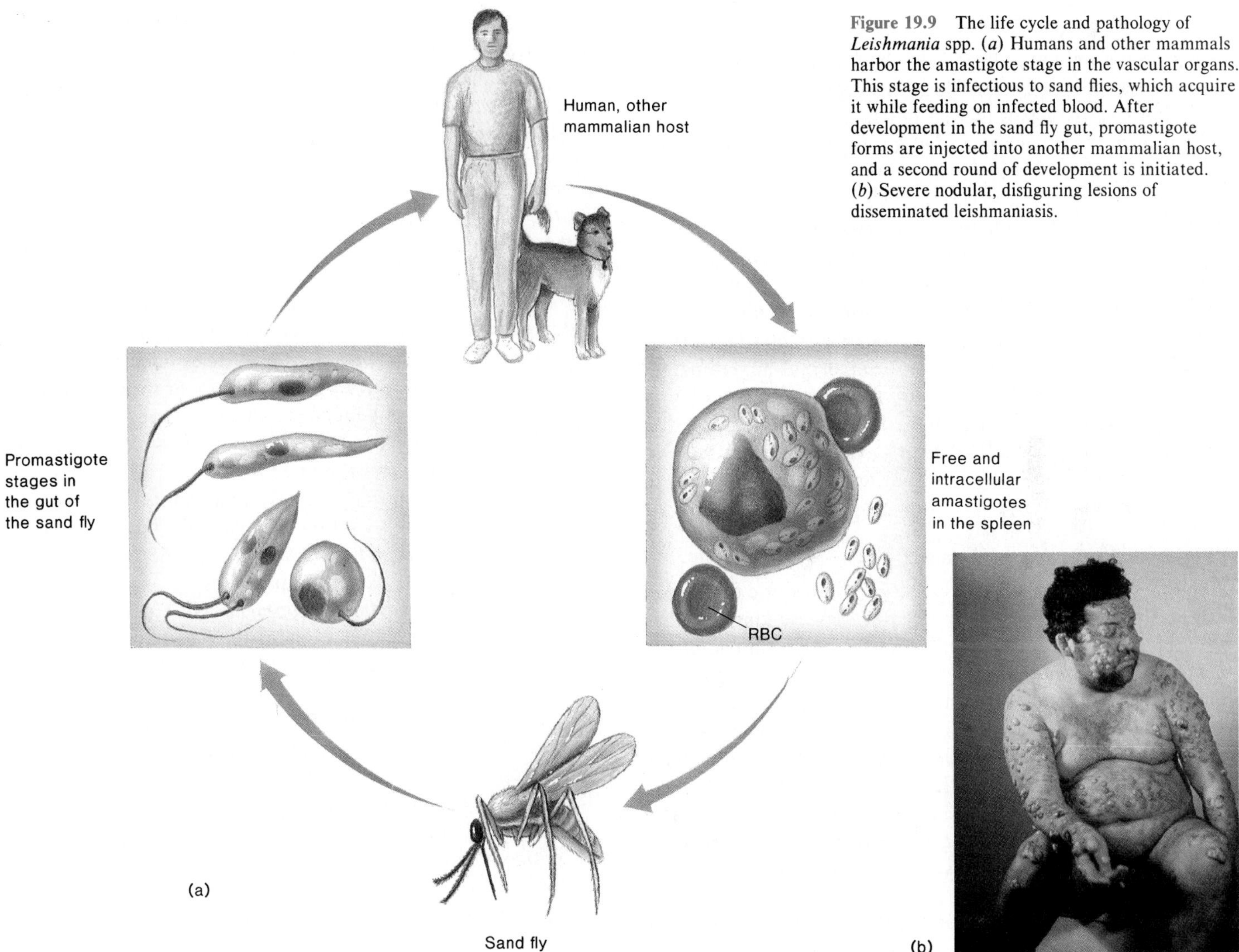

Figure 19.9 The life cycle and pathology of *Leishmania* spp. (*a*) Humans and other mammals harbor the amastigote stage in the vascular organs. This stage is infectious to sand flies, which acquire it while feeding on infected blood. After development in the sand fly gut, promastigote forms are injected into another mammalian host, and a second round of development is initiated. (*b*) Severe nodular, disfiguring lesions of disseminated leishmaniasis.

L. tropica in certain Mediterranean, African, and Indian regions and by *L. mexicana* in Latin America. A single small red papule occurs at the site of a bite and spreads laterally into a large "wet" or "dry" ulcer. A form of mucocutaneous leishmaniasis called espundia, caused by *L. brasiliensis* and endemic to parts of Central and South America, affects both the skin and mucous membranes of the head. The primary skin ulcer develops within a few days after the insect bite, and although it may heal, it often recurs with involvement of the nose, lips, palate, gingiva, and pharynx. When these vulnerable areas develop secondary infections, the disease can be very disfiguring (figure 19.9*b*). Systemic or visceral leishmaniasis that occurs in parts of Africa, Latin America, and China is caused by *L. donovani*. The disease appears gradually after 2 to 18 weeks, with a high, spiking, intermittent fever, weight loss, and enlargement of internal organs, especially the spleen, liver, and lymph nodes. The most deadly form, *kala-azar,* is 75% to 95% fatal in untreated cases. Death is due to complications of infection, including the destruction of blood-forming tissues, anemia, secondary infections, and hemorrhage.

Diagnosis is confounded by the similarity of leishmaniasis to several granulomatous bacterial infections (leprosy, syphilis) and fungal infections. The organism can usually be detected in the cutaneous lesion or in the bone marrow during visceral disease. Additional tests include cultivating the parasite, serological tests, and leishmanin skin testing. Chemotherapy with injected pentavalent antimony is usually effective, though pentamidine or amphotericin B may be necessary in resistant cases. Some disease prevention has been achieved by applying insecticides to sand flies and eliminating their habitats, as well as by destroying infected dogs (an important reservoir of infection in some areas). A type of "vaccination" was practiced in the Middle East for hundreds of years by exposing young children to sand flies or to the lesions of infected persons so that they would acquire the disease and thereafter have immunity. More modern attempts at vaccination have included inoculation with live promastigotes and nonpathogenic strains, but these methods have been generally ineffective. Great hope lies in a newer vaccine strategy using recombinant DNA technology.

The Sporozoan Parasites

One group of unique protozoa are the Class Sporozoea, sometimes called the sporozoans. These tiny organisms parasitize the body fluids or tissues of animal hosts. They are distinct from other protozoans in lacking locomotor organelles in the mature state, though some forms do have a flexing, gliding form of locomotion, and others have flagellated sex cells. Characteristics of the sporozoan life cycle that add greatly to their complexity are alternations between sexual and asexual phases and between different animal hosts. Most members of the group also form specialized spores or sporelike infective bodies that are transmitted by arthropod vectors, food, water, or other means, depending on the parasite. The most important human pathogens are in the genera *Plasmodium* (malaria), *Toxoplasma* (toxoplasmosis), and *Crytosporidium* (cryptosporidiosis). The taxonomic status of the AIDS opportunist *Pneumocystis carinii* has yet to be resolved, but it has traditionally been lumped with the sporozoans.

Plasmodium: The Sporozoan of Malaria

Throughout human history, even back to prehistoric times, malaria has been one of the greatest afflictions, ranking with bubonic plague, influenza, and tuberculosis. It is still one of the most prevalent diseases, killing several million, infecting several hundred million, and threatening more than half the world's population every year. Over time, various regions have had special names for the disease and have given different explanations for its origin, but eventually, the Italian word *malaria,* a combination of *mal,* bad, and *aria,* air, gained universal usage. The superstitions of the Middle Ages explained that evil spirits or mists and vapors arising from swamps caused malaria, because many victims came down with the disease following this sort of exposure. But for about the past 100 years, we have known that the evil spirits were mosquitos breeding in those swamps.

The agent of malaria is an obligate intracellular sporozoan in the genus *Plasmodium,* which contains four species: *P. malariae, P. vivax, P. falciparum,* and *P. ovale.* The human is the primary vertebrate host for these species, which are geographically separate and may show variations in the pattern and severity of disease. All forms are spread primarily by the female *Anopheles* mosquito and occasionally by shared needles, blood transfusions, or from mother to fetus. The detailed study of malaria is a large and engrossing field that is far too complicated for this text. We will cover only the general aspects of its epidemiology, life cycle, and pathology.

Epidemiology, Life Cycle, and Stages of *Plasmodium* Infection Although malaria was once distributed throughout most of the world, the control of mosquitos in temperate areas has successfully restricted it mostly to a belt extending around the equator. Despite this achievement, approximately 300 million cases are still reported each year, about two-thirds of them in Africa. The most frequent victims are children and young adults, of whom at least two million die annually. The total case rate in the United States is about 1,000 new cases per year, most of which occur in immigrants (see figure 11.16).

Because the developmental stages of the malarial parasite are exquisitely timed to coincide with the mosquito's life cycle, and the behavior of the mosquito in turn coincides with human infection, we will present the stages in parasites, mosquitos, and humans as an interactive system (figure 19.10). Development of the malarial parasite is divided into two distinct phases: the **asexual stage,** carried out in the human, and the **sexual stage,** carried out in the mosquito.

The **asexual cycle** and infection begin when a voracious infected female *Anopheles* mosquito lights on an unsuspecting human to extract the blood necessary to develop her eggs. Just prior to feasting, she injects saliva containing anticoagulant into the capillary she has punctured, an event that also inoculates the blood with motile, spindle-shaped, asexual cells called **sporozoites** (Gr. *sporo,* seed, and *zoon,* animal). The sporozoites circulate through the body, but within an hour, they have taken up residence in the liver. Within liver cells, the process of asexual division called *schizogony* (Gr. *schizo,* to divide, and *gone,* seed) generates numerous daughter parasites or *merozoites.* This phase of *exoerythrocytic development* lasts from 5 to 16 days, depending upon the species of parasite. When each packed liver cell finally erupts, it spills from 2,000 to 40,000 merozoites into the circulation.

During the *erythrocytic phase,* merozoites are programmed to infect red blood cells by taking advantage of specific receptors. Inside RBCs, they transform into a circular (ring) form or trophozoite (figure 19.11), which feeds upon hemoglobin, grows, and undergoes multiple divisions to produce a cell or *schizont* filled with more merozoites. When massively infected red cells burst, the liberated merozoites are freed to infect more red cells. As we shall see, these cyclic periods of RBC lysis contribute to some of the symptoms. At some point, certain merozoites differentiate into two types of specialized gametes called *macrogametocytes* (female) and *microgametocytes* (male). Because the human does not provide a suitable environment for completion of fertilization and the next phase of development, this is the end of the cycle in humans.

The **sexual cycle** (sporogony) results when another mosquito sucks human blood containing infected RBCs into her stomach. In the stomach the microgametocyte releases several spermlike gametes that fertilize the larger macrogametes. The resultant diploid cell (oocyst) implants into the stomach wall and undergoes multiple meiotic divisions, releasing haploid sporozoites that migrate to the salivary glands and lodge there. This event completes the sexual cycle and makes the sporozoites available for injection when the mosquito feeds on her next victim.

The clinical signs and symptoms of a classic untreated case of malaria are chiefly due to the body's response to the periodic influx of merozoites, pigment, and debris discharged from infected red cells. After an incubation period of 10 to 16 days, the first prodromal symptoms are malaise, fatigue, vague aches, and nausea. Then, bouts of chills, fever, and sweating appear, at first

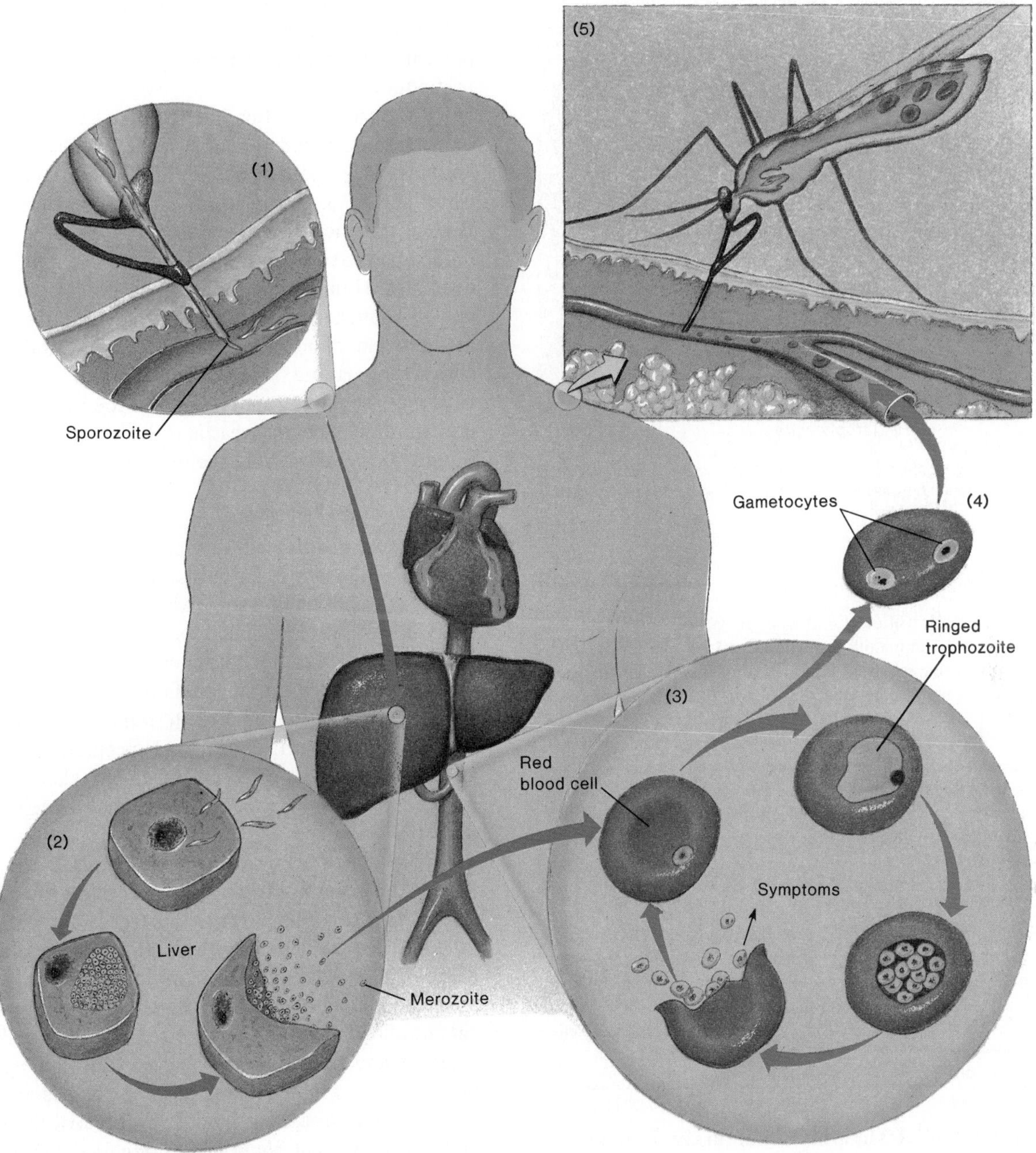

Figure 19.10 The life and transmission cycle of *Plasmodium.* (*1*) In the asexual phase in humans, sporozoites enter a capillary through the saliva of a feeding mosquito. (*2*) Exerythrocytic (liver) phase. Sporozoites invade the liver cells and develop into large numbers of merozoites. (*3*) Erythrocytic phase. Merozoites released into the circulation enter red blood cells. Initial infection is marked by a ring trophozoite; schizogony of the ringed form produces additional merozoites that burst out and infect other red blood cells. (*4*) Gametocytes that develop in certain infected red blood cells are ingested by another mosquito. (*5*) The sexual phase of fertilization and sporozoite formation occurs in the mosquito.

in an irregular pattern, but later recurring at 48- or 72-hour intervals as a result of the synchronous rupturing of RBCs. The interval, length, and regularity of these symptoms reflect the type of malaria. Patients with falciparum malaria, the most malignant type, often manifest persistent fever, rapid pulse, cough, and weakness for weeks without relief. Complications of malaria are hemolytic anemia from lysed blood cells and organ enlargement and rupture due to accumulated excess cellular debris in the spleen, liver, and kidneys. Additional complications are pulmonary failure, cerebral malaria, and gastrointestinal disturbances. Patients with milder, chronic forms of malaria may achieve spontaneous cure after three to five years, but falciparum malaria has a high death rate in the acute phase, especially in children. Certain kinds of malaria are subject to relapses because some infected liver cells harbor dormant infective cells for periods of up to five years.

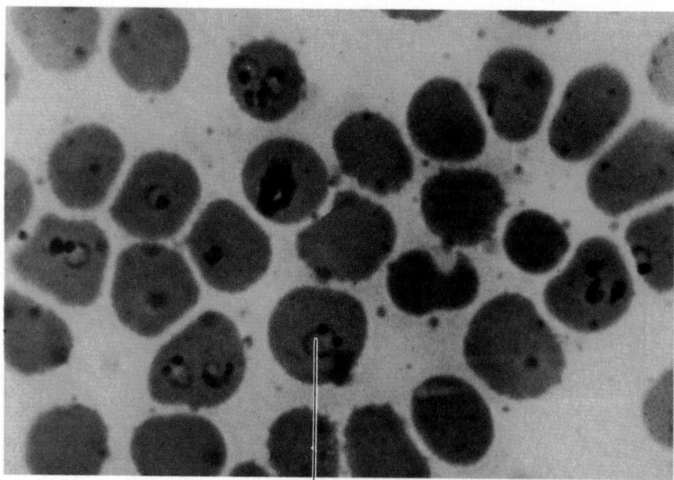

Figure 19.11 The ring trophozoite stage in a *P. falciparum* infection. A smear of peripheral blood shows several ring forms in some red blood cells.

Malaria Resistance: A Marvel of Adaptation In earlier chapters we discussed a type of natural immunity to malaria that does not involve antibodies. It seems that during eons of evolution, certain traits gained prominence in various human populations because they helped their possessors evade infection by *Plasmodium*. One dramatic example occurs in Africans who carry one gene that codes for sickle-cell hemoglobin and another gene that is normal. Although carriers are relatively healthy, they produce some abnormal RBCs that prevent the parasite from deriving sufficient nutrition to multiply. The continued existence of sickle-cell anemia in these geographic areas can be directly attributed to the increased survival of carriers of the gene. Individuals with sickle-cell anemia have even greater protection from malaria; however, the debilitation of sickle-cell disease outweighs its protective effect.

Another example of genetic resistance can be seen in West Africans who lack the Duffy blood receptor on their red blood cells so that one species of the parasite is unable to gain access to the interior of the cells. Other inherited red cell defects, such as persistent fetal hemoglobin, beta thalassemia, glucose-6–dehydrogenase deficiency, and oval-shaped red blood cells, are also believed to interfere with the erythrocytic phase of infection.

Diagnosis and Control of Malaria Malaria can be diagnosed definitively by the discovery of a typical stage of *Plasmodium* in stained blood smears (figure 19.11). Other indications are knowledge of the patient's residence or travel in endemic areas and presenting symptoms of recurring chills, fever, and sweating.

As recently as a decade ago, eradicating malaria seemed possible, but since then, morbidity and mortality have increased in several regions. The two main reasons for this involve the sheer adaptive and survival capacity of both the parasite and vector. Standard chemotherapy with antimalarial drugs has selected for drug-resistant strains of the malarial parasite, and mosquitos have developed resistance to some common insecticides used to control them. The current treatments for malaria include some form of quinine, a drug originally isolated from the cinchona tree (see feature 10.1). Chloroquine, the least toxic one, is used in nonresistant forms of the disease. In areas of the world where resistant strains of *P. falciparum* predominate, a course of quinine sulfate, pyrimethamine, and sulfadiazine is indicated instead. Eliminating the parasite from the liver and preventing relapses can be managed with long-term therapy with primaquine.

Malaria prevention is attempted through long-term mosquito abatement and human prophylaxis. Breeding sites such as swamps, wells, flooded fields, rain barrels, and other standing water should be drained or made inaccessible, and combinations of insecticides may be broadcast into the environment to reduce populations of adult mosquitos, especially near human dwellings. Humans can reduce their risk of infection considerably by using netting, screens, and repellents, remaining indoors at night, and taking weekly doses of prophylactic drugs. Persons with recent malaria must be excluded from blood donation. Even with massive efforts undertaken by the WHO, the prevalence of malaria in endemic areas is still high, thus a vaccine is being sought with great fervor (see feature 19.2).

Other Sporozoan Parasites

***Toxoplasma gondii* and Toxoplasmosis** *Toxoplasma gondii* is an obligate sporozoan parasite with such extensive cosmopolitan distribution that some experts estimate it affects the majority of the world's population at some time in their lives. In most of these cases, toxoplasmosis goes unnoticed, but disease in the fetus and in immunodeficient persons, especially those with AIDS, is severe and often terminal. *T. gondii* is a very successful parasite with so little host specificity that it can attack at least 200 species of birds and mammals. However, its primary reservoir and hosts are members of the feline family, both domestic and wild.

To follow the transmission of toxoplasmosis, we must first look at the general stages of *Toxoplasma's* life cycle in the cat (figure 19.12*a*). The parasite undergoes a sexual phase in the intestine and is released in feces where it becomes an infective *oocyst* that survives in moist soil for several months. The parasite can complete its entire life cycle in the cat population if cats swallow oocysts. The ingested oocysts break open in the intestine and release an invasive asexual tissue phase called a tachyzoite that infects many different tissues and often causes symptomatic disease in the cat. Eventually, these forms enter an asexual cyst state in tissues, also called a pseudocyst, which is capable of infecting animals that prey on cats. Most of the time, the parasite does not cycle in cats alone and is spread to intermediate hosts, usually rodents and birds, that ingest oocysts from the environment or feed on flesh containing pseudocysts. The cycle returns to cats when they eat these infected prey animals.

A wide spectrum of vertebrates gets caught up in this transmission cycle (figure 19.12*b*). The ways that these animals, including humans, become intermediate hosts depend upon their eating habits. Herbivorous animals such as cattle and sheep

Toxoplasma gondii (toks″-oh-plaz′-mah gawn′-dee-eye) L. *toxicum*, poison, and Gr. *plasma*, molded; *gundi*, a small rodent.

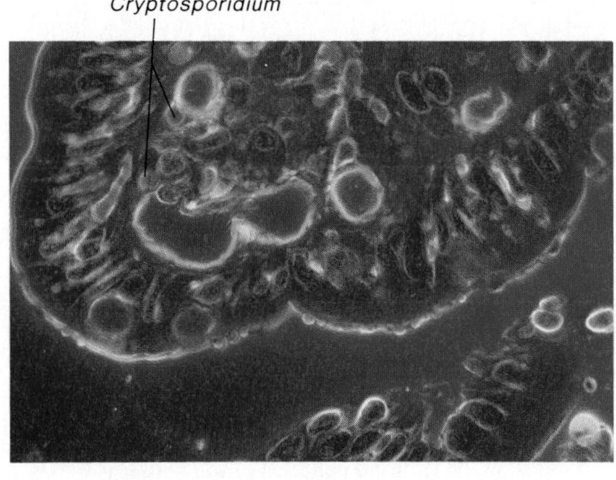

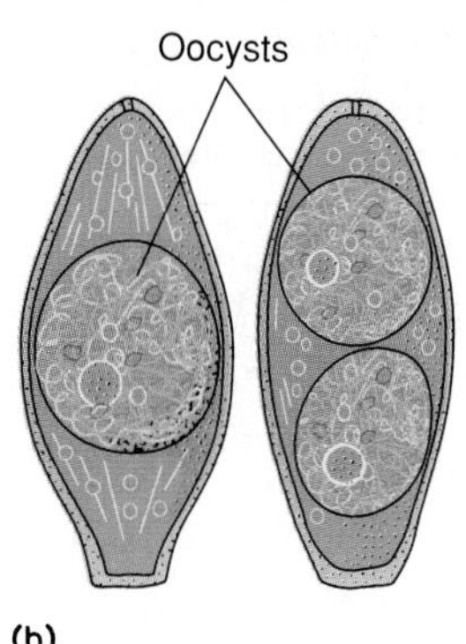

Figure 19.13 Other sporozoan parasites. (*a*) An electron micrograph of a *Cryptosporidium* merozoite penetrating an intestinal cell. (*b*) *Isospora belli,* showing oocysts in two stages of maturation.

where the *Ixodes* tick vector is located. The infection closely resembles malaria in pathology and symptomology, in that red blood cells burst after they are invaded, and it is diagnosed and treated in a similar fashion.

Pneumocystis carinii* and Pneumocystis Pneumonia** Although ***Pneumocystis carinii was discovered in 1942, it remained an obscure protozoan until it was suddenly propelled into clinical prominence as the cause of the most frequent secondary infection in AIDS patients—***Pneumocystis carinii* pneumonia (PCP).** Up to that time, it had been a very rare disease that attacked only severely debilitated, malnourished, elderly, or premature patients. This small unicellular organism has characteristics of both a protozoan and a fungus, and although it has generally been considered a type of protozoan with cystlike and trophozoite phases, recent findings indicate it may actually be a fungus.

Unlike most of the human protozoan pathogens, little is known about the life cycle or epidemiology of *Pneumocystis.* Evidently the parasite is a common resident of the respiratory tracts of animals and humans, where it usually lives with no serious effects. Contact with the agent is so widespread that in some populations, a majority of persons show serological evidence of infection. *Pneumocystis* is probably spread in droplet form from animal to human or from human to human. In healthy persons, it is usually kept in check by phagocytes and lymphocyte immunity, but in those whose immune systems lack these defenses, it multiplies intracellularly and extracellularly. The massive numbers of parasites disrupt the structure of the lung and fill the alveoli with a foamy exudate (figure 19.14). Symptoms are nonspecific and include cough, fever, shallow respiration, and cyanosis. The prognosis for untreated disease ranges from 50% to 100% mortality, but even with treatment, many AIDS patients succumb to this infection.

The primary treatments against PCP are aerosol pentamidine and co-trimoxazole (sulfamethoxazole and trimethoprim) given over a 10-day period. This therapy should be applied even if disease appears mild or is only suspected. The airways of patients in the active stage of infection often must be suctioned to reduce the symptoms, and additional oxygen is sometimes given. Unfortunately, most patients experience numerous recurrences.

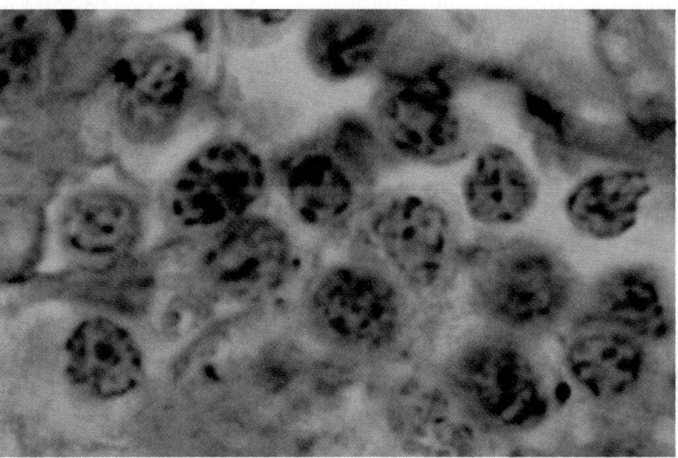

Figure 19.14 *Pneumocystis carinii* cysts in the lung tissue of an AIDS patient (×500).

A Survey of Helminth Parasites

So far our survey has been limited to relatively small infectious agents, but *helminths,* particularly the adults, are comparatively large multicellular animals with specialized tissues and organs similar to the hosts they parasitize. The behavior of these parasites and their adaptations to humans are at once fascinating and grotesque. Imagine having long, squirming white worms suddenly expelled from your nose, feeling and seeing a worm creeping beneath your skin, or having a worm slither across the front of your eye, and you have some idea of the conspicuous nature of these parasites.

Pneumocystis carinii (noo″-moh-sis′-tis kah′-ree-nee-eye) Gr. *pneumo,* lungs, and *cyst,* bladder; and after A. Carini.

helminth Gr. *helmins,* worm.

Table 19.4 Major Helminths of Humans and Their Modes of Transmission

Classification	Life Cycle Requirement	Spread to Humans By
Nematodes (Roundworms)		
Intestinal Nematodes		Ingestion
Infective in egg (embryo) stage		
Trichuris trichiura	Humans	Fecal pollution of soil with eggs
Ascaris lumbricoides	Humans	Fecal pollution of soil with eggs
Enterobius vermicularis	Humans	Close contact
Infective in larval stage		
Hookworms		
Necator americanus	Humans	Fecal pollution of soil with eggs
Ancylostoma duodenale	Humans	Fecal pollution of soil with eggs
Strongyloides stercoralis	Humans; may live free	Fecal pollution of soil with eggs
Trichinella spiralis	Pigs, wild mammals	Consumption of meat containing larvae
Tissue Nematodes		Burrowing of larva into tissue
Wuchereria bancrofti	Humans, mosquitos	Mosquito bite
Onchocerca volvulus	Humans, black flies	Fly bite
Loa loa	Humans, mangrove flies or deer flies	Fly bite
Dracunculus medinensis	Humans and *Cyclops* (an aquatic invertebrate)	Ingestion of water containing *Cyclops*
Trematodes (Flukes)		
Schistosoma japonicum	Humans and snails	Ingestion of fresh water containing larval stage
S. mansoni	Humans and snails	Ingestion of fresh water containing larval stage
S. haematobium	Humans and snails	Ingestion of fresh water containing larval stage
Opisthorchis sinensis	Humans, snails, fish	Consumption of fish
Fasciola hepatica	Herbivores (sheep, cattle)	Consumption of water and water plants
Paragonimus westermani	Humans, mammals, snails, freshwater crabs	Consumption of crabs
Cestodes (Tapeworms)		
Taenia saginata	Humans, cattle	Consumption of undercooked or raw beef
T. solium	Humans, swine	Consumption of undercooked or raw pork
Diphyllobothrium latum	Humans, fish	Consumption of undercooked or raw fish

Worms that parasitize humans are amazingly diverse, ranging from barely visible roundworms (0.3 mm) to huge tapeworms 25 meters long, and varying from oblong- and crescent-shaped to whip- or ribbon-shaped. In the introduction to these organisms in chapter 4, we grouped them into three categories: nematodes (roundworms), trematodes (flukes), and cestodes (tapeworms), and discussed basic characteristics of each group. In this section, we will present major facets of helminth epidemiology, life cycles, pathology, and control, followed by an accounting of the major worm diseases. Table 19.4 presents a selected list of medically significant helminths.

General Epidemiology of Helminth Diseases

The helminths have an impact on human health, economics, and culture that rivals war and famine. It has been said that there are probably more worm infections in humans than there are humans (due to multiple infections in one individual). Although individuals from all societies and regions play host to worms at some time in their lives, most cases occur in children living in rural areas of tropical and subtropical Africa, Asia, the Middle East, South America, and southern Europe. The most prevalent diseases and those creating the highest mortality and morbidity are schistosomiasis, filariasis, hookworm disease, ascariasis, and onchocerciasis. In most developed countries in temperate climates, the major parasitic infections are pinworm and trichinosis, though exotic parasitic worms are often encountered in immigrants and travelers. The continuing cycle between parasites and humans is often the result of poverty and cultural or behavioral practices such as using human excrement for fertilizer, exposing bare feet to soil or standing water, eating undercooked or raw meat and fish, and defecating in open soil. In some areas of the world, parasites are an accepted part of life and have been for thousands of years.

General Life and Transmission Cycles

The complete life cycle of helminths includes the fertilized egg (embryo), larval, and adult stages. In the majority of helminths, adults derive nutrients and reproduce sexually in a host's body. In nematodes, the sexes are separate and usually different in appearance; in trematodes, the male and female sex organs may be separate or in the same worm (hermaphroditic); and cestodes are generally hermaphroditic. For a parasite's continued survival as a species, it must complete the life cycle by transmitting an infective form, usually an egg or larva, to the body of another host, either of the same or a different species. By convention, the host in which the parasite reaches adulthood and mates is the **definitive (final) host,** while larval development occurs in the **intermediate (secondary) host.** A **transport host** is an intermediate host that experiences no parasitic development but is an essential link in the completion of the cycle.

Although individual worms show slight differences in details, most life and transmission cycles can be described by one of five basic patterns (figure 19.15). In general, sources for human infection are contaminated food, soil, and water or infected animals, and routes of infection are by oral intake or penetration of unbroken skin. Humans are the definitive hosts for many of the parasites listed in table 19.4, and in about half the diseases, they are also the sole biological reservoir. In other cases, animals or insect vectors serve as reservoirs or are required to complete worm development. Two important points of departure in helminth infections are that the organisms ordinarily do not complete the entire life cycle without leaving the host and that adults do not increase in numbers by multiplying in the host, as with microparasites. A more accurate term for their presence in the host is an **infestation.**

Pathology of Helminth Infestation

Worms exhibit numerous adaptations to the parasitic habit. They may have specialized mouthparts for attaching to tissues and for feeding, enzymes with which they liquefy and penetrate tissues, and a cuticle or other covering to protect them from the host defenses. As we learned in chapter 4, their organ systems are usually reduced to the essentials: getting food and processing it, moving, and reproducing. The signs and symptoms of infestation depend partly upon the portal of entry. Helminths having an oral portal of entry undergo a period of development in the intestine that is often followed by spread to other organs. Helminths that burrow directly into the skin or are inserted by an insect bite quickly penetrate the lymphatics or bloodstream and often circulate to various organs. In both cases, most worms do not remain in a single location once they have penetrated, and they migrate by various mechanisms through major systems, inflicting injury as they go.

Adult helminths eventually take up final residence in specific anatomical locations such as the intestinal mucosa, blood vessels, lymphatics, subcutaneous tissue, skin, liver, lungs, muscles, brain, or even the eyes. Traumatic damage occurring in the course of migration and organ invasion is primarily due to tissue feeding, blocked ducts and organs, toxic secretions, and pressure within organs due to high worm load. Some consequences are hemorrhage, enlargement and swelling of organs, weight loss, and anemia caused by the parasites feeding on blood. Despite the popular misconception that intestinal worms cause weight loss by competing for food (the person who can eat everything but never gain weight must have a tapeworm!), weight loss is more likely caused by general malnutrition in infested persons, malabsorption of food and vitamins due to intestinal damage, lack of appetite, vomiting, and diarrhea.

The mere presence of a parasite alone does not signify disease. In fact, some persons coexist comfortably with parasites and experience few symptoms. More severe manifestations are the rule if the infectious load is high and the relative health of the patient is poor. One common antihelminthic defense is an increase in eosinophils, blood cells that have a specialized capacity to destroy worms (chapter 12). Although antibodies and sensitized T cells are also formed, their effects are rarely long-term, and reinfection is always a distinct possibility. The inability of the immune system to completely eliminate worm infections appears to be due to their relatively large size, migratory habits, and inaccessibility. The incompleteness of immunity has hampered the development of vaccines for parasitic diseases and has emphasized the use of chemotherapeutic agents instead.

Elements of Diagnosis and Control

Diagnosis of helminthic infestation may come in several stages. A differential blood count showing eosinophilia and serological tests indicating sensitivity to helminths both provide indirect evidence of worm infestation. A history of travel to the tropics or immigration from those regions is also helpful, even if it occurred years ago, because some flukes and nematodes persist for decades. The most definitive evidence is the discovery of eggs, larvae, or adult worms in stools, sputum, urine, gastric washings, blood, or tissue biopsies. The worms are sufficiently distinct in morphology that positive identification may be based on any stage, including eggs.

Although several useful antihelminthic medications exist, the cellular physiology of the eucaryotic parasites resembles that of humans, and drugs toxic to them can also be toxic to humans. Most antihelminthic drugs either suppress a metabolic step or process that is more important to the worm than to the human, or inhibit the worm's movement, thus preventing it from maintaining its position in a certain organ. The rationale for therapy may also take into account a drug's greater toxicity to the relatively smaller and more vulnerable helminths or the action of oral drugs locally on worms in the intestine. Antihelminthic drugs of choice and their effects are given in table 19.5. In some cases, surgery may be necessary to remove worms or larvae, though this can be difficult if the parasite load is high or not confined.

Effective preventive measures aim to minimize human contact with the parasite or to interrupt its life cycle. In cases where the worm is transmitted by fecally contaminated soil, much benefit is achieved by educating the community concerning proper sewage disposal, using sanitary latrines, avoiding human feces as fertilizer, and wearing shoes. If the agent is transmitted by drinking water, disinfection of the water supply by heat or chemicals may be necessary. If the larvae are aquatic and invade through the skin, persons should avoid direct contact with infested water. Food-borne disease may be avoided by thoroughly washing and cooking vegetables and meats. Also,

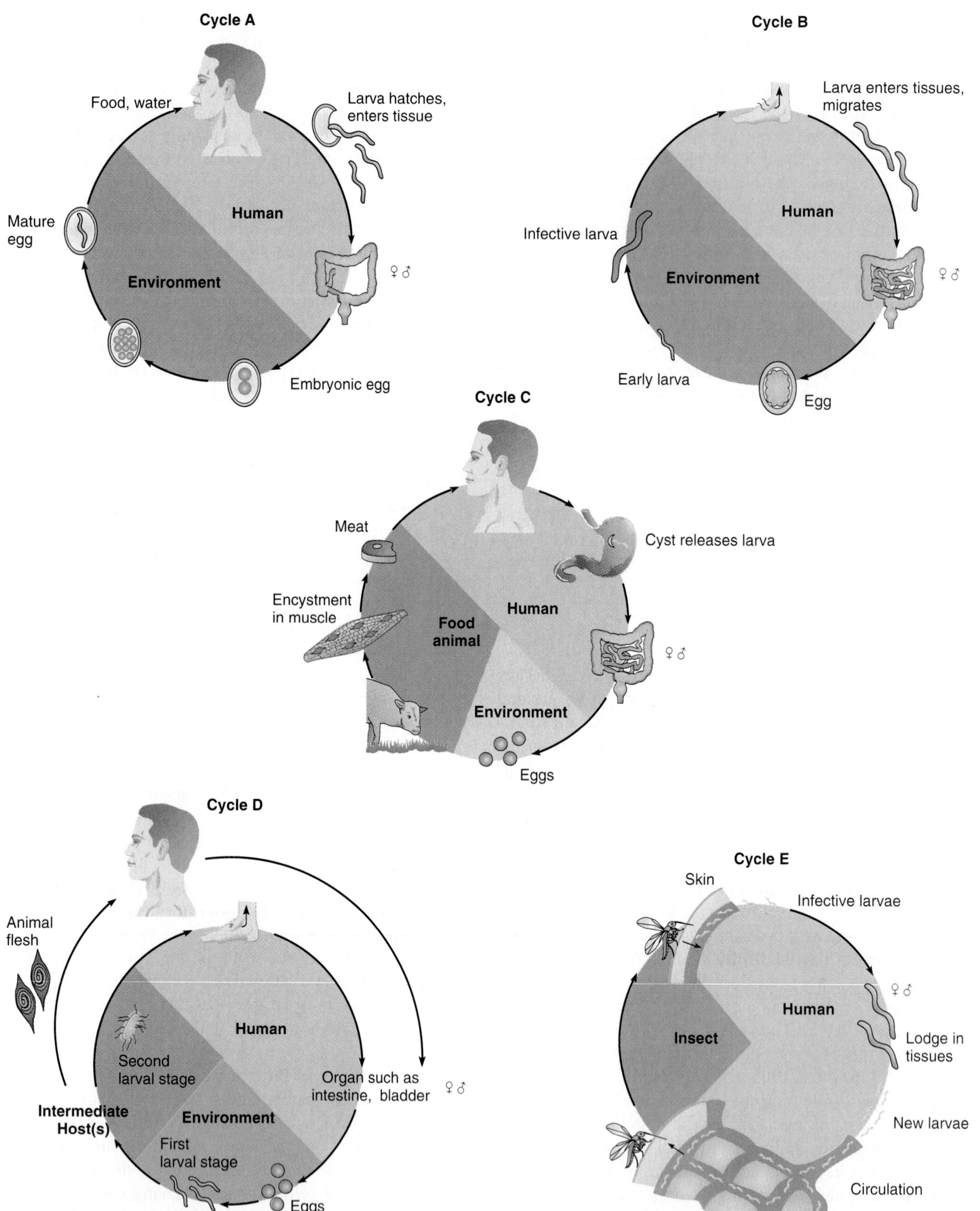

Figure 19.15 Five basic helminth life and transmission cycles. Cycle A occurs between human and environment: Worm develops in intestine; egg is released with feces into environment; eggs are ingested by new host and hatch in intestine. (Examples: *Ascaris, Trichuris.*) Cycle B occurs between human and environment: Worms mature in intestine; eggs are released with feces; larvae hatch and develop in environment; infection occurs through skin penetration by larvae. (Example: hookworms.) Cycle C occurs between humans, environment, and animals used by humans for food: Adult matures in human intestine; eggs are released into environment; eggs are eaten by grazing animals; larval forms encyst in tissue; humans eating animal flesh are infected. (Example: *Taenia.*) Cycle D alternates between human (definitive host) and one or more animal hosts (intermediate hosts) with an intermediate environmental phase: Eggs are released from human; humans are infected through ingestion or direct penetration by larval phase. (Examples: *Opisthorchis* and *Schistosoma.*) In cycle E the entire life cycle occurs in two different animal hosts with no environmental involvement: Human is definitive host and carries larval form in blood; insect vector is intermediate host that picks up and transmits larvae through bite. (Examples: *Wuchereria* and *Onchocerca.*)

Table 19.5 Antihelminthic Therapeutic Agents and Indications for Their Use

Drug	Effect	Used For
Piperazine	Paralyzes worm so it can be expelled in feces	Ascariasis, pinworm, hookworm
Pyrantel	Paralyzes worm so it can be expelled in feces	Ascariasis, pinworm, hookworm, trichinosis
Mebendazole	Blocks key step in worm metabolism	Trichuriasis (whipworm), ascariasis, hookworm, trichinosis
Thiabendazole	Blocks key step in worm metabolism	Strongyloidiasis, guinea worm, hookworm
Biphenium	Paralyzes worm	Hookworm
Diethylcarbamazine	Unknown, but kills microfilariae; does not kill adult worm	Filariasis, loiasis
Niridazole	Alters worm metabolism	Schistosomiasis
Praziquantel	Interferes with worm metabolism	Schistosomiasis, other flukes, tapeworm
Metrifonate	Paralyzes worm	Schistosomiasis
Niclosamide	Inhibits ATP formation in worm; destroys proglottids but not eggs	Tapeworm

because adult worms, larvae, and eggs are sensitive to cold, freezing foods is a highly satisfactory preventative. These methods work best if humans are the sole host of the parasite; if they are not, control of reservoirs or vector populations may be necessary. The complete eradication of helminth diseases is not likely as long as poverty and deeply entrenched cultural and religious practices continue to exist.

Nematode (Roundworm) Infestations

Nematodes are elongate, cylindrical worms that biologists consider one of the most abundant animal groups, since a fistful of soil or mud can easily contain millions of them. The vast majority are free-living soil and freshwater worms, but a few species are parasitic, and about 50 of these affect humans. Nematodes bear a smooth, protective outer covering or cuticle that is periodically shed as the worm grows. They are essentially headless, tapering to a fine point anteriorly. The sexes are distinct, and the male is often smaller than the female, having a hooked posterior containing spicules for coupling with the female (figure 19.16*a*). Beneath the cuticle are longitudinal but not circular muscles, so that a worm placed on a flat surface can only lash about in wild, whiplike motions and cannot crawl.

The human parasites are divided into intestinal nematodes, which develop to some degree in the intestine (for example, *Ascaris, Strongyloides,* and hookworms), and tissue nematodes, which spend their larval and adult phases in various soft tissues other than the intestine (for example, various filarial worms).

Intestinal Nematodes (Cycle A)

Ascaris lumbricoides* and Ascariasis** ***Ascaris lumbricoides is a giant (up to 300 mm long) intestinal roundworm that probably causes more infestations than any worm (estimated at one billion worldwide). Although most cases in the United States probably go unreported, most recorded cases are centered in the southeastern states. The *Ascaris* life cycle comprises larval and adult stages in humans, release of embryonic eggs in feces, and reintroduction of these eggs to humans through food, drink, or soiled objects placed in the mouth (figure 19.16*a, b*). The eggs thrive in warm, moist soils, resist cold and chemical disinfectants, and are killed by sunlight, high temperatures, and drying. Ingested eggs hatch in the human intestine, and the larvae embark upon an odyssey in the tissues. First they penetrate the intestinal wall and enter its lymphatic and circulatory drainage, feeding and growing as they go. Swept along by the bloodstream into the right ventricle of the heart, they are pumped through the pulmonary arteries and eventually arrive at the capillaries of the lungs. From this point, the larvae migrate up the respiratory tree to the glottis. Worms entering the throat are swallowed and returned to the small intestine, where they reach adulthood and reproduce, producing up to 200,000 fertilized eggs per day.

Even as adults, male and female worms are not attached to the intestine and retain some of their exploratory ways. They are known to probe into the bile duct and invade the biliary channels of the liver and gallbladder. On occasion, as a result of their misguided adventures, the worms escape from the mouth, nares, and, rarely, even from an eye. Severe inflammatory reactions mark the migratory route, and allergic reactions such as bronchospasm, asthma, or skin rash may occur. Heavy worm loads can retard the physical and mental development of young, malnourished children (figure 19.16*c*). One possibility with intestinal worm infestations is self-reinoculation due to poor personal hygiene.

Trichuris trichiura* and Whipworm Infection** The common name for ***Trichuris trichiura—whipworm—refers to its likeness to a miniature buggy whip. Although the worm is morphologically different from *Ascaris,* it has a similar transmission cycle, with humans as the sole host. Trichuriasis has a high incidence in the tropics and subtropics, where sanitary management of feces is not observed and children sometimes ingest dirt.

nematode (nem′-ah-tohd) Gr. *nema,* thread.

Ascaris lumbricoides (as′-kah-ris lum″-brih-koy′-dees) Gr. *askaris,* an intestinal worm; L. *lumbricus,* earthworm, and Gr. *eidos,* resemblance.

Trichuris (trik-ur′-is) Gr. *thrix,* hair, and *oura,* tail.

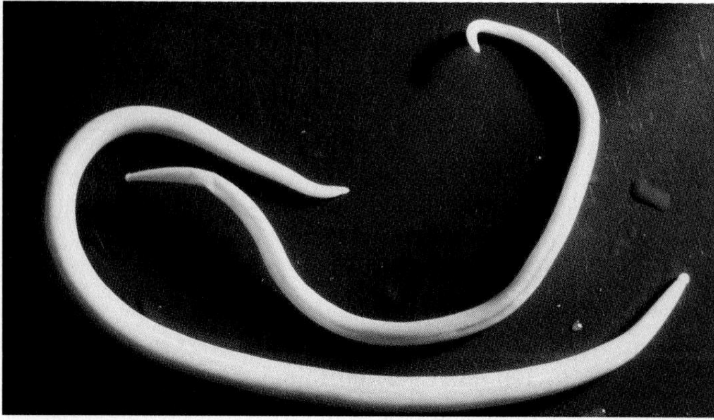

(a)

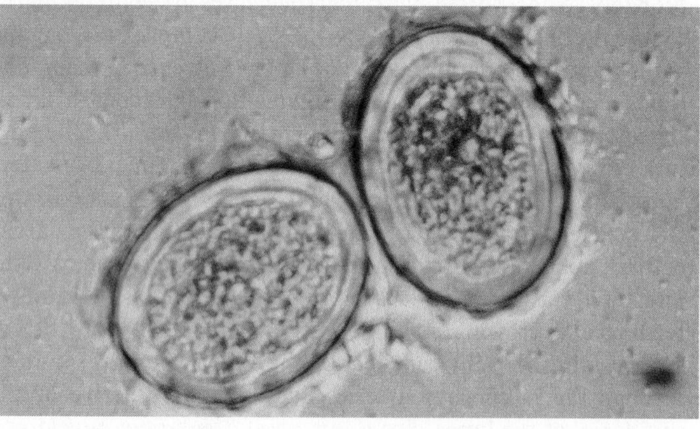

(b)

(c)

Figure 19.16 *Ascaris lumbricoides.* (*a*) Adult worms. The female is larger; the male has a hooked end. (*b*) A microscopic view of *Ascaris lumbricoides* ova from a fecal sample. (*c*) A mass of 800 worms retrieved from the ileum of a child during an autopsy.

Embryonic eggs deposited in the soil are not immediately infective and must first develop for three to six weeks in moist, warm soil. Ingested eggs hatch in the small intestine where the larvae attach, penetrate the outer wall, and go through several molts. From there, the adult nematodes move to the large intestine and grip there by their long, thin heads, while the thicker tail dangles free in the lumen. Following sexual maturation and fertilization, the females eventually lay 3,000 to 5,000 eggs daily into the bowel. The entire cycle requires about 90 days, and the infestation can last up to two years if left untreated.

In anchoring and feeding, the worms burrow into the intestinal mucosa, and some may even pierce capillaries, causing localized hemorrhage and providing a portal of entry for secondary bacterial infection. Light infestation may be asymptomatic, but heavier infestations can cause dysentery, loss of muscle tone, and rectal prolapse that may prove fatal in children.

***Enterobius vermicularis* and Pinworm Infection** The pinworm or seatworm, *Enterobius vermicularis,* was first discussed in chapter 4 (see figure 4.36). Enterobiasis is the most common worm disease of children in temperate zones. The transmission of the pinworm follows a course similar to that of the other enteric parasites. Freshly deposited eggs have a sticky coating that causes them to lodge beneath the fingernails during scratching and to adhere to fomites. Upon drying, the eggs are readily swept by air currents and settle in house dust. Self-inoculation occurs by swallowing eggs from contaminated fingers and by ingesting contaminated food or drink. Eggs hatching in the small intestine release larvae, which migrate to the large intestine, mature, and mate. Itching is brought on when the

---

Enterobius vermicularis (en″-ter-oh′-bee-us ver-mik″-yoo-lah′-ris) Gr. *enteron,* intestine, and *bios,* life; L. *vermiculus,* small worm.

mature female emerges from the anus and lays eggs. Although infestation is not fatal and most cases are asymptomatic, the afflicted child may be irritable, nervous, and pallid from lost or disrupted sleep, and may sometimes suffer nausea, abdominal discomfort, and diarrhea. A rapid, simple, and effective means of diagnosis involves pressing a piece of transparent adhesive tape against the anal skin and then applying it to a slide for microscopic evaluation.

Intestinal Helminths (Cycle B)

The Hookworms The two major agents of human hookworm infestation are ***Necator americanus,*** endemic to the New World, and ***Ancylostoma duodenale,*** endemic to the Old World, though the two species overlap in parts of Latin America. Otherwise, with respect to transmission, life cycle, and pathology, the two are usually lumped together. Humans are sometimes invaded by cat or dog hookworm larvae, sustaining a dermatitis called creeping eruption (see feature 19.3 later in this chapter). The term hookworm was inspired either by the adult's oral cutting plates by which it anchors to the intestinal villi or by its bent anterior end (figure 19.17). What differentiates the hookworm from other intestinal worms is that human infection is acquired when larvae hatched outside the body penetrate the skin.

Hookworm infestation remains endemic to the tropics and subtropics and is absent in arid, cold regions. A generation ago, it outranked all other parasitic diseases for morbidity (WHO estimated about 25% of the world's population was affected). Since that time, it has decreased in incidence, especially in the United States. At one time, hookworm plagued the rural South, often sapping its victims' strength and leaving the unfortunate stigma of being unkempt, irresponsible, ignorant, and shiftless. The decline of the disease in these regions was due to such simple changes as improved sanitation (indoor plumbing) and generally better standards of living, not the least of which included being able to afford shoes.

After being released into warm, moist soil with feces, hookworm eggs hatch into fine *filariform* larvae that instinctively climb onto grass or other vegetation to improve the probability of human contact. Ordinarily, the parasite enters sites on bare feet such as hair follicles, abrasions, or the soft skin between the toes, but cases have also occurred in persons wearing protective footwear due to mud splattering their ankles. Infection has even been reported in persons handling soiled laundry that was left damp for a few days.

Necator americanus (nee-kay'-tor ah-mer"-ih-cah'-nus) L. *necator,* a murderer, and *americanus,* of New World origin.

Ancylostoma duodenale (an"-kih-los'-toh-mah doo-oh-den-ah'-lee) Gr. *ankylos,* curved or hooked, *stoma,* mouth, and *duodenale,* the intestine.

filariform (fih-lar'-ih-form) L. *filum,* a thread.

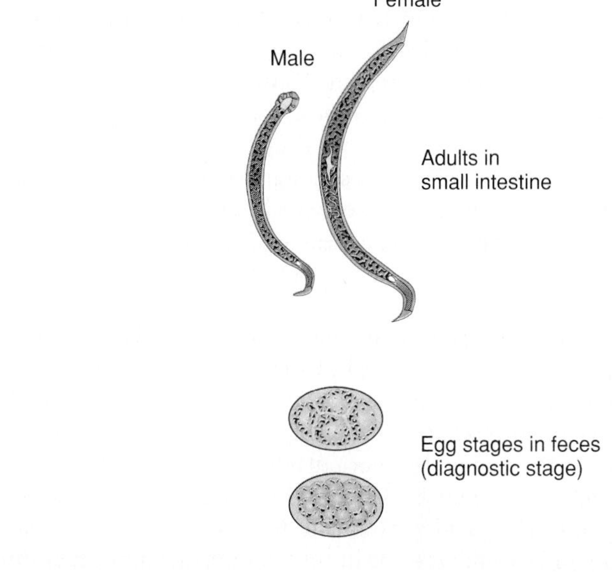

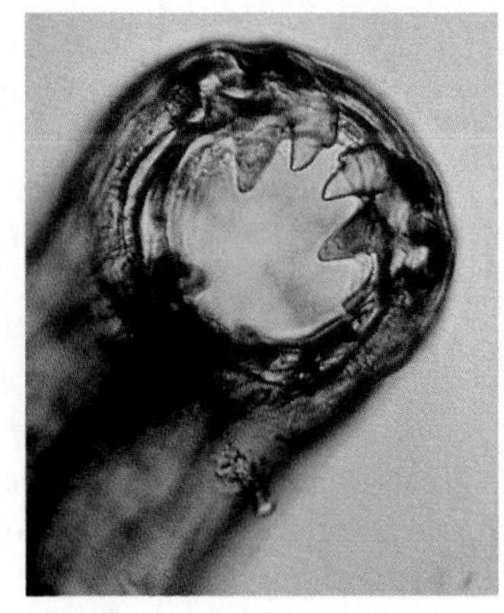

Figure 19.17 (*a*) Adult and egg stages of *Necator americanus.* (*b*) Cutting teeth on the mouths of (left) *N. americanus* and (right) *Ancylostoma duodenale.*

On contact, the hookworm larvae actively burrow into the skin. After several hours they reach the lymphatic or blood circulation and are immediately carried into the heart and lungs. There, the larvae infiltrate the air spaces, and aided by ciliary activity, proceed up the bronchi and trachea to the throat. Although some are ejected by spitting, most larvae are swallowed with sputum and arrive in the small intestine where they anchor, engorge upon blood and villous tissues, and reach the adult stage. Eggs first appear in the stool about six weeks from the time of entry, and the untreated infestation may last about five years.

The symptoms of infestation correspond to the migratory and feeding activities of the hookworm. Initial penetration with larvae may evoke a localized dermatitis called "ground itch."

The eruption usually clears spontaneously in two weeks if secondary bacterial infection is avoided. The transit of the larvae to the lungs is ordinarily brief, but it may cause symptoms of pneumonia and eosinophilia. However, the potential for injury is greatest during the intestinal phase, when heavy worm burdens may cause nausea, vomiting, cramps, pain, and bloody diarrhea. Because blood loss is significant, iron-deficient anemia develops, and infants are especially susceptible to hemorrhagic shock. Chronic fatigue, listlessness, apathy, and anemia worsen with chronic and repeated infections.

Strongyloides stercoralis* and Strongyloidiasis** The agent of strongyloidiasis, or threadworm infection, is ***Strongyloides stercoralis. *S. stercoralis* is an exceptional human nematode due to its minute size and its capacity to complete its life cycle either within the human body or outside in moist soil. This nematode afflicts an estimated 100 to 200 million individuals, shares the same tropical and subtropical distribution as hookworms, and manifests a similar basic life cycle. Infection may occur through penetration of the skin by soil larvae that have developed from fecally deposited or free-living eggs. The worm enters the circulation, is carried to the respiratory tract, is swallowed, and then enters the small intestine to complete development. Although adult *S. stercoralis* lay eggs in the gut just as hookworms do, eggs can hatch into larvae right in the large bowel, penetrate it, and complete their circuitous route through the lungs and back to the small intestine where they mature into adults. Eggs may likewise exit with feces and go through an environmental cycle. These numerous alternative life cycles greatly increase the chance of transmission, and the unique characteristic of larval development makes chronic infection common.

The first hint of threadworm infection is usually a red, intensely itchy skin rash at the site of entry. Mild migratory activity in an otherwise normal person may escape notice, but heavy worm loads can cause symptoms of pneumonitis and eosinophilia. The physical trauma and irritation that result from nematode activities in the intestine produce bloody diarrhea, liver enlargement, bowel obstruction, and malabsorption. In immunocompromised patients there is a risk of severe, disseminated infestation involving numerous organs (figure 19.18). Hardest hit are AIDS patients, transplant patients on immunosuppressant drugs, and cancer patients receiving irradiation therapy, who may die if not treated promptly.

Trichinella spiralis* and Trichinosis** The life cycle of ***Trichinella spiralis is spent entirely within the body of a mammalian host, usually a carnivore or omnivore such as a pig, bear, cat, dog, rat, or whale. In nature, the parasite is maintained in an encapsulated (encysted) larval form in the muscles of these animal reservoirs and is transmitted when other animals prey upon them. Humans acquire **trichinosis** (also called trichinellosis) by eating the meat of an infected animal (usually wild or domestic swine or bears), but humans are essentially dead-end hosts (provided corpses are buried and cannibalism is not practiced!) (figure 19.19).

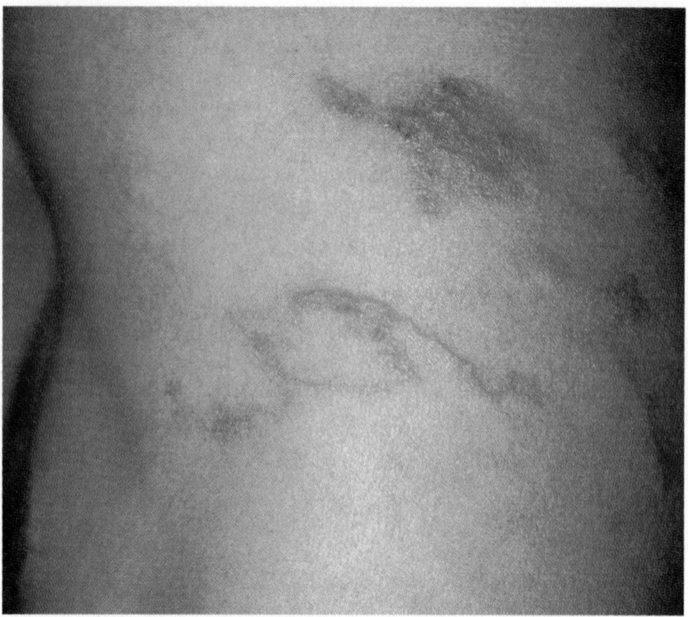

Figure 19.18 A patient with disseminated *Strongyloides* infection. Trails under the skin indicate the migrations of the worms.

Because all wild and domesticated mammals appear to be susceptible to *T. spiralis,* one might expect human trichinellosis to be common worldwide. But in actual fact, its incidence in tropical regions may be lower than in certain temperate regions with a higher standard of living. Relatively high rates of infestation occur in the United States (estimated at 200,000 new cases per year) and in Europe. This appears to be related to regional or ethnic customs of eating raw or rare pork dishes or wild animal meats. Bear meat is the source in up to one-third of the cases in the U.S. Home or small-scale butchering enterprises that do not carefully inspect pork may spread the parasite, although commercial pork may also be a source. Practices such as tasting raw homemade pork sausage for proper seasoning or serving rare pork or pork-beef mixtures have been responsible for sporadic outbreaks.

When meat infested with viable cysts is eaten, the larvae are liberated in the stomach or small intestine by digestive juices that erode the cyst's outer envelope. Following a few molts in the intestine, the larvae reach adulthood and mate. An unusual aspect of this worm's life cycle is that males are expelled in the feces in a harmless form, while females burrow into the intestinal lining. The females incubate eggs and shed live larvae that subsequently penetrate the intestine and enter the lymphatic channels and blood (figure 19.19). All tissues are at risk for destructive invasion, but final development occurs when the coiled larvae are encysted in the skeletal muscles. At maturity, the cyst is about 1 mm long and can be observed by careful inspection

Strongyloides stercoralis (stron″-jih-loy′-deez ster″-kor-ah′-lis) Gr. *strongylos,* round, and *stercoral,* pertaining to feces.

Trichinella spiralis (trik″-ih-nel′-ah spy-ral′-is) A little, spiral-shaped hair worm.

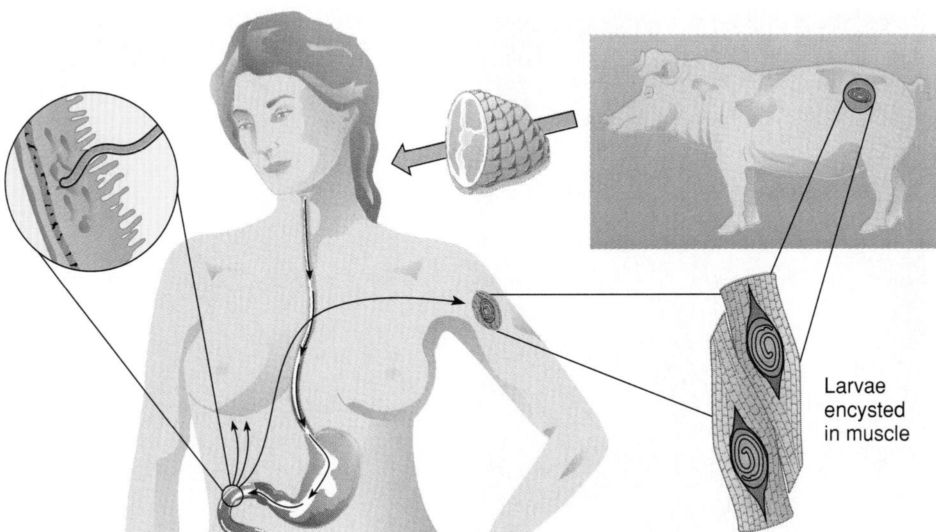

Figure 19.19 The cycle of trichinosis.

of meat. Although larvae may deteriorate over time, they have also been known to survive for years.

In many cases, there is little sign of *T. spiralis* infestation. In clinical illness, however, the first symptoms mimic the flu or viral fevers, with diarrhea, nausea, abdominal pains, fever, and sweating. The second phase, brought on by the migration of numerous larvae and their entrance into muscle, produces puffiness around the eyes, intense muscle and joint pain, shortness of breath, and pronounced eosinophilia. The most serious life-threatening manifestations are heart and brain involvement. In one remarkable case, a woman ate a large amount of raw pork tartare and was dead from massive brain infestation in 48 hours. Although the symptoms eventually subside, a cure is not available once the larvae have encysted in muscles.

Tissue Nematodes

As opposed to the intestinal worms previously described, tissue nematodes parasitize human blood, lymphatics, or skin. While these filarial parasites are responsible for grotesque deformities or loss of sight, they are also wonders of evolution and adaptation. The primary examples of tissue roundworms are the filarial worms, *Loa loa* (the eye worm), and *Dracunculus* (the guinea worm).

Filarial Nematodes and Filariasis (Cycle E)

The filarial parasites are long, threadlike worms that may live in host tissues for years. They are characterized by the production of tiny **microfilarial** larvae that circulate in the blood and lymphatics. Their anatomy is relatively simple, but they have a complex biphasic life cycle that alternates between humans and a blood-sucking mosquito or fly vector. The two species responsible for most **filariases** are *Wuchereria bancrofti,* the cause of elephantiasis, and *Onchocerca volvulus,* the agent of river blindness. Although there are numerous zoonotic species of filarial worms, these two species exist primarily in humans, and there are no other natural vertebrate reservoirs.

***Wuchereria bancrofti* and Bancroftian Filariasis** Filariasis caused by *Wuchereria bancrofti* is a widespread disease afflicting about 90 million persons in the hot, humid regions of the globe native to its intermediate host vectors: female *Culex, Anopheles,* and *Aedes* mosquitos. So intimately attuned are the microfilariae to the mosquito feeding cycle that in night-feeding mosquitos, the blood larvae swarm to the circulation of the skin's surface about two hours before and after midnight. In day-feeding mosquitos that transmit a different strain of the worm, the parasite is more plentiful in the cutaneous blood during the daytime hours.

Infection begins when a mosquito bite deposits the 1.5-mm infective larvae into the skin. These larvae penetrate the lymphatic vessels, where they develop into long-lived, reproducing adults. Mature female worms expel huge numbers of microfilariae into the circulation, a highly accessible location for another mosquito taking blood (figure 19.20*a*). The microfilariae complete larval development in the flight muscles of the mosquito, and the resultant larvae gather in its mouthparts as a ready source of infection for the next human host.

The first symptoms of larval infestation in humans are inflammation and dermatitis in the skin near the mosquito bite, lymphadenitis, and testicular pain (in males). The most conspicuous and bizarre sign of bancroftian filariasis is the swelling of the scrotum or legs called **elephantiasis.** This edema is caused by inflammation and blockage of the main lymphatic channels, which prevents return of lymph to the circulation and causes large amounts of fluid to accumulate in the extremities (figure 19.20*b*). The condition is relatively painless but can predispose to ulceration and infection. In extreme cases, the patient may become physically incapacitated and suffer from the stigma of carrying around a massive growth. Although travelers may

---

Wuchereria bancrofti (voo″-kur-ee′-ree-ah bang-krof′-tee) After Otto Wucherer and J. Bancroft.

elephantiasis (el″-eh-fan-ty′-ah-sis) Gr. *elephas,* elephant.

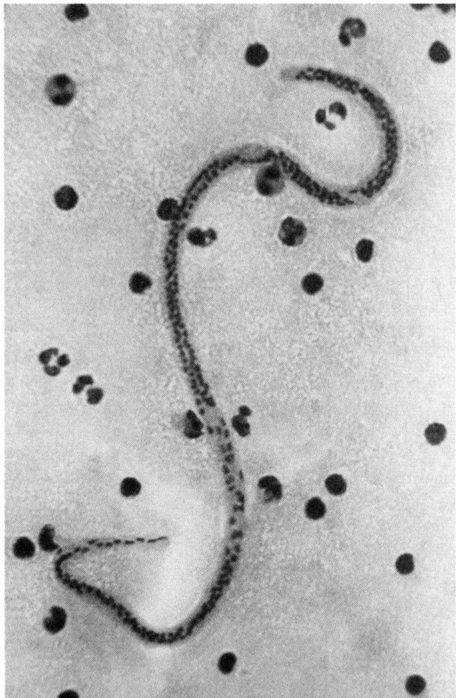

(a)

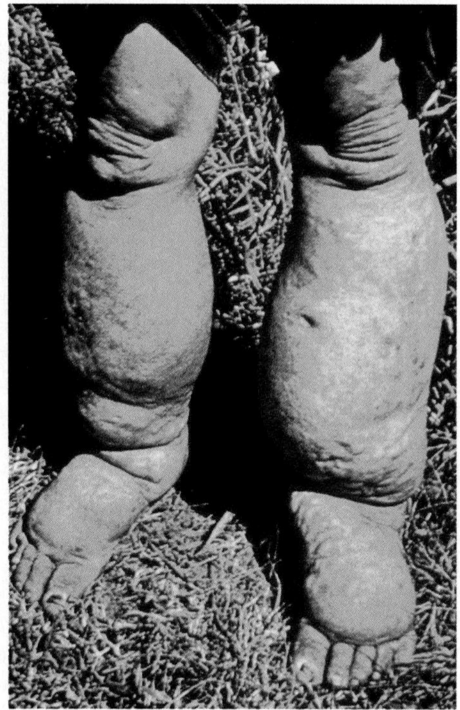

(b)

Figure 19.20 Bancroftian filariasis. (*a*) The appearance of a larva in a blood smear. (*b*) Pronounced elephantiasis in a Costa Rican woman.

regard with horror the thought of developing this condition, it occurs in individuals in endemic regions only after one to five decades of chronic infestation. Acute infection in the young, healthy person is usually subclinical.

***Onchocerca volvulus* and River Blindness** *Onchocerca volvulus* is a filarial worm adapted to subcutaneous tissues and transmitted by small biting vectors called **black flies.** These voracious flies often attack in large numbers, and it is not uncommon to be bitten several hundred times a day. Onchocerciasis is found in rural settlements along turbulent rivers bordered by overhanging vegetation, which are typical breeding sites of the black fly. The life and transmission cycle is similar to that of *Wuchereria,* although the larvae are deposited into the bite wound and develop into adults in the immediate subcutaneous tissues. At the sites of development, disfiguring nodules form within one to two years after initial contact. Microfilariae given off by the adult female migrate into the blood and the eyes. It is the blood phase that infects other feeding black flies.

Some cases of onchocerciasis result in a severe itching rash that may last for years. As worms degenerate, escaping antigens provoke inflammatory and granulomatous lesions. The other, more serious complication is **river blindness,** arising from infestation of the eye. The worms eventually invade the entire eye, producing much inflammation and permanent damage to the retina and optic nerve. In regions of high prevalence, it is not unusual for an ophthalmologist to see microfilariae wiggling in the anterior chamber during a routine eye checkup. Microfilariae die in several months, but adults may endure up to 15 years in nodules.

Worms That Destroy Vision The parasitic diseases provide numerous examples of human suffering, not the least of which is the tragic legacy of river blindness. In some West African villages, half of the adult population is blind and must be cared for and led around by younger family members, who in their time may also fall victim to the disease. River blindness has been so long a part of the lives of these villagers that it is almost fatalistically accepted as inevitable. The disease has caused social problems as well, because many natives thought the disease was spread by the river water and abandoned their villages. In the past 15 years, the WHO has been successful in eradicating the black fly by spraying pesticides along the rivers in some regions. Recently, an American drug company (Merck & Co.) began to distribute **ivermectin** to persons at risk for the disease or already infected. This drug has been used successfully to combat parasites in animals. It kills the blood larvae and prevents the spread of disease. Unfortunately, it does not reverse the blindness or kill the adult worm, so patients must take the drug for up to 15 years.

***Loa loa:* The African Eye Worm** Another nematode that makes disturbing migratory visits across the eye and under the skin is *Loa loa* (loh′-ah loh′-ah). This worm is larger and more apparent than *O. volvulus.* The agent occurs in parts of west and central Africa and is spread by the bites of dipteran flies. Its life cycle is similar to *Onchocerca* in that the larvae and adults remain more or less confined to the subcutaneous tissues. The first signs of disease are itchy, marble-sized lesions called calabar swellings at the site of penetration. These nodules are transient, and the worm lives for only a year or so, but the infestation is often recurrent. Because the nematode is temperature

Onchocerca volvulus (ong″-koh-ser′-kah volv′-yoo-lus) Gr. *onkos,* hook, and *kerkos,* tail; L. *volvere,* to twist.

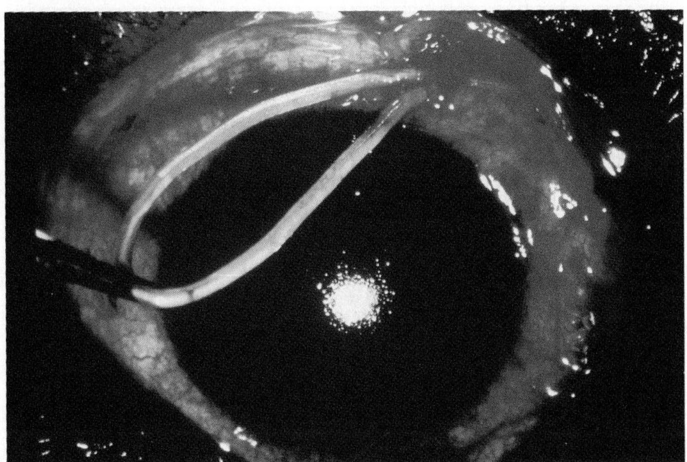
Figure 19.21 *Loa loa* being surgically removed from an eye.

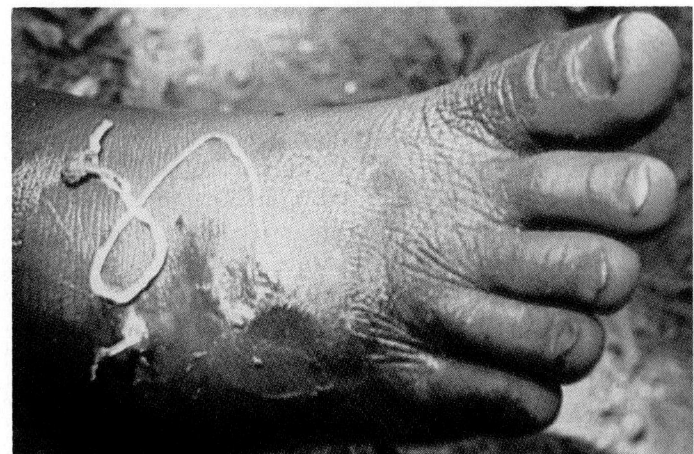
Figure 19.22 A female guinea worm emerging from the ulcer she has formed on a foot.

sensitive, it is drawn to the body surface in response to warmer temperature, and is frequently felt slithering beneath the skin or observed beneath the transparent conjunctiva (figure 19.21). One method of treatment is to carefully pull the worm from a small hole in the conjunctiva following application of local anaesthesia.

***Dracunculus medinensis* and Dracontiasis** In a way, this parasite has earned its name as the dragon worm or guinea worm. Dracontiasis has caused untold misery in desert residents of India, the Middle East, and central Africa for thousands of years. Even today, the disease claims 10 million new victims per year. Infestation begins with ingestion of infected *Cyclops,* an arthropod commonly found in still water such as a community pond or open well that is often the sole source of drinking water. The larvae are freed in the intestine and penetrate into subcutaneous tissues, often leaving no visible signs. Symptoms first occur after a year, coinciding with the maturation of the female worm. Just beneath the skin surface, the worm secretes irritants that provoke an itching and burning blister, hence another common name, fireworm. The worm has remarkably adapted its cycle to a simple but predictable human behavior. When the victim tries to relieve the pain by immersing his feet in well water, the blister erupts and permits the female to expel larvae into the water (figure 19.22). The larvae are then ingested by *Cyclops* that can serve as a continued source of reinfection.

Since antiquity, the "cure" for dracontiasis has involved teasing the worm out of the open blister and gradually winding it around a stick a few centimeters a day to avoid breaking it, which would cause severe irritation. The whole process takes 10 to 14 days, but unfortunately, even with the extraction of the worm, the open sore it leaves is often subject to secondary bacterial infections that may be fatal. An international agency called Global 2000 has currently targeted this disease for complete eradication by 1995. Control efforts are being focussed on water treatment and well-drilling in 18 countries.

Dracunculus medinensis (drah-kung'-kyoo-lus meh-dih-nen'-sis) L. The little dragon of Medina.

The Trematodes or Flukes

The fluke (ME *floke,* flat) is named for the characteristic ovoid or flattened body characteristic of most adult worms, especially liver flukes such as *Fasciola hepatica.* Blood flukes, also called schistosomes, have a more cylindrical body plan. The name **trematode** (Gr. *trema,* a hole) comes from the muscular sucker containing a mouth (hole) at the anterior end of the fluke (see figure 4.35*b*). Flukes have digestive, excretory, neuromuscular, and reproductive systems but lack circulatory and respiratory systems. Each limb of the Y-shaped digestive system repeatedly branches, forming numerous blind pouches to distribute nutrients. Except in the blood flukes, both male and female reproductive systems occur in the same individual and occupy a major portion of the body. Trematodes are adapted to several human practices such as flooding fields for agriculture, using raw sewage as fertilizer, and eating raw fish or aquatic vegetation. In human fluke cycles, animals such as snails or fish are usually the intermediate hosts, and humans are the definitive hosts.

Blood Flukes: The Schistosomes (Cycle D)

Schistosomiasis has afflicted humans for thousands of years. The ancient Egyptian writings that described males "menstruating" were probably referring to blood in the urine caused by renal schistosomiasis. Likewise, petrified *Schistosoma* eggs have been discovered in 3,000-year-old mummies. The disease is caused by *Schistosoma mansoni, S. japonicum,* or *S. haematobium,*[3] species that are morphologically and geographically distinct but share similar life cycles, transmission methods, and general disease manifestations. The disease occurs in 73 countries located in Africa, South America, the Middle East, and the Far East, and is not native to the United States. Schistosomiasis is a major public health problem, probably affecting 200 million persons

Schistosoma (skis''-toh-soh'-mah) Gr. *schisto,* split, and *soma,* body.

3. The species are named after P. Manson; Japan; and Gr. *haem,* blood, and *obe,* to like.

at any one time. Animal species of schistosomes sometimes opportunistically infect humans, with interesting results (see feature 19.3).

The schistosome parasite exhibits the stages of egg, miracidium, cercariae, and adult fluke, and demonstrates intimate adaptations to both the humans and certain species of freshwater snails required to complete its life cycle. The cycle begins when infected humans release eggs into irrigated fields or ponds either by deliberate fertilization with excreta or by defecating or urinating directly into the water. The egg hatches in the water and gives off an actively swimming ciliated larva called a **miracidium** (figure 19.23*a*), which instinctively swims to a snail and burrows into a vulnerable site, shedding its ciliated covering in the process. In the body of the snail, the miracidium multiplies and transforms into a larger, fork-tailed swimming larva called a **cercaria.** Cercariae are given off by the thousands into the water by infected snails.

Upon contact with a human wading or bathing in water, cercariae attach themselves to the skin by ventral suckers and penetrate hair follicles. They pass into small blood and lymphatic vessels, and in about a week, arrive in the liver. Schistosomes achieve sexual maturity in the liver, and the male and female worms remain permanently entwined to facilitate mating (figure 19.23*b*). In time, the pair migrates to and lodges in small blood vessels at specific sites. *S. mansoni* and *S. japonicum* end up in the mesenteric venules of the small intestine, while *S. haematobium* enters the venous plexus of the bladder. While attached to these intravascular sites, the worms feed upon blood, and the female lays spiked eggs that eventually penetrate the intestine or bladder and are voided in feces or urine.

The first symptoms of infestation are itchiness in the vicinity of cercarial entry, followed by fever, chills, diarrhea, and cough. The most severe consequences, associated with chronic infections, are hepatomegaly and liver disease, splenomegaly, bladder obstruction, and blood in the urine. Occasionally, eggs carried into the the central nervous system and heart create a severe granulomatous response. Adult flukes can live for many years, and because they are capable of eluding the host defenses, the immune response is not effective.

Liver and Lung Flukes

Several trematode species that infest humans may be of zoonotic origin. The Chinese liver fluke, ***Opisthorchis (Clonorchis) sinensis,*** completes its sexual development in mammals such as cats, dogs, and swine and has two intermediate hosts in the water phase of development: the snail and the fish (cycle D). Humans ingest cercariae in inadequately cooked or raw freshwater fish and crustaceans. Larvae hatch and crawl into the bile duct where they mature and shed eggs into the intestinal tract. Feces containing eggs are passed into standing water that harbors the intermediate snail host. Infected snails release cercariae that invade fish living in the same water. The cycle is complete when humans ingest the fish containing infectious larvae.

The liver fluke ***Fasciola hepatica*** is a common parasite in sheep, cattle, goats, swine, rabbits, and other mammals, and is

Feature 19.3 When Parasites Attack the "Wrong" Host

Not only do we humans host a number of our own parasites, but occasionally, worms indigenous to other animals attempt to take up residence in us. When this happens, the parasite cannot complete the life cycle in its usual fashion, but in making the attempt, it may cause discomfort and even death in some cases.

One such parasite is *Echinococcus granulosus,* a tiny intestinal tapeworm that causes **hydatid cyst disease.** The parasite's usual definitive host is the dog, though it may be transmitted to other carnivores and herbivores in egg and larval stages. This disease is sometimes an occupational hazard of sheepherders because the parasite cycles back and forth between sheepdogs and sheep. Humans become involved through frequent, intimate contact with fur, paws, and tongue, especially while "kissing" their dog. If a person accidently ingests eggs, the larvae hatch and migrate through the body, eventually coming to rest in the liver and lungs. There they form compact water-filled packets called **hydatid cysts** (L. *hydatos,* water drop). The bursting of these capsules gives rise to an allergic reaction and also proliferates the worm by seeding other sites.

Another disease that arises from invasion by animal worms is **cutaneous larval migrans,** a creeping eruption that occurs when humans walk barefoot through areas where dogs and cats defecate, enabling the larvae of a dog or cat hookworm (genus *Ancylostoma*) to enter the skin. The larva cannot invade completely and so remains just beneath the skin where it meanders about, producing dark red tunnels. These lesions become so itchy and inflamed that the person can't eat or sleep and may have painful secondary infections. A related condition, "swimmer's itch," results when larval schistosomes of birds and small mammals penetrate a short distance into the skin and then die. The disintegrating worm sensitizes the humans and causes intense itching and discomfort.

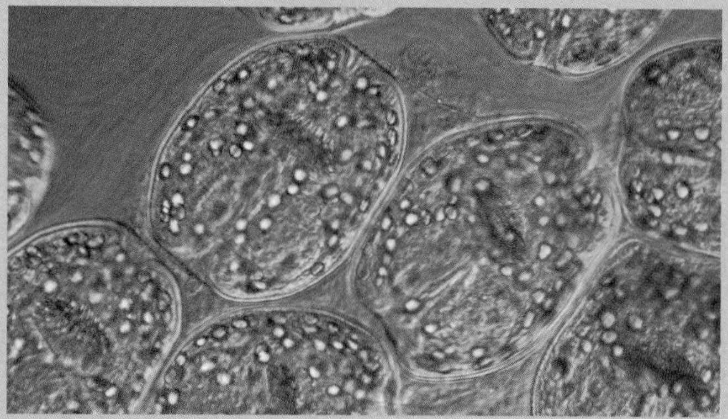

Hydatid cysts from the lung. These fluid-filled spheres are also known as bladder worms.

miracidium (my''-rah-sid'-ee-um) Gr. *meirakidion,* little boy.

cercaria (sir-kair'-ee-uh) Gr. *kerkos,* tail.

Opisthorchis sinensis (oh''-piss-thor'-kis sy-nen'-sis) Gr. *opisthein,* behind, *orchis,* testis, and *sinos,* oriental.

Fasciola hepatica (fah-see'-oh-lah heh-pat'-ih-kah) Gr. *fasciola,* a band, and *hepatos,* liver.

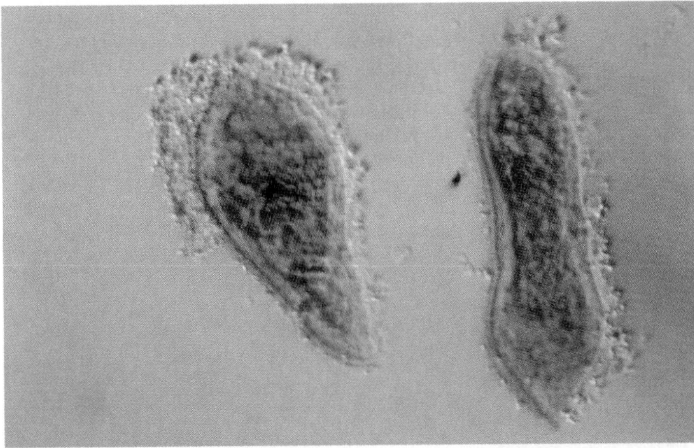

(a)

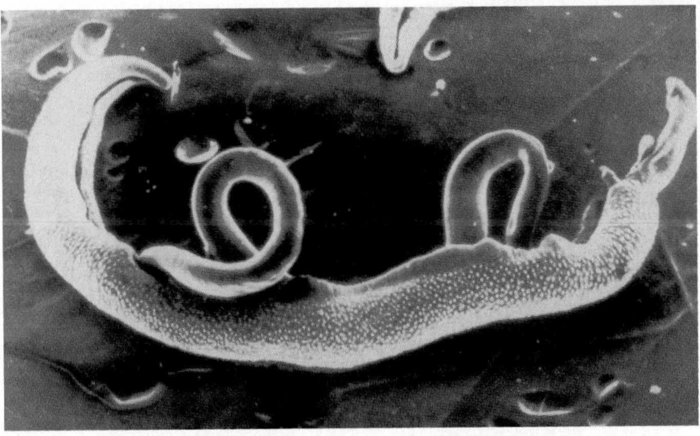

(b)

Figure 19.23 Stages in the life cycle of *Schistosoma*. (*a*) Snail phase or miracidium. (*b*) An electron micrograph of normal mating position. The male worm holds the female in a groove on his ventral surface.

Figure 19.24 *Fasciola hepatica*, the sheep liver fluke (×2).

occasionally transmitted to humans (figure 19.24). Periodic outbreaks in temperate regions of Europe and South America are associated with eating wild watercress. The life cycle is very complex, involving the mammal as the definitive host, the release of eggs in feces, the hatching of the egg in water into miracidia, invasion of freshwater snails, development and release of cercariae, encystment of cercariae on a water plant, and ingestion of the cyst by a mammalian host eating the plant. The cysts release young flukes in the intestine that wander to the liver, lodge in the gallbladder, and develop into the large, leaflike, hermaphroditic adult. Humans develop symptoms of vomiting, diarrhea, hepatomegaly, and bile obstruction only if they are chronically infected by a number of flukes.

The adult lung fluke ***Paragonimus westermani,*** a fluke endemic to the Orient, India, and South America, occupies the pulmonary tissues of various reservoir mammals, usually carnivores such as cats, dogs, foxes, wolves, and weasels. These animals release eggs into water, where they invade the first intermediate host, a snail. Cercariae shed by the snail attack a second intermediate host, a crab or crayfish. Humans accidently contract infection by eating these infected crustaceans in a raw or undercooked state. Following release in the intestine, the worms migrate to the lungs and may cause cough, pleural pain, and lung abscess.

Paragonimus westermani (par″-ah-gon′-ih-mus wes-tur-man′-eye) Gr. *para*, to bear, and *gonimus*, productive; after Westerman, a researcher in flukes.

Cestode or Tapeworm Infestations

The **cestodes,** or tapeworms, are among the most extreme parasitic worms, consisting of little more than a tiny holdfast connected to a chain of flattened sacs that contain the reproductive organs (see feature 4.3). Several anatomical features of the adult emphasize their adaptations to an intestinal existence (figure 19.25). The small **scolex,** or head of the worm, has suckers or hooklets for clinging to the intestinal wall, but it has no mouth because nutrients are absorbed directly through the cuticle covering. Because of its importance in anchoring the worm, the scolex is a potential target for antihelminthic drugs. The head is attached by a **neck** to the **strobila,** a long ribbon composed of individual reproductive segments called **proglottids.** The contents of proglottids are dominated by male and female reproductive organs, though each one also contains a modest

cestode (ses′-tohd) L. *caestus*, to strike.
scolex (skoh′-leks) Gr. *scolos*, worm.
strobila (stroh-bih′-lah) Gr. *strobilos*, twisted.
proglottid (proh-glot′-id) Gr. *pro*, before, and *glottis*.

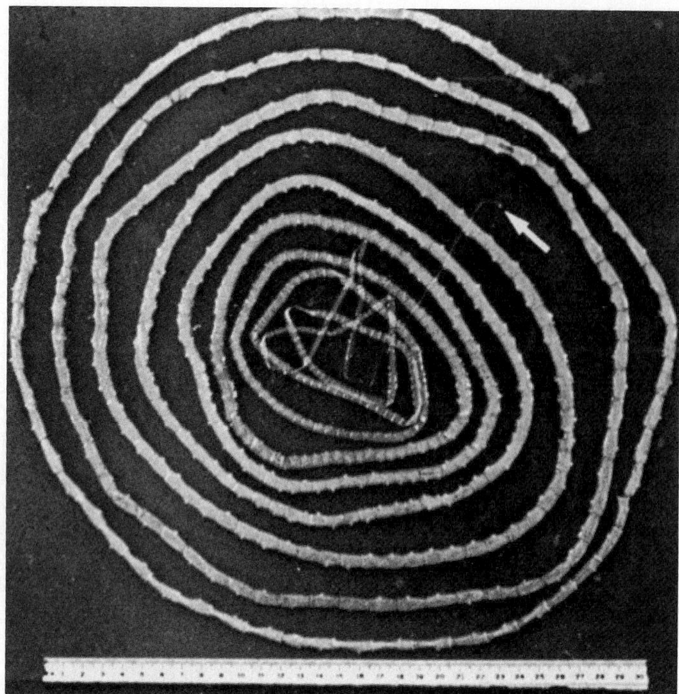

(a)

Figure 19.25 Tapeworm infestation in humans. (*a*) Adult *T. saginata*. The arrow points to the scolex; the remainder of the tape, called the strobila, has a total length of 5 meters. (*b*) Tapeworm scolex. (*c*) A generalized diagram of the life cycle of the beef tapeworm *Taenia saginata*.

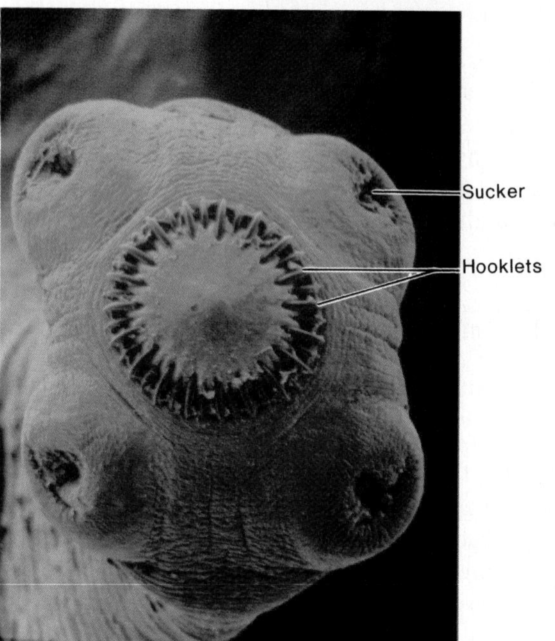

(b)

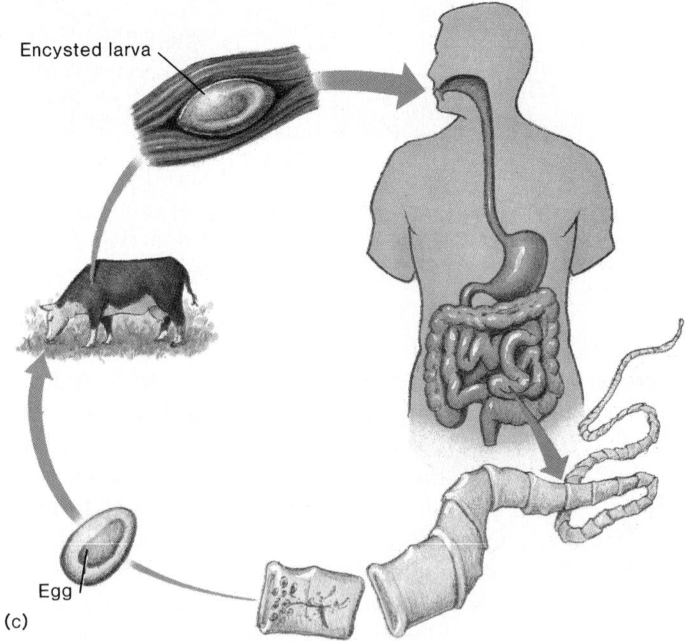

(c)

neuromuscular system that permits feeble movements. The neck generates the proglottids, tapering from the smaller, newly formed, and least mature segments nearest the neck to the larger, older ones containing ripe eggs at the distal end of the worm. In some worms, the proglottids are shed intact in feces after detaching from the strobila, and in others, the proglottids break open in the intestine and expel free eggs into the feces.

All tapeworms for which humans are the definitive hosts require an intermediate animal host. The principal human species are *Taenia saginata,* the beef tapeworm, *Taenia solium,* the swine tapeworm, and *Diphyllobothrium latum,* the fish tapeworm (see feature 19.4). The first two tapeworms cause an important infestation called **taeniasis.** The beef tapeworm follows a cycle C pattern, in which the larval worm is ingested in raw beef (figure 19.25), and the pork tapeworm follows both a cycle C and a modified cycle A, in which the eggs are ingested and hatch in the body.

Feature 19.4 Waiter, There's a Worm in My Fish

Various forms of raw fish or shellfish are part of the cuisine of many cultures. Some of the more popular preparations are Japanese sashimi and sushi. Although these delicacies are definitely an acquired taste, it is possible for humans to acquire much more than flavor from them, because fish flesh often harbors various types of helminths. Usually humans are not natural hosts for these parasites, so the infestation does not develop fully, but some of the symptoms the worms cause during their sojourn in the body can be quite distressing. One of the common fish parasites transmitted this way is ***Anisakis,*** a nematode parasite of marine vertebrates, fish, and mammals. The symptoms of anisakiasis, usually appearing within a few hours after ingesting the half-inch larvae, are due to the attempts of the larvae to burrow into the tissues. The first symptom is often a tingling sensation in the throat, followed by acute gastrointestinal pain, cramping, and vomiting that mimic appendicitis. At times, the cause of these mysterious symptoms is explained when worms are regurgitated or discovered during surgery, but in many cases, they are probably expelled without symptoms. Another worm that can hide in raw or undercooked freshwater fish is the tapeworm *Diphyllobothrium latum,* common to the Great Lakes, Alaska, and Canada. This worm may remain in the intestine for longer periods and continue to develop, because humans serve as its definitive host.

Students always ask about the safety of sushi and seem uncertain about how to eat it and still avoid parasites. One good safeguard is to patronize a reputable sushi bar, since a competent sushi chef carefully examines the fish for the readily visible larvae. Another is to avoid raw salmon and halibut, which are more often involved than tuna and octopus. Fish may also be freed from live parasites by freezing for five days at −20°C.

Taenia saginata is one of the largest tapeworms, being composed of up to 2,000 proglottids and anchored by a scolex with suckers. Taeniasis caused by the beef tapeworm is distributed worldwide, but is mainly concentrated in areas where raw or undercooked beef is eaten and where cattle are permitted to graze in fields or to drink water grossly contaminated with proglottids or eggs. Following their ingestion, the eggs hatch in the small intestine, and the released larvae migrate throughout the organs of the cow. Ultimately, they encyst in the muscles, becoming **cysticerci,** young tapeworms that are the infective stage for humans. When humans ingest a live cysticercus in beef, it is uncoated and flushed into the intestine where it firmly attaches by the scolex and develops into an adult tapeworm. So efficient is this egg-laying machine that a single tapeworm is the usual human load. Humans have not been known to acquire infection by ingesting the eggs. For such a large organism, it is remarkable how inapparent a tapeworm's presence in the intestine can be. Usually inflammation is minimal, and infestations often escape detection until the patient discovers proglottids in his stool. At most, some patients complain of vague abdominal pain and nausea that can be relieved by a meal.

Taenia saginata (tee'-nee-ah saj-ih-nah'-tah) L. *taenia,* a flat band, *sagi,* a pouch, and *natare,* to swim.

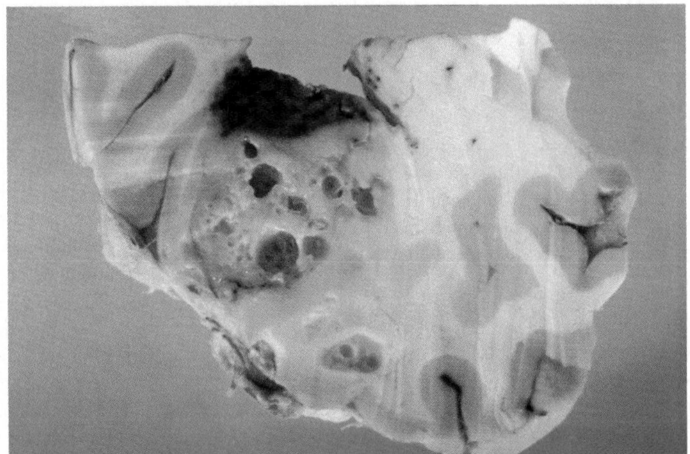

Figure 19.26 Cysticercosis caused by *T. solium* in the brain of a 13-year-old girl.

Taenia solium differs from *T. saginata* by being somewhat smaller, having a scolex with hooklets and suckers to attach to the intestine, and being infective to humans in both the cysticercus and egg stages. The disease is endemic to regions where pigs are allowed to eat fecally contaminated food and humans consume raw or partially cooked pork (Southeast Asia, South America, Mexico, and eastern Europe). The cycle involving ingestion of cysticerci is nearly identical to that shown in figure 19.25.

When humans ingest pork tapeworm eggs that have gotten into their food or water, a different form of the disease, called **cysticercosis,** results. Although humans are not the usual intermediate hosts, the eggs can still hatch in the intestine, releasing tapeworm larvae that migrate to all tissues. They finally settle down to form peculiar cysticerci, or bladder worms, each of which has a small capsule resembling a little bladder. A connective tissue wrapping contributed by the host completes the formation of the bladder worm. The most harmful sites in which they lodge are the heart muscle, the eye, and the brain, especially if a large number of bladder worms occupy tissue space (figure 19.26). It is not uncommon for patients to exhibit seizures, psychiatric disturbances, and other signs of neurologic impairment.

Chapter Review with Key Terms

Parasitology

Parasitology studies **protozoa, helminths** (worms), and arthropods that live on the body of a host. Parasites are spread from human to human directly or indirectly, and from animals and vectors to humans.

Protozoan Pathogens

Protozoans propagate only as **trophozoites** (active feeding stage found in host), or alternate between a trophozoite and a **cyst** (dormant, resistant survival body outside of host). Some have complex life cycles with sexual and asexual phases carried out in more than one host.

Infectious Amebae: Most significant is ***Entamoeba histolytica*** which alternates between a large ameboid trophozoite and a multinucleate cyst that is the infectious stage; human is principal host; **amebiasis** occurs worldwide; affects approximately 500 million persons in the tropics; cysts released in feces of carriers; spread through unsanitary water and food, agricultural practices, and oral-anal or other close contact. Cysts release trophozoites that invade large intestine and feed on tissue and blood; creates ulcers; may penetrate into deeper layers and disseminate to extraintestinal sites (liver, spleen, brain). Symptoms are dysentery, intestinal distress; severe cases may cause fatal intestinal perforation and liver damage; effective drugs are iodoquinol, metronidazole, and chloroquine.

Amebic brain infections are caused by ***Naegleria fowleri*** and ***Acanthamoeba,*** free-living inhabitants of natural waters; **primary acute meningoencephalitis** is acquired through contact with water wherein amebae are flushed into nasal cavity and through traumatic eye damage; infiltration of brain is usually fatal.

Ciliates: The only important pathogen is ***Balantidium coli,*** a common occupant of the intestines of domestic animals such as pigs and cattle; acquired by humans when cyst-containing food and water are ingested; cyst liberates trophozoite, which erodes intestine and creates intestinal symptoms.

Flagellates: Genus ***Trichomonas*** contains pear-shaped flagellated protozoa that are parasites or commensals of animals. Principal human pathogen is ***T. vaginalis***; has no cysts; causes an STD, trichomoniasis; disease spread by direct contact (occasionally through fomites); most common in promiscuous persons; infects vagina, cervix, urethra, producing frothy yellow discharge; both sex partners treated with metronidazole.

Giardia lamblia is an intestinal flagellate with a unique heart shape and four flagella; has cyst stage; causes **giardiasis**; transmission similar to *Entamoeba*; natural reservoir apparently wild and domestic animals; sources of infection are cyst-contaminated fresh water and food; cysts release trophozoites that invade intestinal glands; cause severe abdominal pain, diarrhea; may be severe and chronic in immune-deficient patients; contaminated water must be boiled or filtered as cyst is resistant to chlorine; quinicrine or metronidazole is usual treatment.

Hemoflagellates: Parasites that occur in blood during infection; cause zoonoses spread by insect vectors; most are exotic, tropical; have complex life cycles, with various **mastigote** phases that develop in the insect and human hosts.

Trypanosoma have tapering, crescent-shaped cells with flagellum and undulating membrane; two species cause trypanosomiasis; are geographically isolated and have different blood-feeding vectors; diseases endemic to regions where vectors live.

T. brucei causes **African sleeping sickness**; the two strains are Gambian (a chronic disease) and Rhodesian (an acute disease); parasites may be carried by reservoir hosts such as wild and domestic animals; both are spread by **tsetse fly,** which carries agent in salivary gland and inoculates bite; in the blood, trypanosome multiplies and alters its surface antigen so that immune system cannot control it; enlargement of spleen, lymph nodes, joints involved; brain damage leads to sleepiness, tremors, paralysis, coma, death; treatment with suramin not totally effective.

T. cruzi causes **Chagas' disease**; endemic to Latin America; cycle similar to *T. brucei,* except that the kissing or **reduviid bug** is vector; infection occurs when bug feces inoculate wound or mucous membrane; invasion is marked by inflammation, organ involvement (especially heart and brain); nifurtimox may give some therapeutic benefit.

Leishmania contains several geographically separate species that cause **leishmaniasis,** a zoonosis transmitted by phlebotomous (sand) flies; reservoirs are wild animals; humans enter cycle when fly inoculates them; macrophages engulf infective cells. Disease may be cutaneous, with a local sore on skin; mucocutaneous, with deep granulations on nose, lips, oral cavity; systemic, with fever, enlarged organs, anemia. Most severe and fatal form is **kala-azar**; therapy is antimony, pentamidine.

Sporozoans: Tiny obligate intracellular parasites that lack motility in mature stage and have complex sexual and asexual phases; infection occurs through sporozoites, fecal cysts called oocysts, or tissue cysts; many are zoonotic; some are arthropod vector-borne.

Plasmodium consists of four species that cause **malaria**; human is primary host for asexual phase of parasite; female ***Anopheles*** **mosquito** is vector/host for sexual phase; distributed primarily in malaria belt around the equator where conditions are ideal for mosquito/human interaction; nearly 300 million cases are estimated worldwide. Infective forms for humans (**sporozoites**) enter blood with mosquito saliva; from blood, they penetrate liver cells, multiply, and form hundreds of **merozoites**; merozoites that burst out of liver enter circulation, invade, multiply in, and lyse red blood cells; this cycle continues in RBCs until some RBCs produce sex cells that are picked up by another feeding mosquito; sexual reproduction in mosquito gut produces sporozoites that are harbored in salivary gland. Symptoms are due to lysis of RBCs and include episodes of chill-fever-sweating, anemia, and organ enlargement. Therapy is chloroquine, quinine, primiquine; a vaccine is in development.

Toxoplasma gondii is a sporozoan with wide distribution among birds and mammals; infectious in both tissue cyst and oocyst phases; primary reservoirs are felines harboring oocysts in the GI tract; feces may spread it among other cats, vertebrates (cattle, pigs), and humans. **Toxoplasmosis** is acquired by ingesting raw or rare meats containing tissue cysts or by accidently ingesting or inhaling oocysts from substances contaminated by cat feces. Infection is usually mild and flu-like in older children and adults; more severe in immunodepressed (AIDS) patients or fetuses. **Fetal toxoplasmosis** damages brain, heart, lungs; pregnant women must avoid rare and raw meats and careless handling of cats or cat litter.

Sarcocystis is a zoonotic parasite transmitted to humans from meat containing live tissue cysts; humans develop diarrhea, can transmit cysts to grazing animals.

Cryptosporidium is a common vertebrate pathogen that exists in both tissue and oocyst phases; most cases of **cryptosporidiosis** are caused by handling animals with infections or by drinking water contaminated with animal feces; infection causes diarrhea, abdominal symptoms similar to giardiasis; is a common chronic, debilitating, wasting infection in AIDS patients; can be fatal.

Pneumocystis carinii is a small, funguslike protozoan that appears to be widespread in animals and humans; spread by droplets, close contact. In persons with weakened defenses, it causes acute and dangerous **pneumonia** (**PCP**), in which the lungs fill up with parasites and exudate; is the most prevalent secondary infection in AIDS patients and a common cause of death. Treated with aerosol pentamidine and co-trimoxazole.

Babesia causes **babesiosis,** a zoonosis spread by hard ticks; similar to malaria in symptoms and treatment.

Isospora is an uncommon human intestinal parasite transmitted through fecal contamination; causes coccidiosis.

Helminth Parasites

Helminths, or worms, are relatively larger, multicellular animals with specialized tissues and organs; adaptations to habitat include specialized mouthparts, reduction of organs, protective cuticles, complex life cycles that increase chances of encountering a host. More diseases are caused by worms than by any other group; worms are often long-lived; concentrated in rural semitropics and tropics.

Life cycles involve **adult worms** that mate, produce fertile **eggs** that hatch into **larvae**; larvae mature in several stages to adults; adults live in **definitive host**; eggs and larvae may develop in same host, external environment, or another **intermediate host** (an animal). In some worms, the sexes are separate; others are hermaphroditic, with both sexes in same worm. Because worms do not usually increase in number by multiplying in host, their presence is an **infestation.**

Basic transmission cycles: (1) Egg-laden feces are deposited into soil, water; (2) eggs may be directly infectious when ingested, or (3) go through larval phase in environment that is infectious by direct penetration of skin; (4) animal flesh or plants containing encysted larval worms are ingested; (5) helminth is inoculated into body by insect vector that is intermediate host.

Pathology is due to action of worms on tissues: feeding, breaking down, migrating through tissues, and accumulation of worms and worm products; immune response not completely effective; worms may cause allergies, survive for long periods in the body; **antihelminthic drugs** such as piperazine, pyrantel, and metrifonate control worms by paralyzing their muscles and causing them to be shed; drugs such as mebendazole, diethylcarbamazine, praziquantel interfere with metabolism and kill them; other control measures are improved sanitation (no use of feces as fertilizer), protective clothing, cooking food adequately, and controlling vectors.

Nematodes

Roundworms are filamentous, tapered worms; round in cross section; have protective cuticles, circular muscles, digestive tract; separate sexes with well-developed reproductive organs.

Intestinal Nematodes: These worms spend part of their life cycle in the intestine; adults shed eggs from intestine. ***Ascaris lumbricoides*** is indigenous to humans; very common worldwide; eggs deposited in human feces are the infective stage; ingested eggs hatch in intestine; larvae burrow through intestine into blood, then enter heart, pulmonary circulation, lungs, migrate into pharynx, are swallowed; adult form develops in intestine, produces eggs; symptoms and signs depend upon size of worm burden; may obstruct organs, cause inflammatory reactions.

Trichuris trichiura, the **whipworm,** is a small intestinal pathogen that is transmitted like *Ascaris* but remains in intestine throughout development; symptoms are usually intestinal.

Enterobius vermicularis, the **pinworm,** is a very common childhood worm infection confined to intestine; transmitted by close contact, unclean personal hygiene; eggs swallowed, hatch in intestine, develop into adults, release more eggs; main symptom is anal itching due to egg-laying activities of female; self-inoculation is common.

Hookworms: Worms have curved ends and hooked mouths; ***Necator americanus*** (western hemisphere) and ***Ancylostoma duodenale*** (eastern hemisphere) have similar life cycle, transmission, pathogenesis; very prevalent; human is principal host; spread by eggs shed in feces; infective form is larva that hatches in environment and burrows into skin of lower legs, moves into blood, lymph, ending up in lungs; larvae are coughed up, swallowed, and reach adulthood in intestine; eggs released with feces; symptoms are itching at site of penetration, inflammatory reactions, severe intestinal distress, anemia; can sap strength.

Strongyloides stercoralis, the **threadworm,** is a tiny nematode that can complete its life cycle totally in humans or in the external environment (moist soil); larvae emerging from soil-borne eggs penetrate the skin and migrate to lungs, are swallowed, and develop in intestine; unlike most helminths, can reinfect same host without leaving body; symptoms are pneumonic, intestinal; disease most severe in immunosuppressed or AIDS patients.

Trichinella spiralis causes **trichinosis,** a common zoonosis of hogs and bears, in which human is dead-end host; acquired from eating undercooked or raw meat containing **encysted larvae**; larvae are released in intestine and develop into adults. Female worms produce live larvae that migrate in blood vessels to organs and ultimately become encapsulated in muscle, heart, brain; disease is flu-like, with fever, aches; may be fatal with high worm loads.

Tissue Nematodes: Principal target organs of maturation are nonintestinal (blood, lymphatics, skin). **Filarial worms** are long, threadlike worms with tiny larvae (microfilariae) that circulate in blood and reside in various organs; are spread by various biting insects.

Wuchereria bancrofti causes **bancroftian filariasis,** a tropical infestation spread by various species of mosquitos; insect deposits larvae into bite; larvae move into lymphatics; after development, adult females shed microfilariae into blood; first symptoms of disease are localized inflammation; chronic blockage of lymphatic circulation causes buildup of fluid in lower extremities, or **elephantiasis,** which may manifest as massive swelling.

Onchocerca volvulus causes **river blindness,** African disease spread by black flies and small, river-associated insects that feed on blood; inoculated larvae develop into adults in skin, forming nodules; in long-term infections, microfilariae migrate into eye and produce inflammation that can blind; new treatment is ivermectin.

Loa loa, the African eye worm, is spread by bite of small flies; worm remains just under skin, where it migrates about; often frequents the conjunctiva, where it is best observed.

Dracunculus medinensis or guinea worm causes dracontiasis; acquired by persons drinking water containing a small crustacean, *Cyclops,* that hosts the larvae; ingested larvae penetrate intestine and migrate through body; female lays eggs through blister on leg that opens up when victim stands in water; worm difficult to eradicate.

Trematodes or Flukes

Thin, leaflike worms, with suckers and reduced organs; most are hermaphroditic.

Blood flukes are ***Schistosoma*** species; **schistsomiasis** very prevalent in subtropics and tropics, affecting 200 million persons; adult flukes live in humans and give off eggs that are released into aquatic environment with feces or urine; in water, they hatch into an early larva, the **miracidium,** that infects freshwater snail; snail releases second larva, called a **cercaria,** that is infective to humans wading in water; larvae penetrate skin, move into liver to mature; adults migrate to intestine or bladder and shed eggs. Acute symptoms are fever, diarrhea; chronic symptoms are extreme organ enlargement that can be very destructive.

Zoonotic flukes include the Chinese liver fluke ***Opisthorchis sinensis,*** oriental form whose definitive hosts are cats, dogs, swine; intermediate hosts are snails and fish; humans are infected when eating fish containing larvae; fluke lives in liver. In another liver fluke, ***Fasciola hepatica,*** herbivores (sheep, cattle) are definitive hosts; snails and aquatic plants are intermediate hosts; humans infected by eating raw aquatic plants; fluke lodges in liver.

Cestodes or Tapeworms

Adults are long, very thin, ribbonlike worms that grip host intestine by **scolex**; budding off slender neck is a chain (**strobila**) of sacs (**proglottids**) that make up the worm body; each proglottid is an independent hermaphroditic unit adapted to absorbing food and making eggs; mature proglottids or eggs are shed with feces.

Taenia saginata is the beef tapeworm; humans are definitive hosts; infected by eating raw beef in which the larval form (**cysticercus**) has encysted; in small intestine, larva attaches and becomes adult tapeworm; beef animals are infected by grazing on land contaminated with human feces. ***T. solium*** is the pork tapeworm; humans may be infected with cysticerci as in beef tapeworm or by ingesting eggs in food or drink; disease caused by ingesting eggs is **cysticercosis**; larvae hatch and migrate to encyst in many organs and do not reach tapeworm stage, but damage organs. ***Diphyllobothrium latum,*** the fish tapeworm, matures in animals such as cats, dogs, humans; intermediate freshwater hosts are *Cyclops* and various fish; infection acquired from ingesting raw or rare fish.

True–False Questions

Determine whether the following statements are true (T) or false (F). If you feel a statement is false, explain why, and reword the sentence so that it reads accurately.

____ 1. All protozoan pathogens have a trophozoite phase.

____ 2. All protozoan pathogens have a cyst phase.

____ 3. *Entamoeba* invades primarily the small intestine.

____ 4. Giardiasis is a zoonosis associated with ingesting cysts.

____ 5. Blood flagellates are all transmitted by vectors.

____ 6. *Plasmodium* reproduces both sexually and asexually.

____ 7. In the erythrocytic phase of infection, *Plasmodium* invades the liver cells.

____ 8. An oocyst is the result of fertilization in sporozoans.

____ 9. A person may acquire toxoplasmosis from eating tissue cysts and oocysts.

____ 10. In all helminth life cycles, adults reproduce and form a fertilized egg and larval stages.

____ 11. The definitive host is where the larva develops, and the intermediate host is where the adults produce fertile eggs.

____ 12. Antihelminthic medications aim to paralyze the worm or destroy its metabolism.

____ 13. Worm infestations often cause eosinophilia.

____ 14. Hookworm diseases are spread by the feces of animals.

____ 15. Trichinosis is spread from human to human by fecal-oral contact.

____ 16. Elephantiasis is caused by filarial worm infestation of the lymph nodes that blocks drainage of the lymphatic circulation.

Concept Questions

1. Compare the four major groups of protozoa according to overall cell structure, locomotion, infective state, and mode of transmission.
2. What are the functions of the trophozoite and the cyst in the life cycle?
3. Why does *Entamoeba* require healthy carriers to complete its life cycle? What is the role of night soil? What is the basic pathology of amebiasis? How and where does it invade? Why are so many cases asymptomatic?
4. Briefly describe how primary amebic meningoencephalitis is acquired and its outcome.
5. Compare trichomoniasis and amebiasis with respect to transmission, life cycle, and relative hardiness. Why must both sex partners be treated for trichomoniasis?
6. How are giardiasis and amebic dysentery similar? How are they different?
7. Compare the infective stages and means of vector transfer in the two types of trypanosomiasis and leishmaniasis.
8. What causes the symptoms in trypanosomiasis? In leishmaniasis?
9. What is the infective stage for malaria in humans? In mosquitos? How many times must a female mosquito feed before the parasite can complete the whole cycle? Where in the human does development take place, and what are the results? What is the ring form? What causes the symptoms of malaria? What are some strategies to arrive at an effective vaccine?
10. Give the host range of *Toxoplasma gondii.* How are humans infected? What are the most serious outcomes of infection?
11. Briefly describe the transmission cycle of *Cryptosporidium, Sarcocystis, Babesia,* and *Pneumocystis.*
12. How are helminths different from other parasites, and why are there so many parasitic worm infections worldwide? What is the nature of human defenses against them?
13. Outline the five general cycles, and give examples of helminths that have each cycle. What are the main portals of entry in worm infestations? How is damage done by parasites?
14. What are some ways that worms adapt to parasitism, and how are these adaptations beneficial to them?
15. How are nematodes, trematodes, and cestodes different from one another?
16. Where in the body do the helminth adults ultimately reside and produce fertile eggs?
17. How do adult *Ascaris* get into the intestine? How do adult hookworms get into the intestine? How do microfilariae get into the blood?
18. How is trichinosis different from other worm infections?
19. Which worms can be found in the eye? Which worm can cause blindness?
20. What are the stages in *Schistosoma* development? Which organs are affected by schistosomiasis? What other organs may be invaded by flukes?
21. Describe the structure of a tapeworm. How are tapeworms spread to humans?
22. Which helminths are zoonotic? Which are strictly human parasites? Which are borne by arthropod vectors?

Practical/Thought Questions

1. What special adaptations are needed by parasites that enter through the oral cavity?
2. Why are only female mosquitos involved in malaria and elephantiasis?
3. Which parasitic diseases could conceivably be spread by contaminated blood and needles?
4. For which diseases can one not rely upon chlorination of water as a method of control? Why?
5. If a person returns from traveling afflicted with trypanosomiasis or leishmaniasis, is he or she generally infective to others? Why, or why not?
6. There is no malaria above 6,000 feet in altitude. Why?
7. Why is there no such thing as a safe form of rare or raw meat?
8. Can you think of a possible way to circumvent the antigenic switching of trypanosomes?
9. Which diseases end up in the intestine from swallowing larval worms? From swallowing eggs?
10. What one simple act could in time eradicate *Dracunculus?*
11. To achieve a cure for tapeworm, why must the antihelminthic drug either kill the scolex or slacken its grip? Which tapeworm is pictured in figure 19.25*b*? How can you tell?
12. Can you explain why AIDS patients are so suceptible to certain protozoan diseases?
13. Students sometimes react with horror and distress when they discover that cats and dogs carry parasites to humans. Give an example of a disease for each of these animals that may be spread to humans, and explain how to avoid these diseases and still enjoy your pet.
14. Why must most parasites leave their host to complete the life cycle? What are some ways to prevent this? What is the nature of parasites having numerous hosts? What are the advantages or disadvantages to having more than one host?
15. East Indian natives habitually chew on betel nuts as an alkaloidal stimulant and narcotic. A side effect is phlegm collection, which is eliminated by frequent spitting. The incidence of *Strongyloides* infections is relatively low in this population. Can you account for this? (No, the betel nuts are not effective antihelminthics.)
16. Explain why cystic hydatid disease is very high among women and children in hot, arid Turkana, Kenya, where dogs form an integral part of tribal life.

CHAPTER 20

Introduction to Viral Diseases and DNA Viruses

Detail from "The Cowpock," an 1808 etching that caricatured the worst fears of the English public concerning Edward Jenner's smallpox vaccine.

Chapter Preview

Viruses are probably the most common infectious agents, though no exact figures exist to support this contention. As a group, they are uniformly parasitic and infect not only animals and plants, but other microorganisms as well. Every time a new virus or viral disease is discovered or a connection is made between a virus and a once-unexplained disease, our understanding of these remarkable infectious particles is increased, and we are reminded of their power. Viral diseases are the subjects of chapters 20 (DNA viruses) and 21 (RNA viruses). Chapter coverage will entail those viral groups most significant to humans, with special emphasis on epidemiology, pathology, and methods of control.

Viruses in Infection and Disease

Viruses are the smallest parasites with the simplest biologic structure—essentially small packaged particles of DNA or RNA that depend on the host cell for their qualities of life. Viruses have special adaptations for entering a cell and for reprogramming its genetic and molecular machinery to produce and release new viruses. Animal viruses are divided into families based on the nature of the nucleic acid (DNA or RNA), the type of capsid, and whether or not an envelope is present. All DNA viruses are double-stranded except for the parvoviruses, which have single-stranded DNA. All RNA viruses are single-stranded except for the double-stranded reoviruses (see table 21.1). The genome of RNA viruses may be further described as segmented (consisting of more than one molecule) or nonsegmented (consisting of a single molecule). The envelope is derived in part from host cell membranes as the virus buds off the nuclear envelope or cell membrane, and it often contains spikes that interact with the host cell.

Important Medical Considerations in Viral Diseases

Target Cells

Because receptors on the virus surface react specifically with a molecule on the cell surface, infectiousness is limited to a particular host or cell type in most cases. Various viruses target nearly all types of tissues, including those of the nervous system (polio and rabies), liver (hepatitis), respiratory tract (influenza, colds), intestine (polio), skin and mucous membranes (herpes, pox), and cells of the immune system (AIDS). Most DNA viruses are assembled and budded off the nucleus, and most RNA viruses multiply in and are released from the cytoplasm. The presence of assembling and completed viruses in the cell gives rise to various telltale disruptions called cytopathic effects. Most cells productively infected with viruses are destroyed, which accounts for the sometimes severe pathology and loss of function.

Scope of Infections

Viral diseases range from very mild or asymptomatic infections such as colds to severe, deadly syndromes such as rabies and AIDS. Many are so-called "childhood diseases" that occur primarily in the young and are readily transmitted through droplets. Although we tend to consider these infections self-limited and inevitable, even measles, mumps, chickenpox, and rubella can manifest severe complications. Many diseases are strictly human in origin, but several virulent examples such as yellow fever and viral encephalitis are zoonoses transmitted by arthropod vectors.

The course of viral diseases starts with invasion at the portal of entry by a few virions and a primary infection. In some cases (influenza, colds), the viruses replicate locally and disrupt the tissue, and in others (rabies, polio), the virus enters the blood and creates complications in tissues far from the initial infection. Common manifestations of virus infections are rashes, fever, muscle aches, respiratory involvement, and swollen glands, symptoms that are usually too nonspecific to be of use in diagnosis. Further identification may require isolating the virus in a suitable host such as a cell culture, bird embryo, or animal. Other techniques for detection and diagnosis will be discussed with individual viruses.

Protection in viral infections arises from the combined actions of interferon, neutralizing antibodies, and cytotoxic T cells. Due to the nature of the immune response to viruses, infections frequently result in lifelong immunities, and many viruses are adaptable for vaccines.

Latency/Oncogenicity

Most DNA viruses and a few RNA viruses become permanent residents of the host cell. In some cases, the latent virus alternates between periods of relative inactivity and recurrent infections. In other cases of persistence, the permanent splicing of viral DNA into a site in the host's DNA transforms the cell into a cancer cell. This potential for oncogenesis is more likely with DNA viruses, though retroviruses, which can convert their RNA to DNA, are also important in cancers. This property of combining with host DNA is a drawback in developing live, attenuated DNA virus vaccines.

Teratogenicity/Congenital Defects/Neonatal Infections

Several viruses can cross the placenta from an infected mother to the embryo or fetus. This often causes developmental disturbances and permanent defects in the child that are present at birth (congenital). Among viruses with known *teratogenic* effects are rubella, cytomegalovirus, and adenovirus. Another way that infants may be affected is through infection at the time of birth, as occurs with hepatitis B and herpes simplex.

Survey of DNA Virus Groups

The DNA viruses that cause human diseases are placed into six groups based upon the existence of an envelope, the nature of the genome (double-stranded or single-stranded), size, and target cells.

Enveloped
- Poxviruses
- Herpesviruses
- Hepadnaviruses

Nonenveloped
- Adenoviruses
- Papovaviruses
- Parvoviruses (single-stranded DNA)

---

teratogenic (tur″-ah-toh-jen′-ik) Gr. *teratos,* monster. Any process that causes physical defects in the embryo or fetus.

Enveloped DNA Viruses

Poxviruses: Classification and Structure

Poxviruses produce eruptive skin pustules called **pocks** or **pox** (figure 20.1), which leave small, depressed scars (pockmarks) upon healing. The poxviruses are distinctive because they are the largest and most complex of the animal viruses (see figure 5.9), they have the largest genome of all viruses, and they multiply in the cytoplasm in well-defined sites called factory areas, which appear as inclusion bodies in infected cells.

Among the best known poxviruses are **variola,** the agent of **smallpox,** and vaccinia, a closely related virus adapted for use in vaccination. Other members are of interest to the veterinarian for their role in diseases of domestic or wild animals. A common feature of all poxvirus infections is a specificity for the cytoplasm of epidermal cells and subcutaneous connective tissue. In these sites they produce the typical lesions and also tend to stimulate cell growth that may lead to tumor formation.

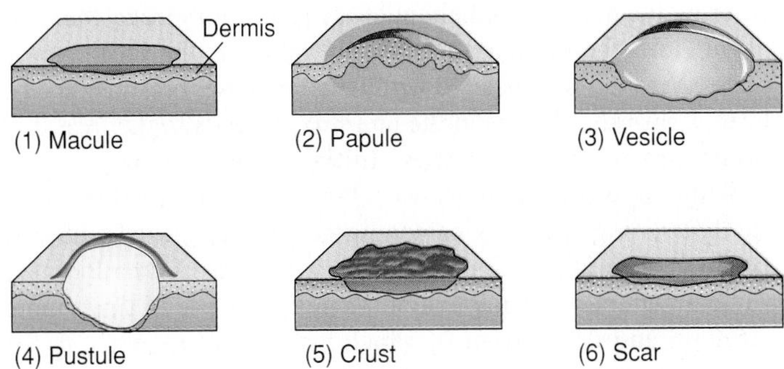

Figure 20.1 Stages in pock development.

Smallpox: A Perspective

Largely through the World Health Organization's comprehensive global efforts, smallpox is now a disease of the past. It is one of the few infectious agents that now exists presumably only in research laboratories. (See feature 20.1 for a discussion of this stunning achievement.) At one time, smallpox was ranked as one of the deadliest infectious diseases, leaving no civilization untouched. Mentions of smallpox-like scourges are found in ancient Chinese, Indian, and Egyptian writings. When introduced by explorers into naive populations (those never before exposed to the disease) such as the Aztecs, Incas, Hawaiians, and Amerindians, smallpox caused terrible losses and had a hand in destroying these civilizations. Even as recently as 1967, approximately 10 to 15 million cases occurred worldwide, with thousands of fatalities.

Disease Manifestations Exposure to smallpox usually occurs through direct inhalation of droplets or skin crusts from an active case, though contaminated fomites may also be a source. After

variola (ver-ee-oh'-lah) L. *varius,* varied, mottled.

Feature 20.1 The End to an Ancient Scourge

The thirty-third World Health Assembly "declares solemnly that the world and all its peoples have won freedom from smallpox . . . an unprecedented achievement in the history of public health. . . ." (Resolution 33–3, May 8, 1980, Geneva, Switzerland).

A concerted vaccination effort against smallpox was first instigated by the United Nations' World Health Organization in 1959. But after seven years, large numbers of cases were still being reported in Africa, Southeast Asia, Indonesia, and Brazil, so in 1966, WHO launched an even more intensive effort. A medical team of 100 thousand persons began a meticulous village-by-village campaign to identify and report population reservoirs and to isolate and vaccinate specific high-risk groups or individuals. Over the next 10 years, this strategy constantly diminished the numbers of cases, and in 1977, the last natural case was reported in Somalia. This program was so successful because (1) unlike many infectious agents, only humans with active cases of smallpox are infectious, (2) variola is not a latent virus, and (3) there is no hidden carrier source. The vaccine is also stable and easy to administer, requiring only one inoculation, and it gives relatively long-term protection. Now only two countries, the United States and the Soviet Union, are known to possess laboratory stocks of variola virus. This virus is maintained because so much remains to be learned about it.

But what of the future? Now that smallpox vaccination for citizens of the world has been suspended, the protective herd effect in populations will gradually wear off, increasing the pool of susceptible individuals. Although the possibility that the laboratory virus will escape accidently is considered remote, surveillance for cases will continue, and armed forces personnel in the United States and Soviet Union continue to receive the vaccination. The CDC also maintains 19 million doses of the vaccine, and Switzerland has stored sufficient doses to administer 300 million vaccinations at the first sign of an outbreak.

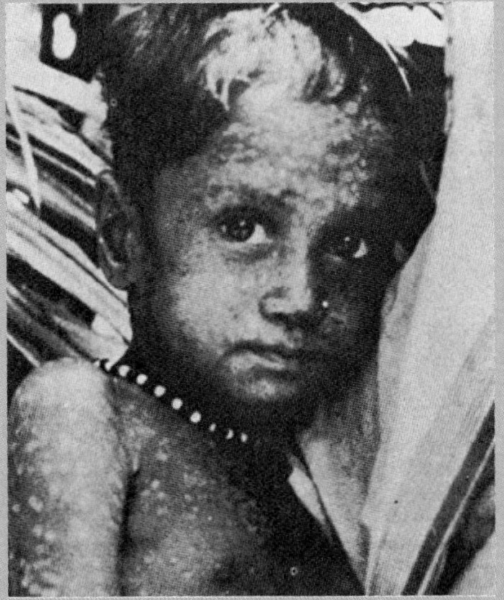

A child in Bangladesh with the last recorded case of variola major (smallpox) in 1975.

Ultimately, the victory over smallpox has demonstrated that cooperative efforts and a program of vaccination can succeed, even on a planet of over five billion people. Hoping for similar success with other agents, WHO has expanded its drive for immunization against six major childhood killers—measles, polio, diphtheria, whooping cough, tetanus, and tuberculosis.

entering the mucous epithelium of the respiratory tract, the virus moves into the circulation and multiplies in the local lymph nodes. Viremia is associated with fever, malaise, prostration, and later, a rash that begins in the pharynx, spreads to the face, and progresses to the extremities. Initially, the rash is macular, evolving in turn to papular, vesicular, and pustular before eventually crusting over, leaving nonpigmented sites pitted with scar tissue (figure 20.1). The two principal forms of smallpox are variola minor and variola major. Variola major is 25 times more virulent, and death from it, which usually occurs early in the illness, is apparently caused by toxemia, shock, and intravascular coagulation. Surviving any form of smallpox nearly always confers lifelong immunity.

Laboratory Diagnosis At the time that smallpox was prevalent, diagnosis was usually based on clinical signs and symptoms. Today, smallpox diagnosis is vested with a special importance and cannot be based on clinical observation. Every year since 1977, apparent cases have been reported that must be carefully differentiated from chickenpox, disseminated herpes, vaccinia, monkeypox, and certain nonviral lesions that mimic smallpox. Scanning stained smears of vesicular fluid for typical elementary bodies and other cytopathic effects can provide useful evidence. The virus may be isolated by inoculating the chorioallantoic membrane of chicken eggs with specimen and looking for pocks (see figure 5.22).

Smallpox Vaccination Edward Jenner's first vaccine to prevent smallpox contained a cowpox virus, which is a close relative of variola (see chapter opening illustration). Now the vaccination involves puncturing a single drop of vaccinia virus into the skin with a double-pronged needle or jetgun. In a successful take, a large, scablike pustule with a reddened periphery develops at the injection site, where it leaves a permanent scar (a pitted emblem that many of us still bear). The vaccine protects persons for up to 10 years against variola as well as other poxvirus infections. Smallpox vaccinations are no longer given and are not required for travel as they were in the past. The decision to stop vaccinating was based not only on the conviction that smallpox no longer poses a threat, but also on the slight risk of vaccinia complications, especially in immunodeficient and allergic persons who may acquire encephalitis, fever, headache, gangrene, rashes, and sometimes paralysis.

Other Poxvirus Diseases

An unclassified poxvirus causes a skin disease called **molluscum contagiosum** (figure 20.2*a*). This disease is distributed throughout the world, with highest incidence in certain Pacific islands. In endemic regions, it is primarily an infection of children and is transmitted by direct contact and fomites. The moist tropical climate appears to favor skin lesions, which take the form of smooth, round, waxy nodules on the face, trunk, and limbs. In adults, the infection is spread by sexual intercourse, and the lesions occur on the pubic and genital regions (figure 20.2*b*). Lesions may be single or multiple, and they follow a progressive 6–36-month course, complicated by itchiness, pain, and self-inoculation.

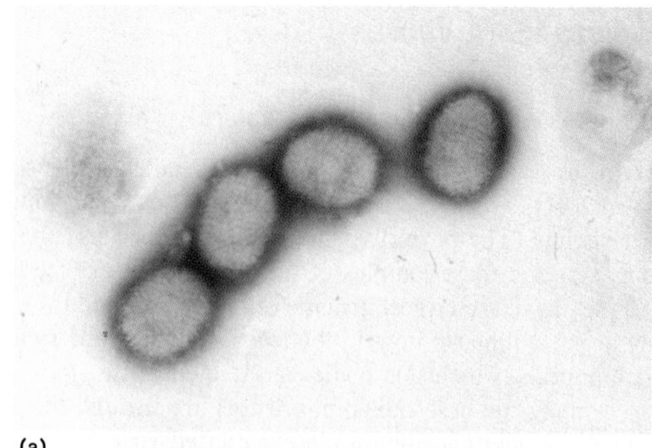

(a)

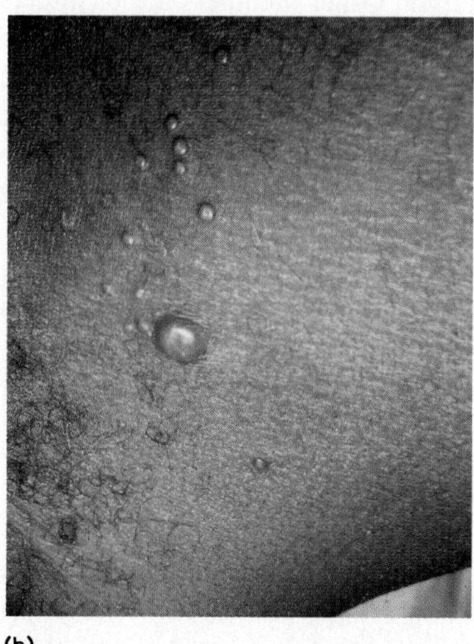

(b)

Figure 20.2 Molluscum contagiosum. (*a*) Virus particles have a typical pox virus structure. (*b*) A sexually acquired infection. The skin eruption takes the form of small waxy papules in the genital region.

Many mammalian groups host some sort of poxvirus infection. Cowpox, rabbitpox, monkeypox, mousepox, camelpox, buffalopox, and elephantpox are among the types. (But chickenpox is a herpesvirus infection, not a poxvirus, and it does not occur in chickens!) Only monkeypox and cowpox appear to be capable of infecting humans symptomatically. Cowpox is an eruptive cutaneous disease that occurs on cows' udders and teats, the usual sources of infection for humans. Human infection is rare and usually confined to the hands, although the face and other cutaneous sites may be involved.

The Herpesviruses: Common, Persistent Human Viruses

Herpesvirus was named for the tendency of some infections to produce a creeping rash. It is the common name for a large family whose members include herpes simplex 1 and 2, the cause

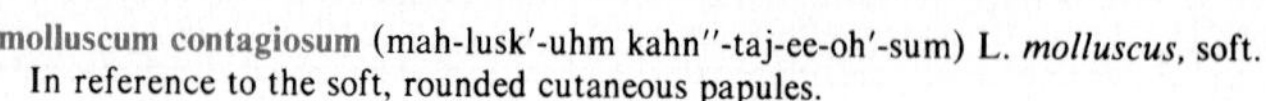

molluscum contagiosum (mah-lusk'-uhm kahn''-taj-ee-oh'-sum) L. *molluscus*, soft. In reference to the soft, rounded cutaneous papules.

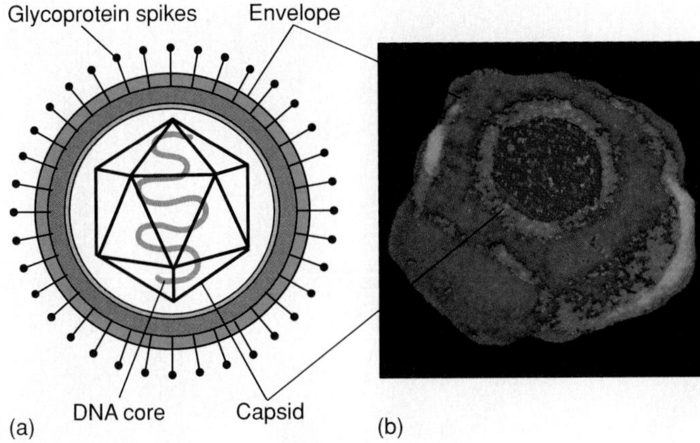

Figure 20.3 (*a*) A schematic of the general structure of herpesvirus alongside (*b*) a color-enhanced view of cytomegalovirus.

of fever blisters and genital infections; herpes zoster, the cause of chickenpox and shingles; cytomegalovirus (CMV), which affects the salivary glands and other viscera; and Epstein-Barr virus (EBV), associated with infection of the lymphoid tissue. One prominent feature of the family is its tendency toward viral latency and recurrent infections. This behavior often involves incorporation of viral nucleic acid into the host genome and carries with it the potential for oncogenesis. Virtually everyone becomes infected with a herpesvirus at some time, usually without adverse effect. The clinical complications of latency and recurrent infections become more severe with advancing age, organ transplants, cancer chemotherapy, or other provoking conditions that compromise the immune defenses.

The herpesviruses are among the larger viruses, with a diameter ranging from about 150 nm to 200 nm (figure 20.3). They are enclosed within a loosely fitting envelope that contains glycoprotein spikes. Being lipid- and protein-based, the herpesviruses, like other enveloped viruses, are prone to deactivation by organic solvents or detergents and are relatively unstable outside the host's body. The icosahedral capsid houses a core of double-stranded DNA that winds around a proteinaceous spindle in some viruses. Replication occurs primarily within the nucleus, and viral release is usually accomplished by cell lysis.

General Properties of Herpes Simplex Viruses

Humans are susceptible to two varieties of **herpes simplex viruses (HSV),** which can be differentiated by several characteristics (table 20.1). Referred to as HSV types 1 and 2, they are ordinarily associated with lesions of the oropharynx and genitalia respectively. Even though laboratory mice, guinea pigs, rabbits, hamsters, and other animals can be experimentally infected, humans appear to be the only natural reservoir for herpes simplex virus.

Epidemiology of Herpes Simplex

The herpes simplex viruses have cosmopolitan distribution. Infection occurs in all seasons and among all age groups, and transmission is promoted by direct exposure to secretions containing the virus. The viruses are relatively sensitive to the environment, but may remain infectious in moist secretions on inanimate objects for a few hours. Persons with active lesions are the most significant source of infection, though asymptomatic carriers can shed small numbers of viruses. Primary infections tend to be age-specific: HSV-1 frequently occurs in infancy and early childhood, and by adulthood, most persons exhibit some serologic evidence of infection. On the other hand, primary infection with HSV-2 occurs most frequently between the ages of 14 and 29, a pattern that reflects the sexual route of transmission. In fact, genital herpes is the third or fourth most common STD in the United States. Instances of HSV-1 genital infections and HSV-2 infections of the oral cavity have recently risen. Such cases are probably caused by autoinoculation with contaminated hands or by oral sexual contact.

Table 20.1 Comparative Epidemiology and Pathology of Herpes Simplex, Types 1 and 2

	HSV-1	HSV-2
Usual Etiologic Agent Of	Herpes labialis Ocular herpes Gingivostomatitis Pharyngitis	Herpes genitalis*
Transmission	Close contact, usually of face	Sexual or close contact
Latency	Occurs in trigeminal ganglion	Occurs primarily in sacral ganglia
Skin Lesions	On face, mouth	On internal, external genitalia, thighs, buttocks
Complications		
Whitlows	Among personnel working on oral cavity	Among obstetric, gynecological personnel
Neonatal Encephalitis	Causes up to 30% of cases**	Causes most cases

*The other herpes simplex type can be involved in this infection, though not as commonly.

**Due to mothers infected genitally by HSV-1 or contamination of the neonate by oral lesions.

The Nature of Latency and Recurrent Attacks In an estimated 20% to 50% of all cases of primary infection, herpes simplex viruses routinely enter the distal regions of sensory neurons and travel to the dorsal root ganglia. Type 1 HSV enters primarily the trigeminal or fifth cranial nerve, which has extensive innervations in the oral region (figure 20.4). Type 2 HSV usually becomes latent in the ganglion of the lumbosacral spinal nerve trunk. The exact details of this unusual mode of latency are not yet fully understood, but the virus remains inside the neuron in a nonproliferative stage for a variable time. Recurrent infection is triggered by various stimuli such as fever, UV radiation, menstruation, stress, or mechanical injury, all of which

herpes (her'-peez) Gr. *herpein,* to creep.

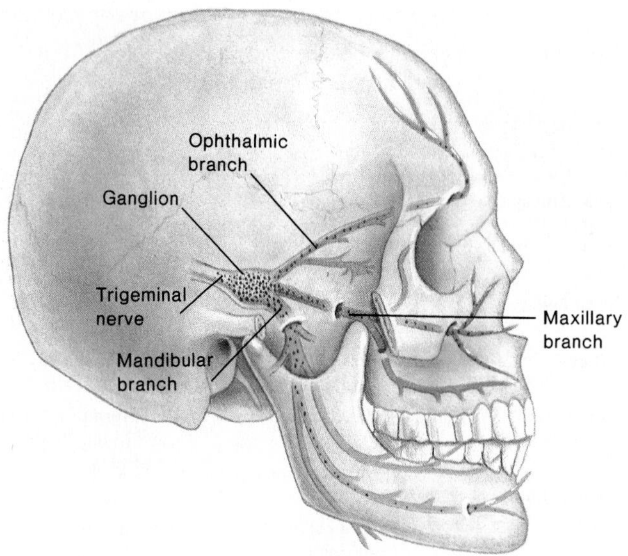

Figure 20.4 The site of latency and routes of recurrence in herpes simplex, type 1. Following primary infection, the virus invades the fifth cranial nerve (trigeminal nerve) and is harbored in its ganglion. Various provocative stimuli cause it to migrate back to the skin surface by the mandibular, maxillary, or ophthalmic branch.

reactivate the virus. The virus migrates to the body surface and produces a local skin or membrane lesion, often in the same site as a previous infection. The number of such attacks can range from one to dozens per year.

The Spectrum of Herpes Infection and Disease

Herpes simplex infections, often called simply "herpes," usually target the mucous membranes. The virus enters cracks or cuts in the membrane surface and multiplies in basal and epithelial cells in the immediate vicinity. An inflammatory response in this region is attended by edema, cell lysis, fusion, degeneration, and a ballooning effect that produces a characteristic thin-walled vesicle. Depending on the portal of entry, herpes simplex viruses are responsible for facial herpes (oral, optic, and pharyngeal), genital herpes, neonatal herpes, and disseminated disease.

Type 1 Herpes Simplex in Children and Adults Although most primary herpetic infections are mild or asymptomatic, an occasional manifestation known as herpetic **gingivostomatitis** may occur (figure 20.5*a*). This infection of the oropharynx, usually due to HSV-1, strikes young children most frequently. Inflammation of the oral mucosa can involve the gums, tongue, soft palate, and lips, sometimes with ulceration and bleeding. A common complication in adolescents is pharyngitis, a syndrome marked by sore throat, fever, chills, headache, swollen lymph nodes, and difficulty in swallowing.

Herpes labialis, otherwise known as **fever blisters** or **coldsores,** is the most common recurrent HSV-1 infection. Vesicles usually crop up on the mucocutaneous junction of the lips or on adjacent skin (figure 20.5*b*). A few hours of tingling or itching precede the formation of one or more vesicles at the site. Pain is most acute in the earlier stages as vesicles ulcerate; lesions crust over in two or three days and heal completely in a week.

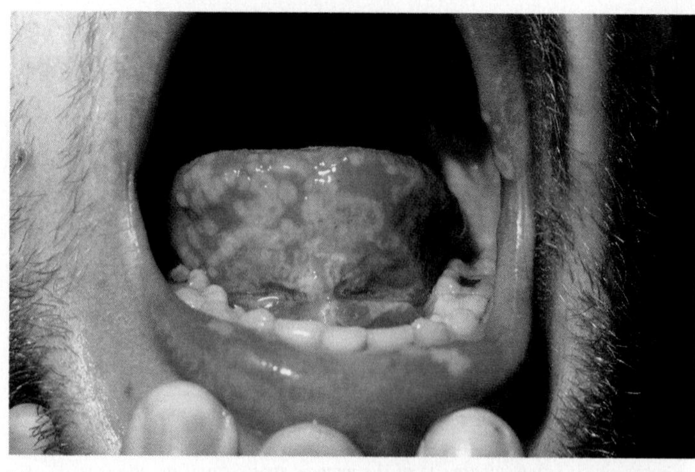

(a)

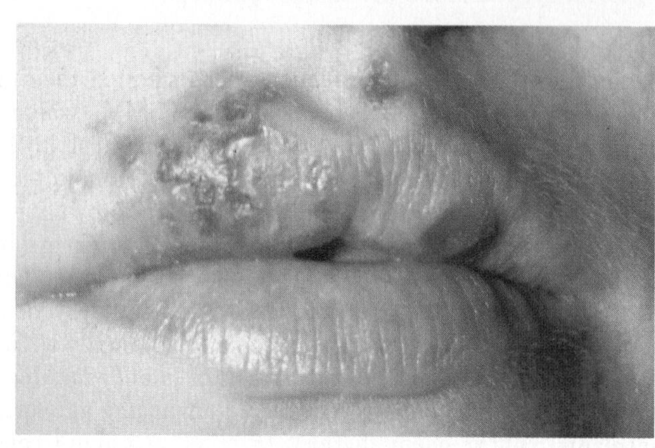

(b)

Figure 20.5 (*a*) Primary herpetic gingivostomatitis in children, involving the entire oral mucosa, tongue, cheeks, and lips. (*b*) Recurrent lesions of herpes labialis (cold sore, fever blister). Tender, itchy papules erupt on the perioral region and progress to vesicles that burst, drain, and scab over. These sores and fluid are highly infectious and should not be handled.

Herpetic keratitis (also called ocular herpes) is an infective inflammation of the eye. In this disease, a latent virus has travelled into the ocular rather than the mandibular branch of the trigeminal nerve. Preliminary symptoms are a gritty feeling, conjunctivitis, sharp pain, and sensitivity to light. Some patients develop characteristic branched or opaque corneal lesions as well. In 25% to 50% of cases, keratitis is recurrent and chronic, and can lead to visual loss requiring surgery to replace the cornea.

Type 2 Herpes Infections Type 2 herpes infection usually coincides with sexual maturation and an increase in sexual encounters. **Genital herpes** (herpes genitalis) is initially attended by malaise, anorexia, fever, and bilateral swelling and tenderness in the groin. Intensely sensitive vesicles break out in clusters on the genitalia, perineum, and buttocks (figure 20.6). The chief symptoms are urethritis, painful urination, cervicitis,

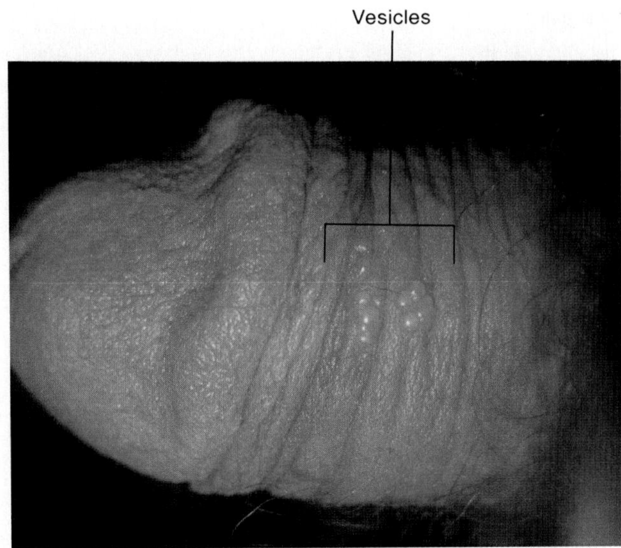

Figure 20.6 Genital herpes. Vesicles start out separate, but become confluent and ulcerate into painful red erosions so tender that inspection is difficult.

itching, and central nervous system involvement (due to HSV in the blood and meninges). After a period of days to weeks, the vesicles ulcerate and develop a light-gray exudate before healing. Bouts of recurrent genital herpes in either sex are usually less severe than the original infection and are brought on by menstruation, stress, and concurrent bacterial infection.

Herpes Genitalis—A New Disease? Actually, genital herpes has been known for several centuries. It was commonly described in patient histories in the eighteenth and nineteenth centuries. Increased reports of this infection during the 1970s and 1980s and greater public awareness and media focus made it seem as though we were seeing a new STD. For reasons that have not been adequately explained, the epidemic struck primarily young, sexually active Caucasians. Perhaps this was due to the sexual revolution and greater promiscuity in this group, or it may be that the ability to diagnose the disease was improved. Regardless, genital herpes is more than a source of pain and discomfort; it often has psychological effects that impair sexual relationships.

Herpes of the Newborn Although HSV infections in adults are annoying, unpleasant, and painful, only rarely are they life-threatening. However, in the neonate and fetus (figure 20.7), HSV infections are very destructive and may even be lethal. Most cases of infection in this group occur when infants are contaminated by the mother's reproductive tract just prior to or during birth (see feature 20.2). Neonatal infections have also been traced to hand transmission from the mother's lesions to the baby. Not unexpectedly, type 2 is more frequently involved, because it is more often associated with genital infection, but type 1 infection has similiar complications. In infants whose disease is confined to the mouth, skin, or eyes, the mortality rate is 30%, but those with disease affecting the central nervous system experience mortality rates ranging from 50% to 80%.

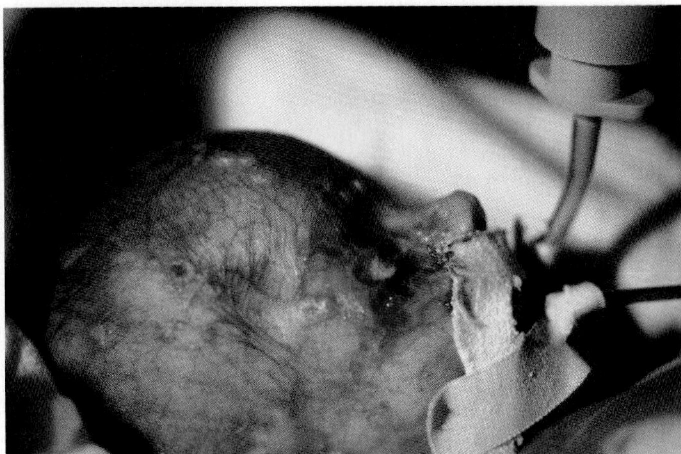

Figure 20.7 Neonatal herpes simplex. This premature infant was born with the classic "cigarette burn" pattern of HSV infection. Babies are either born with the lesions or develop them one to two weeks after birth.

Feature 20.2 Dealing with Pregnancy and Herpes

Because of the danger of herpes to fetuses and newborns and the increase in genital herpes, it is now standard procedure to screen pregnant women for the herpesvirus early in their prenatal care. Women with a history of recurrent infections must be constantly monitored for any signs of viral shedding, especially in the last four weeks of pregnancy. If no evidence of recurrence is seen, vaginal birth is indicated, but any evidence of an outbreak at the time of delivery necessitates a cesarian section. A difficult clinical situation to manage is the woman with active infection and ruptured membranes, a combination that can infect the infant prior to birth and negate the protective effect of a cesarian section.

Very rarely, congenital infections arise when viruses migrate from the infected birth canal to the placenta and fetus. Depending on the time of infection, fetal herpes is marked by spontaneous abortion or disrupted development. The more severe cases show bizarre central nervous system deformities such as microcephaly (small brain) or microphthalmia (small eyes).

Miscellaneous Herpes Infections When either form of herpesvirus enters a break in the skin, a local infection may develop. This route of infection is most often associated with occupational exposure or severely damaged skin. A hazard for health care workers who handle patients or their secretions without hand protection is a disease called herpetic **whitlow.** Workers in the fields of obstetrics, gynecology, dentistry, and respiratory therapy are probably at greatest risk. Whitlows are deepset, usually on one finger, and extremely painful and itchy (figure 20.8). Healing takes place in two to three weeks, but in the interim, afflicted personnel must not work with patients.

whitlow (hwit'-loh) ME *white,* flaw. An abscess on the distal portion of a finger.

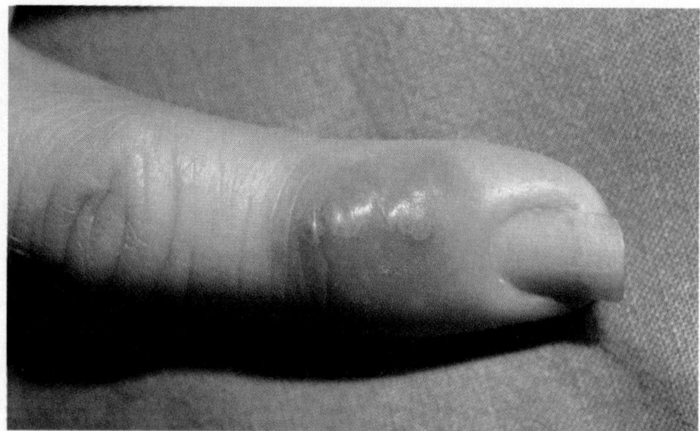

Figure 20.8 Herpetic whitlows. Clusters of these painful, deepset vesicles may develop in dental and medical personnel or any person who carelessly handles his own or others' active lesions.

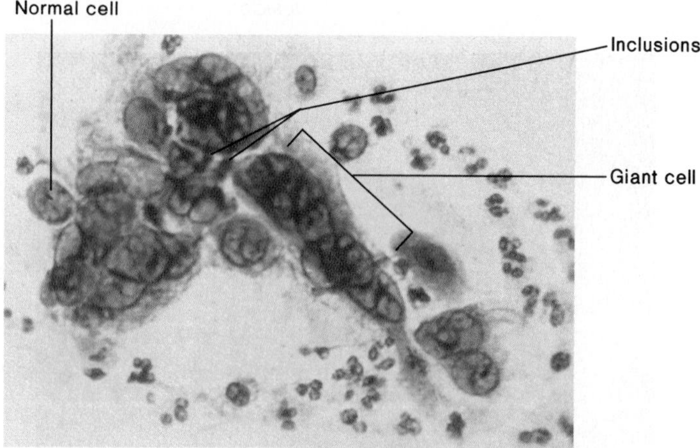

Figure 20.9 Direct cytologic diagnosis of herpesviruses. A direct Papanicolaou (Pap) smear of a cervical scraping shows enlarged (multinucleate giant) cells and intranuclear inclusions typical of herpesvirus infections. This technique is not specific for HSV, but most other herpesviruses do not infect the reproductive mucosa.

Life-Threatening Complications Although herpes simplex encephalitis is a rare complication of type 1 infection, it is probably the most common sporadic form of viral encephalitis in the United States. The infection begins in the epithelium and disseminates along nerve pathways to the brain or spinal cord. The effects on the CNS begin with headache and stiff neck, and may progress to mental disturbances and coma. The fatality rate in untreated cases is 70%. Patients with underlying immunodeficiency are more prone to severe, disseminated herpes infection. Of greatest concern are patients receiving bone marrow, heart, or kidney grafts, cancer patients on immunosuppressive therapy, congenitally immunodeficient persons, and AIDS patients.

For the past 30 years, a considerable amount of evidence has been mounting to implicate type 2 HSV in cervical carcinoma. Although another virus, human papilloma virus (see chapter 21), is known to be involved in this type of cancer, the possible role of HSV is compelling enough for medical practitioners to warn sexually active women about this additional danger from genital herpes.

Diagnosis, Treatment, and Control of Herpes Simplex

Small, painful, vesiculating lesions on the mucous membranes of the mouth or genitalia, lymphadenopathy, and exudate are typical symptoms of primary oral, genital, and recurrent herpes. Further diagnostic support is available by examining scrapings from the base of such lesions stained with Giemsa, Wright, or Papanicolaou (Pap) methods (figure 20.9). The presence of multinuclear cells, giant cells, and intranuclear eosinophilic inclusion bodies can establish herpes infection, though this method will not distinguish among HSV-1, HSV-2, or varicella-zoster viruses (VZV). Subtyping HSV-1, HSV-2, and VZV requires more specific methods (described next).

Laboratory culture and specific tests are essential for diagnosing immunosuppressed and neonatal patients with severe, disseminated herpes infection. A specimen of tissue or fluid is introduced into a primary cell line such as monkey kidney or human embryonic kidney tissue cultures and is observed for cytopathic effects in 24 to 48 hours. Specific confirmatory tests that subtype HSV-1 and HSV-2 use fluorescent antibodies (see figure 1.26) or DNA probes. Serologic analysis is useful for primary infection but inconclusive for recurrent illness, because only 10% of patients show increased antibody titers.

Several agents are available for treatment. Acyclovir (Zovirax) is the most effective therapy developed to date. This drug, in use since the early 1980s, is relatively nontoxic, yet shows high specificity for the herpes simplex virus. Topical medications applied to primary genital and oral lesions cut the length of infection and reduce viral shedding. Systemic therapy is available for more serious complications. The more toxic drugs idoxuridine and trifluridine have been used for topical treatment of herpes keratitis, and vidarabine has been used for complications of disseminated herpes. Injected interferon has shown promise in lessening the symptoms and preventing latency in some herpes infections, but its full value is still being determined. Over-the-counter coldsore medications containing menthol, camphor, and local anaesthetics lessen pain and protect against secondary bacterial infections, but they probably do not affect the progress of the viral infection.

New evidence indicates that a daily dose of oral acyclovir taken for a period of six months to one year is very effective in preventing recurrent genital herpes. Although the barrier protection afforded by condoms can diminish the prospects of spreading genital herpes sexually, persons with active infection should avoid sexual contact. Mothers with coldsores or fever blisters should observe extreme care in handling their newborns; infants should never be kissed on the mouth; and hospital attendants with active oral herpes infection must be barred from the newborn nursery. Medical and dental workers who deal directly with patients can reduce their risk of exposure by wearing gloves when examining lesions.

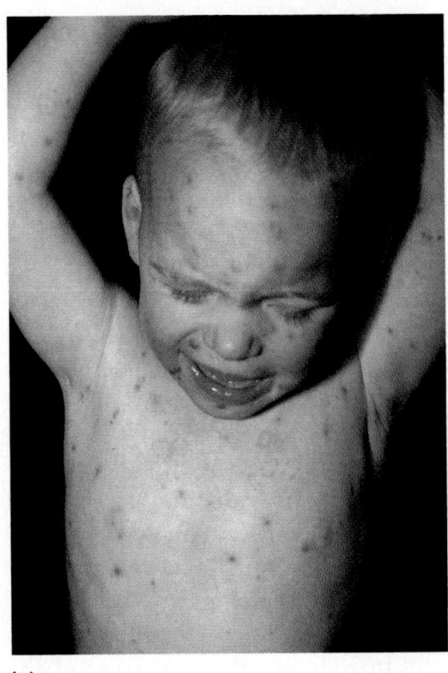

(a)

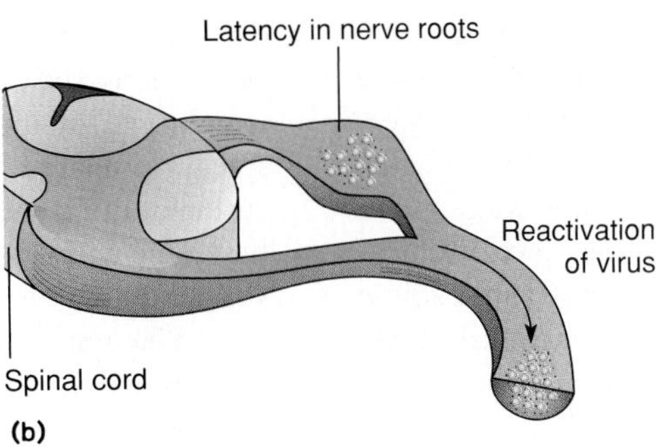

(b)

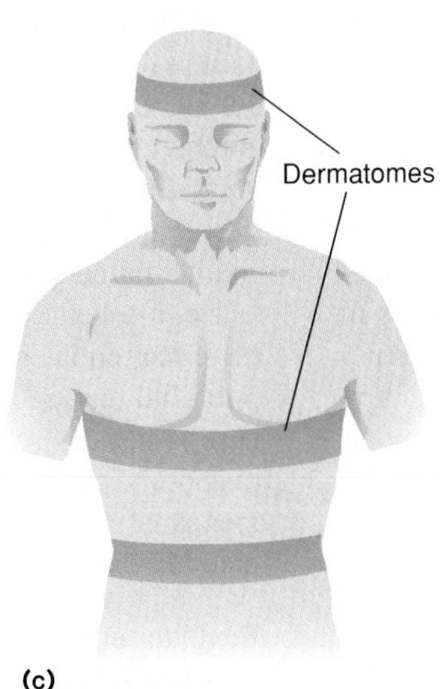

(c)

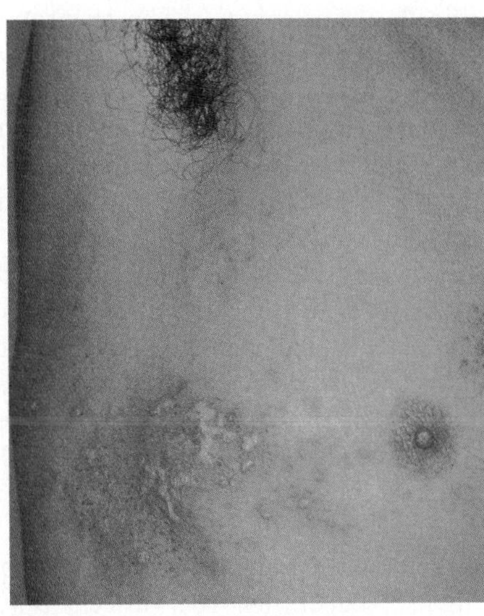

(d)

Figure 20.10 The relationship between varicella (chickenpox) and zoster (shingles) and the clinical appearance of each. (*a*) First contact with the virus (usually in childhood) results in a macular, papular, vesicular rash distributed primarily on the face and trunk. (*b*) The virus becomes latent in the dorsal ganglia of nerves that supply dermatomes in mid-thoracic and facial regions. (*c*) The pattern in thoracic shingles. (*d*) The clinical appearance of shingles. Early symptoms are acute pain in the nerve root and redness of the dermatome, followed by a vesicular-papular rash that may extend completely around the chest and back.

The Biology of Varicella-Zoster Virus

Another herpesvirus causes both **varicella** (commonly known as **chickenpox**) and a recurrent infection called **herpes zoster (shingles)**. Because the same virus causes two forms of disease, it is known by a composite name, **varicella-zoster virus (VZV)**. At one time, the differences in clinical appearance and time separation between varicella and herpes zoster were sufficient to foster the notion that they arose from two separate agents. But about 100 years ago, the observation that young children acquired chickenpox from family members afflicted with shingles gave the first hint that the two diseases were related. More recently, doctors followed the diseases in a single patient using Koch's postulates, and demonstrated that the virus is the same in both diseases, and that zoster is a reactivation of a latent varicella virus (figure 20.10).

Epidemiologic Patterns of VZV Infection Humans are evidently the only natural hosts for varicella-zoster virus. The virus is harbored in the respiratory tract, but it is difficult to isolate from nasal or pharyngeal secretions. Although the virus is unstable and readily loses infectivity when exposed to the environment, it is highly communicable from respiratory droplets

varicella (var″-ih-sel′-ah) The Latin diminutive for smallpox or variola.

zoster (zahs′-tur) Gr. *zoster,* girdle.

shingles (shing′-gulz) L. *cingulus,* cinch. (Both *zoster* and *shingles* refer to the beltlike encircling nature of the rash.)

or from the virus-filled fluid (but not scabs) of skin lesions. Patients with shingles are a source of infection for nonimmune children, but the attack rate is significantly lower than with exposure to chickenpox. Both varicella and zoster occur globally; varicella is more frequent in the winter and spring, while zoster is nonseasonal. Only in rare instances will a child develop chickenpox more than once, and even a subclinical case can result in long-term immunity. Immune persons do not acquire chickenpox or shingles when reexposed to an exogenous source of VZV.

Varicella (Chickenpox) The respiratory epithelium serves as the chief portal of entry and the initial site of viral replication for the chickenpox virus, which is initially without symptoms. Regional lymph nodes and other undetermined body sites probably support secondary multiplication before the virus is detectable in the blood. The first symptoms to appear after an incubation period of 10 to 20 days are fever and an abundant rash that begins on the scalp, face, and trunk and radiates in sparse crops to the extremities. Skin lesions progress quickly from macules and papules to itchy vesicles that encrust and drop off, usually healing completely but sometimes leaving a tiny pit or scar (figure 20.10*a*). Lesions number from a dozen to hundreds and are more abundant in adolescents and adults than in young children.

Herpes Zoster (Shingles) In an unknown percentage of individuals, recuperation from varicella is associated with the entry of the varicella-zoster virus into the sensory endings that innervate dermatomes, regions of the skin supplied by the cutaneous branches of spinal nerves, especially thoracic and trigeminal (figure 20.10*b*). Like HSV, VZV migrates toward the central nervous system and becomes established in the sensory ganglia in latent form. In contrast to herpes simplex, however, zoster generally recurs only once or possibly twice, with a characteristic beltlike distribution after which the disease is named (figure 20.10*c, d*).

Shingles comes on abruptly following reactivation by such stimuli as X-ray treatments, immunosuppressive and other drug therapy, surgery, or developing malignancy. Up to 50% of patients suppressed by radiation therapy, leukemia, or similar malignancies manifest recurrence. The virus is believed to migrate down the ganglion to the skin where multiplication resumes and produces crops of tender skin vesicles typical of varicella, except that they may persist for weeks. Inflammation of the ganglia and the pathways of intercostal nerves may stimulate pain and tenderness (radiculitis) that can last for several months. Involvement of cranial nerves may lead to eye inflammation and ocular and facial paralysis.

Diagnosis, Treatment, and Control The cutaneous manifestations of varicella and zoster are sufficiently characteristic for ready clinical recognition. Supportive evidence for a diagnosis of varicella usually comes from a recent history of close contact with a source or with an active case of varicella or zoster. A diagnosis of zoster is supported by the distribution pattern of skin lesions and a history of varicella. Laboratory confirmation is achieved by identifying multinucleate giant cells in stained smears prepared from vesicle scrapings. Unequivocal identification or differentiation of herpes simplex from herpes zoster is best done with fluorescent antibody detection of viral antigen in skin lesions.

Uncomplicated varicella is self-limited and requires no therapy aside from alleviation of discomfort. To prevent secondary bacterial infection caused by scratching, anaesthetic and antimicrobial ointments are applied. The use of aspirin is contraindicated because of the apparent risk of Reye's syndrome (see feature 21.2). Drugs that may be therapeutically beneficial for systemic disease are intravenous acyclovir, high dose interferon, and vidarabine. Zoster immune globulin (ZIG) and varicella-zoster immune globulin (VZIG) are passive immunotherapies that do not entirely prevent disease but may diminish its complications. Japanese researchers have developed a live, attenuated vaccine that appears to protect children who have leukemia, but it is not in widespread use because of the potential oncogenesis of DNA viruses. In fact, because the disease is usually so mild in normal children, some parents have purposely exposed their children to infected individuals as a means of "immunizing" them. However, this is not recommended due to the potential for viral latency.

The Cytomegalovirus Group

Another group of herpesviruses, the **cytomegaloviruses (CMV)**, got their name from the characteristic cytopathic effects (nuclear and cytoplasmic inclusion bodies) they produce in patient tissues (figure 20.11). At times termed salivary gland virus and cytomegalic inclusion virus, these viruses are among the most ubiquitous pathogens of humans.

Epidemiology of CMV Disease Wherever human populations have been tested, a relatively high percentage (40–100%) show the presence of antibodies to CMV developed during a prior infection. Studies in the United States and the United Kingdom revealed infection in 7.5% of newborns, making CMV the most prevalent viral infection in the fetus. In surveys that include other countries, 20% to 100% of women of childbearing age had CMV antibodies, and about 4% of pregnant women excreted the virus in their urine. In the U.S., 100% of IV drug abusers and 30% of homosexual males test positive for CMV antibodies, and many are persistent shedders of this virus. Although cytomegalovirus is not noted for being highly communicable, it appears to be transmitted in saliva, respiratory mucus, milk, urine, semen, cervical secretions, feces, and lymphocytes. Actual modes of transmission involve intimate exposure as in sexual contact, vaginal birth, transplacental infection, blood transfusion, and

---

cytomegalovirus (sy"-toh-may"-gah-loh-vy'-rus) Gr. *cyto*, cell, and *megale*, big, great. A cytopathic effect that causes the formation of giant cells.

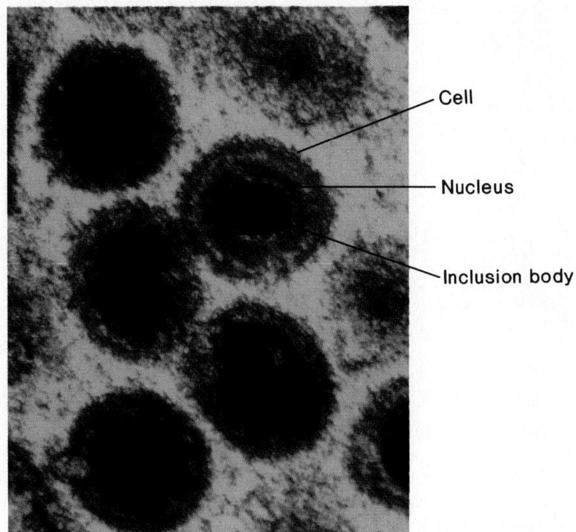

Figure 20.11 Infection of cells with CMV shows the cellular enlargement and distortion of the nucleus by a large inclusion (×66,000).

organ transplantation. As with other herpesviruses, CMV occurs in a latent state in various tissues, and chronic carriage is common.

Infection and Disease Most healthy adults and children with primary CMV infection are asymptomatic. Three groups that develop a more virulent form of disease are fetuses, newborns, and immunodeficient adults. The incidence of **congenital CMV** infection is high, occurring in about 20% of pregnancies in which the mother has concurrent infection. Although most infected newborns are born without signs, a certain number show developmental trauma such as enlarged liver and spleen, jaundice, capillary bleeding, microcephaly, and ocular inflammation. Damage may be so severe and extensive that death follows within a few days or weeks. Babies who survive exhibit long-term neurological sequelae, including hearing and visual disturbances and even mental retardation. In the United States, approximately 5,000 babies a year develop these sequelae. **Perinatal CMV infection,** which occurs following exposure to the mother's vagina, is chiefly asymptomatic, although pneumonitis and a mononucleosis-like syndrome may develop during the first three months after birth.

Cytomegalovirus mononucleosis is a syndrome characterized by fever and lymphocytosis, somewhat similar to the disease caused by Epstein-Barr virus. It may occur in children, but it is chiefly an illness of adults. **Disseminated cytomegalovirus** infection may precipitate a medical crisis. It is a common opportunist of AIDS patients, in whom it produces overwhelming systemic disease, with fever, severe diarrhea, hepatitis, arthritis, pneumonia, and high mortality. Most patients undergoing kidney transplantation and half of those receiving bone marrow develop CMV pneumonitis, hepatitis, myocarditis, meningoencephalitis, hemolytic anemia, and thrombocytopenia. Tumors from patients with Kaposi's sarcoma often contain CMV or its antigens, suggesting its possible role in cancer. But it must be emphasized that isolating the virus from the tumor alone is not sufficient proof and that further evidence is required.

Diagnosis, Treatment, and Prevention In neonates, CMV infection must be differentiated from toxoplasmosis, rubella, herpes simplex, and bacterial sepsis, and in adults, it must be distinguished from Epstein-Barr infection and hepatitis A and B. The cytomegalovirus can be isolated from pathologic specimens taken from virtually all organs as well as from epithelial tissue. Tissues may be observed directly for typical cell enlargement and prominent inclusions in the cytoplasm and nucleus, even though this is not specific. More specific antigen tests involving ELISA and DNA probes are of some benefit. Virus culture is the most effective of all methods, using a rapid monoclonal antibody test against the early nuclear antigen. Serology is unreliable, because it may fail to diagnose infection in neonates and the immunocompromised.

Some antiviral agents give promising results, but most do not modify the course of infection well enough to be clinically useful. Drugs recently approved for therapy are gancyclovir, foscarnet, and hyperimmune CMV immunoglobulin. Interferon appears, at best, to delay excretion of the virus during therapy, but it shows little or no indication of preventing infection or promoting recovery. None of the other anti-herpes drugs is therapeutically active in CMV infections. The development of a vaccine is hampered by the lack of an animal that can be infected with human cytomegalovirus. One crucial concern is whether vaccine-stimulated antibodies would be protective, since patients already seropositive may become reinfected, and another concern is that a vaccine could be oncogenic.

Epstein-Barr Virus: The Newest Herpesvirus

Although the lymphatic disease known as infectious mononucleosis was first described more than a century ago, its cause was finally discovered through a series of accidental events starting in 1958, when Michael Burkitt discovered in African children an unusual malignant tumor (Burkitt's lymphoma) that appeared to be infectious. Later, Michael Epstein and Yvonne Barr cultured a virus from tumors that showed typical herpesvirus morphology. Proof that the two diseases had a common cause was completed when a laboratory technician acquired mononucleosis while working with the Burkitt's lymphoma virus. The **Epstein-Barr** (**EBV**) virus shares morphologic and antigenic features with other herpesviruses, and in addition, it contains a circular form of DNA that is readily spliced into the host cell DNA, thus transforming or immortalizing cultured lymphocytes.

---

mononucleosis (mah″-noh-noo″-klee-oh′-sis) Gr. *mono,* one, *nucleo,* nucleus, and *sis,* state.

Epidemiology of Epstein-Barr Virus EBV is ubiquitous, with a firmly established niche in human lymphoid tissue and salivary glands. Salivary excretion occurs in about 20% of otherwise normal individuals and reaches 50% among the immunosuppressed. Direct oral contact and contamination with saliva are the principal modes of transmission, though fomites and arthropod vectors may play a role. Transmission by means of blood transfusions and organ transplants is also possible.

The nature of infection depends on the patient's age at first exposure, socioeconomic level, geographic region, and genetic predisposition. In less-developed regions of the world, infection rates are generally higher, and infection occurs at an earlier age. Children living in east and west central Africa are predisposed to Burkitt's lymphoma, and residents of parts of China and North Africa show high levels of nasopharyngeal carcinoma, another tumor linked to EBV. In industrialized countries, exposure to EBV is usually postponed until adolescence and young adulthood. This delay is generally ascribed to improved sanitation and less frequent and intimate contact with other children. Because nearly 70% of college-age Americans have never had EBV infection, this population is vulnerable to infectious mononucleosis, often dubbed the "kissing disease" or "mono." By mid-life, 90% to 95% of all persons, regardless of geographic region, show serological evidence of prior or current infection.

Diseases of EBV The epithelium of the oropharynx is the portal of entry for EBV during the primary infection. From this site, the virus moves to the parotid gland, which is the major area of viral replication and the source of viremia. After the initial infection, the virus remains latent in the throat and blood for an extended time, periodically being reactivated by some ill-defined mechanism. In some persons the entire course of infection and latency is asymptomatic.

The symptoms of **infectious mononucleosis** are sore throat, high fever, and cervical lymphadenopathy, which come on after a long incubation period (30–50 days). Many patients also have a gray-white exudate in the pharynx (figure 20.12), a skin rash, and enlarged spleen and liver. A notable sign of mononucleosis is sudden leukocytosis, consisting initially of infected B cells and later of T-suppressor and cytotoxic T cells. The strong, cell-mediated immune response is decisive in controlling the infection and preventing complications. Although most people recover completely with no long-term sequelae, some experts now believe that chronic EBV infection may contribute to a debilitating illness (see feature 20.3).

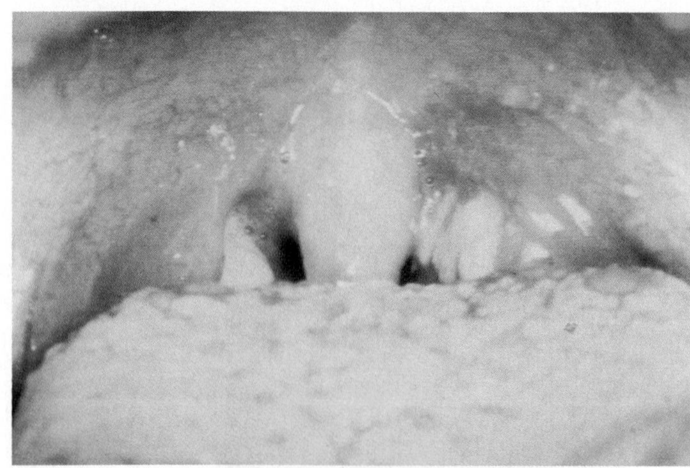

Figure 20.12 Infectious mononucleosis. This view of the pharynx shows the swollen tonsils and throat tissues, a gray-white exudate, and rash on the palate that may be of some use in diagnosis.

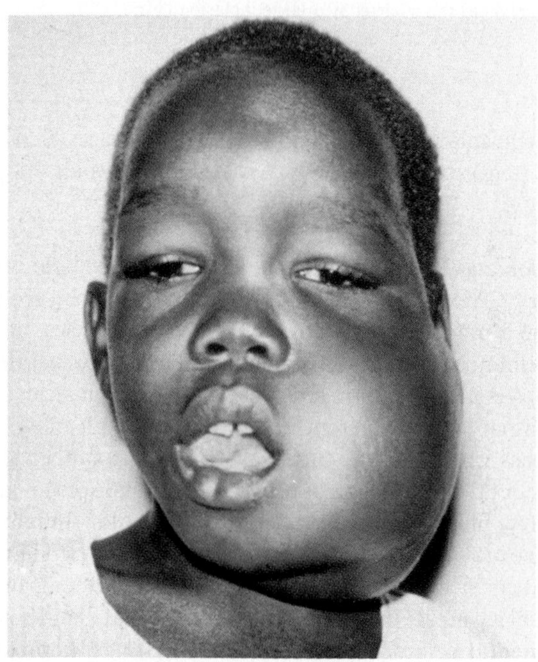

Figure 20.13 Burkitt's lymphoma, a cancer commonly associated with chronic Epstein-Barr infection.

Tumors and Other Complications Associated with EBV Epstein-Barr virus is commonly associated with a B-cell malignancy called **Burkitt's lymphoma.** This cancer usually develops in the jaw and grossly swells the cheek (figure 20.13), though it may also occur in the abdomen. Most cases are reported in central African children four to eight years old, and a small number are seen in North America. The prevalence in Africa may be associated with chronic coinfections with other diseases such as malaria that weaken the immune system and predispose to tumors. One theory to explain the neoplastic effect is that persistent EBV infection in young children eventually immortalizes certain B lymphocytes. When this is coupled with concurrent malaria infection, the T-cell effector response is overwhelmed and unable to control the overgrowth of these malignant B-cell clones (see chapter 14).

Unlike Burkitt's lymphoma, **nasopharyngeal carcinoma** is a malignancy of epithelial cells that occurs in older men living in China and Africa. Although little about this disease is understood, high antibody titer to EBV is a constant feature. It is not known how the virus invades epithelial cells (not its usual target),

Feature 20.3 Viral Infection—The Cause of Chronic Fatigue?

In 1985, two internists in the small resort area of Lake Tahoe, California, witnessed a large cluster of patients with a common group of nonspecific symptoms—disabling fatigue, recurrent colds, memory loss, and enlarged lymph nodes—that lasted for long periods and did not respond to the usual treatments. A series of tests revealed that only one specific sign was shared by all: a high level of antibodies to EBV in the blood. At first, physicians Peter Cheney and Daniel Peterson had difficulty reconciling the signs and symptoms with the usual picture of EBV infection, which usually does not occur in epidemic pattern, causes a relatively mild, short-term illness, and is not highly transmissible by casual contact. Yet, week after week more persons came down with the syndrome, and eventually, nearby regions of California and Nevada also reported outbreaks. Investigators from the CDC were called in to help solve the mystery, and they concluded that EBV infection alone could not account for all aspects of the disease.

By the early 1990s, hundreds of thousands of cases of this mystery disease—now termed **chronic fatigue syndrome (CFS)**—had been diagnosed. One of the earliest large clinical studies at the University of California, San Francisco, made some breakthrough findings that supported the probable connection of CFS to a viral infection. It now appears that CFS patients have chronically overactive immune responses that could readily account for the flu-like nature of the disease.

The causative agent is still a matter of speculation, however. Other possible candidates besides EBV are a newly discovered herpesvirus (human B lymphotropic virus) and a retrovirus (HTLV II). Some hope for control of this devastating syndrome lies in a drug called Ampligen, which appears to diminish many of the symptoms in early trials.

why the incidence of malignancy is so high in certain ethnic groups, or whether dietary carcinogens such as nitrosamines in cured fish are involved.

In general, any person with an immune deficiency (especially of T cells) is highly susceptible to Epstein-Barr virus. Organ transplant patients carry a high risk because their immune systems are weakened by measures to control rejection. The oncogenic potential of the virus is another concern, since it is currently believed to cause some types of lymphatic tumors in kidney transplant patients and is commonly isolated from tonsilar, laryngeal, thymic, and certain AIDS-related cancers. A high correlation also exists between AIDS in male homosexuals and seropositivity for EBV, suggesting it could be a cofactor infection that speeds the onset of AIDS.

Diagnosis, Treatment, and Prevention Because clinical symptoms in EBV infection are common to many other illnesses, laboratory diagnosis is necessary. A differential blood count that shows lymphocytosis, neutropenia, and large atypical lymphocyte cells with lobulated nuclei and vacuolated, basophilic cytoplasm is suggestive of EBV infection (figure 20.14). Serologic assays to detect antibody against the capsid and DNA are helpful, as are direct viral antigen tests using probes and labelled antibodies.

The usual treatments for infectious mononucleosis are nonspecific, directed at symptomatic relief of fever and sore throat. Specific antiviral measures for disseminated and transplant-related disease that have shown some promise are intravenous gammaglobulin, interferon, acyclovir, and monoclonal antibodies. Because Burkitt's lymphoma has usually metastasized by the time of diagnosis, systemic chemotherapy with cyclophosphamide, vincristine, methotrexate, or prednisolone may induce remission. These measures, along with surgical removal of the tumor, have considerably reduced the mortality rate.

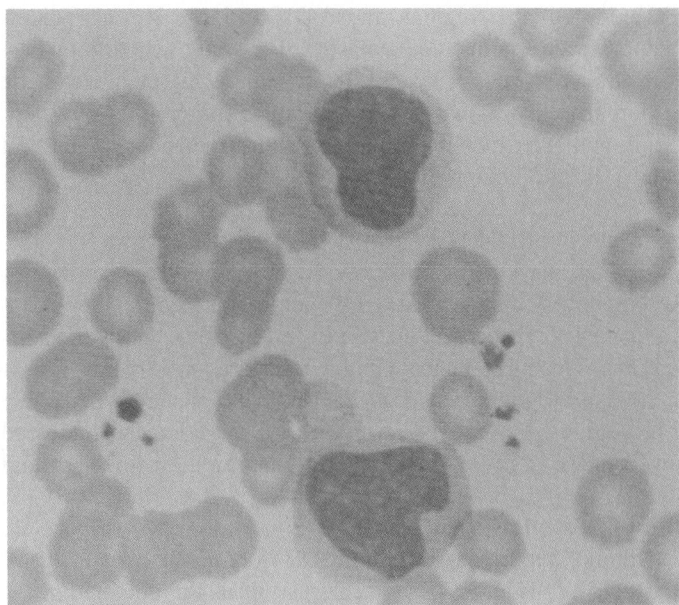

Figure 20.14 Epstein-Barr virus in the blood smear of a patient with infectious mononucleosis. Note the abnormally large lymphocytes containing indented nuclei perforated with holes.

Feature 20.4 The Dangers of Monkey Herpesvirus

Although herpesviruses tend to cause mild infection in their natural mammalian hosts, interspecies infection may be quite severe and even lethal. A herpesvirus called *H. simiae* virus, common among macaque monkeys, produces a localized herpes simplex-like infection in these primates. In that macaques are popular subjects for biomedical research and are maintained in laboratories worldwide, a potential exists for transmission of the virus to humans. Most of the 23 verified human infections have been traced to animal bites or scratches, though contact with saliva, tissues, needles, and even contaminated air is suspected. In all cases, the ensuing infection causes a severe neurological disease that is terminal. Persons who routinely handle these monkeys must take extreme precautions to avoid contact with them or with their secretions.

Hepadnaviruses: Unusual Enveloped DNA Viruses

One group of enveloped DNA viruses, called *hepadnaviruses,* is quite unlike any other so far discovered. They have never been grown in tissue culture and have an unusual genome containing

hepadnavirus (hep″-ah-dee″-en-ay-vy′-rus) Gr. *hepatos,* liver, plus DNA and virus.

both double- and single-stranded DNA (figure 20.15). Hepadnaviruses show a decided tropism for the liver, where they persist and may provoke liver cell carcinoma. One member of the group, **hepatitis B virus (HBV)**, causes a common form of hepatitis, and other members cause hepatitis in woodchucks, ground squirrels, and Peking ducks.

General Considerations: What Is Viral Hepatitis?

When certain viruses infect the liver, they cause **hepatitis**, an inflammatory disease marked by necrosis of hepatocytes and a mononuclear response that swells and disrupts the liver architecture. This pathological change interferes with the liver's excretion of bile pigments such as bilirubin into the intestine. When bilirubin, a greenish-yellow pigment, accumulates in the blood and tissues, it causes **jaundice**, a yellow tinge in the skin and eyes.

Characteristics of the three principal viruses responsible for human hepatitis are summarized in table 20.2. Hepatitis A virus (HAV) is an RNA enterovirus that is the etiologic agent of infectious hepatitis (see chapter 21). In general, HAV disease is far milder, shorter-term, and less virulent than the other forms. Another important agent is hepatitis C virus (HCV; also called non-A, non-B hepatitis virus), a recently discovered RNA virus that causes most cases of infusion hepatitis. These two viruses have no relationship to hepatitis B virus (HBV), a DNA virus that causes serum hepatitis, which will be covered in this chapter.

jaundice (jon'-dis) Fr. *jaunisse*, yellow. Also called icterus.

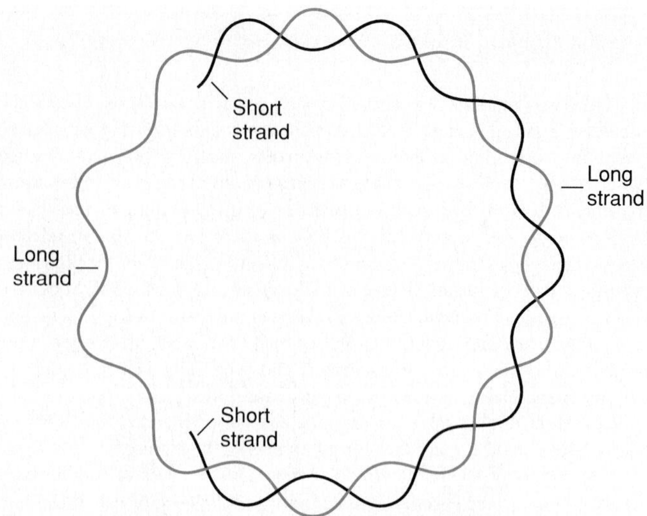

Figure 20.15 The nature of the DNA strand in hepatitis B virus. Like most DNA viruses, the genome of HBV is circular, but unlike them, it is not a completely double-stranded molecule. A section of the molecule is composed of single-stranded DNA.

The incidence of these three forms is presented in figure 20.16. An unusual virus, hepatitis D or delta agent, is a dependovirus, meaning that it is defective and cannot produce infection unless a cell is simultaneously infected by HBV. An astonishing fact about this unique infectious agent is that it appears to be a naked strand of RNA (a viroid) that enters the cell in piggyback fashion on its host virus.

Table 20.2 Principal Morphologic and Pathologic Features of the Hepatitis Viruses

	HAV	HBV	HCV (Non-A, Non-B)
Properties			
Biology			
Nucleic acid	RNA	DNA	RNA
Size	27 nm	42 nm	Various
Protein coat	—	+	Under study
Cell culture	+	—	—
Envelope	—	+	Under study
Disease			
Synonyms	Infectious hepatitis, yellow jaundice	Serum hepatitis, transfusion hepatitis	Post-transfusional hepatitis
Epidemiology	Endemic and epidemic	Endemic	Endemic
Case incidence	Increasing	Increasing	Increasing
Reservoir	Active infections	Chronic carrier	Chronic carrier
Transmission	Oral-fecal; water or food-borne	Overt inoculation from blood, serum; close contact	Inoculation from blood, serum; intimate contact
Incubation Period	2–7 weeks	1–6 months	2–8 weeks
Symptoms	Fever, GI tract disorder	Fever, rash, arthritis	Similar to HBV
Jaundice	1 in 10	Common	Common
Onset/Duration	Acute, short	Gradual, chronic	Acute to chronic
Complications	Uncommon	Chronic active hepatitis, hepatic cancer	Chronic inflammation, cirrhosis
Availability of Vaccine	—	+	—
Diagnostic Tests to Differentiate	+	+	+

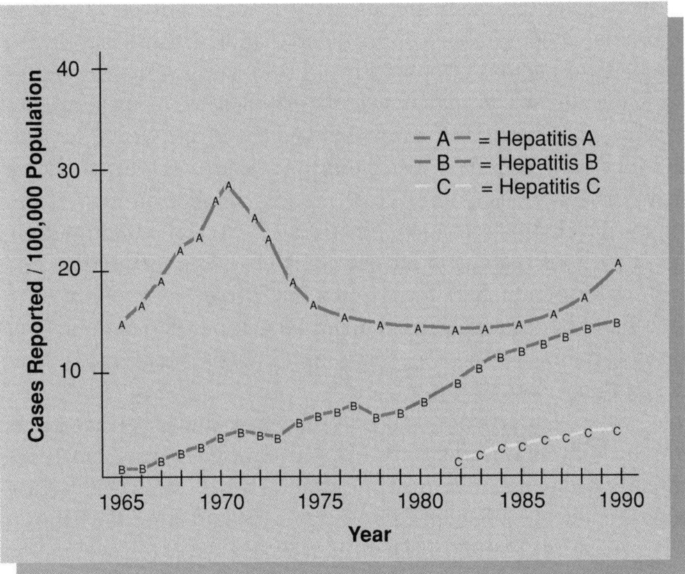

Figure 20.16 The comparative incidence of viral hepatitis in the United States from 1965 to 1990.
Source: Data from the Centers for Disease Control (CDC), Atlanta, GA.

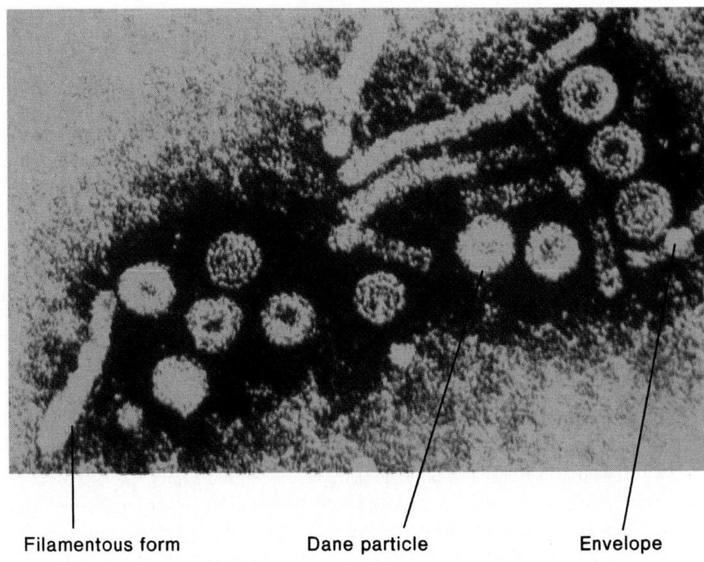

Figure 20.17 Blood from HBV-infected patients presents an array of particles, including Dane particles (the intact virus), the envelope (or surface shell), and filaments that are polymers of the envelope antigen.

Properties of Hepatitis B Virus Because a cell culture system for propagating hepatitis B virus is not yet available, by far the most significant source of information about its morphology and genetics has been obtained from viral fragments, intact virions, and antibodies, which are often abundant in patients' blood. One of these blood-borne pieces, the Dane particle, has been suspected as the infectious virus particle, and modern serologic and electron microscope studies support this view. Blood-borne viral components are shown in figure 20.17.

Epidemiology of Hepatitis B An important factor in the transmission pattern of hepatitis B virus is that it multiplies exclusively in the liver, and that this organ continuously seeds the blood with viruses. Electron microscopic studies have revealed up to 10^7 virions per milliliter of infected blood. Even a minute amount of blood (10^{-6} to 10^{-7} ml) can transmit infection. The abundance of circulating virions is so high and the minimal dose so low that such simple practices as sharing a toothbrush or a razor can transmit infection. Over the past 10 years, HBV has also been detected in semen and vaginal secretions, and it is currently believed to be transmitted by these fluids in certain populations. Spread by means of close family or institutional contact is also well documented. The virus remains infective for days in dried blood, for months when stored in serum at room temperature, and for decades if frozen. It is not inactivated after four hours of exposure to 60°C, but boiling for the same period can destroy it. Disinfectants containing chlorine, iodine, and glutaraldehyde show potent anti-hepatitis B activity.

Hepatitis B is an ancient disease that has been found in all populations, though the incidence and risk are higher among lower socioeconomic classes, drug addicts, the sexually promiscuous, and certain occupations, including persons who conduct medical procedures involving blood or blood products such as serum (see feature 20.5). The incidence of HBV infection and carriage among drug addicts who routinely share needles is very high, especially if blood is pulled back into the syringe and mixed with the drug (see feature 11.2). Homosexual males constitute another high-risk group because such practices as anal intercourse can traumatize membranes and permit intimate contact between broken blood vessels of both participants. Heterosexual intercourse can also spread infection, but is less likely to do so. Infection of the newborn by chronic carrier mothers occurs readily and predisposes to development of the carrier state and increased risk of liver cancer in the child (see feature 20.6). The role of mosquitos, which can harbor the virus for several days after biting infected persons, has been well-documented in the tropics and in some parts of the United States.

The tropism of HBV for the liver becomes a significant epidemiologic complication in some persons (5–10% of infectees). Persistent or chronic carriers harbor the virus for up to six months, and some continue to shed the virus for many years. Worldwide, 300 million persons are estimated to carry the virus, but it is most prevalent in Africa, Asia, and the western Pacific and least prevalent in North America and Europe.

Pathogenesis of Hepatitis B Virus The hepatitis B virus enters the body through a break in the skin or mucous membrane or by injection into the bloodstream. Eventually it reaches the liver cells (hepatocytes), where it multiplies and releases viruses into the blood (viremia). After an incubation period of 4 to 24 weeks

Feature 20.5 The Clinical Connection in Hepatitis B

Hepatitis B is a risk to the patient and the health care worker alike. Obvious gross contact, in the form of blood transfusions or injections of human plasma, clotting factors, serum, and other products, is a classic route of infection. It should be noted, however, that HBV infection among recipients of these substances is quite rare now because of blood tests for HBV antigens and processing techniques that inactivate any HBV present.

Such diverse procedures as kidney dialysis, reused needles, and acupuncture needles are also known to transmit the HBV virus. Outbreaks of hepatitis B recently developed in persons vaccinated with a special needleless gun. These cases were traced to inadequate disinfection of the casing of the gun nozzle, which must be reused continuously between patients. Even cosmetic manipulations such as tattooing and ear or nose piercing may expose a person to infection. The only reliable method for destroying HBV on reusable instruments is through autoclaving.

Insofar as occupational exposure goes, nonimmunized persons working with blood, plasma, sera, or materials contaminated with even small amounts of these substances are very vulnerable to infection. Practically any health occupation may be involved, but hematologists, phlebotomists, dialysis personnel, dental workers, surgical staff, and emergency room attendants must take particular precautions. Accidents involving broken blood vials, needle sticks, improper handling of instruments, and splashed blood and serum are common causes of HBV infection in clinical workers.

Feature 20.6 Hepatitis Testing During Pregnancy

Mothers who are carriers of HBV are highly likely to transmit infection to their newborns at the time of birth. It is not currently known whether or not transplacental HBV infection occurs. Unfortunately, infection with this virus early in life is a known risk factor for chronic liver disease. In light of the increasing incidence of HBV-positive pregnant women, routine testing of all pregnant women is now strongly supported by the CDC and a majority of obstetricians. If adopted, mandatory hepatitis testing would add to the growing number of procedures used to screen prospective mothers for infection with agents that have especially profound risks for fetuses and/or neonates. These diseases, known collectively as STORCH, are syphilis, toxoplasma, other infections, rubella, chlamydia, and herpes. (The "other" infections include CMV, candidiasis, gonorrhea, hepatitis B, group B streptococci, and under some circumstances, AIDS.)

(7 weeks average), some persons experience malaise, fever, chills, anorexia, abdominal discomfort, and diarrhea. Surprisingly, the majority of persons infected exhibit few overt symptoms and go on to develop an immunity to HBV. In those who do experience disease, the severity of symptoms and the aftermath of hepatic damage vary widely. Fever, jaundice, rashes, and arthritis are common reactions (figure 20.18), and a smaller number of patients develop glomerulonephritis and arterial inflammation. Most patients (90%) experience complete liver regeneration and restored function. But for reasons that are not entirely clear, a small number of predisposed older persons develop chronic liver disease in the form of necrosis or cirrhosis (permanent liver scarring and loss of tissue).

The association of HBV with **hepatocellular carcinoma**[1] is based on these observations: (1) The cancer appears most frequently in young men growing up in areas of the world with a high incidence of hepatitis B (Africa and the Far East); (2) persistent carriers of the virus are more likely to develop this cancer; and (3) certain hepatitis B antigens are found in malignant cells and are often detected as integrated components of the host genome. In general, persons with chronic hepatitis are 200 times more likely to develop liver cancer, though the exact role of the virus is still very speculative.

Diagnosis and Management of Hepatitis B The initial stages of most forms of hepatitis are so similar that differential diagnosis on the basis of symptoms is unlikely. A clinician can usually distinguish hepatitis B from hepatitis A by carefully examining the possible risk factors. Some combination of occupational hazard, drug abuse, age (it occurs primarily in adults), relatively long incubation period, and insidious onset strongly suggests HBV infection. Serological tests are increasingly important. Testing procedures fall into two categories: those that detect virus antigen and those that assay for antibodies. Recent developments in RIA and ELISA testing permit detection of the surface antigen very early in the infection. These same tests

1. Also called hepatoma, a primary malignant growth of hepatocytes.

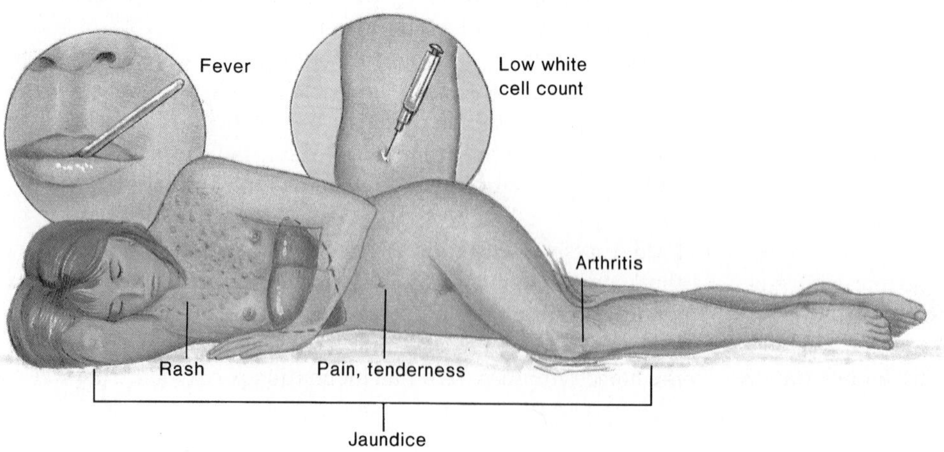

Figure 20.18 The clinical features of hepatitis B. Among the most prominent symptoms are fever, jaundice, and intestinal discomfort. The liver is also enlarged and tender.

are essential for screening blood destined for transfusions, semen in sperm banks, and organs intended for transplant. Antibody tests are most valuable in patients who are negative for the antigen.

As most cases of hepatitis B are self-limited, patient care relies on symptomatic treatment and supportive care. An exciting new treatment for chronic infection is the injection of synthetic interferon, which stops virus multiplication and prevents liver damage in up to 60% of patients. Control measures are currently focussed on prevention. Passive immunization with hepatitis B immune globulin (HBIG) gives significant protection to persons who have been exposed to the virus through needle puncture, broken blood containers, or skin and mucosal contact with blood. Other groups for whom prophylaxis is highly recommended are the sexual contacts of actively infected persons or carriers and neonates born to infected mothers.

The first boon to curbing hepatitis B was a vaccine developed and tested in the late 1970s and committed to general use in 1981. Because HBV cannot be cultivated in tissue cultures or embryos, the source of antigen for this first vaccine was the blood of carriers with high levels of the desired surface antigen. The blood is treated to purify and concentrate the antigen and heated to inactivate any intact viruses. Alternate vaccines (Recombivax, Energix) containing the pure surface antigen cloned in yeasts have largely replaced the original vaccine. Vaccination is a must for medical and dental workers and students, patients receiving multiple transfusions, immunodeficient persons, and cancer patients. Homosexual males, drug addicts, prostitutes, and close contacts of carriers are also strongly advised to seek vaccination.

Nonenveloped DNA Viruses

The Adenoviruses

During a study seeking the cause of the common cold, a new virus was isolated from *adenoids* that had been surgically excised from young children. It turned out that this virus, termed **adenovirus,** was not the sole agent of colds but one of several others (to be described in chapter 21). It was also the first of about 80 strains of nonenveloped, double-stranded DNA viruses to be discovered and eventually classified as adenoviruses. Of this array, about 30 types are associated with human infection, and the rest are animal pathogens. Besides infecting lymphoid tissue, adenoviruses have a preference for the respiratory and intestinal epithelia and the conjunctiva. Adenoviruses produce aggregations of incomplete and assembled viruses that usually lyse the cell (figure 20.19). In cells that do not lyse, the viral DNA may be harbored latently in the nucleus. Experiments show that adenoviruses are oncogenic in animals, but whether this is also true in humans is not known.

adenoid (ad'-eh-noyd) Gr. *adeno,* gland, and *eidos,* form. A lymphoid tissue (tonsil) in the nasopharynx that sometimes becomes enlarged.

Epidemiology of Adenoviruses

Adenoviruses are spread from person to person by means of respiratory and ocular secretions. Most cases of conjunctival infection involve preexisting eye damage and contact with contaminated sources such as swimming pools, dusty working places (shipyards and factories), and unsterilized optical instruments (ocular tonometers used in the glaucoma test). Infection by an adenovirus usually occurs by age 15 in most persons, and in certain individuals, this results in chronic respiratory carriage. Although the reason is obscure, military recruits are at greater risk for infection than the general population, and respiratory epidemics are common in armed services installations. Civilians, even of the same age group and living in close quarters, seldom experience such outbreaks.

Respiratory, Ocular, and Miscellaneous Diseases of Adenoviruses

The patient infected with an adenovirus is typically feverish, with acute rhinitis, cough, and inflammation of the pharynx, enlarged cervical lymph nodes, and a macular rash (somewhat reminiscent of rubella). Occasionally, atypical pneumonia and gastrointestinal disturbances develop. Certain adenoviral strains produce an acute follicular lesion of the conjunctiva. Usually one eye is affected by a watery exudate, redness, and partial closure. A deeper and more serious complication is *keratoconjunctivitis,* an inflammation of the conjunctiva and cornea.

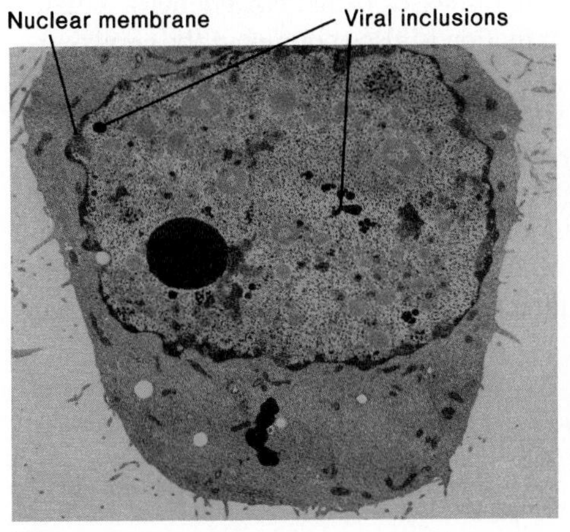

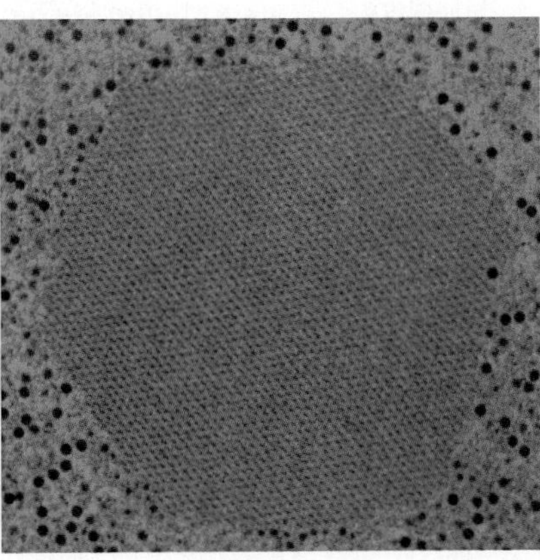

Figure 20.19 (*a*) Cellular and nuclear alterations in adenovirus infection. Granules of early virus parts accumulating in the nucleus cause inclusions to form (×19,000). (*b*) Magnification of the nucleus shows crystalline masses of mature virions (×35,000).

Acute hemorrhagic cystitis in children has been traced to an adenovirus. This self-limiting illness of four to five days duration is marked by hematuria (blood in the urine), frequent and painful urination with occasional episodes of fever, bedwetting, and suprapubic pain. Because the virus apparently infects the intestinal epithelium and can be isolated from some diarrheic stools, adenovirus is also suspected as an agent of infantile gastroenteritis.

Severe cases of adenovirus infection may be treated with interferon in the early stages. An inactivated polyvalent vaccine prepared from viral antigens is an effective preventative measure, especially for military recruits.

Papovaviruses: Papilloma, Polyoma, Vacuolating Viruses

Letters from the terms **pa**pilloma, **po**lyoma, and simian **va**cuolating viruses gave rise to the term **papovavirus.** This family contains two subtypes: the **papillomaviruses,** which are responsible for human and animal papillomas, and the **polyomaviruses,** which include several human (BK and JC viruses) and animal pathogens.

Epidemiology and Pathology of the Human Papilloma Viruses

A papilloma is a benign, squamous epithelial growth commonly referred to as a **wart** or **verruca** and caused by one of 40 different strains of **human papillomavirus (HPV).** Some types are specific for the mucous membranes, while others invade the skin. The appearance and seriousness of the infection vary somewhat from one anatomical region to another. Painless, elevated, rough growths on the fingers and occasionally on other body parts are called **common** or **seed warts** (figure 20.20*a*). These occur routinely in children and young adults. **Plantar warts** are deep, painful papillomas on the soles of the feet, and flat warts are smooth, skin colored lesions that develop on the face, trunk, elbows, and knees. A special form of verruca known as **genital warts** is a prevalent STD and is linked to some types of cancer. Warts are transmissible through direct contact with a wart or indirect contact with a fomite, and they may also spread on the same person by autoinoculation. The incubation period varies anywhere from two weeks to more than a year. Papillomas are common among all sexes, races, and geographical regions.

Genital Warts—An Insidious Papilloma Genital warts are the latest concern among young, sexually active persons. Although this disease is not new, case reports have increased to such an extent that it is now regarded as the most common STD in the United States. Studies show that up to 15% of the population are carriers of one of the five types of HPV associated with genital warts (types 6, 11, 16, 18, and 31). The virus invades the external and internal genital membranes, especially in areas prone to friction in the vagina and the head of the penis. Wart morphology ranges from tiny, flat, inconspicuous bumps to giant, branching, cauliflower-like masses called **condylomata acuminata** (figure 20.20*b*). Infection is also common in the cervix, urethra, and anal skin.

What is most disturbing about the sudden increase in HPV infection is the strong association of the virus with cancer of the reproductive tract, especially the cervix and penis. The virus can be isolated from a majority of female genital cancers and is known to convert some forms of genital warts into malignant tumors. Fear of this infection with its potential for cancer can

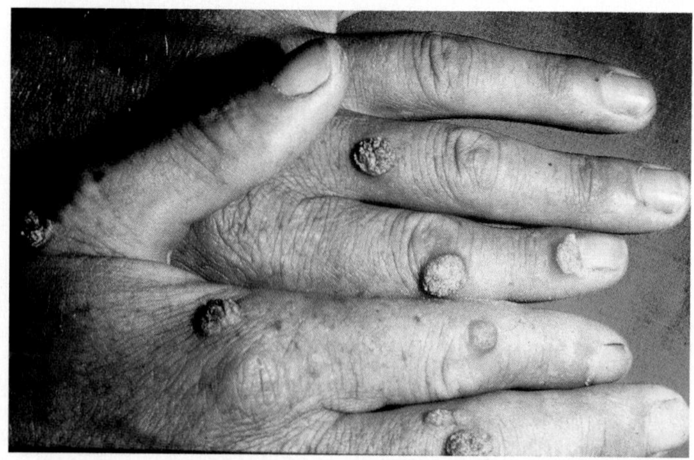

(a)

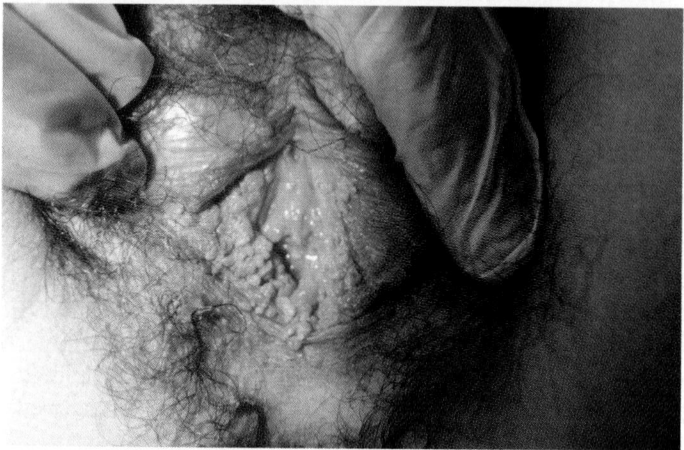

(b)

Figure 20.20 Human papillomas. (*a*) Common warts. Virus infection induces a neoplastic change in the skin and benign growths on the fingers, face, or trunk. One must avoid picking, scratching, or otherwise piercing the lesions to avoid spread and secondary infections. (*b*) Chronic genital warts (condylomata acuminata) that have extended into the labial, perineal, and perianal regions.

papilloma (pap″-il-oh′-mah) L. *papilla,* pimple, and Gr. *oma,* tumor. A benign, nodular, epithelial tumor.

polyoma (paw″-lee-oh′-mah) L. *poly,* many. A glandular tumor caused by a virus.

vacuolating (vak′-yoo-oh-lay″-ting) L. *vacuus,* empty. The process of forming small spaces or vacuoles in the cytoplasm of the cell.

condylomata acuminata (kahn″-dee-loh′-mah-tah ah-kyoo″-min-ah′-tah) Gr. *kondyloma,* knob, and L. *acuminatus,* sharp, pointed. Not to be confused with condyloma lata, a broad, flat lesion of syphilis.

be greatly reduced by early detection and treatment of both the patient and the sexual partner. Such early diagnosis depends on thorough inspection of the genitals and, in women, a Papanicolaou smear to screen for abnormal cervical cells.

Diagnosis, Treatment, and Prevention The warts caused by papillomaviruses are usually distinctive enough to permit reliable clinical diagnosis without much difficulty. However, a biopsy and histologic examination can help clarify any ambiguous cases. A sensitive test such as an RNA probe can detect the HPV type and possibly assess the risk of developing cervical cancer. Although most common warts regress over time, they cause sufficient discomfort and cosmetic concern to require treatment. Treatment strategies for all types of warts include direct chemical application of podophyllin and physical removal of the affected skin or membrane by cauterization, freezing, or laser surgery. Immunotherapy with interferon also shows some promise. Because treatments may not completely destroy the virus, warts may recur.

The Polyomaviruses

In the mid-1950s the search for a leukemia agent in mice led instead to the discovery of a new virus capable of inducing an assortment of tumors—hence its name, polyomavirus. Among the most important human polyomaviruses are the **JC virus** and **BK virus.**[2] Although many polyomaviruses are fully capable of transforming host cells and inducing tumors in experimental animals, none cause cancer in their natural host.

Epidemiology and Pathology On the basis of serological surveys, it appears that infections by JC virus and BK virus are commonplace throughout the world. Because the majority of infections are asymptomatic or mild, not a great deal is known about their mode of transmission, portal of entry, or target cells, though urine and respiratory secretions are suspected as possible sources of the viruses. The main complications of infection occur in patients who have cancer or have been immunosuppressed. **Progressive multifocal leukoencephalopathy (PML)** is an uncommon but generally fatal infection by JC virus that attacks accessory brain cells and gradually demyelinizes certain parts of the cerebrum (figure 20.21). Infection with BK viruses is usually associated with renal and ureter transplants, in which latent virus carried in the transplant becomes active and causes complications in urinary function. The prospects of finding effective therapy for these diseases are bleak. By the time PML is diagnosed, the patient is irreversibly immunocompromised, and extensive brain damage has already occurred. BK infection may be prevented by treating renal transplant patients with human leukocyte interferon.

2. The initials signify the cancer patients from whom the viruses were first isolated.

leukoencephalopathy (loo″-koh-en-sef″-uh-lop′-uh-thee) Gr. *leuko*, white, *cephalos*, brain, and *pathos*, disease.

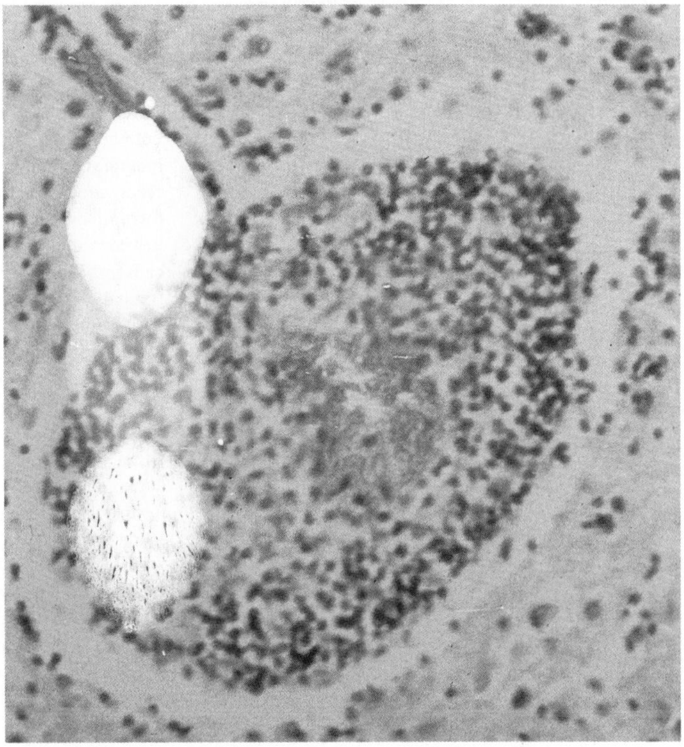

Figure 20.21 Brain pathology in progressive multifocal leukoencephalopathy. Target cells are oligodendrocytes, seen here as an enlarged cell with a giant swollen nucleus and intranuclear inclusions.

Parvoviruses: Nonenveloped Single-Stranded DNA Viruses

The **parvoviruses (PV)** are unique among the viruses in having single-stranded DNA molecules. These resistant viruses are also notable for their extremely small diameter (18–26 nm) and genome size. Parvoviruses are indigenous to and may cause disease in several mammalian groups—for example, distemper in cats, an enteric disease in adult dogs, and a potentially fatal cardiac infection in puppies.

The most important human parvovirus is B19, the cause of erythema contagiosum (fifth disease), a common infection in children. Often the infection goes unnoticed, though the child may have a low-grade fever and a bright red rash on the cheeks. This same virus can be more dangerous in children with immunodeficiency or sickle-cell anemia, because it destroys red blood stem cells. If a pregnant woman contracts infection and transmits it to the fetus, a severe and often fatal anemia can result. One type of parvovirus, the adeno-associated virus (A-AV), is another example of a defective or dependovirus that cannot replicate in the host cell without the helper function of another virus, in this case, an adenovirus. The possible impact of this coinfection on humans is still obscure.

parvovirus (par″-voh-vy′-rus) L. *parvus*, small.

Chapter Review with Key Terms

General Characteristics of Animal Viruses and Human Viral Diseases

Viruses are minute parasitic particles consisting of DNA or RNA genomes packaged within a protein capsid; they invade host cells and appropriate the cell's machinery for mass production of new virions, both in cytoplasm and nucleus; viruses exit the host cell by lysis or budding; budded viruses leave with an envelope; actively infected host cells are usually destroyed.

Viruses attack a variety of host and cell types; individual viruses are relatively host/cell specific, due to the need for receptor recognition; target cells include nervous system, blood, liver, skin, and intestine.

Diseases range from mild and self-limited to fatal in effects; symptoms are local, systemic, and depend on the exact tissue target; identification and diagnosis are by culture, microscopy, genetic probes, and serology; immunity is both humoral and cell-mediated; limited number of drugs available for treatment.

DNA and some RNA viruses persist in inactive state (latent) in host cells; some latent viruses cause recurrent infections, and others integrate into host genome and cause cancerous transformation; some viruses are **teratogenic** and cause congenital defects; zoonotic viruses transmitted by vectors cause severe human disease.

DNA Viruses

DNA viruses are enveloped or nonenveloped nucleocapsids; most have double-stranded DNA; parvoviruses have single-stranded DNA.

Enveloped Viruses

Poxviruses: Large complex viruses; produce skin lesions called **pox** (pocks). **Variola** is the agent of **smallpox,** the first disease to be eliminated from the world through vaccination; last smallpox case reported in 1977; spread by close contact, inhalation of droplets; infection of mucous membrane leads to viremia and large deep pustules that may scar; a highly virulent form was often fatal; vaccination is by scarification with vaccinia virus, a special strain that immunizes against variola; vaccination is no longer done routinely.

Molluscum contagiosum, an unclassified virus, causes waxy nodules on skin; transmitted by direct contact; may be an STD in adults.

Herpesviruses: Persistent, latent viruses that cause recurrent infections and may be involved in neoplastic transformations; attack skin, mucous membranes, and glands.

Herpes simplex virus (HSV): Types 1 and 2 both create lesions of skin and mucous membranes; migrate into nerve ganglia; are reactivated by anatomic and physiologic disruptions; cause serious disseminated disease in newborn infants and immunodeficient patients; cause **whitlows** on fingers of medical and dental personnel. Herpes simplex 1 mainly infects lips (cold sores, fever blisters), eyes, and oropharynx; is transmitted by close contact and droplets. Herpes simplex 2 usually infects genitalia (external, internal) and is sexually transmitted; complications in newborns infected at birth with either virus include encephalitis with severe CNS damage. Treated with acyclovir; controlled by means of barriers and care in handling secretions.

Varicella-Zoster Virus (VZV): This herpesvirus causes both **chickenpox (varicella)** and **zoster (shingles)**; chickenpox is the primary infection, common in children; zoster is later recurrence of infection by latent virus. Chickenpox is transmitted through droplets; symptoms are fever, poxlike papular rash. In zoster, the virus has migrated into spinal nerves, and is reactivated by surgery, cancer, drugs, or other stimuli; travels out to skin, causing painful lesions in a pattern on the trunk or head, also nerve pain; severe disease treated with acyclovir, vidarabine; immune globulin used for prevention; vaccine not in general use.

Cytomegaloviruses (CMV): One of the most common herpesviruses; high infection rate in children, pregnant women, drug abusers, and male homosexuals; spread through close contact with saliva, mucus, urine, and semen; chronically carried; congenital infection causes disturbances of liver, spleen, brain, and eyes; can cause death; in children and adults, causes mononucleosis-like syndrome with fever and leukocytosis; disseminated disease affecting many organs is common in AIDS and transplant patients; antiviral drugs may be of some benefit.

Epstein-Barr Virus (EBV): A herpesvirus of the lymphoid and glandular tissue; transmitted by saliva, oral contact; 95% of all persons develop some form of infection, though often asymptomatic; cause of **mononucleosis,** a "kissing disease" common in college students that is marked by sore throat, fever, swollen lymphoid tissue, leukocytosis; virus is oncogenic; causes **Burkitt's lymphoma,** a malignancy of the B lymphocytes that swells cheek or abdomen and is prevalent in African children; also causes **nasopharyngeal carcinoma,** cancer in older men in China, Africa. Disseminated EBV infection is a complication of AIDS and transplant patients.

Hepadnaviruses: One cause of viral hepatitis, an inflammatory disease of liver cells that may result from several different viruses. **Hepatitis B virus (HBV)** causes **serum hepatitis,** is most severe; virus has strong affinity for liver cells, is carried and shed into blood; only a tiny infectious dose is required; invades skin or mucous membranes; virus resistant to heat and most disinfectants; common clinical risk due to contact with blood; drug addicts, homosexuals are at high risk; may be transmitted from mother to child or by sharing personal articles; infection marked by fever, intestinal disturbance, buildup of bile pigments in blood creates **jaundice**; chronic carriage a problem that may lead to liver disease and liver cancer; managed by serum globulin, interferon; vaccines effective, recommended for health care workers, transfusion patients, other high-risk groups.

Nonenveloped Viruses

Adenoviruses: Common infectious agents of lymphoid organs, respiratory tract, eyes; spread through close contact with secretions, contaminated vehicles, fomites; diseases include common cold with fever and rash; **keratoconjunctivitis,** a severe eye infection; **cystitis,** an acute urinary infection.

Papovaviruses: Papilloma and polyoma viruses.

Human papilloma viruses (HPV) cause skin tumors called **papillomas, verrucas,** and **warts;** spread by close contact with infected skin or fomites. **Common warts** are rough, painless lesions on hands; **plantar warts** are painful, flat, benign tumors on feet, trunk. **Genital warts** are verrucas that start as tiny bumps on membranes or skin of genitals; a very common STD; may progress to large masses called **condyloma acuminata**; chronic infection associated with cervical and penile cancer; treatment by surgery, freezing, cauterization, chemical removal.

Polyomaviruses include animal tumor viruses and human viruses. Human viruses are JC virus and BK virus. **JC virus** causes **progressive multifocal leukoencephalopathy,** a slow destruction of the brain in cancer and immunodeficient patients; **BK virus** causes complications in urinary transplants.

Single-Stranded DNA Viruses: Parvoviruses are the tiniest viruses; cause severe disease in several mammals; **human parvovirus B19** causes erythema contagiosum, a mild respiratory infection of children that can be dangerous to fetus.

True–False Questions

Determine whether the following statements are true (T) or false (F). If you feel a statement is false, explain why, and reword the sentence so that it reads accurately.

___ 1. Live DNA viruses are dangerous to use in vaccines because of their capacity to become latent.

___ 2. Zoonotic viral diseases are generally mild in humans.

___ 3. Smallpox and vaccinia viruses are both causes of smallpox.

___ 4. Herpes simplex 1 causes fever blisters, and herpes simplex 2 causes cold sores.

___ 5. Herpetic whitlows are oral canker sores that occur in children.

___ 6. Infection of neonates with herpes simplex occurs primarily during birth.

___ 7. Acyclovir is an effective treatment for herpes simplex lesions.

___ 8. Varicella and zoster are caused by two different viruses.

___ 9. Hepatitis is a liver disease that often causes jaundice.

___ 10. Hepatitis B causes a chronic liver infection that may lead to cancer.

___ 11. Genital warts are spread by infected fingers.

___ 12. Parvoviruses are unique because they contain a single-stranded DNA genome.

Concept Questions

1. Outline 10 medically important characteristics of viruses.
2. What accounts for the affinity that viruses have for certain hosts or tissues? What accounts for the symptoms of viral diseases? Why are some pathlogical states caused by viruses so much more damaging and life-threatening than others?
3. Outline the target organs and general symptoms of DNA viral infections.
4. What is the general rule governing the cell locations where DNA and RNA viruses multiply? What determines whether latency occurs or not? In what special way can persons carry DNA viruses?
5. Briefly describe the epidemiology and symptoms of smallpox. Why was it such a great killer? What is a pock? What is the basis for the vaccine given for smallpox? How is it protective?
6. What is the disease molluscum contagiosum? What other infection in this chapter could it be mistaken for?
7. What are the common characteristics of herpesviruses?
8. Compare the two types of herpesvirus according to the types of diseases they cause, body areas affected, and complications. Why is neonatal infection accompanied by such severe effects? Is HSV-1 as severe as HSV-2 in this condition? What causes recurrent attacks?
9. Under what circumstances does one get chickenpox and shingles? Where do the lesions of shingles occur, and what causes them to appear there? How are ZIG and VZIG used?
10. What are the main target organs of CMV? How is it transmitted? What group of people is at highest risk for serious disease?
11. What are the main diseases associated with Epstein-Barr virus? What appears to be the pathogenesis involved in Burkitt's lymphoma? Why is mononucleosis a disease primarily of college-age Americans?
12. Define viral hepatitis, and briefly describe the three main types with regard to causative agent, common name, severity, and mode of acquisition.
13. What is unusual about the genome of hepatitis B virus? What is the usual source of the virus for study? What is important about the virus with regard to infectivity and transmission to others? What groups are most at risk for developing hepatitis B?
14. What is the course of hepatitis B infection from portal of entry to portal of exit? What are some serious complications?
15. What is the nature of the vaccines for hepatitis B? Who should receive them?
16. What are the principal diseases of adenoviruses? How are they spread?
17. Compare the types of warts or verrucae. Which type is most serious and why? How are warts cured?
18. Name the most important human parvovirus and the disease it causes.

Practical/Thought Questions

1. Why are DNA viruses more likely to cause cancer than RNA viruses?
2. Why is isolating a virus from a tumor not indicative of its role in disease?
3. What features of variola virus made it possible to eradicate smallpox? Are any other viruses in this chapter possible candidates for this sort of achievement? Why, or why not?
4. Can you think of some reasons that herpes simplex and zoster viruses are carried primarily in nerve trunks? What mechanism might account for reactivation of these viruses by various traumatic events?
5. What measures can health care workers take to avoid whitlows?
6. Why is a baby whose mother has genital herpes not entirely safe even with a cesarian birth?
7. What specific elements do immunodeficient, cancer, and AIDS patients lack that make them so susceptible to the viral diseases in this chapter? Given the ubiquity of most of these viruses, what must a health care worker be on guard for in handling these patients?
8. Can you explain how a dependovirus would function?
9. What are the medical consequences of transfusing blood labelled Dane (+)?
10. A man wants to divorce his wife because he believes her genital herpes could only have been acquired sexually. Her doctor says she has type 1. Can you help them resolve this problem?
11. Explain this statement: One acquires chickenpox from others; one acquires zoster from oneself.
12. A leukemia patient has a high titer of antibodies to EBV. A chronically tired, ill businessman has a high EBV titer. A healthy military recruit has a high EBV titer. An AIDS patient has a high EBV titer. Comment on the probable significance of antibodies to EBV in human serum.
13. What do you think of the practice of exposing children to chickenpox to give them the disease and subsequent immunity?
14. Tell how you would decontaminate a vaccination gun or acupuncture tools to be used between patients. What virus is the greatest risk for infection by this means?

CHAPTER 21

The RNA Virus Diseases

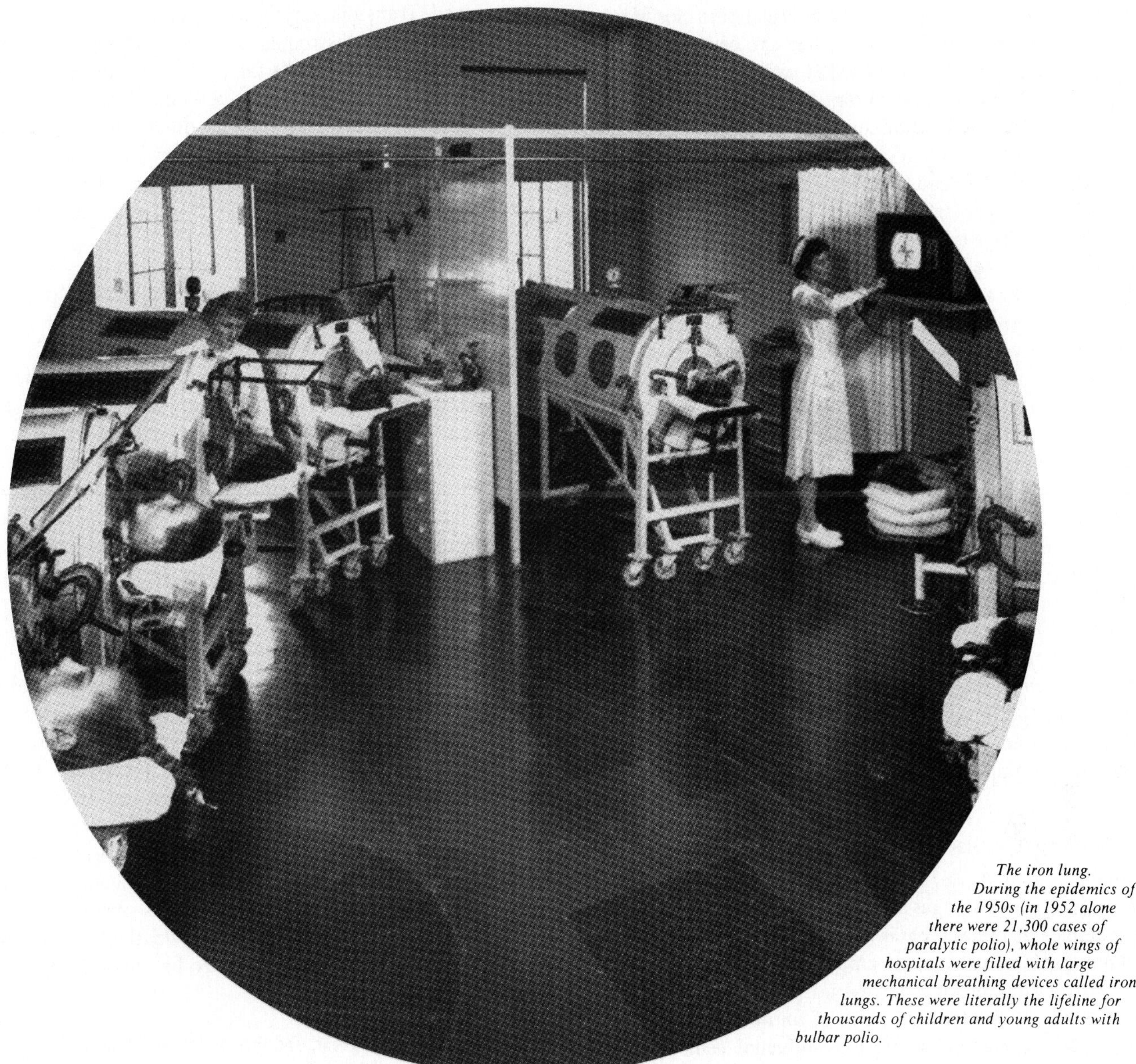

The iron lung. During the epidemics of the 1950s (in 1952 alone there were 21,300 cases of paralytic polio), whole wings of hospitals were filled with large mechanical breathing devices called iron lungs. These were literally the lifeline for thousands of children and young adults with bulbar polio.

Chapter Preview

The RNA viruses are a remarkably diverse group of microbes with a variety of morphologic and genetic adaptations and extreme and novel biological characteristics. Viruses are assigned to one of 11 families on the basis of their envelope, capsid, and the nature of their RNA genome (table 21.1). RNA viruses are etiologic agents in a number of serious and prevalent human diseases, including influenza, AIDS, polio, measles, and rabies.

Enveloped Segmented Single-Stranded RNA Viruses

The Biology of Orthomyxoviruses: Influenza

Orthomyxoviruses are generally spherical particles with an average diameter of 80–120 nm. They are covered with a lipoprotein envelope that is studded with glycoprotein spikes acquired during viral maturation (figure 21.1*a*; see figure 5.19). The two glycoproteins that comprise the spikes of the envelope and contribute to virulence are hemagglutinin (HA) and neuraminidase (NA). Because **hemagglutinin** spikes combine with a specific carbohydrate molecule found in all eucaryotic cell membranes, the virus has the capacity to bind and clump a variety of animal cells. The name hemagglutinin is derived from HA's particular agglutinating action on red blood cells (see figure 13.27). The property of erythrocyte agglutination is the basis for viral assays that have been used to identify several antigenic types. The actual function of HA is to bind to host cell receptors of the respiratory mucosa, thereby facilitating viral adsorption and penetration. The principal activity of **neuraminidase** is to hydrolyze the protective mucous coating of the respiratory tract, assist in viral budding and release, prevent viruses from sticking together, and participate in host cell fusion.

A unique property of the influenza virus genome is its division into eight separate segments capable of reassorting during viral assembly and thus changing the antigenic nature of the proteins for which they code. These and other genetic variations are partly responsible for the *antigenic drift* and *antigenic shift* that occur with the influenza A virus. Antigenic drift is a small change in the antigenicity, while antigenic shift is the accumulation of enough changes to create a new subtype different from the original strain (see feature 21.1).

Epidemiology of Influenza A

Influenza is an acute, highly contagious respiratory illness afflicting people of all ages and marked by seasonal regularity and pandemics at predictable intervals. So convinced were the ancients that this disease was influenced by a celestial alignment of the stars or by some terrestrial power, that one of its first names was an Italian derivative of *un influenza di freddo,* alluding to a vague and mysterious influence. This name has remained and is now contracted to the vernacular "flu." Periodic outbreaks of respiratory illness that were probably influenza have been noted throughout recorded history. Of the five pandemics occurring in 1900, 1918, 1946, 1957, and 1968, the outbreak of 1918 took the greatest toll—at least 50 million deaths. Even today, deaths from influenza or influenzal pneumonia rank among the top 10 causes of death in the United States. Of the three types of influenza viruses, the more virulent type A is most commonly associated with human disease.

Mode of Influenza Transmission

Inhalation of virus-laden aerosols and droplets constitutes the major route of influenza infection, although fomites may play a secondary role. Young children, notorious for their nonchalant hygiene, are the major perpetrators and victims of influenza spread. Transmission is greatly facilitated by crowding and poor ventilation in classrooms, barracks, nursing homes, and military installations, in the late fall and winter. In the tropics, infection occurs throughout the year, without a seasonal distribution. Occupational contact with ducks, other poultry, and swine constitutes a special high-risk category for disease. The overall mortality rate for influenza A is about 0.1% of cases. Deaths are most common among elderly persons and small children.

Infection and Disease

The influenza virus binds primarily to ciliated columnar cells of the respiratory mucosa (figure 21.1). Infection causes the rapid shedding of these cells along with a load of viruses. About one-half of infections remain asymptomatic, while the remainder produce symptoms ranging from inconsequential to lethal. Stripping the trachea and bronchial epithelia to the basal layer eliminates protective ciliary clearance and leads to severe inflammation and irritation. The illness is further aggravated by fever, headache, myalgia, pharyngeal pain, substernal pain, shortness of breath (due to reduced gas exchange), and repetitive bouts of coughing. The viruses tend to remain in the respiratory tract, and viremia is rare. As the normal ciliated columnar epithelium is restored in a week or two, the symptoms usually abate.

Recovery from influenza is related to general health, immunity from previous exposure, vaccination, and age. Patients with emphysema or cardiopulmonary disease, and those who are very young, elderly, or pregnant, are more susceptible to serious complications. Most often, weakened host defenses and immune function predispose patients to secondary bacterial infections, especially pneumonia caused by *Streptococcus pneumoniae* and *Staphylococcus aureus.* Although less common than bacterial pneumonia, primary influenzal virus pneumonia has the ominous reputation for causing rapid death of persons even in their prime. Other complications are bronchitis, myocarditis, meningitis, and neuritis.

Diagnosis, Treatment, and Prevention of Influenza

Except during epidemics, influenza must be differentiated from other acute respiratory diseases. An initial Gram stain of sputum can often differentiate between viral or bacterial infection and thereby form a basis for therapy. If rapid, more definitive identification is necessary, immunofluorescence tests are available to detect influenza antigen in a pharyngeal specimen. Other specific diagnostic methods occasionally used are serological testing for a rising antibody titer and virus isolation in chick embryos or kidney cell culture.

Table 21.1 RNA Virus Families

Enveloped

Segmented, Single-Stranded, Negative-Sense Genome*

Orthomyxoviridae: Influenza
Bunyaviridae: Viral fevers
Arenaviridae: Hemorrhagic fevers

Nonsegmented, Single-Stranded, Negative Sense

Paramyxoviridae: Mumps, measles

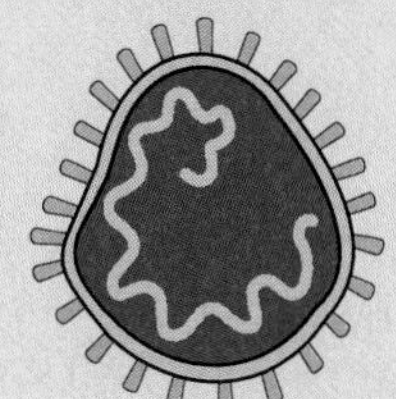

Rhabdoviridae: Rabies

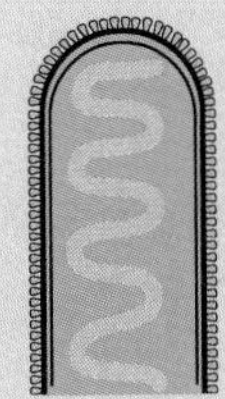

Nonsegmented, Single-Stranded, Positive Sense

Togaviridae: Rubella, arboviral fevers

Coronaviridae: Common cold

Single-Stranded, Positive Sense, Reverse Transcriptase

Retroviridae: AIDS, T-cell leukemia

Nonenveloped

Nonsegmented, Single-Stranded, Positive-Sense Genome

Picornaviridae: Polio, hepatitis A

Caliciviridae: Norwalk agent

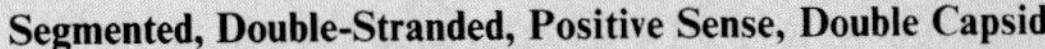

Segmented, Double-Stranded, Positive Sense, Double Capsid

Reoviridae: Tick fever, infant diarrhea

Legend

Nucleocapsid type:

Helical

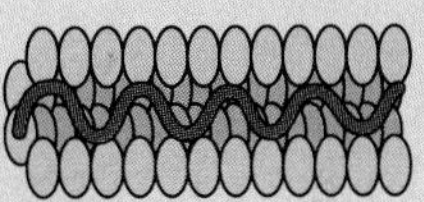

Icosahedral

Genome:

Strandedness

Single (SS)

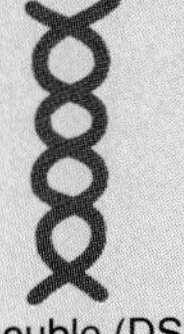

Double (DS)

Number of molecules

Segmented

Nonsegmented

Sense

(+)

(–)

*If the RNA of the virus comes ready to be translated by the host's machinery, it is considered a positive-sense genome, and if it is not directly translatable by the host, it is a negative-sense genome.

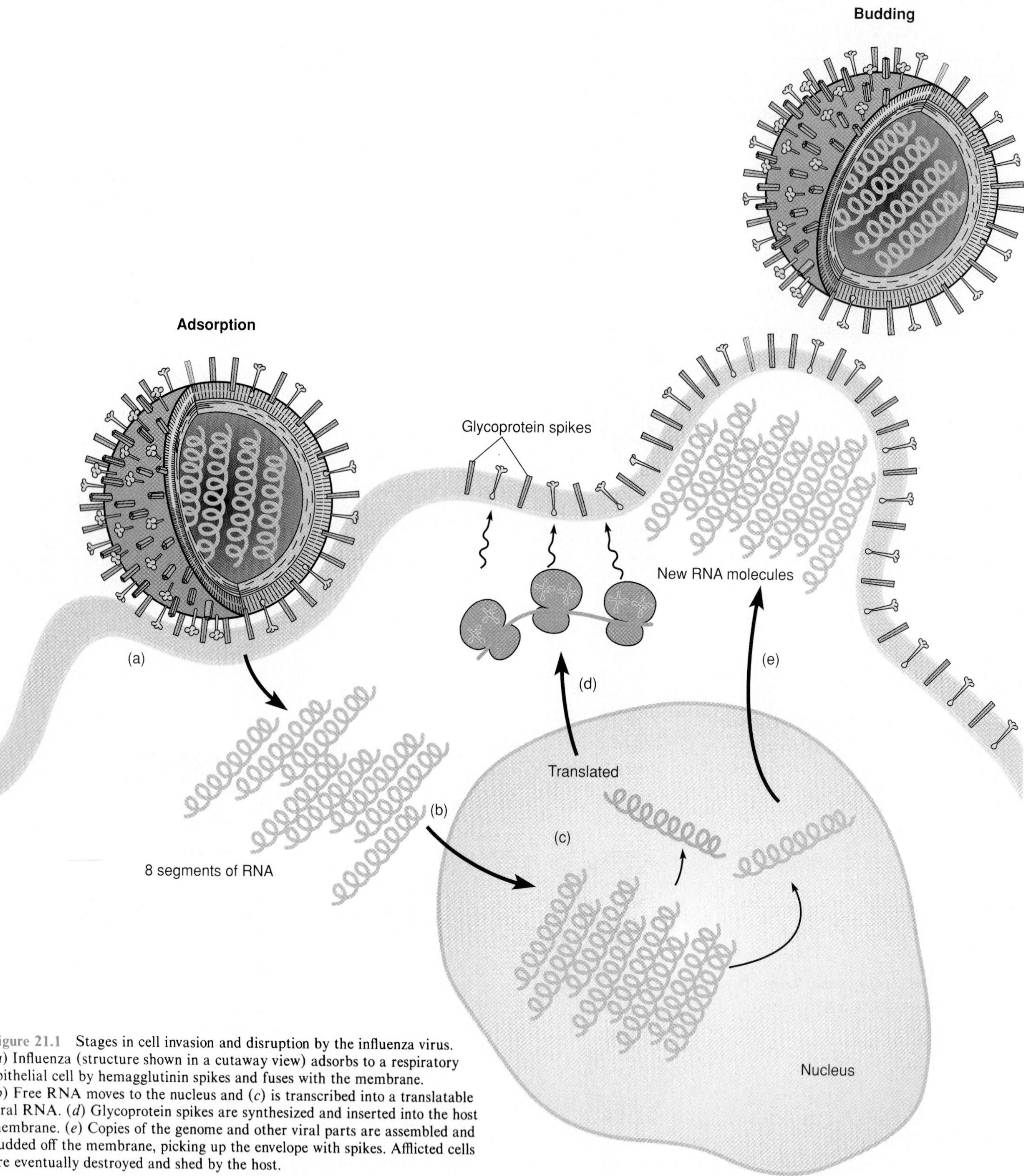

Figure 21.1 Stages in cell invasion and disruption by the influenza virus. (*a*) Influenza (structure shown in a cutaway view) adsorbs to a respiratory epithelial cell by hemagglutinin spikes and fuses with the membrane. (*b*) Free RNA moves to the nucleus and (*c*) is transcribed into a translatable viral RNA. (*d*) Glycoprotein spikes are synthesized and inserted into the host membrane. (*e*) Copies of the genome and other viral parts are assembled and budded off the membrane, picking up the envelope with spikes. Afflicted cells are eventually destroyed and shed by the host.

Feature 21.1 Community Immunity, Genetic Change, and Influenza

Every fall, the word goes out to expect outbreaks of certain types of flu with exotic names such as Hong Kong, swine, or Russian. How can medical experts predict which types of flu to expect in a given year? In order to understand these phenomena, we must first examine how changes in the influenza virus influence its epidemiologic pattern. Four factors in this rather complex interrelationship are (1) the degree of community resistance, (2) viral antigenic drift and shift, (3) animal coinfections, and (4) the success of mutant strains due to natural selection.

Community Immunity The concept of community or herd immunity was introduced in chapter 13. It is defined as the collective resistance of individuals in a population, acquired by active infection, vaccination, and transplacental antibody transfer. Using the logic of herd immunity, the community will be most resistant to a subtype of influenza virus that has most recently passed through it, and it will be most susceptible to a newly developed subtype or one imported from another population or region.

Antigenic Drift and Subtypes The influenza viruses undergo constant genetic variations that challenge the host population with an ever-changing spectrum of viral antigens. Examples of important viral components that are subject to change are the hemagglutinin and neuraminidase spikes that determine infectivity and spread and induce protective antibodies. When shifts result in new viral variants that are so antigenically different that antibodies from a prior infection are no longer protective, there is a lapse in individual and herd immunity. The development of such a new strain through accumulated genetic change may require about 10 years.

Antigenic Shift and Coinfections with Influenza Virus An event with even greater impact takes place when two different influenza virus strains simultaneously infect the same animal. Although the influenza A viruses are common in humans, they have also been found in swine and birds. It is possible for humans to become infected with animal strains and vice versa. Occasionally, coinfection of the same animal by two antigenic variants gives rise to genetically new viral strains quite different from either original. If this totally altered strain is introduced into either population (animal or human), a potential exists for rapid dissemination due to lack of protective antibodies. Experts suspect that the pandemics of 1957, 1968, and 1977, which began in China, arose from a strain developed through coinfections among human, duck, and swine populations.

To characterize each virus strain, virologists have developed a special nomenclature that chronicles the influenza virus type, the presumed animal of origin, the location, and the year of origin. For example, A/duck/Ukraine/63 identifies the strain isolated from domestic or migratory fowl; likewise, A/seal/Mass/1/80 is the causative agent of seal influenza. This terminology may be complicated by the ambiguous origin of some antigenic variants and the appearance of identical strains in widely separated geographic locations. Nevertheless, from this convention come such expressions as "swine flu" and "Hong Kong flu."

Recurrence of Pandemics Influenza pandemics are accelerated by the contagiousness of the virus and widespread migrations or travel, and they are cyclic, with episodes spaced about a decade apart. As we have seen, this synchronization is an expression of declining population immunity and the accumulation of genetic variations in the virus. By compiling extensive serological profiles of populations and the influenza strains to which they are immune, the CDC has a scientific basis for predicting expected epidemic years and the probable strains involved. One advantage of forecasting the expected viral subtypes is that the most vulnerable people can be vaccinated ahead of time with the appropriate virus subtype(s).

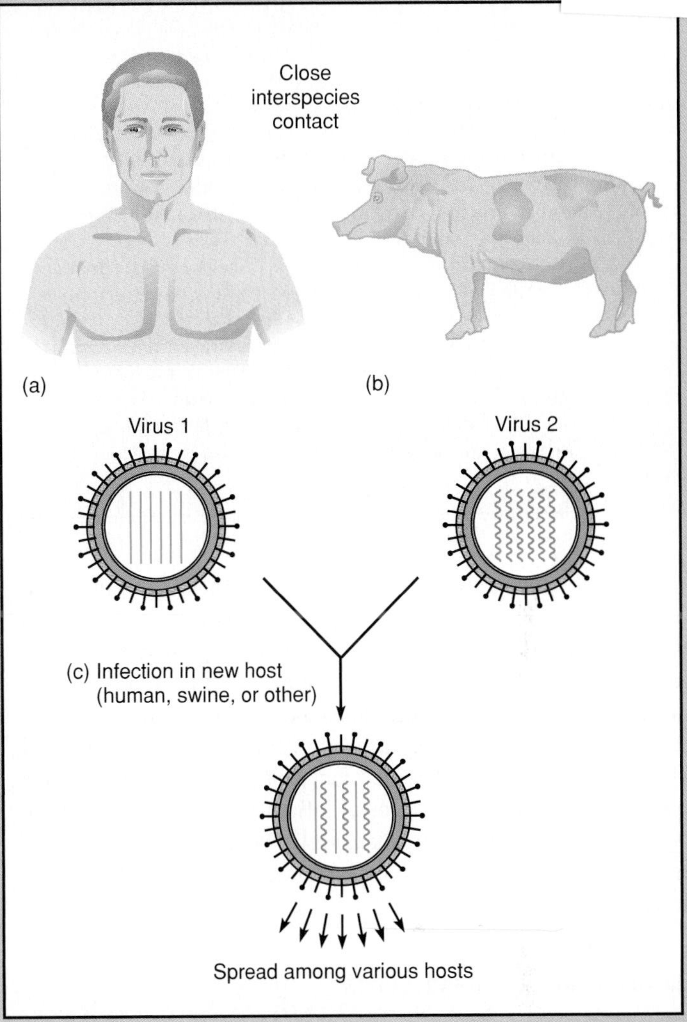

Close interspecies contact
(a)
(b)
Virus 1
Virus 2
(c) Infection in new host (human, swine, or other)
Spread among various hosts

Coinfection of human and pig strains in a single animal. During viral multiplication, the 8-segmented genome of virus 1 may reassort with the 8-segmented genome of virus 2. Theoretically, recombination could produce 256 different genome combinations. Only one is shown here.

In most cases, treatment requires little more than measures to alleviate symptoms. Fluids, bed rest, and nonaspirin drugs (see feature 21.2) to control fever, pain, and inflammation are helpful. Amantadine, which specifically blocks the uncoating of the influenza virus in the cell, is sometimes prescribed to abate the course of influenza, especially in more severe cases. It reduces not only the length of the disease and its symptoms but also the spread of the virus. Amantadine is given orally, though recent studies have shown that aerosol sprays may be beneficial as well. Because the drug shows some toxicity for the CNS, it must be used with caution.

Given the lack of specific therapy and the epidemic potential of influenza, the need for effective prophylaxis is great. Influenza vaccination is not a routine procedure and is recommended primarily for certain high-risk groups, such as the chronically ill, the elderly, and persons with a high degree of

Feature 21.2 Influenza and Long-Term Medical Complications

Two major sequelae are associated with influenza virus or its vaccines. One, **Guillain-Barré syndrome,** is a neurological complication that can arise in a small number of vaccine recipients (1 in 100,000). This syndrome appears to be an autoimmunity induced by viral proteins and marked by varying degrees of demyelination of the peripheral nervous system, leading to weakness and sensory loss. Most patients recover function, but the disease can also be debilitating and fatal. Another deadly neurological complication was associated with the 1970s swine flu vaccinations that killed many recipients. It is now known that such severe reactions are correlated almost exclusively with the H_{swine} antigen, and other vaccines are far safer.

For reasons that are still unknown, a small number of children and young adults infected with influenza virus develop **Reye's syndrome,** a disease that strikes the brain, liver, and kidney and is characterized by the fatty degeneration of these organs. The symptoms are acute and include vomiting, headache, neurological abnormalities, and coma. Fatalities are common. Although the pathological mechanisms underlying the disease are not understood, children given aspirin therapy have a significantly higher risk for Reye's syndrome. The same pathology occurs with other infections such as chickenpox.

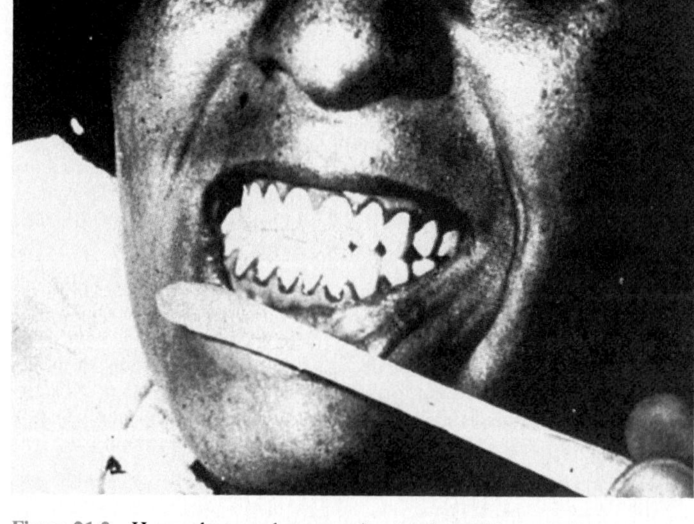

Figure 21.2 Hemorrhage and purpura in a patient with Argentine hemorrhagic fever.

exposure to the public. The standard vaccines, containing inactivated viruses grown in embryonated eggs, have an overall effectiveness of 70%, but they may be accompanied by side effects of chills and fever. Newer subunit vaccines made by disrupting the virus with detergents or solvents produce fewer adverse reactions, but they are more limited in their scope of protection. For complete protection, it is necessary to immunize against all the different strains of influenza viruses occurring in a given year. Much testing is currently being done on an attenuated virus to be administered intranasally, which has greater potential for immunogenicity.

Bunyaviruses and Arenaviruses

Between them, the **bunyavirus** and **arenavirus** groups contain hundreds of viruses whose normal hosts are arthropods or rodents. Although none of the viruses are natural parasites of humans, they may be transmitted zoonotically, occasionally leading to rampant epidemics in human populations. Bunyaviruses will be discussed in a later section along with other arboviruses because they are transmitted primarily by insects and ticks. Some of the more important bunyavirus diseases are California encephalitis, spread by mosquitos; Rift Valley fever, an African disease transmitted by sand flies; and Korean hemorrhagic fever, transmitted by mice, voles, and rats.

Four major arenavirus diseases are Lassa fever, found in parts of Africa; Argentine hemorrhagic fever, typified by severe weakening of the vascular bed, hemorrhage, and shock; Bolivian hemorrhagic fever; and lymphocytic choriomeningitis, a widely distributed infection of the brain and meninges. The viruses are closely associated with a rodent host and are continuously shed by the rodent throughout its lifetime. Transmission of the virus to humans is through aerosols and direct contact with the animal or its excreta. These diseases vary in symptomology and severity from mild fever and malaise to complications such as hemorrhage, renal failure, cardiac damage, and shock syndrome (figure 21.2). The fatality rate in Lassa fever is 25%. So dangerous are these viruses to laboratory workers that the highest level of containment procedures and sterile techniques must be used in handling them.

bunyavirus (bun′-yah-vy″-rus) From Bunyamwera, an area of Africa where this type of virus was first isolated.

arenavirus (ah-ree′-nah-vy″-rus) L. *arena,* sand. In reference to the appearance of virus particles in micrographs.

Enveloped Nonsegmented RNA Viruses

Paramyxoviruses

The important human paramyxoviruses are ***Paramyxovirus*** (parainfluenza and mumps viruses), ***Morbillivirus*** (measles virus), and ***Pneumovirus*** (respiratory syncytial virus), all of which are readily transmitted through respiratory droplets. The envelope of a paramyxovirus possesses HN spikes and F glycoprotein spikes that allow it to infect neighboring cells by a novel mechanism. First the cell membrane of an infected cell is modified by insertion of the spikes. The HN spikes immediately bind an uninfected neighboring cell, and in the presence of F spikes, the two cells permanently fuse. A chain reaction of multiple cell fusions produces a *syncytium* or **multinucleate giant cell** with cytoplasmic inclusion bodies, a diagnostically useful cytopathic effect (figure 21.3).

syncytium (sin-sish′-yum) Gr. *syn,* together, and *kytos,* cell.

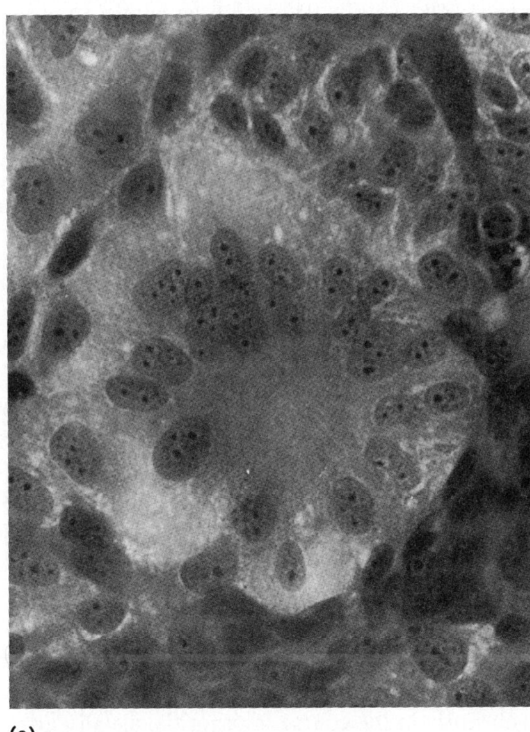

(a)

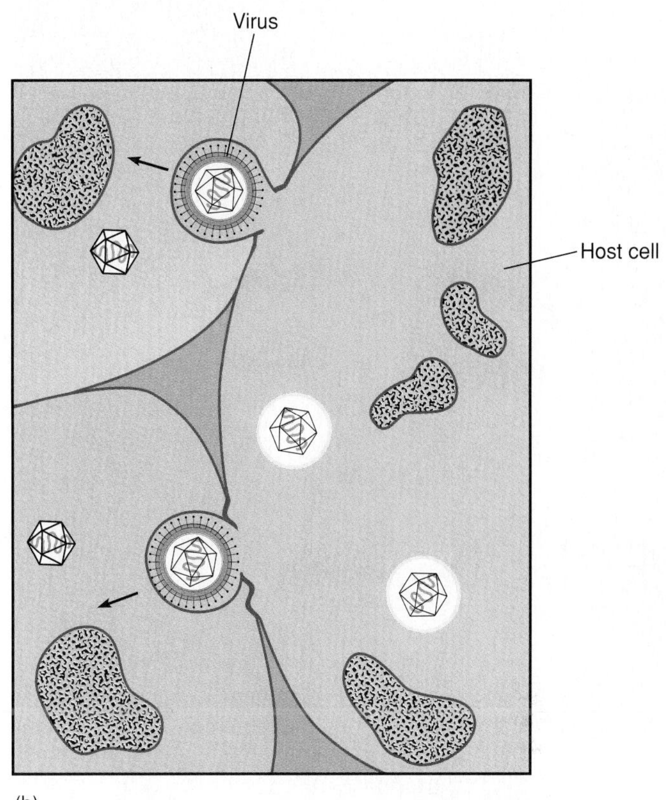

(b)

Figure 21.3 The effects of paramyxoviruses. (*a*) When they infect a host cell, paramyxoviruses induce the cell membranes of adjacent cells to fuse into large, multinucleate giant cells or syncytia. (*b*) This allows direct passage of viruses from an infected cell to uninfected cells by communicating membranes. Through this means, the virus evades antibodies.

Epidemiology and Pathology of Parainfluenza

One type of infection by *Paramyxovirus,* called **parainfluenza,** is as widespread as influenza but usually more benign. Primary infection with parainfluenza virus occurs throughout the year, with seasonal peaks in late fall and winter. Droplets and respiratory secretions effectively disseminate the virus into the air and onto fomites. Inhalation or inoculation of the mucous membranes by contaminated hands are the usual routes of transmission. Parainfluenzal respiratory disease is seen most frequently in children, most of whom have been infected by the age of six. Because protective transplacental immunity is frequently missing, babies in their first year are particularly susceptible, and they develop more severe symptoms. Characteristic symptoms of parainfluenza are minor upper respiratory disease (a cold), bronchitis, bronchopneumonia, and laryngotracheobronchitis (croup). Croup usually afflicts the larynx of infants and young children, causing labored and noisy breathing accompanied by a hoarse cough.

Diagnosis, Treatment, and Prevention of Parainfluenza Presenting symptoms typical of colds are often sufficient to presume respiratory infection of viral origin. Determining the actual viral agent, however, is difficult and usually unnecessary. Whereas in older children and adults infection is usually self-limited and benign, primary infection in infants may be severe enough to be life-threatening. So far no specific chemotherapy is available, but supportive treatment with immune serum globulin or interferon has shown some success. Vaccination of neonates and children has been attempted, but the immunogens used failed to provide enduring protection.

Mumps: Epidemic Parotitis

Another infection caused by *Paramyxovirus* is **mumps** (Old English for lump or bump). So distinctive are the pathologic features of this disease that Hippocrates clearly characterized it several hundred years B.C. as a self-limited, mildly epidemic illness associated with painful swelling at the angle of the jaw (figure 21.4) and occasionally swelling of the testes. The first syndrome, also called **epidemic parotitis,** typically targets the parotid salivary glands, but it is not limited to this region. The mumps virus bears morphologic and antigenic characteristics similar to the parainfluenza virus, and has only a single serological type.

Epidemiology and Pathology of Mumps Humans are the exclusive natural hosts for the mumps virus. Infection occurs worldwide, with epidemic increases in the late winter and early spring in temperate climates. High rates of infection arise among crowded populations or communities with poor herd immunity. Most cases occur in children under the age of 15, and as many

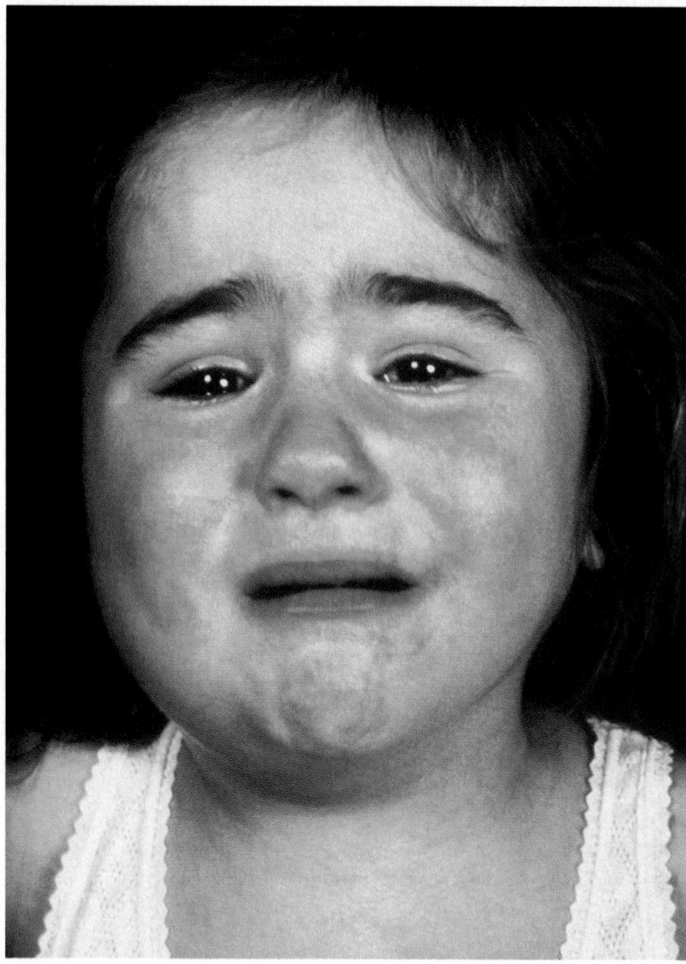

Figure 21.4 The external appearance of swollen parotid glands in mumps (parotitis). Usually both sides are affected, though parotitis affecting one side only occasionally develops.

as 40% are subclinical. Because lasting immunity follows any form of mumps infection, no long-term carrier reservoir exists in the population. The virus is communicated primarily through salivary and respiratory secretions, though it is less contagious than measles or varicella viruses.

After an average incubation period of two to three weeks, symptoms of fever, nasal discharge, muscle pain, and malaise develop. These may be followed by inflammation of the salivary glands (especially the parotids), producing the classic gopher-like swelling of the cheeks on one or both sides (figure 21.4). This swelling of the gland and its duct induces quite a bit of discomfort, especially when the gland is stimulated. Viral multiplication in salivary glands is followed by viremia and invasion of other organs. Such diverse structures as the testes, ovaries, thyroid gland, pancreas, meninges, heart, and kidney can be involved. Despite the invasion of multiple organs, the prognosis for most infections is complete, uncomplicated recovery.

Some Myths About Mumps The popular belief that mumps will cause sterilization of the postpubertal male is still held, despite medical evidence to the contrary. Perhaps this notion has been reinforced by the excruciating pain experienced during testicular infection, by the tenderness that continues afterward, and by the partial atrophy of a testis that occurs in about half the cases. None of these conditions has ever been proved to result in permanent impotence or sterility.

Another frequent assertion is that an individual can develop more than one case of mumps or that, if the swelling of the parotid gland is unilateral, a child can become infected later on the other side. Any case of mumps, even subclinical, leads to permanent immunity. More than a single case of mumps developing in the same person has never been verified, though parotitis arising from a different virus can occur.

Complications in Mumps In 20% to 30% of young adult males, mumps infection localizes in the epididymis and testis, usually on one side only. The resultant syndrome of orchitis and epididymitis may be rather painful, but no permanent damage occurs. In mumps pancreatitis, the virus replicates in beta cells and pancreatic epithelial cells. Viral meningitis, characterized by fever, headache, nausea, vomiting, and stiff neck, is rather common in mumps. It appears 2 to 10 days after the onset of parotitis, lasts for 3 to 5 days, and then dissipates, leaving few or no adverse side effects. On rare occasions, loss of hearing develops abruptly in the course of parotitis, usually affecting only one side. Unfortunately, because the virus replicates in the organ of Corti and causes inflammation and irreversible injury, deafness is usually permanent.

Diagnosis, Treatment, and Prevention of Mumps Mumps can be tentatively diagnosed in a child with swollen parotid glands and a known exposure two or three weeks previously. Because parotitis is not always present and the incubation period may range from 7 to 23 days, a practical diagnostic alternative is to demonstrate a rise in serum antibody titer to mumps virus.

The general pathology of mumps is mild enough that symptomatic treatment to relieve fever, dehydration, and pain is usually adequate. A live, attenuated mumps vaccine given routinely as part of the MMR vaccine at 15 months of age is a powerful and effective control agent. A separate single vaccine is available for adults who require protection. Although the antibody titer is lower than that produced by wild mumps virus, protection often lasts a decade.

Measles: *Morbillivirus* Infection

Measles, an acute disease caused by ***Morbillivirus,*** is also known as **red measles** and **rubeola.** Unfortunately, German measles (rubella), an entirely unrelated viral infection, is also called rubeola in some countries. Some differentiating criteria for these two forms of measles are summarized in table 21.2. Despite a

measles (mee'-zlz) Dutch *maselen*, spotted.
rubeola (roo-bee'-oh-lah) L. *ruber*, red.

Table 21.2 The Two Forms of Measles

	Synonyms	Etiology	Primary Patient	Complications	Skin Rash	Koplik's Spots
Measles	Rubeola, red measles	Paramyxovirus: *Morbillivirus*	Child	SSPE*, pneumonia	Present	Present
German Measles	Rubella, 3-day measles, rubeola	Togavirus: *Rubivirus*	Child/fetus	Fetal deformity	Present	Absent

*See feature 21.4.

tendency to think of measles as a mild childhood illness, a recent epidemic has reminded us that measles can kill babies and young children and is the seventh most frequent cause of death worldwide.

Epidemiology of Measles Measles is one of the most contagious infectious diseases, transmitted principally by direct contact with respiratory aerosols. Epidemic spread is favored by crowding, low levels of herd immunity, a prevalence of nonimmune children, malnutrition, and inadequate medical care. The epidemic of measles that began in 1986 has been linked to the lack of immunization in children or the failure of a single dose of vaccine in many children (see feature 21.3). There is no reservoir other than humans, and although a person is infectious during the periods of incubation, prodromium, and skin rash, the virus is not carried for any length of time during convalescence. Only relatively large, dense populations of susceptible individuals can sustain a continuous chain of infection.

Infection and Disease Infection begins with the invasion of the mucosal lining of the respiratory tract, followed by viremia. The incubation period of nearly two weeks ends with symptoms of sore throat, dry cough, headache, conjunctivitis, lymphadenitis, and fever. In a short time, unusual oral lesions called *Koplik's spots* appear as a prelude to the characteristic red maculopapular *exanthem* that erupts on the head and then progresses to the trunk and extremities, until most of the body is covered (figure 21.5). The rash gradually coalesces into red patches that fade to brown.

Complications of Measles Although in most instances, measles is a self-limited infection, complications may be sufficiently severe to cause death in about 1 in 500 children. Among the complications are laryngitis, bronchopneumonia, bronchitis, pneumonitis, and bacterial secondary infections such as otitis media and sinusitis. Children afflicted with leukemia or thymic deficiency are especially predisposed to pneumonia due to their lack of the natural T-cell defense. Undernourished children frequently have severe diarrhea and abdominal discomfort that adds to their debilitation. The gravest complications involve the central nervous system (see feature 21.4).

Diagnosis, Treatment, and Prevention of Measles The patient's age, a history of recent exposure to measles, and the season of the year are all useful epidemiologic clues for diagnosis. Clinical characteristics such as dry cough, sore throat, conjunctivitis, lymphadenopathy, fever, and especially the appearance of Koplik's spots and a rash are presumptive of measles.

Feature 21.3 Vaccination and the Global Eradication of Measles

Measles vaccination in the United States was at one time so successful that public health authorities envisioned 1982 as a target date for total elimination of measles. Unfortunately, things did not work out that way. After a record low in 1983, there was a sudden resurgence in 1986 that reverted to the case levels of the previous decade. Measles specialists attribute the increases to the breakdown of public vaccination programs and the lack of vaccination in new Southeast Asian and Latin American immigrants. As many as one-third of all American babies are not vaccinated at the most critical times.

Given the elusiveness of measles control even in the United States where vaccination is available to all, how can the disease be exterminated in every corner of the world? One problem is that in less developed countries measles is sometimes viewed as merely a childhood inconvenience or "rite of passage," and this belief may be attended by wariness of meddling with Mother Nature. Other more practical problems involve the proper time to administer the vaccine, how to vaccinate large numbers of people, and the cost of vaccination programs. Because many deaths occur in children before one year of age, it would seem important to vaccinate early in countries where measles persists. Although early residual maternal antibodies can prevent vaccine from taking, vaccinating children at 15 months may be too late to adequately protect them.

Medical teams responsible for vaccination campaigns aim to vaccinate 75% of the world's susceptible children as early as possible, but for this, community support is imperative. Although the vaccine itself is inexpensive ($0.50 per dose), administering it requires time and money. In a few countries, government-supported enticements such as food giveaway programs have helped promote vaccination, and parents are being educated concerning the more severe short- and long-term consequences of measles so that they are willing to participate. Only time will tell if these strategies work.

---

exanthem (eg-zan'-thum) Gr. *exanthema,* to bloom or flower. An eruption or rash of the skin.

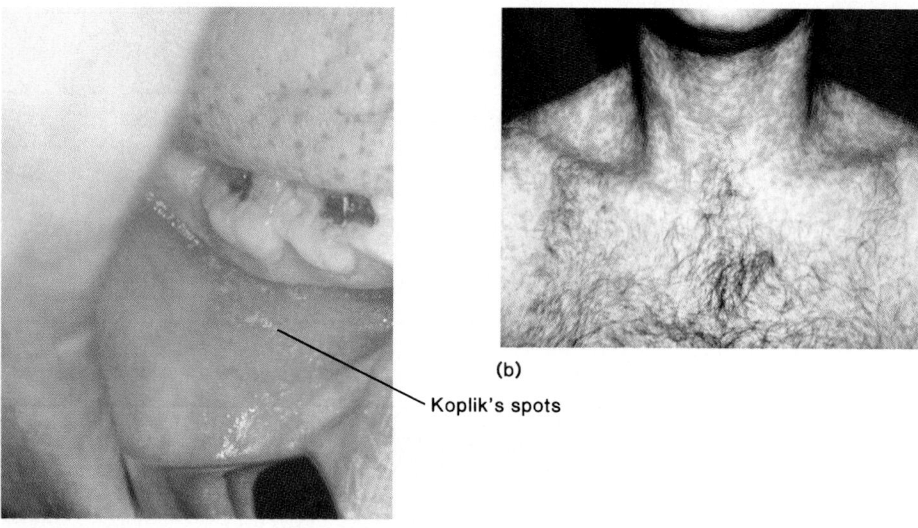

Figure 21.5 Signs and symptoms of measles. (*a*) Koplik's spots, named for the physician who described them in the last century, are tiny white lesions with a red border that form on the buccal mucosa adjacent to the molars just prior to eruption of the skin rash. (*b*) A fully developed measles rash will gradually fade over a week's time.

Feature 21.4 A Rare and Deadly Outgrowth of Measles

Subacute sclerosing panencephalitis (**SSPE**) is a progressive neurologic degeneration of the cerebral cortex, white matter, and brain stem. Its incidence is approximately one case in a million measles infections, and it afflicts primarily male children and adolescents. Its relationship to measles remained unsuspected for three decades because of the prolonged incubation period of SSPE. Eventually, researchers discovered intracellular inclusions in infected brain cells that proved identical to cytological features of measles, and finally they were able to isolate the SSPE agent.

The pathogenesis of SSPE appears to involve a defective virus, one that has lost its ability to form a capsid and be released from an infected cell, though it does continue to spread unchecked through the brain by cell fusion. This sets into motion inflammatory processes that destroy neurons and accessory cells and break down myelin. The disease is known for the profound intellectual and neurological impairment it brings, beginning with such symptoms as decline in memory, judgment, and motor activity. The course of the disease invariably leads to coma and death in a matter of months or years.

subacute sclerosing panencephalitis (sub-uh-kewt' sklair-oh'-sing pan''-en-cef''-uh-ly'-tis) Gr. *skleros,* hard, *pan,* all, and *enkephalos,* brain.

Because specific therapy for primary measles is not available, treatment relies on reducing fever, suppressing cough, and replacing lost fluid. Complications require additional remedies to maintain an open airway, relieve intracranial pressure, control seizures, and sustain nutrient, electrolyte, and fluid levels. Antibiotics may be given for bacterial complications, and large doses of immune globulin may be therapeutic. Vaccination is the most practical, economical, and enduring strategy for combating measles (see feature 21.3). The attenuated viral vaccine is administered by subcutaneous injection and achieves immunity that persists for about 20 years. It must be stressed that since the virus is live, it may cause an atypical infection sometimes accompanied by a rash and fever. Measles immunization is recommended for all healthy children at the age of 15 months (MMR vaccine, with mumps and rubella), and a booster is given prior to entering school. A single antigen vaccine is also available for older patients who require protection against measles alone. As a general rule, anyone who has had the measles is considered protected, but any person who received the vaccine prior to 1980 or whose immunization history is in doubt should be revaccinated.

Respiratory Syncytial Virus: RSV Infections

As its name indicates, **respiratory syncytial virus** (**RSV**), also called *Pneumovirus,* infects the respiratory tract and produces giant multinucleate cells. Outbreaks of droplet-spread RSV disease occur regularly throughout the world, with peak incidence in the winter and early spring. Children six months of age or younger are especially susceptible to serious disease of the respiratory tract. The chance of a youngster being hospitalized for RSV disease before reaching one year of age is estimated to be 5 per 1,000 live births, making RSV the most prevalent cause of respiratory infection in this age group. The infection spreads rapidly among babies in neonatal and nursery units. The mortality rate is highest for children who have complications like prematurity, congenital disease, and immunodeficiency. Because immunity is only partial or transient, infection is usually recurrent and manifests itself in older children and adults as a common cold.

The epithelia of the nose and eye are the principal portals of entry, and the nasopharynx is the main site of RSV replication. The first symptoms of primary infection are fever that lasts for three days, rhinitis, pharyngitis, and otitis. Infection of the bronchial tree and lung parenchyma gives rise to symptoms of *croup* that include acute bouts of coughing, wheezing,

croup (kroop) Scot. *kropan,* to cry aloud.

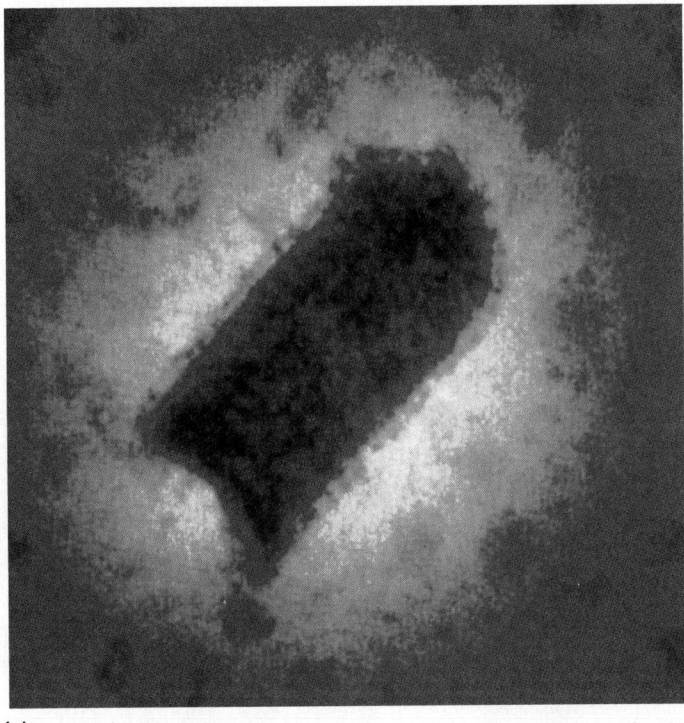

(a)

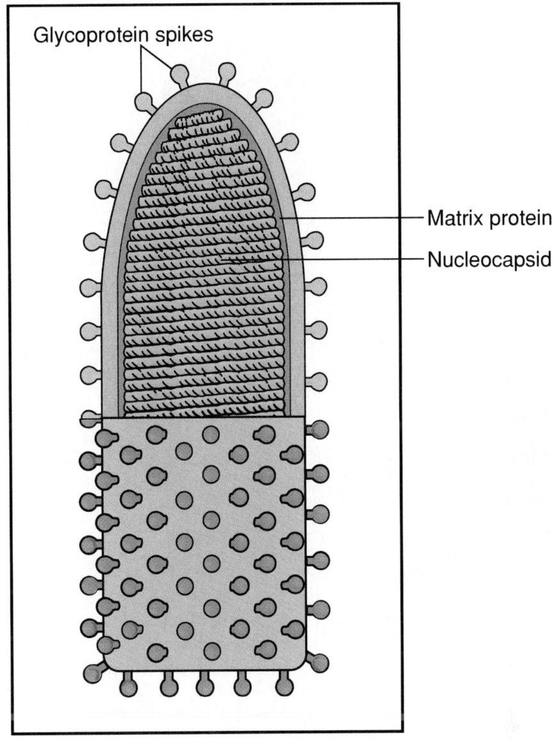

(b)

Figure 21.6 The structure of the rabies virus. (*a*) Color-enhanced virion shows prominent serrations in the core, which represent the tightly coiled nucleocapsid. (*b*) A schematic model of the virus, showing its major features.

dyspnea, and abnormal breathing sounds (rales). Compromised breathing may lead to fatalities. Adults and older children with RSV infection may experience cough and nasal congestion but are frequently asymptomatic.

Diagnosis, Treatment, and Prevention of RSV Diagnosis of RSV infection is more critical in babies than in older children or adults. The afflicted child is conspicuously ill, with signs typical of bronchiolitis and pneumonia. In order to begin the correct treatment, a positive identification of RSV must be made as rapidly as possible. The best diagnostic procedures are those that demonstrate the viral antigen directly from specimens. The sample must be fresh (the virus is very fragile), and it must contain sufficient infected cells. Direct and indirect fluorescent staining, ELISA testing, and DNA probes are useful methods for evaluating a specimen's viral content.

One treatment that appears to be beneficial is ribavirin (virazole), a drug that interferes with the replication of the virus in infected cells. It is administered as an aerosol (inhaled), so it cannot be given to persons whose ventilation is compromised. This relatively new treatment is very costly and does have some adverse side effects. Other supportive measures include drugs to reduce fever, assisting pulmonary ventilation, providing adequate nutrition, and treating secondary bacterial infection if present. In view of the potentially life-threatening nature of RSV infection in youngsters, vaccination seems an attractive alternative. Technologists are working on a live vaccine to be administered intranasally in order to promote formation of antibodies on the mucosal surface.

Rhabdoviruses

The most conspicuous *rhabdovirus* is the **rabies** virus, genus ***Lyssavirus.*** The particles of this virus have a distinctive bullet-like appearance, being round on one end and flat on the other. Additional features are a helical nucleocapsid and spikes that protrude through the envelope (figure 21.6). The family contains approximately 60 different viruses, but only the rabies virus and rarely, some other mammalian lyssaviruses, affect humans.

Epidemiology of Rabies

Rabies is a slow, progressive zoonotic disease characterized by a fatal meningoencephalitis. It is distributed nearly worldwide

dyspnea (dysp'-nee-ah) Gr. *dyspnoia,* difficulty in breathing.

rhabdovirus (rab'-doh-vy''-rus) Gr. *rhabdos,* rod. In reference to its bullet or bacillary form.

rabies (ray'-beez) L. *rabidus,* rage or fury.

Lyssavirus (lye'-suh-vy''-rus) Gr. *lyssa,* madness.

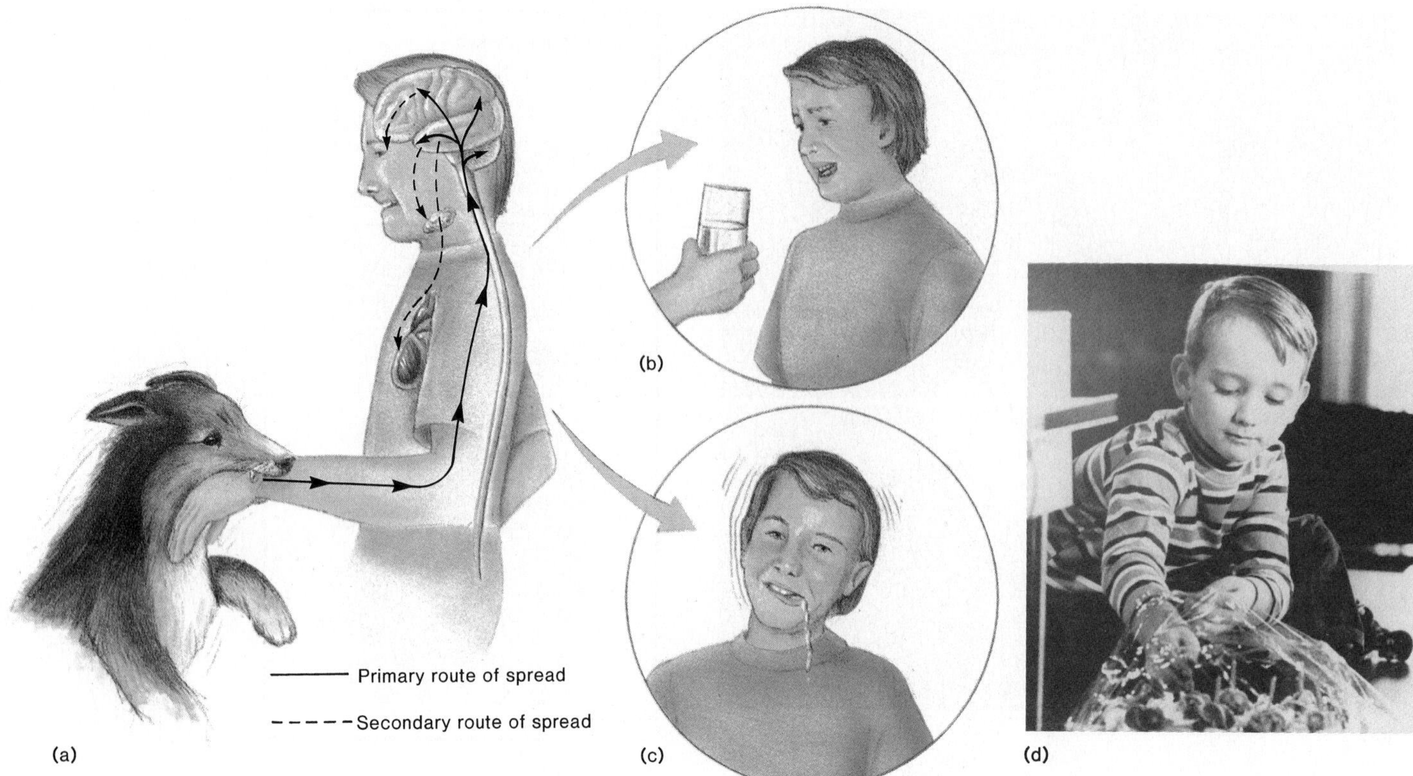

Figure 21.7 A pathological picture of rabies. (*a*) After an animal bite, the virus spreads to the nervous system and multiplies. (*b*) The classic symptom of hydrophobia is elicited by pain in swallowing. (*c*) Severe neurologic symptoms develop, including agitation and twitching. (*d*)A very rare case—a child who survived an attack of rabies.

in various mammals. Carnivores such as skunks, foxes, raccoons, wolves, mongooses, and badgers comprise the major wild reservoirs, and occasionally the virus spreads to dogs and cats. Humans become accidental hosts through dog bites, cat scratches, and contact with sylvan (woodland) animal reservoirs.

Rabies in the World Thirty-four countries, many of which are islands, have remained rabies-free by means of rigorous animal control, quarantine, and isolation. Although the annual number of cases reported worldwide is around 35,000, fewer than 10 per year are reported in the United States, probably due to mandatory rabies vaccination of dogs and other control measures. In some areas, cats are now considered a more likely source of rabies.

Infection and Disease

Infection with rabies virus typically begins when an infected animal's saliva enters a puncture site. Occasionally, the virus is inhaled or inoculated orally. The rabies virus remains up to a week at the trauma site, where it multiplies. The virus gradually enters sensory nerve endings or the neuromuscular junction and advances toward the sensory ganglia, spinal cord, and brain (figure 21.7). Viral multiplication throughout the brain is eventually followed by migration to such diverse sites as the eye, heart, skin, and oral cavity. The infection cycle is completed when the virus replicates in the salivary glands and is shed into the saliva. Clinical rabies proceeds through the identifiable stages of incubation, prodromium, acute neurologic phase (encephalitis), coma, and death. Because rabies may require several months to run its course, it is considered a slow, progressive infection.

Clinical Phases of Rabies

The average incubation period is one to two months, with extremes of a week to more than a year. Its length depends upon the wound site, its severity, and the inoculation dose. The incubation period is shorter in facial, scalp, or neck wounds because of greater proximity to the brain. The prodromal phase begins with fever, anorexia, nausea, vomiting, headache, fatigue, and other nonspecific symptoms. Some patients continue to experience pain, burning, prickling, or tingling sensations at the wound site.

In the form of rabies termed **furious,** the first acute signs of neurological involvement are periods of agitation, disorientation, seizures, and twitching. Spasms in the neck and pharyngeal muscles lead to severe pain upon swallowing. As a result, attempts to swallow or even the sight of liquids bring on *hydrophobia* (fear of water). Throughout this phase, the patient is fully coherent and alert. If afflicted with the **dumb** form of rabies, a patient is not hyperactive but paralyzed, disoriented, and stuporous. Ultimately both forms progress to the coma

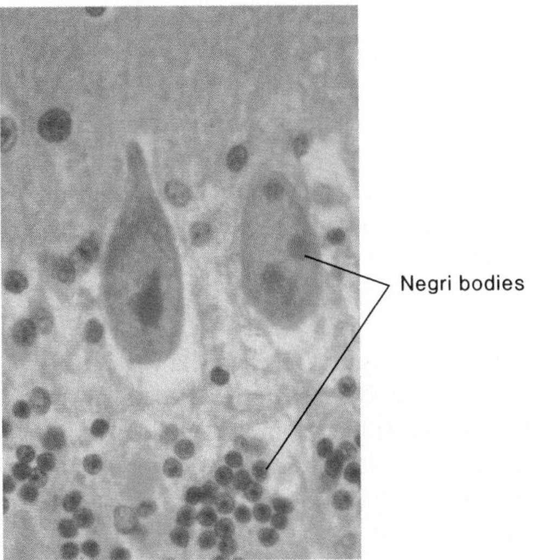

Figure 21.8 Neuronal cells of the cerebellum exhibit distinct intracytoplasmic inclusions called Negri bodies. Though typical of rabies, they are usually not found until after the patient dies, when the brain is examined in an autopsy.

phase, during which death usually occurs from cardiac or respiratory arrest. Up until recently, humans were never known to survive rabies. But three patients have now recovered with minimal aftereffects after receiving intensive, long-term treatment (figure 21.7*d*).

Diagnosis and Management of Rabies

When symptoms appear after a rabid animal attack, the cause-and-effect relationship is vivid, and the disease is readily diagnosed. But the diagnosis can be obscured when contact with an infected animal is not clearly defined or when symptoms are absent or delayed. Anxiety, agitation, and depression may pose as a psychoneurosis; the mental state and pharyngeal spasms resemble tetanus; and encephalitis with convulsions and paralysis mimics a number of other viral infections. Often the disease is diagnosed at autopsy. The laboratory criteria most diagnostic of rabies are intracellular inclusions (Negri bodies) in infected nervous tissue (figure 21.8), isolation and identification of rabies virus in saliva or brain tissue, and demonstration of rabies virus antigens in the brain, serum, or cerebrospinal fluid.

Postexposure Prophylaxis A bite from a wild or stray animal demands prompt action, including assessment of the animal, meticulous care of the wound, and a specific treatment regimen. A wild mammal, especially a skunk, raccoon, fox, or coyote that bites without provocation, is presumed to be rabid, and treatment of the patient is immediately commenced. If the animal is captured, brain samples and other tissue are examined for verification of rabies. The likelihood of a domestic animal becoming rabid is small, so an apparently healthy dog or cat is quarantined 10 days for observation. Preventative therapy is initiated if any signs of rabies appear.

Following an animal bite, the wound should be scrupulously washed with soap or detergent and water, followed by debridement and application of an antiseptic that inactivates the virus such as alcohol or peroxide. Rabies is one of the few infectious diseases for which a combination of passive and active postexposure immunization is therapeutically indicated and successful. Initially the wound is infused with human rabies immune globulin (HRIG) to impede the spread of the virus, and globulin is also injected intramuscularly to provide immediate systemic protection. A full course of human diploid cell vaccine (HDCV) is started simultaneously (see feature 21.5).

Feature 21.5 The Genesis of Rabies Vaccines

The very first attempt to prevent rabies by postexposure vaccination came in 1878 when a boy mauled by a rabid dog was brought to Louis Pasteur for treatment with a vaccine never before tried in humans. Pasteur and others had experimented by infecting rabbit brains with canine rabies virus, harvesting the virus from the rabbits' saliva, and inoculating yet other rabbits. This serial passage repeated a hundred times eventually yielded an attenuated or "fixed" virus. The method of vaccine administration in humans, called the Pasteur treatment, consisted of injecting the patient daily with virus of increasing virulence over a period of two to three weeks. This technique caused many undesirable side effects, such as severe pain at the injection site, allergies, and even neurological damage.

Later replacements for the Pasteur treatment were the phenol-inactivated (Semple) vaccine and the duck embryo vaccine (DEV), prepared from embryonated eggs and given on 21 consecutive days with two boosters. The current vaccine of choice is the **human diploid cell vaccine (HDCV).** This potent inactivated vaccine contains no brain or animal tissue because the virus is cultured in human embryonic fibroblasts. The routine in postexposure vaccination entails intramuscular or intradermal injection on the 1st, 3rd, 7th, 14th, 28th, and 60th days, with two boosters. Although only a few vaccinees develop an adverse reaction such as allergy or local inflammation, the notion of the trauma of rabies vaccination unfortunately persists in the public mind.

Rabies Prevention and Control Measures that effectively prevent and limit rabies are preexposure vaccination and animal control. Three intramuscular injections of HDCV are recommended for high-risk groups such as veterinarians, animal handlers, laboratory personnel, and travelers to endemic areas. Control measures such as vaccination of domestic animals, elimination of strays, and strict quarantine practices have helped reduce the virus reservoir. Several countries have tested a new genetically engineered vaccine made with a vaccinia virus that carries the gene for the rabies virus surface antigen. The vaccine can be incorporated into bait placed in the habitats of wild reservoir species such as skunks and raccoons. It is expected that eating this bait will render the animals immune to rabies.

Other Enveloped RNA Viruses

Coronaviruses

Coronaviruses[1] are relatively large RNA viruses with distinctive, widely spaced spikes on their envelopes. These viruses are

1. Named for the resemblance of the viral spikes to a crown of thorns.

common in domesticated animals and are responsible for epidemic respiratory, enteric, and neurologic diseases in pigs, dogs, cats, and poultry. Thus far, two types of human coronaviruses have been characterized. One of these (HCV) is an etiologic agent of the common cold (see feature 21.13). The same virus is also thought to cause some forms of viral pneumonia and myocarditis. The other human virus is very closely associated with the intestine and may have a role in some human enteric infections. Coronaviruses have also been isolated from the brain, but their role in human neurologic disease is uncertain at this time.

Rubivirus, the Agent of Rubella

Togaviruses are nonsegmented, single-stranded RNA viruses with a loose envelope.[2] There are several important members, including *Rubivirus,* the agent of rubella, and certain arboviruses. **Rubella** or German measles was first recognized as a distinct clinical entity in the mid-eighteenth century, and was considered a benign childhood disease that gave rise to nothing more than mild prodromal symptoms and a brief rash. Then, in 1941, the Australian physician Norman Gregg called attention to the *teratogenic* aspect of rubella virus infection. He discovered that cataracts often developed in neonates born to mothers who had contracted rubella in the first trimester. Epidemiologic investigations over the next generation confirmed the link between rubella and many other congenital defects as well.

Epidemiology of Rubella

Rubella is an endemic disease with worldwide distribution. Infection is initiated much like the paramyxoviruses through contact with respiratory secretions and occasionally urine from infected individuals. The virus is shed from the prodromal phase to a week after the rash appears. Because the virus is only moderately contagious, close living conditions are required for its spread. Although epidemics and pandemics of rubella once regularly occurred in six- to nine-year cycles, the introduction of vaccination has essentially stopped this pattern in the United States. Most cases are reported among adolescents and young adults in military training camps, colleges, and summer camps. The greatest concern is that nonimmune women of childbearing age might be caught up in this cycle, raising the prospect of congenital rubella.

Infection and Disease

Two clinical forms of rubella should be distinguished: postnatal infection, which develops in children or adults, and congenital (prenatal) infection of the fetus, which is expressed in the newborn as various types of birth defects.

Postnatal Rubella During an incubation period of two to three weeks, the virus multiplies in the respiratory epithelium, infiltrates local lymphoid tissue, and enters the bloodstream. Despite widespread invasion of organs, half of infections are asymptomatic. Early symptoms include malaise, mild fever, sore throat, and lymphadenopathy. The rash of pink macules and papules first appears on the face and progresses down the trunk and toward the extremities, advancing and resolving in about three days. Adult rubella is often accompanied by joint inflammation and pain rather than a rash. Except for an occasional complication, postnatal rubella is generally mild and produces lasting immunity.

Congenital Rubella Transmission of the rubella virus to a fetus in utero is a serious complication that gives rise to **congenital rubella.** The mother can transmit the virus even if she is asymptomatic. Although the mechanisms causing embryonic and fetal injury are still under study, the damage appears to come from disruption in the development of the germ layers. Depending upon the timing of the intrauterine invasion, the severity of damage varies. It is generally accepted that infection in the first trimester is most likely to induce spontaneous abortion or multiple permanent defects in the newborn such as cardiac anomalies, ocular lesions, deafness (figure 21.9*b*), and mental and physical retardation. Other effects that may not be immediately noticeable are thrombocytopenic purpura,[3] diabetes mellitus, thyroiditis, and encephalitis. Less drastic sequelae that usually resolve in time are anemia, hepatitis, pneumonia, carditis, and bone infection. Because an infant with congenital rubella teems with the virus and its nasal secretions may shed massive numbers of virus for months or even years, congenitally infected asymptomatic neonates are a concern in the hospital nursery.

Diagnosis of Rubella

Rubella mimics other diseases and is often asymptomatic, so it should never be diagnosed on clinical grounds alone. The confirmatory methods of choice are serologic testing and virus isolation systems. Diagnosis of fetal infection by examining cells and amniotic fluid obtained by amniocentesis is a promising technique that awaits further evaluation. Serologic methods such as hemagglutination inhibition and virus neutralization that demonstrate a fourfold or greater rise in titer are proof of rubella. A quick, simple, miniaturized test for antibody assay (the latex-agglutination card) has great appeal as well.

Prevention of Rubella

Postnatal rubella is generally benign and requires only symptomatic treatment. Because no specific therapy for rubella is available, most control efforts are directed at maintaining herd immunity with attenuated rubella virus vaccine. This vaccine is usually given to children in the combined form (the MMR vaccination) once at 15 months of age. The use of rubella vaccine alone varies according to the age and sex of the individual and whether or not pregnancy is anticipated or already in progress (see feature 21.6).

2. Reminiscent of a toga.

3. Hemorrhage into the tissues due to low blood platelet count.

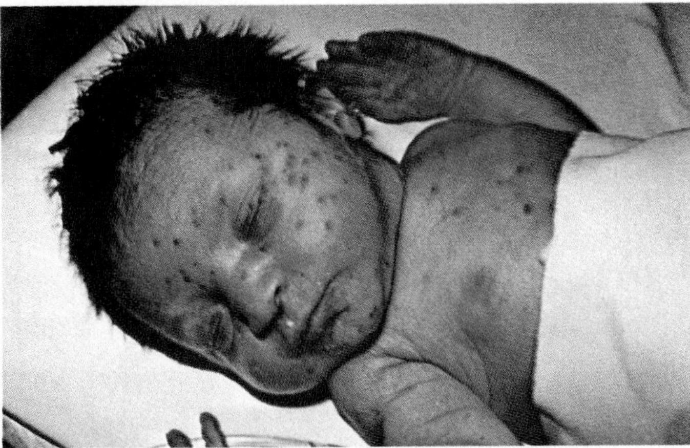

(a)

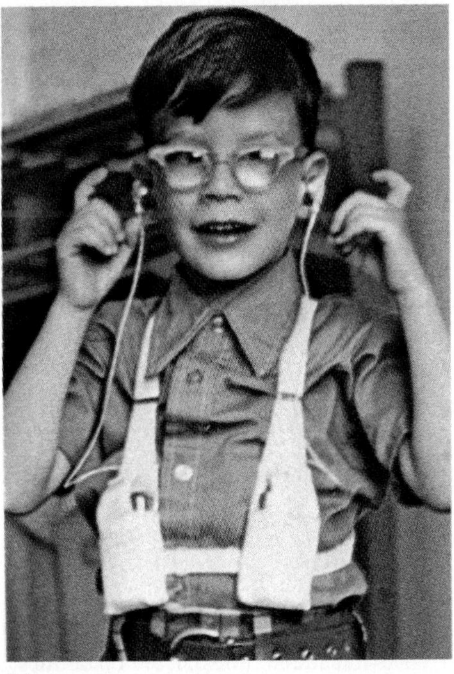

(b)

Figure 21.9 (*a*) An infant born with congenital rubella may manifest a papular and purpurotic rash. (*b*) The sequelae of congenital rubella. Among the permanent defects requiring corrective medical devices are cataracts and deafness. Other children may sustain mental retardation and heart defects.

Feature 21.6 Prevention of Congenital Rubella

Preventing congenital rubella is best accomplished by vaccinating all children at an early age so that girls are already immune by the time of sexual maturity. Unfortunately, the vaccine does not always provide long-term protection, and many children are not vaccinated at all. A second important safety net is the routine testing (mandatory in some states) for rubella antibodies in women prior to marriage. A titer of 1:8 or greater is interpreted as being protective.

The finding of a negative titer in a postpubertal or pregnant woman requires careful evaluation and management. The current recommendation for nonpregnant, antibody-negative women is immediate immunization. Because the vaccine contains live virus and a teratogenic effect is theoretically possible, the vaccine is administered on the condition that the patient uses contraception for three months afterwards. Under no cirumstances is the vaccine to be given to a pregnant woman. The management of verified rubella in a pregnant woman depends upon whether maternal infection occurred early or late. Therapeutic abortion may be indicated if maternal infection occurred in the first trimester and monitoring of the fetus discloses severe damage.

Arboviruses: Viruses Spread by Arthropod Vectors

Vertebrates are hosts to more than four hundred viruses transmitted primarily by arthropods. For simplicity's sake, these viruses are lumped together in a loose grouping called the **arboviruses** (**ar**thropod-**bo**rne **vi**ruses). The major arboviruses pathogenic to humans are togaviruses (*Alphavirus* and *Flavivirus*), some bunyaviruses (*Bunyavirus* and *Phlebovirus*), and reoviruses (*Orbivirus*). The chief vectors are bloodsucking arthropods notorious in their nonselectivity such as mosquitos, ticks, flies, and gnats. Most types of illness caused by these viruses are mild undifferentiated fevers, though some cause severe encephalitides (plural of encephalitis) and life-threatening hemorrhagic fevers. Although these viruses have been assigned taxonomic names, they are more often known by their common names, which are based on geographical location and primary clinical profile (see table 5.3).

Epidemiology of Arbovirus Disease

Wherever there are arthropods there are also arboviruses, so collectively, their distribution is worldwide (figure 21.10). The vectors and viruses tend to be clustered in the tropics and subtropics, but many temperate zones report periodic epidemics. A given arbovirus type may have very restricted distribution, even to a single isolated region, but some range over several continents, and others can spread along with their vectors (see feature 21.7).

The Influence of the Vector As one would expect, the activity and distribution of arboviruses are closely tied to the ecology of the vectors. Factors that weigh most heavily are the longevity of the arthropod, the availability of food and breeding sites, and climatic influences such as temperature and humidity. When vectors take a blood meal to finish their reproductive cycle, they acquire the virus and harbor it for varying periods of time. Infections show a peak incidence when the arthropod is actively feeding and reproducing, usually from late spring through early fall. Warm-blooded vertebrates also maintain the virus during inclement periods of the year (cold and dry seasons). Humans can serve as dead-end, accidental hosts, as in equine encephalitis and Colorado tick fever, or they may be a maintenance reservoir, as in dengue fever and yellow fever. The general patterns of arbovirus transmission involving vectors are illustrated in figure 21.11.

Arboviral diseases have a great impact on humans. Although exact statistics are unavailable, it is accepted knowledge that millions of people acquire infections each year and thousands of them die. The uncertain nature of host and viral cycles

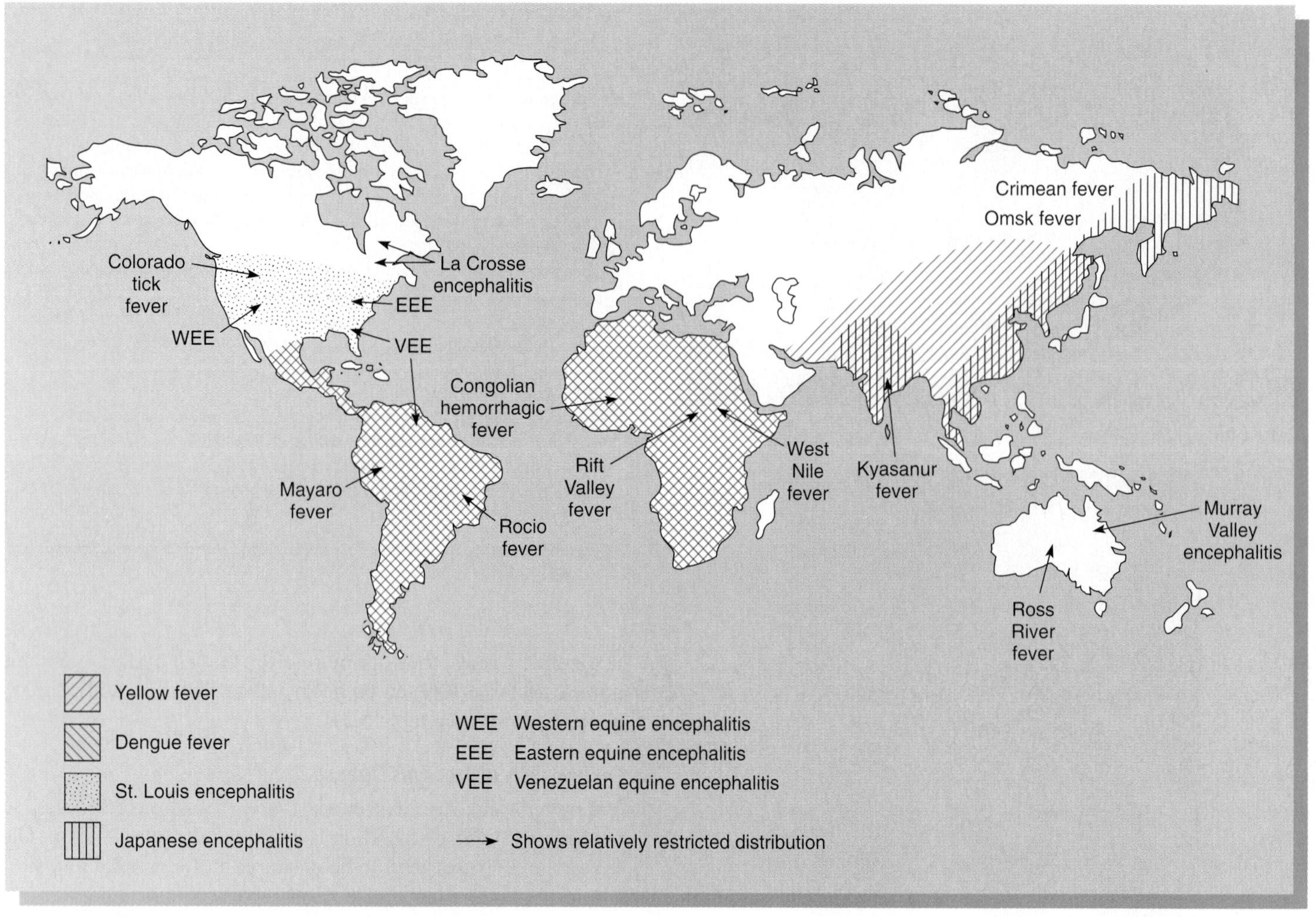

Figure 21.10 Worldwide distribution of major arboviral diseases.

Feature 21.7 Northward Migration of Deadly Dengue Fever

American epidemiologists are currently on guard for the appearance of a deadly form of **dengue fever** caused by a flavivirus and carried by *Aedes* mosquitos. Although mild infection is the usual pattern, a form called dengue hemorrhagic shock syndrome can be lethal. This disease is also called "breakbone fever" because of the severe pain it induces in the muscles and joints. The illness was originally endemic to Southeast Asia, but gradually spread to India, Central America, and the Caribbean. A series of epidemics involving hundreds of thousands of Mexicans, Puerto Ricans, and Cubans spurred investigators from the CDC to survey mosquito populations in states bordering those areas. Eventually they discovered the Asian tiger mosquito *Aedes albopictus,* a recently introduced species that can carry dengue fever virus. These mosquitos currently populate the forests of 12 states, where they breed and maintain themselves in large numbers. Sporadic contact of these mosquitos with infected travelers or immigrants could establish the virus in these mosquito populations. So far, no cases of hemorrhagic dengue fever have been reported in the United States, and it is hoped that continued mosquito abatement will block its spread.

dengue (den′-gha) A corruption of Sp. "dandy" fever. Other synonyms are aden fever and stiffneck fever.

frequently results in sudden, unexpected epidemics, sometimes with previously unreported viruses. Travelers and military personnel entering endemic areas are at special risk because, unlike the natives of that region, they have no immunity to the viruses.

General Characteristics of Arbovirus Infections

Although the arboviruses have many inherently interesting characteristics, our coverage here will be limited to (1) the chief manifestations of disease in humans, (2) aspects of the most important North American viruses, and (3) methods of diagnosis, treatment, and control.

Febrile Illness and Encephalitis One type of disease elicited by arboviruses is an acute, undifferentiated fever, often accompanied by rash. These infections, typified by dengue fever and Colorado tick fever, are usually mild and self-limited, and leave no sequelae. Prominent symptoms are fever up to 40°C, prostration, headache, myalgia, orbital pain, muscle aches, and joint stiffness. About midway through the illness, a maculopapular or petechial rash may erupt over the trunk and limbs.

Figure 21.11 Variations on the arbovirus-vector-human cycle. (*a*) The arthropod vector lives in a habitat separate from humans and spreads the virus among various wild, warm-blooded vertebrates and even to its offspring through transovarial passage. Humans intrude into this cycle by happening on the vector in the wild. (*b*) The infected person returns to human habitation and serves as a source of viruses for related insects living in the community that may act as biological vectors, or (*c*) wild animal reservoirs and vectors bring the virus into human populations, especially through migration or the encroachment of human habitation into wild areas.

When the brain, meninges, and spinal cord are invaded, symptoms of viral encephalitis occur. The most common American types are western equine, eastern equine, St. Louis, and California. The viruses cycle between wild animals (primarily birds) and mosquitos or ticks, and humans are not usually reservoir hosts. The disease begins with an arthropod bite, the release of the virus into the tissues, and its replication in nearby lymphatics. Prolonged viremia establishes the virus in the brain, where inflammation may cause swelling and damage to various nuclei and tracts as well as to the meninges. Symptoms are extremely variable, and may include coma, convulsions, paralysis, tremor, rigidity, loss of coordination, palsies, memory deficits, changes in speech and personality, and heart disorders. Survivors may be left with some degree of permanent brain damage. Young children and very old persons are most sensitive to injury by arboviruses.

Colorado tick fever (**CTF**) is the most common tick-borne viral fever in the United States. Restricted in its distribution to the Rocky Mountain states, it occurs sporadically during the spring and summer months, corresponding to the time of greatest tick activity and human forays into the wilderness. Two hundred to three hundred endemic cases are reported per year, and deaths are rare.

Western equine encephalitis (**WEE**) occurs sporadically in the western United States and Canada, first in horses and later in humans. The mosquito that carries the virus emerges in the early summer when irrigation begins in rural areas and breeding sites are abundant. The disease is extremely dangerous to infants and small children, with a case fatality rate of 3% to 7%. But even during epidemic years, usually no more than one hundred cases are reported.

Eastern equine encephalitis (**EEE**) is endemic to an area along the east coast of North America and Canada. The usual pattern is sporadic cases, but occasional epidemics may occur in humans and horses. High periods of rainfall in the late summer increase the chance of an outbreak, and disease usually appears first in horses and caged birds. The case fatality rate can be very high (70%).

A disease known commonly as **California encephalitis** is caused by two different viral strains. The California strain occurs occasionally in the western states and has little impact on humans. The LaCrosse strain is widely distributed in the eastern United States and Canada, and is a prevalent cause of viral encephalitis in North America. Children living in rural areas are the primary target group, and most of them exhibit mild, transient symptoms. Fatalities are rare.

St. Louis encephalitis (SLE) is the most common of all American viral encephalitides. Cases appear throughout North and South America, but epidemics occur most often in the midwestern and southern states. Inapparent infection is very common, and the total number of cases is probably thousands of times greater than the two hundred or so reported each year. Persons over the age of 55 are most susceptible to life-threatening complications. The seasons of peak activity are spring and summer, depending on the region and species of mosquito. In the east, mosquitos breed in stagnant or polluted water in urban and suburban areas during the summer, while in the west, mosquitos frequent spring flood waters in rural areas.

Hemorrhagic Fevers Certain arboviruses, principally the yellow fever and dengue fever viruses, can disrupt the vascular bed to such an extent that sudden localized bleeding erupts in the tissues, leading to shock and even death. The exact mechanisms of pathology are obscure, but somehow the virus causes capillary fragility and disrupts the blood clotting system. These hemorrhagic syndromes, which are usually attended by fever and pain, are caused by a variety of viruses, are carried by a variety of vectors, and are distributed globally. Reservoir animals are usually small mammals, although yellow fever and dengue fever can be harbored in the human population (see feature 21.7).

As a result of intensive studies begun in the 1920s, more is known about **yellow fever** than any other arboviral disease. Although this disease was at one time cosmopolitan in its distribution, mosquito control measures have eliminated it in many countries, including the United States. Two patterns of transmission occur in nature (figure 21.11). One is an urban cycle between humans and the mosquito *Aedes aegypti,* which reproduces in standing water in cities. The sylvan cycle, on the other hand, is maintained between forest monkeys and jungle species of mosquitos. Most cases in the western hemisphere occur in the jungles of Brazil, Peru, and Colombia during the rainy season. Infection begins acutely with fever, headache, muscle pain, and eye disturbances. While the disease is benign in 80–90% of cases, some patients experience oral hemorrhage, nosebleed, vomiting, jaundice, and liver and kidney damage. The explosive form of the disease leads to death in 50% of cases.

Diagnosis, Treatment, and Control of Arbovirus Infection

Except during epidemics, detecting arboviral infections can be difficult. The patient's history of travel or contact with vectors, along with serum analysis, are highly supportive of a diagnosis. Treatment of the various arbovirus diseases relies completely on support measures to control fever, convulsions, dehydration, shock, and edema.

The most reliable arbovirus vaccine is a live, attenuated one for yellow fever that provides relatively long-lasting immunity. Vaccination is a requirement for travelers in the tropics and may be administered to entire populations during epidemics. Live, attenuated vaccines for WEE and EEE are administered to laboratory workers, veterinarians, stockmen, and horses.

Most of the control safeguards for arbovirus disease are aimed at the arthropod vectors. Mosquito abatement by eliminating breeding sites and broadcasting insecticides have been highly effective in a restricted urban setting, but if the vectors are widely dispersed through a woodland habitat, these techniques are of little value. In the case of sylvan vectors, the only real preventative is to restrict human contact with the vector through insect repellents and protective clothing. Mosquitos may also be avoided by remaining indoors at night, the period of greatest mosquito activity.

Enveloped Single-Stranded RNA Viruses with Reverse Transcriptase: Retroviruses

Retroviruses are among the most unusual and disturbing viruses. They have known oncogenic abilities; they cause dire, often fatal diseases; and they are capable of reprogramming a host's DNA in profound ways. The most familiar of these viruses is the **human immunodeficiency virus** (**HIV**) that causes **acquired immune deficiency syndrome** (**AIDS**), but HIV is only one of a whole array of retroviruses associated with cancer and other diseases. Judging from what is currently known about these viruses, they will continue to influence medicine, research, politics, and economics for decades to come.

The Biology of the Retroviruses

An outstanding characteristic of retroviruses is an unusual enzyme called **reverse transcriptase,** which can convert a single-stranded RNA genome into double-stranded viral DNA. Not only can this newly formed DNA chronically infect host cells, but it may even be incorporated into the host genome as a provirus that can be passed on to progeny cells. Some of the retroviruses are transforming agents—that is, by regulating certain host genes, they convert normal cells into cancer cells (see chapter 14).

The Reversing Viruses Imagine the surprised group of researchers who discovered the revolutionary properties of retroviruses. These viruses do not follow a central thesis of biology, which maintains that the direction of flow of information and function is from DNA to RNA. Previously, all biologic entities, including other viruses, fulfilled this expectation, but this virus has a radical strategy. It comes already equipped with an enzyme that works in the opposite direction by catalyzing the replication of double-stranded DNA from single-stranded RNA. This property gave rise to the name *retro*virus (from Latin for behind or reversal), a property that is apparently found only in retroviruses and some of the DNA viruses.

The association of retroviruses with their hosts may be so intimate that viral genes are permanently integrated into the host genome. In fact, as the technology of DNA probes for detecting retroviral genes is employed, it becomes increasingly evident that retroviruses may be integral parts of chromosomes. So striking is their resemblance to transposons (so-called jumping genes) that some biologists think retroviruses may even have originated from host cells.

Nomenclature and Classification

Most retroviruses isolated so far cause slow infections that lead to leukemia, tumors, anemia, immune dysfunction, and even neurological diseases in birds and mammals. The first human retrovirus (discovered in 1978) was the T-cell leukemia virus I (HTLV I). This was followed by the isolation of HTLV II and HTLV III. Types I and II are associated with human leukemia or lymphoma, while type III is the agent of AIDS. Since it was determined that HTLV III, lymphadenopathy virus (LAV), and AIDS retrovirus (ARV) are all different names for the same virus, a unified name, HIV 1 (human immunodeficiency virus), was assigned to it. Subsequently, a second AIDS virus, HIV 2, was discovered.

Structure and Behavior of Retroviruses

HIV and other retroviruses display structural features typical of enveloped RNA viruses (figure 21.12*a*). The outermost component is a lipid envelope with transmembranous glycoprotein spikes and knobs that mediate viral adsorption to the host cell. Although a few retroviruses are nonspecific, most can infect only host cells that present the required receptors. In the case of the AIDS virus, the cells possess a CD4 receptor that allows viral adsorption and penetration of several types of leukocytes and tissue cells (figure 21.12*b*).

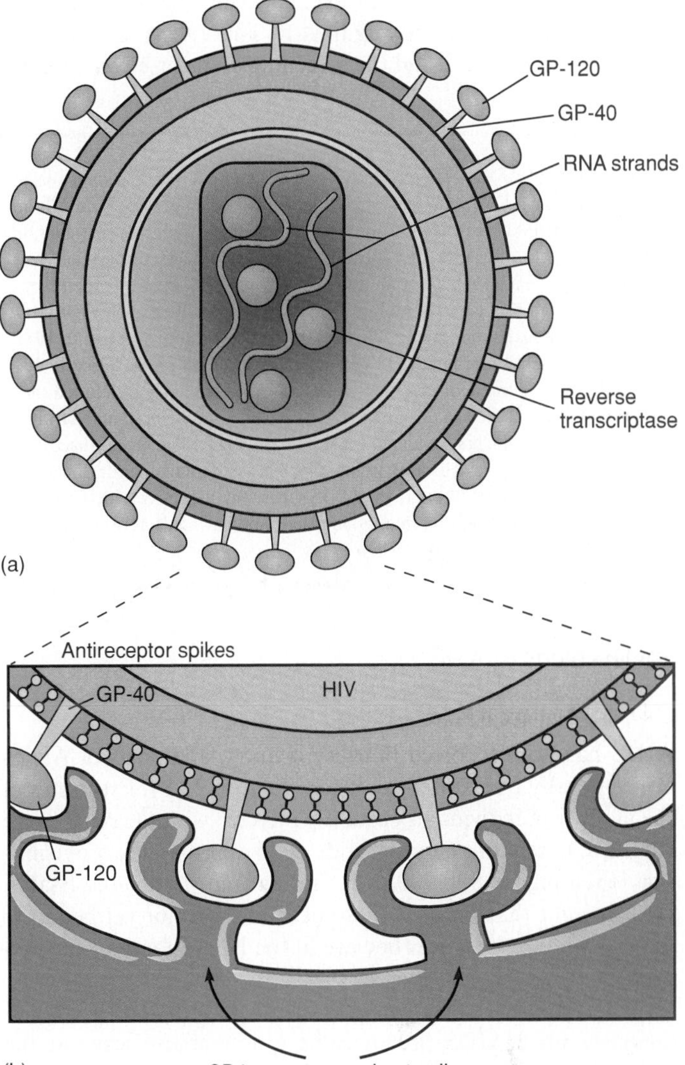

Figure 21.12 (*a*) A cutaway model of HIV. The envelope contains two types of glycoprotein (GP) spikes, two identical RNA strands, and several molecules of reverse transcriptase encased in a protein coating. (*b*) The snug attachment of HIV glycoprotein spikes to their specific receptors on a human cell membrane. This receptor, called CD4, is typically found on T4 lymphocytes and other white blood cells.

A Guide to Acquired Immune Deficiency Syndrome (AIDS)

The explosion of AIDS on the world scene in the early 1980s has altered society's collective view of infectious diseases. In an earlier chapter we emphasized that the AIDS epidemic is not the greatest ever known, but its fatal effects have propelled it almost to the point of a national obsession. At no time in history has a concerted research effort uncovered so much information about a microbe and a disease in so short a time. At no time in history has so much money been spent on treatment, vaccine development, education, and public health for one disease.

Historical Background

To reconstruct the discovery of AIDS, we must go back to the late 1970s, when physicians in Los Angeles, San Francisco, and New York City began to see clusters of patients who were all young males with one or more of a complex of symptoms: severe pneumonia caused by *Pneumocystis carinii* (ordinarily a harmless protozoan), a rare vascular cancer called *Kaposi's sarcoma;* sudden weight loss; swollen lymph nodes; and general loss of immune function. Another mysterious finding was that all of these young men were homosexuals. Early hypotheses attempted to explain the disease as a consequence of the homosexual life-style or a result of immune suppression by chronic drug abuse or infections. Soon, however, cases were reported in nonhomosexual patients who had been transfused with blood or blood products, and from these patients, a novel retrovirus was isolated. That this was clearly a communicable infectious disease left little doubt, and the medical community termed it acquired immune deficiency syndrome or AIDS.

Conclusively linking AIDS to this virus did not happen overnight. Once HIV was isolated from patients' tissues, researchers were hampered by the inability to grow sufficient amounts of the virus for further study and by the lack of accurate tests for differentiating this virus from other retroviruses. Finally in 1984, a functioning line of T cells that remained alive after infection with the new virus became available. This

Kaposi's sarcoma (kah'-poh-seez sahr-koh'-mah) Named for a Viennese dermatologist who first described the condition in the late 1800s.

was soon followed by the creation of sensitive tests for detecting the virus and the antibody against it. These tests in turn made the diagnosis of AIDS and the screening of donor blood a reality.

The Origins of AIDS There are no conclusive answers to the question of where and how HIV originated. Serological examination of stored blood from several countries reveals that, prior to 1970, mostly African samples contain antibodies to the AIDS virus. Initially, it was hypothesized that the virus originated in central Africa from a simian retrovirus carried by green monkeys. Although this virus is benign in monkeys, it could develop high virulence if transmitted to humans. This theory was called into question in 1988 when Japanese researchers showed that HIV 1 is much too genetically different from the simian virus to be a mutated form of it. They concluded that no simian virus studied so far appears to be a recent source of human strains, but they did not rule out an ancient interspecies origin. The recent discovery of an AIDS-like retrovirus among healthy, isolated Amazonian Indians has given credence to the idea that the AIDS virus and its relatives have existed for a long time in remote human populations and have only recently spread as isolated tribes moved into urban areas.

Epidemiology of AIDS

Changing Statistical Patterns

AIDS has been reported in every country, and parts of Africa are especially devastated by it (see feature 21.8). Estimates of the number of individuals currently infected with the virus range from five to ten million worldwide, with approximately one million (the range is 800,000 to 1.3 million) in the United States. It is thought that most of these persons have not yet begun to show symptoms because they are in the latent phase of the disease.

The number of diagnosed AIDS cases has increased continuously since AIDS first became a notifiable disease at the national level in 1984 (figure 21.13). It took nearly seven years for the first 100,000 cases to appear, but only two years for the next 100,000 cases. It is thought that the disease will remain in this epidemic pattern well into the mid- and late 1990s. Although new therapies have lengthened the lives of many patients, the overall death rate is over 50%.

Feature 21.8 The Global Outlook on AIDS

AIDS in the Third World shows the same spectrum of illness as is seen in the United States, but it differs in epidemiology. In Africa, the case rate in men and women is about equal, and in some regions, up to 10% of the population may be carrying the AIDS virus. Furthermore, the mode of transmission there is primarily by sexual contact between men and women and to a lesser extent via blood transfusions and pregnancy. The disease is spreading so rapidly that 10 thousand to 20 thousand new cases and as many deaths are expected each year. Parts of Africa report a virulent form of intestinal AIDS called "slim" that savagely wastes the body. Another complication is the second virus (HIV 2), which is a major cause of AIDS in western Africa. Statistics from the rest of the world point to a growing pandemic that involves large parts of Europe, Latin America, Asia, and the Far East, though inadequate detection probably contributes to underreporting in some areas.

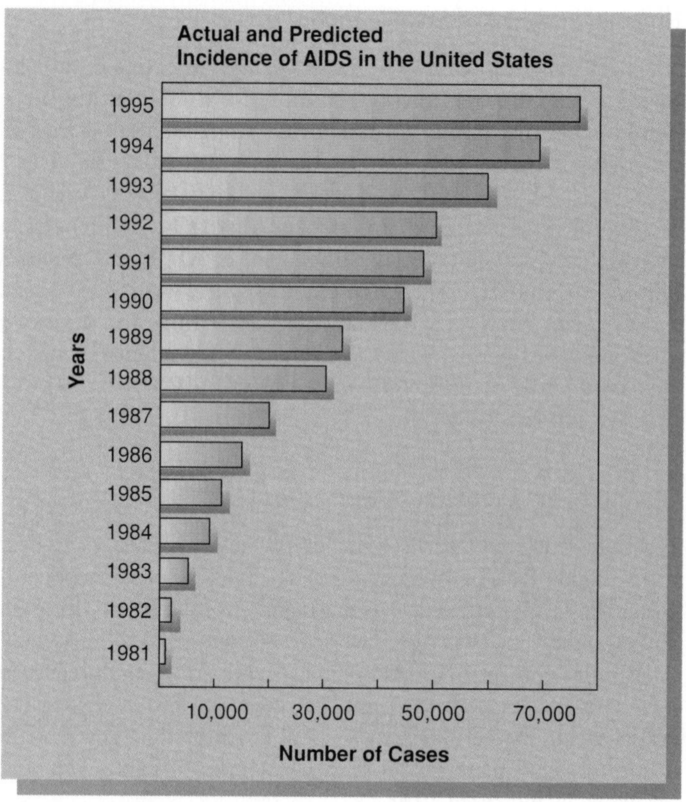

Figure 21.13 Total reported cases of AIDS from 1981 to 1991 and projected occurrence from 1992 to 1995, based on data from the CDC. Source: Data in part from *Summary of Notifiable Diseases, United States 1990* and *Morbidity and Mortality Weekly Report*, December 20, 1991, the Centers for Disease Control, Atlanta, GA.

Although all persons appear to be susceptible to the AIDS virus, epidemiologic statistics show that, overall, it is far more common among males than females, and that ethnically, whites account for the most cases, followed by Hispanics, blacks, and other groups. The average adult patient is about 35 years of age. If one observes specific sociologic groups, however, the sexual, racial, and age statistics can be quite different. A majority of AIDS patients are concentrated in large metropolitan centers of New York, California, Florida, New Jersey, and Texas.

Modes of AIDS Transmission

It is now clear that the mode of HIV transmission is exclusively through two forms of contact: sexual intercourse and transfer of blood or blood products. Each of these can be divided into specific high-risk groups that include persons who become infected through a particular life-style or behavior, or through medical accidents. The mode of transmission is rather similar to that of hepatitis B virus, except that the AIDS virus does not survive for as long outside the host, and it is far more sensitive to heat and disinfectants. In general, AIDS is spread only by direct and rather specific routes, and is not "caught" like influenza or a cold. An infection can take place only if an infectious dose of virus gets past the body's epithelial barriers into the blood. Because the blood of HIV-infected persons often harbors high levels of free virus and infected leukocytes, any form of intimate contact involving transfer of blood (trauma, injection) can be a potential source of infection. Semen and vaginal secretions also harbor the virus, thus they are significant factors in sexual transmission. The virus can be isolated in small amounts from urine, tears, sweat, and saliva, but these fluids are considered an unlikely source of infection. Mother's milk can be a source of neonatal infection, and more surprisingly, babies can spread the virus to their mothers through nursing.

Myths and Misconceptions: How AIDS Is NOT Transmitted So horrifying is the prospect of acquiring a lethal disease like AIDS that inaccurate notions and misconceptions rather than established facts have sometimes abounded. Rumors about the role of bloodsucking insects have been common; however, insects like mosquitos do not serve as biological vectors of the AIDS virus, and the chance of the insect harboring and delivering an infectious dose mechanically is infinitesimally small. At first it was mistakenly believed that airborne droplets, shaking hands, handling fomites, and sharing public facilities, drinking water, food, and swimming pools were possible ways to become infected. However, in-depth studies have indicated that even close contact within families of AIDS patients has not led to infection of healthy family members. One peculiar fear that surfaced during the early stages of the AIDS epidemic was that donating blood could expose the donor to AIDS, but obviously, blood banks use new sterile equipment for each donor. And although kissing is a form of intimate contact, no documented cases have been traced to it. Epidemiologists at the CDC are convinced that if any of these modes of transmission existed, they would have become evident by now.

Feature 21.9 The Odds of Acquiring AIDS Through Heterosexual Contact

A high-risk sexual partner is one who has recently engaged in homosexual activity or intravenous drug use, has lived in a high-prevalence area such as Africa, has had multiple blood transfusions, or has engaged in regular sex with a member of any of these groups. The following estimates on the likelihood of acquiring AIDS come from the Center for AIDS Prevention Studies:

	Chance of Acquiring AIDS
One sexual encounter with an HIV-negative person who does not belong to a risk group (condom used)	1 in 5 billion
One sexual encounter with an HIV-negative person who does not belong to a risk group (no condom used)	1 in 500 million
One sexual encounter with a person who has not been tested but who is not at high risk (condom used)	1 in 50 million
One sexual encounter with a person who has not been tested but who is not at high risk (no condom used)	1 in 5 million
One sexual encounter with an HIV-positive person (no condom)	1 in 500
500 sexual encounters with an HIV-positive person (no condom)	3 in 5

From these data it is apparent that although the number of contacts and the use of protection are important factors, neither is as important as the choice of sex partners. With AIDS, a person's sexual history takes on crucial importance. Epidemiologists cannot overemphasize the need to screen prospective sex partners and to follow a monogamous sexual life-style. This is one point that has the unanimous support of both scientists and moralists.

An Overview of AIDS Cases

The following section will cover definitions of AIDS and ARC, the importance of the serological state of patients, the groups most commonly involved, and behavior or events that increase a person's risk.

The Clinical Definition of AIDS

AIDS is defined as a severe immunodeficiency disease arising from infection with HIV and accompanied by some of the following symptoms: life-threatening opportunistic infections, persistent fever, unusual cancers, chronically swollen lymph nodes, weight loss, diarrhea, and neurological disorders. This definition would also include many patients suffering from what has been termed **AIDS-related complex (ARC)**, which is considered a stage in the development of the disease. At some point during the infection, the patient tests positive for antibodies to HIV. Seropositivity means only that an infection with HIV has taken place, but does not indicate that the disease exists. However, it is presumed that the antibody-positive person is infectious. For further ramifications of this test, see the later section on diagnosis.

Case Descriptions According to Group, in Order of Prevalence

I. Homosexual or Bisexual Males Male homosexuals still account for the majority of AIDS cases (57%), though the rate of infection appears to be declining. Why this demographic group has been so hard hit is apparently due to certain types of homosexual practices. Multiple anonymous sexual encounters were once relatively common among gay men intermingling in bars and bathhouses. Anal sex, which is known to lacerate the rectal mucosa, could certainly provide entry of viruses from semen into the blood. Studies have shown that the passive anal partner in these encounters is the more likely of the two to become infected. In addition, bisexual men provide an avenue for the virus to females and the general heterosexual population.

II. Intravenous Drug Abusers In large metropolitan areas such as New York City, 60% of intravenous drug addicts are thought to be HIV carriers. Needle sharing is most common in the "shooting galleries" of large eastern cities and is especially risky for cocaine users whose rate of drug injection may be more frequent. Infection from contaminated needles is growing more rapidly than for any other mode of transmission (23% of cases), and is another significant factor in the spread of AIDS to the heterosexual population. A fairly large segment of IV users are also homosexual men. So concerned are public health officials about this trend that campaigns to hand out sterile needles or disinfection kits to drug users have begun in some cities.

III. Heterosexual Sex Partners of HIV Carriers About 6% of AIDS cases arise in persons who have had sexual intercourse with an AIDS-infected partner. Individuals in this group include the partners of drug addicts and bisexual men, prostitutes, and spouses of persons who have received transfusions or blood factors. The virus spreads readily in either direction, and multiple sexual encounters increase the chance of infection, although even a single encounter with an AIDS patient has been known to transmit the virus (see feature 21.9). The overall rate of infection in the general heterosexual population has increased dramatically in the past few years in adolescent and young adult women.

IV. Blood Transfusion/Organ Transplant Patients Since routine screening of donated blood for antibodies to the AIDS virus was adopted in 1985, transfusions are no longer considered a serious risk. Before then, approximately 2% of AIDS cases arose in patients receiving multiple transfusions, and most new transfusion-related cases can be traced to transfusions given prior to 1985. Because there may be a lag period of a few weeks to nearly three years before antibodies appear in an infected person (see figure 21.15), it is remotely possible to be infected through donated blood (estimated at 1 in 500,000 transfusions). As added precautions, potential donors are thoroughly screened, and people anticipating surgery are encouraged to stockpile their own blood or to solicit designated donors from their families. A new technique of sterilizing blood with a laser beam shows some promise, but it will not be perfected until the early 1990s. Rarely, organ transplants can carry HIV, so transplant procurement groups are now trying to screen organs for the virus.

V. Patients with Coagulation Disorders Receiving Replacement Factors In this category, which accounts for 1% of all AIDS cases, hemophiliac men predominate. For several years in this country, replacement coagulation products (factor VIII and others) were derived from human plasma and filtered to remove cellular microbial contaminants, but not viruses. Before this route of HIV transmission was discovered, several hundred hemophiliacs became infected. Now replacement infusions are heat-treated to kill any contained viruses, a process that has essentially eliminated risk for this group of people.

VI. Inapparent or Unknown Risk Factors Three percent of all AIDS cases occur in persons who apparently belong to none of the risk groups just described. This is the most troublesome category because it opens up questions about some other, unknown route of spread. When the CDC has carefully followed up such cases, a majority of the patients have admitted to prior contact with a prostitute or a history of other STDs. Factors such as patient denial, unavailability, death, or uncooperativeness make it impossible to explain every case.

VII. Congenital and Neonatal AIDS The chance of an HIV-positive mother infecting her fetus is estimated at 30% to 50%, and in numbers this can mean hundreds of babies infected per year. In some parts of New York City, up to 2% of all newborns test positive for the antibodies. Initial studies show that the population of mothers accounting for this neonatal epidemic are youthful, impoverished IV drug addicts or sex partners of drug addicts. Infected babies develop the disease more rapidly than adults, with half of them dying by the age of three.

VIII. Risks Involving Medical and Dental Personnel Health care personnel are not considered a high-risk group, though about 64 medical and dental workers are known to have acquired AIDS or become antibody-positive as a result of clinical accidents. A health care worker involved in an accident where gross inoculation with blood occurs (a needle stick) has about one chance in five hundred of becoming infected.

The well-publicized case of a Florida dentist who infected several of his patients has prompted a controversial examination of health workers as a *source* of HIV. Statistics from the CDC indicate that at least 6,800 medical and dental personnel have AIDS. With this in mind, lawmakers are considering a federal statute that would force mandatory testing for all health workers. It would also require that HIV-positive physicians and dentists obtain permission from patients and a governing board before performing invasive surgeries.

It must be emphasized that transmission of HIV in either direction will not occur through casual contact or routine nursing procedures, and that basic guidelines and universal precautions for infection control (see appendix D) were designed to give full protection for both worker and patient.

The Evolving Picture of AIDS Pathology

Understanding the pathology of AIDS has been aided to an extent by animal models. Studies of macaque monkeys with a simian form of AIDS that is very similar to the human disease have helped outline some of the mechanisms by which retroviruses suppress the immune system. Although chimpanzees can be infected with HIV, they fail to develop the disease. Much promise lies in studying the colonies of transgenic mice that have been implanted with human immune systems. Preliminary findings show that when these special mice are infected with HIV, they develop some symptoms of AIDS.

Initial Infection of Target Cells

It appears that HIV initially enters special cells on the membranes called dendritic cells. Here it grows and is shed from the cells without killing them. It can also multiply in macrophages in the skin, lymph organs, bone marrow, and blood, and by this means is amplified. One of the great ironies of HIV is that it infects and destroys many of the very cells needed to destroy it, including the helper (T4) class of lymphocytes, monocytes, macrophages, and even B-lymphocytes. The virus enters susceptible cells and transcribes its RNA into DNA. Although initially it may produce a lytic infection, it usually goes into a latent period (figure 21.14), which accounts for the lengthy course of the disease that most often follows.

Stages of HIV Disease

The scope of HIV infection shows a continuous progression from initial contact to full-blown AIDS (figure 21.15). The initial infection is often attended by vague, mononucleosis-like symptoms that soon disappear. This is followed by a period of asymptomatic infection (HIV disease) with an incubation period that varies in length from 2 to 15 years (the average is about 10). The first antibodies to HIV are usually detectable in serum by the third month following infection, but these antibodies are ineffective in neutralizing the virus once it enters the latent period.

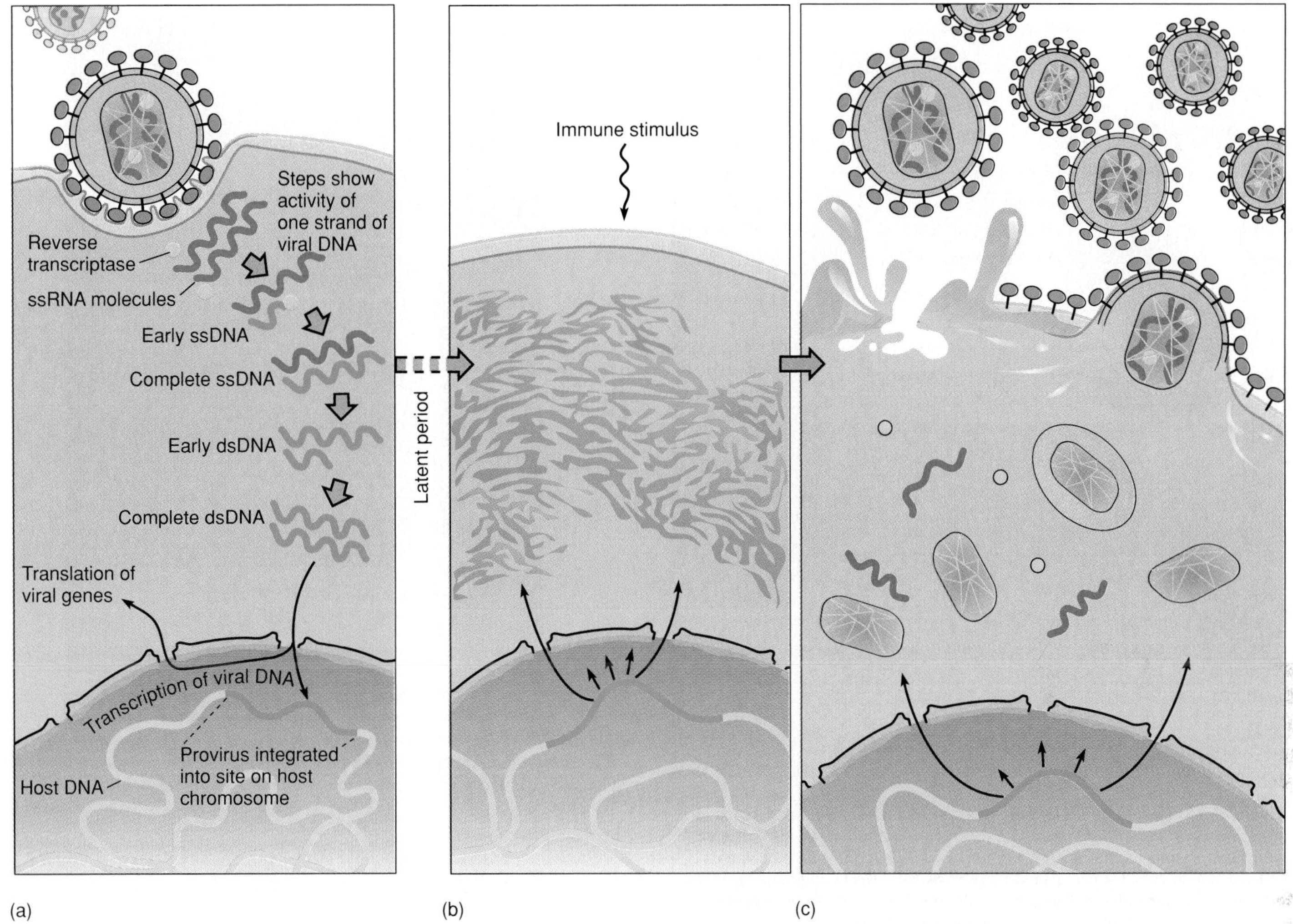

Figure 21.14 The general life cycle of HIV. (*a*) The virus is adsorbed and endocytosed, and the twin RNAs are uncoated. Reverse transcriptase catalyzes the synthesis of a single complementary strand of DNA (ssDNA). This single strand serves as a template for synthesis of a double strand (ds) of DNA. In latency, dsDNA is inserted into the host chromosome as a provirus. (*b*) After an indeterminant period of time, various immune activators stimulate the infected cell, causing reactivation of the provirus genes and production of viral components. (*c*) The shedding of viruses causes lysis of the infected cell.

Disease may initially present itself as ARC, a condition in which patients test positive for the antibody and have some of the symptoms of AIDS, especially fever, swollen glands, fatigue, diarrhea, weight loss, and neurological syndromes, as well as opportunistic infections and neoplasms. ARC may be life-threatening, and it may convert to the full-fledged syndrome, but there are more long-term survivors of ARC than of AIDS. In some persons, the disease bypasses the ARC phase entirely.

Contrary to what is commonly thought, not everyone who becomes infected or is antibody positive develops AIDS. About 15% of antibody-positive persons remain free of disease, indicating that functioning immunity to the virus can develop. The long incubation period makes it difficult to trace the ultimate outcome of infection, but experts feel that 40% to 75% of antibody-positive persons will develop some form of the disease within 6 to 10 years of initial infection. Individuals who acquire the symptoms of full-blown AIDS do not have an optimistic prognosis. There are few long-term survivors of the syndrome, and the death rate has continued to be high even with treatment.

The Primary Effect: Harm to T Cells and the Brain

In order to understand the outcome of AIDS infection, we must think in terms of an impaired immune system. Over the long period of chronic infection, the infected cells play host to the virus in either a latent or an active phase. Cells with active viruses release large quantities into the circulation and are killed in the process. The death of T cells and other white blood cells results in extreme leukopenia and loss of essential T4-memory clones and stem cells (figure 21.16). Evidence indicates that the viruses also cause the formation of giant T-cell and other cell syncytia, which allow the spread of viruses directly from cell to cell, followed by mass destruction of the syncytia (figure 21.17). As the dose of viruses entering the system increases, a vicious

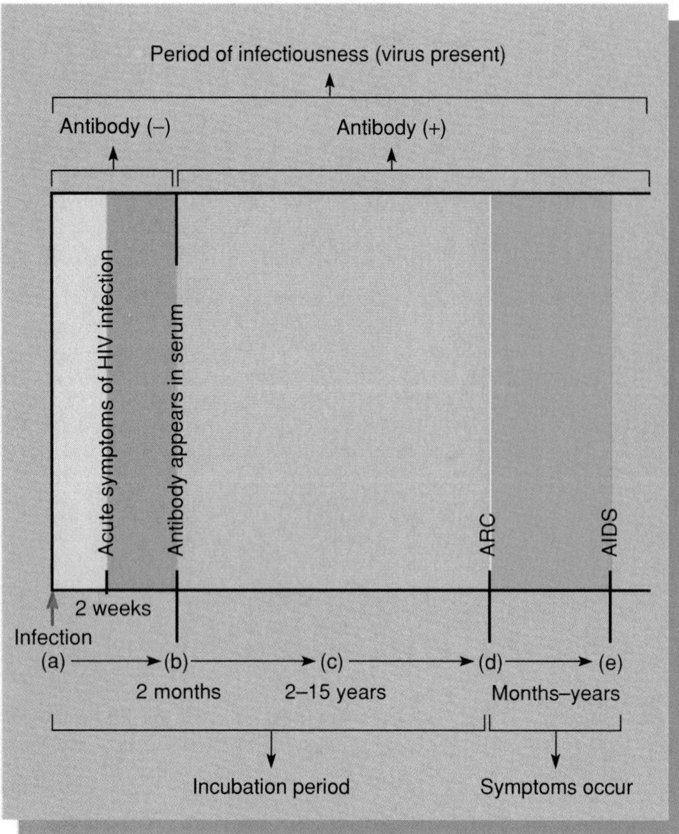

Figure 21.15 Stages in AIDS. (*a*) Infection with virus. (*b*) Appearance of antibodies in standard HIV tests (ELISA). (*c*) Asymptomatic HIV disease, which can encompass an extensive time period. (*d*) AIDS-related complex (ARC), which may not occur in some persons. (*e*) Overt symptoms of AIDS include some combination of opportunistic infections, cancers, and general loss of immune function such as a very low T4 lymphocyte count.

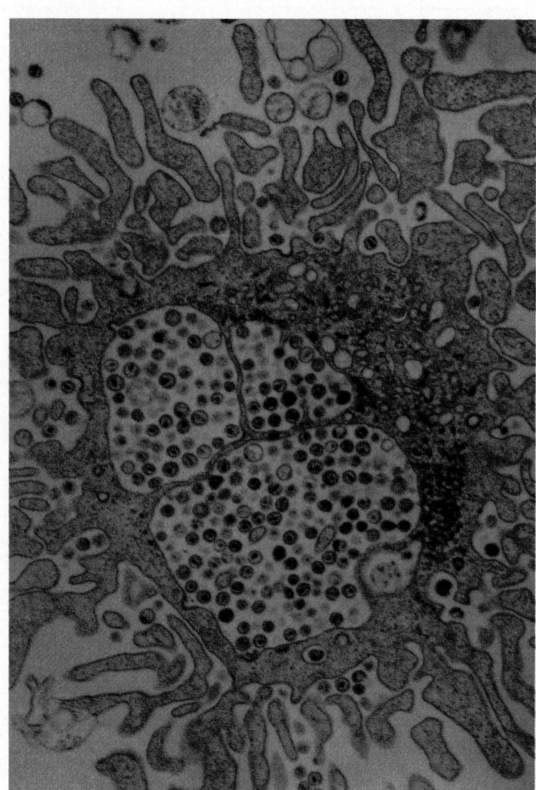

Figure 21.16 Transmission electron micrograph of a cell section during lytic infection by the AIDS virus. The process of viral shedding, which has virtually disintegrated the cell, accounts for the severe decimation of T4 lymphocytes.

cycle occurs, involving a greater infection rate and death of cells and greater viral burden. Because the T4 lymphocytes have multiple cytotoxic, helper, and inducer functions, their destruction or disability paves the way for invasion by opportunistic agents and cancer. When the T4 cell levels fall below 500 cells/mm^3 of blood, overt symptoms of AIDS appear.

Targets of the AIDS virus are not limited to the immune system. The central nervous system is involved in a large proportion of cases. It is believed that infected macrophages cross the blood-brain barrier and release viruses, which then invade nervous tissues. Studies have indicated that some of the viral envelope proteins can have a direct toxic effect on brain's glial and other cells. Other research has shown that some peripheral nerves become demyelinated and the brain becomes inflamed. How the AIDS virus may be involved in these pathologic processes is a topic of intense study and debate.

Secondary Effects That Define AIDS

Opportunistic Infections Debilitating, potentially fatal, and often multiple opportunistic infections are a hallmark of AIDS. In general, the infections involve protozoa (chapter 19), fungi

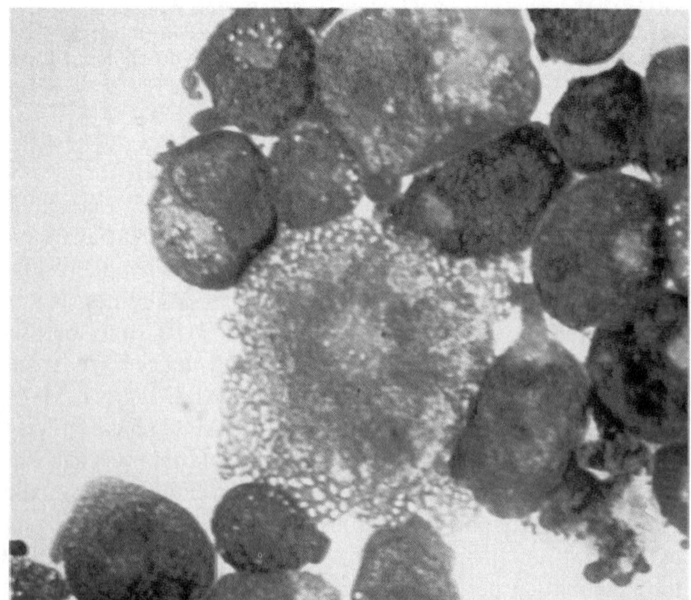

Figure 21.17 The formation of a giant multinucleate syncytium of HIV-infected T cells allows the virus to spread internally from cell to cell (in such sites as the bone marrow) and often results in mass destruction of the syncytium during virus maturation and release.

(chapter 18), viruses, or bacteria (chapters 15, 16, and 17) that establish a foothold in the absence of adequate cytotoxic immunity. The most frequent sites of attack are the respiratory, digestive, and nervous systems.

***Pneumocystis carinii* pneumonia (PCP)**, caused by a common protozoan, is the leading cause of morbidity and mortality in AIDS patients. Its symptoms include fever, cough, shortness of breath, thickening of the lung epithelium, and fluid infiltration that compromises respiration. The course of infection may be acute and rapid, or it may be slow and progressive. Other protozoan infections common in AIDS patients are toxoplasmosis of the brain (caused by *Toxoplasma gondii*) and *Cryptosporidium* diarrhea. Fungi most frequently involved in AIDS complications are *Cryptococcus neoformans,* a cause of meningitis; *Candida albicans,* which attacks the oral, pharyngeal, and esophageal mucosa; and *Aspergillus* species, which infect the lungs.

Common herpesviruses also afflict the AIDS patient. Cytomegalovirus produces a disseminated pneumonia, and may infect the brain, liver and intestine as well. Herpes simplex and herpes zoster can reactivate to produce severe ulceration of the skin, lips, genitalia, and anus. Infections with *Mycobacterium* are particularly problematic and severe. Tuberculosis in AIDS patients is now at epidemic levels. For reasons that are obscure, several nontubercular mycobacteria (MAI) show a strong affinity for AIDS patients, growing in massive numbers in the bone marrow, liver, spleen, and lymph nodes.

AIDS-Associated Cancers The predisposition of AIDS patients for tumors is due to the loss of natural cancer-killing T cells along with the intrusion of microbes that can cause cancer. Kaposi's sarcoma (KS) accounts for most cancers seen clinically. This sarcoma, of unknown etiology (possibly viral), is a nodular purple lesion that develops from endothelial cells in blood vessels of the skin, intestine, and mucous membranes (figure 21.18). Other symptoms that accompany KS are fever, lymphadenopathy, weight loss, and diarrhea. Complications are hemorrhage, perforation, and intestinal obstruction. Additional cancers common to AIDS patients are epithelial carcinomas of the skin, mouth, and rectum, and lymphomas originating from B lymphocytes.

Miscellaneous Conditions AIDS patients experience many nonspecific, disease-related symptoms. Pronounced wasting of body mass that accompanies weight loss, diarrhea, and poor nutrient absorption is a common complaint. Protracted fever, fatigue, sore throat, and night sweats are significant and debilitating. Some of the most virulent complications are neurological. Lesions occur in the brain, meninges, spinal column, and peripheral nerves. Patients with nervous involvement show some degree of withdrawal, persistent memory loss, spasticity, and progressive dementia. Numbness, tingling, loss of reflexes, and pain in the extremities are also common. Both a rash and generalized lymphadenopathy in several chains of lymph nodes are presenting symptoms in many AIDS patients.

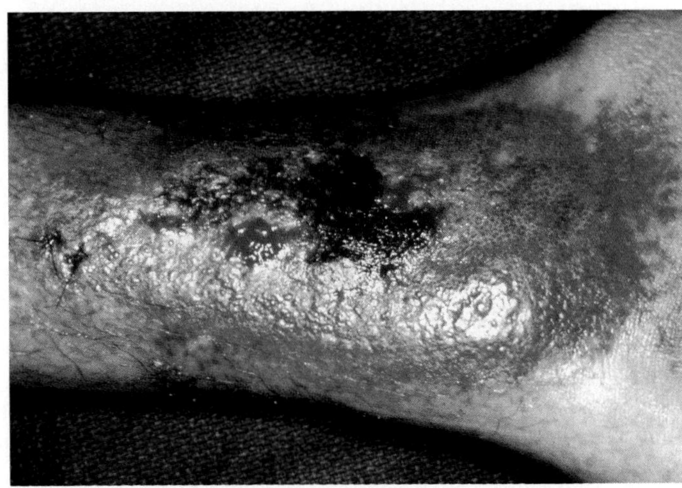

Figure 21.18 Kaposi's sarcoma lesion on the arm. The flat, purple tumors occur in almost any tissue and are frequently multiple.

The Search for Cofactors—The Prevalence of Coinfections

From the very first months of its discovery, medical science sought to identify contributing cofactors in AIDS besides lifestyle or transfusions. One relationship that seems certain is the effect of coinfections, especially STDs. Genital ulcers such as those present in chlamydia, syphilis, and warts increase susceptibility to AIDS because breaks in the mucous membrane promote the shedding and entrance of the virus. AIDS patients frequently have concurrent infections with Epstein-Barr virus (mononucleosis), CMV, syphilis, tuberculosis, hepatitis B, and a recently discovered mycoplasma (see chapter 17). The severe immune suppression accompanying AIDS appears to escalate both diseases, but whether coinfections are required for AIDS to develop is controversial.

Diagnosis of AIDS

Secondary infections and cancers suggestive of AIDS present strong presumptive evidence. When these signs are coupled with a patient's history of predisposing activities and a positive serological test for AIDS antibodies, the diagnosis is confirmed. The enzyme-linked immunosorbent assay for AIDS antibodies (ELISA) is highly sensitive, but variations in the quality of testing and the patient's condition can lead to false-positive results in a small number of tests. An alternate latex agglutination test that is simple, completed in 5 minutes, and 99% accurate has recently been marketed. Positive antibody readings call for the more specific follow-up test known as the Western blot, which detects several different anti-HIV antibodies and can usually rule out a false-positive result (see figure 13.30). False-negative results occur in as many as one in four patients who are indeed infected with HIV. Such silent infections contribute considerably to the unpredictable nature of latency and how to test for it. To rule out the possibility of a false-negative, persons who test negative but who practice high-risk behavior should seek a second test at a later date.

Because an infected person has viruses in his blood from the very earliest stages, a more reliable diagnostic approach would be direct discovery of the AIDS virus, its genes, or antigens in blood and other specimens. This form of testing could eliminate the false tests inherent in antibody analysis and be a particular boon in monitoring transfusion blood. Several rapid, specific, very sensitive tests using monoclonal antibodies are now available for research purposes.

AIDS Treatments

So far, there is no cure for AIDS. None of the therapies do more than prolong life or diminish symptoms. Treatment for AIDS patients focusses on supportive care and drugs to control both opportunistic and HIV infection (see feature 21.10). The respiratory apparatus of pneumocystis patients must be frequently suctioned to remove collected fluids. Drugs used to control PCP are pentamidine, taken in oral or inhaled form; trimethoprim-sulfamethoxazole (Bactrim, Septra); and leucovorin. Drugs that have some control over CMV infections are ganciclovir and foscarnet. Disseminated fungal infections are treated with fluconazole, and Kaposi's sarcoma is relieved by alpha interferon.

AIDS Protection

Measures to prevent direct contact with HIV are an important consideration. There is universal agreement that "safe sex" practices can be very effective. These include the use of condoms with an antimicrobial spermicide, monogamous sex with a long-term sex partner, and avoidance of anal sex. Although avoiding intravenous drugs is an obvious deterrent, many drug addicts do not choose this option. In such cases, risk can be decreased by not sharing syringes and needles or by disinfecting the syringe and needle with hypochlorite solution before use. Anonymous AIDS testing to protect the identity of possible HIV-positive persons is currently underway to screen certain populations, such as newborns and hospital, STD, and drug treatment patients, and some states have adopted AIDS testing as part of the marriage license procedure. Mandatory testing may soon be an expected procedure for medical personnel.

Will There Be an AIDS Vaccine?

From the very first years of the AIDS epidemic, the potential for a vaccine has been regarded warily. The virus presents many seemingly insurmountable problems to vaccine technologists. For example, it becomes latent in cells; its cell surface antigens

Feature 21.10 AIDS Drugs—The Great Hope

Drugs to block the AIDS virus are being actively sought and tested. Of those found so far, only two are currently approved by the Food and Drug Administration. Both azidothymidine (AZT or retrovir) and dedeoxyinosine (DDI or videx) are synthetic nucleotide analogs with broad-spectrum effects. They interrupt the HIV multiplication cycle by being incorporated into the DNA molecule during reverse transcription. Because these drugs lack all of the correct binding sites for further synthesis, this effectively terminates viral DNA synthesis. These drugs are now being used to treat patients with overt AIDS and HIV-positive patients early in the infection, before the viral load is too high and immune system damage is too great. AZT is relatively expensive ($2,200 for one year of treatment), can have highly toxic side effects, and may not be effective. DDI is less expensive and more tolerable by many AIDS patients. Both drugs can improve patient well-being and T_4 cell levels.

An experimental drug related to DDI, dideoxycytidine (DDC), has shown antiviral activity in volunteers. Other drugs, notably CD4, immuthiol, alpha interferon, peptide T, and D-penicillamine, also show promise in clinical trials.

In their urgency for a cure, many AIDS patients are willing to try untested but potentially useful therapies. Feeling they have little to lose, some patients have acquired drugs from countries with less stringent drug policies than the United States. To ease this tremendous demand for alternate treatments, the FDA has liberalized the importation of certain untested drugs for personal use. It is a sad comment that many AIDS patients have been victimized by fraudulent claims of healing by means of everything from urine injections and heat treatments to herbal infusions.

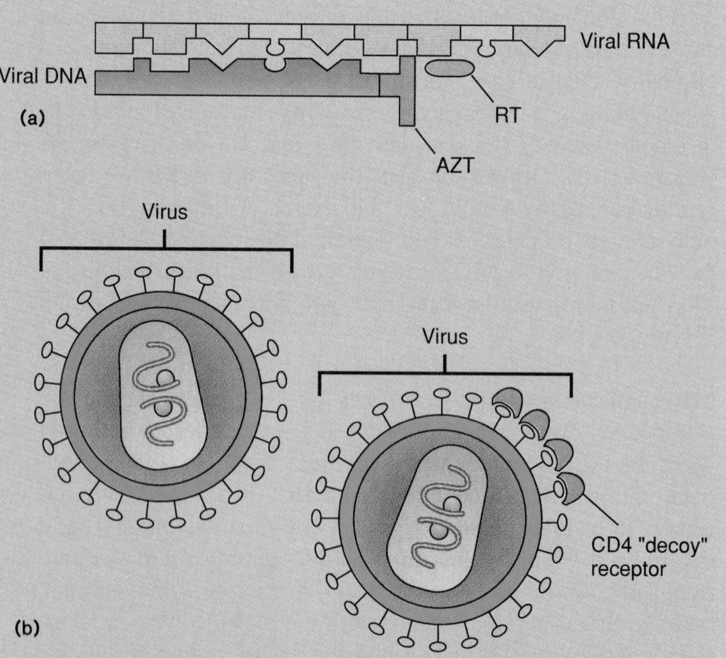

Strategies for AIDS drugs. (*a*) Some drugs block the action of reverse transcriptase (RT). Most of the newer drugs work at this site. AZT, for example, is a molecular mimic of thymine, and when inserted by the transcriptase, it negates further viral DNA synthesis. (*b*) Prevention of HIV adsorption to host cell. CD4 is given as an example. Unfortunately, this therapy will be ineffective on viruses already in the cell.

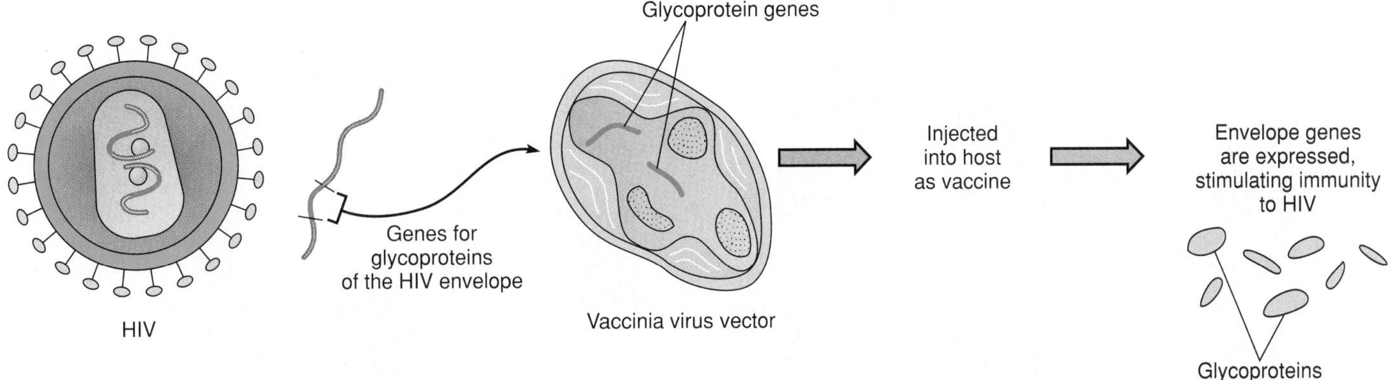

Figure 21.19 "Trojan horse" or viral vector technique, a novel technique for making an AIDS vaccine. The part of the HIV genome coding for envelope glycoproteins is inserted into a carrier virus (vaccinia). This hybrid virus replicates and expresses the HIV genes when it is injected into a host.

mutate rapidly; and although it does elicit immune responses, it is apparently not completely controlled by them. In view of the great need for a vaccine, none of these facts has stopped the medical community from moving ahead.

Because a live, attenuated AIDS virus vaccine could potentially cause cancer, most of the vaccines in various stages of development and testing are based on recombinant viruses and viral envelope or core antigens. In the viral vector technique, vaccinia virus or adenovirus is genetically engineered to carry only the HIV envelope gene (figure 21.19). A French immunologist served as his own guinea pig by repeatedly immunizing himself with a prototype vaccine of this type. He apparently suffered no ill effects and yet developed a fairly strong immune response to the virus. A controversial vaccine, made by inactivating HIV with radiation (HIV immunogen), has been shown to clear the virus from infected chimpanzees and to improve the immune function of HIV-positive human volunteers.

Experimental vaccines await many years of testing in animals and humans. Recent promising experiments are a vaccine tested on chimpanzees that immunized them against one strain of the virus and a vaccine used in mice with human immune systems that protected the human cells against HIV infection. The principal difficulties in human trials lie in determining the safety of the vaccine in volunteers and determining whether it stimulates effective cytotoxic T cells and neutralizing antibodies. Recently the government has approved the use of a recombinant vaccine containing core antigen (HGP-30), but even the most optimistic experts do not expect a vaccine to be generally available before the late 1990s.

Other Retroviral Diseases in Humans

Other human retroviruses are implicated in various transmissible, fatal lymphocyte malignancies. HTLV I causes adult T-cell leukemia (ATL), a type of lymphosarcoma, and a paralytic disease called tropical spastic paraparesis (TSP). HTLV II causes hairy-cell leukemia. The possible mechanisms whereby retroviruses stimulate cancer were covered in chapter 14. One hypothesis says the virus carries an oncogene that is spliced into a host's chromosome during infection and that, when triggered by various carcinogens, immortalizes the cell and frees it from limitations in the number of cell divisions. No oncogenes have been discovered in HTLV I and II, but they do have genes that code for regulatory proteins. One of HTLV's genetic targets seems to be the gene and receptor for interleukin-2, a potent stimulator of T cells.

Epidemiology and Pathology of the Human Leukemia Viruses

Adult T-cell leukemia (ATL) was first described by physicians working with a cluster of patients in southern Japan. Later, a similar clinical disease was described in Caribbean immigrants. In time it was shown that these two diseases and a cancer called Sézary-cell leukemia were one and the same disease. ATL was linked to HTLV I when the virus was isolated from patients' T cells. Although more common in Japan, Europe, and the Caribbean, a small number of cases occur in the United States. The disease is not highly transmissible; studies among families show that repeated close or intimate contact is required. Because the virus is thought to be transferred in infected blood cells, blood transfusions and blood products are potential agents of transmission. Intravenous drug users could spread it through needle sharing. A sexual mode of transmission has also been proposed.

ATL is chronic and invariably fatal in its course. Symptoms are extreme, persistent lymphocytosis, with large, atypical lymphocytes. One variant of ATL, mycosis fungoides or cutaneous T-cell lymphoma, is accompanied by dermatitis, with thickened, scaly, ulcerative, or tumorous skin lesions. Other complications are lymphadenopathy and dissemination of the tumors to the lung, spleen, and liver. Many types of treatments with drugs, radiation, or some combination have been tried, but the long-term prognosis remains poor. Since the discovery of this virus, public health officials have become concerned that it may be present and transmitted in transfused blood, and they are currently screening blood for this virus along with HIV.

Hairy-cell leukemia (HCL) is a rare form of cancer that has been traced to HTLV II infection. HCL derives its name from the appearance of the afflicted lymphocytes, which have

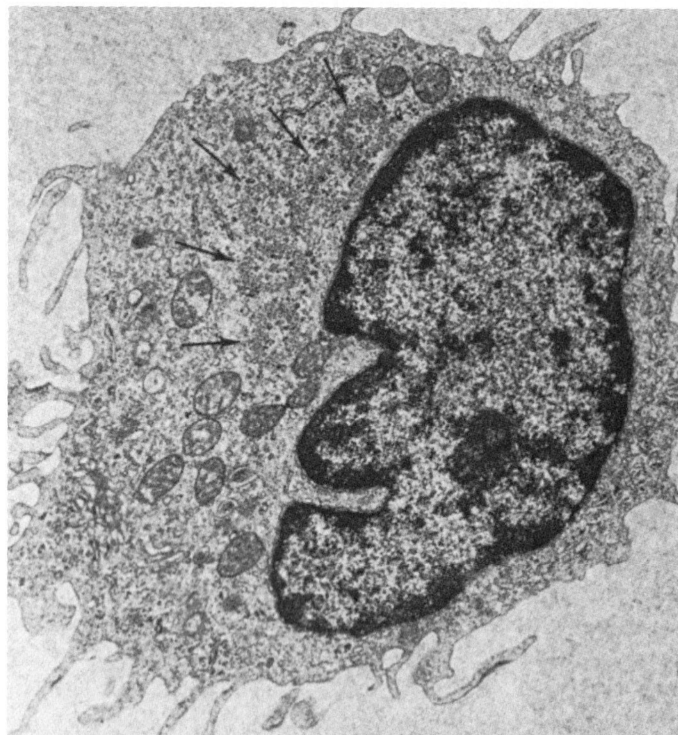

Figure 21.20 Appearance of hairy-cell leukemia, a neoplasm of B lymphocytes caused by HTLV II. Long, hairlike processes are evident in an electron microscope section. Arrows indicate cytopathic effects suggestive of viral infection.

fine cytoplasmic projections or pseudopod-like extensions (figure 21.20). The disease is more common in males, and is probably spread through blood and shared syringes and needles. Unlike ATL, this form of leukemia presents with overall leukopenia but with an increased number of neoplastic B lymphocytes. The most serious complication is the infiltration of the spleen, liver, and bone marrow, which interferes with hematopoeisis. Splenectomy, bone marrow transplants, and antineoplastic drugs have been pursued as treatments, but alpha interferon produces the highest rate of remission.

Other connections have been made between retroviruses and disease. British researchers isolated an apparent retrovirus in the blood cells of 31 of 32 women with breast cancer, and HTLV I has been isolated in the blood of multiple sclerosis and lupus (SLE) patients. Although very preliminary, these findings underline the compelling evidence that retroviruses may indeed be intimately involved in numerous pathologic processes.

Nonenveloped Single-Stranded RNA Viruses: Picornaviruses and Caliciviruses

As suggested by the prefix, *picorna* viruses are named for their small (pico) size and their RNA core (figure 21.21). Important representatives include ***Enterovirus*** and ***Rhinovirus,*** which are responsible for a broad spectrum of human neurologic, enteric, and other illnesses (table 21.3), and *Cardiovirus,* which infects the brain and heart in humans and other mammals. The

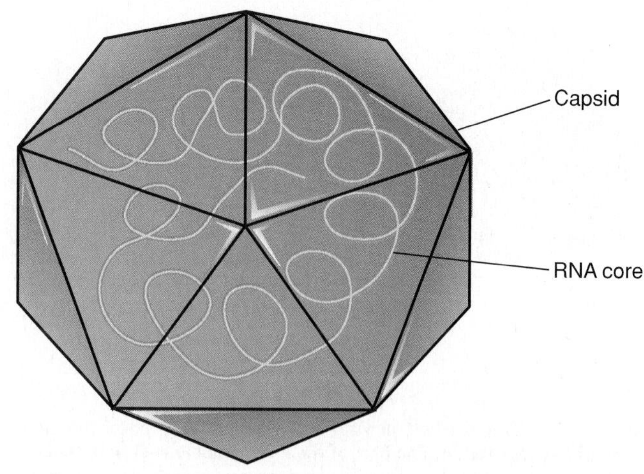

(a)

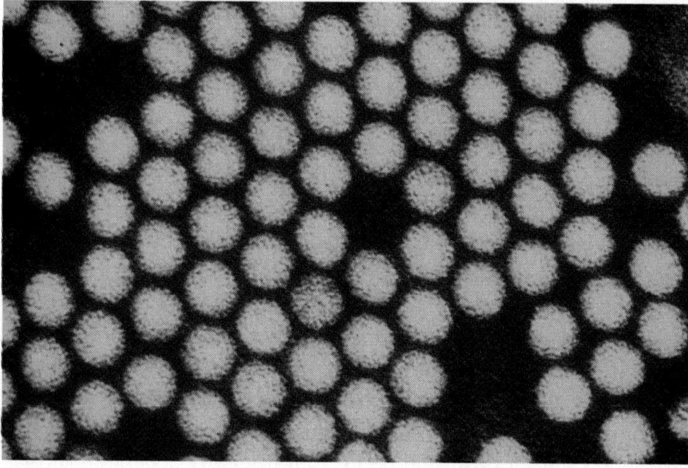

(b)

Figure 21.21 (*a*) A poliovirus, which is a picornavirus and one of the simplest and smallest viruses (30 nm), consists of an icosahedral capsid shell around a tightly packed molecule of RNA. (*b*) A crystalline mass of stacked poliovirus particles in an infected host cell.

Table 21.3 Selected Characteristics of the Human Picornaviruses

Genus	Representative	Primary Diseases
Enterovirus	Poliovirus	Poliomyelitis
	Coxsackievirus A	Focal necrosis, myositis
	Coxsackievirus B	Myocarditis of newborn
	Echovirus	Aseptic meningitis, enteritis, others
	Enterovirus 72	Hepatitis A
Rhinovirus	Rhinovirus	Common cold
Cardiovirus	Cardiovirus	Encephalomyocarditis
Aphthovirus	Aphthovirus	Foot-and-mouth disease (in cloven-foot animals)

Figure 21.22 Polio in an ancient civilization. This stone tablet from the 18th Dynasty in Egypt depicts Siptah bearing the signs of paralytic polio.

Feature 21.11 The Tragic Aftermath of Poliomyelitis

Survivors of the more severe forms of spinal and bulbar polio were left with multiple medical burdens. A patient with respiratory paralysis had to be maintained by a large cylindrical breathing apparatus called the "iron lung" (see chapter opening illustration). Encased and immobilized in these respiratory prisons, many patients could not talk, and their only opportunity to see the world was through a mirror positioned above their beds. Total loss of muscle function meant that even the simplest activities had to be done for them. Maneuvers like bathing and medical examinations were frightening, because the patient had to be disconnected from the machine. In time, these young people were weaned onto air hoses that entered through a tracheostomy tube or were adapted to portable respirators.

Patients with severe paralysis of the skeletal muscles suffered other plights. The unused muscles began to atrophy, growth was slowed, and severe deformities of the trunk and limbs developed. Common sites of deformities were the spine, shoulder, hips, knees, and feet. Because motor function, but not sensation, was compromised, the crippled limbs were often very painful. To prevent this pathology from progressing and even threatening life, the children's spines and limbs were often fused surgically into a rigid state. Physiotherapy was sometimes able to restore partial function if neurons were reparable or if functional muscles could be trained to compensate for dysfunctional ones.

In a desperate attempt to cure polio, patients were treated with gamma globulin, antibiotics, and vitamins, but the epidemic escalated in the early 1950s. As the tragedy and anguish continued to mount, communities were galvanized by a cooperative effort called the March of Dimes. Through the modest donations of millions of schoolchildren and families, enough dimes were collected to underwrite increased research that eventually led to effective vaccines.

What were the prospects for the survivors—those kids in the iron lungs? Many of them died, but due to improved medical care, many others became long-term survivors who have learned to live with their disabilities. As if these impairments were not enough, survivors are also subject to a condition called post-polio syndrome (PPS), an insidious motor neuron disease. PPS manifests as a progressive muscular weakening or sclerosis that comes on many years after the polio attack.

following discussion on human picornaviruses centers upon the poliovirus and other related enteroviruses, the hepatitis A virus, and the human rhinoviruses (HRV).

Poliovirus and Poliomyelitis

Poliomyelitis (polio) is an acute enteroviral infection of the spinal cord that may cause neuromuscular paralysis. Because it often affects small children, it is also called infantile paralysis. No civilization or culture has escaped the devastation of polio. Drawings made by ancient civilizations hint at its long history (figure 21.22), and famous persons such as Sir Walter Scott and Franklin Delano Roosevelt bore its marks. Those of us who experienced the epidemics at the middle of this century will never forget the lasting images of polio: crippling paralysis, braces, the iron lung, and the March of Dimes (see feature 21.11).

Epidemiology of Poliomyelitis

The poliovirus has three serotypes, though types 1 and 3 cause the most severe forms of disease. The protective capsid and lack of an envelope confer chemical stability and resistance to acid, bile, and detergents. As a result, the virus is capable of passing through the gastric environment undamaged, a factor that is instrumental in its transmission.

Sporadic cases of polio can break out at any time of the year, but its incidence is more pronounced during the summer and fall in temperate zones. The virus is passed within the population through food, water, hands, and objects contaminated with feces. Transmission via mechanical vectors such as flies sometimes occurs, but respiratory droplets are only rarely involved. The prevalence of polio varies according to regions. The polio rate in 26 Asian, African, and European countries continues to be high, with an estimated 250 thousand cases of paralytic polio each year and an additional 25 million children afflicted with lesser forms. But thanks to comprehensive vaccination programs, polio has been virtually eliminated in North and South America.

poliomyelitis (poh″-lee-oh-my″-eh-ly′-tis) Gr. *polios,* gray, *myelos,* medulla, and *itis,* inflammation.

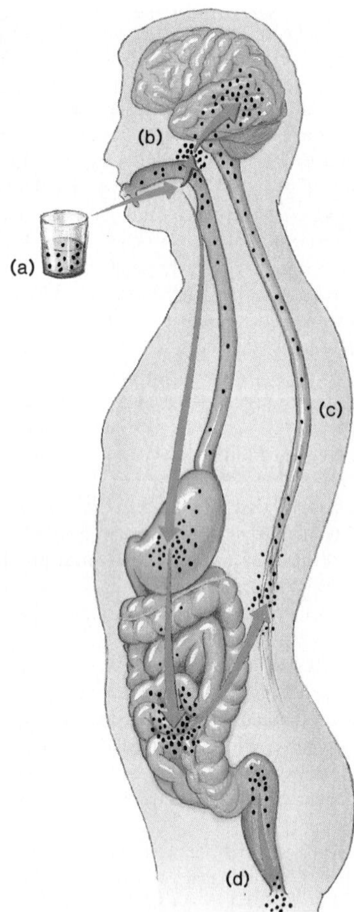

Figure 21.23 The stages of infection and pathogenesis of poliomyelitis. (*a*) The virus is ingested and carried to the throat and intestinal mucosa. (*b*) The virus multiplies in the tonsils. Small numbers of viruses escape to the regional lymph nodes and blood. (*c*) The viruses are further amplified and cross into certain nerve cells of the spinal column and central nervous system. (*d*) The intestine actively sheds viruses.

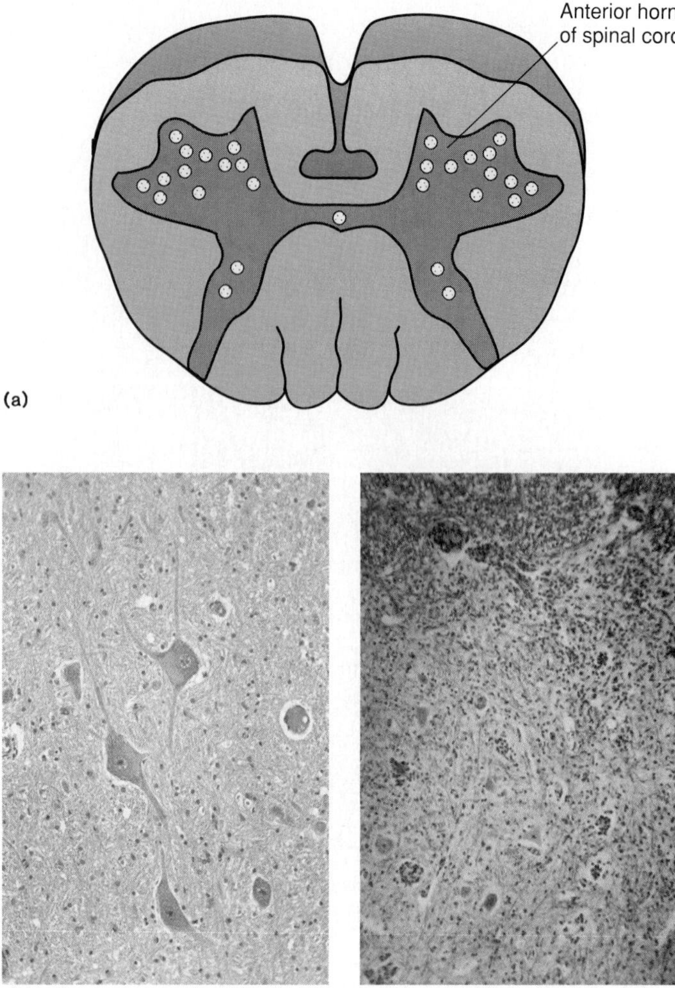

Figure 21.24 Targets of poliovirus. (*a*) A cross section of the spinal column indicates those areas most damaged in spinal poliomyelitis. (*b*) The anterior horn cell of a monkey (left) before and (right) after poliovirus infection.

Infection and Disease

After ingestion of an infectious dose (approximately 10^6 virions), the polioviruses adsorb to receptors of mucosal cells in the oropharynx and intestine (figure 21.23). The precise site of viral multiplication is still unclear, but the likeliest candidates are mucosal epithelia or lymphoid tissue such as tonsils and Peyer's patches. Multiplication of the virus results in large numbers of viruses being shed into the throat and feces, and some of them leaking into the blood. Following an incubation period of one to two weeks, the infection takes one of four courses: (1) subclinical infection, (2) minor disease, (3) aseptic, nonparalytic meningitis, and (4) paralytic disease.

Most infections (95%) are contained as a short-term, asymptomatic viremia. A small number of persons will develop minor disease that includes nonspecific symptoms of fever, headache, nausea, sore throat, and myalgia. If the viremia persists, viruses may gain access to the central nervous system through its blood supply and cross the blood-brain barrier. The virus then spreads along specific pathways in the spinal cord and brain. Being *neurotropic,* the virus infiltrates the motor neurons of the anterior horn of the spinal cord, though it may also attack spinal ganglia, cranial nerves, and motor nuclei (figure 21.24). The so-called major illness, or nonparalytic disease, involves the invasion but not the destruction of nervous tissue. It gives rise to muscle pain and spasm, meningeal inflammation, and vague hypersensitivity.

Paralytic Disease

When the virus is highly virulent or the host is highly susceptible, nervous tissue is destroyed, and various degrees of flaccid paralysis ensue over a period of a few hours to several days. Depending on the level of damage to motor neurons, paralysis of the muscles of the legs, abdomen, back, intercostals, diaphragm, pectoral girdle, and bladder may occur. Less frequently, disintegration of the brain stem, medulla, or even cranial

---

neurotropic (nu″-roh-troh′-pik) Having an affinity for the nervous system.

nerves gives rise to **bulbar poliomyelitis.** In this event, the autonomic cardiorespiratory regulatory centers, palate, pharynx, and vocal cords are affected. The long-term effects of paralytic polio are discussed in feature 21.11. Antibodies to poliovirus occur locally in the intestine and tonsils (secretory antibodies) and in the serum. One exposure confers lifelong immunity.

Diagnosis of Polio

Polio is mainly suspected when epidemics of neuromuscular disease occur in the summer in temperate climates. Conditions requiring differential diagnosis are the Guillain-Barré syndrome, infant botulism, and encephalomyelitis caused by other enteroviruses. Poliovirus can usually be isolated by inoculating cell cultures with stool or throat washings in the early part of the disease. The stage of the patient's infection can also be demonstrated by testing serum samples for the type and amount of antibody.

Measures for Treatment and Control of Polio

In the absence of specific therapy, treatment of polio rests largely on alleviating pain and suffering. During the acute phase, muscle spasm, headache, and associated discomfort can be alleviated by pain-relieving drugs. Respiratory failure may require artificial ventilation maintenance. Because reflex swallowing and coughing may be impaired, drainage or mechanical suction of mucus and secretions is necessary to prevent choking. Impaired ciliary and cough reflexes predispose to secondary pulmonary infection that may require antibiotic treatment. Prompt physical therapy to diminish crippling deformities and to retrain muscles is recommended after the acute febrile phase subsides.

Prevention

The mainstay of prevention is vaccination, a measure that is now taken for granted (see feature 21.12). The two forms of vaccine currently in use are inactivated poliovirus vaccine (IPV), known as the Salk vaccine, and oral poliovirus vaccine (OPV), known as the Sabin vaccine. Both are prepared from animal cell cultures and are trivalent (combinations of the three serotypes). Both vaccines are effective, but one may be favored over the other under certain circumstances. In some tropical countries where exposure of infants to wild virus is common, both vaccines may be given. Polio immunization must be instituted as early in life as possible, usually in four doses starting at about two months of age. Adult candidates are travelers and members of the armed forces.

Feature 21.12 The History of Polio Vaccines

Polio's potential for morbidity and mortality sparked an intensive search for a vaccine as early as the 1930s. The first vaccines were made from inactivated virus extracted from the spinal columns of monkeys, but these preparations turned out to be ineffectual. When a cell culture method for isolating and identifying polioviruses was finally worked out, Jonas Salk and his associates developed an inactivated virus vaccine that they tested in field trials in 1954. Its success led to vaccination on a broad scale. Albert Sabin and his coworkers added a second vaccine in the early 1960s. This oral vaccine, made with attenuated polioviruses, was so successful that it eventually became the vaccine of choice in many countries.

As with most medical blessings, these vaccines did not come without a price. Early incidents cast an unfortunate cloud over the otherwise great promise of the Salk vaccine. Some of the first lots of commercially produced vaccines given in 1955 to children in California and Idaho were faulty, due to incomplete inactivation of some of the viruses. Over 200 children became ill, and of that number, 150 became paralyzed and 11 died. In an episode in Massachusetts in 1959, just the opposite occurred. This time the virus was inactivated by methods that also destroyed its immunogenicity. Sixty-two children receiving this vaccine were later vulnerable to attack by wild virus.

For all of its merits, the oral polio vaccine is not free of medical complications. Being attenuated, it multiplies in vaccinated persons and can be spread to others (some experts consider this an advantage). In very rare instances, the attenuated virus reverts back into a neurovirulent strain and causes disease rather than protects against it. Several instances of paralytic disease have also cropped up among children with hypogammaglobulinemia who have been mistakenly vaccinated. There is a tiny risk (about one case in four million) that an unvaccinated family member will acquire infection and disease from a vaccinated child.

Nonpolio Enteroviruses

Several viruses related to poliovirus commonly cause transient, nonfatal infections. These viruses, **coxsackieviruses** A and B, *echoviruses,* and nonpolio enteroviruses, are like the poliovirus in many of their epidemiologic and infectious characteristics, and are also spread through fecal contamination. The incidence of infection is highest from late spring to early summer in temperate climates and is most frequent in infants and persons living under unhygienic circumstances.

Specific Types of Enterovirus Infection

About 50% to 80% of enteroviral infections are subclinical, and the remainder fall into the category of "undifferentiated febrile illness," characterized by fever, myalgia, and malaise. Symptoms are usually mild and self-limited, and last only a few days. The initial phase of infection is intestinal, after which the virus enters the lymph and blood and disseminates to other organs. The outcome of this infection largely depends on the organ affected. An overview of the more severe complications follows.

Important Complications

Although nonpolio enteroviruses are less virulent than the polioviruses, rare cases of coxsackievirus and echovirus paralysis, aseptic meningitis, and encephalitis occur. Even in severe childhood cases involving seizures, ataxia, coma, and other central

coxsackievirus (kok-sak'-ee-vy"-rus) Named for Coxsackie, New York, where the viruses were first isolated.

echovirus (ek'-oh-vy"-rus) An acronym for enteric cytopathic **h**uman orphan **virus.**

nervous system symptoms, recovery is usually complete. Children are more prone than adults to lower respiratory tract illness, namely bronchitis, bronchiolitis, croup, and pneumonia. All ages, however, are susceptible to the "common cold syndrome" of enteroviruses (see feature 21.13). *Pleurodynia* is an acute disease characterized by recurrent sharp, sudden intercostal or abdominal pain accompanied by fever and sore throat. The pain is often described as stabbing or viselike, hence the name "devil's grip." Recovery is usually spontaneous and complete without specific therapy.

Eruptive skin rashes (exanthems) that resemble the rubella rash are other manifestations of enterovirus infection. Coxsackievirus may cause a peculiar pattern of lesions on the hands, feet, and oral mucosa (hand-foot-mouth disease). Rashes are accompanied by fever, headache, and muscle pain, which usually subside spontaneously with the lesions.

Acute hemorrhagic conjunctivitis is an abrupt inflammation associated with subconjunctival bleeding, serous discharge, painful swelling, and sensitivity to light (figure 21.25). Spread of the virus to the heart in infants can cause extensive damage to the myocardium, leading to heart failure and death in nearly half of the cases. Heart involvement in older children and adults is generally less serious, with symptoms of chest pain, altered heart rhythms, and pericardial inflammation.

Hepatitis A Virus and Infectious Hepatitis

One enterovirus that reacts primarily with the intestinal tract is the **hepatitis A virus (HAV**; enterovirus 72), the cause of infectious or short-term hepatitis. Although this virus is not related to the hepatitis B virus discussed in chapter 20, it shares its tropism for liver cells. Otherwise, the two viruses are different in almost every respect (see table 20.2). The hepatitis A virus is a cubical picornavirus relatively resistant to heat but sensitive to formalin, chlorine, and ultraviolet radiation. There appears to be only one major serotype of this virus.

Epidemiology of Hepatitis A HAV is spread through the oral-fecal route, but the details of transmission vary from one area to another. In general, the disease is associated with deficient personal hygiene and lack of public health measures. In countries with inadequate sewage control, most outbreaks are associated with fecally contaminated water and food. In the United States, the 35,000 cases reported each year are the result of close institutional contact, unhygienic food handling, eating shellfish, sexual transmission, or travel to other countries. In developing countries, children are the most common targets, because exposure to the virus tends to occur early in life, while in North America and Europe, more cases appear in adults. Since the virus is not carried chronically, the principal reservoirs are asymptomatic, short-term carriers or people with clinical disease.

pleurodynia (plur"-oh-din'-ee-ah) Gr. *pleura*, rib, side, and *odyne*, pain.

Feature 21.13 The Common Cold—A Universal Human Condition

The common cold touches the lives of humans more than any other viral infection. Everyone instantly recognizes the sensations of a cold—the nasal stuffiness, scratchy throat, headache, sneezing, and coughing. As an infection, the cold is king, afflicting at least half the population every year and accounting for millions of hours of absenteeism from work and school. The reason for its widespread distribution is not that it is more virulent or transmissible than other infections, but that symptoms of colds are linked to hundreds of different viruses and viral strains. Among the known causative viruses, in order of importance, are rhinoviruses (which cause about half of all colds), paramyxoviruses, enteroviruses, coronaviruses, reoviruses, and adenoviruses. A given cold may be caused by a single virus type or may be due to a mixed infection.

Although the name implies a relationship with cold weather or drafts, studies in which human volunteers with wet heads or feet have been chilled or exposed to moist frigid air have never supported such a link. Most colds occur in the late autumn, winter, and early spring—all periods of colder weather—but this seasonal connection may have more to do with being confined in closed spaces with other people than with temperature. Studies have shown that certain states such as menstruation, extreme fatigue, and allergic rhinitis can facilitate the onset of a cold.

The most significant single factor in the spread of colds are hands contaminated with mucous secretions. The portal of entry is the mucous membranes of the nose and eyes. Multiplication of the virus produces an exudative inflammation with detachment of the epithelium and mild respiratory symptoms. The most common symptom is a nasal discharge, while the least common is fever, except in infants and children. Some people complain of a "head cold," concentrated in the nasal passages and sinuses, while others describe a "chest cold," with sore throat, cough, and chest tightness. Many of the symptoms are highly subjective and depend upon the relative degree of emotional distress or well-being.

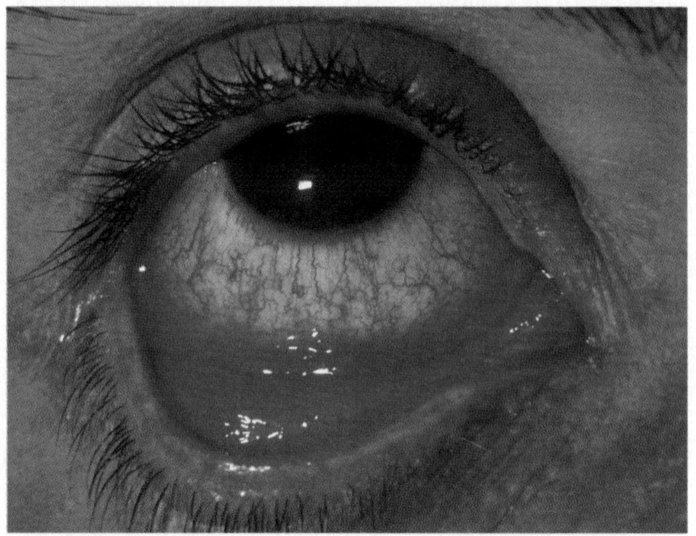

Figure 21.25 Acute hemorrhagic conjunctivitis caused by an enterovirus. In the early phase of disease the eye is severely inflamed, with the sclera bright red due to subconjunctival hemorrhage. Later, edema of the lids causes complete closure of the eye.

The Opportunism of the Hepatitis A Virus In 1988, a massive epidemic of hepatitis A was reported in and around Shanghai, China. Over a period of about two months, more than 400,000 cases were reported, with 11 fatalities. The cause of the epidemic appeared to be contaminated clams eaten without adequate cooking. The clams were taken from local waters that had evidently become polluted by an overload of sewage. Public health officials issued warnings about cleanliness in food procurement and preparation, and advised people not to share utensils. Seafood-associated cases of hepatitis are on the increase in the United States.

Hepatitis A may occasionally be spread by blood or blood products. In 1986, an unusual instance occurred in conjunction with an experimental immunotherapy that uses interleukin-2 to boost a patient's tumor-killing T-lymphocyte population. Therapy involves extracting the patient's T lymphocytes, growing them in cell culture medium, and then replacing the cells in the patient's body. Evidently some of the blood used to make the culture medium was contaminated with HAV. Experiments in some of the treatment centers had to be discontinued during an investigation. But the early findings for this treatment were so encouraging and the needs of these patients so great that the treatments were eventually reinstated, along with measures to screen the donated blood for HAV in the future.

The Course of HAV Infection Swallowed virus is carried to the small intestine, where it multiplies during an incubation period of two to six weeks. Virus is shed in the feces for about two weeks, and at some point, it enters the blood and is carried to the liver. Most infections are either subclinical or accompanied by vague, flu-like symptoms. In more overt cases, the presenting symptoms are loss of appetite, nausea, diarrhea, fever, pain and discomfort in the region of the liver, and darkened urine. Jaundice is present in only about one in 10 cases. Occasionally, hepatitis A occurs as a fulminating disease and causes liver damage, but this is quite rare. Because the virus is not oncogenic and does not predispose to liver cancer, complete uncomplicated cure occurs.

Diagnosis and Control of Hepatitis A A patient's history, liver and blood tests, serological tests, and viral identification all play a role in diagnosing hepatitis A and differentiating it from the other forms of hepatitis. Certain liver enzymes are elevated, and leukopenia is common. The detection of anti-HAV IgM antibodies that are produced early in the infection and tests to identify HA antigen or virus directly in stool samples are of diagnostic value.

There is no specific treatment for hepatitis A once the symptoms begin. Patients receiving immune serum globulin early in the disease usually experience milder symptoms. Prevention of hepatitis A is based primarily on prophylactic immunization with pooled immune serum globulin. Potential globulin recipients are travelers and armed forces personnel planning to enter endemic areas, contacts of known cases, and occupants of day-care centers and other institutions during epidemics. Vaccines in various stages of development and testing include an inactivated viral vaccine and an oral vaccine similar to polio vaccine, based on an attenuated strain of the virus. Control of this disease may be improved by sewage treatment, hygienic food handling and preparation, and adequate cooking of shellfish. Because blood transmission of HAV is rare, blood has not been routinely screened for it (but perhaps it should be under some circumstances).

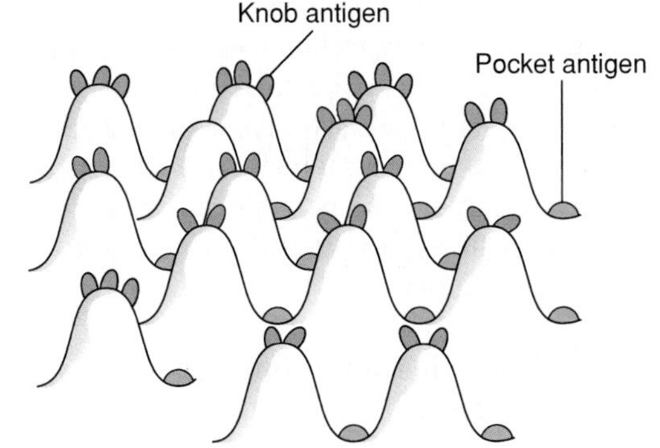

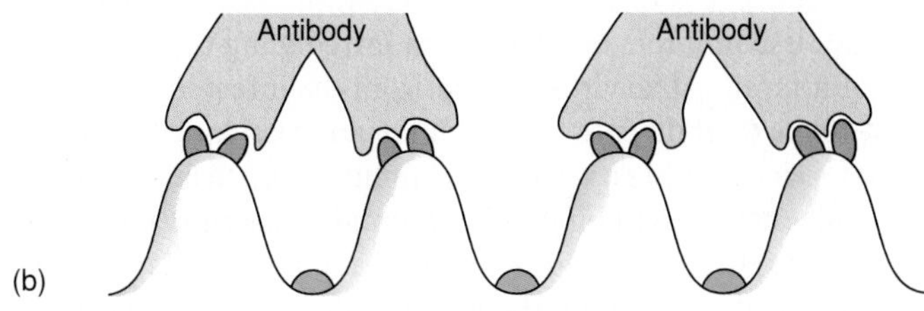

Figure 21.26 (*a*) The surface of a rhinovirus is composed of knobs and pockets. The antigens on the knobs are extremely variable in shape among the scores of viral strains, and those in the pockets are invariable among the strains. (*b*) The knobs are readily accessible to the immune system, and antibodies formed against them will inactivate the virus. But to be fully protected against colds, one would have to produce 100 different kinds of antibodies. The antigens in the pockets are inaccessible and cannot react with antibodies.

Human Rhinovirus (HRV)

The **rhinoviruses,** an extremely large group of picornaviruses having more than 110 serotypes, are associated with the **common cold** (see feature 21.13). Although the majority of characteristics are shared with other picornaviruses, two distinctions set rhinoviruses apart. They are sensitive to acidic environments such as that of the stomach, and their optimum temperature of multiplication is not normal body temperature, but 33°C, the average temperature in the human nose.

Portrait of a Rhinovirus A rhinovirus (type 14) was recently the subject of detailed structural analysis. Over a period of four years, virologists and crystallographers using extremely sophisticated equipment compiled a striking three-dimensional view of its molecular surface that explains why immunity to the rhinovirus has been so elusive. The capsid subunits are of two types: protuberances (knobs), which are antigenically diverse among the rhinoviruses, and indentations (pockets), of which there are only two types (figure 21.26). The antigens on the surface are

rhinovirus (ry′-noh-vy″-rus) Gr. *rhinos,* nose.

the only ones accessible to the immune system, so a successful vaccine would have to contain hundreds of different antigens, and this is not practical. The indented antigens are too deeply situated for either immune surveillance or antibody fit. This strategy is what makes the rhinovirus so successful, and although a vaccine may not be possible, our new knowledge of this virus may spur development of drugs to block the host receptor (see feature 21.14).

Epidemiology and Infection of Rhinoviruses Rhinovirus infections occur in all areas and all age groups at all times of the year. Epidemics caused by a single type arise on occasion, but usually many strains circulate in the population at one time. As mutations increase and herd immunity to a given type is established, newer types predominate. Children are the most successful disseminators of colds, often introducing the virus to the rest of a family. The virus is shed from the infected respiratory tract for several days to weeks, and transmission is linked very closely to inoculation by hands and fomites, and to a lesser extent to droplet nuclei. Although other animals have rhinoviruses, interspecies infections have never been reported. After an incubation period of one to three days, the patient experiences manifestations ranging from headache, chills, fatigue, sore thoat, cough, and a mild nasal drainage to bronchiolitis and atypical pneumonia. Natural host defenses and nasal antibodies have a beneficial local effect on the infection, but immunity is short-lived.

Control of Rhinoviruses The classic therapy is to force fluids and relieve symptoms with various cold remedies and cough syrups that contain nasal decongestants, alcohol, antihistamines, and analgesics. The actual effectiveness of most of these remedies (of which there are hundreds) is rather questionable. See feature 21.14 for other potential treatments. Owing to the extreme diversity of rhinoviruses, prevention of colds through vaccination will probably never be a medical reality. Lacking this, the infection cycle can be interrupted by handwashing and care in handling nasal secretions.

Caliciviruses

Caliciviruses are an ill-defined group of enteric viruses found in humans and mammals. The best known human pathogen is the **Norwalk agent,** named for an outbreak of gastroenteritis that occurred in Norwalk, Ohio, from which a new type of virus was isolated. The Norwalk virus is now believed to cause one-third of all cases of viral gastroenteritis. It is transmitted oral-fecally in schools, camps, cruise ships, and nursing homes, and through contaminated water and shellfish. Infection can occur at any time of the year and in persons of all ages. Onset is acute, accompanied by nausea, vomiting, cramps, diarrhea, and chills; recovery is rapid and complete.

Feature 21.14 Contemplating a Cure for the Common Cold

Finding a cure for the common cold has been such a long-standing hope that it sounds almost like a cliche. Why this mania? It is not motivated by the clinical nature of a cold, which is really a rather benign infection. A more likely reason for the search for a "magic cold bullet" is the prevalence of colds and the money to be made if a truly safe and effective cold drug were discovered.

One nonspecific approach has been to destroy the virus outright and halt its spread. Special facial tissues impregnated with mild acid have been marketed for use during the cold season. One company claims that its aerosol device, which forces hot air and medication to the nose and throat, can nip colds in the bud. Virus multiplication is inhibited by heat, but whether a virus can be killed by heating cells is another question. After years of controversy, a recent study has finally shown that taking megadoses of vitamin C at the onset of a cold can be beneficial.

The important role of natural interferon in controlling many cold viruses has led to the testing and marketing of a nasal spray containing recombinant interferon. Another company is currently developing a novel therapy based on monoclonal antibodies. These antibodies are raised to the site on the human cell (receptor) to which the rhinovirus attaches. In theory, these antibodies should occupy the cell receptor, competitively inhibit viral attachment (receptor blockade), and prevent infection. Experiments in chimpanzees and humans showed that when administered intranasally, this antibody preparation delayed the onset of symptoms and reduced their severity. It is unknown whether any of these products will ever be widely available.

Nonenveloped Double-Stranded RNA Viruses: Reoviruses

Reoviruses have an unusual double-stranded RNA genome and both an inner and outer capsid (see table 21.1). Two of the best-studied viruses of the group are *Rotavirus* and *Reovirus.* Named for its wheel-shaped capsomer, *Rotavirus* (figure 21.27*a*) is a significant cause of diarrhea in newborn humans, calves, and piglets. Because the viruses are transmitted via fecally contaminated food, water, and fomites, disease is prevalent in areas of the world with poor sanitation. Globally, *Rotavirus* is the primary viral cause of mortality and morbidity resulting from diarrhea, accounting for 1 to 1.5 million infant deaths each year. The epidemiology varies with the nutritional state, general health, and living conditions of the infant. Babies from 6 to 24 months of age lacking maternal antibodies have the greatest risk for fatal disease. These children present with symptoms of watery diarrhea, fever, vomiting, dehydration, and shock. The intestinal mucosa may be damaged in a way that chronically compromises nutrition, and long-term or repeated infections can retard growth (figure 21.27*b*). In the United States, rotavirus infection is relatively common, but its course is generally mild. Children are treated as in cholera with oral replacement fluid and electrolytes. A live, attenuated oral vaccine was tested in Venezuela with some success and may be available for general use in the 1990s.

---

calicivirus (kal'-ih-sih-vy''-rus) L. *calix,* the cup of a flower. These viruses have cup-shaped surface depressions.

Rotavirus (roh'-tah-vy''-rus) L. *rota,* wheel.

Reovirus (ree'-oh-vy''-rus) An acronyn of respiratory enteric orphan virus.

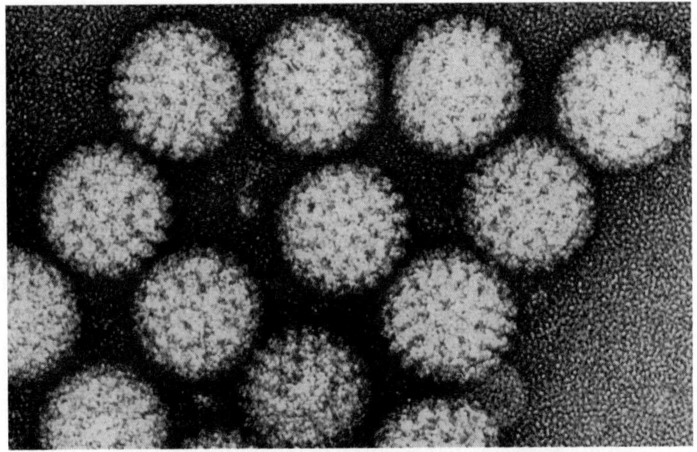

(a)

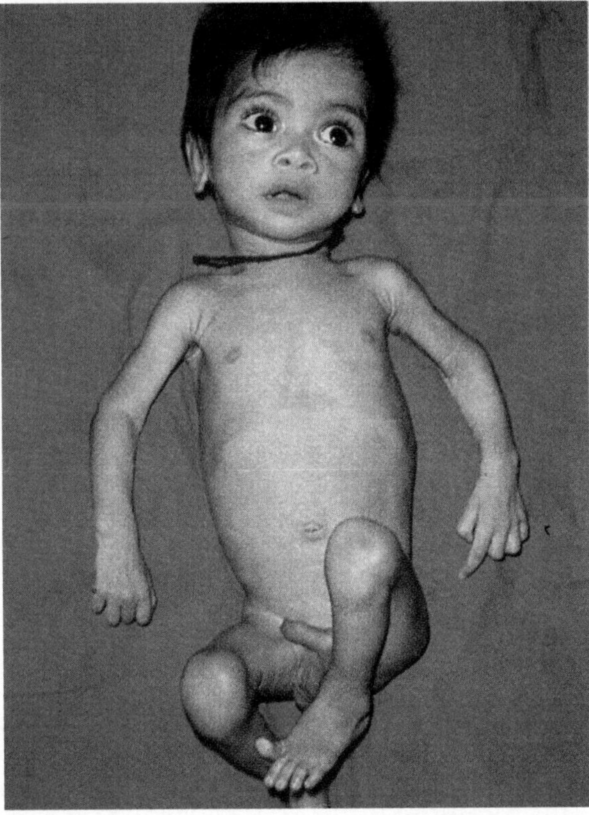

(b)

Figure 21.27 (*a*) A sample of feces from a child with gastroenteritis. So unique is the "spoked-wheel" morphology that identification of a rotavirus is readily made through direct electron microscopy of a fecal specimen. (*b*) An infant suffering from chronic rotavirus gastroenteritis shows evidence of malnutrition, failure to thrive, and stunted growth.

Reovirus is not considered a significant human pathogen. Adults who were voluntarily inoculated with the virus developed symptoms of the common cold. The virus has been isolated from the feces of children with enteritis and from persons suffering from an upper respiratory infection and rash, but most infections are asymptomatic. Although a link to biliary, meningeal, hepatic, and renal syndromes has been suggested, a causal relationship has never been proved.

Table 21.4 Properties of the Agent of Spongiform Encephalopathies

Very resistant to chemicals, radiation, and heat
Does not present virus morphology in electron microscopy of infected brain tissue
Not associated with foreign nucleic acid isolated from infected host cells
Proteinaceous, filterable
Multiplies (slowly) in cell culture; no cytopathic effects
Does not elicit inflammatory reaction in host
Does not elicit antibody formation in host
Responsible for plaques and abnormal fibers forming in brain of host
Very difficult to transmit

Slow Infections by Unconventional Viruslike Agents

Diseases such as subacute sclerosing panencephalitis and progressive multifocal leukoencephalopathy, arising from persistent viral infections of the CNS, have been described in previous sections. They are characterized by a long incubation period, and they can progress over months or years to a state of severe neurological impairment. While these slow infections are caused by known viruses with conventional morphology, there is another group of CNS infections for which no typical virus has been isolated.

The **spongiform encephalopathies** are transmissible, uniformly fatal, chronic infections of the nervous system caused by unusual biological entities that some virologists call unconventional viruses or viroids. Other virologists have concluded that these diseases are caused by agents called *prions*[4] with extremely atypical chemical and physical properties (table 21.4). The agents cause a degeneration in the brain tissue marked by the formation of vacuoles in nerve cells and a spongy appearance of the cerebrum.

Two diseases of humans are known to be caused by unconventional infectious agents. **Kuru** is an endemic disease that was once prevalent among the natives of the New Guinea highlands but has nearly disappeared since cannibalism was discontinued there. Members of these tribes performed mourning rituals that entailed eating the brains of dead relatives. Because women were the principal participants in these ceremonies, they were the most common victims (figure 21.28*a*). It is believed that the kuru agent enters skin cuts and mucosal surfaces and is carried to the brain. After an incubation period of 20–30 years, the parts of the brain controlling gait, posture, speech, and eye movement are adversely affected, leading to symptoms of unsteadiness along with shivering tremor (the tribal word for this

spongiform encephalopathies (spunj'-ih-form en-sef''-uh-lop'-uh-theez)

4. A contraction of **pro**teinaceous **in**fectious particle.

(a)

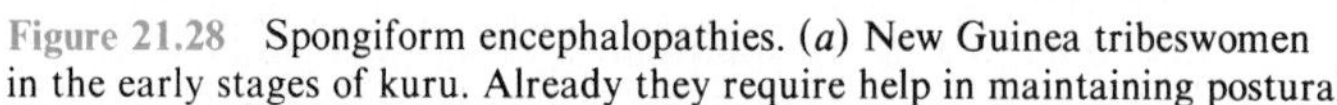

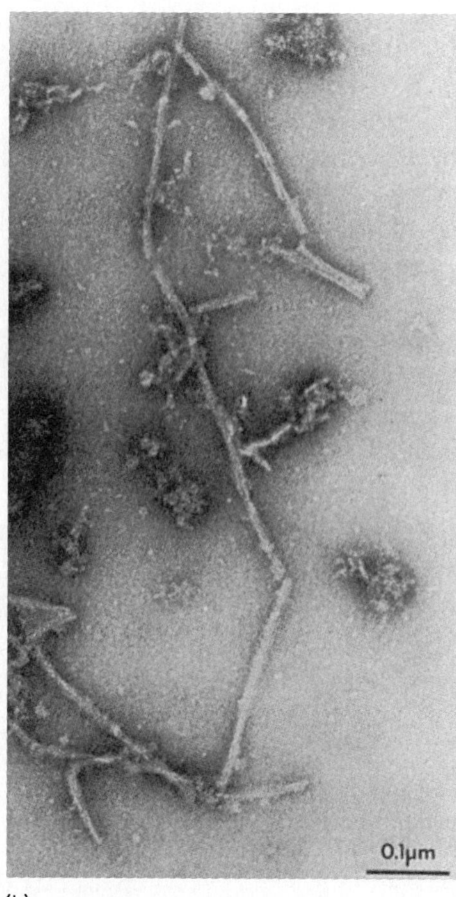

(b)

Figure 21.28 Spongiform encephalopathies. (*a*) New Guinea tribeswomen in the early stages of kuru. Already they require help in maintaining postural stability and exhibit some degree of visual involvement. (*b*) Amyloid fibrils making up the tangles in the brain of a Creutzfeldt-Jakob patient.

is *kuru*). From this point the patient goes downhill toward total loss of locomotion, mental deterioration, and terminal complications.

Creutzfeldt-Jakob disease (CJD) is another encephalopathy with sporadic, worldwide distribution that shows a familial pattern of inheritance and appears to be transmitted by intimate contact with tissues of an infected patient. Persons with CJD have symptoms of altered behavior, memory loss, impaired senses, delirium, and premature senility. Uncontrollable muscle contractions continue until death, which usually occurs within one year of diagnosis. At autopsy the brains show built-up tangled protein fibers (neurofibrillary tangles), enlarged astroglial cells, and vacuoles (figure 21.28*b*). Because the infectious agent is not readily destroyed by usual methods, great care must be taken in handling tissues from CJD patients in a medical setting. Some cases have been transmitted through grafts, inoculation, and handling of brain tissue.

It is becoming clear that a range of neurological diseases have a pathological picture similar to CJD. Examples are scrapie in sheep, amyelotrophic lateral sclerosis, aluminum intoxication, and Alzheimer's disease. It has been suggested that these syndromes be lumped together in the category of transmissible virus dementia (TVD). A common characteristic of these diseases may be that infection or some other factor prevents the transport of a necessary protein neurofilament, which builds up along the neuron and disturbs brain function.

Chapter Review with Key Terms

The RNA Viruses

Enveloped Single-Stranded Viruses

Orthomyxoviruses: Influenza viruses have a segmented genome and glycoprotein spikes for binding to the host (hemagglutinin, neuraminidase); exhibit antigenic shift and drift that creates new viral subtypes. Influenza A is spread in epidemics and pandemics from human to human and from animal (pigs, poultry) to human; virus shed in aerosols in close company; children and elderly most susceptible; clinical disease marked by respiratory symptoms, fever, aches, sore throat; secondary pneumonia is frequent cause of death; long-term complications are Guillain-Barré syndrome, Reye's syndrome in children; controlled with amantadine and vaccination.

Paramyxoviruses: Nonsegmented genome; glycoprotein spikes facilitate development of **multinucleate giant cells**; are spread through droplets.

Paramyxovirus causes **parainfluenza,** a common, mild form of influenza or coldlike illness, mostly in children; **mumps** or **parotitis,** inflammatory infection of salivary glands, sometimes testes, pancreas, and other organs; occurs often in children, creates swelling of jaw and neck; may cause deafness; prevented by live, attenuated vaccine (MMR).

Morbillivirus, the viral agent of **measles** or **rubeola,** is an acute, highly contagious repiratory infection that produces a maculopapular rash over the body and in the oral cavity (**Koplik's spots**); usually mild, but may threaten life in a small number of children; other complication is progressive brain disease called **SSPE;** attenuated vaccine administered in combination (MMR) or alone; booster recommended; recent epidemics due to lapses in community vaccinations.

Pneumovirus or **respiratory syncytial virus (RSV)** causes **croup,** an acute respiratory syndrome in newborns that is a problem in hospital nurseries; treated with aerosol ribavirin.

Rhabdoviruses: The main human pathogen is bullet-shaped ***Lyssavirus,*** the cause of **rabies,** a nearly 100% fatal infection of the brain and meninges; a zoonosis carried by wild carnivores (wild dogs, skunks, raccoons); can spread to domestic animals; usually contracted through an animal bite or scratch but may be inhaled; virus enters nerves and is carried to spinal cord and brain, and eventually to salivary glands; massive infection leads to **furious rabies,** with severe nervous hyperactivity and hydrophobia, or **dumb rabies,** marked by paralysis and loss of consciousness; death results from cardiac or respiratory arrest; treatment after animal contact requires scrupulous cleaning of wound, passive immunization with rabies immune globulin, and active immunization with **human diploid cell vaccine;** vaccine is also given to prevent infection in domestic animals and humans at risk.

Other Enveloped Viruses

Coronaviruses: Have distinctive crown of spikes; human coronavirus (HCV) is one cause of the common cold.

Togaviruses: **Rubella virus** or German measles virus; a **teratogenic** virus that crosses the placenta and creates developmental disruptions; virus spread through close contact, infected mucous droplets; period of respiratory symptoms is followed by rash; usually mild in children; infection during pregnancy may lead to congenital rubella with severe organic (heart, nervous system) damage; control by live, attenuated vaccine (MMR), preferably given in childhood; women should be vaccinated prior to pregnancy.

Zoonotic Viruses

Arenaviruses: Spread by rodent secretions, droppings; example is Lassa fever, a severe and often fatal African hemorrhagic fever.

Arboviruses: Loose grouping of 400 viruses spread by arthropod vectors and harbored by numerous birds and mammals; include togaviruses, bunyaviruses, and reoviruses; diseases occur in conjunction with greatest vector activity; humans usually dead-end host; cause febrile infections (tick-borne fever), mosquito-borne encephalitides in horses and humans (**WEE, EEE**), and in humans and birds (St. Louis encephalitis); hemorrhagic fevers include **yellow fever** and **dengue fever,** sometimes deadly mosquito-borne diseases.

Retroviruses

Enveloped, single-stranded RNA viruses with **reverse transcriptase** that converts their single-stranded RNA into double-stranded DNA genome; insertable into host cell, so viruses are latent and may be oncogenic.

Human Immunodeficiency Virus (HIV, types 1 and 2): Causes **acquired immunodeficiency syndrome (AIDS)**, an HIV antibody-positive state with some combination of opportunistic infections, fever, weight loss, chronic lymphadenitis, neoplasms, diarrhea, brain dysfunction; AIDS-related complex or **ARC** is an HIV-positive intermediate stage of the disease with some combination of the symptoms. HIV appears to have originated in Africa; an estimated 5–10 million persons are infected worldwide; highest prevalence is in United States and Africa.

AIDS is devastating, has high mortality rate; virus occurs in semen, vaginal secretions, and blood; is transmitted through homosexual and heterosexual contact and shared blood; risk groups or behaviors (in order of incidence) are: (1) male homosexuals or bisexuals practicing unsafe anal or oral sex; (2) intravenous drug abusers; (3) heterosexual partners of drug users and bisexuals not using protection; (4) blood transfusion and organ transplant patients; (5) hemophiliacs (less common now); (6) persons with unknown risk factors; (7) fetuses or newborns of infected mothers; (8) health care workers (occasionally), but precautions greatly reduce the risk.

HIV attacks cells with CD4 receptors, starting in macrophages and moving to T cells and other lymphocytes; initial infection creates a flu-like illness; virus becomes latent, gradually destroying T-helper cells, invading brain; after incubation period of 2–15 years, lack of immunities allows infectious agents to take over, including severe intractable **pneumocystis pneumonia (PCP)** and multiple organ infections by protozoa, fungi, mycobacteria, and viruses that would ordinarily be innocuous; **Kaposi's sarcoma** and lymphoma are common cancers; AIDS increases the severity of all other infections, which may serve as cofactors in destroying and weakening immunities; death is caused by secondary factors.

Treatment for AIDS includes drugs for HIV—**azidothymidine** (AZT), dedeoxyinosine (DDI), and many experimental ones—as well as drugs for secondary infections; prevention measures are safe sexual practices; numerous vaccines are in clinical trials.

Other Retroviruses and Diseases: HTLV I causes **adult T-cell leukemia** or mycosis fungoides, a fatal cancer; HTLV II causes another cancer with high mortality rate, **hairy-cell leukemia.**

Nonenveloped RNA Viruses

Picornaviruses are the smallest human viruses.

Enteroviruses: Spread primarily by the oral-fecal route. **Poliovirus,** cause of **polio** or infantile paralysis; common in children; virus is resistant, may be spread through close contact, flies, contaminated food, water; infection usually mild; some cases lead to paralysis caused by infection and destruction of spinal neurons; bulbar type affects most muscles, requires life support, leads to bone and joint deformation; control of disease through vaccination with oral vaccine (OPV) or inactivated vaccine (IPV).

Coxsackievirus causes respiratory infection, hand-foot-mouth disease, conjunctivitis.

Hepatitis A virus causes short-term hepatitis that is a milder form of viral hepatitis; spread through close institutional contact, contaminated food or shellfish; disease is initially flu-like, followed by liver discomfort and diarrhea; jaundice not common; prevented by passive immunization with gamma globulin; no vaccines.

Rhinovirus: Most prominent cause of common cold; spread by droplets; grows in cooler regions of body (nasal membrane); virus exists in over 100 forms, is difficult to control; symptoms are nasal drainage, cough, sneezing, sore throat; treatment symptomatic.

Calicivirus: **Norwalk agent,** a common cause of viral gastroenteritis.

Double-Stranded RNA Viruses

Reoviruses: Most important one is *Rotavirus,* the cause of severe infantile diarrhea; most common cause of viral enteric disease worldwide; children may die from effects of diarrhea, dehydration, shock.

Unconventional Viruslike Agents

Viroids or **prions** have unusual properties: very resistant, proteinaceous, do not produce inflammation but do cause progressive neurological diseases; diseases are the **spongiform encephalopathies,** named for brain's appearance; examples are **kuru,** a disease of cannibals in New Guinea, and **Creutzfeldt-Jakob disease,** a slow deterioration of the brain.

True–False Questions

Determine whether the following statements are true (T) or false (F). If you feel a statement is false, explain why, and reword the sentence so that it reads accurately.

____ 1. Influenza virus, paramyxovirus, mumps virus, and measles virus are all enveloped, single-stranded RNA viruses.

____ 2. Infection with rabies virus causes multinucleate giant cells.

____ 3. Mumps infection affects primarily the salivary glands and rarely other organs.

____ 4. Measles is one of the most contagious diseases of humans.

____ 5. Respiratory syncytial virus causes a life-threatening disease in babies called croup.

____ 6. Many of the symptoms of viral infection are due to viremia.

____ 7. Rabies is caused by a bullet-shaped enveloped virus called *Lyssavirus*.

____ 8. Rabies is treated with both active and passive immunization simultaneously.

____ 9. HIV and other retroviruses cause latent infections in the host cell by means of reverse transcriptase.

____ 10. The organs most affected by the AIDS virus are the lungs, liver, and brain.

____ 11. Polio and hepatitis A viruses are enteric viruses spread by the oral-fecal route.

____ 12. Rhinoviruses are the primary cause of the common cold.

____ 13. The spongiform encephalopathies are brain infections readily spread by droplet nuclei.

Concept Questions

1. How do hemagglutinin and neuraminidase spikes function in influenza virus infections? What are antigenic drift and shift? How do the names for the types of flu originate? What are the complications of influenza? What is the nature of the vaccine?
2. Describe how viruses induce giant cells. Which viruses do this, and what impact does this have on the spread of infection?
3. What is the progression of the measles rash? What is the cause of death in measles, and what are the most severe complications of measles infections? Why has complete global eradication of measles been so difficult, even with vaccination?
4. Describe the epidemiological cycle in rabies. Describe the infection cycle in humans. Why is rabies so uniformly fatal? Why is the latest vaccine less traumatic than earlier ones?
5. What is given in postexposure treatment that is not given in preexposure treatment of rabies?
6. What is a teratogenic virus? Which RNA viruses have this potential? What is the protocol to prevent congenital rubella? How is rubella different from red measles?
7. What are the principal carrier arthropods in arboviruses? How is the cycle of the virus maintained in the wild? Describe the symptoms of the encephalitis types of infections (WEE, EEE) and the hemorrhagic fevers (yellow fever).
8. What are retroviruses, and how are they different from other viruses? Give examples of the three principal human retroviruses and the diseases they cause. Define AIDS.
9. Trace the possible transmission pathways of the AIDS virus from an infected homosexual man to another homosexual man. From one drug abuser to another drug abuser. From a bisexual man to his wife. From a prostitute to a john. From a monogamous man to his monogamous wife of 50 years. From one teenager to another. From a mother to a child.
10. Briefly explain the groups or activities most likely to spread AIDS. How is it not spread? Why is there a concern with bone marrow transplants?
11. Is AIDS an all-or-none thing, or a series of gradual pathological changes? What are the primary target tissues and changes? Why are the T cells so affected? Why is the brain involved in infection? What is an ARC patient? What causes the long latency in AIDS?
12. List the secondary diseases that accompany AIDS. Why are these particular pathogens involved? Why are AIDS patients so susceptible to Kaposi's sarcoma?
13. If a person is seropositive for AIDS, does this mean that the patient has the disease? What does seronegativity mean in a monogamous person who has not had a blood transfusion or taken IV drugs? What does it mean in a person who has participated in high-risk behavior? What is the window for antibody presence in the blood?
14. How does AZT control the AIDS virus? Are there any actual cures for AIDS?
15. Describe the epidemiology and the progress of polio infection and disease. What causes the paralysis and deformity? Compare and contrast the two types of vaccines.
16. Describe the epidemiology of hepatitis A.
17. What exactly is the common cold? Why is control of it so difficult?
18. How are influenza, parainfluenza, mumps, measles, rubella, RSV, and rhinoviruses transmitted? What are some similarities in the symptoms of these infections?
19. How are arboviruses, rabies virus, and arenaviruses spread? What are the roles of nonhuman vertebrates in these diseases?
20. How are poliovirus, hepatitis A virus, rotavirus, caliciviruses, and coxsackieviruses transmitted? What characteristics of the viruses in this group cause them to be readily transmissible?
21. What are the cellular targets of mumps, measles, rabies, and polio viruses?
22. For which of the RNA viruses are vaccines available? For which ones are vaccines in development? Which viruses will probably never have a vaccine developed for them? Why?
23. For which of the RNA viruses are there specific drug treatments?
24. Briefly compare the neurological syndromes caused by measles and AIDS with the slow progressive neurological diseases caused by viruslike agents. How are they different?

Practical/Thought Questions

1. Explain the relationship between herd immunity and the development of influenzal pandemics. Why will herd immunity be lacking with genetic drift or shift?
2. Explain how the antibody content of patient's sera can be used to predict which strain of influenza predominated during a previous epidemic. Offer a way that mixed strain infections in influenza give rise to new and different strains.
3. What is the connection between Guillain-Barré and Reye's syndromes and viral infections?
4. Why do infections such as mumps, measles, polio, rubella, and RSV regularly infect children and not adults?
5. AIDS is thought to have originated from populations of remote people who brought it to civilization where it was amplified. Can you think of a plausible series of events that could explain how it came to be so widespread?
6. Why is there such a discrepancy between the total number of reported cases of AIDS and the projected number of cases estimated by health authorities?
7. Sometimes one reads that AIDS is 100% fatal. Is it? Why, or why not?
8. Why is the AIDS virus such a difficult target for the immune system? For drugs? Why are coinfections with the AIDS virus so devastating?
9. Can you explain why whole live, attenuated (or even dead) AIDS virus vaccines pose a risk? Why is it going to take so long to develop a productive vaccine?
10. What precautions can a person take to prevent himself or herself from getting AIDS? How can a health care worker prevent possible infection?
11. What would be the medical problems of a child being maintained in an iron lung?
12. Various over-the-counter cold remedies control or inhibit inflammation and depress the symptoms of colds. Is this beneficial or not?
13. Three months after having corneal surgery, a patient reported to his physician with fever and numbness of the face. His throat was tight, and he had difficulty swallowing; he was afraid of showering because of the painful spasms it caused. Within hours, he became paralyzed, lapsed into a coma, and died. Physicians were stumped as to the cause. Can you offer some ideas? What therapy should have been given?
14. Late in the spring, a young boy developed fever, loss of memory, difficulty in speech, convulsions, and tremor, and lapsed into a coma. He tested negative for bacterial meningitis and had had no contact with dogs or cats. He survived, but had long-term mental retardation. What are the possible diseases he might have had, and how would he have contracted them?
15. Biopsies from the liver and intestine of an otherwise asymptomatic patient show masses of acid-fast bacilli throughout the tissue. Can you diagnose the illness on the basis of this evidence alone? (Hint: The CDC can.)

CHAPTER 22

Environmental and Applied Microbiology

A complex, computer-controlled machine called a fermentor can manufacture mass quantities of products such as enzymes and drugs. This is just one example of the numerous ways modern industry applies the powers of microbial metabolism to the work of synthesis.

Chapter Preview

The past dozen chapters have focussed on the somewhat narrow field of pathogenic microbiology, necessarily overlooking many other interesting and useful aspects of microbiology and microorganisms. This last chapter emphasizes microbial activities that are not only beneficial to humans and other organisms, but that also help maintain and control the life support systems on earth. This subject will be explored from the standpoints of: (1) the natural roles of microorganisms in the environment and their contributions to the ecological balance, including soil, water, and mineral cycles, and (2) the artificial applications of microbes in the food, medical, biochemical, drug, and agricultural industries. We will finish with a survey of the newest and most explosive area of microbiology: applications in genetic engineering.

Ecology: The Interconnecting Web of Life

The study of microbes in their natural habitats is known as **environmental** or **ecological microbiology,** while the study of the practical uses of microbes in food processing, industrial production, and bioengineering is known as **applied microbiology.** Separating this chapter into two disciplines is an organizational convenience, since the two areas actually overlap to a considerable degree—largely because most natural habitats have been altered by human activities. Human intervention in natural settings has changed the earth's warming and cooling cycles, increased wastes in soil, polluted water, and altered some of the basic relationships between microbial, plant, and animal life. Now that humans are also beginning to release totally new, genetically recombined microbes into the environment and to alter the genes of plants, animals, and even themselves, what does the future hold? Although this question may be imponderable, we know one thing for certain: Silently working in the background will be microbes—the most vast and powerful resource of all.

In chapter 6 we first touched upon the widespread distribution of microorganisms and their adaptations to most habitats of the world, from extreme to temperate. Regardless of their exact location or type of adaptation, microorganisms necessarily are exposed to and interact with their environment in complex and extraordinary ways. The science that studies interactions between microbes and their environment and the effects of these interactions on the earth is **microbial ecology.** Unlike studies of pathogens, which deal with the pathological effects of a single organism or its individual characteristics in the laboratory, ecological studies are aimed at the interactions taking place between organisms and their environment at many levels at any given moment. Therefore, ecology is a broad-based science that merges many subsciences of biology as well as geology, chemistry, and engineering.

Ecological studies take into account both the biotic and abiotic components of an organism's environment. The **biotic** factors include any other living or once-living organisms[1] such as symbionts sharing an organism's habitat, parasites, or food substrates. The **abiotic** factors include any nonliving surroundings such as the atmosphere, soil, water, temperature, and light. A collection of organisms together with its surrounding physical and chemical factors is defined as an **ecosystem** (figure 22.1).

The Organization of Ecosystems

The earth initially may seem like a random, chaotic place, but it is actually an incredibly organized, well-tuned machine. Ecological relationships exist at several levels, ranging from the entire earth all the way down to a single organism (figure 22.1). The most all-encompassing of these levels, the **biosphere,**

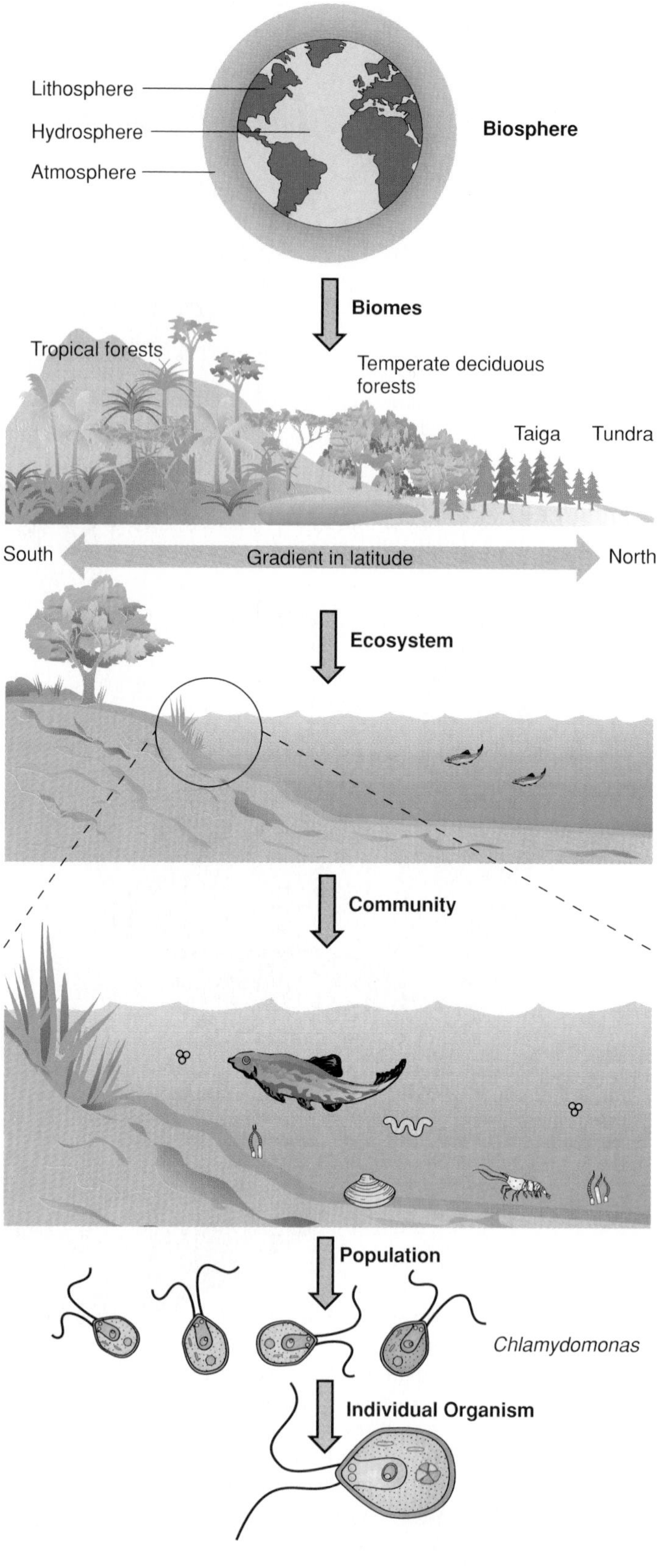

Figure 22.1 The levels of organization in an ecosystem, ranging from the biosphere to the individual organism.

biotic (by-aw′-tik)

1. Biologists make a distinction between nonliving and dead. A nonliving thing has never been alive, whereas a dead thing has been alive but no longer is.

abiotic (ay″-by-aw′-tik)

includes the thin envelope of life (about 14 miles deep) that surrounds the earth's surface. This global ecosystem is comprised of the **hydrosphere** (water), the **lithosphere** (a few miles into the soil), and the **atmosphere** (a few miles into the air). The biosphere maintains or creates the conditions of temperature, light, gases, moisture, and minerals required for life processes. The biosphere may be naturally subdivided into terrestrial and aquatic realms. The terrestrial realm is usually distributed into particular climatic regions called **biomes** (by′-ohmz), each of which is characterized by a dominant plant form, altitude, and latitude. Particular biomes include grassland, desert, mountain, and tropical rain forest (figure 22.1). The aquatic biosphere is generally divisible into freshwater and marine realms.

Biomes and aquatic ecosystems are generally composed of mixed assemblages of organisms that live together at the same place and time, and that usually exhibit well-defined nutritional or behavioral interrelationships. These clustered associations are called **communities.** Although most communities are identified by their easily visualized dominant plants and animals, they also contain a complex assortment of bacteria, fungi, algae, protozoa, and even viruses. The basic units of community structure are **populations,** groups of organisms of the same species. The organizational unit of a population is the individual organism, and each organism in turn has its own levels of organization (organs, tissues, cells).

Ecosystems are generally balanced, with each organism existing in its particular habitat and niche. The **habitat** is the physical location in the environment to which an organism has adapted. In the case of microorganisms, the habitat is frequently a *microenvironment,* where particular qualities of oxygen, light, or nutrient content are somewhat stable. The **niche** is the overall role that a species (or population) serves in a community. This includes such activities as nutritional intake (what it eats), position in the community structure (what eats it), and rate of population growth. A niche can be broad (such as scavengers that feed on nearly any organic food source) or narrow (microbes that decompose cellulose in forest litter). In a subsequent section, we will see that microbes involved in the cycling of nutrients often have narrow, very specialized niches.

Energy and Nutritional Flow in Ecosystems

All living things must obtain nutrients and a usable form of energy from the abiotic and biotic environments. The energy and nutritional relationships in ecosystems may be described in a number of convenient ways. A **food chain** or **energy pyramid** provides a simple summary of the general trophic (feeding) levels, designated as producers, consumers, and decomposers, and traces the flow and quantity of available energy from one level to another (figure 22.2). It is worth noting that microorganisms are the only living things that exist at all three major trophic levels. The nutritional roles of microorganisms in ecosystems are summarized in table 22.1.

Life would not be possible without **producers,** because they provide the fundamental energy source for all levels of the trophic pyramid. Producers are the only organisms in an eco-

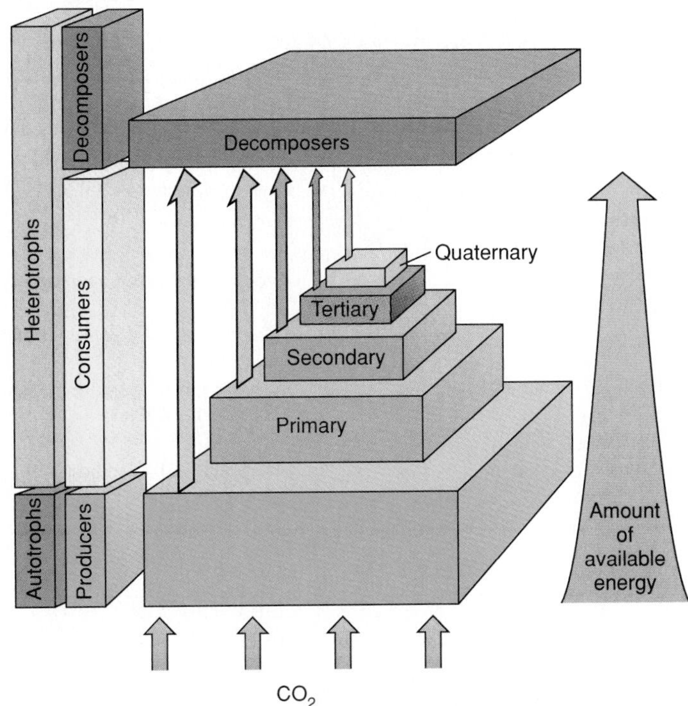

Figure 22.2 A trophic or energy pyramid. The relative size of the blocks indicates the number of individuals that exist at a given trophic level. The arrow on the right indicates the amount of usable energy from producers to top consumers. Both the number of organisms and the amount of usable energy decrease with each trophic level. Decomposers are an exception to this pattern, but only because they can feed from all trophic levels.

system that can produce organic carbon compounds like glucose by assimilating (fixing) inorganic carbon (CO_2) from the atmosphere. Such organisms may also be termed autotrophs. Most producers are photosynthetic organisms such as plants and cyanobacteria that convert the sun's energy into the energy of chemical bonds. Photosynthesis was introduced in chapter 6, and the chemical steps in the process will be discussed in a later section of this chapter. A small but important amount of CO_2 assimilation is brought about by unusual bacteria called lithotrophs. The metabolism of these organisms derives energy from oxidation-reduction reactions of simple inorganic compounds such as sulfides and hydrogen. In certain ecosystems (see thermal vents, feature 6.9), lithotrophs are the sole supporters of the energy pyramid.

Consumers eat the bodies of other living organisms and obtain energy from bonds present in the organic substrates they contain. The category includes animals, protozoa, and a few bacteria and fungi. A pyramid usually has several levels of consumers, ranging from *primary consumers* (grazers), which consume producers; to *secondary consumers* (carnivores), which feed on primary consumers; to *tertiary consumers,* which feed on secondary consumers; and up to *quaternary consumers* (usually the last level), which feed on tertiary consumers. Figure 22.3 shows specific organisms at these levels.

Decomposers, primarily microbes inhabiting soil and water, break down and absorb the organic matter of dead organisms, including plants, animals, and other microorganisms.

Table 22.1 The Major Roles of Microorganisms in Ecosystems

Role	Description of Activity	Examples of Microorganisms Involved
Primary producers	Photosynthesis	Algae, cyanobacteria, sulfur bacteria
	Chemosynthesis	Chemolithotrophic bacteria in thermal vents
Consumers	Predation	Free-living protozoa that feed on algae and bacteria; some fungi that prey upon nematodes
Decomposers	Degradation of plant and animal matter and wastes	Soil saprobes (primarily bacteria and fungi) that degrade cellulose, lignin, and other complex macromolecules
	Mineralization of organic nutrients	Soil saprobes that reduce organic compounds to inorganic compounds such as CO_2 and minerals
Cycling agents for bioelements	Biogeochemical cycles for carbon, nitrogen, phosphorus, sulfur	Specialized bacteria that transform elements into different chemical compounds to keep them cycling from the biotic to the abiotic and back to the biotic phases of the biosphere
Parasites	Living and feeding on hosts	Viruses, bacteria, protozoa, fungi, and worms that play a role in population control

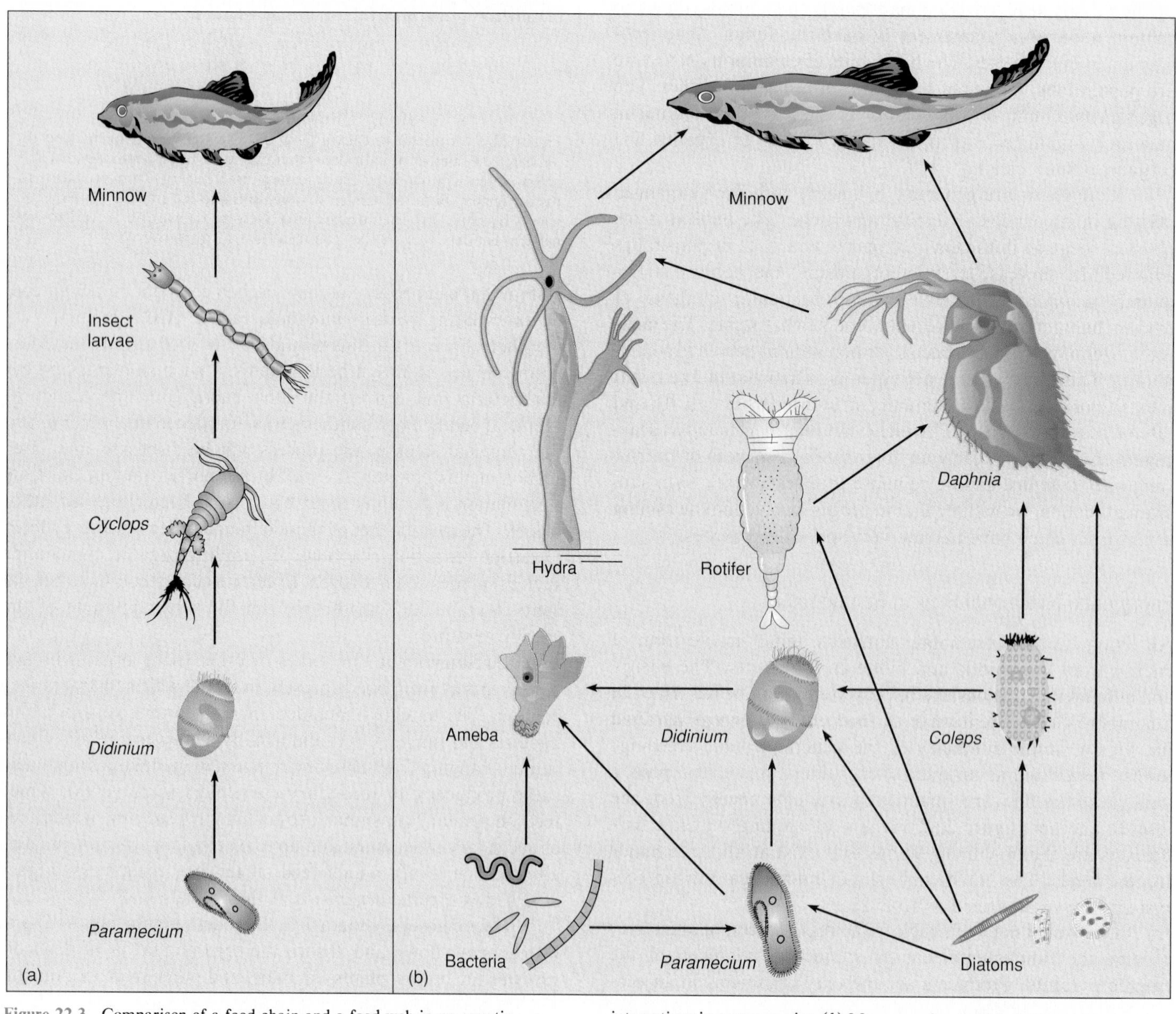

Figure 22.3 Comparison of a food chain and a food web in an aquatic ecosystem. (*a*) A food chain is the simplest way to present specific feeding relationships among organisms, but it may not reflect the total nutritional interactions in a community. (*b*) More complex trophic patterns are accurately depicted by a food web, which traces the multiple feeding options that exist for most organisms. Note: Arrows point toward the consumers.

Because of their biological function, decomposers are active at all levels of the food pyramid. Without this important nutritional class of saprobes, the biosphere would stagnate and die. The work of decomposers is to reduce organic matter into an inorganic form such as minerals and gases that can be cycled back into the ecosystem, especially for the use of primary producers. This process, also termed *mineralization,* is so efficient that almost all biological products can be reduced by some type of decomposer. Numerous bacterial species decompose cellulose and lignin, complex polysaccharides from plant cell walls that account for the vast bulk of detritus in soil and water. Both fungi and bacteria decompose such complex animal compounds as keratin and chitin.

The pyramid illustrates several limitations of ecosystems with regard to energy. Unlike nutrients, which may be passed among trophic levels, recycled, and reused, energy does not cycle. To maintain complex interdependent trophic relationships such as that shown in figure 22.3, a constant input of energy is required at the producer level, and as energy is transferred to the next level, a large proportion (as high as 90%) of the energy will be lost in a form (primarily heat) that cannot be fed back into the system. Thus, the amount of energy available decreases at each successive trophic level. This energy loss also decreases the actual number of individuals that can be supported at each successive level.

A more concrete image of a feeding pathway can be provided by a **food chain.** Although it is a somewhat simplistic way to describe a community feeding pattern, a food chain helps identify the specific types of organisms that may be present at a given trophic level in a natural setting (figure 22.3*a*). Feeding relationships in communities are more accurately represented by a multichannel food chain or **food web** (figure 22.3*b*). This is because, in most ecosystems, a producer group does not ordinarily serve as food for a single type of consumer, but feeds a variety of different consumers. Likewise, the consumer structure is interconnected in such a way that a single organism may feed on a variety of different foods from both producer and consumer levels. These organisms are sometimes called omnivores.

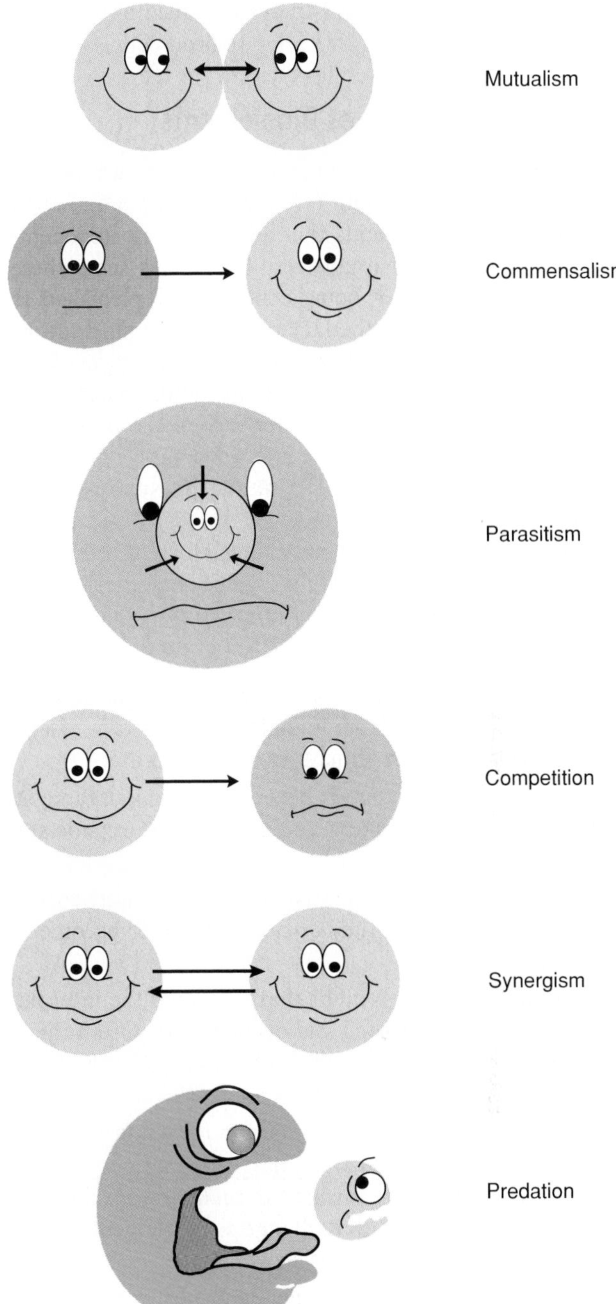

Figure 22.4 Community interactions that have beneficial, harmful, or neutral effects.

Ecological Interactions Between Organisms in a Community

Whenever complex mixtures of organisms associate, they develop various dynamic interrelationships based on nutrition and shared habitat. These relationships, some of which were described in chapter 6, include mutualism, commensalism, parasitism, competition, synergism (cross feeding), predation, and scavenging (figure 22.4).

Mutually beneficial associations (*mutualism*) such as protozoans living in the termite intestine are so well-evolved that the two members require each other for survival. In contrast, *commensalism* is one-sided and independent. Although the action of one microbe favorably affects another, the first microbe receives no benefit. Many commensal unions involve *cometabolism,* meaning that the waste products of the first microbe are useful nutrients for the second one. In *synergism,* two organisms that are usually independent cooperate to break down a nutrient neither one could have metabolized alone. *Parasitism* is an intimate relationship in which the parasite derives its nutrients and habitat from a host that is usually harmed in the process. In *competition,* one microbe gives off antagonistic substances that inhibit or kill susceptible species sharing its habitat. A *predator* is a form of consumer that actively seeks out and ingests live prey. Examples include protozoa that prey on algae and bacteria, and the remarkable *Bdellovibrio,* a bacterium that hunts other bacteria (see feature 3.3). Scavengers are the nutritional jacks-of-all-trades that feed on a variety of food sources, ranging from live cells, to dead cells, to wastes.

The Natural Recycling of Bioelements

Ecosystems are open with regard to energy because the sun is constantly infusing them with a renewable source. In contrast, the bioelements and nutrients that are essential components of protoplasm are supplied exclusively by sources somewhere in the biosphere and are not being continually replenished from outside the earth. In fact, the lack of a required nutrient in the immediate habitat is one of the chief factors limiting organismic and population growth. Because of the finite source of life's building blocks, the long-term sustenance of the biosphere requires continuous **recycling** of elements and nutrients. Essential elements such as carbon, nitrogen, sulfur, phosphorus, oxygen, and iron are cycled through biologic, geologic, and chemical mechanisms called **biogeochemical cycles** (figure 22.5). Although these cycles vary in certain specific characteristics, they share several general qualities, as summarized in the following list:

1. Cycled elements exist in pure or compound form, and ultimately originate from a nonliving, long-term reservoir in the atmosphere, sedimentary rocks, or water.
2. Elements make the rounds between a free, inorganic state in the abiotic environment and a combined, organic state in the biotic environment.
3. Recycling maintains a necessary balance of nutrients in the biosphere so that they do not build up or become unavailable.
4. Cycles are complex systems that rely upon the interplay of producers, consumers, and decomposers. Often the waste products of one organism become a source of energy or building material for another.
5. All organisms participate directly in recycling by removing, adding, or altering nutrients, but only certain categories of microorganisms can make the specific metabolic conversions that change some elements from one nutritional form to another.
6. The net turnover rate for elements is most rapid in the atmosphere and slowest in sedimentary rock.

In the next several sections we shall examine how, jointly and over a period of time, the varied microbial activities affect and are themselves affected by the abiotic environment (see also feature 22.1).

Atmospheric Cycles

The Carbon Cycle

Because carbon is the fundamental atom in all biomolecules and accounts for at least one-half of the dry weight of protoplasm, some form of carbon must be constantly available to living things. As a result, the **carbon cycle** is more intimately associated with

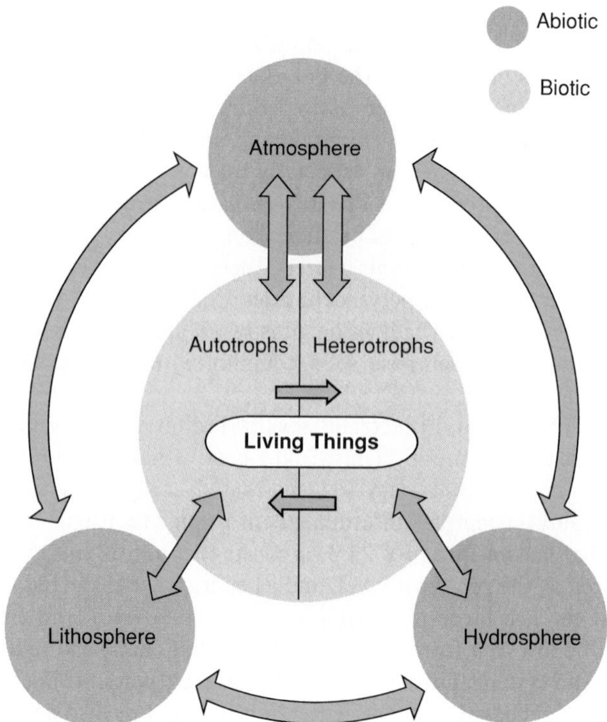

Figure 22.5 The general pattern of biogeochemical cycles. Essential bioelements are in a continuous state of flux, and they constantly recycle between their biotic form in protoplasm and their abiotic (inorganic) form in soil, water, or the atmosphere. Although all living things are involved in these cycles, plants and animals have a more general effect on recycling, while microorganisms act upon and alter the forms of elements in the environment.

Feature 22.1 Gaia: Is the Biosphere a Giant, Self-Sustaining Organism?

Because the earth is so accommodating to life, it occurred to British scientist James Lovelock more than 20 years ago that this was not a random accident. The conventional view has always been that the abiotic realms of the earth existed and evolved independently of living things and that life adapted to the abiotic realm rather than vice versa. But Dr. Lovelock has proposed that the earth contains a diversity of habitats and niches favorable to life because living things have made it that way. In this view, the biosphere is like a super organism that behaves as a single unit, is self-regulating, and maintains an equilibrium.

Dr. Lovelock has called his somewhat controversial idea the **Gaia** (guy'-uh) **Hypothesis** after the mythical Greek goddess of earth. Some scientists have been put off by this unorthodox suggestion of an intent or "wisdom" on the part of the earth, as though it were an organism capable of thought. But Dr. Lovelock has merely used a fanciful name to express a concept that is not really so incredible: As the earth shapes the character of living things, so do living things shape the character of the earth. After all, we know that the chemical composition of the aquatic environment, the atmosphere, and even the soil would not exist as it does without the actions of living things. Organisms are also very active in the recycling of elements, evaporation and precipitation cycles, formation of mineral deposits, and rock weathering. The Gaia Hypothesis has gradually gained in acceptance as scientists have tested some of its predictions and discovered that much of what it says makes sense. It also has great philosophical importance because of the undeniable capacity of humans to purposefully alter the biosphere on a very broad scale.

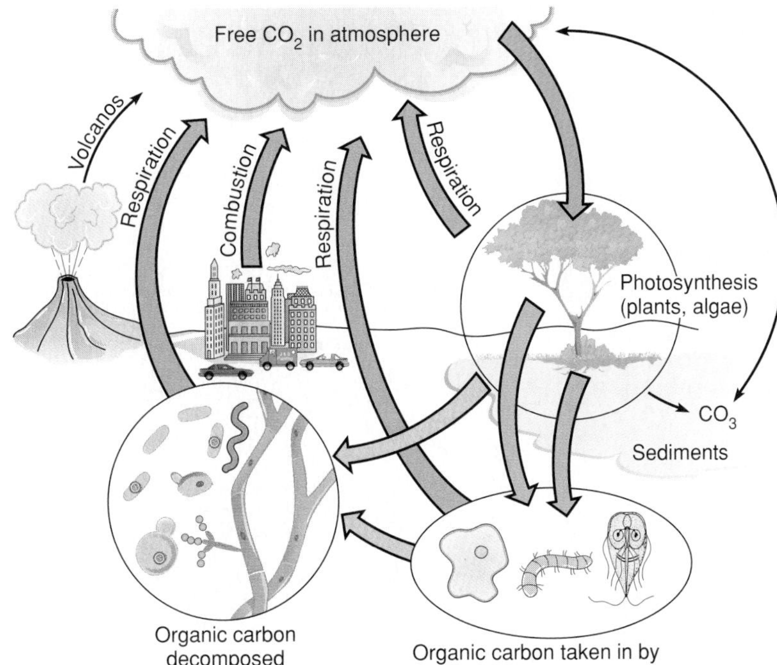

Figure 22.6 The carbon cycle. This cycle traces carbon from the CO_2 pool in the atmosphere to the primary producers, where it is fixed into protoplasm. Organic carbon compounds are taken in by consumers and decomposers that produce CO_2 through respiration and return it to the atmosphere. Combustion of fossil fuels and volcanic eruptions also add to the CO_2 pool. Some of the CO_2 is carried into inorganic sediments by organisms that synthesize carbonate (CO_3) skeletons. In time, natural processes acting on exposed carbonate skeletons may liberate CO_2.

the energy transfers and trophic patterns in the biosphere than are other elements. Besides the enormous organic reservoir in the bodies of organisms, carbon also exists in the gaseous state as carbon dioxide (CO_2) and methane (CH_4) and in the mineral state as carbonate (CO_3). In general, carbon is recycled through ecosystems via photosynthesis (carbon fixation), respiration and fermentation of organic molecules, limestone decomposition, and methane production. A convenient starting point from which to trace the movement of carbon is with carbon dioxide, which occupies a central position in the cycle and represents a large common pool that diffuses into all parts of the ecosystem (figure 22.6). As a general rule, the cycles of oxygen and hydrogen are closely allied to the carbon cycle.

The principal users of the atmospheric carbon dioxide pool are photosynthetic autotrophs (phototrophs) such as plants, algae, and cyanobacteria. An estimated 165 billion tons of organic material per year are produced by terrestrial and aquatic photosynthesis. A smaller amount of CO_2 is used by chemosynthetic autotrophs such as methane bacteria. A review of the general equation for photosynthesis in figure 22.7*a* will reveal that phototrophs use energy from the sun to fix CO_2 into organic compounds such as glucose that can be used in synthesis and respiration. Photosynthesis is also the primary means by which the atmospheric supply of O_2 is regenerated.

Just as photosynthesis removes CO_2 from the atmosphere, other modes of generating energy, such as respiration and fermentation, return it. Recall in the general equation for aerobic respiration (chapter 7) that, in the presence of O_2, organic compounds such as glucose are degraded completely to CO_2 and H_2O, with the release of energy. Carbon dioxide is also released by anaerobic respiration and certain types of fermentation reactions. Heterotrophic organisms, including consumers and decomposers, release CO_2, as do most phototrophs, which must respire in the absence of light.

A small but important phase of the carbon cycle involves certain limestone deposits composed primarily of calcium carbonate ($CaCO_3$). Limestone is produced when marine organisms like molluscs, corals, protozoans, and algae form hardened shells by combining carbon dioxide and calcium ions from the surrounding water. When these organisms die, the durable skeletal components accumulate in marine deposits. As these immense deposits are gradually exposed by geologic upheavals or receding ocean levels, various decomposing agents liberate CO_2 and return it to the CO_2 pool of the water and atmosphere.

The complementary actions of photosynthesis and respiration along with other natural CO_2-releasing processes such as limestone erosion and volcanic activity have maintained a relatively stable atmospheric pool of carbon dioxide. Recent figures show that this balance is being disturbed as humans burn *fossil fuels* and other organic carbon sources. Fossil fuels, including coal, oil, and natural gas, were formed over millions of years through the combined actions of microbes and geologic forces, and so are actually a part of the carbon cycle. Humans are so dependent upon this energy source that within the past 25 years, the proportion of CO_2 in the atmosphere has steadily increased from 0.032% to 0.035%. Although this increase may seem slight and insignificant, many experts now feel it has the potential to profoundly disrupt the delicate temperature balance of the biosphere (see feature 22.2).

Compared to carbon dioxide, **methane** (CH_4) plays a secondary part in the carbon cycle, though it can be a significant product in anaerobic ecosystems dominated by **methanogens**

Feature 22.2 Greenhouse Gases, Fossil Fuels, Cows, Termites, and Global Warming

The sun's radiant energy does more than drive photosynthesis; it also helps maintain the stability of the earth's temperature and climatic conditions. As radiation impinges on the earth's surface, much of it is absorbed, but a large amount of the infrared (heat) radiation bounces back into the upper levels of the atmosphere. For billions of years, the atmosphere has been insulated by a layer of gases (primarily CO_2, CH_4, water vapor, and nitrous oxide) formed by natural processes such as respiration, decomposition, and biogeochemical cycles. This layer traps a certain amount of the reflected heat, yet also allows some of it to escape into space. As long as the amounts of heat entering and leaving are balanced, the mean temperature of the earth will not rise or fall in an erratic or life-threatening way. Although this phenomenon, called the **greenhouse effect,** is popularly viewed in a negative light, it must be emphasized that its function for eons has been primarily to foster life.

The greenhouse effect has recently been a matter of concern because *greenhouse gases* appear to be increasing at a rate that could disrupt the temperature balance. In effect, a denser insulation layer will trap more heat energy and gradually heat the earth. In recent times, 3.5×10^{11} tons per year of CO_2 have been released collectively by respiration, anaerobic microbial activity, fuel combustion, and volcanic activity. By far the greatest increase in CO_2 production results from human activities such as combustion of fossil fuels, burning forests to clear agricultural land, and manufacturing. Deforestation has the added impact of removing large areas of photosynthesizing plants that would otherwise use some of the CO_2.

Originally, experts on the greenhouse effect were concerned primarily about increasing CO_2 levels, but it now appears that the other greenhouse gases combined may have a slightly greater contribution than CO_2, and they, too, are increasing. One of these gases, CH_4, is released from the gastrointestinal tract of ruminant animals such as cattle, goats, and sheep. Anaerobic bacteria in a part of the stomach called the rumen produce large amounts of this gas as they sequentially digest cellulose, a major component of the animals' diet. The gut of termites also harbors wood-digesting and methane-producing bacteria. Although humans cannot digest cellulose, intestinal microenvironments also support methanogens. Other greenhouse gases such as nitrous oxide (N_2O) and sulfur dioxide (SO_2) are also increasing through automobile and industrial pollution.

So far there is no complete agreement as to the extent and effects of global warming. It has been documented that the mean temperature of the earth has increased by 0.5°C since the beginning of the century. In one proposed scenario, by the next century, a rise in the average temperature of 4° to 5°C will begin to melt the polar ice caps and raise the levels of the ocean 2 to 3 feet, but some experts predict more serious effects, including massive flooding of coastal regions, changes in rainfall patterns, expansion of deserts, and long-term climatic disruptions.

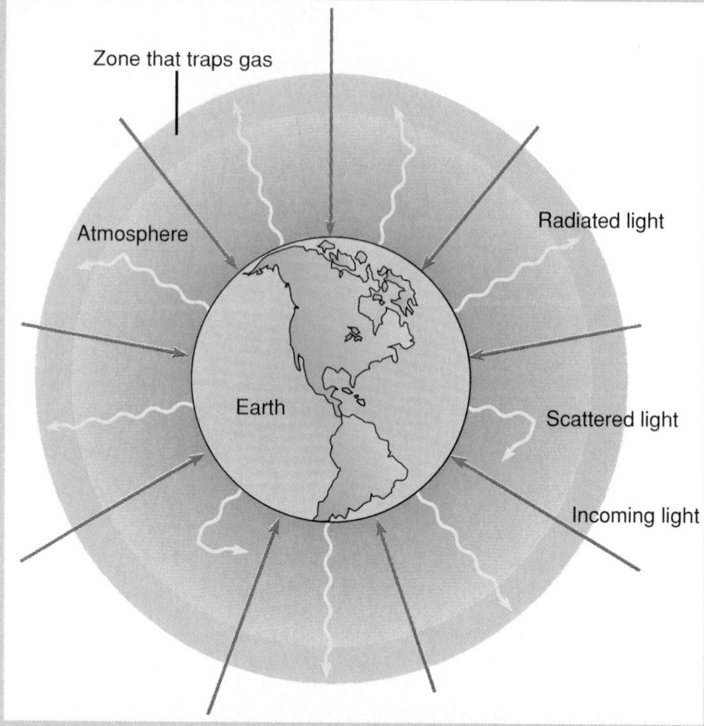

The greenhouse effect. Some of the light radiated or reflected from the earth's surface is trapped in a dense layer of atmospheric gases. Heat radiation trapped in this zone helps maintain the temperature of the atmosphere. Any increase in density of this gas layer could lead to increased entrapment of heat and a gradual rise in the ambient temperature of the earth.

(methane producers). In general, when methanogens reduce CO_2 by means of various oxidizable substrates, they give off CH_4. Lithotrophic methanogens such as *Methanobacterium* and *Methanococcus* use H_2 as an oxidizing agent, and heterotrophic ones use organic acids such as formate or acetate. Typical habitats for methanogens are black muds, marshes, sewage sludge, and gastrointestinal sites of various animals. The practical applications of methanogens are covered in a subsequent section on sewage treatment, and their contribution to the greenhouse effect is discussed in feature 22.2.

Photosynthesis: The Earth's Lifeline

With few exceptions, the energy that drives all living processes comes from the sun, but this source is directly available only to the cells of photosynthesizers. The dominance of photosynthetic organisms in the biosphere is immediately apparent if one observes the brilliant green landscape of a summer day or a magnified sample of pond water. In the terrestrial biosphere, green plants are the primary photosynthesizers, and in aquatic ecosystems, where 80–90% of all photosynthesis occurs, green, golden brown, brown, and red algae, as well as cyanobacteria, fill this role. Other photosynthetic procaryotes are green sulfur, purple sulfur, and purple nonsulfur bacteria.

The anatomy of photosynthetic cells is adapted to trapping sunlight, and their physiology effectively uses this solar energy to produce high-energy glucose from low-energy CO_2 and water. Just how do photosynthetic organisms achieve this remarkable feat? In general, the major steps in photosynthesis include (1) trapping visible light rays by means of special pigments attached to intracellular membranes, (2) excitation of atoms in the pigment molecule and release of electrons, (3) splitting water to release oxygen and hydrogen ions, (4) producing ATP with energy extracted from electrons and hydrogen ions, and (5) fixing CO_2 into carbohydrates using ATP

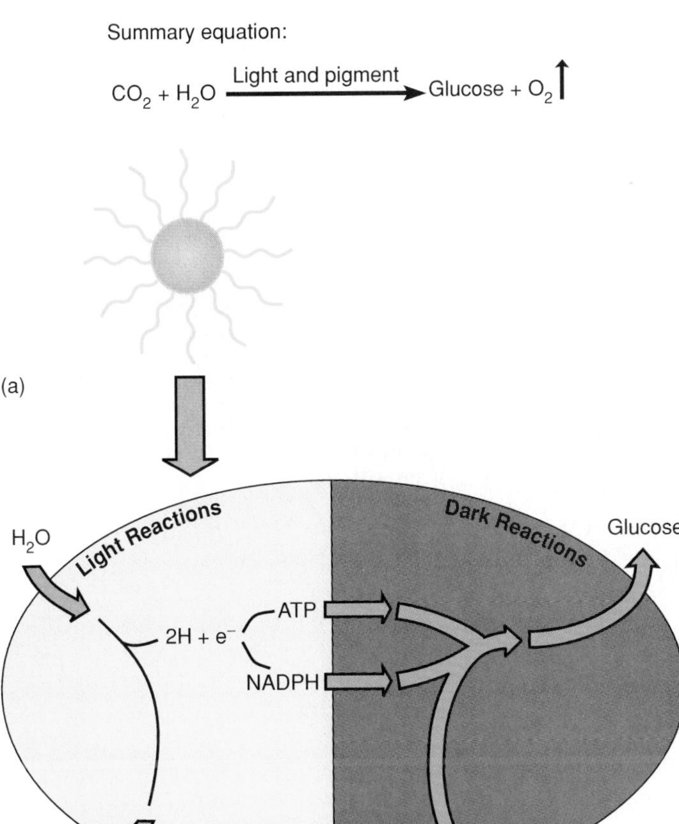

Figure 22.7 (*a*) The equation for photosynthesis. Note that it does not indicate the order of reactions, but merely summarizes the initial reactants and final products. (*b*) The general reactions of photosynthesis. In actuality, photosynthesis is divided into two phases called light reactions and dark reactions. The light reactions require light to activate chlorophyll pigment and use the energy given off during activation to split an H_2O molecule into oxygen and hydrogen, producing ATP and NADPH. The dark reactions, which occur without light, utilize ATP and NADPH produced during the light reactions to fix CO_2 into organic compounds such as glucose.

(figure 22.7). Photosynthesis proceeds in two phases: the **light reactions,** which are dependent on the presence of sunlight, and the **dark reactions,** which proceed in the absence of light.

The Functions of Green Machines Solar energy is delivered in discrete energy packets called **photons** (also called quanta) that travel as waves. The wavelengths of light operating in photosynthesis occur in the visible spectrum between 400 nm (violet) and 700 nm (red). As this light strikes cells, some wavelengths are absorbed, some pass through, and some are reflected. The activity that has greatest impact on photosynthesis is the absorbance of light by a special class of molecules called photosynthetic pigments. These include the **chlorophylls,** which are green; **carotenoids,** which are yellow, orange, or red; and **phycobilins,** which are red or blue-green.[2] By far the most important of these pigments are the chlorophylls, which contain a *photocenter* that consists of a magnesium atom held in the center of a complex ringed structure called a porphyrin. As we will see, the chlorophyll molecule harvests the physical energy of photons and converts it to electron (chemical) energy. The accessory photosynthetic pigments such as carotenes trap light energy and shuttle it to chlorophyll.

The detailed biochemistry of photosynthesis is beyond the scope of this text, but we will provide an overview and account for the major steps as they occur in green plants, algae, and cyanobacteria (figure 22.7). In light reactions, the major reactants are chlorophyll, photons, and H_2O, and the products are ATP, NADPH, and O_2. In dark reactions, the reactants are CO_2, ATP, and NADPH, and the products are glucose and other carbohydrates. Many of the events (coenzyme carriers, electron transport, and phosphorylation) are similar to certain pathways of respiration previously covered in chapter 7.

Light Reactions The systems that carry the photosynthetic pigments and are the sites for the light reactions occur in the *thylakoid* membranes in chloroplasts (figure 22.8*a*) and in specialized lamellae of the cell membranes in procaryotes (see chapter 3). These systems exist as two separate complexes called *photosystem I* (P700) and *photosystem II* (P680)[3] (figure 22.8*c*). Both systems contain chlorophyll and are simultaneously activated by light, but photosystem II performs the reactions that help drive photosystem I. Together the systems are a powerful linear chain for receiving light, transporting electrons, pumping hydrogen ions, and forming ATP.

The influx of photons into the photocenter of the P680 system energizes it and causes electrons to be released from the magnesium atom. The removal of electrons from the photocenter has two major effects. (1) It creates a vacancy in the chlorophyll molecule forceful enough to split an H_2O molecule into hydrogens (as electrons and hydrogen ions) and oxygen. This splitting of water, termed *photolysis,* is the ultimate source of the O_2 gas that is an important product of photosynthesis. The electrons released from the lysed water regenerate the photosystem for the next photon. (2) The electrons released by the first photoevent are immediately boosted through a series of carriers (cytochromes) to the P700 system. At this same time, hydrogen ions accumulate in the internal space of the thylakoid complex, thereby producing an electrochemical gradient.

The P700 system is ready to accept the electrons from the P680 system because its own photosystem has also been activated by light. The electrons it receives are passed along a transport chain to a complex that uses electrons and hydrogen ions to reduce NADP to NADPH. (Recall that reduction in this sense entails the addition of electrons and hydrogens to a substrate.) As noted in chapter 7, a reduced coenzyme is another source of chemical energy and may be used in synthesis. The final event

2. The color of the pigment corresponds to the wavelength of light it reflects.

3. The numbers refer to the principal wavelength of light to which each system is most sensitive.

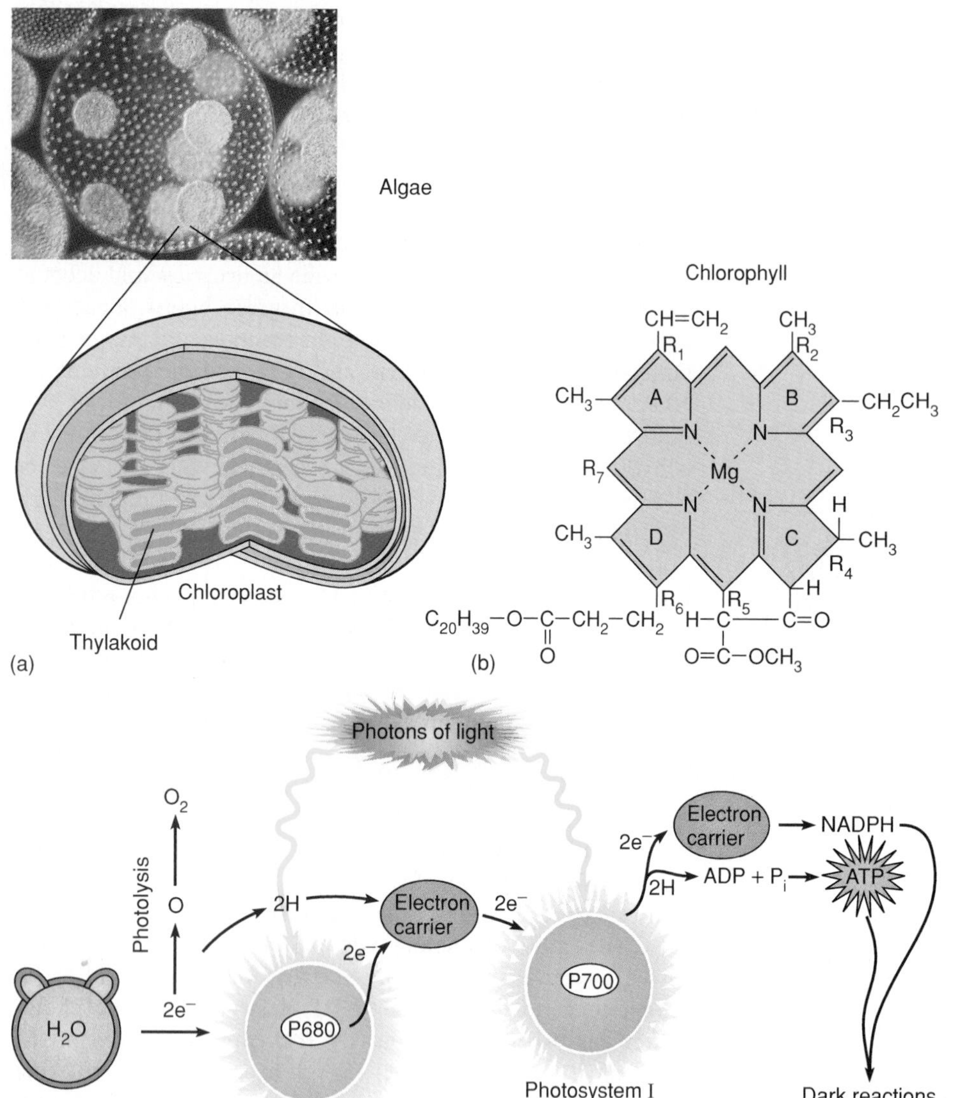

Figure 22.8 (*a*) A colony of the green alga *Volvox* is an aggregate of tiny spherical cells, each of which holds a number of spherical chloroplasts (*inset*). A chloroplast contains membranous compartments called thylakoids where chlorophyll molecules and the photosystems for the light reactions are located. (*b*) A chlorophyll molecule, with a central magnesium atom held by a porphyrin ring. (*c*) The main events of the light reactions. Activation of photosystem II sets up a chain reaction, in which electrons are released from chlorophyll and carried to photosystem I; the empty position in photosystem II is replenished by photolysis of H_2O. Electrons transported to photosystem I are used to synthesize NADPH. Hydrogens released by photolysis are used to synthesize ATP. NADPH and ATP are fed into the stroma of the chloroplast where they function in the dark reactions.

of photosystem I is the synthesis of ATP through a chemiosmotic process. The increased concentration of H^+ ions creates a gradient between the inside of the thylakoid and its immediate surroundings. The passage of these ions through specific membrane enzymes (ATP synthetase) generates enough energy to synthesize an ATP from an ADP and inorganic phosphate. Because it occurs in light, this process is termed *photophosphorylation.*

Dark Reactions The subsequent photosynthetic reactions, not directly dependent upon light, occur in the chloroplast stroma or the cytoplasm of cyanobacteria. During these reactions, the energy produced by the light phase is used to synthesize glucose by means of a cyclic scheme called the *Calvin cycle* (figure 22.9). The cycle begins at the point where CO_2 is combined with a 5-carbon acceptor molecule called ribulose 1, 5 bisphosphate (RuBP). This combination, called carbon fixation, generates a 6-carbon intermediate. This molecule subsequently splits into two 3-carbon molecules of phosphoglycerate (PGA). Two steps prepare the molecules for the final synthesis steps. One of these is the addition of a high-energy phosphate from ATP, which converts PGA to diphosphoglycerate (DPG). The second is the addition of NADPH, which converts DPG to glyceraldehyde phosphate (PGAL). At this point, two high-energy PGAL molecules can combine to form a glucose molecule. Other PGAL molecules are channeled back into the cycle, consuming more ATP to regenerate RuBP.

Other Mechanisms of Photosynthesis The oxygen-releasing (or **oxygenic**) photosynthesis that occurs in plants, algae, and cyanobacteria is the dominant type on earth. Photosynthesis in other types of bacteria differs in several ways. Photosynthetic green and purple bacteria possess bacteriochlorophyll, which is more versatile in capturing light. They have only a cyclic photosystem I, which routes the electrons from the photocenter to the electron carriers and back to the photosystem again. This pathway generates a relatively small amount of ATP, and it may not produce NADPH. Being photolithotrophs, these bacteria

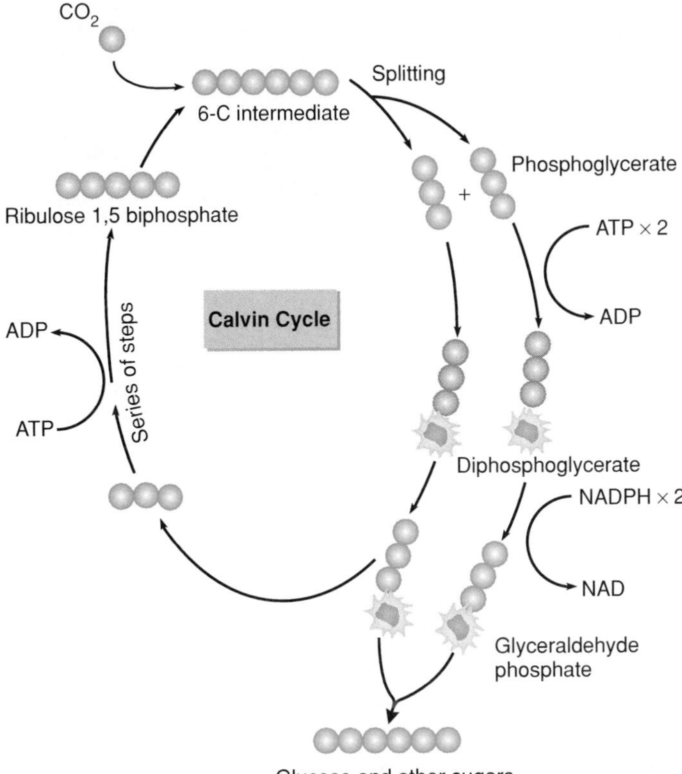

Figure 22.9 The events of the dark reactions, also called the Calvin cycle, are like the TCA (citric acid) cycle running the opposite direction. In this case, CO_2 is added to molecules during the cycle, and the various carbon residues in the scheme are combined rather than split. This cycle also involves the expenditure of ATP and NADPH rather than their release. The glucose produced by these reactions may be stored as complex carbohydrates, or it may be used in various amphibolic pathways to produce other carbohydrate intermediates or amino acids.

use H_2, H_2S, or elemental sulfur rather than H_2O as a source of electrons and reducing power. As a consequence, they are **anoxygenic** (non-oxygen-producing), and many are strict anaerobes. Bacteria that use sulfur compounds like H_2S deposit sulfur residues either intracellularly or extracellularly, depending on the species (*Rhodospirillum,* figure 1.3, and *Chromatium,* figure 3.33). The sulfur bacteria are typically isolated from oxygen-deficient environments such as mud and water at the bottom of shallow ponds and lakes, where they play a central role in the biogeochemical cycle of sulfur.

The Nitrogen Cycle

Nitrogen (N_2) gas is the most abundant component of the atmosphere, accounting for nearly 79% of air volume. As we will see, this extensive reservoir in the air is largely unavailable to most organisms. Only about 0.03% of the earth's nitrogen is combined (or fixed) in some other form such as nitrates (NO_3), nitrites (NO_2), ammonium (NH_4), and organic nitrogen compounds (proteins, nucleic acids).

The nitrogen cycle is relatively more intricate than other cycles because it involves such a diversity of specialized microbes to maintain the flow of the cycle. In many ways, it is actually more of a nitrogen "web" because of the array of

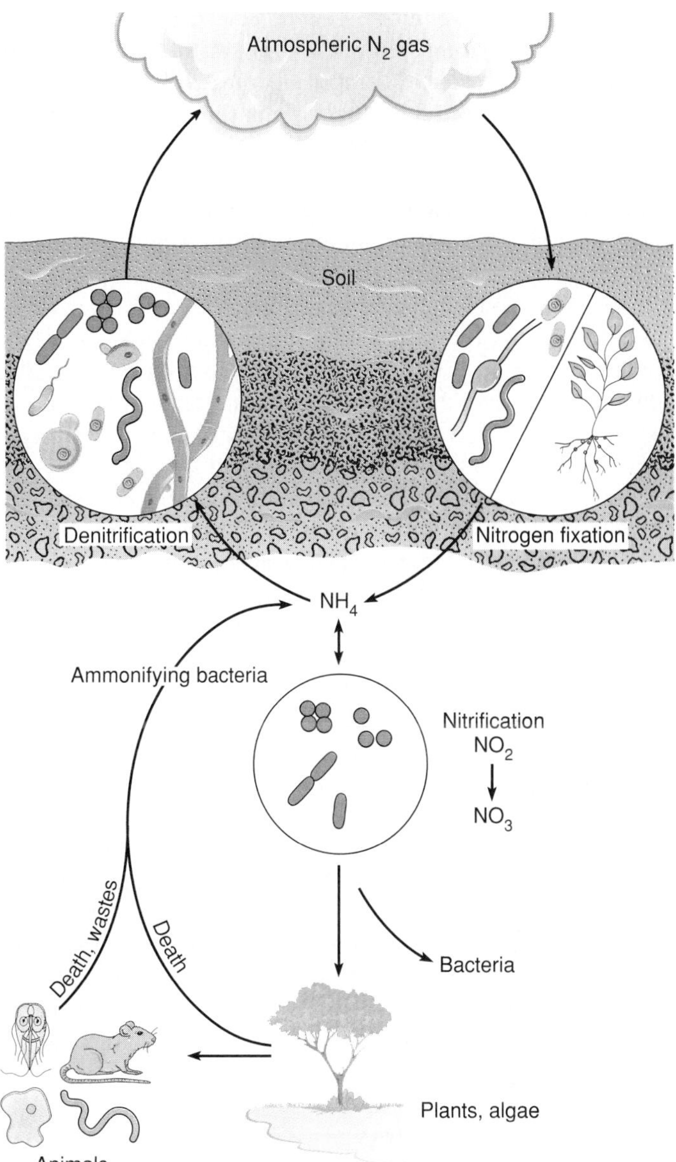

Figure 22.10 The simplified events in the nitrogen cycle. Gaseous nitrogen is acted on by nitrogen-fixing bacteria, which give off ammonia (NH_4). Ammonia is converted to nitrite and nitrate by nitrifying bacteria. Plants, algae, and bacteria use nitrates to synthesize nitrogenous organic compounds (proteins, amino acids, nucleic acids). Organic nitrogen compounds are used by animals and other consumers. Nitrogenous macromolecules from wastes and dead organisms are converted to NH_4 by ammonifying bacteria. This may either be directly recycled into nitrates or returned to the atmospheric N_2 form by denitrifying bacteria.

nitrogen-processing adaptations that occur. Higher plants can utilize NO_3 and NH_4; animals must receive nitrogen in organic form from plants or other animals; and microorganisms vary in their source, using NO_2, NO_3, NH_4, N_2, and organic nitrogen. The cycle includes four basic types of reactions: nitrogen fixation, ammonification, nitrification, and denitrification (figure 22.10).

Nitrogen Fixation The biosphere is most dependent upon the only process that can remove N_2 from the air and fix it into a form usable by living things. This process, called **nitrogen fixation,** is the beginning step in the synthesis of virtually all nitrogenous compounds. Nitrogen fixation is brought about primarily by nitrogen-fixing bacteria in soil and water, though a small amount is formed through nonliving factors such as lightning. The nitrogen-fixers have developed a unique enzyme system capable of breaking the triple bonds of the N_2 molecule and reducing the N atoms, an anaerobic process that requires the expenditure of considerable ATP. The primary product of nitrogen fixation is the ammonium ion, NH_4^+. Nitrogen-fixing bacteria may live free or in a symbiotic relationship with plants. Among the common free-living nitrogen fixers are the aerobic genera *Azotobacter* and *Azospirillum* and certain members of the anaerobic genus *Clostridium.* Other free-living nitrogen fixers are the cyanobacteria *Anabaena* and *Nostoc.*

A significant symbiotic association occurs between *rhizobia* (bacteria in the genera *Rhizobium, Bradyrhizobium,* and *Azorhizobium*) and *legumes* (plants such as soybeans, peas, alfalfa, and clover that characteristically produce seeds in pods). The infection of legume roots by these gram-negative, motile, rod-shaped bacteria causes the formation of special nitrogen-fixing organs, the **root nodules** (figure 22.11). Nodulation begins with rhizobia colonizing specific sites on root hairs. From there, the bacteria invade deeper root cells and induce the cells to form tumorlike masses. Both members of this relationship contribute an essential biochemical activity that leads to N_2 fixation. The bacterium's enzyme system supplies a constant source of reduced nitrogen to the plant, and the plant furnishes nutrients and energy for the activities of the bacterium. The legume uses the NH_4 to aminate (add an amino group to) various carbohydrate intermediates and thereby synthesize amino acids and other nitrogenous compounds that are used in plant and animal synthesis (see figure 7.27). Nodule relationships also occur between the fungus *Frankia* and various woody shrubs.

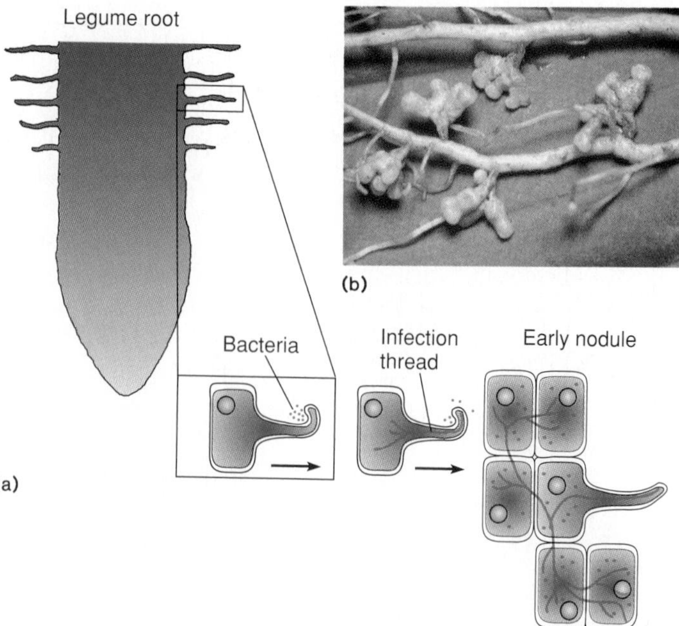

Figure 22.11 Nitrogen fixation through symbiosis. (*a*) Events leading to formation of root nodules. *Rhizobium* cells attach to legume root hair and cause it to curl. Invasion of the legume root proper by *Rhizobium* initiates the formation of an infection thread that spreads into numerous adjacent cells. The presence of bacteria in cells causes nodule formation. (*b*) Mature nodules that have developed in a sweet clover plant.

Figure 22.12 Inoculating legume seeds with *Rhizobium* bacteria increases the plant's capacity to fix nitrogen. The legumes in (*a*) were inoculated and are healthy. The poor growth and yellowish color of the uninoculated legumes in (*b*) indicate a lack of nitrogen.

Natural Fertilizer Factories Plant-bacterial associations have great practical importance in agriculture, since an available source of nitrogen is often a limiting factor in the growth of crops. Because legumes are self-fertilizing and require no additional nitrogen chemicals for growth, they are valuable food plants in areas with poor soils and in countries with limited resources. It has been shown that crop health and yields may be improved by inoculating legume seeds with pure cultures of rhizobia, since the soil is often deficient in the proper strain for forming nodules (figure 22.12). In the past several years, the idea for converting regular, nonleguminous plants into nitrogen-fixers has also been explored. Some researchers attempted to genetically engineer various plants by inserting genes from *Rhizobium* into them, but the plants could not be induced to form nodules or fix nitrogen. A newer, more promising approach has been to artificially fuse *Rhizobium* to the roots of such plants as rice, rapeseed, and wheat. Early results have shown that such manipulations do stimulate nodule formation, but whether the nodules are functional still remains to be seen.

Ammonification In the next phase of the nitrogen cycle, nitrogen-containing organic matter is decomposed by various bacteria (*Clostridium, Proteus*) that live in the soil and water. Organic detritus consists of large amounts of protein and nucleic acids from dead organisms and nitrogenous animal wastes such as urea and uric acid. The decomposition of these substances splits off amino groups and produces NH_4^+. This

rhizobia (ry-zoh'-bee-uh) Gr. *rhiza,* root, and *bios,* to live.
legume (leg'-yoom) L. *legere,* to gather.

process is thus known as **ammonification.** The ammonium released may be reused by certain plants, or it may be acted upon by yet other groups of specialized bacteria.

Nitrification The oxidation of ammonium to NO_2 and NO_3, a process called **nitrification,** is an essential conversion point in the nitrogen cycle because it makes nitrogen available in an additional metabolically useful form. This reaction occurs in two phases and involves lithotrophic aerobic bacteria in soil and water. In the first phase, certain gram-negative genera such as *Nitrosomonas* and *Nitrosococcus* oxidize NH_4^+ to NO_2 as a means of generating energy. Nitrite is not useful to most microbes and is rapidly acted upon by a second group of nitrifiers, including *Nitrobacter* and *Nitrococcus,* which perform the final oxidation of NO_2 to NO_3. Nitrates can be assimilated into protoplasm by a variety of organisms (plants, fungi, and bacteria); a large number of bacteria also use it as a source of oxygen.

Denitrification For the nitrogen cycle to be complete, nitrogen compounds must be returned to the reservoir in the air. This is accomplished by the conversion of NO_3 through intermediate steps to atmospheric nitrogen. This reaction series is essentially the reverse of nitrification. The first step, which involves the reduction of nitrate to nitrite, is so common that hundreds of different bacterial species can do it. Other bacteria such as *Bacillus, Pseudomonas, Spirillum,* and *Thiobacillus* can carry out the denitrification process to completion as follows:

$$NO_3 \rightarrow NO_2 \rightarrow NO \rightarrow N_2O \rightarrow N_2$$

Sedimentary Cycles

The Sulfur Cycle The sulfur cycle resembles the nitrogen cycle, but instead of being drawn from the atmosphere, sulfur originates from natural sedimentary deposits in rocks, oceans, lakes, and swamps. Sulfur exists in the elemental form (S) and as hydrogen sulfide gas (H_2S), sulfate (SO_4), and thiosulfate (S_2O_3). Most of the oxidations and reductions that convert one form of inorganic sulfur compound to another are accomplished by bacteria. Plants and many microorganisms can assimilate only SO_4, and animals must have an organic source. Organic sulfur occurs in the amino acids cystine, cysteine, and methionine, which contain sulfhydryl (—SH) groups and form disulfide (S—S) bonds that contribute to the stability and configuration of proteins.

One of the most remarkable contributors to the cycling of sulfur in the biosphere is the genus *Thiobacillus*. These gram-negative, motile rods are found in mud, sewage, bogs, mining drainage, and brackish springs. Although such environments are forbidding to higher organisms that require complex organic nutrients, they are home to these specialized lithotrophic bacteria. Their metabolism has become adapted to extracting energy by oxidizing elemental sulfur, sulfides, and thiosulfate. One species, *T. thiooxidans,* is so efficient at this process that it secretes large amounts of sulfuric acid into its environment, as shown by the following equation:

$$Na_2S_2O_3 + H_2O + O_2 \rightarrow Na_2SO_4 + H_2SO_4 \text{ (sulfuric acid)} + 4S$$

The marvel of this bacterium is that it can both create and survive in the most acidic habitats on earth. It plays an essential role in the phosphorus cycle, and its relative, *T. ferrooxidans,* participates in the cycling of iron. Other bacteria that can oxidize sulfur to sulfates are the photosynthetic sulfur bacteria mentioned in the section on photosynthesis.

The sulfates formed from oxidation of sulfurous compounds enter the organic phase of the cycle and are assimilated into protoplasm. The sulfur cycle reaches completion when inorganic and organic sulfur compounds are reduced. Bacteria in the genera *Desulfovibrio* and *Desulfuromonas* anaerobically reduce sulfides to H_2S and water as the final step in electron transport. Sites in mud and moist soil where they live usually emanate a strong, rotten egg stench and have a blackened film of iron on the surface. Organic sulfur in amino acids may also be reduced to H_2S by a number of saprobic bacteria.

The Phosphorus Cycle Phosphorus is an integral component of DNA, RNA, and ATP, and all life depends upon a constant supply of it. It cycles between the abiotic and biotic environments in inorganic phosphate (PO_4) rather than in elemental form (figure 22.13). The chief inorganic reservoir is phosphate rock, which contains the insoluble compound $Ca_5(PO_4)_3F$. Before it can enter biological systems, this mineral must be *phosphatized*—converted into more soluble PO_4 by the action of acid. Phosphate is released naturally when the sulfuric acid produced by *Thiobacillus* dissolves phosphate rock. Phosphate fertilizers are made commercially by treating mined phosphate rock with sulfuric acid. Soluble phosphate in the soil and water is the principal source for autotrophs, which fix it onto organic molecules and pass it on to heterotrophs in this form. Organic phosphate is returned to the pool of soluble phosphate by decomposers, and phosphate is finally cycled back to the mineral reservoir by slow geological processes such as sedimentation. Because the low phosphate content of many soils can limit productivity, phosphate is added to soil to increase agricultural yields. The excess runoff of fertilizer into the hydrosphere is often responsible for overgrowth of aquatic pests (see eutrophication in a subsequent section on aquatic habitats).

The involvement of microbes in cycling nutrients has both negative and positive effects. On one hand, they may recycle or create toxic compounds (feature 22.3), while on the other, they may help clean up pesticides (see feature 22.4) and mine minerals.

Thiobacillus (thigh″-oh-bah-sil′-us) Gr. *theion,* sulfur.

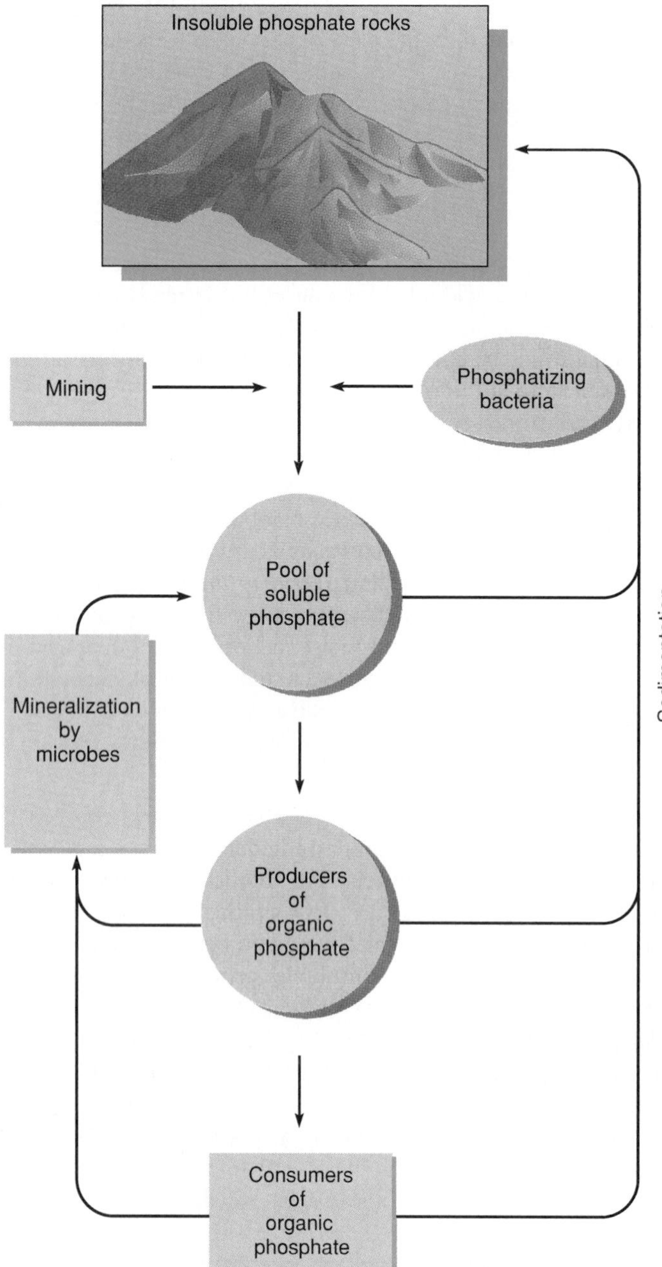

Figure 22.13 The phosphorus cycle. The pool of phosphate existing in sedimentary rocks is released into the ecosystem either naturally by erosion and microbial action or artificially by mining and the use of phosphate fertilizers. Soluble phosphate is cycled through producers, consumers, and decomposers back into the soluble pool of phosphate, or it is returned to sediments in the aquatic biosphere.

Soil Microbiology: The Composition of the Lithosphere

Expressions like "soiled" or "dirty" may suggest that soil is an undesirable, possibly harmful substance. To some persons, soil may also appear to be a somewhat homogenous, inert substance. But at the microscopic level, soil is a dynamic ecosystem that supports complex interactions between numerous geological, chemical, and biological factors. This rich region, called the lithosphere, teems with microbes, serves a dynamic role in biogeochemical cycles, and is an important repository for organic detritus and dead terrestrial organisms.

Figure 22.14 Rocks covered with a thin crust of crustose and fruticose lichens. Such microbial associations begin the process of soil formation by invading and breaking up the rocks into smaller particles.

The abiotic portion of soil is a composite of mineral particles, water, and atmospheric gas. The development of soil begins when geologic sediments are mechanically disturbed and exposed to weather and microbial action. One type of microbe commonly involved in the early colonization and breakdown of rocks are unusual creatures called **lichens** (figure 22.14). Lichens are complex organisms formed from the symbiotic association of a fungus and a cyanobacterium or green alga. This mutual relationship is so interdependent that the two participants cannot live separately. The alga or cyanobacterium feeds itself and the fungus through photosynthesis, while the fungus forms a protective shelter for the alga and supplies water and minerals to it. Most lichens that are early colonists or pioneers of rock are of the compact crustose form.

Rock decomposition releases various-sized particles ranging from rocks, pebbles, and sand grains to microscopic morsels that lie in a loose aggregate (figure 22.15). The porous structure of soil creates various-sized pockets or spaces that provide numerous microhabitats. Some spaces trap moisture and form a liquid phase in which mineral ions and other nutrients are dissolved. Other spaces trap air that will provide gases to soil microbes, plants, and animals. Because both water and air compete for these pockets, the water content of soil is directly related to its oxygen content. Water-saturated soils contain less oxygen, and dry soils have more. Gas tensions in soil may also vary vertically. In general, the concentration of O_2 decreases and that of CO_2 increases with the depth of soil. Aerobic and facultative forms tend to occupy looser, drier soils, whereas anaerobic types are adapted to waterlogged, poorly aerated soils.

---

lichens (ly'-kenz) Gr. *leichen,* a tree moss.

Feature 22.3 Biogeochemical Cycles and Biomagnification of Pollutants

Understanding the action of biogeochemical cycles takes on increased significance when we examine the increasing pollution of the air, soil, and water by humans. The scale of human activity has empowered us to increase the natural cycling rate of such toxic elements as arsenic, chromium, cadmium, lead, and mercury. Many of the 500,000 man-made chemicals introduced into the environment over the past one hundred years can also get caught up in cycles. Some of these chemicals will be converted into less harmful substances, but others, such as DDT and heavy metals, persist and flow along with nutrients into all levels of the biosphere. If a chemical remains stable, accumulates in living tissue, and is not excreted, it can be **biomagnified** or concentrated by living things through the natural trophic flow of the ecosystem. Microscopic producers such as bacteria and algae begin the accumulation process. With each new level of the food chain, the consumers gather an increasing amount of the chemical, until the top consumers may contain toxic levels.

The long-term effects of *bioaccumulation* are best demonstrated using the heavy metal mercury as a model. Mercury compounds are used in household antiseptics and disinfectants, agriculture, and industry. Elemental mercury precipitates proteins by attaching to functional groups, and it interferes with the bonding of uracil and thymine. But the elemental form is not as toxic as the organic mercurials such as ethyl or methyl mercury.

The severity of this toxicity was first demonstrated by an incident in Minamata Bay, Japan, during the 1950s. A local plastics manufacturer had drained large quantities of mercury into the bay. In the aquatic sediments, certain mercury-resistant bacteria (*Desulfvibrio*) metabolized the mercury compounds to the highly toxic organic forms. Motile plankton that fed upon these bacteria accumulated the mercury and passed it on to the next consumer level until eventually, fish and shellfish had accumulated high levels of methyl mercury. These foods were a regular part of the diet of local Japanese fishermen and their families. Many adults and older children in the community suffered nerve and brain damage, and fetuses developed severe birth defects. Of the 130 poisoned persons, 47 died. Since then, no other large-scale cases of toxicity have been traced to eating fish contaminated with methyl mercury. However, recent studies have disclosed increased mercury content in the freshwater lakes of North America, which poses a risk for future toxic tragedies.

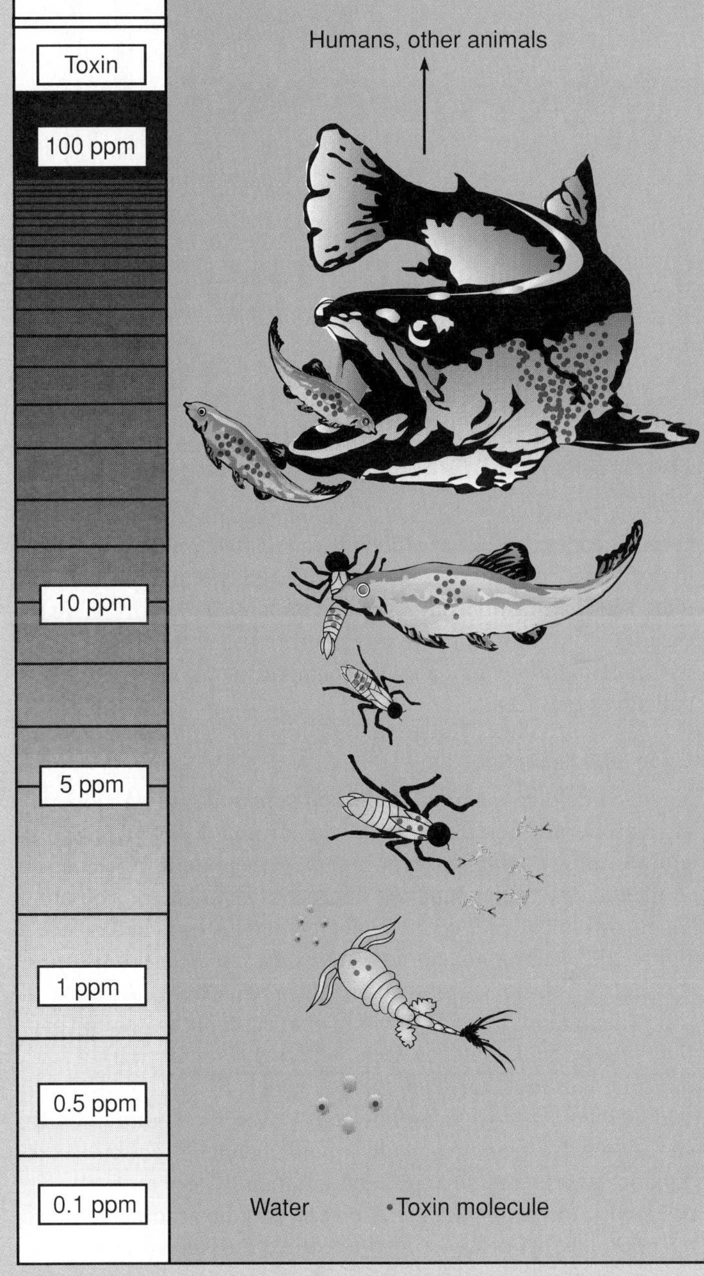

The mechanism for biomagnification of toxic compounds in the food chain.

Within the superstructure of the soil are varying amounts of **humus,** the slowly decaying organic litter from plant and animal tissues. This soft, crumbly mixture holds water like a sponge. It is also an important habitat for microbes that decompose the complex litter and gradually release nutrients. The humus content varies with climate, temperature, moisture and mineral content, and microbial action. Warm tropical soils have a high rate of humus production and microbial decomposition, and because nutrients are swiftly released and used up, they do not accumulate. Fertilized agricultural soils in temperate climates build up humus at a higher rate and are rich in nutrients. The very low content of humus and moisture in desert soils greatly reduces its microbial flora, rate of decomposition, and

humus (hyoo′-mus) L. earth.

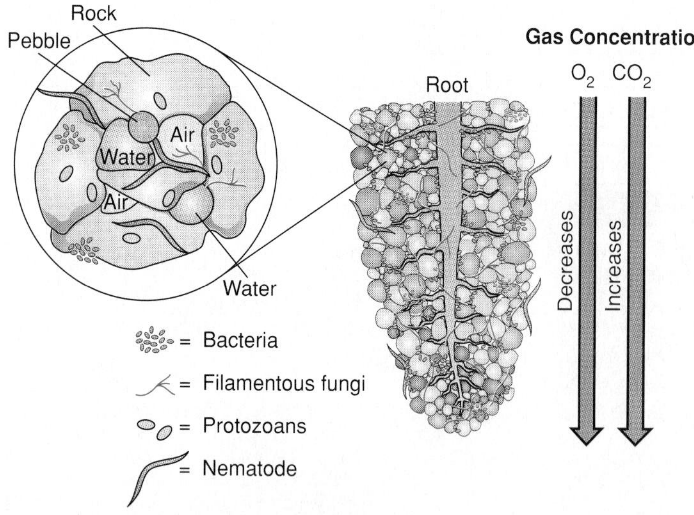

Figure 22.15 The structure of the rhizosphere and the microhabitats that develop in response to soil particles, moisture, air, and gas content.

nutrient content. Bogs are likewise nutrient-poor, but this is due to the dense, oxygen-poor quality of the humus and its acid content, which retard decomposition. Humans artificially increase the amount of humus by mixing plant refuse and animal wastes with soil and allowing natural decompositon to occur, a process called *composting.*

Living Activities in Soil

The rich culture medium of the soil supports a fantastic array of microflora (bacteria, fungi, algae, protozoa, and viruses). A gram of moist loam soil with high humus content may contain as many as 10 billion separate microorganisms, each competing for its own niche. All of the biological interactions described in figure 22.5 occur among organisms in the soil. Some of the most distinctive relationships occur in the **rhizosphere,** the zone of soil surrounding the roots of plants that contains associated bacteria, fungi, and protozoa (figure 22.15). Plants interact with soil microbes in a truly synergistic fashion. Studies have shown that the presence of microbes around the root hairs actually stimulates the plant to exude growth factors such as carbon dioxide, sugars, amino acids, and vitamins. These nutrients are released into fluid spaces where they may be readily captured by microbes. Microbes contribute to the association by converting minerals into forms usable by plants. We saw numerous examples of this in the nitrogen, sulfur, and phosphorus cycles. Certain microbes also produce growth hormones that stimulate plant growth.

We previously observed that plants may form close symbiotic associations with microbes to fix nitrogen. But other mutualistic partnerships that do not involve nitrogen fixation can develop between plant roots and microbes. The most common forms, termed **mycorrhizae,** occur between vascular plants and

Figure 22.16 Mycorrhizae, symbiotic associations between fungi and plant roots that favor the absorption of water and minerals from the soil.

fungi (figure 22.16). As many as 80% of vascular plants live constantly with various species of basidiomycetes, ascomycetes, or zygomycetes. Mycorrhizae involve some degree of fungal attachment to the plant root and have many complex characteristics. The plant feeds the fungus through photosynthesis, which costs the plant energy, but the fungus sustains the relationship in several ways. By extending its mycelium into the rhizosphere, it helps anchor the plant and increases the surface area for capturing water from dry soils and minerals from poor soils. Plants with mycorrhizae can inhabit more severe habitats more successfully than plants without them.

The topsoil, which extends a few inches to a few feet from the surface, supports a host of burrowing animals such as nematodes, termites, and earthworms. Many of these animals are decomposer-reducer organisms that break down organic nutrients through digestion, and also mechanically reduce or fragment the size of particles so that they are more readily mineralized by microbes. Aerobic bacteria initiate the digestion of organic matter into carbon dioxide and water, and generate minerals such as sulfate, phosphate, and nitrate, which may be further degraded by anaerobic bacteria. Fungal enzymes capable of hydrolyzing durable natural substances like cellulose, keratin, lignin, chitin, and paraffin heighten the efficiency of soil decomposition.

The soil may be a source of clinical infection, particularly in connection with fecal contamination in parasitic infections. It is also a repository for agricultural, industrial, and domestic wastes such as insecticides, herbicides, fungicides, manufacturing wastes, and household chemicals. Feature 22.4 explores the problem of soil contamination and the feasibility of harnessing native soil microbes to break down undesirable hydrocarbons and pesticides.

Aquatic Microbiology

Water is the dominant compound on earth. It occupies nearly three-fourths of the earth's surface, and 97% of all water is in the ocean. In the same manner as minerals, the earth's supply

mycorrhizae (my″-koh-ry′-zee) Gr. *mykos,* fungus, and *rhiza,* root.

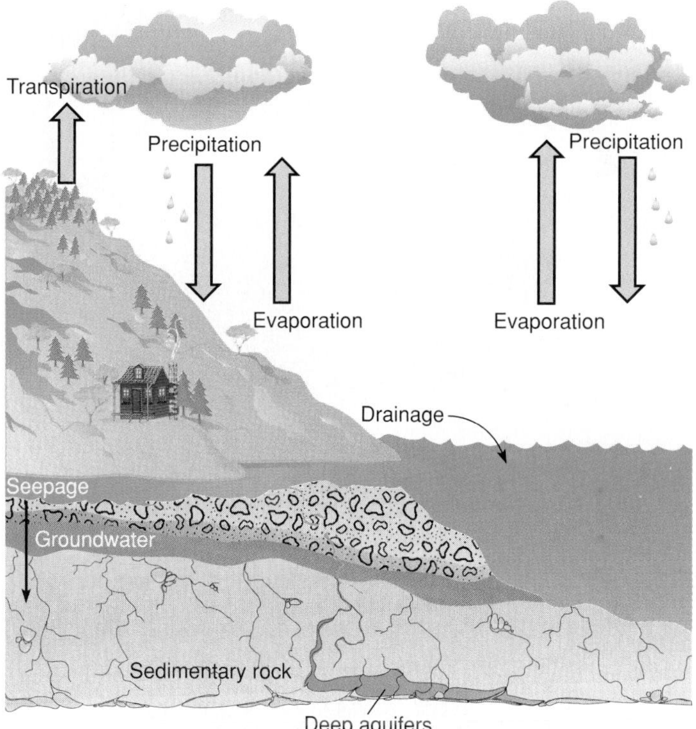

Figure 22.17 The hydrologic cycle. The largest proportion of water cycles through evaporation, transpiration, and precipitation between the hydrosphere and the atmosphere. Other reservoirs of water exist in the groundwater or deep storage aquifers in sedimentary rocks. Plants add to this cycle by releasing water through transpiration, and heterotrophs release it through respiration.

of water is continuously cycled between the hydrosphere, atmosphere, and lithosphere (figure 22.17). The **hydrologic cycle** begins when surface water (lakes, oceans, rivers) exposed to the sun and wind evaporates and enters the vapor phase of the atmosphere. Living things contribute to this reservoir by various activities. Plants lose moisture through transpiration (evaporation through leaves), and all aerobic organisms, from animals to plants to decomposers, give off water during respiration, Airborne moisture accumulates in the atmosphere, most conspicuously as clouds.

Water is returned to the earth through condensation or precipitation (rain, snow). The largest proportion of precipitation falls back into surface waters, where it circulates rapidly between running water (rivers and streams) and standing water (lakes and oceans). Only about 2% of water soaks into the earth or is tied up in ice, but these are very important reservoirs. Surface water seeps into the earth and collects in extensive subterranean pockets produced by the underlying layers of rock, gravel, and sand. This process forms a deep **groundwater** source called an **aquifer.** The water in aquifers circulates very slowly and is an important replenishing source for surface water. It may resurface through springs, geysers, and hot vents, and it is also tapped as the primary supply for one-fourth of all water used by humans.

Although the total amount of water in the hydrologic cycle has not changed over millions of years, its distribution and quality have been greatly altered by human activities. Two serious problems have arisen with aquifers. As a result of increased well-drilling, land development, and persistent local droughts, the aquifers in many areas have not been replenished as rapidly as they have depleted. As these reserves are used up, humans will have to rely on other delivery systems such as pipelines, dams, and reservoirs, which may disrupt the cycling of water even more. Because water picks up materials when falling through air or percolating through the ground, aquifers are also important collection points for pollutants. As we will see, the proper management of water resources is one of the greatest challenges of the next century.

The Structure of Aquatic Ecosystems

Surface waters such as ponds, lakes, oceans, and rivers differ to a considerable extent in size, geographic location, physical makeup, and chemical content. Although an aquatic ecosystem is composed primarily of liquid, it is predictably structured, and it contains significant gradients or local differences in composition. Factors that contribute to the development of zones in aquatic systems are sunlight, temperature, aeration, and dissolved nutrient content. These variations create numerous macro- and microenvironments for communities of organisms. An example of zonation can be seen in the schematic section of a freshwater lake in figure 22.18.

A lake is stratified vertically into three zones or strata. The uppermost region, called the **photic zone,** extends from the surface to the lowest limit of sunlight penetration. Its lower boundary (the compensation depth) is the greatest depth at which photosynthesis can occur. Directly beneath the photic zone lies the **profundal zone,** which extends from the end of the photic zone to the lake sediment. The sediment itself, or **benthic zone,** is composed of organic debris and mud, and it lies directly on the bedrock that forms the lake basin. The horizontal zonation includes the shoreline or **littoral zone,** an area of relatively shallow water. The open, deeper water beyond the littoral is the **limnetic zone.**

Marine Environments The marine profile resembles that of a lake, although special characteristics set it apart. The ocean exhibits extreme variations in salinity, depth, temperature, hydrostatic pressure, and mixing. It contains a unique ecosystem called an *estuary,* an intertidal zone where a river meets the sea. This region fluctuates in salinity, is very high in nutrients, and supports a dense microbial community. It is often dominated by salt-tolerant species of *Pseudomonas* and *Vibrio.* (At least two species of *Vibrio, parahaemolyticus* and *vulnificus,* are pathogens associated with contaminated seafood.) Another important factor is the tidal and wave action that subjects the coastal

---

profundal (proh-fun'-dul) L. *pro,* before, and *fundus,* bottom.
benthic (ben'-thik) Gr. *benthos,* depth of the sea.
littoral (lit'-or-ul) L. *litus,* seashore.
limnetic (lim-neh'-tik) Gr. *limne,* marsh.

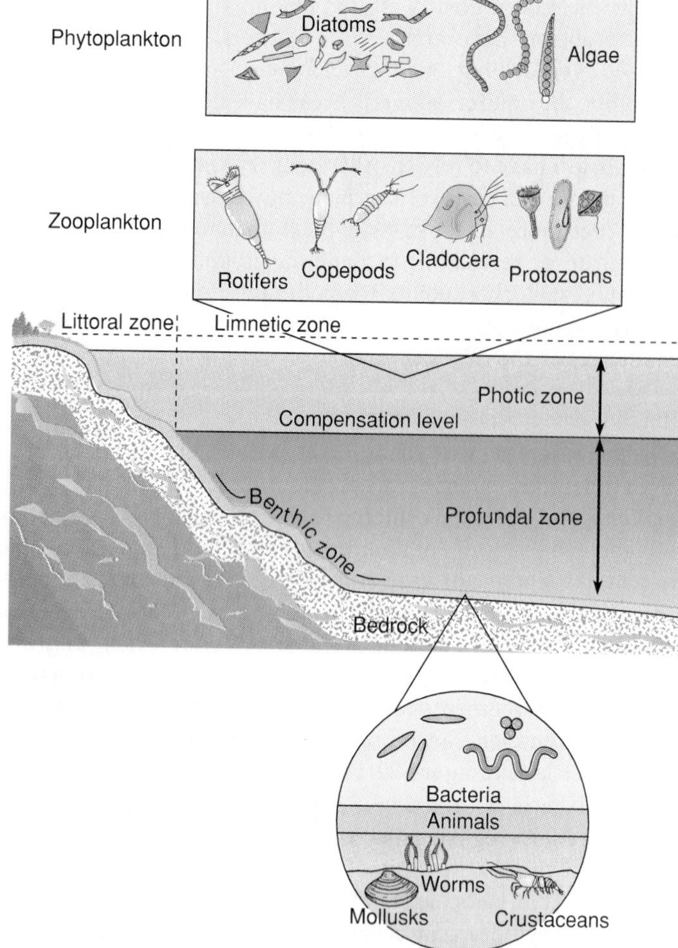

Figure 22.18 Stratification in a freshwater lake. Depth changes in this ecosystem create significant zones that differ in light penetration, temperature, and community structure.

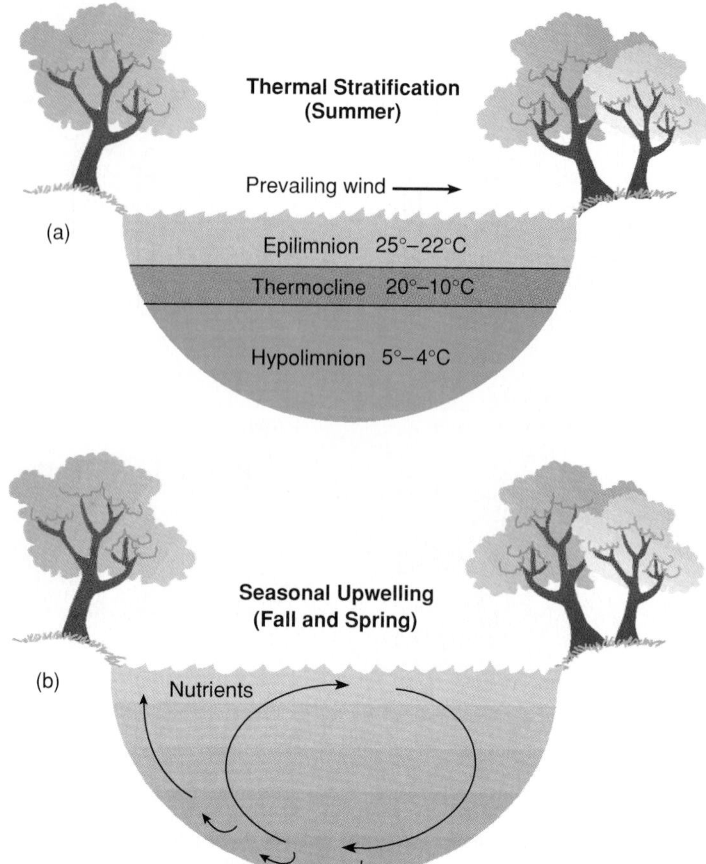

Figure 22.19 Profiles of a lake (*a*) during summer, when it becomes stabilized into three major temperature strata, and (*b*) during fall and spring, when heating or cooling of the water disrupts the temperature strata and causes upwelling of nutrients from the bottom sediments.

habitat to alternate periods of submersion and exposure. The deep ocean, or **abyssal zone,** is poor in nutrients, chiefly due to the absence of sunlight for photosynthesis, and its tremendous depth (up to 10,000 m) makes it oxygen-poor and cold (average temperature 4°C). Although this zone is not totally devoid of life, it supports a community of bacteria with a collection of extreme adaptations. They are halophilic, psychrophilic, barophilic, and anaerobic.

Aquatic Communities Like soil, the freshwater aquatic environment is a site of tremendous microbiological activity. In general, microbial distribution is directly associated with sunlight, temperature, oxygen levels, and nutrient availability. The photic zone, including littoral and limnetic, is the most productive self-sustaining region. This is because it contains large numbers of **plankton,** a floating microbial community that drifts with wave action and currents. The major members of this assemblage are the **phytoplankton,** primarily various photosynthetic algae and cyanobacteria (figure 22.18). The phytoplankton provide nutrition for **zooplankton,** microscopic consumers like protozoa and invertebrates that filter feed, prey, or scavenge. The plankton supports numerous other trophic levels such as larger invertebrates and fish. With its high nutrient content, the benthic zone also supports an extensive variety and concentration of organisms, including aquatic plants, aerobic bacteria (*Bacillus*), and anaerobic bacteria (*Clostridium*) actively involved in recycling organic detritus.

Larger bodies of standing water develop gradients in temperature or thermal stratification, especially during the summer (figure 22.19). The upper region, called the *epilimnion,* is warmest, and the deeper *hypolimnion* is cooler. Between these is a buffer zone, the *thermocline,* that ordinarily prevents the mixing of the two. Twice a year, during the warming cycle of spring and the cooling cycle of fall, temperature changes in the water column break down the thermocline and cause the water from the two strata to mix. This disrupts the stratification and creates currents that bring nutrients up from the sediments. This process, called *upwelling,* is associated with increased activity by certain groups of microbes, and is one explanation for the "red tides" that periodically crop up in oceans (figure 22.20).

plankton (plang′-tun) Gr. *planktos,* wandering.

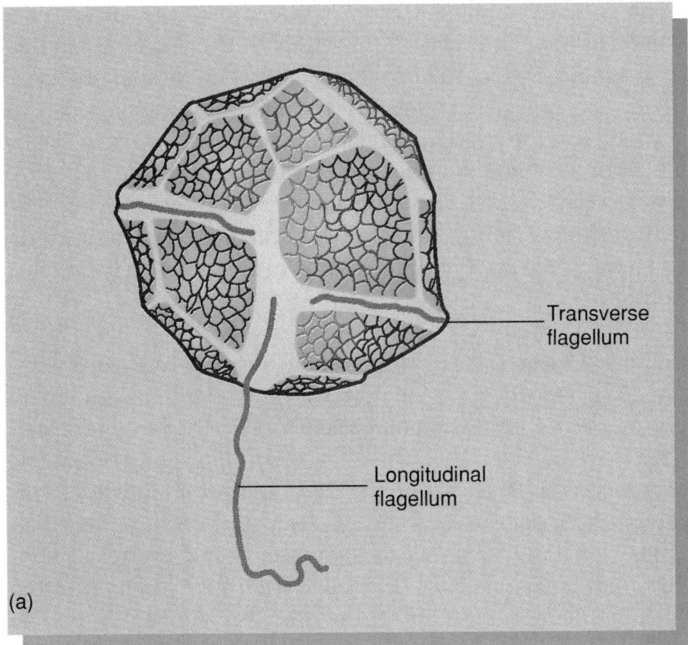

Figure 22.20 Red tides. (*a*) Single-celled red algae called dinoflagellates (*Gymnodinium* shown here) bloom in high-nutrient, warm water and impart a noticeable red color to it, as shown in (*b*). These algae produce a potent muscle toxin that can be concentrated by shellfish through filtration-feeding. When humans eat clams, mussels, or oysters that contain the toxin, they develop paralytic shellfish poisoning. Persons living in coastal areas are cautioned not to eat shellfish during those months of the year associated with red tides (sometimes characterized as months with the letter "r" in them).

Oxygen also forms a gradient that varies with depth. Because it is not very soluble in water and is rapidly used up by the plankton, its concentration varies from highest in the epilimnion to lowest in the benthic zone. In general, the level of dissolved oxygen is inversely related to temperature, thus warm waters have lower levels of this gas. Of all the characteristics of water, the greatest range occurs in nutrient levels. Pure water recently carried from melting snow into cold mountain ponds and lakes is lowest in nutrients. Such nutrient-deficient aquatic ecosystems, called **oligotrophic,** support very few microorganisms, and often they are virtually sterile. Species that can make a living on such starvation rations are *Hyphomicrobium* and *Caulobacter* (see figure 3.35). These bacteria have special stalks that capture even miniscule amounts of hydrocarbons present in this habitat. At one time it was thought that viruses were present only in very low levels in aquatic habitats, but researchers have recently reported finding virus levels of 125 million particles/ml in a water sample from a pristine West German lake. Most of these viruses pose no danger to humans, but as parasites of bacteria, they appear to be a natural control mechanism for bacterial populations.

Figure 22.21 Heavy surface growth of algae and cyanobacteria in a eutrophic pond.

At the other extreme are waters overburdened with organic matter and dissolved nutrients. Some nutrients are added naturally through seasonal upwelling and disasters (floods or typhoons), but the most significant alteration of natural waters comes from effluents from sewage, agriculture, and industry that contain heavy loads of organic debris or nitrate and phosphate fertilizers. The addition of such large quantities of these nutrients to aquatic ecosystems, called *eutrophication,* wreaks havoc on the communities involved. Warm temperatures and the sudden influx of abundant nutrients cause the microscopic algae to flourish (figure 22.21). This creates a heavy surface growth (bloom) that partially or completely shuts off the oxygen supply from the air. At the same time, aerobic heterotrophs actively decompose the organic matter, and this too further depletes the oxygen beneath the surface. The lack of oxygen affects all trophic levels of the community, but it is especially damaging to strict aerobes (fish, invertebrates). If this process continues, the body of water may eventually become a nearly lifeless, anaerobic mire.

---

oligotrophic (ahl″-ih-goh-trof′-ik) Gr. *oligo,* small, and *troph,* to feed.

eutrophication (yoo″-troh-fih-kay′-shun) Gr. *eu,* good.

Microbiology of Drinking Water Supplies We do not have to look far for overwhelming reminders of the importance of safe water. A recent cholera epidemic in Peru killed thousands of persons, and there are constant reports of raw, untreated sewage being drained into waterways that also serve as sources of drinking water. Good health is dependent upon a clean, **potable** (drinkable) water supply. This means the water must be free of pathogens, dissolved toxins, and disagreeable turbidity, odor, color, and taste. As we shall see, water of this quality does not come easily, particularly in developed regions where it must first be treated to render it safe for consumption. Again, we must look to microbes as part of the problem and part of the solution.

Through ordinary exposure to air, soil, and effluents, surface waters usually acquire harmless, saprobic microorganisms. But along its course, water may also pick up pathogenic contaminants. Among the most prominent water-borne pathogens of recent times are the protozoans *Giardia* and *Cryptosporidium,* the bacteria *Campylobacter, Salmonella, Shigella, Vibrio,* and *Mycobacterium,* and the polio, hepatitis A, and Norwalk viruses. Some of these agents (especially encysted protozoans) can survive in natural waters for long periods without a human host, whereas others are present only transiently and will be rapidly lost. To ensure that water is free of infectious agents, its microbial content must be periodically monitored.

Attempting to survey water for specific pathogens can be very difficult and time-consuming, because of their dilution and relatively short survival rates in water. Water-borne pathogens are commonly carried in feces, and this is usually how they get into drinking water supplies like wells or reservoirs. Because high fecal levels mean the water may be unsafe to drink, fecal contamination has become the standard for analyzing water purity. Fecal contamination can be detected by testing the water for the presence of various **indicator bacteria.** These bacteria are normal intestinal flora of humans that live longer in water and are easier to identify than individual pathogens.

Enteric bacteria most useful in the routine monitoring of microbial pollution are *coliforms* and enteric *streptococci.* Because these organisms survive in natural waters but do not multiply there, finding them in high numbers implicates recent or high levels of fecal contamination. Environmental Protection Agency standards for water sanitation are based primarily on the levels of fecal coliforms, which are described as gram-negative, lactose-fermenting, gas-producing enterics such as *Escherichia coli, Enterobacter,* and *Citrobacter.* Fecal contamination of marine waters that poses a risk for gastrointestinal disease is more readily correlated with gram-positive cocci, primarily in the genus *Enterococcus.* Occasionally, coliform bacteriophages and reoviruses (the Norwalk virus) are good indicators of fecal pollution, but their detection is more difficult and more technically demanding.

Water Quality Assays A rapid method for testing the total bacterial levels in water is the **standard plate count.** In this technique, a small sample of water is spread over the surface of a solid medium and observed for numbers of colonies, providing an estimate of the total viable population without differentiating coliforms from other species. This information is particularly helpful in evaluating the effectiveness of various water purification stages. Another general indicator of water quality is the level of dissolved oxygen it contains. It is generally thought that water containing high levels of organic matter and bacteria will have a lower oxygen content due to consumption by aerobic respiration.

Coliform Enumeration Water quality departments employ two standard assays for routine detection and quantification of coliforms. With the **most probable number (MPN)** procedure, coliforms are detected by a series of *presumptive, confirmatory,* and *completed* tests (figure 22.22). The presumptive test involves a series of three subsets of fermentation tubes, each containing different amounts of lactose or lauryl tryptose broth. Each subset contains five tubes, and each tube has an inverted durham tube to collect any gas produced by fermentation. The three subsets are inoculated with water inocula sizes of 10, 1.0, and 0.1 ml, respectively. The scheme for predicting the concentration of coliforms is based upon dilution and statistical standards. After 24 hours of incubation, the tubes are evaluated for gas production. Positive for gas formation is presumptive evidence of coliforms, and negative gas means no coliforms. The number of positive and negative tubes in each subset is tallied, and this set of numbers is applied to a statistical table. From this standard, the most likely or probable concentration of coliforms can be estimated (table 22.2). A confirmatory test for coliforms is made by inoculating another broth from one of the positive tubes. The test is completed by final isolation of the coliform species on selective and differential media, Gram staining the isolate, and reconfirming gas production.

The **membrane filter method** is increasingly preferred for water analysis. It is faster, requires fewer steps and media, is less expensive, is more portable, and can process larger quantities of water. This method is more suitable for dilute fluids like drinking water that are relatively free of particulate matter, and it is less suitable for water containing heavy microbial growth or metal inhibitors that bind to the filter. This technique is related to the method described in chapter 9 for sterilizing fluids by filtering out microbial contaminants, except that in this system, the filter containing the trapped microbes is the desired end product. The steps in membrane filtration are diagrammed in figure 22.23. After filtration, the membrane filter is removed and placed in a small petri dish containing selective broth. After incubation, fecal coliform colonies can be counted and often presumptively identified by their distinctive characteristics on these media (figure 22.23*b*).

When a test is negative for coliforms, the water is considered generally fit for human consumption. But even slight coliform levels are allowable in drinking water under some circumstances. For example, municipal waters may have a maximum of 4 coliforms per 100 ml, while private wells may

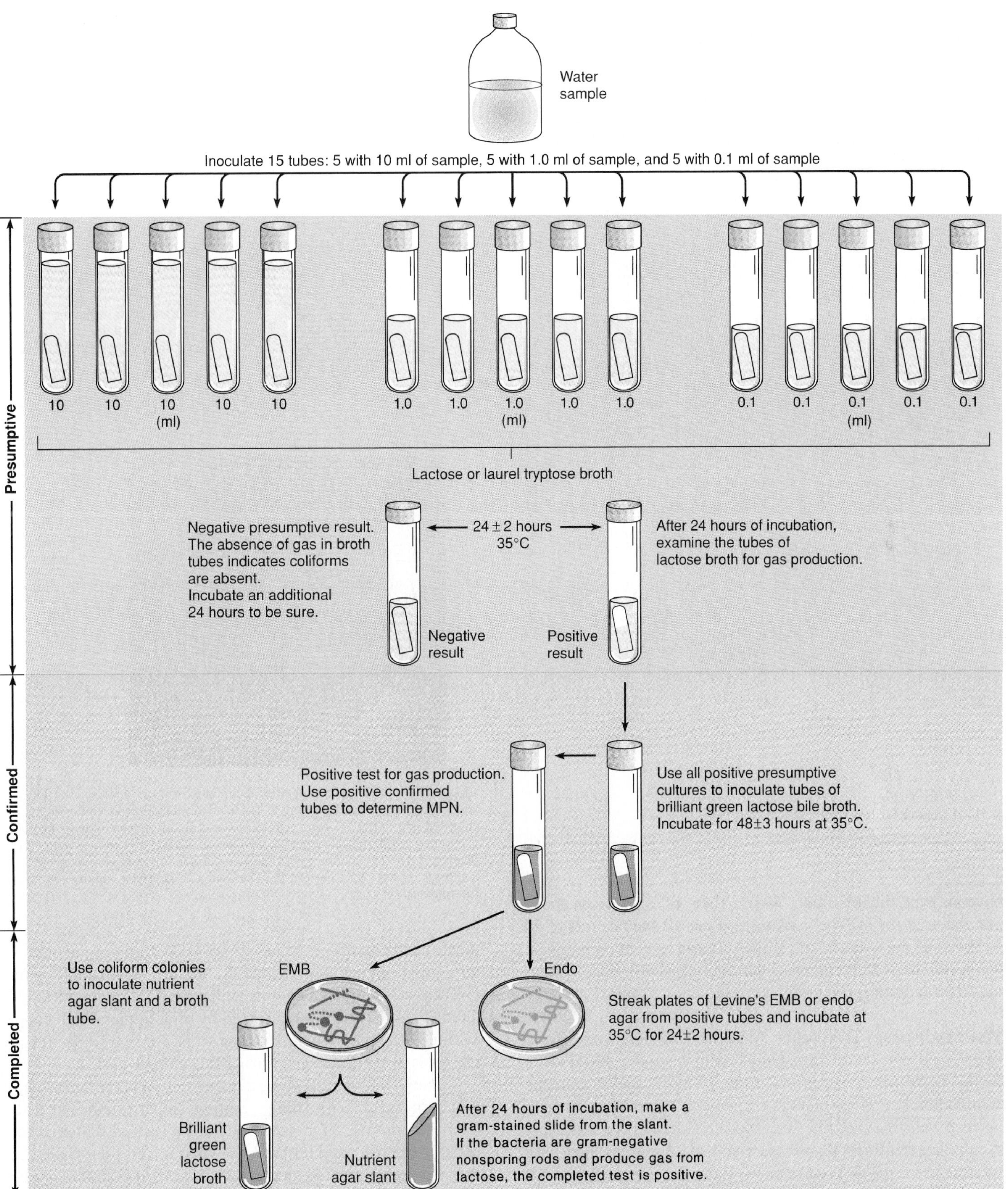

Figure 22.22 The most probable number (MPN) procedure for determining the coliform content of a water sample. In the presumptive test, each set of five tubes of broth is inoculated with a water sample reduced by a factor of 10. After incubation, the series of five tubes is examined and rated for gas production (0 means no tubes with gas, 1 means one tube with gas, 2 means two tubes with gas). Applying this result to an MPN table will indicate the probable number of cells present in 100 ml of the water sample. Confirmation of coliforms may be achieved by confirmatory tests on additional media, and complete identification may be made through selective and differential media and Gram staining.

Table 22.2 Most Probable Number Chart

The Number of Tubes in Series, with Positive Growth at Each Dilution			
10 ml	**1 ml**	**0.1 ml**	**MPN* per 100 ml of Water**
0	1	0	0.18
1	0	0	0.20
1	0	0	0.40
2	0	0	0.45
2	0	1	0.68
2	2	0	0.93
3	0	0	0.78
3	0	1	1.1
3	1	0	1.1
3	2	0	1.4
4	0	0	1.3
4	0	1	1.7
4	1	0	1.7
4	1	1	2.1
4	2	0	2.2
4	2	1	2.6
4	3	0	2.7
5	0	0	2.3
5	0	1	3.1
5	1	0	3.3
5	1	1	4.6
5	2	0	4.9
5	2	1	7.0
5	2	2	9.5
5	3	0	7.9
5	3	1	11.0
5	3	2	14.0
5	4	0	13.0
5	4	1	17.0
5	4	2	22.0
5	4	3	28.0
5	5	0	24.0
5	5	1	35.0
5	5	2	54.0
5	5	3	92.0
5	5	4	160.0

*Most probable number of cells per sample of 100 ml.

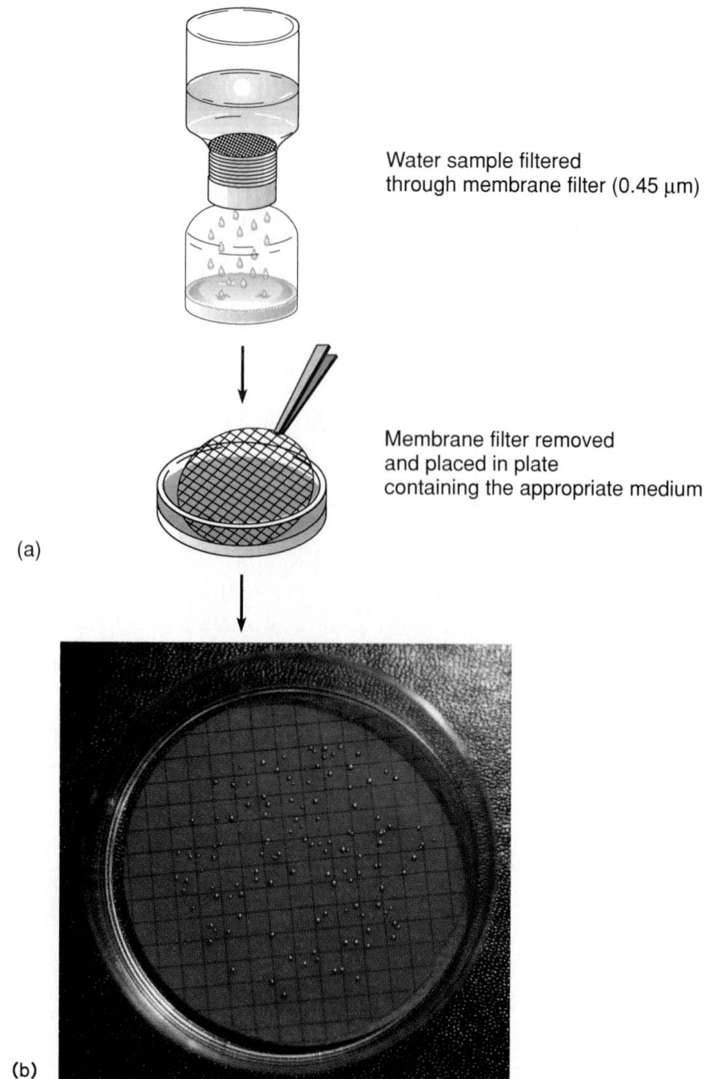

Figure 22.23 The membrane filter technique for water analysis. (*a*) The water sample is filtered through a sterile membrane filter assembly and collected in a flask. The filter is removed and placed in a small petri dish containing a differential, selective medium such as M-FD endo broth and incubated. (*b*) The medium permits easy differentiation of various genera of coliforms, and the grid pattern may be used as a guide for rapidly counting the colonies.

have an even higher count. Waters that will not be consumed but are used for fishing or swimming are allowed counts of 70 to 200 coliforms per 100 ml. If the coliform level of recreational water reaches 1,000 coliforms per 100 ml, health departments usually bar its usage.

Water and Sewage Treatment Most drinking water comes from rivers, aquifers, and springs. Only in remote, undeveloped areas is this water used in its natural form. In most cities, it must be treated before it is supplied to consumers. Water supplies such as deep wells that are relatively clean and free of contaminants require less treatment than those from surface sources laden with wastes. The stepwise process in water purification as carried out by most cities is shown in figure 22.24. Treatment begins with the impoundment of water in a large reservoir such as a dam or catch basin that serves the dual purpose of storage and sedimentation. The access to reservoirs is controlled to avoid contamination by animal carcasses, wastes, and runoff water. Overgrowth of cyanobacteria and algae that add undesirable qualities to the water is prevented by pretreatment with copper sulfate (0.3 ppm). Sedimentation to remove large particulate matter is also encouraged during this storage period.

Next, the water is pumped to holding ponds or tanks, where it undergoes further settling, aeration, and filtration. The water is filtered first through sand beds or pulverized diatomaceous earth to remove residual bacteria, viruses, and protozoans, and then through activated charcoal to remove undesirable organic contaminants. Pipes coming from the filtration beds collect the water in storage tanks. The final step in treatment is chemical disinfection by bubbling chlorine gas through the tank until it

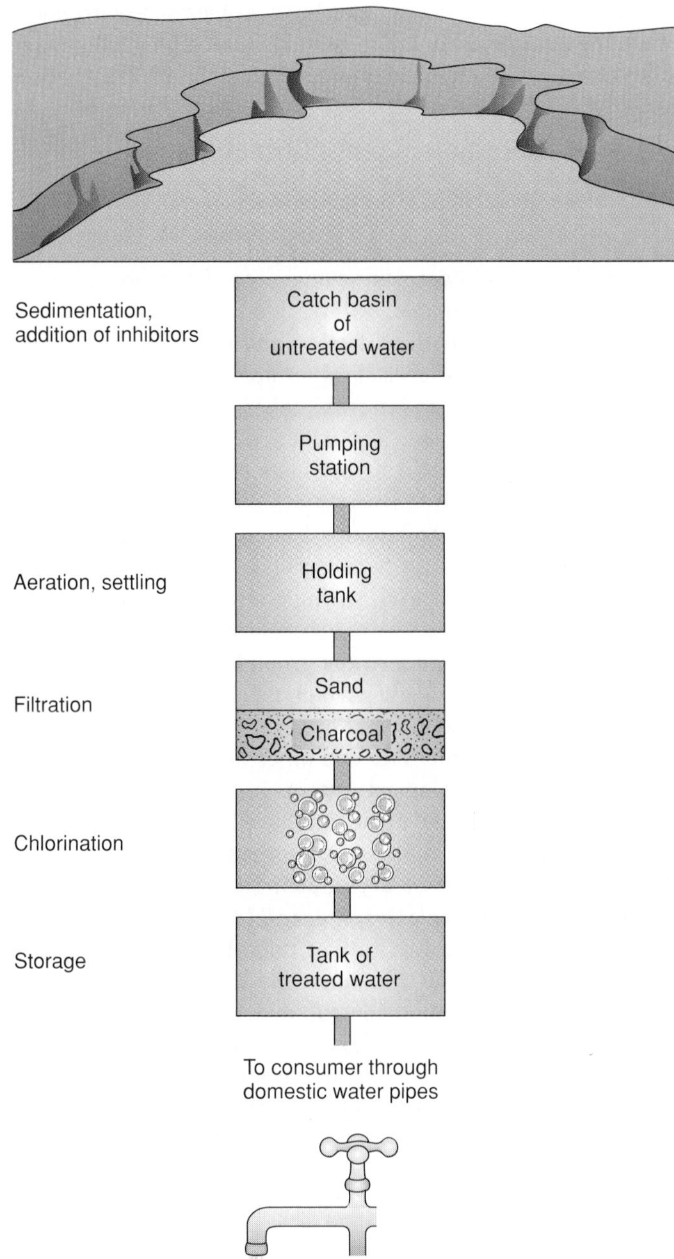

Figure 22.24 The major steps in water purification as carried out by a modern municipal treatment plant.

reaches a concentration of 1–2 ppm, but some municipal plants use chloramines (chapter 9) for this purpose. A few pilot plants in the United States are using ozone or peroxide for final disinfection, but these methods are expensive and cannot sustain an antimicrobial effect over long storage times. The final quality varies, but most tap water has a slight odor or taste from disinfection.

More and more in the developed world, the same water that serves as a source of drinking water is also used as a dump for solid and liquid wastes. This leads to a dilemma, because heavily polluted waters cannot be made safe by normal treatment methods. As the population continues to grow and water resources are spread thin, polluted water may someday be the only source available, and it will have to be salvaged and recycled. One such problem that has already been systematically tackled is the handling of sewage.

Figure 22.25 Aerial photograph of a sewage treatment plant that occupies hundreds of acres and can process several hundred million gallons of water a day.

Sewage is the used wastewater draining out of homes and industries that contains a wide variety of chemicals, debris, and microorganisms. The danger of infection from sewage has been understood for several centuries. Outbreaks of typhoid, cholera, and dysentery were linked to the unsanitary mixing of household water and sewage. But even with this knowledge, sewage effluents are commonly dumped into surface waters. Some sewage is treated to reduce its microbial load before release, but some is emptied raw (untreated). Currently in the U.S. only a small amount of sewage water is reclaimed or recycled. This is largely because heavily contaminated waters require far more stringent and costly methods of treatment than are currently available to most cities. The systems for sewage treatment are massive engineering marvels (figure 22.25).

Another problem is that sewage often contains large amounts of solid wastes, dissolved organic matter, and toxic chemicals that pose a health risk. To remove all potential health hazards, sewage is typically treated in three phases: The primary stage separates out large matter; the secondary stage reduces remaining matter and may remove some toxic substances; and the tertiary stage completes the purification of the water (figure 22.26). Microbial activity is an integral part of the overall process.

In the **primary phase** of treatment, floating bulkier materials such as tires, boxes, bottles, and paper are skimmed off. The remaining smaller, suspended particulates are allowed to settle. Sedimentation in settling tanks usually takes 2 to 10 hours and leaves a mixture rich in organic matter. This aqueous residue is carried into a **secondary phase** of active microbial decomposition or biodegradation. In this phase, a diverse community comprised chiefly of bacteria, algae, and protozoa aerobically digests the remaining particles of wood, paper, fabrics,

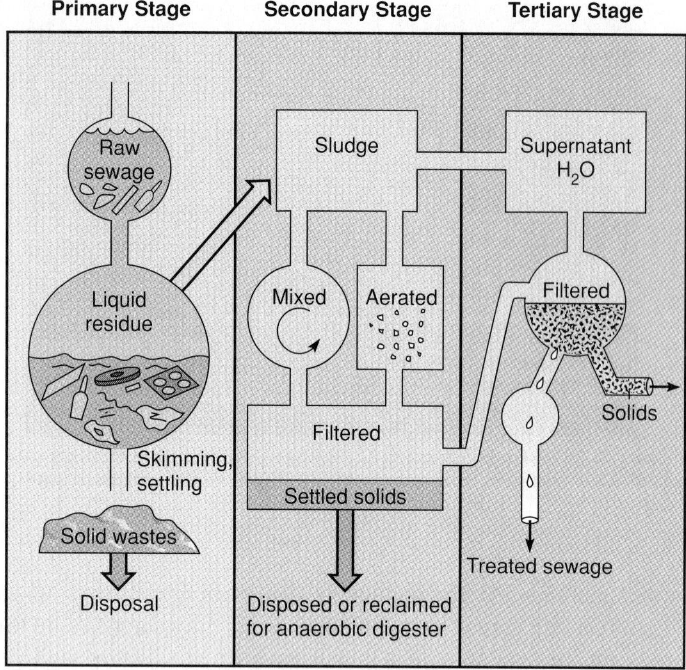

Figure 22.26 The primary, secondary, and tertiary stages in sewage treatment.

petroleum, rubber, and organic molecules. It forms a suspension of material called *sludge* that tends to settle out and slow the process. To hasten aerobic decomposition of the sludge, most processing plants have systems to *activate* the sludge by injecting air, mechanically stirring it, and recirculating it. A large amount of organic matter is mineralized into sulfates, nitrates, phosphates, carbon dioxide, and water. Certain volatile gases such as hydrogen sulfide, ammonia, nitrogen, and methane may also be released. Water from this process is siphoned off and carried to the tertiary stage, which involves further filtering and chlorinating prior to discharge. Such reclaimed sewage water is usually used to water golf courses and parks, rather than for drinking.

In some cases, the solid waste that remains after aerobic decomposition is harvested and reused. Its rich content of nitrogen, potassium, and phosphorus makes it a useful fertilizer. But if the waste contains large amounts of nondegradable or toxic substances, it may be buried in landfills. In many parts of the world, the sludge, which still contains significant amounts of simple, but useful organic matter, is used as a secondary source of energy. Further digestion is carried out by microbes in huge chambers called bioreactors or **anaerobic digesters.** The digesters convert components of the sludge to swamp gas, primarily methane with small amounts of hydrogen, carbon dioxide, nitrogen, and other volatile compounds. Swamp gas may be burned to provide energy to run the sewage processing facility itself or to power small industrial plants. Considering the mounting waste disposal and energy shortage problems, this technology should gain momentum. For another microbiological answer to mounting liquid and solid waste, see feature 22.4.

Applied Microbiology and Biotechnology

"Never underestimate the power of the microbe."
Jackson W. Foster

The profound and sweeping involvement of microbes in the natural world is inescapable. Although our daily encounters with them usually go unnoticed, human and microbial life is clearly intertwined on many levels. It is no wonder that, even long ago, humans realized the power of microbes and harnessed them for specific metabolic tasks. The practical applications of microorganisms in manufacturing products or carrying out a particular decomposition process belong to the large and diverse area of **biotechnology.** Biotechnology has an ancient history, dating back nearly 6,000 years to those first lucky humans who discovered that grape juice left sitting produced wine or that bread dough properly infused with a starter would rise. Today, biotechnology has become a fertile ground for hundreds of applications in industry, medicine, agriculture, food sciences, and environmental protection, and has even come to include the genetic alteration of microbes and other organisms.

Most biotechnological systems involve the actions of bacteria, yeasts, molds, and algae that have been selected or altered to synthesize a certain food, drug, organic acid, alcohol, or vitamin. Many such food and industrial end products are obtained through **fermentation,** a general term used here to refer to the mass, controlled culture of microbes to produce desired organic compounds. It also includes the use of microbes in sewage control, pollution control, and metal mining. A single chapter cannot begin to do justice to this diverse area of microbiology, but we will cover here some of its more important applications in food technology, industrial processes, and bioengineering.

Microorganisms and Food

All human food—from vegetables to caviar to cheese—comes from some other organism, and rarely is it obtained in a sterile, uncontaminated state. Food is but a brief stopover in the overall scheme of biogeochemical cycling. This means that microbes and humans are in direct competition for the nutrients in food, and we must be constantly aware that microbes' fast growth rates give them the winning edge. Somewhere along the route of procurement, processing, or preparation, food becomes contaminated with microbes from the soil, the bodies of plants and animals, water, air, food handlers, or utensils. The final effects depend upon the types and numbers of microbes and whether the food is cooked, preserved, or otherwise processed. In some cases, specific microbes may even be added to food to obtain a

Feature 22.4 Bioremediation: The Pollution Solution?

The soil and water of the earth have long been considered convenient repositories for solid and liquid wastes. Every year, about 264 metric tons of pollutants, industrial wastes, and garbage are deposited into the natural environment. One rationale for this wide-scale discharge of contaminants is an "out-of-sight, out-of-mind" mentality, and another is the often mistaken idea that naturally occurring microbes will eventually biodegrade (break down) waste material.

Humans have been filling landfills with solid wastes for thousands of years, but the process has escalated in the past 50 years. Landfills currently serve as a final resting place for hundreds of castoffs from an affluent society, including yard wastes, paper, glass, plastics, food wastes, wood, textiles, rubber, metal, paints, and solvents. This conglomeration is dumped into holes, covered with soil, and left. Substances that are biodegradable can be readily broken down by microbes, but many materials such as plastics and glass are not biodegradable. Successful biodegradation also requires a compost containing specific types of microorganisms, adequate moisture, and oxygen. The environment surrounding buried trash provides none of these conditions. Large, inaccessible, dry, anaerobic masses of plant materials, paper, and other organic materials will not be readily attacked by the aerobic microorganisms that dominate in biodegradation. A group at the University of Arizona excavating older landfills discovered to their surprise 50-year-old-newspapers that were perfectly readable, and mummified hot dogs and carrots. As we continue to fill up hillsides with waste, the future of these landfills is a prime concern. Should they be capped and allowed to sit unchanged as a legacy to future generations, or should they be artificially degraded, perhaps releasing toxic compounds into the ground and water below?

Another serious concern is pollution of groundwater, the primary source of drinking water for 100 million persons in the United States. Because of the extensive cycling of water through the hydrosphere and lithosphere, groundwater is often the final collection sump for hazardous chemicals released into lakes, streams, oceans, and even garbage dumps. Many of these chemicals are pesticide residues from agriculture (dioxin, selenium, 2,4-D), industrial hydrocarbon wastes (PCBs), and hydrocarbon solvents (benzene, toluene). When millions of Americans turn on their faucets, out flows water containing these and other chemicals, which are often hard to detect, and if detected, are hard to remove.

Once the toxic cat is out of the bag and the environment is polluted, what are the solutions? For many years, polluted soil and water were simply sealed off or dredged and dumped in a different site, without attempting to get rid of the pollutant. But now, with greater awareness of toxic wastes, many Americans are adopting an attitude known as NIMBY (not in my backyard!), and environmentalists are troubled by the long-term effects of contaminating the earth.

The latest method of attacking the pollution problem is through **bioremediation**—using microbes to break down or remove toxic wastes in water and soil. Some of these waste-eating microbes are natural soil and water residents with a surprising capacity to decompose wastes and even man-made toxic substances. Because the natural, unaided process occurs too slowly, most cleanups are accomplished by commercial bioremediation services that treat the contaminated soil with oxygen, nutrients, and water to increase the rate of microbial action. Through these actions, levels of pesticides such as 2,4-D can be reduced to 96% of their original levels, and solvents can be reduced from one million parts per billion (ppb) to 10 ppb or less. Bacteria are also being used to help break up and digest oil spills like the Exxon Valdez disaster of 1989. Among the most important bioremedial microbes are species of *Pseudomonas* and *Bacillus* and various toxin-eating fungi. The current quest in bioremediation is for "super bugs" genetically engineered to convert nasty chemicals into CO_2 or nontoxic residues. Because of their tremendous potential, such microbes will surely become trade secrets, and one has already been patented.

(a)

(b)

Can microbes rescue us from our polluted world? (*a*) A Santa Monica, California, beach receives a heavy runoff of raw sewage. (*b*) Solid wastes collect on a beach in the northeastern United States.

desired effect. The relationships between microorganisms and food can be classified as detrimental, beneficial, or neutral to humans, as summarized by the following outline:

Detrimental Effects
- Food poisoning or food-borne illness
  - Chemical in origin: Pesticides, food additives
  - Biological in origin: Living things or their products
    - Infection: Bacterial, protozoan, worm
    - Intoxication: Bacterial, fungal
- Food spoilage
  - Growth of microbes makes food unfit for consumption; adds undesirable flavors, appearance, and smell; food value is destroyed.

Beneficial Effects
- Food is fermented or otherwise chemically changed by the addition of microbes or microbial products to add or improve flavor, taste, or texture.
- Microbes may serve as food.

Neutral Effects
- The presence or growth of microbes that do not harm or change the nature of the food.

The Philosophy of Food: A Matter of Taste Most of us eat not simply for nutrition but for pleasure as well. It must be noted, however, that taste is an acquired preference and that bouquet to some is decay to others. It is safe to say that, as long as no pathogens are present, the test of whether certain foods are edible is guided by experience. Many ethnic or cultural groups savor delicacies that have flavors, colors, textures, and aromas supplied by bacteria and fungi. Chinese thousand-year eggs (actually only a month old), Norwegian fermented fish, and limburger cheese delight some and offend others. Poi may be viewed as ambrosia or as wallpaper paste. But there is little doubt that most of us enjoy some food that derives its delicious flavor from microbes.

Microbial Involvement in Food-Borne Diseases

Diseases caused by ingesting food are usually referred to as **food poisoning,** and although this term is often used synonymously with microbial food-borne illness, not all food poisoning is caused by microbes or their products. Several illnesses are caused by poisonous plant and animal tissues or by ingesting food contaminated by pesticides or other poisonous substances. Table 22.3 summarizes the major types of microbial food-borne disease.

Table 22.3 Major Forms of Bacterial Food Poisoning

Disease	Microbe	Foods Usually Involved	Comments
Food Intoxications: Caused by Ingestion of Foods Containing Preformed Toxins			
Staphylococcal enteritis	*Staphylococcus aureus*	Custards, cream-filled pastries, ham, dressings	Very common; symptoms come on rapidly; usually nonfatal
Botulism	*Clostridium botulinum*	Home-canned or poorly preserved low-acid foods	Recent cases involve fresh foods; often fatal
Perfringens	*Clostridium perfringens*	Inadequately cooked meats	Spores produce toxin within the intestine
Bacillus cereus enteritis	*Bacillus cereus*	Reheated rice, potatoes, puddings, custards	Mimics staphylococcal enteritis; usually self-limited
Food Infections: Caused by Ingestion of Live Microbes That Invade the Intestine			
Salmonellosis	*Salmonella typhimurium* and *enteriditis*	Poultry, eggs, dairy products, meats	Very common; can be severe and life-threatening
Shigellosis	Various *Shigella* species	Unsanitary cooked food: beans, potatoes	Carriers and flies contaminate food; common in fish, shrimp, salads; cause dysentery
Vibrio enteritis	*Vibrio parahaemolyticus*	Raw or poorly cooked seafoods	Microbe lives naturally on marine animals
Listeriosis	*Listeria monocytogenes*	Poorly pasteurized milk, cheeses	Most severe in fetuses, newborns, and the immunodeficient
Campylobacteriosis	*Campylobacter jejuni*	Raw milk; raw chicken, shellfish, and meats	Very common; animals are carriers of other species
Escherichia enteritis	*Escherichia coli*	Contaminated raw vegetables, cheese	Various strains may produce infantile and traveler's diarrhea

Food poisoning of microbial origin may be divided into two general categories[4] (figure 22.27). **Food intoxication** occurs when microbes growing in food secrete toxin, and the subsequent ingestion of this toxin disrupts a particular target such as the intestine (if an enterotoxin) or the nervous sytem (if a neurotoxin). The symptoms of intoxication vary from bouts of vomiting and diarrhea (staphylococcal intoxication) to severely disrupted muscle function (botulism). **Food infection** occurs when microbial cells multiply in food, are ingested, and act directly on the intestine. In some cases, they infect the surface of the intestine, and in others, they invade the intestine and other body structures. Most food infections manifest some degree of diarrhea and abdominal distress (see feature 16.6). It is important to realize that disease symptoms in food infection can be initiated by toxins, but that the toxins are released by microbes growing in the infected tissue rather than in the food.

Many reported food poisoning outbreaks occur where contaminated food has been served to large groups of people,[5] but most cases probably occur in the home and are not reported. The most common food-borne illness in the United States is staphylococcal food intoxication. Other relatively common agents of food-associated disease include *Campylobacter, Salmonella, Clostridium perfringens,* and *Shigella.* Food poisoning by *Clostridium botulinum, Bacillus cereus,* and *Vibrio* is less common. The detailed pathology of the diseases listed in table 22.3 was covered in chapters 15, 16, and 17.

Prevention Measures for Food Poisoning and Spoilage

It will never be possible to avoid all types of food-borne illness because of the ubiquity of microbes in air, water, food, and the human body. But most types of food poisoning require the growth of microbes in the food. In the case of food infections, an infectious dose (sufficient cells to initiate infection) must be present, and in food intoxication, enough cells to produce the toxin must be present. Thus, food poisoning or spoilage may be prevented by proper food handling, preparation, and storage. The methods shown in figure 22.28 are aimed at preventing the incorporation of microbes into food, removing or destroying microbes in food, and keeping microbes from multiplying.

Preventing the Incorporation of Microbes into Food Most agricultural products such as fruits, vegetables, grains, meats, eggs, and milk are naturally exposed to microbes. To reduce the levels of contaminants, fruits and vegetables may be vigorously washed, while meat, eggs, and milk must be taken from their animal source as aseptically as possible. Aseptic techniques are also essential in the kitchen. Contamination of foods by fingers can be easily remedied by handwashing and proper hygiene, and contamination by flies or other insects can be stopped by covering foods or eliminating pests from the kitchen. Care and common sense also apply in managing utensils. It is important not to cross-contaminate food—for example, by using the same knife for raw meat and salad vegetables without disinfecting it in between, or by using the same spoon to both stir and taste food as it is being prepared.

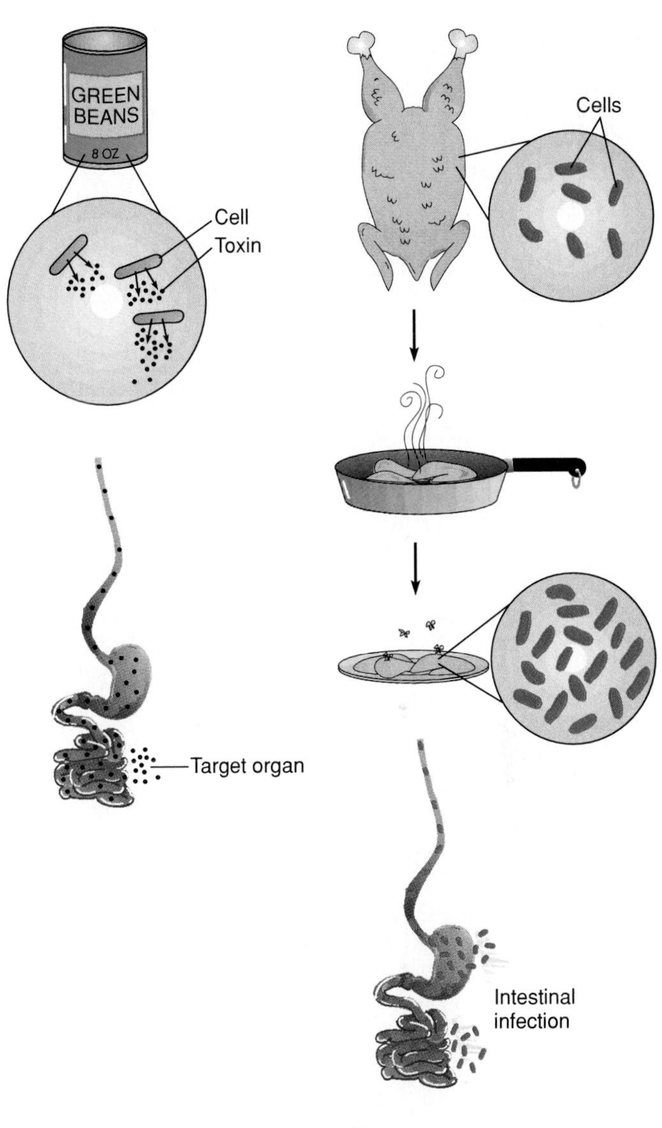

Figure 22.27 Food-borne illnesses of microbial origin. (*a*) Food intoxication. The toxin is produced by microbes growing in the food. After the toxin is ingested, it acts upon its target tissue and causes symptoms. (*b*) Food infection. The microbe itself is ingested and grows in the intestine, which causes symptoms of intestinal inflammation.

4. Although these categories are useful for clarifying the general forms of food poisoning, some diseases are transitional. For example, perfringens intoxication is caused by a toxin, but the toxin is released in the intestinal lumen rather than in the food.
5. One-third of all reported cases result from eating restaurant food.

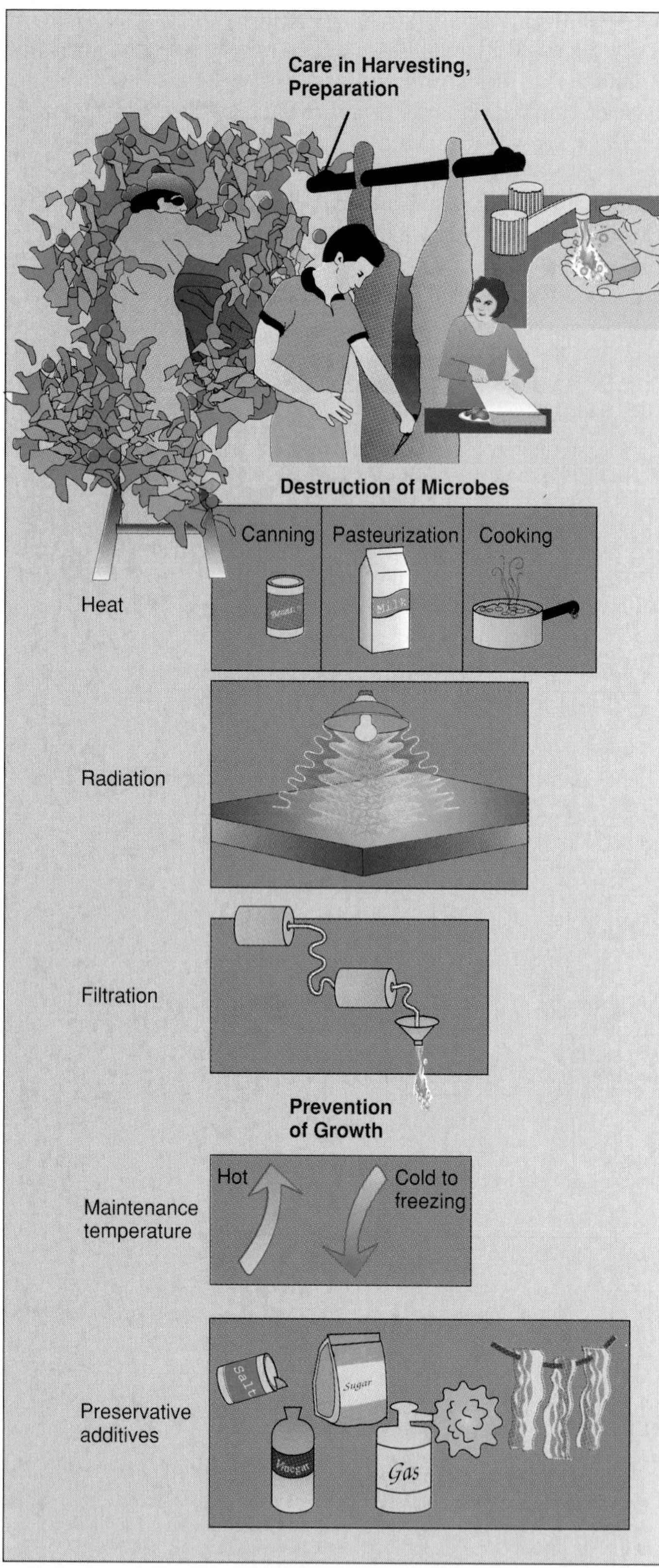

Figure 22.28 The primary methods of preventing food poisoning and food spoilage.

Preventing the Survival or Multiplication of Microbes in Food Even with the most hygienic techniques, it is not possible to eliminate all microbes from certain types of food, so a more efficient approach is to preserve the food by destroying or removing microbial contaminants. Large commercial companies that process and sell bulk foods must ensure that products are free from harmful contaminants. Since management of such mass quantities of food can be difficult, most manufacturers focus their efforts on preservation techniques. Regulations and standards for food processing are administered by two federal agencies—the Food and Drug Administration (FDA) and the United States Department of Agriculture (USDA).

Temperature and Food Preservation Heat is a common way to destroy microbial contaminants or to reduce the load of microorganisms. Commercial canneries preserve food in hermetically sealed containers that have been exposed to high temperatures over a specified time period. The temperature used depends upon the type of food, and it may range from 60°C to 121°C, with exposure times ranging from 20 minutes to 115 minutes. The food is usually processed at a thermal death time (TDT; see chapter 9) that will destroy the main spoilage organisms and pathogens, but not alter the nutrient value or flavor of the food. For example, tomato juice must be heated to between 121°C and 132°C for 20 minutes to destroy the spoilage agent *Bacillus coagulans,* and green beans must be heated to 121°C for 20 minutes to destroy pathogenic *Cl. botulinum.* Most canning methods are rigorous enough to sterilize the food completely, but some only render the food "commercially sterile," which means it contains live bacteria that are unable to grow.

Another use of heat is **pasteurization,** usually defined as the application of heat below 100°C to destroy nonresistant bacteria and yeasts in liquids such as milk, wine, and fruit juices. The heat may be applied in the form of steam, hot water, or even electric current. The most prevalent technology is the *high-temperature short-time (HTST)* or flash method using extensive networks of tubes that expose the liquid to 72°C for 15 seconds (figure 22.29). A newer method, ultrahigh temperature (UHT) pasteurization, steams the product until it reaches a temperature of 134°C for at least one second. Although milk processed this way is not actually sterile, it is often marketed as sterile, with a shelf life of up to three months. Older methods involve large bulk tanks that hold the fluid at a low temperature for a long time, usually 62.3°C for 30 minutes.

In the kitchen, heat is one of the main methods for controlling microbes in food. Temperatures used to boil, roast, or fry foods not only cook them but may also render them free or relatively free of living microbes. In practice, it is necessary to heat foods long enough to destroy any potential pathogens. A quick searing of chicken or an egg will not penetrate deeply enough to kill microbes such as *Salmonella.* In fact, any meat is a potential source of infectious agents and should be adequately cooked. Since most meat-associated food poisoning is caused by nonsporulating bacteria, heating the center of meat to at least 80°C and holding it there for 30 minutes is usually

Figure 22.29 A modern flash pasteurizer, a system used in dairies for high-temperature short-time pasteurization.
Photo taken at Alta Dena Dairy, City of Industry, California.

sufficient to kill pathogens. Roasting or frying food at temperatures of 200°C or above or boiling it will achieve a satisfactory degree of disinfection. Botulism may be prevented in home-canned vegetables or meat by subjecting them to 20 minutes at 100°C to destroy the botulinal toxin.

Any perishable raw or cooked food that could serve as a growth medium must be stored so as to inhibit the multiplication of bacteria that have survived during processing or handling. Because most food-borne bacteria and molds that are agents of spoilage or infection can multiply at room temperature, manipulation of the holding temperature is a useful preservation method (figure 22.30). A good general directive is to store foods at temperatures below 4°C or above 60°C.

Regular refrigeration reduces the growth rate of most mesophilic bacteria by ten times, although some psychrotrophic microbes may continue to grow at a rate that causes spoilage. This factor limits the shelf life of milk, because even at 7°C, a population could go from a few cells to a billion in 10 days. Pathogens such as *Listeria monocytogenes* and *Salmonella* may also continue to grow in refrigerated foods. Freezing is a longer-term method for cold preservation. Foods may be either slow frozen for 3 to 72 hours at −15°C to −23°C or rapidly frozen for 30 minutes at −17°C to −34°C. Because freezing cannot be counted upon to kill microbes, rancid, spoiled, or infectious foods will still be unfit to eat after freezing and defrosting. *Salmonella* is known to survive several months in frozen chicken,

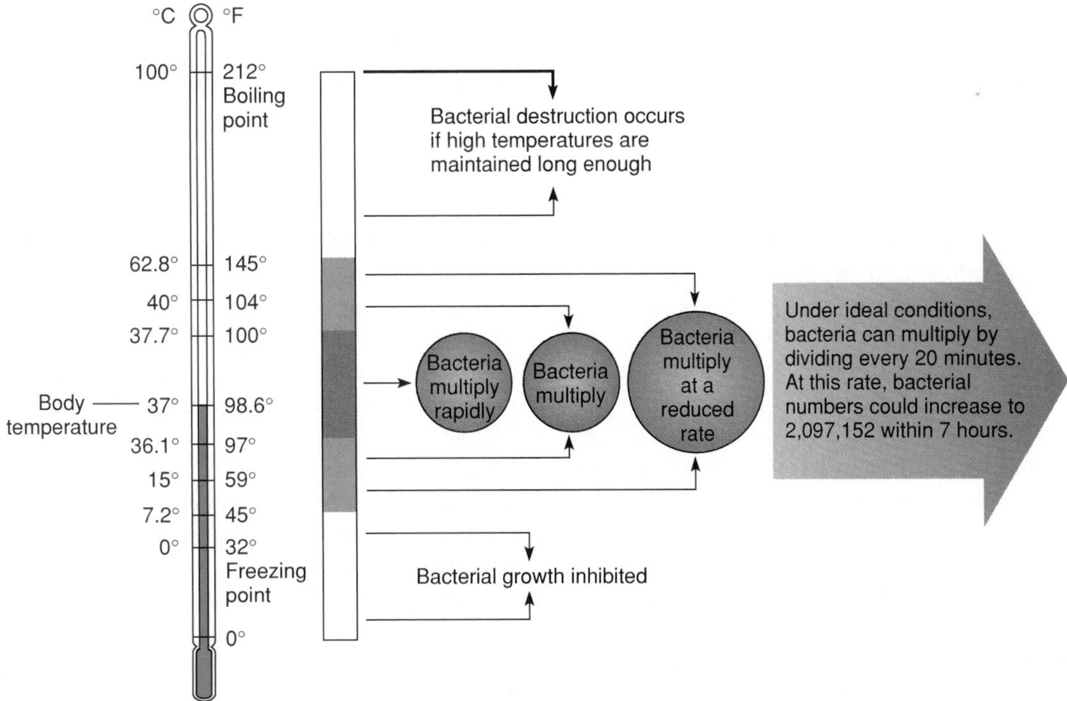

Figure 22.30 Temperatures favoring and inhibiting the growth of microbes in food. Most microbial agents of disease or spoilage grow in the temperature range of 15°–40°C. Preventing unwanted growth in foods in long-term storage is best achieved by refrigeration or freezing (4°C or lower). Preventing microbial growth in foods intended to be consumed warm in a few minutes or hours requires maintaining the foods above 60°C.

and *Vibrio parahaemolyticus* can survive in frozen shellfish. For this reason, frozen foods should be defrosted rapidly and immediately cooked or reheated. Even this practice will not prevent staphylococcal intoxication if the toxin is already present in the food prior to cooking.

Foods such as soups, stews, gravies, meats, and vegetables that are generally eaten hot should not be maintained at warm or room temperatures, especially in settings such as cafeterias, banquets, and picnics. The use of a hot plate, chafing dish, or hot water bath will maintain foods above 60°C, well above the incubation temperature of food poisoning agents.

As a final note about methods to prevent food poisoning, remember the simple axiom: "When in doubt, throw it out."

Radiation Food industries have led the field of radiation sterilization and disinfection (see chapter 9). Ultraviolet (nonionizing) lamps are commonly used to destroy microbes on the surfaces of foods or utensils, but they do not penetrate far enough to sterilize bulky foods or food in packages. Food preparation areas are often equipped with UV radiation devices that are used to destroy spores on the surfaces of cheese, breads, and cakes, to disinfect packaging machines and storage areas, and to treat the air.

Food itself is usually sterilized by gamma or cathode radiation because these ionizing rays are more penetrating. Gamma rays are more dangerous because they originate from a radioactive source, and they are less efficient because they cannot be beamed directly at a target. Because cathode rays are safer and can be focussed, they are the method of choice for most food irradiation applications. Concerns have been raised about the possible secondary effects of radiation that could alter the safety and edibility of foods. Experiments over the past 20 years have demonstrated some side reactions that affect flavor, odor, and vitamin content, but it is currently thought that irradiated foods are relatively free of toxic by-products. The government has currently approved the use of radiation in sterilizing meat, poultry, fish, spices, grain, and some fruits and vegetables.

Microwaves are a form of long-wave radiation used to cook foods. When food is pulsed with microwave radiation, the molecules it contains oscillate rapidly, causing friction that heats the foods. Any antimicrobial effects on the food are due to the action of this heat, not to the radiation.

Other Forms of Preservation Chemical preservatives are added to many foods to prevent the growth of microorganisms that could cause spoilage or disease. These additives may be natural chemicals such as salt (NaCl) or table sugar, or they may be artificial substances such as ethylene oxide. The main classes of preservatives are organic acids, nitrogen salts, sulfur compounds, oxides, salt, and sugar.

Organic acids, including lactic, benzoic, and propionic acids, are among the most widely used preservatives. They are added to baked goods, cheeses, pickles, carbonated beverages, jams, jellies, and dried fruits to reduce spoilage from molds and some bacteria. Nitrites and nitrates are used primarily to maintain the red color of cured meats (hams, bacon, and sausage). They also inhibit the germination of *Clostridium botulinum* spores, but their effects against other microorganisms are limited. Because nitrites and nitrates react with amines in proteins to form carcinogenic nitrosamines, their use is still rather controversial. Sulfite prevents the growth of undesirable molds in dried fruits, juices, and wines, and retards discoloration in various foodstuffs. Many persons are hypersensitive to this preservative, thus any foods containing it must be clearly labelled. Ethylene and propylene oxide gases disinfect various dried foodstuffs. Because they are carcinogenic and can leave a residue in foods, their use is restricted to fruit, cereals, spices, nuts, and cocoa.

The high osmotic pressure contributed by hypertonic levels of salt plasmolyzes bacteria and fungi and removes moisture from food, thereby inhibiting microbial growth. Salt is commonly added to brines, pickled foods, meats, and fish. However, it does not retard the growth of pathogenic halophiles such as *Staphylococcus aureus,* which grows readily even in 7.5% salt solutions. The high sugar concentrations of candies, jellies, and canned fruits also exert an osmotic preservative effect. Other chemical additives that function in preservation are alcohols, antibiotics, and spices. Alcohol is added to flavoring extracts, and antibiotics are approved for treating the carcasses of chickens, fish, and shrimp. Spices such as pepper and cinnamon exert only slight preservative properties, and are usually not sufficiently potent to retard the growth of microorganisms.

Food may also be preserved by **desiccation,** a process that removes moisture needed by microbes for growth by exposing the food to dry, warm air. Solar drying was traditionally used for fruits and vegetables, but modern commercial dehydration is carried out in rapid-evaporation mechanical devices. Drying is not a reliable microbicidal method, however. Numerous resistant microbes such as micrococci, coliforms, staphylococci, salmonellae, and fungi survive in dried milk and eggs, which then spoil rapidly when rehydrated.

Microbial Fermentations in Food Products from Plants

Not only is it difficult to exclude microbes from foods, but in some cases, they are added deliberately. Such common substances as bread, cheese, beer, wine, yogurt, and pickles are the result of microbial **food fermentations.** In these reactions, microbial growth and biochemical activities that impart a particular taste, smell, or appearance to food are actively encouraged. The microbe or microbes may occur naturally on the food substrate as in sauerkraut, or they may be added as pure or mixed samples of known bacteria, molds, or yeasts called **starter cultures.** Many food fermentations are synergistic, with a series of microbes acting in concert to convert the starting substrate to the desired end product. Because large-scale production of fermented milk, butter, cheese, bread, alcoholic brews, and vinegar depends upon inoculation with starter cultures, considerable effort is spent selecting, maintaining, and preparing these cultures and excluding contaminants that can spoil the fermentation. Most starting raw materials are of plant origin (grains, vegetables, beans), and to a lesser extent, of animal origin (milk, meat).

Bread Making Microorganisms accomplish three functions in bread making: (1) **leavening** the flour-based dough, (2) imparting flavor and odor, and (3) conditioning the dough to make it workable. Leavening is achieved primarily through the release of gas to produce a porous and spongy product. Without leavening, bread dough remains dense, flat, and hard. Although various microbes and leavening agents may be used, the most common ones are various strains of the baker's yeast *Saccharomyces cerevisiae*. Other gas-forming microbes such as coliform bacteria, certain *Clostridium* species, heterofermentative lactic acid bacteria, and wild yeasts may be employed, depending on the type of bread desired.

Yeast metabolism requires a source of fermentable sugar, which is produced when amylases act on starch and release maltose. Because the yeast respires aerobically in bread dough, the chief products of maltose fermentation are carbon dioxide and water, rather than alcohol (as is produced in beer and wine). Yeast activity may be modified by adjusting the size of the inoculum, the sugar level, temperature, and salt concentration. Other contributions to bread texture come from kneading, which incorporates air into the dough, and from microbial enzymes, which break down flour proteins (gluten) and give the dough elasticity. Baking the bread expands the trapped gas, causes further rising, and solidifies the structure of the loaf. It also kills the yeast and inactivates the enzymes.

Besides carbon dioxide production, bread fermentation generates other volatile organic acids and alcohols that impart delicate flavors and aromas. These are especially well developed in home-baked bread, which is leavened more slowly than commercial bread. Yeasts and bacteria can also impart unique flavors, depending upon the culture mixture and baking techniques used. The tangy, sour flavor of rye bread, for example, comes in part from starter cultures of lactic acid bacteria such as *Lactobacillus plantarum, L. brevis, L. bulgaricus, Leuconostoc mesenteroides,* and *Streptococcus thermophilus,* while sourdough bread gets its unique tang from *Lactobacillus sanfrancisco* (of course!).

Beer and Other Alcoholic Beverages The production of alcoholic beverages takes advantage of another useful property of yeasts—fermenting carbohydrates in fruits or grains anaerobically to produce ethyl alcohol, as shown by this equation:

$$C_6H_{12}O_6 \rightarrow 2C_2H_5OH + 2CO_2$$
$$\text{(Sugar + Yeast = Ethanol + Carbon dioxide)}$$

Depending upon the starting materials and the processing method, alcoholic beverages vary in alcohol content and flavor. The principal types of fermented beverages are malt liquors, wines, and spirit liquors.

The earliest evidence of brewing appears in ancient tablets left by the Sumerians and Babylonians around 6000 B.C. The starting ingredients for both ancient and present-day versions of beer, ale, stout, porter, and other variations are water, malt (barley grain), hops, and special strains of yeasts. The steps in brewing include malting, mashing, adding hops, fermenting, aging, and finishing (figure 22.31).

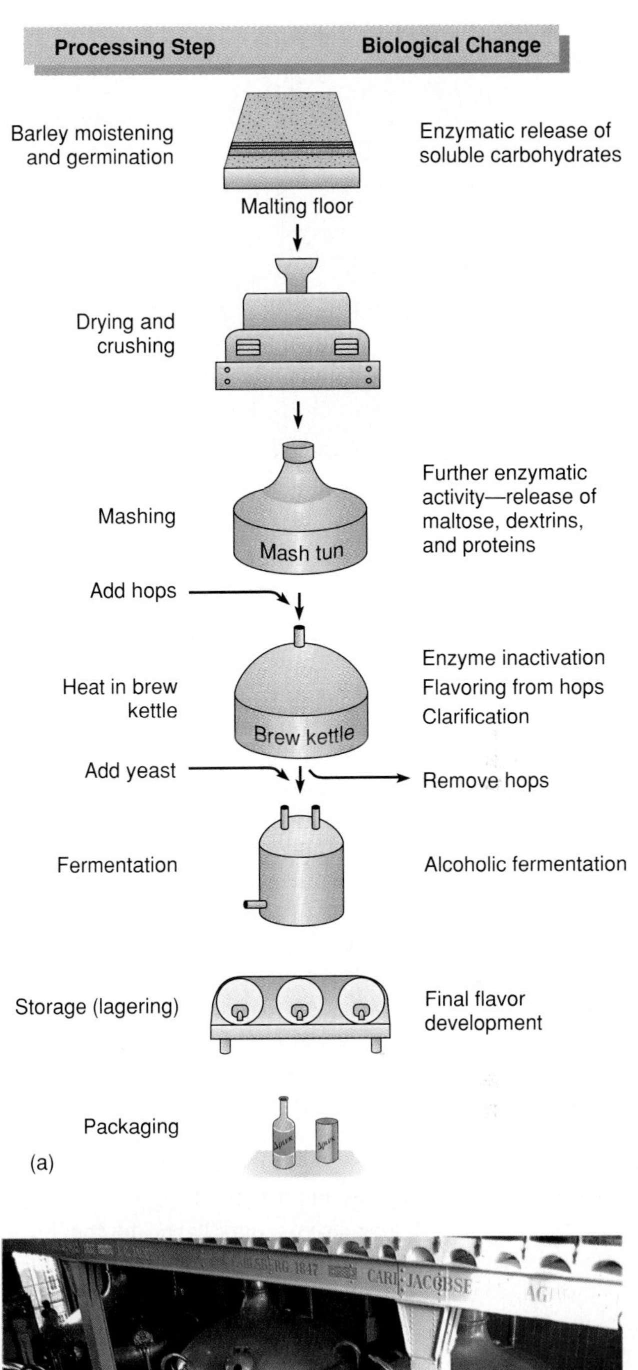

Figure 22.31 (*a*) The general stages in brewing beer, ale, and other malt liquors. (*b*) Fermentation tanks at a commercial brewery.

Figure 22.32 Female hops flowers, the herb that gives beer some of its flavor and aroma.

For brewer's yeast to convert the carbohydrates in grain into ethyl alcohol, the barley must first be sprouted and softened to make its complex nutrients available to microorganisms. This process, called **malting,** releases amylases that convert starch to maltose and dextrins, and proteases that digest proteins. Other sugar and starch supplements that may be added at this point in some forms of beer are corn, rice, wheat, soybeans, potatoes, and sorghum. After separating the sprouts, the remaining malt grain is dried and stored in preparation for mashing.

The malt grain is soaked in warm water and ground up to prepare a **mash.** Sugar and starch supplements are introduced to the mash mixture and heated to a temperature of about 65° to 70°C. During this step, the starch is hydrolyzed by amylase, and simple sugars are released. To stop the activity of the enzymes, this mixture is heated to 75°C. Solid particles are next removed by settling and filtering. **Wort,** the clear fluid that comes off, is rich in dissolved carbohydrates. Wort is boiled for about 2.5 hours with **hops,** the dried scales of the female flower of *Humulus lupulus*, a climbing herb (figure 22.32). This boiling phase serves several purposes: (1) It extracts humulus, the bitter acids and resins that give aroma and flavor to the finished product;[6] (2) it caramelizes the sugar and imparts a golden or brown color; (3) it effectively destroys any bacterial contaminants that can destroy flavor; and (4) it concentrates the mixture. The filtered and cooled supernatant is now ready for the addition of yeasts and fermentation.

Fermentation begins when wort is inoculated with a species of *Saccharomyces* that has been specially developed for beer making. Top yeasts such as *Saccharomyces cerevisiae* function at the surface and are used to produce ales. Bottom yeasts such as *Saccharomyces uvarum* (*carlsbergensis*) function deep in the fermentation vat and are used to make beer. In both cases, the initial inoculum of yeast starter is aerated briefly to promote rapid growth and increase the load of yeast cells. Shortly thereafter, an insulating blanket of foam and carbon dioxide develops on the surface of the vat and promotes anaerobic conditions. During fermentation, which lasts 8 to 14 days, the wort sugar is converted chiefly to ethanol and carbon dioxide. The diversity of flavors in the finished product is partly due to the release of small amounts of glycerol, acetic acid, and esters. Fermentation is self-limited, and it essentially ceases when a concentration of 3% to 6% ethyl alcohol is reached.

Freshly fermented or "green" beer is **lagered,** meaning it is held for several weeks to months in vats at 0°C. During this maturation period, yeast, proteins, resin, and other materials settle, leaving behind a clear, mellow fluid. Lager beer is subjected to a final pasteurization or cold filtration step to remove any residual yeasts that could spoil it. Finally, it is carbonated with carbon dioxide collected during fermentation. The end product is packaged in kegs, bottles, or cans.

Varieties of Malt Liquor Variations in processing and ingredients give rise to numerous types of malt liquors. To develop the pale color, tart taste, and higher alcohol content of *ale,* the top yeasts are continually skimmed off, and more hops are added to the wort. *Stout* and *porter* are heavy, dark ales with a higher sugar content and sweeter flavor. *Malt liquor* and *bock beer* have a higher alcohol content than regular beer, and the latter is aged longer, is darker, and contains more malt and hops. *Pilsener* is a light, lager-type beer with low levels of residual sugars. Light, very low-sugar beers are prepared with wort that has been mixed with fungal enzymes to completely ferment all of the carbohydrates. One example is a Japanese product prepared from fermented yellow rice malt called *sake* (saw'-kay) or rice wine. The starter microbes, *Aspergillus oryzae* mold and *Saccharomyces* yeast, are mixed with rice wort and fermented until the alcohol content reaches 14% to 17%. *Pulque* (pool'-kay) from Latin America is prepared from agave (century plant) that has been fermented by its resident yeast flora to achieve a 6% alcohol content.

Wine is traditionally any alcoholic beverage arising from the fermentation of grape juice, but practically any fruit can be rendered into wine. The essential starting point is the preparation of **must,** the juice given off by crushed fruit that is used as a substrate for fermentation. In general, grape wines are either *white* or *red.* The color comes from the skins of the grapes, so white wine is prepared either from white-skinned grapes or from red-skinned grapes that have had the skin removed. Red wine comes from the red- or purple-skinned varieties, and the intensity of color depends on how long the skins are allowed to remain in the must. Steps in making wine include must preparation (crushing), fermentation, storage, and aging (figure 22.33).

For proper fermentation, a must should contain 12% to 25% glucose or fructose, so the art of wine making begins in the

---

malting (mawlt'-ing) Gr. *malz,* to soften.

wort (wurt) O.E. *wyrt,* a spice.

6. Humulus is an ingredient in beer that provides some rather interesting side effects besides flavor. It is a moderate sedative, a diuretic, and a mild antiseptic.

lagered (law'-gurd) Gr. *lager,* to store or age.

vineyard. Grapes are harvested when their sugar content reaches 15% to 25%, depending on the type of wine to be made. Grapes from the field carry a mixed microflora on their surface called the *bloom* that can serve as a source of wild yeasts. Some wine makers allow these natural yeasts to dominate, but many wineries inoculate the must with a special strain of *Saccharomyces cerevisiae,* variety *ellipsoideus.* To discourage yeasts and bacteria that could spoil the wine, grapes may be treated with sulfur dioxide or potassium metabisulfite. After inoculation, the must is thoroughly aerated and mixed to promote rapid aerobic growth of yeasts, but when the desired level of yeast growth is achieved, these procedures are stopped so that alcoholic fermentation can begin.

During fermentation, the temperature of the vat must be carefully controlled to facilitate alcohol production. The length of fermentation varies from three to five days in red wines and from seven to fourteen days in white wines. The initial fermentation yields ethanol concentrations reaching 7% to 15% by volume, depending upon the type of yeast, the source of the juice, and ambient conditions. From this point, the process varies with the wine. *Dry wines* may undergo a second period of fermentation to reduce the sugar content, whereas *sweet wines* may be immediately placed in storage chambers. The fermented juice (raw wine) is decanted and transferred to large vats to settle and clarify. Before the final aging process, it may be flash-pasteurized to precipitate proteins and filtered to remove any remaining sediments. Wine is aged in wooden casks for varying time periods (months to years), after which it is bottled and stored for further aging. During aging, nonmicrobial changes produce aromas and flavors (the bouquet) characteristic of a particular wine.

Varieties of Wine Like beer, numerous types of wine are available. In sparkling wines such as champagne, CO_2 is incorporated either naturally or artificially during the bottling process. In fortified wines, ethyl alcohol is added to raise the alcohol content to about 20%. Dessert wines are sweet, while table wines are dry and low in alcoholic content. Hard cider is fermented apple wine; perry is from pears; and mead is from honey.

The fermentation processes discussed thus far can only achieve a maximum alcoholic content of 17%, because concentrations above this level inhibit the metabolism of the yeast. To obtain higher concentrations such as those found in liquors, the fermentation product must be **distilled.** During distillation, heating the liquor separates the more volatile alcohol from the less volatile aqueous phase. The alcohol is then condensed and collected. The alcohol content of distilled liquors is rated by *proof,* a measurement that is usually two times the alcohol content. Thus, 80 proof vodka contains 40% ethyl alcohol.

Distilled liquors originate through a process similar to wine making, though the starting substrates can be extremely diverse. In addition to distillation, liquors may be subjected to special treatments such as aging to provide unique flavor or color.

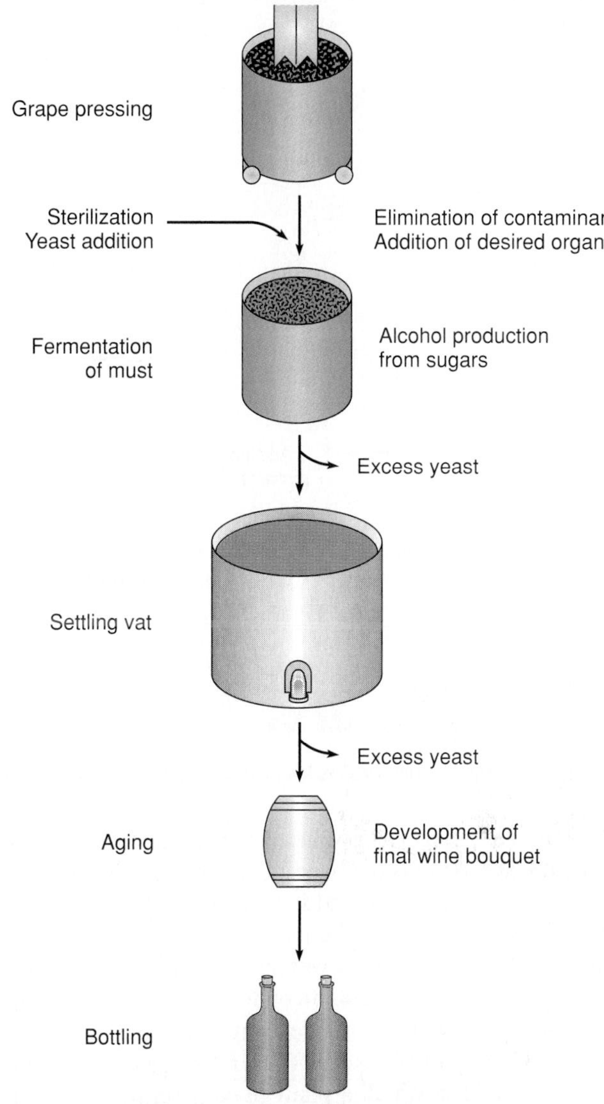

(b)

Figure 22.33 (*a*) General steps in wine making. (*b*) Wine fermentation vats in a larger commercial winery.

Vodka, a colorless liquor, is usually prepared from fermented potatoes, and rum is distilled from fermented sugarcane. Assorted whiskeys are derived from fermented grain mashes; rye whiskey is produced from rye mash, and bourbon from corn mash. Brandy is distilled grape, peach, or apricot wine; tequila is distilled pulque; and applejack is distilled hard cider.

Other Fermented Plant Products Fermentation is not just an effective way to preserve vegetables; the lactic acid produced during this process enhances flavor, and the added salt provides a crisp texture. Two practices are common to virtually all vegetable fermentations: the immersion of vegetables in an aqueous medium to promote anaerobic conditions and the application of salt (brine) to extract sugar and nutrient-laden juices. The salt also disperses bacterial clumps, and its high osmotic pressure inhibits proteolytic bacteria and sporeformers that can spoil the product.

Sauerkraut is the fermentation product of cabbage (figure 22.34). Cabbage is washed, wilted, shredded, salted, and packed tightly into a fermentation vat. Weights are placed on the cabbage mass to squeeze out its juices and to serve as a cover. The fermentation is achieved by cabbage microflora, or a starter culture may be added. The initial agent of fermentation is *Leuconostoc mesenteroides,* which grows rapidy in the brine and produces lactic acid. It is followed by *Lactobacillus plantarum,* which continues to raise the acid content to as high as 2% (pH 3.5) by the end of fermentation. The high acid content restricts the growth of spoilage microbes.

Fermented cucumber pickles come chiefly in salt (sour, sweet-sour, mixed pickles and relishes) and dill varieties. To prepare salt pickles, immature cucumbers are washed, placed in barrels of brine, and allowed to ferment for six to nine weeks. To avoid unfavorable qualities caused by natural microflora and to achieve a more consistent product, the brine can be inoculated with *Pediococcus cerevisiae* and *Lactobacillus plantarum.* Fermented dill pickles are prepared in a somewhat more elaborate fashion, with the addition of dill herb, spices, garlic, onion, and vinegar.

Oriental Specialties The preparation of Oriental fermented vegetables is an ancient art. A combination of starter molds (*Aspergillus soyae, Rhizopus* spp.), yeasts (*Saccharomyces rouxii, Hansenula* spp.), and bacteria (*Lactobacillus delbreuckii, Bacillus subtilis*) furnish the necessary hydrolytic enzymes to digest starch, protein, and lipid in substrates like soybeans, wheat, or rice. Soy sauce, tamari sauce, and ang-khak (Chinese red rice) are used to flavor and color various dishes. Miso, tempeh, and tofu are typical fermented soybean products; minchin is derived from wheat gluten, and idli, an Indian dish, is a combination of rice and black mungo beans. Poi, a Hawaiian staple, is prepared from the root of the taro plant, which is steamed, pounded into a paste, and infused with a starter of bacteria and yeasts. It may be eaten fresh or aged for a few days, which increases its acidity and flavor. Fish, shrimp, and eggs are other novel fermented products from the Orient.

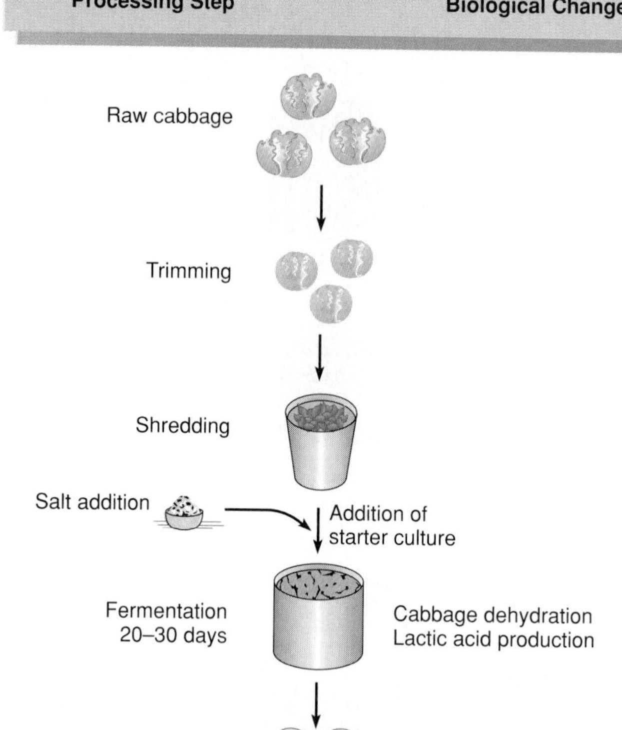

Figure 22.34 Steps in the preparation of sauerkraut.

Natural vinegar is produced when plant juice containing an alcohol substrate is oxidized to acetic acid, which is responsible for the pungent odor and sour taste. Although a reasonable facsimile of vinegar could be made by mixing about 4% acetic acid and a dash of sugar in water, this preparation would lack the traces of various esters, alcohol, glycerin, and volatile oils that give natural vinegar its pleasant character. Vinegar is actually produced in two stages. The first stage is similar to wine or beer making, in which a plant juice is fermented to alcohol by *Saccharomyces.* The second stage involves an aerobic fermentation carried out by acetic acid bacteria in the genera *Acetobacter* and *Gluconobacter.* These bacteria oxidize the ethanol in a two-step process, as shown here:

$$\underset{\text{Ethanol}}{2C_2H_5OH} + \frac{1}{2}O_2 \rightarrow \underset{\text{Acetaldehyde}}{CH_3CHO} + H_2O$$

$$\underset{\text{Acetaldehyde}}{CH_3CHO} + \frac{1}{2}O_2 \rightarrow \underset{\text{Acetic acid}}{CH_3COOH}$$

To furnish the abundance of oxygen necessary in commercial vinegar making, inoculated raw material is exposed to air in thin layers in open trays, allowed to trickle over loosely packed beechwood twigs and shavings, or aerated in a large vat.

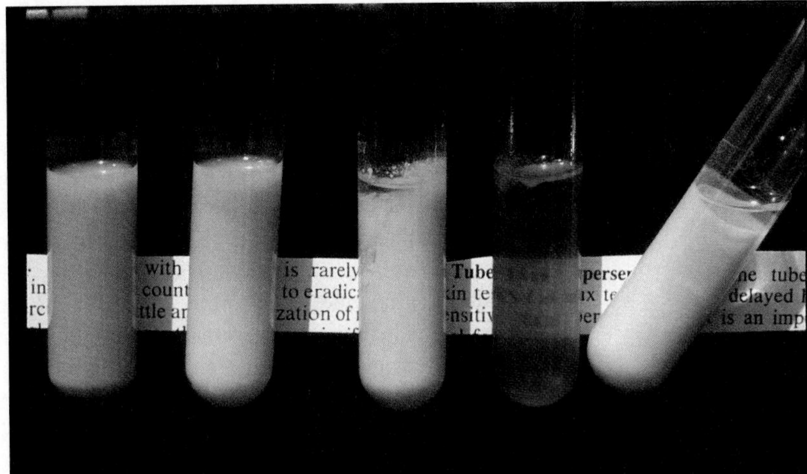

(a)

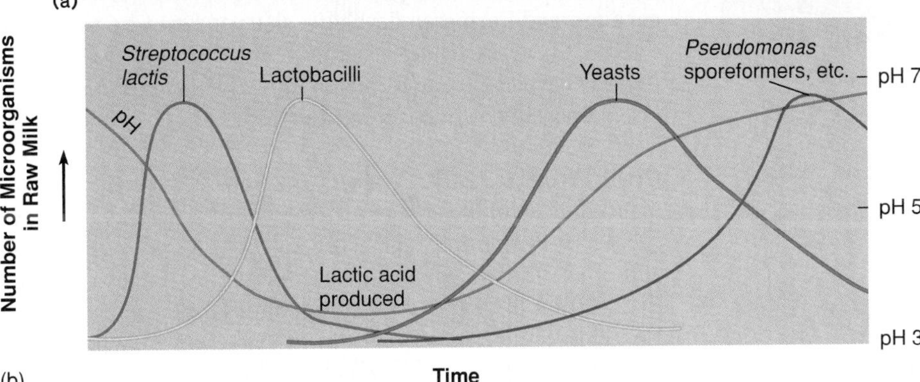

(b)

Figure 22.35 (*a*) Litmus milk is a medium used to indicate pH and consistency changes in milk resulting from microbial action. The first tube is an uninoculated, unchanged control. The second tube has a white, decolorized zone indicative of litmus reduction. The third tube has become acidified (pink), and its proteins have formed a loose curd. In the fourth tube, digestion of milk proteins has caused complete clarification or peptonization of the milk. The fifth tube shows a well-developed solid curd overlaid by a clear fluid, the whey. (*b*) Chart depicting spontaneous changes in the number and type of microorganisms and the pH of raw milk as it incubates.

Different types of vinegar are derived from substrates such as apple cider (cider vinegar), malted grains (malt vinegar), and grape juice (wine vinegar).

Microbes in Milk and Dairy Products

Milk has a highly nutritious composition. It contains an abundance of water and is rich in minerals, protein (chiefly casein), butterfat, sugar (especially lactose), and vitamins. It starts its journey in the udder of a mammal as a sterile substance, but as it passes out of the teat, it is inoculated by the animal's normal flora. Other microbes may be added during milking from the milker, the milking machinery, and collection vessels. Because milk is a nearly perfect culture medium, it is highly susceptible to microbial growth. When raw milk is left at room temperature, a series of bacteria ferment the lactose, produce acid, and alter the milk's content and texture (figure 22.35*a*). This progression can occur naturally, or it can be simulated, as in the production of cheese and yogurt.

In the initial stages of milk fermentation, lactose is rapidly attacked by *Streptococcus lactis* and *Lactobacillus* spp. (figure 22.35*b*). The resultant lactic acid accumulation and lowered pH cause the milk proteins to coagulate into a solid mass called the **curd.** Curdling also causes the separation of a watery liquid called **whey** on the surface. Curd may be produced by microbial action or by an enzyme, **rennin** (casein coagulase), which is isolated from the stomach of unweaned calves.

Cheese Production Since 5000 B.C., various forms of cheese have been produced by spontaneous fermentation of cow, goat, or sheep milk. Present-day, large-scale cheese production is carefully controlled and uses only freeze-dried samples of pure cultures (figure 22.36). These are first inoculated into a small quantity of pasteurized milk to form an active starter culture. This amplified culture is subsequently inoculated into a large vat of milk, where rapid curd development takes place. Such rapid growth is desired because it promotes the overgrowth of the desired inoculum and prevents the activities of undesirable contaminants. Rennin is usually added to increase the rate of curd formation.

After its separation from whey, the curd is rendered to produce one of the 20 major types of soft, semisoft, or hard cheeses. The composition of cheese is varied by adjusting water, fat, acid, and salt content. Cottage and cream cheese are examples of the soft, more perishable variety. After light salting and the optional addition of cream, they are ready for consumption without further processing. Other cheeses acquire their character from "ripening," a complex curing process involving bacterial, mold, and enzyme reactions that develop the final flavor, aroma, and other features characteristic of particular cheeses.

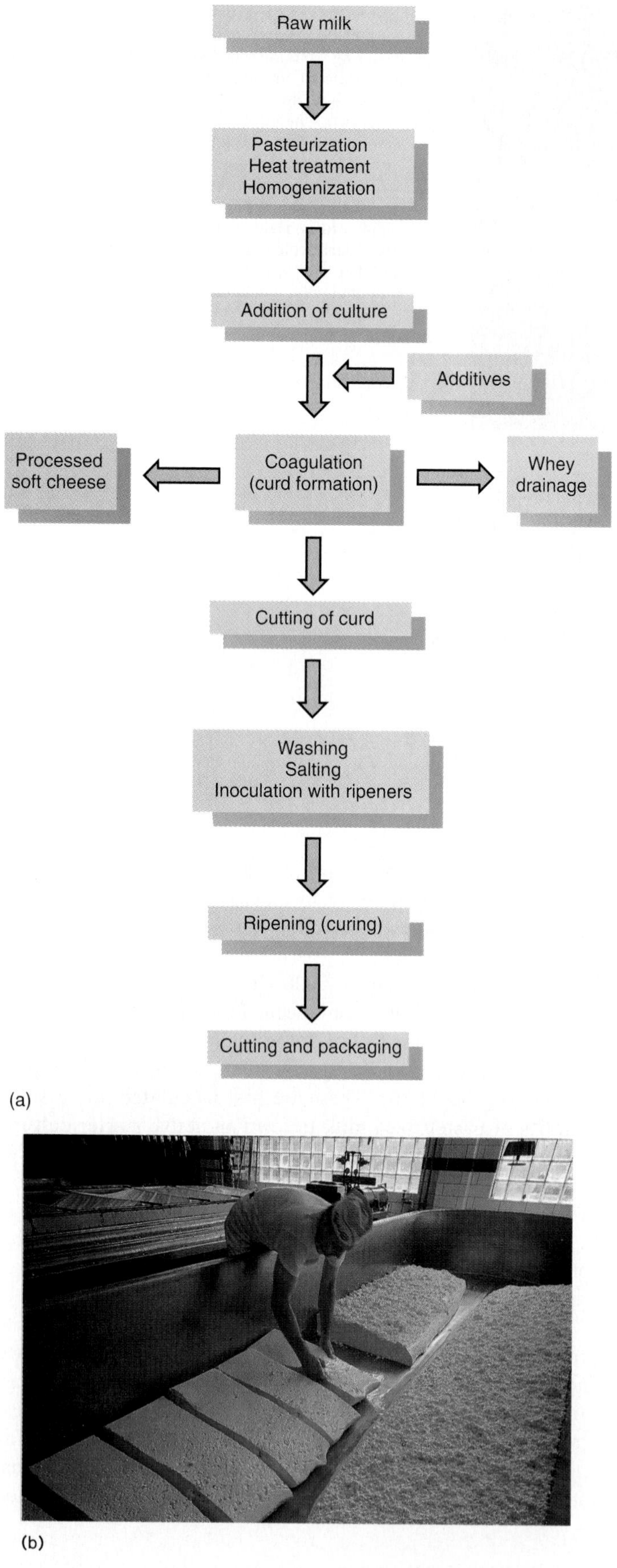

The distinguishing traits of soft cheeses like Limburger, Camembert, and Liederkranz are acquired by ripening with a reddish-brown mucoid coating of yeasts, micrococci, and molds. The microbial enzymes permeate the curd and ferment lipids, proteins, carbohydrates, and other substrates. This process leaves assorted acids, salts, aldehydes, ketones, and other by-products that give the finished cheese powerful aromas and delicate flavors. Semisoft varieties of cheese such as Roquefort, blue, or Gorgonzola are infused and aged with a strain of *Penicillium roqueforti* mold. These cheeses differ in the type of milk used. Roquefort is made with sheep milk, blue cheese with cow milk, and Gorgonzola with goat milk. Hard cheeses like Swiss, cheddar, and Parmesan develop a sharper flavor by aging with selected bacteria. The pockets in Swiss cheese come from entrapped carbon dioxide formed by *Propionibacterium,* which is also responsible for its bittersweet taste.

Other Fermented Milk Products Yogurt is formed by the fermentation of milk by *Lactobacillus bulgaricus* and *Streptococcus thermophilus.* These organisms produce organic acids and other flavor components, and may grow in such numbers that a gram of yogurt regularly contains 100 million bacteria. Live cultures of *Lactobacillus acidophilus* are an important additive to acidophilus milk, which is said to benefit digestion and help maintain the normal flora of the intestine. Fermented milks such as kefir, koumiss, and buttermilk are a basic food source in many cultures.

Microorganisms As Food

At first, the thought of bacteria, molds, algae, and yeasts as food may seem odd or even unappetizing. We do eat their macroscopic relatives such as mushrooms, truffles, and seaweed, but we are used to thinking of the microscopic forms as agents of decay and disease, or at most, as food flavorings. The consumption of microorganisms is not a new concept. In Germany during World War II, it became necessary to supplement the diets of undernourished citizens by adding yeasts and molds to foods. At present, most countries are able to produce enough food for their inhabitants, but in the future, countries with exploding human populations and dwindling arable land may need to consider microbes as a significant source of protein, fat, and vitamins. Several countries already commercially mass produce food yeasts, bacteria, and in a few cases, algae. Although eating microbes has yet to win total public acceptance, their use as feed supplements for livestock is increasing. A technology that shows some promise in increasing world food productivity is **single-cell protein (SCP).** This material is produced from waste products such as molasses from sugar refining, refuse from paper manufacturing, by-products from petroleum processing, and wastes from agriculture and food processing. In England, an animal feed called **pruteen** is produced at a rate of 3,000 tons a month by mass culture of the bacterium *Methylophilus meth-*

Figure 22.36 (*a*) A flow diagram that summarizes the main events in cheese making. (*b*) The curd cutting stage in the making of cheddar cheese.

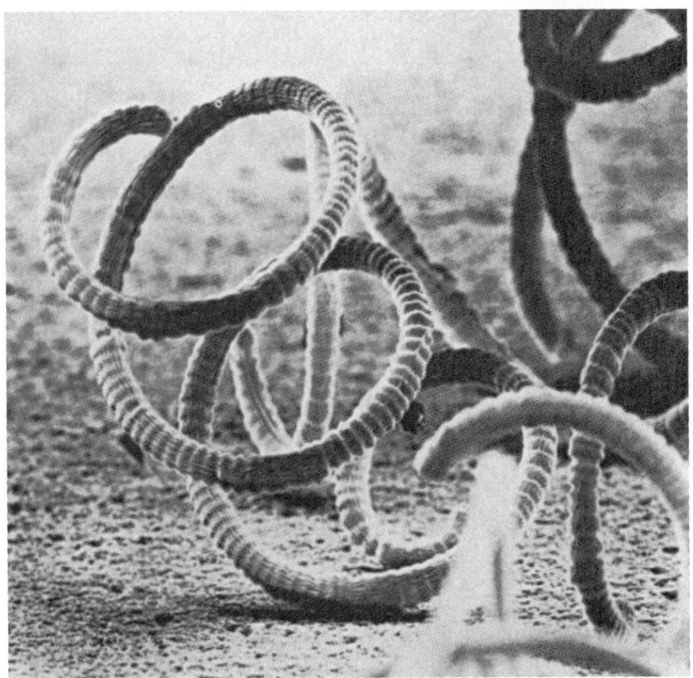

Figure 22.37 Scanning electron micrograph of a single curly filament of *Spirulina plantensis*, the edible cyanobacterium.

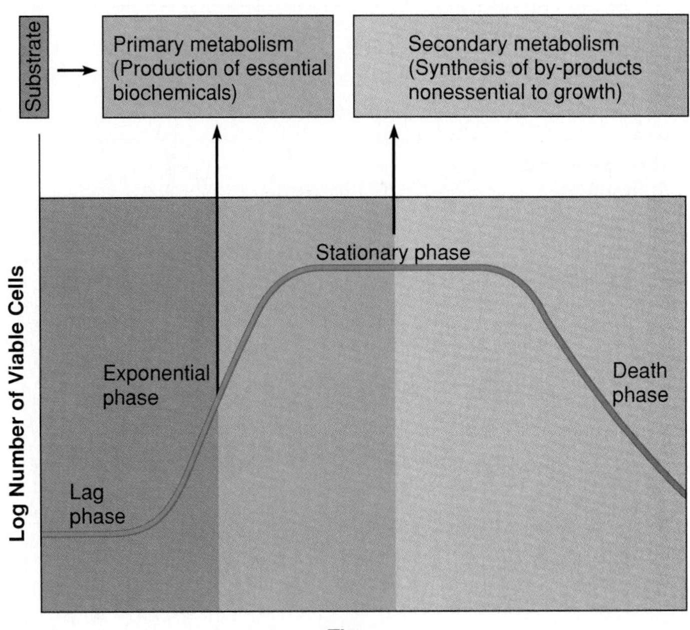

Figure 22.38 The origins of primary and secondary microbial metabolites harvested by industrial processes.

ylotrophus. **Mycoprotein,** a product made from the fungus *Fusarium graminearum,* is also sold there. The filamentous texture of this product makes it a likely candidate for producing meat substitutes for human consumption.

Health food stores carry bottles of dark green pellets or powder that are a pure culture of a spiral-shaped cyanobacterium called *Spirulina* (figure 22.37). This microbe is harvested from the surface of lakes and ponds, where it grows in great mats. In some parts of Africa and Mexico, *Spirulina* has become a viable alternative to green plants as a primary nutrient source, being eaten in its natural form or added to other foods and beverages. The microbe has high productivity and requires very little area or resources to grow. *Spirulina* is nontoxic, high in protein, and has a mild, vegetable-like flavor. Perhaps as the world's population increases, it will replace many traditional food sources.

General Concepts in Industrial Microbiology

Virtually any large-scale commercial enterprise that enlists microorganisms to manufacture consumable materials is part of the realm of **industrial microbiology.** Here the term pertains primarily to bulk production of organic compounds such as antibiotics, hormones, vitamins, acids, solvents (table 22.4), and enzymes (table 22.5). Many of the processing steps involve fermentations similar to those described in food technology, but industrial processes usually occur on a much larger scale, produce a specific compound, and involve numerous complex stages. The aim of industrial microbiology is to produce chemicals that can be purified and packaged for sale or for use in other commercial processes. Thousands of tons of organic chemicals worth several billion dollars are produced by this industry every year. Creating just one of these products requires not only the correct microbe, the correct starting compound, and optimum growth conditions, but also the investment of 10 to 15 years and billions of dollars for research and development.

The microbes used by fermentation industries have traditionally been naturally occurring or wild strains of bacteria or molds that carry out a particular metabolic action on a substrate. At an increasing rate, however, these microbes are genetically altered strains of fungi or bacteria that selectively synthesize large amounts of various metabolic intermediates or **metabolites.** Two basic kinds of metabolic products are harvested by industrial processes: (1) *Primary metabolites* are produced during the major metabolic pathways and are essential to the microbe's function. (2) *Secondary metabolites* are by-products of metabolism that may not be critical to the microbe's function (figure 22.38). In general, primary products are compounds such as amino acids and organic acids synthesized during the logarithmic phase of microbial growth, and secondary products are compounds such as vitamins, antibiotics, and steroids synthesized during the stationary phase (chapter 6). Most strains of industrial microorganisms have been chosen for their high production of a particular primary or secondary metabolite. Certain strains of yeasts and bacteria can produce 20,000 times more metabolite than a wild strain of that same microbe.

Industrial microbiologists have several tricks to increase the amount of the chosen end product. First, they can manipulate the growth environment to increase the synthesis of a metabolite. For instance, adding lactose instead of glucose as the

Table 22.4 Industrial Products of Microorganisms

Chemical	Microbial Source	Substrate	Applications
Pharmaceuticals			
Bacitracin	*Bacillus subtilis*	Glucose	Antibiotic effective against gram-positive bacteria
Cephalosporins	*Cephalosporium*	Glucose	Antibacterial antibiotic, broad spectrum
Penicillins	*Penicillium chrysogenum*	Lactose	Antibacterial antibiotics, broad and narrow spectrum
Erythromycin	*Streptomyces*	Glucose	Antibacterial antibiotic, broad spectrum
Tetracycline	*Streptomyces*	Glucose	Antibacterial antibiotic, broad spectrum
Amphotericin B	*Streptomyces*	Glucose	Antifungal antibiotic
Vitamin B_{12}	*Pseudomonas*	Molasses	Dietary supplement
Riboflavin	*Asbya*	Glucose, corn oil	Animal feed supplement
Steroids (hydrocortisone)	*Rhizopus, Cunninghamella*	Deoxycholic acid	Treatment of inflammation, allergy; hormone replacement therapy
Food Additives and Amino Acids			
Citric acid	*Aspergillus, Candida*	Molasses	Acidifier in soft drinks; used to set jam; candy additive; fish preservative; retards discoloration of crabmeat; delays browning of sliced peaches
Lactic acid	*Lactobacillus, Bacillus*	Whey, corncobs, cottonseed; from maltose, glucose, sucrose	Acidifier of jams, jellies, candies, soft drinks, pickling brine, baking powders
Xanthan	*Xanthomonas*	Glucose medium	Food stabilizer; not digested by humans
Acetic acid	*Acetobacter*	Any ethylene source, ethanol	Food acidifier; used in industrial processes
Glutamic acid	*Corynebacterium, Arthrobacter, Brevibacterium*	Molasses, starch source	Flavor enhancer monosodium glutamate (MSG)
Lysine	*Corynebacterium*	Casein	Dietary supplement (absent in cereals)
Miscellaneous Chemicals			
Ethanol	*Saccharomyces*	Beet, cane, grains, wood, wastes	Additive to gasoline (gasohol)
Acetone	*Clostridium*	Molasses, starch	Solvent for lacquers, resins, rubber, fat, oil
Butanol	*Clostridium*	Molasses, starch	Added to lacquer, rayon, detergent, brake fluid
Gluconic acid	*Aspergillus, Gluconobacter*	Corn steep, any glucose source	Baking powder, glass-bottle washing agent, rust remover, cement mix, pharmaceuticals
Glycerol	Yeast	By-product of alcohol fermentation	Explosive (nitroglycerine)
Dextran	*Klebsiella, Acetobacter, Leuconostoc*	Glucose, molasses, sucrose	Polymer of glucose used as adsorbants, blood expanders, and in burn treatment; a plasma extender; used to stabilize ice cream, sugary syrup, candies
Thuricide	*Bacillus thuringiensis*	Molasses, starch	Used in biocontrol of caterpillars, moths, loopers, and hornworm plant pests

fermentation substrate increases the production of penicillin by *Penicillium.* Another strategy is to select microbial strains that genetically lack a feedback system to regulate the formation of end products, thus encouraging mass accumulation of this product. Many syntheses occur in sequential fashion, wherein the waste products of one organism become the building blocks of the next. During these *biotransformations,* the substrate undergoes a series of slight modifications, each of which gives off a different by-product (figure 22.39). The production of an antibiotic such as tetracycline requires several microorganisms and 72 separate metabolic steps.

From Microbial Factories to Industrial Factories

Industrial fermentations begin with the basic living factory—the single microbial cell. This cell is nurtured to produce trillions of copies of itself, and then coaxed to churn out large volumes of a desired product. This mass fermentation by microbes is the driving force of industrial microbiology. To produce appropriate levels of growth and fermentation, the microbes must be cultivated in a carefully controlled environment. This process is basically similar to culturing bacteria in a test tube of nutrient broth. It requires a sterile medium containing appro-

Table 22.5 Industrial Enzymes and Their Uses

Enzyme	Source	Application
Amylase	*Aspergillus, Bacillus, Rhizopus*	Flour supplement, desizing textiles, mash preparation, syrup manufacture, digestive aid, precooked foods, spot remover in dry cleaning
Catalase	*Micrococcus, Aspergillus*	To prevent oxidation of foods; used in cheese production, cake baking, irradiated foods
Cellulase	*Aspergillus, Trichoderma*	Liquid-coffee concentrates, digestive aid, increase digestibility of animal feed, degradation of wood or wood by-products
Cytochrome *c*	*Candida*	Diagnostic use
Glucose oxidase	*Aspergillus*	Removal of glucose or oxygen that can decolorize or alter flavor in food preparations as in dried egg products; glucose determination in clinical diagnosis
Hyaluronidase	Various bacteria	Medical use in wound cleansing, preventing surgical adhesions
Keratinase	*Streptomyces*	Hair removal from hides in leather preparation
Lipase	*Rhizopus*	Digestive aid, to develop flavors in cheese and milk products
Pectinase	*Aspergillus, Sclerotina*	Clarifies wine, vinegar, syrups and fruit juices by degrading pectin, a gelatinous substance; used in concentrating coffee
Penicillinase	*Bacillus*	Removal of penicillin in research
Proteases	*Aspergillus, Bacillus, Streptomyces*	To clear and flavor sake, process animal feed, remove gelatin from photographic film, recover silver, tenderize meat, unravel silkworm cocoon, remove spots
Rennet	*Mucor*	To curdle milk in cheese making
Streptokinase	*Streptococcus*	Medical use in clot digestion, as a blood thinner
Streptodornase	*Streptococcus*	Promotes healing by removing debris from wounds and burns

priate nutrients, protection from contamination, provisions for introduction of sterile air or total exclusion of air, and a suitable temperature and pH.

Many commercial fermentation processes have been worked out on a small scale in a lab and then *scaled up* to a large commercial venture. An essential component for scaling up is a device called a **fermentor,** in which mass cultures are grown, reactions take place, and product develops. Some fermentors are large tubes, flasks, or vats, but most industrial types are metal cylinders with built-in mechanisms for stirring, cooling, monitoring, and harvesting product (figure 22.40). Fermentors are made of materials that can withstand pressure, and are rust-proof, nontoxic, and leak-proof. Fermentors range in holding capacity from small, 5-gallon systems used in research labs to larger, 5,000- to 100,000-gallon vessels (see chapter opening illustration), and in some industries, to tanks of 250 million to 500 million gallons.

For optimum yield, a fermentor must duplicate the actions occurring in a tiny volume (a test tube) on a massive scale. Most microbes performing fermentations have an aerobic metabolism, and the large volumes make it difficult to provide adequate oxygen. Fermentors have a built-in device called a *sparger* that aerates the medium to promote aerobic growth. To increase the contact between the microbe and the nutrients, paddles (*impellers*) located in the central part of the fermentor vigorously stir the fermentation mixture and maintain its uniformity. The temperature of the chamber is maintained by a cooling jacket.

Substance Production

The general steps in mass production of organic substances in a fermentor are illustrated in figure 22.41. These may be summarized as: (1) introduction of microbes and sterile media into the reaction chamber, (2) fermentation, (3) *downstream processing* (recovery, purification, and packaging of product), and (4) removal of waste. All phases of production must be carried out aseptically and monitored (usually by computer) for rate of flow and quality of product. The starting raw substrates include crude plant residues, molasses, sugars, fish and meat meals, and whey. Additional chemicals may be added to control pH or increase the yield. In *batch fermentation,* the substrate is added to the system all at once and taken through a limited run until product is harvested. In *continuous feed* systems, nutrients are continuously fed into the reactor and the product is siphoned off throughout the run.

Ports in the fermentor allow the raw product and waste materials to be recovered from the reactor chamber when fermentation is complete. The raw product is recovered by settling, precipitation, centrifugation, filtration, or cell lysis. Some products come from this process ready to package, whereas others require further purification, extraction, concentration, or drying. The end product is usually in a powder, cake, granular, or liquid form that is placed in sterilized containers. The waste products may be siphoned off to be used in other processes or discarded, and the residual microbes and nutrients from the fermentation chamber may be recycled back into the system or removed for the next run.

Fermentation technology for large-scale cultivation of microbes and production of microbial products is versatile. Table 22.4 itemizes some of the major pharmaceutical substances, food additives, and solvents produced by microorganisms.

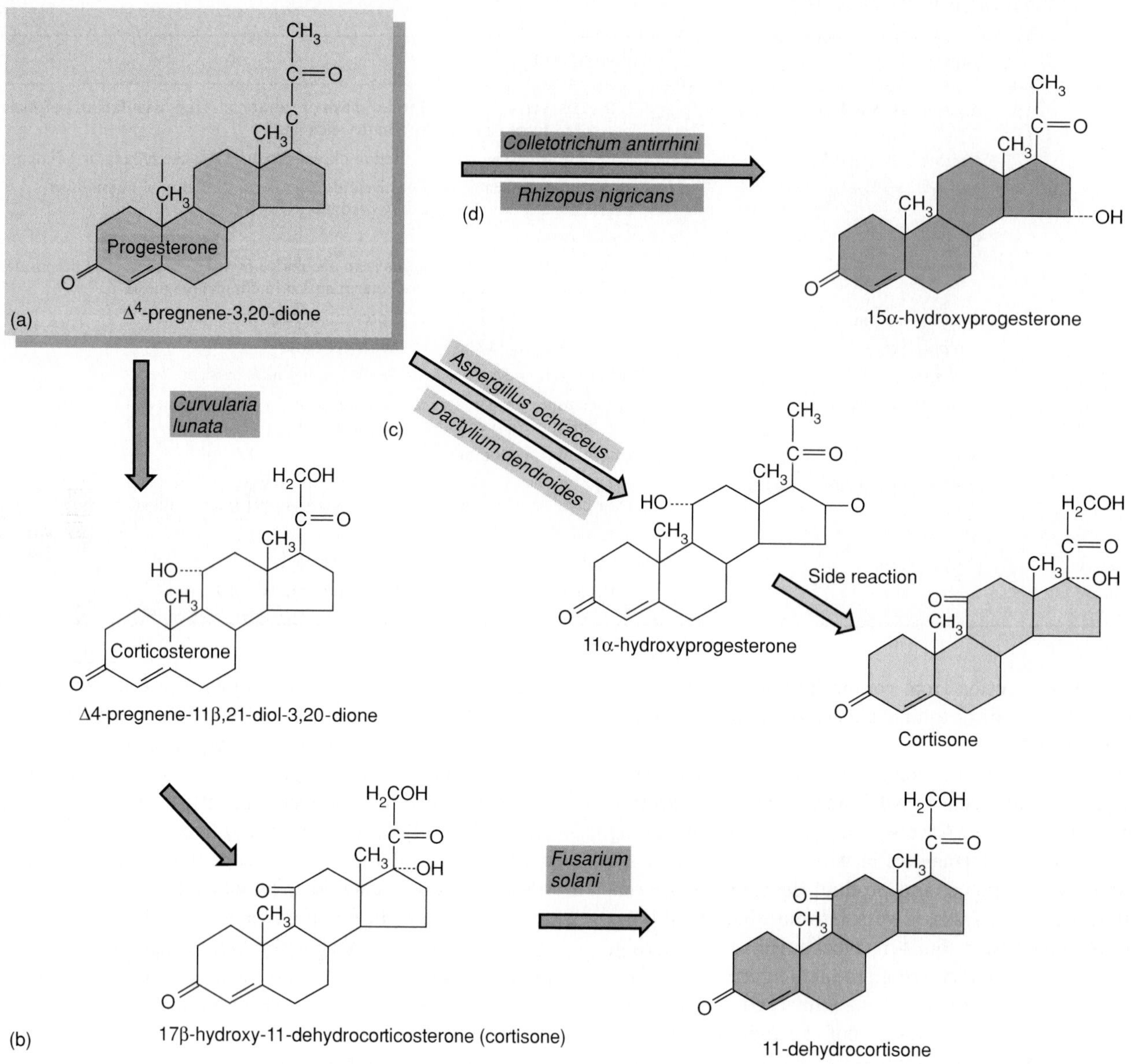

Figure 22.39 An example of biotransformation by microorganisms in the industrial production of steroid hormones. Desired hormones such as cortisone and progesterone may require several steps and microbes, and rarely can a single microbe carry out all of the required synthetic steps. (*a*) Beginning with starting compound Δ 4-pregnene-3,20-dione, (*b*) two species of mold can produce 11-dehydrocortisone in three steps. (*c*) Two other species of mold can convert this starting compound to cortisone in two steps, and (*d*) two species of mold can produce 15α-hydroxyprogesterone in one step.

Pharmaceutical Products Health care products derived from microbial biosynthesis are antibiotics, hormones, vitamins, and vaccines. The first mass-produced antimicrobic was penicillin, which came from *Penicillium chrysogenum,* a mold first isolated from a cantaloupe in Wisconsin. The current strain of this species has gone through 40 years of selective mutation and screening to increase its yield. (The original wild *P. chrysogenum* synthesized 60 mg/ml of medium, and the latest isolate yields 85,000 mg/ml.) To produce the semisynthetic penicillin derivatives (see feature 10.2), the assorted side-chain precursors are introduced to the fermentation vessel during the most appropriate phase of growth. These experiences with penicillin have provided an important model for the manufacture of other antibiotics (see table 22.4).

Several steroid hormones that are clinically therapeutic are produced industrially (figure 22.39). Corticosteroids of the adrenal cortex, cortisone and cortisol (hydrocortisone), are invaluable for treating inflammatory and allergic disorders, and

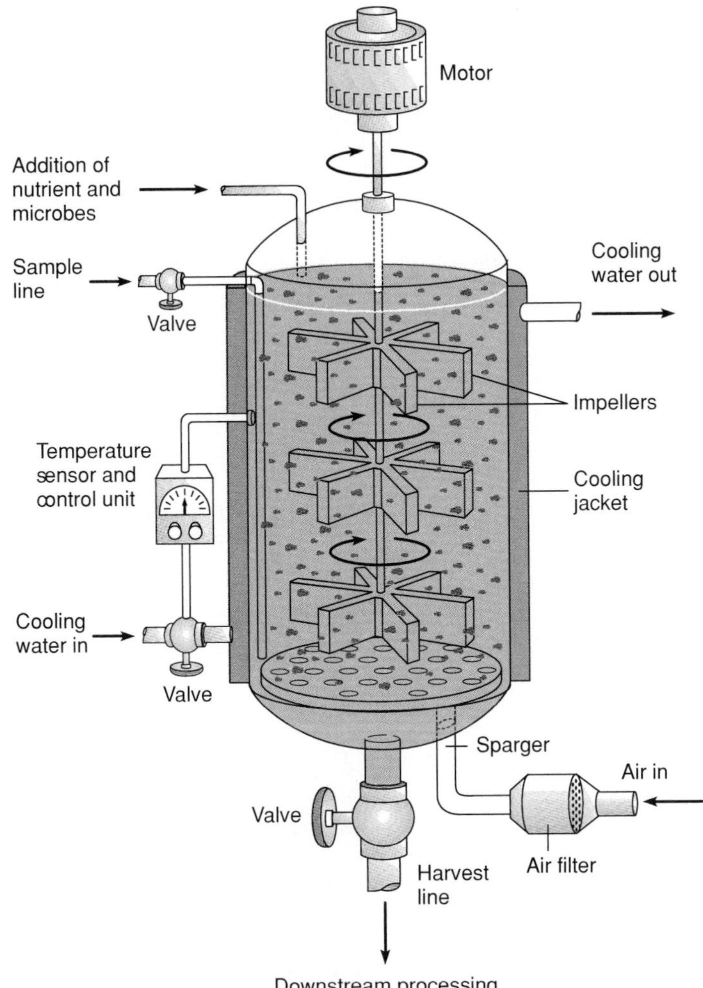

Figure 22.40 A schematic diagram of an industrial fermentor for mass culture of microorganisms. Such instruments are equipped to add nutrients and cultures, to remove product under sterile or aseptic conditions, and to aerate, stir, and cool the mixture automatically.

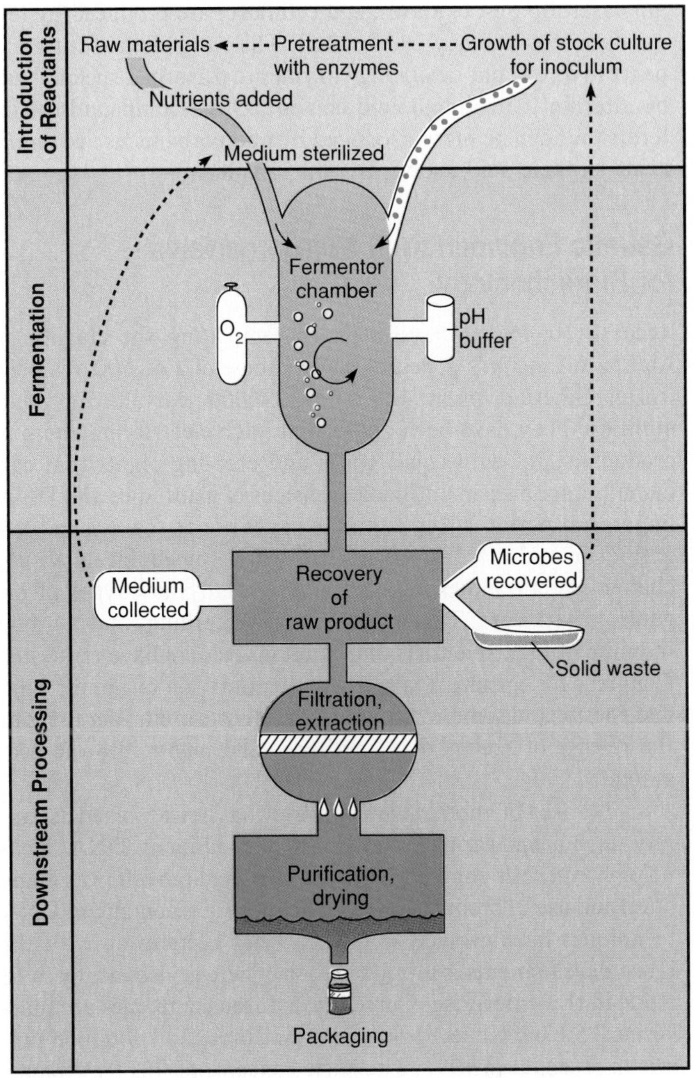

Figure 22.41 The general layout of a fermentation plant for industrial production of drugs, enzymes, fuels, vitamins, and amino acids.

female hormones such as progesterone or estrogens are the active ingredients in birth control pills. For years, the production of these hormones was tedious and expensive because it involved purifying them from slaughterhouse animal glands or chemical syntheses. In time, it was shown that, through biotransformation, various molds could convert a precursor compound called diosgenin taken from the root of an inedible Mexican plant into cortisone. By the same means, stigmasterol from soybean oil could be transformed into progesterone.

Some vaccines are also adaptable to mass production fermentation methods. Vaccines for *Bordetella pertussis, Salmonella typhi, Vibrio cholerae,* and *Mycobacterium tuberculosis* are produced in large batch cultures. *Corynebacterium diphtheriae* and *Clostridium tetani* are propagated for the synthesis of their toxins, from which respective toxoids are prepared.

Miscellaneous Products An exciting innovation has been the development and industrial production of natural *biopesticides* using *Bacillus thuringiensis.* During sporulation, these bacteria produce intracellular crystals that can be toxic to certain insects. When the insect ingests this endotoxin, its digestive tract breaks down and it dies, but the material is relatively nontoxic to other organisms. Commercial dusts are now on the market to suppress caterpillars, moths, and worms on various agricultural crops and trees. A strain of this bacterium is also being considered to control the mosquito vector of malaria and the blackfly vector of onchocercosis.

Enzymes are critical to chemical manufacturing, the agriculture and food industries, textile and paper processing, and even laundry and dry cleaning. The advantage of enzymes is that they are very specific in their activity, and are readily produced and released by microbes. Mass quantities of proteases, amylases, lipases, oxidases, and cellulases are produced by fermentation technology (table 22.5). The wave of the future appears to be custom-designing enzymes to perform a specific task by altering their amino acid content. Other compounds of interest that can be mass-produced by microorganisms are amino acids, organic acids, solvents, and polymers.

Genetic Engineering of Microorganisms for Biotechnology

Recently, biologists have been contemplating the cloning of Abraham Lincoln's genes, as well as those of a 40,000-year-old frozen Siberian mammoth and a 4,000-year-old Egyptian mummy. They have been abuzz with such electrifying ideas as producing pigs with cow's genes and creating plants that can manufacture human antibodies. Visions of giant mice and DNA fingerprint patterns leap from the pages of newspapers and scientific journals. In this futuristic world, bioengineers would change genetic blueprints and create totally new types of organisms with ease. This prospect is so disturbing, that a group of eminent U.S. scientists and religious leaders have requested Congress to legislate against genetic alteration of human eggs and sperm, and some environmental activists are trying to block the release of recombinant microbes and plants into the ecosystem.

No area of applied biotechnology has mushroomed as rapidly as **bioengineering** (also called recombinant DNA technology or genetic engineering), and no other area offers as many novel and useful products. The applications of recombinant DNA techniques have evolved so quickly, that keeping up with the latest developments is a full-time job. There is also money to be made in this enterprise. Currently, a dozen companies are marketing 15 bioengineered products that brought 1.5 billion dollars in revenues in 1991. The basic concepts of this technology, first introduced in the last section of chapter 8, involve the directed, deliberate manipulation (addition or removal) of DNA and other genetic materials to change the genetic character of organisms. Refer to figures 8.30 and 8.31 to refresh your recollection of recombinant DNA technology and cloning techniques.

Most techniques start with the selection of a desired gene, followed by its isolation and excision by restriction endonucleases. This gene is inserted into a cloning vector (plasmid) and then introduced into a host organism, where it is translated into a specific protein. Our discussion of the applications of genetic engineering will be divided into the following sections: genetic engineering of microorganisms; genetic engineering to produce transgenic (genetically recombined) plants, animals, and humans; bioengineered products; and nucleic acid technology.

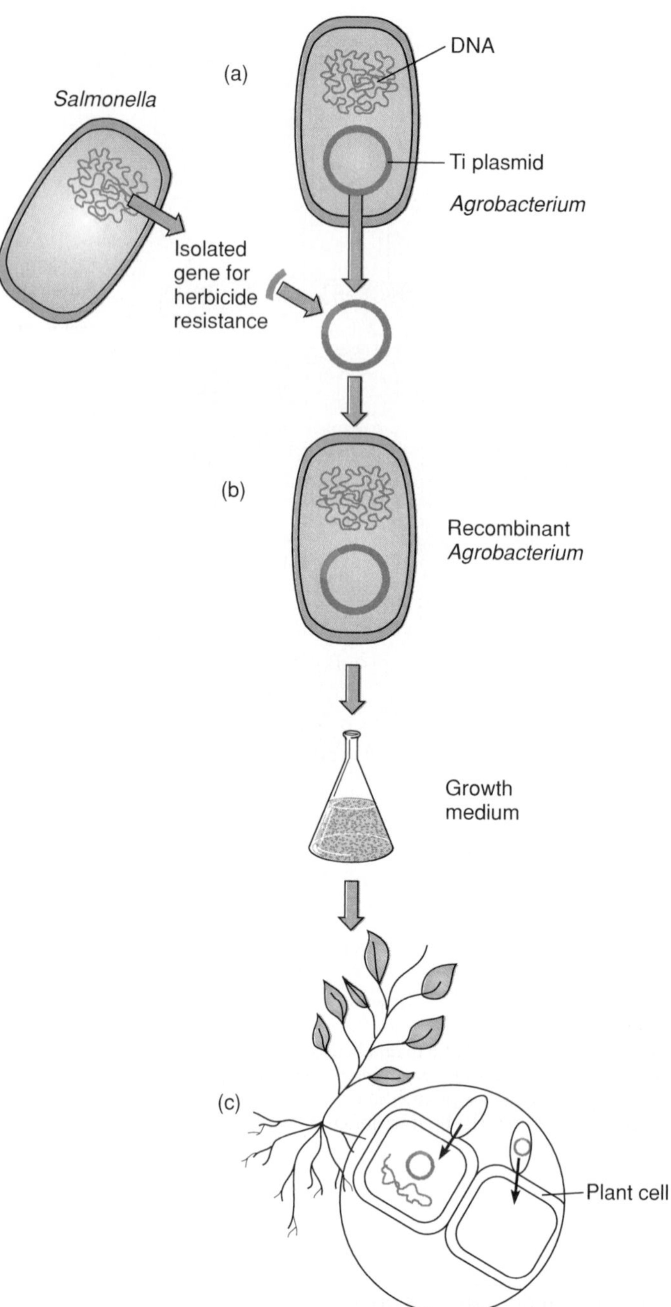

Figure 22.42 Bioengineering of plants using a natural tumor-producing bacterium called *Agrobacterium*. (*a*) The large plasmid (Ti) of this bacterium can be used as a cloning vector for foreign genes that code for herbicide or disease resistance. (*b*) The recombinant plasmids are taken up by the *Agrobacterium* cells, which multiply and copy the foreign gene. (*c*) A plant infected by this recombinant bacterium incorporates the foreign gene and develops the protective traits it provides.

Recombinant Microbes

One of the very first applications of bioengineering was the development of a strain of *Pseudomonas syringae*. The wild strain of this bacterium ordinarily contains an ice nucleation gene that promotes ice or frost formation on cold moist surfaces. Alteration of this gene created a recombinant strain that can actually prevent ice crystals from forming. Manufactured as Frostban, this microbe proved a modest success in preventing frost formation on strawberry and potato plants in the field. In another test, a strain of virus that could kill cabbage looper insects was released into a cabbage patch. It was engineered to

Figure 22.43 Herbicide protection in transgenic plants. Tobacco plants in the upper row have been transformed with a gene that provides protection against buctril, a systemic herbicide. Plants in the lower row are normal and not transformed. Both groups were sprayed with buctril and allowed to sit for six days. (The control plants at the beginning of each row were sprayed with a blank mixture lacking buctril.) Note the relative survival in the transgenic and normal plants.

spontaneously self-destruct after a certain period. *Pseudomonas fluorescens* has been recombined with the genes for producing thuringiensis bioinsecticide. These new bacteria are released to colonize plant roots and help destroy invading insects. All releases of recombinant microbes must be approved by the EPA and must be closely monitored. So far, extensive studies have indicated that such microbes do not survive or proliferate in the environment and probably pose little harm. Genetically engineered viruses have been especially useful for medical applications such as vaccine production (see figure 13.22).

Recombinant Plants

Most bioengineering in plants involves a fascinating bacterium called *Agrobacterium tumefaciens.* This organism is a natural pathogen of plants that invades the roots and causes a tumor called *crown gall disease. A. tumefaciens* contains a large plasmid (Ti) that inserts into the genomes of infected root cells and transforms them. Even after the bacteria in a tumor are dead, the tumor continues to grow. This plasmid is a perfect vector for inserting foreign genes into plant genomes. In most cases, the Ti plasmid is removed, spliced into a selected gene, and returned to *Agrobacterium.* Infection of the plant by the recombinant bacteria automatically transfers the new gene into its root cells (figure 22.42). The insertion of genes in this manner works primarily in dicot plants such as potatoes, tomatoes, tobacco, and cotton. So far, the major recombinant, or **transgenic,** plants have been herbicide-resistant cotton and tobacco plants (figure 22.43), plants with a built-in insecticide, various virus- and fungus-resistant species, and a tomato that does not ripen too soon or break down in transport.

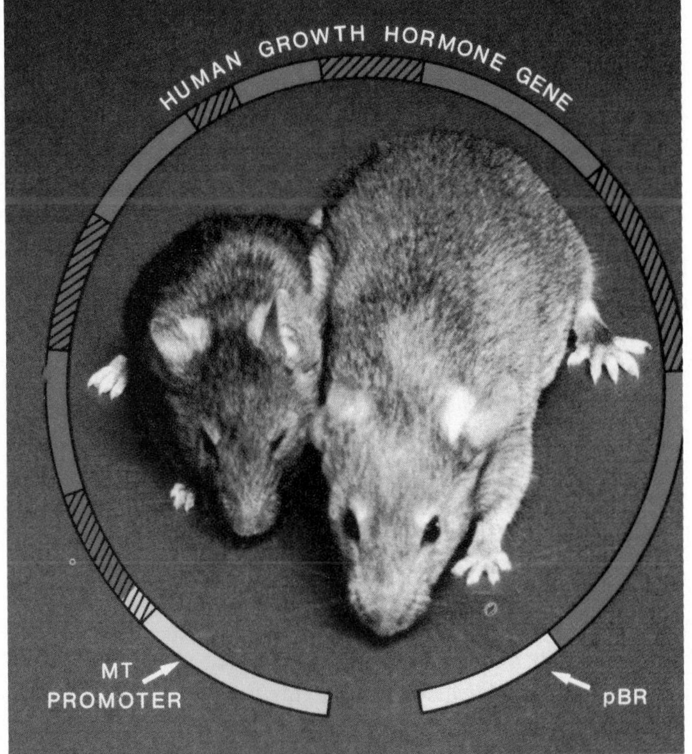

Figure 22.44 These two mice are genetically the same strain, except that the one on the right has been transfected with the human growth hormone gene (shown in its circular plasma vector). The new gene spurred an increase in this mouse's growth rate until it dwarfed its normal-sized companion.

Recombinant Animals

The propensity to be genetically engineered is not exclusive to viruses, bacteria, fungi, and plants. Under some circumstances, animals can also express genes inserted artificially through *transfection.* The first transgenic animals were fruit flies, whose eye color defects were corrected by placing normal genes into defective eggs. Mice have proved particularly amenable to accepting genes through this means. Fertilized mouse embryos have been injected with human genes for growth hormone, producing supermice twice the size of normal mice (figure 22.44). In another experiment, mouse zygotes afflicted with a genetic defect that would have caused them to develop a shivering disease were corrected by having the normal gene added.

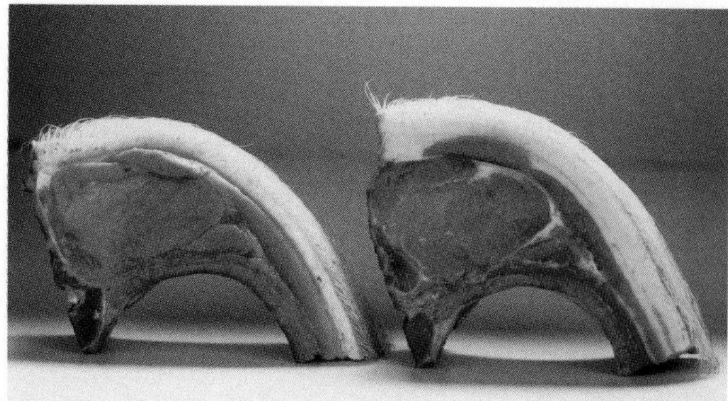

Figure 22.45 Comparison of the meat from a pig into which bovine growth hormone genes were introduced (left) and a normal pig (right). A section taken at the same rib area of the two pigs indicates significantly reduced fat content in the transgenic pig. Both animals weighed approximately the same.

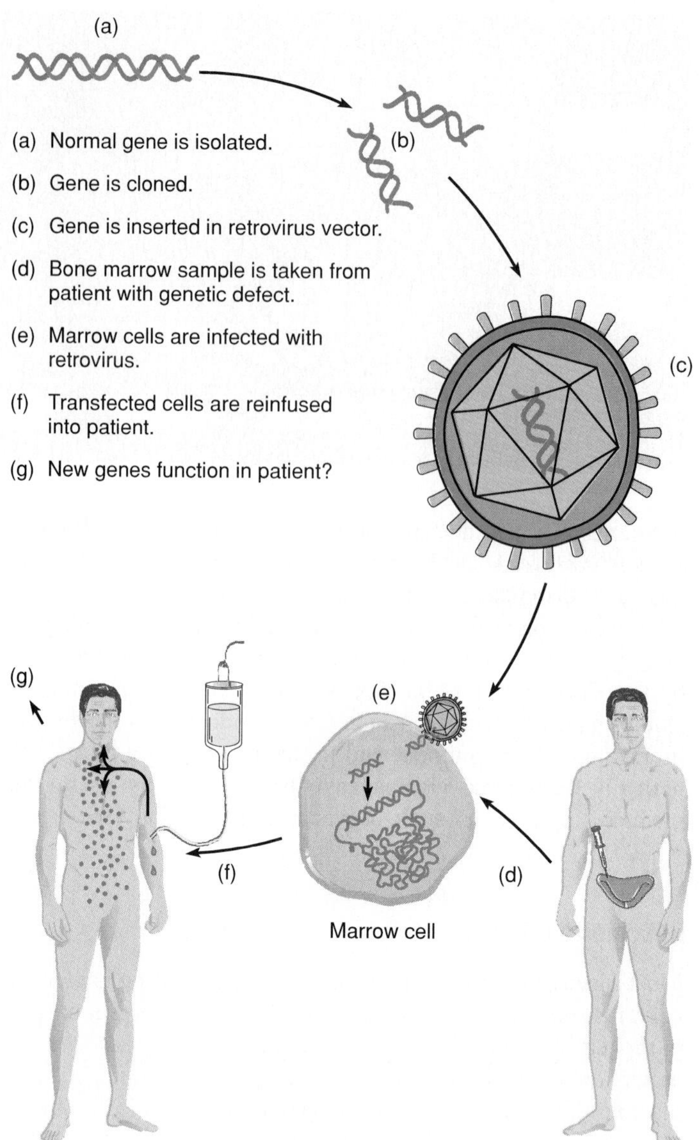

Figure 22.46 Protocol for one type of gene therapy in humans. (*a*) The steps in identifying and isolating a normal human gene. (*b*) Cloning the gene in an appropriate cloning host. (*c*) Inserting the gene into a retrovirus vector. (*d*) Harvesting a bone marrow sample from a patient with a genetic defect. (*e*) Infecting the marrow cells with retrovirus. (*f*) Infusing the transfected cells back into the patient. (*g*) Monitoring the patient for normal function of the new genes.

Animal husbandry has been quick to pick up some of these techniques. In 1987, pig embryos were implanted with the bovine growth hormone gene in an attempt to increase the pig's growth rate and cut down on fat formation in the meat. The first pigs did indeed have less fat (figure 22.45), but unfortunately, they also had some severe health problems (arthritis, heart and kidney disease). The most recent trend in transgenic experiments is to engineer fertilized eggs of domestic animals such as pigs and sheep with human genes. These recombinant animals will serve as living systems to produce large quantities of valuable human proteins such as hemoglobin, clotting factors, and hormones.

Gene Therapy

One outcome of research with transgenic animals is **gene therapy,** a means of replacing a faulty gene with a normal one in children born with fatal or extremely debilitating diseases (figure 22.46). This technique uses a harmless, genetically modified retrovirus vector into which a normal human gene has been spliced. When exposed to the vector, certain cells will pick up and insert the spliced gene into a chromosome. This therapy shows greatest promise in diseases that are treatable with bone marrow transplants. It is possible that, when stem cells take up this replacement gene, they will not only express it, but continue to replicate it for the life of the patient. Because this therapy could be a viable solution for persons with single-gene defects, experimental trials were recently begun for ADA deficiency (a cause of severe combined immunodeficiency disease) and Duchenne's muscular dystrophy. Early evidence shows that children undergoing these forms of therapy do express the recombinant gene, but whether it is a long-term cure is yet to be established. In the future, it may also be attempted as a treatment for sickle-cell anemia, cystic fibrosis, and several other genetic defects.

Miscellaneous Applications

In table 8.8, we listed products that are currently being produced by recombinant DNA technology. This area of bioengineering is gaining the greatest ground. What makes it especially attractive is that the techniques of genetic engineering may be combined with those of fermentation technology to manufacture vast quantities of these substances, just as with antibiotics and steroids. Most of the recombinant proteins currently marketed are useful in medicine. A partial list includes:

Insulin, a replacement hormone for persons with diabetes mellitus.

Human growth hormone, used to treat children with dwarfism or a premature aging disease called progeria.

Interferon, an immune substance used to treat some types of cancer, hepatitis, and genital warts.

Interleukin-2, a T- and B-cell activator used in cancer treatment.

Erythropoeitin (EPO), a stimulant for the marrow cells that give rise to red blood cells; used to treat some forms of anemia.

Tissue plasminogen activating factor (TPA), an enzyme that participates in the dissolution of potentially dangerous blood clots.

Factor VIII, a clotting protein necessary for hemophiliacs.

Recombinant vaccines for hepatitis and *Hemophilus influenzae* B meningitis.

Orthoclone, an immune suppressant in transplant patients.

The Hybridization of Nucleic Acids and Its Applications

A **hybrid** is the offspring of two unlike parents, a product that arises from a mixture of two strains. Although we usually think of organisms as hybridizing, DNA and RNA can be hybridized too. When DNA is melted to form single strands, under controlled conditions, it will automatically renature, or join back into the normal two-stranded form. This is because the complementary sites seek out and reattach to each other. But even more significantly, DNA can renature or hybridize with complementary DNA or RNA from a different source, provided there are complementary areas of attachment (figure 22.47). Knowing this behavior of nucleic acids, it was reasoned that small segments of DNA or RNA of known structure could be used to identify a particular DNA or RNA of unknown sequence by placing them together and determining the degree of hybridization. Short segments of nucleic acid used to characterize genes or RNA are called **gene probes.**

Probes have numerous possibilities. They may be used to detect an infectious agent in a patient's specimen, to identify a culture of an unknown bacterium or virus, to find markers on DNA, and to produce genomic maps. When identification of organisms from unknown specimens is desired, the DNA from the test sample is isolated, denatured, and placed on a special filter (figure 22.47). A DNA probe of known identity is incubated with the sample. Wherever this probe encounters the segment for which it is complementary, it will attach and form a hybrid. Development of the pattern of hybridization is done with photographic film or dyes. A positive hybridization will show up as one or more bands, indicating that the probe is specific for the organism in the sample.

Probes are an exquisitely sensitive and specific way to identify microorganisms. Commercially available diagnostic kits are now on the market for identifying intestinal pathogens such as *Escherichia coli, Salmonella, Campylobacter, Shigella, Clostridium difficile,* rotaviruses, and adenoviruses. Probes are also available for *Mycobacterium* species, *Legionella, Mycoplasma, Chlamydia,* and the viruses of herpes, papilloma, hepatitis A and B, and AIDS, and they are in development for several others. DNA probes have also been developed for human genetic markers and some types of cancer.

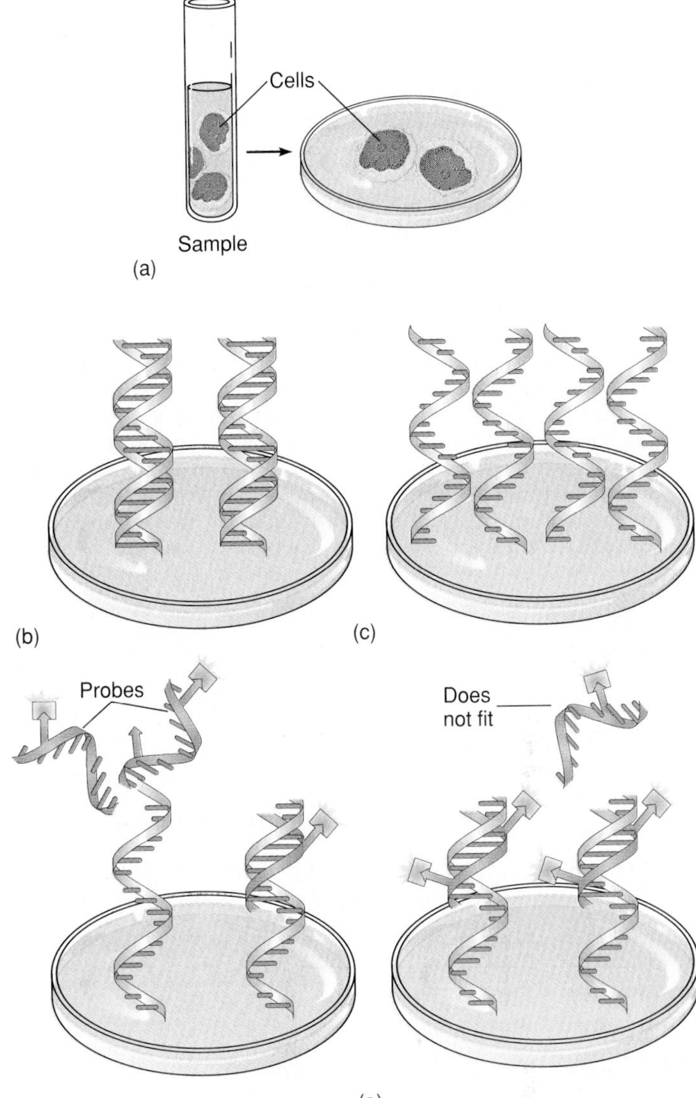

Figure 22.47 Conducting a DNA probe. (Cells and DNA are greatly disproportionate to show reaction.) (*a*) A sample of the test specimen (cells) is bound on a support matrix in a petri plate. (*b*) The cells are disrupted to release their DNA. (*c*) The DNA is denatured to separate into its two strands, which are immobilized on the matrix. (*d*) Labelled probes, consisting of a short single strand of radioactive or fluorescent nucleotides, hybridize with any segments on the unknown test strands to which they are complementary. (*e*) Unreacted probes are rinsed away, leaving only those that have attached. A positive hybridization reaction shows up in the petri dish as bands or blots that fluoresce or can expose a photographic negative.

DNA Fingerprints: A One-of-a-Kind Picture of a Genome

Although DNA is based on a structure of nucleotides, the combination of these nucleotides is unique for each organism. It is now possible to apply the new DNA technology in a manner

Feature 22.5 DNA Fingerprints: The Bar Codes of Life

The elegant technique for producing genetic fingerprints was devised in the mid-1980s by Alex Jeffreys of Great Britain. It involves taking a sample of cells from the organism being tested, breaking the cells open to release the DNA, and exposing the isolated DNA to restriction endonucleases that hydrolyze the DNA at precise locations. Cleavage results in thousands of fragments of DNA that can be separated into a pattern by gel electrophoresis. To make the pattern of bands more visible, the DNA is denatured to produce single strands, and a number of radioactive DNA probes are applied. The probes are known to react only with a given segment on the DNA called a marker. Incubation of the DNA with the probes facilitates attachment of the probe to the precise complementary region (marker) for which it is specific. When this gel is placed against a photographic film, the radioactivity of the probes in place on the DNA exposes the photographic emulsion. The developed emulsion produces a pattern of 20 to 30 dark and light bands that correspond to the separate segments of DNA, a pattern that has been aptly compared to the bar scanning codes on grocery store items.

Each individual's pattern of "bar codes" is shared by no other organism because no two restriction endonucleases cleave DNA in precisely the same way. In humans, only identical twins would be expected to have the exact cleavage pattern; even siblings can have quite different DNA fingerprint patterns. One of the first uses of the technique was in forensic medicine. Besides real fingerprints, a criminal often leaves a bit of his body at the site of a crime—a hair, a piece of skin or fingernail, semen, blood, or saliva. Fingerprinting can be used on almost any specimen, even if only a small amount of DNA is present or if the specimen is old. In the past five years, DNA fingerprinting has been used as evidence in a number of crimes and has helped convict several murderers. A riveting case occurred in an English township, where 4,500 men participated in a "blooding" in hopes of discovering the murderer/rapist of two young girls. The blood of the man who was eventually caught and convicted perfectly matched the pattern of the semen samples taken from the victims. The match was so close that, at minimum, there was only one chance in 30 billion that the prints could have been derived from two different persons! Such specificity promises to revolutionize the use of evidence in trials. The FBI is currently establishing a standardized method to set up a national data bank of fingerprints.

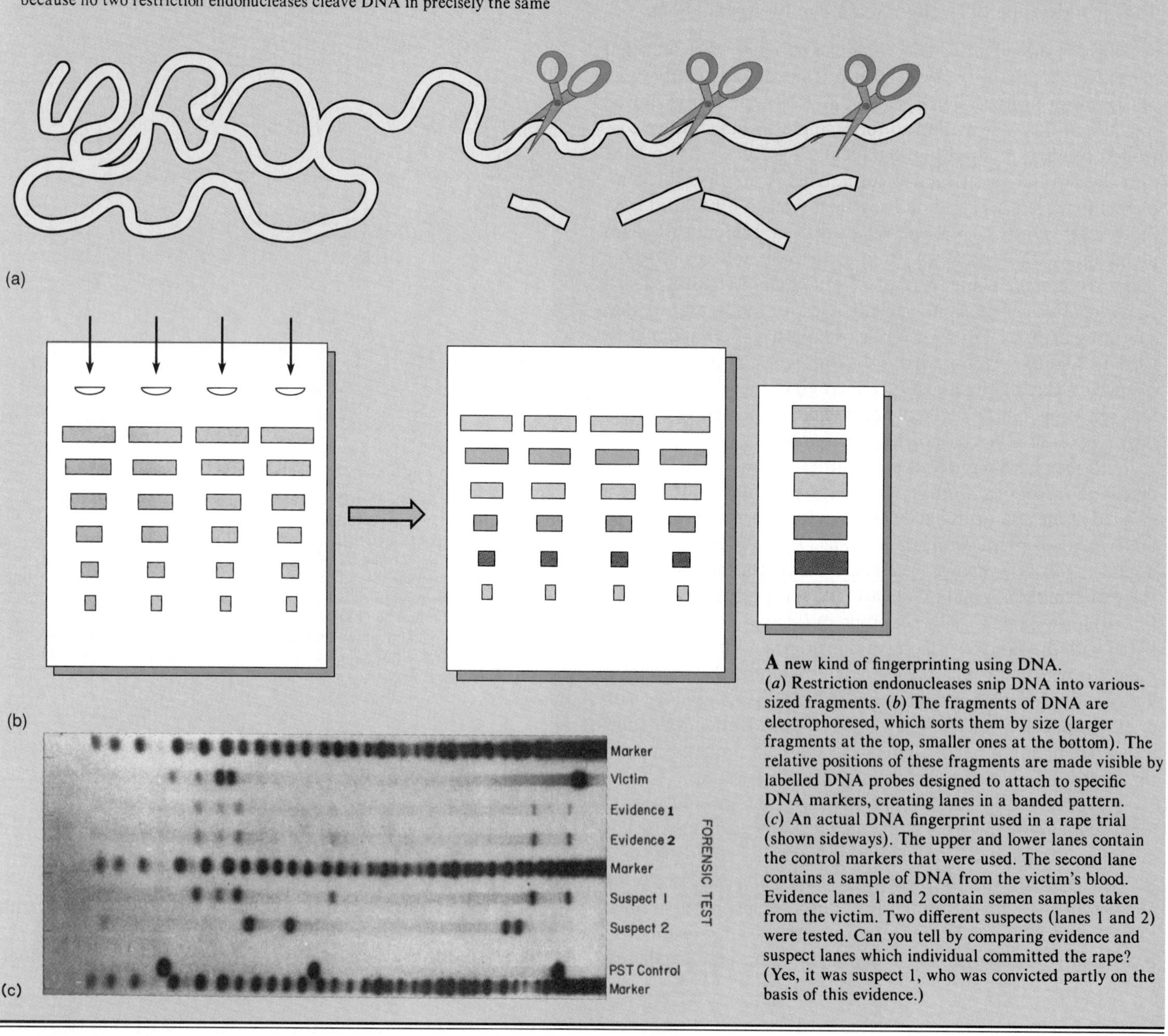

A new kind of fingerprinting using DNA. (*a*) Restriction endonucleases snip DNA into various-sized fragments. (*b*) The fragments of DNA are electrophoresed, which sorts them by size (larger fragments at the top, smaller ones at the bottom). The relative positions of these fragments are made visible by labelled DNA probes designed to attach to specific DNA markers, creating lanes in a banded pattern. (*c*) An actual DNA fingerprint used in a rape trial (shown sideways). The upper and lower lanes contain the control markers that were used. The second lane contains a sample of DNA from the victim's blood. Evidence lanes 1 and 2 contain semen samples taken from the victim. Two different suspects (lanes 1 and 2) were tested. Can you tell by comparing evidence and suspect lanes which individual committed the rape? (Yes, it was suspect 1, who was convicted partly on the basis of this evidence.)

that magnifies these differences and arrays the entire genome in a pattern for comparison. This method of DNA typing, called **DNA fingerprinting,** is a promising new way to decipher the genetic structure of complex organisms, to identify hereditary relationships and hereditary diseases, and even to apprehend criminals (see feature 22.5). DNA fingerprinting has played a part in determining the patterns of inheritance in Huntington's disease, Alzheimer's disease, and cystic fibrosis. It is also used for many other purposes: clarifying paternity, maternity, the pedigree of domestic animals, and the genetic diversity of animals bred in zoos; checking the success of transplants; predicting the likelihood of developing cancer and genetic diseases; and locating mutant genes. Perhaps in the future a DNA fingerprint will be made for every child at birth and kept on file, much as footprints and regular fingerprints have been used up until now.

Genetic Mapping—An Atlas of the Genome

Gene mapping is a process that characterizes the relative order and positions of genes on the chromosomes of living things. Knowing the locus of each gene would make it easier to manipulate genes, locate mutations, understand gene control, and clarify genetic diseases. Because of their relative simplicity, only the genomes of some viruses have been completely mapped. The gene map of *Escherichia coli* has been partly delineated, with the location of more than 900 genes now known.

One of the most exciting prospects is the mapping of the human genome. Approximately 6,500 genes have already been mapped to a given locus, and a larger number have been correlated with a site on one of the chromosomes (figure 22.48). But these are only a drop in the bucket of the 100,000 genes in the DNA library.[7] The federal government has recently underwritten a three-billion-dollar effort to map the human genome. This herculean task is to be carried out in several laboratories and will probably require at least 10 years. So monumental is this project that it promises to forever change the human experience. With a map of the human genome, the genetic background of the 3,000 inherited diseases could eventually be disclosed, clearing the way for diagnosing genetic defects in an embryo or fetus. Genetic profiles of children could possibly predict their predisposition to various organic diseases. Furthermore, knowing the exact sites of genetic defects will help correct or alleviate them, possibly through gene therapy.

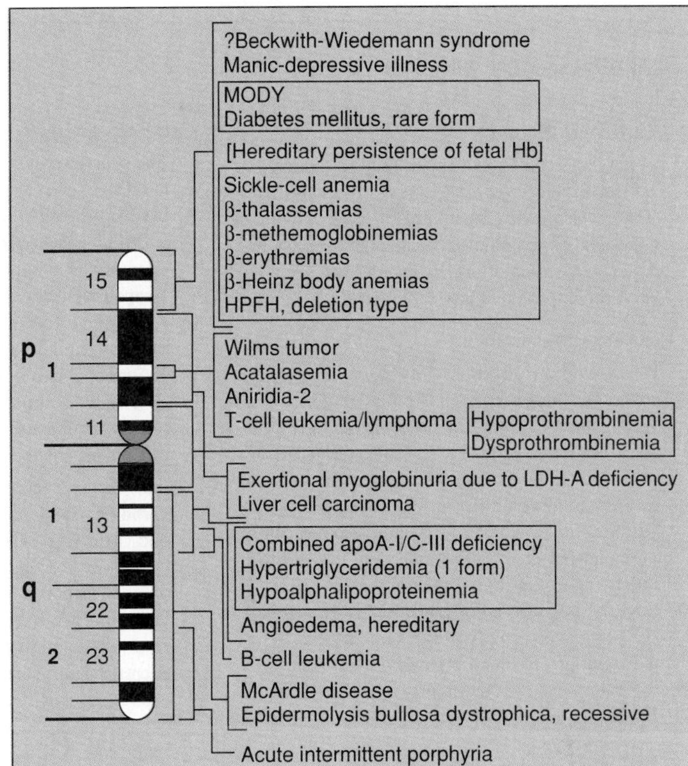

Figure 22.48 The bare beginnings of the human genome map. Chromosome 11, one of the 46 found in human cells, is the location of a relatively large number of genes that cause inherited diseases. So far, many of these loci are only approximate, but as the genome project continues, the map of this and all other chromosomes will become so complex that it will fill several volumes of a book.

Cartoon by J. Chase.

7. This translates to three billion nucleotides. If the DNA library were a book, it would fill a million pages of text—somewhat like an "encyclopedia of humans" composed of 23 volumes, one for each chromosome pair.

Chapter Review with Key Terms

Microbial Ecology

Microbial ecology deals with the interaction between the environment and microorganisms. The environment is comprised of **biotic** (living or once-living) and **abiotic** (nonliving) components. The combination of organisms and the environment make up an **ecosystem.**

Ecosystem Organization

Living things inhabit only that area of the earth called the **biosphere,** which is made up of the **hydrosphere** (water), the **lithosphere** (soil), and the **atmosphere** (air). The biosphere consists of terrestrial ecosystems of **biomes** (characterized by a dominant plant form, elevation, and latitude) and aquatic ecosystems. Biomes contain **communities,** mixed assemblages of organisms coexisting in the same time and space. Communities consist of **populations,** groups of like organisms usually of the same species. The space within which an organism lives is its **habitat**; its role in community dynamics is its **niche.**

Energy and Nutrient Flow

Organisms consume nutrients and derive energy. Their collective trophic status relative to one another is summarized in a **food chain** or **energy pyramid.** At the beginning of the chain or pyramid are **producers**—photosynthetic or lithotrophic organisms that can tap a basic nonliving energy source to synthesize large, complex organic compounds from small, simple inorganic molecules. The levels above producer are occupied by **consumers,** organisms that feed upon other organisms. A primary consumer grazes upon producers; those that prey upon others are ranked as secondary, tertiary, or quaternary consumers, depending upon their predation status. **Decomposers** are consumers that obtain nutrition from the remains of dead organisms from all levels; they help convert organic materials to their inorganic state, a process called **mineralization.** Decomposers are also important nutrient recyclers. Since organisms are usually not related in a linear **food-chain** fashion, a cross-linked **food web** better reflects their ability to use alternate nutritional sources.

Microbial Interrelationships

Mutualism involves a beneficial, dependent, two-way partnership; **commensalism** is a one-sided relationship that benefits one partner while neither benefitting nor harming the other; in **cometabolism,** the by-product of one microbe becomes a nutrient for another; in **synergism,** two microbes degrade a substrate cooperatively; in **parasitism,** a host provides habitat and niche and is harmed by the activities of its passenger. **Competition** involves the release of an antagonistic substance by one organism that inhibits another organism. A **predator** is an organism that seeks and ingests live prey.

Biogeochemical Cycles

The processes by which bioelements and essential building blocks of protoplasm are **recycled** between the biotic and abiotic environments are called **biogeochemical cycles.** They require microorganisms that can remove elements and compounds from a nonliving inorganic reservoir, return them to the food chain, and reconvert them to the inorganic state. Cycles require the actions of producers and decomposers. The **Gaia Hypothesis** is a philosophical explanation for the self-sustaining and self-regulating nature of organisms and the environment.

Atmospheric Cycles: Key compounds in the **carbon cycle** include carbon dioxide, methane, and carbonates. Carbon is **fixed** when autotrophs (photosynthesizers) add carbon dioxide to organic carbon compounds. Most fixed carbon is later returned to the carbon dioxide state by respiration and burning fossil fuels. Some carbon is entrapped for longer periods in the form of calcium carbonate, the compound of limestone deposits from the accumulated shells of marine organisms. Volcanic activity is one of several natural ways in which limestone reverts to carbon dioxide. Anaerobic metabolism of **methane** by **methanogens** also contributes significantly to carbon recycling.

The accumulation of a heat-trapping layer of carbon dioxide, methane, nitrous oxide, and water vapor in the upper atmosphere contributes to the **greenhouse effect.** Human activities such as burning fossil fuels and forests have significantly increased the amounts of these gases, which may cause global warming and have a major climatic impact.

Photosynthesis takes place in two stages—**light reactions** (light-dependent) and **dark reactions** (light-independent). During the light reactions, **photons** of solar energy are absorbed by **chlorophyll, carotenoid,** and **phycobilin** pigments; eucaryotes capture light with **thylakoid** membranes in chloroplasts, and procaryotes have specialized membrane lamellae to trap it. Captured light energy fuels a series of two photosystems that split water by **photolysis** and release oxygen gas and electrons to drive **photophosphorylation.** In this process, released light energy is used to synthesize ATP and NADPH. Dark reactions involve a series of chemical reactions called the **Calvin cycle,** during which ATP is used to fix carbon dioxide to a carrier molecule, ribulose 1,5 bisphosphate. The resultant 6-carbon molecule is converted to glucose in a multistep process. The type of photosynthesis commonly found in plants, algae, and cyanobacteria is called **oxygenic** because it liberates oxygen. **Anoxygenic** photosynthesis occurs in photolithotrophs, which are anaerobic and use hydrogen gas, hydrogen sulfide, or elemental sulfur as an electron source.

The **nitrogen cycle** requires four processes and several types of microbes. In **nitrogen fixation,** atmospheric N_2 gas (the primary reservoir) is combined with oxygen or hydrogen to form NO_2, NO_3, or NH_4 salts that are incorporated into organic compounds (proteins, nucleic acids) by living things. Microbial fixation is achieved both by free-living soil and plant-associated bacteria (*Rhizobium*) that colonize **root nodules** in a mutually beneficial relationship with legumes. **Ammonification** is a stage in the degradation of nitrogenous organic compounds by bacteria that can provide ammonium ions for plants. Some bacteria can extract energy from ammonium ions and in so doing convert it to NO_2. Various **nitrifying** bacteria convert NH_4^+ to NO_2 and then to NO_3, which can be incorporated into protoplasm by still other microbes. **Denitrification** is the multistep process achieved by several microbes that converts various nitrogenous salts back to atmospheric N_2.

Cycles in the Lithosphere: In the **sulfur cycle,** sulfurous compounds in rocks and water are oxidized into useful substrates and returned to the inorganic reservoir through reduction. The major inorganic forms of sulfur are S, H_2S, SO_4, and S_2O_3, and the chief organic forms are in amino acids—cystine, cysteine, and methionine—which contribute to S—S bonds in peptides.

The chief compound in the **phosphorus cycle** occurs as phosphate (PO_4) rock. Microbial action on this reservoir releases it so that it can be incorporated into organic phosphate forms such as DNA, RNA, and ATP. Phosphate is often a limiting factor in ecosystems.

Microorganisms often cycle and help **biomagnify** (accumulate) heavy metals and other toxic pollutants that have been added to habitats by human activities, thereby creating potential hazards in the food chain.

Soil is a dynamic, complex environment that accommodates a vast array of microbes, animals, and plants coexisting among rich organic debris, water and air spaces, and minerals. The breakdown of rocks into various-sized particles and the release of minerals are due in part to the activities of acid-producing bacteria and pioneering **lichens,** symbiotic associations between fungi and cyanobacteria or algae. **Humus** is the rich, moist, porous layer of soil containing plant and animal debris in the process of being decomposed by microorganisms. An example of a microbial habitat is the **rhizosphere,** a thin zone of soil around plant roots, which gives rise to microhabitats having varying gas and moisture contents. **Mycorrhizae** are specialized symbiotic organs formed by the attachment between specialized fungi and certain plant roots.

Cycles in the Hydrosphere: The three environmental compartments (surface water, atmospheric moisture, and groundwater) are linked through a **hydrologic cycle** that involves evaporation and precipitation. Living things contribute to the cycle by releasing metabolic moisture through respiration and transpiration. Water that percolates into the earth accumulates in deep underground **aquifers.**

The diversity and distribution of surface water communities are related to sunlight, temperature, aeration, and dissolved nutrients. In a cross section of a lake, three strata are found along the vertical axis: the **photic zone,** the **profundal zone,** and the **benthic zone**; along the horizontal axis are the **littoral zone** and the **limnetic zone.** Features of the marine environment include tidal and wave action and **estuary, intertidal,** and **abyssal zones** that harbor unique microorganisms. **Phytoplankton** and **zooplankton** living in the photic zone comprise a drifting microbial community that supports larger invertebrates and marine animals. Temperature gradients in still waters cause stratified layers to form; the **epilimnion** and **hypolimnion** meet at an interface called a **thermocline.** Seasonal temperature changes cause **upwelling,** currents that stir the nutrient-rich benthic sediments and account for seasonal outbreaks of abundant microbial growth (red tides). Clear mountain lakes that contain few nutrients and support a sparse population are called **oligotrophic.** Waters spiked with high levels of sewage, fertilizer, or industrial runoff cause **eutrophication,** or explosive aquatic overgrowth of algae and other plants.

Water Analysis: Providing **potable** water is central to prevention of water-borne disease. Water is constantly surveyed for certain **indicator bacteria** (coliforms and streptococci) that signal fecal contamination and are easier to detect and longer-lived than pathogens. Assays for possible water contamination include the **standard plate count** and the coliform enumeration test or **most probable number** (**MPN**) procedure.

Water and Sewage Treatment: Drinking water is rendered safe by a purification process that involves storage, sedimentation, settling, aeration, filtration, and disinfection. Sewage or used wastewater may be processed to remove solid matter and dangerous chemicals and microorganisms before it is released or reclaimed for irrigation. Treatment occurs in three phases. Microbes are used to biodegrade the waste material or sludge. Solid wastes are further processed in **anaerobic digesters** that can provide combustible methane and fertilizer. Future solutions to the environmental waste problem may involve **bioremediation,** the degradation of pollutants by microbial activities.

Applied Microbiology and Biotechnology

Practical application of microbiology in the manufacture of food, industrial chemicals, drugs, and other products is the realm of **biotechnology.** Many of these processes use mass, controlled microbial **fermentations** and bioengineered microorganisms.

Food Microbiology

Microbes and humans compete for the rich nutrients in food. Although some microbes are present on food as harmless contaminants, others can cause either unfavorable or favorable results.

Food-Borne Disease: Some microbes cause spoilage and **food poisoning.** In **food intoxication,** damage is done when the toxin is produced in food, is ingested, and then acts on a specific body target; examples are staphylococcal, clostridial, and fungal poisoning. In **food infection,** the intact microbe does damage when it is ingested and invades the intestine; examples are salmonellosis, shigellosis, listeriosis, and amebic dysentery.

Microbial growth that leads to spoilage and food poisoning can be avoided by maintaining foods at inhibitory temperatures (cold or hot), sterilizing or disinfecting foods, and preserving foods. High temperature and pressure treatment (canning) is commonly used for foods that are not adversely altered by heat, while **pasteurization** (most commonly the flash method) is an alternative to treating fluids (milk, beverages) that are sensitive to heat. For short-term needs, refrigeration and freezing are practical measures that inhibit microbial growth; irradiation can sterilize or disinfect foods for longer-term storage. Alternative preservation methods include additives such as organic acids, nitrites, nitrates, or sulfite, treatment with ethylene and propylene oxide gases, growth inhibition with concentrated sugar, salting, and drying.

Fermentations in Foods: Microbes can impart desirable aroma, flavor, or texture as they grow in foods. Bread, alcoholic beverages, some vegetables, and some dairy products are infused with **starter cultures** of pure microbial strains to yield the necessary fermentation products. Baker's yeast, *Saccharomyces cerevisiae,* is used to **leaven** bread dough by giving off CO_2. Additional flavors may also come from bacterial action.

Beer making involves the following steps: Barley is sprouted (**malted**) to generate digestive enzymes and then dried; warm, wet malt is ground up and spiked with sugar and starch to yield **mash**; the mash mixture is heated with hops to release nutrients and flavor, producing a fraction called **wort**; cooled wort is inoculated with a special strain of yeast that ferments it until a concentration of 3% to 6% alcohol is reached; fermented beer is allowed to mature (**lager**) in large tanks before it is carbonated and packaged.

Wine is usually prepared from grapes, although juice (**must**) from any crushed fruit may be fermented by special strains of yeast to produce it. Wines may be dry or sweet, and may contain from 7% to 15% alcohol. After storage, wine is bottled and aged. Whiskey, vodka, brandy, and other alcoholic beverages are **distilled** to increase their alcohol content. Vegetable products, including sauerkraut, pickles, and a variety of soybean derivatives, may be fermented in the presence of salt or sugar to form an acidic pickling bath. Vinegar is produced by fermenting plant juices first to alcohol and then to acetaldehyde and acetic acid.

Some dairy products are produced by microbes acting on the rich assortment of nutrients in milk. In cheese production, milk proteins are coagulated with **rennin** to form solid **curd** that separates from the watery **whey.** The different kinds of cheeses are obtained by varying such components as water, fat, acid, and salt and by aging or ripening with bacteria and yeasts; yogurt and buttermilk are also processed with live cultures.

Mass-cultured microbes such as yeasts, molds, and algae may serve as food. **Single-cell protein, pruteen,** and **mycoprotein** are currently added to animal feeds; in some countries, large colonies of the cyanobacterium *Spirulina* are harvested to be used as food supplements.

Industrial Microbiology

Industrial microbiology involves the large-scale commercial production of organic compounds such as antibiotics, vitamins, amino acids, enzymes, and hormones using specific microbes in carefully controlled fermentation settings. Microbes are chosen for their production of a desired **metabolite**; several different species may be used to **biotransform** raw materials in a stepwise series of metabolic reactions; fermentations are conducted in massive culture devices called **fermentors** that have special mechanisms for adding nutrients, stirring, oxygenating, altering pH, cooling, monitoring, and harvesting product; plant organization includes downstream processing (purification and packaging).

Bioengineering, or genetic engineering, involves the purposeful manipulation of the genes of microbes and other organisms. Inserting selected genes from one organism into another, also termed recombinant DNA technology, has given rise to numerous applications in industry, agriculture, animal husbandry, and medical therapy. Recombinant microbes have been developed for vaccines, for preventing plant disease, and for biodegrading toxic wastes.

Transgenic plants have been engineered by means of a special cloning host, *Agrobacterium tumefaciens,* to resist herbicides and infections. Transgenic animals have been engineered to contain exogenous growth hormone. Transgenic experiments with humans include **gene therapy,** an attempt to replace a missing or dysfunctional gene.

Genetically engineered proteins for treating disease are produced on a large scale industrially. Genetically engineered DNA and RNA are made to serve as **gene probes** that identify microbes or specific segments on genomes of plants and animals by **hybridization**; offshoots of this technology include **DNA fingerprinting** and genetic maps of chromosomes.

True–False Questions

Determine whether the following statements are true (T) or false (F). If you feel a statement is false, explain why, and reword the sentence so that it reads accurately.

____ 1. The stratosphere is one of the major components of the biosphere.

____ 2. A terrestrial biome is ordinarily defined by its dominant plant form, elevation, and latitude.

____ 3. Interrelated populations sharing a given region constitute a community.

____ 4. The quantity of available nutrients increases from the lower levels of the energy pyramid to the higher ones.

____ 5. Photosynthetic organisms convert the energy of photons into chemical energy.

____ 6. The Calvin cycle is an integral part of the dark reactions of photosynthesis.

____ 7. Although phosphate is commonly found in rock, this form is virtually unavailable for plants because of its insolubility.

____ 8. Plankton exists primarily in the photic and littoral zones.

____ 9. Dissolved oxygen, temperature, hydrostatic pressure, salinity, and sunlight all vary with the depth of the aquatic environment.

____ 10. Water containing coliforms is considered potable.

____ 11. Milk is usually pasteurized by the high-temperature short-time method.

____ 12. Hops are the dried, presprouted grain that is soaked to activate enzymes for beer.

____ 13. Yeasts make the alcohol in wine and the carbon dioxide that leavens bread.

____ 14. In cheese making, rennin is added to facilitate curdling.

____ 15. To prevent food poisoning, foods should be maintained at cold or room temperatures.

____ 16. Secondary metabolites of microbes are formed during the stationary phase of growth.

____ 17. *Agrobacterium* is a pathogen of plant roots that is used as a cloning host.

____ 18. Gene probes are based on the hybridization of complementary strands of nucleic acids.

Concept Questions

1. Present in outline form the levels of organization in the biosphere. Compare autotrophs and heterotrophs, producers and consumers. Where in the energy and trophic schemes do decomposers enter? Compare the concepts of habitat and niche using *Chlamydomonas* (figure 22.1) as an example.
2. Using figure 22.3, point out specific examples of producers, primary consumers, secondary consumers, tertiary consumers, herbivores, primary, secondary, and tertiary carnivores, and omnivores. What is mineralization, and which organisms are responsible for it?
3. Using specific examples, differentiate between commensalism, mutualism, parasitism, competition, parasitism, and scavenging.
4. Outline the general characteristics of a biogeochemical cycle. What are the major sources of carbon, nitrogen, phosphorus, and sulfur?
5. In what major forms is carbon found? Name three ways carbon is returned to the atmosphere; name a way it is fixed into organic compounds. What form is the least available for the majority of living things? Which form is produced during anaerobic reduction of CO_2? Describe the relationship between photosynthesis and respiration with regard to CO_2 and O_2.
6. Tell whether each of the following is produced during the light or dark reactions of photosynthesis: O_2, ATP, NADPH, and glucose. When are water and CO_2 consumed? What is the function of chlorophyll and the photosystems? What is the fate of the ATP and NADH? Compare oxygenic with nonoxygenic photosynthesis.
7. Outline the steps in the nitrogen cycle from N_2 to organic nitrogen and back to N_2. Describe nitrogen fixation, ammonification, nitrification, and denitrification. What form of nitrogen is required by plants? By animals?
8. Outline the phosphorus and sulfur cycles. What are the most important organic phosphorous compounds? The organic sulfur compounds? What are the roles of microorganisms in these cycles?
9. Describe the structure of the soil and the rhizosphere. What is humus? Compare and contrast root nodules with mycorrhizae.
10. Outline the modes of cycling water through the lithosphere, hydrosphere, and atmosphere. What are the roles of precipitation, condensation, respiration, transpiration, surface water, and aquifers?
11. Map a section through the structure of a lake. What types of organisms are found in plankton? Why is phytoplankton not found in the benthic zone? What would you expect to find there?
12. What causes the formation of the epilimnion, hypolimnion, and thermocline? What is upwelling? Compare and contrast red tides and eutrophic algal blooms.
13. Why must water be subjected to analysis? What are the characteristics of good indicator organisms, and why are they monitored rather than pathogens? Give specific examples of indicator organisms and water-borne pathogens. Describe two methods of water analysis. What are the principles behind the most probable number test? Describe the three phases of sewage treatment. What is activated sludge?
14. What is the meaning of fermentation from the standpoint of industrial microbiology? Describe five types of fermentations.
15. What is the main criterion regarding the safety of foods? Differentiate between food infection and food intoxication. When are microbes on food harmless? What are the two most common food intoxications? What are the three most common food infections? Summarize the major methods of preventing food poisoning and spoilage.
16. What microbes are used as starter cultures in bread, beer, wine, cheeses, and sauerkraut? Outline the steps in beer making. List the steps in wine making. What are curds and whey, and what causes them?
17. What are the aims of industrial microbiology? Differentiate between primary and secondary metabolites. Describe a fermentor. How is it scaled up for industrial use? What are specific examples of products produced by these processes?
18. Describe the principles behind recombinant DNA technology, transgenic organisms, gene therapy, DNA hybridization, gene probes, DNA fingerprinting, and gene mapping.

Practical/Thought Questions

1. Why does energy decrease with each trophic level? How is it possible for energy to be lost and the ecosystem to still run efficiently? Are the nutrients on the earth a renewable resource? Why, or why not?
2. Describe the course of events that results in a parasitic relationship evolving into a symbiotic one.
3. Give specific examples that support the Gaia Hypothesis.
4. "Biosphere II" is an experimental, self-contained, self-sustaining ecosystem recently developed in the Arizona desert, completely sealed off from the external environment. Predict what sorts of organisms must be placed in this habitat for it to function in a balanced manner. (Are humans necessary?)
5. Is the greenhouse effect harmful under ordinary circumstances? What has made it dangerous to the global ecosystem? What could each person do on a daily basis to cut down on the potential for disrupting the delicate balance of the earth?
6. Outline a possible genetic engineering method to produce root nodules in nonleguminous plants.
7. If we are to rely on microorganisms to biodegrade wastes in landfills, aquatic habitats, and soil, list some ways that this process could be made more efficient. Since elemental poisons (heavy metals) cannot be further degraded even by microbes, what is a possible fate of these metals? What, if anything, can be done about this form of pollution?
8. Why are organisms in the abyssal zone of the ocean necessarily halophilic, psychrophilic, barophilic, and anaerobic?
9. What happens to the nutrients that run off into the ocean with sewage and other effluents? Why can high mountain communities usually dispense with water treatment?
10. Every year supposedly safe municipal water supplies cause outbreaks of enteric illness. How in the course of water analysis and treatment might these pathogens be missed? What kinds of microbes are they most likely to be?
11. What are some important potential applications of methanogens?
12. Describe four food-preparation and food-maintenance situations in your own kitchen that could expose you and your family to food poisoning. What is the general rule to follow concerning suspicious food?
13. What is the purpose of boiling the wort in beer preparation? What are hops used for? If fermentation of sugars to produce alcohol in wine is anaerobic, why do wine makers make sure that the early phase of yeast growth is aerobic?
14. Predict the differences in the outcome of raw milk that has been incubated for 48 hours versus pasteurized milk that has been incubated for the same length of time.
15. How would you personally feel about eating food products that are partly or wholly composed of algae, fungi, or bacteria?
16. How are cometabolism and biotransformations of microorganisms harnessed in industrial microbiology?
17. Outline the steps in the production of an antibiotic, starting with a newly discovered mold. How can recombinant DNA technology be married to industrial processes?
18. Outline a way that gene therapy could be used to treat a patient with genetic defects due to the presence of extra genes (trisomies as described in chapter 14). How would you regard the release of a bioengineered bacterium in your own back yard? How about eating pork from a transgenic pig?
19. If gene probes, fingerprinting, and mapping could make it possible for you to know of future genetic diseases in you or one of your children, would you wish to use this technology to find out? What if it were used as a screen for employment or insurance?

Appendix A: Important Charts

Greek Letters			
α	alpha	ν	nu
β	beta	ξ	xi
γ	gamma	o	omicron
δ	delta	π	pi
ϵ, ε	epsilon	ρ	rho
ζ	zeta	σ	sigma
η	eta	τ	tau
θ	theta	υ	upsilon
ι	iota	ϕ	phi
κ	kappa	χ	chi
λ	lambda	ψ	psi
μ	mu	ω	omega

Temperature Conversions*

°C	°F	°C	°F	°C	°F	°C	°F	°C	°F	°C	°F
−40	−40.0	9	48.2	57	134.6	−15	5.0	34	93.2	82	179.6
−39	−38.2	10	50.0	58	136.4	−14	6.8	35	95.0	83	181.4
−38	−36.4	11	51.8	59	138.2	−13	8.6	36	96.8	84	183.2
−37	−34.6	12	53.6	60	140.0	−12	10.4	37	98.6	85	185.0
−36	−32.8	13	55.4	61	141.8	−11	12.2	38	100.4	86	186.8
−35	−31.0	14	57.2	62	143.6	−10	14.0	39	102.2	87	188.6
−34	−29.2	15	59.0	63	145.4	−9	15.8	40	104.0	88	190.4
−33	−27.4	16	60.8	64	147.2	−8	17.6	41	105.8	89	192.2
−32	−25.6	17	62.6	65	149.0	−7	19.4	42	107.6	90	194.0
−31	−23.8	18	64.4	66	150.8	−6	21.2	43	109.4	91	195.8
−30	−22.0	19	66.2	67	152.6	−5	23.0	44	111.2	92	197.6
−29	−20.2	20	68.0	68	154.4	−4	24.8	45	113.0	93	199.4
−28	−18.4	21	69.8	69	156.2	−3	26.6	46	114.8	94	201.2
−27	−16.6	22	71.6	70	158.0	−2	28.4	47	116.6	95	203.0
−26	−14.8	23	73.4	71	159.8	−1	30.2	48	118.4	96	204.8
−25	−13.0	24	75.2	72	161.6	0	32.0	49	120.2	97	206.6
−24	−11.2	25	77.0	73	163.4	+1	33.8	50	122.0	98	208.4
−23	−9.4	26	78.8	74	165.2	2	35.6	51	123.8	99	210.2
−22	−7.6	27	80.6	75	167.0	3	37.4	52	125.6	100	212.0
−21	−5.8	28	82.4	76	168.8	4	39.2	53	127.4	101	213.8
−20	−4.0	29	84.2	77	170.6	5	41.0	54	129.2	102	215.6
−19	−2.2	30	86.0	78	172.4	6	42.8	55	131.0	103	217.4
−18	−0.4	31	87.8	79	174.2	7	44.6	56	132.8	104	219.2
−17	+1.4	32	89.6	80	176.0	8	46.4				
−16	3.2	33	91.4	81	177.8						

*General rule of conversion: Degrees Fahrenheit (°F) may be converted to degrees Celsius (°C) by subtracting 32 from °F and multiplying the result by $\frac{5}{9}$. Degrees Celsius may be converted to °F by multiplying °C by $\frac{9}{5}$ and adding 32 to the result.

Appendix B: Exponents

Dealing with concepts such as microbial growth often requires working with numbers in the billions, trillions, and even greater. A mathematical shorthand for expressing such numbers is with exponents. The exponent of a number indicates how many times (designated by a superscript) that number is multiplied by itself. These exponents are also called common *logarithms,* or logs. The following chart, based on multiples of 10, summarizes this system:

Number	Quantity	Exponential Notation*	Number Arrived at By:	One Followed By:
1	One	10^0	Numbers raised to zero power are equal to one	No zeros
10	Ten	10^1**	10×1	One zero
100	Hundred	10^2	10×10	Two zeros
1,000	Thousand	10^3	$10 \times 10 \times 10$	Three zeros
10,000	Ten thousand	10^4	$10 \times 10 \times 10 \times 10$	Four zeros
100,000	Hundred thousand	10^5	$10 \times 10 \times 10 \times 10 \times 10$	Five zeros
1,000,000	Million	10^6	10 times itself 6 times	Six zeros
1,000,000,000	Billion	10^9	10 times itself 9 times	Nine zeros
1,000,000,000,000	Trillion	10^{12}	10 times itself 12 times	Twelve zeros
1,000,000,000,000,000	Quadrillion	10^{15}	10 times itself 15 times	Fifteen zeros
1,000,000,000,000,000,000	Quintillion	10^{18}	10 times itself 18 times	Eighteen zeros

Other large numbers are sextillion (10^{21}), septillion (10^{24}), and octillion (10^{27}).

*The proper way to say the numbers in this column is 10 raised to the *nth* power, where *n* is the exponent. The numbers in this column may also be represented as 1×10^n, but for brevity, the $1\times$ may be omitted.

**The exponent 1 is usually omitted in numbers at this level.

Converting Numbers to Exponent Form

As the chart shows, using exponents to express numbers can be very economical. When working with simple multiples of 10, the exponent is always equal to the number of zeros that follow the 1, but this rule will not work with numbers that are more varied. Other large whole numbers can be converted to exponent form by the following operation: First, move the decimal (which we assume to be at the end of the number) to the left until it sits just behind the first number in the series (example: 3568. = 3.568). Then count the number of spaces (digits) the decimal has moved; that number will be the exponent. (The decimal has moved from 8. to 3., or 3 spaces.) In final notation, the converted number is multiplied by 10 with its appropriate exponent: 3568 is now 3.568×10^3.

Rounding Numbers Off

The notation in the previous example has not actually been shortened, but it can be reduced further by rounding off the decimal fraction to the nearest thousandth (three digits), hundredth (two digits), or tenth (one digit). To round off a number, its last digit is dropped, and the one next to it is either increased or left as is. If the number dropped is 5, 6, 7, 8, or 9, the subsequent digit is increased by one (rounded up); if it is 0, 1, 2, 3, or 4, the subsequent digit remains as is. Using the example of 3.528, removing the 8 rounds off the 2 to a 3, and produces 3.53 (two digits). If further rounding is desired, the same rule of thumb applies, and the number becomes 3.5 (one digit). Other examples of exponential conversions are as follows:

Number	*Is the Same As*	*Rounded Off, Placed in Exponent Form*
16,825.	$1.6825 \times 10 \times 10 \times 10 \times 10$	1.7×10^4
957,654.	$9.57654 \times 10 \times 10 \times 10 \times 10 \times 10$	9.58×10^5
2,855,000.	$2.855000 \times 10 \times 10 \times 10 \times 10 \times 10 \times 10$	2.86×10^6

Negative Exponents

The numbers we have been using so far are greater than 1 and are represented by positive exponents. But the correct notation for numbers less than 1 involves negative exponents (10 raised to a negative power, or 10^{-n}). A negative exponent says that the number has been divided by a certain power of 10 (10, 100, 1,000). This usage is handy when working with concepts such as pH that are based on very small numbers otherwise needing to be represented by large decimal fractions—for example, 0.003528. Converting this and other such numbers to exponential notation is basically similar to converting positive numbers, except that you work from left to right and the exponent is negative. Using the example of 0.003528, first convert the number to a whole integer followed by a decimal fraction and keep track of the number of spaces the decimal point moves (example: 0.003528 = 3.528). Since the decimal has moved three spaces from its original position, the finished product is 3.528×10^{-3}. Other examples include:

Number	*Is the Same As*	*Rounded Off, Expressed with Exponents*
0.0005923	$\frac{5.923}{10 \times 10 \times 10 \times 10}$	5.92×10^{-4}
0.00007295	$\frac{7.295}{10 \times 10 \times 10 \times 10 \times 10}$	7.3×10^{-5}

Appendix C: Methods for Testing Sterilization and Germicidal Processes

Most procedures used for microbial control in the clinical setting are either physical methods (heat, radiation) or chemical methods (disinfectants, antiseptics). These procedures are so crucial to the well-being of patients and staff that their effectiveness must be monitored in a consistent and standardized manner. Particular concerns include the time required for the process, the concentration or intensity of the antimicrobial agent being used, and the nature of the materials being treated. The effectiveness of an agent is frequently established through controlled microbiological analysis. In these tests, a biological indicator (a highly resistant microbe) is exposed to the agent, and its viability is checked. If this known test microbe is destroyed by the treatment, it is assumed that the less resistant microbes have also been destroyed. Growth of the test organism indicates that the sterilization protocol has failed. The general categories of testing include:

Heat The usual form of heat used in microbial control is steam under pressure in an autoclave. Autoclaving materials at a high temperature (121°C) for a sufficient time (15–40 minutes) destroys most bacterial endospores. To monitor the quality of any given industrial or clinical autoclave run, technicians insert a special ampule of *Bacillus stearothermophilus,* a spore-forming bacterium with extreme heat resistance (figure A.1). After autoclaving, the ampule is incubated at 56°C (the optimum temperature for this species) and checked for growth.

Radiation The effectiveness of ionizing radiation in sterilization is determined by placing special strips of dried spores of *Bacillus sphaericus* (a common soil bacterium with extreme resistance to radiation) into a packet of irradiated material.

Filtration The performance of membrane filters used to sterilize liquids may be monitored by adding *Pseudomonas diminuta* to the liquid. Because this species is a tiny bacterium that can escape through the filter if the pore size is not small enough, it is a good indicator of filtrate sterility.

Gas sterilization Ethylene oxide gas, one of the few chemical sterilizing agents, is used on a variety of heat-sensitive medical and laboratory supplies. The most reliable indicator of a successful sterilizing cycle is the sporeformer *Bacillus subtilis,* variety *niger.*

Germicidal assays United States hospitals and clinics commonly employ more than 250 products to control microbes in the environment, on inanimate objects, and on patients. Controlled testing of these products' effectiveness is not usually performed in the clinic itself, but by the chemical or pharmaceutical manufacturer. Twenty-five standardized *in vitro* tests are currently available to assess the effects of germicides. Most of them use a non-spore-forming pathogen as the biological indicator.

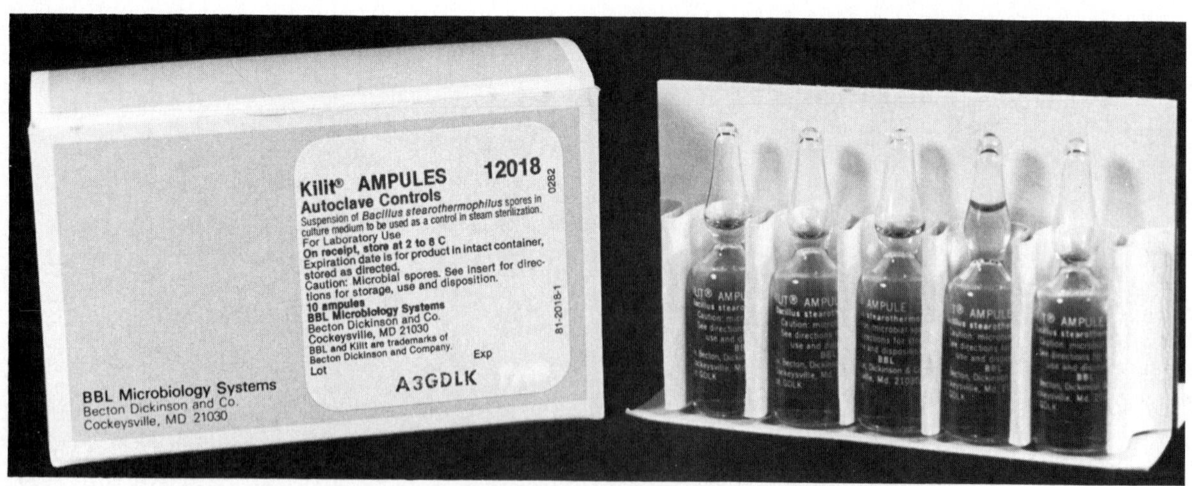

Figure A.1 The sterilizing conditions of autoclave runs may be tested by indicator bacteria in small broth ampules. One ampule is placed in the center of the autoclave chamber along with the regular load. After the process, the ampule is incubated to test for survival.

The Phenol Coefficient (PC)

The older disinfectant phenol has been the traditional standard by which other disinfectants are measured. In the PC test, a water-soluble phenol-based (phenolic) disinfectant (amphyl, lysol) is tested for its bactericidal effectiveness as compared with phenol. These disinfectants are serially diluted in a series of *Salmonella choleraesuis, Staphylococcus aureus,* or *Pseudomonas aeruginosa* cultures. The tubes are left for 5, 10, and 15 minutes and subcultured to test for viability. The PC is a ratio derived by comparing the following data:

$$PC = \frac{\text{Greatest dilution of the phenolic that kills test bacteria in 10 min. but not in 5 min.}}{\text{Greatest dilution of phenol giving same result}}$$

The general interpretation of this ratio is that chemicals with lower phenol coefficients have greater effectiveness. The primary disadvantage of the PC test is that it is restricted to phenolics, which makes it inappropriate for the vast majority of clinical disinfectants.

Use Dilution Test

An alternate test with broader applications is the use dilution test. To perform this test, the test culture (one of those species used in the PC test) is dried onto the surface of small stainless steel cylindrical carriers. These carriers are then exposed to varying concentrations of disinfectant for 10 minutes, removed, rinsed, and placed in tubes of broth. After incubation, the tubes are inspected for growth. The smallest concentration of disinfectant that kills the test organism on 10 carrier pieces is the correct dilution for use.

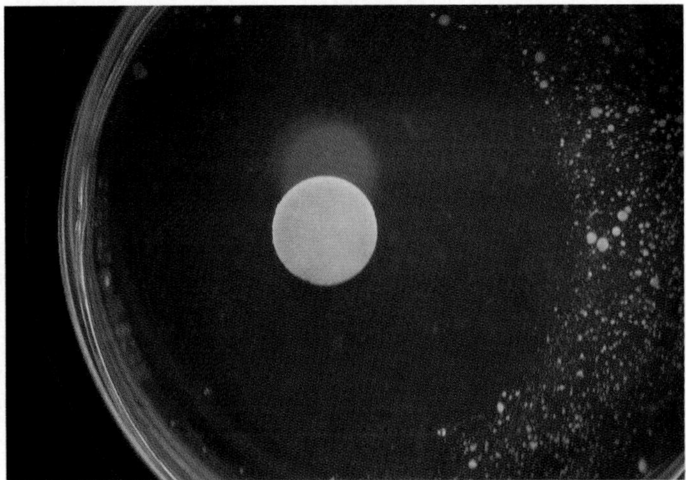

Figure A.2 A filter paper disc containing hydrogen peroxide produces a broad clear zone of inhibition in a culture of *Staphylococcus aureus.*

Filter Paper Disc Method

A quick measure of the inhibitory effects of various disinfectants and antiseptics may be achieved by the filter paper disc method. First, a small (½″) sterile piece of filter paper is dipped into a disinfectant of known concentration. This is placed on an agar medium seeded with a test organism (*S. aureus* or *P. aeruginosa*), and the plate is incubated. A zone devoid of growth (zone of inhibition) around the disc indicates the capacity of the agent to inhibit growth (figure A.2). Like the antimicrobic sensitivity test, this test measures the minimum inhibitory concentration of the chemical. In general, chemicals with wide zones of inhibition are effective even in high dilutions.

Appendix D: Universal Blood and Body Fluid Precautions

Medical and dental settings require stringent measures to prevent the spread of nosocomial (clinically acquired) infections from patient to patient, from patient to worker, and from worker to patient. But even with precautions, the rate of such infections is rather high. Recent evidence indicates that more than one-third of nosocomial infections could be prevented by consistent and rigorous infection control methods.

Previously, control guidelines were disease-specific, and clearly identified infections were managed with particular restrictions and techniques. With this arrangement, personnel tended to handle materials labelled infectious with much greater care than those that were not so labelled. However, the AIDS epidemic spurred a reexamination of this policy. Because of the potential for increased numbers of undiagnosed HIV-infected patients, the Centers for Disease Control laid down more stringent guidelines for handling patients and body substances. These guidelines have been termed **universal precautions (UP)**, because they are based on the assumption that all patient specimens could harbor infectious agents and so must be treated with the same degree of care. They also include body substance isolation techniques (BSI) to be used in known cases of infection.

It is worth mentioning that these precautions are designed to protect all individuals in the clinical setting—patients, workers, and public alike. In general, they include techniques designed to prevent contact with pathogens and contamination, and if this is not possible, to take purposeful measures to decontaminate potentially infectious materials.

The universal precautions recommended for all health care settings are:

1. Barrier precautions, including masks and gloves, should be taken to prevent contact of skin and mucous membranes with patients' blood or other body fluids. Because gloves may have small invisible tears, double gloving decreases the risk further. For protection during surgery, venipuncture, or emergency procedures, gowns, aprons, and other body coverings should be worn. Dental workers should wear eyewear and face shields to protect against splattered blood and saliva.
2. More than 10% of health care personnel are pierced each year by sharp (and usually contaminated) instruments. These accidents carry risks not only for AIDS but also for hepatitis B, hepatitis C, and other diseases. Preventing inoculation infection requires vigilant observance of proper techniques. All disposable needles, scalpels, or sharp devices from invasive procedures must immediately be placed in puncture-proof containers for sterilization and final discard. Under no circumstances should a worker attempt to recap a syringe, remove a needle from a syringe, or leave unprotected used syringes where they pose a risk to others. Reusable needles or other sharp devices must be heat-sterilized in a puncture-proof holder before they are handled. If a needlestick or other injury occurs, immediate attention to the wound such as thorough degermation and application of strong antiseptics, can prevent infection.
3. Dental handpieces should be sterilized between patients, but if this is not possible, they should be thoroughly disinfected with a high-level disinfectant (peroxide, hypochlorite). Blood and saliva should be removed completely from all contaminated dental instruments and intraoral devices prior to sterilization.
4. Hands and other skin surfaces that have been accidentally contaminated with blood or other fluids should be scrubbed immediately with germicidal soap. Hands should likewise be washed after removing rubber gloves, masks, or other barrier devices.
5. Because saliva may be a source of some types of infections, barriers should be used in all mouth-to-mouth resuscitations.
6. Health care workers with active, draining skin or mucous membrane lesions must refrain from handling patients or equipment that will come in contact with patients. Pregnant health care workers risk infecting their fetuses and must pay special attention to these guidelines. Personnel should be protected by vaccination whenever possible.

Isolation procedures for known or suspected infections should still be instituted on a case-by-case basis. The levels of isolation used in clinical settings are summarized in table A.1.

Laboratory Biosafety Levels and Classes of Pathogens

Personnel handling infectious agents in the laboratory must be protected from possible infection through special risk management or containment procedures. This involves (1) carefully observing standard laboratory aseptic and sterile procedures while handling cultures and infectious samples; (2) using large-scale sterilization and disinfection procedures; (3) restricting eating, drinking, and smoking; and (4) wearing personal protective items such as gloves, masks, safety glasses, laboratory coats, boots, and headgear. Some circumstances may also require additional protective equipment such as biological safety cabinets for inoculations and specially engineered facilities to control materials entering and leaving the laboratory in the air and on personnel. Table A.2 summarizes the primary biosafety levels and agents of disease as characterized by the Centers for Disease Control.

Table A.1 Levels of Isolation Used in Clinical Settings

Type of Isolation*	Protective Measures**	To Prevent Spread Of
Enteric precautions	Gowns and gloves must be worn by all persons having direct contact with patient; masks not required; special precautions taken for disposing of feces and urine	Diarrheal diseases; *Shigella, Salmonella,* and *E. coli* gastroenteritis, cholera, hepatitis A, rotavirus, and giardiasis
Respiratory precautions	Private room with closed door is necessary; gowns and gloves not required; masks usually indicated; items contaminated with secretions must be disinfected	Tuberculosis, measles, mumps, meningitis, pertussis, rubella, chickenpox
Wound and skin precautions	Gowns and gloves required for all persons; masks not needed; contaminated instruments and dressings require special precautions	Staphylococcal and streptococcal infections; gas gangrene, herpes zoster, and dermatomycoses
Strict isolation	Private room with closed door required; gowns, masks, and gloves must be worn by all persons; contaminated items must be wrapped and sent to central supply for decontamination	Mostly highly virulent or contagious microbes; includes diphtheria, anthrax, some types of pneumonia, extensive skin and burn infections, disseminated herpes simplex and zoster
Protective isolation (also called reverse isolation)	Same guidelines as for strict isolation; room may be ventilated by unidirectional or laminar airflow filtered through a high-efficiency particulate (HEPA) filter that removes most airborne pathogens; infected persons must be barred	Used to protect patients extremely immunocompromised by cancer, therapy, surgery, genetic defects, burns, prematurity, or AIDS, and therefore vulnerable to opportunistic pathogens

*Precautions are based upon the primary portal of entry and communicability of the pathogen.

**In all cases, visitors to the patient's room must report to the nurses' station before entering the room; all visitors and personnel must wash their hands upon entering and leaving the room.

Table A.2 Primary Biosafety Levels and Agents of Disease

Biosafety Level	Facilities and Practices	Risk of Infection and Class of Pathogens
1	Standard, open bench, no special facilities needed; typical of most microbiology teaching labs; access may be restricted	Low infection hazard; microbes not generally considered pathogens and will not colonize the bodies of healthy persons: *Micrococcus luteus, Bacillus megaterium, Lactobacillus, Saccharomyces*
2	At least Level 1 facilities and practices, plus personnel must be trained in handling pathogens; lab coats and gloves required; safety cabinets may be needed; biohazard signs posted; access restricted	Agents with moderate potential to infect; class 2 pathogens may cause disease in healthy persons but may be contained with proper facilities; most pathogens belong to class 2; includes *Staphylococcus aureus, Escherichia coli, Salmonella* spp., *Corynebacterium diphtheriae;* pathogenic helminths; hepatitis A, B, and rabies viruses; *Cryptococcus* and *Blastomyces*
3	Minimum of Level 2 facilities and practices; all manipulation performed in safety cabinets; lab designed with special containment features; only personnel with special clothing may enter; no unsterilized materials may leave the lab; personnel warned, monitored, and vaccinated against infection dangers	Agents may cause severe or lethal disease especially when inhaled; class 3 microbes include *Mycobacterium tuberculosis, Francisella tularensis, Yersinia pestis, Brucella* spp., *Coxiella burnetii, Coccidioides immitis,* and yellow fever, VEE, and AIDS viruses
4	Minimum of Level 3 facilities and practices; facilities must be isolated with very controlled access; clothing changes and showers required for all persons entering and leaving; materials must be autoclaved or fumigated prior to entering and leaving lab	Agents being handled are highly virulent microbes that pose extreme risk for morbidity and mortality when inhaled in droplet or aerosol form; most are exotic flaviviruses or arenaviruses, including Lassa fever, Ebola, and Marburg viruses

Appendix E: Answers to True–False Questions

Chapter 1

1. F
2. T
3. F
4. F
5. F
6. F
7. F
8. T
9. F
10. F
11. F
12. F
13. F
14. F
15. T

Chapter 2

1. T
2. F
3. T
4. F
5. T
6. F
7. T
8. F
9. T
10. T
11. F
12. T
13. T
14. T
15. T
16. T
17. F
18. T
19. T
20. T

Chapter 3

1. T
2. F
3. F
4. T
5. F
6. F
7. T
8. F
9. T
10. F
11. F
12. F
13. T

Chapter 4

1. T
2. T
3. T
4. F
5. T
6. T
7. F
8. T
9. F
10. T
11. T
12. F
13. T
14. T
15. T

Chapter 5

1. F
2. T
3. T
4. F
5. T
6. T
7. T
8. F
9. F
10. F
11. T
12. T
13. F

Chapter 6

1. T
2. T
3. F
4. F
5. T
6. T
7. F
8. T
9. T
10. T
11. F
12. T
13. T
14. T
15. F
16. T

Chapter 7

1. T
2. F
3. F
4. F
5. T
6. T
7. F
8. T
9. F
10. T
11. T
12. T
13. T
14. T
15. T
16. F
17. T
18. T
19. F
20. F

Chapter 8

1. F
2. T
3. F
4. T
5. T
6. F
7. F
8. T
9. F
10. T
11. F
12. T
13. T

Chapter 9

1. F
2. T
3. T
4. F
5. T
6. T
7. T
8. F
9. T
10. T
11. F
12. F
13. T
14. T

Chapter 10

1. T
2. F
3. F
4. T
5. T
6. T
7. T
8. F
9. T
10. F

Chapter 11

1. F
2. F
3. T
4. F
5. F
6. T
7. T
8. F
9. T
10. F
11. T
12. T
13. F

Chapter 12

1. F
2. T
3. T
4. F
5. F
6. T
7. F
8. T
9. T
10. F

Chapter 13

1. T
2. T
3. F
4. T
5. T
6. F
7. T
8. T
9. F
10. T
11. T
12. F
13. T
14. F
15. T
16. T

Chapter 14

1. T
2. F
3. T
4. T
5. F
6. T
7. T
8. T
9. T
10. T
11. F
12. F
13. T
14. F

Chapter 15

1. F
2. T
3. F
4. F
5. F
6. F
7. T
8. F
9. F
10. T

Chapter 16

1. T
2. T
3. F
4. T
5. F
6. F
7. T
8. F
9. T
10. T
11. T
12. F
13. T
14. T
15. F

Chapter 17

1. F
2. F
3. T
4. F
5. T
6. T
7. F
8. F
9. T
10. F
11. T
12. T

Chapter 18

1. T
2. T
3. F
4. T
5. T
6. T
7. T
8. F
9. F
10. T
11. F
12. F
13. T
14. T

Chapter 19

1. T
2. F
3. F
4. T
5. T
6. T
7. F
8. T
9. T
10. T
11. F
12. T
13. T
14. T
15. F
16. T

Chapter 20

1. T
2. F
3. F
4. F
5. F
6. T
7. T
8. F
9. T
10. T
11. F
12. T

Chapter 21

1. T
2. F
3. F
4. T
5. T
6. T
7. T
8. T
9. T
10. F
11. T
12. T
13. F

Chapter 22

1. F
2. T
3. T
4. F
5. T
6. T
7. T
8. F
9. T
10. F
11. T
12. F
13. T
14. T
15. F
16. T
17. T
18. T

Glossary

A

abyssal zone The deepest region of the ocean, a sunless, high-pressure, cold, anaerobic habitat.

acid-fast A term referring to the property of mycobacteria to retain carbol fuchsin even in the presence of acid alcohol. The staining procedure is used to diagnose tuberculosis.

active immunity Immunity acquired through direct stimulation of the immune system by antigen.

active site The specific region on an apoenzyme that binds substrate. The site for reaction catalysis.

acute Characterized by rapid onset and short duration.

acyclovir A synthetic purine analog that blocks DNA synthesis in certain viruses, particularly the herpes simplex viruses.

adjuvant In immunology, a chemical vehicle that enhances antigenicity, presumably by prolonging antigen retention at the injection site.

aerobe A microorganism that lives and grows in the presence of free gaseous oxygen (O_2).

aflatoxin From *Aspergillus flavus* toxin, a mycotoxin that typically poisons moldy animal feed and may cause liver cancer in humans and other animals.

agammaglobulinemia Also called hypogammaglobulinemia. The absence of or severely reduced levels of antibodies in serum.

agglutination The aggregation by antibodies of suspended cells or like-sized particles (agglutinogens) into clumps that settle.

agglutinin A specific antibody that cross-links agglutinogen, causing it to aggregate.

agglutinogen An antigenic substance on cell surfaces that evokes agglutinin formation against it.

AIDS Acquired immune deficiency syndrome. The complex of signs and symptoms characteristic of the late phase of human immunodeficiency virus (HIV) infection.

allele A gene that occupies the same location as other alternative (allelic) genes on paired chromosomes.

allergen A substance that provokes an allergic response.

allergy The altered, usually exaggerated, immune response to allergen. Also called hypersensitivity.

alloantigen An antigen that is present in some but not all members of the same species. Also called isoantigen.

allograft Relatively compatible tissue exchange between nonidentical members of the same species. Also called homograft.

allosteric Pertaining to the altered activity of an enzyme due to the binding of a molecule to a region other than the enzyme's active site.

amastigote The rounded or ovoid nonflagellated form of the *Leishmania* parasite.

Ames test A method for detecting mutagenic and potentially carcinogenic agents based upon the genetic alteration of nutritionally defective bacteria.

amination The addition of an amine ($—NH_2$) group to a molecule.

aminoglycoside A complex group of drugs derived from soil actinomycetes that impairs ribosome function and has antibiotic potential. Example: streptomycin.

amphibolism Pertaining to the metabolic pathways that serve multiple functions in the breakdown, synthesis, and conversion of metabolites.

amphipathic Relating to a compound that has contrasting characteristics, such as hydrophilic-hydrophobic or acid-base.

amphitrichous Having a single flagellum or a tuft of flagella at opposite poles of a microbial cell.

anabolism The energy-consuming process of incorporating nutrients into protoplasm through biosynthesis.

anaerobe A microorganism that grows best, or exclusively, in the absence of oxygen.

analog In chemistry, a compound that closely resembles another in structure.

anamnestic In immunology, an augmented response or memory related to a prior stimulation of the immune system by antigen. It boosts the levels of immune substances.

anaphylaxis The unusual or exaggerated allergic reaction to antigen that leads to severe respiratory and cardiac complications.

anion A negatively charged ion.

antibiotic A chemical substance from one microorganism that can inhibit or kill another microbe even in minute amounts.

antibody A large protein molecule evoked in response to an antigen that interacts specifically with that antigen.

anticodon The trinucleotide sequence of transfer RNA that is complementary to the trinucleotide sequence of messenger RNA (the codon).

antigen Any cell, particle, or chemical that induces a specific immune response by B cells or T cells and may stimulate resistance to an infection or a toxin. *See* immunogen.

antigenic determinant The precise molecular group of an antigen that defines its specificity and triggers the immune response.

antigenic drift Minor antigenic changes in the influenza A virus due to mutations in the spikes' genes.

antihistamine A drug that counters the action of histamine and is useful in allergy treatment.

anti-idiotype An anti-antibody that reacts specifically with the idiotype (variable region or antigen-binding site) of another antibody.

antimetabolite A substance such as a drug that competes with, substitutes for, or interferes with a normal metabolite.

antiseptic A growth-inhibiting agent used on tissues to prevent infection.

antiserum Antibody-rich serum derived from the blood of animals (deliberately immunized against infectious or toxic antigen) or from persons who have recovered from specific infections.

antitoxin Globulin fraction of serum that neutralizes a specific toxin. Also refers to the specific antitoxin antibody itself.

apoenzyme The protein part of an enzyme, as opposed to the nonprotein or inorganic cofactors.

aquifer A subterranean water-bearing stratum of permeable rock, sand, or gravel.

arthrospore A fungal spore formed by the septation and fragmentation of hyphae.

Arthus reaction An immune complex phenomenon that develops after repeat injection. This localized inflammation results from aggregates of antigen and antibody that bind complement and attract neutrophils.

ascospore A spore formed within a sac-like cell (ascus) of Ascomycetes following nuclear fusion and meiosis.

asepsis A condition free of viable pathogenic microorganisms.

atopy Allergic reaction classified as type I, with a strong familial relationship; caused by allergens such as pollen, insect venom, food, and dander; involves IgE antibody; includes symptoms of hay fever, asthma, and skin rash.

attenuate To reduce the virulence of a pathogenic bacterium or virus by passing it through a non-native host or by long-term subculture.

autoantibody An "anti-self" antibody having an affinity for tissue antigens of the subject in which it is formed.

autoantigen Molecules that are inherently part of self but are perceived by the immune system as foreign.

autograft Tissue or organ surgically transplanted to another site on the same subject.

autoimmune disease The pathologic condition arising from the production of antibodies against autoantigens. Example: rheumatoid arthritis. Also called autoimmunity.

autosome A chromosome of somatic cells as opposed to a sex chromosome of gametes.

autotroph A microorganism that requires only inorganic nutrients and whose sole source of carbon is carbon dioxide.

axenic A sterile state such as a pure culture. An axenic animal is born and raised in a germ-free environment. *See* gnotobiotic.

B

bacteremia The presence of viable bacteria in circulating blood.

bactericide An agent that kills bacteria.

bacteriocin Proteins produced by certain bacteria that are lethal against closely related bacteria and are narrow spectrum compared to antibiotics; these proteins are coded and transferred in plasmids.

bacteriophage A virus that specifically infects bacteria.

bacteriostatic Any process or agent that inhibits bacterial growth.

barophile A microorganism that thrives under high (usually hydrostatic) pressure.

basement membrane A thin layer (1–6 μm) of protein and polysaccharide found at the base of epithelial tissues.

basidiospore A sexual spore that arises from a basidium. Found in basidiomycete fungi.

basidium A reproductive cell created when the swollen terminal cell of a hypha develops filaments (sterigmata) that form spores.

basophil A motile polymorphonuclear leukocyte that binds IgE. The basophilic cytoplasmic granules contain mediators of anaphylaxis and atopy.

bdellovibrio A bacterium that preys on certain other bacteria. It bores a hole into a specific host and inserts itself between the protoplast and the cell wall. There it elongates before subdividing into several cells and devours the host cell.

benthic zone The sedimentary bottom region of a pond, lake, or ocean.

beta-lactamase An enzyme secreted by certain bacteria that cleaves the beta-lactam ring of penicillin and cephalosporin, and thus provides for resistance against that antibiotic. *See* penicillinase.

beta oxidation The degradation of long-chain fatty acids. Two-carbon fragments are formed as a result of enzymatic attack directed against the second or beta carbon of the hydrocarbon chain. Aided by coenzyme A, the fragments enter the Krebs cycle and are processed for ATP synthesis.

binary fission The formation of two new cells of approximately equal size as the result of parent cell division.

bioremediation The use of microbes to reduce or degrade pollutants, industrial wastes, and household garbage.

biosphere Habitable regions comprised of the aquatic (hydrospheric), soil-rock (lithospheric), and air (atmospheric) environments.

blast cell An immature precursor cell of B and T lymphocytes. Also called a lymphoblast.

blocking antibody The IgG class of immunoglobins that competes with IgE antibody for allergens, thus blocking the degranulation of basophils and mast cells.

B lymphocyte One of two main types of lymphocytes, also called B cells. Human B cells (from bursa) mature in the bone marrow. Following antigen stimulation, B cells respond by antibody secretion.

botulin *Clostridium botulinum* toxin. Ingestion of this potent exotoxin leads to flaccid paralysis.

bradykinin An active polypeptide that is a potent vasodilator released from IgE-coated mast cells during anaphylaxis.

broad spectrum A word to denote drugs that affect many different types of bacteria, both gram positive and gram negative.

Brownian movement The passive, erratic, nondirectional motion exhibited by microscopic particles. The jostling comes from being randomly bumped by submicroscopic particles, usually water molecules, in which the visible particles are suspended.

bubo The swelling of one or more lymph nodes due to inflammation.

bulla A large, bubblelike vesicle in a region of separation between the epidermis and the subepidermal layer. The space is usually filled with serum and sometimes with blood.

C

Calvin cycle The recurrent photosynthetic pathway characterized by CO_2 fixation and glucose synthesis. Also called the dark reactions.

cancer Any malignant neoplasm that invades surrounding tissue and may metastasize to other locations. A carcinoma is derived from epithelial tissue, and a sarcoma arises from proliferating mesodermal cells of connective tissue.

capsid The protein covering of a virus's nucleic acid core. Capsids exhibit symmetry due to the regular arrangement of subunits called capsomers. *See* icosahedron.

capsomer A subunit of the virus capsid shaped as a triangle or disc.

capsule In bacteria, the loose, gel-like covering or slime comprised chiefly of simple polysaccharides. This layer is protective, and may be associated with virulence.

carbuncle A deep staphylococcal abscess joining several neighboring hair follicles.

carrier A person who harbors infections and inconspicuously spreads them to others. Also, a chemical agent that can accept an atom, chemical radical, or subatomic particle from one compound and pass it on to another.

caseous lesion Necrotic area of lung tubercle superficially resembling cheese. Typical of tuberculosis.

catabolism The chemical breakdown of complex compounds into simpler units to be used in cell metabolism.

catalyst A substance that alters the rate of a reaction without being consumed or permanently changed by the reaction. In cells, enzymes are catalysts.

cation A positively charged ion.

cavitation The formation of a hollow space such as in the lung in tuberculosis.

cecum The intestinal pocket that forms the first segment of the large intestine. Also called the appendix.

cell-mediated The type of immune responses brought about by T cells, such as cytotoxic, suppressor, and helper effects.

cephalosporins A group of broad-spectrum antibiotics isolated from the fungus *Cephalosporium*.

cercaria The free-swimming larva of the schistosome trematode that emerges from the snail host and may penetrate human skin, causing schistosomiasis.

cestode The common name for tapeworms that parasitize man and domestic animals.

chancre The primary sore of syphilis that forms at the site of penetration by *Treponema pallidum*. It begins as a hard, dull red, painless papule that erodes from the center.

chancroid A lesion that resembles a chancre but is soft and is caused by *Haemophilus ducreyi*.

chemoautotroph An organism that relies upon inorganic chemicals for its energy and carbon dioxide for its carbon. Also called a chemolithotroph.

chemostat A growth chamber with an outflow that is equal to the continuous inflow of nutrient media. This steady-state growth device is used to study such events as cell division, mutation rates, and enzyme regulation.

chemotaxis The tendency of organisms to move in response to a chemical gradient (toward an attractant or to avoid adverse stimuli).

chemotherapy The use of chemical substances or drugs to treat or prevent disease.

chitin A polysaccharide similar to cellulose in chemical structure. This polymer makes up the horny substance of the exoskeletons of arthropods and certain fungi.

chloroplast An organelle containing chlorophyll that is found in photosynthetic eucaryotes.

chromatic aberration Deviant focus or magnification due to refraction of the colored wavelengths that make up white light.

chromatin The genetic material of the nucleus. Chromatin is made up of nucleic acid and stains readily with certain dyes.

chromatin body The bacterial chromosome or nucleoid.

chromosome The tightly coiled bodies in cells that are the primary sites of genes.

chromophore The chemical radical of a dye that is responsible for its color and reactivity.

chronic Any process or disease that persists over a long duration.

clonal selection theory A conceptual explanation for the development of lymphocyte specificity and variety during immune maturation.

clone A colony of cells (or group of organisms) derived from a single cell (or single organism) by asexual reproduction. Each unit shares identical characteristics. Also used as a verb to refer to the process of producing a genetically identical population of cells or genes.

coagulase A plasma-clotting enzyme secreted by *Staphylococcus aureus*. It contributes to virulence and is involved in forming a fibrin wall that surrounds staphylococcal lesions.

coccobacillus An elongated coccus; a short, thick, oval-shaped bacterial rod.

coccus A spherical-shaped bacterial cell.

codon A specific sequence of three nucleotides in mRNA (or the sense strand of DNA) that constitutes the genetic code for a particular amino acid.

coenzyme A complex organic molecule, several of which are derived from vitamins (e.g., nicotinamide, riboflavin). A coenzyme operates in conjunction with an enzyme. Coenzymes serve as transient carriers of specific atoms or functional groups during metabolic reactions.

cofactor An enzyme accessory. It may be organic, such as coenzymes, or inorganic, such as Fe^{2+}, Mn^{2+}, or Zn^{2+} ions.

coliform A collective term that includes normal enteric bacteria that are gram negative and lactose fermenting.

colinear Having corresponding parts of a molecule arranged in the same linear order as another molecule, as in DNA and mRNA.

colony A macroscopic cluster of cells appearing on a solid medium, each arising from the multiplication of a single cell.

colostrum The clear yellow early product of breast milk that is very high in secretory antibodies. Provides passive intestinal protection.

commensalism An unequal relationship in which one species derives benefit without harming the other.

communicable infection Capable of being transmitted from one individual to another.

community The interacting mixture of populations in a given habitat.

complement In immunology, serum protein components that act in a definite sequence when set in motion either by an antigen-antibody complex or by factors of the alternative (properdin) pathway.

conidia Asexual fungal spores shed as free units from the tips of fertile hyphae.

conjugation In bacteria, the contact between donor and recipient cells associated with the transfer of genetic material such as plasmids. May involve special (sex) pili. Also a form of sexual recombination in ciliated protozoans.

constitutive enzyme An enzyme present in bacterial cells in constant amounts, regardless of the presence of substrate. Enzymes of the central catabolic pathways are typical examples.

contagious Communicable; transmissable by direct contact with infected persons and their fresh secretions or excretions.

contaminant An impurity; any undesirable material or organism.

convalescence Recovery; the period between the end of a disease and the complete restoration of health in a patient.

corepressor A molecule that combines with inactive repressor to form active repressor, which attaches to the operator gene site and inhibits the activity of structural genes subordinate to the operator.

crista The infolded inner membrane of a mitochondrion that is the site of the respiratory chain and oxidative phosphorylation.

culture The visible accumulation of microorganisms in or on a nutrient medium. Also, the propagation of microorganisms with various media.

curd The coagulated milk protein used in cheese making.

cyst The resistant, dormant, but infectious form of protozoans. May be important in spread of infectious agents like *Entamoeba histolytica* and *Giardia lamblia*.

cysticercus The larval form of certain *Taenia* species, which typically infest muscles of mammalian intermediate hosts. Also called bladderworm.

cystine An amino acid, $HOOC—CH(NH_2)—CH_2—S—S—CH_2—CH(NH_2)COOH$. An oxidation product of two cysteine molecules in which the —SH (sulfhydryl) groups form a disulfide union. Also called dicysteine.

cytochrome A group of heme protein compounds whose chief role is in electron and/or hydrogen transport occurring in the last phase of aerobic respiration.

cytopathic effect The degenerative changes in cells associated with virus infection. Examples: the formation of multinucleate giant cells or Negri bodies—the prominent cytoplasmic inclusions of nerve cells infected by rabies virus.

cytotoxic Having the capacity to destroy specific cells. One class of T cells attacks cancer cells, virus-infected cells, and eucaryotic pathogens.

D

debridement Trimming away devitalized tissue and foreign matter from a wound.

decontamination The removal or neutralization of an infectious, poisonous, or injurious agent from a site.

definitive host The organism in which a parasite develops into its adult or sexually mature stage. Also called the final host.

degerm The process of physically removing surface oils, debris, and soil from skin to reduce the microbial load.

degranulation The release of cytoplasmic granules, as when chemical mediators are secreted from mast cell granules.

denaturation The loss of normal characteristics resulting from some molecular alteration. Usually in reference to the action of heat or chemicals on proteins whose function depends upon an unaltered tertiary structure.

desquamate To shed the cuticle in scales; to peel off the outer layer of a surface.

diapedesis The migration of intact blood cells between endothelial cells of a blood vessel such as a venule.

differential medium A single substrate that discriminates between groups of microorganisms on the basis of their appearance due to alternate chemical reactions.

differential stain A technique that utilizes two dyes to distinguish between different microbial groups or cell parts by color reaction.

diffusion The dispersal of molecules, ions, or microscopic particles propelled down a concentration gradient by spontaneous random motion to achieve a uniform distribution.

dimorphic In mycology, the tendency of some pathogens to alter their growth form from mold to yeast in response to rising temperature.

diplococcus Spherical or oval-shaped bacteria, typically found in pairs.

diploid Somatic cells having twice the basic chromosome number. One set in the pair is derived from the father, and the other from the mother.

disaccharide A sugar containing two monosaccharides. Example: sucrose (fructose + glucose).

disinfection The destruction of pathogenic nonsporulating microbes or their toxins, usually on inanimate surfaces.

droplet nuclei The dried residue of fine droplets produced by mucus and saliva sprayed while sneezing and coughing. Droplet nuclei are less than 5μm in diameter (large enough to bear a single bacterium and small enough to remain airborne for a long time), and can be carried by air currents. Droplet nuclei are drawn deep into the air passages.

E

ectoplasm The outer, more viscous region of the cytoplasm of a phagocytic cell such as an ameba. It contains microtubules, but not granules or organelles.

eczema An acute or chronic allergy of the skin associated with itching and burning sensations. Typically, red, edematous, vesicular lesions erupt, leaving the skin scaly and sometimes hyperpigmented.

edema The accumulation of excess fluid in cells, tissues, or serous cavities.

electrolyte Any compound that ionizes in solution and conducts current in an electric field.

electromagnetic radiation A form of energy that is emitted as waves and is propagated through space and matter. The spectrum extends from short gamma rays to long radio waves.

electron A negatively charged subatomic particle that is distributed around the nucleus in an atom.

electrostatic Relating to the attraction of opposite charges and the repulsion of like charges. Electrical charge remains stationary as opposed to electrical flow or current.

element A substance comprised of only one kind of atom that cannot be degraded into two or more substances without losing its chemical characteristics.

ELISA Abbreviation for enzyme-linked immunosorbent assay, a very sensitive serological test used to detect antibodies in diseases such as AIDS.

encystment The process of becoming encapsulated by a membranous sac.

endemic disease A native disease that prevails continuously in a geographic region.

endergonic reaction A chemical reaction that occurs with the absorption and storage of surrounding energy.

endocarditis An inflammation of the lining and valves of the heart. Often caused by infection with pyogenic cocci.

endocytosis The process whereby solid and liquid materials are taken into the cell through membrane invagination and engulfment into a vesicle.

endoenzyme An intracellular enzyme, as opposed to enzymes that are secreted.

endogenous Originating or produced within an organism or one of its parts.

endoplasmic reticulum The intracellular network of flattened sacs or tubules with or without ribosomes on their surfaces.

endospore A small, dormant, resistant derivative of a bacterial cell that germinates under favorable growth conditions into a vegetative cell. The bacterial genera *Bacillus* and *Clostridium* are typical sporeformers.

endotoxin A bacterial intracellular toxin that is not ordinarily released (as is exotoxin). Endotoxin is comprised of a phospholipid-polysaccharide complex that is an integral part of gram-negative bacterial cell walls. Endotoxins may cause severe shock and fever.

energy of activation The minimum energy input necessary for reactants to form products in a chemical reaction.

enriched medium A nutrient medium supplemented with blood, serum, or some growth factor to promote the multiplication of fastidious microorganisms.

enteric Pertaining to the intestine.

enteropathogenic Pathogenic to the alimentary canal.

enterotoxin A bacterial toxin that specifically targets intestinal mucous membrane cells. Enterotoxigenic strains of *Escherichia coli* and *Staphylococcus aureus* are typical sources.

enveloped virus A virus whose nucleocapsid is enclosed by a membrane derived in part from the host cell. It usually contains exposed glycoprotein spikes specific for the virus.

enzyme A protein biocatalyst that facilitates metabolic reactions.

enzyme repression The inhibition of enzyme synthesis by the end product of a catabolic pathway.

eosinophil A leukocyte whose cytoplasmic granules readily stain with red eosin dye.

epidemic A sudden and simultaneous outbreak or increase in the number of cases of disease in a community.

epidemiology The study of the factors affecting the prevalence and spread of disease within a community.

epimastigote The trypanosomal form found in the tsetse fly or triatomid bug vector. Its flagellum originates near the nucleus, extends along an undulating membrane, and emerges from the anterior end.

erysipelas An acute, sharply defined inflammatory disease specifically caused by hemolytic *Streptococcus*. The eruption is limited to the skin, but may be complicated by serious systemic symptoms.

erysipeloid An inflammation resembling erysipelas but caused by *Erysipelothrix*, a gram-positive rod. The self-limited cellulitis that appears at the site of an infected wound, usually the hand, comes from handling contaminated fish or meat.

erythema An inflammatory redness of the skin.

erythroblastosis fetalis Hemolytic anemia of the newborn. The anemia comes from hemolysis of Rh-positive fetal erythrocytes by anti-Rh maternal antibodies. Erythroblasts are immature red blood cells prematurely released from the bone marrow.

erythrogenic toxin An exotoxin produced by lysogenized group A strains of β-hemolytic streptococci that is responsible for the severe fever and rash of scarlet fever in the nonimmune individual. Also called a pyrogenic toxin.

eschar A dark, sloughing scab that is the lesion of anthrax and certain rickettsioses.

essential nutrient Any ingredient like a certain amino acid, fatty acid, vitamin, or mineral that cannot be formed by an organism and must be supplied in the diet. *See* growth factor.

ester bond A covalent bond formed by reacting carboxylic acid with an OH group:

$$\left(R - \overset{\overset{\displaystyle O}{\|}}{C} - O - R'\right)$$

Olive and corn oils, lard, and butterfat are examples of triacylglycerols—esters formed between glycerol and three fatty acids.

estuary The intertidal zone where a river empties into the sea.

etiological agent The microbial cause of disease; the pathogen.

ethylene oxide A potent, highly water-soluble gas invaluable for gaseous sterilization of heat-sensitive objects such as plastics, surgical and diagnostic appliances, and spices. Potential hazards are related to its carcinogenic, mutagenic, residual, and explosive nature. Ethylene oxide is rendered nonexplosive by mixing with 90% CO_2 or fluorocarbon.

eucaryotic cell A cell that differs from a procaryotic (bacterial) cell chiefly by having a nuclear membrane (a well-defined nucleus), membrane-bound subcellular organelles, and mitotic cell division.

eutrophication The process whereby dissolved nutrients resulting from natural seasonal enrichment or man-made pollution of water cause overgrowth of algae and cyanobacteria to the detriment of fish and other larger aquatic inhabitants.

exergonic A chemical reaction associated with the release of energy to the surroundings. Antonym: endergonic.

exfoliative toxin A poisonous substance that causes superficial cells of an epithelium to detach and be shed. Example: staphylococcal exfoliatin. Also called an epidermolytic toxin.

exoenzyme An extracellular enzyme chiefly for hydrolysis of nutrient macromolecules that are otherwise impervious to the cell membrane. It functions in saprobic decomposition of organic debris and may be a factor in invasiveness of pathogens.

exon A stretch of eucaryotic DNA coding for a corresponding portion of mRNA that is translated into peptides. Intervening stretches of DNA that are not expressed are called introns. During transcription, exons are separated from introns and spliced together into a continuous mRNA transcript.

exotoxin A toxin (usually protein) that is secreted and acts upon a specific cellular target. Examples: botulin, tetanospasmin, diphtheria toxin, and erythrogenic toxin.

exponential Pertaining to the use of exponents, numbers that are typically written as a superscript to indicate how many times a factor is to be multiplied. Exponents are used in scientific notation to render large cumbersome numbers into small workable quantities.

F

facultative Pertaining to the capacity of microbes to adapt or adjust to variations; not obligatory. Example: A facultative anaerobe can grow without oxygen.

fastidious Requiring special nutritional or environmental conditions for growth. Said of bacteria.

fermentation The extraction of energy through anaerobic degradation of substrates into simpler, reduced metabolites. In large industrial processes, fermentation can mean any use of microbial metabolism to manufacture organic chemicals or other products.

fixation In microscopic slide preparation of tissue sections or bacterial smears, fixation pertains to rapid killing, hardening, and adhesion to the slide, all while retaining as many natural characteristics as possible. Also refers to the assimilation of inorganic molecules into organic ones, as in carbon or nitrogen fixation.

fluid mosaic model A conceptualization of the molecular architecture of cellular membranes. The model suggests a bilipid layer containing proteins. Membrane proteins are embedded to some degree in this bilayer, where they float freely about.

fluorescence The property possessed by certain minerals and dyes to emit visible light when excited by ultraviolet radiation. A fluorescent dye combined with specific antibody provides a sensitive test for the presence of antigen.

folliculitis An inflammatory reaction involving the formation of papules or pustules in clusters of hair follicles.

fomite Virtually any inanimate object an infected individual has contact with that can serve as a vehicle for the spread of disease.

food pyramid A triangular summation of the consumers, producers, and decomposers in a community and how they relate by trophic, number, and energy parameters.

formalin A 37% aqueous solution of formaldehyde gas; a potent chemical fixative and microbicide.

functional group In chemistry, a particular molecular combination that reacts in predictable ways and confers particular properties on a compound. Examples: —COOH, —OH, —CHO.

furuncle A boil; a localized pyogenic infection arising from a hair follicle.

G

Gaia Hypothesis The concept that biotic and abiotic factors sustain suitable conditions for one another simply by their interactions. Named after the mythical Greek goddess of earth.

GALT Abbreviation for **g**ut-**a**ssociated **l**ymphoid **t**issue. Includes Peyer's patches.

gamma globulin The fraction of plasma proteins high in immunoglobulins (antibodies). Preparations from pooled human plasma containing normal antibodies make useful passive immunizing agents against pertussis, polio, measles, and several other diseases.

gene probe Short strands of single-stranded nucleic acid that hybridize specifically with complementary stretches of nucleotides on test samples and thereby serve as a tagging and identification device.

genetic engineering A field involving deliberate alterations (recombinations) of the genomes of microbes, plants, and animals through special technologic processes.

genome The complete set of chromosomes and genes in an organism.

genotype The genetic makeup of an organism. The genotype is ultimately responsible for an organism's phenotype, or expressed characteristics.

germicide An agent lethal to non-endospore-forming pathogens.

gingivitis Inflammation of the gum tissue surrounding the necks of the teeth.

gluconeogenesis The formation of glucose (or glycogen) from noncarbohydrate sources such as protein or fat. Also called glyconeogenesis.

glycan A polysaccharide.

glycocalyx A filamentous network of carbohydrate-rich molecules that coats cells.

glycolysis The energy-yielding breakdown (fermentation) of glucose to pyruvic or lactic acid. It is often called anaerobic glycolysis because no molecular oxygen is consumed in the degradation.

gnotobiotic Referring to experiments performed on germ-free animals.

Gram stain A differential stain for bacteria useful in identification and taxonomy. Gram-positive organisms appear purple from crystal violet-mordant retention, while gram-negative organisms appear red after loss of crystal violet and absorbance of the safranin counterstain.

grana Discrete stacks of chlorophyll-containing thylakoids within chloroplasts.

granulocyte A mature leukocyte that contains noticeable granules in a Wright stain. Examples: neutrophils, eosinophils, and basophils.

granuloma A solid mass or nodule of inflammatory tissue containing modified macrophages and lymphocytes. Usually a chronic pathologic process of diseases such as tuberculosis or syphilis.

greenhouse effect The capacity to retain solar energy by a blanket of atmospheric gases that redirects heat waves back toward the earth.

growth factor Any organic or inorganic substance, such as a mineral, vitamin, or hormone, that must be provided in the diet to facilitate growth.

gumma A nodular, infectious granuloma characteristic of tertiary syphilis.

H

habitat The environment to which an organism is adapted.

halophile A microbe whose growth is either stimulated by salt or requires a high concentration of salt for growth.

H antigen The flagellar antigen of motile bacteria. H comes from the German word *hauch* that denotes the appearance of spreading growth on solid medium.

haploid Having a single set of unpaired chromosomes, such as occurs in gametes and certain microbes.

hapten An incomplete or partial antigen. Although it constitutes the determinative group and can bind antigen, hapten cannot stimulate a full immune response without being carried by a larger protein molecule.

hay fever A form of atopic allergy marked by seasonal acute inflammation of the conjunctiva and mucous membranes of the respiratory passages. Symptoms are irritative itching and rhinitis.

helical Having a spiral or coiled shape. Said of certain virus capsids and bacteria.

helminth A term that designates all parasitic worms.

helper T cell A class of thymus-stimulated lymphocytes that facilitate various immune activities such as assisting B cells and macrophages. Also called a T-helper cell.

hemagglutinin A molecule that causes red blood cells to clump or agglutinate. Often found on the surfaces of viruses.

hemolysin Any biological agent that is capable of destroying red blood cells and causing the release of hemoglobin. Many bacterial pathogens produce exotoxins that act as hemolysins.

hemopoiesis The process by which the various types of blood cells are formed, such as in the bone marrow.

hepatocyte A liver cell.

herd immunity The status of collective acquired immunity in a population that reduces the likelihood that nonimmune individuals will contract and spread infection. One aim of vaccination is to induce herd immunity.

herpes zoster A recurrent infection caused by latent chickenpox virus. Its manifestation on the skin tends to correspond to dermatomes and to occur in patches that "girdle" the trunk. Also called shingles.

herpetic keratitis Corneal or conjunctival inflammation due to herpesvirus type 1.

heterophile antigen An antigen present in a variety of phylogenetically unrelated species. Example: red blood cell antigens and the glycocalyx of bacteria.

heterotroph An organism that relies upon organic compounds for its carbon and energy needs.

histamine A chemical mediator released when mast cells and basophils release their granules. An important mediator of allergy, its effects include smooth muscle contraction, increased vascular permeability, and increased mucous secretion.

histiocyte Another term for macrophage.

histone Proteins associated with eucaryotic DNA. These simple proteins serve as winding spools to compact and condense the chromosomes.

histoplasmin The antigenic extract of *Histoplasma capsulatum,* the causative agent of histoplasmosis. The preparation is used in skin tests for diagnosis and for conducting surveys to determine the geographic distribution of the fungus.

HLA An abbreviation for **h**uman **l**eukocyte **a**ntigens. This closely linked cluster of genes programs for cell surface glycoproteins that control immune interactions between cells and is involved in rejection of allografts. Also called the major histocompatibility complex (MHC).

holoenzyme An enzyme complete with its apoenzyme and cofactors.

hops The ripe, dried fruits of the hop vine (*Humulus lupulus*) that is added to beer wort for flavoring.

hybridoma An artificial cell line that produces monoclonal antibodies. It is formed by fusing (hybridizing) a normal antibody-producing cell with a cancer cell, and it can produce pure antibody indefinitely.

hydatid cyst A sac, usually in liver tissue, containing fluid and larval stages of the echinococcus tapeworm.

hydration The addition of water as in the coating of ions with water molecules as ions enter into aqueous solution.

hydrolase An enzyme that catalyzes the cleavage of a bond with the addition of —H and —OH (parts of a water molecule) at the separation site.

hypertonic Having a greater osmotic pressure than a reference solution.

hyphae The tubular threads that make up filamentous fungi (molds). This web of branched and intertwining fibers is called a mycelium.

hypogammaglobulinemia An inborn disease in which the gamma globulin (antibody) fraction of serum is greatly reduced. The condition is associated with a greater susceptibility to pyogenic infections.

hypotonic Having a lower osmotic pressure than a reference solution.

I

icosahedron A regular geometric figure having 20 surfaces that meet to form 12 corners. Some virions have capsids that resemble icosahedral crystals.

imidazole Five-membered heterocyclic compounds typical of histidine, which are used in antifungal therapy.

immunity An acquired resistance to an infectious agent due to prior contact with that agent.

immunogen Any substance that induces a state of sensitivity or resistance after processing by the immune system of the body.

immunoglobulin The chemical class of proteins to which antibodies belong.

IMViC Abbreviation for four identification tests: **i**ndole production, **m**ethyl red test, **V**oges-Proskauer test, (**i** inserted to concoct a wordlike sound), and **c**itrate as a sole source of carbon. This test was originally developed to distinguish between *Enterobacter aerogenes* (associated with soil) and *Escherichia coli* (a fecal coliform).

incidence In epidemiology, the number of new cases of a disease occurring during a period.

inclusion A relatively inert body in the cytoplasm such as storage granules, glycogen, fat, or some other aggregated metabolic product.

inducible enzyme An enzyme that increases in amount in direct proportion to the amount of substrate present.

infection The entry, establishment, and multiplication of pathogenic organisms within a host.

inflammation A natural, nonspecific response to tissue injury that protects the host from further damage. It stimulates immune reactivity and blocks the spread of an infectious agent.

inoculation The implantation of microorganisms into or upon culture media.

interferon Naturally occurring polypeptides produced by fibroblasts and lymphocytes that can block viral replication and regulate a variety of immune reactions.

interleukin A macrophage agent (interleukin-1 or IL-1) that stimulates lymphocyte function. Stimulated T cells release yet another interleukin (IL-2), which amplifies T-cell response by stimulating additional T cells. T-helper cells stimulated by IL-2 stimulate B-cell proliferation and promote antibody production.

intoxication Poisoning that results from the introduction of a toxin into body tissues through ingestion or injection.

intron The segments on split genes of eucaryotes that do not code for polypeptide. They may have regulatory functions.

in utero Literally means "in the uterus"; pertains to events or developments occurring before birth.

in vitro Literally means "in glass," signifying a process or reaction occurring in an artificial environment, as in a test tube or culture medium.

in vivo Literally means "in a living being," signifying a process or reaction occurring in a living thing.

iodophor A combination of iodine and an organic carrier that is a moderate-level disinfectant and antiseptic.

ionization The aqueous dissociation of an electrolyte into ions.

ionizing radiation Radiant energy consisting of short-wave electromagnetic rays (X ray) or high-speed electrons that cause dislodgment of electrons on target molecules and create ions.

irradiation The application of radiant energy for diagnosis, therapy, disinfection, or sterilization.

isograft Transplanted tissue from one monozygotic twin to the other; transplants between highly inbred animals that are genetically identical.

isolation The separation of microbial cells by serial dilution or mechanical dispersion on solid media to achieve a clone or pure culture.

isotonic Two solutions having the same osmotic pressure such that, when separated by a semipermeable membrane, there is no net movement of solvent in either direction.

isotope A version of an element that is virtually identical in all chemical properties to another version except that their atoms have slightly different atomic masses.

J

jaundice The yellowish pigmentation of skin, mucous membranes, sclera, deeper tissues, and excretions due to abnormal deposition of bile pigments. Jaundice is associated with liver infection, as with hepatitis B virus and leptospirosis.

J chain A small molecule with high sulfhydryl content that secures the heavy chains of IgM to form a pentamer and the heavy chains of IgA to form a dimer.

K

Kaposi's sarcoma A malignant or benign neoplasm that appears as multiple hemorrhagic sites on the skin, lymph nodes, and viscera, and apparently involves the metastasis of abnormal blood vessel cells. It is a clinical feature of AIDS.

keratitis Inflammation of the cornea.

keratoconjunctivitis Inflammation of the conjunctiva and cornea.

killer T cells A T lymphocyte programmed to directly affix cells and kill them. *See* cytotoxic.

Koch's postulates A procedure to establish the specific cause of disease. In all cases of infection, the agent must be found; inoculations of a pure culture must reproduce the same disease in animals; the agent must again be present in the experimental animal; and a pure culture must again be obtained.

Koplik's spots Tiny red blisters with central white specks on the mucosal lining of the cheeks. Symptomatic of measles.

kuru An advancing, fatal form of brain degeneration endemic to certain tribes in New Guinea who practice cannibalism and ingest the causative viruslike agent.

L

labile In chemistry, molecules or compounds that are chemically unstable in the presence of environmental changes.

lacrimation The secretion of tears, especially in profusion.

lager The maturation process of beer, which is allowed to take place in large vats at a reduced temperature.

lagging strand The newly forming 5′ DNA strand that is discontinuously replicated in segments (Okazaki fragments).

lag phase The early phase of population growth during which no signs of growth occur.

latency The state of being inactive. Example: a latent virus or latent infection.

leading strand The newly forming 3′ DNA strand that is replicated in a continuous fashion without segments.

leaven To lighten food material by entrapping gas generated within it. Example: the rising of bread from the CO_2 produced by yeast or baking powder.

lesion A wound, injury, or some other pathological change in tissues.

leukotriene An unsaturated fatty acid derivative of arachidonic acid. Leukotriene functions in chemotactic activity, smooth muscle contractility, mucous secretion, and capillary permeability.

leukocidin A heat-labile substance formed by some pyogenic cocci that impairs and sometimes lyses leukocytes.

leukocytosis An abnormally large number of leukocytes in the blood, which can be indicative of acute infection.

leukopenia A lower than normal leukocyte count in the blood that may be indicative of blood infection or disease.

L form L-phase variants; wall-less forms of some bacteria that are induced by drugs or chemicals. These forms may be involved in infections.

limnetic zone The deep-water region beyond the shoreline.

lipase A fat-splitting enzyme. Example: Triacylglycerol lipase separates the fatty acid chains from the glycerol backbone of triglycerides.

lipopolysaccharide A molecular complex of lipid and carbohydrate found in the bacterial cell wall. The lipopolysaccharide (LPS) of gram-negative bacteria is an endotoxin with generalized pathologic effects such as fever.

lithotroph An autotrophic microbe that derives energy from reduced inorganic compounds such as H_2S.

littoral zone The shallow region along a shoreline.

lophotrichous Bacteria having a tuft of flagella at one or both poles.

lumen The cavity within a tubular organ.

lymphadenitis Inflammation of one or more lymph nodes. Also called lymphadenopathy.

lymphatic system A system of vessels and organs that serve as sites for development of immune cells and immune reactions. It includes the spleen, thymus, nodes, and GALT.

lymphokine A soluble substance secreted by sensitized T lymphocytes upon contact with specific antigen. About 50 types exist, and they stimulate inflammatory cells: macrophages, granulocytes and lymphocytes. Examples: migration inhibitory factor, macrophage activating factor, chemotactic factor.

lyophilization Freeze-drying; the separation of a dissolved solid from the solvent by freezing the solution and evacuating the solvent under vacuum. A means of preserving the viability of cultures.

lysin A complement-fixing antibody that destroys specific targeted cells. Examples: hemolysin and bacteriolysin.

lysis The physical rupture or deterioration of a cell.

lysogeny The indefinite persistence of bacteriophage DNA in a host without bringing about the production of virions. A lysogenic cell can revert to a lytic cycle.

lysosome A cytoplasmic organelle containing lysozyme and other hydrolytic enzymes.

lysozyme An enzyme that attacks the bonds on bacterial peptidoglycan. It is a natural defense found in tears and saliva.

M

macrophage A white blood cell derived from a monocyte that leaves the circulation and enters tissues. These cells are important in nonspecific phagocytosis and in regulating, stimulating, and cleaning up after immune responses.

macule A small, discolored spot or patch of skin that is neither raised above nor depressed below the surrounding surface.

malt The grain, usually barley, that is sprouted to obtain digestive enzymes and dried for making beer.

Mantoux test An intradermal screening test for tuberculin hypersensitivity. A red, firm patch of skin at the injection site greater than 10 mm in diameter after 48 hours is a positive result that indicates current or prior exposure to the TB bacillus.

marker Any trait or factor of a cell, virus, or molecule that makes it distinct and recognizable. Example: a genetic marker.

mash In making beer, this malt grain is steeped in warm water, ground up, and fortified with carbohydrates.

mast cell A nonmotile connective tissue cell implanted along capillaries, especially in the lungs, skin, gastrointestinal tract, and genitourinary tracts. Like a basophil, its granules store mediators of allergy.

mastitis Inflammation of the breast or udder.

matrix The dense ground substance between the cristae of a mitochondrion that serves as a site for metabolic reactions.

meiosis The type of cell division necessary for producing gametes in diploid organisms. Two nuclear divisions in rapid succession produce four gametocytes, each containing a haploid number of chromosomes.

memory cell The long-lived progeny of a sensitized lymphocyte that remains in circulation and is genetically programmed to react rapidly with its antigen.

meningitis An inflammation of the membranes (meninges) that surround and protect the brain. It is often caused by bacteria such as the meningococcus and *Hemophilus influenzae*.

merozoite The motile, infective stage of a sporozoan parasite that comes from a liver or red blood cell undergoing multiple fission.

mesosome The irregular invagination of a bacterial cell membrane that is more prominent in gram-positive than in gram-negative bacteria. Although its function is not definitely known, it appears to participate in DNA replication and cell division, and in certain cells, it appears to play a role in secretion.

messenger RNA A single-stranded transcript that is a copy of the DNA template that corresponds to a gene.

metabolite Small organic molecules that are intermediates in the stepwise biosynthesis or breakdown of macromolecules.

metachromatic Exhibiting a color other than that of the dye used to stain it.

metastasis In cancer, the dissemination of tumor cells so that neoplastic growths appear in other sites away from the primary tumor.

MIC Abbreviation for **m**inimum **i**nhibitory **c**oncentration. The lowest concentration of antibiotic needed to inhibit bacterial growth in a test system.

microaerophile An aerobic bacterium that requires oxygen at a concentration less than that in the atmosphere.

miliary tuberculosis Rapidly fatal tuberculosis due to dissemination of mycobacteria in the blood and formation of tiny granules in various organs and tissues. The term miliary means resembling a millet seed.

mineralization The process by which decomposers (bacteria and fungi) convert organic debris into inorganic and elemental form. It is part of the recycling process.

miracidium The ciliated first-stage larva of a trematode. This form is infective for a corresponding intermediate host snail.

mitosis Somatic cell division that preserves the somatic chromosome number.

monoclonal antibody An antibody produced by a clone of lymphocytes that respond to a particular antigenic determinant and generate identical antibodies only to that determinant. *See* hybridoma.

monocyte A large mononuclear leukocyte normally found in the lymph nodes, spleen, bone marrow, and loose connective tissue. This type of cell makes up 3% to 7% of circulating leukocytes.

monotrichous A microorganism that bears a single flagellum.

morbidity A diseased condition; the relative incidence of disease in a community.

mordant A chemical that fixes a dye in or on cells by forming an insoluble compound and thereby promoting retention of that dye. Example: Gram's iodine in the Gram stain.

must Juices expressed from crushed fruits that are used in fermentation for wine.

mutagen Any agent that induces genetic mutation. Examples: certain chemical substances, ultraviolet light, radioactivity.

mutation A permanent inheritable alteration in the DNA sequence or content of a cell.

mycelium The filamentous mass that makes up a mold. Composed of hyphae.

mycetoma A chronic fungal infection usually afflicting the feet, typified by swelling and multiple draining lesions. Example: maduromycosis or Madura foot.

mycorrhizae Various species of fungi adapted in an intimate, mutualistic relationship to plant roots

mycosis Any disease caused by a fungus.

N

narrow spectrum Denotes drugs that are selective and limited in their effects. For example, they inhibit either gram-negative or gram-positive bacteria, not both.

necrosis A pathologic process in which cells and tissues die and disintegrate.

negative feedback Enzyme regulation of metabolism by the end product of a multienzyme system that blocks the action of a "pacemaker" enzyme at or near the beginning of the pathway.

negative stain A staining technique that renders the background opaque or colored and leaves the object unstained so that it is outlined as a colorless area.

nematode A common name for helminths called roundworms.

neoplasm A synonym for tumor.

neutron An electrically neutral particle in the nuclei of all atoms except hydrogen.

neutrophil A mature granulocyte present in peripheral circulation, exhibiting a multilobular nucleus and numerous cytoplasmic granules that retain a neutral stain. The neutrophil is an active phagocytic cell in bacterial infection.

niche In ecology, an organism's biological role or contribution to its community.

night soil An archaic euphemism for human excrement collected for fertilizing crops.

nitrogen base A ringed compound of which pyrimidines and purines are types.

nitrogen fixation A process occurring in certain bacteria in which atmospheric N_2 gas is converted to a form (NH_4) usable by plants.

nomenclature A set system for scientifically naming organisms, enzymes, anatomic structures, etc.

nonsense codon A triplet of mRNA bases that does not specify an amino acid but signals the end of a polypeptide chain.

normal flora The native microbial forms that an individual harbors.

nosocomial infection An infection not present upon admission to a hospital but incurred while being treated there.

nucleocapsid In viruses, the close physical combination of the nucleic acid with its protective covering.

nucleoid The basophilic nuclear region or nuclear body that contains the bacterial chromosome.

nucleotide A composite of a nitrogen base (purine or pyrimidine), a 5-carbon sugar (ribose or deoxyribose), and a phosphate group; a unit of nucleic acids.

numerical aperture In microscopy, the amount of light passing from the object and into the object in order to maximize optical clarity and resolution.

nutrient Any chemical substance that must be provided to a cell for normal metabolism and growth. Macronutrients are required in large amounts, and micronutrients in small amounts.

O

obligate Without alternative; restricted to a particular characteristic. Example: An obligate parasite survives and grows only in a host; an obligate aerobe must have oxygen to grow; an obligate anaerobe is destroyed by oxygen.

Okazaki fragment In replication of DNA, a segment formed on the lagging strand in which biosynthesis is conducted in a discontinuous manner dictated by the 5′ → 3′ DNA polymerase orientation.

oligodynamic action A chemical having antimicrobial activity in miniscule amounts. Example: Certain heavy metals are effective in a few parts per billion.

oncogene A naturally occurring type of gene that, when activated, can transform a normal cell into a cancer cell.

oncology The study of neoplasms, their cause, disease characteristics, and treatment.

oocyst The encysted form of a fertilized macrogamete or zygote; typical in the life cycles of sporozoan parasites.

operon A genetic operational unit that regulates metabolism by controlling mRNA production. In sequence, the unit consists of a regulatory gene, inducer or repressor control sites, and structural genes.

opportunistic In infection, ordinarily nonpathogenic or weakly pathogenic microbes that cause disease primarily in an immunologically compromised host.

opsonization The process of stimulating phagocytosis by affixing molecules (opsonins such as antibodies and complement) to the surfaces of foreign cells or particles.

osmophile A microorganism that thrives in a medium having high osmotic pressure.

osmosis The diffusion of water across a selectively permeable membrane in the direction of lower water concentration.

oxidation In chemical reactions, the loss of electrons by one reactant.

oxidation-reduction Redox reactions in which paired sets of molecules participate in electron transfers.

oxidative phosphorylation The synthesis of ATP using energy given off during the electron transport phase of respiration.

P

palindrome A word, verse, number, or sentence that reads the same forward or backward. Palindromes of nitrogen bases in DNA have genetic significance as transposable elements, as regulatory protein targets, and in DNA splicing.

palisades The characteristic arrangement of *Corynebacterium* cells resembling a row of fence posts and created by snapping.

pandemic A disease afflicting an increased proportion of the population over a wide geographic area (often worldwide).

pannus The granular membrane occurring on the cornea in trachoma.

papule An elevation of skin that is small, demarcated, firm, and usually conical.

parasite An organism that lives on or within another organism (the host), from which it obtains nutrients and enjoys protection. The parasite produces some degree of harm in the host.

parenteral Administering a substance into a body compartment other than through the gastrointestinal tract, such as via intravenous, subcutaneous, intramuscular, or intramedullary injection.

paroxysmal Events characterized by sharp spasms or convulsions; sudden onset of a symptom such as fever and chills.

passive immunity Specific resistance that is acquired indirectly by donation of preformed immune substances (antibodies) produced in the body of another individual.

pasteurization Heat treatment of perishable fluids like milk, fruit juices, or wine to destroy heat-sensitive vegetative cells, followed by rapid chilling to inhibit growth of survivors and germination of spores. It prevents infection and spoilage.

pathogen Any agent, usually a virus, bacterium, fungus, protozoan, or helminth, that causes disease.

pathogenicity The capacity of microbes to cause disease.

pathology The structural and physiological effects of disease on the body.

pellicle A membranous cover; a thin skin, film, or scum on a liquid surface; a thin film of salivary glycoproteins that forms over newly cleaned tooth enamel when exposed to saliva.

pelvic-inflammatory disease (PID) An infection of the uterus and fallopian tubes that has ascended from the lower reproductive tract. Caused by gonococci and chlamydiae.

penicillinase An enzyme that hydrolyzes penicillin; found in penicillin-resistant strains of bacteria.

penicillins A large group of naturally occurring and synthetic antibiotics produced by *Penicillium* mold and active against the cell wall of bacteria.

pentose A monosaccharide with five carbon atoms per molecule. Examples: arabinose, ribose, xylose.

peptidase An enzyme that can hydrolyze the peptide bonds of a peptide chain.

peptide bond The covalent union between two amino acids that forms between the amine group of one and the carboxyl group of the other. The basic bond of proteins.

peptidoglycan A network of polysaccharide chains cross-linked by short peptides that forms the rigid part of bacterial cell walls. Gram-negative bacteria have a smaller amount of this rigid structure than do gram-positive bacteria.

perinatal In childbirth, occurring before, during, or after delivery.

periodontal Involving the structures that surround the tooth.

periplasmic space The region between the inner and outer membrane of the cell envelopes of gram-negative bacteria.

peritrichous In bacterial morphology, having flagella distributed over the entire cell.

petechiae Minute hemorrhagic spots in the skin that range from pinpoint- to pinhead-sized.

Peyer's patches Oblong lymphoid aggregates of the gut located chiefly in the wall of the terminal small intestine. Along with the tonsils and appendix, Peyer's patches make up the gut-associated-lymphoid tissue that responds to local invasion by infectious agents.

pH The symbol for the negative logarithm of the H ion concentration; p (power) of $[H^+]_{10}$. A system for rating acidity and alkalinity.

phage A bacteriophage; a virus that specifically parasitizes bacteria.

phagocytosis A type of endocytosis in which the cell membrane actively engulfs large particles or cells into vesicles.

phagolysosome A body formed in a phagocyte, consisting of a union between a vesicle containing the ingested particle (the phagosome) and a vacuole of hydrolytic enzymes (the lysosome).

phenotype The observable characteristics of an organism produced by the interaction between its genetic potential (genotype) and the environment.

phlebotomine Pertains to a genus of very small midges or bloodsucking (phlebotomous) sand flies and to diseases associated with those vectors such as kala-azar, Oroya fever, and cutaneous leishmaniasis.

photic zone The aquatic stratum from the surface to the limits of solar light penetration.

photoautotroph An organism that utilizes light for its energy and carbon dioxide chiefly for its carbon needs.

photolysis Literally, splitting water with light. In photosynthesis, this step frees electrons and gives off O_2.

photosynthesis A process occurring in plants, algae, and some bacteria that traps the sun's energy and converts it to the cell. This energy is used to fix CO_2 into organic compounds.

pili Small, stiff, filamentous appendages in gram-negative bacteria that function in DNA exchange during bacterial conjugation.

pinocytosis The engulfment or endocytosis of liquids by extensions of the cell membrane.

plankton Minute animals (zooplankton) or plants (phytoplankton) that float and drift in the limnetic zone of bodies of water.

plaque In virus propagation methods, the clear zone of lysed cells in tissue culture or chick embryo membrane that corresponds to the area containing viruses. In dental application, the filamentous mass of microbes that adheres tenaciously to the tooth and predisposes to caries, calculus, or inflammation.

plasma cell A progeny of an activated B cell that actively produces and secretes antibodies.

plasmids Extrachromosomal genetic units characterized by several features. A plasmid is a circular, double-stranded DNA that is smaller than and replicates independently of the cell chromosome; it bears genes that are not essential for cell growth; it may bear genes that code for adaptive traits; and it is transmissible to other bacteria.

pleomorphism Normal variability of cell shapes in a single species.

pluripotential Stem cells having the developmental plasticity to give rise to more than one type. Example: undifferentiated blood cells in the bone marrow.

pneumonia An inflammation of the lung leading to accumulation of fluid and respiratory compromise.

pneumonic plague The acute, frequently fatal form of pneumonia caused by *Yersinia pestis.*

point mutation A change that involves the loss, substitution, or addition of one or a few nucleotides.

polyclonal In reference to a collection of antibodies with mixed specificities that arose from more than one clone of B cells.

polymer A macromolecule made up of a chain of repeating units. Examples: starch, protein, DNA.

polymerase An enzyme that produces polymers through catalyzing bond formation between building blocks (polymerization).

polymorphonuclear leukocytes (PMNL) White blood cells with variously shaped nuclei. Although this term commonly denotes all granulocytes, it is used especially for the neutrophils.

polymyxin A mixture of antibiotic polypeptides from *Bacillus polymyxa* that are particularly effective against gram-negative bacteria.

polypeptide A relatively large chain of amino acids linked by peptide bonds.

polysaccharide A carbohydrate that can be hydrolyzed into a number of monosaccharides. Examples: cellulose, starch, glycogen.

porin Transmembrane proteins of the outer membrane of gram-negative cells that permit transport of small molecules into the periplasmic space, but bar the penetration of larger molecules.

potable Describing water that is relatively clear, odor-free, and safe to drink.

pox The thick, elevated pustular eruptions of various viral infections. Also called pocks.

precipitin An antibody that combines with and aggregates a soluble antigen to bring about its precipitation from solution.

prevalence The total accumulative number of cases of a disease in a certain area and time period.

prion A concocted word to denote "proteinaceous infectious agent"; a cytopathic protein associated with the slow-virus spongiform encephalopathies of humans and animals.

procaryotic cell A small, simple cell lacking a true nucleus, a nuclear envelope, and membrane-enclosed organelles.

prodromium A short period of mild symptoms occurring at the end of the period of incubation. It indicates the onset of an infection.

proglottid The egg-generating segment of a tapeworm that contains both male and female organs.

promastigote A morphological variation of the trypanosome parasite responsible for leishmaniasis.

promoter region The site comprised of a short signalling DNA sequence that RNA polymerase recognizes and binds to commence transcription.

properdin A normal serum protein involved in the alternate complement pathway that leads to nonspecific lysis of bacterial cells and viruses.

prophage A lysogenized bacteriophage; a phage that is latently incorporated into the host chromosome instead of undergoing viral replication and lysis.

prophylactic Any device, method, or substance used to prevent disease.

prostaglandin A hormonelike substance that regulates many body functions. Prostaglandin comes from a family of organic acids containing 5-carbon rings that are essential to the human diet.

proton An elementary particle that carries the positive charge. It is identical to the nucleus of the hydrogen atom.

protoplast A bacterial cell whose cell wall is completely lacking and is vulnerable to osmotic lysis.

pseudohypha A chain of easily separated, spherical-to-sausage-shaped yeast cells partitioned by constrictions rather than by septa.

pseudomembrane A tenacious, noncellular mucous exudate containing cellular debris that tightly blankets the mucosal surface in infections such as diphtheria and pseudomembranous enterocolitis.

pseudopodium A temporary extension of the protoplasm of an ameboid cell. It serves both in ameboid motion and for food gathering (phagocytosis).

psychrophile A microorganism that thrives at low temperature (0°–20°C), with a temperature optimum of 0°–15°C.

pure culture A container of microbial cells whose identity is known.

purine A nitrogen base that is an important encoding component of DNA and RNA. The two most common purines are adenine and guanine.

purpura A condition characterized by bleeding into skin or mucous membrane, giving rise to patches of red that darken and turn purple.

pus The viscous, opaque, usually yellowish matter formed by an inflammatory infection. It consists of serum exudate, tissue debris, leukocytes, and microorganisms.

pyogenic Pertains to pus-formers, especially the pyogenic cocci: pneumococci, streptococci, staphylococci, and neisseriae.

pyrimidine Nitrogenous bases that help form the genetic code on DNA and RNA. Uracil, thymine, and cytosine are the most important pyrimidines.

pyrimidine dimer The union of two adjacent pyrimidines on the same DNA strand, brought about by exposure to ultraviolet light. It is a form of mutation.

pyrogen A substance that causes a rise in body temperature. It may come from pyrogenic microorganisms or from polymorphonuclear leukocytes (endogenous pyrogens).

Q

Q fever A disease first described in Queensland, Australia, initially dubbed Q for "query" to denote a fever of unknown origin. Q fever is now known to be caused by a rickettsial infection.

quaternary structure In complex proteins made up of more than one polypeptide chain or unit, this term denotes the arrangement of the subunits relative to each other.

quats A byword that pertains to a family of surfactants called quaternary ammonium compounds. These detergents are only weakly microbicidal and are used as sanitizers and preservatives.

Quellung test The capsular swelling phenomenon; a serological test for the presence of pneumococci whose capsules enlarge and become opaque and visible when exposed to specific anticapsular antibodies.

quinolone A new class of synthetic antimicrobic drugs with broad-spectrum effects.

R

rad A unit of measure for absorbed dose of ionizing radiation.

radiation Electromagnetic waves or rays, such as those of light given off from an energy source.

radioactive isotopes Unstable isotopes whose nuclei emit particles of radiation. This emission is called radioactivity or radioactive decay. Three naturally occurring emissions are alpha, beta, and gamma radiation.

radioimmunoassay A highly sensitive laboratory procedure that employs radioisotope-labelled substances to measure the levels of antibodies or antigens in the serum.

real image An image formed at the focal plane of a convex lens. In the compound light microscope, it is the image created by the objective lens.

receptor In intercellular communication, cell surface molecules involved in recognition, binding, and intracellular signalling.

redox Denoting an oxidation-reduction reaction.

reduction In chemistry, the gain of electrons.

refraction In optics, the bending of light as it passes from one medium into another with a different index of refraction.

replication In DNA synthesis, the semiconservative mechanisms that ensure precise duplication of the parent DNA strands.

replication fork The Y-shaped point on a replicating DNA molecule where the DNA polymerase is synthesizing new strands of DNA.

repressor The protein product of a repressor gene that combines with the operator and arrests the transcription and translation of structural genes.

reservoir In disease communication, the natural host or habitat of a pathogen.

resident flora The deeper, more stable microflora that inhabit the skin and exposed mucous membranes, as opposed to the superficial, variable, transient population.

resolving power The capacity of a microscope lens system to accurately distinguish between two separate entities that lie in close proximity. Also called resolution.

respiratory chain In cellular respiration, a series of electron-carrying molecules that transfers energy-rich electrons and protons to molecular oxygen. In transit, energy is extracted and conserved in the form of ATP.

restriction endonuclease An enzyme present naturally in cells that cleaves specific locations on DNA. It is an important means of inactivating viral genomes, and is also used to splice genes in genetic engineering.

reticuloendothelial system Also known as the mononuclear phagocyte system, it pertains to a network of fibers and phagocytic cells (macrophages) that permeates the tissues of all organs. Examples: Kupffer cells in liver sinusoids, alveolar phagocytes in the lung, microglia in nervous tissue.

retrovirus A group of RNA viruses (including HIV) that have the mechanisms for converting their genome into a double strand of DNA that may be inserted on a host's chromosome.

reverse transcriptase The enzyme possessed by retroviruses that carries out the reversion of RNA to DNA—a form of reverse transcription.

Reye's syndrome A sudden, usually fatal unconsciousness in children following a viral infection. Autopsy shows cerebral edema and marked fatty change in the liver and renal tubules.

rhizosphere The zone of soil, complete with microbial inhabitants, in the immediate vicinity of plant roots.

ribose A 5-carbon monosaccharide found in RNA.

ribosome A bilobed macromolecular complex of ribonucleoprotein that coordinates the codons of mRNA with tRNA anticodons, and in so doing, constitutes the peptide assembly site.

ringworm A superficial mycosis caused by various dermatophytic fungi. This common name is actually a misnomer.

rolling circle An intermediate stage in viral replication of circular DNA into linear DNA.

rosette formation A technique for distinguishing surface receptors on T cells by reacting them with sensitized indicator sheep red blood cells. The cluster of red cells around the central white blood cell resembles a little rose blossom and is indicative of the type of receptor.

S

salpingitis Inflammation of the fallopian tubes.

sanitize To clean inanimate objects using soap and degerming agents so that they are safe and free of high levels of microorganisms.

saprobe A microbe that decomposes organic remains from dead organisms. Also known as a saprophyte or saprotroph.

sarcina A cubical packet of eight cells; the cellular arrangement of the genus *Sarcina* in the family Micrococcaceae.

sarcoma A fleshy neoplasm of connective tissue. Growth, usually highly malignant, is of mesodermal origin.

satellite phenomenon A type of commensal relationship in which one microbe produces growth factors that favor the growth of its dependent partner. When plated on solid media, the dependent organism appears as surrounding colonies. Example: *Staphylococcus* surrounded by its *Haemophilus* satellite.

schizogony A process of multiple fission whereby first the nucleus divides several times, and subsequently the cytoplasm is subdivided for each new nucleus during cell division.

scolex The anterior end of a tapeworm characterized by hooks and/or suckers for attachment to the host.

SCP Abbreviation for single-cell protein, a euphemistic expression for microbial protein intended for human and animal consumption.

sebaceous glands The sebum- (oily, fatty) secreting glands of the skin.

secondary immune response The more rapid and heightened response to antigen in a sensitized subject due to memory lymphocytes.

secondary infection An infection that compounds a preexisting one.

secretion The process of actively releasing cellular substances into the extracellular environment.

secretory antibody The immunoglobulin (IgA) that is found in secretions of mucous membranes and serves as a local immediate protection against infection.

selective media Nutrient media designed to favor the growth of certain microbes and to inhibit undesirable competitors.

self-limited Applies to an infection that runs its course without disease or residual effects.

semiconservative replication In DNA replication, the synthesis of paired daughter strands, each retaining a parent strand template.

semisolid media Nutrient media with a firmness midway between that of a broth and an ordinary solid medium; motility media.

sensitizing dose The initial effective exposure to an antigen or an allergen that stimulates an immune response. Often applies to allergies.

sepsis The state of putrefaction; the presence of pathogenic organisms or their toxins in tissue or blood.

septicemia Systemic infection associated with microorganisms multiplying in circulating blood.

septum A partition or cellular crosswall, as in certain fungal hyphae.

sequela A morbid complication that follows a disease.

seropositive Showing the presence of specific antibody in a serologic test. Indicates ongoing infection.

serotonin A vasoconstrictor that inhibits gastric secretion and stimulates smooth muscle.

serotyping The subdivision of a species or subspecies into an immunologic type, based upon antigenic characteristics.

serum The clear fluid expressed from clotted blood that contains dissolved nutrients, antibodies, and hormones, but not cells or clotting factors.

sex pilus A conjugative pilus.

sign Any abnormality uncovered upon physical diagnosis that indicates the presence of disease. A sign is an objective assessment of disease, as opposed to a symptom, which is the subjective assessment perceived by the patient.

sodoku A Japanese term pertaining to rat bite fever.

spheroplast A gram-negative cell whose peptidoglycan, when digested by lysozyme, remains intact but is osmotically vulnerable.

spike A surface receptor on the surface of certain enveloped viruses that facilitates specific attachment to the host cell.

sporangium A fungal cell in which asexual spores are formed by multiple cell cleavage.

sporozoite One of many minute elongated bodies generated by multiple division of the oocyst. This is the infectious form of the malarial parasite that is harbored in the salivary gland of the mosquito and is inoculated into the victim during feeding.

start codon The nucleotide triplet AUG that codes for the first amino acid in protein sequences.

starter culture The sizeable inoculation of pure bacterial, mold, or yeast sample for bulk processing, as in the preparation of fermented foods, beverages, and pharmaceuticals.

stasis A state of rest or inactivity; applied to nongrowing microbial cultures. Also called microbistasis.

sterilization Any process that completely removes or destroys all viable microorganisms, including viruses, from an object or habitat. Material so treated is sterile.

strain In microbiology, a set of descendants cloned from a common ancestor that retain the original characteristics. Any deviation from the original is a different strain.

streptolysin A hemolysin produced by streptococci.

stroma The matrix of the chloroplast that is the site of the dark reactions.

structural gene A gene that codes for the amino acid sequence (peptide structure) of a protein.

subacute Indicates an intermediate status between acute and chronic disease.

subclinical A period of inapparent manifestations that occurs prior to the appearance of symptoms and signs of disease.

substrate The specific molecule upon which an enzyme acts.

subunit vaccine A vaccine against isolated microbial antigens rather than against the entire organism. Example: viral glycoprotein as opposed to the entire virion.

superinfection An infection occurring during antimicrobic therapy that is caused by an overgrowth of drug-resistant microorganisms.

suppressor T cell A class of T cells that inhibits the actions of B cells and other T cells.

surfactant A surface-active agent that forms a water-soluble interface. Examples: detergents, wetting agents, dispersing agents, and surface tension depressants.

sylvatic Denotes the natural presence of disease among wild animal populations. Examples: sylvatic (sylvan) plague, rabies.

symbiosis An intimate association between individuals from two species; used as a synonym for mutualism.

symptom The subjective evidence of infection and disease as perceived by the patient.

syncytium A multinucleated protoplasmic mass formed by consolidation of individual cells.

syndrome The collection of signs and symptoms that, taken together, paint a portrait of the disease.

synergism The coordinated or correlated action by two or more drugs or microbes that results in a heightened response or greater activity.

syngamy Conjugation of the gametes in fertilization.

systemic Occurring throughout the body; said of infections that invade many compartments and organs via the circulation.

T

temperate phage A bacteriophage that enters into a less virulent state by becoming incorporated into the host genome as a prophage instead of in the vegetative or lytic form that eventually destroys the cell.

teratogenic Causing abnormal fetal development.

tetanospasmin The neurotoxin of *Clostridium tetani*, the agent of tetanus. Its chief action is directed upon the inhibitory synapses of the anterior horn motor neurons.

tetracyclines A group of broad-spectrum antibiotics with a complex 4-ring structure.

therapeutic index The ratio of the toxic dose to the effective therapeutic dose that is used to assess the safety and reliability of the drug.

thermal death point The lowest temperature that achieves sterilization in a given quantity of broth culture upon a 10-minute exposure. Examples: 55°C for *Escherichia coli*, 60°C for *Mycobacterium tuberculosis*, and 120°C for spores.

thermal death time The least time required to kill all cells of a culture at a specified temperature.

thermocline A temperature buffer zone in a large body of water that separates the warmer water (the epilimnion) from the colder water (the hypolimnion).

thermoduric Resistant to the harmful effects of high temperature.

thermophile A microorganism that thrives at a temperature of 50°C or higher.

thrush *Candida albicans* infection of the oral cavity.

thylakoid Vesicles of a chloroplast formed by elaborate folding of the inner membrane to form "discs." Solar energy trapped in the thylakoids is used in photosynthesis.

tincture A medicinal substance dissolved in an alcoholic solvent.

tinea Ringworm; a fungal infection of the hair, skin, or nails.

titer In immunochemistry, a measure of antibody level in a patient, determined by agglutination methods.

topoisomerases Enzymes that can add or remove DNA twists and thus regulate the degree of supercoiling.

toxemia An abnormality associated with certain infectious diseases. Toxemia is caused by toxins or other noxious substances released by microorganisms circulating in the blood.

toxigenicity The tendency for a pathogen to produce toxins. It is an important factor in bacterial virulence.

toxoid A toxin that has been rendered nontoxic but is still capable of eliciting the formation of protective antitoxin antibodies; used in vaccines.

tracheostomy A surgically created emergency airway opening into the trachea.

transcription mRNA synthesis; the process by which a strand of RNA is produced against a DNA template.

transduction The transfer of genetic material from one bacterium to another by means of a bacteriophage vector.

transformation In microbial genetics, the transfer of genetic material contained in "naked" DNA fragments from a donor cell to a competent recipient cell.

transgenic technology Introduction of foreign DNA into cells or organisms. Used in genetic engineering to create recombinant plants, animals, and microbes.

translation Protein synthesis; the process of decoding the messenger RNA code into a polypeptide.

transients In normal flora, the assortment of superficial microbes whose numbers and types vary depending upon recent exposure. The deeper-lying residents constitute a more stable population.

transposon A DNA segment with an insertion sequence at each end, enabling it to migrate to another plasmid, to the bacterial chromosome, or to a bacteriophage.

trematode A fluke or flatworm parasite of vertebrates.

triglyceride A type of lipid composed of a glycerol molecule bound to three fatty acids.

trophozoite A vegetative protozoan (feeding form) as opposed to a resting (cyst) form.

trypomastigote The infective morphological stage transmitted by the tsetse fly or the triatomid bug in African trypanosomiasis and Chagas' disease.

tubercle In tuberculosis, the granulomatous, well-defined lung lesion that may serve as a focus for latent infection.

tuberculin A glycerinated broth culture of *Mycobacterium tuberculosis* that is evaporated and filtered. Formerly used to treat tuberculosis, tuberculin is now used chiefly for diagnostic tests.

tyndallization Fractional (discontinuous, intermittent) sterilization designed to destroy spores indirectly. A preparation is exposed to flowing steam for an hour, and then the material is allowed to incubate to permit spore germination. The resultant vegetative cells are destroyed by repeated steaming and incubation.

U

uncoating The process of viral coat removal and release of viral genome by its newly invaded host cell.

universal donor In blood grouping and transfusion, a group O individual whose erythrocytes bear neither agglutinogen A nor B.

uracil A pyrimidine; one of the nitrogenous bases found in ribonucleic acid.

V

vaccine Originally used in reference to inoculation with the cowpox or vaccinia virus to protect against smallpox. In general, the term now pertains to injection of whole microbes (killed or attenuated), toxoids, or parts of microbes as a prevention or cure for disease.

valence The combining power of an atom based upon the number of electrons it can either take on or give up.

variable region The antigen-binding fragment of an immunoglobin molecule, consisting of a combination of heavy and light chains whose molecular conformation is specific for the antigen.

variolation A hazardous, outmoded process of deliberately introducing smallpox material scraped from a victim into the nonimmune subject in the hope of inducing resistance.

VDRL A flocculation test that detects syphilis antibodies. An important screening test. The acronym stands for **V**enereal **D**isease **R**esearch **L**aboratories.

vector An animal that transmits infectious agents from one host to another, usually a biting or piercing arthropod like the tick, mosquito, or fly. Infectious agents may be conveyed mechanically by simple contact or biologically whereby the parasite develops in the vector.

vegetative In describing microbial developmental stages, a metabolically active feeding and dividing form, as opposed to a dormant, seemingly inert, nondividing form. Examples: a bacterial cell versus its spore; a protozoan trophozoite versus its cyst.

vehicle An inanimate material (solid object, liquid, or air) that serves as a transmission agent for pathogens.

verruca A flesh-colored wart. This self-limited tumor arises from accumulations of growing epithelial cells. Example: papilloma warts.

vesicle A blister characterized by a thin-skinned, elevated, superficial pocket inflated with serum.

vibrio A curved, rod-shaped bacterial cell.

viremia The presence of a virus in the bloodstream.

virion An elementary virus particle in its complete morphological and thus infectious form. A virion consists of the nucleic acid core surrounded by a capsid, which may be enclosed in an envelope.

viroid An infectious agent that, unlike a virion, lacks a capsid and consists of a closed circular RNA molecule. Although known viroids are all plant pathogens, it is conceivable that animal versions exist.

virtual image In optics, an image formed by diverging light rays; in the light compound microscope, the second, magnified visual impression formed by the ocular from the real image formed by the objective.

virulence In infection, the relative capacity of a pathogen to invade and harm host cells.

W

wart An epidermal tumor caused by papilloma viruses. Also called a verruca.

Western blot test A procedure for separating and identifying antigen or antibody mixtures by two-dimensional electrophoresis in polyacrylamide gel, followed by immune labelling.

wheal A welt; a marked, slightly red, usually itchy area of the skin that changes in size and shape as it extends to adjacent areas. The reaction is triggered by cutaneous contact or intradermal injection of allergens in sensitive individuals.

whey The residual fluid from milk coagulation that separates from the solidified curd.

white piedra A fungus disease of hair, especially of the scalp, face, and genitals, caused by *Trichosporon beigelii*. The infection is associated with soft, mucilaginous, white-to-light-brown nodules that form within and on the hair shafts.

whitlow A deep inflammation of the finger or toe, especially near the tip or around the nail. Whitlow is a painful herpes simplex virus infection that may last several weeks and is most common among health care personnel who come in contact with the virus in patients.

Widal test An agglutination test for diagnosing typhoid.

wort The clear fluid derived from soaked mash that is fermented for beer.

X

xenograft The transfer of a tissue or an organ from an animal of one species to a recipient of another species.

Y

yaws A tropical disease caused by *Treponema pertenue* that produces granulomatous ulcers on the extremities and occasionally on bone, but not central nervous system or cardiovascular complications.

Z

zoonosis An infectious disease indigenous to animals that humans may acquire through direct or indirect contact with infected animals.

zygospore A thick-walled sexual spore produced by the zygomycete fungi. It develops from the union of two hyphae, each bearing nuclei of opposite mating types.

Credits

ILLUSTRATORS

Hans & Cassady, Inc.: Figures 1.2, 1.3a, 1.4, 1.6a, 1.7, 1.8, 1.10, 1.11, 1.12, 1.13, 1.14, 1.18, 1.19, 1.20, 1.21, 1.22, 1.27a,b, 1.30, 2.1a&b, 2.2, 2.3a-c, 2.4a, 2.4b, 2.4c, 2.5a, 2.5b, 2.6a, 2.6b, 2.7a, 2.7b, 2.8, 2.9, 2.11, 2.12, 2.13a&b, 2.14a, 2.14b, 2.15a, 2.15b, 2.15c, 2.16a, 2.16b, 2.17a&b, 2.18a, 2.18b, 2.19, 2.20, 2.21, 2.22a-d, 2.23a-c, 2.24a, 2.24b&c, 2.25a, 2.26, 3.2b, 3.4, 3.5a&b, 3.6a&b, 3.7a&b, 3.10, 3.11, 3.14a-c, 3.15a 1, 3.15b 1, 3.17, 3.19, 3.21a, 3.22, 3.25, 3.26, 3.28, 3.29a, 3.29c, 3.34, 3.35, 4.3b, 4.4a, 4.7a&b, 4.11, 4.15b, 4.15c, 4.27, 4.28a-d, 4.29a, 4.32a, 5.3a&b, 5.4a-c, 5.5a 1, 5.5b 1, 5.6a&b, 5.9a, 5.9b 2, 5.10, 5.11b 2, 5.12, 5.14, 5.15a&b, 5.16a&b, 5.17a&b, 5.19, 5.21, 6.1, 6.4, 6.5, 6.6a&b, 6.7, 6.8a&b, 6.9, 6.10, 6.12, 6.13a&b, 6.15a-d, 6.16a&b, 6.17, 6.18b, 6.19a&b, 6.20a, 6.21a, 6.22a, 6.24a&b, 7.1, 7.2, 7.3, 7.4a-c, 7.6a-c, 7.7a-d, 7.8a&b, 7.9, 7.10a&b, 7.11, 7.12, 7.13, 7.14, 7.15, 7.16, 7.17, 7.18, 7.19, 7.20, 7.21, 7.22, 7.23, 7.26, 7.27, 8.5, 8.6a-c, 8.7a&b, 8.8a&b, 8.9, 8.10, 8.11, 8.12a&b, 8.13, 8.14a&b, 8.15, 8.16a&b, 8.17, 8.18, 8.19, 8.20a&b, 8.21a&b, 8.22, 8.23a&b, 8.24a-c, 8.25, 8.26a&b, 8.28a-f, 8.29a&b, 8.30, 8.31a-g, 9.1, 9.2a-d, 9.3a&b, 9.4a-d, 9.5b, 9.7, 9.8a-c, 9.11, 9.13a, 9.14, 9.15a&b, 9.16a&b, 9.18, 9.19b, 10.1, 10.2a-d, 10.3a&b, 10.4, 10.5, 10.6, 10.7, 10.8, 10.9a&b, 10.10a-c, 10.11, 10.12, 10.13, 10.15a-c, 10.16a-c, 10.17a-c, 10.19a-c, 10.20a, 10.21a, 11.9, 11.11a-d, 11.12a-c, 11.13a&b, 11.16a, 11.16b, 11.17a-d, 11.18a&b, 11.20, 11.21, 11.24, 12.1, 12.4, 12.5a&b, 12.7a&b, 12.9, 12.12, 12.14a&b, 12.16, 12.21, 12.22, 12.23b&c, 12.24a-e, 12.25, 12.27a&b, 12.28a&b, 13.1, 13.2, 13.3, 13.4, 13.5, 13.6, 13.7, 13.9a-c, 13.10, 13.11a&b, 13.12, 13.13, 13.14a,c, 13.15a&b, 13.16a-d, 13.17, 13.18, 13.19a-e, 13.20, 13.21, 13.22a&b, 13.23, 13.24a&b, 13.25a&b, 13.26a&b, 13.27a-c, 13.28b&c, 13.29, 13.31, 13.32a&b, 13.33a&b, 14.2a, 14.3a&b, 14.4, 14.6c, 14.7, 14.8, 14.9c, 14.10a-c, 14.13, 14.14a&b, 14.16, 14.17, 14.19, 14.21, 14.24a&b, 14.25a,b, 14.26a-d, figure B, p. 472, figure C(a), p. 473, figure D(a,b), p. 474, 15.6b, 15.10, 15.22, 16.3b, 16.12, 16.26, 16.28, 16.30, 16.31, 16.32, 16.35, 16.36, 17.1a, 17.10, 17.14, 17.17, 17.21, 17.25b, 17.28, 17.29a&b, 17.32a-c, 18.1, 18.2, 18.3, 19.4, 19.5a-c, 19.6a, 19.8b, 19.13b, 19.15, 19.17a, 19.19, 20.1, 20.3a, 20.10b,c, 20.15, 20.16, 21.1, 21.3b, 21.6b, 21.10, 21.12a&b, 21.13, 21.14, 21.15, 21.19, 21.21a, 21.24a, 21.26a&b, 22.1, 22.2, 22.3, 22.4, 22.5, 22.6, 22.7, 22.8a-2, 22.8b, 22.8c, 22.9, 22.10, 22.11a, 22.13, 22.15, 22.17, 22.18, 22.19, 22.20a, 22.22, 22.23a, 22.24, 22.26, 22.27, 22.28, 22.30, 22.31a, 22.33a, 22.34, 22.35b, 22.36a, 22.38, 22.39, 22.40, 22.41, 22.42, 22.46, 22.47, 22.48. Box figures 1.1, p. 7; 1.2, p. 15; 2.1, p. 41; 2.2, p. 42; 2.3b, p. 59; 3.1, p. 74; 3.2, p. 91; 4.1, p. 101; 6.5, p. 180; 7.2a-c, p. 203; 8.2, p. 229; 8.3, p. 248; 8.4, p. 251; 8.5a-e, p. 259; 10.2, p. 299; 11.1, p. 340; 12.1, p. 378; 13.1, p. 404; 13.2, p. 414; 14.1, p. 449; 16.2, p. 521; 16.5, p. 531; 17.2a, p. 565; 21.1, p. 675; 21.2, p. 696; 22.1, p. 718; 22.2, p. 725; 22.4a,b, p. 756. Text art figures 1.1, p. 31; 3.1, p. 80; 3.2, p. 83; 3.3, p. 87; 5.1, p. 133; 6.1, p. 169; 13.1, p. 410; 16.1, p. 515; 19.1, p. 620; 21.1, p. 673; 22.1, p. 757.
Molly Babich/Dave Carlson: Figures 11.1, 11.19a&b, 12.19, 16.39. Box figures 7.1a,b, p. 196 and 11.3, p. 350.
Brian Evans: Figures 3.16, 4.6a, 4.9, 4.10, 4.12a, 4.13, 4.14, 4.20, 4.30a,b, 6.3a-d, 7.5a-d, 7.24, 7.25, 8.2, 11.10, 12.17, 12.18, 12.26a, 15.18, 15.23, 16.8a-c, 16.16a,b, 17.27, 19.3, 19.10, 19.25c, 21.23a-d. Box figure 8.1, p. 228.
Phil Sims: Figures 3.1a, 4.2a-c, 4.8a-c, 4.16c, 4.18a-c, 4.19a&b, 4.21, 4.22, 4.26a, 4.31a&b, 4.35a&b, 5.8a-g, 8.1, 8.4, 11.2a&b, 12.10a&b, 12.20a-d, 15.3a, 16.37, 19.1a-c. Box figures 4.2, p. 115; 6.2, p. 160; 11.2, p. 343; 16.1, p. 518.
Molly Babich: Figures 4.33, 4.34a-c, 4.36, 8.27, 11.3a&b, 11.4, 11.6, 11.7a&b, 11.14a-e, 11.15, 12.2, 12.3a, 12.6, 12.8a-d, 12.15a&b, 14.1, 14.11, 14.12, 14.20a&b, 15.4a, figure A, p. 471, 15.12, 15.13a&b, 15.24, 15.27, 16.6a-c, 16.9a-c, 17.11, 17.18, 18.5a&b, 18.7a, 18.8a, 19.7, 19.9a, 19.12a-b, 20.4, 20.18, 21.7a-c, 21.11. Box figures 6.4, p. 176; 7.1a&b, p. 196; 14.2, p. 460; Text art figure 8.1, p. 263.
WCB Graphics: Text art figure 6.2, p. 191, text art figure 10.1, p. 323, figure 9.6b.

ILLUSTRATIONS

Figure 1.20: Figure from *Fundamentals of Microbiology*, Ninth Edition, by M. Frobisher, et al., copyright © 1974 by Saunders College Publishing, reprinted by permission of the publisher.
Figure 1.27a,b: From *Cell Ultrastructure* by William Jensen and Roderic B. Park, © 1967 by Wadsworth Publishing Company, Inc. Reprinted by permission of the publisher.
Figure 3.7: From L. M. Prescott, et al., *Microbiology*. Copyright © 1990 Wm. C. Brown Publishers, Dubuque, Iowa. All Rights Reserved. Reprinted by permission.
Figure 3.34: From L. M. Prescott, et al., *Microbiology*. Copyright © 1990 Wm. C. Brown Publishers, Dubuque, Iowa. All Rights Reserved. Reprinted by permission.
Figure 3.35: Reprinted by permission from *Nature* 196:1189–1192. Copyright © 1962 Macmillan Magazines Ltd.
Figure 5.9a: Reproduced with the permission of the Society for General Microbiology.
Figure 5.9c: Reproduced with permission from F. K. Sanders in *Viruses;* Carolina Biology Reader Series, No. 194. Copyright 1988, Carolina Biological Supply Co., Burlington, North Carolina, USA.
Figure 6.13: From Harold Benson, *Microbiological Applications*, 5th ed. Copyright © 1990 Wm. C. Brown Publishers, Dubuque, Iowa. All Rights Reserved. Reprinted by permission.
Figure 9.16b: From Nolte, et al., *Oral Microbiology*, 4th ed. Copyright © 1982 The C.V. Mosby Company. Reprinted by permission of Gloria-Mae Nolte.
Figure 12.3a: From Kent M. Van De Graaff, *Human Anatomy*, 2d ed. Copyright © 1988 Wm. C. Brown Publishers, Dubuque, Iowa. All Rights Reserved. Reprinted by permission.
Figure 12.9: From John W. Hole, Jr., *Human Anatomy and Physiology*, 5th ed. Copyright © 1990 Wm. C. Brown Publishers, Dubuque, Iowa. All Rights Reserved. Reprinted by permission.
Figure 13.18: From J. A. Bellanti, *Immunology III*. Copyright © 1985 W. B. Saunders, Philadelphia, PA. Reprinted by permission.
Figure 13.27a,b: From L. M. Prescott, et al., *Microbiology*. Copyright © 1990 Wm. C. Brown Publishers, Dubuque, Iowa. All Rights Reserved. Reprinted by permission.
Figure 13.28b,c: From Gillies and Dodds, *Bacteriology Illustrated*. Copyright © Churchill Livingstone, Edinburgh, Scotland. Reprinted by permission.
Figure 13.33: From L. M. Prescott, et al., *Microbiology*. Copyright © 1990 Wm. C. Brown Publishers, Dubuque, Iowa. All Rights Reserved. Reprinted by permission.
Figure 14.2a: Source: Asthma and Allergy Foundation of America, Los Angeles, CA.
Lyrics, page 457: © 1959 Jerry Leiber Music & Mike Stoller Music. All rights reserved. Used by permission.
Excerpt, page 506: From *Bergey's Manual of Systematic Bacteriology*. Copyright © the Williams & Wilkins Co., Baltimore. Reprinted by permission.
Figure 16.31: Copyright © Roche Diagnostic Systems. Reprinted by permission.
Figure 22.22: From L. M. Prescott, et al., *Microbiology*. Copyright © 1990 Wm. C. Brown Publishers, Dubuque, Iowa. All Rights Reserved. Reprinted by permission.
Figure 22.23a: From L. M. Prescott, et al., *Microbiology*. Copyright © 1990 Wm. C. Brown Publishers, Dubuque, Iowa. All Rights Reserved. Reprinted by permission.
Figure 22.30: Reprinted by permission of Macmillan Publishing Company from *Microbiology*, 2d ed., by Ronald M. Atlas. Copyright © 1988 by Macmillan Publishing Company.
Figure 22.31a: From L. M. Prescott, et al., *Microbiology*. Copyright © 1990 Wm. C. Brown Publishers, Dubuque, Iowa. All Rights Reserved. Reprinted by permission.
Figure 22.33a: From L. M. Prescott, et al., *Microbiology*. Copyright © 1990 Wm. C. Brown Publishers, Dubuque, Iowa. All Rights Reserved. Reprinted by permission.
Figure 22.34: From L. M. Prescott, et al., *Microbiology*. Copyright © 1990 Wm. C. Brown Publishers, Dubuque, Iowa. All Rights Reserved. Reprinted by permission.
Figure 22.35b: From M. Frobisher, et al., *Fundamentals of Microbiology*, 9th ed. Copyright © 1974 by W. B. Saunders. Reprinted by permission.
Figure 22.40: From L. M. Prescott, et al., *Microbiology*. Copyright © 1990 Wm. C. Brown Publishers, Dubuque, Iowa. All Rights Reserved. Reprinted by permission.
Figure 22.48: Copyright © Victor A. McKusick, Johns Hopkins Hospital, Baltimore, MD. Reprinted by permission.

PHOTOS

Chapter 1

Opener: The Bettmann Archive; **1.3b:** © George J. Wilder/Visuals Unlimited; **1.3c:** © T. E. Adams/Visuals Unlimited; **1.3d:** © Cabisco/ Visuals Unlimited; **1.5:** The Bettmann Archive; **1.6b:** Science VU/Visuals Unlimited; **1.9:** The Bettmann Archive; **1.15a-d, 1.16a,b:** © K. Talaro/Visuals Unlimited; **1.17:** Courtesy Nikon, Inc.; **1.23a:** © T. E. Adams/Visuals Unlimited; **1.23 b,c:** © M. Abbey/Visuals Unlimited; **1.24:** © E.C.S. Chan/Visuals Unlimited; **1.25a:** © George J. Wilder/Visuals Unlimited; **1.25b:** © M. Abbey/Visuals Unlimited; **1.26:** Science VU-CDC/Visuals Unlimited; **1.27c:** © William Ormerod/Visuals Unlimited; **1.28a:** © David M. Phillips/Visuals Unlimited; **1.28b:** Science VU-CDC/Visuals Unlimited; **1.29:** © Karl Aufderheide/Visuals Unlimited; **p. 31:** Science VU-IBMRL/Visuals Unlimited; **1.31a:** © A. M. Siegelman/Visuals Unlimited; **1.31b:** © John D. Cunningham/ Visuals Unlimited.

Chapter 2

Opener: Courtesy Scott Gould; **2.10a-c:** © John W. Hole, Jr.; **2.25b:** © Austin S. Grindle, Ealing Corporation; **p. 59(a):** © Don Fawcett/Visuals Unlimited.

Chapter 3

Opener: From *ASM News*, 53(2), Feb. 1987, American Society for Microbiology. Photo by H. Kaltwasser; **3.1b:** © Ralph A. Slepecky/ Visuals Unlimited; **3.2a:** Courtesy Julius Adler; **3.3a** (all), **3.3b** (top): © E.C.S. Chan/Visuals Unlimited; **3.3b** (bottom): © Fred Hossler/ Visuals Unlimited; **3.8a:** © David M. Phillips/ Visuals Unlimited; **3.8b:** Science VU-Charles W. Stratton/Visuals Unlimited; **3.9:** Science VU-Fred Marsik/Visuals Unlimited; **3.12a** (both): Photographs by Harriet & Ephrussi-Taylor; **3.12b:** © John D. Cunningham/Visuals Unlimited; **3.13:** Science VU-Charles W. Stratton/Visuals Unlimited; **3.15a,b:** © T. J. Beveridge/Visuals Unlimited; **3.18:** From P. Canumette, *Canadian Journal of Microbiology*, 30:273–84, 1984. Reprinted by permission of National Research Council of Canada; **3.20:** Reprinted by permission from E. J. Laishley et al., *Canadian Journal of Microbiology*, 19:991, 1973. National Research Council of Canada; **3.21b:** © Soad Tabaqchali/Visuals Unlimited; **3.21c:** © Lee D. Simon/Photo Researchers; **3.23a,b:** © David M. Phillips/Visuals Unlimited; **3.23c:** Science VU/Visuals Unlimited; **3.23d:** © R. G. Kessel-C. Y. Shih/Visuals Unlimited; **3.24a:** © A. M. Siegelman/Visuals Unlimited; **3.24b:** From Braude, *Infectious Diseases and Microbiology*, 2/E, Saunders College Publishing Company; **3.27:** Science VU-M.D. Maser/Visuals Unlimited; **3.29b,d:** © K. Talaro/Visuals Unlimited; **3.30:** Science VU-W. Burgdorfer/Visuals Unlimited; **3.31:** © David M. Phillips/Visuals Unlimited; **3.32a:** © George J. Wilder/Visuals Unlimited; **3.32b,c:** © T. E. Adams/Visuals Unlimited; **3.33:** From W. C. Dierksheide and R. M. Pfister, *Canadian Journal of Microbiology*, 19(1):151, 1977. National Research Council of Canada; **3.36a:** From Kessel and Cohen, "Ultrastructure of Square Bacteria From a Brine Pool in Southern Sinai," *Journal of Bacteriology*, 150(2):851–60, 1982. American Society for Microbiology; **3.36b:** From Walter Stoeckenius, *Walsby Square Bacterium: Fine Structure of an Orthogonal Procaryote*.

Chapter 4

Opener: © George J. Wilder/Visuals Unlimited; **4.1:** Reprinted from J. W. Schopf, *Journal of Paleontology*, 46:651–88, 1968; **4.3a:** © Michael Webb/Visuals Unlimited; **4.4b:** Science VU-Charles W. Stratton/Visuals Unlimited; **4.5:** © Don W. Fawcett/Visuals Unlimited; **4.6b:** Science VU/Visuals Unlimited; **4.12b:** © Don W. Fawcett/Visuals Unlimited; **4.15a:** © David M. Phillips/Visuals Unlimited; **4.16a,b:** © Fred Hossler/Visuals Unlimited; **4.17a:** © John D. Cunningham/ Visuals Unlimited; **4.17b:** © T. E. Adams/Peter Arnold, Inc.; **4.17c:** © R. Calentine/Visuals Unlimited; **4.17d:** © Everett S. Beneke/Visuals Unlimited; **4.23:** © K. Talaro/Visuals Unlimited; **4.24a:** © A. M. Siegelman/Visuals Unlimited; **4.24b:** © John D. Cunningham/Visuals Unlimited; **4.25:** © E.C.S. Chan/Visuals Unlimited; **4.26b:** © T. E. Adams/Visuals Unlimited; **4.26c, 4.29b:** © David M. Phillips/ Visuals Unlimited; **4.31c:** © Biophoto Associates/Science Source/Photo Researchers; **4.32b:** Reprinted by permission of Masamichi Aikawa from *Science*, 226:680, 1984. Copyright 1989 by the AAAS.

Chapter 5

Opener: Science VU-Wayside/Visuals Unlimited; **5.1a:** © K. G. Murti/ Visuals Unlimited; **5.1b:** Science VU/Visuals Unlimited; **5.1c:** © A. B. Dowsette/Science Photo Library/Photo Researchers; **5.2a:** Reprinted from Schaffer et al, *Proceedings of the National Academy of Science*, 41:1020, 1955; **5.2b:** © Omikron/Photo Researchers; **5.5a** (top): © Dennis Kunkel/Phototake; **5.5b** (bottom): © K. G. Murti/Visuals Unlimited; **5.6c:** Science VU-NIH, R. Feldman/Visuals Unlimited; **5.7a:** © Boehringer Ingelheim International GMBH; **5.7b:** © Veronika Burmeister/Visuals Unlimited; **5.9b:** © Cabisco/Visuals Unlimited; **5.11a:** © Fred Hossler/Visuals Unlimited; **5.13:** © Lee D.

Simon/Photo Researchers; **5.18:** © K. G. Murti/Visuals Unlimited; **5.20a:** © Patricia Barber/Custom Medical Stock; **5.20b:** Science VU-Charles W. Stratton/Visuals Unlimited; **5.22:** Science VU-Fred Marsik/Visuals Unlimited; **5.23a:** © E.C.S. Chan/Visuals Unlimited; **5.23 b,c:** From J. Prier, *Basic Medical Virology,* 1966. Reprinted by permission of Williams & Wilkins Co., Baltimore. Courtesy of Dr. Jack W. Frankel.

Chapter 6

Opener: From *ASM News,* 53(10): cover, 1987, American Society for Microbiology. Photo by C. W. Sullivan, Marine Biology Research Section, Univ. of Southern California, Los Angeles; **p. 159(a):** © Dieter Blum/Peter Arnold, Inc.; **p. 159(b):** © Martin G. Miller/Visuals Unlimited; **6.2a:** © Ralph Robinson/Visuals Unlimited; **6.2b:** © Friedrich Widdell/Visuals Unlimited; **6.11a:** © Pat Armstrong/Visuals Unlimited; **6.11b:** © Philip Sze/Visuals Unlimited; **p. 177 (a,b):** From *Biology of Paramecium,* 2/e by Ralph Richterman, Plenum Publishing Corp. Courtesy of Dr. Hans-Dieter Gortz, Zoologisches Institut, Munster, Germany; **p. 177(c):** From Orpin and Mathieson, *Applied and Environmental Microbiology,* 52(3):528, 1986. American Society for Microbiology; **6.14:** Science VU-Fred Marsik/Visuals Unlimited; **6.18a:** © K. Talaro/Visuals Unlimited; **6.20b:** From Jean McFaddin, *Biochemical Tests for Identification of Medical Bacteria,* 1980. Reprinted by permission of Williams and Wilkins Co., Baltimore; **6.21b:** © K. Talaro/Visuals Unlimited; **6.22b:** © Raymond B. Otero/Visuals Unlimited; **p. 185:** From *ASM News,* 47(7): 392, 1981. American Society for Microbiology; **6.23a:** © A. M. Siegelman/Visuals Unlimited; **6.23b:** © Raymond B. Otero/Visuals Unlimited; **6.25:** © K. Talaro/Visuals Unlimited; **6.26a:** © Raymond B. Otero/Visuals Unlimited; **6.26b, 6.27:** © K. Talaro/Visuals Unlimited.

Chapter 7

Opener: Reprinted from Per Hellung-Larsen and Jens Dilling Lundgren, *Human Metabolism,* © Munksgaard International Publishers Ltd., Copenhagen, Denmark; **p. 205(a):** © John D. Cunningham/Visuals Unlimited; **p. 205(b):** © Robert F. Myers/ Visuals Unlimited.

Chapter 8

Opener: Lawrence Livermore Laboratory, © Science Photo Library/Custom Medical Stock Photo; **8.3:** © K. G. Murti/Visuals Unlimited; **p. 259(f):** Courtesy of Promega Corporation.

Chapter 9

Opener: The Bettmann Archive; **9.5a:** Science VU/Visuals Unlimited; **9.6a:** © Raymond B. Otero/Visuals Unlimited; **9.9:** Agricultural Research Service, USDA; **9.10:** © K. Talaro/ Visuals Unlimited; **9.12a,b:** Linda Graziadei and Tom Malone, Cold Spring Harbor Laboratory and Heat Systems, Inc.; **9.13b:** Science VU-Pall/Visuals Unlimited; **p. 288(a):** © David M. Phillips/Visuals Unlimited; **p. 288(b), 9.17:** © K. Talaro/Visuals Unlimited; **9.19a:** © Raymond B. Otero/Visuals Unlimited.

Chapter 10

Opener: Courtesy of ICN Pharmaceuticals, Inc.; **p. 299:** © K. Talaro/Visuals Unlimited; **10.2e,f:** Reprinted with permission of Robert L. Perkins from A. S. Klainer and R. L. Perkins, *Journal of Infectious Diseases,* 122:4, 1970, © University of Chicago Press; **10.14:** © Cabisco/Visuals Unlimited; **10.18:** © Kenneth E. Greer/Visuals Unlimited; **10.20b:** Science VU/Visuals Unlimited; **10.21b:** Courtesy of Dr. Josephine Morello.

Chapter 11

Opener: The Bettmann Archive; **11.5:** © David M. Phillips/Visuals Unlimited; **11.8:** Science VU/Visuals Unlimited; **11.22:** Marshall W. Jennison, Syracuse University; **11.23:** Reprinted by permission from Gerald I. Roth and Robert Calmes: *Oral Biology,* St. Louis, 1981, The C.V. Mosby Co.

Chapter 12

Opener: © Manfred Kage/Peter Arnold, Inc.; **12.3b:** © Ellen R. Dirksen/Visuals Unlimited; **12.11:** Reprinted from Cohen and Warren, *Immunology of Parasitic Infections,* 2/E, Blackwell Scientific Publications. Courtesy of D. L. McLaren; **12.13:** © Bill Loncore/Photo Researchers; **12.23a:** © David M. Phillips/ Visuals Unlimited; **12.26b:** Reprinted from Wendell F. Rosse et al., "Immune Lysis of Normal Human and Paroxysmal Nocturnal Hemoglobinuria (PNH) Red Blood Cells," *Journal of Experimental Medicine,* 123:969, 1966. Rockefeller University Press.

Chapter 13

Opener: © Boehringer Ingelheim International GMBH; **13.8:** © David M. Phillips/Visuals Unlimited; **13.14b:** © R. Feldman/Rainbow; **13.28a:** From Gillies and Dodds, *Bacteriology Illustrated,* 5/E, Fig. 14. Reprinted by permissions of Longman Group, Ltd.; **13.30:** From Fields and Knipe, *Fundamentals of Virology,* 1986, Raven Press, Ltd. Courtesy of Philip D. Markham, Advanced BioScience Laboratories, Inc.; **13.34a:** © David M. Phillips/Visuals Unlimited; **13.34b:** From F. Latil et al., *Journal of Clinical Microbiology,* 23(6):1018, American Society of Microbiology.

Chapter 14

Openers (top left) © Stanley Flegler/Visuals Unlimited; (top right, bottom left, bottom right) © David M. Phillips/Visuals Unlimited; **14.2b:** © SPL/Photo Researchers; **14.5:** © Kenneth E. Greer/Visuals Unlimited; **14.6 a,b:** © Custom Medical Stock Photo; **14.9a,b:** © Stuart I. Fox; **14.15a:** © Kenneth E. Greer/ Visuals Unlimited; **14.15b:** © SIU/Visuals Unlimited; **14.18a,b:** © Kenneth E. Greer/ Visuals Unlimited; **14.22:** Reprinted from R. Kretschmer, *New England Journal of Medicine,* 279:1295, 1968. Courtesy of Fred S. Rosen; **14.23:** Courtesy of Baylor College of Medicine, Public Affairs.

Chapter 15

Figure C(b), p. 473: © Fred Marsik/Visuals Unlimited; **Opener:** © Sylvia E. Coleman/ Visuals Unlimited; **15.1:** © David M. Phillips/ Visuals Unlimited; **15.2:** © K. Talaro/Visuals Unlimited; **15.3b,c:** © Carroll H. Weiss/ Camera M.D. Studios; **15.4b:** Science VU-Charles W. Stratton/Visuals Unlimited; **15.5a:** © Charles Stoer/Camera M.D. Studios; **15.5b:** From Braude, *Infectious Diseases and Medical Microbiology,* 2/E, Saunders College Publishing; **15.6a:** © Raymond B. Otero/ Visuals Unlimited; **15.7:** Analytab Products; **15.8:** © A. M. Siegelman/Visuals Unlimited; **15.9a:** © Fred E. Hossler/Visuals Unlimited; **15.9b:** © K. Talaro/Visuals Unlimited; **15.11a,b:** © Kenneth Greer/Visuals Unlimited; **15.14a:** Courtesy Dr. David Schlaes; **15.14b** (all): Diagnostic Products Corporation; **15.15a:** Courtesy of Dr. David Schlaes; **15.15b:** Courtesy Harold Benson; **15.16:** © Carroll H. Weiss/Camera M.D. Studios; **15.17:** © A. M. Siegelman/Visuals Unlimited; **15.19:** © Custom Medical Stock Photo; **15.20:** © Raymond B. Otero/Visuals Unlimited; **15.21:** © G. Musil/Visuals Unlimited; **15.25:** From James Bingham, *Pocket Guide for Clinical Medicine,* 1984. Reprinted by permission of Williams and Wilkins Co., Baltimore; **15.26:** © George J. Wilder/Visuals Unlimited; **15.28:** © Kenneth E. Greer/Visuals Unlimited; **15.29a:** Courtesy of Becton-Dickinson; **15.29b:** © Kathy Talaro/Visuals Unlimited.

Chapter 16

Opener: © Elmer Koneman/Visuals Unlimited; **16.1a:** © A. M. Siegelman/Visuals Unlimited; **16.1b:** © George J. Wilder/Visuals Unlimited; **16.1c:** © John D. Cunningham/Visuals Unlimited; **16.2a:** Science VU-Charles W. Stratton/Visuals Unlimited; **16.2b:** Science VU-Charles W. Stratton/Visuals Unlimited; **16.3a:** From N. A. Boyd et al., *Journal of Medical Microbiology,* 5:459, 1972. Reprinted by permission of Longman Group, Ltd.; **16.4:** Science VU-Charles W. Stratton/Visuals Unlimited; **16.5a:** Reprinted from Farrar and Lambert, *Pocket Guide for Nurses: Infectious Diseases,* 1984. Williams & Wilkins Co., Baltimore. Courtesy of Dr. Fred E. Pittman; **16.5b:** Reprinted from Farrar and Lambert, *Pocket Guide for Nurses: Infectious Diseases,* 1984, Williams and Wilkins Co., Baltimore. Courtesy of Dr. F. Robert Fekety; **16.7:** Reprinted from Farrar and Lambert, *Pocket Guide for Nurses: Infectious Diseases,* 1984, Williams & Wilkins Co., Baltimore. By courtesy of Dr. T. F. Sellers, Jr.; **16.10:** © Kenneth E. Greer/Visuals Unlimited; **16.11:** © George J. Wilder/Visuals Unlimited; **16.13:** Centers for Disease Control; **16.14:** © A. M. Siegelman/Visuals Unlimited; **16.15a:** From Gillies and Dodds, *Bacteriology Illustrated,* 5/E, Fig. 45, p. 58. Reprinted by permission of Churchill Livingstone; **16.15b:** Reprinted from Yegian and Kurung, *American Review of Tuberculosis,* 65:181, 1952; **16.17:** Science VU-Charles W. Stratton/Visuals Unlimited; **p. 521(c):** CNRI/Science Photo Library/Photo Researchers; **p. 521(d):** © Elmer Koneman/ Visuals Unlimited; **16.18a:** © Kenneth E. Greer/Visuals Unlimited; **16.18b:** Science VU/ Visuals Unlimited; **16.19, 16.20a,b:** © Kenneth E. Greer/Visuals Unlimited; **16.21:** © NMSB/ Custom Medical Stock Photo; **16.22:** World Health Organization; **16.23:** © Kenneth E. Greer/Visuals Unlimited; **16.24:** Science VU/ Visuals Unlimited; **16.25:** © K. Talaro/Visuals Unlimited; **16.27:** © Jack M. Bostrack/Visuals Unlimited; **16.29a:** © K. Talaro/Visuals Unlimited; **16.29b:** Science VU-Charles W. Stratton/Visuals Unlimited; **16.33:** © John D. Cunningham/Visuals Unlimited; **16.34:** © A. M. Siegelman/Visuals Unlimited; **16.38, 16.40:** Science VU-Charles W. Stratton/Visuals Unlimited; **16.41:** Reprinted from Gillies and Dodds, *Bacteriology Illustrated,* 5/E, Fig. 78 (right), p. 104. Reprinted by permission of Churchill Livingstone; **16.42:** Reprinted from Farrar and Lambert, *Pocket Guide for Nurses: Infectious Diseases,* 1984, Williams and Wilkins Co., Baltimore. Photo by Dr. M. Tapert.

Chapter 17

Opener: © R. Calentine/Visuals Unlimited; **17.1b-d:** Science VU-Charles W. Stratton/ Visuals Unlimited; **17.2:** © Custom Medical Stock Photo; **17.3:** © Carroll H. Weiss/Camera M.D. Studios; **17.4:** Science VU/Visuals Unlimited; **17.5:** © Kenneth E. Greer/Visuals Unlimited; **17.6a,b, 17.7:** Science VU-CDC/ Visuals Unlimited; **17.8a:** Reprinted from Strickland, *Hunter's Tropical Medicine,* 6/E, Saunders College Publishing; **17.8b:** Science VU/Visuals Unlimited; **17.9:** Science VU-Fred Marsik/Visuals Unlimited; **17.12:** © Centers for Disease Control/Peter Arnold, Inc.; **17.13a-b:** From Jacob S. Teppema, "In Vivo Adherence and Colonization of *Vibrio cholerae* Strains That Differ in Hemagglutinating Activity and Motility," *Journal of Infection and Immunity,* 55(9):2093–2102, Sept. 1987. Reprinted by permission of American Society for Microbiology; **p. 562** (top): AP/Wide World Photos; **17.15:** From R. R. Colwell and D. M. Rollins, "Viable but Nonculturable Stage of *Campylobacter jejuni* and Its Role in Survival in the Natural Aquatic Environment," *Applied and Environmental Microbiology,* 52(3):531–38, 1986. Reprinted by permission of American Society for Microbiology; **17.16a:** From R. L. Anaker et al., "Details of the Ultrastructure of *Rickettsia prowazekii* Grown in the Chick Yolk Sac" *Journal of Bacteriology,* 94(1):260, 1967. Reprinted by permission of American Society of Microbiology; **17.16b:** From Bruce R. Merrell et al., "Morphological and Cell Association Characteristics of *Rochalimaea quintana*: Comparison of the Vole and Fuller Strains," *Journal of Bacteriology,* 135(2):633–640, 1978. Reprinted by permission from American Society of Microbiology; **p. 565(b):** Science VU-PIDC, Richard Endris/Visuals Unlimited; **p. 565(c):** © Richard F. Ashley/Visuals Unlimited; **p. 565(d):** © A. M. Siegelman/ Visuals Unlimited; **17.19:** Dept. of Health and Human Resources, Courtesy of Dr. W. Burgdorfer; **17.20:** From McCaul and Williams, "Development Cycle of *C. burnetti,*" *Journal of Bacteriology,* 147:1063, 1981. Reprinted by permission of American Society for Microbiology; **17.22a:** Armed Forces Institute of Pathology; **17.22b:** Science VU-Bascom Palmer Institute/Visuals Unlimited; **17.23:** © Harvey Blank/Camera M.D. Studios; **17.24a:** Science VU-CDC/Visuals Unlimited; **17.24b:** Science VU-AFIP/Visuals Unlimited; **17.25a:** © David M. Phillips/Visuals Unlimited; **17.26a:** © Kenneth E. Greer/Visuals Unlimited; **17.26b:** From A. W. Margileth et al., *Journal of the American Medical Association,* 252(7):929, Figure 4, 1984. Reprinted by permission of American Medical Association; **17.30a:** © R. Gottsegen/Peter Arnold, Inc.; **17.30b:** © Stanley Flegler/Visuals Unlimited; **17.31a,b, 17.32d:** Science VU-Max A. Listgarten/Visuals Unlimited.

Chapter 18

Opener: © A. M. Siegelman/Visuals Unlimited; **18.4:** From L. K. Green and D. G. Moore, *Laboratory Medicine,* 18(7):cover, July 1987; **18.6a:** © Elmer W. Koneman/Visuals Unlimited; **18.6b:** © Everett S. Beneke/Visuals Unlimited; **18.7b:** © A. M. Siegelman/Visuals Unlimited; **18.8b:** Science VU-Charles Sutton/ Visuals Unlimited; **18.9:** Reprinted from J. Walter Wilson, *Fungous Diseases of Man,* Plate 2, © 1967, Regents of the University of California and University of California Press; **18.10:** From Rippon, *Medical Mycology,* 2/E, Saunders College Publishing; **18.11a:** © Elmer W. Koneman/Visuals Unlimited; **18.11b:** © A. M. Siegelman/Visuals Unlimited; **18.12:** Reprinted by permission of Upjohn Company from E. S. Beneke, et al., *Human Mycoses,* 1984; **18.13:** From Koneman and Roberts, *Practical Laboratory Mycology,* Williams and Wilkins Co., Baltimore; **18.14:** © Harkisan Raj/ Visuals Unlimited; **18.15:** © Everett S. Beneke/Visuals Unlimited; **18.16:** Science VU-Charles W. Stratton/Visuals Unlimited; **18.17:** Reprinted by permission of Upjohn Company from E. S. Beneke, et al., *Human Mycoses,* 1984; **18.18a,b:** From Koneman and Roberts, *Practical Laboratory Mycology,* Williams and Wilkins Co., Baltimore; **18.19a,b:** © Everett S. Beneke/Visuals Unlimited; **18.19c, 18.20a:** © Kenneth E. Greer/Visuals Unlimited; **18.20b:** Reprinted from J. Walter Wilson, *Fungous Diseases of Man,* Plate 42 (middle right), © 1967, Regents of the University of California and University of California Press; **18.21a,b, 18.22a:** © Everett S. Beneke/Visuals Unlimited; **18.22b:** Science VU-CDC/Visuals Unlimited; **18.23a:** © Raymond B. Otero/ Visuals Unlimited; **18.23b:** © Elmer Koneman/ Visuals Unlimited; **18.24:** © A. M. Siegelman/ Visuals Unlimited; **18.25:** Reprinted from J. Walter Wilson, *Fungous Diseases of Man,* Plate 21, © 1967, Regents of the University of California and University of California Press; **18.26a:** © Everett S. Beneke/Visuals Unlimited; **18.26b:** Science VU-AFIP/Visuals Unlimited; **18.27:** © Daniel Snyder/Visuals Unlimited; **18.28:** From Rippon, *Medical Mycology,* 2/E, © Saunders College Publishing; **18.29a,b:** © M. F. Brown/Visuals Unlimited; **18.30a:** © Kenneth E. Greer/Visuals Unlimited; **18.30b:** Science VU-Charles W. Stratton/Visuals Unlimited; **18.31:** Reprinted from J. Walter Wilson, *Fungous Diseases of Man,* Plate 26, © 1967, Regents of University of California and University of California Press.

Chapter 19

Opener: Science VU-David John/Visuals Unlimited; **19.2:** Science VU-Charles W. Stratton/Visuals Unlimited; **19.6b:** © Jerome Paulin/Visuals Unlimited; **19.8a:** Reprinted from Katz, Despommier, and Gwadz, *Parasitic Diseases,* Springer-Verlag. Photo by T. Jones; **19.9b:** Armed Forces Institute of Pathology; **19.11:** Centers for Disease Control; **19.13a:** © Cecil H. Fox/Photo Researchers, Inc.; **19.14:** © A.M. Siegelman/ Visuals Unlimited; **19.16a:** © Lauritz Jensen/ Visuals Unlimited; **19.16b:** © A. M. Siegelman/ Visuals Unlimited; **19.16c:** From Rubin and Farbo, *Pathology.* Reprinted by Permission of J.B. Lippincott Company; **19.17b** (left): © R. Calentine/Visuals Unlimited; **19.17b** (right): Science VU-Fred Marsik/Visuals Unlimited; **19.18:** © Carroll H. Weiss/Camera M.D. Studios; **19.20a,b:** Science VU-Fred Marsik/ Visuals Unlimited; **19.21:** Science VU-AFIP/ Visuals Unlimited; **19.22:** From Farrar and Lambert, *Pocket Guide For Nurses: Infectious Diseases,* 1984, Williams and Wilkins, Co. Baltimore. Photo by Dr. R. Muller; **19.23a:** © Cabisco/Visuals Unlimited; **19.23b:** Science VU/Visuals Unlimited; **19.24:** © A. M. Siegelman/Visuals Unlimited; **19.25a:** From Katz et al., *Parasitic Diseases,* Springer/ Verlag. Photo by U. Martin; **19.25b:** © Stanley Flegler/Visuals Unlimited; **19.26:** Science VU-Charles W. Stratton/Visuals Unlimited; **p. 640:** © Lauritz Jensen/Visuals Unlimited.

Chapter 20

Opener: James Gillroy, British, 1757–1815, *The Cow-Pock,* engraving, 1802, William McCallin McKee Memorial Collection, 1928.1407. © 1991, The Art Institute of Chicago. All Rights Reserved; **p. 651:** World Health Organization; **20.2a:** Science VU-Charles W. Stratton/Visuals Unlimited; **20.2b:** © Charles Stoer/Camera M.D. Studios; **20.3b:** © Boehringer Ingelheim International Gmbh; **20.5a:** © Kenneth E. Greer/Visuals Unlimited; **20.5b, 20.6:** © Carroll H. Weiss/Camera M.D. Studios; **20.7, 20.8:** © Kenneth E. Greer/ Visuals Unlimited; **20.9:** Science VU-Charles W. Stratton/Visuals Unlimited; **20.10a:** © John D. Cunningham/Visuals Unlimited; **20.10d:** © Carroll H. Weiss/Camera M.D. Studios; **20.11:** © Veronika Burmeister/Visuals Unlimited; **20.12:** Science VU-Charles W. Stratton/Visuals Unlimited; **20.13:** Centers for Disease Control; **20.14:** Courtesy Barbara O'Connor; **20.17:** Science VU-NIH/Visuals Unlimited; **20.19a:** © David M. Phillips/Visuals Unlimited; **20.19b:** © K. G. Murti/Visuals Unlimited; **20.20a,b:** © Kenneth E. Greer/ Visuals Unlimited; **20.21:** Science VU-Charles W. Stratton/Visuals Unlimited.

Chapter 21

Opener: The Bettmann Archive; **21.2:** From C. R. Howard, *Perspectives in Medical Virology,* Vol. 2, 1986, Elsevier Science Publishing Company. Photo by J. L. Maiztegui; **21.3a:** From Exeen M. Morgand and Fred Rapp, "Measles Virus and Its Associated Disease," *Bacteriological Reviews,* 41(3):636–666, 1977. Reprinted by permission of American Society for Microbiology; **21.4:** © Biophoto Associates/ Photo Researchers; **21.5a:** Centers for Disease Control; **21.5b:** © Kenneth E. Greer/ Visuals Unlimited; **21.6a:** © Boehringer Ingelheim International GmbH; **21.7d:** AP/ Wide World Photos; **21.8:** Science VU-AFIP/ Visuals Unlimited; **21.9a:** Reprinted by permission of Dr. Kenneth Schiffer from Cooper et al., *American Journal of the Diseased Child,* 110:419, © 1965, American Medical Association; **21.9b:** *Infectious Diseases Magazine,* January 1971; **21.16:** © Hans R. Gelderblom/Visuals Unlimited; **21.17:** From *ASM News,* 53(5): cover, May 1987. American Society for Microbiology; **21.18:** Science VU/Visuals Unlimited; **21.20:** From I. Katayarma, C. Y. Li, and L. T. Yam, "Ultrastructure Characteristics of the Hairy Cells of Leukemic Reticuloendotheliosis," *American Journal of Pathology,* 67:361, 1972. Reprinted by permission of J.B. Lippincott Company; **21.21b:** Science VU-CDC/Visuals Unlimited; **21.22:** The Bettmann Archive; **21.24b (left):** © Cabisco/Visuals Unlimited; **21.24b (right):** Science VU-Charles W. Stratton/Visuals Unlimited; **21.25:** © Carroll H. Weiss/Camera M.D. Studios; **21.27a:** © K. G. Murti/Visuals Unlimited; **21.27b:** From Farrar and Lambert, *Pocket Guide for Nurses: Infectious Diseases,* 1984, Williams and Wilkins, Co., Baltimore. Reprinted by permission of Dr. William Edmund Farrar, Jr.; **21.28a:** Reprinted from Fields and Knipe, *Fundamentals of Virology,* 1986, Raven Press; **21.28b:** From Fields and Knipe, *Fundamentals of Virology,* 1986, Raven Press. Photo by Dr. Patricia Merz.

Chapter 22

Opener: Science VU-SIM/Visuals Unlimited; **22.8a** (top): © Cabisco/Visuals Unlimited; **22.11b:** © John D. Cunningham/Visuals Unlimited; **22.12a,b:** © Sylvan Wittwer/Visuals Unlimited; **22.14, 22.16:** © John D. Cunningham/Visuals Unlimited; **22.20b:** © Carleton Ray/Photo Researchers, Inc.; **22.21:** © John Cunningham/Visuals Unlimited; **22.23b:** © K. Talaro/Visuals Unlimited; **22.25:** Courtesy Donald Klein; **p. 735** (top): © Frank Hanna/Visuals Unlimited; **p. 735** (bottom): © John D. Cunningham/Visuals Unlimited; **22.29:** © K. Talaro/Visuals Unlimited; **22.31b:** © Vance Henry/Taurus Photos; **22.32:** © John D. Cunningham/Visuals Unlimited; **22.33b:** © Kevin Schafer/Peter Arnold, Inc.; **22.35a:** © K. Talaro/Visuals Unlimited; **22.36b:** © Joe Munroe/Photo Researchers, Inc.; **22.37:** Reprinted from O. Ciferri, *Microbiological Reviews,* 47(4):560, 1983. Reprinted by permission of the American Society for Microbiology; **22.43:** Courtesy R. Stalker; **22.44, 22.45:** Courtesy R. L. Brinster; **p. 756(c):** Courtesy of Lifecodes Corporation.

Appendix

A.1: Courtesy of Becton-Dickinson and Company; **A.2:** © Kathy Talaro/Visuals Unlimited.

Index

A

B

C

E

F

H

I

M

Q

R

S

T

U

V

W

X

Y

Z

Primary and Secondary Syphilis
(By state, 1989*)

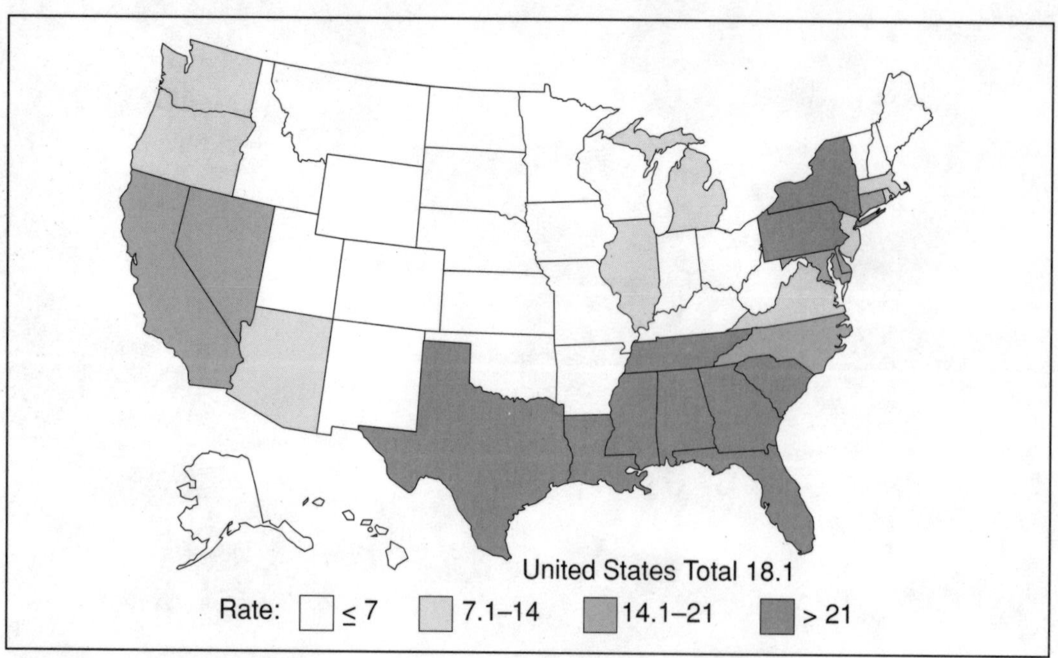

*Based on reported cases per 100,000 population.

Primary and Secondary Syphilis
(By year, United States, 1941–90*)

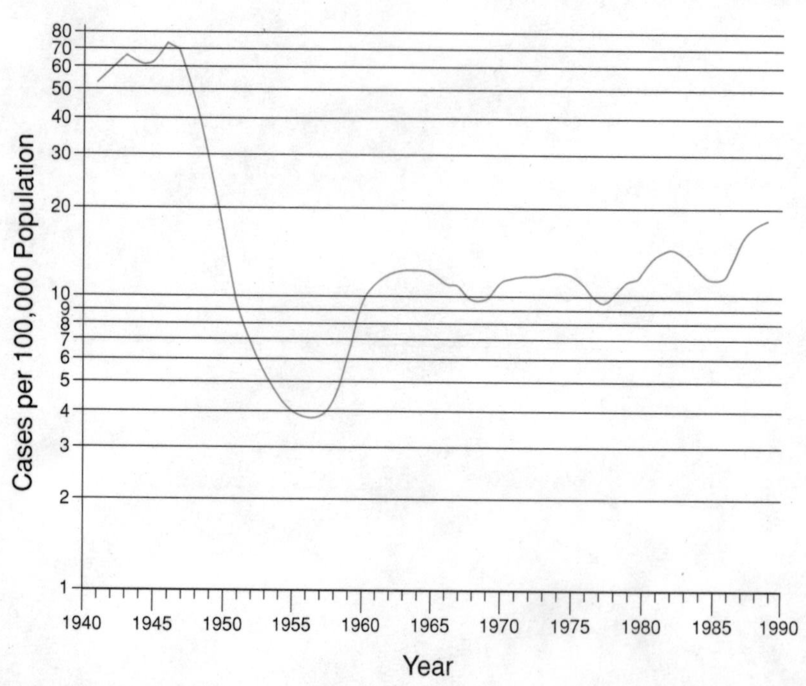

*1941–46 fiscal years (12-month period ending June 30); 1947–90 calendar years.

Source: Map and graph from the CDC, Atlanta, GA.